# WILLIAM SHAKESPEARE

## THE
# COMPLETE
## WORKS

Martin Droeshout's engraving of Shakespeare, first published on the title-page of the First Folio (1623)

*To the Reader*

This figure that thou here seest put,
It was for gentle Shakespeare cut,
Wherein the graver had a strife
With nature to outdo the life.
O, could he but have drawn his wit
As well in brass as he hath hit

His face, the print would then surpass
All that was ever writ in brass!
But since he cannot, reader, look
Not on his picture, but his book.

BEN JONSON

# WILLIAM SHAKESPEARE

# THE COMPLETE WORKS

## COMPACT EDITION

*General Editors*
STANLEY WELLS AND GARY TAYLOR

*Editors*
STANLEY WELLS, GARY TAYLOR
JOHN JOWETT, AND WILLIAM MONTGOMERY

*With Introductions by*
STANLEY WELLS

CLARENDON PRESS · OXFORD

Oxford University Press, Walton Street, Oxford OX2 6DP

Oxford New York Toronto
Delhi Bombay Calcutta Madras Karachi
Petaling Jaya Singapore Hong Kong Tokyo
Nairobi Dar es Salaam Cape Town
Melbourne Auckland

and associated companies in
Berlin Ibadan

OXFORD is a trade mark of Oxford University Press

Published in the United States
by Oxford University Press (USA)

© Oxford University Press 1988

First published 1988
Reprinted 1989, 1990 (twice), 1991 (twice), 1992

British Library Cataloguing in Publication Data
Shakespeare, William, 1564–1616
[Works]. The complete works.
1. Drama in English. Shakespeare, William,
1564–1616. Texts
I. Title   II. Wells, Stanley, 1930–
III. Taylor, Gary
822.3'3
ISBN 0-19-811747-7

Library of Congress Cataloging-in-Publication Data
Shakespeare, William, 1564–1616.
William Shakespeare, the complete works.
Includes index.
I. Wells, Stanley W., 1930–     II. Taylor, Gary.
III. Title.   IV. Title: Complete works.
PR2754.W45 1988   822.3'3   88-5231
ISBN 0-19-811747-7

Printed in Great Britain
by Butler & Tanner Ltd.
Frome, Somerset

# WILLIAM SHAKESPEARE

## THE COMPLETE WORKS

*with a General Introduction, and Introductions to individual works, by*
### STANLEY WELLS

*The Complete Works* has been edited collaboratively under the General Editorship of Stanley Wells and Gary Taylor. Each editor has undertaken prime responsibility for certain works, as follows:

STANLEY WELLS *The Two Gentlemen of Verona*; *The Taming of the Shrew*; *Titus Andronicus*; *Venus and Adonis*; *The Rape of Lucrece*; *Love's Labour's Lost*; *Much Ado About Nothing*; *As You Like It*; *Twelfth Night*; The Sonnets and 'A Lover's Complaint'; Various Poems (printed); *Othello*; *Macbeth*; *Antony and Cleopatra*; *The Winter's Tale*

GARY TAYLOR *1 Henry VI*; *Richard III*; *The Comedy of Errors*; *A Midsummer Night's Dream*; *Henry V*; *Hamlet*; *Troilus and Cressida*; Various Poems (manuscript); *Sir Thomas More* (passages attributed to Shakespeare); *All's Well That Ends Well*; *King Lear*; *Pericles*; *Cymbeline*

JOHN JOWETT *Romeo and Juliet*; *Richard II*; *King John*; *1 Henry IV*; *The Merry Wives of Windsor*; *2 Henry IV*; *Julius Caesar*; *Measure for Measure*; *Timon of Athens*; *Coriolanus*; *The Tempest*

WILLIAM MONTGOMERY *The First Part of the Contention*; *Richard Duke of York*; *The Merchant of Venice*; *All Is True*; *The Two Noble Kinsmen*

*American Advisory Editor · S. Schoenbaum*
*Textual Adviser · G. R. Proudfoot*
*Music Adviser · F. W. Sternfeld*
*Editorial Assistant · Christine Avern-Carr*

# ACKNOWLEDGEMENTS

THE preparation of a volume such as this would be impossible without the generosity that scholars can count on receiving from their colleagues, at home and overseas. Among those to whom we are particularly grateful are: R. E. Alton; John F. Andrews; Peter Beal; Thomas L. Berger; David Bevington; J. W. Binns; Peter W. M. Blayney; Fredson Bowers; A. W. Braunmuller; Alan Brissenden; Susan Brock; J. P. Brockbank; Robert Burchfield; Lou Burnard; Lesley Burnett; John Carey; Janet Clare; Thomas Clayton; T. W. Craik; Norman Davis; Alan Dessen; E. E. Duncan-Jones; K. Duncan-Jones; R. D. Eagleson; Philip Edwards; G. Blakemore Evans; Jean Fuzier; Hans Walter Gabler; Philip Gaskell; A. J. Gurr; Antony Hammond; Richard Hardin; G. R. Hibbard; Myra Hinman; R. V. Holdsworth; E. A. J. Honigmann; T. H. Howard-Hill; MacD. P. Jackson; Harold Jenkins; John Kerrigan; Randall McLeod; Nancy Maguire; Giorgio Melchiori; Peter Milward; Kenneth Muir; Stephen Orgel; Kenneth Palmer; John Pitcher; Eleanor Prosser; S. W. Reid; Marvin Spevack; R. K. Turner; E. M. Waith; Michael Warren; Paul Werstine; G. Walton Williams; Laetitia Yeandle.

We are conscious also of a great debt to the past: to our predecessors R. B. McKerrow and Alice Walker, who did not live to complete an Oxford Shakespeare but whose papers have been of invaluable assistance, and to the long line of editors and other scholars, from Nicholas Rowe onwards, whose work is acknowledged in *William Shakespeare: A Textual Companion*.

We gratefully acknowledge assistance from the staff of the following libraries and institutions: the Beinecke Library, Yale University; the Birmingham Shakespeare Library; the Bodleian Library, Oxford; the British Library; the English Faculty Library, Oxford; the Folger Shakespeare Library; Lambeth Palace Library; St. John's College, Cambridge; the Shakespeare Centre, Stratford-upon-Avon; the Shakespeare Institute, University of Birmingham; Trinity College, Cambridge; the Victoria and Albert Museum; Westminster Abbey Library.

Many debts of gratitude have also been incurred to persons employed in a variety of capacities by Oxford University Press. Among those with whom we have worked especially closely are Linda Agerbak, Sue Dommett, Oonagh Ferrier, Paul Luna, Jamie Mackay, Louise Pengelley, Graham Roberts, Maria Tsoutsos, and Patricia Wilkie. John Bell started it all, Kim Scott Walwyn made sure we finished it, and from beginning to end Christine Avern-Carr's meticulous standards of accuracy have been exemplary.

S.W.W. G.T.
J.J. W.L.M.

# CONTENTS

## THE COMPLETE WORKS

# CONTENTS

# ALPHABETICAL LIST OF CONTENTS

Both full versions and common abbreviations of titles are given

xi

# ILLUSTRATIONS

# GENERAL INTRODUCTION

THIS volume contains all the known plays and poems of William Shakespeare, a writer, actor, and man of the theatre who lived from 1564 to 1616. He was successful and admired in his own time; major literary figures of the subsequent century, such as John Milton, John Dryden, and Alexander Pope, paid tribute to him, and some of his plays continued to be acted during the later seventeenth and earlier eighteenth centuries; but not until the dawn of Romanticism, in the later part of the eighteenth century, did he come to be looked upon as a universal genius who outshone all his fellows and even, some said, partook of the divine. Since then, no other secular imaginative writer has exerted so great an influence over so large a proportion of the world's population. Yet Shakespeare's work is firmly rooted in the circumstances of its conception and development. Its initial success depended entirely on its capacity to please the theatre-goers (and, to a far lesser extent, the readers) of its time; and its later, profound impact is due in great part to that in-built need for constant renewal and adaptation that belongs especially to those works of art that reach full realization only in performance. Shakespeare's power over generations later than his own has been transmitted in part by artists who have drawn on, interpreted, and restructured his texts as others have drawn on the myths of antiquity; but it is the texts as they were originally performed that are the sources of his power, and that we attempt here to present with as much fidelity to his intentions as the circumstances in which they have been preserved will allow.

## Shakespeare's Life: Stratford-upon-Avon and London

Shakespeare's background was commonplace. His father, John, was a glover and wool-dealer in the small Midlands market-town of Stratford-upon-Avon who had married Mary Arden, daughter of a prosperous farmer, in or about 1557. During Shakespeare's childhood his father played a prominent part in local affairs, becoming bailiff (mayor) and justice of the peace in 1568; later his fortunes declined. Of his eight children, four sons and one daughter survived childhood. William, his third child and eldest son, was baptized in Holy Trinity Church, Stratford-upon-Avon, on 26 April 1564; his birthday is traditionally celebrated on 23 April—St. George's Day. The only other member of his family to take up the theatre as a profession was his youngest brother, Edmund, born sixteen years after William. He became an actor and died at the age of twenty-seven: on the last day of 1607 the sexton of St. Saviour's, Southwark, noted 'Edmund Shakspeare A player Buried in yᵉ Church wᵗʰ a forenoone knell of yᵉ great bell, xxs.' The high cost of the funeral suggests that it may have been paid for by his prosperous brother.

John Shakespeare's position in Stratford-upon-Avon would have brought certain privi-leges to his family. When young William was four years old he could have had the excitement of seeing his father, dressed in furred scarlet robes and wearing the alderman's official thumb-ring, regularly attended by two mace-bearing sergeants in buff, presiding at fairs and markets. A little later, he would have begun to attend a 'petty school' to acquire the rudiments of an education that would be continued at the King's New School, an established grammar school with a well-qualified master, assisted by an usher to help with the younger pupils. We have no lists of the school's pupils in Shakespeare's time, but his father's position would have qualified him to attend, and the school offered the kind of education that lies behind the plays and poems. Its boy pupils, aged from about eight to fifteen, endured an arduous routine. Classes began early in the morning: at six, normally; hours were long, holidays infrequent. Education was centred on Latin; in the upper forms, the speaking of English was forbidden. A scene (4.1) in *The Merry Wives of Windsor* showing a schoolmaster taking a boy named William through his Latin grammar draws on the officially approved textbook, William Lily's *Short Introduction of Grammar*, and, no doubt, on Shakespeare's memories of his youth.

From grammar the boys progressed to studying works of classical and neo-classical literature. They might read anthologies of Latin sayings and Aesop's *Fables*, followed by the fairly easy plays of Terence and Plautus (on whose *Menaechmi* Shakespeare was to base *The Comedy of Errors*). They might even act scenes from Latin plays. As they progressed, they would improve their command of language by translating from Latin into English and back, by imitating approved models of style, and by studying manuals of composition, the ancient rules of rhetoric, and modern rules of letter-writing. Putting their training into practice, they would compose formal epistles, orations, and declamations. Their efforts at composition would be stimulated, too, by their reading of the most admired authors. Works that Shakespeare wrote throughout his career show the abiding influence of Virgil's *Aeneid* and of Ovid's *Metamorphoses* (both in the original and in Arthur Golding's translation of 1567). Certainly he developed a taste for books, both classical and modern: his plays show that he continued to read seriously and imaginatively for the whole of his working life.

After Shakespeare died, Ben Jonson accused him of knowing 'small Latin and less Greek'; but Jonson took pride in his classical knowledge: a boy educated at an Elizabethan grammar school would be more thoroughly trained in classical rhetoric and Roman (if not Greek) literature than most present-day holders of a university degree in classics. Modern languages would not normally be on the curriculum. Somehow Shakespeare seems to have picked up a working knowledge of French—which he expected audiences of *Henry V* to understand—and of Italian (the source of *Othello*, for instance, is an Italian tale that had not been published in translation when he wrote his play). We do not know whether he ever travelled outside England.

Shakespeare must have worked hard at school, but there was a life beyond the classroom. He lived in a beautiful and fertile part of the country, with rivers and fields at hand. He had the company of brothers and sisters. Each Sunday the family would go to Stratford's splendid parish church, as the law required; his father, by virtue of his dignified status, would sit in the front pew. There Shakespeare's receptive mind would be impressed by the sonorous phrases of the Bible, in either the Bishops' or the Geneva version, the Homilies, and the Book of Common Prayer. From time to time travelling players would visit Stratford. Shakespeare's father would have the duty of licensing them to perform; probably his son first saw plays professionally acted in the Guildhall below his schoolroom.

Shakespeare would have left school when he was about fifteen. What he did then is not known. One of the earliest legends about him, recorded by John Aubrey around 1681, is that 'he had been in his younger years a schoolmaster in the country'. John Cottom, who was master of the Stratford school between 1579 and 1581 or 1582, and may have taught Shakespeare, was a Lancashire man whose family home was close to that of a landowner, Alexander Houghton. Both Cottom and Houghton were Roman Catholics, and there is some reason to believe that John Shakespeare may have retained loyalties to the old religion. When Houghton died, in 1581, he mentioned in his will one William Shakeshafte, apparently a player. The name is a possible variant of Shakespeare; conceivably Cottom found employment in Lancashire for his talented pupil as a tutor who also acted. If so, Shakespeare was soon back home. On 28 November 1582 a bond was issued permitting him to marry Anne Hathaway of Shottery, a village close to Stratford. She was eight years his senior, and pregnant. Their daughter, Susanna, was baptized on 26 May 1583, and twins, Hamnet and Judith, on 2 February 1585. Though Shakespeare's professional career (described in the next section of this Introduction) was to centre on London, his family remained in Stratford, and he maintained his links with his birthplace till he died and was buried there.

One of the unfounded myths about Shakespeare is that all we know about his life could be written on the back of a postage stamp. In fact we know a lot about some of the less exciting aspects of his life, such as his business dealings and his tax debts (as may be seen from the list of Contemporary Allusions, pp. xxxix–xlii). Though we cannot tell how often he visited Stratford after he moved to London, clearly he felt that he belonged where he

1. The Shakespeare coat of arms, from a draft dated 20 October 1596, prepared by Sir William Dethick, Garter King-of-Arms

was born. His success in his profession may be reflected in his father's application for a grant of arms in 1596, by which John Shakespeare acquired the official status of gentleman. In August of that year William's son, Hamnet, died, aged eleven and a half, and was buried in Stratford. Shakespeare was living in the Bishopsgate area of London, north of the river, in October of the same year, but in the following year showed that he looked on Stratford as his real home by buying a large house, New Place. It was demolished in 1759.

In October 1598 Richard Quiney, whose son was to marry Shakespeare's daughter Judith, travelled to London to plead with the Privy Council on behalf of Stratford Corporation, which was in financial trouble because of fires and bad weather. He wrote the only surviving letter addressed to Shakespeare; as it was found among Quiney's papers, it was presumably never delivered. It requested a loan of £30—a large sum, suggesting confidence in his friend's prosperity. In 1601 Shakespeare's father died, and was buried in Stratford. In May of the following year Shakespeare was able to invest £320 in 107 acres of arable land in Old Stratford. In the same year John Manningham, a London law student, recorded a piece of gossip that gives us a rare contemporary anecdote about the private life of Shakespeare and of Richard Burbage, the leading tragedian of his company:

Upon a time when Burbage played Richard III there was a citizen grew so far in liking with him that before she went from the play she appointed him to come that night unto her by the name of Richard the Third. Shakespeare, overhearing their conclusion, went before, was entertained, and at his game ere Burbage came. Then, message being brought that Richard the Third was at the door, Shakespeare caused return to be made that William the Conqueror was before Richard the Third.

In 1604 Shakespeare was lodging in north London with a Huguenot family called Mountjoy; in 1612 he was to testify in a court case relating to a marriage settlement on

2. Shakespeare's monument, designed by Gheerart Janssen, in Holy Trinity Church, Stratford-upon-Avon

the daughter of the house. The records of the case provide our only transcript of words actually spoken by Shakespeare; they are not characterful. In 1605 Shakespeare invested £440 in the Stratford tithes, which brought him in £60 a year; in June 1607 his elder daughter, Susanna, married a distinguished physician, John Hall, in Stratford, and there his only grandchild, their daughter Elizabeth, was baptized the following February. In 1609 his mother died there, and from about 1610 his increasing involvement with Stratford along with the reduction in his dramatic output suggests that he was withdrawing from his London responsibilities and spending more time at New Place. Perhaps he was deliberately devoting himself to his family's business interests; he was only forty-six years old: an age at which a healthy man was no more likely to retire then than now. If he was ill, he was not totally disabled, as he was in London in 1612 for the Mountjoy lawsuit. In March 1613 he bought a house in the Blackfriars area of London for £140: it seems to have been an investment rather than a home. Also in 1613 the last of his three brothers died. In late 1614 and 1615 he was involved in disputes about the enclosure of the land whose tithes he owned. In February 1616 his second daughter, Judith, married Thomas Quiney, causing William to make alterations to the draft of his will, which he signed on 25 March. His widow was entitled by law and local custom to part of his estate; he left most of the remainder to his elder daughter, Susanna, and her husband. He died on 23 April, and was buried two days later in a prominent position in the chancel of Holy Trinity Church. A monument was commissioned, presumably by members of his family, and was in position by 1623. The work of Gheerart Janssen, a stonemason whose shop was not far from the Globe Theatre, it incorporates a half-length effigy which is one of the only two surviving likenesses of Shakespeare with any strong claim to authenticity.

As this selective survey of the historical records shows, Shakespeare's life is at least as well documented as those of most of his contemporaries who did not belong to great families; we know more about him than about any other dramatist of his time except Ben Jonson. The inscription on the Stratford landowner's memorial links him with Socrates and Virgil; and in the far greater memorial of 1623, the First Folio edition of his plays, Jonson links this 'Star of poets' with his home town as the 'Sweet swan of Avon'. The Folio includes the second reliable likeness of Shakespeare, an engraving by Martin Droeshout which, we must assume, had been commissioned and approved by his friends and colleagues who put the volume together. In the Folio it faces the lines signed 'B.I.' (Ben Jonson) which we print beneath it. Shakespeare's widow died in 1623, and his last surviving descendant, Elizabeth Hall (who inherited New Place and married first a neighbour, Thomas Nash, and secondly John Bernard, knighted in 1661), in 1670.

### Shakespeare's Professional Career

We do not know when Shakespeare joined the theatre after his marriage, or how he was employed in the mean time. In 1587 an actor of the Queen's Men—the most successful company of the 1580s—died as a result of manslaughter shortly before the company visited Stratford. That Shakespeare may have taken his place is an intriguing speculation. Nor do we know when he began to write. It seems likely (though not certain) that he became an actor before starting to write plays; at any rate, none of his extant writings certainly dates from his youth or early manhood. One of his less impressive sonnets—No. 145—apparently plays on the name 'Hathaway' (' "I hate" from hate away she threw'), and may be an early love poem; but this is his only surviving non-dramatic work that seems at all likely to have been written before he became a playwright. Possibly his earliest efforts in verse or drama are lost; just possibly some of them survive anonymously. It would have been very much in keeping with contemporary practice if he had worked in collaboration with other writers at this stage in his career. *1 Henry VI* is the only early play that we feel confident enough to identify as collaborative, but other writers' hands have also been plausibly suspected in *The First Part of the Contention* (*2 Henry VI*), *Richard, Duke of York* (*3 Henry VI*), and the opening scenes, in particular, of *Titus Andronicus*.

The first printed allusion to Shakespeare dates from 1592, in the pamphlet *Greene's Groatsworth of Wit*, published as the work of Robert Greene, writer of plays and prose romances, shortly after he died. Mention of an 'upstart crow' who 'supposes he is as well able to bombast out a blank verse as the best of you' and who 'is in his own conceit the only Shake-scene in a country' suggests rivalry; though parody of a line from *Richard, Duke of York* (*3 Henry VI*) shows that Shakespeare was already known on the London literary scene, the word 'upstart' does not suggest a long-established author.

It seems likely that Shakespeare's earliest surviving plays date from the late 1580s and the early 1590s: they include comedies (*The Two Gentlemen of Verona* and *The Taming of the Shrew*), history plays based on English chronicles (*The First Part of the Contention, Richard, Duke of York*), and a pseudo-classical tragedy (*Titus Andronicus*). We cannot say with any confidence which company (or companies) of players these were written for;

3. Henry Wriothesley, 3rd Earl of Southampton (1573-1624), at the age of twenty: a miniature by Nicholas Hilliard

*Titus Andronicus*, at least, seems to have gone from one company to another, since according to the title-page of the 1594 edition it had been acted by the Earl of Derby's, the Earl of Pembroke's, and the Earl of Sussex's Men. Early in his career, Shakespeare may have worked for more than one company. A watershed in his career was the devastating outbreak of plague which closed London's theatres almost entirely from June 1592 to May 1594. This seems to have turned Shakespeare's thoughts to the possibility of a literary career away from the theatre: in spring 1593 appeared his witty narrative poem *Venus and Adonis*, to be followed in 1594 by its tragic counterpart, *The Rape of Lucrece*. Both carry dedications over Shakespeare's name to Henry Wriothesley, third Earl of Southampton, who, though aged only twenty in 1593, was already making a name for himself as a patron of poets. Patrons could be important to Elizabethan writers; how Southampton rewarded Shakespeare for his dedications we do not know, but the affection with which Shakespeare speaks of him in the dedication to *Lucrece* suggests a strong personal connection and has encouraged the belief that Southampton may be the young man addressed so lovingly in Shakespeare's Sonnets.

Whether Shakespeare began to write the Sonnets at this time is a vexed question. Certainly it is the period at which his plays make most use of the formal characteristics of the sonnet: *Love's Labour's Lost* and *Romeo and Juliet*, for example, both incorporate sonnets into their structure; but *Henry V*, probably dating from 1599, has a sonnet as an Epilogue, and in *All's Well That Er. Well* (c.1604) a letter is cast in this form. Allusions within the

Sonnets suggest that they were written over a period of at least three years. At some later point they seem to have been rearranged into the order in which they were printed. Behind them—if indeed they are autobiographical at all—lies a tantalizingly elusive story of Shakespeare's personal life. Many attempts have been made to identify the poet's friend, the rival poet, and the dark woman who is both the poet's mistress and the seducer of his friend; none has achieved any degree of certainty.

After the epidemic of plague dwindled, a number of actors who had previously belonged to different companies amalgamated to form the Lord Chamberlain's Men. In the first official account that survives, Shakespeare is named, along with the famous comic actor Will Kemp and the tragedian Richard Burbage, as payee for performances at court during the previous Christmas season. The Chamberlain's Men rapidly became the leading dramatic company, though rivalled at first by the Admiral's Men, who had Edward Alleyn as their leading tragedian. Shakespeare stayed with the Chamberlain's (later King's) Men for the rest of his career as actor, playwright, and administrator. He is the only prominent playwright of his time to have had so stable a relationship with a single company.

With the founding of the Lord Chamberlain's Men, Shakespeare's career was placed upon a firm footing. It is not the purpose of this Introduction to describe his development as a dramatist, or to attempt a thorough discussion of the chronology of his writings. The Introductions to individual works state briefly what is known about when they were composed, and also name the principal literary sources on which Shakespeare drew in composing them. The works themselves are arranged in a conjectured order of composition. There are many uncertainties about this, especially in relation to the early plays. The most important single piece of evidence is a passage in a book called *Palladis Tamia: Wit's Treasury*, by a minor writer, Francis Meres, published in 1598. Meres wrote:

As Plautus and Seneca are accounted the best for comedy and tragedy among the Latins, so Shakespeare among the English is the most excellent in both kinds for the stage; for comedy, witness his *Gentlemen of Verona*, his *Errors*, his *Love Labour's Lost*, his *Love Labour's Won*, his *Midsummer's Night Dream*, and his *Merchant of Venice*; for tragedy, his *Richard II*, *Richard III*, *Henry IV*, *King John*, *Titus Andronicus*, and his *Romeo and Juliet*.

Some of the plays that Meres names had already been published or alluded to by 1598; but for others, he supplies a date by which they must have been written. Meres also alludes to Shakespeare's 'sugared sonnets among his private friends', which suggests that some, if not all, of the poems printed in 1609 as Shakespeare's Sonnets were circulating in manuscript by this date. Works not mentioned by Meres that are believed to have been written by 1598 are the three plays concerned with the reign of Henry VI, *The Taming of the Shrew*, and the narrative poems.

Shakespeare seems to have had less success as an actor than as a playwright. We cannot name for certain any of the parts that he played, though seventeenth-century traditions have it that he played Adam in *As You Like It*, and Hamlet's Ghost—and more generally that he had a penchant for 'kingly parts'. Ben Jonson listed him first among the 'principal comedians' in *Every Man in his Humour*, acted in 1598, when he reprinted it in the 1616 Folio, and Shakespeare is also listed among the performers of Jonson's tragedy *Sejanus* in 1603. He was certainly one of the leading administrators of the Chamberlain's Men. Until 1597, when their lease expired, they played mainly in the Theatre, London's first important playhouse, situated north of the River Thames in Shoreditch, outside the jurisdiction of the City fathers, who exercised a repressive influence on the drama. It had been built in 1576 by James Burbage, a joiner, the tragedian's father. Then they seem to have played mainly at the Curtain until some time in 1599. Shakespeare was a member of the syndicate

4. King James I
(1566–1625): a portrait
(1621) by Daniel Mytens

responsible for building the first Globe theatre, in Southwark, on the south bank of the
Thames, out of the dismantled timbers of the Theatre in 1599. Initially he had a ten-
per-cent financial interest in the enterprise, fluctuating as other shareholders joined or
withdrew. It was a valuable share, for the Chamberlain's Men won great acclaim and made
substantial profits. After Queen Elizabeth died, in 1603, they came under the patronage of
the new king, James I; the royal patent of 19 May 1603 names Shakespeare along with
other leaders of the company. London was in the grip of another severe epidemic of plague
which caused a ban on playing till the following spring. The King's processional entry into
London had to be delayed; when at last it took place, on 15 March 1604, each of the
company's leaders was granted four and a half yards of scarlet cloth for his livery as one
of the King's retainers; but the players seem not to have processed. Their association
with the King was far from nominal; during the next thirteen years—up to the time of
Shakespeare's death—they played at court more often than all the other theatre companies
combined. Records are patchy, but we know, for instance, that they gave eleven plays at

court between 1 November 1604 and 31 October 1605, and that seven of them were by Shakespeare: they included older plays—*The Comedy of Errors, Love's Labour's Lost*—and more recent ones—*Othello* and *Measure for Measure*. *The Merchant of Venice* was played twice.

Some measure of Shakespeare's personal success during this period may be gained from the ascription to him of works not now believed to be his; *Locrine* and *Thomas Lord Cromwell* were published in 1595 and 1602 respectively as by 'W.S.'; in 1599 a collection of poems, *The Passionate Pilgrim*, containing some poems certainly by other writers, appeared under his name; so, in 1606 and 1608, did *The London Prodigal* and *A Yorkshire Tragedy*. Since Shakespeare's time, too, many plays of the period, some published, some surviving only in manuscript, have been attributed to him. In modern times, the most plausible case has been made for parts, or all, of *Edward III*, which was entered in the registers of the Stationers' Company (a normal, but not invariable, way of setting in motion the publication process) in 1595 and published in 1596. It was first ascribed to Shakespeare in 1656. Certainly it displays links with some of his writings, but authorship problems are particularly acute during the part of his career when this play seems to have been written, and we cannot feel confident of the attribution.

In August 1608 the King's Men took up the lease of the smaller, 'private' indoor theatre, the Blackfriars; again, Shakespeare was one of the syndicate of owners. The company took possession in 1609; the Blackfriars served as a winter home; in better weather, performances continued to be given at the Globe. By now, Shakespeare was at a late stage in his career. Perhaps he realized it; he seems to have been willing to share his responsibilities as the company's resident dramatist with younger writers. *Timon of Athens*, tentatively dated around 1604–5, seems on internal evidence to be partly the work of Thomas Middleton (*c*.1570–1627). Another collaborative play, very successful in its time, was *Pericles* (*c*.1608), in which Shakespeare probably worked with George Wilkins, an unscrupulous character who gave up his brief career as a writer in favour of a longer one as a tavern (or brothel) keeper. But Shakespeare's most fruitful collaboration was with John Fletcher, his junior by fifteen years. Fletcher was collaborating with Francis Beaumont on plays for the King's Men by about 1608. Beaumont stopped writing plays when he married, in about 1613, and it is at this time that Fletcher seems to have collaborated with Shakespeare. A lost play, *Cardenio*, acted by the King's Men some time before 20 May 1613, was plausibly ascribed to Shakespeare and Fletcher in a document of 1653; *All is True* (*Henry VIII*), first acted about June 1613, is generally agreed on stylistic evidence to be another fruit of the same partnership; and *The Two Noble Kinsmen*, also dated 1613, which seems to be the last play in which Shakespeare had a hand, was ascribed to the pair on its publication in 1634. One of Shakespeare's last professional tasks seems to have been the minor one of devising an *impresa* for the Earl of Rutland to bear at a tournament held on 24 March 1613 to celebrate the tenth anniversary of the King's accession. An *impresa* was a paper or pasteboard shield painted with an emblematic device and motto which would be carried and interpreted for a knight by his squire; such a ceremony is portrayed in *Pericles* (Sc. 6). Shakespeare received forty-four shillings for his share in the work; Richard Burbage was paid the same sum 'for painting and making it'.

## The Drama and Theatre of Shakespeare's Time

Shakespeare came upon the theatrical scene at an auspicious time. English drama and theatre had developed only slowly during the earlier part of the sixteenth century; during Shakespeare's youth they exploded into vigorous life. It was a period of secularization; previously, drama had been largely religious in subject matter and overtly didactic in treatment; as a boy of fifteen, Shakespeare could have seen one of the last performances of a great cycle of plays on religious themes at Coventry, not far from his home town. 1576 saw the building in London of the Theatre, to be rapidly followed by the Curtain: England's first important, custom-built playhouses. There was a sudden spurt in the development of all aspects of theatrical art: acting, production, playwriting, company organization, and administration. Within a few years the twin arts of drama and theatre entered upon a period of achievement whose brilliance remains unequalled.

The new drama was literary and rhetorical rather than scenic and spectacular: but its mainstream was theatrical too. Its writers were poets. Prose was only beginning to be used in plays during Shakespeare's youth; a playwright was often known as a 'poet', and most of the best playwrights of the period wrote with distinction in other forms. Shakespeare's most important predecessors and early contemporaries, from whom he learned much, were John Lyly (c.1554-1606), pre-eminent for courtly comedy and elegant prose, Robert Greene (1558-92), who helped particularly to develop the scope and language of romantic comedy, and Christopher Marlowe (1564-93), whose 'mighty line' put heroism excitingly on the stage and who shares with Shakespeare credit for establishing the English history play as a dramatic mode. As Shakespeare's career progressed, other dramatists displayed their talents and, doubtless, influenced and stimulated him. George Chapman (c.1560-1634) emerged as a dramatist in the mid-1590s and succeeded in both comedy and tragedy. He was deeply interested in classical themes, as was Ben Jonson (1572-1637), who became Shakespeare's chief rival. Jonson was a dominating personality, vocal about his accomplishments (and about Shakespeare, who, he said, 'wanted art'), and biting as a comic satirist. Thomas Dekker (c.1572-1632) wrote comedies that are more akin to Shakespeare's in their romantic warmth; the satirical plays of John Marston (c.1575-1634) are more sensational and cynical than Jonson's. Thomas Middleton brought a sharp wit to the portrayal of contemporary London life, and developed into a great tragic dramatist. Towards the end of Shakespeare's career, Francis Beaumont (1584-1616) and John Fletcher (1579-1625) came upon the scene; the affinity between Shakespeare's late tragicomedies and some of Fletcher's romances is reflected in their collaboration.

The companies for which these dramatists wrote were organized mainly from within. They were led by the sharers: eight in the Lord Chamberlain's Men at first, twelve by the end of Shakespeare's career. Collectively they owned the joint stock of play scripts, costumes, and properties; they shared both expenses and profits. All were working members of the company. Exceptionally, the sharers of Shakespeare's company owned the Globe theatre itself; more commonly, actors rented theatres from financial speculators such as Philip Henslowe, financier of the Admiral's Men. Subordinate to the sharers were the 'hired men'—lesser actors along with prompters ('bookholders'), stagekeepers, wardrobe keepers ('tiremen'), musicians, and money-collectors ('gatherers'). Even those not employed principally as actors might swell a scene at need. The hired men were paid by the week. Companies would need scribes to copy out actors' parts and to make fair copies from the playwrights' foul papers (working manuscripts), but they seem mainly to have been employed part-time. The other important group of company members are the apprentices. These were boys or youths each serving a formal term of apprenticeship to one of the sharers. They played female and juvenile roles.

The success of plays in the Elizabethan theatres depended almost entirely on the actors. They had to be talented, hard-working, and versatile. Plays were given in a repertory system on almost every afternoon of the week except during Lent. Only about two weeks could be allowed for rehearsal of a new play, and during that time the company would be regularly performing a variety of plays. Lacking printed copies, the actors worked from

5. Richard Burbage: reputedly a self-portrait

'parts' written out on scrolls giving only the cue lines from other characters' speeches. The bookholder, or prompter, had to make sure that actors entered at the right moment, properly equipped. Many of them would take several parts in the same play: doubling was a necessary practice. The strain on the memory was great, demanding a high degree of professionalism. Conditions of employment were carefully regulated: a contract of 1614 provides that an actor and sharer, Robert Dawes (not in Shakespeare's company), be fined one shilling for failure to turn up at the beginning of a rehearsal, two shillings for missing a rehearsal altogether, three shillings if he was not 'ready apparelled' for a performance, ten shillings if four other members of the company considered him to be 'overcome with drink' at the time he should be acting, and one pound if he simply failed to turn up for a performance without 'licence or just excuse of sickness'.

There can be no doubt that the best actors of Shakespeare's time would have been greatly admired in any age. English actors became famous abroad; some of the best surviving accounts are in letters written by visitors to England: the actors were literally 'something to write home about', and some of them performed (in English) on the Continent. Edward Alleyn, the leading tragedian of the Admiral's Men, renowned especially for his performances of Marlowe's heroes, made a fortune and founded Dulwich College. All too little is known about the actors of Shakespeare's company and the roles they played, but many testimonies survive to Richard Burbage's excellence in tragic roles. According to an elegy written after he died, in 1619,

> No more young Hamlet, old Hieronimo;
> Kind Lear, the grievèd Moor, and more beside
> That lived in him have now for ever died.

There is no reason to suppose that the boy actors lacked talent and skill; they were highly trained as apprentices to leading actors. Most plays of the period, including Shakespeare's,

6. The hall of the Middle Temple, London

have far fewer female than male roles, but some women's parts—such as Rosalind (in *As You Like It*) and Cleopatra—are long and important; Shakespeare must have had confidence in the boys who played them. Some of them later became sharers themselves.

The playwriting techniques of Shakespeare and his contemporaries were intimately bound up with the theatrical conditions to which they catered. Theatre buildings were virtually confined to London. Plays continued to be given in improvised circumstances when the companies toured the provinces and when they acted at court (that is, wherever the sovereign and his or her entourage happened to be—in London, usually Whitehall or Greenwich). In 1602, *Twelfth Night* was given in the still-surviving hall of one of London's Inns of Court, the Middle Temple. Acting companies could use guildhalls, the halls of great houses or of Oxford and Cambridge colleges, the yards of inns. (In 1608, *Richard II* and *Hamlet* were even performed by ships' crews at sea off the coast of Sierra Leone.) Many plays of the period require no more than an open space and the costumes and properties that the actors carried with them on their travels. Others made more use of the expanding facilities of the professional stage.

Permanent theatres were of two kinds, known now as public and private. Most important to Shakespeare were public theatres such as the Theatre, the Curtain, and the Globe. Unfortunately, the only surviving drawing (reproduced overleaf) portraying the interior of a public theatre in any detail is of the Swan, not used by Shakespeare's company. Though theatres were not uniform in design, they had important features in common. They were large wooden buildings, usually round or polygonal; the Globe, which was about 100 feet in diameter and 36 feet in height, could hold over three thousand spectators. Between the outer and inner walls—a space of about 12 feet—were three levels of tiered benches extending round most of the auditorium and roofed on top; after the Globe burnt down, in 1613, the roof, formerly thatched, was tiled. The surround of benches was broken on the lowest level by the stage, broad and deep, edged with palings, which jutted forth at a height of about 5 feet into the central yard, where spectators ('groundlings') could stand. Actors entered mainly, perhaps entirely, from openings in the wall at the back of the stage. At least two doors, one on each side, could be used; stage directions frequently call for characters to enter simultaneously from different doors, when the dramatic situation requires them to be meeting, and to leave 'severally' (separately) when they are parting. The depth of the stage meant that characters could enter through the stage doors some moments before other characters standing at the front of the stage might be expected to notice them.

Also in the wall at the rear of the stage there appears to have been some kind of central aperture which could be used for the disclosing and putting forth of Desdemona's bed (*Othello*, 5.2) or the concealment of the spying Polonius (*Hamlet*, 3.4) or of the sleeping Lear (*The History of King Lear*, Sc. 20). Behind the stage wall was the tiring-house—the actors' dressing area.

On the second level the seating facilities for spectators seem to have extended even to the back of the stage, forming a balcony which at the Globe was probably divided into five bays. Here was the 'lords' room', which could be taken over by the actors for plays in which action took place 'above' (or 'aloft'), as in Romeo's wooing of Juliet or the death of Mark Antony. It seems to have been possible for actors to move from the main stage to the upper level during the time taken to speak a few lines of verse, as we may see in *The Merchant of Venice* (2.6.51–70) or *Julius Caesar* (5.3.33–5). Somewhere above the lords' room was a window or platform known as 'the top'; Joan la Pucelle appears there briefly in *1 Henry VI* (3.3), and in *The Tempest*, Prospero is seen 'on the top, invisible' (3.3).

Above the stage, at a level higher than the second gallery, was a canopy, probably

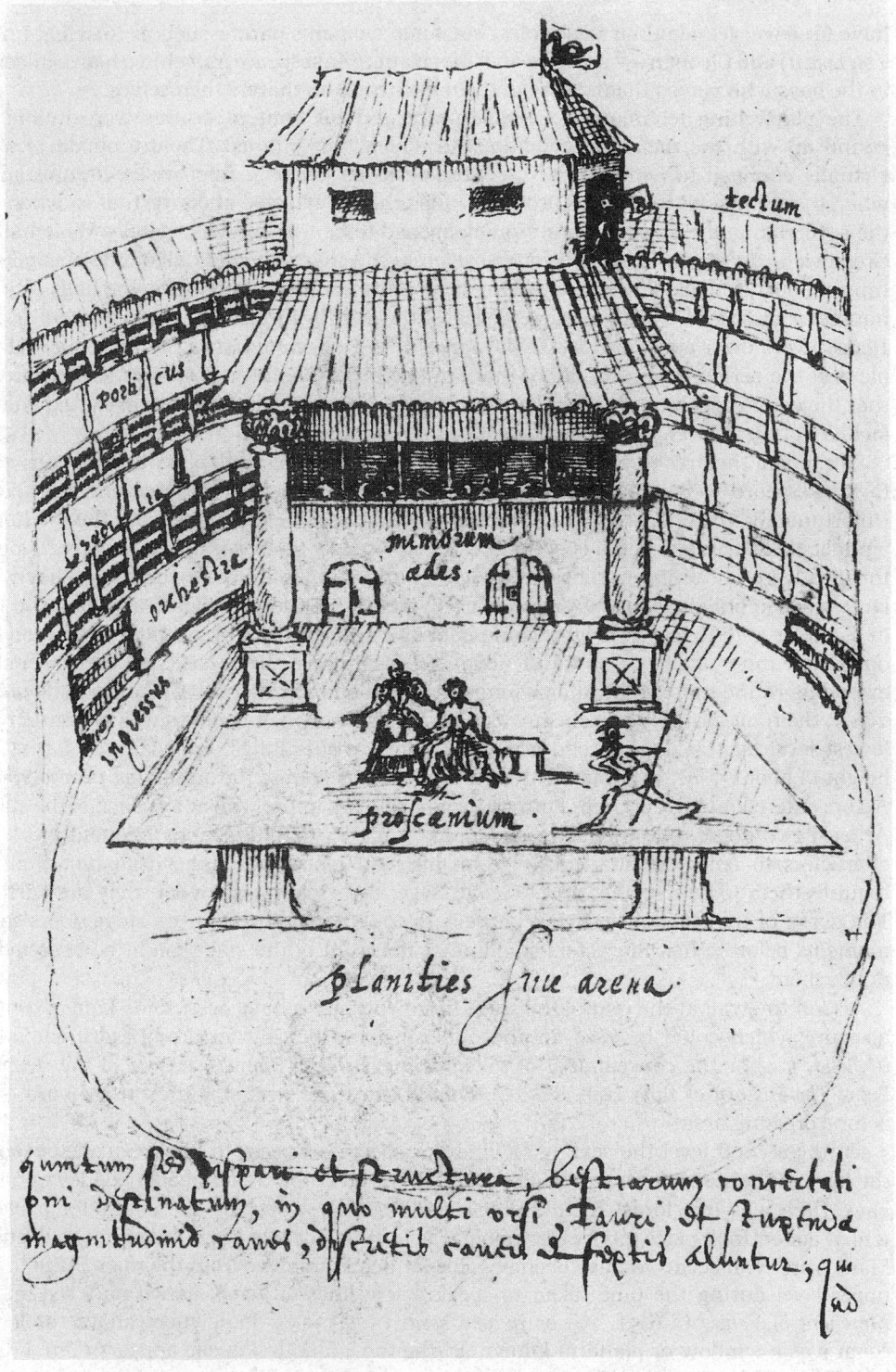

tectum

porticus

orchestra

ingressus

mimorum
ædes

proscænium

planities siue arena

quintum sed dispari et structura, bestiarum concertati
oni destinatum, in quo multi vrsi, tauri, et stupenda
magnitudinis canes, distinctis caueis et septis aluntur, qui
ad

7. The Swan Theatre: a copy, by Aernout van Buchel, of a drawing made about 1596 by
Johannes de Witt, a Dutch visitor to London

supported by two pillars (which could themselves be used as hiding places) rising from the stage. One function of the canopy was to shelter the stage from the weather; it also formed the floor of one or more huts housing the machinery for special effects and its operators. Here cannon-balls could be rolled around a trough to imitate the sound of thunder, and fire crackers could be set off to simulate lightning. And from this area actors could descend in a chair operated by a winch. Shakespeare uses this facility mainly in his late plays: in *Cymbeline* for the descent of Jupiter (5.4), and, probably, in *Pericles* for the descent of Diana (Sc. 21) and in *The Tempest* for Juno's appearance in the masque (4.1). On the stage itself was a trap which could be opened to serve as Ophelia's grave (*Hamlet*, 5.1) or as Malvolio's dungeon (*Twelfth Night*, 4.2).

Somewhere in the backstage area, perhaps in or close to the gallery, must have been a space for the musicians who played a prominent part in many performances. No doubt then, as now, a single musician was capable of playing several instruments. Stringed instruments, plucked (such as the lute) and bowed (such as viols), were needed. Woodwind instruments included recorders (called for in *Hamlet* 3.2) and the stronger, shriller hautboys (ancestors of the modern oboe); trumpets and cornetts were needed for the many flourishes and sennets (more elaborate fanfares) played especially for the comings and goings of royal characters. Sometimes musicians would play on stage: entrances for trumpeters and drummers are common in battle scenes. More often they would be heard but not seen; from behind the stage (as, perhaps, at the opening of *Twelfth Night* or in the concluding dance of *Much Ado About Nothing*), or even occasionally under it (*Antony and Cleopatra*, 4.3). Some actors were themselves musicians: the performers of Feste (in *Twelfth Night*) and Ariel (in *The Tempest*) must sing and, probably, accompany themselves on lute and tabor (a small drum slung around the neck). Though traditional music has survived for some of the songs in Shakespeare's plays (such as Ophelia's mad songs, in *Hamlet*), we have little music which was certainly composed for them in his own time. The principal exception is two songs for *The Tempest* by Robert Johnson, a fine composer who was attached to the King's Men.

Shakespeare's plays require few substantial properties. A 'state', or throne on a dais, is sometimes called for, as are tables and chairs and, occasionally, a bed, a pair of stocks (*King Lear* Sc. 7/2.2), a cauldron (*Macbeth*, 4.1), a rose brier (*1 Henry VI*, 2.4), and a bush (*Two Noble Kinsmen*, 5.3). No doubt these and other such objects were pushed on and off the stage by attendants in full view of the audience. We know that Elizabethan companies spent lavishly on costumes, and some plays require special clothes; at the beginning of *2 Henry IV*, Rumour enters 'painted full of tongues'; regal personages, and supernatural figures such as Hymen in *As You Like It* (5.4) and the goddesses in *The Tempest* (4.1), must have been distinctively costumed; presumably a bear-skin was needed for *The Winter's Tale* (3.3). Probably no serious attempt was made at historical realism. The only surviving contemporary drawing of a scene from a Shakespeare play, illustrating *Titus Andronicus* (reproduced on the following page), shows the characters dressed in a mixture of Elizabethan and classical costumes, and this accords with the often anachronistic references to clothing in plays with a historical setting. The same drawing also illustrates the use of head-dresses, of varied weapons as properties—the guard to the left appears to be wearing a scimitar—of facial and bodily make-up for Aaron, the Moor, and of eloquent gestures. Extended passages of wordless action are not uncommon in Shakespeare's plays. Dumb shows feature prominently in earlier Elizabethan plays, and in Shakespeare the direction 'alarum' or 'alarums and excursions' may stand for lengthy and exciting passages of arms. Even in one of Shakespeare's latest plays, *Cymbeline*, important episodes are conducted in wordless mime (see, for example, 5.2-4).

8. A drawing, attributed to Henry Peacham, illustrating Shakespeare's *Titus Andronicus*

Towards the end of Shakespeare's career his company regularly performed in a private theatre, the Blackfriars, as well as at the Globe. Like other private theatres, this was an enclosed building, using artificial lighting, and so more suitable for winter performances. Private playhouses were smaller than the public ones—the Blackfriars held about 600 spectators—and admission prices were much higher—a minimum of sixpence at the Blackfriars against one penny at the Globe. Facilities at the Blackfriars must have been essentially similar to those at the Globe since some of the same plays were given at both theatres. But the sense of social occasion seems to have been different. Audiences were more elegant (though not necessarily better behaved); music featured more prominently.

It seems to have been under the influence of private-theatre practice that, from about 1609 onwards, performances of plays customarily marked the conventional five-act structure by a pause, graced with music, after each act. Previously, though dramatists often showed awareness of five-act structure (as Shakespeare conspicuously does in *Henry V*, with a Chorus before each act), public performances seem to have been continuous, making the scene the main structural unit. None of the editions of Shakespeare's plays printed in his lifetime (which do not include any written after 1609) marks either act or scene divisions. The innovation of act-pauses threw more emphasis on the act as a unit, and made it possible for dramatists to relax their observance of what has come to be known as 'the law of re-entry', according to which a character who had left the stage at the end of one scene would not normally make an immediate reappearance at the beginning of the next. Thus, if Shakespeare had been writing *The Tempest* before 1609, it is unlikely that Prospero and Ariel, having left the stage at the end of Act 4, would have instantly reappeared at the start of Act 5. We attempt to reflect this feature of Shakespeare's dramaturgy by making no special distinction between scene-breaks and act-breaks except in those later plays in which Shakespeare seems to have observed the new convention (and in *Titus Andronicus*, *Measure for Measure*, and *Macbeth*, since the texts of these plays apparently reflect theatre practice after they were first written, and in *The Comedy of Errors*, a neo-classically structured play in which the act-divisions appear to be authoritative, and to represent a private performance).

Dramatic conventions changed and developed considerably during Shakespeare's career.

Throughout it, they favoured self-evident artifice over naturalism. This is apparent in Shakespeare's dramatic language, with its soliloquies (sometimes addressed directly to the audience), its long, carefully structured speeches, its elaborate use of simile, metaphor, and rhetorical figures of speech (in prose as well as verse), its rhyme, and its patterned dialogue. It is evident in some aspects of behaviour and characterization: Oberon and Prospero have only to declare themselves invisible to become so; disguises can be instantly donned with an appearance of impenetrability, and as rapidly abandoned; some characters—Rumour at the opening of *2 Henry IV*, Time in *The Winter's Tale*, even the Gardeners in *Richard II*—clearly serve a symbolic rather than a realistic function: and supernatural manifestations are common. The calculated positioning of characters on the stage may help to make a dramatic point, as in the scene in *The Two Gentlemen of Verona* (4.2) in which the disguised Julia overhears her faithless lover's serenade to her rival, Silvia; or, more complexly, that in *Troilus and Cressida* (5.2) in which Troilus and Ulysses observe Diomede's courtship of Cressida while they are themselves observed by the cynical Thersites. Not uncommonly, Shakespeare provokes his spectators into consciousness that they are watching a play, as when Cassius, in *Julius Caesar*, looks forward to the time when the conspirators' 'lofty scene' will be 'acted over | In states unborn and accents yet unknown' (3.1.112-14); or, in *Troilus and Cressida*, when Troilus, Cressida, and Pandarus reach out from the past tense of history to the present tense of theatrical performance in a ritualistic anticipation of what their names will come to signify (3.3.169-202).

Techniques such as these are closely related to the non-illusionistic nature of the Elizabethan stage, in which the mechanics of production were frequently visible. Many scenes take place nowhere in particular. Awareness of place was conveyed through dialogue and action rather than through scenery; location could change even within a scene (as, for example, in *2 Henry IV*, where movement of the dying King's bed across the stage establishes the scene as 'some other chamber': 4.3.132). Sometimes Shakespeare uses conflicting reactions to an imagined place as a kind of shorthand guide to character: to the idealistic Gonzalo, Prospero's island is lush, lusty, and green; to the cynical Antonio, 'tawny' (*The Tempest*, 2.1.56-8): such an effect would have been dulled by scenery which proved one or the other right.

In some ways, the changes in Shakespeare's practice as his art develops favour naturalism: his verse becomes freer, metaphor predominates over simile, rhyme and other formalistic elements are reduced, the proportion of prose over verse increases to the middle of his career (but then decreases again), some of his most psychologically complex character portrayals—Coriolanus, Cleopatra—come late. But his drama remains rooted in the conventions of a rhetorical, non-scenic (though not unspectacular) theatre: the supernatural looms largest in his later plays—*Macbeth*, *Pericles*, *Cymbeline*, and *The Tempest*. *The Tempest* draws no less self-consciously on the neo-classical conventions of five-act structure than *The Comedy of Errors*, and Prospero's narration to Miranda (1.2) is as blatant a piece of dramatic exposition as Egeon's tale in the opening scene of the earlier play. Heroines of the late romances—Marina (in *Pericles*), Perdita (in *The Winter's Tale*), and Miranda (in *The Tempest*)—are portrayed with less concern for psychological realism than those of the romantic comedies—Viola (in *Twelfth Night*) and Rosalind (in *As You Like It*)—and the revelation to Leontes at the end of *The Winter's Tale* is both more improbable and more moving than the similar revelation made to Egeon at the end of *The Comedy of Errors*.

The theatre of Shakespeare's time was his most valuable collaborator. Its simplicity was one of its strengths. The actors of his company were the best in their kind. His audiences may not have been learned, or sophisticated, by modern standards; according to some accounts, they could be unruly; but they conferred popularity upon plays which for emotional power, range, and variety, for grandeur of conception and subtlety of execution, are among the most demanding, as well as the most entertaining, ever written. If we value Shakespeare's plays, we must also think well of the theatrical circumstances that permitted, and encouraged, his genius to flourish.

### The Early Printing of Shakespeare's Plays

For all its literary distinction, drama in Shakespeare's time was an art of performance; many plays of the period never got into print: they were published by being acted. It is lucky for us that, so far as we know, all Shakespeare's finished plays except the collaborative *Cardenio* reached print. None of his plays that were printed in his time survives in even a fragment of his own handwriting; the only literary manuscript plausibly ascribed to him is a section of *Sir Thomas More*, a play not printed until the nineteenth century. The only works of Shakespeare that he himself seems to have cared about putting into print are the narrative poems *Venus and Adonis* and *The Rape of Lucrece*. A major reason for this is Shakespeare's exceptionally close involvement with the acting company for which he wrote. There was no effective dramatic copyright; acting companies commonly bought plays from their authors—as a resident playwright, Shakespeare was probably expected to write about two each year—and it was in the companies' interests that their plays should not get into print, when they could be acted by rival companies. Nevertheless, by one means and another, and in one form and another, about half of Shakespeare's plays were printed singly in his lifetime, almost all of them in the flimsy, paperback format of a quarto—a book made from sheets of paper that had been folded twice, and normally costing sixpence. Some of the plays were pirated: printed, that is, in unauthorized editions, from texts that seem to have been put together from memory by actors or even, perhaps, by spectators, perhaps primarily to create scripts for other companies, perhaps purely for publication. These are the so-called 'bad' quartos: bad not because they were, necessarily, badly printed, but because they did not descend in a direct line of written transmission from their author's manuscript. The reported texts of *The First Part of the Contention* and *Richard, Duke of York* (usually known by the titles under which they were printed in the First Folio—*2 Henry VI* and *3 Henry VI*) appeared in 1594 and 1595 respectively; they seem to have been made on the basis of London performances so that the plays could be acted by a company other than the one for which they were written. Also in 1594 appeared *The Taming of A Shrew*, perhaps better described as an imitation of Shakespeare's *The Taming of the Shrew* (the titles may have been regarded as interchangeable) than as a detailed reconstruction of it. The 1597 *Richard III* is perhaps the best of the reported texts; it seems to have been assembled from the company's collective memory, perhaps because they did not have access to the official prompt-book. The text of *Romeo and Juliet* printed in the same year seems to have been put together by a few actors exploiting a popular success. The 1600 quarto of *Henry V* probably presents a text made for a smaller company of actors than that for which it had been written. *The Merry Wives of Windsor* of 1602 seems to derive largely from the memory of the actor who played the Host of the Garter Inn— perhaps a hired man no longer employed by Shakespeare's company. Worst reported of all is the 1603 *Hamlet*, which also appears to derive from the memory of one or more actors in minor roles. Last printed of the 'bad' quartos is *Pericles*, of 1609.

The reported texts have many faults. Frequently they garble the verse and prose of the original—'To be or not to be; ay, there's the point', says the 1603 *Hamlet*; usually they abbreviate—the 1603 *Hamlet* has about 2,200 lines, compared to the 3,800 of the good quarto; sometimes they include lines from plays by other authors (especially Marlowe);

> *Ham.* To be, or not to be, I there's the point,
> To Die, to sleepe, is that all? I all:
> No, to sleepe, to dreame, I mary there it goes,
> For in that dreame of death, when wee awake,
> And borne before an euerlasting Iudge,
> From whence no passenger euer retur'nd,
> The vndiscouered country, at whose sight
> The happy smile, and the accursed damn'd.
> But for this, the ioyfull hope of this,
> Whol'd beare the scornes and flattery of the world,
> Scorned by the right rich, the rich curssed of the poore?
> The widow being opprested the orphan wrong'd,
> The taste of hunger, or a tirants raigne,
> And thousand more calamities besides,
> To grunt and sweate vnder this weary life,
> When that he may his full *Quietus* make,
> With a bare bodkin, who would this indure,
> But for a hope of something after death?
> Which puscles the braine, and doth confound the sence,
> Which makes vs rather beare those euilles we haue,
> Than flie to others that we know not of.
> I that, O this conscience makes cowardes of vs all,
> Lady in thy orizons, be all my sinnes remembred.

9. 'To be or not to be' as it appeared in the 'bad' quarto of 1603

sometimes they include passages clearly cobbled together to supply gaps in the reporter's memory. For all this, they are not without value in helping us to judge how Shakespeare's plays were originally performed. Their stage directions may give us more information about how the plays were staged than is available in other texts: for instance, the reported text of *Hamlet* has the direction '*Enter Ofelia playing on a lute, and her hair down, singing*'—far more vivid than the good Quarto's '*Enter Ophelia*', or even the Folio's '*Enter Ophelia distracted*'. Because these are post-performance texts, they may preserve, in the midst of corruption, authentically Shakespearian changes made to the play after it was first written and not recorded elsewhere. A particularly interesting case is *The Taming of the Shrew*: the play as printed in the Folio, in what is clearly, in general, the more authentic text, abandons early in its action the framework device which makes the story of Katherine and Petruccio a play within the play; the quarto continues this framework through the play, and provides an amusing little episode rounding it off. These passages may derive from ones written by Shakespeare but not printed in the Folio: we print them as Additional Passages at the end of the play. In general, we draw more liberally than most previous editors on the reported texts, in the belief that they can help us to come closer than before to the plays as they were acted by Shakespeare's company as well as by others.

Although in general it was to the advantage of the companies that owned play scripts not to allow them to be printed, some of Shakespeare's plays were printed from authentic manuscripts during his lifetime, and even while they were still being performed by his company; these are the 'good' quartos. First came *Titus Andronicus*, printed in 1594 from Shakespeare's own papers, probably because the company for which he wrote it had been disbanded. In 1597 *Richard II* was printed from Shakespeare's manuscript, minus the politically sensitive episode (4.1) in which Richard gives up his crown to Bolingbroke: a clear instance of censorship, whether self-imposed or not. The first play to be published in

Shakespeare's name is *Love's Labour's Lost*, in 1598. Several other quartos printed from good manuscripts appeared around the same time: *1 Henry IV* (probably from a scribal transcript) in 1598, and *A Midsummer Night's Dream*, *The Merchant of Venice*, *2 Henry IV*, and *Much Ado About Nothing* (all from Shakespeare's papers) in 1600. In 1604 appeared a text of *Hamlet* printed from Shakespeare's own papers and declaring itself to be 'Newly imprinted and enlarged to almost as much again as it was, according to the true and perfect copy': surely an attempt to replace a bad text by a good one. *King Lear* followed, in 1608, in a badly printed quarto whose status has been much disputed, but which we believe to derive from Shakespeare's own manuscript. In 1609 came *Troilus and Cressida*, probably from Shakespeare's own papers, in an edition which in the first-printed copies claims to present the play 'as it was acted by the King's majesty's servants at the Globe', but in later-printed ones declares that it has never been 'staled with the stage'. The only new play to appear between Shakespeare's death and the publication of the Folio in 1623 was *Othello*, printed in 1622 apparently from a transcript of Shakespeare's own papers.

It is clear that publishers found Shakespeare a valuable property—some of these quartos were several times reprinted—and easy to understand why some of his plays were pirated. It is less easy to see why Shakespeare and his colleagues released reliable texts of some plays for publication but not others. As a shareholder in the company to which the plays belonged, Shakespeare himself must have been a partner in its decisions, and it is difficult to believe that he was so lacking in personal vanity that he was happy to be represented in print by garbled texts; but he seems to have taken no interest in the progress of his plays through the press. Even some of those printed from authentic manuscripts—such as the 1604 *Hamlet*—are badly printed, and certainly not proof-read by the author; none of them bears an author's dedication or shows any sign of having been prepared for the press in the way that, for instance, Ben Jonson clearly prepared some of his plays. John Marston, introducing the printed text of his play *The Malcontent* in 1604, wrote: 'Only one thing afflicts me, to think that scenes invented merely to be spoken, should be enforcively published to be read'. Perhaps Shakespeare was similarly afflicted.

In 1616, the year of Shakespeare's death, Ben Jonson published his own collected plays in a handsome Folio. It was the first time that an English writer for the popular stage had been so honoured (or had so honoured himself), and it established a precedent by which Shakespeare's fellows could commemorate their colleague and friend. Principal responsibility for this ambitious enterprise was undertaken by John Heminges and Henry Condell, both long-established actors with Shakespeare's company; latterly, Heminges had been its business manager. They, along with Richard Burbage, had been the colleagues whom Shakespeare remembered in his will: he left each of them 26s. 8d. to buy a mourning ring. Although the Folio did not appear until 1623, they may have started work on it soon after Shakespeare died: big books take a long time to prepare. And they undertook their task with serious care. Most importantly, they printed eighteen plays that had not so far appeared in print, and which might otherwise have vanished. They omitted (so far as we can tell) only *Pericles*, *Cardenio* (now vanished), *The Two Noble Kinsmen*—perhaps because these three were collaborative—and the mysterious *Love's Labour's Won* (see p. 309). And they went to considerable pains to provide good texts. They had no previous experience as editors; they may have had help from others (including Ben Jonson, who wrote commendatory verses for the Folio): anyhow, although printers find it easier to set from print than from manuscript, they were not content simply to reprint quartos whenever they were available. In fact they seem to have made a conscious effort to identify and to avoid making use of the quartos now recognized as bad. In their introductory epistle addressed 'To the Great Variety of Readers', they declare that the public has been 'abused with divers stolen and surreptitious copies, maimed and deformed by the frauds and stealths of injurious impostors'. But now these plays are 'offered to your view cured, and perfect of their limbs; and all the rest absolute in their numbers, as he conceived them'.

None of the quartos believed by modern scholars to be unauthoritative was used unaltered as copy for the Folio. As men of the theatre, Heminges and Condell had access to prompt-books, and they made considerable use of them. For some plays, such as *Titus Andronicus* (which includes a whole scene not present in the quarto), and *A Midsummer Night's Dream*, the printers had a copy of a quarto (not necessarily the first) marked up with alterations made as the result of comparison with a theatre manuscript. For other plays (the first four to be printed in the Folio—*The Tempest*, *The Two Gentlemen of Verona*, *The Merry Wives of Windsor*, and *Measure for Measure*—along with *The Winter's Tale*) they employed a professional scribe, Ralph Crane, to transcribe papers in the theatre's possession. For others, such as *Henry V* and *All's Well That Ends Well*, they seem to have had authorial papers; and for yet others, such as *Macbeth*, a prompt-book. We cannot always be sure of the copy used by the printers, and sometimes it may have been mixed: for *Richard III* they seem to have used pages of the third quarto mixed with pages of the sixth quarto combined with passages in manuscript; a copy of the third quarto of *Richard II*, a copy of the fifth quarto, and a theatre manuscript all contributed to the Folio text of that play; the annotated third quarto of *Titus Andronicus* was supplemented by the 'fly' scene (3.2) which Shakespeare appears to have added after the play was first composed. Dedicating the Folio to the brother Earls of Pembroke and Montgomery, Heminges and Condell claimed that, in collecting Shakespeare's plays together, they had 'done an office to the dead, to procure his orphans guardians' (that is, to provide noble patrons for the works he had left behind), 'without ambition either of self-profit or fame, only to keep the memory of so worthy a friend and fellow alive as was our Shakespeare'. Certainly they deserve our gratitude.

## The Modern Editor's Task

It will be clear from all this that the documents from which even the authoritative early editions of Shakespeare's plays were printed were of a very variable nature. Some were his own papers in a rough state, including loose ends, duplications, inconsistencies, and vaguenesses. At the other extreme were prompt-books representing the play as close to the state in which it appeared in Shakespeare's theatre as we can get; and there were various intermediate states. For those plays of which we have only one text—those first printed in the Folio, along with *Pericles* and *The Two Noble Kinsmen*—the editor is at least not faced with the problem of alternative choices. The surviving text of *Macbeth* gives every sign of being an adaptation: if so, there is no means of recovering what Shakespeare originally wrote. The scribe seems to have entirely expunged Shakespeare's stage directions from *The Two Gentlemen of Verona*: we must make do with what we have. Other plays, however, confront the editor with a problem of choice. Pared down to its essentials, it is this: should he offer his readers a text which is as close as possible to what Shakespeare originally wrote, or should he aim to formulate a text presenting the play as it appeared when performed by the company of which Shakespeare was a principal shareholder in the theatres that he helped to control and on whose success his livelihood depended? The problem exists in two different forms. For some plays, the changes made in the more theatrical text (always the Folio, if we discount the bad quartos) are relatively minor, consisting perhaps in a few reallocations of dialogue, the addition of music cues to the stage directions, and perhaps some cuts. So it is with, for example, *A Midsummer Night's*

*Dream* and *Richard II*. More acute—and more critically exciting—are the problems raised when the more theatrical version appears to represent, not merely the text as originally written after it had been prepared for theatrical use, but a more radical revision of that text made (in some cases) after the first version had been presented in its own terms. At least five of Shakespeare's plays exist in these states: they are *2 Henry IV*, *Hamlet*, *Troilus and Cressida*, *Othello*, and *King Lear*.

The editorial problem is compounded by the existence of conflicting theories to explain the divergences between the surviving texts of these plays. Until recently, it was generally believed that the differences resulted from imperfect transmission: that Shakespeare wrote only one version of each play, and that each variant text represents that original text in a more or less corrupt form. As a consequence of this belief, editors conflated the texts, adding to one passages present only in the other, and selecting among variants in wording in an effort to present what the editor regarded as the most 'Shakespearian' version possible. *Hamlet* provides an example. The 1604 quarto was set from Shakespeare's own papers (with some contamination from the reported text of 1603). The Folio includes about 80 lines that are not in the quarto, but omits about 230 that are there. The Folio was clearly influenced by, if not printed directly from, a theatre manuscript. There are hundreds of local variants. Editors invariably conflate the two texts, assuming that the absence of passages from each was the result either of accidental omission or of cuts made in the theatre against Shakespeare's wishes; they also reject a selection of the variant readings. It is at least arguable that this produces a version that never existed in Shakespeare's time. We believe that the 1604 quarto represents the play as Shakespeare first wrote it, before it was performed, and that the Folio represents a theatrical text of the play after he had revised it. Given this belief, it would be equally logical to base an edition on either text: one the more literary, the other the more theatrical. Both types of edition would be of interest; each would present within its proper context readings which editors who conflate the texts have to abandon.

It would be extravagant in a one-volume edition to present double texts of all the plays that exist in significantly variant form. The theatrical version is, inevitably, that which comes closest to the 'final' version of the play. We have ample testimony from the theatre at all periods, including our own, that play scripts undergo a process of, often, considerable modification on their way from the writing table to the stage. Occasionally, dramatists resent this process; we know that some of Shakespeare's contemporaries resented cuts made in some of their plays. But we know too that plays may be much improved by intelligent cutting, and that dramatists of great literary talent may benefit from the discipline of the theatre. It is, of course, possible that Shakespeare's colleagues occasionally overruled him, forcing him to omit cherished lines, or that practical circumstances—such as the incapacity of a particular actor to do justice to every aspect of his role—necessitated adjustments that Shakespeare would have preferred not to make. But he was himself, supremely, a man of the theatre. We have seen that he displayed no interest in how his plays were printed: in this he is at the opposite extreme from Ben Jonson, who was still in mid-career when he prepared the collected edition of his works. We know that Shakespeare was an actor and shareholder in the leading theatre company of its time, a major financial asset to that company, a man immersed in the life of that theatre and committed to its values. The concept of the director of a play did not exist in his time; but someone must have exercised some, at least, of the functions of the modern director, and there is good reason to believe that that person must have been Shakespeare himself, for his own plays. The very fact that those texts of his plays that contain cuts also give evidence of more 'literary' revision suggests that he was deeply involved in the process by which his plays came to be modified in performance. For these reasons, this edition chooses, when possible, to print the more theatrical version of each play. In some cases, this requires the omission from the body of the text of lines that Shakespeare certainly wrote; there is, of course, no suggestion that these lines are unworthy of their author; merely that, in some if not all performances, he and his company found that the play's overall structure and pace were better without them. All such lines are printed as Additional Passages at the end of the play.

In all but one of Shakespeare's plays the revisions are local—changes in the wording of individual phrases and lines—or else they are effected by additions and cuts. Essentially, then, the story line is not affected. But in *King Lear* the differences between the two texts are more radical. It is not simply that the 1608 quarto lacks over 100 lines that are in the Folio, or that the Folio lacks close on 300 lines that are in the Quarto, or that there are over 850 verbal variants, or that several speeches are assigned to different speakers. It is rather that the sum total of these differences amounts, in this play, to a substantial shift in the presentation and interpretation of the underlying action. The differences are particularly apparent in the military action of the last two acts. We believe, in short, that there are two distinct plays of *King Lear*, not merely two different texts of the same play; so we print edited versions of both the Quarto (*'The History of . . .'*) and the Folio (*'The Tragedy of . . .'*).

Though the editor's selection, when choice is available, of the edition that should form the basis of his text is fundamentally important, many other tasks remain. Elizabethan printers could do meticulously scholarly work, but they rarely expended their best efforts on plays, which—at least in quarto format—they treated as ephemeral publications. Moreover, dramatic manuscripts and heavily annotated quartos must have set them difficult problems. Scribal transcripts would have been easier for the printer, but scribes were themselves liable to introduce error in copying difficult manuscripts, and also had a habit of sophisticating what they copied—for example, by expanding colloquial contractions—in ways that would distort the dramatist's intentions. On the whole, the Folio is a rather well-printed volume; there are not a great many obvious misprints; but for all that, corruption is often discernible. A few quartos—notably *A Midsummer Night's Dream* (1600)—are exceptionally well printed, but others, such as the 1604 *Hamlet*, abound in obvious error, which is a sure sign that they also commit hidden corruptions. Generations of editors have tried to correct the texts; but possible corruptions are still being identified, and new attempts at correction are often made. The preparation of this edition has required a minutely detailed examination of the early texts. At many points we have adopted emendations suggested by previous editors; at other points we offer original readings; and occasionally we revert to the original text at points where it has often been emended.

Stage directions are a special problem, especially in a one-volume edition where some degree of uniformity may be thought desirable. The early editions are often deficient in directions for essential action, even in such basic matters as when characters enter and when they depart. Again, generations of editors have tried to supply such deficiencies, not always systematically. We try to remedy the deficiencies, always bearing in mind the conditions of Shakespeare's stage. At many points the requisite action is apparent from the dialogue; at other points precisely what should happen, or the precise point at which it should happen, is in doubt—and, perhaps, was never clearly determined even by the author. In our edition we use broken brackets—e.g. ⌜*He kneels*⌝—to identify dubious action or placing. Inevitably, this is to some extent a matter of individual interpretation; and, of course, modern directors may, and do, often depart freely from the original directions, both explicit and implicit. Our original-spelling edition, while including the added directions, stays somewhat closer to the wording of the original editions than our modern-spelling edition. Readers interested in the precise directions of the original texts on which ours are based will find them reprinted in *William Shakespeare: A Textual Companion*.

Ever since Shakespeare's plays began to be reprinted, their spelling and punctuation have been modernized. Often, however, this task has been left to the printer; many editors who have undertaken it themselves have merely marked up earlier edited texts, producing a palimpsest; there has been little discussion of the principles involved; and editors have been even less systematic in this area than in that of stage directions. Modernizing the spelling of earlier periods is not the simple business it may appear. Some words are easily handled: 'doe' becomes 'do', 'I' meaning 'yes' becomes 'ay', 'beutie' becomes 'beauty', and so on. But it is not always easy to distinguish between variant spellings and variant forms. It is not our aim to modernize Shakespeare's language: we do not change 'ay' to 'yes', 'ye' to 'you', 'eyne' to 'eyes', or 'hath' to 'has'; we retain obsolete inflections and prefixes. We aim not to make changes that would affect the metre of verse: when the early editions mark an elision—'know'st', 'ha'not', 'i'th'temple'—we do so, too; when scansion requires that an -ed ending be sounded, contrary to modern usage, we mark it with a grave accent—'formèd', 'movèd'. Older forms of words are often preserved when they are required for metre, rhyme, word-play, or characterization. But we do not retain old spellings simply because they may provide a clue to the way words were pronounced by some people in Shakespeare's time, because such clues may be misleading (we know, for instance, that 'boil' was often pronounced as 'bile', 'Rome' as 'room', and 'person' as 'parson'), and, more importantly, because many words which we spell in the same way as the Elizabethans have changed pronunciation in the mean time; it seems pointless to offer in a generally modern context a mere selection of spellings that may convey some of the varied pronunciations available in Shakespeare's time. Many words existed in indifferently variant spellings; we have sometimes preferred the more modern spelling, especially when the older one might mislead: thus, we spell 'beholden', not 'beholding', 'distraught' (when appropriate), not 'distract'.

Similar principles are applied to proper names: it is, for instance, meaningless to preserve the Folio's 'Petruchio' when this is clearly intended to represent the old (as well as the modern) pronunciation of the Italian name 'Petruccio'; failure to modernize adequately here results even in the theatre in the mistaken pedantry of 'Pet-rook-io'. For some words, the arguments for and against modernization are finely balanced. The generally French setting of *As You Like It* has led us to prefer 'Ardenne' to the more familiar 'Arden', though we would not argue that geographical consistency is Shakespeare's strongest point. Problematic too is the military rank of ensign; this appears in early texts of Shakespeare as 'ancient' (or 'aunciant', 'auncient', 'auntient', etc.). 'Ancient' in this sense, in its various forms, was originally a corruption of 'ensign', and from the sixteenth to the eighteenth centuries the forms were interchangeable. Shakespeare himself may well have used both. There is no question that the sense conveyed by modern 'ensign' is overwhelmingly dominant in Shakespeare's designation of Iago (in *Othello*) and Pistol (in *2 Henry IV, Henry V,* and *The Merry Wives of Windsor*), and it is equally clear that 'ancient' could be seriously misleading, so we prefer 'ensign'. This is contrary to the editorial tradition, but a parallel is afforded by the noun 'dolphin', which is the regular spelling in Shakespearian texts for the French 'dauphin'. Here tradition favours 'dauphin', although it did not become common in English until the later seventeenth century. It would be as misleading to imply that Iago and Pistol were ancient as that the Dauphin of France was an aquatic mammal.

Punctuation, too, poses problems. Judging by most of the early, 'good' quartos as well as the fragment of *Sir Thomas More*, Shakespeare himself punctuated lightly. The syntax of his time was in any case more fluid than ours; the imposition upon it of a precisely grammatical system of punctuation reduces ambiguity and imposes definition upon in-definition. But Elizabethan scribes and printers seem to have regarded punctuation as their prerogative; thus, the 1600 quarto of *A Midsummer Night's Dream* is far more precisely punctuated than any Shakespearian manuscript is likely to have been; and Ralph Crane clearly imposed his own system upon the texts he transcribed. So it is impossible to put much faith in the punctuation of the early texts. Additionally, their system is often am-biguous: the question mark could signal an exclamation, and parentheses were idio-syncratically employed. Modern editors, then, may justifiably replace the varying, often conflicting systems of the early texts by one which attempts to convey their sense to the modern reader. Working entirely from the early texts, we have tried to use comparatively light pointing which will not impose certain nuances upon the text at the expense of others. Readers interested in the punctuation of the original texts will find it reproduced with minimal alteration in our original-spelling edition.

Theatre is an endlessly fluid medium. Each performance of a play is unique, differing from others in pace, movement, gesture, audience response, and even—because of the fallibility of human memory—in the words spoken. It is likely too that in Shakespeare's time, as in ours, changes in the texts of plays were consciously made to suit varying circumstances: the characteristics of particular actors, the place in which the play was performed, the anticipated reactions of his audience, and so on. The circumstances by which Shakespeare's plays have been transmitted to us mean that it is impossible to recover exactly the form in which they stood either in his own original manuscripts or in those manuscripts, or transcripts of them, after they had been prepared for use in the theatre. Still less can we hope to pinpoint the words spoken in any particular performance. Never-theless, it is in performance that the plays lived and had their being. Performance is the end to which they were created, and in this edition we have devoted our efforts to recovering and presenting texts of Shakespeare's plays as they were acted in the London playhouses which stood at the centre of his professional life.

# CONTEMPORARY ALLUSIONS TO SHAKESPEARE

MANY contemporary documents, some manuscript, some printed, refer directly to Shakespeare and to members of his family. The following list (which is not exhaustive) briefly indicates the nature of the principal allusions to him and to his closest relatives. It does not include publication records of his plays (given in the *Textual Companion*), the appearances of his name on title-pages, unascribed allusions to his works, commendatory poems, epistles, and dedications printed elsewhere in the edition, or records of performances except for that of 1604–5, in which Shakespeare is named. The principal documents are discussed, and most of them reproduced, in S. Schoenbaum's *William Shakespeare: A Documentary Life* (1975).

| | |
|---|---|
| 26 April 1564 | Baptism of 'Gulielmus filius Iohannes Shakspere' in Stratford-upon-Avon |
| 27 November 1582 | Entry of marriage licence at Worcester 'inter W$^m$ Shaxpere et Annō whateley' |
| 28 November 1582 | Marriage licence bond issued in Worcester for 'Willm Shagspere' and 'Anne Hathwey' |
| 26 May 1583 | Baptism of 'Susanna daughter to William Shakspere' in Stratford-upon-Avon |
| 2 February 1585 | Baptism of 'Hamnet & Iudeth sonne & daughter to Williã Shakspere' in Stratford-upon-Avon |
| 1589 | Reference to William Shakespeare in an earlier lawsuit brought by his parents against John Lambert concerning property at Wilmcote, near Stratford-upon-Avon |
| 1592 | Robert Greene's reference in *Greene's Groatsworth of Wit* to 'an vpstart Crow' who 'is in his owne conceit the onely Shake-scene in a countrey' (see p. xviii); Henry Chettle's (oblique) reference in *Kind-Harts Dreame* to Shakespeare's 'vprightnes of dealing, and fatious [= facetious?] grace in writting'. |
| 1594 | Reference to Shakespeare as author of *Lucrece* in Henry Willobie's *Willobie his Avisa* |
| 1595 | Allusion to 'Sweet Shakespeare' as author of *Lucrece* in William Covell's *Polimanteia* |
| 15 March 1595 | Shakespeare named as joint payee of the Lord Chamberlain's Men for performances at court |
| 1596 | A petition by William Wayte (in London) for sureties of the peace against 'Willm Shakspere' and others |
| 11 August 1596 | Burial of 'Hamnet filius William Shakspere' in Stratford-upon-Avon |
| October, 1596 | Two draft grants of arms to John Shakespeare |
| 4 May 1597, etc. | Documents recording William Shakespeare's purchase of New Place, Stratford-upon-Avon |

| | |
|---|---|
| 15 November 1597 | 'William Shackspeare' of Bishopsgate ward, London, listed as not having paid taxes due in February |
| 1598 | Shakespeare one of the 'principall Comœdians' in Ben Jonson's *Every Man in his Humour* (according to a list printed in the 1616 Jonson Folio) |
| 1598 | Sale by 'mr shaxspere' of a load of stone in Stratford-upon-Avon |
| 1598 | Appearance of Shakespeare's name on title-pages of the second quartos of *Richard II* and *Richard III* and of the first (surviving) quarto of *Love's Labour's Lost*. (Later title-page ascriptions are not listed.) |
| 1598 | Praise of Shakespeare as author of *Venus and Adonis* and *Lucrece* in Richard Barnfield's *Poems in Divers Humours* |
| 1598 | Francis Meres's references in *Palladis Tamia* (see p. xxi) |
| 4 February 1598 | 'Wm Shackespe' listed as holder of corn and malt in Stratford-upon-Avon |
| 1 October 1598 | 'William Shakespeare' named as tax defaulter in Bishopsgate ward, London |
| 15 October 1598 | Letter from Richard Quiney asking 'mr Wm Shackespe' for a loan of £30 |
| 1598–9 | 'Willm Shakespeare' named in Enrolled Subsidy Account (London) |
| 1598–1601 | References to Shakespeare in *The Pilgrimage to Parnassus* and *The Return from Parnassus, Parts 1 and 2*, acted at St. John's College, Cambridge |
| 1599 | Draft of grant permitting John Shakespeare to impale his arms with those of the Arden family |
| 16 May 1599 | The newly built Globe theatre mentioned (in a London inventory) as being in the occupation of 'William Shakespeare and others' |
| 6 October 1599 | 'Willmus Shakspere' named as owing money to the Exchequer (London) |
| 6 October 1600 | 'Willmus Shakspeare' named as owing money to the Exchequer (London) |
| *c.*1601 | References to Shakespeare as author of *Venus and Adonis*, *Lucrece*, and *Hamlet* in a manuscript note by Gabriel Harvey |
| 8 September 1601 | Burial of John Shakespeare in Stratford-upon-Avon |
| 1602 | Reconveyance of New Place to William Shakespeare |
| 1602 | Reference to 'Shakespear ye Player' in York Herald's complaint |
| 13 March 1602 | John Manningham's diary entry of an anecdote about Burbage and Shakespeare (see pp. xvii–xviii) |
| 1 May 1602 | Shakespeare paid William and John Combe £320 for land in Old Stratford |
| 28 September 1602 | Shakespeare bought a cottage in Chapel Lane, Stratford-upon-Avon |
| 1603 | Shakespeare listed among 'The principall Tragœdians' in Ben Jonson's *Sejanus* |
| 1603 | Shakespeare (and others) called on to lament the death of Queen Elizabeth in Henry Chettle's *A Mourneful Dittie, entituled Elizabeths Losse* |
| 17 and 19 May 1603 | Shakespeare named in documents conferring the title of King's Men on the former Lord Chamberlain's Men |
| *c.*1604 | Shakespeare sued Philip Rogers of Stratford-upon-Avon for debt |
| 1604 | Anthony Scoloker, in *Diaphantus, or the Passions of Love*, mentions the popularity of *Hamlet* and refers to '*Friendly Shakespeare's Tragedies*' |
| 15 March 1604 | Shakespeare and his fellows granted scarlet cloth for King James's entry into London (see p. xxii) |
| 24 October 1604 | Shakespeare recorded as owner of a cottage and garden in Rowington Manor |
| 1604–5 | 'Shaxberd' named as author of several plays performed at court |

| | |
|---|---|
| 1605 | Passing reference to Shakespeare in William Camden's *Remaines of a greater Worke concerning Britaine*' |
| 4 May 1605 | Augustine Phillips (actor) bequeathes 'to my fellowe William Shakespeare a xxxˢ peece in gould' |
| 24 July 1605 | Shakespeare pays £440 for an interest in a lease of tithes in the Stratford-upon-Avon area |
| 1607 | Commendatory allusion to Shakespeare in William Barksted's *Myrrha, the Mother of Adonis* |
| 5 June 1607 | Marriage of Susanna Shakespeare to John Hall in Stratford-upon-Avon |
| 9 September 1608 | Burial of Shakespeare's mother, 'Mary Shaxspere wydowe', in Stratford-upon-Avon |
| 1608–9 | Documents recording Shakespeare's lawsuit for debt against John Addenbrooke of Stratford-upon-Avon |
| 11 September 1611 | Shakespeare subscribes to the cost of a bill for repairing highways |
| 1612 | Praise of 'the right happy and copious industry of M. *Shake-speare*' (and others) in John Webster's epistle to *The White Devil* |
| 1612 | Shakespeare testifies in Stephen Belott's suit against Christopher Mountjoy, in London (see pp. xviii–xix) |
| *c.*1613 | Leonard Digges, in a manuscript note, compares Shakespeare to Lope de Vega |
| 28 January 1613 | John Combe of Stratford-upon-Avon bequeathes £5 to 'mr William Shackspere' |
| 10 March 1613 | Shakespeare buys a gatehouse in Blackfriars for £140 |
| 31 March 1613 | Payment to Shakespeare and Richard Burbage of 44 shillings for an *impresa* for the Earl of Rutland (see p. xxiii) |
| 1614 | Passing reference to Shakespeare in an epistle by Richard Carew printed in the second edition of Camden's *Remaines* |
| 5 September 1614, etc. | Documents recording Shakespeare's involvement in disputes concerning enclosures in the Stratford-upon-Avon area |
| *c.*1615 | Complimentary reference to Shakespeare in a manuscript poem by Francis Beaumont |
| 1615 | Passing reference to Shakespeare by Edward Howes in fifth edition of John Stow's *Annales* |
| 26 April 1615 | Shakespeare engaged in litigation concerned with the Blackfriars gatehouse |
| 10 February 1616 | Marriage of Judith Shakespeare to Thomas Quiney in Stratford-upon-Avon |
| 25 March 1616 | Shakespeare's will drawn up in Stratford-upon-Avon |
| 25 April 1616 | Burial of William Shakespeare in Stratford-upon-Avon (his monument records that he died on 23 April) |
| 8 August 1623 | Burial of Anne Shakespeare in Stratford-upon-Avon |
| 16 July 1649 | Burial of Shakespeare's daughter, Susanna Hall, in Stratford-upon-Avon |
| 9 February 1662 | Burial of Shakespeare's daughter, Judith Quiney, in Stratford-upon-Avon |
| 1670 | Death of Shakespeare's last direct descendant, his granddaughter Elizabeth, who married Thomas Nash in 1626 and John (later Sir John) Bernard in 1649 |

# COMMENDATORY POEMS AND
# PREFACES (1599–1640)

### Ad Gulielmum Shakespeare

Honey-tongued Shakespeare, when I saw thine issue
I swore Apollo got them, and none other,
Their rosy-tainted features clothed in tissue,
Some heaven-born goddess said to be their mother.
Rose-cheeked Adonis with his amber tresses,                        5
Fair fire-hot Venus charming him to love her,
Chaste Lucretia virgin-like her dresses,
Proud lust-stung Tarquin seeking still to prove her,
Romeo, Richard, more whose names I know not—
Their sugared tongues and power-attractive beauty     10
Say they are saints although that saints they show not,
For thousands vows to them subjective duty.
They burn in love, thy children; Shakespeare het them;
Go, woo thy muse more nymphish brood beget them.

> John Weever, *Epigrams* (1599)

---

### A never writer to an ever
### reader: news

Eternal reader, you have here a new play never staled
with the stage, never clapper-clawed with the palms of
the vulgar, and yet passing full of the palm comical, for
it is a birth of that brain that never undertook anything
comical vainly; and were but the vain names of comedies
changed for the titles of commodities, or of plays for pleas,
you should see all those grand censors that now style
them such vanities flock to them for the main grace of
their gravities, especially this author's comedies, that are
so framed to the life that they serve for the most common
commentaries of all the actions of our lives, showing such
a dexterity and power of wit that the most displeased
with plays are pleased with his comedies, and all such
dull and heavy-witted worldlings as were never capable
of the wit of a comedy, coming by report of them to his
representations, have found that wit there that they never
found in themselves, and have parted better witted than
they came, feeling an edge of wit set upon them more
than ever they dreamed they had brain to grind it on. So
much and such savoured salt of wit is in his comedies
that they seem, for their height of pleasure, to be born
in that sea that brought forth Venus. Amongst all there
is none more witty than this, and had I time I would
comment upon it, though I know it needs not for so
much as will make you think your testern well bestowed,

but for so much worth as even poor I know to be stuffed
in it. It deserves such a labour as well as the best comedy
in Terence or Plautus. And believe this, that when he is
gone and his comedies out of sale, you will scramble for
them, and set up a new English Inquisition. Take this for
a warning, and at the peril of your pleasure's loss and
judgement's, refuse not, nor like this the less for not being
sullied with the smoky breath of the multitude; but thank
fortune for the scape it hath made amongst you, since
by the grand possessors' wills I believe you should have
prayed for them rather than been prayed. And so I leave
all such to be prayed for, for the states of their        37
wits' healths, that will not praise it.
   *Vale.*

> Anonymous, in *Troilus and Cressida* (1609)

---

### To our English Terence, Master Will Shakespeare

Some say, good Will, which I in sport do sing,
Hadst thou not played some kingly parts in sport
Thou hadst been a companion for a king,
And been a king among the meaner sort.
Some others rail; but rail as they think fit,                5
Thou hast no railing but a reigning wit,
   And honesty thou sow'st, which they do reap,
   So to increase their stock which they do keep.

> John Davies, *The Scourge of Folly* (1610)

---

### To Master William Shakespeare

Shakespeare, that nimble Mercury, thy brain,
Lulls many hundred Argus-eyes asleep,
So fit for all thou fashionest thy vein;
At th'horse-foot fountain thou hast drunk full deep.
Virtue's or vice's theme to thee all one is.                  5
Who loves chaste life, there's Lucrece for a teacher;
Who list read lust, there's Venus and Adonis,
True model of a most lascivious lecher.
Besides, in plays thy wit winds like Meander,
Whence needy new composers borrow more            10
Than Terence doth from Plautus or Menander.
But to praise thee aright, I want thy store.
   Then let thine own works thine own worth upraise,
   And help t'adorn thee with deservèd bays.

> Thomas Freeman, *Run and a Great Cast* (1614)

## Inscriptions upon the Shakespeare monument, Stratford-upon-Avon

*Iudicio Pylium, genio Socratem, arte Maronem,*
*Terra tegit, populus maeret, Olympus habet.*

Stay, passenger, why goest thou by so fast?
Read, if thou canst, whom envious death hath placed
Within this monument: Shakespeare, with whom     5
Quick nature died; whose name doth deck this tomb
Far more than cost, sith all that he hath writ
Leaves living art but page to serve his wit.

> *Obiit anno domini* 1616,
> *aetatis* 53, *die* 23 *Aprilis*

---

## On the death of William Shakespeare

Renownèd Spenser, lie a thought more nigh
To learnèd Chaucer; and rare Beaumont, lie
A little nearer Spenser, to make room
For Shakespeare in your threefold, fourfold tomb.
To lodge all four in one bed make a shift     5
Until doomsday, for hardly will a fifth
Betwixt this day and that by fate be slain
For whom your curtains need be drawn again.
But if precedency in death doth bar
A fourth place in your sacred sepulchre,     10
Under this carvèd marble of thine own,
Sleep, rare tragedian Shakespeare, sleep alone.
Thy unmolested peace, unsharèd cave,
Possess as lord, not tenant, of thy grave,
That unto us or others it may be     15
Honour hereafter to be laid by thee.

> William Basse (*c.*1616–22), in Shakespeare's
> *Poems* (1640)

---

## The Stationer to the Reader
### (in *The Tragedy of Othello*, 1622)

To set forth a book without an epistle were like to the old English proverb, 'A blue coat without a badge', and the author being dead, I thought good to take that piece of work upon me. To commend it I will not, for that which is good, I hope every man will commend without entreaty; and I am the bolder because the author's name is sufficient to vent his work. Thus, leaving everyone to the liberty of judgement, I have ventured to print this play, and leave it to the general censure.     9

> Yours,
> Thomas Walkley.

## The Epistle Dedicatory
### (in *Comedies, Histories, and Tragedies*, 1623)

TO THE MOST NOBLE

AND

INCOMPARABLE PAIR

OF BRETHREN

WILLIAM     5
Earl of Pembroke, &c., Lord Chamberlain to the
*King's most excellent majesty,*

AND

PHILIP
Earl of Montgomery, &c., gentleman of his majesty's
bedchamber; both Knights of the most noble Order
of the Garter, and our singular good
LORDS.

Right Honourable,     14

Whilst we study to be thankful in our particular for the many favours we have received from your lordships, we are fallen upon the ill fortune to mingle two the most diverse things that can be: fear and rashness; rashness in the enterprise, and fear of the success. For when we value the places your highnesses sustain, we cannot but know their dignity greater than to descend to the reading of these trifles; and while we name them trifles we have deprived ourselves of the defence of our dedication. But since your lordships have been pleased to think these trifles something heretofore, and have prosecuted both them and their author, living, with so much favour, we hope that, they outliving him, and he not having the fate, common with some, to be executor to his own writings, you will use the like indulgence toward them you have done unto their parent. There is a great difference whether any book choose his patrons, or find them. This hath done both; for so much were your lordships' likings of the several parts when they were acted as, before they were published, the volume asked to be yours. We have but collected them, and done an office to the dead to procure his orphans guardians, without ambition either of self-profit or fame, only to keep the memory of so worthy a friend and fellow alive as was our Shakespeare, by humble offer of his plays to your most noble patronage. Wherein, as we have justly observed no man to come near your lordships but with a kind of religious address, it hath been the height of our care, who are the presenters, to make the present worthy of your highnesses by the perfection. But there we must also crave our abilities to be considered, my lords. We cannot go beyond our own powers. Country hands reach forth milk, cream, fruits, or what they have; and many nations, we have heard, that had not gums and incense, obtained their requests with a leavened cake. It was no fault to approach their gods by what means they could, and the most, though meanest, of things are made more precious when they are dedicated to temples. In that name, therefore, we most humbly consecrate to your highnesses these remains of your servant Shakespeare, that what delight is in them may be ever your lordships', the reputation his, and the faults ours, if any be committed by a pair so careful to show their gratitude both to the living and the dead as is                                                        58

Your lordships' most bounden,
JOHN HEMINGES.
HENRY CONDELL.

———

## To the Great Variety of Readers

From the most able to him that can but spell: there you are numbered; we had rather you were weighed, especially when the fate of all books depends upon your capacities, and not of your heads alone, but of your purses. Well, it is now public, and you will stand for your privileges, we know: to read and censure. Do so, but buy it first. That doth best commend a book, the stationer says. Then, how odd soever your brains be, or your wisdoms, make your licence the same, and spare not. Judge your six-penn'orth, your shilling's worth, your five shillings' worth at a time, or higher, so you rise to the just rates, and welcome. But whatever you do, buy. Censure will not drive a trade or make the jack go; and though you be a magistrate of wit, and sit on the stage at Blackfriars or the Cockpit to arraign plays daily, know, these plays have had their trial already, and stood out all appeals, and do now come forth quitted rather by a decree of court than any purchased letters of commendation.                                                19

It had been a thing, we confess, worthy to have been wished that the author himself had lived to have set forth and overseen his own writings. But since it hath been ordained otherwise, and he by death departed from that right, we pray you do not envy his friends the office of their care and pain to have collected and published them, and so to have published them as where, before, you were abused with divers stolen and surreptitious copies, maimed and deformed by the frauds and stealths of injurious impostors that exposed them, even those are now offered to your view cured and perfect of their limbs, and all the rest absolute in their numbers, as he conceived them; who, as he was a happy imitator of nature, was a most gentle expresser of it. His mind and hand went together, and what he thought he uttered with that easiness that we have scarce received from him a blot in his papers. But it is not our province, who only gather his works and give them you, to praise him; it is yours, that read him. And there we hope, to your diverse capacities, you will find enough both to draw and hold you; for his wit can no more lie hid than it could be lost. Read him, therefore, and again, and again, and if then you do not like him, surely you are in some manifest danger not to understand him. And so we leave you to other of his friends whom if you need can be your guides;

if you need them not, you can lead yourselves and others. And such readers we wish him.                           46

John Heminges, Henry Condell, in *Comedies, Histories,*
*and Tragedies* (1623)

———

## To the memory of my beloved,
## The AUTHOR
### Master William Shakespeare,
### and
### what he hath left us

To draw no envy, Shakespeare, on thy name
  Am I thus ample to thy book and fame;
While I confess thy writings to be such
  As neither man nor muse can praise too much:
'Tis true, and all men's suffrage. But these ways     5
  Were not the paths I meant unto thy praise,
For silliest ignorance on these may light,
  Which, when it sounds at best, but echoes right;
Or blind affection, which doth ne'er advance
  The truth, but gropes, and urgeth all by chance;   10
Or crafty malice might pretend this praise,
  And think to ruin where it seemed to raise.
These are as some infamous bawd or whore
  Should praise a matron: what could hurt her more?
But thou art proof against them, and indeed          15
  Above th'ill fortune of them, or the need.
I therefore will begin. Soul of the age!
  The applause, delight, the wonder of our stage!
My Shakespeare, rise. I will not lodge thee by
  Chaucer or Spenser, or bid Beaumont lie            20
A little further to make thee a room.
  Thou art a monument without a tomb,
And art alive still while thy book doth live
  And we have wits to read and praise to give.
That I not mix thee so, my brain excuses:            25
  I mean with great but disproportioned muses.
For if I thought my judgement were of years
  I should commit thee surely with thy peers,
And tell how far thou didst our Lyly outshine,
  Or sporting Kyd, or Marlowe's mighty line.         30
And though thou hadst small Latin and less Greek,
  From thence to honour thee I would not seek
For names, but call forth thund'ring Aeschylus,
  Euripides, and Sophocles to us,
Pacuvius, Accius, him of Cordova dead,               35
  To life again, to hear thy buskin tread
And shake a stage; or, when thy socks were on,
  Leave thee alone for the comparison
Of all that insolent Greece or haughty Rome
  Sent forth, or since did from their ashes come.    40
Triumph, my Britain, thou hast one to show
  To whom all scenes of Europe homage owe.
He was not of an age, but for all time,
  And all the muses still were in their prime
When like Apollo he came forth to warm               45
  Our ears, or like a Mercury to charm!

Nature herself was proud of his designs,
　　And joyed to wear the dressing of his lines,
Which were so richly spun, and woven so fit,
　　As since she will vouchsafe no other wit.　　　50
The merry Greek, tart Aristophanes,
　　Neat Terence, witty Plautus, now not please,
But antiquated and deserted lie
　　As they were not of nature's family.
Yet must I not give nature all; thy art,　　　55
　　My gentle Shakespeare, must enjoy a part.
For though the poet's matter nature be,
　　His art doth give the fashion; and that he
Who casts to write a living line must sweat—
　　Such as thine are—and strike the second heat　　　60
Upon the muses' anvil, turn the same,
　　And himself with it that he thinks to frame;
Or for the laurel he may gain a scorn,
　　For a good poet's made as well as born.
And such wert thou. Look how the father's face　　　65
　　Lives in his issue, even so the race
Of Shakespeare's mind and manners brightly shines
　　In his well-turnèd and true-filèd lines,
In each of which he seems to shake a lance,
　　As brandished at the eyes of ignorance.　　　70
Sweet swan of Avon! What a sight it were
　　To see thee in our waters yet appear,
And make those flights upon the banks of Thames
　　That so did take Eliza and our James!
But stay, I see thee in the hemisphere　　　75
　　Advanced, and made a constellation there!
Shine forth, thou star of poets, and with rage
　　Or influence chide or cheer the drooping stage,
Which, since thy flight from hence, hath mourned like
　　night　　　80
　　And despairs day, but for thy volume's light.
　　　　Ben Jonson, in *Comedies, Histories, and Tragedies*
　　　　　　　　(1623)

————

### Upon the Lines and Life of the Famous Scenic Poet, Master William Shakespeare

Those hands which you so clapped go now and wring,
You Britons brave, for done are Shakespeare's days.
His days are done that made the dainty plays
Which made the globe of heav'n and earth to ring.
Dried is that vein, dried is the Thespian spring,　　　5
Turned all to tears, and Phoebus clouds his rays.
That corpse, that coffin now bestick those bays
Which crowned him poet first, then poets' king.
If tragedies might any prologue have,
All those he made would scarce make one to this,　　　10
Where fame, now that he gone is to the grave—
Death's public tiring-house—the *nuntius* is;
　　For though his line of life went soon about,
　　The life yet of his lines shall never out.
　　　　Hugh Holland, in *Comedies, Histories, and Tragedies*
　　　　　　　　(1623)

### TO THE MEMORY of the deceased author Master William Shakespeare

Shakespeare, at length thy pious fellows give
The world thy works, thy works by which outlive
Thy tomb thy name must; when that stone is rent,
And time dissolves thy Stratford monument,
Here we alive shall view thee still. This book,　　　5
When brass and marble fade, shall make thee look
Fresh to all ages. When posterity
Shall loathe what's new, think all is prodigy
That is not Shakespeare's ev'ry line, each verse
Here shall revive, redeem thee from thy hearse.　　　10
Nor fire nor cank'ring age, as Naso said
Of his, thy wit-fraught book shall once invade;
Nor shall I e'er believe or think thee dead—
Though missed—until our bankrupt stage be sped—
Impossible—with some new strain t'outdo　　　15
Passions of Juliet and her Romeo,
Or till I hear a scene more nobly take
Than when thy half-sword parleying Romans spake.
Till these, till any of thy volume's rest
Shall with more fire, more feeling be expressed,　　　20
Be sure, our Shakespeare, thou canst never die,
But crowned with laurel, live eternally.
　　　　Leonard Digges, in *Comedies, Histories, and Tragedies*
　　　　　　　　(1623)

————

### To the memory of Master William Shakespeare

We wondered, Shakespeare, that thou went'st so soon
From the world's stage to the grave's tiring-room.
We thought thee dead, but this thy printed worth
Tells thy spectators that thou went'st but forth
To enter with applause. An actor's art　　　5
Can die, and live to act a second part.
That's but an exit of mortality;
This, a re-entrance to a *plaudite*.
　　　　James Mabbe, in *Comedies, Histories, and Tragedies*
　　　　　　　　(1623)

————

### The Names of the Principal Actors in all these Plays

| | |
|---|---|
| William Shakespeare. | Samuel Gilburn. |
| Richard Burbage. | Robert Armin. |
| John Heminges. | William Ostler. |
| Augustine Phillips. | Nathan Field. |
| William Kempe. | John Underwood.　5 |
| Thomas Pope. | Nicholas Tooley. |
| George Bryan. | William Ecclestone. |
| Henry Condell. | Joseph Taylor. |
| William Sly. | Robert Benfield. |
| Richard Cowley. | Robert Gough.　10 |
| John Lowin. | Richard Robinson. |
| Samuel Cross. | John Shank. |
| Alexander Cook. | John Rice. |

　　　　In *Comedies, Histories, and Tragedies* (1623)

### An Epitaph on the Admirable Dramatic Poet, William Shakespeare

What need my Shakespeare for his honoured bones
The labour of an age in pilèd stones,
Or that his hallowed relics should be hid
Under a star-ypointing pyramid?
Dear son of memory, great heir of fame,　　　　5
What need'st thou such dull witness of thy name?
Thou in our wonder and astonishment
Hast built thyself a lasting monument,
For whilst to th' shame of slow-endeavouring art
Thy easy numbers flow, and that each heart　　　10
Hath from the leaves of thy unvalued book
Those Delphic lines with deep impression took,
Then thou, our fancy of herself bereaving,
Dost make us marble with too much conceiving,
And so sepulchered in such pomp dost lie　　　　15
That kings for such a tomb would wish to die.

John Milton (1630), in *Comedies, Histories, and Tragedies*
(1632)

———

### Upon the Effigies of my Worthy Friend, the Author Master William Shakespeare, and his Works

Spectator, this life's shadow is. To see
The truer image and a livelier he,
Turn reader. But observe his comic vein,
Laugh; and proceed next to a tragic strain,
Then weep. So when thou find'st two contraries,　5
Two different passions from thy rapt soul rise,
Say—who alone effect such wonders could—
Rare Shakespeare to the life thou dost behold.

Anonymous, in *Comedies, Histories, and Tragedies* (1632)

———

### On Worthy Master Shakespeare and his Poems

A mind reflecting ages past, whose clear
And equal surface can make things appear
Distant a thousand years, and represent
Them in their lively colours' just extent;
To outrun hasty time, retrieve the fates,　　　　5
Roll back the heavens, blow ope the iron gates
Of death and Lethe, where confusèd lie
Great heaps of ruinous mortality;
In that deep dusky dungeon to discern
A royal ghost from churls; by art to learn　　　10
The physiognomy of shades, and give
Them sudden birth, wond'ring how oft they live;
What story coldly tells, what poets feign
At second hand, and picture without brain
Senseless and soulless shows; to give a stage,　15
Ample and true with life, voice, action, age,
As Plato's year and new scene of the world
Them unto us or us to them had hurled;
To raise our ancient sovereigns from their hearse,
Make kings his subjects; by exchanging verse　20
Enlive their pale trunks, that the present age

Joys in their joy, and trembles at their rage;
Yet so to temper passion that our ears
Take pleasure in their pain, and eyes in tears
Both weep and smile: fearful at plots so sad,　　25
Then laughing at our fear; abused, and glad
To be abused, affected with that truth
Which we perceive is false; pleased in that ruth
At which we start, and by elaborate play
Tortured and tickled; by a crablike way　　　　30
Time past made pastime, and in ugly sort
Disgorging up his ravin for our sport,
While the plebeian imp from lofty throne
Creates and rules a world, and works upon
Mankind by secret engines; now to move　　　　35
A chilling pity, then a rigorous love;
To strike up and stroke down both joy and ire;
To steer th'affections, and by heavenly fire
Mould us anew; stol'n from ourselves—
　　This, and much more which cannot be expressed　40
But by himself, his tongue and his own breast,
Was Shakespeare's freehold, which his cunning brain
Improved by favour of the ninefold train.
The buskined muse, the comic queen, the grand
And louder tone of Clio; nimble hand　　　　45
And nimbler foot of the melodious pair,
The silver-voicèd lady, the most fair
Calliope, whose speaking silence daunts,
And she whose praise the heavenly body chants.
　　These jointly wooed him, envying one another,　50
Obeyed by all as spouse, but loved as brother,
And wrought a curious robe of sable grave,
Fresh green, and pleasant yellow, red most brave,
And constant blue, rich purple, guiltless white,
The lowly russet, and the scarlet bright,　　　55
Branched and embroidered like the painted spring,
Each leaf matched with a flower, and each string
Of golden wire, each line of silk; there run
Italian works whose thread the sisters spun,
And there did sing, or seem to sing, the choice　60
Birds of a foreign note and various voice.
Here hangs a mossy rock, there plays a fair
But chiding fountain purlèd. Not the air
Nor clouds nor thunder but were living drawn
Not out of common tiffany or lawn,　　　　65
But fine materials which the muses know,
And only know the countries where they grow.
　　Now when they could no longer him enjoy
In mortal garments pent: death may destroy,
They say, his body, but his verse shall live,　　70
And more than nature takes our hands shall give.
In a less volume, but more strongly bound,
Shakespeare shall breathe and speak, with laurel crowned,
Which never fades; fed with Ambrosian meat
In a well-linèd vesture rich and neat.　　　　75
　　So with this robe they clothe him, bid him wear it,
　　For time shall never stain, nor envy tear it.

'The friendly admirer of his endowments', I.M.S.,
in *Comedies, Histories, and Tragedies* (1632)

## Upon Master WILLIAM SHAKESPEARE, the *Deceased Author, and his* POEMS

Poets are born, not made: when I would prove
This truth, the glad remembrance I must love
Of never-dying Shakespeare, who alone
Is argument enough to make that one.
First, that he was a poet none would doubt    5
That heard th'applause of what he sees set out
Imprinted, where thou hast—I will not say,
Reader, his works, for to contrive a play
To him 'twas none—the pattern of all wit,
Art without art unparalleled as yet.    10
Next, nature only helped him, for look thorough
This whole book, thou shalt find he doth not borrow
One phrase from Greeks, nor Latins imitate,
Nor once from vulgar languages translate,
Nor plagiary-like from others glean,    15
Nor begs he from each witty friend a scene
To piece his acts with. All that he doth write
Is pure his own—plot, language exquisite—
But O! what praise more powerful can we give
The dead than that by him the King's men live,    20
His players, which should they but have shared the fate,
All else expired within the short term's date,
How could the Globe have prospered, since through want
Of change the plays and poems had grown scant.
But, happy verse, thou shalt be sung and heard    25
When hungry quills shall be such honour barred.
Then vanish, upstart writers to each stage,
You needy poetasters of this age;
Where Shakespeare lived or spake, vermin, forbear;
Lest with your froth you spot them, come not near.    30
But if you needs must write, if poverty
So pinch that otherwise you starve and die,
On God's name may the Bull or Cockpit have
Your lame blank verse, to keep you from the grave,
Or let new Fortune's younger brethren see    35
What they can pick from your lean industry.
I do not wonder, when you offer at
Blackfriars, that you suffer; 'tis the fate
Of richer veins, prime judgements that have fared
The worse with this deceasèd man compared.    40
So have I seen, when Caesar would appear,
And on the stage at half-sword parley were
Brutus and Cassius; O, how the audience
Were ravished, with what wonder they went thence,
When some new day they would not brook a line    45
Of tedious though well-laboured *Catiline*.
Sejanus too was irksome, they prized more
Honest Iago, or the jealous Moor.
And though the Fox and subtle Alchemist,
Long intermitted, could not quite be missed,    50
Though these have shamed all the ancients, and might
   raise
Their author's merit with a crown of bays,
Yet these, sometimes, even at a friend's desire
Acted, have scarce defrayed the seacoal fire

And doorkeepers; when let but Falstaff come,    55
Hal, Poins, the rest, you scarce shall have a room,
All is so pestered. Let but Beatrice
And Benedick be seen, lo, in a trice
The Cockpit galleries, boxes, all are full
To hear Malvolio, that cross-gartered gull.    60
Brief, there is nothing in his wit-fraught book
Whose sound we would not hear, on whose worth look;
Like old-coined gold, whose lines in every page
Shall pass true current to succeeding age.
But why do I dead Shakespeare's praise recite?    65
Some second Shakespeare must of Shakespeare write;
For me 'tis needless, since an host of men
Will pay to clap his praise, to free my pen.

Leonard Digges (before 1636), in Shakespeare's
*Poems* (1640)

---

## In remembrance of Master *William Shakespeare*.

### ODE

**1.**

Beware, delighted poets, when you sing
To welcome nature in the early spring,
   Your num'rous feet not tread
The banks of Avon; for each flower
(As it ne'er knew a sun or shower)    5
   Hangs there the pensive head.

**2.**

Each tree, whose thick and spreading growth hath made
Rather a night beneath the boughs than shade,
   Unwilling now to grow,
Looks like the plume a captive wears,    10
Whose rifled falls are steeped i'th' tears
   Which from his last rage flow.

**3.**

The piteous river wept itself away
Long since, alas, to such a swift decay
   That, reach the map and look    15
If you a river there can spy,
And for a river your mocked eye
   Will find a shallow brook.

Sir William Davenant, *Madagascar, with other Poems* (1637)

---

## An Elegy on the death of that famous Writer and Actor, Master William Shakspeare

I dare not do thy memory that wrong
Unto our larger griefs to give a tongue;
I'll only sigh in earnest, and let fall
My solemn tears at thy great funeral,
For every eye that rains a show'r for thee    5
Laments thy loss in a sad elegy.
Nor is it fit each humble muse should have
Thy worth his subject, now thou'rt laid in grave;

No, it's a flight beyond the pitch of those
Whose worthless pamphlets are not sense in prose.  10
Let learnèd Jonson sing a dirge for thee,
And fill our orb with mournful harmony;
But we need no remembrancer; thy fame
Shall still accompany thy honoured name
To all posterity, and make us be  15
Sensible of what we lost in losing thee,
Being the age's wonder, whose smooth rhymes
Did more reform than lash the looser times.
Nature herself did her own self admire
As oft as thou wert pleasèd to attire  20
Her in her native lustre, and confess
Thy dressing was her chiefest comeliness.
How can we then forget thee, when the age
Her chiefest tutor, and the widowed stage
Her only favourite, in thee hath lost,  25
And nature's self what she did brag of most?
Sleep, then, rich soul of numbers, whilst poor we
Enjoy the profits of thy legacy,
And think it happiness enough we have
So much of thee redeemèd from the grave  30
As may suffice to enlighten future times
With the bright lustre of thy matchless rhymes.

> Anonymous (before 1638), in Shakespeare's
> *Poems* (1640)

---

## To Shakespeare

Thy muse's sugared dainties seem to us
Like the famed apples of old Tantalus,
For we, admiring, see and hear thy strains,
But none I see or hear those sweets attains.

### To the same

Thou hast so used thy pen, or shook thy spear,
That poets startle, nor thy wit come near.

> Thomas Bancroft, *Two Books of Epigrams and
> Epitaphs* (1639)

---

## To Master William Shakespeare

Shakespeare, we must be silent in thy praise,
'Cause our encomiums will but blast thy bays,
Which envy could not; that thou didst so well,
Let thine own histories prove thy chronicle.

> Anonymous, in *Wit's Recreations* (1640)

## To the Reader

I here presume, under favour, to present to your view some excellent and sweetly composed poems of Master William Shakespeare, which in themselves appear of the same purity the author himself, then living, avouched. They had not the fortune, by reason of their infancy in his death, to have the due accommodation of proportionable glory with the rest of his ever-living works, yet the lines of themselves will afford you a more authentic approbation than my assurance any way can; to invite your allowance, in your perusal you shall find them serene, clear, and elegantly plain, such gentle strains as shall recreate and not perplex your brain, no intricate or cloudy stuff to puzzle intellect, but perfect eloquence, such as will raise your admiration to his praise. This assurance, I know, will not differ from your acknowledgement; and certain I am my opinion will be seconded by the sufficiency of these ensuing lines. I have been somewhat solicitous to bring this forth to the perfect view of all men, and in so doing, glad to be serviceable for the continuance of glory to the deserved author in these his poems.  20

> John Benson, in Shakespeare's *Poems* (1640)

---

## Of Master *William Shakespeare*

What, lofty Shakespeare, art again revived,
And Virbius-like now show'st thyself twice lived?
'Tis Benson's love that thus to thee is shown,
The labour's his, the glory still thine own.
These learnèd poems amongst thine after-birth,  5
That makes thy name immortal on the earth,
Will make the learnèd still admire to see
The muses' gifts so fully infused on thee.
Let carping Momus bark and bite his fill,
And ignorant Davus slight thy learnèd skill,  10
Yet those who know the worth of thy desert,
And with true judgement can discern thy art,
Will be admirers of thy high-tuned strain,
Amongst whose number let me still remain.

> John Warren, in Shakespeare's *Poems* (1640)

# THE COMPLETE WORKS

# THE TWO GENTLEMEN OF VERONA

THE accomplished elegance of the lyrical verse in *The Two Gentlemen of Verona*, as well as the skilful, theatrically effective prose of Lance's monologues, demonstrates that Shakespeare had already developed his writing skills when he composed this play. Nevertheless—and although the earliest mention of it is by Francis Meres in 1598—it may be his first work for the stage; for its dramatic structure is comparatively unambitious, and while some of its scenes are expertly constructed, those involving more than, at the most, four characters betray an uncertainty of technique suggestive of inexperience. It was first printed in the 1623 Folio.

The friendship of the 'two gentlemen'—Valentine and Proteus—is strained when both fall in love with Silvia. Proteus has followed Valentine from Verona to Milan, leaving behind his beloved Julia, who in turn follows him, disguised as a boy. At the climax of the action Valentine displays the depth of his friendship by offering Silvia to Proteus. The conflicting claims of love and friendship illustrated in this plot had been treated in a considerable body of English literature written by the time Shakespeare wrote his play, probably in the late 1580s. John Lyly's didactic fiction *Euphues* (1578) was an immensely popular example; and Lyly's earliest plays, such as *Campaspe* (1584) and *Endimion* (1588), influenced Shakespeare's style as well as his subject matter. Shakespeare was writing in a fashionable mode, but his story of Proteus and Julia is specifically (though perhaps indirectly) indebted to a prose fiction, *Diana*, written in Spanish by the Portuguese Jorge de Montemayor and first published in 1559. Many other influences on the young dramatist may be discerned: his idealized portrayal of Silvia and her relationship with Valentine derives from the medieval tradition of courtly love; Arthur Brooke's long poem *The Tragical History of Romeus and Juliet* (1562) provided some details of the plot; and the comic commentary on the romantic action supplied by the page-boy Speed and the more rustic clown Lance has dramatic antecedents in English plays such as Lyly's early comedies.

Though the play was presumably acted in Shakespeare's time, its first recorded performance is in 1762, in a rewritten version at Drury Lane. Later performances have been sparse, and the play has succeeded best when subjected to adaptation, increasing its musical content, adjusting the emphasis of the last scene so as to reduce the shock of Valentine's donation of Silvia to Proteus, and updating the setting. It can be seen as a dramatic laboratory in which Shakespeare first experimented with conventions of romantic comedy which he would later treat with a more subtle complexity, but it has its own charm. If the whole is not greater than the parts, some of the parts—such as Lance's brilliant monologues, and the delightful scene (4.2) in which Proteus serenades his new love with 'Who is Silvia?' while his disguised old love, Julia, looks wistfully on—are wholly successful. And Lance's dog, Crab, has the most scene-stealing non-speaking role in the canon: this is an experiment that Shakespeare did not repeat.

# THE PERSONS OF THE PLAY

DUKE of Milan
SILVIA, his daughter

PROTEUS, a gentleman of Verona
LANCE, his clownish servant

VALENTINE, a gentleman of Verona
SPEED, his clownish servant
THURIO, a foolish rival to Valentine

ANTONIO, father of Proteus
PANTHINO, his servant

JULIA, beloved of Proteus
LUCETTA, her waiting-woman
HOST, where Julia lodges

EGLAMOUR, agent for Silvia in her escape

OUTLAWS

Servants, musicians

# The Two Gentlemen of Verona

**1.1** *Enter Valentine and Proteus*

VALENTINE
Cease to persuade, my loving Proteus.
Home-keeping youth have ever homely wits.
Were't not affection chains thy tender days
To the sweet glances of thy honoured love,
I rather would entreat thy company                   5
To see the wonders of the world abroad
Than, living dully sluggardized at home,
Wear out thy youth with shapeless idleness.
But since thou lov'st, love still, and thrive therein—
Even as I would, when I to love begin.               10

PROTEUS
Wilt thou be gone? Sweet Valentine, adieu.
Think on thy Proteus when thou haply seest
Some rare noteworthy object in thy travel.
Wish me partaker in thy happiness
When thou dost meet good hap; and in thy danger—
If ever danger do environ thee—                      16
Commend thy grievance to my holy prayers;
For I will be thy beadsman, Valentine.

VALENTINE
And on a love-book pray for my success?

PROTEUS
Upon some book I love I'll pray for thee.            20

VALENTINE
That's on some shallow story of deep love—
How young Leander crossed the Hellespont.

PROTEUS
That's a deep story of a deeper love,
For he was more than over-shoes in love.

VALENTINE
'Tis true, for you are over-boots in love,           25
And yet you never swam the Hellespont.

PROTEUS
Over the boots? Nay, give me not the boots.

VALENTINE
No, I will not; for it boots thee not.

PROTEUS                                  *What?*

VALENTINE
To be in love, where scorn is bought with groans,
Coy looks with heart-sore sighs, one fading moment's
   mirth                                             30
With twenty watchful, weary, tedious nights.
If haply won, perhaps a hapless gain;
If lost, why then a grievous labour won;
However, but a folly bought with wit,
Or else a wit by folly vanquishèd.                   35

PROTEUS
So by your circumstance you call me fool.

VALENTINE
So by your circumstance I fear you'll prove.

PROTEUS
'Tis love you cavil at. I am not love.

VALENTINE
Love is your master, for he masters you,
And he that is so yokèd by a fool                    40
Methinks should not be chronicled for wise.

PROTEUS
Yet writers say 'As in the sweetest bud

The eating canker dwells, so doting love
Inhabits in the finest wits of all.'

VALENTINE
And writers say 'As the most forward bud             45
Is eaten by the canker ere it blow,
Even so by love the young and tender wit
Is turned to folly, blasting in the bud,
Losing his verdure even in the prime,
And all the fair effects of future hopes.'           50
But wherefore waste I time to counsel thee
That art a votary to fond desire?
Once more adieu. My father at the road
Expects my coming, there to see me shipped.

PROTEUS
And thither will I bring thee, Valentine.            55

VALENTINE
Sweet Proteus, no. Now let us take our leave.
To Milan let me hear from thee by letters
Of thy success in love, and what news else
Betideth here in absence of thy friend;
And I likewise will visit thee with mine.            60

PROTEUS
All happiness bechance to thee in Milan.

VALENTINE
As much to you at home; and so farewell.    *Exit*

PROTEUS
He after honour hunts, I after love.
He leaves his friends to dignify them more,
I leave myself, my friends, and all, for love.       65
Thou, Julia, thou hast metamorphosed me,
Made me neglect my studies, lose my time,
War with good counsel, set the world at naught;
Made wit with musing weak, heart sick with thought.
   *Enter Speed*

SPEED
Sir Proteus, save you. Saw you my master?           70

PROTEUS
But now he parted hence to embark for Milan.

SPEED
Twenty to one, then, he is shipped already,
And I have played the sheep in losing him.

PROTEUS
Indeed, a sheep doth very often stray,
An if the shepherd be a while away.                  75

SPEED
You conclude that my master is a shepherd, then,
   and I a sheep?

PROTEUS I do.

SPEED
Why then, my horns are his horns, whether I wake
   or sleep.

PROTEUS A silly answer, and fitting well a sheep.

SPEED This proves me still a sheep.                  80

PROTEUS True, and thy master a shepherd.

SPEED Nay, that I can deny by a circumstance.

PROTEUS It shall go hard but I'll prove it by another.

SPEED The shepherd seeks the sheep, and not the sheep
   the shepherd. But I seek my master, and my master
   seeks not me. Therefore I am no sheep.           86

PROTEUS The sheep for fodder follow the shepherd, the

3

shepherd for food follows not the sheep. Thou for wages
followest thy master, thy master for wages follows not
thee. Therefore thou art a sheep.                                    90

SPEED Such another proof will make me cry 'baa'.

PROTEUS But dost thou hear: gav'st thou my letter to
Julia?

SPEED Ay, sir. I, a lost mutton, gave your letter to her, a
laced mutton, and she, a laced mutton, gave me, a lost
mutton, nothing for my labour.                                       96

PROTEUS Here's too small a pasture for such store of
muttons.

SPEED If the ground be overcharged, you were best stick
her.                                                                100

PROTEUS Nay, in that you are astray. 'Twere best pound
you.

SPEED Nay sir, less than a pound shall serve me for
carrying your letter.

PROTEUS You mistake. I mean the pound, a pinfold.    105

SPEED
    From a pound to a pin? Fold it over and over
    'Tis threefold too little for carrying a letter to your lover.

PROTEUS But what said she?

SPEED (nods, then says) Ay.

PROTEUS Nod-ay? Why, that's 'noddy'.                    110

SPEED You mistook, sir. I say she did nod, and you ask
me if she did nod, and I say 'Ay'.

PROTEUS And that set together is 'noddy'.

SPEED Now you have taken the pains to set it together,
take it for your pains.                                            115

PROTEUS No, no. You shall have it for bearing the letter.

SPEED Well, I perceive I must be fain to bear with you.

PROTEUS Why, sir, how do you bear with me?

SPEED Marry, sir, the letter very orderly, having nothing
but the word 'noddy' for my pains.                            120

PROTEUS Beshrew me but you have a quick wit.

SPEED And yet it cannot overtake your slow purse.

PROTEUS Come, come, open the matter in brief. What said
she?                                                                124

SPEED Open your purse, that the money and the matter
may be both at once delivered.

PROTEUS (giving money) Well, sir, here is for your pains.
What said she?

SPEED Truly, sir, I think you'll hardly win her.         129

PROTEUS Why? Couldst thou perceive so much from her?

SPEED Sir, I could perceive nothing at all from her, no,
not so much as a ducat for delivering your letter. And
being so hard to me, that brought your mind, I fear
she'll prove as hard to you in telling your mind. Give
her no token but stones, for she's as hard as steel.

PROTEUS What said she? Nothing?                         136

SPEED No, not so much as 'Take this for thy pains'. To
testify your bounty, I thank you, you have testerned
me; in requital whereof, henceforth carry your letters
yourself. And so, sir, I'll commend you to my master.
                                                            ⌜Exit⌝

PROTEUS
    Go, go, be gone, to save your ship from wreck,      141
    Which cannot perish having thee aboard,
    Being destined to a drier death on shore.
    I must go send some better messenger.
    I fear my Julia would not deign my lines,            145
    Receiving them from such a worthless post.         Exit

**1.2**    *Enter Julia and Lucetta*

JULIA
    But say, Lucetta, now we are alone—
    Wouldst thou then counsel me to fall in love?

LUCETTA
    Ay, madam, so you stumble not unheedfully.

JULIA
    Of all the fair resort of gentlemen
    That every day with parle encounter me,               5
    In thy opinion which is worthiest love?

LUCETTA
    Please you repeat their names, I'll show my mind
    According to my shallow simple skill.

JULIA
    What think'st thou of the fair Sir Eglamour?

LUCETTA
    As of a knight well spoken, neat, and fine,          10
    But were I you, he never should be mine.

JULIA
    What think'st thou of the rich Mercatio?

LUCETTA
    Well of his wealth, but of himself, so-so.

JULIA
    What think'st thou of the gentle Proteus?

LUCETTA
    Lord, lord, to see what folly reigns in us!          15

JULIA
    How now? What means this passion at his name?

LUCETTA
    Pardon, dear madam, 'tis a passing shame
    That I, unworthy body as I am,
    Should censure thus on lovely gentlemen.

JULIA
    Why not on Proteus, as of all the rest?              20

LUCETTA
    Then thus: of many good, I think him best.

JULIA Your reason?

LUCETTA
    I have no other but a woman's reason:
    I think him so because I think him so.

JULIA
    And wouldst thou have me cast my love on him?        25

LUCETTA
    Ay, if you thought your love not cast away.

JULIA
    Why, he of all the rest hath never moved me.

LUCETTA
    Yet he of all the rest I think best loves ye.

JULIA
    His little speaking shows his love but small.

LUCETTA
    Fire that's closest kept burns most of all.         30

JULIA
    They do not love that do not show their love.

LUCETTA
    O, they love least that let men know their love.

JULIA
    I would I knew his mind.

LUCETTA (giving Proteus' letter)
    Peruse this paper, madam.

JULIA
    'To Julia'—say, from whom?                           35

LUCETTA
    That the contents will show.

JULIA
    Say, say—who gave it thee?

LUCETTA
    Sir Valentine's page; and sent, I think, from Proteus.
    He would have given it you, but I being in the way
    Did in your name receive it. Pardon the fault, I pray.

JULIA
Now, by my modesty, a goodly broker.                    41
Dare you presume to harbour wanton lines?
To whisper, and conspire against my youth?
Now trust me, 'tis an office of great worth,
And you an officer fit for the place.                    45
There. Take the paper.
    *She gives Lucetta the letter*
                        See it be returned,
Or else return no more into my sight.
LUCETTA
To plead for love deserves more fee than hate.
JULIA
Will ye be gone?
LUCETTA              That you may ruminate.    *Exit*
JULIA
And yet I would I had o'erlooked the letter.            50
It were a shame to call her back again
And pray her to a fault for which I chid her.
What fool is she, that knows I am a maid
And would not force the letter to my view,
Since maids in modesty say 'No' to that                 55
Which they would have the profferer construe 'Ay'.
Fie, fie, how wayward is this foolish love
That like a testy babe will scratch the nurse
And presently, all humbled, kiss the rod.
How churlishly I chid Lucetta hence                     60
When willingly I would have had her here.
How angerly I taught my brow to frown
When inward joy enforced my heart to smile.
My penance is to call Lucetta back
And ask remission for my folly past.                    65
What ho! Lucetta!
    *Enter Lucetta*
LUCETTA              What would your ladyship?
JULIA
Is't near dinner-time?
LUCETTA              I would it were,
That you might kill your stomach on your meat
And not upon your maid.
    ⌐*She drops and picks up the letter*⌐
JULIA                   What is't that you
Took up so gingerly?                                     70
LUCETTA Nothing.
JULIA Why didst thou stoop then?
LUCETTA
To take a paper up that I let fall.
JULIA
And is that paper nothing?
LUCETTA
Nothing concerning me.                                  75
JULIA
Then let it lie for those that it concerns.
LUCETTA
Madam, it will not lie where it concerns,
Unless it have a false interpreter.
JULIA
Some love of yours hath writ to you in rhyme.
LUCETTA
That I might sing it, madam, to a tune,                 80
Give me a note. Your ladyship can set.
JULIA
As little by such toys as may be possible.
Best sing it to the tune of 'Light o' love'.
LUCETTA
It is too heavy for so light a tune.

JULIA
Heavy? Belike it hath some burden, then?                85
LUCETTA
Ay, and melodious were it, would you sing it.
JULIA
And why not you?
LUCETTA              I cannot reach so high.
JULIA
Let's see your song.
    ⌐*She tries to take the letter*⌐
                        How now, minion!
LUCETTA
Keep tune there still. So you will sing it out.
And yet methinks I do not like this tune.               90
JULIA You do not?
LUCETTA
No, madam, 'tis too sharp.
JULIA
You, minion, are too saucy.
LUCETTA
Nay, now you are too flat,
And mar the concord with too harsh a descant.           95
There wanteth but a mean to fill your song.
JULIA
The mean is drowned with your unruly bass.
LUCETTA
Indeed, I bid the base for Proteus.
JULIA
This bauble shall not henceforth trouble me.
Here is a coil with protestation.                       100
    *She tears the letter and drops the pieces*
Go, get you gone, and let the papers lie.
You would be fing'ring them to anger me.
LUCETTA (*aside*)
She makes it strange, but she would be best pleased
To be so angered with another letter.    *Exit*
JULIA
Nay, would I were so angered with the same.             105
O hateful hands, to tear such loving words;
Injurious wasps, to feed on such sweet honey
And kill the bees that yield it with your stings.
I'll kiss each several paper for amends.
    *She picks up some of the pieces of paper*
Look, here is writ 'Kind Julia'—unkind Julia,          110
As in revenge of thy ingratitude
I throw thy name against the bruising stones,
Trampling contemptuously on thy disdain.
And here is writ 'Love-wounded Proteus'.
Poor wounded name, my bosom as a bed                    115
Shall lodge thee till thy wound be throughly healed;
And thus I search it with a sovereign kiss.
But twice or thrice was 'Proteus' written down.
Be calm, good wind, blow not a word away
Till I have found each letter in the letter             120
Except mine own name. That, some whirlwind bear
Unto a ragged, fearful, hanging rock
And throw it thence into the raging sea.
Lo, here in one line is his name twice writ:
'Poor forlorn Proteus', 'passionate Proteus',          125
'To the sweet Julia'—that I'll tear away.
And yet I will not, sith so prettily
He couples it to his complaining names.
Thus will I fold them, one upon another.
Now kiss, embrace, contend, do what you will.          130
    *Enter Lucetta*
LUCETTA
Madam, dinner is ready, and your father stays.

JULIA  Well, let us go.

LUCETTA

What, shall these papers lie like telltales here?

JULIA

If you respect them, best to take them up.

LUCETTA

Nay, I was taken up for laying them down.                    135
Yet here they shall not lie, for catching cold.

JULIA

I see you have a month's mind to them.

LUCETTA

Ay, madam, you may say what sights you see.
I see things too, although you judge I wink.

JULIA  Come, come, will't please you go?        *Exeunt*

**1.3**  *Enter Antonio and Panthino*

ANTONIO

Tell me, Panthino, what sad talk was that
Wherewith my brother held you in the cloister?

PANTHINO

'Twas of his nephew Proteus, your son.

ANTONIO

Why, what of him?

PANTHINO                  He wondered that your lordship
Would suffer him to spend his youth at home         5
While other men, of slender reputation,
Put forth their sons to seek preferment out—
Some to the wars, to try their fortune there,
Some to discover islands far away,
Some to the studious universities.                       10
For any or for all these exercises
He said that Proteus your son was meet,
And did request me to importune you
To let him spend his time no more at home,
Which would be great impeachment to his age       15
In having known no travel in his youth.

ANTONIO

Nor need'st thou much importune me to that
Whereon this month I have been hammering.
I have considered well his loss of time,
And how he cannot be a perfect man,                     20
Not being tried and tutored in the world.
Experience is by industry achieved,
And perfected by the swift course of time.
Then tell me, whither were I best to send him?

PANTHINO

I think your lordship is not ignorant                      25
How his companion, youthful Valentine,
Attends the Emperor in his royal court.

ANTONIO  I know it well.

PANTHINO

'Twere good, I think, your lordship sent him thither.
There shall he practise tilts and tournaments,         30
Hear sweet discourse, converse with noblemen,
And be in eye of every exercise
Worthy his youth and nobleness of birth.

ANTONIO

I like thy counsel. Well hast thou advised,
And that thou mayst perceive how well I like it,    35
The execution of it shall make known.
Even with the speediest expedition
I will dispatch him to the Emperor's court.

PANTHINO

Tomorrow, may it please you, Don Alfonso,
With other gentlemen of good esteem,                   40

Are journeying to salute the Emperor
And to commend their service to his will.

ANTONIO

Good company. With them shall Proteus go.
      *Enter Proteus with a letter. He does not see Antonio*
      *and Panthino*
And in good time. Now will we break with him.

PROTEUS  Sweet love, sweet lines, sweet life!        45
Here is her hand, the agent of her heart.
Here is her oath for love, her honour's pawn.
O that our fathers would applaud our loves
To seal our happiness with their consents.
O heavenly Julia!                                             50

ANTONIO

How now, what letter are you reading there?

PROTEUS

May't please your lordship, 'tis a word or two
Of commendations sent from Valentine,
Delivered by a friend that came from him.

ANTONIO

Lend me the letter. Let me see what news.            55

PROTEUS

There is no news, my lord, but that he writes
How happily he lives, how well beloved
And daily gracèd by the Emperor,
Wishing me with him, partner of his fortune.

ANTONIO

And how stand you affected to his wish?             60

PROTEUS

As one relying on your lordship's will,
And not depending on his friendly wish.

ANTONIO

My will is something sorted with his wish.
Muse not that I thus suddenly proceed,
For what I will, I will, and there an end.              65
I am resolved that thou shalt spend some time
With Valentinus in the Emperor's court.
What maintenance he from his friends receives,
Like exhibition thou shalt have from me.
Tomorrow be in readiness to go.                         70
Excuse it not, for I am peremptory.

PROTEUS

My lord, I cannot be so soon provided.
Please you deliberate a day or two.

ANTONIO

Look what thou want'st shall be sent after thee.
No more of stay. Tomorrow thou must go.           75
Come on, Panthino. You shall be employed
To hasten on his expedition.

                            *Exeunt Antonio and Panthino*

PROTEUS

Thus have I shunned the fire for fear of burning
And drenched me in the sea where I am drowned.
I feared to show my father Julia's letter              80
Lest he should take exceptions to my love,
And with the vantage of mine own excuse
Hath he excepted most against my love.
O, how this spring of love resembleth
The uncertain glory of an April day,                   85
Which now shows all the beauty of the sun,
And by and by a cloud takes all away.
      *Enter Panthino*

PANTHINO

Sir Proteus, your father calls for you.
   He is in haste, therefore I pray you go.

PROTEUS
Why, this it is. My heart accords thereto,    90
And yet a thousand times it answers 'No'.     *Exeunt*

**2.1**    *Enter Valentine and Speed*
SPEED (*offering Valentine a glove*)
Sir, your glove.
VALENTINE       Not mine. My gloves are on.
SPEED
Why then, this may be yours, for this is but one.
VALENTINE
Ha, let me see. Ay, give it me, it's mine—
Sweet ornament, that decks a thing divine.
Ah, Silvia, Silvia!      5
SPEED Madam Silvia, Madam Silvia!
VALENTINE How now, sirrah?
SPEED She is not within hearing, sir.
VALENTINE Why, sir, who bade you call her?
SPEED Your worship, sir, or else I mistook.    10
VALENTINE Well, you'll still be too forward.
SPEED And yet I was last chidden for being too slow.
VALENTINE Go to, sir. Tell me, do you know Madam Silvia?
SPEED She that your worship loves?    15
VALENTINE Why, how know you that I am in love?
SPEED Marry, by these special marks: first, you have learned, like Sir Proteus, to wreath your arms, like a malcontent; to relish a love-song, like a robin redbreast; to walk alone, like one that had the pestilence; to sigh, like a schoolboy that had lost his ABC; to weep, like a young wench that had buried her grandam; to fast, like one that takes diet; to watch, like one that fears robbing; to speak puling, like a beggar at Hallowmas. You were wont, when you laughed, to crow like a cock; when you walked, to walk like one of the lions. When you fasted, it was presently after dinner; when you looked sadly, it was for want of money. And now you are metamorphosed with a mistress, that when I look on you I can hardly think you my master.    30
VALENTINE Are all these things perceived in me?
SPEED They are all perceived without ye.
VALENTINE Without me? They cannot.
SPEED Without you? Nay, that's certain, for without you were so simple, none else would. But you are so without these follies that these follies are within you, and shine through you like the water in an urinal, that not an eye that sees you but is a physician to comment on your malady.    39
VALENTINE But tell me, dost thou know my lady Silvia?
SPEED She that you gaze on so as she sits at supper?
VALENTINE Hast thou observed that? Even she I mean.
SPEED Why sir, I know her not.
VALENTINE Dost thou know her by my gazing on her, and yet know'st her not?    45
SPEED Is she not hard-favoured, sir?
VALENTINE Not so fair, boy, as well favoured.
SPEED Sir, I know that well enough.
VALENTINE What dost thou know?
SPEED That she is not so fair as of you well favoured.    50
VALENTINE I mean that her beauty is exquisite but her favour infinite.
SPEED That's because the one is painted and the other out of all count.
VALENTINE How painted? And how out of count?    55
SPEED Marry, sir, so painted to make her fair that no man counts of her beauty.

VALENTINE How esteem'st thou me? I account of her beauty.
SPEED You never saw her since she was deformed.    60
VALENTINE How long hath she been deformed?
SPEED Ever since you loved her.
VALENTINE I have loved her ever since I saw her, and still I see her beautiful.
SPEED If you love her you cannot see her.    65
VALENTINE Why?
SPEED Because love is blind. O that you had mine eyes, or your own eyes had the lights they were wont to have when you chid at Sir Proteus for going ungartered.
VALENTINE What should I see then?    70
SPEED Your own present folly and her passing deformity; for he being in love could not see to garter his hose, and you being in love cannot see to put on your hose.
VALENTINE Belike, boy, then you are in love, for last morning you could not see to wipe my shoes.    75
SPEED True, sir. I was in love with my bed. I thank you, you swinged me for my love, which makes me the bolder to chide you for yours.
VALENTINE In conclusion, I stand affected to her.
SPEED I would you were set. So your affection would cease.    81
VALENTINE Last night she enjoined me to write some lines to one she loves.
SPEED And have you?
VALENTINE I have.    85
SPEED Are they not lamely writ?
VALENTINE No, boy, but as well as I can do them. Peace, here she comes.
    *Enter Silvia*
SPEED (*aside*) O excellent motion! O exceeding puppet! Now will he interpret to her.    90
VALENTINE
Madam and mistress, a thousand good-morrows.
SPEED (*aside*) O, give ye good e'en! Here's a million of manners.
SILVIA
Sir Valentine and servant, to you two thousand.
SPEED (*aside*) He should give her interest, and she gives it him.    96
VALENTINE
As you enjoined me, I have writ your letter
Unto the secret, nameless friend of yours;
Which I was much unwilling to proceed in
But for my duty to your ladyship.    100
    *He gives her a letter*
SILVIA
I thank you, gentle servant. 'Tis very clerkly done.
VALENTINE
Now trust me, madam, it came hardly off;
For being ignorant to whom it goes
I writ at random, very doubtfully.
SILVIA
Perchance you think too much of so much pains?    105
VALENTINE
No, madam. So it stead you I will write—
Please you command—a thousand times as much.
And yet . . .
SILVIA
A pretty period. Well, I guess the sequel.
And yet I will not name it. And yet I care not.    110
And yet, take this again.
    *She offers him the letter*
           And yet I thank you,
Meaning henceforth to trouble you no more.

SPEED (*aside*)
And yet you will, and yet another yet.
VALENTINE
What means your ladyship? Do you not like it?
SILVIA
Yes, yes. The lines are very quaintly writ,                    115
But since unwillingly, take them again.
   *She presses the letter upon him*
Nay, take them.
VALENTINE          Madam, they are for you.
SILVIA
Ay, ay. You writ them, sir, at my request,
But I will none of them. They are for you.
I would have had them writ more movingly.                    120
VALENTINE
Please you, I'll write your ladyship another.
SILVIA
And when it's writ, for my sake read it over,
And if it please you, so. If not, why, so.
VALENTINE
If it please me, madam? What then?
SILVIA
Why, if it please you, take it for your labour.              125
And so good morrow, servant.                        *Exit*
SPEED (*aside*)
O jest unseen, inscrutable, invisible
As a nose on a man's face or a weathercock on a
   steeple.
My master sues to her, and she hath taught her suitor,
He being her pupil, to become her tutor.                    130
O excellent device! Was there ever heard a better?—
That my master, being scribe, to himself should write
   the letter.
VALENTINE How now, sir—what, are you reasoning with
   yourself?
SPEED Nay, I was rhyming. 'Tis you that have the reason.
VALENTINE To do what?                                       136
SPEED To be a spokesman from Madam Silvia.
VALENTINE To whom?
SPEED To yourself. Why, she woos you by a figure.
VALENTINE What figure?                                      140
SPEED By a letter, I should say.
VALENTINE Why, she hath not writ to me.
SPEED What need she, when she hath made you write to
   yourself? Why, do you not perceive the jest?
VALENTINE No, believe me.                                   145
SPEED No believing you indeed, sir. But did you perceive
   her earnest?
VALENTINE She gave me none, except an angry word.
SPEED Why, she hath given you a letter.
VALENTINE That's the letter I writ to her friend.           150
SPEED And that letter hath she delivered, and there an
   end.
VALENTINE I would it were no worse.
SPEED I'll warrant you, 'tis as well.
For often have you writ to her, and she in modesty
Or else for want of idle time could not again reply,        156
Or fearing else some messenger that might her mind
   discover,
Herself hath taught her love himself to write unto her
   lover.
—All this I speak in print, for in print I found it. Why
   muse you, sir? 'Tis dinner-time.                         160
VALENTINE I have dined.
SPEED Ay, but hearken, sir. Though the chameleon love
   can feed on the air, I am one that am nourished by

my victuals, and would fain have meat. O, be not like
your mistress—be moved, be moved!           *Exeunt*

**2.2**  *Enter Proteus and Julia*
PROTEUS
Have patience, gentle Julia.
JULIA
I must where is no remedy.
PROTEUS
When possibly I can I will return.
JULIA
If you turn not, you will return the sooner.
   *She gives him a ring*
Keep this remembrance for thy Julia's sake.                   5
PROTEUS
Why then, we'll make exchange. Here, take you this.
   *He gives her a ring*
JULIA
And seal the bargain with a holy kiss.
   ⌜*They kiss*⌝
PROTEUS
Here is my hand for my true constancy.
And when that hour o'erslips me in the day
Wherein I sigh not, Julia, for thy sake,                     10
The next ensuing hour some foul mischance
Torment me for my love's forgetfulness.
My father stays my coming. Answer not.
The tide is now. (*Julia weeps*) Nay, not thy tide of tears,
That tide will stay me longer than I should.                 15
Julia, farewell.                              *Exit Julia*
        What, gone without a word?
Ay, so true love should do. It cannot speak,
For truth hath better deeds than words to grace it.
   *Enter Panthino*
PANTHINO
Sir Proteus, you are stayed for.
PROTEUS                    Go, I come, I come.—
Alas, this parting strikes poor lovers dumb.    *Exeunt*

**2.3**  *Enter Lance with his dog Crab*
LANCE (*to the audience*) Nay, 'twill be this hour ere I have
   done weeping. All the kind of the Lances have this very
   fault. I have received my proportion, like the prodigious
   son, and am going with Sir Proteus to the Imperial's
   court. I think Crab my dog, be the sourest-natured
   dog that lives. My mother weeping, my father wailing,
   my sister crying, our maid howling, our cat wringing
   her hands, and all our house in a great perplexity, yet
   did not this cruel-hearted cur shed one tear. He is a
   stone, a very pebble-stone, and has no more pity in
   him than a dog. A Jew would have wept to have seen
   our parting. Why, my grandam, having no eyes, look
   you, wept herself blind at my parting. Nay, I'll show
   you the manner of it. This shoe is my father. No, this
   left shoe is my father. No, no, this left shoe is my
   mother. Nay, that cannot be so, neither. Yes, it is so,
   it is so, it hath the worser sole. This shoe with the hole
   in it is my mother, and this my father. A vengeance
   on't, there 'tis. Now, sir, this staff is my sister, for, look
   you, she is as white as a lily and as small as a wand.
   This hat is Nan our maid. I am the dog. No, the dog
   is himself, and I am the dog. O, the dog is me, and I
   am myself. Ay, so, so. Now come I to my father.
   'Father, your blessing.' Now should not the shoe speak
   a word for weeping. Now should I kiss my father. Well,
   he weeps on. Now come I to my mother. O that she

could speak now, like a moved woman. Well, I kiss
her. Why, there 'tis. Here's my mother's breath up and
down. Now come I to my sister. Mark the moan she
makes.—Now the dog all this while sheds not a tear
nor speaks a word. But see how I lay the dust with my
tears.                                                                      32

*Enter Panthino*

PANTHINO Lance, away, away, aboard. Thy master is
shipped, and thou art to post after with oars. What's
the matter? Why weep'st thou, man? Away, ass, you'll
lose the tide if you tarry any longer.                       36

LANCE It is no matter if the tied were lost, for it is the
unkindest tied that ever any man tied.

PANTHINO What's the unkindest tide?

LANCE Why, he that's tied here, Crab my dog.           40

PANTHINO Tut, man, I mean thou'lt lose the flood, and
in losing the flood, lose thy voyage, and in losing thy
voyage, lose thy master, and in losing thy master, lose
thy service, and in losing thy service—

*Lance puts his hand over Panthino's mouth*

Why dost thou stop my mouth?                               45

LANCE For fear thou shouldst lose thy tongue.

PANTHINO Where should I lose my tongue?

LANCE In thy tale.

PANTHINO In thy tail!                                              49

LANCE Lose the tide, and the voyage, and the master, and
the service, and the tied? Why, man, if the river were
dry, I am able to fill it with my tears. If the wind were
down, I could drive the boat with my sighs.

PANTHINO Come, come away, man. I was sent to call
thee.                                                                      55

LANCE Sir, call me what thou darest.

PANTHINO Wilt thou go?

LANCE Well, I will go.                                      *Exeunt*

**2.4**  *Enter Valentine, Silvia, Thurio, and Speed*

SILVIA Servant!

VALENTINE Mistress?

SPEED (*to Valentine*) Master, Sir Thurio frowns on you.

VALENTINE Ay, boy, it's for love.

SPEED Not of you.                                                    5

VALENTINE Of my mistress, then.

SPEED 'Twere good you knocked him.

SILVIA (*to Valentine*) Servant, you are sad.

VALENTINE Indeed, madam, I seem so.

THURIO Seem you that you are not?                        10

VALENTINE Haply I do.

THURIO So do counterfeits.

VALENTINE So do you.

THURIO What seem I that I am not?

VALENTINE Wise.                                                     15

THURIO What instance of the contrary?

VALENTINE Your folly.

THURIO And how quote you my folly?

VALENTINE I quote it in your jerkin.

THURIO My 'jerkin' is a doublet.                            20

VALENTINE Well then, I'll double your folly.

THURIO How!

SILVIA What, angry, Sir Thurio? Do you change colour?

VALENTINE Give him leave, madam, he is a kind of cha-
meleon.                                                                    25

THURIO That hath more mind to feed on your blood than
live in your air.

VALENTINE You have said, sir.

THURIO Ay, sir, and done too, for this time.

VALENTINE I know it well, sir, you always end ere you
begin.                                                                       31

SILVIA A fine volley of words, gentlemen, and quickly
shot off.

VALENTINE 'Tis indeed, madam, we thank the giver.

SILVIA Who is that, servant?                                   35

VALENTINE Yourself, sweet lady, for you gave the fire. Sir
Thurio borrows his wit from your ladyship's looks, and
spends what he borrows kindly in your company.

THURIO Sir, if you spend word for word with me, I shall
make your wit bankrupt.                                          40

VALENTINE I know it well, sir. You have an exchequer of
words, and, I think, no other treasure to give your
followers. For it appears by their bare liveries that they
live by your bare words.

SILVIA No more, gentlemen, no more. Here comes my
father.                                                                       46

*Enter the Duke*

DUKE
Now, daughter Silvia, you are hard beset.
Sir Valentine, your father is in good health,
What say you to a letter from your friends
Of much good news?

VALENTINE                     My lord, I will be thankful   50
To any happy messenger from thence.

DUKE
Know ye Don Antonio, your countryman?

VALENTINE
Ay, my good lord, I know the gentleman
To be of worth, and worthy estimation,
And not without desert so well reputed.             55

DUKE Hath he not a son?

VALENTINE
Ay, my good lord, a son that well deserves
The honour and regard of such a father.

DUKE You know him well?

VALENTINE
I knew him as myself, for from our infancy       60
We have conversed, and spent our hours together.
And though myself have been an idle truant,
Omitting the sweet benefit of time
To clothe mine age with angel-like perfection,
Yet hath Sir Proteus—for that's his name—      65
Made use and fair advantage of his days:
His years but young, but his experience old;
His head unmellowed, but his judgement ripe.
And in a word—for far behind his worth
Comes all the praises that I now bestow—          70
He is complete, in feature and in mind,
With all good grace to grace a gentleman.

DUKE
Beshrew me, sir, but if he make this good
He is as worthy for an empress' love
As meet to be an emperor's counsellor.              75
Well, sir, this gentleman is come to me
With commendation from great potentates,
And here he means to spend his time awhile.
I think 'tis no unwelcome news to you.

VALENTINE
Should I have wished a thing it had been he.     80

DUKE
Welcome him then according to his worth.
Silvia, I speak to you, and you, Sir Thurio;
For Valentine, I need not cite him to it.
I will send him hither to you presently.          *Exit*

VALENTINE
This is the gentleman I told your ladyship          85
Had come along with me, but that his mistress
Did hold his eyes locked in her crystal looks.
SILVIA
Belike that now she hath enfranchised them
Upon some other pawn for fealty.
VALENTINE
Nay, sure, I think she holds them prisoners still.          90
SILVIA
Nay, then he should be blind, and being blind
How could he see his way to seek out you?
VALENTINE
Why, lady, love hath twenty pair of eyes.
THURIO
They say that love hath not an eye at all.
VALENTINE
To see such lovers, Thurio, as yourself.          95
Upon a homely object love can wink.
SILVIA
Have done, have done. Here comes the gentleman.
          Enter Proteus
VALENTINE
Welcome, dear Proteus. Mistress, I beseech you
Confirm his welcome with some special favour.
SILVIA
His worth is warrant for his welcome hither,          100
If this be he you oft have wished to hear from.
VALENTINE
Mistress, it is. Sweet lady, entertain him
To be my fellow-servant to your ladyship.
SILVIA
Too low a mistress for so high a servant.
PROTEUS
Not so, sweet lady, but too mean a servant          105
To have a look of such a worthy mistress.
VALENTINE
Leave off discourse of disability.
Sweet lady, entertain him for your servant.
PROTEUS
My duty will I boast of, nothing else.
SILVIA
And duty never yet did want his meed.          110
Servant, you are welcome to a worthless mistress.
PROTEUS
I'll die on him that says so but yourself.
SILVIA
That you are welcome?
PROTEUS                    That you are worthless.
          ⌈Enter a Servant⌉
⌈SERVANT⌉
Madam, my lord your father would speak with you.
SILVIA
I wait upon his pleasure.          ⌈Exit the Servant⌉
          Come, Sir Thurio,
Go with me. Once more, new servant, welcome.          116
I'll leave you to confer of home affairs.
When you have done, we look to hear from you.
PROTEUS
We'll both attend upon your ladyship.
          Exeunt Silvia and Thurio
VALENTINE
Now tell me, how do all from whence you came?          120
PROTEUS
Your friends are well, and have them much
          commended.

VALENTINE
And how do yours?
PROTEUS          I left them all in health.
VALENTINE
How does your lady, and how thrives your love?
PROTEUS
My tales of love were wont to weary you.
I know you joy not in a love-discourse.          125
VALENTINE
Ay, Proteus, but that life is altered now.
I have done penance for contemning love,
Whose high imperious thoughts have punished me
With bitter fasts, with penitential groans,
With nightly tears and daily heart-sore sighs.          130
For in revenge of my contempt of love
Love hath chased sleep from my enthrallèd eyes,
And made them watchers of mine own heart's sorrow.
O gentle Proteus, love's a mighty lord,
And hath so humbled me as I confess          135
There is no woe to his correction,
Nor to his service no such joy on earth.
Now, no discourse except it be of love.
Now can I break my fast, dine, sup, and sleep
Upon the very naked name of love.          140
PROTEUS
Enough. I read your fortune in your eye.
Was this the idol that you worship so?
VALENTINE
Even she; and is she not a heavenly saint?
PROTEUS
No, but she is an earthly paragon.
VALENTINE
Call her divine.
PROTEUS          I will not flatter her.          145
VALENTINE
O flatter me; for love delights in praises.
PROTEUS
When I was sick you gave me bitter pills,
And I must minister the like to you.
VALENTINE
Then speak the truth by her; if not divine,
Yet let her be a principality,          150
Sovereign to all the creatures on the earth.
PROTEUS
Except my mistress.
VALENTINE          Sweet, except not any,
Except thou wilt except against my love.
PROTEUS
Have I not reason to prefer mine own?
VALENTINE
And I will help thee to prefer her, too.          155
She shall be dignified with this high honour,
To bear my lady's train, lest the base earth
Should from her vesture chance to steal a kiss
And, of so great a favour growing proud,
Disdain to root the summer-swelling flower,          160
And make rough winter everlastingly.
PROTEUS
Why, Valentine, what braggartism is this?
VALENTINE
Pardon me, Proteus, all I can is nothing
To her whose worth makes other worthies nothing.
She is alone.
PROTEUS          Then let her alone.          165
VALENTINE
Not for the world. Why man, she is mine own,

Aiming at Silvia as a sweeter friend.                              30
I cannot now prove constant to myself
Without some treachery used to Valentine.
This night he meaneth with a corded ladder
To climb celestial Silvia's chamber-window,
Myself in counsel his competitor.                                  35
Now presently I'll give her father notice
Of their disguising and pretended flight,
Who, all enraged, will banish Valentine;
For Thurio he intends shall wed his daughter.
But Valentine being gone, I'll quickly cross                       40
By some sly trick blunt Thurio's dull proceeding.
Love, lend me wings to make my purpose swift,
As thou hast lent me wit to plot this drift.            *Exit*

**2.7**    *Enter Julia and Lucetta*
JULIA
Counsel, Lucetta. Gentle girl, assist me,
And e'en in kind love I do conjure thee,
Who art the table wherein all my thoughts
Are visibly charactered and engraved,
To lesson me, and tell me some good mean                            5
How with my honour I may undertake
A journey to my loving Proteus.
LUCETTA
Alas, the way is wearisome and long.
JULIA
A true-devoted pilgrim is not weary
To measure kingdoms with his feeble steps.                         10
Much less shall she that hath love's wings to fly,
And when the flight is made to one so dear,
Of such divine perfection as Sir Proteus.
LUCETTA
Better forbear till Proteus make return.
JULIA
O, know'st thou not his looks are my soul's food?  15
Pity the dearth that I have pinèd in
By longing for that food so long a time.
Didst thou but know the inly touch of love
Thou wouldst as soon go kindle fire with snow
As seek to quench the fire of love with words.                     20
LUCETTA
I do not seek to quench your love's hot fire,
But qualify the fire's extreme rage,
Lest it should burn above the bounds of reason.
JULIA
The more thou damm'st it up, the more it burns.
The current that with gentle murmur glides,                        25
Thou know'st, being stopped, impatiently doth rage.
But when his fair course is not hinderèd
He makes sweet music with th'enamelled stones,
Giving a gentle kiss to every sedge
He overtaketh in his pilgrimage.                                   30
And so by many winding nooks he strays
With willing sport to the wild ocean.
Then let me go, and hinder not my course.
I'll be as patient as a gentle stream,
And make a pastime of each weary step                              35
Till the last step have brought me to my love.
And there I'll rest as after much turmoil
A blessèd soul doth in Elysium.
LUCETTA
But in what habit will you go along?
JULIA
Not like a woman, for I would prevent                              40

The loose encounters of lascivious men.
Gentle Lucetta, fit me with such weeds
As may beseem some well-reputed page.
LUCETTA
Why then, your ladyship must cut your hair.
JULIA
No, girl, I'll knit it up in silken strings                        45
With twenty odd-conceited true-love knots.
To be fantastic may become a youth
Of greater time than I shall show to be.
LUCETTA
What fashion, madam, shall I make your breeches?
JULIA
That fits as well as 'Tell me, good my lord,                       50
What compass will you wear your farthingale?'
Why, e'en what fashion thou best likes, Lucetta.
LUCETTA
You must needs have them with a codpiece, madam.
JULIA
Out, out, Lucetta. That will be ill-favoured.
LUCETTA
A round hose, madam, now's not worth a pin                         55
Unless you have a codpiece to stick pins on.
JULIA
Lucetta, as thou lov'st me let me have
What thou think'st meet and is most mannerly.
But tell me, wench, how will the world repute me
For undertaking so unstaid a journey?                              60
I fear me it will make me scandalized.
LUCETTA
If you think so, then stay at home, and go not.
JULIA Nay, that I will not.
LUCETTA
Then never dream on infamy, but go.
If Proteus like your journey when you come,                        65
No matter who's displeased when you are gone.
I fear me he will scarce be pleased withal.
JULIA
That is the least, Lucetta, of my fear.
A thousand oaths, an ocean of his tears,
And instances of infinite of love                                  70
Warrant me welcome to my Proteus.
LUCETTA
All these are servants to deceitful men.
JULIA
Base men, that use them to so base effect.
But truer stars did govern Proteus' birth.
His words are bonds, his oaths are oracles,                        75
His love sincere, his thoughts immaculate,
His tears pure messengers sent from his heart,
His heart as far from fraud as heaven from earth.
LUCETTA
Pray heaven he prove so when you come to him.
JULIA
Now, as thou lov'st me, do him not that wrong       80
To bear a hard opinion of his truth.
Only deserve my love by loving him,
And presently go with me to my chamber
To take a note of what I stand in need of
To furnish me upon my longing journey.                             85
All that is mine I leave at thy dispose,
My goods, my lands, my reputation;
Only in lieu thereof dispatch me hence.
Come, answer not, but to it presently.
I am impatient of my tarriance.                         *Exeunt*

And I as rich in having such a jewel
As twenty seas, if all their sand were pearl,
The water nectar, and the rocks pure gold.
Forgive me that I do not dream on thee          170
Because thou seest me dote upon my love.
My foolish rival, that her father likes
Only for his possessions are so huge,
Is gone with her along, and I must after;
For love, thou know'st, is full of jealousy.    175
PROTEUS  But she loves you?
VALENTINE
Ay, and we are betrothed. Nay more, our marriage
  hour,
With all the cunning manner of our flight,
Determined of: how I must climb her window,
The ladder made of cords, and all the means     180
Plotted and 'greed on for my happiness.
Good Proteus, go with me to my chamber
In these affairs to aid me with thy counsel.
PROTEUS
Go on before. I shall enquire you forth.
I must unto the road, to disembark              185
Some necessaries that I needs must use,
And then I'll presently attend you.
VALENTINE  Will you make haste?
PROTEUS  I will.                        *Exit Valentine*
Even as one heat another heat expels,           190
Or as one nail by strength drives out another,
So the remembrance of my former love
Is by a newer object quite forgotten.
Is it mine eye, or Valentine's praise,
Her true perfection, or my false transgression   195
That makes me, reasonless, to reason thus?
She is fair, and so is Julia that I love—
That I did love, for now my love is thawed,
Which like a waxen image 'gainst a fire
Bears no impression of the thing it was.         200
Methinks my zeal to Valentine is cold,
And that I love him not as I was wont.
O, but I love his lady too-too much,
And that's the reason I love him so little.
How shall I dote on her with more advice,        205
That thus without advice begin to love her?
'Tis but her picture I have yet beheld,
And that hath dazzled my reason's light.
But when I look on her perfections
There is no reason but I shall be blind.         210
If I can check my erring love I will,
If not, to compass her I'll use my skill.       *Exit*

**2.5**    *Enter Speed, and Lance with his dog Crab*
SPEED  Lance, by mine honesty, welcome to Milan.
LANCE  Forswear not thyself, sweet youth, for I am not
  welcome. I reckon this always, that a man is never
  undone till he be hanged, nor never welcome to a place
  till some certain shot be paid and the hostess say
  'Welcome'.                                          6
SPEED  Come on, you madcap. I'll to the alehouse with
  you presently, where, for one shot of five pence, thou
  shalt have five thousand welcomes. But sirrah, how
  did thy master part with Madam Julia?               10
LANCE  Marry, after they closed in earnest they parted
  very fairly in jest.
SPEED  But shall she marry him?
LANCE  No.
SPEED  How then, shall he marry her?                  15
LANCE  No, neither.

SPEED  What, are they broken?
LANCE  No, they are both as whole as a fish.
SPEED  Why then, how stands the matter with them?
LANCE  Marry, thus: when it stands well with him it
  stands well with her.                               21
SPEED  What an ass art thou! I understand thee not.
LANCE  What a block art thou, that thou canst not! My
  staff understands me.
SPEED  What thou sayst?                               25
LANCE  Ay, and what I do too. Look thee, I'll but lean,
  and my staff under-stands me.
SPEED  It stands under thee indeed.
LANCE  Why, stand-under and under-stand is all one.
SPEED  But tell me true, will't be a match?           30
LANCE  Ask my dog. If he say 'Ay', it will. If he say 'No',
  it will. If he shake his tail and say nothing, it will.
SPEED  The conclusion is, then, that it will.
LANCE  Thou shalt never get such a secret from me but
  by a parable.                                       35
SPEED  'Tis well that I get it so. But Lance, how sayst thou
  that my master is become a notable lover?
LANCE  I never knew him otherwise.
SPEED  Than how?
LANCE  A notable lubber, as thou reportest him to be.  40
SPEED  Why, thou whoreson ass, thou mistak'st me.
LANCE  Why, fool, I meant not thee, I meant thy master.
SPEED  I tell thee my master is become a hot lover.
LANCE  Why, I tell thee I care not, though he burn himself
  in love. If thou wilt, go with me to the alehouse. If not,
  thou art an Hebrew, a Jew, and not worth the name
  of a Christian.                                     47
SPEED  Why?
LANCE  Because thou hast not so much charity in thee as
  to go to the ale with a Christian. Wilt thou go?    50
SPEED  At thy service.                          *Exeunt*

**2.6**    *Enter Proteus*
PROTEUS
To leave my Julia shall I be forsworn;
To love fair Silvia shall I be forsworn;
To wrong my friend I shall be much forsworn.
And e'en that power which gave me first my oath
Provokes me to this threefold perjury.             5
Love bade me swear, and love bids me forswear.
O sweet-suggesting love, if thou hast sinned
Teach me, thy tempted subject, to excuse it.
At first I did adore a twinkling star,
But now I worship a celestial sun.                10
Unheedful vows may heedfully be broken,
And he wants wit that wants resolvèd will
To learn his wit t'exchange the bad for better.
Fie, fie, unreverent tongue, to call her bad
Whose sovereignty so oft thou hast preferred      15
With twenty thousand soul-confirming oaths.
I cannot leave to love, and yet I do.
But there I leave to love where I should love.
Julia I lose, and Valentine I lose.
If I keep them I needs must lose myself.          20
If I lose them, thus find I by their loss
For Valentine, myself, for Julia, Silvia.
I to myself am dearer than a friend,
For love is still most precious in itself,
And Silvia—witness heaven that made her fair—     25
Shows Julia but a swarthy Ethiope.
I will forget that Julia is alive,
Rememb'ring that my love to her is dead,
And Valentine I'll hold an enemy,

**3.1**  *Enter Duke, Thurio, and Proteus*

DUKE
Sir Thurio, give us leave, I pray, awhile.
We have some secrets to confer about.          *Exit Thurio*
Now tell me, Proteus, what's your will with me?

PROTEUS
My gracious lord, that which I would discover
The law of friendship bids me to conceal.          5
But when I call to mind your gracious favours
Done to me, undeserving as I am,
My duty pricks me on to utter that
Which else no worldly good should draw from me.
Know, worthy prince, Sir Valentine my friend          10
This night intends to steal away your daughter.
Myself am one made privy to the plot.
I know you have determined to bestow her
On Thurio, whom your gentle daughter hates,
And should she thus be stol'n away from you          15
It would be much vexation to your age.
Thus, for my duty's sake, I rather chose
To cross my friend in his intended drift
Than by concealing it heap on your head
A pack of sorrows which would press you down,          20
Being unprevented, to your timeless grave.

DUKE
Proteus, I thank thee for thine honest care,
Which to requite command me while I live.
This love of theirs myself have often seen,
Haply, when they have judged me fast asleep,          25
And oftentimes have purposed to forbid
Sir Valentine her company and my court.
But fearing lest my jealous aim might err,
And so unworthily disgrace the man—
A rashness that I ever yet have shunned—          30
I gave him gentle looks, thereby to find
That which thyself hast now disclosed to me.
And that thou mayst perceive my fear of this,
Knowing that tender youth is soon suggested,
I nightly lodge her in an upper tower,          35
The key whereof myself have ever kept;
And thence she cannot be conveyed away.

PROTEUS
Know, noble lord, they have devised a mean
How he her chamber-window will ascend,
And with a corded ladder fetch her down,          40
For which the youthful lover now is gone,
And this way comes he with it presently,
Where, if it please you, you may intercept him.
But, good my lord, do it so cunningly
That my discovery be not aimèd at;          45
For love of you, not hate unto my friend,
Hath made me publisher of this pretence.

DUKE
Upon mine honour, he shall never know
That I had any light from thee of this.

PROTEUS
Adieu, my lord. Sir Valentine is coming.          *Exit*
     *Enter Valentine*

DUKE
Sir Valentine, whither away so fast?          51

VALENTINE
Please it your grace, there is a messenger
That stays to bear my letters to my friends,
And I am going to deliver them.

DUKE Be they of much import?          55

VALENTINE
The tenor of them doth but signify
My health and happy being at your court.

DUKE
Nay then, no matter. Stay with me awhile.
I am to break with thee of some affairs
That touch me near, wherein thou must be secret.          60
'Tis not unknown to thee that I have sought
To match my friend Sir Thurio to my daughter.

VALENTINE
I know it well, my lord; and sure the match
Were rich and honourable. Besides, the gentleman
Is full of virtue, bounty, worth, and qualities          65
Beseeming such a wife as your fair daughter.
Cannot your grace win her to fancy him?

DUKE
No, trust me. She is peevish, sullen, froward,
Proud, disobedient, stubborn, lacking duty,
Neither regarding that she is my child          70
Nor fearing me as if I were her father.
And may I say to thee, this pride of hers
Upon advice hath drawn my love from her,
And where I thought the remnant of mine age
Should have been cherished by her child-like duty,          75
I now am full resolved to take a wife,
And turn her out to who will take her in.
Then let her beauty be her wedding dower,
For me and my possessions she esteems not.

VALENTINE
What would your grace have me to do in this?          80

DUKE
There is a lady of Verona here
Whom I affect, but she is nice, and coy,
And naught esteems my agèd eloquence.
Now therefore would I have thee to my tutor—
For long agone I have forgot to court,          85
Besides, the fashion of the time is changed—
How and which way I may bestow myself
To be regarded in her sun-bright eye.

VALENTINE
Win her with gifts if she respect not words.
Dumb jewels often in their silent kind          90
More than quick words do move a woman's mind.

DUKE
But she did scorn a present that I sent her.

VALENTINE
A woman sometime scorns what best contents her.
Send her another. Never give her o'er,
For scorn at first makes after-love the more.          95
If she do frown, 'tis not in hate of you,
But rather to beget more love in you.
If she do chide, 'tis not to have you gone,
Forwhy the fools are mad if left alone.
Take no repulse, whatever she doth say:          100
For 'Get you gone' she doth not mean 'Away'.
Flatter and praise, commend, extol their graces;
Though ne'er so black, say they have angels' faces.
That man that hath a tongue I say is no man
If with his tongue he cannot win a woman.          105

DUKE
But she I mean is promised by her friends
Unto a youthful gentleman of worth,
And kept severely from resort of men,
That no man hath access by day to her.

VALENTINE
Why then I would resort to her by night.          110

DUKE
Ay, but the doors be locked and keys kept safe,
That no man hath recourse to her by night.
VALENTINE
What lets but one may enter at her window?
DUKE
Her chamber is aloft, far from the ground,
And built so shelving that one cannot climb it          115
Without apparent hazard of his life.
VALENTINE
Why then, a ladder quaintly made of cords
To cast up, with a pair of anchoring hooks,
Would serve to scale another Hero's tower,
So bold Leander would adventure it.          120
DUKE
Now as thou art a gentleman of blood,
Advise me where I may have such a ladder.
VALENTINE
When would you use it? Pray sir, tell me that.
DUKE
This very night; for love is like a child
That longs for everything that he can come by.          125
VALENTINE
By seven o'clock I'll get you such a ladder.
DUKE
But hark thee: I will go to her alone.
How shall I best convey the ladder thither?
VALENTINE
It will be light, my lord, that you may bear it
Under a cloak that is of any length.          130
DUKE
A cloak as long as thine will serve the turn?
VALENTINE
Ay, my good lord.
DUKE                    Then let me see thy cloak,
I'll get me one of such another length.
VALENTINE
Why, any cloak will serve the turn, my lord.
DUKE
How shall I fashion me to wear a cloak?          135
I pray thee let me feel thy cloak upon me.
    *He lifts Valentine's cloak and finds a letter and a*
    *rope-ladder*
What letter is this same? What's here? 'To Silvia'?
And here an engine fit for my proceeding.
I'll be so bold to break the seal for once.
(*Reads*)
'My thoughts do harbour with my Silvia nightly,          140
    And slaves they are to me, that send them flying.
O, could their master come and go as lightly,
    Himself would lodge where, senseless, they are lying.
My herald thoughts in thy pure bosom rest them,
    While I, their king, that thither them importune,
Do curse the grace that with such grace hath blessed
    them,          146
    Because myself do want my servants' fortune.
I curse myself for they are sent by me,
    That they should harbour where their lord should be.'
What's here?          150
'Silvia, this night I will enfranchise thee'?
'Tis so, and here's the ladder for the purpose.
Why, Phaëton, for thou art Merops' son
Wilt thou aspire to guide the heavenly car,
And with thy daring folly burn the world?          155
Wilt thou reach stars because they shine on thee?
Go, base intruder, over-weening slave,

Bestow thy fawning smiles on equal mates,
And think my patience, more than thy desert,
Is privilege for thy departure hence.          160
Thank me for this more than for all the favours
Which, all too much, I have bestowed on thee.
But if thou linger in my territories
Longer than swiftest expedition
Will give thee time to leave our royal court,          165
By heaven, my wrath shall far exceed the love
I ever bore my daughter or thyself.
Be gone. I will not hear thy vain excuse,
But as thou lov'st thy life, make speed from hence.
                                        *Exit*
VALENTINE
And why not death, rather than living torment?          170
To die is to be banished from myself,
And Silvia is my self. Banished from her
Is self from self, a deadly banishment.
What light is light, if Silvia be not seen?
What joy is joy, if Silvia be not by—          175
Unless it be to think that she is by,
And feed upon the shadow of perfection.
Except I be by Silvia in the night
There is no music in the nightingale.
Unless I look on Silvia in the day          180
There is no day for me to look upon.
She is my essence, and I leave to be
If I be not by her fair influence
Fostered, illumined, cherished, kept alive.
I fly not death to fly his deadly doom.          185
Tarry I here I but attend on death,
But fly I hence, I fly away from life.
        *Enter Proteus and Lance*
PROTEUS Run, boy, run, run, and seek him out.
LANCE So-ho, so-ho!
PROTEUS What seest thou?          190
LANCE Him we go to find. There's not a hair on's head
    but 'tis a Valentine.
PROTEUS Valentine?
VALENTINE No.
PROTEUS Who then—his spirit?          195
VALENTINE Neither.
PROTEUS What then?
VALENTINE Nothing.
LANCE Can nothing speak?
        *He threatens Valentine*
Master, shall I strike?          200
PROTEUS Who wouldst thou strike?
LANCE Nothing.
PROTEUS Villain, forbear.
LANCE Why, sir, I'll strike nothing. I pray you—
PROTEUS
Sirrah, I say forbear. Friend Valentine, a word.          205
VALENTINE
My ears are stopped, and cannot hear good news,
So much of bad already hath possessed them.
PROTEUS
Then in dumb silence will I bury mine,
For they are harsh, untuneable, and bad.
VALENTINE
Is Silvia dead?
PROTEUS                No, Valentine.          210
VALENTINE
No Valentine indeed, for sacred Silvia.
Hath she forsworn me?
PROTEUS                No, Valentine.

**VALENTINE**
No Valentine, if Silvia have forsworn me.
What is your news?

**LANCE** Sir, there is a proclamation that you are vanished.

**PROTEUS**
That thou art banished. O that's the news:          216
From hence, from Silvia, and from me thy friend.

**VALENTINE**
O, I have fed upon this woe already,
And now excess of it will make me surfeit.
Doth Silvia know that I am banishèd?          220

**PROTEUS**
Ay, ay; and she hath offered to the doom,
Which unreversed stands in effectual force,
A sea of melting pearl, which some call tears.
Those at her father's churlish feet she tendered,
With them, upon her knees, her humble self,          225
Wringing her hands, whose whiteness so became them
As if but now they waxèd pale, for woe.
But neither bended knees, pure hands held up,
Sad sighs, deep groans, nor silver-shedding tears
Could penetrate her uncompassionate sire,          230
But Valentine, if he be ta'en, must die.
Besides, her intercession chafed him so
When she for thy repeal was suppliant
That to close prison he commanded her,
With many bitter threats of biding there.          235

**VALENTINE**
No more, unless the next word that thou speak'st
Have some malignant power upon my life.
If so I pray thee breathe it in mine ear,
As ending anthem of my endless dolour.

**PROTEUS**
Cease to lament for that thou canst not help,          240
And study help for that which thou lament'st.
Time is the nurse and breeder of all good.
Here if thou stay thou canst not see thy love.
Besides, thy staying will abridge thy life.
Hope is a lover's staff. Walk hence with that,          245
And manage it against despairing thoughts.
Thy letters may be here, though thou art hence,
Which, being writ to me, shall be delivered
Even in the milk-white bosom of thy love.
The time now serves not to expostulate.          250
Come, I'll convey thee through the city gate,
And ere I part with thee confer at large
Of all that may concern thy love affairs.
As thou lov'st Silvia, though not for thyself,
Regard thy danger, and along with me.          255

**VALENTINE**
I pray thee, Lance, an if thou seest my boy
Bid him make haste, and meet me at the North Gate.

**PROTEUS**
Go, sirrah, find him out. Come, Valentine.

**VALENTINE**
O my dear Silvia! Hapless Valentine.          259
*Exeunt Proteus and Valentine*

**LANCE** I am but a fool, look you, and yet I have the wit to think my master is a kind of a knave. But that's all one, if he be but one knave. He lives not now that knows me to be in love, yet I am in love, but a team of horse shall not pluck that from me, nor who 'tis I love; and yet 'tis a woman, but what woman I will not tell myself; and yet 'tis a milkmaid; yet 'tis not a maid, for she hath had gossips; yet 'tis a maid, for she is her master's maid, and serves for wages. She hath more qualities than a water-spaniel, which is much in a bare Christian.          270

*He takes out a paper*
Here is the catalogue of her conditions. '*Imprimis*, she can fetch and carry'—why, a horse can do no more. Nay, a horse cannot fetch, but only carry, therefore is she better than a jade. '*Item*, she can milk.' Look you, a sweet virtue in a maid with clean hands.          275

*Enter Speed*

**SPEED** How now, Signor Lance, what news with your mastership?

**LANCE** With my master's ship? Why, it is at sea.

**SPEED** Well, your old vice still, mistake the word. What news then in your paper?          280

**LANCE** The blackest news that ever thou heard'st.

**SPEED** Why, man, how 'black'?

**LANCE** Why, as black as ink.

**SPEED** Let me read them.

**LANCE** Fie on thee, jolt-head, thou canst not read.          285

**SPEED** Thou liest. I can.

**LANCE** I will try thee. Tell me this: who begot thee?

**SPEED** Marry, the son of my grandfather.

**LANCE** O illiterate loiterer, it was the son of thy grand-mother. This proves that thou canst not read.          290

**SPEED** Come, fool, come. Try me in thy paper.

**LANCE** (*giving Speed the paper*) There: and Saint Nicholas be thy speed.

**SPEED** '*Imprimis*, she can milk.'

**LANCE** Ay, that she can.          295

**SPEED** '*Item*, she brews good ale.'

**LANCE** And thereof comes the proverb 'Blessing of your heart, you brew good ale'.

**SPEED** '*Item*, she can sew.'

**LANCE** That's as much as to say 'Can she so?'          300

**SPEED** '*Item*, she can knit.'

**LANCE** What need a man care for a stock with a wench when she can knit him a stock?

**SPEED** '*Item*, she can wash and scour.'

**LANCE** A special virtue, for then she need not be washed and scoured.          306

**SPEED** '*Item*, she can spin.'

**LANCE** Then may I set the world on wheels, when she can spin for her living.

**SPEED** '*Item*, she hath many nameless virtues.'          310

**LANCE** That's as much as to say 'bastard virtues', that indeed know not their fathers, and therefore have no names.

**SPEED** Here follows her vices.

**LANCE** Close at the heels of her virtues.          315

**SPEED** '*Item*, she is not to be broken with fasting, in respect of her breath.'

**LANCE** Well, that fault may be mended with a breakfast. Read on.

**SPEED** '*Item*, she hath a sweet mouth.'          320

**LANCE** That makes amends for her sour breath.

**SPEED** '*Item*, she doth talk in her sleep.'

**LANCE** It's no matter for that, so she sleep not in her talk.

**SPEED** '*Item*, she is slow in words.'

**LANCE** O villain, that set this down among her vices! To be slow in words is a woman's only virtue. I pray thee out with't, and place it for her chief virtue.          327

**SPEED** '*Item*, she is proud.'

**LANCE** Out with that, too. It was Eve's legacy, and cannot be ta'en from her.          330

**SPEED** '*Item*, she hath no teeth.'

**LANCE** I care not for that, neither, because I love crusts.

SPEED '*Item*, she is curst.'

LANCE Well, the best is, she hath no teeth to bite.

SPEED '*Item*, she will often praise her liquor.'          335

LANCE If her liquor be good, she shall. If she will not, I will; for good things should be praised.

SPEED '*Item*, she is too liberal.'

LANCE Of her tongue she cannot, for that's writ down she is slow of. Of her purse she shall not, for that I'll keep shut. Now of another thing she may, and that cannot I help. Well, proceed.          342

SPEED '*Item*, she hath more hair than wit, and more faults than hairs, and more wealth than faults.'

LANCE Stop there. I'll have her. She was mine and not mine twice or thrice in that last article. Rehearse that once more.          347

SPEED '*Item*, she hath more hair than wit'—

LANCE 'More hair than wit.' It may be. I'll prove it: the cover of the salt hides the salt, and therefore it is more than the salt. The hair that covers the wit is more than the wit, for the greater hides the less. What's next?

SPEED 'And more faults than hairs'—          353

LANCE That's monstrous. O that that were out!

SPEED 'And more wealth than faults.'          355

LANCE Why, that word makes the faults gracious. Well, I'll have her, and if it be a match—as nothing is impossible—

SPEED What then?

LANCE Why then will I tell thee that thy master stays for thee at the North Gate.          361

SPEED For me?

LANCE For thee? Ay, who art thou? He hath stayed for a better man than thee.

SPEED And must I go to him?          365

LANCE Thou must run to him, for thou hast stayed so long that going will scarce serve the turn.

SPEED Why didst not tell me sooner? Pox of your love letters!          *Exit*

LANCE Now will he be swinged for reading my letter. An unmannerly slave, that will thrust himself into secrets. I'll after, to rejoice in the boy's correction.          *Exit*

**3.2**     *Enter the Duke and Thurio*

DUKE

Sir Thurio, fear not but that she will love you
Now Valentine is banished from her sight.

THURIO

Since his exile she hath despised me most,
Forsworn my company, and railed at me,
That I am desperate of obtaining her.          5

DUKE

This weak impress of love is as a figure
Trenchèd in ice, which with an hour's heat
Dissolves to water and doth lose his form.
A little time will melt her frozen thoughts,
And worthless Valentine shall be forgot.          10

*Enter Proteus*

How now, Sir Proteus, is your countryman,
According to our proclamation, gone?

PROTEUS Gone, my good lord.

DUKE

My daughter takes his going grievously?

PROTEUS

A little time, my lord, will kill that grief.          15

DUKE

So I believe, but Thurio thinks not so.
Proteus, the good conceit I hold of thee—

For thou hast shown some sign of good desert—
Makes me the better to confer with thee.

PROTEUS

Longer than I prove loyal to your grace          20
Let me not live to look upon your grace.

DUKE

Thou know'st how willingly I would effect
The match between Sir Thurio and my daughter?

PROTEUS I do, my lord.

DUKE

And also, I think, thou art not ignorant          25
How she opposes her against my will?

PROTEUS

She did, my lord, when Valentine was here.

DUKE

Ay, and perversely she persevers so.
What might we do to make the girl forget
The love of Valentine, and love Sir Thurio?          30

PROTEUS

The best way is to slander Valentine
With falsehood, cowardice, and poor descent,
Three things that women highly hold in hate.

DUKE

Ay, but she'll think that it is spoke in hate.

PROTEUS

Ay, if his enemy deliver it.          35
Therefore it must with circumstance be spoken
By one whom she esteemeth as his friend.

DUKE

Then you must undertake to slander him.

PROTEUS

And that, my lord, I shall be loath to do.
'Tis an ill office for a gentleman,          40
Especially against his very friend.

DUKE

Where your good word cannot advantage him
Your slander never can endamage him.
Therefore the office is indifferent,
Being entreated to it by your friend.          45

PROTEUS

You have prevailed, my lord. If I can do it
By aught that I can speak in his dispraise
She shall not long continue love to him.
But say this weed her love from Valentine,
It follows not that she will love Sir Thurio.          50

THURIO

Therefore, as you unwind her love from him,
Lest it should ravel and be good to none
You must provide to bottom it on me;
Which must be done by praising me as much
As you in worth dispraise Sir Valentine.          55

DUKE

And Proteus, we dare trust you in this kind
Because we know, on Valentine's report,
You are already love's firm votary,
And cannot soon revolt, and change your mind.
Upon this warrant shall you have access          60
Where you with Silvia may confer at large.
For she is lumpish, heavy, melancholy,
And for your friend's sake will be glad of you;
Where you may temper her, by your persuasion,
To hate young Valentine and love my friend.          65

PROTEUS

As much as I can do, I will effect.
But you, Sir Thurio, are not sharp enough.
You must lay lime to tangle her desires

By wailful sonnets, whose composèd rhymes
Should be full-fraught with serviceable vows.                    70
DUKE
Ay, much is the force of heaven-bred poesy.
PROTEUS
Say that upon the altar of her beauty
You sacrifice your tears, your sighs, your heart.
Write till your ink be dry, and with your tears
Moist it again; and frame some feeling line                     75
That may discover such integrity;
For Orpheus' lute was strung with poets' sinews,
Whose golden touch could soften steel and stones,
Make tigers tame, and huge leviathans
Forsake unsounded deeps to dance on sands.                      80
After your dire-lamenting elegies,
Visit by night your lady's chamber-window
With some sweet consort. To their instruments
Tune a deploring dump. The night's dead silence
Will well become such sweet-complaining grievance.
This, or else nothing, will inherit her.                        86
DUKE
This discipline shows thou hast been in love.
THURIO
And thy advice this night I'll put in practice.
Therefore, sweet Proteus, my direction-giver,
Let us into the city presently                                  90
To sort some gentlemen well skilled in music.
I have a sonnet that will serve the turn
To give the onset to thy good advice.
DUKE About it, gentlemen.
PROTEUS
We'll wait upon your grace till after supper,                   95
And afterward determine our proceedings.
DUKE
Even now about it. I will pardon you.
        *Exeunt Thurio and Proteus at one door, and the*
                                        *Duke at another*


4.1    *Enter the Outlaws*
FIRST OUTLAW
Fellows, stand fast. I see a passenger.
SECOND OUTLAW
If there be ten, shrink not, but down with 'em.
        *Enter Valentine and Speed*
THIRD OUTLAW
Stand, sir, and throw us that you have about ye.
If not, we'll make you sit, and rifle you.
SPEED (*to Valentine*)
Sir, we are undone. These are the villains                       5
That all the travellers do fear so much.
VALENTINE (*to the Outlaws*) My friends.
FIRST OUTLAW
That's not so, sir. We are your enemies.
SECOND OUTLAW Peace. We'll hear him.
THIRD OUTLAW  Ay, by my beard will we. For he is a proper
    man.                                                        11
VALENTINE
Then know that I have little wealth to lose.
A man I am, crossed with adversity.
My riches are these poor habiliments,
Of which if you should here disfurnish me                       15
You take the sum and substance that I have.
SECOND OUTLAW  Whither travel you?
VALENTINE  To Verona.
FIRST OUTLAW  Whence came you?
VALENTINE  From Milan.                                          20

THIRD OUTLAW  Have you long sojourned there?
VALENTINE
Some sixteen months, and longer might have stayed
If crooked fortune had not thwarted me.
FIRST OUTLAW
What, were you banished thence?
VALENTINE                       I was.
SECOND OUTLAW                          For what offence?
VALENTINE
For that which now torments me to rehearse.                     25
I killed a man, whose death I much repent,
But yet I slew him manfully, in fight,
Without false vantage or base treachery.
FIRST OUTLAW
Why, ne'er repent it, if it were done so.
But were you banished for so small a fault?                     30
VALENTINE
I was, and held me glad of such a doom.
SECOND OUTLAW  Have you the tongues?
VALENTINE
My youthful travel therein made me happy,
Or else I had been often miserable.
THIRD OUTLAW
By the bare scalp of Robin Hood's fat friar,                    35
This fellow were a king for our wild faction.
FIRST OUTLAW
We'll have him. Sirs, a word.
        *The Outlaws confer*
SPEED (*to Valentine*)          Master, be one of them.
It's an honourable kind of thievery.
VALENTINE  Peace, villain.
SECOND OUTLAW
Tell us this: have you anything to take to?                     40
VALENTINE  Nothing but my fortune.
THIRD OUTLAW
Know, then, that some of us are gentlemen
Such as the fury of ungoverned youth
Thrust from the company of aweful men.
Myself was from Verona banishèd                                 45
For practising to steal away a lady,
An heir, and near allied unto the Duke.
SECOND OUTLAW
And I from Mantua, for a gentleman
Who, in my mood, I stabbed unto the heart.
FIRST OUTLAW
And I, for suchlike petty crimes as these.                      50
But to the purpose, for we cite our faults
That they may hold excused our lawless lives.
And partly seeing you are beautified
With goodly shape, and by your own report
A linguist, and a man of such perfection                        55
As we do in our quality much want—
SECOND OUTLAW
Indeed because you are a banished man,
Therefore above the rest we parley to you.
Are you content to be our general,
To make a virtue of necessity                                   60
And live as we do in this wilderness?
THIRD OUTLAW
What sayst thou? Wilt thou be of our consort?
Say 'Ay', and be the captain of us all.
We'll do thee homage, and be ruled by thee,
Love thee as our commander and our king.                        65
FIRST OUTLAW
But if thou scorn our courtesy, thou diest.
SECOND OUTLAW
Thou shalt not live to brag what we have offered.

**VALENTINE**
I take your offer, and will live with you,
Provided that you do no outrages
On silly women or poor passengers.                        70
**THIRD OUTLAW**
No, we detest such vile, base practices.
Come, go with us. We'll bring thee to our crews
And show thee all the treasure we have got,
Which, with ourselves, all rest at thy dispose.      *Exeunt*

**4.2**   *Enter Proteus*
**PROTEUS**
Already have I been false to Valentine,
And now I must be as unjust to Thurio.
Under the colour of commending him
I have access my own love to prefer.
But Silvia is too fair, too true, too holy              5
To be corrupted with my worthless gifts.
When I protest true loyalty to her
She twits me with my falsehood to my friend.
When to her beauty I commend my vows
She bids me think how I have been forsworn             10
In breaking faith with Julia, whom I loved.
And notwithstanding all her sudden quips,
The least whereof would quell a lover's hope,
Yet, spaniel-like, the more she spurns my love,
The more it grows and fawneth on her still.            15
But here comes Thurio. Now must we to her window,
And give some evening music to her ear.
        *Enter Thurio with Musicians*
**THURIO**
How now, Sir Proteus, are you crept before us?
**PROTEUS**
Ay, gentle Thurio, for you know that love
Will creep in service where it cannot go.              20
**THURIO**
Ay, but I hope, sir, that you love not here.
**PROTEUS**
Sir, but I do, or else I would be hence.
**THURIO**
Who, Silvia?
**PROTEUS**              Ay, Silvia—for your sake.
**THURIO**
I thank you for your own. Now, gentlemen,
Let's tune, and to it lustily awhile.                  25
        *Enter the Host, and Julia dressed as a page-boy.*
        *They talk apart*
**HOST** Now, my young guest, methinks you're allycholly.
I pray you, why is it?
**JULIA** Marry, mine host, because I cannot be merry.
**HOST** Come, we'll have you merry. I'll bring you where
you shall hear music, and see the gentleman that you
asked for.                                             31
**JULIA** But shall I hear him speak?
**HOST** Ay, that you shall.
**JULIA** That will be music.
**HOST** Hark, hark.                                   35
**JULIA** Is he among these?
**HOST** Ay. But peace, let's hear 'em.

            *Song*
    Who is Silvia? What is she,
        That all our swains commend her?
    Holy, fair, and wise is she.                        40
        The heaven such grace did lend her
    That she might admirèd be.

    Is she kind as she is fair?
        For beauty lives with kindness.
    Love doth to her eyes repair                         45
        To help him of his blindness,
    And, being helped, inhabits there.

    Then to Silvia let us sing
        That Silvia is excelling.
    She excels each mortal thing                          50
        Upon the dull earth dwelling.
    To her let us garlands bring.

**HOST** How now, are you sadder than you were before?
How do you, man? The music likes you not.
**JULIA** You mistake. The musician likes me not.        55
**HOST** Why, my pretty youth?
**JULIA** He plays false, father.
**HOST** How, out of tune on the strings?
**JULIA** Not so, but yet so false that he grieves my very
heart-strings.                                          60
**HOST** You have a quick ear.
**JULIA** Ay, I would I were deaf. It makes me have a slow
heart.
**HOST** I perceive you delight not in music.
**JULIA** Not a whit when it jars so.                    65
**HOST** Hark what fine change is in the music.
**JULIA** Ay, that 'change' is the spite.
**HOST** You would have them always play but one thing?
**JULIA** I would always have one play but one thing. But
host, doth this Sir Proteus that we talk on often resort
unto this gentlewoman?                                  71
**HOST** I tell you what Lance his man told me, he loved
her out of all nick.
**JULIA** Where is Lance?
**HOST** Gone to seek his dog, which tomorrow, by his
master's command, he must carry for a present to his
lady.                                                   77
**JULIA** Peace, stand aside. The company parts.
**PROTEUS**
Sir Thurio, fear not you. I will so plead
That you shall say my cunning drift excels.             80
**THURIO**
Where meet we?
**PROTEUS**              At Saint Gregory's well.
**THURIO**                              Farewell.
            *Exeunt Thurio and the Musicians*
        *Enter Silvia, above*
**PROTEUS**
Madam, good even to your ladyship.
**SILVIA**
I thank you for your music, gentlemen.
Who is that that spake?
**PROTEUS**
One, lady, if you knew his pure heart's truth           85
You would quickly learn to know him by his voice.
**SILVIA** Sir Proteus, as I take it.
**PROTEUS**
Sir Proteus, gentle lady, and your servant.
**SILVIA**
What's your will?
**PROTEUS**              That I may compass yours.
**SILVIA**
You have your wish. My will is even this,                90
That presently you hie you home to bed.
Thou subtle, perjured, false, disloyal man,
Think'st thou I am so shallow, so conceitless
To be seducèd by thy flattery,

That hast deceived so many with thy vows?                    95
Return, return, and make thy love amends.
For me—by this pale queen of night I swear—
I am so far from granting thy request
That I despise thee for thy wrongful suit,
And by and by intend to chide myself                         100
Even for this time I spend in talking to thee.
PROTEUS
I grant, sweet love, that I did love a lady,
But she is dead.
JULIA (*aside*)          'Twere false if I should speak it,
For I am sure she is not buried.
SILVIA
Say that she be, yet Valentine, thy friend,                  105
Survives, to whom, thyself art witness,
I am betrothed. And art thou not ashamed
To wrong him with thy importunacy?
PROTEUS
I likewise hear that Valentine is dead.
SILVIA
And so suppose am I, for in his grave,                       110
Assure thyself, my love is buried.
PROTEUS
Sweet lady, let me rake it from the earth.
SILVIA
Go to thy lady's grave and call hers thence,
Or at the least, in hers sepulchre thine.
JULIA (*aside*) He heard not that.                           115
PROTEUS
Madam, if your heart be so obdurate,
Vouchsafe me yet your picture for my love,
The picture that is hanging in your chamber.
To that I'll speak, to that I'll sigh and weep;
For since the substance of your perfect self                 120
Is else devoted, I am but a shadow,
And to your shadow will I make true love.
JULIA (*aside*)
If 'twere a substance, you would sure deceive it
And make it but a shadow, as I am.
SILVIA
I am very loath to be your idol, sir,                        125
But since your falsehood shall become you well
To worship shadows and adore false shapes,
Send to me in the morning, and I'll send it.
And so, good rest.                                  *Exit*
PROTEUS            As wretches have o'ernight,
That wait for execution in the morn.               *Exit*
JULIA Host, will you go?                                     131
HOST By my halidom, I was fast asleep.
JULIA Pray you, where lies Sir Proteus?
HOST Marry, at my house. Trust me, I think 'tis almost
day.                                                         135
JULIA
Not so; but it hath been the longest night
That e'er I watched, and the most heaviest.         *Exeunt*

**4.3**   *Enter Sir Eglamour*
EGLAMOUR
This is the hour that Madam Silvia
Entreated me to call, and know her mind.
There's some great matter she'd employ me in.
Madam, madam!
        *Enter Silvia ⌈above⌉*
SILVIA              Who calls?
EGLAMOUR              Your servant, and your friend.
One that attends your ladyship's command.                    5

SILVIA
Sir Eglamour, a thousand times good morrow!
EGLAMOUR
As many, worthy lady, to yourself.
According to your ladyship's impose
I am thus early come, to know what service
It is your pleasure to command me in.                        10
SILVIA
O Eglamour, thou art a gentleman—
Think not I flatter, for I swear I do not—
Valiant, wise, remorseful, well accomplished.
Thou art not ignorant what dear good will
I bear unto the banished Valentine,                          15
Nor how my father would enforce me marry
Vain Thurio, whom my very soul abhors.
Thyself hast loved, and I have heard thee say
No grief did ever come so near thy heart
As when thy lady and thy true love died,                     20
Upon whose grave thou vowed'st pure chastity.
Sir Eglamour, I would to Valentine,
To Mantua, where I hear he makes abode;
And for the ways are dangerous to pass
I do desire thy worthy company,                              25
Upon whose faith and honour I repose.
Urge not my father's anger, Eglamour,
But think upon my grief, a lady's grief,
And on the justice of my flying hence
To keep me from a most unholy match,                         30
Which heaven and fortune still rewards with plagues.
I do desire thee, even from a heart
As full of sorrows as the sea of sands,
To bear me company and go with me.
If not, to hide what I have said to thee                     35
That I may venture to depart alone.
EGLAMOUR
Madam, I pity much your grievances,
Which, since I know they virtuously are placed,
I give consent to go along with you,
Recking as little what betideth me                           40
As much I wish all good befortune you.
When will you go?
SILVIA                      This evening coming.
EGLAMOUR
Where shall I meet you?
SILVIA                      At Friar Patrick's cell,
Where I intend holy confession.
EGLAMOUR
I will not fail your ladyship.                               45
Good morrow, gentle lady.
SILVIA
Good morrow, kind Sir Eglamour.              *Exeunt*

**4.4**   *Enter Lance and his dog Crab*
LANCE (*to the audience*) When a man's servant shall play
   the cur with him, look you, it goes hard. One that I
   brought up of a puppy, one that I saved from drowning
   when three or four of his blind brothers and sisters
   went to it. I have taught him, even as one would say
   precisely 'Thus I would teach a dog'. I was sent to
   deliver him as a present to Mistress Silvia from my
   master, and I came no sooner into the dining-chamber
   but he steps me to her trencher and steals her capon's
   leg. O, 'tis a foul thing when a cur cannot keep himself
   in all companies. I would have, as one should say, one
   that takes upon him to be a dog indeed, to be, as it
   were, a dog at all things. If I had not had more wit
   than he, to take a fault upon me that he did, I think

verily he had been hanged for't. Sure as I live, he had
suffered for't. You shall judge. He thrusts me himself
into the company of three or four gentleman-like dogs
under the Duke's table. He had not been there—bless
the mark—a pissing-while but all the chamber smelled
him. 'Out with the dog,' says one. 'What cur is that?'
says another. 'Whip him out,' says the third. 'Hang
him up,' says the Duke. I, having been acquainted with
the smell before, knew it was Crab, and goes me to the
fellow that whips the dogs. 'Friend,' quoth I, 'you mean
to whip the dog.' 'Ay, marry do I,' quoth he. 'You do
him the more wrong,' quoth I, ''twas I did the thing
you wot of.' He makes me no more ado, but whips me
out of the chamber. How many masters would do this
for his servant? Nay, I'll be sworn I have sat in the
stocks for puddings he hath stolen, otherwise he had
been executed. I have stood on the pillory for geese he
hath killed, otherwise he had suffered for't. (*To Crab*)
Thou think'st not of this now. Nay, I remember the
trick you served me when I took my leave of Madam
Silvia. Did not I bid thee still mark me, and do as I do?
When didst thou see me heave up my leg and make
water against a gentlewoman's farthingale? Didst thou
ever see me do such a trick?                              38

*Enter Proteus, with Julia dressed as a page-boy*

PROTEUS (*to Julia*)
Sebastian is thy name? I like thee well,
And will employ thee in some service presently.       40
JULIA
In what you please. I'll do what I can.
PROTEUS
I hope thou wilt.—How now, you whoreson peasant,
Where have you been these two days loitering?
LANCE Marry, sir, I carried Mistress Silvia the dog you
bade me.                                                45
PROTEUS And what says she to my little jewel?
LANCE Marry, she says your dog was a cur, and tells you
currish thanks is good enough for such a present.
PROTEUS But she received my dog?
LANCE No indeed did she not. Here have I brought him
back again.                                             51
PROTEUS What, didst thou offer her this from me?
LANCE Ay, sir. The other squirrel was stolen from me by
the hangman boys in the market place, and then I
offered her mine own, who is a dog as big as ten of
yours, and therefore the gift the greater.              56
PROTEUS
Go, get thee hence, and find my dog again,
Or ne'er return again into my sight.
Away, I say. Stay'st thou to vex me here?
*Exit Lance with Crab*
A slave, that still on end turns me to shame.           60
Sebastian, I have entertainèd thee
Partly that I have need of such a youth
That can with some discretion do my business,
For 'tis no trusting to yon foolish lout,
But chiefly for thy face and thy behaviour,             65
Which, if my augury deceive me not,
Witness good bringing up, fortune, and truth.
Therefore know thou, for this I entertain thee.
Go presently, and take this ring with thee.
Deliver it to Madam Silvia.                             70
She loved me well delivered it to me.
JULIA
It seems you loved not her, to leave her token.
She is dead belike?
PROTEUS          Not so. I think she lives.

JULIA
Alas.
PROTEUS Why dost thou cry 'Alas'?
JULIA
I cannot choose but pity her.                           75
PROTEUS
Wherefore shouldst thou pity her?
JULIA
Because methinks that she loved you as well
As you do love your lady Silvia.
She dreams on him that has forgot her love;
You dote on her that cares not for your love.           80
'Tis pity love should be so contrary,
And thinking on it makes me cry 'Alas'.
PROTEUS
Well, give her that ring, and therewithal
This letter. (*Pointing*) That's her chamber. Tell my
lady
I claim the promise for her heavenly picture.           85
Your message done, hie home unto my chamber,
Where thou shalt find me sad and solitary.        *Exit*
JULIA
How many women would do such a message?
Alas, poor Proteus, thou hast entertained
A fox to be the shepherd of thy lambs.                  90
Alas, poor fool, why do I pity him
That with his very heart despiseth me?
Because he loves her, he despiseth me.
Because I love him, I must pity him.
This ring I gave him when he parted from me,            95
To bind him to remember my good will.
And now am I, unhappy messenger,
To plead for that which I would not obtain;
To carry that which I would have refused;
To praise his faith, which I would have dispraised.    100
I am my master's true-confirmèd love,
But cannot be true servant to my master
Unless I prove false traitor to myself.
Yet will I woo for him, but yet so coldly
As, heaven it knows, I would not have him speed.       105
*Enter Silvia*
Gentlewoman, good day. I pray you be my mean
To bring me where to speak with Madam Silvia.
SILVIA
What would you with her, if that I be she?
JULIA
If you be she, I do entreat your patience
To hear me speak the message I am sent on.             110
SILVIA From whom?
JULIA
From my master, Sir Proteus, madam.
SILVIA O, he sends you for a picture?
JULIA Ay, madam.
SILVIA Ursula, bring my picture there.                  115
⌜*An attendant brings a picture*⌝
Go, give your master this. Tell him from me
One Julia, that his changing thoughts forget,
Would better fit his chamber than this shadow.
JULIA
Madam, please you peruse this letter.
*She gives Silvia a letter*
Pardon me, madam, I have unadvised                      120
Delivered you a paper that I should not.
*She takes back the letter and gives Silvia another letter*
This is the letter to your ladyship.
SILVIA
I pray thee, let me look on that again.

JULIA
It may not be. Good madam, pardon me.
SILVIA
There, hold. I will not look upon your master's lines.
I know they are stuffed with protestations,    126
And full of new-found oaths, which he will break
As easily as I do tear his paper.
    *She tears the letter*
JULIA
Madam, he sends your ladyship this ring.
    *She offers Silvia a ring*
SILVIA
The more shame for him, that he sends it me;    130
For I have heard him say a thousand times
His Julia gave it him at his departure.
Though his false finger have profaned the ring,
Mine shall not do his Julia so much wrong.
JULIA She thanks you.    135
SILVIA What sayst thou?
JULIA
I thank you, madam, that you tender her.
Poor gentlewoman, my master wrongs her much.
SILVIA Dost thou know her?
JULIA
Almost as well as I do know myself.    140
To think upon her woes I do protest
That I have wept a hundred several times.
SILVIA
Belike she thinks that Proteus hath forsook her?
JULIA
I think she doth; and that's her cause of sorrow.
SILVIA Is she not passing fair?    145
JULIA
She hath been fairer, madam, than she is.
When she did think my master loved her well
She, in my judgement, was as fair as you.
But since she did neglect her looking-glass,
And threw her sun-expelling mask away,    150
The air hath starved the roses in her cheeks
And pinched the lily tincture of her face,
That now she is become as black as I.
SILVIA How tall was she?
JULIA
About my stature; for at Pentecost,    155
When all our pageants of delight were played,
Our youth got me to play the woman's part,
And I was trimmed in Madam Julia's gown,
Which servèd me as fit, by all men's judgements,
As if the garment had been made for me;    160
Therefore I know she is about my height.
And at that time I made her weep agood,
For I did play a lamentable part.
Madam, 'twas Ariadne, passioning
For Theseus' perjury and unjust flight;    165
Which I so lively acted with my tears
That my poor mistress, movèd therewithal,
Wept bitterly; and would I might be dead
If I in thought felt not her very sorrow.
SILVIA
She is beholden to thee, gentle youth.    170
Alas, poor lady, desolate and left.
I weep myself to think upon thy words.
Here, youth. There is my purse. I give thee this
For thy sweet mistress' sake, because thou lov'st her.
Farewell.    *Exit*
JULIA
And she shall thank you for't, if e'er you know her.—

A virtuous gentlewoman, mild, and beautiful.    177
I hope my master's suit will be but cold,
Since she respects 'my mistress'' love so much.
Alas, how love can trifle with itself.    180
Here is her picture. Let me see, I think
If I had such a tire, this face of mine
Were full as lovely as is this of hers.
And yet the painter flattered her a little,
Unless I flatter with myself too much.    185
Her hair is auburn, mine is perfect yellow.
If that be all the difference in his love,
I'll get me such a coloured periwig.
Her eyes are grey as glass, and so are mine.
Ay, but her forehead's low, and mine's as high.    190
What should it be that he respects in her
But I can make respective in myself,
If this fond love were not a blinded god?
Come, shadow, come, and take this shadow up,
For 'tis thy rival.
    *She picks up the portrait*
                    O thou senseless form,    195
Thou shalt be worshipped, kissed, loved, and adored;
And were there sense in his idolatry
My substance should be statue in thy stead.
I'll use thee kindly, for thy mistress' sake,
That used me so; or else, by Jove I vow,    200
I should have scratched out your unseeing eyes,
To make my master out of love with thee.    *Exit*

**5.1**    *Enter Sir Eglamour*
EGLAMOUR
The sun begins to gild the western sky,
And now it is about the very hour
That Silvia at Friar Patrick's cell should meet me.
She will not fail; for lovers break not hours,
Unless it be to come before their time,    5
So much they spur their expedition.
    *Enter Silvia*
See where she comes. Lady, a happy evening!
SILVIA
Amen, amen. Go on, good Eglamour,
Out at the postern by the abbey wall.
I fear I am attended by some spies.    10
EGLAMOUR
Fear not. The forest is not three leagues off.
If we recover that, we are sure enough.    *Exeunt*

**5.2**    *Enter Thurio, Proteus, and Julia dressed as a pageboy*
THURIO
Sir Proteus, what says Silvia to my suit?
PROTEUS
O sir, I find her milder than she was,
And yet she takes exceptions at your person.
THURIO
What? That my leg is too long?
PROTEUS
No, that it is too little.    5
THURIO
I'll wear a boot, to make it somewhat rounder.
JULIA (*aside*)
But love will not be spurred to what it loathes.
THURIO
What says she to my face?
PROTEUS
She says it is a fair one.
THURIO
Nay, then, the wanton lies. My face is black.    10

21

PROTEUS
But pearls are fair; and the old saying is,
'Black men are pearls in beauteous ladies' eyes'.
JULIA (*aside*)
'Tis true, such pearls as put out ladies' eyes,
For I had rather wink than look on them.
THURIO
How likes she my discourse?                                    15
PROTEUS
Ill, when you talk of war.
THURIO
But well when I discourse of love and peace.
JULIA (*aside*)
But better indeed when you hold your peace.
THURIO
What says she to my valour?
PROTEUS
O sir, she makes no doubt of that.                             20
JULIA (*aside*)
She needs not, when she knows it cowardice.
THURIO
What says she to my birth?
PROTEUS
That you are well derived.
JULIA (*aside*)
True: from a gentleman to a fool.
THURIO
Considers she my possessions?                                  25
PROTEUS
O ay, and pities them.
THURIO Wherefore?
JULIA (*aside*)
That such an ass should owe them.
PROTEUS
That they are out by lease.
JULIA                          Here comes the Duke.
          *Enter the Duke*
DUKE
How now, Sir Proteus. How now, Thurio.
Which of you saw Eglamour of late?                             30
THURIO
Not I.
PROTEUS Nor I.
DUKE          Saw you my daughter?
PROTEUS                          Neither.
DUKE
Why then, she's fled unto that peasant Valentine,
And Eglamour is in her company.
'Tis true, for Friar Laurence met them both          35
As he in penance wandered through the forest.
Him he knew well, and guessed that it was she,
But being masked, he was not sure of it.
Besides, she did intend confession
At Patrick's cell this even, and there she was not.  40
These likelihoods confirm her flight from hence;
Therefore I pray you stand not to discourse,
But mount you presently, and meet with me
Upon the rising of the mountain foot
That leads toward Mantua, whither they are fled.     45
Dispatch, sweet gentlemen, and follow me.        *Exit*
THURIO
Why, this it is to be a peevish girl,
That flies her fortune when it follows her.
I'll after, more to be revenged on Eglamour
Than for the love of reckless Silvia.            ⌐*Exit*⌐

PROTEUS
And I will follow, more for Silvia's love             51
Than hate of Eglamour that goes with her.   ⌐*Exit*⌐
JULIA
And I will follow, more to cross that love
Than hate for Silvia, that is gone for love.   ⌐*Exit*⌐

5.3    *Enter the Outlaws with Silvia captive*
FIRST OUTLAW
Come, come, be patient. We must bring you to our
    captain.
SILVIA
A thousand more mischances than this one
Have learned me how to brook this patiently.
SECOND OUTLAW Come, bring her away.
FIRST OUTLAW
Where is the gentleman that was with her?             5
THIRD OUTLAW
Being nimble-footed he hath outrun us;
But Moses and Valerius follow him.
Go thou with her to the west end of the wood.
There is our captain. We'll follow him that's fled.
The thicket is beset, he cannot scape.               10
          *Exeunt the Second and Third Outlaws*
FIRST OUTLAW (*to Silvia*)
Come, I must bring you to our captain's cave.
Fear not. He bears an honourable mind,
And will not use a woman lawlessly.
SILVIA (*aside*)
O Valentine! This I endure for thee.          *Exeunt*

5.4    *Enter Valentine*
VALENTINE
How use doth breed a habit in a man!
This shadowy desert, unfrequented woods
I better brook than flourishing peopled towns.
Here can I sit alone, unseen of any,
And to the nightingale's complaining notes            5
Tune my distresses and record my woes.
O thou that dost inhabit in my breast,
Leave not the mansion so long tenantless
Lest, growing ruinous, the building fall
And leave no memory of what it was.                  10
Repair me with thy presence, Silvia.
Thou gentle nymph, cherish thy forlorn swain.
What hallooing and what stir is this today?
These are my mates, that make their wills their law,
Have some unhappy passenger in chase.                15
They love me well, yet I have much to do
To keep them from uncivil outrages.
Withdraw thee, Valentine. Who's this comes here?
          *He stands aside.*
          *Enter Proteus, Silvia, and Julia dressed as a pageboy*
PROTEUS
Madam, this service I have done for you—
Though you respect not aught your servant doth—   20
To hazard life, and rescue you from him
That would have forced your honour and your love.
Vouchsafe me for my meed but one fair look.
A smaller boon than this I cannot beg,
And less than this I am sure you cannot give.        25
VALENTINE (*aside*)
How like a dream is this I see and hear!
Love lend me patience to forbear awhile.
SILVIA
O miserable, unhappy that I am!

PROTEUS
Unhappy were you, madam, ere I came.
But by my coming I have made you happy.        30
SILVIA
By thy approach thou mak'st me most unhappy.
JULIA (*aside*)
And me, when he approacheth to your presence.
SILVIA
Had I been seizèd by a hungry lion
I would have been a breakfast to the beast
Rather than have false Proteus rescue me.        35
O heaven be judge how I love Valentine,
Whose life's as tender to me as my soul.
And full as much, for more there cannot be,
I do detest false perjured Proteus.
Therefore be gone, solicit me no more.        40
PROTEUS
What dangerous action, stood it next to death,
Would I not undergo for one calm look!
O, 'tis the curse in love, and still approved,
When women cannot love where they're beloved.
SILVIA
When Proteus cannot love where he's beloved.        45
Read over Julia's heart, thy first, best love,
For whose dear sake thou didst then rend thy faith
Into a thousand oaths, and all those oaths
Descended into perjury to love me.
Thou hast no faith left now, unless thou'dst two,        50
And that's far worse than none. Better have none
Than plural faith, which is too much by one,
Thou counterfeit to thy true friend.
PROTEUS                                    In love
Who respects friend?
SILVIA                    All men but Proteus.
PROTEUS
Nay, if the gentle spirit of moving words        55
Can no way change you to a milder form
I'll woo you like a soldier, at arm's end,
And love you 'gainst the nature of love: force ye.
SILVIA
O heaven!
PROTEUS (*assailing her*) I'll force thee yield to my desire.
VALENTINE (*coming forward*)
Ruffian, let go that rude uncivil touch,        60
Thou friend of an ill fashion.
PROTEUS                              Valentine!
VALENTINE
Thou common friend, that's without faith or love,
For such is a friend now. Treacherous man,
Thou hast beguiled my hopes. Naught but mine eye
Could have persuaded me. Now I dare not say        65
I have one friend alive. Thou wouldst disprove me.
Who should be trusted, when one's right hand
Is perjured to the bosom? Proteus,
I am sorry I must never trust thee more,
But count the world a stranger for thy sake.        70
The private wound is deepest. O time most accursed,
'Mongst all foes that a friend should be the worst!
PROTEUS My shame and guilt confounds me.
Forgive me, Valentine. If hearty sorrow
Be a sufficient ransom for offence,        75
I tender't here. I do as truly suffer
As e'er I did commit.
VALENTINE                Then I am paid,
And once again I do receive thee honest.
Who by repentance is not satisfied

Is nor of heaven nor earth. For these are pleased;        80
By penitence th' Eternal's wrath's appeased.
And that my love may appear plain and free,
All that was mine in Silvia I give thee.
JULIA
O me unhappy!
*She faints*
PROTEUS            Look to the boy.
VALENTINE                          Why, boy!
Why wag, how now? What's the matter? Look up.
Speak.        85
JULIA O good sir, my master charged me to deliver a ring
to Madam Silvia, which out of my neglect was never
done.
PROTEUS Where is that ring, boy?
JULIA Here 'tis. This is it.        90
*She gives Proteus the ring*
PROTEUS How, let me see!
Why, this is the ring I gave to Julia.
JULIA
O, cry you mercy, sir, I have mistook.
*She offers Proteus another ring*
This is the ring you sent to Silvia.
PROTEUS
But how cam'st thou by this ring? At my depart        95
I gave this unto Julia.
JULIA
And Julia herself did give it me,
And Julia herself hath brought it hither.
PROTEUS How? Julia?
JULIA
Behold her that gave aim to all thy oaths        100
And entertained 'em deeply in her heart.
How oft hast thou with perjury cleft the root?
O Proteus, let this habit make thee blush.
Be thou ashamed that I have took upon me
Such an immodest raiment, if shame live        105
In a disguise of love.
It is the lesser blot, modesty finds,
Women to change their shapes than men their minds.
PROTEUS
Than men their minds! 'Tis true. O heaven, were man
But constant, he were perfect. That one error        110
Fills him with faults, makes him run through all th'
sins;
Inconstancy falls off ere it begins.
What is in Silvia's face but I may spy
More fresh in Julia's, with a constant eye?
VALENTINE Come, come, a hand from either.        115
Let me be blessed to make this happy close.
'Twere pity two such friends should be long foes.
*Julia and Proteus join hands*
PROTEUS
Bear witness, heaven, I have my wish for ever.
JULIA
And I mine.
*Enter the Outlaws with the Duke and Thurio as
captives*
OUTLAWS        A prize, a prize, a prize!
VALENTINE
Forbear, forbear, I say. It is my lord the Duke.        120
*The Outlaws release the Duke and Thurio*
(*To the Duke*) Your grace is welcome to a man
disgraced,
Banishèd Valentine.
DUKE                    Sir Valentine!

THURIO
  Yonder is Silvia, and Silvia's mine.
VALENTINE
  Thurio, give back, or else embrace thy death.
  Come not within the measure of my wrath.    125
  Do not name Silvia thine. If once again,
  Verona shall not hold thee. Here she stands.
  Take but possession of her with a touch—
  I dare thee but to breathe upon my love.
THURIO
  Sir Valentine, I care not for her, I.    130
  I hold him but a fool that will endanger
  His body for a girl that loves him not.
  I claim her not, and therefore she is thine.
DUKE
  The more degenerate and base art thou
  To make such means for her as thou hast done,    135
  And leave her on such slight conditions.
  Now by the honour of my ancestry
  I do applaud thy spirit, Valentine,
  And think thee worthy of an empress' love.
  Know then I here forget all former griefs,    140
  Cancel all grudge, repeal thee home again,
  Plead a new state in thy unrivalled merit,
  To which I thus subscribe: Sir Valentine,
  Thou art a gentleman, and well derived.
  Take thou thy Silvia, for thou hast deserved her.    145
VALENTINE
  I thank your grace. The gift hath made me happy.
  I now beseech you, for your daughter's sake,
  To grant one boon that I shall ask of you.

DUKE
  I grant it, for thine own, whate'er it be.
VALENTINE
  These banished men that I have kept withal    150
  Are men endowed with worthy qualities.
  Forgive them what they have committed here,
  And let them be recalled from their exile.
  They are reformèd, civil, full of good,
  And fit for great employment, worthy lord.    155
DUKE
  Thou hast prevailed. I pardon them and thee.
  Dispose of them as thou know'st their deserts.
  Come, let us go. We will include all jars
  With triumphs, mirth, and rare solemnity.
VALENTINE
  And as we walk along I dare be bold    160
  With our discourse to make your grace to smile.
  What think you of this page, my lord?
DUKE
  I think the boy hath grace in him. He blushes.
VALENTINE
  I warrant you, my lord, more grace than boy.
DUKE  What mean you by that saying?    165
VALENTINE
  Please you, I'll tell you as we pass along,
  That you will wonder what hath fortunèd.
  Come, Proteus, 'tis your penance but to hear
  The story of your loves discoverèd.
  That done, our day of marriage shall be yours,    170
  One feast, one house, one mutual happiness.    *Exeunt*

# THE TAMING OF THE SHREW

*The Taming of the Shrew* was first published in the 1623 Folio, but a related play, shorter and simpler, with the title *The Taming of a Shrew*, had appeared in print in 1594. The exact relationship of these plays is disputed. *A Shrew* has sometimes been regarded as the source for *The Shrew*; some scholars have believed that both plays derive independently from an earlier play, now lost; it has even been suggested that Shakespeare wrote both plays. In our view Shakespeare's play was written first, not necessarily on the foundation of an earlier play, and *A Shrew* is an anonymous imitation, written in the hope of capitalizing on the success of Shakespeare's play. The difference between the titles is probably no more significant than the fact that *The Winter's Tale* is even now often loosely referred to as *A Winter's Tale*, or *The Comedy of Errors* as *A Comedy of Errors*.

The plot of *The Taming of the Shrew* has three main strands. First comes the Induction showing how a drunken tinker, Christopher Sly, is made to believe himself a lord for whose entertainment a play is to be presented. This resembles an episode in *The Arabian Nights*, in which Caliph Haroun al Raschid plays a similar trick on Abu Hassan. A Latin version of this story was known in Shakespeare's England; it may also have circulated by word of mouth. Second comes the principal plot of the play performed for Sly, in which the shrewish Katherine is wooed, won, and tamed by the fortune-hunting Petruccio. This is a popular narrative theme; Shakespeare may have known a ballad called 'A merry jest of a shrewd and curst wife lapped in morel's skin for her good behaviour', printed around 1550. The third strand of the play involves Lucentio, Gremio, and Hortensio, all of them suitors for the hand of Katherine's sister, Bianca. This is based on the first English prose comedy, George Gascoigne's *Supposes*, translated from Ludovico Ariosto's *I Suppositi* (1509), acted in 1566, and published in 1573. In *The Taming of the Shrew* as printed in the 1623 Folio Christopher Sly fades out after Act 1, Scene 1; in *A Shrew* he makes other appearances, and rounds off the play. These episodes may derive from a version of Shakespeare's play different from that preserved in the Folio; we print them as Additional Passages.

The adapting of Shakespeare's play that seems to have occurred early in its career foreshadows its later history on the stage. Seven versions appeared during the seventeenth and eighteenth centuries, culminating in David Garrick's *Catharine and Petruchio*, first performed in 1754. This version, omitting Christopher Sly and concentrating on the taming story, held the stage almost unchallenged until late in the nineteenth century. In various incarnations *The Taming of the Shrew* has always been popular on the stage, but its reputation as a robust comedy verging on farce has often obscured its more subtle and imaginative aspects, brutalizing Petruccio and trivializing Kate. The Induction, finely written, establishes a fundamentally serious concern with the powers of persuasion to change not merely appearance but reality, and this theme is acted out at different levels in both strands of the subsequent action.

# THE PERSONS OF THE PLAY

# The Taming of the Shrew

**Induction 1** *Enter Christopher Sly the beggar, and the Hostess*

SLY I'll feeze you, in faith.

HOSTESS A pair of stocks, you rogue.

SLY You're a baggage. The Slys are no rogues. Look in the Chronicles—we came in with Richard Conqueror, therefore *paucas palabras*, let the world slide. Sessa! 5

HOSTESS You will not pay for the glasses you have burst?

SLY No, not a denier. Go by, Saint Jeronimy! Go to thy cold bed and warm thee.

HOSTESS I know my remedy, I must go fetch the headborough. *Exit*

SLY Third or fourth or fifth borough, I'll answer him by law. I'll not budge an inch, boy. Let him come, and kindly. 13

*He falls asleep.*
*Horns sound. Enter a Lord from hunting, with his train*

LORD
Huntsman, I charge thee, tender well my hounds.
Breathe Merriman—the poor cur is embossed— 15
And couple Clowder with the deep-mouthed brach.
Saw'st thou not, boy, how Silver made it good
At the hedge corner, in the coldest fault?
I would not lose the dog for twenty pound.

FIRST HUNTSMAN
Why, Belman is as good as he, my lord. 20
He cried upon it at the merest loss,
And twice today picked out the dullest scent.
Trust me, I take him for the better dog.

LORD
Thou art a fool. If Echo were as fleet
I would esteem him worth a dozen such. 25
But sup them well, and look unto them all.
Tomorrow I intend to hunt again.

FIRST HUNTSMAN I will, my lord.

LORD (*seeing Sly*)
What's here? One dead, or drunk? See, doth he breathe?

SECOND HUNTSMAN
He breathes, my lord. Were he not warmed with ale
This were a bed but cold to sleep so soundly. 31

LORD
O monstrous beast! How like a swine he lies.
Grim death, how foul and loathsome is thine image.
Sirs, I will practise on this drunken man.
What think you: if he were conveyed to bed, 35
Wrapped in sweet clothes, rings put upon his fingers,
A most delicious banquet by his bed,
And brave attendants near him when he wakes—
Would not the beggar then forget himself?

FIRST HUNTSMAN
Believe me, lord, I think he cannot choose. 40

SECOND HUNTSMAN
It would seem strange unto him when he waked.

LORD
Even as a flatt'ring dream or worthless fancy.
Then take him up, and manage well the jest.
Carry him gently to my fairest chamber,
And hang it round with all my wanton pictures. 45

Balm his foul head in warm distillèd waters,
And burn sweet wood to make the lodging sweet.
Procure me music ready when he wakes
To make a dulcet and a heavenly sound,
And if he chance to speak be ready straight, 50
And with a low submissive reverence
Say 'What is it your honour will command?'
Let one attend him with a silver basin
Full of rose-water and bestrewed with flowers;
Another bear the ewer, the third a diaper, 55
And say 'Will't please your lordship cool your hands?'
Someone be ready with a costly suit,
And ask him what apparel he will wear.
Another tell him of his hounds and horse,
And that his lady mourns at his disease. 60
Persuade him that he hath been lunatic,
And when he says he is, say that he dreams,
For he is nothing but a mighty lord.
This do, and do it kindly, gentle sirs.
It will be pastime passing excellent, 65
If it be husbanded with modesty.

FIRST HUNTSMAN
My lord, I warrant you we will play our part
As he shall think by our true diligence
He is no less than what we say he is.

LORD
Take him up gently, and to bed with him; 70
And each one to his office when he wakes.

*Servingmen carry Sly out*
*Trumpets sound*

Sirrah, go see what trumpet 'tis that sounds.

*Exit a Servingman*

Belike some noble gentleman that means,
Travelling some journey, to repose him here.

*Enter a Servingman*

How now? Who is it?

SERVINGMAN An't please your honour, players
That offer service to your lordship. 76

*Enter Players*

LORD
Bid them come near. Now fellows, you are welcome.

PLAYERS We thank your honour.

LORD
Do you intend to stay with me tonight?

A PLAYER
So please your lordship to accept our duty. 80

LORD
With all my heart. This fellow I remember
Since once he played a farmer's eldest son.
'Twas where you wooed the gentlewoman so well.
I have forgot your name, but sure that part
Was aptly fitted and naturally performed. 85

ANOTHER PLAYER
I think 'twas Soto that your honour means.

LORD
'Tis very true. Thou didst it excellent.
Well, you are come to me in happy time,
The rather for I have some sport in hand
Wherein your cunning can assist me much. 90
There is a lord will hear you play tonight;

But I am doubtful of your modesties
Lest, over-eyeing of his odd behaviour—
For yet his honour never heard a play—
You break into some merry passion,                                    95
And so offend him; for I tell you, sirs,
If you should smile he grows impatient.
A PLAYER
Fear not, my lord, we can contain ourselves
Were he the veriest antic in the world.
LORD (*to a Servingman*)
Go, sirrah, take them to the buttery                                  100
And give them friendly welcome every one.
Let them want nothing that my house affords.
                                          *Exit one with the Players*
(*To a Servingman*) Sirrah, go you to Barthol'mew, my
        page,
And see him dressed in all suits like a lady.
That done, conduct him to the drunkard's chamber
And call him 'madam', do him obeisance.                               106
Tell him from me, as he will win my love,
He bear himself with honourable action
Such as he hath observed in noble ladies
Unto their lords by them accomplishèd.                                110
Such duty to the drunkard let him do
With soft low tongue and lowly courtesy,
And say 'What is't your honour will command
Wherein your lady and your humble wife
May show her duty and make known her love?'                           115
And then with kind embracements, tempting kisses,
And with declining head into his bosom
Bid him shed tears, as being overjoyed
To see her noble lord restored to health,
Who for this seven years hath esteemèd him                            120
No better than a poor and loathsome beggar.
And if the boy have not a woman's gift
To rain a shower of commanded tears,
An onion will do well for such a shift,
Which, in a napkin being close conveyed,                              125
Shall in despite enforce a watery eye.
See this dispatched with all the haste thou canst.
Anon I'll give thee more instructions.
                                          *Exit a Servingman*
I know the boy will well usurp the grace,
Voice, gait, and action of a gentlewoman.                             130
I long to hear him call the drunkard husband,
And how my men will stay themselves from laughter
When they do homage to this simple peasant.
I'll in to counsel them. Haply my presence
May well abate the over-merry spleen                                  135
Which otherwise would grow into extremes.    *Exeunt*

**Induction 2**   *Enter aloft Sly, the drunkard, with attendants,*
        *some with apparel, basin, and ewer, and other*
        *appurtenances; and Lord*
SLY For God's sake, a pot of small ale!
FIRST SERVINGMAN
Will't please your lordship drink a cup of sack?
SECOND SERVINGMAN
Will't please your honour taste of these conserves?
THIRD SERVINGMAN
What raiment will your honour wear today?                             4
SLY I am Christophero Sly. Call not me 'honour' nor
'lordship'. I ne'er drank sack in my life, and if you give
me any conserves, give me conserves of beef. Ne'er ask
me what raiment I'll wear, for I have no more doublets
than backs, no more stockings than legs, nor no more

shoes than feet—nay, sometime more feet than shoes,
or such shoes as my toes look through the over-leather.
LORD
Heaven cease this idle humour in your honour.        12
O that a mighty man of such descent,
Of such possessions and so high esteem,
Should be infusèd with so foul a spirit.             15
SLY What, would you make me mad? Am not I
Christopher Sly—old Sly's son of Burton Heath, by
birth a pedlar, by education a cardmaker, by transmuta-
tion a bearherd, and now by present profession a
tinker? Ask Marian Hacket, the fat alewife of Wincot,
if she know me not. If she say I am not fourteen pence
on the score for sheer ale, score me up for the lying'st
knave in Christendom. What, I am not bestraught;
here's—
THIRD SERVINGMAN
O, this it is that makes your lady mourn.            25
SECOND SERVINGMAN
O, this is it that makes your servants droop.
LORD
Hence comes it that your kindred shuns your house,
As beaten hence by your strange lunacy.
O noble lord, bethink thee of thy birth.
Call home thy ancient thoughts from banishment,     30
And banish hence these abject lowly dreams.
Look how thy servants do attend on thee,
Each in his office, ready at thy beck.
Wilt thou have music?                        *Music*
                  Hark, Apollo plays,
And twenty cagèd nightingales do sing.              35
Or wilt thou sleep? We'll have thee to a couch
Softer and sweeter than the lustful bed
On purpose trimmed up for Semiramis.
Say thou wilt walk, we will bestrew the ground.
Or wilt thou ride, thy horses shall be trapped,     40
Their harness studded all with gold and pearl.
Dost thou love hawking? Thou hast hawks will soar
Above the morning lark. Or wilt thou hunt,
Thy hounds shall make the welkin answer them
And fetch shrill echoes from the hollow earth.      45
FIRST SERVINGMAN
Say thou wilt course, thy greyhounds are as swift
As breathèd stags, ay, fleeter than the roe.
SECOND SERVINGMAN
Dost thou love pictures? We will fetch thee straight
Adonis painted by a running brook,
And Cytherea all in sedges hid,                     50
Which seem to move and wanton with her breath
Even as the waving sedges play wi'th' wind.
LORD
We'll show thee Io as she was a maid,
And how she was beguilèd and surprised,
As lively painted as the deed was done.             55
THIRD SERVINGMAN
Or Daphne roaming through a thorny wood,
Scratching her legs that one shall swear she bleeds,
And at that sight shall sad Apollo weep,
So workmanly the blood and tears are drawn.
LORD
Thou art a lord, and nothing but a lord.            60
Thou hast a lady far more beautiful
Than any woman in this waning age.
FIRST SERVINGMAN
And till the tears that she hath shed for thee
Like envious floods o'errun her lovely face

She was the fairest creature in the world;            65
And yet she is inferior to none.

SLY
Am I a lord, and have I such a lady?
Or do I dream? Or have I dreamed till now?
I do not sleep. I see, I hear, I speak.
I smell sweet savours, and I feel soft things.            70
Upon my life, I am a lord indeed,
And not a tinker, nor Christopher Sly.
Well, bring our lady hither to our sight,
And once again a pot o'th' smallest ale.

SECOND SERVINGMAN
Will't please your mightiness to wash your hands?  75
O, how we joy to see your wit restored!
O that once more you knew but what you are!
These fifteen years you have been in a dream,
Or when you waked, so waked as if you slept.

SLY
These fifteen years—by my fay, a goodly nap.            80
But did I never speak of all that time?

FIRST SERVINGMAN
O yes, my lord, but very idle words,
For though you lay here in this goodly chamber
Yet would you say ye were beaten out of door,
And rail upon the hostess of the house,            85
And say you would present her at the leet
Because she brought stone jugs and no sealed quarts.
Sometimes you would call out for Cicely Hacket.

SLY Ay, the woman's maid of the house.

THIRD SERVINGMAN
Why, sir, you know no house, nor no such maid,   90
Nor no such men as you have reckoned up,
As Stephen Sly, and old John Naps of Greet,
And Peter Turf, and Henry Pimpernel,
And twenty more such names and men as these,
Which never were, nor no man ever saw.            95

SLY
Now Lord be thankèd for my good amends.

ALL Amen.

SLY I thank thee. Thou shalt not lose by it.

*Enter Bartholomew the Page, as Lady, with*
*attendants*

BARTHOLOMEW
How fares my noble lord?

SLY                              Marry, I fare well,
For here is cheer enough. Where is my wife?            100

BARTHOLOMEW
Here, noble lord. What is thy will with her?

SLY
Are you my wife, and will not call me husband?
My men should call me lord. I am your goodman.

BARTHOLOMEW
My husband and my lord, my lord and husband;
I am your wife in all obedience.            105

SLY
I know it well. (*To the Lord*) What must I call her?

LORD                              Madam.

SLY Al'ce Madam or Joan Madam?

LORD
Madam, and nothing else. So lords call ladies.

SLY
Madam wife, they say that I have dreamed,
And slept above some fifteen year or more.            110

BARTHOLOMEW
Ay, and the time seems thirty unto me,
Being all this time abandoned from your bed.

SLY
'Tis much. Servants, leave me and her alone.
*Exeunt ⌐Lord and¬ attendants*
Madam, undress you and come now to bed.

BARTHOLOMEW
Thrice-noble lord, let me entreat of you            115
To pardon me yet for a night or two,
Or if not so, until the sun be set,
For your physicians have expressly charged,
In peril to incur your former malady,
That I should yet absent me from your bed.            120
I hope this reason stands for my excuse.

SLY Ay, it stands so that I may hardly tarry so long. But
I would be loath to fall into my dreams again. I will
therefore tarry in despite of the flesh and the blood.
*Enter a Messenger*

MESSENGER
Your honour's players, hearing your amendment,  125
Are come to play a pleasant comedy,
For so your doctors hold it very meet,
Seeing too much sadness hath congealed your blood,
And melancholy is the nurse of frenzy.
Therefore they thought it good you hear a play    130
And frame your mind to mirth and merriment,
Which bars a thousand harms and lengthens life.

SLY
Marry, I will let them play it. Is not a comonty
A Christmas gambol, or a tumbling trick?

BARTHOLOMEW
No, my good lord, it is more pleasing stuff.            135

SLY
What, household stuff?

BARTHOLOMEW                              It is a kind of history.

SLY
Well, we'll see't. Come, madam wife, sit by my side
And let the world slip. We shall ne'er be younger.
*Bartholomew sits*

1.1      *Flourish. Enter Lucentio and his man, Tranio*

LUCENTIO
Tranio, since for the great desire I had
To see fair Padua, nursery of arts,
I am arrived fore fruitful Lombardy,
The pleasant garden of great Italy,
And by my father's love and leave am armed            5
With his good will and thy good company,
My trusty servant, well approved in all,
Here let us breathe, and haply institute
A course of learning and ingenious studies.
Pisa, renownèd for grave citizens,            10
Gave me my being, and my father first—
A merchant of great traffic through the world,
Vincentio, come of the Bentivolii,
Vincentio's son, brought up in Florence,
It shall become to serve all hopes conceived            15
To deck his fortune with his virtuous deeds.
And therefore, Tranio, for the time I study,
Virtue and that part of philosophy
Will I apply that treats of happiness
By virtue specially to be achieved.            20
Tell me thy mind, for I have Pisa left
And am to Padua come as he that leaves
A shallow plash to plunge him in the deep,
And with satiety seeks to quench his thirst.

TRANIO
*Mi perdonate*, gentle master mine.            25

I am in all affected as yourself,
Glad that you thus continue your resolve
To suck the sweets of sweet philosophy.
Only, good master, while we do admire
This virtue and this moral discipline,                          30
Let's be no stoics nor no stocks, I pray,
Or so devote to Aristotle's checks
As Ovid be an outcast quite abjured.
Balk logic with acquaintance that you have,
And practise rhetoric in your common talk.                      35
Music and poesy use to quicken you;
The mathematics and the metaphysics,
Fall to them as you find your stomach serves you.
No profit grows where is no pleasure ta'en.
In brief, sir, study what you most affect.                      40
LUCENTIO
Gramercies, Tranio, well dost thou advise.
If, Biondello, thou wert come ashore,
We could at once put us in readiness
And take a lodging fit to entertain
Such friends as time in Padua shall beget.                      45
But stay a while, what company is this?
TRANIO
Master, some show to welcome us to town.

*Enter Baptista with his two daughters, Katherine*
*and Bianca; Gremio, a pantaloon; Hortensio,*
*suitor to Bianca. Lucentio and Tranio stand by*

BAPTISTA
Gentlemen, importune me no farther,
For how I firmly am resolved you know:
That is, not to bestow my youngest daughter               50
Before I have a husband for the elder.
If either of you both love Katherina,
Because I know you well and love you well
Leave shall you have to court her at your pleasure.
GREMIO
To cart her rather. She's too rough for me.               55
There, there, Hortensio. Will you any wife?
KATHERINE (*to Baptista*)
I pray you, sir, is it your will
To make a stale of me amongst these mates?
HORTENSIO
'Mates', maid? How mean you that? No mates for
    you
Unless you were of gentler, milder mould.                 60
KATHERINE
I'faith, sir, you shall never need to fear.
Iwis it is not half-way to her heart,
But if it were, doubt not her care should be
To comb your noddle with a three-legged stool,
And paint your face, and use you like a fool.             65
HORTENSIO
From all such devils, good Lord deliver us.
GREMIO And me too, good Lord.
TRANIO (*aside to Lucentio*)
Husht, master, here's some good pastime toward.
That wench is stark mad or wonderful froward.
LUCENTIO (*aside to Tranio*)
But in the other's silence do I see                       70
Maid's mild behaviour and sobriety.
Peace, Tranio.
TRANIO (*aside to Lucentio*)
Well said, master. Mum, and gaze your fill.
BAPTISTA
Gentlemen, that I may soon make good
What I have said—Bianca, get you in.                      75

And let it not displease thee, good Bianca,
For I will love thee ne'er the less, my girl.
KATHERINE A pretty peat! It is best
Put finger in the eye, an she knew why.
BIANCA
Sister, content you in my discontent.                     80
(*To Baptista*) Sir, to your pleasure humbly I subscribe.
My books and instruments shall be my company,
On them to look and practise by myself.
LUCENTIO (*aside to Tranio*)
Hark, Tranio, thou mayst hear Minerva speak.
HORTENSIO
Signor Baptista, will you be so strange?                  85
Sorry am I that our good will effects
Bianca's grief.
GREMIO              Why will you mew her up,
Signor Baptista, for this fiend of hell,
And make her bear the penance of her tongue?
BAPTISTA
Gentlemen, content ye. I am resolved.                     90
Go in, Bianca.                                *Exit Bianca*
And for I know she taketh most delight
In music, instruments, and poetry,
Schoolmasters will I keep within my house
Fit to instruct her youth. If you, Hortensio,            95
Or, Signor Gremio, you know any such,
Prefer them hither; for to cunning men
I will be very kind, and liberal
To mine own children in good bringing up.
And so farewell. Katherina, you may stay,               100
For I have more to commune with Bianca.         *Exit*
KATHERINE Why, and I trust I may go too, may I not?
What, shall I be appointed hours, as though belike I
knew not what to take and what to leave? Ha!    *Exit*
GREMIO You may go to the devil's dam. Your gifts are so
good here's none will hold you. Their love is not so
great, Hortensio, but we may blow our nails together
and fast it fairly out. Our cake's dough on both sides.
Farewell. Yet for the love I bear my sweet Bianca, if I
can by any means light on a fit man to teach her that
wherein she delights, I will wish him to her father.
HORTENSIO So will I, Signor Gremio. But a word, I pray.
Though the nature of our quarrel yet never brooked
parle, know now, upon advice, it toucheth us both—
that we may yet again have access to our fair mistress
and be happy rivals in Bianca's love—to labour and
effect one thing specially.                              117
GREMIO What's that, I pray?
HORTENSIO Marry, sir, to get a husband for her sister.
GREMIO A husband?—a devil!                               120
HORTENSIO I say a husband.
GREMIO I say a devil. Think'st thou, Hortensio, though
her father be very rich, any man is so very a fool to
be married to hell?                                      124
HORTENSIO Tush, Gremio. Though it pass your patience
and mine to endure her loud alarums, why, man, there
be good fellows in the world, an a man could light on
them, would take her with all faults, and money
enough.                                                  129
GREMIO I cannot tell, but I had as lief take her dowry
with this condition: to be whipped at the high cross
every morning.                                           132
HORTENSIO Faith, as you say, there's small choice in rotten
apples. But come, since this bar in law makes us friends,
it shall be so far forth friendly maintained till by helping
Baptista's eldest daughter to a husband we set his
youngest free for a husband, and then have to't afresh.

Sweet Bianca! Happy man be his dole. He that runs
fastest gets the ring. How say you, Signor Gremio?
GREMIO I am agreed, and would I had given him the best
horse in Padua to begin his wooing that would
thoroughly woo her, wed her, and bed her, and rid the
house of her. Come on.                                        143
      *Exeunt Hortensio and Gremio. Tranio and Lucentio*
                                                    *remain*
TRANIO
   I pray, sir, tell me: is it possible
   That love should of a sudden take such hold?             145
LUCENTIO
   O Tranio, till I found it to be true
   I never thought it possible or likely.
   But see, while idly I stood looking on
   I found the effect of love in idleness,
   And now in plainness do confess to thee,                 150
   That art to me as secret and as dear
   As Anna to the Queen of Carthage was,
   Tranio, I burn, I pine, I perish, Tranio,
   If I achieve not this young modest girl.
   Counsel me, Tranio, for I know thou canst.               155
   Assist me, Tranio, for I know thou wilt.
TRANIO
   Master, it is no time to chide you now.
   Affection is not rated from the heart.
   If love have touched you, naught remains but so—
   *Redime te captum quam queas minimo.*                    160
LUCENTIO
   Gramercies, lad. Go forward, this contents.
   The rest will comfort, for thy counsel's sound.
TRANIO
   Master, you looked so longly on the maid
   Perhaps you marked not what's the pith of all.
LUCENTIO
   O yes, I saw sweet beauty in her face,                   165
   Such as the daughter of Agenor had,
   That made great Jove to humble him to her hand
   When with his knees he kissed the Cretan strand.
TRANIO
   Saw you no more? Marked you not how her sister
   Began to scold and raise up such a storm                 170
   That mortal ears might hardly endure the din?
LUCENTIO
   Tranio, I saw her coral lips to move,
   And with her breath she did perfume the air.
   Sacred and sweet was all I saw in her.
TRANIO (*aside*)
   Nay, then 'tis time to stir him from his trance.         175
   (*To Lucentio*) I pray, awake, sir. If you love the maid,
   Bend thoughts and wits to achieve her. Thus it
      stands:
   Her elder sister is so curst and shrewd
   That till the father rid his hands of her,
   Master, your love must live a maid at home,              180
   And therefore has he closely mewed her up
   Because she will not be annoyed with suitors.
LUCENTIO
   Ah, Tranio, what a cruel father's he!
   But art thou not advised he took some care
   To get her cunning schoolmasters to instruct her?       185
TRANIO
   Ay, marry am I, sir, and now 'tis plotted.
LUCENTIO
   I have it, Tranio.
TRANIO               Master, for my hand,
   Both our inventions meet and jump in one.

LUCENTIO
   Tell me thine first.
TRANIO                You will be schoolmaster
   And undertake the teaching of the maid.                  190
   That's your device.
LUCENTIO              It is. May it be done?
TRANIO
   Not possible; for who shall bear your part,
   And be in Padua here Vincentio's son,
   Keep house, and ply his book, welcome his friends,
   Visit his countrymen, and banquet them?                  195
LUCENTIO
   *Basta*, content thee, for I have it full.
   We have not yet been seen in any house,
   Nor can we be distinguished by our faces
   For man or master. Then it follows thus:
   Thou shalt be master, Tranio, in my stead;               200
   Keep house, and port, and servants, as I should.
   I will some other be, some Florentine,
   Some Neapolitan, or meaner man of Pisa.
   'Tis hatched, and shall be so. Tranio, at once
   Uncase thee. Take my coloured hat and cloak.             205
   When Biondello comes he waits on thee,
   But I will charm him first to keep his tongue.
TRANIO So had you need.
         ⌈*They exchange clothes*⌉
   In brief, sir, sith it your pleasure is,
   And I am tied to be obedient—                            210
   For so your father charged me at our parting,
   'Be serviceable to my son,' quoth he,
   Although I think 'twas in another sense—
   I am content to be Lucentio
   Because so well I love Lucentio.                         215
LUCENTIO
   Tranio, be so, because Lucentio loves,
   And let me be a slave t'achieve that maid
   Whose sudden sight hath thralled my wounded eye.
         *Enter Biondello*
   Here comes the rogue. Sirrah, where have you been?
BIONDELLO Where have *I* been? Nay, how now, where
   are *you*? Master, has my fellow Tranio stolen your
   clothes, or you stolen his, or both? Pray, what's the
   news?
LUCENTIO
   Sirrah, come hither. 'Tis no time to jest,
   And therefore frame your manners to the time.            225
   Your fellow Tranio here, to save my life
   Puts my apparel and my count'nance on,
   And I for my escape have put on his,
   For in a quarrel since I came ashore
   I killed a man, and fear I was descried.                 230
   Wait you on him, I charge you, as becomes,
   While I make way from hence to save my life.
   You understand me?
BIONDELLO            I sir? Ne'er a whit.
LUCENTIO
   And not a jot of Tranio in your mouth.
   Tranio is changed into Lucentio.                         235
BIONDELLO
   The better for him. Would I were so too.
TRANIO
   So could I, faith, boy, to have the next wish after—
   That Lucentio indeed had Baptista's youngest
      daughter.
   But sirrah, not for my sake but your master's I advise
   You use your manners discreetly in all kind of
      companies.                                            240

When I am alone, why then I am Tranio,
But in all places else your master, Lucentio.
LUCENTIO Tranio, let's go.
One thing more rests that thyself execute—
To make one among these wooers. If thou ask me
    why,                                                    245
Sufficeth my reasons are both good and weighty.
                                                    *Exeunt*

    *The presenters above speak*
FIRST SERVINGMAN
My lord, you nod. You do not mind the play.
SLY Yes, by Saint Anne do I. A good matter, surely.
Comes there any more of it?
BARTHOLOMEW My lord, 'tis but begun.            250
SLY 'Tis a very excellent piece of work, madam lady.
Would 'twere done.
    *They sit and mark*

1.2    *Enter Petruccio and his man, Grumio*
PETRUCCIO
Verona, for a while I take my leave
To see my friends in Padua; but of all
My best-belovèd and approvèd friend
Hortensio, and I trow this is his house.
Here, sirrah Grumio, knock, I say.              5
GRUMIO Knock, sir? Whom should I knock? Is there any
man has rebused your worship?
PETRUCCIO Villain, I say, knock me here soundly.
GRUMIO Knock you here, sir? Why, sir, what am I, sir,
that I should knock you here, sir?              10
PETRUCCIO
Villain, I say, knock me at this gate,
And rap me well or I'll knock your knave's pate.
GRUMIO
My master is grown quarrelsome. I should knock you
    first,
And then I know after who comes by the worst.
PETRUCCIO Will it not be?                       15
Faith, sirrah, an you'll not knock, I'll ring it.
I'll try how you can sol-fa and sing it.
    *He wrings him by the ears.* ⌈*Grumio kneels*⌉
GRUMIO Help, masters, help! My master is mad.
PETRUCCIO Now knock when I bid you, sirrah villain.
    *Enter Hortensio*
HORTENSIO How now, what's the matter? My old friend
Grumio and my good friend Petruccio? How do you all
at Verona?                                      22
PETRUCCIO
Signor Hortensio, come you to part the fray?
*Con tutto il cuore ben trovato,* may I say.
HORTENSIO *Alla nostra casa ben venuto, molto onorato signor
mio Petruccio.*                                 26
Rise, Grumio, rise. We will compound this quarrel.
    *Grumio rises*
GRUMIO Nay, 'tis no matter, sir, what he 'leges in Latin.
If this be not a lawful cause for me to leave his service—
look you, sir: he bid me knock him and rap him
soundly, sir. Well, was it fit for a servant to use his
master so, being perhaps, for aught I see, two-and-
thirty, a pip out?
Whom would to God I had well knocked at first,
Then had not Grumio come by the worst.          35
PETRUCCIO
A senseless villain. Good Hortensio,
I bade the rascal knock upon your gate,
And could not get him for my heart to do it.
GRUMIO Knock at the gate? O heavens, spake you not

these words plain? 'Sirrah, knock me here, rap me
here, knock me well, and knock me soundly'? And
come you now with knocking at the gate?         42
PETRUCCIO
Sirrah, be gone, or talk not, I advise you.
HORTENSIO
Petruccio, patience. I am Grumio's pledge.
Why this' a heavy chance 'twixt him and you,    45
Your ancient, trusty, pleasant servant Grumio.
And tell me now, sweet friend, what happy gale
Blows you to Padua here from old Verona?
PETRUCCIO
Such wind as scatters young men through the world
To seek their fortunes farther than at home,    50
Where small experience grows. But in a few,
Signor Hortensio, thus it stands with me:
Antonio, my father, is deceased,
And I have thrust myself into this maze
Happily to wive and thrive as best I may.       55
Crowns in my purse I have, and goods at home,
And so am come abroad to see the world.
HORTENSIO
Petruccio, shall I then come roundly to thee
And wish thee to a shrewd, ill-favoured wife?
Thou'dst thank me but a little for my counsel,  60
And yet I'll promise thee she shall be rich,
And very rich. But thou'rt too much my friend,
And I'll not wish thee to her.
PETRUCCIO
Signor Hortensio, 'twixt such friends as we
Few words suffice; and therefore, if thou know  65
One rich enough to be Petruccio's wife—
As wealth is burden of my wooing dance—
Be she as foul as was Florentius' love,
As old as Sibyl, and as curst and shrewd
As Socrates' Xanthippe or a worse,              70
She moves me not—or not removes at least
Affection's edge in me, were she as rough
As are the swelling Adriatic seas.
I come to wive it wealthily in Padua;
If wealthily, then happily in Padua.            75
GRUMIO (*to Hortensio*) Nay, look you, sir, he tells you flatly
what his mind is. Why, give him gold enough and
marry him to a puppet or an aglet-baby, or an old trot
with ne'er a tooth in her head, though she have as
many diseases as two-and-fifty horses. Why, nothing
comes amiss so money comes withal.              81
HORTENSIO
Petruccio, since we are stepped thus far in,
I will continue that I broached in jest.
I can, Petruccio, help thee to a wife
With wealth enough, and young and beauteous,    85
Brought up as best becomes a gentlewoman.
Her only fault—and that is faults enough—
Is that she is intolerable curst,
And shrewd and froward so beyond all measure
That, were my state far worser than it is,      90
I would not wed her for a mine of gold.
PETRUCCIO
Hortensio, peace. Thou know'st not gold's effect.
Tell me her father's name and 'tis enough,
For I will board her though she chide as loud
As thunder when the clouds in autumn crack.     95
HORTENSIO
Her father is Baptista Minola,
An affable and courteous gentleman.

Her name is Katherina Minola,
Renowned in Padua for her scolding tongue.

PETRUCCIO
I know her father, though I know not her,          100
And he knew my deceasèd father well.
I will not sleep, Hortensio, till I see her,
And therefore let me be thus bold with you
To give you over at this first encounter,
Unless you will accompany me thither.          105

GRUMIO I pray you, sir, let him go while the humour
lasts. O' my word, an she knew him as well as I do
she would think scolding would do little good upon
him. She may perhaps call him half a score knaves or
so. Why, that's nothing; an he begin once he'll rail in
his rope-tricks. I'll tell you what, sir, an she stand him
but a little he will throw a figure in her face and so
disfigure her with it that she shall have no more eyes
to see withal than a cat. You know him not, sir.

HORTENSIO
Tarry, Petruccio, I must go with thee,          115
For in Baptista's keep my treasure is.
He hath the jewel of my life in hold,
His youngest daughter, beautiful Bianca,
And her withholds from me and other more,
Suitors to her and rivals in my love,          120
Supposing it a thing impossible,
For those defects I have before rehearsed,
That ever Katherina will be wooed.
Therefore this order hath Baptista ta'en:
That none shall have access unto Bianca          125
Till Katherine the curst have got a husband.

GRUMIO Katherine the curst—
A title for a maid of all titles the worst.

HORTENSIO
Now shall my friend Petruccio do me grace,
And offer me disguised in sober robes          130
To old Baptista as a schoolmaster
Well seen in music, to instruct Bianca,
That so I may by this device at least
Have leave and leisure to make love to her,
And unsuspected court her by herself.          135

*Enter Gremio with a paper, and Lucentio disguised
as a schoolmaster*

GRUMIO Here's no knavery. See, to beguile the old folks,
how the young folks lay their heads together. Master,
master, look about you. Who goes there, ha?

HORTENSIO
Peace, Grumio, it is the rival of my love.
Petruccio, stand by a while.          140

GRUMIO
A proper stripling, and an amorous!

*Petruccio, Hortensio, and Grumio stand aside*

GREMIO (*to Lucentio*)
O, very well—I have perused the note.
Hark you, sir, I'll have them very fairly bound—
All books of love, see that at any hand—
And see you read no other lectures to her.          145
You understand me. Over and beside
Signor Baptista's liberality,
I'll mend it with a largess. Take your paper, too,
And let me have them very well perfumed,
For she is sweeter than perfume itself          150
To whom they go to. What will you read to her?

LUCENTIO
Whate'er I read to her, I'll plead for you
As for my patron, stand you so assured,

As firmly as yourself were still in place—
Yea, and perhaps with more successful words          155
Than you, unless you were a scholar, sir.

GREMIO
O this learning, what a thing it is!

GRUMIO (*aside*)
O this woodcock, what an ass it is!

PETRUCCIO Peace, sirrah.

HORTENSIO
Grumio, mum. (*Coming forward*) God save you, Signor
Gremio.          160

GREMIO
And you are well met, Signor Hortensio.
Trow you whither I am going?
To Baptista Minola.
I promised to enquire carefully
About a schoolmaster for the fair Bianca,          165
And by good fortune I have lighted well
On this young man, for learning and behaviour
Fit for her turn, well read in poetry
And other books—good ones, I warrant ye.

HORTENSIO
'Tis well, and I have met a gentleman          170
Hath promised me to help me to another,
A fine musician, to instruct our mistress.
So shall I no whit be behind in duty
To fair Bianca, so beloved of me.

GREMIO
Beloved of me, and that my deeds shall prove.          175

GRUMIO (*aside*) And that his bags shall prove.

HORTENSIO
Gremio, 'tis now no time to vent our love.
Listen to me, and if you speak me fair
I'll tell you news indifferent good for either.
Here is a gentleman whom by chance I met,          180
Upon agreement from us to his liking
Will undertake to woo curst Katherine,
Yea, and to marry her, if her dowry please.

GREMIO So said, so done, is well.
Hortensio, have you told him all her faults?          185

PETRUCCIO
I know she is an irksome brawling scold.
If that be all, masters, I hear no harm.

GREMIO
No, sayst me so, friend? What countryman?

PETRUCCIO
Born in Verona, old Antonio's son.
My father dead, his fortune lives for me,          190
And I do hope good days and long to see.

GREMIO O sir, such a life with such a wife were strange.
But if you have a stomach, to't, a' God's name.
You shall have me assisting you in all.
But will you woo this wildcat?     Will I live!     195

GRUMIO
Will he woo her? Ay, or I'll hang her.

PETRUCCIO
Why came I hither but to that intent?
Think you a little din can daunt mine ears?
Have I not in my time heard lions roar?
Have I not heard the sea, puffed up with winds,          200
Rage like an angry boar chafèd with sweat?
Have I not heard great ordnance in the field,
And heaven's artillery thunder in the skies?
Have I not in a pitchèd battle heard
Loud 'larums, neighing steeds, and trumpets' clang?

And do you tell me of a woman's tongue,                      206
That gives not half so great a blow to hear
As will a chestnut in a farmer's fire?
Tush, tush—fear boys with bugs.
GRUMIO  For he fears none.                                   210
GREMIO  Hortensio, hark.
This gentleman is happily arrived,
My mind presumes, for his own good and ours.
HORTENSIO
I promised we would be contributors,
And bear his charge of wooing, whatsoe'er.                   215
GREMIO
And so we will, provided that he win her.
GRUMIO
I would I were as sure of a good dinner.
    *Enter Tranio, brave, as Lucentio, and Biondello*
TRANIO  Gentlemen, God save you. If I may be bold, tell
me, I beseech you, which is the readiest way to the
house of Signor Baptista Minola?                            220
BIONDELLO  He that has the two fair daughters—is't he
you mean?
TRANIO  Even he, Biondello.
GREMIO
Hark you, sir, you mean not her to—
TRANIO
Perhaps him and her, sir. What have you to do?             225
PETRUCCIO
Not her that chides, sir, at any hand, I pray.
TRANIO
I love no chiders, sir. Biondello, let's away.
LUCENTIO  (*aside*)
Well begun, Tranio.
HORTENSIO                        Sir, a word ere you go.
Are you a suitor to the maid you talk of—yea or no?
TRANIO
And if I be, sir, is it any offence?                        230
GREMIO
No, if without more words you will get you hence.
TRANIO
Why, sir, I pray, are not the streets as free
For me as for you?
GREMIO                        But so is not she.
TRANIO
For what reason, I beseech you?
GREMIO
For this reason, if you'll know—                           235
That she's the choice love of Signor Gremio.
HORTENSIO
That she's the chosen of Signor Hortensio.
TRANIO
Softly, my masters. If you be gentlemen,
Do me this right, hear me with patience.
Baptista is a noble gentleman                               240
To whom my father is not all unknown,
And were his daughter fairer than she is
She may more suitors have, and me for one.
Fair Leda's daughter had a thousand wooers;
Then well one more may fair Bianca have,                    245
And so she shall. Lucentio shall make one,
Though Paris came, in hope to speed alone.
GREMIO
What, this gentleman will out-talk us all!
LUCENTIO
Sir, give him head, I know he'll prove a jade.
PETRUCCIO
Hortensio, to what end are all these words?                 250

HORTENSIO
Sir, let me be so bold as ask you,
Did you yet ever see Baptista's daughter?
TRANIO
No, sir, but hear I do that he hath two,
The one as famous for a scolding tongue
As is the other for beauteous modesty.                      255
PETRUCCIO
Sir, sir, the first's for me. Let her go by.
GREMIO
Yea, leave that labour to great Hercules,
And let it be more than Alcides' twelve.
PETRUCCIO
Sir, understand you this of me in sooth,
The youngest daughter whom you hearken for                  260
Her father keeps from all access of suitors,
And will not promise her to any man
Until the elder sister first be wed.
The younger then is free, and not before.
TRANIO
If it be so, sir, that you are the man                      265
Must stead us all, and me amongst the rest,
And if you break the ice and do this feat,
Achieve the elder, set the younger free
For our access, whose hap shall be to have her
Will not so graceless be to be ingrate.                     270
HORTENSIO
Sir, you say well, and well you do conceive;
And since you do profess to be a suitor
You must, as we do, gratify this gentleman,
To whom we all rest generally beholden.
TRANIO
Sir, I shall not be slack. In sign whereof,                 275
Please ye we may contrive this afternoon,
And quaff carouses to our mistress' health,
And do as adversaries do in law—
Strive mightily, but eat and drink as friends.
GRUMIO *and* BIONDELLO
O excellent motion! Fellows, let's be gone.                 280
HORTENSIO
The motion's good indeed, and be it so.
Petruccio, I shall be your *ben venuto*.          *Exeunt*

**2.1**    *Enter Katherina and Bianca, her hands bound*
BIANCA
Good sister, wrong me not, nor wrong yourself
To make a bondmaid and a slave of me.
That I disdain, but for these other goods,
Unbind my hands, I'll pull them off myself,
Yea, all my raiment to my petticoat,                         5
Or what you will command me will I do,
So well I know my duty to my elders.
KATHERINE
Of all thy suitors here I charge thee tell
Whom thou lov'st best. See thou dissemble not.
BIANCA
Believe me, sister, of all the men alive                    10
I never yet beheld that special face
Which I could fancy more than any other.
KATHERINE
Minion, thou liest. Is't not Hortensio?
BIANCA
If you affect him, sister, here I swear
I'll plead for you myself but you shall have him.           15
KATHERINE
O then, belike you fancy riches more.
You will have Gremio to keep you fair.

BIANCA
Is it for him you do envy me so?
Nay, then, you jest, and now I well perceive
You have but jested with me all this while.          20
I prithee, sister Kate, untie my hands.
KATHERINE (*strikes her*)
If that be jest, then all the rest was so.
        *Enter Baptista*
BAPTISTA
Why, how now, dame, whence grows this insolence?
Bianca, stand aside.—Poor girl, she weeps.—
Go ply thy needle, meddle not with her.            25
(*To Katherine*) For shame, thou hilding of a devilish
        spirit,
Why dost thou wrong her that did ne'er wrong thee?
When did she cross thee with a bitter word?
KATHERINE
Her silence flouts me, and I'll be revenged.
        *She flies after Bianca*
BAPTISTA
What, in my sight? Bianca, get thee in.     *Exit Bianca*
KATHERINE
What, will you not suffer me? Nay, now I see       31
She is your treasure, she must have a husband.
I must dance barefoot on her wedding day,
And for your love to her lead apes in hell.
Talk not to me. I will go sit and weep             35
Till I can find occasion of revenge.              *Exit*
BAPTISTA
Was ever gentleman thus grieved as I?
But who comes here?
        *Enter Gremio, Lucentio as a schoolmaster in the
        habit of a mean man, Petruccio with Hortensio as a
        musician, Tranio as Lucentio, with Biondello his
        boy bearing a lute and books*
GREMIO  Good morrow, neighbour Baptista.
BAPTISTA  Good morrow, neighbour Gremio. God save you,
gentlemen.                                         41
PETRUCCIO
And you, good sir. Pray, have you not a daughter
Called Katherina, fair and virtuous?
BAPTISTA
I have a daughter, sir, called Katherina.
GREMIO
You are too blunt. Go to it orderly.              45
PETRUCCIO
You wrong me, Signor Gremio. Give me leave.
(*To Baptista*) I am a gentleman of Verona, sir,
That hearing of her beauty and her wit,
Her affability and bashful modesty,
Her wondrous qualities and mild behaviour,        50
Am bold to show myself a forward guest
Within your house to make mine eye the witness
Of that report which I so oft have heard,
And for an entrance to my entertainment
I do present you with a man of mine (*presenting
        Hortensio*)                                55
Cunning in music and the mathematics
To instruct her fully in those sciences,
Whereof I know she is not ignorant.
Accept of him, or else you do me wrong.
His name is Licio, born in Mantua.               60
BAPTISTA
You're welcome, sir, and he for your good sake.
But for my daughter, Katherine, this I know:
She is not for your turn, the more my grief.

PETRUCCIO
I see you do not mean to part with her,
Or else you like not of my company.              65
BAPTISTA
Mistake me not, I speak but as I find.
Whence are you, sir? What may I call your name?
PETRUCCIO
Petruccio is my name, Antonio's son,
A man well known throughout all Italy.
BAPTISTA
I know him well. You are welcome for his sake.   70
GREMIO
Saving your tale, Petruccio, I pray
Let us that are poor petitioners speak too.
*Baccare*, you are marvellous forward.
PETRUCCIO
O pardon me, Signor Gremio, I would fain be doing.
GREMIO
I doubt it not, sir. But you will curse your wooing.  75
(*To Baptista*) Neighbour, this is a gift very grateful, I
am sure of it. To express the like kindness, myself, that
have been more kindly beholden to you than any,
freely give unto you this young scholar (*presenting
Lucentio*) that hath been long studying at Rheims, as
cunning in Greek, Latin, and other languages as the
other in music and mathematics. His name is Cambio.
Pray accept his service.                           83
BAPTISTA  A thousand thanks, Signor Gremio. Welcome,
good Cambio. (*To Tranio*) But, gentle sir, methinks you
walk like a stranger. May I be so bold to know the
cause of your coming?                              87
TRANIO
Pardon me, sir, the boldness is mine own
That, being a stranger in this city here,
Do make myself a suitor to your daughter,         90
Unto Bianca, fair and virtuous.
Nor is your firm resolve unknown to me
In the preferment of the eldest sister.
This liberty is all that I request:
That upon knowledge of my parentage               95
I may have welcome 'mongst the rest that woo,
And free access and favour as the rest.
And toward the education of your daughters
I here bestow a simple instrument,
And this small packet of Greek and Latin books.  100
If you accept them, then their worth is great.
BAPTISTA
Lucentio is your name—of whence, I pray?
TRANIO
Of Pisa, sir, son to Vincentio.
BAPTISTA
A mighty man of Pisa. By report
I know him well. You are very welcome, sir.      105
(*To Hortensio*) Take you the lute, (*to Lucentio*) and you
        the set of books.
You shall go see your pupils presently.
Holla, within!
        *Enter a Servant*
                Sirrah, lead these gentlemen
To my daughters, and tell them both
These are their tutors. Bid them use them well.  110
        *Exit Servant with Lucentio and Hortensio,*
                ⌈*Biondello following*⌉
(*To Petruccio*) We will go walk a little in the orchard,
And then to dinner. You are passing welcome—
And so I pray you all to think yourselves.

PETRUCCIO
Signor Baptista, my business asketh haste,
And every day I cannot come to woo.                          115
You knew my father well, and in him me,
Left solely heir to all his lands and goods,
Which I have bettered rather than decreased.
Then tell me, if I get your daughter's love,
What dowry shall I have with her to wife?                    120
BAPTISTA
After my death the one half of my lands,
And in possession twenty thousand crowns.
PETRUCCIO
And for that dowry I'll assure her of
Her widowhood, be it that she survive me,
In all my lands and leases whatsoever.                       125
Let specialties be therefore drawn between us,
That covenants may be kept on either hand.
BAPTISTA
Ay, when the special thing is well obtained—
That is her love, for that is all in all.
PETRUCCIO
Why, that is nothing, for I tell you, father,                130
I am as peremptory as she proud-minded,
And where two raging fires meet together
They do consume the thing that feeds their fury.
Though little fire grows great with little wind,
Yet extreme gusts will blow out fire and all.               135
So I to her, and so she yields to me,
For I am rough, and woo not like a babe.
BAPTISTA
Well mayst thou woo, and happy be thy speed.
But be thou armed for some unhappy words.
PETRUCCIO
Ay, to the proof, as mountains are for winds,               140
That shakes not though they blow perpetually.
*Enter Hortensio with his head broke*
BAPTISTA
How now, my friend, why dost thou look so pale?
HORTENSIO
For fear, I promise you, if I look pale.
BAPTISTA
What, will my daughter prove a good musician?
HORTENSIO
I think she'll sooner prove a soldier.                       145
Iron may hold with her, but never lutes.
BAPTISTA
Why then, thou canst not break her to the lute?
HORTENSIO
Why no, for she hath broke the lute to me.
I did but tell her she mistook her frets,
And bowed her hand to teach her fingering,                   150
When, with a most impatient devilish spirit,
'Frets, call you these?' quoth she, 'I'll fume with
     them,'
And with that word she struck me on the head,
And through the instrument my pate made way,
And there I stood amazèd for a while,                        155
As on a pillory, looking through the lute,
While she did call me rascal, fiddler,
And twangling jack, with twenty such vile terms,
As had she studied to misuse me so.
PETRUCCIO
Now, by the world, it is a lusty wench!                      160
I love her ten times more than e'er I did.
O, how I long to have some chat with her!

BAPTISTA (*to Hortensio*)
Well, go with me, and be not so discomfited.
Proceed in practice with my younger daughter.
She's apt to learn, and thankful for good turns.            165
Signor Petruccio, will you go with us,
Or shall I send my daughter Kate to you?
PETRUCCIO
I pray you, do.                    *Exeunt all but Petruccio*
     I'll attend her here,
And woo her with some spirit when she comes.
Say that she rail, why then I'll tell her plain             170
She sings as sweetly as a nightingale.
Say that she frown, I'll say she looks as clear
As morning roses newly washed with dew.
Say she be mute and will not speak a word,
Then I'll commend her volubility,                           175
And say she uttereth piercing eloquence.
If she do bid me pack, I'll give her thanks
As though she bid me stay by her a week.
If she deny to wed, I'll crave the day
When I shall ask the banns, and when be marrièd.
But here she comes, and now, Petruccio, speak.              181
     *Enter Katherina*
Good morrow, Kate, for that's your name, I hear.
KATHERINE
Well have you heard, but something hard of hearing.
They call me Katherine that do talk of me.
PETRUCCIO
You lie, in faith, for you are called plain Kate,           185
And bonny Kate, and sometimes Kate the curst,
But Kate, the prettiest Kate in Christendom,
Kate of Kate Hall, my super-dainty Kate—
For dainties are all cates, and therefore 'Kate'—
Take this of me, Kate of my consolation:                    190
Hearing thy mildness praised in every town,
Thy virtues spoke of, and thy beauty sounded—
Yet not so deeply as to thee belongs—
Myself am moved to woo thee for my wife.
KATHERINE
Moved? In good time. Let him that moved you hither
Re-move you hence. I knew you at the first                  196
You were a movable.
PETRUCCIO                    Why, what's a movable?
KATHERINE
A joint-stool.
PETRUCCIO          Thou hast hit it. Come, sit on me.
KATHERINE
Asses are made to bear, and so are you.
PETRUCCIO
Women are made to bear, and so are you.                     200
KATHERINE
No such jade as you, if me you mean.
PETRUCCIO
Alas, good Kate, I will not burden thee,
For knowing thee to be but young and light.
KATHERINE
Too light for such a swain as you to catch,
And yet as heavy as my weight should be.                    205
PETRUCCIO
Should be?—should buzz.
KATHERINE                    Well ta'en, and like a buzzard.
PETRUCCIO
O slow-winged turtle, shall a buzzard take thee?
KATHERINE
Ay, for a turtle, as he takes a buzzard.

PETRUCCIO
Come, come, you wasp, i'faith you are too angry.
KATHERINE
If I be waspish, best beware my sting.                    210
PETRUCCIO
My remedy is then to pluck it out.
KATHERINE
Ay, if the fool could find it where it lies.
PETRUCCIO
Who knows not where a wasp does wear his sting?
In his tail.
KATHERINE    In his tongue.
PETRUCCIO                    Whose tongue?
KATHERINE
Yours, if you talk of tales, and so farewell.            215
PETRUCCIO
What, with my tongue in your tail? Nay, come again,
Good Kate, I am a gentleman.
KATHERINE                    That I'll try.
        *She strikes him*
PETRUCCIO
I swear I'll cuff you if you strike again.
KATHERINE  So may you lose your arms.
If you strike me you are no gentleman,                    220
And if no gentleman, why then, no arms.
PETRUCCIO
A herald, Kate? O, put me in thy books.
KATHERINE  What is your crest—a coxcomb?
PETRUCCIO
A combless cock, so Kate will be my hen.
KATHERINE
No cock of mine. You crow too like a craven.             225
PETRUCCIO
Nay, come, Kate, come. You must not look so sour.
KATHERINE
It is my fashion when I see a crab.
PETRUCCIO
Why, here's no crab, and therefore look not sour.
KATHERINE  There is, there is.
PETRUCCIO  Then show it me.                                230
KATHERINE  Had I a glass I would.
PETRUCCIO
What, you mean my face?
KATHERINE                    Well aimed, of such a young one.
PETRUCCIO
Now, by Saint George, I am too young for you.
KATHERINE
Yet you are withered.
PETRUCCIO                    'Tis with cares.
KATHERINE                                    I care not.
PETRUCCIO
Nay, hear you, Kate. In sooth, you scape not so.         235
KATHERINE
I chafe you if I tarry. Let me go.
PETRUCCIO
No, not a whit. I find you passing gentle.
'Twas told me you were rough, and coy, and sullen,
And now I find report a very liar,                        239
For thou art pleasant, gamesome, passing courteous,
But slow in speech, yet sweet as springtime flowers.
Thou canst not frown. Thou canst not look askance,
Nor bite the lip, as angry wenches will,
Nor hast thou pleasure to be cross in talk,
But thou with mildness entertain'st thy wooers,          245
With gentle conference, soft, and affable.
Why does the world report that Kate doth limp?
O sland'rous world! Kate like the hazel twig
Is straight and slender, and as brown in hue
As hazelnuts, and sweeter than the kernels.              250
O let me see thee walk. Thou dost not halt.
KATHERINE
Go, fool, and whom thou keep'st command.
PETRUCCIO
Did ever Dian so become a grove
As Kate this chamber with her princely gait?
O, be thou Dian, and let her be Kate,                    255
And then let Kate be chaste and Dian sportful.
KATHERINE
Where did you study all this goodly speech?
PETRUCCIO
It is extempore, from my mother-wit.
KATHERINE
A witty mother, witless else her son.
PETRUCCIO
Am I not wise?
KATHERINE            Yes, keep you warm.                   260
PETRUCCIO
Marry, so I mean, sweet Katherine, in thy bed.
And therefore setting all this chat aside,
Thus in plain terms: your father hath consented
That you shall be my wife, your dowry 'greed on,
And will you, nill you, I will marry you.                265
Now, Kate, I am a husband for your turn,
For by this light, whereby I see thy beauty—
Thy beauty that doth make me like thee well—
Thou must be married to no man but me,
        *Enter Baptista, Gremio, and Tranio as Lucentio*
For I am he am born to tame you, Kate,                   270
And bring you from a wild Kate to a Kate
Conformable as other household Kates.
Here comes your father. Never make denial.
I must and will have Katherine to my wife.
BAPTISTA  Now, Signor Petruccio, how speed you with my
daughter?                                                276
PETRUCCIO  How but well, sir, how but well?
It were impossible I should speed amiss.
BAPTISTA
Why, how now, daughter Katherine—in your dumps?
KATHERINE
Call you me daughter? Now I promise you               280
You have showed a tender fatherly regard,
To wish me wed to one half-lunatic,
A madcap ruffian and a swearing Jack,
That thinks with oaths to face the matter out.
PETRUCCIO
Father, 'tis thus: yourself and all the world           285
That talked of her have talked amiss of her.
If she be curst, it is for policy,
For she's not froward, but modest as the dove.
She is not hot, but temperate as the morn.
For patience she will prove a second Grissel,           290
And Roman Lucrece for her chastity.
And to conclude, we have 'greed so well together
That upon Sunday is the wedding day.
KATHERINE
I'll see thee hanged on Sunday first.
GREMIO  Hark, Petruccio, she says she'll see thee hanged
first.                                                   296
TRANIO
Is this your speeding? Nay then, goodnight our part.
PETRUCCIO
Be patient, gentlemen. I choose her for myself.
If she and I be pleased, what's that to you?

'Tis bargained 'twixt us twain, being alone,                    300
That she shall still be curst in company.
I tell you, 'tis incredible to believe
How much she loves me. O, the kindest Kate!
She hung about my neck, and kiss on kiss
She vied so fast, protesting oath on oath,                      305
That in a twink she won me to her love.
O, you are novices. 'Tis a world to see
How tame, when men and women are alone,
A meacock wretch can make the curstest shrew.
Give me thy hand, Kate. I will unto Venice,                     310
To buy apparel 'gainst the wedding day.
Provide the feast, father, and bid the guests.
I will be sure my Katherine shall be fine.

BAPTISTA
I know not what to say, but give me your hands.
God send you joy, Petruccio! 'Tis a match.                      315

GREMIO *and* TRANIO
Amen, say we. We will be witnesses.

PETRUCCIO
Father, and wife, and gentlemen, adieu.
I will to Venice. Sunday comes apace.
We will have rings, and things, and fine array;
And kiss me, Kate. We will be married o' Sunday.  320
            *Exeunt Petruccio and Katherine, severally*

GREMIO
Was ever match clapped up so suddenly?

BAPTISTA
Faith, gentlemen, now I play a merchant's part,
And venture madly on a desperate mart.

TRANIO
'Twas a commodity lay fretting by you.
'Twill bring you gain, or perish on the seas.                   325

BAPTISTA
The gain I seek is quiet in the match.

GREMIO
No doubt but he hath got a quiet catch.
But now, Baptista, to your younger daughter.
Now is the day we long have lookèd for.
I am your neighbour, and was suitor first.                      330

TRANIO
And I am one that love Bianca more
Than words can witness, or your thoughts can guess.

GREMIO
Youngling, thou canst not love so dear as I.

TRANIO
Greybeard, thy love doth freeze.

GREMIO                                But thine doth fry.
Skipper, stand back. 'Tis age that nourisheth.                  335

TRANIO
But youth in ladies' eyes that flourisheth.

BAPTISTA
Content you, gentlemen. I will compound this strife.
'Tis deeds must win the prize, and he of both
That can assure my daughter greatest dower
Shall have my Bianca's love.                                    340
Say, Signor Gremio, what can you assure her?

GREMIO
First, as you know, my house within the city
Is richly furnishèd with plate and gold,
Basins and ewers to lave her dainty hands;
My hangings all of Tyrian tapestry.                             345
In ivory coffers I have stuffed my crowns,
In cypress chests my arras counterpoints,
Costly apparel, tents and canopies,
Fine linen, Turkey cushions bossed with pearl,

Valance of Venice gold in needlework,                           350
Pewter, and brass, and all things that belongs
To house or housekeeping. Then at my farm
I have a hundred milch-kine to the pail,
Six score fat oxen standing in my stalls,
And all things answerable to this portion.                      355
Myself am struck in years, I must confess,
And if I die tomorrow this is hers,
If whilst I live she will be only mine.

TRANIO
That 'only' came well in. Sir, list to me.
I am my father's heir and only son.                             360
If I may have your daughter to my wife
I'll leave her houses three or four as good,
Within rich Pisa walls, as any one
Old Signor Gremio has in Padua,
Besides two thousand ducats by the year                         365
Of fruitful land, all which shall be her jointure.
What, have I pinched you, Signor Gremio?

GREMIO
Two thousand ducats by the year of land—
My land amounts not to so much in all.
That she shall have; besides, an argosy                         370
That now is lying in Marseilles road.
What, have I choked you with an argosy?

TRANIO
Gremio, 'tis known my father hath no less
Than three great argosies, besides two galliasses
And twelve tight galleys. These I will assure her,             375
And twice as much whate'er thou off'rest next.

GREMIO
Nay, I have offered all. I have no more,
And she can have no more than all I have.
If you like me, she shall have me and mine.

TRANIO
Why then, the maid is mine from all the world.                 380
By your firm promise Gremio is out-vied.

BAPTISTA
I must confess your offer is the best,
And let your father make her the assurance,
She is your own. Else, you must pardon me,
If you should die before him, where's her dower?  385

TRANIO
That's but a cavil. He is old, I young.

GREMIO
And may not young men die as well as old?

BAPTISTA  Well, gentlemen,
I am thus resolved. On Sunday next, you know,
My daughter Katherine is to be married.                         390
(*To Tranio*) Now, on the Sunday following shall
    Bianca
Be bride to you, if you make this assurance;
If not, to Signor Gremio.
And so I take my leave, and thank you both.

GREMIO
Adieu, good neighbour.                        *Exit Baptista*
                        Now I fear thee not.
Sirrah, young gamester, your father were a fool                396
To give thee all, and in his waning age
Set foot under thy table. Tut, a toy!
An old Italian fox is not so kind, my boy.          *Exit*

TRANIO
A vengeance on your crafty withered hide!                       400
Yet I have faced it with a card of ten.
'Tis in my head to do my master good.
I see no reason but supposed Lucentio

Must get a father called supposed Vincentio—
And that's a wonder; fathers commonly          405
Do get their children, but in this case of wooing
A child shall get a sire, if I fail not of my cunning.
                                                                *Exit*

**3.1**     *Enter Lucentio with books, as Cambio, Hortensio*
            *with a lute, as Licio, and Bianca*

LUCENTIO
Fiddler, forbear. You grow too forward, sir.
Have you so soon forgot the entertainment
Her sister Katherine welcomed you withal?

HORTENSIO
But, wrangling pedant, this Bianca is,
The patroness of heavenly harmony.          5
Then give me leave to have prerogative,
And when in music we have spent an hour
Your lecture shall have leisure for as much.

LUCENTIO
Preposterous ass, that never read so far
To know the cause why music was ordained!          10
Was it not to refresh the mind of man
After his studies or his usual pain?
Then give me leave to read philosophy,
And while I pause, serve in your harmony.

HORTENSIO
Sirrah, I will not bear these braves of thine.          15

BIANCA
Why, gentlemen, you do me double wrong
To strive for that which resteth in my choice.
I am no breeching scholar in the schools.
I'll not be tied to hours nor 'pointed times,
But learn my lessons as I please myself;          20
And to cut off all strife, here sit we down.
(*To Hortensio*) Take you your instrument, play you the
     whiles.
His lecture will be done ere you have tuned.

HORTENSIO
You'll leave his lecture when I am in tune?

LUCENTIO
That will be never. Tune your instrument.          25
     *Hortensio tunes his lute. Lucentio opens a book*

BIANCA Where left we last?

LUCENTIO Here, madam.
     (*Reads*)   'Hic ibat Simois, hic est Sigeia tellus,
     Hic steterat Priami regia celsa senis.'

BIANCA Construe them.          30

LUCENTIO 'Hic ibat', as I told you before—'Simois', I am
Lucentio—'hic est', son unto Vincentio of Pisa—'Sigeia
tellus', disguised thus to get your love—'hic steterat',
and that Lucentio that comes a-wooing—'Priami', is
my man Tranio—'regia', bearing my port—'celsa senis',
that we might beguile the old pantaloon.          36

HORTENSIO Madam, my instrument's in tune.

BIANCA Let's hear. (*Hortensio plays*) O fie, the treble jars.

LUCENTIO Spit in the hole, man, and tune again.
     *Hortensio tunes his lute again*

BIANCA Now let me see if I can construe it. 'Hic ibat
Simois', I know you not—'hic est Sigeia tellus', I trust
you not—'hic steterat Priami', take heed he hear us
not—'regia', presume not—'celsa senis', despair not.  43

HORTENSIO
Madam, 'tis now in tune.

LUCENTIO                    All but the bass.

HORTENSIO
The bass is right, 'tis the base knave that jars.          45

(*Aside*) How fiery and forward our pedant is!
Now, for my life, the knave doth court my love.
*Pedascule*, I'll watch you better yet.

BIANCA (*to Lucentio*)
In time I may believe; yet, I mistrust.

LUCENTIO
Mistrust it not, for sure Aeacides          50
Was Ajax, called so from his grandfather.

BIANCA
I must believe my master, else, I promise you,
I should be arguing still upon that doubt.
But let it rest. Now Licio, to you.
Good master, take it not unkindly, pray,          55
That I have been thus pleasant with you both.

HORTENSIO (*to Lucentio*)
You may go walk and give me leave awhile.
My lessons make no music in three parts.

LUCENTIO
Are you so formal, sir? Well, I must wait.
(*Aside*) And watch withal, for but I be deceived          60
Our fine musician groweth amorous.

HORTENSIO
Madam, before you touch the instrument
To learn the order of my fingering,
I must begin with rudiments of art,
To teach you gamut in a briefer sort,          65
More pleasant, pithy, and effectual
Than hath been taught by any of my trade;
And there it is in writing, fairly drawn.
     *He gives a paper*

BIANCA
Why, I am past my gamut long ago.

HORTENSIO
Yet read the gamut of Hortensio.          70

BIANCA (*reads*)
     'Gam-ut I am, the ground of all accord,
          A—re—to plead Hortensio's passion.
     B—mi—Bianca, take him for thy lord,
          C—fa, ut—that loves with all affection.
     D—sol, re—one clef, two notes have I,          75
     E—la, mi—show pity, or I die.'
Call you this gamut? Tut, I like it not.
Old fashions please me best. I am not so nice
To change true rules for odd inventions.
     *Enter a Messenger*

MESSENGER
Mistress, your father prays you leave your books          80
And help to dress your sister's chamber up.
You know tomorrow is the wedding day.

BIANCA
Farewell, sweet masters both. I must be gone.

LUCENTIO
Faith, mistress, then I have no cause to stay.
     *Exeunt Bianca, Messenger, and Lucentio*

HORTENSIO
But I have cause to pry into this pedant.          85
Methinks he looks as though he were in love.
Yet if thy thoughts, Bianca, be so humble
To cast thy wand'ring eyes on every stale,
Seize thee that list. If once I find thee ranging,
Hortensio will be quit with thee by changing.          *Exit*

**3.2**     *Enter Baptista, Gremio, Tranio as Lucentio,*
            *Katherine, Bianca, and others, attendants*

BAPTISTA (*to Tranio*)
Signor Lucentio, this is the 'pointed day
That Katherine and Petruccio should be married,

39

And yet we hear not of our son-in-law.
What will be said, what mockery will it be,
To want the bridegroom when the priest attends   5
To speak the ceremonial rites of marriage?
What says Lucentio to this shame of ours?

KATHERINE
No shame but mine. I must forsooth be forced
To give my hand opposed against my heart
Unto a mad-brain rudesby full of spleen,   10
Who wooed in haste and means to wed at leisure.
I told you, I, he was a frantic fool,
Hiding his bitter jests in blunt behaviour,
And to be noted for a merry man
He'll woo a thousand, 'point the day of marriage,   15
Make friends, invite them, and proclaim the banns,
Yet never means to wed where he hath wooed.
Now must the world point at poor Katherine
And say 'Lo, there is mad Petruccio's wife,
If it would please him come and marry her.'   20

TRANIO
Patience, good Katherine, and Baptista, too.
Upon my life, Petruccio means but well.
Whatever fortune stays him from his word,
Though he be blunt, I know him passing wise;
Though he be merry, yet withal he's honest.   25

KATHERINE
Would Katherine had never seen him, though.
                      *Exit weeping*

BAPTISTA
Go, girl. I cannot blame thee now to weep.
For such an injury would vex a very saint,
Much more a shrew of thy impatient humour.
    *Enter Biondello*

BIONDELLO Master, master, news—old news, and such
news as you never heard of.   31
BAPTISTA Is it new and old too? How may that be?
BIONDELLO Why, is it not news to hear of Petruccio's
coming?
BAPTISTA Is he come?   35
BIONDELLO Why, no, sir.
BAPTISTA What then?
BIONDELLO He is coming.
BAPTISTA When will he be here?
BIONDELLO When he stands where I am and sees you
there.   41
TRANIO But say, what to thine old news?
BIONDELLO Why, Petruccio is coming in a new hat and
an old jerkin, a pair of old breeches thrice-turned, a
pair of boots that have been candle-cases, one buckled,
another laced, an old rusty sword ta'en out of the town
armoury with a broken hilt, and chapeless, with two
broken points, his horse hipped, with an old mothy
saddle and stirrups of no kindred, besides, possessed
with the glanders and like to mose in the chine, troubled
with the lampass, infected with the fashions, full of
windgalls, sped with spavins, rayed with the yellows,
past cure of the fives, stark spoiled with the staggers,
begnawn with the bots, weighed in the back and
shoulder-shotten, near-legged before and with a half-
cheeked bit and a headstall of sheep's leather which,
being restrained to keep him from stumbling, hath been
often burst and now repaired with knots, one girth six
times pieced, and a woman's crupper of velour which
hath two letters for her name fairly set down in studs,
and here and there pieced with packthread.   61

BAPTISTA Who comes with him?
BIONDELLO O sir, his lackey, for all the world caparisoned
like the horse, with a linen stock on one leg and a
kersey boot-hose on the other, gartered with a red and
blue list; an old hat, and the humour of forty fancies
pricked in't for a feather—a monster, a very monster
in apparel, and not like a Christian footboy or a
gentleman's lackey.

TRANIO
'Tis some odd humour pricks him to this fashion;   70
Yet oftentimes he goes but mean-apparelled.

BAPTISTA
I am glad he's come, howsoe'er he comes.
BIONDELLO Why, sir, he comes not.
BAPTISTA Didst thou not say he comes?
BIONDELLO Who? That Petruccio came?   75
BAPTISTA Ay, that Petruccio came.
BIONDELLO No, sir. I say his horse comes with him on his
back.
BAPTISTA Why, that's all one.
BIONDELLO             Nay, by Saint Jamy,   80
      I hold you a penny,
      A horse and a man
      Is more than one,
      And yet not many.
    *Enter Petruccio and Grumio, fantastically dressed*
PETRUCCIO Come, where be these gallants? Who's at
home?   86
BAPTISTA You are welcome, sir.
PETRUCCIO And yet I come not well.
BAPTISTA And yet you halt not.

TRANIO
Not so well apparelled as I wish you were.   90
PETRUCCIO
Were it not better I should rush in thus—
But where is Kate? Where is my lovely bride?
How does my father? Gentles, methinks you frown.
And wherefore gaze this goodly company
As if they saw some wondrous monument,   95
Some comet or unusual prodigy?

BAPTISTA
Why, sir, you know this is your wedding day.
First were we sad, fearing you would not come;
Now sadder that you come so unprovided.
Fie, doff this habit, shame to your estate,   100
An eyesore to our solemn festival.

TRANIO
And tell us what occasion of import
Hath all so long detained you from your wife
And sent you hither so unlike yourself?

PETRUCCIO
Tedious it were to tell, and harsh to hear.   105
Sufficeth I am come to keep my word,
Though in some part enforcèd to digress,
Which at more leisure I will so excuse
As you shall well be satisfied withal.
But where is Kate? I stay too long from her.   110
The morning wears, 'tis time we were at church.

TRANIO
See not your bride in these unreverent robes,
Go to my chamber, put on clothes of mine.

PETRUCCIO
Not I, believe me. Thus I'll visit her.

BAPTISTA
But thus, I trust, you will not marry her.   115

PETRUCCIO
Good sooth, even thus. Therefore ha' done with
    words.
To me she's married, not unto my clothes.
Could I repair what she will wear in me
As I can change these poor accoutrements,
'Twere well for Kate and better for myself.          120
But what a fool am I to chat with you
When I should bid good morrow to my bride,
And seal the title with a lovely kiss!
                                      *Exit ⌈with Grumio⌉*

TRANIO
He hath some meaning in his mad attire.
We will persuade him, be it possible,                125
To put on better ere he go to church.
                                      ⌈*Exit with Gremio*⌉

BAPTISTA
I'll after him, and see the event of this.     ⌈*Exeunt*⌉

**3.3** ⌈*Enter Lucentio as Cambio, and Tranio as Lucentio*⌉
TRANIO
But, sir, to love concerneth us to add
Her father's liking, which to bring to pass,
As I before imparted to your worship,
I am to get a man—whate'er he be
It skills not much, we'll fit him to our turn—        5
And he shall be Vincentio of Pisa,
And make assurance here in Padua
Of greater sums than I have promisèd.
So shall you quietly enjoy your hope,
And marry sweet Bianca with consent.                  10
LUCENTIO
Were it not that my fellow schoolmaster
Doth watch Bianca's steps so narrowly,
'Twere good, methinks, to steal our marriage,
Which once performed, let all the world say no,
I'll keep mine own, despite of all the world.         15
TRANIO
That by degrees we mean to look into,
And watch our vantage in this business.
We'll overreach the greybeard Gremio,
The narrow-prying father Minola,
The quaint musician, amorous Licio,                   20
All for my master's sake, Lucentio.
        *Enter Gremio*
Signor Gremio, came you from the church?
GREMIO
As willingly as e'er I came from school.
TRANIO
And is the bride and bridegroom coming home?
GREMIO
A bridegroom, say you? 'Tis a groom indeed—          25
A grumbling groom, and that the girl shall find.
TRANIO
Curster than she? Why, 'tis impossible.
GREMIO
Why, he's a devil, a devil, a very fiend.
TRANIO
Why, she's a devil, a devil, the devil's dam.
GREMIO
Tut, she's a lamb, a dove, a fool to him.             30
I'll tell you, Sir Lucentio: when the priest
Should ask if Katherine should be his wife,
'Ay, by Gog's woun's,' quoth he, and swore so loud
That all amazed the priest let fall the book,
And as he stooped again to take it up                 35

This mad-brained bridegroom took him such a cuff
That down fell priest, and book, and book, and priest.
'Now take them up,' quoth he, 'if any list.'
TRANIO
What said the vicar when he rose again?
GREMIO
Trembled and shook, forwhy he stamped and swore
As if the vicar meant to cozen him.                   41
But after many ceremonies done
He calls for wine. 'A health,' quoth he, as if
He had been aboard, carousing to his mates
After a storm; quaffed off the muscatel              45
And threw the sops all in the sexton's face,
Having no other reason
But that his beard grew thin and hungerly
And seemed to ask him sops as he was drinking.
This done, he took the bride about the neck          50
And kissed her lips with such a clamorous smack
That at the parting all the church did echo,
And I seeing this came thence for very shame,
And after me, I know, the rout is coming.
Such a mad marriage never was before.                55
        *Music plays*
Hark, hark, I hear the minstrels play.
        *Enter Petruccio, Katherine, Bianca, Hortensio as*
        *Licio, Baptista, Grumio, and others, attendants*
PETRUCCIO
Gentlemen and friends, I thank you for your pains.
I know you think to dine with me today,
And have prepared great store of wedding cheer.
But so it is my haste doth call me hence,             60
And therefore here I mean to take my leave.
BAPTISTA
Is't possible you will away tonight?
PETRUCCIO
I must away today, before night come.
Make it no wonder. If you knew my business,
You would entreat me rather go than stay.             65
And, honest company, I thank you all
That have beheld me give away myself
To this most patient, sweet, and virtuous wife.
Dine with my father, drink a health to me,
For I must hence; and farewell to you all.            70
TRANIO
Let us entreat you stay till after dinner.
PETRUCCIO
It may not be.
GREMIO               Let me entreat you.
PETRUCCIO
It cannot be.
KATHERINE            Let me entreat you.
PETRUCCIO
I am content.
KATHERINE     Are you content to stay?
PETRUCCIO
I am content you shall entreat me stay,               75
But yet not stay, entreat me how you can.
KATHERINE
Now, if you love me, stay.
PETRUCCIO                Grumio, my horse.
GRUMIO Ay, sir, they be ready. The oats have eaten the
    horses.
KATHERINE
Nay, then, do what thou canst, I will not go today,   80
No, nor tomorrow—not till I please myself.
The door is open, sir, there lies your way.

You may be jogging whiles your boots are green.
For me, I'll not be gone till I please myself.
'Tis like you'll prove a jolly, surly groom,                85
That take it on you at the first so roundly.
PETRUCCIO
O Kate, content thee. Prithee, be not angry.
KATHERINE
I will be angry. What hast thou to do?
Father, be quiet. He shall stay my leisure.
GREMIO
Ay, marry, sir. Now it begins to work.                      90
KATHERINE
Gentlemen, forward to the bridal dinner.
I see a woman may be made a fool
If she had not a spirit to resist.
PETRUCCIO
They shall go forward, Kate, at thy command.
Obey the bride, you that attend on her.                     95
Go to the feast, revel and domineer,
Carouse full measure to her maidenhead.
Be mad and merry, or go hang yourselves.
But for my bonny Kate, she must with me.
Nay, look not big, nor stamp, nor stare, nor fret.          100
I will be master of what is mine own.
She is my goods, my chattels. She is my house,
My household-stuff, my field, my barn,
My horse, my ox, my ass, my anything,
And here she stands, touch her whoever dare.                105
I'll bring mine action on the proudest he
That stops my way in Padua. Grumio,
Draw forth thy weapon, we are beset with thieves.
Rescue thy mistress if thou be a man.
Fear not, sweet wench. They shall not touch thee,
   Kate.                                                     110
I'll buckler thee against a million.
               *Exeunt Petruccio, Katherine, and Grumio*
BAPTISTA
Nay, let them go—a couple of quiet ones!
GREMIO
Went they not quickly I should die with laughing.
TRANIO
Of all mad matches never was the like.
LUCENTIO
Mistress, what's your opinion of your sister?               115
BIANCA
That being mad herself she's madly mated.
GREMIO
I warrant him, Petruccio is Kated.
BAPTISTA
Neighbours and friends, though bride and bridegroom
   wants
For to supply the places at the table,
You know there wants no junkets at the feast.               120
Lucentio, you shall supply the bridegroom's place,
And let Bianca take her sister's room.
TRANIO
Shall sweet Bianca practise how to bride it?
BAPTISTA
She shall, Lucentio. Come, gentlemen, let's go.
                                                  *Exeunt*

**4.1**    *Enter Grumio*
GRUMIO Fie, fie on all tired jades, on all mad masters, and
all foul ways. Was ever man so beaten? Was ever man
so rayed? Was ever man so weary? I am sent before
to make a fire, and they are coming after to warm
them. Now were not I a little pot and soon hot, my

very lips might freeze to my teeth, my tongue to the
roof of my mouth, my heart in my belly ere I should
come by a fire to thaw me. But I with blowing the fire
shall warm myself, for considering the weather, a taller
man than I will take cold. Holla! Hoa, Curtis!          10
   *Enter Curtis*
CURTIS Who is that calls so coldly?
GRUMIO A piece of ice. If thou doubt it, thou mayst slide
from my shoulder to my heel with no greater a run
but my head and my neck. A fire, good Curtis!
CURTIS Is my master and his wife coming, Grumio?        15
GRUMIO O ay, Curtis, ay, and therefore fire, fire! Cast on
no water.
CURTIS Is she so hot a shrew as she's reported?
GRUMIO She was, good Curtis, before this frost; but thou
know'st, winter tames man, woman, and beast, for it
hath tamed my old master, and my new mistress, and
myself, fellow Curtis.                                   22
CURTIS Away, you three-inch fool. I am no beast.
GRUMIO Am I but three inches? Why, thy horn is a foot,
and so long am I, at the least. But wilt thou make a
fire, or shall I complain on thee to our mistress, whose
hand—she being now at hand—thou shalt soon feel to
thy cold comfort, for being slow in thy hot office.
CURTIS I prithee, good Grumio, tell me—how goes the
world?                                                   30
GRUMIO A cold world, Curtis, in every office but thine.
And therefore fire, do thy duty, and have thy duty, for
my master and mistress are almost frozen to death.
CURTIS There's fire ready, and therefore, good Grumio,
the news.                                                35
GRUMIO Why, 'Jack boy, ho boy!', and as much news as
wilt thou.
CURTIS Come, you are so full of cony-catching.
GRUMIO Why, therefore fire, for I have caught extreme
cold. Where's the cook? Is supper ready, the house
trimmed, rushes strewed, cobwebs swept, the servingmen in their new fustian, the white stockings, and
every officer his wedding garment on? Be the Jacks fair
within, the Jills fair without, the carpets laid, and
everything in order?                                     45
CURTIS All ready, and therefore, I pray thee, news.
GRUMIO First, know my horse is tired, my master and
mistress fallen out.
CURTIS How?
GRUMIO Out of their saddles into the dirt, and thereby
hangs a tale.                                            51
CURTIS Let's ha't, good Grumio.
GRUMIO Lend thine ear.
CURTIS Here.
GRUMIO (*cuffing him*) There.                            55
CURTIS This 'tis to feel a tale, not to hear a tale.
GRUMIO And therefore 'tis called a sensible tale, and this
cuff was but to knock at your ear and beseech listening.
Now I begin. *Inprimis*, we came down a foul hill, my
master riding behind my mistress.                        60
CURTIS Both of one horse?
GRUMIO What's that to thee?
CURTIS Why, a horse.
GRUMIO Tell thou the tale. But hadst thou not crossed me
thou shouldst have heard how her horse fell and she
under her horse; thou shouldst have heard in how
miry a place, how she was bemoiled, how he left her
with the horse upon her, how he beat me because her
horse stumbled, how she waded through the dirt to
pluck him off me, how he swore, how she prayed that
never prayed before, how I cried, how the horses ran

away, how her bridle was burst, how I lost my crupper,
with many things of worthy memory which now shall
die in oblivion, and thou return unexperienced to thy
grave.                                                            75
CURTIS By this reckoning he is more shrew than she.
GRUMIO Ay, and that thou and the proudest of you all
  shall find when he comes home. But what talk I of
  this? Call forth Nathaniel, Joseph, Nicholas, Philip,
  Walter, Sugarsop, and the rest. Let their heads be
  sleekly combed, their blue coats brushed, and their
  garters of an indifferent knit. Let them curtsy with their
  left legs and not presume to touch a hair of my master's
  horse-tail till they kiss their hands. Are they all ready?
CURTIS They are.                                                  85
GRUMIO Call them forth.
CURTIS (calling) Do you hear, ho? You must meet my
  master to countenance my mistress.
GRUMIO Why, she hath a face of her own.
CURTIS Who knows not that?                                        90
GRUMIO Thou, it seems, that calls for company to
  countenance her.
CURTIS I call them forth to credit her.
       *Enter four or five servingmen*
GRUMIO Why, she comes to borrow nothing of them.
NATHANIEL Welcome home, Grumio!                                   95
PHILIP How now, Grumio?
JOSEPH What, Grumio?
NICHOLAS Fellow Grumio!
NATHANIEL How now, old lad!
GRUMIO Welcome you, how now you, what you, fellow
  you, and thus much for greeting. Now, my spruce
  companions, is all ready and all things neat?          102
NATHANIEL All things is ready. How near is our master?
GRUMIO E'en at hand, alighted by this, and therefore be
  not—Cock's passion, silence! I hear my master.        105
       *Enter Petruccio and Katherine*
PETRUCCIO
  Where be these knaves? What, no man at door
  To hold my stirrup nor to take my horse?
  Where is Nathaniel, Gregory, Philip?
ALL SERVANTS Here, here sir, here sir.
PETRUCCIO
  Here sir, here sir, here sir, here sir!                         110
  You logger-headed and unpolished grooms,
  What! No attendance! No regard! No duty!
  Where is the foolish knave I sent before?
GRUMIO
  Here, sir, as foolish as I was before.
PETRUCCIO
  You peasant swain, you whoreson, malthorse drudge,
  Did I not bid thee meet me in the park                          116
  And bring along these rascal knaves with thee?
GRUMIO
  Nathaniel's coat, sir, was not fully made,
  And Gabriel's pumps were all unpinked i'th' heel.
  There was no link to colour Peter's hat,                        120
  And Walter's dagger was not come from sheathing.
  There were none fine but Adam, Ralph, and Gregory.
  The rest were ragged, old, and beggarly.
  Yet as they are, here are they come to meet you.
PETRUCCIO
  Go, rascals, go and fetch my supper in.                         125
                                           *Exeunt servants*
  (Sings)     'Where is the life that late I led?
              Where are those—'
  Sit down, Kate, and welcome. Soud, soud, soud, soud.

       *Enter servants with supper*
  Why, when, I say?—Nay, good sweet Kate, be
    merry.—
  Off with my boots, you rogues, you villains. When?
  (Sings)     'It was the friar of orders gray,               131
              As he forth walkèd on his way.'
  Out, you rogue, you pluck my foot awry.
  (Kicking a servant) Take that, and mend the plucking
    of the other.
  Be merry, Kate. (Calling) Some water, here. What,
    hoa!                                                       135
       *Enter one with water*
  Where's my spaniel Troilus? Sirrah, get you hence,
  And bid my cousin Ferdinand come hither—
  One, Kate, that you must kiss and be acquainted
    with.
  (Calling) Where are my slippers? Shall I have some
    water?
  Come, Kate, and wash, and welcome heartily.      140
       ⌈*A servant drops water*⌉
  You whoreson villain, will you let it fall?
KATHERINE
  Patience, I pray you, 'twas a fault unwilling.
PETRUCCIO
  A whoreson, beetle-headed, flap-eared knave.
  Come, Kate, sit down, I know you have a stomach.
  Will you give thanks, sweet Kate, or else shall I?   145
  What's this—mutton?
FIRST SERVINGMAN          Ay.
PETRUCCIO                        Who brought it?
PETER                                                    I.
PETRUCCIO
  'Tis burnt, and so is all the meat.
  What dogs are these? Where is the rascal cook?
  How durst you villains bring it from the dresser
  And serve it thus to me that love it not?        150
  There, (throwing food) take it to you, trenchers, cups,
    and all,
  You heedless jolt-heads and unmannered slaves.
  What, do you grumble? I'll be with you straight.
       *He chases the servants away*
KATHERINE
  I pray you, husband, be not so disquiet.
  The meat was well, if you were so contented.      155
PETRUCCIO
  I tell thee, Kate, 'twas burnt and dried away,
  And I expressly am forbid to touch it,
  For it engenders choler, planteth anger,
  And better 'twere that both of us did fast,
  Since of ourselves ourselves are choleric,        160
  Than feed it with such overroasted flesh.
  Be patient, tomorrow't shall be mended,
  And for this night we'll fast for company.
  Come, I will bring thee to thy bridal chamber.   *Exeunt*
       *Enter servants severally*
NATHANIEL Peter, didst ever see the like?            165
PETER He kills her in her own humour.
       *Enter Curtis, a servant*
GRUMIO Where is he?
CURTIS In her chamber,
  Making a sermon of continency to her,
  And rails, and swears, and rates, that she, poor soul,
  Knows not which way to stand, to look, to speak,  171
  And sits as one new risen from a dream.
  Away, away, for he is coming hither.            *Exeunt*

*Enter Petruccio*

PETRUCCIO
Thus have I politicly begun my reign,
And 'tis my hope to end successfully.                    175
My falcon now is sharp and passing empty,
And till she stoop she must not be full-gorged,
For then she never looks upon her lure.
Another way I have to man my haggard,
To make her come and know her keeper's call—            180
That is, to watch her as we watch these kites
That bate and beat, and will not be obedient.
She ate no meat today, nor none shall eat.
Last night she slept not, nor tonight she shall not.
As with the meat, some undeservèd fault                 185
I'll find about the making of the bed,
And here I'll fling the pillow, there the bolster,
This way the coverlet, another way the sheets,
Ay, and amid this hurly I intend
That all is done in reverent care of her,               190
And in conclusion she shall watch all night,
And if she chance to nod I'll rail and brawl
And with the clamour keep her still awake.
This is a way to kill a wife with kindness,
And thus I'll curb her mad and headstrong humour.
He that knows better how to tame a shrew,               196
Now let him speak. 'Tis charity to show.           *Exit*

**4.2**    *Enter Tranio as Lucentio, and Hortensio as Licio*

TRANIO
Is't possible, friend Licio, that Mistress Bianca
Doth fancy any other but Lucentio?
I tell you, sir, she bears me fair in hand.

HORTENSIO
Sir, to satisfy you in what I have said,
Stand by, and mark the manner of his teaching.          5
      *They stand aside.*
      *Enter Bianca, and Lucentio as Cambio*

LUCENTIO
Now, mistress, profit you in what you read?

BIANCA
What, master, read you? First resolve me that.

LUCENTIO
I read that I profess, *The Art to Love.*

BIANCA
And may you prove, sir, master of your art.

LUCENTIO
While you, sweet dear, prove mistress of my heart.  10
      *They stand aside*

HORTENSIO
Quick proceeders, marry! Now tell me, I pray,
You that durst swear that your mistress Bianca
Loved none in the world so well as Lucentio.

TRANIO
O despiteful love, unconstant womankind!
I tell thee, Licio, this is wonderful.                  15

HORTENSIO
Mistake no more, I am not Licio,
Nor a musician as I seem to be,
But one that scorn to live in this disguise
For such a one as leaves a gentleman
And makes a god of such a cullion.                      20
Know, sir, that I am called Hortensio.

TRANIO
Signor Hortensio, I have often heard
Of your entire affection to Bianca,
And since mine eyes are witness of her lightness

I will with you, if you be so contented,                25
Forswear Bianca and her love for ever.

HORTENSIO
See how they kiss and court. Signor Lucentio,
Here is my hand, and here I firmly vow
Never to woo her more, but do forswear her
As one unworthy all the former favours                  30
That I have fondly flattered her withal.

TRANIO
And here I take the like unfeignèd oath
Never to marry with her, though she would entreat.
Fie on her, see how beastly she doth court him!

HORTENSIO
Would all the world but he had quite forsworn.          35
For me, that I may surely keep mine oath
I will be married to a wealthy widow
Ere three days pass, which hath as long loved me
As I have loved this proud disdainful haggard.
And so farewell, Signor Lucentio.                       40
Kindness in women, not their beauteous looks,
Shall win my love; and so I take my leave,
In resolution as I swore before.               *Exit*

TRANIO
Mistress Bianca, bless you with such grace
As 'longeth to a lover's blessèd case.                  45
Nay, I have ta'en you napping, gentle love,
And have forsworn you with Hortensio.

BIANCA
Tranio, you jest. But have you both forsworn me?

TRANIO
Mistress, we have.

LUCENTIO                    Then we are rid of Licio.

TRANIO
I'faith, he'll have a lusty widow now,                  50
That shall be wooed and wedded in a day.

BIANCA  God give him joy.

TRANIO  Ay, and he'll tame her.

BIANCA  He says so, Tranio.

TRANIO
Faith, he is gone unto the taming-school.               55

BIANCA
The taming-school—what, is there such a place?

TRANIO
Ay, mistress, and Petruccio is the master,
That teacheth tricks eleven-and-twenty long
To tame a shrew and charm her chattering tongue.
      *Enter Biondello*

BIONDELLO
O, master, master, I have watched so long                60
That I am dog-weary, but at last I spied
An ancient angel coming down the hill
Will serve the turn.

TRANIO                    What is he, Biondello?

BIONDELLO
Master, a marcantant or a pedant,
I know not what, but formal in apparel,                   65
In gait and countenance surely like a father.

LUCENTIO  And what of him, Tranio?

TRANIO
If he be credulous and trust my tale,
I'll make him glad to seem Vincentio
And give assurance to Baptista Minola                    70
As if he were the right Vincentio.
Take in your love, and then let me alone.
      *Exeunt Lucentio and Bianca*

*Enter a Pedant*

PEDANT
God save you, sir.

TRANIO                    And you, sir. You are welcome.
Travel you farre on, or are you at the farthest?

PEDANT
Sir, at the farthest for a week or two,                    75
But then up farther and as far as Rome,
And so to Tripoli, if God lend me life.

TRANIO
What countryman, I pray?

PEDANT                         Of Mantua.

TRANIO
Of Mantua, sir? Marry, God forbid,
And come to Padua careless of your life!            80

PEDANT
My life, sir? How, I pray? For that goes hard.

TRANIO
'Tis death for anyone in Mantua
To come to Padua. Know you not the cause?
Your ships are stayed at Venice, and the Duke,
For private quarrel 'twixt your Duke and him,       85
Hath published and proclaimed it openly.
'Tis marvel, but that you are but newly come,
You might have heard it else proclaimed about.

PEDANT
Alas, sir, it is worse for me than so,
For I have bills for money by exchange               90
From Florence, and must here deliver them.

TRANIO
Well, sir, to do you courtesy
This will I do, and this I will advise you.
First tell me, have you ever been at Pisa?

PEDANT
Ay, sir, in Pisa have I often been,                     95
Pisa renownèd for grave citizens.

TRANIO
Among them know you one Vincentio?

PEDANT
I know him not, but I have heard of him,
A merchant of incomparable wealth.

TRANIO
He is my father, sir, and sooth to say,                100
In count'nance somewhat doth resemble you.

BIONDELLO (*aside*) As much as an apple doth an oyster,
and all one.

TRANIO
To save your life in this extremity
This favour will I do you for his sake,               105
And think it not the worst of all your fortunes
That you are like to Sir Vincentio.
His name and credit shall you undertake,
And in my house you shall be friendly lodged.
Look that you take upon you as you should.          110
You understand me, sir? So shall you stay
Till you have done your business in the city.
If this be courtesy, sir, accept of it.

PEDANT
O sir, I do, and will repute you ever
The patron of my life and liberty.                     115

TRANIO
Then go with me to make the matter good.
This, by the way, I let you understand—
My father is here looked for every day
To pass assurance of a dower in marriage

'Twixt me and one Baptista's daughter here.         120
In all these circumstances I'll instruct you.
Go with me to clothe you as becomes you.     *Exeunt*

4.3   *Enter Katherine and Grumio*

GRUMIO
No, no, forsooth. I dare not, for my life.

KATHERINE
The more my wrong, the more his spite appears.
What, did he marry me to famish me?
Beggars that come unto my father's door
Upon entreaty have a present alms,                      5
If not, elsewhere they meet with charity.
But I, who never knew how to entreat,
Nor never needed that I should entreat,
Am starved for meat, giddy for lack of sleep,
With oaths kept waking and with brawling fed,     10
And that which spites me more than all these wants,
He does it under name of perfect love,
As who should say if I should sleep or eat
'Twere deadly sickness, or else present death.
I prithee, go and get me some repast,                 15
I care not what, so it be wholesome food.

GRUMIO What say you to a neat's foot?

KATHERINE
'Tis passing good. I prithee, let me have it.

GRUMIO
I fear it is too choleric a meat.
How say you to a fat tripe finely broiled?           20

KATHERINE
I like it well. Good Grumio, fetch it me.

GRUMIO
I cannot tell, I fear 'tis choleric.
What say you to a piece of beef, and mustard?

KATHERINE
A dish that I do love to feed upon.

GRUMIO
Ay, but the mustard is too hot a little.               25

KATHERINE
Why then, the beef, and let the mustard rest.

GRUMIO
Nay, then I will not. You shall have the mustard,
Or else you get no beef of Grumio.

KATHERINE
Then both, or one, or anything thou wilt.

GRUMIO
Why then, the mustard without the beef.             30

KATHERINE
Go, get thee gone, thou false, deluding slave,
(*Beating him*) That feed'st me with the very name of
       meat.
Sorrow on thee and all the pack of you,
That triumph thus upon my misery.
Go, get thee gone, I say.                                35
       *Enter Petruccio and Hortensio, with meat*

PETRUCCIO
How fares my Kate? What, sweeting, all amort?

HORTENSIO
Mistress, what cheer?

KATHERINE                    Faith, as cold as can be.

PETRUCCIO
Pluck up thy spirits, look cheerfully upon me.
Here, love, thou seest how diligent I am
To dress thy meat myself and bring it thee.         40
I am sure, sweet Kate, this kindness merits thanks.

What, not a word? Nay then, thou lov'st it not,
And all my pains is sorted to no proof.
Here, take away this dish.
KATHERINE                    I pray you, let it stand.
PETRUCCIO
The poorest service is repaid with thanks,                45
And so shall mine before you touch the meat.
KATHERINE I thank you, sir.
HORTENSIO
Signor Petruccio, fie, you are to blame.
Come, Mistress Kate, I'll bear you company.
PETRUCCIO (aside)
Eat it up all, Hortensio, if thou lov'st me.             50
(To Katherine) Much good do it unto thy gentle heart.
Kate, eat apace; and now, my honey love,
Will we return unto thy father's house,
And revel it as bravely as the best,
With silken coats, and caps, and golden rings,           55
With ruffs, and cuffs, and farthingales, and things,
With scarves, and fans, and double change of
    bravery,
With amber bracelets, beads, and all this knavery.
What, hast thou dined? The tailor stays thy leisure,
To deck thy body with his ruffling treasure.             60
        Enter Tailor with a gown
Come, tailor, let us see these ornaments.
Lay forth the gown.
        Enter Haberdasher with a cap
                    What news with you, sir?
HABERDASHER
Here is the cap your worship did bespeak.
PETRUCCIO
Why, this was moulded on a porringer—
A velvet dish. Fie, fie, 'tis lewd and filthy.           65
Why, 'tis a cockle or a walnut-shell,
A knack, a toy, a trick, a baby's cap.
Away with it! Come, let me have a bigger.
KATHERINE
I'll have no bigger. This doth fit the time,
And gentlewomen wear such caps as these.                 70
PETRUCCIO
When you are gentle you shall have one, too,
And not till then.
HORTENSIO (aside)     That will not be in haste.
KATHERINE
Why, sir, I trust I may have leave to speak,
And speak I will. I am no child, no babe.
Your betters have endured me say my mind,                75
And if you cannot, best you stop your ears.
My tongue will tell the anger of my heart,
Or else my heart concealing it will break,
And rather than it shall I will be free
Even to the uttermost as I please in words.              80
PETRUCCIO
Why, thou sayst true. It is a paltry cap,
A custard-coffin, a bauble, a silken pie.
I love thee well in that thou lik'st it not.
KATHERINE
Love me or love me not, I like the cap
And it I will have, or I will have none.                 85
        ⌜Exit Haberdasher⌝
PETRUCCIO
Thy gown? Why, ay. Come, tailor, let us see't.
O mercy, God, what masquing stuff is here?
What's this—a sleeve? 'Tis like a demi-cannon.

What, up and down carved like an apple-tart?
Here's snip, and nip, and cut, and slish and slash,      90
Like to a scissor in a barber's shop.
Why, what o' devil's name, tailor, call'st thou this?
HORTENSIO (aside)
I see she's like to have nor cap nor gown.
TAILOR
You bid me make it orderly and well,
According to the fashion and the time.                   95
PETRUCCIO
Marry, and did, but if you be remembered
I did not bid you mar it to the time.
Go hop me over every kennel home,
For you shall hop without my custom, sir.
I'll none of it. Hence, make your best of it.            100
KATHERINE
I never saw a better fashioned gown,
More quaint, more pleasing, nor more commendable.
Belike you mean to make a puppet of me.
PETRUCCIO
Why true, he means to make a puppet of thee.
TAILOR She says your worship means to make a puppet
    of her.                                              106
PETRUCCIO
O monstrous arrogance! Thou liest, thou thread, thou
    thimble,
Thou yard, three-quarters, half-yard, quarter, nail,
Thou flea, thou nit, thou winter-cricket, thou.
Braved in mine own house with a skein of thread!         110
Away, thou rag, thou quantity, thou remnant,
Or I shall so bemete thee with thy yard
As thou shalt think on prating whilst thou liv'st.
I tell thee, I, that thou hast marred her gown.
TAILOR
Your worship is deceived. The gown is made              115
Just as my master had direction.
Grumio gave order how it should be done.
GRUMIO
I gave him no order, I gave him the stuff.
TAILOR
But how did you desire it should be made?
GRUMIO Marry, sir, with needle and thread.               120
TAILOR
But did you not request to have it cut?
GRUMIO Thou hast faced many things.
TAILOR I have.
GRUMIO Face not me. Thou hast braved many men. Brave
    not me. I will neither be faced nor braved. I say unto
    thee I bid thy master cut out the gown, but I did not
    bid him cut it to pieces. Ergo thou liest.           127
TAILOR (showing a paper) Why, here is the note of the
    fashion, to testify.                                 130
PETRUCCIO Read it.
GRUMIO The note lies in's throat if he say I said so.
TAILOR (reads) 'Imprimis, a loose-bodied gown.'
GRUMIO Master, if ever I said loose-bodied gown, sew me
    in the skirts of it and beat me to death with a bottom
    of brown thread. I said a gown.                      135
PETRUCCIO Proceed.
TAILOR (reads) 'With a small compassed cape.'
GRUMIO I confess the cape.
TAILOR (reads) 'With a trunk sleeve.'
GRUMIO I confess two sleeves.                            140
TAILOR (reads) 'The sleeves curiously cut.'
PETRUCCIO Ay, there's the villany.

GRUMIO Error i'th' bill, sir, error i'th' bill. I commanded
the sleeves should be cut out and sewed up again, and
that I'll prove upon thee though thy little finger be
armed in a thimble.     146

TAILOR This is true that I say. An I had thee in place
where, thou shouldst know it.

GRUMIO I am for thee straight. Take thou the bill, give
me thy mete-yard, and spare not me.     150

HORTENSIO Godamercy, Grumio, then he shall have no
odds.

PETRUCCIO
Well, sir, in brief, the gown is not for me.

GRUMIO You are i'th' right, sir. 'Tis for my mistress.

PETRUCCIO (to the Tailor)
Go, take it up unto thy master's use.     155

GRUMIO (to the Tailor) Villain, not for thy life. Take up my
mistress' gown for thy master's use!

PETRUCCIO Why, sir, what's your conceit in that?

GRUMIO O, sir, the conceit is deeper than you think for.
'Take up my mistress' gown to his master's use'—O
fie, fie, fie!     161

PETRUCCIO (aside)
Hortensio, say thou wilt see the tailor paid.
(To the Tailor) Go, take it hence. Be gone, and say no
more.

HORTENSIO (aside to the Tailor)
Tailor, I'll pay thee for thy gown tomorrow.
Take no unkindness of his hasty words.     165
Away, I say. Commend me to thy master.    *Exit Tailor*

PETRUCCIO
Well, come, my Kate. We will unto your father's
Even in these honest, mean habiliments.
Our purses shall be proud, our garments poor,
For 'tis the mind that makes the body rich,     170
And as the sun breaks through the darkest clouds,
So honour peereth in the meanest habit.
What, is the jay more precious than the lark
Because his feathers are more beautiful?
Or is the adder better than the eel     175
Because his painted skin contents the eye?
O no, good Kate, neither art thou the worse
For this poor furniture and mean array.
If thou account'st it shame, lay it on me,
And therefore frolic; we will hence forthwith     180
To feast and sport us at thy father's house.
Go call my men, and let us straight to him,
And bring our horses unto Long Lane end.
There will we mount, and thither walk on foot.
Let's see, I think 'tis now some seven o'clock,     185
And well we may come there by dinner-time.

KATHERINE
I dare assure you, sir, 'tis almost two,
And 'twill be supper-time ere you come there.

PETRUCCIO
It shall be seven ere I go to horse.
Look what I speak, or do, or think to do,     190
You are still crossing it. Sirs, let't alone.
I will not go today, and ere I do
It shall be what o'clock I say it is.

HORTENSIO (aside)
Why, so this gallant will command the sun.    *Exeunt*

**4.4**    *Enter Tranio as Lucentio, and the Pedant dressed*
*like Vincentio, booted and bare-headed*

TRANIO
Sir, this is the house. Please it you that I call?

PEDANT
Ay, what else. And but I be deceived,
Signor Baptista may remember me
Near twenty years ago in Genoa—

TRANIO
Where we were lodgers at the Pegasus.—     5
'Tis well, and hold your own in any case
With such austerity as 'longeth to a father.
     *Enter Biondello*

PEDANT
I warrant you. But sir, here comes your boy.
'Twere good he were schooled.

TRANIO
Fear you not him. Sirrah Biondello,     10
Now do your duty throughly, I advise you.
Imagine 'twere the right Vincentio.

BIONDELLO Tut, fear not me.

TRANIO
But hast thou done thy errand to Baptista?

BIONDELLO
I told him that your father was at Venice     15
And that you looked for him this day in Padua.

TRANIO (giving money)
Thou'rt a tall fellow. Hold thee that to drink.
Here comes Baptista. Set your countenance, sir.
     *Enter Baptista, and Lucentio as Cambio*

TRANIO
Signor Baptista, you are happily met.
(To the Pedant) Sir, this is the gentleman I told you of.
I pray you stand good father to me now.     21
Give me Bianca for my patrimony.

PEDANT
Soft, son. (To Baptista) Sir, by your leave, having
     come to Padua
To gather in some debts, my son Lucentio
Made me acquainted with a weighty cause     25
Of love between your daughter and himself,
And for the good report I hear of you,
And for the love he beareth to your daughter,
And she to him, to stay him not too long
I am content in a good father's care     30
To have him matched, and if you please to like
No worse than I, upon some agreement
Me shall you find ready and willing
With one consent to have her so bestowed,
For curious I cannot be with you,     35
Signor Baptista, of whom I hear so well.

BAPTISTA
Sir, pardon me in what I have to say.
Your plainness and your shortness please me well.
Right true it is your son Lucentio here
Doth love my daughter, and she loveth him,     40
Or both dissemble deeply their affections.
And therefore if you say no more than this,
That like a father you will deal with him
And pass my daughter a sufficient dower,
The match is made, and all is done.     45
Your son shall have my daughter with consent.

TRANIO
I thank you, sir. Where then do you know best
We be affied, and such assurance ta'en
As shall with either part's agreement stand?

BAPTISTA
Not in my house, Lucentio, for you know     50
Pitchers have ears, and I have many servants.
Besides, old Gremio is heark'ning still,
And happily we might be interrupted.

TRANIO
Then at my lodging, an it like you.
There doth my father lie, and there this night          55
We'll pass the business privately and well.
Send for your daughter by your servant here.
My boy shall fetch the scrivener presently.
The worst is this, that at so slender warning
You are like to have a thin and slender pittance.      60
BAPTISTA
It likes me well. Cambio, hie you home
And bid Bianca make her ready straight,
And if you will, tell what hath happened—
Lucentio's father is arrived in Padua—
And how she's like to be Lucentio's wife.              65
                                    ⌈Exit Lucentio⌉
BIONDELLO
I pray the gods she may with all my heart.
TRANIO
Dally not with the gods, but get thee gone.
                                    ⌈Exit Biondello⌉
Signor Baptista, shall I lead the way?
Welcome. One mess is like to be your cheer.
Come, sir, we will better it in Pisa.                  70
BAPTISTA  I follow you.                          Exeunt

**4.5**   *Enter Lucentio and Biondello*
BIONDELLO  Cambio.
LUCENTIO  What sayst thou, Biondello?
BIONDELLO  You saw my master wink and laugh upon
  you?
LUCENTIO  Biondello, what of that?                    5
BIONDELLO  Faith, nothing, but he's left me here behind to
  expound the meaning or moral of his signs and tokens.
LUCENTIO  I pray thee, moralize them.
BIONDELLO  Then thus: Baptista is safe, talking with the
  deceiving father of a deceitful son.                10
LUCENTIO  And what of him?
BIONDELLO  His daughter is to be brought by you to the
  supper.
LUCENTIO  And then?
BIONDELLO  The old priest at Saint Luke's church is at your
  command at all hours.                               16
LUCENTIO  And what of all this?
BIONDELLO  I cannot tell, except they are busied about a
  counterfeit assurance. Take you assurance of her *cum
  privilegio ad imprimendum solum*—to th' church take the
  priest, clerk, and some sufficient honest witnesses.  21
  If this be not that you look for, I have no more to say,
  But bid Bianca farewell for ever and a day.
LUCENTIO  Hear'st thou, Biondello?
BIONDELLO  I cannot tarry, I knew a wench married in an
  afternoon as she went to the garden for parsley to stuff
  a rabbit, and so may you, sir, and so adieu, sir. My
  master hath appointed me to go to Saint Luke's to bid
  the priest be ready t'attend against you come with your
  appendix.                                      *Exit*
LUCENTIO
I may and will, if she be so contented.               31
She will be pleased, then wherefore should I doubt?
Hap what hap may, I'll roundly go about her.
It shall go hard if Cambio go without her.       *Exit*

**4.6**   *Enter Petruccio, Katherine, Hortensio, and servants*
PETRUCCIO
Come on, i' God's name. Once more toward our
  father's.
Good Lord, how bright and goodly shines the moon!

KATHERINE
The moon?—the sun. It is not moonlight now.
PETRUCCIO
I say it is the moon that shines so bright.
KATHERINE
I know it is the sun that shines so bright.           5
PETRUCCIO
Now, by my mother's son—and that's myself—
It shall be moon, or star, or what I list
Or ere I journey to your father's house.
Go on, and fetch our horses back again.
Evermore crossed and crossed, nothing but crossed.   10
HORTENSIO  (*to Katherine*)
Say as he says or we shall never go.
KATHERINE
Forward, I pray, since we have come so far,
And be it moon or sun or what you please,
And if you please to call it a rush-candle
Henceforth I vow it shall be so for me.               15
PETRUCCIO
I say it is the moon.
KATHERINE
I know it is the moon.
PETRUCCIO
Nay then you lie, it is the blessèd sun.
KATHERINE
Then God be blessed, it is the blessèd sun,
But sun it is not when you say it is not,             20
And the moon changes even as your mind.
What you will have it named, even that it is,
And so it shall be still for Katherine.
HORTENSIO
Petruccio, go thy ways. The field is won.
PETRUCCIO
Well, forward, forward. Thus the bowl should run,    25
And not unluckily against the bias.
But soft, company is coming here.
      *Enter old Vincentio*
(*To Vincentio*) Good morrow, gentle mistress, where
  away?
Tell me, sweet Kate, and tell me truly too,
Hast thou beheld a fresher gentlewoman,              30
Such war of white and red within her cheeks?
What stars do spangle heaven with such beauty
As those two eyes become that heavenly face?
Fair lovely maid, once more good day to thee.
Sweet Kate, embrace her for her beauty's sake.        35
HORTENSIO  A will make the man mad to make the woman
  of him.
KATHERINE
Young budding virgin, fair, and fresh, and sweet,
Whither away, or where is thy abode?
Happy the parents of so fair a child,                 40
Happier the man whom favourable stars
Allots thee for his lovely bedfellow.
PETRUCCIO
Why, how now, Kate, I hope thou art not mad.
This is a man, old, wrinkled, faded, withered,
And not a maiden as thou sayst he is.                 45
KATHERINE
Pardon, old father, my mistaking eyes
That have been so bedazzled with the sun
That everything I look on seemeth green.
Now I perceive thou art a reverend father.
Pardon, I pray thee, for my mad mistaking.            50
PETRUCCIO
Do, good old grandsire, and withal make known

Which way thou travell'st. If along with us,
We shall be joyful of thy company.
VINCENTIO
Fair sir, and you, my merry mistress,
That with your strange encounter much amazed me,
My name is called Vincentio, my dwelling Pisa,          56
And bound I am to Padua, there to visit
A son of mine which long I have not seen.
PETRUCCIO
What is his name?
VINCENTIO          Lucentio, gentle sir.
PETRUCCIO
Happily met, the happier for thy son.          60
And now by law as well as reverend age
I may entitle thee my loving father.
The sister to my wife, this gentlewoman,
Thy son by this hath married. Wonder not,
Nor be not grieved. She is of good esteem,          65
Her dowry wealthy, and of worthy birth,
Beside, so qualified as may beseem
The spouse of any noble gentleman.
Let me embrace with old Vincentio,
And wander we to see thy honest son,          70
Who will of thy arrival be full joyous.
*He embraces Vincentio*
VINCENTIO
But is this true, or is it else your pleasure
Like pleasant travellers to break a jest
Upon the company you overtake?
HORTENSIO
I do assure thee, father, so it is.          75
PETRUCCIO
Come, go along, and see the truth hereof,
For our first merriment hath made thee jealous.
          *Exeunt all but Hortensio*
HORTENSIO
Well, Petruccio, this has put me in heart.
Have to my widow, and if she be froward,
Then hast thou taught Hortensio to be untoward.          80
          *Exit*

**5.1**     *Enter Biondello, Lucentio, and Bianca. Gremio is out*
     *before*
BIONDELLO Softly and swiftly, sir, for the priest is ready.
LUCENTIO I fly, Biondello; but they may chance to need
     thee at home, therefore leave us.
BIONDELLO Nay, faith, I'll see the church a' your back
     and then come back to my master's as soon as I can.
          *Exeunt Lucentio, Bianca, and Biondello*
GREMIO
I marvel Cambio comes not all this while.          6
     *Enter Petruccio, Katherine, Vincentio, Grumio, with*
     *attendants*
PETRUCCIO
Sir, here's the door. This is Lucentio's house.
My father's bears more toward the market-place.
Thither must I, and here I leave you, sir.
VINCENTIO
You shall not choose but drink before you go.          10
I think I shall command your welcome here,
And by all likelihood some cheer is toward.
     *He knocks*
GREMIO They're busy within. You were best knock louder.
     *Vincentio knocks again. The Pedant looks out of the*
     *window*
PEDANT What's he that knocks as he would beat down
·the gate?          15

VINCENTIO Is Signor Lucentio within, sir?
PEDANT He's within, sir, but not to be spoken withal.
VINCENTIO What if a man bring him a hundred pound or
     two to make merry withal?
PEDANT Keep your hundred pounds to yourself. He shall
     need none so long as I live.          21
PETRUCCIO (*to Vincentio*) Nay, I told you your son was well
     beloved in Padua. (*To the Pedant*) Do you hear, sir, to
     leave frivolous circumstances, I pray you tell Signor
     Lucentio that his father is come from Pisa and is here
     at the door to speak with him.          26
PEDANT Thou liest. His father is come from Padua and
     here looking out at the window.
VINCENTIO Art thou his father?
PEDANT Ay, sir, so his mother says, if I may believe her.
PETRUCCIO (*to Vincentio*) Why, how now, gentleman? Why,
     this is flat knavery, to take upon you another man's
     name.          33
PEDANT Lay hands on the villain. I believe a means to
     cozen somebody in this city under my countenance.
     *Enter Biondello*
BIONDELLO (*aside*) I have seen them in the church together,
     God send 'em good shipping. But who is here? Mine
     old master, Vincentio—now we are undone and
     brought to nothing.
VINCENTIO (*to Biondello*) Come hither, crackhemp.          40
BIONDELLO I hope I may choose, sir.
VINCENTIO Come hither, you rogue. What, have you forgot
     me?
BIONDELLO Forgot you? No, sir, I could not forget you,
     for I never saw you before in all my life.          45
VINCENTIO What, you notorious villain, didst thou never
     see thy master's father, Vincentio?
BIONDELLO What, my old worshipful old master? Yes,
     marry, sir, see where he looks out of the window.
VINCENTIO Is't so indeed?          50
     *He beats Biondello*
BIONDELLO Help, help, help! Here's a madman will murder
     me.          *Exit*
PEDANT Help, son! Help, Signor Baptista!          *Exit above*
PETRUCCIO Prithee, Kate, let's stand aside and see the end
     of this controversy.          55
     *They stand aside.*
     *Enter Pedant with servants, Baptista, Tranio as*
     *Lucentio*
TRANIO (*to Vincentio*) Sir, what are you that offer to beat
     my servant?
VINCENTIO What am I, sir? Nay, what are you, sir? O
     immortal gods, O fine villain, a silken doublet, a velvet
     hose, a scarlet cloak, and a copintank hat—O, I am
     undone, I am undone! While I play the good husband
     at home, my son and my servant spend all at the
     university.          63
TRANIO How now, what's the matter?
BAPTISTA What, is the man lunatic?          65
TRANIO Sir, you seem a sober, ancient gentleman by your
     habit, but your words show you a madman. Why sir,
     what 'cerns it you if I wear pearl and gold? I thank
     my good father, I am able to maintain it.
VINCENTIO Thy father! O villain, he is a sailmaker in
     Bergamo.          71
BAPTISTA You mistake, sir, you mistake, sir. Pray what
     do you think is his name?
VINCENTIO His name? As if I knew not his name—I have
     brought him up ever since he was three years old, and
     his name is Tranio.          76

PEDANT Away, away, mad ass. His name is Lucentio, and
  he is mine only son, and heir to the lands of me, Signor
  Vincentio.
VINCENTIO Lucentio? O, he hath murdered his master!
  Lay hold on him, I charge you, in the Duke's name. O
  my son, my son! Tell me, thou villain, where is my
  son Lucentio?                                          83
TRANIO Call forth an officer.
      *Enter an Officer*
  Carry this mad knave to the jail. Father Baptista, I
  charge you see that he be forthcoming.                 86
VINCENTIO Carry me to the jail?
GREMIO Stay, officer, he shall not go to prison.
BAPTISTA Talk not, Signor Gremio. I say he shall go to
  prison.                                                90
GREMIO Take heed, Signor Baptista, lest you be cony-
  catched in this business. I dare swear this is the right
  Vincentio.
PEDANT Swear if thou dar'st.
GREMIO Nay, I dare not swear it.                         95
TRANIO Then thou wert best say that I am not Lucentio.
GREMIO Yes, I know thee to be Signor Lucentio.
BAPTISTA Away with the dotard. To the jail with him.
      *Enter Biondello, Lucentio, and Bianca*
VINCENTIO Thus strangers may be haled and abused. O
  monstrous villain!                                     100
BIONDELLO O, we are spoiled and—yonder he is. Deny
  him, forswear him, or else we are all undone.
          *Exeunt Biondello, Tranio, and Pedant, as fast as*
                                              *may be*
LUCENTIO (*to Vincentio*) Pardon, sweet father.
      *He kneels*
VINCENTIO Lives my sweet son?
BIANCA (*to Baptista*) Pardon, dear father.              105
BAPTISTA
  How hast thou offended? Where is Lucentio?
LUCENTIO
  Here's Lucentio, right son to the right Vincentio,
  That have by marriage made thy daughter mine,
  While counterfeit supposes bleared thine eyne.
GREMIO
  Here's packing with a witness, to deceive us all.     110
VINCENTIO
  Where is that damnèd villain Tranio,
  That faced and braved me in this matter so?
BAPTISTA
  Why, tell me, is not this my Cambio?
BIANCA
  Cambio is changed into Lucentio.
LUCENTIO
  Love wrought these miracles. Bianca's love            115
  Made me exchange my state with Tranio
  While he did bear my countenance in the town,
  And happily I have arrived at the last
  Unto the wishèd haven of my bliss.
  What Tranio did, myself enforced him to.              120
  Then pardon him, sweet father, for my sake.
VINCENTIO I'll slit the villain's nose that would have sent
  me to the jail.
BAPTISTA But do you hear, sir, have you married my
  daughter without asking my good will?                 125
VINCENTIO Fear not, Baptista. We will content you. Go to,
  but I will in to be revenged for this villainy.    *Exit*
BAPTISTA And I to sound the depth of this knavery.   *Exit*
LUCENTIO Look not pale, Bianca. Thy father will not frown.
                              *Exeunt Lucentio and Bianca*

GREMIO
  My cake is dough, but I'll in among the rest,         130
  Out of hope of all but my share of the feast.    *Exit*
KATHERINE (*coming forward*) Husband, let's follow to see
  the end of this ado.
PETRUCCIO First kiss me, Kate, and we will.
KATHERINE What, in the midst of the street?            135
PETRUCCIO What, art thou ashamed of me?
KATHERINE No, sir, God forbid; but ashamed to kiss.
PETRUCCIO
  Why then, let's home again. Come sirrah, let's away.
KATHERINE
  Nay, I will give thee a kiss. Now pray thee love, stay.
      *They kiss*
PETRUCCIO
  Is not this well? Come, my sweet Kate.               140
  Better once than never, for never too late.     *Exeunt*

5.2   *Enter Baptista, Vincentio, Gremio, the Pedant,*
      *Lucentio and Bianca, Petruccio, Katherine, and*
      *Hortensio, Tranio, Biondello, Grumio, and the*
      *Widow, the servingmen with Tranio bringing in a*
      *banquet*
LUCENTIO
  At last, though long, our jarring notes agree,
  And time it is when raging war is done
  To smile at scapes and perils overblown.
  My fair Bianca, bid my father welcome,
  While I with selfsame kindness welcome thine.          5
  Brother Petruccio, sister Katherina,
  And thou, Hortensio, with thy loving widow,
  Feast with the best, and welcome to my house.
  My banquet is to close our stomachs up
  After our great good cheer. Pray you, sit down,       10
  For now we sit to chat as well as eat.
      *They sit*
PETRUCCIO
  Nothing but sit, and sit, and eat, and eat.
BAPTISTA
  Padua affords this kindness, son Petruccio.
PETRUCCIO
  Padua affords nothing but what is kind.
HORTENSIO
  For both our sakes I would that word were true.       15
PETRUCCIO
  Now, for my life, Hortensio fears his widow.
WIDOW
  Then never trust me if I be afeard.
PETRUCCIO
  You are very sensible, and yet you miss my sense.
  I mean Hortensio is afeard of you.
WIDOW
  He that is giddy thinks the world turns round.        20
PETRUCCIO Roundly replied.
KATHERINE Mistress, how mean you that?
WIDOW Thus I conceive by him.
PETRUCCIO
  Conceives by me! How likes Hortensio that?
HORTENSIO
  My widow says thus she conceives her tale.            25
PETRUCCIO Very well mended. Kiss him for that, good
  widow.
KATHERINE
  'He that is giddy thinks the world turns round'—
  I pray you tell me what you meant by that.
WIDOW
  Your husband, being troubled with a shrew,            30

Measures my husband's sorrow by his woe.
And now you know my meaning.

KATHERINE
A very mean meaning.

WIDOW                    Right, I mean you.

KATHERINE
And I am mean indeed respecting you.

PETRUCCIO To her, Kate!                                              35

HORTENSIO To her, widow!

PETRUCCIO
A hundred marks my Kate does put her down.

HORTENSIO That's my office.

PETRUCCIO
Spoke like an officer! Ha' to thee, lad.
            *He drinks to Hortensio*

BAPTISTA
How likes Gremio these quick-witted folks?         40

GREMIO
Believe me, sir, they butt together well.

BIANCA
Head and butt? An hasty-witted body
Would say your head and butt were head and horn.

VINCENTIO
Ay, mistress bride, hath that awakened you?

BIANCA
Ay, but not frighted me, therefore I'll sleep again.   45

PETRUCCIO
Nay, that you shall not. Since you have begun,
Have at you for a better jest or two.

BIANCA
Am I your bird? I mean to shift my bush,
And then pursue me as you draw your bow.
You are welcome all.                                                    50
            *Exit Bianca with Katherine and the Widow*

PETRUCCIO
She hath prevented me here, Signor Tranio.
This bird you aimed at, though you hit her not.
Therefore a health to all that shot and missed.

TRANIO
O sir, Lucentio slipped me like his greyhound,
Which runs himself and catches for his master.     55

PETRUCCIO
A good swift simile, but something currish.

TRANIO
'Tis well, sir, that you hunted for yourself.
'Tis thought your deer does hold you at a bay.

BAPTISTA
O, O, Petruccio, Tranio hits you now.

LUCENTIO
I thank thee for that gird, good Tranio.               60

HORTENSIO
Confess, confess, hath he not hit you here?

PETRUCCIO
A has a little galled me, I confess,
And as the jest did glance away from me,
'Tis ten to one it maimed you two outright.

BAPTISTA
Now in good sadness, son Petruccio,                   65
I think thou hast the veriest shrew of all.

PETRUCCIO
Well, I say no.—And therefore, Sir Assurance,
Let's each one send unto his wife,
And he whose wife is most obedient
To come at first when he doth send for her          70
Shall win the wager which we will propose.

HORTENSIO Content. What's the wager?

LUCENTIO Twenty crowns.

PETRUCCIO Twenty crowns!
I'll venture so much of my hawk or hound,          75
But twenty times so much upon my wife.

LUCENTIO A hundred, then.

HORTENSIO Content.

PETRUCCIO A match, 'tis done.

HORTENSIO Who shall begin?                              80

LUCENTIO That will I.
Go, Biondello, bid your mistress come to me.

BIONDELLO I go.                                    *Exit*

BAPTISTA
Son, I'll be your half Bianca comes.

LUCENTIO
I'll have no halves, I'll bear it all myself.             85
            *Enter Biondello*
How now, what news?

BIONDELLO                    Sir, my mistress sends you word
That she is busy and she cannot come.

PETRUCCIO
How? She's busy and she cannot come?
Is that an answer?

GREMIO                    Ay, and a kind one, too.
Pray God, sir, your wife send you not a worse.    90

PETRUCCIO
I hope, better.

HORTENSIO                    Sirrah Biondello,
Go and entreat my wife to come to me forthwith.
                                        *Exit Biondello*

PETRUCCIO
O ho, 'entreat' her—nay, then she must needs come.

HORTENSIO
I am afraid, sir, do what you can,
            *Enter Biondello*
Yours will not be entreated. Now, where's my wife?

BIONDELLO
She says you have some goodly jest in hand.      96
She will not come. She bids you come to her.

PETRUCCIO
Worse and worse! She will not come—O vile,
Intolerable, not to be endured!
Sirrah Grumio, go to your mistress.                    100
Say I command her come to me.       *Exit Grumio*

HORTENSIO
I know her answer.

PETRUCCIO                    What?

HORTENSIO                    She will not.

PETRUCCIO
The fouler fortune mine, and there an end.
            *Enter Katherine*

BAPTISTA
Now by my halidom, here comes Katherina.

KATHERINE *(to Petruccio)*
What is your will, sir, that you send for me?      105

PETRUCCIO
Where is your sister and Hortensio's wife?

KATHERINE
They sit conferring by the parlour fire.

PETRUCCIO
Go, fetch them hither. If they deny to come,
Swinge me them soundly forth unto their husbands.
Away, I say, and bring them hither straight.         110
            *Exit Katherine*

LUCENTIO
Here is a wonder, if you talk of wonders.

HORTENSIO
And so it is. I wonder what it bodes.

PETRUCCIO

Marry, peace it bodes, and love, and quiet life;
An aweful rule and right supremacy,
And, to be short, what not that's sweet and happy.

BAPTISTA

Now fair befall thee, good Petruccio,                    116
The wager thou hast won, and I will add
Unto their losses twenty thousand crowns,
Another dowry to another daughter,
For she is changed as she had never been.                120

PETRUCCIO

Nay, I will win my wager better yet,
And show more sign of her obedience,
Her new-built virtue and obedience.
    *Enter Katherine, Bianca, and the Widow*
See where she comes, and brings your froward wives
As prisoners to her womanly persuasion.                  125
Katherine, that cap of yours becomes you not.
Off with that bauble, throw it underfoot.
    *Katherine throws down her cap*

WIDOW

Lord, let me never have a cause to sigh
Till I be brought to such a silly pass.

BIANCA

Fie, what a foolish duty call you this?                   130

LUCENTIO

I would your duty were as foolish, too.
The wisdom of your duty, fair Bianca,
Hath cost me a hundred crowns since supper-time.

BIANCA

The more fool you for laying on my duty.

PETRUCCIO

Katherine, I charge thee tell these headstrong women
What duty they do owe their lords and husbands.  136

WIDOW

Come, come, you're mocking. We will have no telling.

PETRUCCIO

Come on, I say, and first begin with her.

WIDOW  She shall not.

PETRUCCIO

I say she shall: and first begin with her.               140

KATHERINE

Fie, fie, unknit that threat'ning, unkind brow,
And dart not scornful glances from those eyes
To wound thy lord, thy king, thy governor.
It blots thy beauty as frosts do bite the meads,
Confounds thy fame as whirlwinds shake fair buds,
And in no sense is meet or amiable.                      146
A woman moved is like a fountain troubled,
Muddy, ill-seeming, thick, bereft of beauty,
And while it is so, none so dry or thirsty
Will deign to sip or touch one drop of it.               150
Thy husband is thy lord, thy life, thy keeper,

Thy head, thy sovereign, one that cares for thee,
And for thy maintenance commits his body
To painful labour both by sea and land,
To watch the night in storms, the day in cold,   155
Whilst thou liest warm at home, secure and safe,
And craves no other tribute at thy hands
But love, fair looks, and true obedience,
Too little payment for so great a debt.
Such duty as the subject owes the prince,        160
Even such a woman oweth to her husband,
And when she is froward, peevish, sullen, sour,
And not obedient to his honest will,
What is she but a foul contending rebel,
And graceless traitor to her loving lord?         165
I am ashamed that women are so simple
To offer war where they should kneel for peace,
Or seek for rule, supremacy, and sway
When they are bound to serve, love, and obey.
Why are our bodies soft, and weak, and smooth,   170
Unapt to toil and trouble in the world,
But that our soft conditions and our hearts
Should well agree with our external parts?
Come, come, you froward and unable worms,
My mind hath been as big as one of yours,        175
My heart as great, my reason haply more,
To bandy word for word and frown for frown;
But now I see our lances are but straws,
Our strength as weak, our weakness past compare,
That seeming to be most which we indeed least are.
Then vail your stomachs, for it is no boot,       181
And place your hands below your husband's foot,
In token of which duty, if he please,
My hand is ready, may it do him ease.

PETRUCCIO

Why, there's a wench! Come on, and kiss me, Kate.
    *They kiss*

LUCENTIO

Well, go thy ways, old lad, for thou shalt ha't.  186

VINCENTIO

'Tis a good hearing when children are toward.

LUCENTIO

But a harsh hearing when women are froward.

PETRUCCIO  Come, Kate, we'll to bed.
We three are married, but you two are sped.       190
'Twas I won the wager, though (*to Lucentio*) you hit
    the white,
And being a winner, God give you good night.
    *Exit Petruccio with Katherine*

HORTENSIO

Now go thy ways, thou hast tamed a curst shrew.

LUCENTIO

'Tis a wonder, by your leave, she will be tamed so.
    *Exeunt*

## ADDITIONAL PASSAGES

*The Taming of A Shrew*, printed in 1594 and believed to derive from Shakespeare's play as performed, contains episodes continuing and rounding off the Christopher Sly framework which may echo passages written by Shakespeare but not printed in the Folio. They are given below.

A. The following exchange occurs at a point for which there is no exact equivalent in Shakespeare's play. It could come at the end of 2.1. The 'fool' of the first line is Sander, the counterpart of Grumio.

> *Then Sly speaks*
SLY  Sim, when will the fool come again?
LORD  He'll come again, my lord, anon.
SLY  Gi's some more drink here. Zounds, where's the
    tapster? Here, Sim, eat some of these things.
LORD  So I do, my lord.        5
SLY  Here, Sim, I drink to thee.
LORD  My lord, here comes the players again.
SLY  O brave, here's two fine gentlewomen.

B. This passage comes between 4.5 and 4.6. If it originates with Shakespeare it implies that Grumio accompanies Petruccio at the beginning of 4.6.

SLY  Sim, must they be married now?
LORD  Ay, my lord.
> *Enter Ferando and Kate and Sander*
SLY  Look, Sim, the fool is come again now.

C. Sly interrupts the action of the play-within-play. This is at 5.1.102 of Shakespeare's play.

> *Phylotus and Valeria runs away.*
> *Then Sly speaks*
SLY  I say we'll have no sending to prison.
LORD  My lord, this is but the play. They're but in jest.
SLY  I tell thee, Sim, we'll have no sending to prison,
    that's flat. Why, Sim, am not I Don Christo Vary?
    Therefore I say they shall not go to prison.    5
LORD  No more they shall not, my lord. They be run away.
SLY  Are they run away, Sim? That's well. Then gi's some
    more drink, and let them play again.
LORD  Here, my lord.
> *Sly drinks and then falls asleep*

D. Sly is carried off between 5.1 and 5.2.

>                                 *Exeunt omnes*
> *Sly sleeps*
LORD
    Who's within there? Come hither, sirs, my lord's
    Asleep again. Go take him easily up
    And put him in his own apparel again,
    And lay him in the place where we did find him
    Just underneath the alehouse side below.    5
    But see you wake him not in any case.
BOY
    It shall be done, my lord. Come help to bear him hence.
>                                         *Exit*

E. The conclusion.

> *Then enter two bearing of Sly in his own apparel*
> *again and leaves him where they found him and then*
> *goes out.*
> *Then enter the Tapster*
TAPSTER
    Now that the darksome night is overpast
    And dawning day appears in crystal sky,
    Now must I haste abroad. But soft, who's this?
    What, Sly! O wondrous, hath he lain here all night?
    I'll wake him. I think he's starved by this,    5
    But that his belly was so stuffed with ale.
    What ho, Sly, awake, for shame!
SLY  Sim, gi's some more wine. What, 's all the players
    gone? Am not I a lord?
TAPSTER
    A lord with a murrain! Come, art thou drunken still?
SLY
    Who's this? Tapster? O Lord, sirrah, I have had    11
    The bravest dream tonight that ever thou
    Heardest in all thy life.
TAPSTER
    Ay, marry, but you had best get you home,
    For your wife will course you for dreaming here tonight.
SLY
    Will she? I know now how to tame a shrew.
    I dreamt upon it all this night till now,
    And thou hast waked me out of the best dream
    That ever I had in my life. But I'll to my
    Wife presently and tame her too,    20
    An if she anger me.
TAPSTER
    Nay, tarry, Sly, for I'll go home with thee
    And hear the rest that thou hast dreamt tonight.
>                                   *Exeunt omnes*

# THE FIRST PART OF THE CONTENTION

## (2 HENRY VI)

WHEN Shakespeare's history plays were gathered together in the 1623 Folio, seven years after he died, they were printed in the order of their historical events, each with a title naming the king in whose reign those events occurred. No one supposes that this is the order in which Shakespeare wrote them; and the Folio titles are demonstrably not, in all cases, those by which the plays were originally known. The three concerned with the reign of Henry VI are listed in the Folio, simply and unappealingly, as the *First*, *Second*, and *Third Parts of King Henry the Sixth*, and these are the names by which they have continued to be known. Versions of the *Second* and *Third* had appeared long before the Folio, in 1594 and 1595; their head titles read *The First Part of the Contention of the two Famous Houses of York and Lancaster with the Death of the Good Duke Humphrey* and *The True Tragedy of Richard, Duke of York, and the Good King Henry the Sixth*. These are, presumably, full versions of the plays' original titles, and we revert to them in preference to the Folio's historical listing.

A variety of internal evidence suggests that the Folio's *Part One* was composed after *The First Part of the Contention* and *Richard, Duke of York*, so we depart from the Folio order, though a reader wishing to read the plays in their narrative sequence will read *Henry VI, Part One* before the other two plays. The dates of all three are uncertain, but *Part One* is alluded to in 1592, when it was probably new. *The First Part of the Contention* probably belongs to 1590-1.

The play draws extensively on English chronicle history for its portrayal of the troubled state of England under Henry VI (1421-71). It dramatizes the touchingly weak King's powerlessness against the machinations of his nobles, especially Richard, Duke of York, himself ambitious for the throne. Richard engineers the Kentish rebellion, led by Jack Cade, which provides some of the play's liveliest episodes; and at the play's end Richard seems poised to take the throne.

Historical events of ten years (1445-55) are dramatized with comparative fidelity within a coherent structure that offers a wide variety of theatrical entertainment. Though the play employs old-fashioned conventions of language (particularly the recurrent classical references) and of dramaturgy (such as the horrors of severed heads), its bold characterization, its fundamentally serious but often ironically comic presentation of moral and political issues, the powerful rhetoric of its verse, and the vivid immediacy of its prose have proved highly effective in its rare modern revivals.

# THE PERSONS OF THE PLAY

### Of the King's Party

KING HENRY VI

QUEEN MARGARET

William de la Pole, Marquis, later Duke, of SUFFOLK, the Queen's lover

Duke Humphrey of GLOUCESTER, the Lord Protector, the King's uncle

Dame Eleanor Cobham, the DUCHESS of Gloucester

CARDINAL BEAUFORT, Bishop of Winchester, Gloucester's uncle and the King's great-uncle

Duke of BUCKINGHAM

Duke of SOMERSET

Old Lord CLIFFORD

YOUNG CLIFFORD, his son

### Of the Duke of York's Party

Duke of YORK

EDWARD, Earl of March ⎫
Crookback RICHARD ⎬ his sons

Earl of SALISBURY

Earl of WARWICK, his son

### The petitions and the combat

Two or three PETITIONERS

Thomas HORNER, an armourer

PETER Thump, his man

Three NEIGHBOURS, who drink to Horner

Three PRENTICES, who drink to Peter

### The conjuration

Sir John HUME ⎫
John SOUTHWELL ⎬ priests

Margery Jordan, a WITCH

Roger BOLINGBROKE, a conjurer

ASNATH, a spirit

### The false miracle

Simon SIMPCOX

SIMPCOX'S WIFE

The MAYOR of Saint Albans

Aldermen of Saint Albans

A BEADLE of Saint Albans

Townsmen of Saint Albans

### Eleanor's penance

Gloucester's SERVANTS

Two SHERIFFS of London

Sir John STANLEY

HERALD

### The murder of Gloucester

Two MURDERERS

COMMONS

### The murder of Suffolk

CAPTAIN of a ship

MASTER of that ship

The Master's MATE

Walter WHITMORE

Two GENTLEMEN

### The Cade Rebellion

Jack CADE, a Kentishman suborned by the Duke of York

Dick the BUTCHER ⎫
Smith the WEAVER ⎪
A Sawyer ⎬ Cade's followers
JOHN ⎪
REBELS ⎭

Emmanuel, the CLERK of Chatham ⎫
Sir Humphrey STAFFORD ⎪
STAFFORD'S BROTHER ⎪
Lord SAYE ⎬ those who die at the rebels' hands
Lord SCALES ⎪
Matthew Gough ⎪
A SERGEANT ⎭

Three or four CITIZENS of London

Alexander IDEN, an esquire of Kent, who kills Cade

### Others

VAUX, a messenger

A POST

MESSENGERS

A SOLDIER

Attendants, guards, servants, soldiers, falconers

# The First Part of the Contention
## of the Two Famous Houses of York and Lancaster

**1.1** *Flourish of trumpets, then hautboys. Enter, at one*
*door, King Henry and Humphrey Duke of*
*Gloucester, the Duke of Somerset, the Duke of*
*Buckingham, Cardinal Beaufort, ⌜and others⌝.*
*Enter, at the other door, the Duke of York, and the*
*Marquis of Suffolk, and Queen Margaret, and the*
*Earls of Salisbury and Warwick*

SUFFOLK (*kneeling before King Henry*)
  As by your high imperial majesty
  I had in charge at my depart for France,
  As Procurator to your excellence,
  To marry Princess Margaret for your grace,
  So, in the famous ancient city Tours,       5
  In presence of the Kings of France and Sicil,
  The Dukes of Orléans, Calaber, Bretagne, and Alençon,
  Seven earls, twelve barons, and twenty reverend
     bishops,
  I have performed my task and was espoused,
  And humbly now upon my bended knee,      10
  In sight of England and her lordly peers,
  Deliver up my title in the Queen
  To your most gracious hands, that are the substance
  Of that great shadow I did represent—
  The happiest gift that ever marquis gave,      15
  The fairest queen that ever king received.
KING HENRY
  Suffolk, arise. Welcome, Queen Margaret.
  I can express no kinder sign of love
  Than this kind kiss.
    *He kisses her*
               O Lord that lends me life,
  Lend me a heart replete with thankfulness!      20
  For thou hast given me in this beauteous face
  A world of earthly blessings to my soul,
  If sympathy of love unite our thoughts.
QUEEN MARGARET
  Th'excess of love I bear unto your grace
  Forbids me to be lavish of my tongue      25
  Lest I should speak more than beseems a woman.
  Let this suffice: my bliss is in your liking,
  And naught can make poor Margaret miserable
  Unless the frown of mighty England's King.
KING HENRY
  Her sight did ravish, but her grace in speech,      30
  Her words yclad with wisdom's majesty,
  Makes me from wond'ring fall to weeping joys,
  Such is the fullness of my heart's content.
  Lords, with one cheerful voice, welcome my love.
LORDS (*kneeling*)
  Long live Queen Margaret, England's happiness.      35
QUEEN MARGARET  We thank you all.
    *Flourish. ⌜They all rise⌝*
SUFFOLK (*to Gloucester*)
  My Lord Protector, so it please your grace,
  Here are the articles of contracted peace
  Between our sovereign and the French King Charles,
  For eighteen months concluded by consent.      40
GLOUCESTER (*reads*) Imprimis: it is agreed between the

French King Charles and William de la Pole, Marquis
of Suffolk, ambassador for Henry, King of England, that
the said Henry shall espouse the Lady Margaret,
daughter unto René, King of Naples, Sicilia, and
Jerusalem, and crown her Queen of England, ere the
thirtieth of May next ensuing.      47
    Item: it is further agreed between them that the
duchy of Anjou and the county of Maine shall be
released and delivered to the King her fa—      50
    ⌜*Gloucester lets the paper fall*⌝
KING HENRY
  Uncle, how now?
GLOUCESTER            Pardon me, gracious lord.
  Some sudden qualm hath struck me at the heart
  And dimmed mine eyes that I can read no further.
KING HENRY (*to Cardinal Beaufort*)
  Uncle of Winchester, I pray read on.      54
CARDINAL BEAUFORT (*reads*) Item: it is further agreed
between them that the duchy of Anjou and the county
of Maine shall be released and delivered to the King
her father, and she sent over of the King of England's
own proper cost and charges, without dowry.
KING HENRY
  They please us well. (*To Suffolk*) Lord Marquis, kneel
    down.      60
    *Suffolk kneels*
  We here create thee first Duke of Suffolk,
  And gird thee with the sword.
    *Suffolk rises*
                      Cousin of York,
  We here discharge your grace from being regent
  I'th' parts of France till term of eighteen months
  Be full expired. Thanks uncle Winchester,      65
  Gloucester, York, and Buckingham, Somerset,
  Salisbury, and Warwick.
  We thank you all for this great favour done
  In entertainment to my princely Queen.
  Come, let us in, and with all speed provide      70
  To see her coronation be performed.
    *Exeunt King Henry, Queen Margaret, and*
    *Suffolk. ⌜Gloucester stays⌝ all the rest*
GLOUCESTER
  Brave peers of England, pillars of the state,
  To you Duke Humphrey must unload his grief,
  Your grief, the common grief of all the land.
  What—did my brother Henry spend his youth,      75
  His valour, coin, and people in the wars?
  Did he so often lodge in open field
  In winter's cold and summer's parching heat
  To conquer France, his true inheritance?
  And did my brother Bedford toil his wits      80
  To keep by policy what Henry got?
  Have you yourselves, Somerset, Buckingham,
  Brave York, Salisbury, and victorious Warwick,
  Received deep scars in France and Normandy?
  Or hath mine uncle Beaufort and myself,      85
  With all the learnèd Council of the realm,
  Studied so long, sat in the Council House

Early and late, debating to and fro,
How France and Frenchmen might be kept in awe,
And had his highness in his infancy                         90
Crownèd in Paris in despite of foes?
And shall these labours and these honours die?
Shall Henry's conquest, Bedford's vigilance,
Your deeds of war, and all our counsel die?
O peers of England, shameful is this league,                95
Fatal this marriage, cancelling your fame,
Blotting your names from books of memory,
Razing the characters of your renown,
Defacing monuments of conquered France,
Undoing all, as all had never been!                        100

CARDINAL BEAUFORT
Nephew, what means this passionate discourse,
This peroration with such circumstance?
For France, 'tis ours; and we will keep it still.

GLOUCESTER
Ay, uncle, we will keep it if we can—
But now it is impossible we should.                         105
Suffolk, the new-made duke that rules the roast,
Hath given the duchy of Anjou and Maine
Unto the poor King René, whose large style
Agrees not with the leanness of his purse.

SALISBURY
Now by the death of Him that died for all,                 110
These counties were the keys of Normandy—
But wherefore weeps Warwick, my valiant son?

WARWICK
For grief that they are past recovery.
For were there hope to conquer them again
My sword should shed hot blood, mine eyes no tears.
Anjou and Maine? Myself did win them both!                 116
Those provinces these arms of mine did conquer—
And are the cities that I got with wounds
Delivered up again with peaceful words?
*Mort Dieu!*                                                120

YORK
For Suffolk's duke, may he be suffocate,
That dims the honour of this warlike isle!
France should have torn and rent my very heart
Before I would have yielded to this league.
I never read but England's kings have had                  125
Large sums of gold and dowries with their wives—
And our King Henry gives away his own,
To match with her that brings no vantages.

GLOUCESTER
A proper jest, and never heard before,
That Suffolk should demand a whole fifteenth               130
For costs and charges in transporting her!
She should have stayed in France and starved in
        France
Before—

CARDINAL BEAUFORT
My lord of Gloucester, now ye grow too hot!
It was the pleasure of my lord the King.                    135

GLOUCESTER
My lord of Winchester, I know your mind.
'Tis not my speeches that you do mislike,
But 'tis my presence that doth trouble ye.
Rancour will out. Proud prelate, in thy face
I see thy fury. If I longer stay                            140
We shall begin our ancient bickerings—
But I'll be gone, and give thee leave to speak.
Lordings, farewell, and say when I am gone,
I prophesied France will be lost ere long.      *Exit*

CARDINAL BEAUFORT
So, there goes our Protector in a rage.                     145
'Tis known to you he is mine enemy;
Nay more, an enemy unto you all,
And no great friend, I fear me, to the King.
Consider, lords, he is the next of blood
And heir apparent to the English crown.                     150
Had Henry got an empire by his marriage,
And all the wealthy kingdoms of the west,
There's reason he should be displeased at it.
Look to it, lords—let not his smoothing words
Bewitch your hearts. Be wise and circumspect.              155
What though the common people favour him,
Calling him 'Humphrey, the good Duke of Gloucester',
Clapping their hands and crying with loud voice
'Jesu maintain your royal excellence!'
With 'God preserve the good Duke Humphrey!'                160
I fear me, lords, for all this flattering gloss,
He will be found a dangerous Protector.

BUCKINGHAM
Why should he then protect our sovereign,
He being of age to govern of himself?
Cousin of Somerset, join you with me,                      165
And all together, with the Duke of Suffolk,
We'll quickly hoist Duke Humphrey from his seat.

CARDINAL BEAUFORT
This weighty business will not brook delay—
I'll to the Duke of Suffolk presently.          *Exit*

SOMERSET
Cousin of Buckingham, though Humphrey's pride  170
And greatness of his place be grief to us,
Yet let us watch the haughty Cardinal;
His insolence is more intolerable
Than all the princes in the land beside.
If Gloucester be displaced, he'll be Protector.            175

BUCKINGHAM
Or thou or I, Somerset, will be Protector,
Despite Duke Humphrey or the Cardinal.
                        *Exeunt Buckingham and Somerset*

SALISBURY
Pride went before, ambition follows him.
While these do labour for their own preferment,
Behoves it us to labour for the realm.                     180
I never saw but Humphrey Duke of Gloucester
Did bear him like a noble gentleman.
Oft have I seen the haughty Cardinal,
More like a soldier than a man o' th' church,
As stout and proud as he were lord of all,                 185
Swear like a ruffian, and demean himself
Unlike the ruler of a commonweal.
Warwick, my son, the comfort of my age,
Thy deeds, thy plainness, and thy housekeeping
Hath won thee greatest favour of the commons,              190
Excepting none but good Duke Humphrey.
And, brother York, thy acts in Ireland,
In bringing them to civil discipline,
Thy late exploits done in the heart of France,
When thou wert Regent for our sovereign,                   195
Have made thee feared and honoured of the people.
The reverence of mine age and Neville's name
Is of no little force if I command.
Join we together for the public good,
In what we can to bridle and suppress                      200
The pride of Suffolk and the Cardinal
With Somerset's and Buckingham's ambition;

And, as we may, cherish Duke Humphrey's deeds
While they do tend the profit of the land.

WARWICK
So God help Warwick, as he loves the land,    205
And common profit of his country!

YORK
And so says York, (*aside*) for he hath greatest cause.

SALISBURY
Then let's away, and look unto the main.

WARWICK
Unto the main? O, father, Maine is lost!
That Maine which by main force Warwick did win,
And would have kept so long as breath did last!    211
Main chance, father, you meant—but I meant Maine,
Which I will win from France or else be slain.

*Exeunt Warwick and Salisbury, leaving only York*

YORK
Anjou and Maine are given to the French,
Paris is lost, the state of Normandy    215
Stands on a tickle point now they are gone;
Suffolk concluded on the articles,
The peers agreed, and Henry was well pleased
To change two dukedoms for a duke's fair daughter.
I cannot blame them all—what is't to them?    220
'Tis thine they give away and not their own!
Pirates may make cheap pennyworths of their pillage,
And purchase friends, and give to courtesans,
Still revelling like lords till all be gone,
Whileas the seely owner of the goods    225
Weeps over them, and wrings his hapless hands,
And shakes his head, and, trembling, stands aloof,
While all is shared and all is borne away,
Ready to starve and dare not touch his own.
So York must sit and fret and bite his tongue,    230
While his own lands are bargained for and sold.
Methinks the realms of England, France, and Ireland
Bear that proportion to my flesh and blood
As did the fatal brand Althaea burnt
Unto the prince's heart of Calydon.    235
Anjou and Maine both given unto the French!
Cold news for me—for I had hope of France,
Even as I have of fertile England's soil.
A day will come when York shall claim his own,
And therefore I will take the Nevilles' parts,    240
And make a show of love to proud Duke Humphrey,
And, when I spy advantage, claim the crown,
For that's the golden mark I seek to hit.
Nor shall proud Lancaster usurp my right,
Nor hold the sceptre in his childish fist,    245
Nor wear the diadem upon his head
Whose church-like humours fits not for a crown.
Then, York, be still a while till time do serve.
Watch thou, and wake when others be asleep,
To pry into the secrets of the state—    250
Till Henry, surfeit in the joys of love
With his new bride and England's dear-bought queen,
And Humphrey with the peers be fall'n at jars.
Then will I raise aloft the milk-white rose,
With whose sweet smell the air shall be perfumed,    255
And in my standard bear the arms of York,
To grapple with the house of Lancaster;
And force perforce I'll make him yield the crown,
Whose bookish rule hath pulled fair England down.

*Exit*

**1.2**     *Enter Duke Humphrey of Gloucester and his wife Eleanor, the Duchess*

DUCHESS
Why droops my lord, like over-ripened corn
Hanging the head at Ceres' plenteous load?
Why doth the great Duke Humphrey knit his brows,
As frowning at the favours of the world?
Why are thine eyes fixed to the sullen earth,    5
Gazing on that which seems to dim thy sight?
What seest thou there? King Henry's diadem,
Enchased with all the honours of the world?
If so, gaze on, and grovel on thy face
Until thy head be circled with the same.    10
Put forth thy hand, reach at the glorious gold.
What, is't too short? I'll lengthen it with mine;
And having both together heaved it up,
We'll both together lift our heads to heaven
And never more abase our sight so low    15
As to vouchsafe one glance unto the ground.

GLOUCESTER
O Nell, sweet Nell, if thou dost love thy lord,
Banish the canker of ambitious thoughts!
And may that hour when I imagine ill
Against my king and nephew, virtuous Henry,    20
Be my last breathing in this mortal world!
My troublous dream this night doth make me sad.

DUCHESS
What dreamed my lord? Tell me and I'll requite it
With sweet rehearsal of my morning's dream.

GLOUCESTER
Methought this staff, mine office-badge in court,    25
Was broke in twain—by whom I have forgot,
But, as I think, it was by th' Cardinal—
And on the pieces of the broken wand
Were placed the heads of Edmund, Duke of Somerset,
And William de la Pole, first Duke of Suffolk.    30
This was my dream—what it doth bode, God knows.

DUCHESS
Tut, this was nothing but an argument
That he that breaks a stick of Gloucester's grove
Shall lose his head for his presumption.
But list to me, my Humphrey, my sweet duke:    35
Methought I sat in seat of majesty
In the cathedral church of Westminster,
And in that chair where kings and queens are crowned,
Where Henry and Dame Margaret kneeled to me,
And on my head did set the diadem.    40

GLOUCESTER
Nay, Eleanor, then must I chide outright.
Presumptuous dame! Ill-nurtured Eleanor!
Art thou not second woman in the realm,
And the Protector's wife beloved of him?
Hast thou not worldly pleasure at command    45
Above the reach or compass of thy thought?
And wilt thou still be hammering treachery
To tumble down thy husband and thyself
From top of honour to disgrace's feet?
Away from me, and let me hear no more!    50

DUCHESS
What, what, my lord? Are you so choleric
With Eleanor for telling but her dream?
Next time I'll keep my dreams unto myself
And not be checked.

**GLOUCESTER**
Nay, be not angry; I am pleased again.     55
*Enter a Messenger*

**MESSENGER**
My Lord Protector, 'tis his highness' pleasure
You do prepare to ride unto Saint Albans,
Whereas the King and Queen do mean to hawk.

**GLOUCESTER**
I go. Come, Nell, thou wilt ride with us?

**DUCHESS**
Yes, my good lord, I'll follow presently.     60
*Exeunt Gloucester and the Messenger*
Follow I must; I cannot go before
While Gloucester bears this base and humble mind.
Were I a man, a duke, and next of blood,
I would remove these tedious stumbling blocks
And smooth my way upon their headless necks.     65
And, being a woman, I will not be slack
To play my part in fortune's pageant.
*(Calling within)* Where are you there? Sir John! Nay,
     fear not man.
We are alone. Here's none but thee and I.
*Enter Sir John Hume*

**HUME**
Jesus preserve your royal majesty.     70

**DUCHESS**
What sayst thou? 'Majesty'? I am but 'grace'.

**HUME**
But by the grace of God and Hume's advice
Your grace's title shall be multiplied.

**DUCHESS**
What sayst thou, man? Hast thou as yet conferred
With Margery Jordan, the cunning witch of Eye,     75
With Roger Bolingbroke, the conjuror?
And will they undertake to do me good?

**HUME**
This they have promisèd: to show your highness
A spirit raised from depth of underground
That shall make answer to such questions     80
As by your Grace shall be propounded him.

**DUCHESS**
It is enough. I'll think upon the questions.
When from Saint Albans we do make return,
We'll see these things effected to the full.
Here, Hume *(giving him money)*, take this reward.
     Make merry, man,     85
With thy confederates in this weighty cause.     *Exit*

**HUME**
Hume must make merry with the Duchess' gold;
Marry, and shall. But how now, Sir John Hume?
Seal up your lips, and give no words but mum;
The business asketh silent secrecy.     90
Dame Eleanor gives gold to bring the witch.
Gold cannot come amiss were she a devil.
Yet have I gold flies from another coast—
I dare not say from the rich Cardinal
And from the great and new-made Duke of Suffolk,     95
Yet I do find it so; for, to be plain,
They, knowing Dame Eleanor's aspiring humour,
Have hired me to undermine the Duchess,
And buzz these conjurations in her brain.
They say 'A crafty knave does need no broker',     100
Yet am I Suffolk and the Cardinal's broker.
Hume, if you take not heed you shall go near
To call them both a pair of crafty knaves.
Well, so it stands; and thus, I fear, at last

Hume's knavery will be the Duchess' wrack,     105
And her attainture will be Humphrey's fall.
Sort how it will, I shall have gold for all.     *Exit*

**1.3**     *Enter Peter, the armourer's man, with two or three
     other Petitioners*

**FIRST PETITIONER** My masters, let's stand close. My Lord
Protector will come this way by and by and then we
may deliver our supplications in the quill.

**SECOND PETITIONER** Marry, the Lord protect him, for he's
a good man, Jesu bless him.     5
*Enter the Duke of Suffolk and Queen Margaret*

⌈**FIRST PETITIONER**⌉ Here a comes, methinks, and the Queen
with him. I'll be the first, sure.
*He goes to meet Suffolk and the Queen*

**SECOND PETITIONER** Come back, fool—this is the Duke of
Suffolk and not my Lord Protector.

**SUFFOLK** *(to the First Petitioner)*
How now, fellow—wouldst anything with me?     10

**FIRST PETITIONER** I pray, my lord, pardon me—I took ye
for my Lord Protector.

**QUEEN MARGARET** ⌈*seeing his supplication, she reads*⌉ 'To my
Lord Protector'—are your supplications to his lordship?
Let me see them.     15
     ⌈*She takes First Petitioner's supplication*⌉
What is thine?

**FIRST PETITIONER** Mine is, an't please your grace, against
John Goodman, my lord Cardinal's man, for keeping
my house and lands and wife and all from me.

**SUFFOLK** Thy wife too? That's some wrong indeed. ⌈*To the
Second Petitioner*⌉ What's yours?     21
     *He takes the supplication*
What's here? *(Reads)* 'Against the Duke of Suffolk for
enclosing the commons of Melford'! ⌈*To the Second
Petitioner*⌉ How now, Sir Knave?

**SECOND PETITIONER** Alas, sir, I am but a poor petitioner of
our whole township.     26

**PETER** ⌈*offering his petition*⌉ Against my master, Thomas
Horner, for saying that the Duke of York was rightful
heir to the crown.

**QUEEN MARGARET** What sayst thou? Did the Duke of York
say he was rightful heir to the crown?     31

**PETER** That my master was? No, forsooth, my master said
that he was and that the King was an usurer.

**QUEEN MARGARET** An usurper thou wouldst say.

**PETER** Ay, forsooth—an usurper.     35

**SUFFOLK** *(calling within)* Who is there?
*Enter a servant*
Take this fellow in and send for his master with a
pursuivant presently. *(To Peter)* We'll hear more of your
matter before the King.     *Exit the servant with Peter*

**QUEEN MARGARET** *(to the Petitioners)*
And as for you that love to be protected     40
Under the wings of our Protector's grace,
Begin your suits anew and sue to him.
     ⌈*She*⌉ *tears the supplication*
Away, base cullions! Suffolk, let them go.

**ALL PETITIONERS** Come, let's be gone.     *Exeunt Petitioners*

**QUEEN MARGARET**
My lord of Suffolk, say, is this the guise?     45
Is this the fashions in the court of England?
Is this the government of Britain's isle,
And this the royalty of Albion's king?
What, shall King Henry be a pupil still
Under the surly Gloucester's governance?     50
Am I a queen in title and in style,
And must be made a subject to a duke?

I tell thee, Pole, when in the city Tours
Thou rann'st a-tilt in honour of my love
And stol'st away the ladies' hearts of France,    55
I thought King Henry had resembled thee
In courage, courtship, and proportion.
But all his mind is bent to holiness,
To number Ave-Maries on his beads.
His champions are the prophets and apostles,    60
His weapons holy saws of sacred writ,
His study is his tilt-yard, and his loves
Are brazen images of canonizèd saints.
I would the college of the cardinals
Would choose him Pope, and carry him to Rome,    65
And set the triple crown upon his head—
That were a state fit for his holiness.

SUFFOLK
Madam, be patient—as I was cause
Your highness came to England, so will I
In England work your grace's full content.    70

QUEEN MARGARET
Beside the haught Protector have we Beaufort
The imperious churchman, Somerset, Buckingham,
And grumbling York; and not the least of these
But can do more in England than the King.

SUFFOLK
And he of these that can do most of all    75
Cannot do more in England than the Nevilles:
Salisbury and Warwick are no simple peers.

QUEEN MARGARET
Not all these lords do vex me half so much
As that proud dame, the Lord Protector's wife.
She sweeps it through the court with troops of ladies
More like an empress than Duke Humphrey's wife.    81
Strangers in court do take her for the queen.
She bears a duke's revenues on her back,
And in her heart she scorns our poverty.
Shall I not live to be avenged on her?    85
Contemptuous base-born callet as she is,
She vaunted 'mongst her minions t'other day
The very train of her worst-wearing gown
Was better worth than all my father's lands,
Till Suffolk gave two dukedoms for his daughter.    90

SUFFOLK
Madam, myself have limed a bush for her,
And placed a choir of such enticing birds
That she will light to listen to their lays,
And never mount to trouble you again.
So let her rest; and, madam, list to me,    95
For I am bold to counsel you in this:
Although we fancy not the Cardinal,
Yet must we join with him and with the lords
Till we have brought Duke Humphrey in disgrace.
As for the Duke of York, this late complaint    100
Will make but little for his benefit.
So one by one we'll weed them all at last,
And you yourself shall steer the happy helm.
    *Sound a sennet.* ⌈*Enter King Henry with the Duke*
    *of York and the Duke of Somerset on either side of*
    *him whispering with him. Also enter Duke*
    *Humphrey of Gloucester, Dame Eleanor the*
    *Duchess of Gloucester, the Duke of Buckingham, the*
    *Earls of Salisbury and Warwick, and Cardinal*
    *Beaufort Bishop of Winchester*⌉

KING HENRY
For my part, noble lords, I care not which:
Or Somerset or York, all's one to me.    105

YORK
If York have ill demeaned himself in France
Then let him be denied the regentship.

SOMERSET
If Somerset be unworthy of the place,
Let York be regent—I will yield to him.

WARWICK
Whether your grace be worthy, yea or no,    110
Dispute not that: York is the worthier.

CARDINAL BEAUFORT
Ambitious Warwick, let thy betters speak.

WARWICK
The Cardinal's not my better in the field.

BUCKINGHAM
All in this presence are thy betters, Warwick.

WARWICK
Warwick may live to be the best of all.    115

SALISBURY
Peace, son; (*to Buckingham*) and show some reason,
    Buckingham,
Why Somerset should be preferred in this.

QUEEN MARGARET
Because the King, forsooth, will have it so.

GLOUCESTER
Madam, the King is old enough himself
To give his censure. These are no women's matters.

QUEEN MARGARET
If he be old enough, what needs your grace    121
To be Protector of his excellence?

GLOUCESTER
Madam, I am Protector of the realm,
And at his pleasure will resign my place.

SUFFOLK
Resign it then, and leave thine insolence.    125
Since thou wert king—as who is king but thou?—
The commonwealth hath daily run to wrack,
The Dauphin hath prevailed beyond the seas,
And all the peers and nobles of the realm
Have been as bondmen to thy sovereignty.    130

CARDINAL BEAUFORT (*to Gloucester*)
The commons hast thou racked, the clergy's bags
Are lank and lean with thy extortions.

SOMERSET (*to Gloucester*)
Thy sumptuous buildings and thy wife's attire
Have cost a mass of public treasury.

BUCKINGHAM (*to Gloucester*)
Thy cruelty in execution    135
Upon offenders hath exceeded law
And left thee to the mercy of the law.

QUEEN MARGARET (*to Gloucester*)
Thy sale of offices and towns in France—
If they were known, as the suspect is great—
Would make thee quickly hop without thy head.    140
    *Exit Gloucester*
    *Queen Margaret lets fall her fan*
(*To the Duchess*)
Give me my fan—what, minion, can ye not?
    *She gives the Duchess a box on the ear*
I cry you mercy, madam! Was it you?

DUCHESS
Was't I? Yea, I it was, proud Frenchwoman!
Could I come near your beauty with my nails,
I'd set my ten commandments in your face.    145

KING HENRY
Sweet aunt, be quiet—'twas against her will.

**DUCHESS**
Against her will? Good King, look to't in time!
She'll pamper thee and dandle thee like a baby.
Though in this place most master wear no breeches,
She shall not strike Dame Eleanor unrevenged!          *Exit*

**BUCKINGHAM** (*aside to Cardinal Beaufort*)
Lord Cardinal, I will follow Eleanor                   151
And listen after Humphrey how he proceeds.
She's tickled now, her fury needs no spurs—
She'll gallop far enough to her destruction.           *Exit*
          *Enter Duke Humphrey of Gloucester*

**GLOUCESTER**
Now, lords, my choler being overblown                  155
With walking once about the quadrangle,
I come to talk of commonwealth affairs.
As for your spiteful false objections,
Prove them, and I lie open to the law.
But God in mercy so deal with my soul                  160
As I in duty love my King and country.
But to the matter that we have in hand—
I say, my sovereign, York is meetest man
To be your regent in the realm of France.

**SUFFOLK**
Before we make election, give me leave                 165
To show some reason of no little force
That York is most unmeet of any man.

**YORK**
I'll tell thee, Suffolk, why I am unmeet:
First, for I cannot flatter thee in pride;
Next, if I be appointed for the place,                 170
My lord of Somerset will keep me here
Without discharge, money, or furniture,
Till France be won into the Dauphin's hands.
Last time I danced attendance on his will
Till Paris was besieged, famished, and lost.           175

**WARWICK**
That can I witness, and a fouler fact
Did never traitor in the land commit.

**SUFFOLK** Peace, headstrong Warwick.

**WARWICK**
Image of pride, why should I hold my peace?
          *Enter, guarded, Horner the armourer and Peter his
          man*

**SUFFOLK**
Because here is a man accused of treason—              180
Pray God the Duke of York excuse himself!

**YORK**
Doth anyone accuse York for a traitor?

**KING HENRY**
What mean'st thou, Suffolk? Tell me, what are these?

**SUFFOLK**
Please it your majesty, this is the man
          *He indicates Peter*
That doth accuse his master (*indicating Horner*) of high
          treason.                                     185
His words were these: that Richard Duke of York
Was rightful heir unto the English crown,
And that your majesty was an usurper.

**KING HENRY** (*to Horner*) Say, man, were these thy words?

**HORNER** An't shall please your majesty, I never said nor
thought any such matter. God is my witness, I am        192
falsely accused by the villain.

**PETER** ⌈*raising his hands*⌉ By these ten bones, my lords, he
did speak them to me in the garret one night as we
were scouring my lord of York's armour.                195

**YORK**
Base dunghill villain and mechanical,
I'll have thy head for this thy traitor's speech!
(*To King Henry*) I do beseech your royal majesty,
Let him have all the rigour of the law.

**HORNER** Alas, my lord, hang me if ever I spake the words.
My accuser is my prentice, and when I did correct him
for his fault the other day, he did vow upon his knees
he would be even with me. I have good witness of this,
therefore, I beseech your majesty, do not cast away an
honest man for a villain's accusation.                 205

**KING HENRY** (*to Gloucester*)
Uncle, what shall we say to this in law?

**GLOUCESTER**
This doom, my lord, if I may judge by case:
Let Somerset be regent o'er the French,
Because in York this breeds suspicion.
(*Indicating Horner and Peter*)
And let these have a day appointed them                210
For single combat in convenient place,
For he (*indicating Horner*) hath witness of his servant's
          malice.
This is the law, and this Duke Humphrey's doom.

**KING HENRY**
Then be it so. (*To Somerset*) My lord of Somerset,
We make you regent o'er the realm of France            215
There to defend our rights 'gainst foreign foes.

**SOMERSET**
I humbly thank your royal majesty.

**HORNER**
And I accept the combat willingly.

**PETER** ⌈*to Gloucester*⌉ Alas, my lord, I cannot fight; for
God's sake, pity my case! The spite of man prevaileth
against me. O Lord, have mercy upon me—I shall never
be able to fight a blow! O Lord, my heart!              222

**GLOUCESTER**
Sirrah, or you must fight or else be hanged.

**KING HENRY**
Away with them to prison, and the day
Of combat be the last of the next month.               225
Come, Somerset, we'll see thee sent away.
                                   *Flourish. Exeunt*

**1.4**    *Enter Margery Jordan, a witch; Sir John Hume and
          John Southwell, two priests; and Roger
          Bolingbroke, a conjuror*

**HUME** Come, my masters, the Duchess, I tell you, expects
performance of your promises.

**BOLINGBROKE** Master Hume, we are therefore provided.
Will her ladyship behold and hear our exorcisms?

**HUME** Ay, what else? Fear you not her courage.        5

**BOLINGBROKE** I have heard her reported to be a woman
of an invincible spirit. But it shall be convenient, Master
Hume, that you be by her, aloft, while we be busy
below. And so, I pray you, go in God's name and leave
us.                                        *Exit Hume*
Mother Jordan, be you prostrate and grovel on the
earth.                                                 12
          *She lies down upon her face.*
          ⌈*Enter Eleanor, the Duchess of Gloucester, aloft*⌉
John Southwell, read you and let us to our work.

**DUCHESS** Well said, my masters, and welcome all. To this
gear the sooner the better.                            15
          ⌈*Enter Hume aloft*⌉

**BOLINGBROKE**
Patience, good lady—wizards know their times.

Deep night, dark night, the silent of the night,
The time of night when Troy was set on fire,
The time when screech-owls cry and bandogs howl,
And spirits walk, and ghosts break up their graves—
That time best fits the work we have in hand.          21
Madam, sit you, and fear not. Whom we raise
We will make fast within a hallowed verge.
          *Here do the ceremonies belonging, and make the*
          *circle. Southwell reads 'Coniuro te', &c. It thunders*
          *and lightens terribly, then the spirit Asnath riseth*
ASNATH *Adsum.*
WITCH Asnath,          25
By the eternal God whose name and power
Thou tremblest at, answer that I shall ask,
For till thou speak, thou shalt not pass from hence.
ASNATH
Ask what thou wilt, that I had said and done.
BOLINGBROKE (*reads*)
'First, of the King: what shall of him become?'          30
ASNATH
The Duke yet lives that Henry shall depose,
But him outlive, and die a violent death.
          *As the spirit speaks,* ⌈*Southwell*⌉ *writes the answer*
BOLINGBROKE (*reads*)
'Tell me what fate awaits the Duke of Suffolk.'
ASNATH
By water shall he die, and take his end.
BOLINGBROKE (*reads*)
'What shall betide the Duke of Somerset?'          35
ASNATH
Let him shun castles. Safer shall he be
Upon the sandy plains than where castles mounted
          stand.
Have done—for more I hardly can endure.
BOLINGBROKE
Descend to darkness and the burning lake!
False fiend, avoid!          40
          *Thunder and lightning. The spirit sinks down again*
          *Enter, breaking in, the Dukes of York and*
          *Buckingham with their guard, among them Sir*
          *Humphrey Stafford*
YORK
Lay hands upon these traitors and their trash.
          ⌈*Bolingbroke, Southwell, and Jordan are taken*
          *prisoner. Buckingham takes the writings from*
          *Bolingbroke and Southwell*⌉
(*To Jordan*) Beldam, I think we watched you at an inch.
(*To the Duchess*) What, madam, are you there? The
          King and common weal
Are deep indebted for this piece of pains.
My lord Protector will, I doubt it not,          45
See you well guerdoned for these good deserts.
DUCHESS
Not half so bad as thine to England's king,
Injurious Duke, that threatest where's no cause.
BUCKINGHAM
True, madam, none at all—
          ⌈*He raises the writings*⌉
                    what call you this?
(*To his men*) Away with them. Let them be clapped up
          close          50
And kept asunder. (*To the Duchess*) You, madam, shall
          with us.
Stafford, take her to thee.
          *Exeunt Stafford* ⌈*and others*⌉ *to the Duchess*
          ⌈*and Hume*⌉ *above*

We'll see your trinkets here all forthcoming.
All away!
          *Exeunt below Jordan, Southwell, and*
          *Bolingbroke, guarded, and, above,* ⌈*Hume and*⌉
          *the Duchess guarded by Stafford* ⌈*and others.*
          *York and Buckingham remain*⌉
YORK
Lord Buckingham, methinks you watched her well.          55
A pretty plot, well chosen to build upon.
Now pray, my lord, let's see the devil's writ.
          ⌈*Buckingham gives him the writings*⌉
What have we here?
          *He reads the writings*
                    Why, this is just
*Aio Aeacidam, Romanos vincere posse.*
These oracles are hardly attained          60
And hardly understood. Come, come, my lord,
The King is now in progress towards Saint Albans;
With him the husband of this lovely lady.
Thither goes these news as fast as horse can carry
          them—
A sorry breakfast for my lord Protector.          65
BUCKINGHAM
Your grace shall give me leave, my lord of York,
To be the post in hope of his reward.
YORK (*returning the writings to Buckingham*)
At your pleasure, my good lord.          ⌈*Exit Buckingham*⌉
(*Calling within*)                    Who's within there, ho!
          *Enter a servingman*
Invite my lords of Salisbury and Warwick
To sup with me tomorrow night. Away.          70
          *Exeunt severally*

2.1          *Enter King Henry, Queen Margaret with her hawk*
          *on her fist, Duke Humphrey of Gloucester, Cardinal*
          *Beaufort, and the Duke of Suffolk, with falconers*
          *hollering*
QUEEN MARGARET
Believe me, lords, for flying at the brook
I saw not better sport these seven years' day;
Yet, by your leave, the wind was very high,
And, ten to one, old Joan had not gone out.
KING HENRY (*to Gloucester*)
But what a point, my lord, your falcon made,          5
And what a pitch she flew above the rest!
To see how God in all his creatures works!
Yea, man and birds are fain of climbing high.
SUFFOLK
No marvel, an it like your majesty,
My Lord Protector's hawks do tower so well;          10
They know their master loves to be aloft,
And bears his thoughts above his falcon's pitch.
GLOUCESTER
My lord, 'tis but a base ignoble mind
That mounts no higher than a bird can soar.
CARDINAL BEAUFORT
I thought as much; he would be above the clouds.          15
GLOUCESTER
Ay, my lord Cardinal, how think you by that?
Were it not good your grace could fly to heaven?
KING HENRY
The treasury of everlasting joy.
CARDINAL BEAUFORT (*to Gloucester*)
Thy heaven is on earth; thine eyes and thoughts
Beat on a crown, the treasure of thy heart,          20
Pernicious Protector, dangerous peer,
That smooth'st it so with King and common weal!

**GLOUCESTER**
What, Cardinal? Is your priesthood grown
  peremptory?
*Tantaene animis caelestibus irae?*
Churchmen so hot? Good uncle, hide such malice    25
With some holiness—can you do it?
**SUFFOLK**
No malice, sir, no more than well becomes
So good a quarrel and so bad a peer.
**GLOUCESTER**
As who, my lord?
**SUFFOLK**                Why, as you, my lord—
An't like your lordly Lord's Protectorship.    30
**GLOUCESTER**
Why, Suffolk, England knows thine insolence.
**QUEEN MARGARET**
And thy ambition, Gloucester.
**KING HENRY**                I prithee peace,
Good Queen, and whet not on these furious peers—
For blessèd are the peacemakers on earth.
**CARDINAL BEAUFORT**
Let me be blessèd for the peace I make    35
Against this proud Protector with my sword.
  ⌈*Gloucester and Cardinal Beaufort speak privately to
  one another*⌉
**GLOUCESTER**
Faith, holy uncle, would't were come to that.
**CARDINAL BEAUFORT**
Marry, when thou dar'st.
**GLOUCESTER**                Dare? I tell thee, priest,
Plantagenets could never brook the dare!
**CARDINAL BEAUFORT**
I am Plantagenet as well as thou,    40
And son to John of Gaunt.
**GLOUCESTER** In bastardy.
**CARDINAL BEAUFORT** I scorn thy words.
**GLOUCESTER**
Make up no factious numbers for the matter,
In thine own person answer thy abuse.    45
**CARDINAL BEAUFORT**
Ay, where thou dar'st not peep; an if thou dar'st,
This evening on the east side of the grove.
**KING HENRY**
How now, my lords?
**CARDINAL BEAUFORT** (*aloud*)
                Believe me, cousin Gloucester,
Had not your man put up the fowl so suddenly,
We had had more sport. (*Aside to Gloucester*) Come
  with thy two-hand sword.    50
**GLOUCESTER** (*aloud*) True, uncle.
(*Aside to Cardinal Beaufort*)
Are ye advised? The east side of the grove.
**CARDINAL BEAUFORT** (*aside to Gloucester*)
I am with you.
**KING HENRY**        Why, how now, uncle Gloucester?
**GLOUCESTER**
Talking of hawking, nothing else, my lord.
(*Aside to the Cardinal*)
Now, by God's mother, priest, I'll shave your crown
  for this,    55
Or all my fence shall fail.
**CARDINAL BEAUFORT** (*aside to Gloucester*)
                Medice, teipsum—
Protector, see to't well; protect yourself.
**KING HENRY**
The winds grow high; so do your stomachs, lords.
How irksome is this music to my heart!

When such strings jar, what hope of harmony?    60
I pray, my lords, let me compound this strife.
  *Enter one crying 'a miracle'*
**GLOUCESTER** What means this noise?
Fellow, what miracle dost thou proclaim?
**ONE**
A miracle, a miracle!
**SUFFOLK**
Come to the King—tell him what miracle.    65
**ONE** (*to King Henry*)
Forsooth, a blind man at Saint Alban's shrine
Within this half-hour hath received his sight—
A man that ne'er saw in his life before.
**KING HENRY**
Now God be praised, that to believing souls
Gives light in darkness, comfort in despair!    70
  *Enter the Mayor and aldermen of Saint Albans,
  with music, bearing the man, Simpcox, between
  two in a chair. Enter Simpcox's Wife ⌈and other
  townsmen⌉ with them*
**CARDINAL BEAUFORT**
Here comes the townsmen on procession
To present your highness with the man.
  ⌈*The townsmen kneel*⌉
**KING HENRY**
Great is his comfort in this earthly vale,
Although by sight his sin be multiplied.
**GLOUCESTER** (*to the townsmen*)
Stand by, my masters, bring him near the King.    75
His highness' pleasure is to talk with him.
  *They ⌈rise and⌉ bear Simpcox before the King*
**KING HENRY** (*to Simpcox*)
Good fellow, tell us here the circumstance,
That we for thee may glorify the Lord.
What, hast thou been long blind and now restored?
**SIMPCOX**
Born blind, an't please your grace.
**SIMPCOX'S WIFE**                Ay, indeed, was he.
**SUFFOLK** What woman is this?    81
**SIMPCOX'S WIFE** His wife, an't like your worship.
**GLOUCESTER** Hadst thou been his mother
Thou couldst have better told.
**KING HENRY** (*to Simpcox*)        Where wert thou born?
**SIMPCOX**
At Berwick, in the north, an't like your grace.    85
**KING HENRY**
Poor soul, God's goodness hath been great to thee.
Let never day nor night unhallowed pass,
But still remember what the Lord hath done.
**QUEEN MARGARET** (*to Simpcox*)
Tell me, good fellow, cam'st thou here by chance,
Or of devotion to this holy shrine?    90
**SIMPCOX**
God knows, of pure devotion, being called
A hundred times and oftener, in my sleep,
By good Saint Alban, who said, 'Simon, come;
Come offer at my shrine and I will help thee.'
**SIMPCOX'S WIFE**
Most true, forsooth, and many time and oft    95
Myself have heard a voice to call him so.
**CARDINAL BEAUFORT** (*to Simpcox*)
What, art thou lame?
**SIMPCOX**                Ay, God almighty help me.
**SUFFOLK**
How cam'st thou so?
**SIMPCOX**                A fall off of a tree.

SIMPCOX'S WIFE (*to Suffolk*)
A plum tree, master.
GLOUCESTER         How long hast thou been blind?
SIMPCOX
O, born so, master.
GLOUCESTER         What, and wouldst climb a tree?
SIMPCOX
But that in all my life, when I was a youth.    101
SIMPCOX'S WIFE (*to Gloucester*)
Too true—and bought his climbing very dear.
GLOUCESTER (*to Simpcox*)
Mass, thou loved'st plums well that wouldst venture
     so.
SIMPCOX
Alas, good master, my wife desired some damsons,
And made me climb with danger of my life.    105
GLOUCESTER ⌈*aside*⌉
A subtle knave, but yet it shall not serve.
(*To Simpcox*) Let me see thine eyes: wink now, now
     open them.
In my opinion yet thou seest not well.
SIMPCOX Yes, master, clear as day, I thank God and Saint
Alban.    110
GLOUCESTER
Sayst thou me so? (*Pointing*) What colour is this cloak
of?
SIMPCOX
Red, master; red as blood.
GLOUCESTER         Why, that's well said.
(*Pointing*) And his cloak?
SIMPCOX         Why, that's green.
GLOUCESTER (*pointing*)         And what colour's
His hose?
SIMPCOX      Yellow, master; yellow as gold.
GLOUCESTER
And what colour's my gown?
SIMPCOX         Black, sir; coal-black, as jet.
KING HENRY
Why, then, thou know'st what colour jet is of?    116
SUFFOLK
And yet I think jet did he never see.
GLOUCESTER
But cloaks and gowns before this day, a many.
SIMPCOX'S WIFE
Never before this day in all his life.
GLOUCESTER Tell me, sirrah, what's my name?    120
SIMPCOX Alas, master, I know not.
GLOUCESTER (*pointing*) What's his name?
SIMPCOX I know not.
GLOUCESTER (*pointing*) Nor his?
SIMPCOX No, truly, sir.    125
GLOUCESTER (*pointing*) Nor his name?
SIMPCOX No indeed, master.
GLOUCESTER What's thine own name?
SIMPCOX
Simon Simpcox, an it please you, master.
GLOUCESTER
Then, Simon, sit thou there the lying'st knave    130
In Christendom. If thou hadst been born blind
Thou mightst as well have known our names as thus
To name the several colours we do wear.
Sight may distinguish colours, but suddenly
To nominate them all—it is impossible.    135
Saint Alban here hath done a miracle.
Would you not think his cunning to be great
That could restore this cripple to his legs again?

SIMPCOX O master, that you could!
GLOUCESTER (*to the Mayor and aldermen*)
My masters of Saint Albans, have you not    140
Beadles in your town, and things called whips?
MAYOR
We have, my lord, an if it please your grace.
GLOUCESTER Then send for one presently.
MAYOR (*to a townsman*)
Sirrah, go fetch the beadle hither straight.      *Exit one*
GLOUCESTER
Bring me a stool.
     *A stool is brought*
(*To Simpcox*)      Now, sirrah, if you mean    145
To save yourself from whipping, leap me o'er
This stool and run away.
SIMPCOX         Alas, master,
I am not able even to stand alone.
You go about to torture me in vain.
     *Enter a Beadle with whips*
GLOUCESTER
Well, sirrah, we must have you find your legs.    150
(*To the Beadle*) Whip him till he leap over that same
     stool.
BEADLE I will, my lord.
(*To Simpcox*) Come on, sirrah, off with your doublet
     quickly.
SIMPCOX Alas, master, what shall I do? I am not able to
stand.    155
     *After the Beadle hath hit him once, he leaps over*
     *the stool and runs away.* ⌈*Some of* ⌉ *the townsmen*
     *follow and cry, 'A miracle! A miracle!'*
KING HENRY
O God, seest thou this and bear'st so long?
QUEEN MARGARET
It made me laugh to see the villain run!
GLOUCESTER ⌈*to the Beadle*⌉
Follow the knave, and take this drab away.
SIMPCOX'S WIFE
Alas, sir, we did it for pure need.
     ⌈*Exit the Beadle with the Wife*⌉
GLOUCESTER ⌈*to the Mayor*⌉
Let them be whipped through every market-town    160
Till they come to Berwick, from whence they came.
     *Exeunt the Mayor* ⌈*and any remaining townsmen*⌉
CARDINAL BEAUFORT
Duke Humphrey has done a miracle today.
SUFFOLK
True: made the lame to leap and fly away.
GLOUCESTER
But you have done more miracles than I—
You made, in a day, my lord, whole towns to fly.    165
     *Enter the Duke of Buckingham*
KING HENRY
What tidings with our cousin Buckingham?
BUCKINGHAM
Such as my heart doth tremble to unfold.
A sort of naughty persons, lewdly bent,
Under the countenance and confederacy
Of Lady Eleanor, the Protector's wife,    170
The ringleader and head of all this rout,
Have practised dangerously against your state,
Dealing with witches and with conjurors,
Whom we have apprehended in the fact,
Raising up wicked spirits from under ground,    175
Demanding of King Henry's life and death

And other of your highness' Privy Council.
And here's the answer the devil did make to them.
*Buckingham gives King Henry the writings*
⌈KING HENRY⌉ (*reads*)
'First of the King: what shall of him become?
The Duke yet lives that Henry shall depose,          180
But him outlive and die a violent death.'
God's will be done in all. Well, to the rest.
(*Reads*) 'Tell me what fate awaits the Duke of Suffolk?
By water shall he die, and take his end.'
SUFFOLK ⌈*aside*⌉
By water must the Duke of Suffolk die?          185
It must be so, or else the devil doth lie.
KING HENRY (*reads*)
'What shall betide the Duke of Somerset?
Let him shun castles. Safer shall he be
Upon the sandy plains than where castles mounted
          stand.'
CARDINAL BEAUFORT (*to Gloucester*)
And so, my Lord Protector, by this means          190
Your lady is forthcoming yet at London.
(*Aside to Gloucester*)
This news, I think, hath turned your weapon's edge.
'Tis like, my lord, you will not keep your hour.
GLOUCESTER
Ambitious churchman, leave to afflict my heart.
Sorrow and grief have vanquished all my powers,          195
And, vanquished as I am, I yield to thee
Or to the meanest groom.
KING HENRY
O God, what mischiefs work the wicked ones,
Heaping confusion on their own heads thereby!
QUEEN MARGARET
Gloucester, see here the tainture of thy nest,          200
And look thyself be faultless, thou wert best.
GLOUCESTER
Madam, for myself, to heaven I do appeal,
How I have loved my King and common weal;
And for my wife, I know not how it stands.
Sorry I am to hear what I have heard.          205
Noble she is, but if she have forgot
Honour and virtue and conversed with such
As, like to pitch, defile nobility,
I banish her my bed and company,
And give her as a prey to law and shame          210
That hath dishonoured Gloucester's honest name.
KING HENRY
Well, for this night we will repose us here;
Tomorrow toward London back again,
To look into this business thoroughly,
And call these foul offenders to their answers,          215
And poise the cause in justice' equal scales,
Whose beam stands sure, whose rightful cause
          prevails.          *Flourish. Exeunt*

**2.2**     *Enter the Duke of York and the Earls of Salisbury*
          *and Warwick*
YORK
Now, my good lords of Salisbury and Warwick,
Our simple supper ended, give me leave
In this close walk to satisfy myself
In craving your opinion of my title,
Which is infallible, to England's crown.          5
SALISBURY
My lord, I long to hear it out at full.

WARWICK
Sweet York, begin, and if thy claim be good,
The Nevilles are thy subjects to command.
YORK Then thus:
Edward the Third, my lords, had seven sons:          10
The first, Edward the Black Prince, Prince of Wales;
The second, William of Hatfield; and the third,
Lionel Duke of Clarence; next to whom
Was John of Gaunt, the Duke of Lancaster;
The fifth was Edmund Langley, Duke of York;          15
The sixth was Thomas of Woodstock, Duke of
          Gloucester;
William of Windsor was the seventh and last.
Edward the Black Prince died before his father
And left behind him Richard, his only son,
Who, after Edward the Third's death, reigned as king
Till Henry Bolingbroke, Duke of Lancaster,          21
The eldest son and heir of John of Gaunt,
Crowned by the name of Henry the Fourth,
Seized on the realm, deposed the rightful king,
Sent his poor queen to France from whence she came,
And him to Pomfret; where, as well you know,          26
Harmless Richard was murdered traitorously.
WARWICK (*to Salisbury*)
Father, the Duke of York hath told the truth;
Thus got the house of Lancaster the crown.
YORK
Which now they hold by force and not by right;          30
For Richard, the first son's heir, being dead,
The issue of the next son should have reigned.
SALISBURY
But William of Hatfield died without an heir.
YORK
The third son, Duke of Clarence, from whose line
I claim the crown, had issue Phillipe, a daughter,          35
Who married Edmund Mortimer, Earl of March;
Edmund had issue, Roger, Earl of March;
Roger had issue, Edmund, Anne and Eleanor.
SALISBURY
This Edmund, in the reign of Bolingbroke,
As I have read, laid claim unto the crown,          40
And, but for Owain Glyndŵr, had been king,
Who kept him in captivity till he died.
But to the rest.
YORK          His eldest sister, Anne,
My mother, being heir unto the crown,
Married Richard, Earl of Cambridge, who was son          45
To Edmund Langley, Edward the Third's fifth son.
By her I claim the kingdom: she was heir
To Roger, Earl of March, who was the son
Of Edmund Mortimer, who married Phillipe,
Sole daughter unto Lionel, Duke of Clarence.          50
So if the issue of the elder son
Succeed before the younger, I am king.
WARWICK
What plain proceedings is more plain than this?
Henry doth claim the crown from John of Gaunt,
The fourth son; York claims it from the third:          55
Till Lionel's issue fails, John's should not reign.
It fails not yet, but flourishes in thee
And in thy sons, fair slips of such a stock.
Then, father Salisbury, kneel we together,
And in this private plot be we the first          60
That shall salute our rightful sovereign
With honour of his birthright to the crown.

SALISBURY *and* WARWICK (*kneeling*)
  Long live our sovereign Richard, England's king!
YORK
  We thank you, lords;
          ⌜*Salisbury and Warwick rise*⌝
                    but I am not your king
  Till I be crowned, and that my sword be stained      65
  With heart-blood of the house of Lancaster—
  And that's not suddenly to be performed,
  But with advice and silent secrecy.
  Do you, as I do, in these dangerous days,
  Wink at the Duke of Suffolk's insolence,             70
  At Beaufort's pride, at Somerset's ambition,
  At Buckingham, and all the crew of them,
  Till they have snared the shepherd of the flock,
  That virtuous prince, the good Duke Humphrey.
  'Tis that they seek, and they, in seeking that,      75
  Shall find their deaths, if York can prophesy.
SALISBURY
  My lord, break off—we know your mind at full.
WARWICK
  My heart assures me that the Earl of Warwick
  Shall one day make the Duke of York a king.
YORK
  And Neville, this I do assure myself—                80
  Richard shall live to make the Earl of Warwick
  The greatest man in England but the King.    *Exeunt*

**2.3**   *Sound trumpets. Enter King Henry and state, with*
          *guard, to banish the Duchess: King Henry and*
          *Queen Margaret, Duke Humphrey of Gloucester, the*
          *Duke of Suffolk* ⌜*and the Duke of Buckingham,*
          *Cardinal Beaufort*⌝, *and, led with officers, Dame*
          *Eleanor Cobham the Duchess, Margery Jordan the*
          *witch, John Southwell and Sir John Hume the two*
          *priests, and Roger Bolingbroke the conjuror;* ⌜*then*
          *enter to them*⌝ *the Duke of York and the Earls of*
          *Salisbury* ⌜*and Warwick*⌝
KING HENRY (*to the Duchess*)
  Stand forth, Dame Eleanor Cobham, Gloucester's wife.
          *She comes forward*
  In sight of God and us your guilt is great;
  Receive the sentence of the law for sins
  Such as by God's book are adjudged to death.
  (*To the Witch, Southwell, Hume, and Bolingbroke*)
  You four, from hence to prison back again;            5
  From thence, unto the place of execution.
  The witch in Smithfield shall be burned to ashes,
  And you three shall be strangled on the gallows.
                    ⌜*Exeunt Witch, Southwell, Hume, and*
                              *Bolingbroke, guarded*⌝
  (*To the Duchess*)
  You, madam, for you are more nobly born,
  Despoilèd of your honour in your life,               10
  Shall, after three days' open penance done,
  Live in your country here in banishment
  With Sir John Stanley in the Isle of Man.
DUCHESS
  Welcome is banishment; welcome were my death.
GLOUCESTER
  Eleanor, the law, thou seest, hath judgèd thee;      15
  I cannot justify whom the law condemns.
                    ⌜*Exit the Duchess, guarded*⌝
  Mine eyes are full of tears, my heart of grief.
  Ah, Humphrey, this dishonour in thine age
  Will bring thy head with sorrow to the grave.

(*To King Henry*)
  I beseech your majesty, give me leave to go.         20
  Sorrow would solace, and mine age would ease.
KING HENRY
  Stay, Humphrey Duke of Gloucester. Ere thou go,
  Give up thy staff. Henry will to himself
  Protector be; and God shall be my hope,
  My stay, my guide, and lantern to my feet.          25
  And go in peace, Humphrey, no less beloved
  Than when thou wert Protector to thy King.
QUEEN MARGARET
  I see no reason why a king of years
  Should be to be protected like a child.
  God and King Henry govern England's helm!            30
  Give up your staff, sir, and the King his realm.
GLOUCESTER
  My staff? Here, noble Henry, is my staff.
  As willingly do I the same resign
  As erst thy father Henry made it mine;
  And even as willing at thy feet I leave it            35
  As others would ambitiously receive it.
          *He lays the staff at King Henry's feet*
  Farewell, good King. When I am dead and gone,
  May honourable peace attend thy throne.    *Exit*
QUEEN MARGARET
  Why, now is Henry King and Margaret Queen,
  And Humphrey Duke of Gloucester scarce himself,       40
  That bears so shrewd a maim; two pulls at once—
  His lady banished and a limb lopped off.
          *She picks up the staff*
  This staff of honour raught, there let it stand
  Where it best fits to be, in Henry's hand.
          *She gives the staff to King Henry*
SUFFOLK
  Thus droops this lofty pine and hangs his sprays;    45
  Thus Eleanor's pride dies in her youngest days.
YORK
  Lords, let him go. Please it your majesty,
  This is the day appointed for the combat,
  And ready are the appellant and defendant—
  The armourer and his man—to enter the lists,          50
  So please your highness to behold the fight.
QUEEN MARGARET
  Ay, good my lord, for purposely therefor
  Left I the court to see this quarrel tried.
KING HENRY
  A God's name, see the lists and all things fit;
  Here let them end it, and God defend the right.       55
YORK
  I never saw a fellow worse bestead,
  Or more afraid to fight, than is the appellant,
  The servant of this armourer, my lords.
          *Enter at one door Horner the armourer and his*
          *Neighbours, drinking to him so much that he is*
          *drunken; and he enters with a drummer before him*
          *and* ⌜*carrying*⌝ *his staff with a sandbag fastened to*
          *it. Enter at the other door Peter his man, also with*
          *a drummer and a staff with sandbag, and Prentices*
          *drinking to him*
FIRST NEIGHBOUR (*offering drink to Horner*) Here, neighbour
  Horner, I drink to you in a cup of sack, and fear not,
  neighbour, you shall do well enough.                  61
SECOND NEIGHBOUR (*offering drink to Horner*) And here,
  neighbour, here's a cup of charneco.
THIRD NEIGHBOUR (*offering drink to Horner*) Here's a pot of
  good double beer, neighbour, drink and be merry, and
  fear not your man.                                    66

HORNER ⌈*accepting the offers of drink*⌉ Let it come, i'faith
I'll pledge you all, and a fig for Peter.

FIRST PRENTICE (*offering drink to Peter*) Here, Peter, I drink
to thee, and be not afeard.                                          70

SECOND PRENTICE (*offering drink to Peter*) Here, Peter, here's
a pint of claret wine for thee.

THIRD PRENTICE (*offering drink to Peter*) And here's a quart
for me, and be merry, Peter, and fear not thy master.
Fight for credit of the prentices!                                   75

PETER ⌈*refusing the offers of drink*⌉ I thank you all. Drink
and pray for me, I pray you, for I think I have taken
my last draught in this world. Here, Robin, an if I die,
I give thee my apron; and, Will, thou shalt have my
hammer; and here, Tom, take all the money that I
have. O Lord bless me, I pray God, for I am never able
to deal with my master, he hath learned so much fence
already.

SALISBURY Come, leave your drinking, and fall to blows.
(*To Peter*) Sirrah, what's thy name?                                 85

PETER Peter, forsooth.

SALISBURY Peter? What more?

PETER Thump.

SALISBURY Thump! Then see that thou thump thy master
well.                                                                 90

HORNER Masters, I am come hither, as it were, upon my
man's instigation, to prove him a knave and myself an
honest man; and touching the Duke of York, I will
take my death I never meant him any ill, nor the King,
nor the Queen; and therefore, Peter, have at thee with
a downright blow.                                                     96

YORK
Dispatch; this knave's tongue begins to double.
     *Sound trumpets an alarum to the combatants. They*
     *fight and Peter hits Horner on the head and strikes*
     *him down*

HORNER Hold, Peter, hold—I confess, I confess treason.
                                                       *He dies*

YORK (*to an attendant, pointing to Horner*) Take away his
weapon. (*To Peter*) Fellow, thank God and the good
wine in thy master's wame.                                           101

PETER ⌈*kneeling*⌉ O God, have I overcome mine enemy in
this presence? O, Peter, thou hast prevailed in right.

KING HENRY (*to attendants, pointing to Horner*)
Go, take hence that traitor from our sight,
For by his death we do perceive his guilt.                           105
And God in justice hath revealed to us
The truth and innocence of this poor fellow,
Which he had thought to have murdered wrongfully.
(*To Peter*) Come, fellow, follow us for thy reward.
                    *Sound a flourish. Exeunt, some carrying*
                                          *Horner's body*

**2.4**     *Enter Duke Humphrey of Gloucester and his men in*
            *mourning cloaks*

GLOUCESTER
Thus sometimes hath the brightest day a cloud;
And after summer evermore succeeds
Barren winter, with his wrathful nipping cold;
So cares and joys abound as seasons fleet.
Sirs, what's o'clock?                                                  5

SERVANT Ten, my lord.

GLOUCESTER
Ten is the hour that was appointed me
To watch the coming of my punished Duchess;
Uneath may she endure the flinty streets,
To tread them with her tender-feeling feet.                           10

Sweet Nell, ill can thy noble mind abrook
The abject people gazing on thy face
With envious looks, laughing at thy shame,
That erst did follow thy proud chariot wheels
When thou didst ride in triumph through the streets.
But soft, I think she comes; and I'll prepare                         16
My tear-stained eyes to see her miseries.
     *Enter the Duchess, Dame Eleanor Cobham, barefoot,*
     *with a white sheet about her, written verses pinned*
     *on her back, and carrying a wax candle in her*
     *hand; she is accompanied by the ⌈two Sheriffs⌉ of*
     *London, and Sir John Stanley, and officers with bills*
     *and halberds*

SERVANT (*to Gloucester*)
So please your grace, we'll take her from the sheriffs.

GLOUCESTER
No, stir not for your lives, let her pass by.

DUCHESS
Come you, my lord, to see my open shame?                              20
Now thou dost penance too. Look how they gaze,
See how the giddy multitude do point
And nod their heads, and throw their eyes on thee.
Ah, Gloucester, hide thee from their hateful looks,
And, in thy closet pent up, rue my shame,                             25
And ban thine enemies—both mine and thine.

GLOUCESTER
Be patient, gentle Nell; forget this grief.

DUCHESS
Ah, Gloucester, teach me to forget myself;
For whilst I think I am thy married wife,
And thou a prince, Protector of this land,                            30
Methinks I should not thus be led along,
Mailed up in shame, with papers on my back,
And followed with a rabble that rejoice
To see my tears and hear my deep-fet groans.
The ruthless flint doth cut my tender feet,                           35
And when I start, the envious people laugh,
And bid me be advisèd how I tread.
Ah, Humphrey, can I bear this shameful yoke?
Trowest thou that e'er I'll look upon the world,
Or count them happy that enjoys the sun?                              40
No, dark shall be my light, and night my day;
To think upon my pomp shall be my hell.
Sometime I'll say I am Duke Humphrey's wife,
And he a prince and ruler of the land;
Yet so he ruled, and such a prince he was,                            45
As he stood by whilst I, his forlorn Duchess,
Was made a wonder and a pointing stock
To every idle rascal follower.
But be thou mild and blush not at my shame,
Nor stir at nothing till the axe of death                             50
Hang over thee, as sure it shortly will.
For Suffolk, he that can do all in all
With her that hateth thee and hates us all,
And York, and impious Beaufort that false priest,
Have all limed bushes to betray thy wings,                            55
And fly thou how thou canst, they'll tangle thee.
But fear not thou until thy foot be snared,
Nor never seek prevention of thy foes.

GLOUCESTER
Ah, Nell, forbear; thou aimest all awry.
I must offend before I be attainted,                                  60
And had I twenty times so many foes,
And each of them had twenty times their power,
All these could not procure me any scathe
So long as I am loyal, true, and crimeless.

Wouldst have me rescue thee from this reproach?     65
Why, yet thy scandal were not wiped away,
But I in danger for the breach of law.
Thy greatest help is quiet, gentle Nell.
I pray thee sort thy heart to patience.
These few days' wonder will be quickly worn.     70
    *Enter a Herald*
HERALD  I summon your grace to his majesty's parliament
  holden at Bury the first of this next month.
GLOUCESTER
And my consent ne'er asked herein before?
This is close dealing. Well, I will be there.     *Exit Herald*
My Nell, I take my leave; and, Master Sheriff,     75
Let not her penance exceed the King's commission.
⌈FIRST⌉ SHERIFF
An't please your grace, here my commission stays,
And Sir John Stanley is appointed now
To take her with him to the Isle of Man.
GLOUCESTER
Must you, Sir John, protect my lady here?     80
STANLEY
So am I given in charge, may't please your grace.
GLOUCESTER
Entreat her not the worse in that I pray
You use her well. The world may laugh again,
And I may live to do you kindness if
You do it her. And so, Sir John, farewell.     85
    ⌈*Gloucester begins to leave*⌉
DUCHESS
What, gone, my lord, and bid me not farewell?
GLOUCESTER
Witness my tears—I cannot stay to speak.
    *Exeunt Gloucester and his men*
DUCHESS
Art thou gone too? All comfort go with thee,
For none abides with me. My joy is death—
Death, at whose name I oft have been afeard,     90
Because I wished this world's eternity.
Stanley, I prithee go and take me hence.
I care not whither, for I beg no favour,
Only convey me where thou art commanded.
STANLEY
Why, madam, that is to the Isle of Man,     95
There to be used according to your state.
DUCHESS
That's bad enough, for I am but reproach;
And shall I then be used reproachfully?
STANLEY
Like to a duchess and Duke Humphrey's lady,
According to that state you shall be used.     100
DUCHESS
Sheriff, farewell, and better than I fare,
Although thou hast been conduct of my shame.
⌈FIRST⌉ SHERIFF
It is my office, and, madam, pardon me.
DUCHESS
Ay, ay, farewell—thy office is discharged.
    ⌈*Exeunt Sheriffs*⌉
Come, Stanley, shall we go?     105
STANLEY
Madam, your penance done, throw off this sheet,
And go we to attire you for our journey.
DUCHESS
My shame will not be shifted with my sheet—
No, it will hang upon my richest robes
And show itself, attire me how I can.     110
Go, lead the way, I long to see my prison.     *Exeunt*

3.1     *Sound a sennet. Enter to the parliament: enter two*
    *heralds before, then the Dukes of Buckingham and*
    *Suffolk, and then the Duke of York and Cardinal*
    *Beaufort, and then King Henry and Queen*
    *Margaret, and then the Earls of Salisbury and*
    *Warwick,* ⌈*with attendants*⌉
KING HENRY
I muse my lord of Gloucester is not come.
'Tis not his wont to be the hindmost man,
Whate'er occasion keeps him from us now.
QUEEN MARGARET
Can you not see, or will ye not observe,
The strangeness of his altered countenance?     5
With what a majesty he bears himself?
How insolent of late he is become?
How proud, how peremptory, and unlike himself?
We know the time since he was mild and affable,
And if we did but glance a far-off look,     10
Immediately he was upon his knee,
That all the court admired him for submission.
But meet him now, and be it in the morn
When everyone will give the time of day,
He knits his brow, and shows an angry eye,     15
And passeth by with stiff unbowèd knee,
Disdaining duty that to us belongs.
Small curs are not regarded when they grin,
But great men tremble when the lion roars—
And Humphrey is no little man in England.     20
First, note that he is near you in descent,
And, should you fall, he is the next will mount.
Meseemeth then it is no policy,
Respecting what a rancorous mind he bears
And his advantage following your decease,     25
That he should come about your royal person,
Or be admitted to your highness' Council.
By flattery hath he won the commons' hearts,
And when he please to make commotion,
'Tis to be feared they all will follow him.     30
Now 'tis the spring, and weeds are shallow-rooted;
Suffer them now, and they'll o'ergrow the garden,
And choke the herbs for want of husbandry.
The reverent care I bear unto my lord
Made me collect these dangers in the Duke.     35
If it be fond, call it a woman's fear;
Which fear, if better reasons can supplant,
I will subscribe and say I wronged the Duke.
My lord of Suffolk, Buckingham, and York,
Reprove my allegation if you can,     40
Or else conclude my words effectual.
SUFFOLK
Well hath your highness seen into this Duke,
And had I first been put to speak my mind,
I think I should have told your grace's tale.
The Duchess by his subornation,     45
Upon my life, began her devilish practices;
Or if he were not privy to those faults,
Yet by reputing of his high descent,
As next the King he was successive heir,
And such high vaunts of his nobility,     50
Did instigate the bedlam brainsick Duchess
By wicked means to frame our sovereign's fall.
Smooth runs the water where the brook is deep,
And in his simple show he harbours treason.
The fox barks not when he would steal the lamb.     55
(*To King Henry*)
No, no, my sovereign, Gloucester is a man
Unsounded yet, and full of deep deceit.

CARDINAL BEAUFORT (*to King Henry*)
Did he not, contrary to form of law,
Devise strange deaths for small offences done?
YORK (*to King Henry*)
And did he not, in his Protectorship,          60
Levy great sums of money through the realm
For soldiers' pay in France, and never sent it,
By means whereof the towns each day revolted?
BUCKINGHAM (*to King Henry*)
Tut, these are petty faults to faults unknown,
Which time will bring to light in smooth Duke
Humphrey.          65
KING HENRY
My lords, at once: the care you have of us
To mow down thorns that would annoy our foot
Is worthy praise, but shall I speak my conscience?
Our kinsman Gloucester is as innocent
From meaning treason to our royal person          70
As is the sucking lamb or harmless dove.
The Duke is virtuous, mild, and too well given
To dream on evil or to work my downfall.
QUEEN MARGARET
Ah, what's more dangerous than this fond affiance?
Seems he a dove? His feathers are but borrowed,          75
For he's disposèd as the hateful raven.
Is he a lamb? His skin is surely lent him,
For he's inclined as is the ravenous wolf.
Who cannot steal a shape that means deceit?
Take heed, my lord, the welfare of us all          80
Hangs on the cutting short that fraudful man.
*Enter the Duke of Somerset*
SOMERSET ⌈*kneeling before King Henry*⌉
All health unto my gracious sovereign.
KING HENRY
Welcome, Lord Somerset. What news from France?
SOMERSET
That all your interest in those territories
Is utterly bereft you—all is lost.          85
KING HENRY
Cold news, Lord Somerset; but God's will be done.
⌈*Somerset rises*⌉
YORK (*aside*)
Cold news for me, for I had hope of France,
As firmly as I hope for fertile England.
Thus are my blossoms blasted in the bud,
And caterpillars eat my leaves away.          90
But I will remedy this gear ere long,
Or sell my title for a glorious grave.
*Enter Duke Humphrey of Gloucester*
GLOUCESTER ⌈*kneeling before King Henry*⌉
All happiness unto my lord the King.
Pardon, my liege, that I have stayed so long.
SUFFOLK
Nay, Gloucester, know that thou art come too soon          95
Unless thou wert more loyal than thou art.
I do arrest thee of high treason here.
GLOUCESTER ⌈*rising*⌉
Well, Suffolk's Duke, thou shalt not see me blush,
Nor change my countenance for this arrest.
A heart unspotted is not easily daunted.          100
The purest spring is not so free from mud
As I am clear from treason to my sovereign.
Who can accuse me? Wherein am I guilty?
YORK
'Tis thought, my lord, that you took bribes of France,
And, being Protector, stayed the soldiers' pay,          105
By means whereof his highness hath lost France.

GLOUCESTER
Is it but thought so? What are they that think it?
I never robbed the soldiers of their pay,
Nor ever had one penny bribe from France.
So help me God, as I have watched the night,          110
Ay, night by night, in studying good for England,
That doit that e'er I wrested from the King,
Or any groat I hoarded to my use,
Be brought against me at my trial day!
No: many a pound of mine own proper store,          115
Because I would not tax the needy commons,
Have I dispursèd to the garrisons,
And never asked for restitution.
CARDINAL BEAUFORT
It serves you well, my lord, to say so much.
GLOUCESTER
I say no more than truth, so help me God.          120
YORK
In your Protectorship you did devise
Strange tortures for offenders, never heard of,
That England was defamed by tyranny.
GLOUCESTER
Why, 'tis well known that whiles I was Protector
Pity was all the fault that was in me,          125
For I should melt at an offender's tears,
And lowly words were ransom for their fault.
Unless it were a bloody murderer,
Or foul felonious thief that fleeced poor passengers,
I never gave them condign punishment.          130
Murder, indeed—that bloody sin—I tortured
Above the felon or what trespass else.
SUFFOLK
My lord, these faults are easy, quickly answerèd,
But mightier crimes are laid unto your charge
Whereof you cannot easily purge yourself.          135
I do arrest you in his highness' name,
And here commit you to my good lord Cardinal
To keep until your further time of trial.
KING HENRY
My lord of Gloucester, 'tis my special hope
That you will clear yourself from all suspense.          140
My conscience tells me you are innocent.
GLOUCESTER
Ah, gracious lord, these days are dangerous.
Virtue is choked with foul ambition,
And charity chased hence by rancour's hand.
Foul subornation is predominant,          145
And equity exiled your highness' land.
I know their complot is to have my life,
And if my death might make this island happy
And prove the period of their tyranny,
I would expend it with all willingness.          150
But mine is made the prologue to their play,
For thousands more that yet suspect no peril
Will not conclude their plotted tragedy.
Beaufort's red sparkling eyes blab his heart's malice,
And Suffolk's cloudy brow his stormy hate;          155
Sharp Buckingham unburdens with his tongue
The envious load that lies upon his heart;
And doggèd York that reaches at the moon,
Whose overweening arm I have plucked back,
By false accuse doth level at my life.          160
(*To Queen Margaret*)
And you, my sovereign lady, with the rest,
Causeless have laid disgraces on my head,
And with your best endeavour have stirred up
My liefest liege to be mine enemy.

Ay, all of you have laid your heads together—        165
Myself had notice of your conventicles—
And all to make away my guiltless life.
I shall not want false witness to condemn me,
Nor store of treasons to augment my guilt.
The ancient proverb will be well effected:        170
'A staff is quickly found to beat a dog'.
CARDINAL BEAUFORT (*to King Henry*)
My liege, his railing is intolerable.
If those that care to keep your royal person
From treason's secret knife and traitor's rage
Be thus upbraided, chid, and rated at,        175
And the offender granted scope of speech,
'Twill make them cool in zeal unto your grace.
SUFFOLK (*to King Henry*)
Hath he not twit our sovereign lady here
With ignominious words, though clerkly couched,
As if she had suborned some to swear        180
False allegations to o'erthrow his state?
QUEEN MARGARET
But I can give the loser leave to chide.
GLOUCESTER
Far truer spoke than meant. I lose indeed;
Beshrew the winners, for they played me false!
And well such losers may have leave to speak.        185
BUCKINGHAM (*to King Henry*)
He'll wrest the sense, and hold us here all day.
Lord Cardinal, he is your prisoner.
CARDINAL BEAUFORT (*to some of his attendants*)
Sirs, take away the Duke and guard him sure.
GLOUCESTER
Ah, thus King Henry throws away his crutch
Before his legs be firm to bear his body.        190
Thus is the shepherd beaten from thy side,
And wolves are gnarling who shall gnaw thee first.
Ah, that my fear were false; ah, that it were!
For, good King Henry, thy decay I fear.
        *Exit Gloucester, guarded by the Cardinal's men*
KING HENRY
My lords, what to your wisdoms seemeth best        195
Do or undo, as if ourself were here.
QUEEN MARGARET
What, will your highness leave the Parliament?
KING HENRY
Ay, Margaret, my heart is drowned with grief,
Whose flood begins to flow within mine eyes,
My body round engirt with misery;        200
For what's more miserable than discontent?
Ah, uncle Humphrey, in thy face I see
The map of honour, truth, and loyalty;
And yet, good Humphrey, is the hour to come
That e'er I proved thee false, or feared thy faith.        205
What louring star now envies thy estate,
That these great lords and Margaret our Queen
Do seek subversion of thy harmless life?
Thou never didst them wrong, nor no man wrong.
And as the butcher takes away the calf,        210
And binds the wretch, and beats it when it strains,
Bearing it to the bloody slaughterhouse,
Even so remorseless have they borne him hence;
And as the dam runs lowing up and down,
Looking the way her harmless young one went,        215
And can do naught but wail her darling's loss;
Even so myself bewails good Gloucester's case
With sad unhelpful tears, and with dimmed eyes
Look after him, and cannot do him good,

So mighty are his vowèd enemies.        220
His fortunes I will weep, and 'twixt each groan,
Say 'Who's a traitor? Gloucester, he is none'.
        *Exit ⌈with Salisbury and Warwick⌉*
QUEEN MARGARET
Free lords, cold snow melts with the sun's hot beams.
Henry my lord is cold in great affairs,
Too full of foolish pity; and Gloucester's show        225
Beguiles him as the mournful crocodile
With sorrow snares relenting passengers,
Or as the snake rolled in a flow'ring bank
With shining chequered slough doth sting a child
That for the beauty thinks it excellent.        230
Believe me, lords, were none more wise than I—
And yet herein I judge mine own wit good—
This Gloucester should be quickly rid the world
To rid us from the fear we have of him.
CARDINAL BEAUFORT
That he should die is worthy policy;        235
But yet we want a colour for his death.
'Tis meet he be condemned by course of law.
SUFFOLK
But, in my mind, that were no policy.
The King will labour still to save his life,
The commons haply rise to save his life;        240
And yet we have but trivial argument
More than mistrust that shows him worthy death.
YORK
So that, by this, you would not have him die?
SUFFOLK
Ah, York, no man alive so fain as I.
YORK (*aside*)
'Tis York that hath more reason for his death.        245
(*Aloud*) But my lord Cardinal, and you my lord of
        Suffolk,
Say as you think, and speak it from your souls.
Were't not all one an empty eagle were set
To guard the chicken from a hungry kite,
As place Duke Humphrey for the King's Protector?
QUEEN MARGARET
So the poor chicken should be sure of death.        251
SUFFOLK
Madam, 'tis true; and were't not madness then
To make the fox surveyor of the fold,
Who being accused a crafty murderer,
His guilt should be but idly posted over        255
Because his purpose is not executed?
No—let him die in that he is a fox,
By nature proved an enemy to the flock,
Before his chaps be stained with crimson blood,
As Humphrey, proved by reasons, to my liege.        260
And do not stand on quillets how to slay him;
Be it by gins, by snares, by subtlety,
Sleeping or waking, 'tis no matter how,
So he be dead; for that is good conceit
Which mates him first that first intends deceit.        265
QUEEN MARGARET
Thrice-noble Suffolk, 'tis resolutely spoke.
SUFFOLK
Not resolute, except so much were done;
For things are often spoke and seldom meant;
But that my heart accordeth with my tongue,
Seeing the deed is meritorious,        270
And to preserve my sovereign from his foe,
Say but the word and I will be his priest.

CARDINAL BEAUFORT
But I would have him dead, my lord of Suffolk,
Ere you can take due orders for a priest.
Say you consent and censure well the deed,          275
And I'll provide his executioner;
I tender so the safety of my liege.

SUFFOLK
Here is my hand; the deed is worthy doing.

QUEEN MARGARET  And so say I.

YORK
And I. And now we three have spoke it,               280
It skills not greatly who impugns our doom.
              *Enter a Post*

POST
Great lord, from Ireland am I come amain
To signify that rebels there are up
And put the Englishmen unto the sword.
Send succours, lords, and stop the rage betime,     285
Before the wound do grow uncurable;
For, being green, there is great hope of help.      ⌈*Exit*⌉

CARDINAL BEAUFORT
A breach that craves a quick expedient stop!
What counsel give you in this weighty cause?

YORK
That Somerset be sent as regent thither.            290
'Tis meet that lucky ruler be employed—
Witness the fortune he hath had in France.

SOMERSET
If York, with all his far-fet policy,
Had been the regent there instead of me,
He never would have stayed in France so long.       295

YORK
No, not to lose it all as thou hast done.
I rather would have lost my life betimes
Than bring a burden of dishonour home
By staying there so long till all were lost.
Show me one scar charactered on thy skin.           300
Men's flesh preserved so whole do seldom win.

QUEEN MARGARET
Nay, then, this spark will prove a raging fire
If wind and fuel be brought to feed it with.
No more, good York; sweet Somerset, be still.
Thy fortune, York, hadst thou been regent there,    305
Might happily have proved far worse than his.

YORK
What, worse than naught? Nay, then a shame take
    all!

SOMERSET
And, in the number, thee that wishest shame.

CARDINAL BEAUFORT
My lord of York, try what your fortune is.
Th'uncivil kerns of Ireland are in arms             310
And temper clay with blood of Englishmen.
To Ireland will you lead a band of men
Collected choicely, from each county some,
And try your hap against the Irishmen?

YORK
I will, my lord, so please his majesty.             315

SUFFOLK
Why, our authority is his consent,
And what we do establish he confirms.
Then, noble York, take thou this task in hand.

YORK
I am content. Provide me soldiers, lords,
Whiles I take order for mine own affairs.           320

SUFFOLK
A charge, Lord York, that I will see performed.
But now return we to the false Duke Humphrey.

CARDINAL BEAUFORT
No more of him—for I will deal with him
That henceforth he shall trouble us no more.
And so, break off; the day is almost spent.         325
Lord Suffolk, you and I must talk of that event.

YORK
My lord of Suffolk, within fourteen days
At Bristol I expect my soldiers;
For there I'll ship them all for Ireland.

SUFFOLK
I'll see it truly done, my lord of York.            330
              *Exeunt all but York*

YORK
Now, York, or never, steel thy fearful thoughts,
And change misdoubt to resolution.
Be that thou hop'st to be, or what thou art
Resign to death; it is not worth th'enjoying.
Let pale-faced fear keep with the mean-born man     335
And find no harbour in a royal heart.
Faster than springtime showers comes thought on
                                         thought,
And not a thought but thinks on dignity.
My brain, more busy than the labouring spider,
Weaves tedious snares to trap mine enemies.         340
Well, nobles, well: 'tis politicly done
To send me packing with an host of men.
I fear me you but warm the starvèd snake,
Who, cherished in your breasts, will sting your hearts.
'Twas men I lacked, and you will give them me.      345
I take it kindly. Yet be well assured
You put sharp weapons in a madman's hands.
Whiles I in Ireland nurse a mighty band,
I will stir up in England some black storm
Shall blow ten thousand souls to heaven or hell,    350
And this fell tempest shall not cease to rage
Until the golden circuit on my head
Like to the glorious sun's transparent beams
Do calm the fury of this mad-bred flaw.
And for a minister of my intent,                    355
I have seduced a headstrong Kentishman,
John Cade of Ashford,
To make commotion, as full well he can,
Under the title of John Mortimer.
In Ireland have I seen this stubborn Cade           360
Oppose himself against a troop of kerns,
And fought so long till that his thighs with darts
Were almost like a sharp-quilled porcupine;
And in the end, being rescued, I have seen
Him caper upright like a wild Morisco,              365
Shaking the bloody darts as he his bells.
Full often like a shag-haired crafty kern
Hath he conversèd with the enemy
And, undiscovered, come to me again
And given me notice of their villainies.            370
This devil here shall be my substitute,
For that John Mortimer, which now is dead,
In face, in gait, in speech, he doth resemble.
By this I shall perceive the commons' mind,
How they affect the house and claim of York.        375
Say he be taken, racked, and torturèd—
I know no pain they can inflict upon him
Will make him say I moved him to those arms.

Say that he thrive, as 'tis great like he will—
Why then from Ireland come I with my strength    380
And reap the harvest which that coistrel sowed.
For Humphrey being dead, as he shall be,
And Henry put apart, the next for me.        *Exit*

**3.2**    ⌈*The curtains are drawn apart, revealing Duke*
*Humphrey of Gloucester in his bed with two men*
*lying on his breast, smothering him in his bed*⌉
FIRST MURDERER (*to the Second Murderer*)
   Run to my lord of Suffolk—let him know
   We have dispatched the Duke as he commanded.
SECOND MURDERER
   O that it were to do! What have we done?
   Didst ever hear a man so penitent?
         *Enter the Duke of Suffolk*
FIRST MURDERER Here comes my lord.       5
SUFFOLK
   Now, sirs, have you dispatched this thing?
FIRST MURDERER Ay, my good lord, he's dead.
SUFFOLK
   Why, that's well said. Go, get you to my house.
   I will reward you for this venturous deed.
   The King and all the peers are here at hand.    10
   Have you laid fair the bed? Is all things well,
   According as I gave directions?
FIRST MURDERER 'Tis, my good lord.
SUFFOLK
   Then draw the curtains close; away, be gone!
       *Exeunt* ⌈*the Murderers, drawing the curtains as*
                      *they leave*⌉
    *Sound trumpets, then enter King Henry and Queen*
    *Margaret, Cardinal Beaufort, the Duke of Somerset,*
    *and attendants*
KING HENRY ⌈*to Suffolk*⌉
   Go call our uncle to our presence straight.    15
   Say we intend to try his grace today
   If he be guilty, as 'tis publishèd.
SUFFOLK
   I'll call him presently, my noble lord.      *Exit*
KING HENRY
   Lords, take your places; and, I pray you all,
   Proceed no straiter 'gainst our uncle Gloucester   20
   Than from true evidence, of good esteem,
   He be approved in practice culpable.
QUEEN MARGARET
   God forbid any malice should prevail
   That faultless may condemn a noble man!
   Pray God he may acquit him of suspicion!    25
KING HENRY
   I thank thee, Meg. These words content me much.
        *Enter Suffolk*
   How now? Why look'st thou pale? Why tremblest
     thou?
   Where is our uncle? What's the matter, Suffolk?
SUFFOLK
   Dead in his bed, my lord—Gloucester is dead.
QUEEN MARGARET Marry, God forfend!     30
CARDINAL BEAUFORT
   God's secret judgement. I did dream tonight
   The Duke was dumb and could not speak a word.
      *King Henry falls to the ground*
QUEEN MARGARET
   How fares my lord? Help, lords—the King is dead!
SOMERSET
   Rear up his body; wring him by the nose.

QUEEN MARGARET
   Run, go, help, help! O Henry, ope thine eyes!    35
SUFFOLK
   He doth revive again. Madam, be patient.
KING HENRY
   O heavenly God!
QUEEN MARGARET    How fares my gracious lord?
SUFFOLK
   Comfort, my sovereign; gracious Henry, comfort.
KING HENRY
   What, doth my lord of Suffolk comfort me?
   Came he right now to sing a raven's note     40
   Whose dismal tune bereft my vital powers;
   And thinks he that the chirping of a wren,
   By crying comfort from a hollow breast
   Can chase away the first-conceivèd sound?
   Hide not thy poison with such sugared words.   45
     ⌈*He begins to rise. Suffolk offers to assist him*⌉
   Lay not thy hands on me—forbear, I say!
   Their touch affrights me as a serpent's sting.
   Thou baleful messenger, out of my sight!
   Upon thy eyeballs murderous tyranny
   Sits in grim majesty to fright the world.     50
   Look not upon me, for thine eyes are wounding—
   Yet do not go away. Come, basilisk,
   And kill the innocent gazer with thy sight.
   For in the shade of death I shall find joy;
   In life, but double death, now Gloucester's dead.   55
QUEEN MARGARET
   Why do you rate my lord of Suffolk thus?
   Although the Duke was enemy to him,
   Yet he most Christian-like laments his death.
   And for myself, foe as he was to me,
   Might liquid tears, or heart-offending groans,    60
   Or blood-consuming sighs recall his life,
   I would be blind with weeping, sick with groans,
   Look pale as primrose with blood-drinking sighs,
   And all to have the noble Duke alive.
   What know I how the world may deem of me?   65
   For it is known we were but hollow friends,
   It may be judged I made the Duke away.
   So shall my name with slander's tongue be wounded
   And princes' courts be filled with my reproach.
   This get I by his death. Ay me, unhappy,     70
   To be a queen, and crowned with infamy.
KING HENRY
   Ah, woe is me for Gloucester, wretched man!
QUEEN MARGARET
   Be woe for me, more wretched than he is.
   What, dost thou turn away and hide thy face?
   I am no loathsome leper—look on me!     75
   What, art thou, like the adder, waxen deaf?
   Be poisonous too and kill thy forlorn queen.
   Is all thy comfort shut in Gloucester's tomb?
   Why, then Queen Margaret was ne'er thy joy.
   Erect his statuë and worship it,       80
   And make my image but an alehouse sign.
   Was I for this nigh wrecked upon the sea,
   And twice by awkward winds from England's bank
   Drove back again unto my native clime?
   What boded this, but well forewarning winds    85
   Did seem to say, 'Seek not a scorpion's nest,
   Nor set no footing on this unkind shore'.
   What did I then, but cursed the gentle gusts
   And he that loosed them forth their brazen caves,
   And bid them blow towards England's blessèd shore,

Or turn our stern upon a dreadful rock.          91
Yet Aeolus would not be a murderer,
But left that hateful office unto thee.
The pretty vaulting sea refused to drown me,
Knowing that thou wouldst have me drowned on
          shore                                  95
With tears as salt as sea through thy unkindness.
The splitting rocks cow'red in the sinking sands,
And would not dash me with their ragged sides,
Because thy flinty heart, more hard than they,
Might in thy palace perish Margaret.             100
As far as I could ken thy chalky cliffs,
When from thy shore the tempest beat us back,
I stood upon the hatches in the storm,
And when the dusky sky began to rob
My earnest-gaping sight of thy land's view,      105
I took a costly jewel from my neck—
A heart it was, bound in with diamonds—
And threw it towards thy land. The sea received it,
And so I wished thy body might my heart.
And even with this I lost fair England's view,   110
And bid mine eyes be packing with my heart,
And called them blind and dusky spectacles
For losing ken of Albion's wishèd coast.
How often have I tempted Suffolk's tongue—
The agent of thy foul inconstancy—               115
To sit and witch me, as Ascanius did,
When he to madding Dido would unfold
His father's acts, commenced in burning Troy!
Am I not witched like her? Or thou not false like him?
Ay me, I can no more. Die, Margaret,             120
For Henry weeps that thou dost live so long.
          *Noise within. Enter the Earls of Warwick and*
          *Salisbury with many commons*

WARWICK (*to King Henry*)
It is reported, mighty sovereign,
That good Duke Humphrey traitorously is murdered
By Suffolk and the Cardinal Beaufort's means.
The commons, like an angry hive of bees          125
That want their leader, scatter up and down
And care not who they sting in his revenge.
Myself have calmed their spleenful mutiny,
Until they hear the order of his death.

KING HENRY
That he is dead, good Warwick, 'tis too true.    130
But how he died God knows, not Henry.
Enter his chamber, view his breathless corpse,
And comment then upon his sudden death.

WARWICK
That shall I do, my liege.—Stay, Salisbury,
With the rude multitude till I return.           135
          *⌈Exeunt Warwick at one door, Salisbury and*
          *commons at another⌉*

KING HENRY
O thou that judgest all things, stay my thoughts,
My thoughts that labour to persuade my soul
Some violent hands were laid on Humphrey's life.
If my suspect be false, forgive me God,
For judgement only doth belong to thee.          140
Fain would I go to chafe his paly lips
With twenty thousand kisses, and to drain
Upon his face an ocean of salt tears,
To tell my love unto his dumb, deaf trunk,
And with my fingers feel his hand unfeeling.     145
But all in vain are these mean obsequies,

          *⌈Enter Warwick who draws apart the curtains and*
          *shows⌉ Gloucester dead in his bed. Bed put forth*
And to survey his dead and earthy image,
What were it but to make my sorrow greater?

WARWICK
Come hither, gracious sovereign, view this body.

KING HENRY
That is to see how deep my grave is made:        150
For with his soul fled all my worldly solace,
For seeing him I see my life in death.

WARWICK
As surely as my soul intends to live
With that dread King that took our state upon Him
To free us from his Father's wrathful curse,     155
I do believe that violent hands were laid
Upon the life of this thrice-famèd Duke.

SUFFOLK
A dreadful oath, sworn with a solemn tongue!
What instance gives Lord Warwick for his vow?

WARWICK
See how the blood is settled in his face.        160
Oft have I seen a timely-parted ghost
Of ashy semblance, meagre, pale, and bloodless,
Being all descended to the labouring heart;
Who, in the conflict that it holds with death,
Attracts the same for aidance 'gainst the enemy; 165
Which, with the heart, there cools, and ne'er returneth
To blush and beautify the cheek again.
But see, his face is black and full of blood;
His eyeballs further out than when he lived,
Staring full ghastly like a strangled man;       170
His hair upreared; his nostrils stretched with
          struggling;
His hands abroad displayed, as one that grasped
And tugged for life and was by strength subdued.
Look on the sheets. His hair, you see, is sticking;
His well-proportioned beard made rough and rugged,
Like to the summer's corn by tempest lodged.     176
It cannot be but he was murdered here.
The least of all these signs were probable.

SUFFOLK
Why, Warwick, who should do the Duke to death?
Myself and Beaufort had him in protection,       180
And we, I hope, sir, are no murderers.

WARWICK
But both of you were vowed Duke Humphrey's foes,
(*To Cardinal Beaufort*)
And you, forsooth, had the good Duke to keep.
'Tis like you would not feast him like a friend;
And 'tis well seen he found an enemy.            185

QUEEN MARGARET
Then you, belike, suspect these noblemen
As guilty of Duke Humphrey's timeless death?

WARWICK
Who finds the heifer dead and bleeding fresh,
And sees fast by a butcher with an axe,
But will suspect 'twas he that made the slaughter?
Who finds the partridge in the puttock's nest    191
But may imagine how the bird was dead,
Although the kite soar with unbloodied beak?
Even so suspicious is this tragedy.

QUEEN MARGARET
Are you the butcher, Suffolk? Where's your knife? 195
Is Beaufort termed a kite? Where are his talons?

SUFFOLK
I wear no knife to slaughter sleeping men.
But here's a vengeful sword, rusted with ease,

That shall be scourèd in his rancorous heart
That slanders me with murder's crimson badge.    200
Say, if thou dar'st, proud Lord of Warwickshire,
That I am faulty in Duke Humphrey's death.
          ⌜Exit Cardinal Beaufort assisted by Somerset⌝

WARWICK
What dares not Warwick, if false Suffolk dare him?

QUEEN MARGARET
He dares not calm his contumelious spirit,
Nor cease to be an arrogant controller,          205
Though Suffolk dare him twenty thousand times.

WARWICK
Madam, be still, with reverence may I say,
For every word you speak in his behalf
Is slander to your royal dignity.

SUFFOLK
Blunt-witted lord, ignoble in demeanour!          210
If ever lady wronged her lord so much,
Thy mother took into her blameful bed
Some stern untutored churl, and noble stock
Was graffed with crabtree slip, whose fruit thou art,
And never of the Nevilles' noble race.            215

WARWICK
But that the guilt of murder bucklers thee
And I should rob the deathsman of his fee,
Quitting thee thereby of ten thousand shames,
And that my sovereign's presence makes me mild,
I would, false murd'rous coward, on thy knee      220
Make thee beg pardon for thy passèd speech,
And say it was thy mother that thou meant'st—
That thou thyself wast born in bastardy!
And after all this fearful homage done,
Give thee thy hire and send thy soul to hell,    225
Pernicious blood-sucker of sleeping men!

SUFFOLK
Thou shalt be waking while I shed thy blood,
If from this presence thou dar'st go with me.

WARWICK
Away, even now, or I will drag thee hence.
Unworthy though thou art, I'll cope with thee,    230
And do some service to Duke Humphrey's ghost.
          Exeunt Suffolk and Warwick

KING HENRY
What stronger breastplate than a heart untainted?
Thrice is he armed that hath his quarrel just;
And he but naked, though locked up in steel,
Whose conscience with injustice is corrupted.     235

COMMONS (within) Down with Suffolk! Down with Suffolk!

QUEEN MARGARET What noise is this?
          Enter Suffolk and Warwick with their weapons
          drawn

KING HENRY
Why, how now, lords? Your wrathful weapons drawn
Here in our presence? Dare you be so bold?
Why, what tumultuous clamour have we here?        240

SUFFOLK
The trait'rous Warwick with the men of Bury
Set all upon me, mighty sovereign!

COMMONS (within) Down with Suffolk! Down with Suffolk!
          Enter from the commons the Earl of Salisbury

SALISBURY (to the commons, within)
Sirs, stand apart. The King shall know your mind.
(To King Henry)
Dread lord, the commons send you word by me       245
Unless Lord Suffolk straight be done to death,

Or banishèd fair England's territories,
They will by violence tear him from your palace
And torture him with grievous ling'ring death.
They say, by him the good Duke Humphrey died;     250
They say, in him they fear your highness' death;
And mere instinct of love and loyalty,
Free from a stubborn opposite intent,
As being thought to contradict your liking,
Makes them thus forward in his banishment.        255
They say, in care of your most royal person,
That if your highness should intend to sleep,
And charge that no man should disturb your rest
In pain of your dislike, or pain of death,
Yet, notwithstanding such a strait edict,         260
Were there a serpent seen with forkèd tongue,
That slily glided towards your majesty,
It were but necessary you were waked,
Lest, being suffered in that harmful slumber,
The mortal worm might make the sleep eternal.     265
And therefore do they cry, though you forbid,
That they will guard you, whe'er you will or no,
From such fell serpents as false Suffolk is,
With whose envenomèd and fatal sting
Your loving uncle, twenty times his worth,        270
They say, is shamefully bereft of life.

COMMONS (within) An answer from the King, my lord of
Salisbury!

SUFFOLK
'Tis like the commons, rude unpolished hinds,
Could send such message to their sovereign.       275
But you, my lord, were glad to be employed,
To show how quaint an orator you are.
But all the honour Salisbury hath won
Is that he was the Lord Ambassador
Sent from a sort of tinkers to the King.          280

COMMONS (within) An answer from the King, or we will
all break in!

KING HENRY
Go, Salisbury, and tell them all from me
I thank them for their tender loving care,
And had I not been 'cited so by them,             285
Yet did I purpose as they do entreat;
For sure my thoughts do hourly prophesy
Mischance unto my state by Suffolk's means.
And therefore by His majesty I swear,
Whose far unworthy deputy I am,                   290
He shall not breathe infection in this air
But three days longer, on the pain of death.
          ⌜Exit Salisbury⌝

QUEEN MARGARET ⌜kneeling⌝
O Henry, let me plead for gentle Suffolk.

KING HENRY
Ungentle Queen, to call him gentle Suffolk.
No more, I say! If thou dost plead for him        295
Thou wilt but add increase unto my wrath.
Had I but said, I would have kept my word;
But when I swear, it is irrevocable.
(To Suffolk) If after three days' space thou here beest
          found
On any ground that I am ruler of,                 300
The world shall not be ransom for thy life.
Come, Warwick; come, good Warwick, go with me.
I have great matters to impart to thee.
          Exeunt King Henry and Warwick with
          attendants ⌜who draw the curtains as they
          leave⌝. Queen Margaret and Suffolk remain

QUEEN MARGARET ⌈rising⌉
Mischance and sorrow go along with you!
Heart's discontent and sour affliction                    305
Be playfellows to keep you company!
There's two of you, the devil make a third,
And threefold vengeance tend upon your steps!

SUFFOLK
Cease, gentle Queen, these execrations,
And let thy Suffolk take his heavy leave.                  310

QUEEN MARGARET
Fie, coward woman and soft-hearted wretch!
Hast thou not spirit to curse thine enemies?

SUFFOLK
A plague upon them! Wherefore should I curse them?
Could curses kill, as doth the mandrake's groan,
I would invent as bitter searching terms,                 315
As curst, as harsh, and horrible to hear,
Delivered strongly through my fixèd teeth,
With full as many signs of deadly hate,
As lean-faced envy in her loathsome cave.
My tongue should stumble in mine earnest words;           320
Mine eyes should sparkle like the beaten flint;
My hair be fixed on end, as one distraught;
Ay, every joint should seem to curse and ban.
And, even now, my burdened heart would break
Should I not curse them. Poison be their drink!           325
Gall, worse than gall, the daintiest that they taste!
Their sweetest shade a grove of cypress trees!
Their chiefest prospect murd'ring basilisks!
Their softest touch as smart as lizards' stings!
Their music frightful as the serpent's hiss,              330
And boding screech-owls make the consort full!
All the foul terrors in dark-seated hell—

QUEEN MARGARET
Enough, sweet Suffolk, thou torment'st thyself,
And these dread curses, like the sun 'gainst glass,
Or like an overchargèd gun, recoil                        335
And turn the force of them upon thyself.

SUFFOLK
You bade me ban, and will you bid me leave?
Now by this ground that I am banished from,
Well could I curse away a winter's night,
Though standing naked on a mountain top,                  340
Where biting cold would never let grass grow,
And think it but a minute spent in sport.

QUEEN MARGARET
O let me entreat thee cease. Give me thy hand,
That I may dew it with my mournful tears;
Nor let the rain of heaven wet this place                 345
To wash away my woeful monuments.
⌈She kisses his palm⌉
O, could this kiss be printed in thy hand
That thou mightst think upon these lips by the seal,
Through whom a thousand sighs are breathed for
    thee!
So get thee gone, that I may know my grief.               350
'Tis but surmised whiles thou art standing by,
As one that surfeits thinking on a want.
I will repeal thee, or, be well assured,
Adventure to be banishèd myself.
And banishèd I am, if but from thee.                      355
Go, speak not to me; even now be gone!
O, go not yet. Even thus two friends condemned
Embrace, and kiss, and take ten thousand leaves,
Loather a hundred times to part than die.
Yet now farewell, and farewell life with thee.            360

SUFFOLK
Thus is poor Suffolk ten times banishèd—
Once by the King, and three times thrice by thee.
'Tis not the land I care for, wert thou thence,
A wilderness is populous enough,
So Suffolk had thy heavenly company.                      365
For where thou art, there is the world itself,
With every several pleasure in the world;
And where thou art not, desolation.
I can no more. Live thou to joy thy life;
Myself no joy in naught but that thou liv'st.             370
    Enter Vaux

QUEEN MARGARET
Whither goes Vaux so fast? What news, I prithee?

VAUX
To signify unto his majesty
That Cardinal Beaufort is at point of death.
For suddenly a grievous sickness took him
That makes him gasp, and stare, and catch the air,
Blaspheming God and cursing men on earth.                 376
Sometime he talks as if Duke Humphrey's ghost
Were by his side; sometime he calls the King,
And whispers to his pillow as to him
The secrets of his over-chargèd soul;                     380
And I am sent to tell his majesty
That even now he cries aloud for him.

QUEEN MARGARET
Go tell this heavy message to the King.        Exit Vaux
Ay me! What is this world? What news are these?
But wherefore grieve I at an hour's poor loss             385
Omitting Suffolk's exile, my soul's treasure?
Why only, Suffolk, mourn I not for thee,
And with the southern clouds contend in tears—
Theirs for the earth's increase, mine for my sorrow's?
Now get thee hence. The King, thou know'st, is
    coming.                                               390
If thou be found by me, thou art but dead.

SUFFOLK
If I depart from thee, I cannot live.
And in thy sight to die, what were it else
But like a pleasant slumber in thy lap?
Here could I breathe my soul into the air,                395
As mild and gentle as the cradle babe
Dying with mother's dug between his lips;
Where, from thy sight, I should be raging mad,
And cry out for thee to close up mine eyes,
To have thee with thy lips to stop my mouth,              400
So shouldst thou either turn my flying soul
Or I should breathe it, so, into thy body—
⌈He kisseth her⌉
And then it lived in sweet Elysium.
By thee to die were but to die in jest;
From thee to die were torture more than death.            405
O, let me stay, befall what may befall!

QUEEN MARGARET
Away. Though parting be a fretful corrosive,
It is applièd to a deathful wound.
To France, sweet Suffolk. Let me hear from thee.
For wheresoe'er thou art in this world's Globe            410
I'll have an Iris that shall find thee out.

SUFFOLK
I go.

QUEEN MARGARET And take my heart with thee.
⌈She kisseth him⌉

SUFFOLK
A jewel, locked into the woefull'st cask

That ever did contain a thing of worth.
Even as a splitted barque, so sunder we—               415
This way fall I to death.
QUEEN MARGARET          This way for me.
                                        *Exeunt severally*

3.3     *Enter King Henry and the Earls of Salisbury and
        Warwick. Then the curtains be drawn revealing
        Cardinal Beaufort in his bed raving and staring as if
        he were mad*
KING HENRY (*to Cardinal Beaufort*)
    How fares my lord? Speak, Beaufort, to thy sovereign.
CARDINAL BEAUFORT
    If thou beest death, I'll give thee England's treasure
    Enough to purchase such another island,
    So thou wilt let me live and feel no pain.
KING HENRY
    Ah, what a sign it is of evil life                      5
    Where death's approach is seen so terrible.
WARWICK
    Beaufort, it is thy sovereign speaks to thee.
CARDINAL BEAUFORT
    Bring me unto my trial when you will.
    Died he not in his bed? Where should he die?
    Can I make men live whe'er they will or no?            10
    O, torture me no more—I will confess.
    Alive again? Then show me where he is.
    I'll give a thousand pound to look upon him.
    He hath no eyes! The dust hath blinded them.
    Comb down his hair—look, look: it stands upright, 15
    Like lime twigs set to catch my wingèd soul.
    Give me some drink, and bid the apothecary
    Bring the strong poison that I bought of him.
KING HENRY
    O Thou eternal mover of the heavens,
    Look with a gentle eye upon this wretch.               20
    O, beat away the busy meddling fiend
    That lays strong siege unto this wretch's soul,
    And from his bosom purge this black despair.
WARWICK
    See how the pangs of death do make him grin.
SALISBURY
    Disturb him not; let him pass peaceably.               25
KING HENRY
    Peace to his soul, if God's good pleasure be.
    Lord Card'nal, if thou think'st on heaven's bliss,
    Hold up thy hand, make signal of thy hope.
                *Cardinal Beaufort dies*
    He dies and makes no sign. O God, forgive him.
WARWICK
    So bad a death argues a monstrous life.                30
KING HENRY
    Forbear to judge, for we are sinners all.
    Close up his eyes and draw the curtain close,
    And let us all to meditation.
                *Exeunt, ⌐drawing the curtains. The bed is
                                            removed⌐*

4.1     *Alarums within, and the chambers be discharged
        like as it were a fight at sea. And then enter the
        Captain of the ship, the Master, the Master's Mate,
        Walter Whitmore, ⌐and others⌐. With them, as
        their prisoners, the Duke of Suffolk, disguised, and
        two Gentlemen*
CAPTAIN
    The gaudy, blabbing, and remorseful day
    Is crept into the bosom of the sea;

And now loud-howling wolves arouse the jades
That drag the tragic melancholy night;
Who, with their drowsy, slow, and flagging wings   5
Clip dead men's graves, and from their misty jaws
Breathe foul contagious darkness in the air.
Therefore bring forth the soldiers of our prize,
For whilst our pinnace anchors in the downs,
Here shall they make their ransom on the sand,     10
Or with their blood stain this discoloured shore.
Master, (*pointing to the First Gentleman*) this prisoner
        freely give I thee,
(*To the Mate*)
And thou, that art his mate, make boot of this.
        *He points to the Second Gentleman*
(*To Walter Whitmore*)
The other (*pointing to Suffolk*), Walter Whitmore, is
        thy share.
FIRST GENTLEMAN (*to the Master*)
    What is my ransom, Master, let me know.            15
MASTER
    A thousand crowns, or else lay down your head.
MATE (*to the Second Gentleman*)
    And so much shall you give, or off goes yours.
CAPTAIN (*to both the Gentlemen*)
    What, think you much to pay two thousand crowns,
    And bear the name and port of gentlemen?
⌐WHITMORE⌐
    Cut both the villains' throats! ⌐To Suffolk⌐ For die you
        shall.                                         20
    The lives of those which we have lost in fight
    ⌐                                               ⌐
    Be counterpoised with such a petty sum.
FIRST GENTLEMAN (*to the Master*)
    I'll give it, sir, and therefore spare my life.
SECOND GENTLEMAN (*to the Mate*)
    And so will I, and write home for it straight.     25
WHITMORE (*to Suffolk*)
    I lost mine eye in laying the prize aboard,
    And therefore to revenge it, shalt thou die—
    And so should these, if I might have my will.
CAPTAIN
    Be not so rash; take ransom; let him live.
SUFFOLK
    Look on my George—I am a gentleman.               30
    Rate me at what thou wilt, thou shalt be paid.
WHITMORE
    And so am I; my name is Walter Whitmore.
                *Suffolk starteth*
    How now—why starts thou? What doth thee affright?
SUFFOLK
    Thy name affrights me, in whose sound is death.
    A cunning man did calculate my birth,             35
    And told me that by 'water' I should die.
    Yet let not this make thee be bloody-minded;
    Thy name is Gualtier, being rightly sounded.
WHITMORE
    Gualtier or Walter—which it is I care not.
    Never yet did base dishonour blur our name        40
    But with our sword we wiped away the blot.
    Therefore, when merchant-like I sell revenge,
    Broke be my sword, my arms torn and defaced,
    And I proclaimed a coward through the world.
SUFFOLK
    Stay, Whitmore; for thy prisoner is a prince,     45
    The Duke of Suffolk, William de la Pole.
WHITMORE
    The Duke of Suffolk muffled up in rags?

SUFFOLK
Ay, but these rags are no part of the Duke.
Jove sometime went disguised, and why not I?
CAPTAIN
But Jove was never slain as thou shalt be.          50
SUFFOLK
Obscure and lousy swain, King Henry's blood,
The honourable blood of Lancaster,
Must not be shed by such a jady groom.
Hast thou not kissed thy hand and held my stirrup?
Bare-headed plodded by my foot-cloth mule          55
And thought thee happy when I shook my head?
How often hast thou waited at my cup,
Fed from my trencher, kneeled down at the board
When I have feasted with Queen Margaret?
Remember it, and let it make thee crestfall'n,          60
Ay, and allay this thy abortive pride,
How in our voiding lobby hast thou stood
And duly waited for my coming forth?
This hand of mine hath writ in thy behalf,
And therefore shall it charm thy riotous tongue.          65
WHITMORE
Speak, Captain—shall I stab the forlorn swain?
CAPTAIN
First let my words stab him as he hath me.
SUFFOLK
Base slave, thy words are blunt and so art thou.
CAPTAIN
Convey him hence and, on our longboat's side,
Strike off his head.
SUFFOLK          Thou dar'st not for thy own.          70
CAPTAIN
Pole—
[SUFFOLK] Pole?
CAPTAIN     Ay, kennel, puddle, sink, whose filth and dirt
Troubles the silver spring where England drinks,
Now will I dam up this thy yawning mouth
For swallowing the treasure of the realm.
Thy lips that kissed the Queen shall sweep the ground,
And thou that smiledst at good Duke Humphrey's
          death          76
Against the senseless winds shalt grin in vain,
Who in contempt shall hiss at thee again.
And wedded be thou to the hags of hell,
For daring to affy a mighty lord          80
Unto the daughter of a worthless king,
Having neither subject, wealth, nor diadem.
By devilish policy art thou grown great,
And like ambitious Sylla, overgorged
With gobbets of thy mother's bleeding heart.          85
By thee Anjou and Maine were sold to France,
The false revolting Normans, thorough thee,
Disdain to call us lord, and Picardy
Hath slain their governors, surprised our forts,
And sent the ragged soldiers, wounded, home.          90
The princely Warwick, and the Nevilles all,
Whose dreadful swords were never drawn in vain,
As hating thee, are rising up in arms;
And now the house of York, thrust from the crown,
By shameful murder of a guiltless king          95
And lofty, proud, encroaching tyranny,
Burns with revenging fire, whose hopeful colours
Advance our half-faced sun, striving to shine,
Under the which is writ, 'Invitis nubibus'.
The commons here in Kent are up in arms,          100
And, to conclude, reproach and beggary
Is crept into the palace of our King,

And all by thee. (To Whitmore) Away, convey him
          hence.
SUFFOLK
O that I were a god, to shoot forth thunder
Upon these paltry, servile, abject drudges.          105
Small things make base men proud. This villain here,
Being captain of a pinnace, threatens more
Than Bargulus, the strong Illyrian pirate.
Drones suck not eagles' blood, but rob beehives.
It is impossible that I should die          110
By such a lowly vassal as thyself.
Thy words move rage, and not remorse in me.
[CAPTAIN]
But my deeds, Suffolk, soon shall stay thy rage.
SUFFOLK
I go of message from the Queen to France—
I charge thee, waft me safely cross the Channel!          115
CAPTAIN  Walter—
WHITMORE
Come, Suffolk, I must waft thee to thy death.
SUFFOLK
Paene gelidus timor occupat artus—
It is thee I fear.
WHITMORE
Thou shalt have cause to fear before I leave thee.          120
What, are ye daunted now? Now will ye stoop?
FIRST GENTLEMAN (to Suffolk)
My gracious lord, entreat him—speak him fair.
SUFFOLK
Suffolk's imperial tongue is stern and rough,
Used to command, untaught to plead for favour.
Far be it we should honour such as these          125
With humble suit. No, rather let my head
Stoop to the block than these knees bow to any
Save to the God of heaven and to my king;
And sooner dance upon a bloody pole
Than stand uncovered to the vulgar groom.          130
True nobility is exempt from fear;
More can I bear than you dare execute.
CAPTAIN
Hale him away, and let him talk no more.
SUFFOLK
Come, 'soldiers', show what cruelty ye can,
That this my death may never be forgot.          135
Great men oft die by vile Besonians;
A Roman sworder and banditto slave
Murdered sweet Tully; Brutus' bastard hand
Stabbed Julius Caesar; savage islanders
Pompey the Great; and Suffolk dies by pirates.          140
          Exit Whitmore with Suffolk
CAPTAIN
And as for these whose ransom we have set,
It is our pleasure one of them depart.
(To the Second Gentleman)
Therefore, come you with us and (to his men, pointing
          to the First Gentleman) let him go.
          Exeunt all but the First Gentleman
          Enter Whitmore with Suffolk's head and body
WHITMORE
There let his head and lifeless body lie,
Until the Queen his mistress bury it.          Exit
FIRST GENTLEMAN
O barbarous and bloody spectacle!          146
His body will I bear unto the King.
If he revenge it not, yet will his friends;
So will the Queen, that living held him dear.
          Exit with Suffolk's head and body

**4.2**    *Enter two Rebels ⌈with long staves⌉*

FIRST REBEL Come and get thee a sword, though made of a lath; they have been up these two days.

SECOND REBEL They have the more need to sleep now then.                                                                                    4

FIRST REBEL I tell thee, Jack Cade the clothier means to dress the commonwealth, and turn it, and set a new nap upon it.

SECOND REBEL So he had need, for 'tis threadbare. Well, I say it was never merry world in England since gentlemen came up.                                                                10

FIRST REBEL O, miserable age! Virtue is not regarded in handicraftsmen.

SECOND REBEL The nobility think scorn to go in leather aprons.

FIRST REBEL Nay more, the King's Council are no good workmen.                                                                              16

SECOND REBEL True; and yet it is said 'Labour in thy vocation'; which is as much to say as 'Let the magistrates be labouring men'; and therefore should we be magistrates.                                                         20

FIRST REBEL Thou hast hit it; for there's no better sign of a brave mind than a hard hand.

SECOND REBEL I see them! I see them! There's Best's son, the tanner of Wingham—

FIRST REBEL He shall have the skins of our enemies to make dog's leather of.                                                         26

SECOND REBEL And Dick the butcher—

FIRST REBEL Then is sin struck down like an ox, and iniquity's throat cut like a calf.

SECOND REBEL And Smith the weaver—                                    30

FIRST REBEL Argo, their thread of life is spun.

SECOND REBEL Come, come, let's fall in with them.
  *Enter Jack Cade, Dick the Butcher, Smith the*
  *Weaver, a sawyer, ⌈and a drummer,⌉ with infinite*
  *numbers, ⌈all with long staves⌉*

CADE We, John Cade, so termed of our supposed father—

BUTCHER (*to his fellows*) Or rather of stealing a cade of herrings.                                                                       35

CADE For our enemies shall fall before us, inspired with the spirit of putting down kings and princes—command silence!

BUTCHER Silence!

CADE My father was a Mortimer—                                          40

BUTCHER (*to his fellows*) He was an honest man and a good bricklayer.

CADE My mother a Plantagenet—

BUTCHER (*to his fellows*) I knew her well, she was a midwife.

CADE My wife descended of the Lacys—                                    45

BUTCHER (*to his fellows*) She was indeed a pedlar's daughter and sold many laces.

WEAVER (*to his fellows*) But now of late, not able to travel with her furred pack, she washes bucks here at home.

CADE Therefore am I of an honourable house.                            50

BUTCHER (*to his fellows*) Ay, by my faith, the field is honourable, and there was he born, under a hedge; for his father had never a house but the cage.

CADE Valiant I am—

WEAVER (*to his fellows*) A must needs, for beggary is valiant.                                                                           56

CADE I am able to endure much—

BUTCHER (*to his fellows*) No question of that, for I have seen him whipped three market days together.

CADE I fear neither sword nor fire.                                     60

WEAVER (*to his fellows*) He need not fear the sword, for his coat is of proof.

BUTCHER (*to his fellows*) But methinks he should stand in fear of fire, being burned i'th' hand for stealing of sheep.                                                                          65

CADE Be brave, then, for your captain is brave and vows reformation. There shall be in England seven halfpenny loaves sold for a penny, the three-hooped pot shall have ten hoops, and I will make it felony to drink small beer. All the realm shall be in common, and in Cheapside shall my palfrey go to grass. And when I am king, as king I will be—                                                         72

ALL CADE'S FOLLOWERS God save your majesty!

CADE I thank you good people!—there shall be no money. All shall eat and drink on my score, and I will apparel them all in one livery that they may agree like brothers, and worship me their lord.                                              77

BUTCHER The first thing we do let's kill all the lawyers.

CADE Nay, that I mean to do. Is not this a lamentable thing that of the skin of an innocent lamb should be made parchment? That parchment, being scribbled o'er, should undo a man? Some say the bee stings, but I say 'tis the bee's wax. For I did but seal once to a thing, and I was never mine own man since. How now? Who's there?                                                         85
  *Enter some bringing forth the Clerk of Chatham*

WEAVER The Clerk of Chatham—he can write and read and cast account.

CADE O, monstrous!

WEAVER We took him setting of boys' copies.

CADE Here's a villain.                                                  90

WEAVER He's a book in his pocket with red letters in't.

CADE Nay, then he is a conjuror!

BUTCHER Nay, he can make obligations and write court hand.                                                                             94

CADE I am sorry for't. The man is a proper man, of mine honour. Unless I find him guilty, he shall not die. Come hither, sirrah, I must examine thee. What is thy name?

CLERK Emmanuel.

BUTCHER They use to write that on the top of letters—'twill go hard with you.                                                     100

CADE Let me alone. (*To the Clerk*) Dost thou use to write thy name? Or hast thou a mark to thyself like an honest plain-dealing man?

CLERK Sir, I thank God I have been so well brought up that I can write my name.                                                    105

ALL CADE'S FOLLOWERS He hath confessed—away with him! He's a villain and a traitor.

CADE Away with him, I say, hang him with his pen and inkhorn about his neck.               *Exit one with the Clerk*
  *Enter a Messenger*

MESSENGER Where's our general?                                         110

CADE Here I am, thou particular fellow.

MESSENGER Fly, fly, fly! Sir Humphrey Stafford and his brother are hard by with the King's forces.

CADE Stand, villain, stand—or I'll fell thee down. He shall be encountered with a man as good as himself. He is but a knight, is a?                                                     116

MESSENGER No.

CADE To equal him I will make myself a knight presently.
  *He kneels and knights himself*
Rise up, Sir John Mortimer.
  *He rises*
Now have at him!                                                       120
  *Enter Sir Humphrey Stafford and his brother, with*
  *a drummer and soldiers*

STAFFORD (*to Cade's followers*)
  Rebellious hinds, the filth and scum of Kent,
  Marked for the gallows, lay your weapons down;

Home to your cottages, forsake this groom.
The King is merciful, if you revolt.
STAFFORD'S BROTHER (*to Cade's followers*)
But angry, wrathful, and inclined to blood,          125
If you go forward. Therefore, yield or die.
CADE (*to his followers*)
As for these silken-coated slaves, I pass not.
It is to you, good people, that I speak,
Over whom, in time to come, I hope to reign—
For I am rightful heir unto the crown.          130
STAFFORD
Villain, thy father was a plasterer
And thou thyself a shearman, art thou not?
CADE
And Adam was a gardener.
STAFFORD'S BROTHER          And what of that?
CADE
Marry, this: Edmund Mortimer, Earl of March,
Married the Duke of Clarence' daughter, did he not?
STAFFORD Ay, sir.          136
CADE
By her he had two children at one birth.
STAFFORD'S BROTHER That's false.
CADE
Ay, there's the question—but I say 'tis true.
The elder of them, being put to nurse,          140
Was by a beggar-woman stol'n away,
And, ignorant of his birth and parentage,
Became a bricklayer when he came to age.
His son am I—deny it an you can.
BUTCHER
Nay, 'tis too true—therefore he shall be king.          145
WEAVER Sir, he made a chimney in my father's house,
and the bricks are alive at this day to testify. Therefore
deny it not.
STAFFORD (*to Cade's followers*)
And will you credit this base drudge's words
That speaks he knows not what?          150
ALL CADE'S FOLLOWERS
Ay, marry, will we—therefore get ye gone.
STAFFORD'S BROTHER
Jack Cade, the Duke of York hath taught you this.
CADE (*aside*)
He lies, for I invented it myself.
(*Aloud*) Go to, sirrah—tell the King from me that for
his father's sake, Henry the Fifth, in whose time boys
went to span-counter for French crowns, I am content
he shall reign; but I'll be Protector over him.          157
BUTCHER And, furthermore, we'll have the Lord Saye's
head for selling the dukedom of Maine.
CADE And good reason, for thereby is England maimed,
and fain to go with a staff, but that my puissance holds
it up. Fellow-kings, I tell you that that Lord Saye hath
gelded the commonwealth, and made it an eunuch,
and, more than that, he can speak French, and
therefore he is a traitor!          165
STAFFORD
O gross and miserable ignorance!
CADE Nay, answer if you can: the Frenchmen are our
enemies; go to, then, I ask but this—can he that speaks
with the tongue of an enemy be a good counsellor or
no?          170
ALL CADE'S FOLLOWERS No, no—and therefore we'll have
his head!
STAFFORD'S BROTHER (*to Stafford*)
Well, seeing gentle words will not prevail,
Assail them with the army of the King.

STAFFORD
Herald, away, and throughout every town          175
Proclaim them traitors that are up with Cade;
That those which fly before the battle ends
May, even in their wives' and children's sight,
Be hanged up for example at their doors.
And you that be the King's friends, follow me!          180
*Exeunt ⌐the Staffords and their soldiers¬*
CADE
And you that love the commons, follow me!
Now show yourselves men—'tis for liberty.
We will not leave one lord, one gentleman—
Spare none but such as go in clouted shoon,
For they are thrifty honest men, and such          185
As would, but that they dare not, take our parts.
BUTCHER They are all in order, and march toward us.
CADE
But then are we in order when we are
Most out of order. Come, march forward!          ⌐*Exeunt*¬

**4.3**  *Alarums to the fight; ⌐excursions,¬ wherein both
the Staffords are slain. Enter Jack Cade, Dick the
Butcher, and the rest*
CADE Where's Dick, the butcher of Ashford?
BUTCHER Here, sir.
CADE They fell before thee like sheep and oxen, and thou
behaved'st thyself as if thou hadst been in thine own
slaughterhouse. Therefore, thus will I reward thee—
the Lent shall be as long again as it is. Thou shalt have
licence to kill for a hundred, lacking one.          7
BUTCHER I desire no more.
CADE And to speak truth, thou deserv'st no less.
⌐*He apparels himself in the Staffords' armour*¬
This monument of the victory will I bear, and the
bodies shall be dragged at my horse heels till I do come
to London, where we will have the Mayor's sword
borne before us.
BUTCHER If we mean to thrive and do good, break open
the jails and let out the prisoners.          15
CADE Fear not that, I warrant thee. Come, let's march
towards London.
*Exeunt, ⌐dragging the Staffords' bodies¬*

**4.4**  *Enter King Henry ⌐reading¬ a supplication, Queen
Margaret carrying Suffolk's head, the Duke of
Buckingham, and the Lord Saye, ⌐with others¬*
QUEEN MARGARET ⌐*aside*¬
Oft have I heard that grief softens the mind,
And makes it fearful and degenerate;
Think, therefore, on revenge, and cease to weep.
But who can cease to weep and look on this?
Here may his head lie on my throbbing breast,          5
But where's the body that I should embrace?
BUCKINGHAM (*to King Henry*)
What answer makes your grace to the rebels'
supplication?
KING HENRY
I'll send some holy bishop to entreat,
For God forbid so many simple souls
Should perish by the sword. And I myself,          10
Rather than bloody war shall cut them short,
Will parley with Jack Cade their general.
But stay, I'll read it over once again.
*He reads*
QUEEN MARGARET (*to Suffolk's head*)
Ah, barbarous villains! Hath this lovely face
Ruled like a wandering planet over me,          15

And could it not enforce them to relent,
That were unworthy to behold the same?

KING HENRY
    Lord Saye, Jack Cade hath sworn to have thy head.

SAYE
    Ay, but I hope your highness shall have his.

KING HENRY (*to Queen Margaret*)
    How now, madam? Still lamenting and mourning    20
    Suffolk's death?
    I fear me, love, if that I had been dead,
    Thou wouldest not have mourned so much for me.

QUEEN MARGARET
    No, my love, I should not mourn, but die for thee.
    *Enter a Messenger, ⌈in haste⌉*

KING HENRY
    How now? What news? Why com'st thou in such
    haste?    25

MESSENGER
    The rebels are in Southwark—fly, my lord!
    Jack Cade proclaims himself Lord Mortimer,
    Descended from the Duke of Clarence' house,
    And calls your grace usurper, openly,
    And vows to crown himself in Westminster.    30
    His army is a ragged multitude
    Of hinds and peasants, rude and merciless.
    Sir Humphrey Stafford and his brother's death
    Hath given them heart and courage to proceed.
    All scholars, lawyers, courtiers, gentlemen,    35
    They call false caterpillars and intend their death.

KING HENRY
    O, graceless men; they know not what they do.

BUCKINGHAM
    My gracious lord, retire to Kenilworth
    Until a power be raised to put them down.

QUEEN MARGARET
    Ah, were the Duke of Suffolk now alive    40
    These Kentish rebels would be soon appeased!

KING HENRY
    Lord Saye, the trait'rous rabble hateth thee—
    Therefore away with us to Kenilworth.

SAYE
    So might your grace's person be in danger.
    The sight of me is odious in their eyes,    45
    And therefore in this city will I stay
    And live alone as secret as I may.
    *Enter another Messenger*

SECOND MESSENGER (*to King Henry*)
    Jack Cade hath almost gotten London Bridge;
    The citizens fly and forsake their houses;
    The rascal people, thirsting after prey,    50
    Join with the traitor; and they jointly swear
    To spoil the city and your royal court.

BUCKINGHAM (*to King Henry*)
    Then linger not, my lord; away, take horse!

KING HENRY
    Come, Margaret. God, our hope, will succour us.

QUEEN MARGARET ⌈*aside*⌉
    My hope is gone, now Suffolk is deceased.    55

KING HENRY (*to Saye*)
    Farewell, my lord. Trust not the Kentish rebels.

BUCKINGHAM (*to Saye*)
    Trust nobody, for fear you be betrayed.

SAYE
    The trust I have is in mine innocence,
    And therefore am I bold and resolute.
    *Exeunt ⌈Saye at one door, the rest at another⌉*

**4.5**    *Enter the Lord Scales upon the Tower, walking.*
        *Enter three or four Citizens below*

SCALES How now? Is Jack Cade slain?

FIRST CITIZEN No, my lord Scales, nor likely to be slain,
    for he and his men have won the bridge, killing all
    those that did withstand them. The Lord Mayor craveth
    aid of your honour from the Tower to defend the city
    from the rebels.    6

SCALES
    Such aid as I can spare you shall command,
    But I am troubled here with them myself.
    The rebels have essayed to win the Tower.
    Get you to Smithfield, there to gather head,    10
    And thither will I send you Matthew Gough.
    Fight for your king, your country, and your lives!
    And so, farewell, for I must hence again.
    *Exeunt, Scales above, the Citizens below*

**4.6**    *Enter Jack Cade, the Weaver, the Butcher, and the*
        *rest. Cade strikes his sword on London Stone*

CADE Now is Mortimer lord of this city. And, here sitting
    upon London Stone, I charge and command that, of
    the city's cost, the Pissing Conduit run nothing but
    claret wine this first year of our reign. And now
    henceforward it shall be treason for any that calls me
    otherwise than Lord Mortimer.    6
    *Enter a Soldier, running*

SOLDIER Jack Cade, Jack Cade!

CADE Zounds, knock him down there!
    *They kill him*

BUTCHER If this fellow be wise, he'll never call ye Jack
    Cade more; I think he hath a very fair warning.    10
    ⌈*He takes a paper from the soldier's body*
    *and reads it*⌉
    My lord, there's an army gathered together in
    Smithfield.

CADE Come then, let's go fight with them—but first, go
    on and set London Bridge afire, and, if you can, burn
    down the Tower too. Come, let's away.    *Exeunt*

**4.7**    *Alarums. ⌈Excursions, wherein⌉ Matthew Gough is*
        *slain, and all the rest of his men with him.*
        *Then enter Jack Cade with his company, among*
        *them the Butcher, the Weaver, and John, a rebel*

CADE So, sirs, now go some and pull down the Savoy;
    others to th' Inns of Court—down with them all.

BUTCHER I have a suit unto your lordship.

CADE Be it a lordship, thou shalt have it for that word.

BUTCHER Only that the laws of England may come out of
    your mouth.    6

JOHN (*aside to his fellows*) Mass, 'twill be sore law then,
    for he was thrust in the mouth with a spear, and 'tis
    not whole yet.

WEAVER (*aside to John*) Nay, John, it will be stinking law,
    for his breath stinks with eating toasted cheese.    11

CADE I have thought upon it—it shall be so. Away! Burn
    all the records of the realm. My mouth shall be the
    Parliament of England.

JOHN (*aside to his fellows*) Then we are like to have biting
    statutes unless his teeth be pulled out.    16

CADE And henceforward all things shall be in common.
    *Enter a Messenger*

MESSENGER My lord, a prize, a prize! Here's the Lord Saye
    which sold the towns in France. He that made us pay
    one-and-twenty fifteens and one shilling to the pound
    the last subsidy.    21

*Enter a rebel with the Lord Saye*

CADE Well, he shall be beheaded for it ten times. (*To Saye*) Ah, thou say, thou serge—nay, thou buckram lord! Now art thou within point-blank of our jurisdiction regal. What canst thou answer to my majesty for giving up of Normandy unto Mounsieur Basimecu, the Dauphin of France? Be it known unto thee by these presence, even the presence of Lord Mortimer, that I am the besom that must sweep the court clean of such filth as thou art. Thou hast most traitorously corrupted the youth of the realm in erecting a grammar school; and, whereas before, our forefathers had no other books but the score and the tally, thou hast caused printing to be used and, contrary to the King his crown and dignity, thou hast built a paper-mill. It will be proved to thy face that thou hast men about thee that usually talk of a noun and a verb and such abominable words as no Christian ear can endure to hear. Thou hast appointed justices of peace to call poor men before them about matters they were not able to answer. Moreover, thou hast put them in prison, and, because they could not read, thou hast hanged them when indeed only for that cause they have been most worthy to live. Thou dost ride on a foot-cloth, dost thou not? 45

SAYE What of that?

CADE Marry, thou ought'st not to let thy horse wear a cloak when honester men than thou go in their hose and doublets.

BUTCHER And work in their shirts, too; as myself, for example, that am a butcher. 50

SAYE You men of Kent.

BUTCHER What say you of Kent?

SAYE
Nothing but this—'tis *bona terra, mala gens*.

CADE *Bonum terrum*—zounds, what's that?

BUTCHER He speaks French. 55

⌈FIRST REBEL⌉ No, 'tis Dutch.

⌈SECOND REBEL⌉ No, 'tis Out-talian, I know it well enough.

SAYE
Hear me but speak, and bear me where you will.
Kent, in the commentaries Caesar writ,
Is termed the civil'st place of all this isle; 60
Sweet is the country, because full of riches;
The people liberal, valiant, active, wealthy;
Which makes me hope you are not void of pity.
I sold not Maine, I lost not Normandy;
Yet to recover them would lose my life. 65
Justice with favour have I always done,
Prayers and tears have moved me—gifts could never.
When have I aught exacted at your hands,
But to maintain the King, the realm, and you?
Large gifts have I bestowed on learnèd clerks 70
Because my book preferred me to the King,
And seeing ignorance is the curse of God,
Knowledge the wing wherewith we fly to heaven.
Unless you be possessed with devilish spirits,
You cannot but forbear to murder me. 75
This tongue hath parleyed unto foreign kings
For your behoof—

CADE Tut, when struck'st thou one blow in the field?

SAYE
Great men have reaching hands. Oft have I struck
Those that I never saw, and struck them dead. 80

REBEL O monstrous coward! What, to come behind folks?

SAYE
These cheeks are pale for watching for your good—

CADE Give him a box o'th' ear, and that will make 'em red again.

⌈*One of the rebels strikes Saye*⌉

SAYE
Long sitting to determine poor men's causes 85
Hath made me full of sickness and diseases.

CADE Ye shall have a hempen caudle, then, and the health o'th' hatchet.

BUTCHER (*to Saye*) Why dost thou quiver, man?

SAYE
The palsy, and not fear, provokes me. 90

CADE Nay, he nods at us as who should say 'I'll be even with you'. I'll see if his head will stand steadier on a pole or no. Take him away, and behead him.

SAYE
Tell me wherein have I offended most?
Have I affected wealth or honour? Speak. 95
Are my chests filled up with extorted gold?
Is my apparel sumptuous to behold?
Whom have I injured, that ye seek my death?
These hands are free from guiltless bloodshedding,
This breast from harbouring foul deceitful thoughts.
O let me live! 101

CADE (*aside*) I feel remorse in myself with his words, but I'll bridle it. He shall die an it be but for pleading so well for his life. (*Aloud*) Away with him—he has a familiar under his tongue; he speaks not a God's name. Go, take him away, I say, to the Standard in Cheapside, and strike off his head presently; and then go to Mile End Green—break into his son-in-law's house, Sir James Cromer, and strike off his head, and bring them both upon two poles hither. 110

ALL CADE'S FOLLOWERS It shall be done!

SAYE
Ah, countrymen, if, when you make your prayers,
God should be so obdurate as yourselves,
How would it fare with your departed souls?
And therefore yet relent and save my life! 115

CADE Away with him, and do as I command ye!

*Exeunt* ⌈*the Butcher and*⌉ *one or two with the Lord Saye*

The proudest peer in the realm shall not wear a head on his shoulders unless he pay me tribute. There shall not a maid be married but she shall pay to me her maidenhead, ere they have it. Married men shall hold of me *in capite*. And we charge and command that their wives be as free as heart can wish or tongue can tell.

*Enter a Rebel*

REBEL O captain, London Bridge is afire!

CADE Run to Billingsgate and fetch pitch and flax and quench it. 125

*Enter the Butcher and a Sergeant*

SERGEANT Justice, justice, I pray you, sir, let me have justice of this fellow here.

CADE Why, what has he done?

SERGEANT Alas, sir, he has ravished my wife.

BUTCHER (*to Cade*) Why, my lord, he would have 'rested me and I went and entered my action in his wife's proper house. 132

CADE Dick, follow thy suit in her common place. (*To the Sergeant*) You whoreson villain, you are a sergeant—you'll take any man by the throat for twelve pence, and 'rest a man when he's at dinner, and have him to prison ere the meat be out of his mouth. (*To the Butcher*) Go, Dick, take him hence: cut out his tongue for cogging, hough him for running, and, to conclude, brain him with his own mace.                                    140

*Exit the Butcher with the Sergeant*

REBEL My lord, when shall we go to Cheapside and take up commodities upon our bills?

CADE Marry, presently. He that will lustily stand to it shall go with me and take up these commodities following—item, a gown, a kirtle, a petticoat, and a smock.                                                            146

ALL CADE'S FOLLOWERS O brave!

*Enter two with the Lord Saye's head and Sir James Cromer's upon two poles*

CADE But is not this braver? Let them kiss one another, for they loved well when they were alive.                   149

⌈*The two heads are made to kiss*⌉

Now part them again, lest they consult about the giving up of some more towns in France. Soldiers, defer the spoil of the city until night. For with these borne before us instead of maces will we ride through the streets, and at every corner have them kiss. Away!        154

⌈*Exeunt two with the heads. The others begin to follow*⌉

Up Fish Street! Down Saint Magnus' Corner! Kill and knock down! Throw them into Thames!

*Sound a parley*

What noise is this? Dare any be so bold to sound retreat or parley when I command them kill?

*Enter the Duke of Buckingham and old Lord Clifford*

BUCKINGHAM

Ay, here they be that dare and will disturb thee!
Know, Cade, we come ambassadors from the King    160
Unto the commons, whom thou hast misled,
And here pronounce free pardon to them all
That will forsake thee and go home in peace.

CLIFFORD

What say ye, countrymen, will ye relent
And yield to mercy whilst 'tis offered you,       165
Or let a rebel lead you to your deaths?
Who loves the King and will embrace his pardon,
Fling up his cap and say 'God save his majesty'.
Who hateth him and honours not his father,
Henry the Fifth, that made all France to quake,    170
Shake he his weapon at us, and pass by.

*They* ⌈*fling up their caps and*⌉ *forsake Cade*

ALL CADE'S FOLLOWERS God save the King! God save the King!

CADE What, Buckingham and Clifford, are ye so brave? (*To the rabble*) And you, base peasants, do ye believe him? Will you needs be hanged with your pardons about your necks? Hath my sword, therefore, broke through London gates that you should leave me at the White Hart in Southwark? I thought ye would never have given out these arms till you had recovered your ancient freedom. But you are all recreants and dastards, and delight to live in slavery to the nobility. Let them break your backs with burdens, take your houses over your heads, ravish your wives and daughters before your faces. For me, I will make shift for one, and so God's curse light upon you all.                      186

ALL CADE'S FOLLOWERS We'll follow Cade! We'll follow Cade!

*They run to Cade again*

CLIFFORD

Is Cade the son of Henry the Fifth
That thus you do exclaim you'll go with him?       190
Will he conduct you through the heart of France

And make the meanest of you earls and dukes?
Alas, he hath no home, no place to fly to,
Nor knows he how to live but by the spoil—
Unless by robbing of your friends and us.          195
Were't not a shame that whilst you live at jar
The fearful French, whom you late vanquishèd,
Should make a start o'er seas and vanquish you?
Methinks already in this civil broil
I see them lording it in London streets,           200
Crying '*Villiago!*' unto all they meet.
Better ten thousand base-born Cades miscarry
Than you should stoop unto a Frenchman's mercy.
To France! To France! And get what you have lost!
Spare England, for it is your native coast.        205
Henry hath money; you are strong and manly;
God on our side, doubt not of victory.

ALL CADE'S FOLLOWERS A Clifford! A Clifford! We'll follow the King and Clifford!

*They forsake Cade*

CADE (*aside*) Was ever feather so lightly blown to and fro as this multitude? The name of Henry the Fifth hales them to an hundred mischiefs, and makes them leave me desolate. I see them lay their heads together to surprise me. My sword make way for me, for here is no staying. (*Aloud*) In despite of the devils and hell, have through the very middest of you! And heavens and honour be witness that no want of resolution in me, but only my followers' base and ignominious treasons, makes me betake me to my heels.

*He runs through them with his staff, and flies away*

BUCKINGHAM

What, is he fled? Go, some, and follow him,        220
And he that brings his head unto the King
Shall have a thousand crowns for his reward.

*Exeunt some of them after Cade*

(*To the remaining rebels*)
Follow us, soldiers, we'll devise a mean
To reconcile you all unto the King.          *Exeunt*

4.8   *Sound trumpets. Enter King Henry, Queen Margaret, and the Duke of Somerset on the terrace*

KING HENRY

Was ever King that joyed an earthly throne
And could command no more content than I?
No sooner was I crept out of my cradle
But I was made a king at nine months old.
Was never subject longed to be a king             5
As I do long and wish to be a subject.

*Enter the Duke of Buckingham and Lord Clifford* ⌈*on the terrace*⌉

BUCKINGHAM (*to King Henry*)
Health and glad tidings to your majesty.

KING HENRY

Why, Buckingham, is the traitor Cade surprised?
Or is he but retired to make him strong?

*Enter, below, multitudes with halters about their necks*

CLIFFORD

He is fled, my lord, and all his powers do yield,  10
And humbly thus with halters on their necks
Expect your highness' doom of life or death.

KING HENRY

Then, heaven, set ope thy everlasting gates
To entertain my vows of thanks and praise.
(*To the multitudes below*)
Soldiers, this day have you redeemed your lives,  15

And showed how well you love your prince and
    country.
Continue still in this so good a mind,
And Henry, though he be infortunate,
Assure yourselves will never be unkind.
And so, with thanks and pardon to you all,    20
I do dismiss you to your several countries.
ALL CADE'S FORMER FOLLOWERS God save the King! God
    save the King!    ⌜*Exeunt multitudes below*⌝
    *Enter a Messenger* ⌜*on the terrace*⌝
MESSENGER (*to King Henry*)
Please it your grace to be advertisèd
The Duke of York is newly come from Ireland,    25
And with a puissant and a mighty power
Of galloglasses and stout Irish kerns
Is marching hitherward in proud array,
And still proclaimeth, as he comes along,
His arms are only to remove from thee    30
The Duke of Somerset, whom he terms a traitor.
KING HENRY
Thus stands my state, 'twixt Cade and York distressed,
Like to a ship that, having scaped a tempest,
Is straightway calmed and boarded with a pirate.
But now is Cade driven back, his men dispersed,    35
And now is York in arms to second him.
I pray thee, Buckingham, go and meet him,
And ask him what's the reason of these arms.
Tell him I'll send Duke Edmund to the Tower;
And, Somerset, we will commit thee thither,    40
Until his army be dismissed from him.
SOMERSET
My lord, I'll yield myself to prison willingly,
Or unto death, to do my country good.
KING HENRY (*to Buckingham*)
In any case, be not too rough in terms,
For he is fierce and cannot brook hard language.    45
BUCKINGHAM
I will, my lord, and doubt not so to deal
As all things shall redound unto your good.
KING HENRY
Come, wife, let's in and learn to govern better;
For yet may England curse my wretched reign.
    *Flourish. Exeunt*

**4.9**    *Enter Jack Cade*
CADE Fie on ambitions; fie on myself that have a sword
and yet am ready to famish. These five days have I hid
me in these woods and durst not peep out, for all the
country is laid for me. But now am I so hungry that if
I might have a lease of my life for a thousand years, I
could stay no longer. Wherefore o'er a brick wall have
I climbed into this garden to see if I can eat grass or
pick a sallet another while, which is not amiss to cool
a man's stomach this hot weather. And I think this
word 'sallet' was born to do me good; for many a time,
but for a sallet, my brain-pan had been cleft with a
brown bill; and many a time, when I have been dry,
and bravely marching, it hath served me instead of a
quart pot to drink in; and now the word 'sallet' must
serve me to feed on.    15
    ⌜*He lies down picking of herbs and eating them.*⌝
    *Enter Sir Alexander Iden* ⌜*and five of his men*⌝
IDEN
Lord, who would live turmoilèd in the court
And may enjoy such quiet walks as these?
This small inheritance my father left me

Contenteth me, and worth a monarchy.
I seek not to wax great by others' waning,    20
Or gather wealth I care not with what envy;
Sufficeth that I have maintains my state,
And sends the poor well pleasèd from my gate.
    ⌜*Cade rises to his knees*⌝
CADE (*aside*) Zounds, here's the lord of the soil come to
seize me for a stray for entering his fee-simple without
leave. (*To Iden*) A villain, thou wilt betray me and get
a thousand crowns of the king by carrying my head
to him; but I'll make thee eat iron like an ostrich and
swallow my sword like a great pin, ere thou and I part.
IDEN
Why, rude companion, whatsoe'er thou be,    30
I know thee not. Why then should I betray thee?
Is't not enough to break into my garden,
And, like a thief, to come to rob my grounds,
Climbing my walls in spite of me the owner,
But thou wilt brave me with these saucy terms?    35
CADE Brave thee? Ay, by the best blood that ever was
broached—and beard thee too! Look on me well—I
have eat no meat these five days, yet come thou and
thy five men, an if I do not leave you all as dead as a
doornail I pray God I may never eat grass more.    40
IDEN
Nay, it shall ne'er be said while England stands
That Alexander Iden, an esquire of Kent,
Took odds to combat a poor famished man.
Oppose thy steadfast gazing eyes to mine—
See if thou canst outface me with thy looks.    45
Set limb to limb, and thou art far the lesser—
Thy hand is but a finger to my fist,
Thy leg a stick comparèd with this truncheon.
My foot shall fight with all the strength thou hast,
And if mine arm be heavèd in the air,    50
Thy grave is digged already in the earth.
As for words, whose greatness answers words,
Let this my sword report what speech forbears.
(*To his men*) Stand you all aside.
CADE By my valour, the most complete champion that
ever I heard. (*To his sword*) Steel, if thou turn the edge
or cut not out the burly-boned clown in chines of beef
ere thou sleep in thy sheath, I beseech God on my
knees thou mayst be turned to hobnails.    59
    ⌜*Cade stands.*⌝ *Here they fight, and Cade falls down*
O, I am slain! Famine and no other hath slain me! Let
ten thousand devils come against me, and give me but
the ten meals I have lost, and I'd defy them all. Wither,
garden, and be henceforth a burying place to all that
do dwell in this house, because the unconquered soul
of Cade is fled.    65
IDEN
Is't Cade that I have slain, that monstrous traitor?
Sword, I will hallow thee for this thy deed
And hang thee o'er my tomb when I am dead.
Ne'er shall this blood be wipèd from thy point
But thou shalt wear it as a herald's coat    70
To emblaze the honour that thy master got.
CADE Iden, farewell, and be proud of thy victory. Tell
Kent from me she hath lost her best man, and exhort
all the world to be cowards. For I, that never feared
any, am vanquished by famine, not by valour.    75
    *He dies*
IDEN
How much thou wrong'st me, heaven be my judge.
Die, damnèd wretch, the curse of her that bore thee!

And ⌐stabbing him again⌐ as I thrust thy body in with
    my sword,
So wish I I might thrust thy soul to hell.
Hence will I drag thee headlong by the heels          80
Unto a dunghill, which shall be thy grave,
And there cut off thy most ungracious head,
Which I will bear in triumph to the King,
Leaving thy trunk for crows to feed upon.
                                    *Exeunt with the body*

5.1    *Enter the Duke of York and his army of Irish with a*
       *drummer and soldiers bearing colours*

YORK
    From Ireland thus comes York to claim his right,
    And pluck the crown from feeble Henry's head.
    Ring, bells, aloud; burn, bonfires, clear and bright,
    To entertain great England's lawful king.
    Ah, *sancta maiestas*! Who would not buy thee dear?   5
    Let them obey that knows not how to rule;
    This hand was made to handle naught but gold.
    I cannot give due action to my words,
    Except a sword or sceptre balance it.
    A sceptre shall it have, have I a sword,              10
    On which I'll toss the fleur-de-lis of France.
                *Enter the Duke of Buckingham*
    (*Aside*) Whom have we here? Buckingham to disturb
        me?
    The King hath sent him sure—I must dissemble.
BUCKINGHAM
    York, if thou meanest well, I greet thee well.
YORK
    Humphrey of Buckingham, I accept thy greeting.       15
    Art thou a messenger, or come of pleasure?
BUCKINGHAM
    A messenger from Henry, our dread liege,
    To know the reason of these arms in peace;
    Or why thou, being a subject as I am,
    Against thy oath and true allegiance sworn,          20
    Should raise so great a power without his leave,
    Or dare to bring thy force so near the court?
YORK (*aside*)
    Scarce can I speak, my choler is so great.
    O, I could hew up rocks and fight with flint,
    I am so angry at these abject terms;                 25
    And now, like Ajax Telamonius,
    On sheep or oxen could I spend my fury.
    I am far better born than is the King,
    More like a king, more kingly in my thoughts;
    But I must make fair weather yet a while,            30
    Till Henry be more weak and I more strong.
    (*Aloud*) Buckingham, I prithee pardon me,
    That I have given no answer all this while;
    My mind was troubled with deep melancholy.
    The cause why I have brought this army hither        35
    Is to remove proud Somerset from the King,
    Seditious to his grace and to the state.
BUCKINGHAM
    That is too much presumption on thy part;
    But if thy arms be to no other end,
    The King hath yielded unto thy demand:               40
    The Duke of Somerset is in the Tower.
YORK
    Upon thine honour, is he prisoner?
BUCKINGHAM
    Upon mine honour, he is prisoner.

YORK
    Then, Buckingham, I do dismiss my powers.
    Soldiers, I thank you all; disperse yourselves;      45
    Meet me tomorrow in Saint George's field.
    You shall have pay and everything you wish.
                                    *Exeunt soldiers*
    (*To Buckingham*) And let my sovereign, virtuous
        Henry,
    Command my eldest son—nay, all my sons—
    As pledges of my fealty and love.                    50
    I'll send them all as willing as I live.
    Lands, goods, horse, armour, anything I have
    Is his to use, so Somerset may die.
BUCKINGHAM
    York, I commend this kind submission.
    We twain will go into his highness' tent.            55
                *Enter King Henry and attendants*
KING HENRY
    Buckingham, doth York intend no harm to us,
    That thus he marcheth with thee arm in arm?
YORK
    In all submission and humility
    York doth present himself unto your highness.
KING HENRY
    Then what intends these forces thou dost bring?      60
YORK
    To heave the traitor Somerset from hence,
    And fight against that monstrous rebel Cade,
    Who since I heard to be discomfited.
                *Enter Iden with Cade's head*
IDEN
    If one so rude and of so mean condition
    May pass into the presence of a king,                65
    ⌐Kneeling⌐ Lo, I present your grace a traitor's head,
    The head of Cade, whom I in combat slew.
KING HENRY
    The head of Cade? Great God, how just art thou!
    O let me view his visage, being dead,
    That living wrought me such exceeding trouble.       70
    Tell me, my friend, art thou the man that slew him?
IDEN ⌐rising⌐
    Iwis, an't like your majesty.
KING HENRY
    How art thou called? And what is thy degree?
IDEN
    Alexander Iden, that's my name;
    A poor esquire of Kent that loves his king.          75
BUCKINGHAM (*to King Henry*)
    So please it you, my lord, 'twere not amiss
    He were created knight for his good service.
KING HENRY
    Iden, kneel down.
                *Iden kneels and King Henry knights him*
                        Rise up a knight.
                *Iden rises*
    We give thee for reward a thousand marks,
    And will that thou henceforth attend on us.          80
IDEN
    May Iden live to merit such a bounty,
    And never live but true unto his liege.        ⌐*Exit*⌐
                *Enter Queen Margaret and the Duke of Somerset*
KING HENRY
    See, Buckingham, Somerset comes wi'th' Queen.
    Go bid her hide him quickly from the Duke.
QUEEN MARGARET
    For thousand Yorks he shall not hide his head,       85
    But boldly stand and front him to his face.

YORK
How now? Is Somerset at liberty?
Then, York, unloose thy long imprisoned thoughts,
And let thy tongue be equal with thy heart.
Shall I endure the sight of Somerset?                    90
False King, why hast thou broken faith with me,
Knowing how hardly I can brook abuse?
'King' did I call thee? No, thou art not king;
Not fit to govern and rule multitudes,
Which dar'st not—no, nor canst not—rule a traitor.
That head of thine doth not become a crown;              96
Thy hand is made to grasp a palmer's staff,
And not to grace an aweful princely sceptre.
That gold must round engird these brows of mine,
Whose smile and frown, like to Achilles' spear,         100
Is able with the change to kill and cure.
Here is a hand to hold a sceptre up,
And with the same to act controlling laws.
Give place! By heaven, thou shalt rule no more
O'er him whom heaven created for thy ruler.             105
SOMERSET
O monstrous traitor! I arrest thee, York,
Of capital treason 'gainst the King and crown.
Obey, audacious traitor; kneel for grace.
YORK (to an attendant)
Sirrah, call in my sons to be my bail.     Exit attendant
I know, ere they will have me go to ward,               110
They'll pawn their swords for my enfranchisement.
QUEEN MARGARET ⌈to Buckingham⌉
Call hither Clifford; bid him come amain,
To say if that the bastard boys of York
Shall be the surety for their traitor father.
                                   Exit ⌈Buckingham⌉
YORK
O blood-bespotted Neapolitan,                           115
Outcast of Naples, England's bloody scourge!
The sons of York, thy betters in their birth,
Shall be their father's bail, and bane to those
That for my surety will refuse the boys.
    Enter ⌈at one door⌉ York's sons Edward and
    crookback Richard ⌈with a drummer and soldiers⌉
See where they come. I'll warrant they'll make it good.
    Enter ⌈at the other door⌉ Clifford ⌈and his son, with
    a drummer and soldiers⌉
QUEEN MARGARET
And here comes Clifford to deny their bail.             121
CLIFFORD (kneeling before King Henry)
Health and all happiness to my lord the King.
    He rises
YORK
I thank thee, Clifford. Say, what news with thee?
Nay, do not fright us with an angry look—
We are thy sovereign, Clifford; kneel again.            125
For thy mistaking so, we pardon thee.
CLIFFORD
This is my king, York; I do not mistake.
But thou mistakes me much to think I do.
(To King Henry)
To Bedlam with him! Is the man grown mad?
KING HENRY
Ay, Clifford, a bedlam and ambitious humour            130
Makes him oppose himself against his king.
CLIFFORD
He is a traitor; let him to the Tower,
And chop away that factious pate of his.

QUEEN MARGARET
He is arrested, but will not obey.
His sons, he says, shall give their words for him.      135
YORK (to Edward and Richard) Will you not, sons?
EDWARD
Ay, noble father, if our words will serve.
RICHARD
And if words will not, then our weapons shall.
CLIFFORD
Why, what a brood of traitors have we here!
YORK
Look in a glass, and call thy image so.                 140
I am thy king, and thou a false-heart traitor.
Call hither to the stake my two brave bears,
That with the very shaking of their chains,
They may astonish these fell-lurking curs.
(To an attendant)
Bid Salisbury and Warwick come to me.                   145
                                Exit attendant
    Enter the Earls of Warwick and Salisbury ⌈with a
    drummer and soldiers⌉
CLIFFORD
Are these thy bears? We'll bait thy bears to death,
And manacle the bearherd in their chains,
If thou dar'st bring them to the baiting place.
RICHARD
Oft have I seen a hot o'erweening cur
Run back and bite, because he was withheld;            150
Who, being suffered with the bear's fell paw,
Hath clapped his tail between his legs and cried;
And such a piece of service will you do,
If you oppose yourselves to match Lord Warwick.
CLIFFORD
Hence, heap of wrath, foul indigested lump,            155
As crooked in thy manners as thy shape!
YORK
Nay, we shall heat you thoroughly anon.
CLIFFORD
Take heed, lest by your heat you burn yourselves.
KING HENRY
Why, Warwick, hath thy knee forgot to bow?
Old Salisbury, shame to thy silver hair,               160
Thou mad misleader of thy brainsick son!
What, wilt thou on thy deathbed play the ruffian,
And seek for sorrow with thy spectacles?
O, where is faith? O, where is loyalty?
If it be banished from the frosty head,                165
Where shall it find a harbour in the earth?
Wilt thou go dig a grave to find out war,
And shame thine honourable age with blood?
Why, art thou old and want'st experience?
Or wherefore dost abuse it if thou hast it?            170
For shame in duty bend thy knee to me,
That bows unto the grave with mickle age.
SALISBURY
My lord, I have considered with myself
The title of this most renownèd Duke,
And in my conscience do repute his grace               175
The rightful heir to England's royal seat.
KING HENRY
Hast thou not sworn allegiance unto me?
SALISBURY I have.
KING HENRY
Canst thou dispense with heaven for such an oath?
SALISBURY
It is great sin to swear unto a sin,                   180
But greater sin to keep a sinful oath.

Who can be bound by any solemn vow
To do a murd'rous deed, to rob a man,
To force a spotless virgin's chastity,
To reave the orphan of his patrimony,                              185
To wring the widow from her customed right,
And have no other reason for this wrong
But that he was bound by a solemn oath?

QUEEN MARGARET
A subtle traitor needs no sophister.

KING HENRY (to an attendant)
Call Buckingham, and bid him arm himself.              190
                                                    Exit attendant

YORK (to King Henry)
Call Buckingham and all the friends thou hast,
I am resolved for death or dignity.

CLIFFORD
The first, I warrant thee, if dreams prove true.

WARWICK
You were best to go to bed and dream again,
To keep you from the tempest of the field.              195

CLIFFORD
I am resolved to bear a greater storm
Than any thou canst conjure up today—
And that I'll write upon thy burgonet
Might I but know thee by thy household badge.

WARWICK
Now by my father's badge, old Neville's crest,         200
The rampant bear chained to the ragged staff,
This day I'll wear aloft my burgonet,
As on a mountain top the cedar shows
That keeps his leaves in spite of any storm,
Even to affright thee with the view thereof.           205

CLIFFORD
And from thy burgonet I'll rend thy bear,
And tread it under foot with all contempt,
Despite the bearherd that protects the bear.

YOUNG CLIFFORD
And so to arms, victorious father,
To quell the rebels and their complices.                210

RICHARD
Fie, charity, for shame! Speak not in spite—
For you shall sup with Jesu Christ tonight.

YOUNG CLIFFORD
Foul stigmatic, that's more than thou canst tell.

RICHARD
If not in heaven, you'll surely sup in hell.
                                                  Exeunt severally

5.2  ⌈An alehouse sign: a castle.⌉ Alarums to the battle.
     Then enter the Duke of Somerset and Richard
     fighting. Richard kills Somerset ⌈under the sign⌉

RICHARD So lie thou there—
For underneath an alehouse' paltry sign,
The Castle in Saint Albans, Somerset
Hath made the wizard famous in his death.
Sword, hold thy temper; heart, be wrathfull still—    5
Priests pray for enemies, but princes kill.
                             Exit ⌈with Somerset's body. The sign is
                                                        removed⌉

5.3  ⌈Alarum again.⌉ Enter the Earl of Warwick

WARWICK
Clifford of Cumberland, 'tis Warwick calls!
An if thou dost not hide thee from the bear,
Now, when the angry trumpet sounds alarum,
And dead men's cries do fill the empty air,
Clifford I say, come forth and fight with me!          5

Proud northern lord, Clifford of Cumberland,
Warwick is hoarse with calling thee to arms!

CLIFFORD (within)
Warwick, stand still; and stir not till I come.
                             Enter the Duke of York

WARWICK
How now, my noble lord? What, all afoot?

YORK
The deadly-handed Clifford slew my steed.              10
But match to match I have encountered him,
And made a prey for carrion kites and crows
Even of the bonny beast he loved so well.
                             Enter Lord Clifford

WARWICK (to Clifford)
Of one or both of us the time is come.

YORK
Hold, Warwick—seek thee out some other chase,         15
For I myself must hunt this deer to death.

WARWICK
Then nobly, York; 'tis for a crown thou fight'st.
(To Clifford) As I intend, Clifford, to thrive today,
It grieves my soul to leave thee unassailed.      Exit

YORK
Clifford, since we are singled here alone,             20
Be this the day of doom to one of us.
For know my heart hath sworn immortal hate
To thee and all the house of Lancaster.

CLIFFORD
And here I stand and pitch my foot to thine,
Vowing not to stir till thou or I be slain.            25
For never shall my heart be safe at rest
Till I have spoiled the hateful house of York.
                  Alarums. They fight. York kills Clifford

YORK
Now, Lancaster, sit sure—thy sinews shrink.
Come, fearful Henry, grovelling on thy face—
Yield up thy crown unto the prince of York.       Exit
                  Alarums, then enter Young Clifford

YOUNG CLIFFORD
Shame and confusion, all is on the rout!               31
Fear frames disorder, and disorder wounds
Where it should guard. O, war, thou son of hell,
Whom angry heavens do make their minister,
Throw in the frozen bosoms of our part                 35
Hot coals of vengeance! Let no soldier fly!
He that is truly dedicate to war
Hath no self-love; nor he that loves himself
Hath not essentially, but by circumstance,
The name of valour.
                  He sees his father's body
                             O, let the vile world end,   40
And the premisèd flames of the last day
Knit earth and heaven together.
Now let the general trumpet blow his blast,
Particularities and petty sounds
To cease! Wast thou ordainèd, dear father,             45
To lose thy youth in peace, and to achieve
The silver livery of advisèd age,
And in thy reverence and thy chair-days, thus
To die in ruffian battle? Even at this sight
My heart is turned to stone, and while 'tis mine       50
It shall be stony. York not our old men spares;
No more will I their babes. Tears virginal
Shall be to me even as the dew to fire,
And beauty that the tyrant oft reclaims
Shall to my flaming wrath be oil and flax.             55

87

Henceforth I will not have to do with pity.
Meet I an infant of the house of York,
Into as many gobbets will I cut it
As wild Medea young Absyrtus did.
In cruelty will I seek out my fame.                              60
Come, thou new ruin of old Clifford's house,
  *He takes his father's body up on his back*
As did Aeneas old Anchises bear,
So bear I thee upon my manly shoulders.
But then Aeneas bare a living load,
Nothing so heavy as these woes of mine.                          65
                                    *Exit with the body*

5.4  ⌈*Alarums again. Then enter three or four bearing*
     *the Duke of Buckingham wounded to his tent.*⌉
     *Alarums still. Enter King Henry, Queen Margaret,*
     *and others*
QUEEN MARGARET
  Away, my lord! You are slow. For shame, away!
KING HENRY
  Can we outrun the heavens? Good Margaret, stay.
QUEEN MARGARET
  What are you made of? You'll nor fight nor fly.
  Now is it manhood, wisdom, and defence,
  To give the enemy way, and to secure us                         5
  By what we can, which can no more but fly.
      *Alarum afar off*
  If you be ta'en, we then should see the bottom
  Of all our fortunes; but if we haply scape—
  As well we may if not through your neglect—
  We shall to London get where you are loved,                    10
  And where this breach now in our fortunes made
  May readily be stopped.
      *Enter Young Clifford*
YOUNG CLIFFORD (*to King Henry*)
  But that my heart's on future mischief set,
  I would speak blasphemy ere bid you fly;
  But fly you must; uncurable discomfit                          15
  Reigns in the hearts of all our present parts.
  Away for your relief, and we will live
  To see their day and them our fortune give.
  Away, my lord, away!                              *Exeunt*

5.5  *Alarum. Retreat. Enter the Duke of York, his sons*
     *Edward and Richard, and soldiers, including a*
     *drummer and some bearing colours*

YORK (*to Edward and Richard*)
  How now, boys! Fortunate this fight hath been,
  I hope, to us and ours for England's good
  And our great honour, that so long we lost
  Whilst faint-heart Henry did usurp our rights.
  Of Salisbury, who can report of him?                            5
  That winter lion who in rage forgets
  Agèd contusions and all brush of time,
  And, like a gallant in the brow of youth,
  Repairs him with occasion. This happy day
  Is not itself, nor have we won one foot                        10
  If Salisbury be lost.
RICHARD              My noble father,
  Three times today I holp him to his horse;
  Three times bestrid him; thrice I led him off,
  Persuaded him from any further act;
  But still where danger was, still there I met him,             15
  And like rich hangings in a homely house,
  So was his will in his old feeble body.
      *Enter the Earls of Salisbury and Warwick*
EDWARD (*to York*)
  See, noble father, where they both do come—
  The only props unto the house of York!
SALISBURY
  Now, by my sword, well hast thou fought today;                 20
  By th' mass, so did we all. I thank you, Richard.
  God knows how long it is I have to live,
  And it hath pleased him that three times today
  You have defended me from imminent death.
  Well, lords, we have not got that which we have—               25
  'Tis not enough our foes are this time fled,
  Being opposites of such repairing nature.
YORK
  I know our safety is to follow them,
  For, as I hear, the King is fled to London,
  To call a present court of Parliament.                         30
  Let us pursue him ere the writs go forth.
  What says Lord Warwick, shall we after them?
WARWICK
  After them? Nay, before them if we can!
  Now by my hand, lords, 'twas a glorious day!
  Saint Albans battle won by famous York                         35
  Shall be eternized in all age to come.
  Sound drums and trumpets, and to London all,
  And more such days as these to us befall!
                              ⌈*Flourish.*⌉ *Exeunt*

## ADDITIONAL PASSAGES

A. We adopt the 1594 Quarto version of the Queen's initial speech, 1.1.24–9; the Folio version, which follows, is probably the author's original draft.

QUEEN MARGARET
Great King of England, and my gracious lord,
The mutual conference that my mind hath had—
By day, by night; waking, and in my dreams;
In courtly company, or at my beads—
With you, mine alder liefest sovereign,    5
Makes me the bolder to salute my king
With ruder terms, such as my wit affords
And overjoy of heart doth minister.

B. For 1.4.39–40.2 the Quarto substitutes the following; it may report a revision made in rehearsal to cover the Spirit's descent.

*The Spirit sinks down again*
BOLINGBROKE
Then down, I say, unto the damnèd pool
Where Pluto in his fiery wagon sits
Riding, amidst the singed and parchèd smokes,
The road of Ditis by the River Styx.
There howl and burn for ever in those flames.    5
Rise, Jordan, rise, and stay thy charming spells—
Zounds, we are betrayed!

C. The entire debate on Duke Humphrey's death in 3.1 is handled differently by the Quarto from the Folio. We retain the Folio version of the debate, but the Quarto version may represent authorial revision. The following Q lines, roughly corresponding to 3.1.310–30.1, are of particular interest because they supply Buckingham with speeches for this latter part of the scene.

[YORK]
Let me have some bands of chosen soldiers,
And York shall try his fortune 'gainst those kerns.
QUEEN MARGARET
York, thou shalt. My lord of Buckingham,
Let it be your charge to muster up such soldiers
As shall suffice him in these needful wars.    5

BUCKINGHAM
Madam, I will, and levy such a band
As soon shall overcome those Irish rebels.
But, York, where shall those soldiers stay for thee?
YORK
At Bristol I will expect them ten days hence.    9
BUCKINGHAM
Then thither shall they come, and so farewell.    *Exit*
YORK
Adieu, my lord of Buckingham.
QUEEN MARGARET
Suffolk, remember what you have to do—
And you, Lord Cardinal—concerning Duke Humphrey.
'Twere good that you did see to it in time.
Come, let us go, that it may be performed.    15
*Exeunt all but York*

D. We adopt the Quarto version of the confrontation between Clifford and York at 5.3.20–30; the Folio version, an edited text of which follows, is probably the author's original draft.

CLIFFORD
What seest thou in me, York? Why dost thou pause?
YORK
With thy brave bearing should I be in love,
But that thou art so fast mine enemy.
CLIFFORD
Nor should thy prowess want praise and esteem,
But that 'tis shown ignobly and in treason.    5
YORK
So let it help me now against thy sword,
As I in justice and true right express it.
CLIFFORD
My soul and body on the action, both.
YORK
A dreadful lay. Address thee instantly.
CLIFFORD
*La fin couronne les oeuvres.*    5
*Alarms. They fight. York kills Clifford*
YORK
Thus war hath given thee peace, for thou art still.
Peace with his soul, heaven, if it be thy will.    *Exit*

# RICHARD DUKE OF YORK
## (3 HENRY VI)

THE play printed in the 1623 Folio as *The Third Part of Henry the Sixth, with the Death of the Duke of York* was described on the title-page of its first, unauthoritative publication in 1595 as *The True Tragedy of Richard, Duke of York, and the Death of Good King Henry the Sixth, with the whole Contention between the two houses Lancaster and York*. It is clearly a continuation of *The First Part of the Contention*, taking up the story where that play had ended, with the aspirations of Richard, Duke of York to the English throne, and was probably composed immediately afterwards.

The final scenes of *The First Part of the Contention* briefly introduce two of York's sons, Edward (the eldest) and Richard (already described as a 'foul, indigested lump, | As crooked in . . . manners as [in] shape'). They, along with their brothers Edmund, Earl of Rutland, and George (later Duke of Clarence), figure more prominently in *Richard Duke of York*. The first scenes show York apparently fulfilling his ambition, as Henry VI weakly cedes his rights to the throne after his death; but Queen Margaret leads an army against York, and, when he is captured, personally taunts him with news of the murder of his youngest son, stabs York to death, and commands that his head be 'set on York gates'. (This powerful scene includes the line 'O tiger's heart wrapped in a woman's hide', paraphrased by Robert Greene before September 1592, which establishes the upward limit of the play's date.)

Though Richard of York dies early in the action, the remainder of the play centres on his sons' efforts (aided by Warwick's politic schemings) to avenge his death and to establish the dominance of Yorkists over Lancastrians. The balance of power shifts frequently, and the brothers' alliance crumbles, but finally Queen Margaret, with her French allies, is defeated and captured, and Richard of York's surviving sons avenge their father's death by killing her son, Edward, before her eyes. Richard of Gloucester starts to clear his way to the throne by murdering 'Good King Henry' in the Tower, and the play ends with the new King Edward IV exulting in his 'country's peace and brothers' loves' while Richard makes clear to the audience that Edward's self-confidence is ill-founded.

Though the play is loud and strife-ridden with war, power politics, and personal ambition, a concern with humane values emerges in the subtle and touching continuing portrayal of the quietist Henry VI, a saintly fool who meditates on the superiority of humble contentment to regal misery in an emblematic scene (2.5) that epitomizes the tragedy of civil strife.

*Richard Duke of York*, like *The First Part of the Contention*, draws extensively on English chronicle history. Historically, the period of the action covers about sixteen years (1455 to 1471), but events are telescoped and rearranged; for instance, the opening scenes move rapidly from the Battle of St Albans (1445) to York's death (1450); the future Richard III was only three years old, and living abroad, at the time of this opening battle in which he takes an active part; and Richard's murder of Henry owes more to legend than to fact.

# THE PERSONS OF THE PLAY

### Of the King's Party

KING HENRY VI

QUEEN MARGARET

PRINCE EDWARD, their son

Duke of SOMERSET

Duke of EXETER

Earl of NORTHUMBERLAND

Earl of WESTMORLAND

Lord CLIFFORD

Lord Stafford

SOMERVILLE

Henry, young Earl of Richmond

A SOLDIER who has killed his father

A HUNTSMAN who guards King Edward

### The Divided House of Neville

Earl of WARWICK, first of York's party, later of Lancaster's

Marquis of MONTAGUE, his brother, of York's party

Earl of OXFORD, their brother-in-law, of Lancaster's party

Lord HASTINGS, their brother-in-law, of York's party

### Of the Duke of York's Party

Richard Plantagenet, Duke of YORK

EDWARD, Earl of March, his son, later Duke of York and KING EDWARD IV

LADY GRAY, a widow, later Edward's wife and queen

Earl RIVERS, her brother

GEORGE, Edward's brother, later Duke OF CLARENCE

RICHARD, Edward's brother, later Duke OF GLOUCESTER

Earl of RUTLAND, Edward's brother

Rutland's TUTOR, a chaplain

SIR JOHN Mortimer, York's uncle

Sir Hugh Mortimer, his brother

Duke of NORFOLK

Sir William Stanley

Earl of Pembroke

Sir John MONTGOMERY

A NOBLEMAN

Two GAMEKEEPERS

Three WATCHMEN, who guard King Edward's tent

LIEUTENANT of the Tower

### The French

KING LOUIS

LADY BONA, his sister-in-law

Lord Bourbon, the French High Admiral

### Others

A SOLDIER who has killed his son

Mayor of Coventry

MAYOR of York

Aldermen of York

Soldiers, messengers, and attendants

# The True Tragedy of Richard Duke of York and the Good King Henry the Sixth

**1.1** *A chair of state. Alarum. Enter Richard*
*Plantagenet, Duke of York, his two sons Edward,*
*Earl of March, and Crookback Richard, the Duke of*
*Norfolk, the Marquis of Montague, and the Earl of*
*Warwick, ⌈with drummers⌉ and soldiers. ⌈They all*
*wear white roses in their hats⌉*

WARWICK
I wonder how the King escaped our hands?

YORK
While we pursued the horsemen of the north,
He slyly stole away and left his men;
Whereat the great lord of Northumberland,
Whose warlike ears could never brook retreat,  5
Cheered up the drooping army; and himself,
Lord Clifford, and Lord Stafford, all abreast,
Charged our main battle's front, and, breaking in,
Were by the swords of common soldiers slain.

EDWARD
Lord Stafford's father, Duke of Buckingham,  10
Is either slain or wounded dangerous.
I cleft his beaver with a downright blow.
That this is true, father, behold his blood.
    *He shows a bloody sword*

MONTAGUE ⌈*to York*⌉
And, brother, here's the Earl of Wiltshire's blood,
    *He shows a bloody sword*
Whom I encountered as the battles joined.  15

RICHARD (*to Somerset's head, which he shows*)
Speak thou for me, and tell them what I did.

YORK
Richard hath best deserved of all my sons.
(*To the head*) But is your grace dead, my lord of
    Somerset?

NORFOLK
Such hap have all the line of John of Gaunt.

RICHARD
Thus do I hope to shake King Henry's head.  20
    ⌈*He holds aloft the head, then throws it down*⌉

WARWICK
And so do I, victorious prince of York.
Before I see thee seated in that throne
Which now the house of Lancaster usurps,
I vow by heaven these eyes shall never close.
This is the palace of the fearful King,  25
And this (*pointing to the chair of state*), the regal
    seat—possess it, York,
For this is thine, and not King Henry's heirs'.

YORK
Assist me then, sweet Warwick, and I will,
For hither we have broken in by force.

NORFOLK
We'll all assist you—he that flies shall die.  30

YORK
Thanks, gentle Norfolk. Stay by me, my lords
And soldiers—stay, and lodge by me this night.
    *They go up upon the state*

WARWICK
And when the King comes, offer him no violence
Unless he seek to thrust you out perforce.
    ⌈*The soldiers withdraw*⌉

YORK
The Queen this day here holds her Parliament,  35
But little thinks we shall be of her council;
By words or blows here let us win our right.

RICHARD
Armed as we are, let's stay within this house.

WARWICK
'The Bloody Parliament' shall this be called,
Unless Plantagenet, Duke of York, be king,  40
And bashful Henry deposed, whose cowardice
Hath made us bywords to our enemies.

YORK
Then leave me not, my lords. Be resolute—
I mean to take possession of my right.

WARWICK
Neither the King nor he that loves him best—  45
The proudest he that holds up Lancaster—
Dares stir a wing if Warwick shake his bells.
I'll plant Plantagenet, root him up who dares.
Resolve thee, Richard—claim the English crown.
    ⌈*York sits in the chair.*⌉
    *Flourish. Enter King Henry, Lord Clifford, the Earls*
    *of Northumberland and Westmorland, the Duke of*
    *Exeter, and the rest. ⌈They all wear red roses in*
    *their hats⌉*

KING HENRY
My lords, look where the sturdy rebel sits—  50
Even in the chair of state! Belike he means,
Backed by the power of Warwick, that false peer,
To aspire unto the crown and reign as king.
Earl of Northumberland, he slew thy father—
And thine, Lord Clifford—and you both have vowed
    revenge  55
On him, his sons, his favourites, and his friends.

NORTHUMBERLAND
If I be not, heavens be revenged on me.

CLIFFORD
The hope thereof makes Clifford mourn in steel.

WESTMORLAND
What, shall we suffer this? Let's pluck him down.
My heart for anger burns—I cannot brook it.  60

KING HENRY
Be patient, gentle Earl of Westmorland.

CLIFFORD
Patience is for poltroons, such as he (*indicating York*).
He durst not sit there had your father lived.
My gracious lord, here in the Parliament
Let us assail the family of York.  65

NORTHUMBERLAND
Well hast thou spoken, cousin, be it so.

KING HENRY
Ah, know you not the city favours them,
And they have troops of soldiers at their beck?

EXETER
But when the Duke is slain, they'll quickly fly.
KING HENRY
Far be the thought of this from Henry's heart,   70
To make a shambles of the Parliament House.
Cousin of Exeter, frowns, words, and threats
Shall be the war that Henry means to use.
(*To York*) Thou factious Duke of York, descend my
    throne
And kneel for grace and mercy at my feet.   75
I am thy sovereign.
YORK                 I am thine.
EXETER
For shame, come down—he made thee Duke of York.
YORK
It was mine inheritance, as the earldom was.
EXETER
Thy father was a traitor to the crown.
WARWICK
Exeter, thou art a traitor to the crown   80
In following this usurping Henry.
CLIFFORD
Whom should he follow but his natural king?
WARWICK
True, Clifford, and that's Richard Duke of York.
KING HENRY (*to York*)
And shall I stand and thou sit in my throne?
YORK
It must and shall be so—content thyself.   85
WARWICK (*to King Henry*)
Be Duke of Lancaster, let him be king.
WESTMORLAND
He is both king and Duke of Lancaster—
And that, the Lord of Westmorland shall maintain.
WARWICK
And Warwick shall disprove it. You forget
That we are those which chased you from the field,   90
And slew your fathers, and, with colours spread,
Marched through the city to the palace gates.
NORTHUMBERLAND
Yes, Warwick, I remember it to my grief,
And, by his soul, thou and thy house shall rue it.
WESTMORLAND (*to York*)
Plantagenet, of thee, and these thy sons,   95
Thy kinsmen, and thy friends, I'll have more lives
Than drops of blood were in my father's veins.
CLIFFORD (*to Warwick*)
Urge it no more, lest that, instead of words,
I send thee, Warwick, such a messenger
As shall revenge his death before I stir.   100
WARWICK ⌈*to York*⌉
Poor Clifford, how I scorn his worthless threats.
YORK ⌈*to King Henry*⌉
Will you we show our title to the crown?
If not, our swords shall plead it in the field.
KING HENRY
What title hast thou, traitor, to the crown?
Thy father was, as thou art, Duke of York;   105
Thy grandfather, Roger Mortimer, Earl of March.
I am the son of Henry the Fifth,
Who made the Dauphin and the French to stoop
And seized upon their towns and provinces.
WARWICK
Talk not of France, sith thou hast lost it all.   110
KING HENRY
The Lord Protector lost it, and not I.
When I was crowned, I was but nine months old.

RICHARD
You are old enough now, and yet, methinks, you lose.
(*To York*) Father, tear the crown from the usurper's
    head.
EDWARD (*to York*)
Sweet father, do so—set it on your head.   115
MONTAGUE (*to York*)
Good brother, as thou lov'st and honour'st arms,
Let's fight it out and not stand cavilling thus.
RICHARD
Sound drums and trumpets, and the King will fly.
YORK Sons, peace!
⌈NORTHUMBERLAND⌉
Peace, thou—and give King Henry leave to speak.   120
KING HENRY
Ah, York, why seekest thou to depose me?
Are we not both Plantagenets by birth,
And from two brothers lineally descent?
Suppose by right and equity thou be king—
Think'st thou that I will leave my kingly throne,   125
Wherein my grandsire and my father sat?
No—first shall war unpeople this my realm;
Ay, and their colours, often borne in France,
And now in England to our heart's great sorrow,
Shall be my winding-sheet. Why faint you, lords?   130
My title's good, and better far than his.
WARWICK
Prove it, Henry, and thou shalt be king.
KING HENRY
Henry the Fourth by conquest got the crown.
YORK
'Twas by rebellion against his king.
KING HENRY ⌈*aside*⌉
I know not what to say—my title's weak.   135
(*To York*) Tell me, may not a king adopt an heir?
YORK What then?
KING HENRY
An if he may, then am I lawful king—
For Richard, in the view of many lords,
Resigned the crown to Henry the Fourth,   140
Whose heir my father was, and I am his.
YORK
He rose against him, being his sovereign,
And made him to resign his crown perforce.
WARWICK
Suppose, my lords, he did it unconstrained—
Think you 'twere prejudicial to his crown?   145
EXETER
No, for he could not so resign his crown
But that the next heir should succeed and reign.
KING HENRY
Art thou against us, Duke of Exeter?
EXETER
His is the right, and therefore pardon me.
YORK
Why whisper you, my lords, and answer not?   150
EXETER ⌈*to King Henry*⌉
My conscience tells me he is lawful king.
KING HENRY ⌈*aside*⌉
All will revolt from me and turn to him.
NORTHUMBERLAND (*to York*)
Plantagenet, for all the claim thou lay'st,
Think not that Henry shall be so deposed.
WARWICK
Deposed he shall be, in despite of all.   155
NORTHUMBERLAND
Thou art deceived—'tis not thy southern power

Of Essex, Norfolk, Suffolk, nor of Kent,
Which makes thee thus presumptuous and proud,
Can set the Duke up in despite of me.
CLIFFORD
King Henry, be thy title right or wrong,                    160
Lord Clifford vows to fight in thy defence.
May that ground gape and swallow me alive
Where I shall kneel to him that slew my father.
KING HENRY
O, Clifford, how thy words revive my heart!
YORK
Henry of Lancaster, resign thy crown.                       165
What mutter you, or what conspire you, lords?
WARWICK
Do right unto this princely Duke of York,
Or I will fill the house with armèd men
And over the chair of state, where now he sits,
Write up his title with usurping blood.                     170
*He stamps with his foot and the soldiers show*
*themselves*
KING HENRY
My lord of Warwick, hear me but one word—
Let me for this my lifetime reign as king.
YORK
Confirm the crown to me and to mine heirs,
And thou shalt reign in quiet while thou liv'st.
KING HENRY
I am content. Richard Plantagenet,                          175
Enjoy the kingdom after my decease.
CLIFFORD
What wrong is this unto the prince your son?
WARWICK
What good is this to England and himself?
WESTMORLAND
Base, fearful, and despairing Henry.
CLIFFORD
How hast thou injured both thyself and us?                  180
WESTMORLAND
I cannot stay to hear these articles.
NORTHUMBERLAND Nor I.
CLIFFORD
Come, cousin, let us tell the Queen these news.
WESTMORLAND (*to King Henry*)
Farewell, faint-hearted and degenerate king,
In whose cold blood no spark of honour bides.               185
                    ⌜*Exit with his soldiers*⌝
NORTHUMBERLAND (*to King Henry*)
Be thou a prey unto the house of York,
And die in bands for this unmanly deed.
                    ⌜*Exit with his soldiers*⌝
CLIFFORD (*to King Henry*)
In dreadful war mayst thou be overcome,
Or live in peace, abandoned and despised.
                    *Exit* ⌜*with his soldiers*⌝
WARWICK (*to King Henry*)
Turn this way, Henry, and regard them not.                  190
EXETER (*to King Henry*)
They seek revenge and therefore will not yield.
KING HENRY
Ah, Exeter.
WARWICK     Why should you sigh, my lord?
KING HENRY
Not for myself, Lord Warwick, but my son,
Whom I unnaturally shall disinherit.
But be it as it may. (*To York*) I here entail              195
The crown to thee and to thine heirs for ever,

Conditionally, that here thou take thine oath
To cease this civil war, and whilst I live
To honour me as thy king and sovereign,
And nor by treason nor hostility                            200
To seek to put me down and reign thyself.
YORK
This oath I willingly take and will perform.
WARWICK
Long live King Henry. (*To York*) Plantagenet, embrace
   him.
     ⌜*York descends.*⌝ *Henry and York embrace*
KING HENRY (*to York*)
And long live thou, and these thy forward sons.
YORK
Now York and Lancaster are reconciled.                      205
EXETER
Accursed be he that seeks to make them foes.
     *Sennet. Here York's train comes down from the*
     *state*
YORK (*to King Henry*)
Farewell, my gracious lord, I'll to my castle.
               *Exeunt York, Edward, and Richard,* ⌜*with*
                                             *soldiers*⌝
WARWICK
And I'll keep London with my soldiers.
                              *Exit* ⌜*with soldiers*⌝
NORFOLK
And I to Norfolk with my followers.
                              *Exit* ⌜*with soldiers*⌝
MONTAGUE
And I unto the sea from whence I came.                      210
                              *Exit* ⌜*with soldiers*⌝
KING HENRY
And I with grief and sorrow to the court.
     ⌜*King Henry and Exeter turn to leave.*⌝
     *Enter Queen Margaret and Prince Edward*
EXETER
Here comes the Queen, whose looks bewray her anger.
I'll steal away.
KING HENRY          Exeter, so will I.
QUEEN MARGARET
Nay, go not from me—I will follow thee.
KING HENRY
Be patient, gentle Queen, and I will stay.                  215
QUEEN MARGARET
Who can be patient in such extremes?
Ah, wretched man, would I had died a maid
And never seen thee, never borne thee son,
Seeing thou hast proved so unnatural a father.
Hath he deserved to lose his birthright thus?               220
Hadst thou but loved him half so well as I,
Or felt that pain which I did for him once,
Or nourished him as I did with my blood,
Thou wouldst have left thy dearest heart-blood there
Rather than have made that savage Duke thine heir
And disinherited thine only son.                            226
PRINCE EDWARD
Father, you cannot disinherit me.
If you be king, why should not I succeed?
KING HENRY
Pardon me, Margaret; pardon me, sweet son—
The Earl of Warwick and the Duke enforced me.               230
QUEEN MARGARET
Enforced thee? Art thou king, and wilt be forced?
I shame to hear thee speak! Ah, timorous wretch,
Thou hast undone thyself, thy son, and me,

And giv'n unto the house of York such head
As thou shalt reign but by their sufferance.                    235
To entail him and his heirs unto the crown—
What is it, but to make thy sepulchre
And creep into it far before thy time?
Warwick is Chancellor and the Lord of Calais;
Stern Falconbridge commands the narrow seas;            240
The Duke is made Protector of the Realm;
And yet shalt thou be safe? Such safety finds
The trembling lamb environèd with wolves.
Had I been there, which am a seely woman,
The soldiers should have tossed me on their pikes   245
Before I would have granted to that act.
But thou preferr'st thy life before thine honour.
And seeing thou dost, I here divorce myself
Both from thy table, Henry, and thy bed,
Until that act of Parliament be repealed                    250
Whereby my son is disinherited.
The northern lords that have forsworn thy colours
Will follow mine, if once they see them spread—
And spread they shall be, to thy foul disgrace
And the utter ruin of the house of York.                    255
Thus do I leave thee. (*To Prince Edward*) Come, son,
       let's away.
Our army is ready—come, we'll after them.
KING HENRY
Stay, gentle Margaret, and hear me speak.
QUEEN MARGARET
Thou hast spoke too much already.
⌈*To Prince Edward*⌉                    Get thee gone.
KING HENRY
Gentle son Edward, thou wilt stay with me?    260
QUEEN MARGARET
Ay, to be murdered by his enemies.
PRINCE EDWARD (*to King Henry*)
When I return with victory from the field,
I'll see your grace. Till then, I'll follow her.
QUEEN MARGARET
Come, son, away—we may not linger thus.
                    *Exit with Prince Edward*
KING HENRY
Poor Queen, how love to me and to her son    265
Hath made her break out into terms of rage.
Revengèd may she be on that hateful Duke,
Whose haughty spirit, wingèd with desire,
Will coast my crown, and, like an empty eagle,
Tire on the flesh of me and of my son.                    270
The loss of those three lords torments my heart.
I'll write unto them and entreat them fair.
Come, cousin, you shall be the messenger.
EXETER
And I, I hope, shall reconcile them all.
                    *Flourish. Exeunt*

**1.2**    *Enter Richard, Edward Earl of March, and the*
          *Marquis of Montague*
RICHARD
Brother, though I be youngest give me leave.
EDWARD
No, I can better play the orator.
MONTAGUE
But I have reasons strong and forcible.
          *Enter the Duke of York*
YORK
Why, how now, sons and brother—at a strife?
What is your quarrel? How began it first?          5

EDWARD
No quarrel, but a slight contention.
YORK About what?
RICHARD
About that which concerns your grace and us—
The crown of England, father, which is yours.
YORK
Mine, boy? Not till King Henry be dead.              10
RICHARD
Your right depends not on his life or death.
EDWARD
Now you are heir—therefore enjoy it now.
By giving the house of Lancaster leave to breathe,
It will outrun you, father, in the end.
YORK
I took an oath that he should quietly reign.        15
EDWARD
But for a kingdom any oath may be broken.
I would break a thousand oaths to reign one year.
RICHARD (*to York*)
No—God forbid your grace should be forsworn.
YORK
I shall be if I claim by open war.
RICHARD
I'll prove the contrary, if you'll hear me speak.   20
YORK
Thou canst not, son—it is impossible.
RICHARD
An oath is of no moment being not took
Before a true and lawful magistrate
That hath authority over him that swears.
Henry had none, but did usurp the place.            25
Then, seeing 'twas he that made you to depose,
Your oath, my lord, is vain and frivolous.
Therefore to arms—and, father, do but think
How sweet a thing it is to wear a crown,
Within whose circuit is Elysium                          30
And all that poets feign of bliss and joy.
Why do we linger thus? I cannot rest
Until the white rose that I wear be dyed
Even in the luke-warm blood of Henry's heart.
YORK
Richard, enough! I will be king or die.              35
(*To Montague*) Brother, thou shalt to London presently
And whet on Warwick to this enterprise.
Thou, Richard, shalt to the Duke of Norfolk
And tell him privily of our intent.
You, Edward, shall to Edmund Brook, Lord Cobham,
With whom the Kentishmen will willingly rise.      41
In them I trust, for they are soldiers
Witty, courteous, liberal, full of spirit.
While you are thus employed, what resteth more
But that I seek occasion how to rise,                   45
And yet the King not privy to my drift,
Nor any of the house of Lancaster.
          *Enter a Messenger*
But stay, what news? Why com'st thou in such post?
MESSENGER
The Queen, with all the northern earls and lords,
Intend here to besiege you in your castle.          50
She is hard by with twenty thousand men,
And therefore fortify your hold, my lord.
YORK
Ay, with my sword. What—think'st thou that we fear
       them?
Edward and Richard, you shall stay with me;

My brother Montague shall post to London.                    55
Let noble Warwick, Cobham, and the rest,
Whom we have left protectors of the King,
With powerful policy strengthen themselves,
And trust not simple Henry nor his oaths.
MONTAGUE
Brother, I go—I'll win them, fear it not.                    60
And thus most humbly I do take my leave.          *Exit*
    *Enter Sir John Mortimer and his brother Sir Hugh*
YORK
Sir John and Sir Hugh Mortimer, mine uncles,
You are come to Sandal in a happy hour.
The army of the Queen mean to besiege us.
SIR JOHN
She shall not need, we'll meet her in the field.           65
YORK  What, with five thousand men?
RICHARD
Ay, with five hundred, father, for a need.
A woman's general—what should we fear?
    *A march sounds afar off*
EDWARD
I hear their drums. Let's set our men in order,
And issue forth and bid them battle straight.              70
YORK ⌜*to Sir John and Sir Hugh*⌝
Five men to twenty—though the odds be great,
I doubt not, uncles, of our victory.
Many a battle have I won in France
Whenas the enemy hath been ten to one—
Why should I not now have the like success?   *Exeunt*

**1.3**  *Alarums, and then enter the young Earl of Rutland*
    *and his Tutor, a chaplain*
RUTLAND
Ah, whither shall I fly to scape their hands?
    *Enter Lord Clifford with soldiers*
Ah, tutor, look where bloody Clifford comes.
CLIFFORD (*to the Tutor*)
Chaplain, away—thy priesthood saves thy life.
As for the brat of this accursèd duke,
Whose father slew my father—he shall die.                   5
TUTOR
And I, my lord, will bear him company.
CLIFFORD  Soldiers, away with him.
TUTOR
Ah, Clifford, murder not this innocent child
Lest thou be hated both of God and man.
                    *Exit, guarded*
    ⌜*Rutland falls to the ground*⌝
CLIFFORD
How now—is he dead already?                                 10
Or is it fear that makes him close his eyes?
I'll open them.
RUTLAND ⌜*reviving*⌝
So looks the pent-up lion o'er the wretch
That trembles under his devouring paws,
And so he walks, insulting o'er his prey,                   15
And so he comes to rend his limbs asunder.
Ah, gentle Clifford, kill me with thy sword
And not with such a cruel threat'ning look.
Sweet Clifford, hear me speak before I die.
I am too mean a subject for thy wrath.                      20
Be thou revenged on men, and let me live.
CLIFFORD
In vain thou speak'st, poor boy. My father's blood
Hath stopped the passage where thy words should
    enter.

RUTLAND
Then let my father's blood open it again.
He is a man, and, Clifford, cope with him.                  25
CLIFFORD
Had I thy brethren here, their lives and thine
Were not revenge sufficient for me.
No—if I digged up thy forefathers' graves,
And hung their rotten coffins up in chains,
It could not slake mine ire nor ease my heart.            30
The sight of any of the house of York
Is as a fury to torment my soul.
And till I root out their accursèd line,
And leave not one alive, I live in hell.
Therefore—                                                  35
RUTLAND
O, let me pray before I take my death.
    ⌜*Kneeling*⌝ To thee I pray: sweet Clifford, pity me.
CLIFFORD
Such pity as my rapier's point affords.
RUTLAND
I never did thee harm—why wilt thou slay me?
CLIFFORD
Thy father hath.
RUTLAND         But 'twas ere I was born.                40
Thou hast one son—for his sake pity me,
Lest in revenge thereof, sith God is just,
He be as miserably slain as I.
Ah, let me live in prison all my days,
And when I give occasion of offence,                       45
Then let me die, for now thou hast no cause.
CLIFFORD
No cause? Thy father slew my father, therefore die.
    *He stabs him*
RUTLAND
*Dii faciant laudis summa sit ista tuae.*      *He dies*
CLIFFORD
Plantagenet—I come, Plantagenet!
And this thy son's blood cleaving to my blade             50
Shall rust upon my weapon till thy blood,
Congealed with this, do make me wipe off both.
    *Exit with Rutland's body* ⌜*and soldiers*⌝

**1.4**  *Alarum. Enter Richard Duke of York*
YORK
The army of the Queen hath got the field;
My uncles both are slain in rescuing me;
And all my followers to the eager foe
Turn back, and fly like ships before the wind,
Or lambs pursued by hunger-starved wolves.                 5
My sons—God knows what hath bechancèd them.
But this I know—they have demeaned themselves
Like men born to renown by life or death.
Three times did Richard make a lane to me,
And thrice cried, 'Courage, father, fight it out!'        10
And full as oft came Edward to my side,
With purple falchion painted to the hilt
In blood of those that had encountered him.
And when the hardiest warriors did retire,
Richard cried, 'Charge and give no foot of ground!'       15
⌜                                    ⌝
And cried 'A crown or else a glorious tomb!
A sceptre or an earthly sepulchre!'
With this, we charged again—but out, alas—
We bodged again, as I have seen a swan                     20
With bootless labour swim against the tide
And spend her strength with over-matching waves.

*A short alarum within*
Ah, hark—the fatal followers do pursue,
And I am faint and cannot fly their fury;
And were I strong, I would not shun their fury. 25
The sands are numbered that makes up my life.
Here must I stay, and here my life must end.
    *Enter Queen Margaret, Lord Clifford, the Earl of*
    *Northumberland, and the young Prince Edward,*
    *with soldiers*
Come bloody Clifford, rough Northumberland—
I dare your quenchless fury to more rage!
I am your butt, and I abide your shot. 30

NORTHUMBERLAND
Yield to our mercy, proud Plantagenet.

CLIFFORD
Ay, to such mercy as his ruthless arm,
With downright payment, showed unto my father.
Now Phaëton hath tumbled from his car,
And made an evening at the noontide prick. 35

YORK
My ashes, as the phoenix, may bring forth
A bird that will revenge upon you all,
And in that hope I throw mine eyes to heaven,
Scorning whate'er you can afflict me with.
Why come you not? What—multitudes, and fear? 40

CLIFFORD
So cowards fight when they can fly no further;
So doves do peck the falcon's piercing talons;
So desperate thieves, all hopeless of their lives,
Breathe out invectives 'gainst the officers.

YORK
O, Clifford, but bethink thee once again, 45
And in thy thought o'errun my former time,
And, if thou canst for blushing, view this face
And bite thy tongue, that slanders him with cowardice
Whose frown hath made thee faint and fly ere this.

CLIFFORD
I will not bandy with thee word for word, 50
But buckle with thee blows twice two for one.
    ⌈*He draws his sword*⌉

QUEEN MARGARET
Hold, valiant Clifford: for a thousand causes
I would prolong a while the traitor's life.
Wrath makes him deaf—speak thou, Northumberland.

NORTHUMBERLAND
Hold, Clifford—do not honour him so much 55
To prick thy finger though to wound his heart.
What valour were it when a cur doth grin
For one to thrust his hand between his teeth
When he might spurn him with his foot away?
It is war's prize to take all vantages, 60
And ten to one is no impeach of valour.
    *They* ⌈*fight and*⌉ *take York*

CLIFFORD
Ay, ay, so strives the woodcock with the gin.

NORTHUMBERLAND
So doth the cony struggle in the net.

YORK
So triumph thieves upon their conquered booty,
So true men yield, with robbers so o'ermatched. 65

NORTHUMBERLAND (*to the Queen*)
What would your grace have done unto him now?

QUEEN MARGARET
Brave warriors, Clifford and Northumberland,
Come make him stand upon this molehill here,
That wrought at mountains with outstretchèd arms

Yet parted but the shadow with his hand. 70
(*To York*) What—was it you that would be England's
    king?
Was't you that revelled in our Parliament,
And made a preachment of your high descent?
Where are your mess of sons to back you now?
The wanton Edward and the lusty George? 75
And where's that valiant crookback prodigy,
Dickie, your boy, that with his grumbling voice
Was wont to cheer his dad in mutinies?
Or with the rest where is your darling Rutland?
Look, York, I stained this napkin with the blood 80
That valiant Clifford with his rapier's point
Made issue from the bosom of thy boy.
And if thine eyes can water for his death,
I give thee this to dry thy cheeks withal.
Alas, poor York, but that I hate thee deadly 85
I should lament thy miserable state.
I prithee, grieve, to make me merry, York.
What—hath thy fiery heart so parched thine entrails
That not a tear can fall for Rutland's death?
Why art thou patient, man? Thou shouldst be mad,
And I, to make thee mad, do mock thee thus. 91
Stamp, rave, and fret, that I may sing and dance.
Thou wouldst be fee'd, I see, to make me sport.
York cannot speak unless he wear a crown.
(*To her men*) A crown for York, and, lords, bow low to
    him. 95
Hold you his hands whilst I do set it on.
    *She puts a paper crown on York's head*
Ay, marry, sir, now looks he like a king,
Ay, this is he that took King Henry's chair,
And this is he was his adopted heir.
But how is it that great Plantagenet 100
Is crowned so soon and broke his solemn oath?
As I bethink me, you should not be king
Till our King Henry had shook hands with death.
And will you pale your head in Henry's glory,
And rob his temples of the diadem 105
Now, in his life, against your holy oath?
O 'tis a fault too, too, unpardonable.
Off with the crown,
    ⌈*She knocks it from his head*⌉
            and with the crown his head,
And whilst we breathe, take time to do him dead.

CLIFFORD
That is my office for my father's sake. 110

QUEEN MARGARET
Nay, stay—let's hear the orisons he makes.

YORK
She-wolf of France, but worse than wolves of France,
Whose tongue more poisons than the adder's tooth—
How ill-beseeming is it in thy sex
To triumph like an Amazonian trull 115
Upon their woes whom fortune captivates!
But that thy face is visor-like, unchanging,
Made impudent with use of evil deeds,
I would essay, proud Queen, to make thee blush.
To tell thee whence thou cam'st, of whom derived,
Were shame enough to shame thee—wert thou not
    shameless. 121
Thy father bears the type of King of Naples,
Of both the Sicils, and Jerusalem—
Yet not so wealthy as an English yeoman.
Hath that poor monarch taught thee to insult? 125
It needs not, nor it boots thee not, proud Queen,

Unless the adage must be verified
That beggars mounted run their horse to death.
'Tis beauty that doth oft make women proud—
But, God he knows, thy share thereof is small;      130
'Tis virtue that doth make them most admired—
The contrary doth make thee wondered at;
'Tis government that makes them seem divine—
The want thereof makes thee abominable.
Thou art as opposite to every good                  135
As the antipodes are unto us,
Or as the south to the septentrion.
O tiger's heart wrapped in a woman's hide!
How couldst thou drain the life-blood of the child
To bid the father wipe his eyes withal,             140
And yet be seen to bear a woman's face?
Women are soft, mild, pitiful, and flexible—
Thou stern, obdurate, flinty, rough, remorseless.
Bidd'st thou me rage? Why, now thou hast thy wish.
Wouldst have me weep? Why, now thou hast thy will.
For raging wind blows up incessant showers,         146
And when the rage allays the rain begins.
These tears are my sweet Rutland's obsequies,
And every drop cries vengeance for his death
'Gainst thee, fell Clifford, and thee, false Frenchwoman.

NORTHUMBERLAND
Beshrew me, but his passions move me so             151
That hardly can I check my eyes from tears.

YORK
That face of his the hungry cannibals
Would not have touched, would not have stained
    with blood—
But you are more inhuman, more inexorable,          155
O, ten times more than tigers of Hyrcania.
See, ruthless Queen, a hapless father's tears.
This cloth thou dipped'st in blood of my sweet boy,
And I with tears do wash the blood away.
Keep thou the napkin and go boast of this,          160
And if thou tell'st the heavy story right,
Upon my soul the hearers will shed tears,
Yea, even my foes will shed fast-falling tears
And say, 'Alas, it was a piteous deed'.
There, take the crown—and with the crown, my
    curse:                                           165
And in thy need such comfort come to thee
As now I reap at thy too cruel hand.
Hard-hearted Clifford, take me from the world.
My soul to heaven, my blood upon your heads.

NORTHUMBERLAND
Had he been slaughter-man to all my kin,            170
I should not, for my life, but weep with him,
To see how inly sorrow gripes his soul.

QUEEN MARGARET
What—weeping-ripe, my lord Northumberland?
Think but upon the wrong he did us all,
And that will quickly dry thy melting tears.        175

CLIFFORD
Here's for my oath, here's for my father's death.
    *He stabs York*

QUEEN MARGARET
And here's to right our gentle-hearted King.
    *She stabs York*

YORK
Open thy gate of mercy, gracious God—
My soul flies through these wounds to seek out thee.
    ⌐He dies⌐

QUEEN MARGARET
Off with his head and set it on York gates,         180
So York may overlook the town of York.
    *Flourish. Exeunt with York's body*

2.1    *A march. Enter Edward Earl of March and Richard,*
       ⌐*with a drummer and soldiers*⌐

EDWARD
I wonder how our princely father scaped,
Or whether he be scaped away or no
From Clifford's and Northumberland's pursuit.
Had he been ta'en we should have heard the news;
Had he been slain we should have heard the news;   5
Or had he scaped, methinks we should have heard
The happy tidings of his good escape.
How fares my brother? Why is he so sad?

RICHARD
I cannot joy until I be resolved
Where our right valiant father is become.           10
I saw him in the battle range about,
And watched him how he singled Clifford forth.
Methought he bore him in the thickest troop,
As doth a lion in a herd of neat;
Or as a bear encompassed round with dogs,           15
Who having pinched a few and made them cry,
The rest stand all aloof and bark at him.
So fared our father with his enemies;
So fled his enemies my warlike father.
Methinks 'tis prize enough to be his son.           20
    ⌐*Three suns appear in the air*⌐
See how the morning opes her golden gates
And takes her farewell of the glorious sun.
How well resembles it the prime of youth,
Trimmed like a younker prancing to his love!

EDWARD
Dazzle mine eyes, or do I see three suns?           25

RICHARD
Three glorious suns, each one a perfect sun;
Not separated with the racking clouds,
But severed in a pale clear-shining sky.
    ⌐*The three suns begin to join*⌐
See, see—they join, embrace, and seem to kiss,
As if they vowed some league inviolable.            30
Now are they but one lamp, one light, one sun.
In this the heaven figures some event.

EDWARD
'Tis wondrous strange, the like yet never heard of.
I think it cites us, brother, to the field,
That we, the sons of brave Plantagenet,             35
Each one already blazing by our meeds,
Should notwithstanding join our lights together
And over-shine the earth as this the world.
Whate'er it bodes, henceforward will I bear
Upon my target three fair-shining suns.             40

RICHARD
Nay, bear three daughters—by your leave I speak it—
You love the breeder better than the male.
    *Enter one blowing*
But what art thou whose heavy looks foretell
Some dreadful story hanging on thy tongue?

MESSENGER
Ah, one that was a woeful looker-on                 45
Whenas the noble Duke of York was slain—
Your princely father and my loving lord.

**EDWARD**

O, speak no more, for I have heard too much.

**RICHARD**

Say how he died, for I will hear it all.

**MESSENGER**

Environèd he was with many foes,                    50
And stood against them as the hope of Troy
Against the Greeks that would have entered Troy.
But Hercules himself must yield to odds;
And many strokes, though with a little axe,
Hews down and fells the hardest-timbered oak.       55
By many hands your father was subdued,
But only slaughtered by the ireful arm
Of unrelenting Clifford and the Queen,
Who crowned the gracious Duke in high despite,
Laughed in his face, and when with grief he wept,   60
The ruthless Queen gave him to dry his cheeks
A napkin steepèd in the harmless blood
Of sweet young Rutland, by rough Clifford slain;
And after many scorns, many foul taunts,
They took his head, and on the gates of York        65
They set the same; and there it doth remain,
The saddest spectacle that e'er I viewed.

**EDWARD**

Sweet Duke of York, our prop to lean upon,
Now thou art gone, we have no staff, no stay.
O Clifford, boist'rous Clifford—thou hast slain     70
The flower of Europe for his chivalry,
And treacherously hast thou vanquished him—
For hand to hand he would have vanquished thee.
Now my soul's palace is become a prison.
Ah, would she break from hence that this my body    75
Might in the ground be closèd up in rest.
For never henceforth shall I joy again—
Never, O never, shall I see more joy.

**RICHARD**

I cannot weep, for all my body's moisture
Scarce serves to quench my furnace-burning heart;   80
Nor can my tongue unload my heart's great burden,
For selfsame wind that I should speak withal
Is kindling coals that fires all my breast,
And burns me up with flames that tears would quench.
To weep is to make less the depth of grief;         85
Tears, then, for babes—blows and revenge for me!
Richard, I bear thy name; I'll venge thy death
Or die renownèd by attempting it.

**EDWARD**

His name that valiant Duke hath left with thee,
His dukedom and his chair with me is left.          90

**RICHARD**

Nay, if thou be that princely eagle's bird,
Show thy descent by gazing 'gainst the sun:
For 'chair and dukedom', 'throne and kingdom' say—
Either that is thine or else thou wert not his.

*March. Enter the Earl of Warwick and the Marquis
of Montague ⌜with drummers, an ensign, and
soldiers⌝*

**WARWICK**

How now, fair lords? What fare? What news abroad?

**RICHARD**

Great lord of Warwick, if we should recount         96
Our baleful news, and at each word's deliverance
Stab poniards in our flesh till all were told,
The words would add more anguish than the wounds.
O valiant lord, the Duke of York is slain.          100

**EDWARD**

O Warwick, Warwick! That Plantagenet,

Which held thee dearly as his soul's redemption,
Is by the stern Lord Clifford done to death.

**WARWICK**

Ten days ago I drowned these news in tears.
And now, to add more measure to your woes,          105
I come to tell you things sith then befall'n.
After the bloody fray at Wakefield fought,
Where your brave father breathed his latest gasp,
Tidings, as swiftly as the posts could run,
Were brought me of your loss and his depart.        110
I then in London, keeper of the King,
Mustered my soldiers, gathered flocks of friends,
And, very well appointed as I thought,
Marched toward Saint Albans to intercept the Queen,
Bearing the King in my behalf along—                115
For by my scouts I was advertisèd
That she was coming with a full intent
To dash our late decree in Parliament
Touching King Henry's oath and your succession.
Short tale to make, we at Saint Albans met,         120
Our battles joined, and both sides fiercely fought;
But whether 'twas the coldness of the King,
Who looked full gently on his warlike queen,
That robbed my soldiers of their heated spleen,
Or whether 'twas report of her success,             125
Or more than common fear of Clifford's rigour—
Who thunders to his captains blood and death—
I cannot judge; but, to conclude with truth,
Their weapons like to lightning came and went;
Our soldiers', like the night-owl's lazy flight,    130
Or like an idle thresher with a flail,
Fell gently down, as if they struck their friends.
I cheered them up with justice of our cause,
With promise of high pay, and great rewards.
But all in vain. They had no heart to fight,        135
And we in them no hope to win the day.
So that we fled—the King unto the Queen,
Lord George your brother, Norfolk, and myself
In haste, post-haste, are come to join with you.
For in the Marches here we heard you were,          140
Making another head to fight again.

**EDWARD**

Where is the Duke of Norfolk, gentle Warwick?
And when came George from Burgundy to England?

**WARWICK**

Some six miles off the Duke is with his soldiers;
And for your brother—he was lately sent             145
From your kind aunt, Duchess of Burgundy,
With aid of soldiers to this needful war.

**RICHARD**

'Twas odd belike when valiant Warwick fled.
Oft have I heard his praises in pursuit,
But ne'er till now his scandal of retire.           150

**WARWICK**

Nor now my scandal, Richard, dost thou hear—
For thou shalt know this strong right hand of mine
Can pluck the diadem from faint Henry's head
And wring the aweful sceptre from his fist,
Were he as famous and as bold in war                155
As he is famed for mildness, peace, and prayer.

**RICHARD**

I know it well, Lord Warwick—blame me not.
'Tis love I bear thy glories make me speak.
But in this troublous time what's to be done?
Shall we go throw away our coats of steel,          160
And wrap our bodies in black mourning gowns,

Numb'ring our Ave-Maries with our beads?
Or shall we on the helmets of our foes
Tell our devotion with revengeful arms?
If for the last, say 'ay', and to it, lords.                    165
WARWICK
Why, therefore Warwick came to seek you out,
And therefore comes my brother Montague.
Attend me, lords. The proud insulting Queen,
With Clifford and the haught Northumberland,
And of their feather many more proud birds,            170
Have wrought the easy-melting King like wax.
(To Edward) He swore consent to your succession,
His oath enrollèd in the Parliament.
And now to London all the crew are gone,
To frustrate both his oath and what beside               175
May make against the house of Lancaster.
Their power, I think, is thirty thousand strong.
Now, if the help of Norfolk and myself,
With all the friends that thou, brave Earl of March,
Amongst the loving Welshmen canst procure,          180
Will but amount to five-and-twenty thousand,
Why, via, to London will we march,
And once again bestride our foaming steeds,
And once again cry 'Charge!' upon our foes—
But never once again turn back and fly.                     185
RICHARD
Ay, now methinks I hear great Warwick speak.
Ne'er may he live to see a sunshine day
That cries 'Retire!' if Warwick bid him stay.
EDWARD
Lord Warwick, on thy shoulder will I lean,
And when thou fail'st—as God forbid the hour—   190
Must Edward fall, which peril heaven forfend!
WARWICK
No longer Earl of March, but Duke of York;
The next degree is England's royal throne—
For King of England shalt thou be proclaimed
In every borough as we pass along,                            195
And he that throws not up his cap for joy,
Shall for the fault make forfeit of his head.
King Edward, valiant Richard, Montague—
Stay we no longer dreaming of renown,
But sound the trumpets and about our task.            200
RICHARD
Then, Clifford, were thy heart as hard as steel,
As thou hast shown it flinty by thy deeds,
I come to pierce it or to give thee mine.
EDWARD
Then strike up drums—God and Saint George for us!
      *Enter a Messenger*
WARWICK How now? What news?                             205
MESSENGER
The Duke of Norfolk sends you word by me
The Queen is coming with a puissant host,
And craves your company for speedy counsel.
WARWICK
Why then it sorts. Brave warriors, let's away.
                              ⌜*March.*⌝ *Exeunt*

**2.2**  ⌜*York's head is thrust out, above.*⌝
      *Flourish. Enter King Henry, Queen Margaret, Lord*
      *Clifford, the Earl of Northumberland, and young*
      *Prince Edward, with a drummer and trumpeters*
QUEEN MARGARET
Welcome, my lord, to this brave town of York.
Yonder's the head of that arch-enemy

That sought to be encompassed with your crown.
Doth not the object cheer your heart, my lord?
KING HENRY
Ay, as the rocks cheer them that fear their wreck.   5
To see this sight, it irks my very soul.
Withhold revenge, dear God—'tis not my fault,
Nor wittingly have I infringed my vow.
CLIFFORD
My gracious liege, this too much lenity
And harmful pity must be laid aside.                          10
To whom do lions cast their gentle looks?
Not to the beast that would usurp their den.
Whose hand is that the forest bear doth lick?
Not his that spoils her young before her face.
Who scapes the lurking serpent's mortal sting?      15
Not he that sets his foot upon her back.
The smallest worm will turn, being trodden on,
And doves will peck in safeguard of their brood.
Ambitious York did level at thy crown,
Thou smiling while he knit his angry brows.            20
He, but a duke, would have his son a king,
And raise his issue like a loving sire;
Thou, being a king, blest with a goodly son,
Didst yield consent to disinherit him,
Which argued thee a most unloving father.             25
Unreasonable creatures feed their young,
And though man's face be fearful to their eyes,
Yet, in protection of their tender ones,
Who hath not seen them, even with those wings
Which sometime they have used with fearful flight,  30
Make war with him that climbed unto their nest,
Offering their own lives in their young's defence?
For shame, my liege, make them your precedent!
Were it not pity that this goodly boy
Should lose his birthright by his father's fault,        35
And long hereafter say unto his child
'What my great-grandfather and grandsire got
My careless father fondly gave away'?
Ah, what a shame were this! Look on the boy,
And let his manly face, which promiseth                  40
Successful fortune, steel thy melting heart
To hold thine own and leave thine own with him.
KING HENRY
Full well hath Clifford played the orator,
Inferring arguments of mighty force.
But, Clifford, tell me—didst thou never hear          45
That things ill got had ever bad success?
And happy always was it for that son
Whose father for his hoarding went to hell?
I'll leave my son my virtuous deeds behind,
And would my father had left me no more.              50
For all the rest is held at such a rate
As brings a thousandfold more care to keep
Than in possession any jot of pleasure.
Ah, cousin York, would thy best friends did know
How it doth grieve me that thy head is here.          55
QUEEN MARGARET
My lord, cheer up your spirits—our foes are nigh,
And this soft courage makes your followers faint.
You promised knighthood to our forward son.
Unsheathe your sword and dub him presently.
Edward, kneel down.                                                 60
      *Prince Edward kneels*
KING HENRY
Edward Plantagenet, arise a knight—
And learn this lesson: draw thy sword in right.

PRINCE EDWARD (*rising*)
  My gracious father, by your kingly leave,
  I'll draw it as apparent to the crown,
  And in that quarrel use it to the death.          65
CLIFFORD
  Why, that is spoken like a toward prince.
          *Enter a Messenger*
MESSENGER
  Royal commanders, be in readiness—
  For with a band of thirty thousand men
  Comes Warwick backing of the Duke of York;
  And in the towns, as they do march along,          70
  Proclaims him king, and many fly to him.
  Darraign your battle, for they are at hand.
CLIFFORD (*to King Henry*)
  I would your highness would depart the field—
  The Queen hath best success when you are absent.
QUEEN MARGARET (*to King Henry*)
  Ay, good my lord, and leave us to our fortune.          75
KING HENRY
  Why, that's my fortune too—therefore I'll stay.
NORTHUMBERLAND
  Be it with resolution then to fight.
PRINCE EDWARD (*to King Henry*)
  My royal father, cheer these noble lords
  And hearten those that fight in your defence.
  Unsheathe your sword, good father; cry 'Saint George!'
          *March. Enter Edward Duke of York, the Earl of*
          *Warwick, Richard, George, the Duke of Norfolk, the*
          *Marquis of Montague, and soldiers*
EDWARD
  Now, perjured Henry, wilt thou kneel for grace,          81
  And set thy diadem upon my head—
  Or bide the mortal fortune of the field?
QUEEN MARGARET
  Go rate thy minions, proud insulting boy!
  Becomes it thee to be thus bold in terms          85
  Before thy sovereign and thy lawful king?
EDWARD
  I am his king, and he should bow his knee.
  I was adopted heir by his consent.
GEORGE (*to Queen Margaret*)
  Since when his oath is broke—for, as I hear,
  You that are king, though he do wear the crown,          90
  Have caused him by new act of Parliament
  To blot our brother out, and put his own son in.
CLIFFORD And reason too—
  Who should succeed the father but the son?
RICHARD
  Are you there, butcher? O, I cannot speak!          95
CLIFFORD
  Ay, crookback, here I stand to answer thee,
  Or any he the proudest of thy sort.
RICHARD
  'Twas you that killed young Rutland, was it not?
CLIFFORD
  Ay, and old York, and yet not satisfied.
RICHARD
  For God's sake, lords, give signal to the fight.          100
WARWICK
  What sayst thou, Henry, wilt thou yield the crown?
QUEEN MARGARET
  Why, how now, long-tongued Warwick, dare you
          speak?
  When you and I met at Saint Albans last,
  Your legs did better service than your hands.

WARWICK
  Then 'twas my turn to fly—and now 'tis thine.          105
CLIFFORD
  You said so much before, and yet you fled.
WARWICK
  'Twas not your valour, Clifford, drove me thence.
NORTHUMBERLAND
  No, nor your manhood that durst make you stay.
RICHARD
  Northumberland, I hold thee reverently.
  Break off the parley, for scarce I can refrain          110
  The execution of my big-swoll'n heart
  Upon that Clifford, that cruel child-killer.
CLIFFORD
  I slew thy father—call'st thou him a child?
RICHARD
  Ay, like a dastard and a treacherous coward,
  As thou didst kill our tender brother Rutland.          115
  But ere sun set I'll make thee curse the deed.
KING HENRY
  Have done with words, my lords, and hear me speak.
QUEEN MARGARET
  Defy them, then, or else hold close thy lips.
KING HENRY
  I prithee give no limits to my tongue—
  I am a king, and privileged to speak.          120
CLIFFORD
  My liege, the wound that bred this meeting here
  Cannot be cured by words—therefore be still.
RICHARD
  Then, executioner, unsheathe thy sword.
  By him that made us all, I am resolved
  That Clifford's manhood lies upon his tongue.          125
EDWARD
  Say, Henry, shall I have my right or no?
  A thousand men have broke their fasts today
  That ne'er shall dine unless thou yield the crown.
WARWICK (*to King Henry*)
  If thou deny, their blood upon thy head;
  For York in justice puts his armour on.          130
PRINCE EDWARD
  If that be right which Warwick says is right,
  There is no wrong, but everything is right.
RICHARD
  Whoever got thee, there thy mother stands—
  For, well I wot, thou hast thy mother's tongue.
QUEEN MARGARET
  But thou art neither like thy sire nor dam,          135
  But like a foul misshapen stigmatic,
  Marked by the destinies to be avoided,
  As venom toads or lizards' dreadful stings.
RICHARD
  Iron of Naples, hid with English gilt,
  Whose father bears the title of a king—          140
  As if a channel should be called the sea—
  Sham'st thou not, knowing whence thou art
          extraught,
  To let thy tongue detect thy base-born heart?
EDWARD
  A wisp of straw were worth a thousand crowns
  To make this shameless callet know herself.          145
  Helen of Greece was fairer far than thou,
  Although thy husband may be Menelaus;
  And ne'er was Agamemnon's brother wronged
  By that false woman, as this king by thee.
  His father revelled in the heart of France,          150

And tamed the King, and made the Dauphin stoop;
And had he matched according to his state,
He might have kept that glory to this day.
But when he took a beggar to his bed,
And graced thy poor sire with his bridal day,        155
Even then that sunshine brewed a shower for him
That washed his father's fortunes forth of France,
And heaped sedition on his crown at home.
For what hath broached this tumult but thy pride?
Hadst thou been meek, our title still had slept,        160
And we, in pity of the gentle King,
Had slipped our claim until another age.

GEORGE (*to Queen Margaret*)
But when we saw our sunshine made thy spring,
And that thy summer bred us no increase,
We set the axe to thy usurping root.        165
And though the edge hath something hit ourselves,
Yet know thou, since we have begun to strike,
We'll never leave till we have hewn thee down,
Or bathed thy growing with our heated bloods.

EDWARD (*to Queen Margaret*)
And in this resolution I defy thee,        170
Not willing any longer conference
Since thou deniest the gentle King to speak.
Sound trumpets—let our bloody colours wave!
And either victory, or else a grave!

QUEEN MARGARET Stay, Edward.        175

EDWARD
No, wrangling woman, we'll no longer stay—
These words will cost ten thousand lives this day.
⌜*Flourish. March. Exeunt Edward and his men
at one door and Queen Margaret and her men
at another door*⌝

**2.3**    *Alarum. Excursions. Enter the Earl of Warwick*

WARWICK
Forespent with toil, as runners with a race,
I lay me down a little while to breathe;
For strokes received, and many blows repaid,
Have robbed my strong-knit sinews of their strength,
And, spite of spite, needs must I rest a while.        5
*Enter Edward, the Duke of York, running*

EDWARD
Smile, gentle heaven, or strike, ungentle death!
For this world frowns, and Edward's sun is clouded.

WARWICK
How now, my lord, what hap? What hope of good?
*Enter George, ⌜running⌝*

GEORGE
Our hap is loss, our hope but sad despair;
Our ranks are broke, and ruin follows us.        10
What counsel give you? Whither shall we fly?

EDWARD
Bootless is flight—they follow us with wings,
And weak we are, and cannot shun pursuit.
*Enter Richard, ⌜running⌝*

RICHARD
Ah, Warwick, why hast thou withdrawn thyself?
Thy brother's blood the thirsty earth hath drunk,        15
Broached with the steely point of Clifford's lance.
And in the very pangs of death he cried,
Like to a dismal clangour heard from far,
'Warwick, revenge—brother, revenge my death!'
So, underneath the belly of their steeds        20
That stained their fetlocks in his smoking blood,
The noble gentleman gave up the ghost.

WARWICK
Then let the earth be drunken with our blood.
I'll kill my horse, because I will not fly.
Why stand we like soft-hearted women here,        25
Wailing our losses, whiles the foe doth rage;
And look upon, as if the tragedy
Were played in jest by counterfeiting actors?
(*Kneeling*) Here, on my knee, I vow to God above
I'll never pause again, never stand still,        30
Till either death hath closed these eyes of mine
Or fortune given me measure of revenge.

EDWARD (*kneeling*)
O, Warwick, I do bend my knee with thine,
And in this vow do chain my soul to thine.
And, ere my knee rise from the earth's cold face,        35
I throw my hands, mine eyes, my heart to Thee,
Thou setter up and plucker down of kings,
Beseeching Thee, if with Thy will it stands
That to my foes this body must be prey,
Yet that Thy brazen gates of heaven may ope        40
And give sweet passage to my sinful soul.
⌜*They rise*⌝
Now, lords, take leave until we meet again,
Where'er it be, in heaven or in earth.

RICHARD
Brother, give me thy hand; and, gentle Warwick,
Let me embrace thee in my weary arms.        45
I, that did never weep, now melt with woe
That winter should cut off our springtime so.

WARWICK
Away, away! Once more, sweet lords, farewell.

GEORGE
Yet let us all together to our troops,
And give them leave to fly that will not stay;        50
And call them pillars that will stand to us;
And, if we thrive, promise them such rewards
As victors wear at the Olympian games.
This may plant courage in their quailing breasts,
For yet is hope of life and victory.        55
Forslow no longer—make we hence amain.        *Exeunt*

**2.4**    ⌜*Alarums.*⌝ *Excursions. Enter Richard* ⌜*at one door*⌝
            *and Lord Clifford* ⌜*at the other*⌝

RICHARD
Now, Clifford, I have singled thee alone.
Suppose this arm is for the Duke of York,
And this for Rutland, both bound to revenge,
Wert thou environed with a brazen wall.

CLIFFORD
Now, Richard, I am with thee here alone.        5
This is the hand that stabbed thy father York,
And this the hand that slew thy brother Rutland,
And here's the heart that triumphs in their death
And cheers these hands that slew thy sire and brother
To execute the like upon thyself—        10
And so, have at thee!
*They fight. The Earl of Warwick comes and rescues
Richard. Lord Clifford flies*

RICHARD
Nay, Warwick, single out some other chase—
For I myself will hunt this wolf to death.        *Exeunt*

**2.5**    *Alarum. Enter King Henry*

KING HENRY
This battle fares like to the morning's war,
    When dying clouds contend with growing light,
    What time the shepherd, blowing of his nails,

Can neither call it perfect day nor night.
Now sways it this way like a mighty sea                         5
  Forced by the tide to combat with the wind,
Now sways it that way like the selfsame sea
  Forced to retire by fury of the wind.
Sometime the flood prevails, and then the wind;
Now one the better, then another best—                        10
Both tugging to be victors, breast to breast,
Yet neither conqueror nor conquerèd.
So is the equal poise of this fell war.
Here on this molehill will I sit me down.
To whom God will, there be the victory.                       15
For Margaret my queen, and Clifford, too,
Have chid me from the battle, swearing both
They prosper best of all when I am thence.
Would I were dead, if God's good will were so—
For what is in this world but grief and woe?                   20
O God! Methinks it were a happy life
To be no better than a homely swain.
To sit upon a hill, as I do now;
To carve out dials quaintly, point by point,
Thereby to see the minutes how they run:                      25
How many makes the hour full complete,
How many hours brings about the day,
How many days will finish up the year,
How many years a mortal man may live.
When this is known, then to divide the times:                 30
So many hours must I tend my flock,
So many hours must I take my rest,
So many hours must I contemplate,
So many hours must I sport myself,
So many days my ewes have been with young,                    35
So many weeks ere the poor fools will ean,
So many years ere I shall shear the fleece.
So minutes, hours, days, weeks, months, and years,
Passed over to the end they were created,
Would bring white hairs unto a quiet grave.                    40
Ah, what a life were this! How sweet! How lovely!
Gives not the hawthorn bush a sweeter shade
To shepherds looking on their seely sheep
Than doth a rich embroidered canopy
To kings that fear their subjects' treachery?                  45
O yes, it doth—a thousandfold it doth.
And to conclude, the shepherd's homely curds,
His cold thin drink out of his leather bottle,
His wonted sleep under a fresh tree's shade,
All which secure and sweetly he enjoys,                        50
Is far beyond a prince's delicates,
His viands sparkling in a golden cup,
His body couchèd in a curious bed,
When care, mistrust, and treason waits on him.
      *Alarum. Enter ⌈at one door⌉ a Soldier with a dead*
      *man in his arms. King Henry stands apart*
SOLDIER
Ill blows the wind that profits nobody.                        55
This man, whom hand to hand I slew in fight,
May be possessèd with some store of crowns;
And I, that haply take them from him now,
May yet ere night yield both my life and them
To some man else, as this dead man doth me.                    60
      ⌈*He removes the dead man's helmet*⌉
Who's this? O God! It is my father's face
Whom in this conflict I, unwares, have killed.
O, heavy times, begetting such events!
From London by the King was I pressed forth;
My father, being the Earl of Warwick's man,                    65
Came on the part of York, pressed by his master;

And I, who at his hands received my life,
Have by my hands of life bereavèd him.
Pardon me, God, I knew not what I did;
And pardon, father, for I knew not thee.                       70
My tears shall wipe away these bloody marks,
And no more words till they have flowed their fill.
      *He weeps*
KING HENRY
O piteous spectacle! O bloody times!
Whiles lions war and battle for their dens,
Poor harmless lambs abide their enmity.                        75
Weep, wretched man, I'll aid thee tear for tear;
And let our hearts and eyes, like civil war,
Be blind with tears, and break, o'ercharged with grief.
      *Enter ⌈at another door⌉ another Soldier with a dead*
      *man ⌈in his arms⌉*
SECOND SOLDIER
Thou that so stoutly hath resisted me,
Give me thy gold, if thou hast any gold—                       80
For I have bought it with an hundred blows.
      ⌈*He removes the dead man's helmet*⌉
But let me see: is this our foeman's face?
Ah, no, no, no—it is mine only son!
Ah, boy, if any life be left in thee,
Throw up thine eye! (*Weeping*) See, see, what showers
      arise,                                                    85
Blown with the windy tempest of my heart,
Upon thy wounds, that kills mine eye and heart!
O, pity, God, this miserable age!
What stratagems, how fell, how butcherly,
Erroneous, mutinous, and unnatural,                            90
This deadly quarrel daily doth beget!
O boy, thy father gave thee life too soon,
And hath bereft thee of thy life too late!
KING HENRY
Woe above woe! Grief more than common grief!
O that my death would stay these ruthful deeds!               95
O, pity, pity, gentle heaven, pity!
The red rose and the white are on his face,
The fatal colours of our striving houses;
The one his purple blood right well resembles,
The other his pale cheeks, methinks, presenteth.             100
Wither one rose, and let the other flourish—
If you contend, a thousand lives must wither.
FIRST SOLDIER
How will my mother for a father's death
Take on with me, and ne'er be satisfied!
SECOND SOLDIER
How will my wife for slaughter of my son                     105
Shed seas of tears, and ne'er be satisfied!
KING HENRY
How will the country for these woeful chances
Misthink the King, and not be satisfied!
FIRST SOLDIER
Was ever son so rued a father's death?
SECOND SOLDIER
Was ever father so bemoaned his son?                         110
KING HENRY
Was ever king so grieved for subjects' woe?
Much is your sorrow, mine ten times so much.
FIRST SOLDIER (*to his father's body*)
I'll bear thee hence where I may weep my fill.
      *Exit ⌈at one door⌉ with the body of his father*
SECOND SOLDIER (*to his son's body*)
These arms of mine shall be thy winding sheet;
My heart, sweet boy, shall be thy sepulchre,                 115

For from my heart thine image ne'er shall go.
My sighing breast shall be thy funeral bell,
And so obsequious will thy father be,
E'en for the loss of thee, having no more,
As Priam was for all his valiant sons.      120
I'll bear thee hence, and let them fight that will—
For I have murdered where I should not kill.
     *Exit ⌈at another door⌉ with the body of his son*

KING HENRY
Sad-hearted men, much overgone with care,
Here sits a king more woeful than you are.
     *Alarums. Excursions. Enter Prince Edward*

PRINCE EDWARD
Fly, father, fly—for all your friends are fled,      125
And Warwick rages like a chafèd bull!
Away—for death doth hold us in pursuit!
     ⌈*Enter Queen Margaret*⌉

QUEEN MARGARET
Mount you, my lord—towards Berwick post amain.
Edward and Richard, like a brace of greyhounds
Having the fearful flying hare in sight,      130
With fiery eyes sparkling for very wrath,
And bloody steel grasped in their ireful hands,
Are at our backs—and therefore hence amain.
     ⌈*Enter Exeter*⌉

EXETER
Away—for vengeance comes along with them!
Nay—stay not to expostulate—make speed—      135
Or else come after. I'll away before.

KING HENRY
Nay, take me with thee, good sweet Exeter.
Not that I fear to stay, but love to go
Whither the Queen intends. Forward, away.      *Exeunt*

**2.6**    *A loud alarum. Enter Lord Clifford, wounded ⌈with*
         *an arrow in his neck⌉*

CLIFFORD
Here burns my candle out—ay, here it dies,
Which, whiles it lasted, gave King Henry light.
O Lancaster, I fear thy overthrow
More than my body's parting with my soul!
My love and fear glued many friends to thee—      5
And, now I fall, thy tough commixture melts,
Impairing Henry, strength'ning misproud York.
The common people swarm like summer flies,
And whither fly the gnats but to the sun?
And who shines now but Henry's enemies?      10
O Phoebus, hadst thou never given consent
That Phaëton should check thy fiery steeds,
Thy burning car never had scorched the earth!
And, Henry, hadst thou swayed as kings should do,
Or as thy father and his father did,      15
Giving no ground unto the house of York,
They never then had sprung like summer flies;
I and ten thousand in this luckless realm
Had left no mourning widows for our death;
And thou this day hadst kept thy chair in peace.      20
For what doth cherish weeds, but gentle air?
And what makes robbers bold, but too much lenity?
Bootless are plaints, and cureless are my wounds;
No way to fly, nor strength to hold out flight;
The foe is merciless and will not pity,      25
For at their hands I have deservèd no pity.
The air hath got into my deadly wounds,
And much effuse of blood doth make me faint.

Come York and Richard, Warwick and the rest—
I stabbed your fathers' bosoms; split my breast.      30
     ⌈*He faints.*⌉
     *Alarum and retreat. Enter Edward Duke of York,*
     *his brothers George and Richard, the Earl of*
     *Warwick, ⌈the Marquis of Montague,⌉ and soldiers*

EDWARD
Now breathe we, lords—good fortune bids us pause,
And smooth the frowns of war with peaceful looks.
Some troops pursue the bloody-minded Queen,
That led calm Henry, though he were a king,
As doth a sail filled with a fretting gust      35
Command an argosy to stem the waves.
But think you, lords, that Clifford fled with them?

WARWICK
No—'tis impossible he should escape;
For, though before his face I speak the words,
Your brother Richard marked him for the grave.      40
And whereso'er he is, he's surely dead.
     *Clifford groans*

⌈EDWARD⌉
Whose soul is that which takes her heavy leave?

⌈RICHARD⌉
A deadly groan, like life and death's departing.

⌈EDWARD⌉ ⌈*to Richard*⌉
See who it is.
     ⌈*Richard goes to Clifford*⌉
               And now the battle's ended,
If friend or foe, let him be gently used.      45

RICHARD
Revoke that doom of mercy, for 'tis Clifford;
Who not contented that he lopped the branch
In hewing Rutland when his leaves put forth,
But set his murd'ring knife unto the root
From whence that tender spray did sweetly spring—     
I mean our princely father, Duke of York.      51

WARWICK
From off the gates of York fetch down the head,
Your father's head, which Clifford placèd there.
Instead whereof let this supply the room—
Measure for measure must be answerèd.      55

EDWARD
Bring forth that fatal screech-owl to our house,
That nothing sung but death to us and ours.
     ⌈*Clifford is dragged forward*⌉
Now death shall stop his dismal threat'ning sound
And his ill-boding tongue no more shall speak.

WARWICK
I think his understanding is bereft.      60
Speak, Clifford, dost thou know who speaks to thee?
Dark cloudy death o'ershades his beams of life,
And he nor sees nor hears us what we say.

RICHARD
O, would he did—and so perhaps he doth.
'Tis but his policy to counterfeit,      65
Because he would avoid such bitter taunts
Which in the time of death he gave our father.

GEORGE
If so thou think'st, vex him with eager words.

RICHARD
Clifford, ask mercy and obtain no grace.

EDWARD
Clifford, repent in bootless penitence.      70

WARWICK
Clifford, devise excuses for thy faults.

GEORGE
  While we devise fell tortures for thy faults.
RICHARD
  Thou didst love York, and I am son to York.
EDWARD
  Thou pitied'st Rutland—I will pity thee.
GEORGE
  Where's Captain Margaret to fence you now?  75
WARWICK
  They mock thee, Clifford—swear as thou wast wont.
RICHARD
  What, not an oath? Nay, then, the world goes hard
  When Clifford cannot spare his friends an oath.
  I know by that he's dead—and, by my soul,
  If this right hand would buy but two hours' life  80
  That I, in all despite, might rail at him,
  This hand should chop it off, and with the issuing
       blood
  Stifle the villain whose unstanchèd thirst
  York and young Rutland could not satisfy.
WARWICK
  Ay, but he's dead. Off with the traitor's head,  85
  And rear it in the place your father's stands.
  And now to London with triumphant march,
  There to be crownèd England's royal king;
  From whence shall Warwick cut the sea to France,
  And ask the Lady Bona for thy queen.  90
  So shalt thou sinew both these lands together.
  And, having France thy friend, thou shalt not dread
  The scattered foe that hopes to rise again,
  For though they cannot greatly sting to hurt,
  Yet look to have them buzz to offend thine ears.  95
  First will I see the coronation,
  And then to Brittany I'll cross the sea
  To effect this marriage, so it please my lord.
EDWARD
  Even as thou wilt, sweet Warwick, let it be.
  For in thy shoulder do I build my seat,  100
  And never will I undertake the thing
  Wherein thy counsel and consent is wanting.
  Richard, I will create thee Duke of Gloucester,
  And George, of Clarence; Warwick, as ourself,
  Shall do and undo as him pleaseth best.  105
RICHARD
  Let me be Duke of Clarence, George of Gloucester—
  For Gloucester's dukedom is too ominous.
WARWICK
  Tut, that's a foolish observation—
  Richard, be Duke of Gloucester. Now to London
  To see these honours in possession.  110
                              Exeunt. ⌈York's head is removed⌉

3.1  *Enter two Gamekeepers, with crossbows in their*
     *hands*
FIRST GAMEKEEPER
  Under this thick-grown brake we'll shroud ourselves,
  For through this laund anon the deer will come,
  And in this covert will we make our stand,
  Culling the principal of all the deer.
SECOND GAMEKEEPER
  I'll stay above the hill, so both may shoot.  5
FIRST GAMEKEEPER
  That cannot be—the noise of thy crossbow
  Will scare the herd, and so my shoot is lost.
  Here stand we both, and aim we at the best.
  And, for the time shall not seem tedious,

  I'll tell thee what befell me on a day  10
  In this self place where now we mean to stand.
FIRST GAMEKEEPER
  Here comes a man—let's stay till he be past.
                              *They stand apart.*
        *Enter King Henry, disguised, carrying a prayer-book*
KING HENRY
  From Scotland am I stolen, even of pure love,
  To greet mine own land with my wishful sight.
  No, Harry, Harry—'tis no land of thine.  15
  Thy place is filled, thy sceptre wrung from thee,
  Thy balm washed off wherewith thou wast anointed.
  No bending knee will call thee Caesar now,
  No humble suitors press to speak for right,
  No, not a man comes for redress of thee—  20
  For how can I help them and not myself?
FIRST GAMEKEEPER (*to the Second Gamekeeper*)
  Ay, here's a deer whose skin's a keeper's fee:
  This is the quondam king—let's seize upon him.
KING HENRY
  Let me embrace thee, sour adversity,
  For wise men say it is the wisest course.  25
SECOND GAMEKEEPER (*to the First Gamekeeper*)
  Why linger we? Let us lay hands upon him.
FIRST GAMEKEEPER (*to the Second Gamekeeper*)
  Forbear awhile—we'll hear a little more.
KING HENRY
  My queen and son are gone to France for aid,
  And, as I hear, the great commanding Warwick
  Is thither gone to crave the French King's sister  30
  To wife for Edward. If this news be true,
  Poor Queen and son, your labour is but lost—
  For Warwick is a subtle orator,
  And Louis a prince soon won with moving words.
  By this account, then, Margaret may win him—  35
  For she's a woman to be pitied much.
  Her sighs will make a batt'ry in his breast,
  Her tears will pierce into a marble heart,
  The tiger will be mild whiles she doth mourn,
  And Nero will be tainted with remorse  40
  To hear and see her plaints, her brinish tears.
  Ay, but she's come to beg; Warwick to give.
  She on his left side, craving aid for Henry;
  He on his right, asking a wife for Edward.
  She weeps and says her Henry is deposed,  45
  He smiles and says his Edward is installed;
  That she, poor wretch, for grief can speak no more,
  Whiles Warwick tells his title, smooths the wrong,
  Inferreth arguments of mighty strength,
  And in conclusion wins the King from her  50
  With promise of his sister and what else
  To strengthen and support King Edward's place.
  O, Margaret, thus 'twill be; and thou, poor soul,
  Art then forsaken, as thou went'st forlorn.
SECOND GAMEKEEPER (*coming forward*)
  Say, what art thou that talk'st of kings and queens?
KING HENRY
  More than I seem, and less than I was born to:  56
  A man at least, for less I should not be;
  And men may talk of kings, and why not I?
SECOND GAMEKEEPER
  Ay, but thou talk'st as if thou wert a king.
KING HENRY
  Why, so I am, in mind—and that's enough.  60
SECOND GAMEKEEPER
  But if thou be a king, where is thy crown?

KING HENRY
My crown is in my heart, not on my head;
Not decked with diamonds and Indian stones,
Nor to be seen. My crown is called content—
A crown it is that seldom kings enjoy.     65

SECOND GAMEKEEPER
Well, if you be a king crowned with content,
Your crown content and you must be contented
To go along with us—for, as we think,
You are the king King Edward hath deposed,
And we his subjects sworn in all allegiance     70
Will apprehend you as his enemy.

KING HENRY
But did you never swear and break an oath?

SECOND GAMEKEEPER
No—never such an oath, nor will not now.

KING HENRY
Where did you dwell when I was King of England?

SECOND GAMEKEEPER
Here in this country, where we now remain.     75

KING HENRY
I was anointed king at nine months old,
My father and my grandfather were kings,
And you were sworn true subjects unto me—
And tell me, then, have you not broke your oaths?

FIRST GAMEKEEPER
No, for we were subjects but while you were king.     80

KING HENRY
Why, am I dead? Do I not breathe a man?
Ah, simple men, you know not what you swear.
Look as I blow this feather from my face,
And as the air blows it to me again,
Obeying with my wind when I do blow,     85
And yielding to another when it blows,
Commanded always by the greater gust—
Such is the lightness of you common men.
But do not break your oaths, for of that sin
My mild entreaty shall not make you guilty.     90
Go where you will, the King shall be commanded;
And be you kings, command, and I'll obey.

FIRST GAMEKEEPER
We are true subjects to the King, King Edward.

KING HENRY
So would you be again to Henry,
If he were seated as King Edward is.     95

FIRST GAMEKEEPER
We charge you, in God's name and in the King's,
To go with us unto the officers.

KING HENRY
In God's name, lead; your king's name be obeyed;
And what God will, that let your king perform;
And what he will I humbly yield unto.     *Exeunt*

3.2     *Enter King Edward, Richard Duke of Gloucester,*
       *George Duke of Clarence, and the Lady Gray*

KING EDWARD
Brother of Gloucester, at Saint Albans field
This lady's husband, Sir Richard Gray, was slain,
His lands then seized on by the conqueror.
Her suit is now to repossess those lands,
Which we in justice cannot well deny,     5
Because in quarrel of the house of York
The worthy gentleman did lose his life.

RICHARD OF GLOUCESTER
Your highness shall do well to grant her suit—
It were dishonour to deny it her.

KING EDWARD
It were no less; but yet I'll make a pause.     10

RICHARD OF GLOUCESTER (*aside to George*) Yea, is it so?
I see the lady hath a thing to grant
Before the King will grant her humble suit.

GEORGE OF CLARENCE (*aside to Richard*)
He knows the game; how true he keeps the wind!

RICHARD OF GLOUCESTER (*aside to George*) Silence.     15

KING EDWARD (*to Lady Gray*)
Widow, we will consider of your suit;
And come some other time to know our mind.

LADY GRAY
Right gracious lord, I cannot brook delay.
May it please your highness to resolve me now,
And what your pleasure is shall satisfy me.     20

RICHARD OF GLOUCESTER (*aside to George*)
Ay, widow? Then I'll warrant you all your lands
An if what pleases him shall pleasure you.
Fight closer, or, good faith, you'll catch a blow.

GEORGE OF CLARENCE (*aside to Richard*)
I fear her not unless she chance to fall.

RICHARD OF GLOUCESTER (*aside to George*)
God forbid that! For he'll take vantages.     25

KING EDWARD (*to Lady Gray*)
How many children hast thou, widow? Tell me.

GEORGE OF CLARENCE (*aside to Richard*)
I think he means to beg a child of her.

RICHARD OF GLOUCESTER (*aside to George*)
Nay, whip me then—he'll rather give her two.

LADY GRAY (*to King Edward*) Three, my most gracious lord.

RICHARD OF GLOUCESTER (*aside*)
You shall have four, an you'll be ruled by him.     30

KING EDWARD (*to Lady Gray*)
'Twere pity they should lose their father's lands.

LADY GRAY
Be pitiful, dread lord, and grant it them.

KING EDWARD (*to Richard and George*)
Lords, give us leave—I'll try this widow's wit.

RICHARD OF GLOUCESTER ⌈*aside to George*⌉
Ay, good leave have you; for you will have leave,
Till youth take leave and leave you to the crutch.     35
      *Richard and George stand apart*

KING EDWARD (*to Lady Gray*)
Now tell me, madam, do you love your children?

LADY GRAY
Ay, full as dearly as I love myself.

KING EDWARD
And would you not do much to do them good?

LADY GRAY
To do them good I would sustain some harm.

KING EDWARD
Then get your husband's lands, to do them good.     40

LADY GRAY
Therefore I came unto your majesty.

KING EDWARD
I'll tell you how these lands are to be got.

LADY GRAY
So shall you bind me to your highness' service.

KING EDWARD
What service wilt thou do me, if I give them?

LADY GRAY
What you command, that rests in me to do.     45

KING EDWARD
But you will take exceptions to my boon.

LADY GRAY
No, gracious lord, except I cannot do it.

KING EDWARD
　Ay, but thou canst do what I mean to ask.
LADY GRAY
　Why, then, I will do what your grace commands.
RICHARD OF GLOUCESTER (*to George*)
　He plies her hard, and much rain wears the marble.
GEORGE OF CLARENCE
　As red as fire! Nay, then her wax must melt.　51
LADY GRAY (*to King Edward*)
　Why stops my lord? Shall I not hear my task?
KING EDWARD
　An easy task—'tis but to love a king.
LADY GRAY
　That's soon performed, because I am a subject.
KING EDWARD
　Why, then, thy husband's lands I freely give thee.　55
LADY GRAY (*curtsies*)
　I take my leave, with many thousand thanks.
RICHARD OF GLOUCESTER (*to George*)
　The match is made—she seals it with a curtsy.
KING EDWARD (*to Lady Gray*)
　But stay thee—'tis the fruits of love I mean.
LADY GRAY
　The fruits of love *I* mean, my loving liege.
KING EDWARD
　Ay, but I fear me in another sense.　60
　What love think'st thou I sue so much to get?
LADY GRAY
　My love till death, my humble thanks, my prayers—
　That love which virtue begs and virtue grants.
KING EDWARD
　No, by my troth, I did not mean such love.
LADY GRAY
　Why, then, you mean not as I thought you did.　65
KING EDWARD
　But now you partly may perceive my mind.
LADY GRAY
　My mind will never grant what I perceive
　Your highness aims at, if I aim aright.
KING EDWARD
　To tell thee plain, I aim to lie with thee.
LADY GRAY
　To tell *you* plain, I had rather lie in prison.　70
KING EDWARD
　Why, then, thou shalt not have thy husband's lands.
LADY GRAY
　Why, then, mine honesty shall be my dower;
　For by that loss I will not purchase them.
KING EDWARD
　Therein thou wrong'st thy children mightily.
LADY GRAY
　Herein your highness wrongs both them and me.　75
　But, mighty lord, this merry inclination
　Accords not with the sadness of my suit.
　Please you dismiss me either with ay or no.
KING EDWARD
　Ay, if thou wilt say 'ay' to my request;
　No, if thou dost say 'no' to my demand.　80
LADY GRAY
　Then, no, my lord—my suit is at an end.
RICHARD OF GLOUCESTER (*to George*)
　The widow likes him not—she knits her brows.
GEORGE OF CLARENCE
　He is the bluntest wooer in Christendom.
KING EDWARD (*aside*)
　Her looks doth argue her replete with modesty;
　Her words doth show her wit incomparable;　85

All her perfections challenge sovereignty.
One way or other, she is for a king;
And she shall be my love or else my queen.
(*To Lady Gray*) Say that King Edward take thee for his
　queen?
LADY GRAY
　'Tis better said than done, my gracious lord.　90
　I am a subject fit to jest withal,
　But far unfit to be a sovereign.
KING EDWARD
　Sweet widow, by my state I swear to thee
　I speak no more than what my soul intends,
　And that is to enjoy thee for my love.　95
LADY GRAY
　And that is more than I will yield unto.
　I know I am too mean to be your queen,
　And yet too good to be your concubine.
KING EDWARD
　You cavil, widow—I did mean my queen.
LADY GRAY
　'Twill grieve your grace my sons should call you father.
KING EDWARD
　No more than when my daughters call thee mother.
　Thou art a widow and thou hast some children;　102
　And, by God's mother, I, being but a bachelor,
　Have other some. Why, 'tis a happy thing
　To be the father unto many sons.　105
　Answer no more, for thou shalt be my queen.
RICHARD OF GLOUCESTER (*to George*)
　The ghostly father now hath done his shrift.
GEORGE OF CLARENCE
　When he was made a shriver, 'twas for shift.
KING EDWARD (*to Richard and George*)
　Brothers, you muse what chat we two have had.
　　　*Richard and George come forward*
RICHARD OF GLOUCESTER
　The widow likes it not, for she looks very sad.　110
KING EDWARD
　You'd think it strange if I should marry her.
GEORGE OF CLARENCE
　To who, my lord?
KING EDWARD　　　　Why, Clarence, to myself.
RICHARD OF GLOUCESTER
　That would be ten days' wonder at the least.
GEORGE OF CLARENCE
　That's a day longer than a wonder lasts.
RICHARD OF GLOUCESTER
　By so much is the wonder in extremes.　115
KING EDWARD
　Well, jest on, brothers—I can tell you both
　Her suit is granted for her husband's lands.
　　　*Enter a Nobleman*
NOBLEMAN
　My gracious lord, Henry your foe is taken
　And brought as prisoner to your palace gate.
KING EDWARD
　See that he be conveyed unto the Tower—　120
　(*To Richard and George*)
　And go we, brothers, to the man that took him,
　To question of his apprehension.
　(*To Lady Gray*) Widow, go you along. ⌈*To Richard and
　　George*⌉ Lords, use her honourably.
　　　　　　　　　*Exeunt all but Richard*
RICHARD OF GLOUCESTER
　Ay, Edward will use women honourably.
　Would he were wasted, marrow, bones, and all,　125
　That from his loins no hopeful branch may spring

To cross me from the golden time I look for.
And yet, between my soul's desire and me—
The lustful Edward's title burièd—
Is Clarence, Henry, and his son young Edward,    130
And all the unlooked-for issue of their bodies,
To take their rooms ere I can place myself.
A cold premeditation for my purpose.
Why, then, I do but dream on sovereignty
Like one that stands upon a promontory    135
And spies a far-off shore where he would tread,
Wishing his foot were equal with his eye,
And chides the sea that sunders him from thence,
Saying he'll lade it dry to have his way—
So do I wish the crown being so far off,    140
And so I chide the means that keeps me from it,
And so I say I'll cut the causes off,
Flattering me with impossibilities.
My eye's too quick, my heart o'erweens too much,
Unless my hand and strength could equal them.    145
Well, say there is no kingdom then for Richard—
What other pleasure can the world afford?
I'll make my heaven in a lady's lap,
And deck my body in gay ornaments,
And 'witch sweet ladies with my words and looks.    150
O, miserable thought! And more unlikely
Than to accomplish twenty golden crowns.
Why, love forswore me in my mother's womb,
And, for I should not deal in her soft laws,
She did corrupt frail nature with some bribe    155
To shrink mine arm up like a withered shrub,
To make an envious mountain on my back—
Where sits deformity to mock my body—
To shape my legs of an unequal size,
To disproportion me in every part,    160
Like to a chaos, or an unlicked bear whelp
That carries no impression like the dam.
And am I then a man to be beloved?
O, monstrous fault, to harbour such a thought!
Then, since this earth affords no joy to me    165
But to command, to check, to o'erbear such
As are of better person than myself,
I'll make my heaven to dream upon the crown,
And whiles I live, t'account this world but hell,
Until my misshaped trunk that bears this head    170
Be round impalèd with a glorious crown.
And yet I know not how to get the crown,
For many lives stand between me and home.
And I—like one lost in a thorny wood,
That rends the thorns and is rent with the thorns,    175
Seeking a way and straying from the way,
Not knowing how to find the open air,
But toiling desperately to find it out—
Torment myself to catch the English crown.
And from that torment I will free myself,    180
Or hew my way out with a bloody axe.
Why, I can smile, and murder whiles I smile,
And cry 'Content!' to that which grieves my heart,
And wet my cheeks with artificial tears,
And frame my face to all occasions.    185
I'll drown more sailors than the mermaid shall;
I'll slay more gazers than the basilisk;
I'll play the orator as well as Nestor,
Deceive more slyly than Ulysses could,
And, like a Sinon, take another Troy.    190
I can add colours to the chameleon,

Change shapes with Proteus for advantages,
And set the murderous Machiavel to school.
Can I do this, and cannot get a crown?
Tut, were it farther off, I'll pluck it down.    *Exit*

3.3    ⌈*Two*⌉ *chairs of state. Flourish. Enter King Louis of*
*France, his sister the Lady Bona, Lord Bourbon his*
*admiral, Prince Edward, Queen Margaret, and the*
*Earl of Oxford. Louis goes up upon the state, sits,*
*and riseth up again*

KING LOUIS
Fair Queen of England, worthy Margaret,
Sit down with us. It ill befits thy state
And birth that thou shouldst stand while Louis
   doth sit.
QUEEN MARGARET
No, mighty King of France, now Margaret
Must strike her sail and learn a while to serve    5
Where kings command. I was, I must confess,
Great Albion's queen in former golden days,
But now mischance hath trod my title down,
And with dishonour laid me on the ground,
Where I must take like seat unto my fortune    10
And to my humble state conform myself.
KING LOUIS
Why, say, fair Queen, whence springs this deep
   despair?
QUEEN MARGARET
From such a cause as fills mine eyes with tears
And stops my tongue, while heart is drowned in cares.
KING LOUIS
Whate'er it be, be thou still like thyself,    15
And sit thee by our side.
   *Seats her by him*
                           Yield not thy neck
To fortune's yoke, but let thy dauntless mind
Still ride in triumph over all mischance.
Be plain, Queen Margaret, and tell thy grief.
It shall be eased if France can yield relief.    20
QUEEN MARGARET
Those gracious words revive my drooping thoughts,
And give my tongue-tied sorrows leave to speak.
Now, therefore, be it known to noble Louis
That Henry, sole possessor of my love,
Is of a king become a banished man,    25
And forced to live in Scotland a forlorn,
While proud ambitious Edward, Duke of York,
Usurps the regal title and the seat
Of England's true-anointed lawful King.
This is the cause that I, poor Margaret,    30
With this my son, Prince Edward, Henry's heir,
Am come to crave thy just and lawful aid.
An if thou fail us all our hope is done.
Scotland hath will to help, but cannot help;
Our people and our peers are both misled,    35
Our treasure seized, our soldiers put to flight,
And, as thou seest, ourselves in heavy plight.
KING LOUIS
Renownèd Queen, with patience calm the storm,
While we bethink a means to break it off.
QUEEN MARGARET
The more we stay, the stronger grows our foe.    40
KING LOUIS
The more I stay, the more I'll succour thee.

QUEEN MARGARET
O, but impatience waiteth on true sorrow.
*Enter the Earl of Warwick*
And see where comes the breeder of my sorrow.
KING LOUIS
What's he approacheth boldly to our presence?
QUEEN MARGARET
Our Earl of Warwick, Edward's greatest friend.          45
KING LOUIS
Welcome, brave Warwick. What brings thee to France?
*He descends. She ariseth*
QUEEN MARGARET (*aside*)
Ay, now begins a second storm to rise,
For this is he that moves both wind and tide.
WARWICK (*to King Louis*)
From worthy Edward, King of Albion,
My lord and sovereign, and thy vowèd friend,          50
I come in kindness and unfeignèd love,
First, to do greetings to thy royal person,
And then, to crave a league of amity,
And lastly, to confirm that amity
With nuptial knot, if thou vouchsafe to grant          55
That virtuous Lady Bona, thy fair sister,
To England's King in lawful marriage.
QUEEN MARGARET (*aside*)
If that go forward, Henry's hope is done.
WARWICK (*to Lady Bona*)
And, gracious madam, in our King's behalf
I am commanded, with your leave and favour,          60
Humbly to kiss your hand, and with my tongue
To tell the passion of my sovereign's heart,
Where fame, late ent'ring at his heedful ears,
Hath placed thy beauty's image and thy virtue.
QUEEN MARGARET
King Louis and Lady Bona, hear me speak          65
Before you answer Warwick. His demand
Springs not from Edward's well-meant honest love,
But from deceit, bred by necessity.
For how can tyrants safely govern home
Unless abroad they purchase great alliance?          70
To prove him tyrant this reason may suffice—
That Henry liveth still; but were he dead,
Yet here Prince Edward stands, King Henry's son.
Look, therefore, Louis, that by this league and
          marriage
Thou draw not on thy danger and dishonour,          75
For though usurpers sway the rule a while,
Yet heav'ns are just and time suppresseth wrongs.
WARWICK
Injurious Margaret.
PRINCE EDWARD          And why not 'Queen'?
WARWICK
Because thy father Henry did usurp,
And thou no more art prince than she is queen.          80
OXFORD
Then Warwick disannuls great John of Gaunt,
Which did subdue the greatest part of Spain;
And, after John of Gaunt, Henry the Fourth,
Whose wisdom was a mirror to the wisest;
And, after that wise prince, Henry the Fifth,          85
Who by his prowess conquerèd all France.
From these our Henry lineally descends.
WARWICK
Oxford, how haps it in this smooth discourse
You told not how Henry the Sixth hath lost
All that which Henry the Fifth had gotten?          90
Methinks these peers of France should smile at that.

But for the rest, you tell a pedigree
Of threescore and two years—a silly time
To make prescription for a kingdom's worth.
OXFORD
Why, Warwick, canst thou speak against thy liege,          95
Whom thou obeyedest thirty and six years,
And not bewray thy treason with a blush?
WARWICK
Can Oxford, that did ever fence the right,
Now buckler falsehood with a pedigree?
For shame—leave Henry, and call Edward king.          100
OXFORD
Call him my king by whose injurious doom
My elder brother, the Lord Aubrey Vere,
Was done to death? And more than so, my father,
Even in the downfall of his mellowed years,
When nature brought him to the door of death?          105
No, Warwick, no—while life upholds this arm,
This arm upholds the house of Lancaster.
WARWICK And I the house of York.
KING LOUIS
Queen Margaret, Prince Edward, and Oxford,
Vouchsafe, at our request, to stand aside          110
While I use further conference with Warwick.
*Queen Margaret ⌜comes down from the state and⌝,
with Prince Edward and Oxford, stands apart*
QUEEN MARGARET
Heavens grant that Warwick's words bewitch him not.
KING LOUIS
Now, Warwick, tell me even upon thy conscience,
Is Edward your true king? For I were loath
To link with him that were not lawful chosen.          115
WARWICK
Thereon I pawn my credit and mine honour.
KING LOUIS
But is he gracious in the people's eye?
WARWICK
The more that Henry was unfortunate.
KING LOUIS
Then further, all dissembling set aside,
Tell me for truth the measure of his love          120
Unto our sister Bona.
WARWICK                    Such it seems
As may beseem a monarch like himself.
Myself have often heard him say and swear
That this his love was an eternal plant,
Whereof the root was fixed in virtue's ground,          125
The leaves and fruit maintained with beauty's sun,
Exempt from envy, but not from disdain,
Unless the Lady Bona quit his pain.
KING LOUIS (*to Lady Bona*)
Now, sister, let us hear your firm resolve.
LADY BONA
Your grant, or your denial, shall be mine.          130
(*To Warwick*) Yet I confess that often ere this day,
When I have heard your king's desert recounted,
Mine ear hath tempted judgement to desire.
KING LOUIS (*to Warwick*)
Then, Warwick, thus—our sister shall be Edward's.
And now, forthwith, shall articles be drawn          135
Touching the jointure that your king must make,
Which with her dowry shall be counterpoised.
(*To Queen Margaret*) Draw near, Queen Margaret, and
          be a witness
That Bona shall be wife to the English king.
*Queen Margaret, Prince Edward, ⌜and Oxford⌝ come
forward*

PRINCE EDWARD
　To Edward, but not to the English king.          140
QUEEN MARGARET
　Deceitful Warwick—it was thy device
　By this alliance to make void my suit!
　Before thy coming Louis was Henry's friend.
KING LOUIS
　And still is friend to him and Margaret.
　But if your title to the crown be weak,          145
　As may appear by Edward's good success,
　Then 'tis but reason that I be released
　From giving aid which late I promisèd.
　Yet shall you have all kindness at my hand
　That your estate requires and mine can yield.     150
WARWICK (to Queen Margaret)
　Henry now lives in Scotland at his ease,
　Where having nothing, nothing can he lose.
　And as for you yourself, our quondam queen,
　You have a father able to maintain you,
　And better 'twere you troubled him than France.   155
QUEEN MARGARET
　Peace, impudent and shameless Warwick, peace!
　Proud setter-up and puller-down of kings!
　I will not hence till, with my talk and tears,
　Both full of truth, I make King Louis behold
　Thy sly conveyance and thy lord's false love,     160
　　　　Post blowing a horn within
　For both of you are birds of selfsame feather.
KING LOUIS
　Warwick, this is some post to us or thee.
　　　　Enter the Post
POST (to Warwick)
　My lord ambassador, these letters are for you,
　Sent from your brother Marquis Montague;
　(To Louis) These from our King unto your majesty;
　(To Queen Margaret)
　And, madam, these for you, from whom I know not.
　　　　They all read their letters
OXFORD (to Prince Edward)
　I like it well that our fair Queen and mistress    167
　Smiles at her news, while Warwick frowns at his.
PRINCE EDWARD
　Nay, mark how Louis stamps as he were nettled.
　I hope all's for the best.                        170
KING LOUIS
　Warwick, what are thy news? And yours, fair Queen?
QUEEN MARGARET
　Mine, such as fill my heart with unhoped joys.
WARWICK
　Mine, full of sorrow and heart's discontent.
KING LOUIS
　What! Has your king married the Lady Gray?
　And now to soothe your forgery and his,           175
　Sends me a paper to persuade me patience?
　Is this th'alliance that he seeks with France?
　Dare he presume to scorn us in this manner?
QUEEN MARGARET
　I told your majesty as much before—
　This proveth Edward's love and Warwick's honesty.
WARWICK
　King Louis, I here protest in sight of heaven     181
　And by the hope I have of heavenly bliss,
　That I am clear from this misdeed of Edward's,
　No more my king, for he dishonours me,
　But most himself, if he could see his shame.      185
　Did I forget that by the house of York

　My father came untimely to his death?
　Did I let pass th'abuse done to my niece?
　Did I impale him with the regal crown?
　Did I put Henry from his native right?           190
　And am I guerdoned at the last with shame?
　Shame on himself, for my desert is honour.
　And to repair my honour, lost for him,
　I here renounce him and return to Henry.
　(To Queen Margaret) My noble Queen, let former
　　　　grudges pass,                              195
　And henceforth I am thy true servitor.
　I will revenge his wrong to Lady Bona
　And replant Henry in his former state.
QUEEN MARGARET
　Warwick, these words have turned my hate to love,
　And I forgive and quite forget old faults,       200
　And joy that thou becom'st King Henry's friend.
WARWICK
　So much his friend, ay, his unfeignèd friend,
　That if King Louis vouchsafe to furnish us
　With some few bands of chosen soldiers,
　I'll undertake to land them on our coast         205
　And force the tyrant from his seat by war.
　'Tis not his new-made bride shall succour him.
　And as for Clarence, as my letters tell me,
　He's very likely now to fall from him
　For matching more for wanton lust than honour,   210
　Or than for strength and safety of our country.
LADY BONA (to King Louis)
　Dear brother, how shall Bona be revenged,
　But by thy help to this distressèd Queen?
QUEEN MARGARET (to King Louis)
　Renownèd Prince, how shall poor Henry live
　Unless thou rescue him from foul despair?        215
LADY BONA (to King Louis)
　My quarrel and this English Queen's are one.
WARWICK
　And mine, fair Lady Bona, joins with yours.
KING LOUIS
　And mine with hers, and thine, and Margaret's.
　Therefore at last I firmly am resolved:
　You shall have aid.                              220
QUEEN MARGARET
　Let me give humble thanks for all at once.
KING LOUIS (to the Post)
　Then, England's messenger, return in post
　And tell false Edward, thy supposèd king,
　That Louis of France is sending over masquers
　To revel it with him and his new bride.          225
　Thou seest what's passed, go fear thy king withal.
LADY BONA (to the Post)
　Tell him, in hope he'll prove a widower shortly,
　I'll wear the willow garland for his sake.
QUEEN MARGARET (to the Post)
　Tell him my mourning weeds are laid aside,
　And I am ready to put armour on.                 230
WARWICK (to the Post)
　Tell him from me that he hath done me wrong,
　And therefore I'll uncrown him ere't be long.
　(Giving money) There's thy reward—be gone.
　　　　　　　　　　　　　　　　　　　Exit Post
KING LOUIS
　But, Warwick, thou and Oxford, with five thousand
　　　　men,
　Shall cross the seas and bid false Edward battle; 235
　And, as occasion serves, this noble Queen
　And Prince shall follow with a fresh supply.

Yet, ere thou go, but answer me one doubt:
What pledge have we of thy firm loyalty?
WARWICK
This shall assure my constant loyalty:    240
That if our Queen and this young Prince agree,
I'll join mine eldest daughter and my joy
To him forthwith in holy wedlock bands.
QUEEN MARGARET
Yes, I agree, and thank you for your motion.
(*To Prince Edward*) Son Edward, she is fair and virtuous,
Therefore delay not. Give thy hand to Warwick,    246
And with thy hand thy faith irrevocable
That only Warwick's daughter shall be thine.
PRINCE EDWARD
Yes, I accept her, for she well deserves it,
And here to pledge my vow I give my hand.    250
*He and Warwick clasp hands*
KING LOUIS
Why stay we now? These soldiers shall be levied,
And thou, Lord Bourbon, our high admiral,
Shall waft them over with our royal fleet.
I long till Edward fall by war's mischance
For mocking marriage with a dame of France.    255
*Exeunt all but Warwick*
WARWICK
I came from Edward as ambassador,
But I return his sworn and mortal foe.
Matter of marriage was the charge he gave me,
But dreadful war shall answer his demand.
Had he none else to make a stale but me?    260
Then none but I shall turn his jest to sorrow.
I was the chief that raised him to the crown,
And I'll be chief to bring him down again.
Not that I pity Henry's misery,
But seek revenge on Edward's mockery.    *Exit*

**4.1**     *Enter Richard Duke of Gloucester, George Duke of*
    *Clarence, the Duke of Somerset, and the Marquis of*
    *Montague*
RICHARD OF GLOUCESTER
Now tell me, brother Clarence, what think you
Of this new marriage with the Lady Gray?
Hath not our brother made a worthy choice?
GEORGE OF CLARENCE
Alas, you know 'tis far from hence to France;
How could he stay till Warwick made return?    5
SOMERSET
My lords, forbear this talk—here comes the King.
    *Flourish. Enter King Edward, the Lady Gray his*
    *Queen, the Earl of Pembroke, and the Lords*
    *Stafford and Hastings. Four stand on one side ⌐of*
    *the King⌐, and four on the other*
RICHARD OF GLOUCESTER   And his well-chosen bride.
GEORGE OF CLARENCE
I mind to tell him plainly what I think.
KING EDWARD
Now, brother of Clarence, how like you our choice,
That you stand pensive, as half-malcontent?    10
GEORGE OF CLARENCE
As well as Louis of France, or the Earl of Warwick,
Which are so weak of courage and in judgement
That they'll take no offence at our abuse.
KING EDWARD
Suppose they take offence without a cause—
They are but Louis and Warwick; I am Edward,    15
Your king and Warwick's, and must have my will.

RICHARD OF GLOUCESTER
And you shall have your will, because our king.
Yet hasty marriage seldom proveth well.
KING EDWARD
Yea, brother Richard, are you offended too?
RICHARD OF GLOUCESTER
Not I, no—God forbid that I should wish them severed
Whom God hath joined together. Ay, and 'twere pity
To sunder them that yoke so well together.    22
KING EDWARD
Setting your scorns and your mislike aside,
Tell me some reason why the Lady Gray
Should not become my wife and England's queen.    25
And you too, Somerset and Montague,
Speak freely what you think.
GEORGE OF CLARENCE
Then this is my opinion: that King Louis
Becomes your enemy for mocking him
About the marriage of the Lady Bona.    30
RICHARD OF GLOUCESTER
And Warwick, doing what you gave in charge,
Is now dishonourèd by this new marriage.
KING EDWARD
What if both Louis and Warwick be appeased
By such invention as I can devise?
MONTAGUE
Yet, to have joined with France in such alliance    35
Would more have strengthened this our
    commonwealth
'Gainst foreign storms than any home-bred marriage.
HASTINGS
Why, knows not Montague that of itself
England is safe, if true within itself?
MONTAGUE
But the safer when 'tis backed with France.    40
HASTINGS
'Tis better using France than trusting France.
Let us be backed with God and with the seas
Which he hath giv'n for fence impregnable,
And with their helps only defend ourselves.
In them and in ourselves our safety lies.    45
GEORGE OF CLARENCE
For this one speech Lord Hastings well deserves
To have the heir of the Lord Hungerford.
KING EDWARD
Ay, what of that? It was my will and grant—
And for this once my will shall stand for law.
RICHARD OF GLOUCESTER
And yet, methinks, your grace hath not done well    50
To give the heir and daughter of Lord Scales
Unto the brother of your loving bride.
She better would have fitted me or Clarence,
But in your bride you bury brotherhood.
GEORGE OF CLARENCE
Or else you would not have bestowed the heir    55
Of the Lord Bonville on your new wife's son,
And leave your brothers to go speed elsewhere.
KING EDWARD
Alas, poor Clarence, is it for a wife
That thou art malcontent? I will provide thee.
GEORGE OF CLARENCE
In choosing for yourself you showed your judgement,
Which being shallow, you shall give me leave    61
To play the broker in mine own behalf,
And to that end I shortly mind to leave you.

KING EDWARD
Leave me, or tarry. Edward will be king,
And not be tied unto his brother's will.    65
LADY GRAY
My lords, before it pleased his majesty
To raise my state to title of a queen,
Do me but right, and you must all confess
That I was not ignoble of descent—
And meaner than myself have had like fortune.    70
But as this title honours me and mine,
So your dislikes, to whom I would be pleasing,
Doth cloud my joys with danger and with sorrow.
KING EDWARD
My love, forbear to fawn upon their frowns.
What danger or what sorrow can befall thee    75
So long as Edward is thy constant friend,
And their true sovereign, whom they must obey?
Nay, whom they shall obey, and love thee too—
Unless they seek for hatred at my hands,
Which if they do, yet will I keep thee safe,    80
And they shall feel the vengeance of my wrath.
RICHARD OF GLOUCESTER (aside)
I hear, yet say not much, but think the more.
     *Enter the Post from France*
KING EDWARD
Now, messenger, what letters or what news from
France?
POST
My sovereign liege, no letters and few words,
But such as I, without your special pardon,    85
Dare not relate.
KING EDWARD
Go to, we pardon thee. Therefore, in brief,
Tell me their words as near as thou canst guess them.
What answer makes King Louis unto our letters?
POST
At my depart these were his very words:    90
'Go tell false Edward, thy supposèd king,
That Louis of France is sending over masquers
To revel it with him and his new bride.'
KING EDWARD
Is Louis so brave? Belike he thinks me Henry.
But what said Lady Bona to my marriage?    95
POST
These were her words, uttered with mild disdain:
'Tell him in hope he'll prove a widower shortly,
I'll wear the willow garland for his sake.'
KING EDWARD
I blame not her, she could say little less;
She had the wrong. But what said Henry's queen?
For I have heard that she was there in place.    101
POST
'Tell him', quoth she, 'my mourning weeds are done,
And I am ready to put armour on.'
KING EDWARD
Belike she minds to play the Amazon.
But what said Warwick to these injuries?    105
POST
He, more incensed against your majesty
Than all the rest, discharged me with these words:
'Tell him from me that he hath done me wrong,
And therefore I'll uncrown him ere't be long.'
KING EDWARD
Ha! Durst the traitor breathe out so proud words?    110
Well, I will arm me, being thus forewarned.
They shall have wars and pay for their presumption.
But say, is Warwick friends with Margaret?

POST
Ay, gracious sovereign, they are so linked in friendship
That young Prince Edward marries Warwick's daughter.
GEORGE OF CLARENCE
Belike the elder; Clarence will have the younger.    116
Now, brother King, farewell, and sit you fast,
For I will hence to Warwick's other daughter,
That, though I want a kingdom, yet in marriage
I may not prove inferior to yourself.    120
You that love me and Warwick, follow me.
     *Exit Clarence, and Somerset follows*
RICHARD OF GLOUCESTER
Not I—⌈*aside*⌉ my thoughts aim at a further matter.
I stay not for the love of Edward, but the crown.
KING EDWARD
Clarence and Somerset both gone to Warwick?
Yet am I armed against the worst can happen,    125
And haste is needful in this desp'rate case.
Pembroke and Stafford, you in our behalf
Go levy men and make prepare for war.
They are already, or quickly will be, landed.
Myself in person will straight follow you.    130
     *Exeunt Pembroke and Stafford*
But ere I go, Hastings and Montague,
Resolve my doubt. You twain, of all the rest,
Are near'st to Warwick by blood and by alliance.
Tell me if you love Warwick more than me.
If it be so, then both depart to him—    135
I rather wish you foes than hollow friends.
But if you mind to hold your true obedience,
Give me assurance with some friendly vow
That I may never have you in suspect.
MONTAGUE
So God help Montague as he proves true.    140
HASTINGS
And Hastings as he favours Edward's cause.
KING EDWARD
Now, brother Richard, will you stand by us?
RICHARD OF GLOUCESTER
Ay, in despite of all that shall withstand you.
KING EDWARD
Why, so. Then am I sure of victory.
Now, therefore, let us hence and lose no hour    145
Till we meet Warwick with his foreign power.    *Exeunt*

**4.2**    *Enter the Earls of Warwick and Oxford in England,
with French soldiers*
WARWICK
Trust me, my lord, all hitherto goes well.
The common sort by numbers swarm to us.
     *Enter the Dukes of Clarence and Somerset*
But see where Somerset and Clarence comes.
Speak suddenly, my lords, are we all friends?
GEORGE OF CLARENCE   Fear not that, my lord.    5
WARWICK
Then, gentle Clarence, welcome unto Warwick—
And welcome, Somerset. I hold it cowardice
To rest mistrustful where a noble heart
Hath pawned an open hand in sign of love,
Else might I think that Clarence, Edward's brother,    10
Were but a feignèd friend to our proceedings.
But come, sweet Clarence, my daughter shall be thine.
And now what rests but, in night's coverture,
Thy brother being carelessly encamped,
His soldiers lurking in the towns about,    15
And but attended by a simple guard,
We may surprise and take him at our pleasure?

Our scouts have found the adventure very easy;
That, as Ulysses and stout Diomed
With sleight and manhood stole to Rhesus' tents          20
And brought from thence the Thracian fatal steeds,
So we, well covered with the night's black mantle,
At unawares may beat down Edward's guard
And seize himself—I say not 'slaughter him',
For I intend but only to surprise him.                   25
You that will follow me to this attempt,
Applaud the name of Henry with your leader.
          *They all cry 'Henry'*
Why, then, let's on our way in silent sort,
For Warwick and his friends, God and Saint George!
                                                *Exeunt*

**4.3**    *Enter three Watchmen, to guard King Edward's tent*
FIRST WATCHMAN
Come on, my masters, each man take his stand.
The King by this is set him down to sleep.
SECOND WATCHMAN What, will he not to bed?
FIRST WATCHMAN
Why, no—for he hath made a solemn vow
Never to lie and take his natural rest                   5
Till Warwick or himself be quite suppressed.
SECOND WATCHMAN
Tomorrow then belike shall be the day,
If Warwick be so near as men report.
THIRD WATCHMAN
But say, I pray, what nobleman is that
That with the King here resteth in his tent?             10
FIRST WATCHMAN
'Tis the Lord Hastings, the King's chiefest friend.
THIRD WATCHMAN
O, is it so? But why commands the King
That his chief followers lodge in towns about him,
While he himself keeps in the cold field?
SECOND WATCHMAN
'Tis the more honour, because more dangerous.            15
THIRD WATCHMAN
Ay, but give me worship and quietness—
I like it better than a dangerous honour.
If Warwick knew in what estate he stands,
'Tis to be doubted he would waken him.
FIRST WATCHMAN
Unless our halberds did shut up his passage.            20
SECOND WATCHMAN
Ay, wherefore else guard we his royal tent
But to defend his person from night-foes?
          *Enter silently the Earl of Warwick, George Duke of*
          *Clarence, the Earl of Oxford, and the Duke of*
          *Somerset, with French soldiers*
WARWICK
This is his tent—and see where stand his guard.
Courage, my masters—honour now or never!
But follow me, and Edward shall be ours.                25
FIRST WATCHMAN Who goes there?
SECOND WATCHMAN Stay or thou diest.
          *Warwick and the rest all cry 'Warwick, Warwick!'*
          *and set upon the guard, who fly, crying 'Arm, arm!'*
          *Warwick and the rest follow them*

**4.4**    *With the drummer playing and trumpeter sounding,*
          *enter the Earl of Warwick, the Duke of Somerset,*
          *and the rest bringing King Edward out in his gown,*
          *sitting in a chair. Richard Duke of Gloucester and*
          *Lord Hastings flies over the stage*
SOMERSET What are they that fly there?

WARWICK
Richard and Hastings—let them go. Here is the Duke.
KING EDWARD
'The Duke'! Why, Warwick, when we parted,
Thou calledst me king.
WARWICK                    Ay, but the case is altered.
When you disgraced me in my embassade,                   5
Then I degraded you from being king,
And come now to create you Duke of York.
Alas, how should you govern any kingdom
That know not how to use ambassadors,
Nor how to be contented with one wife,                   10
Nor how to use your brothers brotherly,
Nor how to study for the people's welfare,
Nor how to shroud yourself from enemies?
KING EDWARD (*seeing George*)
Yea, brother of Clarence, art thou here too?
Nay, then, I see that Edward needs must down.            15
Yet, Warwick, in despite of all mischance,
Of thee thyself and all thy complices,
Edward will always bear himself as king.
Though fortune's malice overthrow my state,
My mind exceeds the compass of her wheel.               20
WARWICK
Then, for his mind, be Edward England's king.
          *Warwick takes off Edward's crown*
But Henry now shall wear the English crown,
And be true king indeed, thou but the shadow.
My lord of Somerset, at my request,
See that, forthwith, Duke Edward be conveyed            25
Unto my brother, Archbishop of York.
When I have fought with Pembroke and his fellows,
I'll follow you, and tell what answer
Louis and the Lady Bona send to him.
Now for a while farewell, good Duke of York.            30
          *They begin to lead Edward out forcibly*
KING EDWARD
What fates impose, that men must needs abide.
It boots not to resist both wind and tide.
                         *Exeunt some with Edward*
OXFORD
What now remains, my lords, for us to do
But march to London with our soldiers?
WARWICK
Ay, that's the first thing that we have to do—          35
To free King Henry from imprisonment
And see him seated in the regal throne.         *Exeunt*

**4.5**    *Enter Earl Rivers and his sister, Lady Gray,*
          *Edward's queen*
RIVERS
Madam, what makes you in this sudden change?
LADY GRAY
Why, brother Rivers, are you yet to learn
What late misfortune is befall'n King Edward?
RIVERS
What? Loss of some pitched battle against Warwick?
LADY GRAY
No, but the loss of his own royal person.               5
RIVERS Then is my sovereign slain?
LADY GRAY
Ay, almost slain—for he is taken prisoner,
Either betrayed by falsehood of his guard
Or by his foe surprised at unawares,
And, as I further have to understand,                   10
Is new committed to the Bishop of York,
Fell Warwick's brother, and by that our foe.

**RIVERS**
These news, I must confess, are full of grief.
Yet, gracious madam, bear it as you may.
Warwick may lose, that now hath won the day.          15
**LADY GRAY**
Till then fair hope must hinder life's decay,
And I the rather wean me from despair
For love of Edward's offspring in my womb.
This is it that makes me bridle passion
And bear with mildness my misfortune's cross.          20
Ay, ay, for this I draw in many a tear
And stop the rising of blood-sucking sighs,
Lest with my sighs or tears I blast or drown
King Edward's fruit, true heir to th'English crown.
**RIVERS**
But, madam, where is Warwick then become?          25
**LADY GRAY**
I am informèd that he comes towards London
To set the crown once more on Henry's head.
Guess thou the rest—King Edward's friends must down.
But to prevent the tyrant's violence—
For trust not him that hath once broken faith—          30
I'll hence forthwith unto the sanctuary,
To save at least the heir of Edward's right.
There shall I rest secure from force and fraud.
Come, therefore, let us fly while we may fly.
If Warwick take us, we are sure to die.          *Exeunt*

**4.6**     *Enter Richard Duke of Gloucester, Lord Hastings,*
           *and Sir William Stanley, ⌈with soldiers⌉*
**RICHARD OF GLOUCESTER**
Now my lord Hastings and Sir William Stanley,
Leave off to wonder why I drew you hither
Into this chiefest thicket of the park.
Thus stands the case: you know our King, my brother,
Is prisoner to the Bishop here, at whose hands          5
He hath good usage and great liberty,
And, often but attended with weak guard,
Comes hunting this way to disport himself.
I have advertised him by secret means
That if about this hour he make this way          10
Under the colour of his usual game,
He shall here find his friends with horse and men
To set him free from his captivity.
           *Enter King Edward and a Huntsman with him*
**HUNTSMAN**
This way, my lord—for this way lies the game.
**KING EDWARD**
Nay, this way, man—see where the huntsmen stand.
Now, brother of Gloucester, Lord Hastings, and the
           rest,          16
Stand you thus close to steal the Bishop's deer?
**RICHARD OF GLOUCESTER**
Brother, the time and case requireth haste.
Your horse stands ready at the park corner.
**KING EDWARD** But whither shall we then?          20
**HASTINGS** To Lynn, my lord,
And shipped from thence to Flanders.
**RICHARD OF GLOUCESTER** ⌈*aside*⌉
Well guessed, believe me—for that was my meaning.
**KING EDWARD**
Stanley, I will requite thy forwardness.
**RICHARD OF GLOUCESTER**
But wherefore stay we? 'Tis no time to talk.          25
**KING EDWARD**
Huntsman, what sayst thou? Wilt thou go along?

**HUNTSMAN**
Better do so than tarry and be hanged.
**RICHARD OF GLOUCESTER**
Come then, away—let's have no more ado.
**KING EDWARD**
Bishop, farewell—shield thee from Warwick's frown,
And pray that I may repossess the crown.          *Exeunt*

**4.7**     *Flourish. Enter the Earl of Warwick and George*
           *Duke of Clarence ⌈with the crown⌉. Then enter*
           *King Henry, the Earl of Oxford, the Duke of*
           *Somerset ⌈with⌉ young Henry Earl of Richmond,*
           *the Marquis of Montague, and the Lieutenant of the*
           *Tower*
**KING HENRY**
Master Lieutenant, now that God and friends
Have shaken Edward from the regal seat
And turned my captive state to liberty,
My fear to hope, my sorrows unto joys,
At our enlargement what are thy due fees?          5
**LIEUTENANT**
Subjects may challenge nothing of their sovereigns—
But if an humble prayer may prevail,
I then crave pardon of your majesty.
**KING HENRY**
For what, Lieutenant? For well using me?
Nay, be thou sure I'll well requite thy kindness,          10
For that it made my prisonment a pleasure—
Ay, such a pleasure as encagèd birds
Conceive when, after many moody thoughts,
At last by notes of household harmony
They quite forget their loss of liberty.          15
But, Warwick, after God, thou sett'st me free,
And chiefly therefore I thank God and thee.
He was the author, thou the instrument.
Therefore, that I may conquer fortune's spite
By living low, where fortune cannot hurt me,          20
And that the people of this blessèd land
May not be punished with my thwarting stars,
Warwick, although my head still wear the crown,
I here resign my government to thee,
For thou art fortunate in all thy deeds.          25
**WARWICK**
Your grace hath still been famed for virtuous,
And now may seem as wise as virtuous
By spying and avoiding fortune's malice,
For few men rightly temper with the stars,
Yet in this one thing let me blame your grace:          30
For choosing me when Clarence is in place.
**GEORGE OF CLARENCE**
No, Warwick, thou art worthy of the sway,
To whom the heav'ns in thy nativity
Adjudged an olive branch and laurel crown,
As likely to be blest in peace and war,          35
And therefore I yield thee my free consent.
**WARWICK**
And I choose Clarence only for Protector.
**KING HENRY**
Warwick and Clarence, give me both your hands.
Now join your hands, and with your hands your
           hearts,
That no dissension hinder government.          40
I make you both Protectors of this land,
While I myself will lead a private life
And in devotion spend my latter days,
To sin's rebuke and my creator's praise.

WARWICK
What answers Clarence to his sovereign's will?    45
GEORGE OF CLARENCE
That he consents, if Warwick yield consent,
For on thy fortune I repose myself.
WARWICK
Why, then, though loath, yet must I be content.
We'll yoke together, like a double shadow
To Henry's body, and supply his place—    50
I mean in bearing weight of government—
While he enjoys the honour and his ease.
And, Clarence, now then it is more than needful
Forthwith that Edward be pronounced a traitor,
And all his lands and goods be confiscate.    55
GEORGE OF CLARENCE
What else? And that succession be determined.
WARWICK
Ay, therein Clarence shall not want his part.
KING HENRY
But with the first of all your chief affairs,
Let me entreat—for I command no more—
That Margaret your queen and my son Edward    60
Be sent for, to return from France with speed.
For, till I see them here, by doubtful fear
My joy of liberty is half eclipsed.
GEORGE OF CLARENCE
It shall be done, my sovereign, with all speed.
KING HENRY
My lord of Somerset, what youth is that    65
Of whom you seem to have so tender care?
SOMERSET
My liege, it is young Henry, Earl of Richmond.
KING HENRY
Come hither, England's hope.
        *King Henry lays his hand on Richmond's head*
                                If secret powers
Suggest but truth to my divining thoughts,
This pretty lad will prove our country's bliss.    70
His looks are full of peaceful majesty,
His head by nature framed to wear a crown,
His hand to wield a sceptre, and himself
Likely in time to bless a regal throne.
Make much of him, my lords, for this is he    75
Must help you more than you are hurt by me.
        *Enter a Post*
WARWICK  What news, my friend?
POST
That Edward is escapèd from your brother
And 'fled, as he hears since, to Burgundy.
WARWICK
Unsavoury news—but how made he escape?    80
POST
He was conveyed by Richard Duke of Gloucester
And the Lord Hastings, who attended him
In secret ambush on the forest side
And from the Bishop's huntsmen rescued him—
For hunting was his daily exercise.    85
WARWICK
My brother was too careless of his charge.
(*To King Henry*) But let us hence, my sovereign, to
        provide
A salve for any sore that may betide.
        *Exeunt all but Somerset, Richmond, and Oxford*
SOMERSET (*to Oxford*)
My lord, I like not of this flight of Edward's,
For doubtless Burgundy will yield him help,    90
And we shall have more wars before't be long.

As Henry's late presaging prophecy
Did glad my heart with hope of this young Richmond,
So doth my heart misgive me, in these conflicts,
What may befall him, to his harm and ours.    95
Therefore, Lord Oxford, to prevent the worst,
Forthwith we'll send him hence to Brittany,
Till storms be past of civil enmity.
OXFORD
Ay, for if Edward repossess the crown,
'Tis like that Richmond with the rest shall down.    100
SOMERSET
It shall be so—he shall to Brittany.
Come, therefore, let's about it speedily.    *Exeunt*

**4.8**    *Flourish. Enter King Edward, Richard Duke of
        Gloucester, and Lord Hastings, ⌈with a troop of
        Hollanders⌉*
KING EDWARD
Now, brother Richard, Lord Hastings, and the rest,
Yet thus far fortune maketh us amends,
And says that once more I shall interchange
My wanèd state for Henry's regal crown.
Well have we passed and now repassed the seas    5
And brought desirèd help from Burgundy.
What then remains, we being thus arrived
From Ravenspurgh haven before the gates of York,
But that we enter, as into our dukedom?
        ⌈*Hastings*⌉ *knocks at the gates of York*
RICHARD OF GLOUCESTER
The gates made fast? Brother, I like not this.    10
For many men that stumble at the threshold
Are well foretold that danger lurks within.
KING EDWARD
Tush, man, abodements must not now affright us.
By fair or foul means we must enter in,
For hither will our friends repair to us.    15
HASTINGS
My liege, I'll knock once more to summon them.
        *He knocks.*
        *Enter, on the walls, the Mayor and aldermen of York*
MAYOR
My lords, we were forewarnèd of your coming,
And shut the gates for safety of ourselves—
For now we owe allegiance unto Henry.
KING EDWARD
But, Master Mayor, if Henry be your king,    20
Yet Edward at the least is Duke of York.
MAYOR
True, my good lord, I know you for no less.
KING EDWARD
Why, and I challenge nothing but my dukedom,
As being well content with that alone.
RICHARD OF GLOUCESTER (*aside*)
But when the fox hath once got in his nose,    25
He'll soon find means to make the body follow.
HASTINGS
Why, Master Mayor, why stand you in a doubt?
Open the gates—we are King Henry's friends.
MAYOR
Ay, say you so? The gates shall then be opened.
        *They descend*
RICHARD OF GLOUCESTER
A wise stout captain, and soon persuaded.    30
HASTINGS
The good old man would fain that all were well,
So 'twere not long of him; but being entered,

I doubt not, I, but we shall soon persuade
Both him and all his brothers unto reason.
    *Enter below the Mayor and two aldermen*
KING EDWARD
So, Master Mayor, these gates must not be shut          35
But in the night or in the time of war.
What—fear not, man, but yield me up the keys,
    *King Edward takes some keys from the Mayor*
For Edward will defend the town and thee,
And all those friends that deign to follow me.
    *March. Enter Sir John Montgomery with a*
    *drummer and soldiers*
RICHARD OF GLOUCESTER
Brother, this is Sir John Montgomery,          40
Our trusty friend, unless I be deceived.
KING EDWARD
Welcome, Sir John—but why come you in arms?
MONTGOMERY
To help King Edward in his time of storm,
As every loyal subject ought to do.
KING EDWARD
Thanks, good Montgomery, but we now forget          45
Our title to the crown, and only claim
Our dukedom till God please to send the rest.
MONTGOMERY
Then fare you well, for I will hence again.
I came to serve a king and not a duke.
Drummer, strike up, and let us march away.          50
    *The drummer begins to sound a march*
KING EDWARD
Nay, stay, Sir John, a while, and we'll debate
By what safe means the crown may be recovered.
MONTGOMERY
What talk you of debating? In few words,
If you'll not here proclaim yourself our king
I'll leave you to your fortune and be gone          55
To keep them back that come to succour you.
Why shall we fight, if you pretend no title?
RICHARD OF GLOUCESTER (*to King Edward*)
Why, brother, wherefore stand you on nice points?
KING EDWARD
When we grow stronger, then we'll make our claim.
Till then 'tis wisdom to conceal our meaning.          60
HASTINGS
Away with scrupulous wit! Now arms must rule.
RICHARD OF GLOUCESTER
And fearless minds climb soonest unto crowns.
Brother, we will proclaim you out of hand,
The bruit thereof will bring you many friends.
KING EDWARD
Then be it as you will, for 'tis my right,          65
And Henry but usurps the diadem.
MONTGOMERY
Ay, now my sovereign speaketh like himself,
And now will I be Edward's champion.
HASTINGS
Sound trumpet, Edward shall be here proclaimed.
    ⌈*To Montgomery*⌉
Come, fellow soldier, make thou proclamation.          70
    *Flourish*
⌈MONTGOMERY⌉ Edward the Fourth, by the grace of God
King of England and France, and Lord of Ireland—
And whosoe'er gainsays King Edward's right,
By this I challenge him to single fight.
    *He throws down his gauntlet*
ALL Long live Edward the Fourth!          75

KING EDWARD
Thanks, brave Montgomery, and thanks unto you all.
If fortune serve me I'll requite this kindness.
Now, for this night, let's harbour here in York;
And when the morning sun shall raise his car
Above the border of this horizon,          80
We'll forward towards Warwick and his mates.
For well I wot that Henry is no soldier.
Ah, froward Clarence, how evil it beseems thee
To flatter Henry and forsake thy brother!
Yet, as we may, we'll meet both thee and Warwick.
Come on, brave soldiers—doubt not of the day          86
And, that once gotten, doubt not of large pay.
    *Exeunt*

**4.9**    *Flourish. Enter King Henry, the Earl of Warwick,*
    *the Marquis of Montague, George Duke of Clarence,*
    *and the Earl of Oxford*
WARWICK
What counsel, lords? Edward from Belgia,
With hasty Germans and blunt Hollanders,
Hath passed in safety through the narrow seas,
And with his troops doth march amain to London,
And many giddy people flock to him.          5
KING HENRY
Let's levy men and beat him back again.
GEORGE OF CLARENCE
A little fire is quickly trodden out,
Which, being suffered, rivers cannot quench.
WARWICK
In Warwickshire I have true-hearted friends,
Not mutinous in peace, yet bold in war.          10
Those will I muster up. And thou, son Clarence,
Shalt stir in Suffolk, Norfolk, and in Kent,
The knights and gentlemen to come with thee.
Thou, brother Montague, in Buckingham,
Northampton, and in Leicestershire shalt find          15
Men well inclined to hear what thou command'st.
And thou, brave Oxford, wondrous well beloved
In Oxfordshire, shalt muster up thy friends.
My sovereign, with the loving citizens,
Like to his island girt in with the ocean,          20
Or modest Dian circled with her nymphs,
Shall rest in London till we come to him.
Fair lords, take leave and stand not to reply.
Farewell, my sovereign.
KING HENRY
Farewell, my Hector, and my Troy's true hope.          25
GEORGE OF CLARENCE
In sign of truth, I kiss your highness' hand.
    *He kisses King Henry's hand*
KING HENRY
Well-minded Clarence, be thou fortunate.
MONTAGUE
Comfort, my lord, and so I take my leave.
    ⌈*He kisses King Henry's hand*⌉
OXFORD
And thus I seal my truth and bid adieu.
    ⌈*He kisses King Henry's hand*⌉
KING HENRY
Sweet Oxford, and my loving Montague,          30
And all at once, once more a happy farewell.          ⌈*Exit*⌉
WARWICK
Farewell, sweet lords—let's meet at Coventry.
    *Exeunt* ⌈*severally*⌉

4.10 ⌈*Enter King Henry and the Duke of Exeter*⌉
KING HENRY
Here at the palace will I rest a while.
Cousin of Exeter, what thinks your lordship?
Methinks the power that Edward hath in field
Should not be able to encounter mine.
EXETER
The doubt is that he will seduce the rest.                    5
KING HENRY
That's not my fear. My meed hath got me fame.
I have not stopped mine ears to their demands,
Nor posted off their suits with slow delays.
My pity hath been balm to heal their wounds,
My mildness hath allayed their swelling griefs,    10
My mercy dried their water-flowing tears.
I have not been desirous of their wealth,
Nor much oppressed them with great subsidies,
Nor forward of revenge, though they much erred.
Then why should they love Edward more than me?  15
No, Exeter, these graces challenge grace;
And when the lion fawns upon the lamb,
The lamb will never cease to follow him.
          *Shout within 'A Lancaster',* ⌈*'A York'*⌉
EXETER
Hark, hark, my lord—what shouts are these?
          *Enter King Edward and Richard Duke of Gloucester,*
          *with soldiers*
KING EDWARD
Seize on the shame-faced Henry—bear him hence,   20
And once again proclaim us King of England.
You are the fount that makes small brooks to flow.
Now stops thy spring—my sea shall suck them dry,
And swell so much the higher by their ebb.
Hence with him to the Tower—let him not speak.   25
          *Exeunt some with King Henry and Exeter*
And lords, towards Coventry bend we our course,
Where peremptory Warwick now remains.
The sun shines hot, and, if we use delay,
Cold biting winter mars our hoped-for hay.
RICHARD OF GLOUCESTER
Away betimes, before his forces join,                    30
And take the great-grown traitor unawares.
Brave warriors, march amain towards Coventry.
                                        *Exeunt*

5.1      *Enter the Earl of Warwick, the Mayor of Coventry,*
          *two Messengers, and others upon the walls*
WARWICK
Where is the post that came from valiant Oxford?
          ⌈*The First Messenger steps forward*⌉
How far hence is thy lord, mine honest fellow?
FIRST MESSENGER
By this at Dunsmore, marching hitherward.
WARWICK
How far off is our brother Montague?
Where is the post that came from Montague?        5
          ⌈*The Second Messenger steps forward*⌉
SECOND MESSENGER
By this at Da'ntry, with a puissant troop.
          *Enter Somerville* ⌈*to them, above*⌉
WARWICK
Say, Somerville—what says my loving son?
And, by thy guess, how nigh is Clarence now?
SOMERVILLE
At Southam I did leave him with his forces,
And do expect him here some two hours hence.    10

                    *A march afar off*
WARWICK
Then Clarence is at hand—I hear his drum.
SOMERVILLE
It is not his, my lord. Here Southam lies.
The drum your honour hears marcheth from Warwick.
WARWICK
Who should that be? Belike, unlooked-for friends.
SOMERVILLE
They are at hand, and you shall quickly know.    15
          *Flourish. Enter below King Edward and Richard*
          *Duke of Gloucester, with soldiers*
KING EDWARD
Go, trumpet, to the walls, and sound a parley.
          ⌈*Sound a parley*⌉
RICHARD OF GLOUCESTER
See how the surly Warwick mans the wall.
WARWICK
O, unbid spite—is sportful Edward come?
Where slept our scouts, or how are they seduced,
That we could hear no news of his repair?         20
KING EDWARD
Now, Warwick, wilt thou ope the city gates,
Speak gentle words, and humbly bend thy knee,
Call Edward king, and at his hands beg mercy?
And he shall pardon thee these outrages.
WARWICK
Nay, rather, wilt thou draw thy forces hence,     25
Confess who set thee up and plucked thee down,
Call Warwick patron, and be penitent?
And thou shalt still remain the Duke of York.
RICHARD OF GLOUCESTER
I thought at least he would have said 'the King'.
Or did he make the jest against his will?          30
WARWICK
Is not a dukedom, sir, a goodly gift?
RICHARD OF GLOUCESTER
Ay, by my faith, for a poor earl to give.
I'll do thee service for so good a gift.
WARWICK
'Twas I that gave the kingdom to thy brother.
KING EDWARD
Why then, 'tis mine, if but by Warwick's gift.    35
WARWICK
Thou art no Atlas for so great a weight;
And, weakling, Warwick takes his gift again;
And Henry is my king, Warwick his subject.
KING EDWARD
But Warwick's king is Edward's prisoner,
And, gallant Warwick, do but answer this:         40
What is the body when the head is off?
RICHARD OF GLOUCESTER
Alas, that Warwick had no more forecast,
But whiles he thought to steal the single ten,
The king was slyly fingered from the deck.
⌈*To Warwick*⌉ You left poor Henry at the Bishop's palace,
And ten to one you'll meet him in the Tower.     46
KING EDWARD
'Tis even so—⌈*to Warwick*⌉ yet you are Warwick still.
RICHARD OF GLOUCESTER
Come, Warwick, take the time—kneel down, kneel
down.
Nay, when? Strike now, or else the iron cools.
WARWICK
I had rather chop this hand off at a blow,          50
And with the other fling it at thy face,
Than bear so low a sail to strike to thee.

KING EDWARD

Sail how thou canst, have wind and tide thy friend,
This hand, fast wound about thy coal-black hair,
Shall, whiles thy head is warm and new cut off,    55
Write in the dust this sentence with thy blood:
'Wind-changing Warwick now can change no more'.
*Enter the Earl of Oxford, with a drummer and*
⌜*soldiers bearing*⌝ *colours*

WARWICK

O cheerful colours! See where Oxford comes.

OXFORD

Oxford, Oxford, for Lancaster!
⌜*Oxford and his men pass over the stage and*
*exeunt into the city*⌝

RICHARD OF GLOUCESTER (*to King Edward*)

The gates are open—let us enter too.    60

KING EDWARD

So other foes may set upon our backs?
Stand we in good array, for they no doubt
Will issue out again and bid us battle.
If not, the city being but of small defence,
We'll quickly rouse the traitors in the same.    65

WARWICK ⌜*to Oxford, within*⌝

O, welcome, Oxford—for we want thy help.
*Enter the Marquis of Montague with a drummer*
*and* ⌜*soldiers bearing*⌝ *colours*

MONTAGUE

Montague, Montague, for Lancaster!
⌜*Montague and his men pass over the stage and*
*exeunt into the city*⌝

RICHARD OF GLOUCESTER

Thou and thy brother both shall bye this treason
Even with the dearest blood your bodies bear.

KING EDWARD

The harder matched, the greater victory.    70
My mind presageth happy gain and conquest.
*Enter the Duke of Somerset with a drummer and*
⌜*soldiers bearing*⌝ *colours*

SOMERSET

Somerset, Somerset, for Lancaster!
⌜*Somerset and his men pass over the stage and*
*exeunt into the city*⌝

RICHARD OF GLOUCESTER

Two of thy name, both dukes of Somerset,
Have sold their lives unto the house of York—
And thou shalt be the third, an this sword hold.    75
*Enter George Duke of Clarence with a drummer and*
⌜*soldiers bearing*⌝ *colours*

WARWICK

And lo, where George of Clarence sweeps along,
Of force enough to bid his brother battle;
With whom an upright zeal to right prevails
More than the nature of a brother's love.

GEORGE OF CLARENCE

Clarence, Clarence, for Lancaster!    80

KING EDWARD

*Et tu, Brute*—wilt thou stab Caesar too?
(*To a trumpeter*) A parley, sirra, to George of Clarence.
*Sound a parley. Richard of Gloucester and George of*
*Clarence whisper together*

WARWICK

Come, Clarence, come—thou wilt if Warwick call.

GEORGE OF CLARENCE

Father of Warwick, know you what this means?
⌜*He takes his red rose out of his hat and throws it at*
*Warwick*⌝

Look—here I throw my infamy at thee!    85
I will not ruinate my father's house,
Who gave his blood to lime the stones together,
And set up Lancaster. Why, trowest thou, Warwick,
That Clarence is so harsh, so blunt, unnatural,
To bend the fatal instruments of war    90
Against his brother and his lawful king?
Perhaps thou wilt object my holy oath.
To keep that oath were more impiety
Than Jephthah, when he sacrificed his daughter.
I am so sorry for my trespass made    95
That, to deserve well at my brothers' hands,
I here proclaim myself thy mortal foe,
With resolution, wheresoe'er I meet thee—
As I will meet thee, if thou stir abroad—
To plague thee for thy foul misleading me.    100
And so, proud-hearted Warwick, I defy thee,
And to my brothers turn my blushing cheeks.
(*To King Edward*)
Pardon me, Edward—I will make amends.
(*To Richard*)
And, Richard, do not frown upon my faults,
For I will henceforth be no more unconstant.    105

KING EDWARD

Now welcome more, and ten times more beloved,
Than if thou never hadst deserved our hate.

RICHARD OF GLOUCESTER (*to George*)

Welcome, good Clarence—this is brother-like.

WARWICK (*to George*)

O, passing traitor—perjured and unjust!

KING EDWARD

What, Warwick, wilt thou leave the town and fight?
Or shall we beat the stones about thine ears?    111

WARWICK ⌜*aside*⌝

Alas, I am not cooped here for defence.
(*To King Edward*)
I will away towards Barnet presently,
And bid thee battle, Edward, if thou dar'st.

KING EDWARD

Yes, Warwick—Edward dares, and leads the way.    115
Lords, to the field—Saint George and victory!
*Exeunt below King Edward and his company.*
*March. The Earl of Warwick and his company*
*descend and follow*

**5.2**    *Alarum and excursions. Enter King Edward bringing*
*forth the Earl of Warwick, wounded*

KING EDWARD

So lie thou there. Die thou, and die our fear—
For Warwick was a bug that feared us all.
Now, Montague, sit fast—I seek for thee
That Warwick's bones may keep thine company.    *Exit*

WARWICK

Ah, who is nigh? Come to me, friend or foe,    5
And tell me who is victor, York or Warwick?
Why ask I that? My mangled body shows,
My blood, my want of strength, my sick heart shows,
That I must yield my body to the earth
And by my fall the conquest to my foe.    10
Thus yields the cedar to the axe's edge,
Whose arms gave shelter to the princely eagle,
Under whose shade the ramping lion slept,
Whose top-branch over-peered Jove's spreading tree
And kept low shrubs from winter's powerful wind.    15
These eyes, that now are dimmed with death's black
veil,

Have been as piercing as the midday sun
To search the secret treasons of the world.
The wrinkles in my brows, now filled with blood,
Were likened oft to kingly sepulchres—                    20
For who lived king, but I could dig his grave?
And who durst smile when Warwick bent his brow?
Lo now my glory smeared in dust and blood.
My parks, my walks, my manors that I had,
Even now forsake me, and of all my lands          25
Is nothing left me but my body's length.
Why, what is pomp, rule, reign, but earth and dust?
And, live we how we can, yet die we must.

*Enter the Earl of Oxford and the Duke of Somerset*

SOMERSET
Ah, Warwick, Warwick—wert thou as we are,
We might recover all our loss again.                       30
The Queen from France hath brought a puissant
      power.
Even now we heard the news. Ah, couldst thou fly!
WARWICK
Why, then I would not fly. Ah, Montague,
If thou be there, sweet brother, take my hand,
And with thy lips keep in my soul a while.          35
Thou lov'st me not—for, brother, if thou didst,
Thy tears would wash this cold congealèd blood
That glues my lips and will not let me speak.
Come quickly, Montague, or I am dead.
SOMERSET
Ah, Warwick—Montague hath breathed his last,   40
And to the latest gasp cried out for Warwick,
And said 'Commend me to my valiant brother.'
And more he would have said, and more he spoke,
Which sounded like a canon in a vault,
That mote not be distinguished; but at last          45
I well might hear, delivered with a groan,
'O, farewell, Warwick.'
WARWICK
Sweet rest his soul. Fly, lords, and save yourselves—
For Warwick bids you all farewell, to meet in heaven.
                                                          *He dies*
OXFORD
Away, away—to meet the Queen's great power!      50
            *Here they bear away Warwick's body. Exeunt*

**5.3**   *Flourish. Enter King Edward in triumph, with*
         *Richard Duke of Gloucester, George Duke of*
         *Clarence, and ⌈soldiers⌉*

KING EDWARD
Thus far our fortune keeps an upward course,
And we are graced with wreaths of victory.
But in the midst of this bright-shining day
I spy a black suspicious threatening cloud
That will encounter with our glorious sun             5
Ere he attain his easeful western bed.
I mean, my lords, those powers that the Queen
Hath raised in Gallia have arrived our coast,
And, as we hear, march on to fight with us.
GEORGE OF CLARENCE
A little gale will soon disperse that cloud,            10
And blow it to the source from whence it came.
Thy very beams will dry those vapours up,
For every cloud engenders not a storm.
RICHARD OF GLOUCESTER
The Queen is valued thirty thousand strong,
And Somerset, with Oxford, fled to her.                15

If she have time to breathe, be well assured,
Her faction will be full as strong as ours.
KING EDWARD
We are advertised by our loving friends
That they do hold their course toward Tewkesbury.
We, having now the best at Barnet field,               20
Will thither straight, for willingness rids way—
And, as we march, our strength will be augmented
In every county as we go along.
Strike up the drum, cry 'Courage!'; and away.
                              ⌈*Flourish. March.*⌉ *Exeunt*

**5.4**   *Flourish. March. Enter Queen Margaret, Prince*
         *Edward, the Duke of Somerset, the Earl of Oxford,*
         *and soldiers*

QUEEN MARGARET
Great lords, wise men ne'er sit and wail their loss,
But cheerly seek how to redress their harms.
What though the mast be now blown overboard,
The cable broke, the holding-anchor lost,
And half our sailors swallowed in the flood?          5
Yet lives our pilot still. Is't meet that he
Should leave the helm and, like a fearful lad,
With tearful eyes add water to the sea,
And give more strength to that which hath too much,
Whiles, in his moan, the ship splits on the rock    10
Which industry and courage might have saved?
Ah, what a shame; ah, what a fault were this.
Say Warwick was our anchor—what of that?
And Montague our top-mast—what of him?
Our slaughtered friends the tackles—what of these? 15
Why, is not Oxford here another anchor?
And Somerset another goodly mast?
The friends of France our shrouds and tacklings?
And, though unskilful, why not Ned and I
For once allowed the skilful pilot's charge?          20
We will not from the helm to sit and weep,
But keep our course, though the rough wind say no,
From shelves and rocks that threaten us with wreck.
As good to chide the waves as speak them fair.
And what is Edward but a ruthless sea?                 25
What Clarence but a quicksand of deceit?
And Richard but a raggèd fatal rock?
All these the enemies to our poor barque.
Say you can swim—alas, 'tis but a while;
Tread on the sand—why, there you quickly sink;    30
Bestride the rock—the tide will wash you off,
Or else you famish. That's a threefold death.
This speak I, lords, to let you understand,
If case some one of you would fly from us,
That there's no hoped-for mercy with the brothers York
More than with ruthless waves, with sands, and rocks.
Why, courage then—what cannot be avoided        37
'Twere childish weakness to lament or fear.
PRINCE EDWARD
Methinks a woman of this valiant spirit
Should, if a coward heard her speak these words,   40
Infuse his breast with magnanimity
And make him, naked, foil a man at arms.
I speak not this as doubting any here—
For did I but suspect a fearful man,
He should have leave to go away betimes,             45
Lest in our need he might infect another
And make him of like spirit to himself.
If any such be here—as God forbid—
Let him depart before we need his help.

OXFORD
Women and children of so high a courage,     50
And warriors faint—why, 'twere perpetual shame!
O brave young Prince, thy famous grandfather
Doth live again in thee! Long mayst thou live
To bear his image and renew his glories!

SOMERSET
And he that will not fight for such a hope,     55
Go home to bed, and like the owl by day,
If he arise, be mocked and wondered at.

QUEEN MARGARET
Thanks, gentle Somerset; sweet Oxford, thanks.

PRINCE EDWARD
And take his thanks that yet hath nothing else.
   *Enter a Messenger*

MESSENGER
Prepare you, lords, for Edward is at hand     60
Ready to fight—therefore be resolute.

OXFORD
I thought no less. It is his policy
To haste thus fast to find us unprovided.

SOMERSET
But he's deceived; we are in readiness.

QUEEN MARGARET
This cheers my heart, to see your forwardness.    65

OXFORD
Here pitch our battle—hence we will not budge.
   *Flourish and march. Enter King Edward, Richard*
   *Duke of Gloucester, and George Duke of Clarence,*
   *with soldiers*

KING EDWARD (*to his followers*)
Brave followers, yonder stands the thorny wood
Which, by the heavens' assistance and your strength,
Must by the roots be hewn up yet ere night.
I need not add more fuel to your fire,     70
For well I wot ye blaze to burn them out.
Give signal to the fight, and to it, lords.

QUEEN MARGARET (*to her followers*)
Lords, knights, and gentlemen—what I should say
My tears gainsay; for every word I speak
Ye see I drink the water of my eye.     75
Therefore, no more but this: Henry your sovereign
Is prisoner to the foe, his state usurped,
His realm a slaughter-house, his subjects slain,
His statutes cancelled, and his treasure spent—
And yonder is the wolf that makes this spoil.    80
You fight in justice; then in God's name, lords,
Be valiant, and give signal to the fight.
   *Alarum, retreat, excursions. Exeunt*

**5.5**  *Flourish. Enter King Edward, Richard Duke of*
   *Gloucester, and George Duke of Clarence with*
   *Queen Margaret, the Earl of Oxford, and the Duke*
   *of Somerset, guarded*

KING EDWARD
Now here a period of tumultuous broils.
Away with Oxford to Hames Castle straight;
For Somerset, off with his guilty head.
Go bear them hence—I will not hear them speak.

OXFORD
For my part, I'll not trouble thee with words.    5
   *Exit, guarded*

SOMERSET
Nor I, but stoop with patience to my fortune.
   *Exit, guarded*

QUEEN MARGARET
So part we sadly in this troublous world
To meet with joy in sweet Jerusalem.

KING EDWARD
Is proclamation made that who finds Edward
Shall have a high reward and he his life?     10

RICHARD OF GLOUCESTER
It is, and lo where youthful Edward comes.
   *Enter Prince Edward, guarded*

KING EDWARD
Bring forth the gallant—let us hear him speak.
What, can so young a thorn begin to prick?
Edward, what satisfaction canst thou make
For bearing arms, for stirring up my subjects,    15
And all the trouble thou hast turned me to?

PRINCE EDWARD
Speak like a subject, proud ambitious York.
Suppose that I am now my father's mouth—
Resign thy chair, and where I stand, kneel thou,
Whilst I propose the self-same words to thee,    20
Which, traitor, thou wouldst have me answer to.

QUEEN MARGARET
Ah, that thy father had been so resolved.

RICHARD OF GLOUCESTER
That you might still have worn the petticoat
And ne'er have stolen the breech from Lancaster.

PRINCE EDWARD
Let Aesop fable in a winter's night—     25
His currish riddles sorts not with this place.

RICHARD OF GLOUCESTER
By heaven, brat, I'll plague ye for that word.

QUEEN MARGARET
Ay, thou wast born to be a plague to men.

RICHARD OF GLOUCESTER
For God's sake take away this captive scold.

PRINCE EDWARD
Nay, take away this scolding crookback rather.    30

KING EDWARD
Peace, wilful boy, or I will charm your tongue.

GEORGE OF CLARENCE (*to Prince Edward*)
Untutored lad, thou art too malapert.

PRINCE EDWARD
I know my duty—you are all undutiful.
Lascivious Edward, and thou, perjured George,
And thou, misshapen Dick—I tell ye all     35
I am your better, traitors as ye are,
And thou usurp'st my father's right and mine.

KING EDWARD
Take that, the likeness of this railer here.
   *King Edward stabs Prince Edward*

RICHARD OF GLOUCESTER
Sprawl'st thou? Take that, to end thy agony.
   *Richard stabs Prince Edward*

GEORGE OF CLARENCE
And there's for twitting me with perjury.     40
   *George stabs Prince Edward, ⌈who dies⌉*

QUEEN MARGARET
O, kill me too!

RICHARD OF GLOUCESTER  Marry, and shall.
   *He offers to kill her*

KING EDWARD
Hold, Richard, hold—for we have done too much.

RICHARD OF GLOUCESTER
Why should she live to fill the world with words?
   *Queen Margaret faints*

KING EDWARD
What—doth she swoon? Use means for her recovery.
RICHARD OF GLOUCESTER (*aside to George*)
Clarence, excuse me to the King my brother.                45
I'll hence to London on a serious matter.
Ere ye come there, be sure to hear some news.
GEORGE OF CLARENCE (*aside to Richard*) What? What?
RICHARD OF GLOUCESTER (*aside to George*)
The Tower, the Tower.                                  *Exit*
QUEEN MARGARET
O Ned, sweet Ned—speak to thy mother, boy.            50
Canst thou not speak? O traitors, murderers!
They that stabbed Caesar shed no blood at all,
Did not offend, nor were not worthy blame,
If this foul deed were by to equal it.
He was a man—this, in respect, a child;              55
And men ne'er spend their fury on a child.
What's worse than murderer that I may name it?
No, no, my heart will burst an if I speak;
And I will speak that so my heart may burst.
Butchers and villains! Bloody cannibals!               60
How sweet a plant have you untimely cropped!
You have no children, butchers; if you had,
The thought of them would have stirred up remorse.
But if you ever chance to have a child,
Look in his youth to have him so cut off             65
As, deathsmen, you have rid this sweet young Prince!
KING EDWARD
Away with her—go, bear her hence perforce.
QUEEN MARGARET
Nay, never bear me hence—dispatch me here.
Here sheathe thy sword—I'll pardon thee my death.
What? Wilt thou not? Then, Clarence, do it thou.    70
GEORGE OF CLARENCE
By heaven, I will not do thee so much ease.
QUEEN MARGARET
Good Clarence, do; sweet Clarence, do thou do it.
GEORGE OF CLARENCE
Didst thou not hear me swear I would not do it?
QUEEN MARGARET
Ay, but thou usest to forswear thyself.
'Twas sin before, but now 'tis charity.             75
What, wilt thou not? Where is that devil's butcher,
Hard-favoured Richard? Richard, where art thou?
Thou art not here. Murder is thy alms-deed—
Petitioners for blood thou ne'er putt'st back.
KING EDWARD
Away, I say—I charge ye, bear her hence.            80
QUEEN MARGARET
So come to you and yours as to this Prince!
                                          *Exit, guarded*
KING EDWARD Where's Richard gone?
GEORGE OF CLARENCE
To London all in post—⌈*aside*⌉ and as I guess,
To make a bloody supper in the Tower.
KING EDWARD
He's sudden if a thing comes in his head.           85
Now march we hence. Discharge the common sort
With pay and thanks, and let's away to London,
And see our gentle Queen how well she fares.
By this I hope she hath a son for me.       *Exeunt*

**5.6**   *Enter on the walls King Henry the Sixth, reading a
          book, Richard Duke of Gloucester, and the
          Lieutenant of the Tower*
RICHARD OF GLOUCESTER
Good day, my lord. What, at your book so hard?

KING HENRY
Ay, my good lord—'my lord', I should say, rather.
'Tis sin to flatter; 'good' was little better.
'Good Gloucester' and 'good devil' were alike,
And both preposterous—therefore not 'good lord'.     5
RICHARD OF GLOUCESTER (*to the Lieutenant*)
Sirrah, leave us to ourselves. We must confer.
                                       *Exit Lieutenant*
KING HENRY
So flies the reckless shepherd from the wolf;
So first the harmless sheep doth yield his fleece,
And next his throat unto the butcher's knife.
What scene of death hath Roscius now to act?        10
RICHARD OF GLOUCESTER
Suspicion always haunts the guilty mind;
The thief doth fear each bush an officer.
KING HENRY
The bird that hath been limèd in a bush
With trembling wings misdoubteth every bush.
And I, the hapless male to one sweet bird,          15
Have now the fatal object in my eye
Where my poor young was limed, was caught and
       killed.
RICHARD OF GLOUCESTER
Why, what a peevish fool was that of Crete,
That taught his son the office of a fowl!
And yet, for all his wings, the fool was drowned.   20
KING HENRY
I, Daedalus; my poor boy, Icarus;
Thy father, Minos, that denied our course;
The sun that seared the wings of my sweet boy,
Thy brother Edward; and thyself, the sea,
Whose envious gulf did swallow up his life.         25
Ah, kill me with thy weapon, not with words!
My breast can better brook thy dagger's point
Than can my ears that tragic history.
But wherefore dost thou come? Is't for my life?
RICHARD OF GLOUCESTER
Think'st thou I am an executioner?                  30
KING HENRY
A persecutor I am sure thou art;
If murdering innocents be executing,
Why, then thou art an executioner.
RICHARD OF GLOUCESTER
Thy son I killed for his presumption.
KING HENRY
Hadst thou been killed when first thou didst presume,
Thou hadst not lived to kill a son of mine.         36
And thus I prophesy: that many a thousand
Which now mistrust no parcel of my fear,
And many an old man's sigh, and many a widow's,
And many an orphan's water-standing eye—            40
Men for their sons', wives for their husbands',
Orphans for their parents' timeless death—
Shall rue the hour that ever thou wast born.
The owl shrieked at thy birth—an evil sign;
The night-crow cried, aboding luckless time;        45
Dogs howled, and hideous tempests shook down trees;
The raven rooked her on the chimney's top;
And chatt'ring pies in dismal discords sung.
Thy mother felt more than a mother's pain,
And yet brought forth less than a mother's hope—    50
To wit, an indigested and deformèd lump,
Not like the fruit of such a goodly tree.
Teeth hadst thou in thy head when thou wast born,
To signify thou cam'st to bite the world;

And if the rest be true which I have heard　　55
Thou cam'st—
RICHARD
I'll hear no more. Die, prophet, in thy speech,
　　*He stabs him*
For this, amongst the rest, was I ordained.
KING HENRY
Ay, and for much more slaughter after this.
O, God forgive my sins, and pardon thee.　　*He dies*
RICHARD OF GLOUCESTER
What—will the aspiring blood of Lancaster　　61
Sink in the ground? I thought it would have mounted.
See how my sword weeps for the poor King's death.
O, may such purple tears be alway shed
From those that wish the downfall of our house!　　65
If any spark of life be yet remaining,
Down, down to hell, and say I sent thee thither—
　　*He stabs him again*
I that have neither pity, love, nor fear.
Indeed, 'tis true that Henry told me of,
For I have often heard my mother say　　70
I came into the world with my legs forward.
Had I not reason, think ye, to make haste,
And seek their ruin that usurped our right?
The midwife wondered and the women cried
'O, Jesus bless us, he is born with teeth!'—　　75
And so I was, which plainly signified
That I should snarl and bite and play the dog.
Then, since the heavens have shaped my body so,
Let hell make crooked my mind to answer it.
I had no father, I am like no father;　　80
I have no brother, I am like no brother;
And this word, 'love', which greybeards call divine,
Be resident in men like one another
And not in me—I am myself alone.
Clarence, beware; thou kept'st me from the light—　　85
But I will sort a pitchy day for thee.
For I will buzz abroad such prophecies
That Edward shall be fearful of his life,
And then, to purge his fear, I'll be thy death.
Henry and his son are gone; thou, Clarence, art next;
And by one and one I will dispatch the rest,　　91
Counting myself but bad till I be best.
I'll throw thy body in another room
And triumph, Henry, in thy day of doom.
　　　　　　　　　　　　*Exit with the body*

5.7　⌈*A chair of state.*⌉ *Flourish. Enter King Edward,*
　　*Lady Gray his Queen, George Duke of Clarence,*
　　*Richard Duke of Gloucester, the Lord Hastings, a*
　　*nurse carrying the infant Prince Edward, and*
　　*attendants*
KING EDWARD
Once more we sit in England's royal throne,
Repurchased with the blood of enemies.
What valiant foemen, like to autumn's corn,
Have we mowed down in tops of all their pride!

Three dukes of Somerset, threefold renowned　　5
For hardy and undoubted champions;
Two Cliffords, as the father and the son;
And two Northumberlands—two braver men
Ne'er spurred their coursers at the trumpet's sound.
With them, the two brave bears, Warwick and
　　Montague,　　10
That in their chains fettered the kingly lion
And made the forest tremble when they roared.
Thus have we swept suspicion from our seat
And made our footstool of security.
*(To Lady Gray)*
Come hither, Bess, and let me kiss my boy.　　15
　　*The nurse brings forth the infant prince. King*
　　*Edward kisses him*
Young Ned, for thee, thine uncles and myself
Have in our armours watched the winter's night,
Went all afoot in summer's scalding heat,
That thou mightst repossess the crown in peace;
And of our labours thou shalt reap the gain.　　20
RICHARD OF GLOUCESTER *(aside)*
I'll blast his harvest, an your head were laid;
For yet I am not looked on in the world.
This shoulder was ordained so thick to heave;
And heave it shall some weight or break my back.
Work thou the way, and thou shalt execute.　　25
KING EDWARD
Clarence and Gloucester, love my lovely queen;
And kiss your princely nephew, brothers, both.
GEORGE OF CLARENCE
The duty that I owe unto your majesty
I seal upon the lips of this sweet babe.
　　*He kisses the infant prince*
LADY GRAY
Thanks, noble Clarence—worthy brother, thanks.　　30
RICHARD OF GLOUCESTER
And that I love the tree from whence thou sprang'st,
Witness the loving kiss I give the fruit.
　　*He kisses the infant prince*
*(Aside)* To say the truth, so Judas kissed his master,
And cried 'All hail!' whenas he meant all harm.
KING EDWARD
Now am I seated as my soul delights,　　35
Having my country's peace and brothers' loves.
GEORGE OF CLARENCE
What will your grace have done with Margaret?
René her father, to the King of France
Hath pawned the Sicils and Jerusalem,
And hither have they sent it for her ransom.　　40
KING EDWARD
Away with her, and waft her hence to France.
And now what rests but that we spend the time
With stately triumphs, mirthful comic shows,
Such as befits the pleasure of the court?
Sound drums and trumpets—farewell, sour annoy!　　45
For here, I hope, begins our lasting joy.
　　　　　　　　　　　　⌈*Flourish.*⌉ *Exeunt*

## ADDITIONAL PASSAGES

A. Our edition adopts the 1595 version of 1.1.120–5 in the belief that it reflects an authorial revision; an edited text of the Folio alternative follows.

KING HENRY
  Peace, thou—and give King Henry leave to speak.
WARWICK
  Plantagenet shall speak first—hear him, lords,
  And be you silent and attentive too,
  For he that interrupts him shall not live.
KING HENRY ⌈*to York*⌉
  Think'st thou that I will leave my kingly throne,    5

B. The 1595 text abridges 5.4.82.1–5.5.17, and may reflect authorial revision. An edited text of the abridged passage follows:

ALL THE LANCASTER PARTY
  Saint George for Lancaster!
      *Alarums to the battle.* ⌈*The house of*⌉ *York flies, then*
      *the chambers are discharged. Then enter King Edward,*
      *George of Clarence, and Richard of Gloucester, and their*
      *followers: they make a great shout, and cry 'For York!*
      *For York!' Then Queen Margaret, Prince Edward,*
      *Oxford and Somerset are all taken prisoner. Flourish,*
      *and enter all again*
KING EDWARD
  Now here a period of tumultuous broils.
  Away with Oxford to Hames Castle straight;
  For Somerset, off with his guilty head.
  Go, bear them hence—I will not hear them speak.    5
OXFORD
  For my part, I'll not trouble thee with words.
                                        *Exit, guarded*
SOMERSET
  Nor I, but stoop with patience to my death.
                                        *Exit, guarded*
KING EDWARD (*to Prince Edward*)
  Edward, what satisfaction canst thou make
  For stirring up my subjects to rebellion?
PRINCE EDWARD
  Speak like a subject, proud ambitious York.    10

# TITUS ANDRONICUS

SHAKESPEARE'S first, most sensation-packed tragedy appeared in print in 1594, and a performance record dating from January of that year appears to indicate that it was then a new play. But according to its title-page it had been acted by three companies, one of which was bankrupt by the summer of 1593; and the play's style, too, suggests that it was written earlier. Shakespeare seems to have added a scene after the play's earliest performances, for Act 3, Scene 2 was first printed in the 1623 Folio. The 1594 performance record may refer to the revised play, not the original, or to the play's first London performance after plague had closed the theatres from June 1592.

By convention, Elizabethan tragedies treated historical subjects, and *Titus Andronicus* is set in Rome during the fourth century AD; but its story (like that of Shakespeare's other early tragedy, *Romeo and Juliet*) is fictitious. Whether Shakespeare invented it is an open question: the same tale is told in both a ballad and a chap-book which survive only in eighteenth-century versions but which could derive from pre-Shakespearian originals. Even if Shakespeare knew these works, they could have supplied only a skeletal narrative. His play's spirit and style owe much to Ovid's *Metamorphoses*, one of his favourite works of classical literature, which he actually brings on stage in Act 4, Scene 1. Ovid's tale of the rape of Philomela was certainly in Shakespeare's mind as he wrote, and the play's more horrific elements owe something to the Roman dramatist Seneca.

In its time, *Titus Andronicus* was popular, perhaps because it combines sensational incident with high-flown rhetoric of a kind that was fashionable around 1590. It tells a story of double revenge. Tamora, Queen of the Goths, seeks revenge on her captor, Titus, for the ritual slaughter of her son Alarbus; she achieves it when her other sons, Chiron and Demetrius, rape and mutilate Titus' daughter, Lavinia. Later, Titus himself seeks revenge on Tamora and her husband, Saturninus, after Tamora's black lover, Aaron, has falsely led him to believe that he can save his sons' lives by allowing his own hand to be chopped off. Though he is driven to madness, Titus, with his brother Marcus and his last surviving son, Lucius, achieves a spectacular sequence of vengeance in which he cuts Tamora's sons' throats, serves their flesh baked in a pie to their mother, kills Lavinia to save her from her shame, and stabs Tamora to death. Then, in rapid succession, Saturninus kills Titus and is himself killed by Lucius, who, as the new Emperor, is left with Marcus to bury the dead, to punish Aaron, and 'To heal Rome's harms and wipe away her woe'.

In *Titus Andronicus*, as in his early history plays, Shakespeare is at his most successful in the expression of grief and the portrayal of vigorously energetic evil. The play's piling of horror upon horror can seem ludicrous, and the reader may be surprised by the apparent disjunction between terrifying events and the measured verse in which characters react; but a few remarkable modern productions have revealed that the play may still arouse pity as well as terror in its audiences.

# THE PERSONS OF THE PLAY

SATURNINUS, eldest son of the late Emperor of Rome; later Emperor

BASSIANUS, his brother

TITUS ANDRONICUS, a Roman nobleman, general against the Goths

LUCIUS ⎫
QUINTUS ⎬ sons of Titus
MARTIUS ⎪
MUTIUS ⎭

LAVINIA, daughter of Titus

YOUNG LUCIUS, a boy, son of Lucius

MARCUS ANDRONICUS, a tribune of the people, Titus' brother

PUBLIUS, his son

SEMPRONIUS ⎫
CAIUS ⎬ kinsmen of Titus
VALENTINE ⎭

A CAPTAIN

AEMILIUS

TAMORA, Queen of the Goths, later wife of Saturninus

ALARBUS ⎫
DEMETRIUS ⎬ her sons
CHIRON ⎭

AARON, a Moor, her lover

A NURSE

A CLOWN

Senators, tribunes, Romans, Goths, soldiers, and attendants

# The Most Lamentable Roman Tragedy of Titus Andronicus

**1.1** ⌜*Flourish.*⌝ *Enter the Tribunes and Senators aloft,*
*and then enter below Saturninus and his followers*
*at one door and Bassianus and his followers* ⌜*at the*
*other, with drummer and colours*⌝

SATURNINUS
Noble patricians, patrons of my right,
Defend the justice of my cause with arms.
And countrymen, my loving followers,
Plead my successive title with your swords.
I am his first-born son that was the last          5
That ware the imperial diadem of Rome.
Then let my father's honours live in me,
Nor wrong mine age with this indignity.

BASSIANUS
Romans, friends, followers, favourers of my right,
If ever Bassianus, Caesar's son,          10
Were gracious in the eyes of royal Rome,
Keep then this passage to the Capitol,
And suffer not dishonour to approach
The imperial seat, to virtue consecrate,
To justice, continence, and nobility;          15
But let desert in pure election shine,
And, Romans, fight for freedom in your choice.

⌜*Enter*⌝ *Marcus Andronicus* ⌜*aloft*⌝ *with the crown*

MARCUS
Princes that strive by factions and by friends
Ambitiously for rule and empery,
Know that the people of Rome, for whom we stand          20
A special party, have by common voice
In election for the Roman empery
Chosen Andronicus, surnamèd *Pius*
For many good and great deserts to Rome.
A nobler man, a braver warrior,          25
Lives not this day within the city walls.
He by the Senate is accited home
From weary wars against the barbarous Goths,
That with his sons, a terror to our foes,
Hath yoked a nation strong, trained up in arms.          30
Ten years are spent since first he undertook
This cause of Rome, and chastisèd with arms
Our enemies' pride. Five times he hath returned
Bleeding to Rome, bearing his valiant sons
In coffins from the field.          35
And now at last, laden with honour's spoils,
Returns the good Andronicus to Rome,
Renownèd Titus, flourishing in arms.
Let us entreat by honour of his name
Whom worthily you would have now succeeded,          40
And in the Capitol and Senate's right,
Whom you pretend to honour and adore,
That you withdraw you and abate your strength,
Dismiss your followers, and, as suitors should,
Plead your deserts in peace and humbleness.          45

SATURNINUS
How fair the Tribune speaks to calm my thoughts.

BASSIANUS
Marcus Andronicus, so I do affy
In thy uprightness and integrity,

And so I love and honour thee and thine,
Thy noble brother Titus and his sons,          50
And her to whom my thoughts are humbled all,
Gracious Lavinia, Rome's rich ornament,
That I will here dismiss my loving friends
And to my fortunes and the people's favour
Commit my cause in balance to be weighed.          55
⌜*Exeunt his soldiers and followers*⌝

SATURNINUS
Friends that have been thus forward in my right,
I thank you all, and here dismiss you all,
And to the love and favour of my country
Commit myself, my person, and the cause.
⌜*Exeunt his soldiers and followers*⌝
(*To the Tribunes and Senators*)
Rome, be as just and gracious unto me          60
As I am confident and kind to thee.
Open the gates and let me in.

BASSIANUS
Tribunes, and me, a poor competitor.
⌜*Flourish.*⌝ *They go up into the Senate House.*
*Enter a Captain*

CAPTAIN
Romans, make way. The good Andronicus,
Patron of virtue, Rome's best champion,          65
Successful in the battles that he fights,
With honour and with fortune is returned
From where he circumscribèd with his sword
And brought to yoke the enemies of Rome.
*Sound drums and trumpets, and then enter Martius*
*and Mutius, two of Titus' sons, and then* ⌜*men*
*bearing coffins*⌝ *covered with black, then Lucius and*
*Quintus, two other sons; then Titus Andronicus* ⌜*in*
*his chariot*⌝ *and then Tamora the Queen of Goths*
*and her sons Alarbus, Chiron, and Demetrius, with*
*Aaron the Moor and others as many as can be.*
*Then set down the* ⌜*coffins*⌝*, and Titus speaks*

TITUS
Hail, Rome, victorious in thy mourning weeds!          70
Lo, as the bark that hath discharged his freight
Returns with precious lading to the bay
From whence at first she weighed her anchorage,
Cometh Andronicus, bound with laurel bows,
To re-salute his country with his tears,          75
Tears of true joy for his return to Rome.
Thou great defender of this Capitol,
Stand gracious to the rites that we intend.
Romans, of five-and-twenty valiant sons,
Half of the number that King Priam had,          80
Behold the poor remains, alive and dead.
These that survive let Rome reward with love;
These that I bring unto their latest home,
With burial amongst their ancestors.
Here Goths have given me leave to sheathe my sword.
Titus unkind, and careless of thine own,          86
Why suffer'st thou thy sons unburied yet
To hover on the dreadful shore of Styx?
Make way to lay them by their brethren.

*They open the tomb*

There greet in silence as the dead are wont,                90
And sleep in peace, slain in your country's wars.
O sacred receptacle of my joys,
Sweet cell of virtue and nobility,
How many sons hast thou of mine in store
That thou wilt never render to me more!                     95

LUCIUS
Give us the proudest prisoner of the Goths,
That we may hew his limbs and on a pile
*Ad manes fratrum* sacrifice his flesh
Before this earthy prison of their bones,
That so the shadows be not unappeased,                     100
Nor we disturbed with prodigies on earth.

TITUS
I give him you, the noblest that survives,
The eldest son of this distressèd Queen.

TAMORA ⌈*kneeling*⌉
Stay, Roman brethren! Gracious conqueror,
Victorious Titus, rue the tears I shed—                    105
A mother's tears in passion for her son—
And if thy sons were ever dear to thee,
O, think my son to be as dear to me!
Sufficeth not that we are brought to Rome
To beautify thy triumphs, and return                       110
Captive to thee and to thy Roman yoke;
But must my sons be slaughtered in the streets
For valiant doings in their country's cause?
O, if to fight for king and commonweal
Were piety in thine, it is in these.                        115
Andronicus, stain not thy tomb with blood.
Wilt thou draw near the nature of the gods?
Draw near them then in being merciful.
Sweet mercy is nobility's true badge.
Thrice-noble Titus, spare my first-born son.               120

TITUS
Patient yourself, madam, and pardon me.
These are their brethren whom your Goths beheld
Alive and dead, and for their brethren slain
Religiously they ask a sacrifice.
To this your son is marked, and die he must                125
T'appease their groaning shadows that are gone.

LUCIUS
Away with him, and make a fire straight,
And with our swords upon a pile of wood
Let's hew his limbs till they be clean consumed.
                *Exeunt Titus' sons with Alarbus*

TAMORA ⌈*rising*⌉
O cruel irreligious piety!                                  130

CHIRON
Was never Scythia half so barbarous.

DEMETRIUS
Oppose not Scythia to ambitious Rome.
Alarbus goes to rest, and we survive
To tremble under Titus' threat'ning look.
Then, madam, stand resolved; but hope withal               135
The selfsame gods that armed the Queen of Troy
With opportunity of sharp revenge
Upon the Thracian tyrant in his tent
May favour Tamora, the Queen of Goths—
When Goths were Goths and Tamora was queen—
To quit her bloody wrongs upon her foes.                   141
        *Enter Quintus, Marcus, Mutius, and Lucius, the*
        *sons of Andronicus, again, with bloody swords*

LUCIUS
See, lord and father, how we have performed

Our Roman rites. Alarbus' limbs are lopped
And entrails feed the sacrificing fire,
Whose smoke like incense doth perfume the sky.             145
Remaineth naught but to inter our brethren
And with loud 'larums welcome them to Rome.

TITUS
Let it be so, and let Andronicus
Make this his latest farewell to their souls.
        ⌈*Flourish.*⌉ *Then sound trumpets and lay the*
        ⌈*coffins*⌉ *in the tomb*
In peace and honour rest you here, my sons;                150
Rome's readiest champions, repose you here in rest,
Secure from worldly chances and mishaps.
Here lurks no treason, here no envy swells,
Here grow no damnèd drugs, here are no storms,
No noise, but silence and eternal sleep.                   155
In peace and honour rest you here, my sons.
        *Enter Lavinia*

LAVINIA
In peace and honour live Lord Titus long,
My noble lord and father, live in fame.
Lo, at this tomb my tributary tears
I render for my brethren's obsequies,                      160
(*Kneeling*) And at thy feet I kneel with tears of joy
Shed on this earth for thy return to Rome.
O, bless me here with thy victorious hand,
Whose fortunes Rome's best citizens applaud.

TITUS
Kind Rome, that hast thus lovingly reserved               165
The cordial of mine age to glad my heart!
Lavinia, live; outlive thy father's days
And fame's eternal date, for virtue's praise.
        ⌈*Lavinia rises*⌉

MARCUS ⌈*aloft*⌉
Long live Lord Titus, my belovèd brother,
Gracious triumpher in the eyes of Rome!                    170

TITUS
Thanks, gentle Tribune, noble brother Marcus.

MARCUS
And welcome, nephews, from successful wars,
You that survive and you that sleep in fame.
Fair lords, your fortunes are alike in all,
That in your country's service drew your swords,           175
But safer triumph is this funeral pomp
That hath aspired to Solon's happiness
And triumphs over chance in honour's bed.
Titus Andronicus, the people of Rome,
Whose friend in justice thou hast ever been,               180
Send thee by me, their tribune and their trust,
This palliament of white and spotless hue,
And name thee in election for the empire
With these our late-deceasèd emperor's sons.
Be *candidatus* then, and put it on,                        185
And help to set a head on headless Rome.

TITUS
A better head her glorious body fits
Than his that shakes for age and feebleness.
What should I don this robe and trouble you?—
Be chosen with proclamations today,                        190
Tomorrow yield up rule, resign my life,
And set abroad new business for you all.
Rome, I have been thy soldier forty years,
And led my country's strength successfully,
And buried one-and-twenty valiant sons                     195
Knighted in field, slain manfully in arms
In right and service of their noble country.

Give me a staff of honour for mine age,
But not a sceptre to control the world.
Upright he held it, lords, that held it last.                    200
MARCUS
Titus, thou shalt obtain and ask the empery.
SATURNINUS
Proud and ambitious Tribune, canst thou tell?
TITUS
Patience, Prince Saturninus.
SATURNINUS                              Romans, do me right.
Patricians, draw your swords, and sheathe them not
Till Saturninus be Rome's emperor.                               205
Andronicus, would thou were shipped to hell
Rather than rob me of the people's hearts!
LUCIUS
Proud Saturnine, interrupter of the good
That noble-minded Titus means to thee.
TITUS
Content thee, Prince. I will restore to thee                     210
The people's hearts, and wean them from themselves.
BASSIANUS
Andronicus, I do not flatter thee
But honour thee, and will do till I die.
My faction if thou strengthen with thy friends
I will most thankful be; and thanks to men                       215
Of noble minds is honourable meed.
TITUS
People of Rome, and people's tribunes here,
I ask your voices and your suffrages.
Will ye bestow them friendly on Andronicus?
TRIBUNES
To gratify the good Andronicus                                   220
And gratulate his safe return to Rome
The people will accept whom he admits.
TITUS
Tribunes, I thank you, and this suit I make:
That you create our emperor's eldest son
Lord Saturnine, whose virtues will, I hope,                      225
Reflect on Rome as Titan's rays on earth,
And ripen justice in this commonweal.
Then if you will elect by my advice,
Crown him and say, 'Long live our Emperor!'
MARCUS
With voices and applause of every sort,                          230
Patricians and plebeians, we create
Lord Saturninus Rome's great emperor,
And say, 'Long live our Emperor Saturnine!'
⎡A long flourish while Marcus and the other
Tribunes, with Saturninus and Bassianus,
come down.
Marcus invests Saturninus in the white
palliament and hands him a sceptre⎤
SATURNINUS
Titus Andronicus, for thy favours done
To us in our election this day                                   235
I give thee thanks in part of thy deserts,
And will with deeds requite thy gentleness.
And for an onset, Titus, to advance
Thy name and honourable family,
Lavinia will I make my empress,                                  240
Rome's royal mistress, mistress of my heart,
And in the sacred Pantheon her espouse.
Tell me, Andronicus, doth this motion please thee?
TITUS
It doth, my worthy lord, and in this match
I hold me highly honoured of your grace,                         245
And here in sight of Rome to Saturnine,

King and commander of our commonweal,
The wide world's emperor, do I consecrate
My sword, my chariot, and my prisoners—
Presents well worthy Rome's imperious lord.                      250
Receive them, then, the tribute that I owe,
Mine honour's ensigns humbled at thy feet.
SATURNINUS
Thanks, noble Titus, father of my life.
How proud I am of thee and of thy gifts
Rome shall record; and when I do forget                          255
The least of these unspeakable deserts,
Romans, forget your fealty to me.
TITUS (to Tamora)
Now, madam, are you prisoner to an emperor,
To him that for your honour and your state
Will use you nobly, and your followers.                          260
SATURNINUS
A goodly lady, trust me, of the hue
That I would choose were I to choose anew.
Clear up, fair queen, that cloudy countenance.
Though chance of war hath wrought this change of
      cheer,
Thou com'st not to be made a scorn in Rome.                      265
Princely shall be thy usage every way.
Rest on my word, and let not discontent
Daunt all your hopes. Madam, he comforts you
Can make you greater than the Queen of Goths.
Lavinia, you are not displeased with this?                       270
LAVINIA
Not I, my lord, sith true nobility
Warrants these words in princely courtesy.
SATURNINUS
Thanks, sweet Lavinia. Romans, let us go.
Ransomless here we set our prisoners free.
Proclaim our honours, lords, with trump and drum.
      ⎡Flourish. Exeunt Saturninus, Tamora,
      Demetrius, Chiron, and Aaron the Moor⎤
BASSIANUS
Lord Titus, by your leave, this maid is mine.                    276
TITUS
How, sir, are you in earnest then, my lord?
BASSIANUS
Ay, noble Titus, and resolved withal
To do myself this reason and this right.
MARCUS
Suum cuique is our Roman justice.                                280
This prince in justice seizeth but his own.
LUCIUS
And that he will and shall, if Lucius live.
TITUS
Traitors, avaunt! Where is the Emperor's guard?
MUTIUS
Brothers, help to convey her hence away,
And with my sword I'll keep this door safe.                      285
      Exeunt Bassianus, Marcus, Quintus, and
      Martius, with Lavinia
(To Titus) My lord, you pass not here.
TITUS                                  What, villain boy,
Barr'st me my way in Rome?
      He attacks Mutius
MUTIUS                                 Help, Lucius, help!
      Titus kills him
LUCIUS (to Titus)
My lord, you are unjust; and more than so,
In wrongful quarrel you have slain your son.
TITUS
Nor thou nor he are any sons of mine.                            290

My sons would never so dishonour me.
Traitor, restore Lavinia to the Emperor.
LUCIUS
Dead, if you will, but not to be his wife
That is another's lawful promised love.

*Exit with Mutius' body*
*Enter aloft Saturninus the Emperor with Tamora*
*and Chiron and Demetrius, her two sons, and*
*Aaron the Moor*

TITUS
Follow, my lord, and I'll soon bring her back.    295
SATURNINUS
No, Titus, no. The Emperor needs her not,
Nor her, nor thee, nor any of thy stock.
I'll trust by leisure him that mocks me once,
Thee never, nor thy traitorous haughty sons,
Confederates all thus to dishonour me.    300
Was none in Rome to make a stale
But Saturnine? Full well, Andronicus,
Agree these deeds with that proud brag of thine
That saidst I begged the empire at thy hands.
TITUS
O monstrous, what reproachful words are these?    305
SATURNINUS
But go thy ways, go give that changing piece
To him that flourished for her with his sword.
A valiant son-in-law thou shalt enjoy,
One fit to bandy with thy lawless sons,
To ruffle in the commonwealth of Rome.    310
TITUS
These words are razors to my wounded heart.
SATURNINUS
And therefore, lovely Tamora, Queen of Goths,
That like the stately Phoebe 'mongst her nymphs
Dost overshine the gallant'st dames of Rome,
If thou be pleased with this my sudden choice,    315
Behold, I choose thee, Tamora, for my bride,
And will create thee Empress of Rome.
Speak, Queen of Goths, dost thou applaud my choice?
And here I swear by all the Roman gods,
Sith priest and holy water are so near,    320
And tapers burn so bright, and everything
In readiness for Hymenaeus stand,
I will not re-salute the streets of Rome,
Or climb my palace, till from forth this place
I lead espoused my bride along with me.    325
TAMORA
And here, in sight of heaven, to Rome I swear
If Saturnine advance the Queen of Goths
She will a handmaid be to his desires,
A loving nurse, a mother to his youth.
SATURNINUS
Ascend, fair Queen, Pantheon. Lords, accompany    330
Your noble emperor and his lovely bride,
Sent by the heavens for Prince Saturnine,
Whose wisdom hath her fortune conquerèd.
There shall we consummate our spousal rites.

*Exeunt all but Titus*

TITUS
I am not bid to wait upon this bride.    335
Titus, when wert thou wont to walk alone,
Dishonoured thus and challengèd of wrongs?

*Enter Marcus and Titus' sons Lucius, Quintus, and*
*Martius, ⌈carrying Mutius' body⌉*

MARCUS
O Titus, see, O see what thou hast done—
In a bad quarrel slain a virtuous son.

TITUS
No, foolish Tribune, no; no son of mine,    340
Nor thou, nor these, confederates in the deed
That hath dishonoured all our family;
Unworthy brother and unworthy sons!
LUCIUS
But let us give him burial as becomes,
Give Mutius burial with our brethren.    345
TITUS
Traitors, away, he rests not in this tomb.
This monument five hundred years hath stood,
Which I have sumptuously re-edified.
Here none but soldiers and Rome's servitors
Repose in fame, none basely slain in brawls.    350
Bury him where you can; he comes not here.
MARCUS
My lord, this is impiety in you.
My nephew Mutius' deeds do plead for him.
He must be buried with his brethren.
⌈QUINTUS *and* MARTIUS⌉
And shall, or him we will accompany.    355
TITUS
'And shall'? What villain was it spake that word?
⌈QUINTUS⌉
He that would vouch it in any place but here.
TITUS
What, would you bury him in my despite?
MARCUS
No, noble Titus, but entreat of thee
To pardon Mutius and to bury him.    360
TITUS
Marcus, even thou hast struck upon my crest,
And with these boys mine honour thou hast
     wounded.
My foes I do repute you every one,
So trouble me no more, but get you gone.
⌈MARTIUS⌉
He is not with himself, let us withdraw.    365
⌈QUINTUS⌉
Not I, till Mutius' bones be burièd.

*Marcus, Lucius, Quintus, and Martius kneel*

MARCUS
Brother, for in that name doth nature plead—
⌈QUINTUS⌉
Father, and in that name doth nature speak—
TITUS
Speak thou no more, if all the rest will speed.
MARCUS
Renownèd Titus, more than half my soul—    370
LUCIUS
Dear father, soul and substance of us all—
MARCUS
Suffer thy brother Marcus to inter
His noble nephew here in virtue's nest,
That died in honour and Lavinia's cause.
Thou art a Roman; be not barbarous.    375
The Greeks upon advice did bury Ajax,
That slew himself; and wise Laertes' son
Did graciously plead for his funerals.
Let not young Mutius then, that was thy joy,
Be barred his entrance here.
TITUS           Rise, Marcus, rise.    380
The dismall'st day is this that e'er I saw,
To be dishonoured by my sons in Rome.
Well, bury him, and bury me the next.

*They put Mutius in the tomb*

LUCIUS
 There lie thy bones, sweet Mutius, with thy friends',
 Till we with trophies do adorn thy tomb.  385
ALL ⌐BUT TITUS⌐ (*kneeling*)
 No man shed tears for noble Mutius;
 He lives in fame, that died in virtue's cause.
      *Exeunt* ⌐*all but Marcus and Titus*⌐
MARCUS
 My lord—to step out of these dreary dumps—
 How comes it that the subtle Queen of Goths
 Is of a sudden thus advanced in Rome?  390
TITUS
 I know not, Marcus, but I know it is—
 Whether by device or no, the heavens can tell.
 Is she not then beholden to the man
 That brought her for this high good turn so far?
⌐MARCUS⌐
 Yes, and will nobly him remunerate.  395
  ⌐*Flourish.*⌐ *Enter the Emperor Saturninus, Tamora,*
  *and her two sons* (*Chiron and Demetrius*), *with*
  *Aaron the Moor at one door.*
  *Enter at the other door Bassianus and Lavinia with*
  ⌐*Lucius, Quintus, and Martius*⌐
SATURNINUS
 So, Bassianus, you have played your prize.
 God give you joy, sir, of your gallant bride.
BASSIANUS
 And you of yours, my lord. I say no more,
 Nor wish no less; and so I take my leave.
SATURNINUS
 Traitor, if Rome have law or we have power,  400
 Thou and thy faction shall repent this rape.
BASSIANUS
 'Rape' call you it, my lord, to seize my own—
 My true betrothèd love, and now my wife?
 But let the laws of Rome determine all;
 Meanwhile am I possessed of that is mine.  405
SATURNINUS
 'Tis good, sir; you are very short with us.
 But if we live we'll be as sharp with you.
BASSIANUS
 My lord, what I have done, as best I may
 Answer I must, and shall do with my life.
 Only thus much I give your grace to know:  410
 By all the duties that I owe to Rome,
 This noble gentleman, Lord Titus here,
 Is in opinion and in honour wronged,
 That, in the rescue of Lavinia,
 With his own hand did slay his youngest son  415
 In zeal to you, and highly moved to wrath
 To be controlled in that he frankly gave.
 Receive him then to favour, Saturnine,
 That hath expressed himself in all his deeds
 A father and a friend to thee and Rome.  420
TITUS
 Prince Bassianus, leave to plead my deeds.
 'Tis thou and those that have dishonoured me.
  ⌐*He kneels*⌐
 Rome and the righteous heavens be my judge
 How I have loved and honoured Saturnine!
TAMORA (*to Saturninus*)
 My worthy lord, if ever Tamora  425
 Were gracious in those princely eyes of thine,
 Then hear me speak indifferently for all;
 And at my suit, sweet, pardon what is past.

SATURNINUS
 What, madam—be dishonoured openly
 And basely put it up without revenge?  430
TAMORA
 Not so, my lord. The gods of Rome forfend
 I should be author to dishonour you.
 But on mine honour dare I undertake
 For good lord Titus' innocence in all,
 Whose fury not dissembled speaks his griefs.  435
 Then at my suit look graciously on him.
 Lose not so noble a friend on vain suppose,
 Nor with sour looks afflict his gentle heart.
 (*Aside to Saturninus*)
 My lord, be ruled by me, be won at last,
 Dissemble all your griefs and discontents.  440
 You are but newly planted in your throne;
 Lest then the people, and patricians too,
 Upon a just survey take Titus' part,
 And so supplant you for ingratitude,
 Which Rome reputes to be a heinous sin,  445
 Yield at entreats; and then let me alone:
 I'll find a day to massacre them all,
 And raze their faction and their family,
 The cruel father and his traitorous sons
 To whom I suèd for my dear son's life,  450
 And make them know what 'tis to let a queen
 Kneel in the streets and beg for grace in vain.
 (*Aloud*) Come, come, sweet Emperor; come,
  Andronicus,
 Take up this good old man, and cheer the heart
 That dies in tempest of thy angry frown.  455
SATURNINUS
 Rise, Titus, rise; my empress hath prevailed.
TITUS (*rising*)
 I thank your majesty and her, my lord,
 These words, these looks, infuse new life in me.
TAMORA
 Titus, I am incorporate in Rome,
 A Roman now adopted happily,  460
 And must advise the Emperor for his good.
 This day all quarrels die, Andronicus;
 And let it be mine honour, good my lord,
 That I have reconciled your friends and you.
 For you, Prince Bassianus, I have passed  465
 My word and promise to the Emperor
 That you will be more mild and tractable.
 And fear not, lords, and you, Lavinia;
 By my advice, all humbled on your knees,
 You shall ask pardon of his majesty.  470
  ⌐*Bassianus*⌐, *Lavinia, Lucius, Quintus, and*
  *Martius kneel*
⌐LUCIUS⌐
 We do, and vow to heaven and to his highness
 That what we did was mildly as we might,
 Tend'ring our sister's honour and our own.
MARCUS ⌐*kneeling*⌐
 That on mine honour here do I protest.
SATURNINUS
 Away, and talk not, trouble us no more.  475
TAMORA
 Nay, nay, sweet Emperor, we must all be friends.
 The Tribune and his nephews kneel for grace.
 I will not be denied; sweetheart, look back.
SATURNINUS
 Marcus, for thy sake and thy brother's here,

And at my lovely Tamora's entreats,    480
I do remit these young men's heinous faults.
Stand up!
    *Marcus, Bassianus, Lavinia, and Titus' sons stand*
         Lavinia, though you left me like a churl,
I found a friend, and sure as death I swore
I would not part a bachelor from the priest.
Come, if the Emperor's court can feast two brides   485
You are my guest, Lavinia, and your friends.
This day shall be a love-day, Tamora.
TITUS
Tomorrow an it please your majesty
To hunt the panther and the hart with me,
With horn and hound we'll give your grace *bonjour*.
SATURNINUS
Be it so, Titus, and gramercy, too.    ⌜*Flourish. Exeunt*⌝

**2.1**    ⌜*Enter Aaron alone*⌝
AARON
Now climbeth Tamora Olympus' top,
Safe out of fortune's shot, and sits aloft,
Secure of thunder's crack or lightning flash,
Advanced above pale envy's threat'ning reach.
As when the golden sun salutes the morn     5
And, having gilt the ocean with his beams,
Gallops the zodiac in his glistering coach
And overlooks the highest-peering hills,
So Tamora.
Upon her wit doth earthly honour wait,     10
And virtue stoops and trembles at her frown.
Then, Aaron, arm thy heart and fit thy thoughts
To mount aloft with thy imperial mistress,
And mount her pitch whom thou in triumph long
Hast prisoner held fettered in amorous chains,    15
And faster bound to Aaron's charming eyes
Than is Prometheus tied to Caucasus.
Away with slavish weeds and servile thoughts!
I will be bright, and shine in pearl and gold
To wait upon this new-made empress.     20
To wait, said I?—to wanton with this queen,
This goddess, this Semiramis, this nymph,
This siren that will charm Rome's Saturnine
And see his shipwreck and his commonweal's.
Hollo, what storm is this?     25
    *Enter Chiron and Demetrius, braving*
DEMETRIUS
Chiron, thy years wants wit, thy wits wants edge
And manners to intrude where I am graced
And may, for aught thou knowest, affected be.
CHIRON
Demetrius, thou dost overween in all,
And so in this, to bear me down with braves.    30
'Tis not the difference of a year or two
Makes me less gracious, or thee more fortunate.
I am as able and as fit as thou
To serve, and to deserve my mistress' grace,
And that my sword upon thee shall approve,    35
And plead my passions for Lavinia's love.
AARON (*aside*)
Clubs, clubs! These lovers will not keep the peace.
DEMETRIUS
Why, boy, although our mother, unadvised,
Gave you a dancing-rapier by your side,
Are you so desperate grown to threat your friends?   40
Go to, have your lath glued within your sheath
Till you know better how to handle it.

CHIRON
Meanwhile, sir, with the little skill I have
Full well shalt thou perceive how much I dare.
DEMETRIUS
Ay, boy, grow ye so brave?
    *They draw*
AARON           Why, how now, lords?   45
So near the Emperor's palace dare ye draw
And maintain such a quarrel openly?
Full well I wot the ground of all this grudge.
I would not for a million of gold
The cause were known to them it most concerns,    50
Nor would your noble mother for much more
Be so dishonoured in the court of Rome.
For shame, put up.
DEMETRIUS       Not I, till I have sheathed
My rapier in his bosom, and withal
Thrust those reproachful speeches down his throat   55
That he hath breathed in my dishonour here.
CHIRON
For that I am prepared and full resolved,
Foul-spoken coward, that thund'rest with thy tongue,
And with thy weapon nothing dar'st perform.
AARON   Away, I say.     60
Now, by the gods that warlike Goths adore,
This petty brabble will undo us all.
Why, lords, and think you not how dangerous
It is to jet upon a prince's right?
What, is Lavinia then become so loose,     65
Or Bassianus so degenerate,
That for her love such quarrels may be broached
Without controlment, justice, or revenge?
Young lords, beware; and should the Empress know
This discord's ground, the music would not please.   70
CHIRON
I care not, I, knew she and all the world,
I love Lavinia more than all the world.
DEMETRIUS
Youngling, learn thou to make some meaner choice.
Lavinia is thine elder brother's hope.
AARON
Why, are ye mad? Or know ye not in Rome    75
How furious and impatient they be,
And cannot brook competitors in love?
I tell you, lords, you do but plot your deaths
By this device.
CHIRON       Aaron, a thousand deaths
Would I propose to achieve her whom I love.    80
AARON
To achieve her how?
DEMETRIUS       Why makes thou it so strange?
She is a woman, therefore may be wooed;
She is a woman, therefore may be won;
She is Lavinia, therefore must be loved.
What, man, more water glideth by the mill    85
Than wots the miller of, and easy it is
Of a cut loaf to steal a shive, we know.
Though Bassianus be the Emperor's brother,
Better than he have worn Vulcan's badge.
AARON (*aside*)
Ay, and as good as Saturninus may.     90
DEMETRIUS
Then why should he despair that knows to court it
With words, fair looks, and liberality?
What, hast not thou full often struck a doe
And borne her cleanly by the keeper's nose?

AARON
  Why then, it seems some certain snatch or so    95
  Would serve your turns.
CHIRON             Ay, so the turn were served.
DEMETRIUS
  Aaron, thou hast hit it.
AARON             Would you had hit it too,
  Then should not we be tired with this ado.
  Why, hark ye, hark ye, and are you such fools
  To square for this? Would it offend you then    100
  That both should speed?
CHIRON  Faith, not me.
DEMETRIUS  Nor me, so I were one.
AARON
  For shame, be friends, and join for that you jar.
  'Tis policy and stratagem must do    105
  That you affect, and so must you resolve
  That what you cannot as you would achieve,
  You must perforce accomplish as you may.
  Take this of me: Lucrece was not more chaste
  Than this Lavinia, Bassianus' love.    110
  A speedier course than ling'ring languishment
  Must we pursue, and I have found the path.
  My lords, a solemn hunting is in hand;
  There will the lovely Roman ladies troop.
  The forest walks are wide and spacious,    115
  And many unfrequented plots there are,
  Fitted by kind for rape and villainy.
  Single you thither then this dainty doe,
  And strike her home by force, if not by words,
  This way or not at all stand you in hope.    120
  Come, come; our Empress, with her sacred wit
  To villainy and vengeance consecrate,
  Will we acquaint with all what we intend,
  And she shall file our engines with advice
  That will not suffer you to square yourselves,    125
  But to your wishes' height advance you both.
  The Emperor's court is like the house of Fame,
  The palace full of tongues, of eyes and ears,
  The woods are ruthless, dreadful, deaf, and dull.
  There speak and strike, brave boys, and take your
    turns.    130
  There serve your lust, shadowed from heaven's eye,
  And revel in Lavinia's treasury.
CHIRON
  Thy counsel, lad, smells of no cowardice.
DEMETRIUS
  *Sit fas aut nefas*, till I find the stream
  To cool this heat, a charm to calm these fits,    135
  *Per Styga, per manes vehor.*           *Exeunt*

**2.2**    *Enter Titus Andronicus and his three sons (Quintus,*
       *Lucius, and Martius), and Marcus, making a noise*
       *with hounds and horns*
TITUS
  The hunt is up, the morn is bright and grey,
  The fields are fragrant and the woods are green.
  Uncouple here, and let us make a bay
  And wake the Emperor and his lovely bride,
  And rouse the Prince, and ring a hunter's peal,    5
  That all the court may echo with the noise.
  Sons, let it be your charge, as it is ours,
  To attend the Emperor's person carefully.
  I have been troubled in my sleep this night,
  But dawning day new comfort hath inspired.    10

*Here a cry of hounds, and wind horns in a peal;*
*then enter Saturninus, Tamora, Bassianus, Lavinia,*
*Chiron, Demetrius, and their attendants*
  Many good-morrows to your majesty.
  Madam, to you as many, and as good.
  I promisèd your grace a hunter's peal.
SATURNINUS
  And you have rung it lustily, my lords,
  Somewhat too early for new-married ladies.    15
BASSIANUS
  Lavinia, how say you?
LAVINIA          I say no.
  I have been broad awake two hours and more.
SATURNINUS
  Come on then, horse and chariots let us have,
  And to our sport. (*To Tamora*) Madam, now shall ye see
  Our Roman hunting.
MARCUS         I have dogs, my lord,    20
  Will rouse the proudest panther in the chase,
  And climb the highest promontory top.
TITUS
  And I have horse will follow where the game
  Makes way, and run like swallows o'er the plain.
DEMETRIUS (*aside*)
  Chiron, we hunt not, we, with horse nor hound,    25
  But hope to pluck a dainty doe to ground.       *Exeunt*

**2.3**    *Enter Aaron alone, with gold*
AARON
  He that had wit would think that I had none,
  To bury so much gold under a tree
  And never after to inherit it.
  Let him that thinks of me so abjectly
  Know that this gold must coin a stratagem    5
  Which, cunningly effected, will beget
  A very excellent piece of villainy.
  And so repose, sweet gold, for their unrest
  That have their alms out of the Empress' chest.
    *He hides the gold.*
    *Enter Tamora alone to the Moor*
TAMORA
  My lovely Aaron, wherefore look'st thou sad    10
  When everything doth make a gleeful boast?
  The birds chant melody on every bush,
  The snakes lies rollèd in the cheerful sun,
  The green leaves quiver with the cooling wind
  And make a chequered shadow on the ground.    15
  Under their sweet shade, Aaron, let us sit,
  And whilst the babbling echo mocks the hounds,
  Replying shrilly to the well-tuned horns,
  As if a double hunt were heard at once,
  Let us sit down and mark their yellowing noise,    20
  And after conflict such as was supposed
  The wand'ring prince and Dido once enjoyed
  When with a happy storm they were surprised,
  And curtained with a counsel-keeping cave,
  We may, each wreathèd in the other's arms,    25
  Our pastimes done, possess a golden slumber
  Whiles hounds and horns and sweet melodious birds
  Be unto us as is a nurse's song
  Of lullaby to bring her babe asleep.
AARON
  Madam, though Venus govern your desires,    30
  Saturn is dominator over mine.
  What signifies my deadly-standing eye,

My silence, and my cloudy melancholy,
My fleece of woolly hair that now uncurls
Even as an adder when she doth unroll                    35
To do some fatal execution?
No, madam, these are no venereal signs.
Vengeance is in my heart, death in my hand,
Blood and revenge are hammering in my head.
Hark, Tamora, the empress of my soul,                    40
Which never hopes more heaven than rests in thee,
This is the day of doom for Bassianus.
His Philomel must lose her tongue today,
Thy sons make pillage of her chastity
And wash their hands in Bassianus' blood.                45
Seest thou this letter? (*Giving a letter*) Take it up, I
    pray thee,
And give the King this fatal-plotted scroll.
Now question me no more. We are espied.
Here comes a parcel of our hopeful booty,
Which dreads not yet their lives' destruction.           50
    *Enter Bassianus and Lavinia*
TAMORA (*aside to Aaron*)
Ah, my sweet Moor, sweeter to me than life!
AARON (*aside to Tamora*)
No more, great Empress; Bassianus comes.
Be cross with him, and I'll go fetch thy sons
To back thy quarrels, whatsoe'er they be.         *Exit*
BASSIANUS
Who have we here? Rome's royal empress              55
Unfurnished of her well-beseeming troop?
Or is it Dian, habited like her
Who hath abandonèd her holy groves
To see the general hunting in this forest?
TAMORA
Saucy controller of my private steps,              60
Had I the power that some say Dian had,
Thy temples should be planted presently
With horns, as was Actaeon's, and the hounds
Should drive upon thy new-transformèd limbs,
Unmannerly intruder as thou art!                   65
LAVINIA
Under your patience, gentle Empress,
'Tis thought you have a goodly gift in horning,
And to be doubted that your Moor and you
Are singled forth to try experiments.
Jove shield your husband from his hounds today—    70
'Tis pity they should take him for a stag.
BASSIANUS
Believe me, Queen, your swart Cimmerian
Doth make your honour of his body's hue,
Spotted, detested, and abominable.
Why are you sequestered from all your train,       75
Dismounted from your snow-white goodly steed,
And wandered hither to an obscure plot,
Accompanied but with a barbarous Moor,
If foul desire had not conducted you?
LAVINIA
And being intercepted in your sport,               80
Great reason that my noble lord be rated
For sauciness. (*To Bassianus*) I pray you, let us hence,
And let her joy her raven-coloured love.
This valley fits the purpose passing well.
BASSIANUS
The King my brother shall have note of this.       85
LAVINIA
Ay, for these slips have made him noted long.
Good King, to be so mightily abused!

TAMORA
Why have I patience to endure all this?
    *Enter Chiron and Demetrius*
DEMETRIUS
How now, dear sovereign and our gracious mother,
Why doth your highness look so pale and wan?       90
TAMORA
Have I not reason, think you, to look pale?
These two have 'ticed me hither to this place.
A barren detested vale you see it is;
The trees, though summer, yet forlorn and lean,
Overcome with moss and baleful mistletoe.          95
Here never shines the sun, here nothing breeds
Unless the nightly owl or fatal raven,
And when they showed me this abhorrèd pit
They told me here at dead time of the night
A thousand fiends, a thousand hissing snakes,      100
Ten thousand swelling toads, as many urchins
Would make such fearful and confusèd cries
As any mortal body hearing it
Should straight fall mad or else die suddenly.
No sooner had they told this hellish tale          105
But straight they told me they would bind me here
Unto the body of a dismal yew
And leave me to this miserable death.
And then they called me foul adulteress,
Lascivious Goth, and all the bitterest terms       110
That ever ear did hear to such effect.
And had you not by wondrous fortune come,
This vengeance on me had they executed.
Revenge it as you love your mother's life,
Or be ye not henceforward called my children.      115
DEMETRIUS
This is a witness that I am thy son.
    *He stabs Bassianus*
CHIRON
And this for me, struck home to show my strength.
    *He stabs Bassianus, who dies.*
    ⌈*Tamora turns to Lavinia*⌉
LAVINIA
Ay, come, Semiramis—nay, barbarous Tamora,
For no name fits thy nature but thy own.
TAMORA (*to Chiron*)
Give me the poniard. You shall know, my boys,      120
Your mother's hand shall right your mother's wrong.
DEMETRIUS
Stay, madam, here is more belongs to her.
First thresh the corn, then after burn the straw.
This minion stood upon her chastity,
Upon her nuptial vow, her loyalty,                 125
And with that quaint hope braves your mightiness.
And shall she carry this unto her grave?
CHIRON
An if she do I would I were an eunuch.
Drag hence her husband to some secret hole,
And make his dead trunk pillow to our lust.        130
TAMORA
But when ye have the honey ye desire
Let not this wasp outlive, us both to sting.
CHIRON
I warrant you, madam, we will make that sure.
Come, mistress, now perforce we will enjoy
That nice-preservèd honesty of yours.              135
LAVINIA
O Tamora, thou bearest a woman's face—
TAMORA
I will not hear her speak. Away with her!

LAVINIA
    Sweet lords, entreat her hear me but a word.
DEMETRIUS (*to Tamora*)
    Listen, fair madam, let it be your glory
    To see her tears, but be your heart to them          140
    As unrelenting flint to drops of rain.
LAVINIA
    When did the tiger's young ones teach the dam?
    O, do not learn her wrath! She taught it thee.
    The milk thou sucked'st from her did turn to marble,
    Even at thy teat thou hadst thy tyranny.              145
    Yet every mother breeds not sons alike.
    (*To Chiron*) Do thou entreat her show a woman's pity.
CHIRON
    What, wouldst thou have me prove myself a bastard?
LAVINIA
    'Tis true, the raven doth not hatch a lark.
    Yet have I heard—O, could I find it now!—            150
    The lion, moved with pity, did endure
    To have his princely paws pared all away.
    Some say that ravens foster forlorn children
    The whilst their own birds famish in their nests.
    O, be to me, though thy hard heart say no,           155
    Nothing so kind, but something pitiful.
TAMORA
    I know not what it means. Away with her!
LAVINIA
    O, let me teach thee for my father's sake,
    That gave thee life when well he might have slain
        thee.
    Be not obdurate, open thy deaf ears.                 160
TAMORA
    Hadst thou in person ne'er offended me
    Even for his sake am I pitiless.
    Remember, boys, I poured forth tears in vain
    To save your brother from the sacrifice,
    But fierce Andronicus would not relent.              165
    Therefore away with her, and use her as you will—
    The worse to her, the better loved of me.
LAVINIA
    O Tamora, be called a gentle queen,
    And with thine own hands kill me in this place;
    For 'tis not life that I have begged so long;        170
    Poor I was slain when Bassianus died.
TAMORA
    What begg'st thou then, fond woman? Let me go.
LAVINIA
    'Tis present death I beg, and one thing more
    That womanhood denies my tongue to tell.
    O, keep me from their worse-than-killing lust,       175
    And tumble me into some loathsome pit
    Where never man's eye may behold my body.
    Do this, and be a charitable murderer.
TAMORA
    So should I rob my sweet sons of their fee.
    No, let them satisfy their lust on thee.             180
DEMETRIUS (*to Lavinia*)
    Away, for thou hast stayed us here too long.
LAVINIA
    No grace, no womanhood—ah, beastly creature,
    The blot and enemy to our general name,
    Confusion fall—
CHIRON
    Nay then, I'll stop your mouth. (*To Demetrius*) Bring
        thou her husband.                                185
    This is the hole where Aaron bid us hide him.

*Demetrius and Chiron cast Bassianus' body into the
pit ⌈and cover the mouth of it with branches⌉, then
exeunt dragging Lavinia*

TAMORA
    Farewell, my sons. See that you make her sure.
    Ne'er let my heart know merry cheer indeed
    Till all the Andronici be made away.
    Now will I hence to seek my lovely Moor,             190
    And let my spleenful sons this trull deflower.    *Exit*
        *Enter Aaron with Quintus and Martius, two of
        Titus' sons*
AARON
    Come on, my lords, the better foot before.
    Straight will I bring you to the loathsome pit
    Where I espied the panther fast asleep.
QUINTUS
    My sight is very dull, whate'er it bodes.            195
MARTIUS
    And mine, I promise you. Were it not for shame,
    Well could I leave our sport to sleep awhile.
        *He falls into the pit*
QUINTUS
    What, art thou fallen? What subtle hole is this,
    Whose mouth is covered with rude-growing briers
    Upon whose leaves are drops of new-shed blood        200
    As fresh as morning dew distilled on flowers?
    A very fatal place it seems to me.
    Speak, brother. Hast thou hurt thee with the fall?
MARTIUS
    O brother, with the dismall'st object hurt
    That ever eye with sight made heart lament.          205
AARON (*aside*)
    Now will I fetch the King to find them here,
    That he thereby may have a likely guess
    How these were they that made away his brother.
                                                    *Exit*
MARTIUS
    Why dost not comfort me and help me out
    From this unhallowed and bloodstainèd hole?         210
QUINTUS
    I am surprisèd with an uncouth fear.
    A chilling sweat o'erruns my trembling joints;
    My heart suspects more than mine eye can see.
MARTIUS
    To prove thou hast a true-divining heart,
    Aaron and thou look down into this den,             215
    And see a fearful sight of blood and death.
QUINTUS
    Aaron is gone, and my compassionate heart
    Will not permit mine eyes once to behold
    The thing whereat it trembles by surmise.
    O, tell me who it is, for ne'er till now             220
    Was I a child to fear I know not what.
MARTIUS
    Lord Bassianus lies berayed in blood
    All on a heap, like to a slaughtered lamb,
    In this detested, dark, blood-drinking pit.
QUINTUS
    If it be dark how dost thou know 'tis he?            225
MARTIUS
    Upon his bloody finger he doth wear
    A precious ring that lightens all this hole,
    Which like a taper in some monument
    Doth shine upon the dead man's earthy cheeks
    And shows the ragged entrails of this pit.           230
    So pale did shine the moon on Pyramus

When he by night lay bathed in maiden blood.
O brother, help me with thy fainting hand—
If fear hath made thee faint, as me it hath—
Out of this fell devouring receptacle, 235
As hateful as Cocytus' misty mouth.

QUINTUS
Reach me thy hand, that I may help thee out,
Or, wanting strength to do thee so much good,
I may be plucked into the swallowing womb
Of this deep pit, poor Bassianus' grave. 240
I have no strength to pluck thee to the brink,

MARTIUS
Nor I no strength to climb without thy help.

QUINTUS
Thy hand once more, I will not loose again
Till thou art here aloft or I below.
Thou canst not come to me; I come to thee. 245
*He falls into the pit.*
*Enter Saturninus the Emperor ⌐with attendants⌐,*
*and Aaron the Moor*

SATURNINUS
Along with me! I'll see what hole is here,
And what he is that now is leapt into it.
*He speaks into the pit*
Say, who art thou that lately didst descend
Into this gaping hollow of the earth?

MARTIUS
The unhappy sons of old Andronicus, 250
Brought hither in a most unlucky hour
To find thy brother Bassianus dead.

SATURNINUS
My brother dead! I know thou dost but jest.
He and his lady both are at the lodge
Upon the north side of this pleasant chase. 255
'Tis not an hour since I left them there.

MARTIUS
We know not where you left them all alive,
But, out alas, here have we found him dead!
*Enter Tamora, Titus Andronicus, and Lucius*

TAMORA Where is my lord the King?

SATURNINUS
Here, Tamora, though gripped with killing grief. 260

TAMORA
Where is thy brother Bassianus?

SATURNINUS
Now to the bottom dost thou search my wound.
Poor Bassianus here lies murderèd.

TAMORA
Then all too late I bring this fatal writ,
The complot of this timeless tragedy, 265
And wonder greatly that man's face can fold
In pleasing smiles such murderous tyranny.
*She giveth Saturnine a letter*

SATURNINUS (*reads*)
'An if we miss to meet him handsomely,
Sweet huntsman—Bassianus 'tis we mean—
Do thou so much as dig the grave for him. 270
Thou know'st our meaning. Look for thy reward
Among the nettles at the elder tree
Which overshades the mouth of that same pit
Where we decreed to bury Bassianus.
Do this, and purchase us thy lasting friends.' 275
O Tamora, was ever heard the like!
This is the pit, and this the elder tree.
Look, sirs, if you can find the huntsman out
That should have murdered Bassianus here.

AARON
My gracious lord, here is the bag of gold. 280

SATURNINUS (*to Titus*)
Two of thy whelps, fell curs of bloody kind,
Have here bereft my brother of his life.
Sirs, drag them from the pit unto the prison.
There let them bide until we have devised
Some never-heard-of torturing pain for them. 285

TAMORA
What, are they in this pit? O wondrous thing!
How easily murder is discoverèd!
*Attendants drag Quintus, Martius, and Bassianus'*
*body from the pit*

TITUS (*kneeling*)
High Emperor, upon my feeble knee
I beg this boon with tears not lightly shed:
That this fell fault of my accursèd sons— 290
Accursèd if the fault be proved in them—

SATURNINUS
If it be proved? You see it is apparent.
Who found this letter? Tamora, was it you?

TAMORA
Andronicus himself did take it up.

TITUS
I did, my lord, yet let me be their bail, 295
For by my father's reverend tomb I vow
They shall be ready at your highness' will
To answer their suspicion with their lives.

SATURNINUS
Thou shalt not bail them. See thou follow me.
Some bring the murdered body, some the murderers.
Let them not speak a word—the guilt is plain; 301
For by my soul, were there worse end than death
That end upon them should be executed. ⌐*Exit*⌐

TAMORA
Andronicus, I will entreat the King.
Fear not thy sons, they shall do well enough. 305

TITUS ⌐*rising*⌐
Come, Lucius, come, stay not to talk with them.
*Exeunt*

2.4　*Enter the Empress' sons, Chiron and Demetrius,*
*with Lavinia, her hands cut off and her tongue cut*
*out, and ravished*

DEMETRIUS
So, now go tell, an if thy tongue can speak,
Who 'twas that cut thy tongue and ravished thee.

CHIRON
Write down thy mind, bewray thy meaning so,
An if thy stumps will let thee play the scribe.

DEMETRIUS
See how with signs and tokens she can scrawl. 5

CHIRON (*to Lavinia*)
Go home, call for sweet water, wash thy hands.

DEMETRIUS
She hath no tongue to call nor hands to wash,
And so let's leave her to her silent walks.

CHIRON
An 'twere my cause I should go hang myself.

DEMETRIUS
If thou hadst hands to help thee knit the cord. 10
*Exeunt Chiron and Demetrius*
⌐*Wind horns.*⌐ *Enter Marcus from hunting to Lavinia*

MARCUS
Who is this—my niece that flies away so fast?

Cousin, a word. Where is your husband?
If I do dream, would all my wealth would wake me.
If I do wake, some planet strike me down
That I may slumber an eternal sleep.     15
Speak, gentle niece, what stern ungentle hands
Hath lopped and hewed and made thy body bare
Of her two branches, those sweet ornaments
Whose circling shadows kings have sought to sleep in,
And might not gain so great a happiness     20
As half thy love. Why dost not speak to me?
Alas, a crimson river of warm blood,
Like to a bubbling fountain stirred with wind,
Doth rise and fall between thy rosèd lips,
Coming and going with thy honey breath.     25
But sure some Tereus hath deflowered thee
And, lest thou shouldst detect him, cut thy tongue.
Ah, now thou turn'st away thy face for shame,
And notwithstanding all this loss of blood,
As from a conduit with three issuing spouts,     30
Yet do thy cheeks look red as Titan's face
Blushing to be encountered with a cloud.
Shall I speak for thee? Shall I say 'tis so?
O that I knew thy heart, and knew the beast,
That I might rail at him to ease my mind!     35
Sorrow concealèd, like an oven stopped,
Doth burn the heart to cinders where it is.
Fair Philomel, why she but lost her tongue
And in a tedious sampler sewed her mind.
But, lovely niece, that mean is cut from thee.     40
A craftier Tereus, cousin, hast thou met,
And he hath cut those pretty fingers off
That could have better sewed than Philomel.
O, had the monster seen those lily hands
Tremble like aspen leaves upon a lute     45
And make the silken strings delight to kiss them,
He would not then have touched them for his life.
Or had he heard the heavenly harmony
Which that sweet tongue hath made,
He would have dropped his knife and fell asleep,     50
As Cerberus at the Thracian poet's feet.
Come, let us go and make thy father blind,
For such a sight will blind a father's eye.
One hour's storm will drown the fragrant meads:
What will whole months of tears thy father's eyes?   55
Do not draw back, for we will mourn with thee.
O, could our mourning ease thy misery!     *Exeunt*

❀

**3.1**   *Enter the Judges, Tribunes, and Senators with Titus'*
*two sons, Martius and Quintus, bound, passing*
⌈*over*⌉ *the stage to the place of execution, and Titus*
*going before, pleading*

TITUS
Hear me, grave fathers; noble Tribunes, stay.
For pity of mine age, whose youth was spent
In dangerous wars whilst you securely slept;
For all my blood in Rome's great quarrel shed;
For all the frosty nights that I have watched,     5
And for these bitter tears which now you see
Filling the agèd wrinkles in my cheeks,
Be pitiful to my condemnèd sons,
Whose souls is not corrupted as 'tis thought.
For two-and-twenty sons I never wept,     10
Because they died in honour's lofty bed.
     *Andronicus lieth down, and the Judges pass by him*
For these two, Tribunes, in the dust I write

My heart's deep languor and my soul's sad tears.
Let my tears stanch the earth's dry appetite;
My sons' sweet blood will make it shame and blush.
                ⌈*Exeunt all but Titus*⌉
O earth, I will befriend thee more with rain     16
That shall distil from these two ancient ruins
Than youthful April shall with all his showers.
In summer's drought I'll drop upon thee still.
In winter with warm tears I'll melt the snow     20
And keep eternal springtime on thy face,
So thou refuse to drink my dear sons' blood.
     *Enter Lucius with his weapon drawn*
O reverend Tribunes, O gentle, agèd men,
Unbind my sons, reverse the doom of death,
And let me say, that never wept before,     25
My tears are now prevailing orators!
LUCIUS
O noble father, you lament in vain.
The Tribunes hear you not. No man is by,
And you recount your sorrows to a stone.
TITUS
Ah Lucius, for thy brothers let me plead.     30
Grave Tribunes, once more I entreat of you—
LUCIUS
My gracious lord, no tribune hears you speak.
TITUS
Why, 'tis no matter, man. If they did hear,
They would not mark me; if they did mark,
They would not pity me; yet plead I must.     35
Therefore I tell my sorrows to the stones,
Who, though they cannot answer my distress,
Yet in some sort they are better than the Tribunes
For that they will not intercept my tale.
When I do weep they humbly at my feet     40
Receive my tears and seem to weep with me,
And were they but attirèd in grave weeds
Rome could afford no tribunes like to these.
A stone is soft as wax, tribunes more hard than stones.
A stone is silent and offendeth not,     45
And tribunes with their tongues doom men to death.
But wherefore stand'st thou with thy weapon drawn?
LUCIUS
To rescue my two brothers from their death,
For which attempt the Judges have pronounced
My everlasting doom of banishment.     50
TITUS ⌈*rising*⌉
O happy man, they have befriended thee!
Why, foolish Lucius, dost thou not perceive
That Rome is but a wilderness of tigers?
Tigers must prey, and Rome affords no prey
But me and mine. How happy art thou then     55
From these devourers to be banishèd!
But who comes with our brother Marcus here?
     *Enter Marcus with Lavinia*
MARCUS
Titus, prepare thy agèd eyes to weep,
Or if not so, thy noble heart to break.
I bring consuming sorrow to thine age.     60
TITUS
Will it consume me? Let me see it then.
MARCUS This was thy daughter.
TITUS Why, Marcus, so she is.
LUCIUS (*falling on his knees*) Ay me, this object kills me.
TITUS
Faint-hearted boy, arise and look upon her.     65
     ⌈*Lucius rises*⌉

Speak, Lavinia, what accursèd hand
Hath made thee handless in thy father's sight?
What fool hath added water to the sea,
Or brought a faggot to bright-burning Troy?
My grief was at the height before thou cam'st,          70
And now like Nilus it disdaineth bounds.
Give me a sword, I'll chop off my hands too,
For they have fought for Rome, and all in vain;
And they have nursed this woe in feeding life;
In bootless prayer have they been held up,              75
And they have served me to effectless use.
Now all the service I require of them
Is that the one will help to cut the other.
'Tis well, Lavinia, that thou hast no hands,
For hands to do Rome service is but vain.               80

LUCIUS
Speak, gentle sister, who hath martyred thee.

MARCUS
O, that delightful engine of her thoughts,
That blabbed them with such pleasing eloquence,
Is torn from forth that pretty hollow cage
Where, like a sweet melodious bird, it sung             85
Sweet varied notes, enchanting every ear.

LUCIUS
O, say thou for her, who hath done this deed?

MARCUS
O, thus I found her, straying in the park,
Seeking to hide herself, as doth the deer
That hath received some unrecuring wound.               90

TITUS
It was my dear, and he that wounded her
Hath hurt me more than had he killed me dead;
For now I stand as one upon a rock
Environed with a wilderness of sea,
Who marks the waxing tide grow wave by wave,            95
Expecting ever when some envious surge
Will in his brinish bowels swallow him.
This way to death my wretched sons are gone.
Here stands my other son, a banished man,
And here my brother, weeping at my woes.                100
But that which gives my soul the greatest spurn
Is dear Lavinia, dearer than my soul.
Had I but seen thy picture in this plight
It would have madded me. What shall I do
Now I behold thy lively body so?                        105
Thou hast no hands to wipe away thy tears,
Nor tongue to tell me who hath martyred thee.
Thy husband he is dead, and for his death
Thy brothers are condemned and dead by this.
Look, Marcus, ah, son Lucius, look on her!              110
When I did name her brothers, then fresh tears
Stood on her cheeks, as doth the honey-dew
Upon a gathered lily almost witherèd.

MARCUS
Perchance she weeps because they killed her
    husband;
Perchance because she knows them innocent.              115

TITUS
If they did kill thy husband, then be joyful,
Because the law hath ta'en revenge on them.
No, no, they would not do so foul a deed;
Witness the sorrow that their sister makes.
Gentle Lavinia, let me kiss thy lips;                   120
Or make some sign how I may do thee ease.
Shall thy good uncle, and thy brother Lucius,
And thou, and I, sit round about some fountain,

Looking all downwards to behold our cheeks
How they are stained, like meadows yet not dry         125
With miry slime left on them by a flood?
And in the fountain shall we gaze so long
Till the fresh taste be taken from that clearness,
And made a brine pit with our bitter tears?
Or shall we cut away our hands like thine?              130
Or shall we bite our tongues, and in dumb shows
Pass the remainder of our hateful days?
What shall we do? Let us that have our tongues
Plot some device of further misery,
To make us wondered at in time to come.                 135

LUCIUS
Sweet father, cease your tears, for at your grief
See how my wretched sister sobs and weeps.

MARCUS
Patience, dear niece. Good Titus, dry thine eyes.

TITUS
Ah, Marcus, Marcus, brother, well I wot
Thy napkin cannot drink a tear of mine,                 140
For thou, poor man, hast drowned it with thine own.

LUCIUS
Ah, my Lavinia, I will wipe thy cheeks.

TITUS
Mark, Marcus, mark. I understand her signs.
Had she a tongue to speak, now would she say
That to her brother which I said to thee.               145
His napkin with his true tears all bewet
Can do no service on her sorrowful cheeks.
O, what a sympathy of woe is this—
As far from help as limbo is from bliss.

*Enter Aaron the Moor, alone*

AARON
Titus Andronicus, my lord the Emperor                   150
Sends thee this word: that, if thou love thy sons,
Let Marcus, Lucius or thyself, old Titus,
Or any one of you, chop off your hand
And send it to the King. He for the same
Will send thee hither both thy sons alive,              155
And that shall be the ransom for their fault.

TITUS
O gracious Emperor! O gentle Aaron,
Did ever raven sing so like a lark
That gives sweet tidings of the sun's uprise?
With all my heart I'll send the Emperor my hand.        160
Good Aaron, wilt thou help to chop it off?

LUCIUS
Stay, father, for that noble hand of thine,
That hath thrown down so many enemies,
Shall not be sent. My hand will serve the turn.
My youth can better spare my blood than you,            165
And therefore mine shall save my brothers' lives.

MARCUS
Which of your hands hath not defended Rome
And reared aloft the bloody battleaxe,
Writing destruction on the enemy's castle?
O, none of both but are of high desert.                 170
My hand hath been but idle; let it serve
To ransom my two nephews from their death,
Then have I kept it to a worthy end.

AARON
Nay, come, agree whose hand shall go along,
For fear they die before their pardon come.             175

MARCUS
My hand shall go.

LUCIUS                    By heaven it shall not go.

TITUS
Sirs, strive no more. Such withered herbs as these
Are meet for plucking up, and therefore mine.
LUCIUS
Sweet father, if I shall be thought thy son,
Let me redeem my brothers both from death.          180
MARCUS
And for our father's sake and mother's care,
Now let me show a brother's love to thee.
TITUS
Agree between you. I will spare my hand.
LUCIUS
Then I'll go fetch an axe.
MARCUS                         But I will use the axe.
                    *Exeunt Lucius and Marcus*
TITUS
Come hither, Aaron. I'll deceive them both.         185
Lend me thy hand, and I will give thee mine.
AARON (*aside*)
If that be called deceit, I will be honest
And never whilst I live deceive men so.
But I'll deceive you in another sort,
And that you'll say ere half an hour pass.          190
      *He cuts off Titus' hand.*
      *Enter Lucius and Marcus again*
TITUS
Now stay your strife. What shall be is dispatched.
Good Aaron, give his majesty my hand.
Tell him it was a hand that warded him
From thousand dangers; bid him bury it.
More hath it merited; that let it have.             195
As for my sons, say I account of them
As jewels purchased at an easy price,
And yet dear too, because I bought mine own.
AARON
I go, Andronicus; and for thy hand
Look by and by to have thy sons with thee.          200
(*Aside*) Their heads, I mean. O, how this villainy
Doth fat me with the very thoughts of it!
Let fools do good, and fair men call for grace:
Aaron will have his soul black like his face.       *Exit*
TITUS
O, here I lift this one hand up to heaven           205
And bow this feeble ruin to the earth.
      *He kneels*
If any power pities wretched tears,
To that I call. (*To Lavinia, who kneels*) What, wouldst
      thou kneel with me?
Do then, dear heart; for heaven shall hear our prayers,
Or with our sighs we'll breathe the welkin dim      210
And stain the sun with fog, as sometime clouds
When they do hug him in their melting bosoms.
MARCUS
O brother, speak with possibility,
And do not break into these deep extremes.
TITUS
Is not my sorrows deep, having no bottom?           215
Then be my passions bottomless with them.
MARCUS
But yet let reason govern thy lament.
TITUS
If there were reason for these miseries,
Then into limits could I bind my woes.
When heaven doth weep, doth not the earth
      o'erflow?                                     220
If the winds rage, doth not the sea wax mad,

Threat'ning the welkin with his big-swoll'n face?
And wilt thou have a reason for this coil?
I am the sea. Hark how her sighs doth blow.
She is the weeping welkin, I the earth.             225
Then must my sea be movèd with her sighs,
Then must my earth with her continual tears
Become a deluge overflowed and drowned,
Forwhy my bowels cannot hide her woes,
But like a drunkard must I vomit them.              230
Then give me leave, for losers will have leave
To ease their stomachs with their bitter tongues.
      *Enter a Messenger with two heads and a hand*
MESSENGER
Worthy Andronicus, ill art thou repaid
For that good hand thou sent'st the Emperor.
Here are the heads of thy two noble sons,           235
And here's thy hand in scorn to thee sent back—
Thy grief their sports, thy resolution mocked,
That woe is me to think upon thy woes
More than remembrance of my father's death.
      ⌐He sets down the heads and hand. Exit⌐
MARCUS
Now let hot Etna cool in Sicily,                    240
And be my heart an ever-burning hell.
These miseries are more than may be borne.
To weep with them that weep doth ease some deal,
But sorrow flouted at is double death.
LUCIUS
Ah, that this sight should make so deep a wound     245
And yet detested life not shrink thereat—
That ever death should let life bear his name
Where life hath no more interest but to breathe!
      *Lavinia kisses Titus*
MARCUS
Alas, poor heart, that kiss is comfortless
As frozen water to a starvèd snake.                 250
TITUS
When will this fearful slumber have an end?
MARCUS
Now farewell, flatt'ry; die, Andronicus.
Thou dost not slumber. See thy two sons' heads,
Thy warlike hand, thy mangled daughter here,
Thy other banished son with this dear sight         255
Struck pale and bloodless, and thy brother, I,
Even like a stony image, cold and numb.
Ah, now no more will I control thy griefs.
Rend off thy silver hair, thy other hand
Gnawing with thy teeth, and be this dismal sight    260
The closing up of our most wretched eyes.
Now is a time to storm. Why art thou still?
TITUS Ha, ha, ha!
MARCUS
Why dost thou laugh? It fits not with this hour.
TITUS
Why, I have not another tear to shed.               265
Besides, this sorrow is an enemy,
And would usurp upon my wat'ry eyes
And make them blind with tributary tears.
Then which way shall I find Revenge's cave?—
For these two heads do seem to speak to me          270
And threat me I shall never come to bliss
Till all these mischiefs be returned again
Even in their throats that hath committed them.
Come, let me see what task I have to do.
      ⌐He and Lavinia rise⌐
You heavy people, circle me about,                  275

That I may turn me to each one of you
And swear unto my soul to right your wrongs.
    *Marcus, Lucius, and Lavinia circle Titus. He*
    *pledges them*
The vow is made. Come, brother, take a head,
And in this hand the other will I bear.
And Lavinia, thou shalt be employed.               280
Bear thou my hand, sweet wench, between thine arms.
As for thee, boy, go get thee from my sight.
Thou art an exile and thou must not stay.
Hie to the Goths, and raise an army there,
And if ye love me, as I think you do,             285
Let's kiss and part, for we have much to do.
    *They kiss. Exeunt all but Lucius*

LUCIUS
Farewell, Andronicus, my noble father,
The woefull'st man that ever lived in Rome.
Farewell, proud Rome, till Lucius come again;
He loves his pledges dearer than his life.         290
Farewell, Lavinia, my noble sister:
O, would thou wert as thou tofore hast been!
But now nor Lucius nor Lavinia lives
But in oblivion and hateful griefs.
If Lucius live he will requite your wrongs     295
And make proud Saturnine and his empress
Beg at the gates like Tarquin and his queen.
Now will I to the Goths and raise a power,
To be revenged on Rome and Saturnine.        *Exit*

**3.2**     *A banquet. Enter Titus Andronicus, Marcus,*
        *Lavinia, and the boy (young Lucius)*
TITUS
So, so, now sit, and look you eat no more
Than will preserve just so much strength in us
As will revenge these bitter woes of ours.
    ⌈*They sit*⌉
Marcus, unknit that sorrow-wreathen knot.
Thy niece and I, poor creatures, want our hands,    5
And cannot passionate our tenfold grief
With folded arms. This poor right hand of mine
Is left to tyrannize upon my breast,
Who, when my heart, all mad with misery,
Beats in this hollow prison of my flesh,         10
Then thus I thump it down.
    *He beats his breast*
(*To Lavinia*) Thou map of woe, that thus dost talk in
    signs,
When thy poor heart beats with outrageous beating
Thou canst not strike it thus to make it still!
Wound it with sighing, girl; kill it with groans,    15
Or get some little knife between thy teeth
And just against thy heart make thou a hole,
That all the tears that thy poor eyes let fall
May run into that sink and, soaking in,
Drown the lamenting fool in sea-salt tears.     20
MARCUS
Fie, brother, fie! Teach her not thus to lay
Such violent hands upon her tender life.
TITUS
How now! Has sorrow made thee dote already?
Why, Marcus, no man should be mad but I.
What violent hands can she lay on her life?     25
Ah, wherefore dost thou urge the name of hands
To bid Aeneas tell the tale twice o'er
How Troy was burnt and he made miserable?
O, handle not the theme, to talk of hands,

Lest we remember still that we have none.     30
Fie, fie, how franticly I square my talk,
As if we should forget we had no hands
If Marcus did not name the word of hands!
Come, let's fall to; and, gentle girl, eat this.
Here is no drink! Hark, Marcus, what she says.    35
I can interpret all her martyred signs.
She says she drinks no other drink but tears,
Brewed with her sorrow, mashed upon her cheeks.
Speechless complainer, I will learn thy thought.
In thy dumb action will I be as perfect     40
As begging hermits in their holy prayers.
Thou shalt not sigh, nor hold thy stumps to heaven,
Nor wink, nor nod, nor kneel, nor make a sign,
But I of these will wrest an alphabet,
And by still practice learn to know thy meaning.    45
YOUNG LUCIUS
Good grandsire, leave these bitter deep laments.
Make my aunt merry with some pleasing tale.
MARCUS
Alas, the tender boy in passion moved
Doth weep to see his grandsire's heaviness.
TITUS
Peace, tender sapling, thou art made of tears,    50
And tears will quickly melt thy life away.
    *Marcus strikes the dish with a knife*
What dost thou strike at, Marcus, with thy knife?
MARCUS
At that that I have killed, my lord—a fly.
TITUS
Out on thee, murderer! Thou kill'st my heart.
Mine eyes are cloyed with view of tyranny.    55
A deed of death done on the innocent
Becomes not Titus' brother. Get thee gone.
I see thou art not for my company.
MARCUS
Alas, my lord, I have but killed a fly.
TITUS
'But'? How if that fly had a father, brother?    60
How would he hang his slender gilded wings
And buzz lamenting dirges in the air!
Poor harmless fly,
That with his pretty buzzing melody     64
Came here to make us merry—and thou hast killed him!
MARCUS
Pardon me, sir, it was a black ill-favoured fly,
Like to the Empress' Moor. Therefore I killed him.
TITUS O, O, O!
Then pardon me for reprehending thee,
For thou hast done a charitable deed.     70
Give me thy knife. I will insult on him,
Flattering myself as if it were the Moor
Come hither purposely to poison me.
    *He takes a knife and strikes*
There's for thyself, and that's for Tamora. Ah, sirrah!
Yet I think we are not brought so low     75
But that between us we can kill a fly
That comes in likeness of a coal-black Moor.
MARCUS
Alas, poor man! Grief has so wrought on him
He takes false shadows for true substances.
TITUS
Come, take away. Lavinia, go with me.     80
I'll to thy closet and go read with thee
Sad stories chancèd in the times of old.

Come, boy, and go with me. Thy sight is young,
And thou shalt read when mine begin to dazzle.
                                                    *Exeunt*

❦

4.1     *Enter Lucius' son and Lavinia running after him,*
        *and the boy flies from her with his books under his*
        *arm. Enter Titus and Marcus*
YOUNG LUCIUS
    Help, grandsire, help! My aunt Lavinia
    Follows me everywhere, I know not why.
    Good uncle Marcus, see how swift she comes.
    Alas, sweet aunt, I know not what you mean.
        ⌈*He drops his books*⌉
MARCUS
    Stand by me, Lucius. Do not fear thine aunt.        5
TITUS
    She loves thee, boy, too well to do thee harm.
YOUNG LUCIUS
    Ay, when my father was in Rome she did.
MARCUS
    What means my niece Lavinia by these signs?
TITUS
    Fear her not, Lucius; somewhat doth she mean.
⌈MARCUS⌉
    See, Lucius, see how much she makes of thee.       10
    Somewhither would she have thee go with her.
    Ah, boy, Cornelia never with more care
    Read to her sons than she hath read to thee
    Sweet poetry and Tully's *Orator.*
    Canst thou not guess wherefore she plies thee thus?
YOUNG LUCIUS
    My lord, I know not, I, nor can I guess,           16
    Unless some fit or frenzy do possess her;
    For I have heard my grandsire say full oft
    Extremity of griefs would make men mad,
    And I have read that Hecuba of Troy                20
    Ran mad for sorrow. That made me to fear,
    Although, my lord, I know my noble aunt
    Loves me as dear as e'er my mother did,
    And would not but in fury fright my youth,
    Which made me down to throw my books and fly,      25
    Causeless, perhaps. But pardon me, sweet aunt;
    And, madam, if my uncle Marcus go
    I will most willingly attend your ladyship.
MARCUS Lucius, I will.
        *Lavinia turns the books over with her stumps*
TITUS
    How now, Lavinia? Marcus, what means this?         30
    Some book there is that she desires to see.
    Which is it, girl, of these?—Open them, boy.
    (*To Lavinia*) But thou art deeper read and better skilled.
    Come and take choice of all my library,
    And so beguile thy sorrow till the heavens         35
    Reveal the damned contriver of this deed.—
    Why lifts she up her arms in sequence thus?
MARCUS
    I think she means that there were more than one
    Confederate in the fact. Ay, more there was,
    Or else to heaven she heaves them for revenge.     40
TITUS
    Lucius, what book is that she tosseth so?
YOUNG LUCIUS
    Grandsire, 'tis Ovid's *Metamorphoses.*
    My mother gave it me.
MARCUS                      For love of her that's gone,
    Perhaps, she culled it from among the rest.

TITUS
    Soft, so busily she turns the leaves.              45
    Help her. What would she find? Lavinia, shall I read?
    This is the tragic tale of Philomel,
    And treats of Tereus' treason and his rape,
    And rape, I fear, was root of thy annoy.
MARCUS
    See, brother, see. Note how she quotes the leaves.  50
TITUS
    Lavinia, wert *thou* thus surprised, sweet girl,
    Ravished and wronged as Philomela was,
    Forced in the ruthless, vast, and gloomy woods?
    See, see. Ay, such a place there is where we did
        hunt—
    O, had we never, never hunted there!—               55
    Patterned by that the poet here describes,
    By nature made for murders and for rapes.
MARCUS
    O, why should nature build so foul a den,
    Unless the gods delight in tragedies?
TITUS
    Give signs, sweet girl, for here are none but friends,
    What Roman lord it was durst do the deed.           61
    Or slunk not Saturnine, as Tarquin erst,
    That left the camp to sin in Lucrece' bed?
MARCUS
    Sit down, sweet niece. Brother, sit down by me.
        *They sit*
    Apollo, Pallas, Jove, or Mercury                    65
    Inspire me, that I may this treason find.
    My lord, look here. Look here, Lavinia.
    This sandy plot is plain. Guide if thou canst
    This after me.
        *He writes his name with his staff, and guides it*
        *with feet and mouth*
                    I here have writ my name
    Without the help of any hand at all.                70
    Cursed be that heart that forced us to this shift!
    Write thou, good niece, and here display at last
    What God will have discovered for revenge.
    Heaven guide thy pen to print thy sorrows plain,
    That we may know the traitors and the truth.        75
        *She takes the staff in her mouth, and guides it with*
        *her stumps, and writes*
    O, do ye read, my lord, what she hath writ?
⌈TITUS⌉ 'Stuprum—Chiron—Demetrius.'
MARCUS
    What, what!—The lustful sons of Tamora
    Performers of this heinous bloody deed?
TITUS
    *Magni dominator poli,*                             80
    *Tam lentus audis scelera, tam lentus vides?*
MARCUS
    O, calm thee, gentle lord, although I know
    There is enough written upon this earth
    To stir a mutiny in the mildest thoughts,
    And arm the minds of infants to exclaims.           85
    My lord, kneel down with me; Lavinia, kneel;
    And kneel, sweet boy, the Roman Hector's hope,
        *All kneel*
    And swear with me—as, with the woeful fere
    And father of that chaste dishonoured dame
    Lord Junius Brutus sware for Lucrece' rape—         90
    That we will prosecute by good advice
    Mortal revenge upon these traitorous Goths,
    And see their blood, or die with this reproach.
        *They rise*

TITUS
'Tis sure enough an you knew how,
But if you hunt these bear-whelps, then beware.          95
The dam will wake, and if she wind ye once
She's with the lion deeply still in league,
And lulls him whilst she playeth on her back,
And when he sleeps will she do what she list.
You are a young huntsman, Marcus. Let alone,          100
And come, I will go get a leaf of brass
And with a gad of steel will write these words,
And lay it by. The angry northern wind
Will blow these sands like Sibyl's leaves abroad,
And where's our lesson then? Boy, what say you?          105
YOUNG LUCIUS
I say, my lord, that if I were a man
Their mother's bedchamber should not be safe
For these base bondmen to the yoke of Rome.
MARCUS
Ay, that's my boy! Thy father hath full oft
For his ungrateful country done the like.          110
YOUNG LUCIUS
And, uncle, so will I, an if I live.
TITUS
Come go with me into mine armoury.
Lucius, I'll fit thee; and withal, my boy,
Shall carry from me to the Empress' sons
Presents that I intend to send them both.          115
Come, come, thou'lt do my message, wilt thou not?
YOUNG LUCIUS
Ay, with my dagger in their bosoms, grandsire.
TITUS
No, boy, not so. I'll teach thee another course.
Lavinia, come. Marcus, look to my house.
Lucius and I'll go brave it at the court.          120
Ay, marry, will we, sir, and we'll be waited on.
                    *Exeunt all but Marcus*
MARCUS
O heavens, can you hear a good man groan
And not relent, or not compassion him?
Marcus, attend him in his ecstasy,
That hath more scars of sorrow in his heart          125
Than foemen's marks upon his battered shield,
But yet so just that he will not revenge.
Revenge the heavens for old Andronicus!          *Exit*

4.2          *Enter Aaron, Chiron, and Demetrius at one door,*
            *and at the other door young Lucius and another*
            *with a bundle of weapons, and verses writ upon*
            *them*
CHIRON
Demetrius, here's the son of Lucius.
He hath some message to deliver us.
AARON
Ay, some mad message from his mad grandfather.
YOUNG LUCIUS
My lords, with all the humbleness I may
I greet your honours from Andronicus          5
(*Aside*) And pray the Roman gods confound you both.
DEMETRIUS
Gramercy, lovely Lucius. What's the news?
YOUNG LUCIUS (*aside*)
That you are both deciphered, that's the news,
For villains marked with rape. (*Aloud*) May it please
    you,
My grandsire, well advised, hath sent by me          10
The goodliest weapons of his armoury

To gratify your honourable youth,
The hope of Rome, for so he bid me say;
        *His attendant gives the weapons*
And so I do, and with his gifts present
Your lordships that, whenever you have need,          15
You may be armèd and appointed well;
And so I leave you both (*aside*) like bloody villains.
                    *Exit with attendant*
DEMETRIUS
What's here—a scroll, and written round about?
Let's see.
'Integer vitae, scelerisque purus,          20
Non eget Mauri iaculis, nec arcu.'
CHIRON
O, 'tis a verse in Horace, I know it well.
I read it in the grammar long ago.
AARON
Ay, just, a verse in Horace; right, you have it.
(*Aside*) Now what a thing it is to be an ass!          25
Here's no sound jest. The old man hath found their
    guilt,
And sends them weapons wrapped about with lines
That wound beyond their feeling to the quick.
But were our witty Empress well afoot
She would applaud Andronicus' conceit.          30
But let her rest in her unrest a while.
(*To Chiron and Demetrius*)
And now, young lords, was't not a happy star
Led us to Rome, strangers and, more than so,
Captives, to be advancèd to this height?
It did me good before the palace gate          35
To brave the Tribune in his brother's hearing.
DEMETRIUS
But me more good to see so great a lord
Basely insinuate and send us gifts.
AARON
Had he not reason, Lord Demetrius?
Did you not use his daughter very friendly?          40
DEMETRIUS
I would we had a thousand Roman dames
At such a bay, by turn to serve our lust.
CHIRON
A charitable wish, and full of love.
AARON
Here lacks but your mother for to say amen.
CHIRON
And that would she, for twenty thousand more.          45
DEMETRIUS
Come, let us go and pray to all the gods
For our belovèd mother in her pains.
AARON
Pray to the devils; the gods have given us over.
                    *Trumpets sound*
DEMETRIUS
Why do the Emperor's trumpets flourish thus?
CHIRON
Belike for joy the Emperor hath a son.          50
DEMETRIUS
Soft, who comes here?
        *Enter Nurse with a blackamoor child*
NURSE                    Good morrow, lords.
O tell me, did you see Aaron the Moor?
AARON
Well, more or less, or ne'er a whit at all,
Here Aaron is; and what with Aaron now?

NURSE
  O gentle Aaron, we are all undone.                           55
  Now help, or woe betide thee evermore!
AARON
  Why, what a caterwauling dost thou keep!
  What dost thou wrap and fumble in thy arms?
NURSE
  O, that which I would hide from heaven's eye,
  Our Empress' shame and stately Rome's disgrace.             60
  She is delivered, lords, she is delivered.
AARON
  To whom?
NURSE               I mean she is brought abed.
AARON
  Well, God give her good rest. What hath he sent her?
NURSE
  A devil.
AARON      Why then, she is the devil's dam.
  A joyful issue!                                              65
NURSE
  A joyless, dismal, black, and sorrowful issue.
  Here is the babe, as loathsome as a toad
  Amongst the fair-faced breeders of our clime.
  The Empress sends it thee, thy stamp, thy seal,
  And bids thee christen it with thy dagger's point.          70
AARON
  Zounds, ye whore, is black so base a hue?
  Sweet blowze, you are a beauteous blossom, sure.
DEMETRIUS  Villain, what hast thou done?
AARON  That which thou canst not undo.
CHIRON  Thou hast undone our mother.                           75
AARON  Villain, I have done thy mother.
DEMETRIUS
  And therein, hellish dog, thou hast undone her.
  Woe to her chance, and damned her loathèd choice,
  Accursed the offspring of so foul a fiend.
CHIRON
  It shall not live.
AARON               It shall not die.                          80
NURSE
  Aaron, it must; the mother wills it so.
AARON
  What, must it, nurse? Then let no man but I
  Do execution on my flesh and blood.
DEMETRIUS
  I'll broach the tadpole on my rapier's point.
  Nurse, give it me. My sword shall soon dispatch it.         85
AARON
  Sooner this sword shall plough thy bowels up.
    *He takes the child and draws his sword*
  Stay, murderous villains, will you kill your brother?
  Now, by the burning tapers of the sky
  That shone so brightly when this boy was got,
  He dies upon my scimitar's sharp point                      90
  That touches this, my first-born son and heir.
  I tell you, younglings, not Enceladus
  With all his threat'ning band of Typhon's brood,
  Nor great Alcides, nor the god of war
  Shall seize this prey out of his father's hands.            95
  What, what, ye sanguine, shallow-hearted boys,
  Ye whitelimed walls, ye alehouse painted signs,
  Coal-black is better than another hue
  In that it scorns to bear another hue;
  For all the water in the ocean                              100
  Can never turn the swan's black legs to white,
  Although she lave them hourly in the flood.

  Tell the Empress from me I am of age
  To keep mine own, excuse it how she can.
DEMETRIUS
  Wilt thou betray thy noble mistress thus?                  105
AARON
  My mistress is my mistress, this myself,
  The figure and the picture of my youth.
  This before all the world do I prefer;
  This maugre all the world will I keep safe,
  Or some of you shall smoke for it in Rome.                 110
DEMETRIUS
  By this our mother is for ever shamed.
CHIRON
  Rome will despise her for this foul escape.
NURSE
  The Emperor in his rage will doom her death.
CHIRON
  I blush to think upon this ignomy.
AARON
  Why, there's the privilege your beauty bears.              115
  Fie, treacherous hue, that will betray with blushing
  The close enacts and counsels of thy heart.
  Here's a young lad framed of another leer.
  Look how the black slave smiles upon the father,
  As who should say 'Old lad, I am thine own.'               120
  He is your brother, lords, sensibly fed
  Of that self blood that first gave life to you,
  And from that womb where you imprisoned were
  He is enfranchisèd and come to light.
  Nay, he is your brother by the surer side,                 125
  Although my seal be stampèd in his face.
NURSE
  Aaron, what shall I say unto the Empress?
DEMETRIUS
  Advise thee, Aaron, what is to be done,
  And we will all subscribe to thy advice.
  Save thou the child, so we may all be safe.                130
AARON
  Then sit we down, and let us all consult.
  My son and I will have the wind of you.
  Keep there; now talk at pleasure of your safety.
    *They sit*
DEMETRIUS (*to the Nurse*)
  How many women saw this child of his?
AARON
  Why, so, brave lords, when we do join in league            135
  I am a lamb; but if you brave the Moor,
  The chafèd boar, the mountain lioness,
  The ocean swells not so as Aaron storms.
  (*To the Nurse*) But say again, how many saw the
      child?
NURSE
  Cornelia the midwife, and myself,                          140
  And no one else but the delivered Empress.
AARON
  The Empress, the midwife, and yourself.
  Two may keep counsel when the third's away.
  Go to the Empress, tell her this I said.                   144
    *He kills her*
  'Wheak, wheak'—so cries a pig preparèd to the spit.
DEMETRIUS
  What mean'st thou, Aaron? Wherefore didst thou this?
AARON
  O Lord, sir, 'tis a deed of policy.                        147
  Shall she live to betray this guilt of ours—
  A long-tongued, babbling gossip? No, lords, no.

And now be it known to you my full intent. 150
Not far, one Muliteus my countryman
His wife but yesternight was brought to bed.
His child is like to her, fair as you are.
Go pack with him, and give the mother gold,
And tell them both the circumstance of all, 155
And how by this their child shall be advanced
And be receivèd for the Emperor's heir,
And substituted in the place of mine,
To calm this tempest whirling in the court;
And let the Emperor dandle him for his own. 160
Hark ye, lords, you see I have given her physic,
And you must needs bestow her funeral.
The fields are near, and you are gallant grooms.
This done, see that you take no longer days,
But send the midwife presently to me. 165
The midwife and the nurse well made away,
Then let the ladies tattle what they please.

CHIRON
Aaron, I see thou wilt not trust the air
With secrets.

DEMETRIUS　　　For this care of Tamora,
Herself and hers are highly bound to thee. 170
　　　　　　　　　Exeunt Chiron and Demetrius with
　　　　　　　　　　　　　　　the Nurse's body
AARON
Now to the Goths, as swift as swallow flies,
There to dispose this treasure in mine arms
And secretly to greet the Empress' friends.
Come on, you thick-lipped slave, I'll bear you hence,
For it is you that puts us to our shifts. 175
I'll make you feed on berries and on roots,
And fat on curds and whey, and suck the goat,
And cabin in a cave, and bring you up
To be a warrior and command a camp.
　　　　　　　　　　　　　Exit with the child

4.3　Enter Titus, old Marcus, his son Publius, young
　　Lucius, and other gentlemen (Sempronius, Caius)
　　with bows; and Titus bears the arrows with letters
　　on the ends of them
TITUS
Come, Marcus, come; kinsmen, this is the way.
Sir boy, let me see your archery.
Look ye draw home enough, and 'tis there straight.
Terras Astraea reliquit.
Be you remembered, Marcus: she's gone, she's fled. 5
Sirs, take you to your tools. You, cousins, shall
Go sound the ocean and cast your nets.
Happily you may catch her in the sea;
Yet there's as little justice as at land.
No, Publius and Sempronius, you must do it. 10
'Tis you must dig with mattock and with spade
And pierce the inmost centre of the earth.
Then, when you come to Pluto's region,
I pray you deliver him this petition.
Tell him it is for justice and for aid, 15
And that it comes from old Andronicus,
Shaken with sorrows in ungrateful Rome.
Ah, Rome! Well, well, I made thee miserable
What time I threw the people's suffrages
On him that thus doth tyrannize o'er me. 20
Go, get you gone, and pray be careful all,
And leave you not a man-of-war unsearched.
This wicked Emperor may have shipped her hence,
And, kinsmen, then we may go pipe for justice.

MARCUS
O, Publius, is not this a heavy case, 25
To see thy noble uncle thus distraught?
PUBLIUS
Therefore, my lords, it highly us concerns
By day and night t'attend him carefully
And feed his humour kindly as we may,
Till time beget some careful remedy. 30
MARCUS
Kinsmen, his sorrows are past remedy,
But ⌈　　　　　　　　　　　　　　　⌉
Join with the Goths, and with revengeful war
Take wreak on Rome for this ingratitude,
And vengeance on the traitor Saturnine. 35
TITUS
Publius, how now? How now, my masters?
What, have you met with her?
PUBLIUS
No, my good lord, but Pluto sends you word
If you will have Revenge from hell, you shall.
Marry, for Justice, she is now employed, 40
He thinks, with Jove, in heaven or somewhere else,
So that perforce you must needs stay a time.
TITUS
He doth me wrong to feed me with delays.
I'll dive into the burning lake below
And pull her out of Acheron by the heels. 45
Marcus, we are but shrubs, no cedars we,
No big-boned men framed of the Cyclops' size,
But metal, Marcus, steel to the very back,
Yet wrung with wrongs more than our backs can
　　bear;
And sith there's no justice in earth nor hell, 50
We will solicit heaven and move the gods
To send down Justice for to wreak our wrongs.
Come, to this gear. You are a good archer, Marcus.
　　　　　　　　He gives them the arrows
'Ad Iovem', that's for you. Here, 'ad Apollinem'.
'Ad Martem', that's for myself. 55
Here, boy, 'to Pallas'. Here 'to Mercury'.
'To Saturn', Caius—not 'to Saturnine'!
You were as good to shoot against the wind.
To it, boy! Marcus, loose when I bid.
Of my word, I have written to effect. 60
There's not a god left unsolicited.
MARCUS
Kinsmen, shoot all your shafts into the court.
We will afflict the Emperor in his pride.
TITUS
Now, masters, draw.
　　　　　They shoot
　　　　　　　　O, well said, Lucius!
Good boy, in Virgo's lap! Give it Pallas. 65
MARCUS
My lord, I aim a mile beyond the moon.
Your letter is with Jupiter by this.
TITUS
Ha, ha! Publius, Publius, what hast thou done?
See, see, thou hast shot off one of Taurus' horns.
MARCUS
This was the sport, my lord. When Publius shot, 70
The Bull, being galled, gave Aries such a knock
That down fell both the Ram's horns in the court,
And who should find them but the Empress' villain!
She laughed, and told the Moor he should not choose
But give them to his master for a present. 75

TITUS
    Why, there it goes. God give his lordship joy.
        *Enter the Clown with a basket and two pigeons in it*
    News, news from heaven; Marcus, the post is come.
    Sirrah, what tidings? Have you any letters?
    Shall I have justice? What says Jupiter?
CLOWN Ho, the gibbet-maker? He says that he hath taken
    them down again, for the man must not be hanged till
    the next week.                                        82
TITUS
    But what says Jupiter, I ask thee?
CLOWN Alas, sir, I know not 'Jupiter'. I never drank with
    him in all my life.                                   85
TITUS
    Why, villain, art not thou the carrier?
CLOWN Ay, of my pigeons, sir; nothing else.
TITUS Why, didst thou not come from heaven?
CLOWN From heaven? Alas, sir, I never came there. God
    forbid I should be so bold to press to heaven in my
    young days. Why, I am going with my pigeons to the
    tribunal plebs to take up a matter of brawl betwixt my
    uncle and one of the Emperal's men.                   93
TITUS
    Sirrah, come hither. Make no more ado,
    But give your pigeons to the Emperor.                 95
    By me thou shalt have justice at his hands.
    Hold, hold—(*giving money*) meanwhile, here's money
        for thy charges.
    Give me pen and ink. Sirrah, can you with a grace
    Deliver up a supplication?
CLOWN Ay, sir.                                           100
TITUS (*writing and giving the Clown a paper*) Then here is
    a supplication for you, and when you come to him, at
    the first approach you must kneel, then kiss his foot,
    then deliver up your pigeons, and then look for your
    reward. I'll be at hand, sir; see you do it bravely.  105
CLOWN I warrant you, sir. Let me alone.
TITUS
    Sirrah, hast thou a knife? Come, let me see it.
    Here, Marcus, fold it in the oration,
    For thou hast made it like an humble suppliant.
    And when thou hast given it to the Emperor,          110
    Knock at my door and tell me what he says.
CLOWN God be with you, sir. I will.                 *Exit*
TITUS
    Come, Marcus, let us go. Publius, follow me.    *Exeunt*

4.4  *Enter Saturninus, the Emperor, and Tamora, the*
     *Empress, and Chiron and Demetrius, her two sons,*
     *and others. The Emperor brings the arrows in his*
     *hand that Titus shot at him*
SATURNINUS
    Why, lords, what wrongs are these! Was ever seen
    An emperor in Rome thus overborne,
    Troubled, confronted thus, and for the extent
    Of egall justice used in such contempt?
    My lords, you know, as know the mightful gods,        5
    However these disturbers of our peace
    Buzz in the people's ears, there naught hath passed
    But even with law against the wilful sons
    Of old Andronicus. And what an if
    His sorrows have so overwhelmed his wits?            10
    Shall we be thus afflicted in his wreaks,
    His fits, his frenzy, and his bitterness?
    And now he writes to heaven for his redress.
    See, here's 'to Jove' and this 'to Mercury',

    This 'to Apollo', this 'to the god of war'—          15
    Sweet scrolls to fly about the streets of Rome!
    What's this but libelling against the Senate
    And blazoning our unjustice everywhere?
    A goodly humour, is it not, my lords?—
    As who would say, in Rome no justice were.            20
    But, if I live, his feignèd ecstasies
    Shall be no shelter to these outrages,
    But he and his shall know that justice lives
    In Saturninus' health, whom if he sleep
    He'll so awake as he in fury shall                    25
    Cut off the proud'st conspirator that lives.
TAMORA
    My gracious lord, my lovely Saturnine,
    Lord of my life, commander of my thoughts,
    Calm thee, and bear the faults of Titus' age,
    Th'effects of sorrow for his valiant sons            30
    Whose loss hath pierced him deep and scarred his
        heart;
    And rather comfort his distressèd plight
    Than prosecute the meanest or the best
    For these contempts. (*Aside*) Why, thus it shall become
    High-witted Tamora to gloze with all.                 35
    But, Titus, I have touched thee to the quick.
    Thy life blood out if Aaron now be wise,
    Then is all safe, the anchor in the port.
        *Enter Clown*
    How now, good fellow, wouldst thou speak with us?
CLOWN Yea, forsooth, an your mistress-ship be Emperial.
TAMORA Empress I am, but yonder sits the Emperor.       41
CLOWN 'Tis he. God and Saint Stephen give you good-
    e'en. I have brought you a letter and a couple of
    pigeons here.
        *Saturninus reads the letter*
SATURNINUS (*to an attendant*)
    Go, take him away, and hang him presently.            45
CLOWN How much money must I have?
TAMORA Come, sirrah, you must be hanged.
CLOWN Hanged, by' Lady? Then I have brought up a
    neck to a fair end.            *Exit ⌈with attendant⌉*
SATURNINUS
    Despiteful and intolerable wrongs!                    50
    Shall I endure this monstrous villainy?
    I know from whence this same device proceeds.
    May this be borne?—As if his traitorous sons,
    That died by law for murder of our brother,
    Have by my means been butchered wrongfully!           55
    Go, drag the villain hither by the hair.
    Nor age nor honour shall shape privilege.
    For this proud mock I'll be thy slaughterman,
    Sly frantic wretch, that holp'st to make me great
    In hope thyself should govern Rome and me.            60
        *Enter Aemilius, a messenger*
SATURNINUS
    What news with thee, Aemilius?
AEMILIUS
    Arm, my lords! Rome never had more cause.
    The Goths have gathered head, and with a power
    Of high-resolvèd men bent to the spoil
    They hither march amain under conduct                 65
    Of Lucius, son to old Andronicus,
    Who threats in course of this revenge to do
    As much as ever Coriolanus did.
SATURNINUS
    Is warlike Lucius general of the Goths?
    These tidings nip me, and I hang the head,            70

As flowers with frost, or grass beat down with storms.
Ay, now begins our sorrows to approach.
'Tis he the common people love so much.
Myself hath often heard them say,
When I have walkèd like a private man,                          75
That Lucius' banishment was wrongfully,
And they have wished that Lucius were their emperor.

TAMORA
Why should you fear? Is not your city strong?

SATURNINUS
Ay, but the citizens favour Lucius,
And will revolt from me to succour him.                         80

TAMORA
King, be thy thoughts imperious like thy name.
Is the sun dimmed, that gnats do fly in it?
The eagle suffers little birds to sing,
And is not careful what they mean thereby,
Knowing that with the shadow of his wings                       85
He can at pleasure stint their melody.
Even so mayst thou the giddy men of Rome.
Then cheer thy spirit; for know thou, Emperor,
I will enchant the old Andronicus
With words more sweet and yet more dangerous                    90
Than baits to fish or honey-stalks to sheep
Whenas the one is wounded with the bait,
The other rotted with delicious feed.

SATURNINUS
But he will not entreat his son for us.

TAMORA
If Tamora entreat him, then he will,                            95
For I can smooth and fill his agèd ears
With golden promises that, were his heart
Almost impregnable, his old ears deaf,
Yet should both ear and heart obey my tongue.
(To Aemilius) Go thou before to be our ambassador.
Say that the Emperor requests a parley                          101
Of warlike Lucius, and appoint the meeting
Even at his father's house, the old Andronicus.

SATURNINUS
Aemilius, do this message honourably,
And if he stand on hostage for his safety,                      105
Bid him demand what pledge will please him best.

AEMILIUS
Your bidding shall I do effectually.                            Exit

TAMORA
Now will I to that old Andronicus,
And temper him with all the art I have
To pluck proud Lucius from the warlike Goths.                   110
And now, sweet Emperor, be blithe again,
And bury all thy fear in my devices.

SATURNINUS
Then go incessantly, and plead to him.
                                            Exeunt severally

                          ✿

5.1   ⌜Flourish.⌝ Enter Lucius with an army of Goths,
       with drummers and soldiers

LUCIUS
Approvèd warriors and my faithful friends,
I have receivèd letters from great Rome
Which signifies what hate they bear their emperor
And how desirous of our sight they are.
Therefore, great lords, be as your titles witness,              5
Imperious, and impatient of your wrongs,
And wherein Rome hath done you any scath
Let him make treble satisfaction.

A GOTH
Brave slip sprung from the great Andronicus,
Whose name was once our terror, now our comfort,
Whose high exploits and honourable deeds                        11
Ingrateful Rome requites with foul contempt,
Be bold in us. We'll follow where thou lead'st,
Like stinging bees in hottest summer's day
Led by their master to the flowered fields,                     15
And be avenged on cursèd Tamora.

GOTHS
And as he saith, so say we all with him.

LUCIUS
I humbly thank him, and I thank you all.
But who comes here, led by a lusty Goth?
          Enter a Goth, leading of Aaron with his child in his
          arms

GOTH
Renownèd Lucius, from our troops I strayed                      20
To gaze upon a ruinous monastery,
And as I earnestly did fix mine eye
Upon the wasted building, suddenly
I heard a child cry underneath a wall.
I made unto the noise, when soon I heard                        25
The crying babe controlled with this discourse:
'Peace, tawny slave, half me and half thy dam!
Did not thy hue bewray whose brat thou art,
Had nature lent thee but thy mother's look,
Villain, thou mightst have been an emperor.                     30
But where the bull and cow are both milk-white
They never do beget a coal-black calf.
Peace, villain, peace!'—even thus he rates the babe—
'For I must bear thee to a trusty Goth
Who, when he knows thou art the Empress' babe,                  35
Will hold thee dearly for thy mother's sake.'
With this, my weapon drawn, I rushed upon him,
Surprised him suddenly, and brought him hither
To use as you think needful of the man.

LUCIUS
O worthy Goth, this is the incarnate devil                      40
That robbed Andronicus of his good hand.
This is the pearl that pleased your Empress' eye,
And here's the base fruit of her burning lust.
(To Aaron) Say, wall-eyed slave, whither wouldst thou
          convey
This growing image of thy fiendlike face?                       45
Why dost not speak? What, deaf? What, not a word?
A halter, soldiers! Hang him on this tree,
And by his side his fruit of bastardy.

AARON
Touch not the boy; he is of royal blood.

LUCIUS
Too like the sire for ever being good.                          50
First hang the child, that he may see it sprawl—
A sight to vex the father's soul withal.
Get me a ladder.
          ⌜A Goth brings a ladder which Aaron climbs⌝
AARON                          Lucius, save the child,
And bear it from me to the Empress.
If thou do this, I'll show thee wondrous things                 55
That highly may advantage thee to hear.
If thou wilt not, befall what may befall,
I'll speak no more but 'Vengeance rot you all!'

LUCIUS
Say on, and if it please me which thou speak'st
Thy child shall live, and I will see it nourished.              60

AARON
  And if it please thee? Why, assure thee, Lucius,
  'Twill vex thy soul to hear what I shall speak;
  For I must talk of murders, rapes, and massacres,
  Acts of black night, abominable deeds,
  Complots of mischief, treason, villainies       65
  Ruthful to hear yet piteously performed,
  And this shall all be buried in my death
  Unless thou swear to me my child shall live.
LUCIUS
  Tell on thy mind. I say thy child shall live.
AARON
  Swear that he shall, and then I will begin.      70
LUCIUS
  Who should I swear by? Thou believest no god.
  That granted, how canst thou believe an oath?
AARON
  What if I do not?—as indeed I do not—
  Yet for I know thou art religious
  And hast a thing within thee callèd conscience,    75
  With twenty popish tricks and ceremonies
  Which I have seen thee careful to observe,
  Therefore I urge thy oath; for that I know
  An idiot holds his bauble for a god,
  And keeps the oath which by that god he swears,  80
  To that I'll urge him, therefore thou shalt vow
  By that same god, what god soe'er it be,
  That thou adorest and hast in reverence,
  To save my boy, to nurse and bring him up,
  Or else I will discover naught to thee.      85
LUCIUS
  Even by my god I swear to thee I will.
AARON
  First know thou I begot him on the Empress.
LUCIUS
  O most insatiate and luxurious woman!
AARON
  Tut, Lucius, this was but a deed of charity
  To that which thou shalt hear of me anon.    90
  'Twas her two sons that murdered Bassianus.
  They cut thy sister's tongue, and ravished her,
  And cut her hands, and trimmed her as thou sawest.
LUCIUS
  O detestable villain! Call'st thou that trimming?
AARON
  Why, she was washed and cut and trimmed, and 'twas
  Trim sport for them which had the doing of it.    96
LUCIUS
  O barbarous beastly villains, like thyself!
AARON
  Indeed, I was their tutor to instruct them.
  That codding spirit had they from their mother,
  As sure a card as ever won the set.    100
  That bloody mind I think they learned of me,
  As true a dog as ever fought at head.
  Well, let my deeds be witness of my worth.
  I trained thy brethren to that guileful hole
  Where the dead corpse of Bassianus lay.    105
  I wrote the letter that thy father found,
  And hid the gold within that letter mentioned,
  Confederate with the Queen and her two sons;
  And what not done that thou hast cause to rue
  Wherein I had no stroke of mischief in it?    110
  I played the cheater for thy father's hand,
  And when I had it drew myself apart,
  And almost broke my heart with extreme laughter.

  I pried me through the crevice of a wall
  When for his hand he had his two sons' heads,    115
  Beheld his tears, and laughed so heartily
  That both mine eyes were rainy like to his;
  And when I told the Empress of this sport
  She swoonèd almost at my pleasing tale,
  And for my tidings gave me twenty kisses.    120
A GOTH
  What, canst thou say all this and never blush?
AARON
  Ay, like a black dog, as the saying is.
LUCIUS
  Art thou not sorry for these heinous deeds?
AARON
  Ay, that I had not done a thousand more.
  Even now I curse the day—and yet I think    125
  Few come within the compass of my curse—
  Wherein I did not some notorious ill,
  As kill a man, or else devise his death;
  Ravish a maid, or plot the way to do it;
  Accuse some innocent and forswear myself;    130
  Set deadly enmity between two friends;
  Make poor men's cattle break their necks;
  Set fire on barns and haystacks in the night,
  And bid the owners quench them with their tears.
  Oft have I digged up dead men from their graves  135
  And set them upright at their dear friends' door,
  Even when their sorrows almost was forgot,
  And on their skins, as on the bark of trees,
  Have with my knife carvèd in Roman letters
  'Let not your sorrow die though I am dead.'    140
  But I have done a thousand dreadful things
  As willingly as one would kill a fly,
  And nothing grieves me heartily indeed
  But that I cannot do ten thousand more.
LUCIUS
  Bring down the devil, for he must not die    145
  So sweet a death as hanging presently.
    *Goths bring Aaron down the ladder*
AARON
  If there be devils, would I were a devil,
  To live and burn in everlasting fire,
  So I might have your company in hell
  But to torment you with my bitter tongue.    150
LUCIUS
  Sirs, stop his mouth, and let him speak no more.
    *Goths gag Aaron.*
    *Enter Aemilius*
A GOTH
  My lord, there is a messenger from Rome
  Desires to be admitted to your presence.
LUCIUS Let him come near.
  Welcome, Aemilius. What's the news from Rome?  155
AEMILIUS
  Lord Lucius, and you princes of the Goths,
  The Roman Emperor greets you all by me,
  And for he understands you are in arms,
  He craves a parley at your father's house,
  Willing you to demand your hostages,    160
  And they shall be immediately delivered.
A GOTH What says our general?
LUCIUS
  Aemilius, let the Emperor give his pledges
  Unto my father and my uncle Marcus,
  And we will come. Away!    165
    ⌈*Flourish.*⌉ *Exeunt* ⌈*marching*⌉

5.2    *Enter Tamora and Chiron and Demetrius, her two*
       *sons, disguised*

TAMORA
Thus, in this strange and sad habiliment,
I will encounter with Andronicus
And say I am Revenge, sent from below
To join with him and right his heinous wrongs.
Knock at his study, where they say he keeps    5
To ruminate strange plots of dire revenge.
Tell him Revenge is come to join with him
And work confusion on his enemies.
       *They knock, and Titus ⌈aloft⌉ opens his study door*

TITUS
Who doth molest my contemplation?
Is it your trick to make me ope the door,    10
That so my sad decrees may fly away
And all my study be to no effect?
You are deceived; for what I mean to do,
See here, in bloody lines I have set down,
And what is written shall be executed.    15

TAMORA
Titus, I am come to talk with thee.

TITUS
No, not a word. How can I grace my talk,
Wanting a hand to give it action?
Thou hast the odds of me, therefore no more.

TAMORA
If thou didst know me thou wouldst talk with me.    20

TITUS
I am not mad, I know thee well enough;
Witness this wretched stump, witness these crimson
   lines,
Witness these trenches made by grief and care,
Witness the tiring day and heavy night,
Witness all sorrow that I know thee well    25
For our proud empress, mighty Tamora.
Is not thy coming for my other hand?

TAMORA
Know, thou sad man, I am not Tamora.
She is thy enemy, and I thy friend.
I am Revenge, sent from th'infernal kingdom    30
To ease the gnawing vulture of thy mind
By working wreakful vengeance on thy foes.
Come down, and welcome me to this world's light.
Confer with me of murder and of death.
There's not a hollow cave or lurking-place,    35
No vast obscurity or misty vale
Where bloody murder or detested rape
Can couch for fear, but I will find them out,
And in their ears tell them my dreadful name,
Revenge, which makes the foul offender quake.    40

TITUS
Art thou Revenge, and art thou sent to me
To be a torment to mine enemies?

TAMORA
I am; therefore come down, and welcome me.

TITUS
Do me some service ere I come to thee.
Lo by thy side where Rape and Murder stands.    45
Now give some surance that thou art Revenge,
Stab them, or tear them on thy chariot wheels,
And then I'll come and be thy wagoner,
And whirl along with thee about the globe,
Provide two proper palfreys, black as jet,    50
To hale thy vengeful wagon swift away
And find out murderers in their guilty caves.
And when thy car is loaden with their heads

I will dismount, and by thy wagon wheel
Trot like a servile footman all day long,    55
Even from Hyperion's rising in the east
Until his very downfall in the sea;
And day by day I'll do this heavy task,
So thou destroy Rapine and Murder there.

TAMORA
These are my ministers, and come with me.    60

TITUS
Are they thy ministers? What are they called?

TAMORA
Rape and Murder, therefore callèd so
'Cause they take vengeance of such kind of men.

TITUS
Good Lord, how like the Empress' sons they are,
And you the Empress! But we worldly men    65
Have miserable, mad, mistaking eyes.
O sweet Revenge, now do I come to thee,
And if one arm's embracement will content thee,
I will embrace thee in it by and by.    *Exit ⌈aloft⌉*

TAMORA
This closing with him fits his lunacy.    70
Whate'er I forge to feed his brainsick humours
Do you uphold and maintain in your speeches,
For now he firmly takes me for Revenge,
And being credulous in this mad thought
I'll make him send for Lucius his son,    75
And whilst I at a banquet hold him sure
I'll find some cunning practice out of hand
To scatter and disperse the giddy Goths,
Or at the least make them his enemies.
See, here he comes, and I must ply my theme.    80
       *Enter Titus, below*

TITUS
Long have I been forlorn, and all for thee.
Welcome, dread Fury, to my woeful house.
Rapine and Murder, you are welcome, too.
How like the Empress and her sons you are!
Well are you fitted, had you but a Moor.    85
Could not all hell afford you such a devil?—
For well I wot the Empress never wags
But in her company there is a Moor,
And would you represent our Queen aright
It were convenient you had such a devil.    90
But welcome as you are. What shall we do?

TAMORA
What wouldst thou have us do, Andronicus?

DEMETRIUS
Show me a murderer, I'll deal with him.

CHIRON
Show me a villain that hath done a rape,
And I am sent to be revenged on him.    95

TAMORA
Show me a thousand that hath done thee wrong,
And I will be revengèd on them all.

TITUS *(to Demetrius)*
Look round about the wicked streets of Rome,
And when thou find'st a man that's like thyself,
Good Murder, stab him; he's a murderer.    100
*(To Chiron)* Go thou with him, and when it is thy hap
To find another that is like to thee,
Good Rapine, stab him; he is a ravisher.
*(To Tamora)* Go thou with them, and in the Emperor's
   court
There is a queen attended by a Moor.    105
Well shalt thou know her by thine own proportion,
For up and down she doth resemble thee.

I pray thee, do on them some violent death;
They have been violent to me and mine.

TAMORA

Well hast thou lessoned us. This shall we do;    110
But would it please thee, good Andronicus,
To send for Lucius, thy thrice-valiant son,
Who leads towards Rome a band of warlike Goths,
And bid him come and banquet at thy house—
When he is here, even at thy solemn feast,    115
I will bring in the Empress and her sons,
The Emperor himself, and all thy foes,
And at thy mercy shall they stoop and kneel,
And on them shalt thou ease thy angry heart.
What says Andronicus to this device?    120

TITUS

Marcus, my brother! 'Tis sad Titus calls.
    *Enter Marcus*
Go, gentle Marcus, to thy nephew Lucius.
Thou shalt enquire him out among the Goths.
Bid him repair to me, and bring with him
Some of the chiefest princes of the Goths.    125
Bid him encamp his soldiers where they are.
Tell him the Emperor and the Empress too
Feast at my house, and he shall feast with them.
This do thou for my love, and so let him,
As he regards his agèd father's life.    130

MARCUS

This will I do, and soon return again.    *Exit*

TAMORA

Now will I hence about thy business,
And take my ministers along with me.

TITUS

Nay, nay, let Rape and Murder stay with me,
Or else I'll call my brother back again,    135
And cleave to no revenge but Lucius.

TAMORA (*aside to her sons*)

What say you, boys, will you abide with him
Whiles I go tell my lord the Emperor
How I have governed our determined jest?
Yield to his humour, smooth and speak him fair,    140
And tarry with him till I turn again.

TITUS (*aside*)

I knew them all, though they supposed me mad,
And will o'erreach them in their own devices—
A pair of cursèd hell-hounds and their dam.

DEMETRIUS

Madam, depart at pleasure. Leave us here.    145

TAMORA

Farewell, Andronicus. Revenge now goes
To lay a complot to betray thy foes.

TITUS

I know thou dost, and sweet Revenge, farewell.
    *Exit Tamora*

CHIRON

Tell us, old man, how shall we be employed?

TITUS

Tut, I have work enough for you to do.    150
Publius, come hither; Caius and Valentine.
    *Enter Publius, Caius, and Valentine*

PUBLIUS

What is your will?

TITUS             Know you these two?

PUBLIUS

The Empress' sons I take them—Chiron, Demetrius.

TITUS

Fie, Publius, fie! Thou art too much deceived.

The one is Murder, and Rape is the other's name.    155
And therefore bind them, gentle Publius;
Caius and Valentine, lay hands on them.
Oft have you heard me wish for such an hour,
And now I find it. Therefore bind them sure,
And stop their mouths if they begin to cry.    *Exit*

CHIRON

Villains, forbear! We are the Empress' sons.    161

PUBLIUS

And therefore do we what we are commanded.
    *Publius, Caius, and Valentine bind and gag Chiron*
    *and Demetrius*
Stop close their mouths. Let them not speak a word.
Is he sure bound? Look that you bind them fast.
    *Enter Titus Andronicus with a knife, and Lavinia*
    *with a basin*

TITUS

Come, come, Lavinia. Look, thy foes are bound.    165
Sirs, stop their mouths. Let them not speak to me,
But let them hear what fearful words I utter.
O villains, Chiron and Demetrius!
Here stands the spring whom you have stained with
    mud,
This goodly summer with your winter mixed.    170
You killed her husband, and for that vile fault
Two of her brothers were condemned to death,
My hand cut off and made a merry jest,
Both her sweet hands, her tongue, and that more
    dear
Than hands or tongue, her spotless chastity,    175
Inhuman traitors, you constrained and forced.
What would you say if I should let you speak?
Villains, for shame. You could not beg for grace.
Hark, wretches, how I mean to martyr you.
This one hand yet is left to cut your throats,    180
Whiles that Lavinia 'tween her stumps doth hold
The basin that receives your guilty blood.
You know your mother means to feast with me,
And calls herself Revenge, and thinks me mad.
Hark, villains, I will grind your bones to dust,    185
And with your blood and it I'll make a paste,
And of the paste a coffin I will rear,
And make two pasties of your shameful heads,
And bid that strumpet, your unhallowed dam,
Like to the earth swallow her own increase.    190
This is the feast that I have bid her to,
And this the banquet she shall surfeit on;
For worse than Philomel you used my daughter,
And worse than Progne I will be revenged.
And now, prepare your throats. Lavinia, come.    195
Receive the blood, and when that they are dead
Let me go grind their bones to powder small,
And with this hateful liquor temper it,
And in that paste let their vile heads be baked.
Come, come, be everyone officious    200
To make this banquet, which I wish may prove
More stern and bloody than the Centaurs' feast.
    *He cuts their throats*
So, now bring them in, for I'll play the cook
And see them ready against their mother comes.
    *Exeunt carrying the bodies*

**5.3**    *Enter Lucius, Marcus, and the Goths, with Aaron,*
    *prisoner, ⌈and an attendant with his child⌉*

LUCIUS

Uncle Marcus, since 'tis my father's mind
That I repair to Rome, I am content.

A GOTH
And ours with thine, befall what fortune will.
LUCIUS
Good uncle, take you in this barbarous Moor,
This ravenous tiger, this accursèd devil. 5
Let him receive no sust'nance, fetter him
Till he be brought unto the Empress' face
For testimony of her foul proceedings,
And see the ambush of our friends be strong.
I fear the Emperor means no good to us. 10
AARON
Some devil whisper curses in my ear
And prompt me, that my tongue may utter forth
The venomous malice of my swelling heart.
LUCIUS
Away, inhuman dog, unhallowed slave!
Sirs, help our uncle to convey him in. 15
⌐Exeunt Goths with Aaron and his child⌐
Flourish
The trumpets show the Emperor is at hand.
Enter Saturninus the Emperor, and Tamora the
Empress, with Aemilius, Tribunes, Senators, and
others
SATURNINUS
What, hath the firmament more suns than one?
LUCIUS
What boots it thee to call thyself a sun?
MARCUS
Rome's emperor and nephew, break the parle.
These quarrels must be quietly debated. 20
The feast is ready which the careful Titus
Hath ordained to an honourable end,
For peace, for love, for league, and good to Rome,
Please you therefore draw nigh, and take your places.
SATURNINUS Marcus, we will. 25
⌐Hautboys. A table brought in.⌐ They sit.
Enter Titus like a cook, placing the dishes, and
Lavinia with a veil over her face; ⌐young Lucius,
and others⌐
TITUS
Welcome, my gracious lord; welcome, dread Queen;
Welcome, ye warlike Goths; welcome, Lucius;
And welcome, all. Although the cheer be poor,
'Twill fill your stomachs. Please you, eat of it.
SATURNINUS
Why art thou thus attired, Andronicus? 30
TITUS
Because I would be sure to have all well
To entertain your highness and your Empress.
TAMORA
We are beholden to you, good Andronicus.
TITUS
An if your highness knew my heart, you were.
My lord the Emperor, resolve me this: 35
Was it well done of rash Virginius
To slay his daughter with his own right hand
Because she was enforced, stained, and deflowered?
SATURNINUS
It was, Andronicus.
TITUS      Your reason, mighty lord?
SATURNINUS
Because the girl should not survive her shame, 40
And by her presence still renew his sorrows.
TITUS
A reason mighty, strong, effectual;
A pattern, precedent, and lively warrant

For me, most wretched, to perform the like.
Die, die, Lavinia, and thy shame with thee, 45
And with thy shame thy father's sorrow die.
⌐He kills her⌐
SATURNINUS
What hast thou done, unnatural and unkind?
TITUS
Killed her for whom my tears have made me blind.
I am as woeful as Virginius was,
And have a thousand times more cause than he 50
To do this outrage, and it now is done.
SATURNINUS
What, was she ravished? Tell who did the deed.
TITUS
Will't please you eat? Will't please your highness
feed?
TAMORA
Why hast thou slain thine only daughter thus?
TITUS
Not I, 'twas Chiron and Demetrius. 55
They ravished her, and cut away her tongue,
And they, 'twas they, that did her all this wrong.
SATURNINUS
Go, fetch them hither to us presently.
TITUS ⌐revealing the heads⌐
Why, there they are, both bakèd in this pie,
Whereof their mother daintily hath fed, 60
Eating the flesh that she herself hath bred.
'Tis true, 'tis true, witness my knife's sharp point.
He stabs the Empress
SATURNINUS
Die, frantic wretch, for this accursèd deed.
He kills Titus
LUCIUS
Can the son's eye behold his father bleed?
There's meed for meed, death for a deadly deed. 65
He kills Saturninus. Confusion follows.
⌐Enter Goths. Lucius, Marcus and others go aloft⌐
MARCUS
You sad-faced men, people and sons of Rome,
By uproars severed, as a flight of fowl
Scattered by winds and high tempestuous gusts,
O, let me teach you how to knit again
This scattered corn into one mutual sheaf, 70
These broken limbs again into one body.
A ROMAN LORD
Let Rome herself be bane unto herself,
And she whom mighty kingdoms curtsy to,
Like a forlorn and desperate castaway,
Do shameful execution on herself 75
But if my frosty signs and chaps of age,
Grave witnesses of true experience,
Cannot induce you to attend my words.
(To Lucius) Speak, Rome's dear friend, as erst our
ancestor
When with his solemn tongue he did discourse 80
To lovesick Dido's sad-attending ear
The story of that baleful-burning night
When subtle Greeks surprised King Priam's Troy.
Tell us what Sinon hath bewitched our ears,
Or who hath brought the fatal engine in 85
That gives our Troy, our Rome, the civil wound.
My heart is not compact of flint nor steel,
Nor can I utter all our bitter grief,
But floods of tears will drown my oratory
And break my utt'rance even in the time 90

When it should move ye to attend me most,
And force you to commiseration.
Here's Rome's young captain. Let him tell the tale,
While I stand by and weep to hear him speak.

LUCIUS
Then, gracious auditory, be it known to you            95
That Chiron and the damnèd Demetrius
Were they that murderèd our Emperor's brother,
And they it were that ravishèd our sister.
For their fell faults our brothers were beheaded,
Our father's tears despised, and basely cozened        100
Of that true hand that fought Rome's quarrel out
And sent her enemies unto the grave.
Lastly myself, unkindly banishèd,
The gates shut on me, and turned weeping out
To beg relief among Rome's enemies,                    105
Who drowned their enmity in my true tears
And oped their arms to embrace me as a friend.
I am the turned-forth, be it known to you,
That have preserved her welfare in my blood,
And from her bosom took the enemy's point,             110
Sheathing the steel in my advent'rous body.
Alas, you know I am no vaunter, I.
My scars can witness, dumb although they are,
That my report is just and full of truth.
But soft, methinks I do digress too much,              115
Citing my worthless praise. O, pardon me,
For when no friends are by, men praise themselves.

MARCUS
Now is my turn to speak. Behold the child.
Of this was Tamora deliverèd,
The issue of an irreligious Moor,                      120
Chief architect and plotter of these woes.
The villain is alive in Titus' house,
And as he is to witness, this is true.
Now judge what cause had Titus to revenge
These wrongs unspeakable, past patience,               125
Or more than any living man could bear.
Now have you heard the truth. What say you,
          Romans?
Have we done aught amiss, show us wherein,
And from the place where you behold us pleading
The poor remainder of Andronici                        130
Will hand in hand all headlong hurl ourselves
And on the ragged stones beat forth our souls
And make a mutual closure of our house.
Speak, Romans, speak, and if you say we shall,
Lo, hand in hand Lucius and I will fall.               135

AEMILIUS
Come, come, thou reverend man of Rome,
And bring our emperor gently in thy hand,
Lucius, our emperor—for well I know
The common voice do cry it shall be so.

ROMANS
Lucius, all hail, Rome's royal emperor!                140

MARCUS (to attendants)
Go, go into old Titus' sorrowful house
And hither hale that misbelieving Moor
To be adjudged some direful slaught'ring death
As punishment for his most wicked life.     Exeunt some
          ⌐Lucius, Marcus, and the others come down⌐
⌐ROMANS⌐
Lucius, all hail, Rome's gracious governor!            145

LUCIUS
Thanks, gentle Romans. May I govern so

To heal Rome's harms and wipe away her woe.
But, gentle people, give me aim awhile,
For nature puts me to a heavy task.
Stand all aloof, but, uncle, draw you near             150
To shed obsequious tears upon this trunk.
(Kissing Titus) O, take this warm kiss on thy pale cold
          lips,
These sorrowful drops upon thy bloodstained face,
The last true duties of thy noble son.

MARCUS (kissing Titus)
Tear for tear, and loving kiss for kiss,               155
Thy brother Marcus tenders on thy lips.
O, were the sum of these that I should pay
Countless and infinite, yet would I pay them.

LUCIUS (to young Lucius)
Come hither, boy, come, come, and learn of us
To melt in showers. Thy grandsire loved thee well.
Many a time he danced thee on his knee,                161
Sung thee asleep, his loving breast thy pillow.
Many a story hath he told to thee,
And bid thee bear his pretty tales in mind,
And talk of them when he was dead and gone.            165

MARCUS
How many thousand times hath these poor lips,
When they were living, warmed themselves on thine!
O now, sweet boy, give them their latest kiss.
Bid him farewell. Commit him to the grave.
Do them that kindness, and take leave of them.         170

YOUNG LUCIUS (kissing Titus)
O grandsire, grandsire, ev'n with all my heart
Would I were dead, so you did live again.
O Lord, I cannot speak to him for weeping.
My tears will choke me if I ope my mouth.
          Enter some with Aaron

A ROMAN
You sad Andronici, have done with woes.                175
Give sentence on this execrable wretch
That hath been breeder of these dire events.

LUCIUS
Set him breast-deep in earth and famish him.
There let him stand, and rave, and cry for food.
If anyone relieves or pities him,                      180
For the offence he dies. This is our doom.
Some stay to see him fastened in the earth.

AARON
Ah, why should wrath be mute and fury dumb?
I am no baby, I, that with base prayers
I should repent the evils I have done.                 185
Ten thousand worse than ever yet I did
Would I perform if I might have my will.
If one good deed in all my life I did
I do repent it from my very soul.

LUCIUS
Some loving friends convey the Emperor hence,          190
And give him burial in his father's grave.
My father and Lavinia shall forthwith
Be closèd in our household's monument.
As for that ravenous tiger, Tamora,
No funeral rite nor man in mourning weed,              195
No mournful bell shall ring her burial;
But throw her forth to beasts and birds to prey.
Her life was beastly and devoid of pity,
And being dead, let birds on her take pity.
          Exeunt with the bodies

## ADDITIONAL PASSAGES

### A. AFTER 1.1.35

The following passage, found in the First Quarto following a comma after 'field' but not included in the Second or Third Quartos or the Folio, conflicts with the subsequent action and presumably should have been deleted. (In the second line, Q1 reads 'of that' for 'of the'.)

> and at this day
> To the monument of the Andronici
> Done sacrifice of expiation,
> And slain the noblest prisoner of the Goths.

### B. AFTER 1.1.283

The following passage found in the quartos and the Folio is difficult to reconcile with the apparent need for Saturninus and his party to leave the stage at 275.1–2 before entering 'above' at 294.2–4. It is omitted from our text in the belief that Shakespeare intended it to be deleted after adding the episode of Mutius' killing to his original draft, and that the printers of Q1 included it by accident.

[TITUS]
  Treason, my lord! Lavinia is surprised.
SATURNINUS
  Surprised, by whom?
BASSIANUS         By him that justly may
  Bear his betrothed from all the world away.

### C. AFTER 4.3.93

The following lines, found in the early texts, appear to be a draft of the subsequent six lines.

MARCUS (*to Titus*) Why, sir, that is as fit as can be to serve for your oration, and let him deliver the pigeons to the Emperor from you.
TITUS (*to the Clown*) Tell me, can you deliver an oration to the Emperor with a grace?
CLOWN Nay, truly, sir, I could never say grace in all my life.

# HENRY VI PART ONE

## BY WILLIAM SHAKESPEARE AND OTHERS

THE play printed here first appeared in the 1623 Folio, as *The First Part of Henry VI*; it tells the beginning of the story that is continued in *The First Part of the Contention* and in *Richard Duke of York*. Although in narrative sequence it belongs before those plays, there is good reason to believe that it was written after them. It is probably the 'new' play referred to as 'harey the vj' in the record of its performance on 3 March 1592 by Lord Strange's Men. The box-office takings of £3 16s. 8d. were a record for the season, and the play was acted another fifteen times during the following ten months. Its success is mentioned in Thomas Nashe's satirical pamphlet *Piers Penniless*, published later in 1592. Defending the drama against moralistic attacks, Nashe claims that plays based on 'our English chronicles' celebrate 'our forefathers' valiant acts' and set them up as a 'reproof to these degenerate effeminate days of ours'. By way of illustration he alludes specifically to the exploits of Lord Talbot, the principal English warrior in *Henry VI Part One*: 'How would it have joyed brave Talbot, the terror of the French, to think that after he had lain two hundred years in his tomb he should triumph again on the stage, and have his bones new-embalmed with the tears of ten thousand spectators at least, at several times, who in the tragedian that represents his person imagine they behold him fresh bleeding!' Nashe may have had personal reasons to puff this play: a variety of evidence suggests that Shakespeare wrote it in collaboration with at least two other authors; Nashe himself was probably responsible for Act 1. The passages most confidently attributed to Shakespeare are Act 2, Scene 4 and Act 4, Scene 2 to the death of Talbot at 4.7.32.

A mass of material, some derived from 'English chronicles', some invented, is packed into this play. It opens impressively with the funeral of Henry V, celebrated for unifying England and subjugating France; but his nobles are at loggerheads even over his coffin, and news rapidly arrives of serious losses in France. The rivalry displayed here between Humphrey, Duke of Gloucester—Protector of the infant Henry VI—and Henry Beaufort, Bishop of Winchester, plays an important part in both this play and *The Contention*, as does the conflict between Richard, Duke of York, and the houses of Somerset and Suffolk; in the Temple Garden scene (2.4), invented by Shakespeare, York's and Somerset's supporters symbolize their respective loyalties by plucking white and red roses. Their dissension weakens England's military strength, but she has a great hero in Lord Talbot, whose nobility as a warrior is pitted against the treachery of the French, led by King Charles and Joan la Pucelle (Joan of Arc), here—following the chronicles—portrayed as a witch and a whore. Historical facts are freely manipulated: Joan was burnt in 1431, though the play's authors have her take part in a battle of 1451 in which Talbot's death is brought forward by two years. The play ends with an uneasy peace between England and France.

# THE PERSONS OF THE PLAY

### The English

KING Henry VI

Duke of GLOUCESTER, Lord Protector, uncle of King Henry

Duke of BEDFORD, Regent of France

Duke of EXETER

Bishop of WINCHESTER (later Cardinal), uncle of King Henry

Duke of SOMERSET

RICHARD PLANTAGENET, later DUKE OF YORK, and Regent of France

Earl of WARWICK

Earl of SALISBURY

Earl of SUFFOLK

Lord TALBOT

JOHN Talbot

Edmund MORTIMER

Sir William GLASDALE

Sir Thomas GARGRAVE

Sir John FASTOLF

Sir William LUCY

WOODVILLE, Lieutenant of the Tower of London

MAYOR of London

VERNON

BASSET

A LAWYER

A LEGATE

Messengers, warders and keepers of the Tower of London, servingmen, officers, captains, soldiers, herald, watch

### The French

CHARLES, Dauphin of France

RENÉ, Duke of Anjou, King of Naples

MARGARET, his daughter

Duke of ALENÇON

BASTARD of Orléans

Duke of BURGUNDY, uncle of King Henry

GENERAL of the French garrison at Bordeaux

COUNTESS of Auvergne

MASTER GUNNER of Orléans

A BOY, his son

JOAN la Pucelle

A SHEPHERD, father of Joan

Porter, French sergeant, French sentinels, French scout, French herald, the Governor of Paris, fiends, and soldiers

# The First Part of Henry the Sixth

**1.1** *Dead march. Enter the funeral of King Henry the Fifth, attended on by the Duke of Bedford (Regent of France), the Duke of Gloucester (Protector), the Duke of Exeter, the Earl of Warwick, the Bishop of Winchester, and the Duke of Somerset*

BEDFORD
Hung be the heavens with black! Yield, day, to night!
Comets, importing change of times and states,
Brandish your crystal tresses in the sky,
And with them scourge the bad revolting stars
That have consented unto Henry's death—          5
King Henry the Fifth, too famous to live long.
England ne'er lost a king of so much worth.

GLOUCESTER
England ne'er had a king until his time.
Virtue he had, deserving to command.
His brandished sword did blind men with his beams.
His arms spread wider than a dragon's wings.          11
His sparkling eyes, replete with wrathful fire,
More dazzled and drove back his enemies
Than midday sun, fierce bent against their faces.
What should I say? His deeds exceed all speech.          15
He ne'er lift up his hand but conquerèd.

EXETER
We mourn in black; why mourn we not in blood?
Henry is dead, and never shall revive.
Upon a wooden coffin we attend,
And death's dishonourable victory          20
We with our stately presence glorify,
Like captives bound to a triumphant car.
What, shall we curse the planets of mishap,
That plotted thus our glory's overthrow?
Or shall we think the subtle-witted French          25
Conjurers and sorcerers, that, afraid of him,
By magic verses have contrived his end?

WINCHESTER
He was a king blest of the King of Kings.
Unto the French, the dreadful judgement day
So dreadful will not be as was his sight.          30
The battles of the Lord of Hosts he fought.
The Church's prayers made him so prosperous.

GLOUCESTER
The Church? Where is it? Had not churchmen prayed,
His thread of life had not so soon decayed.
None do you like but an effeminate prince,          35
Whom like a schoolboy you may overawe.

WINCHESTER
Gloucester, whate'er we like, thou art Protector,
And lookest to command the Prince and realm.
Thy wife is proud: she holdeth thee in awe,
More than God or religious churchmen may.          40

GLOUCESTER
Name not religion, for thou lov'st the flesh,
And ne'er throughout the year to church thou go'st,
Except it be to pray against thy foes.

BEDFORD
Cease, cease these jars, and rest your minds in peace.
Let's to the altar. Heralds, wait on us.          45
⌜*Exeunt Warwick, Somerset, and heralds with coffin*⌝

Instead of gold, we'll offer up our arms—
Since arms avail not, now that Henry's dead.
Posterity, await for wretched years,
When, at their mothers' moistened eyes, babes shall suck,
Our isle be made a marish of salt tears,          50
And none but women left to wail the dead.
Henry the Fifth, thy ghost I invocate:
Prosper this realm; keep it from civil broils;
Combat with adverse planets in the heavens.
A far more glorious star thy soul will make          55
Than Julius Caesar or bright—
*Enter a Messenger*

MESSENGER
My honourable lords, health to you all.
Sad tidings bring I to you out of France,
Of loss, of slaughter, and discomfiture.
Guyenne, Compiègne, Rouen, Rheims, Orléans,          60
Paris, Gisors, Poitiers are all quite lost.

BEDFORD
What sayst thou, man, before dead Henry's corpse?
Speak softly, or the loss of those great towns
Will make him burst his lead and rise from death.

GLOUCESTER (*to the Messenger*)
Is Paris lost? Is Rouen yielded up?          65
If Henry were recalled to life again,
These news would cause him once more yield the ghost.

EXETER (*to the Messenger*)
How were they lost? What treachery was used?

MESSENGER
No treachery, but want of men and money.
Amongst the soldiers this is mutterèd:          70
That here you maintain several factions,
And whilst a field should be dispatched and fought,
You are disputing of your generals.
One would have ling'ring wars, with little cost;
Another would fly swift, but wanteth wings;          75
A third thinks, without expense at all,
By guileful fair words peace may be obtained.
Awake, awake, English nobility!
Let not sloth dim your honours new-begot.
Cropped are the flower-de-luces in your arms;          80
Of England's coat, one half is cut away.          ⌜*Exit*⌝

EXETER
Were our tears wanting to this funeral,
These tidings would call forth her flowing tides.

BEDFORD
Me they concern; Regent I am of France.
Give me my steelèd coat. I'll fight for France.          85
Away with these disgraceful wailing robes!
⌜*He removes his mourning robe*⌝
Wounds will I lend the French, instead of eyes,
To weep their intermissive miseries.
*Enter to them another Messenger with letters*

SECOND MESSENGER
Lords, view these letters, full of bad mischance.
France is revolted from the English quite,          90
Except some petty towns of no import.

The Dauphin Charles is crownèd king in Rheims;
The Bastard of Orléans with him is joined;
René, Duke of Anjou, doth take his part;
The Duke of Alençon flyeth to his side.     *Exit*
EXETER
The Dauphin crownèd King? All fly to him?     96
O whither shall *we* fly from this reproach?
GLOUCESTER
We will not fly, but to our enemies' throats.
Bedford, if thou be slack, I'll fight it out.
BEDFORD
Gloucester, why doubt'st thou of my forwardness?  100
An army have I mustered in my thoughts,
Wherewith already France is overrun.
    *Enter another Messenger*
THIRD MESSENGER
My gracious lords, to add to your laments,
Wherewith you now bedew King Henry's hearse,
I must inform you of a dismal fight     105
Betwixt the stout Lord Talbot and the French.
WINCHESTER
What, wherein Talbot overcame—is't so?
THIRD MESSENGER
O no, wherein Lord Talbot was o'erthrown.
The circumstance I'll tell you more at large.
The tenth of August last, this dreadful lord,   110
Retiring from the siege of Orléans,
Having full scarce six thousand in his troop,
By three-and-twenty thousand of the French
Was round encompassèd and set upon.
No leisure had he to enrank his men.     115
He wanted pikes to set before his archers—
Instead whereof, sharp stakes plucked out of hedges
They pitchèd in the ground confusèdly,
To keep the horsemen off from breaking in.
More than three hours the fight continuèd,   120
Where valiant Talbot above human thought
Enacted wonders with his sword and lance.
Hundreds he sent to hell, and none durst stand him;
Here, there, and everywhere, enraged he slew.
The French exclaimed the devil was in arms:   125
All the whole army stood agazed on him.
His soldiers, spying his undaunted spirit,
'A Talbot! A Talbot!' cried out amain,
And rushed into the bowels of the battle.
Here had the conquest fully been sealed up,   130
If Sir John Fastolf had not played the coward.
He, being in the vanguard placed behind,
With purpose to relieve and follow them,
Cowardly fled, not having struck one stroke.
Hence grew the general wrack and massacre.   135
Enclosèd were they with their enemies.
A base Walloon, to win the Dauphin's grace,
Thrust Talbot with a spear into the back—
Whom all France, with their chief assembled strength,
Durst not presume to look once in the face.   140
BEDFORD
Is Talbot slain then? I will slay myself,
For living idly here in pomp and ease
Whilst such a worthy leader, wanting aid,
Unto his dastard foemen is betrayed.
THIRD MESSENGER
O no, he lives, but is took prisoner,     145
And Lord Scales with him, and Lord Hungerford;
Most of the rest slaughtered, or took likewise.

BEDFORD
His ransom there is none but I shall pay.
I'll hale the Dauphin headlong from his throne;
His crown shall be the ransom of my friend.   150
Four of their lords I'll change for one of ours.
Farewell, my masters; to my task will I.
Bonfires in France forthwith I am to make,
To keep our great Saint George's feast withal.
Ten thousand soldiers with me I will take,   155
Whose bloody deeds shall make all Europe quake.
THIRD MESSENGER
So you had need. Fore Orléans, besieged,
The English army is grown weak and faint.
The Earl of Salisbury craveth supply,
And hardly keeps his men from mutiny,   160
Since they, so few, watch such a multitude.   ⌈*Exit*⌉
EXETER
Remember, lords, your oaths to Henry sworn:
Either to quell the Dauphin utterly,
Or bring him in obedience to your yoke.
BEDFORD
I do remember it, and here take my leave   165
To go about my preparation.     *Exit*
GLOUCESTER
I'll to the Tower with all the haste I can,
To view th'artillery and munition,
And then I will proclaim young Henry king.   *Exit*
EXETER
To Eltham will I, where the young King is,   170
Being ordained his special governor,
And for his safety there I'll best devise.   *Exit*
WINCHESTER
Each hath his place and function to attend;
I am left out; for me, nothing remains.
But long I will not be Jack-out-of-office.   175
The King from Eltham I intend to steal,
And sit at chiefest stern of public weal.   *Exit*

1.2     *Sound a flourish. Enter Charles the Dauphin, the*
    *Duke of Alençon, and René Duke of Anjou,*
    *marching with drummer and soldiers*
CHARLES
Mars his true moving—even as in the heavens,
So in the earth—to this day is not known.
Late did he shine upon the English side;
Now we are victors: upon us he smiles.
What towns of any moment but we have?   5
At pleasure here we lie near Orléans
Otherwhiles the famished English, like pale ghosts,
Faintly besiege us one hour in a month.
ALENÇON
They want their porrage and their fat bull beeves.
Either they must be dieted like mules,   10
And have their provender tied to their mouths,
Or piteous they will look, like drownèd mice.
RENÉ
Let's raise the siege. Why live we idly here?
Talbot is taken, whom we wont to fear.
Remaineth none but mad-brained Salisbury,   15
And he may well in fretting spend his gall:
Nor men nor money hath he to make war.
CHARLES
Sound, sound, alarum! We will rush on them.
Now for the honour of the forlorn French,
Him I forgive my death that killeth me   20
When he sees me go back one foot or flee.   *Exeunt*

**1.3**  *Here alarum. The French are beaten back by the*
*English with great loss. Enter Charles the Dauphin,*
*the Duke of Alençon, and René Duke of Anjou*

CHARLES
Who ever saw the like? What men have I?
Dogs, cowards, dastards! I would ne'er have fled,
But that they left me 'midst my enemies.

RENÉ
Salisbury is a desperate homicide.
He fighteth as one weary of his life.                                        5
The other lords, like lions wanting food,
Do rush upon us as their hungry prey.

ALENÇON
Froissart, a countryman of ours, records
England all Olivers and Rolands bred
During the time Edward the Third did reign.                     10
More truly now may this be verified,
For none but Samsons and Goliases
It sendeth forth to skirmish. One to ten?
Lean raw-boned rascals, who would e'er suppose
They had such courage and audacity?                               15

CHARLES
Let's leave this town, for they are hare-brained slaves,
And hunger will enforce them to be more eager.
Of old I know them: rather with their teeth
The walls they'll tear down, than forsake the siege.

RENÉ
I think by some odd gimmers or device                             20
Their arms are set, like clocks, still to strike on,
Else ne'er could they hold out so as they do.
By my consent we'll even let them alone.

ALENÇON Be it so.
*Enter the Bastard of Orléans*

BASTARD
Where's the Prince Dauphin? I have news for him.   25

CHARLES
Bastard of Orléans, thrice welcome to us.

BASTARD
Methinks your looks are sad, your cheer appalled.
Hath the late overthrow wrought this offence?
Be not dismayed, for succour is at hand.
A holy maid hither with me I bring,                                  30
Which, by a vision sent to her from heaven,
Ordainèd is to raise this tedious siege
And drive the English forth the bounds of France.
The spirit of deep prophecy she hath,
Exceeding the nine sibyls of old Rome.                            35
What's past and what's to come she can descry.
Speak: shall I call her in? Believe my words,
For they are certain and unfallible.

CHARLES
Go call her in.                                    *Exit Bastard*
                    But first, to try her skill,
René stand thou as Dauphin in my place.                        40
Question her proudly; let thy looks be stern.
By this means shall we sound what skill she hath.
*Enter ⌐the Bastard of Orléans with¬ Joan la Pucelle,*
*armed*

RENÉ (*as Charles*)
Fair maid, is't thou wilt do these wondrous feats?

JOAN
René, is't thou that thinkest to beguile me?
Where is the Dauphin? (*To Charles*) Come, come from
behind.                                                                          45
I know thee well, though never seen before.
Be not amazed. There's nothing hid from me.

In private will I talk with thee apart.
Stand back you lords, and give us leave awhile.
*René, Alençon ⌐and Bastard¬ stand apart*

RENÉ ⌐*to Alençon and Bastard*¬
She takes upon her bravely, at first dash.                        50

JOAN
Dauphin, I am by birth a shepherd's daughter,
My wit untrained in any kind of art.
Heaven and our Lady gracious hath it pleased
To shine on my contemptible estate.
Lo, whilst I waited on my tender lambs,                           55
And to sun's parching heat displayed my cheeks,
God's mother deignèd to appear to me,
And in a vision, full of majesty,
Willed me to leave my base vocation
And free my country from calamity.                                 60
Her aid she promised, and assured success.
In complete glory she revealed herself—
And whereas I was black and swart before,
With those clear rays which she infused on me
That beauty am I blest with, which you may see.            65
Ask me what question thou canst possible,
And I will answer unpremeditated.
My courage try by combat, if thou dar'st,
And thou shalt find that I exceed my sex.
Resolve on this: thou shalt be fortunate,                         70
If thou receive me for thy warlike mate.

CHARLES
Thou hast astonished me with thy high terms.
Only this proof I'll of thy valour make:
In single combat thou shalt buckle with me.
An if thou vanquishest, thy words are true;                    75
Otherwise, I renounce all confidence.

JOAN
I am prepared. Here is my keen-edged sword,
Decked with five flower-de-luces on each side—
The which at Touraine, in Saint Katherine's
  churchyard,
Out of a great deal of old iron I chose forth.                   80

CHARLES
Then come a God's name. I fear no woman.

JOAN
And while I live, I'll ne'er fly from a man.
*Here they fight and Joan la Pucelle overcomes*

CHARLES
Stay, stay thy hands! Thou art an Amazon,
And fightest with the sword of Deborah.

JOAN
Christ's mother helps me, else I were too weak.           85

CHARLES
Whoe'er helps thee, 'tis thou that must help me.
Impatiently I burn with thy desire.
My heart and hands thou hast at once subdued.
Excellent Pucelle if thy name be so,
Let me thy servant, and not sovereign be.                       90
'Tis the French Dauphin sueth to thee thus.

JOAN
I must not yield to any rites of love,
For my profession's sacred from above.
When I have chasèd all thy foes from hence,
Then will I think upon a recompense.                              95

CHARLES
Meantime, look gracious on thy prostrate thrall.

RENÉ ⌐*to the other lords apart*¬
My lord, methinks, is very long in talk.

ALENÇON
  Doubtless he shrives this woman to her smock,
  Else ne'er could he so long protract his speech.
RENÉ
  Shall we disturb him, since he keeps no mean?                100
ALENÇON
  He may mean more than we poor men do know.
  These women are shrewd tempters with their tongues.
RENÉ (to Charles)
  My lord, where are you? What devise you on?
  Shall we give o'er Orléans, or no?
JOAN
  Why, no, I say. Distrustful recreants,                      105
  Fight till the last gasp; I'll be your guard.
CHARLES
  What she says, I'll confirm. We'll fight it out.
JOAN
  Assigned am I to be the English scourge.
  This night the siege assurèdly I'll raise.
  Expect Saint Martin's summer, halcyon's days,               110
  Since I have entered into these wars.
  Glory is like a circle in the water,
  Which never ceaseth to enlarge itself
  Till, by broad spreading, it disperse to naught.
  With Henry's death, the English circle ends.                115
  Dispersèd are the glories it included.
  Now am I like that proud insulting ship
  Which Caesar and his fortune bore at once.
CHARLES
  Was Mohammed inspirèd with a dove?
  Thou with an eagle art inspirèd then.                       120
  Helen, the mother of great Constantine,
  Nor yet Saint Philip's daughters were like thee.
  Bright star of Venus, fall'n down on the earth,
  How may I reverently worship thee enough?
ALENÇON
  Leave off delays, and let us raise the siege.               125
RENÉ
  Woman, do what thou canst to save our honours.
  Drive them from Orléans, and be immortalized.
CHARLES
  Presently we'll try. Come, let's away about it.
  No prophet will I trust, if she prove false.        Exeunt

1.4    *Enter the Duke of Gloucester, with his Servingmen*
       *in blue coats*
GLOUCESTER
  I am come to survey the Tower this day.
  Since Henry's death, I fear there is conveyance.
  Where be these warders, that they wait not here?
       *⌈A Servingman⌉ knocketh on the gates*
  Open the gates: 'tis Gloucester that calls.
FIRST WARDER ⌈*within the Tower*⌉
  Who's there that knocketh so imperiously?                   5
GLOUCESTER'S FIRST MAN
  It is the noble Duke of Gloucester.
SECOND WARDER ⌈*within the Tower*⌉
  Whoe'er he be, you may not be let in.
GLOUCESTER'S FIRST MAN
  Villains, answer you so the Lord Protector?
FIRST WARDER ⌈*within the Tower*⌉
  The Lord protect him, so we answer him.
  We do no otherwise than we are willed.                      10
GLOUCESTER
  Who willed you? Or whose will stands, but mine?
  There's none Protector of the realm but I.

(*To Servingmen*) Break up the gates. I'll be your
  warrantize.
  Shall I be flouted thus by dunghill grooms?
       *Gloucester's men rush at the Tower gates*
WOODVILLE ⌈*within the Tower*⌉
  What noise is this? What traitors have we here?             15
GLOUCESTER
  Lieutenant, is it you whose voice I hear?
  Open the gates! Here's Gloucester, that would enter.
WOODVILLE ⌈*within the Tower*⌉
  Have patience, noble duke: I may not open.
  My lord of Winchester forbids.
  From him I have express commandëment                        20
  That thou, nor none of thine, shall be let in.
GLOUCESTER
  Faint-hearted Woodville! Prizest him fore me?—
  Arrogant Winchester, that haughty prelate,
  Whom Henry, our late sovereign, ne'er could brook?
  Thou art no friend to God or to the King.                   25
  Open the gates, or I'll shut thee out shortly.
SERVINGMEN
  Open the gates unto the Lord Protector,
  Or we'll burst them open, if that you come not
  quickly.
       *Enter, to the Lord Protector at the Tower gates, the*
       *Bishop of Winchester and his men in tawny coats*
WINCHESTER
  How now, ambitious vizier! What means this?
GLOUCESTER
  Peeled priest, dost thou command me to be shut out?
WINCHESTER
  I do, thou most usurping proditor,                          31
  And not 'Protector', of the King or realm.
GLOUCESTER
  Stand back, thou manifest conspirator.
  Thou that contrived'st to murder our dead lord,
  Thou that giv'st whores indulgences to sin,                 35
  If thou proceed in this thy insolence—
WINCHESTER
  Nay, stand thou back! I will not budge a foot.
  This be Damascus, be thou cursèd Cain,
  To slay thy brother Abel, if thou wilt.
GLOUCESTER
  I will not slay thee, but I'll drive thee back.             40
  Thy purple robes, as a child's bearing-cloth,
  I'll use to carry thee out of this place.
WINCHESTER
  Do what thou dar'st, I beard thee to thy face.
GLOUCESTER
  What, am I dared and bearded to my face?
  Draw, men, for all this privilegèd place.                   45
       *All draw their swords*
  Blue coats to tawny coats!—Priest, beware your
  beard.
  I mean to tug it, and to cuff you soundly.
  Under my feet I'll stamp thy bishop's mitre.
  In spite of Pope, or dignities of church,
  Here by the cheeks I'll drag thee up and down.              50
WINCHESTER
  Gloucester, thou wilt answer this before the Pope.
GLOUCESTER
  Winchester goose! I cry, 'A rope, a rope!'
  (*To his Servingmen*)
  Now beat them hence. Why do you let them stay?
  (*To Winchester*)
  Thee I'll chase hence, thou wolf in sheep's array.
  Out, tawny coats! Out, cloakèd hypocrite!                   55

*Here Gloucester's men beat out the Bishop's men.*
*Enter in the hurly-burly the Mayor of London and*
*his Officers*

MAYOR
Fie, lords!—that you, being supreme magistrates,
Thus contumeliously should break the peace.

GLOUCESTER
Peace, mayor, thou know'st little of my wrongs.
Here's Beaufort—that regards nor God nor king—
Hath here distrained the Tower to his use.                    60

WINCHESTER (*to Mayor*)
Here's Gloucester—a foe to citizens,
One that still motions war, and never peace,
O'ercharging your free purses with large fines—
That seeks to overthrow religion,
Because he is Protector of the realm,                          65
And would have armour here out of the Tower
To crown himself king and suppress the Prince.

GLOUCESTER
I will not answer thee with words but blows.
*Here the factions skirmish again*

MAYOR
Naught rests for me, in this tumultuous strife,
But to make open proclamation.                                 70
Come, officer, as loud as e'er thou canst, cry.

OFFICER All manner of men, assembled here in arms this
day against God's peace and the King's, we charge and
command you in his highness' name to repair to your
several dwelling places, and not to wear, handle, or
use any sword, weapon, or dagger henceforward, upon
pain of death.                                                 77
*The skirmishes cease*

GLOUCESTER
Bishop, I'll be no breaker of the law.
But we shall meet and break our minds at large.

WINCHESTER
Gloucester, we'll meet to thy cost, be sure.                   80
Thy heart-blood I will have for this day's work.

MAYOR
I'll call for clubs, if you will not away.
(*Aside*) This bishop is more haughty than the devil.

GLOUCESTER
Mayor, farewell. Thou dost but what thou mayst.

WINCHESTER
Abominable Gloucester, guard thy head,                         85
For I intend to have it ere long.
*Exeunt both factions severally*

MAYOR (*to Officers*)
See the coast cleared, and then we will depart.—
Good God, these nobles should such stomachs bear!
I myself fight not once in forty year.            *Exeunt*

**1.5**   *Enter the Master Gunner of Orléans with his Boy*

MASTER GUNNER
Sirrah, thou know'st how Orléans is besieged,
And how the English have the suburbs won.

BOY
Father, I know, and oft have shot at them;
Howe'er, unfortunate, I missed my aim.

MASTER GUNNER
But now thou shalt not. Be thou ruled by me.              5
Chief Master Gunner am I of this town;
Something I must do to procure me grace.
The Prince's spials have informèd me
How the English, in the suburbs close entrenched,
Wont, through a secret grate of iron bars              10

In yonder tower, to overpeer the city,
And thence discover how with most advantage
They may vex us with shot or with assault.
To intercept this inconvenience,
A piece of ordnance 'gainst it I have placed,           15
And even these three days have I watched, if I could
see them.
Now do thou watch, for I can stay no longer.
If thou spy'st any, run and bring me word,
And thou shalt find me at the governor's.

BOY
Father, I warrant you, take you no care—                20
*[Exit Master Gunner at one door]*
I'll never trouble you, if I may spy them.
*Exit [at the other door]*

**1.6**   *Enter the Earl of Salisbury and Lord Talbot above*
*on the turrets with others, among them Sir*
*Thomas Gargrave and Sir William Glasdale*

SALISBURY
Talbot, my life, my joy, again returned?
How wert thou handled, being prisoner?
Or by what means got'st thou to be released?
Discourse, I prithee, on this turret's top.

TALBOT
The Duke of Bedford had a prisoner,                      5
Called the brave Lord Ponton de Santrailles;
For him was I exchanged and ransomèd.
But with a baser man-of-arms by far
Once in contempt they would have bartered me—
Which I, disdaining, scorned, and cravèd death          10
Rather than I would be so pilled esteemed.
In fine, redeemed I was, as I desired.
But O, the treacherous Fastolf wounds my heart,
Whom with my bare fists I would execute
If I now had him brought into my power.                  15

SALISBURY
Yet tell'st thou not how thou wert entertained.

TALBOT
With scoffs and scorns and contumelious taunts.
In open market place produced they me,
To be a public spectacle to all.
'Here', said they, 'is the terror of the French,        20
The scarecrow that affrights our children so.'
Then broke I from the officers that led me
And with my nails digged stones out of the ground
To hurl at the beholders of my shame.
My grisly countenance made others fly.                   25
None durst come near, for fear of sudden death.
In iron walls they deemed me not secure:
So great fear of my name 'mongst them were spread
That they supposed I could rend bars of steel
And spurn in pieces posts of adamant.                    30
Wherefore a guard of chosen shot I had
That walked about me every minute while;
And if I did but stir out of my bed,
Ready they were to shoot me to the heart.
*The Boy [passes over the stage] with a linstock*

SALISBURY
I grieve to hear what torments you endured.              35
But we will be revenged sufficiently.
Now it is supper time in Orléans.
Here, through this grate, I count each one,
And view the Frenchmen how they fortify.
Let us look in: the sight will much delight thee.—       40

Sir Thomas Gargrave and Sir William Glasdale,
Let me have your express opinions
Where is best place to make our batt'ry next.
    ⌈*They look through the grate*⌉

GARGRAVE
I think at the north gate, for there stands Lou.

GLASDALE
And I here, at the bulwark of the Bridge.    45

TALBOT
For aught I see, this city must be famished
Or with light skirmishes enfeeblèd.
    *Here they shoot off chambers* ⌈*within*⌉ *and Salisbury*
    *and Gargrave fall down*

SALISBURY
O Lord have mercy on us, wretched sinners!

GARGRAVE
O Lord have mercy on me, woeful man!

TALBOT
What chance is this that suddenly hath crossed us?
Speak, Salisbury—at least, if thou canst, speak.  51
How far'st thou, mirror of all martial men?
One of thy eyes and thy cheek's side struck off?
Accursèd tower! Accursèd fatal hand
That hath contrived this woeful tragedy!    55
In thirteen battles Salisbury o'ercame;
Henry the Fifth he first trained to the wars;
Whilst any trump did sound or drum struck up
His sword did ne'er leave striking in the field.
Yet liv'st thou, Salisbury? Though thy speech doth
    fail,    60
One eye thou hast to look to heaven for grace.
The sun with one eye vieweth all the world.
Heaven, be thou gracious to none alive
If Salisbury wants mercy at thy hands.—
Sir Thomas Gargrave, hast thou any life?    65
Speak unto Talbot. Nay, look up to him.—
Bear hence his body; I will help to bury it.
    ⌈*Exit one with Gargrave's body*⌉
Salisbury, cheer thy spirit with this comfort:
Thou shalt not die whiles—
He beckons with his hand, and smiles on me,    70
As who should say, 'When I am dead and gone,
Remember to avenge me on the French.'
Plantagenet, I will—and like thee, Nero,
Play on the lute, beholding the towns burn.
Wretched shall France be only in my name.    75
    *Here an alarum, and it thunders and lightens*
What stir is this? What tumult's in the heavens?
Whence cometh this alarum and the noise?
    *Enter a Messenger*

MESSENGER
My lord, my lord, the French have gathered head.
The Dauphin, with one Joan la Pucelle joined,
A holy prophetess new risen up,    80
Is come with a great power to raise the siege.
    *Here Salisbury lifteth himself up and groans*

TALBOT
Hear, hear, how dying Salisbury doth groan!
It irks his heart he cannot be revenged.
Frenchmen, I'll be a Salisbury to you.
*Pucelle* or pucelle, Dauphin or dog-fish,    85
Your hearts I'll stamp out with my horse's heels
And make a quagmire of your mingled brains.—
Convey me Salisbury into his tent,
And then we'll try what these dastard Frenchmen
    dare.    *Alarum. Exeunt carrying Salisbury*

**1.7**    *Here an alarum again, and Lord Talbot pursueth the*
    *Dauphin and driveth him. Then enter Joan la*
    *Pucelle driving Englishmen before her and* ⌈*exeunt*⌉.
    *Then enter Lord Talbot*

TALBOT
Where is my strength, my valour, and my force?
Our English troops retire; I cannot stay them.
A woman clad in armour chaseth men.
    *Enter Joan la Pucelle*
Here, here she comes. (*To Joan*) I'll have a bout with
    thee.
Devil or devil's dam, I'll conjure thee.    5
Blood will I draw on thee—thou art a witch—
And straightway give thy soul to him thou serv'st.

JOAN
Come, come, 'tis only I that must disgrace thee.
    *Here they fight*

TALBOT
Heavens, can you suffer hell so to prevail?
My breast I'll burst with straining of my courage    10
And from my shoulders crack my arms asunder
But I will chastise this high-minded strumpet.
    *They fight again*

JOAN
Talbot, farewell. Thy hour is not yet come.
I must go victual Orléans forthwith.
    *A short alarum, then* ⌈*the French pass over the*
    *stage and*⌉ *enter the town with soldiers*
O'ertake me if thou canst. I scorn thy strength.    15
Go, go, cheer up thy hungry-starvèd men.
Help Salisbury to make his testament.
This day is ours, as many more shall be.
    *Exit into the town*

TALBOT
My thoughts are whirlèd like a potter's wheel.
I know not where I am nor what I do.    20
A witch by fear, not force, like Hannibal
Drives back our troops and conquers as she lists.
So bees with smoke and doves with noisome stench
Are from their hives and houses driven away.
They called us, for our fierceness, English dogs;    25
Now, like to whelps, we crying run away.
    *A short alarum.* ⌈*Enter English soldiers*⌉
Hark, countrymen: either renew the fight
Or tear the lions out of England's coat.
Renounce your style; give sheep in lions' stead.
Sheep run not half so treacherous from the wolf,    30
Or horse or oxen from the leopard,
As you fly from your oft-subduèd slaves.
    *Alarum. Here another skirmish*
It will not be. Retire into your trenches.
You all consented unto Salisbury's death,
For none would strike a stroke in his revenge.    35
    *Pucelle is entered into Orléans*
In spite of us or aught that we could do.
    ⌈*Exeunt Soldiers*⌉
O would I were to die with Salisbury!
The shame hereof will make me hide my head.
    *Exit. Alarum. Retreat*

**1.8**    *Flourish. Enter on the walls Joan la Pucelle, Charles*
    *the Dauphin, René Duke of Anjou, the Duke of*
    *Alençon and French Soldiers* ⌈*with colours*⌉

JOAN
Advance our waving colours on the walls;
Rescued is Orléans from the English.
Thus Joan la Pucelle hath performed her word.

CHARLES
Divinest creature, Astraea's daughter,
How shall I honour thee for this success?                    5
Thy promises are like Adonis' garden,
That one day bloomed and fruitful were the next.
France, triumph in thy glorious prophetess!
Recovered is the town of Orléans.
More blessèd hap did ne'er befall our state.                10

RENÉ
Why ring not out the bells aloud throughout the
    town?
Dauphin, command the citizens make bonfires
And feast and banquet in the open streets
To celebrate the joy that God hath given us.

ALENÇON
All France will be replete with mirth and joy              15
When they shall hear how we have played the men.

CHARLES
'Tis Joan, not we, by whom the day is won—
For which I will divide my crown with her,
And all the priests and friars in my realm
Shall in procession sing her endless praise.               20
A statelier pyramid to her I'll rear
Than Rhodope's of Memphis ever was.
In memory of her, when she is dead
Her ashes, in an urn more precious
Than the rich-jewelled coffer of Darius,                   25
Transported shall be at high festivals
Before the kings and queens of France.
No longer on Saint Denis will we cry,
But Joan la Pucelle shall be France's saint.
Come in, and let us banquet royally                        30
After this golden day of victory.       *Flourish. Exeunt*

**2.1**  *Enter ⌐on the walls⌐ a French Sergeant of a band,*
       *with two Sentinels*

SERGEANT
Sirs, take your places and be vigilant.
If any noise or soldier you perceive
Near to the walls, by some apparent sign
Let us have knowledge at the court of guard.

⌐A SENTINEL⌐
Sergeant, you shall.                           *Exit Sergeant*
       Thus are poor servitors,
When others sleep upon their quiet beds,                    6
Constrained to watch in darkness, rain, and cold.
       *Enter Lord Talbot, the Dukes of Bedford and*
       *Burgundy, and soldiers with scaling ladders, their*
       *drums beating a dead march*

TALBOT
Lord regent, and redoubted Burgundy—
By whose approach the regions of Artois,
Wallon, and Picardy are friends to us—                     10
This happy night the Frenchmen are secure,
Having all day caroused and banqueted.
Embrace we then this opportunity,
As fitting best to quittance their deceit,
Contrived by art and baleful sorcery.                      15

BEDFORD
Coward of France! How much he wrongs his fame,
Despairing of his own arms' fortitude,
To join with witches and the help of hell.

BURGUNDY
Traitors have never other company.
But what's that 'Pucelle' whom they term so pure? 20

TALBOT
A maid, they say.

BEDFORD            A maid? And be so martial?

BURGUNDY
Pray God she prove not masculine ere long.
If underneath the standard of the French
She carry armour as she hath begun—

TALBOT
Well, let them practise and converse with spirits.    25
God is our fortress, in whose conquering name
Let us resolve to scale their flinty bulwarks.

BEDFORD
Ascend, brave Talbot. We will follow thee.

TALBOT
Not all together. Better far, I guess,
That we do make our entrance several ways—            30
That, if it chance the one of us do fail,
The other yet may rise against their force.

BEDFORD
Agreed. I'll to yon corner.

BURGUNDY                     And I to this.
       ⌐*Exeunt severally Bedford and Burgundy with*
                                    *some soldiers*⌐

TALBOT
And here will Talbot mount, or make his grave.
Now, Salisbury, for thee, and for the right           35
Of English Henry, shall this night appear
How much in duty I am bound to both.
       ⌐*Talbot and his soldiers*⌐ *scale the walls*

⌐SENTINELS⌐
Arm! Arm! The enemy doth make assault!

ENGLISH SOLDIERS  Saint George! A Talbot!       *Exeunt above*
       ⌐*Alarum.*⌐ *The French* ⌐*soldiers*⌐ *leap o'er the walls*
       *in their shirts* ⌐*and exeunt*⌐. *Enter several ways the*
       *Bastard of Orléans, the Duke of Alençon, and René*
       *Duke of Anjou, half ready and half unready*

ALENÇON
How now, my lords? What, all unready so?              40

BASTARD
Unready? Ay, and glad we scaped so well.

RENÉ
'Twas time, I trow, to wake and leave our beds,
Hearing alarums at our chamber doors.

ALENÇON
Of all exploits since first I followed arms
Ne'er heard I of a warlike enterprise                 45
More venturous or desperate than this.

BASTARD
I think this Talbot be a fiend of hell.

RENÉ
If not of hell, the heavens sure favour him.

ALENÇON
Here cometh Charles. I marvel how he sped.
       *Enter Charles the Dauphin and Joan la Pucelle*

BASTARD
Tut, holy Joan was his defensive guard.               50

CHARLES (*to Joan*)
Is this thy cunning, thou deceitful dame?
Didst thou at first, to flatter us withal,
Make us partakers of a little gain
That now our loss might be ten times so much?

JOAN
Wherefore is Charles impatient with his friend?      55
At all times will you have my power alike?
Sleeping or waking must I still prevail,

Or will you blame and lay the fault on me?—
Improvident soldiers, had your watch been good,
This sudden mischief never could have fall'n.                60
CHARLES
Duke of Alençon, this was your default,
That, being captain of the watch tonight,
Did look no better to that weighty charge.
ALENÇON
Had all your quarters been as safely kept
As that whereof I had the government,                        65
We had not been thus shamefully surprised.
BASTARD
Mine was secure.
RENÉ                      And so was mine, my lord.
CHARLES
And for myself, most part of all this night
Within her quarter and mine own precinct
I was employed in passing to and fro                         70
About relieving of the sentinels.
Then how or which way should they first break in?
JOAN
Question, my lords, no further of the case,
How or which way. 'Tis sure they found some place
But weakly guarded, where the breach was made.              75
And now there rests no other shift but this—
To gather our soldiers, scattered and dispersed,
And lay new platforms to endamage them.
     *Alarum. Enter an English Soldier*
ENGLISH SOLDIER A Talbot! A Talbot!
     *The French fly, leaving their clothes behind*
ENGLISH SOLDIER
I'll be so bold to take what they have left.                 80
The cry of 'Talbot' serves me for a sword,
For I have loaden me with many spoils,
Using no other weapon but his name.    *Exit with spoils*

2.2    *Enter Lord Talbot, the Dukes of Bedford and*
       *Burgundy, a Captain, ⌈and soldiers⌉*
BEDFORD
The day begins to break and night is fled,
Whose pitchy mantle overveiled the earth.
Here sound retreat and cease our hot pursuit.
     *Retreat is sounded*
TALBOT
Bring forth the body of old Salisbury
And here advance it in the market place,                      5
The middle centre of this cursèd town.
                         ⌈*Exit one or more*⌉
Now have I paid my vow unto his soul:
For every drop of blood was drawn from him
There hath at least five Frenchmen died tonight.
And that hereafter ages may behold                           10
What ruin happened in revenge of him,
Within their chiefest temple I'll erect
A tomb, wherein his corpse shall be interred—
Upon the which, that everyone may read,
Shall be engraved the sack of Orléans,                       15
The treacherous manner of his mournful death,
And what a terror he had been to France.
But, lords, in all our bloody massacre
I muse we met not with the Dauphin's grace,
His new-come champion, virtuous Joan of Arc,                 20
Nor any of his false confederates.
BEDFORD
'Tis thought, Lord Talbot, when the fight began,

Roused on the sudden from their drowsy beds,
They did amongst the troops of armèd men
Leap o'er the walls for refuge in the field.                 25
BURGUNDY
Myself, as far as I could well discern
For smoke and dusky vapours of the night,
Am sure I scared the Dauphin and his trull,
When arm-in-arm they both came swiftly running,
Like to a pair of loving turtle-doves                        30
That could not live asunder day or night.
After that things are set in order here,
We'll follow them with all the power we have.
     *Enter a Messenger*
MESSENGER
All hail, my lords! Which of this princely train
Call ye the warlike Talbot, for his acts                     35
So much applauded through the realm of France?
TALBOT
Here is the Talbot. Who would speak with him?
MESSENGER
The virtuous lady, Countess of Auvergne,
With modesty admiring thy renown,
By me entreats, great lord, thou wouldst vouchsafe  40
To visit her poor castle where she lies,
That she may boast she hath beheld the man
Whose glory fills the world with loud report.
BURGUNDY
Is it even so? Nay, then I see our wars
Will turn unto a peaceful comic sport,                       45
When ladies crave to be encountered with.
You may not, my lord, despise her gentle suit.
TALBOT
Ne'er trust me then, for when a world of men
Could not prevail with all their oratory,
Yet hath a woman's kindness overruled.—                      50
And therefore tell her I return great thanks,
And in submission will attend on her.—
Will not your honours bear me company?
BEDFORD
No, truly, 'tis more than manners will.
And I have heard it said, 'Unbidden guests                   55
Are often welcomest when they are gone'.
TALBOT
Well then, alone—since there's no remedy—
I mean to prove this lady's courtesy.
Come hither, captain.
     *He whispers*
                         You perceive my mind?
CAPTAIN
I do, my lord, and mean accordingly.                         60
                         *Exeunt ⌈severally⌉*

2.3    *Enter the Countess of Auvergne and her Porter*
COUNTESS
Porter, remember what I gave in charge,
And when you have done so, bring the keys to me.
PORTER Madam, I will.                                *Exit*
COUNTESS
The plot is laid. If all things fall out right,
I shall as famous be by this exploit                         5
As Scythian Tomyris by Cyrus' death.
Great is the rumour of this dreadful knight,
And his achievements of no less account.
Fain would mine eyes be witness with mine ears,
To give their censure of these rare reports.                 10

*Enter Messenger and Lord Talbot*

MESSENGER
Madam, according as your ladyship desired,
By message craved, so is Lord Talbot come.

COUNTESS
And he is welcome. What, is this the man?

MESSENGER
Madam, it is.

COUNTESS        Is this the scourge of France?
Is this the Talbot, so much feared abroad          15
That with his name the mothers still their babes?
I see report is fabulous and false.
I thought I should have seen some Hercules,
A second Hector, for his grim aspect
And large proportion of his strong-knit limbs.     20
Alas, this is a child, a seely dwarf.
It cannot be this weak and writhled shrimp
Should strike such terror to his enemies.

TALBOT
Madam, I have been bold to trouble you.
But since your ladyship is not at leisure,          25
I'll sort some other time to visit you.
        *He is going*

COUNTESS *(to Messenger)*
What means he now? Go ask him whither he goes.

MESSENGER
Stay, my Lord Talbot, for my lady craves
To know the cause of your abrupt departure.

TALBOT
Marry, for that she's in a wrong belief,             30
I go to certify her Talbot's here.
        *Enter Porter with keys*

COUNTESS
If thou be he, then art thou prisoner.

TALBOT
Prisoner? To whom?

COUNTESS                To me, bloodthirsty lord;
And for that cause I trained thee to my house.
Long time thy shadow hath been thrall to me,        35
For in my gallery thy picture hangs;
But now the substance shall endure the like,
And I will chain these legs and arms of thine
That hast by tyranny these many years
Wasted our country, slain our citizens,             40
And sent our sons and husbands captivate—

TALBOT Ha, ha, ha!

COUNTESS
Laughest thou, wretch? Thy mirth shall turn to moan.

TALBOT
I laugh to see your ladyship so fond
To think that you have aught but Talbot's shadow     45
Whereon to practise your severity.

COUNTESS Why? Art not thou the man?

TALBOT I am indeed.

COUNTESS Then have I substance too.

TALBOT
No, no, I am but shadow of myself.                   50
You are deceived; my substance is not here.
For what you see is but the smallest part
And least proportion of humanity.
I tell you, madam, were the whole frame here,
It is of such a spacious lofty pitch                 55
Your roof were not sufficient to contain't.

COUNTESS
This is a riddling merchant for the nonce.
He will be here, and yet he is not here.
How can these contrarieties agree?

TALBOT
That will I show you presently.                      60
        *He winds his horn. Within, drums strike up; a peal
        of ordnance. Enter English soldiers*
How say you, madam? Are you now persuaded
That Talbot is but shadow of himself?
These are his substance, sinews, arms, and strength,
With which he yoketh your rebellious necks,
Razeth your cities and subverts your towns,          65
And in a moment makes them desolate.

COUNTESS
Victorious Talbot, pardon my abuse.
I find thou art no less than fame hath bruited,
And more than may be gathered by thy shape.
Let my presumption not provoke thy wrath,            70
For I am sorry that with reverence
I did not entertain thee as thou art.

TALBOT
Be not dismayed, fair lady, nor misconster
The mind of Talbot, as you did mistake
The outward composition of his body.                 75
What you have done hath not offended me;
Nor other satisfaction do I crave
But only, with your patience, that we may
Taste of your wine and see what cates you have:
For soldiers' stomachs always serve them well.       80

COUNTESS
With all my heart; and think me honourèd
To feast so great a warrior in my house.    *Exeunt*

**2.4**   *A rose brier. Enter Richard Plantagenet, the Earl of
        Warwick, the Duke of Somerset, William de la Pole
        (the Earl of Suffolk), Vernon, and a Lawyer*

RICHARD PLANTAGENET
Great lords and gentlemen, what means this silence?
Dare no man answer in a case of truth?

SUFFOLK
Within the Temple hall we were too loud.
The garden here is more convenient.

RICHARD PLANTAGENET
Then say at once if I maintained the truth;          5
Or else was wrangling Somerset in th'error?

SUFFOLK
Faith, I have been a truant in the law,
And never yet could frame my will to it,
And therefore frame the law unto my will.

SOMERSET
Judge you, my lord of Warwick, then between us.      10

WARWICK
Between two hawks, which flies the higher pitch,
Between two dogs, which hath the deeper mouth,
Between two blades, which bears the better temper,
Between two horses, which doth bear him best,
Between two girls, which hath the merriest eye,      15
I have perhaps some shallow spirit of judgement;
But in these nice sharp quillets of the law,
Good faith, I am no wiser than a daw.

RICHARD PLANTAGENET
Tut, tut, here is a mannerly forbearance.
The truth appears so naked on my side                20
That any purblind eye may find it out.

SOMERSET
And on my side it is so well apparelled,
So clear, so shining, and so evident,
That it will glimmer through a blind man's eye.

RICHARD PLANTAGENET
Since you are tongue-tied and so loath to speak,                    25
In dumb significants proclaim your thoughts.
Let him that is a true-born gentleman
And stands upon the honour of his birth,
If he suppose that I have pleaded truth,
From off this briar pluck a white rose with me.                    30
*He plucks a white rose*
SOMERSET
Let him that is no coward nor no flatterer,
But dare maintain the party of the truth,
Pluck a red rose from off this thorn with me.
*He plucks a red rose*
WARWICK
I love no colours, and without all colour
Of base insinuating flattery                                       35
I pluck this white rose with Plantagenet.
SUFFOLK
I pluck this red rose with young Somerset,
And say withal I think he held the right.
VERNON
Stay, lords and gentlemen, and pluck no more
Till you conclude that he upon whose side                          40
The fewest roses from the tree are cropped
Shall yield the other in the right opinion.
SOMERSET
Good Master Vernon, it is well objected.
If I have fewest, I subscribe in silence.
RICHARD PLANTAGENET  And I.                                        45
VERNON
Then for the truth and plainness of the case
I pluck this pale and maiden blossom here,
Giving my verdict on the white rose' side.
SOMERSET
Prick not your finger as you pluck it off,
Lest, bleeding, you do paint the white rose red,                   50
And fall on my side so against your will.
VERNON
If I, my lord, for my opinion bleed,
Opinion shall be surgeon to my hurt
And keep me on the side where still I am.
SOMERSET Well, well, come on! Who else?                            55
LAWYER
Unless my study and my books be false,
The argument you held was wrong in law;
In sign whereof I pluck a white rose too.
RICHARD PLANTAGENET
Now Somerset, where is your argument?
SOMERSET
Here in my scabbard, meditating that                               60
Shall dye your white rose in a bloody red.
RICHARD PLANTAGENET
Meantime your cheeks do counterfeit our roses,
For pale they look with fear, as witnessing
The truth on our side.
SOMERSET                    No, Plantagenet,
'Tis not for fear, but anger, that thy cheeks                      65
Blush for pure shame to counterfeit our roses,
And yet thy tongue will not confess thy error.
RICHARD PLANTAGENET
Hath not thy rose a canker, Somerset?
SOMERSET
Hath not thy rose a thorn, Plantagenet?
RICHARD PLANTAGENET
Ay, sharp and piercing, to maintain his truth,                     70
Whiles thy consuming canker eats his falsehood.

SOMERSET
Well, I'll find friends to wear my bleeding roses,
That shall maintain what I have said is true,
Where false Plantagenet dare not be seen.
RICHARD PLANTAGENET
Now, by this maiden blossom in my hand,                            75
I scorn thee and thy fashion, peevish boy.
SUFFOLK
Turn not thy scorns this way, Plantagenet.
RICHARD PLANTAGENET
Proud Pole, I will, and scorn both him and thee.
SUFFOLK
I'll turn my part thereof into thy throat.
SOMERSET
Away, away, good William de la Pole.                               80
We grace the yeoman by conversing with him.
WARWICK
Now, by God's will, thou wrong'st him, Somerset.
His grandfather was Lionel Duke of Clarence,
Third son to the third Edward, King of England.
Spring crestless yeomen from so deep a root?                       85
RICHARD PLANTAGENET
He bears him on the place's privilege,
Or durst not for his craven heart say thus.
SOMERSET
By him that made me, I'll maintain my words
On any plot of ground in Christendom.
Was not thy father, Richard Earl of Cambridge,                    90
For treason executed in our late king's days?
And by his treason stand'st not thou attainted,
Corrupted, and exempt from ancient gentry?
His trespass yet lives guilty in thy blood,
And till thou be restored thou art a yeoman.                      95
RICHARD PLANTAGENET
My father was attachèd, not attainted;
Condemned to die for treason, but no traitor—
And that I'll prove on better men than Somerset,
Were growing time once ripened to my will.
For your partaker Pole, and you yourself,                         100
I'll note you in my book of memory,
To scourge you for this apprehension.
Look to it well, and say you are well warned.
SOMERSET
Ah, thou shalt find us ready for thee still,
And know us by these colours for thy foes,                        105
For these my friends, in spite of thee, shall wear.
RICHARD PLANTAGENET
And, by my soul, this pale and angry rose,
As cognizance of my blood-drinking hate,
Will I forever, and my faction, wear
Until it wither with me to my grave,                              110
Or flourish to the height of my degree.
SUFFOLK
Go forward, and be choked with thy ambition.
And so farewell until I meet thee next.                *Exit*
SOMERSET
Have with thee, Pole.—Farewell, ambitious Richard.
                                                       *Exit*
RICHARD PLANTAGENET
How I am braved, and must perforce endure it!                     115
WARWICK
This blot that they object against your house
Shall be wiped out in the next parliament,
Called for the truce of Winchester and Gloucester.
An if thou be not then created York,
I will not live to be accounted Warwick.                          120
Meantime, in signal of my love to thee,

Against proud Somerset and William Pole,
Will I upon thy party wear this rose.
And here I prophesy : this brawl today,
Grown to this faction in the Temple garden,          125
Shall send, between the red rose and the white,
A thousand souls to death and deadly night.

RICHARD PLANTAGENET
Good Master Vernon, I am bound to you,
That you on my behalf would pluck a flower.

VERNON
In your behalf still will I wear the same.           130

LAWYER  And so will I.

RICHARD PLANTAGENET  Thanks, gentles.
Come, let us four to dinner. I dare say
This quarrel will drink blood another day.
                    *Exeunt. The rose brier is removed*

**2.5**  *Enter Edmund Mortimer, brought in a chair ⌐by⌐ his
         Keepers*

MORTIMER
Kind keepers of my weak decaying age,
Let dying Mortimer here rest himself.
Even like a man new-halèd from the rack,
So fare my limbs with long imprisonment;
And these grey locks, the pursuivants of death,       5
Argue the end of Edmund Mortimer,
Nestor-like agèd in an age of care.
These eyes, like lamps whose wasting oil is spent,
Wax dim, as drawing to their exigent;
Weak shoulders, overborne with burdening grief,      10
And pithless arms, like to a withered vine
That droops his sapless branches to the ground.
Yet are these feet—whose strengthless stay is numb,
Unable to support this lump of clay—
Swift-wingèd with desire to get a grave,             15
As witting I no other comfort have.
But tell me, keeper, will my nephew come?

KEEPER
Richard Plantagenet, my lord, will come.
We sent unto the Temple, unto his chamber,
And answer was returned that he will come.           20

MORTIMER
Enough. My soul shall then be satisfied.
Poor gentleman, his wrong doth equal mine.
Since Henry Monmouth first began to reign—
Before whose glory I was great in arms—
This loathsome sequestration have I had;             25
And even since then hath Richard been obscured,
Deprived of honour and inheritance.
But now the arbitrator of despairs,
Just Death, kind umpire of men's miseries,
With sweet enlargement doth dismiss me hence.        30
I would his troubles likewise were expired,
That so he might recover what was lost.
                    *Enter Richard Plantagenet*

KEEPER
My lord, your loving nephew now is come.

MORTIMER
Richard Plantagenet, my friend, is he come?

RICHARD PLANTAGENET
Ay, noble uncle, thus ignobly used :                 35
Your nephew, late despisèd Richard, comes.

MORTIMER (*to Keepers*)
Direct mine arms I may embrace his neck
And in his bosom spend my latter gasp.
O tell me when my lips do touch his cheeks,
That I may kindly give one fainting kiss.            40

*He embraces Richard*
And now declare, sweet stem from York's great stock,
Why didst thou say of late thou wert despised?

RICHARD PLANTAGENET
First lean thine agèd back against mine arm,
And in that ease I'll tell thee my dis-ease.
This day in argument upon a case                     45
Some words there grew 'twixt Somerset and me;
Among which terms he used his lavish tongue
And did upbraid me with my father's death;
Which obloquy set bars before my tongue,
Else with the like I had requited him.               50
Therefore, good uncle, for my father's sake,
In honour of a true Plantagenet,
And for alliance' sake, declare the cause
My father, Earl of Cambridge, lost his head.

MORTIMER
That cause, fair nephew, that imprisoned me,         55
And hath detained me all my flow'ring youth
Within a loathsome dungeon, there to pine,
Was cursèd instrument of his decease.

RICHARD PLANTAGENET
Discover more at large what cause that was,
For I am ignorant and cannot guess.                  60

MORTIMER
I will, if that my fading breath permit
And death approach not ere my tale be done.
Henry the Fourth, grandfather to this King,
Deposed his nephew Richard, Edward's son,
The first begotten and the lawful heir               65
Of Edward king, the third of that descent;
During whose reign the Percies of the north,
Finding his usurpation most unjust,
Endeavoured my advancement to the throne.
The reason moved these warlike lords to this         70
Was for that—young King Richard thus removed,
Leaving no heir begotten of his body—
I was the next by birth and parentage,
For by my mother I derivèd am
From Lionel Duke of Clarence, the third son          75
To King Edward the Third—whereas the King
From John of Gaunt doth bring his pedigree,
Being but fourth of that heroic line.
But mark : as in this haughty great attempt
They labourèd to plant the rightful heir,            80
I lost my liberty, and they their lives.
Long after this, when Henry the Fifth,
Succeeding his father Bolingbroke, did reign,
Thy father, Earl of Cambridge then, derived
From famous Edmund Langley, Duke of York,            85
Marrying my sister that thy mother was,
Again, in pity of my hard distress,
Levied an army, weening to redeem
And have installed me in the diadem;
But, as the rest, so fell that noble earl,           90
And was beheaded. Thus the Mortimers,
In whom the title rested, were suppressed.

RICHARD PLANTAGENET
Of which, my lord, your honour is the last.

MORTIMER
True, and thou seest that I no issue have,
And that my fainting words do warrant death.         95
Thou art my heir. The rest I wish thee gather—
But yet be wary in thy studious care.

RICHARD PLANTAGENET
Thy grave admonishments prevail with me.
But yet methinks my father's execution
Was nothing less than bloody tyranny.               100

MORTIMER

With silence, nephew, be thou politic.
Strong-fixèd is the house of Lancaster,
And like a mountain, not to be removed.
But now thy uncle is removing hence,
As princes do their courts, when they are cloyed     105
With long continuance in a settled place.

RICHARD PLANTAGENET

O uncle, would some part of my young years
Might but redeem the passage of your age.

MORTIMER

Thou dost then wrong me, as that slaughterer doth
Which giveth many wounds when one will kill.     110
Mourn not, except thou sorrow for my good.
Only give order for my funeral.
And so farewell, and fair be all thy hopes,
And prosperous be thy life in peace and war.     *Dies*

RICHARD PLANTAGENET

And peace, no war, befall thy parting soul.     115
In prison hast thou spent a pilgrimage,
And like a hermit overpassed thy days.
Well, I will lock his counsel in my breast,
And what I do imagine, let that rest.
Keepers, convey him hence, and I myself     120
Will see his burial better than his life.
                    *Exeunt Keepers with Mortimer's body*
Here dies the dusky torch of Mortimer,
Choked with ambition of the meaner sort.
And for those wrongs, those bitter injuries,
Which Somerset hath offered to my house,     125
I doubt not but with honour to redress.
And therefore haste I to the Parliament,
Either to be restorèd to my blood,
Or make mine ill th'advantage of my good.     *Exit*

**3.1**   *Flourish. Enter young King Henry, the Dukes of*
    *Exeter and Gloucester, the Bishop of Winchester;*
    *the Duke of Somerset and the Earl of Suffolk ⌈with*
    *red roses⌉; the Earl of Warwick and Richard*
    *Plantagenet ⌈with white roses⌉. Gloucester offers to*
    *put up a bill; Winchester snatches it, tears it*

WINCHESTER

Com'st thou with deep premeditated lines?
With written pamphlets studiously devised?
Humphrey of Gloucester, if thou canst accuse,
Or aught intend'st to lay unto my charge,
Do it without invention, suddenly,     5
As I with sudden and extemporal speech
Purpose to answer what thou canst object.

GLOUCESTER

Presumptuous priest, this place commands my
    patience,
Or thou shouldst find thou hast dishonoured me.
Think not, although in writing I preferred     10
The manner of thy vile outrageous crimes,
That therefore I have forged, or am not able
Verbatim to rehearse the method of my pen.
No, prelate, such is thy audacious wickedness,
Thy lewd, pestiferous, and dissentious pranks,     15
As very infants prattle of thy pride.
Thou art a most pernicious usurer,
Froward by nature, enemy to peace,
Lascivious, wanton, more than well beseems
A man of thy profession and degree.     20
And for thy treachery, what's more manifest?—
In that thou laid'st a trap to take my life,
As well at London Bridge as at the Tower.

Beside, I fear me, if thy thoughts were sifted,
The King thy sovereign is not quite exempt     25
From envious malice of thy swelling heart.

WINCHESTER

Gloucester, I do defy thee.—Lords, vouchsafe
To give me hearing what I shall reply.
If I were covetous, ambitious, or perverse,
As he will have me, how am I so poor?     30
Or how haps it I seek not to advance
Or raise myself, but keep my wonted calling?
And for dissension, who preferreth peace
More than I do?—except I be provoked.
No, my good lords, it is not that offends;     35
It is not that that hath incensed the Duke.
It is because no one should sway but he,
No one but he should be about the King—
And that engenders thunder in his breast
And makes him roar these accusations forth.     40
But he shall know I am as good—

GLOUCESTER As good?—
Thou bastard of my grandfather.

WINCHESTER

Ay, lordly sir; for what are you, I pray,
But one imperious in another's throne?     45

GLOUCESTER

Am I not Protector, saucy priest?

WINCHESTER

And am not I a prelate of the Church?

GLOUCESTER

Yes—as an outlaw in a castle keeps
And useth it to patronage his theft.

WINCHESTER

Unreverent Gloucester.

GLOUCESTER                    Thou art reverend     50
Touching thy spiritual function, not thy life.

WINCHESTER

Rome shall remedy this.

⌈GLOUCESTER⌉                    Roam thither then.

⌈WARWICK⌉ (*to Winchester*)
My lord, it were your duty to forbear.

SOMERSET

Ay, so the bishop be not overborne:
Methinks my lord should be religious,     55
And know the office that belongs to such.

WARWICK

Methinks his lordship should be humbler.
It fitteth not a prelate so to plead.

SOMERSET

Yes, when his holy state is touched so near.

WARWICK

State holy or unhallowed, what of that?     60
Is not his grace Protector to the King?

RICHARD PLANTAGENET (*aside*)
Plantagenet, I see, must hold his tongue,
Lest it be said, 'Speak, sirrah, when you should;
Must your bold verdict intertalk with lords?'
Else would I have a fling at Winchester.     65

KING HENRY

Uncles of Gloucester and of Winchester,
The special watchmen of our English weal,
I would prevail, if prayers might prevail,
To join your hearts in love and amity.
O what a scandal is it to our crown     70
That two such noble peers as ye should jar!
Believe me, lords, my tender years can tell
Civil dissension is a viperous worm

That gnaws the bowels of the commonwealth.
  *A noise within*
⌈SERVINGMEN⌉ (*within*) Down with the tawny coats!      75
KING HENRY
  What tumult's this?
WARWICK                    An uproar, I dare warrant,
  Begun through malice of the Bishop's men.
  *A noise again*
⌈SERVINGMEN⌉ (*within*) Stones, stones!
  *Enter the Mayor of London*
MAYOR
  O my good lords, and virtuous Henry,
  Pity the city of London, pity us!                          80
  The Bishop and the Duke of Gloucester's men,
  Forbidden late to carry any weapon,
  Have filled their pockets full of pebble stones
  And, banding themselves in contrary parts,
  Do pelt so fast at one another's pate                      85
  That many have their giddy brains knocked out.
  Our windows are broke down in every street,
  And we for fear compelled to shut our shops.
  *Enter in skirmish. with bloody pates, Winchester's*
  *Servingmen in tawny coats and Gloucester's in blue*
  *coats*
KING HENRY
  We charge you, on allegiance to ourself,
  To hold your slaught'ring hands and keep the peace.
  ⌈*The skirmish ceases*⌉
  Pray, Uncle Gloucester, mitigate this strife.             91
FIRST SERVINGMAN Nay, if we be forbidden stones, we'll
  fall to it with our teeth.
SECOND SERVINGMAN
  Do what ye dare, we are as resolute.
  *Skirmish again*
GLOUCESTER
  You of my household, leave this peevish broil,            95
  And set this unaccustomed fight aside.
THIRD SERVINGMAN
  My lord, we know your grace to be a man
  Just and upright and, for your royal birth,
  Inferior to none but to his majesty;
  And ere that we will suffer such a prince,                100
  So kind a father of the commonweal,
  To be disgracèd by an inkhorn mate,
  We and our wives and children all will fight
  And have our bodies slaughtered by thy foes.
FIRST SERVINGMAN
  Ay, and the very parings of our nails                     105
  Shall pitch a field when we are dead.
  *They begin to skirmish again*
GLOUCESTER                       Stay, stay, I say!
  An if you love me as you say you do,
  Let me persuade you to forbear a while.
KING HENRY
  O how this discord doth afflict my soul!
  Can you, my lord of Winchester, behold                    110
  My sighs and tears, and will not once relent?
  Who should be pitiful if you be not?
  Or who should study to prefer a peace,
  If holy churchmen take delight in broils?
WARWICK
  Yield, my lord Protector; yield, Winchester—              115
  Except you mean with obstinate repulse
  To slay your sovereign and destroy the realm.
  You see what mischief—and what murder, too—
  Hath been enacted through your enmity.
  Then be at peace, except ye thirst for blood.             120

WINCHESTER
  He shall submit, or I will never yield.
GLOUCESTER
  Compassion on the King commands me stoop,
  Or I would see his heart out ere the priest
  Should ever get that privilege of me.
WARWICK
  Behold, my lord of Winchester, the Duke               125
  Hath banished moody discontented fury,
  As by his smoothèd brows it doth appear.
  Why look you still so stern and tragical?
GLOUCESTER
  Here, Winchester, I offer thee my hand.
KING HENRY (*to Winchester*)
  Fie, Uncle Beaufort! I have heard you preach           130
  That malice was a great and grievous sin;
  And will not you maintain the thing you teach,
  But prove a chief offender in the same?
WARWICK
  Sweet King! The Bishop hath a kindly gird.
  For shame, my lord of Winchester, relent.             135
  What, shall a child instruct you what to do?
WINCHESTER
  Well, Duke of Gloucester, I will yield to thee
  Love for thy love, and hand for hand I give.
GLOUCESTER (*aside*)
  Ay, but I fear me with a hollow heart.
  (*To the others*) See here, my friends and loving
    countrymen,                                          140
  This token serveth for a flag of truce
  Betwixt ourselves and all our followers.
  So help me God, as I dissemble not.
WINCHESTER
  So help me God (*aside*) as I intend it not.
KING HENRY
  O loving uncle, kind Duke of Gloucester,              145
  How joyful am I made by this contract!
  (*To Servingmen*) Away, my masters, trouble us no
    more,
  But join in friendship as your lords have done.
FIRST SERVINGMAN Content. I'll to the surgeon's.
SECOND SERVINGMAN And so will I.                         150
THIRD SERVINGMAN And I will see what physic the tavern
  affords.                    *Exeunt the Mayor and Servingmen*
WARWICK
  Accept this scroll, most gracious sovereign,
  Which in the right of Richard Plantagenet
  We do exhibit to your majesty.                         155
GLOUCESTER
  Well urged, my lord of Warwick—for, sweet prince,
  An if your grace mark every circumstance,
  You have great reason to do Richard right,
  Especially for those occasions
  At Eltham Place I told your majesty.                   160
KING HENRY
  And those occasions, uncle, were of force.—
  Therefore, my loving lords, our pleasure is
  That Richard be restorèd to his blood.
WARWICK
  Let Richard be restorèd to his blood.
  So shall his father's wrongs be recompensed.           165
WINCHESTER
  As will the rest, so willeth Winchester.
KING HENRY
  If Richard will be true, not that alone
  But all the whole inheritance I give

That doth belong unto the house of York,
From whence you spring by lineal descent.          170
RICHARD PLANTAGENET
Thy humble servant vows obedience
And humble service till the point of death.
KING HENRY
Stoop then, and set your knee against my foot.
    *Richard kneels*
And in reguerdon of that duty done,
I gird thee with the valiant sword of York.        175
Rise, Richard, like a true Plantagenet,
And rise created princely Duke of York.
RICHARD DUKE OF YORK (*rising*)
And so thrive Richard, as thy foes may fall;
And as my duty springs, so perish they
That grudge one thought against your majesty.      180
ALL BUT RICHARD AND SOMERSET
Welcome, high prince, the mighty Duke of York!
SOMERSET (*aside*)
Perish, base prince, ignoble Duke of York!
GLOUCESTER
Now will it best avail your majesty
To cross the seas and to be crowned in France.
The presence of a king engenders love             185
Amongst his subjects and his loyal friends,
As it disanimates his enemies.
KING HENRY
When Gloucester says the word, King Henry goes,
For friendly counsel cuts off many foes.
GLOUCESTER
Your ships already are in readiness.              190
    *Sennet. Exeunt all but Exeter*
EXETER
Ay, we may march in England or in France,
Not seeing what is likely to ensue.
This late dissension grown betwixt the peers
Burns under feignèd ashes of forged love,
And will at last break out into a flame.          195
As festered members rot but by degree
Till bones and flesh and sinews fall away,
So will this base and envious discord breed.
And now I fear that fatal prophecy
Which, in the time of Henry named the Fifth,      200
Was in the mouth of every sucking babe:
That 'Henry born at Monmouth should win all,
And Henry born at Windsor should lose all'—
Which is so plain that Exeter doth wish
His days may finish, ere that hapless time.       *Exit*

**3.2**    *Enter Joan la Pucelle, disguised, with four French*
         *Soldiers with sacks upon their backs*
JOAN
These are the city gates, the gates of Rouen,
Through which our policy must make a breach.
Take heed. Be wary how you place your words.
Talk like the vulgar sort of market men
That come to gather money for their corn.          5
If we have entrance, as I hope we shall,
And that we find the slothful watch but weak,
I'll by a sign give notice to our friends,
That Charles the Dauphin may encounter them.
A SOLDIER
Our sacks shall be a mean to sack the city,       10
And we be lords and rulers over Rouen.
Therefore we'll knock.
    *They knock*

WATCH (*within*)
Qui là?
JOAN      *Paysans, la pauvre gens de France:*
Poor market folks that come to sell their corn.
WATCH (*opening the gates*)
Enter, go in. The market bell is rung.            15
JOAN (*aside*)
Now, Rouen, I'll shake thy bulwarks to the ground.
    *Exeunt*

**3.3**    *Enter Charles the Dauphin, the Bastard of Orléans,*
         *⌈the Duke of Alençon, René Duke of Anjou, and*
         *French soldiers⌉*
CHARLES
Saint Denis bless this happy stratagem,
And once again we'll sleep secure in Rouen.
BASTARD
Here entered Pucelle and her practisants.
Now she is there, how will she specify
'Here is the best and safest passage in'?          5
RENÉ
By thrusting out a torch from yonder tower—
Which, once discerned, shows that her meaning is:
No way to that, for weakness, which she entered.
    *Enter Joan la Pucelle on the top, thrusting out a*
    *torch burning*
JOAN
Behold, this is the happy wedding torch
That joineth Rouen unto her countrymen,           10
But burning fatal to the Talbonites.
BASTARD
See, noble Charles, the beacon of our friend.
The burning torch in yonder turret stands.
CHARLES
Now shine it like a comet of revenge,
A prophet to the fall of all our foes!            15
RENÉ
Defer no time; delays have dangerous ends.
Enter and cry, 'The Dauphin!', presently,
And then do execution on the watch.    *Alarum. Exeunt*

**3.4**    *An alarum. Enter Lord Talbot in an excursion*
TALBOT
France, thou shalt rue this treason with thy tears,
If Talbot but survive thy treachery.
Pucelle, that witch, that damnèd sorceress,
Hath wrought this hellish mischief unawares,
That hardly we escaped the pride of France.       *Exit*

**3.5**    *An alarum. Excursions. The Duke of Bedford*
         *brought in sick, in a chair. Enter Lord Talbot and*
         *the Duke of Burgundy, without; within, Joan la*
         *Pucelle, Charles the Dauphin, the Bastard of*
         *Orléans, ⌈the Duke of Alençon, and René Duke of*
         *Anjou⌉ on the walls*
JOAN
Good morrow gallants. Want ye corn for bread?
I think the Duke of Burgundy will fast
Before he'll buy again at such a rate.
'Twas full of darnel. Do you like the taste?
BURGUNDY
Scoff on, vile fiend and shameless courtesan.      5
I trust ere long to choke thee with thine own,
And make thee curse the harvest of that corn.
CHARLES
Your grace may starve, perhaps, before that time.

BEDFORD
O let no words, but deeds, revenge this treason.
JOAN
What will you do, good graybeard? Break a lance      10
And run a-tilt at death within a chair?
TALBOT
Foul fiend of France, and hag of all despite,
Encompassed with thy lustful paramours,
Becomes it thee to taunt his valiant age
And twit with cowardice a man half dead?             15
Damsel, I'll have a bout with you again,
Or else let Talbot perish with this shame.
JOAN
Are ye so hot, sir?—Yet, Pucelle, hold thy peace.
If Talbot do but thunder, rain will follow.
*The English whisper together in counsel*
God speed the parliament; who shall be the Speaker?
TALBOT
Dare ye come forth and meet us in the field?         21
JOAN
Belike your lordship takes us then for fools,
To try if that our own be ours or no.
TALBOT
I speak not to that railing Hecate
But unto thee, Alençon, and the rest.                25
Will ye, like soldiers, come and fight it out?
ALENÇON
Seignieur, no.
TALBOT        Seignieur, hang! Base muleteers of France,
Like peasant footboys do they keep the walls
And dare not take up arms like gentlemen.
JOAN
Away, captains, let's get us from the walls,         30
For Talbot means no goodness by his looks.
Goodbye, my lord. We came but to tell you
That we are here.        *Exeunt French from the walls*
TALBOT
And there will we be, too, ere it be long,
Or else reproach be Talbot's greatest fame.          35
Vow Burgundy, by honour of thy house,
Pricked on by public wrongs sustained in France,
Either to get the town again or die.
And I—as sure as English Henry lives,
And as his father here was conqueror;                40
As sure as in this late betrayèd town
Great Cœur-de-lion's heart was buried—
So sure I swear to get the town or die.
BURGUNDY
My vows are equal partners with thy vows.
TALBOT
But ere we go, regard this dying prince,             45
The valiant Duke of Bedford. *(To Bedford)* Come, my
      lord,
We will bestow you in some better place,
Fitter for sickness and for crazy age.
BEDFORD
Lord Talbot, do not so dishonour me.
Here will I sit before the walls of Rouen,           50
And will be partner of your weal or woe.
BURGUNDY
Courageous Bedford, let us now persuade you.
BEDFORD
Not to be gone from hence; for once I read
That stout Pendragon, in his litter sick,
Came to the field and vanquishèd his foes.           55
Methinks I should revive the soldiers' hearts,
Because I ever found them as myself.

TALBOT
Undaunted spirit in a dying breast!
Then be it so; heavens keep old Bedford safe.
And now no more ado, brave Burgundy,                 60
But gather we our forces out of hand,
And set upon our boasting enemy.    *Exit with Burgundy*
      *An alarum. Excursions. Enter Sir John Fastolf and a*
      *Captain*
CAPTAIN
Whither away, Sir John Fastolf, in such haste?
FASTOLF
Whither away? To save myself by flight.
We are like to have the overthrow again.             65
CAPTAIN
What, will you fly, and leave Lord Talbot?
FASTOLF
Ay, all the Talbots in the world, to save my life.    *Exit*
CAPTAIN
Cowardly knight, ill fortune follow thee!            *Exit*
      *Retreat. Excursions. Joan, Alençon, and Charles fly*
BEDFORD
Now, quiet soul, depart when heaven please,
For I have seen our enemies' overthrow.              70
What is the trust or strength of foolish man?
They that of late were daring with their scoffs
Are glad and fain by flight to save themselves.
      *Bedford dies, and is carried in by two in his chair*

3.6    *An alarum. Enter Lord Talbot, the Duke of*
      *Burgundy, and the rest of the English soldiers*
TALBOT
Lost and recovered in a day again!
This is a double honour, Burgundy;
Yet heavens have glory for this victory!
BURGUNDY
Warlike and martial Talbot, Burgundy
Enshrines thee in his heart, and there erects         5
Thy noble deeds as valour's monuments.
TALBOT
Thanks, gentle Duke. But where is Pucelle now?
I think her old familiar is asleep.
Now where's the Bastard's braves, and Charles his
      gleeks?
What, all amort? Rouen hangs her head for grief      10
That such a valiant company are fled.
Now will we take some order in the town,
Placing therein some expert officers,
And then depart to Paris, to the King,
For there young Henry with his nobles lie.           15
BURGUNDY
What wills Lord Talbot pleaseth Burgundy.
TALBOT
But yet, before we go, let's not forget
The noble Duke of Bedford late deceased,
But see his exequies fulfilled in Rouen.
A braver soldier never couchèd lance;                20
A gentler heart did never sway in court.
But kings and mightiest potentates must die,
For that's the end of human misery.        *Exeunt*

3.7    *Enter Charles the Dauphin, the Bastard of Orléans,*
      *the Duke of Alençon, Joan la Pucelle, ⌐and French*
      *soldiers⌐*
JOAN
Dismay not, princes, at this accident,
Nor grieve that Rouen is so recoverèd.

Care is no cure, but rather corrosive,
For things that are not to be remedied.
Let frantic Talbot triumph for a while,                    5
And like a peacock sweep along his tail;
We'll pull his plumes and take away his train,
If Dauphin and the rest will be but ruled.
CHARLES
We have been guided by thee hitherto,
And of thy cunning had no diffidence.                     10
One sudden foil shall never breed distrust.
BASTARD (*to Joan*)
Search out thy wit for secret policies,
And we will make thee famous through the world.
ALENÇON (*to Joan*)
We'll set thy statue in some holy place
And have thee reverenced like a blessèd saint.            15
Employ thee then, sweet virgin, for our good.
JOAN
Then thus it must be; this doth Joan devise:
By fair persuasions mixed with sugared words
We will entice the Duke of Burgundy
To leave the Talbot and to follow us.                     20
CHARLES
Ay, marry, sweeting, if we could do that
France were no place for Henry's warriors,
Nor should that nation boast it so with us,
But be extirpèd from our provinces.
ALENÇON
For ever should they be expulsed from France              25
And not have title of an earldom here.
JOAN
Your honours shall perceive how I will work
To bring this matter to the wishèd end.
        *Drum sounds afar off*
Hark, by the sound of drum you may perceive
Their powers are marching unto Paris-ward.                30
        *Here sound an English march*
There goes the Talbot, with his colours spread,
And all the troops of English after him.
        *Here sound a French march*
Now in the rearward comes the Duke and his;
Fortune in favour makes him lag behind.
Summon a parley. We will talk with him.                   35
        *Trumpets sound a parley*
CHARLES ⌈*calling*⌉
A parley with the Duke of Burgundy.
        ⌈*Enter the Duke of Burgundy*⌉
BURGUNDY
Who craves a parley with the Burgundy?
JOAN
The princely Charles of France, thy countryman.
BURGUNDY
What sayst thou, Charles?—for I am marching hence.
CHARLES
Speak, Pucelle, and enchant him with thy words.           40
JOAN
Brave Burgundy, undoubted hope of France,
Stay. Let thy humble handmaid speak to thee.
BURGUNDY
Speak on, but be not over-tedious.
JOAN
Look on thy country, look on fertile France,
And see the cities and the towns defaced                  45
By wasting ruin of the cruel foe.
As looks the mother on her lowly babe
When death doth close his tender-dying eyes,
See, see the pining malady of France;

Behold the wounds, the most unnatural wounds,             50
Which thou thyself hast given her woeful breast.
O turn thy edgèd sword another way,
Strike those that hurt, and hurt not those that help.
One drop of blood drawn from thy country's bosom
Should grieve thee more than streams of foreign gore.
Return thee, therefore, with a flood of tears,            56
And wash away thy country's stainèd spots.
BURGUNDY ⌈*aside*⌉
Either she hath bewitched me with her words,
Or nature makes me suddenly relent.
JOAN
Besides, all French and France exclaims on thee,          60
Doubting thy birth and lawful progeny,
Who join'st thou with but with a lordly nation
That will not trust thee but for profit's sake?
When Talbot hath set footing once in France
And fashioned thee that instrument of ill,                65
Who then but English Henry will be lord,
And thou be thrust out like a fugitive?
Call we to mind, and mark but this for proof:
Was not the Duke of Orléans thy foe?
And was he not in England prisoner?                       70
But when they heard he was thine enemy
They set him free, without his ransom paid,
In spite of Burgundy and all his friends.
See, then, thou fight'st against thy countrymen,
And join'st with them will be thy slaughtermen.           75
Come, come, return; return, thou wandering lord,
Charles and the rest will take thee in their arms.
BURGUNDY ⌈*aside*⌉
I am vanquishèd. These haughty words of hers
Have battered me like roaring cannon-shot
And made me almost yield upon my knees.                   80
(*To the others*) Forgive me, country, and sweet
        countrymen;
And lords, accept this hearty kind embrace.
My forces and my power of men are yours.
So farewell, Talbot. I'll no longer trust thee.
JOAN
Done like a Frenchman—⌈*aside*⌉ turn and turn again.
CHARLES
Welcome, brave Duke. Thy friendship makes us fresh.
BASTARD
And doth beget new courage in our breasts.                87
ALENÇON
Pucelle hath bravely played her part in this,
And doth deserve a coronet of gold.
CHARLES
Now let us on, my lords, and join our powers,             90
And seek how we may prejudice the foe.          *Exeunt*

3.8    ⌈*Flourish.*⌉ *Enter King Henry, the Duke of
        Gloucester, the Bishop of Winchester, the Duke of
        Exeter; Richard Duke of York, the Earl of
        Warwick, and Vernon* ⌈*with white roses*⌉; *the Earl
        of Suffolk, the Duke of Somerset, and Basset* ⌈*with
        red roses*⌉. *To them, with his soldiers, enter Lord
        Talbot*
TALBOT
My gracious prince and honourable peers,
Hearing of your arrival in this realm
I have a while given truce unto my wars
To do my duty to my sovereign;
In sign whereof, this arm that hath reclaimed             5
To your obedience fifty fortresses,

Twelve cities, and seven walled towns of strength,
Beside five hundred prisoners of esteem,
Lets fall his sword before your highness' feet,          10
And with submissive loyalty of heart
Ascribes the glory of his conquest got
First to my God, and next unto your grace.
    ⌈*He kneels*⌉
KING HENRY
   Is this the Lord Talbot, uncle Gloucester,
   That hath so long been resident in France?
GLOUCESTER
   Yes, if it please your majesty, my liege.          15
KING HENRY (*to Talbot*)
   Welcome, brave captain and victorious lord.
   When I was young—as yet I am not old—
   I do remember how my father said
   A stouter champion never handled sword.
   Long since we were resolvèd of your truth,          20
   Your faithful service and your toil in war,
   Yet never have you tasted our reward,
   Or been reguerdoned with so much as thanks,
   Because till now we never saw your face.
   Therefore stand up,
      *Talbot rises*
              and for these good deserts          25
   We here create you Earl of Shrewsbury;
   And in our coronation take your place.
      *Sennet. Exeunt all but Vernon and Basset*
VERNON
   Now sir, to you that were so hot at sea,
   Disgracing of these colours that I wear
   In honour of my noble lord of York,          30
   Dar'st thou maintain the former words thou spak'st?
BASSET
   Yes, sir, as well as you dare patronage
   The envious barking of your saucy tongue
   Against my lord the Duke of Somerset.
VERNON
   Sirrah, thy lord I honour as he is.          35
BASSET
   Why, what is he?—as good a man as York.
VERNON
   Hark ye, not so. In witness, take ye that.
      *Vernon strikes him*
BASSET
   Villain, thou know'st the law of arms is such
   That whoso draws a sword 'tis present death,
   Or else this blow should broach thy dearest blood.          40
   But I'll unto his majesty and crave
   I may have liberty to venge this wrong,
   When thou shalt see I'll meet thee to thy cost.
VERNON
   Well, miscreant, I'll be there as soon as you,
   And after meet you sooner than you would.    *Exeunt*

**4.1**  ⌈*Flourish.*⌉ *Enter King Henry, the Duke of*
    *Gloucester, the Bishop of Winchester, the Duke of*
    *Exeter; Richard Duke of York, and the Earl of*
    *Warwick with white roses; the Earl of Suffolk and*
    *the Duke of Somerset with red roses; Lord Talbot,*
    *and the Governor of Paris*
GLOUCESTER
   Lord Bishop, set the crown upon his head.
WINCHESTER
   God save King Henry, of that name the sixth!
      *Winchester crowns the King*

GLOUCESTER
   Now, Governor of Paris, take your oath
   That you elect no other king but him;
   Esteem none friends but such as are his friends,          5
   And none your foes but such as shall pretend
   Malicious practices against his state.
   This shall ye do, so help you righteous God.
      *Enter Sir John Fastolf with a letter*
FASTOLF
   My gracious sovereign, as I rode from Calais
   To haste unto your coronation          10
   A letter was delivered to my hands,
      ⌈*He presents the letter*⌉
   Writ to your grace from th' Duke of Burgundy.
TALBOT
   Shame to the Duke of Burgundy and thee!
   I vowed, base knight, when I did meet thee next,
   To tear the Garter from thy craven's leg,          15
      *He tears it off*
   Which I have done because unworthily
   Thou wast installèd in that high degree.—
   Pardon me, princely Henry and the rest.
   This dastard at the battle of Patay
   When but in all I was six thousand strong,          20
   And that the French were almost ten to one,
   Before we met, or that a stroke was given,
   Like to a trusty squire did run away;
   In which assault we lost twelve hundred men.
   Myself and divers gentlemen beside          25
   Were there surprised and taken prisoners.
   Then judge, great lords, if I have done amiss,
   Or whether that such cowards ought to wear
   This ornament of knighthood: yea or no?
GLOUCESTER
   To say the truth, this fact was infamous          30
   And ill beseeming any common man,
   Much more a knight, a captain and a leader.
TALBOT
   When first this order was ordained, my lords,
   Knights of the Garter were of noble birth,
   Valiant and virtuous, full of haughty courage,          35
   Such as were grown to credit by the wars;
   Not fearing death nor shrinking for distress,
   But always resolute in most extremes.
   He then that is not furnished in this sort
   Doth but usurp the sacred name of knight,          40
   Profaning this most honourable order,
   And should—if I were worthy to be judge—
   Be quite degraded, like a hedge-born swain
   That doth presume to boast of gentle blood.
KING HENRY (*to Fastolf*)
   Stain to thy countrymen, thou hear'st thy doom.          45
   Be packing, therefore, thou that wast a knight.
   Henceforth we banish thee on pain of death.
      *Exit Fastolf*
   And now, my Lord Protector, view the letter
   Sent from our uncle, Duke of Burgundy.
GLOUCESTER
   What means his grace that he hath changed his
      style?          50
   No more but plain and bluntly 'To the King'?
   Hath he forgot he is his sovereign?
   Or doth this churlish superscription
   Pretend some alteration in good will?
   What's here? 'I have upon especial cause,          55

Moved with compassion of my country's wrack
Together with the pitiful complaints
Of such as your oppression feeds upon,
Forsaken your pernicious faction
And joined with Charles, the rightful King of France.'
O monstrous treachery! Can this be so?                61
That in alliance, amity, and oaths
There should be found such false dissembling guile?
KING HENRY
What? Doth my uncle Burgundy revolt?
GLOUCESTER
He doth, my lord, and is become your foe.             65
KING HENRY
Is that the worst this letter doth contain?
GLOUCESTER
It is the worst, and all, my lord, he writes.
KING HENRY
Why then, Lord Talbot there shall talk with him
And give him chastisement for this abuse.
(*To Talbot*) How say you, my lord? Are you not
    content?                                          70
TALBOT
Content, my liege? Yes. But that I am prevented,
I should have begged I might have been employed.
KING HENRY
Then gather strength and march unto him straight.
Let him perceive how ill we brook his treason,
And what offence it is to flout his friends.          75
TALBOT
I go, my lord, in heart desiring still
You may behold confusion of your foes.           *Exit*
    *Enter Vernon wearing a white rose, and Basset*
    *wearing a red rose*
VERNON (*to King Henry*)
Grant me the combat, gracious sovereign.
BASSET (*to King Henry*)
And me, my lord; grant me the combat, too.
RICHARD DUKE OF YORK (*to King Henry, pointing to Vernon*)
This is my servant; hear him, noble Prince.           80
SOMERSET (*to King Henry, pointing to Basset*)
And this is mine, sweet Henry; favour him.
KING HENRY
Be patient, lords, and give them leave to speak.
Say, gentlemen, what makes you thus exclaim,
And wherefore crave you combat, or with whom?
VERNON
With him, my lord; for he hath done me wrong.         85
BASSET
And I with him; for he hath done me wrong.
KING HENRY
What is that wrong whereof you both complain?
First let me know, and then I'll answer you.
BASSET
Crossing the sea from England into France,
This fellow here with envious carping tongue          90
Upbraided me about the rose I wear,
Saying the sanguine colour of the leaves
Did represent my master's blushing cheeks
When stubbornly he did repugn the truth
About a certain question in the law                   95
Argued betwixt the Duke of York and him,
With other vile and ignominious terms;
In confutation of which rude reproach,
And in defence of my lord's worthiness,
I crave the benefit of law of arms.                  100

VERNON
And that is my petition, noble lord;
For though he seem with forgèd quaint conceit
To set a gloss upon his bold intent,
Yet know, my lord, I was provoked by him,
And he first took exceptions at this badge,          105
Pronouncing that the paleness of this flower
Bewrayed the faintness of my master's heart.
RICHARD DUKE OF YORK
Will not this malice, Somerset, be left?
SOMERSET
Your private grudge, my lord of York, will out,
Though ne'er so cunningly you smother it.            110
KING HENRY
Good Lord, what madness rules in brainsick men
When for so slight and frivolous a cause
Such factious emulations shall arise?
Good cousins both of York and Somerset,
Quiet yourselves, I pray, and be at peace.           115
RICHARD DUKE OF YORK
Let this dissension first be tried by fight,
And then your highness shall command a peace.
SOMERSET
The quarrel toucheth none but us alone;
Betwixt ourselves let us decide it then.
RICHARD DUKE OF YORK
There is my pledge. Accept it, Somerset.             120
VERNON (*to King Henry*)
Nay, let it rest where it began at first.
BASSET (*to King Henry*)
Confirm it so, mine honourable lord.
GLOUCESTER
Confirm it so? Confounded be your strife,
And perish ye with your audacious prate!
Presumptuous vassals, are you not ashamed            125
With this immodest clamorous outrage
To trouble and disturb the King and us?
And you, my lords, methinks you do not well
To bear with their perverse objections,
Much less to take occasion from their mouths         130
To raise a mutiny betwixt yourselves.
Let me persuade you take a better course.
EXETER
It grieves his highness. Good my lords, be friends.
KING HENRY
Come hither, you that would be combatants.
Henceforth I charge you, as you love our favour,     135
Quite to forget this quarrel and the cause.
And you, my lords, remember where we are—
In France, amongst a fickle wavering nation.
If they perceive dissension in our looks,
And that within ourselves we disagree,               140
How will their grudging stomachs be provoked
To wilful disobedience, and rebel!
Beside, what infamy will there arise
When foreign princes shall be certified
That for a toy, a thing of no regard,                145
King Henry's peers and chief nobility
Destroyed themselves and lost the realm of France!
O, think upon the conquest of my father,
My tender years, and let us not forgo
That for a trifle that was bought with blood.        150
Let me be umpire in this doubtful strife.
I see no reason, if I wear this rose,
    *He takes a red rose*
That anyone should therefore be suspicious

I more incline to Somerset than York.
Both are my kinsmen, and I love them both.    155
As well they may upbraid me with my crown
Because, forsooth, the King of Scots is crowned.
But your discretions better can persuade
Than I am able to instruct or teach,
And therefore, as we hither came in peace,    160
So let us still continue peace and love.
Cousin of York, we institute your grace
To be our regent in these parts of France;
And good my lord of Somerset, unite
Your troops of horsemen with his bands of foot,    165
And like true subjects, sons of your progenitors,
Go cheerfully together and digest
Your angry choler on your enemies.
Ourself, my Lord Protector, and the rest,
After some respite, will return to Calais,    170
From thence to England, where I hope ere long
To be presented by your victories
With Charles, Alençon, and that traitorous rout.
       *Flourish. Exeunt all but York, Warwick,*
                         *Vernon, and Exeter*

WARWICK
My lord of York, I promise you, the King
Prettily, methought, did play the orator.    175
RICHARD DUKE OF YORK
And so he did; but yet I like it not
In that he wears the badge of Somerset.
WARWICK
Tush, that was but his fancy; blame him not.
I dare presume, sweet Prince, he thought no harm.
RICHARD DUKE OF YORK
An if I wist he did—but let it rest.    180
Other affairs must now be managèd.
                  *Exeunt all but Exeter*

EXETER
Well didst thou, Richard, to suppress thy voice;
For had the passions of thy heart burst out
I fear we should have seen deciphered there
More rancorous spite, more furious raging broils,    185
Than yet can be imagined or supposed.
But howsoe'er, no simple man that sees
This jarring discord of nobility,
This shouldering of each other in the court,
This factious bandying of their favourites,    190
But that it doth presage some ill event.
'Tis much when sceptres are in children's hands,
But more when envy breeds unkind division:
There comes the ruin, there begins confusion.    *Exit*

**4.2**    *Enter Lord Talbot with a trumpeter and drummer*
          *and soldiers before Bordeaux*
TALBOT
Go to the gates of Bordeaux, trumpeter.
Summon their general unto the wall.
       *The trumpeter sounds a parley. Enter French*
       *General, aloft*
English John Talbot, captain, calls you forth,
Servant in arms to Harry King of England;
And thus he would: open your city gates,    5
Be humble to us, call my sovereign yours
And do him homage as obedient subjects,
And I'll withdraw me and my bloody power.
But if you frown upon this proffered peace,
You tempt the fury of my three attendants—    10

Lean famine, quartering steel, and climbing fire—
Who in a moment even with the earth
Shall lay your stately and air-braving towers
If you forsake the offer of their love.
GENERAL
Thou ominous and fearful owl of death,    15
Our nation's terror and their bloody scourge,
The period of thy tyranny approacheth.
On us thou canst not enter but by death,
For I protest we are well fortified
And strong enough to issue out and fight.    20
If thou retire, the Dauphin well appointed
Stands with the snares of war to tangle thee.
On either hand thee there are squadrons pitched
To wall thee from the liberty of flight,
And no way canst thou turn thee for redress    25
But death doth front thee with apparent spoil,
And pale destruction meets thee in the face.
Ten thousand French have ta'en the sacrament
To fire their dangerous artillery
Upon no Christian soul but English Talbot.    30
Lo, there thou stand'st, a breathing valiant man
Of an invincible unconquered spirit.
This is the latest glory of thy praise,
That I thy enemy due thee withal,
For ere the glass that now begins to run    35
Finish the process of his sandy hour,
These eyes that see thee now well colourèd
Shall see thee withered, bloody, pale, and dead.
        *Drum afar off*
Hark, hark, the Dauphin's drum, a warning bell,
Sings heavy music to thy timorous soul,    40
And mine shall ring thy dire departure out.    *Exit*
TALBOT
He fables not. I hear the enemy.
Out, some light horsemen, and peruse their wings.
                 ⌈*Exit one or more*⌉
O negligent and heedless discipline,
How are we parked and bounded in a pale!—    45
A little herd of England's timorous deer
Mazed with a yelping kennel of French curs.
If we be English deer, be then in blood,
Not rascal-like to fall down with a pinch,
But rather, moody-mad and desperate stags,    50
Turn on the bloody hounds with heads of steel
And make the cowards stand aloof at bay.
Sell every man his life as dear as mine
And they shall find dear deer of us, my friends.
God and Saint George, Talbot and England's right,    55
Prosper our colours in this dangerous fight!    *Exeunt*

**4.3**    *Enter a Messenger that meets the Duke of York.*
          *Enter Richard Duke of York with a trumpeter and*
          *many soldiers*
RICHARD DUKE OF YORK
Are not the speedy scouts returned again
That dogged the mighty army of the Dauphin?
MESSENGER
They are returned, my lord, and give it out
That he is marched to Bordeaux with his power
To fight with Talbot. As he marched along,    5
By your espials were discoverèd
Two mightier troops than that the Dauphin led,
Which joined with him and made their march for
   Bordeaux.

RICHARD DUKE OF YORK
A plague upon that villain Somerset
That thus delays my promisèd supply                    10
Of horsemen that were levied for this siege!
Renownèd Talbot doth expect my aid,
And I am louted by a traitor villain
And cannot help the noble chevalier.
God comfort him in this necessity;                    15
If he miscarry, farewell wars in France!
    *Enter another messenger, Sir William Lucy*
LUCY
Thou princely leader of our English strength,
Never so needful on the earth of France,
Spur to the rescue of the noble Talbot,
Who now is girdled with a waste of iron            20
And hemmed about with grim destruction.
To Bordeaux, warlike Duke; to Bordeaux, York,
Else farewell Talbot, France, and England's honour.
RICHARD DUKE OF YORK
O God, that Somerset, who in proud heart
Doth stop my cornets, were in Talbot's place!        25
So should we save a valiant gentleman
By forfeiting a traitor and a coward.
Mad ire and wrathful fury makes me weep,
That thus we die while remiss traitors sleep.
LUCY
O, send some succour to the distressed lord.        30
RICHARD DUKE OF YORK
He dies, we lose; I break my warlike word;
We mourn, France smiles; we lose, they daily get,
All 'long of this vile traitor Somerset.
LUCY
Then God take mercy on brave Talbot's soul,
And on his son young John, who two hours since    35
I met in travel toward his warlike father.
This seven years did not Talbot see his son,
And now they meet where both their lives are done.
RICHARD DUKE OF YORK
Alas, what joy shall noble Talbot have
To bid his young son welcome to his grave?        40
Away—vexation almost stops my breath
That sundered friends greet in the hour of death.
Lucy, farewell. No more my fortune can
But curse the cause I cannot aid the man.
Maine, Blois, Poitiers, and Tours are won away      45
'Long all of Somerset and his delay.
    *Exeunt all but Lucy*
LUCY
Thus while the vulture of sedition
Feeds in the bosom of such great commanders,
Sleeping neglection doth betray to loss
The conquest of our scarce-cold conqueror,          50
That ever-living man of memory
Henry the Fifth. Whiles they each other cross,
Lives, honours, lands, and all hurry to loss.      ⌜*Exit*⌝

**4.4**    *Enter the Duke of Somerset with his army*
SOMERSET (*to a Captain*)
It is too late, I cannot send them now.
This expedition was by York and Talbot
Too rashly plotted. All our general force
Might with a sally of the very town
Be buckled with. The over-daring Talbot            5
Hath sullied all his gloss of former honour
By this unheedful, desperate, wild adventure.
York set him on to fight and die in shame
That, Talbot dead, great York might bear the name.

⌜*Enter Lucy*⌝
CAPTAIN
Here is Sir William Lucy, who with me              10
Set from our o'ermatched forces forth for aid.
SOMERSET
How now, Sir William, whither were you sent?
LUCY
Whither, my lord? From bought and sold Lord Talbot,
Who, ringed about with bold adversity,
Cries out for noble York and Somerset              15
To beat assailing death from his weak legions;
And whiles the honourable captain there
Drops bloody sweat from his war-wearied limbs
And, unadvantaged, ling'ring looks for rescue,
You his false hopes, the trust of England's honour,  20
Keep off aloof with worthless emulation.
Let not your private discord keep away
The levied succours that should lend him aid,
While he, renownèd noble gentleman,
Yield up his life unto a world of odds.            25
Orléans the Bastard, Charles, and Burgundy,
Alençon, René, compass him about,
And Talbot perisheth by your default.
SOMERSET
York set him on; York should have sent him aid.
LUCY
And York as fast upon your grace exclaims,          30
Swearing that you withhold his levied horse
Collected for this expedition.
SOMERSET
York lies. He might have sent and had the horse.
I owe him little duty and less love,
And take foul scorn to fawn on him by sending.      35
LUCY
The fraud of England, not the force of France,
Hath now entrapped the noble-minded Talbot.
Never to England shall he bear his life,
But dies betrayed to fortune by your strife.
SOMERSET
Come, go. I will dispatch the horsemen straight.    40
Within six hours they will be at his aid.
LUCY
Too late comes rescue. He is ta'en or slain,
For fly he could not if he would have fled,
And fly would Talbot never, though he might.
SOMERSET
If he be dead, brave Talbot, then adieu.            45
LUCY
His fame lives in the world, his shame in you.
    *Exeunt* ⌜*severally*⌝

**4.5**    *Enter Lord Talbot and his son John*
TALBOT
O young John Talbot, I did send for thee
To tutor thee in stratagems of war,
That Talbot's name might be in thee revived
When sapless age and weak unable limbs
Should bring thy father to his drooping chair.      5
But O—malignant and ill-boding stars!—
Now thou art come unto a feast of death,
A terrible and unavoided danger.
Therefore, dear boy, mount on my swiftest horse,
And I'll direct thee how thou shalt escape          10
By sudden flight. Come, dally not, be gone.
JOHN
Is my name Talbot, and am I your son,
And shall I fly? O, if you love my mother,

Dishonour not her honourable name
To make a bastard and a slave of me.    15
The world will say he is not Talbot's blood
That basely fled when noble Talbot stood.

TALBOT
Fly to revenge my death if I be slain.

JOHN
He that flies so will ne'er return again.

TALBOT
If we both stay, we both are sure to die.   20

JOHN
Then let me stay and, father, do you fly.
Your loss is great; so your regard should be.
My worth unknown, no loss is known in me.
Upon my death the French can little boast;
In yours they will: in you all hopes are lost.  25
Flight cannot stain the honour you have won,
But mine it will, that no exploit have done.
You fled for vantage, everyone will swear,
But if I bow, they'll say it was for fear.
There is no hope that ever I will stay    30
If the first hour I shrink and run away.
Here on my knee I beg mortality
Rather than life preserved with infamy.

TALBOT
Shall all thy mother's hopes lie in one tomb?

JOHN
Ay, rather than I'll shame my mother's womb. 35

TALBOT
Upon my blessing I command thee go.

JOHN
To fight I will, but not to fly the foe.

TALBOT
Part of thy father may be saved in thee.

JOHN
No part of him but will be shamed in me.

TALBOT
Thou never hadst renown, nor canst not lose it. 40

JOHN
Yes, your renownèd name—shall flight abuse it?

TALBOT
Thy father's charge shall clear thee from that stain.

JOHN
You cannot witness for me, being slain.
If death be so apparent, then both fly.

TALBOT
And leave my followers here to fight and die? 45
My age was never tainted with such shame.

JOHN
And shall my youth be guilty of such blame?
No more can I be severed from your side
Than can yourself your self in twain divide.
Stay, go, do what you will: the like do I,   50
For live I will not if my father die.

TALBOT
Then here I take my leave of thee, fair son,
Born to eclipse thy life this afternoon.
Come, side by side together live and die,
And soul with soul from France to heaven fly. *Exeunt*

**4.6** *Alarum. Excursions, wherein Lord Talbot's son John*
   *is hemmed about by French soldiers and Talbot*
   *rescues him.* ⌈*The English drive off the French*⌉

TALBOT
Saint George and victory! Fight, soldiers, fight!
The Regent hath with Talbot broke his word,

And left us to the rage of France his sword.
Where is John Talbot? (*To John*) Pause and take thy
  breath.
I gave thee life, and rescued thee from death.  5

JOHN
O twice my father, twice am I thy son:
The life thou gav'st me first was lost and done
Till with thy warlike sword, despite of fate,
To my determined time thou gav'st new date.

TALBOT
When from the Dauphin's crest thy sword struck fire
It warmed thy father's heart with proud desire  11
Of bold-faced victory. Then leaden age,
Quickened with youthful spleen and warlike rage,
Beat down Alençon, Orléans, Burgundy,
And from the pride of Gallia rescued thee.    15
The ireful Bastard Orléans, that drew blood
From thee, my boy, and had the maidenhood
Of thy first fight, I soon encounterèd,
And interchanging blows, I quickly shed
Some of his bastard blood, and in disgrace   20
Bespoke him thus: 'Contaminated, base,
And misbegotten blood I spill of thine,
Mean and right poor, for that pure blood of mine
Which thou didst force from Talbot, my brave boy.'
Here, purposing the Bastard to destroy,     25
Came in strong rescue. Speak thy father's care:
Art thou not weary, John? How dost thou fare?
Wilt thou yet leave the battle, boy, and fly,
Now thou art sealed the son of chivalry?
Fly to revenge my death when I am dead;   30
The help of one stands me in little stead.
O, too much folly is it, well I wot,
To hazard all our lives in one small boat.
If I today die not with Frenchmen's rage,
Tomorrow I shall die with mickle age.     35
By me they nothing gain, and if I stay
'Tis but the short'ning of my life one day.
In thee thy mother dies, our household's name,
My death's revenge, thy youth, and England's fame.
All these and more we hazard by thy stay;   40
All these are saved if thou wilt fly away.

JOHN
The sword of Orléans hath not made me smart;
These words of yours draw life-blood from my heart.
On that advantage, bought with such a shame,
To save a paltry life and slay bright fame,    45
Before young Talbot from old Talbot fly
The coward horse that bears me fall and die;
And like me to the peasant boys of France,
To be shame's scorn and subject of mischance!
Surely, by all the glory you have won,     50
An if I fly I am not Talbot's son.
Then talk no more of flight; it is no boot.
If son to Talbot, die at Talbot's foot.

TALBOT
Then follow thou thy desp'rate sire of Crete,
Thou Icarus; thy life to me is sweet.     55
If thou wilt fight, fight by thy father's side,
And commendable proved, let's die in pride. *Exeunt*

**4.7** *Alarum. Excursions. Enter old Lord Talbot led by a*
   *Servant*

TALBOT
Where is my other life? Mine own is gone.
O where's young Talbot, where is valiant John?

Triumphant death smeared with captivity,
Young Talbot's valour makes me smile at thee.
When he perceived me shrink and on my knee,    5
His bloody sword he brandished over me,
And like a hungry lion did commence
Rough deeds of rage and stern impatience.
But when my angry guardant stood alone,
Tend'ring my ruin and assailed of none,    10
Dizzy-eyed fury and great rage of heart
Suddenly made him from my side to start
Into the clust'ring battle of the French,
And in that sea of blood my boy did drench
His over-mounting spirit; and there died    15
My Icarus, my blossom, in his pride.
          *Enter English soldiers with John Talbot's body,*
          *borne*
SERVANT
O my dear lord, lo where your son is borne.
TALBOT
Thou antic death, which laugh'st us here to scorn,
Anon from thy insulting tyranny,
Coupled in bonds of perpetuity,    20
Two Talbots wingèd through the lither sky
In thy despite shall scape mortality.
(*To John*) O thou whose wounds become hard-favoured
          death,
Speak to thy father ere thou yield thy breath.
Brave death by speaking, whether he will or no;    25
Imagine him a Frenchman and thy foe.—
Poor boy, he smiles, methinks, as who should say
'Had death been French, then death had died today'.
Come, come, and lay him in his father's arms.
          *Soldiers lay John in Talbot's arms*
My spirit can no longer bear these harms.    30
Soldiers, adieu. I have what I would have,
Now my old arms are young John Talbot's grave.
          *He dies.* ⌈*Alarum.*⌉ *Exeunt soldiers leaving the*
                                                            *bodies*
          *Enter Charles the Dauphin, the dukes of Alençon*
          *and Burgundy, the Bastard of Orléans, and Joan la*
          *Pucelle*
CHARLES
Had York and Somerset brought rescue in,
We should have found a bloody day of this.
BASTARD
How the young whelp of Talbot's, raging wood,    35
Did flesh his puny sword in Frenchmen's blood!
JOAN
Once I encountered him, and thus I said:
'Thou maiden youth, be vanquished by a maid.'
But with a proud, majestical high scorn
He answered thus: 'Young Talbot was not born    40
To be the pillage of a giglot wench.'
So rushing in the bowels of the French,
He left me proudly, as unworthy fight.
BURGUNDY
Doubtless he would have made a noble knight.
See where he lies inhearsèd in the arms    45
Of the most bloody nurser of his harms.
BASTARD
Hew them to pieces, hack their bones asunder,
Whose life was England's glory, Gallia's wonder.
CHARLES
O no, forbear; for that which we have fled
During the life, let us not wrong it dead.    50

          *Enter Sir William Lucy* ⌈*with a French herald*⌉
LUCY
Herald, conduct me to the Dauphin's tent
To know who hath obtained the glory of the day.
CHARLES
On what submissive message art thou sent?
LUCY
Submission, Dauphin? 'Tis a mere French word.
We English warriors wot not what it means.    55
I come to know what prisoners thou hast ta'en,
And to survey the bodies of the dead.
CHARLES
For prisoners ask'st thou? Hell our prison is.
But tell me whom thou seek'st.
LUCY
But where's the great Alcides of the field,    60
Valiant Lord Talbot, Earl of Shrewsbury,
Created for his rare success in arms
Great Earl of Wexford, Waterford, and Valence,
Lord Talbot of Goodrich and Urchinfield,
Lord Strange of Blackmere, Lord Verdun of Alton,    65
Lord Cromwell of Wingfield, Lord Furnival of Sheffield,
The thrice victorious lord of Falconbridge,
Knight of the noble order of Saint George,
Worthy Saint Michael and the Golden Fleece,
Great *Maréchal* to Henry the Sixth    70
Of all his wars within the realm of France?
JOAN
Here's a silly, stately style indeed.
The Turk, that two-and-fifty kingdoms hath,
Writes not so tedious a style as this.
Him that thou magnifi'st with all these titles    75
Stinking and flyblown lies here at our feet.
LUCY
Is Talbot slain, the Frenchmen's only scourge,
Your kingdom's terror and black Nemesis?
O, were mine eye-balls into bullets turned,
That I in rage might shoot them at your faces!    80
O, that I could but call these dead to life!—
It were enough to fright the realm of France.
Were but his picture left amongst you here
It would amaze the proudest of you all.
Give me their bodies, that I may bear them hence    85
And give them burial as beseems their worth.
JOAN (*to Charles*)
I think this upstart is old Talbot's ghost,
He speaks with such a proud commanding spirit.
For God's sake let him have them. To keep them here
They would but stink and putrefy the air.    90
CHARLES  Go, take their bodies hence.
LUCY
I'll bear them hence, but from their ashes shall be
          reared
A phoenix that shall make all France afeard.
CHARLES
So we be rid of them, do with them what thou wilt.
          ⌈*Exeunt Lucy and herald with the bodies*⌉
And now to Paris in this conquering vein.    95
All will be ours, now bloody Talbot's slain.    *Exeunt*

**5.1**    *Sennet. Enter King Henry, the Dukes of Gloucester*
          *and Exeter,* ⌈*and others*⌉
KING HENRY (*to Gloucester*)
Have you perused the letters from the Pope,
The Emperor, and the Earl of Armagnac?

GLOUCESTER
I have, my lord, and their intent is this:
They humbly sue unto your excellence
To have a godly peace concluded of                          5
Between the realms of England and of France.
KING HENRY
How doth your grace affect their motion?
GLOUCESTER
Well, my good lord, and as the only means
To stop effusion of our Christian blood
And 'stablish quietness on every side.                      10
KING HENRY
Ay, marry, uncle; for I always thought
It was both impious and unnatural
That such immanity and bloody strife
Should reign among professors of one faith.
GLOUCESTER
Beside, my lord, the sooner to effect                       15
And surer bind this knot of amity,
The Earl of Armagnac, near knit to Charles—
A man of great authority in France—
Proffers his only daughter to your grace
In marriage, with a large and sumptuous dowry.             20
KING HENRY
Marriage, uncle? Alas, my years are young,
And fitter is my study and my books
Than wanton dalliance with a paramour.
Yet call th'ambassadors,            ⌈Exit one or more⌉
                          and as you please,
So let them have their answers every one.                   25
I shall be well content with any choice
Tends to God's glory and my country's weal.
          Enter the Bishop of Winchester, now in cardinal's
          habit, and three ambassadors, one a Papal Legate
EXETER (aside)
What, is my lord of Winchester installed
And called unto a cardinal's degree?
Then I perceive that will be verified                       30
Henry the Fifth did sometime prophesy:
'If once he come to be a cardinal,
He'll make his cap co-equal with the crown.'
KING HENRY
My lords ambassadors, your several suits
Have been considered and debated on.                        35
Your purpose is both good and reasonable,
And therefore are we certainly resolved
To draw conditions of a friendly peace,
Which by my lord of Winchester we mean
Shall be transported presently to France.                   40
GLOUCESTER ⌈to ambassadors⌉
And for the proffer of my lord your master,
I have informed his highness so at large
As, liking of the lady's virtuous gifts,
Her beauty, and the value of her dower,
He doth intend she shall be England's queen.               45
KING HENRY ⌈to ambassadors⌉
In argument and proof of which contract
Bear her this jewel, pledge of my affection.
(To Gloucester) And so, my lord Protector, see them
          guarded
And safely brought to Dover, wherein shipped,
Commit them to the fortune of the sea.                      50
          Exeunt ⌈severally⌉ all but Winchester and
                                        ⌈Legate⌉

WINCHESTER
Stay, my lord legate; you shall first receive
The sum of money which I promisèd

Should be delivered to his holiness
For clothing me in these grave ornaments.
LEGATE
I will attend upon your lordship's leisure.     ⌈Exit⌉
WINCHESTER
Now Winchester will not submit, I trow,                     56
Or be inferior to the proudest peer.
Humphrey of Gloucester, thou shalt well perceive
That nor in birth or for authority
The Bishop will be overborne by thee.                       60
I'll either make thee stoop and bend thy knee,
Or sack this country with a mutiny.             ⌈Exit⌉

5.2   Enter Charles the Dauphin ⌈reading a letter⌉, the
      Dukes of Burgundy and Alençon, the Bastard of
      Orléans, René Duke of Anjou, and Joan la Pucelle
CHARLES
These news, my lords, may cheer our drooping spirits.
'Tis said the stout Parisians do revolt
And turn again unto the warlike French.
ALENÇON
Then march to Paris, royal Charles of France,
And keep not back your powers in dalliance.                 5
JOAN
Peace be amongst them if they turn to us;
Else, ruin combat with their palaces!
          Enter a Scout
SCOUT
Success unto our valiant general,
And happiness to his accomplices.
CHARLES
What tidings send our scouts? I prithee speak.             10
SCOUT
The English army, that divided was
Into two parties, is now conjoined in one,
And means to give you battle presently.
CHARLES
Somewhat too sudden, sirs, the warning is;
But we will presently provide for them.                     15
BURGUNDY
I trust the ghost of Talbot is not there.
⌈JOAN⌉
Now he is gone, my lord, you need not fear.
Of all base passions, fear is most accursed.
Command the conquest, Charles, it shall be thine;
Let Henry fret and all the world repine.                    20
CHARLES
Then on, my lords; and France be fortunate!    Exeunt

5.3   Alarum. Excursions. Enter Joan la Pucelle
JOAN
The Regent conquers, and the Frenchmen fly.
Now help, ye charming spells and periapts,
And ye choice spirits that admonish me
And give me signs of future accidents.
          Thunder
You speedy helpers, that are substitutes                    5
Under the lordly monarch of the north,
Appear, and aid me in this enterprise.
          Enter Fiends
This speed and quick appearance argues proof
Of your accustomed diligence to me.
Now, ye familiar spirits that are culled                    10
Out of the powerful regions under earth,
Help me this once, that France may get the field.
          They walk and speak not
O, hold me not with silence overlong!

Where I was wont to feed you with my blood,
I'll lop a member off and give it you                    15
In earnest of a further benefit,
So you do condescend to help me now.
    *They hang their heads*
No hope to have redress? My body shall
Pay recompense if you will grant my suit.
    *They shake their heads*
Cannot my body nor blood-sacrifice                    20
Entreat you to your wonted furtherance?
Then take my soul—my body, soul, and all—
Before that England give the French the foil.
    *They depart*
See, they forsake me. Now the time is come
That France must vail her lofty-plumèd crest          25
And let her head fall into England's lap.
My ancient incantations are too weak,
And hell too strong for me to buckle with.
Now, France, thy glory droopeth to the dust.     *Exit*

**5.4**   *Excursions. The Dukes of Burgundy and York fight*
      *hand to hand. The French fly. Joan la Pucelle is*
      *taken*

RICHARD DUKE OF YORK
Damsel of France, I think I have you fast.
Unchain your spirits now with spelling charms,
And try if they can gain your liberty.
A goodly prize, fit for the devil's grace!
[*To his soldiers*] See how the ugly witch doth bend her
    brows,                                              5
As if with Circe she would change my shape.
JOAN
Changed to a worser shape thou canst not be.
RICHARD DUKE OF YORK
O, Charles the Dauphin is a proper man.
No shape but his can please your dainty eye.
JOAN
A plaguing mischief light on Charles and thee,        10
And may ye both be suddenly surprised
By bloody hands in sleeping on your beds!
RICHARD DUKE OF YORK
Fell banning hag, enchantress, hold thy tongue.
JOAN
I prithee give me leave to curse awhile.
RICHARD DUKE OF YORK
Curse, miscreant, when thou comest to the stake.      15
                        *Exeunt*

**5.5**   *Alarum. Enter the Earl of Suffolk with Margaret in*
      *his hand*

SUFFOLK
Be what thou wilt, thou art my prisoner.
    *He gazes on her*
O fairest beauty, do not fear nor fly,
For I will touch thee but with reverent hands,
And lay them gently on thy tender side.
I kiss these fingers for eternal peace.                   5
Who art thou? Say, that I may honour thee.
MARGARET
Margaret my name, and daughter to a king,
The King of Naples, whosoe'er thou art.
SUFFOLK
An earl I am, and Suffolk am I called.
Be not offended, nature's miracle,                       10
Thou art allotted to be ta'en by me.
So doth the swan his downy cygnets save,

Keeping them prisoner underneath his wings.
Yet if this servile usage once offend,
Go, and be free again, as Suffolk's friend.              15
    *She is going*
O stay! (*Aside*) I have no power to let her pass.
My hand would free her, but my heart says no.
As plays the sun upon the glassy stream,
Twinkling another counterfeited beam,
So seems this gorgeous beauty to mine eyes.             20
Fain would I woo her, yet I dare not speak.
I'll call for pen and ink, and write my mind.
Fie, de la Pole, disable not thyself!
Hast not a tongue? Is she not here to hear?
Wilt thou be daunted at a woman's sight?                25
Ay, beauty's princely majesty is such
Confounds the tongue, and makes the senses rough.
MARGARET
Say, Earl of Suffolk—if thy name be so—
What ransom must I pay before I pass?
For I perceive I am thy prisoner.                        30
SUFFOLK (*aside*)
How canst thou tell she will deny thy suit
Before thou make a trial of her love?
MARGARET
Why speak'st thou not? What ransom must I pay?
SUFFOLK (*aside*)
She's beautiful, and therefore to be wooed;
She is a woman, therefore to be won.                     35
MARGARET
Wilt thou accept of ransom, yea or no?
SUFFOLK (*aside*)
Fond man, remember that thou hast a wife;
Then how can Margaret be thy paramour?
MARGARET (*aside*)
I were best to leave him, for he will not hear.
SUFFOLK (*aside*)
There all is marred; there lies a cooling card.          40
MARGARET (*aside*)
He talks at random; sure the man is mad.
SUFFOLK (*aside*)
And yet a dispensation may be had.
MARGARET
And yet I would that you would answer me.
SUFFOLK (*aside*)
I'll win this Lady Margaret. For whom?
Why, for my king—tush, that's a wooden thing.           45
MARGARET (*aside*)
He talks of wood. It is some carpenter.
SUFFOLK (*aside*)
Yet so my fancy may be satisfied,
And peace establishèd between these realms.
But there remains a scruple in that too,
For though her father be the King of Naples,            50
Duke of Anjou and Maine, yet is he poor,
And our nobility will scorn the match.
MARGARET
Hear ye, captain? Are you not at leisure?
SUFFOLK (*aside*)
It shall be so, disdain they ne'er so much.
Henry is youthful, and will quickly yield.               55
(*To Margaret*) Madam, I have a secret to reveal.
MARGARET (*aside*)
What though I be enthralled, he seems a knight
And will not any way dishonour me.
SUFFOLK
Lady, vouchsafe to listen what I say.

MARGARET (*aside*)
 Perhaps I shall be rescued by the French,          60
 And then I need not crave his courtesy.
SUFFOLK
 Sweet madam, give me hearing in a cause.
MARGARET (*aside*)
 Tush, women have been captivate ere now.
SUFFOLK Lady, wherefore talk you so?
MARGARET
 I cry you mercy, 'tis but *quid* for *quo*.        65
SUFFOLK
 Say, gentle Princess, would you not suppose
 Your bondage happy to be made a queen?
MARGARET
 To be a queen in bondage is more vile
 Than is a slave in base servility,
 For princes should be free.
SUFFOLK                    And so shall you,        70
 If happy England's royal king be free.
MARGARET
 Why, what concerns his freedom unto me?
SUFFOLK
 I'll undertake to make thee Henry's queen,
 To put a golden sceptre in thy hand,
 And set a precious crown upon thy head,            75
 If thou wilt condescend to be my—
MARGARET                           What?
SUFFOLK His love.
MARGARET
 I am unworthy to be Henry's wife.
SUFFOLK
 No, gentle madam, I unworthy am
 To woo so fair a dame to be his wife                80
 (*Aside*) And have no portion in the choice myself.—
 How say you, madam; are ye so content?
MARGARET
 An if my father please, I am content.
SUFFOLK
 Then call our captains and our colours forth,
 ⌐Enter captains, colours, and trumpeters⌐
 And, madam, at your father's castle walls          85
 We'll crave a parley to confer with him.
      *Sound a parley. Enter René Duke of Anjou on the
      walls*
 See, René, see thy daughter prisoner.
RENÉ
 To whom?
SUFFOLK     To me.
RENÉ                Suffolk, what remedy?
 I am a soldier, and unapt to weep
 Or to exclaim on fortune's fickleness.              90
SUFFOLK
 Yes, there is remedy enough, my lord.
 Assent, and for thy honour give consent
 Thy daughter shall be wedded to my king,
 Whom I with pain have wooed and won thereto;
 And this her easy-held imprisonment                 95
 Hath gained thy daughter princely liberty.
RENÉ
 Speaks Suffolk as he thinks?
SUFFOLK                    Fair Margaret knows
 That Suffolk doth not flatter, face or feign.
RENÉ
 Upon thy princely warrant I descend
 To give thee answer of thy just demand.            100

SUFFOLK
 And here I will expect thy coming.   ⌐*Exit René above*⌐
      *Trumpets sound. Enter René*
RENÉ
 Welcome, brave Earl, into our territories.
 Command in Anjou what your honour pleases.
SUFFOLK
 Thanks, René, happy for so sweet a child,
 Fit to be made companion with a king.              105
 What answer makes your grace unto my suit?
RENÉ
 Since thou dost deign to woo her little worth
 To be the princely bride of such a lord,
 Upon condition I may quietly
 Enjoy mine own, the countries Maine and Anjou,     110
 Free from oppression or the stroke of war,
 My daughter shall be Henry's, if he please.
SUFFOLK
 That is her ransom. I deliver her,
 And those two counties I will undertake
 Your grace shall well and quietly enjoy.           115
RENÉ
 And I again in Henry's royal name,
 As deputy unto that gracious king,
 Give thee her hand for sign of plighted faith.
SUFFOLK
 René of France, I give thee kingly thanks,
 Because this is in traffic of a king.              120
 (*Aside*) And yet methinks I could be well content
 To be mine own attorney in this case.
 (*To René*) I'll over then to England with this news,
 And make this marriage to be solemnized.
 So farewell, René; set this diamond safe           125
 In golden palaces, as it becomes.
RENÉ
 I do embrace thee as I would embrace
 The Christian prince King Henry, were he here.
MARGARET (*to Suffolk*)
 Farewell, my lord. Good wishes, praise, and prayers
 Shall Suffolk ever have of Margaret.               130
      *She is going*
SUFFOLK
 Farewell, sweet madam; but hark you, Margaret—
 No princely commendations to my king?
MARGARET
 Such commendations as becomes a maid,
 A virgin, and his servant, say to him.
SUFFOLK
 Words sweetly placed, and modestly directed.       135
      ⌐*She is going*⌐
 But madam, I must trouble you again—
 No loving token to his majesty?
MARGARET
 Yes, my good lord: a pure unspotted heart,
 Never yet taint with love, I send the King.
SUFFOLK And this withal.                            140
      *He kisses her*
MARGARET
 That for thyself; I will not so presume
 To send such peevish tokens to a king.
      ⌐*Exeunt René and Margaret*⌐
SUFFOLK ⌐*aside*⌐
 O, wert thou for myself!—but Suffolk, stay.
 Thou mayst not wander in that labyrinth.
 There Minotaurs and ugly treasons lurk.            145

Solicit Henry with her wondrous praise.
Bethink thee on her virtues that surmount,
Mad natural graces that extinguish art.
Repeat their semblance often on the seas,
That when thou com'st to kneel at Henry's feet     150
Thou mayst bereave him of his wits with wonder.
*[Exeunt]*

**5.6**   *Enter Richard Duke of York, the Earl of Warwick,
and a Shepherd*

RICHARD DUKE OF YORK
Bring forth that sorceress condemned to burn.
*[Enter Joan la Pucelle guarded]*
SHEPHERD
Ah, Joan, this kills thy father's heart outright.
Have I sought every country far and near,
And now it is my chance to find thee out
Must I behold thy timeless cruel death?     5
Ah Joan, sweet daughter Joan, I'll die with thee.
JOAN
Decrepit miser, base ignoble wretch,
I am descended of a gentler blood.
Thou art no father nor no friend of mine.
SHEPHERD
Out, out!—My lords, an't please you, 'tis not so.     10
I did beget her, all the parish knows.
Her mother liveth yet, can testify
She was the first fruit of my bach'lorship.
WARWICK *(to Joan)*
Graceless, wilt thou deny thy parentage?
RICHARD DUKE OF YORK
This argues what her kind of life hath been—     15
Wicked and vile; and so her death concludes.
SHEPHERD
Fie, Joan, that thou wilt be so obstacle.
God knows thou art a collop of my flesh,
And for thy sake have I shed many a tear.
Deny me not, I prithee, gentle Joan.     20
JOAN
Peasant, avaunt! *(To the English)* You have suborned
this man
Of purpose to obscure my noble birth.
SHEPHERD *(to the English)*
'Tis true I gave a noble to the priest
The morn that I was wedded to her mother.
*(To Joan)* Kneel down, and take my blessing, good my
girl.     25
Wilt thou not stoop? Now cursèd be the time
Of thy nativity. I would the milk
Thy mother gave thee when thou sucked'st her breast
Had been a little ratsbane for thy sake.
Or else, when thou didst keep my lambs afield,     30
I wish some ravenous wolf had eaten thee.
Dost thou deny thy father, cursèd drab?
*(To the English)* O burn her, burn her! Hanging is too
good.     *Exit*
RICHARD DUKE OF YORK *(to guards)*
Take her away, for she hath lived too long,
To fill the world with vicious qualities.     35
JOAN
First let me tell you whom you have condemned:
Not one begotten of a shepherd swain,
But issued from the progeny of kings;
Virtuous and holy, chosen from above
By inspiration of celestial grace     40

To work exceeding miracles on earth.
I never had to do with wicked spirits;
But you that are polluted with your lusts,
Stained with the guiltless blood of innocents,
Corrupt and tainted with a thousand vices—     45
Because you want the grace that others have,
You judge it straight a thing impossible
To compass wonders but by help of devils.
No, misconceivèd Joan of Arc hath been
A virgin from her tender infancy,     50
Chaste and immaculate in very thought,
Whose maiden-blood thus rigorously effused
Will cry for vengeance at the gates of heaven.
RICHARD DUKE OF YORK
Ay, ay, *(to guards)* away with her to execution.
WARWICK *(to guards)*
And hark ye, sirs: because she is a maid,     55
Spare for no faggots. Let there be enough.
Place barrels of pitch upon the fatal stake,
That so her torture may be shortenèd.
JOAN
Will nothing turn your unrelenting hearts?
Then Joan, discover thine infirmity,     60
That warranteth by law to be thy privilege:
I am with child, ye bloody homicides.
Murder not then the fruit within my womb,
Although ye hale me to a violent death.
RICHARD DUKE OF YORK
Now heaven forfend—the holy maid with child?     65
WARWICK *(to Joan)*
The greatest miracle that e'er ye wrought.
Is all your strict preciseness come to this?
RICHARD DUKE OF YORK
She and the Dauphin have been ingling.
I did imagine what would be her refuge.
WARWICK
Well, go to, we will have no bastards live,     70
Especially since Charles must father it.
JOAN
You are deceived. My child is none of his.
It was Alençon that enjoyed my love.
RICHARD DUKE OF YORK
Alençon, that notorious Machiavel?
It dies an if it had a thousand lives.     75
JOAN
O give me leave, I have deluded you.
'Twas neither Charles nor yet the Duke I named,
But René King of Naples that prevailed.
WARWICK
A married man?—That's most intolerable.
RICHARD DUKE OF YORK
Why, here's a girl; I think she knows not well—     80
There were so many—whom she may accuse.
WARWICK
It's sign she hath been liberal and free.
RICHARD DUKE OF YORK
And yet forsooth she is a virgin pure!
*(To Joan)* Strumpet, thy words condemn thy brat and
thee.
Use no entreaty, for it is in vain.     85
JOAN
Then lead me hence; with whom I leave my curse.
May never glorious sun reflex his beams
Upon the country where you make abode,
But darkness and the gloomy shade of death

Environ you till mischief and despair                                    90
Drive you to break your necks or hang yourselves.
*Enter the Bishop of Winchester, now Cardinal*

RICHARD DUKE OF YORK (*to Joan*)
Break thou in pieces, and consume to ashes,
Thou foul accursèd minister of hell.

⌐*Exit Joan, guarded*⌐

WINCHESTER
Lord Regent, I do greet your excellence
With letters of commission from the King.                                95
For know, my lords, the states of Christendom,
Moved with remorse of these outrageous broils,
Have earnestly implored a general peace
Betwixt our nation and the aspiring French,
And here at hand the Dauphin and his train                              100
Approacheth to confer about some matter.

RICHARD DUKE OF YORK
Is all our travail turned to this effect?
After the slaughter of so many peers,
So many captains, gentlemen, and soldiers
That in this quarrel have been overthrown                               105
And sold their bodies for their country's benefit,
Shall we at last conclude effeminate peace?
Have we not lost most part of all the towns
By treason, falsehood, and by treachery,
Our great progenitors had conquerèd?                                    110
O Warwick, Warwick, I foresee with grief
The utter loss of all the realm of France!

WARWICK
Be patient, York. If we conclude a peace
It shall be with such strict and severe covenants
As little shall the Frenchmen gain thereby.                             115
*Enter Charles the Dauphin, the Duke of Alençon,*
*the Bastard of Orléans, and René Duke of Anjou*

CHARLES
Since, lords of England, it is thus agreed
That peaceful truce shall be proclaimed in France,
We come to be informèd by yourselves
What the conditions of that league must be.

RICHARD DUKE OF YORK
Speak, Winchester; for boiling choler chokes                            120
The hollow passage of my poisoned voice
By sight of these our baleful enemies.

WINCHESTER
Charles and the rest, it is enacted thus:
That, in regard King Henry gives consent,
Of mere compassion and of lenity,                                       125
To ease your country of distressful war
And suffer you to breathe in fruitful peace,
You shall become true liegemen to his crown.
And, Charles, upon condition thou wilt swear
To pay him tribute and submit thyself,                                  130
Thou shalt be placed as viceroy under him,
And still enjoy thy regal dignity.

ALENÇON
Must he be then as shadow of himself?—
Adorn his temples with a coronet,
And yet in substance and authority                                      135
Retain but privilege of a private man?
This proffer is absurd and reasonless.

CHARLES
'Tis known already that I am possessed
With more than half the Gallian territories,
And therein reverenced for their lawful king.                           140
Shall I, for lucre of the rest unvanquished,

Detract so much from that prerogative
As to be called but viceroy of the whole?
No, lord ambassador, I'll rather keep
That which I have than, coveting for more,                              145
Be cast from possibility of all.

RICHARD DUKE OF YORK
Insulting Charles, hast thou by secret means
Used intercession to obtain a league
And, now the matter grows to compromise,
Stand'st thou aloof upon comparison?                                    150
Either accept the title thou usurp'st,
Of benefit proceeding from our king
And not of any challenge of desert,
Or we will plague thee with incessant wars.

RENÉ (*aside to Charles*)
My lord, you do not well in obstinacy                                   155
To cavil in the course of this contract.
If once it be neglected, ten to one
We shall not find like opportunity.

ALENÇON (*aside to Charles*)
To say the truth, it is your policy
To save your subjects from such massacre                                160
And ruthless slaughters as are daily seen
By our proceeding in hostility;
And therefore take this compact of a truce,
Although you break it when your pleasure serves.

WARWICK
How sayst thou, Charles? Shall our condition stand?

CHARLES It shall,                                                       166
Only reserved you claim no interest
In any of our towns of garrison.

RICHARD DUKE OF YORK
Then swear allegiance to his majesty,
As thou art knight, never to disobey                                    170
Nor be rebellious to the crown of England,
Thou nor thy nobles, to the crown of England.
⌐*They swear*⌐
So, now dismiss your army when ye please.
Hang up your ensigns, let your drums be still;
For here we entertain a solemn peace.           *Exeunt*

**5.7**   *Enter the Earl of Suffolk, in conference with King*
        *Henry, and the Dukes of Gloucester and Exeter*

KING HENRY (*to Suffolk*)
Your wondrous rare description, noble Earl,
Of beauteous Margaret hath astonished me.
Her virtues gracèd with external gifts
Do breed love's settled passions in my heart,
And like as rigour of tempestuous gusts                                   5
Provokes the mightiest hulk against the tide,
So am I driven by breath of her renown
Either to suffer shipwreck or arrive
Where I may have fruition of her love.

SUFFOLK
Tush, my good lord, this superficial tale                                10
Is but a preface of her worthy praise.
The chief perfections of that lovely dame,
Had I sufficient skill to utter them,
Would make a volume of enticing lines
Able to ravish any dull conceit;                                         15
And, which is more, she is not so divine,
So full replete with choice of all delights,
But with as humble lowliness of mind
She is content to be at your command—
Command, I mean, of virtuous chaste intents,                             20
To love and honour Henry as her lord.

KING HENRY
And otherwise will Henry ne'er presume.
(*To Gloucester*) Therefore, my lord Protector, give
      consent
That Marg'ret may be England's royal queen.
GLOUCESTER
So should I give consent to flatter sin.                            25
You know, my lord, your highness is betrothed
Unto another lady of esteem.
How shall we then dispense with that contract
And not deface your honour with reproach?
SUFFOLK
As doth a ruler with unlawful oaths,                              30
Or one that, at a triumph having vowed
To try his strength, forsaketh yet the lists
By reason of his adversary's odds.
A poor earl's daughter is unequal odds,
And therefore may be broke without offence.            35
GLOUCESTER
Why, what, I pray, is Margaret more than that?
Her father is no better than an earl,
Although in glorious titles he excel.
SUFFOLK
Yes, my lord; her father is a king,
The King of Naples and Jerusalem,                              40
And of such great authority in France
As his alliance will confirm our peace
And keep the Frenchmen in allegiance.
GLOUCESTER
And so the Earl of Armagnac may do,
Because he is near kinsman unto Charles.                  45
EXETER
Beside, his wealth doth warrant a liberal dower,
Where René sooner will receive than give.
SUFFOLK
A dower, my lords? Disgrace not so your King
That he should be so abject, base, and poor
To choose for wealth and not for perfect love.          50
Henry is able to enrich his queen,
And not to seek a queen to make him rich.
So worthless peasants bargain for their wives,
As market men for oxen, sheep, or horse.
Marriage is a matter of more worth                            55
Than to be dealt in by attorneyship.
Not whom *we* will but whom his grace affects
Must be companion of his nuptial bed.
And therefore, lords, since he affects her most,
That most of all these reasons bindeth us:                  60
In our opinions she should be preferred.
For what is wedlock forcèd but a hell,

An age of discord and continual strife,
Whereas the contrary bringeth bliss,
And is a pattern of celestial peace.                              65
Whom should we match with Henry, being a king,
But Margaret, that is daughter to a king?
Her peerless feature joinèd with her birth
Approves her fit for none but for a king.
Her valiant courage and undaunted spirit,              70
More than in women commonly is seen,
Will answer our hope in issue of a king.
For Henry, son unto a conqueror,
Is likely to beget more conquerors
If with a lady of so high resolve                                75
As is fair Margaret he be linked in love.
Then yield, my lords, and here conclude with me:
That Margaret shall be queen, and none but she.
KING HENRY
Whether it be through force of your report,
My noble lord of Suffolk, or for that                          80
My tender youth was never yet attaint
With any passion of inflaming love,
I cannot tell; but this I am assured:
I feel such sharp dissension in my breast,
Such fierce alarums both of hope and fear,              85
As I am sick with working of my thoughts.
Take therefore shipping; post, my lord, to France;
Agree to any covenants, and procure
That Lady Margaret do vouchsafe to come
To cross the seas to England and be crowned          90
King Henry's faithful and anointed queen.
For your expenses and sufficient charge,
Among the people gather up a tenth.
Be gone, I say; for till you do return
I rest perplexèd with a thousand cares.                    95
(*To Gloucester*) And you, good uncle, banish all offence.
If you do censure me by what you were,
Not what you are, I know it will excuse
This sudden execution of my will.
And so conduct me where from company                100
I may revolve and ruminate my grief.
                                              *Exit* ⌈*with Exeter*⌉
GLOUCESTER
Ay, grief, I fear me, both at first and last.      *Exit*
SUFFOLK
Thus Suffolk hath prevailed, and thus he goes
As did the youthful Paris once to Greece,
With hope to find the like event in love,              105
But prosper better than the Trojan did.
Margaret shall now be queen and rule the King;
But I will rule both her, the King, and realm.     *Exit*

# RICHARD III

IN narrative sequence, *Richard III* follows directly after *Richard Duke of York*, and that play's closing scenes, in which Richard of Gloucester expresses his ambitions for the crown, suggest that Shakespeare had a sequel in mind. But he seems to have gone back to tell the beginning of the story of Henry VI's reign before covering the events from Henry VI's death (in 1471) to the Battle of Bosworth (1485). We have no record of the first performance of *Richard III* (probably in late 1592 or early 1593, outside London); it was printed in 1597, with five reprints before its inclusion in the 1623 Folio.

The principal source of information about Richard III available to Shakespeare was Sir Thomas More's *History of King Richard III* as incorporated in chronicle histories by Edward Hall (1542) and Raphael Holinshed (1577, revised in 1587), both of which Shakespeare seems to have used. His artistic influences include the tragedies of the Roman dramatist Seneca (who was born about 4 BC and died in AD 65), with their ghosts, their rhetorical style, their prominent choruses, and their indirect, highly formal presentation of violent events. (Except for the stabbing of Clarence (1.4) there is no on-stage violence in *Richard III* until the final battle scenes.)

In this play, Shakespeare demonstrates a more complete artistic control of his historical material than in its predecessors: Richard himself is a more dominating central figure than is to be found in any of the earlier plays, historical events are freely manipulated in the interests of an overriding design, and the play's language is more highly patterned and rhetorically unified. That part of the play which shows Richard's bloody progress to the throne is based on the events of some twelve years; the remainder covers the two years of his reign. Shakespeare omits some important events, but invents Richard's wooing of Lady Anne over her father-in-law's coffin, and causes Queen Margaret, who had returned to France in 1476 and who died before Richard became king, to remain in England as a choric figure of grief and retribution. The characterization of Richard as a self-delighting ironist builds upon More. The episodes in which the older women of the play—the Duchess of York, Queen Elizabeth, and Queen Margaret—bemoan their losses, and the climactic procession of ghosts before the final confrontation of Richard with the idealized figure of Richmond, the future Henry VII, help to make *Richard III* the culmination of a tetralogy as well as a masterly poetic drama in its own right. The final speech, in which Richmond, heir to the house of Lancaster and grandfather of Queen Elizabeth I, proclaims the union of 'the white rose and the red' in his marriage to Elizabeth of York, provides a patriotic climax which must have been immensely stirring to the play's early audiences.

Colley Cibber's adaptation (1700) of *Richard III*, incorporating the death of Henry VI, shortening and adapting the play, and making the central role (played by Cibber) even more dominant than it had originally been, held the stage with great success until the late nineteenth century. Since then, Shakespeare's text has been restored (though usually abbreviated—next to *Hamlet*, this is Shakespeare's longest play), and the role of Richard has continued to present a rewarding challenge to leading actors.

# THE PERSONS OF THE PLAY

KING EDWARD IV

DUCHESS OF YORK, his mother

PRINCE EDWARD

Richard, the young Duke of YORK } his sons

George, Duke of CLARENCE

RICHARD, Duke of GLOUCESTER, later KING
  RICHARD } his brothers

Clarence's SON

Clarence's DAUGHTER

QUEEN ELIZABETH, King Edward's wife

Anthony Woodville, Earl RIVERS, her brother

Marquis of DORSET

Lord GRAY } her sons

Sir Thomas VAUGHAN

GHOST OF KING HENRY the Sixth

QUEEN MARGARET, his widow

GHOST OF PRINCE EDWARD, his son

LADY ANNE, Prince Edward's widow

William, LORD HASTINGS, Lord Chamberlain

Lord STANLEY, Earl of Derby, his friend

HENRY EARL OF RICHMOND, later KING HENRY VII, Stanley's
  son-in-law

Earl of OXFORD

Sir James BLUNT } Richmond's followers

Sir Walter HERBERT

Duke of BUCKINGHAM

Duke of NORFOLK

Sir Richard RATCLIFFE

Sir William CATESBY } Richard Gloucester's followers

Sir James TYRREL

Two MURDERERS

A PAGE

CARDINAL

Bishop of ELY

John, a PRIEST

CHRISTOPHER, a Priest

Sir Robert BRACKENBURY, Lieutenant of the Tower of London

Lord MAYOR of London

A SCRIVENER

Hastings, a PURSUIVANT

SHERIFF

Aldermen and Citizens

Attendants, two bishops, messengers, soldiers

# The Tragedy of King Richard the Third

**1.1** *Enter Richard Duke of Gloucester*

RICHARD GLOUCESTER

Now is the winter of our discontent
Made glorious summer by this son of York;
And all the clouds that loured upon our house
In the deep bosom of the ocean buried.
Now are our brows bound with victorious wreaths, 5
Our bruisèd arms hung up for monuments,
Our stern alarums changed to merry meetings,
Our dreadful marches to delightful measures.
Grim-visaged war hath smoothed his wrinkled front,
And now—instead of mounting barbèd steeds 10
To fright the souls of fearful adversaries—
He capers nimbly in a lady's chamber
To the lascivious pleasing of a lute.
But I, that am not shaped for sportive tricks
Nor made to court an amorous looking-glass, 15
I that am rudely stamped and want love's majesty
To strut before a wanton ambling nymph,
I that am curtailed of this fair proportion,
Cheated of feature by dissembling nature,
Deformed, unfinished, sent before my time 20
Into this breathing world scarce half made up—
And that so lamely and unfashionable
That dogs bark at me as I halt by them—
Why, I in this weak piping time of peace
Have no delight to pass away the time, 25
Unless to spy my shadow in the sun
And descant on mine own deformity.
And therefore since I cannot prove a lover
To entertain these fair well-spoken days,
I am determinèd to prove a villain 30
And hate the idle pleasures of these days.
Plots have I laid, inductions dangerous,
By drunken prophecies, libels and dreams
To set my brother Clarence and the King
In deadly hate the one against the other. 35
And if King Edward be as true and just
As I am subtle false and treacherous,
This day should Clarence closely be mewed up
About a prophecy which says that 'G'
Of Edward's heirs the murderer shall be. 40

*Enter George Duke of Clarence, guarded, and Sir
Robert Brackenbury*

Dive, thoughts, down to my soul: here Clarence comes.
Brother, good day. What means this armèd guard
That waits upon your grace?

CLARENCE                                          His majesty,
Tend'ring my person's safety, hath appointed
This conduct to convey me to the Tower. 45

RICHARD GLOUCESTER
Upon what cause?

CLARENCE                    Because my name is George.

RICHARD GLOUCESTER
Alack, my lord, that fault is none of yours.
He should for that commit your godfathers.
Belike his majesty hath some intent
That you should be new-christened in the Tower. 50
But what's the matter, Clarence? May I know?

CLARENCE
Yea, Richard, when I know—for I protest
As yet I do not. But as I can learn
He hearkens after prophecies and dreams,
And from the cross-row plucks the letter 'G' 55
And says a wizard told him that by 'G'
His issue disinherited should be.
And for my name of George begins with 'G',
It follows in his thought that I am he.
These, as I learn, and suchlike toys as these, 60
Hath moved his highness to commit me now.

RICHARD GLOUCESTER
Why, this it is when men are ruled by women.
'Tis not the King that sends you to the Tower;
My Lady Gray, his wife—Clarence, 'tis she
That tempts him to this harsh extremity. 65
Was it not she, and that good man of worship
Anthony Woodeville her brother there,
That made him send Lord Hastings to the Tower,
From whence this present day he is delivered?
We are not safe, Clarence; we are not safe. 70

CLARENCE
By heaven, I think there is no man secure
But the Queen's kindred, and night-walking heralds
That trudge betwixt the King and Mrs Shore.
Heard ye not what an humble suppliant
Lord Hastings was for his delivery? 75

RICHARD GLOUCESTER
Humbly complaining to her deity
Got my Lord Chamberlain his liberty.
I'll tell you what: I think it is our way,
If we will keep in favour with the King,
To be her men and wear her livery. 80
The jealous, o'erworn widow and herself,
Since that our brother dubbed them gentlewomen,
Are mighty gossips in our monarchy.

BRACKENBURY
I beseech your graces both to pardon me.
His majesty hath straitly given in charge 85
That no man shall have private conference,
Of what degree soever, with your brother.

RICHARD GLOUCESTER
Even so. An't please your worship, Brackenbury,
You may partake of anything we say.
We speak no treason, man. We say the King 90
Is wise and virtuous, and his noble Queen
Well struck in years, fair, and not jealous.
We say that Shore's wife hath a pretty foot,
A cherry lip,
A bonny eye, a passing pleasing tongue, 95
And that the Queen's kin are made gentlefolks.
How say you, sir? Can you deny all this?

BRACKENBURY
With this, my lord, myself have naught to do.

RICHARD GLOUCESTER
Naught to do with Mrs Shore? I tell thee, fellow:
He that doth naught with her—excepting one— 100
Were best to do it secretly alone.

BRACKENBURY What one, my lord?

RICHARD GLOUCESTER
    Her husband, knave. Wouldst thou betray me?
BRACKENBURY
    I beseech your grace to pardon me, and do withal
    Forbear your conference with the noble Duke.        105
CLARENCE
    We know thy charge, Brackenbury, and will obey.
RICHARD GLOUCESTER
    We are the Queen's abjects, and must obey.
    Brother, farewell. I will unto the King,
    And whatsoe'er you will employ me in—
    Were it to call King Edward's widow 'sister'—        110
    I will perform it to enfranchise you.
    Meantime, this deep disgrace in brotherhood
    Touches me dearer than you can imagine.
CLARENCE
    I know it pleaseth neither of us well.
RICHARD GLOUCESTER
    Well, your imprisonment shall not be long.        115
    I will deliver you or lie for you.
    Meantime, have patience.
CLARENCE                            I must perforce. Farewell.
        *Exeunt Clarence, Brackenbury, and guard, to*
                                            *the Tower*
RICHARD GLOUCESTER
    Go tread the path that thou shalt ne'er return.
    Simple plain Clarence, I do love thee so
    That I will shortly send thy soul to heaven,        120
    If heaven will take the present at our hands.
    But who comes here? The new-delivered Hastings?
        *Enter Lord Hastings from the Tower*
LORD HASTINGS
    Good time of day unto my gracious lord.
RICHARD GLOUCESTER
    As much unto my good Lord Chamberlain.
    Well are you welcome to the open air.
    How hath your lordship brooked imprisonment?        125
LORD HASTINGS
    With patience, noble lord, as prisoners must.
    But I shall live, my lord, to give them thanks
    That were the cause of my imprisonment.
RICHARD GLOUCESTER
    No doubt, no doubt—and so shall Clarence too,        130
    For they that were your enemies are his,
    And have prevailed as much on him as you.
LORD HASTINGS
    More pity that the eagles should be mewed
    While kites and buzzards prey at liberty.
RICHARD GLOUCESTER What news abroad?        135
LORD HASTINGS
    No news so bad abroad as this at home:
    The King is sickly, weak, and melancholy,
    And his physicians fear him mightily.
RICHARD GLOUCESTER
    Now by Saint Paul, that news is bad indeed.
    O he hath kept an evil diet long,        140
    And overmuch consumed his royal person.
    'Tis very grievous to be thought upon.
    Where is he? In his bed?
LORD HASTINGS                    He is.
RICHARD GLOUCESTER
    Go you before and I will follow you.        *Exit Hastings*
    He cannot live, I hope, and must not die        145
    Till George be packed with post-haste up to heaven.
    I'll in to urge his hatred more to Clarence,
    With lies well steeled with weighty arguments.

    And if I fail not in my deep intent,
    Clarence hath not another day to live—        150
    Which done, God take King Edward to his mercy
    And leave the world for me to bustle in.
    For then I'll marry Warwick's youngest daughter.
    What though I killed her husband and her father?
    The readiest way to make the wench amends        155
    Is to become her husband and her father,
    The which will I: not all so much for love,
    As for another secret close intent,
    By marrying her, which I must reach unto.
    But yet I run before my horse to market.        160
    Clarence still breathes, Edward still lives and reigns;
    When they are gone, then must I count my gains.
                                            *Exit*

1.2        *Enter gentlemen, bearing the corpse of King Henry*
            *the Sixth in an open coffin, with halberdiers to*
            *guard it, Lady Anne being the mourner*
LADY ANNE
    Set down, set down your honourable load,
    If honour may be shrouded in a hearse,
    Whilst I a while obsequiously lament
    Th'untimely fall of virtuous Lancaster.
        *They set the coffin down*
    Poor key-cold figure of a holy king,        5
    Pale ashes of the house of Lancaster,
    Thou bloodless remnant of that royal blood:
    Be it lawful that I invocate thy ghost
    To hear the lamentations of poor Anne,
    Wife to thy Edward, to thy slaughtered son,        10
    Stabbed by the selfsame hand that made these wounds.
    Lo, in these windows that let forth thy life,
    I pour the helpless balm of my poor eyes.
    O cursèd be the hand that made these holes,
    Cursèd the blood that let this blood from hence,        15
    Cursèd the heart that had the heart to do it.
    More direful hap betide that hated wretch
    That makes us wretched by the death of thee
    Than I can wish to wolves, to spiders, toads,
    Or any creeping venomed thing that lives.        20
    If ever he have child, abortive be it,
    Prodigious, and untimely brought to light,
    Whose ugly and unnatural aspect
    May fright the hopeful mother at the view,
    And that be heir to his unhappiness.        25
    If ever he have wife, let her be made
    More miserable by the death of him
    Than I am made by my young lord and thee.—
    Come now towards Chertsey with your holy load,
    Taken from Paul's to be interrèd there,        30
        [*The gentlemen lift the coffin*]
    And still as you are weary of this weight
    Rest you, whiles I lament King Henry's corpse.
        *Enter Richard Duke of Gloucester*
RICHARD GLOUCESTER (*to the gentlemen*)
    Stay, you that bear the corpse, and set it down.
LADY ANNE
    What black magician conjures up this fiend
    To stop devoted charitable deeds?        35
RICHARD GLOUCESTER (*to the gentlemen*)
    Villains, set down the corpse, or by Saint Paul
    I'll make a corpse of him that disobeys.
[HALBERDIER]
    My lord, stand back and let the coffin pass.

RICHARD GLOUCESTER
Unmannered dog, stand thou when I command.
Advance thy halberd higher than my breast,                    40
Or by Saint Paul I'll strike thee to my foot
And spurn upon thee, beggar, for thy boldness.
        *They set the coffin down*
LADY ANNE (*to gentlemen and halberdiers*)
What, do you tremble? Are you all afraid?
Alas, I blame you not, for you are mortal,
And mortal eyes cannot endure the devil.—                     45
Avaunt, thou dreadful minister of hell.
Thou hadst but power over his mortal body;
His soul thou canst not have; therefore be gone.
RICHARD GLOUCESTER
Sweet saint, for charity be not so cursed.
LADY ANNE
Foul devil, for God's sake hence and trouble us not,          50
For thou hast made the happy earth thy hell,
Filled it with cursing cries and deep exclaims.
If thou delight to view thy heinous deeds,
Behold this pattern of thy butcheries.—
O gentlemen, see, see! Dead Henry's wounds                    55
Ope their congealèd mouths and bleed afresh.—
Blush, blush, thou lump of foul deformity,
For 'tis thy presence that ex-hales this blood
From cold and empty veins where no blood dwells.
Thy deed, inhuman and unnatural,                              60
Provokes this deluge supernatural.
O God, which this blood mad'st, revenge his death.
O earth, which this blood drink'st, revenge his death.
Either heav'n with lightning strike the murd'rer dead,
Or earth gape open wide and eat him quick                     65
As thou dost swallow up this good king's blood,
Which his hell-governèd arm hath butcherèd.
RICHARD GLOUCESTER
Lady, you know no rules of charity,
Which renders good for bad, blessings for curses.
LADY ANNE
Villain, thou know'st no law of God nor man.                  70
No beast so fierce but knows some touch of pity.
RICHARD GLOUCESTER
But I know none, and therefore am no beast.
LADY ANNE
O wonderful, when devils tell the truth!
RICHARD GLOUCESTER
More wonderful, when angels are so angry.
Vouchsafe, divine perfection of a woman,                      75
Of these supposèd crimes to give me leave
By circumstance but to acquit myself.
LADY ANNE
Vouchsafe, diffused infection of a man,
Of these known evils but to give me leave
By circumstance t'accuse thy cursèd self.                     80
RICHARD GLOUCESTER
Fairer than tongue can name thee, let me have
Some patient leisure to excuse myself.
LADY ANNE
Fouler than heart can think thee, thou canst make
No excuse current but to hang thyself.
RICHARD GLOUCESTER
By such despair I should accuse myself.                       85
LADY ANNE
And by despairing shalt thou stand excused,
For doing worthy vengeance on thyself
That didst unworthy slaughter upon others.

RICHARD GLOUCESTER
Say that I slew them not.
LADY ANNE                      Then say they were not slain.
But dead they are—and, devilish slave, by thee.               90
RICHARD GLOUCESTER
I did not kill your husband.
LADY ANNE                      Why, then he is alive.
RICHARD GLOUCESTER
Nay, he is dead, and slain by Edward's hand.
LADY ANNE
In thy foul throat thou liest. Queen Margaret saw
Thy murd'rous falchion smoking in his blood,
The which thou once didst bend against her breast,           95
But that thy brothers beat aside the point.
RICHARD GLOUCESTER
I was provokèd by her sland'rous tongue,
That laid their guilt upon my guiltless shoulders.
LADY ANNE
Thou wast provokèd by thy bloody mind,
That never dream'st on aught but butcheries.                 100
Didst thou not kill this king?
RICHARD GLOUCESTER                       I grant ye.
LADY ANNE
Dost grant me, hedgehog? Then God grant me, too,
Thou mayst be damnèd for that wicked deed.
O he was gentle, mild, and virtuous.
RICHARD GLOUCESTER
The better for the King of Heaven that hath him.             105
LADY ANNE
He *is* in heaven, where thou shalt never come.
RICHARD GLOUCESTER
Let him thank me that holp to send him thither,
For he was fitter for that place than earth.
LADY ANNE
And thou unfit for any place but hell.
RICHARD GLOUCESTER
Yes, one place else, if you will hear me name it.            110
LADY ANNE
Some dungeon.
RICHARD GLOUCESTER  Your bedchamber.
LADY ANNE
Ill rest betide the chamber where thou liest.
RICHARD GLOUCESTER
So will it, madam, till I lie with you.
LADY ANNE
I hope so.
RICHARD GLOUCESTER  I know so. But gentle Lady Anne,
To leave this keen encounter of our wits                     115
And fall something into a slower method,
Is not the causer of the timeless deaths
Of these Plantagenets, Henry and Edward,
As blameful as the executioner?
LADY ANNE
Thou wast the cause of that accursed effect.                 120
RICHARD GLOUCESTER
Your beauty was the cause of that effect—
Your beauty that did haunt me in my sleep
To undertake the death of all the world
So I might live one hour in your sweet bosom.
LADY ANNE
If I thought that, I tell thee, homicide,                    125
These nails should rend that beauty from my cheeks.
RICHARD GLOUCESTER
These eyes could not endure sweet beauty's wreck.
You should not blemish it if I stood by.

As all the world is cheerèd by the sun,
So I by that: it is my day, my life.    130
LADY ANNE
Black night o'ershade thy day, and death thy life.
RICHARD GLOUCESTER
Curse not thyself, fair creature: thou art both.
LADY ANNE
I would I were, to be revenged on thee.
RICHARD GLOUCESTER
It is a quarrel most unnatural,
To be revenged on him that loveth you.    135
LADY ANNE
It is a quarrel just and reasonable,
To be revenged on him that killed my husband.
RICHARD GLOUCESTER
He that bereft thee, lady, of thy husband,
Did it to help thee to a better husband.
LADY ANNE
His better doth not breathe upon the earth.    140
RICHARD GLOUCESTER
He lives that loves thee better than he could.
LADY ANNE
Name him.
RICHARD GLOUCESTER   Plantagenet.
LADY ANNE               Why, that was he.
RICHARD GLOUCESTER
The selfsame name, but one of better nature.
LADY ANNE
Where is he?
RICHARD GLOUCESTER  Here.
*She spits at him*
              Why dost thou spit at me?
LADY ANNE
Would it were mortal poison for thy sake.    145
RICHARD GLOUCESTER
Never came poison from so sweet a place.
LADY ANNE
Never hung poison on a fouler toad.
Out of my sight! Thou dost infect mine eyes.
RICHARD GLOUCESTER
Thine eyes, sweet lady, have infected mine.
LADY ANNE
Would they were basilisks to strike thee dead.    150
RICHARD GLOUCESTER
I would they were, that I might die at once,
For now they kill me with a living death.
Those eyes of thine from mine have drawn salt tears,
Shamed their aspects with store of childish drops.
I never sued to friend nor enemy;    155
My tongue could never learn sweet smoothing word;
But now thy beauty is proposed my fee,
My proud heart sues and prompts my tongue to speak.
*She looks scornfully at him*
Teach not thy lip such scorn, for it was made
For kissing, lady, not for such contempt.    160
If thy revengeful heart cannot forgive,
⌈*He kneels and offers her his sword*⌉
Lo, here I lend thee this sharp-pointed sword,
Which if thou please to hide in this true breast
And let the soul forth that adoreth thee,
I lay it naked to the deadly stroke    165
And humbly beg the death upon my knee.
*He lays his breast open; she offers at it with his sword*
Nay, do not pause, for I did kill King Henry;
But 'twas thy beauty that provokèd me.

Nay, now dispatch: 'twas I that stabbed young
   Edward;
But 'twas thy heavenly face that set me on.    170
*She drops the sword*
Take up the sword again, or take up me.
LADY ANNE
Arise, dissembler.
⌈*He rises*⌉
             Though I wish thy death,
I will not be thy executioner.
RICHARD GLOUCESTER
Then bid me kill myself, and I will do it.
LADY ANNE
I have already.
RICHARD GLOUCESTER  That was in thy rage.    175
Speak it again, and even with the word
This hand—which for thy love did kill thy love—
Shall, for thy love, kill a far truer love.
To both their deaths shalt thou be accessary.
LADY ANNE  I would I knew thy heart.    180
RICHARD GLOUCESTER  'Tis figured in my tongue.
LADY ANNE  I fear me both are false.
RICHARD GLOUCESTER  Then never man was true.
LADY ANNE  Well, well, put up your sword.
RICHARD GLOUCESTER  Say then my peace is made.    185
LADY ANNE  That shalt thou know hereafter.
RICHARD GLOUCESTER  But shall I live in hope?
LADY ANNE  All men, I hope, live so.
RICHARD GLOUCESTER  Vouchsafe to wear this ring.
LADY ANNE  To take is not to give.    190
RICHARD GLOUCESTER
Look how my ring encompasseth thy finger;
Even so thy breast encloseth my poor heart.
Wear both of them, for both of them are thine.
And if thy poor devoted servant may
But beg one favour at thy gracious hand,    195
Thou dost confirm his happiness for ever.
LADY ANNE  What is it?
RICHARD GLOUCESTER
That it may please you leave these sad designs
To him that hath most cause to be a mourner,
And presently repair to Crosby House,    200
Where—after I have solemnly interred
At Chertsey monast'ry this noble king,
And wet his grave with my repentant tears—
I will with all expedient duty see you.
For divers unknown reasons, I beseech you    205
Grant me this boon.
LADY ANNE
With all my heart—and much it joys me, too,
To see you are become so penitent.
Tressell and Berkeley, go along with me.
RICHARD GLOUCESTER
Bid me farewell.
LADY ANNE        'Tis more than you deserve.    210
But since you teach me how to flatter you,
Imagine I have said farewell already.
                     *Exeunt two with Anne*
RICHARD GLOUCESTER
Sirs, take up the corpse.
GENTLEMAN        Towards Chertsey, noble lord?
RICHARD GLOUCESTER
No, to Blackfriars; there attend my coming.
          *Exeunt with corpse all but Gloucester*
Was ever woman in this humour wooed?    215
Was ever woman in this humour won?

I'll have her, but I will not keep her long.
What, I that killed her husband and his father,
To take her in her heart's extremest hate,
With curses in her mouth, tears in her eyes,                    220
The bleeding witness of my hatred by,
Having God, her conscience, and these bars against me,
And I no friends to back my suit withal
But the plain devil and dissembling looks—
And yet to win her, all the world to nothing? Ha!    225
Hath she forgot already that brave prince,
Edward her lord, whom I some three months since
Stabbed in my angry mood at Tewkesbury?
A sweeter and a lovelier gentleman,
Framed in the prodigality of nature,                           230
Young, valiant, wise, and no doubt right royal,
The spacious world cannot again afford—
And will she yet abase her eyes on me,
That cropped the golden prime of this sweet prince
And made her widow to a woeful bed?                            235
On me, whose all not equals Edward's moiety?
On me, that halts and am misshapen thus?
My dukedom to a beggarly *denier*,
I do mistake my person all this while.
Upon my life she finds, although I cannot,                     240
Myself to be a marv'lous proper man.
I'll be at charges for a looking-glass
And entertain a score or two of tailors
To study fashions to adorn my body.
Since I am crept in favour with myself,                        245
I will maintain it with some little cost.
But first I'll turn yon fellow in his grave,
And then return lamenting to my love.
Shine out, fair sun, till I have bought a glass,
That I may see my shadow as I pass.                    *Exit*

**1.3**   *Enter Queen Elizabeth, Lord Rivers, ⌈Marquis
         Dorset⌉, and Lord Gray*
RIVERS (*to Elizabeth*)
  Have patience, madam. There's no doubt his majesty
  Will soon recover his accustomed health.
GRAY (*to Elizabeth*)
  In that you brook it ill, it makes him worse.
  Therefore, for God's sake entertain good comfort,
  And cheer his grace with quick and merry eyes.          5
QUEEN ELIZABETH
  If he were dead, what would betide on me?
⌈RIVERS⌉
  No other harm but loss of such a lord.
QUEEN ELIZABETH
  The loss of such a lord includes all harms.
GRAY
  The heavens have blessed you with a goodly son
  To be your comforter when he is gone.                    10
QUEEN ELIZABETH
  Ah, he is young, and his minority
  Is put unto the trust of Richard Gloucester,
  A man that loves not me—nor none of you.
RIVERS
  Is it concluded he shall be Protector?
QUEEN ELIZABETH
  It is determined, not concluded yet;                     15
  But so it must be, if the King miscarry.
         *Enter the Duke of Buckingham and Lord Stanley
         Earl of Derby*
GRAY
  Here come the Lords of Buckingham and Derby.

BUCKINGHAM (*to Elizabeth*)
  Good time of day unto your royal grace.
STANLEY (*to Elizabeth*)
  God make your majesty joyful, as you have been.
QUEEN ELIZABETH
  The Countess Richmond, good my lord of Derby,           20
  To your good prayer will scarcely say 'Amen'.
  Yet, Derby—notwithstanding she's your wife,
  And loves not me—be you, good lord, assured
  I hate not you for her proud arrogance.
STANLEY
  I do beseech you, either not believe                    25
  The envious slanders of her false accusers
  Or, if she be accused on true report,
  Bear with her weakness, which I think proceeds
  From wayward sickness, and no grounded malice.
⌈RIVERS⌉
  Saw you the King today, my lord of Derby?               30
STANLEY
  But now the Duke of Buckingham and I
  Are come from visiting his majesty.
QUEEN ELIZABETH
  With likelihood of his amendment, lords?
BUCKINGHAM
  Madam, good hope: his grace speaks cheerfully.
QUEEN ELIZABETH
  God grant him health. Did you confer with him?         35
BUCKINGHAM
  Ay, madam. He desires to make atonement
  Between the Duke of Gloucester and your brothers,
  And between them and my Lord Chamberlain,
  And sent to warn them to his royal presence.
QUEEN ELIZABETH
  Would all were well! But that will never be.           40
  I fear our happiness is at the height.
         *Enter Richard Duke of Gloucester and Lord Hastings*
RICHARD GLOUCESTER
  They do me wrong, and I will not endure it.
  Who are they that complain unto the King
  That I forsooth am stern and love them not?
  By holy Paul, they love his grace but lightly          45
  That fill his ears with such dissentious rumours.
  Because I cannot flatter and look fair,
  Smile in men's faces, smooth, deceive, and cog,
  Duck with French nods and apish courtesy,
  I must be held a rancorous enemy.                       50
  Cannot a plain man live and think no harm,
  But thus his simple truth must be abused
  With silken, sly, insinuating jacks?
⌈RIVERS⌉
  To whom in all this presence speaks your grace?
RICHARD GLOUCESTER
  To thee, that hast nor honesty nor grace.               55
  When have I injured thee? When done thee wrong?
  Or thee? Or thee? Or any of your faction?
  A plague upon you all! His royal grace—
  Whom God preserve better than you would wish—
  Cannot be quiet scarce a breathing while                60
  But you must trouble him with lewd complaints.
QUEEN ELIZABETH
  Brother of Gloucester, you mistake the matter.
  The King—on his own royal disposition,
  And not provoked by any suitor else—
  Aiming belike at your interior hatred,                  65
  That in your outward action shows itself
  Against my children, brothers, and myself,

Makes him to send, that he may learn the ground
Of your ill will, and thereby to remove it.
RICHARD GLOUCESTER
I cannot tell. The world is grown so bad                    70
That wrens make prey where eagles dare not perch.
Since every jack became a gentleman,
There's many a gentle person made a jack.
QUEEN ELIZABETH
Come, come, we know your meaning, brother
    Gloucester.
You envy my advancement, and my friends'.                    75
God grant we never may have need of you.
RICHARD GLOUCESTER
Meantime, God grants that I have need of you.
Our brother is imprisoned by your means,
Myself disgraced, and the nobility
Held in contempt, while great promotions                    80
Are daily given to ennoble those
That scarce some two days since were worth a noble.
QUEEN ELIZABETH
By him that raised me to this care-full height
From that contented hap which I enjoyed,
I never did incense his majesty                    85
Against the Duke of Clarence, but have been
An earnest advocate to plead for him.
My lord, you do me shameful injury
Falsely to draw me in these vile suspects.
RICHARD GLOUCESTER
You may deny that you were not the mean                    90
Of my Lord Hastings' late imprisonment.
RIVERS She may, my lord, for—
RICHARD GLOUCESTER
She may, Lord Rivers; why, who knows not so?
She may do more, sir, than denying that.
She may help you to many fair preferments,                    95
And then deny her aiding hand therein,
And lay those honours on your high desert.
What may she not? She may—ay, marry, may she.
RIVERS What 'marry, may she'?
RICHARD GLOUCESTER
What marry, may she? Marry with a king:                    100
A bachelor, and a handsome stripling, too.
Iwis your grandam had a worser match.
QUEEN ELIZABETH
My lord of Gloucester, I have too long borne
Your blunt upbraidings and your bitter scoffs.
By heaven, I will acquaint his majesty                    105
Of those gross taunts that oft I have endured.
I had rather be a country servant-maid
Than a great queen, with this condition:
To be so baited, scorned, and stormèd at.
    *Enter old Queen Margaret, unseen behind them*
Small joy have I in being England's queen.                    110
QUEEN MARGARET (*aside*)
And lessened be that small, God I beseech him.
Thy honour, state, and seat is due to me.
RICHARD GLOUCESTER (*to Elizabeth*)
What? Threat you me with telling of the King?
Tell him, and spare not. Look what I have said,
I will avouch't in presence of the King.                    115
I dare adventure to be sent to th' Tower.
'Tis time to speak; my pains are quite forgot.
QUEEN MARGARET (*aside*)
Out, devil! I remember them too well.
Thou killed'st my husband Henry in the Tower,
And Edward, my poor son, at Tewkesbury.                    120

RICHARD GLOUCESTER (*to Elizabeth*)
Ere you were queen—ay, or your husband king—
I was a packhorse in his great affairs,
A weeder-out of his proud adversaries,
A liberal rewarder of his friends.
To royalize his blood, I spent mine own.                    125
QUEEN MARGARET (*aside*)
Ay, and much better blood than his or thine.
RICHARD GLOUCESTER (*to Elizabeth*)
In all which time you and your husband Gray
Were factious for the house of Lancaster;
And Rivers, so were you.—Was not your husband
In Margaret's battle at Saint Albans slain?                    130
Let me put in your minds, if you forget,
What you have been ere this, and what you are;
Withal, what I have been, and what I am.
QUEEN MARGARET (*aside*)
A murd'rous villain, and so still thou art.
RICHARD GLOUCESTER
Poor Clarence did forsake his father Warwick—                    135
Ay, and forswore himself, which Jesu pardon—
QUEEN MARGARET (*aside*) Which God revenge!
RICHARD GLOUCESTER
To fight on Edward's party for the crown,
And for his meed, poor lord, he is mewed up.
I would to God my heart were flint like Edward's,                    140
Or Edward's soft and pitiful like mine.
I am too childish-foolish for this world.
QUEEN MARGARET (*aside*)
Hie thee to hell for shame, and leave this world,
Thou cacodemon; there thy kingdom is.
RIVERS
My lord of Gloucester, in those busy days                    145
Which here you urge to prove us enemies,
We followed then our lord, our sovereign king.
So should we you, if you should be our king.
RICHARD GLOUCESTER
If I should be? I had rather be a pedlar.
Far be it from my heart, the thought thereof.                    150
QUEEN ELIZABETH
As little joy, my lord, as you suppose
You should enjoy, were you this country's king,
As little joy may you suppose in me,
That I enjoy being the queen thereof.
QUEEN MARGARET (*aside*)
Ah, little joy enjoys the queen thereof,                    155
For I am she, and altogether joyless.
I can no longer hold me patient.
    *She comes forward*
Hear me, you wrangling pirates, that fall out
In sharing that which you have pilled from me.
Which of you trembles not that looks on me?                    160
If not that I am Queen, you bow like subjects;
Yet that by you deposed, you quake like rebels.
(*To Richard*) Ah, gentle villain, do not turn away.
RICHARD GLOUCESTER
Foul wrinkled witch, what mak'st thou in my sight?
QUEEN MARGARET
But repetition of what thou hast marred:                    165
That will I make before I let thee go.
A husband and a son thou ow'st to me,
(*To Elizabeth*) And thou a kingdom; (*to the rest*) all of
    you allegiance.
This sorrow that I have by right is yours,
And all the pleasures you usurp are mine.                    170

RICHARD GLOUCESTER
    The curse my noble father laid on thee—
    When thou didst crown his warlike brows with paper,
    And with thy scorns drew'st rivers from his eyes,
    And then, to dry them, gav'st the duke a clout
    Steeped in the faultless blood of pretty Rutland—          175
    His curses then, from bitterness of soul
    Denounced against thee, are all fall'n upon thee,
    And God, not we, hath plagued thy bloody deed.
QUEEN ELIZABETH (to Margaret)
    So just is God to right the innocent.
LORD HASTINGS (to Margaret)
    O 'twas the foulest deed to slay that babe,               180
    And the most merciless that e'er was heard of.
RIVERS (to Margaret)
    Tyrants themselves wept when it was reported.
DORSET (to Margaret)
    No man but prophesied revenge for it.
BUCKINGHAM (to Margaret)
    Northumberland, then present, wept to see it.
QUEEN MARGARET
    What? Were you snarling all before I came,                185
    Ready to catch each other by the throat,
    And turn you all your hatred now on me?
    Did York's dread curse prevail so much with heaven
    That Henry's death, my lovely Edward's death,
    Their kingdom's loss, my woeful banishment,               190
    Should all but answer for that peevish brat?
    Can curses pierce the clouds and enter heaven?
    Why then, give way, dull clouds, to my quick curses!
    Though not by war, by surfeit die your king,
    As ours by murder to make him a king.                     195
    (To Elizabeth) Edward thy son, that now is Prince of
            Wales,
    For Edward my son, that was Prince of Wales,
    Die in his youth by like untimely violence.
    Thyself, a queen, for me that was a queen,
    Outlive thy glory like my wretched self.                  200
    Long mayst thou live—to wail thy children's death,
    And see another, as I see thee now,
    Decked in thy rights, as thou art 'stalled in mine.
    Long die thy happy days before thy death,
    And after many lengthened hours of grief                  205
    Die, neither mother, wife, nor England's queen.—
    Rivers and Dorset, you were standers-by,
    And so wast thou, Lord Hastings, when my son
    Was stabbed with bloody daggers. God I pray him,
    That none of you may live his natural age,                210
    But by some unlooked accident cut off.
RICHARD GLOUCESTER
    Have done thy charm, thou hateful, withered hag.
QUEEN MARGARET
    And leave out thee? Stay, dog, for thou shalt hear me.
    If heaven have any grievous plague in store
    Exceeding those that I can wish upon thee,                215
    O let them keep it till thy sins be ripe,
    And then hurl down their indignation
    On thee, the troubler of the poor world's peace.
    The worm of conscience still begnaw thy soul.
    Thy friends suspect for traitors while thou liv'st,       220
    And take deep traitors for thy dearest friends.
    No sleep close up that deadly eye of thine,
    Unless it be while some tormenting dream
    Affrights thee with a hell of ugly devils.
    Thou elvish-marked, abortive, rooting hog,                225
    Thou that wast sealed in thy nativity
    The slave of nature and the son of hell,

    Thou slander of thy heavy mother's womb,
    Thou loathèd issue of thy father's loins,
    Thou rag of honour, thou detested—                        230
RICHARD GLOUCESTER Margaret.
QUEEN MARGARET
    Richard.
RICHARD GLOUCESTER Ha?
QUEEN MARGARET                    I call thee not.
RICHARD GLOUCESTER
    I cry thee mercy then, for I did think
    That thou hadst called me all these bitter names.
QUEEN MARGARET
    Why so I did, but looked for no reply.                    235
    O let me make the period to my curse.
RICHARD GLOUCESTER
    'Tis done by me, and ends in 'Margaret'.
QUEEN ELIZABETH (to Margaret)
    Thus have you breathed your curse against yourself.
QUEEN MARGARET
    Poor painted Queen, vain flourish of my fortune,
    Why strew'st thou sugar on that bottled spider            240
    Whose deadly web ensnareth thee about?
    Fool, fool, thou whet'st a knife to kill thyself.
    The day will come that thou shalt wish for me
    To help thee curse this poisonous bunch-backed toad.
LORD HASTINGS
    False-boding woman, end thy frantic curse,                245
    Lest to thy harm thou move our patience.
QUEEN MARGARET
    Foul shame upon you, you have all moved mine.
RIVERS
    Were you well served, you would be taught your duty.
QUEEN MARGARET
    To serve me well you all should do me duty.
    Teach me to be your queen, and you my subjects: 250
    O serve me well, and teach yourselves that duty.
DORSET
    Dispute not with her: she is lunatic.
QUEEN MARGARET
    Peace, master Marquis, you are malapert.
    Your fire-new stamp of honour is scarce current.
    O that your young nobility could judge                    255
    What 'twere to lose it and be miserable.
    They that stand high have many blasts to shake them,
    And if they fall they dash themselves to pieces.
RICHARD GLOUCESTER
    Good counsel, marry!—Learn it, learn it, Marquis.
DORSET
    It touches you, my lord, as much as me.                   260
RICHARD GLOUCESTER
    Ay, and much more; but I was born so high.
    Our eyrie buildeth in the cedar's top,
    And dallies with the wind, and scorns the sun.
QUEEN MARGARET
    And turns the sun to shade. Alas, alas!
    Witness my son, now in the shade of death,                265
    Whose bright outshining beams thy cloudy wrath
    Hath in eternal darkness folded up.
    Your eyrie buildeth in our eyrie's nest.—
    O God that seest it, do not suffer it;
    As it was won with blood, lost be it so.                  270
⌈RICHARD GLOUCESTER⌉
    Peace, peace! For shame, if not for charity.
QUEEN MARGARET
    Urge neither charity nor shame to me.
    Uncharitably with me have you dealt,
    And shamefully my hopes by you are butchered.

My charity is outrage; life, my shame;                        275
And in that shame still live my sorrow's rage.
BUCKINGHAM Have done, have done.
QUEEN MARGARET
O princely Buckingham, I'll kiss thy hand
In sign of league and amity with thee.
Now fair befall thee and thy noble house!                     280
Thy garments are not spotted with our blood,
Nor thou within the compass of my curse.
BUCKINGHAM
Nor no one here, for curses never pass
The lips of those that breathe them in the air.
QUEEN MARGARET
I will not think but they ascend the sky                      285
And there awake God's gentle sleeping peace.
O Buckingham, take heed of yonder dog.
        *She points at Richard*
Look when he fawns, he bites; and when he bites,
His venom tooth will rankle to the death.
Have naught to do with him; beware of him;                    290
Sin, death, and hell have set their marks on him,
And all their ministers attend on him.
RICHARD GLOUCESTER
What doth she say, my lord of Buckingham?
BUCKINGHAM
Nothing that I respect, my gracious lord.
QUEEN MARGARET
What, dost thou scorn me for my gentle counsel,               295
And soothe the devil that I warn thee from?
O but remember this another day,
When he shall split thy very heart with sorrow,
And say, 'Poor Margaret was a prophetess'.—
Live each of you the subjects to his hate,                    300
And he to yours, and all of you to God's.              *Exit*
[LORD HASTINGS]
My hair doth stand on end to hear her curses.
RIVERS
And so doth mine. I muse why she's at liberty.
RICHARD GLOUCESTER
I cannot blame her, by God's holy mother.
She hath had too much wrong, and I repent                     305
My part thereof that I have done to her.
QUEEN ELIZABETH
I never did her any, to my knowledge.
RICHARD GLOUCESTER
Yet you have all the vantage of her wrong.
I was too hot to do somebody good,
That is too cold in thinking of it now.                       310
Marry, as for Clarence, he is well repaid:
He is franked up to fatting for his pains.
God pardon them that are the cause thereof.
RIVERS
A virtuous and a Christian-like conclusion,
To pray for them that have done scathe to us.                 315
RICHARD GLOUCESTER
So do I ever— (*speaks to himself*) being well advised:
For had I cursed now, I had cursed myself.
        *Enter Sir William Catesby*
CATESBY
Madam, his majesty doth call for you,
And for your grace, and you my gracious lords.
QUEEN ELIZABETH
Catesby, I come.—Lords, will you go with me?                  320
RIVERS We wait upon your grace.   *Exeunt all but Richard*
RICHARD GLOUCESTER
I do the wrong, and first begin to brawl.
The secret mischiefs that I set abroach

I lay unto the grievous charge of others.
Clarence, whom I indeed have cast in darkness,                325
I do beweep to many simple gulls—
Namely to Derby, Hastings, Buckingham—
And tell them, ''Tis the Queen and her allies
That stir the King against the Duke my brother'.
Now they believe it, and withal whet me                       330
To be revenged on Rivers, Dorset, Gray;
But then I sigh, and with a piece of scripture
Tell them that God bids us do good for evil;
And thus I clothe my naked villainy
With odd old ends, stol'n forth of Holy Writ,                 335
And seem a saint when most I play the devil.
        *Enter two Murderers*
But soft, here come my executioners.—
How now, my hardy, stout, resolvèd mates!
Are you now going to dispatch this thing?
A MURDERER
We are, my lord, and come to have the warrant,                340
That we may be admitted where he is.
RICHARD GLOUCESTER
Well thought upon; I have it here about me.
        *He gives them the warrant*
When you have done, repair to Crosby Place.
But sirs, be sudden in the execution,
Withal obdurate; do not hear him plead,                       345
For Clarence is well spoken, and perhaps
May move your hearts to pity, if you mark him.
A MURDERER
Tut, tut, my lord, we will not stand to prate.
Talkers are no good doers. Be assured,
We go to use our hands, and not our tongues.                  350
RICHARD GLOUCESTER
Your eyes drop millstones when fools' eyes fall tears.
I like you, lads. About your business straight.
Go, go, dispatch.
⌈MURDERERS⌉           We will, my noble lord.
        *Exeunt Richard at one door, the Murderers at
                                            another*

**1.4**    *Enter George Duke of Clarence and* ⌈*Sir Robert
            Brackenbury*⌉
⌈BRACKENBURY⌉
Why looks your grace so heavily today?
CLARENCE
O I have passed a miserable night,
So full of fearful dreams, of ugly sights,
That as I am a Christian faithful man,
I would not spend another such a night                          5
Though 'twere to buy a world of happy days,
So full of dismal terror was the time.
⌈BRACKENBURY⌉
What was your dream, my lord? I pray you, tell me.
CLARENCE
Methoughts that I had broken from the Tower,
And was embarked to cross to Burgundy,                         10
And in my company my brother Gloucester,
Who from my cabin tempted me to walk
Upon the hatches; there we looked toward England,
And cited up a thousand heavy times
During the wars of York and Lancaster                          15
That had befall'n us. As we paced along
Upon the giddy footing of the hatches,
Methought that Gloucester stumbled, and in falling
Struck me—that sought to stay him—overboard
Into the tumbling billows of the main.                         20

O Lord! Methought what pain it was to drown,
What dreadful noise of waters in my ears,
What sights of ugly death within my eyes.
Methoughts I saw a thousand fearful wrecks,
Ten thousand men that fishes gnawed upon,                       25
Wedges of gold, great ouches, heaps of pearl,
Inestimable stones, unvalued jewels,
All scattered in the bottom of the sea.
Some lay in dead men's skulls; and in those holes
Where eyes did once inhabit, there were crept—                  30
As 'twere in scorn of eyes—reflecting gems,
Which wooed the slimy bottom of the deep
And mocked the dead bones that lay scattered by.
⌈BRACKENBURY⌉
Had you such leisure in the time of death,
To gaze upon these secrets of the deep?                         35
CLARENCE
Methought I had, and often did I strive
To yield the ghost, but still the envious flood
Stopped-in my soul and would not let it forth
To find the empty, vast, and wand'ring air,
But smothered it within my panting bulk,                        40
Who almost burst to belch it in the sea.
⌈BRACKENBURY⌉
Awaked you not in this sore agony?
CLARENCE
No, no, my dream was lengthened after life.
O then began the tempest to my soul!
I passed, methought, the melancholy flood,                      45
With that sour ferryman which poets write of,
Unto the kingdom of perpetual night.
The first that there did greet my stranger soul
Was my great father-in-law, renownèd Warwick,
Who cried aloud, 'What scourge for perjury                      50
Can this dark monarchy afford false Clarence?'
And so he vanished. Then came wand'ring by
A shadow like an angel, with bright hair,
Dabbled in blood, and he shrieked out aloud,
'Clarence is come: false, fleeting, perjured Clarence,          55
That stabbed me in the field by Tewkesbury.
Seize on him, furies! Take him unto torment!'
With that, methoughts a legion of foul fiends
Environed me, and howlèd in mine ears
Such hideous cries that with the very noise                     60
I trembling waked, and for a season after
Could not believe but that I was in hell,
Such terrible impression made my dream.
⌈BRACKENBURY⌉
No marvel, lord, though it affrighted you;
I am afraid, methinks, to hear you tell it.                     65
CLARENCE
Ah, Brackenbury, I have done these things,
That now give evidence against my soul,
For Edward's sake; and see how he requites me.
Keeper, I pray thee, sit by me awhile.
My soul is heavy, and I fain would sleep.                       70
⌈BRACKENBURY⌉
I will, my lord. God give your grace good rest.
    *Clarence sleeps*
Sorrow breaks seasons and reposing hours,
Makes the night morning and the noontide night.
Princes have but their titles for their glories,
An outward honour for an inward toil,                           75
And for unfelt imaginations
They often feel a world of restless cares;
So that, between their titles and low name,
There's nothing differs but the outward fame.

    *Enter two Murderers*
FIRST MURDERER Ho, who's here?                                  80
BRACKENBURY
What wouldst thou, fellow? And how cam'st thou
    hither?
SECOND MURDERER I would speak with Clarence, and I
came hither on my legs.
BRACKENBURY What, so brief?
FIRST MURDERER 'Tis better, sir, than to be tedious. (*To
Second Murderer*) Let him see our commission, and talk
no more.                                                        87
    *Brackenbury reads*
BRACKENBURY
I am in this commanded to deliver
The noble Duke of Clarence to your hands.
I will not reason what is meant hereby,                         90
Because I will be guiltless of the meaning.
There lies the Duke asleep, and there the keys.
    ⌈*He throws down the keys*⌉
I'll to the King and signify to him
That thus I have resigned to you my charge.                     94
FIRST MURDERER You may, sir; 'tis a point of wisdom.
Fare you well.                         *Exit Brackenbury*
SECOND MURDERER What, shall I stab him as he sleeps?
FIRST MURDERER No. He'll say 'twas done cowardly, when
he wakes.                                                       99
SECOND MURDERER Why, he shall never wake until the
great judgement day.
FIRST MURDERER Why, then he'll say we stabbed him
sleeping.
SECOND MURDERER The urging of that word 'judgement'
hath bred a kind of remorse in me.                             105
FIRST MURDERER What, art thou afraid?
SECOND MURDERER Not to kill him, having a warrant, but
to be damned for killing him, from the which no
warrant can defend me.
FIRST MURDERER I thought thou hadst been resolute.   110
SECOND MURDERER So I am—to let him live.
FIRST MURDERER I'll back to the Duke of Gloucester and
tell him so.
SECOND MURDERER Nay, I pray thee. Stay a little. I hope
this passionate humour of mine will change. It was
wont to hold me but while one tells twenty.          116
    ⌈*He counts to twenty*⌉
FIRST MURDERER How dost thou feel thyself now?
SECOND MURDERER Some certain dregs of conscience are
yet within me.
FIRST MURDERER Remember our reward, when the deed's
done.                                                          121
SECOND MURDERER 'Swounds, he dies. I had forgot the
reward.
FIRST MURDERER Where's thy conscience now?
SECOND MURDERER O, in the Duke of Gloucester's purse.
FIRST MURDERER When he opens his purse to give us our
reward, thy conscience flies out.                    127
SECOND MURDERER 'Tis no matter. Let it go. There's few
or none will entertain it.
FIRST MURDERER What if it come to thee again?        130
SECOND MURDERER I'll not meddle with it. It makes a man
a coward. A man cannot steal but it accuseth him. A
man cannot swear but it checks him. A man cannot
lie with his neighbour's wife but it detects him. 'Tis a
blushing, shamefaced spirit, that mutinies in a man's
bosom. It fills a man full of obstacles. It made me once
restore a purse of gold that by chance I found. It
beggars any man that keeps it. It is turned out of towns

and cities for a dangerous thing, and every man that
means to live well endeavours to trust to himself and
live without it.                                              141
FIRST MURDERER 'Swounds, 'tis even now at my elbow,
persuading me not to kill the Duke.
SECOND MURDERER Take the devil in thy mind, and believe
him not: he would insinuate with thee but to make
thee sigh.                                                   146
FIRST MURDERER I am strong framed; he cannot prevail
with me.
SECOND MURDERER Spoke like a tall man that respects thy
reputation. Come, shall we fall to work?                     150
FIRST MURDERER Take him on the costard with the hilts
of thy sword, and then throw him into the malmsey
butt in the next room.
SECOND MURDERER O excellent device!—and make a sop
of him.                                                      155
FIRST MURDERER Soft, he wakes.
SECOND MURDERER Strike!
FIRST MURDERER No, we'll reason with him.
CLARENCE
Where art thou, keeper? Give me a cup of wine.
SECOND MURDERER
You shall have wine enough, my lord, anon.                   160
CLARENCE
In God's name, what art thou?
FIRST MURDERER                        A man, as you are.
CLARENCE But not as I am, royal.
FIRST MURDERER Nor you as we are, loyal.
CLARENCE
Thy voice is thunder, but thy looks are humble.
FIRST MURDERER
My voice is now the King's; my looks, mine own.   165
CLARENCE
How darkly and how deadly dost thou speak.
Your eyes do menace me. Why look you pale?
Who sent you hither? Wherefore do you come?
SECOND MURDERER
To, to, to—
CLARENCE       To murder me.
BOTH MURDERERS                Ay, ay.
CLARENCE
You scarcely have the hearts to tell me so,                  170
And therefore cannot have the hearts to do it.
Wherein, my friends, have I offended you?
FIRST MURDERER
Offended us you have not, but the King.
CLARENCE
I shall be reconciled to him again.
SECOND MURDERER
Never, my lord; therefore prepare to die.                    175
CLARENCE
Are you drawn forth among a world of men
To slay the innocent? What is my offence?
Where is the evidence that doth accuse me?
What lawful quest have given their verdict up
Unto the frowning judge, or who pronounced                   180
The bitter sentence of poor Clarence' death?
Before I be convict by course of law,
To threaten me with death is most unlawful.
I charge you, as you hope to have redemption
By Christ's dear blood, shed for our grievous sins,   185
That you depart and lay no hands on me.
The deed you undertake is damnable.
FIRST MURDERER
What we will do, we do upon command.

SECOND MURDERER
And he that hath commanded is our king.
CLARENCE
Erroneous vassals, the great King of Kings                   190
Hath in the table of his law commanded
That thou shalt do no murder. Will you then
Spurn at his edict, and fulfil a man's?
Take heed, for he holds vengeance in his hand
To hurl upon their heads that break his law.                 195
SECOND MURDERER
And that same vengeance doth he hurl on thee,
For false forswearing, and for murder too.
Thou didst receive the sacrament to fight
In quarrel of the house of Lancaster.
FIRST MURDERER
And, like a traitor to the name of God,                      200
Didst break that vow, and with thy treacherous blade
Unripped'st the bowels of thy sov'reign's son.
SECOND MURDERER
Whom thou wast sworn to cherish and defend.
FIRST MURDERER
How canst thou urge God's dreadful law to us,
When thou hast broke it in such dear degree?                 205
CLARENCE
Alas, for whose sake did I that ill deed?
For Edward, for my brother, for his sake.
He sends ye not to murder me for this,
For in that sin he is as deep as I.
If God will be avengèd for the deed,                         210
O know you yet, he doth it publicly.
Take not the quarrel from his pow'rful arm;
He needs no indirect or lawless course
To cut off those that have offended him.
FIRST MURDERER
Who made thee then a bloody minister                         215
When gallant springing brave Plantagenet,
That princely novice, was struck dead by thee?
CLARENCE
My brother's love, the devil, and my rage.
FIRST MURDERER
Thy brother's love, our duty, and thy faults
Provoke us hither now to slaughter thee.                     220
CLARENCE
If you do love my brother, hate not me.
I am his brother, and I love him well.
If you are hired for meed, go back again,
And I will send you to my brother Gloucester,
Who shall reward you better for my life                      225
Than Edward will for tidings of my death.
SECOND MURDERER
You are deceived. Your brother Gloucester hates you.
CLARENCE
O no, he loves me, and he holds me dear.
Go you to him from me.
FIRST MURDERER                Ay, so we will.
CLARENCE
Tell him, when that our princely father York                 230
Blessed his three sons with his victorious arm,
And charged us from his soul to love each other,
He little thought of this divided friendship.
Bid Gloucester think of this, and he will weep.
FIRST MURDERER
Ay, millstones, as he lessoned us to weep.                   235
CLARENCE
O do not slander him, for he is kind.

FIRST MURDERER
  As snow in harvest. Come, you deceive yourself.
  'Tis he that sends us to destroy you here.
CLARENCE
  It cannot be, for he bewept my fortune,
  And hugged me in his arms, and swore with sobs   240
  That he would labour my delivery.
FIRST MURDERER
  Why, so he doth, when he delivers you
  From this earth's thraldom to the joys of heaven.
SECOND MURDERER
  Make peace with God, for you must die, my lord.
CLARENCE
  Have you that holy feeling in your souls          245
  To counsel me to make my peace with God,
  And are you yet to your own souls so blind
  That you will war with God by murd'ring me?
  O sirs, consider: they that set you on
  To do this deed will hate you for the deed.       250
SECOND MURDERER (to First)
  What shall we do?
CLARENCE                Relent, and save your souls.
FIRST MURDERER
  Relent? No. 'Tis cowardly and womanish.
CLARENCE
  Not to relent is beastly, savage, devilish.—
  My friend, I spy some pity in thy looks.
  O if thine eye be not a flatterer,                255
  Come thou on my side, and entreat for me.
  A begging prince, what beggar pities not?
  Which of you, if you were a prince's son,
  Being pent from liberty as I am now,
  If two such murderers as yourselves came to you,  260
  Would not entreat for life? As you would beg
  Were you in my distress—
SECOND MURDERER  Look behind you, my lord!
FIRST MURDERER (stabbing Clarence)
  Take that, and that! If all this will not serve,
  I'll drown you in the malmsey butt within.        265
                              Exit with Clarence's body
SECOND MURDERER
  A bloody deed, and desperately dispatched!
  How fain, like Pilate, would I wash my hands
  Of this most grievous, guilty murder done.
        Enter First Murderer
FIRST MURDERER
  How now? What mean'st thou, that thou help'st me not?
  By heaven, the Duke shall know how slack you have
    been.                                           270
SECOND MURDERER
  I would he knew that I had saved his brother.
  Take thou the fee, and tell him what I say,
  For I repent me that the Duke is slain.           Exit
FIRST MURDERER
  So do not I. Go, coward as thou art.—
  Well, I'll go hide the body in some hole          275
  Till that the Duke give order for his burial.
  And, when I have my meed, I will away,
  For this will out, and then I must not stay.      Exit

2.1   Flourish. Enter King Edward, sick, Queen Elizabeth,
      Lord Marquis Dorset, Lord Rivers, Lord Hastings,
      Sir William Catesby, the Duke of Buckingham ⌈and
      Lord Gray⌉
KING EDWARD
  Why, so! Now have I done a good day's work.

You peers, continue this united league.
I every day expect an embassage
From my redeemer to redeem me hence,
And more in peace my soul shall part to heaven      5
Since I have made my friends at peace on earth.
Hastings and Rivers, take each other's hand.
Dissemble not your hatred; swear your love.
RIVERS
  By heaven, my soul is purged from grudging hate,
  And with my hand I seal my true heart's love.     10
        ⌈He takes Hastings' hand⌉
LORD HASTINGS
  So thrive I, as I truly swear the like.
KING EDWARD
  Take heed you dally not before your king,
  Lest he that is the supreme King of Kings
  Confound your hidden falsehood, and award
  Either of you to be the other's end.              15
LORD HASTINGS
  So prosper I, as I swear perfect love.
RIVERS
  And I, as I love Hastings with my heart.
KING EDWARD (to Elizabeth)
  Madam, yourself is not exempt from this,
  Nor your son Dorset;—Buckingham, nor you.
  You have been factious one against the other.     20
  Wife, love Lord Hastings, let him kiss your hand—
  And what you do, do it unfeignedly.
QUEEN ELIZABETH (giving Hastings her hand to kiss)
  There, Hastings. I will never more remember
  Our former hatred: so thrive I, and mine.
KING EDWARD
  Dorset, embrace him. Hastings, love Lord Marquis.  25
DORSET
  This interchange of love, I here protest,
  Upon my part shall be inviolable.
LORD HASTINGS  And so swear I.
        They embrace
KING EDWARD
  Now, princely Buckingham, seal thou this league
  With thy embracements to my wife's allies,        30
  And make me happy in your unity.
BUCKINGHAM (to Elizabeth)
  Whenever Buckingham doth turn his hate
  Upon your grace, but with all duteous love
  Doth cherish you and yours, God punish me
  With hate in those where I expect most love.       35
  When I have most need to employ a friend,
  And most assurèd that he is a friend,
  Deep, hollow, treacherous, and full of guile
  Be he unto me. This do I beg of heaven,
  When I am cold in love to you or yours.            40
        They embrace
KING EDWARD
  A pleasing cordial, princely Buckingham,
  Is this thy vow unto my sickly heart.
  There wanteth now our brother Gloucester here,
  To make the blessèd period of this peace.
        Enter Sir Richard Ratcliffe and Richard Duke of
        Gloucester
BUCKINGHAM  And in good time,                        45
  Here comes Sir Richard Ratcliffe and the Duke.
RICHARD GLOUCESTER
  Good morrow to my sovereign King and Queen.—
  And princely peers, a happy time of day.

KING EDWARD
Happy indeed, as we have spent the day.
Brother, we have done deeds of charity, 50
Made peace of enmity, fair love of hate,
Between these swelling wrong-incensèd peers.
RICHARD GLOUCESTER
A blessèd labour, my most sovereign lord.
Among this princely heap if any here,
By false intelligence or wrong surmise, 55
Hold me a foe,
If I unwittingly or in my rage
Have aught committed that is hardly borne
By any in this presence, I desire
To reconcile me to his friendly peace. 60
'Tis death to me to be at enmity.
I hate it, and desire all good men's love. —
First, madam, I entreat true peace of you,
Which I will purchase with my duteous service. —
Of you, my noble cousin Buckingham, 65
If ever any grudge were lodged between us. —
Of you, Lord Rivers, and Lord Gray of you,
That all without desert have frowned on me. —
Dukes, earls, lords, gentlemen, indeed of all!
I do not know that Englishman alive 70
With whom my soul is any jot at odds
More than the infant that is born tonight.
I thank my God for my humility.
QUEEN ELIZABETH
A holy day shall this be kept hereafter.
I would to God all strifes were well compounded. — 75
My sovereign lord, I do beseech your highness
To take our brother Clarence to your grace.
RICHARD GLOUCESTER
Why, madam, have I offered love for this,
To be so flouted in this royal presence?
Who knows not that the gentle Duke is dead? 80
    *The others all start*
You do him injury to scorn his corpse.
⌈RIVERS⌉
Who knows not he is dead? Who knows he is?
QUEEN ELIZABETH
All-seeing heaven, what a world is this?
BUCKINGHAM
Look I so pale, Lord Dorset, as the rest?
DORSET
Ay, my good lord, and no one in the presence 85
But his red colour hath forsook his cheeks.
KING EDWARD
Is Clarence dead? The order was reversed.
RICHARD GLOUCESTER
But he, poor man, by your first order died,
And that a wingèd Mercury did bear;
Some tardy cripple bore the countermand, 90
That came too lag to see him buried.
God grant that some, less noble and less loyal,
Nearer in bloody thoughts, but not in blood,
Deserve not worse than wretched Clarence did,
And yet go current from suspicion. 95
    *Enter Lord Stanley Earl of Derby*
STANLEY (*kneeling*)
A boon, my sovereign, for my service done.
KING EDWARD
I pray thee, peace! My soul is full of sorrow.
STANLEY
I will not rise, unless your highness hear me.

KING EDWARD
Then say at once, what is it thou requests?
STANLEY
The forfeit, sovereign, of my servant's life, 100
Who slew today a riotous gentleman,
Lately attendant on the Duke of Norfolk.
KING EDWARD
Have I a tongue to doom my brother's death,
And shall that tongue give pardon to a slave?
My brother slew no man; his fault was thought; 105
And yet his punishment was bitter death.
Who sued to me for him? Who in my wrath
Kneeled at my feet, and bid me be advised?
Who spoke of brotherhood? Who spoke of love?
Who told me how the poor soul did forsake 110
The mighty Warwick and did fight for me?
Who told me, in the field at Tewkesbury,
When Oxford had me down, he rescued me,
And said, 'Dear brother, live, and be a king'?
Who told me, when we both lay in the field, 115
Frozen almost to death, how he did lap me
Even in his garments, and did give himself
All thin and naked to the numb-cold night?
All this from my remembrance brutish wrath
Sinfully plucked, and not a man of you 120
Had so much grace to put it in my mind.
But when your carters or your waiting vassals
Have done a drunken slaughter, and defaced
The precious image of our dear redeemer,
You straight are on your knees for 'Pardon, pardon!' —
And I, unjustly too, must grant it you. 126
But, for my brother, not a man would speak,
Nor I, ungracious, speak unto myself
For him, poor soul. The proudest of you all
Have been beholden to him in his life, 130
Yet none of you would once beg for his life.
O God, I fear thy justice will take hold
On me — and you, and mine, and yours, for this. —
Come, Hastings, help me to my closet.
Ah, poor Clarence! 135
    *Exeunt some with King and Queen*
RICHARD GLOUCESTER
This is the fruits of rashness. Marked you not
How that the guilty kindred of the Queen
Looked pale, when they did hear of Clarence' death?
O, they did urge it still unto the King.
God will revenge it. Come, lords, will you go 140
To comfort Edward with our company?
BUCKINGHAM We wait upon your grace.     *Exeunt*

2.2    *Enter the old Duchess of York with the two children*
      *of Clarence*
BOY
Good grannam, tell us, is our father dead?
DUCHESS OF YORK No, boy.
GIRL
Why do you weep so oft, and beat your breast,
And cry, 'O Clarence, my unhappy son'?
BOY
Why do you look on us and shake your head, 5
And call us orphans, wretches, castaways,
If that our noble father were alive?
DUCHESS OF YORK
My pretty cousins, you mistake me both.
I do lament the sickness of the King,

As loath to lose him, not your father's death.                10
It were lost sorrow to wail one that's lost.
BOY
Then you conclude, my grannam, he is dead.
The King mine uncle is to blame for this.
God will revenge it—whom I will importune
With earnest prayers, all to that effect.                     15
GIRL  And so will I.
DUCHESS OF YORK
Peace, children, peace! The King doth love you well.
Incapable and shallow innocents,
You cannot guess who caused your father's death.
BOY
Grannam, we can. For my good uncle Gloucester         20
Told me the King, provoked to it by the Queen,
Devised impeachments to imprison him,
And when my uncle told me so he wept,
And pitied me, and kindly kissed my cheek,
Bade me rely on him as on my father,                          25
And he would love me dearly as his child.
DUCHESS OF YORK
Ah, that deceit should steal such gentle shapes,
And with a virtuous visor hide deep vice!
He is my son, ay, and therein my shame;
Yet from my dugs he drew not this deceit.                    30
BOY
Think you my uncle did dissemble, grannam?
DUCHESS OF YORK  Ay, boy.
BOY
I cannot think it. Hark, what noise is this?
        *Enter Queen Elizabeth with her hair about her ears*
QUEEN ELIZABETH
Ah, who shall hinder me to wail and weep?
To chide my fortune, and torment myself?                    35
I'll join with black despair against my soul,
And to myself become an enemy.
DUCHESS OF YORK
What means this scene of rude impatience?
QUEEN ELIZABETH
To mark an act of tragic violence.
Edward, my lord, thy son, our king, is dead.               40
Why grow the branches when the root is gone?
Why wither not the leaves that want their sap?
If you will live, lament; if die, be brief,
That our swift-wingèd souls may catch the King's,
Or like obedient subjects follow him                          45
To his new kingdom of ne'er-changing night.
DUCHESS OF YORK
Ah, so much interest have I in thy sorrow
As I had title in thy noble husband.
I have bewept a worthy husband's death,
And lived with looking on his images.                         50
But now two mirrors of his princely semblance
Are cracked in pieces by malignant death,
And I for comfort have but one false glass,
That grieves me when I see my shame in him.
Thou art a widow, yet thou art a mother,                    55
And hast the comfort of thy children left.
But death hath snatched my husband from mine arms
And plucked two crutches from my feeble hands,
Clarence and Edward. O what cause have I,
Thine being but a moiety of my moan,                        60
To overgo thy woes, and drown thy cries?
BOY (*to Elizabeth*)
Ah, aunt, you wept not for our father's death.
How can we aid you with our kindred tears?

DAUGHTER (*to Elizabeth*)
Our fatherless distress was left unmoaned;
Your widow-dolour likewise be unwept.                       65
QUEEN ELIZABETH
Give me no help in lamentation.
I am not barren to bring forth complaints.
All springs reduce their currents to mine eyes,
That I, being governed by the wat'ry moon,
May send forth plenteous tears to drown the world.   70
Ah, for my husband, for my dear Lord Edward!
CHILDREN
Ah, for our father, for our dear Lord Clarence!
DUCHESS OF YORK
Alas, for both, both mine, Edward and Clarence!
QUEEN ELIZABETH
What stay had I but Edward, and he's gone?
CHILDREN
What stay had we but Clarence, and he's gone?       75
DUCHESS OF YORK
What stays had I but they, and they are gone?
QUEEN ELIZABETH
Was never widow had so dear a loss!
CHILDREN
Were never orphans had so dear a loss!
DUCHESS OF YORK
Was never mother had so dear a loss!
Alas, I am the mother of these griefs.                        80
Their woes are parcelled; mine is general.
She for an Edward weeps, and so do I;
I for a Clarence weep, so doth not she.
These babes for Clarence weep, and so do I;
I for an Edward weep, so do not they.                        85
Alas, you three on me, threefold distressed,
Pour all your tears. I am your sorrow's nurse,
And I will pamper it with lamentation.
        *Enter Richard Duke of Gloucester, the Duke of*
        *Buckingham, Lord Stanley Earl of Derby, Lord*
        *Hastings, and Sir Richard Ratcliffe*
RICHARD GLOUCESTER (*to Elizabeth*)
Sister, have comfort. All of us have cause
To wail the dimming of our shining star,                     90
But none can help our harms by wailing them.—
Madam, my mother, I do cry you mercy.
I did not see your grace. Humbly on my knee
I crave your blessing.
DUCHESS OF YORK
God bless thee, and put meekness in thy breast,      95
Love, charity, obedience, and true duty.
RICHARD GLOUCESTER
Amen. (*Aside*) 'And make me die a good old man.'
That is the butt-end of a mother's blessing;
I marvel that her grace did leave it out.
BUCKINGHAM
You cloudy princes and heart-sorrowing peers         100
That bear this heavy mutual load of moan,
Now cheer each other in each other's love.
Though we have spent our harvest of this king,
We are to reap the harvest of his son.
The broken rancour of your high-swoll'n hearts        105
But lately splinted, knit, and joined together,
Must gently be preserved, cherished, and kept.
Meseemeth good that, with some little train,
Forthwith from Ludlow the young Prince be fet
Hither to London to be crowned our king.                  110
RICHARD GLOUCESTER
Then be it so, and go we to determine
Who they shall be that straight shall post to Ludlow.—

Madam, and you my sister, will you go
To give your censures in this weighty business?
QUEEN ELIZABETH *and* DUCHESS OF YORK  With all our hearts.
                    *Exeunt all but Richard and Buckingham*
BUCKINGHAM
My lord, whoever journeys to the Prince,                    116
For God's sake let not us two stay at home,
For by the way I'll sort occasion,
As index to the story we late talked of,
To part the Queen's proud kindred from the Prince.
RICHARD GLOUCESTER
My other self, my counsel's consistory,                    121
My oracle, my prophet, my dear cousin!
I, as a child, will go by thy direction.
Towards Ludlow then, for we'll not stay behind.
                                        *Exeunt*


**2.3**   *Enter one Citizen at one door and another at the*
          *other*
FIRST CITIZEN
Good morrow, neighbour. Whither away so fast?
SECOND CITIZEN
I promise you, I scarcely know myself.
Hear you the news abroad?
FIRST CITIZEN                    Yes, that the King is dead.
SECOND CITIZEN
Ill news, by'r Lady; seldom comes the better.
I fear, I fear, 'twill prove a giddy world.                    5
          *Enter another Citizen*
THIRD CITIZEN
Neighbours, God speed.
FIRST CITIZEN                    Give you good morrow, sir.
THIRD CITIZEN
Doth the news hold of good King Edward's death?
SECOND CITIZEN
Ay, sir, it is too true. God help the while.
THIRD CITIZEN
Then, masters, look to see a troublous world.
FIRST CITIZEN
No, no, by God's good grace his son shall reign.                    10
THIRD CITIZEN
Woe to that land that's governed by a child.
SECOND CITIZEN
In him there is a hope of government,
Which in his nonage council under him,
And in his full and ripened years himself,
No doubt shall then, and till then, govern well.                    15
FIRST CITIZEN
So stood the state when Henry the Sixth
Was crowned in Paris but at nine months old.
THIRD CITIZEN
Stood the state so? No, no, good friends, God wot.
For then this land was famously enriched
With politic, grave counsel; then the King                    20
Had virtuous uncles to protect his grace.
FIRST CITIZEN
Why, so hath this, both by his father and mother.
THIRD CITIZEN
Better it were they all came by his father,
Or by his father there were none at all.
For emulation who shall now be near'st                    25
Will touch us all too near, if God prevent not.
O full of danger is the Duke of Gloucester,
And the Queen's sons and brothers haught and proud.
And were they to be ruled, and not to rule,
This sickly land might solace as before.                    30

FIRST CITIZEN
Come, come, we fear the worst. All will be well.
THIRD CITIZEN
When clouds are seen, wise men put on their cloaks;
When great leaves fall, then winter is at hand;
When the sun sets, who doth not look for night?
Untimely storms make men expect a dearth.                    35
All may be well, but if God sort it so
'Tis more than we deserve, or I expect.
SECOND CITIZEN
Truly the hearts of men are full of fear.
You cannot reason almost with a man
That looks not heavily and full of dread.                    40
THIRD CITIZEN
Before the days of change still is it so.
By a divine instinct men's minds mistrust
Ensuing danger, as by proof we see
The water swell before a boist'rous storm.
But leave it all to God. Whither away?                    45
SECOND CITIZEN
Marry, we were sent for to the justices.
THIRD CITIZEN
And so was I. I'll bear you company.                    *Exeunt*


**2.4**   *Enter* ⌐*Lord Cardinal*⌐, *young Duke of York, Queen*
          *Elizabeth, and the old Duchess of York*
⌐CARDINAL⌐
Last night, I hear, they lay them at Northampton.
At Stony Stratford they do rest tonight.
Tomorrow, or next day, they will be here.
DUCHESS OF YORK
I long with all my heart to see the Prince.
I hope he is much grown since last I saw him.                    5
QUEEN ELIZABETH
But I hear, no. They say my son of York
Has almost overta'en him in his growth.
YORK
Ay, mother, but I would not have it so.
DUCHESS OF YORK
Why, my young cousin, it is good to grow.
YORK
Grandam, one night as we did sit at supper,                    10
My uncle Rivers talked how I did grow
More than my brother. 'Ay', quoth my nuncle
          Gloucester,
'Small herbs have grace; gross weeds do grow apace'.
And since, methinks I would not grow so fast,
Because sweet flow'rs are slow, and weeds make
          haste.
DUCHESS OF YORK
Good faith, good faith, the saying did not hold                    16
In him that did object the same to thee.
He was the wretched'st thing when he was young,
So long a-growing, and so leisurely,
That if his rule were true he should be gracious.                    20
⌐CARDINAL⌐
Why, so no doubt he is, my gracious madam.
DUCHESS OF YORK
I hope he is, but yet let mothers doubt.
YORK
Now, by my troth, if I had been remembered,
I could have given my uncle's grace a flout
To touch his growth, nearer than he touched mine.
DUCHESS OF YORK
How, my young York? I pray thee, let me hear it.                    26

YORK
 Marry, they say my uncle grew so fast
 That he could gnaw a crust at two hours old.
 'Twas full two years ere I could get a tooth.
 Grannam, this would have been a biting jest.          30
DUCHESS OF YORK
 I pray thee, pretty York, who told thee this?
YORK Grannam, his nurse.
DUCHESS OF YORK
 His nurse? Why, she was dead ere thou wast born.
YORK
 If 'twere not she, I cannot tell who told me.
QUEEN ELIZABETH
 A parlous boy! Go to, you are too shrewd.            35
⌈CARDINAL⌉
 Good madam, be not angry with the child.
QUEEN ELIZABETH
 Pitchers have ears.
          Enter ⌈Marquis Dorset⌉
⌈CARDINAL⌉ Here comes your son, Lord Dorset.
 What news, Lord Marquis?
⌈DORSET⌉                   Such news, my lord,
 As grieves me to report.
QUEEN ELIZABETH          How doth the Prince?
⌈DORSET⌉
 Well, madam, and in health.
DUCHESS OF YORK          What is thy news then?
⌈DORSET⌉
 Lord Rivers and Lord Gray are sent to Pomfret,      41
 And with them Thomas Vaughan, prisoners.
DUCHESS OF YORK
 Who hath committed them?
⌈DORSET⌉                    The mighty dukes,
 Gloucester and Buckingham.
⌈CARDINAL⌉                  For what offence?
⌈DORSET⌉
 The sum of all I can, I have disclosed.              45
 Why or for what the nobles were committed
 Is all unknown to me, my gracious lord.
QUEEN ELIZABETH
 Ay me! I see the ruin of our house.
 The tiger now hath seized the gentle hind.
 Insulting tyranny begins to jet                      50
 Upon the innocent and aweless throne.
 Welcome destruction, blood, and massacre!
 I see, as in a map, the end of all.
DUCHESS OF YORK
 Accursèd and unquiet wrangling days,
 How many of you have mine eyes beheld?               55
 My husband lost his life to get the crown,
 And often up and down my sons were tossed,
 For me to joy and weep their gain and loss.
 And being seated, and domestic broils
 Clean overblown, themselves the conquerors          60
 Make war upon themselves, brother to brother,
 Blood to blood, self against self. O preposterous
 And frantic outrage, end thy damnèd spleen,
 Or let me die, to look on death no more.
QUEEN ELIZABETH (to York)
 Come, come, my boy, we will to sanctuary.—           65
 Madam, farewell.
DUCHESS OF YORK   Stay, I will go with you.
QUEEN ELIZABETH
 You have no cause.
⌈CARDINAL⌉ (to Elizabeth) My gracious lady, go,
 And thither bear your treasure and your goods.

 For my part, I'll resign unto your grace
 The seal I keep, and so betide to me                 70
 As well I tender you and all of yours.
 Go, I'll conduct you to the sanctuary.       Exeunt

3.1    The Trumpets sound. Enter young Prince Edward,
       the Dukes of Gloucester and Buckingham, Lord
       Cardinal, with others, including ⌈Lord Stanley Earl
       of Derby and⌉ Sir William Catesby
BUCKINGHAM
 Welcome, sweet Prince, to London, to your chamber.
RICHARD GLOUCESTER (to Prince Edward)
 Welcome, dear cousin, my thoughts' sovereign.
 The weary way hath made you melancholy.
PRINCE EDWARD
 No, uncle, but our crosses on the way
 Have made it tedious, wearisome, and heavy.           5
 I want more uncles here to welcome me.
RICHARD GLOUCESTER
 Sweet Prince, the untainted virtue of your years
 Hath not yet dived into the world's deceit,
 Nor more can you distinguish of a man
 Than of his outward show, which God he knows         10
 Seldom or never jumpeth with the heart.
 Those uncles which you want were dangerous.
 Your grace attended to their sugared words,
 But looked not on the poison of their hearts.
 God keep you from them, and from such false friends.
PRINCE EDWARD
 God keep me from false friends; but they were none.
          Enter Lord Mayor ⌈and his train⌉
RICHARD GLOUCESTER
 My lord, the Mayor of London comes to greet you.     17
MAYOR (kneeling to Prince Edward)
 God bless your grace with health and happy days.
PRINCE EDWARD
 I thank you, good my lord, and thank you all.—
 I thought my mother and my brother York             20
 Would long ere this have met us on the way.
 Fie, what a slug is Hastings, that he hastes not
 To tell us whether they will come or no.
          Enter Lord Hastings
BUCKINGHAM
 In happy time here comes the sweating lord.
PRINCE EDWARD (to Hastings)
 Welcome, my lord. What, will our mother come?        25
LORD HASTINGS
 On what occasion God he knows, not I,
 The Queen your mother, and your brother York,
 Have taken sanctuary. The tender Prince
 Would fain have come with me to meet your grace,
 But by his mother was perforce withheld.             30
BUCKINGHAM
 Fie, what an indirect and peevish course
 Is this of hers!—Lord Cardinal, will your grace
 Persuade the Queen to send the Duke of York
 Unto his princely brother presently?—
 If she deny, Lord Hastings, go with him,             35
 And from her jealous arms pluck him perforce.
CARDINAL
 My lord of Buckingham, if my weak oratory
 Can from his mother win the Duke of York,
 Anon expect him. But if she be obdurate
 To mild entreaties, God in heaven forbid             40
 We should infringe the sacred privilege
 Of blessèd sanctuary. Not for all this land
 Would I be guilty of so deep a sin.

BUCKINGHAM
You are too senseless-obstinate, my lord,
Too ceremonious and traditional.                                45
Weigh it not with the grossness of this age.
You break not sanctuary in seizing him.
The benefit thereof is always granted
To those whose dealings have deserved the place,
And those who have the wit to claim the place.       50
This prince hath neither claimed it nor deserved it,
And therefore, in my mind, he cannot have it.
Then taking him from thence that 'longs not there,
You break thereby no privilege nor charter.
Oft have I heard of 'sanctuary men',                       55
But 'sanctuary children' ne'er till now.
CARDINAL
My lord, you shall o'errule my mind for once.—
Come on, Lord Hastings, will you go with me?
LORD HASTINGS  I come, my lord.
PRINCE EDWARD
Good lords, make all the speedy haste you may.—   60
                    *Exeunt Cardinal and Hastings*
Say, uncle Gloucester, if our brother come,
Where shall we sojourn till our coronation?
RICHARD GLOUCESTER
Where it seems best unto your royal self.
If I may counsel you, some day or two
Your highness shall repose you at the Tower,        65
Then where you please and shall be thought most fit
For your best health and recreation.
PRINCE EDWARD
I do not like the Tower of any place.—
Did Julius Caesar build that place, my lord?
BUCKINGHAM
He did, my gracious lord, begin that place,          70
Which since succeeding ages have re-edified.
PRINCE EDWARD
Is it upon record, or else reported
Successively from age to age, he built it?
BUCKINGHAM
Upon record, my gracious liege.
PRINCE EDWARD
But say, my lord, it were not registered,              75
Methinks the truth should live from age to age,
As 'twere retailed to all posterity
Even to the general all-ending day.
RICHARD GLOUCESTER *(aside)*
So wise so young, they say, do never live long.
PRINCE EDWARD  What say you, uncle?                  80
RICHARD GLOUCESTER
I say, 'Without characters fame lives long'.
*(Aside)* Thus like the formal Vice, Iniquity,
I moralize two meanings in one word.
PRINCE EDWARD
That Julius Caesar was a famous man:
With what his valour did t'enrich his wit,            85
His wit set down to make his valour live.
Death made no conquest of this conqueror,
For yet he lives in fame though not in life.
I'll tell you what, my cousin Buckingham.
BUCKINGHAM  What, my good lord?                      90
PRINCE EDWARD
An if I live until I be a man,
I'll win our ancient right in France again,
Or die a soldier, as I lived a king.
RICHARD GLOUCESTER *(aside)*
Short summers lightly have a forward spring.

                    *Enter young Duke of York, Lord Hastings, and Lord
                    Cardinal*
BUCKINGHAM
Now in good time, here comes the Duke of York.     95
PRINCE EDWARD
Richard of York, how fares our loving brother?
YORK
Well, my dread lord—so must I call you now.
PRINCE EDWARD
Ay, brother, to our grief, as it is yours.
Too late he died that might have kept that title,
Which by his death hath lost much majesty.         100
RICHARD GLOUCESTER
How fares our noble cousin, Lord of York?
YORK
I thank you, gentle uncle, well. O, my lord,
You said that idle weeds are fast in growth;
The Prince, my brother, hath outgrown me far.
RICHARD GLOUCESTER
He hath, my lord.
YORK                        And therefore is he idle?    105
RICHARD GLOUCESTER
O my fair cousin, I must not say so.
YORK
He is more beholden to you then than I.
RICHARD GLOUCESTER
He may command me as my sovereign,
But you have power in me as a kinsman.
YORK
I pray you, uncle, render me this dagger.            110
RICHARD GLOUCESTER
My dagger, little cousin? With all my heart.
PRINCE EDWARD  A beggar, brother?
YORK
Of my kind uncle that I know will give,
It being but a toy which is no grief to give.
RICHARD GLOUCESTER
A greater gift than that I'll give my cousin.        115
YORK
A greater gift? O, that's the sword to it.
RICHARD GLOUCESTER
Ay, gentle cousin, were it light enough.
YORK
O, then I see you will part but with light gifts.
In weightier things you'll say a beggar nay.
RICHARD GLOUCESTER
It is too heavy for your grace to wear.              120
YORK
I'd weigh it lightly, were it heavier.
RICHARD GLOUCESTER
What, would you have my weapon, little lord?
YORK
I would, that I might thank you as you call me.
RICHARD GLOUCESTER  How?
YORK  Little.                                               125
PRINCE EDWARD
My lord of York will still be cross in talk.—
Uncle, your grace knows how to bear with him.
YORK
You mean to bear me, not to bear with me.—
Uncle, my brother mocks both you and me.
Because that I am little like an ape,                   130
He thinks that you should bear me on your shoulders.
BUCKINGHAM
With what a sharp, prodigal wit he reasons.
To mitigate the scorn he gives his uncle,

He prettily and aptly taunts himself.
So cunning and so young is wonderful.                    135
RICHARD GLOUCESTER (to Prince Edward)
My lord, will't please you pass along?
Myself and my good cousin Buckingham
Will to your mother to entreat of her
To meet you at the Tower and welcome you.
YORK (to Prince Edward)
What, will you go unto the Tower, my lord?               140
PRINCE EDWARD
My Lord Protector needs will have it so.
YORK
I shall not sleep in quiet at the Tower.
RICHARD GLOUCESTER Why, what should you fear there?
YORK
Marry, my uncle Clarence' angry ghost.
My grannam told me he was murdered there.               145
PRINCE EDWARD
I fear no uncles dead.
RICHARD GLOUCESTER     Nor none that live, I hope.
PRINCE EDWARD
An if they live, I hope I need not fear.
(To York) But come, my lord, and with a heavy heart,
Thinking on them, go we unto the Tower.
     A Sennet. Exeunt all but Richard, Buckingham,
                                    and Catesby
BUCKINGHAM (to Richard)
Think you, my lord, this little prating York             150
Was not incensèd by his subtle mother
To taunt and scorn you thus opprobriously?
RICHARD GLOUCESTER
No doubt, no doubt. O, 'tis a parlous boy,
Bold, quick, ingenious, forward, capable.
He is all the mother's, from the top to toe.            155
BUCKINGHAM
Well, let them rest.—Come hither, Catesby. Thou art
     sworn
As deeply to effect what we intend
As closely to conceal what we impart.
Thou know'st our reasons, urged upon the way.
What think'st thou? Is it not an easy matter            160
To make Lord William Hastings of our mind,
For the instalment of this noble duke
In the seat royal of this famous isle?
CATESBY
He for his father's sake so loves the Prince
That he will not be won to aught against him.           165
BUCKINGHAM
What think'st thou then of Stanley? Will not he?
CATESBY
He will do all-in-all as Hastings doth.
BUCKINGHAM
Well then, no more but this. Go, gentle Catesby,
And, as it were far off, sound thou Lord Hastings
How he doth stand affected to our purpose.              170
If thou dost find him tractable to us,
Encourage him, and tell him all our reasons.
If he be leaden, icy, cold, unwilling,
Be thou so too, and so break off your talk,
And give us notice of his inclination,                  175
For we tomorrow hold divided counsels,
Wherein thyself shalt highly be employed.
RICHARD GLOUCESTER
Commend me to Lord William. Tell him, Catesby,
His ancient knot of dangerous adversaries

Tomorrow are let blood at Pomfret Castle,               180
And bid my lord, for joy of this good news,
Give Mrs Shore one gentle kiss the more.
BUCKINGHAM
Good Catesby, go effect this business soundly.
CATESBY
My good lords both, with all the heed I can.
RICHARD GLOUCESTER
Shall we hear from you, Catesby, ere we sleep?          185
CATESBY You shall, my lord.
RICHARD GLOUCESTER
At Crosby House, there shall you find us both.
                                        Exit Catesby
BUCKINGHAM
My lord, what shall we do if we perceive
Lord Hastings will not yield to our complots?
RICHARD GLOUCESTER
Chop off his head. Something we will determine.         190
And look when I am king, claim thou of me
The earldom of Hereford, and all the movables
Whereof the King my brother was possessed.
BUCKINGHAM
I'll claim that promise at your grace's hand.
RICHARD GLOUCESTER
And look to have it yielded with all kindness.          195
Come, let us sup betimes, that afterwards
We may digest our complots in some form.      Exeunt

3.2     Enter a Messenger to the door of Lord Hastings
MESSENGER (knocking)
My lord, my lord!
LORD HASTINGS ⌈within⌉ Who knocks?
MESSENGER                          One from Lord Stanley.
     ⌈Enter Lord Hastings⌉
LORD HASTINGS
What is't o'clock?
MESSENGER           Upon the stroke of four.
LORD HASTINGS
Cannot my Lord Stanley sleep these tedious nights?
MESSENGER
So it appears by that I have to say.
First he commends him to your noble self.                 5
LORD HASTINGS What then?
MESSENGER
Then certifies your lordship that this night
He dreamt the boar had razèd off his helm.
Besides, he says there are two councils kept,
And that may be determined at the one                    10
Which may make you and him to rue at th'other.
Therefore he sends to know your lordship's pleasure,
If you will presently take horse with him,
And with all speed post with him toward the north
To shun the danger that his soul divines.                15
LORD HASTINGS
Go, fellow, go, return unto thy lord.
Bid him not fear the separated councils.
His honour and myself are at the one,
And at the other is my good friend Catesby,
Where nothing can proceed that toucheth us               20
Whereof I shall not have intelligence.
Tell him his fears are shallow, without instance.
And for his dreams, I wonder he's so simple,
To trust the mock'ry of unquiet slumbers.
To fly the boar before the boar pursues                  25
Were to incense the boar to follow us,
And make pursuit where he did mean no chase.

Go, bid thy master rise, and come to me,
And we will both together to the Tower,
Where he shall see the boar will use us kindly.　30
MESSENGER
I'll go, my lord, and tell him what you say.　　　*Exit*
　　　*Enter Catesby*
CATESBY
Many good morrows to my noble lord.
LORD HASTINGS
Good morrow, Catesby. You are early stirring.
What news, what news, in this our tott'ring state?
CATESBY
It is a reeling world indeed, my lord,　　　35
And I believe will never stand upright
Till Richard wear the garland of the realm.
LORD HASTINGS
How? 'Wear the garland'? Dost thou mean the crown?
CATESBY Ay, my good lord.
LORD HASTINGS
I'll have this crown of mine cut from my shoulders　40
Before I'll see the crown so foul misplaced.
But canst thou guess that he doth aim at it?
CATESBY
Ay, on my life, and hopes to find you forward
Upon his party for the gain thereof—
And thereupon he sends you this good news:　45
That this same very day your enemies,
The kindred of the Queen, must die at Pomfret.
LORD HASTINGS
Indeed I am no mourner for that news,
Because they have been still my adversaries.
But that I'll give my voice on Richard's side　50
To bar my master's heirs in true descent,
God knows I will not do it, to the death.
CATESBY
God keep your lordship in that gracious mind!
LORD HASTINGS
But I shall laugh at this a twelvemonth hence:
That they which brought me in my master's hate,　55
I live to look upon their tragedy.
Well, Catesby, ere a fortnight make me older,
I'll send some packing that yet think not on't.
CATESBY
'Tis a vile thing to die, my gracious lord,
When men are unprepared, and look not for it.　60
LORD HASTINGS
O monstrous, monstrous! And so falls it out
With Rivers, Vaughan, Gray—and so 'twill do
With some men else, that think themselves as safe
As thou and I, who as thou know'st are dear
To princely Richard and to Buckingham.　65
CATESBY
The Princes both make high account of you—
(*Aside*) For they account his head upon the bridge.
LORD HASTINGS
I know they do, and I have well deserved it.
　　　*Enter Lord Stanley*
Come on, come on, where is your boar-spear, man?
Fear you the boar, and go so unprovided?　70
STANLEY
My lord, good morrow.—Good morrow, Catesby.—
You may jest on, but by the Holy Rood
I do not like these several councils, I.
LORD HASTINGS
My lord, I hold my life as dear as you do yours,
And never in my days, I do protest,　75

Was it so precious to me as 'tis now.
Think you, but that I know our state secure,
I would be so triumphant as I am?
STANLEY
The lords at Pomfret, when they rode from London,
Were jocund, and supposed their states were sure,　80
And they indeed had no cause to mistrust;
But yet you see how soon the day o'ercast.
This sudden stab of rancour I misdoubt.
Pray God, I say, I prove a needless coward.
What, shall we toward the Tower? The day is spent.
LORD HASTINGS
Come, come, have with you! Wot you what, my lord?
Today the lords you talked of are beheaded.　87
STANLEY
They for their truth might better wear their heads
Than some that have accused them wear their hats.
But come, my lord, let us away.　　　90
　　　*Enter a Pursuivant named ⌈Hastings⌉*
LORD HASTINGS
Go on before; I'll follow presently.
　　　　　　　　*Exeunt Stanley and Catesby*
Well met, Hastings. How goes the world with thee?
PURSUIVANT
The better that your lordship please to ask.
LORD HASTINGS
I tell thee, man, 'tis better with me now
Than when I met thee last, where now we meet.　95
Then was I going prisoner to the Tower,
By the suggestion of the Queen's allies;
But now, I tell thee—keep it to thyself—
This day those enemies are put to death,
And I in better state than e'er I was.　　　100
PURSUIVANT
God hold it to your honour's good content.
LORD HASTINGS
Gramercy, Hastings. There, drink that for me.
　　　*He throws him his purse*
PURSUIVANT God save your lordship.　　　*Exit*
　　　*Enter a Priest*
PRIEST
Well met, my lord. I am glad to see your honour.
LORD HASTINGS
I thank thee, good Sir John, with all my heart.　105
I am in your debt for your last exercise.
Come the next sabbath, and I will content you.
　　　⌈*He whispers in his ear.*⌉
　　　*Enter Buckingham*
BUCKINGHAM
What, talking with a priest, Lord Chamberlain?
Your friends at Pomfret, they do need the priest;
Your honour hath no shriving work in hand.　110
LORD HASTINGS
Good faith, and when I met this holy man
The men you talk of came into my mind.
What, go you toward the Tower?
BUCKINGHAM
I do, my lord, but long I cannot stay there;
I shall return before your lordship thence.　115
LORD HASTINGS
Nay, like enough, for I stay dinner there.
BUCKINGHAM (*aside*)
And supper too, although thou know'st it not.
Come, will you go?
LORD HASTINGS　　　I'll wait upon your lordship.
　　　　　　　　　　　　　　　　　　*Exeunt*

**3.3** *Enter Sir Richard Ratcliffe with Halberdiers taking*
*Lord Rivers, Lord Gray, and Sir Thomas Vaughan*
*to death at Pomfret*

RIVERS
Sir Richard Ratcliffe, let me tell thee this:
Today shalt thou behold a subject die
For truth, for duty, and for loyalty.

GRAY (*to Ratcliffe*)
God bless the Prince from all the pack of you!
A knot you are of damnèd bloodsuckers.                    5

VAUGHAN (*to Ratcliffe*)
You live, that shall cry woe for this hereafter.

RATCLIFFE
Dispatch. The limit of your lives is out.

RIVERS
O Pomfret, Pomfret! O thou bloody prison,
Fatal and ominous to noble peers!
Within the guilty closure of thy walls,                   10
Richard the Second here was hacked to death,
And, for more slander to thy dismal seat,
We give to thee our guiltless blood to drink.

GRAY
Now Margaret's curse is fall'n upon our heads,
For standing by when Richard stabbed her son.            15

RIVERS
Then cursed she Hastings; then cursed she
    Buckingham;
Then cursed she Richard. O remember, God,
To hear her prayer for them as now for us.
And for my sister and her princely sons,
Be satisfied, dear God, with our true blood,             20
Which, as thou know'st, unjustly must be spilt.

RATCLIFFE
Make haste: the hour of death is expiate.

RIVERS
Come, Gray; come, Vaughan; let us here embrace.
Farewell, until we meet again in heaven.      *Exeunt*

**3.4** *Enter the Duke of Buckingham, Lord Stanley Earl of*
*Derby, Lord Hastings, Bishop of Ely, the Duke of*
*Norfolk, ⌈Sir William Catesby⌉, with others at a table*

LORD HASTINGS
Now, noble peers, the cause why we are met
Is to determine of the coronation.
In God's name, speak: when is the royal day?

BUCKINGHAM
Is all things ready for that solemn time?

STANLEY
It is, and wants but nomination.                          5

BISHOP OF ELY
Tomorrow, then, I judge a happy day.

BUCKINGHAM
Who knows the Lord Protector's mind herein?
Who is most inward with the noble Duke?

BISHOP OF ELY
Your grace, methinks, should soonest know his mind.

BUCKINGHAM
We know each other's faces. For our hearts,              10
He knows no more of mine than I of yours,
Or I of his, my lord, than you of mine.—
Lord Hastings, you and he are near in love.

LORD HASTINGS
I thank his grace; I know he loves me well.
But for his purpose in the coronation,                    15
I have not sounded him, nor he delivered
His gracious pleasure any way therein.

But you, my honourable lords, may name the time,
And in the Duke's behalf I'll give my voice,
Which I presume he'll take in gentle part.               20
*Enter Richard Duke of Gloucester*

BISHOP OF ELY
In happy time, here comes the Duke himself.

RICHARD GLOUCESTER
My noble lords, and cousins all, good morrow.
I have been long a sleeper, but I trust
My absence doth neglect no great design
Which by my presence might have been concluded.          25

BUCKINGHAM
Had not you come upon your cue, my lord,
William Lord Hastings had pronounced your part—
I mean, your voice, for crowning of the King.

RICHARD GLOUCESTER
Than my Lord Hastings no man might be bolder.
His lordship knows me well, and loves me well.—          30
My lord of Ely, when I was last in Holborn
I saw good strawberries in your garden there.
I do beseech you send for some of them.

BISHOP OF ELY
Marry, and will, my lord, with all my heart.      *Exit*

RICHARD GLOUCESTER
Cousin of Buckingham, a word with you.                   35
(*Aside*) Catesby hath sounded Hastings in our business,
And finds the testy gentleman so hot
That he will lose his head ere give consent
His 'master's child'—as worshipful he terms it—
Shall lose the royalty of England's throne.              40

BUCKINGHAM
Withdraw yourself a while; I'll go with you.
*Exeunt Richard ⌈and Buckingham⌉*

STANLEY
We have not yet set down this day of triumph.
Tomorrow, in my judgement, is too sudden,
For I myself am not so well provided
As else I would be, were the day prolonged.              45
*Enter Bishop of Ely*

BISHOP OF ELY
Where is my lord, the Duke of Gloucester?
I have sent for these strawberries.

LORD HASTINGS
His grace looks cheerfully and smooth this morning.
There's some conceit or other likes him well,
When that he bids good morrow with such spirit.          50
I think there's never a man in Christendom
Can lesser hide his love or hate than he,
For by his face straight shall you know his heart.

STANLEY
What of his heart perceive you in his face
By any likelihood he showed today?                       55

LORD HASTINGS
Marry, that with no man here he is offended—
For were he, he had shown it in his looks.

STANLEY I pray God he be not.
*Enter Richard ⌈and Buckingham⌉*

RICHARD GLOUCESTER
I pray you all, tell me what they deserve
That do conspire my death with devilish plots            60
Of damnèd witchcraft, and that have prevailed
Upon my body with their hellish charms?

LORD HASTINGS
The tender love I bear your grace, my lord,
Makes me most forward in this princely presence
To doom th'offenders, whatsoe'er they be.                65
I say, my lord, they have deservèd death.

RICHARD GLOUCESTER
    Then be your eyes the witness of their evil:
    See how I am bewitched. Behold, mine arm
    Is like a blasted sapling withered up.
    And this is Edward's wife, that monstrous witch,    70
    Consorted with that harlot, strumpet Shore,
    That by their witchcraft thus have markèd me.
LORD HASTINGS
    If they have done this deed, my noble lord—
RICHARD GLOUCESTER
    'If'? Thou protector of this damnèd strumpet,
    Talk'st thou to me of 'ifs'? Thou art a traitor.—    75
    Off with his head. Now, by Saint Paul I swear,
    I will not dine until I see the same.
    Some see it done.
    The rest that love me, rise and follow me.
                            *Exeunt all but* ⌜Catesby⌝ *and Hastings*
LORD HASTINGS
    Woe, woe for England! Not a whit for me,    80
    For I, too fond, might have prevented this.
    Stanley did dream the boar did raze our helms,
    But I did scorn it and disdain to fly.
    Three times today my footcloth horse did stumble,
    And started when he looked upon the Tower,    85
    As loath to bear me to the slaughterhouse.
    O now I need the priest that spake to me.
    I now repent I told the pursuivant,
    As too triumphing, how mine enemies
    Today at Pomfret bloodily were butchered,    90
    And I myself secure in grace and favour.
    O Margaret, Margaret! Now thy heavy curse
    Is lighted on poor Hastings' wretched head.
⌜CATESBY⌝
    Come, come, dispatch: the Duke would be at dinner.
    Make a short shrift; he longs to see your head.    95
LORD HASTINGS
    O momentary grace of mortal men,
    Which we more hunt for than the grace of God.
    Who builds his hope in th'air of your good looks
    Lives like a drunken sailor on a mast,
    Ready with every nod to tumble down    100
    Into the fatal bowels of the deep.
⌜CATESBY⌝
    Come, come, dispatch. 'Tis bootless to exclaim.
LORD HASTINGS
    O bloody Richard! Miserable England!
    I prophesy the fearful'st time to thee
    That ever wretched age hath looked upon.—    105
    Come lead me to the block; bear him my head.
    They smile at me, who shortly shall be dead.    *Exeunt*

**3.5**    *Enter Richard Duke of Gloucester and the Duke of*
        *Buckingham in rotten armour, marvellous ill-*
        *favoured*
RICHARD GLOUCESTER
    Come, cousin, canst thou quake and change thy
        colour?
    Murder thy breath in middle of a word?
    And then again begin, and stop again,
    As if thou wert distraught and mad with terror?
BUCKINGHAM
    Tut, I can counterfeit the deep tragedian,    5
    Tremble and start at wagging of a straw,
    Speak, and look back, and pry on every side,
    Intending deep suspicion; ghastly looks

    Are at my service, like enforcèd smiles,
    And both are ready in their offices    10
    At any time to grace my stratagems.
                            *Enter the Lord Mayor*
RICHARD GLOUCESTER (*aside to Buckingham*)
    Here comes the Mayor.
BUCKINGHAM (*aside to Richard*)
    Let me alone to entertain him.—Lord Mayor—
RICHARD GLOUCESTER ⌜*calling as to one within*⌝
    Look to the drawbridge there!
BUCKINGHAM Hark, a drum!    15
RICHARD GLOUCESTER ⌜*calling as to one within*⌝
    Catesby, o'erlook the walls!
BUCKINGHAM Lord Mayor, the reason we have sent—
RICHARD GLOUCESTER
    Look back, defend thee! Here are enemies.
BUCKINGHAM
    God and our innocence defend and guard us.
                            *Enter* ⌜*Sir William Catesby*⌝ *with Hastings' head*
RICHARD GLOUCESTER
    O, O, be quiet! It is Catesby.    20
CATESBY
    Here is the head of that ignoble traitor,
    The dangerous and unsuspected Hastings.
RICHARD GLOUCESTER
    So dear I loved the man that I must weep.
    I took him for the plainest harmless creature
    That breathed upon the earth, a Christian,    25
    Made him my book wherein my soul recorded
    The history of all her secret thoughts.
    So smooth he daubed his vice with show of virtue
    That, his apparent open guilt omitted—
    I mean, his conversation with Shore's wife—    30
    He lived from all attainture of suspect.
BUCKINGHAM
    The covert'st sheltered traitor that ever lived.
    (*To the Mayor*) Would you imagine, or almost believe—
    Were't not that, by great preservation,
    We live to tell it—that the subtle traitor    35
    This day had plotted in the Council house
    To murder me and my good lord of Gloucester?
MAYOR Had he done so?
RICHARD GLOUCESTER
    What, think you we are Turks or infidels,
    Or that we would against the form of law    40
    Proceed thus rashly in the villain's death
    But that the extreme peril of the case,
    The peace of England, and our persons' safety,
    Enforced us to this execution?
MAYOR
    Now fair befall you, he deserved his death,    45
    And your good graces both have well proceeded,
    To warn false traitors from the like attempts.
    I never looked for better at his hands
    After he once fell in with Mrs Shore.
⌜RICHARD GLOUCESTER⌝
    Yet had not we determined he should die,    50
    Until your lordship came to see his end,
    Which now the loving haste of these our friends—
    Something against our meanings—have prevented;
    Because, my lord, we would have had you hear
    The traitor speak, and timorously confess    55
    The manner and the purpose of his treason,
    That you might well have signified the same
    Unto the citizens, who haply may
    Misconster us in him, and wail his death.

MAYOR
But, my good lord, your graces' word shall serve    60
As well as I had seen and heard him speak.
And do not doubt, right noble princes both,
But I'll acquaint our duteous citizens
With all your just proceedings in this cause.
RICHARD GLOUCESTER
And to that end we wished your lordship here,    65
T'avoid the censures of the carping world.
BUCKINGHAM
Which, since you come too late of our intent,
Yet witness what you hear we did intend,
And so, my good Lord Mayor, we bid farewell.
                *Exit Mayor*
RICHARD GLOUCESTER
Go after; after, cousin Buckingham!    70
The Mayor towards Guildhall hies him in all post;
There, at your meetest vantage of the time,
Infer the bastardy of Edward's children.
Tell them how Edward put to death a citizen    75
Only for saying he would make his son
'Heir to the Crown'—meaning indeed, his house,
Which by the sign thereof was termèd so.
Moreover, urge his hateful luxury
And bestial appetite in change of lust,
Which stretched unto their servants, daughters, wives,
Even where his raging eye, or savage heart,    81
Without control, listed to make a prey.
Nay, for a need, thus far come near my person:
Tell them, when that my mother went with child
Of that insatiate Edward, noble York,    85
My princely father, then had wars in France,
And by true computation of the time
Found that the issue was not his begot—
Which well appearèd in his lineaments,
Being nothing like the noble Duke my father.    90
Yet touch this sparingly, as 'twere far off,
Because, my lord, you know my mother lives.
BUCKINGHAM
Doubt not, my lord, I'll play the orator
As if the golden fee for which I plead
Were for myself. And so, my lord, adieu.    95
                *He starts to go*
RICHARD GLOUCESTER
If you thrive well, bring them to Baynard's Castle,
Where you shall find me well accompanied
With reverend fathers and well-learnèd bishops.
BUCKINGHAM
I go, and towards three or four o'clock
Look for the news that the Guildhall affords.    *Exit*
RICHARD GLOUCESTER
Now will I in, to take some privy order    101
To draw the brats of Clarence out of sight,
And to give notice that no manner person
Have any time recourse unto the Princes.    *Exeunt*

**3.6**    *Enter a Scrivener with a paper in his hand*
SCRIVENER
Here is the indictment of the good Lord Hastings,
Which in a set hand fairly is engrossed,
That it may be today read o'er in Paul's—
And mark how well the sequel hangs together:
Eleven hours I have spent to write it over,    5
For yesternight by Catesby was it sent me;
The precedent was full as long a-doing;
And yet, within these five hours, Hastings lived,

Untainted, unexamined, free, at liberty.
Here's a good world the while! Who is so gross    10
That cannot see this palpable device?
Yet who so bold but says he sees it not?
Bad is the world, and all will come to naught,
When such ill dealing must be seen in thought.    *Exit*

**3.7**    *Enter Richard Duke of Gloucester at one door and
       the Duke of Buckingham at another*
RICHARD GLOUCESTER
How now, how now! What say the citizens?
BUCKINGHAM
Now, by the holy mother of our Lord,
The citizens are mum, say not a word.
RICHARD GLOUCESTER
Touched you the bastardy of Edward's children?
BUCKINGHAM
I did, with his contract with Lady Lucy,    5
And his contract by deputy in France,
Th'insatiate greediness of his desire,
And his enforcement of the city wives,
His tyranny for trifles, his own bastardy—
As being got your father then in France,    10
And his resemblance, being not like the Duke.
Withal, I did infer your lineaments—
Being the right idea of your father
Both in your face and nobleness of mind;
Laid open all your victories in Scotland,    15
Your discipline in war, wisdom in peace,
Your bounty, virtue, fair humility—
Indeed, left nothing fitting for your purpose
Untouched or slightly handled in discourse.
And when mine oratory grew toward end,    20
I bid them that did love their country's good
Cry 'God save Richard, England's royal king!'
RICHARD GLOUCESTER   And did they so?
BUCKINGHAM
No, so God help me. They spake not a word,
But, like dumb statuas or breathing stones,    25
Stared each on other and looked deadly pale—
Which, when I saw, I reprehended them,
And asked the Mayor, what meant this wilful silence?
His answer was, the people were not used
To be spoke to but by the Recorder.    30
Then he was urged to tell my tale again:
'Thus saith the Duke . . . thus hath the Duke
         inferred'—
But nothing spoke in warrant from himself.
When he had done, some followers of mine own,
At lower end of the Hall, hurled up their caps,    35
And some ten voices cried 'God save King Richard!'
And thus I took the vantage of those few:
'Thanks, gentle citizens and friends', quoth I;
'This general applause and cheerful shout
Argues your wisdoms and your love to Richard'—    40
And even here brake off and came away.
RICHARD GLOUCESTER
What tongueless blocks were they! Would they not
     speak?
⌈BUCKINGHAM⌉ No, by my troth, my lord.
RICHARD GLOUCESTER
Will not the Mayor then, and his brethren, come?
BUCKINGHAM
The Mayor is here at hand. Intend some fear;    45
Be not you spoke with, but by mighty suit;
And look you get a prayer book in your hand,

And stand between two churchmen, good my lord,
For on that ground I'll build a holy descant.
And be not easily won to our request.                           50
Play the maid's part: still answer 'nay'—and take it.
RICHARD GLOUCESTER
I go. An if you plead as well for them
As I can say nay to thee for myself,
No doubt we'll bring it to a happy issue.
  *One knocks within*
BUCKINGHAM
Go, go, up to the leads! The Lord Mayor knocks.—  55
         *Exit Richard*
  *Enter the Lord Mayor, aldermen, and citizens*
Welcome, my lord. I dance attendance here.
I think the Duke will not be spoke withal.
  *Enter Catesby*
Now Catesby, what says your lord to my request?
CATESBY
He doth entreat your grace, my noble lord,
To visit him tomorrow, or next day.                             60
He is within with two right reverend fathers,
Divinely bent to meditation,
And in no worldly suits would he be moved,
To draw him from his holy exercise.
BUCKINGHAM
Return, good Catesby, to the gracious Duke.                     65
Tell him myself, the Mayor, and aldermen,
In deep designs, in matter of great moment,
No less importing than our general good,
Are come to have some conference with his grace.
CATESBY
I'll signify so much unto him straight.              *Exit*
BUCKINGHAM
Ah ha! My lord, this prince is not an Edward.                   71
He is not lolling on a lewd day-bed,
But on his knees at meditation;
Not dallying with a brace of courtesans,
But meditating with two deep divines;                           75
Not sleeping to engross his idle body,
But praying to enrich his watchful soul.
Happy were England would this virtuous prince
Take on his grace the sovereignty thereof.
But, sure I fear, we shall not win him to it.                   80
MAYOR
Marry, God defend his grace should say us nay.
BUCKINGHAM
I fear he will. Here Catesby comes again.
  *Enter Catesby*
Now Catesby, what says his grace?
CATESBY
He wonders to what end you have assembled
Such troops of citizens to come to him,                         85
His grace not being warned thereof before.
He fears, my lord, you mean no good to him.
BUCKINGHAM
Sorry I am my noble cousin should
Suspect me that I mean no good to him.
By heaven, we come to him in perfect love,                      90
And so once more return and tell his grace.
        *Exit Catesby*
When holy and devout religious men
Are at their beads, 'tis much to draw them thence.
So sweet is zealous contemplation.
  *Enter Richard aloft, between two bishops. ⌈Enter*
  *Catesby below⌉*
MAYOR
See where his grace stands 'tween two clergymen.    95

BUCKINGHAM
Two props of virtue for a Christian prince,
To stay him from the fall of vanity;
And see, a book of prayer in his hand—
True ornaments to know a holy man.—
Famous Plantagenet, most gracious prince,               100
Lend favourable ear to our request,
And pardon us the interruption
Of thy devotion and right Christian zeal.
RICHARD GLOUCESTER
My lord, there needs no such apology.
I do beseech your grace to pardon me,                   105
Who, earnest in the service of my God,
Deferred the visitation of my friends.
But leaving this, what is your grace's pleasure?
BUCKINGHAM
Even that, I hope, which pleaseth God above,
And all good men of this ungoverned isle.               110
RICHARD GLOUCESTER
I do suspect I have done some offence
That seems disgracious in the city's eye,
And that you come to reprehend my ignorance.
BUCKINGHAM
You have, my lord. Would it might please your grace
On our entreaties to amend your fault.                  115
RICHARD GLOUCESTER
Else wherefore breathe I in a Christian land?
BUCKINGHAM
Know then, it is your fault that you resign
The supreme seat, the throne majestical,
The sceptred office of your ancestors,
Your state of fortune and your due of birth,            120
The lineal glory of your royal house,
To the corruption of a blemished stock,
Whiles in the mildness of your sleepy thoughts—
Which here we waken to our country's good—
The noble isle doth want her proper limbs:              125
Her face defaced with scars of infamy,
Her royal stock graft with ignoble plants
And almost shouldered in the swallowing gulf
Of dark forgetfulness and deep oblivion,
Which to recure we heartily solicit                     130
Your gracious self to take on you the charge
And kingly government of this your land—
Not as Protector, steward, substitute,
Or lowly factor for another's gain,
But as successively, from blood to blood,               135
Your right of birth, your empery, your own.
For this, consorted with the citizens,
Your very worshipful and loving friends,
And by their vehement instigation,
In this just cause come I to move your grace.           140
RICHARD GLOUCESTER
I cannot tell if to depart in silence
Or bitterly to speak in your reproof
Best fitteth my degree or your condition.
Your love deserves my thanks; but my desert,
Unmeritable, shuns your high request.                   145
First, if all obstacles were cut away
And that my path were even to the crown,
As the ripe revenue and due of birth,
Yet so much is my poverty of spirit,
So mighty and so many my defects,                       150
That I would rather hide me from my greatness—
Being a barque to brook no mighty sea—
Than in my greatness covet to be hid,

And in the vapour of my glory smothered.
But God be thanked, there is no need of me,                    155
And much I need to help you, were there need.
The royal tree hath left us royal fruit,
Which, mellowed by the stealing hours of time,
Will well become the seat of majesty
And make, no doubt, us happy by his reign.                     160
On him I lay that you would lay on me,
The right and fortune of his happy stars,
Which God defend that I should wring from him.

BUCKINGHAM
My lord, this argues conscience in your grace,
But the respects thereof are nice and trivial,                 165
All circumstances well considerèd.
You say that Edward is your brother's son;
So say we, too—but not by Edward's wife.
For first was he contract to Lady Lucy—
Your mother lives a witness to his vow—                        170
And afterward, by substitute, betrothed
To Bona, sister to the King of France.
These both put off, a poor petitioner,
A care-crazed mother to a many sons,
A beauty-waning and distressèd widow                           175
Even in the afternoon of her best days,
Made prize and purchase of his wanton eye,
Seduced the pitch and height of his degree
To base declension and loathed bigamy.
By her in his unlawful bed he got                              180
This Edward, whom our manners call the Prince.
More bitterly could I expostulate,
Save that for reverence to some alive
I give a sparing limit to my tongue.
Then, good my lord, take to your royal self                    185
This proffered benefit of dignity—
If not to bless us and the land withal,
Yet to draw forth your noble ancestry
From the corruption of abusing times,
Unto a lineal, true-derivèd course.                            190

MAYOR (to Richard)
Do, good my lord; your citizens entreat you.

BUCKINGHAM (to Richard)
Refuse not, mighty lord, this proffered love.

CATESBY (to Richard)
O make them joyful: grant their lawful suit.

RICHARD GLOUCESTER
Alas, why would you heap this care on me?
I am unfit for state and majesty.                              195
I do beseech you, take it not amiss.
I cannot, nor I will not, yield to you.

BUCKINGHAM
If you refuse it—as, in love and zeal,
Loath to depose the child, your brother's son,
As well we know your tenderness of heart                       200
And gentle, kind, effeminate remorse,
Which we have noted in you to your kindred,
And equally indeed to all estates—
Yet know, whe'er you accept our suit or no,
Your brother's son shall never reign our king,                 205
But we will plant some other in the throne,
To the disgrace and downfall of your house.
And in this resolution here we leave you.—
Come, citizens. 'Swounds, I'll entreat no more.

RICHARD GLOUCESTER
O do not swear, my lord of Buckingham.                         210
        [Exeunt Buckingham and some others]

CATESBY
Call him again, sweet prince. Accept their suit.

[ANOTHER]
If you deny them, all the land will rue it.

RICHARD GLOUCESTER
Will you enforce me to a world of cares?
Call them again.                              Exit one or more
            I am not made of stone,
But penetrable to your kind entreats,                          215
Albeit against my conscience and my soul.
        Enter Buckingham and the rest
Cousin of Buckingham, and sage, grave men,
Since you will buckle fortune on my back,
To bear her burden, whe'er I will or no,
I must have patience to endure the load.                       220
But if black scandal or foul-faced reproach
Attend the sequel of your imposition,
Your mere enforcement shall acquittance me
From all the impure blots and stains thereof;
For God doth know, and you may partly see,                     225
How far I am from the desire of this.

MAYOR
God bless your grace! We see it, and will say it.

RICHARD GLOUCESTER
In saying so, you shall but say the truth.

BUCKINGHAM
Then I salute you with this royal title:
Long live kind Richard, England's worthy king!                 230

[ALL BUT RICHARD] Amen.

BUCKINGHAM
Tomorrow may it please you to be crowned?

RICHARD GLOUCESTER
Even when you please, for you will have it so.

BUCKINGHAM
Tomorrow then, we will attend your grace.
And so, most joyfully, we take our leave.                      235

RICHARD GLOUCESTER (to the bishops)
Come, let us to our holy work again.—
Farewell, my cousin. Farewell, gentle friends.
        Exeunt Richard and bishops above, the rest
                                                 below

4.1     Enter Queen Elizabeth, the old Duchess of York, and
        Marquis Dorset at one door; Lady Anne (Duchess
        of Gloucester) with Clarence's daughter at another
        door

DUCHESS OF YORK
Who meets us here? My niece Plantagenet,
Led in the hand of her kind aunt of Gloucester?
Now for my life, she's wand'ring to the Tower,
On pure heart's love, to greet the tender Prince.—
Daughter, well met.

LADY ANNE                God give your graces both             5
A happy and a joyful time of day.

QUEEN ELIZABETH
As much to you, good sister. Whither away?

LADY ANNE
No farther than the Tower, and—as I guess—
Upon the like devotion as yourselves:
To gratulate the gentle princes there.                         10

QUEEN ELIZABETH
Kind sister, thanks. We'll enter all together—
        Enter from the Tower [Brackenbury] the Lieutenant
And in good time, here the Lieutenant comes.
Master Lieutenant, pray you by your leave,
How doth the Prince, and my young son of York?

BRACKENBURY
Right well, dear madam. By your patience,     15
I may not suffer you to visit them.
The King hath strictly charged the contrary.
QUEEN ELIZABETH
The King? Who's that?
BRACKENBURY          I mean, the Lord Protector.
QUEEN ELIZABETH
The Lord protect him from that kingly title.
Hath he set bounds between their love and me?    20
I am their mother; who shall bar me from them?
DUCHESS OF YORK
I am their father's mother; I will see them.
LADY ANNE
Their aunt I am in law, in love their mother;
Then bring me to their sights. I'll bear thy blame,
And take thy office from thee on my peril.     25
BRACKENBURY
No, madam, no; I may not leave it so.
I am bound by oath, and therefore pardon me.     *Exit*
    *Enter Lord Stanley Earl of Derby*
STANLEY
Let me but meet you ladies one hour hence,
And I'll salute your grace of York as mother
And reverend looker-on of two fair queens.     30
(*To Anne*) Come, madam, you must straight to
    Westminster,
There to be crownèd Richard's royal queen.
QUEEN ELIZABETH
Ah, cut my lace asunder, that my pent heart
May have some scope to beat, or else I swoon
With this dead-killing news.     35
LADY ANNE
Despiteful tidings! O unpleasing news!
DORSET (*to Anne*)
Be of good cheer.—Mother, how fares your grace?
QUEEN ELIZABETH
O Dorset, speak not to me. Get thee gone.
Death and destruction dogs thee at thy heels.
Thy mother's name is ominous to children.     40
If thou wilt outstrip death, go cross the seas,
And live with Richmond from the reach of hell.
Go, hie thee! Hie thee from this slaughterhouse,
Lest thou increase the number of the dead,
And make me die the thrall of Margaret's curses:    45
'Nor mother, wife, nor counted England's Queen'.
STANLEY
Full of wise care is this your counsel, madam.
(*To Dorset*) Take all the swift advantage of the hours.
You shall have letters from me to my son
In your behalf, to meet you on the way.     50
Be not ta'en tardy by unwise delay.
DUCHESS OF YORK
O ill-dispersing wind of misery!
O my accursèd womb, the bed of death!
A cockatrice hast thou hatched to the world,
Whose unavoided eye is murderous.     55
STANLEY (*to Anne*)
Come, madam, come. I in all haste was sent.
LADY ANNE
And I in all unwillingness will go.
O would to God that the inclusive verge
Of golden metal that must round my brow
Were red-hot steel, to sear me to the brains.     60
Anointed let me be with deadly venom,
And die ere men can say 'God save the Queen'.

QUEEN ELIZABETH
Go, go, poor soul. I envy not thy glory.
To feed my humour, wish thyself no harm.
LADY ANNE
No? Why? When he that is my husband now     65
Came to me as I followed Henry's corpse,
When scarce the blood was well washed from his
    hands,
Which issued from my other angel husband
And that dear saint which then I weeping followed—
O when, I say, I looked on Richard's face,     70
This was my wish: 'Be thou', quoth I, 'accursed
For making me, so young, so old a widow,
And when thou wedd'st, let sorrow haunt thy bed;
And be thy wife—if any be so mad—
More miserable made by the life of thee     75
Than thou hast made me by my dear lord's death.'
Lo, ere I can repeat this curse again,
Within so small a time, my woman's heart
Grossly grew captive to his honey words
And proved the subject of mine own soul's curse,     80
Which hitherto hath held mine eyes from rest—
For never yet one hour in his bed
Did I enjoy the golden dew of sleep,
But with his timorous dreams was still awaked.
Besides, he hates me for my father Warwick,     85
And will, no doubt, shortly be rid of me.
QUEEN ELIZABETH
Poor heart, adieu. I pity thy complaining.
LADY ANNE
No more than with my soul I mourn for yours.
DORSET
Farewell, thou woeful welcomer of glory.
LADY ANNE
Adieu, poor soul, that tak'st thy leave of it.     90
DUCHESS OF YORK
Go thou to Richmond, and good fortune guide thee.
           ⌈*Exit Dorset*⌉
Go thou to Richard, and good angels tend thee.
       ⌈*Exeunt Anne, Stanley, and Clarence's daughter*⌉
Go thou to sanctuary, and good thoughts possess thee.
           ⌈*Exit Elizabeth*⌉
I to my grave, where peace and rest lie with me.
Eighty odd years of sorrow have I seen,     95
And each hour's joy racked with a week of teen.
           ⌈*Exit*⌉

**4.2**    *Sound a sennet. Enter King Richard in pomp, the*
       *Duke of Buckingham, Sir William Catesby,* ⌈*other*
       *nobles*⌉, *and a Page*
KING RICHARD
Stand all apart.—Cousin of Buckingham.
BUCKINGHAM My gracious sovereign?
KING RICHARD Give me thy hand.
    *Sound* ⌈*a sennet*⌉. *Here Richard ascendeth the*
    *throne*
Thus high by thy advice
And thy assistance is King Richard seated.     5
But shall we wear these glories for a day?
Or shall they last, and we rejoice in them?
BUCKINGHAM
Still live they, and for ever let them last.
KING RICHARD
Ah, Buckingham, now do I play the touch,
To try if thou be current gold indeed.     10
Young Edward lives. Think now what I would speak.

BUCKINGHAM  Say on, my loving lord.
KING RICHARD
    Why, Buckingham, I say I would be king.
BUCKINGHAM
    Why, so you are, my thrice-renownèd liege.
KING RICHARD
    Ha? Am I king? 'Tis so. But Edward lives.                    15
BUCKINGHAM
    True, noble prince.
KING RICHARD                    O bitter consequence,
    That Edward still should live 'true noble prince'.
    Cousin, thou wast not wont to be so dull.
    Shall I be plain? I wish the bastards dead,
    And I would have it immediately performed.                    20
    What sayst thou now? Speak suddenly, be brief.
BUCKINGHAM  Your grace may do your pleasure.
KING RICHARD
    Tut, tut, thou art all ice. Thy kindness freezes.
    Say, have I thy consent that they shall die?
BUCKINGHAM
    Give me some little breath, some pause, dear lord,    25
    Before I positively speak in this.
    I will resolve you herein presently.                    Exit
CATESBY (to another, aside)
    The King is angry. See, he gnaws his lip.
KING RICHARD (aside)
    I will converse with iron-witted fools
    And unrespective boys. None are for me                    30
    That look into me with considerate eyes.
    High-reaching Buckingham grows circumspect.—
    Boy.
PAGE  My lord?
KING RICHARD
    Know'st thou not any whom corrupting gold                    35
    Will tempt unto a close exploit of death?
PAGE
    I know a discontented gentleman
    Whose humble means match not his haughty spirit.
    Gold were as good as twenty orators,
    And will no doubt tempt him to anything.                    40
KING RICHARD
    What is his name?
PAGE                    His name, my lord, is Tyrrell.
KING RICHARD
    I partly know the man. Go call him hither, boy.
                                            Exit Page
    ⌈Aside⌉ The deep-revolving, witty Buckingham
    No more shall be the neighbour to my counsels.
    Hath he so long held out with me untired,                    45
    And stops he now for breath? Well, be it so.
        Enter Lord Stanley Earl of Derby
    How now, Lord Stanley? What's the news?
STANLEY  Know, my loving lord,
    The Marquis Dorset, as I hear, is fled
    To Richmond, in those parts beyond the seas                    50
    Where he abides.
KING RICHARD
    Come hither, Catesby. (Aside to Catesby) Rumour it
        abroad
    That Anne, my wife, is very grievous sick.
    I will take order for her keeping close.
    Enquire me out some mean-born gentleman,                    55
    Whom I will marry straight to Clarence' daughter.
    The boy is foolish, and I fear not him.
    Look how thou dream'st. I say again, give out
    That Anne, my queen, is sick, and like to die.

About it, for it stands me much upon                    60
    To stop all hopes whose growth may damage me.
                                        ⌈Exit Catesby⌉
    (Aside) I must be married to my brother's daughter,
    Or else my kingdom stands on brittle glass.
    Murder her brothers, and then marry her?
    Uncertain way of gain, but I am in                    65
    So far in blood that sin will pluck on sin.
    Tear-falling pity dwells not in this eye.—
        Enter Sir James Tyrrell; ⌈he kneels⌉
    Is thy name Tyrrell?
TYRRELL
    James Tyrrell, and your most obedient subject.
KING RICHARD
    Art thou indeed?
TYRRELL                    Prove me, my gracious lord.                    70
KING RICHARD
    Dar'st thou resolve to kill a friend of mine?
TYRRELL
    Please you, but I had rather kill two enemies.
KING RICHARD
    Why there thou hast it: two deep enemies,
    Foes to my rest, and my sweet sleep's disturbers,
    Are they that I would have thee deal upon.                    75
    Tyrrell, I mean those bastards in the Tower.
TYRRELL
    Let me have open means to come to them,
    And soon I'll rid you from the fear of them.
KING RICHARD
    Thou sing'st sweet music. Hark, come hither, Tyrrell.
    Go, by this token. Rise, and lend thine ear.                    80
        Richard whispers in his ear
    'Tis no more but so. Say it is done,
    And I will love thee, and prefer thee for it.
TYRRELL  I will dispatch it straight.
⌈KING RICHARD⌉
    Shall we hear from thee, Tyrrell, ere we sleep?
        Enter Buckingham
⌈TYRRELL⌉ Ye shall, my lord.                    Exit
BUCKINGHAM
    My lord, I have considered in my mind                    86
    The late request that you did sound me in.
KING RICHARD
    Well, let that rest. Dorset is fled to Richmond.
BUCKINGHAM  I hear the news, my lord.
KING RICHARD
    Stanley, he is your wife's son. Well, look to it.                    90
BUCKINGHAM
    My lord, I claim the gift, my due by promise,
    For which your honour and your faith is pawned:
    Th'earldom of Hereford, and the movables
    Which you have promisèd I shall possess.
KING RICHARD
    Stanley, look to your wife. If she convey                    95
    Letters to Richmond, you shall answer it.
BUCKINGHAM
    What says your highness to my just request?
KING RICHARD
    I do remember me, Henry the Sixth
    Did prophesy that Richmond should be king,
    When Richmond was a little peevish boy.                    100
    A king . . . perhaps . . . perhaps.
BUCKINGHAM                    My lord?
KING RICHARD
    How chance the prophet could not at that time
    Have told me, I being by, that I should kill him?

BUCKINGHAM
  My lord, your promise for the earldom.
KING RICHARD
  Richmond? When last I was at Exeter,                          105
  The Mayor in courtesy showed me the castle,
  And called it 'Ruge-mount'—at which name I started,
  Because a bard of Ireland told me once
  I should not live long after I saw 'Richmond'.
BUCKINGHAM My lord?                                            110
KING RICHARD Ay? What's o'clock?
BUCKINGHAM
  I am thus bold to put your grace in mind
  Of what you promised me.
KING RICHARD                      But what's o'clock?
BUCKINGHAM Upon the stroke of ten.
KING RICHARD Well, let it strike!                              115
BUCKINGHAM Why 'let it strike'?
KING RICHARD
  Because that, like a jack, thou keep'st the stroke
  Betwixt thy begging and my meditation.
  I am not in the giving vein today.
BUCKINGHAM
  Why then resolve me, whe'er you will or no?                  120
KING RICHARD
  Thou troublest me. I am not in the vein.
            *Exit Richard, followed by all but Buckingham*
BUCKINGHAM
  And is it thus? Repays he my deep service
  With such contempt? Made I him king for this?
  O let me think on Hastings, and be gone
  To Brecon, while my fearful head is on.                      125
                            *Exit ⌈at another door⌉*

**4.3**  *Enter Sir James Tyrrell*
TYRRELL
  The tyrannous and bloody act is done—
  The most arch deed of piteous massacre
  That ever yet this land was guilty of.
  Dighton and Forrest, whom I did suborn
  To do this piece of ruthless butchery,                         5
  Albeit they were fleshed villains, bloody dogs,
  Melted with tenderness and mild compassion,
  Wept like two children in their deaths' sad story.
  'O thus', quoth Dighton, 'lay the gentle babes';
  'Thus, thus', quoth Forrest, 'girdling one another            10
  Within their alabaster innocent arms.
  Their lips were four red roses on a stalk,
  And in their summer beauty kissed each other.
  A book of prayers on their pillow lay,
  Which once', quoth Forrest, 'almost changed my mind.
  But O, the devil'—there the villain stopped,                  16
  When Dighton thus told on, 'We smotherèd
  The most replenishèd sweet work of nature,
  That from the prime creation e'er she framed.'
  Hence both are gone, with conscience and remorse.
  They could not speak, and so I left them both,                21
  To bear this tidings to the bloody king.
            *Enter King Richard*
  And here he comes.—All health, my sovereign lord.
KING RICHARD
  Kind Tyrrell, am I happy in thy news?
TYRRELL
  If to have done the thing you gave in charge                  25
  Beget your happiness, be happy then,
  For it is done.
KING RICHARD    But didst thou see them dead?

TYRRELL
  I did, my lord.
KING RICHARD    And buried, gentle Tyrrell?
TYRRELL
  The chaplain of the Tower hath buried them;
  But where, to say the truth, I do not know.                   30
KING RICHARD
  Come to me, Tyrrell, soon, at after-supper,
  When thou shalt tell the process of their death.
  Meantime, but think how I may do thee good,
  And be inheritor of thy desire.
  Farewell till then.
TYRRELL              I humbly take my leave.         *Exit*
KING RICHARD
  The son of Clarence have I pent up close.                     36
  His daughter meanly have I matched in marriage.
  The sons of Edward sleep in Abraham's bosom,
  And Anne, my wife, hath bid this world goodnight.
  Now, for I know the Breton Richmond aims                      40
  At young Elizabeth, my brother's daughter,
  And by that knot looks proudly o'er the crown,
  To her go I, a jolly thriving wooer—
            *Enter Sir Richard Ratcliffe, ⌈running⌉*
RATCLIFFE My lord.
KING RICHARD
  Good news or bad, that thou com'st in so bluntly?   45
RATCLIFFE
  Bad news, my lord. Ely is fled to Richmond,
  And Buckingham, backed with the hardy Welshmen,
  Is in the field, and still his power increaseth.
KING RICHARD
  Ely with Richmond troubles me more near
  Than Buckingham and his rash-levied strength.       50
  Come, I have learned that fearful commenting
  Is leaden servitor to dull delay.
  Delay leads impotent and snail-paced beggary.
  Then fiery expedition be my wing:
  Jove's Mercury, an herald for a king.               55
  Go, muster men. My counsel is my shield.
  We must be brief, when traitors brave the field.
                                        *Exeunt*

**4.4**  *Enter old Queen Margaret*
QUEEN MARGARET
  So now prosperity begins to mellow
  And drop into the rotten mouth of death.
  Here in these confines slyly have I lurked
  To watch the waning of mine enemies.
  A dire induction am I witness to,                    5
  And will to France, hoping the consequence
  Will prove as bitter, black, and tragical.
            *⌈Enter the old Duchess of York and Queen Elizabeth⌉*
  Withdraw thee, wretched Margaret. Who comes here?
QUEEN ELIZABETH
  Ah, my poor princes! Ah, my tender babes!
  My unblown flowers, new-appearing sweets!           10
  If yet your gentle souls fly in the air,
  And be not fixed in doom perpetual,
  Hover about me with your airy wings
  And hear your mother's lamentation.
QUEEN MARGARET (*aside*)
  Hover about her, say that right for right           15
  Hath dimmed your infant morn to agèd night.
DUCHESS OF YORK
  So many miseries have crazed my voice
  That my woe-wearied tongue is still and mute.
  Edward Plantagenet, why art thou dead?

QUEEN MARGARET (*aside*)
Plantagenet doth quit Plantagenet;                        20
Edward for Edward pays a dying debt.

QUEEN ELIZABETH
Wilt thou, O God, fly from such gentle lambs
And throw them in the entrails of the wolf?
When didst thou sleep, when such a deed was done?

QUEEN MARGARET (*aside*)
When holy Harry died, and my sweet son.                   25

DUCHESS OF YORK
Dead life, blind sight, poor mortal living ghost,
Woe's scene, world's shame, grave's due by life
    usurped,
Brief abstract and record of tedious days,
Rest thy unrest on England's lawful earth,
Unlawfully made drunk with innocents' blood.              30
    ⌜They⌝ sit

QUEEN ELIZABETH
Ah that thou wouldst as soon afford a grave
As thou canst yield a melancholy seat.
Then would I hide my bones, not rest them here.
Ah, who hath any cause to mourn but we?

QUEEN MARGARET (*coming forward*)
If ancient sorrow be most reverend,                       35
Give mine the benefit of seniory,
And let my griefs frown on the upper hand.
If sorrow can admit society,
Tell o'er your woes again by viewing mine.
I had an Edward, till a Richard killed him;               40
I had a husband, till a Richard killed him.
(*To Elizabeth*) Thou hadst an Edward, till a Richard
    killed him;
Thou hadst a Richard, till a Richard killed him.

DUCHESS OF YORK ⌜*rising*⌝
I had a Richard too, and thou didst kill him;
I had a Rutland too, thou holpst to kill him.             45

QUEEN MARGARET
Thou hadst a Clarence too, and Richard killed him.
From forth the kennel of thy womb hath crept
A hell-hound that doth hunt us all to death:
That dog that had his teeth before his eyes,
To worry lambs and lap their gentle blood;               50
That foul defacer of God's handiwork,
That reigns in galled eyes of weeping souls;
That excellent grand tyrant of the earth
Thy womb let loose to chase us to our graves.
O upright, just, and true-disposing God,                  55
How do I thank thee that this charnel cur
Preys on the issue of his mother's body,
And makes her pewfellow with others' moan.

DUCHESS OF YORK
O Harry's wife, triumph not in my woes.
God witness with me, I have wept for thine.               60

QUEEN MARGARET
Bear with me. I am hungry for revenge,
And now I cloy me with beholding it.
Thy Edward, he is dead, that killed my Edward;
Thy other Edward dead, to quite my Edward;
Young York, he is but boot, because both they             65
Matched not the high perfection of my loss;
Thy Clarence, he is dead, that stabbed my Edward,
And the beholders of this frantic play—
Th'adulterate Hastings, Rivers, Vaughan, Gray—
Untimely smothered in their dusky graves.                 70
Richard yet lives, hell's black intelligencer,
Only reserved their factor to buy souls
And send them thither; but at hand, at hand

Ensues his piteous and unpitied end.
Earth gapes, hell burns, fiends roar, saints pray,        75
To have him suddenly conveyed from hence.
Cancel his bond of life, dear God, I plead,
That I may live and say, 'The dog is dead'.

QUEEN ELIZABETH
O thou didst prophesy the time would come
That I should wish for thee to help me curse              80
That bottled spider, that foul bunch-backed toad.

QUEEN MARGARET
I called thee then 'vain flourish of my fortune';
I called thee then, poor shadow, 'painted queen'—
The presentation of but what I was,
The flattering index of a direful pageant,                85
One heaved a-high to be hurled down below,
A mother only mocked with two fair babes,
A dream of what thou wast, a garish flag
To be the aim of every dangerous shot,
A sign of dignity, a breath, a bubble,                    90
A queen in jest, only to fill the scene.
Where is thy husband now? Where be thy brothers?
Where are thy two sons? Wherein dost thou joy?
Who sues, and kneels, and says 'God save the Queen'?
Where be the bending peers that flattered thee?           95
Where be the thronging troops that followed thee?
Decline all this, and see what now thou art:
For happy wife, a most distressèd widow;
For joyful mother, one that wails the name;
For queen, a very caitiff, crowned with care;             100
For one being sued to, one that humbly sues;
For she that scorned at me, now scorned of me;
For she being feared of all, now fearing one;
For she commanding all, obeyed of none.
Thus hath the course of justice whirled about,            105
And left thee but a very prey to time,
Having no more but thought of what thou wert,
To torture thee the more, being what thou art.
Thou didst usurp my place, and dost thou not
Usurp the just proportion of my sorrow?                   110
Now thy proud neck bears half my burdened yoke—
From which, even here, I slip my weary head,
And leave the burden of it all on thee.
Farewell, York's wife, and queen of sad mischance.
These English woes shall make me smile in France.

QUEEN ELIZABETH (*rising*)
O thou, well skilled in curses, stay a while,             116
And teach me how to curse mine enemies.

QUEEN MARGARET
Forbear to sleep the nights, and fast the days;
Compare dead happiness with living woe;
Think that thy babes were sweeter than they were,
And he that slew them fouler than he is.                  121
Bett'ring thy loss makes the bad causer worse.
Revolving this will teach thee how to curse.

QUEEN ELIZABETH
My words are dull. O quicken them with thine!

QUEEN MARGARET
Thy woes will make them sharp and pierce like mine.
                                                    *Exit*

DUCHESS OF YORK
Why should calamity be full of words?                     126

QUEEN ELIZABETH
Windy attorneys to their client woes,
Airy recorders of intestate joys,
Poor breathing orators of miseries.
Let them have scope. Though what they will impart
Help nothing else, yet do they ease the heart.           131

DUCHESS OF YORK
If so, then be not tongue-tied; go with me,
And in the breath of bitter words let's smother
My damnèd son, that thy two sweet sons smothered.
        *A march within*
The trumpet sounds. Be copious in exclaims.          135
        *Enter King Richard and his train ⌈marching with*
        *drummers and trumpeters⌉*
KING RICHARD
Who intercepts me in my expedition?
DUCHESS OF YORK
O, she that might have intercepted thee,
By strangling thee in her accursèd womb,
From all the slaughters, wretch, that thou hast done.
QUEEN ELIZABETH
Hid'st thou that forehead with a golden crown,      140
Where should be branded—if that right were right—
The slaughter of the prince that owed that crown,
And the dire death of my poor sons and brothers?
Tell me, thou villain-slave, where are my children?
DUCHESS OF YORK
Thou toad, thou toad, where is thy brother Clarence?
And little Ned Plantagenet his son?                  146
QUEEN ELIZABETH
Where is the gentle Rivers, Vaughan, Gray?
DUCHESS OF YORK Where is kind Hastings?
KING RICHARD (*to his train*)
A flourish, trumpets! Strike alarum, drums!
Let not the heavens hear these tell-tale women        150
Rail on the Lord's anointed. Strike, I say!
        *Flourish. Alarums*
(*To the women*) Either be patient and entreat me fair,
Or with the clamorous report of war
Thus will I drown your exclamations.
DUCHESS OF YORK Art thou my son?                      155
KING RICHARD
Ay, I thank God, my father, and yourself.
DUCHESS OF YORK
Then patiently hear my impatience.
KING RICHARD
Madam, I have a touch of your condition,
That cannot brook the accent of reproof.
DUCHESS OF YORK
O let me speak!
KING RICHARD        Do, then; but I'll not hear.      160
DUCHESS OF YORK
I will be mild and gentle in my words.
KING RICHARD
And brief, good mother, for I am in haste.
DUCHESS OF YORK
Art thou so hasty? I have stayed for thee,
God knows, in torment and in agony—
KING RICHARD
And came I not at last to comfort you?                165
DUCHESS OF YORK
No, by the Holy Rood, thou know'st it well.
Thou cam'st on earth to make the earth my hell.
A grievous burden was thy birth to me;
Tetchy and wayward was thy infancy;                   169
Thy schooldays frightful, desp'rate, wild, and furious;
Thy prime of manhood daring, bold, and venturous;
Thy age confirmed, proud, subtle, sly, and bloody;   172
More mild, but yet more harmful; kind in hatred.
What comfortable hour canst thou name
That ever graced me in thy company?                   175

KING RICHARD
Faith, none but Humphrey Hewer, that called your
        grace
To breakfast once, forth of my company.
If I be so disgracious in your eye,
Let me march on, and not offend you, madam.—
Strike up the drum.
DUCHESS OF YORK        I pray thee, hear me speak.    180
KING RICHARD
You speak too bitterly.
DUCHESS OF YORK        Hear me a word,
For I shall never speak to thee again.
KING RICHARD So.
DUCHESS OF YORK
Either thou wilt die by God's just ordinance
Ere from this war thou turn a conqueror,              185
Or I with grief and extreme age shall perish,
And never more behold thy face again.
Therefore take with thee my most heavy curse,
Which in the day of battle tire thee more
Than all the complete armour that thou wear'st.      190
My prayers on the adverse party fight,
And there the little souls of Edward's children
Whisper the spirits of thine enemies,
And promise them success and victory.
Bloody thou art, bloody will be thy end;             195
Shame serves thy life, and doth thy death attend.
                                        *Exit*
QUEEN ELIZABETH
Though far more cause, yet much less spirit to curse
Abides in me; I say 'Amen' to all.
KING RICHARD
Stay, madam. I must talk a word with you.
QUEEN ELIZABETH
I have no more sons of the royal blood               200
For thee to slaughter. For my daughters, Richard,
They shall be praying nuns, not weeping queens,
And therefore level not to hit their lives.
KING RICHARD
You have a daughter called Elizabeth,
Virtuous and fair, royal and gracious.               205
QUEEN ELIZABETH
And must she die for this? O let her live,
And I'll corrupt her manners, stain her beauty,
Slander myself as false to Edward's bed,
Throw over her the veil of infamy.
So she may live unscarred of bleeding slaughter,     210
I will confess she was not Edward's daughter.
KING RICHARD
Wrong not her birth. She is a royal princess.
QUEEN ELIZABETH
To save her life I'll say she is not so.
KING RICHARD
Her life is safest only in her birth.
QUEEN ELIZABETH
And only in that safety died her brothers.           215
KING RICHARD
Lo, at their births good stars were opposite.
QUEEN ELIZABETH
No, to their lives ill friends were contrary.
KING RICHARD
All unavoided is the doom of destiny—
QUEEN ELIZABETH
True, when avoided grace makes destiny.
My babes were destined to a fairer death,            220
If grace had blessed thee with a fairer life.

KING RICHARD
  Madam, so thrive I in my enterprise
  And dangerous success of bloody wars,
  As I intend more good to you and yours
  Than ever you or yours by me were harmed.        225
QUEEN ELIZABETH
  What good is covered with the face of heaven,
  To be discovered, that can do me good?
KING RICHARD
  Th'advancement of your children, gentle lady.
QUEEN ELIZABETH
  Up to some scaffold, there to lose their heads.
KING RICHARD
  Unto the dignity and height of fortune,         230
  The high imperial type of this earth's glory.
QUEEN ELIZABETH
  Flatter my sorrow with report of it.
  Tell me what state, what dignity, what honour,
  Canst thou demise to any child of mine?
KING RICHARD
  Even all I have—ay, and myself and all,          235
  Will I withal endow a child of thine,
  So in the Lethe of thy angry soul
  Thou drown the sad remembrance of those wrongs,
  Which thou supposest I have done to thee.
QUEEN ELIZABETH
  Be brief, lest that the process of thy kindness  240
  Last longer telling than thy kindness' date.
KING RICHARD
  Then know that, from my soul, I love thy daughter.
QUEEN ELIZABETH
  My daughter's mother thinks that with her soul.
KING RICHARD What do you think?
QUEEN ELIZABETH
  That thou dost love my daughter *from* thy soul;  245
  So *from* thy soul's love didst thou love her brothers,
  And *from* my heart's love I do thank thee for it.
KING RICHARD
  Be not so hasty to confound my meaning.
  I mean, that *with* my soul I love thy daughter,
  And do intend to make her queen of England.       250
QUEEN ELIZABETH
  Well then, who dost thou mean shall be her king?
KING RICHARD
  Even he that makes her queen. Who else should be?
QUEEN ELIZABETH
  What, thou?
KING RICHARD Even so. How think you of it?
QUEEN ELIZABETH
  How canst thou woo her?
KING RICHARD             That would I learn of you,
  As one being best acquainted with her humour.    255
QUEEN ELIZABETH
  And wilt thou learn of me?
KING RICHARD            Madam, with all my heart.
QUEEN ELIZABETH
  Send to her, by the man that slew her brothers,
  A pair of bleeding hearts; thereon engrave
  'Edward' and 'York'; then haply will she weep.
  Therefore present to her—as sometimes Margaret   260
  Did to thy father, steeped in Rutland's blood—
  A handkerchief which, say to her, did drain
  The purple sap from her sweet brother's body,
  And bid her wipe her weeping eyes withal.
  If this inducement move her not to love,         265
  Send her a letter of thy noble deeds.

  Tell her thou mad'st away her uncle Clarence,
  Her uncle Rivers—ay, and for her sake
  Mad'st quick conveyance with her good aunt Anne.
KING RICHARD
  You mock me, madam. This is not the way          270
  To win your daughter.
QUEEN ELIZABETH         There is no other way,
  Unless thou couldst put on some other shape,
  And not be Richard, that hath done all this.
KING RICHARD
  Infer fair England's peace by this alliance.
QUEEN ELIZABETH
  Which she shall purchase with still-lasting war.  275
KING RICHARD
  Tell her the King, that may command, entreats.
QUEEN ELIZABETH
  That at her hands which the King's King forbids.
KING RICHARD
  Say she shall be a high and mighty queen.
QUEEN ELIZABETH
  To vail the title, as her mother doth.
KING RICHARD
  Say I will love her everlastingly.               280
QUEEN ELIZABETH
  But how long shall that title 'ever' last?
KING RICHARD
  Sweetly in force unto her fair life's end.
QUEEN ELIZABETH
  But how long fairly shall her sweet life last?
KING RICHARD
  As long as heaven and nature lengthens it.
QUEEN ELIZABETH
  As long as hell and Richard likes of it.         285
KING RICHARD
  Say I, her sovereign, am her subject love.
QUEEN ELIZABETH
  But she, your subject, loathes such sovereignty.
KING RICHARD
  Be eloquent in my behalf to her.
QUEEN ELIZABETH
  An honest tale speeds best being plainly told.
KING RICHARD
  Then plainly to her tell my loving tale.         290
QUEEN ELIZABETH
  Plain and not honest is too harsh a style.
KING RICHARD
  Your reasons are too shallow and too quick.
QUEEN ELIZABETH
  O no, my reasons are too deep and dead—
  Too deep and dead, poor infants, in their graves.
KING RICHARD
  Harp not on that string, madam. That is past.    295
QUEEN ELIZABETH
  Harp on it still shall I, till heart-strings break.
KING RICHARD
  Now by my George, my garter, and my crown—
QUEEN ELIZABETH
  Profaned, dishonoured, and the third usurped.
KING RICHARD
  I swear—
QUEEN ELIZABETH By nothing, for this is no oath.
  Thy George, profaned, hath lost his holy honour;  300
  Thy garter, blemished, pawned his lordly virtue;
  Thy crown, usurped, disgraced his kingly glory.
  If something thou wouldst swear to be believed,
  Swear then by something that thou hast not wronged.

KING RICHARD
    Then by myself—
QUEEN ELIZABETH        Thy self is self-misused.                305
KING RICHARD
    Now by the world—
QUEEN ELIZABETH          'Tis full of thy foul wrongs.
KING RICHARD
    My father's death—
QUEEN ELIZABETH        Thy life hath that dishonoured.
KING RICHARD
    Why then, by God—
QUEEN ELIZABETH          God's wrong is most of all.
    If thou didst fear to break an oath with him,
    The unity the King my husband made              310
    Thou hadst not broken, nor my brothers died.
    If thou hadst feared to break an oath by him,
    Th'imperial metal circling now thy head
    Had graced the tender temples of my child,
    And both the princes had been breathing here,       315
    Which now—two tender bedfellows for dust—
    Thy broken faith hath made the prey for worms.
    What canst thou swear by now?
KING RICHARD                    The time to come.
QUEEN ELIZABETH
    That thou hast wrongèd in the time o'erpast,
    For I myself have many tears to wash              320
    Hereafter time, for time past wronged by thee.
    The children live, whose fathers thou hast
            slaughtered—
    Ungoverned youth, to wail it in their age.
    The parents live, whose children thou hast
            butchered—
    Old barren plants, to wail it with their age.        325
    Swear not by time to come, for that thou hast
    Misused ere used, by times ill-used o'erpast.
KING RICHARD
    As I intend to prosper and repent,
    So thrive I in my dangerous affairs
    Of hostile arms—myself myself confound,          330
    Heaven and fortune bar me happy hours,
    Day yield me not thy light nor night thy rest;
    Be opposite, all planets of good luck,
    To my proceeding—if, with dear heart's love,
    Immaculate devotion, holy thoughts,              335
    I tender not thy beauteous, princely daughter.
    In her consists my happiness and thine.
    Without her follows—to myself and thee,
    Herself, the land, and many a Christian soul—
    Death, desolation, ruin, and decay.              340
    It cannot be avoided but by this;
    It will not be avoided but by this.
    Therefore, good-mother—I must call you so—
    Be the attorney of my love to her.
    Plead what I will be, not what I have been;        345
    Not my deserts, but what I will deserve.
    Urge the necessity and state of times,
    And be not peevish-fond in great designs.
QUEEN ELIZABETH
    Shall I be tempted of the devil thus?
KING RICHARD
    Ay, if the devil tempt you to do good.            350
QUEEN ELIZABETH
    Shall I forget myself to be myself?
KING RICHARD
    Ay, if yourself's remembrance wrong yourself.
QUEEN ELIZABETH  Yet thou didst kill my children.

KING RICHARD
    But in your daughter's womb I bury them,
    Where, in that nest of spicery, they will breed        355
    Selves of themselves, to your recomfiture.
QUEEN ELIZABETH
    Shall I go win my daughter to thy will?
KING RICHARD
    And be a happy mother by the deed.
QUEEN ELIZABETH
    I go. Write to me very shortly,
    And you shall understand from me her mind.        360
KING RICHARD
    Bear her my true love's kiss,
        *He kisses her*
                            and so farewell—
                                *Exit Elizabeth*
    Relenting fool, and shallow, changing woman.
        *Enter Sir Richard Ratcliffe*
    How now, what news?
RATCLIFFE
    Most mighty sovereign, on the western coast
    Rideth a puissant navy. To our shores            365
    Throng many doubtful, hollow-hearted friends,
    Unarmed and unresolved, to beat them back.
    'Tis thought that Richmond is their admiral,
    And there they hull, expecting but the aid
    Of Buckingham to welcome them ashore.            370
KING RICHARD
    Some light-foot friend post to the Duke of Norfolk.
    Ratcliffe thyself, or Catesby—where is he?
CATESBY
    Here, my good lord.
KING RICHARD            Catesby, fly to the Duke.
CATESBY
    I will, my lord, with all convenient haste.
KING RICHARD
    Ratcliffe, come hither. Post to Salisbury;          375
    When thou com'st thither— (*to Catesby*) dull,
            unmindful villain,
    Why stay'st thou here, and goest not to the Duke?
CATESBY
    First, mighty liege, tell me your highness' pleasure:
    What from your grace I shall deliver to him?
KING RICHARD
    O true, good Catesby. Bid him levy straight        380
    The greatest strength and power that he can make,
    And meet me suddenly at Salisbury.
CATESBY  I go.                                *Exit*
RATCLIFFE
    What, may it please you, shall I do at Salisbury?
KING RICHARD
    Why, what wouldst thou do there before I go?        385
RATCLIFFE
    Your highness told me I should post before.
KING RICHARD
    My mind is changed.
        *Enter Lord Stanley*
                        Stanley, what news with you?
STANLEY
    None, good my liege, to please you with the hearing,
    Nor none so bad but well may be reported.
KING RICHARD
    Hoyday, a riddle! Neither good nor bad.            390
    Why need'st thou run so many mile about
    When thou mayst tell thy tale the nearest way?
    Once more, what news?
STANLEY                    Richmond is on the seas.

KING RICHARD
There let him sink, and be the seas on him.
White-livered renegade, what doth he there?          395
STANLEY
I know not, mighty sovereign, but by guess.
KING RICHARD Well, as you guess?
STANLEY
Stirred up by Dorset, Buckingham, and Ely,
He makes for England, here to claim the crown.
KING RICHARD
Is the chair empty? Is the sword unswayed?          400
Is the King dead? The empire unpossessed?
What heir of York is there alive but we?
And who is England's king but great York's heir?
Then tell me, what makes he upon the seas?
STANLEY
Unless for that, my liege, I cannot guess.           405
KING RICHARD
Unless for that he comes to be your liege,
You cannot guess wherefore the Welshman comes.
Thou wilt revolt and fly to him, I fear.
STANLEY
No, my good lord, therefore mistrust me not.
KING RICHARD
Where is thy power then? To beat him back,           410
Where be thy tenants and thy followers?
Are they not now upon the western shore,
Safe-conducting the rebels from their ships?
STANLEY
No, my good lord, my friends are in the north.
KING RICHARD
Cold friends to me. What do they in the north,      415
When they should serve their sovereign in the west?
STANLEY
They have not been commanded, mighty King.
Pleaseth your majesty to give me leave,
I'll muster up my friends and meet your grace
Where and what time your majesty shall please.      420
KING RICHARD
Ay, ay, thou wouldst be gone to join with Richmond.
But I'll not trust thee.
STANLEY                    Most mighty sovereign,
You have no cause to hold my friendship doubtful.
I never was, nor never will be, false.
KING RICHARD
Go then and muster men—but leave behind            425
Your son George Stanley. Look your heart be firm,
Or else his head's assurance is but frail.
STANLEY
So deal with him as I prove true to you.             Exit
    Enter a Messenger
MESSENGER
My gracious sovereign, now in Devonshire,
As I by friends am well advisèd,                    430
Sir Edward Courtenay and the haughty prelate,
Bishop of Exeter, his elder brother,
With many more confederates are in arms.
    Enter another Messenger
SECOND MESSENGER
In Kent, my liege, the Guildfords are in arms,
And every hour more competitors                      435
Flock to the rebels, and their power grows strong.
    Enter another Messenger
THIRD MESSENGER
My lord, the army of great Buckingham—

KING RICHARD
Out on ye, owls! Nothing but songs of death?
    He striketh him
There, take thou that, till thou bring better news.
THIRD MESSENGER
The news I have to tell your majesty                 440
Is that, by sudden flood and fall of water,
Buckingham's army is dispersed and scattered,
And he himself wandered away alone,
No man knows whither.
KING RICHARD                    I cry thee mercy.—
Ratcliffe, reward him for the blow I gave him.—      445
Hath any well-advisèd friend proclaimed
Reward to him that brings the traitor in?
THIRD MESSENGER
Such proclamation hath been made, my lord.
    Enter another Messenger
FOURTH MESSENGER
Sir Thomas Lovell and Lord Marquis Dorset—
'Tis said, my liege—in Yorkshire are in arms.        450
But this good comfort bring I to your highness:
The Breton navy is dispersed by tempest.
Richmond in Dorsetshire sent out a boat
Unto the shore, to ask those on the banks
If they were his assistants, yea or no?               455
Who answered him they came from Buckingham
Upon his party. He, mistrusting them,
Hoist sail and made his course again for Bretagne.
KING RICHARD
March on, march on, since we are up in arms,
If not to fight with foreign enemies,                460
Yet to beat down these rebels here at home.
    Enter Catesby
CATESBY
My liege, the Duke of Buckingham is taken.
That is the best news. That the Earl of Richmond
Is with a mighty power landed at Milford
Is colder tidings, yet they must be told.            465
KING RICHARD
Away, towards Salisbury! While we reason here,
A royal battle might be won and lost.
Someone take order Buckingham be brought
To Salisbury. The rest march on with me.
                                    Flourish. Exeunt

4.5    Enter Lord Stanley Earl of Derby and Sir
       Christopher, a priest
STANLEY
Sir Christopher, tell Richmond this from me:
That in the sty of this most deadly boar
My son George Stanley is franked up in hold.
If I revolt, off goes young George's head.
The fear of that holds off my present aid.            5
But tell me, where is princely Richmond now?
SIR CHRISTOPHER
At Pembroke, or at Ha'rfordwest in Wales.
STANLEY
What men of name resort to him?
SIR CHRISTOPHER
Sir Walter Herbert, a renownèd soldier,
Sir Gilbert Talbot, Sir William Stanley,             10
Oxford, redoubted Pembroke, Sir James Blunt,
And Rhys-ap-Thomas with a valiant crew,
And many other of great name and worth—
And towards London do they bend their power,
If by the way they be not fought withal.             15

STANLEY
Well, hie thee to thy lord. Commend me to him.
Tell him the Queen hath heartily consented
He should espouse Elizabeth her daughter.
My letter will resolve him of my mind.
Farewell.                                        *Exeunt severally*

**5.1**   *Enter the Duke of Buckingham with halberdiers, led
by a Sheriff to execution*
BUCKINGHAM
Will not King Richard let me speak with him?
SHERIFF
No, my good lord, therefore be patient.
BUCKINGHAM
Hastings, and Edward's children, Gray and Rivers,
Holy King Henry and thy fair son Edward,
Vaughan, and all that have miscarrièd                    5
By underhand, corrupted, foul injustice:
If that your moody, discontented souls
Do through the clouds behold this present hour,
Even for revenge mock my destruction.
This is All-Souls' day, fellow, is it not?              10
SHERIFF It is.
BUCKINGHAM
Why then All-Souls' day is my body's doomsday.
This is the day which, in King Edward's time,
I wished might fall on me, when I was found
False to his children and his wife's allies.            15
This is the day wherein I wished to fall
By the false faith of him whom most I trusted.
This, this All-Souls' day to my fearful soul
Is the determined respite of my wrongs.
That high all-seer which I dallied with                 20
Hath turned my feignèd prayer on my head,
And given in earnest what I begged in jest.
Thus doth he force the swords of wicked men
To turn their own points in their masters' bosoms.
Thus Margaret's curse falls heavy on my neck.           25
'When he', quoth she, 'shall split thy heart with
      sorrow,
Remember Margaret was a prophetess.'
Come lead me, officers, to the block of shame.
Wrong hath but wrong, and blame the due of blame.
                                                 *Exeunt*

**5.2**   *Enter Henry Earl of Richmond with a letter, the
Earl of Oxford, Sir James Blunt, Sir Walter
Herbert, and others, with drum and colours*
HENRY EARL OF RICHMOND
Fellows in arms, and my most loving friends,
Bruised underneath the yoke of tyranny,
Thus far into the bowels of the land
Have we marched on without impediment,
And here receive we from our father Stanley            5
Lines of fair comfort and encouragement.
The wretched, bloody, and usurping boar,
That spoils your summer fields and fruitful vines,
Swills your warm blood like wash, and makes his
      trough
In your inbowelled bosoms, this foul swine            10
Lies now even in the centry of this isle,
Near to the town of Leicester, as we learn.
From Tamworth thither is but one day's march.
In God's name, cheerly on, courageous friends,
To reap the harvest of perpetual peace                15
By this one bloody trial of sharp war.

OXFORD
Every man's conscience is a thousand swords
To fight against this guilty homicide.
HERBERT
I doubt not but his friends will turn to us.
BLUNT
He hath no friends but what are friends for fear,      20
Which in his dearest need will fly from him.
HENRY EARL OF RICHMOND
All for our vantage. Then, in God's name, march.
True hope is swift, and flies with swallows' wings;
Kings it makes gods, and meaner creatures kings.
                                          *Exeunt ⌜marching⌝*

**5.3**   *Enter King Richard in arms, with the Duke of
Norfolk, Sir Richard Ratcliffe, ⌜Sir William
Catesby, and others⌝*
KING RICHARD
Here pitch our tent, even here in Bosworth field.
      *Soldiers begin to pitch ⌜a tent⌝*
Why, how now, Catesby? Why look you so sad?
⌜CATESBY⌝
My heart is ten times lighter than my looks.
KING RICHARD
My lord of Norfolk.
NORFOLK                     Here, most gracious liege.
KING RICHARD
Norfolk, we must have knocks. Ha, must we not?         5
NORFOLK
We must both give and take, my loving lord.
KING RICHARD
Up with my tent! Here will I lie tonight.
But where tomorrow? Well, all's one for that.
Who hath descried the number of the traitors?
NORFOLK
Six or seven thousand is their utmost power.           10
KING RICHARD
Why, our battalia trebles that account.
Besides, the King's name is a tower of strength,
Which they upon the adverse faction want.
Up with the tent! Come, noble gentlemen,
Let us survey the vantage of the ground.               15
Call for some men of sound direction.
Let's lack no discipline, make no delay—
For, lords, tomorrow is a busy day.
                                          *Exeunt ⌜at one door⌝*

**5.4**   *Enter ⌜at another door⌝ Henry Earl of Richmond,
Sir James Blunt, Sir William Brandon, ⌜the Earl of
Oxford, Marquis Dorset, and others⌝*
HENRY EARL OF RICHMOND
The weary sun hath made a golden set,
And by the bright track of his fiery car
Gives token of a goodly day tomorrow.
Sir William Brandon, you shall bear my standard.
The Earl of Pembroke keeps his regiment;               5
Good Captain Blunt, bear my good night to him,
And by the second hour in the morning
Desire the Earl to see me in my tent.
Yet one thing more, good Captain, do for me:
Where is Lord Stanley quartered, do you know?          10
BLUNT
Unless I have mista'en his colours much,
Which well I am assured I have not done,
His regiment lies half a mile, at least,
South from the mighty power of the King.

HENRY EARL OF RICHMOND
 If without peril it be possible,                        15
 Sweet Blunt, make some good means to speak with
   him,
 And give him from me this most needful note.
BLUNT
 Upon my life, my lord, I'll undertake it.
 And so God give you quiet rest tonight.
HENRY EARL OF RICHMOND
 Good night, good Captain Blunt.          *Exit Blunt*
                           Come, gentlemen.
 Give me some ink and paper in my tent.                  21
 I'll draw the form and model of our battle,
 Limit each leader to his several charge,
 And part in just proportion our small power.
 Let us consult upon tomorrow's business.                25
 Into my tent: the dew is raw and cold.
                         *They withdraw into the tent*

**5.5**   ⌜*A table brought in.*⌝ *Enter King Richard, Sir*
         *Richard Ratcliffe, the Duke of Norfolk, Sir William*
         *Catesby, and others*
KING RICHARD What is't o'clock?
CATESBY
 It's supper-time, my lord. It's nine o'clock.
KING RICHARD
 I will not sup tonight. Give me some ink and paper.
 What, is my beaver easier than it was?
 And all my armour laid into my tent?                     5
CATESBY
 It is, my liege, and all things are in readiness.
KING RICHARD
 Good Norfolk, hie thee to thy charge.
 Use careful watch; choose trusty sentinels.
NORFOLK I go, my lord.
KING RICHARD
 Stir with the lark tomorrow, gentle Norfolk.            10
NORFOLK
 I warrant you, my lord.                            *Exit*
KING RICHARD                      Catesby.
CATESBY                              My lord?
KING RICHARD
 Send out a pursuivant-at-arms
 To Stanley's regiment. Bid him bring his power
 Before sun-rising, lest his son George fall
 Into the blind cave of eternal night.      ⌜*Exit Catesby*⌝
 Fill me a bowl of wine. Give me a watch.                16
 Saddle white Surrey for the field tomorrow.
 Look that my staves be sound, and not too heavy.
 Ratcliffe.
RATCLIFFE My lord?                                       20
KING RICHARD
 Saw'st thou the melancholy Lord Northumberland?
RATCLIFFE
 Thomas the Earl of Surrey and himself,
 Much about cockshut time, from troop to troop
 Went through the army, cheering up the soldiers.
KING RICHARD
 So, I am satisfied. Give me some wine.                  25
 I have not that alacrity of spirit,
 Nor cheer of mind, that I was wont to have.
           *The wine is brought*
 Set it down. Is ink and paper ready?
RATCLIFFE
 It is, my lord.
KING RICHARD   Leave me. Bid my guard watch.

About the mid of night come to my tent,                  30
Ratcliffe, and help to arm me. Leave me, I say.
            *Exit Ratcliffe* ⌜*with others. Richard writes, and*
                                         *later sleeps*⌝
       *Enter Lord Stanley Earl of Derby to Henry Earl of*
       *Richmond and the lords in his tent*
STANLEY
 Fortune and victory sit on thy helm!
HENRY EARL OF RICHMOND
 All comfort that the dark night can afford
 Be to thy person, noble father-in-law.
 Tell me, how fares our loving mother?                   35
STANLEY
 I, by attorney, bless thee from thy mother,
 Who prays continually for Richmond's good.
 So much for that. The silent hours steal on,
 And flaky darkness breaks within the east.
 In brief—for so the season bids us be—                 40
 Prepare thy battle early in the morning,
 And put thy fortune to th'arbitrement
 Of bloody strokes and mortal-staring war.
 I, as I may—that which I would, I cannot—
 With best advantage will deceive the time,              45
 And aid thee in this doubtful shock of arms.
 But on thy side I may not be too forward—
 Lest, being seen, thy brother, tender George,
 Be executed in his father's sight.
 Farewell. The leisure and the fearful time              50
 Cuts off the ceremonious vows of love
 And ample interchange of sweet discourse,
 Which so long sundered friends should dwell upon.
 God give us leisure for these rights of love.
 Once more, adieu. Be valiant, and speed well.           55
HENRY EARL OF RICHMOND
 Good lords, conduct him to his regiment.
 I'll strive with troubled thoughts to take a nap,
 Lest leaden slumber peise me down tomorrow,
 When I should mount with wings of victory.
 Once more, good night, kind lords and gentlemen.        60
            *Exeunt Stanley and the lords*
       ⌜*Richmond kneels*⌝
 O thou, whose captain I account myself,
 Look on my forces with a gracious eye.
 Put in their hands thy bruising irons of wrath,
 That they may crush down with a heavy fall
 Th'usurping helmets of our adversaries.                 65
 Make us thy ministers of chastisement,
 That we may praise thee in the victory.
 To thee I do commend my watchful soul,
 Ere I let fall the windows of mine eyes.
 Sleeping and waking, O defend me still!     *He sleeps*
       *Enter the Ghost of young Prince Edward* ⌜*above*⌝
GHOST OF PRINCE EDWARD (*to Richard*)
 Let me sit heavy on thy soul tomorrow,                  71
 Prince Edward, son to Henry the Sixth.
 Think how thou stabbedst me in my prime of youth
 At Tewkesbury. Despair, therefore, and die.
 (*To Richmond*) Be cheerful, Richmond, for the wrongèd
   souls                                                 75
 Of butchered princes fight in thy behalf.
 King Henry's issue, Richmond, comforts thee.   ⌜*Exit*⌝
       *Enter* ⌜*above*⌝ *the Ghost of King Henry the Sixth*
GHOST OF KING HENRY (*to Richard*)
 When I was mortal, my anointed body
 By thee was punchèd full of deadly holes.
 Think on the Tower and me. Despair and die.             80

Harry the Sixth bids thee despair and die.
(*To Richmond*) Virtuous and holy, be thou conqueror.
Harry that prophesied thou shouldst be king
Comforts thee in thy sleep. Live and flourish!     ⌜*Exit*⌝
          *Enter* ⌜*above*⌝ *the Ghost of George Duke of Clarence*
GHOST OF CLARENCE (*to Richard*)
Let me sit heavy on thy soul tomorrow,                    85
I that was washed to death with fulsome wine,
Poor Clarence, by thy guile betrayed to death.
Tomorrow in the battle think on me,
And fall thy edgeless sword. Despair and die.
(*To Richmond*) Thou offspring of the house of
          Lancaster,                                                          90
The wrongèd heirs of York do pray for thee.
Good angels guard thy battle. Live and flourish!     ⌜*Exit*⌝
          *Enter* ⌜*above*⌝ *the Ghosts of Lord Rivers, Lord Gray,*
          *and Sir Thomas Vaughan*
GHOST OF RIVERS (*to Richard*)
Let me sit heavy on thy soul tomorrow,
Rivers that died at Pomfret. Despair and die.
GHOST OF GRAY (*to Richard*)
Think upon Gray, and let thy soul despair.                95
GHOST OF VAUGHAN (*to Richard*)
Think upon Vaughan, and with guilty fear
Let fall thy pointless lance. Despair and die.
ALL THREE (*to Richmond*)
Awake, and think our wrongs in Richard's bosom
Will conquer him. Awake, and win the day!
                                                  ⌜*Exeunt Ghosts*⌝
          *Enter* ⌜*above*⌝ *the Ghosts of the two young Princes*
⌜GHOSTS OF THE PRINCES⌝ (*to Richard*)
Dream on thy cousins, smothered in the Tower.     100
Let us be lead within thy bosom, Richard,
And weigh thee down to ruin, shame, and death.
Thy nephews' souls bid thee despair and die.
(*To Richmond*) Sleep, Richmond, sleep in peace and
          wake in joy.
Good angels guard thee from the boar's annoy.     105
Live, and beget a happy race of kings!
Edward's unhappy sons do bid thee flourish.
                                                  ⌜*Exeunt Ghosts*⌝
          *Enter* ⌜*above*⌝ *the Ghost of Lord Hastings*
GHOST OF HASTINGS (*to Richard*)
Bloody and guilty, guiltily awake,
And in a bloody battle end thy days.
Think on Lord Hastings, then despair and die.     110
(*To Richmond*) Quiet, untroubled soul, awake, awake!
Arm, fight, and conquer for fair England's sake.     ⌜*Exit*⌝
          *Enter* ⌜*above*⌝ *the Ghost of Lady Anne*
GHOST OF LADY ANNE (*to Richard*)
Richard, thy wife, that wretched Anne thy wife,
That never slept a quiet hour with thee,
Now fills thy sleep with perturbations.                    115
Tomorrow in the battle think on me,
And fall thy edgeless sword. Despair and die.
(*To Richmond*) Thou quiet soul, sleep thou a quiet
          sleep.
Dream of success and happy victory.
Thy adversary's wife doth pray for thee.               ⌜*Exit*⌝
          *Enter* ⌜*above*⌝ *the Ghost of the Duke of Buckingham*
GHOST OF BUCKINGHAM (*to Richard*)
The first was I that helped thee to the crown;     121
The last was I that felt thy tyranny.
O in the battle think on Buckingham,
And die in terror of thy guiltiness!

Dream on, dream on, of bloody deeds and death;     125
Fainting, despair; despairing, yield thy breath.
(*To Richmond*) I died for hope ere I could lend thee aid.
But cheer thy heart, and be thou not dismayed.
God and good angels fight on Richmond's side,
And Richard falls in height of all his pride.     ⌜*Exit*⌝
          *Richard starteth up out of a dream*
KING RICHARD
Give me another horse! Bind up my wounds!     131
Have mercy, Jesu!—Soft, I did but dream.
O coward conscience, how dost thou afflict me?
The lights burn blue. It is now dead midnight.
Cold fearful drops stand on my trembling flesh.     135
What do I fear? Myself? There's none else by.
Richard loves Richard; that is, I am I.
Is there a murderer here? No. Yes, I am.
Then fly! What, from myself? Great reason. Why?
Lest I revenge. Myself upon myself?                    140
Alack, I love myself. Wherefore? For any good
That I myself have done unto myself?
O no, alas, I rather hate myself
For hateful deeds committed by myself.
I am a villain. Yet I lie: I am not.                       145
Fool, of thyself speak well.—Fool, do not flatter.
My conscience hath a thousand several tongues,
And every tongue brings in a several tale,
And every tale condemns me for a villain.
Perjury, perjury, in the high'st degree!               150
Murder, stern murder, in the dir'st degree!
All several sins, all used in each degree,
Throng to the bar, crying all, 'Guilty, guilty!'
I shall despair. There is no creature loves me,
And if I die no soul will pity me.                         155
Nay, wherefore should they?—Since that I myself
Find in myself no pity to myself.
Methought the souls of all that I had murdered
Came to my tent, and every one did threat
Tomorrow's vengeance on the head of Richard.     160
          *Enter Ratcliffe*
RATCLIFFE My lord?
KING RICHARD 'Swounds, who is there?
RATCLIFFE
My lord, 'tis I. The early village cock
Hath twice done salutation to the morn.
Your friends are up, and buckle on their armour.     165
KING RICHARD
O Ratcliffe, I have dreamed a fearful dream.
What thinkest thou, will all our friends prove true?
RATCLIFFE
No doubt, my lord.
KING RICHARD          Ratcliffe, I fear, I fear.
RATCLIFFE
Nay, good my lord, be not afraid of shadows.
KING RICHARD
By the Apostle Paul, shadows tonight                    170
Have struck more terror to the soul of Richard
Than can the substance of ten thousand soldiers
Armèd in proof and led by shallow Richmond.
'Tis not yet near day. Come, go with me.
Under our tents I'll play the eavesdropper,          175
To see if any mean to shrink from me.
                              *Exeunt Richard and Ratcliffe*
          *Enter the lords to Henry Earl of Richmond, sitting*
          *in his tent*
⌜LORDS⌝ Good morrow, Richmond.

HENRY EARL OF RICHMOND
 Cry mercy, lords and watchful gentlemen,
 That you have ta'en a tardy sluggard here.
⌈A LORD⌉ How have you slept, my lord?          180
HENRY EARL OF RICHMOND
 The sweetest sleep and fairest boding dreams
 That ever entered in a drowsy head
 Have I since your departure had, my lords.
 Methought their souls whose bodies Richard murdered
 Came to my tent and cried on victory.          185
 I promise you, my soul is very jocund
 In the remembrance of so fair a dream.
 How far into the morning is it, lords?
⌈A LORD⌉ Upon the stroke of four.
HENRY EARL OF RICHMOND
 Why then, 'tis time to arm, and give direction.   190
       *His oration to his soldiers*
 Much that I could say, loving countrymen,
 The leisure and enforcement of the time
 Forbids to dwell on. Yet remember this:
 God and our good cause fight upon our side.
 The prayers of holy saints and wrongèd souls,   195
 Like high-reared bulwarks, stand before our forces.
 Richard except, those whom we fight against
 Had rather have us win than him they follow.
 For what is he they follow? Truly, friends,
 A bloody tyrant and a homicide;               200
 One raised in blood, and one in blood established;
 One that made means to come by what he hath,
 And slaughtered those that were the means to help
   him;
 A base, foul stone, made precious by the foil
 Of England's chair, where he is falsely set;     205
 One that hath ever been God's enemy.
 Then if you fight against God's enemy,
 God will, in justice, ward you as his soldiers.
 If you do sweat to put a tyrant down,
 You sleep in peace, the tyrant being slain.      210
 If you do fight against your country's foes,
 Your country's foison pays your pains the hire.
 If you do fight in safeguard of your wives,
 Your wives shall welcome home the conquerors.
 If you do free your children from the sword,     215
 Your children's children quites it in your age.
 Then, in the name of God and all these rights,
 Advance your standards! Draw your willing swords!
 For me, the ransom of this bold attempt
 Shall be my cold corpse on the earth's cold face;  220
 But if I thrive, to gain of my attempt,
 The least of you shall share his part thereof.
 Sound, drums and trumpets, bold and cheerfully!
 God and Saint George! Richmond and victory!
       ⌈*Exeunt to the sound of drums and trumpets*⌉

**5.6**   *Enter King Richard, Sir Richard Ratcliffe, Sir*
       *William Catesby, and others*
KING RICHARD
 What said Northumberland, as touching Richmond?
RATCLIFFE
 That he was never trainèd up in arms.
KING RICHARD
 He said the truth. And what said Surrey then?
RATCLIFFE
 He smiled and said, 'The better for our purpose.'
KING RICHARD
 He was in the right, and so indeed it is.        5

       *Clock strikes*
 Tell the clock there. Give me a calendar.
 Who saw the sun today?
       ⌈*A book is brought*⌉
RATCLIFFE                    Not I, my lord.
KING RICHARD
 Then he disdains to shine, for by the book
 He should have braved the east an hour ago.
 A black day will it be to somebody.             10
 Ratcliffe.
RATCLIFFE
 My lord?
KING RICHARD The sun will not be seen today.
 The sky doth frown and lour upon our army.
 I would these dewy tears were from the ground.
 Not shine today—why, what is that to me         15
 More than to Richmond? For the selfsame heaven
 That frowns on me looks sadly upon him.
       *Enter the Duke of Norfolk*
NORFOLK
 Arm, arm, my lord! The foe vaunts in the field.
KING RICHARD
 Come, bustle, bustle! Caparison my horse.
       ⌈*Richard arms*⌉
 Call up Lord Stanley, bid him bring his power.   20
                          *Exit one*
 I will lead forth my soldiers to the plain,
 And thus my battle shall be orderèd.
 My forward shall be drawn out all in length,
 Consisting equally of horse add foot,
 Our archers placèd strongly in the midst.        25
 John Duke of Norfolk, Thomas Earl of Surrey,
 Shall have the leading of this multitude.
 They thus directed, we ourself will follow
 In the main battle, whose puissance on both sides
 Shall be well wingèd with our chiefest horse.    30
 This, and Saint George to boot! What think'st thou,
   Norfolk?
NORFOLK
 A good direction, warlike sovereign.
       *He showeth him a paper*
 This paper found I on my tent this morning.
 (*He reads*)
       'Jackie of Norfolk be not too bold,
       For Dickon thy master is bought and sold.'  35
KING RICHARD
 A thing devisèd by the enemy.—
 Go, gentlemen, each man unto his charge.
 Let not our babbling dreams affright our souls.
 Conscience is but a word that cowards use,
 Devised at first to keep the strong in awe.      40
 Our strong arms be our conscience; swords, our law.
 March on, join bravely! Let us to't, pell mell—
 If not to heaven, then hand in hand to hell.
       *His oration to his army*
 What shall I say, more than I have inferred?
 Remember whom you are to cope withal:          45
 A sort of vagabonds, rascals and runaways,
 A scum of Bretons and base lackey peasants,
 Whom their o'ercloyèd country vomits forth
 To desperate ventures and assured destruction.
 You sleeping safe, they bring to you unrest;     50
 You having lands and blessed with beauteous wives,
 They would distrain the one, distain the other.
 And who doth lead them, but a paltry fellow?

Long kept in Bretagne at our mother's cost;
A milksop; one that never in his life                                55
Felt so much cold as over shoes in snow.
Let's whip these stragglers o'er the seas again,
Lash hence these overweening rags of France,
These famished beggars, weary of their lives,
Who—but for dreaming on this fond exploit—             60
For want of means, poor rats, had hanged themselves.
If we be conquered, let *men* conquer us,
And not these bastard Bretons, whom our fathers
Have in their own land beaten, bobbed, and thumped,
And in record left them the heirs of shame.                       65
Shall these enjoy our lands? Lie with our wives?
Ravish our daughters?
           *Drum afar off*
                                    Hark, I hear their drum.
Fight, gentlemen of England! Fight, bold yeomen!
Draw, archers, draw your arrows to the head!
Spur your proud horses hard, and ride in blood!            70
Amaze the welkin with your broken staves!
           *Enter a Messenger*
What says Lord Stanley? Will he bring his power?
MESSENGER
My lord, he doth deny to come.
KING RICHARD Off with young George's head!
NORFOLK
My lord, the enemy is past the marsh.                              75
After the battle let George Stanley die.
KING RICHARD
A thousand hearts are great within my bosom.
Advance our standards! Set upon our foes!
Our ancient word of courage, fair Saint George,
Inspire us with the spleen of fiery dragons.                      80
Upon them! Victory sits on our helms!            *Exeunt*

**5.7**    *Alarum. Excursions. Enter Sir William Catesby*
CATESBY ⌐*calling*⌐
Rescue, my lord of Norfolk! Rescue, rescue!
⌐*To a soldier*⌐ The King enacts more wonders than a
           man,
Daring an opposite to every danger.
His horse is slain, and all on foot he fights,
Seeking for Richmond in the throat of death.                    5
⌐*Calling*⌐ Rescue, fair lord, or else the day is lost!
           *Alarums. Enter King Richard*
KING RICHARD
A horse! A horse! My kingdom for a horse!
CATESBY
Withdraw, my lord. I'll help you to a horse.
KING RICHARD
Slave, I have set my life upon a cast,
And I will stand the hazard of the die.                            10
I think there be six Richmonds in the field.
Five have I slain today, instead of him.
A horse! A horse! My kingdom for a horse!      *Exeunt*

**5.8**    *Alarum. Enter King Richard ⌐at one door⌐ and*
           *Henry Earl of Richmond ⌐at another⌐. They fight.*
           *Richard is slain. ⌐Exit Richmond.⌐ Retreat and*
           *flourish. Enter Henry Earl of Richmond and Lord*
           *Stanley Earl of Derby, with divers other lords and*
           *soldiers*
HENRY EARL OF RICHMOND
God and your arms be praised, victorious friends!
The day is ours. The bloody dog is dead.
STANLEY (*bearing the crown*)
Courageous Richmond, well hast thou acquit thee.
Lo, here this long usurpèd royalty
From the dead temples of this bloody wretch                 5
Have I plucked off, to grace thy brows withal.
Wear it, enjoy it, and make much of it.
           ⌐*He sets the crown on Henry's head*⌐
KING HENRY THE SEVENTH
Great God of heaven, say 'Amen' to all.
But tell me—young George Stanley, is he living?
STANLEY
He is, my lord, and safe in Leicester town,                  10
Whither, if it please you, we may now withdraw us.
KING HENRY THE SEVENTH
What men of name are slain on either side?
⌐STANLEY⌐ (*reads*)
John Duke of Norfolk, Robert Brackenbury,
Walter Lord Ferrers, and Sir William Brandon.
KING HENRY THE SEVENTH
Inter their bodies as becomes their births.                    15
Proclaim a pardon to the soldiers fled
That in submission will return to us,
And then—as we have ta'en the sacrament—
We will unite the white rose and the red.
Smile, heaven, upon this fair conjunction,                    20
That long have frowned upon their enmity.
What traitor hears me and says not 'Amen'?
England hath long been mad, and scarred herself;
The brother blindly shed the brother's blood;
The father rashly slaughtered his own son;                   25
The son, compelled, been butcher to the sire;
All that divided York and Lancaster,
United in their dire division.
O now let Richmond and Elizabeth,
The true succeeders of each royal house,                      30
By God's fair ordinance conjoin together,
And let their heirs—God, if his will be so—
Enrich the time to come with smooth-faced peace,
With smiling plenty, and fair prosperous days.
Abate the edge of traitors, gracious Lord,                    35
That would reduce these bloody days again
And make poor England weep forth streams of blood.
Let them not live to taste this land's increase,
That would with treason wound this fair land's peace.
Now civil wounds are stopped; peace lives again.      40
That she may long live here, God say 'Amen'.
           ⌐*Flourish.*⌐ *Exeunt*

## ADDITIONAL PASSAGES

The following passages are contained in the Folio text, but not the Quarto; they were apparently omitted from performances.

A. AFTER 1.2.154

These eyes, which never shed remorseful tear—
No, when my father York and Edward wept
To hear the piteous moan that Rutland made
When black-faced Clifford shook his sword at him;
Nor when thy warlike father like a child          5
Told the sad story of my father's death
And twenty times made pause to sob and weep,
That all the standers-by had wet their cheeks
Like trees bedashed with rain. In that sad time
My manly eyes did scorn an humble tear,          10
And what these sorrows could not thence exhale
Thy beauty hath, and made them blind with weeping.

B. AFTER 1.3.166

RICHARD GLOUCESTER
Wert thou not banishèd on pain of death?
QUEEN MARGARET
I was, but I do find more pain in banishment
Than death can yield me here by my abode.

C. AFTER 1.4.68

O God! If my deep prayers cannot appease thee
But thou wilt be avenged on my misdeeds,
Yet execute thy wrath in me alone.
O spare my guiltless wife and my poor children.

D. AFTER 2.2.88

The Folio has Dorset and Rivers enter with Queen Elizabeth at 2.2.33.1.

DORSET
Comfort, dear mother. God is much displeased
That you take with unthankfulness his doing.
In common worldly things 'tis called ungrateful
With dull unwillingness to pay a debt
Which with a bounteous hand was kindly lent;          5
Much more to be thus opposite with heaven
For it requires the royal debt it lent you.
RIVERS
Madam, bethink you like a careful mother
Of the young Prince your son. Send straight for him;
Let him be crowned. In him your comfort lives.          10
Drown desperate sorrow in dead Edward's grave
And plant your joys in living Edward's throne.

E. AFTER 2.2.110

RIVERS
Why with some little train, my lord of Buckingham?
BUCKINGHAM
Marry, my lord, lest by a multitude
The new-healed wound of malice should break out,
Which would be so much the more dangerous
By how much the estate is green and yet ungoverned.
Where every horse bears his commanding rein          6
And may direct his course as please himself,
As well the fear of harm as harm apparent
In my opinion ought to be prevented.

RICHARD GLOUCESTER
I hope the King made peace with all of us,          10
And the compact is firm and true in me.
RIVERS
And so in me, and so I think in all.
Yet since it is but green, it should be put
To no apparent likelihood of breach,
Which haply by much company might be urged.          15
Therefore I say, with noble Buckingham,
That it is meet so few should fetch the Prince.
HASTINGS And so say I.

F. AFTER 3.1.170

And summon him tomorrow to the Tower
To sit about the coronation.

G. AFTER 3.5.100

Beginning Richard Gloucester's speech. The Folio brings on Lovell and Ratcliffe instead of Catesby at 3.5.19.1.

RICHARD GLOUCESTER
Go, Lovell, with all speed to Doctor Shaw;
(To Ratcliffe) Go thou to Friar Penker. Bid them both
Meet me within this hour at Baynard's Castle.
                              Exeunt Lovell and Ratcliffe

H. AFTER 3.7.143

If not to answer, you might haply think
Tongue-tied ambition, not replying, yielded
To bear the golden yoke of sovereignty,
Which fondly you would here impose on me.
If to reprove you for this suit of yours,          5
So seasoned with your faithful love to me,
Then on the other side I checked my friends.
Therefore to speak, and to avoid the first,
And then in speaking not to incur the last,
Definitively thus I answer you.          10

I. AFTER 4.1.96

In the Folio, the characters do not exit during the Duchess of York's preceding speech.

QUEEN ELIZABETH
Stay: yet look back with me unto the Tower.—
Pity, you ancient stones, those tender babes,
Whom envy hath immured within your walls.
Rough cradle for such little pretty ones,
Rude ragged nurse, old sullen playfellow          5
For tender princes: use my babies well.
So foolish sorrow bids your stones farewell.          Exeunt

J. AFTER 4.4.221

KING RICHARD
You speak as if that I had slain my cousins.
QUEEN ELIZABETH
Cousins indeed, and by their uncle cozened
Of comfort, kingdom, kindred, freedom, life.
Whose hand soever lanced their tender hearts,
Thy head all indirectly gave direction.          5
No doubt the murd'rous knife was dull and blunt
Till it was whetted on thy stone-hard heart
To revel in the entrails of my lambs.
But that still use of grief makes wild grief tame,

My tongue should to thy ears not name my boys          10
Till that my nails were anchored in thine eyes—
And I in such a desp'rate bay of death,
Like a poor barque of sails and tackling reft,
Rush all to pieces on thy rocky bosom.

## K. AFTER 4.4.273

KING RICHARD
  Say that I did all this for love of her.
QUEEN ELIZABETH
  Nay, then indeed she cannot choose but hate thee,
  Having bought love with such a bloody spoil.
KING RICHARD
  Look what is done cannot be now amended.
  Men shall deal unadvisedly sometimes,                5
  Which after-hours gives leisure to repent.
  If I did take the kingdom from your sons,
  To make amends I'll give it to your daughter.
  If I have killed the issue of your womb,
  To quicken your increase I will beget               10
  Mine issue of your blood upon your daughter.
  A grandam's name is little less in love
  Than is the doting title of a mother.
  They are as children but one step below,
  Even of your mettall, of your very blood:           15
  Of all one pain, save for a night of groans
  Endured of her for whom you bid like sorrow.
  Your children were vexation to your youth,
  But mine shall be a comfort to your age.
  The loss you have is but a son being king,          20
  And by that loss your daughter is made queen.
  I cannot make you what amends I would,
  Therefore accept such kindness as I can.

  Dorset your son, that with a fearful soul
  Leads discontented steps in foreign soil,           25
  This fair alliance quickly shall call home
  To high promotions and great dignity.
  The king that calls your beauteous daughter wife,
  Familiarly shall call thy Dorset brother.
  Again shall you be mother to a king,                30
  And all the ruins of distressful times
  Repaired with double riches of content.
  What? We have many goodly days to see.
  The liquid drops of tears that you have shed
  Shall come again, transformed to orient pearl,     35
  Advantaging their loan with interest
  Of ten times double gain of happiness.
  Go then, my mother, to thy daughter go.
  Make bold her bashful years with your experience.
  Prepare her ears to hear a wooer's tale.           40
  Put in her tender heart th'aspiring flame
  Of golden sovereignty. Acquaint the Princess
  With the sweet silent hours of marriage joys.
  And when this arm of mine hath chastisèd
  The petty rebel, dull-brained Buckingham,          45
  Bound with triumphant garlands will I come
  And lead thy daughter to a conqueror's bed—
  To whom I will retail my conquest won,
  And she shall be sole victoress: Caesar's Caesar.
QUEEN ELIZABETH
  What were I best to say? Her father's brother      50
  Would be her lord? Or shall I say her uncle?
  Or he that slew her brothers and her uncles?
  Under what title shall I woo for thee,
  That God, the law, my honour, and her love
  Can make seem pleasing to her tender years?        55

# VENUS AND ADONIS

WITH *Venus and Adonis*, Shakespeare made his debut in print: his signature appears at the end of the formal dedication to the Earl of Southampton in which the poem is described as 'the first heir of my invention'—though Shakespeare had already begun to make his mark as a playwright. A terrible outbreak of plague, which was to last for almost two years, began in the summer of 1592, and London's theatres were closed as a precaution against infection. Probably Shakespeare wrote his poem at this time, perhaps seeing a need for an alternative career. It is an early example of the Ovidian erotic narrative poems that were fashionable for about thirty years from 1589; the best known outside Shakespeare is Christopher Marlowe's *Hero and Leander*, written at about the same time.

Ovid, in Book 10 of the *Metamorphoses*, tells the story of Venus and Adonis in about seventy-five lines of verse; Shakespeare's poem—drawing, probably, on both the original Latin and Arthur Golding's English version (1565-7)—is 1,194 lines long. He modified Ovid's tale as well as expanding it. In Ovid, the handsome young mortal Adonis returns the love urged on him by Venus, the goddess of love. Shakespeare turns Adonis into a bashful teenager, unripe for love, who shies away from her advances. In Ovid, the lovers go hunting together (though Venus chases only relatively harmless beasts, and advises Adonis to do the same); in Shakespeare, Adonis takes to the hunt rather as a respite from Venus' remorseless attentions. Whereas Ovid's Venus flies off to Cyprus in her dove-drawn chariot and returns only after Adonis has been mortally wounded, Shakespeare's anxiously awaits the outcome of the chase. She hears the yelping of Adonis' hounds, sees a blood-stained boar, comes upon Adonis' defeated dogs, and at last finds his body. In Ovid, she metamorphoses him into an anemone; in Shakespeare, Adonis' body melts away, and Venus plucks the purple and white flower that springs up in its place.

Shakespeare's only addition to Ovid's narrative is the episode (259-324) in which Adonis' stallion lusts after a mare, so frustrating Adonis' attempt to escape Venus' embraces. But there are many rhetorical elaborations, such as Venus' speech of attempted seduction (95-174), her disquisition on the dangers of boar-hunting (613-714), her metaphysical explanation of why the night is dark (721-68), Adonis' reply (769-810), culminating in his eloquent contrast between lust and love, and Venus' lament over his body (1069-1164).

*Venus and Adonis* is a mythological poem whose landscape is inhabited by none but the lovers and those members of the animal kingdom—the lustful stallion, the timorous hare (679-708), the sensitive snail (1033-6), and the savage boar—which reflect their passions. The boar's disruption of the harmony that existed between Adonis and the animals will, says Venus, result in eternal discord: 'Sorrow on love hereafter shall attend' (1136).

In Shakespeare's own time, *Venus and Adonis* was his most frequently reprinted work, with at least ten editions during his life, and another half-dozen by 1636. After this it fell out of fashion until Coleridge wrote enthusiastically about it in *Biographia Literaria* (1817). Though its conscious artifice may limit its appeal, it is a brilliantly sophisticated erotic comedy, a counterpart in verbal ingenuity to *Love's Labour's Lost*; the comedy of the poem, like that of the play, is darkened and deepened in its later stages by the shadow of sudden death.

*Vilia miretur vulgus ; mihi flavus Apollo*
*Pocula Castalia plena ministret aqua.*

TO THE RIGHT HONOURABLE HENRY WRIOTHESLEY,
EARL OF SOUTHAMPTON, AND BARON OF TITCHFIELD

Right Honourable, I know not how I shall offend in dedicating my unpolished lines to your lordship, nor how the world will censure me for choosing so strong a prop to support so weak a burden. Only, if your honour seem but pleased, I account myself highly praised, and vow to take advantage of all idle hours till I have honoured you with some graver labour. But if the first heir of my invention prove deformed, I shall be sorry it had so noble a godfather, and never after ear so barren a land for fear it yield me still so bad a harvest. I leave it to your honourable survey, and your honour to your heart's content, which I wish may always answer your own wish and the world's hopeful expectation.

Your honour's in all duty,

William Shakespeare

# Venus and Adonis

Even as the sun with purple-coloured face
Had ta'en his last leave of the weeping morn,
Rose-cheeked Adonis hied him to the chase.
Hunting he loved, but love he laughed to scorn.
  Sick-thoughted Venus makes amain unto him,   5
  And like a bold-faced suitor 'gins to woo him.

'Thrice fairer than myself,' thus she began,
'The fields' chief flower, sweet above compare,
Stain to all nymphs, more lovely than a man,
More white and red than doves or roses are—   10
  Nature that made thee with herself at strife
  Saith that the world hath ending with thy life.

'Vouchsafe, thou wonder, to alight thy steed
And rein his proud head to the saddle-bow;
If thou wilt deign this favour, for thy meed   15
A thousand honey secrets shalt thou know.
  Here come and sit where never serpent hisses;
  And, being sat, I'll smother thee with kisses,

'And yet not cloy thy lips with loathed satiety,
But rather famish them amid their plenty,   20
Making them red, and pale, with fresh variety;
Ten kisses short as one, one long as twenty.
  A summer's day will seem an hour but short,
  Being wasted in such time-beguiling sport.'

With this, she seizeth on his sweating palm,   25
The precedent of pith and livelihood,
And, trembling in her passion, calls it balm—
Earth's sovereign salve to do a goddess good.
  Being so enraged, desire doth lend her force
  Courageously to pluck him from his horse.   30

Over one arm, the lusty courser's rein;
Under her other was the tender boy,
Who blushed and pouted in a dull disdain
With leaden appetite, unapt to toy.
  She red and hot as coals of glowing fire;   35
  He red for shame, but frosty in desire.

The studded bridle on a ragged bough
Nimbly she fastens—O, how quick is love!
The steed is stallèd up, and even now
To tie the rider she begins to prove.   40
  Backward she pushed him, as she would be thrust,
  And governed him in strength, though not in lust.

So soon was she along as he was down,
Each leaning on their elbows and their hips.
Now doth she stroke his cheek, now doth he frown   45
And 'gins to chide, but soon she stops his lips,
  And, kissing, speaks, with lustful language broken:
  'If thou wilt chide, thy lips shall never open.'

He burns with bashful shame; she with her tears
Doth quench the maiden burning of his cheeks.   50
Then, with her windy sighs and golden hairs,
To fan and blow them dry again she seeks.

He saith she is immodest, blames her miss;
What follows more she murders with a kiss.

Even as an empty eagle, sharp by fast,   55
Tires with her beak on feathers, flesh, and bone,
Shaking her wings, devouring all in haste
Till either gorge be stuffed or prey be gone,
  Even so she kissed his brow, his cheek, his chin,
  And where she ends she doth anew begin.   60

Forced to content, but never to obey,
Panting he lies and breatheth in her face.
She feedeth on the steam as on a prey
And calls it heavenly moisture, air of grace,
  Wishing her cheeks were gardens full of flowers,   65
  So they were dewed with such distilling showers.

Look how a bird lies tangled in a net,
So fastened in her arms Adonis lies.
Pure shame and awed resistance made him fret,
Which bred more beauty in his angry eyes.   70
  Rain added to a river that is rank
  Perforce will force it overflow the bank.

Still she entreats, and prettily entreats,
For to a pretty ear she tunes her tale.
Still is he sullen, still he lours and frets   75
'Twixt crimson shame and anger ashy-pale.
  Being red, she loves him best; and being white,
  Her best is bettered with a more delight.

Look how he can, she cannot choose but love;
And by her fair immortal hand she swears   80
From his soft bosom never to remove
Till he take truce with her contending tears,
  Which long have rained, making her cheeks all wet;
  And one sweet kiss shall pay this countless debt.

Upon this promise did he raise his chin,   85
Like a divedapper peering through a wave
Who, being looked on, ducks as quickly in—
So offers he to give what she did crave.
  But when her lips were ready for his pay,
  He winks, and turns his lips another way.   90

Never did passenger in summer's heat
More thirst for drink than she for this good turn.
Her help she sees, but help she cannot get.
She bathes in water, yet her fire must burn.
  'O pity,' gan she cry, 'flint-hearted boy!   95
  'Tis but a kiss I beg—why art thou coy?

'I have been wooed as I entreat thee now
Even by the stern and direful god of war,
Whose sinewy neck in battle ne'er did bow,
Who conquers where he comes in every jar.   100
  Yet hath he been my captive and my slave,
  And begged for that which thou unasked shalt have.

'Over my altars hath he hung his lance,
His battered shield, his uncontrollèd crest,
And for my sake hath learned to sport and dance,  105
To toy, to wanton, dally, smile, and jest,
    Scorning his churlish drum and ensign red,
    Making my arms his field, his tent my bed.

'Thus he that over-ruled I overswayed,
Leading him prisoner in a red-rose chain.  110
Strong-tempered steel his stronger strength obeyed,
Yet was he servile to my coy disdain.
    O, be not proud, nor brag not of thy might,
    For mast'ring her that foiled the god of fight.

'Touch but my lips with those fair lips of thine—  115
Though mine be not so fair, yet are they red—
The kiss shall be thine own as well as mine.
What seest thou in the ground? Hold up thy head.
    Look in mine eyeballs: there thy beauty lies.
    Then why not lips on lips, since eyes in eyes?  120

'Art thou ashamed to kiss? Then wink again,
And I will wink. So shall the day seem night.
Love keeps his revels where there are but twain.
Be bold to play—our sport is not in sight.
    These blue-veined violets whereon we lean  125
    Never can blab, nor know not what we mean.

'The tender spring upon thy tempting lip
Shows thee unripe; yet mayst thou well be tasted.
Make use of time; let not advantage slip.
Beauty within itself should not be wasted.  130
    Fair flowers that are not gathered in their prime
    Rot, and consume themselves in little time.

'Were I hard-favoured, foul, or wrinkled-old,
Ill-nurtured, crooked, churlish, harsh in voice,
O'er-worn, despisèd, rheumatic, and cold,  135
Thick-sighted, barren, lean, and lacking juice,
    Then mightst thou pause, for then I were not for
      thee.
    But having no defects, why dost abhor me?

'Thou canst not see one wrinkle in my brow.
Mine eyes are grey, and bright, and quick in turning.
My beauty as the spring doth yearly grow.  141
My flesh is soft and plump, my marrow burning.
    My smooth moist hand, were it with thy hand felt,
    Would in thy palm dissolve, or seem to melt.

'Bid me discourse, I will enchant thine ear;  145
Or like a fairy, trip upon the green;
Or like a nymph, with long, dishevelled hair,
Dance on the sands, and yet no footing seen.
    Love is a spirit all compact of fire,
    Not gross to sink, but light, and will aspire.  150

'Witness this primrose bank whereon I lie:
These forceless flowers like sturdy trees support me.
Two strengthless doves will draw me through the sky
From morn till night, even where I list to sport me.
    Is love so light, sweet boy, and may it be  155
    That thou should think it heavy unto thee?

'Is thine own heart to thine own face affected?
Can thy right hand seize love upon thy left?
Then woo thyself, be of thyself rejected;
Steal thine own freedom, and complain on theft.  160
    Narcissus so himself himself forsook,
    And died to kiss his shadow in the brook.

'Torches are made to light, jewels to wear,
Dainties to taste, fresh beauty for the use,
Herbs for their smell, and sappy plants to bear.  165
Things growing to themselves are growth's abuse.
    Seeds spring from seeds, and beauty breedeth
      beauty:
    Thou wast begot; to get it is thy duty.

'Upon the earth's increase why shouldst thou feed
Unless the earth with thy increase be fed?  170
By law of nature thou art bound to breed,
That thine may live when thou thyself art dead;
    And so in spite of death thou dost survive,
    In that thy likeness still is left alive.'

By this, the lovesick queen began to sweat,  175
For where they lay the shadow had forsook them,
And Titan, tired in the midday heat,
With burning eye did hotly overlook them,
    Wishing Adonis had his team to guide
    So he were like him, and by Venus' side.  180

And now Adonis, with a lazy sprite
And with a heavy, dark, disliking eye,
His louring brows o'erwhelming his fair sight,
Like misty vapours when they blot the sky,
    Souring his cheeks, cries, 'Fie, no more of love!  185
    The sun doth burn my face; I must remove.'

'Ay me,' quoth Venus, 'young, and so unkind?
What bare excuses mak'st thou to be gone?
I'll sigh celestial breath, whose gentle wind
Shall cool the heat of this descending sun.  190
    I'll make a shadow for thee of my hairs;
    If they burn too, I'll quench them with my tears.

'The sun that shines from heaven shines but warm,
And lo, I lie between that sun and thee.
The heat I have from thence doth little harm;  195
Thine eye darts forth the fire that burneth me,
    And were I not immortal, life were done
    Between this heavenly and earthly sun.

'Art thou obdurate, flinty, hard as steel?
Nay, more than flint, for stone at rain relenteth.  200
Art thou a woman's son, and canst not feel
What 'tis to love, how want of love tormenteth?
    O, had thy mother borne so hard a mind,
    She had not brought forth thee, but died unkind.

'What am I, that thou shouldst contemn me this?  205
Or what great danger dwells upon my suit?
What were thy lips the worse for one poor kiss?
Speak, fair; but speak fair words, or else be mute.
    Give me one kiss, I'll give it thee again,
    And one for int'rest, if thou wilt have twain.  210

'Fie, lifeless picture, cold and senseless stone,
Well painted idol, image dull and dead,
Statue contenting but the eye alone,
Thing like a man, but of no woman bred:
    Thou art no man, though of a man's complexion,
    For men will kiss even by their own direction.'  216

This said, impatience chokes her pleading tongue,
And swelling passion doth provoke a pause.
Red cheeks and fiery eyes blaze forth her wrong.
Being judge in love, she cannot right her cause;  220
    And now she weeps, and now she fain would speak,
    And now her sobs do her intendments break.

Sometime she shakes her head, and then his hand;
Now gazeth she on him, now on the ground.
Sometime her arms enfold him like a band;  225
She would, he will not in her arms be bound.
    And when from thence he struggles to be gone,
    She locks her lily fingers one in one.

'Fondling,' she saith, 'since I have hemmed thee here
Within the circuit of this ivory pale,  230
I'll be a park, and thou shalt be my deer.
Feed where thou wilt, on mountain or in dale;
    Graze on my lips, and if those hills be dry,
    Stray lower, where the pleasant fountains lie.

'Within this limit is relief enough,  235
Sweet bottom-grass, and high delightful plain,
Round rising hillocks, brakes obscure and rough,
To shelter thee from tempest and from rain.
    Then be my deer, since I am such a park;
    No dog shall rouse thee, though a thousand bark.'

At this Adonis smiles as in disdain,  241
That in each cheek appears a pretty dimple.
Love made those hollows, if himself were slain,
He might be buried in a tomb so simple,
    Foreknowing well, if there he came to lie,  245
    Why, there love lived, and there he could not die.

These lovely caves, these round enchanting pits,
Opened their mouths to swallow Venus' liking.
Being mad before, how doth she now for wits?
Struck dead at first, what needs a second striking?  250
    Poor queen of love, in thine own law forlorn,
    To love a cheek that smiles at thee in scorn!

Now which way shall she turn? What shall she say?
Her words are done, her woes the more increasing.
The time is spent; her object will away,  255
And from her twining arms doth urge releasing.
    'Pity,' she cries; 'some favour, some remorse!'
    Away he springs, and hasteth to his horse.

But lo, from forth a copse that neighbours by
A breeding jennet, lusty, young, and proud,  260
Adonis' trampling courser doth espy,
And forth she rushes, snorts, and neighs aloud.
    The strong-necked steed, being tied unto a tree,
    Breaketh his rein, and to her straight goes he.

Imperiously he leaps, he neighs, he bounds,  265
And now his woven girths he breaks asunder.
The bearing earth with his hard hoof he wounds,
Whose hollow womb resounds like heaven's thunder.
    The iron bit he crusheth 'tween his teeth,
    Controlling what he was controllèd with.  270

His ears up-pricked, his braided hanging mane
Upon his compassed crest now stand on end;
His nostrils drink the air, and forth again,
As from a furnace, vapours doth he send.
    His eye, which scornfully glisters like fire,  275
    Shows his hot courage and his high desire.

Sometime he trots, as if he told the steps,
With gentle majesty and modest pride.
Anon he rears upright, curvets, and leaps,
As who should say, 'Lo, thus my strength is tried,  280
    And this I do to captivate the eye
    Of the fair breeder that is standing by.'

What recketh he his rider's angry stir,
His flattering 'Holla', or his 'Stand, I say!'?
What cares he now for curb or pricking spur,  285
For rich caparisons or trappings gay?
    He sees his love, and nothing else he sees,
    For nothing else with his proud sight agrees.

Look when a painter would surpass the life
In limning out a well proportioned steed,  290
His art with nature's workmanship at strife,
As if the dead the living should exceed:
    So did this horse excel a common one
    In shape, in courage, colour, pace, and bone.

Round-hoofed, short-jointed, fetlocks shag and long,
Broad breast, full eye, small head, and nostril wide,
High crest, short ears, straight legs, and passing
    strong;  297
Thin mane, thick tail, broad buttock, tender hide—
    Look what a horse should have he did not lack,
    Save a proud rider on so proud a back.  300

Sometime he scuds far off, and there he stares;
Anon he starts at stirring of a feather.
To bid the wind a base he now prepares,
And whe'er he run or fly they know not whether;
    For through his mane and tail the high wind sings,
    Fanning the hairs, who wave like feathered wings.

He looks upon his love, and neighs unto her;  307
She answers him as if she knew his mind.
Being proud, as females are, to see him woo her,
She puts on outward strangeness, seems unkind,  310
    Spurns at his love, and scorns the heat he feels,
    Beating his kind embracements with her heels.

Then, like a melancholy malcontent,
He vails his tail that, like a falling plume,
Cool shadow to his melting buttock lent.  315
He stamps, and bites the poor flies in his fume.
    His love, perceiving how he was enraged,
    Grew kinder, and his fury was assuaged.

His testy master goeth about to take him,
When lo, the unbacked breeder, full of fear,　320
Jealous of catching, swiftly doth forsake him,
With her the horse, and left Adonis there.
　As they were mad unto the wood they hie them,
　Outstripping crows that strive to overfly them.

All swoll'n with chafing, down Adonis sits,　325
Banning his boist'rous and unruly beast;
And now the happy season once more fits
That lovesick love by pleading may be blessed;
　For lovers say the heart hath treble wrong
　When it is barred the aidance of the tongue.　330

An oven that is stopped, or river stayed,
Burneth more hotly, swelleth with more rage.
So of concealèd sorrow may be said
Free vent of words love's fire doth assuage.
　But when the heart's attorney once is mute,　335
　The client breaks, as desperate in his suit.

He sees her coming, and begins to glow,
Even as a dying coal revives with wind,
And with his bonnet hides his angry brow,
Looks on the dull earth with disturbèd mind,　340
　Taking no notice that she is so nigh,
　For all askance he holds her in his eye.

O, what a sight it was wistly to view
How she came stealing to the wayward boy,
To note the fighting conflict of her hue,　345
How white and red each other did destroy!
　But now her cheek was pale; and by and by
　It flashed forth fire, as lightning from the sky.

Now was she just before him as he sat,
And like a lowly lover down she kneels;　350
With one fair hand she heaveth up his hat;
Her other tender hand his fair cheek feels.
　His tend'rer cheek receives her soft hand's print
　As apt as new-fall'n snow takes any dint.

O, what a war of looks was then between them,　355
Her eyes petitioners to his eyes suing!
His eyes saw her eyes as they had not seen them;
Her eyes wooed still; his eyes disdained the wooing;
　And all this dumb play had his acts made plain
　With tears which, chorus-like, her eyes did rain.

Full gently now she takes him by the hand,　361
A lily prisoned in a jail of snow,
Or ivory in an alabaster band;
So white a friend engirds so white a foe.
　This beauteous combat, wilful and unwilling,　365
　Showed like two silver doves that sit a-billing.

Once more the engine of her thoughts began:
'O fairest mover on this mortal round,
Would thou wert as I am, and I a man,
My heart all whole as thine, thy heart my wound;
　For one sweet look thy help I would assure thee,
　Though nothing but my body's bane would cure
　　thee.'
　　　　372

'Give me my hand,' saith he. 'Why dost thou feel it?'
'Give me my heart,' saith she, 'and thou shalt have it.
O, give it me, lest thy hard heart do steel it,　375
And, being steeled, soft sighs can never grave it;
　Then love's deep groans I never shall regard,
　Because Adonis' heart hath made mine hard.'

'For shame,' he cries, 'let go, and let me go!
My day's delight is past; my horse is gone,　380
And 'tis your fault I am bereft him so.
I pray you hence, and leave me here alone;
　For all my mind, my thought, my busy care
　Is how to get my palfrey from the mare.'

Thus she replies: 'Thy palfrey, as he should,　385
Welcomes the warm approach of sweet desire.
Affection is a coal that must be cooled,
Else, suffered, it will set the heart on fire.
　The sea hath bounds, but deep desire hath none;
　Therefore no marvel though thy horse be gone.　390

'How like a jade he stood tied to the tree,
Servilely mastered with a leathern rein!
But when he saw his love, his youth's fair fee,
He held such petty bondage in disdain,
　Throwing the base thong from his bending crest,
　Enfranchising his mouth, his back, his breast.　396

'Who sees his true-love in her naked bed,
Teaching the sheets a whiter hue than white,
But when his glutton eye so full hath fed
His other agents aim at like delight?　400
　Who is so faint that dares not be so bold
　To touch the fire, the weather being cold?

'Let me excuse thy courser, gentle boy;
And learn of him, I heartily beseech thee,
To take advantage on presented joy.　405
Though I were dumb, yet his proceedings teach thee.
　O, learn to love! The lesson is but plain,
　And, once made perfect, never lost again.'

'I know not love,' quoth he, 'nor will not know it,
Unless it be a boar, and then I chase it.　410
'Tis much to borrow, and I will not owe it.
My love to love is love but to disgrace it;
　For I have heard it is a life in death,
　That laughs and weeps, and all but with a breath.

'Who wears a garment shapeless and unfinished?　415
Who plucks the bud before one leaf put forth?
If springing things be any jot diminished,
They wither in their prime, prove nothing worth.
　The colt that's backed and burdened being young,
　Loseth his pride, and never waxeth strong.　420

'You hurt my hand with wringing. Let us part,
And leave this idle theme, this bootless chat.
Remove your siege from my unyielding heart;
To love's alarms it will not ope the gate.
　Dismiss your vows, your feignèd tears, your
　　flatt'ry;　425
　For where a heart is hard they make no batt'ry.'

'What, canst thou talk?' quoth she. 'Hast thou a tongue?
O, would thou hadst not, or I had no hearing!
Thy mermaid's voice hath done me double wrong.
I had my load before, now pressed with bearing:   430
  Melodious discord, heavenly tune harsh sounding,
  Ears' deep-sweet music, and heart's deep-sore
    wounding.

'Had I no eyes but ears, my ears would love
That inward beauty and invisible;
Or were I deaf, thy outward parts would move   435
Each part in me that were but sensible.
  Though neither eyes nor ears to hear nor see,
  Yet should I be in love by touching thee.

'Say that the sense of feeling were bereft me,
And that I could not see, nor hear, nor touch,   440
And nothing but the very smell were left me,
Yet would my love to thee be still as much;
  For from the stillitory of thy face excelling
  Comes breath perfumed, that breedeth love by
    smelling.

'But O, what banquet wert thou to the taste,   445
Being nurse and feeder of the other four!
Would they not wish the feast might ever last
And bid suspicion double-lock the door
  Lest jealousy, that sour unwelcome guest,
  Should by his stealing-in disturb the feast?'   450

Once more the ruby-coloured portal opened
Which to his speech did honey passage yield,
Like a red morn that ever yet betokened
Wrack to the seaman, tempest to the field,
  Sorrow to shepherds, woe unto the birds,   455
  Gusts and foul flaws to herdmen and to herds.

This ill presage advisedly she marketh.
Even as the wind is hushed before it raineth,
Or as the wolf doth grin before he barketh,
Or as the berry breaks before it staineth,   460
  Or like the deadly bullet of a gun,
  His meaning struck her ere his words begun,

And at his look she flatly falleth down,
For looks kill love, and love by looks reviveth;
A smile recures the wounding of a frown,   465
But blessèd bankrupt that by loss so thriveth!
  The silly boy, believing she is dead,
  Claps her pale cheek till clapping makes it red,

And, all amazed, brake off his late intent,
For sharply he did think to reprehend her,   470
Which cunning love did wittily prevent.
Fair fall the wit that can so well defend her!
  For on the grass she lies as she were slain,
  Till his breath breatheth life in her again.

He wrings her nose, he strikes her on the cheeks,   475
He bends her fingers, holds her pulses hard;
He chafes her lips; a thousand ways he seeks
To mend the hurt that his unkindness marred.
  He kisses her; and she, by her good will,
  Will never rise, so he will kiss her still.   480

The night of sorrow now is turned to day.
Her two blue windows faintly she upheaveth,
Like the fair sun when, in his fresh array,
He cheers the morn, and all the earth relieveth;
  And as the bright sun glorifies the sky,   485
  So is her face illumined with her eye,

Whose beams upon his hairless face are fixed,
As if from thence they borrowed all their shine.
Were never four such lamps together mixed,
Had not his clouded with his brow's repine.   490
  But hers, which through the crystal tears gave light,
  Shone like the moon in water seen by night.

'O, where am I?' quoth she; 'in earth or heaven,
Or in the ocean drenched, or in the fire?
What hour is this: or morn or weary even?   495
Do I delight to die, or life desire?
  But now I lived, and life was death's annoy;
  But now I died, and death was lively joy.

'O, thou didst kill me; kill me once again!
Thy eyes' shrewd tutor, that hard heart of thine,   500
Hath taught them scornful tricks, and such disdain
That they have murdered this poor heart of mine,
  And these mine eyes, true leaders to their queen,
  But for thy piteous lips no more had seen.

'Long may they kiss each other, for this cure!   505
O, never let their crimson liveries wear,
And as they last, their verdure still endure
To drive infection from the dangerous year,
  That the star-gazers, having writ on death,
  May say the plague is banished by thy breath!   510

'Pure lips, sweet seals in my soft lips imprinted,
What bargains may I make still to be sealing?
To sell myself I can be well contented,
So thou wilt buy, and pay, and use good dealing;
  Which purchase if thou make, for fear of slips   515
  Set thy seal manual on my wax-red lips.

'A thousand kisses buys my heart from me;
And pay them at thy leisure, one by one.
What is ten hundred touches unto thee?
Are they not quickly told, and quickly gone?   520
  Say for non-payment that the debt should double,
  Is twenty hundred kisses such a trouble?'

'Fair queen,' quoth he, 'if any love you owe me,
Measure my strangeness with my unripe years.
Before I know myself, seek not to know me.   525
No fisher but the ungrown fry forbears.
  The mellow plum doth fall, the green sticks fast,
  Or, being early plucked, is sour to taste.

'Look, the world's comforter with weary gait
His day's hot task hath ended in the west.   530
The owl, night's herald, shrieks 'tis very late;
The sheep are gone to fold, birds to their nest,
  And coal-black clouds, that shadow heaven's light,
  Do summon us to part and bid good night.

'Now let me say good night, and so say you.     535
If you will say so, you shall have a kiss.'
'Good night,' quoth she; and ere he says adieu
The honey fee of parting tendered is.
    Her arms do lend his neck a sweet embrace.
    Incorporate then they seem; face grows to face,     540

Till breathless he disjoined, and backward drew
The heavenly moisture, that sweet coral mouth,
Whose precious taste her thirsty lips well knew,
Whereon they surfeit, yet complain on drought.
    He with her plenty pressed, she faint with dearth,
    Their lips together glued, fall to the earth.     546

Now quick desire hath caught the yielding prey,
And glutton-like she feeds, yet never filleth.
Her lips are conquerors, his lips obey,
Paying what ransom the insulter willeth,     550
    Whose vulture thought doth pitch the price so high
    That she will draw his lips' rich treasure dry,

And, having felt the sweetness of the spoil,
With blindfold fury she begins to forage.
Her face doth reek and smoke, her blood doth boil,
And careless lust stirs up a desperate courage,     556
    Planting oblivion, beating reason back,
    Forgetting shame's pure blush and honour's wrack.

Hot, faint, and weary with her hard embracing,
Like a wild bird being tamed with too much handling,
Or as the fleet-foot roe that's tired with chasing,     561
Or like the froward infant stilled with dandling,
    He now obeys, and now no more resisteth,
    While she takes all she can, not all she listeth.

What wax so frozen but dissolves with temp'ring     565
And yields at last to every light impression?
Things out of hope are compassed oft with vent'ring,
Chiefly in love, whose leave exceeds commission.
    Affection faints not, like a pale-faced coward,
    But then woos best when most his choice is froward.

When he did frown, O, had she then gave over,     571
Such nectar from his lips she had not sucked.
Foul words and frowns must not repel a lover.
What though the rose have prickles, yet 'tis plucked!
    Were beauty under twenty locks kept fast,     575
    Yet love breaks through, and picks them all at last.

For pity now she can no more detain him.
The poor fool prays her that he may depart.
She is resolved no longer to restrain him,
Bids him farewell, and look well to her heart,     580
    The which, by Cupid's bow she doth protest,
    He carries thence encagèd in his breast.

'Sweet boy,' she says, 'this night I'll waste in sorrow,
For my sick heart commands mine eyes to watch.
Tell me, love's master, shall we meet tomorrow?     585
Say, shall we, shall we? Wilt thou make the match?'
    He tells her no, tomorrow he intends
    To hunt the boar with certain of his friends.

'The boar!' quoth she; whereat a sudden pale,
Like lawn being spread upon the blushing rose,     590
Usurps her cheek. She trembles at his tale,
And on his neck her yoking arms she throws.
    She sinketh down, still hanging by his neck.
    He on her belly falls, she on her back.

Now is she in the very lists of love,     595
Her champion mounted for the hot encounter.
All is imaginary she doth prove.
He will not manage her, although he mount her,
    That worse than Tantalus' is her annoy,
    To clip Elysium, and to lack her joy.     600

Even so poor birds, deceived with painted grapes,
Do surfeit by the eye, and pine the maw;
Even so languisheth in her mishaps
As those poor birds that helpless berries saw.
    The warm effects which she in him finds missing
    She seeks to kindle with continual kissing.     606

But all in vain, good queen! It will not be.
She hath assayed as much as may be proved;
Her pleading hath deserved a greater fee:
She's Love; she loves; and yet she is not loved.     610
    'Fie, fie,' he says, 'you crush me. Let me go.
    You have no reason to withhold me so.'

'Thou hadst been gone,' quoth she, 'sweet boy, ere
    this,
But that thou told'st me thou wouldst hunt the boar.
O, be advised; thou know'st not what it is     615
With javelin's point a churlish swine to gore,
    Whose tushes, never sheathed, he whetteth still,
    Like to a mortal butcher, bent to kill.

'On his bow-back he hath a battle set
Of bristly pikes that ever threat his foes.     620
His eyes like glow-worms shine; when he doth fret
His snout digs sepulchres where'er he goes.
    Being moved, he strikes, whate'er is in his way,
    And whom he strikes his crooked tushes slay.

'His brawny sides with hairy bristles armed     625
Are better proof than thy spear's point can enter.
His short thick neck cannot be easily harmed.
Being ireful, on the lion he will venture.
    The thorny brambles and embracing bushes,
    As fearful of him, part; through whom he rushes.

'Alas, he naught esteems that face of thine,     631
To which love's eyes pays tributary gazes,
Nor thy soft hands, sweet lips, and crystal eyne,
Whose full perfection all the world amazes;
    But having thee at vantage—wondrous dread!—
    Would root these beauties as he roots the mead.     636

'O, let him keep his loathsome cabin still.
Beauty hath naught to do with such foul fiends.
Come not within his danger by thy will.
They that thrive well take counsel of their friends.     640
    When thou didst name the boar, not to dissemble,
    I feared thy fortune, and my joints did tremble.

'Didst thou not mark my face? Was it not white?
Sawest thou not signs of fear lurk in mine eye?
Grew I not faint, and fell I not downright?        645
Within my bosom, whereon thou dost lie,
    My boding heart pants, beats, and takes no rest,
    But like an earthquake shakes thee on my breast.

'For where love reigns, disturbing jealousy
Doth call himself affection's sentinel,             650
Gives false alarms, suggesteth mutiny,
And in a peaceful hour doth cry, "Kill, kill!",
    Distemp'ring gentle love in his desire,
    As air and water do abate the fire.

'This sour informer, this bate-breeding spy,        655
This canker that eats up love's tender spring,
This carry-tale, dissentious jealousy,
That sometime true news, sometime false doth bring,
    Knocks at my heart, and whispers in mine ear
    That if I love thee, I thy death should fear;    660

'And, more than so, presenteth to mine eye
The picture of an angry chafing boar,
Under whose sharp fangs on his back doth lie
An image like thyself, all stained with gore,
    Whose blood upon the fresh flowers being shed    665
    Doth make them droop with grief, and hang the
        head.

'What should I do, seeing thee so indeed,
That tremble at th'imagination?
The thought of it doth make my faint heart bleed,
And fear doth teach it divination.                  670
    I prophesy thy death, my living sorrow,
    If thou encounter with the boar tomorrow.

'But if thou needs wilt hunt, be ruled by me:
Uncouple at the timorous flying hare,
Or at the fox which lives by subtlety,              675
Or at the roe which no encounter dare.
    Pursue these fearful creatures o'er the downs,
    And on thy well-breathed horse keep with thy
        hounds.

'And when thou hast on foot the purblind hare,
Mark the poor wretch, to overshoot his troubles,    680
How he outruns the wind, and with what care
He cranks and crosses with a thousand doubles.
    The many musits through the which he goes
    Are like a labyrinth to amaze his foes.

'Sometime he runs among a flock of sheep            685
To make the cunning hounds mistake their smell,
And sometime where earth-delving conies keep,
To stop the loud pursuers in their yell;
    And sometime sorteth with a herd of deer.
    Danger deviseth shifts; wit waits on fear.       690

'For there his smell with others being mingled,
The hot scent-snuffing hounds are driven to doubt,
Ceasing their clamorous cry till they have singled,
With much ado, the cold fault cleanly out.
    Then do they spend their mouths. Echo replies,   695
    As if another chase were in the skies.

'By this, poor Wat, far off upon a hill,
Stands on his hinder legs with list'ning ear,
To hearken if his foes pursue him still.
Anon their loud alarums he doth hear,               700
    And now his grief may be comparèd well
    To one sore sick that hears the passing-bell.

'Then shalt thou see the dew-bedabbled wretch
Turn, and return, indenting with the way.
Each envious brier his weary legs do scratch;       705
Each shadow makes him stop, each murmur stay;
    For misery is trodden on by many,
    And, being low, never relieved by any.

'Lie quietly, and hear a little more;
Nay, do not struggle, for thou shalt not rise.      710
To make thee hate the hunting of the boar
Unlike myself thou hear'st me moralize,
    Applying this to that, and so to so,
    For love can comment upon every woe.

'Where did I leave?' 'No matter where,' quoth he;   715
'Leave me, and then the story aptly ends.
The night is spent.' 'Why what of that?' quoth she.
'I am,' quoth he, 'expected of my friends,
    And now 'tis dark, and going I shall fall.'
    'In night,' quoth she, 'desire sees best of all.  720

'But if thou fall, O, then imagine this:
The earth, in love with thee, thy footing trips,
And all is but to rob thee of a kiss.
Rich preys make true men thieves; so do thy lips
    Make modest Dian cloudy and forlorn             725
    Lest she should steal a kiss, and die forsworn.

'Now of this dark night I perceive the reason.
Cynthia, for shame, obscures her silver shine
Till forging nature be condemned of treason
For stealing moulds from heaven, that were divine,
    Wherein she framed thee, in high heaven's despite,
    To shame the sun by day and her by night.        732

'And therefore hath she bribed the destinies
To cross the curious workmanship of nature,
To mingle beauty with infirmities,                  735
And pure perfection with impure defeature,
    Making it subject to the tyranny
    Of mad mischances and much misery;

'As burning fevers, agues pale and faint,
Life-poisoning pestilence, and frenzies wood,       740
The marrow-eating sickness whose attaint
Disorder breeds by heating of the blood;
    Surfeits, impostumes, grief, and damned despair
    Swear nature's death for framing thee so fair.

'And not the least of all these maladies            745
But in one minute's fight brings beauty under.
Both favour, savour, hue, and qualities,
Whereat th'impartial gazer late did wonder,
    Are on the sudden wasted, thawed, and done,
    As mountain snow melts with the midday sun.      750

'Therefore, despite of fruitless chastity,
Love-lacking vestals and self-loving nuns,
That on the earth would breed a scarcity
And barren dearth of daughters and of sons,
   Be prodigal. The lamp that burns by night    755
   Dries up his oil to lend the world his light.

'What is thy body but a swallowing grave,
Seeming to bury that posterity
Which, by the rights of time, thou needs must have
If thou destroy them not in dark obscurity?    760
   If so, the world will hold thee in disdain,
   Sith in thy pride so fair a hope is slain.

'So in thyself thyself art made away,
A mischief worse than civil, home-bred strife,
Or theirs whose desperate hands themselves do slay,
Or butcher sire that reaves his son of life.    766
   Foul cank'ring rust the hidden treasure frets,
   But gold that's put to use more gold begets.'

'Nay, then,' quoth Adon, 'You will fall again
Into your idle, over-handled theme.    770
The kiss I gave you is bestowed in vain,
And all in vain you strive against the stream;
   For, by this black-faced night, desire's foul nurse,
   Your treatise makes me like you worse and worse.

'If love have lent you twenty thousand tongues,    775
And every tongue more moving than your own,
Bewitching like the wanton mermaid's songs,
Yet from mine ear the tempting tune is blown;
   For know, my heart stands armèd in mine ear,
   And will not let a false sound enter there,    780

'Lest the deceiving harmony should run
Into the quiet closure of my breast,
And then my little heart were quite undone,
In his bedchamber to be barred of rest.
   No, lady, no. My heart longs not to groan,    785
   But soundly sleeps, while now it sleeps alone.

'What have you urged that I cannot reprove?
The path is smooth that leadeth on to danger.
I hate not love, but your device in love,
That lends embracements unto every stranger.    790
   You do it for increase—O strange excuse,
   When reason is the bawd to lust's abuse!

'Call it not love, for love to heaven is fled
Since sweating lust on earth usurped his name,
Under whose simple semblance he hath fed    795
Upon fresh beauty, blotting it with blame;
   Which the hot tyrant stains, and soon bereaves,
   As caterpillars do the tender leaves.

'Love comforteth, like sunshine after rain,
But lust's effect is tempest after sun.    800
Love's gentle spring doth always fresh remain;
Lust's winter comes ere summer half be done.
   Love surfeits not; lust like a glutton dies.
   Love is all truth, lust full of forgèd lies.

'More I could tell, but more I dare not say;    805
The text is old, the orator too green.
Therefore in sadness now I will away;
My face is full of shame, my heart of teen.
   Mine ears that to your wanton talk attended
   Do burn themselves for having so offended.'    810

With this he breaketh from the sweet embrace
Of those fair arms which bound him to her breast,
And homeward through the dark laund runs apace,
Leaves love upon her back, deeply distressed.
   Look how a bright star shooteth from the sky,    815
   So glides he in the night from Venus' eye,

Which after him she darts, as one on shore
Gazing upon a late-embarkèd friend
Till the wild waves will have him seen no more,
Whose ridges with the meeting clouds contend.    820
   So did the merciless and pitchy night
   Fold in the object that did feed her sight.

Whereat amazed, as one that unaware
Hath dropped a precious jewel in the flood,
Or stonished, as night wand'rers often are,    825
Their light blown out in some mistrustful wood:
   Even so, confounded in the dark she lay,
   Having lost the fair discovery of her way.

And now she beats her heart, whereat it groans,
That all the neighbour caves, as seeming troubled, 830
Make verbal repetition of her moans;
Passion on passion deeply is redoubled.
   'Ay me,' she cries, and twenty times 'Woe, woe!'
   And twenty echoes twenty times cry so.

She, marking them, begins a wailing note,    835
And sings extemporally a woeful ditty,
How love makes young men thrall, and old men dote,
How love is wise in folly, foolish-witty.
   Her heavy anthem still concludes in woe,
   And still the choir of echoes answer so.    840

Her song was tedious, and outwore the night;
For lovers' hours are long, though seeming short.
If pleased themselves, others, they think, delight
In such-like circumstance, with such-like sport.
   Their copious stories oftentimes begun    845
   End without audience, and are never done.

For who hath she to spend the night withal
But idle sounds resembling parasites,
Like shrill-tongued tapsters answering every call,
Soothing the humour of fantastic wits?    850
   She says ''Tis so'; they answer all ''Tis so',
   And would say after her, if she said 'No'.

Lo, here the gentle lark, weary of rest,
From his moist cabinet mounts up on high
And wakes the morning, from whose silver breast 855
The sun ariseth in his majesty,
   Who doth the world so gloriously behold
   That cedar tops and hills seem burnished gold.

Venus salutes him with this fair good-morrow:
'O thou clear god, and patron of all light,  860
From whom each lamp and shining star doth borrow
The beauteous influence that makes him bright:
    There lives a son that sucked an earthly mother
    May lend thee light, as thou dost lend to other.'

This said, she hasteth to a myrtle grove,  865
Musing the morning is so much o'erworn
And yet she hears no tidings of her love.
She hearkens for his hounds, and for his horn.
    Anon she hears them chant it lustily,
    And all in haste she coasteth to the cry.  870

And as she runs, the bushes in the way
Some catch her by the neck, some kiss her face,
Some twine about her thigh to make her stay.
She wildly breaketh from their strict embrace,
    Like a milch doe whose swelling dugs do ache,  875
    Hasting to feed her fawn hid in some brake.

By this she hears the hounds are at a bay,
Whereat she starts, like one that spies an adder
Wreathed up in fatal folds just in his way,
The fear whereof doth make him shake and shudder;
    Even so the timorous yelping of the hounds  881
    Appals her senses, and her spirit confounds.

For now she knows it is no gentle chase,
But the blunt boar, rough bear, or lion proud,
Because the cry remaineth in one place,  885
Where fearfully the dogs exclaim aloud.
    Finding their enemy to be so curst,
    They all strain court'sy who shall cope him first.

This dismal cry rings sadly in her ear,
Through which it enters to surprise her heart,  890
Who, overcome by doubt and bloodless fear,
With cold-pale weakness numbs each feeling part;
    Like soldiers when their captain once doth yield,
    They basely fly, and dare not stay the field.

Thus stands she in a trembling ecstasy,  895
Till, cheering up her senses all dismayed,
She tells them 'tis a causeless fantasy
And childish error that they are afraid;
    Bids them leave quaking, bids them fear no more;
    And with that word she spied the hunted boar,  900

Whose frothy mouth, bepainted all with red,
Like milk and blood being mingled both together,
A second fear through all her sinews spread,
Which madly hurries her, she knows not whither:
    This way she runs, and now she will no further,
    But back retires to rate the boar for murder.  906

A thousand spleens bear her a thousand ways.
She treads the path that she untreads again.
Her more than haste is mated with delays,
Like the proceedings of a drunken brain,  910
    Full of respects, yet naught at all respecting;
    In hand with all things, naught at all effecting.

Here kennelled in a brake she finds a hound,
And asks the weary caitiff for his master;
And there another licking of his wound,  915
'Gainst venomed sores the only sovereign plaster.
    And here she meets another, sadly scowling,
    To whom she speaks; and he replies with howling.

When he hath ceased his ill-resounding noise,
Another flap-mouthed mourner, black and grim,  920
Against the welkin volleys out his voice.
Another, and another, answer him,
    Clapping their proud tails to the ground below,
    Shaking their scratched ears, bleeding as they go.

Look how the world's poor people are amazed  925
At apparitions, signs, and prodigies,
Whereon with fearful eyes they long have gazed,
Infusing them with dreadful prophecies:
    So she at these sad signs draws up her breath,
    And, sighing it again, exclaims on death.  930

'Hard-favoured tyrant, ugly, meagre, lean,
Hateful divorce of love'—thus chides she death;
'Grim-grinning ghost, earth's worm: what dost thou
    mean
To stifle beauty, and to steal his breath
    Who, when he lived, his breath and beauty set  935
    Gloss on the rose, smell to the violet?

'If he be dead—O no, it cannot be,
Seeing his beauty, thou shouldst strike at it.
O yes, it may; thou hast no eyes to see,
But hatefully, at random dost thou hit.  940
    Thy mark is feeble age; but thy false dart
    Mistakes that aim, and cleaves an infant's heart.

'Hadst thou but bid beware, then he had spoke,
And, hearing him, thy power had lost his power.
The destinies will curse thee for this stroke.  945
They bid thee crop a weed; thou pluck'st a flower.
    Love's golden arrow at him should have fled,
    And not death's ebon dart to strike him dead.

'Dost thou drink tears, that thou provok'st such
    weeping?
What may a heavy groan advantage thee?  950
Why hast thou cast into eternal sleeping
Those eyes that taught all other eyes to see?
    Now nature cares not for thy mortal vigour,
    Since her best work is ruined with thy rigour.'

Here overcome, as one full of despair,  955
She vailed her eyelids, who like sluices stopped
The crystal tide that from her two cheeks fair
In the sweet channel of her bosom dropped.
    But through the flood-gates breaks the silver rain,
    And with his strong course opens them again.  960

O, how her eyes and tears did lend and borrow!
Her eye seen in the tears, tears in her eye,
Both crystals, where they viewed each other's sorrow:
Sorrow, that friendly sighs sought still to dry,
    But, like a stormy day, now wind, now rain,  965
    Sighs dry her cheeks, tears make them wet again.

Variable passions throng her constant woe,
As striving who should best become her grief.
All entertained, each passion labours so
That every present sorrow seemeth chief,  970
    But none is best. Then join they all together,
    Like many clouds consulting for foul weather.

By this, far off she hears some huntsman hollo;
A nurse's song ne'er pleased her babe so well.
The dire imagination she did follow  975
This sound of hope doth labour to expel;
    For now reviving joy bids her rejoice
    And flatters her it is Adonis' voice.

Whereat her tears began to turn their tide,
Being prisoned in her eye like pearls in glass;  980
Yet sometimes falls an orient drop beside,
Which her cheek melts, as scorning it should pass
    To wash the foul face of the sluttish ground,
    Who is but drunken when she seemeth drowned.

O hard-believing love—how strange it seems  985
Not to believe, and yet too credulous!
Thy weal and woe are both of them extremes.
Despair, and hope, makes thee ridiculous.
    The one doth flatter thee in thoughts unlikely;
    In likely thoughts the other kills thee quickly.  990

Now she unweaves the web that she hath wrought.
Adonis lives, and death is not to blame.
It was not she that called him all to naught.
Now she adds honours to his hateful name.
    She clepes him king of graves, and grave for kings,
    Imperious supreme of all mortal things.  996

'No, no,' quoth she, 'sweet death, I did but jest.
Yet pardon me, I felt a kind of fear
Whenas I met the boar, that bloody beast,
Which knows no pity, but is still severe.  1000
    Then, gentle shadow—truth I must confess—
    I railed on thee, fearing my love's decease.

''Tis not my fault; the boar provoked my tongue.
Be wreaked on him, invisible commander.
'Tis he, foul creature, that hath done thee wrong.  1005
I did but act; he's author of thy slander.
    Grief hath two tongues, and never woman yet
    Could rule them both, without ten women's wit.'

Thus, hoping that Adonis is alive,
Her rash suspect she doth extenuate,  1010
And, that his beauty may the better thrive,
With death she humbly doth insinuate,
    Tells him of trophies, statues, tombs; and stories
    His victories, his triumphs, and his glories.

'O Jove,' quoth she, 'how much a fool was I  1015
To be of such a weak and silly mind
To wail his death who lives, and must not die
Till mutual overthrow of mortal kind!
    For he being dead, with him is beauty slain,
    And beauty dead, black chaos comes again.  1020

'Fie, fie, fond love, thou art as full of fear
As one with treasure laden, hemmed with thieves.
Trifles unwitnessèd with eye or ear
Thy coward heart with false bethinking grieves.'
    Even at this word she hears a merry horn,  1025
    Whereat she leaps, that was but late forlorn.

As falcons to the lure, away she flies.
The grass stoops not, she treads on it so light;
And in her haste unfortunately spies
The foul boar's conquest on her fair delight;  1030
    Which seen, her eyes, as murdered with the view,
    Like stars ashamed of day, themselves withdrew.

Or as the snail, whose tender horns being hit
Shrinks backward in his shelly cave with pain,
And there, all smothered up, in shade doth sit,  1035
Long after fearing to creep forth again;
    So at his bloody view her eyes are fled
    Into the deep dark cabins of her head,

Where they resign their office and their light
To the disposing of her troubled brain,  1040
Who bids them still consort with ugly night,
And never wound the heart with looks again,
    Who, like a king perplexèd in his throne,
    By their suggestion gives a deadly groan,

Whereat each tributary subject quakes,  1045
As when the wind, imprisoned in the ground,
Struggling for passage, earth's foundation shakes,
Which with cold terror doth men's minds confound.
    This mutiny each part doth so surprise
    That from their dark beds once more leap her eyes,.

And, being opened, threw unwilling light  1051
Upon the wide wound that the boar had trenched
In his soft flank, whose wonted lily-white
With purple tears that his wound wept was drenched.
    No flower was nigh, no grass, herb, leaf, or weed,
    But stole his blood, and seemed with him to bleed.

This solemn sympathy poor Venus noteth.  1057
Over one shoulder doth she hang her head.
Dumbly she passions, franticly she doteth.
She thinks he could not die, he is not dead.  1060
    Her voice is stopped, her joints forget to bow,
    Her eyes are mad that they have wept till now.

Upon his hurt she looks so steadfastly
That her sight, dazzling, makes the wound seem three;
And then she reprehends her mangling eye,  1065
That makes more gashes where no breach should be.
    His face seems twain; each several limb is doubled;
    For oft the eye mistakes, the brain being troubled.

'My tongue cannot express my grief for one,
And yet,' quoth she, 'behold two Adons dead!  1070
My sighs are blown away, my salt tears gone,
Mine eyes are turned to fire, my heart to lead.
    Heavy heart's lead, melt at mine eyes' red fire!
    So shall I die by drops of hot desire.

'Alas, poor world, what treasure hast thou lost,      1075
What face remains alive that's worth the viewing?
Whose tongue is music now? What canst thou boast
Of things long since, or anything ensuing?
    The flowers are sweet, their colours fresh and trim;
    But true sweet beauty lived and died with him.      1080

'Bonnet nor veil henceforth no creature wear:
Nor sun nor wind will ever strive to kiss you.
Having no fair to lose, you need not fear.
The sun doth scorn you, and the wind doth hiss you.
    But when Adonis lived, sun and sharp air      1085
    Lurked like two thieves to rob him of his fair;

'And therefore would he put his bonnet on,
Under whose brim the gaudy sun would peep.
The wind would blow it off, and, being gone,
Play with his locks; then would Adonis weep,      1090
    And straight, in pity of his tender years,
    They both would strive who first should dry his tears.

'To see his face the lion walked along
Behind some hedge, because he would not fear him.
To recreate himself when he hath sung,      1095
The tiger would be tame, and gently hear him.
    If he had spoke, the wolf would leave his prey,
    And never fright the silly lamb that day.

'When he beheld his shadow in the brook,
The fishes spread on it their golden gills.      1100
When he was by, the birds such pleasure took
That some would sing, some other in their bills
    Would bring him mulberries and ripe-red cherries.
    He fed them with his sight, they him with berries.

'But this foul, grim, and urchin-snouted boar,      1105
Whose downward eye still looketh for a grave,
Ne'er saw the beauteous livery that he wore:
Witness the entertainment that he gave.
    If he did see his face, why then, I know
    He thought to kiss him, and hath killed him so.      1110

''Tis true, 'tis true; thus was Adonis slain;
He ran upon the boar with his sharp spear,
Who did not whet his teeth at him again,
But by a kiss thought to persuade him there,
    And, nuzzling in his flank, the loving swine      1115
    Sheathed unaware the tusk in his soft groin.

'Had I been toothed like him, I must confess
With kissing him I should have killed him first;
But he is dead, and never did he bless
My youth with his, the more am I accursed.'      1120
    With this she falleth in the place she stood,
    And stains her face with his congealèd blood.

She looks upon his lips, and they are pale.
She takes him by the hand, and that is cold.
She whispers in his ears a heavy tale,      1125
As if they heard the woeful words she told.
    She lifts the coffer-lids that close his eyes,
    Where lo, two lamps burnt out in darkness lies;

Two glasses, where herself herself beheld
A thousand times, and now no more reflect,      1130
Their virtue lost, wherein they late excelled,
And every beauty robbed of his effect.
    'Wonder of time,' quoth she, 'this is my spite,
    That, thou being dead, the day should yet be light.

'Since thou art dead, lo, here I prophesy      1135
Sorrow on love hereafter shall attend.
It shall be waited on with jealousy,
Find sweet beginning, but unsavoury end;
    Ne'er settled equally, but high or low,
    That all love's pleasure shall not match his woe.

'It shall be fickle, false, and full of fraud,      1141
Bud, and be blasted, in a breathing-while:
The bottom poison, and the top o'erstrawed
With sweets that shall the truest sight beguile.
    The strongest body shall it make most weak,      1145
    Strike the wise dumb, and teach the fool to speak.

'It shall be sparing, and too full of riot,
Teaching decrepit age to tread the measures.
The staring ruffian shall it keep in quiet,
Pluck down the rich, enrich the poor with treasures;
    It shall be raging-mad, and silly-mild;      1151
    Make the young old, the old become a child.

'It shall suspect where is no cause of fear;
It shall not fear where it should most mistrust.
It shall be merciful, and too severe,      1155
And most deceiving when it seems most just.
    Perverse it shall be where it shows most toward,
    Put fear to valour, courage to the coward.

'It shall be cause of war and dire events,
And set dissension 'twixt the son and sire;      1160
Subject and servile to all discontents,
As dry combustious matter is to fire.
    Sith in his prime death doth my love destroy,
    They that love best their loves shall not enjoy.'

By this, the boy that by her side lay killed      1165
Was melted like a vapour from her sight,
And in his blood that on the ground lay spilled
A purple flower sprung up, chequered with white,
    Resembling well his pale cheeks, and the blood
    Which in round drops upon their whiteness stood.

She bows her head the new-sprung flower to smell,
Comparing it to her Adonis' breath,      1172
And says within her bosom it shall dwell,
Since he himself is reft from her by death.
    She crops the stalk, and in the breach appears      1175
    Green-dropping sap, which she compares to tears.

'Poor flower,' quoth she, 'this was thy father's guise—
Sweet issue of a more sweet-smelling sire—
For every little grief to wet his eyes.
To grow unto himself was his desire,      1180
    And so 'tis thine; but know it is as good
    To wither in my breast as in his blood.

'Here was thy father's bed, here in my breast.
Thou art the next of blood, and 'tis thy right.
Lo, in this hollow cradle take thy rest;        1185
My throbbing heart shall rock thee day and night.
   There shall not be one minute in an hour
   Wherein I will not kiss my sweet love's flower.'

Thus, weary of the world, away she hies,
And yokes her silver doves, by whose swift aid    1190
Their mistress, mounted, through the empty skies
In her light chariot quickly is conveyed,
   Holding their course to Paphos, where their queen
   Means to immure herself, and not be seen.

# THE RAPE OF LUCRECE

DEDICATING *Venus and Adonis* to the Earl of Southampton in 1593, Shakespeare promised, if the poem pleased, to 'take advantage of all idle hours' to honour the Earl with 'some graver labour'. *The Rape of Lucrece*, also dedicated to Southampton, was entered in the Stationers' Register on 9 May 1594, and printed in the same year. The warmth of the dedication suggests that the Earl was by then a friend as well as a patron.

Like *Venus and Adonis*, *The Rape of Lucrece* is an erotic narrative based on Ovid, but this time the subject matter is historical, the tone tragic. The events took place in 509 BC, and were already legendary at the time of the first surviving account, by Livy in his history of Rome published between 27 and 25 BC. Shakespeare's main source was Ovid's *Fasti*, but he seems also to have known Livy's and other accounts.

Historically, Lucretia's rape had political consequences. Her ravisher, Tarquin, was a member of the tyrannical ruling family of Rome. During the siege of Ardea, a group of noblemen boasted of their wives' virtue, and rode home to test them; only Collatine's wife, Lucretia, lived up to her husband's claims, and Sextus Tarquinius was attracted to her. Failing to seduce her, he raped her and returned to Rome. Lucretia committed suicide, and her husband's friend, Lucius Junius Brutus, used the occasion as an opportunity to rouse the Roman people against Tarquinius' rule and to constitute themselves a republic.

Shakespeare concentrates on the private side of the story; Tarquin is lusting after Lucrece in the poem's opening lines, and the ending devotes only a few lines to the consequence of her suicide. As in *Venus and Adonis*, Shakespeare makes a little narrative material go a long way. At first, the focus is on Tarquin; after he has threatened Lucrece, it swings over to her. The opening sequence, with its marvellously dramatic account of Tarquin's tormented state of mind as he approaches Lucrece's chamber, is the more intense. Tarquin disappears from the action soon after the rape, when Lucrece delivers herself of a long complaint, apostrophizing night, opportunity, and time, and cursing Tarquin with rhetorical fervour, before deciding to kill herself. After summoning her husband, she seeks consolation in a painting of Troy which is described (1373-1442) in lines indebted to the first and second books of Virgil's *Aeneid* and to Book 13 of Ovid's *Metamorphoses*. After she dies, her husband and father mourn, but Brutus calls for deeds not words, and determines on revenge. The last lines of the poem look forward to the banishment of the Tarquins, but nothing is said of the establishment of a republic.

Like *Venus and Adonis*, *Lucrece*, initially popular (with six editions in Shakespeare's lifetime and another three by 1655), was later neglected. Coleridge admired it, and more recent criticism has recognized in it a profoundly dramatic quality combined with, if sometimes dissipated by, a remarkable force of rhetoric. The writing of the poem seems to have been a formative experience for Shakespeare. In it he not only laid the basis for his later plays on Roman history, but also explored themes that were to figure prominently in his later work. This is especially apparent in the portrayal of a man who 'still pursues his fear' (308), the relentless power of self-destructive evil that Shakespeare remembered when he made Macbeth, on his way to murder Duncan, speak of 'withered murder' which, 'With Tarquin's ravishing strides, towards his design | Moves like a ghost'.

The love I dedicate to your lordship is without end, whereof this pamphlet without beginning is but a superfluous moiety. The warrant I have of your honourable disposition, not the worth of my untutored lines, makes it assured of acceptance. What I have done is yours; what I have to do is yours, being part in all I have, devoted yours. Were my worth greater my duty would show greater, meantime, as it is, it is bound to your lordship, to whom I wish long life still lengthened with all happiness.

Your lordship's in all duty,

William Shakespeare

## THE ARGUMENT

Lucius Tarquinius (for his excessive pride surnamed Superbus), after he had caused his own father-in-law Servius Tullius to be cruelly murdered, and, contrary to the Roman laws and customs, not requiring or staying for the people's suffrages had possessed himself of the kingdom, went accompanied with his sons and other noblemen of Rome to besiege Ardea, during which siege the principal men of the army meeting one evening at the tent of Sextus Tarquinius, the King's son, in their discourses after supper everyone commended the virtues of his own wife, among whom Collatinus extolled the incomparable chastity of his wife, Lucretia. In that pleasant humour they all posted to Rome, and, intending by their secret and sudden arrival to make trial of that which everyone had before avouched, only Collatinus finds his wife (though it were late in the night) spinning amongst her maids. The other ladies were all found dancing, and revelling, or in several disports. Whereupon the noblemen yielded Collatinus the victory and his wife the fame. At that time Sextus Tarquinius, being enflamed with Lucrece' beauty, yet smothering his passions for the present, departed with the rest back to the camp, from whence he shortly after privily withdrew himself and was, according to his estate, royally entertained and lodged by Lucrece at Collatium. The same night he treacherously stealeth into her chamber, violently ravished her, and early in the morning speedeth away. Lucrece, in this lamentable plight, hastily dispatcheth messengers—one to Rome for her father, another to the camp for Collatine. They came, the one accompanied with Junius Brutus, the other with Publius Valerius, and, finding Lucrece attired in mourning habit, demanded the cause of her sorrow. She, first taking an oath of them for her revenge, revealed the actor and whole manner of his dealing, and withal suddenly stabbed herself. Which done, with one consent they all vowed to root out the whole hated family of the Tarquins, and, bearing the dead body to Rome, Brutus acquainted the people with the doer and manner of the vile deed, with a bitter invective against the tyranny of the King; wherewith the people were so moved that with one consent and a general acclamation the Tarquins were all exiled and the state government changed from kings to consuls.

# The Rape of Lucrece

From the besieged Ardea all in post,
Borne by the trustless wings of false desire,
Lust-breathèd Tarquin leaves the Roman host
And to Collatium bears the lightless fire
Which, in pale embers hid, lurks to aspire     5
    And girdle with embracing flames the waist
    Of Collatine's fair love, Lucrece the chaste.

Haply that name of chaste unhapp'ly set
This bateless edge on his keen appetite,
When Collatine unwisely did not let     10
To praise the clear unmatchèd red and white
Which triumphed in that sky of his delight,
    Where mortal stars as bright as heaven's beauties
    With pure aspects did him peculiar duties.

For he the night before in Tarquin's tent     15
Unlocked the treasure of his happy state,
What priceless wealth the heavens had him lent
In the possession of his beauteous mate,
Reck'ning his fortune at such high-proud rate
    That kings might be espousèd to more fame,     20
    But king nor peer to such a peerless dame.

O happiness enjoyed but of a few,
And, if possessed, as soon decayed and done
As is the morning's silver melting dew
Against the golden splendour of the sun,     25
An expired date cancelled ere well begun!
    Honour and beauty in the owner's arms
    Are weakly fortressed from a world of harms.

Beauty itself doth of itself persuade
The eyes of men without an orator.     30
What needeth then apology be made
To set forth that which is so singular?
Or why is Collatine the publisher
    Of that rich jewel he should keep unknown
    From thievish ears, because it is his own?     35

Perchance his boast of Lucrece' sov'reignty
Suggested this proud issue of a king,
For by our ears our hearts oft tainted be.
Perchance that envy of so rich a thing,
Braving compare, disdainfully did sting     40
    His high-pitched thoughts, that meaner men should
       vaunt
    That golden hap which their superiors want.

But some untimely thought did instigate
His all-too-timeless speed, if none of those.
His honour, his affairs, his friends, his state     45
Neglected all, with swift intent he goes
To quench the coal which in his liver glows.
    O rash false heat, wrapped in repentant cold,
    Thy hasty spring still blasts and ne'er grows old!

When at Collatium this false lord arrived,     50
Well was he welcomed by the Roman dame,
Within whose face beauty and virtue strived
Which of them both should underprop her fame.

When virtue bragged, beauty would blush for shame;
    When beauty boasted blushes, in despite     55
    Virtue would stain that or with silver white.

But beauty, in that white entitulèd
From Venus' doves, doth challenge that fair field.
Then virtue claims from beauty beauty's red,
Which virtue gave the golden age to gild     60
Their silver cheeks, and called it then their shield,
    Teaching them thus to use it in the fight:
    When shame assailed, the red should fence the
       white.

This heraldry in Lucrece' face was seen,
Argued by beauty's red and virtue's white.     65
Of either's colour was the other queen,
Proving from world's minority their right.
Yet their ambition makes them still to fight,
    The sovereignty of either being so great
    That oft they interchange each other's seat.     70

This silent war of lilies and of roses
Which Tarquin viewed in her fair face's field
In their pure ranks his traitor eye encloses,
Where, lest between them both it should be killed,
The coward captive vanquishèd doth yield     75
    To those two armies that would let him go
    Rather than triumph in so false a foe.

Now thinks he that her husband's shallow tongue,
The niggard prodigal that praised her so,
In that high task hath done her beauty wrong,     80
Which far exceeds his barren skill to show.
Therefore that praise which Collatine doth owe
    Enchanted Tarquin answers with surmise
    In silent wonder of still-gazing eyes.

This earthly saint adorèd by this devil     85
Little suspecteth the false worshipper,
For unstained thoughts do seldom dream on evil.
Birds never limed no secret bushes fear,
So guiltless she securely gives good cheer
    And reverent welcome to her princely guest,     90
    Whose inward ill no outward harm expressed,

For that he coloured with his high estate,
Hiding base sin in pleats of majesty,
That nothing in him seemed inordinate
Save sometime too much wonder of his eye,     95
Which, having all, all could not satisfy,
    But poorly rich so wanteth in his store
    That, cloyed with much, he pineth still for more.

But she that never coped with stranger eyes
Could pick no meaning from their parling looks,     100
Nor read the subtle shining secrecies
Writ in the glassy margins of such books.
She touched no unknown baits nor feared no hooks,
    Nor could she moralize his wanton sight
    More than his eyes were opened to the light.     105

He stories to her ears her husband's fame
Won in the fields of fruitful Italy,
And decks with praises Collatine's high name
Made glorious by his manly chivalry
With bruisèd arms and wreaths of victory.          110
    Her joy with heaved-up hand she doth express,
    And wordless so greets heaven for his success.

Far from the purpose of his coming thither
He makes excuses for his being there.
No cloudy show of stormy blust'ring weather       115
Doth yet in his fair welkin once appear
Till sable night, mother of dread and fear,
    Upon the world dim darkness doth display
    And in her vaulty prison stows the day.

For then is Tarquin brought unto his bed,         120
Intending weariness with heavy sprite;
For after supper long he questionèd
With modest Lucrece, and wore out the night.
Now leaden slumber with life's strength doth fight,
    And everyone to rest himself betakes            125
    Save thieves, and cares, and troubled minds that
      wakes.

As one of which doth Tarquin lie revolving
The sundry dangers of his will's obtaining,
Yet ever to obtain his will resolving,
Though weak-built hopes persuade him to abstaining.
Despair to gain doth traffic oft for gaining,     131
    And when great treasure is the meed proposed,
    Though death be adjunct, there's no death supposed.

Those that much covet are with gain so fond
That what they have not, that which they possess,
They scatter and unloose it from their bond,      136
And so by hoping more they have but less,
Or, gaining more, the profit of excess
    Is but to surfeit and such griefs sustain
    That they prove bankrupt in this poor-rich gain.

The aim of all is but to nurse the life           141
With honour, wealth, and ease in waning age,
And in this aim there is such thwarting strife
That one for all, or all for one, we gage,
As life for honour in fell battle's rage,         145
    Honour for wealth; and oft that wealth doth cost
    The death of all, and all together lost.

So that, in vent'ring ill, we leave to be
The things we are for that which we expect,
And this ambitious foul infirmity                 150
In having much, torments us with defect
Of that we have; so then we do neglect
    The thing we have, and all for want of wit
    Make something nothing by augmenting it.

Such hazard now must doting Tarquin make,         155
Pawning his honour to obtain his lust,
And for himself himself he must forsake.
Then where is truth if there be no self-trust?
When shall he think to find a stranger just
    When he himself himself confounds, betrays     160
    To sland'rous tongues and wretched hateful days?

Now stole upon the time the dead of night
When heavy sleep had closed up mortal eyes.
No comfortable star did lend his light,
No noise but owls' and wolves' death-boding cries 165
Now serves the season, that they may surprise
    The silly lambs. Pure thoughts are dead and still,
    While lust and murder wakes to stain and kill.

And now this lustful lord leapt from his bed,
Throwing his mantle rudely o'er his arm,          170
Is madly tossed between desire and dread.
Th'one sweetly flatters, th'other feareth harm,
But honest fear, bewitched with lust's foul charm,
    Doth too-too oft betake him to retire,
    Beaten away by brainsick rude desire.           175

His falchion on a flint he softly smiteth,
That from the cold stone sparks of fire do fly,
Whereat a waxen torch forthwith he lighteth,
Which must be lodestar to his lustful eye,
And to the flame thus speaks advisedly:           180
    'As from this cold flint I enforced this fire,
    So Lucrece must I force to my desire.'

Here pale with fear he doth premeditate
The dangers of his loathsome enterprise,
And in his inward mind he doth debate            185
What following sorrow may on this arise.
Then, looking scornfully, he doth despise
    His naked armour of still-slaughtered lust,
    And justly thus controls his thoughts unjust:

'Fair torch, burn out thy light, and lend it not  190
To darken her whose light excelleth thine;
And die, unhallowed thoughts, before you blot
With your uncleanness that which is divine.
Offer pure incense to so pure a shrine.
    Let fair humanity abhor the deed               195
    That spots and stains love's modest snow-white weed.

'O shame to knighthood and to shining arms!
O foul dishonour to my household's grave!
O impious act including all foul harms!
A martial man to be soft fancy's slave!           200
True valour still a true respect should have;
    Then my digression is so vile, so base,
    That it will live engraven in my face.

'Yea, though I die the scandal will survive
And be an eyesore in my golden coat.              205
Some loathsome dash the herald will contrive
To cipher me how fondly I did dote,
That my posterity, shamed with the note,
    Shall curse my bones and hold it for no sin
    To wish that I their father had not been.       210

'What win I if I gain the thing I seek?
A dream, a breath, a froth of fleeting joy.
Who buys a minute's mirth to wail a week,
Or sells eternity to get a toy?
For one sweet grape who will the vine destroy?    215
    Or what fond beggar, but to touch the crown,
    Would with the sceptre straight be strucken down?

'If Collatinus dream of my intent
Will he not wake, and in a desp'rate rage
Post hither this vile purpose to prevent?—          220
This siege that hath engirt his marriage,
This blur to youth, this sorrow to the sage,
  This dying virtue, this surviving shame,
  Whose crime will bear an ever-during blame.

'O what excuse can my invention make          225
When thou shalt charge me with so black a deed?
Will not my tongue be mute, my frail joints shake,
Mine eyes forgo their light, my false heart bleed?
The guilt being great, the fear doth still exceed,
  And extreme fear can neither fight nor fly,          230
  But coward-like with trembling terror die.

'Had Collatinus killed my son or sire,
Or lain in ambush to betray my life,
Or were he not my dear friend, this desire
Might have excuse to work upon his wife          235
As in revenge or quittal of such strife.
  But as he is my kinsman, my dear friend,
  The shame and fault finds no excuse nor end.

'Shameful it is—ay, if the fact be known.
Hateful it is—there is no hate in loving.          240
I'll beg her love—but she is not her own.
The worst is but denial and reproving;
My will is strong past reason's weak removing.
  Who fears a sentence or an old man's saw
  Shall by a painted cloth be kept in awe.'          245

Thus graceless holds he disputation
'Tween frozen conscience and hot-burning will,
And with good thoughts makes dispensation,
Urging the worser sense for vantage still;
Which in a moment doth confound and kill          250
  All pure effects, and doth so far proceed
  That what is vile shows like a virtuous deed.

Quoth he, 'She took me kindly by the hand,
And gazed for tidings in my eager eyes,
Fearing some hard news from the warlike band          255
Where her belovèd Collatinus lies.
O how her fear did make her colour rise!
  First red as roses that on lawn we lay,
  Then white as lawn, the roses took away.

'And how her hand, in my hand being locked,          260
Forced it to tremble with her loyal fear,
Which struck her sad, and then it faster rocked
Until her husband's welfare she did hear,
Whereat she smilèd with so sweet a cheer
  That had Narcissus seen her as she stood          265
  Self-love had never drowned him in the flood.

'Why hunt I then for colour or excuses?
All orators are dumb when beauty pleadeth.
Poor wretches have remorse in poor abuses;
Love thrives not in the heart that shadows dreadeth;
Affection is my captain, and he leadeth,          271
  And when his gaudy banner is displayed,
  The coward fights, and will not be dismayed.

'Then childish fear avaunt, debating die,
Respect and reason wait on wrinkled age!          275
My heart shall never countermand mine eye,
Sad pause and deep regard beseems the sage.
My part is youth, and beats these from the stage.
  Desire my pilot is, beauty my prize.
  Then who fears sinking where such treasure lies?'

As corn o'ergrown by weeds, so heedful fear          281
Is almost choked by unresisted lust.
Away he steals, with open list'ning ear,
Full of foul hope and full of fond mistrust,
Both which as servitors to the unjust          285
  So cross him with their opposite persuasion
  That now he vows a league, and now invasion.

Within his thought her heavenly image sits,
And in the selfsame seat sits Collatine.
That eye which looks on her confounds his wits,          290
That eye which him beholds, as more divine,
Unto a view so false will not incline,
  But with a pure appeal seeks to the heart,
  Which once corrupted, takes the worser part,

And therein heartens up his servile powers          295
Who, flattered by their leader's jocund show,
Stuff up his lust as minutes fill up hours,
And as their captain, so their pride doth grow,
Paying more slavish tribute than they owe.
  By reprobate desire thus madly led          300
  The Roman lord marcheth to Lucrece' bed.

The locks between her chamber and his will,
Each one by him enforced, retires his ward;
But as they open they all rate his ill,
Which drives the creeping thief to some regard.          305
The threshold grates the door to have him heard,
  Night-wand'ring weasels shriek to see him there.
  They fright him, yet he still pursues his fear.

As each unwilling portal yields him way,
Through little vents and crannies of the place          310
The wind wars with his torch to make him stay,
And blows the smoke of it into his face,
Extinguishing his conduct in this case.
  But his hot heart, which fond desire doth scorch,
  Puffs forth another wind that fires the torch,          315

And being lighted, by the light he spies
Lucretia's glove wherein her needle sticks.
He takes it from the rushes where it lies,
And gripping it, the needle his finger pricks,
As who should say 'This glove to wanton tricks          320
  Is not inured. Return again in haste.
  Thou seest our mistress' ornaments are chaste.'

But all these poor forbiddings could not stay him;
He in the worst sense consters their denial.
The doors, the wind, the glove that did delay him          325
He takes for accidental things of trial,
Or as those bars which stop the hourly dial,
  Who with a ling'ring stay his course doth let
  Till every minute pays the hour his debt.

'So, so,' quoth he, 'these lets attend the time,          330
Like little frosts that sometime threat the spring
To add a more rejoicing to the prime,
And give the sneapèd birds more cause to sing.
Pain pays the income of each precious thing.
 Huge rocks, high winds, strong pirates, shelves, and
   sands                                              335
 The merchant fears, ere rich at home he lands.'

Now is he come unto the chamber door
That shuts him from the heaven of his thought,
Which with a yielding latch, and with no more,
Hath barred him from the blessèd thing he sought.
So from himself impiety hath wrought              341
 That for his prey to pray he doth begin,
 As if the heavens should countenance his sin.

But in the midst of his unfruitful prayer
Having solicited th'eternal power                          345
That his foul thoughts might compass his fair fair,
And they would stand auspicious to the hour,
Even there he starts. Quoth he, 'I must deflower.
 The powers to whom I pray abhor this fact;
 How can they then assist me in the act?              350

'Then love and fortune be my gods, my guide!
My will is backed with resolution.
Thoughts are but dreams till their effects be tried;
The blackest sin is cleared with absolution.
Against love's fire fear's frost hath dissolution.        355
 The eye of heaven is out, and misty night
 Covers the shame that follows sweet delight.'

This said, his guilty hand plucked up the latch,
And with his knee the door he opens wide.
The dove sleeps fast that this night-owl will catch.   360
Thus treason works ere traitors be espied.
Who sees the lurking serpent steps aside,
 But she, sound sleeping, fearing no such thing,
 Lies at the mercy of his mortal sting.

Into the chamber wickedly he stalks,                      365
And gazeth on her yet-unstainèd bed.
The curtains being close, about he walks,
Rolling his greedy eye-balls in his head.
By their high treason is his heart misled,
 Which gives the watchword to his hand full soon
 To draw the cloud that hides the silver moon.       371

Look as the fair and fiery-pointed sun
Rushing from forth a cloud bereaves our sight,
Even so, the curtain drawn, his eyes begun
To wink, being blinded with a greater light.             375
Whether it is that she reflects so bright
 That dazzleth them, or else some shame supposed,
 But blind they are, and keep themselves enclosed.

O had they in that darksome prison died,
Then had they seen the period of their ill.              380
Then Collatine again by Lucrece' side
In his clear bed might have reposèd still.
But they must ope, this blessèd league to kill,
 And holy-thoughted Lucrece to their sight
 Must sell her joy, her life, her world's delight.    385

Her lily hand her rosy cheek lies under,
Coz'ning the pillow of a lawful kiss,
Who therefore angry seems to part in sunder,
Swelling on either side to want his bliss;
Between whose hills her head entombèd is,                390
 Where like a virtuous monument she lies
 To be admired of lewd unhallowed eyes.

Without the bed her other fair hand was,
On the green coverlet, whose perfect white
Showed like an April daisy on the grass,                 395
With pearly sweat resembling dew of night,
Her eyes like marigolds had sheathed their light,
 And canopied in darkness sweetly lay
 Till they might open to adorn the day.

Her hair like golden threads played with her breath—
O modest wantons, wanton modesty!—                      401
Showing life's triumph in the map of death,
And death's dim look in life's mortality.
Each in her sleep themselves so beautify
 As if between them twain there were no strife,       405
 But that life lived in death, and death in life.

Her breasts like ivory globes circled with blue,
A pair of maiden worlds unconquerèd,
Save of their lord no bearing yoke they knew,
And him by oath they truly honourèd.                     410
These worlds in Tarquin new ambition bred,
 Who like a foul usurper went about
 From this fair throne to heave the owner out.

What could he see but mightily he noted?
What did he note but strongly he desired?                415
What he beheld, on that he firmly doted,
And in his will his wilful eye he tired.
With more than admiration he admired
 Her azure veins, her alabaster skin,
 Her coral lips, her snow-white dimpled chin.        420

As the grim lion fawneth o'er his prey,
Sharp hunger by the conquest satisfied,
So o'er this sleeping soul doth Tarquin stay,
His rage of lust by gazing qualified,
Slaked not suppressed for standing by her side.         425
 His eye which late this mutiny restrains
 Unto a greater uproar tempts his veins,

And they like straggling slaves for pillage fighting,
Obdurate vassals fell exploits effecting,
In bloody death and ravishment delighting,              430
Nor children's tears nor mothers' groans respecting,
Swell in their pride, the onset still expecting.
 Anon his beating heart, alarum striking,
 Gives the hot charge, and bids them do their liking.

His drumming heart cheers up his burning eye,           435
His eye commends the leading to his hand.
His hand, as proud of such a dignity,
Smoking with pride marched on to make his stand
On her bare breast, the heart of all her land,
 Whose ranks of blue veins as his hand did scale   440
 Left their round turrets destitute and pale.

They, must'ring to the quiet cabinet
Where their dear governess and lady lies,
Do tell her she is dreadfully beset,
And fright her with confusion of their cries.    445
  She much amazed breaks ope her locked-up eyes,
    Who, peeping forth this tumult to behold,
    Are by his flaming torch dimmed and controlled.

Imagine her as one in dead of night
From forth dull sleep by dreadful fancy waking,    450
That thinks she hath beheld some ghastly sprite
Whose grim aspect sets every joint a-shaking.
  What terror 'tis! But she in worser taking,
    From sleep disturbèd, heedfully doth view
    The sight which makes supposèd terror true.    455

Wrapped and confounded in a thousand fears,
Like to a new-killed bird she trembling lies.
She dares not look, yet, winking, there appears
Quick-shifting antics, ugly in her eyes.
  Such shadows are the weak brain's forgeries,    460
    Who, angry that the eyes fly from their lights,
    In darkness daunts them with more dreadful sights.

His hand that yet remains upon her breast—
Rude ram, to batter such an ivory wall—
May feel her heart, poor citizen, distressed,    465
Wounding itself to death, rise up and fall,
Beating her bulk, that his hand shakes withal.
  This moves in him more rage and lesser pity
  To make the breach and enter this sweet city.

First like a trumpet doth his tongue begin    470
To sound a parley to his heartless foe,
Who o'er the white sheet peers her whiter chin,
The reason of this rash alarm to know,
Which he by dumb demeanour seeks to show.
  But she with vehement prayers urgeth still    475
  Under what colour he commits this ill.

Thus he replies: 'The colour in thy face,
That even for anger makes the lily pale
And the red rose blush at her own disgrace,
Shall plead for me and tell my loving tale.    480
Under that colour am I come to scale
  Thy never-conquered fort. The fault is thine,
  For those thine eyes betray thee unto mine.

'Thus I forestall thee, if thou mean to chide:
Thy beauty hath ensnared thee to this night,    485
Where thou with patience must my will abide,
My will that marks thee for my earth's delight,
Which I to conquer sought with all my might.
  But as reproof and reason beat it dead,
  By thy bright beauty was it newly bred.    490

'I see what crosses my attempt will bring,
I know what thorns the growing rose defends;
I think the honey guarded with a sting;
All this beforehand counsel comprehends.
But will is deaf, and hears no heedful friends.    495
  Only he hath an eye to gaze on beauty,
  And dotes on what he looks, 'gainst law or duty.

'I have debated even in my soul
What wrong, what shame, what sorrow I shall breed;
But nothing can affection's course control,    500
Or stop the headlong fury of his speed.
I know repentant tears ensue the deed,
  Reproach, disdain, and deadly enmity,
  Yet strive I to embrace mine infamy.'

This said, he shakes aloft his Roman blade,    505
Which like a falcon tow'ring in the skies
Coucheth the fowl below with his wings' shade
Whose crooked beak threats, if he mount he dies.
So under his insulting falchion lies
  Harmless Lucretia, marking what he tells    510
  With trembling fear, as fowl hear falcons' bells.

'Lucrece,' quoth he, 'this night I must enjoy thee.
If thou deny, then force must work my way,
For in thy bed I purpose to destroy thee.
That done, some worthless slave of thine I'll slay    515
To kill thine honour with thy life's decay;
  And in thy dead arms do I mean to place him,
  Swearing I slew him seeing thee embrace him.

'So thy surviving husband shall remain
The scornful mark of every open eye,    520
Thy kinsmen hang their heads at this disdain,
Thy issue blurred with nameless bastardy,
And thou, the author of their obloquy,
  Shalt have thy trespass cited up in rhymes
  And sung by children in succeeding times.    525

'But if thou yield, I rest thy secret friend.
The fault unknown is as a thought unacted.
A little harm done to a great good end
For lawful policy remains enacted.
The poisonous simple sometime is compacted    530
  In a pure compound; being so applied,
  His venom in effect is purified.

'Then for thy husband and thy children's sake
Tender my suit; bequeath not to their lot
The shame that from them no device can take,    535
The blemish that will never be forgot,
Worse than a slavish wipe or birth-hour's blot;
  For marks descried in men's nativity
  Are nature's faults, not their own infamy.'

Here with a cockatrice' dead-killing eye    540
He rouseth up himself, and makes a pause,
While she, the picture of pure piety,
Like a white hind under the gripe's sharp claws,
Pleads in a wilderness where are no laws
  To the rough beast that knows no gentle right,    545
  Nor aught obeys but his foul appetite.

But when a black-faced cloud the world doth threat,
In his dim mist th'aspiring mountains hiding,
From earth's dark womb some gentle gust doth get
Which blows these pitchy vapours from their biding,    551
  Hind'ring their present fall by this dividing;
    So his unhallowed haste her words delays,
    And moody Pluto winks while Orpheus plays.

Yet, foul night-waking cat, he doth but dally
While in his holdfast foot the weak mouse panteth.
Her sad behaviour feeds his vulture folly, 556
A swallowing gulf that even in plenty wanteth.
His ear her prayers admits, but his heart granteth
   No penetrable entrance to her plaining.
   Tears harden lust, though marble wear with raining.

Her pity-pleading eyes are sadly fixed 561
In the remorseless wrinkles of his face.
Her modest eloquence with sighs is mixed,
Which to her oratory adds more grace.
She puts the period often from his place, 565
   And midst the sentence so her accent breaks
   That twice she doth begin ere once she speaks.

She conjures him by high almighty Jove,
By knighthood, gentry, and sweet friendship's oath,
By her untimely tears, her husband's love, 570
By holy human law and common troth,
By heaven and earth and all the power of both,
   That to his borrowed bed he make retire,
   And stoop to honour, not to foul desire.

Quoth she, 'Reward not hospitality 575
With such black payment as thou hast pretended.
Mud not the fountain that gave drink to thee;
Mar not the thing that cannot be amended;
End thy ill aim before thy shoot be ended.
   He is no woodman that doth bend his bow 580
   To strike a poor unseasonable doe.

'My husband is thy friend; for his sake spare me.
Thyself art mighty; for thine own sake leave me;
Myself a weakling; do not then ensnare me.
Thou look'st not like deceit; do not deceive me. 585
My sighs like whirlwinds labour hence to heave thee.
   If ever man were moved with woman's moans,
   Be movèd with my tears, my sighs, my groans.

'All which together, like a troubled ocean,
Beat at thy rocky and wreck-threat'ning heart 590
To soften it with their continual motion,
For stones dissolved to water do convert.
O, if no harder than a stone thou art,
   Melt at my tears, and be compassionate.
   Soft pity enters at an iron gate. 595

'In Tarquin's likeness I did entertain thee.
Hast thou put on his shape to do him shame?
To all the host of heaven I complain me.
Thou wrong'st his honour, wound'st his princely name.
Thou art not what thou seem'st, and if the same, 600
   Thou seem'st not what thou art, a god, a king,
   For kings like gods should govern everything.

'How will thy shame be seeded in thine age
When thus thy vices bud before thy spring?
If in thy hope thou dar'st do such outrage, 605
What dar'st thou not when once thou art a king?
O be remembered, no outrageous thing
   From vassal actors can be wiped away;
   Then kings' misdeeds cannot be hid in clay.

'This deed will make thee only loved for fear, 610
But happy monarchs still are feared for love.
With foul offenders thou perforce must bear
When they in thee the like offences prove.
If but for fear of this, thy will remove;
   For princes are the glass, the school, the book 615
   Where subjects' eyes do learn, do read, do look.

'And wilt thou be the school where lust shall learn?
Must he in thee read lectures of such shame?
Wilt thou be glass wherein it shall discern
Authority for sin, warrant for blame, 620
To privilege dishonour in thy name?
   Thou back'st reproach against long-living laud,
   And mak'st fair reputation but a bawd.

'Hast thou command? By him that gave it thee,
From a pure heart command thy rebel will. 625
Draw not thy sword to guard iniquity,
For it was lent thee all that brood to kill.
Thy princely office how canst thou fulfil
   When, patterned by thy fault, foul sin may say
   He learned to sin, and thou didst teach the way?

'Think but how vile a spectacle it were 631
To view thy present trespass in another.
Men's faults do seldom to themselves appear;
Their own transgressions partially they smother.
This guilt would seem death-worthy in thy brother.
   O, how are they wrapped in with infamies 636
   That from their own misdeeds askance their eyes!

'To thee, to thee my heaved-up hands appeal,
Not to seducing lust, thy rash relier.
I sue for exiled majesty's repeal; 640
Let him return, and flatt'ring thoughts retire.
His true respect will prison false desire,
   And wipe the dim mist from thy doting eyne,
   That thou shalt see thy state, and pity mine.'

'Have done,' quoth he; 'my uncontrollèd tide 645
Turns not, but swells the higher by this let.
Small lights are soon blown out; huge fires abide,
And with the wind in greater fury fret.
The petty streams, that pay a daily debt
   To their salt sovereign, with their fresh falls' haste
   Add to his flow, but alter not his taste.' 651

'Thou art,' quoth she, 'a sea, a sovereign king,
And lo, there falls into thy boundless flood
Black lust, dishonour, shame, misgoverning,
Who seek to stain the ocean of thy blood. 655
If all these petty ills shall change thy good,
   Thy sea within a puddle's womb is hearsed,
   And not the puddle in thy sea dispersed.

'So shall these slaves be king, and thou their slave;
Thou nobly base, they basely dignified; 660
Thou their fair life, and they thy fouler grave;
Thou loathèd in their shame, they in thy pride.
The lesser thing should not the greater hide.
   The cedar stoops not to the base shrub's foot,
   But low shrubs wither at the cedar's root. 665

'So let thy thoughts, low vassals to thy state'-
'No more,' quoth he, 'by heaven, I will not hear thee.
Yield to my love. If not, enforcèd hate
Instead of love's coy touch shall rudely tear thee.
That done, despitefully I mean to bear thee     670
    Unto the base bed of some rascal groom
    To be thy partner in this shameful doom.'

This said, he sets his foot upon the light;
For light and lust are deadly enemies.
Shame folded up in blind concealing night     675
When most unseen, then most doth tyrannize.
The wolf hath seized his prey, the poor lamb cries,
    Till with her own white fleece her voice controlled
    Entombs her outcry in her lips' sweet fold.

For with the nightly linen that she wears     680
He pens her piteous clamours in her head,
Cooling his hot face in the chastest tears
That ever modest eyes with sorrow shed.
O that prone lust should stain so pure a bed,
    The spots whereof could weeping purify,     685
    Her tears should drop on them perpetually!

But she hath lost a dearer thing than life,
And he hath won what he would lose again.
This forcèd league doth force a further strife,
This momentary joy breeds months of pain;     690
This hot desire converts to cold disdain.
    Pure chastity is rifled of her store,
    And lust, the thief, far poorer than before.

Look as the full-fed hound or gorgèd hawk,
Unapt for tender smell or speedy flight,     695
Make slow pursuit, or altogether balk
The prey wherein by nature they delight,
So surfeit-taking Tarquin fares this night.
    His taste delicious, in digestion souring,
    Devours his will that lived by foul devouring.     700

O deeper sin than bottomless conceit
Can comprehend in still imagination!
Drunken desire must vomit his receipt
Ere he can see his own abomination.
While lust is in his pride, no exclamation     705
    Can curb his heat or rein his rash desire,
    Till like a jade self-will himself doth tire.

And then with lank and lean discoloured cheek,
With heavy eye, knit brow, and strengthless pace,
Feeble desire, all recreant, poor, and meek,     710
Like to a bankrupt beggar wails his case.
The flesh being proud, desire doth fight with grace,
    For there it revels, and when that decays,
    The guilty rebel for remission prays.

So fares it with this faultful lord of Rome     715
Who this accomplishment so hotly chased;
For now against himself he sounds this doom,
That through the length of times he stands disgraced.
Besides, his soul's fair temple is defaced,
    To whose weak ruins muster troops of cares     720
    To ask the spotted princess how she fares.

She says her subjects with foul insurrection
Have battered down her consecrated wall,
And by their mortal fault brought in subjection
Her immortality, and made her thrall     725
To living death and pain perpetual,
    Which in her prescience she controllèd still,
    But her foresight could not forestall their will.

Ev'n in this thought through the dark night he
    stealeth,
A captive victor that hath lost in gain,     730
Bearing away the wound that nothing healeth,
The scar that will, despite of cure, remain;
Leaving his spoil perplexed in greater pain.
    She bears the load of lust he left behind,
    And he the burden of a guilty mind.     735

He like a thievish dog creeps sadly thence;
She like a wearied lamb lies panting there.
He scowls, and hates himself for his offence;
She, desperate, with her nails her flesh doth tear.
He faintly flies, sweating with guilty fear;     740
    She stays, exclaiming on the direful night.
    He runs, and chides his vanished loathed delight.

He thence departs, a heavy convertite;
She there remains, a hopeless castaway.
He in his speed looks for the morning light;     745
She prays she never may behold the day.
'For day,' quoth she, 'night's scapes doth open lay,
    And my true eyes have never practised how
    To cloak offences with a cunning brow.

'They think not but that every eye can see     750
The same disgrace which they themselves behold,
And therefore would they still in darkness be,
To have their unseen sin remain untold.
For they their guilt with weeping will unfold,
    And grave, like water that doth eat in steel,     755
    Upon my cheeks what helpless shame I feel.'

Here she exclaims against repose and rest,
And bids her eyes hereafter still be blind.
She wakes her heart by beating on her breast,
And bids it leap from thence where it may find     760
Some purer chest to close so pure a mind.
    Frantic with grief, thus breathes she forth her spite
    Against the unseen secrecy of night:

'O comfort-killing night, image of hell,
Dim register and notary of shame,     765
Black stage for tragedies and murders fell,
Vast sin-concealing chaos, nurse of blame!
Blind muffled bawd, dark harbour for defame,
    Grim cave of death, whisp'ring conspirator
    With close-tongued treason and the ravisher!     770

'O hateful, vaporous, and foggy night,
Since thou art guilty of my cureless crime,
Muster thy mists to meet the eastern light,
Make war against proportioned course of time.
Or if thou wilt permit the sun to climb     775
    His wonted height, yet ere he go to bed
    Knit poisonous clouds about his golden head.

'With rotten damps ravish the morning air,
Let their exhaled unwholesome breaths make sick
The life of purity, the supreme fair,                               780
Ere he arrive his weary noon-tide prick;
And let thy musty vapours march so thick
    That in their smoky ranks his smothered light
    May set at noon, and make perpetual night.

'Were Tarquin night, as he is but night's child,     785
The silver-shining queen he would distain;
Her twinkling handmaids too, by him defiled,
Through night's black bosom should not peep again.
So should I have co-partners in my pain,
    And fellowship in woe doth woe assuage,          790
    As palmers' chat makes short their pilgrimage.

'Where now I have no one to blush with me,
To cross their arms and hang their heads with mine,
To mask their brows and hide their infamy,
But I alone, alone must sit and pine,                             795
Seasoning the earth with showers of silver brine,
    Mingling my talk with tears, my grief with groans,
    Poor wasting monuments of lasting moans.

'O night, thou furnace of foul reeking smoke,
Let not the jealous day behold that face             800
Which underneath thy black all-hiding cloak
Immodestly lies martyred with disgrace!
Keep still possession of thy gloomy place,
    That all the faults which in thy reign are made
    May likewise be sepulchred in thy shade.         805

'Make me not object to the tell-tale day:
The light will show charactered in my brow
The story of sweet chastity's decay,
The impious breach of holy wedlock vow.
Yea, the illiterate that know not how                 810
    To cipher what is writ in learnèd books
    Will quote my loathsome trespass in my looks.

'The nurse to still her child will tell my story,
And fright her crying babe with Tarquin's name.
The orator to deck his oratory                        815
Will couple my reproach to Tarquin's shame.
Feast-finding minstrels tuning my defame
    Will tie the hearers to attend each line,
    How Tarquin wrongèd me, I Collatine.

'Let my good name, that senseless reputation,       820
For Collatine's dear love be kept unspotted;
If that be made a theme for disputation,
The branches of another root are rotted,
And undeserved reproach to him allotted
    That is as clear from this attaint of mine       825
    As I ere this was pure to Collatine.

'O unseen shame, invisible disgrace!
O unfelt sore, crest-wounding private scar!
Reproach is stamped in Collatinus' face,
And Tarquin's eye may read the mot afar,             830
How in peace is wounded, not in war.
    Alas, how many bear such shameful blows,
    Which not themselves but he that gives them knows!

'If, Collatine, thine honour lay in me,
From me by strong assault it is bereft;              835
My honey lost, and I, a drone-like bee,
Have no perfection of my summer left,
But robbed and ransacked by injurious theft.
    In thy weak hive a wandering wasp hath crept,
    And sucked the honey which thy chaste bee kept.

'Yet am I guilty of thy honour's wrack;              841
Yet for thy honour did I entertain him.
Coming from thee, I could not put him back,
For it had been dishonour to disdain him.
Besides, of weariness he did complain him,           845
    And talked of virtue—O unlooked-for evil,
    When virtue is profaned in such a devil!

'Why should the worm intrude the maiden bud,
Or hateful cuckoos hatch in sparrows' nests,
Or toads infect fair founts with venom mud,          850
Or tyrant folly lurk in gentle breasts,
Or kings be breakers of their own behests?
    But no perfection is so absolute
    That some impurity doth not pollute.

'The agèd man that coffers up his gold               855
Is plagued with cramps, and gouts, and painful fits,
And scarce hath eyes his treasure to behold,
But like still-pining Tantalus he sits,
And useless barns the harvest of his wits,
    Having no other pleasure of his gain             860
    But torment that it cannot cure his pain.

'So then he hath it when he cannot use it,
And leaves it to be mastered by his young,
Who in their pride do presently abuse it.
Their father was too weak and they too strong        865
To hold their cursèd-blessèd fortune long.
    The sweets we wish for turn to loathèd sours
    Even in the moment that we call them ours.

'Unruly blasts wait on the tender spring,
Unwholesome weeds take root with precious flowers,
The adder hisses where the sweet birds sing,         871
What virtue breeds, iniquity devours.
We have no good that we can say is ours
    But ill-annexèd opportunity
    Or kills his life or else his quality.            875

'O opportunity, thy guilt is great!
'Tis thou that execut'st the traitor's treason;
Thou sets the wolf where he the lamb may get;
Whoever plots the sin, thou points the season.
'Tis thou that spurn'st at right, at law, at reason;  880
    And in thy shady cell where none may spy him
    Sits sin, to seize the souls that wander by him.

'Thou mak'st the vestal violate her oath,
Thou blow'st the fire when temperance is thawed,
Thou smother'st honesty, thou murd'rest troth,       885
Thou foul abettor, thou notorious bawd;
Thou plantest scandal and displacest laud.
    Thou ravisher, thou traitor, thou false thief,
    Thy honey turns to gall, thy joy to grief.

'Thy secret pleasure turns to open shame,            890
Thy private feasting to a public fast,
Thy smoothing titles to a ragged name,
Thy sugared tongue to bitter wormwood taste.
Thy violent vanities can never last.
    How comes it then, vile opportunity,            895
    Being so bad, such numbers seek for thee?

'When wilt thou be the humble suppliant's friend,
And bring him where his suit may be obtained?
When wilt thou sort an hour great strifes to end,
Or free that soul which wretchedness hath chained,
Give physic to the sick, ease to the pained?            901
    The poor, lame, blind, halt, creep, cry out for thee,
    But they ne'er meet with opportunity.

'The patient dies while the physician sleeps,
The orphan pines while the oppressor feeds,            905
Justice is feasting while the widow weeps,
Advice is sporting while infection breeds.
Thou grant'st no time for charitable deeds.
    Wrath, envy, treason, rape, and murder's rages,
    Thy heinous hours wait on them as their pages. 910

'When truth and virtue have to do with thee
A thousand crosses keep them from thy aid.
They buy thy help, but sin ne'er gives a fee;
He gratis comes, and thou art well appaid
As well to hear as grant what he hath said.            915
    My Collatine would else have come to me
    When Tarquin did, but he was stayed by thee.

'Guilty thou art of murder and of theft,
Guilty of perjury and subornation,
Guilty of treason, forgery, and shift,            920
Guilty of incest, that abomination:
An accessory by thine inclination
    To all sins past and all that are to come
    From the creation to the general doom.

'Misshapen time, copesmate of ugly night,            925
Swift subtle post, carrier of grisly care,
Eater of youth, false slave to false delight,
Base watch of woes, sin's pack-horse, virtue's snare,
Thou nursest all, and murd'rest all that are.
    O hear me then, injurious shifting time;            930
    Be guilty of my death, since of my crime.

'Why hath thy servant opportunity
Betrayed the hours thou gav'st me to repose,
Cancelled my fortunes, and enchainèd me
To endless date of never-ending woes?            935
Time's office is to fine the hate of foes,
    To eat up errors by opinion bred,
    Not spend the dowry of a lawful bed.

'Time's glory is to calm contending kings,
To unmask falsehood and bring truth to light,            940
To stamp the seal of time in agèd things,
To wake the morn and sentinel the night,
To wrong the wronger till he render right,
    To ruinate proud buildings with thy hours
    And smear with dust their glitt'ring golden towers;

'To fill with worm-holes stately monuments,            946
To feed oblivion with decay of things,
To blot old books and alter their contents,
To pluck the quills from ancient ravens' wings,
To dry the old oak's sap and blemish springs,            950
    To spoil antiquities of hammered steel,
    And turn the giddy round of fortune's wheel;

'To show the beldame daughters of her daughter,
To make the child a man, the man a child,
To slay the tiger that doth live by slaughter,            955
To tame the unicorn and lion wild,
To mock the subtle in themselves beguiled,
    To cheer the ploughman with increaseful crops,
    And waste huge stones with little water drops.

'Why work'st thou mischief in thy pilgrimage,            960
Unless thou couldst return to make amends?
One poor retiring minute in an age
Would purchase thee a thousand thousand friends,
Lending him wit that to bad debtors lends.
    O this dread night, wouldst thou one hour come
        back,            965
    I could prevent this storm and shun thy wrack!

'Thou ceaseless lackey to eternity,
With some mischance cross Tarquin in his flight.
Devise extremes beyond extremity
To make him curse this cursèd crimeful night.            970
Let ghastly shadows his lewd eyes affright,
    And the dire thought of his committed evil
    Shape every bush a hideous shapeless devil.

'Disturb his hours of rest with restless trances;
Afflict him in his bed with bedrid groans;            975
Let there bechance him pitiful mischances
To make him moan, but pity not his moans.
Stone him with hardened hearts harder than stones,
    And let mild women to him lose their mildness,
    Wilder to him than tigers in their wildness.            980

'Let him have time to tear his curlèd hair,
Let him have time against himself to rave,
Let him have time of time's help to despair,
Let him have time to live a loathèd slave,
Let him have time a beggar's orts to crave,            985
    And time to see one that by alms doth live
    Disdain to him disdainèd scraps to give.

'Let him have time to see his friends his foes,
And merry fools to mock at him resort.
Let him have time to mark how slow time goes            990
In time of sorrow, and how swift and short
His time of folly and his time of sport;
    And ever let his unrecalling crime
    Have time to wail th'abusing of his time.

'O time, thou tutor both to good and bad,            995
Teach me to curse him that thou taught'st this ill;
At his own shadow let the thief run mad,
Himself himself seek every hour to kill;
Such wretched hands such wretched blood should spill,
    For who so base would such an office have            1000
    As sland'rous deathsman to so base a slave?

'The baser is he, coming from a king,
To shame his hope with deeds degenerate.
The mightier man, the mightier is the thing
That makes him honoured or begets him hate,    1005
For greatest scandal waits on greatest state.
    The moon being clouded presently is missed,
    But little stars may hide them when they list.

'The crow may bathe his coal-black wings in mire
And unperceived fly with the filth away,    1010
But if the like the snow-white swan desire,
The stain upon his silver down will stay.
Poor grooms are sightless night, kings glorious day.
    Gnats are unnoted wheresoe'er they fly,
    But eagles gazed upon with every eye.    1015

'Out, idle words, servants to shallow fools,
Unprofitable sounds, weak arbitrators!
Busy yourselves in skill-contending schools,
Debate where leisure serves with dull debaters,
To trembling clients be you mediators;    1020
    For me, I force not argument a straw,
    Since that my case is past the help of law.

'In vain I rail at opportunity,
At time, at Tarquin, and uncheerful night.
In vain I cavil with mine infamy,    1025
In vain I spurn at my confirmed despite.
This helpless smoke of words doth me no right;
    The remedy indeed is to do me good
    Is to let forth my foul defilèd blood.

'Poor hand, why quiver'st thou at this decree?    1030
Honour thyself to rid me of this shame,
For if I die, my honour lives in thee,
But if I live, thou liv'st in my defame.
Since thou couldst not defend thy loyal dame,
    And wast afeard to scratch her wicked foe,    1035
    Kill both thyself and her for yielding so.'

This said, from her betumbled couch she starteth,
To find some desp'rate instrument of death.
But this, no slaughterhouse, no tool imparteth
To make more vent for passage of her breath,    1040
Which thronging through her lips so vanisheth
    As smoke from Etna that in air consumes,
    Or that which from dischargèd cannon fumes.

'In vain,' quoth she, 'I live, and seek in vain
Some happy mean to end a hapless life.    1045
I feared by Tarquin's falchion to be slain,
Yet for the selfsame purpose seek a knife.
But when I feared I was a loyal wife:
    So am I now—O no, that cannot be,
    Of that true type hath Tarquin rifled me.    1050

'O, that is gone for which I sought to live,
And therefore now I need not fear to die.
To clear this spot by death, at least I give
A badge of fame to slander's livery,
A dying life to living infamy,    1055
    Poor helpless help, the treasure stol'n away,
    To burn the guiltless casket where it lay!

'Well, well, dear Collatine, thou shalt not know
The stainèd taste of violated troth.
I will not wrong thy true affection so    1060
To flatter thee with an infringèd oath.
This bastard graft shall never come to growth.
    He shall not boast, who did thy stock pollute,
    That thou art doting father of his fruit,

'Nor shall he smile at thee in secret thought,    1065
Nor laugh with his companions at thy state.
But thou shalt know thy int'rest was not bought
Basely with gold, but stol'n from forth thy gate.
For me, I am the mistress of my fate,
    And with my trespass never will dispense    1070
    Till life to death acquit my forced offence.

'I will not poison thee with my attaint,
Nor fold my fault in cleanly coined excuses.
My sable ground of sin I will not paint
To hide the truth of this false night's abuses.    1075
My tongue shall utter all; mine eyes, like sluices,
    As from a mountain spring that feeds a dale
    Shall gush pure streams to purge my impure tale.'

By this, lamenting Philomel had ended
The well-tuned warble of her nightly sorrow,    1080
And solemn night with slow sad gait descended
To ugly hell, when lo, the blushing morrow
Lends light to all fair eyes that light will borrow.
    But cloudy Lucrece shames herself to see,
    And therefore still in night would cloistered be.    1085

Revealing day through every cranny spies,
And seems to point her out where she sits weeping;
To whom she sobbing speaks, 'O eye of eyes,
Why pry'st thou through my window? Leave thy
    peeping,
Mock with thy tickling beams eyes that are sleeping,
    Brand not my forehead with thy piercing light,    1091
    For day hath naught to do what's done by night.'

Thus cavils she with everything she sees:
True grief is fond and testy as a child
Who, wayward once, his mood with naught agrees;
Old woes, not infant sorrows, bear them mild.    1096
Continuance tames the one; the other wild,
    Like an unpractised swimmer plunging still,
    With too much labour drowns for want of skill.

So she, deep drenchèd in a sea of care,    1100
Holds disputation with each thing she views,
And to herself all sorrow doth compare;
No object but her passion's strength renews,
And as one shifts, another straight ensues.
    Sometime her grief is dumb and hath no words, 1105
    Sometime 'tis mad and too much talk affords.

The little birds that tune their morning's joy
Make her moans mad with their sweet melody,
For mirth doth search the bottom of annoy;
Sad souls are slain in merry company;    1110
Grief best is pleased with grief's society.
    True sorrow then is feelingly sufficed
    When with like semblance it is sympathized.

'Tis double death to drown in ken of shore;
He ten times pines that pines beholding food; 1115
To see the salve doth make the wound ache more;
Great grief grieves most at that would do it good;
Deep woes roll forward like a gentle flood
    Who, being stopped, the bounding banks o'erflows.
    Grief dallied with nor law nor limit knows. 1120

'You mocking birds,' quoth she, 'your tunes entomb
Within your hollow-swelling feathered breasts,
And in my hearing be you mute and dumb;
My restless discord loves no stops nor rests;
A woeful hostess brooks not merry guests. 1125
    Relish your nimble notes to pleasing ears;
    Distress likes dumps when time is kept with tears.

'Come, Philomel, that sing'st of ravishment,
Make thy sad grove in my dishevelled hair.
As the dank earth weeps at thy languishment, 1130
So I at each sad strain will strain a tear,
And with deep groans the diapason bear;
    For burden-wise I'll hum on Tarquin still,
    While thou on Tereus descants better skill.

'And whiles against a thorn thou bear'st thy part 1135
To keep thy sharp woes waking, wretched I,
To imitate thee well, against my heart
Will fix a sharp knife to affright mine eye,
Who if it wink shall thereon fall and die.
    These means, as frets upon an instrument, 1140
    Shall tune our heart-strings to true languishment.

'And for, poor bird, thou sing'st not in the day,
As shaming any eye should thee behold,
Some dark deep desert seated from the way,
That knows not parching heat nor freezing cold, 1145
Will we find out, and there we will unfold
    To creatures stern sad tunes to change their kinds.
    Since men prove beasts, let beasts bear gentle minds.'

As the poor frighted deer that stands at gaze,
Wildly determining which way to fly, 1150
Or one encompassed with a winding maze,
That cannot tread the way out readily,
So with herself is she in mutiny,
    To live or die which of the twain were better
    When life is shamed and death reproach's debtor.

'To kill myself,' quoth she, 'alack, what were it 1156
But with my body my poor soul's pollution?
They that lose half with greater patience bear it
Than they whose whole is swallowed in confusion.
That mother tries a merciless conclusion 1160
    Who, having two sweet babes, when death takes one
    Will slay the other and be nurse to none.

'My body or my soul, which was the dearer,
When the one pure the other made divine?
Whose love of either to myself was nearer, 1165
When both were kept for heaven and Collatine?
Ay me, the bark peeled from the lofty pine
    His leaves will wither and his sap decay;
    So must my soul, her bark being peeled away.

'Her house is sacked, her quiet interrupted, 1170
Her mansion battered by the enemy,
Her sacred temple spotted, spoiled, corrupted,
Grossly engirt with daring infamy.
Then let it not be called impiety
    If in this blemished fort I make some hole 1175
    Through which I may convey this troubled soul.

'Yet die I will not till my Collatine
Have heard the cause of my untimely death,
That he may vow in that sad hour of mine
Revenge on him that made me stop my breath. 1180
My stainèd blood to Tarquin I'll bequeath,
    Which by him tainted shall for him be spent,
    And as his due writ in my testament.

'My honour I'll bequeath unto the knife
That wounds my body so dishonourèd; 1185
'Tis honour to deprive dishonoured life;
The one will live, the other being dead.
So of shame's ashes shall my fame be bred,
    For in my death I murder shameful scorn;
    My shame so dead, mine honour is new born. 1190

'Dear lord of that dear jewel I have lost,
What legacy shall I bequeath to thee?
My resolution, love, shall be thy boast,
By whose example thou revenged mayst be.
How Tarquin must be used, read it in me. 1195
    Myself, thy friend, will kill myself, thy foe;
    And for my sake serve thou false Tarquin so.

'This brief abridgement of my will I make:
My soul and body to the skies and ground;
My resolution, husband, do thou take; 1200
Mine honour be the knife's that makes my wound;
My shame be his that did my fame confound;
    And all my fame that lives disbursèd be
    To those that live and think no shame of me.

'Thou, Collatine, shalt oversee this will. 1205
How was I overseen that thou shalt see it!
My blood shall wash the slander of mine ill;
My life's foul deed my life's fair end shall free it.
Faint not, faint heart, but stoutly say "So be it".
    Yield to my hand, my hand shall conquer thee; 1210
    Thou dead, both die, and both shall victors be.'

This plot of death when sadly she had laid,
And wiped the brinish pearl from her bright eyes,
With untuned tongue she hoarsely calls her maid,
Whose swift obedience to her mistress hies; 1215
For fleet-winged duty with thought's feathers flies.
    Poor Lucrece' cheeks unto her maid seem so
    As winter meads when sun doth melt their snow.

Her mistress she doth give demure good-morrow
With soft slow tongue, true mark of modesty, 1220
And sorts a sad look to her lady's sorrow,
For why her face wore sorrow's livery;
But durst not ask of her audaciously
    Why her two suns were cloud-eclipsèd so,
    Nor why her fair cheeks over-washed with woe. 1225

But as the earth doth weep, the sun being set,
Each flower moistened like a melting eye,
Even so the maid with swelling drops gan wet
Her circled eyne, enforced by sympathy
Of those fair suns set in her mistress' sky, 1230
 Who in a salt-waved ocean quench their light;
 Which makes the maid weep like the dewy night.

A pretty while these pretty creatures stand,
Like ivory conduits coral cisterns filling.
One justly weeps, the other takes in hand 1235
No cause but company of her drops' spilling.
Their gentle sex to weep are often willing,
 Grieving themselves to guess at others' smarts,
 And then they drown their eyes or break their hearts.

For men have marble, women waxen minds, 1240
And therefore are they formed as marble will.
The weak oppressed, th'impression of strange kinds
Is formed in them by force, by fraud, or skill.
Then call them not the authors of their ill,
 No more than wax shall be accounted evil 1245
 Wherein is stamped the semblance of a devil.

Their smoothness like a goodly champaign plain
Lays open all the little worms that creep;
In men as in a rough-grown grove remain
Cave-keeping evils that obscurely sleep. 1250
Through crystal walls each little mote will peep;
 Though men can cover crimes with bold stern looks,
 Poor women's faces are their own faults' books.

No man inveigh against the withered flower,
But chide rough winter that the flower hath killed.
Not that devoured, but that which doth devour 1256
Is worthy blame. O, let it not be held
Poor women's faults that they are so full-filled
 With men's abuses. Those proud lords, to blame,
 Make weak-made women tenants to their shame.

The precedent whereof in Lucrece view, 1261
Assailed by night with circumstances strong
Of present death, and shame that might ensue
By that her death, to do her husband wrong.
Such danger to resistance did belong 1265
 That dying fear through all her body spread;
 And who cannot abuse a body dead?

By this, mild patience bid fair Lucrece speak
To the poor counterfeit of her complaining.
'My girl,' quoth she, 'on what occasion break 1270
Those tears from thee that down thy cheeks are
  raining?
If thou dost weep for grief of my sustaining,
 Know, gentle wench, it small avails my mood.
 If tears could help, mine own would do me good.

'But tell me, girl, when went'—and there she stayed,
Till after a deep groan—'Tarquin from hence?' 1276
'Madam, ere I was up,' replied the maid,
'The more to blame my sluggard negligence.
Yet with the fault I thus far can dispense:
 Myself was stirring ere the break of day, 1280
 And ere I rose was Tarquin gone away.

'But lady, if your maid may be so bold,
She would request to know your heaviness.'
'O, peace,' quoth Lucrece, 'if it should be told,
The repetition cannot make it less; 1285
For more it is than I can well express,
 And that deep torture may be called a hell
 When more is felt than one hath power to tell.

'Go, get me hither paper, ink, and pen;
Yet save that labour, for I have them here. 1290
What should I say? One of my husband's men
Bid thou be ready by and by to bear
A letter to my lord, my love, my dear.
 Bid him with speed prepare to carry it;
 The cause craves haste, and it will soon be writ.'

Her maid is gone, and she prepares to write, 1296
First hovering o'er the paper with her quill.
Conceit and grief an eager combat fight;
What wit sets down is blotted straight with will;
This is too curious-good, this blunt and ill. 1300
 Much like a press of people at a door
 Throng her inventions, which shall go before.

At last she thus begins: 'Thou worthy lord
Of that unworthy wife that greeteth thee,
Health to thy person! Next, vouchsafe t'afford— 1305
If ever, love, thy Lucrece thou wilt see—
Some present speed to come and visit me.
 So I commend me, from our house in grief;
 My woes are tedious, though my words are brief.'

Here folds she up the tenor of her woe, 1310
Her certain sorrow writ uncertainly.
By this short schedule Collatine may know
Her grief, but not her grief's true quality.
She dares not thereof make discovery,
 Lest he should hold it her own gross abuse, 1315
 Ere she with blood had stained her stain's excuse.

Besides, the life and feeling of her passion
She hoards, to spend when he is by to hear her,
When sighs and groans and tears may grace the
  fashion
Of her disgrace, the better so to clear her 1320
From that suspicion which the world might bear her.
 To shun this blot she would not blot the letter
 With words, till action might become them better.

To see sad sights moves more than hear them told,
For then the eye interprets to the ear 1325
The heavy motion that it doth behold,
When every part a part of woe doth bear.
'Tis but a part of sorrow that we hear;
 Deep sounds make lesser noise than shallow fords,
 And sorrow ebbs, being blown with wind of words.

Her letter now is sealed, and on it writ 1331
'At Ardea to my lord with more than haste'.
The post attends, and she delivers it,
Charging the sour-faced groom to hie as fast
As lagging fowls before the northern blast. 1335
 Speed more than speed but dull and slow she deems;
 Extremity still urgeth such extremes.

The homely villain curtsies to her low,
And blushing on her with a steadfast eye
Receives the scroll without or yea or no,          1340
And forth with bashful innocence doth hie.
But they whose guilt within their bosoms lie
  Imagine every eye beholds their blame,
  For Lucrece thought he blushed to see her shame,

When, silly groom, God wot, it was defect          1345
Of spirit, life, and bold audacity.
Such harmless creatures have a true respect
To talk in deeds, while others saucily
Promise more speed, but do it leisurely.
  Even so this pattern of the worn-out age          1350
  Pawned honest looks, but laid no words to gage.

His kindled duty kindled her mistrust,
That two red fires in both their faces blazed.
She thought he blushed as knowing Tarquin's lust,
And blushing with him, wistly on him gazed.          1355
Her earnest eye did make him more amazed.
  The more she saw the blood his cheeks replenish,
  The more she thought he spied in her some blemish.

But long she thinks till he return again,
And yet the duteous vassal scarce is gone.          1360
The weary time she cannot entertain,
For now 'tis stale to sigh, to weep, and groan.
So woe hath wearied woe, moan tired moan,
  That she her plaints a little while doth stay,
  Pausing for means to mourn some newer way.          1365

At last she calls to mind where hangs a piece
Of skilful painting made for Priam's Troy,
Before the which is drawn the power of Greece,
For Helen's rape the city to destroy,
Threat'ning cloud-kissing Ilion with annoy;          1370
  Which the conceited painter drew so proud
  As heaven, it seemed, to kiss the turrets bowed.

A thousand lamentable objects there,
In scorn of nature, art gave lifeless life.
Many a dry drop seemed a weeping tear          1375
Shed for the slaughtered husband by the wife.
The red blood reeked to show the painter's strife,
  And dying eyes gleamed forth their ashy lights
  Like dying coals burnt out in tedious nights.

There might you see the labouring pioneer          1380
Begrimed with sweat and smeared all with dust,
And from the towers of Troy there would appear
The very eyes of men through loop-holes thrust,
Gazing upon the Greeks with little lust.
  Such sweet observance in this work was had          1385
  That one might see those far-off eyes look sad.

In great commanders grace and majesty
You might behold, triumphing in their faces;
In youth, quick bearing and dexterity;
And here and there the painter interlaces          1390
Pale cowards marching on with trembling paces,
  Which heartless peasants did so well resemble
  That one would swear he saw them quake and
    tremble.

In Ajax and Ulysses, O what art
Of physiognomy might one behold!          1395
The face of either ciphered either's heart;
Their face their manners most expressly told.
In Ajax' eyes blunt rage and rigour rolled,
  But the mild glance that sly Ulysses lent
  Showed deep regard and smiling government.          1400

There pleading might you see grave Nestor stand,
As 'twere encouraging the Greeks to fight,
Making such sober action with his hand
That it beguiled attention, charmed the sight.
In speech it seemed his beard all silver-white          1405
  Wagged up and down, and from his lips did fly
  Thin winding breath which purled up to the sky.

About him were a press of gaping faces
Which seemed to swallow up his sound advice,
All jointly list'ning, but with several graces,          1410
As if some mermaid did their ears entice;
Some high, some low, the painter was so nice.
  The scalps of many, almost hid behind,
  To jump up higher seemed, to mock the mind.

Here one man's hand leaned on another's head,          1415
His nose being shadowed by his neighbour's ear;
Here one being thronged bears back, all boll'n and red;
Another, smothered, seems to pelt and swear,
And in their rage such signs of rage they bear
  As but for loss of Nestor's golden words          1420
  It seemed they would debate with angry swords.

For much imaginary work was there;
Conceit deceitful, so compact, so kind,
That for Achilles' image stood his spear
Gripped in an armèd hand; himself behind          1425
Was left unseen save to the eye of mind;
  A hand, a foot, a face, a leg, a head,
  Stood for the whole to be imaginèd.

And from the walls of strong-besiegèd Troy
When their brave hope, bold Hector, marched to field,
Stood many Trojan mothers sharing joy          1431
To see their youthful sons bright weapons wield;
And to their hope they such odd action yield
  That through their light joy seemèd to appear,
  Like bright things stained, a kind of heavy fear.          1435

And from the strand of Dardan where they fought
To Simois' reedy banks the red blood ran,
Whose waves to imitate the battle sought
With swelling ridges, and their ranks began
To break upon the gallèd shore, and then          1440
  Retire again, till meeting greater ranks
  They join, and shoot their foam at Simois' banks.

To this well painted piece is Lucrece come,
To find a face where all distress is stelled.
Many she sees where cares have carvèd some,          1445
But none where all distress and dolour dwelled
Till she despairing Hecuba beheld
  Staring on Priam's wounds with her old eyes,
  Which bleeding under Pyrrhus' proud foot lies.

In her the painter had anatomized 1450
Time's ruin, beauty's wreck, and grim care's reign.
Her cheeks with chaps and wrinkles were disguised;
Of what she was no semblance did remain.
Her blue blood changed to black in every vein,
  Wanting the spring that those shrunk pipes had fed,
  Showed life imprisoned in a body dead. 1456

On this sad shadow Lucrece spends her eyes,
And shapes her sorrow to the beldame's woes,
Who nothing wants to answer her but cries
And bitter words to ban her cruel foes. 1460
The painter was no god to lend her those,
  And therefore Lucrece swears he did her wrong
  To give her so much grief, and not a tongue.

'Poor instrument,' quoth she, 'without a sound,
I'll tune thy woes with my lamenting tongue, 1465
And drop sweet balm in Priam's painted wound,
And rail on Pyrrhus that hath done him wrong,
And with my tears quench Troy that burns so long,
  And with my knife scratch out the angry eyes
  Of all the Greeks that are thine enemies. 1470

'Show me the strumpet that began this stir,
That with my nails her beauty I may tear.
Thy heat of lust, fond Paris, did incur
This load of wrath that burning Troy doth bear;
Thine eye kindled the fire that burneth here, 1475
  And here in Troy, for trespass of thine eye,
  The sire, the son, the dame and daughter die.

'Why should the private pleasure of someone
Become the public plague of many moe?
Let sin alone committed light alone 1480
Upon his head that hath transgressèd so;
Let guiltless souls be freed from guilty woe.
  For one's offence why should so many fall,
  To plague a private sin in general?

'Lo, here weeps Hecuba, here Priam dies, 1485
Here manly Hector faints, here Troilus swoons,
Here friend by friend in bloody channel lies,
And friend to friend gives unadvisèd wounds,
And one man's lust these many lives confounds.
  Had doting Priam checked his son's desire, 1490
  Troy had been bright with fame, and not with fire.'

Here feelingly she weeps Troy's painted woes;
For sorrow, like a heavy hanging bell
Once set on ringing, with his own weight goes;
Then little strength rings out the doleful knell. 1495
So Lucrece, set a-work, sad tales doth tell
  To pencilled pensiveness and coloured sorrow.
  She lends them words, and she their looks doth
    borrow.

She throws her eyes about the painting round,
And who she finds forlorn she doth lament. 1500
At last she sees a wretched image bound,
That piteous looks to Phrygian shepherds lent.
His face, though full of cares, yet showed content.
  Onward to Troy with the blunt swains he goes,
  So mild that patience seemed to scorn his woes. 1505

In him the painter laboured with his skill
To hide deceit and give the harmless show
An humble gait, calm looks, eyes wailing still,
A brow unbent that seemed to welcome woe;
Cheeks neither red nor pale, but mingled so 1510
  That blushing red no guilty instance gave,
  Nor ashy pale the fear that false hearts have.

But like a constant and confirmèd devil
He entertained a show so seeming just,
And therein so ensconced his secret evil 1515
That jealousy itself could not mistrust
False creeping craft and perjury should thrust
  Into so bright a day such blackfaced storms,
  Or blot with hell-born sin such saint-like forms.

The well skilled workman this mild image drew 1520
For perjured Sinon, whose enchanting story
The credulous old Priam after slew;
Whose words like wildfire burnt the shining glory
Of rich-built Ilion, that the skies were sorry,
  And little stars shot from their fixèd places 1525
  When their glass fell wherein they viewed their faces.

This picture she advisedly perused,
And chid the painter for his wondrous skill,
Saying some shape in Sinon's was abused,
So fair a form lodged not a mind so ill; 1530
And still on him she gazed, and gazing still,
  Such signs of truth in his plain face she spied
  That she concludes the picture was belied.

'It cannot be,' quoth she, 'that so much guile'—
She would have said 'can lurk in such a look', 1535
But Tarquin's shape came in her mind the while,
And from her tongue 'can lurk' from 'cannot' took.
'It cannot be' she in that sense forsook,
  And turned it thus: 'It cannot be, I find,
  But such a face should bear a wicked mind. 1540

'For even as subtle Sinon here is painted,
So sober-sad, so weary, and so mild,
As if with grief or travail he had fainted,
To me came Tarquin armèd, too beguiled
With outward honesty, but yet defiled 1545
  With inward vice. As Priam him did cherish,
  So did I Tarquin, so my Troy did perish.

'Look, look, how list'ning Priam wets his eyes
To see those borrowed tears that Sinon sheds.
Priam, why art thou old and yet not wise? 1550
For every tear he falls a Trojan bleeds.
His eye drops fire, no water thence proceeds.
  Those round clear pearls of his that move thy pity
  Are balls of quenchless fire to burn thy city.

'Such devils steal effects from lightless hell, 1555
For Sinon in his fire doth quake with cold,
And in that cold hot-burning fire doth dwell.
These contraries such unity do hold
Only to flatter fools and make them bold;
  So Priam's trust false Sinon's tears doth flatter 1560
  That he finds means to burn his Troy with water.'

Here, all enraged, such passion her assails
That patience is quite beaten from her breast.
She tears the senseless Sinon with her nails,
Comparing him to that unhappy guest          1565
Whose deed hath made herself herself detest.
    At last she smilingly with this gives o'er:
    'Fool, fool,' quoth she, 'his wounds will not be sore.'

Thus ebbs and flows the current of her sorrow,
And time doth weary time with her complaining.  1570
She looks for night, and then she longs for morrow,
And both she thinks too long with her remaining.
Short time seems long in sorrow's sharp sustaining.
    Though woe be heavy, yet it seldom sleeps,
    And they that watch see time how slow it creeps.

Which all this time hath overslipped her thought  1576
That she with painted images hath spent,
Being from the feeling of her own grief brought
By deep surmise of others' detriment,
Losing her woes in shows of discontent.          1580
    It easeth some, though none it ever cured,
    To think their dolour others have endured.

But now the mindful messenger come back
Brings home his lord and other company,
Who finds his Lucrece clad in mourning black,    1585
And round about her tear-distainèd eye
Blue circles streamed, like rainbows in the sky.
    These water-galls in her dim element
    Foretell new storms to those already spent.

Which when her sad beholding husband saw,        1590
Amazedly in her sad face he stares.
Her eyes, though sod in tears, looked red and raw,
Her lively colour killed with deadly cares.
He hath no power to ask her how she fares.
    Both stood like old acquaintance in a trance,  1595
    Met far from home, wond'ring each other's chance.

At last he takes her by the bloodless hand,
And thus begins: 'What uncouth ill event
Hath thee befall'n, that thou dost trembling stand?
Sweet love, what spite hath thy fair colour spent?  1600
Why art thou thus attired in discontent?
    Unmask, dear dear, this moody heaviness,
    And tell thy grief, that we may give redress.'

Three times with sighs she gives her sorrow fire
Ere once she can discharge one word of woe.      1605
At length addressed to answer his desire,
She modestly prepares to let them know
Her honour is ta'en prisoner by the foe,
    While Collatine and his consorted lords
    With sad attention long to hear her words.    1610

And now this pale swan in her wat'ry nest
Begins the sad dirge of her certain ending.
'Few words,' quoth she, 'shall fit the trespass best,
Where no excuse can give the fault amending.
In me more woes than words are now depending,    1615
    And my laments would be drawn out too long
    To tell them all with one poor tired tongue.

'Then be this all the task it hath to say:
Dear husband, in the interest of thy bed
A stranger came, and on that pillow lay          1620
Where thou wast wont to rest thy weary head;
And what wrong else may be imaginèd
    By foul enforcement might be done to me,
    From that, alas, thy Lucrece is not free.

'For in the dreadful dead of dark midnight       1625
With shining falchion in my chamber came
A creeping creature with a flaming light,
And softly cried, "Awake, thou Roman dame,
And entertain my love; else lasting shame
    On thee and thine this night I will inflict,  1630
    If thou my love's desire do contradict.

' "For some hard-favoured groom of thine," quoth he,
"Unless thou yoke thy liking to my will,
I'll murder straight, and then I'll slaughter thee,
And swear I found you where you did fulfil       1635
The loathsome act of lust, and so did kill
    The lechers in their deed. This act will be
    My fame, and thy perpetual infamy."

'With this I did begin to start and cry,
And then against my heart he set his sword,      1640
Swearing unless I took all patiently
I should not live to speak another word.
So should my shame still rest upon record,
    And never be forgot in mighty Rome
    Th'adulterate death of Lucrece and her groom.  1645

'Mine enemy was strong, my poor self weak,
And far the weaker with so strong a fear.
My bloody judge forbade my tongue to speak;
No rightful plea might plead for justice there.
His scarlet lust came evidence to swear          1650
    That my poor beauty had purloined his eyes;
    And when the judge is robbed, the prisoner dies.

'O teach me how to make mine own excuse,
Or at the least this refuge let me find:
Though my gross blood be stained with this abuse,
Immaculate and spotless is my mind.              1656
That was not forced, that never was inclined
    To accessory yieldings, but still pure
    Doth in her poisoned closet yet endure.'

Lo, here the hopeless merchant of this loss,     1660
With head declined and voice dammed up with woe,
With sad set eyes and wreathèd arms across,
From lips new waxen pale begins to blow
The grief away that stops his answer so;
    But wretched as he is, he strives in vain.    1665
    What he breathes out, his breath drinks up again.

As through an arch the violent roaring tide
Outruns the eye that doth behold his haste,
Yet in the eddy boundeth in his pride
Back to the strait that forced him on so fast,   1670
In rage sent out, recalled in rage being past;
    Even so his sighs, his sorrows, make a saw,
    To push grief on, and back the same grief draw.

Which speechless woe of his poor she attendeth,
And his untimely frenzy thus awaketh:          1675
'Dear lord, thy sorrow to my sorrow lendeth
Another power; no flood by raining slaketh.
My woe too sensible thy passion maketh,
    More feeling-painful. Let it then suffice
    To drown on woe one pair of weeping eyes.          1680

'And for my sake, when I might charm thee so,
For she that was thy Lucrece, now attend me.
Be suddenly revengèd on my foe—
Thine, mine, his own. Suppose thou dost defend me
From what is past. The help that thou shalt lend me
    Comes all too late, yet let the traitor die,          1686
    For sparing justice feeds iniquity.

'But ere I name him, you fair lords,' quoth she,
Speaking to those that came with Collatine,
'Shall plight your honourable faiths to me          1690
With swift pursuit to venge this wrong of mine;
For 'tis a meritorious fair design
    To chase injustice with revengeful arms.
    Knights, by their oaths, should right poor ladies'
        harms.'

At this request with noble disposition          1695
Each present lord began to promise aid,
As bound in knighthood to her imposition,
Longing to hear the hateful foe bewrayed.
But she that yet her sad task hath not said
    The protestation stops. 'O speak,' quoth she;          1700
    'How may this forcèd stain be wiped from me?

'What is the quality of my offence,
Being constrained with dreadful circumstance?
May my pure mind with the foul act dispense,
My low-declinèd honour to advance?          1705
May any terms acquit me from this chance?
    The poisoned fountain clears itself again,
    And why not I from this compellèd stain?'

With this they all at once began to say
Her body's stain her mind untainted clears,          1710
While with a joyless smile she turns away
The face, that map which deep impression bears
Of hard misfortune, carved in it with tears.
    'No, no,' quoth she, 'no dame hereafter living
    By my excuse shall claim excuse's giving.'          1715

Here with a sigh as if her heart would break
She throws forth Tarquin's name. 'He, he,' she says—
But more than he her poor tongue could not speak,
Till after many accents and delays,
Untimely breathings, sick and short essays,          1720
    She utters this: 'He, he, fair lords, 'tis he
    That guides this hand to give this wound to me.'

Even here she sheathèd in her harmless breast
A harmful knife, that thence her soul unsheathed.
That blow did bail it from the deep unrest          1725
Of that polluted prison where it breathed.
Her contrite sighs unto the clouds bequeathed
    Her wingèd sprite, and through her wounds doth fly
    Life's lasting date from cancelled destiny.

Stone-still, astonished with this deadly deed          1730
Stood Collatine and all his lordly crew,
Till Lucrece' father that beholds her bleed
Himself on her self-slaughtered body threw;
And from the purple fountain Brutus drew
    The murd'rous knife; and as it left the place          1735
    Her blood in poor revenge held it in chase,

And bubbling from her breast it doth divide
In two slow rivers, that the crimson blood
Circles her body in on every side,
Who like a late-sacked island vastly stood,          1740
Bare and unpeopled in this fearful flood.
    Some of her blood still pure and red remained,
    And some looked black, and that false Tarquin-
        stained.

About the mourning and congealèd face
Of that black blood a wat'ry rigol goes,          1745
Which seems to weep upon the tainted place;
And ever since, as pitying Lucrece' woes,
Corrupted blood some watery token shows,
    And blood untainted still doth red abide,
    Blushing at that which is so putrefied.          1750

'Daughter, dear daughter,' old Lucretius cries,
'That life was mine which thou hast here deprived.
If in the child the father's image lies,
Where shall I live now Lucrece is unlived?
Thou wast not to this end from me derived.          1755
    If children predecease progenitors,
    We are their offspring, and they none of ours.

'Poor broken glass, I often did behold
In thy sweet semblance my old age new born;
But now that fair fresh mirror, dim and old,          1760
Shows me a bare-boned death by time outworn.
O, from thy cheeks my image thou hast torn,
    And shivered all the beauty of my glass,
    That I no more can see what once I was.

'O time, cease thou thy course and last no longer,          1765
If they surcease to be that should survive!
Shall rotten death make conquest of the stronger,
And leave the falt'ring feeble souls alive?
The old bees die, the young possess their hive.
    Then live, sweet Lucrece, live again and see          1770
    Thy father die, and not thy father thee.'

By this starts Collatine as from a dream,
And bids Lucretius give his sorrow place;
And then in key-cold Lucrece' bleeding stream
He falls, and bathes the pale fear in his face,          1775
And counterfeits to die with her a space,
    Till manly shame bids him possess his breath,
    And live to be revengèd on her death.

The deep vexation of his inward soul
Hath served a dumb arrest upon his tongue,          1780
Who, mad that sorrow should his use control,
Or keep him from heart-easing words so long,
Begins to talk; but through his lips do throng
    Weak words, so thick come in his poor heart's aid
    That no man could distinguish what he said.          1785

Yet sometime 'Tarquin' was pronouncèd plain,
But through his teeth, as if the name he tore.
This windy tempest, till it blow up rain,
Held back his sorrow's tide to make it more.
At last it rains, and busy winds give o'er.          1790
  Then son and father weep with equal strife
  Who should weep most, for daughter or for wife.

The one doth call her his, the other his,
Yet neither may possess the claim they lay.
The father says 'She's mine'; 'O, mine she is,'       1795
Replies her husband, 'do not take away
My sorrow's interest; let no mourner say
  He weeps for her, for she was only mine,
  And only must be wailed by Collatine.'

'O,' quoth Lucretius, 'I did give that life            1800
Which she too early and too late hath spilled.'
'Woe, woe,' quoth Collatine, 'she was my wife.
I owed her, and 'tis mine that she hath killed.'
'My daughter' and 'my wife' with clamours filled
  The dispersed air, who, holding Lucrece' life,    1805
  Answered their cries, 'my daughter' and 'my wife'.

Brutus, who plucked the knife from Lucrece' side,
Seeing such emulation in their woe
Began to clothe his wit in state and pride,
Burying in Lucrece' wound his folly's show.           1810
He with the Romans was esteemèd so
  As silly jeering idiots are with kings,
  For sportive words and utt'ring foolish things.

But now he throws that shallow habit by
Wherein deep policy did him disguise,                 1815
And armed his long-hid wits advisedly
To check the tears in Collatinus' eyes.
'Thou wrongèd lord of Rome,' quoth he, 'arise.
  Let my unsounded self, supposed a fool,
  Now set thy long-experienced wit to school.       1820

'Why, Collatine, is woe the cure for woe?
Do wounds help wounds, or grief help grievous deeds?
Is it revenge to give thyself a blow
For his foul act by whom thy fair wife bleeds?
Such childish humour from weak minds proceeds;        1825
  Thy wretched wife mistook the matter so
  To slay herself, that should have slain her foe.

'Courageous Roman, do not steep thy heart
In such relenting dew of lamentations,
But kneel with me, and help to bear thy part          1830
To rouse our Roman gods with invocations
That they will suffer these abominations—
  Since Rome herself in them doth stand disgraced—
  By our strong arms from forth her fair streets chased.

'Now by the Capitol that we adore,                    1835
And by this chaste blood so unjustly stained,
By heaven's fair sun that breeds the fat earth's store,
By all our country rights in Rome maintained,
And by chaste Lucrece' soul that late complained
  Her wrongs to us, and by this bloody knife,       1840
  We will revenge the death of this true wife.'

This said, he struck his hand upon his breast,
And kissed the fatal knife to end his vow,
And to his protestation urged the rest,
Who, wond'ring at him, did his words allow.           1845
Then jointly to the ground their knees they bow,
  And that deep vow which Brutus made before
  He doth again repeat, and that they swore.

When they had sworn to this advisèd doom
They did conclude to bear dead Lucrece thence,        1850
To show her bleeding body thorough Rome,
And so to publish Tarquin's foul offence;
Which being done with speedy diligence,
  The Romans plausibly did give consent
  To Tarquin's everlasting banishment.              1855

# THE COMEDY OF ERRORS

ON the night of 28 December 1594, the Christmas revels at Gray's Inn—one of London's law schools—became so uproarious that one performance planned for the occasion had to be abandoned. Eventually 'it was thought good not to offer anything of account saving dancing and revelling with gentlewomen; and after such sports a comedy of errors (like to Plautus his *Menaechmus*) was played by the players. So that night was begun, and continued to the end, in nothing but confusion and errors; whereupon it was ever afterwards called "The Night of Errors".'

This sounds like a reference to Shakespeare's play, first printed in the 1623 Folio, which is certainly based in large part on the Roman dramatist Plautus' comedy *Menaechmi*. As Shakespeare's shortest play, it would have been especially suited to late-night performance; there is no evidence that it was written for the occasion, but it may well have been new in 1594.

The comedy in *Menaechmi* derives from the embarrassment experienced by a man in search of his long-lost twin brother when various people intimately acquainted with that twin—including his wife, his mistress, and his father—mistake the one for the other. Shakespeare greatly increases the possibilities of comic confusion by giving the brothers (both called Antipholus) servants (both called Dromio) who themselves are long-separated twins. An added episode in which Antipholus of Ephesus' wife, Adriana, bars him from his own house in which she is entertaining his brother is based on another play by Plautus, *Amphitruo*. Shakespeare sets the comic action within a more serious framework, opening with a scene in which the twin masters' old father, Egeon, who has arrived at Ephesus in search of them, is shown under imminent sentence of death unless he finds someone to redeem him. This strand of the plot, as well as the surprising revelation that brings about the resolution of the action, is based on the story of Apollonius of Tyre which Shakespeare was to use again, many years later, in *Pericles*.

*The Comedy of Errors* is a kind of diploma piece, as if Shakespeare were displaying his ability to outshine both his classical progenitors and their English imitators. Along with *The Tempest*, it is his most classically constructed play: all the action takes place within a few hours and in a single place. Moreover, it seems to make use of the conventionalized arcade setting of academic drama, with three 'houses'—the Phoenix, the Porcupine, and the Priory—represented by doors and signs on stage. The working out of the complexities inherent in the basic situation represents a considerable intellectual feat. But the comedy is humanized by the interweaving of romantic elements, such as Egeon's initial plight, the love between the visiting Antipholus and his twin brother's wife's sister, Luciana, and the entirely serious portrayal of Egeon's suffering when his own son fails to recognize him at the moment of his greatest need. From time to time the comic tension is relaxed by the presence of discursive set pieces, none more memorable than Dromio of Syracuse's description of Nell, the kitchen wench who is 'spherical, like a globe'.

# THE PERSONS OF THE PLAY

Solinus, DUKE of Ephesus

EGEON, a merchant of Syracuse, father of the Antipholus twins

ANTIPHOLUS OF EPHESUS ⎫
ANTIPHOLUS OF SYRACUSE ⎭ twin brothers, sons of Egeon

DROMIO OF EPHESUS ⎫ twin brothers, and bondmen of the
DROMIO OF SYRACUSE ⎭ Antipholus twins

ADRIANA, wife of Antipholus of Ephesus

LUCIANA, her sister

NELL, Adriana's kitchen-maid

ANGELO, a goldsmith

BALTHASAR, a merchant

A COURTESAN

Doctor PINCH, a schoolmaster and exorcist

MERCHANT OF EPHESUS, a friend of Antipholus of Syracuse

SECOND MERCHANT, Angelo's creditor

EMILIA, an abbess at Ephesus

Jailer, messenger, headsman, officers, and other attendants

# The Comedy of Errors

*Enter Solinus, the Duke of Ephesus, with Egeon the*
*Merchant of Syracuse, Jailer, and other attendants*

EGEON
Proceed, Solinus, to procure my fall,
And by the doom of death end woes and all.

DUKE
Merchant of Syracusa, plead no more.
I am not partial to infringe our laws.
The enmity and discord which of late                    5
Sprung from the rancorous outrage of your Duke
To merchants, our well-dealing countrymen,
Who, wanting guilders to redeem their lives,
Have sealed his rigorous statutes with their bloods,
Excludes all pity from our threat'ning looks.          10
For since the mortal and intestine jars
'Twixt thy seditious countrymen and us,
It hath in solemn synods been decreed,
Both by the Syracusians and ourselves,
To admit no traffic to our adverse towns.              15
Nay more: if any born at Ephesus
Be seen at Syracusian marts and fairs;
Again, if any Syracusian born
Come to the bay of Ephesus—he dies,
His goods confiscate to the Duke's dispose,            20
Unless a thousand marks be levièd
To quit the penalty and ransom him.
Thy substance, valued at the highest rate,
Cannot amount unto a hundred marks.
Therefore by law thou art condemned to die.            25

EGEON
Yet this my comfort: when your words are done,
My woes end likewise with the evening sun.

DUKE
Well, Syracusian, say in brief the cause
Why thou departed'st from thy native home,
And for what cause thou cam'st to Ephesus.             30

EGEON
A heavier task could not have been imposed
Than I to speak my griefs unspeakable.
Yet, that the world may witness that my end
Was wrought by nature, not by vile offence,
I'll utter what my sorrow gives me leave.              35
In Syracusa was I born, and wed
Unto a woman happy but for me,
And by me happy, had not our hap been bad.
With her I lived in joy, our wealth increased
By prosperous voyages I often made                     40
To Epidamnum, till my factor's death,
And the great care of goods at random left,
Drew me from kind embracements of my spouse,
From whom my absence was not six months old
Before herself—almost at fainting under               45
The pleasing punishment that women bear—
Had made provision for her following me,
And soon and safe arrivèd where I was.
There had she not been long but she became
A joyful mother of two goodly sons;                    50
And, which was strange, the one so like the other
As could not be distinguished but by names.
That very hour, and in the selfsame inn,

A mean-born woman was deliverèd
Of such a burden male, twins both alike.               55
Those, for their parents were exceeding poor,
I bought, and brought up to attend my sons.
My wife, not meanly proud of two such boys,
Made daily motions for our home return.
Unwilling, I agreed. Alas! Too soon                    60
We came aboard.
A league from Epidamnum had we sailed
Before the always-wind-obeying deep
Gave any tragic instance of our harm.
But longer did we not retain much hope,                65
For what obscurèd light the heavens did grant
Did but convey unto our fearful minds
A doubtful warrant of immediate death,
Which though myself would gladly have embraced,
Yet the incessant weepings of my wife—                70
Weeping before for what she saw must come—
And piteous plainings of the pretty babes,
That mourned for fashion, ignorant what to fear,
Forced me to seek delays for them and me.
And this it was—for other means was none:              75
The sailors sought for safety by our boat,
And left the ship, then sinking-ripe, to us.
My wife, more careful for the latter-born,
Had fastened him unto a small spare mast
Such as seafaring men provide for storms.              80
To him one of the other twins was bound,
Whilst I had been like heedful of the other.
The children thus disposed, my wife and I,
Fixing our eyes on whom our care was fixed,
Fastened ourselves at either end the mast,             85
And floating straight, obedient to the stream,
Was carried towards Corinth, as we thought.
At length the sun, gazing upon the earth,
Dispersed those vapours that offended us,
And by the benefit of his wishèd light                 90
The seas waxed calm, and we discoverèd
Two ships from far, making amain to us:
Of Corinth that, of Epidaurus this.
But ere they came—O let me say no more!
Gather the sequel by that went before.                 95

DUKE
Nay, forward, old man; do not break off so,
For we may pity though not pardon thee.

EGEON
O, had the gods done so, I had not now
Worthily termed them merciless to us.
For, ere the ships could meet by twice five leagues,
We were encountered by a mighty rock,                 101
Which being violently borne upon,
Our helpful ship was splitted in the midst,
So that in this unjust divorce of us
Fortune had left to both of us alike                  105
What to delight in, what to sorrow for.
Her part, poor soul, seeming as burdenèd
With lesser weight but not with lesser woe,
Was carried with more speed before the wind,
And in our sight they three were taken up             110
By fishermen of Corinth, as we thought.

At length another ship had seized on us,
And, knowing whom it was their hap to save,
Gave healthful welcome to their shipwrecked guests,
And would have reft the fishers of their prey          115
Had not their barque been very slow of sail;
And therefore homeward did they bend their course.
Thus have you heard me severed from my bliss,
That by misfortunes was my life prolonged
To tell sad stories of my own mishaps.          120

DUKE
And for the sake of them thou sorrow'st for,
Do me the favour to dilate at full
What have befall'n of them and thee till now.

EGEON
My youngest boy, and yet my eldest care,
At eighteen years became inquisitive          125
After his brother, and importuned me
That his attendant—so his case was like,
Reft of his brother, but retained his name—
Might bear him company in the quest of him;
Whom whilst I laboured of a love to see,          130
I hazarded the loss of whom I loved.
Five summers have I spent in farthest Greece,
Roaming clean through the bounds of Asia,
And coasting homeward came to Ephesus,
Hopeless to find, yet loath to leave unsought          135
Or that or any place that harbours men.
But here must end the story of my life,
And happy were I in my timely death
Could all my travels warrant me they live.

DUKE
Hapless Egeon, whom the fates have marked          140
To bear the extremity of dire mishap,
Now trust me, were it not against our laws—
Which princes, would they, may not disannul—
Against my crown, my oath, my dignity,
My soul should sue as advocate for thee.          145
But though thou art adjudgèd to the death,
And passèd sentence may not be recalled
But to our honour's great disparagement,
Yet will I favour thee in what I can.
Therefore, merchant, I'll limit thee this day          150
To seek thy health by beneficial help.
Try all the friends thou hast in Ephesus:
Beg thou or borrow to make up the sum,
And live. If no, then thou art doomed to die.
Jailer, take him to thy custody.          155

JAILER I will, my lord.

EGEON
Hopeless and helpless doth Egeon wend,
But to procrastinate his lifeless end.          *Exeunt*

**1.2**     *Enter ⌐from the bay⌐ Antipholus of Syracuse,*
        *Merchant ⌐of Ephesus⌐, and Dromio of Syracuse*

MERCHANT ⌐OF EPHESUS⌐
Therefore give out you are of Epidamnum,
Lest that your goods too soon be confiscate.
This very day a Syracusian merchant
Is apprehended for arrival here,
And, not being able to buy out his life,          5
According to the statute of the town
Dies ere the weary sun set in the west.
There is your money that I had to keep.

ANTIPHOLUS OF SYRACUSE (*to Dromio*)
Go bear it to the Centaur, where we host,
And stay there, Dromio, till I come to thee.          10

Within this hour it will be dinner-time.
Till that I'll view the manners of the town,
Peruse the traders, gaze upon the buildings,
And then return and sleep within mine inn;
For with long travel I am stiff and weary.          15
Get thee away.

DROMIO OF SYRACUSE
Many a man would take you at your word,
And go indeed, having so good a mean.          *Exit*

ANTIPHOLUS OF SYRACUSE
A trusty villain, sir, that very oft,
When I am dull with care and melancholy,          20
Lightens my humour with his merry jests.
What, will you walk with me about the town,
And then go to my inn and dine with me?

MERCHANT ⌐OF EPHESUS⌐
I am invited, sir, to certain merchants
Of whom I hope to make much benefit.          25
I crave your pardon. Soon at five o'clock,
Please you, I'll meet with you upon the mart,
And afterward consort you till bedtime.
My present business calls me from you now.

ANTIPHOLUS OF SYRACUSE
Farewell till then. I will go lose myself,          30
And wander up and down to view the city.

MERCHANT ⌐OF EPHESUS⌐
Sir, I commend you to your own content.          *Exit*

ANTIPHOLUS OF SYRACUSE
He that commends me to mine own content
Commends me to the thing I cannot get.
I to the world am like a drop of water          35
That in the ocean seeks another drop,
Who, falling there to find his fellow forth,
Unseen, inquisitive, confounds himself.
So I, to find a mother and a brother,
In quest of them, unhappy, lose myself.          40

            *Enter Dromio of Ephesus*
Here comes the almanac of my true date.
What now? How chance thou art returned so soon?

DROMIO OF EPHESUS
Returned so soon? Rather approached too late.
The capon burns, the pig falls from the spit.
The clock hath strucken twelve upon the bell;          45
My mistress made it one upon my cheek.
She is so hot because the meat is cold.
The meat is cold because you come not home.
You come not home because you have no stomach.
You have no stomach, having broke your fast;          50
But we that know what 'tis to fast and pray
Are penitent for your default today.

ANTIPHOLUS OF SYRACUSE
Stop in your wind, sir. Tell me this, I pray:
Where have you left the money that I gave you?

DROMIO OF EPHESUS
O—sixpence that I had o' Wednesday last          55
To pay the saddler for my mistress' crupper?
The saddler had it, sir; I kept it not.

ANTIPHOLUS OF SYRACUSE
I am not in a sportive humour now.
Tell me, and dally not: where is the money?
We being strangers here, how dar'st thou trust          60
So great a charge from thine own custody?

DROMIO OF EPHESUS
I pray you, jest, sir, as you sit at dinner.
I from my mistress come to you in post.
If I return I shall be post indeed,

For she will scour your fault upon my pate. 65
Methinks your maw, like mine, should be your clock,
And strike you home without a messenger.

ANTIPHOLUS OF SYRACUSE
Come, Dromio, come, these jests are out of season.
Reserve them till a merrier hour than this.
Where is the gold I gave in charge to thee? 70

DROMIO OF EPHESUS
To me, sir? Why, you gave no gold to me.

ANTIPHOLUS OF SYRACUSE
Come on, sir knave, have done your foolishness,
And tell me how thou hast disposed thy charge.

DROMIO OF EPHESUS
My charge was but to fetch you from the mart
Home to your house, the Phoenix, sir, to dinner. 75
My mistress and her sister stays for you.

ANTIPHOLUS OF SYRACUSE
Now, as I am a Christian, answer me
In what safe place you have bestowed my money,
Or I shall break that merry sconce of yours
That stands on tricks when I am undisposed. 80
Where is the thousand marks thou hadst of me?

DROMIO OF EPHESUS
I have some marks of yours upon my pate,
Some of my mistress' marks upon my shoulders,
But not a thousand marks between you both.
If I should pay your worship those again, 85
Perchance you will not bear them patiently.

ANTIPHOLUS OF SYRACUSE
Thy mistress' marks? What mistress, slave, hast thou?

DROMIO OF EPHESUS
Your worship's wife, my mistress, at the Phoenix:
She that doth fast till you come home to dinner,
And prays that you will hie you home to dinner. 90

ANTIPHOLUS OF SYRACUSE
What, wilt thou flout me thus unto my face,
Being forbid? There, take you that, sir knave!
    *He beats Dromio*

DROMIO OF EPHESUS
What mean you, sir? For God's sake, hold your hands!
Nay, an you will not, sir, I'll take my heels.    *Exit*

ANTIPHOLUS OF SYRACUSE
Upon my life, by some device or other 95
The villain is o'er-raught of all my money.
They say this town is full of cozenage,
As nimble jugglers that deceive the eye,
Dark-working sorcerers that change the mind,
Soul-killing witches that deform the body, 100
Disguisèd cheaters, prating mountebanks,
And many suchlike libertines of sin.
If it prove so, I will be gone the sooner.
I'll to the Centaur to go seek this slave.
I greatly fear my money is not safe.    *Exit*

❧

**2.1** *Enter ⌈from the Phoenix⌉ Adriana, wife of*
    *Antipholus of Ephesus, with Luciana, her sister*

ADRIANA
Neither my husband nor the slave returned
That in such haste I sent to seek his master?
Sure, Luciana, it is two o'clock.

LUCIANA
Perhaps some merchant hath invited him,
And from the mart he's somewhere gone to dinner. 5
Good sister, let us dine, and never fret.
A man is master of his liberty.

Time is their mistress, and when they see time
They'll go or come. If so, be patient, sister.

ADRIANA
Why should their liberty than ours be more? 10

LUCIANA
Because their business still lies out o' door.

ADRIANA
Look when I serve him so, he takes it ill.

LUCIANA
O, know he is the bridle of your will.

ADRIANA
There's none but asses will be bridled so.

LUCIANA
Why, headstrong liberty is lashed with woe. 15
There's nothing situate under heaven's eye
But hath his bound in earth, in sea, in sky.
The beasts, the fishes, and the wingèd fowls
Are their males' subjects and at their controls.
Man, more divine, the master of all these, 20
Lord of the wide world and wild wat'ry seas,
Indued with intellectual sense and souls,
Of more pre-eminence than fish and fowls,
Are masters to their females, and their lords.
Then let your will attend on their accords. 25

ADRIANA
This servitude makes you to keep unwed.

LUCIANA
Not this, but troubles of the marriage bed.

ADRIANA
But were you wedded, you would bear some sway.

LUCIANA
Ere I learn love, I'll practise to obey.

ADRIANA
How if your husband start some otherwhere? 30

LUCIANA
Till he come home again, I would forbear.

ADRIANA
Patience unmoved! No marvel though she pause:
They can be meek that have no other cause.
A wretched soul, bruised with adversity,
We bid be quiet when we hear it cry. 35
But were we burdened with like weight of pain,
As much or more we should ourselves complain.
So thou, that hast no unkind mate to grieve thee,
With urging helpless patience would relieve me.
But if thou live to see like right bereft, 40
This fool-begged patience in thee will be left.

LUCIANA
Well, I will marry one day, but to try.
    *Enter Dromio of Ephesus*
Here comes your man. Now is your husband nigh.

ADRIANA
Say, is your tardy master now at hand?

DROMIO OF EPHESUS Nay, he's at two hands with me, and
    that my two ears can witness. 46

ADRIANA
Say, didst thou speak with him? Know'st thou his
    mind?

DROMIO OF EPHESUS
I? Ay, he told his mind upon mine ear.
Beshrew his hand, I scarce could understand it.

LUCIANA
Spake he so doubtfully thou couldst not feel his
    meaning? 50

DROMIO OF EPHESUS Nay, he struck so plainly I could too
    well feel his blows, and withal so doubtfully that I
    could scarce under-stand them.

ADRIANA
But say, I prithee, is he coming home?
It seems he hath great care to please his wife.          55
DROMIO OF EPHESUS
Why, mistress, sure my master is horn-mad.
ADRIANA Horn-mad, thou villain?
DROMIO OF EPHESUS
I mean not cuckold-mad, but sure he is stark mad.
When I desired him to come home to dinner,
He asked me for a thousand marks in gold.          60
''Tis dinner-time,' quoth I. 'My gold,' quoth he.
'Your meat doth burn,' quoth I. 'My gold,' quoth he.
'Will you come home?' quoth I. 'My gold,' quoth he;
'Where is the thousand marks I gave thee, villain?'
'The pig', quoth I, 'is burned.' 'My gold!' quoth he.          65
'My mistress, sir—' quoth I. 'Hang up thy mistress!
I know thy mistress not. Out on thy mistress!'
LUCIANA Quoth who?
DROMIO OF EPHESUS Quoth my master.
'I know', quoth he, 'no house, no wife, no mistress.'
So that my errand, due unto my tongue,          71
I thank him, I bare home upon my shoulders;
For, in conclusion, he did beat me there.
ADRIANA
Go back again, thou slave, and fetch him home.
DROMIO OF EPHESUS
Go back again and be new beaten home?          75
For God's sake, send some other messenger.
ADRIANA
Back, slave, or I will break thy pate across.
DROMIO OF EPHESUS
An he will bless that cross with other beating,
Between you I shall have a holy head.
ADRIANA
Hence, prating peasant. Fetch thy master home.          80
          She beats Dromio
DROMIO OF EPHESUS
Am I so round with you as you with me,
That like a football you do spurn me thus?
You spurn me hence, and he will spurn me hither.
If I last in this service, you must case me in leather.
          Exit
LUCIANA (to Adriana)
Fie, how impatience loureth in your face!          85
ADRIANA
His company must do his minions grace,
Whilst I at home starve for a merry look.
Hath homely age th'alluring beauty took
From my poor cheek? Then he hath wasted it.
Are my discourses dull? Barren my wit?          90
If voluble and sharp discourse be marred,
Unkindness blunts it more than marble hard.
Do their gay vestments his affections bait?
That's not my fault: he's master of my state.
What ruins are in me that can be found          95
By him not ruined? Then is he the ground
Of my defeatures. My decayèd fair
A sunny look of his would soon repair.
But, too unruly deer, he breaks the pale,
And feeds from home. Poor I am but his stale.          100
LUCIANA
Self-harming jealousy! Fie, beat it hence.
ADRIANA
Unfeeling fools can with such wrongs dispense.
I know his eye doth homage otherwhere,
Or else what lets it but he would be here?

Sister, you know he promised me a chain.          105
Would that alone o' love he would detain,
So he would keep fair quarter with his bed.
I see the jewel best enamellèd
Will lose her beauty. Yet the gold bides still
That others touch; and often touching will          110
Wear gold, and yet no man that hath a name
By falsehood and corruption doth it shame.
Since that my beauty cannot please his eye,
I'll weep what's left away, and weeping die.
LUCIANA
How many fond fools serve mad jealousy!          115
          ⌈Exeunt into the Phoenix⌉

2.2          Enter Antipholus of Syracuse
ANTIPHOLUS OF SYRACUSE
The gold I gave to Dromio is laid up
Safe at the Centaur, and the heedful slave
Is wandered forth in care to seek me out.
By computation and mine host's report,
I could not speak with Dromio since at first          5
I sent him from the mart! See, here he comes.
          Enter Dromio of Syracuse
How now, sir, is your merry humour altered?
As you love strokes, so jest with me again.
You know no Centaur? You received no gold?
Your mistress sent to have me home to dinner?          10
My house was at the Phoenix?—Wast thou mad,
That thus so madly thou didst answer me?
DROMIO OF SYRACUSE
What answer, sir? When spake I such a word?
ANTIPHOLUS OF SYRACUSE
Even now, even here, not half an hour since.
DROMIO OF SYRACUSE
I did not see you since you sent me hence          15
Home to the Centaur with the gold you gave me.
ANTIPHOLUS OF SYRACUSE
Villain, thou didst deny the gold's receipt,
And told'st me of a mistress and a dinner,
For which I hope thou felt'st I was displeased.
DROMIO OF SYRACUSE
I am glad to see you in this merry vein.          20
What means this jest? I pray you, master, tell me.
ANTIPHOLUS OF SYRACUSE
Yea, dost thou jeer and flout me in the teeth?
Think'st thou I jest? Hold, take thou that, and that.
          He beats Dromio
DROMIO OF SYRACUSE
Hold, sir, for God's sake—now your jest is earnest!
Upon what bargain do you give it me?          25
ANTIPHOLUS OF SYRACUSE
Because that I familiarly sometimes
Do use you for my fool, and chat with you,
Your sauciness will jest upon my love,
And make a common of my serious hours.
When the sun shines, let foolish gnats make sport,          30
But creep in crannies when he hides his beams.
If you will jest with me, know my aspect,
And fashion your demeanour to my looks,
Or I will beat this method in your sconce.
DROMIO OF SYRACUSE 'Sconce' call you it? So you would
leave battering, I had rather have it a head. An you
use these blows long, I must get a sconce for my head,
and ensconce it too, or else I shall seek my wit in my
shoulders. But I pray, sir, why am I beaten?
ANTIPHOLUS OF SYRACUSE Dost thou not know?          40

DROMIO OF SYRACUSE Nothing, sir, but that I am beaten.

ANTIPHOLUS OF SYRACUSE Shall I tell you why?

DROMIO OF SYRACUSE Ay, sir, and wherefore; for they say every why hath a wherefore.

ANTIPHOLUS OF SYRACUSE
'Why' first: for flouting me; and then 'wherefore': 45
For urging it the second time to me.

DROMIO OF SYRACUSE
Was there ever any man thus beaten out of season,
When in the why and the wherefore is neither rhyme
    nor reason?—
Well, sir, I thank you.

ANTIPHOLUS OF SYRACUSE Thank me, sir, for what?

DROMIO OF SYRACUSE Marry, sir, for this something that you gave me for nothing. 51

ANTIPHOLUS OF SYRACUSE I'll make you amends next, to give you nothing for something. But say, sir, is it dinner-time?

DROMIO OF SYRACUSE No, sir, I think the meat wants that I have. 56

ANTIPHOLUS OF SYRACUSE In good time, sir. What's that?

DROMIO OF SYRACUSE Basting.

ANTIPHOLUS OF SYRACUSE Well, sir, then 'twill be dry.

DROMIO OF SYRACUSE If it be, sir, I pray you eat none of it.

ANTIPHOLUS OF SYRACUSE Your reason? 61

DROMIO OF SYRACUSE Lest it make you choleric and purchase me another dry basting.

ANTIPHOLUS OF SYRACUSE Well, sir, learn to jest in good time. There's a time for all things. 65

DROMIO OF SYRACUSE I durst have denied that before you were so choleric.

ANTIPHOLUS OF SYRACUSE By what rule, sir?

DROMIO OF SYRACUSE Marry, sir, by a rule as plain as the plain bald pate of Father Time himself. 70

ANTIPHOLUS OF SYRACUSE Let's hear it.

DROMIO OF SYRACUSE There's no time for a man to recover his hair that grows bald by nature.

ANTIPHOLUS OF SYRACUSE May he not do it by fine and recovery? 75

DROMIO OF SYRACUSE Yes, to pay a fine for a periwig, and recover the lost hair of another man.

ANTIPHOLUS OF SYRACUSE Why is Time such a niggard of hair, being, as it is, so plentiful an excrement?

DROMIO OF SYRACUSE Because it is a blessing that he bestows on beasts, and what he hath scanted men in hair he hath given them in wit. 82

ANTIPHOLUS OF SYRACUSE Why, but there's many a man hath more hair than wit.

DROMIO OF SYRACUSE Not a man of those but he hath the wit to lose his hair. 86

ANTIPHOLUS OF SYRACUSE Why, thou didst conclude hairy men plain dealers, without wit.

DROMIO OF SYRACUSE The plainer dealer, the sooner lost. Yet he loseth it in a kind of jollity. 90

ANTIPHOLUS OF SYRACUSE For what reason?

DROMIO OF SYRACUSE For two, and sound ones too.

ANTIPHOLUS OF SYRACUSE Nay, not sound, I pray you.

DROMIO OF SYRACUSE Sure ones, then.

ANTIPHOLUS OF SYRACUSE Nay, not sure, in a thing falsing.

DROMIO OF SYRACUSE Certain ones, then. 96

ANTIPHOLUS OF SYRACUSE Name them.

DROMIO OF SYRACUSE The one, to save the money that he spends in tiring; the other, that at dinner they should not drop in his porridge. 100

ANTIPHOLUS OF SYRACUSE You would all this time have proved there is no time for all things.

DROMIO OF SYRACUSE Marry, and did, sir: namely, e'en no time to recover hair lost by nature.

ANTIPHOLUS OF SYRACUSE But your reason was not substantial, why there is no time to recover. 106

DROMIO OF SYRACUSE Thus I mend it: Time himself is bald, and therefore to the world's end will have bald followers.

ANTIPHOLUS OF SYRACUSE I knew 'twould be a bald conclusion. 111

     *Enter [from the Phoenix] Adriana and Luciana*
But soft—who wafts us yonder?

ADRIANA
Ay, ay, Antipholus, look strange and frown:
Some other mistress hath thy sweet aspects.
I am not Adriana, nor thy wife. 115
The time was once when thou unurged wouldst vow
That never words were music to thine ear,
That never object pleasing in thine eye,
That never touch well welcome to thy hand,
That never meat sweet-savoured in thy taste, 120
Unless I spake, or looked, or touched, or carved to
    thee.
How comes it now, my husband, O how comes it
That thou art then estrangèd from thyself?—
Thy 'self' I call it, being strange to me
That, undividable, incorporate, 125
Am better than thy dear self's better part.
Ah, do not tear away thyself from me;
For know, my love, as easy mayst thou fall
A drop of water in the breaking gulf,
And take unmingled thence that drop again 130
Without addition or diminishing,
As take from me thyself, and not me too.
How dearly would it touch thee to the quick
Shouldst thou but hear I were licentious,
And that this body, consecrate to thee, 135
By ruffian lust should be contaminate?
Wouldst thou not spit at me, and spurn at me,
And hurl the name of husband in my face,
And tear the stained skin off my harlot brow,
And from my false hand cut the wedding ring, 140
And break it with a deep-divorcing vow?
I know thou canst, and therefore see thou do it!
I am possessed with an adulterate blot;
My blood is mingled with the crime of lust.
For if we two be one, and thou play false, 145
I do digest the poison of thy flesh,
Being strumpeted by thy contagion.
Keep then fair league and truce with thy true bed,
I live unstained, thou undishonourèd.

ANTIPHOLUS OF SYRACUSE
Plead you to *me*, fair dame? I know you not. 150
In Ephesus I am but two hours old,
As strange unto your town as to your talk,
Who, every word by all my wit being scanned,
Wants wit in all one word to understand.

LUCIANA
Fie, brother, how the world is changed with you! 155
When were you wont to use my sister thus?
She sent for you by Dromio home to dinner.

ANTIPHOLUS OF SYRACUSE By Dromio?

DROMIO OF SYRACUSE By me?

ADRIANA
By thee; and this thou didst return from him— 160
That he did buffet thee, and in his blows
Denied my house for his, me for his wife.

ANTIPHOLUS OF SYRACUSE
  Did you converse, sir, with this gentlewoman?
  What is the course and drift of your compact?
DROMIO OF SYRACUSE
  I, sir? I never saw her till this time.                    165
ANTIPHOLUS OF SYRACUSE
  Villain, thou liest; for even her very words
  Didst thou deliver to me on the mart.
DROMIO OF SYRACUSE
  I never spake with her in all my life.
ANTIPHOLUS OF SYRACUSE
  How can she thus then call us by our names?—
  Unless it be by inspiration.                               170
ADRIANA
  How ill agrees it with your gravity
  To counterfeit thus grossly with your slave,
  Abetting him to thwart me in my mood!
  Be it my wrong you are from me exempt,
  But wrong not that wrong with a more contempt.  175
  Come, I will fasten on this sleeve of thine.
  Thou art an elm, my husband; I a vine,
  Whose weakness, married to thy stronger state,
  Makes me with thy strength to communicate.
  If aught possess thee from me, it is dross,             180
  Usurping ivy, brier, or idle moss,
  Who, all for want of pruning, with intrusion
  Infect thy sap, and live on thy confusion.
ANTIPHOLUS OF SYRACUSE (aside)
  To me she speaks, she moves me for her theme.
  What, was I married to her in my dream?               185
  Or sleep I now, and think I hear all this?
  What error drives our eyes and ears amiss?
  Until I know this sure uncertainty,
  I'll entertain the offered fallacy.
LUCIANA
  Dromio, go bid the servants spread for dinner.        190
DROMIO OF SYRACUSE (aside)
  O, for my beads! I cross me for a sinner.
  This is the fairy land. O spite of spites,
  We talk with goblins, oafs, and sprites.
  If we obey them not, this will ensue:
  They'll suck our breath or pinch us black and blue.
LUCIANA
  Why prat'st thou to thyself, and answer'st not?      196
  Dromio, thou drone, thou snail, thou slug, thou sot.
DROMIO OF SYRACUSE (to Antipholus)
  I am transformèd, master, am not I?
ANTIPHOLUS OF SYRACUSE
  I think thou art in mind, and so am I.
DROMIO OF SYRACUSE
  Nay, master, both in mind and in my shape.          200
ANTIPHOLUS OF SYRACUSE
  Thou hast thine own form.
DROMIO OF SYRACUSE          No, I am an ape.
LUCIANA
  If thou art changed to aught, 'tis to an ass.
DROMIO OF SYRACUSE ⌈to Antipholus⌉
  'Tis true she rides me, and I long for grass.
  'Tis so, I am an ass; else it could never be
  But I should know her as well as she knows me.     205
ADRIANA
  Come, come, no longer will I be a fool,
  To put the finger in the eye and weep
  Whilst man and master laughs my woes to scorn.
  (To Antipholus) Come, sir, to dinner.—Dromio, keep
      the gate.—

  Husband, I'll dine above with you today,               210
  And shrive you of a thousand idle pranks.—
  Sirrah, if any ask you for your master,
  Say he dines forth, and let no creature enter.—
  Come, sister.—Dromio, play the porter well.
ANTIPHOLUS OF SYRACUSE (aside)
  Am I in earth, in heaven, or in hell?                   215
  Sleeping or waking? Mad or well advised?
  Known unto these, and to myself disguised!
  I'll say as they say, and persever so,
  And in this mist at all adventures go.
DROMIO OF SYRACUSE
  Master, shall I be porter at the gate?                  220
ADRIANA
  Ay, and let none enter, lest I break your pate.
LUCIANA
  Come, come, Antipholus, we dine too late.
                          Exeunt ⌈into the Phoenix⌉

                              ❀

3.1     Enter Antipholus of Ephesus, his man Dromio,
        Angelo the goldsmith, and Balthasar the merchant
ANTIPHOLUS OF EPHESUS
  Good Signor Angelo, you must excuse us all.
  My wife is shrewish when I keep not hours.
  Say that I lingered with you at your shop
  To see the making of her carcanet,
  And that tomorrow you will bring it home.—        5
  But here's a villain that would face me down
  He met me on the mart, and that I beat him,
  And charged him with a thousand marks in gold,
  And that I did deny my wife and house.
  Thou drunkard, thou, what didst thou mean by this?
DROMIO OF EPHESUS
  Say what you will, sir, but I know what I know—   11
  That you beat me at the mart I have your hand to
      show.
  If the skin were parchment, and the blows you gave
      were ink,
  Your own handwriting would tell you what I think.
ANTIPHOLUS OF EPHESUS
  I think thou art an ass.
DROMIO OF EPHESUS          Marry, so it doth appear   15
  By the wrongs I suffer and the blows I bear.
  I should kick being kicked, and, being at that pass,
  You would keep from my heels, and beware of an ass.
ANTIPHOLUS OF EPHESUS
  You're sad, Signor Balthasar. Pray God our cheer
  May answer my good will, and your good welcome
      here.                                               20
BALTHASAR
  I hold your dainties cheap, sir, and your welcome dear.
ANTIPHOLUS OF EPHESUS
  O, Signor Balthasar, either at flesh or fish
  A table full of welcome makes scarce one dainty dish.
BALTHASAR
  Good meat, sir, is common; that every churl affords.
ANTIPHOLUS OF EPHESUS
  And welcome more common, for that's nothing but
      words.                                             25
BALTHASAR
  Small cheer and great welcome makes a merry feast.
ANTIPHOLUS OF EPHESUS
  Ay, to a niggardly host and more sparing guest.
  But though my cates be mean, take them in good part.

Better cheer may you have, but not with better heart.
But soft, my door is locked. (*To Dromio*) Go bid them
   let us in.   30
DROMIO OF EPHESUS (*calling*)
  Maud, Bridget, Marian, Cicely, Gillian, Ginn!
  ⌐*Enter Dromio of Syracuse within the Phoenix*⌐
DROMIO OF SYRACUSE (*within the Phoenix*)
  Mome, malt-horse, capon, coxcomb, idiot, patch!
  Either get thee from the door or sit down at the hatch.
  Dost thou conjure for wenches, that thou call'st for
   such store
  When one is one too many? Go, get thee from the
   door.
DROMIO OF EPHESUS
  What patch is made our porter? My master stays in
   the street.   36
DROMIO OF SYRACUSE (*within*)
  Let him walk from whence he came, lest he catch
   cold on's feet.
ANTIPHOLUS OF EPHESUS
  Who talks within there? Ho, open the door!
DROMIO OF SYRACUSE (*within the Phoenix*)
  Right, sir, I'll tell you when, an you'll tell me wherefore.
ANTIPHOLUS OF EPHESUS
  Wherefore? For my dinner—I have not dined today.
DROMIO OF SYRACUSE (*within the Phoenix*)
  Nor today here you must not. Come again when you
   may.   41
ANTIPHOLUS OF EPHESUS
  What art thou that keep'st me out from the house I
   owe?
DROMIO OF SYRACUSE (*within the Phoenix*)
  The porter for this time, sir, and my name is Dromio.
DROMIO OF EPHESUS
  O villain, thou hast stol'n both mine office and my
   name.
  The one ne'er got me credit, the other mickle blame.
  If thou hadst been Dromio today in my place,   46
  Thou wouldst have changed thy pate for an aim, or
   thy name for an ass.
  *Enter Nell within the Phoenix*
NELL (*within the Phoenix*)
  What a coil is there, Dromio? Who are those at the
   gate?
DROMIO OF EPHESUS
  Let my master in, Nell.
NELL (*within the Phoenix*) Faith no, he comes too late;
  And so tell your master.
DROMIO OF EPHESUS   O Lord, I must laugh.   50
  Have at you with a proverb: 'Shall I set in my staff?'
NELL (*within the Phoenix*)
  Have at you with another—that's 'When? Can you
   tell?'
DROMIO OF SYRACUSE (*within the Phoenix*)
  If thy name be called Nell, Nell, thou hast answered
   him well.
  ⌐               ⌐
ANTIPHOLUS OF EPHESUS (*to Nell*)
  Do you hear, you minion? You'll let us in, I hope?  55
NELL (*within the Phoenix*)
  I thought to have asked you.
DROMIO OF SYRACUSE (*within*)   And you said no.
DROMIO OF EPHESUS
  So, come help.
  ⌐*He and Antipholus beat the door*⌐
           Well struck! There was blow for blow.

ANTIPHOLUS OF EPHESUS (*to Nell*)
  Thou baggage, let me in.
NELL (*within the Phoenix*)   Can you tell for whose sake?
DROMIO OF EPHESUS
  Master, knock the door hard.
NELL (*within the Phoenix*)   Let him knock till it ache.
ANTIPHOLUS OF EPHESUS
  You'll cry for this, minion, if I beat the door down.  60
NELL (*within the Phoenix*)
  What needs all that, and a pair of stocks in the town?
  *Enter Adriana within the Phoenix*
ADRIANA (*within the Phoenix*)
  Who is that at the door that keeps all this noise?
DROMIO OF SYRACUSE (*within the Phoenix*)
  By my troth, your town is troubled with unruly boys.
ANTIPHOLUS OF EPHESUS (*to Adriana*)
  Are you there, wife? You might have come before.
ADRIANA (*within the Phoenix*)
  Your wife, sir knave? Go, get you from the door.  65
                          *Exit with Nell*
DROMIO OF EPHESUS (*to Antipholus*)
  If you went in pain, master, this knave would go sore.
ANGELO (*to Antipholus*)
  Here is neither cheer, sir, nor welcome; we would
   fain have either.
BALTHASAR
  In debating which was best, we shall part with neither.
DROMIO OF EPHESUS (*to Antipholus*)
  They stand at the door, master. Bid them welcome
   hither.
ANTIPHOLUS OF EPHESUS
  There is something in the wind, that we cannot get in.
DROMIO OF EPHESUS
  You would say so, master, if your garments were thin.
  Your cake here is warm within: you stand here in the
   cold.   72
  It would make a man mad as a buck to be so bought
   and sold.
ANTIPHOLUS OF EPHESUS
  Go fetch me something. I'll break ope the gate.
DROMIO OF SYRACUSE (*within the Phoenix*)
  Break any breaking here, and I'll break your knave's
   pate.   75
DROMIO OF EPHESUS
  A man may break a word with you, sir, and words
   are but wind;
  Ay, and break it in your face, so he break it not
   behind.
DROMIO OF SYRACUSE (*within the Phoenix*)
  It seems thou want'st breaking. Out upon thee, hind!
DROMIO OF EPHESUS
  Here's too much 'Out upon thee!' I pray thee, let me
   in.
DROMIO OF SYRACUSE (*within the Phoenix*)
  Ay, when fowls have no feathers, and fish have no fin.
ANTIPHOLUS OF EPHESUS
  Well, I'll break in.—Go borrow me a crow.  81
DROMIO OF EPHESUS
  A crow without feather? Master, mean you so?
  For a fish without a fin, there's a fowl without a feather.
  (*To Dromio of Syracuse*)
  If a crow help us in, sirrah, we'll pluck a crow together.
ANTIPHOLUS OF EPHESUS
  Go, get thee gone. Fetch me an iron crow.  85
BALTHASAR
  Have patience, sir. O, let it not be so!
  Herein you war against your reputation,

And draw within the compass of suspect
Th'unviolated honour of your wife.
Once this: your long experience of her wisdom,               90
Her sober virtue, years, and modesty,
Plead on her part some cause to you unknown;
And doubt not, sir, but she will well excuse
Why at this time the doors are made against you.
Be ruled by me. Depart in patience,                          95
And let us to the Tiger all to dinner,
And about evening come yourself alone
To know the reason of this strange restraint.
If by strong hand you offer to break in
Now in the stirring passage of the day,                     100
A vulgar comment will be made of it,
And that supposèd by the common rout
Against your yet ungallèd estimation,
That may with foul intrusion enter in
And dwell upon your grave when you are dead.                105
For slander lives upon succession,
For ever housed where once it gets possession.

ANTIPHOLUS OF EPHESUS
You have prevailed. I will depart in quiet,
And in despite of mirth mean to be merry.
I know a wench of excellent discourse,                      110
Pretty and witty; wild, and yet, too, gentle.
There will we dine. This woman that I mean,
My wife—but, I protest, without desert—
Hath oftentimes upbraided me withal.
To her will we to dinner. (To Angelo) Get you home
And fetch the chain. By this, I know, 'tis made.            116
Bring it, I pray you, to the Porcupine,
For there's the house. That chain will I bestow—
Be it for nothing but to spite my wife—
Upon mine hostess there. Good sir, make haste:              120
Since mine own doors refuse to entertain me,
I'll knock elsewhere, to see if they'll disdain me.

ANGELO
I'll meet you at that place some hour hence.

ANTIPHOLUS OF EPHESUS
Do so.                                         [Exit Angelo]
        This jest shall cost me some expense.
                    Exeunt [Dromio of Syracuse within the
                    Phoenix, and the others into the Porcupine]

3.2    Enter [from the Phoenix] Luciana with Antipholus
       of Syracuse

LUCIANA
And may it be that you have quite forgot
A husband's office? Shall, Antipholus,
Even in the spring of love thy love-springs rot?
    Shall love, in building, grow so ruinous?
If you did wed my sister for her wealth,                      5
    Then for her wealth's sake use her with more
        kindness;
Or if you like elsewhere, do it by stealth:
    Muffle your false love with some show of blindness.
Let not my sister read it in your eye.
    Be not thy tongue thy own shame's orator.               10
Look sweet, speak fair, become disloyalty;
    Apparel vice like virtue's harbinger.
Bear a fair presence, though your heart be tainted:
    Teach sin the carriage of a holy saint.
Be secret-false. What need she be acquainted?               15
    What simple thief brags of his own attaint?
'Tis double wrong to truant with your bed,
    And let her read it in thy looks at board.
Shame hath a bastard fame, well managèd;

Ill deeds is doubled with an evil word.                      20
Alas, poor women, make us but believe—
    Being compact of credit—that you love us.
Though others have the arm, show us the sleeve.
    We in your motion turn, and you may move us.
Then, gentle brother, get you in again.                      25
    Comfort my sister, cheer her, call her wife:
'Tis holy sport to be a little vain
    When the sweet breath of flattery conquers strife.

ANTIPHOLUS OF SYRACUSE
Sweet mistress—what your name is else I know not,
    Nor by what wonder you do hit of mine.                   30
Less in your knowledge and your grace you show not
    Than our earth's wonder, more than earth divine.
Teach me, dear creature, how to think and speak.
    Lay open to my earthy gross conceit,
Smothered in errors, feeble, shallow, weak,                  35
    The folded meaning of your words' deceit.
Against my soul's pure truth why labour you
    To make it wander in an unknown field?
Are you a god? Would you create me new?
    Transform me, then, and to your power I'll yield.
But if that I am I, then well I know                         41
    Your weeping sister is no wife of mine,
Nor to her bed no homage do I owe.
    Far more, far more, to you do I decline.
O, train me not, sweet mermaid, with thy note               45
    To drown me in thy sister's flood of tears.
Sing, siren, for thyself, and I will dote.
    Spread o'er the silver waves thy golden hairs,
And as a bed I'll take them, and there lie,
    And in that glorious supposition think                  50
He gains by death that hath such means to die.
    Let love, being light, be drownèd if she sink.

LUCIANA
What, are you mad, that you do reason so?

ANTIPHOLUS OF SYRACUSE
Not mad, but mated—how, I do not know.

LUCIANA
It is a fault that springeth from your eye.                  55

ANTIPHOLUS OF SYRACUSE
For gazing on your beams, fair sun, being by.

LUCIANA
Gaze where you should, and that will clear your
        sight.

ANTIPHOLUS OF SYRACUSE
As good to wink, sweet love, as look on night.

LUCIANA
Why call you me 'love'? Call my sister so.

ANTIPHOLUS OF SYRACUSE
Thy sister's sister.

LUCIANA                That's my sister.

ANTIPHOLUS OF SYRACUSE                      No,              60
It is thyself, mine own self's better part,
Mine eye's clear eye, my dear heart's dearer heart,
My food, my fortune, and my sweet hope's aim,
My sole earth's heaven, and my heaven's claim.

LUCIANA
All this my sister is, or else should be.                   65

ANTIPHOLUS OF SYRACUSE
Call thyself sister, sweet, for I am thee.
Thee will I love, and with thee lead my life.
Thou hast no husband yet, nor I no wife.
Give me thy hand.

LUCIANA                O soft, sir, hold you still;
I'll fetch my sister to get her good will.                  70
                              Exit [into the Phoenix]

*Enter ⌜from the Phoenix⌝ Dromio of Syracuse*

ANTIPHOLUS OF SYRACUSE Why, how now, Dromio! Where runn'st thou so fast?

DROMIO OF SYRACUSE Do you know me, sir? Am I Dromio? Am I your man? Am I myself?

ANTIPHOLUS OF SYRACUSE Thou art Dromio, thou art my man, thou art thyself.                                    76

DROMIO OF SYRACUSE I am an ass, I am a woman's man, and besides myself.

ANTIPHOLUS OF SYRACUSE What woman's man? And how besides thyself?                                              80

DROMIO OF SYRACUSE Marry, sir, besides myself I am due to a woman: one that claims me, one that haunts me, one that will have me.

ANTIPHOLUS OF SYRACUSE What claim lays she to thee?

DROMIO OF SYRACUSE Marry, sir, such claim as you would lay to your horse; and she would have me as a beast— not that, I being a beast, she would have me, but that she, being a very beastly creature, lays claim to me.

ANTIPHOLUS OF SYRACUSE What is she?                       89

DROMIO OF SYRACUSE A very reverend body; ay, such a one as a man may not speak of without he say 'sir-reverence'. I have but lean luck in the match, and yet is she a wondrous fat marriage.

ANTIPHOLUS OF SYRACUSE How dost thou mean, a fat marriage?                                                    95

DROMIO OF SYRACUSE Marry, sir, she's the kitchen wench, and all grease; and I know not what use to put her to but to make a lamp of her, and run from her by her own light. I warrant her rags and the tallow in them will burn a Poland winter. If she lives till doomsday, she'll burn a week longer than the whole world.    101

ANTIPHOLUS OF SYRACUSE What complexion is she of?

DROMIO OF SYRACUSE Swart like my shoe, but her face nothing like so clean kept. For why?—She sweats a man may go overshoes in the grime of it.            105

ANTIPHOLUS OF SYRACUSE That's a fault that water will mend.

DROMIO OF SYRACUSE No, sir, 'tis in grain. Noah's flood could not do it.

ANTIPHOLUS OF SYRACUSE What's her name?                  110

DROMIO OF SYRACUSE Nell, sir. But her name and three-quarters—that's an ell and three-quarters—will not measure her from hip to hip.

ANTIPHOLUS OF SYRACUSE Then she bears some breadth?

DROMIO OF SYRACUSE No longer from head to foot than from hip to hip. She is spherical, like a globe. I could find out countries in her.                           117

ANTIPHOLUS OF SYRACUSE In what part of her body stands Ireland?

DROMIO OF SYRACUSE Marry, sir, in her buttocks. I found it out by the bogs.                                    121

ANTIPHOLUS OF SYRACUSE Where Scotland?

DROMIO OF SYRACUSE I found it by the barrenness, hard in the palm of her hand.

ANTIPHOLUS OF SYRACUSE Where France?                     125

DROMIO OF SYRACUSE In her forehead, armed and reverted, making war against her heir.

ANTIPHOLUS OF SYRACUSE Where England?

DROMIO OF SYRACUSE I looked for the chalky cliffs, but I could find no whiteness in them. But I guess it stood in her chin, by the salt rheum that ran between France and it.

ANTIPHOLUS OF SYRACUSE Where Spain?

DROMIO OF SYRACUSE Faith, I saw it not, but I felt it hot in her breath.                                        135

ANTIPHOLUS OF SYRACUSE Where America, the Indies?

DROMIO OF SYRACUSE O, sir, upon her nose, all o'er embellished with rubies, carbuncles, sapphires, declining their rich aspect to the hot breath of Spain, who sent whole armadas of carracks to be ballast at her nose.                                            141

ANTIPHOLUS OF SYRACUSE Where stood Belgia, the Netherlands?

DROMIO OF SYRACUSE O, sir, I did not look so low. To conclude, this drudge or diviner laid claim to me, called me Dromio, swore I was assured to her, told me what privy marks I had about me—as the mark of my shoulder, the mole in my neck, the great wart on my left arm—that I, amazed, ran from her as a witch. And I think if my breast had not been made of faith, and my heart of steel, she had transformed me to a curtal dog, and made me turn i'th' wheel.             152

ANTIPHOLUS OF SYRACUSE
Go, hie thee presently. Post to the road.
An if the wind blow any way from shore,
I will not harbour in this town tonight.            155
If any barque put forth, come to the mart,
Where I will walk till thou return to me.
If everyone knows us, and we know none,
'Tis time, I think, to trudge, pack, and be gone.

DROMIO OF SYRACUSE
As from a bear a man would run for life,             160
So fly I from her that would be my wife.
                                    *Exit ⌜to the bay⌝*

ANTIPHOLUS OF SYRACUSE
There's none but witches do inhabit here,
And therefore 'tis high time that I were hence.
She that doth call me husband, even my soul
Doth for a wife abhor. But her fair sister,          165
Possessed with such a gentle sovereign grace,
Of such enchanting presence and discourse,
Hath almost made me traitor to myself.
But lest myself be guilty to self-wrong,
I'll stop mine ears against the mermaid's song.      170
                    *Enter Angelo with the chain*

ANGELO
Master Antipholus.

ANTIPHOLUS OF SYRACUSE Ay, that's my name.

ANGELO
I know it well, sir. Lo, here's the chain.
I thought to have ta'en you at the Porcupine.
The chain unfinished made me stay thus long.

ANTIPHOLUS OF SYRACUSE *(taking the chain)*
What is your will that I shall do with this?         175

ANGELO
What please yourself, sir. I have made it for you.

ANTIPHOLUS OF SYRACUSE
Made it for me, sir? I bespoke it not.

ANGELO
Not once, nor twice, but twenty times you have.
Go home with it, and please your wife withal,
And soon at supper-time I'll visit you,              180
And then receive my money for the chain.

ANTIPHOLUS OF SYRACUSE
I pray you, sir, receive the money now,
For fear you ne'er see chain nor money more.

ANGELO
You are a merry man, sir. Fare you well.        *Exit*

ANTIPHOLUS OF SYRACUSE
What I should think of this I cannot tell.           185
But this I think: there's no man is so vain
That would refuse so fair an offered chain.

I see a man here needs not live by shifts,
When in the streets he meets such golden gifts.
I'll to the mart, and there for Dromio stay.          190
If any ship put out, then straight away!          *Exit*

❁

**4.1**     *Enter Second Merchant, Angelo the goldsmith, and*
          *an Officer*
SECOND MERCHANT (*to Angelo*)
     You know since Pentecost the sum is due,
     And since I have not much importuned you;
     Nor now I had not, but that I am bound
     To Persia, and want guilders for my voyage.
     Therefore make present satisfaction,          5
     Or I'll attach you by this officer.
ANGELO
     Even just the sum that I do owe to you
     Is growing to me by Antipholus,
     And in the instant that I met with you
     He had of me a chain. At five o'clock          10
     I shall receive the money for the same.
     Pleaseth you walk with me down to his house,
     I will discharge my bond, and thank you too.
          *Enter Antipholus of Ephesus and Dromio of Ephesus*
          *from the Courtesan's house (the Porcupine)*
OFFICER
     That labour may you save. See where he comes.
ANTIPHOLUS OF EPHESUS (*to Dromio*)
     While I go to the goldsmith's house, go thou          15
     And buy a rope's end. That will I bestow
     Among my wife and her confederates
     For locking me out of my doors by day.
     But soft, I see the goldsmith. Get thee gone.
     Buy thou a rope, and bring it home to me.          20
DROMIO OF EPHESUS
     I buy a thousand pound a year, I buy a rope.     *Exit*
ANTIPHOLUS OF EPHESUS (*to Angelo*)
     A man is well holp up that trusts to you!
     I promisèd your presence and the chain,
     But neither chain nor goldsmith came to me.
     Belike you thought our love would last too long          25
     If it were chained together, and therefore came not.
ANGELO
     Saving your merry humour, here's the note
     How much your chain weighs to the utmost carat,
     The fineness of the gold, and chargeful fashion,
     Which doth amount to three odd ducats more          30
     Than I stand debted to this gentleman.
     I pray you see him presently discharged,
     For he is bound to sea, and stays but for it.
ANTIPHOLUS OF EPHESUS
     I am not furnished with the present money.
     Besides, I have some business in the town.          35
     Good signor, take the stranger to my house,
     And with you take the chain, and bid my wife
     Disburse the sum on the receipt thereof.
     Perchance I will be there as soon as you.
ANGELO
     Then you will bring the chain to her yourself?          40
ANTIPHOLUS OF EPHESUS
     No, bear it with you, lest I come not time enough.
ANGELO
     Well, sir, I will. Have you the chain about you?
ANTIPHOLUS OF EPHESUS
     An if I have not, sir, I hope you have;
     Or else you may return without your money.

ANGELO
     Nay, come, I pray you, sir, give me the chain.          45
     Both wind and tide stays for this gentleman,
     And I, to blame, have held him here too long.
ANTIPHOLUS OF EPHESUS
     Good Lord! You use this dalliance to excuse
     Your breach of promise to the Porcupine.
     I should have chid you for not bringing it,          50
     But like a shrew you first begin to brawl.
SECOND MERCHANT (*to Angelo*)
     The hour steals on. I pray you, sir, dispatch.
ANGELO (*to Antipholus*)
     You hear how he importunes me. The chain!
ANTIPHOLUS OF EPHESUS
     Why, give it to my wife, and fetch your money.
ANGELO
     Come, come, you know I gave it you even now.          55
     Either send the chain, or send me by some token.
ANTIPHOLUS OF EPHESUS
     Fie, now you run this humour out of breath.
     Come, where's the chain? I pray you let me see it.
SECOND MERCHANT
     My business cannot brook this dalliance.
     Good sir, say whe'er you'll answer me or no;          60
     If not, I'll leave him to the officer.
ANTIPHOLUS OF EPHESUS
     I answer you? What should I answer you?
ANGELO
     The money that you owe me for the chain.
ANTIPHOLUS OF EPHESUS
     I owe you none till I receive the chain.
ANGELO
     You know I gave it you half an hour since.          65
ANTIPHOLUS OF EPHESUS
     You gave me none. You wrong me much to say so.
ANGELO
     You wrong me more, sir, in denying it.
     Consider how it stands upon my credit.
SECOND MERCHANT
     Well, officer, arrest him at my suit.
OFFICER (*to Angelo*)
     I do, and charge you in the Duke's name to obey me.
ANGELO (*to Antipholus*)
     This touches me in reputation.          71
     Either consent to pay this sum for me,
     Or I attach you by this officer.
ANTIPHOLUS OF EPHESUS
     Consent to pay thee that I never had?
     Arrest me, foolish fellow, if thou dar'st.          75
ANGELO
     Here is thy fee: arrest him, officer.
     I would not spare my brother in this case
     If he should scorn me so apparently.
OFFICER (*to Antipholus*)
     I do arrest you, sir. You hear the suit.
ANTIPHOLUS OF EPHESUS
     I do obey thee till I give thee bail.          80
     (*To Angelo*) But, sirrah, you shall buy this sport as dear
     As all the metal in your shop will answer.
ANGELO
     Sir, sir, I shall have law in Ephesus,
     To your notorious shame, I doubt it not.
          *Enter Dromio of Syracuse, from the bay*
DROMIO OF SYRACUSE
     Master, there's a barque of Epidamnum          85
     That stays but till her owner comes aboard,

And then she bears away. Our freightage, sir,
I have conveyed aboard, and I have bought
The oil, the balsamum, and aqua-vitae.
The ship is in her trim; the merry wind          90
Blows fair from land. They stay for naught at all
But for their owner, master, and yourself.
ANTIPHOLUS OF EPHESUS
How now? A madman? Why, thou peevish sheep,
What ship of Epidamnum stays for me?
DROMIO OF SYRACUSE
A ship you sent me to, to hire waftage.          95
ANTIPHOLUS OF EPHESUS
Thou drunken slave, I sent thee for a rope,
And told thee to what purpose and what end.
DROMIO OF SYRACUSE
You sent me for a ropë's end as soon.
You sent me to the bay, sir, for a barque.
ANTIPHOLUS OF EPHESUS
I will debate this matter at more leisure,       100
And teach your ears to list me with more heed.
To Adriana, villain, hie thee straight.
Give her this key, and tell her in the desk
That's covered o'er with Turkish tapestry
There is a purse of ducats. Let her send it.     105
Tell her I am arrested in the street,
And that shall bail me. Hie thee, slave. Be gone!—
On, officer, to prison, till it come.
                    *Exeunt all but Dromio of Syracuse*
DROMIO OF SYRACUSE
To Adriana. That is where we dined,
Where Dowsabel did claim me for her husband.     110
She is too big, I hope, for me to compass.
Thither I must, although against my will;
For servants must their masters' minds fulfil.   *Exit*

**4.2**   *Enter ⌜from the Phoenix⌝ Adriana and Luciana*
ADRIANA
Ah, Luciana, did he tempt thee so?
   Mightst thou perceive austerely in his eye
That he did plead in earnest, yea or no?
   Looked he or red or pale, or sad or merrily?
What observation mad'st thou in this case        5
Of his heart's meteors tilting in his face?
LUCIANA
First he denied you had in him no right.
ADRIANA
He meant he did me none, the more my spite.
LUCIANA
Then swore he that he was a stranger here.
ADRIANA
And true he swore, though yet forsworn he were.  10
LUCIANA
Then pleaded I for you.
ADRIANA                   And what said he?
LUCIANA
That love I begged for you, he begged of me.
ADRIANA
With what persuasion did he tempt thy love?
LUCIANA
With words that in an honest suit might move.
First he did praise my beauty, then my speech.   15
ADRIANA
Didst speak him fair?
LUCIANA               Have patience, I beseech.
ADRIANA
I cannot, nor I will not, hold me still.

My tongue, though not my heart, shall have his will.
He is deformèd, crookèd, old, and sere,
Ill-faced, worse-bodied, shapeless everywhere,   20
Vicious, ungentle, foolish, blunt, unkind,
Stigmatical in making, worse in mind.
LUCIANA
Who would be jealous, then, of such a one?
No evil lost is wailed when it is gone.
ADRIANA
Ah, but I think him better than I say,           25
   And yet would herein others' eyes were worse.
Far from her nest the lapwing cries away.
   My heart prays for him, though my tongue do curse.
                    *Enter Dromio of Syracuse running*
DROMIO OF SYRACUSE
Here, go—the desk, the purse! Sweet now, make haste!
LUCIANA
How? Hast thou lost thy breath?
DROMIO OF SYRACUSE            By running fast.   30
ADRIANA
Where is thy master, Dromio? Is he well?
DROMIO OF SYRACUSE
No, he's in Tartar limbo, worse than hell.
A devil in an everlasting garment hath him,
One whose hard heart is buttoned up with steel;
A fiend, a fairy, pitiless and rough;            35
A wolf, nay worse, a fellow all in buff;
A back-friend, a shoulder-clapper, one that
   countermands
The passages of alleys, creeks, and narrow launds;
A hound that runs counter, and yet draws dryfoot
   well;
One that before the Judgement carries poor souls to
   hell.                                         40
ADRIANA Why, man, what is the matter?
DROMIO OF SYRACUSE
I do not know the matter, he is 'rested on the case.
ADRIANA
What, is he arrested? Tell me at whose suit.
DROMIO OF SYRACUSE
I know not at whose suit he is arrested well,
But is in a suit of buff which 'rested him, that can I
   tell.
Will you send him, mistress, redemption—the money
   in his desk?                                  46
ADRIANA
Go fetch it, sister.        *Exit Luciana ⌜into the Phoenix⌝*
                    This I wonder at,
That he unknown to me should be in debt.
Tell me, was he arrested on a bond?
DROMIO OF SYRACUSE
Not on a bond but on a stronger thing:           50
A chain, a chain—do you not hear it ring?
ADRIANA
What, the chain?
DROMIO OF SYRACUSE
              No, no, the bell. 'Tis time that I were gone:
It was two ere I left him, and now the clock strikes
   one.
ADRIANA
The hours come back! That did I never hear.
DROMIO OF SYRACUSE
O yes, if any hour meet a sergeant, a turns back for
   very fear.                                    55
ADRIANA
As if time were in debt. How fondly dost thou reason!

DROMIO OF SYRACUSE
 Time is a very bankrupt, and owes more than he's
 worth to season.
 Nay, he's a thief too. Have you not heard men say
 That time comes stealing on by night and day?
 If a be in debt and theft, and a sergeant in the way,
 Hath he not reason to turn back an hour in a day? 61
  *Enter Luciana ⌈from the Phoenix⌉ with the money*
ADRIANA
 Go, Dromio, there's the money. Bear it straight,
  And bring thy master home immediately.
            *⌈Exit Dromio⌉*
 Come, sister, I am pressed down with conceit:
  Conceit, my comfort and my injury.          65
            *Exeunt ⌈into the Phoenix⌉*

**4.3**  *Enter Antipholus of Syracuse, wearing the chain*
ANTIPHOLUS OF SYRACUSE
 There's not a man I meet but doth salute me
 As if I were their well-acquainted friend,
 And everyone doth call me by my name.
 Some tender money to me, some invite me,
 Some other give me thanks for kindnesses.      5
 Some offer me commodities to buy.
 Even now a tailor called me in his shop,
 And showed me silks that he had bought for me,
 And therewithal took measure of my body.
 Sure, these are but imaginary wiles,           10
 And Lapland sorcerers inhabit here.
   *Enter Dromio of Syracuse with the money*
DROMIO OF SYRACUSE Master, here's the gold you sent me
 for. What, have you got redemption from the picture
 of old Adam new apparelled?
ANTIPHOLUS OF SYRACUSE
 What gold is this? What Adam dost thou mean?  15
DROMIO OF SYRACUSE Not that Adam that kept the
 Paradise, but that Adam that keeps the prison—he that
 goes in the calf's skin, that was killed for the Prodigal;
 he that came behind you, sir, like an evil angel, and
 bid you forsake your liberty.               20
ANTIPHOLUS OF SYRACUSE I understand thee not.
DROMIO OF SYRACUSE No? Why, 'tis a plain case: he that
 went like a bass viol in a case of leather, the man, sir,
 that when gentlemen are tired gives them a sob and
 'rests them; he, sir, that takes pity on decayed men
 and gives them suits of durance; he that sets up his
 rest to do more exploits with his mace than a Moorish
 pike.                                  28
ANTIPHOLUS OF SYRACUSE What, thou mean'st an officer?
DROMIO OF SYRACUSE Ay, sir, the sergeant of the band: he
 that brings any man to answer it that breaks his bond;
 one that thinks a man always going to bed, and says
 'God give you good rest.'                   33
ANTIPHOLUS OF SYRACUSE Well, sir, there rest in your
 foolery. Is there any ships puts forth tonight? May we
 be gone?
DROMIO OF SYRACUSE Why, sir, I brought you word an
 hour since that the barque *Expedition* put forth tonight,
 and then were you hindered by the sergeant to tarry
 for the hoy *Delay*. Here are the angels that you sent
 for to deliver you.                        41
ANTIPHOLUS OF SYRACUSE
 The fellow is distraught, and so am I,
 And here we wander in illusions.
 Some blessèd power deliver us from hence.

 *Enter a Courtesan ⌈from the Porcupine⌉*
COURTESAN
 Well met, well met, Master Antipholus.        45
 I see, sir, you have found the goldsmith now.
 Is that the chain you promised me today?
ANTIPHOLUS OF SYRACUSE
 Satan, avoid! I charge thee, tempt me not!
DROMIO OF SYRACUSE Master, is this Mistress Satan?
ANTIPHOLUS OF SYRACUSE It is the devil.        50
DROMIO OF SYRACUSE Nay, she is worse, she is the devil's
 dam; and here she comes in the habit of a light wench.
 And thereof comes that the wenches say 'God damn
 me'—that's as much to say, 'God make me a light
 wench.' It is written they appear to men like angels of
 light. Light is an effect of fire, and fire will burn. Ergo,
 light wenches will burn. Come not near her.   57
COURTESAN
 Your man and you are marvellous merry, sir.
 Will you go with me? We'll mend our dinner here.
DROMIO OF SYRACUSE Master, if you do, expect spoon-meat,
 and bespeak a long spoon.                   61
ANTIPHOLUS OF SYRACUSE Why, Dromio?
DROMIO OF SYRACUSE Marry, he must have a long spoon
 that must eat with the devil.
ANTIPHOLUS OF SYRACUSE (*to Courtesan*)
 Avoid, thou fiend! What tell'st thou me of supping?
 Thou art, as you are all, a sorceress.        66
 I conjure thee to leave me and be gone.
COURTESAN
 Give me the ring of mine you had at dinner,
 Or for my diamond the chain you promised,
 And I'll be gone, sir, and not trouble you.    70
DROMIO OF SYRACUSE
 Some devils ask but the parings of one's nail,
 A rush, a hair, a drop of blood, a pin,
 A nut, a cherry-stone;
 But she, more covetous, would have a chain.
 Master, be wise; an if you give it her,        75
 The devil will shake her chain, and fright us with it.
COURTESAN (*to Antipholus*)
 I pray you, sir, my ring, or else the chain.
 I hope you do not mean to cheat me so?
ANTIPHOLUS OF SYRACUSE
 Avaunt, thou witch!—Come, Dromio, let us go.
DROMIO OF SYRACUSE
 'Fly pride' says the peacock. Mistress, that you know.
      *Exeunt Antipholus of Syracuse
           and Dromio of Syracuse*
COURTESAN
 Now, out of doubt, Antipholus is mad;        81
 Else would he never so demean himself.
 A ring he hath of mine worth forty ducats,
 And for the same he promised me a chain.
 Both one and other he denies me now.          85
 The reason that I gather he is mad,
 Besides this present instance of his rage,
 Is a mad tale he told today at dinner
 Of his own doors being shut against his entrance.
 Belike his wife, acquainted with his fits,     90
 On purpose shut the doors against his way.
 My way is now to hie home to his house,
 And tell his wife that, being lunatic,
 He rushed into my house, and took perforce
 My ring away. This course I fittest choose,    95
 For forty ducats is too much to lose.      *Exit*

**4.4** *Enter Antipholus of Ephesus with the Officer*
ANTIPHOLUS OF EPHESUS
Fear me not, man, I will not break away.
I'll give thee ere I leave thee so much money
To warrant thee as I am 'rested for.
My wife is in a wayward mood today,
And will not lightly trust the messenger          5
That I should be attached in Ephesus.
I tell you 'twill sound harshly in her ears.
*Enter Dromio of Ephesus with a rope's end*
Here comes my man. I think he brings the money.—
How now, sir? Have you that I sent you for?
DROMIO OF EPHESUS
Here's that, I warrant you, will pay them all.    10
ANTIPHOLUS OF EPHESUS But where's the money?
DROMIO OF EPHESUS
Why, sir, I gave the money for the rope.
ANTIPHOLUS OF EPHESUS
Five hundred ducats, villain, for a rope?
DROMIO OF EPHESUS
I'll serve you, sir, five hundred at the rate.
ANTIPHOLUS OF EPHESUS
To what end did I bid thee hie thee home?         15
DROMIO OF EPHESUS To a rope's end, sir, and to that end
am I returned.
ANTIPHOLUS OF EPHESUS
And to that end, sir, I will welcome you.
*He beats Dromio*
OFFICER Good sir, be patient.
DROMIO OF EPHESUS Nay, 'tis for me to be patient: I am
in adversity.                                     21
OFFICER Good now, hold thy tongue.
DROMIO OF EPHESUS Nay, rather persuade *him* to hold his
hands.
ANTIPHOLUS OF EPHESUS Thou whoreson, senseless villain!
DROMIO OF EPHESUS I would I were senseless, sir, that I
might not feel your blows.                        27
ANTIPHOLUS OF EPHESUS Thou art sensible in nothing but
blows, and so is an ass.
DROMIO OF EPHESUS I am an ass indeed. You may prove
it by my long ears.—I have served him from the hour
of my nativity to this instant, and have nothing at his
hands for my service but blows. When I am cold, he
heats me with beating. When I am warm, he cools me
with beating. I am waked with it when I sleep, raised
with it when I sit, driven out of doors with it when I
go from home, welcomed home with it when I return.
Nay, I bear it on my shoulders, as a beggar wont her
brat, and I think when he hath lamed me I shall beg
with it from door to door.                        40
*Enter Adriana, Luciana, Courtesan, and a
schoolmaster called Pinch*
ANTIPHOLUS OF EPHESUS
Come, go along: my wife is coming yonder.
DROMIO OF EPHESUS (*to Adriana*) Mistress, *respice finem*—
respect your end—or rather, to prophesy like the parrot,
'Beware the rope's end'.
ANTIPHOLUS OF EPHESUS Wilt thou still talk?         45
*He beats Dromio*
COURTESAN (*to Adriana*)
How say you now? Is not your husband mad?
ADRIANA
His incivility confirms no less.—
Good Doctor Pinch, you are a conjurer.
Establish him in his true sense again,
And I will please you what you will demand.        50

LUCIANA
Alas, how fiery and how sharp he looks!
COURTESAN
Mark how he trembles in his ecstasy.
PINCH (*to Antipholus*)
Give me your hand, and let me feel your pulse.
ANTIPHOLUS OF EPHESUS
There is my hand, and let it feel your ear.
*He strikes Pinch*
PINCH
I charge thee, Satan, housed within this man,      55
To yield possession to my holy prayers,
And to thy state of darkness hie thee straight:
I conjure thee by all the saints in heaven.
ANTIPHOLUS OF EPHESUS
Peace, doting wizard, peace! I am not mad.
ADRIANA
O that thou wert not, poor distressèd soul.        60
ANTIPHOLUS OF EPHESUS
You minion, you, are these your customers?
Did this companion with the saffron face
Revel and feast it at my house today,
Whilst upon me the guilty doors were shut,
And I denied to enter in my house?                 65
ADRIANA
O husband, God doth know you dined at home,
Where would you had remained until this time,
Free from these slanders and this open shame.
ANTIPHOLUS OF EPHESUS
Dined at home?
(*To Dromio*)     Thou villain, what sayst thou?
DROMIO OF EPHESUS
Sir, sooth to say, you did not dine at home.       70
ANTIPHOLUS OF EPHESUS
Were not my doors locked up, and I shut out?
DROMIO OF EPHESUS
Pardie, your doors were locked, and you shut out.
ANTIPHOLUS OF EPHESUS
And did not she herself revile me there?
DROMIO OF EPHESUS
Sans fable, she herself reviled you there.
ANTIPHOLUS OF EPHESUS
Did not her kitchen-maid rail, taunt, and scorn me?
DROMIO OF EPHESUS
Certes she did. The kitchen vestal scorned you.    76
ANTIPHOLUS OF EPHESUS
And did not I in rage depart from thence?
DROMIO OF EPHESUS
In verity you did. My bones bears witness,
That since have felt the vigour of his rage.
ADRIANA (*aside to Pinch*)
Is't good to soothe him in these contraries?       80
PINCH (*aside to Adriana*)
It is no shame. The fellow finds his vein,
And, yielding to him, humours well his frenzy.
ANTIPHOLUS OF EPHESUS (*to Adriana*)
Thou hast suborned the goldsmith to arrest me.
ADRIANA
Alas, I sent you money to redeem you,
By Dromio here, who came in haste for it.          85
DROMIO OF EPHESUS
Money by me? Heart and good will you might,
But surely, master, not a rag of money.
ANTIPHOLUS OF EPHESUS
Went'st not thou to her for a purse of ducats?

**ADRIANA**
He came to me, and I delivered it.

**LUCIANA**
And I am witness with her that she did.      90

**DROMIO OF EPHESUS**
God and the ropemaker bear me witness
That I was sent for nothing but a rope.

**PINCH** (*aside to Adriana*)
Mistress, both man and master is possessed.
I know it by their pale and deadly looks.
They must be bound and laid in some dark room.   95

**ANTIPHOLUS OF EPHESUS** (*to Adriana*)
Say wherefore didst thou lock me forth today,
(*To Dromio*) And why dost thou deny the bag of gold?

**ADRIANA**
I did not, gentle husband, lock thee forth.

**DROMIO OF EPHESUS**
And, gentle master, I received no gold.
But I confess, sir, that we were locked out.     100

**ADRIANA**
Dissembling villain, thou speak'st false in both.

**ANTIPHOLUS OF EPHESUS**
Dissembling harlot, thou art false in all,
And art confederate with a damnèd pack
To make a loathsome abject scorn of me.
But with these nails I'll pluck out those false eyes,  105
That would behold in me this shameful sport.
⌈*He reaches for Adriana; she shrieks.*⌉
*Enter three or four, and offer to bind him. He strives*

**ADRIANA**
O, bind him, bind him. Let him not come near me.

**PINCH**
More company! The fiend is strong within him.

**LUCIANA**
Ay me, poor man, how pale and wan he looks.

**ANTIPHOLUS OF EPHESUS**
What, will you murder me?—Thou, jailer, thou,   110
I am thy prisoner. Wilt thou suffer them
To make a rescue?

**OFFICER**             Masters, let him go.
He is my prisoner, and you shall not have him.

**PINCH**
Go, bind his man, for he is frantic too.
*They bind Dromio*

**ADRIANA**
What wilt thou do, thou peevish officer?     115
Hast thou delight to see a wretched man
Do outrage and displeasure to himself?

**OFFICER**
He is my prisoner. If I let him go,
The debt he owes will be required of me.

**ADRIANA**
I will discharge thee ere I go from thee.     120
Bear me forthwith unto his creditor,
And, knowing how the debt grows, I will pay it.—
Good Master Doctor, see him safe conveyed
Home to my house. O most unhappy day!

**ANTIPHOLUS OF EPHESUS** O most unhappy strumpet!   125

**DROMIO OF EPHESUS**
Master, I am here entered in bond for you.

**ANTIPHOLUS OF EPHESUS**
Out on thee, villain! Wherefore dost thou mad me?

**DROMIO OF EPHESUS**
Will you be bound for nothing? Be mad, good master—
Cry, 'The devil!'

**LUCIANA**
God help, poor souls, how idly do they talk!   130

**ADRIANA**
Go bear him hence. Sister, go you with me.
*Exeunt ⌈into the Phoenix⌉, Pinch and others*
*carrying off Antipholus of Ephesus and Dromio of*
*Ephesus. The Officer, Adriana, Luciana, and the*
*Courtesan remain*
(*To the Officer*) Say now, whose suit is he arrested at?

**OFFICER**
One Angelo, a goldsmith. Do you know him?

**ADRIANA**
I know the man. What is the sum he owes?

**OFFICER**
Two hundred ducats.

**ADRIANA**          Say, how grows it due?    135

**OFFICER**
Due for a chain your husband had of him.

**ADRIANA**
He did bespeak a chain for me, but had it not.

**COURTESAN**
Whenas your husband all in rage today
Came to my house, and took away my ring—
The ring I saw upon his finger now—     140
Straight after did I meet him with a chain.

**ADRIANA**
It may be so, but I did never see it.
Come, jailer, bring me where the goldsmith is.
I long to know the truth hereof at large.
*Enter Antipholus of Syracuse (wearing the chain)*
*and Dromio of Syracuse with their rapiers drawn*

**LUCIANA**
God, for thy mercy, they are loose again!    145

**ADRIANA**
And come with naked swords. Let's call more help
To have them bound again.

**OFFICER**           Away, they'll kill us!
*All but Antipholus and Dromio run out, as fast as*
*may be, frighted*

**ANTIPHOLUS OF SYRACUSE**
I see these witches are afraid of swords.

**DROMIO OF SYRACUSE**
She that would be your wife now ran from you.

**ANTIPHOLUS OF SYRACUSE**
Come to the Centaur. Fetch our stuff from thence.  150
I long that we were safe and sound aboard.

**DROMIO OF SYRACUSE** Faith, stay here this night. They will
surely do us no harm. You saw they speak us fair, give
us gold. Methinks they are such a gentle nation that,
but for the mountain of mad flesh that claims marriage
of me, I could find in my heart to stay here still, and
turn witch.    157

**ANTIPHOLUS OF SYRACUSE**
I will not stay tonight for all the town.
Therefore away, to get our stuff aboard.       *Exeunt*

               ❦

**5.1**    *Enter Second Merchant and Angelo the goldsmith*

**ANGELO**
I am sorry, sir, that I have hindered you,
But I protest he had the chain of me,
Though most dishonestly he doth deny it.

**SECOND MERCHANT**
How is the man esteemed here in the city?

**ANGELO**
Of very reverend reputation, sir,             5
Of credit infinite, highly beloved,
Second to none that lives here in the city.
His word might bear my wealth at any time.

SECOND MERCHANT
  Speak softly. Yonder, as I think, he walks.
    *Enter Antipholus of Syracuse, wearing the chain,*
    *and Dromio of Syracuse again*
ANGELO
  'Tis so, and that self chain about his neck    10
  Which he forswore most monstrously to have.
  Good sir, draw near to me. I'll speak to him.—
  Signor Antipholus, I wonder much
  That you would put me to this shame and trouble,
  And not without some scandal to yourself,    15
  With circumstance and oaths so to deny
  This chain, which now you wear so openly.
  Beside the charge, the shame, imprisonment,
  You have done wrong to this my honest friend,
  Who, but for staying on our controversy,    20
  Had hoisted sail and put to sea today.
  This chain you had of me. Can you deny it?
ANTIPHOLUS OF SYRACUSE
  I think I had. I never did deny it.
SECOND MERCHANT
  Yes, that you did, sir, and forswore it too.
ANTIPHOLUS OF SYRACUSE
  Who heard me to deny it or forswear it?    25
SECOND MERCHANT
  These ears of mine, thou know'st, did hear thee.
  Fie on thee, wretch! 'Tis pity that thou liv'st
  To walk where any honest men resort.
ANTIPHOLUS OF SYRACUSE
  Thou art a villain to impeach me thus.
  I'll prove mine honour and mine honesty    30
  Against thee presently, if thou dar'st stand.
SECOND MERCHANT
  I dare, and do defy thee for a villain.
    *They draw. Enter Adriana, Luciana, Courtesan,*
    *and others ⌈from the Phoenix⌉*
ADRIANA
  Hold, hurt him not, for God's sake; he is mad.
  Some get within him, take his sword away.
  Bind Dromio too, and bear them to my house.    35
DROMIO OF SYRACUSE
  Run, master, run! For God's sake take a house.
  This is some priory—in, or we are spoiled.
    *Exeunt Antipholus of Syracuse and*
    *Dromio of Syracuse to the priory*
  *Enter ⌈from the priory⌉ the Lady Abbess*
ABBESS
  Be quiet, people. Wherefore throng you hither?
ADRIANA
  To fetch my poor distracted husband hence.
  Let us come in, that we may bind him fast,    40
  And bear him home for his recovery.
ANGELO
  I knew he was not in his perfect wits.
SECOND MERCHANT
  I am sorry now that I did draw on him.
ABBESS
  How long hath this possession held the man?
ADRIANA
  This week he hath been heavy, sour, sad,    45
  And much, much different from the man he was;
  But till this afternoon his passion
  Ne'er brake into extremity of rage.
ABBESS
  Hath he not lost much wealth by wreck at sea?
  Buried some dear friend? Hath not else his eye    50

  Strayed his affection in unlawful love—
  A sin prevailing much in youthful men,
  Who give their eyes the liberty of gazing?
  Which of these sorrows is he subject to?
ADRIANA
  To none of these, except it be the last,    55
  Namely some love that drew him oft from home.
ABBESS
  You should for that have reprehended him.
ADRIANA
  Why, so I did.
ABBESS        Ay, but not rough enough.
ADRIANA
  As roughly as my modesty would let me.
ABBESS Haply in private.    60
ADRIANA And in assemblies too.
ABBESS Ay, but not enough.
ADRIANA
  It was the copy of our conference.
  In bed he slept not for my urging it.
  At board he fed not for my urging it.    65
  Alone, it was the subject of my theme.
  In company I often glancèd it.
  Still did I tell him it was vile and bad.
ABBESS
  And thereof came it that the man was mad.
  The venom clamours of a jealous woman    70
  Poisons more deadly than a mad dog's tooth.
  It seems his sleeps were hindered by thy railing,
  And thereof comes it that his head is light.
  Thou sayst his meat was sauced with thy upbraidings.
  Unquiet meals make ill digestions.    75
  Thereof the raging fire of fever bred,
  And what's a fever but a fit of madness?
  Thou sayst his sports were hindered by thy brawls.
  Sweet recreation barred, what doth ensue
  But moody and dull melancholy,    80
  Kinsman to grim and comfortless despair,
  And at her heels a huge infectious troop
  Of pale distemperatures and foes to life?
  In food, in sport, and life-preserving rest
  To be disturbed would mad or man or beast.    85
  The consequence is, then, thy jealous fits
  Hath scared thy husband from the use of wits.
LUCIANA
  She never reprehended him but mildly
  When he demeaned himself rough, rude, and wildly.
  (*To Adriana*) Why bear you these rebukes, and answer
    not?    90
ADRIANA
  She did betray me to my own reproof.—
  Good people, enter, and lay hold on him.
ABBESS
  No, not a creature enters in my house.
ADRIANA
  Then let your servants bring my husband forth.
ABBESS
  Neither. He took this place for sanctuary,    95
  And it shall privilege him from your hands
  Till I have brought him to his wits again,
  Or lose my labour in essaying it.
ADRIANA
  I will attend my husband, be his nurse,
  Diet his sickness, for it is my office,    100
  And will have no attorney but myself.
  And therefore let me have him home with me.

ABBESS
  Be patient, for I will not let him stir
  Till I have used the approvèd means I have,
  With wholesome syrups, drugs, and holy prayers   105
  To make of him a formal man again.
  It is a branch and parcel of mine oath,
  A charitable duty of my order.
  Therefore depart, and leave him here with me.
ADRIANA
  I will not hence, and leave my husband here;   110
  And ill it doth beseem your holiness
  To separate the husband and the wife.
ABBESS
  Be quiet and depart. Thou shalt not have him.
              ⌈*Exit into the priory*⌉
LUCIANA (*to Adriana*)
  Complain unto the Duke of this indignity.
ADRIANA
  Come, go, I will fall prostrate at his feet,   115
  And never rise until my tears and prayers
  Have won his grace to come in person hither
  And take perforce my husband from the Abbess.
SECOND MERCHANT
  By this, I think, the dial point's at five.
  Anon, I'm sure, the Duke himself in person   120
  Comes this way to the melancholy vale,
  The place of death and sorry execution,
  Behind the ditches of the abbey here.
ANGELO Upon what cause?
SECOND MERCHANT
  To see a reverend Syracusian merchant,   125
  Who put unluckily into this bay
  Against the laws and statutes of this town,
  Beheaded publicly for his offence.
ANGELO
  See where they come. We will behold his death.
LUCIANA
  Kneel to the Duke before he pass the abbey.   130
      *Enter Solinus Duke of Ephesus, and Egeon the*
      *merchant of Syracuse, bareheaded, with the*
      *headsman and other officers*
DUKE
  Yet once again proclaim it publicly:
  If any friend will pay the sum for him,
  He shall not die, so much we tender him.
ADRIANA (*kneeling*)
  Justice, most sacred Duke, against the Abbess!
DUKE
  She is a virtuous and a reverend lady.   135
  It cannot be that she hath done thee wrong.
ADRIANA
  May it please your grace, Antipholus my husband,
  Who I made lord of me and all I had
  At your important letters—this ill day
  A most outrageous fit of madness took him,   140
  That desp'rately he hurried through the street,
  With him his bondman, all as mad as he,
  Doing displeasure to the citizens
  By rushing in their houses, bearing thence
  Rings, jewels, anything his rage did like.   145
  Once did I get him bound, and sent him home,
  Whilst to take order for the wrongs I went
  That here and there his fury had committed.
  Anon, I wot not by what strong escape,
  He broke from those that had the guard of him,   150

  And with his mad attendant and himself,
  Each one with ireful passion, with drawn swords,
  Met us again, and, madly bent on us,
  Chased us away; till, raising of more aid,
  We came again to bind them. Then they fled   155
  Into this abbey, whither we pursued them,
  And here the Abbess shuts the gates on us,
  And will not suffer us to fetch him out,
  Nor send him forth that we may bear him hence.
  Therefore, most gracious Duke, with thy command
  Let him be brought forth, and borne hence for help.
DUKE ⌈*raising Adriana*⌉
  Long since, thy husband served me in my wars,   162
  And I to thee engaged a prince's word,
  When thou didst make him master of thy bed,
  To do him all the grace and good I could.—   165
  Go, some of you, knock at the abbey gate,
  And bid the Lady Abbess come to me.
  I will determine this before I stir.
      *Enter a Messenger* ⌈*from the Phoenix*⌉
MESSENGER (*to Adriana*)
  O mistress, mistress, shift and save yourself!
  My master and his man are both broke loose,   170
  Beaten the maids a-row, and bound the Doctor,
  Whose beard they have singed off with brands of fire,
  And ever as it blazed they threw on him
  Great pails of puddled mire to quench the hair.
  My master preaches patience to him, and the while
  His man with scissors nicks him like a fool;   176
  And sure—unless you send some present help—
  Between them they will kill the conjurer.
ADRIANA
  Peace, fool. Thy master and his man are here,
  And that is false thou dost report to us.   180
MESSENGER
  Mistress, upon my life I tell you true.
  I have not breathed almost since I did see it.
  He cries for you, and vows, if he can take you,
  To scorch your face and to disfigure you.
      *Cry within*
  Hark, hark, I hear him, mistress. Fly, be gone!   185
DUKE (*to Adriana*)
  Come stand by me. Fear nothing. Guard with halberds!
      *Enter Antipholus of Ephesus and Dromio of Ephesus*
      ⌈*from the Phoenix*⌉
ADRIANA
  Ay me, it is my husband! Witness you
  That he is borne about invisible.
  Even now we housed him in the abbey here,
  And now he's there, past thought of human reason.
ANTIPHOLUS OF EPHESUS
  Justice, most gracious Duke, O grant me justice,   191
  Even for the service that long since I did thee,
  When I bestrid thee in the wars, and took
  Deep scars to save thy life; even for the blood
  That then I lost for thee, now grant me justice!   195
EGEON (*aside*)
  Unless the fear of death doth make me dote,
  I see my son Antipholus, and Dromio.
ANTIPHOLUS OF EPHESUS
  Justice, sweet prince, against that woman there,
  She whom thou gav'st to me to be my wife,
  That hath abusèd and dishonoured me   200
  Even in the strength and height of injury.
  Beyond imagination is the wrong
  That she this day hath shameless thrown on me.

DUKE
  Discover how, and thou shalt find me just.
ANTIPHOLUS OF EPHESUS
  This day, great Duke, she shut the doors upon me  205
  While she with harlots feasted in my house.
DUKE
  A grievous fault!—Say, woman, didst thou so?
ADRIANA
  No, my good lord. Myself, he, and my sister
  Today did dine together. So befall my soul
  As this is false he burdens me withal.  210
LUCIANA
  Ne'er may I look on day nor sleep on night
  But she tells to your highness simple truth.
ANGELO (aside)
  O perjured woman! They are both forsworn.
  In this the madman justly chargeth them.
ANTIPHOLUS OF EPHESUS
  My liege, I am advisèd what I say,  215
  Neither disturbed with the effect of wine,
  Nor heady-rash provoked with raging ire,
  Albeit my wrongs might make one wiser mad.
  This woman locked me out this day from dinner.
  That goldsmith there, were he not packed with her,
  Could witness it, for he was with me then,  221
  Who parted with me to go fetch a chain,
  Promising to bring it to the Porcupine,
  Where Balthasar and I did dine together.
  Our dinner done, and he not coming thither,  225
  I went to seek him. In the street I met him,
  And in his company that gentleman.
      He points to the Second Merchant
  There did this perjured goldsmith swear me down
  That I this day of him received the chain,
  Which, God he knows, I saw not. For the which  230
  He did arrest me with an officer.
  I did obey, and sent my peasant home
  For certain ducats. He with none returned.
  Then fairly I bespoke the officer
  To go in person with me to my house.  235
  By th' way, we met my wife, her sister, and a rabble
      more
  Of vile confederates. Along with them
  They brought one Pinch, a hungry lean-faced villain,
  A mere anatomy, a mountebank,
  A threadbare juggler, and a fortune-teller,  240
  A needy, hollow-eyed, sharp-looking wretch,
  A living dead man. This pernicious slave,
  Forsooth, took on him as a conjurer,
  And gazing in mine eyes, feeling my pulse,
  And with no face, as 'twere, outfacing me,  245
  Cries out I was possessed. Then all together
  They fell upon me, bound me, bore me thence,
  And in a dark and dankish vault at home
  There left me and my man, both bound together,
  Till, gnawing with my teeth my bonds in sunder,  250
  I gained my freedom, and immediately
  Ran hither to your grace, whom I beseech
  To give me ample satisfaction
  For these deep shames and great indignities.
ANGELO
  My lord, in truth, thus far I witness with him:  255
  That he dined not at home, but was locked out.
DUKE
  But had he such a chain of thee, or no?

ANGELO
  He had, my lord, and when he ran in here
  These people saw the chain about his neck.
SECOND MERCHANT (to Antipholus)
  Besides, I will be sworn these ears of mine  260
  Heard you confess you had the chain of him,
  After you first forswore it on the mart,
  And thereupon I drew my sword on you;
  And then you fled into this abbey here,
  From whence I think you are come by miracle.  265
ANTIPHOLUS OF EPHESUS
  I never came within these abbey walls,
  Nor ever didst thou draw thy sword on me.
  I never saw the chain, so help me heaven,
  And this is false you burden me withal.
DUKE
  Why, what an intricate impeach is this!  270
  I think you all have drunk of Circe's cup.
  If here you housed him, here he would have been.
  If he were mad, he would not plead so coldly.
  (To Adriana) You say he dined at home, the goldsmith
      here
  Denies that saying. (To Dromio) Sirrah, what say you?
DROMIO OF EPHESUS (pointing out the Courtesan)
  Sir, he dined with her there, at the Porcupine.  276
COURTESAN
  He did, and from my finger snatched that ring.
ANTIPHOLUS OF EPHESUS
  'Tis true, my liege, this ring I had of her.
DUKE (to Courtesan)
  Saw'st thou him enter at the abbey here?
COURTESAN
  As sure, my liege, as I do see your grace.  280
DUKE
  Why, this is strange. Go call the Abbess hither.
  I think you are all mated, or stark mad.
                      Exit one to the priory
EGEON (coming forward)
  Most mighty Duke, vouchsafe me speak a word.
  Haply I see a friend will save my life,
  And pay the sum that may deliver me.  285
DUKE
  Speak freely, Syracusian, what thou wilt.
EGEON (to Antipholus)
  Is not your name, sir, called Antipholus?
  And is not that your bondman Dromio?
DROMIO OF EPHESUS
  Within this hour I was his bondman, sir,
  But he, I thank him, gnawed in two my cords.  290
  Now am I Dromio, and his man, unbound.
EGEON
  I am sure you both of you remember me.
DROMIO OF EPHESUS
  Ourselves we do remember, sir, by you;
  For lately we were bound as you are now.
  You are not Pinch's patient, are you, sir?  295
EGEON
  Why look you strange on me? You know me well.
ANTIPHOLUS OF EPHESUS
  I never saw you in my life till now.
EGEON
  O, grief hath changed me since you saw me last,
  And careful hours with time's deformèd hand
  Have written strange defeatures in my face.  300
  But tell me yet, dost thou not know my voice?
ANTIPHOLUS OF EPHESUS Neither.
EGEON Dromio, nor thou?

DROMIO OF EPHESUS No, trust me sir, nor I.

EGEON I am sure thou dost.                                    305

DROMIO OF EPHESUS Ay, sir, but I am sure I do not, and
whatsoever a man denies, you are now bound to believe
him.

EGEON
Not know my voice? O time's extremity,
Hast thou so cracked and splitted my poor tongue  310
In seven short years that here my only son
Knows not my feeble key of untuned cares?
Though now this grainèd face of mine be hid
In sap-consuming winter's drizzled snow,
And all the conduits of my blood froze up,        315
Yet hath my night of life some memory,
My wasting lamps some fading glimmer left,
My dull deaf ears a little use to hear.
All these old witnesses, I cannot err,
Tell me thou art my son Antipholus.               320

ANTIPHOLUS OF EPHESUS
I never saw my father in my life.

EGEON
But seven years since, in Syracusa bay,
Thou know'st we parted. But perhaps, my son,
Thou sham'st to acknowledge me in misery.

ANTIPHOLUS OF EPHESUS
The Duke, and all that know me in the city,       325
Can witness with me that it is not so.
I ne'er saw Syracusa in my life.

DUKE (to Egeon)
I tell thee, Syracusian, twenty years
Have I been patron to Antipholus,
During which time he ne'er saw Syracusa.          330
I see thy age and dangers make thee dote.

*Enter ⌜from the priory⌝ the Abbess, with Antipholus
of Syracuse, wearing the chain, and Dromio of
Syracuse*

ABBESS
Most mighty Duke, behold a man much wronged.

*All gather to see them*

ADRIANA
I see two husbands, or mine eyes deceive me.

DUKE
One of these men is *genius* to the other:
And so of these, which is the natural man,        335
And which the spirit? Who deciphers them?

DROMIO OF SYRACUSE
I, sir, am Dromio. Command him away.

DROMIO OF EPHESUS
I, sir, am Dromio. Pray let me stay.

ANTIPHOLUS OF SYRACUSE
Egeon, art thou not? Or else his ghost.

DROMIO OF SYRACUSE
O, my old master, who hath bound him here?        340

ABBESS
Whoever bound him, I will loose his bonds,
And gain a husband by his liberty.
Speak, old Egeon, if thou beest the man
That hadst a wife once called Emilia,
That bore thee at a burden two fair sons.         345
O, if thou beest the same Egeon, speak,
And speak unto the same Emilia.

DUKE
Why, here begins his morning story right:
These two Antipholus', these two so like,
And these two Dromios, one in semblance—          350
Besides his urging of her wreck at sea.

These are the parents to these children,
Which accidentally are met together.

EGEON
If I dream not, thou art Emilia.
If thou art she, tell me, where is that son        355
That floated with thee on the fatal raft?

ABBESS
By men of Epidamnum he and I
And the twin Dromio all were taken up.
But, by and by, rude fishermen of Corinth
By force took Dromio and my son from them,         360
And me they left with those of Epidamnum.
What then became of them I cannot tell;
I, to this fortune that you see me in.

DUKE (to Antipholus of Syracuse)
Antipholus, thou cam'st from Corinth first.

ANTIPHOLUS OF SYRACUSE
No, sir, not I. I came from Syracuse.              365

DUKE
Stay, stand apart. I know not which is which.

ANTIPHOLUS OF EPHESUS
I came from Corinth, my most gracious lord.

DROMIO OF EPHESUS And I with him.

ANTIPHOLUS OF EPHESUS
Brought to this town by that most famous warrior,
Duke Menaphon, your most renownèd uncle.           370

ADRIANA
Which of you two did dine with me today?

ANTIPHOLUS OF SYRACUSE I, gentle mistress.

ADRIANA And are not you my husband?

ANTIPHOLUS OF EPHESUS No, I say nay to that.

ANTIPHOLUS OF SYRACUSE
And so do I. Yet did she call me so;               375
And this fair gentlewoman, her sister here,
Did call me brother. (*To Luciana*) What I told you then
I hope I shall have leisure to make good,
If this be not a dream I see and hear.

ANGELO
That is the chain, sir, which you had of me.       380

ANTIPHOLUS OF SYRACUSE
I think it be, sir. I deny it not.

ANTIPHOLUS OF EPHESUS (*to Angelo*)
And you, sir, for this chain arrested me.

ANGELO
I think I did, sir. I deny it not.

ADRIANA (*to Antipholus of Ephesus*)
I sent you money, sir, to be your bail,
By Dromio, but I think he brought it not.          385

DROMIO OF EPHESUS No, none by me.

ANTIPHOLUS OF SYRACUSE (*to Adriana*)
This purse of ducats I received from you,
And Dromio my man did bring them me.
I see we still did meet each other's man,
And I was ta'en for him, and he for me,            390
And thereupon these errors are arose.

ANTIPHOLUS OF EPHESUS
These ducats pawn I for my father here.

DUKE
It shall not need. Thy father hath his life.

COURTESAN
Sir, I must have that diamond from you.

ANTIPHOLUS OF EPHESUS
There, take it, and much thanks for my good cheer.

ABBESS
Renownèd Duke, vouchsafe to take the pains        396
To go with us into the abbey here,

And hear at large discoursèd all our fortunes,
And all that are assembled in this place,
That by this sympathizèd one day's error          400
Have suffered wrong. Go, keep us company,
And we shall make full satisfaction.
Thirty-three years have I but gone in travail
Of you, my sons, and till this present hour
My heavy burden ne'er deliverèd.          405
The Duke, my husband, and my children both,
And you the calendars of their nativity,
Go to a gossips' feast, and joy with me.
After so long grief, such festivity!

DUKE
With all my heart I'll gossip at this feast.          410
          *Exeunt ⌜into the priory⌝ all but the two*
          *Dromios and two brothers Antipholus*

DROMIO OF SYRACUSE (*to Antipholus of Ephesus*)
Master, shall I fetch your stuff from shipboard?

ANTIPHOLUS OF EPHESUS
Dromio, what stuff of mine hast thou embarked?

DROMIO OF SYRACUSE
Your goods that lay at host, sir, in the Centaur.

ANTIPHOLUS OF SYRACUSE
He speaks to me.—I am your master, Dromio.
Come, go with us. We'll look to that anon.          415
Embrace thy brother there; rejoice with him.
          *Exeunt the brothers Antipholus*

DROMIO OF SYRACUSE
There is a fat friend at your master's house,
That kitchened me for you today at dinner.
She now shall be my sister, not my wife.

DROMIO OF EPHESUS
Methinks you are my glass and not my brother.          420
I see by you I am a sweet-facèd youth.
Will you walk in to see their gossiping?

DROMIO OF SYRACUSE Not I, sir, you are my elder.

DROMIO OF EPHESUS That's a question. How shall we try
it?          425

DROMIO OF SYRACUSE We'll draw cuts for the senior. Till
then, lead thou first.

DROMIO OF EPHESUS Nay, then thus:
We came into the world like brother and brother,
And now let's go hand in hand, not one before
          another.          *Exeunt ⌜to the priory⌝*

# LOVE'S LABOUR'S LOST

THE 1598 edition of *Love's Labour's Lost* is the first play text to carry Shakespeare's name on the title-page, which also refers to performance before the Queen 'this last Christmas'. The play is said to be 'Newly corrected and augmented', so perhaps an earlier edition has failed to survive. Even so, the text shows every sign of having been printed from Shakespeare's working papers, since it includes some passages in draft as well as in revised form. We print the drafts as Additional Passages. The play was probably written some years before publication, in 1593 or 1594.

The setting is Navarre—a kingdom straddling the border between Spain and France—where the young King and three of his friends vow to devote the following three years to austere self-improvement, forgoing the company of women. But they have forgotten the imminent arrival on a diplomatic mission of the Princess of France with, as it happens, three of her ladies; much comedy derives from, first, the men's embarrassed attempts to conceal from one another that they are falling in love, and second, the girls' practical joke in exchanging identities when the men, disguised as Russians, come to entertain and to woo them. Shakespeare seems to have picked up the King's friends' names—Biron, Dumaine, and Longueville—from leading figures in contemporary France, but to have invented the plot himself. He counterpoints the main action with events involving characters based in part on the type-figures of Italian commedia dell'arte who reflect facets of the lords' personalities. Costard, an unsophisticated, open-hearted yokel, and his girl-friend Jaquenetta are sexually uninhibited; Don Adriano de Armado, 'a refinèd traveller of Spain' who also, though covertly, loves Jaquenetta, is full of pompous affectation; and Holofernes, a schoolmaster (seen always with his admiring companion, the curate Sir Nathaniel), demonstrates the avid pedantry into which the young men's verbal brilliance could degenerate. Much of the play's language is highly sophisticated (this is, as the title-page claims, a 'conceited comedy'), in keeping with its subject matter. But the action reaches its climax when a messenger brings news which is communicated entirely without verbal statement. This is a theatrical masterstroke which also signals Shakespeare's most daring experiment with comic form. 'The scene begins to cloud'; in the play's closing minutes the lords and ladies seek to readjust themselves to the new situation, and the play ends in subdued fashion with a third entertainment, the songs of the owl and the cuckoo.

*Love's Labour's Lost* was for long regarded as a play of excessive verbal sophistication, of interest mainly because of a series of supposed topical allusions; but a number of distinguished twentieth-century productions have revealed its theatrical mastery.

# THE PERSONS OF THE PLAY

Ferdinand, KING of Navarre

BIRON  
LONGUEVILLE  } lords attending on the King  
DUMAINE

Don Adriano de ARMADO, an affected Spanish braggart  
MOTE, his page

PRINCESS of France

ROSALINE  
CATHERINE  } ladies attending on the Princess  
MARIA

BOYET  
Two other LORDS  } attending on the Princess

COSTARD, a Clown  
JAQUENETTA, a country wench

Sir NATHANIEL, a curate  
HOLOFERNES, a schoolmaster  
Anthony DULL, a constable

MERCADÉ, a messenger

A FORESTER

# Love's Labour's Lost

**1.1** *Enter Ferdinand, King of Navarre, Biron,*
*Longueville, and Dumaine*

KING

Let fame, that all hunt after in their lives,
Live registered upon our brazen tombs,
And then grace us in the disgrace of death
When, spite of cormorant devouring time,
Th'endeavour of this present breath may buy    5
That honour which shall bate his scythe's keen edge
And make us heirs of all eternity.
Therefore, brave conquerors—for so you are,
That war against your own affections
And the huge army of the world's desires—    10
Our late edict shall strongly stand in force.
Navarre shall be the wonder of the world.
Our court shall be a little academe,
Still and contemplative in living art.
You three—Biron, Dumaine, and Longueville—    15
Have sworn for three years' term to live with me
My fellow scholars, and to keep those statutes
That are recorded in this schedule here.
Your oaths are passed; and now subscribe your
    names,
That his own hand may strike his honour down    20
That violates the smallest branch herein.
If you are armed to do as sworn to do,
Subscribe to your deep oaths, and keep it, too.

LONGUEVILLE

I am resolved. 'Tis but a three years' fast.
The mind shall banquet, though the body pine.    25
Fat paunches have lean pates, and dainty bits
Make rich the ribs but bankrupt quite the wits.
    *He signs*

DUMAINE

My loving lord, Dumaine is mortified.
The grosser manner of these world's delights
He throws upon the gross world's baser slaves.    30
To love, to wealth, to pomp I pine and die,
With all these living in philosophy.
    *He signs*

BIRON

I can but say their protestation over.
So much, dear liege, I have already sworn:
That is, to live and study here three years.    35
But there are other strict observances,
As not to see a woman in that term,
Which I hope well is not enrollèd there;
And one day in a week to touch no food,
And but one meal on every day beside,    40
The which I hope is not enrollèd there;
And then to sleep but three hours in the night,
And not be seen to wink of all the day,
When I was wont to think no harm all night,
And make a dark night too of half the day,    45
Which I hope well is not enrollèd there.
O, these are barren tasks, too hard to keep—
Not to see ladies, study, fast, not sleep.

KING

Your oath is passed to pass away from these.

BIRON

Let me say no, my liege, an if you please.    50
I only swore to study with your grace,
And stay here in your court, for three years' space.

LONGUEVILLE

You swore to that, Biron, and to the rest.

BIRON

By yea and nay, sir, then I swore in jest.
What is the end of study, let me know?    55

KING

Why, that to know which else we should not know.

BIRON

Things hid and barred, you mean, from common
    sense.

KING

Ay, that is study's god-like recompense.

BIRON

Come on, then, I will swear to study so
To know the thing I am forbid to know,    60
As thus: to study where I well may dine
    When I to feast expressly am forbid;
Or study where to meet some mistress fine
    When mistresses from common sense are hid;
Or having sworn too hard a keeping oath,    65
Study to break it and not break my troth.
If study's gain be thus, and this be so,
Study knows that which yet it doth not know.
Swear me to this, and I will ne'er say no.

KING

These be the stops that hinder study quite,    70
And train our intellects to vain delight.

BIRON

Why, all delights are vain, but that most vain
Which, with pain purchased, doth inherit pain;
As painfully to pore upon a book
    To seek the light of truth while truth the while    75
Doth falsely blind the eyesight of his look.
    Light, seeking light, doth light of light beguile;
So ere you find where light in darkness lies
Your light grows dark by losing of your eyes.
Study me how to please the eye indeed    80
    By fixing it upon a fairer eye,
Who dazzling so, that eye shall be his heed,
    And give him light that it was blinded by.
Study is like the heavens' glorious sun,
    That will not be deep searched with saucy looks.    85
Small have continual plodders ever won
    Save base authority from others' books.
These earthly godfathers of heaven's lights,
    That give a name to every fixèd star,
Have no more profit of their shining nights    90
    Than those that walk and wot not what they are.
Too much to know is to know naught but fame,
And every godfather can give a name.

KING

How well he's read, to reason against reading!

DUMAINE

Proceeded well, to stop all good proceeding.    95

LONGUEVILLE

He weeds the corn and still lets grow the weeding.

BIRON
    The spring is near when green geese are a-breeding.
DUMAINE
    How follows that?
BIRON              Fit in his place and time.
DUMAINE
    In reason nothing.
BIRON           Something then in rhyme.
KING
    Biron is like an envious sneaping frost,     100
      That bites the first-born infants of the spring.
BIRON
    Well, say I am! Why should proud summer boast
      Before the birds have any cause to sing?
    Why should I joy in any abortive birth?
    At Christmas I no more desire a rose     105
    Than wish a snow in May's new-fangled shows,
    But like of each thing that in season grows.
    So you to study, now it is too late,
    Climb o'er the house to unlock the little gate.
KING
    Well, sit you out. Go home, Biron. Adieu.     110
BIRON
    No, my good lord, I have sworn to stay with you.
    And though I have for barbarism spoke more
      Than for that angel knowledge you can say,
    Yet confident I'll keep what I have sworn,
      And bide the penance of each three years' day.     115
    Give me the paper. Let me read the same,
    And to the strict'st decrees I'll write my name.
KING (*giving a paper*)
    How well this yielding rescues thee from shame!
BIRON (*reads*) 'Item: that no woman shall come within a
    mile of my court.' Hath this been proclaimed?     120
LONGUEVILLE Four days ago.
BIRON Let's see the penalty. 'On pain of losing her tongue.'
    Who devised this penalty?
LONGUEVILLE Marry, that did I.
BIRON Sweet lord, and why?     125
LONGUEVILLE
    To fright them hence with that dread penalty.
BIRON
    A dangerous law against gentility.
    'Item: if any man be seen to talk with a woman within
    the term of three years, he shall endure such public
    shame as the rest of the court can possible devise.'     130
    This article, my liege, yourself must break;
      For well you know here comes in embassy
    The French King's daughter with yourself to speak—
      A maid of grace and complete majesty—
    About surrender-up of Aquitaine     135
      To her decrepit, sick, and bedrid father.
    Therefore this article is made in vain,
      Or vainly comes th'admirèd Princess hither.
KING
    What say you, lords? Why, this was quite forgot.
BIRON
    So study evermore is overshot.     140
    While it doth study to have what it would,
    It doth forget to do the thing it should;
    And when it hath the thing it hunteth most,
    'Tis won as towns with fire—so won, so lost.
KING
    We must of force dispense with this decree.     145
    She must lie here, on mere necessity.

BIRON
    Necessity will make us all forsworn
      Three thousand times within this three years' space;
    For every man with his affects is born,
      Not by might mastered, but by special grace.     150
    If I break faith, this word shall speak for me:
    I am forsworn on mere necessity.
    So to the laws at large I write my name,
      And he that breaks them in the least degree
    Stands in attainder of eternal shame.     155
      *He signs*
    Suggestions are to other as to me,
    But I believe, although I seem so loath,
    I am the last that will last keep his oath.
    But is there no quick recreation granted?
KING
    Ay, that there is. Our court, you know, is haunted
      With a refinèd traveller of Spain,     161
    A man in all the world's new fashion planted,
      That hath a mint of phrases in his brain.
    One who the music of his own vain tongue
      Doth ravish like enchanting harmony;     165
    A man of complements, whom right and wrong
      Have chose as umpire of their mutiny.
    This child of fancy, that Armado hight,
      For interim to our studies shall relate
    In high-borne words the worth of many a knight     170
      From tawny Spain lost in the world's debate.
    How you delight, my lords, I know not, I;
    But I protest I love to hear him lie,
    And I will use him for my minstrelsy.
BIRON
    Armado is a most illustrious wight,     175
    A man of fire-new words, fashion's own knight.
LONGUEVILLE
    Costard the swain and he shall be our sport,
    And so to study three years is but short.
      *Enter a constable, Anthony Dull, with Costard with*
      *a letter*
DULL Which is the Duke's own person?
BIRON This, fellow. What wouldst?     180
DULL I myself reprehend his own person, for I am his
    grace's farborough. But I would see his own person in
    flesh and blood.
BIRON This is he.
DULL Señor Arm—Arm—commends you. There's villainy
    abroad. This letter will tell you more.     186
COSTARD Sir, the contempts thereof are as touching me.
KING A letter from the magnificent Armado.
BIRON How low soever the matter, I hope in God for high
    words.     190
LONGUEVILLE A high hope for a low heaven. God grant
    us patience.
BIRON To hear, or forbear laughing?
LONGUEVILLE To hear meekly, sir, and to laugh moder-
    ately, or to forbear both.     195
BIRON Well, sir, be it as the style shall give us cause to
    climb in the merriness.
COSTARD The matter is to me, sir, as concerning
    Jaquenetta. The manner of it is, I was taken with the
    manner.     200
BIRON In what manner?
COSTARD In manner and form following, sir—all those
    three. I was seen with her in the manor house, sitting
    with her upon the form, and taken following her into
    the park; which put together is 'in manner and form

following'. Now, sir, for the manner: it is the manner of a man to speak to a woman. For the form: in some form.

BIRON For the 'following', sir?                              209

COSTARD As it shall follow in my correction; and God defend the right.

KING Will you hear this letter with attention?

BIRON As we would hear an oracle.

COSTARD Such is the simplicity of man to hearken after the flesh.                                              215

KING (reads) 'Great deputy, the welkin's vicegerent and sole dominator of Navarre, my soul's earth's god, and body's fostering patron'—

COSTARD Not a word of Costard yet.

KING 'So it is'—                                             220

COSTARD It may be so; but if he say it is so, he is, in telling true, but so.

KING Peace!

COSTARD Be to me and every man that dares not fight.

KING No words!                                               225

COSTARD Of other men's secrets, I beseech you.

KING 'So it is, besieged with sable-coloured melancholy, I did commend the black-oppressing humour to the most wholesome physic of thy health-giving air, and, as I am a gentleman, betook myself to walk. The time when? About the sixth hour, when beasts most graze, birds best peck, and men sit down to that nourishment which is called supper. So much for the time when. Now for the ground which—which, I mean, I walked upon. It is yclept thy park. Then for the place where—where, I mean, I did encounter that obscene and most preposterous event that draweth from my snow-white pen the ebon-coloured ink which here thou viewest, beholdest, surveyest, or seest. But to the place where. It standeth north-north-east and by east from the west corner of thy curious-knotted garden. There did I see that low-spirited swain, that base minnow of thy mirth'—

COSTARD Me?

KING 'That unlettered, small-knowing soul'—                 245

COSTARD Me?

KING 'That shallow vassal'—

COSTARD Still me?

KING 'Which, as I remember, hight Costard'—

COSTARD O, me!                                               250

KING 'Sorted and consorted, contrary to thy established proclaimed edict and continent canon, with, with, O with—but with this I passion to say wherewith'—

COSTARD With a wench.

KING 'With a child of our grandmother Eve, a female, or for thy more sweet understanding a woman. Him I, as my ever-esteemed duty pricks me on, have sent to thee, to receive the meed of punishment, by thy sweet grace's officer Anthony Dull, a man of good repute, carriage, bearing, and estimation.'                                260

DULL Me, an't shall please you. I am Anthony Dull.

KING 'For Jaquenetta—so is the weaker vessel called—which I apprehended with the aforesaid swain, I keep her as a vessel of thy law's fury, and shall at the least of thy sweet notice bring her to trial. Thine in all compliments of devoted and heartburning heat of duty, Don Adriano de Armado.'

BIRON This is not so well as I looked for, but the best that ever I heard.                                             269

KING Ay, the best for the worst. (To Costard) But, sirrah, what say you to this?

COSTARD Sir, I confess the wench.

KING Did you hear the proclamation?

COSTARD I do confess much of the hearing it, but little of the marking of it.                                        275

KING It was proclaimed a year's imprisonment to be taken with a wench.

COSTARD I was taken with none, sir. I was taken with a damsel.

KING Well, it was proclaimed 'damsel'.                       280

COSTARD This was no damsel, neither, sir. She was a virgin.

⌜KING⌝ It is so varied, too, for it was proclaimed 'virgin'.

COSTARD If it were, I deny her virginity. I was taken with a maid.                                                  285

KING This 'maid' will not serve your turn, sir.

COSTARD This maid will serve my turn, sir.

KING Sir, I will pronounce your sentence. You shall fast a week with bran and water.

COSTARD I had rather pray a month with mutton and porridge.                                                   291

KING
    And Don Armado shall be your keeper.
My lord Biron, see him delivered o'er,
And go we, lords, to put in practice that
    Which each to other hath so strongly sworn.      295
        *Exeunt the King, Longueville, and Dumaine*

BIRON
I'll lay my head to any good man's hat
    These oaths and laws will prove an idle scorn.
Sirrah, come on.

COSTARD I suffer for the truth, sir; for true it is I was taken with Jaquenetta, and Jaquenetta is a true girl, and therefore, welcome the sour cup of prosperity, affliction may one day smile again; and till then, sit thee down, sorrow.                                    *Exeunt*

**1.2** *Enter Armado and Mote, his page*

ARMADO Boy, what sign is it when a man of great spirit grows melancholy?

MOTE A great sign, sir, that he will look sad.

ARMADO Why, sadness is one and the selfsame thing, dear imp.                                                     5

MOTE No, no, O Lord, sir, no.

ARMADO How canst thou part sadness and melancholy, my tender juvenal?

MOTE By a familiar demonstration of the working, my tough señor.                                                10

ARMADO Why 'tough señor'? Why 'tough señor'?

MOTE Why 'tender juvenal'? Why 'tender juvenal'?

ARMADO I spoke it, tender juvenal, as a congruent epitheton appertaining to thy young days, which we may nominate 'tender'.                                    15

MOTE And I, tough señor, as an appertinent title to your old time, which we may name 'tough'.

ARMADO Pretty and apt.

MOTE How mean you, sir? I 'pretty' and my saying 'apt'? Or I 'apt' and my saying 'pretty'?                       20

ARMADO Thou 'pretty', because little.

MOTE Little pretty, because little. Wherefore 'apt'?

ARMADO And therefore 'apt' because quick.

MOTE Speak you this in my praise, master?

ARMADO In thy condign praise.                                25

MOTE I will praise an eel with the same praise.

ARMADO What—that an eel is ingenious?

MOTE That an eel is quick.

ARMADO I do say thou art quick in answers. Thou heatest my blood.                                               30

MOTE I am answered, sir.

ARMADO I love not to be crossed.

MOTE *(aside)* He speaks the mere contrary—crosses love not him.

ARMADO I have promised to study three years with the Duke.                                                              36

MOTE You may do it in an hour, sir.

ARMADO Impossible.

MOTE How many is one, thrice told?

ARMADO I am ill at reckoning; it fitteth the spirit of a tapster.                                                          41

MOTE You are a gentleman and a gamester, sir.

ARMADO I confess both. They are both the varnish of a complete man.

MOTE Then I am sure you know how much the gross sum of deuce-ace amounts to.                                         46

ARMADO It doth amount to one more than two.

MOTE Which the base vulgar do call three.

ARMADO True.

MOTE Why, sir, is this such a piece of study? Now here is 'three' studied ere ye'll thrice wink, and how easy it is to put 'years' to the word 'three' and study 'three years' in two words, the dancing horse will tell you.

ARMADO A most fine figure.

MOTE *(aside)* To prove you a cipher.                                 55

ARMADO I will hereupon confess I am in love; and as it is base for a soldier to love, so am I in love with a base wench. If drawing my sword against the humour of affection would deliver me from the reprobate thought of it, I would take desire prisoner and ransom him to any French courtier for a new-devised curtsy. I think scorn to sigh. Methinks I should outswear Cupid. Comfort me, boy. What great men have been in love?

MOTE Hercules, master.                                              64

ARMADO Most sweet Hercules! More authority, dear boy. Name more—and, sweet my child, let them be men of good repute and carriage.

MOTE Samson, master; he was a man of good carriage, great carriage, for he carried the town-gates on his back like a porter, and he was in love.                         70

ARMADO O well-knit Samson, strong-jointed Samson! I do excel thee in my rapier as much as thou didst me in carrying gates. I am in love, too. Who was Samson's love, my dear Mote?

MOTE A woman, master.                                              75

ARMADO Of what complexion?

MOTE Of all the four, or the three, or the two, or one of the four.

ARMADO Tell me precisely of what complexion?

MOTE Of the sea-water green, sir.                                  80

ARMADO Is that one of the four complexions?

MOTE As I have read, sir; and the best of them, too.

ARMADO Green indeed is the colour of lovers, but to have a love of that colour, methinks Samson had small reason for it. He surely affected her for her wit.       85

MOTE It was so, sir, for she had a green wit.

ARMADO My love is most immaculate white and red.

MOTE Most maculate thoughts, master, are masked under such colours.

ARMADO Define, define, well-educated infant.                     90

MOTE My father's wit and my mother's tongue assist me!

ARMADO Sweet invocation of a child!—most pretty and pathetical.

MOTE      If she be made of white and red
            Her faults will ne'er be known,                        95
          For blushing cheeks by faults are bred
            And fears by pale white shown.

Then if she fear or be to blame,
  By this you shall not know;
For still her cheeks possess the same               100
  Which native she doth owe.
A dangerous rhyme, master, against the reason of white and red.

ARMADO Is there not a ballad, boy, of the King and the Beggar?                                                         105

MOTE The world was very guilty of such a ballad some three ages since, but I think now 'tis not to be found; or if it were, it would neither serve for the writing nor the tune.

ARMADO I will have that subject newly writ o'er, that I may example my digression by some mighty precedent. Boy, I do love that country girl that I took in the park with the rational hind Costard. She deserves well.

MOTE *(aside)* To be whipped—and yet a better love than my master.                                                        115

ARMADO Sing, boy. My spirit grows heavy in love.

MOTE And that's great marvel, loving a light wench.

ARMADO I say, sing.

MOTE Forbear till this company be past.

*Enter Costard the clown, Constable Dull, and Jaquenetta, a wench*

DULL *(to Armado)* Sir, the Duke's pleasure is that you keep Costard safe, and you must suffer him to take no delight, nor no penance, but a must fast three days a week. For this damsel, I must keep her at the park. She is allowed for the dey-woman. Fare you well.        124

ARMADO *(aside)* I do betray myself with blushing.—Maid.

JAQUENETTA Man.

ARMADO I will visit thee at the lodge.

JAQUENETTA That's hereby.

ARMADO I know where it is situate.

JAQUENETTA Lord, how wise you are!                                 130

ARMADO I will tell thee wonders.

JAQUENETTA With that face?

ARMADO I love thee.

JAQUENETTA So I heard you say.

ARMADO And so farewell.                                            135

JAQUENETTA Fair weather after you.

⌈DULL⌉ Come, Jaquenetta, away.

                              ⌈*Exeunt Dull and Jaquenetta*⌉

ARMADO Villain, thou shalt fast for thy offences ere thou be pardoned.

COSTARD Well, sir, I hope when I do it I shall do it on a full stomach.                                                     141

ARMADO Thou shalt be heavily punished.

COSTARD I am more bound to you than your fellows, for they are but lightly rewarded.

ARMADO Take away this villain. Shut him up.                       145

MOTE Come, you transgressing slave. Away!

COSTARD Let me not be pent up, sir. I will fast, being loose.

MOTE No, sir. That were fast and loose. Thou shalt to prison.                                                           150

COSTARD Well, if ever I do see the merry days of desolation that I have seen, some shall see.

MOTE What shall some see?

COSTARD Nay, nothing, Master Mote, but what they look upon. It is not for prisoners to be too silent in their words, and therefore I will say nothing. I thank God I have as little patience as another man, and therefore I can be quiet.                    *Exeunt Mote and Costard*

ARMADO I do affect the very ground—which is base—where her shoe—which is baser—guided by her foot—which is basest—doth tread. I shall be forsworn—which

is a great argument of falsehood—if I love. And how
can that be true love which is falsely attempted? Love
is a familiar; love is a devil. There is no evil angel but
love. Yet was Samson so tempted, and he had an
excellent strength. Yet was Solomon so seduced, and
he had a very good wit. Cupid's butt-shaft is too hard
for Hercules' club, and therefore too much odds for a
Spaniard's rapier. The first and second cause will not
serve my turn: the passado he respects not, the duello
he regards not. His disgrace is to be called boy, but his
glory is to subdue men. Adieu, valour; rust, rapier; be
still, drum: for your manager is in love; yea, he loveth.
Assist me, some extemporal god of rhyme, for I am
sure I shall turn sonnet. Devise wit, write pen, for I
am for whole volumes, in folio.                    *Exit*

2.1    *Enter the Princess of France with three attending*
       *ladies—Maria, Catherine, and Rosaline—and three*
       *lords, one named Boyet*
BOYET
  Now, madam, summon up your dearest spirits.
  Consider who the King your father sends,
  To whom he sends, and what's his embassy:
  Yourself, held precious in the world's esteem,
  To parley with the sole inheritor                    5
  Of all perfections that a man may owe,
  Matchless Navarre; the plea of no less weight
  Than Aquitaine, a dowry for a queen.
  Be now as prodigal of all dear grace
  As nature was in making graces dear                  10
  When she did starve the general world beside
  And prodigally gave them all to you.
PRINCESS
  Good Lord Boyet, my beauty, though but mean,
  Needs not the painted flourish of your praise.
  Beauty is bought by judgement of the eye,            15
  Not uttered by base sale of chapmen's tongues.
  I am less proud to hear you tell my worth
  Than you much willing to be counted wise
  In spending your wit in the praise of mine.
  But now to task the tasker: good Boyet,              20
  You are not ignorant all-telling fame
  Doth noise abroad Navarre hath made a vow
  Till painful study shall outwear three years
  No woman may approach his silent court.
  Therefore to's seemeth it a needful course,          25
  Before we enter his forbidden gates,
  To know his pleasure; and in that behalf,
  Bold of your worthiness, we single you
  As our best-moving fair solicitor.
  Tell him the daughter of the King of France          30
  On serious business, craving quick dispatch,
  Importunes personal conference with his grace.
  Haste, signify so much while we attend,
  Like humble-visaged suitors, his high will.
BOYET
  Proud of employment, willingly I go.                 35
PRINCESS
  All pride is willing pride, and yours is so.    *Exit Boyet*
  Who are the votaries, my loving lords,
  That are vow-fellows with this virtuous duke?
A LORD
  Lord Longueville is one.
PRINCESS                    Know you the man?
MARIA
  I know him, madam. At a marriage feast               40
  Between Lord Périgord and the beauteous heir

Of Jaques Fauconbridge solemnizèd
In Normandy saw I this Longueville.
A man of sovereign parts he is esteemed,
Well fitted in arts, glorious in arms.                 45
Nothing becomes him ill that he would well.
The only soil of his fair virtue's gloss—
If virtue's gloss will stain with any soil—
Is a sharp wit matched with too blunt a will,
Whose edge hath power to cut, whose will still wills
It should none spare that come within his power.       51
PRINCESS
  Some merry mocking lord, belike—is't so?
MARIA
  They say so most that most his humours know.
PRINCESS
  Such short-lived wits do wither as they grow.
  Who are the rest?                                    55
CATHERINE
  The young Dumaine, a well-accomplished youth,
  Of all that virtue love for virtue loved.
  Most power to do most harm, least knowing ill,
  For he hath wit to make an ill shape good,
  And shape to win grace, though he had no wit.        60
  I saw him at the Duke Alençon's once,
  And much too little of that good I saw
  Is my report to his great worthiness.
ROSALINE
  Another of these students at that time
  Was there with him, if I have heard a truth.         65
  Biron they call him, but a merrier man,
  Within the limit of becoming mirth,
  I never spent an hour's talk withal.
  His eye begets occasion for his wit,
  For every object that the one doth catch             70
  The other turns to a mirth-moving jest,
  Which his fair tongue, conceit's expositor,
  Delivers in such apt and gracious words
  That agèd ears play truant at his tales,
  And younger hearings are quite ravishèd,             75
  So sweet and voluble is his discourse.
PRINCESS
  God bless my ladies, are they all in love,
  That every one her own hath garnishèd
  With such bedecking ornaments of praise?
A LORD
  Here comes Boyet.
       *Enter Boyet*
PRINCESS                    Now, what admittance, lord?    80
BOYET
  Navarre had notice of your fair approach,
  And he and his competitors in oath
  Were all addressed to meet you, gentle lady,
  Before I came. Marry, thus much I have learnt:
  He rather means to lodge you in the field,           85
  Like one that comes here to besiege his court,
  Than seek a dispensation for his oath
  To let you enter his unpeopled house.
       *Enter Navarre, Longueville, Dumaine, and Biron*
  Here comes Navarre.
KING  Fair Princess, welcome to the court of Navarre.   90
PRINCESS  'Fair' I give you back again, and welcome I
  have not yet. The roof of this court is too high to be
  yours, and welcome to the wide fields too base to be
  mine.
KING
  You shall be welcome, madam, to my court.            95

PRINCESS
   I will be welcome, then. Conduct me thither.
KING
   Hear me, dear lady. I have sworn an oath—
PRINCESS
   Our Lady help my lord! He'll be forsworn.
KING
   Not for the world, fair madam, by my will.
PRINCESS
   Why, will shall break it—will and nothing else.   100
KING
   Your ladyship is ignorant what it is.
PRINCESS
   Were my lord so his ignorance were wise,
   Where now his knowledge must prove ignorance.
   I hear your grace hath sworn out housekeeping.
   'Tis deadly sin to keep that oath, my lord,   105
   And sin to break it.
   But pardon me, I am too sudden-bold.
   To teach a teacher ill beseemeth me.
   Vouchsafe to read the purpose of my coming,
   And suddenly resolve me in my suit.   110
     *She gives him a paper*
KING
   Madam, I will, if suddenly I may.
PRINCESS
   You will the sooner that I were away,
   For you'll prove perjured if you make me stay.
     *Navarre reads the paper*
BIRON (*to Rosaline*)
   Did not I dance with you in Brabant once?
⌈ROSALINE⌉
   Did not I dance with you in Brabant once?   115
BIRON
   I know you did.
⌈ROSALINE⌉       How needless was it then
   To ask the question!
BIRON       You must not be so quick.
⌈ROSALINE⌉
   'Tis 'long of you, that spur me with such questions.
BIRON
   Your wit's too hot, it speeds too fast, 'twill tire.
⌈ROSALINE⌉
   Not till it leave the rider in the mire.   120
BIRON
   What time o' day?
⌈ROSALINE⌉
   The hour that fools should ask.
BIRON
   Now fair befall your mask.
⌈ROSALINE⌉
   Fair fall the face it covers.
BIRON
   And send you many lovers.   125
⌈ROSALINE⌉
   Amen, so you be none.
BIRON
   Nay, then will I be gone.
KING (*to the Princess*)
   Madam, your father here doth intimate
   The payment of a hundred thousand crowns,
   Being but the one-half of an entire sum   130
   Disbursèd by my father in his wars.
   But say that he or we—as neither have—
   Received that sum, yet there remains unpaid
   A hundred thousand more, in surety of the which

   One part of Aquitaine is bound to us,   135
   Although not valued to the money's worth.
   If then the King your father will restore
   But that one half which is unsatisfied,
   We will give up our right in Aquitaine
   And hold fair friendship with his majesty.   140
   But that, it seems, he little purposeth,
   For here he doth demand to have repaid
   A hundred thousand crowns, and not demands,
   On payment of a hundred thousand crowns,
   To have his title live in Aquitaine,   145
   Which we much rather had depart withal,
   And have the money by our father lent,
   Than Aquitaine, so gelded as it is.
   Dear Princess, were not his requests so far
   From reason's yielding, your fair self should make   150
   A yielding 'gainst some reason in my breast,
   And go well satisfied to France again.
PRINCESS
   You do the King my father too much wrong,
   And wrong the reputation of your name,
   In so unseeming to confess receipt   155
   Of that which hath so faithfully been paid.
KING
   I do protest I never heard of it,
   And if you prove it I'll repay it back
   Or yield up Aquitaine.
PRINCESS       We arrest your word.
   Boyet, you can produce acquittances   160
   For such a sum from special officers
   Of Charles, his father.
KING       Satisfy me so.
BOYET
   So please your grace, the packet is not come
   Where that and other specialties are bound.
   Tomorrow you shall have a sight of them.   165
KING
   It shall suffice me, at which interview
   All liberal reason I will yield unto.
   Meantime receive such welcome at my hand
   As honour, without breach of honour, may
   Make tender of to thy true worthiness.   170
   You may not come, fair princess, within my gates,
   But here without you shall be so received
   As you shall deem yourself lodged in my heart,
   Though so denied fair harbour in my house.
   Your own good thoughts excuse me, and farewell.   175
   Tomorrow shall we visit you again.
PRINCESS
   Sweet health and fair desires consort your grace.
KING
   Thy own wish wish I thee in every place.
     *Exit with Longueville and Dumaine*
BIRON (*to Rosaline*) Lady, I will commend you to mine
   own heart.   180
ROSALINE Pray you, do my commendations. I would be
   glad to see it.
BIRON I would you heard it groan.
ROSALINE Is the fool sick?
BIRON Sick at the heart.   185
ROSALINE
   Alack, let it blood.
BIRON
   Would that do it good?
ROSALINE
   My physic says 'Ay'.

BIRON
  Will you prick't with your eye?
ROSALINE
  *Non point*, with my knife.                190
BIRON
  Now God save thy life.
ROSALINE
  And yours, from long living.
BIRON
  I cannot stay thanksgiving.          *Exit*
    *Enter Dumaine*
DUMAINE (*to Boyet*)
  Sir, I pray you a word. What lady is that same?
BOYET
  The heir of Alençon, Catherine her name.    195
DUMAINE
  A gallant lady. Monsieur, fare you well.    *Exit*
    *Enter Longueville*
LONGUEVILLE (*to Boyet*)
  I beseech you a word, what is she in the white?
BOYET
  A woman sometimes, an you saw her in the light.
LONGUEVILLE
  Perchance light in the light. I desire her name.
BOYET
  She hath but one for herself; to desire that were a
    shame.                           200
LONGUEVILLE
  Pray you, sir, whose daughter?
BOYET
  Her mother's, I have heard.
LONGUEVILLE
  God's blessing on your beard!
BOYET
  Good sir, be not offended.
  She is an heir of Fauconbridge.         205
LONGUEVILLE
  Nay, my choler is ended.
  She is a most sweet lady.
BOYET
  Not unlike, sir. That may be.    *Exit Longueville*
    *Enter Biron*
BIRON
  What's her name in the cap?
BOYET
  Rosaline, by good hap.              210
BIRON
  Is she wedded or no?
BOYET
  To her will, sir, or so.
BIRON
  O, you are welcome, sir. Adieu.
BOYET
  Farewell to me, sir, and welcome to you.    *Exit Biron*
MARIA
  That last is Biron, the merry madcap lord.    215
  Not a word with him but a jest.
BOYET              And every jest but a word.
PRINCESS
  It was well done of you to take him at his word.
BOYET
  I was as willing to grapple as he was to board.
⌈CATHERINE⌉
  Two hot sheeps, marry.
BOYET          And wherefore not ships?
  No sheep, sweet lamb, unless we feed on your lips.

⌈CATHERINE⌉
  You sheep and I pasture—shall that finish the jest?
BOYET
  So you grant pasture for me.
⌈CATHERINE⌉         Not so, gentle beast.  222
  My lips are no common, though several they be.
BOYET
  Belonging to whom?
⌈CATHERINE⌉         To my fortunes and me.
PRINCESS
  Good wits will be jangling; but, gentles, agree.  225
  This civil war of wits were much better used
  On Navarre and his bookmen, for here 'tis abused.
BOYET
  If my observation, which very seldom lies,
  By the heart's still rhetoric disclosèd with eyes,
  Deceive me not now, Navarre is infected.    230
PRINCESS With what?
BOYET
  With that which we lovers entitle 'affected'.
PRINCESS Your reason?
BOYET
  Why, all his behaviours did make their retire
  To the court of his eye, peeping thorough desire.  235
  His heart like an agate with your print impressed,
  Proud with his form, in his eye pride expressed.
  His tongue, all impatient to speak and not see,
  Did stumble with haste in his eyesight to be.
  All senses to that sense did make their repair,  240
  To feel only looking on fairest of fair.
  Methought all his senses were locked in his eye,
  As jewels in crystal, for some prince to buy,
  Who, tendering their own worth from where they
    were glassed,
  Did point you to buy them along as you passed.  245
  His face's own margin did quote such amazes
  That all eyes saw his eyes enchanted with gazes.
  I'll give you Aquitaine and all that is his
  An you give him for my sake but one loving kiss.
PRINCESS
  Come, to our pavilion. Boyet is disposed.    250
BOYET
  But to speak that in words which his eye hath
    disclosed.
  I only have made a mouth of his eye
  By adding a tongue, which I know will not lie.
⌈ROSALINE⌉
  Thou art an old love-monger, and speak'st skilfully.
⌈MARIA⌉
  He is Cupid's grandfather, and learns news of him.
⌈CATHERINE⌉
  Then was Venus like her mother, for her father is but
    grim.                        256
BOYET
  Do you hear, my mad wenches?
⌈MARIA⌉            No.
BOYET           What then, do you see?
⌈CATHERINE⌉
  Ay—our way to be gone.
BOYET         You are too hard for me.
    *Exeunt*

**3.1**    *Enter Armado the braggart, and Mote his boy*
ARMADO Warble, child; make passionate my sense of
  hearing.
MOTE (*sings*) Concolinel.

ARMADO Sweet air! Go, tenderness of years, take this key.
Give enlargement to the swain. Bring him festinately
hither. I must employ him in a letter to my love.     6
MOTE Master, will you win your love with a French
brawl?
ARMADO How meanest thou—brawling in French?
MOTE No, my complete master; but to jig off a tune at
the tongue's end, canary to it with your feet, humour
it with turning up your eyelids, sigh a note and sing
a note, sometime through the throat as if you swallowed
love with singing love, sometime through the nose as
if you snuffed up love by smelling love, with your hat
penthouse-like o'er the shop of your eyes, with your
arms crossed on your thin-belly doublet like a rabbit
on a spit, or your hands in your pocket like a man
after the old painting, and keep not too long in one
tune, but a snip and away. These are complements,
these are humours; these betray nice wenches that
would be betrayed without these, and make them men
of note—do you note? men—that most are affected to
these.
ARMADO How hast thou purchased this experience?     25
MOTE By my penny of observation.
ARMADO But O, but O—
MOTE 'The hobby-horse is forgot.'
ARMADO Call'st thou my love hobby-horse?
MOTE No, master, the hobby-horse is but a colt, and your
love perhaps a hackney. But have you forgot your love?
ARMADO Almost I had.                                 32
MOTE Negligent student, learn her by heart.
ARMADO By heart and in heart, boy.
MOTE And out of heart, master. All those three I will
prove.                                               36
ARMADO What wilt thou prove?
MOTE A man, if I live; and this, 'by', 'in', and 'without',
upon the instant: 'by' heart you love her because your
heart cannot come *by* her; 'in' heart you love her
because your heart is *in* love with her; and 'out' of
heart you love her, being *out* of heart that you cannot
enjoy her.
ARMADO I am all these three.                         44
MOTE (*aside*) And three times as much more, and yet
nothing at all.
ARMADO Fetch hither the swain. He must carry me a
letter.
MOTE (*aside*) A message well sympathized—a horse to be
ambassador for an ass.                               50
ARMADO Ha, ha! What sayst thou?
MOTE Marry, sir, you must send the ass upon the horse,
for he is very slow-gaited. But I go.
ARMADO The way is but short. Away!
MOTE As swift as lead, sir.                          55
ARMADO The meaning, pretty ingenious?
Is not lead a metal heavy, dull, and slow?
MOTE
*Minime*, honest master—or rather, master, no.
ARMADO
I say lead is slow.
MOTE                          You are too swift, sir, to say so.
Is that lead slow which is fired from a gun?         60
ARMADO Sweet smoke of rhetoric!
He reputes me a cannon, and the bullet, that's he.
I shoot thee at the swain.
MOTE                          Thump, then, and I flee.
                                                *Exit*

ARMADO
A most acute juvenal—voluble and free of grace.
By thy favour, sweet welkin, I must sigh in thy face.
Most rude melancholy, valour gives thee place.       66
My herald is returned.
        *Enter Mote the page, and Costard the clown*
MOTE
A wonder, master—here's a costard broken in a shin.
ARMADO
Some enigma, some riddle; come, thy *l'envoi*. Begin.
COSTARD No egma, no riddle, no *l'envoi*, no salve in the
mail, sir. O sir, plantain, a plain plantain—no *l'envoi*,
no *l'envoi*, no salve, sir, but a plantain.          72
ARMADO By virtue, thou enforcest laughter—thy silly
thought my spleen. The heaving of my lungs provokes
me to ridiculous smiling. O pardon me, my stars! Doth
the inconsiderate take salve for *l'envoi*, and the word
*l'envoi* for a salve?                                77
MOTE
Do the wise think them other? Is not *l'envoi* a salve?
ARMADO
No, page, it is an epilogue or discourse to make plain
Some obscure precedence that hath tofore been sain.
I will example it.                                   81
        The fox, the ape, and the humble-bee
        Were still at odds, being but three.
There's the moral. Now the *l'envoi*.
MOTE I will add the *l'envoi*. Say the moral again.   85
ARMADO        The fox, the ape, and the humble-bee
              Were still at odds, being but three.
MOTE          Until the goose came out of door
              And stayed the odds by adding four.
Now will I begin your moral, and do you follow with
my *l'envoi*.                                         91
        The fox, the ape, and the humble-bee
        Were still at odds, being but three.
ARMADO        Until the goose came out of door,
              Staying the odds by adding four.        95
MOTE A good *l'envoi*, ending in the goose. Would you
desire more?
COSTARD
The boy hath sold him a bargain—a goose, that's flat.
Sir, your pennyworth is good an your goose be fat.
To sell a bargain well is as cunning as fast and loose.
Let me see, a fat *l'envoi*—ay, that's a fat goose.   101
ARMADO
Come hither, come hither. How did this argument
        begin?
MOTE
By saying that a costard was broken in a shin.
Then called you for the *l'envoi*.
COSTARD True, and I for a plantain. Thus came your
argument in. Then the boy's fat *l'envoi*, the goose that
you bought, and he ended the market.                 107
ARMADO But tell me, how was there a costard broken in
a shin?
MOTE I will tell you sensibly.                        110
COSTARD Thou hast no feeling of it. Mote, I will speak
that *l'envoi*.
I, Costard, running out, that was safely within,
Fell over the threshold and broke my shin.
ARMADO We will talk no more of this matter.           115
COSTARD Till there be more matter in the shin.
ARMADO Sirrah Costard, I will enfranchise thee.
COSTARD O, marry me to one Frances! I smell some *l'envoi*,
some goose, in this.

ARMADO By my sweet soul, I mean setting thee at liberty,
enfreedoming thy person. Thou wert immured,
restrained, captivated, bound.                          122
COSTARD True, true, and now you will be my purgation
and let me loose.
ARMADO I give thee thy liberty, set thee from durance,
and in lieu thereof impose on thee nothing but this:
bear this significant to the country maid, Jaquenetta.
(*Giving him a letter*) There is remuneration (*giving him
money*), for the best ward of mine honour is rewarding
my dependants. Mote, follow.                          *Exit*
MOTE
Like the sequel, I. Signor Costard, adieu.            *Exit*
COSTARD
My sweet ounce of man's flesh, my incony Jew!    132
Now will I look to his remuneration. Remuneration—
O, that's the Latin word for three-farthings. Three-
farthings—remuneration. 'What's the price of this
inkle?' 'One penny.' 'No, I'll give you a remuneration.'
Why, it carries it! Remuneration! Why, it is a fairer
name than French crown. I will never buy and sell out
of this word.                                          139
    *Enter Biron*
BIRON My good knave Costard, exceedingly well met.
COSTARD Pray you, sir, how much carnation ribbon may
a man buy for a remuneration?
BIRON What is a remuneration?
COSTARD Marry, sir, halfpenny-farthing.
BIRON Why, then, three-farthing-worth of silk.    145
COSTARD I thank your worship. God be wi' you.
BIRON Stay, slave, I must employ thee.
    As thou wilt win my favour, good my knave,
    Do one thing for me that I shall entreat.
COSTARD When would you have it done, sir?    150
BIRON This afternoon.
COSTARD Well, I will do it, sir. Fare you well.
BIRON Thou knowest not what it is.
COSTARD I shall know, sir, when I have done it.
BIRON Why, villain, thou must know first.    155
COSTARD I will come to your worship tomorrow morning.
BIRON
    It must be done this afternoon. Hark, slave,
    It is but this:
    The Princess comes to hunt here in the park,
    And in her train there is a gentle lady.    160
    When tongues speak sweetly, then they name her
        name,
    And Rosaline they call her. Ask for her,
    And to her white hand see thou do commend
    This sealed-up counsel. There's thy guerdon (*giving
        him a letter and money*), go.
COSTARD Guerdon! O sweet guerdon!—better than
remuneration, elevenpence-farthing better—most sweet
guerdon! I will do it, sir, in print. Guerdon—
remuneration.                                          *Exit*
BIRON
    And I, forsooth, in love—I that have been love's whip,
    A very beadle to a humorous sigh,    170
    A critic, nay, a night-watch constable,
    A domineering pedant o'er the boy,
    Than whom no mortal so magnificent.
    This wimpled, whining, purblind, wayward boy,
    This Signor Junior, giant dwarf, Dan Cupid,    175
    Regent of love-rhymes, lord of folded arms,
    Th'anointed sovereign of sighs and groans,

    Liege of all loiterers and malcontents,
    Dread prince of plackets, king of codpieces,
    Sole imperator and great general    180
    Of trotting paritors—O my little heart!—
    And I to be a corporal of his field,
    And wear his colours like a tumbler's hoop!
    What? I love, I sue, I seek a wife?—
    A woman, that is like a German clock,    185
    Still a-repairing, ever out of frame,
    And never going aright, being a watch,
    But being watched that it may still go right.
    Nay, to be perjured, which is worst of all,
    And among three to love the worst of all—    190
    A whitely wanton with a velvet brow,
    With two pitch-balls stuck in her face for eyes—
    Ay, and, by heaven, one that will do the deed
    Though Argus were her eunuch and her guard.
    And I to sigh for her, to watch for her,    195
    To pray for her—go to, it is a plague
    That Cupid will impose for my neglect
    Of his almighty dreadful little might.
    Well, I will love, write, sigh, pray, sue, groan:
    Some men must love my lady, and some Joan.    *Exit*

**4.1**    *Enter the Princess, a Forester, her ladies—Rosaline,
        Maria, and Catherine—and her lords, among them
        Boyet*
PRINCESS
    Was that the King that spurred his horse so hard
    Against the steep uprising of the hill?
⌈BOYET⌉
    I know not, but I think it was not he.
PRINCESS
    Whoe'er a was, a showed a mounting mind.
    Well, lords, today we shall have our dispatch.    5
    Ere Saturday we will return to France.
    Then, forester my friend, where is the bush
    That we must stand and play the murderer in?
FORESTER
    Hereby, upon the edge of yonder coppice—
    A stand where you may make the fairest shoot.    10
PRINCESS
    I thank my beauty, I am fair that shoot,
    And thereupon thou speak'st 'the fairest shoot'.
FORESTER
    Pardon me, madam, for I meant not so.
PRINCESS
    What, what? First praise me, and again say no?
    O short-lived pride! Not fair? Alack, for woe!    15
FORESTER
    Yes, madam, fair.
PRINCESS                    Nay, never paint me now.
    Where fair is not, praise cannot mend the brow.
    Here, good my glass, take this for telling true.
        *She gives him money*
    Fair payment for foul words is more than due.
FORESTER
    Nothing but fair is that which you inherit.    20
PRINCESS
    See, see, my beauty will be saved by merit!
    O heresy in fair, fit for these days—
    A giving hand, though foul, shall have fair praise.
    But come, the bow. Now mercy goes to kill,
    And shooting well is then accounted ill.    25
    Thus will I save my credit in the shoot,
    Not wounding—pity would not let me do't.

If wounding, then it was to show my skill,
That more for praise than purpose meant to kill.
And, out of question, so it is sometimes—                    30
Glory grows guilty of detested crimes
When for fame's sake, for praise, an outward part,
We bend to that the working of the heart,
As I for praise alone now seek to spill
The poor deer's blood that my heart means no ill.    35
BOYET
Do not curst wives hold that self-sovereignty
Only for praise' sake when they strive to be
Lords o'er their lords?
PRINCESS
Only for praise, and praise we may afford
To any lady that subdues a lord.                    40
          *Enter Costard the clown*
BOYET
Here comes a member of the commonwealth.
COSTARD God dig-you-de'en, all. Pray you, which is the
head lady?
PRINCESS Thou shalt know her, fellow, by the rest that
have no heads.                    45
COSTARD Which is the greatest lady, the highest?
PRINCESS The thickest and the tallest.
COSTARD
The thickest and the tallest—it is so, truth is truth.
An your waist, mistress, were as slender as my wit
One o' these maids' girdles for your waist should be fit.
Are not you the chief woman? You are the thickest
here.                    51
PRINCESS What's your will, sir? What's your will?
COSTARD
I have a letter from Monsieur Biron to one Lady
Rosaline.
PRINCESS
O, thy letter, thy letter! (*She takes it*) He's a good
friend of mine.
(*To Costard*) Stand aside, good bearer. Boyet, you can
carve.                    55
Break up this capon.
          *She gives the letter to Boyet*
BOYET                    I am bound to serve.
This letter is mistook. It importeth none here.
It is writ to Jaquenetta.
PRINCESS                    We will read it, I swear.
Break the neck of the wax, and everyone give ear.    59
BOYET (*reads*) 'By heaven, that thou art fair is most
infallible, true that thou art beauteous, truth itself that
thou art lovely. More fairer than fair, beautiful than
beauteous, truer than truth itself, have commiseration
on thy heroical vassal. The magnanimous and most
illustrate King Cophetua set's eye upon the penurious
and indubitate beggar Zenelophon, and he it was that
might rightly say "*Veni, vidi, vici*", which to
annothanize in the vulgar—O base and obscure
vulgar!—*videlicet* "He came, see, and overcame." He
came, one; see, two; overcame, three. Who came? The
King. Why did he come? To see. Why did he see? To
overcome. To whom came he? To the beggar. What
saw he? The beggar. Who overcame he? The beggar.
The conclusion is victory. On whose side? The King's.
The captive is enriched. On whose side? The beggar's.
The catastrophe is a nuptial. On whose side? The
King's—no, on both in one, or one in both. I am the
King—for so stands the comparison—thou the beggar,
for so witnesseth thy lowliness. Shall I command thy

love? I may. Shall I enforce thy love? I could. Shall I
entreat thy love? I will. What shalt thou exchange for
rags? Robes. For tittles? Titles. For thyself? Me. Thus,
expecting thy reply, I profane my lips on thy foot, my
eyes on thy picture, and my heart on thy every part.
          Thine in the dearest design of industry,    85
                    Don Adriano de Armado.
Thus dost thou hear the Nemean lion roar
   'Gainst thee, thou lamb, that standest as his prey.
Submissive fall his princely feet before,
   And he from forage will incline to play.                    90
But if thou strive, poor soul, what art thou then?
Food for his rage, repasture for his den.'
PRINCESS
What plume of feathers is he that indited this letter?
What vane? What weathercock? Did you ever hear
   better?
BOYET
I am much deceived but I remember the style.    95
PRINCESS
Else your memory is bad, going o'er it erewhile.
BOYET
This Armado is a Spaniard that keeps here in court,
A phantasim, a Monarcho, and one that makes sport
To the Prince and his bookmates.
PRINCESS (*to Costard*)                    Thou, fellow, a word.
Who gave thee this letter?
COSTARD                    I told you—my lord.    100
PRINCESS
To whom shouldst thou give it?
COSTARD                    From my lord to my lady.
PRINCESS
From which lord to which lady?
COSTARD
From my lord Biron, a good master of mine,
To a lady of France that he called Rosaline.
PRINCESS
Thou hast mistaken his letter. Come, lords, away.    105
(*To Rosaline, giving her the letter*)
Here, sweet, put up this, 'twill be thine another day.
                    *Exit attended*
BOYET
Who is the suitor? Who is the suitor?
ROSALINE                    Shall I teach you to know?
BOYET
Ay, my continent of beauty.
ROSALINE                    Why, she that bears the bow.
Finely put off.
BOYET
My lady goes to kill horns, but if thou marry,    110
Hang me by the neck if horns that year miscarry.
Finely put on.
ROSALINE
Well then, I am the shooter.
BOYET                    And who is your deer?
ROSALINE
If we choose by the horns, yourself come not near.
Finely put on indeed!                    115
MARIA
You still wrangle with her, Boyet, and she strikes at
   the brow.
BOYET
But she herself is hit lower—have I hit her now?
ROSALINE Shall I come upon thee with an old saying that
was a man when King Pépin of France was a little boy,
as touching the hit it?                    120

BOYET So I may answer thee with one as old that was a
   woman when Queen Guinevere of Britain was a little
   wench, as touching the hit it.
ROSALINE (sings)
            Thou canst not hit it, hit it, hit it,
            Thou canst not hit it, my good man.    125
BOYET (sings)
            An I cannot, cannot, cannot,
            An I cannot, another can.        Exit Rosaline
COSTARD
   By my troth, most pleasant! How both did fit it!
MARIA
   A mark marvellous well shot, for they both did hit it.
BOYET
   A mark—O mark but that mark! A mark, says my
      lady.                                        130
   Let the mark have a prick in't to mete at, if it may be.
MARIA
   Wide o' the bow hand—i'faith, your hand is out.
COSTARD
   Indeed, a must shoot nearer, or he'll ne'er hit the clout.
BOYET
   An if my hand be out, then belike your hand is in.
COSTARD
   Then will she get the upshoot by cleaving the pin.   135
MARIA
   Come, come, you talk greasily, your lips grow foul.
COSTARD
   She's too hard for you at pricks, sir. Challenge her to
      bowl.
BOYET
   I fear too much rubbing. Goodnight, my good owl.
            Exeunt Boyet, Maria, ⌈and Catherine⌉
COSTARD
   By my soul, a swain, a most simple clown.
   Lord, Lord, how the ladies and I have put him down!
   O' my troth, most sweet jests, most incony vulgar
      wit,                                         141
   When it comes so smoothly off, so obscenely, as it
      were, so fit!
   Armado o'th' t'other side—O, a most dainty man!—
   To see him walk before a lady and to bear her fan!
   To see him kiss his hand, and how most sweetly a
      will swear,                                  145
   And his page o' t'other side, that handful of wit—
   Ah heavens, it is a most pathetical nit!
      Shout within
   Sola, sola!                                        Exit

4.2    Enter Dull, Holofernes the pedant, and Nathaniel the
       curate
NATHANIEL Very reverend sport, truly, and done in the
   testimony of a good conscience.
HOLOFERNES The deer was, as you know—sanguis—in
   blood, ripe as the pomewater who now hangeth like a
   jewel in the ear of caelo, the sky, the welkin, the heaven,
   and anon falleth like a crab on the face of terra, the
   soil, the land, the earth.                          7
NATHANIEL Truly, Master Holofernes, the epithets are
   sweetly varied, like a scholar at the least. But, sir, I
   assure ye it was a buck of the first head.          10
HOLOFERNES Sir Nathaniel, haud credo.
DULL 'Twas not a 'auld grey doe', 'twas a pricket.
HOLOFERNES Most barbarous intimation! Yet a kind of
   insinuation, as it were in via, in way, of explication,
   facere, as it were, replication, or rather ostentare, to

show, as it were, his inclination after his undressed,
   unpolished, uneducated, unpruned, untrained, or
   rather unlettered, or ratherest unconfirmed, fashion, to
   insert again my 'haud credo' for a deer.             19
DULL I said the deer was not a 'auld grey doe', 'twas a
   pricket.
HOLOFERNES Twice-sod simplicity, bis coctus!
   O thou monster ignorance, how deformed dost thou
      look!
NATHANIEL
   Sir, he hath never fed of the dainties that are bred in
      a book.                                         24
   He hath not eat paper, as it were, he hath not drunk
   ink. His intellect is not replenished, he is only an
   animal, only sensible in the duller parts,
   And such barren plants are set before us that we
      thankful should be,
   Which we of taste and feeling are, for those parts that
      do fructify in us more than he.
   For as it would ill become me to be vain, indiscreet,
      or a fool,                                      30
   So were there a patch set on learning to see him in a
      school.
   But omne bene say I, being of an old father's mind:
   'Many can brook the weather that love not the wind.'
DULL
   You two are bookmen. Can you tell me by your wit
   What was a month old at Cain's birth that's not five
      weeks old as yet?                              35
HOLOFERNES Dictynna, Goodman Dull, Dictynna, Goodman
   Dull.
DULL What is 'Dictima'?
NATHANIEL A title to Phoebe, to luna, to the moon.
HOLOFERNES
   The moon was a month old when Adam was no
      more,                                          40
   And raught not to five weeks when he came to five
      score.
   Th'allusion holds in the exchange.
DULL 'Tis true, indeed, the collusion holds in the
   exchange.
HOLOFERNES God comfort thy capacity, I say th'allusion
   holds in the exchange.                              46
DULL And I say the pollution holds in the exchange, for
   the moon is never but a month old—and I say beside
   that 'twas a pricket that the Princess killed.
HOLOFERNES Sir Nathaniel, will you hear an extemporal
   epitaph on the death of the deer? And to humour the
   ignorant call I the deer the Princess killed a pricket.
NATHANIEL Perge, good Master Holofernes, perge, so it
   shall please you to abrogate scurrility.             54
HOLOFERNES I will something affect the letter, for it argues
   facility.
   The preyful Princess pierced and pricked a pretty
      pleasing pricket.
      Some say a sore, but not a sore till now made sore
         with shooting.
   The dogs did yell; put 'l' to 'sore', then 'sorel' jumps
      from thicket—
      Or pricket sore, or else sorel. The people fall a-
         hooting.                                     60
   If sore be sore, then 'l' to 'sore' makes fifty sores—O
      sore 'l'!
   Of one sore I an hundred make by adding but one
      more 'l'.
NATHANIEL A rare talent!

DULL If a talent be a claw, look how he claws him with a talent. 65

HOLOFERNES This is a gift that I have, simple, simple—a foolish extravagant spirit, full of forms, figures, shapes, objects, ideas, apprehensions, motions, revolutions. These are begot in the ventricle of memory, nourished in the womb of *pia mater*, and delivered upon the mellowing of occasion. But the gift is good in those in whom it is acute, and I am thankful for it. 72

NATHANIEL Sir, I praise the Lord for you, and so may my parishioners; for their sons are well tutored by you, and their daughters profit very greatly under you. You are a good member of the commonwealth. 76

HOLOFERNES *Mehercle*, if their sons be ingenious they shall want no instruction; if their daughters be capable, I will put it to them. But *Vir sapit qui pauca loquitur*; a soul feminine saluteth us. 80

*Enter Jaquenetta, and Costard the clown*

JAQUENETTA God give you good-morrow, Master Parson.

HOLOFERNES Master Parson, *quasi* 'pierce one'? And if one should be pierced, which is the one?

COSTARD Marry, Master Schoolmaster, he that is likeliest to a hogshead. 85

HOLOFERNES 'Of piercing a hogshead'—a good lustre of conceit in a turf of earth, fire enough for a flint, pearl enough for a swine—'tis pretty, it is well.

JAQUENETTA Good Master Parson, be so good as read me this letter. It was given me by Costard, and sent me from Don Armado. I beseech you read it. 91

*She gives the letter to Nathaniel, who reads it*

HOLOFERNES (*to himself*) '*Facile precor gelida quando pecas omnia sub umbra ruminat*', and so forth. Ah, good old Mantuan! I may speak of thee as the traveller doth of Venice: 95

    *Venezia, Venezia,*
    *Chi non ti vede, chi non ti prezia.*

Old Mantuan, old Mantuan—who understandeth thee not, loves thee not. (*He sings*) Ut, re, sol, la, mi, fa. (*To Nathaniel*) Under pardon, sir, what are the contents? Or rather, as Horace says in his—what, my soul—verses? 102

NATHANIEL Ay, sir, and very learned.

HOLOFERNES Let me hear a staff, a stanza, a verse. *Lege, domine.* 105

NATHANIEL (*reads*)
'If love make me forsworn, how shall I swear to love?
  Ah, never faith could hold, if not to beauty vowed.
Though to myself forsworn, to thee I'll faithful prove.
  Those thoughts to me were oaks, to thee like osiers bowed.
Study his bias leaves, and makes his book thine eyes,
  Where all those pleasures live that art would comprehend. 111
If knowledge be the mark, to know thee shall suffice.
  Well learnèd is that tongue that well can thee commend;
All ignorant that soul that sees thee without wonder;
  Which is to me some praise that I thy parts admire.
Thy eye Jove's lightning bears, thy voice his dreadful thunder, 116
  Which, not to anger bent, is music and sweet fire.
Celestial as thou art, O pardon, love, this wrong,
  That singeth heaven's praise with such an earthly tongue.'

HOLOFERNES You find not the apostrophus, and so miss the accent. Let me supervise the canzonet. Here are

only numbers ratified, but for the elegancy, facility, and golden cadence of poesy—*caret*. Ovidius Naso was the man. And why indeed 'Naso' but for smelling out the odoriferous flowers of fancy, the jerks of invention? *Imitari* is nothing. So doth the hound his master, the ape his keeper, the tired horse his rider. But *domicella*—virgin—was this directed to you? 128

JAQUENETTA Ay, sir.

HOLOFERNES I will overglance the superscript. 'To the snow-white hand of the most beauteous Lady Rosaline.' I will look again on the intellect of the letter for the nomination of the party writing to the person written unto. 'Your ladyship's in all desired employment, Biron.' Sir Nathaniel, this Biron is one of the votaries with the King, and here he hath framed a letter to a sequent of the stranger Queen's, which, accidentally or by the way of progression, hath miscarried. (*To Jaquenetta*) Trip and go, my sweet, deliver this paper into the royal hand of the King. It may concern much. Stay not thy compliment, I forgive thy duty. Adieu.

JAQUENETTA Good Costard, go with me.—Sir, God save your life. 143

COSTARD Have with thee, my girl.     *Exit with Jaquenetta*

NATHANIEL Sir, you have done this in the fear of God very religiously, and, as a certain father saith—

HOLOFERNES Sir, tell not me of the father; I do fear colourable colours. But to return to the verses—did they please you, Sir Nathaniel?

NATHANIEL Marvellous well for the pen. 150

HOLOFERNES I do dine today at the father's of a certain pupil of mine where, if before repast it shall please you to gratify the table with a grace, I will on my privilege I have with the parents of the foresaid child or pupil undertake your *ben venuto*, where I will prove those verses to be very unlearned, neither savouring of poetry, wit, nor invention. I beseech your society. 157

NATHANIEL And thank you too, for society, saith the text, is the happiness of life.

HOLOFERNES And certes the text most infallibly concludes it. (*To Dull*) Sir, I do invite you too. You shall not say me nay. *Pauca verba.* Away, the gentles are at their game, and we will to our recreation.     *Exeunt*

**4.3**     *Enter Biron with a paper in his hand, alone*

BIRON The King, he is hunting the deer. I am coursing myself. They have pitched a toil, I am toiling in a pitch—pitch that defiles. Defile—a foul word. Well, set thee down, sorrow; for so they say the fool said, and so say I, and I the fool. Well proved, wit! By the Lord, this love is as mad as Ajax, it kills sheep, it kills me, I a sheep—well proved again o' my side. I will not love. If I do, hang me; i'faith, I will not. O, but her eye! By this light, but for her eye I would not love her. Yes, for her two eyes. Well, I do nothing in the world but lie, and lie in my throat. By heaven, I do love, and it hath taught me to rhyme and to be melancholy, and here (*showing a paper*) is part of my rhyme, and here (*touching his breast*) my melancholy. Well, she hath one o' my sonnets already. The clown bore it, the fool sent it, and the lady hath it. Sweet clown, sweeter fool, sweetest lady. By the world, I would not care a pin if the other three were in. Here comes one with a paper. God give him grace to groan.

*He stands aside. The King entereth with a paper*

KING Ay me! 20

BIRON (*aside*) Shot, by heaven! Proceed, sweet Cupid, thou
    hast thumped him with thy birdbolt under the left pap.
    In faith, secrets.
KING (*reads*)
    'So sweet a kiss the golden sun gives not
      To those fresh morning drops upon the rose    25
    As thy eyebeams when their fresh rays have smote
      The night of dew that on my cheeks down flows.
    Nor shines the silver moon one-half so bright
      Through the transparent bosom of the deep
    As doth thy face through tears of mine give light.   30
      Thou shin'st in every tear that I do weep.
    No drop but as a coach doth carry thee,
      So ridest thou triumphing in my woe.
    Do but behold the tears that swell in me
      And they thy glory through my grief will show.
    But do not love thyself; then thou wilt keep    36
    My tears for glasses, and still make me weep.
    O Queen of queens, how far dost thou excel,
    No thought can think nor tongue of mortal tell.'
How shall she know my griefs? I'll drop the paper.   40
Sweet leaves, shade folly. Who is he comes here?
    *Enter Longueville with papers. The King steps aside*
What, Longueville, and reading—listen, ear!
BIRON (*aside*)
    Now in thy likeness one more fool appear!
LONGUEVILLE Ay me! I am forsworn.
BIRON (*aside*)
    Why, he comes in like a perjure, wearing papers.   45
KING (*aside*)
    In love, I hope! Sweet fellowship in shame.
BIRON (*aside*)
    One drunkard loves another of the name.
LONGUEVILLE
    Am I the first that have been perjured so?
BIRON (*aside*)
    I could put thee in comfort, not by two that I know.
    Thou makest the triumviry, the corner-cap of society,
    The shape of love's Tyburn, that hangs up simplicity.
LONGUEVILLE
    I fear these stubborn lines lack power to move.    52
    O sweet Maria, empress of my love,
    These numbers will I tear, and write in prose.
BIRON (*aside*)
    O, rhymes are guards on wanton Cupid's hose,    55
    Disfigure not his slop.
LONGUEVILLE           This same shall go.
    *He reads the sonnet*
    'Did not the heavenly rhetoric of thine eye,
      'Gainst whom the world cannot hold argument,
    Persuade my heart to this false perjury?
      Vows for thee broke deserve not punishment.    60
    A woman I forswore, but I will prove,
      Thou being a goddess, I forswore not thee.
    My vow was earthly, thou a heavenly love.
      Thy grace being gained cures all disgrace in me.
    Vows are but breath, and breath a vapour is.    65
      Then thou, fair sun, which on my earth dost shine,
    Exhal'st this vapour-vow; in thee it is.
      If broken then, it is no fault of mine.
    If by me broke, what fool is not so wise
    To lose an oath to win a paradise?'    70
BIRON (*aside*)
    This is the liver vein, which makes flesh a deity,
    A green goose a goddess, pure, pure idolatry.

God amend us, God amend: we are much out o'th'
    way.
    *Enter Dumaine with a paper*
LONGUEVILLE (*aside*)
    By whom shall I send this? Company? Stay.
    *He steps aside*
BIRON (*aside*)
    All hid, all hid—an old infant play.    75
    Like a demigod here sit I in the sky,
    And wretched fools' secrets heedfully o'er-eye.
    More sacks to the mill! O heavens, I have my wish.
    Dumaine transformed—four woodcocks in a dish!
DUMAINE O most divine Kate!    80
BIRON (*aside*) O most profane coxcomb!
DUMAINE
    By heaven, the wonder in a mortal eye!
BIRON (*aside*)
    By earth, she is not, corporal; there you lie.
DUMAINE
    Her amber hairs for foul hath amber quoted.
BIRON (*aside*)
    An amber-coloured raven was well noted.    85
DUMAINE
    As upright as the cedar.
BIRON (*aside*)          Stoop, I say.
    Her shoulder is with child.
DUMAINE           As fair as day.
BIRON (*aside*)
    Ay, as some days; but then no sun must shine.
DUMAINE O that I had my wish!
LONGUEVILLE (*aside*) And I had mine!    90
KING (*aside*) And I mine too, good Lord!
BIRON (*aside*)
    Amen, so I had mine. Is not that a good word?
DUMAINE
    I would forget her, but a fever she
    Reigns in my blood and will remembered be.
BIRON (*aside*)
    A fever in your blood—why then, incision    95
    Would let her out in saucers—sweet misprision.
DUMAINE
    Once more I'll read the ode that I have writ.
BIRON (*aside*)
    Once more I'll mark how love can vary wit.
    *Dumaine reads his sonnet*
DUMAINE
    'On a day—alack the day—
    Love, whose month is ever May,    100
    Spied a blossom passing fair
    Playing in the wanton air.
    Through the velvet leaves the wind
    All unseen can passage find,
    That the lover, sick to death,    105
    Wished himself the heavens' breath.
    "Air", quoth he, "thy cheeks may blow;
    Air, would I might triumph so.
    But, alack, my hand is sworn
    Ne'er to pluck thee from thy thorn—    110
    Vow, alack, for youth unmeet,
    Youth so apt to pluck a sweet.
    Do not call it sin in me
    That I am forsworn for thee,
    Thou for whom great Jove would swear    115
    Juno but an Ethiop were,
    And deny himself for Jove,
    Turning mortal for thy love." '

This will I send, and something else more plain,
That shall express my true love's fasting pain.          120
O, would the King, Biron, and Longueville
Were lovers too! Ill to example ill
Would from my forehead wipe a perjured note,
For none offend where all alike do dote.

LONGUEVILLE (coming forward)
Dumaine, thy love is far from charity,          125
That in love's grief desir'st society.
You may look pale, but I should blush, I know,
To be o'erheard and taken napping so.

KING (coming forward)
Come, sir, you blush. As his, your case is such.
You chide at him, offending twice as much.          130
You do not love Maria? Longueville
Did never sonnet for her sake compile,
Nor never lay his wreathèd arms athwart
His loving bosom to keep down his heart?
I have been closely shrouded in this bush,          135
And marked you both, and for you both did blush.
I heard your guilty rhymes, observed your fashion,
Saw sighs reek from you, noted well your passion.
'Ay me!' says one, 'O Jove!' the other cries.
One, her hairs were gold; crystal the other's eyes.          140
(To Longueville) You would for paradise break faith and
          troth,
(To Dumaine) And Jove for your love would infringe an
          oath.
What will Biron say when that he shall hear
Faith so infringèd, which such zeal did swear?
How will he scorn, how will he spend his wit!          145
How will he triumph, leap, and laugh at it!
For all the wealth that ever I did see
I would not have him know so much by me.

BIRON (coming forward)
Now step I forth to whip hypocrisy.
Ah, good my liege, I pray thee pardon me.          150
Good heart, what grace hast thou thus to reprove
These worms for loving, that art most in love?
Your eyes do make no coaches. In your tears
There is no certain princess that appears.
You'll not be perjured, 'tis a hateful thing;          155
Tush, none but minstrels like of sonneting!
But are you not ashamed, nay, are you not,
All three of you, to be thus much o'ershot?
(To Longueville) You found his mote, the King your
          mote did see,
But I a beam do find in each of three.          160
O, what a scene of fool'ry have I seen,
Of sighs, of groans, of sorrow, and of teen!
O me, with what strict patience have I sat,
To see a king transformèd to a gnat!
To see great Hercules whipping a gig,          165
And profound Solomon to tune a jig,
And Nestor play at pushpin with the boys,
And critic Timon laugh at idle toys!
Where lies thy grief, O tell me, good Dumaine?
And, gentle Longueville, where lies thy pain?          170
And where my liege's? All about the breast.
A caudle, ho!

KING          Too bitter is thy jest.
Are we betrayed thus to thy over-view?

BIRON
Not you to me, but I betrayed by you.
I that am honest, I that hold it sin          175
To break the vow I am engagèd in.

I am betrayed by keeping company
With men like you, men of inconstancy.
When shall you see me write a thing in rhyme,
Or groan for Joan, or spend a minute's time          180
In pruning me? When shall you hear that I
Will praise a hand, a foot, a face, an eye,
A gait, a state, a brow, a breast, a waist,
A leg, a limb?

KING          Soft, whither away so fast?
A true man or a thief, that gallops so?          185

BIRON
I post from love; good lover, let me go.
          Enter Jaquenetta with a letter, and Costard the
          clown

JAQUENETTA
God bless the King!

KING          What present hast thou there?

COSTARD
Some certain treason.

KING          What makes treason here?

COSTARD
Nay, it makes nothing, sir.

KING          If it mar nothing neither,
The treason and you go in peace away together!          190

JAQUENETTA
I beseech your grace, let this letter be read.
Our parson misdoubts it; 'twas treason, he said.

KING Biron, read it over.
          Biron takes and reads the letter
(To Jaquenetta) Where hadst thou it?

JAQUENETTA Of Costard.          195

KING (to Costard) Where hadst thou it?

COSTARD Of Dun Adramadio, Dun Adramadio.
          Biron tears the letter

KING (to Biron)
How now, what is in you? Why dost thou tear it?

BIRON
A toy, my liege, a toy. Your grace needs not fear it.

LONGUEVILLE
It did move him to passion, and therefore let's hear it.

DUMAINE (taking up a piece of the letter)
It is Biron's writing, and here is his name.          201

BIRON (to Costard)
Ah, you whoreson loggerhead, you were born to do
          me shame!
Guilty, my lord, guilty! I confess, I confess.

KING What?

BIRON
That you three fools lacked me fool to make up the
          mess.          205
He, he, and you—e'en you, my liege—and I
Are pickpurses in love, and we deserve to die.
O, dismiss this audience, and I shall tell you more.

DUMAINE
Now the number is even.

BIRON          True, true; we are four.
Will these turtles be gone?

KING          Hence, sirs; away.          210

COSTARD
Walk aside the true folk, and let the traitors stay.
          Exeunt Costard and Jaquenetta

BIRON
Sweet lords, sweet lovers!—O, let us embrace.
As true we are as flesh and blood can be.
The sea will ebb and flow, heaven show his face.
Young blood doth not obey an old decree.          215

We cannot cross the cause why we were born,
Therefore of all hands must we be forsworn.

KING
What, did these rent lines show some love of thine?

BIRON
'Did they', quoth you? Who sees the heavenly
        Rosaline
That, like a rude and savage man of Ind                    220
    At the first op'ning of the gorgeous east,
Bows not his vassal head and, strucken blind,
    Kisses the base ground with obedient breast?
What peremptory eagle-sighted eye
    Dares look upon the heaven of her brow                 225
That is not blinded by her majesty?

KING
What zeal, what fury hath inspired thee now?
My love, her mistress, is a gracious moon,
    She an attending star, scarce seen a light.

BIRON
My eyes are then no eyes, nor I Biron.                     230
O, but for my love, day would turn to night.
Of all complexions the culled sovereignty
    Do meet as at a fair in her fair cheek,
Where several worthies make one dignity,
    Where nothing wants that want itself doth seek.
Lend me the flourish of all gentle tongues—               236
    Fie, painted rhetoric! O, she needs it not.
To things of sale a seller's praise belongs.
    She passes praise—then praise too short doth blot.
A withered hermit fivescore winters worn                  240
    Might shake off fifty, looking in her eye.
Beauty doth varnish age as if new-born,
    And gives the crutch the cradle's infancy.
O, 'tis the sun that maketh all things shine.

KING
By heaven, thy love is black as ebony.                    245

BIRON
Is ebony like her? O word divine!
A wife of such wood were felicity.
O, who can give an oath? Where is a book,
    That I may swear beauty doth beauty lack
If that she learn not of her eye to look?                 250
    No face is fair that is not full so black.

KING
O paradox! Black is the badge of hell,
    The hue of dungeons and the style of night,
And beauty's crest becomes the heavens well.

BIRON
Devils soonest tempt, resembling spirits of light.        255
O, if in black my lady's brows be decked,
    It mourns that painting and usurping hair
Should ravish doters with a false aspect,
    And therefore is she born to make black fair.
Her favour turns the fashion of the days,                 260
    For native blood is counted painting now,
And therefore red that would avoid dispraise
    Paints itself black to imitate her brow.

DUMAINE
To look like her are chimney-sweepers black.

LONGUEVILLE
And since her time are colliers counted bright.           265

KING
And Ethiops of their sweet complexion crack.

DUMAINE
Dark needs no candles now, for dark is light.

BIRON
Your mistresses dare never come in rain,
    For fear their colours should be washed away.

KING
'Twere good yours did; for, sir, to tell you plain,       270
    I'll find a fairer face not washed today.

BIRON
I'll prove her fair, or talk till doomsday here.

KING
No devil will fright thee then so much as she.

DUMAINE
I never knew man hold vile stuff so dear.

LONGUEVILLE (showing his foot)
    Look, here's thy love—my foot and her face see.       275

BIRON
O, if the streets were pavèd with thine eyes
    Her feet were much too dainty for such tread.

DUMAINE
O vile! Then as she goes, what upward lies
    The street should see as she walked overhead.

KING
But what of this? Are we not all in love?                 280

BIRON
    Nothing so sure, and thereby all forsworn.

KING
Then leave this chat and, good Biron, now prove
    Our loving lawful and our faith not torn.

DUMAINE
Ay, marry there, some flattery for this evil.

LONGUEVILLE
O, some authority how to proceed,                         285
Some tricks, some quillets how to cheat the devil.

DUMAINE
Some salve for perjury.

BIRON                    O, 'tis more than need.
Have at you, then, affection's men-at-arms.
Consider what you first did swear unto:
To fast, to study, and to see no woman—                   290
Flat treason 'gainst the kingly state of youth.
Say, can you fast? Your stomachs are too young,
And abstinence engenders maladies.
O, we have made a vow to study, lords,
And in that vow we have forsworn our books;               295
For when would you, my liege, or you, or you
In leaden contemplation have found out
Such fiery numbers as the prompting eyes
Of beauty's tutors have enriched you with?
Other slow arts entirely keep the brain,                  300
And therefore, finding barren practisers,
Scarce show a harvest of their heavy toil.
But love, first learnèd in a lady's eyes,
Lives not alone immurèd in the brain,
But with the motion of all elements                       305
Courses as swift as thought in every power,
And gives to every power a double power
Above their functions and their offices.
It adds a precious seeing to the eye—
A lover's eyes will gaze an eagle blind.                  310
A lover's ear will hear the lowest sound
When the suspicious head of theft is stopped.
Love's feeling is more soft and sensible
Than are the tender horns of cockled snails.
Love's tongue proves dainty Bacchus gross in taste.
For valour, is not love a Hercules,                       316
Still climbing trees in the Hesperides?
Subtle as Sphinx, as sweet and musical

As bright Apollo's lute strung with his hair;
And when love speaks, the voice of all the gods    320
Make heaven drowsy with the harmony.
Never durst poet touch a pen to write
Until his ink were tempered with love's sighs.
O, then his lines would ravish savage ears,
And plant in tyrants mild humility.    325
From women's eyes this doctrine I derive.
They sparkle still the right Promethean fire.
They are the books, the arts, the academes
That show, contain, and nourish all the world,
Else none at all in aught proves excellent.    330
Then fools you were these women to forswear,
Or keeping what is sworn, you will prove fools.
For wisdom's sake—a word that all men love—
Or for love's sake—a word that loves all men—
Or for men's sake—the authors of these women—    335
Or women's sake—by whom we men are men—
Let us once lose our oaths to find ourselves,
Or else we lose ourselves to keep our oaths.
It is religion to be thus forsworn,
For charity itself fulfils the law,    340
And who can sever love from charity?
KING
Saint Cupid, then, and, soldiers, to the field!
BIRON
Advance your standards, and upon them, lords.
Pell-mell, down with them; but be first advised
In conflict that you get the sun of them.    345
LONGUEVILLE
Now to plain dealing. Lay these glozes by.
Shall we resolve to woo these girls of France?
KING
And win them, too! Therefore let us devise
Some entertainment for them in their tents.
BIRON
First, from the park let us conduct them thither;    350
Then homeward every man attach the hand
Of his fair mistress. In the afternoon
We will with some strange pastime solace them,
Such as the shortness of the time can shape,
For revels, dances, masques, and merry hours    355
Forerun fair love, strewing her way with flowers.
KING
Away, away, no time shall be omitted
That will be time, and may by us be fitted.
BIRON
*Allons, allons*! Sowed cockle reaped no corn,
And justice always whirls in equal measure.    360
Light wenches may prove plagues to men forsworn.
If so, our copper buys no better treasure.    *Exeunt*

**5.1**    *Enter Holofernes the pedant, Nathaniel the curate,
and Anthony Dull*
HOLOFERNES *Satis quid sufficit.*
NATHANIEL I praise God for you, sir. Your reasons at
dinner have been sharp and sententious, pleasant
without scurrility, witty without affection, audacious
without impudency, learned without opinion, and
strange without heresy. I did converse this quondam
day with a companion of the King's who is intituled,
nominated, or called Don Adriano de Armado.    8
HOLOFERNES *Novi hominum tanquam te.* His humour is
lofty, his discourse peremptory, his tongue filed, his eye
ambitious, his gait majestical, and his general
behaviour vain, ridiculous, and thrasonical. He is too

picked, too spruce, too affected, too odd, as it were, too
peregrinate, as I may call it.
NATHANIEL A most singular and choice epithet.    15
*He draws out his table-book*
HOLOFERNES He draweth out the thread of his verbosity
finer than the staple of his argument. I abhor such
fanatical phantasims, such insociable and point-device
companions, such rackers of orthography as to speak
'dout', *sine* 'b', when he should say 'doubt'; 'det' when
he should pronounce 'debt'—'d, e, b, t', not 'd, e, t'.
He clepeth a calf 'cauf', half 'hauf', neighbour
*vocatur* 'nebour'—'neigh' abbreviated 'ne'. This is
abhominable—which he would call 'abominable'. It
insinuateth me of *insanire—ne intelligis, domine?*—to
make frantic, lunatic.    26
NATHANIEL *Laus deo, bone intelligo.*
HOLOFERNES *Bone? Bon, fort bon*—Priscian a little
scratched—'twill serve.
*Enter Armado the braggart, Mote his boy, and
Costard the clown*
NATHANIEL *Videsne quis venit?*    30
HOLOFERNES *Video, et gaudio.*
ARMADO (*to Mote*) Chirrah.
HOLOFERNES (*to Nathaniel*) *Quare* 'chirrah', not 'sirrah'?
ARMADO Men of peace, well encountered.
HOLOFERNES Most military sir, salutation!    35
MOTE (*aside to Costard*) They have been at a great feast of
languages and stolen the scraps.
COSTARD (*aside to Mote*) O, they have lived long on the
alms-basket of words. I marvel thy master hath not
eaten thee for a word, for thou art not so long by the
head as *honorificabilitudinitatibus*. Thou art easier
swallowed than a flapdragon.    42
MOTE (*aside to Costard*) Peace, the peal begins.
ARMADO (*to Holofernes*) Monsieur, are you not lettered?
MOTE Yes, yes, he teaches boys the horn-book. What is
'a, b' spelled backward, with the horn on his head?
HOLOFERNES Ba, *pueritia*, with a horn added.    47
MOTE Ba, most silly sheep, with a horn! You hear his
learning.
HOLOFERNES *Quis, quis*, thou consonant?    50
MOTE The last of the five vowels if you repeat them, or
the fifth if I.
HOLOFERNES I will repeat them: a, e, i—
MOTE The sheep. The other two concludes it: o, u.    54
ARMADO Now by the salt wave of the *Mediterraneum* a
sweet touch, a quick venue of wit; snip, snap, quick,
and home. It rejoiceth my intellect—true wit.
MOTE Offered by a child to an old man, which is
'wit-old'.
HOLOFERNES What is the figure? What is the figure?    60
MOTE Horns.
HOLOFERNES Thou disputes like an infant. Go whip thy
gig.
MOTE Lend me your horn to make one, and I will whip
about your infamy *circum circa*—a gig of a cuckold's
horn.    66
COSTARD An I had but one penny in the world, thou
shouldst have it to buy gingerbread. (*Giving money*)
Hold, there is the very remuneration I had of thy
master, thou halfpenny purse of wit, thou pigeon-egg
of discretion. O, an the heavens were so pleased that
thou wert but my bastard, what a joyful father wouldst
thou make me! Go to, thou hast it *ad dunghill*, at the
fingers' ends, as they say.    74

HOLOFERNES O, I smell false Latin—'dunghill' for *unguem*.

ARMADO Arts-man, *preambulate*. We will be singled from the barbarous. Do you not educate youth at the charge-house on the top of the mountain?

HOLOFERNES Or *mons*, the hill.

ARMADO At your sweet pleasure, for the mountain.    80

HOLOFERNES I do, sans question.

ARMADO Sir, it is the King's most sweet pleasure and affection to congratulate the Princess at her pavilion in the posteriors of this day, which the rude multitude call the afternoon.    85

HOLOFERNES The posterior of the day, most generous sir, is liable, congruent, and measurable for the afternoon. The word is well culled, choice, sweet, and apt, I do assure you, sir, I do assure.

ARMADO Sir, the King is a noble gentleman, and my familiar, I do assure ye, very good friend. For what is inward between us, let it pass. I do beseech thee, remember thy courtesy. I beseech thee, apparel thy head. And, among other important and most serious designs, and of great import indeed, too—but let that pass, for I must tell thee it will please his grace, by the world, sometime to lean upon my poor shoulder and with his royal finger thus dally with my excrement, with my mustachio. But, sweetheart, let that pass. By the world, I recount no fable. Some certain special honours it pleaseth his greatness to impart to Armado, a soldier, a man of travel, that hath seen the world. But let that pass. The very all of all is—but, sweetheart, I do implore secrecy—that the King would have me present the Princess—sweet chuck—with some delightful ostentation, or show, or pageant, or antic, or firework. Now, understanding that the curate and your sweet self are good at such eruptions and sudden breaking-out of mirth, as it were, I have acquainted you withal to the end to crave your assistance.    110

HOLOFERNES Sir, you shall present before her the Nine Worthies. Sir Nathaniel, as concerning some entertainment of time, some show in the posterior of this day to be rendered by our assistance, the King's command, and this most gallant, illustrate, and learned gentleman before the Princess, I say none so fit as to present the Nine Worthies.    117

NATHANIEL Where will you find men worthy enough to present them?

HOLOFERNES Joshua, yourself; myself, Judas Maccabeus; and this gallant gentleman, Hector. This swain, because of his great limb or joint, shall pass Pompey the Great; the page, Hercules.    123

ARMADO Pardon, sir, error! He is not quantity enough for that Worthy's thumb. He is not so big as the end of his club.

HOLOFERNES Shall I have audience? He shall present Hercules in minority. His enter and exit shall be strangling a snake, and I will have an apology for that purpose.    130

MOTE An excellent device! So, if any of the audience hiss, you may cry 'Well done, Hercules, now thou crushest the snake!'—that is the way to make an offence gracious, though few have the grace to do it.

ARMADO For the rest of the Worthies?    135

HOLOFERNES I will play three myself.

MOTE Thrice-worthy gentleman!

ARMADO Shall I tell you a thing?

HOLOFERNES We attend.

ARMADO We will have, if this fadge not, an antic. I beseech you, follow.    141

HOLOFERNES *Via*, goodman Dull! Thou hast spoken no word all this while.

DULL Nor understood none neither, sir.

HOLOFERNES *Allons!* We will employ thee.    145

DULL I'll make one in a dance or so, or I will play on the tabor to the Worthies, and let them dance the hay.

HOLOFERNES Most dull, honest Dull! To our sport, away.
                                                   *Exeunt*

5.2    *Enter the Princess and her ladies: Rosaline, Maria, and Catherine*

PRINCESS
Sweethearts, we shall be rich ere we depart,
If fairings come thus plentifully in.
A lady walled about with diamonds—
Look you what I have from the loving King.

ROSALINE
Madam, came nothing else along with that?    5

PRINCESS
Nothing but this?—yes, as much love in rhyme
As would be crammed up in a sheet of paper
Writ o' both sides the leaf, margin and all,
That he was fain to seal on Cupid's name.

ROSALINE
That was the way to make his godhead wax,    10
For he hath been five thousand year a boy.

CATHERINE
Ay, and a shrewd unhappy gallows, too.

ROSALINE
You'll ne'er be friends with him, a killed your sister.

CATHERINE
He made her melancholy, sad, and heavy,
And so she died. Had she been light like you,    15
Of such a merry, nimble, stirring spirit,
She might ha' been a grandam ere she died;
And so may you, for a light heart lives long.

ROSALINE
What's your dark meaning, mouse, of this light word?

CATHERINE
A light condition in a beauty dark.    20

ROSALINE
We need more light to find your meaning out.

CATHERINE
You'll mar the light by taking it in snuff,
Therefore I'll darkly end the argument.

ROSALINE
Look what you do, you do it still i'th' dark.

CATHERINE
So do not you, for you are a light wench.    25

ROSALINE
Indeed I weigh not you, and therefore light.

CATHERINE
You weigh me not? O, that's you care not for me.

ROSALINE
Great reason, for past care is still past cure.

PRINCESS
Well bandied, both; a set of wit well played.
But Rosaline, you have a favour, too.    30
Who sent it? And what is it?

ROSALINE                    I would you knew.
An if my face were but as fair as yours
My favour were as great, be witness this.
Nay, I have verses, too, I thank Biron,

The numbers true, and were the numb'ring, too,          35
I were the fairest goddess on the ground.
I am compared to twenty thousand fairs.
O, he hath drawn my picture in his letter.
PRINCESS  Anything like?
ROSALINE
Much in the letters, nothing in the praise.          40
PRINCESS
Beauteous as ink—a good conclusion.
CATHERINE
Fair as a text B in a copy-book.
ROSALINE
Ware pencils, ho! Let me not die your debtor,
My red dominical, my golden letter.
O, that your face were not so full of O's!          45
PRINCESS
A pox of that jest; I beshrew all shrews.
But Catherine, what was sent to you from fair
          Dumaine?
CATHERINE
Madam, this glove.
PRINCESS                    Did he not send you twain?
CATHERINE  Yes, madam; and moreover,
Some thousand verses of a faithful lover.          50
A huge translation of hypocrisy
Vilely compiled, profound simplicity.
MARIA
This and these pearls to me sent Longueville.
The letter is too long by half a mile.
PRINCESS
I think no less. Dost thou not wish in heart          55
The chain were longer and the letter short?
MARIA
Ay, or I would these hands might never part.
PRINCESS
We are wise girls to mock our lovers so.
ROSALINE
They are worse fools to purchase mocking so.
That same Biron I'll torture ere I go.          60
O that I knew he were but in by th' week!—
How I would make him fawn, and beg, and seek,
And wait the season, and observe the times,
And spend his prodigal wits in bootless rhymes,
And shape his service wholly to my hests,          65
And make him proud to make me proud that jests!
So pursuivant-like would I o'ersway his state
That he should be my fool, and I his fate.
PRINCESS
None are so surely caught when they are catched
As wit turned fool. Folly in wisdom hatched          70
Hath wisdom's warrant, and the help of school,
And wit's own grace, to grace a learnèd fool.
ROSALINE
The blood of youth burns not with such excess
As gravity's revolt to wantonness.
MARIA
Folly in fools bears not so strong a note          75
As fool'ry in the wise when wit doth dote,
Since all the power thereof it doth apply
To prove, by wit, worth in simplicity.
          *Enter Boyet*
PRINCESS
Here comes Boyet, and mirth is in his face.
BOYET
O, I am stabbed with laughter! Where's her grace?          80

PRINCESS
Thy news, Boyet?
BOYET                    Prepare, madam, prepare.
Arm, wenches, arm. Encounters mounted are
Against your peace. Love doth approach disguised,
Armèd in arguments. You'll be surprised.
Muster your wits, stand in your own defence,          85
Or hide your heads like cowards and fly hence.
PRINCESS
Saint Denis to Saint Cupid! What are they
That charge their breath against us? Say, scout, say.
BOYET
Under the cool shade of a sycamore
I thought to close mine eyes some half an hour          90
When lo, to interrupt my purposed rest
Toward that shade I might behold addressed
The King and his companions. Warily
I stole into a neighbour thicket by
And overheard what you shall overhear:          95
That by and by disguised they will be here.
Their herald is a pretty knavish page
That well by heart hath conned his embassage.
Action and accent did they teach him there.
'Thus must thou speak', and 'thus thy body bear'.          100
And ever and anon they made a doubt
Presence majestical would put him out,
'For', quoth the King, 'an angel shalt thou see,
Yet fear not thou, but speak audaciously.'
The boy replied, 'An angel is not evil.          105
I should have feared her had she been a devil.'
With that all laughed and clapped him on the
          shoulder,
Making the bold wag by their praises bolder.
One rubbed his elbow thus, and fleered, and swore
A better speech was never spoke before.          110
Another with his finger and his thumb
Cried '*Via*, we will do't, come what will come!'
The third he capered and cried 'All goes well!'
The fourth turned on the toe and down he fell.
With that they all did tumble on the ground          115
With such a zealous laughter, so profound,
That in this spleen ridiculous appears,
To check their folly, passion's solemn tears.
PRINCESS
But what, but what—come they to visit us?
BOYET
They do, they do, and are apparelled thus          120
⌐                                                    ¬
Like Muscovites or Russians, as I guess.
Their purpose is to parley, to court and dance,
And every one his love-suit will advance
Unto his several mistress, which they'll know          125
By favours several which they did bestow.
PRINCESS
And will they so? The gallants shall be tasked,
For, ladies, we will every one be masked,
And not a man of them shall have the grace,
Despite of suit, to see a lady's face.          130
(*To Rosaline*) Hold, take thou this, my sweet, and give
          me thine.
So shall Biron take me for Rosaline.
          *She changes favours with Rosaline*
(*To Catherine and Maria*)
And change you favours, too. So shall your loves
Woo contrary, deceived by these removes.
          *Catherine and Maria change favours*

ROSALINE
  Come on, then, wear the favours most in sight.   135
CATHERINE
  But in this changing what is your intent?
PRINCESS
  The effect of my intent is to cross theirs.
  They do it but in mockery-merriment,
  And mock for mock is only my intent.
  Their several counsels they unbosom shall   140
  To loves mistook, and so be mocked withal
  Upon the next occasion that we meet
  With visages displayed to talk and greet.
ROSALINE
  But shall we dance if they desire us to't?
PRINCESS
  No, to the death we will not move a foot,   145
  Nor to their penned speech render we no grace,
  But while 'tis spoke each turn away her face.
BOYET
  Why, that contempt will kill the speaker's heart,
  And quite divorce his memory from his part.
PRINCESS
  Therefore I do it; and I make no doubt   150
  The rest will ne'er come in if he be out.
  There's no such sport as sport by sport o'erthrown,
  To make theirs ours, and ours none but our own.
  So shall we stay, mocking intended game,
  And they well mocked depart away with shame.   155
    *A trumpet sounds*
BOYET
  The trumpet sounds, be masked, the masquers come.
    *The ladies mask.*
    *Enter blackamoors with music; the boy Mote with*
    *a speech; the King and his lords, disguised as*
    *Russians*
MOTE
  All hail, the richest beauties on the earth!
BIRON (*aside*)
  Beauties no richer than rich taffeta.
MOTE
  A holy parcel of the fairest dames—
    *The ladies turn their backs to him*
  That ever turned their—backs to mortal views.   160
BIRON 'Their eyes', villain, 'their eyes'!
MOTE
  That ever turned their eyes to mortal views.
  Out . . .
BOYET True, out indeed!
MOTE
  Out of your favours, heavenly spirits, vouchsafe   165
  Not to behold—
BIRON 'Once to behold', rogue!
MOTE
  Once to behold with your sun-beamèd eyes—
  With your sun-beamèd eyes—
BOYET
  They will not answer to that epithet.   170
  You were best call it 'daughter-beamèd' eyes.
MOTE
  They do not mark me, and that brings me out.
BIRON
  Is this your perfectness? Be gone, you rogue!
    *Exit Mote*
ROSALINE (*as the Princess*)
  What would these strangers? Know their minds, Boyet.

  If they do speak our language, 'tis our will   175
  That some plain man recount their purposes.
  Know what they would.
BOYET         What would you with the Princess?
BIRON
  Nothing but peace and gentle visitation.
ROSALINE What would they, say they?
BOYET
  Nothing but peace and gentle visitation.   180
ROSALINE
  Why, that they have, and bid them so be gone.
BOYET
  She says you have it, and you may be gone.
KING
  Say to her we have measured many miles
  To tread a measure with her on this grass.
BOYET
  They say that they have measured many a mile   185
  To tread a measure with you on this grass.
ROSALINE
  It is not so. Ask them how many inches
  Is in one mile. If they have measured many,
  The measure then of one is easily told.
BOYET
  If to come hither you have measured miles,   190
  And many miles, the Princess bids you tell
  How many inches doth fill up one mile.
BIRON
  Tell her we measure them by weary steps.
BOYET
  She hears herself.
ROSALINE         How many weary steps
  Of many weary miles you have o'ergone   195
  Are numbered in the travel of one mile?
BIRON
  We number nothing that we spend for you.
  Our duty is so rich, so infinite,
  That we may do it still without account.
  Vouchsafe to show the sunshine of your face   200
  That we, like savages, may worship it.
ROSALINE
  My face is but a moon, and clouded, too.
KING
  Blessèd are clouds to do as such clouds do.
  Vouchsafe, bright moon, and these thy stars, to shine,
  Those clouds removed, upon our watery eyne.   205
ROSALINE
  O vain petitioner, beg a greater matter.
  Thou now requests but moonshine in the water.
KING
  Then in our measure do but vouchsafe one change.
  Thou bid'st me beg; this begging is not strange.
ROSALINE
  Play, music, then.
    ⌜*Music plays*⌝
                Nay, you must do it soon.   210
  Not yet?—no dance! Thus change I like the moon.
KING
  Will you not dance? How come you thus estranged?
ROSALINE
  You took the moon at full, but now she's changed.
KING
  Yet still she is the moon, and I the man.
  ⌜                      ⌝   215
  The music plays, vouchsafe some motion to it.

ROSALINE
Our ears vouchsafe it.
KING                                 But your legs should do it.
ROSALINE
Since you are strangers and come here by chance
We'll not be nice. Take hands. We will not dance.
KING
Why take we hands, then?
ROSALINE                          Only to part friends.          220
Curtsy, sweethearts, and so the measure ends.
KING
More measure of this measure, be not nice.
ROSALINE
We can afford no more at such a price.
KING
Price you yourselves. What buys your company?
ROSALINE
Your absence only.
KING                            That can never be.          225
ROSALINE
Then cannot we be bought, and so adieu—
Twice to your visor, and half once to you.
KING
If you deny to dance, let's hold more chat.
ROSALINE
In private, then.
KING                          I am best pleased with that.
          *The King and Rosaline talk apart*
BIRON (*to the Princess, taking her for Rosaline*)
White-handed mistress, one sweet word with thee.   230
PRINCESS
Honey and milk and sugar—there is three.
BIRON
Nay then, two treys, an if you grow so nice—
Metheglin, wort, and malmsey—well run, dice!
There's half-a-dozen sweets.
PRINCESS                          Seventh sweet, adieu.
Since you can cog, I'll play no more with you.      235
BIRON
One word in secret.
PRINCESS                          Let it not be sweet.
BIRON
Thou griev'st my gall.
PRINCESS                          Gall—bitter!
BIRON                                    Therefore meet.
          *Biron and the Princess talk apart*
DUMAINE (*to Maria, taking her for Catherine*)
Will you vouchsafe with me to change a word?
MARIA
Name it.
DUMAINE   Fair lady—
MARIA                          Say you so? Fair lord—
Take that for your 'fair lady'.
DUMAINE                          Please it you,      240
As much in private, and I'll bid adieu.
          *Dumaine and Maria talk apart*
CATHERINE
What, was your visor made without a tongue?
LONGUEVILLE (*taking Catherine for Maria*)
I know the reason, lady, why you ask.
CATHERINE
O, for your reason! Quickly, sir, I long.
LONGUEVILLE
You have a double tongue within your mask,        245
And would afford my speechless visor half.
CATHERINE
'Veal', quoth the Dutchman. Is not veal a calf?

LONGUEVILLE
A calf, fair lady?
CATHERINE                  No, a fair lord calf.
LONGUEVILLE
Let's part the word.
CATHERINE                  No, I'll not be your half.
Take all and wean it, it may prove an ox.          250
LONGUEVILLE
Look how you butt yourself in these sharp mocks!
Will you give horns, chaste lady? Do not so.
CATHERINE
Then die a calf before your horns do grow.
LONGUEVILLE
One word in private with you ere I die.
CATHERINE
Bleat softly, then. The butcher hears you cry.     255
          *Longueville and Catherine talk apart*
BOYET
The tongues of mocking wenches are as keen
     As is the razor's edge invisible,
Cutting a smaller hair than may be seen,
     Above the sense of sense; so sensible
Seemeth their conference. Their conceits have wings
Fleeter than arrows, bullets, wind, thought, swifter
     things.                                         261
ROSALINE
Not one word more, my maids. Break off, break off.
BIRON
By heaven, all dry-beaten with pure scoff!
KING
Farewell, mad wenches, you have simple wits.
          *Exeunt the King, lords, and blackamoors*
          ⌈*The ladies unmask*⌉
PRINCESS
Twenty adieus, my frozen Muscovites.                265
Are these the breed of wits so wondered at?
BOYET
     Tapers they are, with your sweet breaths puffed
          out.
ROSALINE
Well-liking wits they have; gross, gross; fat, fat.
PRINCESS
     O poverty in wit, kingly-poor flout!
Will they not, think you, hang themselves tonight,
     Or ever but in visors show their faces?        271
This pert Biron was out of count'nance quite.
ROSALINE
     Ah, they were all in lamentable cases.
The King was weeping-ripe for a good word.
PRINCESS
     Biron did swear himself out of all suit.       275
MARIA
Dumaine was at my service, and his sword.
     'Non point,' quoth I. My servant straight was mute.
CATHERINE
Lord Longueville said I came o'er his heart,
     And trow you what he called me?
PRINCESS                          'Qualm', perhaps.
CATHERINE
Yes, in good faith.
PRINCESS                  Go, sickness as thou art.   280
ROSALINE
Well, better wits have worn plain statute-caps.
But will you hear? The King is my love sworn.
PRINCESS
And quick Biron hath plighted faith to me.

CATHERINE
  And Longueville was for my service born.
MARIA
    Dumaine is mine, as sure as bark on tree.          285
BOYET
  Madam, and pretty mistresses, give ear.
  Immediately they will again be here
  In their own shapes, for it can never be
  They will digest this harsh indignity.
PRINCESS
  Will they return?
BOYET                    They will, they will, God knows,   290
  And leap for joy, though they are lame with blows.
  Therefore change favours, and when they repair,
  Blow like sweet roses in this summer air.
PRINCESS
  How 'blow'? How 'blow'? Speak to be understood.
BOYET
  Fair ladies masked are roses in their bud;          295
  Dismasked, their damask sweet commixture shown,
  Are angels vailing clouds, or roses blown.
PRINCESS
  Avaunt, perplexity! What shall we do
  If they return in their own shapes to woo?
ROSALINE
  Good madam, if by me you'll be advised,             300
  Let's mock them still, as well known as disguised.
  Let us complain to them what fools were here,
  Disguised like Muscovites in shapeless gear,
  And wonder what they were, and to what end
  Their shallow shows, and prologue vilely penned,    305
  And their rough carriage so ridiculous,
  Should be presented at our tent to us.
BOYET
  Ladies, withdraw. The gallants are at hand.
PRINCESS
  Whip, to our tents, as roes run over land!
                              *Exeunt the ladies*
  *Enter the King, Biron, Dumaine, and Longueville, as*
                              *themselves*
KING
  Fair sir, God save you. Where's the Princess?      310
BOYET
  Gone to her tent. Please it your majesty
  Command me any service to her thither?
KING
  That she vouchsafe me audience for one word.
BOYET
  I will, and so will she, I know, my lord.        *Exit*
BIRON
  This fellow pecks up wit as pigeons peas,          315
  And utters it again when God doth please.
  He is wit's pedlar, and retails his wares
  At wakes and wassails, meetings, markets, fairs.
  And we that sell by gross, the Lord doth know,
  Have not the grace to grace it with such show.     320
  This gallant pins the wenches on his sleeve.
  Had he been Adam, he had tempted Eve.
  A can carve too, and lisp, why, this is he
  That kissed his hand away in courtesy.
  This is the ape of form, Monsieur the Nice,        325
  That when he plays at tables chides the dice
  In honourable terms. Nay, he can sing
  A mean most meanly, and in ushering
  Mend him who can. The ladies call him sweet.
  The stairs as he treads on them kiss his feet.     330

  This is the flower that smiles on everyone
  To show his teeth as white as whalës bone,
  And consciences that will not die in debt
  Pay him the due of 'honey-tongued' Boyet.
KING
  A blister on his sweet tongue with my heart,       335
  That put Armado's page out of his part!
                        *Enter the ladies and Boyet*
BIRON
  See where it comes. Behaviour, what wert thou
  Till this madman showed thee, and what art thou
      now?
KING
  All hail, sweet madam, and fair time of day!
PRINCESS
  'Fair' in 'all hail' is foul, as I conceive.       340
KING
  Construe my speeches better, if you may.
PRINCESS
    Then wish me better. I will give you leave.
KING
  We came to visit you, and purpose now
  To lead you to our court. Vouchsafe it, then.
PRINCESS
  This field shall hold me, and so hold your vow.    345
  Nor God nor I delights in perjured men.
KING
  Rebuke me not for that which you provoke.
  The virtue of your eye must break my oath.
PRINCESS
  You nickname virtue. 'Vice' you should have spoke,
  For virtue's office never breaks men's troth.      350
  Now by my maiden honour, yet as pure
    As the unsullied lily, I protest,
  A world of torments though I should endure,
    I would not yield to be your house's guest,
  So much I hate a breaking cause to be             355
  Of heavenly oaths, vowed with integrity.
KING
  O, you have lived in desolation here,
    Unseen, unvisited, much to our shame.
PRINCESS
  Not so, my lord. It is not so, I swear.
    We have had pastimes here, and pleasant game.
  A mess of Russians left us but of late.            361
KING
    How, madam? Russians?
PRINCESS                          Ay, in truth, my lord.
  Trim gallants, full of courtship and of state.
ROSALINE
    Madam, speak true.—It is not so, my lord.
  My lady, to the manner of the days,                365
  In courtesy gives undeserving praise.
  We four indeed confronted were with four
  In Russian habit. Here they stayed an hour,
  And talked apace, and in that hour, my lord,
  They did not bless us with one happy word.         370
  I dare not call them fools, but this I think:
  When they are thirsty, fools would fain have drink.
BIRON
  This jest is dry to me. Gentle sweet,
  Your wits makes wise things foolish. When we greet,
  With eyes' best seeing, heaven's fiery eye,        375
  By light we lose light. Your capacity
  Is of that nature that to your huge store
  Wise things seem foolish, and rich things but poor.

ROSALINE
This proves you wise and rich, for in my eye—
BIRON
I am a fool, and full of poverty.                                  380
ROSALINE
But that you take what doth to you belong
It were a fault to snatch words from my tongue.
BIRON
O, I am yours, and all that I possess.
ROSALINE
All the fool mine!
BIRON                    I cannot give you less.
ROSALINE
Which of the visors was it that you wore?        385
BIRON
Where? When? What visor? Why demand you this?
ROSALINE
There, then, that visor, that superfluous case,
That hid the worse and showed the better face.
KING (aside to the lords)
We were descried. They'll mock us now, downright.
DUMAINE (aside to the King)
Let us confess, and turn it to a jest.                      390
PRINCESS
Amazed, my lord? Why looks your highness sad?
ROSALINE
Help, hold his brows, he'll swoon. Why look you
     pale?
Seasick, I think, coming from Muscovy.
BIRON
Thus pour the stars down plagues for perjury.
Can any face of brass hold longer out?             395
Here stand I, lady. Dart thy skill at me—
Bruise me with scorn, confound me with a flout,
Thrust thy sharp wit quite through my ignorance,
Cut me to pieces with thy keen conceit,
And I will wish thee nevermore to dance,          400
Nor nevermore in Russian habit wait.
O, never will I trust to speeches penned,
Nor to the motion of a schoolboy's tongue,
Nor never come in visor to my friend,
Nor woo in rhyme, like a blind harper's song.    405
Taffeta phrases, silken terms precise,
Three-piled hyperboles, spruce affectation,
Figures pedantical—these summer flies
Have blown me full of maggot ostentation.
I do forswear them, and I here protest,            410
By this white glove—how white the hand, God
     knows!—
Henceforth my wooing mind shall be expressed
In russet yeas, and honest kersey noes.
And to begin, wench, so God help me, law!
My love to thee is sound, sans crack or flaw.   415
ROSALINE
Sans 'sans', I pray you.
BIRON                    Yet I have a trick
Of the old rage. Bear with me, I am sick.
I'll leave it by degrees. Soft, let us see.
Write 'Lord have mercy on us' on those three.
They are infected, in their hearts it lies.          420
They have the plague, and caught it of your eyes.
These lords are visited, you are not free;
For the Lord's tokens on you do I see.
PRINCESS
No, they are free that gave these tokens to us.
BIRON
Our states are forfeit. Seek not to undo us.      425

ROSALINE
It is not so, for how can this be true,
That you stand forfeit, being those that sue?
BIRON
Peace, for I will not have to do with you.
ROSALINE
Nor shall not, if I do as I intend.
BIRON (to the lords)
Speak for yourselves. My wit is at an end.         430
KING
Teach us, sweet madam, for our rude transgression
Some fair excuse.
PRINCESS                    The fairest is confession.
Were not you here but even now disguised?
KING
Madam, I was.
PRINCESS                    And were you well advised?
KING
I was, fair madam.
PRINCESS                    When you then were here,   435
What did you whisper in your lady's ear?
KING
That more than all the world I did respect her.
PRINCESS
When she shall challenge this, you will reject her.
KING
Upon mine honour, no.
PRINCESS                    Peace, peace, forbear.
Your oath once broke, you force not to forswear.   440
KING
Despise me when I break this oath of mine.
PRINCESS
I will, and therefore keep it. Rosaline,
What did the Russian whisper in your ear?
ROSALINE
Madam, he swore that he did hold me dear
As precious eyesight, and did value me             445
Above this world, adding thereto moreover
That he would wed me, or else die my lover.
PRINCESS
God give thee joy of him! The noble lord
Most honourably doth uphold his word.
KING
What mean you, madam? By my life, my troth,   450
I never swore this lady such an oath.
ROSALINE
By heaven, you did, and to confirm it plain,
You gave me this. But take it, sir, again.
KING
My faith and this the Princess I did give.
I knew her by this jewel on her sleeve.            455
PRINCESS
Pardon me, sir, this jewel did she wear,
And Lord Biron, I thank him, is my dear.
(To Biron) What, will you have me, or your pearl again?
BIRON
Neither of either. I remit both twain.
I see the trick on't. Here was a consent,          460
Knowing aforehand of our merriment,
To dash it like a Christmas comedy.
Some carry-tale, some please-man, some slight zany,
Some mumble-news, some trencher-knight, some Dick
That smiles his cheek in years, and knows the trick
To make my lady laugh when she's disposed,    466
Told our intents before, which once disclosed,
The ladies did change favours, and then we,
Following the signs, wooed but the sign of she.

Now, to our perjury to add more terror,       470
We are again forsworn, in will and error.
Much upon this 'tis, (*to Boyet*) and might not you
Forestall our sport, to make us thus untrue?
Do not you know my lady's foot by th' square,
    And laugh upon the apple of her eye,       475
And stand between her back, sir, and the fire,
    Holding a trencher, jesting merrily?
You put our page out. Go, you are allowed.
Die when you will, a smock shall be your shroud.
You leer upon me, do you? There's an eye     480
Wounds like a leaden sword.
BOYET               Full merrily
Hath this brave manège, this career been run.
BIRON
Lo, he is tilting straight. Peace, I have done.
    *Enter Costard the clown*
Welcome, pure wit. Thou partest a fair fray.
COSTARD
O Lord, sir, they would know          485
Whether the three Worthies shall come in or no.
BIRON
What, are there but three?
COSTARD             No, sir, but it is vara fine,
For everyone pursents three.
BIRON               And three times thrice is nine.
COSTARD
Not so, sir, under correction, sir, I hope it is not so.
You cannot beg us, sir. I can assure you, sir, we
    know what we know.          490
I hope, sir, three times thrice, sir—
BIRON               Is not nine?
COSTARD Under correction, sir, we know whereuntil it
doth amount.
BIRON By Jove, I always took three threes for nine.
COSTARD O Lord, sir, it were pity you should get your
living by reck'ning, sir.          496
BIRON How much is it?
COSTARD O Lord, sir, the parties themselves, the actors,
sir, will show whereuntil it doth amount. For mine
own part, I am, as they say, but to parfect one man
in one poor man, Pompion the Great, sir.    501
BIRON Art thou one of the Worthies?
COSTARD It pleased them to think me worthy of Pompey
the Great. For mine own part, I know not the degree
of the Worthy, but I am to stand for him.    505
BIRON Go, bid them prepare.
COSTARD
We will turn it finely off, sir. We will take some care.
                                 *Exit*
KING
Biron, they will shame us. Let them not approach.
BIRON
We are shame-proof, my lord, and 'tis some policy
To have one show worse than the King's and his
    company.          510
KING I say they shall not come.
PRINCESS
Nay, my good lord, let me o'errule you now.
That sport best pleases that doth least know how.
Where zeal strives to content, and the contents
Dies in the zeal of that which it presents,    515
There form confounded makes most form in mirth,
When great things labouring perish in their birth.
BIRON
A right description of our sport, my lord.

    *Enter Armado the braggart*
ARMADO (*to the King*) Anointed, I implore so much expense
of thy royal sweet breath as will utter a brace of words.
    ⌈*Armado and the King speak apart*⌉
PRINCESS Doth this man serve God?          521
BIRON Why ask you?
PRINCESS
A speaks not like a man of God his making.
ARMADO That is all one, my fair sweet honey monarch,
for, I protest, the schoolmaster is exceeding fantastical,
too-too vain, too-too vain. But we will put it, as they
say, to *fortuna de la guerra*. I wish you the peace of
mind, most royal couplement.          *Exit*
KING Here is like to be a good presence of Worthies. He
presents Hector of Troy, the swain Pompey the Great,
the parish curate Alexander, Armado's page Hercules,
the pedant Judas Maccabeus,          532
And if these four Worthies in their first show thrive,
These four will change habits and present the other
    five.
BIRON
There is five in the first show.          535
KING
You are deceived, 'tis not so.
BIRON
The pedant, the braggart, the hedge-priest, the fool,
    and the boy,
Abate throw at novum and the whole world again
Cannot pick out five such, take each one in his vein.
KING
The ship is under sail, and here she comes amain.  540
    *Enter Costard the clown as Pompey*
COSTARD (*as Pompey*)
I Pompey am—
BIRON            You lie, you are not he.
COSTARD (*as Pompey*)
I Pompey am—
BOYET            With leopard's head on knee.
BIRON
Well said, old mocker. I must needs be friends with
    thee.
COSTARD (*as Pompey*)
I Pompey am, Pompey surnamed the Big.
DUMAINE 'The Great'.          545
COSTARD It is 'Great', sir—
    (*As Pompey*) Pompey surnamed the Great,
That oft in field with targe and shield did make my
    foe to sweat,
And travelling along this coast I here am come by
    chance,
And lay my arms before the legs of this sweet lass of
    France.—          550
If your ladyship would say 'Thanks, Pompey', I had
    done.
⌈PRINCESS⌉ Great thanks, great Pompey.
COSTARD 'Tis not so much worth, but I hope I was perfect.
I made a little fault in 'great'.          555
BIRON My hat to a halfpenny Pompey proves the best
    Worthy.
    *Costard stands aside.*
    *Enter Nathaniel the curate as Alexander*
NATHANIEL (*as Alexander*)
When in the world I lived I was the world's
    commander.
    By east, west, north, and south, I spread my
      conquering might.
My scutcheon plain declares that I am Alisander.   560

BOYET
  Your nose says no, you are not, for it stands too
    right.
BIRON (to Boyet)
  Your nose smells 'no' in this, most tender-smelling
    knight.
PRINCESS
  The conqueror is dismayed. Proceed, good Alexander.
NATHANIEL (as Alexander)
  When in the world I lived I was the world's
    commander.
BOYET
  Most true, 'tis right, you were so, Alisander.      565
BIRON (to Costard) Pompey the Great.
COSTARD Your servant, and Costard.
BIRON Take away the conqueror, take away Alisander.
COSTARD (to Nathaniel) O, sir, you have overthrown
  Alisander the Conqueror. You will be scraped out of
  the painted cloth for this. Your lion that holds his pole-
  axe sitting on a close-stool will be given to Ajax. He
  will be the ninth Worthy. A conqueror and afeard to
  speak? Run away for shame, Alisander.      574
                    ⌜Exit Nathaniel the curate⌝
  There, an't shall please you, a foolish mild man, an
  honest man, look you, and soon dashed. He is a
  marvellous good neighbour, faith, and a very good
  bowler, but for Alisander—alas, you see how 'tis—a
  little o'erparted. But there are Worthies a-coming will
  speak their mind in some other sort.      580
PRINCESS Stand aside, good Pompey.
      Enter Holofernes the pedant as Judas, and the boy
      Mote as Hercules
HOLOFERNES
  Great Hercules is presented by this imp,
    Whose club killed Cerberus, that three-headed
      canus,
  And when he was a babe, a child, a shrimp,
    Thus did he strangle serpents in his manus.      585
  Quoniam he seemeth in minority,
  Ergo I come with this apology.
  (To Mote) Keep some state in thy exit, and vanish.
                                        Exit Mote
HOLOFERNES (as Judas)
  Judas I am—
DUMAINE A Judas?      590
HOLOFERNES Not Iscariot, sir.
  (As Judas) Judas I am, yclept Maccabeus.
DUMAINE Judas Maccabeus clipped is plain Judas.
BIRON A kissing traitor. How art thou proved Judas?
HOLOFERNES (as Judas)
  Judas I am—      595
DUMAINE The more shame for you, Judas.
HOLOFERNES What mean you, sir?
BOYET To make Judas hang himself.
HOLOFERNES Begin, sir. You are my elder.
BIRON Well followed—Judas was hanged on an elder.
HOLOFERNES I will not be put out of countenance.      601
BIRON Because thou hast no face.
HOLOFERNES What is this?
BOYET A cittern-head.
DUMAINE The head of a bodkin.      605
BIRON A death's face in a ring.
LONGUEVILLE The face of an old Roman coin, scarce seen.
BOYET The pommel of Caesar's falchion.
DUMAINE The carved-bone face on a flask.
BIRON Saint George's half-cheek in a brooch.      610
DUMAINE Ay, and in a brooch of lead.

BIRON Ay, and worn in the cap of a tooth-drawer. And
  now forward, for we have put thee in countenance.
HOLOFERNES You have put me out of countenance.
BIRON False, we have given thee faces.      615
HOLOFERNES But you have outfaced them all.
BIRON
  An thou wert a lion, we would do so.
BOYET
  Therefore, as he is an ass, let him go.
  And so adieu, sweet Jude. Nay, why dost thou stay?
DUMAINE For the latter end of his name.      620
BIRON
  For the ass to the Jude. Give it him. Jud-as, away.
HOLOFERNES
  This is not generous, not gentle, not humble.
BOYET
  A light for Monsieur Judas. It grows dark, he may
    stumble.                          Exit Holofernes
PRINCESS Alas, poor Maccabeus, how hath he been baited!
      Enter Armado the braggart as Hector
BIRON Hide thy head, Achilles, here comes Hector in
  arms.      626
DUMAINE Though my mocks come home by me, I will
  now be merry.
KING Hector was but a Trojan in respect of this.
BOYET But is this Hector?      630
KING I think Hector was not so clean-timbered.
LONGUEVILLE His leg is too big for Hector's.
DUMAINE More calf, certain.
BOYET No, he is best endowed in the small.
BIRON This cannot be Hector.      635
DUMAINE He's a god, or a painter, for he makes faces.
ARMADO (as Hector)
  The armipotent Mars, of lances the almighty,
    Gave Hector a gift—
DUMAINE A gilt nutmeg.
BIRON A lemon.      640
LONGUEVILLE Stuck with cloves.
DUMAINE No, cloven.
ARMADO Peace!
  (As Hector) The armipotent Mars, of lances the
    almighty,
    Gave Hector a gift, the heir of Ilion,      645
  A man so breathèd that certain he would fight, yea,
    From morn till night, out of his pavilion.
  I am that flower—
DUMAINE                    That mint.
LONGUEVILLE                          That colombine.
ARMADO Sweet Lord Longueville, rein thy tongue.
LONGUEVILLE I must rather give it the rein, for it runs
  against Hector.      651
DUMAINE Ay, and Hector's a greyhound.
ARMADO The sweet war-man is dead and rotten. Sweet
  chucks, beat not the bones of the buried. When he
  breathed he was a man. But I will forward with my
  device. (To the Princess) Sweet royalty, bestow on me
  the sense of hearing.      657
      Biron steps forth
PRINCESS
  Speak, brave Hector, we are much delighted.
ARMADO I do adore thy sweet grace's slipper.
BOYET Loves her by the foot.      660
DUMAINE He may not by the yard.
ARMADO (as Hector)
  This Hector far surmounted Hannibal.
⌜
                                                    ⌝
ARMADO The party is gone.

COSTARD Fellow Hector, she is gone, she is two months
on her way.    666
ARMADO What meanest thou?
COSTARD Faith, unless you play the honest Trojan the
poor wench is cast away. She's quick. The child brags
in her belly already. 'Tis yours.    670
ARMADO Dost thou infamonize me among potentates?
Thou shalt die.
COSTARD Then shall Hector be whipped for Jaquenetta
that is quick by him, and hanged for Pompey that is
dead by him.    675
DUMAINE Most rare Pompey!
BOYET Renowned Pompey!
BIRON Greater than great—great, great, great Pompey,
Pompey the Huge.
DUMAINE Hector trembles.    680
BIRON Pompey is moved. More Ates, more Ates—stir them
on, stir them on!
DUMAINE Hector will challenge him.
BIRON Ay, if a have no more man's blood in his belly
than will sup a flea.    685
ARMADO By the North Pole, I do challenge thee.
COSTARD I will not fight with a pole, like a northern man.
I'll slash, I'll do it by the sword. I bepray you, let me
borrow my arms again.
DUMAINE Room for the incensed Worthies.    690
COSTARD I'll do it in my shirt.
DUMAINE Most resolute Pompey.
MOTE (*aside to Armado*) Master, let me take you a button-
hole lower. Do you not see Pompey is uncasing for the
combat? What mean you? You will lose your
reputation.    696
ARMADO Gentlemen and soldiers, pardon me. I will not
combat in my shirt.
DUMAINE You may not deny it, Pompey hath made the
challenge.    700
ARMADO Sweet bloods, I both may and will.
BIRON What reason have you for't?
ARMADO The naked truth of it is, I have no shirt. I go
woolward for penance.
⌈MOTE⌉ True, and it was enjoined him in Rome for want
of linen, since when I'll be sworn he wore none but a
dish-clout of Jaquenetta's, and that a wears next his
heart, for a favour.    708
*Enter a messenger, Monsieur Mercadé*
MERCADÉ
God save you, madam.
PRINCESS      Welcome, Mercadé,
But that thou interrupt'st our merriment.    710
MERCADÉ
I am sorry, madam, for the news I bring
Is heavy in my tongue. The King your father—
PRINCESS
Dead, for my life.
MERCADÉ      Even so. My tale is told.
BIRON
Worthies, away. The scene begins to cloud.    714
ARMADO For mine own part, I breathe free breath. I have
seen the day of wrong through the little hole of
discretion, and I will right myself like a soldier.
     *Exeunt the Worthies*
KING How fares your majesty?
QUEEN
Boyet, prepare. I will away tonight.
KING
Madam, not so, I do beseech you stay.    720

QUEEN
Prepare, I say. I thank you, gracious lords,
For all your fair endeavours, and entreat,
Out of a new-sad soul, that you vouchsafe
In your rich wisdom to excuse or hide
The liberal opposition of our spirits.    725
If overboldly we have borne ourselves
In the converse of breath, your gentleness
Was guilty of it. Farewell, worthy lord.
A heavy heart bears not a nimble tongue.
Excuse me so coming too short of thanks,    730
For my great suit so easily obtained.
KING
The extreme parts of time extremely forms
All causes to the purpose of his speed,
And often at his very loose decides
That which long process could not arbitrate.    735
And though the mourning brow of progeny
Forbid the smiling courtesy of love
The holy suit which fain it would convince,
Yet since love's argument was first on foot,
Let not the cloud of sorrow jostle it    740
From what it purposed, since to wail friends lost
Is not by much so wholesome-profitable
As to rejoice at friends but newly found.
QUEEN
I understand you not. My griefs are double.
BIRON
Honest plain words best pierce the ear of grief,    745
And by these badges understand the King.
For your fair sakes have we neglected time,
Played foul play with our oaths. Your beauty, ladies,
Hath much deformed us, fashioning our humours
Even to the opposèd end of our intents,    750
And what in us hath seemed ridiculous—
As love is full of unbefitting strains,
All wanton as a child, skipping and vain,
Formed by the eye and therefore like the eye,
Full of strange shapes, of habits and of forms,    755
Varying in subjects as the eye doth roll
To every varied object in his glance;
Which parti-coated presence of loose love
Put on by us, if in your heavenly eyes
Have misbecomed our oaths and gravities,    760
Those heavenly eyes that look into these faults
Suggested us to make them. Therefore, ladies,
Our love being yours, the error that love makes
Is likewise yours. We to ourselves prove false
By being once false for ever to be true    765
To those that make us both—fair ladies, you.
And even that falsehood, in itself a sin,
Thus purifies itself and turns to grace.
QUEEN
We have received your letters full of love,
Your favours the ambassadors of love,    770
And in our maiden council rated them
At courtship, pleasant jest, and courtesy,
As bombast and as lining to the time.
But more devout than this in our respects
Have we not been, and therefore met your loves    775
In their own fashion, like a merriment.
DUMAINE
Our letters, madam, showed much more than jest.
LONGUEVILLE
So did our looks.
ROSALINE      We did not quote them so.

KING
Now, at the latest minute of the hour,
Grant us your loves.
QUEEN                         A time, methinks, too short    780
To make a world-without-end bargain in.
No, no, my lord, your grace is perjured much,
Full of dear guiltiness, and therefore this:
If for my love—as there is no such cause—
You will do aught, this shall you do for me:    785
Your oath I will not trust, but go with speed
To some forlorn and naked hermitage
Remote from all the pleasures of the world.
There stay until the twelve celestial signs
Have brought about the annual reckoning.    790
If this austere, insociable life
Change not your offer made in heat of blood;
If frosts and fasts, hard lodging and thin weeds
Nip not the gaudy blossoms of your love,
But that it bear this trial and last love,    795
Then at the expiration of the year
Come challenge me, challenge me by these deserts,
And, by this virgin palm now kissing thine,
I will be thine, and till that instance shut
My woeful self up in a mourning house,    800
Raining the tears of lamentation
For the remembrance of my father's death.
If this thou do deny, let our hands part,
Neither entitled in the other's heart.
KING
If this, or more than this, I would deny,    805
    To flatter up these powers of mine with rest
The sudden hand of death close up mine eye.
    Hence, hermit, then. My heart is in thy breast.
        *They talk apart*
DUMAINE (*to Catherine*)
But what to me, my love? But what to me?
A wife?
CATHERINE  A beard, fair health, and honesty.    810
With three-fold love I wish you all these three.
DUMAINE
O, shall I say 'I thank you, gentle wife'?
CATHERINE
Not so, my lord. A twelvemonth and a day
I'll mark no words that smooth-faced wooers say.
Come when the King doth to my lady come;    815
Then if I have much love, I'll give you some.
DUMAINE
I'll serve thee true and faithfully till then.
CATHERINE
Yet swear not, lest ye be forsworn again.
        *They talk apart*
LONGUEVILLE
What says Maria?
MARIA                     At the twelvemonth's end
I'll change my black gown for a faithful friend.    820
LONGUEVILLE
I'll stay with patience; but the time is long.
MARIA
The liker you—few taller are so young.
        *They talk apart*
BIRON (*to Rosaline*)
Studies my lady? Mistress, look on me.
Behold the window of my heart, mine eye,
What humble suit attends thy answer there.    825
Impose some service on me for thy love.
ROSALINE
Oft have I heard of you, my lord Biron,

Before I saw you; and the world's large tongue
Proclaims you for a man replete with mocks,
Full of comparisons and wounding flouts,    830
Which you on all estates will execute
That lie within the mercy of your wit.
To weed this wormwood from your fruitful brain,
And therewithal to win me if you please,
Without the which I am not to be won,    835
You shall this twelvemonth term from day to day
Visit the speechless sick and still converse
With groaning wretches, and your task shall be
With all the fierce endeavour of your wit
To enforce the painèd impotent to smile.    840
BIRON
To move wild laughter in the throat of death?—
It cannot be, it is impossible.
Mirth cannot move a soul in agony.
ROSALINE
Why, that's the way to choke a gibing spirit,
Whose influence is begot of that loose grace    845
Which shallow laughing hearers give to fools.
A jest's prosperity lies in the ear
Of him that hears it, never in the tongue
Of him that makes it. Then if sickly ears,
Deafed with the clamours of their own dear groans,
Will hear your idle scorns, continue then,    851
And I will have you and that fault withal.
But if they will not, throw away that spirit,
And I shall find you empty of that fault,
Right joyful of your reformation.    855
BIRON
A twelvemonth? Well, befall what will befall,
I'll jest a twelvemonth in an hospital.
QUEEN (*to the King*)
Ay, sweet my lord, and so I take my leave.
KING
No, madam, we will bring you on your way.
BIRON
Our wooing doth not end like an old play.    860
Jack hath not Jill. These ladies' courtesy
Might well have made our sport a comedy.
KING
Come, sir, it wants a twelvemonth an' a day,
And then 'twill end.
BIRON                         That's too long for a play.
        *Enter Armado the braggart*
ARMADO (*to the King*) Sweet majesty, vouchsafe me.    865
QUEEN Was not that Hector?
DUMAINE The worthy knight of Troy.
ARMADO
I will kiss thy royal finger and take leave.
I am a votary, I have vowed to Jaquenetta
To hold the plough for her sweet love three year.    870
But, most esteemed greatness, will you hear the
dialogue that the two learned men have compiled in
praise of the owl and the cuckoo? It should have
followed in the end of our show.
KING Call them forth quickly, we will do so.    875
ARMADO
Holla, approach!
        *Enter Holofernes, Nathaniel, Costard, Mote, Dull,*
        *Jaquenetta, and others*
                         This side is Hiems, winter,
This Ver, the spring, the one maintained by the owl,
The other by the cuckoo. Ver, begin.

SPRING (sings)

When daisies pied and violets blue,
   And lady-smocks, all silver-white,     880
And cuckoo-buds of yellow hue
   Do paint the meadows with delight,
The cuckoo then on every tree
Mocks married men, for thus sings he:
     Cuckoo!     885
Cuckoo, cuckoo—O word of fear,
Unpleasing to a married ear.

When shepherds pipe on oaten straws,
   And merry larks are ploughmen's clocks;
When turtles tread, and rooks and daws,     890
   And maidens bleach their summer smocks,
The cuckoo then on every tree
Mocks married men, for thus sings he:
     Cuckoo!
Cuckoo, cuckoo—O word of fear,     895
Unpleasing to a married ear.

WINTER (sings)

When icicles hang by the wall,
   And Dick the shepherd blows his nail,
And Tom bears logs into the hall,
   And milk comes frozen home in pail;     900
When blood is nipped, and ways be foul,
Then nightly sings the staring owl:
Tu-whit, tu-whoo!—a merry note,
While greasy Joan doth keel the pot.

When all aloud the wind doth blow,     905
   And coughing drowns the parson's saw,
And birds sit brooding in the snow,
   And Marian's nose looks red and raw;
When roasted crabs hiss in the bowl,
Then nightly sings the staring owl:     910
Tu-whit, tu-whoo!—a merry note,
While greasy Joan doth keel the pot.

⌈ARMADO⌉ The words of Mercury are harsh after the songs
of Apollo. You that way, we this way.   *Exeunt, severally*

## ADDITIONAL PASSAGES

A. The following lines found after 4.3.293 in the First
Quarto represent an unrevised version of parts of Biron's
long speech, 4.3.287–341. The first six lines form the
basis of 4.3.294–9; the next three are revised at 4.3.326–
30; the next four at 4.3.300–2; the last nine are less
directly related to the revised version.

And where that you have vowed to study, lords,
In that each of you have forsworn his book,
Can you still dream, and pore, and thereon look?
For when would you, my lord, or you, or you,
Have found the ground of study's excellence     5
Without the beauty of a woman's face?
From women's eyes this doctrine I derive.
They are the ground, the books, the academes,
From whence doth spring the true Promethean fire.
Why, universal plodding poisons up     10
The nimble spirits in the arteries,
As motion and long-during action tires
The sinewy vigour of the traveller.
Now, for not looking on a woman's face
You have in that forsworn the use of eyes,     15
And study, too, the causer of your vow.
For where is any author in the world

Teaches such beauty as a woman's eye?
Learning is but an adjunct to ourself,
And where we are, our learning likewise is.     20
Then when ourselves we see in ladies' eyes
With ourselves.
Do we not likewise see our learning there?

B. The following two lines, spoken by the Princess and
found after 5.2.130 in the First Quarto, seem to represent
a first draft of 5.2.131–2.

Hold, Rosaline. This favour thou shalt wear,
And then the King will court thee for his dear.

C. The following lines found after 5.2.809 in the First
Quarto represent a draft version of 5.2.824–41.

BIRON
   And what to me, my love? And what to me?
ROSALINE
   You must be purgèd, too. Your sins are rank.
   You are attaint with faults and perjury.
   Therefore if you my favour mean to get
   A twelvemonth shall you spend, and never rest     5
   But seek the weary beds of people sick.

# LOVE'S LABOUR'S WON
## A BRIEF ACCOUNT

IN 1598, Francis Meres called as witnesses to Shakespeare's excellence in comedy 'his *Gētlemē of Verona*, his *Errors*, his *Loue labors lost*, his *Loue labours wonne*, his *Midsummers night dreame*, & his *Merchant of Venice*'. This was the only evidence that Shakespeare wrote a play called *Love's Labour's Won* until the discovery in 1953 of a fragment of a bookseller's list that had been used in the binding of a volume published in 1637/8. The fragment itself appears to record items sold from 9 to 17 August 1603 by a book dealer in the south of England. Among items headed '[inte]rludes & tragedyes' are

> marchant of vennis
> taming of a shrew
> knak to know a knave
> knak to know an honest man
> loves labor lost
> loves labor won

No author is named for any of the items. All the plays named in the list except *Love's Labour's Won* are known to have been printed by 1600; all were written by 1596-7. Taken together, Meres's reference in 1598 and the 1603 fragment appear to demonstrate that a play by Shakespeare called *Love's Labour's Won* had been performed by the time Meres wrote and was in print by August 1603. Conceivably the phrase served as an alternative title for one of Shakespeare's other comedies, though the only one believed to have been written by 1598 but not listed by Meres is *The Taming of the Shrew*, which is named (as *The Taming of A Shrew*) in the bookseller's fragment. Otherwise we must suppose that *Love's Labour's Won* is the title of a lost play by Shakespeare, that no copy of the edition mentioned in the bookseller's list is extant, and that Heminges and Condell failed to include it in the 1623 Folio.

None of these suppositions is implausible. We know of at least one other lost play attributed to Shakespeare (see *Cardenio*, below), and of many lost works by contemporary playwrights. No copy of the first edition of *Titus Andronicus* was known until 1904; for *1 Henry IV* and *The Passionate Pilgrim* only a fragment of the first edition survives. And we now know that *Troilus and Cressida* was almost omitted from the 1623 Folio (probably for copyright reasons) despite its evident authenticity. It is also possible that, like most of the early editions of Shakespeare's plays, the lost edition of *Love's Labour's Won* did not name him on the title-page, and this omission might go some way to explaining the failure of the edition to survive, or (if it does still survive) to be noticed. *Love's Labour's Won* stands a much better chance of having survived, somewhere, than *Cardenio*: because it was printed, between 500 and 1,500 copies were once in circulation, whereas for *Cardenio* we know of only a single manuscript.

The evidence for the existence of the lost play (unlike that for *Cardenio*) gives us little indication of its content. Meres explicitly states, and the title implies, that it was a comedy. Its titular pairing with *Love's Labour's Lost* suggests that they may have been written at about the same time. Both Meres and the bookseller's catalogue place it after *Love's Labour's Lost*; although neither list is necessarily chronological, Meres's does otherwise agree with our own view of the order of composition of Shakespeare's comedies.

# A MIDSUMMER NIGHT'S DREAM

FRANCIS MERES mentions *A Midsummer Night's Dream* in his *Palladis Tamia*, of 1598, and it was first printed in 1600. It has often been thought that Shakespeare wrote the play for an aristocratic wedding, but there is no evidence to support this speculation, and the 1600 title-page states that it had been 'sundry times publicly acted' by the Lord Chamberlain's Men. In stylistic variation it resembles *Love's Labour's Lost*: both plays employ a wide variety of verse measures and rhyme schemes, along with prose that is sometimes (as in Bottom's account of his dream, 4.1.202-15) rhetorically patterned. Probably it was written in 1594 or 1595, either just before or just after *Romeo and Juliet*.

Shakespeare built his own plot from diverse elements of literature, drama, legend, and folklore, supplemented by his imagination and observation. There are four main strands. One, which forms the basis of the action, shows the preparations for the marriage of Theseus, Duke of Athens, to Hippolyta, Queen of the Amazons, and (in the last act) its celebration. This is indebted to Chaucer's *Knight's Tale*, as is the play's second strand, the love story of Lysander and Hermia (who elope to escape her father's opposition) and of Demetrius. In Chaucer, two young men fall in love with the same girl and quarrel over her; Shakespeare adds the comic complication of another girl (Helena) jilted by, but still loving, one of the young men. A third strand shows the efforts of a group of Athenian workmen—the 'mechanicals'—led by Bottom the Weaver to prepare a play, *Pyramus and Thisbe* (based mainly on Arthur Golding's translation of Ovid's *Metamorphoses*) for performance at the Duke's wedding. The mechanicals themselves belong rather to Elizabethan England than to ancient Greece. Bottom's partial transformation into an ass has many literary precedents. Fourthly, Shakespeare depicts a quarrel between Oberon and Titania, King and Queen of the Fairies. Oberon's attendant, Robin Goodfellow, a puck (or pixie), interferes mischievously in the workmen's rehearsals and the affairs of the lovers. The fairy part of the play owes something to both folklore and literature; Robin Goodfellow was a well-known figure about whom Shakespeare could have read in Reginald Scot's *Discovery of Witchcraft* (1586).

*A Midsummer Night's Dream* offers a glorious celebration of the powers of the human imagination while also making comic capital out of its limitations. It is one of Shakespeare's most polished achievements, a poetic drama of exquisite grace, wit, and humanity. In performance, its imaginative unity has sometimes been violated, but it has become one of Shakespeare's most popular plays, with a special appeal for the young.

# THE PERSONS OF THE PLAY

THESEUS, Duke of Athens

HIPPOLYTA, Queen of the Amazons, betrothed to Theseus

PHILOSTRATE, Master of the Revels to Theseus

EGEUS, father of Hermia

HERMIA, daughter of Egeus, in love with Lysander

LYSANDER, loved by Hermia

DEMETRIUS, suitor to Hermia

HELENA, in love with Demetrius

OBERON, King of Fairies

TITANIA, Queen of Fairies

ROBIN GOODFELLOW, a puck

PEASEBLOSSOM
COBWEB
MOTE
MUSTARDSEED
} fairies

Peter QUINCE, a carpenter

Nick BOTTOM, a weaver

Francis FLUTE, a bellows-mender

Tom SNOUT, a tinker

SNUG, a joiner

Robin STARVELING, a tailor

Attendant lords and fairies

# A Midsummer Night's Dream

**1.1**  *Enter Theseus, Hippolyta, and Philostrate, with
others*

THESEUS
  Now, fair Hippolyta, our nuptial hour
  Draws on apace. Four happy days bring in
  Another moon—but O, methinks how slow
  This old moon wanes! She lingers my desires
  Like to a stepdame or a dowager                                       5
  Long withering out a young man's revenue.
HIPPOLYTA
  Four days will quickly steep themselves in night,
  Four nights will quickly dream away the time;
  And then the moon, like to a silver bow
  New bent in heaven, shall behold the night        10
  Of our solemnities.
THESEUS                              Go, Philostrate,
  Stir up the Athenian youth to merriments.
  Awake the pert and nimble spirit of mirth.
  Turn melancholy forth to funerals—
  The pale companion is not for our pomp.           15
                                    ⌜*Exit Philostrate*⌝
  Hippolyta, I wooed thee with my sword,
  And won thy love doing thee injuries.
  But I will wed thee in another key—
  With pomp, with triumph, and with revelling.
      *Enter Egeus and his daughter Hermia, and Lysander
      and Demetrius*
EGEUS
  Happy be Theseus, our renownèd Duke.              20
THESEUS
  Thanks, good Egeus. What's the news with thee?
EGEUS
  Full of vexation come I, with complaint
  Against my child, my daughter Hermia.—
  Stand forth Demetrius.—My noble lord,
  This man hath my consent to marry her.—           25
  Stand forth Lysander.—And, my gracious Duke,
  This hath bewitched the bosom of my child.
  Thou, thou, Lysander, thou hast given her rhymes,
  And interchanged love tokens with my child.
  Thou hast by moonlight at her window sung         30
  With feigning voice verses of feigning love,
  And stol'n the impression of her fantasy
  With bracelets of thy hair, rings, gauds, conceits,
  Knacks, trifles, nosegays, sweetmeats—messengers
  Of strong prevailment in unhardened youth.        35
  With cunning hast thou filched my daughter's heart,
  Turned her obedience which is due to me
  To stubborn harshness. And, my gracious Duke,
  Be it so she will not here before your grace
  Consent to marry with Demetrius,                  40
  I beg the ancient privilege of Athens:
  As she is mine, I may dispose of her,
  Which shall be either to this gentleman
  Or to her death, according to our law
  Immediately provided in that case.                45
THESEUS
  What say you, Hermia? Be advised, fair maid.
  To you your father should be as a god,
  One that composed your beauties, yea, and one

To whom you are but as a form in wax,
By him imprinted, and within his power             50
To leave the figure or disfigure it.
Demetrius is a worthy gentleman.
HERMIA
  So is Lysander.
THESEUS                    In himself he is,
  But in this kind, wanting your father's voice,
  The other must be held the worthier.              55
HERMIA
  I would my father looked but with my eyes.
THESEUS
  Rather your eyes must with his judgement look.
HERMIA
  I do entreat your grace to pardon me.
  I know not by what power I am made bold,
  Nor how it may concern my modesty                 60
  In such a presence here to plead my thoughts,
  But I beseech your grace that I may know
  The worst that may befall me in this case
  If I refuse to wed Demetrius.
THESEUS
  Either to die the death, or to abjure             65
  For ever the society of men.
  Therefore, fair Hermia, question your desires.
  Know of your youth, examine well your blood,
  Whether, if you yield not to your father's choice,
  You can endure the livery of a nun,               70
  For aye to be in shady cloister mewed,
  To live a barren sister all your life,
  Chanting faint hymns to the cold fruitless moon.
  Thrice blessèd they that master so their blood
  To undergo such maiden pilgrimage;                75
  But earthlier happy is the rose distilled
  Than that which, withering on the virgin thorn,
  Grows, lives, and dies in single blessedness.
HERMIA
  So will I grow, so live, so die, my lord,
  Ere I will yield my virgin patent up              80
  Unto his lordship whose unwishèd yoke
  My soul consents not to give sovereignty.
THESEUS
  Take time to pause, and by the next new moon—
  The sealing day betwixt my love and me
  For everlasting bond of fellowship—               85
  Upon that day either prepare to die
  For disobedience to your father's will,
  Or else to wed Demetrius, as he would,
  Or on Diana's altar to protest
  For aye austerity and single life.                90
DEMETRIUS
  Relent, sweet Hermia; and, Lysander, yield
  Thy crazèd title to my certain right.
LYSANDER
  You have her father's love, Demetrius;
  Let me have Hermia's. Do you marry him.
EGEUS
  Scornful Lysander! True, he hath my love;         95
  And what is mine my love shall render him,
  And she is mine, and all my right of her
  I do estate unto Demetrius.

LYSANDER [to Theseus]
I am, my lord, as well derived as he,
As well possessed. My love is more than his,    100
My fortunes every way as fairly ranked,
If not with vantage, as Demetrius;
And—which is more than all these boasts can be—
I am beloved of beauteous Hermia.
Why should not I then prosecute my right?    105
Demetrius—I'll avouch it to his head—
Made love to Nedar's daughter, Helena,
And won her soul, and she, sweet lady, dotes,
Devoutly dotes, dotes in idolatry
Upon this spotted and inconstant man.    110

THESEUS
I must confess that I have heard so much,
And with Demetrius thought to have spoke thereof;
But, being over-full of self affairs,
My mind did lose it. But, Demetrius, come;
And come, Egeus. You shall go with me.    115
I have some private schooling for you both.
For you, fair Hermia, look you arm yourself
To fit your fancies to your father's will,
Or else the law of Athens yields you up—
Which by no means we may extenuate—    120
To death or to a vow of single life.
Come, my Hippolyta; what cheer, my love?—
Demetrius and Egeus, go along.
I must employ you in some business
Against our nuptial, and confer with you    125
Of something nearly that concerns yourselves.

EGEUS
With duty and desire we follow you.
        *Exeunt all but Lysander and Hermia*

LYSANDER
How now, my love? Why is your cheek so pale?
How chance the roses there do fade so fast?

HERMIA
Belike for want of rain, which I could well    130
Beteem them from the tempest of my eyes.

LYSANDER
Ay me, for aught that I could ever read,
Could ever hear by tale or history,
The course of true love never did run smooth,
But either it was different in blood—    135

HERMIA
O cross!—too high to be enthralled to low.

LYSANDER
Or else misgrafted in respect of years—

HERMIA
O spite!—too old to be engaged to young.

LYSANDER
Or merit stood upon the choice of friends—

HERMIA
O hell!—to choose love by another's eyes.    140

LYSANDER
Or if there were a sympathy in choice,
War, death, or sickness did lay siege to it,
Making it momentany as a sound,
Swift as a shadow, short as any dream,
Brief as the lightning in the collied night,    145
That, in a spleen, unfolds both heaven and earth,
And, ere a man hath power to say 'Behold!',
The jaws of darkness do devour it up.
So quick bright things come to confusion.

HERMIA
If then true lovers have been ever crossed,    150
It stands as an edict in destiny.

Then let us teach our trial patience,
Because it is a customary cross,
As due to love as thoughts, and dreams, and sighs,
Wishes, and tears, poor fancy's followers.    155

LYSANDER
A good persuasion. Therefore hear me, Hermia.
I have a widow aunt, a dowager
Of great revenue, and she hath no child,
And she respects me as her only son.
From Athens is her house remote seven leagues.    160
There, gentle Hermia, may I marry thee,
And to that place the sharp Athenian law
Cannot pursue us. If thou lov'st me then,
Steal forth thy father's house tomorrow night,
And in the wood, a league without the town,    165
Where I did meet thee once with Helena
To do observance to a morn of May,
There will I stay for thee.

HERMIA             My good Lysander,
I swear to thee by Cupid's strongest bow,
By his best arrow with the golden head,    170
By the simplicity of Venus' doves,
By that which knitteth souls and prospers loves,
And by that fire which burned the Carthage queen
When the false Trojan under sail was seen;
By all the vows that ever men have broke—    175
In number more than ever women spoke—
In that same place thou hast appointed me
Tomorrow truly will I meet with thee.

LYSANDER
Keep promise, love. Look, here comes Helena.
        *Enter Helena*

HERMIA
God speed, fair Helena. Whither away?    180

HELENA
Call you me fair? That 'fair' again unsay.
Demetrius loves your fair—O happy fair!
Your eyes are lodestars, and your tongue's sweet air
More tuneable than lark to shepherd's ear
When wheat is green, when hawthorn buds appear.
Sickness is catching. O, were favour so!    186
Your words I catch, fair Hermia; ere I go,
My ear should catch your voice, my eye your eye,
My tongue should catch your tongue's sweet melody.
Were the world mine, Demetrius being bated,    190
The rest I'd give to be to you translated.
O, teach me how you look, and with what art
You sway the motion of Demetrius' heart.

HERMIA
I frown upon him, yet he loves me still.

HELENA
O that your frowns would teach my smiles such skill!

HERMIA
I give him curses, yet he gives me love.    196

HELENA
O that my prayers could such affection move!

HERMIA
The more I hate, the more he follows me.

HELENA
The more I love, the more he hateth me.

HERMIA
His folly, Helen, is no fault of mine.    200

HELENA
None but your beauty; would that fault were mine!

HERMIA
Take comfort. He no more shall see my face.
Lysander and myself will fly this place.

Before the time I did Lysander see
Seemed Athens as a paradise to me.                              205
O then, what graces in my love do dwell,
That he hath turned a heaven unto a hell?

LYSANDER
Helen, to you our minds we will unfold.
Tomorrow night, when Phoebe doth behold
Her silver visage in the wat'ry glass,                          210
Decking with liquid pearl the bladed grass—
A time that lovers' sleights doth still conceal—
Through Athens' gates have we devised to steal.

HERMIA
And in the wood where often you and I
Upon faint primrose beds were wont to lie,                      215
Emptying our bosoms of their counsel sweet,
There my Lysander and myself shall meet,
And thence from Athens turn away our eyes
To seek new friends and stranger companies.
Farewell, sweet playfellow. Pray thou for us,                   220
And good luck grant thee thy Demetrius.—
Keep word, Lysander. We must starve our sight
From lovers' food till morrow deep midnight.

LYSANDER
I will, my Hermia.                              *Exit Hermia*
        Helena, adieu.
As you on him, Demetrius dote on you.                  *Exit*

HELENA
How happy some o'er other some can be!                          226
Through Athens I am thought as fair as she.
But what of that? Demetrius thinks not so.
He will not know what all but he do know.
And as he errs, doting on Hermia's eyes,                        230
So I, admiring of his qualities.
Things base and vile, holding no quantity,
Love can transpose to form and dignity.
Love looks not with the eyes, but with the mind,
And therefore is winged Cupid painted blind.                    235
Nor hath love's mind of any judgement taste;
Wings and no eyes figure unheedy haste.
And therefore is love said to be a child
Because in choice he is so oft beguiled.
As waggish boys in game themselves forswear,                    240
So the boy Love is perjured everywhere.
For ere Demetrius looked on Hermia's eyne
He hailed down oaths that he was only mine,
And when this hail some heat from Hermia felt,
So he dissolved, and showers of oaths did melt.                 245
I will go tell him of fair Hermia's flight.
Then to the wood will he tomorrow night
Pursue her, and for this intelligence
If I have thanks it is a dear expense.
But herein mean I to enrich my pain,                            250
To have his sight thither and back again.              *Exit*

1.2    *Enter Quince the carpenter, and Snug the joiner,*
       *and Bottom the weaver, and Flute the bellows-*
       *mender, and Snout the tinker, and Starveling the*
       *tailor*

QUINCE Is all our company here?
BOTTOM You were best to call them generally, man by
man, according to the scrip.
QUINCE Here is the scroll of every man's name which is
thought fit through all Athens to play in our interlude
before the Duke and the Duchess on his wedding day
at night.                                                        7

BOTTOM First, good Peter Quince, say what the play treats
on; then read the names of the actors; and so grow to
a point.                                                         10
QUINCE Marry, our play is *The Most Lamentable Comedy*
*and Most Cruel Death of Pyramus and Thisbe.*
BOTTOM A very good piece of work, I assure you, and a
merry. Now, good Peter Quince, call forth your actors
by the scroll. Masters, spread yourselves.                       15
QUINCE Answer as I call you. Nick Bottom, the weaver?
BOTTOM Ready. Name what part I am for, and proceed.
QUINCE You, Nick Bottom, are set down for Pyramus.
BOTTOM What is Pyramus? A lover or a tyrant?
QUINCE A lover, that kills himself most gallant for love.
BOTTOM That will ask some tears in the true performing
of it. If I do it, let the audience look to their eyes. I will
move stones. I will condole, in some measure. To the
rest.—Yet my chief humour is for a tyrant. I could play
'erc'les rarely, or a part to tear a cat in, to make all
split.                                                           26

                        The raging rocks
                        And shivering shocks
                        Shall break the locks
                            Of prison gates,                     30
                        And Phibus' car
                        Shall shine from far
                        And make and mar
                            The foolish Fates.

This was lofty. Now name the rest of the players.—
This is 'erc'les' vein, a tyrant's vein. A lover is more
condoling.                                                       37
QUINCE Francis Flute, the bellows-mender?
FLUTE Here, Peter Quince.
QUINCE Flute, you must take Thisbe on you.                       40
FLUTE What is Thisbe? A wand'ring knight?
QUINCE It is the lady that Pyramus must love.
FLUTE Nay, faith, let not me play a woman. I have a
beard coming.
QUINCE That's all one. You shall play it in a mask, and
you may speak as small as you will.                              46
BOTTOM An I may hide my face, let me play Thisbe too.
I'll speak in a monstrous little voice: 'Thisne, Thisne!'—
'Ah Pyramus, my lover dear, thy Thisbe dear and lady
dear.'                                                           50
QUINCE No, no, you must play Pyramus; and Flute, you
Thisbe.
BOTTOM Well, proceed.
QUINCE Robin Starveling, the tailor?
STARVELING Here, Peter Quince.                                   55
QUINCE Robin Starveling, you must play Thisbe's mother.
Tom Snout, the tinker?
SNOUT Here, Peter Quince.
QUINCE You, Pyramus' father; myself, Thisbe's father.
Snug the joiner, you the lion's part; and I hope here
is a play fitted.                                                61
SNUG Have you the lion's part written? Pray you, if it be,
give it me; for I am slow of study.
QUINCE You may do it extempore, for it is nothing but
roaring.                                                         65
BOTTOM Let me play the lion too. I will roar that I will
do any man's heart good to hear me. I will roar that I
will make the Duke say 'Let him roar again; let him
roar again'.
QUINCE An you should do it too terribly you would fright
the Duchess and the ladies that they would shriek, and
that were enough to hang us all.                                 72

ALL THE REST That would hang us, every mother's son.

BOTTOM I grant you, friends, if you should fright the ladies out of their wits they would have no more discretion but to hang us, but I will aggravate my voice so that I will roar you as gently as any sucking dove. I will roar you an 'twere any nightingale.                                   78

QUINCE You can play no part but Pyramus; for Pyramus is a sweet-faced man; a proper man as one shall see in a summer's day; a most lovely, gentlemanlike man. Therefore you must needs play Pyramus.

BOTTOM Well, I will undertake it. What beard were I best to play it in?

QUINCE Why, what you will.                                   85

BOTTOM I will discharge it in either your straw-colour beard, your orange-tawny beard, your purple-in-grain beard, or your French-crown-colour beard, your perfect yellow.                                   89

QUINCE Some of your French crowns have no hair at all, and then you will play bare faced. But masters, here are your parts, and I am to entreat you, request you, and desire you to con them by tomorrow night, and meet me in the palace wood a mile without the town by moonlight. There will we rehearse; for if we meet in the city we shall be dogged with company, and our devices known. In the meantime I will draw a bill of properties such as our play wants. I pray you fail me not.                                   99

BOTTOM We will meet, and there we may rehearse most obscenely and courageously. Take pains; be perfect. Adieu.

QUINCE At the Duke's oak we meet.

BOTTOM Enough. Hold, or cut bowstrings.          *Exeunt*

**2.1**    *Enter a Fairy at one door and Robin Goodfellow, a puck, at another*

ROBIN

How now, spirit, whither wander you?

FAIRY

Over hill, over dale,
    Thorough bush, thorough brier,
Over park, over pale,
    Thorough flood, thorough fire:                   5
I do wander everywhere
Swifter than the moonës sphere,
And I serve the Fairy Queen
To dew her orbs upon the green.
The cowslips tall her pensioners be.              10
In their gold coats spots you see;
Those be rubies, fairy favours.
In those freckles live their savours.
I must go seek some dewdrops here,
And hang a pearl in every cowslip's ear.          15
Farewell, thou lob of spirits; I'll be gone.
Our Queen and all her elves come here anon.

ROBIN

The King doth keep his revels here tonight.
Take heed the Queen come not within his sight,
For Oberon is passing fell and wroth               20
Because that she, as her attendant, hath
A lovely boy stol'n from an Indian king.
She never had so sweet a changeling;
And jealous Oberon would have the child
Knight of his train, to trace the forests wild.   25
But she perforce withholds the lovèd boy,
Crowns him with flowers, and makes him all her joy.
And now they never meet in grove, or green,
By fountain clear, or spangled starlight sheen,

But they do square, that all their elves for fear   30
Creep into acorn cups, and hide them there.

FAIRY

Either I mistake your shape and making quite
Or else you are that shrewd and knavish sprite
Called Robin Goodfellow. Are not you he
That frights the maidens of the villag'ry,          35
Skim milk, and sometimes labour in the quern,
And bootless make the breathless housewife churn,
And sometime make the drink to bear no barm—
Mislead night wanderers, laughing at their harm?
Those that 'hobgoblin' call you, and 'sweet puck',  40
You do their work, and they shall have good luck.
Are not you he?

ROBIN                     Thou speak'st aright;
I am that merry wanderer of the night.
I jest to Oberon, and make him smile
When I a fat and bean-fed horse beguile,            45
Neighing in likeness of a filly foal;
And sometime lurk I in a gossip's bowl
In very likeness of a roasted crab,
And when she drinks, against her lips I bob,
And on her withered dewlap pour the ale.            50
The wisest aunt telling the saddest tale
Sometime for three-foot stool mistaketh me;
Then slip I from her bum. Down topples she,
And 'tailor' cries, and falls into a cough,
And then the whole choir hold their hips, and laugh,
And waxen in their mirth, and sneeze, and swear     56
A merrier hour was never wasted there.—
    *Enter Oberon the King of Fairies at one door, with*
    *his train, and Titania the Queen at another, with hers*
But make room, fairy: here comes Oberon.

FAIRY

And here my mistress. Would that he were gone.

OBERON

Ill met by moonlight, proud Titania.                60

TITANIA

What, jealous Oberon?—Fairies, skip hence.
I have forsworn his bed and company.

OBERON

Tarry, rash wanton. Am not I thy lord?

TITANIA

Then I must be thy lady; but I know
When thou hast stol'n away from fairyland           65
And in the shape of Corin sat all day,
Playing on pipes of corn, and versing love
To amorous Phillida. Why art thou here
Come from the farthest step of India,
But that, forsooth, the bouncing Amazon,            70
Your buskined mistress and your warrior love,
To Theseus must be wedded, and you come
To give their bed joy and prosperity?

OBERON

How canst thou thus for shame, Titania,
Glance at my credit with Hippolyta,                 75
Knowing I know thy love to Theseus?
Didst not thou lead him through the glimmering night
From Perigouna whom he ravishèd,
And make him with fair Aegles break his faith,
With Ariadne and Antiopa?                           80

TITANIA

These are the forgeries of jealousy,
And never since the middle summer's spring
Met we on hill, in dale, forest, or mead,
By pavèd fountain or by rushy brook,
Or in the beachèd margin of the sea                 85

To dance our ringlets to the whistling wind,
But with thy brawls thou hast disturbed our sport.
Therefore the winds, piping to us in vain,
As in revenge have sucked up from the sea
Contagious fogs which, falling in the land,     90
Hath every pelting river made so proud
That they have overborne their continents.
The ox hath therefore stretched his yoke in vain,
The ploughman lost his sweat, and the green corn
Hath rotted ere his youth attained a beard.     95
The fold stands empty in the drownèd field,
And crows are fatted with the murrain flock.
The nine men's morris is filled up with mud,
And the quaint mazes in the wanton green
For lack of tread are undistinguishable.     100
The human mortals want their winter cheer.
No night is now with hymn or carol blessed.
Therefore the moon, the governess of floods,
Pale in her anger washes all the air,
That rheumatic diseases do abound;     105
And thorough this distemperature we see
The seasons alter: hoary-headed frosts
Fall in the fresh lap of the crimson rose,
And on old Hiems' thin and icy crown
An odorous chaplet of sweet summer buds     110
Is, as in mock'ry, set. The spring, the summer,
The childing autumn, angry winter change
Their wonted liveries, and the mazèd world
By their increase now knows not which is which;
And this same progeny of evils comes     115
From our debate, from our dissension.
We are their parents and original.

OBERON
Do you amend it, then. It lies in you.
Why should Titania cross her Oberon?
I do but beg a little changeling boy     120
To be my henchman.

TITANIA              Set your heart at rest.
The fairyland buys not the child of me.
His mother was a vot'ress of my order,
And in the spicèd Indian air by night
Full often hath she gossiped by my side,     125
And sat with me on Neptune's yellow sands,
Marking th'embarkèd traders on the flood,
When we have laughed to see the sails conceive
And grow big-bellied with the wanton wind,
Which she with pretty and with swimming gait     130
Following, her womb then rich with my young squire,
Would imitate, and sail upon the land
To fetch me trifles, and return again
As from a voyage, rich with merchandise.
But she, being mortal, of that boy did die;     135
And for her sake do I rear up her boy;
And for her sake I will not part with him.

OBERON
How long within this wood intend you stay?

TITANIA
Perchance till after Theseus' wedding day.
If you will patiently dance in our round,     140
And see our moonlight revels, go with us.
If not, shun me, and I will spare your haunts.

OBERON
Give me that boy and I will go with thee.

TITANIA
Not for thy fairy kingdom.—Fairies, away.
We shall chide downright if I longer stay.     145
             *Exeunt Titania and her train*

OBERON
Well, go thy way. Thou shalt not from this grove
Till I torment thee for this injury.—
My gentle puck, come hither. Thou rememb'rest
Since once I sat upon a promontory
And heard a mermaid on a dolphin's back     150
Uttering such dulcet and harmonious breath
That the rude sea grew civil at her song
And certain stars shot madly from their spheres
To hear the sea-maid's music?

ROBIN              I remember.

OBERON
That very time I saw, but thou couldst not,     155
Flying between the cold moon and the earth
Cupid, all armed. A certain aim he took
At a fair vestal thronèd by the west,
And loosed his love-shaft smartly from his bow
As it should pierce a hundred thousand hearts.     160
But I might see young Cupid's fiery shaft
Quenched in the chaste beams of the wat'ry moon,
And the imperial vot'ress passèd on,
In maiden meditation, fancy-free.
Yet marked I where the bolt of Cupid fell.     165
It fell upon a little western flower—
Before, milk-white; now, purple with love's wound—
And maidens call it love-in-idleness.
Fetch me that flower; the herb I showed thee once.
The juice of it on sleeping eyelids laid     170
Will make or man or woman madly dote
Upon the next live creature that it sees.
Fetch me this herb, and be thou here again
Ere the leviathan can swim a league.

ROBIN
I'll put a girdle round about the earth     175
In forty minutes.              *Exit*

OBERON          Having once this juice
I'll watch Titania when she is asleep,
And drop the liquor of it in her eyes.
The next thing then she waking looks upon—
Be it on lion, bear, or wolf, or bull,     180
On meddling monkey, or on busy ape—
She shall pursue it with the soul of love.
And ere I take this charm from off her sight—
As I can take it with another herb—
I'll make her render up her page to me.     185
But who comes here? I am invisible,
And I will overhear their conference.
         *Enter Demetrius, Helena following him*

DEMETRIUS
I love thee not, therefore pursue me not.
Where is Lysander, and fair Hermia?
The one I'll slay, the other slayeth me.     190
Thou told'st me they were stol'n unto this wood,
And here am I, and wood within this wood
Because I cannot meet my Hermia.
Hence, get thee gone, and follow me no more.

HELENA
You draw me, you hard-hearted adamant,     195
But yet you draw not iron; for my heart
Is true as steel. Leave you your power to draw,
And I shall have no power to follow you.

DEMETRIUS
Do I entice you? Do I speak you fair?
Or rather do I not in plainest truth     200
Tell you I do not nor I cannot love you?

HELENA
And even for that do I love you the more.

I am your spaniel, and, Demetrius,
The more you beat me I will fawn on you.
Use me but as your spaniel: spurn me, strike me,          205
Neglect me, lose me; only give me leave,
Unworthy as I am, to follow you.
What worser place can I beg in your love—
And yet a place of high respect with me—
Than to be usèd as you use your dog?          210

DEMETRIUS
Tempt not too much the hatred of my spirit;
For I am sick when I do look on thee.

HELENA
And I am sick when I look not on you.

DEMETRIUS
You do impeach your modesty too much,
To leave the city and commit yourself          215
Into the hands of one that loves you not;
To trust the opportunity of night,
And the ill counsel of a desert place,
With the rich worth of your virginity.

HELENA
Your virtue is my privilege, for that          220
It is not night when I do see your face;
Therefore I think I am not in the night,
Nor doth this wood lack worlds of company;
For you in my respect are all the world.
Then how can it be said I am alone,          225
When all the world is here to look on me?

DEMETRIUS
I'll run from thee, and hide me in the brakes,
And leave thee to the mercy of wild beasts.

HELENA
The wildest hath not such a heart as you.
Run when you will. The story shall be changed:          230
Apollo flies, and Daphne holds the chase.
The dove pursues the griffin, the mild hind
Makes speed to catch the tiger: bootless speed,
When cowardice pursues, and valour flies.

DEMETRIUS
I will not stay thy questions. Let me go;          235
Or if thou follow me, do not believe
But I shall do thee mischief in the wood.

HELENA
Ay, in the temple, in the town, the field,
You do me mischief. Fie, Demetrius,
Your wrongs do set a scandal on my sex.          240
We cannot fight for love as men may do;
We should be wooed, and were not made to woo.
I'll follow thee, and make a heaven of hell,
To die upon the hand I love so well.
        ⌈*Exit Demetrius, Helena following him*⌉

OBERON
Fare thee well, nymph. Ere he do leave this grove          245
Thou shalt fly him, and he shall seek thy love.
        *Enter Robin Goodfellow the puck*
Hast thou the flower there? Welcome, wanderer.

ROBIN
Ay, there it is.

OBERON          I pray thee give it me.
I know a bank where the wild thyme blows,
Where oxlips and the nodding violet grows,          250
Quite overcanopied with luscious woodbine,
With sweet musk-roses, and with eglantine.
There sleeps Titania sometime of the night,
Lulled in these flowers with dances and delight;

And there the snake throws her enamelled skin,          255
Weed wide enough to wrap a fairy in;
And with the juice of this I'll streak her eyes,
And make her full of hateful fantasies.
Take thou some of it, and seek through this grove.
A sweet Athenian lady is in love          260
With a disdainful youth. Anoint his eyes;
But do it when the next thing he espies
May be the lady. Thou shalt know the man
By the Athenian garments he hath on.
Effect it with some care, that he may prove          265
More fond on her than she upon her love;
And look thou meet me ere the first cock crow.

ROBIN
Fear not, my lord. Your servant shall do so.
        *Exeunt severally*

**2.2**     *Enter Titania, Queen of Fairies, with her train*
TITANIA
Come, now a roundel and a fairy song,
Then for the third part of a minute hence:
Some to kill cankers in the musk-rose buds,
Some war with reremice for their leathern wings
To make my small elves coats, and some keep back          5
The clamorous owl, that nightly hoots and wonders
At our quaint spirits. Sing me now asleep;
Then to your offices, and let me rest.
        *She lies down. Fairies sing*

⌈FIRST FAIRY⌉
You spotted snakes with double tongue,
    Thorny hedgehogs, be not seen;          10
Newts and blindworms, do no wrong;
    Come not near our Fairy Queen.

⌈CHORUS⌉ ⌈*dancing*⌉
    Philomel with melody,
    Sing in our sweet lullaby;
Lulla, lulla, lullaby; lulla, lulla, lullaby.          15
    Never harm
    Nor spell nor charm
Come our lovely lady nigh.
So good night, with lullaby.

FIRST FAIRY
Weaving spiders, come not here;          20
    Hence, you long-legged spinners, hence;
Beetles black, approach not near;
    Worm nor snail do no offence.

⌈CHORUS⌉ ⌈*dancing*⌉
    Philomel with melody,
    Sing in our sweet lullaby;          25
Lulla, lulla, lullaby; lulla, lulla, lullaby.
    Never harm
    Nor spell nor charm
Come our lovely lady nigh.
So good night, with lullaby.          30

        *Titania sleeps*
SECOND FAIRY
Hence, away. Now all is well.
One aloof stand sentinel.
        *Exeunt all but Titania ⌈and the sentinel⌉*
        *Enter Oberon. He drops the juice on Titania's*
        *eyelids*
OBERON
What thou seest when thou dost wake,
Do it for thy true love take;

Love and languish for his sake.                    35
Be it ounce, or cat, or bear,
Pard, or boar with bristled hair,
In thy eye that shall appear
When thou wak'st, it is thy dear.
Wake when some vile thing is near.              *Exit*
    *Enter Lysander and Hermia*
LYSANDER
Fair love, you faint with wand'ring in the wood,     41
    And, to speak truth, I have forgot our way.
We'll rest us, Hermia, if you think it good,
    And tarry for the comfort of the day.
HERMIA
Be it so, Lysander. Find you out a bed;             45
For I upon this bank will rest my head.
    ⌈*She lies down*⌉
LYSANDER
One turf shall serve as pillow for us both;
One heart, one bed; two bosoms, and one troth.
HERMIA
Nay, good Lysander; for my sake, my dear,
Lie further off yet; do not lie so near.            50
LYSANDER
O, take the sense, sweet, of my innocence!
Love takes the meaning in love's conference—
I mean that my heart unto yours is knit,
So that but one heart we can make of it.
Two bosoms interchainèd with an oath;               55
So, then, two bosoms and a single troth.
Then by your side no bed-room me deny;
For lying so, Hermia, I do not lie.
HERMIA
Lysander riddles very prettily.
Now much beshrew my manners and my pride           60
If Hermia meant to say Lysander lied.
But, gentle friend, for love and courtesy,
Lie further off, in humane modesty.
Such separation as may well be said
Becomes a virtuous bachelor and a maid,            65
So far be distant; and good night, sweet friend.
Thy love ne'er alter till thy sweet life end.
LYSANDER
Amen, amen, to that fair prayer say I;
And then end life when I end loyalty.
Here is my bed; sleep give thee all his rest.      70
    *He lies down*
HERMIA
With half that wish the wisher's eyes be pressed.
    *They sleep apart.*
    *Enter Robin Goodfellow the puck*
ROBIN
Through the forest have I gone,
But Athenian found I none
On whose eyes I might approve
This flower's force in stirring love.              75
Night and silence. Who is here?
Weeds of Athens he doth wear.
This is he my master said
Despisèd the Athenian maid—
And here the maiden, sleeping sound                80
On the dank and dirty ground.
Pretty soul, she durst not lie
Near this lack-love, this kill-courtesy.
Churl, upon thy eyes I throw
All the power this charm doth owe.                 85
    *He drops the juice on Lysander's eyelids*

When thou wak'st, let love forbid
Sleep his seat on thy eyelid.
So, awake when I am gone.
For I must now to Oberon.                          *Exit*
    *Enter Demetrius and Helena, running*
HELENA
Stay, though thou kill me, sweet Demetrius.        90
DEMETRIUS
I charge thee hence, and do not haunt me thus.
HELENA
O, wilt thou darkling leave me? Do not so.
DEMETRIUS
Stay, on thy peril; I alone will go.               *Exit*
HELENA
O, I am out of breath in this fond chase.
The more my prayer, the lesser is my grace.        95
Happy is Hermia, wheresoe'er she lies;
For she hath blessèd and attractive eyes.
How came her eyes so bright? Not with salt tears—
If so, my eyes are oft'ner washed than hers.
No, no; I am as ugly as a bear,                   100
For beasts that meet me run away for fear.
Therefore no marvel though Demetrius
Do, as a monster, fly my presence thus.
What wicked and dissembling glass of mine
Made me compare with Hermia's sphery eyne!        105
But who is here? Lysander, on the ground?
Dead, or asleep? I see no blood, no wound.
Lysander, if you live, good sir, awake.
LYSANDER (*awaking*)
And run through fire I will for thy sweet sake.
Transparent Helena, nature shows art              110
That through thy bosom makes me see thy heart.
Where is Demetrius? O, how fit a word
Is that vile name to perish on my sword!
HELENA
Do not say so, Lysander; say not so.
What though he love your Hermia? Lord, what
    though?                                        115
Yet Hermia still loves you; then be content.
LYSANDER
Content with Hermia? No, I do repent
The tedious minutes I with her have spent.
Not Hermia but Helena I love.
Who will not change a raven for a dove?           120
The will of man is by his reason swayed,
And reason says you are the worthier maid.
Things growing are not ripe until their season,
So I, being young, till now ripe not to reason.
And, touching now the point of human skill,       125
Reason becomes the marshal to my will,
And leads me to your eyes, where I o'erlook
Love's stories written in love's richest book.
HELENA
Wherefore was I to this keen mockery born?
When at your hands did I deserve this scorn?       130
Is't not enough, is't not enough, young man,
That I did never—no, nor never can—
Deserve a sweet look from Demetrius' eye,
But you must flout my insufficiency?
Good troth, you do me wrong; good sooth, you do,
In such disdainful manner me to woo.               136
But fare you well. Perforce I must confess
I thought you lord of more true gentleness.
O, that a lady of one man refused
Should of another therefore be abused!            *Exit*

LYSANDER
She sees not Hermia. Hermia, sleep thou there, 141
And never mayst thou come Lysander near;
For as a surfeit of the sweetest things
The deepest loathing to the stomach brings,
Or as the heresies that men do leave 145
Are hated most of those they did deceive,
So thou, my surfeit and my heresy,
Of all be hated, but the most of me;
And all my powers, address your love and might
To honour Helen, and to be her knight. *Exit*

HERMIA (*awaking*)
Help me, Lysander, help me! Do thy best 151
To pluck this crawling serpent from my breast!
Ay me, for pity. What a dream was here?
Lysander, look how I do quake with fear.
Methought a serpent ate my heart away, 155
And you sat smiling at his cruel prey.
Lysander—what, removed? Lysander, lord—
What, out of hearing, gone? No sound, no word?
Alack, where are you? Speak an if you hear,
Speak, of all loves. I swoon almost with fear. 160
No? Then I well perceive you are not nigh.
Either death or you I'll find immediately. *Exit*

**3.1**   *Enter the clowns: Quince, Snug, Bottom, Flute,
Snout, and Starveling*

BOTTOM Are we all met?
QUINCE Pat, pat; and here's a marvellous convenient
place for our rehearsal. This green plot shall be our
stage, this hawthorn brake our tiring-house, and we
will do it in action as we will do it before the Duke. 5
BOTTOM Peter Quince?
QUINCE What sayst thou, bully Bottom?
BOTTOM There are things in this comedy of Pyramus and
Thisbe that will never please. First, Pyramus must draw
a sword to kill himself, which the ladies cannot abide.
How answer you that? 11
SNOUT By'r la'kin, a parlous fear.
STARVELING I believe we must leave the killing out, when
all is done.
BOTTOM Not a whit. I have a device to make all well.
Write me a prologue, and let the prologue seem to say
we will do no harm with our swords, and that Pyramus
is not killed indeed; and for the more better assurance,
tell them that I, Pyramus, am not Pyramus, but Bottom
the weaver. This will put them out of fear. 20
QUINCE Well, we will have such a prologue; and it shall
be written in eight and six.
BOTTOM No, make it two more: let it be written in eight
and eight.
SNOUT Will not the ladies be afeard of the lion? 25
STARVELING I fear it, I promise you.
BOTTOM Masters, you ought to consider with yourself, to
bring in—God shield us—a lion among ladies is a most
dreadful thing; for there is not a more fearful wild fowl
than your lion living, and we ought to look to't. 30
SNOUT Therefore another prologue must tell he is not a
lion.
BOTTOM Nay, you must name his name, and half his face
must be seen through the lion's neck, and he himself
must speak through, saying thus or to the same defect:
'ladies', or 'fair ladies, I would wish you' or 'I would
request you' or 'I would entreat you not to fear, not
to tremble. My life for yours. If you think I come hither

as a lion, it were pity of my life. No, I am no such
thing. I am a man, as other men are'—and there,
indeed, let him name his name, and tell them plainly
he is Snug the joiner. 42
QUINCE Well, it shall be so; but there is two hard things:
that is, to bring the moonlight into a chamber—for
you know Pyramus and Thisbe meet by moonlight.
⌜SNOUT⌝ Doth the moon shine that night we play our
play? 47
BOTTOM A calendar, a calendar—look in the almanac,
find out moonshine, find out moonshine.
⌜*Enter Robin Goodfellow the puck, invisible*⌝
QUINCE ⌜*with a book*⌝ Yes, it doth shine that night. 50
BOTTOM Why, then may you leave a casement of the great
chamber window where we play open, and the moon
may shine in at the casement.
QUINCE Ay, or else one must come in with a bush of
thorns and a lantern and say he comes to disfigure, or
to present, the person of Moonshine. Then there is
another thing: we must have a wall in the great
chamber; for Pyramus and Thisbe, says the story, did
talk through the chink of a wall. 59
SNOUT You can never bring in a wall. What say you,
Bottom?
BOTTOM Some man or other must present Wall; and let
him have some plaster, or some loam, or some rough-
cast about him, to signify 'wall'; and let him hold his
fingers thus, and through that cranny shall Pyramus
and Thisbe whisper. 66
QUINCE If that may be, then all is well. Come, sit down
every mother's son, and rehearse your parts. Pyramus,
you begin. When you have spoken your speech, enter
into that brake; and so everyone according to his cue.
ROBIN (*aside*)
What hempen homespuns have we swagg'ring here
So near the cradle of the Fairy Queen? 72
What, a play toward? I'll be an auditor—
An actor, too, perhaps, if I see cause.
QUINCE Speak, Pyramus. Thisbe, stand forth. 75
BOTTOM (*as Pyramus*)
Thisbe, the flowers of odious savours sweet.
QUINCE Odours, odours.
BOTTOM (*as Pyramus*)    Odours savours sweet.
So hath thy breath, my dearest Thisbe dear.
But hark, a voice. Stay thou but here a while, 80
And by and by I will to thee appear. *Exit*
⌜ROBIN⌝ (*aside*)
A stranger Pyramus than e'er played here. *Exit*
FLUTE Must I speak now?
QUINCE Ay, marry must you. For you must understand
he goes but to see a noise that he heard, and is to
come again. 86
FLUTE (*as Thisbe*)
Most radiant Pyramus, most lily-white of hue,
Of colour like the red rose on triumphant brier;
Most bristly juvenile, and eke most lovely Jew,
As true as truest horse that yet would never tire: 90
I'll meet thee, Pyramus, at Ninny's tomb.
QUINCE Ninus' tomb, man!—Why, you must not speak
that yet. That you answer to Pyramus. You speak all
your part at once, cues and all.—Pyramus, enter: your
cue is past; it is 'never tire'. 95
FLUTE O.
(*As Thisbe*) As true as truest horse that yet would
never tire.

*Enter ⌐Robin leading⌐ Bottom with the ass-head*

BOTTOM (*as Pyramus*)
  If I were fair, Thisbe, I were only thine.

QUINCE O monstrous! O strange! We are haunted. Pray,
  masters; fly, masters: help!    ⌐*The clowns all exeunt*⌐

ROBIN
  I'll follow you, I'll lead you about a round,    101
    Through bog, through bush, through brake,
      through brier.
  Sometime a horse I'll be, sometime a hound,
    A hog, a headless bear, sometime a fire,    104
  And neigh, and bark, and grunt, and roar, and burn,
  Like horse, hound, hog, bear, fire, at every turn.    *Exit*
    ⌐*Enter Bottom again, with the ass-head*⌐

BOTTOM Why do they run away? This is a knavery of
  them to make me afeard.
    *Enter Snout*

SNOUT O Bottom, thou art changed. What do I see on
  thee?

BOTTOM What do you see? You see an ass-head of your
  own, do you?    ⌐*Exit Snout*⌐
    *Enter Quince*

QUINCE Bless thee, Bottom, bless thee. Thou art translated.
    *Exit*

BOTTOM I see their knavery. This is to make an ass of me,
  to fright me, if they could; but I will not stir from this
  place, do what they can. I will walk up and down here,
  and I will sing, that they shall hear I am not afraid.
    (*Sings*)
      The ousel cock so black of hue,
        With orange-tawny bill;
      The throstle with his note so true,    120
        The wren with little quill.

TITANIA (*awaking*)
  What angel wakes me from my flow'ry bed?

BOTTOM (*sings*)
      The finch, the sparrow, and the lark,
        The plainsong cuckoo grey,
      Whose note full many a man doth mark,    125
        And dares not answer 'Nay'—
  for indeed, who would set his wit to so foolish a bird?
  Who would give a bird the lie, though he cry 'Cuckoo'
  never so?

TITANIA
  I pray thee, gentle mortal, sing again.    130
  Mine ear is much enamoured of thy note;
  So is mine eye enthrallèd to thy shape;
  And thy fair virtue's force perforce doth move me
  On the first view to say, to swear, I love thee.

BOTTOM Methinks, mistress, you should have little reason
  for that. And yet, to say the truth, reason and love
  keep little company together nowadays—the more the
  pity that some honest neighbours will not make them
  friends. Nay, I can gleek upon occasion.

TITANIA
  Thou art as wise as thou art beautiful.    140

BOTTOM Not so, neither; but if I had wit enough to get
  out of this wood, I have enough to serve mine own
  turn.

TITANIA
  Out of this wood do not desire to go.
  Thou shalt remain here, whether thou wilt or no.    145
  I am a spirit of no common rate:
  The summer still doth tend upon my state;
  And I do love thee. Therefore go with me.
  I'll give thee fairies to attend on thee,

  And they shall fetch thee jewels from the deep,    150
  And sing while thou on pressèd flowers dost sleep;
  And I will purge thy mortal grossness so
  That thou shalt like an airy spirit go.
  Peaseblossom, Cobweb, Mote, and Mustardseed!
    *Enter four fairies: Peaseblossom, Cobweb, Mote, and*
      *Mustardseed*

A FAIRY
  Ready.

ANOTHER  And I.

ANOTHER        And I.

ANOTHER              And I.

⌐ALL FOUR⌐              Where shall we go?    155

TITANIA
  Be kind and courteous to this gentleman.
  Hop in his walks, and gambol in his eyes.
  Feed him with apricots and dewberries,
  With purple grapes, green figs, and mulberries;
  The honeybags steal from the humble-bees,    160
  And for night tapers crop their waxen thighs
  And light them at the fiery glow-worms' eyes
  To have my love to bed, and to arise;
  And pluck the wings from painted butterflies
  To fan the moonbeams from his sleeping eyes.    165
  Nod to him, elves, and do him courtesies.

A FAIRY Hail, mortal.

⌐ANOTHER⌐ Hail.

ANOTHER Hail.

ANOTHER Hail.    170

BOTTOM I cry your worships mercy, heartily.—I beseech
  your worship's name.

COBWEB Cobweb.

BOTTOM I shall desire you of more acquaintance, good
  Master Cobweb. If I cut my finger, I shall make bold
  with you.—Your name, honest gentleman?    176

PEASEBLOSSOM Peaseblossom.

BOTTOM I pray you commend me to Mistress Squash, your
  mother, and to Master Peascod, your father. Good
  Master Peaseblossom, I shall desire you of more
  acquaintance, too.—Your name, I beseech you, sir?

MUSTARDSEED Mustardseed.    182

BOTTOM Good Master Mustardseed, I know your patience
  well. That same cowardly giantlike ox-beef hath
  devoured many a gentleman of your house. I promise
  you your kindred hath made my eyes water ere now.
  I desire you of more acquaintance, good Master
  Mustardseed.

TITANIA (*to the Fairies*)
  Come, wait upon him, lead him to my bower.
    The moon, methinks, looks with a wat'ry eye,    190
  And when she weeps, weeps every little flower,
    Lamenting some enforcèd chastity.
  Tie up my love's tongue; bring him silently.    *Exeunt*

**3.2**    *Enter Oberon, King of Fairies*

OBERON
  I wonder if Titania be awaked,
  Then what it was that next came in her eye,
  Which she must dote on in extremity.
    *Enter Robin Goodfellow*
  Here comes my messenger. How now, mad spirit?
  What nightrule now about this haunted grove?    5

ROBIN
  My mistress with a monster is in love.
  Near to her close and consecrated bower
  While she was in her dull and sleeping hour

A crew of patches, rude mechanicals
That work for bread upon Athenian stalls,                    10
Were met together to rehearse a play
Intended for great Theseus' nuptial day.
The shallowest thickskin of that barren sort,
Who Pyramus presented, in their sport
Forsook his scene and entered in a brake,                    15
When I did him at this advantage take.
An ass's nole I fixèd on his head.
Anon his Thisbe must be answerèd,
And forth my mimic comes. When they him spy—
As wild geese that the creeping fowler eye,                  20
Or russet-pated choughs, many in sort,
Rising and cawing at the gun's report,
Sever themselves and madly sweep the sky—
So, at his sight, away his fellows fly,
And at our stamp here o'er and o'er one falls.              25
He 'Murder' cries, and help from Athens calls.
Their sense thus weak, lost with their fears thus
      strong,
Made senseless things begin to do them wrong.
For briers and thorns at their apparel snatch;
Some sleeves, some hats—from yielders all things catch.
I led them on in this distracted fear,                       31
And left sweet Pyramus translated there;
When in that moment, so it came to pass,
Titania waked and straightway loved an ass.

OBERON
This falls out better than I could devise.                   35
But hast thou yet latched the Athenian's eyes
With the love juice, as I did bid thee do?

ROBIN
I took him sleeping; that is finished, too;
And the Athenian woman by his side,
That when he waked of force she must be eyed.               40
      *Enter Demetrius and Hermia*

OBERON
Stand close. This is the same Athenian.

ROBIN
This is the woman, but not this the man.
      ⌈*They stand apart*⌉

DEMETRIUS
O, why rebuke you him that loves you so?
Lay breath so bitter on your bitter foe.

HERMIA
Now I but chide, but I should use thee worse;                45
For thou, I fear, hast given me cause to curse.
If thou hast slain Lysander in his sleep,
Being o'er shoes in blood, plunge in the deep,
And kill me too.
The sun was not so true unto the day                         50
As he to me. Would he have stolen away
From sleeping Hermia? I'll believe as soon
This whole earth may be bored, and that the moon
May through the centre creep, and so displease
Her brother's noontide with th'Antipodes.                    55
It cannot be but thou hast murdered him.
So should a murderer look—so dead, so grim.

DEMETRIUS
So should the murdered look, and so should I,
Pierced through the heart with your stern cruelty.
Yet you, the murderer, look as bright, as clear              60
As yonder Venus in her glimmering sphere.

HERMIA
What's this to my Lysander? Where is he?
Ah, good Demetrius, wilt thou give him me?

DEMETRIUS
I had rather give his carcass to my hounds.

HERMIA
Out, dog; out, cur. Thou driv'st me past the bounds
Of maiden's patience. Hast thou slain him then?             66
Henceforth be never numbered among men.
O, once tell true; tell true, even for my sake.
Durst thou have looked upon him being awake,
And hast thou killed him sleeping? O brave touch!           70
Could not a worm, an adder do so much?—
An adder did it, for with doubler tongue
Than thine, thou serpent, never adder stung.

DEMETRIUS
You spend your passion on a misprised mood.
I am not guilty of Lysander's blood,                         75
Nor is he dead, for aught that I can tell.

HERMIA
I pray thee, tell me then that he is well.

DEMETRIUS
And if I could, what should I get therefor?

HERMIA
A privilege never to see me more;
And from thy hated presence part I so.                       80
See me no more, whether he be dead or no.        *Exit*

DEMETRIUS
There is no following her in this fierce vein.
Here therefore for a while I will remain.
So sorrow's heaviness doth heavier grow
For debt that bankrupt sleep doth sorrow owe,               85
Which now in some slight measure it will pay,
If for his tender here I make some stay.
      *He lies down and sleeps*

OBERON (*to Robin*)
What hast thou done? Thou hast mistaken quite,
And laid the love juice on some true love's sight.
Of thy misprision must perforce ensue                        90
Some true love turned, and not a false turned true.

ROBIN
Then fate o'errules that, one man holding troth,
A million fail, confounding oath on oath.

OBERON
About the wood go swifter than the wind,
And Helena of Athens look thou find.                         95
All fancy-sick she is, and pale of cheer
With sighs of love that costs the fresh blood dear.
By some illusion see thou bring her here.
I'll charm his eyes against she do appear.

ROBIN
I go, I go—look how I go,                                    100
Swifter than arrow from the Tartar's bow.        *Exit*

OBERON
Flower of this purple dye,
Hit with Cupid's archery,
Sink in apple of his eye.
      *He drops the juice on Demetrius' eyelids*
When his love he doth espy,                                  105
Let her shine as gloriously
As the Venus of the sky.
When thou wak'st, if she be by,
Beg of her for remedy.
      *Enter Robin Goodfellow, the puck*

ROBIN
Captain of our fairy band,                                   110
Helena is here at hand,
And the youth mistook by me,
Pleading for a lover's fee.

Shall we their fond pageant see?
Lord, what fools these mortals be!                              115
OBERON
Stand aside. The noise they make
Will cause Demetrius to awake.
ROBIN
Then will two at once woo one.
That must needs be sport alone;
And those things do best please me                             120
That befall prepost'rously.
　　⌐They stand apart.⌐
　　Enter Helena, Lysander ⌐following her⌐
LYSANDER
Why should you think that I should woo in scorn?
Scorn and derision never come in tears.
Look when I vow, I weep; and vows so born,
In their nativity all truth appears.                           125
How can these things in me seem scorn to you,
Bearing the badge of faith to prove them true?
HELENA
You do advance your cunning more and more,
When truth kills truth—O devilish holy fray!
These vows are Hermia's. Will you give her o'er?              130
Weigh oath with oath, and you will nothing weigh.
Your vows to her and me put in two scales
Will even weigh, and both as light as tales.
LYSANDER
I had no judgement when to her I swore.
HELENA
Nor none, in my mind, now you give her o'er.                  135
LYSANDER
Demetrius loves her, and he loves not you.
⌐HELENA⌐
⌐                                                         ⌐
DEMETRIUS (awaking)
O Helen, goddess, nymph, perfect, divine!
To what, my love, shall I compare thine eyne?
Crystal is muddy. O, how ripe in show                         140
Thy lips, those kissing cherries, tempting grow!
That pure congealèd white—high Taurus' snow,
Fanned with the eastern wind—turns to a crow
When thou hold'st up thy hand. O, let me kiss
This princess of pure white, this seal of bliss!             145
HELENA
O spite! O hell! I see you all are bent
To set against me for your merriment.
If you were civil, and knew courtesy,
You would not do me thus much injury.
Can you not hate me—as I know you do—                         150
But you must join in souls to mock me too?
If you were men, as men you are in show,
You would not use a gentle lady so,
To vow and swear and superpraise my parts
When I am sure you hate me with your hearts.                 155
You both are rivals and love Hermia,
And now both rivals to mock Helena.
A trim exploit, a manly enterprise—
To conjure tears up in a poor maid's eyes
With your derision. None of noble sort                        160
Would so offend a virgin, and extort
A poor soul's patience, all to make you sport.
LYSANDER
You are unkind, Demetrius. Be not so.
For you love Hermia; this you know I know.
And here with all good will, with all my heart,              165
In Hermia's love I yield you up my part;

And yours of Helena to me bequeath,
Whom I do love, and will do till my death.
HELENA
Never did mockers waste more idle breath.
DEMETRIUS
Lysander, keep thy Hermia. I will none.                       170
If e'er I loved her, all that love is gone.
My heart to her but as guestwise sojourned
And now to Helen is it home returned,
There to remain.
LYSANDER　　　　　　Helen, it is not so.
DEMETRIUS
Disparage not the faith thou dost not know,                   175
Lest to thy peril thou aby it dear.
　　Enter Hermia
Look where thy love comes; yonder is thy dear.
HERMIA
Dark night, that from the eye his function takes,
The ear more quick of apprehension makes.
Wherein it doth impair the seeing sense,                      180
It pays the hearing double recompense.
Thou art not by mine eye, Lysander, found;
Mine ear, I thank it, brought me to thy sound.
But why unkindly didst thou leave me so?
LYSANDER
Why should he stay whom love doth press to go? 185
HERMIA
What love could press Lysander from my side?
LYSANDER
Lysander's love, that would not let him bide:
Fair Helena, who more engilds the night
Than all yon fiery O's and eyes of light.
Why seek'st thou me? Could not this make thee know
The hate I bare thee made me leave thee so?                   191
HERMIA
You speak not as you think. It cannot be.
HELENA ⌐aside⌐
Lo, she is one of this confederacy.
Now I perceive they have conjoined all three
To fashion this false sport in spite of me.—                  195
Injurious Hermia, most ungrateful maid,
Have you conspired, have you with these contrived
To bait me with this foul derision?
Is all the counsel that we two have shared—
The sisters' vows, the hours that we have spent               200
When we have chid the hasty-footed time
For parting us—O, is all quite forgot?
All schooldays' friendship, childhood innocence?
We, Hermia, like two artificial gods
Have with our needles created both one flower,                205
Both on one sampler, sitting on one cushion,
Both warbling of one song, both in one key,
As if our hands, our sides, voices, and minds
Had been incorporate. So we grew together,
Like to a double cherry: seeming parted,                      210
But yet an union in partition,
Two lovely berries moulded on one stem.
So, with two seeming bodies but one heart,
Two of the first—like coats in heraldry,
Due but to one and crownèd with one crest.                    215
And will you rend our ancient love asunder,
To join with men in scorning your poor friend?
It is not friendly, 'tis not maidenly.
Our sex as well as I may chide you for it,
Though I alone do feel the injury.                            220

HERMIA
I am amazèd at your passionate words.
I scorn you not. It seems that you scorn me.
HELENA
Have you not set Lysander, as in scorn,
To follow me, and praise my eyes and face?
And made your other love, Demetrius—                      225
Who even but now did spurn me with his foot—
To call me goddess, nymph, divine, and rare,
Precious, celestial? Wherefore speaks he this
To her he hates? And wherefore doth Lysander
Deny your love so rich within his soul,                   230
And tender me, forsooth, affection,
But by your setting on, by your consent?
What though I be not so in grace as you,
So hung upon with love, so fortunate,
But miserable most, to love unloved—                      235
This you should pity rather than despise.
HERMIA
I understand not what you mean by this.
HELENA
Ay, do. Persever, counterfeit sad looks,
Make mouths upon me when I turn my back,
Wink each at other, hold the sweet jest up.               240
This sport well carried shall be chronicled.
If you have any pity, grace, or manners,
You would not make me such an argument.
But fare ye well. 'Tis partly my own fault,
Which death or absence soon shall remedy.                 245
LYSANDER
Stay, gentle Helena, hear my excuse,
My love, my life, my soul, fair Helena.
HELENA
O excellent!
HERMIA (*to Lysander*) Sweet, do not scorn her so.
DEMETRIUS (*to Lysander*)
If she cannot entreat I can compel.
LYSANDER
Thou canst compel no more than she entreat.              250
Thy threats have no more strength than her weak
      prayers.—
Helen, I love thee; by my life I do.
I swear by that which I will lose for thee
To prove him false that says I love thee not.
DEMETRIUS (*to Helena*)
I say I love thee more than he can do.                    255
LYSANDER
If thou say so, withdraw, and prove it too.
DEMETRIUS
Quick, come.
HERMIA          Lysander, whereto tends all this?
⌈*She takes him by the arm*⌉
LYSANDER
Away, you Ethiope.
DEMETRIUS            No, no, sir, yield.
Seem to break loose, take on as you would follow,
But yet come not. You are a tame man; go.                 260
LYSANDER (*to Hermia*)
Hang off, thou cat, thou burr; vile thing, let loose,
Or I will shake thee from me like a serpent.
HERMIA
Why are you grown so rude? What change is this,
Sweet love?
LYSANDER      Thy love? Out, tawny Tartar, out;
Out, loathèd med'cine; O hated potion, hence.             265

HERMIA
Do you not jest?
HELENA           Yes, sooth, and so do you.
LYSANDER
Demetrius, I will keep my word with thee.
DEMETRIUS
I would I had your bond, for I perceive
A weak bond holds you. I'll not trust your word.
LYSANDER
What, should I hurt her, strike her, kill her dead?      270
Although I hate her, I'll not harm her so.
HERMIA
What, can you do me greater harm than hate?
Hate me—wherefore? O me, what news, my love?
Am not I Hermia? Are not you Lysander?
I am as fair now as I was erewhile.                       275
Since night you loved me, yet since night you left me.
Why then, you left me—O, the gods forbid—
In earnest, shall I say?
LYSANDER              Ay, by my life,
And never did desire to see thee more.
Therefore be out of hope, of question, doubt.            280
Be certain, nothing truer; 'tis no jest
That I do hate thee and love Helena.
HERMIA (*to Helena*)
O me, you juggler, you canker blossom,
You thief of love—what, have you come by night
And stol'n my love's heart from him?
HELENA                            Fine, i'faith.      285
Have you no modesty, no maiden shame,
No touch of bashfulness? What, will you tear
Impatient answers from my gentle tongue?
Fie, fie, you counterfeit, you puppet, you!
HERMIA
Puppet? Why, so! Ay, that way goes the game.             290
Now I perceive that she hath made compare
Between our statures; she hath urged her height,
And with her personage, her tall personage,
Her height, forsooth, she hath prevailed with him—
And are you grown so high in his esteem                   295
Because I am so dwarfish and so low?
How low am I, thou painted maypole? Speak,
How low am I? I am not yet so low
But that my nails can reach unto thine eyes.
HELENA (*to Demetrius and Lysander*)
I pray you, though you mock me, gentlemen,                300
Let her not hurt me. I was never curst.
I have no gift at all in shrewishness.
I am a right maid for my cowardice.
Let her not strike me. You perhaps may think
Because she is something lower than myself                305
That I can match her—
HERMIA              Lower? Hark again.
HELENA
Good Hermia, do not be so bitter with me.
I evermore did love you, Hermia,
Did ever keep your counsels, never wronged you—
Save that in love unto Demetrius                          310
I told him of your stealth unto this wood.
He followed you; for love I followed him.
But he hath chid me hence, and threatened me
To strike me, spurn me, nay, to kill me too.
And now, so you will let me quiet go,                     315
To Athens will I bear my folly back,
And follow you no further. Let me go.
You see how simple and how fond I am.

HERMIA
Why, get you gone. Who is't that hinders you?
HELENA
A foolish heart that I leave here behind.          320
HERMIA
What, with Lysander?
HELENA                    With Demetrius.
LYSANDER
Be not afraid; she shall not harm thee, Helena.
DEMETRIUS
No, sir, she shall not, though you take her part.
HELENA
O, when she is angry she is keen and shrewd.
She was a vixen when she went to school,          325
And though she be but little, she is fierce.
HERMIA
Little again? Nothing but 'low' and 'little'?—
Why will you suffer her to flout me thus?
Let me come to her.
LYSANDER          Get you gone, you dwarf,
You *minimus* of hind'ring knot-grass made,          330
You bead, you acorn.
DEMETRIUS                    You are too officious
In her behalf that scorns your services.
Let her alone. Speak not of Helena.
Take not her part. For if thou dost intend
Never so little show of love to her,          335
Thou shalt aby it.
LYSANDER          Now she holds me not.
Now follow, if thou dar'st, to try whose right,
Of thine or mine, is most in Helena.
DEMETRIUS
Follow? Nay, I'll go with thee, cheek by jowl.
                    *Exeunt Lysander and Demetrius*
HERMIA
You, mistress, all this coil is long of you.          340
Nay, go not back.
HELENA          I will not trust you, I,
Nor longer stay in your curst company.
Your hands than mine are quicker for a fray;
My legs are longer, though, to run away.          *Exit*
HERMIA
I am amazed, and know not what to say.          *Exit*
          ⌐Oberon and Robin come forward¬
OBERON
This is thy negligence. Still thou mistak'st,          346
Or else commit'st thy knaveries wilfully.
ROBIN
Believe me, king of shadows, I mistook.
Did not you tell me I should know the man
By the Athenian garments he had on?—          350
And so far blameless proves my enterprise
That I have 'nointed an Athenian's eyes;
And so far am I glad it so did sort
As this their jangling I esteem a sport.
OBERON
Thou seest these lovers seek a place to fight.          355
Hie therefore, Robin, overcast the night;
The starry welkin cover thou anon
With drooping fog as black as Acheron,
And lead these testy rivals so astray
As one come not within another's way.          360
Like to Lysander sometime frame thy tongue,
Then stir Demetrius up with bitter wrong;
And sometime rail thou like Demetrius,
And from each other look thou lead them thus

Till o'er their brows death-counterfeiting sleep          365
With leaden legs and batty wings doth creep.
Then crush this herb into Lysander's eye—
Whose liquor hath this virtuous property,
To take from thence all error with his might,
And make his eyeballs roll with wonted sight.          370
When they next wake, all this derision
Shall seem a dream and fruitless vision,
And back to Athens shall the lovers wend
With league whose date till death shall never end.
Whiles I in this affair do thee employ,          375
I'll to my queen and beg her Indian boy;
And then I will her charmèd eye release
From monster's view, and all things shall be peace.
ROBIN
My fairy lord, this must be done with haste,
For night's swift dragons cut the clouds full fast,          380
And yonder shines Aurora's harbinger,
At whose approach ghosts, wand'ring here and there,
Troop home to churchyards; damnèd spirits all
That in cross-ways and floods have burial
Already to their wormy beds are gone,          385
For fear lest day should look their shames upon.
They wilfully themselves exiled from light,
And must for aye consort with black-browed night.
OBERON
But we are spirits of another sort.
I with the morning's love have oft made sport,          390
And like a forester the groves may tread
Even till the eastern gate, all fiery red,
Opening on Neptune with fair blessèd beams
Turns into yellow gold his salt green streams.
But notwithstanding, haste, make no delay;          395
We may effect this business yet ere day.          *Exit*
ROBIN
Up and down, up and down,
I will lead them up and down.
I am feared in field and town.
Goblin, lead them up and down.          400
Here comes one.
          *Enter Lysander*
LYSANDER
Where art thou, proud Demetrius? Speak thou now.
ROBIN ⌐*shifting place*¬
Here, villain, drawn and ready. Where art thou?
LYSANDER
I will be with thee straight.
ROBIN ⌐*shifting place*¬          Follow me then
To plainer ground.          ⌐*Exit Lysander*¬
          *Enter Demetrius*
DEMETRIUS ⌐*shifting place*¬ Lysander, speak again.          405
Thou runaway, thou coward, art thou fled?
Speak! In some bush? Where dost thou hide thy head?
ROBIN ⌐*shifting place*¬
Thou coward, art thou bragging to the stars,
Telling the bushes that thou look'st for wars,
And wilt not come? Come, recreant; come, thou child,
I'll whip thee with a rod. He is defiled          411
That draws a sword on thee.
DEMETRIUS ⌐*shifting place*¬          Yea, art thou there?
ROBIN ⌐*shifting place*¬
Follow my voice; we'll try no manhood here.          *Exeunt*

**3.3** ⌐*Enter Lysander*¬
LYSANDER
He goes before me, and still dares me on;
When I come where he calls, then he is gone.

The villain is much lighter heeled than I;
I followed fast, but faster he did fly,
That fallen am I in dark uneven way,                                    5
And here will rest me.
    *He lies down*
             Come, thou gentle day;
For if but once thou show me thy grey light,
I'll find Demetrius, and revenge this spite.   *He sleeps*
    *Enter Robin Goodfellow and Demetrius*
ROBIN ⌈*shifting place*⌉
Ho, ho, ho, coward, why com'st thou not?
DEMETRIUS
Abide me if thou dar'st, for well I wot                                 10
Thou runn'st before me, shifting every place,
And dar'st not stand nor look me in the face.
Where art thou now?
ROBIN ⌈*shifting place*⌉    Come hither, I am here.
DEMETRIUS
Nay, then thou mock'st me. Thou shalt buy this
    dear
If ever I thy face by daylight see.                                     15
Now go thy way. Faintness constraineth me
To measure out my length on this cold bed.
    *He lies down*
By day's approach look to be visited.    *He sleeps*
    *Enter Helena*
HELENA
O weary night, O long and tedious night,
   Abate thy hours; shine comforts from the east    20
That I may back to Athens by daylight
   From these that my poor company detest;
And sleep, that sometimes shuts up sorrow's eye,
Steal me a while from mine own company.
    *She lies down and sleeps*
ROBIN
Yet but three? Come one more,                                          25
Two of both kinds makes up four.
    ⌈*Enter Hermia*⌉
Here she comes, curst and sad.
Cupid is a knavish lad
Thus to make poor females mad.
HERMIA
Never so weary, never so in woe,                                       30
   Bedabbled with the dew, and torn with briers,
I can no further crawl, no further go.
My legs can keep no pace with my desires.
Here will I rest me till the break of day.
    *She lies down*
Heavens shield Lysander, if they mean a fray.    35
    *She sleeps*
ROBIN    On the ground sleep sound.
    I'll apply to your eye,
    Gentle lover, remedy.
    *He drops the juice on Lysander's eyelids*
    When thou wak'st thou tak'st
    True delight in the sight                                    40
    Of thy former lady's eye,
   And the country proverb known,
   That 'every man should take his own',
   In your waking shall be shown.
    Jack shall have Jill,                                         45
    Naught shall go ill,

the man shall have his mare again, and all shall be
well.                                     *Exit*

**4.1**    *Enter Titania, Queen of Fairies, and Bottom the*
    *clown with the ass-head, and fairies: Peaseblossom,*
    *Cobweb, Mote, and Mustardseed*
TITANIA (*to Bottom*)
Come, sit thee down upon this flow'ry bed,
   While I thy amiable cheeks do coy,
And stick musk-roses in thy sleek smooth head,
   And kiss thy fair large ears, my gentle joy.
BOTTOM Where's Peaseblossom?                                            5
PEASEBLOSSOM Ready.
BOTTOM Scratch my head, Peaseblossom. Where's Mon-
sieur Cobweb?
COBWEB Ready.
BOTTOM Monsieur Cobweb, good monsieur, get you your
weapons in your hand and kill me a red-hipped humble-
bee on the top of a thistle; and, good monsieur, bring
me the honeybag. Do not fret yourself too much in the
action, monsieur; and, good monsieur, have a care the
honeybag break not. I would be loath to have you
overflowen with a honeybag, signor.   ⌈*Exit Cobweb*⌉
Where's Monsieur Mustardseed?                                          17
MUSTARDSEED Ready.
BOTTOM Give me your neaf, Monsieur Mustardseed. Pray
you, leave your courtesy, good monsieur.                               20
MUSTARDSEED What's your will?
BOTTOM Nothing, good monsieur, but to help Cavaliery
Peaseblossom to scratch. I must to the barber's,
monsieur, for methinks I am marvellous hairy about
the face; and I am such a tender ass, if my hair do but
tickle me I must scratch.                                              26
TITANIA
What, wilt thou hear some music, my sweet love?
BOTTOM I have a reasonable good ear in music. Let's have
the tongs and the bones.
    ⌈*Rural music*⌉
TITANIA
Or say, sweet love, what thou desir'st to eat.    30
BOTTOM Truly, a peck of provender. I could munch your
good dry oats. Methinks I have a great desire to a bottle
of hay. Good hay, sweet hay, hath no fellow.
TITANIA
I have a venturous fairy that shall seek
The squirrel's hoard, and fetch thee off new nuts.    35
BOTTOM I had rather have a handful or two of dried peas.
But I pray you, let none of your people stir me. I have
an exposition of sleep come upon me.
TITANIA
Sleep thou, and I will wind thee in my arms.
Fairies, be gone, and be all ways away.                                40
                       *Exeunt Fairies*
So doth the woodbine the sweet honeysuckle
Gently entwist; the female ivy so
Enrings the barky fingers of the elm.
O how I love thee, how I dote on thee!
    *They sleep*
    *Enter Robin Goodfellow* ⌈*and Oberon, meeting*⌉
OBERON
Welcome, good Robin. Seest thou this sweet sight?    45
Her dotage now I do begin to pity,
For meeting her of late behind the wood,
Seeking sweet favours for this hateful fool,
I did upbraid her and fall out with her,
For she his hairy temples then had rounded                             50
With coronet of fresh and fragrant flowers,
And that same dew which sometime on the buds
Was wont to swell like round and orient pearls

Stood now within the pretty flow'rets' eyes,
Like tears that did their own disgrace bewail.    55
When I had at my pleasure taunted her,
And she in mild terms begged my patience,
I then did ask of her her changeling child,
Which straight she gave me, and her fairy sent
To bear him to my bower in fairyland.    60
And now I have the boy, I will undo
This hateful imperfection of her eyes.
And, gentle puck, take this transformèd scalp
From off the head of this Athenian swain,
That he, awaking when the other do,    65
May all to Athens back again repair,
And think no more of this night's accidents
But as the fierce vexation of a dream.
But first I will release the Fairy Queen.
        *He drops the juice on Titania's eyelids*
        Be as thou wast wont to be,    70
        See as thou wast wont to see.
        Dian's bud o'er Cupid's flower
        Hath such force and blessèd power.
Now, my Titania, wake you, my sweet queen.

TITANIA (*awaking*)
My Oberon, what visions have I seen!    75
Methought I was enamoured of an ass.

OBERON
There lies your love.

TITANIA                How came these things to pass?
O, how mine eyes do loathe his visage now!

OBERON
Silence a while.—Robin, take off this head.—
Titania, music call, and strike more dead    80
Than common sleep of all these five the sense.

TITANIA
Music, ho—music such as charmeth sleep.
        ⌜*Still music*⌝

ROBIN (*taking the ass-head off Bottom*)
Now when thou wak'st with thine own fool's eyes
        peep.

OBERON
Sound music.
        ⌜*The music changes*⌝
                Come, my queen, take hands with me,
And rock the ground whereon these sleepers be.    85
        *Oberon and Titania dance*
Now thou and I are new in amity,
And will tomorrow midnight solemnly
Dance in Duke Theseus' house, triumphantly,
And bless it to all fair prosperity.
There shall the pairs of faithful lovers be    90
Wedded with Theseus, all in jollity.

ROBIN
        Fairy King, attend and mark.
        I do hear the morning lark.

OBERON
        Then, my queen, in silence sad
        Trip we after nightës shade.    95
        We the globe can compass soon,
        Swifter than the wand'ring moon.

TITANIA
        Come, my lord, and in our flight
        Tell me how it came this night
        That I sleeping here was found    100
        With these mortals on the ground.
        *Exeunt Oberon, Titania, and*
        *Robin. The sleepers lie still*

*Wind horns within. Enter Theseus with Egeus,*
*Hippolyta, and all his train*

THESEUS
Go, one of you, find out the forester,
For now our observation is performed;
And since we have the vanguard of the day,
My love shall hear the music of my hounds.    105
Uncouple in the western valley; let them go.
Dispatch, I say, and find the forester.    *Exit one*
We will, fair Queen, up to the mountain's top,
And mark the musical confusion
Of hounds and echo in conjunction.    110

HIPPOLYTA
I was with Hercules and Cadmus once
When in a wood of Crete they bayed the bear
With hounds of Sparta. Never did I hear
Such gallant chiding; for besides the groves,
The skies, the fountains, every region near    115
Seemed all one mutual cry. I never heard
So musical a discord, such sweet thunder.

THESEUS
My hounds are bred out of the Spartan kind,
So flewed, so sanded; and their heads are hung
With ears that sweep away the morning dew,    120
Crook-kneed, and dewlapped like Thessalian bulls,
Slow in pursuit, but matched in mouth like bells,
Each under each. A cry more tuneable
Was never holla'd to nor cheered with horn
In Crete, in Sparta, nor in Thessaly.    125
Judge when you hear. But soft: what nymphs are
        these?

EGEUS
My lord, this is my daughter here asleep,
And this Lysander; this Demetrius is;
This Helena, old Nedar's Helena.
I wonder of their being here together.    130

THESEUS
No doubt they rose up early to observe
The rite of May, and, hearing our intent,
Came here in grace of our solemnity.
But speak, Egeus: is not this the day
That Hermia should give answer of her choice?    135

EGEUS It is, my lord.

THESEUS
Go bid the huntsmen wake them with their horns.
        ⌜*Exit one*⌝
        *Shout within: wind horns. The lovers all start up*
Good morrow, friends. Saint Valentine is past.
Begin these wood-birds but to couple now?

LYSANDER
Pardon, my lord.
        *The lovers kneel*

THESEUS                I pray you all stand up.    140
        *The lovers stand*
(*To Demetrius and Lysander*) I know you two are rival
        enemies.
How comes this gentle concord in the world,
That hatred is so far from jealousy
To sleep by hate, and fear no enmity?

LYSANDER
My lord, I shall reply amazèdly,    145
Half sleep, half waking. But as yet, I swear,
I cannot truly say how I came here,
But as I think—for truly would I speak,
And, now I do bethink me, so it is—
I came with Hermia hither. Our intent    150

Was to be gone from Athens where we might,
Without the peril of the Athenian law—

EGEUS (*to Theseus*)
Enough, enough, my lord, you have enough.
I beg the law, the law upon his head.—
They would have stol'n away, they would, Demetrius,　156
Thereby to have defeated you and me—
You of your wife, and me of my consent,
Of my consent that she should be your wife.

DEMETRIUS (*to Theseus*)
My lord, fair Helen told me of their stealth,
Of this their purpose hither to this wood,　160
And I in fury hither followed them,
Fair Helena in fancy following me.
But, my good lord, I wot not by what power—
But by some power it is—my love to Hermia,
Melted as the snow, seems to me now　165
As the remembrance of an idle gaud
Which in my childhood I did dote upon,
And all the faith, the virtue of my heart,
The object and the pleasure of mine eye
Is only Helena. To her, my lord,　170
Was I betrothed ere I see Hermia.
But like in sickness did I loathe this food;
But, as in health come to my natural taste,
Now I do wish it, love it, long for it,
And will for evermore be true to it.　175

THESEUS
Fair lovers, you are fortunately met.
Of this discourse we more will hear anon.—
Egeus, I will overbear your will,
For in the temple by and by with us
These couples shall eternally be knit.—　180
And, for the morning now is something worn,
Our purposed hunting shall be set aside.
Away with us to Athens. Three and three,
We'll hold a feast in great solemnity.
Come, Hippolyta.　185

*Exit Duke Theseus with Hippolyta, Egeus,*
*and all his train*

DEMETRIUS
These things seem small and undistinguishable,
Like far-off mountains turnèd into clouds.

HERMIA
Methinks I see these things with parted eye,
When everything seems double.

HELENA　　　　　　　　　　　So methinks,
And I have found Demetrius like a jewel,　190
Mine own and not mine own.

DEMETRIUS　　　　　　　　It seems to me
That yet we sleep, we dream. Do not you think
The Duke was here and bid us follow him?

HERMIA
Yea, and my father.

HELENA　　　　　　　And Hippolyta.

LYSANDER
And he did bid us follow to the temple.　195

DEMETRIUS
Why then, we are awake. Let's follow him,
And by the way let us recount our dreams.

*Exeunt the lovers*

*Bottom wakes*

BOTTOM When my cue comes, call me, and I will answer.
My next is 'most fair Pyramus'. Heigh-ho. Peter Quince?
Flute the bellows-mender? Snout the tinker?
Starveling? God's my life! Stolen hence, and left me
asleep?—I have had a most rare vision. I have had a

dream past the wit of man to say what dream it was.
Man is but an ass if he go about t'expound this dream.
Methought I was—there is no man can tell what.
Methought I was, and methought I had—but man is
but a patched fool if he will offer to say what methought
I had. The eye of man hath not heard, the ear of man
hath not seen, man's hand is not able to taste, his
tongue to conceive, nor his heart to report what my
dream was. I will get Peter Quince to write a ballad of
this dream. It shall be called 'Bottom's Dream', because
it hath no bottom, and I will sing it in the latter end
of a play, before the Duke. Peradventure, to make it
the more gracious, I shall sing it at her death.　*Exit*

**4.2**　*Enter Quince, Flute, Snout, and Starveling*
QUINCE Have you sent to Bottom's house? Is he come
home yet?
STARVELING He cannot be heard of. Out of doubt he is
transported.
FLUTE If he come not, then the play is marred. It goes
not forward. Doth it?　6
QUINCE It is not possible. You have not a man in all
Athens able to discharge Pyramus but he.
FLUTE No, he hath simply the best wit of any handicraft-
man in Athens.　10
QUINCE Yea, and the best person, too; and he is a very
paramour for a sweet voice.
FLUTE You must say 'paragon'. A paramour is, God bless
us, a thing of naught.　14

*Enter Snug the joiner*
SNUG Masters, the Duke is coming from the temple, and
there is two or three lords and ladies more married. If
our sport had gone forward we had all been made men.
FLUTE O sweet bully Bottom! Thus hath he lost sixpence
a day during his life. He could not have scaped sixpence
a day. An the Duke had not given him sixpence a day
for playing Pyramus, I'll be hanged. He would have
deserved it. Sixpence a day in Pyramus, or nothing.

*Enter Bottom*
BOTTOM Where are these lads? Where are these hearts?
QUINCE Bottom! O most courageous day! O most happy
hour!　25
BOTTOM Masters, I am to discourse wonders; but ask me
not what. For if I tell you, I am no true Athenian. I
will tell you everything right as it fell out.
QUINCE Let us hear, sweet Bottom.　29
BOTTOM Not a word of me. All that I will tell you is that
the Duke hath dined. Get your apparel together, good
strings to your beards, new ribbons to your pumps.
Meet presently at the palace; every man look o'er his
part. For the short and the long is, our play is preferred.
In any case let Thisbe have clean linen, and let not
him that plays the lion pare his nails, for they shall
hang out for the lion's claws. And, most dear actors,
eat no onions nor garlic, for we are to utter sweet
breath, and I do not doubt but to hear them say it is
a sweet comedy. No more words. Away, go, away!
*Exeunt*

**5.1**　*Enter Theseus, Hippolyta, ⌈Egeus⌉, and attendant*
*lords*
HIPPOLYTA
'Tis strange, my Theseus, that these lovers speak of.
THESEUS
More strange than true. I never may believe
These antique fables, nor these fairy toys.
Lovers and madmen have such seething brains,

Such shaping fantasies, that apprehend    5
More than cool reason ever comprehends.
The lunatic, the lover, and the poet
Are of imagination all compact.
One sees more devils than vast hell can hold:
That is the madman. The lover, all as frantic,    10
Sees Helen's beauty in a brow of Egypt.
The poet's eye, in a fine frenzy rolling,
Doth glance from heaven to earth, from earth to
     heaven,
And as imagination bodies forth
The forms of things unknown, the poet's pen    15
Turns them to shapes, and gives to airy nothing
A local habitation and a name.
Such tricks hath strong imagination
That if it would but apprehend some joy
It comprehends some bringer of that joy;    20
Or in the night, imagining some fear,
How easy is a bush supposed a bear!

HIPPOLYTA
But all the story of the night told over,
And all their minds transfigured so together,
More witnesseth than fancy's images,    25
And grows to something of great constancy;
But howsoever, strange and admirable.
     *Enter the lovers: Lysander, Demetrius, Hermia,*
     *and Helena*

THESEUS
Here come the lovers, full of joy and mirth.
Joy, gentle friends—joy and fresh days of love
Accompany your hearts.

LYSANDER            More than to us    30
Wait in your royal walks, your board, your bed.

THESEUS
Come now, what masques, what dances shall we have
To wear away this long age of three hours
Between our after-supper and bed-time?
Where is our usual manager of mirth?    35
What revels are in hand? Is there no play
To ease the anguish of a torturing hour?
Call Egeus.

⌈EGEUS⌉       Here, mighty Theseus.

THESEUS
Say, what abridgement have you for this evening?
What masque, what music? How shall we beguile    40
The lazy time if not with some delight?

⌈EGEUS⌉
There is a brief how many sports are ripe.
Make choice of which your highness will see first.

⌈LYSANDER⌉ (*reads*)
'The battle with the centaurs, to be sung
By an Athenian eunuch to the harp.'    45

THESEUS
We'll none of that. That have I told my love
In glory of my kinsman Hercules.

⌈LYSANDER⌉ (*reads*)
'The riot of the tipsy bacchanals
Tearing the Thracian singer in their rage.'

THESEUS
That is an old device, and it was played    50
When I from Thebes came last a conqueror.

⌈LYSANDER⌉ (*reads*)
'The thrice-three muses mourning for the death
Of learning, late deceased in beggary.'

THESEUS
That is some satire, keen and critical,
Not sorting with a nuptial ceremony.    55

⌈LYSANDER⌉ (*reads*)
'A tedious brief scene of young Pyramus
And his love Thisbe: very tragical mirth.'

THESEUS
'Merry' *and* 'tragical'? 'Tedious' *and* 'brief'?—
That is, hot ice and wondrous strange black snow.
How shall we find the concord of this discord?    60

⌈EGEUS⌉
A play there is, my lord, some ten words long,
Which is as 'brief' as I have known a play;
But by ten words, my lord, it is too long,
Which makes it 'tedious'; for in all the play
There is not one word apt, one player fitted.    65
And 'tragical', my noble lord, it is,
For Pyramus therein doth kill himself;
Which when I saw rehearsed, I must confess,
Made mine eyes water; but more merry tears
The passion of loud laughter never shed.    70

THESEUS What are they that do play it?

⌈EGEUS⌉
Hard-handed men that work in Athens here,
Which never laboured in their minds till now,
And now have toiled their unbreathed memories
With this same play against your nuptial.    75

THESEUS
And we will hear it.

⌈EGEUS⌉         No, my noble lord,
It is not for you. I have heard it over,
And it is nothing, nothing in the world,
Unless you can find sport in their intents
Extremely stretched, and conned with cruel pain    80
To do you service.

THESEUS        I will hear that play;
For never anything can be amiss
When simpleness and duty tender it.
Go, bring them in; and take your places, ladies.
                       *Exit* ⌈*Egeus*⌉

HIPPOLYTA
I love not to see wretchedness o'ercharged,    85
And duty in his service perishing.

THESEUS
Why, gentle sweet, you shall see no such thing.

HIPPOLYTA
He says they can do nothing in this kind.

THESEUS
The kinder we, to give them thanks for nothing.
Our sport shall be to take what they mistake,    90
And what poor duty cannot do,
Noble respect takes it in might, not merit.
Where I have come, great clerks have purposèd
To greet me with premeditated welcomes,
Where I have seen them shiver and look pale,    95
Make periods in the midst of sentences,
Throttle their practised accent in their fears,
And in conclusion dumbly have broke off,
Not paying me a welcome. Trust me, sweet,
Out of this silence yet I picked a welcome,    100
And in the modesty of fearful duty
I read as much as from the rattling tongue
Of saucy and audacious eloquence.
Love, therefore, and tongue-tied simplicity
In least speak most, to my capacity.    105
         *Enter* ⌈*Egeus*⌉

⌈EGEUS⌉
So please your grace, the Prologue is addressed.

THESEUS Let him approach.

*⌐Flourish trumpets.⌐ Enter ⌐Quince as⌐ the Prologue*
⌐QUINCE⌐ (*as Prologue*)
    If we offend, it is with our good will.
        That you should think: we come not to offend
    But with good will. To show our simple skill,       110
        That is the true beginning of our end.
    Consider then we come but in despite.
        We do not come as minding to content you,
    Our true intent is. All for your delight
        We are not here. That you should here repent you
    The actors are at hand, and by their show       116
    You shall know all that you are like to know.
THESEUS This fellow doth not stand upon points.
LYSANDER He hath rid his prologue like a rough colt: he
    knows not the stop. A good moral, my lord: it is not
    enough to speak, but to speak true.       121
HIPPOLYTA Indeed, he hath played on this prologue like
    a child on a recorder—a sound, but not in government.
THESEUS His speech was like a tangled chain—nothing
    impaired, but all disordered. Who is next?       125
        *Enter ⌐with a trumpeter before them⌐ Bottom as*
        *Pyramus, Flute as Thisbe, Snout as Wall, Starveling*
        *as Moonshine, and Snug as Lion, for the dumb show*
⌐QUINCE⌐ (*as Prologue*)
    Gentles, perchance you wonder at this show,
        But wonder on, till truth make all things plain.
    This man is Pyramus, if you would know;
        This beauteous lady Thisbe is, certain.
    This man with lime and roughcast doth present       130
        Wall, that vile wall which did these lovers sunder;
    And through Wall's chink, poor souls, they are content
        To whisper; at the which let no man wonder.
    This man, with lantern, dog, and bush of thorn,
        Presenteth Moonshine. For if you will know,       135
    By moonshine did these lovers think no scorn
        To meet at Ninus' tomb, there, there to woo.
    This grizzly beast, which 'Lion' hight by name,
        The trusty Thisbe coming first by night
    Did scare away, or rather did affright;       140
    And as she fled, her mantle she did fall,
        Which Lion vile with bloody mouth did stain.
    Anon comes Pyramus, sweet youth and tall,
        And finds his trusty Thisbe's mantle slain;
    Whereat with blade—with bloody, blameful blade—
        He bravely broached his boiling bloody breast;       146
    And Thisbe, tarrying in mulberry shade,
        His dagger drew and died. For all the rest,
    Let Lion, Moonshine, Wall, and lovers twain
    At large discourse, while here they do remain.       150
        *⌐Exeunt all the clowns but Snout as Wall⌐*
THESEUS I wonder if the lion be to speak.
DEMETRIUS No wonder, my lord—one lion may when
    many asses do.
⌐SNOUT⌐ (*as Wall*)
    In this same interlude it doth befall
    That I, one Snout by name, present a wall;       155
    And such a wall as I would have you think
    That had in it a crannied hole or chink,
    Through which the lovers Pyramus and Thisbe
    Did whisper often, very secretly.
    This loam, this roughcast, and this stone doth show
    That I am that same wall; the truth is so.       161
    And this the cranny is, right and sinister,
    Through which the fearful lovers are to whisper.
THESEUS Would you desire lime and hair to speak better?
DEMETRIUS It is the wittiest partition that ever I heard
    discourse, my lord.       166

*Enter Bottom as Pyramus*
THESEUS Pyramus draws near the wall. Silence.
BOTTOM (*as Pyramus*)
    O grim-looked night, O night with hue so black,
        O night which ever art when day is not;
    O night, O night, alack, alack, alack,       170
        I fear my Thisbe's promise is forgot.
    And thou, O wall, O sweet O lovely wall,
        That stand'st between her father's ground and mine,
    Thou wall, O wall, O sweet and lovely wall,
        Show me thy chink, to blink through with mine
        eyne.       175
        *Wall shows his chink*
    Thanks, courteous wall. Jove shield thee well for this.
        But what see I? No Thisbe do I see.
    O wicked wall, through whom I see no bliss,
        Cursed be thy stones for thus deceiving me.
THESEUS The wall methinks, being sensible, should curse
    again.       181
BOTTOM (*to Theseus*) No, in truth, sir, he should not.
    'Deceiving me' is Thisbe's cue. She is to enter now,
    and I am to spy her through the wall. You shall see,
    it will fall pat as I told you.       185
        *Enter Flute as Thisbe*
    Yonder she comes.
FLUTE (*as Thisbe*)
    O wall, full often hast thou heard my moans
        For parting my fair Pyramus and me.
    My cherry lips have often kissed thy stones,
        Thy stones with lime and hair knit up in thee.       190
BOTTOM (*as Pyramus*)
    I see a voice. Now will I to the chink
        To spy an I can hear my Thisbe's face.
    Thisbe?
FLUTE (*as Thisbe*) My love—thou art my love, I think.
BOTTOM (*as Pyramus*)
    Think what thou wilt, I am thy lover's grace,
    And like Lemander am I trusty still.       195
FLUTE (*as Thisbe*)
    And I like Helen, till the fates me kill.
BOTTOM (*as Pyramus*)
    Not Shaphalus to Procrus was so true.
FLUTE (*as Thisbe*)
    As Shaphalus to Procrus, I to you.
BOTTOM (*as Pyramus*)
    O kiss me through the hole of this vile wall.
FLUTE (*as Thisbe*)
    I kiss the wall's hole, not your lips at all.       200
BOTTOM (*as Pyramus*)
    Wilt thou at Ninny's tomb meet me straightway?
FLUTE (*as Thisbe*)
    Tide life, tide death, I come without delay.
        *Exeunt Bottom and Flute severally*
SNOUT (*as Wall*)
    Thus have I, Wall, my part dischargèd so;
    And being done, thus Wall away doth go.       *Exit*
THESEUS Now is the wall down between the two
    neighbours.       206
DEMETRIUS No remedy, my lord, when walls are so wilful
    to hear without warning.
HIPPOLYTA This is the silliest stuff that ever I heard.
THESEUS The best in this kind are but shadows, and the
    worst are no worse if imagination amend them.       211
HIPPOLYTA It must be your imagination, then, and not
    theirs.
THESEUS If we imagine no worse of them than they of

themselves, they may pass for excellent men. Here
come two noble beasts in: a man and a lion.   216
   *Enter Snug as Lion, and Starveling as Moonshine*
   *with a lantern, thorn bush, and dog*
SNUG (*as Lion*)
You, ladies, you whose gentle hearts do fear
   The smallest monstrous mouse that creeps on floor,
May now perchance both quake and tremble here
   When lion rough in wildest rage doth roar.   220
Then know that I as Snug the joiner am
A lion fell, nor else no lion's dam.
For if I should as Lion come in strife
Into this place, 'twere pity on my life.
THESEUS A very gentle beast, and of a good conscience.
DEMETRIUS The very best at a beast, my lord, that e'er I
saw.   227
LYSANDER This lion is a very fox for his valour.
THESEUS True, and a goose for his discretion.
DEMETRIUS Not so, my lord, for his valour cannot carry
his discretion, and the fox carries the goose.   231
THESEUS His discretion, I am sure, cannot carry his valour,
for the goose carries not the fox. It is well. Leave it to
his discretion, and let us listen to the moon.
STARVELING (*as Moonshine*)
This lantern doth the hornèd moon present.   235
DEMETRIUS He should have worn the horns on his head.
THESEUS He is no crescent, and his horns are invisible
within the circumference.
STARVELING (*as Moonshine*)
This lantern doth the hornèd moon present.
   Myself the man i'th' moon do seem to be.   240
THESEUS This is the greatest error of all the rest—the man
should be put into the lantern. How is it else the man
i'th' moon?
DEMETRIUS He dares not come there for the candle; for
you see it is already in snuff.   245
HIPPOLYTA I am aweary of this moon. Would he would
change.
THESEUS It appears by his small light of discretion that he
is in the wane; but yet in courtesy, in all reason, we
must stay the time.   250
LYSANDER Proceed, Moon.
STARVELING All that I have to say is to tell you that the
lantern is the moon, I the man i'th' moon, this thorn
bush my thorn bush, and this dog my dog.
DEMETRIUS Why, all these should be in the lantern, for
all these are in the moon. But silence; here comes
Thisbe.   257
   *Enter Flute as Thisbe*
FLUTE (*as Thisbe*)
This is old Ninny's tomb. Where is my love?
SNUG (*as Lion*) O.
   *Lion roars. Thisbe drops her mantle and runs off*
DEMETRIUS Well roared, Lion.   260
THESEUS Well run, Thisbe.
HIPPOLYTA Well shone, Moon.—Truly, the moon shines
with a good grace.
   *Lion worries Thisbe's mantle*
THESEUS Well moused, Lion.
DEMETRIUS And then came Pyramus.   265
   ⌈*Enter Bottom as Pyramus*⌉
LYSANDER And so the lion vanished.   ⌈*Exit Lion*⌉
BOTTOM (*as Pyramus*)
Sweet moon, I thank thee for thy sunny beams.
   I thank thee, moon, for shining now so bright;
For by thy gracious, golden, glittering gleams
   I trust to take of truest Thisbe sight.   270

But stay, O spite!
But mark, poor knight,
What dreadful dole is here?
   Eyes, do you see?
   How can it be?   275
O dainty duck, O dear!
   Thy mantle good,
   What, stained with blood?
Approach, ye furies fell.
   O fates, come, come,   280
   Cut thread and thrum,
Quail, crush, conclude, and quell.
THESEUS This passion—and the death of a dear friend—
would go near to make a man look sad.
HIPPOLYTA Beshrew my heart, but I pity the man.   285
BOTTOM (*as Pyramus*)
O wherefore, nature, didst thou lions frame,
   Since lion vile hath here deflowered my dear?—
Which is—no, no, which *was*—the fairest dame
   That lived, that loved, that liked, that looked, with
      cheer.
   Come tears, confound;   290
   Out sword, and wound
The pap of Pyramus.
   Ay, that left pap,
   Where heart doth hop.
Thus die I: thus, thus, thus.   295
   *He stabs himself*
   Now am I dead,
   Now am I fled,
My soul is in the sky.
   Tongue, lose thy light;   299
   Moon, take thy flight.   ⌈*Exit Moonshine*⌉
Now die, die, die, die, die.   *He dies*
DEMETRIUS No die but an ace for him; for he is but one.
LYSANDER Less than an ace, man; for he is dead; he is
nothing.
THESEUS With the help of a surgeon he might yet recover
and prove an ass.   306
HIPPOLYTA How chance Moonshine is gone before Thisbe
comes back and finds her lover?
THESEUS She will find him by starlight.
   ⌈*Enter Flute as Thisbe*⌉
Here she comes, and her passion ends the play.   310
HIPPOLYTA Methinks she should not use a long one for
such a Pyramus. I hope she will be brief.
DEMETRIUS A mote will turn the balance which Pyramus,
which Thisbe, is the better—he for a man, God warrant
us; she for a woman, God bless us.   315
LYSANDER She hath spied him already with those sweet
eyes.
DEMETRIUS And thus she means, videlicet:
FLUTE (*as Thisbe*)
   Asleep, my love?
   What, dead, my dove?   320
O Pyramus, arise.
   Speak, speak. Quite dumb?
   Dead, dead? A tomb
Must cover thy sweet eyes.
   These lily lips,   325
   This cherry nose,
These yellow cowslip cheeks
   Are gone, are gone.
   Lovers, make moan.
His eyes were green as leeks.   330
   O sisters three,
   Come, come to me

With hands as pale as milk.
  Lay them in gore,
  Since you have shore                                    335
With shears his thread of silk.
  Tongue, not a word.
  Come, trusty sword,
Come, blade, my breast imbrue.
  *She stabs herself*
  And farewell friends,                                   340
  Thus Thisbe ends.
Adieu, adieu, adieu.                                  *She dies*
THESEUS Moonshine and Lion are left to bury the dead.
DEMETRIUS Ay, and Wall too.
⌈BOTTOM⌉ No, I assure you, the wall is down that parted
their fathers. Will it please you to see the epilogue or
to hear a bergamask dance between two of our
company?                                                        348
THESEUS No epilogue, I pray you; for your play needs no
excuse. Never excuse; for when the players are all dead
there need none to be blamed. Marry, if he that writ
it had played Pyramus and hanged himself in Thisbe's
garter it would have been a fine tragedy; and so it is,
truly, and very notably discharged. But come, your
bergamask. Let your epilogue alone.                             355
  ⌈*Bottom and Flute*⌉ *dance a bergamask, then exeunt*
The iron tongue of midnight hath told twelve.
Lovers, to bed; 'tis almost fairy time.
I fear we shall outsleep the coming morn
As much as we this night have overwatched.
This palpable-gross play hath well beguiled                      360
The heavy gait of night. Sweet friends, to bed.
A fortnight hold we this solemnity
In nightly revels and new jollity.                          *Exeunt*

**5.2**   *Enter Robin Goodfellow with a broom*
ROBIN
  Now the hungry lion roars,
    And the wolf behowls the moon,
  Whilst the heavy ploughman snores,
    All with weary task fordone.
  Now the wasted brands do glow                                   5
    Whilst the screech-owl, screeching loud,
  Puts the wretch that lies in woe
    In remembrance of a shroud.
  Now it is the time of night
    That the graves, all gaping wide,                             10
  Every one lets forth his sprite
    In the churchway paths to glide;
  And we fairies that do run
    By the triple Hecate's team
  From the presence of the sun,                                  15
    Following darkness like a dream,
  Now are frolic. Not a mouse
  Shall disturb this hallowed house.
  I am sent with broom before
  To sweep the dust behind the door.                             20

*Enter Oberon and Titania, King and Queen of
Fairies, with all their train*
OBERON
  Through the house give glimmering light.
    By the dead and drowsy fire
  Every elf and fairy sprite
    Hop as light as bird from brier,
  And this ditty after me                                        25
  Sing, and dance it trippingly.
TITANIA
  First rehearse your song by rote,
  To each word a warbling note.
  Hand in hand with fairy grace
  Will we sing and bless this place.                             30
    ⌈*The song. The fairies dance*⌉
OBERON
  Now until the break of day
  Through this house each fairy stray.
  To the best bride bed will we,
  Which by us shall blessèd be,
  And the issue there create                                     35
  Ever shall be fortunate.
  So shall all the couples three
  Ever true in loving be,
  And the blots of nature's hand
  Shall not in their issue stand.                                40
  Never mole, harelip, nor scar,
  Nor mark prodigious such as are
  Despisèd in nativity
  Shall upon their children be.
  With this field-dew consecrate                                 45
  Every fairy take his gait
  And each several chamber bless
  Through this palace with sweet peace;
  And the owner of it blessed
  Ever shall in safety rest.                                     50
  Trip away, make no stay,
  Meet me all by break of day.        *Exeunt all but Robin*

**Epilogue**
ROBIN
  If we shadows have offended,
  Think but this, and all is mended:
  That you have but slumbered here,
  While these visions did appear;
  And this weak and idle theme,                                   5
  No more yielding but a dream,
  Gentles, do not reprehend.
  If you pardon, we will mend.
  And as I am an honest puck,
  If we have unearnèd luck                                       10
  Now to 'scape the serpent's tongue,
  We will make amends ere long,
  Else the puck a liar call.
  So, good night unto you all.
  Give me your hands, if we be friends,                          15
  And Robin shall restore amends.

## ADDITIONAL PASSAGES

An unusual quantity and kind of mislineation in the first edition has persuaded most scholars that the text at the beginning of 5.1 was revised, with new material written in the margins. We here offer a reconstruction of the passage as originally drafted, which can be compared with 5.1.1–86 of the edited text.

**5.1**    *Enter Theseus, Hippolyta, and Philostrate*

HIPPOLYTA
'Tis strange, my Theseus, that these lovers speak of.

THESEUS
More strange than true. I never may believe
These antique fables, nor these fairy toys.
Lovers and mad men have such seething brains.
One sees more devils than vast hell can hold:          5
That is the madman. The lover, all as frantic,
Sees Helen's beauty in a brow of Egypt.
Such tricks hath strong imagination
That if it would but apprehend some joy
It comprehends some bringer of that joy;          10
Or in the night, imagining some fear,
How easy is a bush supposed a bear!

HIPPOLYTA
But all the story of the night told over,
And all their minds transfigured so together,
More witnesseth than fancy's images,          15
And grows to something of great constancy;
But howsoever, strange and admirable.
        *Enter the lovers: Lysander, Demetrius, Hermia, and*
        *Helena*

THESEUS
Here come the lovers, full of joy and mirth.
Come now, what masques, what dances shall we
    have
To ease the anguish of a torturing hour?          20
Call Philostrate.

PHILOSTRATE          Here mighty Theseus.

THESEUS
Say, what abridgement have you for this evening?
What masque, what music? How shall we beguile
The lazy time if not with some delight?

PHILOSTRATE
There is a brief how many sports are ripe.          25
Make choice of which your highness will see first.

THESEUS
'The battle with the centaurs to be sung
By an Athenian eunuch to the harp.'
We'll none of that. That have I told my love
In glory of my kinsman Hercules.          30
'The riot of the tipsy Bacchanals
Tearing the Thracian singer in their rage.'
That is an old device, and it was played
When I from Thebes came last a conquerer.
'The thrice-three Muses mourning for the death          35
Of learning, late deceased in beggary.'
That is some satire, keen and critical,
Not sorting with a nuptial ceremony.
'A tedious brief scene of young Pyramus
And his love Thisby.' 'Tedious' *and* 'brief'?          40

PHILOSTRATE
A play there is, my lord, some ten words long,
Which is as 'brief' as I have known a play;
But by ten words, my lord, it is too long,
Which makes it 'tedious'; for in all the play
There is not one word apt, one player fitted.          45

THESEUS What are they that do play it?

PHILOSTRATE
Hard-handed men that work in Athens here,
Which never laboured in their minds till now,
And now have toiled their unbreathed memories
With this same play against your nuptial.          50

THESEUS
Go, bring them in; and take your places, ladies.
                    *Exit Philostrate*

HIPPOLYTA
I love not to see wretchedness o'ercharged
And duty in his service perishing.

# ROMEO AND JULIET

ON its first appearance in print, in 1597, *Romeo and Juliet* was described as 'An excellent conceited tragedy' that had 'been often (with great applause) played publicly'; its popularity is witnessed by the fact that this is a pirated version, put together from actors' memories as a way of cashing in on its success. A second printing, two years later, offered a greatly superior text apparently printed from Shakespeare's working papers. Probably he wrote it in 1594 or 1595.

The story was already well known, in Italian, French, and English. Shakespeare owes most to Arthur Brooke's long poem *The Tragical History of Romeus and Juliet* (1562), which had already supplied hints for *The Two Gentlemen of Verona*; he may also have looked at some of the other versions. In his address 'To the Reader', Brooke says that he has seen 'the same argument lately set forth on stage with more commendation than I can look for', but no earlier play survives.

Shakespeare's Prologue neatly sketches the plot of the two star-crossed lovers born of feuding families whose deaths 'bury their parents' strife'; and the formal verse structure of the Prologue—a sonnet—is matched by the carefully patterned layout of the action. At the climax of the first scene, Prince Escalus stills a brawl between representatives of the houses of Montague (Romeo's family) and Capulet (Juliet's); at the end of Act 3, Scene 1, he passes judgement on another, more serious brawl, banishing Romeo for killing Juliet's cousin Tybalt after Tybalt had killed Romeo's friend and the Prince's kinsman, Mercutio; and at the end of Act 5, the Prince presides over the reconciliation of Montagues and Capulets. Within this framework of public life Romeo and Juliet act out their brief tragedy: in the first act they meet and declare their love—in another sonnet; in the second they arrange to marry in secret; in the third, after Romeo's banishment, they consummate their marriage and part; in the fourth, Juliet drinks a sleeping draught prepared by Friar Laurence so that she may escape marriage to Paris and, after waking in the family tomb, run off with Romeo; in the fifth, after Romeo, believing her to be dead, has taken poison, she stabs herself to death.

The play's structural formality is offset by an astonishing fertility of linguistic invention, showing itself no less in the comic bawdiness of the servants, the Nurse, and (on a more sophisticated level) Mercutio than in the rapt and impassioned poetry of the lovers. Shakespeare's mastery over a wide range of verbal styles combines with his psychological perceptiveness to create a richer gallery of memorable characters than in any of his earlier plays; and his theatrical imagination compresses Brooke's leisurely narrative into a dramatic masterpiece.

# THE PERSONS OF THE PLAY

CHORUS

ROMEO

MONTAGUE, his father

MONTAGUE'S WIFE

BENVOLIO, Montague's nephew

ABRAHAM, Montague's servingman

BALTHASAR, Romeo's man

JULIET

CAPULET, her father

CAPULET'S WIFE

TYBALT, her nephew

His page

PETRUCCIO

CAPULET'S COUSIN

Juliet's NURSE

PETER  ⎫
SAMSON ⎬ servingmen of the Capulets
GREGORY ⎭

Other SERVINGMEN

MUSICIANS

Escalus, PRINCE of Verona

MERCUTIO      ⎫
County PARIS  ⎬ his kinsmen

PAGE to Paris

FRIAR LAURENCE

FRIAR JOHN

An APOTHECARY

CHIEF WATCHMAN

Other CITIZENS OF THE WATCH

Masquers, guests, gentlewomen, followers of the Montague and
    Capulet factions

# The Most Excellent and Lamentable Tragedy of Romeo and Juliet

**Prologue** *Enter Chorus*

CHORUS

Two households, both alike in dignity
  In fair Verona, where we lay our scene,
From ancient grudge break to new mutiny,
  Where civil blood makes civil hands unclean.
From forth the fatal loins of these two foes     5
  A pair of star-crossed lovers take their life,
Whose misadventured piteous overthrows
  Doth with their death bury their parents' strife.
The fearful passage of their death-marked love
  And the continuance of their parents' rage—   10
Which but their children's end, naught could remove—
  Is now the two-hours' traffic of our stage;
The which if you with patient ears attend,
What here shall miss, our toil shall strive to mend.
                              *Exit*

**1.1**   *Enter Samson and Gregory, of the house of Capulet,*
       *with swords and bucklers*

SAMSON Gregory, on my word, we'll not carry coals.

GREGORY No, for then we should be colliers.

SAMSON I mean an we be in choler, we'll draw.

GREGORY Ay, while you live, draw your neck out of collar.

SAMSON I strike quickly, being moved.     5

GREGORY But thou art not quickly moved to strike.

SAMSON A dog of the house of Montague moves me.

GREGORY To move is to stir, and to be valiant is to stand,
  therefore if thou art moved, thou runn'st away.

SAMSON A dog of that house shall move me to stand. I
  will take the wall of any man or maid of Montague's.

GREGORY That shows thee a weak slave, for the weakest
  goes to the wall.     13

SAMSON 'Tis true, and therefore women, being the weaker
  vessels, are ever thrust to the wall; therefore I will
  push Montague's men from the wall, and thrust his
  maids to the wall.

GREGORY The quarrel is between our masters and us their
  men.     19

SAMSON 'Tis all one. I will show myself a tyrant: when I
  have fought with the men I will be civil with the
  maids—I will cut off their heads.

GREGORY The heads of the maids?

SAMSON Ay, the heads of the maids, or their maidenheads,
  take it in what sense thou wilt.     25

GREGORY They must take it in sense that feel it.

SAMSON Me they shall feel while I am able to stand, and
  'tis known I am a pretty piece of flesh.

GREGORY 'Tis well thou art not fish. If thou hadst, thou
  hadst been poor-john.     30
    *Enter Abraham and another servingman of the*
    *Montagues*
  Draw thy tool. Here comes of the house of Montagues.

SAMSON My naked weapon is out. Quarrel, I will back
  thee.

GREGORY How—turn thy back and run?

SAMSON Fear me not.     35

GREGORY No, marry—I fear thee!

SAMSON Let us take the law of our side. Let them begin.

GREGORY I will frown as I pass by, and let them take it
  as they list.

SAMSON Nay, as they dare. I will bite my thumb at them,
  which is disgrace to them if they bear it.     41
    *He bites his thumb*

ABRAHAM Do you bite your thumb at us, sir?

SAMSON I do bite my thumb, sir.

ABRAHAM Do you bite your thumb at us, sir?

SAMSON (*to Gregory*) Is the law of our side if I say 'Ay'?

GREGORY No.     46

SAMSON (*to Abraham*) No, sir, I do not bite my thumb at
  you, sir, but I bite my thumb, sir.

GREGORY (*to Abraham*) Do you quarrel, sir?

ABRAHAM Quarrel, sir? No, sir.     50

SAMSON But if you do, sir, I am for you. I serve as good
  a man as you.

ABRAHAM No better.

SAMSON Well, sir.
    *Enter Benvolio*

GREGORY Say 'better'. Here comes one of my master's
  kinsmen.     56

SAMSON (*to Abraham*) Yes, better, sir.

ABRAHAM You lie.

SAMSON Draw, if you be men. Gregory, remember thy
  washing blow.     60
    *They draw and fight*

BENVOLIO (*drawing*) Part, fools. Put up your swords. You
  know not what you do.
    *Enter Tybalt*

TYBALT (*drawing*)
What, art thou drawn among these heartless hinds?
Turn thee, Benvolio. Look upon thy death.

BENVOLIO
I do but keep the peace. Put up thy sword,     65
Or manage it to part these men with me.

TYBALT
What, drawn and talk of peace? I hate the word
As I hate hell, all Montagues, and thee.
Have at thee, coward.
    *They fight. Enter three or four Citizens ⌜of the*
    *watch⌝, with clubs or partisans*

⌜CITIZENS OF THE WATCH⌝
Clubs, bills and partisans! Strike! Beat them down!  70
Down with the Capulets. Down with the Montagues.
    *Enter Capulet in his gown, and his Wife*

CAPULET
What noise is this? Give me my long sword, ho!

CAPULET'S WIFE
A crutch, a crutch—why call you for a sword?
    *Enter Montague ⌜with his sword drawn⌝, and his*
    *Wife*

CAPULET
My sword, I say. Old Montague is come,
And flourishes his blade in spite of me.     75

337

MONTAGUE
Thou villain Capulet!
[*His Wife holds him back*]
                    Hold me not, let me go.
MONTAGUE'S WIFE
Thou shalt not stir one foot to seek a foe.
[*The Citizens of the watch attempt to part
the factions.*]
*Enter Prince Escalus with his train*
PRINCE
Rebellious subjects, enemies to peace,
Profaners of this neighbour-stainèd steel—
Will they not hear? What ho, you men, you beasts,
That quench the fire of your pernicious rage        81
With purple fountains issuing from your veins:
On pain of torture, from those bloody hands
Throw your mistempered weapons to the ground,
And hear the sentence of your movèd Prince.        85
[*Montague, Capulet, and their followers throw down
their weapons*]
Three civil brawls bred of an airy word
By thee, old Capulet, and Montague,
Have thrice disturbed the quiet of our streets
And made Verona's ancient citizens
Cast by their grave-beseeming ornaments        90
To wield old partisans in hands as old,
Cankered with peace, to part your cankered hate.
If ever you disturb our streets again
Your lives shall pay the forfeit of the peace.
For this time all the rest depart away.        95
You, Capulet, shall go along with me;
And Montague, come you this afternoon
To know our farther pleasure in this case
To old Freetown, our common judgement-place.
Once more, on pain of death, all men depart.        100
                    *Exeunt all but Montague,*
                    *his Wife, and Benvolio*
MONTAGUE
Who set this ancient quarrel new abroach?
Speak, nephew: were you by when it began?
BENVOLIO
Here were the servants of your adversary
And yours, close fighting ere I did approach.
I drew to part them. In the instant came        105
The fiery Tybalt with his sword prepared,
Which, as he breathed defiance to my ears,
He swung about his head and cut the winds
Who, nothing hurt withal, hissed him in scorn.
While we were interchanging thrusts and blows,        110
Came more and more, and fought on part and part
Till the Prince came, who parted either part.
MONTAGUE'S WIFE
O where is Romeo—saw you him today?
Right glad I am he was not at this fray.
BENVOLIO
Madam, an hour before the worshipped sun        115
Peered forth the golden window of the east,
A troubled mind drive me to walk abroad,
Where, underneath the grove of sycamore
That westward rooteth from this city side,
So early walking did I see your son.        120
Towards him I made, but he was ware of me,
And stole into the covert of the wood.
I, measuring his affections by my own—
Which then most sought where most might not be
found,

Being one too many by my weary self—        125
Pursued my humour not pursuing his,
And gladly shunned who gladly fled from me.
MONTAGUE
Many a morning hath he there been seen,
With tears augmenting the fresh morning's dew,
Adding to clouds more clouds with his deep sighs.        130
But all so soon as the all-cheering sun
Should in the farthest east begin to draw
The shady curtains from Aurora's bed,
Away from light steals home my heavy son,
And private in his chamber pens himself,        135
Shuts up his windows, locks fair daylight out,
And makes himself an artificial night.
Black and portentous must this humour prove,
Unless good counsel may the cause remove.
BENVOLIO
My noble uncle, do you know the cause?        140
MONTAGUE
I neither know it nor can learn of him.
BENVOLIO
Have you importuned him by any means?
MONTAGUE
Both by myself and many other friends,
But he, his own affection's counsellor,
Is to himself—I will not say how true,        145
But to himself so secret and so close,
So far from sounding and discovery,
As is the bud bit with an envious worm
Ere he can spread his sweet leaves to the air
Or dedicate his beauty to the sun.        150
Could we but learn from whence his sorrows grow
We would as willingly give cure as know.
                    *Enter Romeo*
BENVOLIO
See where he comes. So please you step aside,
I'll know his grievance or be much denied.
MONTAGUE
I would thou wert so happy by thy stay        155
To hear true shrift. Come, madam, let's away.
                    *Exeunt Montague and his Wife*
BENVOLIO
Good morrow, cousin.
ROMEO                    Is the day so young?
BENVOLIO
But new struck nine.
ROMEO                    Ay me, sad hours seem long.
Was that my father that went hence so fast?
BENVOLIO
It was. What sadness lengthens Romeo's hours?        160
ROMEO
Not having that which, having, makes them short.
BENVOLIO In love.
ROMEO Out.
BENVOLIO Of love?
ROMEO
Out of her favour where I am in love.        165
BENVOLIO
Alas that love, so gentle in his view,
Should be so tyrannous and rough in proof.
ROMEO
Alas that love, whose view is muffled still,
Should without eyes see pathways to his will.
Where shall we dine? [*Seeing blood*] O me! What fray
was here?        170

Yet tell me not, for I have heard it all.
Here's much to do with hate, but more with love.
Why then, O brawling love, O loving hate,
O anything of nothing first create;
O heavy lightness, serious vanity,                                    175
Misshapen chaos of well-seeming forms,
Feather of lead, bright smoke, cold fire, sick health,
Still-waking sleep, that is not what it is!
This love feel I, that feel no love in this.
Dost thou not laugh?
BENVOLIO                       No, coz, I rather weep.          180
ROMEO
Good heart, at what?
BENVOLIO                       At thy good heart's oppression.
ROMEO Why, such is love's transgression.
Griefs of mine own lie heavy in my breast,
Which thou wilt propagate to have it pressed
With more of thine. This love that thou hast shown
Doth add more grief to too much of mine own.     186
Love is a smoke made with the fume of sighs,
Being purged, a fire sparkling in lovers' eyes,
Being vexed, a sea nourished with lovers' tears.
What is it else? A madness most discreet,              190
A choking gall and a preserving sweet.
Farewell, my coz.
BENVOLIO             Soft, I will go along;
An if you leave me so, you do me wrong.
ROMEO
Tut, I have lost myself. I am not here.
This is not Romeo; he's some other where.           195
BENVOLIO
Tell me in sadness, who is that you love?
ROMEO What, shall I groan and tell thee?
BENVOLIO
Groan? Why no; but sadly tell me who.
ROMEO
Bid a sick man in sadness make his will,
A word ill urged to one that is so ill.                      200
In sadness, cousin, I do love a woman.
BENVOLIO
I aimed so near when I supposed you loved.
ROMEO
A right good markman; and she's fair I love.
BENVOLIO
A right fair mark, fair coz, is soonest hit.
ROMEO
Well, in that hit you miss. She'll not be hit         205
With Cupid's arrow; she hath Dian's wit,
And, in strong proof of chastity well armed,
From love's weak childish bow she lives unharmed.
She will not stay the siege of loving terms,
Nor bide th'encounter of assailing eyes,             210
Nor ope her lap to saint-seducing gold.
O, she is rich in beauty, only poor
That when she dies, with beauty dies her store.
BENVOLIO
Then she hath sworn that she will still live chaste?
ROMEO
She hath, and in that sparing makes huge waste;  215
For beauty starved with her severity
Cuts beauty off from all posterity.
She is too fair, too wise, wisely too fair,
To merit bliss by making me despair.
She hath forsworn to love, and in that vow       220
Do I live dead, that live to tell it now.

BENVOLIO
Be ruled by me; forget to think of her.
ROMEO
O, teach me how I should forget to think!
BENVOLIO
By giving liberty unto thine eyes.
Examine other beauties.
ROMEO                       'Tis the way                              225
To call hers, exquisite, in question more.
These happy masks that kiss fair ladies' brows,
Being black, puts us in mind they hide the fair.
He that is strucken blind cannot forget
The precious treasure of his eyesight lost.          230
Show me a mistress that is passing fair,
What doth her beauty serve but as a note
Where I may read who passed that passing fair?
Farewell, thou canst not teach me to forget.
BENVOLIO
I'll pay that doctrine, or else die in debt.      *Exeunt*

**1.2**     *Enter Capulet, Paris, and ⌈Peter,⌉ a servingman*
CAPULET
But Montague is bound as well as I,
In penalty alike, and 'tis not hard, I think,
For men so old as we to keep the peace.
PARIS
Of honourable reckoning are you both,
And pity 'tis you lived at odds so long.               5
But now, my lord: what say you to my suit?
CAPULET
But saying o'er what I have said before.
My child is yet a stranger in the world;
She hath not seen the change of fourteen years.
Let two more summers wither in their pride        10
Ere we may think her ripe to be a bride.
PARIS
Younger than she are happy mothers made.
CAPULET
And too soon marred are those so early made.
But woo her, gentle Paris, get her heart;
My will to her consent is but a part,                    15
And, she agreed, within her scope of choice
Lies my consent and fair-according voice.
This night I hold an old-accustomed feast
Whereto I have invited many a guest
Such as I love, and you among the store,            20
One more most welcome, makes my number more.
At my poor house look to behold this night
Earth-treading stars that make dark heaven light.
Such comfort as do lusty young men feel
When well-apparelled April on the heel              25
Of limping winter treads—even such delight
Among fresh female buds shall you this night
Inherit at my house; hear all, all see,
And like her most whose merit most shall be,
Which on more view of many, mine, being one,   30
May stand in number, though in reck'ning none.
Come, go with me. (*Giving ⌈Peter⌉ a paper*) Go, sirrah,
           trudge about;
Through fair Verona find those persons out
Whose names are written there, and to them say
My house and welcome on their pleasure stay.    35
                       *Exeunt Capulet and Paris*
⌈PETER⌉ Find them out whose names are written here? It
is written that the shoemaker should meddle with his
yard and the tailor with his last, the fisher with his

pencil and the painter with his nets; but I am sent to
find those persons whose names are here writ, and can
never find what names the writing person hath here
writ. I must to the learned.                                    42
    *Enter Benvolio and Romeo*
In good time.
BENVOLIO *(to Romeo)*
  Tut, man, one fire burns out another's burning,
    One pain is lessened by another's anguish.         45
  Turn giddy, and be holp by backward turning.
    One desperate grief cures with another's languish.
  Take thou some new infection to thy eye,
  And the rank poison of the old will die.
ROMEO
  Your plantain leaf is excellent for that.            50
BENVOLIO For what, I pray thee?
ROMEO For your broken shin.
BENVOLIO Why, Romeo, art thou mad?
ROMEO
  Not mad, but bound more than a madman is;
  Shut up in prison, kept without my food,             55
  Whipped and tormented and— *(to ⌈Peter⌉)* Good e'en,
    good fellow.
⌈PETER⌉
  God gi'good e'en. I pray, sir, can you read?
ROMEO
  Ay, mine own fortune in my misery.
⌈PETER⌉ Perhaps you have learned it without book. But I
  pray, can you read anything you see?                  60
ROMEO
  Ay, if I know the letters and the language.
⌈PETER⌉ Ye say honestly. Rest you merry.
ROMEO Stay, fellow, I can read.
    *He reads the letter*
  'Signor Martino and his wife and daughters,
  County Anselme and his beauteous sisters,            65
  The lady widow of Vitruvio,
  Signor Placentio and his lovely nieces,
  Mercutio and his brother Valentine,
  Mine uncle Capulet, his wife and daughters,
  My fair niece Rosaline and Livia,                    70
  Signor Valentio and his cousin Tybalt,
  Lucio and the lively Helena.'
  A fair assembly. Whither should they come?
⌈PETER⌉ Up.
ROMEO Whither?                                               75
⌈PETER⌉ To supper to our house.
ROMEO Whose house?
⌈PETER⌉ My master's.
ROMEO
  Indeed, I should have asked thee that before.
⌈PETER⌉ Now I'll tell you without asking. My master is
  the great rich Capulet, and if you be not of the house
  of Montagues, I pray come and crush a cup of wine.
  Rest you merry.                                 *Exit*
BENVOLIO
  At this same ancient feast of Capulet's
  Sups the fair Rosaline, whom thou so loves,          85
  With all the admirèd beauties of Verona.
  Go thither, and with unattainted eye
  Compare her face with some that I shall show,
  And I will make thee think thy swan a crow.
ROMEO
  When the devout religion of mine eye                 90
    Maintains such falsehood, then turn tears to fires;
  And these who, often drowned, could never die,
    Transparent heretics, be burnt for liars.

  One fairer than my love!—the all-seeing sun
  Ne'er saw her match since first the world begun.      95
BENVOLIO
  Tut, you saw her fair, none else being by,
  Herself poised with herself in either eye;
  But in that crystal scales let there be weighed
  Your lady's love against some other maid
  That I will show you shining at this feast,           100
  And she shall scant show well that now seems best.
ROMEO
  I'll go along, no such sight to be shown,
  But to rejoice in splendour of mine own.      *Exeunt*

**1.3**    *Enter Capulet's Wife and the Nurse*
CAPULET'S WIFE
  Nurse, where's my daughter? Call her forth to me.
NURSE
  Now, by my maidenhead at twelve year old,
  I bade her come. What, lamb, what, ladybird—
  God forbid—where is this girl? What, Juliet!
    *Enter Juliet*
JULIET How now, who calls?                                    5
NURSE Your mother.
JULIET
  Madam, I am here. What is your will?
CAPULET'S WIFE
  This is the matter.—Nurse, give leave a while.
  We must talk in secret.—Nurse, come back again.
  I have remembered me, thou s' hear our counsel.      10
  Thou knowest my daughter's of a pretty age.
NURSE
  Faith, I can tell her age unto an hour.
CAPULET'S WIFE She's not fourteen.
NURSE I'll lay fourteen of my teeth—and yet, to my teen
  be it spoken, I have but four—she's not fourteen. How
  long is it now to Lammastide?                         16
CAPULET'S WIFE A fortnight and odd days.
NURSE
  Even or odd, of all days in the year
  Come Lammas Eve at night shall she be fourteen.
  Susan and she—God rest all Christian souls!—         20
  Were of an age. Well, Susan is with God;
  She was too good for me. But, as I said,
  On Lammas Eve at night shall she be fourteen,
  That shall she, marry, I remember it well.
  'Tis since the earthquake now eleven years,          25
  And she was weaned—I never shall forget it—
  Of all the days of the year upon that day,
  For I had then laid wormwood to my dug,
  Sitting in the sun under the dovehouse wall.
  My lord and you were then at Mantua.                 30
  Nay, I do bear a brain! But, as I said,
  When it did taste the wormwood on the nipple
  Of my dug and felt it bitter, pretty fool,
  To see it tetchy and fall out wi'th' dug!
  'Shake', quoth the dove-house! 'Twas no need, I trow,
  To bid me trudge;                                    36
  And since that time it is eleven years,
  For then she could stand high-lone. Nay, by th' rood,
  She could have run and waddled all about,
  For even the day before, she broke her brow,         40
  And then my husband—God be with his soul,
  A was a merry man!—took up the child.
  'Yea,' quoth he, 'dost thou fall upon thy face?
  Thou wilt fall backward when thou hast more wit,
  Wilt thou not, Jule?' And, by my halidom,           45

The pretty wretch left crying and said 'Ay'.
To see now how a jest shall come about!
I warrant an I should live a thousand years
I never should forget it. 'Wilt thou not, Jule?' quoth he,
And, pretty fool, it stinted and said 'Ay'.                50

CAPULET'S WIFE
Enough of this. I pray thee hold thy peace.

NURSE
Yes, madam. Yet I cannot choose but laugh
To think it should leave crying and say 'Ay'.
And yet, I warrant, it had upon it brow
A bump as big as a young cock'rel's stone.                55
A perilous knock, and it cried bitterly.
'Yea,' quoth my husband, 'fall'st upon thy face?
Thou wilt fall backward when thou com'st to age,
Wilt thou not, Jule?' It stinted and said 'Ay'.

JULIET
And stint thou too, I pray thee, Nurse, say I.            60

NURSE
Peace, I have done. God mark thee to his grace,
Thou wast the prettiest babe that e'er I nursed.
An I might live to see thee married once,
I have my wish.

CAPULET'S WIFE
Marry, that 'marry' is the very theme                     65
I came to talk of. Tell me, daughter Juliet,
How stands your dispositions to be married?

JULIET
It is an honour that I dream not of.

NURSE
'An honour'! Were not I thine only nurse,
I would say thou hadst sucked wisdom from thy teat.

CAPULET'S WIFE
Well, think of marriage now. Younger than you            71
Here in Verona, ladies of esteem,
Are made already mothers. By my count
I was your mother much upon these years
That you are now a maid. Thus then, in brief:            75
The valiant Paris seeks you for his love.

NURSE
A man, young lady, lady, such a man
As all the world—why, he's a man of wax.

CAPULET'S WIFE
Verona's summer hath not such a flower.

NURSE
Nay, he's a flower, in faith, a very flower.             80

CAPULET'S WIFE (to Juliet)
What say you? Can you love the gentleman?
This night you shall behold him at our feast.
Read o'er the volume of young Paris' face,
And find delight writ there with beauty's pen.
Examine every married lineament,                          85
And see how one another lends content;
And what obscured in this fair volume lies
Find written in the margin of his eyes.
This precious book of love, this unbound lover,
To beautify him only lacks a cover.                       90
The fish lives in the sea, and 'tis much pride
For fair without the fair within to hide.
That book in many's eyes doth share the glory
That in gold clasps locks in the golden story.
So shall you share all that he doth possess              95
By having him, making yourself no less.

NURSE
No less, nay, bigger. Women grow by men.

CAPULET'S WIFE (to Juliet)
Speak briefly: can you like of Paris' love?

JULIET
I'll look to like, if looking liking move;
But no more deep will I endart mine eye                  100
Than your consent gives strength to make it fly.
        Enter [Peter]

[PETER] Madam, the guests are come, supper served up,
you called, my young lady asked for, the Nurse cursed
in the pantry, and everything in extremity. I must hence
to wait. I beseech you follow straight.                  105

CAPULET'S WIFE
We follow thee.                            Exit [Peter]
        Juliet, the County stays.

NURSE
Go, girl; seek happy nights to happy days.       Exeunt

**1.4**   *Enter Romeo, Mercutio, and Benvolio, as masquers,
        with five or six other masquers, [bearing a drum and
        torches]*

ROMEO
What, shall this speech be spoke for our excuse,
Or shall we on without apology?

BENVOLIO
The date is out of such prolixity.
We'll have no Cupid hoodwinked with a scarf,
Bearing a Tartar's painted bow of lath,                   5
Scaring the ladies like a crowkeeper,
Nor no without-book Prologue faintly spoke
After the prompter for our entrance.
But let them measure us by what they will,
We'll measure them a measure, and be gone.               10

ROMEO
Give me a torch. I am not for this ambling;
Being but heavy, I will bear the light.

MERCUTIO
Nay, gentle Romeo, we must have you dance.

ROMEO
Not I, believe me. You have dancing shoes
With nimble soles; I have a soul of lead                  15
So stakes me to the ground I cannot move.

MERCUTIO
You are a lover; borrow Cupid's wings,
And soar with them above a common bound.

ROMEO
I am too sore empiercèd with his shaft
To soar with his light feathers, and so bound            20
I cannot bound a pitch above dull woe;
Under love's heavy burden do I sink.

MERCUTIO
And to sink in it should you burden love—
Too great oppression for a tender thing.

ROMEO
Is love a tender thing? It is too rough,                 25
Too rude, too boist'rous, and it pricks like thorn.

MERCUTIO
If love be rough with you, be rough with love.
Prick love for pricking, and you beat love down.
Give me a case to put my visage in,
A visor for a visor. What care I                          30
What curious eye doth quote deformity?
Here are the beetle brows shall blush for me.
        [They put on visors]

BENVOLIO
Come, knock and enter, and no sooner in
But every man betake him to his legs.

**ROMEO**
A torch for me. Let wantons light of heart            35
Tickle the sense-less rushes with their heels,
For I am proverbed with a grandsire phrase.
I'll be a candle-holder and look on.
The game was ne'er so fair, and I am done.
⌜*He takes a torch*⌝

**MERCUTIO**
Tut, dun's the mouse, the constable's own word.       40
If thou art dun we'll draw thee from the mire
Of—save your reverence—love, wherein thou stickest
Up to the ears. Come, we burn daylight, ho!

**ROMEO**
Nay, that's not so.

**MERCUTIO**                I mean, sir, in delay
We waste our lights in vain, like lights by day.      45
Take our good meaning, for our judgement sits
Five times in that ere once in our five wits.

**ROMEO**
And we mean well in going to this masque,
But 'tis no wit to go.

**MERCUTIO**            Why, may one ask?

**ROMEO**
I dreamt a dream tonight.

**MERCUTIO**            And so did I.                  50

**ROMEO**
Well, what was yours?

**MERCUTIO**            That dreamers often lie.

**ROMEO**
In bed asleep while they do dream things true.

**MERCUTIO**
O, then I see Queen Mab hath been with you.

**BENVOLIO** Queen Mab, what's she?

**MERCUTIO**
She is the fairies' midwife, and she comes            55
In shape no bigger than an agate stone
On the forefinger of an alderman,
Drawn with a team of little atomi
Athwart men's noses as they lie asleep.
Her wagon spokes made of long spinners' legs;        60
The cover, of the wings of grasshoppers;
Her traces, of the moonshine's wat'ry beams;
Her collars, of the smallest spider web;
Her whip, of cricket's bone, the lash of film;
Her wagoner, a small grey-coated gnat                 65
Not half so big as a round little worm
Pricked from the lazy finger of a maid.
Her chariot is an empty hazelnut
Made by the joiner squirrel or old grub,
Time out o' mind the fairies' coachmakers.           70
And in this state she gallops night by night
Through lovers' brains, and then they dream of love;
O'er courtiers' knees, that dream on curtsies straight;
O'er ladies' lips, who straight on kisses dream,
Which oft the angry Mab with blisters plagues        75
Because their breaths with sweetmeats tainted are.
Sometime she gallops o'er a lawyer's lip,
And then dreams he of smelling out a suit;
And sometime comes she with a tithe-pig's tail
Tickling a parson's nose as a lies asleep;           80
Then dreams he of another benefice.
Sometime she driveth o'er a soldier's neck,
And then dreams he of cutting foreign throats,
Of breaches, ambuscados, Spanish blades,
Of healths five fathom deep; and then anon           85
Drums in his ear, at which he starts and wakes,

And being thus frighted, swears a prayer or two,
And sleeps again. This is that very Mab
That plaits the manes of horses in the night,
And bakes the elf-locks in foul sluttish hairs,      90
Which once untangled much misfortune bodes.
This is the hag, when maids lie on their backs,
That presses them and learns them first to bear,
Making them women of good carriage.
This is she—

**ROMEO**       Peace, peace, Mercutio, peace!        95
Thou talk'st of nothing.

**MERCUTIO**            True. I talk of dreams,
Which are the children of an idle brain,
Begot of nothing but vain fantasy,
Which is as thin of substance as the air,
And more inconstant than the wind, who woos         100
Even now the frozen bosom of the north,
And, being angered, puffs away from thence,
Turning his face to the dew-dropping south.

**BENVOLIO**
This wind you talk of blows us from ourselves.
Supper is done, and we shall come too late.         105

**ROMEO**
I fear too early, for my mind misgives
Some consequence yet hanging in the stars
Shall bitterly begin his fearful date
With this night's revels, and expire the term
Of a despisèd life, closed in my breast,            110
By some vile forfeit of untimely death.
But he that hath the steerage of my course
Direct my sail! On, lusty gentlemen.

**BENVOLIO** Strike, drum.
            *They march about the stage and* ⌜*exeunt*⌝

**1.5**     ⌜*Peter*⌝ *and other Servingmen come forth with*
            *napkins*

⌜**PETER**⌝ Where's Potpan, that he helps not to take away?
He shift a trencher, he scrape a trencher!

**FIRST SERVINGMAN** When good manners shall lie all in one
or two men's hands, and they unwashed too, 'tis a foul
thing.                                                5

⌜**PETER**⌝ Away with the joint-stools, remove the court-
cupboard, look to the plate. Good thou, save me a piece
of marzipan, and, as thou loves me, let the porter let in
Susan Grindstone and Nell. Anthony and Potpan!

**SECOND SERVINGMAN** Ay, boy, ready.                  10

⌜**PETER**⌝ You are looked for and called for, asked for and
sought for, in the great chamber.

⌜**FIRST**⌝ **SERVINGMAN** We cannot be here and there too.
Cheerly, boys! Be brisk a while, and the longest liver
take all.                                             15

            ⌜*They come and go, setting forth tables and chairs.*⌝
            *Enter* ⌜*Musicians, then*⌝ *at one door Capulet,* ⌜*his*
            *Wife,*⌝ *his Cousin, Juliet,* ⌜*the Nurse,*⌝ *Tybalt, his*
            *page, Petruccio, and all the guests and gentlewomen;*
            *at another door, the masquers:* ⌜*Romeo, Benvolio and*
            *Mercutio*⌝

**CAPULET** (*to the masquers*)
Welcome, gentlemen. Ladies that have their toes
Unplagued with corns will walk a bout with you.
Aha, my mistresses, which of you all
Will now deny to dance? She that makes dainty,
She, I'll swear, hath corns. Am I come near ye now?
Welcome, gentlemen. I have seen the day               21
That I have worn a visor, and could tell
A whispering tale in a fair lady's ear

Such as would please. 'Tis gone, 'tis gone, 'tis gone.
You are welcome, gentlemen. Come, musicians, play.
   *Music plays, and the masquers, guests, and*
   *gentlewomen dance. ⌜Romeo stands apart⌝*
A hall, a hall! Give room, and foot it, girls.         26
(*To Servingmen*) More light, you knaves, and turn the
   tables up,
And quench the fire, the room is grown too hot.
(*To his Cousin*) Ah sirrah, this unlooked-for sport comes
   well.
Nay, sit, nay, sit, good cousin Capulet,               30
For you and I are past our dancing days.
   ⌜*Capulet and his Cousin sit*⌝
How long is't now since last yourself and I
Were in a masque?
CAPULET'S COUSIN         By'r Lady, thirty years.
CAPULET
What, man, 'tis not so much, 'tis not so much.
'Tis since the nuptial of Lucentio,                    35
Come Pentecost as quickly as it will,
Some five-and-twenty years; and then we masqued.
CAPULET'S COUSIN
'Tis more, 'tis more. His son is elder, sir.
His son is thirty.
CAPULET            Will you tell me that?
His son was but a ward two years ago.                  40
ROMEO (*to a Servingman*)
What lady's that which doth enrich the hand
Of yonder knight?
SERVINGMAN            I know not, sir.
ROMEO
O, she doth teach the torches to burn bright!
It seems she hangs upon the cheek of night
As a rich jewel in an Ethiope's ear—                   45
Beauty too rich for use, for earth too dear.
So shows a snowy dove trooping with crows
As yonder lady o'er her fellows shows.
The measure done, I'll watch her place of stand,
And, touching hers, make blessèd my rude hand.         50
Did my heart love till now? Forswear it, sight,
For I ne'er saw true beauty till this night.
TYBALT
This, by his voice, should be a Montague.
Fetch me my rapier, boy.              ⌜*Exit page*⌝
                 What, dares the slave
Come hither, covered with an antic face,               55
To fleer and scorn at our solemnity?
Now, by the stock and honour of my kin,
To strike him dead I hold it not a sin.
CAPULET ⌜*standing*⌝
Why, how now, kinsman? Wherefore storm you so?
TYBALT
Uncle, this is a Montague, our foe,                    60
A villain that is hither come in spite
To scorn at our solemnity this night.
CAPULET
Young Romeo, is it?
TYBALT            'Tis he, that villain Romeo.
CAPULET
Content thee, gentle coz, let him alone.
A bears him like a portly gentleman,                   65
And, to say truth, Verona brags of him
To be a virtuous and well-governed youth.
I would not for the wealth of all this town
Here in my house do him disparagement.

Therefore be patient, take no note of him.             70
It is my will, the which if thou respect,
Show a fair presence and put off these frowns,
An ill-beseeming semblance for a feast.
TYBALT
It fits when such a villain is a guest.
I'll not endure him.
CAPULET            He shall be endured.                75
What, goodman boy, I say he shall. Go to,
Am I the master here or you? Go to—
You'll not endure him! God shall mend my soul.
You'll make a mutiny among my guests,
You will set cock-a-hoop! You'll be the man!           80
TYBALT
Why, uncle, 'tis a shame.
CAPULET                   Go to, go to,
You are a saucy boy. Is't so, indeed?
This trick may chance to scathe you. I know what,
You must contrary me. Marry, 'tis time—
   ⌜*A dance ends. Juliet retires to her place of stand,*
   *where Romeo awaits her*⌝
(*To the guests*) Well said, my hearts! (*To Tybalt*) You are
   a princox, go.                                      85
Be quiet, or— (*to Servingmen*) more light, more light!—
   (*to Tybalt*) for shame,
I'll make you quiet. (*To the guests*) What, cheerly, my
   hearts!
   ⌜*The music plays again, and the guests dance*⌝
TYBALT
Patience perforce with wilful choler meeting
Makes my flesh tremble in their different greeting.
I will withdraw, but this intrusion shall,             90
Now seeming sweet, convert to bitt'rest gall.     *Exit*
ROMEO (*to Juliet, touching her hand*)
If I profane with my unworthiest hand
   This holy shrine, the gentler sin is this:
My lips, two blushing pilgrims, ready stand
   To smooth that rough touch with a tender kiss.      95
JULIET
Good pilgrim, you do wrong your hand too much,
   Which mannerly devotion shows in this.
For saints have hands that pilgrims' hands do touch,
   And palm to palm is holy palmers' kiss.
ROMEO
Have not saints lips, and holy palmers, too?          100
JULIET
Ay, pilgrim, lips that they must use in prayer.
ROMEO
O then, dear saint, let lips do what hands do:
   They pray; grant thou, lest faith turn to despair.
JULIET
Saints do not move, though grant for prayers' sake.
ROMEO
Then move not while my prayer's effect I take.        105
   *He kisses her*
Thus from my lips, by thine my sin is purged.
JULIET
Then have my lips the sin that they have took.
ROMEO
Sin from my lips? O trespass sweetly urged!
   Give me my sin again.
   *He kisses her*
JULIET                   You kiss by th' book.
NURSE
Madam, your mother craves a word with you.            110
   ⌜*Juliet departs to her mother*⌝

ROMEO
What is her mother?
NURSE                          Marry, bachelor,
Her mother is the lady of the house,
And a good lady, and a wise and virtuous.
I nursed her daughter that you talked withal.
I tell you, he that can lay hold of her          115
Shall have the chinks.
ROMEO (aside)                Is she a Capulet?
O dear account! My life is my foe's debt.
BENVOLIO
Away, be gone, the sport is at the best.
ROMEO
Ay, so I fear, the more is my unrest.
CAPULET
Nay, gentlemen, prepare not to be gone.          120
We have a trifling foolish banquet towards.
    ⌈They whisper in his ear⌉
Is it e'en so? Why then, I thank you all.
I thank you, honest gentlemen. Good night.
More torches here! Come on then, let's to bed.
(To his Cousin) Ah, sirrah, by my fay, it waxes late.  125
I'll to my rest.
            Exeunt Capulet, ⌈his Wife,⌉ and his Cousin. The
            guests, gentlewomen, masquers, musicians, and
                  servingmen begin to leave
JULIET
Come hither, Nurse. What is yon gentleman?
NURSE
The son and heir of old Tiberio.
JULIET
What's he that now is going out of door?
NURSE
Marry, that, I think, be young Petruccio.        130
JULIET
What's he that follows here, that would not dance?
NURSE I know not.
JULIET
Go ask his name.
    The Nurse goes
                If he be married,
My grave is like to be my wedding bed.
NURSE (returning)
His name is Romeo, and a Montague,               135
The only son of your great enemy.
JULIET ⌈aside⌉
My only love sprung from my only hate!
Too early seen unknown, and known too late!
Prodigious birth of love it is to me
That I must love a loathèd enemy.                140
NURSE
What's tis? what's tis?
JULIET                    A rhyme I learnt even now
Of one I danced withal.
    One calls within 'Juliet!'
NURSE                       Anon, anon.
Come, let's away. The strangers all are gone.    Exeunt

2.0    Enter Chorus
CHORUS
Now old desire doth in his deathbed lie,
    And young affection gapes to be his heir.
That fair for which love groaned for and would die,
    With tender Juliet matched, is now not fair.
Now Romeo is beloved and loves again,              5
    Alike bewitchèd by the charm of looks;

But to his foe supposed he must complain,
    And she steal love's sweet bait from fearful hooks.
Being held a foe, he may not have access
    To breathe such vows as lovers use to swear,    10
And she as much in love, her means much less
    To meet her new belovèd anywhere.
But passion lends them power, time means, to meet,
Temp'ring extremities with extreme sweet.        Exit

2.1    Enter Romeo
ROMEO
Can I go forward when my heart is here?
Turn back, dull earth, and find thy centre out.
    ⌈He turns back and withdraws.⌉
    Enter Benvolio with Mercutio
BENVOLIO (calling)
Romeo, my cousin Romeo, Romeo!
MERCUTIO
He is wise, and, on my life, hath stol'n him home to
    bed.
BENVOLIO
He ran this way, and leapt this orchard wall.     5
Call, good Mercutio.
⌈MERCUTIO⌉                   Nay, I'll conjure too.
Romeo! Humours! Madman! Passion! Lover!
Appear thou in the likeness of a sigh.
Speak but one rhyme and I am satisfied.
Cry but 'Ay me!' Pronounce but 'love' and 'dove'.  10
Speak to my gossip Venus one fair word,
One nickname for her purblind son and heir,
Young Adam Cupid, he that shot so trim
When King Cophetua loved the beggar maid.—
He heareth not, he stirreth not, he moveth not.   15
The ape is dead, and I must conjure him.—
I conjure thee by Rosaline's bright eyes,
By her high forehead and her scarlet lip,
By her fine foot, straight leg, and quivering thigh,
And the demesnes that there adjacent lie,          20
That in thy likeness thou appear to us.
BENVOLIO
An if he hear thee, thou wilt anger him.
MERCUTIO
This cannot anger him. 'Twould anger him
To raise a spirit in his mistress' circle
Of some strange nature, letting it there stand     25
Till she had laid it and conjured it down.
That were some spite. My invocation
Is fair and honest. In his mistress' name,
I conjure only but to raise up him.
BENVOLIO
Come, he hath hid himself among these trees        30
To be consorted with the humorous night.
Blind is his love, and best befits the dark.
MERCUTIO
If love be blind, love cannot hit the mark.
Now will he sit under a medlar tree
And wish his mistress were that kind of fruit       35
As maids call medlars when they laugh alone.
O Romeo, that she were, O that she were
An open-arse, and thou a popp'rin' pear.
Romeo, good night. I'll to my truckle-bed.
This field-bed is too cold for me to sleep.        40
Come, shall we go?
BENVOLIO              Go then, for 'tis in vain
To seek him here that means not to be found.
                  Exeunt Benvolio and Mercutio

ROMEO [*coming forward*]
He jests at scars that never felt a wound.
But soft, what light through yonder window breaks?
It is the east, and Juliet is the sun.                                    45
Arise, fair sun, and kill the envious moon,
Who is already sick and pale with grief
That thou, her maid, art far more fair than she.
Be not her maid, since she is envious.
Her vestal livery is but sick and green,                                  50
And none but fools do wear it; cast it off.
        [*Enter Juliet aloft*]
It is my lady, O, it is my love.
O that she knew she were!
She speaks, yet she says nothing. What of that?
Her eye discourses; I will answer it.                                     55
I am too bold. 'Tis not to me she speaks.
Two of the fairest stars in all the heaven,
Having some business, do entreat her eyes
To twinkle in their spheres till they return.
What if her eyes were there, they in her head?—                          60
The brightness of her cheek would shame those stars
As daylight doth a lamp; her eye in heaven
Would through the airy region stream so bright
That birds would sing and think it were not night.
See how she leans her cheek upon her hand.                               65
O, that I were a glove upon that hand,
That I might touch that cheek!

JULIET                                      Ay me.
ROMEO (*aside*)                        She speaks.
O, speak again, bright angel; for thou art
As glorious to this night, being o'er my head,
As is a wingèd messenger of heaven                                       70
Unto the white upturnèd wond'ring eyes
Of mortals that fall back to gaze on him
When he bestrides the lazy-passing clouds
And sails upon the bosom of the air.

JULIET (*not knowing Romeo hears her*)
O Romeo, Romeo, wherefore art thou Romeo?                                75
Deny thy father and refuse thy name,
Or if thou wilt not, be but sworn my love,
And I'll no longer be a Capulet.

ROMEO (*aside*)
Shall I hear more, or shall I speak at this?

JULIET
'Tis but thy name that is my enemy.                                      80
Thou art thyself, though not a Montague.
What's Montague? It is nor hand, nor foot,
Nor arm, nor face, nor any other part
Belonging to a man. O, be some other name!
What's in a name? That which we call a rose                              85
By any other word would smell as sweet.
So Romeo would, were he not Romeo called,
Retain that dear perfection which he owes
Without that title. Romeo, doff thy name,
And for thy name—which is no part of thee—                              90
Take all myself.

ROMEO (*to Juliet*)   I take thee at thy word.
Call me but love and I'll be new baptized.
Henceforth I never will be Romeo.

JULIET
What man art thou that, thus bescreenèd in night,
So stumblest on my counsel?

ROMEO                                 By a name                          95
I know not how to tell thee who I am.
My name, dear saint, is hateful to myself
Because it is an enemy to thee.
Had I it written, I would tear the word.

JULIET
My ears have yet not drunk a hundred words                              100
Of thy tongue's uttering, yet I know the sound.
Art thou not Romeo, and a Montague?

ROMEO
Neither, fair maid, if either thee dislike.

JULIET
How cam'st thou hither, tell me, and wherefore?
The orchard walls are high and hard to climb,                           105
And the place death, considering who thou art,
If any of my kinsmen find thee here.

ROMEO
With love's light wings did I o'erperch these walls,
For stony limits cannot hold love out,
And what love can do, that dares love attempt.                          110
Therefore thy kinsmen are no stop to me.

JULIET
If they do see thee, they will murder thee.

ROMEO
Alack, there lies more peril in thine eye
Than twenty of their swords. Look thou but sweet,
And I am proof against their enmity.                                     115

JULIET
I would not for the world they saw thee here.

ROMEO
I have night's cloak to hide me from their eyes,
And but thou love me, let them find me here.
My life were better ended by their hate
Than death proroguèd, wanting of thy love.                              120

JULIET
By whose direction found'st thou out this place?

ROMEO
By love, that first did prompt me to enquire.
He lent me counsel, and I lent him eyes.
I am no pilot, yet wert thou as far
As that vast shore washed with the farthest sea,                        125
I should adventure for such merchandise.

JULIET
Thou knowest the mask of night is on my face,
Else would a maiden blush bepaint my cheek
For that which thou hast heard me speak tonight.
Fain would I dwell on form, fain, fain deny                             130
What I have spoke; but farewell, compliment.
Dost thou love me? I know thou wilt say 'Ay',
And I will take thy word. Yet if thou swear'st
Thou mayst prove false. At lovers' perjuries,
They say, Jove laughs. O gentle Romeo,                                  135
If thou dost love, pronounce it faithfully;
Or if thou think'st I am too quickly won,
I'll frown, and be perverse, and say thee nay,
So thou wilt woo; but else, not for the world.
In truth, fair Montague, I am too fond,                                 140
And therefore thou mayst think my 'haviour light.
But trust me, gentleman, I'll prove more true
Than those that have more cunning to be strange.
I should have been more strange, I must confess,
But that thou overheard'st, ere I was ware,                             145
My true-love passion. Therefore pardon me,
And not impute this yielding to light love,
Which the dark night hath so discoverèd.

ROMEO
Lady, by yonder blessèd moon I vow,
That tips with silver all these fruit-tree tops—                        150

JULIET
O swear not by the moon, th'inconstant moon
That monthly changes in her circled orb,
Lest that thy love prove likewise variable.

ROMEO
  What shall I swear by?
JULIET               Do not swear at all,
  Or if thou wilt, swear by thy gracious self,     155
  Which is the god of my idolatry,
  And I'll believe thee.
ROMEO           If my heart's dear love—
JULIET
  Well, do not swear. Although I joy in thee,
  I have no joy of this contract tonight.
  It is too rash, too unadvised, too sudden,     160
  Too like the lightning which doth cease to be
  Ere one can say it lightens. Sweet, good night.
  This bud of love by summer's ripening breath
  May prove a beauteous flower when next we meet.
  Good night, good night. As sweet repose and rest  165
  Come to thy heart as that within my breast.
ROMEO
  O, wilt thou leave me so unsatisfied?
JULIET
  What satisfaction canst thou have tonight?
ROMEO
  Th'exchange of thy love's faithful vow for mine.
JULIET
  I gave thee mine before thou didst request it,    170
  And yet I would it were to give again.
ROMEO
  Wouldst thou withdraw it? For what purpose, love?
JULIET
  But to be frank and give it thee again.
  And yet I wish but for the thing I have.
  My bounty is as boundless as the sea,     175
  My love as deep. The more I give to thee
  The more I have, for both are infinite.
    *Nurse calls within*
  I hear some noise within. Dear love, adieu.—
  Anon, good Nurse!—Sweet Montague, be true.
  Stay but a little; I will come again.            *Exit*
ROMEO
  O blessèd, blessèd night! I am afeard,    181
  Being in night, all this is but a dream,
  Too flattering-sweet to be substantial.
    *Enter Juliet aloft*
JULIET
  Three words, dear Romeo, and good night indeed.
  If that thy bent of love be honourable,    185
  Thy purpose marriage, send me word tomorrow,
  By one that I'll procure to come to thee,
  Where and what time thou wilt perform the rite,
  And all my fortunes at thy foot I'll lay,
  And follow thee, my lord, throughout the world.  190
⌐NURSE⌐ (*within*)
  Madam!
JULIET
  I come, anon. (*To Romeo*) But if thou mean'st not well,
  I do beseech thee—
⌐NURSE⌐ (*within*) Madam!
JULIET By and by I come.—              195
  To cease thy strife and leave me to my grief.
  Tomorrow will I send.
ROMEO So thrive my soul—
JULIET A thousand times good night.          *Exit*
ROMEO
  A thousand times the worse to want thy light.  200
  Love goes toward love as schoolboys from their books,
  But love from love, toward school with heavy looks.
    ⌐*He is going.*⌐

    *Enter Juliet aloft again*
JULIET
  Hist, Romeo! Hist! O for a falconer's voice
  To lure this tassel-gentle back again.
  Bondage is hoarse, and may not speak aloud,    205
  Else would I tear the cave where Echo lies,
  And make her airy tongue more hoarse than mine
  With repetition of my Romeo's name. Romeo!
ROMEO
  It is my soul that calls upon my name.
  How silver-sweet sound lovers' tongues by night,  210
  Like softest music to attending ears!
JULIET
  Romeo!
ROMEO     My nyas?
JULIET                What o'clock tomorrow
  Shall I send to thee?
ROMEO              By the hour of nine.
JULIET
  I will not fail; 'tis twenty year till then.
  I have forgot why I did call thee back.     215
ROMEO
  Let me stand here till thou remember it.
JULIET
  I shall forget, to have thee still stand there,
  Rememb'ring how I love thy company.
ROMEO
  And I'll still stay, to have thee still forget,
  Forgetting any other home but this.     220
JULIET
  'Tis almost morning. I would have thee gone—
  And yet no farther than a wanton's bird,
  That lets it hop a little from his hand,
  Like a poor prisoner in his twisted gyves,
  And with a silk thread plucks it back again,    225
  So loving-jealous of his liberty.
ROMEO
  I would I were thy bird.
JULIET              Sweet, so would I.
  Yet I should kill thee with much cherishing.
  Good night, good night. Parting is such sweet sorrow
  That I shall say good night till it be morrow.    230
⌐ROMEO⌐
  Sleep dwell upon thine eyes, peace in thy breast.
                          *Exit Juliet*
  Would I were sleep and peace, so sweet to rest.
  Hence will I to my ghostly sire's close cell,
  His help to crave, and my dear hap to tell.
                              *Exit*

2.2    *Enter Friar Laurence, with a basket*
FRIAR LAURENCE
  The grey-eyed morn smiles on the frowning night,
  Chequ'ring the eastern clouds with streaks of light,
  And fleckled darkness like a drunkard reels
  From forth day's path and Titan's fiery wheels.
  Now, ere the sun advance his burning eye    5
  The day to cheer and night's dank dew to dry,
  I must up-fill this osier cage of ours
  With baleful weeds and precious-juicèd flowers.
  The earth, that's nature's mother, is her tomb,
  What is her burying grave, that is her womb,    10
  And from her womb children of divers kind
  We sucking on her natural bosom find,
  Many for many virtues excellent,
  None but for some, and yet all different.

O mickle is the powerful grace that lies      15
In plants, herbs, stones, and their true qualities,
For naught so vile that on the earth doth live
But to the earth some special good doth give;
Nor aught so good but, strained from that fair use,
Revolts from true birth, stumbling on abuse.      20
Virtue itself turns vice being misapplied,
And vice sometime's by action dignified.
    *Enter Romeo*
Within the infant rind of this weak flower
Poison hath residence, and medicine power,
For this, being smelt, with that part cheers each part;
Being tasted, slays all senses with the heart.      26
Two such opposèd kings encamp them still
In man as well as herbs—grace and rude will;
And where the worser is predominant,
Full soon the canker death eats up that plant.      30

ROMEO
Good morrow, father.

FRIAR LAURENCE     *Benedicite.*
What early tongue so sweet saluteth me?
Young son, it argues a distempered head
So soon to bid good morrow to thy bed.
Care keeps his watch in every old man's eye,      35
And where care lodges, sleep will never lie,
But where unbruisèd youth with unstuffed brain
Doth couch his limbs, there golden sleep doth reign.
Therefore thy earliness doth me assure
Thou art uproused with some distemp'rature;      40
Or if not so, then here I hit it right:
Our Romeo hath not been in bed tonight.

ROMEO
That last is true; the sweeter rest was mine.

FRIAR LAURENCE
God pardon sin!—Wast thou with Rosaline?

ROMEO
With Rosaline, my ghostly father? No,      45
I have forgot that name and that name's woe.

FRIAR LAURENCE
That's my good son; but where hast thou been then?

ROMEO
I'll tell thee ere thou ask it me again.
I have been feasting with mine enemy,
Where on a sudden one hath wounded me      50
That's by me wounded. Both our remedies
Within thy help and holy physic lies.
I bear no hatred, blessèd man, for lo,
My intercession likewise steads my foe.

FRIAR LAURENCE
Be plain, good son, and homely in thy drift.      55
Riddling confession finds but riddling shrift.

ROMEO
Then plainly know my heart's dear love is set
On the fair daughter of rich Capulet.
As mine on hers, so hers is set on mine,
And all combined save what thou must combine      60
By holy marriage. When and where and how
We met, we wooed, and made exchange of vow
I'll tell thee as we pass; but this I pray,
That thou consent to marry us today.

FRIAR LAURENCE
Holy Saint Francis, what a change is here!      65
Is Rosaline, that thou didst love so dear,
So soon forsaken? Young men's love then lies
Not truly in their hearts, but in their eyes.

Jesu Maria, what a deal of brine
Hath washed thy sallow cheeks for Rosaline!      70
How much salt water thrown away in waste
To season love, that of it doth not taste!
The sun not yet thy sighs from heaven clears.
Thy old groans yet ring in mine ancient ears.
Lo, here upon thy cheek the stain doth sit      75
Of an old tear that is not washed off yet.
If e'er thou wast thyself, and these woes thine,
Thou and these woes were all for Rosaline.
And art thou changed? Pronounce this sentence then:
Women may fall when there's no strength in men.      80

ROMEO
Thou chidd'st me oft for loving Rosaline.

FRIAR LAURENCE
For doting, not for loving, pupil mine.

ROMEO
And bad'st me bury love.

FRIAR LAURENCE      Not in a grave
To lay one in, another out to have.

ROMEO
I pray thee, chide me not. Her I love now      85
Doth grace for grace and love for love allow.
The other did not so.

FRIAR LAURENCE      O, she knew well
Thy love did read by rote, that could not spell.
But come, young waverer, come, go with me.
In one respect I'll thy assistant be;      90
For this alliance may so happy prove
To turn your households' rancour to pure love.

ROMEO
O, let us hence! I stand on sudden haste.

FRIAR LAURENCE
Wisely and slow. They stumble that run fast.      *Exeunt*

2.3     *Enter Benvolio and Mercutio*

MERCUTIO Where the devil should this Romeo be? Came
he not home tonight?

BENVOLIO
Not to his father's. I spoke with his man.

MERCUTIO
Why, that same pale hard-hearted wench, that Rosaline,
Torments him so that he will sure run mad.      5

BENVOLIO
Tybalt, the kinsman to old Capulet,
Hath sent a letter to his father's house.

MERCUTIO
A challenge, on my life.

BENVOLIO      Romeo will answer it.

MERCUTIO Any man that can write may answer a letter.

BENVOLIO Nay, he will answer the letter's master, how he
dares, being dared.      11

MERCUTIO Alas, poor Romeo, he is already dead—stabbed
with a white wench's black eye, run through the ear
with a love song, the very pin of his heart cleft with the
blind bow-boy's butt-shaft; and is he a man to encounter
Tybalt?      16

⌜BENVOLIO⌝ Why, what is Tybalt?

MERCUTIO More than Prince of Cats. O, he's the courageous
captain of compliments. He fights as you sing pricksong:
keeps time, distance, and proportion. He rests his minim
rests: one, two, and the third in your bosom; the very
butcher of a silk button. A duellist, a duellist; a gentleman
of the very first house of the first and second cause. Ah,
the immortal *passado*, the *punto reverso*, the *hai*.

BENVOLIO The what?      25

MERCUTIO The pox of such antic, lisping, affecting phantasims, these new tuners of accent! 'By Jesu, a very good blade, a very tall man, a very good whore.' Why is not this a lamentable thing, grandsire, that we should be thus afflicted with these strange flies, these fashionmongers, these 'pardon-me's', who stand so much on the new form that they cannot sit at ease on the old bench? O, their bones, their bones! 33

*Enter Romeo*

BENVOLIO Here comes Romeo, here comes Romeo!

MERCUTIO Without his roe, like a dried herring. O flesh, flesh, how art thou fishified! Now is he for the numbers that Petrarch flowed in. Laura to his lady was a kitchen wench—marry, she had a better love to berhyme her—Dido a dowdy, Cleopatra a gypsy, Helen and Hero hildings and harlots, Thisbe a grey eye or so, but not to the purpose. Signor Romeo, *bonjour*. There's a French salutation to your French slop. You gave us the counterfeit fairly last night.

ROMEO Good morrow to you both. What counterfeit did I give you? 45

MERCUTIO The slip, sir, the slip. Can you not conceive?

ROMEO Pardon, good Mercutio. My business was great, and in such a case as mine a man may strain courtesy.

MERCUTIO That's as much as to say such a case as yours constrains a man to bow in the hams. 50

ROMEO Meaning to curtsy.

MERCUTIO Thou hast most kindly hit it.

ROMEO A most courteous exposition.

MERCUTIO Nay, I am the very pink of courtesy.

ROMEO Pink for flower. 55

MERCUTIO Right.

ROMEO Why, then is my pump well flowered.

MERCUTIO Sure wit, follow me this jest now till thou hast worn out thy pump, that when the single sole of it is worn, the jest may remain, after the wearing, solely singular. 61

ROMEO O single-soled jest, solely singular for the singleness!

MERCUTIO Come between us, good Benvolio. My wits faints.

ROMEO Switch and spurs, switch and spurs, or I'll cry a match. 65

MERCUTIO Nay, if our wits run the wild-goose chase, I am done, for thou hast more of the wild goose in one of thy wits than I am sure I have in my whole five. Was I with you there for the goose?

ROMEO Thou wast never with me for anything when thou wast not there for the goose. 71

MERCUTIO I will bite thee by the ear for that jest.

ROMEO Nay, good goose, bite not.

MERCUTIO Thy wit is very bitter sweeting, it is a most sharp sauce. 75

ROMEO And is it not then well served in to a sweet goose?

MERCUTIO O, here's a wit of cheveril, that stretches from an inch narrow to an ell broad.

ROMEO I stretch it out for that word 'broad', which, added to the goose, proves thee far and wide a broad goose.

MERCUTIO Why, is not this better now than groaning for love? Now art thou sociable, now art thou Romeo, now art thou what thou art by art as well as by nature, for this drivelling love is like a great natural that runs lolling up and down to hide his bauble in a hole. 85

BENVOLIO Stop there, stop there.

MERCUTIO Thou desirest me to stop in my tale against the hair.

BENVOLIO Thou wouldst else have made thy tale large.

MERCUTIO O, thou art deceived, I would have made it short, for I was come to the whole depth of my tale, and meant indeed to occupy the argument no longer. 92

*Enter the Nurse, and Peter, her man*

ROMEO Here's goodly gear.

⌈BENVOLIO⌉ A sail, a sail!

MERCUTIO Two, two—a shirt and a smock. 95

NURSE Peter.

PETER Anon.

NURSE My fan, Peter.

MERCUTIO Good Peter, to hide her face, for her fan's the fairer face. 100

NURSE God ye good morrow, gentlemen.

MERCUTIO God ye good e'en, fair gentlewoman.

NURSE Is it good e'en?

MERCUTIO 'Tis no less, I tell ye: for the bawdy hand of the dial is now upon the prick of noon. 105

NURSE Out upon you, what a man are you!

ROMEO One, gentlewoman, that God hath made for himself to mar.

NURSE By my troth, it is well said. 'For himself to mar', quoth a? Gentlemen, can any of you tell me where I may find the young Romeo? 111

ROMEO I can tell you, but young Romeo will be older when you have found him than he was when you sought him. I am the youngest of that name, for fault of a worse.

NURSE You say well. 115

MERCUTIO Yea, is the worst well? Very well took, i'faith, wisely, wisely.

NURSE (*to Romeo*) If you be he, sir, I desire some confidence with you.

BENVOLIO She will endite him to some supper. 120

MERCUTIO A bawd, a bawd, a bawd. So ho!

ROMEO What hast thou found?

MERCUTIO No hare, sir, unless a hare, sir, in a lenten pie, that is something stale and hoar ere it be spent.

⌈*He walks by them and*⌉ *sings*

    An old hare hoar 125
    And an old hare hoar
      Is very good meat in Lent.
    But a hare that is hoar
    Is too much for a score
      When it hoars ere it be spent. 130

Romeo, will you come to your father's? We'll to dinner thither.

ROMEO I will follow you.

MERCUTIO Farewell, ancient lady. Farewell, ⌈*sings*⌉ 'lady, lady, lady'. *Exeunt Mercutio and Benvolio*

NURSE I pray you, sir, what saucy merchant was this that was so full of his ropery? 137

ROMEO A gentleman, Nurse, that loves to hear himself talk, and will speak more in a minute than he will stand to in a month. 140

NURSE An a speak anything against me, I'll take him down an a were lustier than he is, and twenty such jacks; an if I cannot, I'll find those that shall. Scurvy knave! I am none of his flirt-jills, I am none of his skeans-mates. (*To Peter*) And thou must stand by, too, and suffer every knave to use me at his pleasure. 146

PETER I saw no man use you at his pleasure. If I had, my weapon should quickly have been out; I warrant you, I dare draw as soon as another man if I see occasion in a good quarrel, and the law on my side. 150

NURSE Now, afore God, I am so vexed that every part about me quivers. Scurvy knave! (*To Romeo*) Pray you, sir, a word; and, as I told you, my young lady bid me enquire you out. What she bid me say I will keep to myself, but first let me tell ye if ye should lead her in a fool's paradise, as they say, it were a very gross kind of behaviour, as they say, for the gentlewoman is young; and therefore if you should deal double with her, truly it were an ill thing to be offered to any gentlewoman, and very weak dealing.                                     160

ROMEO Nurse, commend me to thy lady and mistress. I protest unto thee—

NURSE Good heart, and i'faith I will tell her as much. Lord, Lord, she will be a joyful woman.

ROMEO What wilt thou tell her, Nurse? Thou dost not mark me.                                                          166

NURSE I will tell her, sir, that you do protest; which as I take it is a gentlemanlike offer.

ROMEO Bid her devise
Some means to come to shrift this afternoon,        170
And there she shall at Friar Laurence' cell
Be shrived and married. (*Offering money*) Here is for thy pains.

NURSE No, truly, sir, not a penny.

ROMEO Go to, I say, you shall.

NURSE ⌈*taking the money*⌉
This afternoon, sir. Well, she shall be there.       175

ROMEO
And stay, good Nurse, behind the abbey wall.
Within this hour my man shall be with thee
And bring thee cords made like a tackled stair,
Which to the high topgallant of my joy
Must be my convoy in the secret night.              180
Farewell. Be trusty, and I'll quit thy pains.
Farewell. Commend me to thy mistress.

NURSE
Now God in heaven bless thee! Hark you, sir.

ROMEO What sayst thou, my dear Nurse?

NURSE
Is your man secret? Did you ne'er hear say          185
'Two may keep counsel, putting one away'?

ROMEO
I warrant thee my man's as true as steel.

NURSE
Well, sir, my mistress is the sweetest lady.
Lord, Lord, when 'twas a little prating thing—
O, there is a nobleman in town, one Paris,          190
That would fain lay knife aboard; but she, good soul,
Had as lief see a toad, a very toad,
As see him. I anger her sometimes,
And tell her that Paris is the properer man;
But I'll warrant you, when I say so she looks        195
As pale as any clout in the versal world.
Doth not rosemary and Romeo begin
Both with a letter?

ROMEO
Ay, Nurse, what of that? Both with an 'R'.          199

NURSE Ah, mocker—that's the dog's name. 'R' is for the—no, I know it begins with some other letter, and she hath the prettiest sententious of it, of you and rosemary, that it would do you good to hear it.

ROMEO Commend me to thy lady.

NURSE Ay, a thousand times. Peter!                   205

PETER Anon.

NURSE ⌈*giving Peter her fan*⌉ Before, and apace.

                    *Exeunt* ⌈*Peter and Nurse at one door,*
                              *Romeo at another door*⌉

**2.4**    *Enter Juliet*

JULIET
The clock struck nine when I did send the Nurse.
In half an hour she promised to return.
Perchance she cannot meet him. That's not so.
O, she is lame! Love's heralds should be thoughts,
Which ten times faster glides than the sun's beams   5
Driving back shadows over louring hills.
Therefore do nimble-pinioned doves draw Love,
And therefore hath the wind-swift Cupid wings.
Now is the sun upon the highmost hill
Of this day's journey, and from nine till twelve     10
Is three long hours, yet she is not come.
Had she affections and warm youthful blood
She would be as swift in motion as a ball.
My words would bandy her to my sweet love,
And his to me.                                       15
But old folks, many feign as they were dead—
Unwieldy, slow, heavy, and pale as lead.
        *Enter the Nurse and Peter*
O God, she comes! O honey Nurse, what news?
Hast thou met with him? Send thy man away.

NURSE Peter, stay at the gate.                *Exit Peter*

JULIET
Now, good sweet Nurse—O Lord, why look'st thou sad?
Though news be sad, yet tell them merrily;           22
If good, thou sham'st the music of sweet news
By playing it to me with so sour a face.

NURSE
I am a-weary. Give me leave a while.                 25
Fie, how my bones ache. What a jaunce have I!

JULIET
I would thou hadst my bones and I thy news.
Nay, come, I pray thee speak, good, good Nurse, speak.

NURSE
Jesu, what haste! Can you not stay a while?
Do you not see that I am out of breath?              30

JULIET
How art thou out of breath when thou hast breath
To say to me that thou art out of breath?
The excuse that thou dost make in this delay
Is longer than the tale thou dost excuse.
Is thy news good or bad? Answer to that.            35
Say either, and I'll stay the circumstance.
Let me be satisfied: is't good or bad?

NURSE Well, you have made a simple choice. You know not how to choose a man. Romeo? No, not he; though his face be better than any man's, yet his leg excels all men's, and for a hand and a foot and a body, though they be not to be talked on, yet they are past compare. He is not the flower of courtesy, but, I'll warrant him, as gentle as a lamb. Go thy ways, wench. Serve God. What, have you dined at home?                      45

JULIET
No, no. But all this did I know before.
What says he of our marriage—what of that?

NURSE
Lord, how my head aches! What a head have I!
It beats as it would fall in twenty pieces.
My back—
        ⌈*Juliet rubs her back*⌉
                a' t'other side—ah, my back, my back!   50
Beshrew your heart for sending me about
To catch my death with jauncing up and down.

JULIET
I'faith, I am sorry that thou art not well.
Sweet, sweet, sweet Nurse, tell me, what says my love?

NURSE Your love says, like an honest gentleman, and a
courteous, and a kind, and a handsome, and, I warrant,
a virtuous—where is your mother?      57

JULIET
Where is my mother? Why, she is within.
Where should she be? How oddly thou repliest!
'Your love says like an honest gentleman      60
"Where is your mother?"'

NURSE          O, God's Lady dear!
Are you so hot? Marry come up, I trow.
Is this the poultice for my aching bones?
Henceforward do your messages yourself.

JULIET
Here's such a coil! Come, what says Romeo?      65

NURSE
Have you got leave to go to shrift today?

JULIET I have.

NURSE
Then hie you hence to Friar Laurence' cell.
There stays a husband to make you a wife.
Now comes the wanton blood up in your cheeks.      70
They'll be in scarlet straight at any news.
Hie you to church. I must another way,
To fetch a ladder by the which your love
Must climb a bird's nest soon, when it is dark.
I am the drudge, and toil in your delight,      75
But you shall bear the burden soon at night.
Go, I'll to dinner. Hie you to the cell.

JULIET
Hie to high fortune! Honest Nurse, farewell.
*Exeunt ⌈severally⌉*

**2.5**    *Enter Friar Laurence and Romeo*

FRIAR LAURENCE
So smile the heavens upon this holy act
That after-hours with sorrow chide us not!

ROMEO
Amen, amen. But come what sorrow can,
It cannot countervail the exchange of joy
That one short minute gives me in her sight.      5
Do thou but close our hands with holy words,
Then love-devouring death do what he dare—
It is enough I may but call her mine.

FRIAR LAURENCE
These violent delights have violent ends,
And in their triumph die like fire and powder,      10
Which as they kiss consume. The sweetest honey
Is loathsome in his own deliciousness,
And in the taste confounds the appetite.
Therefore love moderately. Long love doth so.
Too swift arrives as tardy as too slow.      15
     *Enter Juliet ⌈somewhat fast, and embraceth Romeo⌉*
Here comes the lady. O, so light a foot
Will ne'er wear out the everlasting flint.
A lover may bestride the gossamers
That idles in the wanton summer air,
And yet not fall, so light is vanity.      20

JULIET
Good even to my ghostly confessor.

FRIAR LAURENCE
Romeo shall thank thee, daughter, for us both.

JULIET
As much to him, else is his thanks too much.

ROMEO
Ah, Juliet, if the measure of thy joy
Be heaped like mine, and that thy skill be more      25
To blazon it, then sweeten with thy breath

This neighbour air, and let rich music's tongue
Unfold the imagined happiness that both
Receive in either by this dear encounter.

JULIET
Conceit, more rich in matter than in words,      30
Brags of his substance, not of ornament.
They are but beggars that can count their worth,
But my true love is grown to such excess
I cannot sum up some of half my wealth.

FRIAR LAURENCE
Come, come with me, and we will make short work, 35
For, by your leaves, you shall not stay alone
Till Holy Church incorporate two in one.     *Exeunt*

**3.1**    *Enter Mercutio with his page, Benvolio, and men*

BENVOLIO
I pray thee, good Mercutio, let's retire.
The day is hot, the Capels are abroad,
And if we meet we shall not scape a brawl,
For now, these hot days, is the mad blood stirring.    4

MERCUTIO Thou art like one of these fellows that, when he
enters the confines of a tavern, claps me his sword upon
the table and says 'God send me no need of thee', and
by the operation of the second cup, draws him on the
drawer when indeed there is no need.

BENVOLIO Am I like such a fellow?      10

MERCUTIO Come, come, thou art as hot a jack in thy mood
as any in Italy, and as soon moved to be moody, and
as soon moody to be moved.

BENVOLIO And what to?

MERCUTIO Nay, an there were two such, we should have
none shortly, for one would kill the other. Thou—why,
thou wilt quarrel with a man that hath a hair more or
a hair less in his beard than thou hast. Thou wilt quarrel
with a man for cracking nuts, having no other reason
but because thou hast hazel eyes. What eye but such
an eye would spy out such a quarrel? Thy head is as
full of quarrels as an egg is full of meat, and yet thy
head hath been beaten as addle as an egg for quarrelling.
Thou hast quarrelled with a man for coughing in the
street because he hath wakened thy dog that hath lain
asleep in the sun. Didst thou not fall out with a tailor
for wearing his new doublet before Easter; with another
for tying his new shoes with old ribbon? And yet thou
wilt tutor me from quarrelling!      29

BENVOLIO An I were so apt to quarrel as thou art, any
man should buy the fee-simple of my life for an hour
and a quarter.

MERCUTIO The fee simple? O, simple!
     *Enter Tybalt, Petruccio, and others*

BENVOLIO By my head, here comes the Capulets.

MERCUTIO By my heel, I care not.      35

TYBALT (*to Petruccio and the others*)
Follow me close, for I will speak to them.
(*To the Montagues*) Gentlemen, good e'en. A word with
one of you.

MERCUTIO And but one word with one of us? Couple it
with something: make it a word and a blow.

TYBALT You shall find me apt enough to that, sir, an you
will give me occasion.      41

MERCUTIO Could you not take some occasion without
giving?

TYBALT
Mercutio, thou consort'st with Romeo.

MERCUTIO 'Consort'? What, dost thou make us minstrels?
An thou make minstrels of us, look to hear nothing but

discords. ⌈*Touching his rapier*⌉ Here's my fiddlestick; here's
that shall make you dance. Zounds—'Consort'!

BENVOLIO
We talk here in the public haunt of men.
Either withdraw unto some private place,                    50
Or reason coldly of your grievances,
Or else depart. Here all eyes gaze on us.

MERCUTIO
Men's eyes were made to look, and let them gaze.
I will not budge for no man's pleasure, I.
    *Enter Romeo*

TYBALT
Well, peace be with you, sir. Here comes my man.    55

MERCUTIO
But I'll be hanged, sir, if he wear your livery.
Marry, go before to field, he'll be your follower.
Your worship in that sense may call him 'man'.

TYBALT
Romeo, the love I bear thee can afford
No better term than this: thou art a villain.              60

ROMEO
Tybalt, the reason that I have to love thee
Doth much excuse the appertaining rage
To such a greeting. Villain am I none.
Therefore, farewell. I see thou knowest me not.

TYBALT
Boy, this shall not excuse the injuries                     65
That thou hast done me. Therefore turn and draw.

ROMEO
I do protest I never injured thee,
But love thee better than thou canst devise
Till thou shalt know the reason of my love.
And so, good Capulet—which name I tender               70
As dearly as mine own—be satisfied.

MERCUTIO ⌈*drawing*⌉
O calm, dishonourable, vile submission!
*Alla stoccado* carries it away.
Tybalt, you ratcatcher, come, will you walk?

TYBALT  What wouldst thou have with me?            75

MERCUTIO  Good King of Cats, nothing but one of your nine
lives. That I mean to make bold withal, and, as you
shall use me hereafter, dry-beat the rest of the eight.
Will you pluck your sword out of his pilcher by the ears?
Make haste, lest mine be about your ears ere it be out.

TYBALT  (*drawing*) I am for you.                           81

ROMEO
Gentle Mercutio, put thy rapier up.

MERCUTIO  (*to Tybalt*) Come, sir, your *passado.*
    *They fight*

ROMEO ⌈*drawing*⌉
Draw, Benvolio. Beat down their weapons.
Gentlemen, for shame forbear this outrage.                 85
Tybalt, Mercutio, the Prince expressly hath
Forbid this bandying in Verona streets.
Hold, Tybalt, good Mercutio.
    ⌈*Romeo beats down their points and rushes between*
    *them. Tybalt under Romeo's arm thrusts Mercutio in*⌉

⌈PETRUCCIO⌉ Away, Tybalt!
    *Exeunt Tybalt, Petruccio, and their followers*

MERCUTIO I am hurt.                                          90
A plague o' both your houses. I am sped.
Is he gone, and hath nothing?

BENVOLIO                          What, art thou hurt?

MERCUTIO
Ay, ay, a scratch, a scratch; marry, 'tis enough.
Where is my page? Go, villain. Fetch a surgeon.
    *Exit page*

ROMEO
Courage, man. The hurt cannot be much.                     95

MERCUTIO  No, 'tis not so deep as a well, nor so wide as a
church door, but 'tis enough. 'Twill serve. Ask for me
tomorrow, and you shall find me a grave man. I am
peppered, I warrant, for this world. A plague o' both
your houses! Zounds, a dog, a rat, a mouse, a cat, to
scratch a man to death! A braggart, a rogue, a villain,
that fights by the book of arithmetic! Why the devil
came you between us? I was hurt under your arm.

ROMEO  I thought all for the best.

MERCUTIO
Help me into some house, Benvolio,                        105
Or I shall faint. A plague o' both your houses.
They have made worms' meat of me.
I have it, and soundly, too. Your houses!
    *Exeunt all but Romeo*

ROMEO
This gentleman, the Prince's near ally,
My very friend, hath got this mortal hurt                 110
In my behalf, my reputation stained
With Tybalt's slander—Tybalt, that an hour
Hath been my cousin! O sweet Juliet,
Thy beauty hath made me effeminate,
And in my temper softened valour's steel.                 115
    *Enter Benvolio*

BENVOLIO
O Romeo, Romeo, brave Mercutio is dead!
That gallant spirit hath aspired the clouds,
Which too untimely here did scorn the earth.

ROMEO
This day's black fate on more days doth depend.
This but begins the woe others must end.                  120
    *Enter Tybalt*

BENVOLIO
Here comes the furious Tybalt back again.

ROMEO
He gad in triumph, and Mercutio slain?
Away to heaven, respective lenity,
And fire-eyed fury be my conduct now.
Now, Tybalt, take the 'villain' back again                125
That late thou gav'st me, for Mercutio's soul
Is but a little way above our heads,
Staying for thine to keep him company.
Either thou, or I, or both must go with him.

TYBALT
Thou, wretched boy, that didst consort him here,          130
Shalt with him hence.

ROMEO                      This shall determine that.
    *They fight. Tybalt is wounded. He falls and dies*

BENVOLIO  Romeo, away, be gone.
The citizens are up, and Tybalt slain.
Stand not amazed. The Prince will doom thee death
If thou art taken. Hence, be gone, away.                  135

ROMEO
O, I am fortune's fool!

BENVOLIO                      Why dost thou stay?
    *Exit Romeo*

    *Enter Citizens* ⌈*of the watch*⌉

CITIZEN ⌈OF THE WATCH⌉
Which way ran he that killed Mercutio?
Tybalt, that murderer, which way ran he?

BENVOLIO
There lies that Tybalt.

CITIZEN ⌈OF THE WATCH⌉ (*to Tybalt*) Up, sir, go with me.
I charge thee in the Prince's name, obey.                 140

*Enter the Prince, old Montague, Capulet, their Wives, and all*

PRINCE
Where are the vile beginners of this fray?

BENVOLIO
O noble Prince, I can discover all
The unlucky manage of this fatal brawl.
There lies the man, slain by young Romeo,
That slew thy kinsman, brave Mercutio.                145

CAPULET'S WIFE
Tybalt, my cousin, O, my brother's child!
O Prince, O cousin, husband! O, the blood is spilled
Of my dear kinsman! Prince, as thou art true,
For blood of ours shed blood of Montague!
O cousin, cousin!

PRINCE                          Benvolio, who began this fray?    150

BENVOLIO
Tybalt, here slain, whom Romeo's hand did slay.
Romeo, that spoke him fair, bid him bethink
How nice the quarrel was, and urged withal
Your high displeasure. All this—utterèd
With gentle breath, calm look, knees humbly bowed—
Could not take truce with the unruly spleen           156
Of Tybalt deaf to peace, but that he tilts
With piercing steel at bold Mercutio's breast,
Who, all as hot, turns deadly point to point,
And, with a martial scorn, with one hand beats        160
Cold death aside, and with the other sends
It back to Tybalt, whose dexterity
Retorts it. Romeo, he cries aloud,
'Hold, friends, friends, part!' and swifter than his
    tongue
His agent arm beats down their fatal points,          165
And 'twixt them rushes, underneath whose arm
An envious thrust from Tybalt hit the life
Of stout Mercutio, and then Tybalt fled,
But by and by comes back to Romeo,
Who had but newly entertained revenge,                170
And to't they go like lightning; for ere I
Could draw to part them was stout Tybalt slain,
And as he fell did Romeo turn and fly.
This is the truth, or let Benvolio die.

CAPULET'S WIFE
He is a kinsman to the Montague.                      175
Affection makes him false; he speaks not true.
Some twenty of them fought in this black strife,
And all those twenty could but kill one life.
I beg for justice, which thou, Prince, must give.
Romeo slew Tybalt; Romeo must not live.               180

PRINCE
Romeo slew him, he slew Mercutio.
Who now the price of his dear blood doth owe?

⌜MONTAGUE⌝
Not Romeo, Prince. He was Mercutio's friend.
His fault concludes but what the law should end,
The life of Tybalt.

PRINCE                    And for that offence        185
Immediately we do exile him hence.
I have an interest in your hate's proceeding;
My blood for your rude brawls doth lie a-bleeding;
But I'll amerce you with so strong a fine
That you shall all repent the loss of mine.           190
I will be deaf to pleading and excuses.
Nor tears nor prayers shall purchase out abuses.
Therefore use none. Let Romeo hence in haste,
Else, when he is found, that hour is his last.

Bear hence this body, and attend our will.            195
Mercy but murders, pardoning those that kill.
                              *Exeunt with the body*

3.2    *Enter Juliet*

JULIET
Gallop apace, you fiery-footed steeds,
Towards Phoebus' lodging. Such a waggoner
As Phaëton would whip you to the west
And bring in cloudy night immediately.
Spread thy close curtain, love-performing night,       5
That runaways' eyes may wink, and Romeo
Leap to these arms untalked of and unseen.
Lovers can see to do their amorous rites
By their own beauties; or, if love be blind,
It best agrees with night. Come, civil night,         10
Thou sober-suited matron all in black,
And learn me how to lose a winning match
Played for a pair of stainless maidenhoods.
Hood my unmanned blood, bating in my cheeks,
With thy black mantle till strange love grown bold    15
Think true love acted simple modesty.
Come night, come Romeo; come, thou day in night,
For thou wilt lie upon the wings of night
Whiter than new snow on a raven's back.
Come, gentle night; come, loving, black-browed night,
Give me my Romeo, and when I shall die                21
Take him and cut him out in little stars,
And he will make the face of heaven so fine
That all the world will be in love with night
And pay no worship to the garish sun.                 25
O, I have bought the mansion of a love
But not possessed it, and though I am sold,
Not yet enjoyed. So tedious is this day
As is the night before some festival
To an impatient child that hath new robes             30
And may not wear them.
    *Enter the Nurse, ⌜wringing her hands,⌝ with the
    ladder of cords ⌜in her lap⌝*
                                O, here comes my Nurse,
And she brings news, and every tongue that speaks
But Romeo's name speaks heavenly eloquence.
Now, Nurse, what news? What, hast thou there
The cords that Romeo bid thee fetch?

NURSE ⌜*putting down the cords*⌝            Ay, ay, the cords.

JULIET
Ay me, what news? Why dost thou wring thy hands?

NURSE
Ah, welladay! He's dead, he's dead, he's dead!        37
We are undone, lady, we are undone.
Alack the day, he's gone, he's killed, he's dead!

JULIET
Can heaven be so envious?

NURSE                        Romeo can,               40
Though heaven cannot. O Romeo, Romeo,
Who ever would have thought it Romeo?

JULIET
What devil art thou that dost torment me thus?
This torture should be roared in dismal hell.
Hath Romeo slain himself? Say thou but 'Ay',          45
And that bare vowel 'I' shall poison more
Than the death-darting eye of cockatrice.
I am not I if there be such an 'Ay',
Or those eyes shut that makes thee answer 'Ay'.
If he be slain, say 'Ay'; or if not, 'No'.            50
Brief sounds determine of my weal or woe.

NURSE
I saw the wound, I saw it with mine eyes,
God save the mark, here on his manly breast—
A piteous corpse, a bloody, piteous corpse—
Pale, pale as ashes, all bedaubed in blood,                55
All in gore blood; I swoonèd at the sight.
JULIET
O, break, my heart, poor bankrupt, break at once!
To prison, eyes; ne'er look on liberty.
Vile earth, to earth resign; end motion here,
And thou and Romeo press one heavy bier!                60
NURSE
O Tybalt, Tybalt, the best friend I had!
O courteous Tybalt, honest gentleman,
That ever I should live to see thee dead!
JULIET
What storm is this that blows so contrary?
Is Romeo slaughtered, and is Tybalt dead?                65
My dearest cousin and my dearer lord?
Then, dreadful trumpet, sound the general doom,
For who is living if those two are gone?
NURSE
Tybalt is gone and Romeo banishèd.
Romeo that killed him—he is banishèd.                70
JULIET
O God, did Romeo's hand shed Tybalt's blood?
⌜NURSE⌝
It did, it did, alas the day, it did.
⌜JULIET⌝
O serpent heart hid with a flow'ring face!
Did ever dragon keep so fair a cave?
Beautiful tyrant, fiend angelical!                75
Dove-feathered raven, wolvish-ravening lamb!
Despisèd substance of divinest show!
Just opposite to what thou justly seem'st—
A damnèd saint, an honourable villain.
O nature, what hadst thou to do in hell                80
When thou didst bower the spirit of a fiend
In mortal paradise of such sweet flesh?
Was ever book containing such vile matter
So fairly bound? O, that deceit should dwell
In such a gorgeous palace!                85
NURSE
There's no trust, no faith, no honesty in men;
All perjured, all forsworn, all naught, dissemblers all.
Ah, where's my man? Give me some aqua vitae.
These griefs, these woes, these sorrows make me old.
Shame come to Romeo!
JULIET                    Blistered be thy tongue                90
For such a wish! He was not born to shame.
Upon his brow shame is ashamed to sit,
For 'tis a throne where honour may be crowned
Sole monarch of the universal earth.
O, what a beast was I to chide at him!                95
NURSE
Will you speak well of him that killed your cousin?
JULIET
Shall I speak ill of him that is my husband?
Ah, poor my lord, what tongue shall smooth thy name
When I, thy three-hours wife, have mangled it?
But wherefore, villain, didst thou kill my cousin?                100
That villain cousin would have killed my husband.
Back, foolish tears, back to your native spring!
Your tributary drops belong to woe,
Which you, mistaking, offer up to joy.
My husband lives, that Tybalt would have slain;                105

And Tybalt's dead, that would have slain my husband.
All this is comfort. Wherefore weep I then?
Some word there was, worser than Tybalt's death,
That murdered me. I would forget it fain,
But O, it presses to my memory                110
Like damnèd guilty deeds to sinners' minds!
'Tybalt is dead, and Romeo banishèd.'
That 'banishèd', that one word 'banishèd'
Hath slain ten thousand Tybalts. Tybalt's death
Was woe enough, if it had ended there;                115
Or, if sour woe delights in fellowship
And needly will be ranked with other griefs,
Why followed not, when she said 'Tybalt's dead',
'Thy father', or 'thy mother', nay, or both,
Which modern lamentation might have moved?                120
But with a rearward following Tybalt's death,
'Romeo is banishèd'—to speak that word
Is father, mother, Tybalt, Romeo, Juliet,
All slain, all dead. 'Romeo is banishèd'—
There is no end, no limit, measure, bound,                125
In that word's death. No words can that woe sound.
Where is my father and my mother, Nurse?
NURSE
Weeping and wailing over Tybalt's corpse.
Will you go to them? I will bring you thither.
JULIET
Wash they his wounds with tears; mine shall be spent
When theirs are dry, for Romeo's banishment.                131
Take up those cords. Poor ropes, you are beguiled,
Both you and I, for Romeo is exiled.
He made you for a highway to my bed,
But I, a maid, die maiden-widowèd.                135
Come, cords; come, Nurse; I'll to my wedding bed,
And death, not Romeo, take my maidenhead!
NURSE (taking up the cords)
Hie to your chamber. I'll find Romeo
To comfort you. I wot well where he is.
Hark ye, your Romeo will be here at night.                140
I'll to him. He is hid at Laurence' cell.
JULIET (giving her a ring)
O, find him! Give this ring to my true knight,
And bid him come to take his last farewell.
                    Exeunt ⌜severally⌝

3.3    Enter Friar Laurence
FRIAR LAURENCE
Romeo, come forth, come forth, thou fear-full man.
Affliction is enamoured of thy parts,
And thou art wedded to calamity.
        Enter Romeo
ROMEO
Father, what news? What is the Prince's doom?
What sorrow craves acquaintance at my hand                5
That I yet know not?
FRIAR LAURENCE            Too familiar
Is my dear son with such sour company.
I bring thee tidings of the Prince's doom.
ROMEO
What less than doomsday is the Prince's doom?
FRIAR LAURENCE
A gentler judgement vanished from his lips:                10
Not body's death, but body's banishment.
ROMEO
Ha, banishment? Be merciful, say 'death',
For exile hath more terror in his look,
Much more than death. Do not say 'banishment'.

FRIAR LAURENCE
Hence from Verona art thou banishèd.                                     15
Be patient, for the world is broad and wide.
ROMEO
There is no world without Verona walls
But purgatory, torture, hell itself.
Hence banishèd is banished from the world,
And world's exile is death. Then 'banishèd'                              20
Is death mistermed. Calling death 'banishèd'
Thou cutt'st my head off with a golden axe,
And smil'st upon the stroke that murders me.
FRIAR LAURENCE
O deadly sin, O rude unthankfulness!
Thy fault our law calls death, but the kind Prince,                      25
Taking thy part, hath rushed aside the law
And turned that black word 'death' to banishment.
This is dear mercy, and thou seest it not.
ROMEO
'Tis torture, and not mercy. Heaven is here
Where Juliet lives, and every cat and dog                                30
And little mouse, every unworthy thing,
Live here in heaven and may look on her,
But Romeo may not. More validity,
More honourable state, more courtship lives
In carrion flies than Romeo. They may seize                              35
On the white wonder of dear Juliet's hand,
And steal immortal blessing from her lips,
Who, even in pure and vestal modesty,
Still blush, as thinking their own kisses sin.
But Romeo may not, he is banishèd.                                       40
Flies may do this, but I from this must fly.
They are free men, but I am banishèd.
And sayst thou yet that exile is not death?
Hadst thou no poison mixed, no sharp-ground knife,
No sudden mean of death, though ne'er so mean,                           45
But 'banishèd' to kill me—'banishèd'?
O friar, the damnèd use that word in hell.
Howling attends it. How hast thou the heart,
Being a divine, a ghostly confessor,
A sin-absolver and my friend professed,                                  50
To mangle me with that word 'banishèd'?
FRIAR LAURENCE
Thou fond mad man, hear me a little speak.
ROMEO
O, thou wilt speak again of banishment.
FRIAR LAURENCE
I'll give thee armour to keep off that word—
Adversity's sweet milk, philosophy,                                      55
To comfort thee though thou art banishèd.
ROMEO
Yet 'banishèd'? Hang up philosophy!
Unless philosophy can make a Juliet,
Displant a town, reverse a prince's doom,
It helps not, it prevails not. Talk no more.                             60
FRIAR LAURENCE
O, then I see that madmen have no ears.
ROMEO
How should they, when that wise men have no eyes?
FRIAR LAURENCE
Let me dispute with thee of thy estate.
ROMEO
Thou canst not speak of that thou dost not feel.
Wert thou as young as I, Juliet thy love,                                65
An hour but married, Tybalt murderèd,
Doting like me, and like me banishèd,
Then mightst thou speak, then mightst thou tear thy
hair,

And fall upon the ground, as I do now,
*He falls upon the ground*
Taking the measure of an unmade grave.                                   70
*Knock within*
FRIAR LAURENCE
Arise, one knocks. Good Romeo, hide thyself.
ROMEO
Not I, unless the breath of heartsick groans
Mist-like enfold me from the search of eyes.
*Knocking within*
FRIAR LAURENCE
Hark, how they knock!—Who's there?—Romeo, arise.
Thou wilt be taken.—Stay a while.—Stand up.                             75
*Still knock within*
Run to my study.—By and by!—God's will,
What simpleness is this?
*Knock within*
                              I come, I come.
Who knocks so hard? Whence come you? What's your
will?
NURSE (*within*)
Let me come in, and you shall know my errand.
I come from Lady Juliet.
FRIAR LAURENCE ⌈*opening the door*⌉ Welcome then.                        80
*Enter the Nurse*
NURSE
O holy friar, O tell me, holy friar,
Where is my lady's lord? Where's Romeo?
FRIAR LAURENCE
There on the ground, with his own tears made drunk.
NURSE
O, he is even in my mistress' case,
Just in her case! O woeful sympathy,                                     85
Piteous predicament! Even so lies she,
Blubb'ring and weeping, weeping and blubb'ring.
(*To Romeo*) Stand up, stand up, stand an you be a man,
For Juliet's sake, for her sake, rise and stand.
Why should you fall into so deep an O?                                   90
ROMEO (*rising*)
Nurse.
NURSE    Ah sir, ah sir, death's the end of all.
ROMEO
Spak'st thou of Juliet? How is it with her?
Doth not she think me an old murderer,
Now I have stained the childhood of our joy
With blood removed but little from her own?                              95
Where is she, and how doth she, and what says
My concealed lady to our cancelled love?
NURSE
O, she says nothing, sir, but weeps and weeps,
And now falls on her bed, and then starts up,
And 'Tybalt' calls, and then on Romeo cries,                            100
And then down falls again.
ROMEO               As if that name
Shot from the deadly level of a gun
Did murder her as that name's cursèd hand
Murdered her kinsman. O tell me, friar, tell me,
In what vile part of this anatomy                                       105
Doth my name lodge? Tell me, that I may sack
The hateful mansion.
⌈*He offers to stab himself, and the Nurse snatches the
dagger away*⌉
FRIAR LAURENCE            Hold thy desperate hand.
Art thou a man? Thy form cries out thou art.
Thy tears are womanish, thy wild acts denote
The unreasonable fury of a beast.                                       110

Unseemly woman in a seeming man,
And ill-beseeming beast in seeming both!
Thou hast amazed me. By my holy order,
I thought thy disposition better tempered.
Hast thou slain Tybalt? Wilt thou slay thyself,          115
And slay thy lady that in thy life lives
By doing damnèd hate upon thyself?
Why rail'st thou on thy birth, the heaven, and earth,
Since birth and heaven and earth, all three, do meet
In thee at once, which thou at once wouldst lose?          120
Fie, fie, thou sham'st thy shape, thy love, thy wit,
Which like a usurer abound'st in all,
And usest none in that true use indeed
Which should bedeck thy shape, thy love, thy wit.
Thy noble shape is but a form of wax,                    125
Digressing from the valour of a man;
Thy dear love sworn but hollow perjury,
Killing that love which thou hast vowed to cherish;
Thy wit, that ornament to shape and love,
Misshapen in the conduct of them both,                  130
Like powder in a skilless soldier's flask
Is set afire by thine own ignorance,
And thou dismembered with thine own defence.
What, rouse thee, man! Thy Juliet is alive,
For whose dear sake thou wast but lately dead:          135
There art thou happy. Tybalt would kill thee,
But thou slewest Tybalt: there art thou happy.
The law that threatened death becomes thy friend,
And turns it to exile: there art thou happy.
A pack of blessings light upon thy back,                140
Happiness courts thee in her best array,
But, like a mishavèd and sullen wench,
Thou pout'st upon thy fortune and thy love.
Take heed, take heed, for such die miserable.
Go, get thee to thy love, as was decreed.               145
Ascend her chamber; hence and comfort her.
But look thou stay not till the watch be set,
For then thou canst not pass to Mantua,
Where thou shalt live till we can find a time
To blaze your marriage, reconcile your friends,         150
Beg pardon of the Prince, and call thee back
With twenty hundred thousand times more joy
Than thou went'st forth in lamentation.
Go before, Nurse. Commend me to thy lady,
And bid her hasten all the house to bed,                155
Which heavy sorrow makes them apt unto.
Romeo is coming.

NURSE
O Lord, I could have stayed here all the night
To hear good counsel! O, what learning is!
My lord, I'll tell my lady you will come.               160

ROMEO
Do so, and bid my sweet prepare to chide.
          ⌐Nurse offers to go in, and turns again⌐

NURSE (giving the ring)
Here, sir, a ring she bid me give you, sir.
Hie you, make haste, for it grows very late.

ROMEO
How well my comfort is revived by this.     Exit Nurse

FRIAR LAURENCE
Go hence, good night, and here stands all your state.
Either be gone before the watch be set,                 166
Or by the break of day disguised from hence.
Sojourn in Mantua. I'll find out your man,
And he shall signify from time to time

Every good hap to you that chances here.                170
Give me thy hand. 'Tis late. Farewell. Good night.

ROMEO
But that a joy past joy calls out on me,
It were a grief so brief to part with thee.
Farewell.                          Exeunt ⌐severally⌐

3.4    Enter Capulet, his Wife, and Paris

CAPULET
Things have fall'n out, sir, so unluckily
That we have had no time to move our daughter.
Look you, she loved her kinsman Tybalt dearly,
And so did I. Well, we were born to die.
'Tis very late. She'll not come down tonight.             5
I promise you, but for your company
I would have been abed an hour ago.

PARIS
These times of woe afford no times to woo.
Madam, good night. Commend me to your daughter.

CAPULET'S WIFE
I will, and know her mind early tomorrow.                10
Tonight she's mewed up to her heaviness.
          ⌐Paris offers to go in, and Capulet calls him again⌐

CAPULET
Sir Paris, I will make a desperate tender
Of my child's love. I think she will be ruled
In all respects by me. Nay, more, I doubt it not.
Wife, go you to her ere you go to bed.                    15
Acquaint her here of my son Paris' love,
And bid her—mark you me?—on Wednesday next—
But soft—what day is this?

PARIS                             Monday, my lord.

CAPULET
Monday. Ha, ha! Well, Wednesday is too soon.
O' Thursday let it be. O' Thursday, tell her,            20
She shall be married to this noble earl.
Will you be ready? Do you like this haste?
We'll keep no great ado—a friend or two.
For hark you, Tybalt being slain so late,
It may be thought we held him carelessly,                25
Being our kinsman, if we revel much.
Therefore we'll have some half a dozen friends,
And there an end. But what say you to Thursday?

PARIS
My lord, I would that Thursday were tomorrow.

CAPULET
Well, get you gone. O' Thursday be it, then.             30
(To his Wife) Go you to Juliet ere you go to bed.
Prepare her, wife, against this wedding day.—
Farewell, my lord.—Light to my chamber, ho!—
Afore me, it is so very late that we
May call it early by and by. Good night.                 35
          Exeunt ⌐Capulet and his wife at
               one door, Paris at another door⌐

3.5    Enter Romeo and Juliet aloft ⌐with the ladder of cords⌐

JULIET
Wilt thou be gone? It is not yet near day.
It was the nightingale, and not the lark,
That pierced the fear-full hollow of thine ear.
Nightly she sings on yon pom'granate tree.
Believe me, love, it was the nightingale.                 5

ROMEO
It was the lark, the herald of the morn,
No nightingale. Look, love, what envious streaks
Do lace the severing clouds in yonder east.

Night's candles are burnt out, and jocund day
Stands tiptoe on the misty mountain tops.                    10
I must be gone and live, or stay and die.
JULIET
Yon light is not daylight; I know it, I.
It is some meteor that the sun exhaled
To be to thee this night a torchbearer
And light thee on thy way to Mantua.                        15
Therefore stay yet. Thou need'st not to be gone.
ROMEO
Let me be ta'en, let me be put to death.
I am content, so thou wilt have it so.
I'll say yon grey is not the morning's eye,
'Tis but the pale reflex of Cynthia's brow;                 20
Nor that is not the lark whose notes do beat
The vaulty heaven so high above our heads.
I have more care to stay than will to go.
Come, death, and welcome; Juliet wills it so.
How is't, my soul? Let's talk. It is not day.               25
JULIET
It is, it is. Hie hence, be gone, away.
It is the lark that sings so out of tune,
Straining harsh discords and unpleasing sharps.
Some say the lark makes sweet division;
This doth not so, for she divideth us.                      30
Some say the lark and loathèd toad changed eyes.
O, now I would they had changed voices, too,
Since arm from arm that voice doth us affray,
Hunting thee hence with hunt's-up to the day.
O, now be gone! More light and light it grows.             35
ROMEO
More light and light, more dark and dark our woes.
            *Enter the Nurse ⌐hastily⌐*
NURSE Madam.
JULIET Nurse.
NURSE
Your lady mother is coming to your chamber.
The day is broke; be wary, look about.                 *Exit*
JULIET
Then, window, let day in, and let life out.                 41
ROMEO
Farewell, farewell! One kiss, and I'll descend.
    ⌐*He lets down the ladder of cords and goes down*⌐
JULIET
Art thou gone so, love, lord, my husband, friend?
I must hear from thee every day in the hour,
For in a minute there are many days.                        45
O, by this count I shall be much in years
Ere I again behold my Romeo.
ROMEO Farewell.
I will omit no opportunity
That may convey my greetings, love, to thee.               50
JULIET
O, think'st thou we shall ever meet again?
ROMEO
I doubt it not, and all these woes shall serve
For sweet discourses in our times to come.
⌐JULIET⌐
O God, I have an ill-divining soul!
Methinks I see thee, now thou art so low,                   55
As one dead in the bottom of a tomb.
Either my eyesight fails, or thou look'st pale.
ROMEO
And trust me, love, in my eye so do you.
Dry sorrow drinks our blood. Adieu, adieu.            *Exit*

JULIET ⌐*pulling up the ladder and weeping*⌐
O fortune, fortune, all men call thee fickle.               60
If thou art fickle, what dost thou with him
That is renowned for faith? Be fickle, fortune,
For then I hope thou wilt not keep him long,
But send him back.
            *Enter Capulet's Wife ⌐below⌐*
CAPULET'S WIFE            Ho, daughter, are you up?
JULIET
Who is't that calls? It is my lady mother.                  65
Is she not down so late, or up so early?
What unaccustomed cause procures her hither?
        ⌐*She goes down and enters below*⌐
CAPULET'S WIFE
Why, how now, Juliet?
JULIET                   Madam, I am not well.
CAPULET'S WIFE
Evermore weeping for your cousin's death?
What, wilt thou wash him from his grave with tears?
An if thou couldst, thou couldst not make him live,        71
Therefore have done. Some grief shows much of love,
But much of grief shows still some want of wit.
JULIET
Yet let me weep for such a feeling loss.
CAPULET'S WIFE
So shall you feel the loss, but not the friend             75
Which you so weep for.
JULIET                   Feeling so the loss,
I cannot choose but ever weep the friend.
CAPULET'S WIFE
Well, girl, thou weep'st not so much for his death
As that the villain lives which slaughtered him.
JULIET
What villain, madam?
CAPULET'S WIFE            That same villain Romeo.           80
JULIET (*aside*)
Villain and he be many miles asunder.
(*To her mother*) God pardon him—I do, with all my
            heart,
And yet no man like he doth grieve my heart.
CAPULET'S WIFE
That is because the traitor murderer lives.
JULIET
Ay, madam, from the reach of these my hands.               85
Would none but I might venge my cousin's death.
CAPULET'S WIFE
We will have vengeance for it, fear thou not.
Then weep no more. I'll send to one in Mantua,
Where that same banished runagate doth live,
Shall give him such an unaccustomed dram                    90
That he shall soon keep Tybalt company;
And then I hope thou wilt be satisfied.
JULIET
Indeed, I never shall be satisfied
With Romeo till I behold him, dead,
Is my poor heart so for a kinsman vexed.                    95
Madam, if you could find out but a man
To bear a poison, I would temper it
That Romeo should, upon receipt thereof,
Soon sleep in quiet. O, how my heart abhors
To hear him named and cannot come to him                  100
To wreak the love I bore my cousin
Upon his body that hath slaughtered him!
CAPULET'S WIFE
Find thou the means, and I'll find such a man.
But now I'll tell thee joyful tidings, girl.

JULIET
And joy comes well in such a needy time.          105
What are they, I beseech your ladyship?
CAPULET'S WIFE
Well, well, thou hast a careful father, child;
One who, to put thee from thy heaviness,
Hath sorted out a sudden day of joy
That thou expect'st not, nor I looked not for.    110
JULIET
Madam, in happy time. What day is that?
CAPULET'S WIFE
Marry, my child, early next Thursday morn
The gallant, young, and noble gentleman
The County Paris at Saint Peter's Church
Shall happily make thee there a joyful bride.     115
JULIET
Now, by Saint Peter's Church, and Peter too,
He shall not make me there a joyful bride.
I wonder at this haste, that I must wed
Ere he that should be husband comes to woo.
I pray you, tell my lord and father, madam,       120
I will not marry yet; and when I do, I swear
It shall be Romeo—whom you know I hate—
Rather than Paris. These are news indeed.
      *Enter Capulet and the Nurse*
CAPULET'S WIFE
Here comes your father. Tell him so yourself,
And see how he will take it at your hands.        125
CAPULET
When the sun sets, the earth doth drizzle dew,
But for the sunset of my brother's son
It rains downright.
How now, a conduit, girl? What, still in tears?
Evermore show'ring? In one little body           130
Thou counterfeit'st a barque, a sea, a wind,
For still thy eyes—which I may call the sea—
Do ebb and flow with tears. The barque thy body is,
Sailing in this salt flood; the winds thy sighs,
Who, raging with thy tears and they with them,   135
Without a sudden calm will overset
Thy tempest-tossèd body.—How now, wife?
Have you delivered to her our decree?
CAPULET'S WIFE
Ay, sir, but she will none, she gives you thanks.
I would the fool were married to her grave.      140
CAPULET
Soft, take me with you, take me with you, wife.
How, will she none? Doth she not give us thanks?
Is she not proud? Doth she not count her blest,
Unworthy as she is, that we have wrought
So worthy a gentleman to be her bride?           145
JULIET
Not proud you have, but thankful that you have.
Proud can I never be of what I hate,
But thankful even for hate that is meant love.
CAPULET
How, how, how, how—chopped logic? What is this?
'Proud', and 'I thank you', and 'I thank you not',  150
And yet 'not proud'? Mistress minion, you,
Thank me no thankings, nor proud me no prouds,
But fettle your fine joints 'gainst Thursday next
To go with Paris to Saint Peter's Church,
Or I will drag thee on a hurdle thither.         155
Out, you green-sickness carrion! Out, you baggage,
You tallow-face!
CAPULET'S WIFE        Fie, fie, what, are you mad?

JULIET *(kneeling)*
Good father, I beseech you on my knees,
Hear me with patience but to speak a word.
CAPULET
Hang thee, young baggage, disobedient wretch!    160
I tell thee what: get thee to church o' Thursday,
Or never after look me in the face.
Speak not, reply not, do not answer me.
      ⌜*Juliet rises*⌝
My fingers itch. Wife, we scarce thought us blest
That God had lent us but this only child,        165
But now I see this one is one too much,
And that we have a curse in having her.
Out on her, hilding!
NURSE                God in heaven bless her!
You are to blame, my lord, to rate her so.
CAPULET
And why, my lady Wisdom? Hold your tongue,       170
Good Prudence. Smatter with your gossips, go!
NURSE
I speak no treason.
⌜CAPULET⌝            O, God-i'-good-e'en!
⌜NURSE⌝
May not one speak?
CAPULET              Peace, you mumbling fool,
Utter your gravity o'er a gossip's bowl,
For here we need it not.
CAPULET'S WIFE       You are too hot.             175
CAPULET
God's bread, it makes me mad. Day, night; work, play;
Alone, in company, still my care hath been
To have her matched; and having now provided
A gentleman of noble parentage,
Of fair demesnes, youthful, and nobly lined,     180
Stuffed, as they say, with honourable parts,
Proportioned as one's thought would wish a man—
And then to have a wretched puling fool,
A whining maumet, in her fortune's tender,
To answer 'I'll not wed, I cannot love;          185
I am too young, I pray you pardon me'!
But an you will not wed, I'll pardon you!
Graze where you will, you shall not house with me.
Look to't, think on't. I do not use to jest.
Thursday is near. Lay hand on heart. Advise.     190
An you be mine, I'll give you to my friend.
An you be not, hang, beg, starve, die in the streets,
For, by my soul, I'll ne'er acknowledge thee,
Nor what is mine shall never do thee good.
Trust to't. Bethink you. I'll not be forsworn.  *Exit*
JULIET
Is there no pity sitting in the clouds            196
That sees into the bottom of my grief?
O sweet my mother, cast me not away!
Delay this marriage for a month, a week;
Or if you do not, make the bridal bed            200
In that dim monument where Tybalt lies.
CAPULET'S WIFE
Talk not to me, for I'll not speak a word.
Do as thou wilt, for I have done with thee.      *Exit*
JULIET
O, God—O Nurse, how shall this be prevented?
My husband is on earth, my faith in heaven.       205
How shall that faith return again to earth
Unless that husband send it me from heaven
By leaving earth? Comfort me, counsel me.
Alack, alack, that heaven should practise stratagems

Upon so soft a subject as myself!     210
What sayst thou? Hast thou not a word of joy?
Some comfort, Nurse.
NURSE           Faith, here it is: Romeo
Is banishèd, and all the world to nothing
That he dares ne'er come back to challenge you,
Or if he do, it needs must be by stealth.    215
Then, since the case so stands as now it doth,
I think it best you married with the County.
O, he's a lovely gentleman!
Romeo's a dishclout to him. An eagle, madam,
Hath not so green, so quick, so fair an eye    220
As Paris hath. Beshrew my very heart,
I think you are happy in this second match,
For it excels your first; or if it did not,
Your first is dead, or 'twere as good he were
As living hence and you no use of him.    225
JULIET Speak'st thou from thy heart?
NURSE
And from my soul, too, else beshrew them both.
JULIET Amen.
NURSE What?
JULIET
Well, thou hast comforted me marvellous much.    230
Go in; and tell my lady I am gone,
Having displeased my father, to Laurence' cell
To make confession and to be absolved.
NURSE
Marry, I will; and this is wisely done.    ⌜Exit⌝
JULIET (watching her go)
Ancient damnation! O most wicked fiend!    235
Is it more sin to wish me thus forsworn,
Or to dispraise my lord with that same tongue
Which she hath praised him with above compare
So many thousand times? Go, counsellor!
Thou and my bosom henceforth shall be twain.    240
I'll to the friar, to know his remedy.
If all else fail, myself have power to die.    Exit

**4.1**    *Enter Friar Laurence and Paris*
FRIAR LAURENCE
On Thursday, sir? The time is very short.
PARIS
My father Capulet will have it so,
And I am nothing slow to slack his haste.
FRIAR LAURENCE
You say you do not know the lady's mind?
Uneven is the course. I like it not.    5
PARIS
Immoderately she weeps for Tybalt's death,
And therefore have I little talked of love,
For Venus smiles not in a house of tears.
Now, sir, her father counts it dangerous
That she do give her sorrow so much sway,    10
And in his wisdom hastes our marriage
To stop the inundation of her tears,
Which, too much minded by herself alone,
May be put from her by society.
Now do you know the reason of this haste.    15
FRIAR LAURENCE (aside)
I would I knew not why it should be slowed.—
    *Enter Juliet*
Look, sir, here comes the lady toward my cell.
PARIS
Happily met, my lady and my wife.

JULIET
That may be, sir, when I may be a wife.
PARIS
That 'may be' must be, love, on Thursday next.    20
JULIET
What must be shall be.
FRIAR LAURENCE        That's a certain text.
PARIS
Come you to make confession to this father?
JULIET
To answer that, I should confess to you.
PARIS
Do not deny to him that you love me.
JULIET
I will confess to you that I love him.    25
PARIS
So will ye, I am sure, that you love me.
JULIET
If I do so, it will be of more price,
Being spoke behind your back, than to your face.
PARIS
Poor soul, thy face is much abused with tears.
JULIET
The tears have got small victory by that,    30
For it was bad enough before their spite.
PARIS
Thou wrong'st it more than tears with that report.
JULIET
That is no slander, sir, which is a truth,
And what I spake, I spake it to my face.
PARIS
Thy face is mine, and thou hast slandered it.    35
JULIET
It may be so, for it is not mine own.—
Are you at leisure, holy father, now,
Or shall I come to you at evening mass?
FRIAR LAURENCE
My leisure serves me, pensive daughter, now.
My lord, we must entreat the time alone.    40
PARIS
God shield I should disturb devotion!—
Juliet, on Thursday early will I rouse ye.
(Kissing her) Till then, adieu, and keep this holy kiss.    Exit
JULIET
O, shut the door, and when thou hast done so,
Come weep with me, past hope, past cure, past help!
FRIAR LAURENCE
O Juliet, I already know thy grief.    46
It strains me past the compass of my wits.
I hear thou must, and nothing may prorogue it,
On Thursday next be married to this County.
JULIET
Tell me not, friar, that thou hear'st of this,    50
Unless thou tell me how I may prevent it.
If in thy wisdom thou canst give no help,
Do thou but call my resolution wise,
    *She draws a knife*
And with this knife I'll help it presently.
God joined my heart and Romeo's, thou our hands,    55
And ere this hand, by thee to Romeo's sealed,
Shall be the label to another deed,
Or my true heart with treacherous revolt
Turn to another, this shall slay them both.
Therefore, out of thy long-experienced time,    60
Give me some present counsel; or, behold,

'Twixt my extremes and me this bloody knife
Shall play the umpire, arbitrating that
Which the commission of thy years and art
Could to no issue of true honour bring.                    65
Be not so long to speak. I long to die
If what thou speak'st speak not of remedy.
FRIAR LAURENCE
Hold, daughter, I do spy a kind of hope
Which craves as desperate an execution
As that is desperate which we would prevent.              70
If, rather than to marry County Paris,
Thou hast the strength of will to slay thyself,
Then is it likely thou wilt undertake
A thing like death to chide away this shame,
That cop'st with death himself to scape from it;          75
And, if thou dar'st, I'll give thee remedy.
JULIET
O, bid me leap, rather than marry Paris,
From off the battlements of any tower,
Or walk in thievish ways, or bid me lurk
Where serpents are. Chain me with roaring bears,          80
Or hide me nightly in a charnel house,
O'ercovered quite with dead men's rattling bones,
With reeky shanks and yellow chapless skulls;
Or bid me go into a new-made grave
And hide me with a dead man in his tomb—                  85
Things that, to hear them told, have made me
        tremble—
And I will do it without fear or doubt,
To live an unstained wife to my sweet love.
FRIAR LAURENCE
Hold, then; go home, be merry, give consent
To marry Paris. Wednesday is tomorrow.                    90
Tomorrow night look that thou lie alone.
Let not the Nurse lie with thee in thy chamber.
Take thou this vial, being then in bed,
And this distilling liquor drink thou off,
When presently through all thy veins shall run           95
A cold and drowsy humour; for no pulse
Shall keep his native progress, but surcease.
No warmth, no breath shall testify thou livest.
The roses in thy lips and cheeks shall fade
To wanny ashes, thy eyes' windows fall                    100
Like death when he shuts up the day of life.
Each part, deprived of supple government,
Shall, stiff and stark and cold, appear like death;
And in this borrowed likeness of shrunk death
Thou shalt continue two-and-forty hours,                  105
And then awake as from a pleasant sleep.
Now, when the bridegroom in the morning comes
To rouse thee from thy bed, there art thou dead.
Then, as the manner of our country is,
In thy best robes, uncovered on the bier                  110
Thou shalt be borne to that same ancient vault
Where all the kindred of the Capulets lie.
In the meantime, against thou shalt awake,
Shall Romeo by my letters know our drift,
And hither shall he come, and he and I                    115
Will watch thy waking, and that very night
Shall Romeo bear thee hence to Mantua.
And this shall free thee from this present shame,
If no inconstant toy nor womanish fear
Abate thy valour in the acting it.                        120
JULIET
Give me, give me! O, tell not me of fear!

FRIAR LAURENCE (giving her the vial)
Hold, get you gone. Be strong and prosperous
In this resolve. I'll send a friar with speed
To Mantua with my letters to thy lord.
JULIET
Love give me strength, and strength shall help afford.
Farewell, dear father.              Exeunt ⌈severally⌉

**4.2**    Enter Capulet, his Wife, the Nurse, and ⌈two⌉
        Servingmen
CAPULET (giving a Servingman a paper)
So many guests invite as here are writ.
                                    ⌈Exit Servingman⌉
(To the other Servingman) Sirrah, go hire me twenty
        cunning cooks.
SERVINGMAN You shall have none ill, sir, for I'll try if they
    can lick their fingers.
CAPULET How canst thou try them so?               5
SERVINGMAN Marry, sir, 'tis an ill cook that cannot lick his
    own fingers, therefore he that cannot lick his fingers
    goes not with me.
CAPULET Go, be gone.              ⌈Exit Servingman⌉
We shall be much unfurnished for this time.       10
(To the Nurse) What, is my daughter gone to Friar
        Laurence?
NURSE Ay, forsooth.
CAPULET
Well, he may chance to do some good on her.
A peevish, self-willed harlotry it is.
        Enter Juliet
NURSE
See where she comes from shrift with merry look.  15
CAPULET (to Juliet)
How now, my headstrong, where have you been
        gadding?
JULIET
Where I have learned me to repent the sin
Of disobedient opposition
To you and your behests, and am enjoined
By holy Laurence to fall prostrate here           20
To beg your pardon. (Kneeling) Pardon, I beseech you.
Henceforward I am ever ruled by you.
CAPULET ⌈to the Nurse⌉
Send for the County; go tell him of this.
I'll have this knot knit up tomorrow morning.
JULIET
I met the youthful lord at Laurence' cell,        25
And gave him what becoming love I might,
Not stepping o'er the bounds of modesty.
CAPULET
Why, I am glad on't. This is well. Stand up.
        Juliet rises
This is as't should be. Let me see the County.
⌈To Nurse⌉ Ay, marry, go, I say, and fetch him hither.
Now, afore God, this reverend holy friar,         31
All our whole city is much bound to him.
JULIET
Nurse, will you go with me into my closet
To help me sort such needful ornaments
As you think fit to furnish me tomorrow?          35
CAPULET'S WIFE
No, not till Thursday. There is time enough.
CAPULET
Go, Nurse, go with her. We'll to church tomorrow.
                        Exeunt Juliet and Nurse

**CAPULET'S WIFE**
We shall be short in our provision.
'Tis now near night.
**CAPULET**                    Tush, I will stir about,
And all things shall be well, I warrant thee, wife.                    40
Go thou to Juliet, help to deck up her.
I'll not to bed tonight. Let me alone.
I'll play the housewife for this once. What, ho!
They are all forth. Well, I will walk myself
To County Paris to prepare up him                    45
Against tomorrow. My heart is wondrous light,
Since this same wayward girl is so reclaimed.
                    *Exeunt ⌈severally⌉*

**4.3**    *Enter Juliet and the Nurse ⌈with garments⌉*
**JULIET**
Ay, those attires are best. But, gentle Nurse,
I pray thee leave me to myself tonight,
For I have need of many orisons
To move the heavens to smile upon my state,
Which—well thou knowest—is cross and full of sin.    5
                    *Enter Capulet's Wife*
**CAPULET'S WIFE**
What, are you busy, ho? Need you my help?
**JULIET**
No, madam, we have culled such necessaries
As are behoveful for our state tomorrow.
So please you, let me now be left alone,
And let the Nurse this night sit up with you,    10
For I am sure you have your hands full all
In this so sudden business.
**CAPULET'S WIFE**                    Good night.
Get thee to bed, and rest, for thou hast need.
                    *Exeunt Capulet's Wife ⌈and Nurse⌉*
**JULIET**
Farewell. God knows when we shall meet again.
I have a faint cold fear thrills through my veins    15
That almost freezes up the heat of life.
I'll call them back again to comfort me.
Nurse!—What should she do here?
⌈*She opens curtains, behind which is seen her bed*⌉
My dismal scene I needs must act alone.
Come, vial. What if this mixture do not work at all?    20
Shall I be married then tomorrow morning?
No, no, this shall forbid it. Lie thou there.
                    *She lays down a knife*
What if it be a poison which the friar
Subtly hath ministered to have me dead,
Lest in this marriage he should be dishonoured    25
Because he married me before to Romeo?
I fear it is—and yet methinks it should not,
For he hath still been tried a holy man.
How if, when I am laid into the tomb,
I wake before the time that Romeo    30
Come to redeem me? There's a fearful point.
Shall I not then be stifled in the vault,
To whose foul mouth no healthsome air breathes in,
And there die strangled ere my Romeo comes?
Or, if I live, is it not very like    35
The horrible conceit of death and night,
Together with the terror of the place—
As in a vault, an ancient receptacle
Where for this many hundred years the bones
Of all my buried ancestors are packed;    40
Where bloody Tybalt, yet but green in earth,
Lies fest'ring in his shroud; where, as they say,
At some hours in the night spirits resort—

Alack, alack, is it not like that I,
So early waking—what with loathsome smells,    45
And shrieks like mandrakes torn out of the earth,
That living mortals, hearing them, run mad—
O, if I wake, shall I not be distraught,
Environèd with all these hideous fears,
And madly play with my forefathers' joints,    50
And pluck the mangled Tybalt from his shroud,
And, in this rage, with some great kinsman's bone
As with a club dash out my desp'rate brains?
O, look! Methinks I see my cousin's ghost
Seeking out Romeo that did spit his body    55
Upon a rapier's point. Stay, Tybalt, stay!
Romeo, Romeo, Romeo! Here's drink. I drink to thee.
                    *She drinks from the vial and falls upon the bed,*
                    ⌈*pulling closed the curtains*⌉

**4.4**    *Enter Capulet's Wife, and the Nurse ⌈with herbs⌉*
**CAPULET'S WIFE**
Hold, take these keys, and fetch more spices, Nurse.
**NURSE**
They call for dates and quinces in the pastry.
                    *Enter Capulet*
**CAPULET**
Come, stir, stir, stir! The second cock hath crowed.
The curfew bell hath rung. 'Tis three o'clock.
Look to the baked meats, good Angelica.    5
Spare not for cost.
**NURSE**                    Go, you cot-quean, go.
Get you to bed. Faith, you'll be sick tomorrow
For this night's watching.
**CAPULET**
No, not a whit. What, I have watched ere now
All night for lesser cause, and ne'er been sick.    10
**CAPULET'S WIFE**
Ay, you have been a mouse-hunt in your time,
But I will watch you from such watching now.
                    *Exeunt Capulet's Wife and Nurse*
**CAPULET**
A jealous-hood, a jealous-hood!
                    *Enter three or four Servingmen, with spits and
                    logs and baskets*
                    Now, fellow, what is there?
**FIRST SERVINGMAN**
Things for the cook, sir, but I know not what.
**CAPULET**
Make haste, make haste.
                    *Exit First Servingman ⌈and one or two others⌉*
                    Sirrah, fetch drier logs.
Call Peter. He will show thee where they are.    16
**SECOND SERVINGMAN**
I have a head, sir, that will find out logs
And never trouble Peter for the matter.
**CAPULET**
Mass, and well said! A merry whoreson, ha!
Thou shalt be loggerhead.    *Exit Second Servingman*
                    Good faith, 'tis day.
The County will be here with music straight,    21
For so he said he would.
                    *Music plays within*
                    I hear him near.
Nurse! Wife! What ho, what, Nurse, I say!
                    *Enter the Nurse*
Go waken Juliet. Go and trim her up.
I'll go and chat with Paris. Hie, make haste,    25
Make haste, the bridegroom he is come already.
Make haste, I say.                    *Exit*

NURSE
  Mistress, what, mistress! Juliet! Fast, I warrant her, she.
  Why, lamb, why, lady! Fie, you slug-abed!
  Why, love, I say, madam, sweetheart, why, bride!     30
  What, not a word? You take your pennyworths now.
  Sleep for a week, for the next night, I warrant,
  The County Paris hath set up his rest
  That you shall rest but little. God forgive me!
  Marry, and amen. How sound is she asleep!     35
  I needs must wake her. Madam, madam, madam!
  Ay, let the County take you in your bed.
  He'll fright you up, i'faith. Will it not be?
        ⌜She draws back the curtains⌝
  What, dressed and in your clothes, and down again?
  I must needs wake you. Lady, lady, lady!     40
  Alas, alas! Help, help! My lady's dead.
  O welladay, that ever I was born!
  Some aqua-vitae, ho! My lord, my lady!
        Enter Capulet's Wife
CAPULET'S WIFE
  What noise is here?
NURSE                          O lamentable day!
CAPULET'S WIFE
  What is the matter?
NURSE                        Look, look. O heavy day!     45
CAPULET'S WIFE
  O me, O me, my child, my only life!
  Revive, look up, or I will die with thee.
  Help, help, call help!
        Enter Capulet
CAPULET
  For shame, bring Juliet forth. Her lord is come.
NURSE
  She's dead, deceased. She's dead, alack the day!     50
CAPULET'S WIFE
  Alack the day, she's dead, she's dead, she's dead!
CAPULET
  Ha, let me see her! Out, alas, she's cold.
  Her blood is settled, and her joints are stiff.
  Life and these lips have long been separated.
  Death lies on her like an untimely frost     55
  Upon the sweetest flower of all the field.
NURSE
  O lamentable day!
CAPULET'S WIFE         O woeful time!
CAPULET
  Death, that hath ta'en her hence to make me wail,
  Ties up my tongue, and will not let me speak.
        Enter Friar Laurence and Paris, with Musicians
FRIAR LAURENCE
  Come, is the bride ready to go to church?     60
CAPULET
  Ready to go, but never to return.
  (To Paris) O son, the night before thy wedding day
  Hath death lain with thy wife. See, there she lies,
  Flower as she was, deflowerèd by him.
  Death is my son-in-law, death is my heir.     65
  My daughter he hath wedded. I will die,
  And leave him all. Life, living, all is death's.
        ⌜Paris, Capulet and his Wife, and the Nurse all at
        once wring their hands and cry out together:⌝
PARIS
  Have I thought long to see this morning's face,
  And doth it give me such a sight as this?
  Beguiled, divorcèd, wrongèd, spited, slain!     70
  Most detestable death, by thee beguiled,

By cruel, cruel thee quite overthrown.
  O love, O life: not life, but love in death.
CAPULET'S WIFE
  Accursed, unhappy, wretched, hateful day!
  Most miserable hour that e'er time saw     75
  In lasting labour of his pilgrimage!
  But one, poor one, one poor and loving child,
  But one thing to rejoice and solace in,
  And cruel death hath catched it from my sight!
NURSE
  O woe! O woeful, woeful, woeful day!     80
  Most lamentable day! Most woeful day
  That ever, ever, I did yet behold!
  O day, O day, O day, O hateful day,
  Never was seen so black a day as this!
  O woeful day, O woeful day!     85
CAPULET
  Despised, distressèd, hated, martyred, killed!
  Uncomfortable time, why cam'st thou now
  To murder, murder our solemnity?
  O child, O child, my soul and not my child!
  Dead art thou, alack, my child is dead,     90
  And with my child my joys are burièd.
FRIAR LAURENCE
  Peace, ho, for shame! Confusion's cure lives not
  In these confusions. Heaven and yourself
  Had part in this fair maid. Now heaven hath all,
  And all the better is it for the maid.     95
  Your part in her you could not keep from death,
  But heaven keeps his part in eternal life.
  The most you sought was her promotion,
  For 'twas your heaven she should be advanced,
  And weep ye now, seeing she is advanced     100
  Above the clouds as high as heaven itself?
  O, in this love you love your child so ill
  That you run mad, seeing that she is well.
  She's not well married that lives married long,
  But she's best married that dies married young.     105
  Dry up your tears, and stick your rosemary
  On this fair corpse, and, as the custom is,
  All in her best array bear her to church;
  For though fond nature bids us all lament,
  Yet nature's tears are reason's merriment.     110
CAPULET
  All things that we ordainèd festival
  Turn from their office to black funeral.
  Our instruments to melancholy bells,
  Our wedding cheer to a sad burial feast,
  Our solemn hymns to sullen dirges change;     115
  Our bridal flowers serve for a buried corpse,
  And all things change them to the contrary.
FRIAR LAURENCE
  Sir, go you in; and madam, go with him,
  And go, Sir Paris. Everyone prepare
  To follow this fair corpse unto her grave.     120
  The heavens do lour upon you for some ill.
  Move them no more by crossing their high will.
        ⌜They cast rosemary on Juliet, and shut the curtains.⌝
        Exeunt all but the Nurse and Musicians
⌜FIRST⌝ MUSICIAN Faith, we may put up our pipes and be
  gone.
NURSE
  Honest good fellows, ah, put up, put up,     125
  For well you know this is a pitiful case.
⌜FIRST⌝ MUSICIAN
  Ay, by my troth, the case may be amended.
        Exit Nurse

*Enter Peter*

PETER Musicians, O, musicians! 'Heart's ease', 'Heart's
ease'; O, an you will have me live, play 'Heart's ease'.

⌈FIRST⌉ MUSICIAN Why 'Heart's ease'?                    130

PETER O, musicians, because my heart itself plays 'My heart
is full of woe'. O, play me some merry dump to comfort
me.

⌈FIRST⌉ MUSICIAN Not a dump, we. 'Tis no time to play
now.                                                    135

PETER You will not then?

FIRST MUSICIAN No.

PETER I will then give it you soundly.

FIRST MUSICIAN What will you give us?

PETER No money, on my faith, but the gleek. I will give
you the minstrel.                                        141

FIRST MUSICIAN Then will I give you the serving-creature.

PETER (*drawing his dagger*) Then will I lay the serving-
creature's dagger on your pate. I will carry no crochets.
I'll re you, I'll fa you. Do you note me?                145

FIRST MUSICIAN An you re us and fa us, you note us.

SECOND MUSICIAN Pray you, put up your dagger and put
out your wit.

⌈PETER⌉ Then have at you with my wit. I will dry-beat you
with an iron wit, and put up my iron dagger. Answer
me like men.                                             151

⌈*Sings*⌉

> When griping grief the heart doth wound,
>     And doleful dumps the mind oppress,
>     Then music with her silver sound—

Why 'silver sound', why 'music with her silver sound'?
What say you, Matthew Minikin?                           156

FIRST MUSICIAN Marry, sir, because silver hath a sweet
sound.

PETER Prates! What say you, Hugh Rebec?

SECOND MUSICIAN I say 'silver sound' because musicians
sound for silver.                                        161

PETER Prates too! What say you, Simon Soundpost?

THIRD MUSICIAN Faith, I know not what to say.

PETER O, I cry you mercy, you are the singer. I will say
for you. It is 'music with her silver sound' because
musicians have no gold for sounding.                     166

⌈*Sings*⌉

> Then music with her silver sound
>     With speedy help doth lend redress.        *Exit*

FIRST MUSICIAN What a pestilent knave is this same!

SECOND MUSICIAN Hang him, jack! Come, we'll in here,
tarry for the mourners, and stay dinner.        *Exeunt*

**5.1**    *Enter Romeo*

ROMEO

If I may trust the flattering truth of sleep,
My dreams presage some joyful news at hand.
My bosom's lord sits lightly in his throne,
And all this day an unaccustomed spirit
Lifts me above the ground with cheerful thoughts.     5
I dreamt my lady came and found me dead—
Strange dream, that gives a dead man leave to
    think!—
And breathed such life with kisses in my lips
That I revived and was an emperor.
Ah me, how sweet is love itself possessed            10
When but love's shadows are so rich in joy!

*Enter Balthasar, Romeo's man, ⌈booted⌉*

News from Verona! How now, Balthasar?
Dost thou not bring me letters from the friar?
How doth my lady? Is my father well?
How fares my Juliet? That I ask again,               15
For nothing can be ill if she be well.

BALTHASAR

Then she is well, and nothing can be ill.
Her body sleeps in Capel's monument,
And her immortal part with angels lives.
I saw her laid low in her kindred's vault,           20
And presently took post to tell it you.
O, pardon me for bringing these ill news,
Since you did leave it for my office, sir.

ROMEO

Is it e'en so? Then I defy you, stars.
Thou knowest my lodging. Get me ink and paper,       25
And hire posthorses. I will hence tonight.

BALTHASAR

I do beseech you, sir, have patience.
Your looks are pale and wild, and do import
Some misadventure.

ROMEO                    Tush, thou art deceived.
Leave me, and do the thing I bid thee do.            30
Hast thou no letters to me from the friar?

BALTHASAR

No, my good lord.

ROMEO                No matter. Get thee gone,
And hire those horses. I'll be with thee straight.

*Exit Balthasar*

Well, Juliet, I will lie with thee tonight.
Let's see for means. O mischief, thou art swift      35
To enter in the thoughts of desperate men!
I do remember an apothecary,
And hereabouts a dwells, which late I noted,
In tattered weeds, with overwhelming brows,
Culling of simples. Meagre were his looks.           40
Sharp misery had worn him to the bones,
And in his needy shop a tortoise hung,
An alligator stuffed, and other skins
Of ill-shaped fishes; and about his shelves
A beggarly account of empty boxes,                   45
Green earthen pots, bladders, and musty seeds,
Remnants of packthread, and old cakes of roses
Were thinly scattered to make up a show.
Noting this penury, to myself I said
'An if a man did need a poison now,                  50
Whose sale is present death in Mantua,
Here lives a caitiff wretch would sell it him.'
O, this same thought did but forerun my need,
And this same needy man must sell it me.
As I remember, this should be the house.             55
Being holiday, the beggar's shop is shut.
What ho, apothecary!

*Enter Apothecary*

APOTHECARY                Who calls so loud?

ROMEO

Come hither, man. I see that thou art poor.

*He offers money*

Hold, there is forty ducats. Let me have
A dram of poison—such soon-speeding gear            60
As will disperse itself through all the veins,
That the life-weary taker may fall dead,
And that the trunk may be discharged of breath
As violently as hasty powder fired
Doth hurry from the fatal cannon's womb.             65

APOTHECARY

Such mortal drugs I have, but Mantua's law
Is death to any he that utters them.

ROMEO
  Art thou so bare and full of wretchedness,
  And fear'st to die? Famine is in thy cheeks,
  Need and oppression starveth in thy eyes,                70
  Contempt and beggary hangs upon thy back.
  The world is not thy friend, nor the world's law.
  The world affords no law to make thee rich.
  Then be not poor, but break it, and take this.
APOTHECARY
  My poverty but not my will consents.                     75
ROMEO
  I pay thy poverty and not thy will.
APOTHECARY (handing Romeo poison)
  Put this in any liquid thing you will
  And drink it off, and if you had the strength
  Of twenty men it would dispatch you straight.
ROMEO (giving money)
  There is thy gold—worse poison to men's souls,           80
  Doing more murder in this loathsome world,
  Than these poor compounds that thou mayst not sell.
  I sell thee poison; thou hast sold me none.
  Farewell, buy food, and get thyself in flesh.
                              ⌈Exit Apothecary⌉
  Come, cordial and not poison, go with me                 85
  To Juliet's grave, for there must I use thee.       Exit

5.2    Enter Friar John at one door
FRIAR JOHN
  Holy Franciscan friar, brother, ho!
      Enter Friar Laurence at another door
FRIAR LAURENCE
  This same should be the voice of Friar John.
  Welcome from Mantua! What says Romeo?
  Or if his mind be writ, give me his letter.
FRIAR JOHN
  Going to find a barefoot brother out—                    5
  One of our order—to associate me
  Here in this city visiting the sick,
  And finding him, the searchers of the town,
  Suspecting that we both were in a house
  Where the infectious pestilence did reign,               10
  Sealed up the doors, and would not let us forth,
  So that my speed to Mantua there was stayed.
FRIAR LAURENCE
  Who bare my letter then to Romeo?
FRIAR JOHN
  I could not send it—here it is again—
  Nor get a messenger to bring it thee,                    15
  So fearful were they of infection.
FRIAR LAURENCE
  Unhappy fortune! By my brotherhood,
  The letter was not nice, but full of charge,
  Of dear import, and the neglecting it
  May do much danger. Friar John, go hence.               20
  Get me an iron crow, and bring it straight
  Unto my cell.
FRIAR JOHN        Brother, I'll go and bring it thee.  Exit
FRIAR LAURENCE
  Now must I to the monument alone.
  Within this three hours will fair Juliet wake.
  She will beshrew me much that Romeo                      25
  Hath had no notice of these accidents.
  But I will write again to Mantua,
  And keep her at my cell till Romeo come.
  Poor living corpse, closed in a dead man's tomb!    Exit

5.3    Enter Paris and his Page, with flowers, sweet water,
       and a torch
PARIS
  Give me thy torch, boy. Hence, and stand aloof.
  Yet put it out, for I would not be seen.
          ⌈His Page puts out the torch⌉
  Under yon yew trees lay thee all along,
  Holding thy ear close to the hollow ground.
  So shall no foot upon the churchyard tread,              5
  Being loose, unfirm, with digging up of graves,
  But thou shalt hear it. Whistle then to me
  As signal that thou hear'st something approach.
  Give me those flowers. Do as I bid thee. Go.
PAGE ⌈aside⌉
  I am almost afraid to stand alone                        10
  Here in the churchyard, yet I will adventure.
          He hides himself at a distance from Paris
PARIS (strewing flowers)
  Sweet flower, with flowers thy bridal bed I strew.
          He sprinkles water
    O woe! Thy canopy is dust and stones,
  Which with sweet water nightly I will dew,
    Or, wanting that, with tears distilled by moans.       15
  The obsequies that I for thee will keep
  Nightly shall be to strew thy grave and weep.
          The Page whistles
  The boy gives warning. Something doth approach.
  What cursèd foot wanders this way tonight
  To cross my obsequies and true love's rite?             20
          Enter Romeo and ⌈Balthasar⌉ his man, with a torch,
          a mattock, and a crow of iron
  What, with a torch? Muffle me, night, a while.
          He stands aside
ROMEO
  Give me that mattock and the wrenching iron.
  Hold, take this letter. Early in the morning
  See thou deliver it to my lord and father.
  Give me the light. Upon thy life I charge thee,         25
  Whate'er thou hear'st or seest, stand all aloof,
  And do not interrupt me in my course.
  Why I descend into this bed of death
  Is partly to behold my lady's face,
  But chiefly to take thence from her dead finger         30
  A precious ring, a ring that I must use
  In dear employment. Therefore hence, be gone.
  But if thou, jealous, dost return to pry
  In what I farther shall intend to do,
  By heaven, I will tear thee joint by joint,             35
  And strew this hungry churchyard with thy limbs.
  The time and my intents are savage-wild,
  More fierce and more inexorable far
  Than empty tigers or the roaring sea.
⌈BALTHASAR⌉
  I will be gone, sir, and not trouble ye.                40
ROMEO
  So shalt thou show me friendship. Take thou that.
          He gives money
  Live and be prosperous, and farewell, good fellow.
⌈BALTHASAR⌉ (aside)
  For all this same, I'll hide me hereabout.
  His looks I fear, and his intents I doubt.
          He hides himself at a distance from Romeo.
          ⌈Romeo begins to force open the tomb⌉
ROMEO
  Thou detestable maw, thou womb of death,                45
  Gorged with the dearest morsel of the earth,

Thus I enforce thy rotten jaws to open,
And in despite I'll cram thee with more food.

PARIS (aside)
This is that banished haughty Montague
That murdered my love's cousin, with which grief   50
It is supposèd the fair creature died;
And here is come to do some villainous shame
To the dead bodies. I will apprehend him.
⌜Drawing⌝ Stop thy unhallowed toil, vile Montague!
Can vengeance be pursued further than death?   55
Condemnèd villain, I do apprehend thee.
Obey and go with me, for thou must die.

ROMEO
I must indeed, and therefore came I hither.
Good gentle youth, tempt not a desp'rate man.
Fly hence, and leave me. Think upon these gone.   60
Let them affright thee. I beseech thee, youth,
Put not another sin upon my head
By urging me to fury. O, be gone.
By heaven, I love thee better than myself,
For I come hither armed against myself.   65
Stay not, be gone. Live, and hereafter say
A madman's mercy bid thee run away.

PARIS
I do defy thy conjuration,
And apprehend thee for a felon here.

ROMEO (drawing)
Wilt thou provoke me? Then have at thee, boy.   70
    They fight

⌜PAGE⌝
O Lord, they fight! I will go call the watch.   Exit

PARIS
O, I am slain! If thou be merciful,
Open the tomb, lay me with Juliet.

ROMEO
In faith, I will.                   Paris dies
        Let me peruse this face.
Mercutio's kinsman, noble County Paris!   75
What said my man when my betossèd soul
Did not attend him as we rode? I think
He told me Paris should have married Juliet.
Said he not so? Or did I dream it so?
Or am I mad, hearing him talk of Juliet,   80
To think it was so? O, give me thy hand,
One writ with me in sour misfortune's book.
I'll bury thee in a triumphant grave.
    ⌜He opens the tomb, revealing Juliet⌝
A grave—O no, a lantern, slaughtered youth,
For here lies Juliet, and her beauty makes   85
This vault a feasting presence full of light.
    ⌜He bears the body of Paris to the tomb⌝
Death, lie thou there, by a dead man interred.
How oft, when men are at the point of death,
Have they been merry, which their keepers call
A lightning before death! O, how may I   90
Call this a lightning? O my love, my wife!
Death, that hath sucked the honey of thy breath,
Hath had no power yet upon thy beauty.
Thou art not conquered. Beauty's ensign yet
Is crimson in thy lips and in thy cheeks,   95
And death's pale flag is not advancèd there.
Tybalt, liest thou there in thy bloody sheet?
O, what more favour can I do to thee
Than with that hand that cut thy youth in twain
To sunder his that was thine enemy?   100
Forgive me, cousin. Ah, dear Juliet,

Why art thou yet so fair? Shall I believe
That unsubstantial death is amorous,
And that the lean abhorrèd monster keeps
Thee here in dark to be his paramour?   105
For fear of that I still will stay with thee,
And never from this pallet of dim night
Depart again. Here, here will I remain
With worms that are thy chambermaids. O, here   110
Will I set up my everlasting rest,
And shake the yoke of inauspicious stars
From this world-wearied flesh. Eyes, look your last.
Arms, take your last embrace, and lips, O you
The doors of breath, seal with a righteous kiss
A dateless bargain to engrossing death.   115
    ⌜He kisses Juliet, then pours poison into the cup⌝
Come, bitter conduct, come, unsavoury guide,
Thou desperate pilot, now at once run on
The dashing rocks thy seasick weary barque!
Here's to my love.
    He drinks the poison
               O true apothecary,
Thy drugs are quick! Thus with a kiss I die.   120
    He kisses Juliet, falls, and dies.
    Enter Friar Laurence with lantern, crow, and spade

FRIAR LAURENCE
Saint Francis be my speed! How oft tonight
Have my old feet stumbled at graves? Who's there?

BALTHASAR
Here's one, a friend, and one that knows you well.

FRIAR LAURENCE
Bliss be upon you. Tell me, good my friend,
What torch is yon that vainly lends his light   125
To grubs and eyeless skulls? As I discern,
It burneth in the Capels' monument.

BALTHASAR
It doth so, holy sir, and there's my master,
One that you love.

FRIAR LAURENCE       Who is it?

BALTHASAR               Romeo.

FRIAR LAURENCE
How long hath he been there?

BALTHASAR            Full half an hour.   130

FRIAR LAURENCE
Go with me to the vault.

BALTHASAR           I dare not, sir.
My master knows not but I am gone hence,
And fearfully did menace me with death
If I did stay to look on his intents.

FRIAR LAURENCE
Stay then, I'll go alone. Fear comes upon me.   135
O, much I fear some ill unthrifty thing.

BALTHASAR
As I did sleep under this yew tree here
I dreamt my master and another fought,
And that my master slew him.

FRIAR LAURENCE          Romeo!
    He ⌜stoops and⌝ looks on the blood and weapons
Alack, alack, what blood is this which stains   140
The stony entrance of this sepulchre?
What mean these masterless and gory swords
To lie discoloured by this place of peace?
Romeo! O, pale! Who else? What, Paris, too,
And steeped in blood? Ah, what an unkind hour   145
Is guilty of this lamentable chance!
    Juliet awakes ⌜and rises⌝
The lady stirs.

JULIET
O comfortable friar, where is my lord?
I do remember well where I should be,
And there I am. Where is my Romeo?                            150
FRIAR LAURENCE
I hear some noise. Lady, come from that nest
Of death, contagion, and unnatural sleep.
A greater power than we can contradict
Hath thwarted our intents. Come, come away.
Thy husband in thy bosom there lies dead,                      155
And Paris, too. Come, I'll dispose of thee
Among a sisterhood of holy nuns.
Stay not to question, for the watch is coming.
Come, go, good Juliet. I dare no longer stay.            *Exit*
JULIET
Go, get thee hence, for I will not away.                       160
What's here? A cup closed in my true love's hand?
Poison, I see, hath been his timeless end.
O churl!—drunk all, and left no friendly drop
To help me after? I will kiss thy lips.
Haply some poison yet doth hang on them,                       165
To make me die with a restorative.
        *She kisses Romeo's lips*
Thy lips are warm.
CHIEF WATCHMAN ⌈*within*⌉ Lead, boy. Which way?
JULIET
Yea, noise? Then I'll be brief.
        *She takes Romeo's dagger*
                                O happy dagger,
This is thy sheath! There rust, and let me die.
        *She stabs herself, falls, and dies.*
        *Enter the Page and Watchmen*
⌈PAGE⌉
This is the place, there where the torch doth burn.    170
CHIEF WATCHMAN
The ground is bloody. Search about the churchyard.
Go, some of you. Whoe'er you find, attach.
                        *Exeunt some Watchmen*
Pitiful sight! Here lies the County slain,
And Juliet bleeding, warm, and newly dead,
Who here hath lain this two days burièd.               175
Go tell the Prince. Run to the Capulets,
Raise up the Montagues. Some others search.
                *Exeunt other Watchmen* ⌈*severally*⌉
We see the ground whereon these woes do lie,
But the true ground of all these piteous woes
We cannot without circumstance descry.                 180
        *Enter* ⌈*Watchmen*⌉ *with Balthasar*
⌈SECOND⌉ WATCHMAN
Here's Romeo's man. We found him in the
        churchyard.
CHIEF WATCHMAN
Hold him in safety till the Prince come hither.
        *Enter another Watchman with Friar Laurence*
THIRD WATCHMAN
Here is a friar that trembles, sighs, and weeps.
We took this mattock and this spade from him
As he was coming from this churchyard's side.          185
CHIEF WATCHMAN
A great suspicion. Stay the friar, too.
        *Enter the Prince* ⌈*with others*⌉
PRINCE
What misadventure is so early up,
That calls our person from our morning rest?
        *Enter Capulet and his Wife*
CAPULET
What should it be that is so shrieked abroad?

CAPULET'S WIFE
O, the people in the street cry 'Romeo',               190
Some 'Juliet', and some 'Paris', and all run
With open outcry toward our monument.
PRINCE
What fear is this which startles in our ears?
CHIEF WATCHMAN
Sovereign, here lies the County Paris slain,
And Romeo dead, and Juliet, dead before,               195
Warm, and new killed.
PRINCE
Search, seek, and know how this foul murder comes.
CHIEF WATCHMAN
Here is a friar, and slaughtered Romeo's man,
With instruments upon them fit to open
These dead men's tombs.                                200
CAPULET
O heavens! O wife, look how our daughter bleeds!
This dagger hath mista'en, for lo, his house
Is empty on the back of Montague,
And it mis-sheathèd in my daughter's bosom.
CAPULET'S WIFE
O me, this sight of death is as a bell                 205
That warns my old age to a sepulchre.
        *Enter Montague*
PRINCE
Come, Montague, for thou art early up
To see thy son and heir more early down.
MONTAGUE
Alas, my liege, my wife is dead tonight.
Grief of my son's exile hath stopped her breath.       210
What further woe conspires against mine age?
PRINCE Look, and thou shalt see.
MONTAGUE (*seeing Romeo's body*)
O thou untaught! What manners is in this,
To press before thy father to a grave?
PRINCE
Seal up the mouth of outrage for a while,              215
Till we can clear these ambiguities
And know their spring, their head, their true descent;
And then will I be general of your woes,
And lead you even to death. Meantime, forbear,
And let mischance be slave to patience.                220
Bring forth the parties of suspicion.
FRIAR LAURENCE
I am the greatest, able to do least,
Yet most suspected, as the time and place
Doth make against me, of this direful murder;
And here I stand, both to impeach and purge           225
Myself condemnèd and myself excused.
PRINCE
Then say at once what thou dost know in this.
FRIAR LAURENCE
I will be brief, for my short date of breath
Is not so long as is a tedious tale.
Romeo, there dead, was husband to that Juliet,         230
And she, there dead, that Romeo's faithful wife.
I married them, and their stol'n marriage day
Was Tybalt's doomsday, whose untimely death
Banished the new-made bridegroom from this city,
For whom, and not for Tybalt, Juliet pined.           235
You, to remove that siege of grief from her,
Betrothed and would have married her perforce
To County Paris. Then comes she to me,
And with wild looks bid me devise some mean
To rid her from this second marriage,                  240
Or in my cell there would she kill herself.

Then gave I her—so tutored by my art—
A sleeping potion, which so took effect
As I intended, for it wrought on her
The form of death. Meantime I writ to Romeo          245
That he should hither come as this dire night
To help to take her from her borrowed grave,
Being the time the potion's force should cease.
But he which bore my letter, Friar John,
Was stayed by accident, and yesternight              250
Returned my letter back. Then all alone,
At the prefixèd hour of her waking,
Came I to take her from her kindred's vault,
Meaning to keep her closely at my cell
Till I conveniently could send to Romeo.             255
But when I came, some minute ere the time
Of her awakening, here untimely lay
The noble Paris and true Romeo dead.
She wakes, and I entreated her come forth
And bear this work of heaven with patience.          260
But then a noise did scare me from the tomb,
And she, too desperate, would not go with me,
But, as it seems, did violence on herself.
All this I know, and to the marriage
Her nurse is privy; and if aught in this             265
Miscarried by my fault, let my old life
Be sacrificed, some hour before his time,
Unto the rigour of severest law.

PRINCE
We still have known thee for a holy man.
Where's Romeo's man? What can he say to this?        270

BALTHASAR
I brought my master news of Juliet's death,
And then in post he came from Mantua
To this same place, to this same monument.
This letter he early bid me give his father,
And threatened me with death, going in the vault,    275
If I departed not and left him there.

PRINCE
Give me the letter. I will look on it.
        *He takes the letter*

Where is the County's page that raised the watch?
Sirrah, what made your master in this place?

PAGE
He came with flowers to strew his lady's grave,      280
And bid me stand aloof, and so I did.
Anon comes one with light to ope the tomb,
And by and by my master drew on him,
And then I ran away to call the watch.

PRINCE
This letter doth make good the friar's words,        285
Their course of love, the tidings of her death;
And here he writes that he did buy a poison
Of a poor 'pothecary, and therewithal
Came to this vault to die, and lie with Juliet.
Where be these enemies? Capulet, Montague,           290
See what a scourge is laid upon your hate,
That heaven finds means to kill your joys with love.
And I, for winking at your discords, too
Have lost a brace of kinsmen. All are punishèd.

CAPULET
O brother Montague, give me thy hand.                295
This is my daughter's jointure, for no more
Can I demand.

MONTAGUE          But I can give thee more,
For I will raise her statue in pure gold,
That whiles Verona by that name is known
There shall no figure at such rate be set            300
As that of true and faithful Juliet.

CAPULET
As rich shall Romeo's by his lady's lie,
Poor sacrifices of our enmity.

PRINCE
A glooming peace this morning with it brings.
The sun for sorrow will not show his head.           305
Go hence, to have more talk of these sad things.
Some shall be pardoned, and some punishèd;
For never was a story of more woe
Than this of Juliet and her Romeo.
                        ⌜*The tomb is closed.*⌝ *Exeunt*

# RICHARD II

THE subject-matter of *Richard II* seemed inflammatorily topical to Shakespeare's contemporaries. Richard, who had notoriously indulged his favourites, had been compelled to yield his throne to Henry Bolingbroke, Earl of Hereford: like Richard, the ageing Queen Elizabeth had no obvious successor, and she too encouraged favourites—such as the Earl of Essex—who might aspire to the throne. When Shakespeare's play first appeared in print (in 1597), and in the two succeeding editions printed during Elizabeth's life, the episode (4.1.145–308) showing Richard yielding the crown was omitted; and in 1601, on the day before Essex led his ill-fated rebellion against Elizabeth, his fellow conspirators commissioned a special performance in the hope of arousing popular support, even though the play was said to be 'long out of use'—surprisingly, since it was probably written no earlier than 1595.

But Shakespeare introduced no obvious topicality into his dramatization of Richard's reign, for which he read widely while using Raphael Holinshed's *Chronicles* (1577, revised and enlarged in 1587) as his main source of information. In choosing to write about Richard II (1367–1400) he was returning to the beginning of the story whose ending he had staged in *Richard III*; for Bolingbroke's usurpation of the throne to which Richard's hereditary right was indisputable had set in train the series of events finally expiated only in the union of the houses of York and Lancaster celebrated in the last speech of *Richard III*. Like *Richard III*, this is a tragical history, focusing on a single character; but Richard II is a far more introverted and morally ambiguous figure than Richard III. In this play, written entirely in verse, Shakespeare forgoes stylistic variety in favour of an intense, plangent lyricism.

Our early impressions of Richard are unsympathetic. Having banished Mowbray and Bolingbroke, he behaves callously to Bolingbroke's father, John of Gaunt, a stern upholder of the old order to whose warning against his irresponsible behaviour he pays no attention, and upon Gaunt's death confiscates his property with no regard for Bolingbroke's rights. During Richard's absence on an Irish campaign, Bolingbroke returns to England and gains support in his efforts to claim his inheritance. Gradually, as the balance of power shifts, Richard makes deeper claims on the audience's sympathy. When he confronts Bolingbroke at Flint Castle (3.1) he eloquently laments his imminent deposition even though Bolingbroke insists that he comes only to claim what is his; soon afterwards (4.1.98–103) the Duke of York announces Richard's abdication. The transference of power is effected in a scene of lyrical expansiveness, and Richard becomes a pitiable figure as he is led to imprisonment in Pomfret (Pontefract) Castle while his former queen is banished to France. Richard's self-exploration reaches its climax in his soliloquy spoken shortly before his murder at the hands of Piers Exton; at the end of the play, Henry, anxious and guilt-laden, denies responsibility for the murder and plans an expiatory pilgrimage to the Holy Land.

# THE PERSONS OF THE PLAY

KING RICHARD II

The QUEEN, his wife

JOHN OF GAUNT, Duke of Lancaster, Richard's uncle

Harry BOLINGBROKE, Duke of Hereford, John of Gaunt's son, later
  KING HENRY IV

DUCHESS OF GLOUCESTER, widow of Gaunt's and York's brother

Duke of YORK, King Richard's uncle

DUCHESS OF YORK

Duke of AUMERLE, their son

Thomas MOWBRAY, Duke of Norfolk

GREEN
BAGOT      } followers of King Richard
BUSHY

Percy, Earl of NORTHUMBERLAND
HARRY PERCY, his son
Lord ROSS                          } of Bolingbroke's party
Lord WILLOUGHBY

Earl of SALISBURY
BISHOP OF CARLISLE   } of King Richard's party
Sir Stephen SCROPE

Lord BERKELEY

Lord FITZWALTER

Duke of SURREY

ABBOT OF WESTMINSTER

Sir Piers EXTON

LORD MARSHAL

HERALDS

CAPTAIN of the Welsh army

LADIES attending the Queen

GARDENER

Gardener's MEN

Exton's MEN

KEEPER of the prison at Pomfret

GROOM of King Richard's stable

Lords, soldiers, attendants

# The Tragedy of King Richard the Second

**1.1** *Enter King Richard and John of Gaunt, with the*
*Lord Marshal, other nobles, and attendants*

KING RICHARD
Old John of Gaunt, time-honoured Lancaster,
Hast thou according to thy oath and bond
Brought hither Henry Hereford, thy bold son,
Here to make good the boist'rous late appeal,
Which then our leisure would not let us hear, 5
Against the Duke of Norfolk, Thomas Mowbray?

JOHN OF GAUNT I have, my liege.

KING RICHARD
Tell me moreover, hast thou sounded him
If he appeal the Duke on ancient malice
Or worthily, as a good subject should, 10
On some known ground of treachery in him?

JOHN OF GAUNT
As near as I could sift him on that argument,
On some apparent danger seen in him
Aimed at your highness, no inveterate malice.

KING RICHARD
Then call them to our presence. ⌈*Exit one or more*⌉
                                    Face to face
And frowning brow to brow, ourselves will hear 16
The accuser and the accusèd freely speak.
High-stomached are they both and full of ire;
In rage, deaf as the sea, hasty as fire.
    *Enter Bolingbroke Duke of Hereford, and Mowbray*
    *Duke of Norfolk*

BOLINGBROKE
Many years of happy days befall 20
My gracious sovereign, my most loving liege!

MOWBRAY
Each day still better others' happiness,
Until the heavens, envying earth's good hap,
Add an immortal title to your crown!

KING RICHARD
We thank you both. Yet one but flatters us, 25
As well appeareth by the cause you come,
Namely, to appeal each other of high treason.
Cousin of Hereford, what dost thou object
Against the Duke of Norfolk, Thomas Mowbray?

BOLINGBROKE
First—heaven be the record to my speech— 30
In the devotion of a subject's love,
Tend'ring the precious safety of my Prince,
And free from other misbegotten hate,
Come I appellant to this princely presence.
Now, Thomas Mowbray, do I turn to thee; 35
And mark my greeting well, for what I speak
My body shall make good upon this earth,
Or my divine soul answer it in heaven.
Thou art a traitor and a miscreant,
Too good to be so, and too bad to live, 40
Since the more fair and crystal is the sky,
The uglier seem the clouds that in it fly.
Once more, the more to aggravate the note,
With a foul traitor's name stuff I thy throat,
And wish, so please my sovereign, ere I move 45
What my tongue speaks my right-drawn sword may
prove.

MOWBRAY
Let not my cold words here accuse my zeal.
'Tis not the trial of a woman's war,
The bitter clamour of two eager tongues,
Can arbitrate this cause betwixt us twain. 50
The blood is hot that must be cooled for this.
Yet can I not of such tame patience boast
As to be hushed and naught at all to say.
First, the fair reverence of your highness curbs me
From giving reins and spurs to my free speech, 55
Which else would post until it had returned
These terms of treason doubled down his throat.
Setting aside his high blood's royalty,
And let him be no kinsman to my liege,
I do defy him, and I spit at him, 60
Call him a slanderous coward and a villain;
Which to maintain I would allow him odds,
And meet him, were I tied to run afoot
Even to the frozen ridges of the Alps,
Or any other ground inhabitable, 65
Wherever Englishman durst set his foot.
Meantime let this defend my loyalty:
By all my hopes, most falsely doth he lie.

BOLINGBROKE (*throwing down his gage*)
Pale trembling coward, there I throw my gage,
Disclaiming here the kindred of the King, 70
And lay aside my high blood's royalty,
Which fear, not reverence, makes thee to except.
If guilty dread have left thee so much strength
As to take up mine honour's pawn, then stoop.
By that, and all the rites of knighthood else, 75
Will I make good against thee, arm to arm,
What I have spoke or thou canst worse devise.

MOWBRAY (*taking up the gage*)
I take it up, and by that sword I swear
Which gently laid my knighthood on my shoulder,
I'll answer thee in any fair degree 80
Or chivalrous design of knightly trial;
And when I mount, alive may I not light
If I be traitor or unjustly fight!

KING RICHARD (*to Bolingbroke*)
What doth our cousin lay to Mowbray's charge?
It must be great that can inherit us 85
So much as of a thought of ill in him.

BOLINGBROKE
Look what I speak, my life shall prove it true:
That Mowbray hath received eight thousand nobles
In name of lendings for your highness' soldiers,
The which he hath detained for lewd employments, 90
Like a false traitor and injurious villain.
Besides I say, and will in battle prove,
Or here or elsewhere, to the furthest verge
That ever was surveyed by English eye,
That all the treasons for these eighteen years 95
Complotted and contrivèd in this land
Fetch from false Mowbray their first head and spring.
Further I say, and further will maintain
Upon his bad life, to make all this good,
That he did plot the Duke of Gloucester's death, 100
Suggest his soon-believing adversaries,

And consequently, like a traitor-coward,
Sluiced out his innocent soul through streams of blood;
Which blood, like sacrificing Abel's, cries
Even from the tongueless caverns of the earth          105
To me for justice and rough chastisement.
And, by the glorious worth of my descent,
This arm shall do it or this life be spent.

KING RICHARD
How high a pitch his resolution soars!
Thomas of Norfolk, what sayst thou to this?          110

MOWBRAY
O, let my sovereign turn away his face,
And bid his ears a little while be deaf,
Till I have told this slander of his blood
How God and good men hate so foul a liar!

KING RICHARD
Mowbray, impartial are our eyes and ears.          115
Were he my brother, nay, my kingdom's heir,
As he is but my father's brother's son,
Now by my sceptre's awe I make a vow
Such neighbour-nearness to our sacred blood
Should nothing privilege him, nor partialize          120
The unstooping firmness of my upright soul.
He is our subject, Mowbray; so art thou.
Free speech and fearless I to thee allow.

MOWBRAY
Then, Bolingbroke, as low as to thy heart
Through the false passage of thy throat thou liest!          125
Three parts of that receipt I had for Calais
Disbursed I duly to his highness' soldiers.
The other part reserved I by consent,
For that my sovereign liege was in my debt
Upon remainder of a dear account          130
Since last I went to France to fetch his queen.
Now swallow down that lie. For Gloucester's death,
I slew him not, but to my own disgrace
Neglected my sworn duty in that case.
For you, my noble lord of Lancaster,          135
The honourable father to my foe,
Once did I lay an ambush for your life,
A trespass that doth vex my grievèd soul;
But ere I last received the Sacrament
I did confess it, and exactly begged          140
Your grace's pardon, and I hope I had it.
This is my fault. As for the rest appealed,
It issues from the rancour of a villain,
A recreant and most degenerate traitor,
Which in myself I boldly will defend,          145
    *He throws down his gage*
And interchangeably hurl down my gage
Upon this overweening traitor's foot,
To prove myself a loyal gentleman
Even in the best blood chambered in his bosom;
In haste whereof most heartily I pray          150
Your highness to assign our trial day.
    ⌈*Bolingbroke takes up the gage*⌉

KING RICHARD
Wrath-kindled gentlemen, be ruled by me.
Let's purge this choler without letting blood.
This we prescribe, though no physician:
Deep malice makes too deep incision;          155
Forget, forgive, conclude, and be agreed;
Our doctors say this is no time to bleed.
Good uncle, let this end where it begun.
We'll calm the Duke of Norfolk, you your son.

JOHN OF GAUNT
To be a make-peace shall become my age.          160
Throw down, my son, the Duke of Norfolk's gage.

KING RICHARD
And, Norfolk, throw down his.

JOHN OF GAUNT                    When, Harry, when?
Obedience bids I should not bid again.

KING RICHARD
Norfolk, throw down! We bid; there is no boot.

MOWBRAY (*kneeling*)
Myself I throw, dread sovereign, at thy foot.          165
My life thou shalt command, but not my shame.
The one my duty owes, but my fair name,
Despite of death that lives upon my grave,
To dark dishonour's use thou shalt not have.
I am disgraced, impeached, and baffled here,          170
Pierced to the soul with slander's venomed spear,
The which no balm can cure but his heart blood
Which breathed this poison.

KING RICHARD                    Rage must be withstood.
Give me his gage. Lions make leopards tame.

MOWBRAY ⌈*standing*⌉
Yea, but not change his spots. Take but my shame,
And I resign my gage. My dear dear lord,          176
The purest treasure mortal times afford
Is spotless reputation; that away,
Men are but gilded loam, or painted clay.
A jewel in a ten-times barred-up chest          180
Is a bold spirit in a loyal breast.
Mine honour is my life. Both grow in one.
Take honour from me, and my life is done.
Then, dear my liege, mine honour let me try.
In that I live, and for that will I die.          185

KING RICHARD
Cousin, throw down your gage. Do you begin.

BOLINGBROKE
O God defend my soul from such deep sin!
Shall I seem crest-fallen in my father's sight?
Or with pale beggar-fear impeach my height
Before this out-dared dastard? Ere my tongue          190
Shall wound my honour with such feeble wrong,
Or sound so base a parle, my teeth shall tear
The slavish motive of recanting fear,
And spit it bleeding in his high disgrace
Where shame doth harbour, even in Mowbray's face.
    ⌈*Exit John of Gaunt*⌉

KING RICHARD
We were not born to sue, but to command;          196
Which since we cannot do to make you friends,
Be ready, as your lives shall answer it,
At Coventry upon Saint Lambert's day.
There shall your swords and lances arbitrate          200
The swelling difference of your settled hate.
Since we cannot atone you, we shall see
Justice design the victor's chivalry.
Lord Marshal, command our officers-at-arms
Be ready to direct these home alarms.          *Exeunt*

1.2          *Enter John of Gaunt, Duke of Lancaster, with the*
             *Duchess of Gloucester*

JOHN OF GAUNT
Alas, the part I had in Gloucester's blood
Doth more solicit me than your exclaims
To stir against the butchers of his life.
But since correction lieth in those hands
Which made the fault that we cannot correct,          5

Put we our quarrel to the will of heaven,
Who, when they see the hours ripe on earth,
Will rain hot vengeance on offenders' heads.

DUCHESS OF GLOUCESTER
Finds brotherhood in thee no sharper spur?
Hath love in thy old blood no living fire?                10
Edward's seven sons, whereof thyself art one,
Were as seven vials of his sacred blood,
Or seven fair branches springing from one root.
Some of those seven are dried by nature's course,
Some of those branches by the destinies cut;              15
But Thomas, my dear lord, my life, my Gloucester,
One vial full of Edward's sacred blood,
One flourishing branch of his most royal root,
Is cracked, and all the precious liquor spilt;
Is hacked down, and his summer leaves all faded           20
By envy's hand and murder's bloody axe.
Ah, Gaunt, his blood was thine! That bed, that womb,
That mettle, that self mould that fashioned thee,
Made him a man; and though thou liv'st and
       breathest,
Yet art thou slain in him. Thou dost consent             25
In some large measure to thy father's death
In that thou seest thy wretched brother die,
Who was the model of thy father's life.
Call it not patience, Gaunt, it is despair.
In suff'ring thus thy brother to be slaughtered          30
Thou show'st the naked pathway to thy life,
Teaching stern murder how to butcher thee.
That which in mean men we entitle patience
Is pale cold cowardice in noble breasts.
What shall I say? To safeguard thine own life            35
The best way is to venge my Gloucester's death.

JOHN OF GAUNT
God's is the quarrel; for God's substitute,
His deputy anointed in his sight,
Hath caused his death; the which if wrongfully,
Let heaven revenge, for I may never lift                 40
An angry arm against his minister.

DUCHESS OF GLOUCESTER
Where then, alas, may I complain myself?

JOHN OF GAUNT
To God, the widow's champion and defence.

DUCHESS OF GLOUCESTER
Why then, I will. Farewell, old Gaunt.
Thou goest to Coventry, there to behold                  45
Our cousin Hereford and fell Mowbray fight.
O, set my husband's wrongs on Hereford's spear,
That it may enter butcher Mowbray's breast!
Or if misfortune miss the first career,
Be Mowbray's sins so heavy in his bosom                  50
That they may break his foaming courser's back
And throw the rider headlong in the lists,
A caitiff, recreant to my cousin Hereford!
Farewell, old Gaunt. Thy sometimes brother's wife
With her companion, grief, must end her life.            55

JOHN OF GAUNT
Sister, farewell. I must to Coventry.
As much good stay with thee as go with me.

DUCHESS OF GLOUCESTER
Yet one word more. Grief boundeth where it falls,
Not with the empty hollowness, but weight.
I take my leave before I have begun,                     60
For sorrow ends not when it seemeth done.
Commend me to thy brother, Edmund York.
Lo, this is all.—Nay, yet depart not so!
Though this be all, do not so quickly go.

I shall remember more. Bid him—ah, what?—               65
With all good speed at Pleshey visit me.
Alack, and what shall good old York there see
But empty lodgings and unfurnished walls,
Unpeopled offices, untrodden stones,
And what hear there for welcome but my groans?          70
Therefore commend me; let him not come there
To seek out sorrow that dwells everywhere.
Desolate, desolate will I hence and die.
The last leave of thee takes my weeping eye.
                              *Exeunt ⌈severally⌉*

1.3     *Enter Lord Marshal ⌈with officers setting out*
         *chairs⌉, and the Duke of Aumerle*

LORD MARSHAL
My lord Aumerle, is Harry Hereford armed?

AUMERLE
Yea, at all points, and longs to enter in.

LORD MARSHAL
The Duke of Norfolk, sprightfully and bold,
Stays but the summons of the appellant's trumpet.

AUMERLE
Why then, the champions are prepared, and stay         5
For nothing but his majesty's approach.
       *The trumpets sound, and King Richard enters, with*
       *John of Gaunt, Duke of Lancaster, ⌈Bushy, Bagot,*
       *Green,⌉ and other nobles. When they are set, enter*
       *Mowbray Duke of Norfolk, defendant, in arms, ⌈and*
       *a Herald⌉*

KING RICHARD
Marshal, demand of yonder champion
The cause of his arrival here in arms.
Ask him his name, and orderly proceed
To swear him in the justice of his cause.              10

LORD MARSHAL (*to Mowbray*)
In God's name and the King's, say who thou art,
And why thou com'st thus knightly clad in arms,
Against what man thou com'st, and what thy
       quarrel.
Speak truly on thy knighthood and thy oath,
As so defend thee heaven and thy valour!               15

MOWBRAY
My name is Thomas Mowbray, Duke of Norfolk,
Who hither come engagèd by my oath—
Which God defend a knight should violate—
Both to defend my loyalty and truth
To God, my king, and my succeeding issue,              20
Against the Duke of Hereford that appeals me;
And by the grace of God and this mine arm
To prove him, in defending of myself,
A traitor to my God, my king, and me.
And as I truly fight, defend me heaven!                 25
       ⌈*He sits.*⌉
       *The trumpets sound. Enter Bolingbroke Duke of*
       *Hereford, appellant, in armour, ⌈and a Herald⌉*

KING RICHARD
Marshal, ask yonder knight in arms
Both who he is and why he cometh hither
Thus plated in habiliments of war;
And formally, according to our law,
Depose him in the justice of his cause.                30

LORD MARSHAL (*to Bolingbroke*)
What is thy name? And wherefore com'st thou hither
Before King Richard in his royal lists?
Against whom comest thou? And what's thy quarrel?
Speak like a true knight, so defend thee heaven!

BOLINGBROKE
Harry of Hereford, Lancaster, and Derby        35
Am I, who ready here do stand in arms
To prove by God's grace and my body's valour
In lists on Thomas Mowbray, Duke of Norfolk,
That he is a traitor foul and dangerous
To God of heaven, King Richard, and to me.        40
And as I truly fight, defend me heaven!
    ⌈He sits⌉
LORD MARSHAL
On pain of death, no person be so bold
Or daring-hardy as to touch the lists
Except the Marshal and such officers
Appointed to direct these fair designs.        45
BOLINGBROKE ⌈standing⌉
Lord Marshal, let me kiss my sovereign's hand
And bow my knee before his majesty,
For Mowbray and myself are like two men
That vow a long and weary pilgrimage;
Then let us take a ceremonious leave        50
And loving farewell of our several friends.
LORD MARSHAL (to King Richard)
The appellant in all duty greets your highness,
And craves to kiss your hand and take his leave.
KING RICHARD
We will descend and fold him in our arms.
    He descends from his seat and embraces Bolingbroke
Cousin of Hereford, as thy cause is just,        55
So be thy fortune in this royal fight.
Farewell, my blood, which if today thou shed,
Lament we may, but not revenge thee dead.
BOLINGBROKE
O, let no noble eye profane a tear
For me if I be gored with Mowbray's spear.        60
As confident as is the falcon's flight
Against a bird do I with Mowbray fight.
(To the Lord Marshal) My loving lord, I take my leave
    of you;
(To Aumerle) Of you, my noble cousin, Lord Aumerle;
Not sick, although I have to do with death,        65
But lusty, young, and cheerly drawing breath.
Lo, as at English feasts, so I regreet
The daintiest last, to make the end most sweet.
(To Gaunt, ⌈kneeling⌉) O thou, the earthly author of my
    blood,
Whose youthful spirit in me regenerate        70
Doth with a two-fold vigour lift me up
To reach at victory above my head,
Add proof unto mine armour with thy prayers,
And with thy blessings steel my lance's point,
That it may enter Mowbray's waxen coat        75
And furbish new the name of John a Gaunt
Even in the lusty haviour of his son.
JOHN OF GAUNT
God in thy good cause make thee prosperous!
Be swift like lightning in the execution,
And let thy blows, doubly redoublèd,        80
Fall like amazing thunder on the casque
Of thy adverse pernicious enemy.
Rouse up thy youthful blood, be valiant, and live.
BOLINGBROKE ⌈standing⌉
Mine innocence and Saint George to thrive!
MOWBRAY ⌈standing⌉
However God or fortune cast my lot,        85
There lives or dies, true to King Richard's throne,
A loyal, just, and upright gentleman.

Never did captive with a freer heart
Cast off his chains of bondage and embrace
His golden uncontrolled enfranchisement        90
More than my dancing soul doth celebrate
This feast of battle with mine adversary.
Most mighty liege, and my companion peers,
Take from my mouth the wish of happy years.
As gentle and as jocund as to jest        95
Go I to fight. Truth hath a quiet breast.
KING RICHARD
Farewell, my lord. Securely I espy
Virtue with valour couchèd in thine eye.—
Order the trial, Marshal, and begin.
LORD MARSHAL
Harry of Hereford, Lancaster, and Derby,        100
Receive thy lance; and God defend the right!
    ⌈An officer bears a lance to Bolingbroke⌉
BOLINGBROKE
Strong as a tower in hope, I cry 'Amen!'
LORD MARSHAL (to an officer)
Go bear this lance to Thomas, Duke of Norfolk.
    ⌈An officer bears a lance to Mowbray⌉
FIRST HERALD
Harry of Hereford, Lancaster, and Derby
Stands here for God, his sovereign, and himself,        105
On pain to be found false and recreant,
To prove the Duke of Norfolk, Thomas Mowbray,
A traitor to his God, his king, and him,
And dares him to set forward to the fight.
SECOND HERALD
Here standeth Thomas Mowbray, Duke of Norfolk,        110
On pain to be found false and recreant,
Both to defend himself and to approve
Henry of Hereford, Lancaster, and Derby
To God his sovereign and to him disloyal,
Courageously and with a free desire        115
Attending but the signal to begin.
LORD MARSHAL
Sound trumpets, and set forward combatants!
    ⌈A charge is sounded.⌉
    King Richard throws down his warder
Stay, the King hath thrown his warder down.
KING RICHARD
Let them lay by their helmets and their spears,
And both return back to their chairs again.        120
    ⌈Bolingbroke and Mowbray disarm and sit⌉
(To the nobles) Withdraw with us, and let the trumpets
    sound
While we return these dukes what we decree.
    A long flourish, during which King Richard and his
    nobles withdraw and hold council, ⌈then come
    forward⌉. King Richard addresses Bolingbroke and
    Mowbray
Draw near, and list what with our council we have
    done.
For that our kingdom's earth should not be soiled
With that dear blood which it hath fosterèd,        125
And for our eyes do hate the dire aspect
Of civil wounds ploughed up with neighbours' swords,
Which, so roused up with boist'rous untuned drums,
With harsh-resounding trumpets' dreadful bray,
And grating shock of wrathful iron arms,        130
Might from our quiet confines fright fair peace
And make us wade even in our kindred's blood,
Therefore we banish you our territories.
You, cousin Hereford, upon pain of life,

Till twice five summers have enriched our fields    135
Shall not regreet our fair dominions,
But tread the stranger paths of banishment.
BOLINGBROKE
Your will be done. This must my comfort be:
That sun that warms you here shall shine on me,
And those his golden beams to you here lent    140
Shall point on me and gild my banishment.
KING RICHARD
Norfolk, for thee remains a heavier doom,
Which I with some unwillingness pronounce.
The sly slow hours shall not determinate
The dateless limit of thy dear exile.    145
The hopeless word of 'never to return'
Breathe I against thee, upon pain of life.
MOWBRAY
A heavy sentence, my most sovereign liege,
And all unlooked-for from your highness' mouth.
A dearer merit, not so deep a maim    150
As to be cast forth in the common air,
Have I deservèd at your highness' hands.
The language I have learnt these forty years,
My native English, now I must forgo,
And now my tongue's use is to me no more    155
Than an unstringèd viol or a harp,
Or like a cunning instrument cased up,
Or, being open, put into his hands
That knows no touch to tune the harmony.
Within my mouth you have enjailed my tongue,    160
Doubly portcullised with my teeth and lips,
And dull unfeeling barren ignorance
Is made my jailer to attend on me.
I am too old to fawn upon a nurse,
Too far in years to be a pupil now.    165
What is thy sentence then but speechless death,
Which robs my tongue from breathing native breath?
KING RICHARD
It boots thee not to be compassionate.
After our sentence, plaining comes too late.
MOWBRAY
Then thus I turn me from my country's light,    170
To dwell in solemn shades of endless night.
KING RICHARD
Return again, and take an oath with thee.
(To both) Lay on our royal sword your banished hands.
Swear by the duty that you owe to God—
Our part therein we banish with yourselves—    175
To keep the oath that we administer.
You never shall, so help you truth and God,
Embrace each other's love in banishment,
Nor never look upon each other's face,
Nor never write, regreet, nor reconcile    180
This low'ring tempest of your home-bred hate,
Nor never by advisèd purpose meet
To plot, contrive, or complot any ill
'Gainst us, our state, our subjects, or our land.
BOLINGBROKE
I swear.
MOWBRAY And I, to keep all this.    185
BOLINGBROKE
Norfolk, so far as to mine enemy:
By this time, had the King permitted us,
One of our souls had wandered in the air,
Banished this frail sepulchre of our flesh,
As now our flesh is banished from this land.    190
Confess thy treasons ere thou fly the realm.

Since thou hast far to go, bear not along
The clogging burden of a guilty soul.
MOWBRAY
No, Bolingbroke, if ever I were traitor,
My name be blotted from the book of life,    195
And I from heaven banished as from hence.
But what thou art, God, thou, and I do know,
And all too soon I fear the King shall rue.
Farewell, my liege. Now no way can I stray:
Save back to England, all the world's my way.    Exit
KING RICHARD
Uncle, even in the glasses of thine eyes    201
I see thy grievèd heart. Thy sad aspect
Hath from the number of his banished years
Plucked four away. (To Bolingbroke) Six frozen winters
    spent,
Return with welcome home from banishment.    205
BOLINGBROKE
How long a time lies in one little word!
Four lagging winters and four wanton springs
End in a word: such is the breath of kings.
JOHN OF GAUNT
I thank my liege that in regard of me
He shortens four years of my son's exile.    210
But little vantage shall I reap thereby,
For ere the six years that he hath to spend
Can change their moons and bring their times about,
My oil-dried lamp and time-bewasted light
Shall be extinct with age and endless night.    215
My inch of taper will be burnt and done,
And blindfold death not let me see my son.
KING RICHARD
Why, uncle, thou hast many years to live.
JOHN OF GAUNT
But not a minute, King, that thou canst give.
Shorten my days thou canst with sudden sorrow,    220
And pluck nights from me, but not lend a morrow.
Thou canst help time to furrow me with age,
But stop no wrinkle in his pilgrimage.
Thy word is current with him for my death,
But dead, thy kingdom cannot buy my breath.    225
KING RICHARD
Thy son is banished upon good advice,
Whereto thy tongue a party verdict gave.
Why at our justice seem'st thou then to lour?
JOHN OF GAUNT
Things sweet to taste prove in digestion sour.
You urged me as a judge, but I had rather    230
You would have bid me argue like a father.
Alas, I looked when some of you should say
I was too strict to make mine own away,
But you gave leave to my unwilling tongue
Against my will to do myself this wrong.    235
KING RICHARD
Cousin, farewell; and uncle, bid him so.
Six years we banish him, and he shall go.
    [Flourish.] Exeunt all but Aumerle, the Lord
        Marshal, John of Gaunt, and Bolingbroke
AUMERLE (to Bolingbroke)
Cousin, farewell. What presence must not know,
From where you do remain let paper show.    [Exit]
LORD MARSHAL (to Bolingbroke)
My lord, no leave take I, for I will ride    240
As far as land will let me by your side.
JOHN OF GAUNT (to Bolingbroke)
O, to what purpose dost thou hoard thy words,
That thou return'st no greeting to thy friends?

**BOLINGBROKE**
I have too few to take my leave of you,
When the tongue's office should be prodigal    245
To breathe the abundant dolour of the heart.
**JOHN OF GAUNT**
Thy grief is but thy absence for a time.
**BOLINGBROKE**
Joy absent, grief is present for that time.
**JOHN OF GAUNT**
What is six winters? They are quickly gone.
**BOLINGBROKE**
To men in joy, but grief makes one hour ten.    250
**JOHN OF GAUNT**
Call it a travel that thou tak'st for pleasure.
**BOLINGBROKE**
My heart will sigh when I miscall it so,
Which finds it an enforcèd pilgrimage.
**JOHN OF GAUNT**
The sullen passage of thy weary steps
Esteem as foil wherein thou art to set    255
The precious jewel of thy home return.
**BOLINGBROKE**
O, who can hold a fire in his hand
By thinking on the frosty Caucasus,
Or cloy the hungry edge of appetite
By bare imagination of a feast,    260
Or wallow naked in December snow
By thinking on fantastic summer's heat?
O no, the apprehension of the good
Gives but the greater feeling to the worse.
Fell sorrow's tooth doth never rankle more    265
Than when he bites, but lanceth not the sore.
**JOHN OF GAUNT**
Come, come, my son, I'll bring thee on thy way.
Had I thy youth and cause, I would not stay.
**BOLINGBROKE**
Then England's ground, farewell. Sweet soil, adieu,
My mother and my nurse that bears me yet!    270
Where'er I wander, boast of this I can:
Though banished, yet a trueborn Englishman.    *Exeunt*

**1.4**    *Enter King Richard with ⌜Green and Bagot⌝ at one*
      *door, and the Lord Aumerle at another*
**KING RICHARD**
We did observe.—Cousin Aumerle,
How far brought you high Hereford on his way?
**AUMERLE**
I brought high Hereford, if you call him so,
But to the next highway, and there I left him.
**KING RICHARD**
And say, what store of parting tears were shed?    5
**AUMERLE**
Faith, none for me, except the north-east wind,
Which then grew bitterly against our faces,
Awaked the sleeping rheum, and so by chance
Did grace our hollow parting with a tear.
**KING RICHARD**
What said our cousin when you parted with him?    10
**AUMERLE**
'Farewell.' And for my heart disdainèd that my tongue
Should so profane the word, that taught me craft
To counterfeit oppression of such grief
That words seemed buried in my sorrow's grave.
Marry, would the word 'farewell' have lengthened
     hours    15
And added years to his short banishment,

He should have had a volume of farewells;
But since it would not, he had none of me.
**KING RICHARD**
He is our cousin, cousin; but 'tis doubt,
When time shall call him home from banishment,    20
Whether our kinsman come to see his friends.
Ourself and Bushy, Bagot here, and Green
Observed his courtship to the common people,
How he did seem to dive into their hearts
With humble and familiar courtesy,    25
What reverence he did throw away on slaves,
Wooing poor craftsmen with the craft of smiles
And patient underbearing of his fortune,
As 'twere to banish their affects with him.
Off goes his bonnet to an oysterwench,    30
A brace of draymen bid God speed him well,
And had the tribute of his supple knee
With 'Thanks, my countrymen, my loving friends',
As were our England in reversion his,
And he our subjects' next degree in hope.    35
**GREEN**
Well, he is gone, and with him go these thoughts.
Now for the rebels which stand out in Ireland.
Expedient manage must be made, my liege,
Ere further leisure yield them further means
For their advantage and your highness' loss.    40
**KING RICHARD**
We will ourself in person to this war,
And for our coffers with too great a court
And liberal largess are grown somewhat light,
We are enforced to farm our royal realm,
The revenue whereof shall furnish us    45
For our affairs in hand. If that come short,
Our substitutes at home shall have blank charters,
Whereto, when they shall know what men are rich,
They shall subscribe them for large sums of gold,
And send them after to supply our wants;    50
For we will make for Ireland presently.
     *Enter Bushy*
Bushy, what news?
**BUSHY**
Old John of Gaunt is grievous sick, my lord,
Suddenly taken, and hath sent post-haste
To entreat your majesty to visit him.    55
**KING RICHARD** Where lies he?
**BUSHY** At Ely House.
**KING RICHARD**
Now put it, God, in his physician's mind
To help him to his grave immediately.
The lining of his coffers shall make coats    60
To deck our soldiers for these Irish wars.
Come, gentlemen, let's all go visit him.
Pray God we may make haste and come too late!
     *Exeunt*

**2.1**    *Enter John of Gaunt, Duke of Lancaster, sick,*
      *⌜carried in a chair,⌝ with the Duke of York*
**JOHN OF GAUNT**
Will the King come, that I may breathe my last
In wholesome counsel to his unstaid youth?
**YORK**
Vex not yourself, nor strive not with your breath,
For all in vain comes counsel to his ear.
**JOHN OF GAUNT**
O, but they say the tongues of dying men    5
Enforce attention, like deep harmony.

Where words are scarce they are seldom spent in
  vain,
For they breathe truth that breathe their words in
  pain.
He that no more must say is listened more
  Than they whom youth and ease have taught to
  glose.                                                    10
More are men's ends marked than their lives before.
  The setting sun, and music at the close,
As the last taste of sweets, is sweetest last,
Writ in remembrance more than things long past.
Though Richard my life's counsel would not hear,    15
My death's sad tale may yet undeaf his ear.
YORK
No, it is stopped with other, flattering sounds,
As praises of whose taste the wise are feared,
Lascivious metres to whose venom sound
The open ear of youth doth always listen,           20
Report of fashions in proud Italy,
Whose manners still our tardy-apish nation
Limps after in base imitation.
Where doth the world thrust forth a vanity—
So it be new there's no respect how vile—           25
That is not quickly buzzed into his ears?
Then all too late comes counsel, to be heard
Where will doth mutiny with wit's regard.
Direct not him whose way himself will choose:
'Tis breath thou lack'st, and that breath wilt thou lose.
JOHN OF GAUNT
Methinks I am a prophet new-inspired,               31
And thus, expiring, do foretell of him.
His rash, fierce blaze of riot cannot last,
For violent fires soon burn out themselves.
Small showers last long, but sudden storms are short.
He tires betimes that spurs too fast betimes.       36
With eager feeding food doth choke the feeder.
Light vanity, insatiate cormorant,
Consuming means, soon preys upon itself.
This royal throne of kings, this sceptred isle,     40
This earth of majesty, this seat of Mars,
This other Eden, demi-paradise,
This fortress built by nature for herself
Against infection and the hand of war,
This happy breed of men, this little world,         45
This precious stone set in the silver sea,
Which serves it in the office of a wall,
Or as a moat defensive to a house
Against the envy of less happier lands;
This blessèd plot, this earth, this realm, this England,
This nurse, this teeming womb of royal kings,       51
Feared by their breed and famous by their birth,
Renownèd for their deeds as far from home
For Christian service and true chivalry
As is the sepulchre, in stubborn Jewry,             55
Of the world's ransom, blessèd Mary's son;
This land of such dear souls, this dear dear land,
Dear for her reputation through the world,
Is now leased out—I die pronouncing it—
Like to a tenement or pelting farm.                 60
England, bound in with the triumphant sea,
Whose rocky shore beats back the envious siege
Of wat'ry Neptune, is now bound in with shame,
With inky blots and rotten parchment bonds.
That England that was wont to conquer others        65
Hath made a shameful conquest of itself.
Ah, would the scandal vanish with my life,
How happy then were my ensuing death!

*Enter King Richard and the Queen; ⌈the Duke of
Aumerle,⌉ Bushy, ⌈Green, Bagot,⌉ Lord Ross, and
Lord Willoughby*
YORK
The King is come. Deal mildly with his youth,
For young hot colts, being reined, do rage the more.
QUEEN
How fares our noble uncle Lancaster?                71
KING RICHARD
What comfort, man? How is't with agèd Gaunt?
JOHN OF GAUNT
O, how that name befits my composition!
Old Gaunt indeed, and gaunt in being old.
Within me grief hath kept a tedious fast,           75
And who abstains from meat that is not gaunt?
For sleeping England long time have I watched.
Watching breeds leanness, leanness is all gaunt.
The pleasure that some fathers feed upon
Is my strict fast: I mean my children's looks.      80
And therein fasting, hast thou made me gaunt.
Gaunt am I for the grave, gaunt as a grave,
Whose hollow womb inherits naught but bones.
KING RICHARD
Can sick men play so nicely with their names?
JOHN OF GAUNT
No, misery makes sport to mock itself.              85
Since thou dost seek to kill my name in me,
I mock my name, great King, to flatter thee.
KING RICHARD
Should dying men flatter with those that live?
JOHN OF GAUNT
No, no, men living flatter those that die.
KING RICHARD
Thou now a-dying sayst thou flatt'rest me.          90
JOHN OF GAUNT
O no: thou diest, though I the sicker be.
KING RICHARD
I am in health; I breathe, and see thee ill.
JOHN OF GAUNT
Now He that made me knows I see thee ill:
Ill in myself to see, and in thee seeing ill.
Thy deathbed is no lesser than thy land,            95
Wherein thou liest in reputation sick;
And thou, too careless patient as thou art,
Committ'st thy anointed body to the cure
Of those physicians that first wounded thee.
A thousand flatterers sit within thy crown,         100
Whose compass is no bigger than thy head,
And yet, encagèd in so small a verge,
The waste is no whit lesser than thy land.
O, had thy grandsire with a prophet's eye
Seen how his son's son should destroy his sons,     105
From forth thy reach he would have laid thy shame,
Deposing thee before thou wert possessed,
Which art possessed now to depose thyself.
Why, cousin, wert thou regent of the world
It were a shame to let this land by lease.          110
But, for thy world, enjoying but this land,
Is it not more than shame to shame it so?
Landlord of England art thou now, not king.
Thy state of law is bondslave to the law,
And—                                                115
KING RICHARD
And thou, a lunatic lean-witted fool,
Presuming on an ague's privilege,
Dar'st with thy frozen admonition

Make pale our cheek, chasing the royal blood
With fury from his native residence.                              120
Now by my seat's right royal majesty,
Wert thou not brother to great Edward's son,
This tongue that runs so roundly in thy head
Should run thy head from thy unreverent shoulders.

JOHN OF GAUNT
O, spare me not, my brother Edward's son,                    125
For that I was his father Edward's son.
That blood already, like the pelican,
Hast thou tapped out and drunkenly caroused.
My brother Gloucester, plain well-meaning soul—
Whom fair befall in heaven 'mongst happy souls—
May be a precedent and witness good                            131
That thou respect'st not spilling Edward's blood.
Join with the present sickness that I have,
And thy unkindness be like crookèd age,
To crop at once a too-long withered flower.                    135
Live in thy shame, but die not shame with thee.
These words hereafter thy tormentors be.
(To attendants) Convey me to my bed, then to my
    grave.
Love they to live that love and honour have.
                        Exit, ⌈carried in the chair⌉
KING RICHARD
And let them die that age and sullens have,                    140
For both hast thou, and both become the grave.
YORK
I do beseech your majesty impute his words
To wayward sickliness and age in him.
He loves you, on my life, and holds you dear
As Harry Duke of Hereford, were he here.                       145
KING RICHARD
Right, you say true: as Hereford's love, so his.
As theirs, so mine; and all be as it is.
        Enter the Earl of Northumberland
NORTHUMBERLAND
My liege, old Gaunt commends him to your majesty.
KING RICHARD
What says he?
NORTHUMBERLAND Nay, nothing: all is said.
His tongue is now a stringless instrument.                      150
Words, life, and all, old Lancaster hath spent.
YORK
Be York the next that must be bankrupt so!
Though death be poor, it ends a mortal woe.
KING RICHARD
The ripest fruit first falls, and so doth he.
His time is spent; our pilgrimage must be.                      155
So much for that. Now for our Irish wars.
We must supplant those rough rug-headed kerns,
Which live like venom where no venom else
But only they have privilege to live.
And for these great affairs do ask some charge,                160
Towards our assistance we do seize to us
The plate, coin, revenues, and movables
Whereof our uncle Gaunt did stand possessed.
YORK
How long shall I be patient? Ah, how long
Shall tender duty make me suffer wrong?                         165
Not Gloucester's death, nor Hereford's banishment,
Nor Gaunt's rebukes, nor England's private wrongs,
Nor the prevention of poor Bolingbroke
About his marriage, nor my own disgrace,
Have ever made me sour my patient cheek,                        170
Or bend one wrinkle on my sovereign's face.

I am the last of noble Edward's sons,
Of whom thy father, Prince of Wales, was first.
In war was never lion raged more fierce,
In peace was never gentle lamb more mild,                      175
Than was that young and princely gentleman.
His face thou hast, for even so looked he,
Accomplished with the number of thy hours.
But when he frowned it was against the French,
And not against his friends. His noble hand                    180
Did win what he did spend, and spent not that
Which his triumphant father's hand had won.
His hands were guilty of no kindred blood,
But bloody with the enemies of his kin.
O, Richard, York is too far gone with grief,                   185
Or else he never would compare between.
KING RICHARD
Why uncle, what's the matter?
YORK                                    O my liege,
Pardon me if you please; if not, I, pleased
Not to be pardoned, am content withal.
Seek you to seize and grip into your hands                     190
The royalties and rights of banished Hereford?
Is not Gaunt dead? And doth not Hereford live?
Was not Gaunt just? And is not Harry true?
Did not the one deserve to have an heir?
Is not his heir a well-deserving son?                          195
Take Hereford's rights away, and take from Time
His charters and his customary rights:
Let not tomorrow then ensue today;
Be not thyself, for how art thou a king
But by fair sequence and succession?                           200
Now afore God—God forbid I say true!—
If you do wrongfully seize Hereford's rights,
Call in the letters patents that he hath
By his attorneys general to sue
His livery, and deny his offered homage,                       205
You pluck a thousand dangers on your head,
You lose a thousand well-disposèd hearts,
And prick my tender patience to those thoughts
Which honour and allegiance cannot think.
KING RICHARD
Think what you will, we seize into our hands                   210
His plate, his goods, his money, and his lands.
YORK
I'll not be by the while. My liege, farewell.
What will ensue hereof there's none can tell.
But by bad courses may be understood
That their events can never fall out good.            Exit
KING RICHARD
Go, Bushy, to the Earl of Wiltshire straight.                  216
Bid him repair to us to Ely House
To see this business. Tomorrow next
We will for Ireland, and 'tis time, I trow.
And we create, in absence of ourself,                          220
Our uncle York Lord Governor of England;
For he is just and always loved us well.—
Come on, our Queen; tomorrow must we part.
Be merry, for our time of stay is short.
        ⌈Flourish.⌉ Exeunt ⌈Bushy at one door; King
                Richard, the Queen, Aumerle, Green, and
                Bagot at another door⌉. Northumberland,
                        Willoughby, and Ross remain
NORTHUMBERLAND
Well, lords, the Duke of Lancaster is dead.                    225
ROSS
And living too, for now his son is Duke.

WILLOUGHBY
Barely in title, not in revenues.
NORTHUMBERLAND
Richly in both, if justice had her right.
ROSS
My heart is great, but it must break with silence
Ere't be disburdened with a liberal tongue.    230
NORTHUMBERLAND
Nay, speak thy mind, and let him ne'er speak more
That speaks thy words again to do thee harm.
WILLOUGHBY
Tends that that thou wouldst speak to the Duke of
    Hereford?
If it be so, out with it boldly, man.
Quick is mine ear to hear of good towards him.    235
ROSS
No good at all that I can do for him,
Unless you call it good to pity him,
Bereft and gelded of his patrimony.
NORTHUMBERLAND
Now afore God, 'tis shame such wrongs are borne
In him, a royal prince, and many more    240
Of noble blood in this declining land.
The King is not himself, but basely led
By flatterers; and what they will inform
Merely in hate 'gainst any of us all,
That will the King severely prosecute    245
'Gainst us, our lives, our children, and our heirs.
ROSS
The commons hath he pilled with grievous taxes,
And quite lost their hearts. The nobles hath he fined
For ancient quarrels, and quite lost their hearts.
WILLOUGHBY
And daily new exactions are devised,    250
As blanks, benevolences, and I wot not what.
But what, a' God's name, doth become of this?
NORTHUMBERLAND
Wars hath not wasted it; for warred he hath not,
But basely yielded upon compromise
That which his ancestors achieved with blows.    255
More hath he spent in peace than they in wars.
ROSS
The Earl of Wiltshire hath the realm in farm.
WILLOUGHBY
The King's grown bankrupt like a broken man.
NORTHUMBERLAND
Reproach and dissolution hangeth over him.
ROSS
He hath not money for these Irish wars,    260
His burdenous taxations notwithstanding,
But by the robbing of the banished Duke.
NORTHUMBERLAND
His noble kinsman. Most degenerate King!
But, lords, we hear this fearful tempest sing,
Yet seek no shelter to avoid the storm.    265
We see the wind sit sore upon our sails,
And yet we strike not, but securely perish.
ROSS
We see the very wreck that we must suffer,
And unavoided is the danger now
For suffering so the causes of our wreck.    270
NORTHUMBERLAND
Not so: even through the hollow eyes of death
I spy life peering; but I dare not say
How near the tidings of our comfort is.

WILLOUGHBY
Nay, let us share thy thoughts, as thou dost ours.
ROSS
Be confident to speak, Northumberland.    275
We three are but thyself, and, speaking so,
Thy words are but as thoughts. Therefore be bold.
NORTHUMBERLAND
Then thus. I have from Port le Blanc,
A bay in Brittaine, received intelligence
That Harry Duke of Hereford, Reinold Lord Cobham,
Thomas son and heir to the Earl of Arundel    281
That late broke from the Duke of Exeter,
His brother, Archbishop late of Canterbury,
Sir Thomas Erpingham, Sir Thomas Ramston,
Sir John Norbery,    285
Sir Robert Waterton, and Francis Coint,
All these well furnished by the Duke of Brittaine
With eight tall ships, three thousand men of war,
Are making hither with all due expedience,
And shortly mean to touch our northern shore.    290
Perhaps they had ere this, but that they stay
The first departing of the King for Ireland.
If then we shall shake off our slavish yoke,
Imp out our drooping country's broken wing,
Redeem from broking pawn the blemished crown,    295
Wipe off the dust that hides our sceptre's gilt,
And make high majesty look like itself,
Away with me in post to Ravenspurgh.
But if you faint, as fearing to do so,
Stay, and be secret, and myself will go.    300
ROSS
To horse, to horse! Urge doubts to them that fear.
WILLOUGHBY
Hold out my horse, and I will first be there.    *Exeunt*

2.2    *Enter the Queen, Bushy, and Bagot*
BUSHY
Madam, your majesty is too much sad.
You promised when you parted with the King
To lay aside life-harming heaviness
And entertain a cheerful disposition.
QUEEN
To please the King I did; to please myself    5
I cannot do it. Yet I know no cause
Why I should welcome such a guest as grief,
Save bidding farewell to so sweet a guest
As my sweet Richard. Yet again, methinks
Some unborn sorrow, ripe in fortune's womb,    10
Is coming towards me; and my inward soul
At nothing trembles. With something it grieves
More than with parting from my lord the King.
BUSHY
Each substance of a grief hath twenty shadows
Which shows like grief itself but is not so.    15
For sorrow's eye, glazèd with blinding tears,
Divides one thing entire to many objects—
Like perspectives, which, rightly gazed upon,
Show nothing but confusion; eyed awry,
Distinguish form. So your sweet majesty,    20
Looking awry upon your lord's departure,
Find shapes of grief more than himself to wail,
Which, looked on as it is, is naught but shadows
Of what it is not. Then, thrice-gracious Queen,
More than your lord's departure weep not: more is
    not seen,    25

Or if it be, 'tis with false sorrow's eye,
Which for things true weeps things imaginary.

QUEEN
It may be so, but yet my inward soul
Persuades me it is otherwise. Howe'er it be,
I cannot but be sad: so heavy-sad                          30
As thought—on thinking on no thought I think—
Makes me with heavy nothing faint and shrink.

BUSHY
'Tis nothing but conceit, my gracious lady.

QUEEN
'Tis nothing less: conceit is still derived
From some forefather grief; mine is not so;               35
For nothing hath begot my something grief—
Or something hath the nothing that I grieve—
'Tis in reversion that I do possess—
But what it is that is not yet known what,
I cannot name; 'tis nameless woe, I wot.                  40
            *Enter Green*

GREEN
God save your majesty, and well met, gentlemen.
I hope the King is not yet shipped for Ireland.

QUEEN
Why hop'st thou so? 'Tis better hope he is,
For his designs crave haste, his haste good hope.
Then wherefore dost thou hope he is not shipped?         45

GREEN
That he, our hope, might have retired his power,
And driven into despair an enemy's hope,
Who strongly hath set footing in this land.
The banished Bolingbroke repeals himself,
And with uplifted arms is safe arrived                    50
At Ravenspurgh.

QUEEN              Now God in heaven forbid!

GREEN
Ah madam, 'tis too true! And, that is worse,
The Lord Northumberland, his son young Harry Percy,
The Lords of Ross, Beaumont, and Willoughby,
With all their powerful friends, are fled to him.         55

BUSHY
Why have you not proclaimed Northumberland,
And all the rest, revolted faction-traitors?

GREEN
We have; whereupon the Earl of Worcester
Hath broke his staff, resigned his stewardship,
And all the household servants fled with him              60
To Bolingbroke.

QUEEN
So, Green, thou art the midwife to my woe,
And Bolingbroke my sorrow's dismal heir.
Now hath my soul brought forth her prodigy,
And I, a gasping new-delivered mother,                    65
Have woe to woe, sorrow to sorrow joined.

BUSHY
Despair not, madam.

QUEEN                   Who shall hinder me?
I will despair, and be at enmity
With cozening hope. He is a flatterer,
A parasite, a keeper-back of death,                       70
Who gently would dissolve the bonds of life,
Which false hope lingers in extremity.
            *Enter the Duke of York,* ⌈*wearing a gorget*⌉

GREEN Here comes the Duke of York.

QUEEN
With signs of war about his agèd neck.
O, full of careful business are his looks!                75
Uncle, for God's sake speak comfortable words.

YORK
Should I do so, I should belie my thoughts.
Comfort's in heaven, and we are on the earth,
Where nothing lives but crosses, cares, and grief.
Your husband, he is gone to save far off,                 80
Whilst others come to make him lose at home.
Here am I, left to underprop his land,
Who, weak with age, cannot support myself.
Now comes the sick hour that his surfeit made.
Now shall he try his friends that flattered him.          85
            *Enter a Servingman*

SERVINGMAN
My lord, your son was gone before I came.

YORK
He was? Why so, go all which way it will.
The nobles they are fled. The commons they are cold,
And will, I fear, revolt on Hereford's side.
Sirrah, get thee to Pleshey, to my sister Gloucester.     90
Bid her send me presently a thousand pound—
Hold; take my ring.

SERVINGMAN
My lord, I had forgot to tell your lordship,
Today as I came by I callèd there—
But I shall grieve you to report the rest.                95

YORK What is't, knave?

SERVINGMAN
An hour before I came, the Duchess died.

YORK
God for his mercy, what a tide of woes
Comes rushing on this woeful land at once!
I know not what to do. I would to God,                   100
So my untruth had not provoked him to it,
The King had cut off my head with my brother's.
What, are there no posts dispatched for Ireland?
How shall we do for money for these wars?
(*To the Queen*) Come, sister—cousin, I would say; pray
            pardon me.                                     105
(*To the Servingman*) Go, fellow, get thee home. Provide
            some carts,
And bring away the armour that is there.
                        ⌈*Exit Servingman*⌉
Gentlemen, will you go muster men?
If I know how or which way to order these affairs
Thus disorderly thrust into my hands,                    110
Never believe me. Both are my kinsmen.
T'one is my sovereign, whom both my oath
And duty bids defend; t'other again
Is my kinsman, whom the King hath wronged,
Whom conscience and my kindred bids to right.            115
Well, somewhat we must do. (*To the Queen*) Come,
            cousin,
I'll dispose of you.—
Gentlemen, go muster up your men,
And meet me presently at Berkeley Castle.
I should to Pleshey too, but time will not permit.       120
All is uneven,
And everything is left at six and seven.
            *Exeunt the Duke of York and the Queen. Bushy,*
                        *Bagot, and Green remain*

BUSHY
The wind sits fair for news to go for Ireland,
But none returns. For us to levy power
Proportionable to the enemy                              125
Is all unpossible.

GREEN
Besides, our nearness to the King in love
Is near the hate of those love not the King.

BAGOT
And that is the wavering commons; for their love
Lies in their purses, and whoso empties them        130
By so much fills their hearts with deadly hate.
BUSHY
Wherein the King stands generally condemned.
BAGOT
If judgement lie in them, then so do we,
Because we ever have been near the King.
GREEN
Well, I will for refuge straight to Bristol Castle.   135
The Earl of Wiltshire is already there.
BUSHY
Thither will I with you; for little office
Will the hateful commoners perform for us,
Except like curs to tear us all to pieces.
(To Bagot) Will you go along with us?               140
BAGOT
No, I will to Ireland, to his majesty.
Farewell: if heart's presages be not vain
We three here part that ne'er shall meet again.
BUSHY
That's as York thrives to beat back Bolingbroke.
GREEN
Alas, poor Duke, the task he undertakes             145
Is numb'ring sands and drinking oceans dry.
Where one on his side fights, thousands will fly.
⌈BAGOT⌉
Farewell at once, for once, for all and ever.
BUSHY
Well, we may meet again.
BAGOT                              I fear me never.
             Exeunt ⌈Bushy and Green at one door, and
                                Bagot at another door⌉

2.3   Enter Bolingbroke Duke of Lancaster and Hereford,
      and the Earl of Northumberland
BOLINGBROKE
How far is it, my lord, to Berkeley now?
NORTHUMBERLAND Believe me, noble lord,
I am a stranger here in Gloucestershire.
These high wild hills and rough uneven ways
Draws out our miles and makes them wearisome;      5
And yet your fair discourse hath been as sugar,
Making the hard way sweet and delectable.
But I bethink me what a weary way
From Ravenspurgh to Cotswold will be found
In Ross and Willoughby, wanting your company,      10
Which I protest hath very much beguiled
The tediousness and process of my travel.
But theirs is sweetened with the hope to have
The present benefit which I possess;
And hope to joy is little less in joy              15
Than hope enjoyed. By this the weary lords
Shall make their way seem short as mine hath done
By sight of what I have: your noble company.
BOLINGBROKE
Of much less value is my company
Than your good words.
      Enter Harry Percy
                        But who comes here?          20
NORTHUMBERLAND
It is my son, young Harry Percy,
Sent from my brother Worcester, whencesoever.
Harry, how fares your uncle?

HARRY PERCY
I had thought, my lord, to have learned his health of
   you.
NORTHUMBERLAND Why, is he not with the Queen?     25
HARRY PERCY
No, my good lord; he hath forsook the court,
Broken his staff of office, and dispersed
The household of the King.
NORTHUMBERLAND                What was his reason?
He was not so resolved when last we spake together.
HARRY PERCY
Because your lordship was proclaimèd traitor.      30
But he, my lord, is gone to Ravenspurgh
To offer service to the Duke of Hereford,
And sent me over by Berkeley to discover
What power the Duke of York had levied there,
Then with directions to repair to Ravenspurgh.     35
NORTHUMBERLAND
Have you forgot the Duke of Hereford, boy?
HARRY PERCY
No, my good lord, for that is not forgot
Which ne'er I did remember. To my knowledge,
I never in my life did look on him.
NORTHUMBERLAND
Then learn to know him now. This is the Duke.       40
HARRY PERCY
My gracious lord, I tender you my service,
Such as it is, being tender, raw, and young,
Which elder days shall ripen and confirm
To more approvèd service and desert.
BOLINGBROKE
I thank thee, gentle Percy, and be sure             45
I count myself in nothing else so happy
As in a soul rememb'ring my good friends;
And as my fortune ripens with thy love,
It shall be still thy true love's recompense.
My heart this covenant makes; my hand thus seals it.
      He gives Percy his hand
NORTHUMBERLAND
How far is it to Berkeley, and what stir            51
Keeps good old York there with his men of war?
HARRY PERCY
There stands the castle, by yon tuft of trees,
Manned with three hundred men, as I have heard,
And in it are the Lords of York, Berkeley, and
   Seymour,                                          55
None else of name and noble estimate.
      Enter Lord Ross and Lord Willoughby
NORTHUMBERLAND
Here come the Lords of Ross and Willoughby,
Bloody with spurring, fiery red with haste.
BOLINGBROKE
Welcome, my lords. I wot your love pursues
A banished traitor. All my treasury                 60
Is yet but unfelt thanks, which, more enriched,
Shall be your love and labour's recompense.
ROSS
Your presence makes us rich, most noble lord.
WILLOUGHBY
And far surmounts our labour to attain it.
BOLINGBROKE
Evermore thank's the exchequer of the poor,         65
Which till my infant fortune comes to years
Stands for my bounty.
      Enter Berkeley
                        But who comes here?

NORTHUMBERLAND
It is my lord of Berkeley, as I guess.
BERKELEY
My lord of Hereford, my message is to you.
BOLINGBROKE
My lord, my answer is to 'Lancaster',                    70
And I am come to seek that name in England,
And I must find that title in your tongue
Before I make reply to aught you say.
BERKELEY
Mistake me not, my lord, 'tis not my meaning
To raze one title of your honour out.                    75
To you, my lord, I come—what lord you will—
From the most gracious regent of this land,
The Duke of York, to know what pricks you on
To take advantage of the absent time
And fright our native peace with self-borne arms.        80
    *Enter the Duke of York*
BOLINGBROKE
I shall not need transport my words by you.
Here comes his grace in person.—My noble uncle!
    *He kneels*
YORK
Show me thy humble heart, and not thy knee,
Whose duty is deceivable and false.
BOLINGBROKE My gracious uncle—                          85
YORK
Tut, tut, grace me no grace, nor uncle me no uncle.
I am no traitor's uncle, and that word 'grace'
In an ungracious mouth is but profane.
Why have those banished and forbidden legs
Dared once to touch a dust of England's ground?          90
But then more 'why': why have they dared to march
So many miles upon her peaceful bosom,
Frighting her pale-faced villages with war
And ostentation of despisèd arms?
Com'st thou because the anointed King is hence?          95
Why, foolish boy, the King is left behind,
And in my loyal bosom lies his power.
Were I but now the lord of such hot youth
As when brave Gaunt, thy father, and myself
Rescued the Black Prince, that young Mars of men,
From forth the ranks of many thousand French,           101
O then how quickly should this arm of mine,
Now prisoner to the palsy, chastise thee
And minister correction to thy fault!
BOLINGBROKE
My gracious uncle, let me know my fault.                 105
On what condition stands it and wherein?
YORK
Even in condition of the worst degree:
In gross rebellion and detested treason.
Thou art a banished man, and here art come
Before the expiration of thy time                        110
In braving arms against thy sovereign.
BOLINGBROKE ⌈*standing*⌉
As I was banished, I was banished Hereford;
But as I come, I come for Lancaster.
And, noble uncle, I beseech your grace,
Look on my wrongs with an indifferent eye.               115
You are my father, for methinks in you
I see old Gaunt alive. O then, my father,
Will you permit that I shall stand condemned
A wandering vagabond, my rights and royalties
Plucked from my arms perforce and given away            120
To upstart unthrifts? Wherefore was I born?

If that my cousin King be King in England,
It must be granted I am Duke of Lancaster.
You have a son, Aumerle my noble kinsman.
Had you first died and he been thus trod down,           125
He should have found his uncle Gaunt a father
To rouse his wrongs and chase them to the bay.
I am denied to sue my livery here,
And yet my letters patents give me leave.
My father's goods are all distrained and sold,          130
And these and all are all amiss employed.
What would you have me do? I am a subject,
And I challenge law; attorneys are denied me;
And therefore personally I lay my claim
To my inheritance of free descent.                      135
NORTHUMBERLAND
The noble Duke hath been too much abused.
ROSS
It stands your grace upon to do him right.
WILLOUGHBY
Base men by his endowments are made great.
YORK
My lords of England, let me tell you this.
I have had feeling of my cousin's wrongs,               140
And laboured all I could to do him right.
But in this kind to come, in braving arms,
Be his own carver, and cut out his way
To find out right with wrong—it may not be.
And you that do abet him in this kind                    145
Cherish rebellion, and are rebels all.
NORTHUMBERLAND
The noble Duke hath sworn his coming is
But for his own, and for the right of that
We all have strongly sworn to give him aid;
And let him never see joy that breaks that oath.        150
YORK
Well, well, I see the issue of these arms.
I cannot mend it, I must needs confess,
Because my power is weak and all ill-left.
But if I could, by Him that gave me life,
I would attach you all, and make you stoop              155
Unto the sovereign mercy of the King.
But since I cannot, be it known to you
I do remain as neuter. So fare you well—
Unless you please to enter in the castle
And there repose you for this night.                    160
BOLINGBROKE
An offer, uncle, that we will accept.
But we must win your grace to go with us
To Bristol Castle, which they say is held
By Bushy, Bagot, and their complices,
The caterpillars of the commonwealth,                   165
Which I have sworn to weed and pluck away.
YORK
It may be I will go with you—but yet I'll pause,
For I am loath to break our country's laws.
Nor friends nor foes, to me welcome you are.            169
Things past redress are now with me past care. *Exeunt*

2.4    *Enter the Earl of Salisbury and a Welsh Captain*
WELSH CAPTAIN
My lord of Salisbury, we have stayed ten days,
And hardly kept our countrymen together,
And yet we hear no tidings from the King.
Therefore we will disperse ourselves. Farewell.
SALISBURY
Stay yet another day, thou trusty Welshman.             5
The King reposeth all his confidence in thee.

WELSH CAPTAIN
'Tis thought the King is dead. We will not stay.
The bay trees in our country are all withered,
And meteors fright the fixèd stars of heaven.
The pale-faced moon looks bloody on the earth,                    10
And lean-looked prophets whisper fearful change.
Rich men look sad, and ruffians dance and leap;
The one in fear to lose what they enjoy,
The other to enjoy by rage and war.
These signs forerun the death or fall of kings.                    15
Farewell. Our countrymen are gone and fled,
As well assured Richard their king is dead.                    *Exit*

SALISBURY
Ah, Richard! With the eyes of heavy mind
I see thy glory, like a shooting star,
Fall to the base earth from the firmament.                    20
Thy sun sets weeping in the lowly west,
Witnessing storms to come, woe, and unrest.
Thy friends are fled to wait upon thy foes,
And crossly to thy good all fortune goes.                    *Exit*

3.1    *Enter Bolingbroke Duke of Lancaster and Hereford,*
       *the Duke of York, the Earl of Northumberland,*
       *⌐Lord Ross, Harry Percy, and Lord Willoughby⌐*

BOLINGBROKE Bring forth these men.
       *Enter Bushy and Green, guarded as prisoners*
Bushy and Green, I will not vex your souls,
Since presently your souls must part your bodies,
With too much urging your pernicious lives,
For 'twere no charity. Yet to wash your blood                    5
From off my hands, here in the view of men
I will unfold some causes of your deaths.
You have misled a prince, a royal king,
A happy gentleman in blood and lineaments,
By you unhappied and disfigured clean.                    10
You have, in manner, with your sinful hours
Made a divorce betwixt his queen and him,
Broke the possession of a royal bed,
And stained the beauty of a fair queen's cheeks
With tears drawn from her eyes by your foul wrongs.
Myself—a prince by fortune of my birth,                    16
Near to the King in blood, and near in love
Till you did make him misinterpret me—
Have stooped my neck under your injuries,
And sighed my English breath in foreign clouds,                    20
Eating the bitter bread of banishment,
Whilst you have fed upon my signories,
Disparked my parks and felled my forest woods,
From my own windows torn my household coat,
Razed out my imprese, leaving me no sign,                    25
Save men's opinions and my living blood,
To show the world I am a gentleman.
This and much more, much more than twice all this,
Condemns you to the death.—See them delivered over
To execution and the hand of death.                    30

BUSHY
More welcome is the stroke of death to me
Than Bolingbroke to England.

GREEN
My comfort is that heaven will take our souls,
And plague injustice with the pains of hell.

BOLINGBROKE
My lord Northumberland, see them dispatched.                    35
       *Exit Northumberland, with Bushy and Green,*
                                              *guarded*

Uncle, you say the Queen is at your house.
For God's sake, fairly let her be intreated.
Tell her I send to her my kind commends.
Take special care my greetings be delivered.

YORK
A gentleman of mine I have dispatched                    40
With letters of your love to her at large.

BOLINGBROKE
Thanks, gentle uncle.—Come, lords, away,
To fight with Glyndŵr and his complices.
A while to work, and after, holiday.                    *Exeunt*

3.2    *⌐Flourish.⌐ Enter King Richard, the Duke of*
       *Aumerle, the Bishop of Carlisle, and ⌐soldiers, with*
       *drum and colours⌐*

KING RICHARD
Harlechly Castle call they this at hand?

AUMERLE
Yea, my lord. How brooks your grace the air
After your late tossing on the breaking seas?

KING RICHARD
Needs must I like it well. I weep for joy
To stand upon my kingdom once again.                    5
       *He touches the ground*
Dear earth, I do salute thee with my hand,
Though rebels wound thee with their horses' hoofs.
As a long-parted mother with her child
Plays fondly with her tears, and smiles in meeting,
So, weeping, smiling, greet I thee my earth,                    10
And do thee favours with my royal hands.
Feed not thy sovereign's foe, my gentle earth,
Nor with thy sweets comfort his ravenous sense;
But let thy spiders that suck up thy venom
And heavy-gaited toads lie in their way,                    15
Doing annoyance to the treacherous feet
Which with usurping steps do trample thee.
Yield stinging nettles to mine enemies,
And when they from thy bosom pluck a flower
Guard it, I pray thee, with a lurking adder,                    20
Whose double tongue may with a mortal touch
Throw death upon thy sovereign's enemies.—
Mock not my senseless conjuration, lords.
This earth shall have a feeling, and these stones
Prove armèd soldiers, ere her native king                    25
Shall falter under foul rebellion's arms.

BISHOP OF CARLISLE
Fear not, my lord. That power that made you king
Hath power to keep you king in spite of all.

AUMERLE
He means, my lord, that we are too remiss,
Whilst Bolingbroke, through our security,                    30
Grows strong and great in substance and in friends.

KING RICHARD
Discomfortable cousin, know'st thou not
That when the searching eye of heaven is hid
Behind the globe, that lights the lower world,
Then thieves and robbers range abroad unseen                    35
In murders and in outrage bloody here;
But when from under this terrestrial ball
He fires the proud tops of the eastern pines,
And darts his light through every guilty hole,
Then murders, treasons, and detested sins,                    40
The cloak of night being plucked from off their backs,
Stand bare and naked, trembling at themselves?
So when this thief, this traitor, Bolingbroke,

Who all this while hath revelled in the night
Whilst we were wand'ring with the Antipodes,                    45
Shall see us rising in our throne, the east,
His treasons will sit blushing in his face,
Not able to endure the sight of day,
But, self-affrighted, tremble at his sin.
Not all the water in the rough rude sea                    50
Can wash the balm from an anointed king.
The breath of worldly men cannot depose
The deputy elected by the Lord.
For every man that Bolingbroke hath pressed
To lift shrewd steel against our golden crown,                    55
God for his Richard hath in heavenly pay
A glorious angel. Then if angels fight,
Weak men must fall; for heaven still guards the right.
    *Enter the Earl of Salisbury*
Welcome, my lord. How far off lies your power?
SALISBURY
Nor nea'er nor farther off, my gracious lord,                    60
Than this weak arm. Discomfort guides my tongue,
And bids me speak of nothing but despair.
One day too late, I fear me, noble lord,
Hath clouded all thy happy days on earth.
O, call back yesterday, bid time return,                    65
And thou shalt have twelve thousand fighting men.
Today, today, unhappy day too late,
Overthrows thy joys, friends, fortune, and thy state;
For all the Welshmen, hearing thou wert dead,
Are gone to Bolingbroke, dispersed, and fled.                    70
AUMERLE
Comfort, my liege. Why looks your grace so pale?
KING RICHARD
But now the blood of twenty thousand men
   Did triumph in my face, and they are fled;
And till so much blood thither come again
   Have I not reason to look pale and dead?                    75
All souls that will be safe fly from my side,
For time hath set a blot upon my pride.
AUMERLE
Comfort, my liege. Remember who you are.
KING RICHARD
I had forgot myself. Am I not King?
Awake, thou sluggard majesty, thou sleep'st!                    80
Is not the King's name forty thousand names?
Arm, arm, my name! A puny subject strikes
At thy great glory. Look not to the ground,
Ye favourites of a king: are we not high?
High be our thoughts. I know my uncle York                    85
Hath power enough to serve our turn.
    *Enter Scrope*
                    But who comes here?
SCROPE
More health and happiness betide my liege
Than can my care-tuned tongue deliver him.
KING RICHARD
Mine ear is open and my heart prepared.
The worst is worldly loss thou canst unfold.                    90
Say, is my kingdom lost? Why 'twas my care,
And what loss is it to be rid of care?
Strives Bolingbroke to be as great as we?
Greater he shall not be. If he serve God
We'll serve Him too, and be his fellow so.                    95
Revolt our subjects? That we cannot mend.
They break their faith to God as well as us.
Cry woe, destruction, ruin, loss, decay:
The worst is death, and death will have his day.

SCROPE
Glad am I that your highness is so armed                    100
To bear the tidings of calamity.
Like an unseasonable stormy day,
Which makes the silver rivers drown their shores
As if the world were all dissolved to tears,
So high above his limits swells the rage                    105
Of Bolingbroke, covering your fearful land
With hard bright steel, and hearts harder than steel.
Whitebeards have armed their thin and hairless scalps
Against thy majesty. Boys with women's voices
Strive to speak big, and clap their female joints                    110
In stiff unwieldy arms against thy crown.
Thy very beadsmen learn to bend their bows
Of double-fatal yew against thy state.
Yea, distaff-women manage rusty bills
Against thy seat. Both young and old rebel,                    115
And all goes worse than I have power to tell.
KING RICHARD
Too well, too well thou tell'st a tale so ill.
Where is the Earl of Wiltshire? Where is Bagot?
What is become of Bushy, where is Green,
That they have let the dangerous enemy                    120
Measure our confines with such peaceful steps?
If we prevail, their heads shall pay for it.
I warrant they have made peace with Bolingbroke.
SCROPE
Peace have they made with him indeed, my lord.
KING RICHARD
O villains, vipers damned without redemption!                    125
Dogs easily won to fawn on any man!
Snakes in my heart-blood warmed, that sting my
    heart!
Three Judases, each one thrice-worse than Judas!
Would they make peace? Terrible hell make war
Upon their spotted souls for this offence!                    130
SCROPE
Sweet love, I see, changing his property,
Turns to the sourest and most deadly hate.
Again uncurse their souls. Their peace is made
With heads, and not with hands. Those whom you
    curse
Have felt the worst of death's destroying wound,                    135
And lie full low, graved in the hollow ground.
AUMERLE
Is Bushy, Green, and the Earl of Wiltshire dead?
SCROPE
Ay, all of them at Bristol lost their heads.
AUMERLE
Where is the Duke my father, with his power?
KING RICHARD
No matter where. Of comfort no man speak.                    140
Let's talk of graves, of worms and epitaphs,
Make dust our paper, and with rainy eyes
Write sorrow on the bosom of the earth.
Let's choose executors and talk of wills—
And yet not so, for what can we bequeath                    145
Save our deposèd bodies to the ground?
Our lands, our lives, and all are Bolingbroke's;
And nothing can we call our own but death,
And that small model of the barren earth
Which serves as paste and cover to our bones.                    150
⌜*Sitting*⌝ For God's sake, let us sit upon the ground,
And tell sad stories of the death of kings—
How some have been deposed, some slain in war,
Some haunted by the ghosts they have deposed,

Some poisoned by their wives, some sleeping killed,
All murdered. For within the hollow crown          156
That rounds the mortal temples of a king
Keeps Death his court; and there the antic sits,
Scoffing his state and grinning at his pomp,
Allowing him a breath, a little scene,             160
To monarchize, be feared, and kill with looks,
Infusing him with self and vain conceit,
As if this flesh which walls about our life
Were brass impregnable; and humoured thus,
Comes at the last, and with a little pin           165
Bores through his castle wall; and farewell, king.
Cover your heads, and mock not flesh and blood
With solemn reverence. Throw away respect,
Tradition, form, and ceremonious duty,
For you have but mistook me all this while.        170
I live with bread, like you; feel want,
Taste grief, need friends. Subjected thus,
How can you say to me I am a king?

BISHOP OF CARLISLE
My lord, wise men ne'er wail their present woes,
But presently prevent the ways to wail.            175
To fear the foe, since fear oppresseth strength,
Gives in your weakness strength unto your foe;
And so your follies fight against yourself.
Fear, and be slain. No worse can come to fight;
And fight and die is death destroying death,       180
Where fearing dying pays death servile breath.

AUMERLE
My father hath a power. Enquire of him,
And learn to make a body of a limb.

KING RICHARD ⌈standing⌉
Thou chid'st me well. Proud Bolingbroke, I come
To change blows with thee for our day of doom.     185
This ague-fit of fear is overblown.
An easy task it is to win our own.
Say, Scrope, where lies our uncle with his power?
Speak sweetly, man, although thy looks be sour.

SCROPE
Men judge by the complexion of the sky             190
   The state and inclination of the day.
So may you by my dull and heavy eye
   My tongue hath but a heavier tale to say.
I play the torturer by small and small
To lengthen out the worst that must be spoken.     195
Your uncle York is joined with Bolingbroke,
And all your northern castles yielded up,
And all your southern gentlemen in arms
Upon his faction.

KING RICHARD              Thou hast said enough.
(To Aumerle) Beshrew thee, cousin, which didst lead
   me forth                                        200
Of that sweet way I was in to despair.
What say you now? What comfort have we now?
By heaven, I'll hate him everlastingly
That bids me be of comfort any more.
Go to Flint Castle; there I'll pine away.          205
A king, woe's slave, shall kingly woe obey.
That power I have, discharge, and let them go
To ear the land that hath some hope to grow;
For I have none. Let no man speak again
To alter this, for counsel is but vain.            210

AUMERLE
My liege, one word.

KING RICHARD              He does me double wrong
That wounds me with the flatteries of his tongue.

Discharge my followers. Let them hence away
From Richard's night to Bolingbroke's fair day.
                                        *Exeunt*

3.3     *Enter Bolingbroke Duke of Lancaster and Hereford,*
        *the Duke of York, the Earl of Northumberland,*
        ⌈*and soldiers, with drum and colours*⌉
BOLINGBROKE
So that by this intelligence we learn
The Welshmen are dispersed, and Salisbury
Is gone to meet the King, who lately landed
With some few private friends upon this coast.

NORTHUMBERLAND
The news is very fair and good, my lord.           5
Richard not far from hence hath hid his head.

YORK
It would beseem the Lord Northumberland
To say 'King Richard'. Alack the heavy day
When such a sacred king should hide his head!

NORTHUMBERLAND
Your grace mistakes. Only to be brief              10
Left I his title out.

YORK                    The time hath been,
Would you have been so brief with him, he would
Have been so brief with you to shorten you,
For taking so the head, your whole head's length.

BOLINGBROKE
Mistake not, uncle, further than you should.       15

YORK
Take not, good cousin, further than you should,
Lest you mistake the heavens are over our heads.

BOLINGBROKE
I know it, uncle, and oppose not myself
Against their will.
        *Enter Harry Percy* ⌈*and a trumpeter*⌉
                    But who comes here?
Welcome, Harry. What, will not this castle yield?  20

HARRY PERCY
The castle royally is manned, my lord,
Against thy entrance.

BOLINGBROKE                    Royally?
Why, it contains no king.

HARRY PERCY                    Yes, my good lord,
It doth contain a king. King Richard lies
Within the limits of yon lime and stone,           25
And with him are the Lord Aumerle, Lord Salisbury,
Sir Stephen Scrope, besides a clergyman
Of holy reverence; who, I cannot learn.

NORTHUMBERLAND
O, belike it is the Bishop of Carlisle.

BOLINGBROKE (*to Northumberland*) Noble lord,      30
Go to the rude ribs of that ancient castle;
Through brazen trumpet send the breath of parley
Into his ruined ears, and thus deliver.
Henry Bolingbroke
Upon his knees doth kiss King Richard's hand,      35
And sends allegiance and true faith of heart
To his most royal person, hither come
Even at his feet to lay my arms and power,
Provided that my banishment repealed
And lands restored again be freely granted.        40
If not, I'll use the advantage of my power,
And lay the summer's dust with showers of blood
Rained from the wounds of slaughtered Englishmen;
The which how far off from the mind of Bolingbroke

It is such crimson tempest should bedrench          45
The fresh green lap of fair King Richard's land,
My stooping duty tenderly shall show.
Go, signify as much, while here we march
Upon the grassy carpet of this plain.
Let's march without the noise of threat'ning drum,   50
That from this castle's tottered battlements
Our fair appointments may be well perused.
Methinks King Richard and myself should meet
With no less terror than the elements
Of fire and water when their thund'ring shock        55
At meeting tears the cloudy cheeks of heaven.
Be he the fire, I'll be the yielding water.
The rage be his, whilst on the earth I rain
My waters: on the earth, and not on him.—
March on, and mark King Richard, how he looks.      60
    ⌈They march about the stage; then Bolingbroke,
    York, Percy, and soldiers stand at a distance from
    the walls; Northumberland and trumpeter advance
    to the walls.⌉ The trumpets sound ⌈a parley
    without, and an answer within; then a flourish
    within⌉. King Richard appeareth on the walls, with
    the Bishop of Carlisle, the Duke of Aumerle,
    ⌈Scrope, and the Earl of Salisbury⌉
See, see, King Richard doth himself appear,
As doth the blushing discontented sun
From out the fiery portal of the east
When he perceives the envious clouds are bent
To dim his glory and to stain the track              65
Of his bright passage to the occident.
YORK
Yet looks he like a king. Behold, his eye,
As bright as is the eagle's, lightens forth
Controlling majesty. Alack, alack for woe
That any harm should stain so fair a show!           70
KING RICHARD (to Northumberland)
We are amazed; and thus long have we stood
To watch the fearful bending of thy knee,
Because we thought ourself thy lawful king.
An if we be, how dare thy joints forget
To pay their aweful duty to our presence?            75
If we be not, show us the hand of God
That hath dismissed us from our stewardship.
For well we know no hand of blood and bone
Can grip the sacred handle of our sceptre,
Unless he do profane, steal, or usurp.               80
And though you think that all—as you have done—
Have torn their souls by turning them from us,
And we are barren and bereft of friends,
Yet know my master, God omnipotent,
Is mustering in his clouds on our behalf             85
Armies of pestilence; and they shall strike
Your children yet unborn and unbegot,
That lift your vassal hands against my head
And threat the glory of my precious crown.
Tell Bolingbroke, for yon methinks he is,            90
That every stride he makes upon my land
Is dangerous treason. He is come to open
The purple testament of bleeding war;
But ere the crown he looks for live in peace
Ten thousand bloody crowns of mothers' sons          95
Shall ill become the flower of England's face,
Change the complexion of her maid-pale peace
To scarlet indignation, and bedew
Her pastures' grass with faithful English blood.

NORTHUMBERLAND ⌈kneeling⌉
The King of heaven forbid our lord the King          100
Should so with civil and uncivil arms
Be rushed upon. Thy thrice-noble cousin
Harry Bolingbroke doth humbly kiss thy hand,
And by the honourable tomb he swears,
That stands upon your royal grandsire's bones,       105
And by the royalties of both your bloods,
Currents that spring from one most gracious head,
And by the buried hand of warlike Gaunt,
And by the worth and honour of himself,
Comprising all that may be sworn or said,            110
His coming hither hath no further scope
Than for his lineal royalties, and to beg
Enfranchisement immediate on his knees;
Which on thy royal party granted once,
His glittering arms he will commend to rust,         115
His barbèd steeds to stables, and his heart
To faithful service of your majesty.
This swears he as he is a prince and just,
And as I am a gentleman I credit him.
KING RICHARD
Northumberland, say thus the King returns:           120
His noble cousin is right welcome hither,
And all the number of his fair demands
Shall be accomplished without contradiction.
With all the gracious utterance thou hast,
Speak to his gentle hearing kind commends.           125
    Northumberland and the trumpeter return to
    Bolingbroke
(To Aumerle) We do debase ourself, cousin, do we not,
To look so poorly and to speak so fair?
Shall we call back Northumberland, and send
Defiance to the traitor, and so die?
AUMERLE
No, good my lord, let's fight with gentle words      130
Till time lend friends, and friends their helpful swords.
KING RICHARD
O God, O God, that e'er this tongue of mine,
That laid the sentence of dread banishment
On yon proud man, should take it off again
With words of sooth! O, that I were as great         135
As is my grief, or lesser than my name,
Or that I could forget what I have been,
Or not remember what I must be now!
Swell'st thou, proud heart? I'll give thee scope to
    beat,
Since foes have scope to beat both thee and me.      140
    Northumberland advances to the walls
AUMERLE
Northumberland comes back from Bolingbroke.
KING RICHARD
What must the King do now? Must he submit?
The King shall do it. Must he be deposed?
The King shall be contented. Must he lose
The name of King? A God's name, let it go.           145
I'll give my jewels for a set of beads,
My gorgeous palace for a hermitage,
My gay apparel for an almsman's gown,
My figured goblets for a dish of wood,
My sceptre for a palmer's walking staff,             150
My subjects for a pair of carvèd saints,
And my large kingdom for a little grave,
A little, little grave, an obscure grave;
Or I'll be buried in the King's highway,

Some way of common trade where subjects' feet   155
May hourly trample on their sovereign's head,
For on my heart they tread now, whilst I live,
And buried once, why not upon my head?
Aumerle, thou weep'st, my tender-hearted cousin.
We'll make foul weather with despisèd tears.   160
Our sighs and they shall lodge the summer corn,
And make a dearth in this revolting land.
Or shall we play the wantons with our woes,
And make some pretty match with shedding tears;
As thus to drop them still upon one place   165
Till they have fretted us a pair of graves
Within the earth, and therein laid? 'There lies
Two kinsmen digged their graves with weeping eyes.'
Would not this ill do well? Well, well, I see
I talk but idly and you mock at me.   170
Most mighty prince, my lord Northumberland,
What says King Bolingbroke? Will his majesty
Give Richard leave to live till Richard die?
You make a leg, and Bolingbroke says 'Ay'.

NORTHUMBERLAND
My lord, in the base court he doth attend   175
To speak with you. May it please you to come down?

KING RICHARD
Down, down I come like glist'ring Phaethon,
Wanting the manage of unruly jades.
In the base court: base court where kings grow base
To come at traitors' calls, and do them grace.   180
In the base court, come down: down court, down
    King,
For night-owls shriek where mounting larks should
    sing.         *Exeunt King Richard and his party*
    *Northumberland returns to Bolingbroke*

BOLINGBROKE
What says his majesty?

NORTHUMBERLAND         Sorrow and grief of heart
Makes him speak fondly, like a frantic man.
    *Enter King Richard ⌐and his party⌐ below*
Yet he is come.

BOLINGBROKE         Stand all apart,   185
And show fair duty to his majesty.
    *He kneels down*
My gracious lord.

KING RICHARD
Fair cousin, you debase your princely knee
To make the base earth proud with kissing it.
Me rather had my heart might feel your love   190
Than my unpleased eye see your courtesy.
Up, cousin, up. Your heart is up, I know,
Thus high at least, although your knee be low.

BOLINGBROKE
My gracious lord, I come but for mine own.

KING RICHARD
Your own is yours, and I am yours, and all.   195

BOLINGBROKE
So far be mine, my most redoubted lord,
As my true service shall deserve your love.

KING RICHARD
Well you deserve. They well deserve to have
That know the strong'st and surest way to get.
    ⌐*Bolingbroke rises*⌐
(*To York*) Uncle, give me your hands. Nay, dry your
    eyes.   200
Tears show their love, but want their remedies.
(*To Bolingbroke*) Cousin, I am too young to be your
    father,
Though you are old enough to be my heir.

What you will have I'll give, and willing too;
For do we must what force will have us do.   205
Set on towards London, cousin: is it so?

BOLINGBROKE
Yea, my good lord.

KING RICHARD         Then I must not say no.
    *Flourish. Exeunt*

**3.4**    *Enter the Queen, with her two Ladies*

QUEEN
What sport shall we devise here in this garden,
To drive away the heavy thought of care?

⌐FIRST⌐ LADY Madam, we'll play at bowls.

QUEEN
'Twill make me think the world is full of rubs,
And that my fortune runs against the bias.   5

⌐SECOND⌐ LADY Madam, we'll dance.

QUEEN
My legs can keep no measure in delight
When my poor heart no measure keeps in grief;
Therefore no dancing, girl. Some other sport.

⌐FIRST⌐ LADY Madam, we'll tell tales.   10

QUEEN Of sorrow or of joy?

⌐FIRST⌐ LADY Of either, madam.

QUEEN Of neither, girl.
For if of joy, being altogether wanting,
It doth remember me the more of sorrow.   15
Or if of grief, being altogether had,
It adds more sorrow to my want of joy.
For what I have I need not to repeat,
And what I want it boots not to complain.

⌐SECOND⌐ LADY
Madam, I'll sing.

QUEEN         'Tis well that thou hast cause;   20
But thou shouldst please me better wouldst thou
    weep.

⌐SECOND⌐ LADY
I could weep, madam, would it do you good.

QUEEN
And I could sing, would weeping do me good,
And never borrow any tear of thee.
    *Enter a Gardener and two Men*
But stay; here come the gardeners.   25
Let's step into the shadow of these trees.
My wretchedness unto a row of pins
They will talk of state, for everyone doth so
Against a change. Woe is forerun with woe.
    *The Queen and her Ladies stand apart*

GARDENER ⌐*to First Man*⌐
Go, bind thou up young dangling apricots   30
Which, like unruly children, make their sire
Stoop with oppression of their prodigal weight.
Give some supportance to the bending twigs.
⌐*To Second Man*⌐ Go thou, and, like an executioner,
Cut off the heads of too fast-growing sprays   35
That look too lofty in our commonwealth.
All must be even in our government.
You thus employed, I will go root away
The noisome weeds which without profit suck
The soil's fertility from wholesome flowers.   40

⌐FIRST⌐ MAN
Why should we, in the compass of a pale,
Keep law and form and due proportion,
Showing as in a model our firm estate,
When our sea-wallèd garden, the whole land,
Is full of weeds, her fairest flowers choked up,   45

Her fruit trees all unpruned, her hedges ruined,
Her knots disordered, and her wholesome herbs
Swarming with caterpillars?

GARDENER            Hold thy peace.
He that hath suffered this disordered spring
Hath now himself met with the fall of leaf.      50
The weeds which his broad spreading leaves did
     shelter,
That seemed in eating him to hold him up,
Are plucked up, root and all, by Bolingbroke—
I mean the Earl of Wiltshire, Bushy, Green.

⌈SECOND⌉ MAN
What, are they dead?

GARDENER          They are; and Bolingbroke    55
Hath seized the wasteful King. O, what pity is it
That he had not so trimmed and dressed his land
As we this garden! We at time of year
Do wound the bark, the skin of our fruit trees,
Lest, being over-proud in sap and blood,      60
With too much riches it confound itself.
Had he done so to great and growing men,
They might have lived to bear, and he to taste,
Their fruits of duty. Superfluous branches
We lop away, that bearing boughs may live.      65
Had he done so, himself had borne the crown,
Which waste of idle hours hath quite thrown down.

⌈FIRST⌉ MAN
What, think you then the King shall be deposed?

GARDENER
Depressed he is already, and deposed
'Tis doubt he will be. Letters came last night      70
To a dear friend of the good Duke of York's
That tell black tidings.

QUEEN
O, I am pressed to death through want of speaking!
     *She comes forward*
Thou, old Adam's likeness, set to dress this garden,
How dares thy harsh rude tongue sound this
     unpleasing news?      75
What Eve, what serpent hath suggested thee
To make a second fall of cursèd man?
Why dost thou say King Richard is deposed?
Dar'st thou, thou little better thing than earth,
Divine his downfall? Say where, when, and how    80
Cam'st thou by this ill tidings? Speak, thou wretch!

GARDENER
Pardon me, madam. Little joy have I
To breathe this news, yet what I say is true.
King Richard he is in the mighty hold
Of Bolingbroke. Their fortunes both are weighed.    85
In your lord's scale is nothing but himself
And some few vanities that make him light.
But in the balance of great Bolingbroke,
Besides himself, are all the English peers,
And with that odds he weighs King Richard down.   90
Post you to London and you will find it so.
I speak no more than everyone doth know.

QUEEN
Nimble mischance that art so light of foot,
Doth not thy embassage belong to me,
And am I last that knows it? O, thou think'st      95
To serve me last, that I may longest keep
Thy sorrow in my breast. Come, ladies, go
To meet at London London's king in woe.
What, was I born to this, that my sad look
Should grace the triumph of great Bolingbroke?    100

Gard'ner, for telling me these news of woe,
Pray God the plants thou graft'st may never grow.
     *Exit with her Ladies*

GARDENER
Poor Queen, so that thy state might be no worse
I would my skill were subject to thy curse.
Here did she fall a tear. Here in this place      105
I'll set a bank of rue, sour herb-of-grace.
Rue even for ruth here shortly shall be seen
In the remembrance of a weeping queen.      *Exeunt*

4.1      *Enter, as to Parliament, Bolingbroke Duke of*
        *Lancaster and Hereford, the Duke of Aumerle, the*
        *Earl of Northumberland, Harry Percy, Lord*
        *Fitzwalter, the Duke of Surrey, the Bishop of*
        *Carlisle, and the Abbot of Westminster*

BOLINGBROKE
Call forth Bagot.
     *Enter Bagot, with officers*
            Now, Bagot, freely speak thy mind:
What thou dost know of noble Gloucester's death,
Who wrought it with the King, and who performed
The bloody office of his timeless end.

BAGOT
Then set before my face the Lord Aumerle.      5

BOLINGBROKE (*to Aumerle*)
Cousin, stand forth, and look upon that man.
     *Aumerle stands forth*

BAGOT
My lord Aumerle, I know your daring tongue
Scorns to unsay what once it hath delivered.
In that dead time when Gloucester's death was plotted
I heard you say 'Is not my arm of length,      10
That reacheth from the restful English court
As far as Calais, to mine uncle's head?'
Amongst much other talk that very time
I heard you say that you had rather refuse
The offer of an hundred thousand crowns      15
Than Bolingbroke's return to England,
Adding withal how blest this land would be
In this your cousin's death.

AUMERLE         Princes and noble lords,
What answer shall I make to this base man?
Shall I so much dishonour my fair stars      20
On equal terms to give him chastisement?
Either I must, or have mine honour soiled
With the attainder of his slanderous lips.
     *He throws down his gage*
There is my gage, the manual seal of death
That marks thee out for hell. I say thou liest,      25
And will maintain what thou hast said is false
In thy heart blood, though being all too base
To stain the temper of my knightly sword.

BOLINGBROKE
Bagot, forbear. Thou shalt not take it up.

AUMERLE
Excepting one, I would he were the best      30
In all this presence that hath moved me so.

FITZWALTER
If that thy valour stand on sympathy,
There is my gage, Aumerle, in gage to thine.
     *He throws down his gage*
By that fair sun which shows me where thou stand'st,
I heard thee say, and vauntingly thou spak'st it,    35
That thou wert cause of noble Gloucester's death.

If thou deny'st it twenty times, thou liest,
And I will turn thy falsehood to thy heart,
Where it was forgèd, with my rapier's point.

AUMERLE
Thou dar'st not, coward, live to see that day.          40

FITZWALTER
Now by my soul, I would it were this hour.

AUMERLE
Fitzwalter, thou art damned to hell for this.

HARRY PERCY
Aumerle, thou liest. His honour is as true
In this appeal as thou art all unjust;
And that thou art so, there I throw down my gage          45
  *He throws down his gage*
To prove it on thee to the extremest point
Of mortal breathing. Seize it if thou dar'st.

AUMERLE
An if I do not, may my hands rot off,
And never brandish more revengeful steel
Over the glittering helmet of my foe.          50

SURREY
My lord Fitzwalter, I do remember well
The very time Aumerle and you did talk.

FITZWALTER
'Tis very true. You were in presence then,
And you can witness with me this is true.

SURREY
As false, by heaven, as heaven itself is true.          55

FITZWALTER
Surrey, thou liest.

SURREY                     Dishonourable boy,
That lie shall lie so heavy on my sword
That it shall render vengeance and revenge,
Till thou, the lie-giver, and that lie do lie
In earth as quiet as thy father's skull;          60
In proof whereof, there is my honour's pawn.
  *He throws down his gage*
Engage it to the trial if thou dar'st.

FITZWALTER
How fondly dost thou spur a forward horse!
If I dare eat, or drink, or breathe, or live,
I dare meet Surrey in a wilderness          65
And spit upon him whilst I say he lies,
And lies, and lies. There is my bond of faith
To tie thee to my strong correction.
As I intend to thrive in this new world,
Aumerle is guilty of my true appeal.          70
Besides, I heard the banished Norfolk say
That thou, Aumerle, didst send two of thy men
To execute the noble Duke at Calais.

AUMERLE
Some honest Christian trust me with a gage.
  *He takes another's gage and throws it down*
That Norfolk lies, here do I throw down this,          75
If he may be repealed, to try his honour.

BOLINGBROKE
These differences shall all rest under gage
Till Norfolk be repealed. Repealed he shall be,
And, though mine enemy, restored again
To all his lands and signories. When he is returned,
Against Aumerle we will enforce his trial.          81

BISHOP OF CARLISLE
That honourable day shall never be seen.
Many a time hath banished Norfolk fought
For Jesu Christ in glorious Christian field,
Streaming the ensign of the Christian cross          85
Against black pagans, Turks, and Saracens;
And, toiled with works of war, retired himself
To Italy, and there at Venice gave
His body to that pleasant country's earth,
And his pure soul unto his captain, Christ,          90
Under whose colours he had fought so long.

BOLINGBROKE
Why, Bishop of Carlisle, is Norfolk dead?

BISHOP OF CARLISLE
As surely as I live, my lord.

BOLINGBROKE
Sweet peace conduct his sweet soul to the bosom
Of good old Abraham! Lords appellants,          95
Your differences shall all rest under gage
Till we assign you to your days of trial.
  *Enter the Duke of York*

YORK
Great Duke of Lancaster, I come to thee
From plume-plucked Richard, who with willing soul
Adopts thee heir, and his high sceptre yields          100
To the possession of thy royal hand.
Ascend his throne, descending now from him,
And long live Henry, of that name the fourth!

BOLINGBROKE
In God's name I'll ascend the regal throne.

BISHOP OF CARLISLE Marry, God forbid!          105
Worst in this royal presence may I speak,
Yet best beseeming me to speak the truth.
Would God that any in this noble presence
Were enough noble to be upright judge
Of noble Richard. Then true noblesse would          110
Learn him forbearance from so foul a wrong.
What subject can give sentence on his king?
And who sits here that is not Richard's subject?
Thieves are not judged but they are by to hear,
Although apparent guilt be seen in them;          115
And shall the figure of God's majesty,
His captain, steward, deputy elect,
Anointed, crownèd, planted many years,
Be judged by subject and inferior breath,
And he himself not present? O, forfend it, God,          120
That in a Christian climate souls refined
Should show so heinous, black, obscene a deed!
I speak to subjects, and a subject speaks
Stirred up by God thus boldly for his king.
My lord of Hereford here, whom you call king,          125
Is a foul traitor to proud Hereford's king;
And, if you crown him, let me prophesy
The blood of English shall manure the ground,
And future ages groan for this foul act.
Peace shall go sleep with Turks and infidels,          130
And in this seat of peace tumultuous wars
Shall kin with kin and kind with kind confound.
Disorder, horror, fear, and mutiny
Shall here inhabit, and this land be called
The field of Golgotha and dead men's skulls.          135
O, if you rear this house against this house
It will the woefullest division prove
That ever fell upon this cursèd earth!
Prevent, resist it; let it not be so,
Lest child, child's children, cry against you woe.          140

NORTHUMBERLAND
Well have you argued, sir, and for your pains
Of capital treason we arrest you here.
My lord of Westminster, be it your charge
To keep him safely till his day of trial.
May it please you, lords, to grant the Commons' suit?

BOLINGBROKE

Fetch hither Richard, that in common view          146
He may surrender. So we shall proceed
Without suspicion.

YORK                          I will be his conduct.          *Exit*

BOLINGBROKE

Lords, you that here are under our arrest,
Procure your sureties for your days of answer.          150
Little are we beholden to your love,
And little looked for at your helping hands.

*Enter Richard and the Duke of York, ⌈with
attendants bearing the crown and sceptre⌉*

RICHARD

Alack, why am I sent for to a king
Before I have shook off the regal thoughts
Wherewith I reigned? I hardly yet have learned          155
To insinuate, flatter, bow, and bend my knee.
Give sorrow leave awhile to tutor me
To this submission. Yet I well remember
The favours of these men. Were they not mine?
Did they not sometime cry 'All haill' to me?          160
So Judas did to Christ. But He in twelve
Found truth in all but one; I, in twelve thousand, none.
God save the King! Will no man say 'Amen'?
Am I both priest and clerk? Well then, Amen.
God save the King, although I be not he.          165
And yet Amen, if heaven do think him me.
To do what service am I sent for hither?

YORK

To do that office of thine own good will
Which tired majesty did make thee offer:
The resignation of thy state and crown          170
To Henry Bolingbroke.

RICHARD *(to an attendant)*

Give me the crown. *(To Bolingbroke)* Here, cousin,
          seize the crown.
Here, cousin. On this side my hand, on that side thine.
Now is this golden crown like a deep well
That owes two buckets filling one another,          175
The emptier ever dancing in the air,
The other down, unseen, and full of water.
That bucket down and full of tears am I,
Drinking my griefs, whilst you mount up on high.

BOLINGBROKE

I thought you had been willing to resign.          180

RICHARD

My crown I am, but still my griefs are mine.
You may my glories and my state depose,
But not my griefs; still am I king of those.

BOLINGBROKE

Part of your cares you give me with your crown.

RICHARD

Your cares set up do not pluck my cares down.          185
My care is loss of care by old care done;
Your care is gain of care by new care won.
The cares I give I have, though given away;
They 'tend the crown, yet still with me they stay.

BOLINGBROKE

Are you contented to resign the crown?          190

RICHARD

Ay, no; no, ay; for I must nothing be;
Therefore no, no, for I resign to thee.
Now mark me how I will undo myself.
I give this heavy weight from off my head,

⌈*Bolingbroke accepts the crown*⌉

And this unwieldy sceptre from my hand,          195

⌈*Bolingbroke accepts the sceptre*⌉

The pride of kingly sway from out my heart.
With mine own tears I wash away my balm,
With mine own hands I give away my crown,
With mine own tongue deny my sacred state,
With mine own breath release all duteous oaths.          200
All pomp and majesty I do forswear.
My manors, rents, revenues I forgo.
My acts, decrees, and statutes I deny.
God pardon all oaths that are broke to me.
God keep all vows unbroke are made to thee.          205
Make me, that nothing have, with nothing grieved,
And thou with all pleased, that hast all achieved.
Long mayst thou live in Richard's seat to sit,
And soon lie Richard in an earthy pit.
'God save King Henry,' unkinged Richard says,          210
'And send him many years of sunshine days.'
What more remains?

NORTHUMBERLAND *(giving Richard papers)*
                    No more but that you read
These accusations and these grievous crimes
Committed by your person and your followers
Against the state and profit of this land,          215
That by confessing them, the souls of men
May deem that you are worthily deposed.

RICHARD

Must I do so? And must I ravel out
My weaved-up follies? Gentle Northumberland,
If thy offences were upon record,          220
Would it not shame thee in so fair a troop
To read a lecture of them? If thou wouldst,
There shouldst thou find one heinous article
Containing the deposing of a king
And cracking the strong warrant of an oath,          225
Marked with a blot, damned in the book of heaven.
Nay, all of you that stand and look upon
Whilst that my wretchedness doth bait myself,
Though some of you, with Pilate, wash your hands,
Showing an outward pity, yet you Pilates          230
Have here delivered me to my sour cross,
And water cannot wash away your sin.

NORTHUMBERLAND

My lord, dispatch. Read o'er these articles.

RICHARD

Mine eyes are full of tears; I cannot see.
And yet salt water blinds them not so much          235
But they can see a sort of traitors here.
Nay, if I turn mine eyes upon myself
I find myself a traitor with the rest,
For I have given here my soul's consent
T'undeck the pompous body of a king,          240
Made glory base and sovereignty a slave,
Proud majesty a subject, state a peasant.

NORTHUMBERLAND My lord—

RICHARD

No lord of thine, thou haught-insulting man,
Nor no man's lord. I have no name, no title,          245
No, not that name was given me at the font,
But 'tis usurped. Alack the heavy day,
That I have worn so many winters out
And know not now what name to call myself!
O, that I were a mockery king of snow,          250
Standing before the sun of Bolingbroke
To melt myself away in water-drops!
Good king, great king—and yet not greatly good—
An if my word be sterling yet in England,
Let it command a mirror hither straight,          255

That it may show me what a face I have,
Since it is bankrupt of his majesty.

BOLINGBROKE
Go some of you and fetch a looking-glass.
*Exit one or more*

NORTHUMBERLAND
Read o'er this paper while the glass doth come.

RICHARD
Fiend, thou torment'st me ere I come to hell.        260

BOLINGBROKE
Urge it no more, my lord Northumberland.

NORTHUMBERLAND
The Commons will not then be satisfied.

RICHARD
They shall be satisfied. I'll read enough
When I do see the very book indeed
Where all my sins are writ, and that's myself.       265
*Enter one with a glass*
Give me that glass, and therein will I read.
*Richard takes the glass and looks in it*
No deeper wrinkles yet? Hath sorrow struck
So many blows upon this face of mine
And made no deeper wounds? O flatt'ring glass,
Like to my followers in prosperity,                  270
Thou dost beguile me! Was this face the face
That every day under his household roof
Did keep ten thousand men? Was this the face
That like the sun did make beholders wink?
Is this the face which faced so many follies,        275
That was at last outfaced by Bolingbroke?
A brittle glory shineth in this face.
As brittle as the glory is the face,
*He shatters the glass*
For there it is, cracked in an hundred shivers.
Mark, silent King, the moral of this sport:          280
How soon my sorrow hath destroyed my face.

BOLINGBROKE
The shadow of your sorrow hath destroyed
The shadow of your face.

RICHARD                            Say that again:
'The shadow of my sorrow'—ha, let's see.
'Tis very true: my grief lies all within,            285
And these external manner of laments
Are merely shadows to the unseen grief
That swells with silence in the tortured soul.
There lies the substance, and I thank thee, King,
For thy great bounty that not only giv'st            290
Me cause to wail, but teachest me the way
How to lament the cause. I'll beg one boon,
And then be gone and trouble you no more.
Shall I obtain it?

BOLINGBROKE               Name it, fair cousin.

RICHARD
Fair cousin? I am greater than a king;               295
For when I was a king my flatterers
Were then but subjects; being now a subject,
I have a king here to my flatterer.
Being so great, I have no need to beg.

BOLINGBROKE Yet ask.                                 300

RICHARD And shall I have?

BOLINGBROKE You shall.

RICHARD Then give me leave to go.

BOLINGBROKE Whither?

RICHARD
Whither you will, so I were from your sights.        305

BOLINGBROKE
Go some of you, convey him to the Tower.

RICHARD
O good, 'convey'! Conveyors are you all,
That rise thus nimbly by a true king's fall.
⌈*Exit, guarded*⌉

BOLINGBROKE
On Wednesday next we solemnly set down
Our coronation. Lords, prepare yourselves.           310
*Exeunt all but the Abbot of Westminster, the
Bishop of Carlisle, and Aumerle*

ABBOT OF WESTMINSTER
A woeful pageant have we here beheld.

BISHOP OF CARLISLE
The woe's to come, the children yet unborn
Shall feel this day as sharp to them as thorn.

AUMERLE
You holy clergymen, is there no plot
To rid the realm of this pernicious blot?            315

ABBOT OF WESTMINSTER
My lord, before I freely speak my mind herein,
You shall not only take the sacrament
To bury mine intents, but also to effect
Whatever I shall happen to devise.
I see your brows are full of discontent,             320
Your hearts of sorrow, and your eyes of tears.
Come home with me to supper. I will lay
A plot shall show us all a merry day.      *Exeunt*

**5.1**  *Enter the Queen, with her Ladies*

QUEEN
This way the King will come. This is the way
To Julius Caesar's ill-erected Tower,
To whose flint bosom my condemnèd lord
Is doomed a prisoner by proud Bolingbroke.
Here let us rest, if this rebellious earth          5
Have any resting for her true king's queen.
*Enter Richard ⌈and guard⌉*
But soft, but see—or rather do not see—
My fair rose wither. Yet look up, behold,
That you in pity may dissolve to dew,
And wash him fresh again with true-love tears.—     10
Ah, thou the model where old Troy did stand!
Thou map of honour, thou King Richard's tomb,
And not King Richard! Thou most beauteous inn:
Why should hard-favoured grief be lodged in thee,
When triumph is become an alehouse guest?           15

RICHARD
Join not with grief, fair woman, do not so,
To make my end too sudden. Learn, good soul,
To think our former state a happy dream,
From which awaked, the truth of what we are
Shows us but this. I am sworn brother, sweet,       20
To grim necessity, and he and I
Will keep a league till death. Hie thee to France,
And cloister thee in some religious house.
Our holy lives must win a new world's crown,
Which our profane hours here have stricken down.    25

QUEEN
What, is my Richard both in shape and mind
Transformed and weakenèd? Hath Bolingbroke
Deposed thine intellect? Hath he been in thy heart?
The lion dying thrusteth forth his paw
And wounds the earth, if nothing else, with rage    30
To be o'erpowered; and wilt thou, pupil-like,

Take the correction, mildly kiss the rod,
And fawn on rage with base humility,
Which art a lion and the king of beasts?
RICHARD
A king of beasts indeed! If aught but beasts,                    35
I had been still a happy king of men.
Good sometimes Queen, prepare thee hence for
     France.
Think I am dead, and that even here thou tak'st,
As from my death-bed, thy last living leave.
In winter's tedious nights, sit by the fire                    40
With good old folks, and let them tell thee tales
Of woeful ages long ago betid;
And ere thou bid goodnight, to quit their griefs
Tell thou the lamentable fall of me,
And send the hearers weeping to their beds;                    45
Forwhy the senseless brands will sympathize
The heavy accent of thy moving tongue,
And in compassion weep the fire out;
And some will mourn in ashes, some coal black,
For the deposing of a rightful king.                    50
     *Enter the Earl of Northumberland*
NORTHUMBERLAND
My lord, the mind of Bolingbroke is changed.
You must to Pomfret, not unto the Tower.
And, madam, there is order ta'en for you.
With all swift speed you must away to France.
RICHARD
Northumberland, thou ladder wherewithal                    55
The mounting Bolingbroke ascends my throne,
The time shall not be many hours of age
More than it is ere foul sin, gathering head,
Shall break into corruption. Thou shalt think,
Though he divide the realm and give thee half,                    60
It is too little helping him to all.
He shall think that thou, which know'st the way
To plant unrightful kings, wilt know again,
Being ne'er so little urged another way,
To pluck him headlong from the usurpèd throne.                    65
The love of wicked friends converts to fear,
That fear to hate, and hate turns one or both
To worthy danger and deservèd death.
NORTHUMBERLAND
My guilt be on my head, and there an end.
Take leave and part, for you must part forthwith.                    70
RICHARD
Doubly divorced! Bad men, you violate
A twofold marriage: 'twixt my crown and me,
And then betwixt me and my married wife.
(*To the Queen*) Let me unkiss the oath 'twixt thee and
     me—
And yet not so, for with a kiss 'twas made.                    75
Part us, Northumberland: I towards the north,
Where shivering cold and sickness pines the clime;
My queen to France, from whence set forth in pomp
She came adornèd hither like sweet May,
Sent back like Hallowmas or short'st of day.                    80
QUEEN
And must we be divided? Must we part?
RICHARD
Ay, hand from hand, my love, and heart from heart.
QUEEN
Banish us both, and send the King with me.
⌈NORTHUMBERLAND⌉
That were some love, but little policy.
QUEEN
Then whither he goes, thither let me go.                    85

RICHARD
So two together weeping make one woe.
Weep thou for me in France, I for thee here.
Better far off than, near, be ne'er the nea'er.
Go count thy way with sighs, I mine with groans.
QUEEN
So longest way shall have the longest moans.                    90
RICHARD
Twice for one step I'll groan, the way being short,
And piece the way out with a heavy heart.
Come, come, in wooing sorrow let's be brief,
Since, wedding it, there is such length in grief.
One kiss shall stop our mouths, and dumbly part.                    95
Thus give I mine, and thus take I thy heart.
     *They kiss*
QUEEN
Give me mine own again. 'Twere no good part
To take on me to keep and kill thy heart.
     *They kiss*
So now I have mine own again, be gone,
That I may strive to kill it with a groan.                    100
RICHARD
We make woe wanton with this fond delay.
Once more, adieu. The rest let sorrow say.
     *Exeunt ⌈Richard, guarded, and Northumberland*
     *at one door, the Queen and her Ladies at*
     *another door⌉*

**5.2**     *Enter the Duke and Duchess of York*
DUCHESS OF YORK
My lord, you told me you would tell the rest,
When weeping made you break the story off,
Of our two cousins' coming into London.
YORK
Where did I leave?
DUCHESS OF YORK     At that sad stop, my lord,
Where rude misgoverned hands from windows' tops                    5
Threw dust and rubbish on King Richard's head.
YORK
Then, as I said, the Duke, great Bolingbroke,
Mounted upon a hot and fiery steed,
Which his aspiring rider seemed to know,
With slow but stately pace kept on his course,                    10
Whilst all tongues cried 'God save thee, Bolingbroke!'
You would have thought the very windows spake,
So many greedy looks of young and old
Through casements darted their desiring eyes
Upon his visage, and that all the walls                    15
With painted imagery had said at once,
'Jesu preserve thee! Welcome, Bolingbroke!'
Whilst he, from the one side to the other turning,
Bare-headed, lower than his proud steed's neck,
Bespake them thus: 'I thank you, countrymen',                    20
And thus still doing, thus he passed along.
DUCHESS OF YORK
Alack, poor Richard! Where rode he the whilst?
YORK
As in a theatre the eyes of men,
After a well-graced actor leaves the stage,
Are idly bent on him that enters next,                    25
Thinking his prattle to be tedious,
Even so, or with much more contempt, men's eyes
Did scowl on gentle Richard. No man cried 'God save
     him!'
No joyful tongue gave him his welcome home;
But dust was thrown upon his sacred head,                    30
Which with such gentle sorrow he shook off,

His face still combating with tears and smiles,
The badges of his grief and patience,
That had not God for some strong purpose steeled
The hearts of men, they must perforce have melted,
And barbarism itself have pitied him.                                36
But heaven hath a hand in these events,
To whose high will we bound our calm contents.
To Bolingbroke are we sworn subjects now,
Whose state and honour I for aye allow.                              40
     *Enter the Duke of Aumerle*
DUCHESS OF YORK
Here comes my son Aumerle.
YORK                                Aumerle that was;
But that is lost for being Richard's friend,
And, madam, you must call him 'Rutland' now.
I am in Parliament pledge for his truth
And lasting fealty to the new-made King.                             45
DUCHESS OF YORK
Welcome, my son. Who are the violets now
That strew the green lap of the new-come spring?
AUMERLE
Madam, I know not, nor I greatly care not.
God knows I had as lief be none as one.
YORK
Well, bear you well in this new spring of time,                      50
Lest you be cropped before you come to prime.
What news from Oxford? Hold these jousts and
    triumphs?
AUMERLE
For aught I know, my lord, they do.
YORK You will be there, I know.
AUMERLE
If God prevent it not, I purpose so.                                 55
YORK
What seal is that that hangs without thy bosom?
Yea, look'st thou pale? Let me see the writing.
AUMERLE
My lord, 'tis nothing.
YORK                                No matter, then, who see it.
I will be satisfied. Let me see the writing.
AUMERLE
I do beseech your grace to pardon me.                                60
It is a matter of small consequence,
Which for some reasons I would not have seen.
YORK
Which for some reasons, sir, I mean to see.
I fear, I fear!
DUCHESS OF YORK What should you fear?
'Tis nothing but some bond that he is entered into               65
For gay apparel 'gainst the triumph day.
YORK
Bound to himself? What doth he with a bond
That he is bound to? Wife, thou art a fool.
Boy, let me see the writing.
AUMERLE
I do beseech you, pardon me. I may not show it.                     70
YORK
I will be satisfied. Let me see it, I say.
    *He plucks it out of Aumerle's bosom, and reads it*
Treason, foul treason! Villain, traitor, slave!
DUCHESS OF YORK What is the matter, my lord?
YORK
Ho, who is within there? Saddle my horse.—
God for his mercy, what treachery is here!                          75
DUCHESS OF YORK Why, what is it, my lord?
YORK
Give me my boots, I say. Saddle my horse.—

Now by mine honour, by my life, my troth,
I will appeach the villain.
DUCHESS OF YORK What is the matter?                                  80
YORK Peace, foolish woman.
DUCHESS OF YORK
I will not peace. What is the matter, son?
AUMERLE
Good mother, be content. It is no more
Than my poor life must answer.
DUCHESS OF YORK                          Thy life answer?
YORK
Bring me my boots. I will unto the King.                            85
    *His man enters with his boots*
DUCHESS OF YORK
Strike him, Aumerle! Poor boy, thou art amazed.
(*To York's man*) Hence, villain! Never more come in
    my sight.
YORK
Give me my boots, I say.
DUCHESS OF YORK               Why, York, what wilt thou do?
Wilt thou not hide the trespass of thine own?
Have we more sons? Or are we like to have?                          90
Is not my teeming date drunk up with time?
And wilt thou pluck my fair son from mine age,
And rob me of a happy mother's name?
Is he not like thee? Is he not thine own?
YORK Thou fond, mad woman,                                          95
Wilt thou conceal this dark conspiracy?
A dozen of them here have ta'en the sacrament,
And interchangeably set down their hands
To kill the King at Oxford.
DUCHESS OF YORK               He shall be none.
We'll keep him here, then what is that to him?                     100
YORK
Away, fond woman! Were he twenty times my son
I would appeach him.
DUCHESS OF YORK          Hadst thou groaned for him
As I have done thou wouldst be more pitiful.
But now I know thy mind: thou dost suspect
That I have been disloyal to thy bed,                              105
And that he is a bastard, not thy son.
Sweet York, sweet husband, be not of that mind.
He is as like thee as a man may be,
Not like to me or any of my kin,
And yet I love him.
YORK                     Make way, unruly woman.                    110
    *Exit ⌜with his man⌝*
DUCHESS OF YORK
After, Aumerle! Mount thee upon his horse.
Spur, post, and get before him to the King,
And beg thy pardon ere he do accuse thee.
I'll not be long behind—though I be old,
I doubt not but to ride as fast as York—                           115
And never will I rise up from the ground
Till Bolingbroke have pardoned thee. Away, be gone!
    *Exeunt ⌜severally⌝*

**5.3**    *Enter Bolingbroke, crowned King Henry, with*
    *Harry Percy, and other nobles*
KING HENRY
Can no man tell of my unthrifty son?
'Tis full three months since I did see him last.
If any plague hang over us, 'tis he.
I would to God, my lords, he might be found.
Enquire at London 'mongst the taverns there,                        5
For there, they say, he daily doth frequent
With unrestrainèd loose companions—

Even such, they say, as stand in narrow lanes
And beat our watch and rob our passengers—
Which he, young wanton and effeminate boy,           10
Takes on the point of honour to support
So dissolute a crew.
HARRY PERCY
My lord, some two days since, I saw the Prince,
And told him of these triumphs held at Oxford.
KING HENRY And what said the gallant?                15
HARRY PERCY
His answer was he would unto the stews,
And from the common'st creature pluck a glove,
And wear it as a favour, and with that
He would unhorse the lustiest challenger.
KING HENRY
As dissolute as desperate. Yet through both         20
I see some sparks of better hope, which elder days
May happily bring forth.
          *Enter the Duke of Aumerle, amazed*
                    But who comes here?
AUMERLE Where is the King?
KING HENRY
What means our cousin that he stares and looks so
     wildly?
AUMERLE (*kneeling*)
God save your grace! I do beseech your majesty      25
To have some conference with your grace alone.
KING HENRY (*to lords*)
Withdraw yourselves, and leave us here alone.
          *Exeunt all but King Henry and Aumerle*
What is the matter with our cousin now?
AUMERLE
For ever may my knees grow to the earth,
My tongue cleave to the roof within my mouth,       30
Unless a pardon ere I rise or speak.
KING HENRY
Intended or committed was this fault?
If on the first, how heinous e'er it be,
To win thy after-love I pardon thee.
AUMERLE (*rising*)
Then give me leave that I may turn the key,         35
That no man enter till my tale be done.
KING HENRY
Have thy desire.
          *Aumerle locks the door.*
          *The Duke of York knocks at the door and crieth*
YORK (*within*)      My liege, beware! Look to thyself!
Thou hast a traitor in thy presence there.
          *King Henry draws his sword*
KING HENRY (*to Aumerle*) Villain, I'll make thee safe.
AUMERLE
Stay thy revengeful hand! Thou hast no cause to fear.
YORK (*knocking within*)
Open the door, secure foolhardy King!               41
Shall I for love speak treason to thy face?
Open the door, or I will break it open.
          ⌜*King Henry*⌝ *opens the door. Enter the Duke of*
          *York*
KING HENRY
What is the matter, uncle? Speak,
Recover breath, tell us how near is danger,         45
That we may arm us to encounter it.
YORK
Peruse this writing here, and thou shalt know
The treason that my haste forbids me show.
          *He gives King Henry the paper*

AUMERLE
Remember, as thou read'st, thy promise past.
I do repent me. Read not my name there.             50
My heart is not confederate with my hand.
YORK
It was, villain, ere thy hand did set it down.
I tore it from the traitor's bosom, King.
Fear, and not love, begets his penitence.
Forget to pity him, lest pity prove                 55
A serpent that will sting thee to the heart.
KING HENRY
O, heinous, strong, and bold conspiracy!
O loyal father of a treacherous son!
Thou sheer, immaculate, and silver fountain,
From whence this stream through muddy passages      60
Hath held his current and defiled himself,
Thy overflow of good converts to bad,
And thy abundant goodness shall excuse
This deadly blot in thy digressing son.
YORK
So shall my virtue be his vice's bawd,              65
And he shall spend mine honour with his shame,
As thriftless sons their scraping fathers' gold.
Mine honour lives when his dishonour dies,
Or my shamed life in his dishonour lies.
Thou kill'st me in his life: giving him breath      70
The traitor lives, the true man's put to death.
DUCHESS OF YORK (*within*)
What ho, my liege, for God's sake let me in!
KING HENRY
What shrill-voiced suppliant makes this eager cry?
DUCHESS OF YORK (*within*)
A woman, and thy aunt, great King; 'tis I.
Speak with me, pity me! Open the door!              75
A beggar begs that never begged before.
KING HENRY
Our scene is altered from a serious thing,
And now changed to 'The Beggar and the King'.
My dangerous cousin, let your mother in.
I know she is come to pray for your foul sin.       80
          *Aumerle opens the door. Enter the Duchess of York*
YORK
If thou do pardon, whosoever pray,
More sins for this forgiveness prosper may.
This festered joint cut off, the rest rest sound.
This let alone will all the rest confound.
DUCHESS OF YORK (*kneeling*)
O King, believe not this hard-hearted man.          85
Love loving not itself, none other can.
YORK
Thou frantic woman, what dost thou make here?
Shall thy old dugs once more a traitor rear?
DUCHESS OF YORK
Sweet York, be patient.—Hear me, gentle liege.
KING HENRY
Rise up, good aunt.
DUCHESS OF YORK          Not yet, I thee beseech.     90
Forever will I kneel upon my knees,
And never see day that the happy sees,
Till thou give joy, until thou bid me joy
By pardoning Rutland, my transgressing boy.
AUMERLE (*kneeling*)
Unto my mother's prayers I bend my knee.            95
YORK (*kneeling*)
Against them both my true joints bended be.
Ill mayst thou thrive if thou grant any grace.

DUCHESS OF YORK
Pleads he in earnest? Look upon his face.
His eyes do drop no tears, his prayers are in jest.
His words come from his mouth; ours from our
    breast.                                              100
He prays but faintly, and would be denied;
We pray with heart and soul, and all beside.
His weary joints would gladly rise, I know;
Our knees shall kneel till to the ground they grow.
His prayers are full of false hypocrisy;                 105
Ours of true zeal and deep integrity.
Our prayers do outpray his; then let them have
That mercy which true prayer ought to have.
⌈KING HENRY⌉
Good aunt, stand up.
DUCHESS OF YORK            Nay, do not say 'Stand up'.
Say 'Pardon' first, and afterwards 'Stand up'.          110
An if I were thy nurse, thy tongue to teach,
'Pardon' should be the first word of thy speech.
I never longed to hear a word till now.
Say 'Pardon', King. Let pity teach thee how.
The word is short, but not so short as sweet;           115
No word like 'Pardon' for kings' mouths so meet.
YORK
Speak it in French, King: say 'Pardonnez-moi'.
DUCHESS OF YORK
Dost thou teach pardon pardon to destroy?
Ah, my sour husband, my hard-hearted lord
That sets the word itself against the word!             120
Speak 'Pardon' as 'tis current in our land;
The chopping French we do not understand.
Thine eye begins to speak; set thy tongue there;
Or in thy piteous heart plant thou thine ear,
That hearing how our plaints and prayers do pierce,
Pity may move thee 'Pardon' to rehearse.                126
KING HENRY
Good aunt, stand up.
DUCHESS OF YORK            I do not sue to stand.
Pardon is all the suit I have in hand.
KING HENRY
I pardon him as God shall pardon me.
    ⌈York and Aumerle rise⌉
DUCHESS OF YORK
O, happy vantage of a kneeling knee!                    130
Yet am I sick for fear. Speak it again.
Twice saying pardon doth not pardon twain,
But makes one pardon strong.
KING HENRY                    I pardon him
With all my heart.
DUCHESS OF YORK (rising) A god on earth thou art.
KING HENRY
But for our trusty brother-in-law and the Abbot,       135
With all the rest of that consorted crew,
Destruction straight shall dog them at the heels.
Good uncle, help to order several powers
To Oxford, or where'er these traitors are.
They shall not live within this world, I swear,         140
But I will have them if I once know where.
Uncle, farewell; and cousin, so adieu.
Your mother well hath prayed; and prove you true.
DUCHESS OF YORK
Come, my old son. I pray God make thee new.
    Exeunt ⌈King Henry at one door; York,
    the Duchess of York, and Aumerle at
    another door⌉

5.4    Enter Sir Piers Exton, and his Men
EXTON
Didst thou not mark the King, what words he spake?
'Have I no friend will rid me of this living fear?'
Was it not so?
⌈FIRST⌉ MAN        Those were his very words.
EXTON
'Have I no friend?' quoth he. He spake it twice,
And urged it twice together, did he not?                5
⌈SECOND⌉ MAN He did.
EXTON
And speaking it, he wishtly looked on me,
As who should say 'I would thou wert the man
That would divorce this terror from my heart',
Meaning the King at Pomfret. Come, let's go.            10
I am the King's friend, and will rid his foe.    Exeunt

5.5    Enter Richard, alone
RICHARD
I have been studying how I may compare
This prison where I live unto the world;
And for because the world is populous,
And here is not a creature but myself,
I cannot do it. Yet I'll hammer it out.                 5
My brain I'll prove the female to my soul,
My soul the father, and these two beget
A generation of still-breeding thoughts;
And these same thoughts people this little world
In humours like the people of this world.               10
For no thought is contented. The better sort,
As thoughts of things divine, are intermixed
With scruples, and do set the faith itself
Against the faith, as thus: 'Come, little ones',
And then again,                                          15
'It is as hard to come as for a camel
To thread the postern of a small needle's eye.'
Thoughts tending to ambition, they do plot
Unlikely wonders: how these vain weak nails
May tear a passage through the flinty ribs              20
Of this hard world, my ragged prison walls;
And for they cannot, die in their own pride.
Thoughts tending to content flatter themselves
That they are not the first of fortune's slaves,
Nor shall not be the last—like seely beggars,          25
Who, sitting in the stocks, refuge their shame
That many have, and others must, set there;
And in this thought they find a kind of ease,
Bearing their own misfortunes on the back
Of such as have before endured the like.                30
Thus play I in one person many people,
And none contented. Sometimes am I king;
Then treason makes me wish myself a beggar,
And so I am. Then crushing penury
Persuades me I was better when a king.                  35
Then am I kinged again, and by and by
Think that I am unkinged by Bolingbroke,
And straight am nothing. But whate'er I be,
Nor I, nor any man that but man is,
With nothing shall be pleased till he be eased          40
With being nothing.
    The music plays
                Music do I hear.
Ha, ha; keep time! How sour sweet music is
When time is broke and no proportion kept.
So is it in the music of men's lives.

And here have I the daintiness of ear                                45
To check time broke in a disordered string;
But for the concord of my state and time
Had not an ear to hear my true time broke.
I wasted time, and now doth time waste me,
For now hath time made me his numb'ring clock.       50
My thoughts are minutes, and with sighs they jar
Their watches on unto mine eyes, the outward watch
Whereto my finger, like a dial's point,
Is pointing still in cleansing them from tears.
Now, sir, the sounds that tell what hour it is          55
Are clamorous groans that strike upon my heart,
Which is the bell. So sighs, and tears, and groans
Show minutes, hours, and times. But my time
Runs posting on in Bolingbroke's proud joy,
While I stand fooling here, his jack of the clock.      60
This music mads me. Let it sound no more,
For though it have holp madmen to their wits,
In me it seems it will make wise men mad.
⌈The music ceases⌉
Yet blessing on his heart that gives it me,
For 'tis a sign of love, and love to Richard            65
Is a strange brooch in this all-hating world.
*Enter a Groom of the stable*
GROOM
Hail, royal Prince!
RICHARD                    Thanks, noble peer.
The cheapest of us is ten groats too dear.
What art thou, and how com'st thou hither,
Where no man never comes but that sad dog               70
That brings me food to make misfortune live?
GROOM
I was a poor groom of thy stable, King,
When thou wert king; who, travelling towards York,
With much ado at length have gotten leave
To look upon my sometimes royal master's face.         75
O, how it erned my heart when I beheld
In London streets, that coronation day,
When Bolingbroke rode on roan Barbary,
That horse that thou so often hast bestrid,
That horse that I so carefully have dressed!            80
RICHARD
Rode he on Barbary? Tell me, gentle friend,
How went he under him?
GROOM
So proudly as if he disdained the ground.
RICHARD
So proud that Bolingbroke was on his back.
That jade hath eat bread from my royal hand;           85
This hand hath made him proud with clapping him.
Would he not stumble, would he not fall down—
Since pride must have a fall—and break the neck
Of that proud man that did usurp his back?
Forgiveness, horse! Why do I rail on thee,             90
Since thou, created to be awed by man,
Wast born to bear? I was not made a horse,
And yet I bear a burden like an ass,
Spur-galled and tired by jauncing Bolingbroke.
*Enter Keeper to Richard, with meat*
KEEPER (*to Groom*)
Fellow, give place. Here is no longer stay.            95
RICHARD (*to Groom*)
If thou love me, 'tis time thou wert away.
GROOM
What my tongue dares not, that my heart shall say.
*Exit*

KEEPER
My lord, will't please you to fall to?
RICHARD
Taste of it first, as thou art wont to do.
KEEPER
My lord, I dare not. Sir Piers of Exton,               100
Who lately came from the King, commands the
    contrary.
RICHARD (*striking the Keeper*)
The devil take Henry of Lancaster and thee!
Patience is stale, and I am weary of it.
KEEPER Help, help, help!
*Exton and his men rush in*
RICHARD
How now! What means death in this rude assault?
*He seizes a weapon from a man, and kills him*
Villain, thy own hand yields thy death's instrument.
*He kills another*
Go thou, and fill another room in hell.                107
*Here Exton strikes him down*
RICHARD
That hand shall burn in never-quenching fire
That staggers thus my person. Exton, thy fierce hand
Hath with the King's blood stained the King's own
    land.                                              110
Mount, mount, my soul; thy seat is up on high,
Whilst my gross flesh sinks downward, here to die.
*He dies*

EXTON
As full of valour as of royal blood.
Both have I spilt. O, would the deed were good!
For now the devil that told me I did well              115
Says that this deed is chronicled in hell.
This dead King to the living King I'll bear.
Take hence the rest, and give them burial here.
*Exeunt ⌈Exton with Richard's body at one door,
    and his men with the other bodies at another
    door⌉*

5.6   ⌈*Flourish.*⌉ *Enter King Henry and the Duke of York,*
      ⌈*with other lords and attendants*⌉
KING HENRY
Kind uncle York, the latest news we hear
Is that the rebels have consumed with fire
Our town of Ci'cester in Gloucestershire;
But whether they be ta'en or slain we hear not.
*Enter the Earl of Northumberland*
Welcome, my lord. What is the news?                    5
NORTHUMBERLAND
First, to thy sacred state wish I all happiness.
The next news is, I have to London sent
The heads of Salisbury, Spencer, Blunt, and Kent.
The manner of their taking may appear
At large discoursèd in this paper here.                10
*He gives the paper to King Henry*
KING HENRY
We thank thee, gentle Percy, for thy pains,
And to thy worth will add right worthy gains.
*Enter Lord Fitzwalter*
FITZWALTER
My lord, I have from Oxford sent to London
The heads of Brocas and Sir Bennet Seely,
Two of the dangerous consorted traitors                15
That sought at Oxford thy dire overthrow.
KING HENRY
Thy pains, Fitzwalter, shall not be forgot.
Right noble is thy merit, well I wot.

*Enter Harry Percy, with the Bishop of Carlisle,*
*guarded*

HARRY PERCY
  The grand conspirator Abbot of Westminster,
  With clog of conscience and sour melancholy,    20
  Hath yielded up his body to the grave.
  But here is Carlisle living, to abide
  Thy kingly doom and sentence of his pride.
KING HENRY Carlisle, this is your doom.
  Choose out some secret place, some reverent room  25
  More than thou hast, and with it joy thy life.
  So as thou liv'st in peace, die free from strife.
  For though mine enemy thou hast ever been,
  High sparks of honour in thee have I seen.
      *Enter Exton with ⌈his men bearing⌉ the coffin*
EXTON
  Great King, within this coffin I present    30
  Thy buried fear. Herein all breathless lies
  The mightiest of thy greatest enemies,
  Richard of Bordeaux, by me hither brought.
KING HENRY
  Exton, I thank thee not, for thou hast wrought

  A deed of slander with thy fatal hand    35
  Upon my head and all this famous land.
EXTON
  From your own mouth, my lord, did I this deed.
KING HENRY
  They love not poison that do poison need;
  Nor do I thee. Though I did wish him dead,
  I hate the murderer, love him murderèd.    40
  The guilt of conscience take thou for thy labour,
  But neither my good word nor princely favour.
  With Cain go wander through the shades of night,
  And never show thy head by day nor light.
      ⌈*Exeunt Exton and his men*⌉
  Lords, I protest my soul is full of woe    45
  That blood should sprinkle me to make me grow.
  Come mourn with me for what I do lament,
  And put on sullen black incontinent.
  I'll make a voyage to the Holy Land
  To wash this blood off from my guilty hand.    50
  March sadly after. Grace my mournings here
  In weeping after this untimely bier.
      *Exeunt ⌈with the coffin⌉*

## ADDITIONAL PASSAGES

The following passages of four lines or more appear in
the 1597 Quarto but not the Folio; Shakespeare probably
deleted them as part of his limited revisions to the text.

A. AFTER 1.3.127

  And for we think the eagle-wingèd pride
  Of sky-aspiring and ambitious thoughts
  With rival-hating envy set on you
  To wake our peace, which in our country's cradle
  Draws the sweet infant breath of gentle sleep,    5

B. AFTER 1.3.235

  O, had't been a stranger, not my child,
  To smooth his fault I should have been more mild.
  A partial slander sought I to avoid,
  And in the sentence my own life destroyed.

C. AFTER 1.3.256

BOLINGBROKE
  Nay, rather every tedious stride I make
  Will but remember what a deal of world
  I wander from the jewels that I love.
  Must I not serve a long apprenticehood
  To foreign passages, and in the end,    5
  Having my freedom, boast of nothing else
  But that I was a journeyman to grief?
JOHN OF GAUNT
  All places that the eye of heaven visits
  Are to a wise man ports and happy havens.
  Teach thy necessity to reason thus:    10
  There is no virtue like necessity.
  Think not the King did banish thee,
  But thou the King. Woe doth the heavier sit

  Where it perceives it is but faintly borne.
  Go, say I sent thee forth to purchase honour,    15
  And not the King exiled thee; or suppose
  Devouring pestilence hangs in our air
  And thou art flying to a fresher clime.
  Look what thy soul holds dear, imagine it
  To lie that way thou goest, not whence thou com'st.
      20
  Suppose the singing birds musicians,
  The grass whereon thou tread'st the presence strewed,
  The flowers fair ladies, and thy steps no more
  Than a delightful measure or a dance;
  For gnarling sorrow hath less power to bite    25
  The man that mocks at it and sets it light.

D. AFTER 3.2.28

  The means that heavens yield must be embraced
  And not neglected; else heaven would,
  And we will not: heaven's offer we refuse,
  The proffered means of succour and redress.

E. AFTER 4.1.50

ANOTHER LORD
  I task the earth to the like, forsworn Aumerle,
  And spur thee on with full as many lies
  As may be hollowed in thy treacherous ear
  From sun to sun. There is my honour's pawn.
  Engage it to the trial if thou darest.    5
      *He throws down his gage*
AUMERLE
  Who sets me else? By heaven, I'll throw at all.
  I have a thousand spirits in one breast
  To answer twenty thousand such as you.

# KING JOHN

A PLAY called *The Troublesome Reign of John, King of England*, published anonymously in 1591, has sometimes been thought to be a derivative version of Shakespeare's *King John*, first published in the 1623 Folio; more probably Shakespeare wrote his play in 1595 or 1596, using *The Troublesome Reign*—itself based on Holinshed's *Chronicles* and John Foxe's *Book of Martyrs* (1563)—as his principal source. Like *Richard II*, *King John* is written entirely in verse.

King John (c.1167-1216) was famous as the opponent of papal tyranny, and *The Troublesome Reign* is a violently anti-Catholic play; but Shakespeare is more moderate. He portrays selected events from John's reign—like *The Troublesome Reign*, making no mention of Magna Carta—and ends with John's death, but John is not so dominant a figure in his play as Richard II or Richard III in theirs. Indeed, the longest—and liveliest—role is that of Richard Cœur-de-lion's illegitimate son, Philip Falconbridge, the Bastard.

King John's reign was troublesome initially because of his weak claim to his brother Richard Cœur-de-lion's throne. Prince Arthur, son of John's elder brother Geoffrey, had no less strong a claim, which is upheld by his mother, Constance, and by King Philip of France. The waste and futility of the consequent war between power-hungry leaders is satirically demonstrated in the dispute over the French town of Angers, which is resolved by a marriage between John's niece, Lady Blanche of Spain, and Louis, the French Dauphin. The moral is strikingly drawn by the Bastard—the man best fitted to be king, but debarred by accident of birth—in his speech (2.1.562-99) on 'commodity' (self-interest). King Philip breaks his treaty with England, and in the ensuing battle Prince Arthur is captured. He becomes the play's touchstone of humanity as he persuades John's agent, Hubert, to disobey John's orders to blind him, only to kill himself while trying to escape. John's noblemen, thinking the King responsible for the boy's death, defect to the French, but return to their allegiance on learning that the Dauphin intends to kill them after conquering England. John dies, poisoned by a monk; the play ends with the reunited noblemen swearing allegiance to John's son, the young Henry III, and with the Bastard's boast that

> This England never did, nor never shall,
> Lie at the proud foot of a conqueror
> But when it first did help to wound itself.

Twentieth-century revivals of *King John* have been infrequent, but it was popular in the nineteenth century, when the roles of the King, the Bastard, and Constance all appealed to successful actors; a production of 1823 at Covent Garden inaugurated a trend for historically accurate settings and costumes which led to a number of spectacular revivals.

# THE PERSONS OF THE PLAY

KING JOHN of England
QUEEN ELEANOR, his mother

LADY FALCONBRIDGE
Philip the BASTARD, later knighted as Sir Richard Plantagenet,
  her illegitimate son by King Richard I (Coeur-de-lion)
Robert FALCONBRIDGE, her legitimate son
James GURNEY, her attendant

Lady BLANCHE of Spain, niece of King John
PRINCE HENRY, son of King John
HUBERT, a follower of King John
Earl of SALISBURY
Earl of PEMBROKE
Earl of ESSEX
Lord BIGOT

KING PHILIP of France
LOUIS THE DAUPHIN, his son
ARTHUR, Duke of Brittaine, nephew of King John
Lady CONSTANCE, his mother
Duke of AUSTRIA (Limoges)
CHÂTILLON, ambassador from France to England
Count MELUN

A CITIZEN of Angers
Cardinal PANDOLF, a legate from the Pope
PETER OF POMFRET, a prophet

HERALDS
EXECUTIONERS
MESSENGERS
SHERIFF
Lords, soldiers, attendants

# The Life and Death of King John

**1.1** ⌈*Flourish.*⌉ *Enter King John, Queen Eleanor, and the*
*Earls of Pembroke, Essex, and Salisbury; with*
*them Châtillon of France*

KING JOHN
Now say, Châtillon, what would France with us?

CHÂTILLON
Thus, after greeting, speaks the King of France,
In my behaviour, to the majesty—
The borrowed majesty—of England here.

QUEEN ELEANOR
A strange beginning: 'borrowed majesty'? 5

KING JOHN
Silence, good mother, hear the embassy.

CHÂTILLON
Philip of France, in right and true behalf
Of thy deceasèd brother Geoffrey's son,
Arthur Plantagenet, lays most lawful claim
To this fair island and the territories, 10
To Ireland, Poitou, Anjou, Touraine, Maine;
Desiring thee to lay aside the sword
Which sways usurpingly these several titles,
And put the same into young Arthur's hand,
Thy nephew and right royal sovereign. 15

KING JOHN
What follows if we disallow of this?

CHÂTILLON
The proud control of fierce and bloody war,
To enforce these rights so forcibly withheld—

KING JOHN
Here have we war for war, and blood for blood,
Controlment for controlment: so answer France. 20

CHÂTILLON
Then take my king's defiance from my mouth,
The farthest limit of my embassy.

KING JOHN
Bear mine to him, and so depart in peace.
Be thou as lightning in the eyes of France,
For ere thou canst report, I will be there; 25
The thunder of my cannon shall be heard.
So hence. Be thou the trumpet of our wrath,
And sullen presage of your own decay.—
An honourable conduct let him have;
Pembroke, look to't.—Farewell, Châtillon. 30
*Exeunt Châtillon and Pembroke*

QUEEN ELEANOR
What now, my son? Have I not ever said
How that ambitious Constance would not cease
Till she had kindled France and all the world
Upon the right and party of her son?
This might have been prevented and made whole 35
With very easy arguments of love,
Which now the manage of two kingdoms must
With fearful-bloody issue arbitrate.

KING JOHN
Our strong possession and our right for us.

QUEEN ELEANOR (*aside to King John*)
Your strong possession much more than your right,
Or else it must go wrong with you and me: 41
So much my conscience whispers in your ear,
Which none but heaven and you and I shall hear.

*Enter a Sheriff,* ⌈*who whispers to Essex*⌉

ESSEX
My liege, here is the strangest controversy,
Come from the country to be judged by you, 45
That e'er I heard. Shall I produce the men?

KING JOHN Let them approach.— ⌈*Exit Sheriff*⌉
Our abbeys and our priories shall pay
This expeditious charge.
*Enter Robert Falconbridge and Philip the Bastard*
⌈*with the Sheriff*⌉
What men are you?

BASTARD
Your faithful subject I, a gentleman 50
Born in Northamptonshire, and eldest son,
As I suppose, to Robert Falconbridge,
A soldier, by the honour-giving hand
Of Cœur-de-lion knighted in the field.

KING JOHN What art thou? 55

FALCONBRIDGE
The son and heir to that same Falconbridge.

KING JOHN
Is that the elder, and art thou the heir?
You came not of one mother then, it seems.

BASTARD
Most certain of one mother, mighty King—
That is well known—and, as I think, one father. 60
But for the certain knowledge of that truth
I put you o'er to heaven, and to my mother.
Of that I doubt as all men's children may.

QUEEN ELEANOR
Out on thee, rude man! Thou dost shame thy mother
And wound her honour with this diffidence. 65

BASTARD
I, Madam? No, I have no reason for it.
That is my brother's plea and none of mine,
The which if he can prove, a pops me out
At least from fair five hundred pound a year.
Heaven guard my mother's honour, and my land! 70

KING JOHN
A good blunt fellow.—Why, being younger born,
Doth he lay claim to thine inheritance?

BASTARD
I know not why, except to get the land;
But once he slandered me with bastardy.
But whe'er I be as true begot or no, 75
That still I lay upon my mother's head;
But that I am as well begot, my liege—
Fair fall the bones that took the pains for me—
Compare our faces and be judge yourself.
If old Sir Robert did beget us both 80
And were our father, and this son like him,
O old Sir Robert, father, on my knee
I give heaven thanks I was not like to thee.

KING JOHN
Why, what a madcap hath heaven lent us here!

QUEEN ELEANOR
He hath a trick of Cœur-de-lion's face; 85
The accent of his tongue affecteth him.
Do you not read some tokens of my son
In the large composition of this man?

KING JOHN
Mine eye hath well examinèd his parts,
And finds them perfect Richard.
(*To Robert Falconbridge*)       Sirrah, speak:     90
What doth move you to claim your brother's land?
BASTARD
Because he hath a half-face like my father!
With half that face would he have all my land,
A half-faced groat five hundred pound a year.
FALCONBRIDGE
My gracious liege, when that my father lived,     95
Your brother did employ my father much—
BASTARD
Well, sir, by this you cannot get my land.
Your tale must be how he employed my mother.
FALCONBRIDGE
And once dispatched him in an embassy
To Germany, there with the Emperor     100
To treat of high affairs touching that time.
Th'advantage of his absence took the King,
And in the meantime sojourned at my father's,
Where how he did prevail I shame to speak.
But truth is truth: large lengths of seas and shores
Between my father and my mother lay,     106
As I have heard my father speak himself,
When this same lusty gentleman was got.
Upon his deathbed he by will bequeathed
His lands to me, and took it on his death     110
That this my mother's son was none of his;
And if he were, he came into the world
Full fourteen weeks before the course of time.
Then, good my liege, let me have what is mine,
My father's land, as was my father's will.     115
KING JOHN
Sirrah, your brother is legitimate.
Your father's wife did after wedlock bear him,
And if she did play false, the fault was hers,
Which fault lies on the hazards of all husbands
That marry wives. Tell me, how if my brother,     120
Who, as you say, took pains to get this son,
Had of your father claimed this son for his?
In sooth, good friend, your father might have kept
This calf, bred from his cow, from all the world;
In sooth he might. Then if he were my brother's,     125
My brother might not claim him, nor your father,
Being none of his, refuse him. This concludes:
My mother's son did get your father's heir;
Your father's heir must have your father's land.
FALCONBRIDGE
Shall then my father's will be of no force     130
To dispossess that child which is not his?
BASTARD
Of no more force to dispossess me, sir,
Than was his will to get me, as I think.
QUEEN ELEANOR
Whether hadst thou rather be: a Falconbridge,
And like thy brother to enjoy thy land,     135
Or the reputed son of Cœur-de-lion,
Lord of thy presence, and no land beside?
BASTARD
Madam, an if my brother had my shape,
And I had his, Sir Robert's his like him,
And if my legs were two such riding-rods,     140
My arms such eel-skins stuffed, my face so thin
That in mine ear I durst not stick a rose
Lest men should say 'Look where three-farthings
    goes!',

And, to his shape, were heir to all this land,
Would I might never stir from off this place.     145
I would give it every foot to have this face;
It would not be Sir Nob in any case.
QUEEN ELEANOR
I like thee well. Wilt thou forsake thy fortune,
Bequeath thy land to him, and follow me?
I am a soldier and now bound to France.     150
BASTARD
Brother, take you my land; I'll take my chance.
Your face hath got five hundred pound a year,
Yet sell your face for fivepence and 'tis dear.—
Madam, I'll follow you unto the death.
QUEEN ELEANOR
Nay, I would have you go before me thither.     155
BASTARD
Our country manners give our betters way.
KING JOHN What is thy name?
BASTARD
Philip, my liege, so is my name begun:
Philip, good old Sir Robert's wife's eldest son.
KING JOHN
From henceforth bear his name whose form thou
    bear'st.     160
Kneel thou down Philip, but arise more great:
       *He knights the Bastard*
Arise Sir Richard and Plantagenet.
BASTARD
Brother by th' mother's side, give me your hand.
My father gave me honour, yours gave land.
Now blessèd be the hour, by night or day,     165
When I was got, Sir Robert was away.
QUEEN ELEANOR
The very spirit of Plantagenet!
I am thy grandam, Richard; call me so.
BASTARD
Madam, by chance, but not by truth; what though?
Something about, a little from the right,     170
    In at the window, or else o'er the hatch;
Who dares not stir by day must walk by night,
    And have is have, however men do catch.
Near or far off, well won is still well shot,
And I am I, howe'er I was begot.     175
KING JOHN
Go, Falconbridge, now hast thou thy desire:
A landless knight makes thee a landed squire.—
Come, madam, and come, Richard; we must speed
For France; for France, for it is more than need.
BASTARD
Brother, adieu. Good fortune come to thee,     180
For thou wast got i'th' way of honesty.
       *Exeunt all but the Bastard*
A foot of honour better than I was,
But many a many foot of land the worse.
Well, now can I make any Joan a lady.
'Good e'en, Sir Richard'—'God-a-mercy fellow';     185
And if his name be George I'll call him Peter,
For new-made honour doth forget men's names;
'Tis too respective and too sociable
For your conversion. Now your traveller,
He and his toothpick at my worship's mess;     190
And when my knightly stomach is sufficed,
Why then I suck my teeth and catechize
My pickèd man of countries. 'My dear sir,'
Thus leaning on mine elbow I begin,
'I shall beseech you—'. That is Question now;     195

And then comes Answer like an Absey book.
'O sir,' says Answer, 'at your best command,
At your employment, at your service, sir.'
'No sir,' says Question, 'I, sweet sir, at yours.'
And so, ere Answer knows what Question would,     200
Saving in dialogue of compliment,
And talking of the Alps and Apennines,
The Pyrenean and the River Po,
It draws toward supper in conclusion so.
But this is worshipful society,                    205
And fits the mounting spirit like myself;
For he is but a bastard to the time
That doth not smack of observation;
And so am I—whether I smack or no,
And not alone in habit and device,                 210
Exterior form, outward accoutrement,
But from the inward motion—to deliver
Sweet, sweet, sweet poison for the age's tooth;
Which, though I will not practise to deceive,
Yet to avoid deceit I mean to learn;              215
For it shall strew the footsteps of my rising.
          *Enter Lady Falconbridge and James Gurney*
But who comes in such haste in riding-robes?
What woman-post is this? Hath she no husband
That will take pains to blow a horn before her?
O me, 'tis my mother! How now, good lady?         220
What brings you here to court so hastily?

LADY FALCONBRIDGE
Where is that slave thy brother? Where is he
That holds in chase mine honour up and down?

BASTARD
My brother Robert, old Sir Robert's son?
Colbrand the Giant, that same mighty man?         225
Is it Sir Robert's son that you seek so?

LADY FALCONBRIDGE
Sir Robert's son, ay, thou unreverent boy,
Sir Robert's son. Why scorn'st thou at Sir Robert?
He is Sir Robert's son, and so art thou.

BASTARD
James Gurney, wilt thou give us leave awhile?     230

GURNEY
Good leave, good Philip.

BASTARD                    Philip Sparrow, James!
There's toys abroad; anon I'll tell thee more.
                              *Exit James Gurney*
Madam, I was not old Sir Robert's son.
Sir Robert might have eat his part in me
Upon Good Friday, and ne'er broke his fast.       235
Sir Robert could do well, marry to confess,
Could a get me! Sir Robert could not do it:
We know his handiwork. Therefore, good mother,
To whom am I beholden for these limbs?
Sir Robert never holp to make this leg.           240

LADY FALCONBRIDGE
Hast thou conspirèd with thy brother too,
That for thine own gain shouldst defend mine honour?
What means this scorn, thou most untoward knave?

BASTARD
Knight, knight, good mother, Basilisco-like!
What! I am dubbed; I have it on my shoulder.      245
But, mother, I am not Sir Robert's son.
I have disclaimed Sir Robert; and my land,
Legitimation, name, and all is gone.
Then, good my mother, let me know my father;
Some proper man, I hope; who was it, mother?      250

LADY FALCONBRIDGE
Hast thou denied thyself a Falconbridge?

BASTARD
As faithfully as I deny the devil.

LADY FALCONBRIDGE
King Richard Cœur-de-lion was thy father.
By long and vehement suit I was seduced
To make room for him in my husband's bed.         255
Heaven lay not my transgression to my charge!
Thou art the issue of my dear offence,
Which was so strongly urged past my defence.

BASTARD
Now by this light, were I to get again,
Madam, I would not wish a better father.          260
Some sins do bear their privilege on earth,
And so doth yours; your fault was not your folly.
Needs must you lay your heart at his dispose,
Subjected tribute to commanding love,
Against whose fury and unmatchèd force            265
The aweless lion could not wage the fight,
Nor keep his princely heart from Richard's hand.
He that perforce robs lions of their hearts
May easily win a woman's. Ay, my mother,
With all my heart I thank thee for my father.     270
Who lives and dares but say thou didst not well
When I was got, I'll send his soul to hell.
Come, lady, I will show thee to my kin,
    And they shall say, when Richard me begot,
If thou hadst said him nay, it had been sin.      275
    Who says it was, he lies: I say 'twas not. *Exeunt*

2.1    ⌐*Flourish.*⌐ *Enter before Angers* ⌐*at one door*⌐ *Philip*
       *King of France, Louis the Dauphin, Lady*
       *Constance, and Arthur Duke of Brittaine, with*
       *soldiers;* ⌐*at another door*⌐ *the Duke of Austria,*
       *wearing a lion's hide, with soldiers*

⌐KING PHILIP⌐
Before Angers well met, brave Austria.—
Arthur, that great forerunner of thy blood,
Richard that robbed the lion of his heart
And fought the holy wars in Palestine,
By this brave duke came early to his grave;        5
And, for amends to his posterity,
At our importance hither is he come
To spread his colours, boy, in thy behalf,
And to rebuke the usurpation
Of thy unnatural uncle, English John.              10
Embrace him, love him, give him welcome hither.

ARTHUR (*to Austria*)
God shall forgive you Cœur-de-lion's death,
The rather that you give his offspring life,
Shadowing their right under your wings of war.
I give you welcome with a powerless hand,          15
But with a heart full of unstainèd love.
Welcome before the gates of Angers, Duke.

⌐KING PHILIP⌐
A noble boy. Who would not do thee right?

AUSTRIA (*kissing Arthur*)
Upon thy cheek lay I this zealous kiss
As seal to this indenture of my love:              20
That to my home I will no more return
Till Angers and the right thou hast in France,
Together with that pale, that white-faced shore,
Whose foot spurns back the ocean's roaring tides
And coops from other lands her islanders,          25
Even till that England, hedged in with the main,

That water-wallèd bulwark, still secure
And confident from foreign purposes,
Even till that utmost corner of the west
Salute thee for her king. Till then, fair boy,          30
Will I not think of home, but follow arms.
CONSTANCE
O, take his mother's thanks, a widow's thanks,
Till your strong hand shall help to give him strength
To make a more requital to your love.
AUSTRIA
The peace of heaven is theirs that lift their swords     35
In such a just and charitable war.
KING PHILIP
Well then, to work! Our cannon shall be bent
Against the brows of this resisting town.
Call for our chiefest men of discipline
To cull the plots of best advantages.                    40
We'll lay before this town our royal bones,
Wade to the market-place in Frenchmen's blood,
But we will make it subject to this boy.
CONSTANCE
Stay for an answer to your embassy,
Lest unadvised you stain your swords with blood.         45
My lord Châtillon may from England bring
That right in peace which here we urge in war,
And then we shall repent each drop of blood
That hot rash haste so indirectly shed.
        *Enter Châtillon*
KING PHILIP
A wonder, lady: lo upon thy wish                         50
Our messenger Châtillon is arrived.—
What England says, say briefly, gentle lord;
We coldly pause for thee. Châtillon, speak.
CHÂTILLON
Then turn your forces from this paltry siege,
And stir them up against a mightier task.                55
England, impatient of your just demands,
Hath put himself in arms. The adverse winds,
Whose leisure I have stayed, have given him time
To land his legions all as soon as I.
His marches are expedient to this town,                  60
His forces strong, his soldiers confident.
With him along is come the Mother-Queen,
An Ate stirring him to blood and strife;
With her her niece, the Lady Blanche of Spain;
With them a bastard of the King's deceased;              65
And all th'unsettled humours of the land—
Rash, inconsiderate, fiery voluntaries,
With ladies' faces and fierce dragons' spleens—
Have sold their fortunes at their native homes,
Bearing their birthrights proudly on their backs,        70
To make a hazard of new fortunes here.
In brief, a braver choice of dauntless spirits
Than now the English bottoms have waft o'er
Did never float upon the swelling tide
To do offence and scathe in Christendom.                 75
        *Drum beats*
The interruption of their churlish drums
Cuts off more circumstance. They are at hand;
To parley or to fight therefore prepare.
KING PHILIP
How much unlooked-for is this expedition!
AUSTRIA
By how much unexpected, by so much                       80
We must awake endeavour for defence,

For courage mounteth with occasion.
Let them be welcome then: we are prepared.
        *Enter, ⌈marching,⌉ King John of England, the
        Bastard, Queen Eleanor, Lady Blanche, the Earl of
        Pembroke, and soldiers*
KING JOHN
Peace be to France, if France in peace permit
Our just and lineal entrance to our own.                 85
If not, bleed France, and peace ascend to heaven,
Whiles we, God's wrathful agent, do correct
Their proud contempt that beats his peace to heaven.
KING PHILIP
Peace be to England, if that war return
From France to England, there to live in peace.          90
England we love, and for that England's sake
With burden of our armour here we sweat.
This toil of ours should be a work of thine;
But thou from loving England art so far
That thou hast underwrought his lawful king,             95
Cut off the sequence of posterity,
Outfacèd infant state, and done a rape
Upon the maiden virtue of the crown.
*(Pointing to Arthur)*
Look here upon thy brother Geoffrey's face.
These eyes, these brows, were moulded out of his;       100
This little abstract doth contain that large
Which died in Geoffrey; and the hand of time
Shall draw this brief into as huge a volume.
That Geoffrey was thy elder brother born,
And this his son; England was Geoffrey's right,          105
And this is Geoffrey's. In the name of God,
How comes it then that thou art called a king,
When living blood doth in these temples beat,
Which owe the crown that thou o'ermasterest?
KING JOHN
From whom hast thou this great commission, France,
To draw my answer from thy articles?                    111
KING PHILIP
From that supernal judge that stirs good thoughts
In any breast of strong authority
To look into the blots and stains of right.
That judge hath made me guardian to this boy,           115
Under whose warrant I impeach thy wrong,
And by whose help I mean to chastise it.
KING JOHN
Alack, thou dost usurp authority.
KING PHILIP
Excuse it is to beat usurping down.
QUEEN ELEANOR
Who is it thou dost call usurper, France?               120
CONSTANCE
Let me make answer: thy usurping son.
QUEEN ELEANOR
Out, insolent! Thy bastard shall be king
That thou mayst be a queen and check the world.
CONSTANCE
My bed was ever to thy son as true
As thine was to thy husband; and this boy               125
Liker in feature to his father Geoffrey
Than thou and John in manners, being as like
As rain to water, or devil to his dam.
My boy a bastard? By my soul I think
His father never was so true begot.                     130
It cannot be, an if thou wert his mother.
QUEEN ELEANOR *(to Arthur)*
There's a good mother, boy, that blots thy father.

CONSTANCE (*to Arthur*)
There's a good grandam, boy, that would blot thee.
AUSTRIA
Peace!
BASTARD Hear the crier!
AUSTRIA            What the devil art thou?
BASTARD
One that will play the devil, sir, with you,    135
An a may catch your hide and you alone.
You are the hare of whom the proverb goes,
Whose valour plucks dead lions by the beard.
I'll smoke your skin-coat an I catch you right—
Sirrah, look to't—i'faith I will, i'faith!    140
BLANCHE
O, well did he become that lion's robe
That did disrobe the lion of that robe!
BASTARD
It lies as sightly on the back of him
As great Alcides' shows upon an ass.
But, ass, I'll take that burden from your back,    145
Or lay on that shall make your shoulders crack.
AUSTRIA
What cracker is this same that deafs our ears
With this abundance of superfluous breath?—
King Philip, determine what we shall do straight.
⌈KING PHILIP⌉
Women and fools, break off your conference.—    150
King John, this is the very sum of all:
England and Ireland, Anjou, Touraine, Maine,
In right of Arthur do I claim of thee.
Wilt thou resign them and lay down thy arms?
KING JOHN
My life as soon. I do defy thee, France.—    155
Arthur of Brittaine, yield thee to my hand,
And out of my dear love I'll give thee more
Than e'er the coward hand of France can win.
Submit thee, boy.
QUEEN ELEANOR (*to Arthur*) Come to thy grandam, child.
CONSTANCE (*to Arthur*)
Do, child, go to it grandam, child.    160
Give grandam kingdom, and it grandam will
Give it a plum, a cherry, and a fig.
There's a good grandam.
ARTHUR           Good my mother, peace.
I would that I were low laid in my grave.
I am not worth this coil that's made for me.    165
*He weeps*
QUEEN ELEANOR
His mother shames him so, poor boy, he weeps.
CONSTANCE
Now shame upon you, whe'er she does or no!
His grandam's wrongs, and not his mother's shames,
Draw those heaven-moving pearls from his poor eyes,
Which heaven shall take in nature of a fee;    170
Ay, with these crystal beads heaven shall be bribed
To do him justice and revenge on you.
QUEEN ELEANOR
Thou monstrous slanderer of heaven and earth!
CONSTANCE
Thou monstrous injurer of heaven and earth!
Call not me slanderer. Thou and thine usurp    175
The dominations, royalties and rights
Of this oppressèd boy. This is thy eld'st son's son,
Infortunate in nothing but in thee.
Thy sins are visited in this poor child;
The canon of the law is laid on him,    180

Being but the second generation
Removèd from thy sin-conceiving womb.
KING JOHN
Bedlam, have done.
CONSTANCE        I have but this to say:
That he is not only plaguèd for her sin,
But God hath made her sin and her the plague    185
On this removèd issue, plagued for her
And with her plague; her sin his injury,
Her injury the beadle to her sin;
All punished in the person of this child,
And all for her. A plague upon her!    190
QUEEN ELEANOR
Thou unadvisèd scold, I can produce
A will that bars the title of thy son.
CONSTANCE
Ay, who doubts that? A will, a wicked will,
A woman's will, a cankered grandam's will!
KING PHILIP
Peace, lady; pause or be more temperate.    195
It ill beseems this presence to cry aim
To these ill-tunèd repetitions.—
Some trumpet summon hither to the walls
These men of Angers. Let us hear them speak
Whose title they admit, Arthur's or John's.    200
*Trumpet sounds. Enter a Citizen upon the walls*
CITIZEN
Who is it that hath warned us to the walls?
KING PHILIP
'Tis France for England.
KING JOHN        England for itself.
You men of Angers and my loving subjects—
KING PHILIP
You loving men of Angers, Arthur's subjects,
Our trumpet called you to this gentle parle—    205
KING JOHN
For our advantage; therefore hear us first.
These flags of France that are advancèd here
Before the eye and prospect of your town,
Have hither marched to your endamagement.
The cannons have their bowels full of wrath,    210
And ready mounted are they to spit forth
Their iron indignation 'gainst your walls.
All preparation for a bloody siege
And merciless proceeding by these French
Confront your city's eyes, your winking gates;    215
And but for our approach, those sleeping stones
That as a waist doth girdle you about,
By the compulsion of their ordinance,
By this time from their fixèd beds of lime
Had been dishabited, and wide havoc made    220
For bloody power to rush upon your peace.
But on the sight of us your lawful king,
Who painfully, with much expedient march,
Have brought a countercheck before your gates
To save unscratched your city's threatened cheeks,
Behold the French, amazed, vouchsafe a parle;    226
And now instead of bullets wrapped in fire
To make a shaking fever in your walls,
They shoot but calm words folded up in smoke
To make a faithless error in your ears;    230
Which trust accordingly, kind citizens,
And let us in, your king, whose laboured spirits,
Forwearied in this action of swift speed,
Craves harbourage within your city walls.

KING PHILIP
  When I have said, make answer to us both.       235
    *He takes Arthur's hand*
  Lo, in this right hand, whose protection
  Is most divinely vowed upon the right
  Of him it holds, stands young Plantagenet,
  Son to the elder brother of this man
  And king o'er him and all that he enjoys.       240
  For this downtrodden equity we tread
  In warlike march these greens before your town,
  Being no further enemy to you
  Than the constraint of hospitable zeal
  In the relief of this oppressèd child           245
  Religiously provokes. Be pleasèd then
  To pay that duty which you truly owe
  To him that owes it, namely this young prince;
  And then our arms, like to a muzzled bear,
  Save in aspect, hath all offence sealed up:     250
  Our cannons' malice vainly shall be spent
  Against th'invulnerable clouds of heaven,
  And with a blessèd and unvexed retire,
  With unhacked swords and helmets all unbruised,
  We will bear home that lusty blood again        255
  Which here we came to spout against your town,
  And leave your children, wives, and you in peace.
  But if you fondly pass our proffered offer,
  'Tis not the roundure of your old-faced walls
  Can hide you from our messengers of war,        260
  Though all these English and their discipline
  Were harboured in their rude circumference.
  Then tell us, shall your city call us lord
  In that behalf which we have challenged it,
  Or shall we give the signal to our rage,        265
  And stalk in blood to our possession?
CITIZEN
  In brief, we are the King of England's subjects.
  For him and in his right we hold this town.
KING JOHN
  Acknowledge then the King, and let me in.
CITIZEN
  That can we not; but he that proves the king,   270
  To him will we prove loyal; till that time
  Have we rammed up our gates against the world.
KING JOHN
  Doth not the crown of England prove the king?
  And if not that, I bring you witnesses:
  Twice fifteen thousand hearts of England's breed—
BASTARD (*aside*) Bastards and else.              276
KING JOHN
  To verify our title with their lives.
KING PHILIP
  As many and as well-born bloods as those—
BASTARD (*aside*) Some bastards too.
KING PHILIP
  Stand in his face to contradict his claim.      280
CITIZEN
  Till you compound whose right is worthiest,
  We for the worthiest hold the right from both.
KING JOHN
  Then God forgive the sin of all those souls
  That to their everlasting residence,
  Before the dew of evening fall, shall fleet      285
  In dreadful trial of our kingdom's king.
KING PHILIP
  Amen, Amen! Mount, chevaliers! To arms!

BASTARD
  Saint George that swinged the dragon, and e'er since
  Sits on's horseback at mine hostess' door,
  Teach us some fence! (*To Austria*) Sirrah, were I at
    home                                          290
  At your den, sirrah, with your lioness,
  I would set an ox-head to your lion's hide
  And make a monster of you.
AUSTRIA                        Peace, no more.
BASTARD
  O tremble, for you hear the lion roar!
KING JOHN
  Up higher to the plain, where we'll set forth   295
  In best appointment all our regiments.
BASTARD
  Speed then, to take advantage of the field.
KING PHILIP
  It shall be so, and at the other hill
  Command the rest to stand. God and our right!
    *Exeunt ⌈severally⌉ King John and King Philip*
                    *with their powers. The Citizen remains*
                                    *on the walls*
      ⌈*Alarum.*⌉ *Here, after excursions, enter ⌈at one door⌉*
      *the French Herald, with ⌈a trumpeter⌉, to the gates*
FRENCH HERALD
  You men of Angers, open wide your gates          300
  And let young Arthur Duke of Brittaine in,
  Who by the hand of France this day hath made
  Much work for tears in many an English mother,
  Whose sons lie scattered on the bleeding ground;
  Many a widow's husband grovelling lies,          305
  Coldly embracing the discoloured earth;
  And victory with little loss doth play
  Upon the dancing banners of the French,
  Who are at hand, triumphantly displayed,
  To enter conquerors, and to proclaim             310
  Arthur of Brittaine England's king and yours.
    *Enter ⌈at another door⌉ the English Herald, with a*
    *trumpeter*
ENGLISH HERALD
  Rejoice, you men of Angers, ring your bells!
  King John, your king and England's, doth approach,
  Commander of this hot malicious day.
  Their armours that marched hence so silver-bright 315
  Hither return all gilt with Frenchmen's blood.
  There stuck no plume in any English crest
  That is removèd by a staff of France;
  Our colours do return in those same hands
  That did display them when we first marched forth;
  And like a jolly troop of huntsmen come         321
  Our lusty English, all with purpled hands
  Dyed in the dying slaughter of their foes.
  Open your gates and give the victors way.
⌈CITIZEN⌉
  Heralds, from off our towers we might behold     325
  From first to last the onset and retire
  Of both your armies, whose equality
  By our best eyes cannot be censurèd.
  Blood hath bought blood and blows have answered
    blows,
  Strength matched with strength and power confronted
    power.                                        330
  Both are alike, and both alike we like.
  One must prove greatest. While they weigh so even,
  We hold our town for neither, yet for both.

*Enter at one door King John, the Bastard, Queen
Eleanor and Lady Blanche, with soldiers; at
another door King Philip, Louis the Dauphin, and
the Duke of Austria with soldiers*

KING JOHN

France, hast thou yet more blood to cast away?

Say, shall the current of our right run on,      335

Whose passage, vexed with thy impediment,

Shall leave his native channel and o'erswell

With course disturbed even thy confining shores,

Unless thou let his silver water keep

A peaceful progress to the ocean?      340

KING PHILIP

England, thou hast not saved one drop of blood

In this hot trial more than we of France;

Rather, lost more. And by this hand I swear,

That sways the earth this climate overlooks,

Before we will lay down our just-borne arms,      345

We'll put thee down 'gainst whom these arms we
     bear,

Or add a royal number to the dead,

Gracing the scroll that tells of this war's loss

With slaughter coupled to the name of kings.

BASTARD

Ha, majesty! How high thy glory towers      350

When the rich blood of kings is set on fire!

O, now doth Death line his dead chaps with steel;

The swords of soldiers are his teeth, his fangs;

And now he feasts, mousing the flesh of men

In undetermined differences of kings.      355

Why stand these royal fronts amazèd thus?

Cry havoc, Kings! Back to the stainèd field,

You equal potents, fiery-kindled spirits!

Then let confusion of one part confirm

The other's peace; till then, blows, blood, and death!

KING JOHN

Whose party do the townsmen yet admit?      361

KING PHILIP

Speak, citizens, for England: who's your king?

⌈CITIZEN⌉

The King of England, when we know the King.

KING PHILIP

Know him in us, that here hold up his right.

KING JOHN

In us, that are our own great deputy      365

And bear possession of our person here,

Lord of our presence, Angers, and of you.

⌈CITIZEN⌉

A greater power than we denies all this,

And, till it be undoubted, we do lock

Our former scruple in our strong-barred gates,      370

Kinged of our fear, until our fears resolved

Be by some certain king, purged and deposed.

BASTARD

By heaven, these scroyles of Angers flout you, Kings,

And stand securely on their battlements

As in a theatre, whence they gape and point      375

At your industrious scenes and acts of death.

Your royal presences be ruled by me.

Do like the mutines of Jerusalem:

Be friends awhile, and both conjointly bend

Your sharpest deeds of malice on this town.      380

By east and west let France and England mount

Their battering cannon, chargèd to the mouths,

Till their soul-fearing clamours have brawled down

The flinty ribs of this contemptuous city.

I'd play incessantly upon these jades,      385

Even till unfencèd desolation

Leave them as naked as the vulgar air.

That done, dissever your united strengths,

And part your mingled colours once again;

Turn face to face, and bloody point to point.      390

Then in a moment Fortune shall cull forth

Out of one side her happy minion,

To whom in favour she shall give the day,

And kiss him with a glorious victory.

How like you this wild counsel, mighty states?      395

Smacks it not something of the policy?

KING JOHN

Now, by the sky that hangs above our heads,

I like it well.—France, shall we knit our powers,

And lay this Angers even with the ground,

Then after fight who shall be king of it?      400

BASTARD (*to King Philip*)

An if thou hast the mettle of a king,

Being wronged as we are by this peevish town,

Turn thou the mouth of thy artillery,

As we will ours, against these saucy walls;

And when that we have dashed them to the ground,

Why, then defy each other, and pell-mell      406

Make work upon ourselves, for heaven or hell.

KING PHILIP

Let it be so.—Say, where will you assault?

KING JOHN

We from the west will send destruction

Into this city's bosom.      410

AUSTRIA I from the north.

KING PHILIP Our thunder from the south

Shall rain their drift of bullets on this town.

BASTARD ⌈*to King John*⌉

O prudent discipline! From north to south

Austria and France shoot in each other's mouth.      415

I'll stir them to it. Come, away, away!

⌈CITIZEN⌉

Hear us, great Kings, vouchsafe a while to stay,

And I shall show you peace and fair-faced league.

Win you this city without stroke or wound;

Rescue those breathing lives to die in beds,      420

That here come sacrifices for the field.

Persever not, but hear me, mighty Kings.

KING JOHN

Speak on with favour; we are bent to hear.

⌈CITIZEN⌉

That daughter there of Spain, the Lady Blanche,

Is niece to England. Look upon the years      425

Of Louis the Dauphin and that lovely maid.

If lusty love should go in quest of beauty,

Where should he find it fairer than in Blanche?

If zealous love should go in search of virtue,

Where should he find it purer than in Blanche?      430

If love ambitious sought a match of birth,

Whose veins bound richer blood than Lady Blanche?

Such as she is in beauty, virtue, birth,

Is the young Dauphin every way complete;

If not complete, O, say he is not she;      435

And she again wants nothing—to name want—

If want it be not that she is not he.

He is the half part of a blessèd man,

Left to be finishèd by such as she;

And she a fair divided excellence,      440

Whose fullness of perfection lies in him.

O, two such silver currents when they join

Do glorify the banks that bound them in,
And two such shores to two such streams made one,
Two such controlling bounds, shall you be, Kings,   445
To these two princes if you marry them.
This union shall do more than battery can
To our fast-closèd gates, for at this match,
With swifter spleen than powder can enforce,
The mouth of passage shall we fling wide ope,   450
And give you entrance. But without this match
The sea enragèd is not half so deaf,
Lions more confident, mountains and rocks
More free from motion, no, not Death himself
In mortal fury half so peremptory,   455
As we to keep this city.

BASTARD ⌈aside⌉          Here's a stay
That shakes the rotten carcass of old Death
Out of his rags. Here's a large mouth, indeed,
That spits forth Death and mountains, rocks and seas,
Talks as familiarly of roaring lions   460
As maids of thirteen do of puppy-dogs.
What cannoneer begot this lusty blood?
He speaks plain cannon: fire, and smoke, and bounce;
He gives the bastinado with his tongue;
Our ears are cudgelled; not a word of his   465
But buffets better than a fist of France.
Zounds! I was never so bethumped with words
Since I first called my brother's father Dad.

QUEEN ELEANOR (aside to King John)
Son, list to this conjunction, make this match,
Give with our niece a dowry large enough;   470
For, by this knot, thou shalt so surely tie
Thy now unsured assurance to the crown
That yon green boy shall have no sun to ripe
The bloom that promiseth a mighty fruit.
I see a yielding in the looks of France;   475
Mark how they whisper. Urge them while their souls
Are capable of this ambition,
Lest zeal, now melted by the windy breath
Of soft petitions, pity, and remorse,
Cool and congeal again to what it was.   480

⌈CITIZEN⌉
Why answer not the double majesties
This friendly treaty of our threatened town?

KING PHILIP
Speak England first, that hath been forward first
To speak unto this city: what say you?

KING JOHN
If that the Dauphin there, thy princely son,   485
Can in this book of beauty read 'I love',
Her dowry shall weigh equal with a queen;
For Anjou and fair Touraine, Maine, Poitou,
And all that we upon this side the sea—
Except this city now by us besieged—   490
Find liable to our crown and dignity,
Shall gild her bridal bed, and make her rich
In titles, honours, and promotions,
As she in beauty, education, blood,
Holds hand with any princess of the world.   495

KING PHILIP
What sayst thou, boy? Look in the lady's face.

LOUIS THE DAUPHIN
I do, my lord, and in her eye I find
A wonder, or a wondrous miracle,
The shadow of myself formed in her eye;
Which, being but the shadow of your son,   500

Becomes a sun and makes your son a shadow.
I do protest I never loved myself
Till now enfixèd I beheld myself
Drawn in the flattering table of her eye.
    *He whispers with Blanche*

BASTARD (aside)
Drawn in the flattering table of her eye,   505
  Hanged in the frowning wrinkle of her brow,
And quartered in her heart: he doth espy
  Himself love's traitor. This is pity now,
That hanged and drawn and quartered there should be
In such a love so vile a lout as he.   510

BLANCHE (to Louis the Dauphin)
My uncle's will in this respect is mine.
If he see aught in you that makes him like,
That anything he sees which moves his liking
I can with ease translate it to my will;
Or if you will, to speak more properly,   515
I will enforce it easily to my love.
Further I will not flatter you, my lord,
That all I see in you is worthy love,
Than this: that nothing do I see in you,
Though churlish thoughts themselves should be your
    judge,   520
That I can find should merit any hate.

KING JOHN
What say these young ones? What say you, my niece?

BLANCHE
That she is bound in honour still to do
What you in wisdom shall vouchsafe to say.

KING JOHN
Speak then, Prince Dauphin, can you love this lady?

LOUIS THE DAUPHIN
Nay, ask me if I can refrain from love,   526
For I do love her most unfeignedly.

KING JOHN
Then do I give Volquessen, Touraine, Maine,
Poitou, and Anjou, these five provinces,
With her to thee, and this addition more:   530
Full thirty thousand marks of English coin.
Philip of France, if thou be pleased withal,
Command thy son and daughter to join hands.

KING PHILIP
It likes us well.—Young princes, close your hands.

AUSTRIA
And your lips too, for I am well assured   535
That I did so when I was first assured.
    ⌈*Louis the Dauphin and Lady Blanche join hands
    and kiss*⌉

KING PHILIP
Now citizens of Angers, ope your gates.
Let in that amity which you have made,
For at Saint Mary's chapel presently
The rites of marriage shall be solemnized.—   540
Is not the Lady Constance in this troop?
(Aside) I know she is not, for this match made up
Her presence would have interrupted much.
  (Aloud) Where is she and her son? Tell me who knows.

LOUIS THE DAUPHIN
She is sad and passionate at your highness' tent.   545

KING PHILIP
And by my faith this league that we have made
Will give her sadness very little cure.—
Brother of England, how may we content
This widow lady? In her right we came,

Which we, God knows, have turned another way 550
To our own vantage.
KING JOHN                    We will heal up all,
For we'll create young Arthur Duke of Brittaine
And Earl of Richmond, and this rich fair town
We make him lord of. Call the Lady Constance.
Some speedy messenger bid her repair          555
To our solemnity. I trust we shall,
If not fill up the measure of her will,
Yet in some measure satisfy her so
That we shall stop her exclamation.
Go we as well as haste will suffer us          560
To this unlooked-for, unpreparèd pomp.
                    ⌈Flourish.⌉ Exeunt all but the Bastard
BASTARD
Mad world, mad kings, mad composition!
John, to stop Arthur's title in the whole,
Hath willingly departed with a part;
And France, whose armour conscience buckled on,
Whom zeal and charity brought to the field    566
As God's own soldier, rounded in the ear
With that same purpose-changer, that sly devil,
That broker that still breaks the pate of faith,
That daily break-vow, he that wins of all,     570
Of kings, of beggars, old men, young men, maids,—
Who having no external thing to lose
But the word 'maid', cheats the poor maid of that—
That smooth-faced gentleman, tickling commodity;
Commodity, the bias of the world,              575
The world who of itself is peisèd well,
Made to run even upon even ground,
Till this advantage, this vile-drawing bias,
This sway of motion, this commodity,
Makes it take head from all indifferency,      580
From all direction, purpose, course, intent;
And this same bias, this commodity,
This bawd, this broker, this all-changing word,
Clapped on the outward eye of fickle France,
Hath drawn him from his own determined aid,    585
From a resolved and honourable war,
To a most base and vile-concluded peace.
And why rail I on this commodity?
But for because he hath not wooed me yet—
Not that I have the power to clutch my hand    590
When his fair angels would salute my palm,
But for my hand, as unattempted yet,
Like a poor beggar raileth on the rich.
Well, whiles I am a beggar I will rail,
And say there is no sin but to be rich,        595
And being rich, my virtue then shall be
To say there is no vice but beggary.
Since kings break faith upon commodity,
Gain, be my lord, for I will worship thee.     Exit

2.2    Enter Lady Constance, Arthur Duke of Brittaine,
       and the Earl of Salisbury
CONSTANCE (to Salisbury)
Gone to be married? Gone to swear a peace?
False blood to false blood joined! Gone to be friends?
Shall Louis have Blanche, and Blanche those
       provinces?
It is not so, thou hast misspoke, misheard.
Be well advised, tell o'er thy tale again.      5
It cannot be, thou dost but say 'tis so.
I trust I may not trust thee, for thy word
Is but the vain breath of a common man.
Believe me, I do not believe thee, man;

I have a king's oath to the contrary.           10
Thou shalt be punished for thus frighting me;
For I am sick and capable of fears;
Oppressed with wrongs, and therefore full of fears;
A widow husbandless, subject to fears;
A woman naturally born to fears;                15
And though thou now confess thou didst but jest,
With my vexed spirits I cannot take a truce,
But they will quake and tremble all this day.
What dost thou mean by shaking of thy head?
Why dost thou look so sadly on my son?          20
What means that hand upon that breast of thine?
Why holds thine eye that lamentable rheum,
Like a proud river peering o'er his bounds?
Be these sad signs confirmers of thy words?
Then speak again—not all thy former tale,       25
But this one word: whether thy tale be true.
SALISBURY
As true as I believe you think them false
That give you cause to prove my saying true.
CONSTANCE
O, if thou teach me to believe this sorrow,
Teach thou this sorrow how to make me die;      30
And let belief and life encounter so
As doth the fury of two desperate men
Which in the very meeting fall and die.
Louis marry Blanche! (To Arthur) O boy, then where
       art thou?
France friend with England!—What becomes of me?
(To Salisbury) Fellow, be gone, I cannot brook thy
       sight;                                   36
This news hath made thee a most ugly man.
SALISBURY
What other harm have I, good lady, done,
But spoke the harm that is by others done?
CONSTANCE
Which harm within itself so heinous is          40
As it makes harmful all that speak of it.
ARTHUR
I do beseech you, madam, be content.
CONSTANCE
If thou that bidd'st me be content wert grim,
Ugly and sland'rous to thy mother's womb,
Full of unpleasing blots and sightless stains,  45
Lame, foolish, crooked, swart, prodigious,
Patched with foul moles and eye-offending marks,
I would not care, I then would be content,
For then I should not love thee, no, nor thou
Become thy great birth, nor deserve a crown.    50
But thou art fair, and at thy birth, dear boy,
Nature and Fortune joined to make thee great.
Of Nature's gifts thou mayst with lilies boast,
And with the half-blown rose. But Fortune, O,
She is corrupted, changed, and won from thee;   55
Sh'adulterates hourly with thine uncle John,
And with her golden hand hath plucked on France
To tread down fair respect of sovereignty,
And made his majesty the bawd to theirs.
France is a bawd to Fortune and King John,      60
That strumpet Fortune, that usurping John.
(To Salisbury) Tell me, thou fellow, is not France
       forsworn?
Envenom him with words, or get thee gone
And leave those woes alone, which I alone
Am bound to underbear.
SALISBURY             Pardon me, madam,        65
I may not go without you to the Kings.

CONSTANCE
Thou mayst, thou shalt; I will not go with thee.
I will instruct my sorrows to be proud,
For grief is proud and makes his owner stoop.
⌈*She sits upon the ground*⌉
To me and to the state of my great grief                      70
Let kings assemble, for my grief's so great
That no supporter but the huge firm earth
Can hold it up. Here I and sorrows sit;
Here is my throne; bid kings come bow to it.
⌈*Exeunt Salisbury and Arthur*⌉

3.1  ⌈*Flourish.*⌉ *Enter King John and King Philip* ⌈*hand
in hand*⌉; *Louis the Dauphin and Lady Blanche,*
⌈*married*⌉; *Queen Eleanor, the Bastard, and the
Duke of Austria*
KING PHILIP (*to Blanche*)
'Tis true, fair daughter, and this blessèd day
Ever in France shall be kept festival.
To solemnize this day, the glorious sun
Stays in his course and plays the alchemist,
Turning with splendour of his precious eye             5
The meagre cloddy earth to glittering gold.
The yearly course that brings this day about
Shall never see it but a holy day.
CONSTANCE (*rising*)
A wicked day, and not a holy day!
What hath this day deserved? What hath it done,    10
That it in golden letters should be set
Among the high tides in the calendar?
Nay, rather turn this day out of the week,
This day of shame, oppression, perjury.
Or if it must stand still, let wives with child          15
Pray that their burdens may not fall this day,
Lest that their hopes prodigiously be crossed;
But on this day let seamen fear no wreck;
No bargains break that are not this day made;
This day all things begun come to ill end,               20
Yea, faith itself to hollow falsehood change.
KING PHILIP
By heaven, lady, you shall have no cause
To curse the fair proceedings of this day.
Have I not pawned to you my majesty?
CONSTANCE
You have beguiled me with a counterfeit                 25
Resembling majesty, which being touched and tried
Proves valueless. You are forsworn, forsworn.
You came in arms to spill mine enemies' blood,
But now in arms you strengthen it with yours.
The grappling vigour and rough frown of war      30
Is cold in amity and painted peace,
And our oppression hath made up this league.
Arm, arm, you heavens, against these perjured Kings!
A widow cries, be husband to me, God!
Let not the hours of this ungodly day                    35
Wear out the day in peace, but ere sun set
Set armèd discord 'twixt these perjured Kings.
Hear me, O hear me!
AUSTRIA                         Lady Constance, peace.
CONSTANCE
War, war, no peace! Peace is to me a war.
O Limoges, O Austria, thou dost shame              40
That bloody spoil. Thou slave, thou wretch, thou
      coward!
Thou little valiant, great in villainy;
Thou ever strong upon the stronger side;

Thou Fortune's champion, that dost never fight
But when her humorous ladyship is by                  45
To teach thee safety. Thou art perjured too,
And sooth'st up greatness. What a fool art thou,
A ramping fool, to brag and stamp, and swear
Upon my party! Thou cold-blooded slave,
Hast thou not spoke like thunder on my side,      50
Been sworn my soldier, bidding me depend
Upon thy stars, thy fortune, and thy strength?
And dost thou now fall over to my foes?
Thou wear a lion's hide! Doff it, for shame,
And hang a calf's-skin on those recreant limbs.   55
AUSTRIA
O, that a man should speak those words to me!
BASTARD
And hang a calf's-skin on those recreant limbs.
AUSTRIA
Thou dar'st not say so, villain, for thy life.
BASTARD
And hang a calf's-skin on those recreant limbs.
KING JOHN (*to the Bastard*)
We like not this. Thou dost forget thyself.           60
      *Enter Cardinal Pandolf*
KING PHILIP
Here comes the holy legate of the Pope.
PANDOLF
Hail, you anointed deputies of God.—
To thee, King John, my holy errand is.
I Pandolf, of fair Milan Cardinal,
And from Pope Innocent the legate here,            65
Do in his name religiously demand
Why thou against the Church, our Holy Mother,
So wilfully dost spurn, and force perforce
Keep Stephen Langton, chosen Archbishop
Of Canterbury, from that holy see.                       70
This, in our foresaid Holy Father's name,
Pope Innocent, I do demand of thee.
KING JOHN
What earthy name to interrogatories
Can task the free breath of a sacred king?
Thou canst not, Cardinal, devise a name            75
So slight, unworthy, and ridiculous
To charge me to an answer, as the Pope.
Tell him this tale, and from the mouth of England
Add thus much more: that no Italian priest
Shall tithe or toll in our dominions;                      80
But as we, under God, are supreme head,
So, under him, that great supremacy
Where we do reign we will alone uphold
Without th'assistance of a mortal hand.
So tell the Pope, all reverence set apart             85
To him and his usurped authority.
KING PHILIP
Brother of England, you blaspheme in this.
KING JOHN
Though you and all the kings of Christendom
Are led so grossly by this meddling priest,
Dreading the curse that money may buy out,     90
And by the merit of vile gold, dross, dust,
Purchase corrupted pardon of a man,
Who in that sale sells pardon from himself;
Though you and all the rest so grossly led
This juggling witchcraft with revenue cherish;    95
Yet I alone, alone do me oppose
Against the Pope, and count his friends my foes.

PANDOLF
　　Then by the lawful power that I have
　　Thou shalt stand cursed and excommunicate;
　　And blessèd shall he be that doth revolt      100
　　From his allegiance to an heretic;
　　And meritorious shall that hand be called,
　　Canonizèd and worshipped as a saint,
　　That takes away by any secret course
　　Thy hateful life.
CONSTANCE　　　　　O lawful let it be      105
　　That I have room with Rome to curse awhile.
　　Good Father Cardinal, cry thou 'Amen'
　　To my keen curses, for without my wrong
　　There is no tongue hath power to curse him right.
PANDOLF
　　There's law and warrant, lady, for my curse.      110
CONSTANCE
　　And for mine too. When law can do no right,
　　Let it be lawful that law bar no wrong.
　　Law cannot give my child his kingdom here,
　　For he that holds his kingdom holds the law.
　　Therefore, since law itself is perfect wrong,      115
　　How can the law forbid my tongue to curse?
PANDOLF
　　Philip of France, on peril of a curse,
　　Let go the hand of that arch-heretic,
　　And raise the power of France upon his head,
　　Unless he do submit himself to Rome.      120
QUEEN ELEANOR
　　Look'st thou pale, France? Do not let go thy hand.
CONSTANCE [to King John]
　　Look to it, devil, lest that France repent,
　　And by disjoining hands hell lose a soul.
AUSTRIA
　　King Philip, listen to the Cardinal.
BASTARD
　　And hang a calf's-skin on his recreant limbs.      125
AUSTRIA
　　Well, ruffian, I must pocket up these wrongs,
　　Because—
BASTARD　　　Your breeches best may carry them.
KING JOHN
　　Philip, what sayst thou to the Cardinal?
CONSTANCE
　　What should he say, but as the Cardinal?
LOUIS THE DAUPHIN
　　Bethink you, Father, for the difference      130
　　Is purchase of a heavy curse from Rome,
　　Or the light loss of England for a friend.
　　Forgo the easier.
BLANCHE　　　　　　That's the curse of Rome.
CONSTANCE
　　O Louis, stand fast; the devil tempts thee here
　　In likeness of a new untrimmèd bride.      135
BLANCHE
　　The Lady Constance speaks not from her faith,
　　But from her need.
CONSTANCE [to King Philip] O if thou grant my need,
　　Which only lives but by the death of faith,
　　That need must needs infer this principle:
　　That faith would live again by death of need.      140
　　O, then tread down my need, and faith mounts up;
　　Keep my need up, and faith is trodden down.
KING JOHN
　　The King is moved, and answers not to this.
CONSTANCE (to King Philip)
　　O, be removed from him, and answer well.

AUSTRIA
　　Do so, King Philip, hang no more in doubt.      145
BASTARD
　　Hang nothing but a calf's-skin, most sweet lout.
KING PHILIP
　　I am perplexed, and know not what to say.
PANDOLF
　　What canst thou say but will perplex thee more,
　　If thou stand excommunicate and cursed?
KING PHILIP
　　Good Reverend Father, make my person yours,      150
　　And tell me how you would bestow yourself.
　　This royal hand and mine are newly knit,
　　And the conjunction of our inward souls
　　Married in league, coupled and linked together
　　With all religious strength of sacred vows;      155
　　The latest breath that gave the sound of words
　　Was deep-sworn faith, peace, amity, true love,
　　Between our kingdoms and our royal selves;
　　And even before this truce, but new before,
　　No longer than we well could wash our hands      160
　　To clap this royal bargain up of peace,
　　God knows, they were besmeared and over-stained
　　With slaughter's pencil, where Revenge did paint
　　The fearful difference of incensèd kings;
　　And shall these hands, so lately purged of blood,      165
　　So newly joined in love, so strong in both,
　　Unyoke this seizure and this kind regreet,
　　Play fast and loose with faith, so jest with heaven,
　　Make such unconstant children of ourselves,
　　As now again to snatch our palm from palm,      170
　　Unswear faith sworn, and on the marriage-bed
　　Of smiling peace to march a bloody host,
　　And make a riot on the gentle brow
　　Of true sincerity? O holy sir,
　　My Reverend Father, let it not be so.      175
　　Out of your grace, devise, ordain, impose
　　Some gentle order, and then we shall be blessed
　　To do your pleasure and continue friends.
PANDOLF
　　All form is formless, order orderless,
　　Save what is opposite to England's love.      180
　　Therefore to arms, be champion of our Church,
　　Or let the Church, our mother, breathe her curse,
　　A mother's curse, on her revolting son.
　　France, thou mayst hold a serpent by the tongue,
　　A crazèd lion by the mortal paw,      185
　　A fasting tiger safer by the tooth,
　　Than keep in peace that hand which thou dost hold.
KING PHILIP
　　I may disjoin my hand, but not my faith.
PANDOLF
　　So mak'st thou faith an enemy to faith,
　　And like a civil war, sett'st oath to oath,      190
　　Thy tongue against thy tongue. O, let thy vow,
　　First made to heaven, first be to heaven performed;
　　That is, to be the champion of our Church.
　　What since thou swor'st is sworn against thyself,
　　And may not be performèd by thyself;      195
　　For that which thou hast sworn to do amiss
　　Is not amiss when it is truly done;
　　And being not done where doing tends to ill,
　　The truth is then most done not doing it.
　　The better act of purposes mistook      200
　　Is to mistake again; though indirect,
　　Yet indirection thereby grows direct,

And falsehood falsehood cures, as fire cools fire
Within the scorchèd veins of one new burned.
It is religion that doth make vows kept;                    205
But thou hast sworn against religion;
By what thou swear'st, against the thing thou
    swear'st;
And mak'st an oath the surety for thy troth:
Against an oath, the truth. Thou art unsure
To swear: swear'st only not to be forsworn—    210
Else what a mockery should it be to swear!—
But thou dost swear only to be forsworn,
And most forsworn to keep what thou dost swear;
Therefore thy later vows against thy first
Is in thyself rebellion to thyself,                        215
And better conquest never canst thou make
Than arm thy constant and thy nobler parts
Against these giddy loose suggestions;
Upon which better part our prayers come in
If thou vouchsafe them. But if not, then know    220
The peril of our curses light on thee
So heavy as thou shalt not shake them off,
But in despair die under their black weight.

AUSTRIA
Rebellion, flat rebellion!

BASTARD                              Wilt not be?
Will not a calf's-skin stop that mouth of thine?    225

LOUIS THE DAUPHIN
Father, to arms!

BLANCHE                    Upon thy wedding day?
Against the blood that thou hast married?
What, shall our feast be kept with slaughtered men?
Shall braying trumpets and loud churlish drums,
Clamours of hell, be measures to our pomp?    230
    She kneels
O husband, hear me! Ay, alack, how new
Is 'husband' in my mouth! Even for that name
Which till this time my tongue did ne'er pronounce,
Upon my knee I beg, go not to arms
Against mine uncle.

CONSTANCE (kneeling)    O, upon my knee    235
Made hard with kneeling, I do pray to thee,
Thou virtuous Dauphin, alter not the doom
Forethought by heaven.

BLANCHE (to Louis the Dauphin)
Now shall I see thy love: what motive may
Be stronger with thee than the name of wife?    240

CONSTANCE
That which upholdeth him that thee upholds:
His honour—O thine honour, Louis, thine honour!

LOUIS THE DAUPHIN (to King Philip)
I muse your majesty doth seem so cold
When such profound respects do pull you on.

PANDOLF
I will denounce a curse upon his head.    245

KING PHILIP
Thou shalt not need.—England, I will fall from thee.
    [He takes his hand from King John's hand. Blanche
    and Constance rise]

CONSTANCE
O, fair return of banished majesty!

QUEEN ELEANOR
O, foul revolt of French inconstancy!

KING JOHN
France, thou shalt rue this hour within this hour.

BASTARD
Old Time the clock-setter, that bald sexton Time,    250
Is it as he will?—Well then, France shall rue.

BLANCHE
The sun's o'ercast with blood; fair day, adieu!
Which is the side that I must go withal?
I am with both, each army hath a hand,
And in their rage, I having hold of both,    255
They whirl asunder and dismember me.
Husband, I cannot pray that thou mayst win.—
Uncle, I needs must pray that thou must lose.—
Father, I may not wish the fortune thine.—
Grandam, I will not wish thy wishes thrive.    260
Whoever wins, on that side shall I lose,
Assurèd loss before the match be played.

LOUIS THE DAUPHIN
Lady, with me, with me thy fortune lies.

BLANCHE
There where my fortune lives, there my life dies.

KING JOHN (to the Bastard)
Cousin, go draw our puissance together.—    265
    [Exit the Bastard]
France, I am burned up with inflaming wrath,
A rage whose heat hath this condition:
That nothing can allay, nothing but blood,
The blood, and dearest-valued blood, of France.

KING PHILIP
Thy rage shall burn thee up, and thou shalt turn    270
To ashes ere our blood shall quench that fire.
Look to thyself, thou art in jeopardy.

KING JOHN
No more than he that threats.—To arms let's hie!
    Exeunt [severally]

3.2    Alarum; excursions. Enter the Bastard, with the
       Duke of Austria's head

BASTARD
Now, by my life, this day grows wondrous hot;
Some airy devil hovers in the sky
And pours down mischief. Austria's head lie there,
While Philip breathes.
    Enter King John, Arthur Duke of Brittaine, and
    Hubert

KING JOHN
Hubert, keep this boy.—Philip, make up!    5
My mother is assailèd in our tent,
And ta'en I fear.

BASTARD                    My lord, I rescued her;
Her highness is in safety; fear you not.
But on, my liege, for very little pains
Will bring this labour to an happy end.    10
    Exeunt [King John and the Bastard at one door,
    Hubert and Arthur at another door]

3.3    Alarum; excursions; retreat. Enter King John,
       Queen Eleanor, Arthur Duke of Brittaine, the
       Bastard, Hubert, lords, [with soldiers]

KING JOHN (to Queen Eleanor)
So shall it be; your grace shall stay behind
So strongly guarded. (To Arthur) Cousin, look not sad;
Thy grandam loves thee, and thy uncle will
As dear be to thee as thy father was.

ARTHUR
O, this will make my mother die with grief.    5

KING JOHN (to the Bastard)
Cousin, away for England! Haste before,
And ere our coming, see thou shake the bags
Of hoarding abbots. The fat ribs of peace
Must by the hungry now be fed upon.

Imprisoned angels set at liberty.                                    10
Use our commission in his utmost force.
BASTARD
Bell, book, and candle shall not drive me back
When gold and silver becks me to come on.
I leave your highness.—Grandam, I will pray,
If ever I remember to be holy,                                       15
For your fair safety. So I kiss your hand.
QUEEN ELEANOR
Farewell, gentle cousin.
KING JOHN                     Coz, farewell.     *Exit the Bastard*
QUEEN ELEANOR
Come hither, little kinsman. Hark, a word.
          *She takes Arthur aside*
KING JOHN
Come hither, Hubert.
          *He takes Hubert aside*
                    O my gentle Hubert,
We owe thee much. Within this wall of flesh               20
There is a soul counts thee her creditor,
And with advantage means to pay thy love;
And, my good friend, thy voluntary oath
Lives in this bosom, dearly cherishèd.
Give me thy hand.
          *He takes Hubert's hand*
                    I had a thing to say,                  25
But I will fit it with some better tune.
By heaven, Hubert, I am almost ashamed
To say what good respect I have of thee.
HUBERT
I am much bounden to your majesty.
KING JOHN
Good friend, thou hast no cause to say so yet,            30
But thou shalt have; and creep time ne'er so slow,
Yet it shall come for me to do thee good.
I had a thing to say—but let it go.
The sun is in the heaven, and the proud day,
Attended with the pleasures of the world,                 35
Is all too wanton and too full of gauds
To give me audience. If the midnight bell
Did with his iron tongue and brazen mouth
Sound on into the drowsy race of night;
If this same were a churchyard where we stand,           40
And thou possessèd with a thousand wrongs;
Or if that surly spirit, melancholy,
Had baked thy blood and made it heavy, thick,
Which else runs tickling up and down the veins,
Making that idiot, laughter, keep men's eyes             45
And strain their cheeks to idle merriment—
A passion hateful to my purposes—
Or if that thou couldst see me without eyes,
Hear me without thine ears, and make reply
Without a tongue, using conceit alone,                   50
Without eyes, ears, and harmful sound of words;
Then in despite of broad-eyed watchful day
I would into thy bosom pour my thoughts.
But, ah, I will not. Yet I love thee well,
And by my troth, I think thou lov'st me well.            55
HUBERT
So well that what you bid me undertake,
Though that my death were adjunct to my act,
By heaven, I would do it.
KING JOHN                     Do not I know thou wouldst?
Good Hubert, Hubert, Hubert, throw thine eye
On yon young boy. I'll tell thee what, my friend,        60
He is a very serpent in my way,

And wheresoe'er this foot of mine doth tread,
He lies before me. Dost thou understand me?
Thou art his keeper.
HUBERT                     And I'll keep him so
That he shall not offend your majesty.                   65
KING JOHN
Death.
HUBERT  My lord.
KING JOHN          A grave.
HUBERT                     He shall not live.
KING JOHN                                  Enough.
I could be merry now. Hubert, I love thee.
Well, I'll not say what I intend for thee.
Remember. (*To Queen Eleanor*) Madam, fare you well.
I'll send those powers o'er to your majesty.             70
QUEEN ELEANOR
My blessing go with thee.
KING JOHN (*to Arthur*)       For England, cousin, go.
Hubert shall be your man, attend on you
With all true duty.—On toward Calais, ho!
          *Exeunt ⌐Queen Eleanor, attended, at one door,*
                          *the rest at another door⌐*

**3.4**    *Enter King Philip, Louis the Dauphin, Cardinal*
          *Pandolf, and attendants*
KING PHILIP
So, by a roaring tempest on the flood,
A whole armada of convicted sail
Is scattered and disjoined from fellowship.
PANDOLF
Courage and comfort; all shall yet go well.
KING PHILIP
What can go well when we have run so ill?                 5
Are we not beaten? Is not Angers lost,
Arthur ta'en prisoner, divers dear friends slain,
And bloody England into England gone,
O'erbearing interruption, spite of France?
LOUIS THE DAUPHIN
What he hath won, that hath he fortified.                10
So hot a speed, with such advice disposed,
Such temperate order in so fierce a cause,
Doth want example. Who hath read or heard
Of any kindred action like to this?
KING PHILIP
Well could I bear that England had this praise,          15
So we could find some pattern of our shame.
          *Enter Constance, distracted, with her hair about*
          *her ears*
Look who comes here! A grave unto a soul,
Holding th'eternal spirit against her will
In the vile prison of afflicted breath.—
I prithee, lady, go away with me.                        20
CONSTANCE
Lo, now, now see the issue of your peace!
KING PHILIP
Patience, good lady; comfort, gentle Constance.
CONSTANCE
No, I defy all counsel, all redress,
But that which ends all counsel, true redress:
Death, Death, O amiable, lovely Death!                   25
Thou odoriferous stench, sound rottenness!
Arise forth from the couch of lasting night,
Thou hate and terror to prosperity,
And I will kiss thy detestable bones,
And put my eyeballs in thy vaulty brows,                 30
And ring these fingers with thy household worms,
And stop this gap of breath with fulsome dust,

And be a carrion monster like thyself.
Come grin on me, and I will think thou smil'st,
And buss thee as thy wife. Misery's love,
O, come to me!                                                    35
KING PHILIP          O fair affliction, peace!
CONSTANCE
No, no, I will not, having breath to cry.
O, that my tongue were in the thunder's mouth!
Then with a passion would I shake the world,
And rouse from sleep that fell anatomy,                           40
Which cannot hear a lady's feeble voice,
Which scorns a modern invocation.
PANDOLF
Lady, you utter madness, and not sorrow.
CONSTANCE
Thou art not holy to belie me so.
I am not mad: this hair I tear is mine;                           45
My name is Constance; I was Geoffrey's wife;
Young Arthur is my son; and he is lost.
I am not mad; I would to God I were,
For then 'tis like I should forget myself.
O, if I could, what grief should I forget!                        50
Preach some philosophy to make me mad,
And thou shalt be canonized, Cardinal.
For, being not mad, but sensible of grief,
My reasonable part produces reason
How I may be delivered of these woes,                             55
And teaches me to kill or hang myself.
If I were mad I should forget my son,
Or madly think a babe of clouts were he.
I am not mad; too well, too well I feel
The different plague of each calamity.                            60
KING PHILIP
Bind up those tresses. O, what love I note
In the fair multitude of those her hairs!
Where but by chance a silver drop hath fallen,
Even to that drop ten thousand wiry friends
Do glue themselves in sociable grief,                             65
Like true, inseparable, faithful loves,
Sticking together in calamity.
CONSTANCE
To England, if you will.
KING PHILIP                    Bind up your hairs.
CONSTANCE
Yes, that I will. And wherefore will I do it?
I tore them from their bonds, and cried aloud,                    70
'O that these hands could so redeem my son,
As they have given these hairs their liberty!'
But now I envy at their liberty,
And will again commit them to their bonds,
Because my poor child is a prisoner.                              75
        *She binds up her hair*
And Father Cardinal, I have heard you say
That we shall see and know our friends in heaven.
If that be true, I shall see my boy again;
For since the birth of Cain, the first male child,
To him that did but yesterday suspire,                            80
There was not such a gracious creature born.
But now will canker-sorrow eat my bud,
And chase the native beauty from his cheek;
And he will look as hollow as a ghost,
As dim and meagre as an ague's fit,                               85
And so he'll die; and rising so again,
When I shall meet him in the court of heaven,
I shall not know him; therefore never, never
Must I behold my pretty Arthur more.

PANDOLF
You hold too heinous a respect of grief.                          90
CONSTANCE
He talks to me that never had a son.
KING PHILIP
You are as fond of grief as of your child.
CONSTANCE
Grief fills the room up of my absent child,
Lies in his bed, walks up and down with me,
Puts on his pretty looks, repeats his words,                      95
Remembers me of all his gracious parts,
Stuffs out his vacant garments with his form;
Then have I reason to be fond of grief.
Fare you well. Had you such a loss as I,
I could give better comfort than you do.                          100
        *She unbinds her hair*
I will not keep this form upon my head
When there is such disorder in my wit.
O Lord, my boy, my Arthur, my fair son,
My life, my joy, my food, my all the world,                       104
My widow-comfort, and my sorrow's cure!          *Exit*
KING PHILIP
I fear some outrage, and I'll follow her.    *Exit ⌐attended⌐*
LOUIS THE DAUPHIN
There's nothing in this world can make me joy.
Life is as tedious as a twice-told tale,
Vexing the dull ear of a drowsy man;
And bitter shame hath spoiled the sweet world's taste,
That it yields naught but shame and bitterness.                   111
PANDOLF
Before the curing of a strong disease,
Even in the instant of repair and health,
The fit is strongest. Evils that take leave,
On their departure most of all show evil.                         115
What have you lost by losing of this day?
LOUIS THE DAUPHIN
All days of glory, joy, and happiness.
PANDOLF
If you had won it, certainly you had.
No, no; when Fortune means to men most good,
She looks upon them with a threat'ning eye.                       120
'Tis strange to think how much King John hath lost
In this which he accounts so clearly won.
Are not you grieved that Arthur is his prisoner?
LOUIS THE DAUPHIN
As heartily as he is glad he hath him.
PANDOLF
Your mind is all as youthful as your blood.                       125
Now hear me speak with a prophetic spirit,
For even the breath of what I mean to speak
Shall blow each dust, each straw, each little rub,
Out of the path which shall directly lead
Thy foot to England's throne. And therefore mark.                 130
John hath seized Arthur, and it cannot be
That whiles warm life plays in that infant's veins
The misplaced John should entertain an hour,
One minute, nay, one quiet breath of rest.
A sceptre snatched with an unruly hand                            135
Must be as boisterously maintained as gained;
And he that stands upon a slipp'ry place
Makes nice of no vile hold to stay him up.
That John may stand, then Arthur needs must fall;
So be it, for it cannot be but so.                                140
LOUIS THE DAUPHIN
But what shall I gain by young Arthur's fall?

PANDOLF
You, in the right of Lady Blanche your wife,
May then make all the claim that Arthur did.

LOUIS THE DAUPHIN
And lose it, life and all, as Arthur did.

PANDOLF
How green you are, and fresh in this old world!    145
John lays you plots; the times conspire with you;
For he that steeps his safety in true blood
Shall find but bloody safety and untrue.
This act, so vilely born, shall cool the hearts
Of all his people, and freeze up their zeal,        150
That none so small advantage shall step forth
To check his reign but they will cherish it;
No natural exhalation in the sky,
No scope of nature, no distempered day,
No common wind, no customèd event,                  155
But they will pluck away his natural cause,
And call them meteors, prodigies, and signs,
Abortives, presages, and tongues of heaven
Plainly denouncing vengeance upon John.

LOUIS THE DAUPHIN
Maybe he will not touch young Arthur's life,        160
But hold himself safe in his prisonment.

PANDOLF
O sir, when he shall hear of your approach,
If that young Arthur be not gone already,
Even at that news he dies; and then the hearts
Of all his people shall revolt from him,            165
And kiss the lips of unacquainted change,
And pick strong matter of revolt and wrath
Out of the bloody fingers' ends of John.
Methinks I see this hurly all on foot,
And O, what better matter breeds for you            170
Than I have named! The Bastard Falconbridge
Is now in England, ransacking the Church,
Offending charity. If but a dozen French
Were there in arms, they would be as a call
To train ten thousand English to their side,        175
Or as a little snow tumbled about
Anon becomes a mountain. O noble Dauphin,
Go with me to the King. 'Tis wonderful
What may be wrought out of their discontent
Now that their souls are top-full of offence.       180
For England, go! I will whet on the King.

LOUIS THE DAUPHIN
Strong reasons make strange actions. Let us go.
If you say ay, the King will not say no.      Exeunt

**4.1**  *Enter Hubert, and Executioners with a rope and*
        *irons*

HUBERT
Heat me these irons hot, and look thou stand
Within the arras. When I strike my foot
Upon the bosom of the ground, rush forth
And bind the boy which you shall find with me
Fast to the chair. Be heedful. Hence, and watch!    5

EXECUTIONER
I hope your warrant will bear out the deed.

HUBERT
Uncleanly scruples: fear not you. Look to't!
    ⌈*The Executioners withdraw behind the arras*⌉
Young lad, come forth, I have to say with you.
    *Enter Arthur Duke of Brittaine*

ARTHUR
Good morrow, Hubert.

HUBERT                     Good morrow, little Prince.

ARTHUR
As little prince, having so great a title           10
To be more prince, as may be. You are sad.

HUBERT
Indeed I have been merrier.

ARTHUR                      Mercy on me!
Methinks nobody should be sad but I.
Yet I remember, when I was in France,
Young gentlemen would be as sad as night            15
Only for wantonness. By my christendom,
So I were out of prison and kept sheep,
I should be as merry as the day is long;
And so I would be here, but that I doubt
My uncle practises more harm to me.                 20
He is afraid of me, and I of him.
Is it my fault that I was Geoffrey's son?
No, indeed is't not, and I would to God
I were your son, so you would love me, Hubert.

HUBERT (*aside*)
If I talk to him, with his innocent prate           25
He will awake my mercy, which lies dead;
Therefore I will be sudden, and dispatch.

ARTHUR
Are you sick, Hubert? You look pale today.
In sooth, I would you were a little sick,
That I might sit all night and watch with you.      30
I warrant I love you more than you do me.

HUBERT (*aside*)
His words do take possession of my bosom.
    *He shows Arthur a paper*
Read here, young Arthur. (*Aside*) How now: foolish
    rheum,
Turning dispiteous torture out of door?
I must be brief, lest resolution drop               35
Out at mine eyes in tender womanish tears.
(*To Arthur*) Can you not read it? Is it not fair writ?

ARTHUR
Too fairly, Hubert, for so foul effect.
Must you with hot irons burn out both mine eyes?

HUBERT
Young boy, I must.

ARTHUR                  And will you?

HUBERT                                  And I will.     40

ARTHUR
Have you the heart? When your head did but ache
I knit my handkerchief about your brows,
The best I had—a princess wrought it me,
And I did never ask it you again—
And with my hand at midnight held your head,        45
And like the watchful minutes to the hour
Still and anon cheered up the heavy time,
Saying 'What lack you?' and 'Where lies your grief?'
Or 'What good love may I perform for you?'
Many a poor man's son would have lain still         50
And ne'er have spoke a loving word to you,
But you at your sick service had a prince.
Nay, you may think my love was crafty love,
And call it cunning. Do, an if you will.
If heaven be pleased that you must use me ill,      55
Why then you must. Will you put out mine eyes,
These eyes that never did, nor never shall,
So much as frown on you?

HUBERT                        I have sworn to do it,
And with hot irons must I burn them out.

ARTHUR
Ah, none but in this iron age would do it.          60
The iron of itself, though heat red hot,

Approaching near these eyes would drink my tears,
And quench his fiery indignation
Even in the matter of mine innocence;
Nay, after that, consume away in rust,                              65
But for containing fire to harm mine eye.
Are you more stubborn-hard than hammered iron?
An if an angel should have come to me
And told me Hubert should put out mine eyes,
I would not have believed him; no tongue but
      Hubert's.                                                      70
      *Hubert stamps his foot*

HUBERT
Come forth!
      *The Executioners come forth*
                        Do as I bid you do.

ARTHUR
O, save me, Hubert, save me! My eyes are out
Even with the fierce looks of these bloody men.

HUBERT (*to the Executioners*)
Give me the iron, I say, and bind him here.
      *He takes the iron*

ARTHUR
Alas, what need you be so boisterous-rough?                         75
I will not struggle; I will stand stone-still.
For God's sake, Hubert, let me not be bound.
Nay, hear me, Hubert! Drive these men away,
And I will sit as quiet as a lamb;
I will not stir, nor wince, nor speak a word,                      80
Nor look upon the iron angerly.
Thrust but these men away, and I'll forgive you,
Whatever torment you do put me to.

HUBERT (*to the Executioners*)
Go stand within. Let me alone with him.

EXECUTIONER
I am best pleased to be from such a deed.                           85
      *Exeunt Executioners*

ARTHUR
Alas, I then have chid away my friend!
He hath a stern look, but a gentle heart.
Let him come back, that his compassion may
Give life to yours.

HUBERT                  Come, boy, prepare yourself.

ARTHUR
Is there no remedy?

HUBERT                  None but to lose your eyes.                 90

ARTHUR
O God, that there were but a mote in yours,
A grain, a dust, a gnat, a wandering hair,
Any annoyance in that precious sense,
Then, feeling what small things are boisterous there,
Your vile intent must needs seem horrible.                         95

HUBERT
Is this your promise? Go to, hold your tongue!

ARTHUR
Hubert, the utterance of a brace of tongues
Must needs want pleading for a pair of eyes.
Let me not hold my tongue, let me not, Hubert;
Or, Hubert, if you will, cut out my tongue,                        100
So I may keep mine eyes. O, spare mine eyes,
Though to no use but still to look on you.
Lo, by my troth, the instrument is cold
And would not harm me.

HUBERT                  I can heat it, boy.

ARTHUR
No, in good sooth: the fire is dead with grief,                    105
Being create for comfort, to be used

In undeserved extremes. See else yourself.
There is no malice in this burning coal;
The breath of heaven hath blown his spirit out,
And strewed repentant ashes on his head.                           110

HUBERT
But with my breath I can revive it, boy.

ARTHUR
An if you do, you will but make it blush
And glow with shame of your proceedings, Hubert.
Nay, it perchance will sparkle in your eyes,
And like a dog that is compelled to fight,                         115
Snatch at his master that doth tarre him on.
All things that you should use to do me wrong
Deny their office; only you do lack
That mercy which fierce fire and iron extends,
Creatures of note for mercy-lacking uses.                          120

HUBERT
Well, see to live. I will not touch thine eye
For all the treasure that thine uncle owes.
Yet am I sworn, and I did purpose, boy,
With this same very iron to burn them out.

ARTHUR
O, now you look like Hubert. All this while                        125
You were disguisèd.

HUBERT                  Peace, no more. Adieu.
Your uncle must not know but you are dead.
I'll fill these doggèd spies with false reports;
And, pretty child, sleep doubtless and secure
That Hubert, for the wealth of all the world,                      130
Will not offend thee.

ARTHUR                  O God! I thank you, Hubert.

HUBERT
Silence, no more. Go closely in with me.
Much danger do I undergo for thee.          *Exeunt*

4.2    ⌈*Flourish.*⌉ *Enter King John, the Earls of Pembroke*
       *and Salisbury, and other lords. King John ascends*
       *the throne*

KING JOHN
Here once again we sit, once again crowned,
And looked upon, I hope, with cheerful eyes.

PEMBROKE
This 'once again', but that your highness pleased,
Was once superfluous. You were crowned before,
And that high royalty was ne'er plucked off,                        5
The faiths of men ne'er stainèd with revolt;
Fresh expectation troubled not the land
With any longed-for change or better state.

SALISBURY
Therefore to be possessed with double pomp,
To guard a title that was rich before,                             10
To gild refinèd gold, to paint the lily,
To throw a perfume on the violet,
To smooth the ice, or add another hue
Unto the rainbow, or with taper-light
To seek the beauteous eye of heaven to garnish,                    15
Is wasteful and ridiculous excess.

PEMBROKE
But that your royal pleasure must be done,
This act is as an ancient tale new-told,
And in the last repeating troublesome,
Being urgèd at a time unseasonable.                                20

SALISBURY
In this the antique and well-noted face
Of plain old form is much disfigurèd,
And like a shifted wind unto a sail,

It makes the course of thoughts to fetch about,
Startles and frights consideration,                                25
Makes sound opinion sick, and truth suspected
For putting on so new a fashioned robe.
PEMBROKE
When workmen strive to do better than well,
They do confound their skill in covetousness;
And oftentimes excusing of a fault                                30
Doth make the fault the worser by th'excuse;
As patches set upon a little breach
Discredit more in hiding of the fault
Than did the fault before it was so patched.
SALISBURY
To this effect: before you were new-crowned      35
We breathed our counsel, but it pleased your
    highness
To overbear it; and we are all well pleased,
Since all and every part of what we would
Doth make a stand at what your highness will.
KING JOHN
Some reasons of this double coronation            40
I have possessed you with, and think them strong.
And more, more strong, when lesser is my fear
I shall endue you with. Meantime but ask
What you would have reformed that is not well,
And well shall you perceive how willingly          45
I will both hear and grant you your requests.
PEMBROKE
Then I, as one that am the tongue of these
To sound the purposes of all their hearts,
Both for myself and them, but chief of all
Your safety, for the which myself and them      50
Bend their best studies, heartily request
Th'enfranchisement of Arthur, whose restraint
Doth move the murmuring lips of discontent
To break into this dangerous argument:
If what in rest you have, in right you hold,          55
Why then your fears—which, as they say, attend
The steps of wrong—should move you to mew up
Your tender kinsman, and to choke his days
With barbarous ignorance, and deny his youth
The rich advantage of good exercise?               60
That the time's enemies may not have this
To grace occasions, let it be our suit
That you have bid us ask, his liberty;
Which for our goods we do no further ask
Than whereupon our weal, on you depending,   65
Counts it your weal he have his liberty.
        *Enter Hubert*
KING JOHN
Let it be so. I do commit his youth
To your direction.—Hubert, what news with you?
        *He takes Hubert aside*
PEMBROKE
This is the man should do the bloody deed:
He showed his warrant to a friend of mine.        70
The image of a wicked heinous fault
Lives in his eye; that close aspect of his
Does show the mood of a much troubled breast;
And I do fearfully believe 'tis done
What we so feared he had a charge to do.          75
SALISBURY
The colour of the King doth come and go
Between his purpose and his conscience,
Like heralds 'twixt two dreadful battles set.
His passion is so ripe it needs must break.

PEMBROKE
And when it breaks, I fear will issue thence        80
The foul corruption of a sweet child's death.
KING JOHN (*coming forward*)
We cannot hold mortality's strong hand.
Good lords, although my will to give is living,
The suit which you demand is gone and dead.
He tells us Arthur is deceased tonight.                85
SALISBURY
Indeed we feared his sickness was past cure.
PEMBROKE
Indeed we heard how near his death he was,
Before the child himself felt he was sick.
This must be answered, either here or hence.
KING JOHN
Why do you bend such solemn brows on me?    90
Think you I bear the shears of destiny?
Have I commandment on the pulse of life?
SALISBURY
It is apparent foul play, and 'tis shame
That greatness should so grossly offer it.
So thrive it in your game; and so, farewell.          95
PEMBROKE
Stay yet, Lord Salisbury; I'll go with thee,
And find th'inheritance of this poor child,
His little kingdom of a forcèd grave.
That blood which owed the breadth of all this isle
Three foot of it doth hold. Bad world the while.   100
This must not be thus borne. This will break out
To all our sorrows; and ere long, I doubt.
        *Exeunt Pembroke, Salisbury, ⌐and other lords⌐*
KING JOHN
They burn in indignation. I repent.
There is no sure foundation set on blood,
No certain life achieved by others' death.           105
        *Enter a Messenger*
A fearful eye thou hast. Where is that blood
That I have seen inhabit in those cheeks?
So foul a sky clears not without a storm:
Pour down thy weather: how goes all in France?
MESSENGER
From France to England. Never such a power     110
For any foreign preparation
Was levied in the body of a land.
The copy of your speed is learned by them,
For when you should be told they do prepare,
The tidings comes that they are all arrived.        115
KING JOHN
O, where hath our intelligence been drunk?
Where hath it slept? Where is my mother's ear,
That such an army could be drawn in France,
And she not hear of it?
MESSENGER                          My liege, her ear
Is stopped with dust. The first of April died      120
Your noble mother. And as I hear, my lord,
The Lady Constance in a frenzy died
Three days before; but this from rumour's tongue
I idly heard; if true or false I know not.
KING JOHN
Withhold thy speed, dreadful Occasion;            125
O, make a league with me till I have pleased
My discontented peers. What, Mother dead?—
How wildly then walks my estate in France!—
Under whose conduct came those powers of France
That thou for truth giv'st out are landed here?   130

MESSENGER
   Under the Dauphin.
     *Enter the Bastard and Peter of Pomfret*
KING JOHN          Thou hast made me giddy
   With these ill tidings. (*To the Bastard*) Now, what says
     the world
   To your proceedings? Do not seek to stuff
   My head with more ill news, for it is full.
BASTARD
   But if you be afeard to hear the worst,     135
   Then let the worst, unheard, fall on your head.
KING JOHN
   Bear with me, cousin, for I was amazed
   Under the tide; but now I breathe again
   Aloft the flood, and can give audience
   To any tongue, speak it of what it will.     140
BASTARD
   How I have sped among the clergymen
   The sums I have collected shall express.
   But as I travelled hither through the land,
   I find the people strangely fantasied,
   Possessed with rumours, full of idle dreams,     145
   Not knowing what they fear, but full of fear.
   And here's a prophet that I brought with me
   From forth the streets of Pomfret, whom I found
   With many hundreds treading on his heels;
   To whom he sung, in rude, harsh-sounding rhymes,
   That ere the next Ascension Day at noon     151
   Your highness should deliver up your crown.
KING JOHN
   Thou idle dreamer, wherefore didst thou so?
PETER OF POMFRET
   Foreknowing that the truth will fall out so.
KING JOHN
   Hubert, away with him! Imprison him,     155
   And on that day, at noon, whereon he says
   I shall yield up my crown, let him be hanged.
   Deliver him to safety, and return,
   For I must use thee.
     *Exeunt Hubert and Peter of Pomfret*
     O my gentle cousin,
   Hear'st thou the news abroad, who are arrived?     160
BASTARD
   The French, my lord: men's mouths are full of it.
   Besides, I met Lord Bigot and Lord Salisbury
   With eyes as red as new-enkindled fire,
   And others more, going to seek the grave
   Of Arthur, whom they say is killed tonight     165
   On your suggestion.
KING JOHN          Gentle kinsman, go
   And thrust thyself into their companies.
   I have a way to win their loves again.
   Bring them before me.
BASTARD          I will seek them out.
KING JOHN
   Nay, but make haste, the better foot before.     170
   O, let me have no subject enemies
   When adverse foreigners affright my towns
   With dreadful pomp of stout invasion!
   Be Mercury, set feathers to thy heels,
   And fly like thought from them to me again.     175
BASTARD
   The spirit of the time shall teach me speed.     *Exit*
KING JOHN
   Spoke like a sprightful noble gentleman!—
   Go after him, for he perhaps shall need

Some messenger betwixt me and the peers,
   And be thou he.     180
MESSENGER  With all my heart, my liege.     *Exit*
KING JOHN  My mother dead!
     *Enter Hubert*
HUBERT
   My lord, they say five moons were seen tonight,
   Four fixèd, and the fifth did whirl about
   The other four in wondrous motion.     185
KING JOHN
   Five moons?
HUBERT          Old men and beldams in the streets
   Do prophesy upon it dangerously.
   Young Arthur's death is common in their mouths,
   And when they talk of him they shake their heads,
   And whisper one another in the ear;     190
   And he that speaks doth grip the hearer's wrist,
   Whilst he that hears makes fearful action,
   With wrinkled brows, with nods, with rolling eyes.
   I saw a smith stand with his hammer, thus,
   The whilst his iron did on the anvil cool,     195
   With open mouth swallowing a tailor's news,
   Who, with his shears and measure in his hand,
   Standing on slippers which his nimble haste
   Had falsely thrust upon contrary feet,
   Told of a many thousand warlike French     200
   That were embattailèd and ranked in Kent.
   Another lean unwashed artificer
   Cuts off his tale, and talks of Arthur's death.
KING JOHN
   Why seek'st thou to possess me with these fears?
   Why urgest thou so oft young Arthur's death?     205
   Thy hand hath murdered him. I had a mighty cause
   To wish him dead, but thou hadst none to kill him.
HUBERT
   No had, my lord? Why, did you not provoke me?
KING JOHN
   It is the curse of kings to be attended
   By slaves that take their humours for a warrant     210
   To break within the bloody house of life,
   And on the winking of authority
   To understand a law, to know the meaning
   Of dangerous majesty, when perchance it frowns
   More upon humour than advised respect.     215
HUBERT
   Here is your hand and seal for what I did.
     *He shows a paper*
KING JOHN
   O, when the last account 'twixt heaven and earth
   Is to be made, then shall this hand and seal
   Witness against us to damnation!
   How oft the sight of means to do ill deeds     220
   Make deeds ill done! Hadst not thou been by,
   A fellow by the hand of nature marked,
   Quoted, and signed to do a deed of shame,
   This murder had not come into my mind.
   But taking note of thy abhorred aspect,     225
   Finding thee fit for bloody villainy,
   Apt, liable to be employed in danger,
   I faintly broke with thee of Arthur's death;
   And thou, to be endearèd to a king,
   Made it no conscience to destroy a prince.     230
HUBERT  My lord—
KING JOHN
   Hadst thou but shook thy head or made a pause
   When I spake darkly what I purposèd,
   Or turned an eye of doubt upon my face,

As bid me tell my tale in express words,                  235
Deep shame had struck me dumb, made me break off,
And those thy fears might have wrought fears in me.
But thou didst understand me by my signs,
And didst in signs again parley with sin;
Yea, without stop, didst let thy heart consent,          240
And consequently thy rude hand to act
The deed which both our tongues held vile to name.
Out of my sight, and never see me more!
My nobles leave me, and my state is braved,
Even at my gates, with ranks of foreign powers;         245
Nay, in the body of this fleshly land,
This kingdom, this confine of blood and breath,
Hostility and civil tumult reigns
Between my conscience and my cousin's death.

HUBERT
Arm you against your other enemies;                      250
I'll make a peace between your soul and you.
Young Arthur is alive. This hand of mine
Is yet a maiden and an innocent hand,
Not painted with the crimson spots of blood.
Within this bosom never entered yet                      255
The dreadful motion of a murderous thought;
And you have slandered nature in my form,
Which, howsoever rude exteriorly,
Is yet the cover of a fairer mind
Than to be butcher of an innocent child.                 260

KING JOHN
Doth Arthur live? O, haste thee to the peers;
Throw this report on their incensèd rage,
And make them tame to their obedience.
Forgive the comment that my passion made
Upon thy feature, for my rage was blind,                 265
And foul imaginary eyes of blood
Presented thee more hideous than thou art.
O, answer not, but to my closet bring
The angry lords with all expedient haste.
I conjure thee but slowly; run more fast.                270
                            *Exeunt ⌐severally⌐*

**4.3** *Enter Arthur Duke of Brittaine on the walls,*
      *disguised as a ship-boy*

ARTHUR
The wall is high, and yet will I leap down.
Good ground, be pitiful, and hurt me not.
There's few or none do know me; if they did,
This ship-boy's semblance hath disguised me quite.
I am afraid, and yet I'll venture it.                      5
If I get down and do not break my limbs,
I'll find a thousand shifts to get away.
As good to die and go, as die and stay.
                    *He leaps down*
O me! My uncle's spirit is in these stones.
Heaven take my soul, and England keep my bones!  10
                                    *He dies*
      *Enter the Earls of Pembroke and Salisbury, and Lord*
      *Bigot*

SALISBURY
Lords, I will meet him at Saint Edmundsbury.
It is our safety, and we must embrace
This gentle offer of the perilous time.

PEMBROKE
Who brought that letter from the Cardinal?

SALISBURY
The Count Melun, a noble lord of France,                 15
Who's private with me of the Dauphin's love;
'Tis much more general than these lines import.

BIGOT
Tomorrow morning let us meet him then.

SALISBURY
Or rather, then set forward, for 'twill be
Two long days' journey, lords, or ere we meet.           20
                    *Enter the Bastard*

BASTARD
Once more today well met, distempered lords.
The King by me requests your presence straight.

SALISBURY
The King hath dispossessed himself of us.
We will not line his thin bestainèd cloak
With our pure honours, nor attend the foot               25
That leaves the print of blood where'er it walks.
Return and tell him so; we know the worst.

BASTARD
Whate'er you think, good words I think were best.

SALISBURY
Our griefs and not our manners reason now.

BASTARD
But there is little reason in your grief.                 30
Therefore 'twere reason you had manners now.

PEMBROKE
Sir, sir, impatience hath his privilege.

BASTARD
'Tis true—to hurt his master, no man else.

SALISBURY
This is the prison.
            *He sees Arthur's body*
                        What is he lies here?

PEMBROKE
O death, made proud with pure and princely beauty!
The earth had not a hole to hide this deed.              36

SALISBURY
Murder, as hating what himself hath done,
Doth lay it open to urge on revenge.

BIGOT
Or when he doomed this beauty to a grave,
Found it too precious-princely for a grave.              40

SALISBURY (*to the Bastard*)
Sir Richard, what think you? You have beheld.
Or have you read or heard; or could you think,
Or do you almost think, although you see,
That you do see? Could thought, without this object,
Form such another? This is the very top,                 45
The height, the crest, or crest unto the crest,
Of murder's arms; this is the bloodiest shame,
The wildest savagery, the vilest stroke
That ever wall-eyed wrath or staring rage
Presented to the tears of soft remorse.                  50

PEMBROKE
All murders past do stand excused in this,
And this, so sole and so unmatchable,
Shall give a holiness, a purity,
To the yet-unbegotten sin of times,
And prove a deadly bloodshed but a jest,                 55
Exampled by this heinous spectacle.

BASTARD
It is a damnèd and a bloody work,
The graceless action of a heavy hand—
If that it be the work of any hand.

SALISBURY
If that it be the work of any hand?                      60
We had a kind of light what would ensue:
It is the shameful work of Hubert's hand,
The practice and the purpose of the King;

From whose obedience I forbid my soul,
Kneeling before this ruin of sweet life,                         65
And breathing to his breathless excellence
The incense of a vow, a holy vow,
Never to taste the pleasures of the world,
Never to be infected with delight,
Nor conversant with ease and idleness,                           70
Till I have set a glory to this hand
By giving it the worship of revenge.

PEMBROKE *and* BIGOT
Our souls religiously confirm thy words.
          *Enter Hubert*

HUBERT
Lords, I am hot with haste in seeking you.
Arthur doth live; the King hath sent for you.                    75

SALISBURY
O, he is bold, and blushes not at death!—
Avaunt, thou hateful villain, get thee gone!

HUBERT
I am no villain.

SALISBURY                Must I rob the law?
     *He draws his sword*

BASTARD
Your sword is bright, sir; put it up again.

SALISBURY
Not till I sheathe it in a murderer's skin.                      80

HUBERT (*drawing his sword*)
Stand back, Lord Salisbury, stand back, I say!
By heaven, I think my sword's as sharp as yours.
I would not have you, lord, forget yourself,
Nor tempt the danger of my true defence,
Lest I, by marking of your rage, forget                          85
Your worth, your greatness and nobility.

BIGOT
Out, dunghill! Dar'st thou brave a nobleman?

HUBERT
Not for my life; but yet I dare defend
My innocent life against an emperor.

SALISBURY
Thou art a murderer.

HUBERT                Do not prove me so;                        90
Yet I am none. Whose tongue soe'er speaks false,
Not truly speaks; who speaks not truly, lies.

PEMBROKE
Cut him to pieces!

BASTARD (*drawing his sword*) Keep the peace, I say!

SALISBURY
Stand by, or I shall gall you, Falconbridge.

BASTARD
Thou wert better gall the devil, Salisbury.                      95
If thou but frown on me, or stir thy foot,
Or teach thy hasty spleen to do me shame,
I'll strike thee dead. Put up thy sword betime,
Or I'll so maul you and your toasting-iron
That you shall think the devil is come from hell.               100

BIGOT
What wilt thou do, renownèd Falconbridge,
Second a villain and a murderer?

HUBERT
Lord Bigot, I am none.

BIGOT                Who killed this prince?

HUBERT
'Tis not an hour since I left him well.
I honoured him, I loved him, and will weep                      105
My date of life out for his sweet life's loss.

SALISBURY
Trust not those cunning waters of his eyes,
For villainy is not without such rheum,
And he, long traded in it, makes it seem
Like rivers of remorse and innocency.                           110
Away with me, all you whose souls abhor
Th'uncleanly savours of a slaughter-house,
For I am stifled with this smell of sin.

BIGOT
Away toward Bury, to the Dauphin there.

PEMBROKE
There, tell the King, he may enquire us out.                    115
          *Exeunt Pembroke, Salisbury, and Bigot*

BASTARD
Here's a good world! Knew you of this fair work?
Beyond the infinite and boundless reach
Of mercy, if thou didst this deed of death
Art thou damned, Hubert.

HUBERT Do but hear me, sir.                                     120

BASTARD Ha! I'll tell thee what:
Thou'rt damned as black—nay nothing is so black—
Thou art more deep damned than Prince Lucifer;
There is not yet so ugly a fiend of hell
As thou shalt be if thou didst kill this child.                 125

HUBERT
Upon my soul—

BASTARD                If thou didst but consent
To this most cruel act, do but despair;
And if thou want'st a cord, the smallest thread
That ever spider twisted from her womb
Will serve to strangle thee; a rush will be a beam              130
To hang thee on; or wouldst thou drown thyself,
Put but a little water in a spoon
And it shall be, as all the ocean,
Enough to stifle such a villain up.
I do suspect thee very grievously.                              135

HUBERT
If I in act, consent, or sin of thought
Be guilty of the stealing that sweet breath
Which was embounded in this beauteous clay,
Let hell want pains enough to torture me.
I left him well.

BASTARD                Go bear him in thine arms.               140
I am amazed, methinks, and lose my way
Among the thorns and dangers of this world.
          *Hubert takes up Arthur in his arms*
How easy dost thou take all England up!
From forth this morsel of dead royalty,
The life, the right, and truth of all this realm                145
Is fled to heaven, and England now is left
To tug and scramble, and to part by th' teeth
The unowed interest of proud swelling state.
Now for the bare-picked bone of majesty
Doth doggèd war bristle his angry crest,                        150
And snarleth in the gentle eyes of peace;
Now powers from home and discontents at home
Meet in one line, and vast confusion waits,
As doth a raven on a sick-fall'n beast,
The imminent decay of wrested pomp.                             155
Now happy he whose cloak and cincture can
Hold out this tempest. Bear away that child,
And follow me with speed. I'll to the King.
A thousand businesses are brief in hand,
And heaven itself doth frown upon the land.                     160
          *Exeunt ⌜severally⌝*

**5.1** ⌈*Flourish.*⌉ *Enter King John and Cardinal Pandolf,*
     *with attendants*

KING JOHN ⌈*giving Pandolf the crown*⌉
    Thus have I yielded up into your hand
    The circle of my glory.

PANDOLF (*giving back the crown*) Take again
    From this my hand, as holding of the Pope,
    Your sovereign greatness and authority.

KING JOHN
    Now keep your holy word: go meet the French,       5
    And from his Holiness use all your power
    To stop their marches 'fore we are enflamed.
    Our discontented counties do revolt,
    Our people quarrel with obedience,
    Swearing allegiance and the love of soul           10
    To stranger blood, to foreign royalty.
    This inundation of mistempered humour
    Rests by you only to be qualified.
    Then pause not, for the present time's so sick
    That present med'cine must be ministered,          15
    Or overthrow incurable ensues.

PANDOLF
    It was my breath that blew this tempest up,
    Upon your stubborn usage of the Pope,
    But since you are a gentle convertite,
    My tongue shall hush again this storm of war        20
    And make fair weather in your blust'ring land.
    On this Ascension Day, remember well,
    Upon your oath of service to the Pope,
    Go I to make the French lay down their arms.
                        ⌈*Exeunt all but King John*⌉

KING JOHN
    Is this Ascension Day? Did not the prophet          25
    Say that before Ascension Day at noon
    My crown I should give off? Even so I have.
    I did suppose it should be on constraint,
    But, heaven be thanked, it is but voluntary.
                        *Enter Bastard*

BASTARD
    All Kent hath yielded; nothing there holds out       30
    But Dover Castle. London hath received,
    Like a kind host, the Dauphin and his powers.
    Your nobles will not hear you, but are gone
    To offer service to your enemy;
    And wild amazement hurries up and down              35
    The little number of your doubtful friends.

KING JOHN
    Would not my lords return to me again
    After they heard young Arthur was alive?

BASTARD
    They found him dead and cast into the streets,
    An empty casket, where the jewel of life            40
    By some damned hand was robbed and ta'en away.

KING JOHN
    That villain Hubert told me he did live.

BASTARD
    So on my soul he did, for aught he knew.
    But wherefore do you droop? Why look you sad?
    Be great in act as you have been in thought.        45
    Let not the world see fear and sad distrust
    Govern the motion of a kingly eye.
    Be stirring as the time, be fire with fire;
    Threaten the threat'ner, and outface the brow
    Of bragging horror. So shall inferior eyes,          50
    That borrow their behaviours from the great,

    Grow great by your example, and put on
    The dauntless spirit of resolution.
    Away, and glisten like the god of war
    When he intendeth to become the field.              55
    Show boldness and aspiring confidence.
    What, shall they seek the lion in his den
    And fright him there, and make him tremble there?
    O, let it not be said! Forage, and run
    To meet displeasure farther from the doors,          60
    And grapple with him ere he come so nigh.

KING JOHN
    The legate of the Pope hath been with me,
    And I have made a happy peace with him,
    And he hath promised to dismiss the powers
    Led by the Dauphin.

BASTARD                     O inglorious league!         65
    Shall we, upon the footing of our land,
    Send fair-play orders, and make compromise,
    Insinuation, parley, and base truce
    To arms invasive? Shall a beardless boy,
    A cockered silken wanton, brave our fields           70
    And flesh his spirit in a warlike soil,
    Mocking the air with colours idly spread,
    And find no check? Let us, my liege, to arms!
    Perchance the Cardinal cannot make your peace,
    Or if he do, let it at least be said                 75
    They saw we had a purpose of defence.

KING JOHN
    Have thou the ordering of this present time.

BASTARD
    Away, then, with good courage! ⌈*Aside*⌉ Yet I know
    Our party may well meet a prouder foe.    *Exeunt*

**5.2**  *Enter,* ⌈*marching*⌉ *in arms, Louis the Dauphin, the*
      *Earl of Salisbury, Count Melun, the Earl of*
      *Pembroke, and Lord Bigot, with soldiers*

LOUIS THE DAUPHIN
    My Lord Melun, let this be copied out,
    And keep it safe for our remembrance.
    Return the precedent to these lords again,
    That having our fair order written down,
    Both they and we, perusing o'er these notes,          5
    May know wherefore we took the sacrament
    And keep our faiths firm and inviolable.

SALISBURY
    Upon our sides it never shall be broken.
    And, noble Dauphin, albeit we swear
    A voluntary zeal and an unurgèd faith                10
    To your proceedings, yet believe me, Prince,
    I am not glad that such a sore of time
    Should seek a plaster by contemnèd revolt,
    And heal the inveterate canker of one wound
    By making many. O, it grieves my soul                15
    That I must draw this metal from my side
    To be a widow-maker! O, and there
    Where honourable rescue and defence
    Cries out upon the name of Salisbury!
    But such is the infection of the time,               20
    That for the health and physic of our right,
    We cannot deal but with the very hand
    Of stern injustice and confusèd wrong.
    And is't not pity, O my grievèd friends,
    That we the sons and children of this isle           25
    Was born to see so sad an hour as this,
    Wherein we step after a stranger, march

Upon her gentle bosom, and fill up
Her enemies' ranks? I must withdraw and weep
Upon the spot of this enforcèd cause—                    30
To grace the gentry of a land remote,
And follow unacquainted colours here.
What, here? O nation, that thou couldst remove;
That Neptune's arms who clippeth thee about
Would bear thee from the knowledge of thyself          35
And gripple thee unto a pagan shore,
Where these two Christian armies might combine
The blood of malice in a vein of league,
And not to spend it so unneighbourly.

LOUIS THE DAUPHIN
A noble temper dost thou show in this,                    40
And great affections, wrestling in thy bosom,
Doth make an earthquake of nobility.
O, what a noble combat hast thou fought
Between compulsion and a brave respect!
Let me wipe off this honourable dew                       45
That silverly doth progress on thy cheeks.
My heart hath melted at a lady's tears,
Being an ordinary inundation;
But this effusion of such manly drops,
This shower blown up by tempest of the soul,            50
Startles mine eyes, and makes me more amazed
Than had I seen the vaulty top of heaven
Figured quite o'er with burning meteors.
Lift up thy brow, renownèd Salisbury,
And with a great heart heave away this storm;           55
Commend these waters to those baby eyes
That never saw the giant world enraged,
Nor met with Fortune other than at feasts,
Full warm of blood, of mirth, of gossiping.
Come, come, for thou shalt thrust thy hand as deep
Into the purse of rich prosperity                         61
As Louis himself. So, nobles, shall you all
That knit your sinews to the strength of mine.
⌈A trumpet sounds⌉
And even there methinks an angel spake!
Enter Cardinal Pandolf
Look where the holy legate comes apace,                  65
To give us warrant from the hand of heaven,
And on our actions set the name of right
With holy breath.

PANDOLF                    Hail, noble prince of France!
The next is this. King John hath reconciled
Himself to Rome; his spirit is come in                   70
That so stood out against the Holy Church,
The great metropolis and See of Rome;
Therefore thy threat'ning colours now wind up,
And tame the savage spirit of wild war,
That like a lion fostered up at hand                     75
It may lie gently at the foot of peace,
And be no further harmful than in show.

LOUIS THE DAUPHIN
Your grace shall pardon me: I will not back.
I am too high-born to be propertied,
To be a secondary at control,                             80
Or useful serving-man and instrument
To any sovereign state throughout the world.
Your breath first kindled the dead coal of wars
Between this chastisèd kingdom and myself,
And brought in matter that should feed this fire;        85
And now 'tis far too huge to be blown out
With that same weak wind which enkindled it.

You taught me how to know the face of right,
Acquainted me with interest to this land,
Yea, thrust this enterprise into my heart;               90
And come ye now to tell me John hath made
His peace with Rome? What is that peace to me?
I, by the honour of my marriage bed,
After young Arthur, claim this land for mine;
And now it is half conquered, must I back                95
Because that John hath made his peace with Rome?
Am I Rome's slave? What penny hath Rome borne,
What men provided, what munition sent
To underprop this action? Is't not I
That undergo this charge? Who else but I,               100
And such as to my claim are liable,
Sweat in this business and maintain this war?
Have I not heard these islanders shout out
'Vive le Roi!' as I have banked their towns?
Have I not here the best cards for the game,            105
To win this easy match played for a crown?
And shall I now give o'er the yielded set?
No, no, on my soul, it never shall be said.

PANDOLF
You look but on the outside of this work.

LOUIS THE DAUPHIN
Outside or inside, I will not return                     110
Till my attempt so much be glorified
As to my ample hope was promisèd
Before I drew this gallant head of war,
And culled these fiery spirits from the world
To outlook conquest and to win renown                   115
Even in the jaws of danger and of death.
A trumpet sounds
What lusty trumpet thus doth summon us?
Enter the Bastard

BASTARD
According to the fair play of the world,
Let me have audience; I am sent to speak.
My holy lord of Milan, from the King                    120
I come to learn how you have dealt for him,
And as you answer I do know the scope
And warrant limited unto my tongue.

PANDOLF
The Dauphin is too wilful-opposite,
And will not temporize with my entreaties.              125
He flatly says he'll not lay down his arms.

BASTARD
By all the blood that ever fury breathed,
The youth says well. Now hear our English king,
For thus his royalty doth speak in me.
He is prepared, and reason too he should.               130
This apish and unmannerly approach,
This harnessed masque and unadvisèd revel,
This unhaired sauciness and boyish troops,
The King doth smile at, and is well prepared
To whip this dwarfish war, these pigmy arms,            135
From out the circle of his territories.
That hand which had the strength even at your door
To cudgel you and make you take the hatch,
To dive like buckets in concealèd wells,
To crouch in litter of your stable planks,              140
To lie like pawns locked up in chests and trunks,
To hug with swine, to seek sweet safety out
In vaults and prisons, and to thrill and shake
Even at the crying of your nation's crow,
Thinking his voice an armèd Englishman;                 145

Shall that victorious hand be feebled here
That in your chambers gave you chastisement?
No! Know the gallant monarch is in arms,
And like an eagle o'er his eyrie towers
To souse annoyance that comes near his nest.          150
*(To the English lords)*
And you degenerate, you ingrate revolts,
You bloody Neros, ripping up the womb
Of your dear mother England, blush for shame;
For your own ladies and pale-visaged maids
Like Amazons come tripping after drums;               155
Their thimbles into armèd gauntlets change,
Their needles to lances, and their gentle hearts
To fierce and bloody inclination.

LOUIS THE DAUPHIN
There end thy brave, and turn thy face in peace.
We grant thou canst outscold us. Fare thee well:      160
We hold our time too precious to be spent
With such a brabbler.

PANDOLF                        Give me leave to speak.

BASTARD
No, I will speak.

LOUIS THE DAUPHIN  We will attend to neither.—
Strike up the drums, and let the tongue of war
Plead for our interest and our being here.            165

BASTARD
Indeed your drums, being beaten, will cry out;
And so shall you, being beaten. Do but start
An echo with the clamour of thy drum,
And even at hand a drum is ready braced
That shall reverberate all as loud as thine.          170
Sound but another, and another shall
As loud as thine rattle the welkin's ear,
And mock the deep-mouthed thunder; for at hand,
Not trusting to this halting legate here,
Whom he hath used rather for sport than need,         175
Is warlike John; and in his forehead sits
A bare-ribbed Death, whose office is this day
To feast upon whole thousands of the French.

LOUIS THE DAUPHIN
Strike up our drums to find this danger out.

BASTARD
And thou shalt find it, Dauphin, do not doubt.        180
⌜*Drums beat.*⌝ *Exeunt the Bastard* ⌜*at one door*⌝,
*all the rest,* ⌜*marching, at another door*⌝

5.3    *Alarum. Enter King John* ⌜*at one door*⌝ *and Hubert*
⌜*at another door*⌝

KING JOHN
How goes the day with us? O, tell me, Hubert.

HUBERT
Badly, I fear. How fares your majesty?

KING JOHN
This fever that hath troubled me so long
Lies heavy on me. O, my heart is sick!
*Enter a Messenger*

MESSENGER
My lord, your valiant kinsman Falconbridge
Desires your majesty to leave the field,               5
And send him word by me which way you go.

KING JOHN
Tell him toward Swineshead, to the abbey there.

MESSENGER
Be of good comfort, for the great supply
That was expected by the Dauphin here
Are wrecked three nights ago on Goodwin Sands.        10

This news was brought to Richard, but even now
The French fight coldly and retire themselves.

KING JOHN
Ay me, this tyrant fever burns me up,
And will not let me welcome this good news.           15
Set on toward Swineshead. To my litter straight;
Weakness possesseth me, and I am faint.      *Exeunt*

5.4    ⌜*Alarum.*⌝ *Enter the Earls of Salisbury and*
*Pembroke, and Lord Bigot*

SALISBURY
I did not think the King so stored with friends.

PEMBROKE
Up once again; put spirit in the French.
If they miscarry, we miscarry too.

SALISBURY
That misbegotten devil Falconbridge,
In spite of spite, alone upholds the day.              5

PEMBROKE
They say King John, sore sick, hath left the field.
*Enter Count Melun, wounded,* ⌜*led by a soldier*⌝

MELUN
Lead me to the revolts of England here.

SALISBURY
When we were happy, we had other names.

PEMBROKE
It is the Count Melun.

SALISBURY                      Wounded to death.

MELUN
Fly, noble English, you are bought and sold.          10
Unthread the rude eye of rebellion,
And welcome home again discarded faith;
Seek out King John and fall before his feet,
For if the French be lords of this loud day
He means to recompense the pains you take             15
By cutting off your heads. Thus hath he sworn,
And I with him, and many more with me,
Upon the altar at Saint Edmundsbury,
Even on that altar where we swore to you
Dear amity and everlasting love.                      20

SALISBURY
May this be possible? May this be true?

MELUN
Have I not hideous death within my view,
Retaining but a quantity of life,
Which bleeds away, even as a form of wax
Resolveth from his figure 'gainst the fire?           25
What in the world should make me now deceive,
Since I must lose the use of all deceit?
Why should I then be false, since it is true
That I must die here, and live hence by truth?
I say again, if Louis do win the day,                 30
He is forsworn if e'er those eyes of yours
Behold another daybreak in the east;
But even this night, whose black contagious breath
Already smokes about the burning cresset
Of the old, feeble, and day-wearied sun,              35
Even this ill night your breathing shall expire,
Paying the fine of rated treachery
Even with a treacherous fine of all your lives,
If Louis by your assistance win the day.
Commend me to one Hubert with your king.              40
The love of him, and this respect besides,
For that my grandsire was an Englishman,
Awakes my conscience to confess all this;

In lieu whereof, I pray you bear me hence
From forth the noise and rumour of the field,          45
Where I may think the remnant of my thoughts
In peace, and part this body and my soul
With contemplation and devout desires.
SALISBURY
We do believe thee; and beshrew my soul
But I do love the favour and the form                  50
Of this most fair occasion; by the which
We will untread the steps of damnèd flight,
And like a bated and retirèd flood,
Leaving our rankness and irregular course,
Stoop low within those bounds we have o'erlooked, 55
And calmly run on in obedience
Even to our ocean, to our great King John.
My arm shall give thee help to bear thee hence,
For I do see the cruel pangs of death
Right in thine eye.—Away, my friends! New flight, 60
And happy newness that intends old right.    *Exeunt*

5.5   ⌜*Alarum; retreat.*⌝ *Enter Louis the Dauphin, and*
      *his train*
LOUIS THE DAUPHIN
The sun of heaven, methought, was loath to set,
But stayed and made the western welkin blush,
When English measured backward their own ground
In faint retire. O, bravely came we off,
When with a volley of our needless shot,                5
After such bloody toil, we bid good night,
And wound our tatt'ring colours clearly up,
Last in the field and almost lords of it.
        *Enter a Messenger*
MESSENGER
Where is my prince the Dauphin?
LOUIS THE DAUPHIN              Here. What news?
MESSENGER
The Count Melun is slain; the English lords            10
By his persuasion are again fall'n off;
And your supply which you have wished so long
Are cast away and sunk on Goodwin Sands.
LOUIS THE DAUPHIN
Ah, foul shrewd news! Beshrew thy very heart!
I did not think to be so sad tonight                   15
As this hath made me. Who was he that said
King John did fly an hour or two before
The stumbling night did part our weary powers?
MESSENGER
Whoever spoke it, it is true, my lord.
LOUIS THE DAUPHIN
Well, keep good quarter and good care tonight.          20
The day shall not be up so soon as I,
To try the fair adventure of tomorrow.    *Exeunt*

5.6   *Enter the Bastard* ⌜*with a light*⌝ *and Hubert* ⌜*with a*
      *pistol*⌝, *severally*
HUBERT
Who's there? Speak, ho! Speak quickly, or I shoot.
BASTARD
A friend. What art thou?
HUBERT                        Of the part of England.
BASTARD
Whither dost thou go?
HUBERT                   What's that to thee?
Why may not I demand of thine affairs
As well as thou of mine?                                5

BASTARD Hubert, I think.
HUBERT Thou hast a perfect thought.
I will upon all hazards well believe
Thou art my friend that know'st my tongue so well.
Who art thou?
BASTARD            Who thou wilt. An if thou please,     10
Thou mayst befriend me so much as to think
I come one way of the Plantagenets.
HUBERT
Unkind remembrance! Thou and eyeless night
Have done me shame. Brave soldier, pardon me
That any accent breaking from thy tongue               15
Should 'scape the true acquaintance of mine ear.
BASTARD
Come, come, sans compliment. What news abroad?
HUBERT
Why, here walk I in the black brow of night
To find you out.
BASTARD            Brief, then, and what's the news?
HUBERT
O my sweet sir, news fitting to the night:             20
Black, fearful, comfortless, and horrible.
BASTARD
Show me the very wound of this ill news;
I am no woman, I'll not swoon at it.
HUBERT
The King, I fear, is poisoned by a monk.
I left him almost speechless, and broke out            25
To acquaint you with this evil, that you might
The better arm you to the sudden time
Than if you had at leisure known of this.
BASTARD
How did he take it? Who did taste to him?
HUBERT
A monk, I tell you, a resolvèd villain,                30
Whose bowels suddenly burst out. The King
Yet speaks, and peradventure may recover.
BASTARD
Who didst thou leave to tend his majesty?
HUBERT
Why, know you not? The lords are all come back,
And brought Prince Henry in their company,             35
At whose request the King hath pardoned them,
And they are all about his majesty.
BASTARD
Withhold thine indignation, mighty heaven,
And tempt us not to bear above our power.
I'll tell thee, Hubert, half my power this night,      40
Passing these flats, are taken by the tide.
These Lincoln Washes have devourèd them;
Myself, well mounted, hardly have escaped.
Away before! Conduct me to the King.
I doubt he will be dead or ere I come.    *Exeunt*

5.7   *Enter Prince Henry, the Earl of Salisbury, and Lord*
      *Bigot*
PRINCE HENRY
It is too late. The life of all his blood
Is touched corruptibly, and his pure brain,
Which some suppose the soul's frail dwelling-house,
Doth by the idle comments that it makes
Foretell the ending of mortality.                      5
        *Enter the Earl of Pembroke*
PEMBROKE
His highness yet doth speak, and holds belief
That being brought into the open air,

It would allay the burning quality
Of that fell poison which assaileth him.

PRINCE HENRY
Let him be brought into the orchard here.—                    10
                              *Exit Lord Bigot*
Doth he still rage?

PEMBROKE                    He is more patient
Than when you left him. Even now, he sung.

PRINCE HENRY
O, vanity of sickness! Fierce extremes
In their continuance will not feel themselves.
Death, having preyed upon the outward parts,                    15
Leaves them invincible, and his siege is now .
Against the mind; the which he pricks and wounds
With many legions of strange fantasies,
Which in their throng and press to that last hold
Confound themselves. 'Tis strange that death should
       sing.                    20
I am the cygnet to this pale faint swan,
Who chants a doleful hymn to his own death,
And from the organ-pipe of frailty sings
His soul and body to their lasting rest.

SALISBURY
Be of good comfort, Prince, for you are born                    25
To set a form upon that indigest
Which he hath left so shapeless and so rude.
       *King John is brought in, ⌈with Lord Bigot attending⌉*

KING JOHN
Ay marry, now my soul hath elbow-room;
It would not out at windows nor at doors.
There is so hot a summer in my bosom                    30
That all my bowels crumble up to dust;
I am a scribbled form, drawn with a pen
Upon a parchment, and against this fire
Do I shrink up.

PRINCE HENRY       How fares your majesty?

KING JOHN
Poisoned, ill fare! Dead, forsook, cast off;                    35
And none of you will bid the winter come
To thrust his icy fingers in my maw,
Nor let my kingdom's rivers take their course
Through my burned bosom, nor entreat the north
To make his bleak winds kiss my parchèd lips                    40
And comfort me with cold. I do not ask you much;
I beg cold comfort, and you are so strait
And so ingrateful you deny me that.

PRINCE HENRY
O, that there were some virtue in my tears
That might relieve you!

KING JOHN                    The salt in them is hot.                    45
Within me is a hell, and there the poison
Is, as a fiend, confined to tyrannize
On unreprievable condemnèd blood.
       *Enter the Bastard*

BASTARD
O, I am scalded with my violent motion
And spleen of speed to see your majesty!                    50

KING JOHN
O cousin, thou art come to set mine eye.
The tackle of my heart is cracked and burnt,
And all the shrouds wherewith my life should sail
Are turnèd to one thread, one little hair;
My heart hath one poor string to stay it by,                    55
Which holds but till thy news be utterèd,
And then all this thou seest is but a clod
And module of confounded royalty.

BASTARD
The Dauphin is preparing hitherward,
Where God He knows how we shall answer him;                    60
For in a night the best part of my power,
As I upon advantage did remove,
Were in the Washes all unwarily
Devourèd by the unexpected flood.
       *King John dies*

SALISBURY
You breathe these dead news in as dead an ear.                    65
(*To King John*) My liege, my lord!—But now a king,
       now thus.

PRINCE HENRY
Even so must I run on, and even so stop.
What surety of the world, what hope, what stay,
When this was now a king and now is clay?

BASTARD (*to King John*)
Art thou gone so? I do but stay behind                    70
To do the office for thee of revenge,
And then my soul shall wait on thee to heaven,
As it on earth hath been thy servant still.
(*To the lords*) Now, now, you stars that move in your
       right spheres,
Where be your powers? Show now your mended
       faiths,                    75
And instantly return with me again,
To push destruction and perpetual shame
Out of the weak door of our fainting land.
Straight let us seek, or straight we shall be sought.
The Dauphin rages at our very heels.                    80

SALISBURY
It seems you know not, then, so much as we.
The Cardinal Pandolf is within at rest,
Who half an hour since came from the Dauphin,
And brings from him such offers of our peace
As we with honour and respect may take,                    85
With purpose presently to leave this war.

BASTARD
He will the rather do it when he sees
Ourselves well-sinewed to our own defence.

SALISBURY
Nay, 'tis in a manner done already,
For many carriages he hath dispatched                    90
To the sea-side, and put his cause and quarrel
To the disposing of the Cardinal,
With whom yourself, myself, and other lords,
If you think meet, this afternoon will post
To consummate this business happily.                    95

BASTARD
Let it be so.—And you, my noble prince,
With other princes that may best be spared,
Shall wait upon your father's funeral.

PRINCE HENRY
At Worcester must his body be interred,
For so he willed it.

BASTARD                    Thither shall it then,                    100
And happily may your sweet self put on
The lineal state and glory of the land,
To whom with all submission, on my knee,
I do bequeath my faithful services
And true subjection everlastingly.                    105
       *He kneels*

SALISBURY
And the like tender of our love we make,
To rest without a spot for evermore.
       *Salisbury, Pembroke and Bigot kneel*

PRINCE HENRY
  I have a kind of soul that would give thanks,
  And knows not how to do it but with tears.
     *He weeps*
BASTARD ⌜*rising*⌝
  O, let us pay the time but needful woe,      110
  Since it hath been beforehand with our griefs.
  This England never did, nor never shall,
Lie at the proud foot of a conqueror
But when it first did help to wound itself.
Now these her princes are come home again,    115
Come the three corners of the world in arms
And we shall shock them. Naught shall make us rue
If England to itself do rest but true.
     ⌜*Flourish.*⌝ *Exeunt* ⌜*with the body*⌝

# THE MERCHANT OF VENICE

ENTRY of 'a book of *The Merchant of Venice* or otherwise called *The Jew of Venice*' in the Stationers' Register on 22 July 1598 probably represents an attempt by Shakespeare's company to prevent the unauthorized printing of a popular play: it eventually appeared in print as '*The Comical History of the Merchant of Venice*' in 1600, when it was said to have 'been divers times acted by the Lord Chamberlain his servants'; probably Shakespeare wrote it in 1596 or 1597. The alternative title—*The Jew of Venice*—probably reflects Shylock's impact on the play's first audiences.

The play is constructed on the basis of two romantic tales using motifs well known to sixteenth-century readers. The story of Giannetto (Shakespeare's Bassanio) and the Lady (Portia) of Belmont comes from an Italian collection of fifty stories published under the title of *Il Pecorone* ('the big sheep', or 'dunce') and attributed to one Ser Giovanni of Florence. Written in the later part of the fourteenth century, the volume did not appear until 1558. No sixteenth-century translation is known, so (unless there was a lost intermediary) Shakespeare must have read it in Italian. It gave him the main outline of the plot involving Antonio (the merchant), Bassanio (the wooer), Portia, and the Jew (Shylock). The pound of flesh motif was available also in other versions, one of which, in Alexander Silvayn's *The Orator* (translated 1596), influenced the climactic scene (4.1) in which Shylock attempts to exact the full penalty of his bond.

In the story from *Il Pecorone* the lady (a widow) challenges her suitors to seduce her, on pain of the forfeiture of their wealth, and thwarts them by drugging their wine. Shakespeare more romantically shows a maiden required by her father's will to accept only a wooer who will forswear marriage if he fails to make the right choice among caskets of gold, silver, and lead. The story of the caskets was readily available in versions by John Gower (in his *Confessio Amantis*) and Giovanni Boccaccio (in his *Decameron*), and in an anonymous anthology (the *Gesta Romanorum*). Shakespeare added the character of Jessica, Shylock's daughter who elopes with the Christian Lorenzo—perhaps influenced by episodes in Christopher Marlowe's play *The Jew of Malta* (c.1589)—and made many adjustments to the stories from which he borrowed.

*The Merchant of Venice* is a natural development from Shakespeare's earlier comedies, especially *The Two Gentlemen of Verona*, with its heroine disguised as a boy and its portrayal of the competing demands of love and friendship. But Portia is the first of his great romantic heroines, and Shylock his first great comic antagonist. Though the play grew out of fairy tales, its moral scheme is not entirely clear cut: the Christians are open to criticism, the Jew is true to his own code of conduct. The response of twentieth-century audiences has been complicated by racial issues; in any case, the role of Shylock affords such strong opportunities for an actor capable of arousing an undercurrent of sympathy for a vindictive character that it has sometimes unbalanced the play in performance. But the so-called trial scene (4.1) is unfailing in its impact on audiences, and the closing episodes modulate skilfully from romantic lyricism to high comedy, while sustaining the play's concern with true and false values.

# THE PERSONS OF THE PLAY

ANTONIO, a merchant of Venice

BASSANIO, his friend and Portia's suitor

LEONARDO, Bassanio's servant

LORENZO
GRAZIANO   friends of Antonio and Bassanio
SALERIO
SOLANIO

SHYLOCK, a Jew

JESSICA, his daughter

TUBAL, a Jew

LANCELOT, a clown, first Shylock's servant and then Bassanio's

GOBBO, his father

PORTIA, an heiress

NERISSA, her waiting-gentlewoman

BALTHASAR
            Portia's servants
STEFANO

Prince of MOROCCO
                  Portia's suitors
Prince of ARAGON

DUKE of Venice

Magnificoes of Venice

A jailer, attendants, and servants

# The Comical History of the Merchant of Venice, or Otherwise Called the Jew of Venice

**1.1** *Enter Antonio, Salerio, and Solanio*

ANTONIO
In sooth, I know not why I am so sad.
It wearies me, you say it wearies you,
But how I caught it, found it, or came by it,
What stuff 'tis made of, whereof it is born,
I am to learn;                                                    5
And such a want-wit sadness makes of me
That I have much ado to know myself.

SALERIO
Your mind is tossing on the ocean,
There where your argosies with portly sail,
Like signors and rich burghers on the flood—        10
Or as it were the pageants of the sea—
Do overpeer the petty traffickers
That curtsy to them, do them reverence,
As they fly by them with their woven wings.

SOLANIO (*to Antonio*)
Believe me, sir, had I such venture forth            15
The better part of my affections would
Be with my hopes abroad. I should be still
Plucking the grass to know where sits the wind,
Peering in maps for ports and piers and roads,
And every object that might make me fear            20
Misfortune to my ventures out of doubt
Would make me sad.

SALERIO                    My wind cooling my broth
Would blow me to an ague when I thought
What harm a wind too great might do at sea.
I should not see the sandy hour-glass run            25
But I should think of shallows and of flats,
And see my wealthy Andrew, decks in sand,
Vailing her hightop lower than her ribs
To kiss her burial. Should I go to church
And see the holy edifice of stone                    30
And not bethink me straight of dangerous rocks
Which, touching but my gentle vessel's side,
Would scatter all her spices on the stream,
Enrobe the roaring waters with my silks,
And, in a word, but even now worth this,            35
And now worth nothing? Shall I have the thought
To think on this, and shall I lack the thought
That such a thing bechanced would make me sad?
But tell not me. I know Antonio
Is sad to think upon his merchandise.                40

ANTONIO
Believe me, no. I thank my fortune for it,
My ventures are not in one bottom trusted,
Nor to one place; nor is my whole estate
Upon the fortune of this present year.
Therefore my merchandise makes me not sad.          45

SOLANIO
Why then, you are in love.

ANTONIO                            Fie, fie.

SOLANIO
Not in love neither? Then let us say you are sad
Because you are not merry, and 'twere as easy
For you to laugh, and leap, and say you are merry
Because you are not sad. Now, by two-headed Janus,
Nature hath framed strange fellows in her time:     51
Some that will evermore peep through their eyes
And laugh like parrots at a bagpiper,
And other of such vinegar aspect
That they'll not show their teeth in way of smile   55
Though Nestor swear the jest be laughable.
    *Enter Bassanio, Lorenzo, and Graziano*
Here comes Bassanio, your most noble kinsman,
Graziano, and Lorenzo. Fare ye well.
We leave you now with better company.

SALERIO
I would have stayed till I had made you merry       60
If worthier friends had not prevented me.

ANTONIO
Your worth is very dear in my regard.
I take it your own business calls on you,
And you embrace th'occasion to depart.

SALERIO Good morrow, my good lords.                 65

BASSANIO
Good signors both, when shall we laugh? Say, when?
You grow exceeding strange. Must it be so?

SALERIO
We'll make our leisures to attend on yours.
                        *Exeunt Salerio and Solanio*

LORENZO
My lord Bassanio, since you have found Antonio,
We two will leave you; but at dinner-time           70
I pray you have in mind where we must meet.

BASSANIO I will not fail you.

GRAZIANO
You look not well, Signor Antonio.
You have too much respect upon the world.
They lose it that do buy it with much care.         75
Believe me, you are marvellously changed.

ANTONIO
I hold the world but as the world, Graziano—
A stage where every man must play a part,
And mine a sad one.

GRAZIANO                    Let me play the fool.
With mirth and laughter let old wrinkles come,      80
And let my liver rather heat with wine
Than my heart cool with mortifying groans.
Why should a man whose blood is warm within
Sit like his grandsire cut in alabaster,
Sleep when he wakes, and creep into the jaundice    85
By being peevish? I tell thee what, Antonio—
I love thee, and 'tis my love that speaks—
There are a sort of men whose visages
Do cream and mantle like a standing pond,
And do a wilful stillness entertain                 90
With purpose to be dressed in an opinion
Of wisdom, gravity, profound conceit,
As who should say 'I am Sir Oracle,
And when I ope my lips, let no dog bark.'
O my Antonio, I do know of these                     95

That therefore only are reputed wise
For saying nothing, when I am very sure,
If they should speak, would almost damn those ears
Which, hearing them, would call their brothers fools.
I'll tell thee more of this another time.                    100
But fish not with this melancholy bait
For this fool gudgeon, this opinion.—
Come, good Lorenzo.—Fare ye well a while.
I'll end my exhortation after dinner.

LORENZO (to Antonio and Bassanio)
Well, we will leave you then till dinner-time.               105
I must be one of these same dumb wise men,
For Graziano never lets me speak.

GRAZIANO
Well, keep me company but two years more
Thou shalt not know the sound of thine own tongue.

ANTONIO
Fare you well. I'll grow a talker for this gear.             110

GRAZIANO
Thanks, i'faith, for silence is only commendable
In a neat's tongue dried and a maid not vendible.
                              *Exeunt Graziano and Lorenzo*

ANTONIO Yet is that anything now?
BASSANIO Graziano speaks an infinite deal of nothing,
more than any man in all Venice. His reasons are as
two grains of wheat hid in two bushels of chaff: you
shall seek all day ere you find them, and when you
have them they are not worth the search.

ANTONIO
Well, tell me now what lady is the same
To whom you swore a secret pilgrimage,                       120
That you today promised to tell me of.

BASSANIO
'Tis not unknown to you, Antonio,
How much I have disabled mine estate
By something showing a more swelling port
Than my faint means would grant continuance,                125
Nor do I now make moan to be abridged
From such a noble rate; but my chief care
Is to come fairly off from the great debts
Wherein my time, something too prodigal,
Hath left me gaged. To you, Antonio,                         130
I owe the most in money and in love,
And from your love I have a warranty
To unburden all my plots and purposes
How to get clear of all the debts I owe.

ANTONIO
I pray you, good Bassanio, let me know it,                   135
And if it stand as you yourself still do,
Within the eye of honour, be assured
My purse, my person, my extremest means
Lie all unlocked to your occasions.

BASSANIO
In my schooldays, when I had lost one shaft,                 140
I shot his fellow of the selfsame flight
The selfsame way, with more advisèd watch,
To find the other forth; and by adventuring both,
I oft found both. I urge this childhood proof
Because what follows is pure innocence.                      145
I owe you much, and, like a wilful youth,
That which I owe is lost; but if you please
To shoot another arrow that self way
Which you did shoot the first, I do not doubt,
As I will watch the aim, or to find both                     150
Or bring your latter hazard back again,
And thankfully rest debtor for the first.

ANTONIO
You know me well, and herein spend but time
To wind about my love with circumstance;
And out of doubt you do me now more wrong                    155
In making question of my uttermost
Than if you had made waste of all I have.
Then do but say to me what I should do
That in your knowledge may by me be done,
And I am pressed unto it. Therefore speak.                   160

BASSANIO
In Belmont is a lady richly left,
And she is fair, and, fairer than that word,
Of wondrous virtues. Sometimes from her eyes
I did receive fair speechless messages.
Her name is Portia, nothing undervalued                      165
To Cato's daughter, Brutus' Portia;
Nor is the wide world ignorant of her worth,
For the four winds blow in from every coast
Renownèd suitors, and her sunny locks
Hang on her temples like a golden fleece,                    170
Which makes her seat of Belmont Colchis' strand,
And many Jasons come in quest of her.
O my Antonio, had I but the means
To hold a rival place with one of them,
I have a mind presages me such thrift                        175
That I should questionless be fortunate.

ANTONIO
Thou know'st that all my fortunes are at sea,
Neither have I money nor commodity
To raise a present sum. Therefore go forth—
Try what my credit can in Venice do;                         180
That shall be racked even to the uttermost
To furnish thee to Belmont, to fair Portia.
Go presently enquire, and so will I,
Where money is; and I no question make
To have it of my trust or for my sake.                       185
                              *Exeunt ⌜severally⌝*

**1.2**  *Enter Portia with Nerissa, her waiting-woman*
PORTIA By my troth, Nerissa, my little body is aweary of
this great world.
NERISSA You would be, sweet madam, if your miseries
were in the same abundance as your good fortunes
are; and yet, for aught I see, they are as sick that
surfeit with too much as they that starve with nothing.
It is no mean happiness, therefore, to be seated in the
mean. Superfluity comes sooner by white hairs, but
competency lives longer.
PORTIA Good sentences, and well pronounced.                  10
NERISSA They would be better if well followed.
PORTIA If to do were as easy as to know what were good
to do, chapels had been churches, and poor men's
cottages princes' palaces. It is a good divine that follows
his own instructions. I can easier teach twenty what
were good to be done than to be one of the twenty to
follow mine own teaching. The brain may devise laws
for the blood, but a hot temper leaps o'er a cold decree.
Such a hare is madness, the youth, to skip o'er the
meshes of good counsel, the cripple. But this reasoning
is not in the fashion to choose me a husband. O me,
the word 'choose'! I may neither choose who I would
nor refuse who I dislike; so is the will of a living
daughter curbed by the will of a dead father. Is it not
hard, Nerissa, that I cannot choose one nor refuse
none?                                                        26

NERISSA Your father was ever virtuous, and holy men at their death have good inspirations; therefore the lottery that he hath devised in these three chests of gold, silver, and lead, whereof who chooses his meaning chooses you, will no doubt never be chosen by any rightly but one who you shall rightly love. But what warmth is there in your affection towards any of these princely suitors that are already come?                           34

PORTIA I pray thee overname them, and as thou namest them I will describe them; and according to my description, level at my affection.

NERISSA First there is the Neapolitan prince.

PORTIA Ay, that's a colt indeed, for he doth nothing but talk of his horse, and he makes it a great appropriation to his own good parts that he can shoe him himself. I am much afeard my lady his mother played false with a smith.                                                43

NERISSA Then is there the County Palatine.

PORTIA He doth nothing but frown, as who should say 'An you will not have me, choose'. He hears merry tales and smiles not. I fear he will prove the weeping philosopher when he grows old, being so full of unmannerly sadness in his youth. I had rather be married to a death's-head with a bone in his mouth than to either of these. God defend me from these two!

NERISSA How say you by the French lord, Monsieur le Bon?                                                    53

PORTIA God made him, and therefore let him pass for a man. In truth, I know it is a sin to be a mocker, but he—why, he hath a horse better than the Neapolitan's, a better bad habit of frowning than the Count Palatine. He is every man in no man. If a throstle sing, he falls straight a-cap'ring. He will fence with his own shadow. If I should marry him, I should marry twenty husbands. If he would despise me, I would forgive him, for if he love me to madness, I shall never requite him.    62

NERISSA What say you then to Falconbridge, the young baron of England?

PORTIA You know I say nothing to him, for he understands not me, nor I him. He hath neither Latin, French, nor Italian, and you will come into the court and swear that I have a poor pennyworth in the English. He is a proper man's picture, but alas, who can converse with a dumb show? How oddly he is suited! I think he bought his doublet in Italy, his round hose in France, his bonnet in Germany, and his behaviour everywhere.

NERISSA What think you of the Scottish lord, his neighbour?                                                75

PORTIA That he hath a neighbourly charity in him, for he borrowed a box of the ear of the Englishman and swore he would pay him again when he was able. I think the Frenchman became his surety, and sealed under for another.                                        80

NERISSA How like you the young German, the Duke of Saxony's nephew?

PORTIA Very vilely in the morning when he is sober, and most vilely in the afternoon when he is drunk. When he is best he is a little worse than a man, and when he is worst he is little better than a beast. An the worst fall that ever fell, I hope I shall make shift to go without him.                                                    88

NERISSA If he should offer to choose, and choose the right casket, you should refuse to perform your father's will if you should refuse to accept him.

PORTIA Therefore, for fear of the worst, I pray thee set a deep glass of Rhenish wine on the contrary casket; for if the devil be within and that temptation without, I know he will choose it. I will do anything, Nerissa, ere I will be married to a sponge.                           96

NERISSA You need not fear, lady, the having any of these lords. They have acquainted me with their determinations, which is indeed to return to their home and to trouble you with no more suit unless you may be won by some other sort than your father's imposition depending on the caskets.                              102

PORTIA If I live to be as old as Sibylla I will die as chaste as Diana unless I be obtained by the manner of my father's will. I am glad this parcel of wooers are so reasonable, for there is not one among them but I dote on his very absence; and I pray God grant them a fair departure.

NERISSA Do you not remember, lady, in your father's time, a Venetian, a scholar and a soldier, that came hither in company of the Marquis of Montferrat?   111

PORTIA Yes, yes, it was Bassanio—as I think, so was he called.

NERISSA True, madam. He of all the men that ever my foolish eyes looked upon was the best deserving a fair lady.

PORTIA I remember him well, and I remember him worthy of thy praise.                                          118

*Enter a Servingman*

How now, what news?

SERVINGMAN The four strangers seek for you, madam, to take their leave, and there is a forerunner come from a fifth, the Prince of Morocco, who brings word the Prince his master will be here tonight.              123

PORTIA If I could bid the fifth welcome with so good heart as I can bid the other four farewell, I should be glad of his approach. If he have the condition of a saint and the complexion of a devil, I had rather he should shrive me than wive me.

Come, Nerissa. (*To the Servingman*) Sirrah, go before. Whiles we shut the gate upon one wooer,         130
Another knocks at the door.                *Exeunt*

**1.3** *Enter Bassanio with Shylock the Jew*

SHYLOCK Three thousand ducats. Well.

BASSANIO Ay, sir, for three months.

SHYLOCK For three months. Well.

BASSANIO For the which, as I told you, Antonio shall be bound.                                                 5

SHYLOCK Antonio shall become bound. Well.

BASSANIO May you stead me? Will you pleasure me? Shall I know your answer?

SHYLOCK Three thousand ducats for three months, and Antonio bound.                                           10

BASSANIO Your answer to that.

SHYLOCK Antonio is a good man.

BASSANIO Have you heard any imputation to the contrary?

SHYLOCK Ho, no, no, no, no! My meaning in saying he is a good man is to have you understand me that he is sufficient. Yet his means are in supposition. He hath an argosy bound to Tripolis, another to the Indies. I understand moreover upon the Rialto he hath a third at Mexico, a fourth for England, and other ventures he hath squandered abroad. But ships are but boards, sailors but men. There be land rats and water rats, water thieves and land thieves—I mean pirates—and

then there is the peril of waters, winds, and rocks. The
man is, notwithstanding, sufficient. Three thousand
ducats. I think I may take his bond.                          26

BASSANIO Be assured you may.

SHYLOCK I will be assured I may, and that I may be
assured, I will bethink me. May I speak with Antonio?

BASSANIO If it please you to dine with us.                   30

SHYLOCK ⌈aside⌉ Yes, to smell pork, to eat of the habitation
which your prophet the Nazarite conjured the devil
into! I will buy with you, sell with you, talk with you,
walk with you, and so following, but I will not eat
with you, drink with you, nor pray with you.                 35

    *Enter Antonio*

⌈*To Antonio*⌉ What news on the Rialto? ⌈*To Bassanio*⌉
Who is he comes here?

BASSANIO This is Signor Antonio.

    ⌈*Bassanio and Antonio speak silently to one another*⌉

SHYLOCK (*aside*)
How like a fawning publican he looks.
I hate him for he is a Christian;                            40
But more, for that in low simplicity
He lends out money gratis, and brings down
The rate of usance here with us in Venice.
If I can catch him once upon the hip
I will feed fat the ancient grudge I bear him.               45
He hates our sacred nation, and he rails,
Even there where merchants most do congregate,
On me, my bargains, and my well-won thrift—
Which he calls interest. Cursèd be my tribe
If I forgive him.

BASSANIO        Shylock, do you hear?          50

SHYLOCK
I am debating of my present store,
And by the near guess of my memory
I cannot instantly raise up the gross
Of full three thousand ducats. What of that?
Tubal, a wealthy Hebrew of my tribe,                         55
Will furnish me. But soft—how many months
Do you desire? ⌈*To Antonio*⌉ Rest you fair, good signor.
Your worship was the last man in our mouths.

ANTONIO
Shylock, albeit I neither lend nor borrow
By taking nor by giving of excess,                           60
Yet to supply the ripe wants of my friend
I'll break a custom. (*To Bassanio*) Is he yet possessed
How much ye would?

SHYLOCK Ay, ay, three thousand ducats.

ANTONIO And for three months.                                65

SHYLOCK
I had forgot—three months. (*To Bassanio*) You told me
    so.—
Well then, your bond; and let me see—but hear you,
Methoughts you said you neither lend nor borrow
Upon advantage.

ANTONIO        I do never use it.

SHYLOCK
When Jacob grazed his uncle Laban's sheep—                   70
This Jacob from our holy Abram was,
As his wise mother wrought in his behalf,
The third possessor; ay, he was the third—

ANTONIO
And what of him? Did he take interest?

SHYLOCK
No, not take interest, not, as you would say,               75
Directly int'rest. Mark what Jacob did:
When Laban and himself were compromised
That all the eanlings which were streaked and pied

Should fall as Jacob's hire, the ewes, being rank,
In end of autumn turnèd to the rams,                         80
And when the work of generation was
Between these woolly breeders in the act,
The skilful shepherd peeled me certain wands,
And in the doing of the deed of kind
He stuck them up before the fulsome ewes                     85
Who, then conceiving, did in eaning time
Fall parti-coloured lambs; and those were Jacob's.
This was a way to thrive; and he was blest;
And thrift is blessing, if men steal it not.

ANTONIO
This was a venture, sir, that Jacob served for—              90
A thing not in his power to bring to pass,
But swayed and fashioned by the hand of heaven.
Was this inserted to make interest good,
Or is your gold and silver ewes and rams?

SHYLOCK
I cannot tell. I make it breed as fast.                      95
But note me, signor—

ANTONIO        Mark you this, Bassanio?
The devil can cite Scripture for his purpose.
An evil soul producing holy witness
Is like a villain with a smiling cheek,
A goodly apple rotten at the heart.                         100
O, what a goodly outside falsehood hath!

SHYLOCK
Three thousand ducats. 'Tis a good round sum.
Three months from twelve—then let me see the rate.

ANTONIO
Well, Shylock, shall we be beholden to you?

SHYLOCK
Signor Antonio, many a time and oft                         105
In the Rialto you have rated me
About my moneys and my usances.
Still have I borne it with a patient shrug,
For suff'rance is the badge of all our tribe.
You call me misbeliever, cut-throat, dog,                   110
And spit upon my Jewish gaberdine,
And all for use of that which is mine own.
Well then, it now appears you need my help.
Go to, then. You come to me, and you say
'Shylock, we would have moneys'—you say so,                 115
You, that did void your rheum upon my beard,
And foot me as you spurn a stranger cur
Over your threshold. Moneys is your suit.
What should I say to you? Should I not say
'Hath a dog money? Is it possible                           120
A cur can lend three thousand ducats?' Or
Shall I bend low, and in a bondman's key,
With bated breath and whisp'ring humbleness
Say this: 'Fair sir, you spat on me on Wednesday last;
You spurned me such a day; another time                     125
You called me dog; and for these courtesies
I'll lend you thus much moneys'?

ANTONIO
I am as like to call thee so again,
To spit on thee again, to spurn thee too.
If thou wilt lend this money, lend it not                   130
As to thy friends; for when did friendship take
A breed for barren metal of his friend?
But lend it rather to thine enemy,
Who if he break, thou mayst with better face
Exact the penalty.

SHYLOCK      Why, look you, how you storm!
I would be friends with you, and have your love,   136

Forget the shames that you have stained me with,
Supply your present wants, and take no doit
Of usance for my moneys; and you'll not hear me.
This is kind I offer.                                           140
BASSANIO                This were kindness.
SHYLOCK This kindness will I show.
Go with me to a notary, seal me there
Your single bond, and, in a merry sport,
If you repay me not on such a day,                             145
In such a place, such sum or sums as are
Expressed in the condition, let the forfeit
Be nominated for an equal pound
Of your fair flesh to be cut off and taken
In what part of your body pleaseth me.                         150
ANTONIO
Content, in faith. I'll seal to such a bond,
And say there is much kindness in the Jew.
BASSANIO
You shall not seal to such a bond for me.
I'll rather dwell in my necessity.
ANTONIO
Why, fear not, man; I will not forfeit it.                     155
Within these two months—that's a month before
This bond expires—I do expect return
Of thrice three times the value of this bond.
SHYLOCK
O father Abram, what these Christians are,
Whose own hard dealings teaches them suspect    160
The thoughts of others! (*To Bassanio*) Pray you tell me
   this:
If he should break his day, what should I gain
By the exaction of the forfeiture?
A pound of man's flesh taken from a man
Is not so estimable, profitable neither,                       165
As flesh of muttons, beeves, or goats. I say,
To buy his favour I extend this friendship.
If he will take it, so. If not, adieu,
And, for my love, I pray you wrong me not.
ANTONIO
Yes, Shylock, I will seal unto this bond.                      170
SHYLOCK
Then meet me forthwith at the notary's.
Give him direction for this merry bond,
And I will go and purse the ducats straight,
See to my house—left in the fearful guard
Of an unthrifty knave—and presently                            175
I'll be with you.
ANTONIO                Hie thee, gentle Jew.    *Exit Shylock*
The Hebrew will turn Christian; he grows kind.
BASSANIO
I like not fair terms and a villain's mind.
ANTONIO
Come on. In this there can be no dismay.
My ships come home a month before the day.    *Exeunt*

2.1   ⌈*Flourish of cornetts.*⌉ *Enter the Prince of Morocco,*
      *a tawny Moor all in white, and three or four*
      *followers accordingly, with Portia, Nerissa, and*
      *their train*
MOROCCO (*to Portia*)
Mislike me not for my complexion,
The shadowed livery of the burnished sun,
To whom I am a neighbour and near bred.
Bring me the fairest creature northward born,
Where Phoebus' fire scarce thaws the icicles,                  5
And let us make incision for your love

To prove whose blood is reddest, his or mine.
I tell thee, lady, this aspect of mine
Hath feared the valiant. By my love I swear,
The best regarded virgins of our clime                         10
Have loved it too. I would not change this hue
Except to steal your thoughts, my gentle queen.
PORTIA
In terms of choice I am not solely led
By nice direction of a maiden's eyes.
Besides, the lott'ry of my destiny                             15
Bars me the right of voluntary choosing.
But if my father had not scanted me,
And hedged me by his wit to yield myself
His wife who wins me by that means I told you,
Yourself, renownèd Prince, then stood as fair                  20
As any comer I have looked on yet
For my affection.
MOROCCO                Even for that I thank you.
Therefore I pray you lead me to the caskets
To try my fortune. By this scimitar,
That slew the Sophy and a Persian prince                       25
That won three fields of Sultan Suleiman,
I would o'erstare the sternest eyes that look,
Outbrave the heart most daring on the earth,
Pluck the young sucking cubs from the she-bear,
Yea, mock the lion when a roars for prey,                      30
To win the lady. But alas the while,
If Hercules and Lichas play at dice
Which is the better man, the greater throw
May turn by fortune from the weaker hand.
So is Alcides beaten by his rage,                              35
And so may I, blind Fortune leading me,
Miss that which one unworthier may attain,
And die with grieving.
PORTIA                You must take your chance,
And either not attempt to choose at all,
Or swear before you choose, if you choose wrong     40
Never to speak to lady afterward
In way of marriage. Therefore be advised.
MOROCCO
Nor will not. Come, bring me unto my chance.
PORTIA
First, forward to the temple. After dinner
Your hazard shall be made.
MOROCCO                Good fortune then,           45
To make me blest or cursèd'st among men.
                       ⌈*Flourish of cornetts.*⌉ *Exeunt*

2.2   *Enter Lancelot the clown*
LANCELOT Certainly my conscience will serve me to run
   from this Jew my master. The fiend is at mine elbow
   and tempts me, saying to me 'Gobbo, Lancelot Gobbo,
   good Lancelot,' or 'good Gobbo,' or 'good Lancelot
   Gobbo—use your legs, take the start, run away.' My
   conscience says 'No, take heed, honest Lancelot, take
   heed, honest Gobbo,' or, as aforesaid, 'honest Lancelot
   Gobbo—do not run, scorn running with thy heels.'
   Well, the most courageous fiend bids me pack. '*Via!*'
   says the fiend; 'Away!' says the fiend. 'For the heavens,
   rouse up a brave mind,' says the fiend, 'and run.' Well,
   my conscience hanging about the neck of my heart
   says very wisely to me, 'My honest friend Lancelot'—
   being an honest man's son, or rather an honest
   woman's son, for indeed my father did something
   smack, something grow to; he had a kind of taste—
   well, my conscience says, 'Lancelot, budge not';

'Budge!' says the fiend; 'Budge not', says my conscience. 'Conscience,' say I, 'you counsel well'; 'Fiend,' say I, 'you counsel well.' To be ruled by my conscience I should stay with the Jew my master who, God bless the mark, is a kind of devil; and to run away from the Jew I should be ruled by the fiend who, saving your reverence, is the devil himself. Certainly the Jew is the very devil incarnation; and in my conscience, my conscience is but a kind of hard conscience to offer to counsel me to stay with the Jew. The fiend gives the more friendly counsel. I will run, fiend. My heels are at your commandment. I will run.     29

*Enter old Gobbo, ⌐blind,⌐ with a basket*

GOBBO Master young man, you, I pray you, which is the way to Master Jew's?

LANCELOT (*aside*) O heavens, this is my true-begotten father who, being more than sand-blind—high-gravel-blind—knows me not. I will try confusions with him.    34

GOBBO Master young gentleman, I pray you which is the way to Master Jew's?

LANCELOT Turn up on your right hand at the next turning, but at the next turning of all on your left, marry at the very next turning, turn of no hand but turn down indirectly to the Jew's house.    40

GOBBO By God's sonties, 'twill be a hard way to hit. Can you tell me whether one Lancelot that dwells with him dwell with him or no?

LANCELOT Talk you of young Master Lancelot? (*Aside*) Mark me now, now will I raise the waters. (*To Gobbo*) Talk you of young Master Lancelot?    46

GOBBO No master, sir, but a poor man's son. His father, though I say't, is an honest exceeding poor man, and, God be thanked, well to live.

LANCELOT Well, let his father be what a will, we talk of young Master Lancelot.    51

GOBBO Your worship's friend, and Lancelot, sir.

LANCELOT But I pray you, *ergo* old man, *ergo* I beseech you, talk you of young Master Lancelot?

GOBBO Of Lancelot, an't please your mastership.    55

LANCELOT *Ergo* Master Lancelot. Talk not of Master Lancelot, father, for the young gentleman, according to fates and destinies and such odd sayings—the sisters three and such branches of learning—is indeed deceased; or, as you would say in plain terms, gone to heaven.    61

GOBBO Marry, God forbid! The boy was the very staff of my age, my very prop.

LANCELOT ⌐aside⌐ Do I look like a cudgel or a hovel-post, a staff or a prop? (*To Gobbo*) Do you know me, father?

GOBBO Alack the day, I know you not, young gentleman. But I pray you tell me, is my boy—God rest his soul—alive or dead?

LANCELOT Do you not know me, father?

GOBBO Alack, sir, I am sand-blind. I know you not.    70

LANCELOT Nay, indeed, if you had your eyes you might fail of the knowing me. It is a wise father that knows his own child. Well, old man, I will tell you news of your son. (*Kneeling*) Give me your blessing. Truth will come to light; murder cannot be hid long—a man's son may, but in the end truth will out.    76

GOBBO Pray you, sir, stand up. I am sure you are not Lancelot, my boy.

LANCELOT Pray you, let's have no more fooling about it, but give me your blessing. I am Lancelot, your boy that was, your son that is, your child that shall be.    81

GOBBO I cannot think you are my son.

LANCELOT I know not what I shall think of that, but I am Lancelot the Jew's man, and I am sure Margery your wife is my mother.    85

GOBBO Her name is Margery indeed. I'll be sworn, if thou be Lancelot thou art mine own flesh and blood.

*He feels Lancelot's head*

Lord worshipped might he be, what a beard hast thou got! Thou hast got more hair on thy chin than Dobbin my fill-horse has on his tail.    90

LANCELOT It should seem then that Dobbin's tail grows backward. I am sure he had more hair of his tail than I have of my face when I last saw him.

GOBBO Lord, how art thou changed! How dost thou and thy master agree? I have brought him a present. How 'gree you now?    96

LANCELOT Well, well; but for mine own part, as I have set up my rest to run away, so I will not rest till I have run some ground. My master's a very Jew. Give him a present?—give him a halter! I am famished in his service. You may tell every finger I have with my ribs. Father, I am glad you are come. Give me your present to one Master Bassanio, who indeed gives rare new liveries. If I serve not him, I will run as far as God has any ground.    105

*Enter Bassanio with Leonardo and followers*

O rare fortune! Here comes the man. To him, father, for I am a Jew if I serve the Jew any longer.

BASSANIO (*to one of his men*) You may do so, but let it be so hasted that supper be ready at the farthest by five of the clock. See these letters delivered, put the liveries to making, and desire Graziano to come anon to my lodging.         *Exit one*

LANCELOT (*to Gobbo*) To him, father.

GOBBO (*to Bassanio*) God bless your worship.

BASSANIO Gramercy. Wouldst thou aught with me?    115

GOBBO Here's my son, sir, a poor boy—

LANCELOT (*to Bassanio*) Not a poor boy, sir, but the rich Jew's man that would, sir, as my father shall specify.

GOBBO (*to Bassanio*) He hath a great infection, sir, as one would say, to serve—    120

LANCELOT Indeed, the short and the long is, I serve the Jew, and have a desire as my father shall specify.

GOBBO (*to Bassanio*) His master and he, saving your worship's reverence, are scarce cater-cousins.

LANCELOT (*to Bassanio*) To be brief, the very truth is that the Jew, having done me wrong, doth cause me, as my father—being, I hope, an old man—shall frutify unto you.    128

GOBBO (*to Bassanio*) I have here a dish of doves that I would bestow upon your worship, and my suit is—

LANCELOT (*to Bassanio*) In very brief, the suit is impertinent to myself, as your worship shall know by this honest old man; and though I say it, though old man, yet, poor man, my father.

BASSANIO One speak for both. What would you?    135

LANCELOT Serve you, sir.

GOBBO (*to Bassanio*) That is the very defect of the matter, sir.

BASSANIO (*to Lancelot*)
I know thee well. Thou hast obtained thy suit.
Shylock thy master spoke with me this day,    140
And hath preferred thee, if it be preferment
To leave a rich Jew's service to become
The follower of so poor a gentleman.

LANCELOT The old proverb is very well parted between my master Shylock and you, sir: you have the grace of God, sir, and he hath enough.    146

BASSANIO
    Thou speak'st it well. (*To Gobbo*) Go, father, with thy son.
    (*To Lancelot*) Take leave of thy old master and enquire
    My lodging out. (*To one of his men*) Give him a livery
    More guarded than his fellows'. See it done.    150
LANCELOT (*to Gobbo*) Father, in. I cannot get a service, no,
    I have ne'er a tongue in my head—well!
        *He looks at his palm*
    If any man in Italy have a fairer table which doth offer
    to swear upon a book, I shall have good fortune. Go
    to, here's a simple line of life, here's a small trifle of
    wives—alas, fifteen wives is nothing. Eleven widows
    and nine maids is a simple coming-in for one man, and
    then to scape drowning thrice, and to be in peril of my
    life with the edge of a featherbed—here are simple
    scapes. Well, if Fortune be a woman, she's a good
    wench for this gear. Father, come. I'll take my leave
    of the Jew in the twinkling.        *Exit with old Gobbo*
BASSANIO
    I pray thee, good Leonardo, think on this.
    These things being bought and orderly bestowed,
    Return in haste, for I do feast tonight        165
    My best-esteemed acquaintance. Hie thee. Go.
LEONARDO
    My best endeavours shall be done herein.
        *He begins to leave. Enter Graziano*
GRAZIANO (*to Leonardo*)
    Where's your master?
LEONARDO            Yonder, sir, he walks.    *Exit*
GRAZIANO
    Signor Bassanio.
BASSANIO        Graziano.
GRAZIANO
    I have a suit to you.
BASSANIO            You have obtained it.    170
GRAZIANO
    You must not deny me. I must go with you to
    Belmont.
BASSANIO
    Why then, you must. But hear thee, Graziano,
    Thou art too wild, too rude and bold of voice—
    Parts that become thee happily enough,
    And in such eyes as ours appear not faults;    175
    But where thou art not known, why, there they show
    Something too liberal. Pray thee, take pain
    To allay with some cold drops of modesty
    Thy skipping spirit, lest through thy wild behaviour
    I be misconstered in the place I go to,        180
    And lose my hopes.
GRAZIANO        Signor Bassanio, hear me.
    If I do not put on a sober habit,
    Talk with respect, and swear but now and then,
    Wear prayer books in my pocket, look demurely—
    Nay more, while grace is saying hood mine eyes    185
    Thus with my hat, and sigh, and say 'Amen',
    Use all the observance of civility,
    Like one well studied in a sad ostent
    To please his grandam, never trust me more.
BASSANIO Well, we shall see your bearing.        190
GRAZIANO
    Nay, but I bar tonight. You shall not gauge me
    By what we do tonight.
BASSANIO            No, that were pity.
    I would entreat you rather to put on
    Your boldest suit of mirth, for we have friends

    That purpose merriment. But fare you well.    195
    I have some business.
GRAZIANO
    And I must to Lorenzo and the rest.
    But we will visit you at supper-time.    *Exeunt severally*

2.3    *Enter Jessica and Lancelot, the clown*
JESSICA
    I am sorry thou wilt leave my father so.
    Our house is hell, and thou, a merry devil,
    Didst rob it of some taste of tediousness.
    But fare thee well. There is a ducat for thee.
    And, Lancelot, soon at supper shalt thou see    5
    Lorenzo, who is thy new master's guest.
    Give him this letter, do it secretly;
    And so farewell. I would not have my father
    See me in talk with thee.
LANCELOT Adieu. Tears exhibit my tongue, most beautiful
    pagan; most sweet Jew; if a Christian do not play the
    knave and get thee, I am much deceived. But adieu.
    These foolish drops do something drown my manly
    spirit. Adieu.                14
JESSICA Farewell, good Lancelot.    *Exit Lancelot*
    Alack, what heinous sin is it in me
    To be ashamed to be my father's child!
    But though I am a daughter to his blood,
    I am not to his manners. O Lorenzo,
    If thou keep promise I shall end this strife,    20
    Become a Christian and thy loving wife.    *Exit*

2.4    *Enter Graziano, Lorenzo, Salerio, and Salanio*
LORENZO
    Nay, we will slink away in supper-time,
    Disguise us at my lodging, and return
    All in an hour.
GRAZIANO
    We have not made good preparation.
SALERIO
    We have not spoke as yet of torchbearers.    5
SOLANIO
    'Tis vile, unless it may be quaintly ordered,
    And better in my mind not undertook.
LORENZO
    'Tis now but four o'clock. We have two hours
    To furnish us.
        *Enter Lancelot with a letter*
            Friend Lancelot, what's the news?
LANCELOT (*presenting the letter*) An it shall please you to
    break up this, it shall seem to signify.    11
LORENZO (*taking the letter*)
    I know the hand. In faith, 'tis a fair hand,
    And whiter than the paper it writ on
    Is the fair hand that writ.
GRAZIANO            Love-news, in faith.
LANCELOT ⌈*to Lorenzo*⌉ By your leave, sir.    15
LORENZO Whither goest thou?
LANCELOT Marry, sir, to bid my old master the Jew to sup
    tonight with my new master the Christian.
LORENZO
    Hold, here, take this. (*Giving money*) Tell gentle Jessica
    I will not fail her. Speak it privately.        20
    Go.                *Exit Lancelot*
        Gentlemen,
    Will you prepare you for this masque tonight?
    I am provided of a torchbearer.

SALERIO
Ay, marry, I'll be gone about it straight.

SOLANIO
And so will I.

LORENZO        Meet me and Graziano     25
At Graziano's lodging some hour hence.

SALERIO 'Tis good we do so.       *Exit with Solanio*

GRAZIANO
Was not that letter from fair Jessica?

LORENZO
I must needs tell thee all. She hath directed
How I shall take her from her father's house,    30
What gold and jewels she is furnished with,
What page's suit she hath in readiness.
If e'er the Jew her father come to heaven
It will be for his gentle daughter's sake;
And never dare misfortune cross her foot     35
Unless she do it under this excuse:
That she is issue to a faithless Jew.
Come, go with me. Peruse this as thou goest.
     *He gives Graziano the letter*
Fair Jessica shall be my torchbearer.      *Exeunt*

**2.5**    *Enter Shylock the Jew and his man that was,*
       *Lancelot the clown*

SHYLOCK
Well, thou shalt see, thy eyes shall be thy judge,
The difference of old Shylock and Bassanio.
(*Calling*) What, Jessica! (*To Lancelot*) Thou shalt not
     gormandize
As thou hast done with me. (*Calling*) What, Jessica!
(*To Lancelot*) And sleep and snore and rend apparel
     out.     5
(*Calling*) Why, Jessica, I say! ·

LANCELOT (*calling*)       Why, Jessica!

SHYLOCK
Who bids thee call? I do not bid thee call.

LANCELOT Your worship was wont to tell me I could do
nothing without bidding.
     *Enter Jessica*

JESSICA (*to Shylock*) Call you? What is your will?    10

SHYLOCK
I am bid forth to supper, Jessica.
There are my keys. But wherefore should I go?
I am not bid for love. They flatter me,
But yet I'll go in hate, to feed upon
The prodigal Christian. Jessica, my girl,     15
Look to my house. I am right loath to go.
There is some ill a-brewing towards my rest,
For I did dream of money-bags tonight.

LANCELOT I beseech you, sir, go. My young master doth
expect your reproach.     20

SHYLOCK So do I his.

LANCELOT And they have conspired together. I will not
say you shall see a masque, but if you do, then it was
not for nothing that my nose fell a-bleeding on Black
Monday last at six o'clock i'th' morning, falling out
that year on Ash Wednesday was four year in
th'afternoon.     27

SHYLOCK
What, are there masques? Hear you me, Jessica,
Lock up my doors; and when you hear the drum
And the vile squealing of the wry-necked fife,    30
Clamber not you up to the casements then,
Nor thrust your head into the public street

To gaze on Christian fools with varnished faces,
But stop my house's ears—I mean my casements.
Let not the sound of shallow fopp'ry enter     35
My sober house. By Jacob's staff I swear
I have no mind of feasting forth tonight.
But I will go. (*To Lancelot*) Go you before me, sirrah.
Say I will come.

LANCELOT       I will go before, sir.
(*Aside to Jessica*)
Mistress, look out at window for all this.     40
     There will come a Christian by
     Will be worth a Jewës eye.       *Exit*

SHYLOCK (*to Jessica*)
What says that fool of Hagar's offspring, ha?

JESSICA
His words were 'Farewell, mistress'; nothing else.

SHYLOCK
The patch is kind enough, but a huge feeder,    45
Snail-slow in profit, and he sleeps by day
More than the wildcat. Drones hive not with me;
Therefore I part with him, and part with him
To one that I would have him help to waste
His borrowed purse. Well, Jessica, go in.     50
Perhaps I will return immediately.
Do as I bid you. Shut doors after you.
Fast bind, fast find—
A proverb never stale in thrifty mind.
              *Exit at one door*

JESSICA
Farewell; and if my fortune be not crossed,     55
I have a father, you a daughter lost.
           *Exit at another door*

**2.6**    *Enter the masquers, Graziano and Salerio, ⌜with*
       *torchbearers⌝*

GRAZIANO
This is the penthouse under which Lorenzo
Desired us to make stand.

SALERIO          His hour is almost past.

GRAZIANO
And it is marvel he outdwells his hour,
For lovers ever run before the clock.

SALERIO
O, ten times faster Venus' pigeons fly     5
To seal love's bonds new made than they are wont
To keep obligèd faith unforfeited.

GRAZIANO
That ever holds. Who riseth from a feast
With that keen appetite that he sits down?
Where is the horse that doth untread again     10
His tedious measures with the unbated fire
That he did pace them first? All things that are
Are with more spirit chasèd than enjoyed.
How like a younker or a prodigal
The scarfèd barque puts from her native bay,    15
Hugged and embracèd by the strumpet wind!
How like the prodigal doth she return,
With over-weathered ribs and raggèd sails,
Lean, rent, and beggared by the strumpet wind!
     *Enter Lorenzo, ⌜with a torch⌝*

SALERIO
Here comes Lorenzo. More of this hereafter.     20

LORENZO
Sweet friends, your patience for my long abode.
Not I but my affairs have made you wait.

When you shall please to play the thieves for wives
I'll watch as long for you therein. Approach.
Here dwells my father Jew. (*Calling*) Ho, who's
    within?                                                        25
    *Enter Jessica above in boy's apparel*
JESSICA
    Who are you? Tell me for more certainty,
    Albeit I'll swear that I do know your tongue.
LORENZO Lorenzo, and thy love.
JESSICA
    Lorenzo, certain, and my love indeed,
    For who love I so much? And now who knows        30
    But you, Lorenzo, whether I am yours?
LORENZO
    Heaven and thy thoughts are witness that thou art.
JESSICA
    Here, catch this casket. It is worth the pains.
    I am glad 'tis night, you do not look on me,
    For I am much ashamed of my exchange;            35
    But love is blind, and lovers cannot see
    The pretty follies that themselves commit;
    For if they could, Cupid himself would blush
    To see me thus transformèd to a boy.
LORENZO
    Descend, for you must be my torchbearer.         40
JESSICA
    What, must I hold a candle to my shames?
    They in themselves, good sooth, are too too light.
    Why, 'tis an office of discovery, love,
    And I should be obscured.
LORENZO                          So are you, sweet,
    Even in the lovely garnish of a boy.             45
    But come at once,
    For the close night doth play the runaway,
    And we are stayed for at Bassanio's feast.
JESSICA
    I will make fast the doors, and gild myself
    With some more ducats, and be with you straight.  50
                                      *Exit above*
GRAZIANO
    Now, by my hood, a gentile, and no Jew.
LORENZO
    Beshrew me but I love her heartily,
    For she is wise, if I can judge of her;
    And fair she is, if that mine eyes be true;
    And true she is, as she hath proved herself;      55
    And therefore like herself, wise, fair, and true,
    Shall she be placèd in my constant soul.
        *Enter Jessica below*
    What, art thou come? On, gentlemen, away.
    Our masquing mates by this time for us stay.
                        *Exit with Jessica and Salerio*
        *Enter Antonio*
ANTONIO
    Who's there?
GRAZIANO          Signor Antonio?                      60
ANTONIO
    Fie, fie, Graziano, where are all the rest?
    'Tis nine o'clock. Our friends all stay for you.
    No masque tonight. The wind is come about.
    Bassanio presently will go aboard.
    I have sent twenty out to seek for you.            65
GRAZIANO
    I am glad on't. I desire no more delight
    Than to be under sail and gone tonight.   *Exeunt*

2.7   ⌜*Flourish of cornetts.*⌝ *Enter Portia with Morocco
      and both their trains*
PORTIA
    Go, draw aside the curtains, and discover
    The several caskets to this noble prince.
        *The curtains are drawn aside, revealing three caskets*
    (*To Morocco*) Now make your choice.
MOROCCO
    This first of gold, who this inscription bears:
    'Who chooseth me shall gain what many men desire.'
    The second silver, which this promise carries:     6
    'Who chooseth me shall get as much as he deserves.'
    This third dull lead, with warning all as blunt:
    'Who chooseth me must give and hazard all he hath.'
    How shall I know if I do choose the right?         10
PORTIA
    The one of them contains my picture, Prince.
    If you choose that, then I am yours withal.
MOROCCO
    Some god direct my judgement! Let me see.
    I will survey th'inscriptions back again.
    What says this leaden casket?                      15
    'Who chooseth me must give and hazard all he hath.'
    Must give, for what? For lead? Hazard for lead?
    This casket threatens. Men that hazard all
    Do it in hope of fair advantages.
    A golden mind stoops not to shows of dross.        20
    I'll then nor give nor hazard aught for lead.
    What says the silver with her virgin hue?
    'Who chooseth me shall get as much as he deserves.'
    'As much as he deserves': pause there, Morocco,
    And weigh thy value with an even hand.             25
    If thou beest rated by thy estimation
    Thou dost deserve enough, and yet 'enough'
    May not extend so far as to the lady.
    And yet to be afeard of my deserving
    Were but a weak disabling of myself.               30
    As much as I deserve—why, that's the lady!
    I do in birth deserve her, and in fortunes,
    In graces, and in qualities of breeding;
    But more than these, in love I do deserve.
    What if I strayed no farther, but chose here?      35
    Let's see once more this saying graved in gold:
    'Who chooseth me shall gain what many men desire.'
    Why, that's the lady! All the world desires her.
    From the four corners of the earth they come
    To kiss this shrine, this mortal breathing saint.  40
    The Hyrcanian deserts and the vasty wilds
    Of wide Arabia are as throughfares now
    For princes to come view fair Portia.
    The watery kingdom, whose ambitious head
    Spits in the face of heaven, is no bar             45
    To stop the foreign spirits, but they come
    As o'er a brook to see fair Portia.
    One of these three contains her heavenly picture.
    Is't like that lead contains her? 'Twere damnation
    To think so base a thought. It were too gross      50
    To rib her cerecloth in the obscure grave.
    Or shall I think in silver she's immured,
    Being ten times undervalued to tried gold?
    O sinful thought! Never so rich a gem
    Was set in worse than gold. They have in England   55
    A coin that bears the figure of an angel
    Stamped in gold, but that's insculped upon;
    But here an angel in a golden bed

Lies all within. Deliver me the key.
Here do I choose, and thrive I as I may.                    60
            *He is given a key*
PORTIA
  There, take it, Prince; and if my form lie there,
  Then I am yours.
            *Morocco opens the golden casket*
MOROCCO                    O hell! What have we here?
  A carrion death, within whose empty eye
  There is a written scroll. I'll read the writing.
    'All that glisters is not gold;                         65
    Often have you heard that told.
    Many a man his life hath sold
    But my outside to behold.
    Gilded tombs do worms infold.
    Had you been as wise as bold,                           70
    Young in limbs, in judgement old,
    Your answer had not been enscrolled.
    Fare you well; your suit is cold.'
    Cold indeed, and labour lost.
    Then farewell heat, and welcome frost.                  75
  Portia, adieu. I have too grieved a heart
  To take a tedious leave. Thus losers part.
            ⌈*Flourish of cornetts.*⌉ *Exit with his train*
PORTIA
  A gentle riddance. Draw the curtains, go.
  Let all of his complexion choose me so.
            *The curtains are drawn. Exeunt*

**2.8**  *Enter Salerio and Solanio*
SALERIO
  Why, man, I saw Bassanio under sail.
  With him is Graziano gone along,
  And in their ship I am sure Lorenzo is not.
SOLANIO
  The villain Jew with outcries raised the Duke,
  Who went with him to search Bassanio's ship.          5
SALERIO
  He came too late. The ship was under sail.
  But there the Duke was given to understand
  That in a gondola were seen together
  Lorenzo and his amorous Jessica.
  Besides, Antonio certified the Duke                   10
  They were not with Bassanio in his ship.
SOLANIO
  I never heard a passion so confused,
  So strange, outrageous, and so variable
  As the dog Jew did utter in the streets.
  'My daughter! O, my ducats! O, my daughter!          15
  Fled with a Christian! O, my Christian ducats!
  Justice! The law! My ducats and my daughter!
  A sealèd bag, two sealèd bags of ducats,
  Of double ducats, stol'n from me by my daughter!
  And jewels, two stones, two rich and precious stones,
  Stol'n by my daughter! Justice! Find the girl!       21
  She hath the stones upon her, and the ducats!'
SALERIO
  Why, all the boys in Venice follow him,
  Crying, 'His stones, his daughter, and his ducats!'
SOLANIO
  Let good Antonio look he keep his day,               25
  Or he shall pay for this.
SALERIO                    Marry, well remembered.
  I reasoned with a Frenchman yesterday,
  Who told me in the narrow seas that part
  The French and English there miscarrièd

A vessel of our country, richly fraught.                30
  I thought upon Antonio when he told me,
  And wished in silence that it were not his.
SOLANIO
  You were best to tell Antonio what you hear—
  Yet do not suddenly, for it may grieve him.
SALERIO
  A kinder gentleman treads not the earth.              35
  I saw Bassanio and Antonio part.
  Bassanio told him he would make some speed
  Of his return. He answered, 'Do not so.
  Slubber not business for my sake, Bassanio,
  But stay the very riping of the time;                 40
  And for the Jew's bond which he hath of me,
  Let it not enter in your mind of love.
  Be merry, and employ your chiefest thoughts
  To courtship and such fair ostents of love
  As shall conveniently become you there.'              45
  And even there, his eye being big with tears,
  Turning his face, he put his hand behind him
  And, with affection wondrous sensible,
  He wrung Bassanio's hand; and so they parted.
SOLANIO
  I think he only loves the world for him.              50
  I pray thee let us go and find him out,
  And quicken his embracèd heaviness
  With some delight or other.
SALERIO                    Do we so.        *Exeunt*

**2.9**  *Enter Nerissa and a servitor*
NERISSA
  Quick, quick, I pray thee, draw the curtain straight.
  The Prince of Aragon hath ta'en his oath,
  And comes to his election presently.
        *The servitor draws aside the curtain, revealing the*
        *three caskets.* ⌈*Flourish of cornetts.*⌉ *Enter Aragon,*
        *his train, and Portia*
PORTIA
  Behold, there stand the caskets, noble Prince.
  If you choose that wherein I am contained,             5
  Straight shall our nuptial rites be solemnized.
  But if you fail, without more speech, my lord,
  You must be gone from hence immediately.
ARAGON
  I am enjoined by oath to observe three things:
  First, never to unfold to anyone                      10
  Which casket 'twas I chose. Next, if I fail
  Of the right casket, never in my life
  To woo a maid in way of marriage.
  Lastly, if I do fail in fortune of my choice,
  Immediately to leave you and be gone.                 15
PORTIA
  To these injunctions everyone doth swear
  That comes to hazard for my worthless self.
ARAGON
  And so have I addressed me. Fortune now
  To my heart's hope! Gold, silver, and base lead.
        *He reads the leaden casket*
  'Who chooseth me must give and hazard all he hath.'
  You shall look fairer ere I give or hazard.           21
  What says the golden chest? Ha, let me see.
  'Who chooseth me shall gain what many men desire.'
  'What many men desire'—that 'many' may be meant
  By the fool multitude, that choose by show,           25
  Not learning more than the fond eye doth teach,
  Which pries not to th'interior but, like the martlet,

Builds in the weather on the outward wall
Even in the force and road of casualty.
I will not choose what many men desire, 30
Because I will not jump with common spirits
And rank me with the barbarous multitudes.
Why then, to thee, thou silver treasure-house.
Tell me once more what title thou dost bear.
'Who chooseth me shall get as much as he deserves'—
And well said too, for who shall go about 36
To cozen fortune, and be honourable
Without the stamp of merit? Let none presume
To wear an undeservèd dignity.
O, that estates, degrees, and offices 40
Were not derived corruptly, and that clear honour
Were purchased by the merit of the wearer!
How many then should cover that stand bare,
How many be commanded that command?
How much low peasantry would then be gleaned 45
From the true seed of honour, and how much honour
Picked from the chaff and ruin of the times
To be new varnished? Well; but to my choice.
'Who chooseth me shall get as much as he deserves.'
I will assume desert. Give me a key for this, 50
And instantly unlock my fortunes here.
*He is given a key. ⌈He⌉ opens the silver casket*

PORTIA
Too long a pause for that which you find there.

ARAGON
What's here? The portrait of a blinking idiot
Presenting me a schedule. I will read it.
How much unlike art thou to Portia! 55
How much unlike my hopes and my deservings!
'Who chooseth me shall have as much as he deserves.'
Did I deserve no more than a fool's head?
Is that my prize? Are my deserts no better?

PORTIA
To offend and judge are distinct offices, 60
And of opposèd natures.

ARAGON                    What is here?
*He reads the schedule*
'The fire seven times tried this;
Seven times tried that judgement is
That did never choose amiss.
Some there be that shadows kiss; 65
Such have but a shadow's bliss.
There be fools alive, iwis,
Silvered o'er; and so was this.
Take what wife you will to bed,
I will ever be your head. 70
So be gone; you are sped.'
Still more fool I shall appear
By the time I linger here.
With one fool's head I came to woo,
But I go away with two. 75
Sweet, adieu. I'll keep my oath
Patiently to bear my wroth.
⌈*Flourish of cornetts.*⌉ *Exit with his train*

PORTIA
Thus hath the candle singed the moth.
O, these deliberate fools! When they do choose
They have the wisdom by their wit to lose. 80

NERISSA
The ancient saying is no heresy:
Hanging and wiving goes by destiny.

PORTIA
Come, draw the curtain, Nerissa.
*Nerissa draws the curtain.*

*Enter a Messenger*
MESSENGER
Where is my lady?

PORTIA                    Here. What would my lord?
MESSENGER
Madam, there is alighted at your gate 85
A young Venetian, one that comes before
To signify th'approaching of his lord,
From whom he bringeth sensible regreets,
To wit, besides commends and courteous breath,
Gifts of rich value. Yet I have not seen 90
So likely an ambassador of love.
A day in April never came so sweet
To show how costly summer was at hand
As this fore-spurrer comes before his lord.

PORTIA
No more, I pray thee, I am half afeard 95
Thou wilt say anon he is some kin to thee,
Thou spend'st such high-day wit in praising him.
Come, come, Nerissa, for I long to see
Quick Cupid's post that comes so mannerly.

NERISSA
Bassanio, Lord Love, if thy will it be!    *Exeunt*

**3.1**    *Enter Solanio and Salerio*
SOLANIO
Now, what news on the Rialto?

SALERIO Why, yet it lives there unchecked that Antonio
hath a ship of rich lading wrecked on the narrow
seas—the Goodwins I think they call the place—a very
dangerous flat, and fatal, where the carcasses of many
a tall ship lie buried, as they say, if my gossip Report
be an honest woman of her word. 7

SOLANIO I would she were as lying a gossip in that as
ever knapped ginger or made her neighbours believe
she wept for the death of a third husband. But it is
true, without any slips of prolixity or crossing the plain
highway of talk, that the good Antonio, the honest
Antonio—O that I had a title good enough to keep his
name company—

SALERIO Come, the full stop. 15

SOLANIO Ha, what sayst thou? Why, the end is he hath
lost a ship.

SALERIO I would it might prove the end of his losses.

SOLANIO Let me say amen betimes, lest the devil cross my
prayer— 20
*Enter Shylock*
for here he comes in the likeness of a Jew. How now,
Shylock, what news among the merchants?

SHYLOCK You knew, none so well, none so well as you,
of my daughter's flight.

SALERIO That's certain. I for my part knew the tailor that
made the wings she flew withal. 26

SOLANIO And Shylock for his own part knew the bird was
fledge, and then it is the complexion of them all to
leave the dam.

SHYLOCK She is damned for it. 30

SALERIO That's certain, if the devil may be her judge.

SHYLOCK My own flesh and blood to rebel!

SOLANIO Out upon it, old carrion, rebels it at these years?

SHYLOCK I say my daughter is my flesh and my blood.

SALERIO There is more difference between thy flesh and
hers than between jet and ivory; more between your
bloods than there is between red wine and Rhenish.
But tell us, do you hear whether Antonio have had
any loss at sea or no? 39

SHYLOCK There I have another bad match. A bankrupt, a

prodigal, who dare scarce show his head on the Rialto; a beggar, that was used to come so smug upon the mart. Let him look to his bond. He was wont to call me usurer: let him look to his bond. He was wont to lend money for a Christian courtesy: let him look to his bond.                                                          46

SALERIO Why, I am sure if he forfeit thou wilt not take his flesh. What's that good for?

SHYLOCK To bait fish withal. If it will feed nothing else it will feed my revenge. He hath disgraced me, and hindered me half a million; laughed at my losses, mocked at my gains, scorned my nation, thwarted my bargains, cooled my friends, heated mine enemies, and what's his reason?—I am a Jew. Hath not a Jew eyes? Hath not a Jew hands, organs, dimensions, senses, affections, passions; fed with the same food, hurt with the same weapons, subject to the same diseases, healed by the same means, warmed and cooled by the same winter and summer as a Christian is? If you prick us do we not bleed? If you tickle us do we not laugh? If you poison us do we not die? And if you wrong us shall we not revenge? If we are like you in the rest, we will resemble you in that. If a Jew wrong a Christian, what is his humility? Revenge. If a Christian wrong a Jew, what should his sufferance be by Christian example? Why, revenge. The villainy you teach me I will execute, and it shall go hard but I will better the instruction.

*Enter a Man from Antonio*

MAN (*to Solanio and Salerio*) Gentlemen, my master Antonio is at his house and desires to speak with you both.   70

SALERIO We have been up and down to seek him.

*Enter Tubal*

SOLANIO Here comes another of the tribe. A third cannot be matched unless the devil himself turn Jew.

*Exeunt Solanio and Salerio, with Antonio's Man*

SHYLOCK How now, Tubal? What news from Genoa? Hast thou found my daughter?                                          75

TUBAL I often came where I did hear of her, but cannot find her.

SHYLOCK Why, there, there, there, there. A diamond gone cost me two thousand ducats in Frankfurt. The curse never fell upon our nation till now—I never felt it till now. Two thousand ducats in that and other precious, precious jewels. I would my daughter were dead at my foot and the jewels in her ear! Would she were hearsed at my foot and the ducats in her coffin! No news of them? Why, so. And I know not what's spent in the search. Why thou, loss upon loss: the thief gone with so much, and so much to find the thief, and no satisfaction, no revenge, nor no ill luck stirring but what lights o' my shoulders, no sighs but o' my breathing, no tears but o' my shedding.            90

TUBAL Yes, other men have ill luck too. Antonio, as I heard in Genoa—

SHYLOCK What, what, what? Ill luck, ill luck?

TUBAL Hath an argosy cast away coming from Tripolis.

SHYLOCK I thank God, I thank God! Is it true, is it true?

TUBAL I spoke with some of the sailors that escaped the wreck.                                                              97

SHYLOCK I thank thee, good Tubal. Good news, good news! Ha, ha—heard in Genoa?

TUBAL Your daughter spent in Genoa, as I heard, one night fourscore ducats.                                            101

SHYLOCK Thou stick'st a dagger in me. I shall never see my gold again. Fourscore ducats at a sitting? Fourscore ducats?

TUBAL There came divers of Antonio's creditors in my company to Venice that swear he cannot choose but break.

SHYLOCK I am very glad of it. I'll plague him, I'll torture him. I am glad of it.                                           109

TUBAL One of them showed me a ring that he had of your daughter for a monkey.

SHYLOCK Out upon her! Thou torturest me, Tubal. It was my turquoise. I had it of Leah when I was a bachelor. I would not have given it for a wilderness of monkeys.

TUBAL But Antonio is certainly undone.                     115

SHYLOCK Nay, that's true, that's very true. Go, Tubal, fee me an officer. Bespeak him a fortnight before. I will have the heart of him if he forfeit, for were he out of Venice I can make what merchandise I will. Go, Tubal, and meet me at our synagogue. Go, good Tubal; at our synagogue, Tubal.                        *Exeunt severally*

3.2   *Enter Bassanio, Portia, Nerissa, Graziano, and all their trains. [The curtains are drawn aside revealing the three caskets]*

PORTIA (*to Bassanio*)
  I pray you tarry. Pause a day or two
  Before you hazard, for in choosing wrong
  I lose your company. Therefore forbear a while.
  There's something tells me—but it is not love—
  I would not lose you; and you know yourself        5
  Hate counsels not in such a quality.
  But lest you should not understand me well—
  And yet a maiden hath no tongue but thought—
  I would detain you here some month or two
  Before you venture for me. I could teach you      10
  How to choose right, but then I am forsworn.
  So will I never be; so may you miss me.
  But if you do, you'll make me wish a sin,
  That I had been forsworn. Beshrew your eyes,
  They have o'erlooked me and divided me.            15
  One half of me is yours, the other half yours—
  Mine own, I would say, but if mine, then yours,
  And so all yours. O, these naughty times
  Puts bars between the owners and their rights;
  And so, though yours, not yours. Prove it so,     20
  Let fortune go to hell for it, not I.
  I speak too long, but 'tis to piece the time,
  To eke it, and to draw it out in length
  To stay you from election.

BASSANIO                            Let me choose,
  For as I am, I live upon the rack.                 25

PORTIA
  Upon the rack, Bassanio? Then confess
  What treason there is mingled with your love.

BASSANIO
  None but that ugly treason of mistrust
  Which makes me fear th'enjoying of my love.
  There may as well be amity and life               30
  'Tween snow and fire as treason and my love.

PORTIA
  Ay, but I fear you speak upon the rack,
  Where men enforcèd do speak anything.

BASSANIO
  Promise me life and I'll confess the truth.

PORTIA
  Well then, confess and live.

BASSANIO                            'Confess and love'   35
  Had been the very sum of my confession.
  O happy torment, when my torturer

Doth teach me answers for deliverance!
But let me to my fortune and the caskets.
PORTIA
Away then. I am locked in one of them.                    40
If you do love me, you will find me out.
Nerissa and the rest, stand all aloof.
Let music sound while he doth make his choice.
Then if he lose he makes a swanlike end,
Fading in music. That the comparison                      45
May stand more proper, my eye shall be the stream
And wat'ry deathbed for him. He may win,
And what is music then? Then music is
Even as the flourish when true subjects bow
To a new-crownèd monarch. Such it is                      50
As are those dulcet sounds in break of day
That creep into the dreaming bridegroom's ear
And summon him to marriage. Now he goes,
With no less presence but with much more love
Than young Alcides when he did redeem                     55
The virgin tribute paid by howling Troy
To the sea-monster. I stand for sacrifice.
The rest aloof are the Dardanian wives,
With blearèd visages come forth to view
The issue of th'exploit. Go, Hercules.                    60
Live thou, I live. With much more more dismay
I view the fight than thou that mak'st the fray.
       ⌈Here music.⌉ A song the whilst Bassanio comments
       on the caskets to himself

⌈ONE FROM PORTIA'S TRAIN⌉
          Tell me where is fancy bred,
          Or in the heart, or in the head?
          How begot, how nourishèd?                       65
⌈ALL⌉          Reply, reply.

⌈ONE FROM PORTIA'S TRAIN⌉
          It is engendered in the eyes,
          With gazing fed; and fancy dies
          In the cradle where it lies.
          Let us all ring fancy's knell.                  70
          I'll begin it: ding, dong, bell.
ALL          Ding, dong, bell.

BASSANIO (aside)
So may the outward shows be least themselves.
The world is still deceived with ornament.
In law, what plea so tainted and corrupt                  75
But, being seasoned with a gracious voice,
Obscures the show of evil? In religion,
What damnèd error but some sober brow
Will bless it and approve it with a text,
Hiding the grossness with fair ornament?                  80
There is no vice so simple but assumes
Some mark of virtue on his outward parts.
How many cowards whose hearts are all as false
As stairs of sand, wear yet upon their chins
The beards of Hercules and frowning Mars,                 85
Who, inward searched, have livers white as milk?
And these assume but valour's excrement
To render them redoubted. Look on beauty
And you shall see 'tis purchased by the weight,
Which therein works a miracle in nature,                  90
Making them lightest that wear most of it.
So are those crispèd, snaky, golden locks
Which makes such wanton gambols with the wind
Upon supposèd fairness, often known
To be the dowry of a second head,                         95

The skull that bred them in the sepulchre.
Thus ornament is but the guilèd shore
To a most dangerous sea, the beauteous scarf
Veiling an Indian beauty; in a word,
The seeming truth which cunning times put on            100
To entrap the wisest. (Aloud) Therefore, thou gaudy
      gold,
Hard food for Midas, I will none of thee.
(To the silver casket) Nor none of thee, thou pale and
      common drudge
'Tween man and man. But thou, thou meagre lead,
Which rather threaten'st than dost promise aught,
Thy paleness moves me more than eloquence,              106
And here choose I. Joy be the consequence!
PORTIA (aside)
How all the other passions fleet to air,
As doubtful thoughts, and rash-embraced despair,
And shudd'ring fear, and green-eyed jealousy.           110
O love, be moderate! Allay thy ecstasy.
In measure rain thy joy; scant this excess.
I feel too much thy blessing: make it less,
For fear I surfeit.
       Bassanio opens the leaden casket
BASSANIO          What find I here?
Fair Portia's counterfeit. What demi-god                115
Hath come so near creation? Move these eyes?
Or whether, riding on the balls of mine,
Seem they in motion? Here are severed lips
Parted with sugar breath. So sweet a bar
Should sunder such sweet friends. Here in her hairs
The painter plays the spider, and hath woven            121
A golden mesh t'untrap the hearts of men
Faster than gnats in cobwebs. But her eyes—
How could he see to do them? Having made one,
Methinks it should have power to steal both his         125
And leave itself unfurnished. Yet look how far
The substance of my praise doth wrong this shadow
In underprizing it, so far this shadow
Doth limp behind the substance. Here's the scroll,
The continent and summary of my fortune.                130
      'You that choose not by the view
      Chance as fair and choose as true.
      Since this fortune falls to you,
      Be content, and seek no new.
      If you be well pleased with this,                  135
      And hold your fortune for your bliss,
      Turn you where your lady is,
      And claim her with a loving kiss.'
A gentle scroll. Fair lady, by your leave,
I come by note to give and to receive,                  140
Like one of two contending in a prize,
That thinks he hath done well in people's eyes,
Hearing applause and universal shout,
Giddy in spirit, still gazing in a doubt
Whether those peals of praise be his or no.             145
So, thrice-fair lady, stand I even so,
As doubtful whether what I see be true
Until confirmed, signed, ratified by you.
PORTIA
You see me, Lord Bassanio, where I stand,
Such as I am. Though for myself alone                   150
I would not be ambitious in my wish
To wish myself much better, yet for you
I would be trebled twenty times myself,
A thousand times more fair, ten thousand times more
      rich,

That only to stand high in your account
I might in virtues, beauties, livings, friends,
Exceed account. But the full sum of me
Is sum of something which, to term in gross,
Is an unlessoned girl, unschooled, unpractisèd,
Happy in this, she is not yet so old
But she may learn; happier than this,
She is not bred so dull but she can learn;
Happiest of all is that her gentle spirit
Commits itself to yours to be directed
As from her lord, her governor, her king.
Myself and what is mine to you and yours
Is now converted. But now I was the lord·
Of this fair mansion, master of my servants,
Queen o'er myself; and even now, but now,
This house, these servants, and this same myself
Are yours, my lord's. I give them with this ring,
Which when you part from, lose, or give away,
Let it presage the ruin of your love,
And be my vantage to exclaim on you.

BASSANIO
Madam, you have bereft me of all words.
Only my blood speaks to you in my veins,
And there is such confusion in my powers
As after some oration fairly spoke
By a belovèd prince there doth appear
Among the buzzing pleasèd multitude,
Where every something being blent together
Turns to a wild of nothing save of joy,
Expressed and not expressed. But when this ring
Parts from this finger, then parts life from hence.
O, then be bold to say Bassanio's dead.

NERISSA
My lord and lady, it is now our time
That have stood by and seen our wishes prosper
To cry 'Good joy, good joy, my lord and lady!'

GRAZIANO
My lord Bassanio, and my gentle lady,
I wish you all the joy that you can wish,
For I am sure you can wish none from me.
And when your honours mean to solemnize
The bargain of your faith, I do beseech you
Even at that time I may be married too.

BASSANIO
With all my heart, so thou canst get a wife.

GRAZIANO
I thank your lordship, you have got me one.
My eyes, my lord, can look as swift as yours.
You saw the mistress, I beheld the maid.
You loved, I loved; for intermission
No more pertains to me, my lord, than you.
Your fortune stood upon the caskets there,
And so did mine too, as the matter falls;
For wooing here until I sweat again,
And swearing till my very roof was dry
With oaths of love, at last—if promise last—
I got a promise of this fair one here
To have her love, provided that your fortune
Achieved her mistress.

PORTIA                          Is this true, Nerissa?

NERISSA
Madam, it is, so you stand pleased withal.

BASSANIO
And do you, Graziano, mean good faith?

GRAZIANO Yes, faith, my lord.

BASSANIO
Our feast shall be much honoured in your marriage.

GRAZIANO (to Nerissa)
We'll play with them the first boy for a thousand
      ducats.

NERISSA What, and stake down?

GRAZIANO
No, we shall ne'er win at that sport and stake down.
      *Enter Lorenzo, Jessica, and Salerio, a messenger*
      *from Venice*
But who comes here? Lorenzo and his infidel!
What, and my old Venetian friend Salerio!

BASSANIO
Lorenzo and Salerio, welcome hither,
If that the youth of my new int'rest here
Have power to bid you welcome. (*To Portia*) By your
      leave,
I bid my very friends and countrymen,
Sweet Portia, welcome.

PORTIA
So do I, my lord. They are entirely welcome.

LORENZO
I thank your honour. For my part, my lord,
My purpose was not to have seen you here,
But meeting with Salerio by the way
He did entreat me past all saying nay
To come with him along.

SALERIO                          I did, my lord,
And I have reason for it. Signor Antonio
Commends him to you.
      *He gives Bassanio a letter*

BASSANIO                          Ere I ope his letter
I pray you tell me how my good friend doth.

SALERIO
Not sick, my lord, unless it be in mind;
Nor well, unless in mind. His letter there
Will show you his estate.
      *Bassanio opens the letter and reads*

GRAZIANO
Nerissa, (*indicating Jessica*) cheer yon stranger. Bid her
      welcome.
Your hand, Salerio. What's the news from Venice?
How doth that royal merchant good Antonio?
I know he will be glad of our success.
We are the Jasons; we have won the fleece.

SALERIO
I would you had won the fleece that he hath lost.

PORTIA
There are some shrewd contents in yon same paper
That steals the colour from Bassanio's cheek.
Some dear friend dead, else nothing in the world
Could turn so much the constitution
Of any constant man. What, worse and worse?
With leave, Bassanio, I am half yourself,
And I must freely have the half of anything
That this same paper brings you.

BASSANIO                          O sweet Portia,
Here are a few of the unpleasant'st words
That ever blotted paper. Gentle lady,
When I did first impart my love to you
I freely told you all the wealth I had
Ran in my veins: I was a gentleman;
And then I told you true; and yet, dear lady,
Rating myself at nothing, you shall see
How much I was a braggart. When I told you
My state was nothing, I should then have told you

155

160

165

170

175

180

185

190

195

200

205

210

216

220

225

230

235

240

245

250

255

That I was worse than nothing, for indeed
I have engaged myself to a dear friend,
Engaged my friend to his mere enemy,                           260
To feed my means. Here is a letter, lady,
The paper as the body of my friend,
And every word in it a gaping wound
Issuing life-blood. But is it true, Salerio?
Hath all his ventures failed? What, not one hit?              265
From Tripolis, from Mexico, and England,
From Lisbon, Barbary, and India,
And not one vessel scape the dreadful touch
Of merchant-marring rocks?

SALERIO                                    Not one, my lord.
Besides, it should appear that if he had                       270
The present money to discharge the Jew
He would not take it. Never did I know
A creature that did bear the shape of man
So keen and greedy to confound a man.
He plies the Duke at morning and at night,                     275
And doth impeach the freedom of the state
If they deny him justice. Twenty merchants,
The Duke himself, and the magnificoes
Of greatest port, have all persuaded with him,
But none can drive him from the envious plea                   280
Of forfeiture, of justice, and his bond.

JESSICA
When I was with him I have heard him swear
To Tubal and to Cush, his countrymen,
That he would rather have Antonio's flesh
Than twenty times the value of the sum                         285
That he did owe him; and I know, my lord,
If law, authority, and power deny not,
It will go hard with poor Antonio.

PORTIA (to Bassanio)
Is it your dear friend that is thus in trouble?

BASSANIO
The dearest friend to me, the kindest man,                     290
The best-conditioned and unwearied spirit
In doing courtesies, and one in whom
The ancient Roman honour more appears
Than any that draws breath in Italy.

PORTIA What sum owes he the Jew?                               295

BASSANIO
For me, three thousand ducats.

PORTIA                             What, no more?
Pay him six thousand and deface the bond.
Double six thousand, and then treble that,
Before a friend of this description
Shall lose a hair thorough Bassanio's fault.                   300
First go with me to church and call me wife,
And then away to Venice to your friend;
For never shall you lie by Portia's side
With an unquiet soul. You shall have gold
To pay the petty debt twenty times over.                       305
When it is paid, bring your true friend along.
My maid Nerissa and myself meantime
Will live as maids and widows. Come, away,
For you shall hence upon your wedding day.
Bid your friends welcome, show a merry cheer.                  310
Since you are dear bought, I will love you dear.
But let me hear the letter of your friend.

[BASSANIO] (reads) 'Sweet Bassanio, my ships have all
miscarried, my creditors grow cruel, my estate is very
low, my bond to the Jew is forfeit, and since in paying
it, it is impossible I should live, all debts are cleared

between you and I if I might but see you at my death.
Notwithstanding, use your pleasure. If your love do
not persuade you to come, let not my letter.'

PORTIA
O, love! Dispatch all business, and be gone.                  320

BASSANIO
Since I have your good leave to go away
I will make haste, but till I come again
No bed shall e'er be guilty of my stay
Nor rest be interposer 'twixt us twain.          Exeunt

3.3    Enter Shylock the Jew, Solanio, Antonio, and the
       jailer

SHYLOCK
Jailer, look to him. Tell not me of mercy.
This is the fool that lent out money gratis.
Jailer, look to him.

ANTONIO                       Hear me yet, good Shylock.

SHYLOCK
I'll have my bond. Speak not against my bond.
I have sworn an oath that I will have my bond.                 5
Thou called'st me dog before thou hadst a cause,
But since I am a dog, beware my fangs.
The Duke shall grant me justice. I do wonder,
Thou naughty jailer, that thou art so fond
To come abroad with him at his request.                       10

ANTONIO I pray thee hear me speak.

SHYLOCK
I'll have my bond. I will not hear thee speak.
I'll have my bond, and therefore speak no more.
I'll not be made a soft and dull-eyed fool
To shake the head, relent, and sigh, and yield                15
To Christian intercessors. Follow not.
I'll have no speaking. I will have my bond.          Exit

SOLANIO
It is the most impenetrable cur
That ever kept with men.

ANTONIO                    Let him alone.
I'll follow him no more with bootless prayers.                20
He seeks my life. His reason well I know:
I oft delivered from his forfeitures
Many that have at times made moan to me.
Therefore he hates me.

SOLANIO                   I am sure the Duke
Will never grant this forfeiture to hold.                     25

ANTONIO
The Duke cannot deny the course of law,
For the commodity that strangers have
With us in Venice, if it be denied,
Will much impeach the justice of the state,
Since that the trade and profit of the city                   30
Consisteth of all nations. Therefore go.
These griefs and losses have so bated me
That I shall hardly spare a pound of flesh
Tomorrow to my bloody creditor.
Well, jailer, on. Pray God Bassanio come                      35
To see me pay his debt, and then I care not.        Exeunt

3.4    Enter Portia, Nerissa, Lorenzo, Jessica, and
       Balthasar, a man of Portia's

LORENZO (to Portia)
Madam, although I speak it in your presence,
You have a noble and a true conceit
Of godlike amity, which appears most strongly
In bearing thus the absence of your lord.

But if you knew to whom you show this honour,    5
How true a gentleman you send relief,
How dear a lover of my lord your husband,
I know you would be prouder of the work
Than customary bounty can enforce you.
PORTIA
I never did repent for doing good,    10
Nor shall not now; for in companions
That do converse and waste the time together,
Whose souls do bear an equal yoke of love,
There must be needs a like proportion
Of lineaments, of manners, and of spirit,    15
Which makes me think that this Antonio,
Being the bosom lover of my lord,
Must needs be like my lord. If it be so,
How little is the cost I have bestowed
In purchasing the semblance of my soul    20
From out the state of hellish cruelty.
This comes too near the praising of myself,
Therefore no more of it. Hear other things:
Lorenzo, I commit into your hands
The husbandry and manage of my house    25
Until my lord's return. For mine own part,
I have toward heaven breathed a secret vow
To live in prayer and contemplation,
Only attended by Nerissa here,
Until her husband and my lord's return.    30
There is a monastery two miles off,
And there we will abide. I do desire you
Not to deny this imposition,
The which my love and some necessity
Now lays upon you.
LORENZO             Madam, with all my heart,    35
I shall obey you in all fair commands.
PORTIA
My people do already know my mind,
And will acknowledge you and Jessica
In place of Lord Bassanio and myself.
So fare you well till we shall meet again.    40
LORENZO
Fair thoughts and happy hours attend on you!
JESSICA
I wish your ladyship all heart's content.
PORTIA
I thank you for your wish, and am well pleased
To wish it back on you. Fare you well, Jessica.
            *Exeunt Lorenzo and Jessica*
Now, Balthasar,    45
As I have ever found thee honest-true,
So let me find thee still. Take this same letter,
And use thou all th'endeavour of a man
In speed to Padua. See thou render this
Into my cousin's hands, Doctor Bellario,    50
And look what notes and garments he doth give
    thee,
Bring them, I pray thee, with imagined speed
Unto the traject, to the common ferry
Which trades to Venice. Waste no time in words,
But get thee gone. I shall be there before thee.    55
BALTHASAR
Madam, I go with all convenient speed.    *Exit*
PORTIA
Come on, Nerissa. I have work in hand
That you yet know not of. We'll see our husbands
Before they think of us.
NERISSA            Shall they see us?

PORTIA
They shall, Nerissa, but in such a habit    60
That they shall think we are accomplishèd
With that we lack. I'll hold thee any wager,
When we are both accoutered like young men
I'll prove the prettier fellow of the two,
And wear my dagger with the braver grace,    65
And speak between the change of man and boy
With a reed voice, and turn two mincing steps
Into a manly stride, and speak of frays
Like a fine bragging youth, and tell quaint lies
How honourable ladies sought my love,    70
Which I denying, they fell sick and died.
I could not do withal. Then I'll repent,
And wish for all that that I had not killed them;
And twenty of these puny lies I'll tell,
That men shall swear I have discontinued school    75
Above a twelvemonth. I have within my mind
A thousand raw tricks of these bragging Jacks
Which I will practise.
NERISSA Why, shall we turn to men?
PORTIA Fie, what a question's that    80
If thou wert near a lewd interpreter!
But come, I'll tell thee all my whole device
When I am in my coach, which stays for us
At the park gate; and therefore haste away,
For we must measure twenty miles today.    *Exeunt*

**3.5**    *Enter Lancelot the clown, and Jessica*
LANCELOT Yes, truly; for look you, the sins of the father
are to be laid upon the children, therefore I promise
you I fear you. I was always plain with you, and so
now I speak my agitation of the matter, therefore be
o' good cheer, for truly I think you are damned. There
is but one hope in it that can do you any good, and
that is but a kind of bastard hope, neither.    7
JESSICA And what hope is that, I pray thee?
LANCELOT Marry, you may partly hope that your father
got you not, that you are not the Jew's daughter.    10
JESSICA That were a kind of bastard hope indeed. So the
sins of my mother should be visited upon me.
LANCELOT Truly then, I fear you are damned both by
father and mother. Thus, when I shun Scylla your
father, I fall into Charybdis your mother. Well, you are
gone both ways.    16
JESSICA I shall be saved by my husband. He hath made
me a Christian.
LANCELOT Truly, the more to blame he! We were
Christians enough before, e'en as many as could well
live one by another. This making of Christians will
raise the price of hogs. If we grow all to be pork-eaters
we shall not shortly have a rasher on the coals for
money.
    *Enter Lorenzo*
JESSICA I'll tell my husband, Lancelot, what you say. Here
he comes.    26
LORENZO I shall grow jealous of you shortly, Lancelot, if
you thus get my wife into corners.
JESSICA Nay, you need not fear us, Lorenzo. Lancelot and
I are out. He tells me flatly there's no mercy for me in
heaven because I am a Jew's daughter, and he says
you are no good member of the commonwealth, for in
converting Jews to Christians you raise the price of
pork.    34
LORENZO (*to Lancelot*) I shall answer that better to the
commonwealth than you can the getting up of the
Negro's belly. The Moor is with child by you, Lancelot.

LANCELOT It is much that the Moor should be more than
  reason, but if she be less than an honest woman, she
  is indeed more than I took her for.                    40
LORENZO How every fool can play upon the word! I think
  the best grace of wit will shortly turn into silence, and
  discourse grow commendable in none only but parrots.
  Go in, sirrah, bid them prepare for dinner.
LANCELOT That is done, sir. They have all stomachs.    45
LORENZO Goodly Lord, what a wit-snapper are you! Then
  bid them prepare dinner.
LANCELOT That is done too, sir; only 'cover' is the word.
LORENZO Will you cover then, sir?
LANCELOT Not so, sir, neither. I know my duty.          50
LORENZO Yet more quarrelling with occasion! Wilt thou
  show the whole wealth of thy wit in an instant? I pray
  thee understand a plain man in his plain meaning. Go
  to thy fellows; bid them cover the table, serve in the
  meat, and we will come in to dinner.                   55
LANCELOT For the table, sir, it shall be served in. For the
  meat, sir, it shall be covered. For your coming in to
  dinner, sir, why, let it be as humours and conceits
  shall govern.                                    *Exit*

LORENZO
  O dear discretion, how his words are suited!          60
  The fool hath planted in his memory
  An army of good words, and I do know
  A many fools that stand in better place,
  Garnished like him, that for a tricksy word
  Defy the matter. How cheer'st thou, Jessica?          65
  And now, good sweet, say thy opinion:
  How dost thou like the Lord Bassanio's wife?

JESSICA
  Past all expressing. It is very meet
  The Lord Bassanio live an upright life,
  For, having such a blessing in his lady,              70
  He finds the joys of heaven here on earth,
  And if on earth he do not merit it,
  In reason he should never come to heaven.
  Why, if two gods should play some heavenly match
  And on the wager lay two earthly women,               75
  And Portia one, there must be something else
  Pawned with the other; for the poor rude world
  Hath not her fellow.

LORENZO                    Even such a husband
  Hast thou of me as she is for a wife.

JESSICA
  Nay, but ask my opinion too of that!                  80

LORENZO
  I will anon. First let us go to dinner.

JESSICA
  Nay, let me praise you while I have a stomach.

LORENZO
  No, pray thee, let it serve for table-talk.
  Then, howsome'er thou speak'st, 'mong other things
  I shall digest it.

JESSICA               Well, I'll set you forth.    *Exeunt*

**4.1**   *Enter the Duke, the magnificoes, Antonio, Bassanio,*
          *Graziano, and Salerio*

DUKE
  What, is Antonio here?

ANTONIO                    Ready, so please your grace.

DUKE
  I am sorry for thee. Thou art come to answer
  A stony adversary, an inhuman wretch
  Uncapable of pity, void and empty

  From any dram of mercy.

ANTONIO                    I have heard            5
  Your grace hath ta'en great pains to qualify
  His rigorous course, but since he stands obdurate,
  And that no lawful means can carry me
  Out of his envy's reach, I do oppose
  My patience to his fury, and am armed              10
  To suffer with a quietness of spirit
  The very tyranny and rage of his.

DUKE
  Go one, and call the Jew into the court.

SALERIO
  He is ready at the door. He comes, my lord.
          *Enter Shylock*

DUKE
  Make room, and let him stand before our face.    15
  Shylock, the world thinks—and I think so too—
  That thou but lead'st this fashion of thy malice
  To the last hour of act, and then 'tis thought
  Thou'lt show thy mercy and remorse more strange
  Than is thy strange apparent cruelty,             20
  And where thou now exacts the penalty—
  Which is a pound of this poor merchant's flesh—
  Thou wilt not only loose the forfeiture,
  But, touched with human gentleness and love,
  Forgive a moiety of the principal,                25
  Glancing an eye of pity on his losses,
  That have of late so huddled on his back
  Enough to press a royal merchant down
  And pluck commiseration of his state
  From brassy bosoms and rough hearts of flint,     30
  From stubborn Turks and Tartars never trained
  To offices of tender courtesy.
  We all expect a gentle answer, Jew.

SHYLOCK
  I have possessed your grace of what I purpose,
  And by our holy Sabbath have I sworn              35
  To have the due and forfeit of my bond.
  If you deny it, let the danger light
  Upon your charter and your city's freedom.
  You'll ask me why I rather choose to have
  A weight of carrion flesh than to receive         40
  Three thousand ducats. I'll not answer that,
  But say it is my humour. Is it answered?
  What if my house be troubled with a rat,
  And I be pleased to give ten thousand ducats
  To have it baned? What, are you answered yet?     45
  Some men there are love not a gaping pig,
  Some that are mad if they behold a cat,
  And others when the bagpipe sings i'th' nose
  Cannot contain their urine; for affection,
  Mistress of passion, sways it to the mood         50
  Of what it likes or loathes. Now for your answer:
  As there is no firm reason to be rendered
  Why he cannot abide a gaping pig,
  Why he a harmless necessary cat,
  Why he a woollen bagpipe, but of force            55
  Must yield to such inevitable shame
  As to offend himself being offended,
  So can I give no reason, nor I will not,
  More than a lodged hate and a certain loathing
  I bear Antonio, that I follow thus                60
  A losing suit against him. Are you answered?

BASSANIO
  This is no answer, thou unfeeling man,
  To excuse the current of thy cruelty.

SHYLOCK
I am not bound to please thee with my answers.

BASSANIO
Do all men kill the things they do not love?          65

SHYLOCK
Hates any man the thing he would not kill?

BASSANIO
Every offence is not a hate at first.

SHYLOCK
What, wouldst thou have a serpent sting thee twice?

ANTONIO
I pray you think you question with the Jew.
You may as well go stand upon the beach          70
And bid the main flood bate his usual height;
You may as well use question with the wolf
Why he hath made the ewe bleat for the lamb;
You may as well forbid the mountain pines
To wag their high tops and to make no noise          75
When they are fretten with the gusts of heaven,
You may as well do anything most hard
As seek to soften that—than which what's harder?—
His Jewish heart. Therefore, I do beseech you,
Make no more offers, use no farther means,          80
But with all brief and plain conveniency
Let me have judgement and the Jew his will.

BASSANIO (to Shylock)
For thy three thousand ducats here is six.

SHYLOCK
If every ducat in six thousand ducats
Were in six parts, and every part a ducat,          85
I would not draw them. I would have my bond.

DUKE
How shalt thou hope for mercy, rend'ring none?

SHYLOCK
What judgement shall I dread, doing no wrong?
You have among you many a purchased slave
Which, like your asses and your dogs and mules,          90
You use in abject and in slavish parts
Because you bought them. Shall I say to you
'Let them be free, marry them to your heirs.
Why sweat they under burdens? Let their beds
Be made as soft as yours, and let their palates          95
Be seasoned with such viands.' You will answer
'The slaves are ours.' So do I answer you.
The pound of flesh which I demand of him
Is dearly bought. 'Tis mine, and I will have it.
If you deny me, fie upon your law:          100
There is no force in the decrees of Venice.
I stand for judgement. Answer: shall I have it?

DUKE
Upon my power I may dismiss this court
Unless Bellario, a learnèd doctor
Whom I have sent for to determine this,          105
Come here today.

SALERIO          My lord, here stays without
A messenger with letters from the doctor,
New come from Padua.

DUKE
Bring us the letters. Call the messenger.  [Exit Salerio]

BASSANIO
Good cheer, Antonio. What, man, courage yet!          110
The Jew shall have my flesh, blood, bones, and all
Ere thou shalt lose for me one drop of blood.

ANTONIO
I am a tainted wether of the flock,
Meetest for death. The weakest kind of fruit

Drops earliest to the ground; and so let me.          115
You cannot better be employed, Bassanio,
Than to live still and write mine epitaph.
          Enter ⌈Salerio, with⌉ Nerissa apparelled as a judge's
          clerk

DUKE
Came you from Padua, from Bellario?

NERISSA
From both, my lord. Bellario greets your grace.
          She gives a letter to the Duke.
          Shylock whets his knife on his shoe

BASSANIO (to Shylock)
Why dost thou whet thy knife so earnestly?          120

SHYLOCK
To cut the forfeit from that bankrupt there.

GRAZIANO
Not on thy sole but on thy soul, harsh Jew,
Thou mak'st thy knife keen. But no metal can,
No, not the hangman's axe, bear half the keenness
Of thy sharp envy. Can no prayers pierce thee?          125

SHYLOCK
No, none that thou hast wit enough to make.

GRAZIANO
O, be thou damned, inexorable dog,
And for thy life let justice be accused!
Thou almost mak'st me waver in my faith
To hold opinion with Pythagoras          130
That souls of animals infuse themselves
Into the trunks of men. Thy currish spirit
Governed a wolf who, hanged for human slaughter,
Even from the gallows did his fell soul fleet,
And, whilst thou lay'st in thy unhallowed dam,          135
Infused itself in thee; for thy desires
Are wolvish, bloody, starved, and ravenous.

SHYLOCK
Till thou canst rail the seal from off my bond
Thou but offend'st thy lungs to speak so loud.
Repair thy wit, good youth, or it will fall          140
To cureless ruin. I stand here for law.

DUKE
This letter from Bellario doth commend
A young and learnèd doctor to our court.
Where is he?

NERISSA          He attendeth here hard by
To know your answer, whether you'll admit him.          145

DUKE
With all my heart. Some three or four of you
Go give him courteous conduct to this place.
          Exeunt three or four
Meantime the court shall hear Bellario's letter.
(Reads) 'Your grace shall understand that at the receipt
of your letter I am very sick, but in the instant that
your messenger came, in loving visitation was with me
a young doctor of Rome; his name is Balthasar. I
acquainted him with the cause in controversy between
the Jew and Antonio, the merchant. We turned o'er
many books together. He is furnished with my opinion
which, bettered with his own learning—the greatness
whereof I cannot enough commend—comes with him
at my importunity to fill up your grace's request in my
stead. I beseech you let his lack of years be no
impediment to let him lack a reverend estimation, for
I never knew so young a body with so old a head. I
leave him to your gracious acceptance, whose trial
shall better publish his commendation.'          163

*Enter ⌈three or four with⌉ Portia as Balthasar*
You hear the learn'd Bellario, what he writes;
And here, I take it, is the doctor come.                165
(*To Portia*) Give me your hand. Come you from old
        Bellario?
PORTIA
I did, my lord.
DUKE            You are welcome. Take your place.
Are you acquainted with the difference
That holds this present question in the court?
PORTIA
I am informèd throughly of the cause.                  170
Which is the merchant here, and which the Jew?
DUKE
Antonio and old Shylock, both stand forth.
        *Antonio and Shylock stand forth*
PORTIA
Is your name Shylock?
SHYLOCK            Shylock is my name.
PORTIA
Of a strange nature is the suit you follow,
Yet in such rule that the Venetian law              175
Cannot impugn you as you do proceed.
(*To Antonio*) You stand within his danger, do you not?
ANTONIO
Ay, so he says.
PORTIA            Do you confess the bond?
ANTONIO
I do.
PORTIA  Then must the Jew be merciful.
SHYLOCK
On what compulsion must I? Tell me that.            180
PORTIA
The quality of mercy is not strained.
It droppeth as the gentle rain from heaven
Upon the place beneath. It is twice blest:
It blesseth him that gives, and him that takes.
'Tis mightiest in the mightiest. It becomes          185
The thronèd monarch better than his crown.
His sceptre shows the force of temporal power,
The attribute to awe and majesty,
Wherein doth sit the dread and fear of kings;
But mercy is above this sceptred sway.               190
It is enthronèd in the hearts of kings;
It is an attribute to God himself,
And earthly power doth then show likest God's
When mercy seasons justice. Therefore, Jew,
Though justice be thy plea, consider this:           195
That in the course of justice none of us
Should see salvation. We do pray for mercy,
And that same prayer doth teach us all to render
The deeds of mercy. I have spoke thus much
To mitigate the justice of thy plea,                 200
Which if thou follow, this strict court of Venice
Must needs give sentence 'gainst the merchant there.
SHYLOCK
My deeds upon my head! I crave the law,
The penalty and forfeit of my bond.
PORTIA
Is he not able to discharge the money?              205
BASSANIO
Yes, here I tender it for him in the court,
Yea, twice the sum. If that will not suffice
I will be bound to pay it ten times o'er
On forfeit of my hands, my head, my heart.
If this will not suffice, it must appear            210

That malice bears down truth. And, I beseech you,
Wrest once the law to your authority.
To do a great right, do a little wrong,
And curb this cruel devil of his will.
PORTIA
It must not be. There is no power in Venice         215
Can alter a decree establishèd.
'Twill be recorded for a precedent,
And many an error by the same example
Will rush into the state. It cannot be.
SHYLOCK
A Daniel come to judgement, yea, a Daniel!          220
O wise young judge, how I do honour thee!
PORTIA
I pray you let me look upon the bond.
SHYLOCK
Here 'tis, most reverend doctor, here it is.
PORTIA
Shylock, there's thrice thy money offered thee.
SHYLOCK
An oath, an oath! I have an oath in heaven.         225
Shall I lay perjury upon my soul?
No, not for Venice.
PORTIA            Why, this bond is forfeit,
And lawfully by this the Jew may claim
A pound of flesh, to be by him cut off
Nearest the merchant's heart. (*To Shylock*) Be merciful.
Take thrice thy money. Bid me tear the bond.        231
SHYLOCK
When it is paid according to the tenor.
It doth appear you are a worthy judge.
You know the law. Your exposition
Hath been most sound. I charge you, by the law      235
Whereof you are a well-deserving pillar,
Proceed to judgement. By my soul I swear
There is no power in the tongue of man
To alter me. I stay here on my bond.
ANTONIO
Most heartily I do beseech the court                240
To give the judgement.
PORTIA            Why, then thus it is:
You must prepare your bosom for his knife—
SHYLOCK
O noble judge, O excellent young man!
PORTIA
For the intent and purpose of the law
Hath full relation to the penalty                   245
Which here appeareth due upon the bond.
SHYLOCK
'Tis very true. O wise and upright judge!
How much more elder art thou than thy looks!
PORTIA (*to Antonio*)
Therefore lay bare your bosom.
SHYLOCK            Ay, his breast.
So says the bond, doth it not, noble judge?         250
'Nearest his heart'—those are the very words.
PORTIA
It is so. Are there balance here to weigh the flesh?
SHYLOCK  I have them ready.
PORTIA
Have by some surgeon, Shylock, on your charge
To stop his wounds, lest he do bleed to death.      255
SHYLOCK
Is it so nominated in the bond?
PORTIA
It is not so expressed, but what of that?
'Twere good you do so much for charity.

SHYLOCK
I cannot find it. 'Tis not in the bond.
PORTIA (*to Antonio*)
You, merchant, have you anything to say?    260
ANTONIO
But little. I am armed and well prepared.
Give me your hand, Bassanio; fare you well.
Grieve not that I am fall'n to this for you,
For herein Fortune shows herself more kind
Than is her custom; it is still her use    265
To let the wretched man outlive his wealth
To view with hollow eye and wrinkled brow
An age of poverty, from which ling'ring penance
Of such misery doth she cut me off.
Commend me to your honourable wife.    270
Tell her the process of Antonio's end.
Say how I loved you. Speak me fair in death,
And when the tale is told, bid her be judge
Whether Bassanio had not once a love.
Repent but you that you shall lose your friend,    275
And he repents not that he pays your debt;
For if the Jew do cut but deep enough,
I'll pay it instantly, with all my heart.
BASSANIO
Antonio, I am married to a wife
Which is as dear to me as life itself,    280
But life itself, my wife, and all the world
Are not with me esteemed above thy life.
I would lose all, ay, sacrifice them all
Here to this devil, to deliver you.
PORTIA ⌈*aside*⌉
Your wife would give you little thanks for that    285
If she were by to hear you make the offer.
GRAZIANO
I have a wife who, I protest, I love.
I would she were in heaven so she could
Entreat some power to change this currish Jew.
NERISSA ⌈*aside*⌉
'Tis well you offer it behind her back;    290
The wish would make else an unquiet house.
SHYLOCK ⌈*aside*⌉
These be the Christian husbands. I have a daughter.
Would any of the stock of Barabbas
Had been her husband rather than a Christian.
(*Aloud*) We trifle time. I pray thee pursue sentence.
PORTIA
A pound of that same merchant's flesh is thine.    296
The court awards it, and the law doth give it.
SHYLOCK Most rightful judge!
PORTIA
And you must cut this flesh from off his breast.
The law allows it, and the court awards it.    300
SHYLOCK
Most learnèd judge! A sentence: (*to Antonio*) come,
    prepare.
PORTIA
Tarry a little. There is something else.
This bond doth give thee here no jot of blood.
The words expressly are 'a pound of flesh'.
Take then thy bond. Take thou thy pound of flesh.
But in the cutting it, if thou dost shed    306
One drop of Christian blood, thy lands and goods
Are by the laws of Venice confiscate
Unto the state of Venice.
GRAZIANO                O upright judge!
Mark, Jew! O learnèd judge!    310

SHYLOCK Is that the law?
PORTIA Thyself shalt see the act;
For as thou urgest justice, be assured
Thou shalt have justice more than thou desir'st.
GRAZIANO
O learnèd judge! Mark, Jew—a learnèd judge!    315
SHYLOCK
I take this offer, then. Pay the bond thrice,
And let the Christian go.
BASSANIO                Here is the money.
PORTIA
Soft, the Jew shall have all justice. Soft, no haste.
He shall have nothing but the penalty.
GRAZIANO
O Jew, an upright judge, a learnèd judge!    320
PORTIA (*to Shylock*)
Therefore prepare thee to cut off the flesh.
Shed thou no blood, nor cut thou less nor more
But just a pound of flesh. If thou tak'st more
Or less than a just pound, be it but so much
As makes it light or heavy in the substance    325
Or the division of the twentieth part
Of one poor scruple—nay, if the scale do turn
But in the estimation of a hair,
Thou diest, and all thy goods are confiscate.
GRAZIANO
A second Daniel, a Daniel, Jew!    330
Now, infidel, I have you on the hip.
PORTIA
Why doth the Jew pause? Take thy forfeiture.
SHYLOCK
Give me my principal, and let me go.
BASSANIO
I have it ready for thee. Here it is.
PORTIA
He hath refused it in the open court.    335
He shall have merely justice and his bond.
GRAZIANO
A Daniel, still say I, a second Daniel!
I thank thee, Jew, for teaching me that word.
SHYLOCK
Shall I not have barely my principal?
PORTIA
Thou shalt have nothing but the forfeiture    340
To be so taken at thy peril, Jew.
SHYLOCK
Why then, the devil give him good of it.
I'll stay no longer question.
PORTIA                Tarry, Jew.
The law hath yet another hold on you.
It is enacted in the laws of Venice,    345
If it be proved against an alien
That by direct or indirect attempts
He seek the life of any citizen,
The party 'gainst the which he doth contrive
Shall seize one half his goods; the other half    350
Comes to the privy coffer of the state,
And the offender's life lies in the mercy
Of the Duke only, 'gainst all other voice—
In which predicament I say thou stand'st,
For it appears by manifest proceeding    355
That indirectly, and directly too,
Thou hast contrived against the very life
Of the defendant, and thou hast incurred
The danger formerly by me rehearsed.
Down, therefore, and beg mercy of the Duke.    360

GRAZIANO (*to Shylock*)
Beg that thou mayst have leave to hang thyself—
And yet, thy wealth being forfeit to the state,
Thou hast not left the value of a cord.
Therefore thou must be hanged at the state's charge.

DUKE (*to Shylock*)
That thou shalt see the difference of our spirit,      365
I pardon thee thy life before thou ask it.
For half thy wealth, it is Antonio's.
The other half comes to the general state,
Which humbleness may drive unto a fine.

PORTIA
Ay, for the state, not for Antonio.      370

SHYLOCK
Nay, take my life and all, pardon not that.
You take my house when you do take the prop
That doth sustain my house; you take my life
When you do take the means whereby I live.

PORTIA
What mercy can you render him, Antonio?      375

GRAZIANO
A halter, gratis. Nothing else, for God's sake.

ANTONIO
So please my lord the Duke and all the court
To quit the fine for one half of his goods,
I am content, so he will let me have
The other half in use, to render it      380
Upon his death unto the gentleman
That lately stole his daughter.
Two things provided more: that for this favour
He presently become a Christian;
The other, that he do record a gift      385
Here in the court of all he dies possessed
Unto his son, Lorenzo, and his daughter.

DUKE
He shall do this, or else I do recant
The pardon that I late pronouncèd here.

PORTIA
Art thou contented, Jew? What dost thou say?      390

SHYLOCK
I am content.

PORTIA (*to Nerissa*) Clerk, draw a deed of gift.

SHYLOCK
I pray you give me leave to go from hence.
I am not well. Send the deed after me,
And I will sign it.

DUKE                    Get thee gone, but do it.

GRAZIANO (*to Shylock*)
In christ'ning shalt thou have two godfathers.      395
Had I been judge thou shouldst have had ten more,
To bring thee to the gallows, not the font.
                                        *Exit Shylock*

DUKE (*to Portia*)
Sir, I entreat you home with me to dinner.

PORTIA
I humbly do desire your grace of pardon.
I must away this night toward Padua,      400
And it is meet I presently set forth.

DUKE
I am sorry that your leisure serves you not.
Antonio, gratify this gentleman,
For in my mind you are much bound to him.
                              *Exit Duke and his train*

BASSANIO (*to Portia*)
Most worthy gentleman, I and my friend      405

Have by your wisdom been this day acquitted
Of grievous penalties, in lieu whereof
Three thousand ducats due unto the Jew
We freely cope your courteous pains withal.

ANTONIO
And stand indebted over and above      410
In love and service to you evermore.

PORTIA
He is well paid that is well satisfied,
And I, delivering you, am satisfied,
And therein do account myself well paid.
My mind was never yet more mercenary.      415
I pray you know me when we meet again.
I wish you well; and so I take my leave.

BASSANIO
Dear sir, of force I must attempt you further.
Take some remembrance of us as a tribute,
Not as fee. Grant me two things, I pray you:      420
Not to deny me, and to pardon me.

PORTIA
You press me far, and therefore I will yield.
⌐*To Antonio*⌐ Give me your gloves. I'll wear them for
        your sake.
(*To Bassanio*) And for your love I'll take this ring from
        you.
Do not draw back your hand. I'll take no more,      425
And you in love shall not deny me this.

BASSANIO
This ring, good sir? Alas, it is a trifle.
I will not shame myself to give you this.

PORTIA
I will have nothing else, but only this;
And now, methinks, I have a mind to it.      430

BASSANIO
There's more depends on this than on the value.
The dearest ring in Venice will I give you,
And find it out by proclamation.
Only for this, I pray you pardon me.

PORTIA
I see, sir, you are liberal in offers.      435
You taught me first to beg, and now methinks
You teach me how a beggar should be answered.

BASSANIO
Good sir, this ring was given me by my wife,
And when she put it on she made me vow
That I should neither sell, nor give, nor lose it.      440

PORTIA
That 'scuse serves many men to save their gifts.
An if your wife be not a madwoman,
And know how well I have deserved this ring,
She would not hold out enemy for ever
For giving it to me. Well, peace be with you.      445
                              *Exeunt Portia and Nerissa*

ANTONIO
My lord Bassanio, let him have the ring.
Let his deservings and my love withal
Be valued 'gainst your wife's commandëment.

BASSANIO
Go, Graziano, run and overtake him.
Give him the ring, and bring him, if thou canst,      450
Unto Antonio's house. Away, make haste.
                                        *Exit Graziano*
Come, you and I will thither presently,
And in the morning early will we both
Fly toward Belmont. Come, Antonio.      *Exeunt*

**4.2** *Enter Portia and Nerissa, still disguised*

PORTIA
Enquire the Jew's house out, give him this deed,
And let him sign it. We'll away tonight,
And be a day before our husbands home.
This deed will be well welcome to Lorenzo.
*Enter Graziano*

GRAZIANO Fair sir, you are well o'erta'en.    5
My lord Bassanio upon more advice
Hath sent you here this ring, and doth entreat
Your company at dinner.

PORTIA              That cannot be.
His ring I do accept most thankfully,
And so I pray you tell him. Furthermore,    10
I pray you show my youth old Shylock's house.

GRAZIANO
That will I do.

NERISSA         Sir, I would speak with you.
*(Aside to Portia)* I'll see if I can get my husband's ring
Which I did make him swear to keep for ever.

PORTIA *(aside to Nerissa)*
Thou mayst; I warrant we shall have old swearing  15
That they did give the rings away to men.
But we'll outface them, and outswear them too.
Away, make haste. Thou know'st where I will tarry.
*Exit ⌐at one door¬*

NERISSA *(to Graziano)*
Come, good sir, will you show me to this house?
*Exeunt ⌐at another door¬*

**5.1** *Enter Lorenzo and Jessica*

LORENZO
The moon shines bright. In such a night as this,
When the sweet wind did gently kiss the trees
And they did make no noise—in such a night
Troilus, methinks, mounted the Trojan walls,
And sighed his soul toward the Grecian tents    5
Where Cressid lay that night.

JESSICA           In such a night
Did Thisbe fearfully o'ertrip the dew
And saw the lion's shadow ere himself,
And ran dismayed away.

LORENZO          In such a night
Stood Dido with a willow in her hand    10
Upon the wild sea banks, and waft her love
To come again to Carthage.

JESSICA           In such a night
Medea gatherèd the enchanted herbs
That did renew old Aeson.

LORENZO         In such a night
Did Jessica steal from the wealthy Jew,    15
And with an unthrift love did run from Venice
As far as Belmont.

JESSICA         In such a night
Did young Lorenzo swear he loved her well,
Stealing her soul with many vows of faith,
And ne'er a true one.

LORENZO         In such a night    20
Did pretty Jessica, like a little shrew,
Slander her love, and he forgave it her.

JESSICA
I would outnight you, did nobody come.
But hark, I hear the footing of a man.
*Enter Stefano, a messenger*

LORENZO
Who comes so fast in silence of the night?    25

STEFANO A friend.

LORENZO
A friend—what friend? Your name, I pray you, friend?

STEFANO
Stefano is my name, and I bring word
My mistress will before the break of day
Be here at Belmont. She doth stray about    30
By holy crosses, where she kneels and prays
For happy wedlock hours.

LORENZO         Who comes with her?

STEFANO
None but a holy hermit and her maid.
I pray you, is my master yet returned?

LORENZO
He is not, nor we have not heard from him.    35
But go we in, I pray thee, Jessica,
And ceremoniously let us prepare
Some welcome for the mistress of the house.
*Enter Lancelot, the clown*

LANCELOT *(calling)* Sola, sola! Wo, ha, ho! Sola, sola!

LORENZO Who calls?    40

LANCELOT *(calling)* Sola!—Did you see Master Lorenzo?
*(Calling)* Master Lorenzo! Sola, sola!

LORENZO Leave hollering, man: here.

LANCELOT *(calling)* Sola! —Where, where?

LORENZO Here.    45

LANCELOT Tell him there's a post come from my master
with his horn full of good news. My master will be here
ere morning.    *Exit*

LORENZO *(to Jessica)*
Sweet soul, let's in, and there expect their coming.
And yet no matter. Why should we go in?    50
My friend Stefano, signify, I pray you,
Within the house your mistress is at hand,
And bring your music forth into the air.    *Exit Stefano*
How sweet the moonlight sleeps upon this bank!
Here will we sit, and let the sounds of music    55
Creep in our ears. Soft stillness and the night
Become the touches of sweet harmony.
Sit, Jessica.
*⌐They¬ sit*
           Look how the floor of heaven
Is thick inlaid with patens of bright gold.
There's not the smallest orb which thou behold'st  60
But in his motion like an angel sings,
Still choiring to the young-eyed cherubins.
Such harmony is in immortal souls,
But whilst this muddy vesture of decay
Doth grossly close it in, we cannot hear it.    65
*⌐Enter Musicians¬*
*(To the Musicians)* Come, ho, and wake Diana with a
hymn.
With sweetest touches pierce your mistress' ear,
And draw her home with music.
*The Musicians play*

JESSICA
I am never merry when I hear sweet music.

LORENZO
The reason is your spirits are attentive,    70
For do but note a wild and wanton herd
Or race of youthful and unhandled colts,
Fetching mad bounds, bellowing and neighing loud,
Which is the hot condition of their blood,
If they but hear perchance a trumpet sound,    75
Or any air of music touch their ears,
You shall perceive them make a mutual stand,

Their savage eyes turned to a modest gaze
By the sweet power of music. Therefore the poet
Did feign that Orpheus drew trees, stones, and floods,
Since naught so stockish, hard, and full of rage          81
But music for the time doth change his nature.
The man that hath no music in himself,
Nor is not moved with concord of sweet sounds,
Is fit for treasons, stratagems, and spoils.          85
The motions of his spirit are dull as night,
And his affections dark as Erebus.
Let no such man be trusted. Mark the music.
      *Enter Portia and Nerissa, as themselves*

PORTIA
That light we see is burning in my hall.
How far that little candle throws his beams—          90
So shines a good deed in a naughty world.

NERISSA
When the moon shone we did not see the candle.

PORTIA
So doth the greater glory dim the less.
A substitute shines brightly as a king
Until a king be by, and then his state          95
Empties itself as doth an inland brook
Into the main of waters. Music, hark.

NERISSA
It is your music, madam, of the house.

PORTIA
Nothing is good, I see, without respect.
Methinks it sounds much sweeter than by day.          100

NERISSA
Silence bestows that virtue on it, madam.

PORTIA
The crow doth sing as sweetly as the lark
When neither is attended, and I think
The nightingale, if she should sing by day,
When every goose is cackling, would be thought          105
No better a musician than the wren.
How many things by season seasoned are
To their right praise and true perfection!
      ⌈*She sees Lorenzo and Jessica*⌉
Peace, ho!
      ⌈*Music ceases*⌉
            The moon sleeps with Endymion,
And would not be awaked.

LORENZO ⌈*rising*⌉            That is the voice,          110
Or I am much deceived, of Portia.

PORTIA
He knows me as the blind man knows the cuckoo—
By the bad voice.

LORENZO            Dear lady, welcome home.

PORTIA
We have been praying for our husbands' welfare,
Which speed we hope the better for our words.          115
Are they returned?

LORENZO            Madam, they are not yet,
But there is come a messenger before
To signify their coming.

PORTIA            Go in, Nerissa.
Give order to my servants that they take
No note at all of our being absent hence;          120
Nor you, Lorenzo; Jessica, nor you.
      ⌈*A tucket sounds*⌉

LORENZO
Your husband is at hand. I hear his trumpet.
We are no tell-tales, madam. Fear you not.

PORTIA
This night, methinks, is but the daylight sick.
It looks a little paler. 'Tis a day          125
Such as the day is when the sun is hid.
      *Enter Bassanio, Antonio, Graziano, and their*
      *followers. Graziano and Nerissa speak silently to*
      *one another*

BASSANIO
We should hold day with the Antipodes
If you would walk in absence of the sun.

PORTIA
Let me give light, but let me not be light;
For a light wife doth make a heavy husband,          130
And never be Bassanio so for me.
But God sort all. You are welcome home, my lord.

BASSANIO
I thank you, madam. Give welcome to my friend.
This is the man, this is Antonio,
To whom I am so infinitely bound.          135

PORTIA
You should in all sense be much bound to him,
For as I hear he was much bound for you.

ANTONIO
No more than I am well acquitted of.

PORTIA
Sir, you are very welcome to our house.
It must appear in other ways than words,          140
Therefore I scant this breathing courtesy.

GRAZIANO (*to Nerissa*)
By yonder moon I swear you do me wrong.
In faith, I gave it to the judge's clerk.
Would he were gelt that had it for my part,
Since you do take it, love, so much at heart.          145

PORTIA
A quarrel, ho, already! What's the matter?

GRAZIANO
About a hoop of gold, a paltry ring
That she did give me, whose posy was
For all the world like cutlers' poetry
Upon a knife—'Love me and leave me not'.          150

NERISSA
What talk you of the posy or the value?
You swore to me when I did give it you
That you would wear it till your hour of death,
And that it should lie with you in your grave.
Though not for me, yet for your vehement oaths          155
You should have been respective and have kept it.
Gave it a judge's clerk?—no, God's my judge,
The clerk will ne'er wear hair on's face that had it.

GRAZIANO
He will an if he live to be a man.

NERISSA
Ay, if a woman live to be a man.          160

GRAZIANO
Now by this hand, I gave it to a youth,
A kind of boy, a little scrubbèd boy
No higher than thyself, the judge's clerk,
A prating boy that begged it as a fee.
I could not for my heart deny it him.          165

PORTIA
You were to blame, I must be plain with you,
To part so slightly with your wife's first gift,
A thing stuck on with oaths upon your finger,
And so riveted with faith unto your flesh.
I gave my love a ring, and made him swear          170
Never to part with it; and here he stands.

I dare be sworn for him he would not leave it,
Nor pluck it from his finger for the wealth
That the world masters. Now, in faith, Graziano,
You give your wife too unkind a cause of grief.          175
An 'twere to me, I should be mad at it.

BASSANIO (aside)
Why, I were best to cut my left hand off
And swear I lost the ring defending it.

GRAZIANO ⌈to Portia⌉
My lord Bassanio gave his ring away
Unto the judge that begged it, and indeed          180
Deserved it, too, and then the boy his clerk,
That took some pains in writing, he begged mine,
And neither man nor master would take aught
But the two rings.

PORTIA (to Bassanio)     What ring gave you, my lord?
Not that, I hope, which you received of me.          185

BASSANIO
If I could add a lie unto a fault
I would deny it; but you see my finger
Hath not the ring upon it. It is gone.

PORTIA
Even so void is your false heart of truth.
By heaven, I will ne'er come in your bed          190
Until I see the ring.

NERISSA (to Graziano)     Nor I in yours
Till I again see mine.

BASSANIO               Sweet Portia,
If you did know to whom I gave the ring,
If you did know for whom I gave the ring,
And would conceive for what I gave the ring,          195
And how unwillingly I left the ring
When naught would be accepted but the ring,
You would abate the strength of your displeasure.

PORTIA
If you had known the virtue of the ring,
Or half her worthiness that gave the ring,          200
Or your own honour to contain the ring,
You would not then have parted with the ring.
What man is there so much unreasonable,
If you had pleased to have defended it
With any terms of zeal, wanted the modesty          205
To urge the thing held as a ceremony?
Nerissa teaches me what to believe.
I'll die for't but some woman had the ring.

BASSANIO
No, by my honour, madam, by my soul,
No woman had it, but a civil doctor          210
Which did refuse three thousand ducats of me,
And begged the ring, the which I did deny him,
And suffered him to go displeased away,
Even he that had held up the very life
Of my dear friend. What should I say, sweet lady?          215
I was enforced to send it after him.
I was beset with shame and courtesy.
My honour would not let ingratitude
So much besmear it. Pardon me, good lady,
For by these blessèd candles of the night,          220
Had you been there I think you would have begged
The ring of me to give the worthy doctor.

PORTIA
Let not that doctor e'er come near my house.
Since he hath got the jewel that I loved,
And that which you did swear to keep for me,          225
I will become as liberal as you.
I'll not deny him anything I have,

No, not my body nor my husband's bed.
Know him I shall, I am well sure of it.
Lie not a night from home. Watch me like Argus.          230
If you do not, if I be left alone,
Now by mine honour, which is yet mine own,
I'll have that doctor for my bedfellow.

NERISSA (to Graziano)
And I his clerk, therefore be well advised
How you do leave me to mine own protection.          235

GRAZIANO
Well, do you so. Let not me take him then,
For if I do, I'll mar the young clerk's pen.

ANTONIO
I am th'unhappy subject of these quarrels.

PORTIA
Sir, grieve not you. You are welcome notwithstanding.

BASSANIO
Portia, forgive me this enforcèd wrong,          240
And in the hearing of these many friends
I swear to thee, even by thine own fair eyes,
Wherein I see myself—

PORTIA                    Mark you but that?
In both my eyes he doubly sees himself,
In each eye one. Swear by your double self,          245
And there's an oath of credit.

BASSANIO               Nay, but hear me.
Pardon this fault, and by my soul I swear
I never more will break an oath with thee.

ANTONIO (to Portia)
I once did lend my body for his wealth
Which, but for him that had your husband's ring,          250
Had quite miscarried. I dare be bound again,
My soul upon the forfeit, that your lord
Will never more break faith advisedly.

PORTIA
Then you shall be his surety. Give him this,
And bid him keep it better than the other.          255

ANTONIO
Here, Lord Bassanio, swear to keep this ring.

BASSANIO
By heaven, it is the same I gave the doctor!

PORTIA
I had it of him. Pardon me, Bassanio,
For by this ring, the doctor lay with me.

NERISSA
And pardon me, my gentle Graziano,          260
For that same scrubbèd boy, the doctor's clerk,
In lieu of this last night did lie with me.

GRAZIANO
Why, this is like the mending of highways
In summer where the ways are fair enough!
What, are we cuckolds ere we have deserved it?          265

PORTIA
Speak not so grossly. You are all amazed.
Here is a letter. Read it at your leisure.
It comes from Padua, from Bellario.
There you shall find that Portia was the doctor,
Nerissa there her clerk. Lorenzo here          270
Shall witness I set forth as soon as you,
And even but now returned. I have not yet
Entered my house. Antonio, you are welcome,
And I have better news in store for you
Than you expect. Unseal this letter soon.          275
There you shall find three of your argosies
Are richly come to harbour suddenly.

You shall not know by what strange accident
I chancèd on this letter.

ANTONIO          I am dumb!

BASSANIO (*to Portia*)
Were you the doctor and I knew you not?          280

GRAZIANO (*to Nerissa*)
Were you the clerk that is to make me cuckold?

NERISSA
Ay, but the clerk that never means to do it
Unless he live until he be a man.

BASSANIO (*to Portia*)
Sweet doctor, you shall be my bedfellow.
When I am absent, then lie with my wife.          285

ANTONIO (*to Portia*)
Sweet lady, you have given me life and living,
For here I read for certain that my ships
Are safely come to road.

PORTIA          How now, Lorenzo?
My clerk hath some good comforts, too, for you.

NERISSA
Ay, and I'll give them him without a fee.          290

There do I give to you and Jessica
From the rich Jew a special deed of gift,
After his death, of all he dies possessed of.

LORENZO
Fair ladies, you drop manna in the way
Of starvèd people.

PORTIA          It is almost morning,          295
And yet I am sure you are not satisfied
Of these events at full. Let us go in,
And charge us there upon inter'gatories,
And we will answer all things faithfully.

GRAZIANO
Let it be so. The first inter'gatory          300
That my Nerissa shall be sworn on is
Whether till the next night she had rather stay,
Or go to bed now, being two hours to day.
But were the day come, I should wish it dark
Till I were couching with the doctor's clerk.          305
Well, while I live I'll fear no other thing
So sore as keeping safe Nerissa's ring.          *Exeunt*

# 1 HENRY IV

THE play described in the 1623 Folio as *The First Part of Henry the Fourth* had been entered on the Stationers' Register on 25 February 1598 as *The History of Henry the Fourth*, and that is the title of the first surviving edition, of the same year. An earlier edition, doubtless also printed in 1598, is known only from a single, eight-page fragment. Five more editions appeared before the Folio.

The printing of at least two editions within a few months, and the fact that one of them was read almost out of existence, reflect a matter of exceptional topical interest. The earliest title-page advertises the play's portrayal of 'the humorous conceits of Sir John Falstaff'; but when it was first acted, probably in 1596, this character bore the name of his historical counterpart, the Protestant martyr Sir John Oldcastle. Shakespeare changed his surname as the result of protests from Oldcastle's descendants, the influential Cobham family, one of whom—William Brooke, 7th Lord Cobham—was Elizabeth I's Lord Chamberlain from August 1596 till he died on 5 March 1597. Our edition restores Sir John's original surname for the first time in printed texts (though there is reason to believe that even after 1596 the name 'Oldcastle' was sometimes used on the stage), and also restores Russell and Harvey, names Shakespeare was probably obliged to alter to Bardolph and Peto.

Shakespeare had already shown Henry IV's rise to power, and his troubled state of mind on achieving power, in *Richard II*; that play also shows Henry's dissatisfaction with his wayward son, Prince Harry, later Henry V. *1 Henry IV* continues the story, but in a very different dramatic style. A play called *The Famous Victories of Henry V*, entered in the Stationers' Register in 1594, was published anonymously, in a debased and shortened text, in 1598. This text—which also features Oldcastle as a reprobate—gives a sketchy version of the events portrayed in *1* and *2 Henry IV* and *Henry V*. Shakespeare must have known the original play, but in the absence of a full text we cannot tell how much he depended on it. The surviving version contains nothing about the rebellions against Henry IV, for which Shakespeare seems to have gone to Holinshed's, and perhaps other, *Chronicles*; he draws also on Samuel Daniel's poem *The First Four Books of the Civil Wars* (1595).

*1 Henry IV* is the first of Shakespeare's history plays to make extensive use of the techniques of comedy. On a national level, the play shows the continuing problems of Henry Bolingbroke, insecure in his hold on the throne, and the victim of rebellions led by Worcester, Hotspur (Harry Percy), and Glyndŵr. These scenes are counterpointed by others, written mainly in prose, which, in the manner of a comic sub-plot, provide humorous diversion while also reflecting and extending the concerns of the main plot. Henry suffers not only public insurrection but the personal rebellion of Prince Harry, in his unprincely exploits with the reprobate old knight, Oldcastle. Sir John has become Shakespeare's most famous comic character, but Shakespeare shows that the Prince's treatment of him as a surrogate father who must eventually be abandoned has an intensely serious side.

# THE PERSONS OF THE PLAY

KING HENRY IV

PRINCE HARRY, Prince of Wales,
    familiarly known as Hal  } King Henry's sons

Lord JOHN OF LANCASTER

Earl of WESTMORLAND

Sir Walter BLUNT

Earl of WORCESTER

Percy, Earl of NORTHUMBERLAND, his brother

Henry Percy, known as HOTSPUR,
    Northumberland's son

Kate, LADY PERCY, Hotspur's wife

Lord Edmund MORTIMER, called Earl of March,
    Lady Percy's brother

LADY MORTIMER, his wife

Owain GLYNDŴR, Lady Mortimer's father

Earl of DOUGLAS

Sir Richard VERNON

Scrope, ARCHBISHOP of York

SIR MICHAEL, a member of the Archbishop's
    household

} rebels against King Henry

SIR JOHN Oldcastle

Edward (Ned) POINS

RUSSELL

HARVEY

Mistress Quickly, HOSTESS of a tavern
    in Eastcheap

FRANCIS, a drawer

VINTNER

} associates of Prince Harry

GADSHILL

CARRIERS

CHAMBERLAIN

OSTLER

TRAVELLERS

SHERIFF

MESSENGERS

SERVANT

Lords, soldiers

# The History of Henry the Fourth

**1.1** *Enter King Henry, Lord John of Lancaster, and the Earl of Westmorland, with other ⌈lords⌉*

KING HENRY

So shaken as we are, so wan with care,
Find we a time for frighted peace to pant
And breathe short-winded accents of new broils
To be commenced in strands afar remote.
No more the thirsty entrance of this soil                        5
Shall daub her lips with her own children's blood.
No more shall trenching war channel her fields,
Nor bruise her flow'rets with the armèd hoofs
Of hostile paces. Those opposèd eyes,
Which, like the meteors of a troubled heaven,           10
All of one nature, of one substance bred,
Did lately meet in the intestine shock
And furious close of civil butchery,
Shall now in mutual well-beseeming ranks
March all one way, and be no more opposed            15
Against acquaintance, kindred, and allies.
The edge of war, like an ill-sheathèd knife,
No more shall cut his master. Therefore, friends,
As far as to the sepulchre of Christ—
Whose soldier now, under whose blessèd cross        20
We are impressèd and engaged to fight—
Forthwith a power of English shall we levy,
Whose arms were moulded in their mothers' womb
To chase these pagans in those holy fields
Over whose acres walked those blessèd feet              25
Which fourteen hundred years ago were nailed,
For our advantage, on the bitter cross.
But this our purpose now is twelve month old,
And bootless 'tis to tell you we will go.
Therefor we meet not now. Then let me hear           30
Of you, my gentle cousin Westmorland,
What yesternight our Council did decree
In forwarding this dear expedience.

WESTMORLAND

My liege, this haste was hot in question,
And many limits of the charge set down              35
But yesternight, when all athwart there came
A post from Wales, loaden with heavy news,
Whose worst was that the noble Mortimer,
Leading the men of Herefordshire to fight
Against the irregular and wild Glyndŵr,             40
Was by the rude hands of that Welshman taken,
A thousand of his people butcherèd,
Upon whose dead corpse' there was such misuse,
Such beastly shameless transformation,
By those Welshwomen done as may not be           45
Without much shame retold or spoken of.

KING HENRY

It seems then that the tidings of this broil
Brake off our business for the Holy Land.

WESTMORLAND

This matched with other did, my gracious lord,
For more uneven and unwelcome news                 50
Came from the north, and thus it did import:
On Holy-rood day the gallant Hotspur there—
Young Harry Percy—and brave Archibald,
That ever valiant and approvèd Scot,

At Holmedon met,                                          55
Where they did spend a sad and bloody hour,
As by discharge of their artillery
And shape of likelihood the news was told;
For he that brought them in the very heat
And pride of their contention did take horse,           60
Uncertain of the issue any way.

KING HENRY

Here is a dear, a true industrious friend,
Sir Walter Blunt, new lighted from his horse,
Stained with the variation of each soil
Betwixt that Holmedon and this seat of ours;           65
And he hath brought us smooth and welcome news.
The Earl of Douglas is discomfited.
Ten thousand bold Scots, two-and-twenty knights,
Balked in their own blood did Sir Walter see
On Holmedon's plains. Of prisoners Hotspur took      70
Mordake the Earl of Fife and eldest son
To beaten Douglas, and the Earl of Athol,
Of Moray, Angus, and Menteith;
And is not this an honourable spoil,
A gallant prize? Ha, cousin, is it not?                     75

WESTMORLAND

In faith, it is a conquest for a prince to boast of.

KING HENRY

Yea, there thou mak'st me sad, and mak'st me sin
In envy that my lord Northumberland
Should be the father to so blest a son—
A son who is the theme of honour's tongue,           80
Amongst a grove the very straightest plant,
Who is sweet Fortune's minion and her pride—
Whilst I by looking on the praise of him
See riot and dishonour stain the brow
Of my young Harry. O, that it could be proved      85
That some night-tripping fairy had exchanged
In cradle clothes our children where they lay,
And called mine Percy, his Plantagenet!
Then would I have his Harry, and he mine.
But let him from my thoughts. What think you, coz,
Of this young Percy's pride? The prisoners            91
Which he in this adventure hath surprised
To his own use he keeps, and sends me word
I shall have none but Mordake Earl of Fife.

WESTMORLAND

This is his uncle's teaching. This is Worcester,        95
Malevolent to you in all aspects,
Which makes him prune himself, and bristle up
The crest of youth against your dignity.

KING HENRY

But I have sent for him to answer this;
And for this cause awhile we must neglect            100
Our holy purpose to Jerusalem.
Cousin, on Wednesday next our Council we
Will hold at Windsor. So inform the lords.
But come yourself with speed to us again,
For more is to be said and to be done                  105
Than out of anger can be utterèd.

WESTMORLAND I will, my liege.

*Exeunt ⌈King Henry, Lancaster, and other lords at one door; Westmorland at another door⌉*

**1.2**   *Enter Harry Prince of Wales and Sir John Oldcastle*

SIR JOHN Now, Hal, what time of day is it, lad?

PRINCE HARRY Thou art so fat-witted with drinking of old sack, and unbuttoning thee after supper, and sleeping upon benches after noon, that thou hast forgotten to demand that truly which thou wouldst truly know. What a devil hast thou to do with the time of the day? Unless hours were cups of sack, and minutes capons, and clocks the tongues of bawds, and dials the signs of leaping-houses, and the blessed sun himself a fair hot wench in flame-coloured taffeta, I see no reason why thou shouldst be so superfluous to demand the time of the day.                    12

SIR JOHN Indeed you come near me now, Hal, for we that take purses go by the moon and the seven stars, and not 'By Phoebus, he, that wand'ring knight so fair'. And I prithee, sweet wag, when thou art a king, as God save thy grace—'majesty' I should say, for grace thou wilt have none—

PRINCE HARRY What, none?                    19

SIR JOHN No, by my troth, not so much as will serve to be prologue to an egg and butter.

PRINCE HARRY Well, how then? Come, roundly, roundly.

SIR JOHN Marry then, sweet wag, when thou art king let not us that are squires of the night's body be called thieves of the day's beauty. Let us be 'Diana's foresters', 'gentlemen of the shade', 'minions of the moon', and let men say we be men of good government, being governed, as the sea is, by our noble and chaste mistress the moon, under whose countenance we steal.                    29

PRINCE HARRY Thou sayst well, and it holds well too, for the fortune of us that are the moon's men doth ebb and flow like the sea, being governed as the sea is by the moon. As for proof now: a purse of gold most resolutely snatched on Monday night, and most dissolutely spent on Tuesday morning; got with swearing 'lay by!', and spent with crying 'bring in!'; now in as low an ebb as the foot of the ladder, and by and by in as high a flow as the ridge of the gallows.

SIR JOHN By the Lord, thou sayst true, lad; and is not my Hostess of the tavern a most sweet wench?                    40

PRINCE HARRY As the honey of Hybla, my old lad of the castle; and is not a buff jerkin a most sweet robe of durance?

SIR JOHN How now, how now, mad wag? What, in thy quips and thy quiddities? What a plague have I to do with a buff jerkin?                    46

PRINCE HARRY Why, what a pox have I to do with my Hostess of the tavern?

SIR JOHN Well, thou hast called her to a reckoning many a time and oft.                    50

PRINCE HARRY Did I ever call for thee to pay thy part?

SIR JOHN No, I'll give thee thy due, thou hast paid all there.

PRINCE HARRY Yea, and elsewhere so far as my coin would stretch; and where it would not, I have used my credit.

SIR JOHN Yea, and so used it that were it not here apparent that thou art heir apparent—but I prithee, sweet wag, shall there be gallows standing in England when thou art king, and resolution thus fubbed as it is with the rusty curb of old father Antic the law? Do not thou when thou art king hang a thief.                    61

PRINCE HARRY No, thou shalt.

SIR JOHN Shall I? O, rare! By the Lord, I'll be a brave judge!

PRINCE HARRY Thou judgest false already. I mean thou shalt have the hanging of the thieves, and so become a rare hangman.                    67

SIR JOHN Well, Hal, well; and in some sort it jumps with my humour as well as waiting in the court, I can tell you.                    70

PRINCE HARRY For obtaining of suits?

SIR JOHN Yea, for obtaining of suits, whereof the hangman hath no lean wardrobe. 'Sblood, I am as melancholy as a gib cat, or a lugged bear.

PRINCE HARRY Or an old lion, or a lover's lute.                    75

SIR JOHN Yea, or the drone of a Lincolnshire bagpipe.

PRINCE HARRY What sayst thou to a hare, or the melancholy of Moor-ditch?

SIR JOHN Thou hast the most unsavoury similes, and art indeed the most comparative, rascalliest sweet young Prince. But Hal, I prithee trouble me no more with vanity. I would to God thou and I knew where a commodity of good names were to be bought. An old lord of the Council rated me the other day in the street about you, sir, but I marked him not; and yet he talked very wisely, but I regarded him not; and yet he talked wisely, and in the street too.                    87

PRINCE HARRY Thou didst well, for wisdom cries out in the streets, and no man regards it.

SIR JOHN O, thou hast damnable iteration, and art indeed able to corrupt a saint. Thou hast done much harm upon me, Hal, God forgive thee for it. Before I knew thee, Hal, I knew nothing; and now am I, if a man should speak truly, little better than one of the wicked. I must give over this life, and I will give it over. By the Lord, an I do not, I am a villain. I'll be damned for never a king's son in Christendom.                    97

PRINCE HARRY Where shall we take a purse tomorrow, Jack?

SIR JOHN Zounds, where thou wilt, lad! I'll make one; an I do not, call me villain and baffle me.                    101

PRINCE HARRY I see a good amendment of life in thee, from praying to purse-taking.

SIR JOHN Why, Hal, 'tis my vocation, Hal. 'Tis no sin for a man to labour in his vocation.                    105

*Enter Poins*

Poins! Now shall we know if Gadshill have set a match. O, if men were to be saved by merit, what hole in hell were hot enough for him? This is the most omnipotent villain that ever cried 'Stand!' to a true man.

PRINCE HARRY Good morrow, Ned.                    110

POINS Good morrow, sweet Hal. (*To Sir John*) What says Monsieur Remorse? What says Sir John, sack-and-sugar Jack? How agrees the devil and thee about thy soul, that thou soldest him on Good Friday last, for a cup of Madeira and a cold capon's leg?                    115

PRINCE HARRY Sir John stands to his word, the devil shall have his bargain, for he was never yet a breaker of proverbs: he will give the devil his due.

POINS (*to Sir John*) Then art thou damned for keeping thy word with the devil.                    120

PRINCE HARRY Else he had been damned for cozening the devil.

POINS But my lads, my lads, tomorrow morning by four o'clock early, at Gads Hill, there are pilgrims going to Canterbury with rich offerings, and traders riding to London with fat purses. I have visors for you all; you have horses for yourselves. Gadshill lies tonight in Rochester. I have bespoke supper tomorrow night in

Eastcheap. We may do it as secure as sleep. If you will
go, I will stuff your purses full of crowns; if you will
not, tarry at home and be hanged.                      131

SIR JOHN Hear ye, Edward, if I tarry at home and go not,
I'll hang you for going.

POINS You will, chops?

SIR JOHN Hal, wilt thou make one?                      135

PRINCE HARRY Who, I rob? I a thief? Not I, by my faith.

SIR JOHN There's neither honesty, manhood, nor good
fellowship in thee, nor thou camest not of the blood
royal, if thou darest not stand for ten shillings.

PRINCE HARRY Well then, once in my days I'll be a
madcap.                                                141

SIR JOHN Why, that's well said.

PRINCE HARRY Well, come what will, I'll tarry at home.

SIR JOHN By the Lord, I'll be a traitor then, when thou
art king.                                              145

PRINCE HARRY I care not.

POINS Sir John, I prithee leave the Prince and me alone.
I will lay him down such reasons for this adventure
that he shall go.

SIR JOHN Well, God give thee the spirit of persuasion and
him the ears of profiting, that what thou speakest may
move and what he hears may be believed, that the true
prince may, for recreation' sake, prove a false thief; for
the poor abuses of the time want countenance. Farewell.
You shall find me in Eastcheap.                        155

PRINCE HARRY Farewell, the latter spring; farewell, All-
hallown summer.                          *Exit Sir John*

POINS Now, my good sweet honey lord, ride with us
tomorrow. I have a jest to execute that I cannot manage
alone. Oldcastle, Harvey, Russell, and Gadshill shall rob
those men that we have already waylaid—yourself and
I will not be there—and when they have the booty, if
you and I do not rob them, cut this head off from my
shoulders.                                             164

PRINCE HARRY But how shall we part with them in setting
forth?

POINS Why, we will set forth before or after them and
appoint them a place of meeting, wherein it is at our
pleasure to fail. And then will they adventure upon the
exploit themselves, which they shall have no sooner
achieved but we'll set upon them.                      171

PRINCE HARRY Ay, but 'tis like that they will know us by
our horses, by our habits, and by every other
appointment, to be ourselves.

POINS Tut, our horses they shall not see—I'll tie them in
the wood; our visors we will change after we leave
them; and, sirrah, I have cases of buckram for the
nonce, to immask our noted outward garments.      178

PRINCE HARRY But I doubt they will be too hard for us.

POINS Well, for two of them, I know them to be as true-
bred cowards as ever turned back; and for the third,
if he fight longer than he sees reason, I'll forswear
arms. The virtue of this jest will be the incomprehensible
lies that this same fat rogue will tell us when we meet
at supper: how thirty at least he fought with, what
wards, what blows, what extremities he endured; and
in the reproof of this lives the jest.               187

PRINCE HARRY Well, I'll go with thee. Provide us all things
necessary, and meet me tomorrow night in Eastcheap;
there I'll sup. Farewell.                              190

POINS Farewell, my lord.                              *Exit*

PRINCE HARRY
I know you all, and will a while uphold
The unyoked humour of your idleness.

Yet herein will I imitate the sun,
Who doth permit the base contagious clouds        195
To smother up his beauty from the world,
That when he please again to be himself,
Being wanted he may be more wondered at
By breaking through the foul and ugly mists
Of vapours that did seem to strangle him.          200
If all the year were playing holidays,
To sport would be as tedious as to work;
But when they seldom come, they wished-for come,
And nothing pleaseth but rare accidents.
So when this loose behaviour I throw off           205
And pay the debt I never promisèd,
By how much better than my word I am,
By so much shall I falsify men's hopes;
And like bright metal on a sullen ground,
My reformation, glitt'ring o'er my fault,           210
Shall show more goodly and attract more eyes
Than that which hath no foil to set it off.
I'll so offend to make offence a skill,
Redeeming time when men think least I will.     *Exit*

1.3     *Enter the King, the Earls of Northumberland and
        Worcester, Hotspur, Sir Walter Blunt, with other
        ⌈lords⌉*

KING HENRY (*to Hotspur, Northumberland, and Worcester*)
My blood hath been too cold and temperate,
Unapt to stir at these indignities,
And you have found me, for accordingly
You tread upon my patience; but be sure
I will from henceforth rather be myself,                5
Mighty and to be feared, than my condition,
Which hath been smooth as oil, soft as young down,
And therefore lost that title of respect
Which the proud soul ne'er pays but to the proud.

WORCESTER
Our house, my sovereign liege, little deserves        10
The scourge of greatness to be used on it,
And that same greatness too, which our own hands
Have holp to make so portly.

NORTHUMBERLAND (*to the King*)  My lord—

KING HENRY
Worcester, get thee gone, for I do see
Danger and disobedience in thine eye.                 15
O sir, your presence is too bold and peremptory,
And majesty might never yet endure
The moody frontier of a servant brow.
You have good leave to leave us. When we need
Your use and counsel we shall send for you.           20
                                        *Exit Worcester*
You were about to speak.

NORTHUMBERLAND          Yea, my good lord.
Those prisoners in your highness' name demanded,
Which Harry Percy here at Holmedon took,
Were, as he says, not with such strength denied
As was delivered to your majesty,                     25
Who either through envy or misprision
Was guilty of this fault, and not my son.

HOTSPUR (*to the King*)
My liege, I did deny no prisoners;
But I remember, when the fight was done,
When I was dry with rage and extreme toil,            30
Breathless and faint, leaning upon my sword,
Came there a certain lord, neat and trimly dressed,
Fresh as a bridegroom, and his chin, new-reaped,
Showed like a stubble-land at harvest-home.

He was perfumèd like a milliner,                                    35
And 'twixt his finger and his thumb he held
A pouncet-box, which ever and anon
He gave his nose and took't away again—
Who therewith angry, when it next came there
Took it in snuff—and still he smiled and talked;   40
And as the soldiers bore dead bodies by,
He called them untaught knaves, unmannerly
To bring a slovenly unhandsome corpse
Betwixt the wind and his nobility.
With many holiday and lady terms                        45
He questioned me; amongst the rest demanded
My prisoners in your majesty's behalf.
I then, all smarting with my wounds being cold—
To be so pestered with a popinjay!—
Out of my grief and my impatience                        50
Answered neglectingly, I know not what—
He should, or should not—for he made me mad
To see him shine so brisk, and smell so sweet,
And talk so like a waiting gentlewoman
Of guns, and drums, and wounds, God save the mark!
And telling me the sovereign'st thing on earth   56
Was parmacity for an inward bruise,
And that it was great pity, so it was,
This villainous saltpetre should be digged
Out of the bowels of the harmless earth,               60
Which many a good tall fellow had destroyed
So cowardly, and but for these vile guns
He would himself have been a soldier.
This bald unjointed chat of his, my lord,
Made me to answer indirectly, as I said,                 65
And I beseech you, let not his report
Come current for an accusation
Betwixt my love and your high majesty.

BLUNT (to the King)
Why, yet he doth deny his prisoners,
Whate'er Lord Harry Percy then had said        70
To such a person, and in such a place,
At such a time, with all the rest retold,
May reasonably die, and never rise
To do him wrong or any way impeach
What then he said, so he unsay it now.            75

KING HENRY
Why, yet he doth deny his prisoners,
But with proviso and exception
That we at our own charge shall ransom straight
His brother-in-law the foolish Mortimer,
Who, on my soul, hath wilfully betrayed          80
The lives of those that he did lead to fight
Against that great magician, damned Glyndŵr—
Whose daughter, as we hear, the Earl of March
Hath lately married. Shall our coffers, then,
Be emptied to redeem a traitor home?            85
Shall we buy treason, and indent with fears
When they have lost and forfeited themselves?
No, on the barren mountains let him starve;
For I shall never hold that man my friend
Whose tongue shall ask me for one penny cost   90
To ransom home revolted Mortimer—

HOTSPUR Revolted Mortimer?
He never did fall off, my sovereign liege,
But by the chance of war. To prove that true
Needs no more but one tongue for all those wounds,
Those mouthèd wounds, which valiantly he took   96
When on the gentle Severn's sedgy bank,
In single opposition, hand to hand,

He did confound the best part of an hour
In changing hardiment with great Glyndŵr.      100
Three times they breathed, and three times did they
   drink,
Upon agreement, of swift Severn's flood,
Who, then affrighted with their bloody looks,
Ran fearfully among the trembling reeds,
And hid his crisp head in the hollow bank,       105
Bloodstainèd with these valiant combatants.
Never did bare and rotten policy
Colour her working with such deadly wounds,
Nor never could the noble Mortimer
Receive so many, and all willingly.                    110
Then let not him be slandered with revolt.

KING HENRY
Thou dost belie him, Percy, thou dost belie him.
He never did encounter with Glyndŵr. I tell thee,
He durst as well have met the devil alone
As Owain Glyndŵr for an enemy.                    115
Art thou not ashamed? But, sirrah, henceforth
Let me not hear you speak of Mortimer.
Send me your prisoners with the speediest means,
Or you shall hear in such a kind from me
As will displease you.—My lord Northumberland,   120
We license your departure with your son.
(To Hotspur) Send us your prisoners, or you'll hear of it.
                   Exeunt all but Hotspur and Northumberland
HOTSPUR
An if the devil come and roar for them
I will not send them. I will after straight
And tell him so, for I will ease my heart,          125
Although it be with hazard of my head.
NORTHUMBERLAND
What, drunk with choler? Stay and pause awhile.
            Enter the Earl of Worcester
Here comes your uncle.
HOTSPUR                              Speak of Mortimer?
Zounds, I will speak of him, and let my soul
Want mercy if I do not join with him.            130
In his behalf I'll empty all these veins,
And shed my dear blood drop by drop in the dust,
But I will lift the downfall Mortimer
As high in the air as this unthankful King,
As this ingrate and cankered Bolingbroke.     135
NORTHUMBERLAND (to Worcester)
Brother, the King hath made your nephew mad.
WORCESTER
Who struck this heat up after I was gone?
HOTSPUR
He will forsooth have all my prisoners;
And when I urged the ransom once again
Of my wife's brother, then his cheek looked pale,   140
And on my face he turned an eye of death,
Trembling even at the name of Mortimer.
WORCESTER
I cannot blame him: was not he proclaimed
By Richard, that dead is, the next of blood?
NORTHUMBERLAND
He was; I heard the proclamation.               145
And then it was when the unhappy King,
Whose wrongs in us God pardon, did set forth
Upon his Irish expedition,
From whence he, intercepted, did return
To be deposed, and shortly murderèd.          150
WORCESTER
And for whose death we in the world's wide mouth
Live scandalized and foully spoken of.

HOTSPUR
But soft, I pray you; did King Richard then
Proclaim my brother Edmund Mortimer
Heir to the crown?
NORTHUMBERLAND    He did; myself did hear it.    155
HOTSPUR
Nay, then I cannot blame his cousin King
That wished him on the barren mountains starve.
But shall it be that you that set the crown
Upon the head of this forgetful man,
And for his sake wear the detested blot    160
Of murderous subornation, shall it be
That you a world of curses undergo,
Being the agents or base second means,
The cords, the ladder, or the hangman, rather?
O, pardon me that I descend so low    165
To show the line and the predicament
Wherein you range under this subtle King!
Shall it for shame be spoken in these days,
Or fill up chronicles in time to come,
That men of your nobility and power    170
Did gage them both in an unjust behalf,
As both of you, God pardon it, have done:
To put down Richard, that sweet lovely rose,
And plant this thorn, this canker, Bolingbroke?
And shall it in more shame be further spoken    175
That you are fooled, discarded, and shook off
By him for whom these shames ye underwent?
No; yet time serves wherein you may redeem
Your banished honours, and restore yourselves
Into the good thoughts of the world again,    180
Revenge the jeering and disdained contempt
Of this proud King, who studies day and night
To answer all the debt he owes to you
Even with the bloody payment of your deaths.
Therefore, I say—
WORCESTER    Peace, cousin, say no more.    185
And now I will unclasp a secret book,
And to your quick-conceiving discontents
I'll read you matter deep and dangerous,
As full of peril and adventurous spirit
As to o'erwalk a current roaring loud    190
On the unsteadfast footing of a spear.
HOTSPUR
If he fall in, good night, or sink or swim.
Send danger from the east unto the west,
So honour cross it from the north to south;
And let them grapple. O, the blood more stirs    195
To rouse a lion than to start a hare!
NORTHUMBERLAND (to Worcester)
Imagination of some great exploit
Drives him beyond the bounds of patience.
⌜HOTSPUR⌝
By heaven, methinks it were an easy leap
To pluck bright honour from the pale-faced moon,    200
Or dive into the bottom of the deep,
Where fathom-line could never touch the ground,
And pluck up drownèd honour by the locks,
So he that doth redeem her thence might wear,
Without corrival, all her dignities.    205
But out upon this half-faced fellowship!
WORCESTER (to Northumberland)
He apprehends a world of figures here,
But not the form of what he should attend.
(To Hotspur) Good cousin, give me audience for a while,
And list to me.    210

HOTSPUR
I cry you mercy.
WORCESTER    Those same noble Scots
That are your prisoners—
HOTSPUR    I'll keep them all.
By God, he shall not have a Scot of them;
No, if a scot would save his soul he shall not.
I'll keep them, by this hand.
WORCESTER    You start away,    215
And lend no ear unto my purposes.
Those prisoners you shall keep.
HOTSPUR    Nay, I will; that's flat.
He said he would not ransom Mortimer,
Forbade my tongue to speak of Mortimer;
But I will find him when he lies asleep,    220
And in his ear I'll hollo 'Mortimer!'
Nay, I'll have a starling shall be taught to speak
Nothing but 'Mortimer', and give it him
To keep his anger still in motion.
WORCESTER Hear you, cousin, a word.    225
HOTSPUR
All studies here I solemnly defy,
Save how to gall and pinch this Bolingbroke.
And that same sword-and-buckler Prince of Wales—
But that I think his father loves him not
And would be glad he met with some mischance—
I would have him poisoned with a pot of ale.    231
WORCESTER
Farewell, kinsman. I'll talk to you
When you are better tempered to attend.
NORTHUMBERLAND (to Hotspur)
Why, what a wasp-stung and impatient fool
Art thou to break into this woman's mood,    235
Tying thine ear to no tongue but thine own!
HOTSPUR
Why, look you, I am whipped and scourged with rods,
Nettled and stung with pismires, when I hear
Of this vile politician Bolingbroke.
In Richard's time—what d'ye call the place?    240
A plague upon't, it is in Gloucestershire.
'Twas where the madcap Duke his uncle kept—
His uncle York—where I first bowed my knee
Unto this king of smiles, this Bolingbroke.
'Sblood, when you and he came back from
    Ravenspurgh.    245
NORTHUMBERLAND
At Berkeley castle.
HOTSPUR    You say true.
Why, what a candy deal of courtesy
This fawning greyhound then did proffer me!
'Look when his infant fortune came to age',
And 'gentle Harry Percy', and 'kind cousin'.    250
O, the devil take such cozeners!—God forgive me.
Good uncle, tell your tale; I have done.
WORCESTER
Nay, if you have not, to't again.
We'll stay your leisure.
HOTSPUR    I have done, i'faith.
WORCESTER
Then once more to your Scottish prisoners.    255
Deliver them up without their ransom straight;
And make the Douglas' son your only mean
For powers in Scotland, which, for divers reasons
Which I shall send you written, be assured
Will easily be granted. (To Northumberland) You, my
    lord,    260

Your son in Scotland being thus employed,
Shall secretly into the bosom creep
Of that same noble prelate well-beloved,
The Archbishop.
HOTSPUR                    Of York, is't not?
WORCESTER                              True, who bears hard
His brother's death at Bristol, the Lord Scrope.    265
I speak not this in estimation,
As what I think might be, but what I know
Is ruminated, plotted, and set down,
And only stays but to behold the face
Of that occasion that shall bring it on.    270
HOTSPUR
I smell it; upon my life, it will do well!
NORTHUMBERLAND
Before the game is afoot thou still lett'st slip.
HOTSPUR
Why, it cannot choose but be a noble plot—
And then the power of Scotland and of York
To join with Mortimer, ha?
WORCESTER                    And so they shall.    275
HOTSPUR
In faith, it is exceedingly well aimed.
WORCESTER
And 'tis no little reason bids us speed
To save our heads by raising of a head;
For, bear ourselves as even as we can,
The King will always think him in our debt,    280
And think we think ourselves unsatisfied
Till he hath found a time to pay us home.
And see already how he doth begin
To make us strangers to his looks of love.
HOTSPUR
He does, he does. We'll be revenged on him.    285
WORCESTER
Cousin, farewell. No further go in this
Than I by letters shall direct your course.
When time is ripe, which will be suddenly,
I'll steal to Glyndŵr and Lord Mortimer,
Where you and Douglas and our powers at once,    290
As I will fashion it, shall happily meet,
To bear our fortunes in our own strong arms,
Which now we hold at much uncertainty.
NORTHUMBERLAND
Farewell, good brother. We shall thrive, I trust.
HOTSPUR (to Worcester)
Uncle, adieu. O, let the hours be short    295
Till fields and blows and groans applaud our sport!
*Exeunt ⌈Worcester at one door, Northumberland*
*and Hotspur at another door⌉*

**2.1**    *Enter a Carrier, with a lantern in his hand*
FIRST CARRIER Heigh-ho! An't be not four by the day, I'll
be hanged. Charles's Wain is over the new chimney,
and yet our horse not packed. What, ostler!
OSTLER (*within*) Anon, anon!
FIRST CARRIER I prithee, Tom, beat cut's saddle, put a few
flocks in the point. Poor jade is wrung in the withers,
out of all cess.    7
*Enter another Carrier*
SECOND CARRIER Peas and beans are as dank here as a
dog, and that is the next way to give poor jades the
bots. This house is turned upside down since Robin
Ostler died.    11
FIRST CARRIER Poor fellow never joyed since the price of
oats rose; it was the death of him.

SECOND CARRIER I think this be the most villainous house
in all London road for fleas. I am stung like a tench.
FIRST CARRIER Like a tench? By the mass, there is ne'er
a king christen could be better bit than I have been
since the first cock.    18
SECOND CARRIER Why, they will allow us ne'er a jordan,
and then we leak in your chimney, and your chamber-
lye breeds fleas like a loach.    21
FIRST CARRIER What, ostler! Come away, and be hanged,
come away!
SECOND CARRIER I have a gammon of bacon and two races
of ginger to be delivered as far as Charing Cross.    25
FIRST CARRIER God's body, the turkeys in my pannier are
quite starved! What, ostler! A plague on thee, hast
thou never an eye in thy head? Canst not hear? An
'twere not as good deed as drink to break the pate on
thee, I am a very villain. Come, and be hanged! Hast
no faith in thee?    31
*Enter Gadshill*
GADSHILL Good morrow, carriers. What's o'clock?
FIRST CARRIER I think it be two o'clock.
GADSHILL I prithee lend me thy lantern to see my gelding
in the stable.    35
FIRST CARRIER Nay, by God, soft. I know a trick worth
two of that, i'faith.
GADSHILL (*to Second Carrier*) I pray thee, lend me thine.
SECOND CARRIER Ay, when? Canst tell? 'Lend me thy
lantern,' quoth a. Marry, I'll see thee hanged first.    40
GADSHILL Sirrah carrier, what time do you mean to come
to London?
SECOND CARRIER Time enough to go to bed with a candle,
I warrant thee.—Come, neighbour Mugs, we'll call up
the gentlemen. They will along with company, for they
have great charge.    *Exeunt Carriers*
GADSHILL What ho, chamberlain!    47
*Enter Chamberlain*
CHAMBERLAIN 'At hand' quoth Pickpurse.
GADSHILL That's even as fair as ' "At hand" quoth the
chamberlain', for thou variest no more from picking of
purses than giving direction doth from labouring: thou
layest the plot how.    52
CHAMBERLAIN Good morrow, Master Gadshill. It holds
current that I told you yesternight. There's a franklin
in the Weald of Kent hath brought three hundred
marks with him in gold. I heard him tell it to one of
his company last night at supper—a kind of auditor,
one that hath abundance of charge too, God knows
what. They are up already, and call for eggs and butter;
they will away presently.    60
GADSHILL Sirrah, if they meet not with Saint Nicholas's
clerks, I'll give thee this neck.
CHAMBERLAIN No, I'll none of it; I pray thee keep that
for the hangman, for I know thou worshippest Saint
Nicholas as truly as a man of falsehood may.    65
GADSHILL What talkest thou to me of the hangman? If I
hang, I'll make a fat pair of gallows, for if I hang, old
Sir John hangs with me, and thou knowest he's no
starveling. Tut, there are other Trojans that thou
dreamest not of, the which for sport' sake are content
to do the profession some grace, that would, if matters
should be looked into, for their own credit' sake make
all whole. I am joined with no foot-landrakers, no long-
staff sixpenny strikers, none of these mad mustachio
purple-hued maltworms, but with nobility and tran-
quillity, burgomasters and great 'oyez'-ers; such as can

hold in, such as will strike sooner than speak, and speak sooner than drink, and drink sooner than pray. And yet, zounds, I lie, for they pray continually to their saint the commonwealth; or rather, not pray to her, but prey on her; for they ride up and down on her and make her their boots.

CHAMBERLAIN What, the commonwealth their boots? Will she hold out water in foul way?                         84

GADSHILL She will, she will, justice hath liquored her. We steal as in a castle, cocksure; we have the recipe of fern-seed, we walk invisible.

CHAMBERLAIN Nay, by my faith, I think you are more beholden to the night than to fern-seed for your walking invisible.                                                   90

GADSHILL Give me thy hand; thou shalt have a share in our purchase, as I am a true man.

CHAMBERLAIN Nay, rather let me have it as you are a false thief.

GADSHILL Go to, 'homo' is a common name to all men. Bid the ostler bring my gelding out of the stable. Farewell, you muddy knave.          *Exeunt ⌈severally⌉*

**2.2**  *Enter Prince Harry, Poins, Harvey, ⌈and Russell⌉*
POINS Come, shelter, shelter!
          *⌈Exeunt Harvey and Russell at another door⌉*
I have removed Oldcastle's horse, and he frets like a gummed velvet.
PRINCE HARRY Stand close!                        *⌈Exit Poins⌉*
     *Enter Sir John Oldcastle*
SIR JOHN Poins! Poins, and be hanged! Poins!        5
PRINCE HARRY Peace, ye fat-kidneyed rascal! What a brawling dost thou keep!
SIR JOHN Where's Poins, Hal?
PRINCE HARRY He is walked up to the top of the hill. I'll go seek him.                                      *⌈Exit⌉*
SIR JOHN I am accursed to rob in that thief's company. The rascal hath removed my horse and tied him I know not where. If I travel but four foot by the square further afoot, I shall break my wind. Well, I doubt not but to die a fair death, for all this—if I scape hanging for killing that rogue. I have forsworn his company hourly any time this two-and-twenty years, and yet I am bewitched with the rogue's company. If the rascal have not given me medicines to make me love him, I'll be hanged. It could not be else: I have drunk medicines. Poins! Hal! A plague upon you both! Russell! Harvey! I'll starve ere I'll rob a foot further. An 'twere not as good a deed as drink to turn true man and to leave these rogues, I am the veriest varlet that ever chewed with a tooth. Eight yards of uneven ground is threescore and ten miles afoot with me, and the stony-hearted villains know it well enough. A plague upon't when thieves cannot be true one to another!             28
     *They whistle. ⌈Enter Prince Harry, Poins, Harvey, and Russell⌉*
Whew! A plague upon you all! Give me my horse, you rogues, give me my horse, and be hanged!         30
PRINCE HARRY Peace, ye fat-guts. Lie down, lay thine ear close to the ground, and list if thou canst hear the tread of travellers.
SIR JOHN Have you any levers to lift me up again, being down? 'Sblood, I'll not bear my own flesh so far afoot again for all the coin in thy father's exchequer. What a plague mean ye to colt me thus?              37
PRINCE HARRY Thou liest: thou art not colted, thou art uncolted.

SIR JOHN I prithee, good Prince Hal, help me to my horse, good king's son.                                     41
PRINCE HARRY Out, ye rogue, shall I be your ostler?
SIR JOHN Hang thyself in thine own heir-apparent garters! If I be ta'en, I'll peach for this. An I have not ballads made on you all and sung to filthy tunes, let a cup of sack be my poison. When a jest is so forward, and afoot too! I hate it.                                       47
     *Enter Gadshill ⌈visored⌉*
GADSHILL Stand!
SIR JOHN So I do, against my will.
POINS O, 'tis our setter, I know his voice. Gadshill, what news?                                                      51
⌈GADSHILL⌉ Case ye, case ye, on with your visors! There's money of the King's coming down the hill; 'tis going to the King's exchequer.
SIR JOHN You lie, ye rogue, 'tis going to the King's tavern.
GADSHILL There's enough to make us all.          56
SIR JOHN To be hanged.
     *⌈They put on visors⌉*
PRINCE HARRY Sirs, you four shall front them in the narrow lane. Ned Poins and I will walk lower. If they scape from your encounter, then they light on us.  60
HARVEY How many be there of them?
GADSHILL Some eight or ten.
SIR JOHN Zounds, will they not rob us?
PRINCE HARRY What, a coward, Sir John Paunch?
SIR JOHN Indeed I am not John of Gaunt your grandfather, but yet no coward, Hal.                           66
PRINCE HARRY Well, we leave that to the proof.
POINS Sirrah Jack, thy horse stands behind the hedge. When thou needest him, there thou shalt find him. Farewell, and stand fast.                              70
SIR JOHN Now cannot I strike him if I should be hanged.
PRINCE HARRY *(aside to Poins)* Ned, where are our disguises?
POINS *(aside to the Prince)* Here, hard by. Stand close.
     *⌈Exeunt the Prince and Poins⌉*
SIR JOHN Now, my masters, happy man be his dole, say I; every man to his business.                         75
     *⌈They stand aside.⌉*
     *Enter the Travellers, ⌈amongst them the Carriers⌉*
⌈FIRST⌉ TRAVELLER Come, neighbour, the boy shall lead our horses down the hill. We'll walk afoot a while, and ease their legs.
THIEVES ⌈coming forward⌉ Stand!
⌈SECOND⌉ TRAVELLER Jesus bless us!               80
SIR JOHN Strike, down with them, cut the villains' throats! Ah, whoreson caterpillars, bacon-fed knaves! They hate us youth. Down with them, fleece them!
⌈FIRST⌉ TRAVELLER O, we are undone, both we and ours for ever!                                             85
SIR JOHN Hang ye, gorbellied knaves, are ye undone? No, ye fat chuffs; I would your store were here. On, bacons, on! What, ye knaves! Young men must live. You are grand-jurors, are ye? We'll jure ye, faith.
     *Here they rob them and bind them. Exeunt the thieves with the travellers*

**2.3**  *Enter Prince Harry and Poins, disguised in buckram suits*
PRINCE HARRY The thieves have bound the true men; now could thou and I rob the thieves, and go merrily to London. It would be argument for a week, laughter for a month, and a good jest for ever.
POINS Stand close; I hear them coming.             5
     *They stand aside.*

*Enter Sir John Oldcastle, Russell, Harvey, and*
*Gadshill, with the travellers' money*

SIR JOHN Come, my masters, let us share, and then to
horse before day. An the Prince and Poins be not two
arrant cowards, there's no equity stirring. There's no
more valour in that Poins than in a wild duck.

*As they are sharing, the Prince and Poins set upon*
*them*

PRINCE HARRY Your money!                                    10
POINS Villains!

*Gadshill, Russell, and Harvey run away ⌐severally⌐,*
*and Oldcastle, after a blow or two, ⌐roars and⌐*
*runs away too, leaving the booty behind them*

PRINCE HARRY
Got with much ease. Now merrily to horse.
The thieves are all scattered, and possessed with fear
So strongly that they dare not meet each other.
Each takes his fellow for an officer.                       15
Away, good Ned. Oldcastle sweats to death,
And lards the lean earth as he walks along.
Were't not for laughing, I should pity him.

POINS
How the fat rogue roared!          *Exeunt with the booty*

**2.4**    *Enter Hotspur, reading a letter*

HOTSPUR 'But for mine own part, my lord, I could be well
contented to be there, in respect of the love I bear your
house.'—He could be contented; why is he not then?
In respect of the love he bears our house! He shows in
this he loves his own barn better than he loves our
house. Let me see some more.—'The purpose you
undertake is dangerous'—Why, that's certain: 'tis
dangerous to take a cold, to sleep, to drink; but I tell
you, my lord fool, out of this nettle danger we pluck
this flower safety.—'The purpose you undertake is
dangerous, the friends you have named uncertain, the
time itself unsorted, and your whole plot too light for
the counterpoise of so great an opposition.'—Say you
so, sáy you so? I say unto you again, you are a shallow,
cowardly hind, and you lie. What a lack-brain is this!
By the Lord, our plot is a good plot as ever was laid,
our friends true and constant; a good plot, good friends,
and full of expectation; an excellent plot, very good
friends. What a frosty-spirited rogue is this! Why, my
lord of York commends the plot and the general course
of the action. Zounds, an I were now by this rascal, I
could brain him with his lady's fan! Is there not my
father, my uncle, and myself? Lord Edmund Mortimer,
my lord of York, and Owain Glyndŵr? Is there not
besides the Douglas? Have I not all their letters, to
meet me in arms by the ninth of the next month? And
are they not some of them set forward already? What
a pagan rascal is this, an infidel! Ha, you shall see
now, in very sincerity of fear and cold heart will he to
the King, and lay open all our proceedings! O, I could
divide myself and go to buffets for moving such a dish
of skim-milk with so honourable an action! Hang him!
Let him tell the King we are prepared; I will set forward
tonight.                                                    34

*Enter Lady Percy*
How now, Kate? I must leave you within these two
hours.

LADY PERCY
O my good lord, why are you thus alone?
For what offence have I this fortnight been
A banished woman from my Harry's bed?

Tell me, sweet lord, what is't that takes from thee    40
Thy stomach, pleasure, and thy golden sleep?
Why dost thou bend thine eyes upon the earth,
And start so often when thou sitt'st alone?
Why hast thou lost the fresh blood in thy cheeks,
And given my treasures and my rights of thee          45
To thick-eyed musing and curst melancholy?
In thy faint slumbers I by thee have watched,
And heard thee murmur tales of iron wars,
Speak terms of manège to thy bounding steed,
Cry 'Courage! To the field!' And thou hast talked     50
Of sallies and retires, of trenches, tents,
Of palisadoes, frontiers, parapets,
Of basilisks, of cannon, culverin,
Of prisoners ransomed, and of soldiers slain,
And all the currents of a heady fight.                 55
Thy spirit within thee hath been so at war,
And thus hath so bestirred thee in thy sleep,
That beads of sweat have stood upon thy brow
Like bubbles in a late-disturbèd stream;
And in thy face strange motions have appeared,        60
Such as we see when men restrain their breath
On some great sudden hest. O, what portents are
      these?
Some heavy business hath my lord in hand,
And I must know it, else he loves me not.

HOTSPUR
What ho!
      *Enter Servant*
                  Is Gilliams with the packet gone?    65

SERVANT
He is, my lord, an hour ago.

HOTSPUR
Hath Butler brought those horses from the sheriff?

SERVANT
One horse, my lord, he brought even now.

HOTSPUR
What horse? A roan, a crop-ear, is it not?

SERVANT
It is, my lord.

HOTSPUR            That roan shall be my throne.       70
Well, I will back him straight.—O, *Esperance*!—
Bid Butler lead him forth into the park.

LADY PERCY
But hear you, my lord.

HOTSPUR               What sayst thou, my lady?

LADY PERCY
What is it carries you away?

HOTSPUR                        Why, my horse,
My love, my horse.

LADY PERCY            Out, you mad-headed ape!          75
A weasel hath not such a deal of spleen
As you are tossed with. In faith, I'll know your business, Harry, that I will.
I fear my brother Mortimer doth stir
About his title, and hath sent for you                 80
To line his enterprise; but if you go—

HOTSPUR
So far afoot? I shall be weary, love.

LADY PERCY
Come, come, you paraquito, answer me
Directly to this question that I ask.
In faith, I'll break thy little finger, Harry,         85
An if thou wilt not tell me all things true.

HOTSPUR
Away, away, you trifler! Love? I love thee not,
I care not for thee, Kate. This is no world

To play with maumets and to tilt with lips.
We must have bloody noses and cracked crowns,    90
And pass them current, too. God's me, my horse!—
What sayst thou, Kate? What wouldst thou have
    with me?

LADY PERCY
Do you not love me? Do you not indeed?
Well, do not, then, for since you love me not
I will not love myself. Do you not love me?    95
Nay, tell me if you speak in jest or no.

HOTSPUR Come, wilt thou see me ride?
And when I am a-horseback, I will swear
I love thee infinitely. But hark you, Kate.
I must not have you henceforth question me    100
Whither I go, nor reason whereabout.
Whither I must, I must; and, to conclude,
This evening must I leave you, gentle Kate.
I know you wise, but yet no farther wise
Than Harry Percy's wife; constant you are,    105
But yet a woman; and for secrecy
No lady closer, for I well believe
Thou wilt not utter what thou dost not know.
And so far will I trust thee, gentle Kate.

LADY PERCY How, so far?    110

HOTSPUR
Not an inch further. But hark you, Kate,
Whither I go, thither shall you go too.
Today will I set forth, tomorrow you.
Will this content you, Kate?

LADY PERCY            It must, of force.    *Exeunt*

**2.5**    *Enter Prince Harry*

PRINCE HARRY Ned, prithee come out of that fat room,
and lend me thy hand to laugh a little.
     *Enter Poins ⌈at another door⌉*

POINS Where hast been, Hal?

PRINCE HARRY With three or four loggerheads, amongst
three or fourscore hogsheads. I have sounded the very
bass-string of humility. Sirrah, I am sworn brother to
a leash of drawers, and can call them all by their
christen names, as 'Tom', 'Dick', and 'Francis'. They
take it already, upon their salvation, that though I be
but Prince of Wales yet I am the king of courtesy, and
tell me flatly I am no proud jack like Oldcastle, but a
Corinthian, a lad of mettle, a good boy—by the Lord,
so they call me; and when I am King of England I shall
command all the good lads in Eastcheap. They call
drinking deep 'dyeing scarlet', and when you breathe
in your watering they cry 'Hem!' and bid you 'Play it
off!' To conclude, I am so good a proficient in one
quarter of an hour that I can drink with any tinker in
his own language during my life. I tell thee, Ned, thou
hast lost much honour that thou wert not with me in
this action. But, sweet Ned—to sweeten which name
of Ned I give thee this pennyworth of sugar, clapped
even now into my hand by an underskinker, one that
never spake other English in his life than 'Eight shillings
and sixpence', and 'You are welcome', with this shrill
addition, 'Anon, anon, sir! Score a pint of bastard in
the Half-moon!' or so. But, Ned, to drive away the time
till Oldcastle come, I prithee do thou stand in some
by-room, while I question my puny drawer to what
end he gave me the sugar, and do thou never leave
calling 'Francis!', that his tale to me may be nothing
but 'Anon!' Step aside, and I'll show thee a precedent.
     *Exit Poins*

POINS (*within*) Francis!

PRINCE HARRY Thou art perfect.

POINS (*within*) Francis!    35
     *Enter Francis, a drawer*

FRANCIS Anon, anon, sir!—Look down into the Pomegranate, Ralph!

PRINCE HARRY Come hither, Francis.

FRANCIS My lord.

PRINCE HARRY How long hast thou to serve, Francis?    40

FRANCIS Forsooth, five years, and as much as to—

POINS (*within*) Francis!

FRANCIS Anon, anon, sir!

PRINCE HARRY Five year! By'r Lady, a long lease for the
clinking of pewter. But Francis, darest thou be so
valiant as to play the coward with thy indenture, and
show it a fair pair of heels, and run from it?    47

FRANCIS O Lord, sir, I'll be sworn upon all the books in
England, I could find in my heart—

POINS (*within*) Francis!    50

FRANCIS Anon, sir!

PRINCE HARRY How old art thou, Francis?

FRANCIS Let me see, about Michaelmas next I shall be—

POINS (*within*) Francis!

FRANCIS Anon, sir! (*To the Prince*) Pray, stay a little, my
lord.    56

PRINCE HARRY Nay, but hark you, Francis. For the sugar
thou gavest me, 'twas a pennyworth, was't not?

FRANCIS O Lord, I would it had been two!

PRINCE HARRY I will give thee for it a thousand pound.
Ask me when thou wilt, and thou shalt have it—    61

POINS (*within*) Francis!

FRANCIS Anon, anon!

PRINCE HARRY Anon, Francis? No, Francis, but tomorrow,
Francis; or, Francis, o' Thursday; or, indeed, Francis,
when thou wilt. But Francis.    66

FRANCIS My lord.

PRINCE HARRY Wilt thou rob this leathern-jerkin, crystal-
button, knot-pated, agate-ring, puke-stocking, caddis-
garter, smooth-tongue, Spanish-pouch?    70

FRANCIS O Lord, sir, who do you mean?

PRINCE HARRY Why, then, your brown bastard is your
only drink! For look you, Francis, your white canvas
doublet will sully. In Barbary, sir, it cannot come to so
much.    75

FRANCIS What, sir?

POINS (*within*) Francis!

PRINCE HARRY Away, you rogue! Dost thou not hear them
call?
     ⌈*As he departs*⌉ *Poins and the Prince both call him.*
     *The Drawer stands amazed, not knowing which*
     *way to go.*
     *Enter Vintner*

VINTNER What, standest thou still, and hearest such a
calling? Look to the guests within.    *Exit Francis*
My lord, old Sir John with half a dozen more are at
the door. Shall I let them in?    83

PRINCE HARRY Let them alone a while, and then open the
door.    *Exit Vintner*
Poins!

POINS ⌈*within*⌉ Anon, anon, sir!
     *Enter Poins*

PRINCE HARRY Sirrah, Oldcastle and the rest of the thieves
are at the door. Shall we be merry?    89

POINS As merry as crickets, my lad. But hark ye, what
cunning match have you made with this jest of the
drawer? Come, what's the issue?

PRINCE HARRY I am now of all humours that have showed themselves humours since the old days of goodman Adam to the pupil age of this present twelve o'clock at midnight.                                                                    96

*[Enter Francis]*

What's o'clock, Francis?

FRANCIS Anon, anon, sir!                    *[Exit at another door]*

PRINCE HARRY That ever this fellow should have fewer words than a parrot, and yet the son of a woman! His industry is upstairs and downstairs, his eloquence the parcel of a reckoning. I am not yet of Percy's mind, the Hotspur of the North—he that kills me some six or seven dozen of Scots at a breakfast, washes his hands, and says to his wife, 'Fie upon this quiet life! I want work.' 'O my sweet Harry,' says she, 'how many hast thou killed today?' Give my roan horse a drench,' says he, and answers, 'Some fourteen,' an hour after; 'a trifle, a trifle.' I prithee call in Oldcastle. I'll play Percy, and that damned brawn shall play Dame Mortimer his wife. 'Rivo!' says the drunkard. Call in Ribs, call in Tallow.                                                                    112

*Enter Sir John Oldcastle, with sword and buckler, Russell, Harvey, and Gadshill, [followed by] Francis, with wine*

POINS Welcome, Jack. Where hast thou been?

SIR JOHN A plague of all cowards, I say, and a vengeance too, marry and amen!—Give me a cup of sack, boy.— Ere I lead this life long, I'll sew netherstocks and mend them and foot them too. A plague of all cowards!— Give me a cup of sack, rogue. Is there no virtue extant?

*He drinketh*

PRINCE HARRY Didst thou never see Titan kiss a dish of butter—pitiful hearted Titan—that melted at the sweet tale of the sun's? If thou didst, then behold that compound.                                                                    122

SIR JOHN *(to Francis)* You rogue, here's lime in this sack too. There is nothing but roguery to be found in villainous man, yet a coward is worse than a cup of sack with lime in it.                    *[Exit Francis]*

A villainous coward! Go thy ways, old Jack, die when thou wilt. If manhood, good manhood, be not forgot upon the face of the earth, then am I a shotten herring. There lives not three good men unhanged in England, and one of them is fat and grows old, God help the while. A bad world, I say. I would I were a weaver—I could sing psalms, or anything. A plague of all cowards, I say still.                                                                    134

PRINCE HARRY How now, woolsack, what mutter you?

SIR JOHN A king's son! If I do not beat thee out of thy kingdom with a dagger of lath, and drive all thy subjects afore thee like a flock of wild geese, I'll never wear hair on my face more. You, Prince of Wales!                    139

PRINCE HARRY Why, you whoreson round man, what's the matter?

SIR JOHN Are not you a coward? Answer me to that. And Poins there?

POINS Zounds, ye fat paunch, an ye call me coward, by the Lord I'll stab thee.                                                                    145

SIR JOHN I call thee coward? I'll see thee damned ere I call thee coward, but I would give a thousand pound I could run as fast as thou canst. You are straight enough in the shoulders; you care not who sees your back. Call you that backing of your friends? A plague upon such backing! Give me them that will face me. Give me a cup of sack. I am a rogue if I drunk today.

PRINCE HARRY O villain, thy lips are scarce wiped since thou drunkest last.

SIR JOHN All is one for that.                                                                    155

*He drinketh*

A plague of all cowards, still say I.

PRINCE HARRY What's the matter?

SIR JOHN What's the matter? There be four of us here have ta'en a thousand pound this day morning.

PRINCE HARRY Where is it, Jack, where is it?                    160

SIR JOHN Where is it? Taken from us it is. A hundred upon poor four of us.

PRINCE HARRY What, a hundred, man?

SIR JOHN I am a rogue if I were not at half-sword with a dozen of them, two hours together. I have scaped by miracle. I am eight times thrust through the doublet, four through the hose, my buckler cut through and through, my sword hacked like a handsaw. *Ecce signum.*

*[He shows his sword]*

I never dealt better since I was a man. All would not do. A plague of all cowards! *(Pointing to Gadshill, Harvey, and Russell)* Let them speak. If they speak more or less than truth, they are villains and the sons of darkness.

*[*PRINCE HARRY*]* Speak, sirs, how was it?

*[*GADSHILL*]* We four set upon some dozen—                    175

SIR JOHN *(to the Prince)* Sixteen at least, my lord.

*[*GADSHILL*]* And bound them.

HARVEY No, no, they were not bound.

SIR JOHN You rogue, they were bound every man of them, or I am a Jew else, an Hebrew Jew.                    180

*[*GADSHILL*]* As we were sharing, some six or seven fresh men set upon us.

SIR JOHN And unbound the rest; and then come in the other.

PRINCE HARRY What, fought you with them all?                    185

SIR JOHN All? I know not what you call all, but if I fought not with fifty of them, I am a bunch of radish. If there were not two- or three-and-fifty upon poor old Jack, then am I no two-legged creature.

PRINCE HARRY Pray God you have not murdered some of them.                                                                    191

SIR JOHN Nay, that's past praying for. I have peppered two of them. Two I am sure I have paid—two rogues in buckram suits. I tell thee what, Hal, if I tell thee a lie, spit in my face, call me horse. Thou knowest my old ward—                                                                    196

*[He stands as to fight]*

here I lay, and thus I bore my point. Four rogues in buckram let drive at me.

PRINCE HARRY What, four? Thou saidst but two even now.                                                                    200

SIR JOHN Four, Hal, I told thee four.

POINS Ay, ay, he said four.

SIR JOHN These four came all afront, and mainly thrust at me. I made me no more ado, but took all their seven points in my target, thus.                                                                    205

*[He wards himself with his buckler]*

PRINCE HARRY Seven? Why, there were but four even now.

SIR JOHN In buckram?

POINS Ay, four in buckram suits.

SIR JOHN Seven, by these hilts, or I am a villain else.

PRINCE HARRY *(aside to Poins)* Prithee, let him alone. We shall have more anon.                                                                    212

SIR JOHN Dost thou hear me, Hal?

PRINCE HARRY Ay, and mark thee too, Jack.

SIR JOHN Do so, for it is worth the listening to. These nine in buckram that I told thee of— 216

PRINCE HARRY (aside to Poins) So, two more already.

SIR JOHN Their points being broken—

POINS ⌈aside to the Prince⌉ Down fell their hose.

SIR JOHN Began to give me ground. But I followed me close, came in foot and hand, and, with a thought, seven of the eleven I paid. 222

PRINCE HARRY (aside to Poins) O monstrous! Eleven buckram men grown out of two!

SIR JOHN But, as the devil would have it, three misbegotten knaves in Kendal green came at my back and let drive at me; for it was so dark, Hal, that thou couldst not see thy hand. 228

PRINCE HARRY These lies are like their father that begets them—gross as a mountain, open, palpable. Why, thou clay-brained guts, thou knotty-pated fool, thou whoreson obscene greasy tallow-catch—

SIR JOHN What, art thou mad? Art thou mad? Is not the truth the truth? 234

PRINCE HARRY Why, how couldst thou know these men in Kendal green when it was so dark thou couldst not see thy hand? Come, tell us your reason. What sayst thou to this?

POINS Come, your reason, Jack, your reason. 239

SIR JOHN What, upon compulsion? Zounds, an I were at the strappado, or all the racks in the world, I would not tell you on compulsion. Give you a reason on compulsion? If reasons were as plentiful as blackberries, I would give no man a reason upon compulsion, I.

PRINCE HARRY I'll be no longer guilty of this sin. This sanguine coward, this bed-presser, this horse-back-breaker, this huge hill of flesh— 247

SIR JOHN 'Sblood, you starveling, you elf-skin, you dried neat's tongue, you bull's pizzle, you stock-fish—O, for breath to utter what is like thee!—you tailor's yard, you sheath, you bow-case, you vile standing tuck—

PRINCE HARRY Well, breathe awhile, and then to't again, and when thou hast tired thyself in base comparisons, hear me speak but this.

POINS Mark, Jack. 255

PRINCE HARRY We two saw you four set on four, and bound them, and were masters of their wealth.—Mark now how a plain tale shall put you down.—Then did we two set on you four, and, with a word, outfaced you from your prize, and have it; yea, and can show it you here in the house. And Oldcastle, you carried your guts away as nimbly, with as quick dexterity, and roared for mercy, and still run and roared, as ever I heard bull-calf. What a slave art thou, to hack thy sword as thou hast done, and then say it was in fight! What trick, what device, what starting-hole canst thou now find out to hide thee from this open and apparent shame? 268

POINS Come, let's hear, Jack; what trick hast thou now?

SIR JOHN By the Lord, I knew ye as well as he that made ye. Why, hear you, my masters. Was it for me to kill the heir-apparent? Should I turn upon the true prince? Why, thou knowest I am as valiant as Hercules; but beware instinct. The lion will not touch the true prince—instinct is a great matter. I was now a coward on instinct. I shall think the better of myself and thee during my life—I for a valiant lion, and thou for a true prince. But by the Lord, lads, I am glad you have the money.—(Calling) Hostess, clap to the doors.—Watch tonight, pray tomorrow. Gallants, lads, boys, hearts of

gold, all the titles of good fellowship come to you! What, shall we be merry, shall we have a play extempore?

PRINCE HARRY Content, and the argument shall be thy running away. 285

SIR JOHN Ah, no more of that, Hal, an thou lovest me.

Enter Hostess

HOSTESS O Jesu, my lord the Prince!

PRINCE HARRY How now, my lady the Hostess, what sayst thou to me? 289

HOSTESS Marry, my lord, there is a nobleman of the court at door would speak with you. He says he comes from your father.

PRINCE HARRY Give him as much as will make him a royal man, and send him back again to my mother.

SIR JOHN What manner of man is he? 295

HOSTESS An old man.

SIR JOHN What doth gravity out of his bed at midnight? Shall I give him his answer?

PRINCE HARRY Prithee do, Jack.

SIR JOHN Faith, and I'll send him packing. Exit

PRINCE HARRY Now, sirs; (to Gadshill) by'r Lady, you fought fair—so did you, Harvey, so did you, Russell. You are lions too—you ran away upon instinct, you will not touch the true prince; no, fie!

RUSSELL Faith, I ran when I saw others run. 305

PRINCE HARRY Faith, tell me now in earnest, how came Oldcastle's sword so hacked?

HARVEY Why, he hacked it with his dagger, and said he would swear truth out of England but he would make you believe it was done in fight, and persuaded us to do the like. 311

RUSSELL Yea, and to tickle our noses with speargrass, to make them bleed; and then to beslubber our garments with it, and swear it was the blood of true men. I did that I did not this seven year before—I blushed to hear his monstrous devices. 316

PRINCE HARRY O villain, thou stolest a cup of sack eighteen years ago, and wert taken with the manner, and ever since thou hast blushed extempore. Thou hadst fire and sword on thy side, and yet thou rannest away. What instinct hadst thou for it? 321

RUSSELL (indicating his face) My lord, do you see these meteors? Do you behold these exhalations?

PRINCE HARRY I do.

RUSSELL What think you they portend? 325

PRINCE HARRY Hot livers, and cold purses.

RUSSELL Choler, my lord, if rightly taken. ⌈Exit⌉

PRINCE HARRY No, if rightly taken, halter.

Enter Sir John Oldcastle

Here comes lean Jack; here comes bare-bone. How now, my sweet creature of bombast? How long is't ago, Jack, since thou sawest thine own knee? 331

SIR JOHN My own knee? When I was about thy years, Hal, I was not an eagle's talon in the waist; I could have crept into any alderman's thumb-ring. A plague of sighing and grief—it blows a man up like a bladder. There's villainous news abroad. Here was Sir John Bracy from your father; you must to the court in the morning. That same mad fellow of the North, Percy, and he of Wales that gave Amamon the bastinado, and made Lucifer cuckold, and swore the devil his true liegeman upon the cross of a Welsh hook—what a plague call you him? 342

POINS Owain Glyndŵr.

SIR JOHN Owain, Owain, the same; and his son-in-law
Mortimer, and old Northumberland, and that sprightly
Scot of Scots Douglas, that runs a-horseback up a hill
perpendicular—                                             347
PRINCE HARRY He that rides at high speed and with his
pistol kills a sparrow flying.
SIR JOHN You have hit it.                                  350
PRINCE HARRY So did he never the sparrow.
SIR JOHN Well, that rascal hath good mettle in him; he
will not run.
PRINCE HARRY Why, what a rascal art thou, then, to
praise him so for running!                                 355
SIR JOHN A-horseback, ye cuckoo, but afoot he will not
budge a foot.
PRINCE HARRY Yes, Jack, upon instinct.
SIR JOHN I grant ye, upon instinct. Well, he is there too,
and one Mordake, and a thousand blue-caps more.
Worcester is stolen away tonight. Thy father's beard is
turned white with the news. You may buy land now
as cheap as stinking mackerel.                             363
PRINCE HARRY Why then, it is like, if there come a hot
June and this civil buffeting hold, we shall buy
maidenheads as they buy hobnails: by the hundreds.
SIR JOHN By the mass, lad, thou sayst true; it is like we
shall have good trading that way. But tell me, Hal, art
not thou horrible afeard? Thou being heir-apparent,
could the world pick thee out three such enemies again
as that fiend Douglas, that spirit Percy, and that devil
Glyndŵr? Art thou not horribly afraid? Doth not thy
blood thrill at it?
PRINCE HARRY Not a whit, i'faith. I lack some of thy
instinct.                                                  375
SIR JOHN Well, thou wilt be horribly chid tomorrow when
thou comest to thy father. If thou love me, practise an
answer.
PRINCE HARRY Do thou stand for my father, and examine
me upon the particulars of my life.                        380
SIR JOHN Shall I? Content. This chair shall be my state,
this dagger my sceptre, and this cushion my crown.
        *He sits*
PRINCE HARRY Thy state is taken for a joint-stool, thy
golden sceptre for a leaden dagger, and thy precious
rich crown for a pitiful bald crown.                       385
SIR JOHN Well, an the fire of grace be not quite out of
thee, now shalt thou be moved. Give me a cup of sack
to make my eyes look red, that it may be thought I
have wept; for I must speak in passion, and I will do
it in King Cambyses' vein.                                 390
PRINCE HARRY (*bowing*) Well, here is my leg.
SIR JOHN And here is my speech. (*To Harvey, Poins, and
Gadshill*) Stand aside, nobility.
HOSTESS O Jesu, this is excellent sport, i'faith.          394
SIR JOHN
    Weep not, sweet Queen, for trickling tears are vain.
HOSTESS O the Father, how he holds his countenance!
SIR JOHN
    For God's sake, lords, convey my tristful Queen,
    For tears do stop the floodgates of her eyes.
HOSTESS O Jesu, he doth it as like one of these harlotry
players as ever I see!                                     400
SIR JOHN
    Peace, good pint-pot; peace, good tickle-brain.—
Harry, I do not only marvel where thou spendest thy
time, but also how thou art accompanied. For though
the camomile, the more it is trodden on, the faster it
grows, yet youth, the more it is wasted, the sooner it

wears. That thou art my son I have partly thy mother's
word, partly my own opinion, but chiefly a villainous
trick of thine eye, and a foolish hanging of thy nether
lip, that doth warrant me. If then thou be son to me,
here lies the point. Why, being son to me, art thou so
pointed at? Shall the blessed sun of heaven prove a
micher, and eat blackberries?—A question not to be
asked. Shall the son of England prove a thief, and take
purses?—A question to be asked. There is a thing,
Harry, which thou hast often heard of, and it is known
to many in our land by the name of pitch. This pitch,
as ancient writers do report, doth defile. So doth the
company thou keepest. For Harry, now I do not speak
to thee in drink, but in tears; not in pleasure, but in
passion; not in words only, but in woes also. And yet
there is a virtuous man whom I have often noted in
thy company, but I know not his name.                     422
PRINCE HARRY What manner of man, an it like your
majesty?
SIR JOHN A goodly, portly man, i'faith, and a corpulent;
of a cheerful look, a pleasing eye, and a most noble
carriage; and, as I think, his age some fifty, or, by'r
Lady, inclining to threescore. And now I remember me,
his name is Oldcastle. If that man should be lewdly
given, he deceiveth me; for, Harry, I see virtue in his
looks. If, then, the tree may be known by the fruit, as
the fruit by the tree, then peremptorily I speak it—
there is virtue in that Oldcastle. Him keep with; the
rest banish. And tell me now, thou naughty varlet, tell
me, where hast thou been this month?                      435
PRINCE HARRY Dost thou speak like a king? Do thou stand
for me, and I'll play my father.
SIR JOHN (*standing*) Depose me. If thou dost it half so
gravely, so majestically both in word and matter, hang
me up by the heels for a rabbit sucker, or a poulter's
hare.                                                     441
PRINCE HARRY (*sitting*) Well, here I am set.
SIR JOHN And here I stand. (*To the others*) Judge, my
masters.
PRINCE HARRY Now, Harry, whence come you?                 445
SIR JOHN My noble lord, from Eastcheap.
PRINCE HARRY The complaints I hear of thee are grievous.
SIR JOHN 'Sblood, my lord, they are false. [*To the others*]
Nay, I'll tickle ye for a young prince, i'faith.
PRINCE HARRY Swearest thou, ungracious boy? Hence-
forth ne'er look on me. Thou art violently carried away
from grace. There is a devil haunts thee in the likeness
of an old fat man; a tun of man is thy companion.
Why dost thou converse with that trunk of humours,
that bolting-hutch of beastliness, that swollen parcel of
dropsies, that huge bombard of sack, that stuffed cloak-
bag of guts, that roasted Manningtree ox with the
pudding in his belly, that reverend Vice, that grey
Iniquity, that father Ruffian, that Vanity in Years?
Wherein is he good, but to taste sack and drink it?
Wherein neat and cleanly, but to carve a capon and
eat it? Wherein cunning, but in craft? Wherein crafty,
but in villainy? Wherein villainous, but in all things?
Wherein worthy, but in nothing?                           464
SIR JOHN I would your grace would take me with you.
Whom means your grace?
PRINCE HARRY That villainous, abominable misleader of
youth, Oldcastle; that old white-bearded Satan.
SIR JOHN My lord, the man I know.
PRINCE HARRY I know thou dost.                            470

SIR JOHN But to say I know more harm in him than in myself were to say more than I know. That he is old, the more the pity, his white hairs do witness it. But that he is, saving your reverence, a whoremaster, that I utterly deny. If sack and sugar be a fault, God help the wicked. If to be old and merry be a sin, then many an old host that I know is damned. If to be fat be to be hated, then Pharaoh's lean kine are to be loved. No, my good lord, banish Harvey, banish Russell, banish Poins, but for sweet Jack Oldcastle, kind Jack Oldcastle, true Jack Oldcastle, and therefore more valiant being, as he is, old Jack Oldcastle,     482
Banish not him thy Harry's company,
Banish not him thy Harry's company.
Banish plump Jack, and banish all the world.     485
PRINCE HARRY I do; I will.
          *Knocking within.* ⌈*Exit Hostess.*⌉
          *Enter Russell, running*
RUSSELL O my lord, my lord, the sheriff with a most monstrous watch is at the door.
SIR JOHN Out, ye rogue! Play out the play! I have much to say in the behalf of that Oldcastle.     490
          *Enter the Hostess*
HOSTESS O Jesu! My lord, my lord!
PRINCE HARRY Heigh, heigh, the devil rides upon a fiddlestick! What's the matter?
HOSTESS The sheriff and all the watch are at the door. They are come to search the house. Shall I let them in?     496
SIR JOHN Dost thou hear, Hal? Never call a true piece of gold a counterfeit—thou art essentially made, without seeming so.
PRINCE HARRY And thou a natural coward without instinct.     501
SIR JOHN I deny your major. If you will deny the sheriff, so. If not, let him enter. If I become not a cart as well as another man, a plague on my bringing up. I hope I shall as soon be strangled with a halter as another.
PRINCE HARRY Go, hide thee behind the arras. The rest walk up above. Now, my masters, for a true face and good conscience.     *Exeunt Poins, Russell, and Gadshill*
SIR JOHN Both which I have had, but their date is out; and therefore I'll hide me.     510
          *He withdraws behind the arras*
PRINCE HARRY (*to Hostess*) Call in the sheriff.     *Exit Hostess*
          *Enter Sheriff and a Carrier*
Now, master sheriff, what is your will with me?
SHERIFF
First, pardon me, my lord. A hue and cry
Hath followed certain men unto this house.
PRINCE HARRY What men?     515
SHERIFF
One of them is well known, my gracious lord,
A gross, fat man.
CARRIER          As fat as butter.
PRINCE HARRY
The man, I do assure you, is not here,
For I myself at this time have employed him.
And, sheriff, I will engage my word to thee     520
That I will by tomorrow dinner-time
Send him to answer thee, or any man,
For anything he shall be charged withal.
And so let me entreat you leave the house.
SHERIFF
I will, my lord. There are two gentlemen     525
Have in this robbery lost three hundred marks.

PRINCE HARRY
It may be so. If he have robbed these men,
He shall be answerable. And so, farewell.
SHERIFF Good night, my noble lord.
PRINCE HARRY
I think it is good morrow, is it not?     530
SHERIFF
Indeed, my lord, I think it be two o'clock.
                    *Exeunt Sheriff and Carrier*
PRINCE HARRY
This oily rascal is known as well as Paul's.
Go call him forth.
HARVEY          Oldcastle!
          ⌈*He draws back the arras, revealing Sir John asleep*⌉
                              Fast asleep
Behind the arras, and snorting like a horse.
PRINCE HARRY
Hark how hard he fetches breath. Search his pockets.
          *Harvey searcheth his pocket and findeth certain*
          *papers. He* ⌈*closeth the arras and*⌉ *cometh forward*
What hast thou found?
HARVEY          Nothing but papers, my lord.
PRINCE HARRY Let's see what they be. Read them.     537
⌈HARVEY⌉ (*reads*)
Item: a capon.                              2s. 2d.
Item: sauce.                                      4d.
Item: sack, two gallons.                    5s. 8d.
Item: anchovies and sack after supper.      2s. 6d.
Item: bread.                                        ob.
⌈PRINCE HARRY⌉ O monstrous! But one halfpennyworth of bread to this intolerable deal of sack! What there is else, keep close; we'll read it at more advantage. There let him sleep till day. I'll to the court in the morning. We must all to the wars, and thy place shall be honourable. I'll procure this fat rogue a charge of foot, and I know his death will be a march of twelve score. The money shall be paid back again, with advantage. Be with me betimes in the morning; and so good morrow, Harvey.     552
HARVEY Good morrow, good my lord.     *Exeunt* ⌈*severally*⌉

3.1     *Enter Hotspur, the Earl of Worcester, Lord*
          *Mortimer, and Owain Glyndŵr, with a map*
MORTIMER
These promises are fair, the parties sure,
And our induction full of prosperous hope.
HOTSPUR
Lord Mortimer and cousin Glyndŵr,
Will you sit down? And uncle Worcester?
          ⌈*Mortimer, Glyndŵr, and Worcester sit*⌉
A plague upon it, I have forgot the map!     5
GLYNDŴR
No, here it is. Sit, cousin Percy, sit,
Good cousin Hotspur;
          ⌈*Hotspur sits*⌉
                    For by that name
As oft as Lancaster doth speak of you,
His cheek looks pale, and with a rising sigh
He wisheth you in heaven.
HOTSPUR          And you in hell,     10
As oft as he hears Owain Glyndŵr spoke of.
GLYNDŴR
I cannot blame him. At my nativity
The front of heaven was full of fiery shapes,
Of burning cressets; and at my birth

The frame and huge foundation of the earth          15
Shaked like a coward.
HOTSPUR                    Why, so it would have done
At the same season if your mother's cat
Had but kittened, though yourself had never been
    born.
GLYNDŴR
I say the earth did shake when I was born.
HOTSPUR
And I say the earth was not of my mind          20
If you suppose as fearing you it shook.
GLYNDŴR
The heavens were all on fire, the earth did tremble—
HOTSPUR
O, then the earth shook to see the heavens on fire,
And not in fear of your nativity.
Diseasèd nature oftentimes breaks forth          25
In strange eruptions; oft the teeming earth
Is with a kind of colic pinched and vexed
By the imprisoning of unruly wind
Within her womb, which for enlargement striving
Shakes the old beldam earth, and topples down          30
Steeples and moss-grown towers. At your birth
Our grandam earth, having this distemp'rature,
In passion shook.
GLYNDŴR                    Cousin, of many men
I do not bear these crossings. Give me leave
To tell you once again that at my birth          35
The front of heaven was full of fiery shapes,
The goats ran from the mountains, and the herds
Were strangely clamorous to the frighted fields.
These signs have marked me extraordinary,
And all the courses of my life do show          40
I am not in the roll of commen men.
Where is he living, clipped in with the sea
That chides the banks of England, Scotland, Wales,
Which calls me pupil or hath read to me?
And bring him out that is but woman's son          45
Can trace me in the tedious ways of art,
And hold me pace in deep experiments.
HOTSPUR ⌈standing⌉
I think there's no man speaketh better Welsh.
I'll to dinner.
MORTIMER
Peace, cousin Percy, you will make him mad.          50
GLYNDŴR
I can call spirits from the vasty deep.
HOTSPUR
Why, so can I, or so can any man;
But will they come when you do call for them?
GLYNDŴR
Why, I can teach you, cousin, to command the devil.
HOTSPUR
And I can teach thee, coz, to shame the devil,          55
By telling truth: 'Tell truth, and shame the devil'.
If thou have power to raise him, bring him hither,
And I'll be sworn I have power to shame him hence.
O, while you live, tell truth and shame the devil.
MORTIMER
Come, come, no more of this unprofitable chat.          60
GLYNDŴR
Three times hath Henry Bolingbroke made head
Against my power; thrice from the banks of Wye
And sandy-bottomed Severn have I sent him
Bootless home, and weather-beaten back.

HOTSPUR
Home without boots, and in foul weather too!          65
How scapes he agues, in the devil's name?
GLYNDŴR
Come, here's the map. Shall we divide our right,
According to our threefold order ta'en?
MORTIMER
The Archdeacon hath divided it
Into three limits very equally.          70
England from Trent and Severn hitherto
By south and east is to my part assigned;
All westward—Wales beyond the Severn shore
And all the fertile land within that bound—
To Owain Glyndŵr; (to Hotspur) and, dear coz, to you
The remnant northward lying off from Trent.          76
And our indentures tripartite are drawn,
Which, being sealèd interchangeably—
A business that this night may execute—
Tomorrow, cousin Percy, you and I          80
And my good lord of Worcester will set forth
To meet your father and the Scottish power,
As is appointed us, at Shrewsbury.
My father, Glyndŵr, is not ready yet,
Nor shall we need his help these fourteen days.          85
Within that space you may have drawn together
Your tenants, friends, and neighbouring gentlemen.
GLYNDŴR
A shorter time shall send me to you, lords;
And in my conduct shall your ladies come,
From whom you now must steal and take no leave;
For there will be a world of water shed          91
Upon the parting of your wives and you.
HOTSPUR
Methinks my moiety north from Burton here
In quantity equals not one of yours.
See how this river comes me cranking in,          95
And cuts me from the best of all my land
A huge half-moon, a monstrous cantle, out.
I'll have the current in this place dammed up,
And here the smug and silver Trent shall run
In a new channel fair and evenly.          100
It shall not wind with such a deep indent,
To rob me of so rich a bottom here.
GLYNDŴR
Not wind? It shall, it must; you see it doth.
MORTIMER
Yea, but mark how he bears his course, and runs
    me up
With like advantage on the other side,          105
Gelding the opposèd continent as much
As on the other side it takes from you.
WORCESTER
Yea, but a little charge will trench him here,
And on this north side win this cape of land,
And then he runs straight and even.          110
HOTSPUR
I'll have it so; a little charge will do it.
GLYNDŴR I'll not have it altered.
HOTSPUR Will not you?
GLYNDŴR No, nor you shall not.
HOTSPUR Who shall say me nay?          115
GLYNDŴR Why, that will I.
HOTSPUR
Let me not understand you, then: speak it in Welsh.
GLYNDŴR
I can speak English, lord, as well as you;

For I was trained up in the English court,
Where, being but young, I framèd to the harp          120
Many an English ditty lovely well,
And gave the tongue a helpful ornament—
A virtue that was never seen in you.

HOTSPUR
Marry, and I am glad of it, with all my heart.
I had rather be a kitten and cry 'mew'          125
Than one of these same metre ballad-mongers.
I had rather hear a brazen canstick turned,
Or a dry wheel grate on the axle-tree,
And that would set my teeth nothing on edge,
Nothing so much as mincing poetry.          130
'Tis like the forced gait of a shuffling nag.

GLYNDŴR Come, you shall have Trent turned.

HOTSPUR
I do not care. I'll give thrice so much land
To any well-deserving friend;
But in the way of bargain—mark ye me—          135
I'll cavil on the ninth part of a hair.
Are the indentures drawn? Shall we be gone?

GLYNDŴR
The moon shines fair. You may away by night.
I'll haste the writer, and withal
Break with your wives of your departure hence.          140
I am afraid my daughter will run mad,
So much she doteth on her Mortimer.          *Exit*

MORTIMER
Fie, cousin Percy, how you cross my father!

HOTSPUR
I cannot choose. Sometime he angers me
With telling me of the moldwarp and the ant,          145
Of the dreamer Merlin and his prophecies,
And of a dragon and a finless fish,
A clip-winged griffin and a moulten raven,
A couching lion and a ramping cat,
And such a deal of skimble-skamble stuff          150
As puts me from my faith. I tell you what,
He held me last night at the least nine hours
In reckoning up the several devils' names
That were his lackeys. I cried, 'Hum!' and, 'Well,
   go to!',
But marked him not a word. O, he is as tedious          155
As a tired horse, a railing wife,
Worse than a smoky house. I had rather live
With cheese and garlic, in a windmill, far,
Than feed on cates and have him talk to me
In any summer house in Christendom.          160

MORTIMER
In faith, he is a worthy gentleman,
Exceedingly well read, and profited
In strange concealments, valiant as a lion,
And wondrous affable, and as bountiful
As mines of India. Shall I tell you, cousin?          165
He holds your temper in a high respect,
And curbs himself even of his natural scope
When you come 'cross his humour; faith, he does.
I warrant you, that man is not alive
Might so have tempted him as you have done          170
Without the taste of danger and reproof.
But do not use it oft, let me entreat you.

WORCESTER (*to Hotspur*)
In faith, my lord, you are too wilful-blame,
And since your coming hither have done enough
To put him quite besides his patience.          175
You must needs learn, lord, to amend this fault.

Though sometimes it show greatness, courage, blood—
And that's the dearest grace it renders you—
Yet oftentimes it doth present harsh rage,
Defect of manners, want of government,          180
Pride, haughtiness, opinion, and disdain,
The least of which haunting a nobleman
Loseth men's hearts, and leaves behind a stain
Upon the beauty of all parts besides,
Beguiling them of commendation.          185

HOTSPUR
Well, I am schooled. Good manners be your speed!
   *Enter Glyndŵr with Lady Percy and Mortimer's
   wife*
Here come our wives, and let us take our leave.
   ⌜*Mortimer's wife weeps, and speaks to him in
   Welsh*⌝

MORTIMER
This is the deadly spite that angers me:
My wife can speak no English, I no Welsh.

GLYNDŴR
My daughter weeps she'll not part with you.          190
She'll be a soldier, too; she'll to the wars.

MORTIMER
Good father, tell her that she and my aunt Percy
Shall follow in your conduct speedily.
   *Glyndŵr speaks to her in Welsh, and she answers
   him in the same*

GLYNDŴR
She is desperate here, a peevish self-willed harlotry,
One that no persuasion can do good upon.          195
   *The lady speaks in Welsh*

MORTIMER
I understand thy looks. That pretty Welsh
Which thou down pourest from these swelling
   heavens
I am too perfect in, and but for shame
In such a parley should I answer thee.
   *The lady kisses him, and speaks again in Welsh*

MORTIMER
I understand thy kisses, and thou mine,          200
And that's a feeling disputation;
But I will never be a truant, love,
Till I have learnt thy language, for thy tongue
Makes Welsh as sweet as ditties highly penned,
Sung by a fair queen in a summer's bower          205
With ravishing division, to her lute.

GLYNDŴR
Nay, if you melt, then will she run mad.
   *The lady* ⌜*sits on the rushes and*⌝ *speaks again in
   Welsh*

MORTIMER
O, I am ignorance itself in this!

GLYNDŴR
She bids you on the wanton rushes lay you down
And rest your gentle head upon her lap,          210
And she will sing the song that pleaseth you,
And on your eyelids crown the god of sleep,
Charming your blood with pleasing heaviness,
Making such difference 'twixt wake and sleep
As is the difference betwixt day and night          215
The hour before the heavenly-harnessed team
Begins his golden progress in the east.

MORTIMER
With all my heart, I'll sit and hear her sing.
By that time will our book, I think, be drawn.
   *He sits,* ⌜*resting his head on the Welsh lady's lap*⌝

GLYNDŴR
  Do so, and those musicians that shall play to you  220
  Hang in the air a thousand leagues from hence,
  And straight they shall be here. Sit and attend.
HOTSPUR
  Come, Kate, thou art perfect in lying down.
  Come, quick, quick, that I may lay my head in thy
    lap.
LADY PERCY (sitting) Go, ye giddy goose!  225
  Hotspur sits, resting his head on Lady Percy's lap.
  The music plays
HOTSPUR
  Now I perceive the devil understands Welsh;
  And 'tis no marvel, he is so humorous.
  By'r Lady, he's a good musician.
LADY PERCY
  Then should you be nothing but musical,
  For you are altogether governed by humours.  230
  Lie still, ye thief, and hear the lady sing in Welsh.
HOTSPUR I had rather hear Lady my brach howl in Irish.
LADY PERCY Wouldst thou have thy head broken?
HOTSPUR No.
LADY PERCY Then be still.  235
HOTSPUR Neither—'tis a woman's fault.
LADY PERCY Now God help thee!
HOTSPUR To the Welsh lady's bed.
LADY PERCY What's that?
HOTSPUR Peace; she sings.  240
  Here the lady sings a Welsh song
HOTSPUR Come, Kate, I'll have your song too.
LADY PERCY Not mine, in good sooth.
HOTSPUR Not yours, in good sooth! Heart, you swear like
  a comfit-maker's wife: 'Not you, in good sooth!' and
  'As true as I live!' and  245
  'As God shall mend me!' and 'As sure as day!';
  And giv'st such sarcenet surety for thy oaths
  As if thou never walk'st further than Finsbury.
  Swear me, Kate, like a lady as thou art,
  A good mouth-filling oath, and leave 'in sooth'  250
  And such protest of pepper gingerbread
  To velvet-guards and Sunday citizens.
  Come, sing.
LADY PERCY I will not sing.
HOTSPUR 'Tis the next way to turn tailor, or be redbreast
  teacher. (Rising) An the indentures be drawn, I'll away
  within these two hours; and so come in when ye will.
                                                      Exit
GLYNDŴR
  Come, come, Lord Mortimer. You are as slow
  As hot Lord Percy is on fire to go.
  By this our book is drawn. We'll but seal,  260
  And then to horse immediately.
MORTIMER (rising)              With all my heart.
  The ladies rise, and all exeunt

3.2  Enter King Henry, Prince Harry, and lords
KING HENRY
  Lords, give us leave—the Prince of Wales and I
  Must have some private conference—but be near at
    hand,
  For we shall presently have need of you.
                                          Exeunt Lords
  I know not whether God will have it so
  For some displeasing service I have done,  5
  That in his secret doom out of my blood
  He'll breed revengement and a scourge for me,

  But thou dost in thy passages of life
  Make me believe that thou art only marked
  For the hot vengeance and the rod of heaven  10
  To punish my mistreadings. Tell me else,
  Could such inordinate and low desires,
  Such poor, such bare, such lewd, such mean attempts,
  Such barren pleasures, rude society,
  As thou art matched withal and grafted to,  15
  Accompany the greatness of thy blood,
  And hold their level with thy princely heart?
PRINCE HARRY
  So please your majesty, I would I could
  Quit all offences with as clear excuse
  As well as I am doubtless I can purge  20
  Myself of many I am charged withal;
  Yet such extenuation let me beg
  As, in reproof of many tales devised—
  Which oft the ear of greatness needs must hear
  By smiling pickthanks and base newsmongers—  25
  I may, for some things true wherein my youth
  Hath faulty wandered and irregular,
  Find pardon on my true submission.
KING HENRY
  God pardon thee! Yet let me wonder, Harry,
  At thy affections, which do hold a wing  30
  Quite from the flight of all thy ancestors.
  Thy place in Council thou hast rudely lost—
  Which by thy younger brother is supplied—
  And art almost an alien to the hearts
  Of all the court and princes of my blood.  35
  The hope and expectation of thy time
  Is ruined, and the soul of every man
  Prophetically do forethink thy fall.
  Had I so lavish of my presence been,
  So common-hackneyed in the eyes of men,  40
  So stale and cheap to vulgar company,
  Opinion, that did help me to the crown,
  Had still kept loyal to possession,
  And left me in reputeless banishment,
  A fellow of no mark nor likelihood.  45
  By being seldom seen, I could not stir
  But, like a comet, I was wondered at,
  That men would tell their children 'This is he.'
  Others would say 'Where, which is Bolingbroke?'
  And then I stole all courtesy from heaven,  50
  And dressed myself in such humility
  That I did pluck allegiance from men's hearts,
  Loud shouts and salutations from their mouths,
  Even in the presence of the crownèd King.
  Thus did I keep my person fresh and new,  55
  My presence like a robe pontifical—
  Ne'er seen but wondered at—and so my state,
  Seldom but sumptuous, showed like a feast,
  And won by rareness such solemnity.
  The skipping King, he ambled up and down  60
  With shallow jesters and rash bavin wits,
  Soon kindled and soon burnt, carded his state,
  Mingled his royalty with cap'ring fools,
  Had his great name profanèd with their scorns,
  And gave his countenance, against his name,  65
  To laugh at gibing boys, and stand the push
  Of every beardless vain comparative;
  Grew a companion to the common streets,
  Enfeoffed himself to popularity,
  That, being daily swallowed by men's eyes,  70
  They surfeited with honey, and began

To loathe the taste of sweetness, whereof a little
More than a little is by much too much.
So when he had occasion to be seen,
He was but as the cuckoo is in June,       75
Heard, not regarded, seen but with such eyes
As, sick and blunted with community,
Afford no extraordinary gaze
Such as is bent on sun-like majesty
When it shines seldom in admiring eyes,      80
But rather drowsed and hung their eyelids down,
Slept in his face, and rendered such aspect
As cloudy men use to their adversaries,
Being with his presence glutted, gorged, and full.
And in that very line, Harry, standest thou;    85
For thou hast lost thy princely privilege
With vile participation. Not an eye
But is a-weary of thy common sight,
Save mine, which hath desired to see thee more,
Which now doth that I would not have it do—   90
Make blind itself with foolish tenderness.

    *He weeps*

PRINCE HARRY
I shall hereafter, my thrice-gracious lord,
Be more myself.

KING HENRY         For all the world,
As thou art to this hour was Richard then,
When I from France set foot at Ravenspurgh,   95
And even as I was then is Percy now.
Now by my sceptre, and my soul to boot,
He hath more worthy interest to the state
Than thou, the shadow of succession;
For, of no right, nor colour like to right,     100
He doth fill fields with harness in the realm,
Turns head against the lion's armèd jaws,
And, being no more in debt to years than thou,
Leads ancient lords and reverend bishops on
To bloody battles, and to bruising arms.      105
What never-dying honour hath he got
Against renownèd Douglas!—whose high deeds,
Whose hot incursions and great name in arms,
Holds from all soldiers chief majority
And military title capital             110
Through all the kingdoms that acknowledge Christ.
Thrice hath this Hotspur, Mars in swaddling-clothes,
This infant warrior, in his enterprises
Discomfited great Douglas; ta'en him once;
Enlargèd him; and made a friend of him     115
To fill the mouth of deep defiance up,
And shake the peace and safety of our throne.
And what say you to this? Percy, Northumberland,
The Archbishop's grace of York, Douglas, Mortimer,
Capitulate against us, and are up.         120
But wherefore do I tell these news to thee?
Why, Harry, do I tell thee of my foes,
Which art my near'st and dearest enemy?—
Thou that art like enough, through vassal fear,
Base inclination, and the start of spleen,     125
To fight against me under Percy's pay,
To dog his heels, and curtsy at his frowns,
To show how much thou art degenerate.

PRINCE HARRY
Do not think so; you shall not find it so.
And God forgive them that so much have swayed 130
Your majesty's good thoughts away from me.
I will redeem all this on Percy's head,
And in the closing of some glorious day

Be bold to tell you that I am your son;
When I will wear a garment all of blood,     135
And stain my favours in a bloody mask,
Which, washed away, shall scour my shame with it.
And that shall be the day, whene'er it lights,
That this same child of honour and renown,
This gallant Hotspur, this all-praisèd knight,   140
And your unthought-of Harry chance to meet.
For every honour sitting on his helm,
Would they were multitudes, and on my head
My shames redoubled; for the time will come
That I shall make this northern youth exchange 145
His glorious deeds for my indignities.
Percy is but my factor, good my lord,
To engross up glorious deeds on my behalf;
And I will call him to so strict account
That he shall render every glory up,       150
Yea, even the slightest worship of his time,
Or I will tear the reckoning from his heart.
This, in the name of God, I promise here,
The which if he be pleased I shall perform,
I do beseech your majesty may salve       155
The long-grown wounds of my intemperature;
If not, the end of life cancels all bonds,
And I will die a hundred thousand deaths
Ere break the smallest parcel of this vow.

KING HENRY
A hundred thousand rebels die in this.      160
Thou shalt have charge and sovereign trust herein.
    *Enter Sir Walter Blunt*
How now, good Blunt? Thy looks are full of speed.

BLUNT
So hath the business that I come to speak of.
Lord Mortimer of Scotland hath sent word
That Douglas and the English rebels met     165
The eleventh of this month at Shrewsbury.
A mighty and a fearful head they are,
If promises be kept on every hand,
As ever offered foul play in a state.

KING HENRY
The Earl of Westmorland set forth today,    170
With him my son Lord John of Lancaster,
For this advertisement is five days old.
On Wednesday next, Harry, you shall set forward.
On Thursday we ourselves will march.
Our meeting is Bridgnorth, and, Harry, you   175
Shall march through Gloucestershire, by which
    account,
Our business valuèd, some twelve days hence
Our general forces at Bridgnorth shall meet.
Our hands are full of business; let's away.
Advantage feeds him fat while men delay.   *Exeunt*

**3.3**   *Enter Sir John Oldcastle ⌜with a truncheon at his
    waist⌝, and Russell*

SIR JOHN Russell, am I not fallen away vilely since this
last action? Do I not bate? Do I not dwindle? Why,
my skin hangs about me like an old lady's loose gown.
I am withered like an old apple-john. Well, I'll repent,
and that suddenly, while I am in some liking. I shall
be out of heart shortly, and then I shall have no
strength to repent. An I have not forgotten what the
inside of a church is made of, I am a peppercorn, a
brewer's horse—the inside of a church! Company,
villainous company, hath been the spoil of me.    10
RUSSELL Sir John, you are so fretful you cannot live long.

SIR JOHN Why, there is it. Come, sing me a bawdy song, make me merry. I was as virtuously given as a gentleman need to be: virtuous enough; swore little; diced not—above seven times a week; went to a bawdy-house not—above once in a quarter—of an hour; paid money that I borrowed—three or four times; lived well, and in good compass. And now I live out of all order, out of all compass.                                                   19

RUSSELL Why, you are so fat, Sir John, that you must needs be out of all compass, out of all reasonable compass, Sir John.

SIR JOHN Do thou amend thy face, and I'll amend my life. Thou art our admiral, thou bearest the lantern in the poop—but 'tis in the nose of thee. Thou art the Knight of the Burning Lamp.                                              26

RUSSELL Why, Sir John, my face does you no harm.

SIR JOHN No, I'll be sworn; I make as good use of it as many a man doth of a death's head, or a *memento mori*. I never see thy face but I think upon hell-fire and Dives that lived in purple—for there he is in his robes, burning, burning. If thou wert any way given to virtue, I would swear by thy face; my oath should be 'By this fire that's God's angel!' But thou art altogether given over, and wert indeed, but for the light in thy face, the son of utter darkness. When thou rannest up Gads Hill in the night to catch my horse, if I did not think thou hadst been an *ignis fatuus* or a ball of wildfire, there's no purchase in money. O, thou art a perpetual triumph, an everlasting bonfire-light! Thou hast saved me a thousand marks in links and torches, walking with thee in the night betwixt tavern and tavern—but the sack that thou hast drunk me would have bought me lights as good cheap at the dearest chandler's in Europe. I have maintained that salamander of yours with fire any time this two-and-thirty years, God reward me for it.                                                              47

RUSSELL 'Sblood, I would my face were in your belly!

SIR JOHN God-a-mercy! So should I be sure to be heart-burnt.                                                            50

*Enter Hostess*

How now, Dame Partlet the hen, have you enquired yet who picked my pocket?

HOSTESS Why, Sir John, what do you think, Sir John? Do you think I keep thieves in my house? I have searched, I have enquired; so has my husband, man by man, boy by boy, servant by servant. The tithe of a hair was never lost in my house before.                                  57

SIR JOHN Ye lie, Hostess: Russell was shaved and lost many a hair, and I'll be sworn my pocket was picked. Go to, you are a woman, go.                                    60

HOSTESS Who, I? No, I defy thee! God's light, I was never called so in mine own house before.

SIR JOHN Go to, I know you well enough.

HOSTESS No, Sir John, you do not know me, Sir John; I know you, Sir John. You owe me money, Sir John, and now you pick a quarrel to beguile me of it. I bought you a dozen of shirts to your back.

SIR JOHN Dowlas, filthy dowlas. I have given them away to bakers' wives; they have made bolters of them.      69

HOSTESS Now as I am a true woman, holland of eight shillings an ell. You owe money here besides, Sir John: for your diet, and by-drinkings, and money lent you, four-and-twenty pound.

SIR JOHN (*pointing at Russell*) He had his part of it. Let him pay.                                                         75

HOSTESS He? Alas, he is poor; he hath nothing.

SIR JOHN How, poor? Look upon his face. What call you rich? Let them coin his nose, let them coin his cheeks, I'll not pay a denier. What, will you make a younker of me? Shall I not take mine ease in mine inn, but I shall have my pocket picked? I have lost a seal-ring of my grandfather's worth forty mark.                              82

HOSTESS O Jesu, (*to Russell*) I have heard the Prince tell him, I know not how oft, that that ring was copper.

SIR JOHN How? The Prince is a jack, a sneak-up. ⌈*Raising his truncheon*⌉ 'Sblood, an he were here I would cudgel him like a dog if he would say so.                          87

*Enter Prince Harry and Harvey, marching; and Sir John Oldcastle meets them, playing upon his truncheon like a fife*

How now, lad, is the wind in that door, i'faith? Must we all march?

RUSSELL Yea, two and two, Newgate fashion.          90

HOSTESS My lord, I pray you hear me.

PRINCE HARRY
What sayst thou, Mistress Quickly? How doth thy husband?
I love him well; he is an honest man.

HOSTESS Good my lord, hear me!

SIR JOHN Prithee, let her alone, and list to me.      95

PRINCE HARRY What sayst thou, Jack?

SIR JOHN The other night I fell asleep here behind the arras, and had my pocket picked. This house is turned bawdy-house: they pick pockets.

PRINCE HARRY What didst thou lose, Jack?               100

SIR JOHN Wilt thou believe me, Hal, three or four bonds of forty pound apiece, and a seal-ring of my grand-father's.

PRINCE HARRY A trifle, some eightpenny matter.

HOSTESS So I told him, my lord; and I said I heard your grace say so; and, my lord, he speaks most vilely of you, like a foul-mouthed man as he is, and said he would cudgel you.                                              108

PRINCE HARRY What? He did not!

HOSTESS There's neither faith, truth, nor womanhood in me else.

SIR JOHN There's no more faith in thee than in a stewed prune, nor no more truth in thee than in a drawn fox; and, for womanhood, Maid Marian may be the deputy's wife of the ward to thee. Go, you thing, go!          115

HOSTESS Say, what thing, what thing?

SIR JOHN What thing? Why, a thing to thank God on.

HOSTESS I am no thing to thank God on. I would thou shouldst know it, I am an honest man's wife; and setting thy knighthood aside, thou art a knave to call me so.                                                             121

SIR JOHN Setting thy womanhood aside, thou art a beast to say otherwise.

HOSTESS Say, what beast, thou knave, thou?

SIR JOHN What beast? Why, an otter.                    125

PRINCE HARRY An otter, Sir John? Why an otter?

SIR JOHN Why? She's neither fish nor flesh; a man knows not where to have her.

HOSTESS Thou art an unjust man in saying so. Thou or any man knows where to have me, thou knave, thou.

PRINCE HARRY Thou sayst true, Hostess, and he slanders thee most grossly.                                           132

HOSTESS So he doth you, my lord, and said this other day you owed him a thousand pound.

PRINCE HARRY (*to Sir John*) Sirrah, do I owe you a thousand pound?                                              136

SIR JOHN A thousand pound, Hal? A million! Thy love is worth a million; thou owest me thy love.

HOSTESS Nay, my lord, he called you 'jack' and said he
    would cudgel you.                                      140
SIR JOHN Did I, Russell?
RUSSELL Indeed, Sir John, you said so.
SIR JOHN Yea, if he said my ring was copper.
PRINCE HARRY I say 'tis copper; darest thou be as good
    as thy word now?                                      145
SIR JOHN Why, Hal, thou knowest as thou art but man I
    dare, but as thou art prince, I fear thee as I fear the
    roaring of the lion's whelp.
PRINCE HARRY And why not as the lion?
SIR JOHN The King himself is to be feared as the lion. Dost
    thou think I'll fear thee as I fear thy father? Nay, an I
    do, I pray God my girdle break.                       152
PRINCE HARRY O, if it should, how would thy guts fall
    about thy knees! But sirrah, there's no room for faith,
    truth, nor honesty in this bosom of thine; it is all filled
    up with guts and midriff. Charge an honest woman
    with picking thy pocket? Why, thou whoreson
    impudent embossed rascal, if there were anything in
    thy pocket but tavern reckonings, memorandums of
    bawdy-houses, and one poor pennyworth of sugar-
    candy to make thee long-winded—if thy pocket were
    enriched with any other injuries but these, I am a
    villain. And yet you will stand to it, you will not pocket
    up wrong. Art thou not ashamed?                        164
SIR JOHN Dost thou hear, Hal? Thou knowest in the state
    of innocency Adam fell, and what should poor Jack
    Oldcastle do in the days of villainy? Thou seest I have
    more flesh than another man, and therefore more
    frailty. You confess, then, you picked my pocket.
PRINCE HARRY It appears so by the story.                  170
SIR JOHN Hostess, I forgive thee. Go make ready breakfast.
    Love thy husband, look to thy servants, cherish thy
    guests. Thou shalt find me tractable to any honest
    reason; thou seest I am pacified still. Nay, prithee, be
    gone.                                       Exit Hostess
    Now, Hal, to the news at court. For the robbery, lad,
    how is that answered?                                  177
PRINCE HARRY O, my sweet beef, I must still be good angel
    to thee. The money is paid back again.
SIR JOHN O, I do not like that paying back; 'tis a double
    labour.                                                181
PRINCE HARRY I am good friends with my father, and may
    do anything.
SIR JOHN Rob me the exchequer the first thing thou dost,
    and do it with unwashed hands too.                     185
RUSSELL Do, my lord.
PRINCE HARRY I have procured thee, Jack, a charge of
    foot.
SIR JOHN I would it had been of horse! Where shall I find
    one that can steal well? O, for a fine thief of the age
    of two-and-twenty or thereabouts! I am heinously
    unprovided. Well, God be thanked for these rebels—
    they offend none but the virtuous. I laud them, I praise
    them.
PRINCE HARRY Russell.                                     195
RUSSELL My lord?
PRINCE HARRY (giving letters)
    Go bear this letter to Lord John of Lancaster,
    To my brother John; this to my lord of Westmorland.
                                              Exit Russell
    Go, Harvey, to horse, to horse, for thou and I
    Have thirty miles to ride yet ere dinner time.        200
                                              Exit Harvey

    Jack, meet me tomorrow in the Temple Hall
    At two o'clock in the afternoon.
    There shalt thou know thy charge, and there receive
    Money and order for their furniture.
    The land is burning, Percy stands on high,            205
    And either we or they must lower lie.          Exit
SIR JOHN
    Rare words! Brave world! (Calling) Hostess, my
        breakfast, come!—
    O, I could wish this tavern were my drum!        Exit

4.1    Enter Hotspur and the Earls of Worcester and
        Douglas
HOTSPUR
    Well said, my noble Scot! If speaking truth
    In this fine age were not thought flattery,
    Such attribution should the Douglas have
    As not a soldier of this season's stamp
    Should go so general current through the world.       5
    By God, I cannot flatter, I do defy
    The tongues of soothers, but a braver place
    In my heart's love hath no man than yourself.
    Nay, task me to my word, approve me, lord.
DOUGLAS Thou art the king of honour.                      10
    No man so potent breathes upon the ground
    But I will beard him.
HOTSPUR                 Do so, and 'tis well.
    Enter a Messenger with letters
    What letters hast thou there? I can but thank you.
MESSENGER These letters come from your father.
HOTSPUR
    Letters from him? Why comes he not himself?           15
MESSENGER
    He cannot come, my lord, he is grievous sick.
HOTSPUR
    Zounds, how has he the leisure to be sick
    In such a jostling time? Who leads his power?
    Under whose government come they along?
MESSENGER
    His letters bears his mind, not I, my lord.           20
    Hotspur reads the letter
WORCESTER
    I prithee tell me, doth he keep his bed?
MESSENGER
    He did, my lord, four days ere I set forth;
    And at the time of my departure thence
    He was much feared by his physicians.
WORCESTER
    I would the state of time had first been whole         25
    Ere he by sickness had been visited.
    His health was never better worth than now.
HOTSPUR
    Sick now? Droop now? This sickness doth infect
    The very life-blood of our enterprise.
    'Tis catching hither, even to our camp.                30
    He writes me here that inward sickness stays him,
    And that his friends by deputation
    Could not so soon be drawn; nor did he think it meet
    To lay so dangerous and dear a trust
    On any soul removed but on his own.                    35
    Yet doth he give us bold advertisement
    That with our small conjunction we should on,
    To see how fortune is disposed to us;
    For, as he writes, there is no quailing now,
    Because the King is certainly possessed                40
    Of all our purposes. What say you to it?

WORCESTER
Your father's sickness is a maim to us.
HOTSPUR
A perilous gash, a very limb lopped off.
And yet, in faith, it is not. His present want
Seems more than we shall find it. Were it good　45
To set the exact wealth of all our states
All at one cast, to set so rich a main
On the nice hazard of one doubtful hour?
It were not good, for therein should we read
The very bottom and the sole of hope,　50
The very list, the very utmost bound,
Of all our fortunes.
DOUGLAS
Faith, and so we should, where now remains
A sweet reversion—we may boldly spend
Upon the hope of what is to come in.　55
A comfort of retirement lives in this.
HOTSPUR
A rendezvous, a home to fly unto,
If that the devil and mischance look big
Upon the maidenhead of our affairs.
WORCESTER
But yet I would your father had been here.　60
The quality and hair of our attempt
Brooks no division. It will be thought
By some that know not why he is away
That wisdom, loyalty, and mere dislike
Of our proceedings kept the Earl from hence;　65
And think how such an apprehension
May turn the tide of fearful faction,
And breed a kind of question in our cause.
For, well you know, we of the off'ring side
Must keep aloof from strict arbitrement,　70
And stop all sight-holes, every loop from whence
The eye of reason may pry in upon us.
This absence of your father's draws a curtain
That shows the ignorant a kind of fear
Before not dreamt of.
HOTSPUR　　　　You strain too far.　75
I rather of his absence make this use:
It lends a lustre, and more great opinion,
A larger dare to our great enterprise,
Than if the Earl were here; for men must think
If we without his help can make a head　80
To push against a kingdom, with his help
We shall o'erturn it topsy-turvy down.
Yet all goes well, yet all our joints are whole.
DOUGLAS
As heart can think, there is not such a word
Spoke of in Scotland as this term of fear.　85
*Enter Sir Richard Vernon*
HOTSPUR
My cousin Vernon! Welcome, by my soul!
VERNON
Pray God my news be worth a welcome, lord.
The Earl of Westmorland, seven thousand strong,
Is marching hitherwards; with him Prince John.
HOTSPUR
No harm. What more?
VERNON　　　　And further I have learned　90
The King himself in person is set forth,
Or hitherwards intended speedily,
With strong and mighty preparation.
HOTSPUR
He shall be welcome too. Where is his son,
The nimble-footed madcap Prince of Wales,　95

And his comrades that daffed the world aside
And bid it pass?
VERNON　　　　All furnished, all in arms,
All plumed like ostriches, that with the wind
⌈　　　　　　　　　　　　　　　　　　　⌉
Baiting like eagles having lately bathed,　100
Glittering in golden coats like images,
As full of spirit as the month of May,
And gorgeous as the sun at midsummer;
Wanton as youthful goats, wild as young bulls.
I saw young Harry with his beaver on,　105
His cuishes on his thighs, gallantly armed,
Rise from the ground like feathered Mercury,
And vaulted with such ease into his seat
As if an angel dropped down from the clouds
To turn and wind a fiery Pegasus,　110
And witch the world with noble horsemanship.
HOTSPUR
No more, no more! Worse than the sun in March,
This praise doth nourish agues. Let them come!
They come like sacrifices in their trim,
And to the fire-eyed maid of smoky war　115
All hot and bleeding will we offer them.
The mailèd Mars shall on his altar sit
Up to the ears in blood. I am on fire
To hear this rich reprisal is so nigh,
And yet not ours! Come, let me taste my horse,　120
Who is to bear me like a thunderbolt
Against the bosom of the Prince of Wales.
Harry to Harry shall, hot horse to horse,
Meet and ne'er part till one drop down a corpse.
O, that Glyndŵr were come!
VERNON　　　　There is more news.　125
I learned in Worcester, as I rode along,
He cannot draw his power this fourteen days.
DOUGLAS
That's the worst tidings that I hear of yet.
WORCESTER
Ay, by my faith, that bears a frosty sound.
HOTSPUR
What may the King's whole battle reach unto?　130
VERNON
To thirty thousand.
HOTSPUR　　　　Forty let it be.
My father and Glyndŵr being both away,
The powers of us may serve so great a day.
Come, let us take a muster speedily.
Doomsday is near: die all, die merrily.　135
DOUGLAS
Talk not of dying; I am out of fear
Of death or death's hand for this one half year.
*Exeunt*

**4.2**　*Enter Sir John Oldcastle and Russell*
SIR JOHN Russell, get thee before to Coventry; fill me a
bottle of sack. Our soldiers shall march through. We'll
to Sutton Coldfield tonight.
RUSSELL Will you give me money, captain?
SIR JOHN Lay out, lay out.　5
RUSSELL This bottle makes an angel.
SIR JOHN ⌈*giving Russell money*⌉ An if it do, take it for thy
labour; an if it make twenty, take them all; I'll answer
the coinage. Bid my lieutenant Harvey meet me at
town's end.　10
RUSSELL I will, captain. Farewell.　*Exit*
SIR JOHN If I be not ashamed of my soldiers, I am a soused
gurnet. I have misused the King's press damnably. I

have got in exchange of one hundred and fifty soldiers three hundred and odd pounds. I press me none but good householders, yeomen's sons, enquire me out contracted bachelors, such as had been asked twice on the banns, such a commodity of warm slaves as had as lief hear the devil as a drum, such as fear the report of a caliver worse than a struck fowl or a hurt wild duck. I pressed me none but such toasts and butter, with hearts in their bellies no bigger than pins' heads, and they have bought out their services; and now my whole charge consists of ensigns, corporals, lieutenants, gentlemen of companies—slaves as ragged as Lazarus in the painted cloth, where the glutton's dogs licked his sores—and such as indeed were never soldiers, but discarded unjust servingmen, younger sons to younger brothers, revolted tapsters, and ostlers trade-fallen, the cankers of a calm world and a long peace, ten times more dishonourable-ragged than an old feazed ensign; and such have I to fill up the rooms of them as have bought out their services, that you would think that I had a hundred and fifty tattered prodigals lately come from swine-keeping, from eating draff and husks. A mad fellow met me on the way and told me I had unloaded all the gibbets and pressed the dead bodies. No eye hath seen such scarecrows. I'll not march through Coventry with them, that's flat. Nay, and the villains march wide betwixt the legs, as if they had gyves on, for indeed I had the most of them out of prison. There's not a shirt and a half in all my company; and the half-shirt is two napkins tacked together and thrown over the shoulders like a herald's coat without sleeves; and the shirt, to say the truth, stolen from my host at Saint Albans, or the red-nose innkeeper of Daventry. But that's all one; they'll find linen enough on every hedge.                                                    48

*Enter Prince Harry and the Earl of Westmorland*

PRINCE HARRY How now, blown Jack? How now, quilt?

SIR JOHN What, Hal! How now, mad wag? What a devil dost thou in Warwickshire? My good lord of Westmorland, I cry you mercy! I thought your honour had already been at Shrewsbury.                                       53

WESTMORLAND Faith, Sir John, 'tis more than time that I were there, and you too; but my powers are there already. The King, I can tell you, looks for us all. We must away all night.

SIR JOHN Tut, never fear me. I am as vigilant as a cat to steal cream.                                                             59

PRINCE HARRY I think to steal cream indeed, for thy theft hath already made thee butter. But tell me, Jack, whose fellows are these that come after?

SIR JOHN Mine, Hal, mine.

PRINCE HARRY I did never see such pitiful rascals.         64

SIR JOHN Tut, tut, good enough to toss, food for powder, food for powder. They'll fill a pit as well as better. Tush, man, mortal men, mortal men.

WESTMORLAND Ay, but Sir John, methinks they are exceeding poor and bare, too beggarly.                         69

SIR JOHN Faith, for their poverty, I know not where they had that, and for their bareness, I am sure they never learned that of me.

PRINCE HARRY No, I'll be sworn, unless you call three fingers in the ribs bare. But sirrah, make haste. Percy is already in the field.                                           *Exit*

SIR JOHN What, is the King encamped?                         76

WESTMORLAND He is, Sir John. I fear we shall stay too long.                                                            ⌈*Exit*⌉

SIR JOHN
Well, to the latter end of a fray
And the beginning of a feast                                          80
Fits a dull fighter and a keen guest.                        *Exit*

4.3    *Enter Hotspur, the Earls of Worcester and Douglas, and Sir Richard Vernon*

HOTSPUR
We'll fight with him tonight.

WORCESTER                                      It may not be.

DOUGLAS
You give him then advantage.

VERNON                                         Not a whit.

HOTSPUR
Why say you so? Looks he not for supply?

VERNON
So do we.

HOTSPUR      His is certain; ours is doubtful.

WORCESTER
Good cousin, be advised. Stir not tonight.                   5

VERNON (*to Hotspur*)
Do not, my lord.

DOUGLAS                        You do not counsel well.
You speak it out of fear and cold heart.

VERNON
Do me no slander, Douglas. By my life—
And I dare well maintain it with my life—
If well-respected honour bid me on,                          10
I hold as little counsel with weak fear
As you, my lord, or any Scot that this day lives.
Let it be seen tomorrow in the battle
Which of us fears.

DOUGLAS Yea, or tonight.                                      15

VERNON Content.

HOTSPUR Tonight, say I.

VERNON
Come, come, it may not be. I wonder much,
Being men of such great leading as you are,
That you foresee not what impediments                        20
Drag back our expedition. Certain horse
Of my cousin Vernon's are not yet come up.
Your uncle Worcester's horse came but today,
And now their pride and mettle is asleep,
Their courage with hard labour tame and dull,                25
That not a horse is half the half himself.

HOTSPUR
So are the horses of the enemy
In general journey-bated and brought low.
The better part of ours are full of rest.

WORCESTER
The number of the King exceedeth our.                        30
For God's sake, cousin, stay till all come in.
*The trumpet sounds a parley* ⌈*within*⌉. *Enter Sir Walter Blunt*

BLUNT
I come with gracious offers from the King,
If you vouchsafe me hearing and respect.

HOTSPUR
Welcome, Sir Walter Blunt; and would to God
You were of our determination.                               35
Some of us love you well, and even those some
Envy your great deservings and good name,
Because you are not of our quality,
But stand against us like an enemy.

BLUNT
And God defend but still I should stand so,                  40
So long as out of limit and true rule

You stand against anointed majesty.
But to my charge. The King hath sent to know
The nature of your griefs, and whereupon
You conjure from the breast of civil peace          45
Such bold hostility, teaching his duteous land
Audacious cruelty. If that the King
Have any way your good deserts forgot,
Which he confesseth to be manifold,
He bids you name your griefs, and with all speed     50
You shall have your desires, with interest,
And pardon absolute for yourself and these
Herein misled by your suggestion.

HOTSPUR
The King is kind, and well we know the King
Knows at what time to promise, when to pay.          55
My father and my uncle and myself
Did give him that same royalty he wears;
And when he was not six-and-twenty strong,
Sick in the world's regard, wretched and low,
A poor unminded outlaw sneaking home,                60
My father gave him welcome to the shore;
And when he heard him swear and vow to God
He came but to be Duke of Lancaster,
To sue his livery, and beg his peace
With tears of innocency and terms of zeal,           65
My father, in kind heart and pity moved,
Swore him assistance, and performed it too.
Now when the lords and barons of the realm
Perceived Northumberland did lean to him,
The more and less came in with cap and knee,         70
Met him in boroughs, cities, villages,
Attended him on bridges, stood in lanes,
Laid gifts before him, proffered him their oaths,
Gave him their heirs as pages, followed him,
Even at the heels, in golden multitudes.             75
He presently, as greatness knows itself,
Steps me a little higher than his vow
Made to my father while his blood was poor
Upon the naked shore at Ravenspurgh,
And now forsooth takes on him to reform              80
Some certain edicts and some strait decrees
That lie too heavy on the commonwealth,
Cries out upon abuses, seems to weep
Over his country's wrongs; and by this face,
This seeming brow of justice, did he win             85
The hearts of all that he did angle for;
Proceeded further, cut me off the heads
Of all the favourites that the absent King
In deputation left behind him here
When he was personal in the Irish war.               90

BLUNT
Tut, I came not to hear this.

HOTSPUR                        Then to the point.
In short time after, he deposed the King,
Soon after that deprived him of his life,
And in the neck of that tasked the whole state;
To make that worse, suffered his kinsman March—      95
Who is, if every owner were well placed,
Indeed his king—to be engaged in Wales,
There without ransom to lie forfeited;
Disgraced me in my happy victories,
Sought to entrap me by intelligence,                 100
Rated mine uncle from the Council-board,
In rage dismissed my father from the court,
Broke oath on oath, committed wrong on wrong,
And in conclusion drove us to seek out

This head of safety, and withal to pry               105
Into his title, the which we find
Too indirect for long continuance.

BLUNT
Shall I return this answer to the King?

HOTSPUR
Not so, Sir Walter. We'll withdraw awhile.
Go to the King, and let there be impawned            110
Some surety for a safe return again;
And in the morning early shall mine uncle
Bring him our purposes. And so, farewell.

BLUNT
I would you would accept of grace and love.

HOTSPUR
And maybe so we shall.

BLUNT                        Pray God you do.          115
*Exeunt ⌈Hotspur, Worcester, Douglas, and*
*Vernon at one door, Blunt at another door⌉*

**4.4**    *Enter the Archbishop of York, and Sir Michael*

ARCHBISHOP *(giving letters)*
Hie, good Sir Michael, bear this sealèd brief
With wingèd haste to the Lord Marshal,
This to my cousin Scrope, and all the rest
To whom they are directed. If you knew
How much they do import, you would make haste.        5

SIR MICHAEL My good lord,
I guess their tenor.

ARCHBISHOP                Like enough you do.
Tomorrow, good Sir Michael, is a day
Wherein the fortune of ten thousand men
Must bide the touch; for, sir, at Shrewsbury,         10
As I am truly given to understand,
The King with mighty and quick-raisèd power
Meets with Lord Harry. And I fear, Sir Michael,
What with the sickness of Northumberland,
Whose power was in the first proportion,             15
And what with Owain Glyndŵr's absence thence,
Who with them was a rated sinew too,
And comes not in, overruled by prophecies,
I fear the power of Percy is too weak
To wage an instant trial with the King.               20

SIR MICHAEL
Why, my good lord, you need not fear; there is
    Douglas
And Lord Mortimer.

ARCHBISHOP            No, Mortimer is not there.

SIR MICHAEL
But there is Mordake, Vernon, Lord Harry Percy;
And there is my lord of Worcester, and a head
Of gallant warriors, noble gentlemen.                 25

ARCHBISHOP
And so there is; but yet the King hath drawn
The special head of all the land together—
The Prince of Wales, Lord John of Lancaster,
The noble Westmorland, and warlike Blunt,
And many more corrivals, and dear men                 30
Of estimation and command in arms.

SIR MICHAEL
Doubt not, my lord, they shall be well opposed.

ARCHBISHOP
I hope no less, yet needful 'tis to fear;
And to prevent the worst, Sir Michael, speed.
For if Lord Percy thrive not, ere the King            35
Dismiss his power he means to visit us,
For he hath heard of our confederacy,

And 'tis but wisdom to make strong against him;
Therefore make haste. I must go write again
To other friends; and so farewell, Sir Michael.          40

*Exeunt ⌐severally⌐*

5.1     *Enter King Henry, Prince Harry, Lord John of*
        *Lancaster, the Earl of Westmorland, Sir Walter*
        *Blunt, and Sir John Oldcastle*

KING HENRY
How bloodily the sun begins to peer
Above yon bulky hill! The day looks pale
At his distemp'rature.

PRINCE HARRY          The southern wind
Doth play the trumpet to his purposes,
And by his hollow whistling in the leaves                    5
Foretells a tempest and a blust'ring day.

KING HENRY
Then with the losers let it sympathize,
For nothing can seem foul to those that win.

*The trumpet sounds ⌐a parley within⌐. Enter the*
*Earl of Worcester ⌐and Sir Richard Vernon⌐*

How now, my lord of Worcester? 'Tis not well
That you and I should meet upon such terms                  10
As now we meet. You have deceived our trust,
And made us doff our easy robes of peace
To crush our old limbs in ungentle steel.
This is not well, my lord, this is not well.
What say you to it? Will you again unknit                   15
This churlish knot of all-abhorrèd war,
And move in that obedient orb again
Where you did give a fair and natural light,
And be no more an exhaled meteor,
A prodigy of fear, and a portent                            20
Of broachèd mischief to the unborn times?

WORCESTER Hear me, my liege.
For mine own part, I could be well content
To entertain the lag-end of my life
With quiet hours; for I protest,                            25
I have not sought the day of this dislike.

KING HENRY
You have not sought it? How comes it, then?

SIR JOHN Rebellion lay in his way, and he found it.

PRINCE HARRY Peace, chewet, peace!

WORCESTER (*to the King*)
It pleased your majesty to turn your looks                  30
Of favour from myself and all our house;
And yet I must remember you, my lord,
We were the first and dearest of your friends.
For you my staff of office did I break
In Richard's time, and posted day and night                35
To meet you on the way and kiss your hand
When yet you were in place and in account
Nothing so strong and fortunate as I.
It was myself, my brother, and his son
That brought you home, and boldly did outdare               40
The dangers of the time. You swore to us,
And you did swear that oath at Doncaster,
That you did nothing purpose 'gainst the state,
Nor claim no further than your new-fall'n right,
The seat of Gaunt, dukedom of Lancaster.                    45
To this we swore our aid, but in short space
It rained down fortune show'ring on your head,
And such a flood of greatness fell on you,
What with our help, what with the absent King,
What with the injuries of a wanton time,                    50
The seeming sufferances that you had borne,

And the contrarious winds that held the King
So long in his unlucky Irish wars
That all in England did repute him dead;
And from this swarm of fair advantages                      55
You took occasion to be quickly wooed
To gripe the general sway into your hand,
Forgot your oath to us at Doncaster,
And being fed by us, you used us so
As that ungentle gull, the cuckoo's bird,                   60
Useth the sparrow—did oppress our nest,
Grew by our feeding to so great a bulk
That even our love durst not come near your sight
For fear of swallowing. But with nimble wing
We were enforced for safety' sake to fly                    65
Out of your sight, and raise this present head,
Whereby we stand opposèd by such means
As you yourself have forged against yourself,
By unkind usage, dangerous countenance,
And violation of all faith and troth                        70
Sworn to us in your younger enterprise.

KING HENRY
These things indeed you have articulate,
Proclaimed at market crosses, read in churches,
To face the garment of rebellion
With some fine colour that may please the eye               75
Of fickle changelings and poor discontents,
Which gape and rub the elbow at the news
Of hurly-burly innovation;
And never yet did insurrection want
Such water-colours to impaint his cause,                    80
Nor moody beggars starving for a time
Of pell-mell havoc and confusion.

PRINCE HARRY
In both our armies there is many a soul
Shall pay full dearly for this encounter
If once they join in trial. Tell your nephew                85
The Prince of Wales doth join with all the world
In praise of Henry Percy. By my hopes,
This present enterprise set off his head,
I do not think a braver gentleman,
More active-valiant or more valiant-young,                  90
More daring, or more bold, is now alive
To grace this latter age with noble deeds.
For my part, I may speak it to my shame,
I have a truant been to chivalry;
And so I hear he doth account me too.                       95
Yet this, before my father's majesty:
I am content that he shall take the odds
Of his great name and estimation,
And will, to save the blood on either side,
Try fortune with him in a single fight.                     100

KING HENRY
And, Prince of Wales, so dare we venture thee,
Albeit considerations infinite
Do make against it. No, good Worcester, no.
We love our people well; even those we love
That are misled upon your cousin's part;                    105
And will they take the offer of our grace,
Both he and they and you, yea, every man
Shall be my friend again, and I'll be his.
So tell your cousin, and bring me word
What he will do. But if he will not yield,                  110
Rebuke and dread correction wait on us,
And they shall do their office. So be gone.
We will not now be troubled with reply.
We offer fair; take it advisedly.

*Exeunt Worcester ⌐and Vernon⌐*

PRINCE HARRY
It will not be accepted, on my life.                              115
The Douglas and the Hotspur both together
Are confident against the world in arms.
KING HENRY
Hence, therefore, every leader to his charge,
For on their answer will we set on them,
And God befriend us as our cause is just!          120
             *Exeunt all but Prince Harry and Oldcastle*
SIR JOHN Hal, if thou see me down in the battle, and
bestride me, so. 'Tis a point of friendship.
PRINCE HARRY Nothing but a colossus can do thee that
friendship. Say thy prayers, and farewell.
SIR JOHN I would 'twere bed-time, Hal, and all well.    125
PRINCE HARRY Why, thou owest God a death.          *Exit*
SIR JOHN 'Tis not due yet. I would be loath to pay him
before his day. What need I be so forward with him
that calls not on me? Well, 'tis no matter; honour
pricks me on. Yea, but how if honour prick me off
when I come on? How then? Can honour set-to a leg?
No. Or an arm? No. Or take away the grief of a wound?
No. Honour hath no skill in surgery, then? No. What
is honour? A word. What is in that word 'honour'?
What is that 'honour'? Air. A trim reckoning! Who
hath it? He that died o' Wednesday. Doth he feel it?
No. Doth he hear it? No. 'Tis insensible then? Yea, to
the dead. But will it not live with the living? No. Why?
Detraction will not suffer it. Therefore I'll none of it.
Honour is a mere scutcheon. And so ends my catechism.
                                                                    *Exit*


**5.2**    *Enter the Earl of Worcester and Sir Richard Vernon*
WORCESTER
O no, my nephew must not know, Sir Richard,
The liberal and kind offer of the King.
VERNON
'Twere best he did.
WORCESTER          Then are we all undone.
It is not possible, it cannot be,
The King should keep his word in loving us.          5
He will suspect us still, and find a time
To punish this offence in other faults.
Supposition all our lives shall be stuck full of eyes,
For treason is but trusted like the fox,
Who, ne'er so tame, so cherished, and locked up,     10
Will have a wild trick of his ancestors.
Look how we can, or sad or merrily,
Interpretation will misquote our looks,
And we shall feed like oxen at a stall,
The better cherished still the nearer death.          15
My nephew's trespass may be well forgot;
It hath the excuse of youth and heat of blood,
And an adopted name of privilege—
A hare-brained Hotspur, governed by a spleen.
All his offences live upon my head,                      20
And on his father's. We did train him on,
And, his corruption being ta'en from us,
We as the spring of all shall pay for all.
Therefore, good cousin, let not Harry know
In any case the offer of the King.                       25
VERNON
Deliver what you will; I'll say 'tis so.
           *Enter Hotspur and the Earl of Douglas*
Here comes your cousin.
HOTSPUR                    My uncle is returned.
Deliver up my lord of Westmorland.
Uncle, what news?

WORCESTER
The King will bid you battle presently.               30
DOUGLAS
Defy him by the Lord of Westmorland.
HOTSPUR
Lord Douglas, go you and tell him so.
DOUGLAS
Marry, and shall, and very willingly.          *Exit*
WORCESTER
There is no seeming mercy in the King.
HOTSPUR
Did you beg any? God forbid!                            35
WORCESTER
I told him gently of our grievances,
Of his oath-breaking, which he mended thus:
By now forswearing that he is forsworn.
He calls us 'rebels', 'traitors', and will scourge
With haughty arms this hateful name in us.       40
           *Enter the Earl of Douglas*
DOUGLAS
Arm, gentlemen, to arms, for I have thrown
A brave defiance in King Henry's teeth—
And Westmorland that was engaged did bear it—
Which cannot choose but bring him quickly on.
WORCESTER (*to Hotspur*)
The Prince of Wales stepped forth before the King   45
And, nephew, challenged you to single fight.
HOTSPUR
O, would the quarrel lay upon our heads,
And that no man might draw short breath today
But I and Harry Monmouth! Tell me, tell me,
How showed his tasking? Seemed it in contempt?   50
VERNON
No, by my soul, I never in my life
Did hear a challenge urged more modestly,
Unless a brother should a brother dare
To gentle exercise and proof of arms.
He gave you all the duties of a man,                   55
Trimmed up your praises with a princely tongue,
Spoke your deservings like a chronicle,
Making you ever better than his praise
By still dispraising praise valued with you;
And, which became him like a prince indeed,       60
He made a blushing cital of himself,
And chid his truant youth with such a grace
As if he mastered there a double spirit
Of teaching and of learning instantly.
There did he pause; but let me tell the world,      65
If he outlive the envy of this day,
England did never owe so sweet a hope,
So much misconstrued in his wantonness.
HOTSPUR
Cousin, I think thou art enamourèd
On his follies. Never did I hear                         70
Of any prince so wild a liberty.
But be he as he will, yet once ere night
I will embrace him with a soldier's arm,
That he shall shrink under my courtesy.
Arm, arm, with speed! And fellows, soldiers, friends,
Better consider what you have to do                    76
Than I, that have not well the gift of tongue,
Can lift your blood up with persuasion.
           *Enter a Messenger*
MESSENGER My lord, here are letters for you.
HOTSPUR I cannot read them now.   ⌈*Exit Messenger*⌉
O gentlemen, the time of life is short.               81

To spend that shortness basely were too long
If life did ride upon a dial's point,
Still ending at the arrival of an hour.
An if we live, we live to tread on kings; 85
If die, brave death when princes die with us!
Now for our consciences: the arms are fair
When the intent of bearing them is just.
    *Enter another Messenger*

MESSENGER
My lord, prepare; the King comes on apace. ⌈*Exit*⌉

HOTSPUR
I thank him that he cuts me from my tale, 90
For I profess not talking, only this:
Let each man do his best. And here draw I
A sword whose temper I intend to stain
With the best blood that I can meet withal
In the adventure of this perilous day. 95
Now *Esperance*! Percy! And set on!
Sound all the lofty instruments of war,
And by that music let us all embrace,
For, heaven to earth, some of us never shall
A second time do such a courtesy. 100
    *The trumpets sound. Here they embrace. Exeunt*

**5.3**  *King Henry enters with his power. Alarum, and*
    *exeunt to the battle. Then enter the Earl of*
    *Douglas, and Sir Walter Blunt, disguised as the*
    *King*

BLUNT
What is thy name, that in the battle thus
Thou crossest me? What honour dost thou seek
Upon my head?

DOUGLAS        Know then my name is Douglas,
And I do haunt thee in the battle thus
Because some tell me that thou art a king. 5

BLUNT They tell thee true.

DOUGLAS
The Lord of Stafford dear today hath bought
Thy likeness, for instead of thee, King Harry,
This sword hath ended him. So shall it thee,
Unless thou yield thee as my prisoner. 10

BLUNT
I was not born a yielder, thou proud Scot,
And thou shalt find a king that will revenge
Lord Stafford's death.
    *They fight. Douglas kills Blunt. Then enter Hotspur*

HOTSPUR
O Douglas, hadst thou fought at Holmedon thus,
I never had triumphed upon a Scot. 15

DOUGLAS
All's done, all's won: here breathless lies the King.

HOTSPUR Where?

DOUGLAS Here.

HOTSPUR
This, Douglas? No, I know this face full well.
A gallant knight he was; his name was Blunt— 20
Semblably furnished like the King himself.

DOUGLAS (*to Blunt's body*)
A fool go with thy soul, whither it goes!
A borrowed title hast thou bought too dear.
Why didst thou tell me that thou wert a king?

HOTSPUR
The king hath many marching in his coats. 25

DOUGLAS
Now by my sword, I will kill all his coats.

I'll murder all his wardrobe, piece by piece,
Until I meet the King.

HOTSPUR        Up and away!
Our soldiers stand full fairly for the day.
    *Exeunt, leaving Blunt's body*
    *Alarum. Enter Sir John Oldcastle*

SIR JOHN Though I could scape shot-free at London, I fear
the shot here. Here's no scoring but upon the pate.—
Soft, who are you?—Sir Walter Blunt. There's honour
for you. Here's no vanity. I am as hot as molten lead,
and as heavy too. God keep lead out of me; I need no
more weight than mine own bowels. I have led my
ragamuffins where they are peppered; there's not three
of my hundred and fifty left alive, and they are for the
town's end, to beg during life. 38
    *Enter Prince Harry*
But who comes here?

PRINCE HARRY
What, stand'st thou idle here? Lend me thy sword. 40
Many a noble man lies stark and stiff
Under the hoofs of vaunting enemies,
Whose deaths as yet are unrevenged. I prithee
Lend me thy sword.

SIR JOHN O Hal, I prithee give me leave to breathe awhile. 45
Turk Gregory never did such deeds in arms
As I have done this day. I have paid Percy,
I have made him sure.

PRINCE HARRY        He is indeed,
And living to kill thee. I prithee
Lend me thy sword.

SIR JOHN        Nay, before God, Hal, 50
If Percy be alive thou gett'st not my sword;
But take my pistol if thou wilt.

PRINCE HARRY
Give it me. What, is it in the case?

SIR JOHN        Ay, Hal;
'Tis hot, 'tis hot. There's that will sack a city.
    *The Prince draws it out, and finds it to be a bottle of*
    *sack*

PRINCE HARRY
What, is it a time to jest and dally now? 55
    *He throws the bottle at him. Exit*

SIR JOHN Well, if Percy be alive, I'll pierce him. If he do
come in my way, so; if he do not, if I come in his
willingly, let him make a carbonado of me. I like not
such grinning honour as Sir Walter hath. Give me life,
which if I can save, so; if not, honour comes unlooked
for, and there's an end.    *Exit* ⌈*with Blunt's body*⌉

**5.4**  *Alarum. Excursions. Enter King Henry, Prince*
    *Harry, wounded, Lord John of Lancaster, and the*
    *Earl of Westmorland*

KING HENRY
I prithee, Harry, withdraw thyself, thou bleed'st too
much.
Lord John of Lancaster, go you with him.

JOHN OF LANCASTER
Not I, my lord, unless I did bleed too.

PRINCE HARRY (*to the King*)
I beseech your majesty, make up,
Lest your retirement do amaze your friends. 5

KING HENRY
I will do so. My lord of Westmorland,
Lead him to his tent.

WESTMORLAND (*to the Prince*)
  Come, my lord, I'll lead you to your tent.
PRINCE HARRY
  Lead me, my lord? I do not need your help,
  And God forbid a shallow scratch should drive          10
  The Prince of Wales from such a field as this,
  Where stained nobility lies trodden on,
  And rebels' arms triumph in massacres.
JOHN OF LANCASTER
  We breathe too long. Come, cousin Westmorland,
  Our duty this way lies. For God's sake, come.          15
              *Exeunt Lancaster and Westmorland*
PRINCE HARRY
  By God, thou hast deceived me, Lancaster;
  I did not think thee lord of such a spirit.
  Before I loved thee as a brother, John,
  But now I do respect thee as my soul.
KING HENRY
  I saw him hold Lord Percy at the point               20
  With lustier maintenance than I did look for
  Of such an ungrown warrior.
PRINCE HARRY
  O, this boy lends mettle to us all!                     *Exit*
              *Enter the Earl of Douglas*
DOUGLAS
  Another king! They grow like Hydra's heads.
  I am the Douglas, fatal to all those                 25
  That wear those colours on them. What art thou
  That counterfeit'st the person of a king?
KING HENRY
  The King himself, who, Douglas, grieves at heart
  So many of his shadows thou hast met
  And not the very King. I have two boys               30
  Seek Percy and thyself about the field;
  But seeing thou fall'st on me so luckily,
  I will assay thee; and defend thyself.
DOUGLAS
  I fear thou art another counterfeit;
  And yet, in faith, thou bear'st thee like a king.    35
  But mine I am sure thou art, whoe'er thou be,
  And thus I win thee.
         *They fight. The King being in danger, enter Prince
         Harry*
PRINCE HARRY
  Hold up thy head, vile Scot, or thou art like
  Never to hold it up again. The spirits
  Of valiant Shirley, Stafford, Blunt, are in my arms.  40
  It is the Prince of Wales that threatens thee,
  Who never promiseth but he means to pay.
         *They fight. Douglas flieth*
  Cheerly, my lord! How fares your grace?
  Sir Nicholas Gawsey hath for succour sent,
  And so hath Clifton. I'll to Clifton straight.       45
KING HENRY Stay and breathe awhile.
  Thou hast redeemed thy lost opinion,
  And showed thou mak'st some tender of my life,
  In this fair rescue thou hast brought to me.
PRINCE HARRY
  O God, they did me too much injury                   50
  That ever said I hearkened for your death.
  If it were so, I might have let alone
  The insulting hand of Douglas over you,
  Which would have been as speedy in your end
  As all the poisonous potions in the world,          55
  And saved the treacherous labour of your son.

KING HENRY
  Make up to Clifton; I'll to Sir Nicholas Gawsey.    *Exit*
              *Enter Hotspur*
HOTSPUR
  If I mistake not, thou art Harry Monmouth.
PRINCE HARRY
  Thou speak'st as if I would deny my name.
HOTSPUR
  My name is Harry Percy.
PRINCE HARRY                    Why then, I see        60
  A very valiant rebel of the name.
  I am the Prince of Wales; and think not, Percy,
  To share with me in glory any more.
  Two stars keep not their motion in one sphere,
  Nor can one England brook a double reign            65
  Of Harry Percy and the Prince of Wales.
HOTSPUR
  Nor shall it, Harry, for the hour is come
  To end the one of us, and would to God
  Thy name in arms were now as great as mine.
PRINCE HARRY
  I'll make it greater ere I part from thee,          70
  And all the budding honours on thy crest
  I'll crop to make a garland for my head.
HOTSPUR
  I can no longer brook thy vanities.
         *They fight.*
         *Enter Sir John Oldcastle*
SIR JOHN Well said, Hal! To it, Hal! Nay, you shall find
  no boy's play here, I can tell you.                  75
         *Enter Douglas. He fighteth with Sir John, who falls
         down as if he were dead. Exit Douglas. The Prince
         killeth Hotspur*
HOTSPUR
  O Harry, thou hast robbed me of my youth.
  I better brook the loss of brittle life
  Than those proud titles thou hast won of me.
  They wound my thoughts worse than thy sword my
     flesh.
  But thoughts, the slaves of life, and life, time's fool, 80
  And time, that takes survey of all the world,
  Must have a stop. O, I could prophesy,
  But that the earthy and cold hand of death
  Lies on my tongue. No, Percy, thou art dust,
  And food for—                                  *He dies*
PRINCE HARRY
  For worms, brave Percy. Fare thee well, great heart.
  Ill-weaved ambition, how much art thou shrunk!      87
  When that this body did contain a spirit,
  A kingdom for it was too small a bound,
  But now two paces of the vilest earth              90
  Is room enough. This earth that bears thee dead
  Bears not alive so stout a gentleman.
  If thou wert sensible of courtesy,
  I should not make so dear a show of zeal;
  But let my favours hide thy mangled face,          95
         *He covers Hotspur's face*
  And even in thy behalf I'll thank myself
  For doing these fair rites of tenderness.
  Adieu, and take thy praise with thee to heaven.
  Thy ignominy sleep with thee in the grave,
  But not remembered in thy epitaph.                 100
         *He spieth Sir John on the ground*
  What, old acquaintance! Could not all this flesh
  Keep in a little life? Poor Jack, farewell.
  I could have better spared a better man.

O, I should have a heavy miss of thee,
If I were much in love with vanity. 105
Death hath not struck so fat a deer today,
Though many dearer in this bloody fray.
Embowelled will I see thee by and by.
Till then, in blood by noble Percy lie.      *Exit*
     *Sir John riseth up*

SIR JOHN Embowelled? If thou embowel me today, I'll give
you leave to powder me, and eat me too, tomorrow.
'Sblood, 'twas time to counterfeit, or that hot termagant
Scot had paid me, scot and lot too. Counterfeit? I lie, I
am no counterfeit. To die is to be a counterfeit, for he
is but the counterfeit of a man who hath not the life
of a man. But to counterfeit dying when a man thereby
liveth is to be no counterfeit, but the true and perfect
image of life indeed. The better part of valour is
discretion, in the which better part I have saved my
life. Zounds, I am afraid of this gunpowder Percy,
though he be dead. How if he should counterfeit too,
and rise? By my faith, I am afraid he would prove the
better counterfeit. Therefore I'll make him sure; yea,
and I'll swear I killed him. Why may not he rise as
well as I? Nothing confutes me but eyes, and nobody
sees me. Therefore, sirrah, (*stabbing Hotspur*) with a
new wound in your thigh, come you along with me.
     *He takes up Hotspur on his back.*
     *Enter Prince Harry and Lord John of Lancaster*

PRINCE HARRY
Come, brother John. Full bravely hast thou fleshed 128
Thy maiden sword.

JOHN OF LANCASTER      But soft; whom have we here?
Did you not tell me this fat man was dead? 130

PRINCE HARRY I did; I saw him dead,
Breathless and bleeding on the ground.
(*To Sir John*)      Art thou alive?
Or is it fantasy that plays upon our eyesight?
I prithee speak; we will not trust our eyes
Without our ears. Thou art not what thou seem'st.

SIR JOHN No, that's certain: I am not a double man. But
if I be not Jack Oldcastle, then am I a jack. There is
Percy. If your father will do me any honour, so; if not,
let him kill the next Percy himself. I look to be either
earl or duke, I can assure you. 140

PRINCE HARRY
Why, Percy I killed myself, and saw thee dead.

SIR JOHN Didst thou? Lord, Lord, how this world is given
to lying! I grant you I was down and out of breath,
and so was he; but we rose both at an instant, and
fought a long hour by Shrewsbury clock. If I may be
believed, so; if not, let them that should reward valour
bear the sin upon their own heads. I'll take't on my
death I gave him this wound in the thigh. If the man
were alive and would deny it, zounds, I would make
him eat a piece of my sword. 150

JOHN OF LANCASTER
This is the strangest tale that e'er I heard.

PRINCE HARRY
This is the strangest fellow, brother John.
(*To Sir John*) Come, bring your luggage nobly on your
back.
For my part, if a lie may do thee grace,
I'll gild it with the happiest terms I have. 155
     *A retreat is sounded*
The trumpet sounds retreat; the day is our.
Come, brother, let us to the highest of the field
To see what friends are living, who are dead.
     *Exeunt the Prince and Lancaster*

SIR JOHN I'll follow, as they say, for reward. He that
rewards me, God reward him. If I do grow great, I'll
grow less; for I'll purge, and leave sack, and live
cleanly, as a nobleman should do. 162
     *Exit, bearing Hotspur's body*

**5.5**      *The trumpets sound. Enter King Henry, Prince*
     *Harry, Lord John of Lancaster, the Earl of*
     *Westmorland, with the Earl of Worcester and Sir*
     *Richard Vernon, prisoners, ⌈and soldiers⌉*

KING HENRY
Thus ever did rebellion find rebuke.
Ill-spirited Worcester, did not we send grace,
Pardon, and terms of love to all of you?
And wouldst thou turn our offers contrary,
Misuse the tenor of thy kinsman's trust? 5
Three knights upon our party slain today,
A noble earl, and many a creature else,
Had been alive this hour
If like a Christian thou hadst truly borne
Betwixt our armies true intelligence. 10

WORCESTER
What I have done my safety urged me to,
And I embrace this fortune patiently,
Since not to be avoided it falls on me.

KING HENRY
Bear Worcester to the death, and Vernon too.
Other offenders we will pause upon. 15
     *Exeunt Worcester and Vernon, guarded*
How goes the field?

PRINCE HARRY
The noble Scot Lord Douglas, when he saw
The fortune of the day quite turned from him,
The noble Percy slain, and all his men
Upon the foot of fear, fled with the rest; 20
And falling from a hill he was so bruised
That the pursuers took him. At my tent
The Douglas is, and I beseech your grace
I may dispose of him.

KING HENRY With all my heart. 25

PRINCE HARRY
Then, brother John of Lancaster,
To you this honourable bounty shall belong.
Go to the Douglas, and deliver him
Up to his pleasure ransomless and free.
His valours shown upon our crests today 30
Have taught us how to cherish such high deeds
Even in the bosom of our adversaries.

JOHN OF LANCASTER
I thank your grace for this high courtesy,
Which I shall give away immediately.

KING HENRY
Then this remains, that we divide our power. 35
You, son John, and my cousin Westmorland,
Towards York shall bend you with your dearest speed
To meet Northumberland and the prelate Scrope,
Who, as we hear, are busily in arms.
Myself and you, son Harry, will towards Wales, 40
To fight with Glyndŵr and the Earl of March.
Rebellion in this land shall lose his sway,
Meeting the check of such another day;
And since this business so fair is done,
Let us not leave till all our own be won. 45
     *Exeunt ⌈the King, the Prince, and their power*
     *at one door, Lancaster, Westmorland, and ⌈their⌉*
     *power at another door⌉*

# THE MERRY WIVES OF WINDSOR

A LEGEND dating from 1702 claims that Shakespeare wrote *The Merry Wives of Windsor* in fourteen days and by command of Queen Elizabeth; in 1709 she was said to have wished particularly to see Falstaff in love. Whether or not this is true, a passage towards the end of the play alluding directly to the ceremonies of the Order of the Garter, Britain's highest order of chivalry, encourages the belief that the play has a direct connection with a specific occasion. In 1597 George Carey, Lord Hunsdon, Lord Chamberlain and patron of Shakespeare's company, was installed at Windsor as a Knight of the Garter. The Queen was not present at the installation but had attended the Garter Feast at the Palace of Westminster on St George's Day (23 April). Shakespeare's play was probably performed in association with this occasion, and may have been written especially for it. It was first printed, in a corrupt text, in 1602; a better text appears in the 1623 Folio.

Some of the characters—Sir John Falstaff, Mistress Quickly, Pistol, Nim, Justice Shallow—appear also in *1* and *2 Henry IV* and *Henry V*, but in spite of a reference to 'the wild Prince and Poins' at 3.2.66-7, this is essentially an Elizabethan comedy, the only one that Shakespeare set firmly in England. The play is full of details that would have been familiar to Elizabethan Londoners, and the language is colloquial and up to date. The plot, however, is made up of conventional situations whose ancestry is literary rather than realistic. There are many analogues in medieval and other tales to Shakespeare's basic plot situations, some in books that he probably or certainly knew. The central story, of Sir John's unsuccessful attempts to seduce Mistress Page and Mistress Ford, and of Master Ford's unfounded jealousy, is in the tradition of the Italian *novella*, and may have been suggested by Ser Giovanni Fiorentino's *Il Pecorone* (1558). Alongside it Shakespeare places the comical but finally romantic love story of Anne Page, wooed by the foolish but rich Abraham Slender and the irascible French Doctor Caius, but won by the young and handsome Fenton. The play contains a higher proportion of prose to verse than any other play by Shakespeare, and the action is often broadly comic; but it ends, after the midnight scene in Windsor Forest during which Sir John is frightened out of his lechery, in forgiveness and love.

The play is known to have been acted for James I on 4 November 1604, and for Charles I in 1638. It was revived soon after the theatres reopened, in 1660; at first it was not particularly popular, but since 1720 it has consistently pleased audiences. Many artists have illustrated it, and it forms the basis for a number of operas, including Otto Nicolai's *Die lustigen Weiben von Windsor* (1848) and Giuseppe Verdi's comic masterpiece, *Falstaff* (1893).

# THE PERSONS OF THE PLAY

MISTRESS Margaret PAGE

Master George PAGE, her husband

ANNE Page, their daughter

WILLIAM Page, their son

MISTRESS Alice FORD

Master Frank FORD, her husband

JOHN

ROBERT } their servants

citizens of Windsor

SIR JOHN Falstaff

BARDOLPH

PISTOL } Sir John's followers

NIM

ROBIN, Sir John's page

The HOST of the Garter Inn

Sir Hugh EVANS, a Welsh parson

Doctor CAIUS, a French physician

MISTRESS QUICKLY, his housekeeper

John RUGBY, his servant

Master FENTON, a young gentleman, in love with Anne Page

Master Abraham SLENDER

Robert SHALLOW, his uncle, a Justice

Peter SIMPLE, Slender's servant

Children of Windsor, appearing as fairies

# The Merry Wives of Windsor

**1.1** *Enter Justice Shallow, Master Slender, and Sir Hugh Evans*

SHALLOW Sir Hugh, persuade me not. I will make a Star Chamber matter of it. If he were twenty Sir John Falstaffs, he shall not abuse Robert Shallow, Esquire.

SLENDER In the county of Gloucester, Justice of Peace and Coram. 5

SHALLOW Ay, cousin Slender, and Custalorum.

SLENDER Ay, and Ratolorum too; and a gentleman born, Master Parson, who writes himself 'Armigero' in any bill, warrant, quittance, or obligation: 'Armigero'.

SHALLOW Ay, that I do, and have done any time these three hundred years. 11

SLENDER All his successors gone before him hath done't, and all his ancestors that come after him may. They may give the dozen white luces in their coat.

SHALLOW It is an old coat. 15

EVANS The dozen white louses do become an old coad well. It agrees well passant: it is a familiar beast to man, and signifies love.

SHALLOW The luce is the fresh fish; the salt fish is an old cod. 20

SLENDER I may quarter, coz.

SHALLOW You may, by marrying.

EVANS It is marring indeed if he quarter it.

SHALLOW Not a whit.

EVANS Yes, py'r Lady. If he has a quarter of your coat, there is but three skirts for yourself, in my simple conjectures. But that is all one. If Sir John Falstaff have committed disparagements unto you, I am of the Church, and will be glad to do my benevolence to make atonements and compromises between you. 30

SHALLOW The Council shall hear it; it is a riot.

EVANS It is not meet the Council hear a riot. There is no fear of Got in a riot. The Council, look you, shall desire to hear the fear of Got, and not to hear a riot. Take your 'visaments in that. 35

SHALLOW Ha! O' my life, if I were young again, the sword should end it.

EVANS It is petter that friends is the sword and end it. And there is also another device in my prain, which peradventure prings goot discretions with it. There is Anne Page which is daughter to Master George Page, which is pretty virginity. 42

SLENDER Mistress Anne Page? She has brown hair, and speaks small like a woman?

EVANS It is that fery person for all the 'orld, as just as you will desire. And seven hundred pounds of moneys, and gold and silver, is her grandsire upon his death's-bed—Got deliver to a joyful resurrections—give, when she is able to overtake seventeen years old. It were a goot motion if we leave our pribbles and prabbles, and desire a marriage between Master Abraham and Mistress Anne Page. 52

SLENDER Did her grandsire leave her seven hundred pound?

EVANS Ay, and her father is make her a petter penny.

⌈SHALLOW⌉ I know the young gentlewoman. She has good gifts. 57

EVANS Seven hundred pounds and possibilities is goot gifts.

SHALLOW Well, let us see honest Master Page. Is Falstaff there? 61

EVANS Shall I tell you a lie? I do despise a liar as I do despise one that is false, or as I despise one that is not true. The knight Sir John is there, and I beseech you be ruled by your well-willers. I will peat the door for Master Page. 66

*He knocks on the door*

What ho! Got pless your house here!

PAGE ⌈*within*⌉ Who's there?

EVANS Here is Got's plessing and your friend, and Justice Shallow, and here young Master Slender, that peradventures shall tell you another tale if matters grow to your likings. 72

⌈*Enter Master Page*⌉

PAGE I am glad to see your worships well. I thank you for my venison, Master Shallow.

SHALLOW Master Page, I am glad to see you. Much good do it your good heart! I wished your venison better; it was ill killed.—How doth good Mistress Page?—And I thank you always with my heart, la, with my heart.

PAGE Sir, I thank you.

SHALLOW Sir, I thank you. By yea and no, I do. 80

PAGE I am glad to see you, good Master Slender.

SLENDER How does your fallow greyhound, sir? I heard say he was outrun on Cotswold.

PAGE It could not be judged, sir.

SLENDER You'll not confess, you'll not confess. 85

SHALLOW That he will not. 'Tis your fault, 'tis your fault. (*To Page*) 'Tis a good dog.

PAGE A cur, sir.

SHALLOW Sir, he's a good dog and a fair dog. Can there be more said? He is good and fair. Is Sir John Falstaff here? 91

PAGE Sir, he is within; and I would I could do a good office between you.

EVANS It is spoke as a Christians ought to speak.

SHALLOW He hath wronged me, Master Page. 95

PAGE Sir, he doth in some sort confess it.

SHALLOW If it be confessed, it is not redressed. Is not that so, Master Page? He hath wronged me; indeed he hath; at a word, he hath. Believe me, Robert Shallow, Esquire, saith he is wronged. 100

*Enter Sir John Falstaff, Bardolph, Nim, and Pistol*

PAGE Here comes Sir John.

SIR JOHN Now, Master Shallow, you'll complain of me to the King?

SHALLOW Knight, you have beaten my men, killed my deer, and broke open my lodge. 105

SIR JOHN But not kissed your keeper's daughter?

SHALLOW Tut, a pin. This shall be answered.

SIR JOHN I will answer it straight: I have done all this. That is now answered.

SHALLOW The Council shall know this. 110

SIR JOHN 'Twere better for you if it were known in counsel. You'll be laughed at.

EVANS *Pauca verba*, Sir John, good worts.

SIR JOHN Good worts? Good cabbage!—Slender, I broke
your head. What matter have you against me?    115
SLENDER Marry, sir, I have matter in my head against
you, and against your cony-catching rascals, Bardolph,
Nim, and Pistol.
BARDOLPH You Banbury cheese!
SLENDER Ay, it is no matter.    120
PISTOL How now, Mephistopheles?
SLENDER Ay, it is no matter.
NIM Slice, I say *pauca, pauca*. Slice, that's my humour.
SLENDER (*to Shallow*) Where's Simple, my man? Can you
tell, cousin?    125
EVANS Peace, I pray you. Now let us understand. There
is three umpires in this matter, as I understand: that
is, Master Page, *fidelicet* Master Page; and there is
myself, *fidelicet* myself; and the three party is, lastly
and finally, mine Host of the Garter.    130
PAGE We three to hear it, and end it between them.
EVANS Fery goot. I will make a prief of it in my notebook,
and we will afterwards 'ork upon the cause with as
great discreetly as we can.
SIR JOHN Pistol.    135
PISTOL He hears with ears.
EVANS The tevil and his tam! What phrase is this? 'He
hears with ear'! Why, it is affectations.
SIR JOHN Pistol, did you pick Master Slender's purse?
SLENDER Ay, by these gloves did he—or I would I might
never come in mine own great chamber again else—
of seven groats in mill-sixpences, and two Edward
shovel-boards that cost me two shilling and twopence
apiece of Ed Miller. By these gloves.
SIR JOHN Is this true, Pistol?    145
EVANS No, it is false, if it is a pickpurse.
PISTOL
Ha, thou mountain-foreigner! Sir John and master
mine,
I combat challenge of this latten bilbo.—
Word of denial in thy *labras* here,
Word of denial: froth and scum, thou liest.    150
SLENDER (*pointing to Nim*) By these gloves, then, 'twas he.
NIM Be advised, sir, and pass good humours. I will say
'marry, trap with you' if you run the nuthook's humour
on me. That is the very note of it.
SLENDER By this hat, then, he in the red face had it. For
though I cannot remember what I did when you made
me drunk, yet I am not altogether an ass.    157
SIR JOHN (*to Bardolph*) What say you, Scarlet and John?
BARDOLPH Why, sir, for my part I say the gentleman had
drunk himself out of his five sentences.    160
EVANS It is 'his five senses'. Fie, what the ignorance is!
BARDOLPH And being fap, sir, was, as they say, cashiered.
And so conclusions passed the careers.
SLENDER Ay, you spake in Latin then, too. But 'tis no
matter. I'll ne'er be drunk, whilst I live, again, but in
honest, civil, godly company, for this trick. If I be
drunk, I'll be drunk with those that have the fear of
God, and not with drunken knaves.    168
EVANS So Got 'udge me, that is a virtuous mind.
SIR JOHN You hear all these matters denied, gentlemen,
you hear it.
*Enter Anne Page, with wine*
PAGE Nay, daughter, carry the wine in; we'll drink
within.    *Exit Anne*
SLENDER O heaven, this is Mistress Anne Page!
⌜*Enter at another door Mistress Ford and Mistress
Page*⌝
PAGE How now, Mistress Ford?    175

SIR JOHN Mistress Ford, by my troth, you are very well
met. By your leave, good mistress.
⌜*He kisses her*⌝
PAGE Wife, bid these gentlemen welcome.—Come, we
have a hot venison pasty to dinner. Come, gentlemen,
I hope we shall drink down all unkindness.    180
*Exeunt all but Slender*
SLENDER I had rather than forty shillings I had my book
of songs and sonnets here.
*Enter Simple*
How now, Simple, where have you been? I must wait
on myself, must I? You have not the book of riddles
about you, have you?    185
SIMPLE Book of riddles? Why, did you not lend it to Alice
Shortcake upon Allhallowmas last, a fortnight afore
Michaelmas?
*Enter Shallow and Evans*
SHALLOW (*to Slender*) Come, coz; come, coz; we stay for
you. (*Aside to him*) A word with you, coz.    190
*He draws Slender aside*
Marry, this, coz: there is, as 'twere, a tender, a kind
of tender, made afar off by Sir Hugh here. Do you
understand me?
SLENDER Ay, sir, you shall find me reasonable. If it be so,
I shall do that that is reason.    195
SHALLOW Nay, but understand me.
SLENDER So I do, sir.
EVANS Give ear to his motions. Master Slender, I will
description the matter to you, if you be capacity of it.
SLENDER Nay, I will do as my cousin Shallow says. I pray
you pardon me. He's a Justice of Peace in his country,
simple though I stand here.    202
EVANS But that is not the question. The question is
concerning your marriage.
SHALLOW Ay, there's the point, sir.    205
EVANS Marry, is it, the very point of it—to Mistress Anne
Page.
SLENDER Why, if it be so, I will marry her upon any
reasonable demands.
EVANS But can you affection the 'oman? Let us command
to know that of your mouth, or of your lips—for divers
philosophers hold that the lips is parcel of the mouth.
Therefore, precisely, can you carry your good will to
the maid?    214
SHALLOW Cousin Abraham Slender, can you love her?
SLENDER I hope, sir, I will do as it shall become one that
would do reason.
EVANS Nay, Got's lords and his ladies, you must speak
positable if you can carry her your desires towards her.
SHALLOW That you must. Will you, upon good dowry,
marry her?    221
SLENDER I will do a greater thing than that upon your
request, cousin, in any reason.
SHALLOW Nay, conceive me, conceive me, sweet coz.
What I do is to pleasure you, coz. Can you love the
maid?    226
SLENDER I will marry her, sir, at your request. But if there
be no great love in the beginning, yet heaven may
decrease it upon better acquaintance, when we are
married and have more occasion to know one another.
I hope upon familiarity will grow more contempt. But
if you say 'marry her', I will marry her. That I am
freely dissolved, and dissolutely.    233
EVANS It is a fery discretion answer, save the faul' is in
the 'ord 'dissolutely'. The 'ort is, according to our
meaning, 'resolutely'. His meaning is good.
SHALLOW Ay, I think my cousin meant well.

SLENDER Ay, or else I would I might be hanged, la.
*Enter Anne Page*
SHALLOW Here comes fair Mistress Anne.—Would I were
young for your sake, Mistress Anne.                    240
ANNE The dinner is on the table. My father desires your
worships' company.
SHALLOW I will wait on him, fair Mistress Anne.
EVANS 'Od's plessed will, I will not be absence at the
grace.                         *Exeunt Shallow and Evans*
ANNE (*to Slender*) Will't please your worship to come in,
sir?                                                   247
SLENDER No, I thank you, forsooth, heartily; I am very
well.
ANNE The dinner attends you, sir.                      250
SLENDER I am not a-hungry, I thank you, forsooth. (*To
Simple*) Go, sirrah; for all you are my man, go wait
upon my cousin Shallow.                *Exit Simple*
A Justice of Peace sometime may be beholden to his
friend for a man. I keep but three men and a boy yet,
till my mother be dead. But what though? Yet I live
like a poor gentleman born.                            257
ANNE I may not go in without your worship. They will
not sit till you come.
SLENDER I'faith, I'll eat nothing. I thank you as much as
though I did.                                          261
ANNE I pray you, sir, walk in.
⌈*Dogs bark within*⌉
SLENDER I had rather walk here, I thank you. I bruised
my shin th'other day, with playing at sword and dagger
with a master of fence—three veneys for a dish of
stewed prunes—and, by my troth, I cannot abide the
smell of hot meat since. Why do your dogs bark so?
Be there bears i'th' town?                             268
ANNE I think there are, sir. I heard them talked of.
SLENDER I love the sport well—but I shall as soon quarrel
at it as any man in England. You are afraid if you see
the bear loose, are you not?
ANNE Ay, indeed, sir.                                  273
SLENDER That's meat and drink to me, now. I have seen
Sackerson loose twenty times, and have taken him by
the chain. But I warrant you, the women have so cried
and shrieked at it that it passed. But women, indeed,
cannot abide 'em. They are very ill-favoured, rough
things.                                                279
*Enter Page*
PAGE Come, gentle Master Slender, come. We stay for
you.
SLENDER I'll eat nothing, I thank you, sir.
PAGE By cock and pie, you shall not choose, sir. Come,
come.
SLENDER Nay, pray you lead the way.                    285
PAGE Come on, sir.
SLENDER Mistress Anne, yourself shall go first.
ANNE Not I, sir. Pray you keep on.
SLENDER Truly, I will not go first, truly, la. I will not do
you that wrong.                                        290
ANNE I pray you, sir.
SLENDER I'll rather be unmannerly than troublesome. You
do yourself wrong, indeed, la.
*Exeunt ⌈Slender first, the others following⌉*

**1.2**     *Enter Sir Hugh Evans and Simple, ⌈from dinner⌉*
EVANS Go your ways, and ask of Doctor Caius' house
which is the way. And there dwells one Mistress
Quickly, which is in the manner of his 'oman, or his
dry-nurse, or his cook, or his laundry, his washer, and
his wringer.                                             5

SIMPLE Well, sir.
EVANS Nay, it is petter yet. Give her this letter, for it is a
'oman that altogethers acquaintance with Mistress
Anne Page. And the letter is to desire and require her
to solicit your master's desires to Mistress Anne Page.
I pray you be gone.                      ⌈*Exit Simple*⌉
I will make an end of my dinner; there's pippins and
cheese to come.                                    *Exit*

**1.3**     *Enter Sir John Falstaff, Bardolph, Nim, Pistol, and
Robin*
SIR JOHN Mine Host of the Garter!
*Enter the Host of the Garter*
HOST What says my bully rook? Speak scholarly and
wisely.
SIR JOHN Truly, mine Host, I must turn away some of my
followers.                                               5
HOST Discard, bully Hercules, cashier. Let them wag. Trot,
trot.
SIR JOHN I sit at ten pounds a week.
HOST Thou'rt an emperor: Caesar, kaiser, and pheezer. I
will entertain Bardolph. He shall draw, he shall tap.
Said I well, bully Hector?                              11
SIR JOHN Do so, good mine Host.
HOST I have spoke; let him follow. (*To Bardolph*) Let me
see thee froth and lime. I am at a word: follow.    *Exit*
SIR JOHN Bardolph, follow him. A tapster is a good trade.
An old cloak makes a new jerkin; a withered
servingman a fresh tapster. Go; adieu.                 17
BARDOLPH It is a life that I have desired. I will thrive.
⌈*Exit*⌉
PISTOL
O base Hungarian wight, wilt thou the spigot wield?
NIM He was gotten in drink; his mind is not heroic. Is
not the humour conceited?                              21
SIR JOHN I am glad I am so acquit of this tinderbox. His
thefts were too open. His filching was like an unskilful
singer: he kept not time.
NIM The good humour is to steal at a minute's rest.     25
PISTOL
'Convey' the wise it call. 'Steal'? Foh, a fico for the
phrase!
SIR JOHN Well, sirs, I am almost out at heels.
PISTOL Why then, let kibes ensue.
SIR JOHN There is no remedy: I must cony-catch, I must
shift.                                                 30
PISTOL Young ravens must have food.
SIR JOHN Which of you know Ford of this town?
PISTOL I ken the wight. He is of substance good.
SIR JOHN My honest lads, I will tell you what I am about.
PISTOL Two yards and more.                             35
SIR JOHN No quips now, Pistol. Indeed, I am in the waist
two yards about. But I am now about no waste; I am
about thrift. Briefly, I do mean to make love to Ford's
wife. I spy entertainment in her. She discourses, she
carves, she gives the leer of invitation. I can construe
the action of her familiar style; and the hardest voice
of her behaviour, to be Englished rightly, is 'I am Sir
John Falstaff's'.
PISTOL He hath studied her well, and translated her will:
out of honesty, into English.                          45
NIM The anchor is deep. Will that humour pass?
SIR JOHN Now, the report goes, she has all the rule of her
husband's purse; he hath a legion of angels.
PISTOL
As many devils entertain, and 'To her, boy!' say I.

NIM The humour rises; it is good. Humour me the angels!

SIR JOHN (*showing letters*) I have writ me here a letter to her—and here another to Page's wife, who even now gave me good eyes too, examined my parts with most judicious oeillades; sometimes the beam of her view gilded my foot, sometimes my portly belly. 55

PISTOL
Then did the sun on dunghill shine.

NIM I thank thee for that humour.

SIR JOHN O, she did so course o'er my exteriors, with such a greedy intention, that the appetite of her eye did seem to scorch me up like a burning-glass! Here's another letter to her. She bears the purse too. She is a region in Guiana, all gold and bounty. I will be cheaters to them both, and they shall be exchequers to me. They shall be my East and West Indies, and I will trade to them both. (*Giving a letter to Pistol*) Go bear thou this letter to Mistress Page, (*giving a letter to Nim*) and thou this to Mistress Ford. We will thrive, lads, we will thrive.

PISTOL (*returning the letter*)
Shall I Sir Pandarus of Troy become,
And by my side wear steel? Then Lucifer take all. 70

NIM (*returning the letter*) I will run no base humour. Here, take the humour-letter. I will keep the haviour of reputation.

SIR JOHN (*to Robin*)
Hold, sirrah. Bear you these letters tightly.
Sail like my pinnace to these golden shores. 75
*He gives Robin the letters*
Rogues, hence, avaunt! Vanish like hailstones! Go!
Trudge, plod, away o'th' hoof, seek shelter, pack!
Falstaff will learn the humour of the age:
French thrift, you rogues—myself and skirted page.
*Exeunt Sir John and Robin*

PISTOL
Let vultures gripe thy guts!—for gourd and fullam holds, 80
And high and low beguiles the rich and poor.
Tester I'll have in pouch when thou shalt lack,
Base Phrygian Turk!

NIM
I have operations which be humours of revenge.

PISTOL
Wilt thou revenge?

NIM                    By welkin and her stars! 85

PISTOL
With wit or steel?

NIM                    With both the humours, I.
I will discuss the humour of this love to Ford.

PISTOL
And I to Page shall eke unfold
    How Falstaff, varlet vile,
His dove will prove, his gold will hold, 90
    And his soft couch defile.

NIM My humour shall not cool. I will incense Ford to deal with poison; I will possess him with yellowness; for this revolt of mine is dangerous. That is my true humour. 95

PISTOL
Thou art the Mars of malcontents.
I second thee. Troop on.                    *Exeunt*

**1.4** *Enter Mistress Quickly and Simple*
MISTRESS QUICKLY What, John Rugby!
    *Enter John Rugby*
I pray thee, go to the casement and see if you can see

my master, Master Doctor Caius, coming. If he do, i'faith, and find anybody in the house, here will be an old abusing of God's patience and the King's English.

RUGBY I'll go watch. 6

MISTRESS QUICKLY Go; and we'll have a posset for't soon at night, in faith, at the latter end of a seacoal fire.
    *Exit Rugby*
An honest, willing, kind fellow as ever servant shall come in house withal; and, I warrant you, no telltale, nor no breedbate. His worst fault is that he is given to prayer; he is something peevish that way—but nobody but has his fault. But let that pass. Peter Simple you say your name is?

SIMPLE Ay, for fault of a better. 15

MISTRESS QUICKLY And Master Slender's your master?

SIMPLE Ay, forsooth.

MISTRESS QUICKLY Does he not wear a great round beard, like a glover's paring-knife?

SIMPLE No, forsooth; he hath but a little whey face, with a little yellow beard, a Cain-coloured beard. 21

MISTRESS QUICKLY A softly spirited man, is he not?

SIMPLE Ay, forsooth; but he is as tall a man of his hands as any is between this and his head. He hath fought with a warrener. 25

MISTRESS QUICKLY How say you?—O, I should remember him: does he not hold up his head, as it were, and strut in his gait?

SIMPLE Yes, indeed does he.

MISTRESS QUICKLY Well, heaven send Anne Page no worse fortune! Tell Master Parson Evans I will do what I can for your master. Anne is a good girl, and I wish— 32
    *Enter Rugby*

RUGBY Out, alas, here comes my master!    ⌜*Exit*⌝

MISTRESS QUICKLY We shall all be shent. Run in here, good young man; for God's sake, go into this closet. He will not stay long. 36
    *Simple steps into the closet*
What, John Rugby! John! What, John, I say!
    ⌜*Enter Rugby*⌝
⌜*Speaking loudly*⌝ Go, John, go enquire for my master. I doubt he be not well, that he comes not home.
    ⌜*Exit Rugby*⌝
(*Singing*) 'And down, down, adown-a' (*etc.*) 40
    *Enter Doctor Caius*

CAIUS Vat is you sing? I do not like dese toys. Pray you go and vetch me in my closet *un boîtier vert*—a box, a green-a box. Do intend vat I speak? A green-a box.

MISTRESS QUICKLY Ay, forsooth, I'll fetch it you. (*Aside*) I am glad he went not in himself. If he had found the young man, he would have been horn-mad. 46
    *She goes to fetch the box*

CAIUS *Fe, fe, fe, fe! Ma foi, il fait fort chaud! Je m'en vais à la cour. La grande affaire.*

MISTRESS QUICKLY Is it this, sir?

CAIUS *Oui. Mets-le à ma pochette. Dépêche*, quickly! Vere is dat knave Rugby? 51

MISTRESS QUICKLY What, John Rugby! John!
    ⌜*Enter Rugby*⌝

RUGBY Here, sir.

CAIUS You are John Rugby, and you are Jack Rugby. Come, take-a your rapier, and come after my heel to the court. 56

RUGBY 'Tis ready, sir, here in the porch.
    *He fetches the rapier*

CAIUS By my trot, I tarry too long. 'Od's me, *qu'ai-j' oublié?* Dere is some simples in my closet dat I vill not for the varld I shall leave behind. 60

MISTRESS QUICKLY (*aside*) Ay me, he'll find the young man there, and be mad.

CAIUS (*discovering Simple*) O *diable, diable*! Vat is in my closet? Villainy, *larron*! Rugby, my rapier!

*He takes the rapier*

MISTRESS QUICKLY Good master, be content.        65

CAIUS Wherefore shall I be content-a?

MISTRESS QUICKLY The young man is an honest man.

CAIUS What shall de honest man do in my closet? Dere is no honest man dat shall come in my closet.

MISTRESS QUICKLY I beseech you, be not so phlegmatic. Hear the truth of it. He came of an errand to me from Parson Hugh.        72

CAIUS Vell.

SIMPLE Ay, forsooth, to desire her to—

MISTRESS QUICKLY Peace, I pray you.        75

CAIUS Peace-a your tongue. (*To Simple*) Speak-a your tale.

SIMPLE To desire this honest gentlewoman, your maid, to speak a good word to Mistress Anne Page for my master in the way of marriage.

MISTRESS QUICKLY This is all, indeed, la; but I'll ne'er put my finger in the fire an need not.        81

CAIUS Sir Hugh send-a you?—Rugby, *baile* me some paper.

*Rugby brings paper*

(*To Simple*) Tarry you a little-a while.

*Caius writes*

MISTRESS QUICKLY (*aside to Simple*) I am glad he is so quiet. If he had been throughly moved, you should have heard him so loud and so melancholy. But notwithstanding, man, I'll do your master what good I can. And the very yea and the no is, the French doctor, my master—I may call him my master, look you, for I keep his house, and I wash, wring, brew, bake, scour, dress meat and drink, make the beds, and do all myself—

SIMPLE (*aside to Mistress Quickly*) 'Tis a great charge to come under one body's hand.        95

MISTRESS QUICKLY (*aside to Simple*) Are you advised o' that? You shall find it a great charge—and to be up early, and down late. But notwithstanding, to tell you in your ear—I would have no words of it—my master himself is in love with Mistress Anne Page. But notwithstanding that, I know Anne's mind: that's neither here nor there.        102

CAIUS (*giving the letter to Simple*) You, jack'nape, give-a this letter to Sir Hugh. By Gar, it is a shallenge. I will cut his troat in de Park, and I will teach a scurvy jackanape priest to meddle or make. You may be gone. It is not good you tarry here. By Gar, I will cut all his two stones. By Gar, he shall not have a stone to throw at his dog.        *Exit Simple*

MISTRESS QUICKLY Alas, he speaks but for his friend.        110

CAIUS It is no matter-a ver dat. Do not you tell-a me dat I shall have Anne Page for myself? By Gar, I vill kill de jack-priest. And I have appointed mine Host of de Jarteer to measure our weapon. By Gar, I will myself have Anne Page.        115

MISTRESS QUICKLY Sir, the maid loves you, and all shall be well. We must give folks leave to prate, what the goodyear!

CAIUS Rugby, come to the court with me. (*To Mistress Quickly*) By Gar, if I have not Anne Page, I shall turn your head out of my door. Follow my heels, Rugby.

MISTRESS QUICKLY You shall have Anne—        122

*Exeunt Caius and Rugby*

—ass-head of your own. No, I know Anne's mind for that. Never a woman in Windsor knows more of Anne's mind than I do, nor can do more than I do with her, I thank heaven.        126

FENTON (*within*) Who's within there, ho!

MISTRESS QUICKLY Who's there, I trow?—Come near the house, I pray you.

*Enter Master Fenton*

FENTON How now, good woman, how dost thou?        130

MISTRESS QUICKLY The better that it pleases your good worship to ask.

FENTON What news? How does pretty Mistress Anne?

MISTRESS QUICKLY In truth, sir, and she is pretty, and honest, and gentle, and one that is your friend. I can tell you that by the way, I praise heaven for it.        136

FENTON Shall I do any good, thinkest thou? Shall I not lose my suit?

MISTRESS QUICKLY Troth, sir, all is in His hands above. But notwithstanding, Master Fenton, I'll be sworn on a book she loves you. Have not your worship a wart above your eye?        142

FENTON Yes, marry, have I. What of that?

MISTRESS QUICKLY Well, thereby hangs a tale. Good faith, it is such another Nan!—But I detest, an honest maid as ever broke bread.—We had an hour's talk of that wart. I shall never laugh but in that maid's company.— But indeed she is given too much to allicholy and musing.—But for you—well—go to!        149

FENTON Well, I shall see her today. Hold, there's money for thee. Let me have thy voice in my behalf. If thou seest her before me, commend me.

MISTRESS QUICKLY Will I? I'faith, that I will. And I will tell your worship more of the wart the next time we have confidence, and of other wooers.        155

FENTON Well, farewell. I am in great haste now.

MISTRESS QUICKLY Farewell to your worship.

*Exit Fenton*

Truly, an honest gentleman; but Anne loves him not, for I know Anne's mind as well as another does.—Out upon't, what have I forgot?        *Exit*

**2.1        *Enter Mistress Page, with a letter***

MISTRESS PAGE What, have I scaped love-letters in the holiday time of my beauty, and am I now a subject for them? Let me see.

*She reads*

'Ask me no reason why I love you, for though Love use Reason for his precision, he admits him not for his counsellor. You are not young; no more am I. Go to, then, there's sympathy. You are merry; so am I. Ha, ha, then, there's more sympathy. You love sack, and so do I. Would you desire better sympathy? Let it suffice thee, Mistress Page, at the least if the love of soldier can suffice, that I love thee. I will not say "pity me"— 'tis not a soldier-like phrase—but I say "love me".        12

> By me, thine own true knight,
> By day or night,
> Or any kind of light,        15
> With all his might
> For thee to fight,
>
> John Falstaff.'

What a Herod of Jewry is this! O, wicked, wicked world! One that is well-nigh worn to pieces with age, to show himself a young gallant! What an unweighed behaviour hath this Flemish drunkard picked, i'th' devil's name, out of my conversation, that he dares in this manner assay me? Why, he hath not been thrice in my company. What should I say to him? I was then frugal of my mirth, heaven forgive me. Why, I'll exhibit a bill in the Parliament for the putting down of men. O God, that I knew how to be revenged on him! For revenged I will be, as sure as his guts are made of puddings.                                                                      30

*Enter Mistress Ford*

MISTRESS FORD Mistress Page! By my faith, I was going to your house.

MISTRESS PAGE And by my faith, I was coming to you. You look very ill.

MISTRESS FORD Nay, I'll ne'er believe that: I have to show to the contrary.                                                          36

MISTRESS PAGE Faith, but you do, in my mind.

MISTRESS FORD Well, I do, then. Yet I say I could show you to the contrary. O Mistress Page, give me some counsel.                                                                     40

MISTRESS PAGE What's the matter, woman?

MISTRESS FORD O woman, if it were not for one trifling respect, I could come to such honour!

MISTRESS PAGE Hang the trifle, woman; take the honour. What is it? Dispense with trifles. What is it?           45

MISTRESS FORD If I would but go to hell for an eternal moment or so, I could be knighted.

MISTRESS PAGE What? Thou liest! Sir Alice Ford? These knights will hack, and so thou shouldst not alter the article of thy gentry.                                                   50

MISTRESS FORD We burn daylight. Here: read, read.

*She gives Mistress Page a letter*

Perceive how I might be knighted.

*Mistress Page reads*

I shall think the worse of fat men as long as I have an eye to make difference of men's liking. And yet he would not swear, praised women's modesty, and gave such orderly and well-behaved reproof to all uncomeliness that I would have sworn his disposition would have gone to the truth of his words. But they do no more adhere and keep place together than the hundred and fifty psalms to the tune of 'Greensleeves'. What tempest, I trow, threw this whale, with so many tuns of oil in his belly, ashore at Windsor? How shall I be revenged on him? I think the best way were to entertain him with hope, till the wicked fire of lust have melted him in his own grease. Did you ever hear the like?                                                                         66

MISTRESS PAGE Letter for letter, but that the name of Page and Ford differs.

*She gives Mistress Ford her letter*

To thy great comfort in this mystery of ill opinions, here's the twin brother of thy letter. But let thine inherit first, for I protest mine never shall. I warrant he hath a thousand of these letters, writ with blank space for different names—sure, more, and these are of the second edition. He will print them, out of doubt—for he cares not what he puts into the press when he would put us two. I had rather be a giantess, and lie under Mount Pelion. Well, I will find you twenty lascivious turtles ere one chaste man.

MISTRESS FORD Why, this is the very same: the very hand, the very words. What doth he think of us?           80

MISTRESS PAGE Nay, I know not. It makes me almost ready to wrangle with mine own honesty. I'll entertain myself like one that I am not acquainted withal; for, sure,

unless he know some strain in me that I know not myself, he would never have boarded me in this fury.

MISTRESS FORD 'Boarding' call you it? I'll be sure to keep him above deck.                                                              87

MISTRESS PAGE So will I. If he come under my hatches, I'll never to sea again. Let's be revenged on him. Let's appoint him a meeting, give him a show of comfort in his suit, and lead him on with a fine baited delay till he hath pawned his horses to mine Host of the Garter.

MISTRESS FORD Nay, I will consent to act any villainy against him that may not sully the chariness of our honesty. O that my husband saw this letter! It would give eternal food to his jealousy.                                    96

*Enter Master Ford with Pistol, and Master Page with Nim*

MISTRESS PAGE Why, look where he comes, and my goodman too. He's as far from jealousy as I am from giving him cause; and that, I hope, is an unmeasurable distance.                                                                          100

MISTRESS FORD You are the happier woman.

MISTRESS PAGE Let's consult together against this greasy knight. Come hither.

*They withdraw.*

FORD Well, I hope it be not so.

PISTOL

Hope is a curtal dog in some affairs.                          105
Sir John affects thy wife.

FORD Why, sir, my wife is not young.

PISTOL

He woos both high and low, both rich and poor,
Both young and old, one with another, Ford.
He loves the gallimaufry, Ford. Perpend.            110

FORD Love my wife?

PISTOL

With liver burning hot. Prevent,
Or go thou like Sir Actaeon, he,
With Ringwood at thy heels.
O, odious is the name!                                             115

FORD What name, sir?

PISTOL The horn, I say. Farewell.
Take heed; have open eye; for thieves do foot by night.
Take heed ere summer comes, or cuckoo-birds do
     sing.—
Away, Sir Corporal Nim!—Believe it, Page; he speaks
     sense.                                                                  *Exit*

FORD (*aside*) I will be patient. I will find out this.    121

NIM (*to Page*) And this is true. I like not the humour of lying. He hath wronged me in some humours. I should have borne the humoured letter to her; but I have a sword, and it shall bite upon my necessity. He loves your wife. There's the short and the long.                 126
My name is Corporal Nim. I speak and I avouch 'tis true.
My name is Nim, and Falstaff loves your wife. Adieu.
I love not the humour of bread and cheese. Adieu.

*Exit*

PAGE (*aside*) The humour of it, quoth a? Here's a fellow frights English out of his wits.                                    131

FORD (*aside*) I will seek out Falstaff.

PAGE (*aside*) I never heard such a drawling, affecting rogue.

FORD (*aside*) If I do find it—well.                               135

PAGE (*aside*) I will not believe such a Cathayan though the priest o'th' town commended him for a true man.

FORD (*aside*) 'Twas a good, sensible fellow. Well.

*Mistress Page and Mistress Ford come forward*

PAGE How now, Meg?

MISTRESS PAGE Whither go you, George? Hark you. 140
*They talk apart*

MISTRESS FORD How now, sweet Frank? Why art thou melancholy?

FORD I melancholy? I am not melancholy. Get you home, go.

MISTRESS FORD Faith, thou hast some crotchets in thy head now. Will you go, Mistress Page? 146

MISTRESS PAGE Have with you.—You'll come to dinner, George?
*Enter Mistress Quickly*
(*Aside to Mistress Ford*) Look who comes yonder. She shall be our messenger to this paltry knight. 150

MISTRESS FORD (*aside to Mistress Page*) Trust me, I thought on her. She'll fit it.

MISTRESS PAGE (*to Mistress Quickly*) You are come to see my daughter Anne?

MISTRESS QUICKLY Ay, forsooth; and I pray how does good Mistress Anne? 156

MISTRESS PAGE Go in with us and see. We have an hour's talk with you.
*Exeunt Mistress Page, Mistress Ford, and Mistress Quickly*

PAGE How now, Master Ford?

FORD You heard what this knave told me, did you not?

PAGE Yes, and you heard what the other told me? 161

FORD Do you think there is truth in them?

PAGE Hang 'em, slaves! I do not think the knight would offer it. But these that accuse him in his intent towards our wives are a yoke of his discarded men—very rogues, now they be out of service. 166

FORD Were they his men?

PAGE Marry, were they.

FORD I like it never the better for that. Does he lie at the Garter? 170

PAGE Ay, marry, does he. If he should intend this voyage toward my wife, I would turn her loose to him; and what he gets more of her than sharp words, let it lie on my head.

FORD I do not misdoubt my wife, but I would be loath to turn them together. A man may be too confident. I would have nothing lie on my head. I cannot be thus satisfied. 178
*Enter the Host of the Garter*

PAGE Look where my ranting Host of the Garter comes. There is either liquor in his pate or money in his purse when he looks so merrily.—How now, mine Host?

HOST God bless you, bully rook, God bless you! Thou'rt a gentleman. 183
*Enter Shallow*
Cavaliero Justice, I say!

SHALLOW I follow, mine Host, I follow.—Good even and twenty, good Master Page. Master Page, will you go with us? We have sport in hand.

HOST Tell him, Cavaliero Justice, tell him, bully rook.

SHALLOW Sir, there is a fray to be fought between Sir Hugh, the Welsh priest, and Caius, the French doctor.

FORD Good mine Host o'th' Garter, a word with you.

HOST What sayst thou, my bully rook? 192
*They talk apart*

SHALLOW (*to Page*) Will you go with us to behold it? My merry Host hath had the measuring of their weapons, and, I think, hath appointed them contrary places. For,

believe me, I hear the parson is no jester. Hark, I will tell you what our sport shall be. 197
*They talk apart*

HOST (*to Ford*) Hast thou no suit against my knight, my guest cavaliero?

⌈FORD⌉ None, I protest. But I'll give you a pottle of burnt sack to give me recourse to him and tell him my name is Brooke—only for a jest. 202

HOST My hand, bully. Thou shalt have egress and regress—said I well?—and thy name shall be Brooke. It is a merry knight. (*To Shallow and Page*) Will you go, mijn'heers? 206

SHALLOW Have with you, mine Host.

PAGE I have heard the Frenchman hath good skill in his rapier.

SHALLOW Tut, sir, I could have told you more. In these times you stand on distance—your passes, stoccados, and I know not what. 'Tis the heart, Master Page; ⌈*showing his rapier-passes*⌉ 'tis here, 'tis here. I have seen the time with my long sword I would have made you four tall fellows skip like rats. 215

HOST Here, boys; here, here! Shall we wag?

PAGE Have with you. I had rather hear them scold than fight.                    *Exeunt Host, Shallow, and Page*

FORD Though Page be a secure fool and stands so firmly on his wife's frailty, yet I cannot put off my opinion so easily. She was in his company at Page's house, and what they made there I know not. Well, I will look further into't; and I have a disguise to sound Falstaff. If I find her honest, I lose not my labour. If she be otherwise, 'tis labour well bestowed.                    *Exit*

**2.2**    *Enter Sir John Falstaff and Pistol*

SIR JOHN I will not lend thee a penny.

PISTOL
I will retort the sum in equipage.

SIR JOHN Not a penny.

PISTOL ⌈*drawing his sword*⌉ Why then, the world's mine oyster, which I with sword will open. 5

SIR JOHN Not a penny. I have been content, sir, you should lay my countenance to pawn. I have grated upon my good friends for three reprieves for you and your coach-fellow Nim, or else you had looked through the grate like a gemini of baboons. I am damned in hell for swearing to gentlemen my friends you were good soldiers and tall fellows. And when Mistress Bridget lost the handle of her fan, I took't upon mine honour thou hadst it not. 14

PISTOL
Didst not thou share? Hadst thou not fifteen pence?

SIR JOHN Reason, you rogue, reason. Thinkest thou I'll endanger my soul gratis? At a word, hang no more about me. I am no gibbet for you. Go, a short knife and a throng, to your manor of Pickt-hatch, go. You'll not bear a letter for me, you rogue? You stand upon your honour? Why, thou unconfinable baseness, it is as much as I can do to keep the terms of my honour precise. Ay, ay, I myself sometimes, leaving the fear of God on the left hand and hiding mine honour in my necessity, am fain to shuffle, to hedge, and to lurch; and yet you, you rogue, will ensconce your rags, your cat-a-mountain looks, your red-lattice phrases, and your bold beating oaths, under the shelter of your honour! You will not do it, you?

PISTOL ⌈*sheathing his sword*⌉
I do relent. What wouldst thou more of man? 30

*Enter Robin*
ROBIN Sir, here's a woman would speak with you.
SIR JOHN Let her approach.
    *Enter Mistress Quickly*
MISTRESS QUICKLY Give your worship good morrow.
SIR JOHN Good morrow, goodwife.
MISTRESS QUICKLY Not so, an't please your worship.    35
SIR JOHN Good maid, then.
MISTRESS QUICKLY I'll be sworn: as my mother was the
first hour I was born.
SIR JOHN I do believe the swearer. What with me?
MISTRESS QUICKLY Shall I vouchsafe your worship a word
or two?                                              41
SIR JOHN Two thousand, fair woman, and I'll vouchsafe
thee the hearing.
MISTRESS QUICKLY There is one Mistress Ford, sir—I pray
come a little nearer this ways.                      45
    *She draws Sir John aside*
I myself dwell with Master Doctor Caius—
SIR JOHN Well, on. Mistress Ford, you say.
MISTRESS QUICKLY Your worship says very true. I pray
your worship come a little nearer this ways.
SIR JOHN I warrant thee nobody hears. Mine own people,
mine own people.                                     51
MISTRESS QUICKLY Are they so? God bless them and make
them His servants!
SIR JOHN Well, Mistress Ford: what of her?
MISTRESS QUICKLY Why, sir, she's a good creature. Lord,
Lord, your worship's a wanton! Well, heaven forgive
you, and all of us, I pray—                          57
SIR JOHN Mistress Ford; come, Mistress Ford.
MISTRESS QUICKLY Marry, this is the short and the long of
it. You have brought her into such a canaries as 'tis
wonderful. The best courtier of them all, when the
court lay at Windsor, could never have brought her to
such a canary. Yet there has been knights, and lords,
and gentlemen, with their coaches; I warrant you,
coach after coach, letter after letter, gift after gift,
smelling so sweetly, all musk; and so rustling, I warrant
you, in silk and gold, and in such aligant terms, and
in such wine and sugar of the best and the fairest, that
would have won any woman's heart; and, I warrant
you, they could never get an eye-wink of her. I had
myself twenty angels given me this morning—but I
defy all angels, in any such sort, as they say, but in
the way of honesty. And, I warrant you, they could
never get her so much as sip on a cup with the proudest
of them all. And yet there has been earls, nay, which
is more, pensioners. But, I warrant you, all is one with
her.                                                 77
SIR JOHN But what says she to me? Be brief, my good
she-Mercury.
MISTRESS QUICKLY Marry, she hath received your letter,
for the which she thanks you a thousand times, and
she gives you to notify that her husband will be absence
from his house between ten and eleven.
SIR JOHN Ten and eleven.                             84
MISTRESS QUICKLY Ay, forsooth, and then you may come
and see the picture, she says, that you wot of. Master
Ford, her husband, will be from home. Alas, the sweet
woman leads an ill life with him. He's a very jealousy
man. She leads a very frampold life with him, good
heart.                                               90
SIR JOHN Ten and eleven. Woman, commend me to her.
I will not fail her.

MISTRESS QUICKLY Why, you say well. But I have another
messenger to your worship. Mistress Page hath her
hearty commendations to you too; and, let me tell you
in your ear, she's as fartuous a civil modest wife, and
one, I tell you, that will not miss you morning nor
evening prayer, as any is in Windsor, whoe'er be the
other; and she bade me tell your worship that her
husband is seldom from home, but she hopes there will
come a time. I never knew a woman so dote upon a
man. Surely I think you have charms, la; yes, in truth.
SIR JOHN Not I, I assure thee. Setting the attraction of my
good parts aside, I have no other charms.
MISTRESS QUICKLY Blessing on your heart for't!       105
SIR JOHN But I pray thee tell me this: has Ford's wife and
Page's wife acquainted each other how they love me?
MISTRESS QUICKLY O God no, sir; that were a jest indeed!
They have not so little grace, I hope. That were a trick
indeed! But Mistress Page would desire you to send her
your little page of all loves. Her husband has a
marvellous infection to the little page; and, truly,
Master Page is an honest man. Never a wife in Windsor
leads a better life than she does. Do what she will; say
what she will; take all, pay all; go to bed when she
list; rise when she list; all is as she will. And, truly,
she deserves it, for if there be a kind woman in Windsor,
she is one. You must send her your page, no remedy.
SIR JOHN Why, I will.                                119
MISTRESS QUICKLY Nay, but do so, then; and, look you,
he may come and go between you both. And in any
case have a nay-word, that you may know one
another's mind, and the boy never need to understand
anything—for 'tis not good that children should know
any wickedness. Old folks, you know, have discretion,
as they say, and know the world.                     126
SIR JOHN Fare thee well. Commend me to them both.
There's my purse; I am yet thy debtor.—Boy, go along
with this woman.    *Exeunt Mistress Quickly and Robin*
*(Aside)* This news distracts me.                     130
PISTOL *(aside)*
This punk is one of Cupid's carriers.
Clap on more sails! Pursue! Up with your sights!
Give fire! She is my prize, or ocean whelm them all!
                                                *Exit*
SIR JOHN Sayst thou so, old Jack? Go thy ways! I'll make
more of thy old body than I have done. Will they yet
look after thee? Wilt thou, after the expense of so much
money, be now a gainer? Good body, I thank thee. Let
them say 'tis grossly done; so it be fairly done, no
matter.                                              139
    *Enter Bardolph, ⌈with sack⌉*
BARDOLPH Sir John, there's one Master Brooke below
would fain speak with you and be acquainted with
you, and hath sent your worship a morning's draught
of sack.
SIR JOHN Brooke is his name?
BARDOLPH Ay, sir.                                    145
SIR JOHN Call him in. ⌈*Drinking sack*⌉ Such Brookes are
welcome to me, that o'erflows such liquor.
                                          *Exit Bardolph*
Aha, Mistress Ford and Mistress Page, have I
encompassed you? ⌈*Drinking*⌉ Go to. Via!
    *Enter Bardolph, and Master Ford disguised as*
    *Brooke*
FORD God bless you, sir.                             150
SIR JOHN And you, sir. Would you speak with me?

FORD I make bold to press with so little preparation upon you.

SIR JOHN You're welcome. What's your will? *(To Bardolph)* Give us leave, drawer.    *Exit Bardolph*

FORD Sir, I am a gentleman that have spent much. My name is Brooke.    157

SIR JOHN Good Master Brooke, I desire more acquaintance of you.

FORD Good Sir John, I sue for yours—not to charge you, for I must let you understand I think myself in better plight for a lender than you are; the which hath something emboldened me to this unseasoned intrusion; for they say if money go before, all ways do lie open.    165

SIR JOHN Money is a good soldier, sir, and will on.

FORD Troth, and I have a bag of money here troubles me. If you will help to bear it, Sir John, take half, or all, for easing me of the carriage.

SIR JOHN Sir, I know not how I may deserve to be your porter.    171

FORD I will tell you, sir, if you will give me the hearing.

SIR JOHN Speak, good Master Brooke. I shall be glad to be your servant.

FORD Sir, I hear you are a scholar—I will be brief with you—and you have been a man long known to me, though I had never so good means as desire to make myself acquainted with you. I shall discover a thing to you wherein I must very much lay open mine own imperfection; but, good Sir John, as you have one eye upon my follies, as you hear them unfolded, turn another into the register of your own, that I may pass with a reproof the easier, sith you yourself know how easy it is to be such an offender.

SIR JOHN Very well, sir, proceed.    185

FORD There is a gentlewoman in this town; her husband's name is Ford.

SIR JOHN Well, sir.

FORD I have long loved her, and, I protest to you, bestowed much on her, followed her with a doting observance, engrossed opportunities to meet her, fee'd every slight occasion that could but niggardly give me sight of her; not only bought many presents to give her, but have given largely to many to know what she would have given. Briefly, I have pursued her as love hath pursued me, which hath been on the wing of all occasions. But, whatsoever I have merited, either in my mind or in my means, meed I am sure I have received none, unless experience be a jewel. That I have purchased at an infinite rate, and that hath taught me to say this: 'Love like a shadow flies when substance love pursues, Pursuing that that flies, and flying what pursues.'    202

SIR JOHN Have you received no promise of satisfaction at her hands?

FORD Never.    205

SIR JOHN Have you importuned her to such a purpose?

FORD Never.

SIR JOHN Of what quality was your love then?

FORD Like a fair house built on another man's ground, so that I have lost my edifice by mistaking the place where I erected it.    211

SIR JOHN To what purpose have you unfolded this to me?

FORD When I have told you that, I have told you all. Some say that though she appear honest to me, yet in other places she enlargeth her mirth so far that there is shrewd construction made of her. Now, Sir John, here is the heart of my purpose. You are a gentleman of excellent breeding, admirable discourse, of great admittance, authentic in your place and person, generally allowed for your many warlike, court-like, and learned preparations.    221

SIR JOHN O sir!

FORD Believe it, for you know it. There is money.

⌈*He offers money*⌉

Spend it, spend it; spend more; spend all I have; only give me so much of your time in exchange of it as to lay an amiable siege to the honesty of this Ford's wife. Use your art of wooing, win her to consent to you. If any man may, you may as soon as any.    228

SIR JOHN Would it apply well to the vehemency of your affection that I should win what you would enjoy? Methinks you prescribe to yourself very preposterously.

FORD O, understand my drift. She dwells so securely on the excellency of her honour that the folly of my soul dares not present itself. She is too bright to be looked against. Now, could I come to her with any detection in my hand, my desires had instance and argument to commend themselves. I could drive her then from the ward of her purity, her reputation, her marriage vow, and a thousand other her defences which now are too too strongly embattled against me. What say you to't, Sir John?    241

SIR JOHN Master Brooke, I will first make bold with your money.

⌈*He takes the money*⌉

Next, give me your hand.

*He takes his hand*

And last, as I am a gentleman, you shall, if you will, enjoy Ford's wife.    246

FORD O, good sir!

SIR JOHN I say you shall.

FORD Want no money, Sir John, you shall want none.

SIR JOHN Want no Mistress Ford, Master Brooke, you shall want none. I shall be with her, I may tell you, by her own appointment. Even as you came in to me, her spokesmate, or go-between, parted from me. I say I shall be with her between ten and eleven, for at that time the jealous rascally knave her husband will be forth. Come you to me at night; you shall know how I speed.

FORD I am blessed in your acquaintance. Do you know Ford, sir?    259

SIR JOHN Hang him, poor cuckoldly knave, I know him not. Yet I wrong him to call him poor. They say the jealous wittolly knave hath masses of money, for the which his wife seems to me well favoured. I will use her as the key of the cuckoldly rogue's coffer, and there's my harvest-home.    265

FORD I would you knew Ford, sir, that you might avoid him if you saw him.

SIR JOHN Hang him, mechanical salt-butter rogue! I will stare him out of his wits. I will awe him with my cudgel; it shall hang like a meteor o'er the cuckold's horns. Master Brooke, thou shalt know I will predominate over the peasant, and thou shalt lie with his wife. Come to me soon at night. Ford's a knave, and I will aggravate his style: thou, Master Brooke, shalt know him for knave and cuckold. Come to me soon at night.    *Exit*

FORD What a damned epicurean rascal is this! My heart is ready to crack with impatience. Who says this is improvident jealousy? My wife hath sent to him, the hour is fixed, the match is made. Would any man have thought this? See the hell of having a false woman!

My bed shall be abused, my coffers ransacked, my reputation gnawn at, and I shall not only receive this villainous wrong, but stand under the adoption of abominable terms, and by him that does me this wrong. Terms! Names! 'Amaimon' sounds well, 'Lucifer' well, 'Barbason' well; yet they are devils' additions, the names of fiends. But 'cuckold', 'wittol'! 'Cuckold'—the devil himself hath not such a name. Page is an ass, a secure ass. He will trust his wife, he will not be jealous. I will rather trust a Fleming with my butter, Parson Hugh the Welshman with my cheese, an Irishman with my aqua-vitae bottle, or a thief to walk my ambling gelding, than my wife with herself. Then she plots, then she ruminates, then she devises; and what they think in their hearts they may effect, they will break their hearts but they will effect. God be praised for my jealousy! Eleven o'clock the hour. I will prevent this, detect my wife, be revenged on Falstaff, and laugh at Page. I will about it. Better three hours too soon than a minute too late. God's my life: cuckold, cuckold, cuckold!                                                      *Exit*

**2.3**  *Enter Doctor Caius and John Rugby, with rapiers*
CAIUS Jack Rugby!
RUGBY Sir.
CAIUS Vat is the clock, Jack?
RUGBY 'Tis past the hour, sir, that Sir Hugh promised to meet.                                                                      5
CAIUS By Gar, he has save his soul dat he is no come; he has pray his Pible well dat he is no come. By Gar, Jack Rugby, he is dead already if he be come.
RUGBY He is wise, sir, he knew your worship would kill him if he came.                                                          10
CAIUS ⌐drawing his rapier⌐ By Gar, de herring is no dead so as I vill kill him. Take your rapier, Jack. I vill tell you how I vill kill him.
RUGBY Alas, sir, I cannot fence.
CAIUS Villainy, take your rapier.                                   15
RUGBY Forbear: here's company.
        ⌐Caius sheathes his rapier.⌐
        *Enter the Host of the Garter, Justice Shallow,*
        *Master Page, and Master Slender*
HOST God bless thee, bully Doctor.
SHALLOW God save you, Master Doctor Caius.
PAGE Now, good Master Doctor.
SLENDER Give you good morrow, sir.                                  20
CAIUS Vat be all you, one, two, tree, four, come for?
HOST To see thee fight, to see thee foin, to see thee traverse, to see thee here, to see thee there; to see thee pass thy punto, thy stock, thy reverse, thy distance, thy montant. Is he dead, my Ethiopian? Is he dead, my Francisco? Ha, bully? What says my Aesculapius, my Galen, my heart of elder, ha? Is he dead, bully stale? Is he dead?
CAIUS By Gar, he is de coward jack-priest of de vorld. He is not show his face.                                              30
HOST Thou art a Castalian King Urinal, Hector of Greece, my boy.
CAIUS I pray you bear witness that me have stay six or seven, two, tree hours for him, and he is no come.
SHALLOW He is the wiser man, Master Doctor. He is a curer of souls, and you a curer of bodies. If you should fight you go against the hair of your professions. Is it not true, Master Page?
PAGE Master Shallow, you have yourself been a great fighter, though now a man of peace.                            40

SHALLOW Bodykins, Master Page, though I now be old and of the peace, if I see a sword out my finger itches to make one. Though we are justices and doctors and churchmen, Master Page, we have some salt of our youth in us. We are the sons of women, Master Page.
PAGE 'Tis true, Master Shallow.                                    46
SHALLOW It will be found so, Master Page.—Master Doctor Caius, I am come to fetch you home. I am sworn of the peace. You have showed yourself a wise physician, and Sir Hugh hath shown himself a wise and patient churchman. You must go with me, Master Doctor.   51
HOST Pardon, guest Justice. (*To Caius*) A word, Monsieur Mockwater.
CAIUS Mockvater? Vat is dat?
HOST Mockwater, in our English tongue, is valour, bully.
CAIUS By Gar, then I have as much mockvater as de Englishman. Scurvy jack-dog priest! By Gar, me vill cut his ears.
HOST He will clapper-claw thee tightly, bully.
CAIUS Clapper-de-claw? Vat is dat?                                  60
HOST That is, he will make thee amends.
CAIUS By Gar, me do look he shall clapper-de-claw me, for, by Gar, me vill have it.
HOST And I will provoke him to't, or let him wag.
CAIUS Me tank you for dat.                                          65
HOST And moreover, bully— (*Aside to the others*) But first, master guest and Master Page, and eke Cavaliero Slender, go you through the town to Frogmore.
PAGE Sir Hugh is there, is he?
HOST He is there. See what humour he is in, and I will bring the Doctor about by the fields. Will it do well?
SHALLOW We will do it.                                             72
⌐PAGE, SHALLOW, *and* SLENDER⌐ Adieu, good Master Doctor.
                          *Exeunt Page, Shallow, and Slender*
CAIUS ⌐drawing his rapier⌐ By Gar, me vill kill de priest, for he speak for a jackanape to Anne Page.                   75
HOST Let him die. Sheathe thy impatience; throw cold water on thy choler. Go about the fields with me through Frogmore. I will bring thee where Mistress Anne Page is, at a farmhouse a-feasting; and thou shalt woo her. Cried game? Said I well?            80
CAIUS ⌐sheathing his rapier⌐ By Gar, me dank you vor dat. By Gar, I love you, and I shall procure-a you de good guest: de earl, de knight, de lords, de gentlemen, my patiences.
HOST For the which I will be thy adversary toward Anne Page. Said I well?                                              86
CAIUS By Gar, 'tis good. Vell said.
HOST Let us wag, then.
CAIUS Come at my heels, Jack Rugby.                    *Exeunt*

**3.1**  *Enter Sir Hugh Evans ⌐with a rapier, and bearing a book⌐ and Simple ⌐bearing Evans's gown⌐*
EVANS I pray you now, good Master Slender's servingman, and friend Simple by your name, which way have you looked for Master Caius, that calls himself Doctor of Physic?                                                          4
SIMPLE Marry, sir, the Petty Ward, the Park Ward, every way; old Windsor way, and every way but the town way.
EVANS I most fehemently desire you you will also look that way.                                                         9
SIMPLE I will, sir.                                          ⌐*Exit*⌐
EVANS ⌐opening the book⌐ Jeshu pless me, how full of cholers I am, and trempling of mind! I shall be glad if he have deceived me. How melancholies I am! I will knog his

urinals about his knave's costard when I have good
opportunities for the 'ork. Pless my soul!—     15

(*Singing*)
> To shallow rivers, to whose falls
> Melodious birds sings madrigals.
> There will we make our peds of roses,
> And a thousand fragrant posies.
> To shallow—     20

Mercy on me! I have a great dispositions to cry.—

(*Singing*)
> Melodious birds sing madrigals.—
> When as I sat in Pabylon—
> And a thousand vagram posies.
> To shallow (*etc.*)     25

⌜*Enter Simple*⌝

SIMPLE Yonder he is coming. This way, Sir Hugh.
EVANS He's welcome.
(*Singing*) 'To shallow rivers to whose falls—'
God prosper the right! What weapons is he?     29
SIMPLE No weapons, sir. There comes my master, Master
Shallow, and another gentleman, from Frogmore, over
the stile this way.
EVANS Pray you give me my gown—or else keep it in
your arms.     34
⌜*He reads.*⌝
*Enter Justice Shallow, Master Slender, and Master
Page*
SHALLOW How now, Master Parson? Good morrow, good
Sir Hugh. Keep a gamester from the dice and a good
student from his book, and it is wonderful.     37
SLENDER (*aside*) Ah, sweet Anne Page!
PAGE God save you, good Sir Hugh.
EVANS God pless you from his mercy sake, all of you.     40
SHALLOW What, the sword and the Word? Do you study
them both, Master Parson?
PAGE And youthful still: in your doublet and hose this
raw, rheumatic day!
EVANS There is reasons and causes for it.     45
PAGE We are come to you to do a good office, Master
Parson.
EVANS Fery well. What is it?
PAGE Yonder is a most reverend gentleman, who, belike
having received wrong by some person, is at most odds
with his own gravity and patience that ever you saw.
SHALLOW I have lived fourscore years and upward; I
never heard a man of his place, gravity, and learning
so wide of his own respect.
EVANS What is it?     55
PAGE I think you know-him: Master Doctor Caius, the
renowned French physician.
EVANS Got's will and his passion of my heart! I had as
lief you would tell me of a mess of pottage.
PAGE Why?     60
EVANS He has no more knowledge in Hibbocrates and
Galen, and he is a knave besides—a cowardly knave
as you would desires to be acquainted withal.
PAGE ⌜*to Shallow*⌝ I warrant you, he's the man should
fight with him.     65
SLENDER (*aside*) O sweet Anne Page!
SHALLOW It appears so by his weapons.
*Enter the Host of the Garter, Doctor Caius, and
John Rugby*
Keep them asunder—here comes Doctor Caius.
*Evans and Caius draw and offer to fight*
PAGE Nay, good Master Parson, keep in your weapon.

SHALLOW So do you, good Master Doctor.     70
HOST Disarm them and let them question. Let them keep
their limbs whole, and hack our English.
*Shallow and Page take Caius's and Evans's rapiers*
CAIUS (*to Evans*) I pray you let-a me speak a word with
your ear. Wherefore vill you not meet-a me?
EVANS ⌜*aside to Caius*⌝ Pray you use your patience. ⌜*Aloud*⌝
In good time!     76
CAIUS By Gar, you are de coward, de jack-dog, john-ape.
EVANS (*aside to Caius*) Pray you let us not be laughing-
stocks to other men's humours. I desire you in
friendship, and I will one way or other make you
amends. (*Aloud*) By Jeshu, I will knog your urinal about
your knave's cogscomb.     82
CAIUS *Diable*! Jack Rugby, mine Host de Jarteer, have I
not stay for him to kill him? Have I not, at de place I
did appoint?     85
EVANS As I am a Christians soul, now look you, this is
the place appointed. I'll be judgement by mine Host of
the Garter.
HOST Peace, I say, Gallia and Gaul, French and Welsh,
soul-curer and body-curer.     90
CAIUS Ay, dat is very good, *excellent*.
HOST Peace, I say. Hear mine Host of the Garter. Am I
politic? Am I subtle? Am I a Machiavel? Shall I lose
my doctor? No, he gives me the potions and the
motions. Shall I lose my parson, my priest, my Sir
Hugh? No, he gives me the Proverbs and the No-verbs.
(*To Caius*) Give me thy hand terrestrial—so. (*To Evans*)
Give me thy hand celestial—so. Boys of art, I have
deceived you both, I have directed you to wrong places.
Your hearts are mighty, your skins are whole, and let
burnt sack be the issue. (*To Shallow and Page*) Come,
lay their swords to pawn. (*To Caius and Evans*) Follow
me, lads of peace, follow, follow, follow.     *Exit*
SHALLOW Afore God, a mad host! Follow, gentlemen,
follow.     *Exeunt Shallow and Page*
SLENDER (*aside*) O sweet Anne Page!     *Exit*
CAIUS Ha, do I perceive dat? Have you make-a de sot of
us, ha, ha?     108
EVANS This is well: he has made us his vlouting-stog. I
desire you that we may be friends, and let us knog our
prains together to be revenge on this same scall, scurvy,
cogging companion, the Host of the Garter.
CAIUS By Gar, with all my heart. He promise to bring me
where is Anne Page. By Gar, he deceive me too.     114
EVANS Well, I will smite his noddles. Pray you follow.
    *Exeunt*

**3.2**     *Enter Robin, followed by Mistress Page*
MISTRESS PAGE Nay, keep your way, little gallant. You
were wont to be a follower, but now you are a leader.
Whether had you rather, lead mine eyes, or eye your
master's heels?
ROBIN I had rather, forsooth, go before you like a man
than follow him like a dwarf.     6
MISTRESS PAGE O, you are a flattering boy! Now I see
you'll be a courtier.
*Enter Master Ford*
FORD
Well met, Mistress Page. Whither go you?
MISTRESS PAGE Truly, sir, to see your wife. Is she at home?
FORD Ay, and as idle as she may hang together, for want
of company. I think if your husbands were dead you
two would marry.     13
MISTRESS PAGE Be sure of that—two other husbands.

FORD Where had you this pretty weathercock?    15

MISTRESS PAGE I cannot tell what the dickens his name is my husband had him of.—What do you call your knight's name, sirrah?

ROBIN Sir John Falstaff.

FORD Sir John Falstaff?    20

MISTRESS PAGE He, he; I can never hit on's name. There is such a league between my goodman and he! Is your wife at home indeed?

FORD Indeed she is.

MISTRESS PAGE By your leave, sir, I am sick till I see her.
*Exeunt Robin and Mistress Page*

FORD Has Page any brains? Hath he any eyes? Hath he any thinking? Sure they sleep; he hath no use of them. Why, this boy will carry a letter twenty mile, as easy as a cannon will shoot point-blank twelve score. He pieces out his wife's inclination; he gives her folly motion and advantage. And now she's going to my wife, and Falstaff's boy with her. A man may hear this shower sing in the wind. And Falstaff's boy with her. Good plots—they are laid; and our revolted wives share damnation together. Well, I will take him; then torture my wife, pluck the borrowed veil of modesty from the so-seeming Mistress Page, divulge Page himself for a secure and wilful Actaeon, and to these violent proceedings all my neighbours shall cry aim.    39
⌜*Clock strikes*⌝
The clock gives me my cue, and my assurance bids me search. There I shall find Falstaff. I shall be rather praised for this than mocked, for it is as positive as the earth is firm that Falstaff is there. I will go.
*Enter Master Page, Justice Shallow, Master Slender, the Host of the Garter, Sir Hugh Evans, Doctor Caius, and John Rugby*

SHALLOW, PAGE, *etc.* Well met, Master Ford.    44

FORD (*aside*) By my faith, a good knot! (*To them*) I have good cheer at home, and I pray you all go with me.

SHALLOW I must excuse myself, Master Ford.

SLENDER And so must I, sir. We have appointed to dine with Mistress Anne, and I would not break with her for more money than I'll speak of.    50

SHALLOW We have lingered about a match between Anne Page and my cousin Slender, and this day we shall have our answer.

SLENDER I hope I have your good will, father Page.

PAGE You have, Master Slender: I stand wholly for you. (*To Caius*) But my wife, Master Doctor, is for you altogether.    57

CAIUS Ay, be Gar, and de maid is love-a me. My nursh-a Quickly tell me so mush.

HOST (*to Page*) What say you to young Master Fenton? He capers, he dances, he has eyes of youth; he writes verses, he speaks holiday, he smells April and May. He will carry't, he will carry't; 'tis in his buttons he will carry't.    64

PAGE Not by my consent, I promise you. The gentleman is of no having. He kept company with the wild Prince and Poins. He is of too high a region; he knows too much. No, he shall not knit a knot in his fortunes with the finger of my substance. If he take her, let him take her simply: the wealth I have waits on my consent, and my consent goes not that way.    71

FORD I beseech you heartily, some of you go home with me to dinner. Besides your cheer, you shall have sport: I will show you a monster. Master Doctor, you shall go. So shall you, Master Page, and you, Sir Hugh.    75

SHALLOW Well, God be with you! ⌜*Aside to Slender*⌝ We shall have the freer wooing at Master Page's.
*Exeunt Shallow and Slender*

CAIUS Go home, John Rugby; I come anon.    *Exit Rugby*

HOST Farewell, my hearts. I will to my honest knight Falstaff, and drink canary with him.    *Exit*

FORD (*aside*) I think I shall drink in pipe-wine first with him: I'll make him dance. (*To Page, Caius, and Evans*) Will you go, gentles?    83

⌜PAGE, CAIUS, *and* EVANS⌝ Have with you to see this monster.    *Exeunt*

**3.3**    *Enter Mistress Ford and Mistress Page*

MISTRESS FORD What, John! What, Robert!

MISTRESS PAGE Quickly, quickly! Is the buck-basket—

MISTRESS FORD I warrant.—What, Robert, I say!

MISTRESS PAGE Come, come, come!
*Enter John and Robert, with a buck-basket*

MISTRESS FORD Here, set it down.    5

MISTRESS PAGE Give your men the charge. We must be brief.

MISTRESS FORD Marry, as I told you before, John and Robert, be ready here hard by in the brew-house; and when I suddenly call you, come forth, and without any pause or staggering take this basket on your shoulders. That done, trudge with it in all haste, and carry it among the whitsters in Datchet Mead, and there empty it in the muddy ditch close by the Thames' side.

MISTRESS PAGE (*to John and Robert*) You will do it?    15

MISTRESS FORD I ha' told them over and over; they lack no direction.—Be gone, and come when you are called.
*Exeunt John and Robert*
*Enter Robin*

MISTRESS PAGE Here comes little Robin.

MISTRESS FORD How now, my eyas-musket, what news with you?    20

ROBIN My master Sir John is come in at your back door, Mistress Ford, and requests your company.

MISTRESS PAGE You little Jack-a-Lent, have you been true to us?

ROBIN Ay, I'll be sworn. My master knows not of your being here, and hath threatened to put me into everlasting liberty if I tell you of it; for he swears he'll turn me away.    28

MISTRESS PAGE Thou'rt a good boy. This secrecy of thine shall be a tailor to thee, and shall make thee a new doublet and hose.—I'll go hide me.

MISTRESS FORD Do so. (*To Robin*) Go tell thy master I am alone.    *Exit Robin*
Mistress Page, remember you your cue.    34

MISTRESS PAGE I warrant thee. If I do not act it, hiss me.

MISTRESS FORD Go to, then.    ⌜*Exit Mistress Page*⌝
We'll use this unwholesome humidity, this gross watery pumpkin. We'll teach him to know turtles from jays.
*Enter Sir John Falstaff*

SIR JOHN Have I caught thee, my heavenly jewel? Why, now let me die, for I have lived long enough. This is the period of my ambition. O, this blessed hour!    41

MISTRESS FORD O sweet Sir John!

SIR JOHN Mistress Ford, I cannot cog; I cannot prate, Mistress Ford. Now shall I sin in my wish: I would thy husband were dead. I'll speak it before the best lord. I would make thee my lady.    46

MISTRESS FORD I your lady, Sir John? Alas, I should be a pitiful lady.

SIR JOHN Let the court of France show me such another. I see how thine eye would emulate the diamond. Thou

hast the right arched beauty of the brow that becomes
the ship-tire, the tire-valiant, or any tire of Venetian
admittance.

MISTRESS FORD A plain kerchief, Sir John—my brows
become nothing else, nor that well neither.        55

SIR JOHN By the Lord, thou art a tyrant to say so. Thou
wouldst make an absolute courtier, and the firm fixture
of thy foot would give an excellent motion to thy gait
in a semicircled farthingale. I see what thou wert if
fortune, thy foe, were, with nature, thy friend. Come,
thou canst not hide it.        61

MISTRESS FORD Believe me, there's no such thing in me.

SIR JOHN What made me love thee? Let that persuade
thee there's something extraordinary in thee. Come, I
cannot cog and say thou art this and that, like a-many
of these lisping hawthorn-buds that come like women
in men's apparel and smell like Bucklersbury in simple
time; I cannot. But I love thee, none but thee; and
thou deservest it.        69

MISTRESS FORD Do not betray me, sir. I fear you love
Mistress Page.

SIR JOHN Thou mightst as well say I love to walk by the
Counter gate, which is as hateful to me as the reek of
a lime-kiln.        74

MISTRESS FORD Well, heaven knows how I love you; and
you shall one day find it.

SIR JOHN Keep in that mind. I'll deserve it.

MISTRESS FORD Nay, I must tell you, so you do; or else I
could not be in that mind.        79

*Enter Robin*

ROBIN Mistress Ford, Mistress Ford! Here's Mistress Page
at the door, sweating and blowing, and looking wildly,
and would needs speak with you presently.

SIR JOHN She shall not see me. I will ensconce me behind
the arras.        84

MISTRESS FORD Pray you do so; she's a very tattling
woman.

*Sir John hides behind the arras.*
*Enter Mistress Page*

What's the matter? How now?

MISTRESS PAGE O Mistress Ford, what have you done?
You're shamed, you're overthrown, you're undone for
ever.        90

MISTRESS FORD What's the matter, good Mistress Page?

MISTRESS PAGE O well-a-day, Mistress Ford! Having an
honest man to your husband, to give him such cause
of suspicion!

MISTRESS FORD What cause of suspicion?        95

MISTRESS PAGE What cause of suspicion? Out upon you!
How am I mistook in you!

MISTRESS FORD Why, alas, what's the matter?

MISTRESS PAGE Your husband's coming hither, woman,
with all the officers in Windsor, to search for a
gentleman that he says is here now in the house, by
your consent, to take an ill advantage of his absence.
You are undone.

MISTRESS FORD 'Tis not so, I hope.        104

MISTRESS PAGE Pray heaven it be not so that you have
such a man here! But 'tis most certain your husband's
coming, with half Windsor at his heels, to search for
such a one. I come before to tell you. If you know
yourself clear, why, I am glad of it; but if you have a
friend here, convey, convey him out. Be not amazed.
Call all your senses to you. Defend your reputation, or
bid farewell to your good life for ever.        112

MISTRESS FORD What shall I do? There is a gentleman,
my dear friend; and I fear not mine own shame so
much as his peril. I had rather than a thousand pound
he were out of the house.        116

MISTRESS PAGE For shame, never stand 'you had rather'
and 'you had rather'. Your husband's here at hand.
Bethink you of some conveyance: in the house you
cannot hide him. O, how have you deceived me! Look,
here is a basket. If he be of any reasonable stature, he
may creep in here; and throw foul linen upon him as
if it were going to bucking. Or—it is whiting time—
send him by your two men to Datchet Mead.        124

MISTRESS FORD He's too big to go in there. What shall I
do?

SIR JOHN (*coming forward*) Let me see't, let me see't, O let
me see't! I'll in, I'll in. Follow your friend's counsel;
I'll in.        129

MISTRESS PAGE What, Sir John Falstaff! (*Aside to him*) Are
these your letters, knight?

SIR JOHN (*aside to Mistress Page*) I love thee. Help me
away. Let me creep in here.
*He goes into the basket*

I'll never—        134
*Mistress Page and Mistress Ford put foul clothes
over him*

MISTRESS PAGE (*to Robin*) Help to cover your master, boy.—
Call your men, Mistress Ford. ⌐Aside to Sir John⌐ You
dissembling knight!

MISTRESS FORD What, John! Robert, John!
*Enter John and Robert*

Go take up these clothes here quickly. Where's the
cowl-staff?        140
*John and Robert fit the cowl-staff*

Look how you drumble! Carry them to the laundress
in Datchet Mead. Quickly, come!
*They lift the basket and start to leave.*
*Enter Master Ford, Master Page, Doctor Caius, and
Sir Hugh Evans*

FORD (*to Page, Caius, and Evans*) Pray you come near. If I
suspect without cause, why then, make sport at me;
then let me be your jest—I deserve it. (*To John and
Robert*) How now? Whither bear you this?        146

⌐JOHN⌐ To the laundress, forsooth.

MISTRESS FORD Why, what have you to do whither they
bear it? You were best meddle with buck-washing!

FORD Buck? I would I could wash myself of the buck!
Buck, buck, buck? Ay, buck, I warrant you, buck. And
of the season too, it shall appear.        152
⌐*Exeunt John and Robert, with the basket*⌐

Gentlemen, I have dreamt tonight. I'll tell you my
dream. Here, here, here be my keys. Ascend my
chambers, search, seek, find out. I'll warrant we'll
unkennel the fox. Let me stop this way first.        156
*He locks the door*

So, now, uncoop.

PAGE Good Master Ford, be contented. You wrong yourself
too much.

FORD True, Master Page.—Up, gentlemen! You shall see
sport anon. Follow me, gentlemen.        *Exit*

EVANS This is fery fantastical humours and jealousies.

CAIUS By Gar, 'tis no the fashion of France; it is not
jealous in France.        164

PAGE Nay, follow him, gentlemen. See the issue of his
search.        *Exeunt Caius, Evans, and Page*

MISTRESS PAGE Is there not a double excellency in this?

MISTRESS FORD I know not which pleases me better: that my husband is deceived, or Sir John. 169

MISTRESS PAGE What a taking was he in when your husband asked what was in the basket!

MISTRESS FORD I am half afraid he will have need of washing, so throwing him into the water will do him a benefit. 174

MISTRESS PAGE Hang him, dishonest rascal! I would all of the same strain were in the same distress.

MISTRESS FORD I think my husband hath some special suspicion of Falstaff's being here, for I never saw him so gross in his jealousy till now. 179

MISTRESS PAGE I will lay a plot to try that, and we will yet have more tricks with Falstaff. His dissolute disease will scarce obey this medicine.

MISTRESS FORD Shall we send that foolish carrion Mistress Quickly to him, and excuse his throwing into the water, and give him another hope, to betray him to another punishment? 186

MISTRESS PAGE We will do it. Let him be sent for tomorrow eight o'clock, to have amends.

*Enter Ford, Page, Caius, and Evans*

FORD I cannot find him. Maybe the knave bragged of that he could not compass. 190

MISTRESS PAGE (*aside to Mistress Ford*) Heard you that?

MISTRESS FORD You use me well, Master Ford, do you?

FORD Ay, I do so.

MISTRESS FORD Heaven make me better than your thoughts! 195

FORD Amen.

MISTRESS PAGE You do yourself mighty wrong, Master Ford.

FORD Ay, ay, I must bear it.

EVANS If there be anypody in the house, and in the chambers, and in the coffers, and in the presses, heaven forgive my sins at the day of judgement! 202

CAIUS Be Gar, nor I too. There is nobodies.

PAGE Fie, fie, Master Ford, are you not ashamed? What spirit, what devil suggests this imagination? I would not ha' your distemper in this kind for the wealth of Windsor Castle. 207

FORD 'Tis my fault, Master Page. I suffer for it.

EVANS You suffer for a pad conscience. Your wife is as honest a 'omans as I will desires among five thousand, and five hundred too. 211

CAIUS By Gar, I see 'tis an honest woman.

FORD Well, I promised you a dinner. Come, come, walk in the park. I pray you pardon me. I will hereafter make known to you why I have done this.—Come, wife; come, Mistress Page. I pray you pardon me. Pray heartily pardon me. 217

PAGE (*to Caius and Evans*) Let's go in, gentlemen. (*Aside to them*) But trust me, we'll mock him. (*To Ford, Caius, and Evans*) I do invite you tomorrow morning to my house to breakfast. After, we'll a-birding together. I have a fine hawk for the bush. Shall it be so? 222

FORD Anything.

EVANS If there is one, I shall make two in the company.

CAIUS If there be one or two, I shall make-a the turd.

FORD Pray you go, Master Page. 226

*Exeunt* [*all but Evans and Caius*]

EVANS I pray you now, remembrance tomorrow on the lousy knave mine Host.

CAIUS Dat is good, by Gar; with all my heart.

EVANS A lousy knave, to have his gibes and his mockeries.

*Exeunt*

**3.4**   *Enter Master Fenton and Anne Page*

FENTON
I see I cannot get thy father's love;
Therefore no more turn me to him, sweet Nan.

ANNE
Alas, how then?

FENTON      Why, thou must be thyself.
He doth object I am too great of birth,
And that, my state being galled with my expense, 5
I seek to heal it only by his wealth.
Besides these, other bars he lays before me—
My riots past, my wild societies;
And tells me 'tis a thing impossible
I should love thee but as a property. 10

ANNE Maybe he tells you true.

[FENTON]
No, heaven so speed me in my time to come!
Albeit I will confess thy father's wealth
Was the first motive that I wooed thee, Anne,
Yet, wooing thee, I found thee of more value 15
Than stamps in gold or sums in sealèd bags;
And 'tis the very riches of thyself
That now I aim at.

ANNE      Gentle Master Fenton,
Yet seek my father's love, still seek it, sir.
If opportunity and humblest suit 20
Cannot attain it, why then—

*Enter Justice Shallow, Master Slender* [*richly dressed*]*, and Mistress Quickly*

               Hark you hither.

*They talk apart*

SHALLOW Break their talk, Mistress Quickly. My kinsman shall speak for himself.

SLENDER I'll make a shaft or a bolt on't. 'Slid, 'tis but venturing. 25

SHALLOW
Be not dismayed.

SLENDER      No, she shall not dismay me.
I care not for that, but that I am afeard.

MISTRESS QUICKLY (*to Anne*) Hark ye, Master Slender would speak a word with you.

ANNE
I come to him. (*To Fenton*) This is my father's choice.
O, what a world of vile ill-favoured faults 31
Looks handsome in three hundred pounds a year!

MISTRESS QUICKLY And how does good Master Fenton? Pray you, a word with you.

*She draws Fenton aside*

SHALLOW She's coming. To her, coz! O boy, thou hadst a father! 36

SLENDER I had a father, Mistress Anne; my uncle can tell you good jests of him.—Pray you, uncle, tell Mistress Anne the jest how my father stole two geese out of a pen, good uncle. 40

SHALLOW Mistress Anne, my cousin loves you.

SLENDER Ay, that I do, as well as I love any woman in Gloucestershire.

SHALLOW He will maintain you like a gentlewoman.

SLENDER Ay, by God, that I will, come cut and long-tail, under the degree of a squire. 46

SHALLOW He will make you a hundred and fifty pounds jointure.

ANNE Good Master Shallow, let him woo for himself.

SHALLOW Marry, I thank you for it, I thank you for that good comfort.—She calls you, coz. I'll leave you. 51

*He stands aside*

ANNE Now, Master Slender.

SLENDER Now, good Mistress Anne.

ANNE What is your will?

SLENDER My will? 'Od's heartlings, that's a pretty jest indeed! I ne'er made my will yet, I thank God; I am not such a sickly creature, I give God praise. 57

ANNE I mean, Master Slender, what would you with me?

SLENDER Truly, for mine own part, I would little or nothing with you. Your father and my uncle hath made motions. If it be my luck, so. If not, happy man be his dole. They can tell you how things go better than I can.

*Enter Master Page and Mistress Page*

You may ask your father: here he comes. 64

PAGE

Now, Master Slender.—Love him, daughter Anne.—
Why, how now? What does Master Fenton here?
You wrong me, sir, thus still to haunt my house.
I told you, sir, my daughter is disposed of.

FENTON

Nay, Master Page, be not impatient.

MISTRESS PAGE

Good Master Fenton, come not to my child. 70

PAGE She is no match for you.

FENTON Sir, will you hear me?

PAGE No, good Master Fenton.—
Come, Master Shallow; come, son Slender, in.—
Knowing my mind, you wrong me, Master Fenton. 75

*Exeunt Page, Shallow, and Slender*

MISTRESS QUICKLY (*to Fenton*) Speak to Mistress Page.

FENTON

Good Mistress Page, for that I love your daughter
In such a righteous fashion as I do,
Perforce against all checks, rebukes, and manners
I must advance the colours of my love, 80
And not retire. Let me have your good will.

ANNE Good mother, do not marry me to yon fool.

MISTRESS PAGE I mean it not; I seek you a better husband.

MISTRESS QUICKLY ⌈*aside to Anne*⌉ That's my master, Master Doctor. 85

ANNE

Alas, I had rather be set quick i'th' earth
And bowled to death with turnips.

MISTRESS PAGE

Come, trouble not yourself, good Master Fenton.
I will not be your friend nor enemy.
My daughter will I question how she loves you, 90
And as I find her, so am I affected.
Till then, farewell, sir. She must needs go in.
Her father will be angry.

FENTON

Farewell, gentle mistress.—Farewell, Nan.

*Exeunt Mistress Page and Anne*

MISTRESS QUICKLY This is my doing now. 'Nay', said I, 'will you cast away your child on a fool and a physician? Look on Master Fenton.' This is my doing.

FENTON

I thank thee, (*giving her a ring*) and I pray thee, once tonight 98
Give my sweet Nan this ring. (*Giving money*) There's for thy pains.

MISTRESS QUICKLY Now heaven send thee good fortune!

*Exit Fenton*

A kind heart he hath. A woman would run through fire and water for such a kind heart. But yet I would my master had Mistress Anne; or I would Master Slender had her; or, in sooth, I would Master Fenton had her. I will do what I can for them all three, for so I have promised, and I'll be as good as my word—but speciously for Master Fenton. Well, I must of another errand to Sir John Falstaff from my two mistresses. What a beast am I to slack it! *Exit*

**3.5** *Enter Sir John Falstaff*

SIR JOHN Bardolph, I say!

*Enter Bardolph*

BARDOLPH Here, sir.

SIR JOHN Go fetch me a quart of sack; put a toast in't.

*Exit Bardolph*

Have I lived to be carried in a basket like a barrow of butcher's offal, and to be thrown in the Thames? Well, if I be served such another trick, I'll have my brains ta'en out and buttered, and give them to a dog for a New Year's gift. 'Sblood, the rogues slighted me into the river with as little remorse as they would have drowned a blind bitch's puppies, fifteen i'th' litter! And you may know by my size that I have a kind of alacrity in sinking. If the bottom were as deep as hell, I should down. I had been drowned, but that the shore was shelvy and shallow—a death that I abhor, for the water swells a man, and what a thing should I have been when I had been swelled? By the Lord, a mountain of mummy! 17

*Enter Bardolph, with ⌈two large cups of⌉ sack*

BARDOLPH Here's Mistress Quickly, sir, to speak with you.

SIR JOHN Come, let me pour in some sack to the Thames' water, for my belly's as cold as if I had swallowed snowballs for pills to cool the reins. 21

*He drinks*

Call her in.

BARDOLPH Come in, woman!

*Enter Mistress Quickly*

MISTRESS QUICKLY (*to Sir John*) By your leave; I cry you mercy. Give your worship good morrow! 25

SIR JOHN (⌈*drinking, then*⌉ *speaking to Bardolph*) Take away these chalices. Go brew me a pottle of sack, finely.

BARDOLPH With eggs, sir?

SIR JOHN Simple of itself. I'll no pullet-sperms in my brewage. *Exit Bardolph,* ⌈*with cups*⌉
How now? 31

MISTRESS QUICKLY Marry, sir, I come to your worship from Mistress Ford.

SIR JOHN Mistress Ford? I have had ford enough: I was thrown into the ford, I have my belly full of ford. 35

MISTRESS QUICKLY Alas the day, good heart, that was not her fault. She does so take on with her men; they mistook their erection.

SIR JOHN So did I mine, to build upon a foolish woman's promise. 40

MISTRESS QUICKLY Well, she laments, sir, for it, that it would yearn your heart to see it. Her husband goes this morning a-birding. She desires you once more to come to her, between eight and nine. I must carry her word quickly. She'll make you amends, I warrant you.

SIR JOHN Well, I will visit her. Tell her so, and bid her think what a man is; let her consider his frailty, and then judge of my merit.

MISTRESS QUICKLY I will tell her.

SIR JOHN Do so. Between nine and ten, sayst thou? 50

MISTRESS QUICKLY Eight and nine, sir.

SIR JOHN Well, be gone. I will not miss her.

MISTRESS QUICKLY Peace be with you, sir. *Exit*

SIR JOHN I marvel I hear not of Master Brooke; he sent
me word to stay within. I like his money well.          55
    *Enter Master Ford, disguised as Brooke*
By the mass, here he comes.

FORD God bless you, sir.

SIR JOHN Now, Master Brooke, you come to know what
hath passed between me and Ford's wife.

FORD That indeed, Sir John, is my business.          60

SIR JOHN Master Brooke, I will not lie to you. I was at
her house the hour she appointed me.

FORD And sped you, sir?

SIR JOHN Very ill-favouredly, Master Brooke.

FORD How so, sir? Did she change her determination?

SIR JOHN No, Master Brooke, but the peaking cornuto her
husband, Master Brooke, dwelling in a continual 'larum
of jealousy, comes me in the instant of our encounter—
after we had embraced, kissed, protested, and, as it
were, spoke the prologue of our comedy—and at his
heels a rabble of his companions, thither provoked and
instigated by his distemper, and, forsooth, to search his
house for his wife's love.

FORD What, while you were there?

SIR JOHN While I was there.          75

FORD And did he search for you, and could not find you?

SIR JOHN You shall hear. As God would have it, comes
in one Mistress Page, gives intelligence of Ford's
approach, and, by her invention and Ford's wife's
distraction, they conveyed me into a buck-basket—

FORD A buck-basket?          81

SIR JOHN By the Lord, a buck-basket!—rammed me in
with foul shirts and smocks, socks, foul stockings,
greasy napkins, that, Master Brooke, there was the
rankest compound of villainous smell that ever offended
nostril.          86

FORD And how long lay you there?

SIR JOHN Nay, you shall hear, Master Brooke, what I have
suffered to bring this woman to evil, for your good.
Being thus crammed in the basket, a couple of Ford's
knaves, his hinds, were called forth by their mistress,
to carry me, in the name of foul clothes, to Datchet
Lane. They took me on their shoulders, met the jealous
knave their master in the door, who asked them once
or twice what they had in their basket. I quaked for
fear lest the lunatic knave would have searched it, but
fate, ordaining he should be a cuckold, held his hand.
Well, on went he for a search, and away went I for
foul clothes. But mark the sequel, Master Brooke. I
suffered the pangs of three several deaths. First, an
intolerable fright, to be detected with a jealous rotten
bell-wether. Next, to be compassed like a good bilbo in
the circumference of a peck, hilt to point, heel to head.
And then, to be stopped in, like a strong distillation,
with stinking clothes that fretted in their own grease.
Think of that—a man of my kidney—think of that—
that am as subject to heat as butter, a man of continual
dissolution and thaw. It was a miracle to scape
suffocation. And in the height of this bath, when I was
more than half stewed in grease like a Dutch dish, to
be thrown into the Thames and cooled, glowing-hot,
in that surge, like a horseshoe. Think of that—hissing
hot—think of that, Master Brooke!          113

FORD In good sadness, sir, I am sorry that for my sake
you have suffered all this. My suit then is desperate.
You'll undertake her no more?

SIR JOHN Master Brooke, I will be thrown into Etna as I
have been into Thames ere I will leave her thus. Her
husband is this morning gone a-birding. I have received
from her another embassy of meeting. 'Twixt eight and
nine is the hour, Master Brooke.          121

FORD 'Tis past eight already, sir.

SIR JOHN Is it? I will then address me to my appointment.
Come to me at your convenient leisure, and you shall
know how I speed; and the conclusion shall be crowned
with your enjoying her. Adieu. You shall have her,
Master Brooke; Master Brooke, you shall cuckold Ford.
    *Exit*

FORD Hum! Ha! Is this a vision? Is this a dream? Do I
sleep? Master Ford, awake! Awake, Master Ford!
There's a hole made in your best coat, Master Ford.
This 'tis to be married! This 'tis to have linen and
buck-baskets! Well, I will proclaim myself what I am.
I will now take the lecher. He is at my house. He
cannot scape me; 'tis impossible he should. He cannot
creep into a halfpenny purse, nor into a pepperbox.
But lest the devil that guides him should aid him, I will
search impossible places. Though what I am I cannot
avoid, yet to be what I would not shall not make me
tame. If I have horns to make one mad, let the proverb
go with me: I'll be horn-mad.          *Exit*

**4.1**    *Enter Mistress Page, Mistress Quickly, and William
    Page*

MISTRESS PAGE Is he at Mistress Ford's already, thinkest
thou?

MISTRESS QUICKLY Sure he is by this, or will be presently.
But truly he is very courageous-mad about his throwing
into the water. Mistress Ford desires you to come
suddenly.          6

MISTRESS PAGE I'll be with her by and by. I'll but bring
my young man here to school.
    *Enter Sir Hugh Evans*
Look where his master comes. 'Tis a playing day, I
see.—How now, Sir Hugh, no school today?          10

EVANS No, Master Slender is let the boys leave to play.

MISTRESS QUICKLY Blessing of his heart!

MISTRESS PAGE Sir Hugh, my husband says my son profits
nothing in the world at his book. I pray you ask him
some questions in his accidence.          15

EVANS Come hither, William. Hold up your head. Come.

MISTRESS PAGE Come on, sirrah. Hold up your head.
Answer your master; be not afraid.

EVANS William, how many numbers is in nouns?

WILLIAM Two.          20

MISTRESS QUICKLY Truly, I thought there had been one
number more, because they say ''Od's nouns'.

EVANS Peace your tattlings!—What is 'fair', William?

WILLIAM '*Pulcher*'.

MISTRESS QUICKLY Polecats? There are fairer things than
polecats, sure.          26

EVANS You are a very simplicity 'oman. I pray you
peace.—What is '*lapis*', William?

WILLIAM A stone.

EVANS And what is 'a stone', William?          30

WILLIAM A pebble.

EVANS No, it is '*lapis*'. I pray you remember in your prain.

WILLIAM '*Lapis*'.

EVANS That is a good William. What is he, William, that
does lend articles?          35

WILLIAM Articles are borrowed of the pronoun, and be
thus declined. *Singulariter nominativo*: '*hic, haec, hoc*'.

EVANS *Nominativo*: '*hig, hag, hog*'. Pray you mark:
*genitivo*: '*huius*'. Well, what is your accusative case?

WILLIAM *Accusativo*: 'hinc'—                                   40
EVANS I pray you have your remembrance, child.
    *Accusativo*: 'hing, hang, hog'.
MISTRESS QUICKLY 'Hang-hog' is Latin for bacon, I warrant
    you.
EVANS Leave your prabbles, 'oman!—What is the focative
    case, William?                                            46
WILLIAM O—*vocativo*, O—
EVANS Remember, William, focative is *caret*.
MISTRESS QUICKLY And that's a good root.
EVANS 'Oman, forbear.                                         50
MISTRESS PAGE (*to Mistress Quickly*) Peace.
EVANS What is your genitive case plural, William?
WILLIAM Genitive case?
EVANS Ay.
WILLIAM *Genitivo*: 'horum, harum, horum'.                    55
MISTRESS QUICKLY Vengeance of Jenny's case! Fie on her!
    Never name her, child, if she be a whore.
EVANS For shame, 'oman!
MISTRESS QUICKLY You do ill to teach the child such words.
    He teaches him to hick and to hack, which they'll do
    fast enough of themselves, and to call 'whorum'. Fie
    upon you!                                                 62
EVANS 'Oman, art thou lunatics? Hast thou no
    understandings for thy cases, and the numbers of the
    genders? Thou art as foolish Christian creatures as I
    would desires.                                            66
MISTRESS PAGE (*to Mistress Quickly*) Prithee, hold thy peace.
EVANS Show me now, William, some declensions of your
    pronouns.
WILLIAM Forsooth, I have forgot.                              70
EVANS It is '*qui, que, quod*'. If you forget your '*qui*'s, your
    '*que*'s, and your '*quod*'s, you must be preeches. Go your
    ways and play; go.
MISTRESS PAGE He is a better scholar than I thought he
    was.                                                      75
EVANS He is a good sprag memory. Farewell, Mistress
    Page.
MISTRESS PAGE Adieu, good Sir Hugh.              *Exit Evans*
    Get you home, boy.                         *Exit William*
    (*To Mistress Quickly*) Come, we stay too long.   *Exeunt*

**4.2**    *Enter Sir John Falstaff and Mistress Ford*
SIR JOHN Mistress Ford, your sorrow hath eaten up my
    sufferance. I see you are obsequious in your love, and
    I profess requital to a hair's breadth: not only, Mistress
    Ford, in the simple office of love, but in all the
    accoutrement, complement, and ceremony of it. But
    are you sure of your husband now?                          6
MISTRESS FORD He's a-birding, sweet Sir John.
MISTRESS PAGE (*within*) What ho, gossip Ford, what ho!
MISTRESS FORD Step into th' chamber, Sir John.
                    *Sir John steps into the chamber*
        *Enter Mistress Page*
MISTRESS PAGE How now, sweetheart, who's at home
    besides yourself?                                         11
MISTRESS FORD Why, none but mine own people.
MISTRESS PAGE Indeed?
MISTRESS FORD No, certainly. (*Aside to her*) Speak louder.
MISTRESS PAGE Truly, I am so glad you have nobody here.
MISTRESS FORD Why?                                            16
MISTRESS PAGE Why, woman, your husband is in his old
    lines again. He so takes on yonder with my husband,
    so rails against all married mankind, so curses all Eve's
    daughters of what complexion soever, and so buffets
    himself on the forehead, crying 'Peer out, peer out!',

that any madness I ever yet beheld seemed but
tameness, civility, and patience to this his distemper
he is in now. I am glad the fat knight is not here.
MISTRESS FORD Why, does he talk of him?                       25
MISTRESS PAGE Of none but him; and swears he was
    carried out, the last time he searched for him, in a
    basket, protests to my husband he is now here, and
    hath drawn him and the rest of their company from
    their sport to make another experiment of his suspicion.
    But I am glad the knight is not here. Now he shall see
    his own foolery.                                          32
MISTRESS FORD How near is he, Mistress Page?
MISTRESS PAGE Hard by at street end. He will be here
    anon.                                                     35
MISTRESS FORD I am undone: the knight is here.
MISTRESS PAGE Why then, you are utterly shamed, and
    he's but a dead man. What a woman are you! Away
    with him, away with him! Better shame than murder.
MISTRESS FORD Which way should he go? How should I
    bestow him? Shall I put him into the basket again?
            *Sir John comes forth from the chamber*
SIR JOHN No, I'll come no more i'th' basket. May I not
    go out ere he come?                                       43
MISTRESS PAGE Alas, three of Master Ford's brothers watch
    the door with pistols, that none shall issue out.
    Otherwise you might slip away ere he came. But what
    make you here?
SIR JOHN What shall I do? I'll creep up into the chimney.
MISTRESS FORD There they always use to discharge their
    birding-pieces.                                           50
⌈MISTRESS PAGE⌉ Creep into the kiln-hole.
SIR JOHN Where is it?
MISTRESS FORD He will seek there, on my word. Neither
    press, coffer, chest, trunk, well, vault, but he hath an
    abstract for the remembrance of such places, and goes
    to them by his note. There is no hiding you in the
    house.                                                    57
SIR JOHN I'll go out, then.
MISTRESS ⌈PAGE⌉ If you go out in your own semblance,
    you die, Sir John—unless you go out disguised.            60
MISTRESS FORD How might we disguise him?
MISTRESS PAGE Alas the day, I know not. There is no
    woman's gown big enough for him; otherwise he might
    put on a hat, a muffler, and a kerchief, and so escape.
SIR JOHN Good hearts, devise something. Any extremity
    rather than a mischief.                                   66
MISTRESS FORD My maid's aunt, the fat woman of Brent-
    ford, has a gown above.
MISTRESS PAGE On my word, it will serve him; she's as
    big as he is; and there's her thrummed hat, and her
    muffler too.—Run up, Sir John.                            71
MISTRESS FORD Go, go, sweet Sir John. Mistress Page and
    I will look some linen for your head.
MISTRESS PAGE Quick, quick! We'll come dress you
    straight. Put on the gown the while.       *Exit Sir John*
MISTRESS FORD I would my husband would meet him in
    this shape. He cannot abide the old woman of Brentford.
    He swears she's a witch, forbade her my house, and
    hath threatened to beat her.                              79
MISTRESS PAGE Heaven guide him to thy husband's cudgel,
    and the devil guide his cudgel afterwards!
MISTRESS FORD But is my husband coming?
MISTRESS PAGE Ay, in good sadness is he, and talks of the
    basket too, howsoever he hath had intelligence.          84
MISTRESS FORD We'll try that, for I'll appoint my men to
    carry the basket again, to meet him at the door with
    it as they did last time.

MISTRESS PAGE Nay, but he'll be here presently. Let's go
dress him like the witch of Brentford. 89
MISTRESS FORD I'll first direct my men what they shall do
with the basket. Go up; I'll bring linen for him straight.
MISTRESS PAGE Hang him, dishonest varlet! We cannot
misuse him enough. ⌈*Exit Mistress Ford*⌉
We'll leave a proof by that which we will do,
Wives may be merry, and yet honest, too. 95
We do not act that often jest and laugh.
'Tis old but true: 'Still swine eats all the draff'. *Exit*
*Enter* ⌈*Mistress Ford, with*⌉ *John and Robert*
MISTRESS FORD Go, sirs, take the basket again on your
shoulders. Your master is hard at door. If he bid you
set it down, obey him. Quickly, dispatch! *Exit*
⌈JOHN⌉ Come, come, take it up. 101
⌈ROBERT⌉ Pray heaven it be not full of knight again.
⌈JOHN⌉ I hope not; I had as lief bear so much lead.
*They lift the basket.*
*Enter Master Ford, Master Page, Doctor Caius, Sir*
*Hugh Evans, and Justice Shallow*
FORD Ay, but if it prove true, Master Page, have you any
way then to unfool me again? (*To John and Robert*) Set
down the basket, villains. 106
*John and Robert set down the basket*
Somebody call my wife. Youth in a basket! O, you
panderly rascals! There's a knot, a gang, a pack, a
conspiracy against me. Now shall the devil be
shamed.—What, wife, I say! Come, come forth! Behold
what honest clothes you send forth to bleaching. 111
PAGE Why, this passes, Master Ford. You are not to go
loose any longer; you must be pinioned.
EVANS Why, this is lunatics; this is mad as a mad dog.
SHALLOW Indeed, Master Ford, this is not well, indeed.
FORD So say I too, sir. 116
*Enter Mistress Ford*
Come hither, Mistress Ford! Mistress Ford, the honest
woman, the modest wife, the virtuous creature, that
hath the jealous fool to her husband! I suspect without
cause, mistress, do I? 120
MISTRESS FORD God be my witness you do, if you suspect
me in any dishonesty.
FORD Well said, brazen-face; hold it out.
*He opens the basket and starts to take out clothes*
Come forth, sirrah!
PAGE This passes. 125
MISTRESS FORD (*to Ford*) Are you not ashamed? Let the
clothes alone.
FORD I shall find you anon.
EVANS 'Tis unreasonable: will you take up your wife's
clothes? Come, away. 130
FORD ⌈*to John and Robert*⌉ Empty the basket, I say.
⌈PAGE⌉ Why, man, why?
FORD Master Page, as I am a man, there was one conveyed
out of my house yesterday in this basket. Why may
not he be there again? In my house I am sure he is.
My intelligence is true, my jealousy is reasonable. ⌈*To*
*John and Robert*⌉ Pluck me out all the linen. 137
*He takes out clothes*
MISTRESS FORD If you find a man there, he shall die a
flea's death.
PAGE Here's no man. 140
SHALLOW By my fidelity, this is not well, Master Ford.
This wrongs you.
EVANS Master Ford, you must pray, and not follow the
imaginations of your own heart. This is jealousies.

FORD Well, he's not here I seek for. 145
PAGE No, nor nowhere else but in your brain.
FORD Help to search my house this one time. If I find not
what I seek, show no colour for my extremity; let me
for ever be your table-sport; let them say of me, 'As
jealous as Ford, that searched a hollow walnut for his
wife's leman'. Satisfy me once more; once more search
with me. ⌈*Exeunt John and Robert with the basket*⌉
MISTRESS FORD What ho, Mistress Page! Come you and
the old woman down. My husband will come into the
chamber. 155
FORD Old woman? What old woman's that?
MISTRESS FORD Why, it is my maid's Aunt of Brentford.
FORD A witch, a quean, an old, cozening quean! Have I
not forbid her my house? She comes of errands, does
she? We are simple men; we do not know what's
brought to pass under the profession of fortune-telling.
She works by charms, by spells, by th' figure, and such
daubery as this is, beyond our element. We know
nothing.—Come down, you witch, you hag, you! Come
down, I say! 165
⌈*Enter Mistress Page, and Sir John Falstaff,*
*disguised as an old woman.*⌉
⌈*Ford makes towards them*⌉
MISTRESS FORD Nay, good sweet husband!—Good gentlemen, let him not strike the old woman.
MISTRESS PAGE (*to Sir John*) Come, Mother Prat. Come,
give me your hand.
FORD I'll prat her! 170
*He beats Sir John*
Out of my door, you witch, you rag, you baggage, you
polecat, you runnion! Out, out! I'll conjure you, I'll
fortune-tell you! *Exit Sir John*
MISTRESS PAGE Are you not ashamed? I think you have
killed the poor woman. 175
MISTRESS FORD Nay, he will do it.—'Tis a goodly credit
for you!
FORD Hang her, witch!
EVANS By Jeshu, I think the 'oman is a witch indeed. I
like not when a 'oman has a great peard. I spy a great
peard under his muffler. 181
FORD Will you follow, gentlemen? I beseech you, follow.
See but the issue of my jealousy. If I cry out thus upon
no trail, never trust me when I open again.
PAGE Let's obey his humour a little further. Come,
gentlemen. *Exeunt the men*
MISTRESS PAGE By my troth, he beat him most pitifully.
MISTRESS FORD Nay, by th' mass, that he did not—he beat
him most unpitifully, methought. 189
MISTRESS PAGE I'll have the cudgel hallowed and hung
o'er the altar. It hath done meritorious service.
MISTRESS FORD What think you—may we, with the
warrant of womanhood and the witness of a good
conscience, pursue him with any further revenge?
MISTRESS PAGE The spirit of wantonness is sure scared out
of him. If the devil have him not in fee-simple, with
fine and recovery, he will never, I think, in the way of
waste attempt us again.
MISTRESS FORD Shall we tell our husbands how we have
served him? 200
MISTRESS PAGE Yes, by all means, if it be but to scrape
the figures out of your husband's brains. If they can
find in their hearts the poor, unvirtuous, fat knight
shall be any further afflicted, we two will still be the
ministers. 205

MISTRESS FORD I'll warrant they'll have him publicly shamed, and methinks there would be no period to the jest should he not be publicly shamed.

MISTRESS PAGE Come, to the forge with it, then shape it. I would not have things cool.    *Exeunt*

**4.3**    *Enter the Host of the Garter and Bardolph*

BARDOLPH Sir, the Germans desire to have three of your horses. The Duke himself will be tomorrow at court, and they are going to meet him.

HOST What duke should that be comes so secretly? I hear not of him in the court. Let me speak with the gentlemen. They speak English?    6

BARDOLPH Ay, sir. I'll call them to you.

HOST They shall have my horses, but I'll make them pay; I'll sauce them. They have had my house a week at command; I have turned away my other guests. They must come off: I'll sauce them. Come.    *Exeunt*

**4.4**    *Enter Master Page, Master Ford, Mistress Page, Mistress Ford, and Sir Hugh Evans*

EVANS 'Tis one of the best discretions of a 'oman as ever I did look upon.

PAGE And did he send you both these letters at an instant?

MISTRESS PAGE Within a quarter of an hour.

FORD
Pardon me, wife. Henceforth do what thou wilt.    5
I rather will suspect the sun with cold
Than thee with wantonness. Now doth thy honour stand,
In him that was of late an heretic,
As firm as faith.

PAGE    'Tis well, 'tis well; no more.
Be not as extreme in submission    10
As in offence.
But let our plot go forward. Let our wives
Yet once again, to make us public sport,
Appoint a meeting with this old fat fellow,
Where we may take him and disgrace him for it.    15

FORD
There is no better way than that they spoke of.

PAGE
How, to send him word they'll meet him in the Park
At midnight? Fie, fie, he'll never come.

EVANS You say he has been thrown in the rivers, and has been grievously peaten as an old 'oman. Methinks there should be terrors in him, that he should not come. Methinks his flesh is punished; he shall have no desires.

PAGE So think I too.

⌈MISTRESS⌉ FORD
Devise but how you'll use him when he comes,    25
And let us two devise to bring him thither.

MISTRESS PAGE
There is an old tale goes that Herne the hunter,
Sometime a keeper here in Windsor Forest,
Doth all the winter time at still midnight
Walk round about an oak with great ragg'd horns;    30
And there he blasts the trees, and takes the cattle,
And makes milch-kine yield blood, and shakes a chain
In a most hideous and dreadful manner.
You have heard of such a spirit, and well you know
The superstitious idle-headed eld    35
Received, and did deliver to our age,
This tale of Herne the hunter for a truth.

PAGE
Why, yet there want not many that do fear
In deep of night to walk by this Herne's Oak.
But what of this?

MISTRESS FORD    Marry, this is our device:    40
That Falstaff at that oak shall meet with us,
Disguised like Herne, with huge horns on his head.

PAGE
Well, let it not be doubted but he'll come,
And in this shape. When you have brought him thither
What shall be done with him? What is your plot?    45

MISTRESS PAGE
That likewise have we thought upon, and thus.
Nan Page my daughter, and my little son,
And three or four more of their growth, we'll dress
Like urchins, oafs, and fairies, green and white,
With rounds of waxen tapers on their heads,    50
And rattles in their hands. Upon a sudden,
As Falstaff, she, and I are newly met,
Let them from forth a saw-pit rush at once,
With some diffusèd song. Upon their sight
We two in great amazèdness will fly.    55
Then let them all encircle him about,
And, fairy-like, to pinch the unclean knight,
And ask him why, that hour of fairy revel,
In their so sacred paths he dares to tread
In shape profane.

⌈MISTRESS⌉ FORD    And till he tell the truth,    60
Let the supposèd fairies pinch him sound,
And burn him with their tapers.

MISTRESS PAGE    The truth being known,
We'll all present ourselves, dis-horn the spirit,
And mock him home to Windsor.

FORD    The children must
Be practised well to this, or they'll ne'er do't.    65

EVANS I will teach the children their behaviours, and I will be like a jackanapes also, to burn the knight with my taber.

FORD
That will be excellent. I'll go buy them vizors.

MISTRESS PAGE
My Nan shall be the Queen of all the Fairies,    70
Finely attirèd in a robe of white.

PAGE
That silk will I go buy— (*aside*) and in that tire
Shall Master Slender steal my Nan away,
And marry her at Eton. (*To Mistress Page*) Go send to Falstaff straight.

FORD
Nay, I'll to him again in name of Brooke.    75
He'll tell me all his purpose. Sure he'll come.

MISTRESS PAGE
Fear not you that. (*To Page, Ford, and Evans*) Go get us properties
And tricking for our fairies.

EVANS Let us about it. It is admirable pleasures, and fery honest knaveries.    *Exeunt Ford, Page, and Evans*

MISTRESS PAGE    Go, Mistress Ford,
Send quickly to Sir John, to know his mind.    81
*Exit Mistress Ford*
I'll to the Doctor. He hath my good will,
And none but he, to marry with Nan Page.
That Slender, though well landed, is an idiot;
And he my husband best of all affects.    85

The Doctor is well moneyed, and his friends
Potent at court. He, none but he, shall have her,
Though twenty thousand worthier come to crave her.

*Exit*

**4.5**   *Enter the Host of the Garter and Simple*

HOST What wouldst thou have, boor? What, thick-skin?
Speak, breathe, discuss. Brief, short, quick, snap.

SIMPLE Marry, sir, I come to speak with Sir John Falstaff,
from Master Slender.                                               4

HOST There's his chamber, his house, his castle, his
standing-bed and truckle-bed. 'Tis painted about with
the story of the Prodigal, fresh and new. Go knock and
call. He'll speak like an Anthropophaginian unto thee.
Knock, I say.                                                          9

SIMPLE There's an old woman, a fat woman, gone up
into his chamber. I'll be so bold as stay, sir, till she
come down. I come to speak with her, indeed.

HOST Ha, a fat woman? The knight may be robbed. I'll
call.—Bully knight, bully Sir John! Speak from thy
lungs military! Art thou there? It is thine Host, thine
Ephesian, calls.                                                     16

SIR JOHN (*within*) How now, mine Host?

HOST Here's a Bohemian Tartar tarries the coming down
of thy fat woman. Let her descend, bully, let her
descend. My chambers are honourable. Fie, privacy!
Fie!                                                                    21

*Enter Sir John Falstaff*

SIR JOHN There was, mine Host, an old fat woman even
now with me; but she's gone.

SIMPLE Pray you, sir, was't not the wise woman of
Brentford?                                                            25

SIR JOHN Ay, marry was it, mussel-shell. What would you
with her?

SIMPLE My master, sir, my master Slender, sent to her,
seeing her go through the streets, to know, sir, whether
one Nim, sir, that beguiled him of a chain, had the
chain or no.                                                          31

SIR JOHN I spake with the old woman about it.

SIMPLE And what says she, I pray, sir?

SIR JOHN Marry, she says that the very same man that
beguiled Master Slender of his chain cozened him of it.

SIMPLE I would I could have spoken with the woman
herself. I had other things to have spoken with her,
too, from him.

SIR JOHN What are they? Let us know.

HOST Ay, come, quick.                                           40

⌈SIMPLE⌉ I may not conceal them, sir.

HOST Conceal them, or thou diest.

SIMPLE Why, sir, they were nothing but about Mistress
Anne Page, to know if it were my master's fortune to
have her or no.                                                      45

SIR JOHN 'Tis, 'tis his fortune.

SIMPLE What, sir?

SIR JOHN To have her or no. Go say the woman told me
so.

SIMPLE May I be bold to say so, sir?                         50

SIR JOHN Ay, Sir Tike; who more bold?

SIMPLE I thank your worship. I shall make my master
glad with these tidings.                                    *Exit*

HOST Thou art clerkly, thou art clerkly, Sir John. Was
there a wise woman with thee?                                55

SIR JOHN Ay, that there was, mine Host, one that hath
taught me more wit than ever I learned before in my
life. And I paid nothing for it, neither, but was paid for
my learning.

*Enter Bardolph, ⌈muddy⌉*

BARDOLPH O Lord, sir, cozenage, mere cozenage!     60

HOST Where be my horses? Speak well of them, varletto.

BARDOLPH Run away with the cozeners. For so soon as I
came beyond Eton, they threw me off from behind one
of them, in a slough of mire, and set spurs and away,
like three German devils, three Doctor Faustuses.   65

HOST They are gone but to meet the Duke, villain. Do not
say they be fled. Germans are honest men.

*Enter Sir Hugh Evans*

EVANS Where is mine Host?

HOST What is the matter, sir?

EVANS Have a care of your entertainments. There is a
friend of mine come to town tells me there is three
cozen Garmombles that has cozened all the hosts of
Reading, of Maidenhead, of Colnbrook, of horses and
money. I tell you for good will, look you. You are wise,
and full of gibes and vlouting-stocks, and 'tis not
convenient you should be cozened. Fare you well.    76

*Exit*

*Enter Doctor Caius*

CAIUS Vere is mine Host de Jarteer?

HOST Here, Master Doctor, in perplexity and doubtful
dilemma.

CAIUS I cannot tell vat is dat, but it is tell-a me dat you
make grand preparation for a duke de Jamany. By my
trot, der is no duke that the court is know to come. I
tell you for good will. Adieu.                          *Exit*

HOST (*to Bardolph*) Hue and cry, villain, go! (*To Sir John*)
Assist me, knight. I am undone. (*To Bardolph*) Fly, run,
hue and cry, villain. I am undone.                        86

*Exeunt Host and Bardolph ⌈severally⌉*

SIR JOHN I would all the world might be cozened, for I
have been cozened, and beaten too. If it should come
to the ear of the court how I have been transformed,
and how my transformation hath been washed and
cudgelled, they would melt me out of my fat, drop by
drop, and liquor fishermen's boots with me. I warrant
they would whip me with their fine wits till I were as
crestfallen as a dried pear. I never prospered since I
forswore myself at primero. Well, if my wind were but
long enough, I would repent.                                96

*Enter Mistress Quickly*

Now; whence come you?

MISTRESS QUICKLY From the two parties, forsooth.

SIR JOHN The devil take one party, and his dam the other,
and so they shall be both bestowed. I have suffered
more for their sakes, more than the villainous
inconstancy of man's disposition is able to bear.   102

MISTRESS QUICKLY O Lord, sir, and have not they suffered?
Yes, I warrant, speciously one of them. Mistress Ford,
good heart, is beaten black and blue, that you cannot
see a white spot about her.                                106

SIR JOHN What tellest thou me of black and blue? I was
beaten myself into all the colours of the rainbow, and
I was like to be apprehended for the witch of Brentford.
But that my admirable dexterity of wit, my
counterfeiting the action of an old woman, delivered
me, the knave constable had set me i'th' stocks, i'th'
common stocks, for a witch.                               113

MISTRESS QUICKLY Sir, let me speak with you in your
chamber. You shall hear how things go, and, I warrant,
to your content. Here is a letter will say somewhat.
Good hearts, what ado here is to bring you together!
Sure one of you does not serve heaven well, that you
are so crossed.                                                119

SIR JOHN Come up into my chamber.                *Exeunt*

**4.6** *Enter Master Fenton and the Host of the Garter*

HOST Master Fenton, talk not to me. My mind is heavy.
I will give over all.

FENTON
Yet hear me speak. Assist me in my purpose,
And, as I am a gentleman, I'll give thee
A hundred pound in gold more than your loss.  5

HOST I will hear you, Master Fenton, and I will at the
least keep your counsel.

FENTON
From time to time I have acquainted you
With the dear love I bear to fair Anne Page,
Who mutually hath answered my affection,  10
So far forth as herself might be her chooser,
Even to my wish. I have a letter from her
Of such contents as you will wonder at,
The mirth whereof so larded with my matter
That neither singly can be manifested  15
Without the show of both. Fat Falstaff
Hath a great scene. The image of the jest
I'll show you here at large. Hark, good mine Host.
Tonight at Herne's Oak, just 'twixt twelve and one,
Must my sweet Nan present the Fairy Queen—  20
⌈*Showing the letter*⌉
The purpose why is here—in which disguise,
While other jests are something rank on foot,
Her father hath commanded her to slip
Away with Slender, and with him at Eton
Immediately to marry. She hath consented.  25
Now, sir, her mother, ever strong against that match
And firm for Doctor Caius, hath appointed
That he shall likewise shuffle her away,
While other sports are tasking of their minds,
And at the dean'ry, where a priest attends,  30
Straight marry her. To this her mother's plot
She, seemingly obedient, likewise hath
Made promise to the Doctor. Now, thus it rests.
Her father means she shall be all in white;
And in that habit, when Slender sees his time  35
To take her by the hand and bid her go,
She shall go with him. Her mother hath intended,
The better to denote her to the Doctor—
For they must all be masked and visorèd—
That quaint in green she shall be loose enrobed,  40
With ribbons pendant flaring 'bout her head;
And when the Doctor spies his vantage ripe,
To pinch her by the hand, and on that token
The maid hath given consent to go with him.

HOST
Which means she to deceive, father or mother?  45

FENTON
Both, my good Host, to go along with me.
And here it rests: that you'll procure the vicar
To stay for me at church 'twixt twelve and one,
And, in the lawful name of marrying,
To give our hearts united ceremony.  50

HOST
Well, husband your device. I'll to the vicar.
Bring you the maid, you shall not lack a priest.

FENTON
So shall I evermore be bound to thee.
Besides, I'll make a present recompense.
*Exeunt* ⌈*severally*⌉

**5.1** *Enter Sir John Falstaff and Mistress Quickly*

SIR JOHN Prithee, no more prattling; go; I'll hold. This is
the third time; I hope good luck lies in odd numbers.

Away, go! They say there is divinity in odd numbers,
either in nativity, chance, or death. Away!

MISTRESS QUICKLY I'll provide you a chain, and I'll do
what I can to get you a pair of horns.  6

SIR JOHN Away, I say! Time wears. Hold up your head,
and mince.  *Exit Mistress Quickly*
*Enter Master Ford, disguised as Brooke*
How now, Master Brooke? Master Brooke, the matter
will be known tonight or never. Be you in the Park
about midnight at Herne's Oak, and you shall see
wonders.  12

FORD Went you not to her yesterday, sir, as you told me
you had appointed?

SIR JOHN I went to her, Master Brooke, as you see, like
a poor old man; but I came from her, Master Brooke,
like a poor old woman. That same knave Ford, her
husband, hath the finest mad devil of jealousy in him,
Master Brooke, that ever governed frenzy. I will tell
you, he beat me grievously in the shape of a woman—
for in the shape of man, Master Brooke, I fear not
Goliath with a weaver's beam, because I know also life
is a shuttle. I am in haste. Go along with me; I'll tell
you all, Master Brooke. Since I plucked geese, played
truant, and whipped top, I knew not what 'twas to be
beaten till lately. Follow me. I'll tell you strange things
of this knave Ford, on whom tonight I will be revenged,
and I will deliver his wife into your hand. Follow.
Strange things in hand, Master Brooke. Follow.  29
*Exeunt*

**5.2** *Enter Master Page, Justice Shallow, and Master Slender*

PAGE Come, come, we'll couch i'th' Castle ditch till we
see the light of our fairies. Remember, son Slender, my
daughter.

SLENDER Ay, forsooth. I have spoke with her, and we
have a nay-word how to know one another. I come to
her in white and cry 'mum'; she cries 'budget'; and
by that we know one another.  7

SHALLOW That's good, too. But what needs either your
'mum' or her 'budget'? The white will decipher her
well enough. (*To Page*) It hath struck ten o'clock.  10

PAGE The night is dark; lights and spirits will become it
well. God prosper our sport! No man means evil but
the devil, and we shall know him by his horns. Let's
away. Follow me.  *Exeunt*

**5.3** *Enter Mistress Page, Mistress Ford, and Doctor Caius*

MISTRESS PAGE Master Doctor, my daughter is in green.
When you see your time, take her by the hand, away
with her to the deanery, and dispatch it quickly. Go
before into the Park. We two must go together.

CAIUS I know vat I have to do. Adieu.  5

MISTRESS PAGE Fare you well, sir.  *Exit Caius*
My husband will not rejoice so much at the abuse of
Falstaff as he will chafe at the doctor's marrying my
daughter. But 'tis no matter. Better a little chiding than
a great deal of heartbreak.  10

MISTRESS FORD Where is Nan now, and her troop of fairies,
and the Welsh devil Hugh?

MISTRESS PAGE They are all couched in a pit hard by
Herne's Oak, with obscured lights, which, at the very
instant of Falstaff's and our meeting, they will at once
display to the night.  16

MISTRESS FORD That cannot choose but amaze him.

MISTRESS PAGE If he be not amazed, he will be mocked. If he be amazed, he will every way be mocked.

MISTRESS FORD We'll betray him finely.          20

MISTRESS PAGE
Against such lewdsters and their lechery
Those that betray them do no treachery.

MISTRESS FORD The hour draws on. To the Oak, to the Oak!          *Exeunt*

**5.4**    *Enter Sir Hugh Evans, ⌐disguised as a satyr,⌐ and*
*⌐William Page and other⌐ children, disguised as fairies*

EVANS Trib, trib, fairies! Come! And remember your parts. Be pold, I pray you. Follow me into the pit, and when I give the watch'ords, do as I pid you. Come, come; trib, trib!          *Exeunt*

**5.5**    *Enter Sir John Falstaff, disguised as Herne, ⌐with*
*horns on his head, and bearing a chain⌐*

SIR JOHN The Windsor bell hath struck twelve; the minute draws on. Now the hot-blooded gods assist me! Remember, Jove, thou wast a bull for thy Europa; love set on thy horns. O powerful love, that in some respects makes a beast a man; in some other, a man a beast! You were also, Jupiter, a swan, for the love of Leda. O omnipotent love! How near the god drew to the complexion of a goose! A fault done first in the form of a beast—O Jove, a beastly fault!—and then another fault in the semblance of a fowl—think on't, Jove, a foul fault! When gods have hot backs, what shall poor men do? For me, I am here a Windsor stag, and the fattest, I think, i'th' forest. Send me a cool rut-time, Jove, or who can blame me to piss my tallow?

*Enter Mistress Ford ⌐followed by⌐ Mistress Page*
Who comes here? My doe!          15

MISTRESS FORD Sir John! Art thou there, my deer, my male deer?

SIR JOHN My doe with the black scut! Let the sky rain potatoes, let it thunder to the tune of 'Greensleeves', hail kissing-comfits, and snow eringoes; let there come a tempest of provocation, I will shelter me here.          21
⌐*He embraces her*⌐

MISTRESS FORD Mistress Page is come with me, sweetheart.

SIR JOHN Divide me like a bribed buck, each a haunch. I will keep my sides to myself, my shoulders for the fellow of this walk, and my horns I bequeath your husbands. Am I a woodman, ha? Speak I like Herne the hunter? Why, now is Cupid a child of conscience; he makes restitution. As I am a true spirit, welcome!
⌐*A noise within*⌐

MISTRESS PAGE Alas, what noise?

MISTRESS FORD God forgive our sins!          30

SIR JOHN What should this be?

MISTRESS FORD *and* MISTRESS PAGE Away, away!
*Exeunt Mistress Ford and Mistress Page,*
⌐*running*⌐

SIR JOHN I think the devil will not have me damned, lest the oil that's in me should set hell on fire. He would never else cross me thus.          35

*Enter Sir Hugh Evans, ⌐William Page,⌐ and*
*children, disguised as before, with tapers; Mistress*
*Quickly, disguised as the Fairy Queen; Anne Page,*
*disguised as a fairy; and one disguised as*
*Hobgoblin*

MISTRESS QUICKLY
Fairies black, grey, green, and white,
You moonshine revellers, and shades of night,

You orphan heirs of fixèd destiny,
Attend your office and your quality.—
Crier hobgoblin, make the fairy oyes.          40

⌐HOBGOBLIN⌐
Elves, list your names. Silence, you airy toys.
Cricket, to Windsor chimneys shalt thou leap.
Where fires thou find'st unraked and hearths unswept,
There pinch the maids as blue as bilberry.
Our radiant Queen hates sluts and sluttery.          45

SIR JOHN (*aside*)
They are fairies. He that speaks to them shall die.
I'll wink and couch; no man their works must eye.
*He lies down, and hides his face*

EVANS
Where's Bead? Go you, and, where you find a maid
That ere she sleep has thrice her prayers said,
Raise up the organs of her fantasy,          50
Sleep she as sound as careless infancy.
But those as sleep and think not on their sins,
Pinch them, arms, legs, backs, shoulders, sides, and shins.

MISTRESS QUICKLY About, about!
Search Windsor Castle, elves, within and out.          55
Strew good luck, oafs, on every sacred room,
That it may stand till the perpetual doom
In state as wholesome as in state 'tis fit,
Worthy the owner, and the owner it.
The several chairs of order look you scour          60
With juice of balm and every precious flower.
Each fair instalment, coat, and sev'ral crest
With loyal blazon evermore be blessed;
And nightly, meadow-fairies, look you sing,
Like to the Garter's compass, in a ring.          65
Th'expressure that it bears, green let it be,
More fertile-fresh than all the field to see;
And 'Honi soit qui mal y pense' write
In em'rald tufts, flowers purple, blue, and white,
Like sapphire, pearl, and rich embroidery,          70
Buckled below fair knighthood's bending knee—
Fairies use flowers for their charactery.
Away, disperse!—But till 'tis one o'clock
Our dance of custom, round about the oak
Of Herne the hunter, let us not forget.          75

EVANS
Pray you, lock hand in hand; yourselves in order set;
And twenty glow-worms shall our lanterns be
To guide our measure round about the tree.—
But stay; I smell a man of middle earth.

SIR JOHN (*aside*)
God defend me from that Welsh fairy,          80
Lest he transform me to a piece of cheese!

⌐HOBGOBLIN⌐ (*to Sir John*)
Vile worm, thou wast o'erlooked even in thy birth.

MISTRESS QUICKLY (*to fairies*)
With trial-fire, touch me his finger-end.
If he be chaste, the flame will back descend,
And turn him to no pain; but if he start,          85
It is the flesh of a corrupted heart.

⌐HOBGOBLIN⌐
A trial, come!

EVANS          Come, will this wood take fire?
*They burn Sir John with tapers*

SIR JOHN O, O, O!

MISTRESS QUICKLY
Corrupt, corrupt, and tainted in desire.

About him, fairies; sing a scornful rhyme;    90
And, as you trip, still pinch him to your time.

*They dance around Sir John, pinching him and singing:*

FAIRIES   Fie on sinful fantasy!
     Fie on lust and luxury!
     Lust is but a bloody fire,
     Kindled with unchaste desire,    95
     Fed in heart, whose flames aspire,
     As thoughts do blow them, higher and higher.
     Pinch him, fairies, mutually.
     Pinch him for his villainy.
     Pinch him, and burn him, and turn him about,    100
     Till candles and starlight and moonshine be out.

*During the song, enter Doctor Caius one way, and exit stealing away a fairy in green; enter Master Slender another way, and exit stealing away a fairy in white; enter Master Fenton, and exit stealing away Anne Page. After the song, a noise of hunting within. Exeunt Mistress Quickly, Evans, Hobgoblin, and fairies, running. Sir John rises, and starts to run away. Enter Master Page, Master Ford, Mistress Page, and Mistress Ford*

PAGE
Nay, do not fly. I think we have watched you now.
Will none but Herne the hunter serve your turn?

MISTRESS PAGE
I pray you, come, hold up the jest no higher.
Now, good Sir John, how like you Windsor wives?   105
*(Pointing to Falstaff's horns)*
See you these, husband? Do not these fair yokes
Become the forest better than the town?

FORD *(to Sir John)* Now, sir, who's a cuckold now? Master Brooke, Falstaff's a knave, a cuckoldly knave. Here are his horns, Master Brooke. And, Master Brooke, he hath enjoyed nothing of Ford's but his buck-basket, his cudgel, and twenty pounds of money which must be paid to Master Brooke; his horses are arrested for it, Master Brooke.    114

MISTRESS FORD Sir John, we have had ill luck. We could never mate. I will never take you for my love again, but I will always count you my deer.

SIR JOHN I do begin to perceive that I am made an ass.
*He takes off the horns*

FORD Ay, and an ox, too. Both the proofs are extant.

SIR JOHN And these are not fairies? By the Lord, I was three or four times in the thought they were not fairies, and yet the guiltiness of my mind, the sudden surprise of my powers, drove the grossness of the foppery into a received belief—in despite of the teeth of all rhyme and reason—that they were fairies. See now how wit may be made a Jack-a-Lent when 'tis upon ill employment!    127

EVANS Sir John Falstaff, serve Got and leave your desires, and fairies will not pinse you.

FORD Well said, Fairy Hugh.    130

EVANS And leave you your jealousies too, I pray you.

FORD I will never mistrust my wife again till thou art able to woo her in good English.

SIR JOHN Have I laid my brain in the sun and dried it, that it wants matter to prevent so gross o'er-reaching as this? Am I ridden with a Welsh goat too? Shall I have a coxcomb of frieze? 'Tis time I were choked with a piece of toasted cheese.

EVANS Seese is not good to give putter; your belly is all putter.    140

SIR JOHN 'Seese' and 'putter'? Have I lived to stand at the taunt of one that makes fritters of English? This is enough to be the decay of lust and late walking through the realm.

MISTRESS PAGE Why, Sir John, do you think, though we would have thrust virtue out of our hearts by the head and shoulders, and have given ourselves without scruple to hell, that ever the devil could have made you our delight?

FORD What, a hodge-pudding, a bag of flax?    150

MISTRESS PAGE A puffed man?

PAGE Old, cold, withered, and of intolerable entrails?

FORD And one that is as slanderous as Satan?

PAGE And as poor as Job?

FORD And as wicked as his wife?    155

EVANS And given to fornications, and to taverns, and sack, and wine, and metheglins; and to drinkings, and swearings, and starings, pribbles and prabbles?

SIR JOHN Well, I am your theme; you have the start of me. I am dejected. I am not able to answer the Welsh flannel. Ignorance itself is a plummet o'er me. Use me as you will.    162

FORD Marry, sir, we'll bring you to Windsor, to one Master Brooke, that you have cozened of money, to whom you should have been a pander. Over and above that you have suffered, I think to repay that money will be a biting affliction.    167

PAGE Yet be cheerful, knight. Thou shalt eat a posset tonight at my house, where I will desire thee to laugh at my wife that now laughs at thee. Tell her Master Slender hath married her daughter.    171

MISTRESS PAGE *(aside)* Doctors doubt that! If Anne Page be my daughter, she is, by this, Doctor Caius's wife.

*Enter Master Slender*

SLENDER Whoa, ho, ho, father Page!

PAGE Son, how now? How now, son? Have you dispatched?    176

SLENDER Dispatched? I'll make the best in Gloucestershire know on't; would I were hanged, la, else.

PAGE Of what, son?

SLENDER I came yonder at Eton to marry Mistress Anne Page, and she's a great lubberly boy. If it had not been i'th' church, I would have swinged him, or he should have swinged me. If I did not think it had been Anne Page, would I might never stir; and 'tis a postmaster's boy.    185

PAGE Upon my life, then, you took the wrong.

SLENDER What need you tell me that? I think so, when I took a boy for a girl. If I had been married to him, for all he was in woman's apparel, I would not have had him.    190

PAGE Why, this is your own folly. Did not I tell you how you should know my daughter by her garments?

SLENDER I went to her in white and cried 'mum', and she cried 'budget', as Anne and I had appointed; and yet it was not Anne, but a postmaster's boy.    195

MISTRESS PAGE Good George, be not angry. I knew of your purpose, turned my daughter into green, and indeed she is now with the Doctor at the deanery, and there married.

*Enter Doctor Caius*

CAIUS Ver is Mistress Page? By Gar, I am cozened! I ha' married *un garçon*, a boy, *un paysan*, by Gar. A boy! It is not Anne Page, by Gar. I am cozened.    202

PAGE Why, did you take her in green?

CAIUS Ay, be Gar, and 'tis a boy. Be Gar, I'll raise all Windsor.    205

FORD This is strange. Who hath got the right Anne?
 *Enter Master Fenton and Anne*
PAGE
 My heart misgives me: here comes Master Fenton.—
 How now, Master Fenton?
ANNE
 Pardon, good father. Good my mother, pardon.
PAGE
 Now, mistress, how chance you went not with Master
  Slender?           210
⌈MISTRESS⌉ PAGE
 Why went you not with Master Doctor, maid?
FENTON
 You do amaze her. Hear the truth of it.
 You would have married her, most shamefully,
 Where there was no proportion held in love.
 The truth is, she and I, long since contracted,  215
 Are now so sure that nothing can dissolve us.
 Th'offence is holy that she hath committed,
 And this deceit loses the name of craft,
 Of disobedience, or unduteous title,
 Since therein she doth evitate and shun   220

 A thousand irreligious cursèd hours
 Which forcèd marriage would have brought upon her.
FORD (*to Page and Mistress Page*)
 Stand not amazed. Here is no remedy.
 In love the heavens themselves do guide the state;
 Money buys lands, and wives are sold by fate. 225
SIR JOHN I am glad, though you have ta'en a special
 stand to strike at me, that your arrow hath glanced.
PAGE
 Well, what remedy? Fenton, heaven give thee joy!
 What cannot be eschewed must be embraced.
SIR JOHN
 When night-dogs run, all sorts of deer are chased. 230
MISTRESS PAGE
 Well, I will muse no further. Master Fenton,
 Heaven give you many, many merry days!
 Good husband, let us every one go home,
 And laugh this sport o'er by a country fire,
 Sir John and all.
FORD     Let it be so, Sir John.  235
 To Master Brooke you yet shall hold your word,
 For he tonight shall lie with Mistress Ford.  *Exeunt*

# 2 HENRY IV

*2 Henry IV*, printed in 1600 as *The Second Part of Henry the Fourth*, was not reprinted until it was included in somewhat revised form in the 1623 Folio, with the same title. Shakespeare may have started to write it late in 1596, or in 1597, directly after *1 Henry IV*, but have laid it aside while he composed *The Merry Wives of Windsor*. As in *1 Henry IV*, he drew on *The Famous Victories of Henry the Fifth*, Holinshed's *Chronicles*, and Samuel Daniel's *Four Books of the Civil Wars*, along with other, minor sources; but the play contains a greater proportion of non-historical material apparently invented by Shakespeare. In this play Shakespeare seems from the start to have accepted the change of Sir John's surname to Falstaff.

Like *1 Henry IV*, Part Two draws on the techniques of comedy, but its overall tone is more sombre. At its start, the Prince seems to have regressed from his reformed state at the end of Part One; his father still has many causes for anxiety, has not made his expiatory pilgrimage to the Holy Land, and is again the victim of rebellion, led this time by the Earl of Northumberland, the Archbishop of York, and the Lords Hastings and Mowbray. Again Henry's public responsibilities are exacerbated by anxieties about Prince Harry's behaviour; the climax of their relationship comes after Harry, discovering his sick father asleep and thinking him dead, tries on his crown; after bitterly upbraiding him, Henry accepts his son's assertions of good faith, and, recalling the devious means by which he himself came to the throne, warns Harry that he may need to protect himself against civil strife by pursuing 'foreign quarrels'—the campaigning against France depicted in *Henry V*. The King dies in the Jerusalem Chamber of Westminster Abbey, the closest he will get to the Holy Land.

In this play the Prince spends less time than in Part One with Sir John, who is shown much in the company of Mistress Quickly and Doll Tearsheet at the Boar's Head tavern in Eastcheap and later in Gloucestershire on his way to and from the place of battle. Shakespeare never excelled the bitter-sweet comedy of the passages involving Falstaff and his old comrade Justice Shallow. The play ends in a counterpointing of major and minor keys as the newly crowned Henry V rejects Sir John and all that he has stood for.

# THE PERSONS OF THE PLAY

RUMOUR, the Presenter

EPILOGUE

KING HENRY IV

PRINCE HARRY, later crowned King Henry V ⎤

PRINCE JOHN of Lancaster ⎟ sons of King

Humphrey, Duke of GLOUCESTER ⎟ Henry IV

Thomas, Duke of CLARENCE ⎦

Percy, Earl of NORTHUMBERLAND, of the rebels' party

LADY NORTHUMBERLAND

KATE, their son Hotspur's widow

TRAVERS, Northumberland's servant

MORTON, a bearer of news from Shrewsbury

Scrope, ARCHBISHOP of York ⎤

LORD BARDOLPH ⎟ rebels against

Thomas, Lord MOWBRAY, the Earl Marshal ⎬ King Henry

Lord HASTINGS ⎟ IV

Sir John COLEVILLE ⎦

LORD CHIEF JUSTICE

His SERVANT

GOWER, a Messenger

SIR JOHN Falstaff ⎤

His PAGE ⎟

BARDOLPH ⎟

POINS ⎬ 'irregular humorists'

Ensign PISTOL ⎟

PETO ⎦

MISTRESS QUICKLY, hostess of a tavern

DOLL TEARSHEET, a whore

SNARE ⎤
⎬ sergeants
FANG ⎦

Neville, Earl of WARWICK ⎤

Earl of SURREY ⎟

Earl of WESTMORLAND ⎬ supporters of King Henry IV

HARCOURT ⎟

Sir John Blunt ⎦

Robert SHALLOW ⎤
⎬ country justices
SILENCE ⎦

DAVY, Shallow's servant

Ralph MOULDY ⎤

Simon SHADOW ⎟

Thomas WART ⎬ men levied to fight for King Henry IV

Francis FEEBLE ⎟

Peter BULLCALF ⎦

PORTER of Northumberland's household

DRAWERS

BEADLES

GROOMS

MESSENGER

Sneak and other musicians

Lord Chief Justice's men, soldiers and attendants

# The Second Part of Henry the Fourth

**Induction** *Enter Rumour ⌈in a robe⌉ painted full of*
    *tongues*

RUMOUR
Open your ears; for which of you will stop
The vent of hearing when loud Rumour speaks?
I from the orient to the drooping west,
Making the wind my post-horse, still unfold
The acts commencèd on this ball of earth. 5
Upon my tongues continual slanders ride,
The which in every language I pronounce,
Stuffing the ears of men with false reports.
I speak of peace, while covert enmity
Under the smile of safety wounds the world; 10
And who but Rumour, who but only I,
Make fearful musters and prepared defence
Whiles the big year, swoll'n with some other griefs,
Is thought with child by the stern tyrant war,
And no such matter? Rumour is a pipe 15
Blown by surmises, Jealousy's conjectures,
And of so easy and so plain a stop
That the blunt monster with uncounted heads,
The still-discordant wav'ring multitude,
Can play upon it. But what need I thus 20
My well-known body to anatomize
Among my household? Why is Rumour here?
I run before King Harry's victory,
Who in a bloody field by Shrewsbury
Hath beaten down young Hotspur and his troops, 25
Quenching the flame of bold rebellion
Even with the rebels' blood. But what mean I
To speak so true at first? My office is
To noise abroad that Harry Monmouth fell
Under the wrath of noble Hotspur's sword, 30
And that the King before the Douglas' rage
Stooped his anointed head as low as death.
This have I rumoured through the peasant towns
Between that royal field of Shrewsbury
And this worm-eaten hold of raggèd stone, 35
Where Hotspur's father, old Northumberland,
Lies crafty-sick. The posts come tiring on,
And not a man of them brings other news
Than they have learnt of me. From Rumour's
    tongues
They bring smooth comforts false, worse than true
    wrongs. *Exit*

**1.1** *Enter Lord Bardolph at one door. ⌈He crosses the*
    *stage to another door⌉*

LORD BARDOLPH
Who keeps the gate here, ho?
    *Enter Porter ⌈above⌉*
                      Where is the Earl?
PORTER
What shall I say you are?
LORD BARDOLPH         Tell thou the Earl
That the Lord Bardolph doth attend him here.
PORTER
His lordship is walked forth into the orchard.
Please it your honour knock but at the gate, 5

And he himself will answer.
    *Enter the Earl Northumberland ⌈at the other door⌉,*
    *as sick, with a crutch and coif*
LORD BARDOLPH         Here comes the Earl.
                           *⌈Exit Porter⌉*
NORTHUMBERLAND
What news, Lord Bardolph? Every minute now
Should be the father of some stratagem.
The times are wild; contention, like a horse
Full of high feeding, madly hath broke loose, 10
And bears down all before him.
LORD BARDOLPH          Noble Earl,
I bring you certain news from Shrewsbury.
NORTHUMBERLAND
Good, an God will.
LORD BARDOLPH     As good as heart can wish.
The King is almost wounded to the death;
And, in the fortune of my lord your son, 15
Prince Harry slain outright; and both the Blunts
Killed by the hand of Douglas; young Prince John
And Westmorland and Stafford fled the field;
And Harry Monmouth's brawn, the hulk Sir John,
Is prisoner to your son. O, such a day, 20
So fought, so followed, and so fairly won,
Came not till now to dignify the times
Since Caesar's fortunes!
NORTHUMBERLAND      How is this derived?
Saw you the field? Came you from Shrewsbury?
LORD BARDOLPH
I spake with one, my lord, that came from thence, 25
A gentleman well bred and of good name,
That freely rendered me these news for true.
    *Enter Travers*
NORTHUMBERLAND
Here comes my servant Travers, who I sent
On Tuesday last to listen after news.
LORD BARDOLPH
My lord, I overrode him on the way, 30
And he is furnished with no certainties
More than he haply may retail from me.
NORTHUMBERLAND
Now, Travers, what good tidings comes with you?
TRAVERS
My lord, Lord Bardolph turned me back
With joyful tidings, and being better horsed 35
Outrode me. After him came spurring hard
A gentleman almost forspent with speed,
That stopped by me to breathe his bloodied horse.
He asked the way to Chester, and of him
I did demand what news from Shrewsbury. 40
He told me that rebellion had ill luck,
And that young Harry Percy's spur was cold.
With that he gave his able horse the head,
And, bending forward, struck his armèd heels
Against the panting sides of his poor jade 45
Up to the rowel-head; and starting so,
He seemed in running to devour the way,
Staying no longer question.
NORTHUMBERLAND      Ha? Again:
Said he young Harry Percy's spur was cold?

Of Hotspur, 'Coldspur'? that rebellion                    50
Had met ill luck?
LORD BARDOLPH          My lord, I'll tell you what:
If my young lord your son have not the day,
Upon mine honour, for a silken point
I'll give my barony. Never talk of it.
NORTHUMBERLAND
Why should the gentleman that rode by Travers          55
Give then such instances of loss?
LORD BARDOLPH                    Who, he?
He was some hilding fellow that had stol'n
The horse he rode on, and, upon my life,
Spoke at a venture.
          *Enter Morton*
                    Look, here comes more news.
NORTHUMBERLAND
Yea, this man's brow, like to a title leaf,          60
Foretells the nature of a tragic volume.
So looks the strand whereon the imperious flood
Hath left a witnessed usurpation.
Say, Morton, didst thou come from Shrewsbury?
MORTON
I ran from Shrewsbury, my noble lord,          65
Where hateful death put on his ugliest mask
To fright our party.
NORTHUMBERLAND          How doth my son and brother?
Thou tremblest, and the whiteness in thy cheek
Is apter than thy tongue to tell thy errand.
Even such a man, so faint, so spiritless,          70
So dull, so dead in look, so woebegone,
Drew Priam's curtain in the dead of night,
And would have told him half his Troy was burnt;
But Priam found the fire ere he his tongue,
And I my Percy's death ere thou report'st it.          75
This thou wouldst say: 'Your son did thus and thus,
Your brother thus; so fought the noble Douglas',
Stopping my greedy ear with their bold deeds;
But in the end, to stop my ear indeed,
Thou hast a sigh to blow away this praise,          80
Ending with 'Brother, son, and all are dead.'
MORTON
Douglas is living, and your brother yet;
But for my lord your son—
NORTHUMBERLAND          Why, he is dead.
See what a ready tongue suspicion hath!
He that but fears the thing he would not know          85
Hath by instinct knowledge from others' eyes
That what he feared is chanced. Yet speak, Morton.
Tell thou an earl his divination lies,
And I will take it as a sweet disgrace,
And make thee rich for doing me such wrong.          90
MORTON
You are too great to be by me gainsaid,
Your spirit is too true, your fears too certain.
NORTHUMBERLAND
Yet for all this, say not that Percy's dead.
I see a strange confession in thine eye—
Thou shak'st thy head, and hold'st it fear or sin          95
To speak a truth. If he be slain, say so.
The tongue offends not that reports his death;
And he doth sin that doth belie the dead,
Not he which says the dead is not alive.
Yet the first bringer of unwelcome news          100
Hath but a losing office, and his tongue
Sounds ever after as a sullen bell
Remembered knolling a departing friend.

LORD BARDOLPH
I cannot think, my lord, your son is dead.
MORTON (*to Northumberland*)
I am sorry I should force you to believe          105
That which I would to God I had not seen;
But these mine eyes saw him in bloody state,
Rend'ring faint quittance, wearied and out-breathed,
To Harry Monmouth, whose swift wrath beat down
The never-daunted Percy to the earth,          110
From whence with life he never more sprung up.
In few, his death, whose spirit lent a fire
Even to the dullest peasant in his camp,
Being bruited once, took fire and heat away
From the best-tempered courage in his troops;          115
For from his metal was his party steeled,
Which once in him abated, all the rest
Turned on themselves, like dull and heavy lead;
And, as the thing that's heavy in itself
Upon enforcement flies with greatest speed,          120
So did our men, heavy in Hotspur's loss,
Lend to this weight such lightness with their fear
That arrows fled not swifter toward their aim
Than did our soldiers, aiming at their safety,
Fly from the field. Then was that noble Worcester          125
Too soon ta'en prisoner; and that furious Scot
The bloody Douglas, whose well-labouring sword
Had three times slain th'appearance of the King,
Gan vail his stomach, and did grace the shame
Of those that turned their backs, and in his flight,          130
Stumbling in fear, was took. The sum of all
Is that the King hath won, and hath sent out
A speedy power to encounter you, my lord,
Under the conduct of young Lancaster
And Westmorland. This is the news at full.          135
NORTHUMBERLAND
For this I shall have time enough to mourn.
In poison there is physic; and these news,
Having been well, that would have made me sick,
Being sick, have in some measure made me well;
And, as the wretch whose fever-weakened joints,          140
Like strengthless hinges, buckle under life,
Impatient of his fit, breaks like a fire
Out of his keeper's arms, even so my limbs,
Weakened with grief, being now enraged with grief,
Are thrice themselves.
          ⌜*He casts away his crutch*⌝
                    Hence therefore, thou nice crutch!
A scaly gauntlet now with joints of steel          146
Must glove this hand.
          ⌜*He snatches off his coif*⌝
                    And hence, thou sickly coif!
Thou art a guard too wanton for the head
Which princes fleshed with conquest aim to hit.
Now bind my brows with iron, and approach          150
The ragged'st hour that time and spite dare bring
To frown upon th'enraged Northumberland!
Let heaven kiss earth! Now let not nature's hand
Keep the wild flood confined! Let order die!
And let this world no longer be a stage          155
To feed contention in a ling'ring act;
But let one spirit of the first-born Cain
Reign in all bosoms, that each heart being set
On bloody courses, the rude scene may end,
And darkness be the burier of the dead!          160
LORD BARDOLPH
Sweet Earl, divorce not wisdom from your honour.

MORTON

The lives of all your loving complices
Lean on your health, the which, if you give o'er
To stormy passion, must perforce decay.
You cast th'event of war, my noble lord,                165
And summed the account of chance, before you said
'Let us make head'. It was your presurmise
That in the dole of blows your son might drop.
You knew he walked o'er perils on an edge,
More likely to fall in than to get o'er.                170
You were advised his flesh was capable
Of wounds and scars, and that his forward spirit
Would lift him where most trade of danger ranged.
Yet did you say, 'Go forth'; and none of this,
Though strongly apprehended, could restrain          175
The stiff-borne action. What hath then befall'n?
Or what doth this bold enterprise bring forth,
More than that being which was like to be?

LORD BARDOLPH

We all that are engagèd to this loss
Knew that we ventured on such dangerous seas          180
That if we wrought out life was ten to one;
And yet we ventured for the gain proposed,
Choked the respect of likely peril feared;
And since we are o'erset, venture again.
Come, we will all put forth body and goods.           185

MORTON

'Tis more than time; and, my most noble lord,
I hear for certain, and dare speak the truth,
The gentle Archbishop of York is up
With well-appointed powers. He is a man
Who with a double surety binds his followers.         190
My lord, your son had only but the corpse,
But shadows and the shows of men, to fight;
For that same word 'rebellion' did divide
The action of their bodies from their souls,
And they did fight with queasiness, constrained,      195
As men drink potions, that their weapons only
Seemed on our side; but, for their spirits and souls,
This word 'rebellion', it had froze them up,
As fish are in a pond. But now the Bishop
Turns insurrection to religion.                       200
Supposed sincere and holy in his thoughts,
He's followed both with body and with mind,
And doth enlarge his rising with the blood
Of fair King Richard, scraped from Pomfret stones;
Derives from heaven his quarrel and his cause;        205
Tells them he doth bestride a bleeding land
Gasping for life under great Bolingbroke;
And more and less do flock to follow him.

NORTHUMBERLAND

I knew of this before, but, to speak truth,
This present grief had wiped it from my mind.         210
Go in with me, and counsel every man
The aptest way for safety and revenge.
Get posts and letters, and make friends with speed.
Never so few, and never yet more need.      *Exeunt*

1.2     *Enter Sir John Falstaff, ⌈followed by⌉ his Page*
        *bearing his sword and buckler*

SIR JOHN Sirrah, you giant, what says the doctor to my
   water?
PAGE He said, sir, the water itself was a good healthy
   water, but, for the party that owed it, he might have
   more diseases than he knew for.                        5
SIR JOHN Men of all sorts take a pride to gird at me. The
   brain of this foolish-compounded clay, man, is not able

to invent anything that tends to laughter more than I
invent, or is invented on me. I am not only witty in
myself, but the cause that wit is in other men. I do
here walk before thee like a sow that hath o'erwhelmed
all her litter but one. If the Prince put thee into my
service for any other reason than to set me off, why
then, I have no judgement. Thou whoreson mandrake,
thou art fitter to be worn in my cap than to wait at
my heels. I was never manned with an agate till now;
but I will set you neither in gold nor silver, but in vile
apparel, and send you back again to your master for
a jewel—the juvenal the Prince your master, whose
chin is not yet fledge. I will sooner have a beard grow
in the palm of my hand than he shall get one off his
cheek; and yet he will not stick to say his face is a
face-royal. God may finish it when he will; 'tis not a
hair amiss yet. He may keep it still at a face-royal, for
a barber shall never earn sixpence out of it. And yet
he'll be crowing as if he had writ man ever since his
father was a bachelor. He may keep his own grace, but
he's almost out of mine, I can assure him. What said
Master Dumbleton about the satin for my short cloak
and slops?                                              30
PAGE He said, sir, you should procure him better
   assurance than Bardolph. He would not take his bond
   and yours; he liked not the security.
SIR JOHN Let him be damned like the glutton! Pray God
   his tongue be hotter! A whoreson Achitophel, a rascally
   yea-forsooth knave, to bear a gentleman in hand and
   then stand upon security! The whoreson smooth-pates
   do now wear nothing but high shoes and bunches of
   keys at their girdles; and if a man is through with
   them in honest taking-up, then they must stand upon
   security. I had as lief they would put ratsbane in my
   mouth as offer to stop it with security. I looked a should
   have sent me two-and-twenty yards of satin, as I am
   a true knight, and he sends me 'security'! Well, he
   may sleep in security, for he hath the horn of
   abundance, and the lightness of his wife shines through
   it; and yet cannot he see, though he have his own
   lanthorn to light him. Where's Bardolph?
PAGE He's gone in Smithfield to buy your worship a
   horse.                                               50
SIR JOHN I bought him in Paul's, and he'll buy me a horse
   in Smithfield. An I could get me but a wife in the stews,
   I were manned, horsed, and wived.

   *Enter the Lord Chief Justice and his Servant*

PAGE Sir, here comes the nobleman that committed the
   Prince for striking him about Bardolph.              55
SIR JOHN ⌈*moving away*⌉ Wait close; I will not see him.
LORD CHIEF JUSTICE (*to his Servant*) What's he that goes
   there?
SERVANT Falstaff, an't please your lordship.
LORD CHIEF JUSTICE He that was in question for the
   robbery?                                             61
SERVANT He, my lord; but he hath since done good service
   at Shrewsbury, and, as I hear, is now going with some
   charge to the Lord John of Lancaster.
LORD CHIEF JUSTICE What, to York? Call him back again.
SERVANT Sir John Falstaff!                              66
SIR JOHN Boy, tell him I am deaf.
PAGE (*to the Servant*) You must speak louder; my master
   is deaf.
LORD CHIEF JUSTICE I am sure he is to the hearing of
   anything good. (*To the Servant*) Go pluck him by the
   elbow; I must speak with him.                        72

SERVANT Sir John!

SIR JOHN What, a young knave and begging! Is there not wars? Is there not employment? Doth not the King lack subjects? Do not the rebels want soldiers? Though it be a shame to be on any side but one, it is worse shame to beg than to be on the worst side, were it worse than the name of rebellion can tell how to make it.                                                         80

SERVANT You mistake me, sir.

SIR JOHN Why, sir, did I say you were an honest man? Setting my knighthood and my soldiership aside, I had lied in my throat if I had said so.

SERVANT I pray you, sir, then set your knighthood and your soldiership aside, and give me leave to tell you you lie in your throat if you say I am any other than an honest man.                                              88

SIR JOHN I give thee leave to tell me so? I lay aside that which grows to me? If thou gettest any leave of me, hang me. If thou takest leave, thou wert better be hanged. You hunt counter. Hence, avaunt!

SERVANT Sir, my lord would speak with you.

LORD CHIEF JUSTICE Sir John Falstaff, a word with you. 94

SIR JOHN My good lord! God give your lordship good time of day. I am glad to see your lordship abroad. I heard say your lordship was sick. I hope your lordship goes abroad by advice. Your lordship, though not clean past your youth, have yet some smack of age in you, some relish of the saltness of time in you; and I most humbly beseech your lordship to have a reverent care of your health.                                               102

LORD CHIEF JUSTICE Sir John, I sent for you before your expedition to Shrewsbury.

SIR JOHN An't please your lordship, I hear his majesty is returned with some discomfort from Wales.      106

LORD CHIEF JUSTICE I talk not of his majesty. You would not come when I sent for you.

SIR JOHN And I hear, moreover, his highness is fallen into this same whoreson apoplexy.                     110

LORD CHIEF JUSTICE Well, God mend him! I pray you, let me speak with you.

SIR JOHN This apoplexy is, as I take it, a kind of lethargy, an't please your lordship, a kind of sleeping in the blood, a whoreson tingling.                            115

LORD CHIEF JUSTICE What tell you me of it? Be it as it is.

SIR JOHN It hath it original from much grief, from study, and perturbation of the brain. I have read the cause of his effects in Galen. It is a kind of deafness.

LORD CHIEF JUSTICE I think you are fallen into the disease, for you hear not what I say to you.             121

SIR JOHN Very well, my lord, very well. Rather, an't please you, it is the disease of not listening, the malady of not marking, that I am troubled withal.

LORD CHIEF JUSTICE To punish you by the heels would amend the attention of your ears, and I care not if I do become your physician.                             127

SIR JOHN I am as poor as Job, my lord, but not so patient. Your lordship may minister the potion of imprisonment to me in respect of poverty; but how I should be your patient to follow your prescriptions, the wise may make some dram of a scruple, or indeed a scruple itself.

LORD CHIEF JUSTICE I sent for you, when there were matters against you for your life, to come speak with me.  134

SIR JOHN As I was then advised by my learned counsel in the laws of this land-service, I did not come.

LORD CHIEF JUSTICE Well, the truth is, Sir John, you live in great infamy.

SIR JOHN He that buckles himself in my belt cannot live in less.                                              140

LORD CHIEF JUSTICE Your means are very slender, and your waste is great.

SIR JOHN I would it were otherwise; I would my means were greater and my waist slenderer.

LORD CHIEF JUSTICE You have misled the youthful Prince.

SIR JOHN The young Prince hath misled me. I am the fellow with the great belly, and he my dog.       147

LORD CHIEF JUSTICE Well, I am loath to gall a new-healed wound. Your day's service at Shrewsbury hath a little gilded over your night's exploit on Gads Hill. You may thank th'unquiet time for your quiet o'erposting that action.                                              152

SIR JOHN My lord—

LORD CHIEF JUSTICE But since all is well, keep it so. Wake not a sleeping wolf.                                 155

SIR JOHN To wake a wolf is as bad as smell a fox.

LORD CHIEF JUSTICE What! You are as a candle, the better part burnt out.

SIR JOHN A wassail candle, my lord, all tallow—if I did say of wax, my growth would approve the truth.  160

LORD CHIEF JUSTICE There is not a white hair in your face but should have his effect of gravity.

SIR JOHN His effect of gravy, gravy, gravy.

LORD CHIEF JUSTICE You follow the young Prince up and down like his ill angel.                             165

SIR JOHN Not so, my lord; your ill angel is light, but I hope he that looks upon me will take me without weighing. And yet in some respects, I grant, I cannot go. I cannot tell, virtue is of so little regard in these costermongers' times that true valour is turned bearherd; pregnancy is made a tapster, and his quick wit wasted in giving reckonings; all the other gifts appertinent to man, as the malice of this age shapes them, are not worth a gooseberry. You that are old consider not the capacities of us that are young. You do measure the heat of our livers with the bitterness of your galls. And we that are in the vanguard of our youth, I must confess, are wags too.            178

LORD CHIEF JUSTICE Do you set down your name in the scroll of youth, that are written down old with all the characters of age? Have you not a moist eye, a dry hand, a yellow cheek, a white beard, a decreasing leg, an increasing belly? Is not your voice broken, your wind short, your chin double, your wit single, and every part about you blasted with antiquity? And will you yet call yourself young? Fie, fie, fie, Sir John! 186

SIR JOHN My lord, I was born about three of the clock in the afternoon with a white head, and something a round belly. For my voice, I have lost it with hallowing and singing of anthems. To approve my youth further, I will not. The truth is, I am only old in judgement and understanding; and he that will caper with me for a thousand marks, let him lend me the money, and have at him! For the box of th'ear that the Prince gave you, he gave it like a rude prince, and you took it like a sensible lord. I have checked him for it, and the young lion repents—⌈aside⌉ marry, not in ashes and sackcloth, but in new silk and old sack.

LORD CHIEF JUSTICE Well, God send the Prince a better companion!                                            200

SIR JOHN God send the companion a better prince! I cannot rid my hands of him.

LORD CHIEF JUSTICE Well, the King hath severed you and Prince Harry. I hear you are going with Lord John of

Lancaster against the Archbishop and the Earl of Northumberland. 206

SIR JOHN Yea, I thank your pretty sweet wit for it. But look you pray, all you that kiss my lady Peace at home, that our armies join not in a hot day; for, by the Lord, I take but two shirts out with me, and I mean not to sweat extraordinarily. If it be a hot day and I brandish anything but my bottle, would I might never spit white again. There is not a dangerous action can peep out his head but I am thrust upon it. Well, I cannot last ever. But it was alway yet the trick of our English nation, if they have a good thing, to make it too common. If ye will needs say I am an old man, you should give me rest. I would to God my name were not so terrible to the enemy as it is. I were better to be eaten to death with a rust than to be scoured to nothing with perpetual motion. 221

LORD CHIEF JUSTICE Well, be honest, be honest, and God bless your expedition.

SIR JOHN Will your lordship lend me a thousand pound to furnish me forth? 225

LORD CHIEF JUSTICE Not a penny, not a penny. You are too impatient to bear crosses. Fare you well. Commend me to my cousin Westmorland.

*Exeunt Lord Chief Justice and his Servant*

SIR JOHN If I do, fillip me with a three-man beetle. A man can no more separate age and covetousness than a can part young limbs and lechery; but the gout galls the one and the pox pinches the other, and so both the degrees prevent my curses. Boy!

PAGE Sir.

SIR JOHN What money is in my purse? 235

PAGE Seven groats and two pence.

SIR JOHN I can get no remedy against this consumption of the purse. Borrowing only lingers and lingers it out, but the disease is incurable. (*Giving letters*) Go bear this letter to my lord of Lancaster; this to the Prince; this to the Earl of Westmorland; and this to old Mistress Ursula, whom I have weekly sworn to marry since I perceived the first white hair of my chin. About it. You know where to find me. ⌈*Exit Page*⌉

A pox of this gout!—or a gout of this pox!—for the one or the other plays the rogue with my great toe. 'Tis no matter if I do halt; I have the wars for my colour, and my pension shall seem the more reasonable. A good wit will make use of anything. I will turn diseases to commodity. *Exit*

**1.3** *Enter the Archbishop of York, Thomas Mowbray the Earl Marshal, Lord Hastings, and Lord Bardolph*

ARCHBISHOP OF YORK
Thus have you heard our cause and known our means,
And, my most noble friends, I pray you all
Speak plainly your opinions of our hopes.
And first, Lord Marshal, what say you to it?

MOWBRAY
I well allow the occasion of our arms, 5
But gladly would be better satisfied
How in our means we should advance ourselves
To look with forehead bold and big enough
Upon the power and puissance of the King.

HASTINGS
Our present musters grow upon the file 10
To five-and-twenty thousand men of choice,
And our supplies live largely in the hope

Of great Northumberland, whose bosom burns
With an incensèd fire of injuries.

LORD BARDOLPH
The question then, Lord Hastings, standeth thus: 15
Whether our present five-and-twenty thousand
May hold up head without Northumberland.

HASTINGS
With him we may.

LORD BARDOLPH          Yea, marry, there's the point;
But if without him we be thought too feeble,
My judgement is, we should not step too far 20
Till we had his assistance by the hand;
For in a theme so bloody-faced as this,
Conjecture, expectation, and surmise
Of aids uncertain should not be admitted.

ARCHBISHOP OF YORK
'Tis very true, Lord Bardolph, for indeed 25
It was young Hotspur's case at Shrewsbury.

LORD BARDOLPH
It was, my lord; who lined himself with hope,
Eating the air on promise of supply,
Flatt'ring himself with project of a power
Much smaller than the smallest of his thoughts; 30
And so, with great imagination
Proper to madmen, led his powers to death,
And winking leapt into destruction.

HASTINGS
But by your leave, it never yet did hurt
To lay down likelihoods and forms of hope. 35

LORD BARDOLPH
Yes, if this present quality of war—
Indeed the instant action, a cause on foot—
Lives so in hope; as in an early spring
We see th'appearing buds, which to prove fruit
Hope gives not so much warrant as despair 40
That frosts will bite them. When we mean to build
We first survey the plot, then draw the model;
And when we see the figure of the house,
Then must we rate the cost of the erection,
Which if we find outweighs ability, 45
What do we then but draw anew the model
In fewer offices, or, at least, desist
To build at all? Much more in this great work—
Which is almost to pluck a kingdom down
And set another up—should we survey 50
The plot of situation and the model,
Consent upon a sure foundation,
Question surveyors, know our own estate,
How able such a work to undergo,
To weigh against his opposite; or else 55
We fortify in paper and in figures,
Using the names of men instead of men,
Like one that draws the model of an house
Beyond his power to build it, who, half-through,
Gives o'er, and leaves his part-created cost 60
A naked subject to the weeping clouds,
And waste for churlish winter's tyranny.

HASTINGS
Grant that our hopes, yet likely of fair birth,
Should be stillborn, and that we now possessed
The utmost man of expectation, 65
I think we are a body strong enough,
Even as we are, to equal with the King.

LORD BARDOLPH
What, is the King but five-and-twenty thousand?

HASTINGS
To us no more, nay, not so much, Lord Bardolph;
For his divisions, as the times do brawl,                70
Are in three heads: one power against the French,
And one against Glyndŵr, perforce a third
Must take up us. So is the unfirm King
In three divided, and his coffers sound
With hollow poverty and emptiness.                75
ARCHBISHOP OF YORK
That he should draw his several strengths together
And come against us in full puissance
Need not be dreaded.
HASTINGS                    If he should do so,
He leaves his back unarmed, the French and Welsh
Baying him at the heels. Never fear that.                80
LORD BARDOLPH
Who is it like should lead his forces hither?
HASTINGS
The Duke of Lancaster and Westmorland;
Against the Welsh, himself and Harry Monmouth;
But who is substituted 'gainst the French
I have no certain notice.
ARCHBISHOP OF YORK            Let us on,                85
And publish the occasion of our arms.
The commonwealth is sick of their own choice;
Their over-greedy love hath surfeited.
An habitation giddy and unsure
Hath he that buildeth on the vulgar heart.                90
O thou fond many, with what loud applause
Didst thou beat heaven with blessing Bolingbroke,
Before he was what thou wouldst have him be!
And being now trimmed in thine own desires,
Thou, beastly feeder, art so full of him                95
That thou provok'st thyself to cast him up.
So, so, thou common dog, didst thou disgorge
Thy glutton bosom of the royal Richard;
And now thou wouldst eat thy dead vomit up,
And howl'st to find it. What trust is in these times?
They that when Richard lived would have him die    101
Are now become enamoured on his grave.
Thou that threw'st dust upon his goodly head,
When through proud London he came sighing on
After th'admirèd heels of Bolingbroke,                105
Cri'st now, 'O earth, yield us that king again,
And take thou this!' O thoughts of men accursed!
Past and to come seems best; things present, worst.
⌜MOWBRAY⌝
Shall we go draw our numbers and set on?
HASTINGS
We are time's subjects, and time bids be gone.    110
                                        Exeunt

**2.1**    *Enter Mistress Quickly (the hostess of a tavern),*
*and an officer, Fang ⌜followed at a distance by⌝*
*another officer, Snare*
MISTRESS QUICKLY Master Fang, have you entered the
action?
FANG It is entered.
MISTRESS QUICKLY Where's your yeoman? Is't a lusty
yeoman? Will a stand to't?                5
FANG Sirrah!—Where's Snare?
MISTRESS QUICKLY O Lord, ay, good Master Snare.
SNARE ⌜coming forward⌝ Here, here.
FANG Snare, we must arrest Sir John Falstaff.
MISTRESS QUICKLY Yea, good Master Snare, I have entered
him and all.                11

SNARE It may chance cost some of us our lives, for he
will stab.
MISTRESS QUICKLY Alas the day, take heed of him; he
stabbed me in mine own house, most beastly, in good
faith. A cares not what mischief he does; if his weapon
be out, he will foin like any devil, he will spare neither
man, woman, nor child.                18
FANG If I can close with him, I care not for his thrust.
MISTRESS QUICKLY No, nor I neither. I'll be at your elbow.
FANG An I but fist him once, an a come but within my
vice—
MISTRESS QUICKLY I am undone by his going, I warrant
you; he's an infinitive thing upon my score. Good
Master Fang, hold him sure. Good Master Snare, let
him not scape. A comes continuantly to Pie Corner—
saving your manhoods—to buy a saddle, and he is
indited to dinner to the Lubber's Head in Lombard
Street, to Master Smooth's the silkman. I pray you,
since my exion is entered, and my case so openly
known to the world, let him be brought in to his
answer. A hundred mark is a long one for a poor lone
woman to bear; and I have borne, and borne, and
borne, and have been fobbed off, and fobbed off, and
fobbed off, from this day to that day, that it is a shame
to be thought on. There is no honesty in such dealing,
unless a woman should be made an ass and a beast,
to bear every knave's wrong.                38
        *Enter Sir John Falstaff, Bardolph, and the Page*
Yonder he comes, and that arrant malmsey-nose knave
Bardolph with him. Do your offices, do your offices,
Master Fang and Master Snare; do me, do me, do me
your offices.
SIR JOHN How now, whose mare's dead? What's the
matter?                44
FANG Sir John, I arrest you at the suit of Mistress Quickly.
SIR JOHN ⌜drawing⌝ Away, varlets! Draw, Bardolph! Cut
me off the villain's head! Throw the quean in the
channel!
        ⌜Bardolph draws⌝
MISTRESS QUICKLY Throw me in the channel? I'll throw
thee in the channel!                50
        *A brawl*
Wilt thou, wilt thou, thou bastardly rogue? Murder,
murder! Ah, thou honeysuckle villain, wilt thou kill
God's officers, and the King's? Ah, thou honeyseed
rogue! Thou art a honeyseed, a man-queller, and a
woman-queller.                55
SIR JOHN Keep them off, Bardolph!
FANG A rescue, a rescue!
MISTRESS QUICKLY Good people, bring a rescue or two.
Thou wot, wot thou? Thou wot, wot'a? Do, do, thou
rogue, do, thou hempseed!                60
PAGE Away, you scullion, you rampallian, you
fustilarian! I'll tickle your catastrophe!
        *Enter the Lord Chief Justice and his men*
LORD CHIEF JUSTICE
What is the matter? Keep the peace here, ho!
        *Brawl ends. ⌜Fang⌝ seizes Sir John*
MISTRESS QUICKLY Good my lord, be good to me; I beseech
you, stand to me.                65
LORD CHIEF JUSTICE
How now, Sir John? What, are you brawling here?
Doth this become your place, your time and business?
You should have been well on your way to York.
        ⌜To Fang⌝ Stand from him, fellow. Wherefore hang'st
        thou upon him?

MISTRESS QUICKLY O my most worshipful lord, an't please your grace, I am a poor widow of Eastcheap, and he is arrested at my suit. 72

LORD CHIEF JUSTICE For what sum?

MISTRESS QUICKLY It is more than for some, my lord, it is for all, all I have. He hath eaten me out of house and home. He hath put all my substance into that fat belly of his; (to Sir John) but I will have some of it out again, or I will ride thee a-nights like the mare.

SIR JOHN I think I am as like to ride the mare, if I have any vantage of ground to get up. 80

LORD CHIEF JUSTICE How comes this, Sir John? Fie, what man of good temper would endure this tempest of exclamation? Are you not ashamed, to enforce a poor widow to so rough a course to come by her own?

SIR JOHN (to the Hostess) What is the gross sum that I owe thee? 86

MISTRESS QUICKLY Marry, if thou wert an honest man, thyself, and the money too. Thou didst swear to me upon a parcel-gilt goblet, sitting in my Dolphin chamber, at the round table, by a sea-coal fire, upon Wednesday in Wheeson week, when the Prince broke thy head for liking his father to a singing-man of Windsor—thou didst swear to me then, as I was washing thy wound, to marry me, and make me my lady thy wife. Canst thou deny it? Did not goodwife Keech the butcher's wife come in then, and call me 'Gossip Quickly'—coming in to borrow a mess of vinegar, telling us she had a good dish of prawns, whereby thou didst desire to eat some, whereby I told thee they were ill for a green wound? And didst thou not, when she was gone downstairs, desire me to be no more so familiarity with such poor people, saying that ere long they should call me 'madam'? And didst thou not kiss me, and bid me fetch thee thirty shillings? I put thee now to thy book-oath; deny it if thou canst. *She weeps*

SIR JOHN My lord, this is a poor mad soul, and she says up and down the town that her eldest son is like you. She hath been in good case, and the truth is, poverty hath distracted her. But for these foolish officers, I beseech you I may have redress against them. 110

LORD CHIEF JUSTICE Sir John, Sir John, I am well acquainted with your manner of wrenching the true cause the false way. It is not a confident brow, nor the throng of words that come with such more than impudent sauciness from you, can thrust me from a level consideration. You have, as it appears to me, practised upon the easy-yielding spirit of this woman, and made her serve your uses both in purse and in person. 118

MISTRESS QUICKLY Yea, in truth, my lord.

LORD CHIEF JUSTICE Pray thee, peace. (To Sir John) Pay her the debt you owe her, and unpay the villainy you have done with her. The one you may do with sterling money, and the other with current repentance. 123

SIR JOHN My lord, I will not undergo this sneap without reply. You call honourable boldness 'impudent sauciness'; if a man will make curtsy and say nothing, he is virtuous. No, my lord, my humble duty remembered, I will not be your suitor. I say to you I do desire deliverance from these officers, being upon hasty employment in the King's affairs. 130

LORD CHIEF JUSTICE You speak as having power to do wrong; but answer in th'effect of your reputation, and satisfy the poor woman.

SIR JOHN (*drawing apart*) Come hither, hostess. *She goes to him.*

*Enter Master Gower, a messenger*

LORD CHIEF JUSTICE Now, Master Gower, what news?

GOWER
The King, my lord, and Harry Prince of Wales 136
Are near at hand; the rest the paper tells.
*Lord Chief Justice reads the paper, and converses apart with Gower*

SIR JOHN As I am a gentleman!

MISTRESS QUICKLY Faith, you said so before.

SIR JOHN As I am a gentleman! Come, no more words of it. 141

MISTRESS QUICKLY By this heavenly ground I tread on, I must be fain to pawn both my plate and the tapestry of my dining-chambers.

SIR JOHN Glasses, glasses, is the only drinking; and for thy walls, a pretty slight drollery, or the story of the Prodigal, or the German hunting in waterwork, is worth a thousand of these bed-hangers and these fly-bitten tapestries. Let it be ten pound if thou canst. Come, an 'twere not for thy humours, there's not a better wench in England. Go, wash thy face, and draw the action. Come, thou must not be in this humour with me. Dost not know me? Come, I know thou wast set on to this. 154

MISTRESS QUICKLY Pray thee, Sir John, let it be but twenty nobles. I'faith, I am loath to pawn my plate, so God save me, la!

SIR JOHN Let it alone; I'll make other shift. You'll be a fool still. 159

MISTRESS QUICKLY Well, you shall have it, though I pawn my gown. I hope you'll come to supper. You'll pay me altogether?

SIR JOHN Will I live? *To Bardolph and the Page* Go with her, with her. Hook on, hook on! 164

MISTRESS QUICKLY Will you have Doll Tearsheet meet you at supper?

SIR JOHN No more words; let's have her.
*Exeunt Mistress Quickly, Bardolph, the Page, Fang and Snare*

LORD CHIEF JUSTICE (*to Gower*) I have heard better news.

SIR JOHN What's the news, my good lord? 169

LORD CHIEF JUSTICE (*to Gower*) Where lay the King tonight?

GOWER At Basingstoke, my lord.

SIR JOHN (*to Lord Chief Justice*) I hope, my lord, all's well. What is the news, my lord?

LORD CHIEF JUSTICE (*to Gower*) Come all his forces back?

GOWER
No; fifteen hundred foot, five hundred horse, 175
Are marched up to my lord of Lancaster
Against Northumberland and the Archbishop.

SIR JOHN (*to Lord Chief Justice*)
Comes the King back from Wales, my noble lord?

LORD CHIEF JUSTICE (*to Gower*)
You shall have letters of me presently.
Come, go along with me, good Master Gower. 180
*They are going*

SIR JOHN My lord!

LORD CHIEF JUSTICE What's the matter?

SIR JOHN Master Gower, shall I entreat you with me to dinner?

GOWER I must wait upon my good lord here, I thank you, good Sir John. 186

LORD CHIEF JUSTICE Sir John, you loiter here too long, being you are to take soldiers up in counties as you go.

SIR JOHN Will you sup with me, Master Gower? 190

LORD CHIEF JUSTICE What foolish master taught you these manners, Sir John?
SIR JOHN Master Gower, if they become me not, he was a fool that taught them me. (*To Lord Chief Justice*) This is the right fencing grace, my lord—tap for tap, and so part fair.                                                              196
LORD CHIEF JUSTICE Now the Lord lighten thee; thou art a great fool. *Exeunt ⌈Lord Chief Justice and Gower at one door, Sir John at another⌉*

**2.2**   *Enter Prince Harry and Poins*
PRINCE HARRY Before God, I am exceeding weary.
POINS Is't come to that? I had thought weariness durst not have attached one of so high blood.
PRINCE HARRY Faith, it does me, though it discolours the complexion of my greatness to acknowledge it. Doth it not show vilely in me to desire small beer?              6
POINS Why, a prince should not be so loosely studied as to remember so weak a composition.
PRINCE HARRY Belike then my appetite was not princely got; for, by my troth, I do now remember the poor creature small beer. But indeed, these humble considerations make me out of love with my greatness. What a disgrace is it to me to remember thy name! Or to know thy face tomorrow! Or to take note how many pair of silk stockings thou hast—videlicet these, and those that were thy peach-coloured ones! Or to bear the inventory of thy shirts—as one for superfluity, and another for use. But that the tennis-court keeper knows better than I, for it is a low ebb of linen with thee when thou keepest not racket there; as thou hast not done a great while, because the rest of thy low countries have made a shift to eat up thy holland.              22
POINS How ill it follows, after you have laboured so hard, you should talk so idly! Tell me, how many good young princes would do so, their fathers lying so sick as yours is?                                                                              26
PRINCE HARRY Shall I tell thee one thing, Poins?
POINS Yes, faith, and let it be an excellent good thing.
PRINCE HARRY It shall serve among wits of no higher breeding than thine.                                                     30
POINS Go to, I stand the push of your one thing that you'll tell.
PRINCE HARRY Marry, I tell thee, it is not meet that I should be sad now my father is sick; albeit I could tell to thee, as to one it pleases me, for fault of a better, to call my friend, I could be sad; and sad indeed too.   36
POINS Very hardly, upon such a subject.
PRINCE HARRY By this hand, thou thinkest me as far in the devil's book as thou and Falstaff, for obduracy and persistency. Let the end try the man. But I tell thee, my heart bleeds inwardly that my father is so sick; and keeping such vile company as thou art hath, in reason, taken from me all ostentation of sorrow.
POINS The reason?                                                     44
PRINCE HARRY What wouldst thou think of me if I should weep?
POINS I would think thee a most princely hypocrite.
PRINCE HARRY It would be every man's thought, and thou art a blessed fellow to think as every man thinks. Never a man's thought in the world keeps the roadway better than thine. Every man would think me an hypocrite indeed. And what accites your most worshipful thought to think so?
POINS Why, because you have been so lewd, and so much engrafted to Falstaff.                                                   55

PRINCE HARRY And to thee.
POINS By this light, I am well spoke on; I can hear it with mine own ears. The worst that they can say of me is that I am a second brother, and that I am a proper fellow of my hands; and those two things I confess I cannot help.                                              61
*Enter Bardolph ⌈followed by⌉ the Page*
By the mass, here comes Bardolph.
PRINCE HARRY And the boy that I gave Falstaff. A had him from me Christian, and look if the fat villain have not transformed him ape.                                           65
BARDOLPH God save your grace!
PRINCE HARRY And yours, most noble Bardolph!
POINS (*to Bardolph*) Come, you virtuous ass, you bashful fool, must you be blushing? Wherefore blush you now? What a maidenly man at arms are you become! Is't such a matter to get a pottle-pot's maidenhead?     71
PAGE A calls me e'en now, my lord, through a red lattice, and I could discern no part of his face from the window. At last I spied his eyes, and methought he had made two holes in the ale-wife's red petticoat, and so peeped through.                                                          76
PRINCE HARRY (*to Poins*) Has not the boy profited?
BARDOLPH (*to the Page*) Away, you whoreson upright rabbit, away!
PAGE Away, you rascally Althea's dream, away!          80
PRINCE HARRY Instruct us, boy; what dream, boy?
PAGE Marry, my lord, Althea dreamt she was delivered of a firebrand, and therefore I call him her dream.
PRINCE HARRY (*giving him money*) A crown's-worth of good interpretation! There 'tis, boy.                                  85
POINS O, that this good blossom could be kept from cankers! (*Giving the Page money*) Well, there is sixpence to preserve thee.
BARDOLPH An you do not make him hanged among you, the gallows shall be wronged.                                    90
PRINCE HARRY And how doth thy master, Bardolph?
BARDOLPH Well, my good lord. He heard of your grace's coming to town. There's a letter for you.
POINS Delivered with good respect. And how doth the Martlemas your master?                                              95
BARDOLPH In bodily health, sir.
*Prince Harry reads the letter*
POINS Marry, the immortal part needs a physician, but that moves not him. Though that be sick, it dies not.
PRINCE HARRY I do allow this wen to be as familiar with me as my dog; and he holds his place, for look you how he writes.                                                         101
⌈*He gives Poins the letter*⌉
POINS 'John Falstaff, knight'.—Every man must know that, as oft as he has occasion to name himself; even like those that are kin to the King, for they never prick their finger but they say 'There's some of the King's blood spilt.' 'How comes that?' says he that takes upon him not to conceive. The answer is as ready as a borrower's cap: 'I am the King's poor cousin, sir.'
PRINCE HARRY Nay, they will be kin to us, or they will fetch it from Japhet. (*Taking the letter*) But the letter. 'Sir John Falstaff, knight, to the son of the King nearest his father, Harry Prince of Wales, greeting.'          112
POINS Why, this is a certificate.
PRINCE HARRY Peace!—'I will imitate the honourable Romans in brevity.'                                                    115
POINS (*taking the letter*) Sure he means brevity in breath, short winded. (*Reads*) 'I commend me to thee, I commend thee, and I leave thee. Be not too familiar

with Poins, for he misuses thy favours so much that
he swears thou art to marry his sister Nell. Repent at
idle times as thou mayst. And so, farewell.          121
     Thine by yea and no—which is as much as to
     say, as thou usest him—Jack Falstaff with my
     familiars, John with my brothers and sisters,
     and Sir John with all Europe.'          125
My lord, I'll steep this letter in sack and make him eat
it.

PRINCE HARRY That's to make him eat twenty of his
    words. But do you use me thus, Ned? Must I marry
    your sister?          130

POINS God send the wench no worse fortune, but I never
    said so.

PRINCE HARRY Well, thus we play the fools with the time,
    and the spirits of the wise sit in the clouds and mock
    us. (To Bardolph) Is your master here in London?   135

BARDOLPH Yea, my lord.

PRINCE HARRY Where sups he? Doth the old boar feed in
    the old frank?

BARDOLPH At the old place, my lord, in Eastcheap.

PRINCE HARRY What company?          140

PAGE Ephesians, my lord, of the old church.

PRINCE HARRY Sup any women with him?

PAGE None, my lord, but old Mistress Quickly and
    Mistress Doll Tearsheet.

PRINCE HARRY What pagan may that be?          145

PAGE A proper gentlewoman, sir, and a kinswoman of
    my master's.

PRINCE HARRY Even such kin as the parish heifers are to
    the town bull. Shall we steal upon them, Ned, at
    supper?          150

POINS I am your shadow, my lord; I'll follow you.

PRINCE HARRY Sirrah, you, boy, and Bardolph, no word
    to your master that I am yet come to town. (Giving
    money) There's for your silence.

BARDOLPH I have no tongue, sir.          155

PAGE And for mine, sir, I will govern it.

PRINCE HARRY Fare you well; go.
                Exeunt Bardolph and the Page
This Doll Tearsheet should be some road.

POINS I warrant you, as common as the way between
    Saint Albans and London.          160

PRINCE HARRY How might we see Falstaff bestow himself
    tonight in his true colours, and not ourselves be seen?

POINS Put on two leathern jerkins and aprons, and wait
    upon him at his table like drawers.

PRINCE HARRY From a god to a bull—a heavy declension—
    it was Jove's case. From a prince to a prentice—a low
    transformation—that shall be mine; for in everything
    the purpose must weigh with the folly. Follow me, Ned.
                   Exeunt

**2.3**    *Enter the Earl of Northumberland, Lady*
       *Northumberland, and Lady Percy*

NORTHUMBERLAND
    I pray thee, loving wife and gentle daughter,
    Give even way unto my rough affairs.
    Put not you on the visage of the times
    And be like them to Percy troublesome.

LADY NORTHUMBERLAND
    I have given over; I will speak no more.          5
    Do what you will; your wisdom be your guide.

NORTHUMBERLAND
    Alas, sweet wife, my honour is at pawn,
    And, but my going, nothing can redeem it.

LADY PERCY
    O yet, for God's sake, go not to these wars!
    The time was, father, that you broke your word          10
    When you were more endeared to it than now—
    When your own Percy, when my heart's dear Harry,
    Threw many a northward look to see his father
    Bring up his powers; but he did long in vain.
    Who then persuaded you to stay at home?          15
    There were two honours lost, yours and your son's.
    For yours, the God of heaven brighten it!
    For his, it stuck upon him as the sun
    In the grey vault of heaven, and by his light
    Did all the chivalry of England move          20
    To do brave acts. He was indeed the glass
    Wherein the noble youth did dress themselves.
    He had no legs that practised not his gait;
    And speaking thick, which nature made his blemish,
    Became the accents of the valiant;          25
    For those that could speak low and tardily
    Would turn their own perfection to abuse
    To seem like him. So that in speech, in gait,
    In diet, in affections of delight,
    In military rules, humours of blood,          30
    He was the mark and glass, copy and book,
    That fashioned others. And him—O wondrous him!
    O miracle of men!—him did you leave,
    Second to none, unseconded by you,
    To look upon the hideous god of war          35
    In disadvantage, to abide a field
    Where nothing but the sound of Hotspur's name
    Did seem defensible; so you left him.
    Never, O never do his ghost the wrong
    To hold your honour more precise and nice          40
    With others than with him. Let them alone.
    The Marshal and the Archbishop are strong.
    Had my sweet Harry had but half their numbers,
    Today might I, hanging on Hotspur's neck,
    Have talked of Monmouth's grave.

NORTHUMBERLAND            Beshrew your heart,
    Fair daughter, you do draw my spirits from me          46
    With new lamenting ancient oversights.
    But I must go and meet with danger there,
    Or it will seek me in another place,
    And find me worse provided.

LADY NORTHUMBERLAND        O fly to Scotland,          50
    Till that the nobles and the armèd commons
    Have of their puissance made a little taste.

LADY PERCY
    If they get ground and vantage of the King,
    Then join you with them like a rib of steel,
    To make strength stronger; but, for all our loves,          55
    First let them try themselves. So did your son.
    He was so suffered. So came I a widow,
    And never shall have length of life enough
    To rain upon remembrance with mine eyes,
    That it may grow and sprout as high as heaven          60
    For recordation to my noble husband.

NORTHUMBERLAND
    Come, come, go in with me. 'Tis with my mind
    As with the tide swelled up unto his height,
    That makes a still stand, running neither way.
    Fain would I go to meet the Archbishop,          65
    But many thousand reasons hold me back.
    I will resolve for Scotland. There am I
    Till time and vantage crave my company.    *Exeunt*

2.4 ⌜*A table and chairs set forth.*⌝ *Enter a Drawer*
⌜*with wine*⌝ *and another Drawer* ⌜*with a dish of apple-johns*⌝

⌜FIRST DRAWER⌝ What the devil hast thou brought there—apple-johns? Thou knowest Sir John cannot endure an apple-john.

⌜SECOND DRAWER⌝ Mass, thou sayst true. The Prince once set a dish of apple-johns before him; and told him, there were five more Sir Johns; and, putting off his hat, said 'I will now take my leave of these six dry, round, old, withered knights.' It angered him to the heart. But he hath forgot that.                               9

⌜FIRST DRAWER⌝ Why then, cover, and set them down; and see if thou canst find out Sneak's noise. Mistress Tearsheet would fain hear some music.
                                        ⌜*Exit the Second Drawer*⌝
⌜*The First Drawer covers the table.*⌝
⌜*Enter the Second Drawer*⌝

⌜SECOND DRAWER⌝ Sirrah, here will be the Prince and Master Poins anon, and they will put on two of our jerkins and aprons, and Sir John must not know of it. Bardolph hath brought word.                        16

⌜FIRST DRAWER⌝ By the mass, here will be old utis! It will be an excellent stratagem.

⌜SECOND DRAWER⌝ I'll see if I can find out Sneak.   *Exeunt*
*Enter Mistress Quickly and Doll Tearsheet, drunk*

MISTRESS QUICKLY I'faith, sweetheart, methinks now you are in an excellent good temperality. Your pulsidge beats as extraordinarily as heart would desire, and your colour, I warrant you, is as red as any rose, in good truth, la; but i'faith, you have drunk too much canaries, and that's a marvellous searching wine, and it perfumes the blood ere we can say 'What's this?' How do you now?                                     27

DOLL TEARSHEET Better than I was.—Hem!

MISTRESS QUICKLY Why, that's well said! A good heart's worth gold.                                          30
*Enter Sir John Falstaff*
Lo, here comes Sir John.

SIR JOHN (*sings*) 'When Arthur first in court'—⌜*Calls*⌝ Empty the jordan!—(*Sings*) 'And was a worthy king'—How now, Mistress Doll?

MISTRESS QUICKLY Sick of a qualm, yea, good faith.       35

SIR JOHN So is all her sect; an they be once in a calm, they are sick.

DOLL TEARSHEET A pox damn you, you muddy rascal! Is that all the comfort you give me?

SIR JOHN You make fat rascals, Mistress Doll.          40

DOLL TEARSHEET I make them? Gluttony and diseases make them; I make them not.

SIR JOHN If the cook help to make the gluttony, you help to make the diseases, Doll. We catch of you, Doll, we catch of you; grant that, my poor virtue, grant that.

DOLL TEARSHEET Yea, Jesu, our chains and our jewels.   46

SIR JOHN 'Your brooches, pearls, and ouches'—for to serve bravely is to come halting off, you know; to come off the breach with his pike bent bravely, and to surgery bravely; to venture upon the charged chambers bravely.                                               51

MISTRESS QUICKLY By my troth, this is the old fashion. You two never meet but you fall to some discord. You are both, i' good truth, as rheumatic as two dry toasts; you cannot one bear with another's confirmities. What the goodyear, one must bear, (*to Doll*) and that must be you. You are the weaker vessel, as they say, the emptier vessel.                                        58

DOLL TEARSHEET Can a weak empty vessel bear such a huge full hogshead? There's a whole merchant's venture of Bordeaux stuff in him; you have not seen a hulk better stuffed in the hold.—Come, I'll be friends with thee, Jack. Thou art going to the wars, and whether I shall ever see thee again or no there is nobody cares.                                        65
*Enter a Drawer*

DRAWER Sir, Ensign Pistol's below, and would speak with you.

DOLL TEARSHEET Hang him, swaggering rascal, let him not come hither. It is the foul-mouthedest rogue in England.                                               70

MISTRESS QUICKLY If he swagger, let him not come here. No, by my faith! I must live among my neighbours; I'll no swaggerers. I am in good name and fame with the very best. Shut the door; there comes no swaggerers here. I have not lived all this while to have swaggering now. Shut the door, I pray you.                        76

SIR JOHN Dost thou hear, hostess?

MISTRESS QUICKLY Pray ye pacify yourself, Sir John. There comes no swaggerers here.

SIR JOHN Dost thou hear? It is mine ensign.             80

MISTRESS QUICKLY Tilly-fally, Sir John, ne'er tell me. Your ensign-swaggerer comes not in my doors. I was before Master Tisick the debuty t'other day, and, as he said to me—'twas no longer ago than Wed'sday last, i' good faith—'Neighbour Quickly,' says he—Master Dumb our minister was by then—'Neighbour Quickly,' says he, 'receive those that are civil, for,' said he, 'you are in an ill name.' Now a said so, I can tell whereupon. 'For,' says he, 'you are an honest woman, and well thought on; therefore take heed what guests you receive. Receive,' says he, 'no swaggering companions.' There comes none here. You would bless you to hear what he said. No, I'll no swaggerers.                 93

SIR JOHN He's no swaggerer, hostess—a tame cheater, i'faith. You may stroke him as gently as a puppy greyhound. He'll not swagger with a Barbary hen, if her feathers turn back in any show of resistance.—Call him up, drawer.                          ⌜*Exit Drawer*⌝

MISTRESS QUICKLY Cheater call you him? I will bar no honest man my house, nor no cheater, but I do not love swaggering, by my troth, I am the worse when one says 'swagger'. Feel, masters, how I shake, look you, I warrant you.

DOLL TEARSHEET So you do, hostess.                    104

MISTRESS QUICKLY Do I? Yea, in very truth do I, an 'twere an aspen leaf. I cannot abide swaggerers.
*Enter Pistol, Bardolph, and the Page*

PISTOL God save you, Sir John.

SIR JOHN Welcome, Ensign Pistol. Here, Pistol, I charge you with a cup of sack. Do you discharge upon mine hostess.                                             110

PISTOL I will discharge upon her, Sir John, with two bullets.

SIR JOHN She is pistol-proof, sir, you shall not hardly offend her.

MISTRESS QUICKLY Come, I'll drink no proofs, nor no bullets. I'll drink no more than will do me good, for no man's pleasure, I.                                    117

PISTOL Then to you, Mistress Dorothy! I will charge you.

DOLL TEARSHEET Charge me? I scorn you, scurvy companion. What, you poor, base, rascally, cheating, lack-linen mate! Away, you mouldy rogue, away! I am meat for your master.                                122

PISTOL I know you, Mistress Dorothy.

DOLL TEARSHEET Away, you cutpurse rascal, you filthy
bung, away! By this wine, I'll thrust my knife in your
mouldy chaps an you play the saucy cuttle with me!
　　*She brandishes a knife*
Away, you bottle-ale rascal, you basket-hilt stale
juggler, you!
　　*Pistol draws his sword*
Since when, I pray you, sir? God's light, with two
points on your shoulder! Much!　　　　　　　　130

PISTOL God let me not live, but I will murder your ruff
for this.

MISTRESS QUICKLY No, good Captain Pistol; not here, sweet
captain.

DOLL TEARSHEET Captain? Thou abominable damned
cheater, art thou not ashamed to be called 'captain'?
An captains were of my mind, they would truncheon
you out, for taking their names upon you before you
have earned them. You a captain? You slave! For
what? For tearing a poor whore's ruff in a bawdy-
house! He a captain? Hang him, rogue, he lives upon
mouldy stewed prunes and dried cakes. A captain?
God's light, these villains will make the word 'captain'
odious; therefore captains had need look to't.

BARDOLPH Pray thee, go down, good ensign.　　　145

SIR JOHN Hark thee hither, Mistress Doll.
　　*He takes her aside*

PISTOL Not I! I tell thee what, Corporal Bardolph, I could
tear her! I'll be revenged of her.

PAGE Pray thee, go down.

PISTOL I'll see her damned first　　　　　　　150
To Pluto's damned lake, by this hand,
To th'infernal deep,
Where Erebus, and tortures vile also.
'Hold hook and line!' say I.
Down, down, dogs; down, Fates.　　　　　　155
Have we not Hiren here?

MISTRESS QUICKLY Good Captain Pizzle, be quiet. 'Tis very
late, i'faith. I beseek you now, aggravate your choler.

PISTOL These be good humours indeed!
Shall pack-horses　　　　　　　　　　　160
And hollow pampered jades of Asia,
Which cannot go but thirty mile a day,
Compare with Caesars and with cannibals,
And Trojan Greeks?
Nay, rather damn them with King Cerberus,　　165
And let the welkin roar. Shall we fall foul for toys?

MISTRESS QUICKLY By my troth, captain, these are very
bitter words.

BARDOLPH Be gone, good ensign; this will grow to a brawl
anon.　　　　　　　　　　　　　　170

PISTOL
Die men like dogs! Give crowns like pins!
Have we not Hiren here?

MISTRESS QUICKLY O' my word, captain, there's none such
here. What the goodyear, do you think I would deny
her? For God's sake, be quiet.　　　　　　175

PISTOL
Then feed and be fat, my fair Calipolis.
Come, give's some sack.
*Si fortune me tormente, sperato me contento.*
Fear we broadsides? No; let the fiend give fire!
Give me some sack; and, sweetheart, lie thou there.
　　*He lays down his sword*
Come we to full points here? And are etceteras nothings?
　　*He drinks*

SIR JOHN Pistol, I would be quiet.

PISTOL Sweet knight, I kiss thy neaf. What, we have seen
the seven stars!

DOLL TEARSHEET For God's sake, thrust him downstairs. I
cannot endure such a fustian rascal.　　　186

PISTOL Thrust him downstairs? Know we not Galloway
nags?

SIR JOHN Quoit him down, Bardolph, like a shove-groat
shilling. Nay, an a do nothing but speak nothing, a
shall be nothing here.　　　　　　　　191

BARDOLPH (*to Pistol*) Come, get you downstairs.

PISTOL *taking up his sword*
What, shall we have incision? Shall we imbrue?
Then death rock me asleep, abridge my doleful days.
Why then, let grievous, ghastly, gaping wounds　195
Untwine the Sisters Three. Come, Atropos, I say!

MISTRESS QUICKLY Here's goodly stuff toward!

SIR JOHN Give me my rapier, boy.

DOLL TEARSHEET I pray thee, Jack, I pray thee, do not
draw.　　　　　　　　　　　　　　200

SIR JOHN (*taking his rapier and speaking to Pistol*) Get you
downstairs.
　　*Sir John, Bardolph, and Pistol brawl*

MISTRESS QUICKLY Here's a goodly tumult! I'll forswear
keeping house afore I'll be in these tirrits and frights!
　　*Sir John thrusts at Pistol*
So!　　　　　　　　　　　　　　205
　　*Pistol thrusts at Sir John*
Murder, I warrant now! Alas, alas, put up your naked
weapons, put up your naked weapons!
　　　　　*Exit Pistol, pursued by Bardolph*

DOLL TEARSHEET I pray thee, Jack, be quiet; the rascal's
gone. Ah, you whoreson little valiant villain, you!

MISTRESS QUICKLY (*to Sir John*) Are you not hurt i'th'
groin? Methought a made a shrewd thrust at your
belly.　　　　　　　　　　　　　212
　　*Enter Bardolph*

SIR JOHN Have you turned him out o'doors?

BARDOLPH Yea, sir. The rascal's drunk. You have hurt
him, sir, i'th' shoulder.　　　　　　　215

SIR JOHN A rascal, to brave me!

DOLL TEARSHEET Ah, you sweet little rogue, you! Alas,
poor ape, how thou sweatest! Come, let me wipe thy
face; come on, you whoreson chops. Ah rogue, i'faith,
I love thee. Thou art as valorous as Hector of Troy,
worth five of Agamemnon, and ten times better than
the Nine Worthies. Ah, villain!　　　　　222

SIR JOHN A rascally slave! I will toss the rogue in a
blanket.

DOLL TEARSHEET Do, an thou darest for thy heart. An thou
dost, I'll canvas thee between a pair of sheets.　226
　　*Enter musicians*

PAGE The music is come, sir.

SIR JOHN Let them play.—Play, sirs!
　　*Music plays*
Sit on my knee, Doll. A rascal bragging slave! The
rogue fled from me like quicksilver.　　　230

DOLL TEARSHEET I'faith, and thou followed'st him like a
church. Thou whoreson little tidy Bartholomew boar-
pig, when wilt thou leave fighting o'days, and foining
o'nights, and begin to patch up thine old body for
heaven?　　　　　　　　　　　　235
　　*Enter Prince Harry and Poins, disguised as drawers*

SIR JOHN Peace, good Doll, do not speak like a death's-
head, do not bid me remember mine end.

DOLL TEARSHEET Sirrah, what humour's the Prince of?

SIR JOHN A good shallow young fellow. A would have made a good pantler; a would ha' chipped bread well.

DOLL TEARSHEET They say Poins has a good wit.                241

SIR JOHN He a good wit? Hang him, baboon! His wit's as thick as Tewkesbury mustard; there's no more conceit in him than is in a mallet.

DOLL TEARSHEET Why does the Prince love him so, then?

SIR JOHN Because their legs are both of a bigness, and a plays at quoits well, and eats conger and fennel, and drinks off candles' ends for flap-dragons, and rides the wild mare with the boys, and jumps upon joint-stools, and swears with a good grace, and wears his boot very smooth like unto the sign of the leg, and breeds no bate with telling of discreet stories, and such other gambol faculties a has that show a weak mind and an able body; for the which the Prince admits him; for the Prince himself is such another—the weight of a hair will turn the scales between their avoirdupois.

PRINCE HARRY (aside to Poins) Would not this nave of a wheel have his ears cut off?                258

POINS Let's beat him before his whore.

PRINCE HARRY Look whe'er the withered elder hath not his poll clawed like a parrot.

POINS Is it not strange that desire should so many years outlive performance?                263

SIR JOHN Kiss me, Doll.

*They kiss*

PRINCE HARRY (aside to Poins) Saturn and Venus this year in conjunction! What says th'almanac to that?

POINS And look whether the fiery Trigon his man be not lisping to his master's old tables, his note-book, his counsel-keeper!                269

SIR JOHN (to Doll) Thou dost give me flattering busses.

DOLL TEARSHEET By my troth, I kiss thee with a most constant heart.

SIR JOHN I am old, I am old.

DOLL TEARSHEET I love thee better than I love e'er a scurvy young boy of them all.                275

SIR JOHN What stuff wilt have a kirtle of? I shall receive money o'Thursday; shalt have a cap tomorrow.—A merry song!

⌈*The music plays again*⌉

Come, it grows late; we'll to bed. Thou'lt forget me when I am gone.                280

DOLL TEARSHEET By my troth, thou'lt set me a-weeping an thou sayst so. Prove that ever I dress myself handsome till thy return—well, hearken a'th' end.

SIR JOHN Some sack, Francis.

PRINCE *and* POINS (coming forward) Anon, anon, sir.   285

SIR JOHN Ha, a bastard son of the King's!—And art not thou Poins his brother?

PRINCE HARRY Why, thou globe of sinful continents, what a life dost thou lead!

SIR JOHN A better than thou: I am a gentleman, thou art a drawer.                291

PRINCE HARRY Very true, sir, and I come to draw you out by the ears.

MISTRESS QUICKLY O, the Lord preserve thy grace! By my troth, welcome to London! Now the Lord bless that sweet face of thine! O Jesu, are you come from Wales?

SIR JOHN (to Prince Harry) Thou whoreson mad compound of majesty! By this light—flesh and corrupt blood, thou art welcome.

DOLL TEARSHEET How, you fat fool? I scorn you.        300

POINS (to Prince Harry) My lord, he will drive you out of your revenge and turn all to a merriment, if you take not the heat.

PRINCE HARRY (to Sir John) You whoreson candlemine you, how vilely did you speak of me now, before this honest, virtuous, civil gentlewoman!                306

MISTRESS QUICKLY God's blessing of your good heart, and so she is, by my troth!

SIR JOHN (to Prince Harry) Didst thou hear me?

PRINCE HARRY Yea, and you knew me as you did when you ran away by Gads Hill; you knew I was at your back, and spoke it on purpose to try my patience.

SIR JOHN No, no, no, not so, I did not think thou wast within hearing.                314

PRINCE HARRY I shall drive you, then, to confess the wilful abuse, and then I know how to handle you.

SIR JOHN No abuse, Hal; o'mine honour, no abuse.

PRINCE HARRY Not? To dispraise me, and call me 'pantler' and 'bread-chipper' and I know not what?

SIR JOHN No abuse, Hal.                320

POINS No abuse?

SIR JOHN No abuse, Ned, i'th' world, honest Ned, none. I dispraised him before the wicked, that the wicked might not fall in love with him; (to Prince Harry) in which doing I have done the part of a careful friend and a true subject, and thy father is to give me thanks for it. No abuse, Hal; none, Ned, none; no, faith, boys, none.                328

PRINCE HARRY See now whether pure fear and entire cowardice doth not make thee wrong this virtuous gentlewoman to close with us. Is she of the wicked? Is thine hostess here of the wicked? Or is thy boy of the wicked? Or honest Bardolph, whose zeal burns in his nose, of the wicked?

POINS (to Sir John) Answer, thou dead elm, answer.   335

SIR JOHN The fiend hath pricked down Bardolph irrecoverable, and his face is Lucifer's privy kitchen, where he doth nothing but roast malt-worms. For the boy, there is a good angel about him, but the devil outbids him, too.                340

PRINCE HARRY For the women?

SIR JOHN For one of them, she's in hell already, and burns poor souls. For th'other, I owe her money, and whether she be damned for that I know not.

MISTRESS QUICKLY No, I warrant you.                345

SIR JOHN No, I think thou art not; I think thou art quit for that. Marry, there is another indictment upon thee, for suffering flesh to be eaten in thy house, contrary to the law, for the which I think thou wilt howl.

MISTRESS QUICKLY All victuallers do so. What's a joint of mutton or two in a whole Lent?                351

PRINCE HARRY You, gentlewoman—

DOLL TEARSHEET What says your grace?

SIR JOHN His grace says that which his flesh rebels against.

*Peto knocks at door within*

MISTRESS QUICKLY Who knocks so loud at door? (Calls) Look to th' door there, Francis.                356

*Enter Peto*

PRINCE HARRY Peto, how now, what news?

PETO

The King your father is at Westminster;
And there are twenty weak and wearied posts
Come from the north; and as I came along          360
I met and overtook a dozen captains,
Bareheaded, sweating, knocking at the taverns,
And asking every one for Sir John Falstaff.

PRINCE HARRY
By heaven, Poins, I feel me much to blame
So idly to profane the precious time,                              365
When tempest of commotion, like the south
Borne with black vapour, doth begin to melt
And drop upon our bare unarmèd heads.—
Give me my sword and cloak.—Falstaff, good night.
                    *Exeunt Prince Harry and Poins*
SIR JOHN  Now comes in the sweetest morsel of the night,
and we must hence and leave it unpicked.                  371
                    *Knocking within.* ⌈*Exit Bardolph*⌉
More knocking at the door!
    *Enter Bardolph*
How now, what's the matter?
BARDOLPH
You must away to court, sir, presently.
A dozen captains stay at door for you.                          375
SIR JOHN ⌈*to the Page*⌉ Pay the musicians, sirrah. Farewell,
hostess; farewell, Doll. You see, my good wenches, how
men of merit are sought after. The undeserver may
sleep, when the man of action is called on. Farewell,
good wenches. If I be not sent away post, I will see
you again ere I go.                            ⌈*Exeunt musicians*⌉
DOLL TEARSHEET ⌈*weeping*⌉ I cannot speak. If my heart be
not ready to burst—well, sweet Jack, have a care of
thyself.
SIR JOHN  Farewell, farewell!                                     385
            *Exit* ⌈*with Bardolph, Peto, and the Page*⌉
MISTRESS QUICKLY  Well, fare thee well. I have known thee
these twenty-nine years come peascod-time, but an
honester and truer-hearted man—well, fare thee well.
    ⌈*Enter Bardolph*⌉
BARDOLPH  Mistress Tearsheet!
MISTRESS QUICKLY  What's the matter?                    390
BARDOLPH  Bid Mistress Tearsheet come to my master.
                                                          ⌈*Exit*⌉
MISTRESS QUICKLY  O run, Doll; run, run, good Doll!
                    *Exeunt* ⌈*Doll at one door, Mistress Quickly at
                    another door*⌉

3.1    *Enter King Henry in his nightgown, with a page*
KING HENRY (*giving letters*)
Go call the Earls of Surrey and of Warwick.
But ere they come, bid them o'er-read these letters
And well consider of them. Make good speed.
                                        *Exit page*
How many thousand of my poorest subjects
Are at this hour asleep? O sleep, O gentle sleep,         5
Nature's soft nurse, how have I frighted thee,
That thou no more wilt weigh my eyelids down
And steep my senses in forgetfulness?
Why rather, sleep, liest thou in smoky cribs,
Upon uneasy pallets stretching thee,                            10
And hushed with buzzing night-flies to thy slumber,
Than in the perfumed chambers of the great,
Under the canopies of costly state,
And lulled with sound of sweetest melody?
O thou dull god, why li'st thou with the vile              15
In loathsome beds, and leav'st the kingly couch
A watch-case, or a common 'larum-bell?
Wilt thou upon the high and giddy mast
Seal up the ship-boy's eyes, and rock his brains
In cradle of the rude imperious surge,                          20
And in the visitation of the winds,
Who take the ruffian billows by the top,

Curling their monstrous heads, and hanging them
With deafing clamour in the slippery clouds,
That, with the hurly, death itself awakes?                  25
Canst thou, O partial sleep, give thy repose
To the wet sea-boy in an hour so rude,
And in the calmest and most stillest night,
With all appliances and means to boot,
Deny it to a king? Then happy low, lie down.           30
Uneasy lies the head that wears a crown.
    *Enter the Earls of Warwick and Surrey*
WARWICK
Many good morrows to your majesty!
KING HENRY
Is it good morrow, lords?
WARWICK                                  'Tis one o'clock, and past.
KING HENRY
Why then, good morrow to you all, my lords.
Have you read o'er the letter that I sent you?           35
WARWICK  We have, my liege.
KING HENRY
Then you perceive the body of our kingdom,
How foul it is, what rank diseases grow,
And with what danger near the heart of it.
WARWICK
It is but as a body yet distempered,                            40
Which to his former strength may be restored
With good advice and little medicine.
My lord Northumberland will soon be cooled.
KING HENRY
O God, that one might read the book of fate,
And see the revolution of the times                            45
Make mountains level, and the continent,
Weary of solid firmness, melt itself
Into the sea; and other times to see
The beachy girdle of the ocean
Too wide for Neptune's hips; how chance's mocks  50
And changes fill the cup of alteration
With divers liquors! 'Tis not ten years gone
Since Richard and Northumberland, great friends,
Did feast together; and in two year after
Were they at wars. It is but eight years since          55
This Percy was the man nearest my soul,
Who like a brother toiled in my affairs,
And laid his love and life under my foot,
Yea, for my sake, even to the eyes of Richard
Gave him defiance. But which of you was by—      60
(*To Warwick*) You, cousin Neville, as I may
    remember—
When Richard, with his eye brimful of tears,
Then checked and rated by Northumberland,
Did speak these words, now proved a prophecy?—
'Northumberland, thou ladder by the which               65
My cousin Bolingbroke ascends my throne'—
Though then, God knows, I had no such intent,
But that necessity so bowed the state
That I and greatness were compelled to kiss—
'The time shall come'—thus did he follow it—         70
'The time will come that foul sin, gathering head,
Shall break into corruption'; so went on,
Foretelling this same time's condition,
And the division of our amity.
WARWICK
There is a history in all men's lives                            75
Figuring the natures of the times deceased;
The which observed, a man may prophesy,
With a near aim, of the main chance of things

As yet not come to life, who in their seeds
And weak beginnings lie intreasurèd.                    80
Such things become the hatch and brood of time;
And by the necessary form of this
King Richard might create a perfect guess
That great Northumberland, then false to him,
Would of that seed grow to a greater falseness,        85
Which should not find a ground to root upon
Unless on you.
KING HENRY          Are these things then necessities?
Then let us meet them like necessities;
And that same word even now cries out on us.
They say the Bishop and Northumberland                  90
Are fifty thousand strong.
WARWICK                    It cannot be, my lord.
Rumour doth double, like the voice and echo,
The numbers of the feared. Please it your grace
To go to bed? Upon my soul, my lord,
The powers that you already have sent forth            95
Shall bring this prize in very easily.
To comfort you the more, I have received
A certain instance that Glyndŵr is dead.
Your majesty hath been this fortnight ill,
And these unseasoned hours perforce must add          100
Unto your sickness.
KING HENRY          I will take your counsel.
And were these inward wars once out of hand,
We would, dear lords, unto the Holy Land.    *Exeunt*

**3.2**    *Enter Justice Shallow and Justice Silence*
SHALLOW Come on, come on, come on! Give me your
hand, sir, give me your hand, sir. An early stirrer, by
the rood! And how doth my good cousin Silence?
SILENCE Good morrow, good cousin Shallow.              4
SHALLOW And how doth my cousin your bedfellow? And
your fairest daughter and mine, my god-daughter Ellen?
SILENCE Alas, a black ouzel, cousin Shallow.
SHALLOW By yea and no, sir, I dare say my cousin William
is become a good scholar. He is at Oxford still, is he
not?                                                   10
SILENCE Indeed, sir, to my cost.
SHALLOW A must then to the Inns o' Court shortly. I was
once of Clement's Inn, where I think they will talk of
mad Shallow yet.
SILENCE You were called 'lusty Shallow' then, cousin.  15
SHALLOW By the mass, I was called anything; and I would
have done anything indeed, too, and roundly, too.
There was I, and little John Doit of Staffordshire, and
black George Barnes, and Francis Pickbone, and Will
Squeal, a Cotswold man; you had not four such swinge-
bucklers in all the Inns o' Court again. And I may say
to you, we knew where the bona-robas were, and had
the best of them all at commandment. Then was Jack
Falstaff, now Sir John, a boy, and page to Thomas
Mowbray, Duke of Norfolk.                              25
SILENCE This Sir John, cousin, that comes hither anon
about soldiers?
SHALLOW The same Sir John, the very same. I see him
break Scoggin's head at the court gate when a was a
crack, not thus high. And the very same day did I fight
with one Samson Stockfish, a fruiterer, behind Gray's
Inn. Jesu, Jesu, the mad days that I have spent! And
to see how many of my old acquaintance are dead.
SILENCE We shall all follow, cousin.                   34
SHALLOW Certain, 'tis certain; very sure, very sure. Death,
as the Psalmist saith, is certain to all; all shall die.
How a good yoke of bullocks at Stamford fair?

SILENCE By my troth, I was not there.
SHALLOW Death is certain. Is old Double of your town
living yet?                                            40
SILENCE Dead, sir.
SHALLOW Jesu, Jesu, dead! A drew a good bow; and dead!
A shot a fine shoot. John o' Gaunt loved him well, and
betted much money on his head. Dead! A would have
clapped i'th' clout at twelve score, and carried you a
forehand shaft a fourteen and fourteen and a half, that
it would have done a man's heart good to see. How a
score of ewes now?
SILENCE Thereafter as they be. A score of good ewes may
be worth ten pounds.                                   50
SHALLOW And is old Double dead?
         *Enter Bardolph and ⌈the Page⌉*
SILENCE Here come two of Sir John Falstaff's men, as I
think.
⌈SHALLOW⌉ Good morrow, honest gentlemen.
BARDOLPH I beseech you, which is Justice Shallow?     55
SHALLOW I am Robert Shallow, sir, a poor esquire of this
county, and one of the King's Justices of the Peace.
What is your good pleasure with me?
BARDOLPH My captain, sir, commends him to you—my
captain Sir John Falstaff, a tall gentleman, by heaven,
and a most gallant leader.                             61
SHALLOW He greets me well, sir. I knew him a good
backsword man. How doth the good knight? May I ask
how my lady his wife doth?
BARDOLPH Sir, pardon, a soldier is better accommodated
than with a wife.                                      66
SHALLOW It is well said, in faith, sir, and it is well said
indeed, too. 'Better accommodated'—it is good; yea,
indeed is it. Good phrases are surely, and ever were,
very commendable. 'Accommodated'—it comes of
'accommodo'. Very good, a good phrase.                71
BARDOLPH Pardon, sir, I have heard the word—'phrase'
call you it?—By this day, I know not the phrase; but
I will maintain the word with my sword to be a soldier-
like word, and a word of exceeding good command, by
heaven. 'Accommodated'; that is, when a man is, as
they say, accommodated; or when a man is being
whereby a may be thought to be accommodated; which
is an excellent thing.                                 79
         *Enter Sir John Falstaff*
SHALLOW It is very just. Look, here comes good Sir John.
(*To Sir John*) Give me your hand, give me your worship's
good hand. By my troth, you like well, and bear your
years very well. Welcome, good Sir John.
SIR JOHN I am glad to see you well, good Master Robert
Shallow. (*To Silence*) Master Surecard, as I think.    85
SHALLOW No, Sir John, it is my cousin Silence, in
commission with me.
SIR JOHN Good Master Silence, it well befits you should
be of the peace.
SILENCE Your good worship is welcome.                  90
SIR JOHN Fie, this is hot weather, gentlemen. Have you
provided me here half a dozen sufficient men?
SHALLOW Marry, have we, sir. Will you sit?
SIR JOHN Let me see them, I beseech you.
         ⌈*He sits*⌉
SHALLOW Where's the roll, where's the roll, where's the
roll? Let me see, let me see, let me see; so, so, so, so,
so. Yea, marry, sir: 'Ralph Mouldy'. ⌈*To Silence*⌉ Let
them appear as I call, let them do so, let them do so.
Let me see, (*calls*) where is Mouldy?
         ⌈*Enter Mouldy*⌉
MOULDY Here, an't please you.                         100

SHALLOW What think you, Sir John? A good-limbed fellow, young, strong, and of good friends.

SIR JOHN Is thy name Mouldy?

MOULDY Yea, an't please you.                                    104

SIR JOHN 'Tis the more time thou wert used.           105

SHALLOW Ha, ha, ha, most excellent, i'faith! Things that are mouldy lack use. Very singular good, in faith, well said, Sir John, very well said.

SIR JOHN Prick him.

MOULDY I was pricked well enough before, an you could have let me alone. My old dame will be undone now for one to do her husbandry and her drudgery. You need not to have pricked me; there are other men fitter to go out than I.                                           114

SIR JOHN Go to, peace, Mouldy. You shall go, Mouldy; it is time you were spent.

MOULDY Spent?

SHALLOW Peace, fellow, peace. Stand aside; know you where you are?                                             119

⌈*Mouldy stands aside*⌉

For th'other, Sir John, let me see: 'Simon Shadow'—

SIR JOHN Yea, marry, let me have him to sit under. He's like to be a cold soldier.

SHALLOW (*calls*) Where's Shadow?

⌈*Enter Shadow*⌉

SHADOW Here, sir.

SIR JOHN Shadow, whose son art thou?                    125

SHADOW My mother's son, sir.

SIR JOHN Thy mother's son! Like enough, and thy father's shadow. So the son of the female is the shadow of the male—it is often so indeed—but not of the father's substance.                                               130

SHALLOW Do you like him, Sir John?

SIR JOHN Shadow will serve for summer. Prick him, for we have a number of shadows fill up the muster book.

⌈*Shadow stands aside*⌉

SHALLOW (*calls*) 'Thomas Wart.'

SIR JOHN Where's he?                                            135

⌈*Enter Wart*⌉

WART Here, sir.

SIR JOHN Is thy name Wart?

WART Yea, sir.

SIR JOHN Thou art a very ragged wart.

SHALLOW Shall I prick him, Sir John?                    140

SIR JOHN It were superfluous, for his apparel is built upon his back, and the whole frame stands upon pins. Prick him no more.

SHALLOW Ha, ha, ha, you can do it, sir, you can do it! I commend you well.                                         145

⌈*Wart stands aside*⌉

(*Calls*) 'Francis Feeble.'

⌈*Enter Feeble*⌉

FEEBLE Here, sir.

SHALLOW What trade art thou, Feeble?

FEEBLE A woman's tailor, sir.

SHALLOW Shall I prick him, sir?                            150

SIR JOHN You may, but if he had been a man's tailor, he'd ha' pricked you. (*To Feeble*) Wilt thou make as many holes in an enemy's battle as thou hast done in a woman's petticoat?

FEEBLE I will do my good will, sir; you can have no more.

SIR JOHN Well said, good woman's tailor; well said, courageous Feeble! Thou wilt be as valiant as the wrathful dove or most magnanimous mouse. Prick the woman's tailor. Well, Master Shallow; deep, Master Shallow.                                               160

FEEBLE I would Wart might have gone, sir.

SIR JOHN I would thou wert a man's tailor, that thou mightst mend him and make him fit to go. I cannot put him to a private soldier that is the leader of so many thousands. Let that suffice, most forcible Feeble.

FEEBLE It shall suffice, sir.                                      166

SIR JOHN I am bound to thee, reverend Feeble.

⌈*Feeble stands aside*⌉

Who is next?

SHALLOW (*calls*) 'Peter Bullcalf o'th' green.'

SIR JOHN Yea, marry, let's see Bullcalf.                 170

⌈*Enter Bullcalf*⌉

BULLCALF Here, sir.

SIR JOHN Fore God, a likely fellow! Come, prick Bullcalf till he roar again.

BULLCALF O Lord, good my lord captain!

SIR JOHN What, dost thou roar before thou'rt pricked?

BULLCALF O Lord, sir, I am a diseased man.             176

SIR JOHN What disease hast thou?

BULLCALF A whoreson cold, sir; a cough, sir, which I caught with ringing in the King's affairs upon his coronation day, sir.                                          180

SIR JOHN Come, thou shalt go to the wars in a gown. We will have away thy cold, and I will take such order that thy friends shall ring for thee.

⌈*Bullcalf stands aside*⌉

Is here all?                                                       184

SHALLOW There is two more called than your number. You must have but four here, sir, and so I pray you go in with me to dinner.

SIR JOHN Come, I will go drink with you, but I cannot tarry dinner. I am glad to see you, by my troth, Master Shallow.                                                      190

SHALLOW O, Sir John, do you remember since we lay all night in the Windmill in Saint George's Field?

SIR JOHN No more of that, good Master Shallow, no more of that.

SHALLOW Ha, 'twas a merry night! And is Jane Nightwork alive?                                                         196

SIR JOHN She lives, Master Shallow.

SHALLOW She never could away with me.

SIR JOHN Never, never. She would always say she could not abide Master Shallow.                               200

SHALLOW By the mass, I could anger her to th' heart. She was then a bona-roba. Doth she hold her own well?

SIR JOHN Old, old, Master Shallow.

SHALLOW Nay, she must be old; she cannot choose but be old; certain she's old; and had Robin Nightwork by old Nightwork before I came to Clement's Inn.       206

SILENCE That's fifty-five year ago.

SHALLOW Ha, cousin Silence, that thou hadst seen that that this knight and I have seen! Ha, Sir John, said I well?                                                           210

SIR JOHN We have heard the chimes at midnight, Master Shallow.

SHALLOW That we have, that we have; in faith, Sir John, we have. Our watchword was 'Hem boys!' Come, let's to dinner; come, let's to dinner. Jesus, the days that we have seen! Come, come.                               216

*Exeunt Shallow, Silence, and Sir John*

BULLCALF ⌈*coming forward*⌉ Good Master Corporate Bardolph, stand my friend, and here's four Harry ten shillings in French crowns for you. In very truth, sir, I had as lief be hanged, sir, as go. And yet for mine own part, sir, I do not care; but rather because I am unwilling, and, for mine own part, have a desire to stay with my friends. Else, sir, I did not care, for mine own part, so much.

BARDOLPH ⌈taking the money⌉ Go to; stand aside.    225
⌈Bullcalf stands aside⌉

MOULDY ⌈coming forward⌉ And, good Master Corporal Captain, for my old dame's sake stand my friend. She has nobody to do anything about her when I am gone, and she is old and cannot help herself. You shall have forty, sir.    230

BARDOLPH Go to; stand aside.
⌈Mouldy stands aside⌉

FEEBLE By my troth, I care not. A man can die but once. We owe God a death. I'll ne'er bear a base mind. An't be my destiny, so; an't be not, so. No man's too good to serve's prince. And let it go which way it will, he that dies this year is quit for the next.    236

BARDOLPH Well said; thou'rt a good fellow.

FEEBLE Faith, I'll bear no base mind.
Enter Sir John Falstaff, Shallow, and Silence

SIR JOHN Come, sir, which men shall I have?

SHALLOW Four of which you please.    240

BARDOLPH (to Sir John) Sir, a word with you. (Aside to him) I have three pound to free Mouldy and Bullcalf.

SIR JOHN Go to, well.

SHALLOW Come, Sir John, which four will you have?

SIR JOHN Do you choose for me.    245

SHALLOW Marry, then: Mouldy, Bullcalf, Feeble, and Shadow.

SIR JOHN Mouldy and Bullcalf. For you, Mouldy, stay at home till you are past service; and for your part, Bullcalf, grow till you come unto it. I will none of you.
⌈Exeunt Bullcalf and Mouldy⌉

SHALLOW Sir John, Sir John, do not yourself wrong. They are your likeliest men, and I would have you served with the best.    253

SIR JOHN Will you tell me, Master Shallow, how to choose a man? Care I for the limb, the thews, the stature, bulk, and big assemblance of a man? Give me the spirit, Master Shallow. Here's Wart; you see what a ragged appearance it is? A shall charge you and discharge you with the motion of a pewterer's hammer, come off and on swifter than he that gibbets on the brewer's bucket. And this same half-faced fellow Shadow; give me this man. He presents no mark to the enemy; the foeman may with as great aim level at the edge of a penknife. And for a retreat, how swiftly will this Feeble the woman's tailor run off! O, give me the spare men, and spare me the great ones.—Put me a caliver into Wart's hand, Bardolph.

BARDOLPH (giving Wart a caliver) Hold, Wart. Traverse— thas, thas, thas!    269
⌈Wart marches⌉

SIR JOHN (to Wart) Come, manage me your caliver. So; very well. Go to, very good, exceeding good. O, give me always a little, lean, old, chapped, bald shot! Well said, i'faith, Wart; thou'rt a good scab. Hold; (giving a coin) there's a tester for thee.    274

SHALLOW He is not his craft's master; he doth not do it right. I remember at Mile-End Green, when I lay at Clement's Inn—I was then Sir Dagonet in Arthur's show—there was a little quiver fellow, and a would manage you his piece thus, and a would about and about, and come you in and come you in. 'Ra-ta-ta!' would a say; 'Bounce!' would a say; and away again would a go; and again would a come. I shall ne'er see such a fellow.    283

SIR JOHN These fellows will do well, Master Shallow. God keep you, Master Silence; I will not use many words with you. Fare you well, gentlemen both; I thank you. I must a dozen mile tonight.—Bardolph, give the soldiers coats.    288

SHALLOW Sir John, the Lord bless you; God prosper your affairs! God send us peace! As you return, visit my house; let our old acquaintance be renewed. Peradventure I will with ye to the court.

SIR JOHN Fore God, would you would!

SHALLOW Go to, I have spoke at a word. God keep you!

SIR JOHN Fare you well, gentle gentlemen.    295
Exeunt Shallow and Silence
On, Bardolph, lead the men away.
Exeunt Bardolph, Wart, Shadow, and Feeble
As I return, I will fetch off these justices. I do see the bottom of Justice Shallow. Lord, Lord, how subject we old men are to this vice of lying! This same starved justice hath done nothing but prate to me of the wildness of his youth and the feats he hath done about Turnbull Street; and every third word a lie, duer paid to the hearer than the Turk's tribute. I do remember him at Clement's Inn, like a man made after supper of a cheese paring. When a was naked, he was for all the world like a forked radish, with a head fantastically carved upon it with a knife. A was so forlorn that his dimensions, to any thick sight, were invisible. A was the very genius of famine. And now is this Vice's dagger become a squire, and talks as familiarly of John o' Gaunt as if he had been sworn brother to him, and I'll be sworn a ne'er saw him but once, in the Tilt-yard, and then he burst his head for crowding among the marshal's men. I saw it, and told John o' Gaunt he beat his own name; for you might have trussed him and all his apparel into an eel-skin. The case of a treble hautboy was a mansion for him, a court. And now has he land and beeves. Well, I'll be acquainted with him if I return; and't shall go hard but I'll make him a philosopher's two stones to me. If the young dace be a bait for the old pike, I see no reason in the law of nature but I may snap at him. Let time shape, and there an end.    Exit

4.1    Enter ⌈in arms⌉ the Archbishop of York, Thomas Mowbray, Lord Hastings, and ⌈Coleville⌉, within the Forest of Gaultres

ARCHBISHOP OF YORK What is this forest called?

HASTINGS
'Tis Gaultres Forest, an't shall please your grace.

ARCHBISHOP OF YORK
Here stand, my lords, and send discoverers forth
To know the numbers of our enemies.

HASTINGS
We have sent forth already.

ARCHBISHOP OF YORK    'Tis well done.    5
My friends and brethren in these great affairs,
I must acquaint you that I have received
New-dated letters from Northumberland,
Their cold intent, tenor, and substance, thus:
Here doth he wish his person, with such powers    10
As might hold sortance with his quality,
The which he could not levy; whereupon
He is retired to ripe his growing fortunes
To Scotland, and concludes in hearty prayers
That your attempts may overlive the hazard    15
And fearful meeting of their opposite.

MOWBRAY
  Thus do the hopes we have in him touch ground
  And dash themselves to pieces.
      *Enter a Messenger*
HASTINGS              Now, what news?
MESSENGER
  West of this forest, scarcely off a mile,
  In goodly form comes on the enemy;        20
  And, by the ground they hide, I judge their number
  Upon or near the rate of thirty thousand.
MOWBRAY
  The just proportion that we gave them out.
  Let us sway on, and face them in the field.
      *Enter the Earl of Westmorland*
ARCHBISHOP OF YORK
  What well-appointed leader fronts us here?     25
MOWBRAY
  I think it is my lord of Westmorland.
WESTMORLAND
  Health and fair greeting from our general,
  The Prince, Lord John and Duke of Lancaster.
ARCHBISHOP OF YORK
  Say on, my lord of Westmorland, in peace,
  What doth concern your coming.
WESTMORLAND           Then, my lord,   30
  Unto your grace do I in chief address
  The substance of my speech. If that rebellion
  Came like itself, in base and abject routs,
  Led on by bloody youth, guarded with rags,
  And countenanced by boys and beggary;      35
  I say, if damned commotion so appeared
  In his true native and most proper shape,
  You, reverend father, and these noble lords
  Had not been here to dress the ugly form
  Of base and bloody insurrection          40
  With your fair honours. You, Lord Archbishop,
  Whose see is by a civil peace maintained,
  Whose beard the silver hand of peace hath touched,
  Whose learning and good letters peace hath tutored,
  Whose white investments figure innocence,     45
  The dove and very blessèd spirit of peace,
  Wherefore do you so ill translate yourself
  Out of the speech of peace that bears such grace
  Into the harsh and boist'rous tongue of war,
  Turning your books to graves, your ink to blood,   50
  Your pens to lances, and your tongue divine
  To a loud trumpet and a point of war?
ARCHBISHOP OF YORK
  Wherefore do I this? So the question stands.
  Briefly, to this end: we are all diseased,
  And with our surfeiting and wanton hours     55
  Have brought ourselves into a burning fever,
  And we must bleed for it—of which disease
  Our late King Richard, being infected, died.
  But, my most noble lord of Westmorland,
  I take not on me here as a physician,        60
  Nor do I as an enemy to peace
  Troop in the throngs of military men;
  But rather show a while like fearful war
  To diet rank minds, sick of happiness,
  And purge th'obstructions which begin to stop   65
  Our very veins of life. Hear me more plainly.
  I have in equal balance justly weighed
  What wrongs our arms may do, what wrongs we suffer,
  And find our griefs heavier than our offences.
  We see which way the stream of time doth run,   70

  And are enforced from our most quiet shore
  By the rough torrent of occasion;
  And have the summary of all our griefs,
  When time shall serve, to show in articles,
  Which long ere this we offered to the King,     75
  And might by no suit gain our audience.
  When we are wronged, and would unfold our griefs,
  We are denied access unto his person
  Even by those men that most have done us wrong.
  The dangers of the days but newly gone,      80
  Whose memory is written on the earth
  With yet appearing blood, and the examples
  Of every minute's instance, present now,
  Hath put us in these ill-beseeming arms,
  Not to break peace, or any branch of it,      85
  But to establish here a peace indeed,
  Concurring both in name and quality.
WESTMORLAND
  Whenever yet was your appeal denied?
  Wherein have you been gallèd by the King?
  What peer hath been suborned to grate on you,   90
  That you should seal this lawless bloody book
  Of forged rebellion with a seal divine?
ARCHBISHOP OF YORK
  My brother general, the commonwealth
  I make my quarrel in particular.
WESTMORLAND
  There is no need of any such redress;       95
  Or if there were, it not belongs to you.
MOWBRAY
  Why not to him in part, and to us all
  That feel the bruises of the days before,
  And suffer the condition of these times
  To lay a heavy and unequal hand          100
  Upon our honours?
WESTMORLAND       O my good Lord Mowbray,
  Construe the times to their necessities,
  And you shall say indeed it is the time,
  And not the King, that doth you injuries.
  Yet for your part, it not appears to me,      105
  Either from the King or in the present time,
  That you should have an inch of any ground
  To build a grief on. Were you not restored
  To all the Duke of Norfolk's signories,
  Your noble and right well-remembered father's?   110
MOWBRAY
  What thing in honour had my father lost
  That need to be revived and breathed in me?
  The King that loved him, as the state stood then,
  Was force perforce compelled to banish him;
  And then that Henry Bolingbroke and he,      115
  Being mounted and both rousèd in their seats,
  Their neighing coursers daring of the spur,
  Their armèd staves in charge, their beavers down,
  Their eyes of fire sparkling through sights of steel,
  And the loud trumpet blowing them together,   120
  Then, then, when there was nothing could have stayed
  My father from the breast of Bolingbroke—
  O, when the King did throw his warder down,
  His own life hung upon the staff he threw;
  Then threw he down himself and all their lives   125
  That by indictment and by dint of sword
  Have since miscarried under Bolingbroke.
WESTMORLAND
  You speak, Lord Mowbray, now you know not what.
  The Earl of Hereford was reputed then

In England the most valiant gentleman.                    130
Who knows on whom fortune would then have
    smiled?
But if your father had been victor there,
He ne'er had borne it out of Coventry;
For all the country in a general voice
Cried hate upon him, and all their prayers and love
Were set on Hereford, whom they doted on          136
And blessed and graced, indeed, more than the King.
But this is mere digression from my purpose.
Here come I from our princely general
To know your griefs, to tell you from his grace     140
That he will give you audience; and wherein
It shall appear that your demands are just,
You shall enjoy them, everything set off
That might so much as think you enemies.

MOWBRAY
But he hath forced us to compel this offer,          145
And it proceeds from policy, not love.

WESTMORLAND
Mowbray, you overween to take it so.
This offer comes from mercy, not from fear;
For lo, within a ken our army lies,
Upon mine honour, all too confident                     150
To give admittance to a thought of fear.
Our battle is more full of names than yours,
Our men more perfect in the use of arms,
Our armour all as strong, our cause the best.
Then reason will our hearts should be as good.     155
Say you not then our offer is compelled.

MOWBRAY
Well, by my will we shall admit no parley.

WESTMORLAND
That argues but the shame of your offence.
A rotten case abides no handling.

HASTINGS
Hath the Prince John a full commission,               160
In very ample virtue of his father,
To hear and absolutely to determine
Of what conditions we shall stand upon?

WESTMORLAND
That is intended in the general's name.
I muse you make so slight a question.                   165

ARCHBISHOP OF YORK
Then take, my lord of Westmorland, this schedule;
For this contains our general grievances.
Each several article herein redressed,
All members of our cause, both here and hence,
That are ensinewed to this action                        170
Acquitted by a true substantial form,
And present execution of our wills
To us and to our purposes consigned,
We come within our awe-full banks again,
And knit our powers to the arm of peace.          175

WESTMORLAND (taking the schedule)
This will I show the general. Please you, lords,
In sight of both our battles we may meet,
And either end in peace—which God so frame—
Or to the place of diff'rence call the swords
Which must decide it.

ARCHBISHOP OF YORK     My lord, we will do so.     180
                        Exit Westmorland

MOWBRAY
There is a thing within my bosom tells me
That no conditions of our peace can stand.

HASTINGS
Fear you not that. If we can make our peace
Upon such large terms and so absolute
As our conditions shall consist upon,               185
Our peace shall stand as firm as rocky mountains.

MOWBRAY
Yea, but our valuation shall be such
That every slight and false-derivèd cause,
Yea, every idle, nice, and wanton reason,
Shall to the King taste of this action,                 190
That, were our royal faiths martyrs in love,
We shall be winnowed with so rough a wind
That even our corn shall seem as light as chaff,
And good from bad find no partition.

ARCHBISHOP OF YORK
No, no, my lord; note this. The King is weary     195
Of dainty and such picking grievances,
For he hath found to end one doubt by death
Revives two greater in the heirs of life;
And therefore will he wipe his tables clean,
And keep no tell-tale to his memory                   200
That may repeat and history his loss
To new remembrance; for full well he knows
He cannot so precisely weed this land
As his misdoubts present occasion.
His foes are so enrooted with his friends             205
That, plucking to unfix an enemy,
He doth unfasten so and shake a friend;
So that this land, like an offensive wife
That hath enragèd him on to offer strokes,
As he is striking, holds his infant up,                 210
And hangs resolved correction in the arm
That was upreared to execution.

HASTINGS
Besides, the King hath wasted all his rods
On late offenders, that he now doth lack
The very instruments of chastisement;              215
So that his power, like to a fangless lion,
May offer, but not hold.

ARCHBISHOP OF YORK            'Tis very true.
And therefore be assured, my good Lord Marshal,
If we do now make our atonement well,
Our peace will, like a broken limb united,          220
Grow stronger for the breaking.

MOWBRAY                        Be it so.
        Enter Westmorland
Here is returned my lord of Westmorland.

WESTMORLAND
The Prince is here at hand. Pleaseth your lordship
To meet his grace just distance 'tween our armies?

MOWBRAY
Your grace of York, in God's name then set forward.

ARCHBISHOP OF YORK
Before, and greet his grace!—My lord, we come.   226
        [They march over the stage.]
        Enter Prince John [with one or more soldiers
        carrying wine]

PRINCE JOHN
You are well encountered here, my cousin Mowbray.
Good day to you, gentle lord Archbishop;
And so to you, Lord Hastings, and to all.
My lord of York, it better showed with you          230
When that your flock, assembled by the bell,
Encircled you to hear with reverence
Your exposition on the holy text,
Than now to see you here an iron man,

Cheering a rout of rebels with your drum,                235
Turning the word to sword, and life to death.
That man that sits within a monarch's heart
And ripens in the sunshine of his favour,
Would he abuse the countenance of the King,
Alack, what mischiefs might he set abroach               240
In shadow of such greatness! With you, Lord Bishop,
It is even so. Who hath not heard it spoken
How deep you were within the books of God—
To us, the speaker in his parliament,
To us, th'imagined voice of God himself,                 245
The very opener and intelligencer
Between the grace, the sanctities of heaven
And our dull workings? O, who shall believe
But you misuse the reverence of your place,
Employ the countenance and grace of heav'n              250
As a false favourite doth his prince's name
In deeds dishonourable? You have ta'en up,
Under the counterfeited zeal of God,
The subjects of his substitute, my father;
And, both against the peace of heaven and him,          255
Have here upswarmèd them.
ARCHBISHOP OF YORK              Good my lord of Lancaster,
I am not here against your father's peace;
But, as I told my lord of Westmorland,
The time misordered doth, in common sense,
Crowd us and crush us to this monstrous form,           260
To hold our safety up. I sent your grace
The parcels and particulars of our grief,
The which hath been with scorn shoved from the
    court,
Whereon this Hydra son of war is born;
Whose dangerous eyes may well be charmed asleep
With grant of our most just and right desires,          266
And true obedience, of this madness cured,
Stoop tamely to the foot of majesty.
MOWBRAY
If not, we ready are to try our fortunes
To the last man.
HASTINGS              And though we here fall down,      270
We have supplies to second our attempt.
If they miscarry, theirs shall second them;
And so success of mischief shall be born,
And heir from heir shall hold this quarrel up,
Whiles England shall have generation.                   275
PRINCE JOHN
You are too shallow, Hastings, much too shallow,
To sound the bottom of the after-times.
WESTMORLAND
Pleaseth your grace to answer them directly
How far forth you do like their articles?
PRINCE JOHN
I like them all, and do allow them well,                280
And swear here, by the honour of my blood,
My father's purposes have been mistook,
And some about him have too lavishly
Wrested his meaning and authority.
(To the Archbishop)
My lord, these griefs shall be with speed redressed;
Upon my soul they shall. If this may please you,        286
Discharge your powers unto their several counties,
As we will ours; and here between the armies
Let's drink together friendly and embrace,
That all their eyes may bear those tokens home          290
Of our restorèd love and amity.
ARCHBISHOP OF YORK
I take your princely word for these redresses.

[PRINCE JOHN]
I give it you, and will maintain my word;
And thereupon I drink unto your grace.
    He drinks
[HASTINGS] [to Coleville]
Go, captain, and deliver to the army                    295
This news of peace. Let them have pay, and part.
I know it will well please them. Hie thee, captain.
    Exit [Coleville]
ARCHBISHOP OF YORK
To you, my noble lord of Westmorland!
    He drinks
WESTMORLAND (drinking)
I pledge your grace. An if you knew what pains
I have bestowed to breed this present peace,            300
You would drink freely; but my love to ye
Shall show itself more openly hereafter.
ARCHBISHOP OF YORK
I do not doubt you.
WESTMORLAND              I am glad of it.
(Drinking) Health to my lord and gentle cousin
    Mowbray!
MOWBRAY
You wish me health in very happy season,                305
For I am on the sudden something ill.
ARCHBISHOP OF YORK
Against ill chances men are ever merry;
But heaviness foreruns the good event.
WESTMORLAND
Therefore be merry, coz, since sudden sorrow
Serves to say thus: some good thing comes tomorrow.
ARCHBISHOP OF YORK
Believe me, I am passing light in spirit.               311
MOWBRAY
So much the worse, if your own rule be true.
    Shout within
PRINCE JOHN
The word of peace is rendered. Hark how they shout.
MOWBRAY
This had been cheerful after victory.
ARCHBISHOP OF YORK
A peace is of the nature of a conquest,                 315
For then both parties nobly are subdued,
And neither party loser.
PRINCE JOHN (to Westmorland) Go, my lord,
And let our army be dischargèd too.
    Exit Westmorland
(To the Archbishop) And, good my lord, so please you,
    let our trains
March by us, that we may peruse the men                 320
We should have coped withal.
ARCHBISHOP OF YORK              Go, good Lord Hastings,
And ere they be dismissed, let them march by.
    Exit Hastings
PRINCE JOHN
I trust, lords, we shall lie tonight together.
    Enter the Earl of Westmorland, [with captains]
Now, cousin, wherefore stands our army still?
WESTMORLAND
The leaders, having charge from you to stand,           325
Will not go off until they hear you speak.
PRINCE JOHN
They know their duties.
    Enter Lord Hastings
HASTINGS [to the Archbishop] Our army is dispersed.
Like youthful steers unyoked, they take their courses,

East, west, north, south; or, like a school broke up,
Each hurries toward his home and sporting place.    330
WESTMORLAND
Good tidings, my lord Hastings, for the which
I do arrest thee, traitor, of high treason;
And you, Lord Archbishop, and you, Lord Mowbray,
Of capital treason I attach you both.
            ⌈*The captains guard Hastings, the Archbishop, and*
            *Mowbray*⌉
MOWBRAY
Is this proceeding just and honourable?    335
WESTMORLAND Is your assembly so?
ARCHBISHOP OF YORK Will you thus break your faith?
PRINCE JOHN I pawned thee none.
I promised you redress of these same grievances
Whereof you did complain; which, by mine honour,
I will perform with a most Christian care.    341
But for you rebels, look to taste the due
Meet for rebellion and such acts as yours.
Most shallowly did you these arms commence,
Fondly brought here, and foolishly sent hence.—    345
Strike up our drums, pursue the scattered stray.
God, and not we, hath safely fought today.
Some guard these traitors to the block of death,
Treason's true bed and yielder up of breath.    *Exeunt*

**4.2**    *Alarum. Excursions. Enter Sir John Falstaff and*
        *Coleville*

SIR JOHN What's your name, sir, of what condition are
you, and of what place, I pray?
COLEVILLE I am a knight, sir, and my name is Coleville
of the Dale.    4
SIR JOHN Well then, Coleville is your name, a knight is
your degree, and your place the Dale. Coleville shall be
still your name, a traitor your degree, and the dungeon
your place—a place deep enough, so shall you be still
Coleville of the Dale.
COLEVILLE Are not you Sir John Falstaff?    10
SIR JOHN As good a man as he, sir, whoe'er I am. Do ye
yield, sir, or shall I sweat for you? If I do sweat, they
are the drops of thy lovers, and they weep for thy
death; therefore rouse up fear and trembling, and do
observance to my mercy.    15
COLEVILLE (*kneeling*) I think you are Sir John Falstaff, and
in that thought yield me.
SIR JOHN (*aside*) I have a whole school of tongues in this
belly of mine, and not a tongue of them all speaks any
other word but my name. An I had but a belly of any
indifferency, I were simply the most active fellow in
Europe. My womb, my womb, my womb undoes me.
        *Enter Prince John, the Earl of Westmorland, Sir*
        *John Blunt, and other lords and soldiers*
Here comes our general.    23
PRINCE JOHN
The heat is past; follow no further now.
        *A retreat is sounded*
Call in the powers, good cousin Westmorland.    25
                        *Exit Westmorland*
Now, Falstaff, where have you been all this while?
When everything is ended, then you come.
These tardy tricks of yours will, on my life,
One time or other break some gallows' back.
SIR JOHN I would be sorry, my lord, but it should be thus.
I never knew yet but rebuke and check was the reward
of valour. Do you think me a swallow, an arrow, or a
bullet? Have I in my poor and old motion the expedition

of thought? I have speeded hither with the very
extremest inch of possibility; I have foundered nine-
score and odd posts; and here, travel-tainted as I am,
have in my pure and immaculate valour taken Sir John
Coleville of the Dale, a most furious knight and valorous
enemy. But what of that? He saw me, and yielded, that
I may justly say, with the hook-nosed fellow of Rome,
'I came, saw, and overcame.'    41
PRINCE JOHN It was more of his courtesy than your
deserving.
SIR JOHN I know not. Here he is, and here I yield him;
and I beseech your grace, let it be booked with the rest
of this day's deeds; or, by the Lord, I will have it in a
particular ballad else, with mine own picture on the
top on't, Coleville kissing my foot; to the which course
if I be enforced, if you do not all show like gilt twopences
to me, and I in the clear sky of fame o'ershine you as
much as the full moon doth the cinders of the element,
which show like pins' heads to her, believe not the
word of the noble. Therefore let me have right, and let
desert mount.
PRINCE JOHN Thine's too heavy to mount.    55
SIR JOHN Let it shine then.
PRINCE JOHN Thine's too thick to shine.
SIR JOHN Let it do something, my good lord, that may do
me good, and call it what you will.
PRINCE JOHN
Is thy name Coleville?
COLEVILLE                        It is, my lord.    60
PRINCE JOHN
A famous rebel art thou, Coleville.
SIR JOHN And a famous true subject took him.
COLEVILLE
I am, my lord, but as my betters are
That led me hither. Had they been ruled by me,
You should have won them dearer than you have.    65
SIR JOHN
I know not how—they sold themselves, but thou
Like a kind fellow gav'st thyself away,
And I thank thee for thee.
        *Enter the Earl of Westmorland*
PRINCE JOHN                        Have you left pursuit?
WESTMORLAND
Retreat is made, and execution stayed.
PRINCE JOHN
Send Coleville with his confederates    70
To York, to present execution.
Blunt, lead him hence, and see you guard him sure.
                        *Exit Blunt, with Coleville*
And now dispatch we toward the court, my lords.
I hear the King my father is sore sick.
(*To Westmorland*) Our news shall go before us to his
        majesty,    75
Which, cousin, you shall bear to comfort him;
And we with sober speed will follow you.
SIR JOHN
My lord, I beseech you give me leave to go
Through Gloucestershire, and when you come to court
Stand, my good lord, pray, in your good report.    80
PRINCE JOHN
Fare you well, Falstaff. I in my condition
Shall better speak of you than you deserve.
                        *Exeunt all but Sir John*
SIR JOHN I would you had but the wit; 'twere better than
your dukedom. Good faith, this same young sober-
blooded boy doth not love me, nor a man cannot make

him laugh. But that's no marvel; he drinks no wine. There's never none of these demure boys come to any proof; for thin drink doth so overcool their blood, and making many fish meals, that they fall into a kind of male green-sickness; and then when they marry, they get wenches. They are generally fools and cowards— which some of us should be too, but for inflammation. A good sherry-sack hath a two-fold operation in it. It ascends me into the brain, dries me there all the foolish and dull and crudy vapours which environ it, makes it apprehensive, quick, forgetive, full of nimble, fiery, and delectable shapes, which, delivered o'er to the voice, the tongue, which is the birth, becomes excellent wit. The second property of your excellent sherry is the warming of the blood, which, before cold and settled, left the liver white and pale, which is the badge of pusillanimity and cowardice. But the sherry warms it, and makes it course from the inwards to the parts' extremes; it illuminateth the face, which, as a beacon, gives warning to all the rest of this little kingdom, man, to arm; and then the vital commoners and inland petty spirits muster me all to their captain, the heart; who, great and puffed up with his retinue, doth any deed of courage. And this valour comes of sherry. So that skill in the weapon is nothing without sack, for that sets it a-work; and learning a mere hoard of gold kept by a devil, till sack commences it and sets it in act and use. Hereof comes it that Prince Harry is valiant; for the cold blood he did naturally inherit of his father he hath, like lean, sterile, and bare land, manured, husbanded, and tilled, with excellent endeavour of drinking good, and good store of fertile sherry, that he is become very hot and valiant. If I had a thousand sons, the first human principle I would teach them should be to forswear thin potations, and to addict themselves to sack.                                                    121
    *Enter Bardolph*
How now, Bardolph?
BARDOLPH
    The army is dischargèd all and gone.
SIR JOHN  Let them go. I'll through Gloucestershire, and there will I visit Master Robert Shallow, Esquire. I have him already tempering between my finger and my thumb, and shortly will I seal with him. Come, away!
                                                    *Exeunt*

**4.3**  *Enter King Henry ⌈in his bed⌉, attended by the Earl of Warwick, Thomas Duke of Clarence, Humphrey Duke of Gloucester, ⌈and others⌉*

KING HENRY
    Now, lords, if God doth give successful end
    To this debate that bleedeth at our doors,
    We will our youth lead on to higher fields,
    And draw no swords but what are sanctified.
    Our navy is addressed, our power collected,          5
    Our substitutes in absence well invested,
    And everything lies level to our wish;
    Only we want a little personal strength,
    And pause us till these rebels now afoot
    Come underneath the yoke of government.              10
WARWICK
    Both which we doubt not but your majesty
    Shall soon enjoy.
KING HENRY              Humphrey, my son of Gloucester,
    Where is the Prince your brother?

GLOUCESTER
    I think he's gone to hunt, my lord, at Windsor.
KING HENRY
    And how accompanied?
GLOUCESTER                    I do not know, my lord.    15
KING HENRY
    Is not his brother Thomas of Clarence with him?
GLOUCESTER
    No, my good lord, he is in presence here.
CLARENCE  What would my lord and father?
KING HENRY
    Nothing but well to thee, Thomas of Clarence.
    How chance thou art not with the Prince thy brother?
    He loves thee, and thou dost neglect him, Thomas.    21
    Thou hast a better place in his affection
    Than all thy brothers. Cherish it, my boy,
    And noble offices thou mayst effect
    Of mediation, after I am dead,                       25
    Between his greatness and thy other brethren.
    Therefore omit him not, blunt not his love,
    Nor lose the good advantage of his grace
    By seeming cold or careless of his will;
    For he is gracious, if he be observed;               30
    He hath a tear for pity, and a hand
    Open as day for melting charity.
    Yet notwithstanding, being incensed, he is flint,
    As humorous as winter, and as sudden
    As flaws congealèd in the spring of day.             35
    His temper therefore must be well observed.
    Chide him for faults, and do it reverently,
    When you perceive his blood inclined to mirth;
    But being moody, give him line and scope
    Till that his passions, like a whale on ground,      40
    Confound themselves with working. Learn this,
        Thomas,
    And thou shalt prove a shelter to thy friends,
    A hoop of gold to bind thy brothers in,
    That the united vessel of their blood,
    Mingled with venom of suggestion—                    45
    As force perforce the age will pour it in—
    Shall never leak, though it do work as strong
    As aconitum or rash gunpowder.
CLARENCE
    I shall observe him with all care and love.
KING HENRY
    Why art thou not at Windsor with him, Thomas?       50
CLARENCE
    He is not there today; he dines in London.
KING HENRY
    And how accompanied? Canst thou tell that?
CLARENCE
    With Poins and other his continual followers.
KING HENRY
    Most subject is the fattest soil to weeds,
    And he, the noble image of my youth,                 55
    Is overspread with them; therefore my grief
    Stretches itself beyond the hour of death.
    The blood weeps from my heart when I do shape
    In forms imaginary th'unguided days
    And rotten times that you shall look upon             60
    When I am sleeping with my ancestors;
    For when his headstrong riot hath no curb,
    When rage and hot blood are his counsellors,
    When means and lavish manners meet together,
    O, with what wings shall his affections fly          65
    Towards fronting peril and opposed decay?

WARWICK
My gracious lord, you look beyond him quite.
The Prince but studies his companions,
Like a strange tongue, wherein, to gain the language,
'Tis needful that the most immodest word          70
Be looked upon and learnt, which once attained,
Your highness knows, comes to no further use
But to be known and hated; so, like gross terms,
The Prince will in the perfectness of time
Cast off his followers, and their memory          75
Shall as a pattern or a measure live
By which his grace must mete the lives of other,
Turning past evils to advantages.
KING HENRY
'Tis seldom when the bee doth leave her comb
In the dead carrion.
          *Enter the Earl of Westmorland*
                              Who's here? Westmorland?     80
WESTMORLAND
Health to my sovereign, and new happiness
Added to that that I am to deliver!
Prince John your son doth kiss your grace's hand.
Mowbray, the Bishop Scrope, Hastings, and all
Are brought to the correction of your law.        85
There is not now a rebel's sword unsheathed,
But peace puts forth her olive everywhere.
The manner how this action hath been borne
Here at more leisure may your highness read,
With every course in his particular.              90
          *He gives the King papers*
KING HENRY
O Westmorland, thou art a summer bird
Which ever in the haunch of winter sings
The lifting up of day.
          *Enter Harcourt*
                    Look, here's more news.
HARCOURT
From enemies heaven keep your majesty;
And when they stand against you, may they fall    95
As those that I am come to tell you of!
The Earl Northumberland and the Lord Bardolph,
With a great power of English and of Scots,
Are by the sheriff of Yorkshire overthrown.
The manner and true order of the fight            100
This packet, please it you, contains at large.
          *He gives the King papers*
KING HENRY
And wherefore should these good news make me sick?
Will fortune never come with both hands full,
But write her fair words still in foulest letters?
She either gives a stomach and no food—           105
Such are the poor in health—or else a feast,
And takes away the stomach—such are the rich,
That have abundance and enjoy it not.
I should rejoice now at this happy news,
And now my sight fails, and my brain is giddy.    110
O me! Come near me now; I am much ill.
          *He swoons*
GLOUCESTER
Comfort, your majesty!
CLARENCE                    O my royal father!
WESTMORLAND
My sovereign lord, cheer up yourself, look up.
WARWICK
Be patient, princes; you do know these fits
Are with his highness very ordinary.              115
Stand from him, give him air; he'll straight be well.

CLARENCE
No, no, he cannot long hold out these pangs.
Th'incessant care and labour of his mind
Hath wrought the mure that should confine it in
So thin that life looks through and will break out.   120
GLOUCESTER
The people fear me, for they do observe
Unfathered heirs and loathly births of nature.
The seasons change their manners, as the year
Had found some months asleep and leaped them over.
CLARENCE
The river hath thrice flowed, no ebb between,     125
And the old folk, time's doting chronicles,
Say it did so a little time before
That our great grandsire Edward sicked and died.
WARWICK
Speak lower, princes, for the King recovers.
GLOUCESTER
This apoplexy will certain be his end.            130
KING HENRY
I pray you take me up and bear me hence
Into some other chamber; softly, pray.
          ⌈*The King is carried over the stage in his bed*⌉
Let there be no noise made, my gentle friends,
Unless some dull and favourable hand
Will whisper music to my weary spirit.            135
WARWICK
Call for the music in the other room.
          ⌈*Exit one or more. Still music within*⌉
KING HENRY
Set me the crown upon my pillow here.
          ⌈*Clarence*⌉ *takes the crown* ⌈*from the King's head*⌉,
          *and sets it on his pillow*
CLARENCE
His eye is hollow, and he changes much.
          ⌈*A noise within*⌉
WARWICK
Less noise, less noise!
          *Enter Prince Harry*
PRINCE HARRY                    Who saw the Duke of Clarence?
CLARENCE
I am here, brother, full of heaviness.            140
PRINCE HARRY
How now, rain within doors, and none abroad?
How doth the King?
GLOUCESTER                    Exceeding ill.
PRINCE HARRY
Heard he the good news yet? Tell it him.
GLOUCESTER
He altered much upon the hearing it.
PRINCE HARRY If he be sick with joy, he'll recover without
physic.                                           146
WARWICK
Not so much noise, my lords! Sweet prince, speak low.
The King your father is disposed to sleep.
CLARENCE
Let us withdraw into the other room.
WARWICK
Will't please your grace to go along with us?     150
PRINCE HARRY
No, I will sit and watch here by the King.
          *Exeunt all but the King and Prince Harry*
Why doth the crown lie there upon his pillow,
Being so troublesome a bedfellow?
O polished perturbation, golden care,
That keep'st the ports of slumber open wide       155

To many a watchful night!—Sleep with it now;
Yet not so sound, and half so deeply sweet,
As he whose brow with homely biggen bound
Snores out the watch of night. O majesty,
When thou dost pinch thy bearer, thou dost sit      160
Like a rich armour worn in heat of day,
That scald'st with safety.—By his gates of breath
There lies a downy feather which stirs not.
Did he suspire, that light and weightless down
Perforce must move.—My gracious lord, my father!—
This sleep is sound indeed. This is a sleep      166
That from this golden rigol hath divorced
So many English kings.—Thy due from me
Is tears and heavy sorrows of the blood,
Which nature, love, and filial tenderness      170
Shall, O dear father, pay thee plenteously.
My due from thee is this imperial crown,
Which, as immediate from thy place and blood,
Derives itself to me.
            *He puts the crown on his head*
                        Lo where it sits,
Which God shall guard; and put the world's whole
    strength      175
Into one giant arm, it shall not force
This lineal honour from me. This from thee
Will I to mine leave, as 'tis left to me.            *Exit*
    ⌈*Music ceases.*⌉ *The King awakes*
KING HENRY
Warwick, Gloucester, Clarence!
    *Enter the Earl of Warwick, and the Dukes of
    Gloucester and Clarence*
CLARENCE                        Doth the King call?
WARWICK
What would your majesty? How fares your grace?
KING HENRY
Why did you leave me here alone, my lords?      181
CLARENCE
We left the Prince my brother here, my liege,
Who undertook to sit and watch by you.
KING HENRY
The Prince of Wales? Where is he? Let me see him.
WARWICK
This door is open; he is gone this way.      185
GLOUCESTER
He came not through the chamber where we stayed.
KING HENRY
Where is the crown? Who took it from my pillow?
WARWICK
When we withdrew, my liege, we left it here.
KING HENRY
The Prince hath ta'en it hence. Go seek him out.
Is he so hasty that he doth suppose      190
My sleep my death?
Find him, my lord of Warwick; chide him hither.
                        *Exit Warwick*
This part of his conjoins with my disease,
And helps to end me. See, sons, what things you are,
How quickly nature falls into revolt      195
When gold becomes her object!
For this the foolish over-careful fathers
Have broke their sleep with thoughts, their brains with
    care,
Their bones with industry; for this they have
Engrossèd and piled up the cankered heaps      200
Of strange-achievèd gold; for this they have
Been thoughtful to invest their sons with arts

And martial exercises; when, like the bee
Culling from every flower the virtuous sweets,
Our thighs packed with wax, our mouths with honey,
We bring it to the hive; and, like the bees,      206
Are murdered for our pains. This bitter taste
Yields his engrossments to the ending father.
        *Enter the Earl of Warwick*
Now where is he that will not stay so long
Till his friend sickness have determined me?      210
WARWICK
My lord, I found the Prince in the next room,
Washing with kindly tears his gentle cheeks
With such a deep demeanour, in great sorrow,
That tyranny, which never quaffed but blood,
Would, by beholding him, have washed his knife      215
With gentle eye-drops. He is coming hither.
KING HENRY
But wherefore did he take away the crown?
        *Enter Prince Harry with the crown*
Lo where he comes.—Come hither to me, Harry.
(*To the others*) Depart the chamber; leave us here
    alone.      *Exeunt all but the King and Prince Harry*
PRINCE HARRY
I never thought to hear you speak again.      220
KING HENRY
Thy wish was father, Harry, to that thought.
I stay too long by thee, I weary thee.
Dost thou so hunger for mine empty chair
That thou wilt needs invest thee with my honours
Before thy hour be ripe? O foolish youth,      225
Thou seek'st the greatness that will overwhelm thee!
Stay but a little, for my cloud of dignity
Is held from falling with so weak a wind
That it will quickly drop. My day is dim.
Thou hast stol'n that which after some few hours      230
Were thine without offence, and at my death
Thou hast sealed up my expectation.
Thy life did manifest thou loved'st me not,
And thou wilt have me die assured of it.
Thou hid'st a thousand daggers in thy thoughts,      235
Whom thou hast whetted on thy stony heart
To stab at half an hour of my life.
What, canst thou not forbear me half an hour?
Then get thee gone and dig my grave thyself,
And bid the merry bells ring to thine ear      240
That thou art crownèd, not that I am dead.
Let all the tears that should bedew my hearse
Be drops of balm to sanctify thy head.
Only compound me with forgotten dust.
Give that which gave thee life unto the worms.      245
Pluck down my officers, break my decrees;
For now a time is come to mock at form—
Harry the Fifth is crowned. Up, vanity!
Down, royal state! All you sage counsellors, hence!
And to the English court assemble now      250
From every region, apes of idleness!
Now, neighbour confines, purge you of your scum!
Have you a ruffian that will swear, drink, dance,
Revel the night, rob, murder, and commit
The oldest sins the newest kind of ways?      255
Be happy; he will trouble you no more.
England shall double gild his treble guilt,
England shall give him office, honour, might;
For the fifth Harry from curbed licence plucks
The muzzle of restraint, and the wild dog      260
Shall flesh his tooth on every innocent.

O my poor kingdom, sick with civil blows!
When that my care could not withhold thy riots,
What wilt thou do when riot is thy care?
O, thou wilt be a wilderness again,
Peopled with wolves, thy old inhabitants.                265
PRINCE HARRY
O pardon me, my liege! But for my tears,
The moist impediments unto my speech,
I had forestalled this dear and deep rebuke
Ere you with grief had spoke and I had heard            270
The course of it so far. There is your crown;
        ⌈He returns the crown and kneels⌉
And He that wears the crown immortally
Long guard it yours! If I affect it more
Than as your honour and as your renown,
Let me no more from this obedience rise,                275
Which my most true and inward duteous spirit
Teacheth this prostrate and exterior bending.
God witness with me, when I here came in
And found no course of breath within your majesty,
How cold it struck my heart. If I do feign,             280
O, let me in my present wildness die,
And never live to show th'incredulous world
The noble change that I have purposèd.
Coming to look on you, thinking you dead,
And dead almost, my liege, to think you were,           285
I spake unto this crown as having sense,
And thus upbraided it: 'The care on thee depending
Hath fed upon the body of my father;
Therefore thou best of gold art worst of gold.
Other, less fine in carat, is more precious,            290
Preserving life in medicine potable;
But thou, most fine, most honoured, most renowned,
Hast eat thy bearer up.' Thus, my royal liege,
Accusing it, I put it on my head,
To try with it, as with an enemy                        295
That had before my face murdered my father,
The quarrel of a true inheritor.
But if it did infect my blood with joy
Or swell my thoughts to any strain of pride,
If any rebel or vain spirit of mine                     300
Did with the least affection of a welcome
Give entertainment to the might of it,
Let God for ever keep it from my head,
And make me as the poorest vassal is,
That doth with awe and terror kneel to it.              305
KING HENRY O my son,
God put it in thy mind to take it hence,
That thou mightst win the more thy father's love,
Pleading so wisely in excuse of it!
Come hither, Harry; sit thou by my bed,                 310
And hear, I think, the very latest counsel
That ever I shall breathe.
        Prince Harry ⌈rises from kneeling and⌉ sits by
        the bed
                        God knows, my son,
By what bypaths and indirect crook'd ways
I met this crown; and I myself know well
How troublesome it sat upon my head.                    315
To thee it shall descend with better quiet,
Better opinion, better confirmation;
For all the soil of the achievement goes
With me into the earth. It seemed in me
But as an honour snatched with boist'rous hand;         320
And I had many living to upbraid

My gain of it by their assistances,
Which daily grew to quarrel and to bloodshed,
Wounding supposèd peace. All these bold fears
Thou seest with peril I have answerèd;                  325
For all my reign hath been but as a scene
Acting that argument. And now my death
Changes the mood, for what in me was purchased
Falls upon thee in a more fairer sort,
So thou the garland wear'st successively.               330
Yet though thou stand'st more sure than I could do,
Thou art not firm enough, since griefs are green,
And all thy friends—which thou must make thy
        friends—
Have but their stings and teeth newly ta'en out,
By whose fell working I was first advanced,             335
And by whose power I well might lodge a fear
To be again displaced; which to avoid
I cut them off, and had a purpose now
To lead out many to the Holy Land,
Lest rest and lying still might make them look          340
Too near unto my state. Therefore, my Harry,
Be it thy course to busy giddy minds
With foreign quarrels, that action hence borne out
May waste the memory of the former days.
More would I, but my lungs are wasted so               345
That strength of speech is utterly denied me.
How I came by the crown, O God forgive,
And grant it may with thee in true peace live!
PRINCE HARRY My gracious liege,
You won it, wore it, kept it, gave it me;                350
Then plain and right must my possession be,
Which I with more than with a common pain
'Gainst all the world will rightfully maintain.
        Enter Prince John of Lancaster ⌈followed by⌉ the
        Earl of Warwick ⌈and others⌉
KING HENRY
Look, look, here comes my John of Lancaster.
PRINCE JOHN
Health, peace, and happiness to my royal father!   355
KING HENRY
Thou bring'st me happiness and peace, son John;
But health, alack, with youthful wings is flown
From this bare withered trunk. Upon thy sight
My worldly business makes a period.
Where is my lord of Warwick?
PRINCE HARRY                    My lord of Warwick!
        ⌈Warwick comes forward to the King⌉
KING HENRY
Doth any name particular belong                          361
Unto the lodging where I first did swoon?
WARWICK
'Tis called Jerusalem, my noble lord.
KING HENRY
Laud be to God! Even there my life must end.
It hath been prophesied to me many years                365
I should not die but in Jerusalem,
Which vainly I supposed the Holy Land;
But bear me to that chamber; there I'll lie;
In that Jerusalem shall Harry die.
                Exeunt, bearing the King in his bed

5.1     Enter Shallow, ⌈Silence,⌉ Sir John Falstaff,
        Bardolph, and the Page
SHALLOW (to Sir John ) By cock and pie, you shall not
        away tonight.—What, Davy, I say!
SIR JOHN You must excuse me, Master Robert Shallow.

SHALLOW I will not excuse you; you shall not be excused;
excuses shall not be admitted; there is no excuse shall
serve; you shall not be excused.—Why, Davy!    6
    *Enter Davy*
DAVY Here, sir.
SHALLOW Davy, Davy, Davy; let me see, Davy; let me
see. William Cook—bid him come hither.—Sir John,
you shall not be excused.    10
DAVY Marry, sir, thus: those precepts cannot be served.
And again, sir: shall we sow the headland with wheat?
SHALLOW With red wheat, Davy. But for William Cook;
are there no young pigeons?
DAVY Yes, sir. Here is now the smith's note for shoeing
and plough-irons.    16
SHALLOW Let it be cast and paid. Sir John, you shall not
be excused.
DAVY Sir, a new link to the bucket must needs be had;
and, sir, do you mean to stop any of William's wages,
about the sack he lost at Hinkley Fair?    21
SHALLOW A shall answer it. Some pigeons, Davy, a couple
of short-legged hens, a joint of mutton, and any pretty
little tiny kickshaws, tell William Cook.
DAVY Doth the man of war stay all night, sir?    25
SHALLOW Yea, Davy. I will use him well; a friend i'th'
court is better than a penny in purse. Use his men well,
Davy, for they are arrant knaves, and will backbite.
DAVY No worse than they are back-bitten, sir, for they
have marvellous foul linen.    30
SHALLOW Well conceited, Davy. About thy business, Davy.
DAVY I beseech you, sir, to countenance William Visor of
Wo'ncot against Clement Perks o'th' Hill.
SHALLOW There is many complaints, Davy, against that
Visor. That Visor is an arrant knave, on my knowledge.
DAVY I grant your worship that he is a knave, sir; but
yet God forbid, sir, but a knave should have some
countenance at his friend's request. An honest man,
sir, is able to speak for himself, when a knave is not. I
have served your worship truly, sir, this eight years.
An I cannot once or twice in a quarter bear out a
knave against an honest man, I have little credit with
your worship. The knave is mine honest friend, sir;
therefore I beseech you let him be countenanced.    44
SHALLOW Go to; I say he shall have no wrong. Look
about, Davy.    *⌜Exit Davy⌝*
Where are you, Sir John? Come, off with your boots.—
Give me your hand, Master Bardolph.
BARDOLPH I am glad to see your worship.    49
SHALLOW I thank thee with all my heart, kind Master
Bardolph. *⌜To the Page⌝* And welcome, my tall fellow.—
Come, Sir John.
SIR JOHN I'll follow you, good Master Robert Shallow.
    *Exit Shallow ⌜with Silence⌝*
Bardolph, look to our horses.    54
    *Exit Bardolph ⌜with the Page⌝*
If I were sawed into quantities, I should make four
dozen of such bearded hermits' staves as Master
Shallow. It is a wonderful thing to see the semblable
coherence of his men's spirits and his. They, by
observing him, do bear themselves like foolish justices;
he, by conversing with them, is turned into a justice-
like servingman. Their spirits are so married in
conjunction, with the participation of society, that they
flock together in consent like so many wild geese. If I
had a suit to Master Shallow, I would humour his men
with the imputation of being near their master; if to
his men, I would curry with Master Shallow that no

man could better command his servants. It is certain
that either wise bearing or ignorant carriage is caught
as men take diseases, one of another; therefore let men
take heed of their company. I will devise matter enough
out of this Shallow to keep Prince Harry in continual
laughter the wearing out of six fashions—which is four
terms, or two actions—and a shall laugh without
intervallums. O, it is much that a lie with a slight oath,
and a jest with a sad brow, will do with a fellow that
never had the ache in his shoulders! O, you shall see
him laugh till his face be like a wet cloak ill laid up!
SHALLOW (*within*) Sir John!    78
SIR JOHN I come, Master Shallow; I come, Master Shallow.
    *Exit*

**5.2**    *Enter the Earl of Warwick ⌜at one door⌝, and the
    Lord Chief Justice ⌜at another door⌝*
WARWICK
How now, my Lord Chief Justice, whither away?
LORD CHIEF JUSTICE How doth the King?
WARWICK
Exceeding well: his cares are now all ended.
LORD CHIEF JUSTICE
I hope not dead.
WARWICK        He's walked the way of nature,
And to our purposes he lives no more.    5
LORD CHIEF JUSTICE
I would his majesty had called me with him.
The service that I truly did his life
Hath left me open to all injuries.
WARWICK
Indeed I think the young King loves you not.
LORD CHIEF JUSTICE
I know he doth not, and do arm myself    10
To welcome the condition of the time,
Which cannot look more hideously upon me
Than I have drawn it in my fantasy.
    *Enter Prince John of Lancaster, and the Dukes of
    Clarence and Gloucester*
WARWICK
Here come the heavy issue of dead Harry.
O, that the living Harry had the temper    15
Of he the worst of these three gentlemen!
How many nobles then should hold their places,
That must strike sail to spirits of vile sort!
LORD CHIEF JUSTICE
O God, I fear all will be overturned.
PRINCE JOHN
Good morrow, cousin Warwick, good morrow.    20
GLOUCESTER *and* CLARENCE Good morrow, cousin.
PRINCE JOHN
We meet like men that had forgot to speak.
WARWICK
We do remember, but our argument
Is all too heavy to admit much talk.
PRINCE JOHN
Well, peace be with him that hath made us heavy!    25
LORD CHIEF JUSTICE
Peace be with us, lest we be heavier!
GLOUCESTER
O good my lord, you have lost a friend indeed;
And I dare swear you borrow not that face
Of seeming sorrow—it is sure your own.
PRINCE JOHN (*to Lord Chief Justice*)
Though no man be assured what grace to find,    30
You stand in coldest expectation.
I am the sorrier; would 'twere otherwise.

CLARENCE (*to Lord Chief Justice*)
    Well, you must now speak Sir John Falstaff fair,
    Which swims against your stream of quality.
LORD CHIEF JUSTICE
    Sweet princes, what I did I did in honour,                    35
    Led by th'impartial conduct of my soul;
    And never shall you see that I will beg
    A raggèd and forestalled remission.
    If truth and upright innocency fail me,
    I'll to the King my master, that is dead,                     40
    And tell him who hath sent me after him.
            *Enter Prince Harry, as King*
WARWICK Here comes the Prince.
LORD CHIEF JUSTICE
    Good morrow, and God save your majesty!
PRINCE HARRY
    This new and gorgeous garment, majesty,
    Sits not so easy on me as you think.                          45
    Brothers, you mix your sadness with some fear.
    This is the English not the Turkish court;
    Not Amurath an Amurath succeeds,
    But Harry Harry. Yet be sad, good brothers,
    For, by my faith, it very well becomes you.                   50
    Sorrow so royally in you appears
    That I will deeply put the fashion on,
    And wear it in my heart. Why then, be sad;
    But entertain no more of it, good brothers,
    Than a joint burden laid upon us all.                         55
    For me, by heaven, I bid you be assured
    I'll be your father and your brother too.
    Let me but bear your love, I'll bear your cares.
    Yet weep that Harry's dead, and so will I;
    But Harry lives that shall convert those tears               60
    By number into hours of happiness.
PRINCE JOHN, GLOUCESTER, *and* CLARENCE
    We hope no other from your majesty.
PRINCE HARRY
    You all look strangely on me, (*to Lord Chief Justice*)
        and you most.
    You are, I think, assured I love you not.
LORD CHIEF JUSTICE
    I am assured, if I be measured rightly,                       65
    Your majesty hath no just cause to hate me.
PRINCE HARRY
    No? How might a prince of my great hopes forget
    So great indignities you laid upon me?
    What—rate, rebuke, and roughly send to prison
    Th'immediate heir of England? Was this easy?                  70
    May this be washed in Lethe and forgotten?
LORD CHIEF JUSTICE
    I then did use the person of your father.
    The image of his power lay then in me;
    And in th'administration of his law,
    Whiles I was busy for the commonwealth,                       75
    Your highness pleasèd to forget my place,
    The majesty and power of law and justice,
    The image of the King whom I presented,
    And struck me in my very seat of judgement;
    Whereon, as an offender to your father,                       80
    I gave bold way to my authority
    And did commit you. If the deed were ill,
    Be you contented, wearing now the garland,
    To have a son set your decrees at naught—
    To pluck down justice from your awe-full bench,               85
    To trip the course of law, and blunt the sword
    That guards the peace and safety of your person,

    Nay, more, to spurn at your most royal image,
    And mock your workings in a second body?
    Question your royal thoughts, make the case yours,
    Be now the father, and propose a son;                         91
    Hear your own dignity so much profaned,
    See your most dreadful laws so loosely slighted,
    Behold yourself so by a son disdained;
    And then imagine me taking your part,                         95
    And in your power soft silencing your son.
    After this cold considerance, sentence me;
    And, as you are a king, speak in your state
    What I have done that misbecame my place,
    My person, or my liege's sovereignty.                        100
PRINCE HARRY
    You are right Justice, and you weigh this well.
    Therefore still bear the balance and the sword;
    And I do wish your honours may increase
    Till you do live to see a son of mine
    Offend you and obey you as I did.                            105
    So shall I live to speak my father's words:
    'Happy am I that have a man so bold
    That dares do justice on my proper son,
    And not less happy having such a son
    That would deliver up his greatness so                       110
    Into the hands of justice.' You did commit me,
    For which I do commit into your hand
    Th'unstainèd sword that you have used to bear,
    With this remembrance: that you use the same
    With the like bold, just, and impartial spirit               115
    As you have done 'gainst me. There is my hand.
    You shall be as a father to my youth;
    My voice shall sound as you do prompt mine ear,
    And I will stoop and humble my intents
    To your well-practisèd wise directions.—                     120
    And princes all, believe me, I beseech you,
    My father is gone wild into his grave,
    For in his tomb lie my affections;
    And with his spirits sadly I survive
    To mock the expectation of the world,                        125
    To frustrate prophecies, and to raze out
    Rotten opinion, who hath writ me down
    After my seeming. The tide of blood in me
    Hath proudly flowed in vanity till now.
    Now doth it turn, and ebb back to the sea,                   130
    Where it shall mingle with the state of floods,
    And flow henceforth in formal majesty.
    Now call we our high court of Parliament,
    And let us choose such limbs of noble counsel
    That the great body of our state may go                      135
    In equal rank with the best-governed nation;
    That war, or peace, or both at once, may be
    As things acquainted and familiar to us;
    (*To Lord Chief Justice*)
    In which you, father, shall have foremost hand.
    (*To all*) Our coronation done, we will accite,             140
    As I before remembered, all our state;
    And, God consigning to my good intents,
    No prince nor peer shall have just cause to say,
    'God shorten Harry's happy life one day.'        *Exeunt*

5.3    ⌈*A table and chairs set forth.*⌉ *Enter Sir John*
       *Falstaff, Shallow, Silence, Davy* ⌈*with vessels for*
       *the table*⌉, *Bardolph, and the Page*
SHALLOW (*to Sir John*) Nay, you shall see my orchard,
    where, in an arbour, we will eat a last year's pippin
    of mine own grafting, with a dish of caraways, and so
    forth—come, cousin Silence—and then to bed.

536

SIR JOHN Fore God, you have here a goodly dwelling and
a rich.   6

SHALLOW Barren, barren, barren; beggars all, beggars all,
Sir John. Marry, good air.—Spread, Davy; spread, Davy.
    ⌈*Davy begins to spread the table*⌉
Well said, Davy.

SIR JOHN This Davy serves you for good uses; he is your
serving-man and your husband.   11

SHALLOW A good varlet, a good varlet, a very good varlet,
Sir John.—By the mass, I have drunk too much sack
at supper.—A good varlet. Now sit down, now sit
down. (*To Silence*) Come, cousin.   15

SILENCE Ah, sirrah, quoth-a, we shall
   (*sings*)
      Do nothing but eat, and make good cheer,
      And praise God for the merry year,
      When flesh is cheap and females dear,
      And lusty lads roam here and there   20
      So merrily,
      And ever among so merrily.

SIR JOHN There's a merry heart, good Master Silence! I'll
give you a health for that anon.

SHALLOW Good Master Bardolph!—Some wine, Davy.   25

DAVY ⌈*to Sir John*⌉ Sweet sir, sit. ⌈*To Bardolph*⌉ I'll be with
you anon. ⌈*To Sir John*⌉ Most sweet sir, sit. Master page,
good master page, sit.
    ⌈*All but Davy sit. Davy pours wine*⌉
Proface! What you want in meat, we'll have in drink;
but you must bear; the heart's all.   30

SHALLOW Be merry, Master Bardolph and my little soldier
there, be merry.

SILENCE (*sings*)
      Be merry, be merry, my wife has all,
      For women are shrews, both short and tall,
      'Tis merry in hall when beards wags all,   35
      And welcome merry shrovetide.
Be merry, be merry.

SIR JOHN I did not think Master Silence had been a man
of this mettle.

SILENCE Who, I? I have been merry twice and once ere
now.   41
    *Enter Davy* ⌈*with a dish of apples*⌉

DAVY There's a dish of leather-coats for you.

SHALLOW Davy!

DAVY Your worship! I'll be with you straight. ⌈*To Sir
John*⌉ A cup of wine, sir?   45

SILENCE ⌈*sings*⌉
      A cup of wine
      That's brisk and fine,
      And drink unto thee, leman mine,
      And a merry heart lives long-a.

SIR JOHN Well said, Master Silence.   50

SILENCE And we shall be merry; now comes in the sweet
o'th' night.

SIR JOHN Health and long life to you, Master Silence!
    *He drinks*

SILENCE Fill the cup and let it come. I'll pledge you a mile
to th' bottom.   55

SHALLOW Honest Bardolph, welcome! If thou want'st
anything and wilt not call, beshrew thy heart! (*To the
Page*) Welcome, my little tiny thief, and welcome indeed,
too!—I'll drink to Master Bardolph, and to all the
cavalieros about London.   60
    *He drinks*

DAVY I hope to see London once ere I die.

BARDOLPH An I might see you there, Davy!

SHALLOW By the mass, you'll crack a quart together, ha,
will you not, Master Bardolph?

BARDOLPH Yea, sir, in a pottle-pot.   65

SHALLOW By God's liggens, I thank thee. The knave will
stick by thee, I can assure thee that; a will not out;
'tis true-bred.

BARDOLPH And I'll stick by him, sir.

SHALLOW Why, there spoke a king! Lack nothing, be
merry!   71
    *One knocks at the door within*
Look who's at door there, ho! Who knocks?
    ⌈*Exit Davy*⌉
    ⌈*Silence drinks*⌉

SIR JOHN ⌈*to Silence*⌉ Why, now you have done me right!

SILENCE ⌈*sings*⌉      Do me right,
      And dub me knight—   75
      Samingo.
Is't not so?

SIR JOHN 'Tis so.

SILENCE Is't so?—Why then, say an old man can do
somewhat.   80
    ⌈*Enter Davy*⌉

DAVY An't please your worship, there's one Pistol come
from the court with news.

SIR JOHN From the court? Let him come in.
    *Enter Pistol*
How now, Pistol?

PISTOL Sir John, God save you.   85

SIR JOHN What wind blew you hither, Pistol?

PISTOL
Not the ill wind which blows no man to good.
Sweet knight, thou art now one of the greatest men in
this realm.

SILENCE By'r Lady, I think a be—but goodman Puff of
Bar'son.   91

PISTOL Puff?
Puff in thy teeth, most recreant coward base!—
Sir John, I am thy Pistol and thy friend,
And helter-skelter have I rode to thee,   95
And tidings do I bring, and lucky joys,
And golden times, and happy news of price.

SIR JOHN I pray thee now, deliver them like a man of this
world.

PISTOL
A foutre for the world and worldlings base!   100
I speak of Africa and golden joys.

SIR JOHN
O base Assyrian knight, what is thy news?
Let King Cophetua know the truth thereof.

SILENCE ⌈*singing*⌉
'And Robin Hood, Scarlet, and John.'

PISTOL
Shall dunghill curs confront the Helicons?   105
And shall good news be baffled?
Then Pistol lay thy head in Furies' lap.

SHALLOW Honest gentleman, I know not your breeding.

PISTOL Why then, lament therefor.

SHALLOW Give me pardon, sir. If, sir, you come with news
from the court, I take it there's but two ways: either
to utter them, or conceal them. I am, sir, under the
King in some authority.

PISTOL
Under which king, besonian? Speak, or die.

SHALLOW
Under King Harry.

PISTOL          Harry the Fourth, or Fifth?   115

SHALLOW
  Harry the Fourth.
PISTOL                          A foutre for thine office!
  Sir John, thy tender lambkin now is king.
  Harry the Fifth's the man. I speak the truth.
  When Pistol lies, do this, (*making the fig*) and fig me,
  Like the bragging Spaniard.
SIR JOHN                        What, is the old King dead?
PISTOL
  As nail in door. The things I speak are just.      121
SIR JOHN Away, Bardolph, saddle my horse! Master Robert
  Shallow, choose what office thou wilt in the land; 'tis
  thine. Pistol, I will double-charge thee with dignities.
BARDOLPH O joyful day!                               125
  I would not take a knighthood for my fortune.
PISTOL What, I do bring good news?
SIR JOHN (*to Davy*) Carry Master Silence to bed.
                           ⌐Exit Davy with Silence⌐
  Master Shallow—my lord Shallow—be what thou wilt,
  I am fortune's steward—get on thy boots; we'll ride
  all night.—O sweet Pistol!—Away, Bardolph!     131
                           ⌐Exit Bardolph⌐
  Come, Pistol, utter more to me, and withal devise
  something to do thyself good. Boot, boot, Master
  Shallow! I know the young King is sick for me. Let us
  take any man's horses—the laws of England are at my
  commandment. Blessed are they that have been my
  friends, and woe to my Lord Chief Justice.        137
PISTOL
  Let vultures vile seize on his lungs also!
  'Where is the life that late I led?' say they.
  Why, here it is. Welcome these pleasant days.    *Exeunt*

**5.4**    *Enter Beadles, dragging in Mistress Quickly and*
           *Doll Tearsheet*

MISTRESS QUICKLY No, thou arrant knave! I would to God
  that I might die, that I might have thee hanged. Thou
  hast drawn my shoulder out of joint.
FIRST BEADLE The constables have delivered her over to
  me; and she shall have whipping-cheer, I warrant her.
  There hath been a man or two killed about her.     6
DOLL TEARSHEET Nut-hook, nut-hook, you lie! Come on,
  I'll tell thee what, thou damned tripe-visaged rascal,
  an the child I go with do miscarry, thou wert better
  thou hadst struck thy mother, thou paper-faced villain.
MISTRESS QUICKLY O the Lord, that Sir John were come!
  He would make this a bloody day to somebody. But I
  pray God the fruit of her womb miscarry!         13
FIRST BEADLE If it do, you shall have a dozen of cushions
  again; you have but eleven now. Come, I charge you
  both go with me, for the man is dead that you and
  Pistol beat amongst you.
DOLL TEARSHEET I'll tell you what, you thin man in a
  censer, I will have you as soundly swinged for this,
  you bluebottle rogue, you filthy famished correctioner!
  If you be not swinged, I'll forswear half-kirtles.    21
FIRST BEADLE Come, come, you she knight-errant, come!
MISTRESS QUICKLY O God, that right should thus o'ercome
  might! Well, of sufferance comes ease.
DOLL TEARSHEET Come, you rogue, come; bring me to a
  justice.                                              26
MISTRESS QUICKLY Ay, come, you starved bloodhound.
DOLL TEARSHEET Goodman death, goodman bones!
MISTRESS QUICKLY Thou atomy, thou!
DOLL TEARSHEET Come, you thin thing; come, you rascal.
FIRST BEADLE Very well.                              *Exeunt*

**5.5**    *Enter* ⌐two⌐ *Grooms, strewing rushes*
FIRST GROOM More rushes, more rushes!
SECOND GROOM The trumpets have sounded twice.
⌐FIRST⌐ GROOM 'Twill be two o'clock ere they come from
  the coronation.                                 *Exeunt*
           *Enter Sir John Falstaff, Shallow, Pistol, Bardolph,*
           *and the Page*
SIR JOHN Stand here by me, Master Robert Shallow. I will
  make the King do you grace. I will leer upon him as a
  comes by, and do but mark the countenance that he
  will give me.                                        8
PISTOL God bless thy lungs, good knight.
SIR JOHN Come here, Pistol; stand behind me. (*To Shallow*)
  O, if I had had time to have made new liveries, I would
  have bestowed the thousand pound I borrowed of you!
  But 'tis no matter; this poor show doth better; this
  doth infer the zeal I had to see him.
⌐SHALLOW⌐ It doth so.                                 15
SIR JOHN It shows my earnestness of affection—
PISTOL It doth so.
SIR JOHN My devotion—
PISTOL It doth, it doth, it doth.
SIR JOHN As it were, to ride day and night, and not to
  deliberate, not to remember, not to have patience to
  shift me—                                           22
SHALLOW It is most certain.
⌐SIR JOHN⌐ But to stand stained with travel and sweating
  with desire to see him, thinking of nothing else, putting
  all affairs in oblivion, as if there were nothing else to
  be done but to see him.                            27
PISTOL 'Tis *semper idem*, for *absque hoc nihil est*: 'tis all in
  every part.
SHALLOW 'Tis so indeed.                              30
PISTOL
  My knight, I will inflame thy noble liver,
  And make thee rage.
  Thy Doll, and Helen of thy noble thoughts,
  Is in base durance and contagious prison,
  Haled thither                                      35
  By most mechanical and dirty hand.
  Rouse up Revenge from ebon den with fell Alecto's
           snake,
  For Doll is in. Pistol speaks naught but truth.
SIR JOHN I will deliver her.
           ⌐Shouts within.⌐ Trumpets sound
PISTOL
  There roared the sea, and trumpet-clangour sounds!
           *Enter King Harry the Fifth, Prince John of*
           *Lancaster, the Dukes of Clarence and Gloucester,*
           *the Lord Chief Justice,* ⌐and others⌐
SIR JOHN
  God save thy grace, King Hal, my royal Hal!       41
PISTOL
  The heavens thee guard and keep, most royal imp of
           fame!
SIR JOHN God save thee, my sweet boy!
KING HARRY
  My Lord Chief Justice, speak to that vain man.
LORD CHIEF JUSTICE (*to Sir John*)
  Have you your wits? Know you what 'tis you speak?
SIR JOHN
  My king, my Jove, I speak to thee, my heart!      46
KING HARRY
  I know thee not, old man. Fall to thy prayers.
  How ill white hairs becomes a fool and jester!
  I have long dreamt of such a kind of man,
  So surfeit-swelled, so old, and so profane;        50

But being awake, I do despise my dream.
Make less thy body hence, and more thy grace.
Leave gormandizing; know the grave doth gape
For thee thrice wider than for other men.
Reply not to me with a fool-born jest.    55
Presume not that I am the thing I was,
For God doth know, so shall the world perceive,
That I have turned away my former self;
So will I those that kept me company.
When thou dost hear I am as I have been,    60
Approach me, and thou shalt be as thou wast,
The tutor and the feeder of my riots.
Till then I banish thee, on pain of death,
As I have done the rest of my misleaders,
Not to come near our person by ten mile.    65
For competence of life I will allow you,
That lack of means enforce you not to evils;
And as we hear you do reform yourselves,
We will, according to your strengths and qualities,
Give you advancement. (*To Lord Chief Justice*) Be it
     your charge, my lord,    70
To see performed the tenor of our word. (*To his train*)
     Set on!       *Exeunt King Harry and his train*
SIR JOHN Master Shallow, I owe you a thousand pound.
SHALLOW Yea, marry, Sir John; which I beseech you to
let me have home with me.
SIR JOHN That can hardly be, Master Shallow. Do not you
grieve at this. I shall be sent for in private to him. Look
you, he must seem thus to the world. Fear not your
advancements. I will be the man yet that shall make
you great.    79
SHALLOW I cannot perceive how, unless you give me your
doublet and stuff me out with straw. I beseech you,
good Sir John, let me have five hundred of my thousand.
SIR JOHN Sir, I will be as good as my word. This that you
heard was but a colour.    84
SHALLOW A colour I fear that you will die in, Sir John.
SIR JOHN Fear no colours. Go with me to dinner. Come,
Lieutenant Pistol; come, Bardolph. I shall be sent for
soon at night.
     *Enter the Lord Chief Justice and Prince John,*
       *with officers*
LORD CHIEF JUSTICE (*to officers*)
Go carry Sir John Falstaff to the Fleet.
Take all his company along with him.    90
SIR JOHN My lord, my lord!
LORD CHIEF JUSTICE
I cannot now speak. I will hear you soon.—
Take them away.
PISTOL
*Si fortuna me tormenta, spero me contenta.*
             *Exeunt all but Prince John*
               *and Lord Chief Justice*

PRINCE JOHN
I like this fair proceeding of the King's.    95
He hath intent his wonted followers
Shall all be very well provided for,
But all are banished till their conversations
Appear more wise and modest to the world.
LORD CHIEF JUSTICE And so they are.    100
PRINCE JOHN
The King hath called his parliament, my lord.
LORD CHIEF JUSTICE He hath.
PRINCE JOHN
I will lay odds that, ere this year expire,
We bear our civil swords and native fire
As far as France. I heard a bird so sing,    105
Whose music, to my thinking, pleased the King.
Come, will you hence?        *Exeunt*

**Epilogue**    *Enter Epilogue*
EPILOGUE First my fear, then my curtsy, last my speech.
     My fear is your displeasure; my curtsy, my duty;
and my speech to beg your pardons. If you look for a
good speech now, you undo me; for what I have to
say is of mine own making, and what indeed I should
say will, I doubt, prove mine own marring. But to the
purpose, and so to the venture. Be it known to you, as
it is very well, I was lately here in the end of a
displeasing play, to pray your patience for it, and to
promise you a better. I did mean indeed to pay you
with this; which, if like an ill venture it come unluckily
home, I break, and you, my gentle creditors, lose. Here
I promised you I would be, and here I commit my body
to your mercies. Bate me some, and I will pay you
some, and, as most debtors do, promise you infinitely.
     If my tongue cannot entreat you to acquit me, will
you command me to use my legs? And yet that were
but light payment, to dance out of your debt. But a
good conscience will make any possible satisfaction,
and so would I. All the gentlewomen here have forgiven
me; if the gentlemen will not, then the gentlemen do
not agree with the gentlewomen, which was never
seen before in such an assembly.    23
     One word more, I beseech you. If you be not too
much cloyed with fat meat, our humble author will
continue the story with Sir John in it, and make you
merry with fair Catherine of France; where, for
anything I know, Falstaff shall die of a sweat—unless
already a be killed with your hard opinions. For
Oldcastle died a martyr, and this is not the man. My
tongue is weary; when my legs are too, I will bid you
good night, and so kneel down before you—but, indeed,
to pray for the Queen.    33
     ⌈*He dances, then kneels for applause.*⌉ *Exit*

## ADDITIONAL PASSAGES

Along with some substantial additions, Shakespeare probably made a number of short excisions when preparing the finished version of the play. The following, present in the Quarto but entirely or substantially omitted in the later Folio text, are the most significant:

A. AFTER 2.2.22

And God knows whether those that bawl out the ruins of thy linen shall inherit his kingdom—but the midwives say the children are not in the fault, whereupon the world increases, and kindreds are mightily strengthened.

B. AFTER 'LIQUORS!', 3.1.52

O, if this were seen,
The happiest youth, viewing his progress through,
What perils past, what crosses to ensue,
Would shut the book and sit him down and die.

C. AFTER 'FAMINE.', 3.2.309

yet lecherous as a monkey; and the whores called him 'mandrake'. A came ever in the rearward of the fashion, and sung those tunes to the overscutched hussies that he heard the carmen whistle, and sware they were his fancies or his good-nights.

# MUCH ADO ABOUT NOTHING

*Much Ado About Nothing* is not mentioned in the list of plays by Shakespeare given by Francis Meres in his *Palladis Tamia*, published in the autumn of 1598. Certain speech-prefixes of the first edition, published in 1600, suggest that as Shakespeare wrote he had in mind for the role of Dogberry the comic actor Will Kemp, who is believed to have left the Lord Chamberlain's Men during 1599. Probably Shakespeare wrote the play between summer 1598 and spring 1599.

The action is set in Sicily, where Don Pedro, Prince of Aragon, has recently defeated his half-brother, the bastard Don John, in a military engagement. Apparently reconciled, they return to the capital, Messina, as guests of the Governor, Leonato. There Count Claudio, a young nobleman serving in Don Pedro's army, falls in love with Hero, Leonato's daughter, whom Don Pedro woos on his behalf. The play's central plot, written mainly in verse, shows how Don John maliciously deceives Claudio into believing that Hero has taken a lover on the eve of her marriage, causing Claudio to repudiate her publicly, at the altar. This is a variation on an old tale that existed in many versions; it had been told in Italian verse by Ariosto, in his *Orlando Furioso* (1516, translated into English verse by Sir John Harington, 1591), in Italian prose by Matteo Bandello in his *Novelle* (1554, adapted into French by P. de Belleforest, 1569), in English prose by George Whetstone (*The Rock of Regard*, 1576), in English verse by Edmund Spenser (*The Faerie Queene*, Book 2, canto 4, 1590), and in a number of plays including Luigi Pasqualigo's *Il Fedele* (1579), adapted into English—perhaps by Anthony Munday—as *Fedele and Fortunio* (published in 1583). Shakespeare, whose plot is an independent reworking of the traditional story, seems to owe most to Ariosto and Bandello, perhaps indirectly.

Don John's deception, with its tragicomical resolution, is offset by a parallel plot written mainly in prose, portraying another, more light-hearted deception, by which Hero's cousin, Beatrice, and Benedick—friend of Don Pedro and Claudio—are tricked into acknowledging, first to themselves and then to each other, that they are in love. This part of the play seems to be of Shakespeare's invention: the juxtaposition of this clever, sophisticated, apparently unillusioned pair with the more naïve Claudio and Hero recalls Shakespeare's earlier contrast of romantic and antiromantic attitudes to love and marriage in *The Taming of the Shrew*. The play's third main strand is provided by Constable Dogberry, his partner Verges, and the Watchmen, clearly English rather than Sicilian in origin. Although Benedick and Beatrice are, technically, subordinate characters, they have dominated the imagination of both readers and playgoers.

# THE PERSONS OF THE PLAY

DON PEDRO, Prince of Aragon

BENEDICK, of Padua  
CLAUDIO, of Florence  } lords, companions of Don Pedro

BALTHASAR, attendant on Don Pedro, a singer

DON JOHN, the bastard brother of Don Pedro

BORACHIO  
CONRAD  } followers of Don John

LEONATO, Governor of Messina

HERO, his daughter

BEATRICE, an orphan, his niece

ANTONIO, an old man, brother of Leonato

MARGARET  
URSULA  } waiting-gentlewomen attendant on Hero

FRIAR Francis

DOGBERRY, the Constable in charge of the Watch

VERGES, the Headborough, Dogberry's partner

A SEXTON

WATCHMEN

A BOY, serving Benedick

Attendants and messengers

# Much Ado About Nothing

**1.1** *Enter Leonato, governor of Messina, Hero his*
*daughter, and Beatrice his niece, with a Messenger*

LEONATO I learn in this letter that Don Pedro of Aragon comes this night to Messina.

MESSENGER He is very near by this. He was not three leagues off when I left him.

LEONATO How many gentlemen have you lost in this action? 6

MESSENGER But few of any sort, and none of name.

LEONATO A victory is twice itself when the achiever brings home full numbers. I find here that Don Pedro hath bestowed much honour on a young Florentine called Claudio. 11

MESSENGER Much deserved on his part, and equally remembered by Don Pedro. He hath borne himself beyond the promise of his age, doing in the figure of a lamb the feats of a lion. He hath indeed better bettered expectation than you must expect of me to tell you how. 17

LEONATO He hath an uncle here in Messina will be very much glad of it.

MESSENGER I have already delivered him letters, and there appears much joy in him—even so much that joy could not show itself modest enough without a badge of bitterness.

LEONATO Did he break out into tears?

MESSENGER In great measure. 25

LEONATO A kind overflow of kindness, there are no faces truer than those that are so washed. How much better is it to weep at joy than to joy at weeping!

BEATRICE I pray you, is Signor Montanto returned from the wars, or no? 30

MESSENGER I know none of that name, lady. There was none such in the army, of any sort.

LEONATO What is he that you ask for, niece?

HERO My cousin means Signor Benedick of Padua.

MESSENGER O, he's returned, and as pleasant as ever he was. 36

BEATRICE He set up his bills here in Messina, and challenged Cupid at the flight; and my uncle's fool, reading the challenge, subscribed for Cupid and challenged him at the bird-bolt. I pray you, how many hath he killed and eaten in these wars? But how many hath he killed? For indeed I promised to eat all of his killing.

LEONATO Faith, niece, you tax Signor Benedick too much. But he'll be meet with you, I doubt it not. 45

MESSENGER He hath done good service, lady, in these wars.

BEATRICE You had musty victual, and he hath holp to eat it. He is a very valiant trencherman, he hath an excellent stomach. 50

MESSENGER And a good soldier too, lady.

BEATRICE And a good soldier to a lady, but what is he to a lord?

MESSENGER A lord to a lord, a man to a man, stuffed with all honourable virtues. 55

BEATRICE It is so, indeed. He is no less than a stuffed man. But for the stuffing—well, we are all mortal.

LEONATO You must not, sir, mistake my niece. There is a kind of merry war betwixt Signor Benedick and her.

They never meet but there's a skirmish of wit between them. 61

BEATRICE Alas, he gets nothing by that. In our last conflict four of his five wits went halting off, and now is the whole man governed with one, so that if he have wit enough to keep himself warm, let him bear it for a difference between himself and his horse, for it is all the wealth that he hath left to be known a reasonable creature. Who is his companion now? He hath every month a new sworn brother.

MESSENGER Is't possible? 70

BEATRICE Very easily possible. He wears his faith but as the fashion of his hat, it ever changes with the next block.

MESSENGER I see, lady, the gentleman is not in your books.

BEATRICE No. An he were, I would burn my study. But I pray you, who is his companion? Is there no young squarer now that will make a voyage with him to the devil?

MESSENGER He is most in the company of the right noble Claudio. 80

BEATRICE O Lord, he will hang upon him like a disease. He is sooner caught than the pestilence, and the taker runs presently mad. God help the noble Claudio. If he have caught the Benedick, it will cost him a thousand pound ere a be cured. 85

MESSENGER I will hold friends with you, lady.

BEATRICE Do, good friend.

LEONATO You will never run mad, niece.

BEATRICE No, not till a hot January.

MESSENGER Don Pedro is approached. 90

*Enter Don Pedro, Claudio, Benedick, Balthasar, and*
*Don John the bastard*

DON PEDRO Good Signor Leonato, are you come to meet your trouble? The fashion of the world is to avoid cost, and you encounter it.

LEONATO Never came trouble to my house in the likeness of your grace; for trouble being gone, comfort should remain, but when you depart from me, sorrow abides and happiness takes his leave. 97

DON PEDRO You embrace your charge too willingly. I think this is your daughter.

LEONATO Her mother hath many times told me so. 100

BENEDICK Were you in doubt, sir, that you asked her?

LEONATO Signor Benedick, no, for then were you a child.

DON PEDRO You have it full, Benedick. We may guess by this what you are, being a man. Truly, the lady fathers herself. Be happy, lady, for you are like an honourable father. 106

BENEDICK If Signor Leonato be her father, she would not have his head on her shoulders for all Messina, as like him as she is.

BEATRICE I wonder that you will still be talking, Signor Benedick. Nobody marks you. 111

BENEDICK What, my dear Lady Disdain! Are you yet living?

BEATRICE Is it possible disdain should die while she hath such meet food to feed it as Signor Benedick? Courtesy itself must convert to disdain if you come in her presence. 117

BENEDICK Then is courtesy a turncoat. But it is certain I am loved of all ladies, only you excepted. And I would I could find in my heart that I had not a hard heart, for truly I love none. 121

BEATRICE A dear happiness to women. They would else have been troubled with a pernicious suitor. I thank God and my cold blood I am of your humour for that. I had rather hear my dog bark at a crow than a man swear he loves me. 126

BENEDICK God keep your ladyship still in that mind. So some gentleman or other shall scape a predestinate scratched face.

BEATRICE Scratching could not make it worse an 'twere such a face as yours were. 131

BENEDICK Well, you are a rare parrot-teacher.

BEATRICE A bird of my tongue is better than a beast of yours.

BENEDICK I would my horse had the speed of your tongue, and so good a continuer. But keep your way, o' God's name. I have done. 137

BEATRICE You always end with a jade's trick. I know you of old.

DON PEDRO That is the sum of all, Leonato. Signor Claudio and Signor Benedick, my dear friend Leonato hath invited you all. I tell him we shall stay here at the least a month, and he heartily prays some occasion may detain us longer. I dare swear he is no hypocrite, but prays from his heart. 145

LEONATO If you swear, my lord, you shall not be forsworn. (To Don John) Let me bid you welcome, my lord. Being reconciled to the Prince your brother, I owe you all duty.

DON JOHN I thank you. I am not of many words, but I thank you. 151

LEONATO (to Don Pedro) Please it your grace lead on?

DON PEDRO Your hand, Leonato. We will go together.

*Exeunt all but Benedick and Claudio*

CLAUDIO Benedick, didst thou note the daughter of Signor Leonato? 155

BENEDICK I noted her not, but I looked on her.

CLAUDIO Is she not a modest young lady?

BENEDICK Do you question me as an honest man should do, for my simple true judgement, or would you have me speak after my custom, as being a professed tyrant to their sex? 161

CLAUDIO No, I pray thee speak in sober judgement.

BENEDICK Why, i'faith, methinks she's too low for a high praise, too brown for a fair praise, and too little for a great praise. Only this commendation I can afford her, that were she other than she is she were unhandsome, and being no other but as she is, I do not like her.

CLAUDIO Thou thinkest I am in sport. I pray thee tell me truly how thou likest her. 169

BENEDICK Would you buy her, that you enquire after her?

CLAUDIO Can the world buy such a jewel?

BENEDICK Yea, and a case to put it into. But speak you this with a sad brow, or do you play the flouting jack, to tell us Cupid is a good hare-finder and Vulcan a rare carpenter? Come, in what key shall a man take you to go in the song? 176

CLAUDIO In mine eye she is the sweetest lady that ever I looked on.

BENEDICK I can see yet without spectacles, and I see no such matter. There's her cousin, an she were not possessed with a fury, exceeds her as much in beauty as the first of May doth the last of December. But I hope you have no intent to turn husband, have you?

CLAUDIO I would scarce trust myself though I had sworn the contrary, if Hero would be my wife. 185

BENEDICK Is't come to this? In faith, hath not the world one man but he will wear his cap with suspicion? Shall I never see a bachelor of three-score again? Go to, i'faith, an thou wilt needs thrust thy neck into a yoke, wear the print of it, and sigh away Sundays. Look, Don Pedro is returned to seek you. 191

*Enter Don Pedro*

DON PEDRO What secret hath held you here that you followed not to Leonato's?

BENEDICK I would your grace would constrain me to tell.

DON PEDRO I charge thee on thy allegiance. 195

BENEDICK You hear, Count Claudio? I can be secret as a dumb man, I would have you think so. But on my allegiance, mark you this, on my allegiance! He is in love. With who? Now that is your grace's part. Mark how short his answer is: with Hero, Leonato's short daughter. 201

CLAUDIO If this were so, so were it uttered.

BENEDICK Like the old tale, my lord—it is not so, nor 'twas not so, but indeed, God forbid it should be so.

CLAUDIO If my passion change not shortly, God forbid it should be otherwise. 206

DON PEDRO Amen, if you love her, for the lady is very well worthy.

CLAUDIO You speak this to fetch me in, my lord.

DON PEDRO By my troth, I speak my thought. 210

CLAUDIO And in faith, my lord, I spoke mine.

BENEDICK And by my two faiths and troths, my lord, I spoke mine.

CLAUDIO That I love her, I feel.

DON PEDRO That she is worthy, I know. 215

BENEDICK That I neither feel how she should be loved nor know how she should be worthy is the opinion that fire cannot melt out of me. I will die in it at the stake.

DON PEDRO Thou wast ever an obstinate heretic in the despite of beauty. 220

CLAUDIO And never could maintain his part but in the force of his will.

BENEDICK That a woman conceived me, I thank her. That she brought me up, I likewise give her most humble thanks. But that I will have a recheat winded in my forehead, or hang my bugle in an invisible baldric, all women shall pardon me. Because I will not do them the wrong to mistrust any, I will do myself the right to trust none. And the fine is—for the which I may go the finer—I will live a bachelor. 230

DON PEDRO I shall see thee ere I die look pale with love.

BENEDICK With anger, with sickness, or with hunger, my lord; not with love. Prove that ever I lose more blood with love than I will get again with drinking, pick out mine eyes with a ballad-maker's pen and hang me up at the door of a brothel house for the sign of blind Cupid.

DON PEDRO Well, if ever thou dost fall from this faith thou wilt prove a notable argument. 239

BENEDICK If I do, hang me in a bottle like a cat, and shoot at me, and he that hits me, let him be clapped on the shoulder and called Adam.

DON PEDRO Well, as time shall try. 'In time the savage bull doth bear the yoke.' 244

BENEDICK The savage bull may, but if ever the sensible Benedick bear it, pluck off the bull's horns and set them in my forehead, and let me be vilely painted, and in

such great letters as they write 'Here is good horse to
hire' let them signify under my sign 'Here you may see
Benedick, the married man'.                                    250

CLAUDIO If this should ever happen thou wouldst be horn-
mad.

DON PEDRO Nay, if Cupid have not spent all his quiver in
Venice thou wilt quake for this shortly.

BENEDICK I look for an earthquake too, then.                   255

DON PEDRO Well, you will temporize with the hours. In
the mean time, good Signor Benedick, repair to
Leonato's, commend me to him, and tell him I will not
fail him at supper, for indeed he hath made great
preparation.                                                    260

BENEDICK I have almost matter enough in me for such an
embassage. And so I commit you—

CLAUDIO To the tuition of God, from my house if I had
it—

DON PEDRO The sixth of July,                                    265
                    Your loving friend,
                                   Benedick.

BENEDICK Nay, mock not, mock not. The body of your
discourse is sometime guarded with fragments, and the
guards are but slightly basted on neither. Ere you flout
old ends any further, examine your conscience. And
so I leave you.                                          *Exit*

CLAUDIO
My liege, your highness now may do me good.

DON PEDRO
My love is thine to teach. Teach it but how
And thou shalt see how apt it is to learn                       275
Any hard lesson that may do thee good.

CLAUDIO
Hath Leonato any son, my lord?

DON PEDRO
No child but Hero. She's his only heir.
Dost thou affect her, Claudio?

CLAUDIO                              O my lord,
When you went onward on this ended action                       280
I looked upon her with a soldier's eye,
That liked, but had a rougher task in hand
Than to drive liking to the name of love.
But now I am returned, and that war-thoughts
Have left their places vacant, in their rooms                   285
Come thronging soft and delicate desires,
All prompting me how fair young Hero is,
Saying I liked her ere I went to wars.

DON PEDRO
Thou wilt be like a lover presently,
And tire the hearer with a book of words.                       290
If thou dost love fair Hero, cherish it,
And I will break with her, and with her father,
And thou shalt have her. Was't not to this end
That thou began'st to twist so fine a story?

CLAUDIO
How sweetly you do minister to love,                            295
That know love's grief by his complexion!
But lest my liking might too sudden seem
I would have salved it with a longer treatise.

DON PEDRO
What need the bridge much broader than the flood?
The fairest grant is the necessity.                             300
Look what will serve is fit. 'Tis once: thou lovest,
And I will fit thee with the remedy.
I know we shall have revelling tonight.
I will assume thy part in some disguise,
And tell fair Hero I am Claudio.                                305

And in her bosom I'll unclasp my heart
And take her hearing prisoner with the force
And strong encounter of my amorous tale.
Then after to her father will I break,
And the conclusion is, she shall be thine.                      310
In practice let us put it presently.              *Exeunt*

**1.2**    *Enter Leonato and Antonio, an old man brother to*
           *Leonato, severally*

LEONATO How now, brother, where is my cousin, your
son? Hath he provided this music?

ANTONIO He is very busy about it. But brother, I can tell
you strange news that you yet dreamt not of.

LEONATO Are they good?                                           5

ANTONIO As the event stamps them. But they have a good
cover, they show well outward. The Prince and Count
Claudio, walking in a thick-pleached alley in mine
orchard, were thus much overheard by a man of mine:
the Prince discovered to Claudio that he loved my niece,
your daughter, and meant to acknowledge it this night
in a dance, and if he found her accordant he meant to
take the present time by the top and instantly break
with you of it.

LEONATO Hath the fellow any wit that told you this?    15

ANTONIO A good sharp fellow. I will send for him, and
question him yourself.

LEONATO No, no. We will hold it as a dream till it appear
itself. But I will acquaint my daughter withal, that
she may be the better prepared for an answer if
peradventure this be true. Go you and tell her of it.   21
        ⌐Enter attendants⌐
Cousins, you know what you have to do. O, I cry you
mercy, friend. Go you with me and I will use your
skill.—Good cousin, have a care this busy time.  *Exeunt*

**1.3**    *Enter Don John the bastard and Conrad, his*
           *companion*

CONRAD What the goodyear, my lord, why are you thus
out of measure sad?

DON JOHN There is no measure in the occasion that breeds
it, therefore the sadness is without limit.

CONRAD You should hear reason.                                   5

DON JOHN And when I have heard it, what blessing brings
it?

CONRAD If not a present remedy, at least a patient
sufferance.

DON JOHN I wonder that thou—being, as thou sayst thou
art, born under Saturn—goest about to apply a moral
medicine to a mortifying mischief. I cannot hide what
I am. I must be sad when I have cause, and smile at
no man's jests; eat when I have stomach, and wait for
no man's leisure; sleep when I am drowsy, and tend
on no man's business; laugh when I am merry, and
claw no man in his humour.                                      17

CONRAD Yea, but you must not make the full show of this
till you may do it without controlment. You have of
late stood out against your brother, and he hath ta'en
you newly into his grace, where it is impossible you
should take true root but by the fair weather that you
make yourself. It is needful that you frame the season
for your own harvest.                                           24

DON JOHN I had rather be a canker in a hedge than a
rose in his grace, and it better fits my blood to be
disdained of all than to fashion a carriage to rob love
from any. In this, though I cannot be said to be a

flattering honest man, it must not be denied but I am
a plain-dealing villain. I am trusted with a muzzle, and
enfranchised with a clog. Therefore I have decreed not
to sing in my cage. If I had my mouth I would bite. If
I had my liberty I would do my liking. In the mean
time, let me be that I am, and seek not to alter me.

CONRAD Can you make no use of your discontent?  35

DON JOHN I make all use of it, for I use it only. Who
comes here?

    *Enter Borachio*

What news, Borachio?

BORACHIO I came yonder from a great supper. The Prince
your brother is royally entertained by Leonato, and I
can give you intelligence of an intended marriage.  41

DON JOHN Will it serve for any model to build mischief
on? What is he for a fool that betroths himself to
unquietness?

BORACHIO Marry, it is your brother's right hand.  45

DON JOHN Who, the most exquisite Claudio?

BORACHIO Even he.

DON JOHN A proper squire. And who, and who? Which
way looks he?

BORACHIO Marry, on Hero, the daughter and heir of
Leonato.  51

DON JOHN A very forward March chick. How came you
to this?

BORACHIO Being entertained for a perfumer, as I was
smoking a musty room comes me the Prince and
Claudio hand in hand, in sad conference. I whipped
me behind the arras, and there heard it agreed upon
that the Prince should woo Hero for himself and, having
obtained her, give her to Count Claudio.  59

DON JOHN Come, come, let us thither. This may prove
food to my displeasure. That young start-up hath all
the glory of my overthrow. If I can cross him any way
I bless myself every way. You are both sure, and will
assist me?

CONRAD To the death, my lord.  65

DON JOHN Let us to the great supper. Their cheer is the
greater that I am subdued. Would the cook were o' my
mind. Shall we go prove what's to be done?

BORACHIO We'll wait upon your lordship.  *Exeunt*

**2.1**    *Enter Leonato, Antonio his brother, Hero his*
    *daughter, Beatrice his niece,* ⌜*Margaret, and Ursula*⌝

LEONATO Was not Count John here at supper?

ANTONIO I saw him not.

BEATRICE How tartly that gentleman looks. I never can
see him but I am heartburned an hour after.

HERO He is of a very melancholy disposition.  5

BEATRICE He were an excellent man that were made just
in the midway between him and Benedick. The one is
too like an image and says nothing, and the other too
like my lady's eldest son, evermore tattling.

LEONATO Then half Signor Benedick's tongue in Count
John's mouth, and half Count John's melancholy in
Signor Benedick's face—  12

BEATRICE With a good leg and a good foot, uncle, and
money enough in his purse—such a man would win
any woman in the world, if a could get her good will.

LEONATO By my troth, niece, thou wilt never get thee a
husband if thou be so shrewd of thy tongue.  17

ANTONIO In faith, she's too curst.

BEATRICE Too curst is more than curst. I shall lessen God's
sending that way, for it is said God sends a curst cow
short horns, but to a cow too curst he sends none.  21

LEONATO So, by being too curst, God will send you no
horns.

BEATRICE Just, if he send me no husband, for the which
blessing I am at him upon my knees every morning
and evening. Lord, I could not endure a husband with
a beard on his face. I had rather lie in the woollen.

LEONATO You may light on a husband that hath no beard.

BEATRICE What should I do with him—dress him in my
apparel and make him my waiting gentlewoman? He
that hath a beard is more than a youth, and he that
hath no beard is less than a man; and he that is more
than a youth is not for me, and he that is less than a
man, I am not for him. Therefore I will even take
sixpence in earnest of the bearherd and lead his apes
into hell.  36

LEONATO Well then, go you into hell?

BEATRICE No, but to the gate, and there will the devil
meet me like an old cuckold with horns on his head,
and say, 'Get you to heaven, Beatrice, get you to
heaven. Here's no place for you maids.' So deliver I up
my apes and away to Saint Peter fore the heavens. He
shows me where the bachelors sit, and there live we
as merry as the day is long.  44

ANTONIO (*to Hero*) Well, niece, I trust you will be ruled
by your father.

BEATRICE Yes, faith, it is my cousin's duty to make curtsy
and say, 'Father, as it please you.' But yet for all that,
cousin, let him be a handsome fellow, or else make
another curtsy and say, 'Father, as it please me.'  50

LEONATO Well, niece, I hope to see you one day fitted
with a husband.

BEATRICE Not till God make men of some other mettle
than earth. Would it not grieve a woman to be
overmastered with a piece of valiant dust?—to make
an account of her life to a clod of wayward marl? No,
uncle, I'll none. Adam's sons are my brethren, and
truly I hold it a sin to match in my kindred.  58

LEONATO (*to Hero*) Daughter, remember what I told you.
If the Prince do solicit you in that kind, you know your
answer.

BEATRICE The fault will be in the music, cousin, if you be
not wooed in good time. If the Prince be too important,
tell him there is measure in everything, and so dance
out the answer. For hear me, Hero, wooing, wedding,
and repenting is as a Scotch jig, a measure, and a
cinquepace. The first suit is hot and hasty, like a Scotch
jig—and full as fantastical; the wedding mannerly
modest, as a measure, full of state and ancientry. And
then comes repentance, and with his bad legs falls into
the cinquepace faster and faster till he sink into his
grave.  72

LEONATO Cousin, you apprehend passing shrewdly.

BEATRICE I have a good eye, uncle. I can see a church by
daylight.  75

LEONATO The revellers are entering, brother. Make good
room.

    *Enter Don Pedro, Claudio, Benedick, and Balthasar,*
    *all masked, Don John, and Borachio,* ⌜*with a*
    *drummer*⌝

DON PEDRO (*to Hero*) Lady, will you walk a bout with your
friend?

HERO So you walk softly, and look sweetly, and say
nothing, I am yours for the walk; and especially when
I walk away.  82

DON PEDRO With me in your company?

HERO I may say so when I please.

DON PEDRO And when please you to say so?    85

HERO When I like your favour; for God defend the lute should be like the case.

DON PEDRO

My visor is Philemon's roof. Within the house is Jove.

HERO

Why, then, your visor should be thatched.

DON PEDRO            Speak low if you speak love.

    *They move aside*

⌈BALTHASAR⌉ (*to Margaret*) Well, I would you did like me.

MARGARET So would not I, for your own sake, for I have many ill qualities.

⌈BALTHASAR⌉ Which is one?

MARGARET I say my prayers aloud.    94

⌈BALTHASAR⌉ I love you the better—the hearers may cry amen.

MARGARET God match me with a good dancer.

BALTHASAR Amen.

MARGARET And God keep him out of my sight when the dance is done. Answer, clerk.    100

BALTHASAR No more words. The clerk is answered.

    *They move aside*

URSULA (*to Antonio*) I know you well enough, you are Signor Antonio.

ANTONIO At a word, I am not.

URSULA I know you by the waggling of your head.    105

ANTONIO To tell you true, I counterfeit him.

URSULA You could never do him so ill-well unless you were the very man. Here's his dry hand up and down. You are he, you are he.

ANTONIO At a word, I am not.    110

URSULA Come, come, do you think I do not know you by your excellent wit? Can virtue hide itself? Go to, mum, you are he. Graces will appear, and there's an end.

    *They move aside*

BEATRICE (*to Benedick*) Will you not tell me who told you so?    115

BENEDICK No, you shall pardon me.

BEATRICE Nor will you not tell me who you are?

BENEDICK Not now.

BEATRICE That I was disdainful, and that I had my good wit out of the Hundred Merry Tales—well, this was Signor Benedick that said so.    121

BENEDICK What's he?

BEATRICE I am sure you know him well enough.

BENEDICK Not I, believe me.

BEATRICE Did he never make you laugh?    125

BENEDICK I pray you, what is he?

BEATRICE Why, he is the Prince's jester, a very dull fool. Only his gift is in devising impossible slanders. None but libertines delight in him, and the commendation is not in his wit but in his villainy, for he both pleases men and angers them, and then they laugh at him, and beat him. I am sure he is in the fleet. I would he had boarded me.

BENEDICK When I know the gentleman, I'll tell him what you say.    135

BEATRICE Do, do. He'll but break a comparison or two on me, which peradventure not marked, or not laughed at, strikes him into melancholy, and then there's a partridge wing saved, for the fool will eat no supper that night.    140

    ⌈*Music*⌉

We must follow the leaders.

BENEDICK In every good thing.

BEATRICE Nay, if they lead to any ill I will leave them at the next turning.

    *Dance. Exeunt all but Don John, Borachio, and Claudio*

DON JOHN (*aside to Borachio*) Sure my brother is amorous on Hero, and hath withdrawn her father to break with him about it. The ladies follow her, and but one visor remains.

BORACHIO (*aside to Don John*) And that is Claudio. I know him by his bearing.    150

DON JOHN Are not you Signor Benedick?

CLAUDIO You know me well. I am he.

DON JOHN Signor, you are very near my brother in his love. He is enamoured on Hero. I pray you dissuade him from her. She is no equal for his birth. You may do the part of an honest man in it.    156

CLAUDIO How know you he loves her?

DON JOHN I heard him swear his affection.

BORACHIO So did I, too, and he swore he would marry her tonight.    160

DON JOHN Come, let us to the banquet.

               *Exeunt all but Claudio*

CLAUDIO

Thus answer I in name of Benedick,

But hear these ill news with the ears of Claudio.

'Tis certain so, the Prince woos for himself.

Friendship is constant in all other things    165

Save in the office and affairs of love.

Therefore all hearts in love use their own tongues.

Let every eye negotiate for itself,

And trust no agent; for beauty is a witch

Against whose charms faith melteth into blood.    170

This is an accident of hourly proof,

Which I mistrusted not. Farewell, therefore, Hero.

    *Enter Benedick*

BENEDICK Count Claudio?

CLAUDIO Yea, the same.

BENEDICK Come, will you go with me?    175

CLAUDIO Whither?

BENEDICK Even to the next willow, about your own business, County. What fashion will you wear the garland of? About your neck, like an usurer's chain? Or under your arm, like a lieutenant's scarf? You must wear it one way, for the Prince hath got your Hero.

CLAUDIO I wish him joy of her.    182

BENEDICK Why, that's spoken like an honest drover; so they sell bullocks. But did you think the Prince would have served you thus?    185

CLAUDIO I pray you leave me.

BENEDICK Ho, now you strike like the blind man—'twas the boy that stole your meat, and you'll beat the post.

CLAUDIO If it will not be, I'll leave you.        *Exit*

BENEDICK Alas, poor hurt fowl, now will he creep into sedges. But that my Lady Beatrice should know me, and not know me! The Prince's fool! Ha, it may be I go under that title because I am merry. Yea, but so I am apt to do myself wrong. I am not so reputed. It is the base, though bitter, disposition of Beatrice that puts the world into her person, and so gives me out. Well, I'll be revenged as I may.    197

    *Enter Don Pedro the Prince*

DON PEDRO Now, signor, where's the Count? Did you see him?

BENEDICK Troth, my lord, I have played the part of Lady Fame. I found him here as melancholy as a lodge in a warren. I told him—and I think I told him true—that your grace had got the good will of this young lady, and I offered him my company to a willow tree, either to make him a garland, as being forsaken, or to bind him up a rod, as being worthy to be whipped.    206

DON PEDRO To be whipped—what's his fault?

BENEDICK The flat transgression of a schoolboy who, being overjoyed with finding a bird's nest, shows it his companion, and he steals it.    210

DON PEDRO Wilt thou make a trust a transgression? The transgression is in the stealer.

BENEDICK Yet it had not been amiss the rod had been made, and the garland too, for the garland he might have worn himself, and the rod he might have bestowed on you, who, as I take it, have stolen his bird's nest.

DON PEDRO I will but teach them to sing, and restore them to the owner.

BENEDICK If their singing answer your saying, by my faith you say honestly.    220

DON PEDRO The Lady Beatrice hath a quarrel to you. The gentleman that danced with her told her she is much wronged by you.

BENEDICK O, she misused me past the endurance of a block. An oak but with one green leaf on it would have answered her. My very visor began to assume life and scold with her. She told me—not thinking I had been myself—that I was the Prince's jester, that I was duller than a great thaw, huddling jest upon jest with such impossible conveyance upon me that I stood like a man at a mark, with a whole army shooting at me. She speaks poniards, and every word stabs. If her breath were as terrible as her terminations, there were no living near her, she would infect to the North Star. I would not marry her though she were endowed with all that Adam had left him before he transgressed. She would have made Hercules have turned spit, yea, and have cleft his club to make the fire, too. Come, talk not of her. You shall find her the infernal Ate in good apparel. I would to God some scholar would conjure her, for certainly, while she is here a man may live as quiet in hell as in a sanctuary, and people sin upon purpose because they would go thither, so indeed all disquiet, horror, and perturbation follows her.

*Enter Claudio and Beatrice, ⌈and Leonato with Hero⌉*

DON PEDRO Look, here she comes.    245

BENEDICK Will your grace command me any service to the world's end? I will go on the slightest errand now to the Antipodes that you can devise to send me on. I will fetch you a tooth-picker now from the furthest inch of Asia, bring you the length of Prester John's foot, fetch you a hair off the Great Cham's beard, do you any embassage to the pigmies, rather than hold three words' conference with this harpy. You have no employment for me?

DON PEDRO None but to desire your good company.    255

BENEDICK O God, sir, here's a dish I love not. I cannot endure my Lady Tongue.      *Exit*

DON PEDRO Come, lady, come, you have lost the heart of Signor Benedick.

BEATRICE Indeed, my lord, he lent it me a while, and I gave him use for it, a double heart for his single one. Marry, once before he won it of me, with false dice. Therefore your grace may well say I have lost it.

DON PEDRO You have put him down, lady, you have put him down.    265

BEATRICE So I would not he should do me, my lord, lest I should prove the mother of fools. I have brought Count Claudio, whom you sent me to seek.

DON PEDRO Why, how now, Count, wherefore are you sad?    270

CLAUDIO Not sad, my lord.

DON PEDRO How then? Sick?

CLAUDIO Neither, my lord.

BEATRICE The Count is neither sad, nor sick, nor merry, nor well, but civil count, civil as an orange, and something of that jealous complexion.    276

DON PEDRO I'faith, lady, I think your blazon to be true, though I'll be sworn, if he be so, his conceit is false. Here, Claudio, I have wooed in thy name, and fair Hero is won. I have broke with her father and his good will obtained. Name the day of marriage, and God give thee joy.    282

LEONATO Count, take of me my daughter, and with her my fortunes. His grace hath made the match, and all grace say amen to it.    285

BEATRICE Speak, Count, 'tis your cue.

CLAUDIO Silence is the perfectest herald of joy. I were but little happy if I could say how much. (*To Hero*) Lady, as you are mine, I am yours. I give away myself for you, and dote upon the exchange.    290

BEATRICE (*to Hero*) Speak, cousin. Or, if you cannot, stop his mouth with a kiss, and let not him speak, neither.

DON PEDRO In faith, lady, you have a merry heart.

BEATRICE Yea, my lord, I thank it. Poor fool, it keeps on the windy side of care.—My cousin tells him in his ear that he is in her heart.    296

CLAUDIO And so she doth, cousin.

BEATRICE Good Lord, for alliance! Thus goes everyone to the world but I, and I am sunburnt. I may sit in a corner and cry 'Heigh-ho for a husband'.    300

DON PEDRO Lady Beatrice, I will get you one.

BEATRICE I would rather have one of your father's getting. Hath your grace ne'er a brother like you? Your father got excellent husbands if a maid could come by them.

DON PEDRO Will you have me, lady?    305

BEATRICE No, my lord, unless I might have another for working days. Your grace is too costly to wear every day. But I beseech your grace, pardon me. I was born to speak all mirth and no matter.    309

DON PEDRO Your silence most offends me, and to be merry best becomes you; for out o' question, you were born in a merry hour.

BEATRICE No, sure, my lord, my mother cried. But then there was a star danced, and under that was I born. (*To Hero and Claudio*) Cousins, God give you joy.    315

LEONATO Niece, will you look to those things I told you of?

BEATRICE I cry you mercy, uncle. (*To Don Pedro*) By your grace's pardon.      *Exit Beatrice*

DON PEDRO By my troth, a pleasant-spirited lady.    320

LEONATO There's little of the melancholy element in her, my lord. She is never sad but when she sleeps, and not ever sad then; for I have heard my daughter say she hath often dreamt of unhappiness and waked herself with laughing.    325

DON PEDRO She cannot endure to hear tell of a husband.

LEONATO O, by no means. She mocks all her wooers out of suit.

DON PEDRO She were an excellent wife for Benedick.

LEONATO O Lord, my lord, if they were but a week married they would talk themselves mad.    331

DON PEDRO County Claudio, when mean you to go to church?

CLAUDIO Tomorrow, my lord. Time goes on crutches till love have all his rites.                                    335

LEONATO Not till Monday, my dear son, which is hence a just sevennight, and a time too brief, too, to have all things answer my mind.

DON PEDRO Come, you shake the head at so long a breathing, but I warrant thee, Claudio, the time shall not go dully by us. I will in the interim undertake one of Hercules' labours, which is to bring Signor Benedick and the Lady Beatrice into a mountain of affection th'one with th'other. I would fain have it a match, and I doubt not but to fashion it, if you three will but minister such assistance as I shall give you direction.

LEONATO My lord, I am for you, though it cost me ten nights' watchings.

CLAUDIO And I, my lord.

DON PEDRO And you too, gentle Hero?                       350

HERO I will do any modest office, my lord, to help my cousin to a good husband.

DON PEDRO And Benedick is not the unhopefullest husband that I know. Thus far can I praise him: he is of a noble strain, of approved valour and confirmed honesty. I will teach you how to humour your cousin that she shall fall in love with Benedick, and I, with your two helps, will so practise on Benedick that, in despite of his quick wit and his queasy stomach, he shall fall in love with Beatrice. If we can do this, Cupid is no longer an archer; his glory shall be ours, for we are the only love-gods. Go in with me, and I will tell you my drift.
                                                   *Exeunt*

**2.2    Enter Don John and Borachio**

DON JOHN It is so. The Count Claudio shall marry the daughter of Leonato.

BORACHIO Yea, my lord, but I can cross it.

DON JOHN Any bar, any cross, any impediment will be medicinable to me. I am sick in displeasure to him, and whatsoever comes athwart his affection ranges evenly with mine. How canst thou cross this marriage?       7

BORACHIO Not honestly, my lord, but so covertly that no dishonesty shall appear in me.

DON JOHN Show me briefly how.                             10

BORACHIO I think I told your lordship a year since how much I am in the favour of Margaret, the waiting gentlewoman to Hero.

DON JOHN I remember.

BORACHIO I can at any unseasonable instant of the night appoint her to look out at her lady's chamber window.

DON JOHN What life is in that to be the death of this marriage?                                            18

BORACHIO The poison of that lies in you to temper. Go you to the Prince your brother. Spare not to tell him that he hath wronged his honour in marrying the renowned Claudio—whose estimation do you mightily hold up—to a contaminated stale, such a one as Hero.

DON JOHN What proof shall I make of that?                  24

BORACHIO Proof enough to misuse the Prince, to vex Claudio, to undo Hero, and kill Leonato. Look you for any other issue?

DON JOHN Only to despite them I will endeavour anything.

BORACHIO Go then. Find me a meet hour to draw Don Pedro and the Count Claudio alone. Tell them that you know that Hero loves me. Intend a kind of zeal both to the Prince and Claudio as in love of your brother's honour who hath made this match, and his friend's reputation who is thus like to be cozened with the semblance of a maid, that you have discovered thus. They will scarcely believe this without trial. Offer them instances, which shall bear no less likelihood than to see me at her chamber window, hear me call Margaret Hero, hear Margaret term me Claudio. And bring them to see this the very night before the intended wedding, for in the mean time I will so fashion the matter that Hero shall be absent, and there shall appear such seeming truth of Hero's disloyalty that jealousy shall be called assurance, and all the preparation over-thrown.                                              45

DON JOHN Grow this to what adverse issue it can, I will put it in practice. Be cunning in the working this, and thy fee is a thousand ducats.

BORACHIO Be you constant in the accusation, and my cunning shall not shame me.                              50

DON JOHN I will presently go learn their day of marriage.
                                                   *Exeunt*

**2.3    Enter Benedick**

BENEDICK Boy!

    ⌈*Enter Boy*⌉

BOY Signor?

BENEDICK In my chamber window lies a book. Bring it hither to me in the orchard.

BOY I am here already, sir.                                5

BENEDICK I know that, but I would have thee hence and here again.                               ⌈*Exit Boy*⌉
    I do much wonder that one man, seeing how much another man is a fool when he dedicates his behaviours to love, will, after he hath laughed at such shallow follies in others, become the argument of his own scorn by falling in love. And such a man is Claudio. I have known when there was no music with him but the drum and the fife, and now had he rather hear the tabor and the pipe. I have known when he would have walked ten mile afoot to see a good armour, and now will he lie ten nights awake carving the fashion of a new doublet. He was wont to speak plain and to the purpose, like an honest man and a soldier, and now is he turned orthography. His words are a very fantastical banquet, just so many strange dishes. May I be so converted, and see with these eyes? I cannot tell. I think not. I will not be sworn but love may transform me to an oyster, but I'll take my oath on it, till he have made an oyster of me he shall never make me such a fool. One woman is fair, yet I am well. Another is wise, yet I am well. Another virtuous, yet I am well. But till all graces be in one woman, one woman shall not come in my grace. Rich she shall be, that's certain. Wise, or I'll none. Virtuous, or I'll never cheapen her. Fair, or I'll never look on her. Mild, or come not near me. Noble, or not I for an angel. Of good discourse, an excellent musician, and her hair shall be of what colour it please God. Ha! The Prince and Monsieur Love. I will hide me in the arbour.                               35

    *He hides.*
    *Enter Don Pedro the Prince, Leonato, and Claudio*

DON PEDRO Come, shall we hear this music?

CLAUDIO
    Yea, my good lord. How still the evening is,
    As hushed on purpose to grace harmony.

DON PEDRO (*aside*)
    See you where Benedick hath hid himself?

CLAUDIO (*aside*)
　　O, very well, my lord. The music ended,　　　　40
　　We'll fit the hid-fox with a pennyworth.
　　　　*Enter Balthasar with music*
DON PEDRO
　　Come, Balthasar, we'll hear that song again.
BALTHASAR
　　O good my lord, tax not so bad a voice
　　To slander music any more than once.
DON PEDRO
　　It is the witness still of excellency　　　　　　45
　　To put a strange face on his own perfection.
　　I pray thee sing, and let me woo no more.
BALTHASAR
　　Because you talk of wooing I will sing,
　　Since many a wooer doth commence his suit
　　To her he thinks not worthy, yet he woos,　　50
　　Yet will he swear he loves.
DON PEDRO　　　　　　　　　　Nay pray thee, come;
　　Or if thou wilt hold longer argument,
　　Do it in notes.
BALTHASAR　　　　Note this before my notes:
　　There's not a note of mine that's worth the noting.
DON PEDRO
　　Why, these are very crotchets that he speaks—　55
　　Note notes, forsooth, and nothing!
　　　　*The accompaniment begins*
BENEDICK Now, divine air! Now is his soul ravished. Is it
　　not strange that sheep's guts should hale souls out of
　　men's bodies? Well, a horn for my money, when all's
　　done.　　　　　　　　　　　　　　　　　　　　60

BALTHASAR (*sings*)
　　　　Sigh no more, ladies, sigh no more.
　　　　　　Men were deceivers ever,
　　　　One foot in sea, and one on shore,
　　　　　　To one thing constant never.
　　　　Then sigh not so, but let them go,　　　　65
　　　　　　And be you blithe and bonny,
　　　　Converting all your sounds of woe
　　　　　　Into hey nonny, nonny.

　　　　Sing no more ditties, sing no more
　　　　　　Of dumps so dull and heavy.
　　　　The fraud of men was ever so　　　　　　70
　　　　　　Since summer first was leafy.
　　　　Then sigh not so, but let them go,
　　　　　　And be you blithe and bonny,
　　　　Converting all your sounds of woe　　　　75
　　　　　　Into hey nonny, nonny.

DON PEDRO By my troth, a good song.
BALTHASAR And an ill singer, my lord.
DON PEDRO Ha, no, no, faith. Thou singest well enough
　　for a shift.　　　　　　　　　　　　　　　　80
BENEDICK (*aside*) An he had been a dog that should have
　　howled thus, they would have hanged him; and I pray
　　God his bad voice bode no mischief. I had as lief have
　　heard the night-raven, come what plague could have
　　come after it.　　　　　　　　　　　　　　　　85
DON PEDRO Yea, marry, dost thou hear, Balthasar? I pray
　　thee get us some excellent music, for tomorrow night
　　we would have it at the Lady Hero's chamber window.
BALTHASAR The best I can, my lord.　　　　　　*Exit*
DON PEDRO Do so. Farewell. Come hither, Leonato. What
　　was it you told me of today, that your niece Beatrice
　　was in love with Signor Benedick?　　　　　　92

CLAUDIO (*aside*) O, ay, stalk on, stalk on. The fowl sits.—
　　I did never think that lady would have loved any man.
LEONATO No, nor I neither. But most wonderful that she
　　should so dote on Signor Benedick, whom she hath in
　　all outward behaviours seemed ever to abhor.　　97
BENEDICK (*aside*) Is't possible? Sits the wind in that corner?
LEONATO By my troth, my lord, I cannot tell what to
　　think of it. But that she loves him with an enraged
　　affection, it is past the infinite of thought.　　　101
DON PEDRO Maybe she doth but counterfeit.
CLAUDIO Faith, like enough.
LEONATO O God! Counterfeit? There was never counterfeit
　　of passion came so near the life of passion as she
　　discovers it.　　　　　　　　　　　　　　　　106
DON PEDRO Why, what effects of passion shows she?
CLAUDIO (*aside*) Bait the hook well. This fish will bite.
LEONATO What effects, my lord? She will sit you—you
　　heard my daughter tell you how.　　　　　　110
CLAUDIO She did indeed.
DON PEDRO How, how, I pray you? You amaze me. I
　　would have thought her spirit had been invincible
　　against all assaults of affection.
LEONATO I would have sworn it had, my lord, especially
　　against Benedick.　　　　　　　　　　　　116
BENEDICK (*aside*) I should think this a gull, but that the
　　white-bearded fellow speaks it. Knavery cannot, sure,
　　hide himself in such reverence.
CLAUDIO (*aside*) He hath ta'en th'infection. Hold it up.
DON PEDRO Hath she made her affection known to
　　Benedick?　　　　　　　　　　　　　　　　122
LEONATO No, and swears she never will. That's her
　　torment.
CLAUDIO 'Tis true, indeed, so your daughter says. 'Shall
　　I,' says she, 'that have so oft encountered him with
　　scorn, write to him that I love him?'　　　　127
LEONATO This says she now when she is beginning to
　　write to him, for she'll be up twenty times a night, and
　　there will she sit in her smock till she have writ a sheet
　　of paper. My daughter tells us all.　　　　131
CLAUDIO Now you talk of a sheet of paper, I remember a
　　pretty jest your daughter told us of.
LEONATO O, when she had writ it and was reading it over,
　　she found Benedick and Beatrice between the sheet.
CLAUDIO That.　　　　　　　　　　　　　136
LEONATO O, she tore the letter into a thousand halfpence,
　　railed at herself that she should be so immodest to
　　write to one that she knew would flout her. 'I measure
　　him,' says she, 'by my own spirit, for I should flout
　　him if he writ to me, yea, though I love him I should.'
CLAUDIO Then down upon her knees she falls, weeps,
　　sobs, beats her heart, tears her hair, prays, curses, 'O
　　sweet Benedick, God give me patience.'　　　144
LEONATO She doth indeed, my daughter says so, and the
　　ecstasy hath so much overborne her that my daughter
　　is sometime afeard she will do a desperate outrage to
　　herself. It is very true.
DON PEDRO It were good that Benedick knew of it by some
　　other, if she will not discover it.　　　　　150
CLAUDIO To what end? He would make but a sport of it
　　and torment the poor lady worse.
DON PEDRO An he should, it were an alms to hang him.
　　She's an excellent sweet lady, and, out of all suspicion,
　　she is virtuous.　　　　　　　　　　　　155
CLAUDIO And she is exceeding wise.
DON PEDRO In everything but in loving Benedick.

LEONATO O my lord, wisdom and blood combating in so tender a body, we have ten proofs to one that blood hath the victory. I am sorry for her, as I have just cause, being her uncle and her guardian.          161

DON PEDRO I would she had bestowed this dotage on me. I would have doffed all other respects and made her half myself. I pray you tell Benedick of it, and hear what a will say.          165

LEONATO Were it good, think you?

CLAUDIO Hero thinks surely she will die, for she says she will die if he love her not, and she will die ere she make her love known, and she will die if he woo her, rather than she will bate one breath of her accustomed crossness.          171

DON PEDRO She doth well. If she should make tender of her love 'tis very possible he'll scorn it, for the man, as you know all, hath a contemptible spirit.

CLAUDIO He is a very proper man.          175

DON PEDRO He hath indeed a good outward happiness.

CLAUDIO Before God; and in my mind, very wise.

DON PEDRO He doth indeed show some sparks that are like wit.

CLAUDIO And I take him to be valiant.          180

DON PEDRO As Hector, I assure you; and in the managing of quarrels you may say he is wise, for either he avoids them with great discretion or undertakes them with a most Christianlike fear.

LEONATO If he do fear God, a must necessarily keep peace. If he break the peace, he ought to enter into a quarrel with fear and trembling.          187

DON PEDRO And so will he do, for the man doth fear God, howsoever it seems not in him by some large jests he will make. Well, I am sorry for your niece. Shall we go seek Benedick and tell him of her love?          191

CLAUDIO Never tell him, my lord. Let her wear it out with good counsel.

LEONATO Nay, that's impossible. She may wear her heart out first.          195

DON PEDRO Well, we will hear further of it by your daughter. Let it cool the while. I love Benedick well, and I could wish he would modestly examine himself to see how much he is unworthy so good a lady.

LEONATO My lord, will you walk? Dinner is ready.          200

CLAUDIO (aside) If he do not dote on her upon this, I will never trust my expectation.

DON PEDRO (aside) Let there be the same net spread for her, and that must your daughter and her gentlewomen carry. The sport will be when they hold one an opinion of another's dotage, and no such matter. That's the scene that I would see, which will be merely a dumb show. Let us send her to call him in to dinner.          208

*Exeunt Don Pedro, Claudio, and Leonato*

BENEDICK (coming forward) This can be no trick. The conference was sadly borne. They have the truth of this from Hero. They seem to pity the lady. It seems her affections have their full bent. Love me! Why, it must be requited. I hear how I am censured. They say I will bear myself proudly if I perceive the love come from her. They say too that she will rather die than give any sign of affection. I did never think to marry. I must not seem proud. Happy are they that hear their detractions and can put them to mending. They say the lady is fair. 'Tis a truth, I can bear them witness. And virtuous—'tis so, I cannot reprove it. And wise, but for loving me. By my troth, it is no addition to her wit—nor no great argument of her folly, for I will be horribly in love with her. I may chance have some odd quirks and remnants of wit broken on me because I have railed so long against marriage; but doth not the appetite alter? A man loves the meat in his youth that he cannot endure in his age. Shall quips and sentences and these paper bullets of the brain awe a man from the career of his humour? No. The world must be peopled. When I said I would die a bachelor, I did not think I should live till I were married. Here comes Beatrice.          232

*Enter Beatrice*

By this day, she's a fair lady. I do spy some marks of love in her.

BEATRICE Against my will I am sent to bid you come in to dinner.          236

BENEDICK

Fair Beatrice, I thank you for your pains.

BEATRICE I took no more pains for those thanks than you take pains to thank me. If it had been painful I would not have come.          240

BENEDICK You take pleasure, then, in the message?

BEATRICE Yea, just so much as you may take upon a knife's point and choke a daw withal. You have no stomach, signor? Fare you well.          *Exit*

BENEDICK Ha! 'Against my will I am sent to bid you come in to dinner.' There's a double meaning in that. 'I took no more pains for those thanks than you took pains to thank me.' That's as much as to say 'Any pains that I take for you is as easy as thanks.'—If I do not take pity of her I am a villain. If I do not love her I am a Jew. I will go get her picture.          *Exit*

3.1     *Enter Hero and two gentlewomen, Margaret and Ursula*

HERO

Good Margaret, run thee to the parlour.
There shalt thou find my cousin Beatrice
Proposing with the Prince and Claudio.
Whisper her ear, and tell her I and Ursula
Walk in the orchard, and our whole discourse          5
Is all of her. Say that thou overheard'st us,
And bid her steal into the pleachèd bower
Where honeysuckles, ripened by the sun,
Forbid the sun to enter—like favourites
Made proud by princes, that advance their pride          10
Against that power that bred it. There will she hide her
To listen our propose. This is thy office.
Bear thee well in it, and leave us alone.

MARGARET

I'll make her come, I warrant you, presently.          *Exit*

HERO

Now, Ursula, when Beatrice doth come,          15
As we do trace this alley up and down
Our talk must only be of Benedick.
When I do name him, let it be thy part
To praise him more than ever man did merit.
My talk to thee must be how Benedick          20
Is sick in love with Beatrice. Of this matter
Is little Cupid's crafty arrow made,
That only wounds by hearsay.

*Enter Beatrice*

                              Now begin,
For look where Beatrice like a lapwing runs
Close by the ground to hear our conference.          25

URSULA

The pleasant'st angling is to see the fish
Cut with her golden oars the silver stream
And greedily devour the treacherous bait.

So angle we for Beatrice, who even now
Is couchèd in the woodbine coverture.                              30
Fear you not my part of the dialogue.

HERO
Then go we near her, that her ear lose nothing
Of the false-sweet bait that we lay for it.—
*They approach Beatrice's hiding-place*
No, truly, Ursula, she is too disdainful.
I know her spirits are as coy and wild                             35
As haggards of the rock.

URSULA                              But are you sure
That Benedick loves Beatrice so entirely?

HERO
So says the Prince and my new trothèd lord.

URSULA
And did they bid you tell her of it, madam?

HERO
They did entreat me to acquaint her of it,                         40
But I persuaded them, if they loved Benedick,
To wish him wrestle with affection
And never to let Beatrice know of it.

URSULA
Why did you so? Doth not the gentleman
Deserve as full as fortunate a bed                                 45
As ever Beatrice shall couch upon?

HERO
O god of love! I know he doth deserve
As much as may be yielded to a man.
But nature never framed a woman's heart
Of prouder stuff than that of Beatrice.                            50
Disdain and scorn ride sparkling in her eyes,
Misprising what they look on, and her wit
Values itself so highly that to her
All matter else seems weak. She cannot love,
Nor take no shape nor project of affection,                        55
She is so self-endearèd.

URSULA                              Sure, I think so.
And therefore certainly it were not good
She knew his love, lest she'll make sport at it.

HERO
Why, you speak truth. I never yet saw man,
How wise, how noble, young, how rarely featured,    60
But she would spell him backward. If fair-faced,
She would swear the gentleman should be her sister.
If black, why nature, drawing of an antic,
Made a foul blot. If tall, a lance ill headed;
If low, an agate very vilely cut;                                  65
If speaking, why, a vane blown with all winds;
If silent, why, a block movèd with none.
So turns she every man the wrong side out,
And never gives to truth and virtue that
Which simpleness and merit purchaseth.                             70

URSULA
Sure, sure, such carping is not commendable.

HERO
No, not to be so odd and from all fashions
As Beatrice is cannot be commendable.
But who dare tell her so? If I should speak
She would mock me into air, O, she would laugh me
Out of myself, press me to death with wit.                         76
Therefore let Benedick, like covered fire,
Consume away in sighs, waste inwardly.
It were a better death than die with mocks,
Which is as bad as die with tickling.                              80

URSULA
Yet tell her of it, hear what she will say.

HERO
No. Rather I will go to Benedick
And counsel him to fight against his passion.
And truly, I'll devise some honest slanders
To stain my cousin with. One doth not know           85
How much an ill word may empoison liking.

URSULA
O, do not do your cousin such a wrong.
She cannot be so much without true judgement,
Having so swift and excellent a wit
As she is prized to have, as to refuse                             90
So rare a gentleman as Signor Benedick.

HERO
He is the only man of Italy,
Always excepted my dear Claudio.

URSULA
I pray you be not angry with me, madam,
Speaking my fancy. Signor Benedick,                                95
For shape, for bearing, argument, and valour
Goes foremost in report through Italy.

HERO
Indeed, he hath an excellent good name.

URSULA
His excellence did earn it ere he had it.
When are you married, madam?                                       100

HERO
Why, every day, tomorrow. Come, go in.
I'll show thee some attires and have thy counsel
Which is the best to furnish me tomorrow.

URSULA *(aside)*
She's limed, I warrant you. We have caught her,
madam.

HERO *(aside)*
If it prove so, then loving goes by haps.                          105
Some Cupid kills with arrows, some with traps.
*Exeunt Hero and Ursula*

BEATRICE *(coming forward)*
What fire is in mine ears? Can this be true?
Stand I condemned for pride and scorn so much?
Contempt, farewell; and maiden pride, adieu.
No glory lives behind the back of such.                            110
And, Benedick, love on. I will requite thee,
Taming my wild heart to thy loving hand.
If thou dost love, my kindness shall incite thee
To bind our loves up in a holy band.
For others say thou dost deserve, and I                            115
Believe it better than reportingly.              *Exit*

**3.2**   *Enter Don Pedro the Prince, Claudio, Benedick, and*
         *Leonato*

DON PEDRO I do but stay till your marriage be consummate, and then go I toward Aragon.

CLAUDIO I'll bring you thither, my lord, if you'll vouchsafe
me.

DON PEDRO Nay, that would be as great a soil in the new
gloss of your marriage as to show a child his new coat
and forbid him to wear it. I will only be bold with
Benedick for his company, for from the crown of his
head to the sole of his foot he is all mirth. He hath
twice or thrice cut Cupid's bow-string, and the little
hangman dare not shoot at him. He hath a heart as
sound as a bell, and his tongue is the clapper, for what
his heart thinks his tongue speaks.

BENEDICK Gallants, I am not as I have been.

LEONATO So say I. Methinks you are sadder.                         15

CLAUDIO I hope he be in love.

DON PEDRO Hang him, truant! There's no true drop of
blood in him to be truly touched with love. If he be
sad, he wants money.

BENEDICK I have the toothache.                              20

DON PEDRO Draw it.

BENEDICK Hang it.

CLAUDIO You must hang it first and draw it afterwards.

DON PEDRO What? Sigh for the toothache?

LEONATO Where is but a humour or a worm.                    25

BENEDICK Well, everyone can master a grief but he that
has it.

CLAUDIO Yet say I he is in love.

DON PEDRO There is no appearance of fancy in him, unless
it be a fancy that he hath to strange disguises, as to
be a Dutchman today, a Frenchman tomorrow, or in
the shape of two countries at once, as a German from
the waist downward, all slops, and a Spaniard from
the hip upward, no doublet. Unless he have a fancy to
this foolery, as it appears he hath, he is no fool for
fancy, as you would have it appear he is.                   36

CLAUDIO If he be not in love with some woman there is
no believing old signs. A brushes his hat o' mornings,
what should that bode?

DON PEDRO Hath any man seen him at the barber's?           40

CLAUDIO No, but the barber's man hath been seen with
him, and the old ornament of his cheek hath already
stuffed tennis balls.

LEONATO Indeed, he looks younger than he did by the
loss of a beard.                                            45

DON PEDRO Nay, a rubs himself with civet. Can you smell
him out by that?

CLAUDIO That's as much as to say the sweet youth's in
love.

DON PEDRO The greatest note of it is his melancholy.       50

CLAUDIO And when was he wont to wash his face?

DON PEDRO Yea, or to paint himself?—for the which I
hear what they say of him.

CLAUDIO Nay, but his jesting spirit, which is now crept
into a lute-string, and now governed by stops.             55

DON PEDRO Indeed, that tells a heavy tale for him.
Conclude, conclude, he is in love.

CLAUDIO Nay, but I know who loves him.

DON PEDRO That would I know, too. I warrant, one that
knows him not.                                              60

CLAUDIO Yes, and his ill conditions, and in despite of all,
dies for him.

DON PEDRO She shall be buried with her face upwards.

BENEDICK Yet is this no charm for the toothache. Old
signor, walk aside with me. I have studied eight or
nine wise words to speak to you which these hobby-
horses must not hear.        Exeunt Benedick and Leonato

DON PEDRO For my life, to break with him about Beatrice.

CLAUDIO 'Tis even so. Hero and Margaret have by this
played their parts with Beatrice, and then the two bears
will not bite one another when they meet.                   71

    Enter Don John the bastard

DON JOHN My lord, and brother, God save you.

DON PEDRO Good-e'en, brother.

DON JOHN If your leisure served I would speak with you.

DON PEDRO In private?                                       75

DON JOHN If it please you. Yet Count Claudio may hear,
for what I would speak of concerns him.

DON PEDRO What's the matter?

DON JOHN (to Claudio) Means your lordship to be married
tomorrow?                                                   80

DON PEDRO You know he does.

DON JOHN I know not that when he knows what I know.

CLAUDIO If there be any impediment, I pray you discover
it.

DON JOHN You may think I love you not. Let that appear
hereafter, and aim better at me by that I now will
manifest. For my brother, I think he holds you well
and in dearness of heart hath holp to effect your
ensuing marriage—surely suit ill spent, and labour ill
bestowed.                                                   90

DON PEDRO Why, what's the matter?

DON JOHN I came hither to tell you, and, circumstances
shortened—for she has been too long a-talking of—the
lady is disloyal.

CLAUDIO Who, Hero?                                          95

DON JOHN Even she. Leonato's Hero, your Hero, every
man's Hero.

CLAUDIO Disloyal?

DON JOHN The word is too good to paint out her
wickedness. I could say she were worse. Think you of
a worse title, and I will fit her to it. Wonder not till
further warrant. Go but with me tonight, you shall see
her chamber window entered, even the night before
her wedding day. If you love her then, tomorrow wed
her. But it would better fit your honour to change your
mind.                                                      106

CLAUDIO May this be so?

DON PEDRO I will not think it.

DON JOHN If you dare not trust that you see, confess not
that you know. If you will follow me I will show you
enough, and when you have seen more and heard
more, proceed accordingly.                                 112

CLAUDIO If I see anything tonight why I should not marry
her, tomorrow, in the congregation where I should
wed, there will I shame her.                               115

DON PEDRO And as I wooed for thee to obtain her, I will
join with thee to disgrace her.

DON JOHN I will disparage her no farther till you are my
witnesses. Bear it coldly but till midnight, and let the
issue show itself.                                         120

DON PEDRO O day untowardly turned!

CLAUDIO O mischief strangely thwarting!

DON JOHN O plague right well prevented!—So will you
say when you have seen the sequel.            Exeunt

3.3    *Enter Dogberry and his compartner Verges, with the*
       *Watch*

DOGBERRY Are you good men and true?

VERGES Yea, or else it were pity but they should suffer
salvation, body and soul.

DOGBERRY Nay, that were a punishment too good for them
if they should have any allegiance in them, being
chosen for the Prince's watch.

VERGES Well, give them their charge, neighbour Dogberry.

DOGBERRY First, who think you the most desertless man
to be constable?                                            9

SECOND WATCHMAN Hugh Oatcake, sir, or George Seacoal,
for they can write and read.

DOGBERRY Come hither, neighbour Seacoal, God hath blest
you with a good name. To be a well-favoured man is
the gift of fortune, but to write and read comes by
nature.                                                    15

FIRST WATCHMAN Both which, Master Constable—

DOGBERRY You have. I knew it would be your answer.
Well, for your favour, sir, why, give God thanks, and
make no boast of it. And for your writing and reading,
let that appear when there is no need of such vanity.

You are thought here to be the most senseless and fit man for the constable of the watch, therefore bear you the lantern. This is your charge: you shall comprehend all vagrom men. You are to bid any man stand, in the Prince's name. 25

FIRST WATCHMAN How if a will not stand?

DOGBERRY Why then take no note of him, but let him go, and presently call the rest of the watch together, and thank God you are rid of a knave.

VERGES If he will not stand when he is bidden he is none of the Prince's subjects. 31

DOGBERRY True, and they are to meddle with none but the Prince's subjects.—You shall also make no noise in the streets, for for the watch to babble and to talk is most tolerable and not to be endured. 35

A WATCHMAN We will rather sleep than talk. We know what belongs to a watch.

DOGBERRY Why, you speak like an ancient and most quiet watchman, for I cannot see how sleeping should offend. Only have a care that your bills be not stolen. Well, you are to call at all the alehouses and bid those that are drunk get them to bed. 42

A WATCHMAN How if they will not?

DOGBERRY Why then, let them alone till they are sober. If they make you not then the better answer, you may say they are not the men you took them for. 46

A WATCHMAN Well, sir.

DOGBERRY If you meet a thief you may suspect him, by virtue of your office, to be no true man; and for such kind of men, the less you meddle or make with them why, the more is for your honesty. 51

A WATCHMAN If we know him to be a thief, shall we not lay hands on him?

DOGBERRY Truly, by your office you may, but I think they that touch pitch will be defiled. The most peaceable way for you if you do take a thief is to let him show himself what he is, and steal out of your company.

VERGES You have been always called a merciful man, partner. 59

DOGBERRY Truly, I would not hang a dog by my will, much more a man who hath any honesty in him.

VERGES If you hear a child cry in the night you must call to the nurse and bid her still it.

A WATCHMAN How if the nurse be asleep and will not hear us? 65

DOGBERRY Why then, depart in peace and let the child wake her with crying, for the ewe that will not hear her lamb when it baes will never answer a calf when he bleats.

VERGES 'Tis very true. 70

DOGBERRY This is the end of the charge. You, constable, are to present the Prince's own person. If you meet the Prince in the night you may stay him.

VERGES Nay, by'r Lady, that I think a cannot.

DOGBERRY Five shillings to one on't with any man that knows the statutes he may stay him. Marry, not without the Prince be willing, for indeed the watch ought to offend no man, and it is an offence to stay a man against his will.

VERGES By'r Lady, I think it be so. 80

DOGBERRY Ha ha ha! Well, masters, good night. An there be any matter of weight chances, call up me. Keep your fellows' counsels, and your own, and good night. Come, neighbour. 84

⌜FIRST⌝ WATCHMAN Well, masters, we hear our charge. Let us go sit here upon the church bench till two, and then all to bed.

DOGBERRY One word more, honest neighbours. I pray you watch about Signor Leonato's door, for the wedding being there tomorrow, there is a great coil tonight. Adieu. Be vigitant, I beseech you. 91

     *Exeunt Dogberry and Verges.* ⌜*The Watch sit*⌝
     *Enter Borachio and Conrad*

BORACHIO What, Conrad!

⌜FIRST⌝ WATCHMAN (*aside*) Peace, stir not.

BORACHIO Conrad, I say.

CONRAD Here, man, I am at thy elbow. 95

BORACHIO Mass, an my elbow itched, I thought there would a scab follow.

CONRAD I will owe thee an answer for that. And now, forward with thy tale.

BORACHIO Stand thee close, then, under this penthouse, for it drizzles rain, and I will, like a true drunkard, utter all to thee.

A WATCHMAN (*aside*) Some treason, masters. Yet stand close. 104

BORACHIO Therefore, know I have earned of Don John a thousand ducats.

CONRAD Is it possible that any villainy should be so dear?

BORACHIO Thou shouldst rather ask if it were possible any villainy should be so rich. For when rich villains have need of poor ones, poor ones may make what price they will. 111

CONRAD I wonder at it.

BORACHIO That shows thou art unconfirmed. Thou knowest that the fashion of a doublet, or a hat, or a cloak is nothing to a man. 115

CONRAD Yes, it is apparel.

BORACHIO I mean the fashion.

CONRAD Yes, the fashion is the fashion.

BORACHIO Tush, I may as well say the fool's the fool. But seest thou not what a deformed thief this fashion is?

A WATCHMAN (*aside*) I know that Deformed. A has been a vile thief this seven year. A goes up and down like a gentleman. I remember his name.

BORACHIO Didst thou not hear somebody?

CONRAD No, 'twas the vane on the house. 125

BORACHIO Seest thou not, I say, what a deformed thief this fashion is, how giddily a turns about all the hot-bloods between fourteen and five-and-thirty, sometimes fashioning them like Pharaoh's soldiers in the reechy painting, sometime like god Bel's priests in the old church window, sometime like the shaven Hercules in the smirched, worm-eaten tapestry, where his codpiece seems as massy as his club? 133

CONRAD All this I see, and I see that the fashion wears out more apparel than the man. But art not thou thyself giddy with the fashion, too, that thou hast shifted out of thy tale into telling me of the fashion?

BORACHIO Not so, neither. But know that I have tonight wooed Margaret, the Lady Hero's gentlewoman, by the name of Hero. She leans me out at her mistress' chamber window, bids me a thousand times good night—I tell this tale vilely, I should first tell thee how the Prince, Claudio, and my master, planted and placed and possessed by my master, Don John, saw afar off in the orchard this amiable encounter. 145

CONRAD And thought they Margaret was Hero?

BORACHIO Two of them did, the Prince and Claudio, but the devil my master knew she was Margaret, and partly by his oaths, which first possessed them, partly by the dark night, which did deceive them, but chiefly by my villainy, which did confirm any slander that Don John

had made, away went Claudio enraged, swore he would meet her as he was appointed next morning at the temple, and there, before the whole congregation, shame her with what he saw o'ernight, and send her home again without a husband.                156
⌈FIRST⌉ WATCHMAN (*coming forward*) We charge you in the Prince's name.
⌈A WATCHMAN⌉ Call up the right Master Constable. We have here recovered the most dangerous piece of lechery that ever was known in the commonwealth.      161
⌈FIRST⌉ WATCHMAN And one Deformed is one of them. I know him—a wears a lock.
CONRAD Masters, masters!
⌈A WATCHMAN⌉ You'll be made bring Deformed forth, I warrant you.                166
⌈CONRAD⌉ Masters—
⌈A WATCHMAN⌉ Never speak. We charge you. Let us obey you to go with us.
BORACHIO (*to Conrad*) We are like to prove a goodly commodity, being taken up of these men's bills.   171
CONRAD A commodity in question, I warrant you. Come, we'll obey you.                *Exeunt*

3.4    *Enter Hero, Margaret, and Ursula*
HERO Good Ursula, wake my cousin Beatrice, and desire her to rise.
URSULA I will, lady.
HERO And bid her come hither.
URSULA Well.                *Exit*
MARGARET Troth, I think your other rebato were better.
HERO No, pray thee, good Meg, I'll wear this.      7
MARGARET By my troth, 's not so good, and I warrant your cousin will say so.
HERO My cousin's a fool, and thou art another: I'll wear none but this.                11
MARGARET I like the new tire within excellently, if the hair were a thought browner. And your gown's a most rare fashion, i'faith. I saw the Duchess of Milan's gown that they praise so.                15
HERO O, that exceeds, they say.
MARGARET By my troth, 's but a night-gown in respect of yours—cloth o' gold, and cuts, and laced with silver, set with pearls, down sleeves, side sleeves, and skirts round underborne with a bluish tinsel. But for a fine, quaint, graceful, and excellent fashion, yours is worth ten on't.                22
HERO God give me joy to wear it, for my heart is exceeding heavy.
MARGARET 'Twill be heavier soon by the weight of a man.
HERO Fie upon thee, art not ashamed?
MARGARET Of what, lady? Of speaking honourably? Is not marriage honourable in a beggar? Is not your lord honourable without marriage? I think you would have me say 'saving your reverence, a husband'. An bad thinking do not wrest true speaking, I'll offend nobody. Is there any harm in 'the heavier for a husband'? None, I think, an it be the right husband and the right wife—otherwise 'tis light and not heavy. Ask my Lady Beatrice else. Here she comes.                35
        *Enter Beatrice*
HERO Good morrow, coz.
BEATRICE Good morrow, sweet Hero.
HERO Why, how now? Do you speak in the sick tune?
BEATRICE I am out of all other tune, methinks.
MARGARET Clap 's into 'Light o' love'. That goes without a burden. Do you sing it, and I'll dance it.      41

BEATRICE Ye light o' love with your heels. Then if your husband have stables enough, you'll see he shall lack no barns.
MARGARET O illegitimate construction! I scorn that with my heels.                46
BEATRICE (*to Hero*) 'Tis almost five o'clock, cousin. 'Tis time you were ready. By my troth, I am exceeding ill. Heigh-ho!
MARGARET For a hawk, a horse, or a husband?      50
BEATRICE For the letter that begins them all—h.
MARGARET Well, an you be not turned Turk, there's no more sailing by the star.
BEATRICE What means the fool, trow?
MARGARET Nothing, I. But God send everyone their heart's desire.                56
HERO These gloves the Count sent me, they are an excellent perfume.
BEATRICE I am stuffed, cousin. I cannot smell.
MARGARET A maid, and stuffed! There's goodly catching of cold.                61
BEATRICE O, God help me, God help me. How long have you professed apprehension?
MARGARET Ever since you left it. Doth not my wit become me rarely?                65
BEATRICE It is not seen enough. You should wear it in your cap. By my troth, I am sick.
MARGARET Get you some of this distilled *carduus benedictus*, and lay it to your heart. It is the only thing for a qualm.                70
HERO There thou prickest her with a thistle.
BEATRICE Benedictus—why Benedictus? You have some moral in this Benedictus.
MARGARET Moral? No, by my troth, I have no moral meaning. I meant plain holy-thistle. You may think perchance that I think you are in love. Nay, by'r Lady, I am not such a fool to think what I list, nor I list not to think what I can, nor indeed I cannot think, if I would think my heart out of thinking, that you are in love, or that you will be in love, or that you can be in love. Yet Benedick was such another, and now is he become a man. He swore he would never marry, and yet now in despite of his heart he eats his meat without grudging. And how you may be converted I know not, but methinks you look with your eyes, as other women do.                86
BEATRICE What pace is this that thy tongue keeps?
MARGARET Not a false gallop.
        *Enter Ursula*
URSULA (*to Hero*) Madam, withdraw. The Prince, the Count, Signor Benedick, Don John, and all the gallants of the town are come to fetch you to church.      91
HERO Help to dress me, good coz, good Meg, good Ursula.
                *Exeunt*

3.5    *Enter Leonato, and Dogberry the constable, and Verges the headborough*
LEONATO What would you with me, honest neighbour?
DOGBERRY Marry, sir, I would have some confidence with you that decerns you nearly.
LEONATO Brief I pray you, for you see it is a busy time with me.                5
DOGBERRY Marry, this it is, sir.
VERGES Yes, in truth it is, sir.
LEONATO What is it, my good friends?
DOGBERRY Goodman Verges, sir, speaks a little off the matter—an old man, sir, and his wits are not so blunt

as, God help, I would desire they were. But in faith, honest as the skin between his brows.   12

VERGES Yes, I thank God, I am as honest as any man living that is an old man and no honester than I.

DOGBERRY Comparisons are odorous. Palabras, neighbour Verges.   16

LEONATO Neighbours, you are tedious.

DOGBERRY It pleases your worship to say so, but we are the poor Duke's officers. But truly, for mine own part, if I were as tedious as a king I could find in my heart to bestow it all of your worship.   21

LEONATO All thy tediousness on me, ah?

DOGBERRY Yea, an 'twere a thousand pound more than 'tis, for I hear as good exclamation on your worship as of any man in the city, and though I be but a poor man, I am glad to hear it.   26

VERGES And so am I.

LEONATO I would fain know what you have to say.

VERGES Marry, sir, our watch tonight, excepting your worship's presence, ha' ta'en a couple of as arrant knaves as any in Messina.   31

DOGBERRY A good old man, sir. He will be talking. As they say, when the age is in, the wit is out. God help us, it is a world to see. Well said, i'faith, neighbour Verges. Well, God's a good man. An two men ride of a horse, one must ride behind. An honest soul, i'faith, sir, by my troth he is, as ever broke bread. But, God is to be worshipped, all men are not alike, alas, good neighbour.   39

LEONATO Indeed, neighbour, he comes too short of you.

DOGBERRY Gifts that God gives!

LEONATO I must leave you.

DOGBERRY One word, sir. Our watch, sir, have indeed comprehended two auspicious persons, and we would have them this morning examined before your worship.

LEONATO Take their examination yourself, and bring it me. I am now in great haste, as it may appear unto you.

DOGBERRY It shall be suffigance.

LEONATO Drink some wine ere you go. Fare you well.   50

*Enter a Messenger*

MESSENGER My lord, they stay for you to give your daughter to her husband.

LEONATO I'll wait upon them, I am ready.

*Exeunt Leonato and Messenger*

DOGBERRY Go, good partner, go get you to Francis Seacoal, bid him bring his pen and inkhorn to the jail. We are now to examination these men.   56

VERGES And we must do it wisely.

DOGBERRY We will spare for no wit, I warrant you. Here's that shall drive some of them to a non-com. Only get the learned writer to set down our excommunication, and meet me at the jail.   *Exeunt*

**4.1**   *Enter Don Pedro the Prince, Don John the bastard, Leonato, Friar Francis, Claudio, Benedick, Hero, and Beatrice*

LEONATO Come, Friar Francis, be brief. Only to the plain form of marriage, and you shall recount their particular duties afterwards.

FRIAR (*to Claudio*) You come hither, my lord, to marry this lady?   5

CLAUDIO No.

LEONATO To be married to her. Friar, you come to marry her.

FRIAR (*to Hero*) Lady, you come hither to be married to this count?   10

HERO I do.

FRIAR If either of you know any inward impediment why you should not be conjoined, I charge you on your souls to utter it.

CLAUDIO Know you any, Hero?   15

HERO None, my lord.

FRIAR Know you any, Count?

LEONATO I dare make his answer—none.

CLAUDIO O, what men dare do! What men may do! What men daily do, not knowing what they do!   20

BENEDICK How now! Interjections? Why then, some be of laughing, as 'ah, ha, he!'

CLAUDIO
Stand thee by, Friar. Father, by your leave,
Will you with free and unconstrainèd soul
Give me this maid, your daughter?   25

LEONATO
As freely, son, as God did give her me.

CLAUDIO
And what have I to give you back whose worth
May counterpoise this rich and precious gift?

DON PEDRO
Nothing, unless you render her again.

CLAUDIO
Sweet Prince, you learn me noble thankfulness.   30
There, Leonato, take her back again.
Give not this rotten orange to your friend.
She's but the sign and semblance of her honour.
Behold how like a maid she blushes here!
O, what authority and show of truth   35
Can cunning sin cover itself withal!
Comes not that blood as modest evidence
To witness simple virtue? Would you not swear,
All you that see her, that she were a maid,
By these exterior shows? But she is none.   40
She knows the heat of a luxurious bed.
Her blush is guiltiness, not modesty.

LEONATO
What do you mean, my lord?

CLAUDIO      Not to be married,
Not to knit my soul to an approvèd wanton.

LEONATO
Dear my lord, if you in your own proof   45
Have vanquished the resistance of her youth
And made defeat of her virginity—

CLAUDIO
I know what you would say. If I have known her,
You will say she did embrace me as a husband,
And so extenuate the forehand sin.   50
No, Leonato,
I never tempted her with word too large,
But as a brother to his sister showed
Bashful sincerity and comely love.

HERO
And seemed I ever otherwise to you?   55

CLAUDIO
Out on thee, seeming! I will write against it.
You seem to me as Dian in her orb,
As chaste as is the bud ere it be blown.
But you are more intemperate in your blood
Than Venus or those pampered animals   60
That rage in savage sensuality.

HERO
Is my lord well that he doth speak so wide?

LEONATO
Sweet Prince, why speak not you?
DON PEDRO                      What should I speak?
I stand dishonoured, that have gone about
To link my dear friend to a common stale.          65
LEONATO
Are these things spoken, or do I but dream?
DON JOHN
Sir, they are spoken, and these things are true.
BENEDICK This looks not like a nuptial.
HERO 'True'! O God!
CLAUDIO Leonato, stand I here?          70
Is this the Prince? Is this the Prince's brother?
Is this face Hero's? Are our eyes our own?
LEONATO
All this is so. But what of this, my lord?
CLAUDIO
Let me but move one question to your daughter,
And by that fatherly and kindly power          75
That you have in her, bid her answer truly.
LEONATO (to Hero)
I charge thee do so, as thou art my child.
HERO
O God defend me, how am I beset!
What kind of catechizing call you this?
CLAUDIO
To make you answer truly to your name.          80
HERO
Is it not Hero? Who can blot that name
With any just reproach?
CLAUDIO                      Marry, that can Hero.
Hero itself can blot out Hero's virtue.
What man was he talked with you yesternight
Out at your window betwixt twelve and one?          85
Now if you are a maid, answer to this.
HERO
I talked with no man at that hour, my lord.
DON PEDRO
Why, then are you no maiden. Leonato,
I am sorry you must hear. Upon mine honour,
Myself, my brother, and this grievèd Count          90
Did see her, hear her, at that hour last night
Talk with a ruffian at her chamber window,
Who hath indeed, most like a liberal villain,
Confessed the vile encounters they have had
A thousand times in secret.
DON JOHN                      Fie, fie, they are          95
Not to be named, my lord, not to be spoke of.
There is not chastity enough in language
Without offence to utter them. Thus, pretty lady,
I am sorry for thy much misgovernment.
CLAUDIO
O Hero! What a Hero hadst thou been          100
If half thy outward graces had been placed
About thy thoughts and counsels of thy heart!
But fare thee well, most foul, most fair, farewell
Thou pure impiety and impious purity.
For thee I'll lock up all the gates of love,          105
And on my eyelids shall conjecture hang
To turn all beauty into thoughts of harm,
And never shall it more be gracious.
LEONATO
Hath no man's dagger here a point for me?
    *Hero falls to the ground*
BEATRICE
Why, how now, cousin, wherefore sink you down?

DON JOHN
Come. Let us go. These things come thus to light          111
Smother her spirits up.
    *Exeunt Don Pedro, Don John, and Claudio*
BENEDICK
How doth the lady?
BEATRICE                      Dead, I think. Help, uncle.
Hero, why Hero! Uncle, Signor Benedick, Friar—
LEONATO
O fate, take not away thy heavy hand.          115
Death is the fairest cover for her shame
That may be wished for.
BEATRICE                      How now, cousin Hero?
FRIAR (to Hero) Have comfort, lady.
LEONATO (to Hero) Dost thou look up?
FRIAR Yea, wherefore should she not?          120
LEONATO
Wherefore? Why, doth not every earthly thing
Cry shame upon her? Could she here deny
The story that is printed in her blood?
Do not live, Hero, do not ope thine eyes,
For did I think thou wouldst not quickly die,          125
Thought I thy spirits were stronger than thy shames,
Myself would on the rearward of reproaches
Strike at thy life. Grieved I I had but one?
Chid I for that at frugal nature's frame?
O one too much by thee! Why had I one?          130
Why ever wast thou lovely in my eyes?
Why had I not with charitable hand
Took up a beggar's issue at my gates,
Who smirchèd thus and mired with infamy,
I might have said 'No part of it is mine,          135
This shame derives itself from unknown loins.'
But mine, and mine I loved, and mine I praised,
And mine that I was proud on, mine so much
That I myself was to myself not mine,
Valuing of her—why she, O she is fallen          140
Into a pit of ink, that the wide sea
Hath drops too few to wash her clean again,
And salt too little which may season give
To her foul tainted flesh.
BENEDICK                      Sir, sir, be patient.
For my part, I am so attired in wonder          145
I know not what to say.
BEATRICE
O, on my soul, my cousin is belied.
BENEDICK
Lady, were you her bedfellow last night?
BEATRICE
No, truly not, although until last night
I have this twelvemonth been her bedfellow.          150
LEONATO
Confirmed, confirmed. O, that is stronger made
Which was before barred up with ribs of iron.
Would the two princes lie? And Claudio lie,
Who loved her so that, speaking of her foulness,
Washed it with tears? Hence from her, let her die.          155
FRIAR Hear me a little,
For I have only been silent so long
And given way unto this course of fortune
⌜                                              ⌝
By noting of the lady. I have marked          160
A thousand blushing apparitions
To start into her face, a thousand innocent shames
In angel whiteness beat away those blushes,
And in her eye there hath appeared a fire

To burn the errors that these princes hold     165
Against her maiden truth. Call me a fool,
Trust not my reading nor my observations,
Which with experimental seal doth warrant
The tenor of my book. Trust not my age,
My reverence, calling, nor my divinity,     170
If this sweet lady lie not guiltless here
Under some biting error.

LEONATO                Friar, it cannot be.
Thou seest that all the grace that she hath left
Is that she will not add to her damnation
A sin of perjury. She not denies it.     175
Why seek'st thou then to cover with excuse
That which appears in proper nakedness?

FRIAR (to Hero)
Lady, what man is he you are accused of?

HERO
They know that do accuse me. I know none.
If I know more of any man alive     180
Than that which maiden modesty doth warrant,
Let all my sins lack mercy. O my father,
Prove you that any man with me conversed
At hours unmeet, or that I yesternight
Maintained the change of words with any creature,
Refuse me, hate me, torture me to death.     186

FRIAR
There is some strange misprision in the princes.

BENEDICK
Two of them have the very bent of honour,
And if their wisdoms be misled in this
The practice of it lives in John the bastard,     190
Whose spirits toil in frame of villainies.

LEONATO
I know not. If they speak but truth of her
These hands shall tear her. If they wrong her honour
The proudest of them shall well hear of it.
Time hath not yet so dried this blood of mine,     195
Nor age so eat up my invention,
Nor fortune made such havoc of my means,
Nor my bad life reft me so much of friends,
But they shall find awaked in such a kind
Both strength of limb and policy of mind,     200
Ability in means, and choice of friends,
To quit me of them throughly.

FRIAR               Pause awhile,
And let my counsel sway you in this case.
Your daughter here the princes left for dead,
Let her a while be secretly kept in,     205
And publish it that she is dead indeed.
Maintain a mourning ostentation,
And on your family's old monument
Hang mournful epitaphs, and do all rites
That appertain unto a burial.     210

LEONATO
What shall become of this? What will this do?

FRIAR
Marry, this, well carried, shall on her behalf
Change slander to remorse. That is some good.
But not for that dream I on this strange course,
But on this travail look for greater birth.     215
She—dying, as it must be so maintained,
Upon the instant that she was accused—
Shall be lamented, pitied, and excused
Of every hearer. For it so falls out
That what we have, we prize not to the worth     220
Whiles we enjoy it, but, being lacked and lost,

Why then we rack the value, then we find
The virtue that possession would not show us
Whiles it was ours. So will it fare with Claudio.
When he shall hear she died upon his words,     225
Th'idea of her life shall sweetly creep
Into his study of imagination,
And every lovely organ of her life
Shall come apparelled in more precious habit,
More moving-delicate, and full of life,     230
Into the eye and prospect of his soul
Than when she lived indeed. Then shall he mourn,
If ever love had interest in his liver,
And wish he had not so accusèd her,
No, though he thought his accusation true.     235
Let this be so, and doubt not but success
Will fashion the event in better shape
Than I can lay it down in likelihood.
But if all aim but this be levelled false,
The supposition of the lady's death     240
Will quench the wonder of her infamy.
And if it sort not well, you may conceal her,
As best befits her wounded reputation,
In some reclusive and religious life,
Out of all eyes, tongues, minds, and injuries.     245

BENEDICK
Signor Leonato, let the Friar advise you.
And though you know my inwardness and love
Is very much unto the Prince and Claudio,
Yet, by mine honour, I will deal in this
As secretly and justly as your soul     250
Should with your body.

LEONATO Being that I flow in grief,
The smallest twine may lead me.

FRIAR
'Tis well consented. Presently away,
For to strange sores strangely they strain the cure.
(To Hero) Come, lady, die to live. This wedding day
Perhaps is but prolonged. Have patience, and endure.

           *Exeunt all but Beatrice and Benedick*

BENEDICK Lady Beatrice, have you wept all this while?
BEATRICE Yea, and I will weep a while longer.
BENEDICK I will not desire that.     260
BEATRICE You have no reason, I do it freely.
BENEDICK Surely I do believe your fair cousin is wronged.
BEATRICE Ah, how much might the man deserve of me
    that would right her!
BENEDICK Is there any way to show such friendship?
BEATRICE A very even way, but no such friend.     266
BENEDICK May a man do it?
BEATRICE It is a man's office, but not yours.
BENEDICK I do love nothing in the world so well as you.
    Is not that strange?     270
BEATRICE As strange as the thing I know not. It were as
    possible for me to say I loved nothing so well as you,
    but believe me not, and yet I lie not. I confess nothing
    nor I deny nothing. I am sorry for my cousin.
BENEDICK By my sword, Beatrice, thou lovest me.     275
BEATRICE Do not swear and eat it.
BENEDICK I will swear by it that you love me, and I will
    make him eat it that says I love not you.
BEATRICE Will you not eat your word?
BENEDICK With no sauce that can be devised to it. I protest
    I love thee.     281
BEATRICE Why then, God forgive me.
BENEDICK What offence, sweet Beatrice?
BEATRICE You have stayed me in a happy hour. I was
    about to protest I loved you.     285

BENEDICK  And do it with all thy heart.

BEATRICE  I love you with so much of my heart that none is left to protest.

BENEDICK  Come, bid me do anything for thee.

BEATRICE  Kill Claudio.                                                    290

BENEDICK  Ha! Not for the wide world.

BEATRICE  You kill me to deny it. Farewell.

BENEDICK  Tarry, sweet Beatrice.

BEATRICE  I am gone though I am here. There is no love in you.—Nay, I pray you, let me go.                        295

BENEDICK  Beatrice.

BEATRICE  In faith, I will go.

BENEDICK  We'll be friends first.

BEATRICE  You dare easier be friends with me than fight with mine enemy.                                              300

BENEDICK  Is Claudio thine enemy?

BEATRICE  Is a not approved in the height a villain, that hath slandered, scorned, dishonoured my kinswoman? O that I were a man! What, bear her in hand until they come to take hands, and then with public accusation, uncovered slander, unmitigated rancour— O God that I were a man! I would eat his heart in the market place.

BENEDICK  Hear me, Beatrice.                                      309

BEATRICE  Talk with a man out at a window—a proper saying!

BENEDICK  Nay, but Beatrice.

BEATRICE  Sweet Hero, she is wronged, she is slandered, she is undone.

BENEDICK  Beat—                                                   315

BEATRICE  Princes and counties! Surely a princely testimony, a goodly count, Count Comfit, a sweet gallant, surely. O that I were a man for his sake! Or that I had any friend would be a man for my sake! But manhood is melted into courtesies, valour into compliment, and men are only turned into tongue, and trim ones, too. He is now as valiant as Hercules that only tells a lie and swears it. I cannot be a man with wishing, therefore I will die a woman with grieving.                          324

BENEDICK  Tarry, good Beatrice. By this hand, I love thee.

BEATRICE  Use it for my love some other way than swearing by it.

BENEDICK  Think you in your soul the Count Claudio hath wronged Hero?

BEATRICE  Yea, as sure as I have a thought or a soul.

BENEDICK  Enough, I am engaged, I will challenge him. I will kiss your hand, and so I leave you. By this hand, Claudio shall render me a dear account. As you hear of me, so think of me. Go comfort your cousin. I must say she is dead. And so, farewell.                    *Exeunt*

**4.2**  *Enter Dogberry and Verges the constables, and the Sexton, in gowns, and the Watch, with Conrad and Borachio*

DOGBERRY  Is our whole dissembly appeared?

VERGES  O, a stool and a cushion for the Sexton.

SEXTON  ⌈*sits*⌉ Which be the malefactors?

DOGBERRY  Marry, that am I, and my partner.

VERGES  Nay, that's certain, we have the exhibition to examine.                                                          6

SEXTON  But which are the offenders that are to be examined? Let them come before Master Constable.

DOGBERRY  Yea, marry, let them come before me. What is your name, friend?                                          10

BORACHIO  Borachio.

DOGBERRY  (*to the Sexton*) Pray write down 'Borachio'. (*To Conrad*) Yours, sirrah?

CONRAD  I am a gentleman, sir, and my name is Conrad.

DOGBERRY  Write down 'Master Gentleman Conrad'.— Masters, do you serve God?                                  16

CONRAD *and* BORACHIO  Yea, sir, we hope.

DOGBERRY  Write down that they hope they serve God. And write 'God' first, for God defend but God should go before such villains. Masters, it is proved already that you are little better than false knaves, and it will go near to be thought so shortly. How answer you for yourselves?                                                      23

CONRAD  Marry, sir, we say we are none.

DOGBERRY  A marvellous witty fellow, I assure you, but I will go about with him. Come you hither, sirrah. A word in your ear, sir. I say to you it is thought you are false knaves.                                                 28

BORACHIO  Sir, I say to you we are none.

DOGBERRY  Well, stand aside. Fore God, they are both in a tale. Have you writ down that they are none?

SEXTON  Master Constable, you go not the way to examine. You must call forth the watch that are their accusers.

DOGBERRY  Yea, marry, that's the eftest way. Let the watch come forth. Masters, I charge you in the Prince's name accuse these men.                                            36

FIRST WATCHMAN  This man said, sir, that Don John, the Prince's brother, was a villain.

DOGBERRY  Write down Prince John a villain. Why, this is flat perjury, to call a prince's brother villain.       40

BORACHIO  Master Constable.

DOGBERRY  Pray thee, fellow, peace. I do not like thy look, I promise thee.

SEXTON  What heard you him say else?                      44

SECOND WATCHMAN  Marry, that he had received a thousand ducats of Don John for accusing the Lady Hero wrongfully.

DOGBERRY  Flat burglary, as ever was committed.

VERGES  Yea, by mass, that it is.

SEXTON  What else, fellow?                                       50

FIRST WATCHMAN  And that Count Claudio did mean upon his words to disgrace Hero before the whole assembly, and not marry her.

DOGBERRY  O villain! Thou wilt be condemned into everlasting redemption for this.                                   55

SEXTON  What else?

WATCH  This is all.

SEXTON  And this is more, masters, than you can deny. Prince John is this morning secretly stolen away. Hero was in this manner accused, in this very manner refused, and upon the grief of this suddenly died. Master Constable, let these men be bound and brought to Leonato's. I will go before and show him their examination.                                                    *Exit*

DOGBERRY  Come, let them be opinioned.                 65

VERGES  Let them be, in the hands—

⌈CONRAD⌉ Off, coxcomb!

DOGBERRY  God's my life, where's the Sexton? Let him write down the Prince's officer coxcomb. Come, bind them. Thou naughty varlet!                               70

CONRAD  Away, you are an ass, you are an ass.

DOGBERRY  Dost thou not suspect my place? Dost thou not suspect my years? O that he were here to write me down an ass! But masters, remember that I am an ass. Though it be not written down, yet forget not that I am an ass. No, thou villain, thou art full of piety, as shall be proved upon thee by good witness. I am a wise fellow, and which is more, an officer, and which is

more, a householder, and which is more, as pretty a
piece of flesh as any is in Messina, and one that knows
the law, go to, and a rich fellow enough, go to, and a
fellow that hath had losses, and one that hath two
gowns, and everything handsome about him. Bring
him away. O that I had been writ down an ass!   84
                                    *Exeunt*

**5.1**   *Enter Leonato and Antonio his brother*
ANTONIO
If you go on thus, you will kill yourself,
And 'tis not wisdom thus to second grief
Against yourself.
LEONATO          I pray thee cease thy counsel,
Which falls into mine ears as profitless
As water in a sieve. Give not me counsel,   5
Nor let no comforter delight mine ear
But such a one whose wrongs do suit with mine.
Bring me a father that so loved his child,
Whose joy of her is overwhelmed like mine,
And bid him speak of patience.   10
Measure his woe the length and breadth of mine,
And let it answer every strain for strain,
As thus for thus, and such a grief for such,
In every lineament, branch, shape, and form.
If such a one will smile and stroke his beard,   15
Bid sorrow wag, cry 'hem' when he should groan,
Patch grief with proverbs, make misfortune drunk
With candle-wasters, bring him yet to me,
And I of him will gather patience.
But there is no such man, for, brother, men   20
Can counsel and speak comfort to that grief
Which they themselves not feel, but tasting it
Their counsel turns to passion, which before
Would give preceptial medicine to rage,
Fetter strong madness in a silken thread,   25
Charm ache with air and agony with words.
No, no, 'tis all men's office to speak patience
To those that wring under the load of sorrow,
But no man's virtue nor sufficiency
To be so moral when he shall endure   30
The like himself. Therefore give me no counsel.
My griefs cry louder than advertisement.
ANTONIO
Therein do men from children nothing differ.
LEONATO
I pray thee peace, I will be flesh and blood,
For there was never yet philosopher   35
That could endure the toothache patiently,
However they have writ the style of gods,
And made a pish at chance and sufferance.
ANTONIO
Yet bend not all the harm upon yourself.
Make those that do offend you suffer, too.   40
LEONATO
There thou speak'st reason, nay I will do so.
My soul doth tell me Hero is belied,
And that shall Claudio know, so shall the Prince,
And all of them that thus dishonour her.
    *Enter Don Pedro the Prince and Claudio*
ANTONIO
Here comes the Prince and Claudio hastily.   45
DON PEDRO
Good e'en, good e'en.
CLAUDIO            Good day to both of you.

LEONATO
Hear you, my lords?
DON PEDRO          We have some haste, Leonato.
LEONATO
Some haste, my lord! Well, fare you well, my lord.
Are you so hasty now? Well, all is one.
DON PEDRO
Nay, do not quarrel with us, good old man.   50
ANTONIO
If he could right himself with quarrelling,
Some of us would lie low.
CLAUDIO           Who wrongs him?
LEONATO
Marry, thou dost wrong me, thou dissembler, thou.
Nay, never lay thy hand upon thy sword,
I fear thee not.
CLAUDIO        Marry, beshrew my hand   55
If it should give your age such cause of fear.
In faith, my hand meant nothing to my sword.
LEONATO
Tush, tush, man, never fleer and jest at me.
I speak not like a dotard nor a fool,
As under privilege of age to brag   60
What I have done being young, or what would do
Were I not old. Know Claudio to thy head,
Thou hast so wronged mine innocent child and me
That I am forced to lay my reverence by
And with grey hairs and bruise of many days   65
Do challenge thee to trial of a man.
I say thou hast belied mine innocent child.
Thy slander hath gone through and through her heart,
And she lies buried with her ancestors,
O, in a tomb where never scandal slept   70
Save this of hers, framed by thy villainy.
CLAUDIO
My villainy?
LEONATO        Thine, Claudio, thine I say.
DON PEDRO
You say not right, old man.
LEONATO           My lord, my lord,
I'll prove it on his body if he dare,
Despite his nice fence and his active practice,   75
His May of youth and bloom of lustihood.
CLAUDIO
Away, I will not have to do with you.
LEONATO
Canst thou so doff me? Thou hast killed my child.
If thou kill'st me, boy, thou shalt kill a man.
ANTONIO
He shall kill two of us, and men indeed.   80
But that's no matter, let him kill one first.
Win me and wear me. Let him answer me.
Come follow me boy, come sir boy, come follow me,
Sir boy, I'll whip you from your foining fence.
Nay, as I am a gentleman, I will.   85
LEONATO Brother.
ANTONIO
Content yourself. God knows, I loved my niece,
And she is dead, slandered to death by villains
That dare as well answer a man indeed
As I dare take a serpent by the tongue.   90
Boys, apes, braggarts, jacks, milksops!
LEONATO Brother Antony—
ANTONIO
Hold you content. What, man, I know them, yea
And what they weigh, even to the utmost scruple.

Scambling, outfacing, fashion-monging boys, 95
That lie, and cog, and flout, deprave, and slander,
Go anticly, and show an outward hideousness,
And speak off half a dozen dangerous words,
How they might hurt their enemies, if they durst,
And this is all. 100

LEONATO But brother Antony—

ANTONIO Come, 'tis no matter,
Do not you meddle, let me deal in this.

DON PEDRO
Gentlemen both, we will not wake your patience.
My heart is sorry for your daughter's death, 105
But on my honour she was charged with nothing
But what was true and very full of proof.

LEONATO
My lord, my lord—

DON PEDRO I will not hear you.

LEONATO
No? Come brother, away. I will be heard.

ANTONIO
And shall, or some of us will smart for it. 110

*Exeunt Leonato and Antonio*

*Enter Benedick*

DON PEDRO
See, see, here comes the man we went to seek.

CLAUDIO Now signor, what news?

BENEDICK (*to Don Pedro*) Good day, my lord.

DON PEDRO Welcome, signor. You are almost come to part
almost a fray. 115

CLAUDIO We had liked to have had our two noses snapped
off with two old men without teeth.

DON PEDRO Leonato and his brother. What thinkest thou?
Had we fought, I doubt we should have been too young
for them. 120

BENEDICK In a false quarrel there is no true valour. I came
to seek you both.

CLAUDIO We have been up and down to seek thee, for we
are high-proof melancholy and would fain have it
beaten away. Wilt thou use thy wit? 125

BENEDICK It is in my scabbard. Shall I draw it?

DON PEDRO Dost thou wear thy wit by thy side?

CLAUDIO Never any did so, though very many have been
beside their wit. I will bid thee draw as we do the
minstrels, draw to pleasure us. 130

DON PEDRO As I am an honest man he looks pale. Art
thou sick, or angry?

CLAUDIO What, courage, man. What though care killed
a cat, thou hast mettle enough in thee to kill care.

BENEDICK Sir, I shall meet your wit in the career an you
charge it against me. I pray you choose another subject.

CLAUDIO Nay then, give him another staff. This last was
broke cross.

DON PEDRO By this light, he changes more and more. I
think he be angry indeed. 140

CLAUDIO If he be, he knows how to turn his girdle.

BENEDICK (*aside to Claudio*) Shall I speak a word in your
ear?

CLAUDIO God bless me from a challenge.

BENEDICK You are a villain. I jest not. I will make it good
how you dare, with what you dare, and when you
dare. Do me right, or I will protest your cowardice.
You have killed a sweet lady, and her death shall fall
heavy on you. Let me hear from you. 149

CLAUDIO Well, I will meet you, so I may have good cheer.

DON PEDRO What, a feast, a feast?

CLAUDIO I'faith, I thank him, he hath bid me to a calf's
head and a capon, the which if I do not carve most

curiously, say my knife's naught. Shall I not find a
woodcock too? 155

BENEDICK Sir, your wit ambles well, it goes easily.

DON PEDRO I'll tell thee how Beatrice praised thy wit the
other day. I said thou hadst a fine wit. 'True,' said she,
'a fine little one.' 'No,' said I, 'a great wit.' 'Right,'
says she, 'a great gross one.' 'Nay,' said I, 'a good wit.'
'Just,' said she, 'it hurts nobody.' 'Nay,' said I, 'the
gentleman is wise.' 'Certain,' said she, 'a wise gentleman.' 'Nay,' said I, 'he hath the tongues.' 'That I
believe,' said she, 'for he swore a thing to me on
Monday night which he forswore on Tuesday morning.
There's a double tongue, there's two tongues.' Thus
did she an hour together trans-shape thy particular
virtues, yet at last she concluded with a sigh thou wast
the properest man in Italy. 169

CLAUDIO For the which she wept heartily and said she
cared not.

DON PEDRO Yea, that she did. But yet for all that, an if
she did not hate him deadly she would love him dearly.
The old man's daughter told us all. 174

CLAUDIO All, all. And moreover, God saw him when he
was hid in the garden.

DON PEDRO But when shall we set the savage bull's horns
on the sensible Benedick's head?

CLAUDIO Yea, and text underneath, 'Here dwells Benedick
the married man'. 180

BENEDICK Fare you well, boy, you know my mind. I will
leave you now to your gossip-like humour. You break
jests as braggarts do their blades which, God be
thanked, hurt not. (*To Don Pedro*) My lord, for your
many courtesies I thank you. I must discontinue your
company. Your brother the bastard is fled from Messina.
You have among you killed a sweet and innocent lady.
For my lord Lackbeard there, he and I shall meet, and
till then, peace be with him. *Exit*

DON PEDRO He is in earnest. 190

CLAUDIO In most profound earnest, and, I'll warrant you,
for the love of Beatrice.

DON PEDRO And hath challenged thee.

CLAUDIO Most sincerely.

DON PEDRO What a pretty thing man is when he goes in
his doublet and hose and leaves off his wit! 196

*Enter Dogberry and Verges the constables, the
Watch, Conrad, and Borachio*

CLAUDIO He is then a giant to an ape. But then is an ape
a doctor to such a man.

DON PEDRO But soft you, let me be. Pluck up, my heart,
and be sad. Did he not say my brother was fled? 200

DOGBERRY Come you sir, if justice cannot tame you, she
shall ne'er weigh more reasons in her balance. Nay, an
you be a cursing hypocrite once, you must be looked to.

DON PEDRO How now, two of my brother's men bound?
Borachio one. 205

CLAUDIO Hearken after their offence, my lord.

DON PEDRO Officers, what offence have these men done?

DOGBERRY Marry, sir, they have committed false report,
moreover they have spoken untruths, secondarily they
are slanders, sixth and lastly they have belied a lady,
thirdly they have verified unjust things, and to
conclude, they are lying knaves. 212

DON PEDRO First I ask thee what they have done, thirdly
I ask thee what's their offence, sixth and lastly why
they are committed, and to conclude, what you lay to
their charge. 216

CLAUDIO Rightly reasoned, and in his own division. And
by my troth there's one meaning well suited.

DON PEDRO (*to Conrad and Borachio*) Who have you
offended, masters, that you are thus bound to your
answer? This learned constable is too cunning to be
understood. What's your offence?                                222

BORACHIO Sweet Prince, let me go no farther to mine
answer. Do you hear me, and let this Count kill me. I
have deceived even your very eyes. What your wisdoms
could not discover, these shallow fools have brought
to light, who in the night overheard me confessing to
this man how Don John your brother incensed me to
slander the Lady Hero, how you were brought into the
orchard and saw me court Margaret in Hero's garments,
how you disgraced her when you should marry her.
My villainy they have upon record, which I had rather
seal with my death than repeat over to my shame. The
lady is dead upon mine and my master's false
accusation, and briefly, I desire nothing but the reward
of a villain.                                                             236

DON PEDRO (*to Claudio*)
Runs not this speech like iron through your blood?

CLAUDIO
I have drunk poison whiles he uttered it.

DON PEDRO (*to Borachio*)
But did my brother set thee on to this?

BORACHIO
Yea, and paid me richly for the practice of it.           240

DON PEDRO
He is composed and framed of treachery,
And fled he is upon this villainy.

CLAUDIO
Sweet Hero, now thy image doth appear
In the rare semblance that I loved it first.               244

DOGBERRY Come, bring away the plaintiffs. By this time
our Sexton hath reformed Signor Leonato of the matter.
And masters, do not forget to specify, when time and
place shall serve, that I am an ass.

VERGES Here, here comes Master Signor Leonato, and the
Sexton, too.                                                            250

*Enter Leonato, Antonio his brother, and the Sexton*

LEONATO
Which is the villain? Let me see his eyes,
That when I note another man like him
I may avoid him. Which of these is he?

BORACHIO
If you would know your wronger, look on me.

LEONATO
Art thou the slave that with thy breath hast killed    255
Mine innocent child?

BORACHIO                        Yea, even I alone.

LEONATO
No, not so, villain, thou beliest thyself.
Here stand a pair of honourable men.
A third is fled that had a hand in it.
I thank you, Princes, for my daughter's death.          260
Record it with your high and worthy deeds.
'Twas bravely done, if you bethink you of it.

CLAUDIO
I know not how to pray your patience,
Yet I must speak. Choose your revenge yourself,
Impose me to what penance your invention               265
Can lay upon my sin. Yet sinned I not
But in mistaking.

DON PEDRO                  By my soul, nor I,
And yet to satisfy this good old man

I would bend under any heavy weight
That he'll enjoin me to.                                             270

LEONATO
I cannot bid you bid my daughter live—
That were impossible—but I pray you both
Possess the people in Messina here
How innocent she died, and if your love
Can labour aught in sad invention,                           275
Hang her an epitaph upon her tomb
And sing it to her bones, sing it tonight.
Tomorrow morning come you to my house,
And since you could not be my son-in-law,
Be yet my nephew. My brother hath a daughter,       280
Almost the copy of my child that's dead,
And she alone is heir to both of us.
Give her the right you should have giv'n her cousin,
And so dies my revenge.

CLAUDIO                        O noble sir!
Your overkindness doth wring tears from me.            285
I do embrace your offer; and dispose
For henceforth of poor Claudio.

LEONATO
Tomorrow then I will expect your coming.
Tonight I take my leave. This naughty man
Shall face to face be brought to Margaret,               290
Who I believe was packed in all this wrong,
Hired to it by your brother.

BORACHIO                        No, by my soul, she was not,
Nor knew not what she did when she spoke to me,
But always hath been just and virtuous
In anything that I do know by her.                            295

DOGBERRY (*to Leonato*) Moreover, sir, which indeed is not
under white and black, this plaintiff here, the offender,
did call me ass. I beseech you let it be remembered in
his punishment. And also the watch heard them talk
of one Deformed. They say he wears a key in his ear
and a lock hanging by it, and borrows money in God's
name, the which he hath used so long and never paid
that now men grow hard-hearted and will lend nothing
for God's sake. Pray you examine him upon that point.

LEONATO
I thank thee for thy care and honest pains.              305

DOGBERRY Your worship speaks like a most thankful and
reverend youth, and I praise God for you.

LEONATO (*giving him money*) There's for thy pains.

DOGBERRY God save the foundation.

LEONATO Go. I discharge thee of thy prisoner, and I thank
thee.                                                                        311

DOGBERRY I leave an arrant knave with your worship,
which I beseech your worship to correct yourself, for
the example of others. God keep your worship, I wish
your worship well. God restore you to health. I humbly
give you leave to depart, and if a merry meeting may
be wished, God prohibit it. Come, neighbour.          317

*Exeunt Dogberry and Verges*

LEONATO
Until tomorrow morning, lords, farewell.

ANTONIO
Farewell, my lords. We look for you tomorrow.

DON PEDRO
We will not fail.

CLAUDIO                        Tonight I'll mourn with Hero.      320

LEONATO (*to the Watch*)
Bring you these fellows on.—We'll talk with Margaret
How her acquaintance grew with this lewd fellow.

*Exeunt*

**5.2**    *Enter Benedick and Margaret*

BENEDICK Pray thee, sweet Mistress Margaret, deserve well at my hands by helping me to the speech of Beatrice.

MARGARET Will you then write me a sonnet in praise of my beauty?

BENEDICK In so high a style, Margaret, that no man living shall come over it, for in most comely truth, thou deservest it. 7

MARGARET To have no man come over me—why, shall I always keep below stairs?

BENEDICK Thy wit is as quick as the greyhound's mouth, it catches. 11

MARGARET And yours as blunt as the fencer's foils, which hit but hurt not.

BENEDICK A most manly wit, Margaret, it will not hurt a woman. And so I pray thee call Beatrice. I give thee the bucklers. 16

MARGARET Give us the swords. We have bucklers of our own.

BENEDICK If you use them, Margaret, you must put in the pikes with a vice—and they are dangerous weapons for maids. 21

MARGARET Well, I will call Beatrice to you, who I think hath legs. *Exit*

BENEDICK And therefore will come.

(*Sings*)     The god of love 25
       That sits above,
       And knows me, and knows me,
       How pitiful I deserve—

I mean in singing; but in loving, Leander the good swimmer, Troilus the first employer of panders, and a whole book full of these quondam carpet-mongers whose names yet run smoothly in the even road of a blank verse, why they were never so truly turned over and over as my poor self in love. Marry, I cannot show it in rhyme. I have tried. I can find out no rhyme to 'lady' but 'baby', an innocent rhyme; for 'scorn' 'horn', a hard rhyme; for 'school' 'fool', a babbling rhyme. Very ominous endings. No, I was not born under a rhyming planet, nor I cannot woo in festival terms.

*Enter Beatrice*

Sweet Beatrice, wouldst thou come when I called thee?

BEATRICE Yea, signor, and depart when you bid me. 41

BENEDICK O, stay but till then.

BEATRICE 'Then' is spoken. Fare you well now. And yet ere I go, let me go with that I came for, which is with knowing what hath passed between you and Claudio.

BENEDICK Only foul words, and thereupon I will kiss thee.

BEATRICE Foul words is but foul wind, and foul wind is but foul breath, and foul breath is noisome, therefore I will depart unkissed. 49

BENEDICK Thou hast frighted the word out of his right sense, so forcible is thy wit. But I must tell thee plainly, Claudio undergoes my challenge, and either I must shortly hear from him or I will subscribe him a coward. And I pray thee now tell me, for which of my bad parts didst thou first fall in love with me? 55

BEATRICE For them all together, which maintain so politic a state of evil that they will not admit any good part to intermingle with them. But for which of my good parts did you first suffer love for me?

BENEDICK Suffer love—a good epithet. I do suffer love indeed, for I love thee against my will. 61

BEATRICE In spite of your heart, I think. Alas, poor heart. If you spite it for my sake I will spite it for yours, for I will never love that which my friend hates.

BENEDICK Thou and I are too wise to woo peaceably. 65

BEATRICE It appears not in this confession. There's not one wise man among twenty that will praise himself.

BENEDICK An old, an old instance, Beatrice, that lived in the time of good neighbours. If a man do not erect in this age his own tomb ere he dies, he shall live no longer in monument than the bell rings and the widow weeps. 72

BEATRICE And how long is that, think you?

BENEDICK Question—why, an hour in clamour and a quarter in rheum. Therefore is it most expedient for the wise, if Don Worm—his conscience—find no impediment to the contrary, to be the trumpet of his own virtues, as I am to myself. So much for praising myself who, I myself will bear witness, is praiseworthy. And now tell me, how doth your cousin? 80

BEATRICE Very ill.

BENEDICK And how do you?

BEATRICE Very ill too.

BENEDICK Serve God, love me, and mend. There will I leave you too, for here comes one in haste. 85

*Enter Ursula*

URSULA Madam, you must come to your uncle. Yonder's old coil at home. It is proved my lady Hero hath been falsely accused, and the Prince and Claudio mightily abused, and Don John is the author of all, who is fled and gone. Will you come presently? 90

BEATRICE Will you go hear this news, signor?

BENEDICK I will live in thy heart, die in thy lap, and be buried in thy eyes. And moreover, I will go with thee to thy uncle's. *Exeunt*

**5.3**    *Enter Claudio, Don Pedro the Prince, and three or four with tapers, all in black*

CLAUDIO
Is this the monument of Leonato?

A LORD
It is, my lord.

⌈CLAUDIO (*reading from a scroll*)⌉
     Done to death by slanderous tongues
       Was the Hero that here lies.
     Death in guerdon of her wrongs 5
       Gives her fame which never dies.
     So the life that died with shame
     Lives in death with glorious fame.

*He hangs the epitaph on the tomb*

     Hang thou there upon the tomb,
     Praising her when I am dumb. 10

Now music sound, and sing your solemn hymn.

*Song*

     Pardon, goddess of the night,
     Those that slew thy virgin knight,
     For the which with songs of woe
     Round about her tomb they go. 15
     Midnight, assist our moan,
     Help us to sigh and groan,
       Heavily, heavily.
     Graves yawn, and yield your dead
     Till death be utterèd, 20
       Heavily, heavily.

⌈CLAUDIO⌉
Now, unto thy bones good night.
Yearly will I do this rite.

DON PEDRO
  Good morrow, masters, put your torches out.
    The wolves have preyed, and look, the gentle day
  Before the wheels of Phoebus round about        26
    Dapples the drowsy east with spots of grey.
  Thanks to you all, and leave us. Fare you well.
CLAUDIO
  Good morrow, masters. Each his several way.
DON PEDRO
  Come, let us hence, and put on other weeds,        30
    And then to Leonato's we will go.
CLAUDIO
  And Hymen now with luckier issue speed 's
    Than this for whom we rendered up this woe.
                                        *Exeunt*

**5.4**    *Enter Leonato, Antonio, Benedick, Beatrice,*
          *Margaret, Ursula, Friar Francis, and Hero*
FRIAR
  Did I not tell you she was innocent?
LEONATO
  So are the Prince and Claudio who accused her
  Upon the error that you heard debated.
  But Margaret was in some fault for this,
  Although against her will as it appears        5
  In the true course of all the question.
ANTONIO
  Well, I am glad that all things sorts so well.
BENEDICK
  And so am I, being else by faith enforced
  To call young Claudio to a reckoning for it.
LEONATO
  Well, daughter, and you gentlewomen all,        10
  Withdraw into a chamber by yourselves,
  And when I send for you come hither masked.
          *Exeunt Beatrice, Hero, Margaret, and Ursula*
  The Prince and Claudio promised by this hour
  To visit me. You know your office, brother,
  You must be father to your brother's daughter,        15
  And give her to young Claudio.
ANTONIO
  Which I will do with confirmed countenance.
BENEDICK
  Friar, I must entreat your pains, I think.
FRIAR  To do what, signor?
BENEDICK
  To bind me or undo me, one of them.        20
  Signor Leonato, truth it is, good signor,
  Your niece regards me with an eye of favour.
LEONATO
  That eye my daughter lent her, 'tis most true.
BENEDICK
  And I do with an eye of love requite her.
LEONATO
  The sight whereof I think you had from me,        25
  From Claudio and the Prince. But what's your will?
BENEDICK
  Your answer, sir, is enigmatical.
  But for my will, my will is your good will
  May stand with ours this day to be conjoined
  In the state of honourable marriage,        30
  In which, good Friar, I shall desire your help.
LEONATO
  My heart is with your liking.
FRIAR                          And my help.
  Here comes the Prince and Claudio.

          *Enter Don Pedro and Claudio with attendants*
DON PEDRO
  Good morrow to this fair assembly.
LEONATO
  Good morrow, Prince. Good morrow, Claudio.        35
  We here attend you. Are you yet determined
  Today to marry with my brother's daughter?
CLAUDIO
  I'll hold my mind, were she an Ethiope.
LEONATO
  Call her forth, brother, here's the Friar ready.
                                        *Exit Antonio*
DON PEDRO
  Good morrow, Benedick. Why, what's the matter        40
  That you have such a February face,
  So full of frost, of storm and cloudiness?
CLAUDIO
  I think he thinks upon the savage bull.
  Tush, fear not, man, we'll tip thy horns with gold,
  And all Europa shall rejoice at thee        45
  As once Europa did at lusty Jove
  When he would play the noble beast in love.
BENEDICK
  Bull Jove, sir, had an amiable low,
  And some such strange bull leapt your father's cow
  And got a calf in that same noble feat        50
  Much like to you, for you have just his bleat.
          *Enter Antonio with Hero, Beatrice, Margaret, and*
          *Ursula, masked*
CLAUDIO
  For this I owe you. Here comes other reck'nings.
  Which is the lady I must seize upon?
⌈ANTONIO⌉
  This same is she, and I do give you her.
CLAUDIO
  Why then, she's mine. Sweet, let me see your face.  55
LEONATO
  No, that you shall not till you take her hand
  Before this Friar and swear to marry her.
CLAUDIO (*to Hero*)
  Give me your hand before this holy friar.
  I am your husband if you like of me.
HERO (*unmasking*)
  And when I lived I was your other wife;        60
  And when you loved, you were my other husband.
CLAUDIO
  Another Hero!
HERO                Nothing certainer.
  One Hero died defiled, but I do live,
  And surely as I live, I am a maid.
DON PEDRO
  The former Hero, Hero that is dead!        65
LEONATO
  She died, my lord, but whiles her slander lived.
FRIAR
  All this amazement can I qualify
  When after that the holy rites are ended
  I'll tell you largely of fair Hero's death.
  Meantime, let wonder seem familiar,        70
  And to the chapel let us presently.
BENEDICK
  Soft and fair, Friar, which is Beatrice?
BEATRICE (*unmasking*)
  I answer to that name, what is your will?
BENEDICK
  Do not you love me?
BEATRICE                Why no, no more than reason.

BENEDICK
    Why then, your uncle and the Prince and Claudio   75
    Have been deceived. They swore you did.

BEATRICE
    Do not you love me?

BENEDICK              Troth no, no more than reason.

BEATRICE
    Why then, my cousin, Margaret, and Ursula
    Are much deceived, for they did swear you did.

BENEDICK
    They swore that you were almost sick for me.     80

BEATRICE
    They swore that you were wellnigh dead for me.

BENEDICK
    'Tis no such matter. Then you do not love me?

BEATRICE
    No, truly, but in friendly recompense.

LEONATO
    Come, cousin, I am sure you love the gentleman.

CLAUDIO
    And I'll be sworn upon't that he loves her,     85
    For here's a paper written in his hand,
    A halting sonnet of his own pure brain,
    Fashioned to Beatrice.

HERO             And here's another,
    Writ in my cousin's hand, stol'n from her pocket,
    Containing her affection unto Benedick.     90

BENEDICK A miracle! Here's our own hands against our
    hearts. Come, I will have thee, but by this light, I take
    thee for pity.

BEATRICE I would not deny you, but by this good day, I
    yield upon great persuasion, and partly to save your
    life, for I was told you were in a consumption.     96

BENEDICK (kissing her) Peace, I will stop your mouth.

DON PEDRO
    How dost thou, Benedick the married man?

BENEDICK I'll tell thee what, Prince: a college of wit-
crackers cannot flout me out of my humour. Dost thou
think I care for a satire or an epigram? No, if a man
will be beaten with brains, a shall wear nothing
handsome about him. In brief, since I do purpose to
marry, I will think nothing to any purpose that the
world can say against it, and therefore never flout at
me for what I have said against it. For man is a giddy
thing, and this is my conclusion. For thy part, Claudio,
I did think to have beaten thee, but in that thou art
like to be my kinsman, live unbruised, and love my
cousin.     110

CLAUDIO I had well hoped thou wouldst have denied
Beatrice, that I might have cudgelled thee out of thy
single life to make thee a double dealer, which out of
question thou wilt be, if my cousin do not look
exceeding narrowly to thee.     115

BENEDICK Come, come, we are friends, let's have a dance
ere we are married, that we may lighten our own
hearts and our wives' heels.

LEONATO We'll have dancing afterward.

BENEDICK First, of my word. Therefore play, music. (To
Don Pedro) Prince, thou art sad, get thee a wife, get
thee a wife. There is no staff more reverend than one
tipped with horn.

    Enter Messenger

MESSENGER
    My lord, your brother John is ta'en in flight,
    And brought with armèd men back to Messina.     125

BENEDICK Think not on him till tomorrow, I'll devise thee
brave punishments for him. Strike up, pipers.

    Dance, and exeunt

# HENRY V

THE Chorus to Act 5 of *Henry V* contains an uncharacteristic, direct topical reference:

> Were now the General of our gracious Empress—
> As in good time he may—from Ireland coming,
> Bringing rebellion broachèd on his sword,
> How many would the peaceful city quit
> To welcome him!

'The General' must be the Earl of Essex, whose 'Empress'—Queen Elizabeth—had sent him on an Irish campaign on 27 March 1599; he returned, disgraced, on 28 September. Plans for his campaign had been known at least since the previous November; the idea that he might return in triumph would have been meaningless after September 1599, and it seems likely that Shakespeare wrote his play during 1599, probably in the spring. It appeared in print, in a debased text, in (probably) August 1600, when it was said to have 'been sundry times played by the Right Honourable the Lord Chamberlain his servants'. Although this text seems to have been put together from memory by actors playing in an abbreviated adaptation, the Shakespearian text behind it appears to have been in a later state than the generally superior text printed from Shakespeare's own papers in the 1623 Folio. Our edition draws on the 1600 quarto in the attempt to represent the play as acted by Shakespeare's company. The principal difference is the reversion to historical authenticity in the substitution at Agincourt of the Duke of Bourbon for the Dauphin.

As in the two plays about Henry IV, Shakespeare is indebted to *The Famous Victories of Henry the Fifth* (printed 1598). Other Elizabethan plays about Henry V, now lost, may have influenced him; he certainly used the chronicle histories of Edward Hall (1542) and Holinshed (1577, revised and enlarged in 1587).

From the 'civil broils' of the earlier history plays, Shakespeare turns to portray a country united in war against France. Each act is prefaced by a Chorus, speaking some of the play's finest poetry, and giving it an epic quality. Henry V, 'star of England', is Shakespeare's most heroic warrior king, but (like his predecessors) has an introspective side, and is aware of the crime by which his father came to the throne. We are reminded of his 'wilder days', and see that the transition from 'madcap prince' to the 'mirror of all Christian kings' involves loss: although the epilogue to *2 Henry IV* had suggested that Sir John would reappear, he is only, though poignantly, an off-stage presence. Yet Shakespeare's infusion of comic form into historical narrative reaches its natural conclusion in this play. Sir John's cronies, Pistol, Bardolph, Nim, and Mistress Quickly, reappear to provide a counterpart to the heroic action, and Shakespeare invents comic episodes involving an Englishman (Gower), a Welshman (Fluellen), an Irishman (MacMorris), and a Scot (Jamy). The play also has romance elements, in the almost incredible extent of the English victory over the French and in the disguised Henry's comradely mingling with his soldiers, as well as in his courtship of the French princess. The play's romantic and heroic aspects have made it popular especially in times of war and have aroused accusations of jingoism, but the horrors of war are vividly depicted, and the Chorus's closing speech reminds us that Henry died young, and that his son's protector 'lost France and made his England bleed'.

# THE PERSONS OF THE PLAY

CHORUS

KING HARRY V of England, claimant to the French throne

Duke of GLOUCESTER  } his brothers
Duke of CLARENCE

Duke of EXETER, his uncle

Duke of YORK

SALISBURY

WESTMORLAND

WARWICK

Archbishop of CANTERBURY
Bishop of ELY

Richard, Earl of CAMBRIDGE  }
Henry, Lord SCROPE of Masham  } traitors
Sir Thomas GREY

PISTOL  }
NIM  } formerly Falstaff's companions
BARDOLPH

BOY, formerly Falstaff's page

HOSTESS, formerly Mistress Quickly, now Pistol's wife

Captain GOWER, an Englishman

Captain FLUELLEN, a Welshman

Captain MACMORRIS, an Irishman

Captain JAMY, a Scot

Sir Thomas ERPINGHAM

John BATES  }
Alexander COURT  } English soldiers
Michael WILLIAMS

HERALD

KING CHARLES VI of France

ISABEL, his wife and queen

The DAUPHIN, their son and heir

CATHERINE, their daughter

ALICE, an old gentlewoman

The CONSTABLE of France  }
Duke of BOURBON
Duke of ORLÉANS
Duke of BERRI  } French noblemen at Agincourt
Lord RAMBURES
Lord GRANDPRÉ

Duke of BURGUNDY

MONTJOY, the French Herald

GOVERNOR of Harfleur

French AMBASSADORS to England

# The Life of Henry the Fifth

**Prologue** *Enter Chorus as Prologue*

CHORUS

O for a muse of fire, that would ascend
The brightest heaven of invention:
A kingdom for a stage, princes to act,
And monarchs to behold the swelling scene.
Then should the warlike Harry, like himself,          5
Assume the port of Mars, and at his heels,
Leashed in like hounds, should famine, sword, and fire
Crouch for employment. But pardon, gentles all,
The flat unraisèd spirits that hath dared
On this unworthy scaffold to bring forth          10
So great an object. Can this cock-pit hold
The vasty fields of France? Or may we cram
Within this wooden O the very casques
That did affright the air at Agincourt?
O pardon: since a crookèd figure may          15
Attest in little place a million,
And let us, ciphers to this great account,
On your imaginary forces work.
Suppose within the girdle of these walls
Are now confined two mighty monarchies,          20
Whose high uprearèd and abutting fronts
The perilous narrow ocean parts asunder.
Piece out our imperfections with your thoughts:
Into a thousand parts divide one man,
And make imaginary puissance.          25
Think, when we talk of horses, that you see them,
Printing their proud hoofs i'th' receiving earth;
For 'tis your thoughts that now must deck our kings,
Carry them here and there, jumping o'er times,
Turning th'accomplishment of many years          30
Into an hourglass—for the which supply,
Admit me Chorus to this history,
Who Prologue-like your humble patience pray
Gently to hear, kindly to judge, our play.          *Exit*

**1.1** *Enter the Archbishop of Canterbury and the Bishop
of Ely*

CANTERBURY

My lord, I'll tell you. That self bill is urged
Which in th'eleventh year of the last king's reign
Was like, and had indeed against us passed,
But that the scrambling and unquiet time
Did push it out of farther question.          5

ELY

But how, my lord, shall we resist it now?

CANTERBURY

It must be thought on. If it pass against us,
We lose the better half of our possession,
For all the temporal lands which men devout
By testament have given to the Church          10
Would they strip from us—being valued thus:
As much as would maintain, to the King's honour,
Full fifteen earls and fifteen hundred knights,
Six thousand and two hundred good esquires;
And, to relief of lazars and weak age,          15
Of indigent faint souls past corporal toil,
A hundred almshouses right well supplied;
And to the coffers of the King beside
A thousand pounds by th' year. Thus runs the bill.

ELY  This would drink deep.          20
CANTERBURY  'Twould drink the cup and all.
ELY  But what prevention?
CANTERBURY
The King is full of grace and fair regard.
ELY
And a true lover of the holy Church.
CANTERBURY
The courses of his youth promised it not.          25
The breath no sooner left his father's body
But that his wildness, mortified in him,
Seemed to die too. Yea, at that very moment
Consideration like an angel came
And whipped th'offending Adam out of him,          30
Leaving his body as a paradise
T'envelop and contain celestial spirits.
Never was such a sudden scholar made;
Never came reformation in a flood
With such a heady currance scouring faults;          35
Nor never Hydra-headed wilfulness
So soon did lose his seat—and all at once—
As in this king.
ELY                                We are blessèd in the change.
CANTERBURY
Hear him but reason in divinity
And, all-admiring, with an inward wish          40
You would desire the King were made a prelate;
Hear him debate of commonwealth affairs,
You would say it hath been all-in-all his study;
List his discourse of war, and you shall hear
A fearful battle rendered you in music;          45
Turn him to any cause of policy,
The Gordian knot of it he will unloose,
Familiar as his garter—that when he speaks,
The air, a chartered libertine, is still,
And the mute wonder lurketh in men's ears          50
To steal his sweet and honeyed sentences:
So that the art and practic part of life
Must be the mistress to this theoric.
Which is a wonder how his grace should glean it,
Since his addiction was to courses vain,          55
His companies unlettered, rude, and shallow,
His hours filled up with riots, banquets, sports,
And never noted in him any study,
Any retirement, any sequestration
From open haunts and popularity.          60
ELY
The strawberry grows underneath the nettle,
And wholesome berries thrive and ripen best
Neighboured by fruit of baser quality;
And so the Prince obscured his contemplation
Under the veil of wildness—which, no doubt,          65
Grew like the summer grass, fastest by night,
Unseen, yet crescive in his faculty.
CANTERBURY
It must be so, for miracles are ceased,
And therefore we must needs admit the means
How things are perfected.
ELY                                But, my good lord,          70
How now for mitigation of this bill

Urged by the Commons? Doth his majesty
Incline to it, or no?
CANTERBURY                    He seems indifferent,
Or rather swaying more upon our part
Than cherishing th'exhibitors against us;                    75
For I have made an offer to his majesty,
Upon our spiritual convocation
And in regard of causes now in hand,
Which I have opened to his grace at large:
As touching France, to give a greater sum                    80
Than ever at one time the clergy yet
Did to his predecessors part withal.
ELY
How did this offer seem received, my lord?
CANTERBURY
With good acceptance of his majesty,
Save that there was not time enough to hear,                    85
As I perceived his grace would fain have done,
The severals and unhidden passages
Of his true titles to some certain dukedoms,
And generally to the crown and seat of France,
Derived from Edward, his great-grandfather.                    90
ELY
What was th'impediment that broke this off?
CANTERBURY
The French ambassador upon that instant
Craved audience—and the hour I think is come
To give him hearing. Is it four o'clock?
ELY It is.                    95
CANTERBURY
Then go we in, to know his embassy—
Which I could with a ready guess declare
Before the Frenchman speak a word of it.
ELY
I'll wait upon you, and I long to hear it.                    *Exeunt*

**1.2**    *Enter King Harry, the Dukes of Gloucester,*
        *⌜Clarence⌝, and Exeter, and the Earls of Warwick*
        *and Westmorland*
KING HARRY
Where is my gracious lord of Canterbury?
EXETER
Not here in presence.
KING HARRY                    Send for him, good uncle.
WESTMORLAND
Shall we call in th'ambassador, my liege?
KING HARRY
Not yet, my cousin. We would be resolved,
Before we hear him, of some things of weight                    5
That task our thoughts, concerning us and France.
        *Enter the Archbishop of Canterbury and the Bishop*
        *of Ely*
CANTERBURY
God and his angels guard your sacred throne,
And make you long become it.
KING HARRY                    Sure we thank you.
My learnèd lord, we pray you to proceed,
And justly and religiously unfold                    10
Why the law Salic that they have in France
Or should or should not bar us in our claim.
And God forbid, my dear and faithful lord,
That you should fashion, wrest, or bow your reading,
Or nicely charge your understanding soul                    15
With opening titles miscreate, whose right
Suits not in native colours with the truth;

For God doth know how many now in health
Shall drop their blood in approbation
Of what your reverence shall incite us to.                    20
Therefore take heed how you impawn our person,
How you awake our sleeping sword of war;
We charge you in the name of God take heed.
For never two such kingdoms did contend
Without much fall of blood, whose guiltless drops    25
Are every one a woe, a sore complaint
'Gainst him whose wrongs gives edge unto the swords
That makes such waste in brief mortality.
Under this conjuration speak, my lord,
For we will hear, note, and believe in heart                    30
That what you speak is in your conscience washed
As pure as sin with baptism.
CANTERBURY
Then hear me, gracious sovereign, and you peers
That owe your selves, your lives, and services
To this imperial throne. There is no bar                    35
To make against your highness' claim to France
But this, which they produce from Pharamond:
'*In terram Salicam mulieres ne succedant*'—
'No woman shall succeed in Salic land'—
Which 'Salic land' the French unjustly gloss    40
To be the realm of France, and Pharamond
The founder of this law and female bar.
Yet their own authors faithfully affirm
That the land Salic is in Germany,
Between the floods of Saale and of Elbe,                    45
Where, Charles the Great having subdued the Saxons,
There left behind and settled certain French
Who, holding in disdain the German women
For some dishonest manners of their life,
Established there this law: to wit, no female    50
Should be inheritrix in Salic land—
Which Salic, as I said, 'twixt Elbe and Saale,
Is at this day in Germany called Meissen.
Then doth it well appear the Salic Law
Was not devisèd for the realm of France.                    55
Nor did the French possess the Salic land
Until four hundred one-and-twenty years
After defunction of King Pharamond,
Idly supposed the founder of this law,
Who died within the year of our redemption    60
Four hundred twenty-six; and Charles the Great
Subdued the Saxons, and did seat the French
Beyond the river Saale, in the year
Eight hundred five. Besides, their writers say,
King Pépin, which deposèd Childéric,                    65
Did, as heir general—being descended
Of Blithild, which was daughter to King Clotaire—
Make claim and title to the crown of France.
Hugh Capet also—who usurped the crown
Of Charles the Duke of Lorraine, sole heir male    70
Of the true line and stock of Charles the Great—
To fine his title with some shows of truth,
Though in pure truth it was corrupt and naught,
Conveyed himself as heir to th' Lady Lingard,
Daughter to Charlemain, who was the son    75
To Louis the Emperor, and Louis the son
Of Charles the Great. Also, King Louis the Ninth,
Who was sole heir to the usurper Capet,
Could not keep quiet in his conscience,
Wearing the crown of France, till satisfied    80
That fair Queen Isabel, his grandmother,

Was lineal of the Lady Ermengarde,
Daughter to Charles, the foresaid Duke of Lorraine;
By the which marriage, the line of Charles the Great
Was reunited to the crown of France.                    85
So that, as clear as is the summer's sun,
King Pépin's title and Hugh Capet's claim,
King Louis his satisfaction, all appear
To hold in right and title of the female;
So do the kings of France unto this day,                90
Howbeit they would hold up this Salic Law
To bar your highness claiming from the female,
And rather choose to hide them in a net
Than amply to embar their crookèd titles,
Usurped from you and your progenitors.                  95
KING HARRY
May I with right and conscience make this claim?
CANTERBURY
The sin upon my head, dread sovereign.
For in the Book of Numbers is it writ,
'When the son dies, let the inheritance
Descend unto the daughter.' Gracious lord,             100
Stand for your own; unwind your bloody flag;
Look back into your mighty ancestors.
Go, my dread lord, to your great-grandsire's tomb,
From whom you claim; invoke his warlike spirit,
And your great-uncle's, Edward the Black Prince,       105
Who on the French ground played a tragedy,
Making defeat on the full power of France,
Whiles his most mighty father on a hill
Stood smiling to behold his lion's whelp
Forage in blood of French nobility.                    110
O noble English, that could entertain
With half their forces the full pride of France,
And let another half stand laughing by,
All out of work, and cold for action.
ELY
Awake remembrance of those valiant dead,               115
And with your puissant arm renew their feats.
You are their heir, you sit upon their throne,
The blood and courage that renownèd them
Runs in your veins—and my thrice-puissant liege
Is in the very May-morn of his youth,                  120
Ripe for exploits and mighty enterprises.
EXETER
Your brother kings and monarchs of the earth
Do all expect that you should rouse yourself
As did the former lions of your blood.
WESTMORLAND
They know your grace hath cause; and means and
          might,                                        125
So hath your highness. Never king of England
Had nobles richer and more loyal subjects,
Whose hearts have left their bodies here in England
And lie pavilioned in the fields of France.
CANTERBURY
O let their bodies follow, my dear liege,              130
With blood and sword and fire, to win your right.
In aid whereof, we of the spiritualty
Will raise your highness such a mighty sum
As never did the clergy at one time
Bring in to any of your ancestors.                     135
KING HARRY
We must not only arm t'invade the French,
But lay down our proportions to defend
Against the Scot, who will make raid upon us
With all advantages.

CANTERBURY
They of those marches, gracious sovereign,             140
Shall be a wall sufficient to defend
Our inland from the pilfering borderers.
KING HARRY
We do not mean the coursing snatchers only,
But fear the main intendment of the Scot,
Who hath been still a giddy neighbour to us.           145
For you shall read that my great-grandfather
Never unmasked his power unto France
But that the Scot on his unfurnished kingdom
Came pouring like the tide into a breach
With ample and brim fullness of his force              150
Galling the gleanèd land with hot assays,
Girding with grievous siege castles and towns,
That England, being empty of defence,
Hath shook and trembled at the bruit thereof.
CANTERBURY
She hath been then more feared than harmed, my
          liege.                                        155
For hear her but exampled by herself:
When all her chivalry hath been in France
And she a mourning widow of her nobles,
She hath herself not only well defended
But taken and impounded as a stray                     160
The King of Scots, whom she did send to France
To fill King Edward's fame with prisoner kings
And make your chronicle as rich with praise
As is the ooze and bottom of the sea
With sunken wrack and sumless treasuries.              165
⌜A LORD⌝
But there's a saying very old and true:
          'If that you will France win,
          Then with Scotland first begin.'
For once the eagle England being in prey,
To her unguarded nest the weasel Scot                  170
Comes sneaking, and so sucks her princely eggs,
Playing the mouse in absence of the cat,
To 'tame and havoc more than she can eat.
EXETER
It follows then the cat must stay at home.
Yet that is but a crushed necessity,                   175
Since we have locks to safeguard necessaries
And pretty traps to catch the petty thieves.
While that the armèd hand doth fight abroad,
Th'advisèd head defends itself at home.
For government, though high and low and lower,  180
Put into parts, doth keep in one consent,
Congreeing in a full and natural close,
Like music.
CANTERBURY  True. Therefore doth heaven divide
The state of man in divers functions,
Setting endeavour in continual motion;                 185
To which is fixèd, as an aim or butt,
Obedience. For so work the honey-bees,
Creatures that by a rule in nature teach
The act of order to a peopled kingdom.
They have a king, and officers of sorts,               190
Where some like magistrates correct at home;
Others like merchants venture trade abroad;
Others like soldiers, armèd in their stings,
Make boot upon the summer's velvet buds,
Which pillage they with merry march bring home  195
To the tent royal of their emperor,
Who busied in his majesty surveys
The singing masons building roofs of gold,

The civil citizens lading up the honey,
The poor mechanic porters crowding in                              200
Their heavy burdens at his narrow gate,
The sad-eyed justice with his surly hum
Delivering o'er to executors pale
The lazy yawning drone. I this infer:
That many things, having full reference                           205
To one consent, may work contrariously.
As many arrows, loosèd several ways,
Fly to one mark, as many ways meet in one town,
As many fresh streams meet in one salt sea,
As many lines close in the dial's centre,                         210
So may a thousand actions once afoot
End in one purpose, and be all well borne
Without defect. Therefore to France, my liege.
Divide your happy England into four,
Whereof take you one quarter into France,                         215
And you withal shall make all Gallia shake.
If we with thrice such powers left at home
Cannot defend our own doors from the dog,
Let us be worried, and our nation lose
The name of hardiness and policy.                                 220

KING HARRY
Call in the messengers sent from the Dauphin.
*Exit one or more*
Now are we well resolved, and by God's help
And yours, the noble sinews of our power,
France being ours we'll bend it to our awe,
Or break it all to pieces. Or there we'll sit,                    225
Ruling in large and ample empery
O'er France and all her almost kingly dukedoms,
Or lay these bones in an unworthy urn,
Tombless, with no remembrance over them.
Either our history shall with full mouth                          230
Speak freely of our acts, or else our grave,
Like Turkish mute, shall have a tongueless mouth,
Not worshipped with a waxen epitaph.
*Enter Ambassadors of France, with a tun*
Now are we well prepared to know the pleasure
Of our fair cousin Dauphin, for we hear                           235
Your greeting is from him, not from the King.

AMBASSADOR
May't please your majesty to give us leave
Freely to render what we have in charge,
Or shall we sparingly show you far off
The Dauphin's meaning and our embassy?                            240

KING HARRY
We are no tyrant, but a Christian king,
Unto whose grace our passion is as subject
As is our wretches fettered in our prisons.
Therefore with frank and with uncurbèd plainness
Tell us the Dauphin's mind.

AMBASSADOR                          Thus then in few:              245
Your highness lately sending into France
Did claim some certain dukedoms, in the right
Of your great predecessor, King Edward the Third.
In answer of which claim, the Prince our master
Says that you savour too much of your youth,                      250
And bids you be advised, there's naught in France
That can be with a nimble galliard won:
You cannot revel into dukedoms there.
He therefore sends you, meeter for your spirit,
This tun of treasure, and in lieu of this                         255
Desires you let the dukedoms that you claim
Hear no more of you. This the Dauphin speaks.

KING HARRY
What treasure, uncle?

EXETER (*opening the tun*) Tennis balls, my liege.

KING HARRY
We are glad the Dauphin is so pleasant with us.
His present and your pains we thank you for.                      260
When we have matched our rackets to these balls,
We will in France, by God's grace, play a set
Shall strike his father's crown into the hazard.
Tell him he hath made a match with such a wrangler
That all the courts of France will be disturbed                   265
With chases. And we understand him well,
How he comes o'er us with our wilder days,
Not measuring what use we made of them.
We never valued this poor seat of England,
And therefore, living hence, did give ourself                     270
To barbarous licence—as 'tis ever common
That men are merriest when they are from home.
But tell the Dauphin I will keep my state,
Be like a king, and show my sail of greatness
When I do rouse me in my throne of France.                        275
For that have I laid by my majesty
And plodded like a man for working days,
But I will rise there with so full a glory
That I will dazzle all the eyes of France,
Yea strike the Dauphin blind to look on us.                       280
And tell the pleasant Prince this mock of his
Hath turned his balls to gunstones, and his soul
Shall stand sore chargèd for the wasteful vengeance
That shall fly from them—for many a thousand
widows                                                            284
Shall this his mock mock out of their dear husbands,
Mock mothers from their sons, mock castles down;
Ay, some are yet ungotten and unborn
That shall have cause to curse the Dauphin's scorn.
But this lies all within the will of God,
To whom I do appeal, and in whose name                            290
Tell you the Dauphin I am coming on
To venge me as I may, and to put forth
My rightful hand in a well-hallowed cause.
So get you hence in peace. And tell the Dauphin
His jest will savour but of shallow wit                           295
When thousands weep more than did laugh at it.—
Convey them with safe conduct.—Fare you well.
*Exeunt Ambassadors*

EXETER This was a merry message.

KING HARRY
We hope to make the sender blush at it.
Therefore, my lords, omit no happy hour                           300
That may give furth'rance to our expedition;
For we have now no thought in us but France,
Save those to God, that run before our business.
Therefore let our proportions for these wars
Be soon collected, and all things thought upon                   305
That may with reasonable swiftness add
More feathers to our wings; for, God before,
We'll chide this Dauphin at his father's door.
Therefore let every man now task his thought,
That this fair action may on foot be brought.                     310
⌜*Flourish.*⌝ *Exeunt*

2.0     *Enter Chorus*

CHORUS
Now all the youth of England are on fire,
And silken dalliance in the wardrobe lies;

Now thrive the armourers, and honour's thought
Reigns solely in the breast of every man.
They sell the pasture now to buy the horse,                    5
Following the mirror of all Christian kings
With wingèd heels, as English Mercuries.
For now sits expectation in the air
And hides a sword from hilts unto the point
With crowns imperial, crowns and coronets,                    10
Promised to Harry and his followers.
The French, advised by good intelligence
Of this most dreadful preparation,
Shake in their fear, and with pale policy
Seek to divert the English purposes.                          15
O England!—model to thy inward greatness,
Like little body with a mighty heart,
What mightst thou do, that honour would thee do,
Were all thy children kind and natural?
But see, thy fault France hath in thee found out:             20
A nest of hollow bosoms, which he fills
With treacherous crowns; and three corrupted men—
One, Richard, Earl of Cambridge; and the second
Henry, Lord Scrope of Masham; and the third
Sir Thomas Grey, knight, of Northumberland—                   25
Have, for the gilt of France—O guilt indeed!—
Confirmed conspiracy with fearful France;
And by their hands this grace of kings must die,
If hell and treason hold their promises,
Ere he take ship for France, and in Southampton.              30
Linger your patience on, and we'll digest
Th'abuse of distance, force—perforce—a play.
The sum is paid, the traitors are agreed,
The King is set from London, and the scene
Is now transported, gentles, to Southampton.                  35
There is the playhouse now, there must you sit,
And thence to France shall we convey you safe,
And bring you back, charming the narrow seas
To give you gentle pass—for if we may
We'll not offend one stomach with our play.                   40
But till the King come forth, and not till then,
Unto Southampton do we shift our scene.            *Exit*

**2.1**    *Enter Corporal Nim and Lieutenant Bardolph*
BARDOLPH Well met, Corporal Nim.
NIM Good morrow, Lieutenant Bardolph.
BARDOLPH What, are Ensign Pistol and you friends yet?
NIM For my part, I care not. I say little, but when time
   shall serve, there shall be smiles—but that shall be as
   it may. I dare not fight, but I will wink and hold out
   mine iron. It is a simple one, but what though? It will
   toast cheese, and it will endure cold, as another man's
   sword will—and there's an end.                             9
BARDOLPH I will bestow a breakfast to make you friends,
   and we'll be all three sworn brothers to France. Let't
   be so, good Corporal Nim.
NIM Faith, I will live so long as I may, that's the certain
   of it, and when I cannot live any longer, I will do as I
   may. That is my rest, that is the rendezvous of it.       15
BARDOLPH It is certain, corporal, that he is married to
   Nell Quickly, and certainly she did you wrong, for you
   were troth-plight to her.
NIM I cannot tell. Things must be as they may. Men may
   sleep, and they may have their throats about them at
   that time, and some say knives have edges. It must be
   as it may. Though Patience be a tired mare, yet she
   will plod. There must be conclusions. Well, I cannot
   tell.                                                      24

*Enter Ensign Pistol and Hostess Quickly*
BARDOLPH Good morrow, Ensign Pistol. (*To Nim*) Here
   comes Ensign Pistol and his wife. Good Corporal, be
   patient here.
⌈NIM⌉ How now, mine host Pistol?
PISTOL
   Base tick, call'st thou me host? Now by Gad's lugs
   I swear I scorn the term. Nor shall my Nell keep
       lodgers.                                               30
HOSTESS No, by my troth, not long, for we cannot lodge
   and board a dozen or fourteen gentlewomen that live
   honestly by the prick of their needles, but it will be
   thought we keep a bawdy-house straight.
       ⌈*Nim draws his sword*⌉
   O well-a-day, Lady! If he be not hewn now, we shall
   see wilful adultery and murder committed.                 36
       ⌈*Pistol draws his sword*⌉
BARDOLPH Good lieutenant, good corporal, offer nothing
   here.
NIM Pish.
PISTOL
   Pish for thee, Iceland dog. Thou prick-eared cur of
       Iceland.                                               40
HOSTESS Good Corporal Nim, show thy valour, and put
   up your sword.
       *They sheathe their swords*
NIM Will you shog off? I would have you *solus*.
PISTOL
   'Solus', egregious dog? O viper vile!
   The *solus* in thy most marvellous face,                  45
   The *solus* in thy teeth, and in thy throat,
   And in thy hateful lungs, yea in thy maw pardie—
   And which is worse, within thy nasty mouth.
   I do retort the *solus* in thy bowels,
   For I can take, and Pistol's cock is up,                  50
   And flashing fire will follow.
NIM I am not Barbason, you cannot conjure me. I have
   an humour to knock you indifferently well. If you grow
   foul with me, Pistol, I will scour you with my rapier,
   as I may, in fair terms. If you would walk off, I would
   prick your guts a little, in good terms, as I may, and
   that's the humour of it.                                  57
PISTOL
   O braggart vile, and damnèd furious wight!
   The grave doth gape and doting death is near.
   Therefore ex-hale.                                        60
       *Pistol and Nim draw their swords*
BARDOLPH Hear me, hear me what I say.
       ⌈*He draws his sword*⌉
   He that strikes the first stroke, I'll run him up to the
   hilts, as I am a soldier.
PISTOL
   An oath of mickle might, and fury shall abate.
       ⌈*They sheathe their swords*⌉
   (*To Nim*) Give me thy fist, thy forefoot to me give.     65
   Thy spirits are most tall.
NIM I will cut thy throat one time or other, in fair terms,
   that is the humour of it.
PISTOL *Couple a gorge*,
   That is the word. I thee defy again.                      70
   O hound of Crete, think'st thou my spouse to get?
   No, to the spital go,
   And from the powd'ring tub of infamy
   Fetch forth the lazar kite of Cressid's kind,
   Doll Tearsheet she by name, and her espouse.              75
   I have, and I will hold, the quondam Quickly
   For the only she, and—*pauca*, there's enough. Go to.

*Enter the Boy ⌐running⌐*

BOY Mine host Pistol, you must come to my master, and
you, hostess. He is very sick, and would to bed.—Good
Bardolph, put thy face between his sheets, and do the
office of a warming-pan.—Faith, he's very ill.            81

BARDOLPH Away, you rogue!

HOSTESS By my troth, he'll yield the crow a pudding one
of these days. The King has killed his heart. Good
husband, come home presently.            *Exit ⌐with Boy⌐*

BARDOLPH Come, shall I make you two friends? We must
to France together. Why the devil should we keep
knives to cut one another's throats?

PISTOL
Let floods o'erswell, and fiends for food howl on!     89

NIM You'll pay me the eight shillings I won of you at
betting?

PISTOL Base is the slave that pays.

NIM That now I will have. That's the humour of it.

PISTOL
As manhood shall compound. Push home.            94
*Pistol and Nim draw their swords*

BARDOLPH ⌐*drawing his sword*⌐ By this sword, he that makes
the first thrust, I'll kill him. By this sword, I will.

PISTOL
Sword is an oath, and oaths must have their course.
⌐*He sheathes his sword*⌐

BARDOLPH Corporal Nim, an thou wilt be friends, be
friends. An thou wilt not, why then be enemies with
me too. Prithee, put up.            100

NIM I shall have my eight shillings?

PISTOL
A noble shalt thou have, and present pay,
And liquor likewise will I give to thee,
And friendship shall combine, and brotherhood.
I'll live by Nim, and Nim shall live by me.            105
Is not this just? For I shall sutler
Unto the camp, and profits will accrue.
Give me thy hand.

NIM I shall have my noble?

PISTOL In cash, most justly paid.            110

NIM Well then, that's the humour of't.
⌐*Nim and Bardolph sheathe their swords.*⌐
*Enter Hostess Quickly*

HOSTESS As ever you come of women, come in quickly to
Sir John. Ah, poor heart, he is so shaked of a burning
quotidian-tertian, that it is most lamentable to behold.
Sweet men, come to him.            ⌐*Exit*⌐

NIM The King hath run bad humours on the knight,
that's the even of it.            117

PISTOL Nim, thou hast spoke the right.
His heart is fracted and corroborate.

NIM The King is a good king, but it must be as it may.
He passes some humours and careers.            121

PISTOL
Let us condole the knight—for, lambkins, we will live.
*Exeunt*

**2.2** *Enter the Dukes of Exeter and ⌐Gloucester⌐, and the
Earl of Westmorland*

⌐GLOUCESTER⌐
Fore God, his grace is bold to trust these traitors.

EXETER
They shall be apprehended by and by.

WESTMORLAND
How smooth and even they do bear themselves,
As if allegiance in their bosoms sat,
Crownèd with faith and constant loyalty.            5

⌐GLOUCESTER⌐
The King hath note of all that they intend,
By interception which they dream not of.

EXETER
Nay, but the man that was his bedfellow,
Whom he hath dulled and cloyed with gracious
favours—
That he should for a foreign purse so sell            10
His sovereign's life to death and treachery.
*Sound trumpets. Enter King Harry, Lord Scrope, the
Earl of Cambridge, and Sir Thomas Grey*

KING HARRY
Now sits the wind fair, and we will aboard.
My lord of Cambridge, and my kind lord of Masham,
And you, my gentle knight, give me your thoughts.
Think you not that the powers we bear with us            15
Will cut their passage through the force of France,
Doing the execution and the act
For which we have in head assembled them?

SCROPE
No doubt, my liege, if each man do his best.

KING HARRY
I doubt not that, since we are well persuaded            20
We carry not a heart with us from hence
That grows not in a fair consent with ours,
Nor leave not one behind that doth not wish
Success and conquest to attend on us.

CAMBRIDGE
Never was monarch better feared and loved            25
Than is your majesty. There's not, I think, a subject
That sits in heart-grief and uneasiness
Under the sweet shade of your government.

GREY
True. Those that were your father's enemies
Have steeped their galls in honey, and do serve you
With hearts create of duty and of zeal.            31

KING HARRY
We therefore have great cause of thankfulness,
And shall forget the office of our hand
Sooner than quittance of desert and merit,
According to their weight and worthiness.            35

SCROPE
So service shall with steelèd sinews toil,
And labour shall refresh itself with hope,
To do your grace incessant services.

KING HARRY
We judge no less.—Uncle of Exeter,
Enlarge the man committed yesterday            40
That railed against our person. We consider
It was excess of wine that set him on,
And on his more advice we pardon him.

SCROPE
That's mercy, but too much security.
Let him be punished, sovereign, lest example            45
Breed, by his sufferance, more of such a kind.

KING HARRY
O let us yet be merciful.

CAMBRIDGE
So may your highness, and yet punish too.

GREY
Sir, you show great mercy if you give him life,
After the taste of much correction.            50

KING HARRY
Alas, your too much love and care of me
Are heavy orisons 'gainst this poor wretch.
If little faults proceeding on distemper

Shall not be winked at, how shall we stretch our eye
When capital crimes, chewed, swallowed, and
    digested,                                               55
Appear before us? We'll yet enlarge that man,
Though Cambridge, Scrope, and Grey, in their dear
    care
And tender preservation of our person,
Would have him punished. And now to our French
    causes.
Who are the late commissioners?
CAMBRIDGE                          I one, my lord.       60
Your highness bade me ask for it today.
SCROPE
So did you me, my liege.
GREY                      And I, my royal sovereign.
KING HARRY
Then Richard, Earl of Cambridge, there is yours;
There yours, Lord Scrope of Masham, and sir knight,
Grey of Northumberland, this same is yours.            65
Read them, and know I know your worthiness.—
My lord of Westmorland, and Uncle Exeter,
We will aboard tonight.—Why, how now, gentlemen?
What see you in those papers, that you lose
So much complexion?—Look ye how they change:          70
Their cheeks are paper.—Why, what read you there
That have so cowarded and chased your blood
Out of appearance?
CAMBRIDGE              I do confess my fault,
And do submit me to your highness' mercy.
GREY and SCROPE To which we all appeal.                75
KING HARRY
The mercy that was quick in us but late
By your own counsel is suppressed and killed.
You must not dare, for shame, to talk of mercy,
For your own reasons turn into your bosoms,
As dogs upon their masters, worrying you.—            80
See you, my princes and my noble peers,
These English monsters? My lord of Cambridge here,
You know how apt our love was to accord
To furnish him with all appurtenants
Belonging to his honour; and this vile man             85
Hath for a few light crowns lightly conspired
And sworn unto the practices of France
To kill us here in Hampton. To the which
This knight, no less for bounty bound to us
Than Cambridge is, hath likewise sworn. But O          90
What shall I say to thee, Lord Scrope, thou cruel,
Ingrateful, savage, and inhuman creature?
Thou that didst bear the key of all my counsels,
That knew'st the very bottom of my soul,
That almost mightst ha' coined me into gold             95
Wouldst thou ha' practised on me for thy use:
May it be possible that foreign hire
Could out of thee extract one spark of evil
That might annoy my finger? 'Tis so strange
That though the truth of it stands off as gross        100
As black on white, my eye will scarcely see it.
Treason and murder ever kept together,
As two yoke-devils sworn to either's purpose,
Working so grossly in a natural cause
That admiration did not whoop at them;                 105
But thou, 'gainst all proportion, didst bring in
Wonder to wait on treason and on murder.
And whatsoever cunning fiend it was
That wrought upon thee so preposterously
Hath got the voice in hell for excellence.            110

And other devils that suggest by treasons
Do botch and bungle up damnation
With patches, colours, and with forms, being fetched
From glist'ring semblances of piety;
But he that tempered thee, bade thee stand up,        115
Gave thee no instance why thou shouldst do treason,
Unless to dub thee with the name of traitor.
If that same demon that hath gulled thee thus
Should with his lion gait walk the whole world,
He might return to vasty Tartar back                  120
And tell the legions, 'I can never win
A soul so easy as that Englishman's.'
O how hast thou with jealousy infected
The sweetness of affiance. Show men dutiful?
Why so didst thou. Seem they grave and learned?       125
Why so didst thou. Come they of noble family?
Why so didst thou. Seem they religious?
Why so didst thou. Or are they spare in diet,
Free from gross passion, or of mirth or anger,
Constant in spirit, not swerving with the blood,      130
Garnished and decked in modest complement,
Not working with the eye without the ear,
And but in purgèd judgement trusting neither?
Such, and so finely boulted, didst thou seem.
And thus thy fall hath left a kind of blot            135
To mark the full-fraught man, and best endowed,
With some suspicion. I will weep for thee,
For this revolt of thine methinks is like
Another fall of man.—Their faults are open.
Arrest them to the answer of the law,                 140
And God acquit them of their practices.
EXETER I arrest thee of high treason, by the name of
    Richard, Earl of Cambridge.—I arrest thee of high
    treason, by the name of Henry, Lord Scrope of
    Masham.—I arrest thee of high treason, by the name
    of Thomas Grey, knight, of Northumberland.        146
SCROPE
Our purposes God justly hath discovered,
And I repent my fault more than my death,
Which I beseech your highness to forgive
Although my body pay the price of it.                 150
CAMBRIDGE
For me, the gold of France did not seduce,
Although I did admit it as a motive
The sooner to effect what I intended.
But God be thankèd for prevention,
Which heartily in sufferance will rejoice,            155
Beseeching God and you to pardon me.
GREY
Never did faithful subject more rejoice
At the discovery of most dangerous treason
Than I do at this hour joy o'er myself,
Prevented from a damnèd enterprise.                   160
My fault, but not my body, pardon, sovereign.
KING HARRY
God 'quit you in his mercy. Hear your sentence.
You have conspired against our royal person,
Joined with an enemy proclaimed and fixed,
And from his coffers                                  165
Received the golden earnest of our death,
Wherein you would have sold your king to slaughter,
His princes and his peers to servitude,
His subjects to oppression and contempt,
And his whole kingdom into desolation.                170
Touching our person seek we no revenge,
But we our kingdom's safety must so tender,

575

Whose ruin you have sought, that to her laws
We do deliver you. Get ye therefore hence,
Poor miserable wretches, to your death;                    175
The taste whereof, God of his mercy give
You patience to endure, and true repentance
Of all your dear offences.—Bear them hence.
                    *Exeunt the traitors, guarded*
Now lords for France, the enterprise whereof
Shall be to you, as us, like glorious.                    180
We doubt not of a fair and lucky war,
Since God so graciously hath brought to light
This dangerous treason lurking in our way
To hinder our beginnings. We doubt not now
But every rub is smoothèd on our way.                    185
Then forth, dear countrymen. Let us deliver
Our puissance into the hand of God,
Putting it straight in expedition.
Cheerly to sea, the signs of war advance:
No king of England, if not king of France.                    190
                    *Flourish. Exeunt*

**2.3**    *Enter Ensign Pistol, Corporal Nim, Lieutenant*
        *Bardolph, Boy, and Hostess Quickly*

HOSTESS Prithee, honey, sweet husband, let me bring thee
  to Staines.
PISTOL
  No, for my manly heart doth erne. Bardolph,
  Be blithe; Nim, rouse thy vaunting veins; boy, bristle
  Thy courage up. For Falstaff he is dead,                    5
  And we must earn therefore.
BARDOLPH Would I were with him, wheresome'er he is,
  either in heaven or in hell.
HOSTESS Nay, sure he's not in hell. He's in Arthur's
  bosom, if ever man went to Arthur's bosom. A made
  a finer end, and went away an it had been any christom
  child. A parted ev'n just between twelve and one, ev'n
  at the turning o'th' tide—for after I saw him fumble
  with the sheets, and play with flowers, and smile upon
  his finger's end, I knew there was but one way. For
  his nose was as sharp as a pen, and a babbled of green
  fields. 'How now, Sir John?' quoth I. 'What, man! Be
  o' good cheer.' So a cried out, 'God, God, God', three
  or four times. Now I, to comfort him, bid him a should
  not think of God; I hoped there was no need to trouble
  himself with any such thoughts yet. So a bade me lay
  more clothes on his feet. I put my hand into the bed
  and felt them, and they were as cold as any stone.
  Then I felt to his knees, and so up'ard and up'ard, and
  all was as cold as any stone.                    25
NIM They say he cried out of sack.
HOSTESS Ay, that a did.
BARDOLPH And of women.
HOSTESS Nay, that a did not.                    29
BOY Yes, that a did, and said they were devils incarnate.
HOSTESS A could never abide carnation, 'twas a colour he
  never liked.
BOY A said once the devil would have him about women.
HOSTESS A did in some sort, indeed, handle women—but
  then he was rheumatic, and talked of the Whore of
  Babylon.                    36
BOY Do you not remember, a saw a flea stick upon
  Bardolph's nose, and a said it was a black soul burning
  in hell-fire.
BARDOLPH Well, the fuel is gone that maintained that fire.
  That's all the riches I got in his service.                    41

NIM Shall we shog? The King will be gone from
  Southampton.
PISTOL
  Come, let's away.—My love, give me thy lips.
      *He kisses her*
  Look to my chattels and my movables.                    45
  Let senses rule. The word is 'Pitch and pay'.
  Trust none, for oaths are straws, men's faiths are
    wafer-cakes,
  And Holdfast is the only dog, my duck.
  Therefore *caveto* be thy counsellor.
  Go, clear thy crystals.—Yokefellows in arms,                    50
  Let us to France, like horseleeches, my boys,
  To suck, to suck, the very blood to suck!
BOY (*aside*) And that's but unwholesome food, they say.
PISTOL Touch her soft mouth, and march.
BARDOLPH Farewell, hostess.                    55
      *He kisses her*
NIM I cannot kiss, that is the humour of it, but adieu.
PISTOL (*to Hostess*)
  Let housewifery appear. Keep close, I thee command.
HOSTESS Farewell! Adieu!                    *Exeunt severally*

**2.4**    *Flourish. Enter King Charles the Sixth of France,*
        *the Dauphin, the Constable, and the Dukes of Berri*
        *and ⌈Bourbon⌉*

KING CHARLES
  Thus comes the English with full power upon us,
  And more than carefully it us concerns
  To answer royally in our defences.
  Therefore the Dukes of Berri and of Bourbon,
  Of Brabant and of Orléans shall make forth,                    5
  And you Prince Dauphin, with all swift dispatch
  To line and new-repair our towns of war
  With men of courage and with means defendant.
  For England his approaches makes as fierce
  As waters to the sucking of a gulf.                    10
  It fits us then to be as provident
  As fear may teach us, out of late examples
  Left by the fatal and neglected English
  Upon our fields.
DAUPHIN          My most redoubted father,
  It is most meet we arm us 'gainst the foe,                    15
  For peace itself should not so dull a kingdom—
  Though war, nor no known quarrel, were in
    question—
  But that defences, musters, preparations
  Should be maintained, assembled, and collected
  As were a war in expectation.                    20
  Therefore, I say, 'tis meet we all go forth
  To view the sick and feeble parts of France.
  And let us do it with no show of fear,
  No, with no more than if we heard that England
  Were busied with a Whitsun morris dance.                    25
  For, my good liege, she is so idly kinged,
  Her sceptre so fantastically borne
  By a vain, giddy, shallow, humorous youth,
  That fear attends her not.
CONSTABLE         O peace, Prince Dauphin.
  You are too much mistaken in this king.                    30
  Question your grace the late ambassadors
  With what great state he heard their embassy,
  How well supplied with agèd counsellors,
  How modest in exception, and withal
  How terrible in constant resolution,                    35
  And you shall find his vanities forespent

Were but the outside of the Roman Brutus,
Covering discretion with a coat of folly,
As gardeners do with ordure hide those roots
That shall first spring and be most delicate.                    40
DAUPHIN
Well, 'tis not so, my Lord High Constable.
But though we think it so, it is no matter.
In cases of defence 'tis best to weigh
The enemy more mighty than he seems.
So the proportions of defence are filled—                        45
Which, of a weak and niggardly projection,
Doth like a miser spoil his coat with scanting
A little cloth.
KING CHARLES   Think we King Harry strong.
And princes, look you strongly arm to meet him.
The kindred of him hath been fleshed upon us,                    50
And he is bred out of that bloody strain
That haunted us in our familiar paths.
Witness our too-much-memorable shame
When Crécy battle fatally was struck,
And all our princes captived by the hand                         55
Of that black name, Edward, Black Prince of Wales,
Whiles that his mountant sire, on mountain standing,
Up in the air, crowned with the golden sun,
Saw his heroical seed and smiled to see him
Mangle the work of nature and deface                             60
The patterns that by God and by French fathers
Had twenty years been made. This is a stem
Of that victorious stock, and let us fear
The native mightiness and fate of him.
        *Enter a Messenger*
MESSENGER
Ambassadors from Harry, King of England,                         65
Do crave admittance to your majesty.
KING CHARLES
We'll give them present audience. Go and bring them.
                                    *Exit Messenger*
You see this chase is hotly followed, friends.
DAUPHIN
Turn head and stop pursuit. For coward dogs
Most spend their mouths when what they seem to
        threaten                                                 70
Runs far before them. Good my sovereign,
Take up the English short, and let them know
Of what a monarchy you are the head.
Self-love, my liege, is not so vile a sin
As self-neglecting.
        *Enter the Duke of Exeter, ⌈attended⌉*
KING CHARLES        From our brother England?                    75
EXETER
From him, and thus he greets your majesty:
He wills you, in the name of God Almighty,
That you divest yourself and lay apart
The borrowed glories that by gift of heaven,
By law of nature and of nations, 'longs                          80
To him and to his heirs, namely the crown,
And all wide-stretchèd honours that pertain
By custom and the ordinance of times
Unto the crown of France. That you may know
'Tis no sinister nor no awkward claim,                           85
Picked from the worm-holes of long-vanished days,
Nor from the dust of old oblivion raked,
He sends you this most memorable line,
In every branch truly demonstrative,
Willing you over-look this pedigree,                             90
And when you find him evenly derived

From his most famed of famous ancestors,
Edward the Third, he bids you then resign
Your crown and kingdom, indirectly held
From him, the native and true challenger.                        95
KING CHARLES  Or else what follows?
EXETER
Bloody constraint. For if you hide the crown
Even in your hearts, there will he rake for it.
Therefore in fierce tempest is he coming,
In thunder and in earthquake, like a Jove,                      100
That if requiring fail, he will compel;
And bids you, in the bowels of the Lord,
Deliver up the crown, and to take mercy
On the poor souls for whom this hungry war
Opens his vasty jaws; and on your head                          105
Turns he the widows' tears, the orphans' cries,
The dead men's blood, the pining maidens' groans,
For husbands, fathers, and betrothèd lovers
That shall be swallowed in this controversy.
This is his claim, his threat'ning, and my message—
Unless the Dauphin be in presence here,                         111
To whom expressly I bring greeting too.
KING CHARLES
For us, we will consider of this further.
Tomorrow shall you bear our full intent
Back to our brother England.
DAUPHIN                     For the Dauphin,                     115
I stand here for him. What to him from England?
EXETER
Scorn and defiance, slight regard, contempt;
And anything that may not misbecome
The mighty sender, doth he prize you at.
Thus says my king: an if your father's highness                 120
Do not, in grant of all demands at large,
Sweeten the bitter mock you sent his majesty,
He'll call you to so hot an answer for it
That caves and womby vaultages of France
Shall chide your trespass and return your mock                  125
In second accent of his ordinance.
DAUPHIN
Say if my father render fair return
It is against my will, for I desire
Nothing but odds with England. To that end,
As matching to his youth and vanity,                            130
I did present him with the Paris balls.
EXETER
He'll make your Paris Louvre shake for it,
Were it the mistress court of mighty Europe.
And be assured, you'll find a diff'rence,
As we his subjects have in wonder found,                        135
Between the promise of his greener days
And these he masters now: now he weighs time
Even to the utmost grain. That you shall read
In your own losses, if he stay in France.
KING CHARLES ⌈*rising*⌉
Tomorrow shall you know our mind at full.                       140
        *Flourish*
EXETER
Dispatch us with all speed, lest that our king
Come here himself to question our delay—
For he is footed in    is land already.
KING CHARLES
You shall be soon dispatched with fair conditions.
A night is but small breath and little pause                    145
To answer matters of this consequence.
                                    ⌈*Flourish.*⌉ *Exeunt*

**3.0**   *Enter Chorus*

CHORUS

Thus with imagined wing our swift scene flies
In motion of no less celerity
Than that of thought. Suppose that you have seen
The well-appointed king at Dover pier
Embark his royalty, and his brave fleet    5
With silken streamers the young Phoebus fanning.
Play with your fancies, and in them behold
Upon the hempen tackle ship-boys climbing;
Hear the shrill whistle, which doth order give
To sounds confused; behold the threaden sails,    10
Borne with th'invisible and creeping wind,
Draw the huge bottoms through the furrowed sea,
Breasting the lofty surge. O do but think
You stand upon the rivage and behold
A city on th'inconstant billows dancing—    15
For so appears this fleet majestical,
Holding due course to Harfleur. Follow, follow!
Grapple your minds to sternage of this navy,
And leave your England, as dead midnight still,
Guarded with grandsires, babies, and old women,    20
Either past or not arrived to pith and puissance.
For who is he, whose chin is but enriched
With one appearing hair, that will not follow
These culled and choice-drawn cavaliers to France?
Work, work your thoughts, and therein see a siege.    25
Behold the ordnance on their carriages,
With fatal mouths gaping on girded Harfleur.
Suppose th'ambassador from the French comes back,
Tells Harry that the King doth offer him
Catherine his daughter, and with her, to dowry,    30
Some petty and unprofitable dukedoms.
The offer likes not, and the nimble gunner
With linstock now the devilish cannon touches,
    *Alarum, and chambers go off*
And down goes all before them. Still be kind,
And eke out our performance with your mind.    *Exit*

**3.1**   *Alarum. Enter King Harry ⌈and the English army,*
    *with⌉ scaling ladders*

KING HARRY

Once more unto the breach, dear friends, once more,
Or close the wall up with our English dead.
In peace there's nothing so becomes a man
As modest stillness and humility,
But when the blast of war blows in our ears,    5
Then imitate the action of the tiger.
Stiffen the sinews, conjure up the blood,
Disguise fair nature with hard-favoured rage.
Then lend the eye a terrible aspect,
Let it pry through the portage of the head    10
Like the brass cannon, let the brow o'erwhelm it
As fearfully as doth a gallèd rock
O'erhang and jutty his confounded base,
Swilled with the wild and wasteful ocean.
Now set the teeth and stretch the nostril wide,    15
Hold hard the breath, and bend up every spirit
To his full height. On, on, you noblest English,
Whose blood is fet from fathers of war-proof,
Fathers that like so many Alexanders
Have in these parts from morn till even fought,    20
And sheathed their swords for lack of argument.
Dishonour not your mothers; now attest
That those whom you called fathers did beget you.
Be copy now to men of grosser blood,
And teach them how to war. And you, good yeomen,

Whose limbs were made in England, show us here    26
The mettle of your pasture; let us swear
That you are worth your breeding—which I doubt not,
For there is none of you so mean and base
That hath not noble lustre in your eyes.    30
I see you stand like greyhounds in the slips,
Straining upon the start. The game's afoot.
Follow your spirit, and upon this charge
Cry, 'God for Harry! England and Saint George!'
    *Alarum, and chambers go off. Exeunt*

**3.2**   *Enter Nim, Bardolph, Ensign Pistol, and Boy*

BARDOLPH On, on, on, on, oh! To the breach, to the
breach!
NIM Pray thee corporal, stay. The knocks are too hot,
and for mine own part I have not a case of lives. The
humour of it is too hot, that is the very plainsong of
it.    6
PISTOL
'The plainsong' is most just, for humours do abound.
Knocks go and come, God's vassals drop and die,
⌈*sings*⌉    And sword and shield
           In bloody field    10
           Doth win immortal fame.
BOY Would I were in an alehouse in London. I would
give all my fame for a pot of ale, and safety.
PISTOL ⌈*sings*⌉ And I.
           If wishes would prevail with me    15
           My purpose should not fail with me
           But thither would I hie.
BOY ⌈*sings*⌉ As duly
           But not as truly
           As bird doth sing on bough.    20
    *Enter Captain Fluellen and beats them in*
FLUELLEN God's plud! Up to the breaches, you dogs!
Avaunt, you cullions!
PISTOL
Be merciful, great duke, to men of mould.
Abate thy rage, abate thy manly rage,
Abate thy rage, great duke. Good bawcock, bate    25
Thy rage. Use lenity, sweet chuck.
NIM These be good humours!
    ⌈*Fluellen begins to beat Nim*⌉
Your honour runs bad humours.
                   *Exeunt all but ⌈the Boy⌉*
BOY As young as I am, I have observed these three
swashers. I am boy to them all three, but all they three,
though they should serve me, could not be man to me,
for indeed three such antics do not amount to a man.
For Bardolph, he is white-livered and red-faced—by the
means whereof a faces it out, but fights not. For Pistol,
he hath a killing tongue and a quiet sword—by the
means whereof a breaks words, and keeps whole
weapons. For Nim, he hath heard that men of few
words are the best men, and therefore he scorns to say
his prayers, lest a should be thought a coward. But his
few bad words are matched with as few good deeds—
for a never broke any man's head but his own, and
that was against a post, when he was drunk. They will
steal anything, and call it 'purchase'. Bardolph stole a
lute case, bore it twelve leagues, and sold it for three
halfpence. Nim and Bardolph are sworn brothers in
filching, and in Calais they stole a fire shovel. I knew
by that piece of service the men would carry coals.
They would have me as familiar with men's pockets as
their gloves or their handkerchiefs—which makes much

against my manhood, if I should take from another's
pocket to put into mine, for it is plain pocketing up of
wrongs. I must leave them, and seek some better
service. Their villainy goes against my weak stomach,
and therefore I must cast it up.                    *Exit*

**3.3**    *Enter Captain Gower ⌈and Captain Fluellen,*
        *meeting⌉*
GOWER  Captain Fluellen, you must come presently to the
mines. The Duke of Gloucester would speak with you.
FLUELLEN  To the mines? Tell you the Duke it is not so
good to come to the mines. For look you, the mines is
not according to the disciplines of the war. The concavi-
ties of it is not sufficient. For look you, th'athversary,
you may discuss unto the Duke, look you, is digt
himself, four yard under, the countermines. By Cheshu,
I think a will plow up all, if there is not better directions.
GOWER  The Duke of Gloucester, to whom the order of the
siege is given, is altogether directed by an Irishman, a
very valiant gentleman, i'faith.
FLUELLEN  It is Captain MacMorris, is it not?
GOWER  I think it be.                                    14
FLUELLEN  By Cheshu, he is an ass, as in the world. I will
verify as much in his beard. He has no more directions
in the true disciplines of the wars, look you—of the
*Roman* disciplines—than is a puppy dog.
        *Enter Captain MacMorris and Captain Jamy*
GOWER  Here a comes, and the Scots captain, Captain
Jamy, with him.                                          20
FLUELLEN  Captain Jamy is a marvellous falorous gentle-
man, that is certain, and of great expedition and
knowledge in th'anciant wars, upon my particular
knowledge of his directions. By Cheshu, he will
maintain his argument as well as any military man in
the world, in the disciplines of the pristine wars of the
Romans.                                                  27
JAMY  I say gud day, Captain Fluellen.
FLUELLEN  Good e'en to your worship, good Captain James.
GOWER  How now, Captain MacMorris, have you quit the
mines? Have the pioneers given o'er?                     31
MACMORRIS  By Chrish law, 'tish ill done. The work ish
give over, the trumpet sound the retreat. By my hand
I swear, and my father's soul, the work ish ill done, it
ish give over. I would have blowed up the town, so
Chrish save me law, in an hour. O 'tish ill done, 'tish
ill done, by my hand 'tish ill done.                     37
FLUELLEN  Captain MacMorris, I beseech you now, will
you vouchsafe me, look you, a few disputations with
you, as partly touching or concerning the disciplines
of the war, the Roman wars, in the way of argument,
look you, and friendly communication? Partly to satisfy
my opinion and partly for the satisfaction, look you, of
my mind. As touching the direction of the military
discipline, that is the point.                           45
JAMY  It sall be vary gud, gud feith, gud captains bath,
and I sall quite you with gud leve, as I may pick
occasion. That sall I, marry.
MACMORRIS  It is no time to discourse, so Chrish save me.
The day is hot, and the weather and the wars and the
King and the dukes. It is no time to discourse. The
town is besieched, and the trumpet call us to the breach,
and we talk and, be Chrish, do nothing, 'tis shame for
us all. So God sa' me, 'tis shame to stand still, it is
shame by my hand. And there is throats to be cut, and
works to be done, and there ish nothing done, so Chrish
sa' me law.                                              57

JAMY  By the mess, ere these eyes of mine take themselves
to slumber, ay'll de gud service, or I'll lig i'th' grund
for it. Ay owe Got a death, and I'll pay't as valorously
as I may, that sall I suirely do, that is the brief and
the long. Marry, I wad full fain heard some question
'tween you twae.                                         63
FLUELLEN  Captain MacMorris, I think, look you, under
your correction, there is not many of your nation—
MACMORRIS  Of my nation? What ish my nation? Ish a
villain and a bastard and a knave and a rascal? What
ish my nation? Who talks of my nation?                   68
FLUELLEN  Look you, if you take the matter otherwise than
is meant, Captain MacMorris, peradventure I shall think
you do not use me with that affability as in discretion
you ought to use me, look you, being as good a man
as yourself, both in the disciplines of war and in the
derivation of my birth, and in other particularities.    74
MACMORRIS  I do not know you so good a man as myself.
So Chrish save me, I will cut off your head.
GOWER  Gentlemen both, you will mistake each other.
JAMY  Ah, that's a foul fault.
        *A parley is sounded*
GOWER  The town sounds a parley.                         79
FLUELLEN  Captain MacMorris, when there is more better
opportunity to be required, look you, I will be so bold
as to tell you I know the disciplines of war. And there
is an end.                                              *Exit*
        ⌈*Flourish.*⌉ *Enter King Harry and all his train before*
        *the gates*
KING HARRY
How yet resolves the Governor of the town?
This is the latest parle we will admit.                  85
Therefore to our best mercy give yourselves,
Or like to men proud of destruction
Defy us to our worst. For as I am a soldier,
A name that in my thoughts becomes me best,
If I begin the batt'ry once again                        90
I will not leave the half-achievèd Harfleur
Till in her ashes she lie buried.
The gates of mercy shall be all shut up,
And the fleshed soldier, rough and hard of heart,
In liberty of bloody hand shall range                    95
With conscience wide as hell, mowing like grass
Your fresh fair virgins and your flow'ring infants.
What is it then to me if impious war
Arrayed in flames like to the prince of fiends
Do with his smirched complexion all fell feats          100
Enlinked to waste and desolation?
What is't to me, when you yourselves are cause,
If your pure maidens fall into the hand
Of hot and forcing violation?
What rein can hold licentious wickedness                 105
When down the hill he holds his fierce career?
We may as bootless spend our vain command
Upon th'enragèd soldiers in their spoil
As send precepts to the leviathan
To come ashore. Therefore, you men of Harfleur,         110
Take pity of your town and of your people
Whiles yet my soldiers are in my command,
Whiles yet the cool and temperate wind of grace
O'erblows the filthy and contagious clouds
Of heady murder, spoil, and villainy.                    115
If not—why, in a moment look to see
The blind and bloody soldier with foul hand
Defile the locks of your shrill-shrieking daughters;
Your fathers taken by the silver beards,

And their most reverend heads dashed to the walls;
Your naked infants spitted upon pikes,                    121
Whiles the mad mothers with their howls confused
Do break the clouds, as did the wives of Jewry
At Herod's bloody-hunting slaughtermen.
What say you? Will you yield, and this avoid?             125
Or, guilty in defence, be thus destroyed?
          *Enter Governor* ⌜*on the wall*⌝
GOVERNOR
Our expectation hath this day an end.
The Dauphin, whom of succours we entreated,
Returns us that his powers are yet not ready
To raise so great a siege. Therefore, dread King,        130
We yield our town and lives to thy soft mercy.
Enter our gates, dispose of us and ours,
For we no longer are defensible.
KING HARRY
Open your gates.               ⌜*Exit Governor*⌝
          Come, Uncle Exeter,
Go you and enter Harfleur. There remain,                 135
And fortify it strongly 'gainst the French.
Use mercy to them all. For us, dear uncle,
The winter coming on, and sickness growing
Upon our soldiers, we will retire to Calais.
Tonight in Harfleur will we be your guest;               140
Tomorrow for the march are we addressed.
          ⌜*The gates are opened.*⌝ *Flourish, and they enter*
                                              *the town*

**3.4**  *Enter Princess Catherine and Alice, an old*
          *gentlewoman*
CATHERINE Alice, tu as été en Angleterre, et tu bien parles
le langage.
ALICE Un peu, madame.
CATHERINE Je te prie, m'enseignez. Il faut que j'apprenne
à parler. Comment appelez-vous la main en anglais?
ALICE La main? Elle est appelée *de hand*.                    6
CATHERINE *De hand*. Et les doigts?
ALICE Les doigts? Ma foi, j'oublie les doigts, mais je me
souviendrai. Les doigts—je pense qu'ils sont appelés *de
fingres*. Oui, *de fingres*.                              10
CATHERINE La main, *de hand*; les doigts, *de fingres*. Je pense
que je suis la bonne écolière; j'ai gagné deux mots
d'anglais vitement. Comment appelez-vous les ongles?
ALICE Les ongles? Nous les appelons *de nails*.
CATHERINE *De nails*. Écoutez—dites-moi si je parle bien:
*de hand, de fingres*, et *de nails*.                     16
ALICE C'est bien dit, madame. Il est fort bon anglais.
CATHERINE Dites-moi l'anglais pour le bras.
ALICE *De arma*, madame.
CATHERINE Et le coude?                                    20
ALICE *D'elbow*.
CATHERINE *D'elbow*. Je m'en fais la répétition de tous les
mots que vous m'avez appris dès à présent.
ALICE Il est trop difficile, madame, comme je pense.
CATHERINE Excusez-moi, Alice. Écoutez: *d'hand, de fingre,
de nails, d'arma, de bilbow*.                             26
ALICE *D'elbow*, madame.
CATHERINE O Seigneur Dieu, je m'en oublie! *D'elbow*.
Comment appelez-vous le col?
ALICE *De nick*, madame.                                  30
CATHERINE *De nick*. Et le menton?
ALICE *De chin*.
CATHERINE *De sin*. Le col, *de nick*; le menton, *de sin*.
ALICE Oui. Sauf votre honneur, en vérité vous prononcez
les mots aussi droit que les natifs d'Angleterre.        35

CATHERINE Je ne doute point d'apprendre, par la grâce de
Dieu, et en peu de temps.
ALICE N'avez-vous y déjà oublié ce que je vous ai
enseigné?
CATHERINE Non, et je réciterai à vous promptement:
*d'hand, de fingre, de mailès*—                          41
ALICE *De nails*, madame.
CATHERINE *De nails, de arma, de ilbow*—
ALICE Sauf votre honneur, *d'elbow*.
CATHERINE Ainsi dis-je. *D'elbow, de nick*, et *de sin*. Comment
appelez-vous les pieds et la robe?                       46
ALICE *De foot*, madame, et *de cown*.
CATHERINE *De foot* et *de cown*? O Seigneur Dieu! Ils sont
les mots de son mauvais, corruptible, gros, et
impudique, et non pour les dames d'honneur d'user. Je
ne voudrais prononcer ces mots devant les seigneurs
de France pour tout le monde. Foh! *De foot* et *de cown*!
Néanmoins, je réciterai une autre fois ma leçon
ensemble. *D'hand, de fingre, de nails, d'arma, d'elbow, de
nick, de sin, de foot, de cown*.                         55
ALICE Excellent, madame!
CATHERINE C'est assez pour une fois. Allons-nous à dîner.
                                              *Exeunt*

**3.5**  *Enter King Charles the Sixth of France, the
          Dauphin, the Constable, the Duke of* ⌜*Bourbon*⌝,
          *and others*
KING CHARLES
'Tis certain he hath passed the River Somme.
CONSTABLE
And if he be not fought withal, my lord,
Let us not live in France; let us quit all
And give our vineyards to a barbarous people.
DAUPHIN
*O Dieu vivant!* Shall a few sprays of us,                   5
The emptying of our fathers' luxury,
Our scions, put in wild and savage stock,
Spirt up so suddenly into the clouds
And over-look their grafters?
⌜BOURBON⌝
Normans, but bastard Normans, Norman bastards!          10
*Mort de ma vie*, if they march along
Unfought withal, but I will sell my dukedom
To buy a slobb'ry and a dirty farm
In that nook-shotten isle of Albion.
CONSTABLE
*Dieu de batailles!* Where have they this mettle?           15
Is not their climate foggy, raw, and dull,
On whom as in despite the sun looks pale,
Killing their fruit with frowns? Can sodden water,
A drench for sur-reined jades—their barley-broth—
Decoct their cold blood to such valiant heat?           20
And shall our quick blood, spirited with wine,
Seem frosty? O for honour of our land
Let us not hang like roping icicles
Upon our houses' thatch, whiles a more frosty people
Sweat drops of gallant youth in our rich fields—        25
'Poor' may we call them, in their native lords.
DAUPHIN By faith and honour,
Our madams mock at us and plainly say
Our mettle is bred out, and they will give
Their bodies to the lust of English youth,              30
To new-store France with bastard warriors.
⌜BOURBON⌝
They bid us, 'To the English dancing-schools,
And teach lavoltas high and swift corantos'—

Saying our grace is only in our heels,
And that we are most lofty runaways.                    35
KING CHARLES
    Where is Montjoy the herald? Speed him hence.
    Let him greet England with our sharp defiance.
    Up, princes, and with spirit of honour edged
    More sharper than your swords, hie to the field.
    Charles Delabret, High Constable of France,         40
    You Dukes of Orléans, Bourbon, and of Berri,
    Alençon, Brabant, Bar, and Burgundy,
    Jaques Châtillion, Rambures, Vaudemont,
    Beaumont, Grandpré, Roussi, and Fauconbridge,
    Foix, Lestrelles, Boucicault, and Charolais,        45
    High dukes, great princes, barons, lords, and knights,
    For your great seats now quit you of great shames.
    Bar Harry England, that sweeps through our land
    With pennons painted in the blood of Harfleur;
    Rush on his host, as doth the melted snow           50
    Upon the valleys, whose low vassal seat
    The Alps doth spit and void his rheum upon.
    Go down upon him, you have power enough,
    And in a captive chariot into Rouen
    Bring him our prisoner.
CONSTABLE                       This becomes the great.  55
    Sorry am I his numbers are so few,
    His soldiers sick and famished in their march,
    For I am sure when he shall see our army
    He'll drop his heart into the sink of fear
    And, fore achievement, offer us his ransom.         60
KING CHARLES
    Therefore, Lord Constable, haste on Montjoy,
    And let him say to England that we send
    To know what willing ransom he will give.—
    Prince Dauphin, you shall stay with us in Rouen.
DAUPHIN
    Not so, I do beseech your majesty.                  65
KING CHARLES
    Be patient, for you shall remain with us.—
    Now forth, Lord Constable, and princes all,
    And quickly bring us word of England's fall.
                                        *Exeunt severally*

**3.6**  *Enter Captains Gower and Fluellen, meeting*
GOWER How now, Captain Fluellen, come you from the
    bridge?
FLUELLEN I assure you there is very excellent services
    committed at the bridge.
GOWER Is the Duke of Exeter safe?                       5
FLUELLEN The Duke of Exeter is as magnanimous as
    Agamemnon, and a man that I love and honour with
    my soul and my heart and my duty and my live and
    my living and my uttermost power. He is not, God be
    praised and blessed, any hurt in the world, but keeps
    the bridge most valiantly, with excellent discipline.
    There is an ensign lieutenant there at the pridge, I
    think in my very conscience he is as valiant a man as
    Mark Antony, and he is a man of no estimation in the
    world, but I did see him do as gallant service.     15
GOWER What do you call him?
FLUELLEN He is called Ensign Pistol.
GOWER I know him not.
            *Enter Ensign Pistol*
FLUELLEN Here is the man.
PISTOL
    Captain, I thee beseech to do me favours.           20
    The Duke of Exeter doth love thee well.

FLUELLEN Ay, I praise God, and I have merited some love
    at his hands.
PISTOL
    Bardolph, a soldier firm and sound of heart,
    Of buxom valour, hath by cruel fate                 25
    And giddy Fortune's furious fickle wheel,
    That goddess blind that stands upon the rolling
        restless stone—
FLUELLEN By your patience, Ensign Pistol: Fortune is
    painted blind, with a muffler afore her eyes, to signify
    to you that Fortune is blind. And she is painted also
    with a wheel, to signify to you—which is the moral of
    it—that she is turning and inconstant and mutability
    and variation. And her foot, look you, is fixed upon a
    spherical stone, which rolls and rolls and rolls. In good
    truth, the poet makes a most excellent description of
    it; Fortune is an excellent moral.                  36
PISTOL
    Fortune is Bardolph's foe and frowns on him,
    For he hath stol'n a pax, and hangèd must a be.
    A damnèd death—
    Let gallows gape for dog, let man go free,          40
    And let not hemp his windpipe suffocate.
    But Exeter hath given the doom of death
    For pax of little price.
    Therefore go speak, the Duke will hear thy voice,
    And let not Bardolph's vital thread be cut          45
    With edge of penny cord and vile reproach.
    Speak, captain, for his life, and I will thee requite.
FLUELLEN Ensign Pistol, I do partly understand your
    meaning.
PISTOL Why then rejoice therefor.                       50
FLUELLEN Certainly, ensign, it is not a thing to rejoice at.
    For if, look you, he were my brother, I would desire
    the Duke to use his good pleasure, and put him to
    executions. For discipline ought to be used.
PISTOL
    Die and be damned! and *fico* for thy friendship.  55
FLUELLEN It is well.
PISTOL The fig of Spain.
FLUELLEN Very good.
PISTOL
    I say the fig within thy bowels and thy dirty maw.
                                                *Exit*
FLUELLEN Captain Gower, cannot you hear it lighten and
    thunder?                                            61
GOWER Why, is this the ensign you told me of? I remember
    him now. A bawd, a cutpurse.
FLUELLEN I'll assure you, a uttered as prave words at the
    pridge as you shall see in a summer's day. But it is
    very well. What he has spoke to me, that is well, I
    warrant you, when time is serve.                    67
GOWER Why 'tis a gull, a fool, a rogue, that now and
    then goes to the wars, to grace himself at his return
    into London under the form of a soldier. And such
    fellows are perfect in the great commanders' names,
    and they will learn you by rote where services were
    done—at such and such a sconce, at such a breach,
    at such a convoy, who came off bravely, who was shot,
    who disgraced, what terms the enemy stood on—and
    this they con perfectly in the phrase of war, which they
    trick up with new-tuned oaths. And what a beard of
    the General's cut and a horrid suit of the camp will do
    among foaming bottles and ale-washed wits is
    wonderful to be thought on. But you must learn to
    know such slanders of the age, or else you may be
    marvellously mistook.                               82

FLUELLEN I tell you what, Captain Gower, I do perceive he is not the man that he would gladly make show to the world he is. If I find a hole in his coat, I will tell him my mind.    86

*A drum is heard*

Hark you, the King is coming, and I must speak with him from the pridge.

*Enter King Harry and his poor soldiers, with drum and colours*

God pless your majesty.

KING HARRY
How now, Fluellen, com'st thou from the bridge?    90

FLUELLEN Ay, so please your majesty. The Duke of Exeter has very gallantly maintained the pridge. The French is gone off, look you, and there is gallant and most prave passages. Marry, th'athversary was have possession of the pridge, but he is enforced to retire, and the Duke of Exeter is master of the pridge. I can tell your majesty, the Duke is a prave man.    97

KING HARRY What men have you lost, Fluellen?

FLUELLEN The perdition of th'athversary hath been very great, reasonable great. Marry, for my part I think the Duke hath lost never a man, but one that is like to be executed for robbing a church, one Bardolph, if your majesty know the man. His face is all bubuncles and whelks and knobs and flames o' fire, and his lips blows at his nose, and it is like a coal of fire, sometimes plue and sometimes red. But his nose is executed, and his fire's out.    107

KING HARRY We would have all such offenders so cut off, and we here give express charge that in our marches through the country there be nothing compelled from the villages, nothing taken but paid for, none of the French upbraided or abused in disdainful language. For when lenity and cruelty play for a kingdom, the gentler gamester is the soonest winner.

*Tucket. Enter Montjoy*

MONTJOY You know me by my habit.    115

KING HARRY
Well then, I know thee. What shall I know of thee?

MONTJOY
My master's mind.

KING HARRY        Unfold it.

MONTJOY             Thus says my King:
'Say thou to Harry of England, though we seemed dead, we did but sleep. Advantage is a better soldier than rashness. Tell him, we could have rebuked him at Harfleur, but that we thought not good to bruise an injury till it were full ripe. Now we speak upon our cue, and our voice is imperial. England shall repent his folly, see his weakness, and admire our sufferance. Bid him therefore consider of his ransom, which must proportion the losses we have borne, the subjects we have lost, the disgrace we have digested—which in weight to re-answer, his pettiness would bow under. For our losses, his exchequer is too poor; for th'effusion of our blood, the muster of his kingdom too faint a number; and for our disgrace, his own person kneeling at our feet but a weak and worthless satisfaction. To this add defiance, and tell him for conclusion he hath betrayed his followers, whose condemnation is pronounced.'    135
So far my King and master; so much my office.

KING HARRY
What is thy name? I know thy quality.

MONTJOY Montjoy.

KING HARRY
Thou dost thy office fairly. Turn thee back
And tell thy king I do not seek him now,    140
But could be willing to march on to Calais
Without impeachment, for to say the sooth—
Though 'tis no wisdom to confess so much
Unto an enemy of craft and vantage—
My people are with sickness much enfeebled,    145
My numbers lessened, and those few I have
Almost no better than so many French;
Who when they were in health—I tell thee herald,
I thought upon one pair of English legs
Did march three Frenchmen. Yet forgive me, God,    150
That I do brag thus. This your air of France
Hath blown that vice in me. I must repent.
Go, therefore, tell thy master here I am;
My ransom is this frail and worthless trunk,
My army but a weak and sickly guard.    155
Yet, God before, tell him we will come on,
Though France himself and such another neighbour
Stand in our way. There's for thy labour, Montjoy.
Go bid thy master well advise himself.
If we may pass, we will; if we be hindered,    160
We shall your tawny ground with your red blood
Discolour. And so, Montjoy, fare you well.
The sum of all our answer is but this:
We would not seek a battle as we are,
Nor as we are we say we will not shun it.    165
So tell your master.

MONTJOY
I shall deliver so. Thanks to your highness.      *Exit*

GLOUCESTER
I hope they will not come upon us now.

KING HARRY
We are in God's hand, brother, not in theirs.
March to the bridge. It now draws toward night.    170
Beyond the river we'll encamp ourselves,
And on tomorrow bid them march away.      *Exeunt*

**3.7**    *Enter the Constable, Lord Rambures, the Dukes of Orléans and ⌈Bourbon⌉, with others*

CONSTABLE Tut, I have the best armour of the world. Would it were day.

ORLÉANS You have an excellent armour. But let my horse have his due.

CONSTABLE It is the best horse of Europe.    5

ORLÉANS Will it never be morning?

⌈BOURBON⌉ My lord of Orléans and my Lord High Constable, you talk of horse and armour?

ORLÉANS You are as well provided of both as any prince in the world.    10

⌈BOURBON⌉ What a long night is this! I will not change my horse with any that treads but on four pasterns. Ah ha! He bounds from the earth as if his entrails were hares—*le cheval volant*, the Pegasus, *qui a les narines de feu!* When I bestride him, I soar, I am a hawk; he trots the air, the earth sings when he touches it, the basest horn of his hoof is more musical than the pipe of Hermes.

ORLÉANS He's of the colour of the nutmeg.    19

⌈BOURBON⌉ And of the heat of the ginger. It is a beast for Perseus. He is pure air and fire, and the dull elements of earth and water never appear in him, but only in patient stillness while his rider mounts him. He is indeed a horse, and all other jades you may call beasts.

CONSTABLE Indeed, my lord, it is a most absolute and excellent horse.

⌈BOURBON⌉ It is the prince of palfreys. His neigh is like the bidding of a monarch, and his countenance enforces homage.

ORLÉANS No more, cousin. 30

⌈BOURBON⌉ Nay, the man hath no wit, that cannot from the rising of the lark to the lodging of the lamb vary deserved praise on my palfrey. It is a theme as fluent as the sea. Turn the sands into eloquent tongues, and my horse is argument for them all. 'Tis a subject for a sovereign to reason on, and for a sovereign's sovereign to ride on, and for the world, familiar to us and unknown, to lay apart their particular functions, and wonder at him. I once writ a sonnet in his praise, and began thus: 'Wonder of nature!—' 40

ORLÉANS I have heard a sonnet begin so to one's mistress.

⌈BOURBON⌉ Then did they imitate that which I composed to my courser, for my horse is my mistress.

ORLÉANS Your mistress bears well.

⌈BOURBON⌉ Me well, which is the prescribed praise and perfection of a good and particular mistress. 46

CONSTABLE Nay, for methought yesterday your mistress shrewdly shook your back.

⌈BOURBON⌉ So perhaps did yours.

CONSTABLE Mine was not bridled. 50

⌈BOURBON⌉ O then belike she was old and gentle, and you rode like a kern of Ireland, your French hose off, and in your strait strossers.

CONSTABLE You have good judgement in horsemanship.

⌈BOURBON⌉ Be warned by me then: they that ride so, and ride not warily, fall into foul bogs. I had rather have my horse to my mistress. 57

CONSTABLE I had as lief have my mistress a jade.

⌈BOURBON⌉ I tell thee, Constable, my mistress wears his own hair. 60

CONSTABLE I could make as true a boast as that, if I had a sow to my mistress.

⌈BOURBON⌉ 'Le chien est retourné à son propre vomissement, et la truie lavée au bourbier.' Thou makest use of anything. 65

CONSTABLE Yet do I not use my horse for my mistress, or any such proverb so little kin to the purpose.

RAMBURES My Lord Constable, the armour that I saw in your tent tonight, are those stars or suns upon it?

CONSTABLE Stars, my lord. 70

⌈BOURBON⌉ Some of them will fall tomorrow, I hope.

CONSTABLE And yet my sky shall not want.

⌈BOURBON⌉ That may be, for you bear a many superfluously, and 'twere more honour some were away.

CONSTABLE Even as your horse bears your praises, who would trot as well were some of your brags dismounted.

⌈BOURBON⌉ Would I were able to load him with his desert! Will it never be day? I will trot tomorrow a mile, and my way shall be paved with English faces. 79

CONSTABLE I will not say so, for fear I should be faced out of my way. But I would it were morning, for I would fain be about the ears of the English.

RAMBURES Who will go to hazard with me for twenty prisoners? 84

CONSTABLE You must first go yourself to hazard, ere you have them.

⌈BOURBON⌉ 'Tis midnight. I'll go arm myself.       *Exit*

ORLÉANS The Duke of Bourbon longs for morning.

RAMBURES He longs to eat the English.

CONSTABLE I think he will eat all he kills. 90

ORLÉANS By the white hand of my lady, he's a gallant prince.

CONSTABLE Swear by her foot, that she may tread out the oath.

ORLÉANS He is simply the most active gentleman of France.

CONSTABLE Doing is activity, and he will still be doing.

ORLÉANS He never did harm that I heard of. 97

CONSTABLE Nor will do none tomorrow. He will keep that good name still.

ORLÉANS I know him to be valiant. 100

CONSTABLE I was told that by one that knows him better than you.

ORLÉANS What's he?

CONSTABLE Marry, he told me so himself, and he said he cared not who knew it. 105

ORLÉANS He needs not; it is no hidden virtue in him.

CONSTABLE By my faith, sir, but it is. Never anybody saw it but his lackey. 'Tis a hooded valour, and when it appears it will bate.

ORLÉANS 'Ill will never said well.' 110

CONSTABLE I will cap that proverb with 'There is flattery in friendship.'

ORLÉANS And I will take up that with 'Give the devil his due.'

CONSTABLE Well placed! There stands your friend for the devil. Have at the very eye of that proverb with 'A pox of the devil!' 117

ORLÉANS You are the better at proverbs by how much 'a fool's bolt is soon shot'.

CONSTABLE You have shot over. 120

ORLÉANS 'Tis not the first time you were overshot.

    *Enter a Messenger*

MESSENGER My Lord High Constable, the English lie within fifteen hundred paces of your tents.

CONSTABLE Who hath measured the ground?

MESSENGER The Lord Grandpré. 125

CONSTABLE A valiant and most expert gentleman.

                    ⌈*Exit Messenger*⌉

Would it were day! Alas, poor Harry of England. He longs not for the dawning as we do.

ORLÉANS What a wretched and peevish fellow is this King of England, to mope with his fat-brained followers so far out of his knowledge. 131

CONSTABLE If the English had any apprehension, they would run away.

ORLÉANS That they lack—for if their heads had any intellectual armour, they could never wear such heavy headpieces. 136

RAMBURES That island of England breeds very valiant creatures. Their mastiffs are of unmatchable courage.

ORLÉANS Foolish curs, that run winking into the mouth of a Russian bear, and have their heads crushed like rotten apples. You may as well say, 'That's a valiant flea that dare eat his breakfast on the lip of a lion.'

CONSTABLE Just, just. And the men do sympathize with the mastiffs in robustious and rough coming on, leaving their wits with their wives. And then, give them great meals of beef, and iron and steel, they will eat like wolves and fight like devils. 147

ORLÉANS Ay, but these English are shrewdly out of beef.

CONSTABLE Then shall we find tomorrow they have only stomachs to eat, and none to fight. Now is it time to arm. Come, shall we about it? 151

ORLÉANS

It is now two o'clock. But let me see—by ten
We shall have each a hundred Englishmen.     *Exeunt*

**4.0**   *Enter Chorus*

CHORUS

Now entertain conjecture of a time
When creeping murmur and the poring dark
Fills the wide vessel of the universe.
From camp to camp through the foul womb of night
The hum of either army stilly sounds,              5
That the fixed sentinels almost receive
The secret whispers of each other's watch.
Fire answers fire, and through their paly flames
Each battle sees the other's umbered face.
Steed threatens steed, in high and boastful neighs    10
Piercing the night's dull ear, and from the tents
The armourers, accomplishing the knights,
With busy hammers closing rivets up,
Give dreadful note of preparation.
The country cocks do crow, the clocks do toll      15
And the third hour of drowsy morning name.
Proud of their numbers and secure in soul,
The confident and overlusty French
Do the low-rated English play at dice,
And chide the cripple tardy-gaited night,         20
Who like a foul and ugly witch doth limp
So tediously away. The poor condemnèd English,
Like sacrifices, by their watchful fires
Sit patiently and inly ruminate
The morning's danger; and their gesture sad,     25
Investing lank lean cheeks and war-worn coats,
Presented them unto the gazing moon
So many horrid ghosts. O now, who will behold
The royal captain of this ruined band
Walking from watch to watch, from tent to tent,   30
Let him cry, 'Praise and glory on his head!'
For forth he goes and visits all his host,
Bids them good morrow with a modest smile
And calls them brothers, friends, and countrymen.
Upon his royal face there is no note          35
How dread an army hath enrounded him;
Nor doth he dedicate one jot of colour
Unto the weary and all-watchèd night,
But freshly looks and overbears attaint
With cheerful semblance and sweet majesty,     40
That every wretch, pining and pale before,
Beholding him, plucks comfort from his looks.
A largess universal, like the sun,
His liberal eye doth give to everyone,
Thawing cold fear, that mean and gentle all    45
Behold, as may unworthiness define,
A little touch of Harry in the night.
And so our scene must to the battle fly,
Where O for pity, we shall much disgrace,
With four or five most vile and ragged foils,    50
Right ill-disposed in brawl ridiculous,
The name of Agincourt. Yet sit and see,
Minding true things by what their mock'ries be.   *Exit*

**4.1**   *Enter King Harry and the Duke of Gloucester, then
the Duke of ⌐Clarence⌐*

KING HARRY

Gloucester, 'tis true that we are in great danger;
The greater therefore should our courage be.
Good morrow, brother Clarence. God Almighty!
There is some soul of goodness in things evil,
Would men observingly distil it out—         5
For our bad neighbour makes us early stirrers,
Which is both healthful and good husbandry.

Besides, they are our outward consciences,
And preachers to us all, admonishing
That we should dress us fairly for our end.     10
Thus may we gather honey from the weed
And make a moral of the devil himself.

    *Enter Sir Thomas Erpingham*

Good morrow, old Sir Thomas Erpingham.
A good soft pillow for that good white head
Were better than a churlish turf of France.     15

ERPINGHAM

Not so, my liege. This lodging likes me better,
Since I may say, 'Now lie I like a king.'

KING HARRY

'Tis good for men to love their present pains
Upon example. So the spirit is eased,
And when the mind is quickened, out of doubt    20
The organs, though defunct and dead before,
Break up their drowsy grave and newly move
With casted slough and fresh legerity.
Lend me thy cloak, Sir Thomas.

    *He puts on Erpingham's cloak*

                      Brothers both,
Commend me to the princes in our camp.     25
Do my good morrow to them, and anon
Desire them all to my pavilion.

GLOUCESTER We shall, my liege.

ERPINGHAM Shall I attend your grace?

KING HARRY No, my good knight.         30
Go with my brothers to my lords of England.
I and my bosom must debate awhile,
And then I would no other company.

ERPINGHAM

The Lord in heaven bless thee, noble Harry.

KING HARRY

God-a-mercy, old heart, thou speak'st cheerfully.   35

    *Exeunt all but King Harry*

    *Enter Pistol ⌐to him⌐*

PISTOL *Qui vous là?*

KING HARRY A friend.

PISTOL

Discuss unto me: art thou officer,
Or art thou base, common, and popular?

KING HARRY I am a gentleman of a company.     40

PISTOL Trail'st thou the puissant pike?

KING HARRY Even so. What are you?

PISTOL

As good a gentleman as the Emperor.

KING HARRY Then you are a better than the King.

PISTOL

The King's a bawcock and a heart-of-gold,     45
A lad of life, an imp of fame,
Of parents good, of fist most valiant.
I kiss his dirty shoe, and from heartstring
I love the lovely bully. What is thy name?

KING HARRY Harry *le roi*.           50

PISTOL

Leroi? A Cornish name. Art thou of Cornish crew?

KING HARRY No, I am a Welshman.

PISTOL Know'st thou Fluellen?

KING HARRY Yes.

PISTOL

Tell him I'll knock his leek about his pate     55
Upon Saint Davy's day.

KING HARRY Do not you wear your dagger in your cap
that day, lest he knock that about yours.

PISTOL  Art thou his friend?
KING HARRY  And his kinsman too.                                60
PISTOL  The *fico* for thee then.
KING HARRY  I thank you. God be with you.
PISTOL  My name is Pistol called.
KING HARRY  It sorts well with your fierceness.
*Exit Pistol*
*Enter Captains Fluellen and Gower ⌈severally⌉. King Harry stands apart*
GOWER  Captain Fluellen!                                        65
FLUELLEN  So! In the name of Jesu Christ, speak fewer. It is the greatest admiration in the universal world, when the true and ancient prerogatifs and laws of the wars is not kept. If you would take the pains but to examine the wars of Pompey the Great, you shall find, I warrant you, that there is no tiddle-taddle nor pibble-babble in Pompey's camp. I warrant you, you shall find the ceremonies of the wars, and the cares of it, and the forms of it, and the sobriety of it, and the modesty of it, to be otherwise.                                        75
GOWER  Why, the enemy is loud. You hear him all night.
FLUELLEN  If the enemy is an ass and a fool and a prating coxcomb, is it meet, think you, that we should also, look you, be an ass and a fool and a prating coxcomb? In your own conscience now?                              80
GOWER  I will speak lower.
FLUELLEN  I pray you and beseech you that you will.
*Exeunt Fluellen and Gower*
KING HARRY
  Though it appear a little out of fashion,
  There is much care and valour in this Welshman.
*Enter three soldiers: John Bates, Alexander Court, and Michael Williams*
COURT  Brother John Bates, is not that the morning which breaks yonder?                                           86
BATES  I think it be. But we have no great cause to desire the approach of day.
WILLIAMS  We see yonder the beginning of the day, but I think we shall never see the end of it.—Who goes there?                                                        91
KING HARRY  A friend.
WILLIAMS  Under what captain serve you?
KING HARRY  Under Sir Thomas Erpingham.                         94
WILLIAMS  A good old commander and a most kind gentleman. I pray you, what thinks he of our estate?
KING HARRY  Even as men wrecked upon a sand, that look to be washed off the next tide.
BATES  He hath not told his thought to the King?             99
KING HARRY  No, nor it is not meet he should. For though I speak it to you, I think the King is but a man, as I am. The violet smells to him as it doth to me; the element shows to him as it doth to me. All his senses have but human conditions. His ceremonies laid by, in his nakedness he appears but a man, and though his affections are higher mounted than ours, yet when they stoop, they stoop with the like wing. Therefore, when he sees reason of fears, as we do, his fears, out of doubt, be of the same relish as ours are. Yet, in reason, no man should possess him with any appearance of fear, lest he, by showing it, should dishearten his army.                                       112
BATES  He may show what outward courage he will, but I believe, as cold a night as 'tis, he could wish himself in Thames up to the neck. And so I would he were, and I by him, at all adventures, so we were quit here.

KING HARRY  By my troth, I will speak my conscience of the King. I think he would not wish himself anywhere but where he is.                                            119
BATES  Then I would he were here alone. So should he be sure to be ransomed, and a many poor men's lives saved.
KING HARRY  I dare say you love him not so ill to wish him here alone, howsoever you speak this to feel other men's minds. Methinks I could not die anywhere so contented as in the King's company, his cause being just and his quarrel honourable.                           127
WILLIAMS  That's more than we know.
BATES  Ay, or more than we should seek after. For we know enough if we know we are the King's subjects. If his cause be wrong, our obedience to the King wipes the crime of it out of us.                                  132
WILLIAMS  But if the cause be not good, the King himself hath a heavy reckoning to make, when all those legs and arms and heads chopped off in a battle shall join together at the latter day, and cry all, 'We died at such a place'—some swearing, some crying for a surgeon, some upon their wives left poor behind them, some upon the debts they owe, some upon their children rawly left. I am afeard there are few die well that die in a battle, for how can they charitably dispose of anything, when blood is their argument? Now, if these men do not die well, it will be a black matter for the King that led them to it—who to disobey were against all proportion of subjection.                             145
KING HARRY  So, if a son that is by his father sent about merchandise do sinfully miscarry upon the sea, the imputation of his wickedness, by your rule, should be imposed upon his father, that sent him. Or if a servant, under his master's command transporting a sum of money, be assailed by robbers, and die in many irreconciled iniquities, you may call the business of the master the author of the servant's damnation. But this is not so. The King is not bound to answer the particular endings of his soldiers, the father of his son, nor the master of his servant, for they purpose not their deaths when they propose their services. Besides, there is no king, be his cause never so spotless, if it come to the arbitrament of swords, can try it out with all unspotted soldiers. Some, peradventure, have on them the guilt of premeditated and contrived murder; some, of beguiling virgins with the broken seals of perjury; some, making the wars their bulwark, that have before gored the gentle bosom of peace with pillage and robbery. Now, if these men have defeated the law and outrun native punishment, though they can outstrip men, they have no wings to fly from God. War is his beadle. War is his vengeance. So that here men are punished for before-breach of the King's laws, in now the King's quarrel. Where they feared the death, they have borne life away; and where they would be safe, they perish. Then if they die unprovided, no more is the King guilty of their damnation than he was before guilty of those impieties for the which they are now visited. Every subject's duty is the King's, but every subject's soul is his own. Therefore should every soldier in the wars do as every sick man in his bed: wash every mote out of his conscience. And dying so, death is to him advantage; or not dying, the time was blessedly lost wherein such preparation was gained. And in him that escapes, it were not sin to think that, making God so free an offer, he let him outlive that

day to see his greatness and to teach others how they
should prepare.                                                    184
⌈BATES⌉ 'Tis certain, every man that dies ill, the ill upon
his own head. The King is not to answer it. I do not
desire he should answer for me, and yet I determine to
fight lustily for him.
KING HARRY I myself heard the King say he would not be
ransomed.                                                          190
WILLIAMS Ay, he said so, to make us fight cheerfully, but
when our throats are cut he may be ransomed, and
we ne'er the wiser.
KING HARRY If I live to see it, I will never trust his word
after.                                                             195
WILLIAMS You pay him then! That's a perilous shot out
of an elder-gun, that a poor and a private displeasure
can do against a monarch. You may as well go about
to turn the sun to ice with fanning in his face with a
peacock's feather. You'll never trust his word after!
Come, 'tis a foolish saying.                                       201
KING HARRY Your reproof is something too round. I should
be angry with you, if the time were convenient.
WILLIAMS Let it be a quarrel between us, if you live.
KING HARRY I embrace it.                                           205
WILLIAMS How shall I know thee again?
KING HARRY Give me any gage of thine, and I will wear
it in my bonnet. Then if ever thou darest acknowledge
it, I will make it my quarrel.
WILLIAMS Here's my glove. Give me another of thine.
KING HARRY There.                                                  211
    *They exchange gloves*
WILLIAMS This will I also wear in my cap. If ever thou
come to me and say, after tomorrow, 'This is my glove',
by this hand I will take thee a box on the ear.
KING HARRY If ever I live to see it, I will challenge it.
WILLIAMS Thou darest as well be hanged.                            216
KING HARRY Well, I will do it, though I take thee in the
King's company.
WILLIAMS Keep thy word. Fare thee well.
BATES Be friends, you English fools, be friends. We have
French quarrels enough, if you could tell how to reckon.
KING HARRY Indeed, the French may lay twenty French
crowns to one they will beat us, for they bear them on
their shoulders. But it is no English treason to cut
French crowns, and tomorrow the King himself will be
a clipper.                                        *Exeunt soldiers*
Upon the King.                                                     227
'Let us our lives, our souls, our debts, our care-full
    wives,
Our children, and our sins, lay on the King.'
We must bear all. O hard condition,                                230
Twin-born with greatness: subject to the breath
Of every fool, whose sense no more can feel
But his own wringing. What infinite heartsease
Must kings neglect that private men enjoy?
And what have kings that privates have not too,                    235
Save ceremony, save general ceremony?
And what art thou, thou idol ceremony?
What kind of god art thou, that suffer'st more
Of mortal griefs than do thy worshippers?
What are thy rents? What are thy comings-in?                       240
O ceremony, show me but thy worth.
What is thy soul of adoration?
Art thou aught else but place, degree, and form,
Creating awe and fear in other men?
Wherein thou art less happy, being feared,                         245
Than they in fearing.
What drink'st thou oft, instead of homage sweet,
But poisoned flattery? O be sick, great greatness,
And bid thy ceremony give thee cure.
Think'st thou the fiery fever will go out                          250
With titles blown from adulation?
Will it give place to flexure and low bending?
Canst thou, when thou command'st the beggar's knee,
Command the health of it? No, thou proud dream
That play'st so subtly with a king's repose;                       255
I am a king that find thee, and I know
'Tis not the balm, the sceptre, and the ball,
The sword, the mace, the crown imperial,
The intertissued robe of gold and pearl,
The farcèd title running fore the king,                            260
The throne he sits on, nor the tide of pomp
That beats upon the high shore of this world—
No, not all these, thrice-gorgeous ceremony,
Not all these, laid in bed majestical,
Can sleep so soundly as the wretched slave                         265
Who with a body filled and vacant mind
Gets him to rest, crammed with distressful bread;
Never sees horrid night, the child of hell,
But like a lackey from the rise to set
Sweats in the eye of Phoebus, and all night                        270
Sleeps in Elysium; next day, after dawn
Doth rise and help Hyperion to his horse,
And follows so the ever-running year
With profitable labour to his grave.
And but for ceremony such a wretch,                                275
Winding up days with toil and nights with sleep,
Had the forehand and vantage of a king.
The slave, a member of the country's peace,
Enjoys it, but in gross brain little wots
What watch the King keeps to maintain the peace,
Whose hours the peasant best advantages.                           281
    *Enter Sir Thomas Erpingham*
ERPINGHAM
My lord, your nobles, jealous of your absence,
Seek through your camp to find you.
KING HARRY                                        Good old knight,
Collect them all together at my tent.
I'll be before thee.
ERPINGHAM            I shall do't, my lord.                  *Exit*
KING HARRY
O God of battles, steel my soldiers' hearts.                       286
Possess them not with fear. Take from them now
The sense of reck'ning, ere th'opposèd numbers
Pluck their hearts from them. Not today, O Lord,
O not today, think not upon the fault                              290
My father made in compassing the crown.
I Richard's body have interrèd new,
And on it have bestowed more contrite tears
Than from it issued forcèd drops of blood.
Five hundred poor have I in yearly pay                             295
Who twice a day their withered hands hold up
Toward heaven to pardon blood. And I have built
Two chantries, where the sad and solemn priests
Sing still for Richard's soul. More will I do,
Though all that I can do is nothing worth,                         300
Since that my penitence comes after ill,
Imploring pardon.
    *Enter the Duke of Gloucester*
GLOUCESTER
My liege.
KING HARRY My brother Gloucester's voice? Ay.
I know thy errand, I will go with thee.
The day, my friends, and all things stay for me.                   305
                                                          *Exeunt*

**4.2** *Enter the Dukes of* ⌐Bourbon⌐ *and Orléans, and Lord*
*Rambures*

ORLÉANS
The sun doth gild our armour. Up, my lords!
⌐BOURBON⌐ *Monte cheval!* My horse! *Varlet, lacquais!* Ha!
ORLÉANS  O brave spirit!
⌐BOURBON⌐ *Via les eaux et terre!*
ORLÉANS *Rien plus? L'air et feu!*                                5
⌐BOURBON⌐ *Cieux,* cousin Orléans!
          *Enter the Constable*
Now, my Lord Constable!
CONSTABLE
Hark how our steeds for present service neigh.
⌐BOURBON⌐
Mount them and make incision in their hides,
That their hot blood may spin in English eyes          10
And dout them with superfluous courage. Ha!
RAMBURES
What, will you have them weep our horses' blood?
How shall we then behold their natural tears?
          *Enter a Messenger*
MESSENGER
The English are embattled, you French peers.
CONSTABLE
To horse, you gallant princes, straight to horse!      15
Do but behold yon poor and starvèd band,
And your fair show shall suck away their souls,
Leaving them but the shells and husks of men.
There is not work enough for all our hands,
Scarce blood enough in all their sickly veins          20
To give each naked curtal-axe a stain
That our French gallants shall today draw out
And sheathe for lack of sport. Let us but blow on
     them,
The vapour of our valour will o'erturn them.
'Tis positive 'gainst all exceptions, lords,            25
That our superfluous lackeys and our peasants,
Who in unnecessary action swarm
About our squares of battle, were enough
To purge this field of such a hilding foe,
Though we upon this mountain's basis by                30
Took stand for idle speculation,
But that our honours must not. What's to say?
A very little little let us do
And all is done. Then let the trumpets sound
The tucket sonance and the note to mount,              35
For our approach shall so much dare the field
That England shall couch down in fear and yield.
          *Enter Lord Grandpré*
GRANDPRÉ
Why do you stay so long, my lords of France?
Yon island carrions, desperate of their bones,
Ill-favouredly become the morning field.               40
Their ragged curtains poorly are let loose
And our air shakes them passing scornfully.
Big Mars seems bankrupt in their beggared host
And faintly through a rusty beaver peeps.
The horsemen sit like fixèd candlesticks               45
With torchstaves in their hands, and their poor jades
Lob down their heads, drooping the hides and hips,
The gum down-roping from their pale dead eyes,
And in their palled dull mouths the gimmaled bit
Lies foul with chewed grass, still and motionless.     50
And their executors, the knavish crows,
Fly o'er them all impatient for their hour.
Description cannot suit itself in words

To demonstrate the life of such a battle
In life so lifeless as it shows itself.                55
CONSTABLE
They have said their prayers, and they stay for death.
⌐BOURBON⌐
Shall we go send them dinners and fresh suits
And give their fasting horses provender,
And after fight with them?
CONSTABLE
I stay but for my guidon. To the field!                60
I will the banner from a trumpet take
And use it for my haste. Come, come away!
The sun is high, and we outwear the day.      *Exeunt*

**4.3** *Enter the Dukes of Gloucester,* ⌐Clarence⌐, *and*
*Exeter, the Earls of Salisbury and* ⌐Warwick⌐, *and*
*Sir Thomas Erpingham, with all* ⌐the⌐ *host*

GLOUCESTER  Where is the King?
⌐CLARENCE⌐
The King himself is rode to view their battle.
⌐WARWICK⌐
Of fighting men they have full threescore thousand.
EXETER
There's five to one. Besides, they all are fresh.
SALISBURY
God's arm strike with us! 'Tis a fearful odds.          5
God b'wi' you, princes all. I'll to my charge.
If we no more meet till we meet in heaven,
Then joyfully, my noble Lord of Clarence,
My dear Lord Gloucester, and my good Lord Exeter,
And (*to Warwick*) my kind kinsman, warriors all,
     adieu.                                             10
⌐CLARENCE⌐
Farewell, good Salisbury, and good luck go with thee.
EXETER
Farewell, kind lord. Fight valiantly today—
And yet I do thee wrong to mind thee of it,
For thou art framed of the firm truth of valour.
               *Exit Salisbury*
⌐CLARENCE⌐
He is as full of valour as of kindness,                15
Princely in both.
          *Enter King Harry, behind*
⌐WARWICK⌐               O that we now had here
But one ten thousand of those men in England
That do no work today.
KING HARRY               What's he that wishes so?
My cousin Warwick? No, my fair cousin.
If we are marked to die, we are enough                 20
To do our country loss; and if to live,
The fewer men, the greater share of honour.
God's will, I pray thee wish not one man more.
By Jove, I am not covetous for gold,
Nor care I who doth feed upon my cost;                 25
It ernes me not if men my garments wear;
Such outward things dwell not in my desires.
But if it be a sin to covet honour
I am the most offending soul alive.
No, faith, my coz, wish not a man from England.        30
God's peace, I would not lose so great an honour
As one man more methinks would share from me
For the best hope I have. O do not wish one more.
Rather proclaim it presently through my host
That he which hath no stomach to this fight,           35
Let him depart. His passport shall be made
And crowns for convoy put into his purse.

We would not die in that man's company
That fears his fellowship to die with us.
This day is called the Feast of Crispian.                    40
He that outlives this day and comes safe home
Will stand a-tiptoe when this day is named
And rouse him at the name of Crispian.
He that shall see this day and live t'old age
Will yearly on the vigil feast his neighbours          45
And say, 'Tomorrow is Saint Crispian.'
Then will he strip his sleeve and show his scars
And say, 'These wounds I had on Crispin's day.'
Old men forget; yet all shall be forgot,
But he'll remember, with advantages,                         50
What feats he did that day. Then shall our names,
Familiar in his mouth as household words—
Harry the King, Bedford and Exeter,
Warwick and Talbot, Salisbury and Gloucester—
Be in their flowing cups freshly remembered.           55
This story shall the good man teach his son,
And Crispin Crispian shall ne'er go by
From this day to the ending of the world
But we in it shall be rememberèd,
We few, we happy few, we band of brothers.           60
For he today that sheds his blood with me
Shall be my brother; be he ne'er so vile,
This day shall gentle his condition.
And gentlemen in England now abed
Shall think themselves accursed they were not here,
And hold their manhoods cheap whiles any speaks  66
That fought with us upon Saint Crispin's day.

*Enter the Earl of Salisbury*

SALISBURY
My sovereign lord, bestow yourself with speed.
The French are bravely in their battles set
And will with all expedience charge on us.               70

KING HARRY
All things are ready if our minds be so.

⌈WARWICK⌉
Perish the man whose mind is backward now.

KING HARRY
Thou dost not wish more help from England, coz?

⌈WARWICK⌉
God's will, my liege, would you and I alone,
Without more help, could fight this royal battle.     75

KING HARRY
Why now thou hast unwished five thousand men,
Which likes me better than to wish us one.—
You know your places. God be with you all.

*Tucket. Enter Montjoy*

MONTJOY
Once more I come to know of thee, King Harry,
If for thy ransom thou wilt now compound          80
Before thy most assurèd overthrow.
For certainly thou art so near the gulf
Thou needs must be englutted. Besides, in mercy
The Constable desires thee thou wilt mind
Thy followers of repentance, that their souls          85
May make a peaceful and a sweet retire
From off these fields where, wretches, their poor
      bodies
Must lie and fester.

KING HARRY Who hath sent thee now?

MONTJOY The Constable of France.                           90

KING HARRY
I pray thee bear my former answer back.
Bid them achieve me, and then sell my bones.

Good God, why should they mock poor fellows thus?
The man that once did sell the lion's skin
While the beast lived, was killed with hunting him.  95
A many of our bodies shall no doubt
Find native graves, upon the which, I trust,
Shall witness live in brass of this day's work.
And those that leave their valiant bones in France,
Dying like men, though buried in your dunghills   100
They shall be famed. For there the sun shall greet
      them
And draw their honours reeking up to heaven,
Leaving their earthly parts to choke your clime,
The smell whereof shall breed a plague in France.
Mark then abounding valour in our English,          105
That, being dead, like to the bullets grazing
Break out into a second course of mischief,
Killing in relapse of mortality.
Let me speak proudly. Tell the Constable
We are but warriors for the working day.               110
Our gayness and our gilt are all besmirched
With rainy marching in the painful field.
There's not a piece of feather in our host—
Good argument, I hope, we will not fly—
And time hath worn us into slovenry.                       115
But by the mass, our hearts are in the trim.
And my poor soldiers tell me, yet ere night
They'll be in fresher robes, as they will pluck
The gay new coats o'er your French soldiers' heads,
And turn them out of service. If they do this—     120
As if God please, they shall—my ransom then
Will soon be levied. Herald, save thou thy labour.
Come thou no more for ransom, gentle herald.
They shall have none, I swear, but these my joints—
Which if they have as I will leave 'em them,          125
Shall yield them little. Tell the Constable.

MONTJOY
I shall, King Harry. And so fare thee well.
Thou never shalt hear herald any more.

KING HARRY
I fear thou wilt once more come for a ransom.
                                                           *Exit Montjoy*

*Enter the Duke of York*

YORK
My lord, most humbly on my knee I beg                 130
The leading of the vanguard.

KING HARRY
Take it, brave York.—Now soldiers, march away,
And how thou pleasest, God, dispose the day.  *Exeunt*

**4.4**    *Alarum. Excursions. Enter Pistol, a French soldier,
      and the Boy*

PISTOL Yield, cur.

FRENCH SOLDIER *Je pense que vous êtes le gentilhomme de
bon qualité.*

PISTOL
*Qualité? 'Calin o custure me!'*
Art thou a gentleman? What is thy name? Discuss.   5

FRENCH SOLDIER *O Seigneur Dieu!*

PISTOL ⌈*aside*⌉
O Seigneur Dew should be a gentleman.—
Perpend my words, O Seigneur Dew, and mark:
O Seigneur Dew, thou diest, on point of fox,
Except, O Seigneur, thou do give to me                  10
Egregious ransom.

FRENCH SOLDIER *O prenez miséricorde! Ayez pitié de moi!*

PISTOL
  'Moy' shall not serve, I will have forty 'moys',
  Or I will fetch thy rim out at thy throat
  In drops of crimson blood.                              15
FRENCH SOLDIER *Est-il impossible d'échapper la force de ton*
  *bras?*
PISTOL
  Brass, cur? Thou damnèd and luxurious mountain
    goat,
  Offer'st me brass?
FRENCH SOLDIER *O pardonne-moi!*                          20
PISTOL
  Sayst thou me so? Is that a ton of moys?—
  Come hither boy. Ask me this slave in French
  What is his name.
BOY *Écoutez: comment êtes-vous appelé?*
FRENCH SOLDIER *Monsieur le Fer.*                         25
BOY He says his name is Master Fer.
PISTOL Master Fer? I'll fer him, and firk him, and ferret
  him.
  Discuss the same in French unto him.
BOY I do not know the French for fer and ferret and firk.
PISTOL
  Bid him prepare, for I will cut his throat.             31
FRENCH SOLDIER *Que dit-il, monsieur?*
BOY *Il me commande à vous dire que vous faites vous prêt,*
  *car ce soldat ici est disposé tout à cette heure de couper*
  *votre gorge.*                                          35
PISTOL
  *Oui, couper la gorge, par ma foi,*
  Peasant, unless thou give me crowns, brave crowns;
  Or mangled shalt thou be by this my sword.
FRENCH SOLDIER *O je vous supplie, pour l'amour de Dieu, me*
  *pardonner. Je suis le gentilhomme de bonne maison. Gardez*
  *ma vie, et je vous donnerai deux cents écus.*          41
PISTOL What are his words?
BOY He prays you to save his life. He is a gentleman of a
  good house, and for his ransom he will give you two
  hundred crowns.                                         45
PISTOL
  Tell him, my fury shall abate, and I the crowns will
    take.
FRENCH SOLDIER *Petit monsieur, que dit-il?*
BOY *Encore qu'il est contre son jurement de pardonner aucun*
  *prisonnier; néanmoins, pour les écus que vous lui ci*
  *promettez, il est content à vous donner la liberté, le*
  *franchisement.*                                        51
FRENCH SOLDIER (*kneeling to Pistol*) *Sur mes genoux je vous*
  *donne mille remerciements, et je m'estime heureux que j'ai*
  *tombé entre les mains d'un chevalier, comme je pense, le*
  *plus brave, vaillant, et treis-distingué seigneur d'Angleterre.*
PISTOL Expound unto me, boy.                              56
BOY He gives you upon his knees a thousand thanks, and
  he esteems himself happy that he hath fallen into the
  hands of one, as he thinks, the most brave, valorous,
  and thrice-worthy seigneur of England.                 60
PISTOL
  As I suck blood, I will some mercy show.
  Follow me.
BOY *Suivez-vous le grand capitaine.*
                    *Exeunt Pistol and French Soldier*
  I did never know so full a voice issue from so empty a
  heart. But the saying is true: 'The empty vessel makes
  the greatest sound.' Bardolph and Nim had ten times
  more valour than this roaring devil i'th' old play, that
everyone may pare his nails with a wooden dagger,
and they are both hanged, and so would this be, if he
durst steal anything adventurously. I must stay with
the lackeys with the luggage of our camp. The French
might have a good prey of us, if he knew of it, for there
is none to guard it but boys.                        *Exit*

4.5    *Enter the Constable, the Dukes of Orléans and*
       ⌈*Bourbon*⌉, *and Lord Rambures*
CONSTABLE *O diable!*
ORLÉANS *O Seigneur! Le jour est perdu, tout est perdu!*
⌈BOURBON⌉
  *Mort de ma vie!* All is confounded, all.
  Reproach and everlasting shame
  Sits mocking in our plumes.                             5
           *A short alarum*
  *O méchante fortune!*— (*To Rambures*) Do not run away.
⌈ORLÉANS⌉
  We are enough yet living in the field
  To smother up the English in our throngs,
  If any order might be thought upon.
BOURBON
  The devil take order. Once more back again!            10
  And he that will not follow Bourbon now,
  Let him go home, and with his cap in hand
  Like a base leno hold the chamber door
  Whilst by a slave no gentler than my dog
  His fairest daughter is contaminated.                  15
CONSTABLE
  Disorder that hath spoiled us friend us now.
  Let us on heaps go offer up our lives.
BOURBON I'll to the throng.
  Let life be short, else shame will be too long.  *Exeunt*

4.6    *Alarum. Enter King Harry and his train, with*
       *prisoners*
KING HARRY
  Well have we done, thrice-valiant countrymen.
  But all's not done; yet keep the French the field.
           ⌈*Enter the Duke of Exeter*⌉
EXETER
  The Duke of York commends him to your majesty.
KING HARRY
  Lives he, good uncle? Thrice within this hour
  I saw him down, thrice up again and fighting.          5
  From helmet to the spur, all blood he was.
EXETER
  In which array, brave soldier, doth he lie,
  Larding the plain. And by his bloody side,
  Yokefellow to his honour-owing wounds,
  The noble Earl of Suffolk also lies.                   10
  Suffolk first died, and York, all haggled over,
  Comes to him, where in gore he lay insteeped,
  And takes him by the beard, kisses the gashes
  That bloodily did yawn upon his face,
  And cries aloud, 'Tarry, dear cousin Suffolk.          15
  My soul shall thine keep company to heaven.
  Tarry, sweet soul, for mine, then fly abreast,
  As in this glorious and well-foughten field
  We kept together in our chivalry.'
  Upon these words I came and cheered him up.            20
  He smiled me in the face, raught me his hand,
  And with a feeble grip says, 'Dear my lord,
  Commend my service to my sovereign.'
  So did he turn, and over Suffolk's neck
  He threw his wounded arm, and kissed his lips,         25

And so espoused to death, with blood he sealed
A testament of noble-ending love.
The pretty and sweet manner of it forced
Those waters from me which I would have stopped.
But I had not so much of man in me,                                    30
And all my mother came into mine eyes
And gave me up to tears.
KING HARRY                              I blame you not,
For hearing this I must perforce compound
With mistful eyes, or they will issue too.
        *Alarum*
But hark, what new alarum is this same?                                35
The French have reinforced their scattered men.
Then every soldier kill his prisoners.
        ⌈*The soldiers kill their prisoners*⌉
Give the word through.
⌈PISTOL⌉ *Coup' la gorge.*                              *Exeunt*

**4.7**   *Enter Captains Fluellen and Gower*
FLUELLEN Kill the poys and the luggage! 'Tis expressly
against the law of arms. 'Tis as arrant a piece of
knavery, mark you now, as can be offert. In your
conscience now, is it not?
GOWER 'Tis certain there's not a boy left alive. And the
cowardly rascals that ran from the battle ha' done this
slaughter. Besides, they have burned and carried away
all that was in the King's tent; wherefore the King
most worthily hath caused every soldier to cut his
prisoner's throat. O 'tis a gallant king.                              10
FLUELLEN Ay, he was porn at Monmouth. Captain Gower,
what call you the town's name where Alexander the
Pig was born?
GOWER Alexander the Great.
FLUELLEN Why I pray you, is not 'pig' great? The pig or
the great or the mighty or the huge or the
magnanimous are all one reckonings, save the phrase
is a little variations.                                                18
GOWER I think Alexander the Great was born in Macedon.
His father was called Philip of Macedon, as I take it.
FLUELLEN I think it is e'en Macedon where Alexander is
porn. I tell you, captain, if you look in the maps of the
world I warrant you sall find, in the comparisons
between Macedon and Monmouth, that the situations,
look you, is both alike. There is a river in Macedon,
and there is also moreover a river at Monmouth. It is
called Wye at Monmouth, but it is out of my prains
what is the name of the other river—but 'tis all one,
'tis alike as my fingers is to my fingers, and there is
salmons in both. If you mark Alexander's life well,
Harry of Monmouth's life is come after it indifferent
well. For there is figures in all things. Alexander, God
knows, and you know, in his rages and his furies and
his wraths and his cholers and his moods and his
displeasures and his indignations, and also being a little
intoxicates in his prains, did in his ales and his angers,
look you, kill his best friend Cleitus—
GOWER Our King is not like him in that. He never killed
any of his friends.                                                    39
FLUELLEN It is not well done, mark you now, to take the
tales out of my mouth ere it is made an end and
finished. I speak but in the figures and comparisons of
it. As Alexander killed his friend Cleitus, being in his
ales and his cups, so also Harry Monmouth, being in
his right wits and his good judgements, turned away
the fat knight with the great-belly doublet—he was full
of jests and gipes and knaveries and mocks—I have
forgot his name.                                                       48

GOWER Sir John Falstaff.
FLUELLEN That is he. I'll tell you, there is good men porn
at Monmouth.
GOWER Here comes his majesty.
        *Alarum. Enter King Harry* ⌈*and the English army*⌉*,
        with the Duke of Bourbon,* ⌈*the Duke of Orléans,*⌉
        *and other prisoners. Flourish*
KING HARRY
I was not angry since I came to France
Until this instant. Take a trumpet, herald;
Ride thou unto the horsemen on yon hill.                               55
If they will fight with us, bid them come down,
Or void the field: they do offend our sight.
If they'll do neither, we will come to them,
And make them skirr away as swift as stones
Enforcèd from the old Assyrian slings.                                 60
Besides, we'll cut the throats of those we have,
And not a man of them that we shall take
Shall taste our mercy. Go and tell them so.
        *Enter Montjoy*
EXETER
Here comes the herald of the French, my liege.
GLOUCESTER
His eyes are humbler than they used to be.                             65
KING HARRY
How now, what means this, herald? Know'st thou
        not
That I have fined these bones of mine for ransom?
Com'st thou again for ransom?
MONTJOY                              No, great King.
I come to thee for charitable licence,
That we may wander o'er this bloody field                              70
To book our dead and then to bury them,
To sort our nobles from our common men—
For many of our princes, woe the while,
Lie drowned and soaked in mercenary blood.
So do our vulgar drench their peasant limbs                            75
In blood of princes, and our wounded steeds
Fret fetlock-deep in gore, and with wild rage
Jerk out their armèd heels at their dead masters,
Killing them twice. O give us leave, great King,
To view the field in safety, and dispose                               80
Of their dead bodies.
KING HARRY                 I tell thee truly, herald,
I know not if the day be ours or no,
For yet a many of your horsemen peer
And gallop o'er the field.
MONTJOY                          The day is yours.
KING HARRY
Praisèd be God, and not our strength, for it.                          85
What is this castle called that stands hard by?
MONTJOY They call it Agincourt.
KING HARRY
Then call we this the field of Agincourt,
Fought on the day of Crispin Crispian.
FLUELLEN Your grandfather of famous memory, an't
please your majesty, and your great-uncle Edward the
Plack Prince of Wales, as I have read in the chronicles,
fought a most prave pattle here in France.
KING HARRY They did, Fluellen.                                         94
FLUELLEN Your majesty says very true. If your majesties
is remembered of it, the Welshmen did good service in
a garden where leeks did grow, wearing leeks in their
Monmouth caps, which your majesty know to this
hour is an honourable badge of the service. And I do
believe your majesty takes no scorn to wear the leek
upon Saint Tavy's day.                                                101

KING HARRY
I wear it for a memorable honour,
For I am Welsh, you know, good countryman.

FLUELLEN All the water in Wye cannot wash your majesty's Welsh plood out of your pody, I can tell you that. God pless it and preserve it, as long as it pleases his grace, and his majesty too. 107

KING HARRY Thanks, good my countryman.

FLUELLEN By Jeshu, I am your majesty's countryman. I care not who know it, I will confess it to all the world. I need not to be ashamed of your majesty, praised be God, so long as your majesty is an honest man. 112

KING HARRY
God keep me so.
    *Enter Williams with a glove in his cap*
Our heralds go with him.
Bring me just notice of the numbers dead
On both our parts.
    *Exeunt Montjoy, ⌈Gower,⌉ and an English
                herald*
Call yonder fellow hither.

EXETER (*to Williams*) Soldier, you must come to the King.

KING HARRY Soldier, why wearest thou that glove in thy cap?

WILLIAMS An't please your majesty, 'tis the gage of one that I should fight withal, if he be alive. 120

KING HARRY An Englishman?

WILLIAMS An't please your majesty, a rascal, that swaggered with me last night—who, if a live, and ever dare to challenge this glove, I have sworn to take him a box o'th' ear; or if I can see my glove in his cap—which he swore, as he was a soldier, he would wear if a lived—I will strike it out soundly. 127

KING HARRY What think you, Captain Fluellen? Is it fit this soldier keep his oath?

FLUELLEN He is a craven and a villain else, an't please your majesty, in my conscience. 131

KING HARRY It may be his enemy is a gentleman of great sort, quite from the answer of his degree.

FLUELLEN Though he be as good a gentleman as the devil is, as Lucifer and Beelzebub himself, it is necessary, look your grace, that he keep his vow and his oath. If he be perjured, see you now, his reputation is as arrant a villain and a Jack-sauce as ever his black shoe trod upon God's ground and his earth, in my conscience, law. 140

KING HARRY Then keep thy vow, sirrah, when thou meetest the fellow.

WILLIAMS So I will, my liege, as I live.

KING HARRY Who serv'st thou under?

WILLIAMS Under Captain Gower, my liege. 145

FLUELLEN Gower is a good captain, and is good knowledge and literatured in the wars.

KING HARRY Call him hither to me, soldier.

WILLIAMS I will, my liege.           *Exit*

KING HARRY (*giving him Williams's other glove*) Here, Fluellen, wear thou this favour for me and stick it in thy cap. When Alençon and myself were down together, I plucked this glove from his helm. If any man challenge this, he is a friend to Alençon and an enemy to our person. If thou encounter any such, apprehend him, an thou dost me love. 156

FLUELLEN Your grace does me as great honours as can be desired in the hearts of his subjects. I would fain see the man that has but two legs that shall find himself aggriefed at this glove, that is all; but I would fain see it once. An't please God of his grace, that I would see.

KING HARRY Know'st thou Gower? 162

FLUELLEN He is my dear friend, an't please you.

KING HARRY Pray thee, go seek him and bring him to my tent. 165

FLUELLEN I will fetch him.          *Exit*

KING HARRY
My lord of Warwick and my brother Gloucester,
Follow Fluellen closely at the heels.
The glove which I have given him for a favour
May haply purchase him a box o'th' ear. 170
It is the soldier's. I by bargain should
Wear it myself. Follow, good cousin Warwick.
If that the soldier strike him, as I judge
By his blunt bearing he will keep his word,
Some sudden mischief may arise of it, 175
For I do know Fluellen valiant
And touched with choler, hot as gunpowder,
And quickly will return an injury.
Follow, and see there be no harm between them.
Go you with me, uncle of Exeter.    *Exeunt severally*

**4.8**    *Enter Captain Gower and Williams*

WILLIAMS I warrant it is to knight you, captain.
    *Enter Captain Fluellen*

FLUELLEN God's will and his pleasure, captain, I beseech you now, come apace to the King. There is more good toward you, peradventure, than is in your knowledge to dream of. 5

WILLIAMS Sir, know you this glove?

FLUELLEN Know the glove? I know the glove is a glove.

WILLIAMS ⌈*plucking the glove from Fluellen's cap*⌉ I know this, and thus I challenge it. 9
    *He strikes Fluellen*

FLUELLEN God's plood, and his! An arrant traitor as any's in the universal world, or in France, or in England.

GOWER (*to Williams*) How now, sir? You villain!

WILLIAMS Do you think I'll be forsworn?

FLUELLEN Stand away, Captain Gower. I will give treason his payment into plows, I warrant you. 15

WILLIAMS I am no traitor.

FLUELLEN That's a lie in thy throat. I charge you in his majesty's name, apprehend him. He's a friend of the Duke Alençon's.
    *Enter the Earl of Warwick and the Duke of
    Gloucester*

WARWICK How now, how now, what's the matter? 20

FLUELLEN My lord of Warwick, here is—praised be God for it—a most contagious treason come to light, look you, as you shall desire in a summer's day.
    *Enter King Harry and the Duke of Exeter*
Here is his majesty.

KING HARRY How now, what is the matter? 25

FLUELLEN My liege, here is a villain and a traitor that, look your grace, has struck the glove which your majesty is take out of the helmet of Alençon.

WILLIAMS My liege, this was my glove—here is the fellow of it—and he that I gave it to in change promised to wear it in his cap. I promised to strike him, if he did. I met this man with my glove in his cap, and I have been as good as my word. 33

FLUELLEN Your majesty hear now, saving your majesty's manhood, what an arrant rascally beggarly lousy knave it is. I hope your majesty is pear me testimony and witness, and will avouchment that this is the glove of Alençon that your majesty is give me, in your conscience now. 39

KING HARRY Give me thy glove, soldier. Look, here is the
fellow of it.
    'Twas I indeed thou promisèd'st to strike,
    And thou hast given me most bitter terms.
FLUELLEN An't please your majesty, let his neck answer
for it, if there is any martial law in the world.        45
KING HARRY
    How canst thou make me satisfaction?
WILLIAMS All offences, my lord, come from the heart.
Never came any from mine that might offend your
majesty.
KING HARRY
    It was ourself thou didst abuse.                      50
WILLIAMS Your majesty came not like yourself. You
appeared to me but as a common man. Witness the
night, your garments, your lowliness. And what your
highness suffered under that shape, I beseech you take
it for your own fault, and not mine, for had you been
as I took you for, I made no offence. Therefore I beseech
your highness pardon me.                                 57
KING HARRY
    Here, Uncle Exeter, fill this glove with crowns
    And give it to this fellow.—Keep it, fellow,
    And wear it for an honour in thy cap                  60
    Till I do challenge it.—Give him the crowns.
    —And captain, you must needs be friends with him.
FLUELLEN By this day and this light, the fellow has mettle
enough in his belly.—Hold, there is twelve pence for
you, and I pray you to serve God, and keep you out of
prawls and prabbles and quarrels and dissensions, and
I warrant you it is the better for you.                  67
WILLIAMS I will none of your money.
FLUELLEN It is with a good will. I can tell you, it will
serve you to mend your shoes. Come, wherefore should
you be so pashful? Your shoes is not so good. 'Tis a
good shilling, I warrant you, or I will change it.       72
    *Enter ⌈an English⌉ Herald*
KING HARRY Now, herald, are the dead numbered?
HERALD
    Here is the number of the slaughtered French.
KING HARRY
    What prisoners of good sort are taken, uncle?        75
EXETER
    Charles, Duke of Orléans, nephew to the King;
    Jean, Duke of Bourbon, and Lord Boucicault;
    Of other lords and barons, knights and squires,
    Full fifteen hundred, besides common men.
KING HARRY
    This note doth tell me of ten thousand French        80
    That in the field lie slain. Of princes in this number
    And nobles bearing banners, there lie dead
    One hundred twenty-six; added to these,
    Of knights, esquires, and gallant gentlemen,
    Eight thousand and four hundred, of the which        85
    Five hundred were but yesterday dubbed knights.
    So that in these ten thousand they have lost
    There are but sixteen hundred mercenaries;
    The rest are princes, barons, lords, knights, squires,
    And gentlemen of blood and quality.                  90
    The names of those their nobles that lie dead:
    Charles Delabret, High Constable of France;
    Jaques of Châtillon, Admiral of France;
    The Master of the Crossbows, Lord Rambures;
    Great-Master of France, the brave Sir Guiscard
        Dauphin;                                         95

    Jean, Duke of Alençon; Antony, Duke of Brabant,
    The brother to the Duke of Burgundy;
    And Édouard, Duke of Bar; of lusty earls,
    Grandpré and Roussi, Fauconbridge and Foix,
    Beaumont and Marle, Vaudemont and Lestrelles.       100
    Here was a royal fellowship of death.
    Where is the number of our English dead?
        *He is given another paper*
    Edward the Duke of York, the Earl of Suffolk,
    Sir Richard Keighley, Davy Gam Esquire;
    None else of name, and of all other men             105
    But five-and-twenty. O God, thy arm was here,
    And not to us, but to thy arm alone
    Ascribe we all. When, without stratagem,
    But in plain shock and even play of battle,
    Was ever known so great and little loss             110
    On one part and on th'other? Take it God,
    For it is none but thine.
EXETER                                'Tis wonderful.
KING HARRY
    Come, go we in procession to the village,
    And be it death proclaimèd through our host
    To boast of this, or take that praise from God      115
    Which is his only.
FLUELLEN Is it not lawful, an't please your majesty, to tell
how many is killed?
KING HARRY
    Yes, captain, but with this acknowledgement,
    That God fought for us.                              120
FLUELLEN Yes, in my conscience, he did us great good.
KING HARRY Do we all holy rites:
    Let there be sung *Non nobis* and *Te Deum*,
    The dead with charity enclosed in clay;
    And then to Calais, and to England then,            125
    Where ne'er from France arrived more-happy men.
                                              *Exeunt*

**5.0**    *Enter Chorus*
CHORUS
    Vouchsafe to those that have not read the story
    That I may prompt them—and of such as have,
    I humbly pray them to admit th'excuse
    Of time, of numbers, and due course of things,
    Which cannot in their huge and proper life            5
    Be here presented. Now we bear the King
    Toward Calais. Grant him there; there seen,
    Heave him away upon your wingèd thoughts
    Athwart the sea. Behold, the English beach
    Pales-in the flood, with men, maids, wives, and boys,
    Whose shouts and claps out-voice the deep-mouthed
        sea,                                             11
    Which like a mighty whiffler fore the King
    Seems to prepare his way. So let him land,
    And solemnly see him set on to London.
    So swift a pace hath thought, that even now          15
    You may imagine him upon Blackheath,
    Where that his lords desire him to have borne
    His bruisèd helmet and his bended sword
    Before him through the city; he forbids it,
    Being free from vainness and self-glorious pride,    20
    Giving full trophy, signal, and ostent
    Quite from himself, to God. But now behold,
    In the quick forge and working-house of thought,
    How London doth pour out her citizens.
    The Mayor and all his brethren, in best sort,        25

Like to the senators of th'antique Rome
With the plebeians swarming at their heels,
Go forth and fetch their conqu'ring Caesar in—
As, by a lower but high-loving likelihood,
Were now the General of our gracious Empress—          30
As in good time he may—from Ireland coming,
Bringing rebellion broachèd on his sword,
How many would the peaceful city quit
To welcome him! Much more, and much more cause,
Did they this Harry. Now in London place him;          35
As yet the lamentation of the French
Invites the King of England's stay at home.
The Emperor's coming in behalf of France,
To order peace between them ⌐
                            ⌐ and omit              40
All the occurrences, whatever chanced,
Till Harry's back-return again to France.
There must we bring him, and myself have played
The interim by remem'bring you 'tis past.
Then brook abridgement, and your eyes advance,         45
After your thoughts, straight back again to France.
                                              *Exit*

**5.1**  *Enter Captain Gower and Captain Fluellen, with a*
         *leek in his cap and a cudgel*
GOWER  Nay, that's right. But why wear you your leek
today? Saint Davy's day is past.
FLUELLEN  There is occasions and causes why and
wherefore in all things. I will tell you, ass my friend,
Captain Gower. The rascally scald beggarly lousy
pragging knave Pistol—which you and yourself and all
the world know to be no petter than a fellow, look you
now, of no merits—he is come to me, and prings me
pread and salt yesterday, look you, and bid me eat my
leek. It was in a place where I could not breed no
contention with him, but I will be so bold as to wear
it in my cap till I see him once again, and then I will
tell him a little piece of my desires.               13
         *Enter Ensign Pistol*
GOWER  Why, here a comes, swelling like a turkey-cock.
FLUELLEN  'Tis no matter for his swellings nor his turkey-
cocks.—God pless you Ensign Pistol, you scurvy lousy
knave, God pless you.
PISTOL
Ha, art thou bedlam? Dost thou thirst, base Trojan,
To have me fold up Parca's fatal web?
Hence! I am qualmish at the smell of leek.           20
FLUELLEN  I peseech you heartily, scurvy lousy knave, at
my desires and my requests and my petitions, to eat,
look you, this leek. Because, look you, you do not love
it, nor your affections and your appetites and your
digestions does not agree with it, I would desire you
to eat it.                                           26
PISTOL
Not for Cadwallader and all his goats.
FLUELLEN  There is one goat for you. (*He strikes Pistol*) Will
you be so good, scald knave, as eat it?
PISTOL  Base Trojan, thou shalt die.                 30
FLUELLEN  You say very true, scald knave, when God's
will is. I will desire you to live in the mean time, and
eat your victuals. Come, there is sauce for it. (*He strikes
him*) You called me yesterday 'mountain-squire', but I
will make you today a 'squire of low degree'. I pray
you, fall to. If you can mock a leek you can eat a leek.
         ⌐*He strikes him*⌐
GOWER  Enough, captain, you have astonished him.      37

FLUELLEN  By Jesu, I will make him eat some part of my
leek, or I will peat his pate four days and four nights.—
Bite, I pray you. It is good for your green wound and
your ploody coxcomb.                                 41
PISTOL  Must I bite?
FLUELLEN  Yes, certainly, and out of doubt and out of
question too, and ambiguities.
PISTOL  By this leek, I will most horribly revenge—   45
         ⌐*Fluellen threatens him*⌐
I eat and eat—I swear—
FLUELLEN  Eat, I pray you. Will you have some more sauce
to your leek? There is not enough leek to swear by.
PISTOL
Quiet thy cudgel, thou dost see I eat.
FLUELLEN  Much good do you, scald knave, heartily. Nay,
pray you throw none away. The skin is good for your
broken coxcomb. When you take occasions to see leeks
hereafter, I pray you mock at 'em, that is all.       53
PISTOL  Good.
FLUELLEN  Ay, leeks is good. Hold you, there is a groat to
heal your pate.
PISTOL  Me, a groat?
FLUELLEN  Yes, verily, and in truth you shall take it, or I
have another leek in my pocket which you shall eat.
PISTOL
I take thy groat in earnest of revenge.              60
FLUELLEN  If I owe you anything, I will pay you in cudgels.
You shall be a woodmonger, and buy nothing of me
but cudgels. God b'wi' you, and keep you, and heal
your pate.                                        *Exit*
PISTOL  All hell shall stir for this.                 65
GOWER  Go, go, you are a counterfeit cowardly knave.
Will you mock at an ancient tradition, begun upon an
honourable respect and worn as a memorable trophy
of predeceased valour, and dare not avouch in your
deeds any of your words? I have seen you gleeking and
galling at this gentleman twice or thrice. You thought,
because he could not speak English in the native garb,
he could not therefore handle an English cudgel. You
find it otherwise. And henceforth let a Welsh correction
teach you a good English condition. Fare ye well.     75
                                              *Exit*
PISTOL
Doth Fortune play the hussy with me now?
News have I that my Nell is dead
I'th' spital of a malady of France,
And there my rendezvous is quite cut off.
Old I do wax, and from my weary limbs                80
Honour is cudgelled. Well, bawd I'll turn,
And something lean to cutpurse of quick hand.
To England will I steal, and there I'll steal,
And patches will I get unto these cudgelled scars,
And swear I got them in the Gallia wars.       *Exit*

**5.2**  *Enter at one door King Harry, the Dukes of Exeter*
         *and ⌐Clarence⌐, the Earl of Warwick, and other*
         *lords; at another, King Charles the Sixth of France,*
         *Queen Isabel, the Duke of Burgundy, and other*
         *French, among them Princess Catherine and Alice*
KING HARRY
Peace to this meeting, wherefor we are met.
Unto our brother France and to our sister,
Health and fair time of day. Joy and good wishes
To our most fair and princely cousin Catherine;
And as a branch and member of this royalty,          5

By whom this great assembly is contrived,
We do salute you, Duke of Burgundy.
And princes French, and peers, health to you all.

KING CHARLES
Right joyous are we to behold your face.
Most worthy brother England, fairly met.　　　10
So are you, princes English, every one.

QUEEN ISABEL
So happy be the issue, brother England,
Of this good day and of this gracious meeting,
As we are now glad to behold your eyes—
Your eyes which hitherto have borne in them,　　　15
Against the French that met them in their bent,
The fatal balls of murdering basilisks.
The venom of such looks we fairly hope
Have lost their quality, and that this day
Shall change all griefs and quarrels into love.　　　20

KING HARRY
To cry amen to that, thus we appear.

QUEEN ISABEL
You English princes all, I do salute you.

BURGUNDY
My duty to you both, on equal love,
Great Kings of France and England. That I have
　　　laboured
With all my wits, my pains, and strong endeavours,
To bring your most imperial majesties　　　26
Unto this bar and royal interview,
Your mightiness on both parts best can witness.
Since, then, my office hath so far prevailed
That face to face and royal eye to eye　　　30
You have congreeted, let it not disgrace me
If I demand, before this royal view,
What rub or what impediment there is
Why that the naked, poor, and mangled peace,
Dear nurse of arts, plenties, and joyful births,　　　35
Should not in this best garden of the world,
Our fertile France, put up her lovely visage?
Alas, she hath from France too long been chased,
And all her husbandry doth lie on heaps,
Corrupting in it own fertility.　　　40
Her vine, the merry cheerer of the heart,
Unprunèd dies; her hedges even-plashed
Like prisoners wildly overgrown with hair
Put forth disordered twigs; her fallow leas
The darnel, hemlock, and rank fumitory　　　45
Doth root upon, while that the coulter rusts
That should deracinate such savagery.
The even mead—that erst brought sweetly forth
The freckled cowslip, burnet, and green clover—
Wanting the scythe, all uncorrected, rank,　　　50
Conceives by idleness, and nothing teems
But hateful docks, rough thistles, kecksies, burs,
Losing both beauty and utility.
An all our vineyards, fallows, meads, and hedges,
Defective in their natures, grow to wildness,　　　55
Even so our houses and ourselves and children
Have lost, or do not learn for want of time,
The sciences that should become our country,
But grow like savages—as soldiers will
That nothing do but meditate on blood—　　　60
To swearing and stern looks, diffused attire,
And everything that seems unnatural.
Which to reduce into our former favour
You are assembled, and my speech entreats

That I may know the let why gentle peace　　　65
Should not expel these inconveniences
And bless us with her former qualities.

KING HARRY
If, Duke of Burgundy, you would the peace
Whose want gives growth to th'imperfections
Which you have cited, you must buy that peace　　　70
With full accord to all our just demands,
Whose tenors and particular effects
You have enscheduled briefly in your hands.

BURGUNDY
The King hath heard them, to the which as yet
There is no answer made.

KING HARRY　　　　　　Well then, the peace,　　　75
Which you before so urged, lies in his answer.

KING CHARLES
I have but with a cursitory eye
O'erglanced the articles. Pleaseth your grace
To appoint some of your council presently
To sit with us once more, with better heed　　　80
To re-survey them, we will suddenly
Pass our accept and peremptory answer.

KING HARRY
Brother, we shall.—Go, Uncle Exeter
And brother Clarence, and you, brother Gloucester;
Warwick and Huntingdon, go with the King,　　　85
And take with you free power to ratify,
Augment, or alter, as your wisdoms best
Shall see advantageable for our dignity,
Anything in or out of our demands,
And we'll consign thereto.—Will you, fair sister,　　　90
Go with the princes, or stay here with us?

QUEEN
Our gracious brother, I will go with them.
Haply a woman's voice may do some good
When articles too nicely urged be stood on.

KING HARRY
Yet leave our cousin Catherine here with us.　　　95
She is our capital demand, comprised
Within the fore-rank of our articles.

QUEEN
She hath good leave.
　　　　　　　　　*Exeunt all but King Harry, Catherine, and Alice*

KING HARRY　　　　　　Fair Catherine, and most fair,
Will you vouchsafe to teach a soldier terms
Such as will enter at a lady's ear　　　100
And plead his love-suit to her gentle heart?

CATHERINE Your majesty shall mock at me. I cannot speak
your England.

KING HARRY O fair Catherine, if you will love me soundly
with your French heart, I will be glad to hear you
confess it brokenly with your English tongue. Do you
like me, Kate?　　　107

CATHERINE *Pardonnez-moi*, I cannot tell vat is 'like me'.

KING HARRY An angel is like you, Kate, and you are like
an angel.　　　110

CATHERINE (*to Alice*) *Que dit-il?—que je suis semblable à les
anges?*

ALICE *Oui, vraiment—sauf votre grâce—ainsi dit-il.*

KING HARRY I said so, dear Catherine, and I must not
blush to affirm it.　　　115

CATHERINE *O bon Dieu! Les langues des hommes sont pleines
de tromperies.*

KING HARRY What says she, fair one? That the tongues
of men are full of deceits?

ALICE *Oui*, dat de tongeus of de mans is be full of deceits— dat is de Princess.    121

KING HARRY The Princess is the better Englishwoman. I'faith, Kate, my wooing is fit for thy understanding. I am glad thou canst speak no better English, for if thou couldst, thou wouldst find me such a plain king that thou wouldst think I had sold my farm to buy my crown. I know no ways to mince it in love, but directly to say, 'I love you'; then if you urge me farther than to say, 'Do you in faith?', I wear out my suit. Give me your answer, i'faith do, and so clap hands and a bargain. How say you, lady?    131

CATHERINE *Sauf votre honneur*, me understand well.

KING HARRY Marry, if you would put me to verses, or to dance for your sake, Kate, why, you undid me. For the one I have neither words nor measure, and for the other I have no strength in measure—yet a reasonable measure in strength. If I could win a lady at leap-frog, or by vaulting into my saddle with my armour on my back, under the correction of bragging be it spoken, I should quickly leap into a wife. Or if I might buffet for my love, or bound my horse for her favours, I could lay on like a butcher, and sit like a jackanapes, never off. But before God, Kate, I cannot look greenly, nor gasp out my eloquence, nor I have no cunning in protestation—only downright oaths, which I never use till urged, nor never break for urging. If thou canst love a fellow of this temper, Kate, whose face is not worth sunburning, that never looks in his glass for love of anything he sees there, let thine eye be thy cook. I speak to thee plain soldier: if thou canst love me for this, take me. If not, to say to thee that I shall die, is true—but for thy love, by the Lord, no. Yet I love thee, too. And while thou livest, dear Kate, take a fellow of plain and uncoined constancy, for he perforce must do thee right, because he hath not the gift to woo in other places. For these fellows of infinite tongue, that can rhyme themselves into ladies' favours, they do always reason themselves out again. What! A speaker is but a prater, a rhyme is but a ballad; a good leg will fall, a straight back will stoop, a black beard will turn white, a curled pate will grow bald, a fair face will wither, a full eye will wax hollow, but a good heart, Kate, is the sun and the moon—or rather the sun and not the moon, for it shines bright and never changes, but keeps his course truly. If thou would have such a one, take me; and take me, take a soldier; take a soldier, take a king. And what sayst thou then to my love? Speak, my fair—and fairly, I pray thee.

CATHERINE Is it possible dat I sould love de *ennemi* of France?    170

KING HARRY No, it is not possible you should love the enemy of France, Kate. But in loving me, you should love the friend of France, for I love France so well that I will not part with a village of it, I will have it all mine; and Kate, when France is mine, and I am yours, then yours is France, and you are mine.    176

CATHERINE I cannot tell vat is dat.

KING HARRY No, Kate? I will tell thee in French—which I am sure will hang upon my tongue like a new-married wife about her husband's neck, hardly to be shook off. *Je quand suis le possesseur de France, et quand vous avez le possession de moi*—let me see, what then? Saint Denis be my speed!—*donc vôtre est France, et vous êtes mienne.* It is as easy for me, Kate, to conquer the kingdom as

to speak so much more French. I shall never move thee in French, unless it be to laugh at me.    186

CATHERINE *Sauf votre honneur, le français que vous parlez, il est meilleur que l'anglais lequel je parle.*

KING HARRY No, faith, is't not, Kate. But thy speaking of my tongue, and I thine, most truly-falsely, must needs be granted to be much at one. But Kate, dost thou understand thus much English? Canst thou love me?

CATHERINE I cannot tell.    193

KING HARRY Can any of your neighbours tell, Kate? I'll ask them. Come, I know thou lovest me, and at night when you come into your closet you'll question this gentlewoman about me, and I know, Kate, you will to her dispraise those parts in me that you love with your heart. But good Kate, mock me mercifully—the rather, gentle princess, because I love thee cruelly. If ever thou be'st mine, Kate—as I have a saving faith within me tells me thou shalt—I get thee with scrambling, and thou must therefore needs prove a good soldier-breeder. Shall not thou and I, between Saint Denis and Saint George, compound a boy, half-French half-English, that shall go to Constantinople and take the Turk by the beard? Shall we not? What sayst thou, my fair flower-de-luce?    208

CATHERINE I do not know dat.

KING HARRY No, 'tis hereafter to know, but now to promise. Do but now promise, Kate, you will endeavour for your French part of such a boy, and for my English moiety take the word of a king and a bachelor. How answer you, *la plus belle Catherine du monde, mon très chère et divine déesse?*    215

CATHERINE Your *majesté* 'ave *faux* French enough to deceive de most sage *demoiselle* dat is *en France.*

KING HARRY Now fie upon my false French! By mine honour, in true English, I love thee, Kate. By which honour I dare not swear thou lovest me, yet my blood begins to flatter me that thou dost, notwithstanding the poor and untempering effect of my visage. Now beshrew my father's ambition! He was thinking of civil wars when he got me; therefore was I created with a stubborn outside, with an aspect of iron, that when I come to woo ladies I fright them. But in faith, Kate, the elder I wax the better I shall appear. My comfort is that old age, that ill layer-up of beauty, can do no more spoil upon my face. Thou hast me, if thou hast me, at the worst, and thou shalt wear me, if thou wear me, better and better; and therefore tell me, most fair Catherine, will you have me? Put off your maiden blushes, avouch the thoughts of your heart with the looks of an empress, take me by the hand and say, 'Harry of England, I am thine'—which word thou shalt no sooner bless mine ear withal, but I will tell thee aloud, 'England is thine, Ireland is thine, France is thine, and Henry Plantagenet is thine'—who, though I speak it before his face, if he be not fellow with the best king, thou shalt find the best king of good fellows. Come, your answer in broken music—for thy voice is music and thy English broken. Therefore, queen of all, Catherine, break thy mind to me in broken English: wilt thou have me?

CATHERINE Dat is as it shall please de *roi mon père.*    245

KING HARRY Nay, it will please him well, Kate. It shall please him, Kate.

CATHERINE Den it sall also content me.

KING HARRY Upon that I kiss your hand, and I call you my queen.    250

CATHERINE *Laissez, mon seigneur, laissez, laissez! Ma foi, je ne veux point que vous abbaissez votre grandeur en baisant la main d'une de votre seigneurie indigne serviteur. Excusezmoi, je vous supplie, mon treis-puissant seigneur.*

KING HARRY Then I will kiss your lips, Kate.     255

CATHERINE *Les dames et demoiselles pour être baisées devant leurs noces, il n'est pas la coutume de France.*

KING HARRY *(to Alice)* Madam my interpreter, what says she?

ALICE Dat it is not be de *façon pour les* ladies of France— I cannot tell vat is *baiser en* Anglish.     261

KING HARRY To kiss.

ALICE Your *majesté entend* bettre *que moi.*

KING HARRY It is not a fashion for the maids in France to kiss before they are married, would she say?     265

ALICE *Oui, vraiment.*

KING HARRY O Kate, nice customs curtsy to great kings. Dear Kate, you and I cannot be confined within the weak list of a country's fashion. We are the makers of manners, Kate, and the liberty that follows our places stops the mouth of all find-faults, as I will do yours, for upholding the nice fashion of your country in denying me a kiss. Therefore, patiently and yielding. *(He kisses her)* You have witchcraft in your lips, Kate. There is more eloquence in a sugar touch of them than in the tongues of the French Council, and they should sooner persuade Harry of England than a general petition of monarchs. Here comes your father.     278

*Enter King Charles, Queen Isabel, the Duke of Burgundy, and the French and English lords*

BURGUNDY God save your majesty. My royal cousin, teach you our princess English?     280

KING HARRY I would have her learn, my fair cousin, how perfectly I love her, and that is good English.

BURGUNDY Is she not apt?

KING HARRY Our tongue is rough, coz, and my condition is not smooth, so that having neither the voice nor the heart of flattery about me I cannot so conjure up the spirit of love in her that he will appear in his true likeness.     288

BURGUNDY Pardon the frankness of my mirth, if I answer you for that. If you would conjure in her, you must make a circle; if conjure up love in her in his true likeness, he must appear naked and blind. Can you blame her then, being a maid yet rosed over with the virgin crimson of modesty, if she deny the appearance of a naked blind boy in her naked seeing self? It were, my lord, a hard condition for a maid to consign to.

KING HARRY Yet they do wink and yield, as love is blind and enforces.

BURGUNDY They are then excused, my lord, when they see not what they do.     300

KING HARRY Then, good my lord, teach your cousin to consent winking.

BURGUNDY I will wink on her to consent, my lord, if you will teach her to know my meaning. For maids, well summered and warm kept, are like flies at Bartholomew-tide: blind, though they have their eyes. And then they will endure handling, which before would not abide looking on.

KING HARRY This moral ties me over to time and a hot summer, and so I shall catch the fly, your cousin, in the latter end, and she must be blind too.     311

BURGUNDY As love is, my lord, before that it loves.

KING HARRY It is so. And you may, some of you, thank love for my blindness, who cannot see many a fair French city for one fair French maid that stands in my way.     316

KING CHARLES Yes, my lord, you see them perspectively, the cities turned into a maid—for they are all girdled with maiden walls that war hath never entered.

KING HARRY Shall Kate be my wife?     320

KING CHARLES So please you.

KING HARRY I am content, so the maiden cities you talk of may wait on her: so the maid that stood in the way for my wish shall show me the way to my will.

KING CHARLES
We have consented to all terms of reason.     325

KING HARRY Is't so, my lords of England?

⌈WARWICK⌉
The King hath granted every article:
His daughter first, and so in sequel all,
According to their firm proposèd natures.

EXETER
Only he hath not yet subscribèd this:     330
where your majesty demands that the King of France,
having any occasion to write for matter of grant, shall
name your highness in this form and with this addition:
⌈*reads*⌉ in French, *Notre très cher fils Henri, Roi
d'Angleterre, Héritier de France,* and thus in Latin,
*Praeclarissimus filius noster Henricus, Rex Angliae et
Haeres Franciae.*

KING CHARLES
Nor this I have not, brother, so denied,
But your request shall make me let it pass.

KING HARRY
I pray you then, in love and dear alliance,     340
Let that one article rank with the rest,
And thereupon give me your daughter.

KING CHARLES
Take her, fair son, and from her blood raise up
Issue to me, that the contending kingdoms
Of France and England, whose very shores look pale
With envy of each other's happiness,     346
May cease their hatred, and this dear conjunction
Plant neighbourhood and Christian-like accord
In their sweet bosoms, that never war advance
His bleeding sword 'twixt England and fair France.

⌈ALL⌉ Amen.     351

KING HARRY
Now welcome, Kate, and bear me witness all
That here I kiss her as my sovereign Queen.
*Flourish*

QUEEN ISABEL
God, the best maker of all marriages,
Combine your hearts in one, your realms in one.     355
As man and wife, being two, are one in love,
So be there 'twixt your kingdoms such a spousal
That never may ill office or fell jealousy,
Which troubles oft the bed of blessèd marriage,
Thrust in between the paction of these kingdoms     360
To make divorce of their incorporate league;
That English may as French, French Englishmen,
Receive each other, God speak this 'Amen'.

ALL Amen.

KING HARRY
Prepare we for our marriage. On which day,     365
My lord of Burgundy, we'll take your oath,
And all the peers', for surety of our leagues.
Then shall I swear to Kate, and you to me,
And may our oaths well kept and prosp'rous be.
*Sennet. Exeunt*

**Epilogue** *Enter Chorus*
CHORUS
    Thus far with rough and all-unable pen
      Our bending author hath pursued the story,
    In little room confining mighty men,
      Mangling by starts the full course of their glory.
    Small time, but in that small most greatly lived    5
      This star of England. Fortune made his sword,

By which the world's best garden he achieved,
  And of it left his son imperial lord.
Henry the Sixth, in infant bands crowned king
  Of France and England, did this king succeed,    10
Whose state so many had the managing
  That they lost France and made his England bleed,
Which oft our stage hath shown—and, for their sake,
In your fair minds let this acceptance take.    *Exit*

## ADDITIONAL PASSAGES

The Dauphin/Bourbon variant, which usually involves only the alteration of speech-prefixes, has several consequences for the dialogue and structure of 4.5. There follow edited texts of the Folio and Quarto versions of this scene.

### A. FOLIO

      *Enter the Constable, Orléans, Bourbon, the Dauphin,*
      *and Rambures*
CONSTABLE  *O diable!*
ORLÉANS  *O Seigneur! Le jour est perdu, tout est perdu.*
DAUPHIN
  *Mort de ma vie!* All is confounded, all.
  Reproach and everlasting shame
  Sits mocking in our plumes.    5
      *A short alarum*
  *O méchante fortune!* Do not run away.
             ⌜*Exit Rambures*⌝
CONSTABLE  Why, all our ranks are broke.
DAUPHIN
  O perdurable shame! Let's stab ourselves:
  Be these the wretches that we played at dice for?
ORLÉANS
  Is this the king we sent to for his ransom?    10
BOURBON
  Shame, an eternall shame, nothing but shame!
  Let us die in pride. In once more, back again!
  And he that will not follow Bourbon now,
  Let him go home, and with his cap in hand
  Like a base leno hold the chamber door,    15
  Whilst by a slave no gentler than my dog
  His fairest daughter is contaminated.
CONSTABLE
  Disorder that hath spoiled us, friend us now,
  Let us on heaps go offer up our lives.

ORLÉANS
  We are enough yet living in the field    20
  To smother up the English in our throngs,
  If any order might be thought upon.
BOURBON
  The devil take order now. I'll to the throng.
  Let life be short, else shame will be too long.    *Exeunt*

### B. QUARTO

      *Enter the four French lords: the Constable, Orléans,*
      *Bourbon, and Gebon*
GEBON  *O diabello!*
CONSTABLE  *Mort de ma vie!*
ORLÉANS  O what a day is this!
BOURBON
  *O jour de honte,* all is gone, all is lost.
CONSTABLE  We are enough yet living in the field    5
  To smother up the English,
  If any order might be thought upon.
BOURBON
  A plague of order! Once more to the field!
  And he that will not follow Bourbon now,
  Let him go home, and with his cap in hand,    10
  Like a base leno hold the chamber door,
  Whilst by a slave no gentler than my dog
  His fairest daughter is contaminated.
CONSTABLE
  Disorder that hath spoiled us, right us now.
  Come we in heaps, we'll offer up our lives    15
  Unto these English, or else die with fame.
⌜BOURBON⌝  Come, come along.
  Let's die with honour, our shame doth last too long.
                    *Exeunt*

# JULIUS CAESAR

ON 21 September 1599 a Swiss doctor, Thomas Platter, saw what can only have been Shakespeare's *Julius Caesar* 'very pleasingly performed' in the newly built Globe Theatre— 'the straw-thatched house'—on the south side of the Thames. Francis Meres does not mention the play in *Palladis Tamia* of 1598, and minor resemblances with works printed in the early part of 1599 suggest that Shakespeare wrote it during that year. It was first printed in the 1623 Folio.

*Julius Caesar* shows Shakespeare turning from English to Roman history, which he had last used in *Titus Andronicus* and *The Rape of Lucrece*. Caesar was regarded as perhaps the greatest ruler in the history of the world, and his murder by Brutus as one of the foulest crimes: but it was also recognized that Caesar had faults and Brutus virtues. Other plays, some now lost, had been written about Caesar and may have influenced Shakespeare; but there is no question that he made extensive use (for the first time in this play) of Sir Thomas North's great translation (based on Jacques Amyot's French version and published in 1579) of *Lives of the Noble Grecians and Romans* by the Greek historian Plutarch, who lived from about AD 50 to 130.

Shakespeare was interested in the aftermath of Caesar's death as well as in the events leading up to it, and in the public and private motives of those responsible for it. So, although the Folio calls the play *The Tragedy of Julius Caesar*, Caesar is dead before the play is half over; Brutus, Cassius, and Antony have considerably longer roles, and Brutus is portrayed with a degree of introspection which links him more closely to Shakespeare's other tragic heroes. Shakespeare draws mainly on the last quarter of Plutarch's Life of Caesar, showing his fall; he also uses the Lives of Antony and Brutus for the play's first sweep of action, showing the rise of the conspiracy against Caesar, its leaders' efforts to persuade Brutus to join them, the assassination itself, and its immediate aftermath as Antony incites the citizens to revenge. The second part, showing the formation of the triumvirate of Antony, Lepidus, and Octavius Caesar, the uneasy alliance of Brutus and Cassius, and the battles in which Caesar's spirit revenges itself, depends mainly on the Life of Brutus. Facts are often altered and rearranged in the interests of dramatic economy and effectiveness.

Although Shakespeare wrote the play at a point in his career at which he was tending to use a high proportion of prose, *Julius Caesar* is written mainly in verse; as if to suit the subject matter, the style is classical in its lucidity and eloquence, reaching a climax of rhetorical effectiveness in the speeches over Caesar's body (3.2). The play's stageworthiness has been repeatedly demonstrated; it offers excellent opportunities in all its main roles, and the quarrel between Brutus and Cassius (4.2) has been admired ever since Leonard Digges, a contemporary of Shakespeare, praised it at the expense of Ben Jonson:

> So have I seen, when Caesar would appear,
> And on the stage at half-sword parley were
> Brutus and Cassius; O, how the audience
> Were ravished, with what wonder they went thence,
> When some new day they would not brook a line
> Of tedious though well-laboured *Catiline*.

# THE PERSONS OF THE PLAY

Julius CAESAR
CALPURNIA, his wife

Marcus BRUTUS, a noble Roman, opposed to Caesar
PORTIA, his wife
LUCIUS, his servant

Caius CASSIUS
CASCA
TREBONIUS
DECIUS Brutus } opposed to Caesar
METELLUS Cimber
CINNA
Caius LIGARIUS

Mark ANTONY
OCTAVIUS Caesar } rulers of Rome after Caesar's death
LEPIDUS

FLAVIUS
MURELLUS } tribunes of the people

CICERO
PUBLIUS } senators
POPILLIUS Laena

A SOOTHSAYER
ARTEMIDORUS

CINNA the Poet

PINDARUS, Cassius' bondman
TITINIUS, an officer in Cassius' army
LUCILLIUS
MESSALA
VARRUS
CLAUDIO
YOUNG CATO
STRATO } officers and soldiers in Brutus' army
VOLUMNIUS
FLAVIUS
DARDANIUS
CLITUS

A POET
GHOST of Caesar

A COBBLER
A CARPENTER
Other PLEBEIANS
A MESSENGER
SERVANTS
Senators, soldiers, and attendants

# The Tragedy of Julius Caesar

**1.1**   *Enter Flavius, Murellus, and certain commoners*
       *over the stage*

FLAVIUS
Hence, home, you idle creatures, get you home!
Is this a holiday? What, know you not,
Being mechanical, you ought not walk
Upon a labouring day without the sign
Of your profession?—Speak, what trade art thou?   5

CARPENTER Why, sir, a carpenter.

MURELLUS
Where is thy leather apron and thy rule?
What dost thou with thy best apparel on?—
You, sir, what trade are you?

COBBLER Truly, sir, in respect of a fine workman I am
but, as you would say, a cobbler.   11

MURELLUS
But what trade art thou? Answer me directly.

COBBLER A trade, sir, that I hope I may use with a safe
conscience, which is indeed, sir, a mender of bad soles.

FLAVIUS
What trade, thou knave? Thou naughty knave, what
    trade?   15

COBBLER Nay, I beseech you, sir, be not out with me. Yet
if you be out, sir, I can mend you.

MURELLUS
What mean'st thou by that? Mend me, thou saucy
    fellow?

COBBLER Why, sir, cobble you.

FLAVIUS Thou art a cobbler, art thou?   20

COBBLER Truly, sir, all that I live by is with the awl. I
meddle with no tradesman's matters, nor women's
matters, but withal I am indeed, sir, a surgeon to old
shoes: when they are in great danger I recover them.
As proper men as ever trod upon neat's leather have
gone upon my handiwork.   26

FLAVIUS
But wherefore art not in thy shop today?
Why dost thou lead these men about the streets?

COBBLER Truly, sir, to wear out their shoes to get myself
into more work. But indeed, sir, we make holiday to
see Caesar, and to rejoice in his triumph.   31

MURELLUS
Wherefore rejoice? What conquest brings he home?
What tributaries follow him to Rome
To grace in captive bonds his chariot wheels?
You blocks, you stones, you worse than senseless
    things!   35
O, you hard hearts, you cruel men of Rome,
Knew you not Pompey? Many a time and oft
Have you climbed up to walls and battlements,
To towers and windows, yea to chimney-tops,
Your infants in your arms, and there have sat   40
The livelong day with patient expectation
To see great Pompey pass the streets of Rome.
And when you saw his chariot but appear,
Have you not made an universal shout,
That Tiber trembled underneath her banks   45
To hear the replication of your sounds
Made in her concave shores?
And do you now put on your best attire?
And do you now cull out a holiday?
And do you now strew flowers in his way   50
That comes in triumph over Pompey's blood?
Be gone!
Run to your houses, fall upon your knees,
Pray to the gods to intermit the plague
That needs must light on this ingratitude.   55

FLAVIUS
Go, go, good countrymen, and for this fault
Assemble all the poor men of your sort;
Draw them to Tiber banks, and weep your tears
Into the channel, till the lowest stream
Do kiss the most exalted shores of all.   60
               *Exeunt all the commoners*
See whe'er their basest mettle be not moved.
They vanish tongue-tied in their guiltiness.
Go you down that way towards the Capitol;
This way will I. Disrobe the images
If you do find them decked with ceremonies.   65

MURELLUS May we do so?
You know it is the Feast of Lupercal.

FLAVIUS
It is no matter. Let no images
Be hung with Caesar's trophies. I'll about,
And drive away the vulgar from the streets;   70
So do you too where you perceive them thick.
These growing feathers plucked from Caesar's wing
Will make him fly an ordinary pitch,
Who else would soar above the view of men
And keep us all in servile fearfulness.      *Exeunt*

**1.2**   ⌈*Loud music.*⌉ *Enter Caesar, Antony stripped for the*
      *course, Calpurnia, Portia, Decius, Cicero, Brutus,*
      *Cassius, Casca, a Soothsayer,* ⌈*a throng of citizens*⌉;
      *after them, Murellus and Flavius*

CAESAR Calpurnia.

CASCA Peace, ho! Caesar speaks.
    ⌈*Music ceases*⌉

CAESAR Calpurnia.

CALPURNIA Here, my lord.

CAESAR
Stand you directly in Antonio's way   5
When he doth run his course.—Antonio.

ANTONY Caesar, my lord.

CAESAR
Forget not in your speed, Antonio,
To touch Calpurnia, for our elders say
The barren, touchèd in this holy chase,   10
Shake off their sterile curse.

ANTONY                I shall remember:
When Caesar says 'Do this', it is performed.

CAESAR
Set on, and leave no ceremony out.
    ⌈*Music*⌉

SOOTHSAYER Caesar!

CAESAR Ha! Who calls?   15

CASCA
Bid every noise be still. Peace yet again.
    ⌈*Music ceases*⌉

CAESAR
Who is it in the press that calls on me?
I hear a tongue shriller than all the music
Cry 'Caesar!' Speak. Caesar is turned to hear.
SOOTHSAYER
Beware the ides of March.
CAESAR                              What man is that?          20
BRUTUS
A soothsayer bids you beware the ides of March.
CAESAR
Set him before me; let me see his face.
CASSIUS
Fellow, come from the throng; look upon Caesar.
      *The Soothsayer comes forward*
CAESAR
What sayst thou to me now? Speak once again.
SOOTHSAYER Beware the ides of March.                          25
CAESAR
He is a dreamer. Let us leave him. Pass!
            *Sennet. Exeunt all but Brutus and Cassius*
CASSIUS
Will you go see the order of the course?
BRUTUS Not I.
CASSIUS I pray you, do.
BRUTUS
I am not gamesome; I do lack some part          30
Of that quick spirit that is in Antony.
Let me not hinder, Cassius, your desires.
I'll leave you.
CASSIUS
Brutus, I do observe you now of late.
I have not from your eyes that gentleness          35
And show of love as I was wont to have.
You bear too stubborn and too strange a hand
Over your friend that loves you.
BRUTUS                              Cassius,
Be not deceived. If I have veiled my look,
I turn the trouble of my countenance          40
Merely upon myself. Vexèd I am
Of late with passions of some difference,
Conceptions only proper to myself,
Which give some soil, perhaps, to my behaviours.
But let not therefore my good friends be grieved—          45
Among which number, Cassius, be you one—
Nor construe any further my neglect
Than that poor Brutus, with himself at war,
Forgets the shows of love to other men.
CASSIUS
Then, Brutus, I have much mistook your passion,          50
By means whereof this breast of mine hath buried
Thoughts of great value, worthy cogitations.
Tell me, good Brutus, can you see your face?
BRUTUS
No, Cassius, for the eye sees not itself
But by reflection, by some other things.          55
CASSIUS 'Tis just;
And it is very much lamented, Brutus,
That you have no such mirrors as will turn
Your hidden worthiness into your eye,
That you might see your shadow. I have heard          60
Where many of the best respect in Rome—
Except immortal Caesar—speaking of Brutus,
And groaning underneath this age's yoke,
Have wished that noble Brutus had his eyes.
BRUTUS
Into what dangers would you lead me, Cassius,          65

That you would have me seek into myself
For that which is not in me?
CASSIUS
Therefor, good Brutus, be prepared to hear.
And since you know you cannot see yourself
So well as by reflection, I, your glass,          70
Will modestly discover to yourself
That of yourself which you yet know not of.
And be not jealous on me, gentle Brutus.
Were I a common laughter, or did use
To stale with ordinary oaths my love          75
To every new protester; if you know
That I do fawn on men and hug them hard,
And after scandal them; or if you know
That I profess myself in banqueting
To all the rout: then hold me dangerous.          80
      *Flourish and shout within*
BRUTUS
What means this shouting? I do fear the people
Choose Caesar for their king.
CASSIUS                              Ay, do you fear it?
Then must I think you would not have it so.
BRUTUS
I would not, Cassius; yet I love him well.
But wherefore do you hold me here so long?          85
What is it that you would impart to me?
If it be aught toward the general good,
Set honour in one eye and death i'th' other,
And I will look on both indifferently;
For let the gods so speed me as I love          90
The name of honour more than I fear death.
CASSIUS
I know that virtue to be in you, Brutus,
As well as I do know your outward favour.
Well, honour is the subject of my story.
I cannot tell what you and other men          95
Think of this life; but for my single self,
I had as lief not be, as live to be
In awe of such a thing as I myself.
I was born free as Caesar, so were you.
We both have fed as well, and we can both          100
Endure the winter's cold as well as he.
For once upon a raw and gusty day,
The troubled Tiber chafing with her shores,
Said Caesar to me 'Dar'st thou, Cassius, now
Leap in with me into this angry flood,          105
And swim to yonder point?' Upon the word,
Accoutred as I was I plungèd in,
And bade him follow. So indeed he did.
The torrent roared, and we did buffet it
With lusty sinews, throwing it aside,          110
And stemming it with hearts of controversy.
But ere we could arrive the point proposed,
Caesar cried 'Help me, Cassius, or I sink!'
Ay, as Aeneas our great ancestor
Did from the flames of Troy upon his shoulder          115
The old Anchises bear, so from the waves of Tiber
Did I the tirèd Caesar. And this man
Is now become a god, and Cassius is
A wretched creature, and must bend his body
If Caesar carelessly but nod on him.          120
He had a fever when he was in Spain,
And when the fit was on him, I did mark
How he did shake. 'Tis true, this god did shake.
His coward lips did from their colour fly;

And that same eye whose bend doth awe the world
Did lose his lustre. I did hear him groan, 126
Ay, and that tongue of his that bade the Romans
Mark him and write his speeches in their books,
'Alas!' it cried, 'Give me some drink, Titinius',
As a sick girl. Ye gods, it doth amaze me 130
A man of such a feeble temper should
So get the start of the majestic world,
And bear the palm alone!
        *Flourish and shout within*
BRUTUS                    Another general shout!
I do believe that these applauses are
For some new honours that are heaped on Caesar. 135
CASSIUS
Why, man, he doth bestride the narrow world
Like a Colossus, and we petty men
Walk under his huge legs, and peep about
To find ourselves dishonourable graves.
Men at sometime were masters of their fates. 140
The fault, dear Brutus, is not in our stars,
But in ourselves, that we are underlings.
Brutus and Caesar: what should be in that 'Caesar'?
Why should that name be sounded more than yours?
Write them together: yours is as fair a name. 145
Sound them: it doth become the mouth as well.
Weigh them: it is as heavy. Conjure with 'em:
'Brutus' will start a spirit as soon as 'Caesar'.
Now in the names of all the gods at once,
Upon what meat doth this our Caesar feed 150
That he is grown so great? Age, thou art shamed.
Rome, thou hast lost the breed of noble bloods.
When went there by an age since the great flood,
But it was famed with more than with one man?
When could they say till now, that talked of Rome,
That her wide walls encompassed but one man? 156
Now is it Rome indeed, and room enough
When there is in it but one only man.
O, you and I have heard our fathers say
There was a Brutus once that would have brooked
Th'eternal devil to keep his state in Rome 161
As easily as a king.
BRUTUS
That you do love me I am nothing jealous.
What you would work me to I have some aim.
How I have thought of this and of these times 165
I shall recount hereafter. For this present,
I would not, so with love I might entreat you,
Be any further moved. What you have said
I will consider. What you have to say
I will with patience hear, and find a time 170
Both meet to hear and answer such high things.
Till then, my noble friend, chew upon this:
Brutus had rather be a villager
Than to repute himself a son of Rome
Under these hard conditions as this time 175
Is like to lay upon us.
CASSIUS            I am glad
That my weak words have struck but thus much show
Of fire from Brutus.
        *⌜Music.⌝ Enter Caesar and his train*
BRUTUS
The games are done, and Caesar is returning.
CASSIUS
As they pass by, pluck Casca by the sleeve, 180
And he will, after his sour fashion, tell you
What hath proceeded worthy note today.

BRUTUS
I will do so. But look you, Cassius,
The angry spot doth glow on Caesar's brow,
And all the rest look like a chidden train. 185
Calpurnia's cheek is pale, and Cicero
Looks with such ferret and such fiery eyes
As we have seen him in the Capitol
Being crossed in conference by some senators.
CASSIUS
Casca will tell us what the matter is. 190
CAESAR Antonio.
ANTONY Caesar.
CAESAR
Let me have men about me that are fat,
Sleek-headed men, and such as sleep a-nights.
Yon Cassius has a lean and hungry look. 195
He thinks too much. Such men are dangerous.
ANTONY
Fear him not, Caesar, he's not dangerous.
He is a noble Roman, and well given.
CAESAR
Would he were fatter! But I fear him not.
Yet if my name were liable to fear, 200
I do not know the man I should avoid
So soon as that spare Cassius. He reads much,
He is a great observer, and he looks
Quite through the deeds of men. He loves no plays,
As thou dost, Antony; he hears no music. 205
Seldom he smiles, and smiles in such a sort
As if he mocked himself, and scorned his spirit
That could be moved to smile at anything.
Such men as he be never at heart's ease
Whiles they behold a greater than themselves, 210
And therefore are they very dangerous.
I rather tell thee what is to be feared
Than what I fear, for always I am Caesar.
Come on my right hand, for this ear is deaf,
And tell me truly what thou think'st of him. 215
        *Sennet. Exeunt Caesar and his train. Brutus,*
                        *Cassius, and Casca remain*
CASCA (*to Brutus*) You pulled me by the cloak. Would you
        speak with me?
BRUTUS
Ay, Casca. Tell us what hath chanced today,
That Caesar looks so sad.
CASCA Why, you were with him, were you not? 220
BRUTUS
I should not then ask Casca what had chanced.
CASCA Why, there was a crown offered him; and being
offered him, he put it by with the back of his hand,
thus; and then the people fell a-shouting.
BRUTUS What was the second noise for? 225
CASCA Why, for that too.
CASSIUS
They shouted thrice. What was the last cry for?
CASCA Why, for that too.
BRUTUS Was the crown offered him thrice?
CASCA Ay, marry, was't; and he put it by thrice, every
time gentler than other; and at every putting by, mine
honest neighbours shouted. 232
CASSIUS
Who offered him the crown?
CASCA                    Why, Antony.
BRUTUS
Tell us the manner of it, gentle Casca.

CASCA I can as well be hanged as tell the manner of it. It was mere foolery, I did not mark it. I saw Mark Antony offer him a crown—yet 'twas not a crown neither, 'twas one of these coronets—and as I told you he put it by once; but for all that, to my thinking he would fain have had it. Then he offered it to him again; then he put it by again—but to my thinking he was very loath to lay his fingers off it. And then he offered it the third time; he put it the third time by. And still as he refused it, the rabblement hooted, and clapped their chapped hands, and threw up their sweaty nightcaps, and uttered such a deal of stinking breath because Caesar refused the crown that it had almost choked Caesar; for he swooned and fell down at it. And for mine own part, I durst not laugh for fear of opening my lips and receiving the bad air.      250

CASSIUS
But soft, I pray you. What, did Caesar swoon?

CASCA He fell down in the market-place, and foamed at mouth, and was speechless.

BRUTUS
'Tis very like: he hath the falling sickness.

CASSIUS
No, Caesar hath it not; but you and I           255
And honest Casca, we have the falling sickness.

CASCA I know not what you mean by that, but I am sure Caesar fell down. If the tag-rag people did not clap him and hiss him, according as he pleased and displeased them, as they use to do the players in the theatre, I am no true man.                                    261

BRUTUS
What said he when he came unto himself?

CASCA Marry, before he fell down, when he perceived the common herd was glad he refused the crown, he plucked me ope his doublet and offered them his throat to cut. An I had been a man of any occupation, if I would not have taken him at a word, I would I might go to hell among the rogues. And so he fell. When he came to himself again, he said, if he had done or said anything amiss, he desired their worships to think it was his infirmity. Three or four wenches where I stood cried 'Alas, good soul!' and forgave him with all their hearts. But there's no heed to be taken of them: if Caesar had stabbed their mothers they would have done no less.                                    275

BRUTUS
And after that he came thus sad away?

CASCA Ay.

CASSIUS Did Cicero say anything?

CASCA Ay, he spoke Greek.

CASSIUS To what effect?                                    280

CASCA Nay, an I tell you that, I'll ne'er look you i'th' face again. But those that understood him smiled at one another, and shook their heads. But for mine own part, it was Greek to me. I could tell you more news, too. Murellus and Flavius, for pulling scarves off Caesar's images, are put to silence. Fare you well. There was more foolery yet, if I could remember it.      287

CASSIUS Will you sup with me tonight, Casca?

CASCA No, I am promised forth.

CASSIUS Will you dine with me tomorrow?           290

CASCA Ay, if I be alive, and your mind hold, and your dinner worth the eating.

CASSIUS Good; I will expect you.

CASCA Do so. Farewell both.                    *Exit*

BRUTUS
What a blunt fellow is this grown to be!           295
He was quick mettle when he went to school.

CASSIUS
So is he now, in execution
Of any bold or noble enterprise,
However he puts on this tardy form.
This rudeness is a sauce to his good wit,           300
Which gives men stomach to digest his words
With better appetite.

BRUTUS
And so it is. For this time I will leave you.
Tomorrow, if you please to speak with me,
I will come home to you; or if you will,           305
Come home to me and I will wait for you.

CASSIUS
I will do so. Till then, think of the world.   *Exit Brutus*
Well, Brutus, thou art noble; yet I see
Thy honourable mettle may be wrought
From that it is disposed. Therefore it is meet      310
That noble minds keep ever with their likes;
For who so firm that cannot be seduced?
Caesar doth bear me hard, but he loves Brutus.
If I were Brutus now, and he were Cassius,
He should not humour me. I will this night      315
In several hands in at his windows throw—
As if they came from several citizens—
Writings, all tending to the great opinion
That Rome holds of his name, wherein obscurely
Caesar's ambition shall be glancèd at.           320
And after this, let Caesar seat him sure,
For we will shake him, or worse days endure.   *Exit*

**1.3**   *Thunder and lightning. Enter Casca, ⌈at one door,*
        *with his sword drawn,⌉ and Cicero ⌈at another⌉*

CICERO
Good even, Casca. Brought you Caesar home?
Why are you breathless, and why stare you so?

CASCA
Are not you moved, when all the sway of earth
Shakes like a thing unfirm? O Cicero,
I have seen tempests when the scolding winds      5
Have rived the knotty oaks, and I have seen
Th'ambitious ocean swell and rage and foam
To be exalted with the threat'ning clouds;
But never till tonight, never till now,
Did I go through a tempest dropping fire.           10
Either there is a civil strife in heaven,
Or else the world, too saucy with the gods,
Incenses them to send destruction.

CICERO
Why, saw you anything more wonderful?

CASCA
A common slave—you know him well by sight—      15
Held up his left hand, which did flame and burn
Like twenty torches joined; and yet his hand,
Not sensible of fire, remained unscorched.
Besides—I ha' not since put up my sword—
Against the Capitol I met a lion                    20
Who glazed upon me, and went surly by
Without annoying me. And there were drawn
Upon a heap a hundred ghastly women,
Transformèd with their fear, who swore they saw
Men all in fire walk up and down the streets.      25
And yesterday the bird of night did sit
Even at noonday upon the market-place,

Hooting and shrieking. When these prodigies
Do so conjointly meet, let not men say
'These are their reasons', 'they are natural',     30
For I believe they are portentous things
Unto the climate that they point upon.

CICERO
Indeed it is a strange-disposèd time;
But men may construe things after their fashion,
Clean from the purpose of the things themselves.   35
Comes Caesar to the Capitol tomorrow?

CASCA
He doth, for he did bid Antonio
Send word to you he would be there tomorrow.

CICERO
Good night then, Casca. This disturbèd sky
Is not to walk in.

CASCA             Farewell, Cicero.      *Exit Cicero*
    *Enter Cassius,* ⌈*unbraced*⌉

CASSIUS
Who's there?

CASCA          A Roman.

CASSIUS          Casca, by your voice.    41

CASCA
Your ear is good. Cassius, what night is this?

CASSIUS
A very pleasing night to honest men.

CASCA
Who ever knew the heavens menace so?

CASSIUS
Those that have known the earth so full of faults.  45
For my part, I have walked about the streets,
Submitting me unto the perilous night;
And thus unbracèd, Casca, as you see,
Have bared my bosom to the thunder-stone;
And when the cross blue lightning seemed to open  50
The breast of heaven, I did present myself
Even in the aim and very flash of it.

CASCA
But wherefore did you so much tempt the heavens?
It is the part of men to fear and tremble
When the most mighty gods by tokens send     55
Such dreadful heralds to astonish us.

CASSIUS
You are dull, Casca, and those sparks of life
That should be in a Roman you do want,
Or else you use not. You look pale, and gaze,
And put on fear, and cast yourself in wonder,   60
To see the strange impatience of the heavens;
But if you would consider the true cause
Why all these fires, why all these gliding ghosts,
Why birds and beasts from quality and kind—
Why old men, fools, and children calculate—   65
Why all these things change from their ordinance,
Their natures, and preformèd faculties,
To monstrous quality—why, you shall find
That heaven hath infused them with these spirits
To make them instruments of fear and warning  70
Unto some monstrous state. Now could I, Casca,
Name to thee a man most like this dreadful night,
That thunders, lightens, opens graves, and roars
As doth the lion in the Capitol;
A man no mightier than thyself or me     75
In personal action, yet prodigious grown,
And fearful, as these strange eruptions are.

CASCA
'Tis Caesar that you mean, is it not, Cassius?

CASSIUS
Let it be who it is; for Romans now
Have thews and limbs like to their ancestors.   80
But woe the while! Our fathers' minds are dead,
And we are governed with our mothers' spirits.
Our yoke and sufferance show us womanish.

CASCA
Indeed they say the senators tomorrow
Mean to establish Caesar as a king,     85
And he shall wear his crown by sea and land
In every place save here in Italy.

CASSIUS (*drawing his dagger*)
I know where I will wear this dagger then:
Cassius from bondage will deliver Cassius.
Therein, ye gods, you make the weak most strong;  90
Therein, ye gods, you tyrants do defeat.
Nor stony tower, nor walls of beaten brass,
Nor airless dungeon, nor strong links of iron,
Can be retentive to the strength of spirit;
But life, being weary of these worldly bars,   95
Never lacks power to dismiss itself.
If I know this, know all the world besides,
That part of tyranny that I do bear
I can shake off at pleasure.
    *Thunder still*

CASCA          So can I.
So every bondman in his own hand bears    100
The power to cancel his captivity.

CASSIUS
And why should Caesar be a tyrant then?
Poor man, I know he would not be a wolf
But that he sees the Romans are but sheep.
He were no lion, were not Romans hinds.    105
Those that with haste will make a mighty fire
Begin it with weak straws. What trash is Rome,
What rubbish, and what offal, when it serves
For the base matter to illuminate
So vile a thing as Caesar! But, O grief,    110
Where hast thou led me? I perhaps speak this
Before a willing bondman; then I know
My answer must be made. But I am armed,
And dangers are to me indifferent.

CASCA
You speak to Casca, and to such a man    115
That is no fleering tell-tale. Hold. My hand.
Be factious for redress of all these griefs,
And I will set this foot of mine as far
As who goes farthest.
    *They join hands*

CASSIUS          There's a bargain made.
Now know you, Casca, I have moved already   120
Some certain of the noblest-minded Romans
To undergo with me an enterprise
Of honourable-dangerous consequence.
And I do know by this they stay for me
In Pompey's Porch; for now, this fearful night,  125
There is no stir or walking in the streets,
And the complexion of the element
In favour's like the work we have in hand,
Most bloody, fiery, and most terrible.
    *Enter Cinna*

CASCA
Stand close a while, for here comes one in haste.  130

CASSIUS
'Tis Cinna; I do know him by his gait.
He is a friend.—Cinna, where haste you so?

CINNA
  To find out you. Who's that? Metellus Cimber?
CASSIUS
  No, it is Casca, one incorporate
  To our attempts. Am I not stayed for, Cinna?    135
CINNA
  I am glad on't. What a fearful night is this!
  There's two or three of us have seen strange sights.
CASSIUS Am I not stayed for? Tell me.
CINNA Yes, you are.
  O Cassius, if you could                         140
  But win the noble Brutus to our party—
CASSIUS
  Be you content. Good Cinna, take this paper,
    *He gives Cinna letters*
  And look you lay it in the Praetor's Chair,
  Where Brutus may but find it; and throw this
  In at his window. Set this up with wax          145
  Upon old Brutus' statue. All this done,
  Repair to Pompey's Porch where you shall find us.
  Is Decius Brutus and Trebonius there?
CINNA
  All but Metellus Cimber, and he's gone
  To seek you at your house. Well, I will hie,    150
  And so bestow these papers as you bade me.
CASSIUS
  That done, repair to Pompey's Theatre.    *Exit Cinna*
  Come, Casca, you and I will yet ere day
  See Brutus at his house. Three parts of him
  Is ours already, and the man entire            155
  Upon the next encounter yields him ours.
CASCA
  O, he sits high in all the people's hearts,
  And that which would appear offence in us
  His countenance, like richest alchemy,
  Will change to virtue and to worthiness.        160
CASSIUS
  Him and his worth, and our great need of him,
  You have right well conceited. Let us go,
  For it is after midnight, and ere day
  We will awake him and be sure of him.     *Exeunt*

**2.1**    *Enter Brutus in his orchard*
BRUTUS What, Lucius, ho!—
  I cannot by the progress of the stars
  Give guess how near to day.—Lucius, I say!—
  I would it were my fault to sleep so soundly.—
  When, Lucius, when? Awake, I say! What, Lucius!   5
    *Enter Lucius*
LUCIUS Called you, my lord?
BRUTUS
  Get me a taper in my study, Lucius.
  When it is lighted, come and call me here.
LUCIUS I will, my lord.                              *Exit*
BRUTUS
  It must be by his death. And for my part        10
  I know no personal cause to spurn at him,
  But for the general. He would be crowned.
  How that might change his nature, there's the
      question.
  It is the bright day that brings forth the adder,
  And that craves wary walking. Crown him: that!  15
  And then I grant we put a sting in him
  That at his will he may do danger with.
  Th'abuse of greatness is when it disjoins
  Remorse from power. And to speak truth of Caesar,
  I have not known when his affections swayed     20

More than his reason. But 'tis a common proof
  That lowliness is young ambition's ladder,
  Whereto the climber-upward turns his face;
  But when he once attains the upmost round,
  He then unto the ladder turns his back,         25
  Looks in the clouds, scorning the base degrees
  By which he did ascend. So Caesar may.
  Then lest he may, prevent. And since the quarrel
  Will bear no colour for the thing he is,
  Fashion it thus: that what he is, augmented,     30
  Would run to these and these extremities;
  And therefore think him as a serpent's egg,
  Which, hatched, would as his kind grow mischievous,
  And kill him in the shell.
    *Enter Lucius, with a letter*
LUCIUS
  The taper burneth in your closet, sir.          35
  Searching the window for a flint, I found
  This paper, thus sealed up, and I am sure
  It did not lie there when I went to bed.
    *He gives him the letter*
BRUTUS
  Get you to bed again; it is not day.
  Is not tomorrow, boy, the ides of March?         40
LUCIUS I know not, sir.
BRUTUS
  Look in the calendar and bring me word.
LUCIUS I will, sir.                                  *Exit*
BRUTUS
  The exhalations whizzing in the air
  Give so much light that I may read by them.      45
    *He opens the letter and reads*
  'Brutus, thou sleep'st. Awake, and see thyself.
  Shall Rome, et cetera? Speak, strike, redress.'—
  'Brutus, thou sleep'st. Awake.'
  Such instigations have been often dropped
  Where I have took them up.                        50
  'Shall Rome, et cetera?' Thus must I piece it out:
  Shall Rome stand under one man's awe? What,
      Rome?
  My ancestors did from the streets of Rome
  The Tarquin drive when he was called a king.
  'Speak, strike, redress.' Am I entreated          55
  To speak and strike? O Rome, I make thee promise,
  If the redress will follow, thou receivest
  Thy full petition at the hand of Brutus.
    *Enter Lucius*
LUCIUS
  Sir, March is wasted fifteen days.
    *Knock within*
BRUTUS
  'Tis good. Go to the gate; somebody knocks.      60
                                          *Exit Lucius*
  Since Cassius first did whet me against Caesar
  I have not slept.
  Between the acting of a dreadful thing
  And the first motion, all the interim is
  Like a phantasma or a hideous dream.             65
  The genius and the mortal instruments
  Are then in counsel, and the state of man,
  Like to a little kingdom, suffers then
  The nature of an insurrection.
    *Enter Lucius*
LUCIUS
  Sir, 'tis your brother Cassius at the door,      70
  Who doth desire to see you.
BRUTUS                            Is he alone?

LUCIUS
No, sir, there are more with him.
BRUTUS                              Do you know them?
LUCIUS
No, sir; their hats are plucked about their ears,
And half their faces buried in their cloaks,
That by no means I may discover them          75
By any mark of favour.
BRUTUS                    Let 'em enter.      *Exit Lucius*
They are the faction. O conspiracy,
Sham'st thou to show thy dang'rous brow by night,
When evils are most free? O then by day
Where wilt thou find a cavern dark enough      80
To mask thy monstrous visage? Seek none, conspiracy.
Hide it in smiles and affability;
For if thou put thy native semblance on,
Not Erebus itself were dim enough
To hide thee from prevention.                  85
    *Enter the conspirators, muffled: Cassius, Casca,*
    *Decius, Cinna, Metellus, and Trebonius*
CASSIUS
I think we are too bold upon your rest.
Good morrow, Brutus. Do we trouble you?
BRUTUS
I have been up this hour, awake all night.
Know I these men that come along with you?
CASSIUS
Yes, every man of them; and no man here        90
But honours you; and every one doth wish
You had but that opinion of yourself
Which every noble Roman bears of you.
This is Trebonius.
BRUTUS               He is welcome hither.
CASSIUS
This, Decius Brutus.
BRUTUS                He is welcome too.        95
CASSIUS
This, Casca; Cinna, this; and this, Metellus Cimber.
BRUTUS They are all welcome.
What watchful cares do interpose themselves
Betwixt your eyes and night?
CASSIUS                      Shall I entreat a word?
    *Cassius and Brutus ⌈stand aside and⌉ whisper*
DECIUS
Here lies the east. Doth not the day break here?   100
CASCA  No.
CINNA
O pardon, sir, it doth; and yon grey lines
That fret the clouds are messengers of day.
CASCA
You shall confess that you are both deceived.
    *He points his sword*
Here, as I point my sword, the sun arises,     105
Which is a great way growing on the south,
Weighing the youthful season of the year.
Some two months hence up higher toward the north
He first presents his fire, and the high east
Stands, as the Capitol, directly here.         110
    *He points his sword.*
    ⌈*Brutus and Cassius join the other conspirators*⌉
BRUTUS
Give me your hands all over, one by one.
    *He shakes their hands*
CASSIUS
And let us swear our resolution.

BRUTUS
No, not an oath. If not the face of men,
The sufferance of our souls, the time's abuse—
If these be motives weak, break off betimes,   115
And every man hence to his idle bed.
So let high-sighted tyranny range on
Till each man drop by lottery. But if these,
As I am sure they do, bear fire enough
To kindle cowards and to steel with valour     120
The melting spirits of women, then, countrymen,
What need we any spur but our own cause
To prick us to redress? What other bond
Than secret Romans, that have spoke the word
And will not palter? And what other oath       125
Than honesty to honesty engaged
That this shall be or we will fall for it?
Swear priests and cowards and men cautelous,
Old feeble carrions, and such suffering souls
That welcome wrongs; unto bad causes swear     130
Such creatures as men doubt; but do not stain
The even virtue of our enterprise,
Nor th'insuppressive mettle of our spirits,
To think that or our cause or our performance
Did need an oath, when every drop of blood     135
That every Roman bears, and nobly bears,
Is guilty of a several bastardy
If he do break the smallest particle
Of any promise that hath passed from him.
CASSIUS
But what of Cicero? Shall we sound him?         140
I think he will stand very strong with us.
CASCA
Let us not leave him out.
CINNA                      No, by no means.
METELLUS
O, let us have him, for his silver hairs
Will purchase us a good opinion,
And buy men's voices to commend our deeds.      145
It shall be said his judgement ruled our hands.
Our youths and wildness shall no whit appear,
But all be buried in his gravity.
BRUTUS
O, name him not! Let us not break with him,
For he will never follow anything               150
That other men begin.
CASSIUS Then leave him out.
CASCA Indeed he is not fit.
DECIUS
Shall no man else be touched, but only Caesar?
CASSIUS
Decius, well urged. I think it is not meet      155
Mark Antony, so well beloved of Caesar,
Should outlive Caesar. We shall find of him
A shrewd contriver. And you know his means,
If he improve them, may well stretch so far
As to annoy us all; which to prevent,           160
Let Antony and Caesar fall together.
BRUTUS
Our course will seem too bloody, Caius Cassius,
To cut the head off and then hack the limbs,
Like wrath in death and envy afterwards—
For Antony is but a limb of Caesar.             165
Let's be sacrificers, but not butchers, Caius.
We all stand up against the spirit of Caesar,
And in the spirit of men there is no blood.

O, that we then could come by Caesar's spirit,
And not dismember Caesar! But, alas,          170
Caesar must bleed for it. And, gentle friends,
Let's kill him boldly, but not wrathfully.
Let's carve him as a dish fit for the gods,
Not hew him as a carcass fit for hounds.
And let our hearts, as subtle masters do,          175
Stir up their servants to an act of rage,
And after seem to chide 'em. This shall make
Our purpose necessary, and not envious;
Which so appearing to the common eyes,
We shall be called purgers, not murderers.          180
And for Mark Antony, think not of him,
For he can do no more than Caesar's arm
When Caesar's head is off.

CASSIUS                              Yet I fear him;
For in the engrafted love he bears to Caesar—

BRUTUS
Alas, good Cassius, do not think of him.          185
If he love Caesar, all that he can do
Is to himself: take thought, and die for Caesar.
And that were much he should, for he is given
To sports, to wildness, and much company.

TREBONIUS
There is no fear in him. Let him not die;          190
For he will live, and laugh at this hereafter.
          *Clock strikes*

BRUTUS
Peace, count the clock.

CASSIUS                              The clock hath stricken three.

TREBONIUS
'Tis time to part.

CASSIUS                    But it is doubtful yet
Whether Caesar will come forth today or no;
For he is superstitious grown of late,          195
Quite from the main opinion he held once
Of fantasy, of dreams and ceremonies.
It may be these apparent prodigies,
The unaccustomed terror of this night,
And the persuasion of his augurers,          200
May hold him from the Capitol today.

DECIUS
Never fear that. If he be so resolved
I can o'ersway him; for he loves to hear
That unicorns may be betrayed with trees,
And bears with glasses, elephants with holes,          205
Lions with toils, and men with flatterers;
But when I tell him he hates flatterers;
He says he does, being then most flattered. Let me work,
For I can give his humour the true bent,
And I will bring him to the Capitol.          210

CASSIUS
Nay, we will all of us be there to fetch him.

BRUTUS
By the eighth hour. Is that the uttermost?

CINNA
Be that the uttermost, and fail not then.

METELLUS
Caius Ligarius doth bear Caesar hard,
Who rated him for speaking well of Pompey.          215
I wonder none of you have thought of him.

BRUTUS
Now good Metellus, go along by him.
He loves me well, and I have given him reasons.
Send him but hither, and I'll fashion him.          219

CASSIUS
The morning comes upon's. We'll leave you, Brutus.
And, friends, disperse yourselves; but all remember
What you have said, and show yourselves true Romans.

BRUTUS
Good gentlemen, look fresh and merrily.
Let not our looks put on our purposes;
But bear it as our Roman actors do,          225
With untired spirits and formal constancy.
And so good morrow to you every one.
                              *Exeunt all but Brutus*
Boy, Lucius!—Fast asleep? It is no matter.
Enjoy the honey-heavy dew of slumber.
Thou hast no figures nor no fantasies          230
Which busy care draws in the brains of men;
Therefore thou sleep'st so sound.
          *Enter Portia*

PORTIA                              Brutus, my lord.

BRUTUS
Portia, what mean you? Wherefore rise you now?
It is not for your health thus to commit
Your weak condition to the raw cold morning.          235

PORTIA
Nor for yours neither. You've ungently, Brutus,
Stole from my bed; and yesternight at supper
You suddenly arose, and walked about
Musing and sighing, with your arms across;
And when I asked you what the matter was,          240
You stared upon me with ungentle looks.
I urged you further; then you scratched your head,
And too impatiently stamped with your foot.
Yet I insisted; yet you answered not,
But with an angry wafture of your hand          245
Gave sign for me to leave you. So I did,
Fearing to strengthen that impatience
Which seemed too much enkindled, and withal
Hoping it was but an effect of humour,
Which sometime hath his hour with every man.          250
It will not let you eat, nor talk, nor sleep;
And could it work so much upon your shape
As it hath much prevailed on your condition,
I should not know you Brutus. Dear my lord,
Make me acquainted with your cause of grief.          255

BRUTUS
I am not well in health, and that is all.

PORTIA
Brutus is wise, and were he not in health
He would embrace the means to come by it.

BRUTUS
Why, so I do. Good Portia, go to bed.

PORTIA
Is Brutus sick? And is it physical          260
To walk unbracèd and suck up the humours
Of the dank morning? What, is Brutus sick?
And will he steal out of his wholesome bed
To dare the vile contagion of the night,
And tempt the rheumy and unpurgèd air          265
To add unto his sickness? No, my Brutus,
You have some sick offence within your mind,
Which by the right and virtue of my place
I ought to know of. (*Kneeling*) And upon my knees,
I charm you by my once-commended beauty,          270
By all your vows of love, and that great vow
Which did incorporate and make us one,
That you unfold to me, your self, your half,
Why you are heavy, and what men tonight

Have had resort to you—for here have been       275
Some six or seven, who did hide their faces
Even from darkness.
BRUTUS                          Kneel not, gentle Portia.
PORTIA ⌈*rising*⌉
I should not need if you were gentle Brutus.
Within the bond of marriage, tell me, Brutus,
Is it excepted I should know no secrets       280
That appertain to you? Am I your self
But as it were in sort or limitation?
To keep with you at meals, comfort your bed,
And talk to you sometimes? Dwell I but in the
       suburbs
Of your good pleasure? If it be no more,       285
Portia is Brutus' harlot, not his wife.
BRUTUS
You are my true and honourable wife,
As dear to me as are the ruddy drops
That visit my sad heart.
PORTIA
If this were true, then should I know this secret.       290
I grant I am a woman, but withal
A woman that Lord Brutus took to wife.
I grant I am a woman, but withal
A woman well reputed, Cato's daughter.
Think you I am no stronger than my sex,       295
Being so fathered and so husbanded?
Tell me your counsels; I will not disclose 'em.
I have made strong proof of my constancy,
Giving myself a voluntary wound
Here in the thigh. Can I bear that with patience,       300
And not my husband's secrets?
BRUTUS                          O ye gods,
Render me worthy of this noble wife!
       *Knocking within*
Hark, hark, one knocks. Portia, go in a while,
And by and by thy bosom shall partake
The secrets of my heart.       305
All my engagements I will construe to thee,
All the charactery of my sad brows.
Leave me with haste.       *Exit Portia*
       Lucius, who's that knocks?
       *Enter Lucius, and Ligarius, with a kerchief ⌈round
       his head⌉*
LUCIUS
Here is a sick man that would speak with you.
BRUTUS
Caius Ligarius, that Metellus spake of.—       310
Boy, stand aside.       ⌈*Exit*⌉ *Lucius*
       Caius Ligarius, how?
LIGARIUS
Vouchsafe good morrow from a feeble tongue.
BRUTUS
O, what a time have you chose out, brave Caius,
To wear a kerchief! Would you were not sick!
LIGARIUS
I am not sick if Brutus have in hand       315
Any exploit worthy the name of honour.
BRUTUS
Such an exploit have I in hand, Ligarius,
Had you a healthful ear to hear of it.
LIGARIUS
By all the gods that Romans bow before,
I here discard my sickness.
       *He pulls off his kerchief*
       Soul of Rome,       320
Brave son derived from honourable loins,

Thou like an exorcist hast conjured up
My mortifièd spirit. Now bid me run,
And I will strive with things impossible,
Yea, get the better of them. What's to do?       325
BRUTUS
A piece of work that will make sick men whole.
LIGARIUS
But are not some whole that we must make sick?
BRUTUS
That must we also. What it is, my Caius,
I shall unfold to thee as we are going
To whom it must be done.
LIGARIUS                          Set on your foot,       330
And with a heart new-fired I follow you
To do I know not what; but it sufficeth
That Brutus leads me on.
BRUTUS                          Follow me then.       *Exeunt*

2.2       *Thunder and lightning.*
       *Enter Julius Caesar in his nightgown*
CAESAR
Nor heaven nor earth have been at peace tonight.
Thrice hath Calpurnia in her sleep cried out
'Help, ho! They murder Caesar!'—Who's within?
       *Enter a Servant*
SERVANT My lord.
CAESAR
Go bid the priests do present sacrifice,       5
And bring me their opinions of success.
SERVANT I will, my lord.       *Exit*
       *Enter Calpurnia*
CALPURNIA
What mean you, Caesar? Think you to walk forth?
You shall not stir out of your house today.
CAESAR
Caesar shall forth. The things that threatened me       10
Ne'er looked but on my back; when they shall see
The face of Caesar, they are vanishèd.
CALPURNIA
Caesar, I never stood on ceremonies,
Yet now they fright me. There is one within,
Besides the things that we have heard and seen,       15
Recounts most horrid sights seen by the watch.
A lioness hath whelpèd in the streets,
And graves have yawned and yielded up their dead.
Fierce fiery warriors fight upon the clouds,
In ranks and squadrons and right form of war,       20
Which drizzled blood upon the Capitol.
The noise of battle hurtled in the air.
Horses do neigh, and dying men did groan,
And ghosts did shriek and squeal about the streets.
O Caesar, these things are beyond all use,       25
And I do fear them.
CAESAR                          What can be avoided
Whose end is purposed by the mighty gods?
Yet Caesar shall go forth, for these predictions
Are to the world in general as to Caesar.
CALPURNIA
When beggars die there are no comets seen;       30
The heavens themselves blaze forth the death of
       princes.
CAESAR
Cowards die many times before their deaths;
The valiant never taste of death but once.
Of all the wonders that I yet have heard,
It seems to me most strange that men should fear,       35

Seeing that death, a necessary end,
Will come when it will come.
*Enter Servant*
                              What say the augurers?
SERVANT
They would not have you to stir forth today.
Plucking the entrails of an offering forth,
They could not find a heart within the beast.        40
CAESAR
The gods do this in shame of cowardice.
Caesar should be a beast without a heart
If he should stay at home today for fear.
No, Caesar shall not. Danger knows full well
That Caesar is more dangerous than he.              45
We are two lions littered in one day,
And I the elder and more terrible.
And Caesar shall go forth.
CALPURNIA                    Alas, my lord,
Your wisdom is consumed in confidence.
Do not go forth today. Call it my fear              50
That keeps you in the house, and not your own.
We'll send Mark Antony to the Senate House,
And he shall say you are not well today.
Let me upon my knee prevail in this.
*She kneels*
CAESAR
Mark Antony shall say I am not well,               55
And for thy humour I will stay at home.
*Enter Decius*
Here's Decius Brutus; he shall tell them so.
⌜*Calpurnia rises*⌝
DECIUS
Caesar, all hail! Good morrow, worthy Caesar.
I come to fetch you to the Senate House.
CAESAR
And you are come in very happy time               60
To bear my greeting to the senators
And tell them that I will not come today.
Cannot is false, and that I dare not, falser.
I will not come today; tell them so, Decius.
CALPURNIA
Say he is sick.
CAESAR          Shall Caesar send a lie?           65
Have I in conquest stretched mine arm so far,
To be afeard to tell greybeards the truth?
Decius, go tell them Caesar will not come.
DECIUS
Most mighty Caesar, let me know some cause,
Lest I be laughed at when I tell them so.          70
CAESAR
The cause is in my will; I will not come.
That is enough to satisfy the Senate.
But for your private satisfaction,
Because I love you, I will let you know.
Calpurnia here, my wife, stays me at home.         75
She dreamt tonight she saw my statue,
Which like a fountain with an hundred spouts
Did run pure blood; and many lusty Romans
Came smiling and did bathe their hands in it.
And these does she apply for warnings and portents
Of evils imminent, and on her knee                 81
Hath begged that I will stay at home today.
DECIUS
This dream is all amiss interpreted.
It was a vision fair and fortunate.

Your statue spouting blood in many pipes,          85
In which so many smiling Romans bathed,
Signifies that from you great Rome shall suck
Reviving blood, and that great men shall press
For tinctures, stains, relics, and cognizance.
This by Calpurnia's dream is signified.            90
CAESAR
And this way have you well expounded it.
DECIUS
I have, when you have heard what I can say.
And know it now: the Senate have concluded
To give this day a crown to mighty Caesar.
If you shall send them word you will not come,      95
Their minds may change. Besides, it were a mock
Apt to be rendered for someone to say
'Break up the Senate till another time,
When Caesar's wife shall meet with better dreams.'
If Caesar hide himself, shall they not whisper      100
'Lo, Caesar is afraid'?
Pardon me, Caesar; for my dear dear love
To your proceeding bids me tell you this,
And reason to my love is liable.
CAESAR
How foolish do your fears seem now, Calpurnia!     105
I am ashamèd I did yield to them.
Give me my robe, for I will go.
*Enter* ⌜*Cassius,*⌝ *Brutus, Ligarius, Metellus, Casca,*
*Trebonius, and Cinna*
And look where Cassius is come to fetch me.
⌜CASSIUS⌝
Good morrow, Caesar.
CAESAR                    Welcome, Cassius.—
What, Brutus, are you stirred so early too?—       110
Good morrow, Casca.—Caius Ligarius,
Caesar was ne'er so much your enemy
As that same ague which hath made you lean.
What is't o'clock?
BRUTUS          Caesar, 'tis strucken eight.
CAESAR
I thank you for your pains and courtesy.           115
*Enter Antony*
See, Antony that revels long a-nights
Is notwithstanding up. Good morrow, Antony.
ANTONY
So to most noble Caesar.
CAESAR ⌜*to Calpurnia*⌝        Bid them prepare within.
I am to blame to be thus waited for.   ⌜*Exit Calpurnia*⌝
Now, Cinna.—Now, Metellus.—What, Trebonius!   120
I have an hour's talk in store for you.
Remember that you call on me today.
Be near me, that I may remember you.
TREBONIUS
Caesar, I will, ⌜*aside*⌝ and so near will I be
That your best friends shall wish I had been further.
CAESAR
Good friends, go in and taste some wine with me,   126
And we, like friends, will straightway go together.
BRUTUS (*aside*)
That every like is not the same, O Caesar,
The heart of Brutus ernes to think upon.   *Exeunt*

**2.3**   *Enter Artemidorus, reading a letter*
ARTEMIDORUS 'Caesar, beware of Brutus. Take heed of
Cassius. Come not near Casca. Have an eye to Cinna.
Trust not Trebonius. Mark well Metellus Cimber. Decius

Brutus loves thee not. Thou hast wronged Caius
Ligarius. There is but one mind in all these men, and
it is bent against Caesar. If thou beest not immortal,
look about you. Security gives way to conspiracy.       7
The mighty gods defend thee!
                                         Thy lover,
                                         Artemidorus.'
Here will I stand till Caesar pass along,               11
And as a suitor will I give him this.
My heart laments that virtue cannot live
Out of the teeth of emulation.
If thou read this, O Caesar, thou mayst live.           15
If not, the fates with traitors do contrive.       Exit

**2.4**   *Enter Portia and Lucius*
PORTIA
I prithee, boy, run to the Senate House.
Stay not to answer me, but get thee gone.—
Why dost thou stay?
LUCIUS                To know my errand, madam.
PORTIA
I would have had thee there and here again
Ere I can tell thee what thou shouldst do there.        5
(*Aside*) O constancy, be strong upon my side;
Set a huge mountain 'tween my heart and tongue.
I have a man's mind, but a woman's might.
How hard it is for women to keep counsel!
(*To Lucius*) Art thou here yet?
LUCIUS                 Madam, what should I do?
Run to the Capitol, and nothing else?                   11
And so return to you, and nothing else?
PORTIA
Yes, bring me word, boy, if thy lord look well,
For he went sickly forth; and take good note
What Caesar doth, what suitors press to him.            15
Hark, boy, what noise is that?
LUCIUS  I hear none, madam.
PORTIA  Prithee, listen well.
I heard a bustling rumour, like a fray,
And the wind brings it from the Capitol.                20
LUCIUS  Sooth, madam, I hear nothing.
        *Enter the Soothsayer*
PORTIA
Come hither, fellow. Which way hast thou been?
SOOTHSAYER
At mine own house, good lady.
PORTIA  What is't o'clock?
SOOTHSAYER  About the ninth hour, lady.                 25
PORTIA
Is Caesar yet gone to the Capitol?
SOOTHSAYER
Madam, not yet. I go to take my stand
To see him pass on to the Capitol.
PORTIA
Thou hast some suit to Caesar, hast thou not?
SOOTHSAYER
That I have, lady. If it will please Caesar             30
To be so good to Caesar as to hear me,
I shall beseech him to befriend himself.
PORTIA
Why, know'st thou any harms intended towards him?
SOOTHSAYER
None that I know will be; much that I fear may chance.
Good morrow to you.
        ⌈*He moves away*⌉
                        Here the street is narrow.      35

The throng that follows Caesar at the heels,
Of senators, of praetors, common suitors,
Will crowd a feeble man almost to death.
I'll get me to a place more void, and there
Speak to great Caesar as he comes along.          *Exit*
PORTIA (*aside*)
I must go in. Ay me! How weak a thing               41
The heart of woman is! O Brutus,
The heavens speed thee in thine enterprise!—
Sure the boy heard me. (*To Lucius*) Brutus hath a suit
That Caesar will not grant. (*Aside*) O, I grow faint!   45
(*To Lucius*) Run, Lucius, and commend me to my lord.
Say I am merry. Come to me again,
And bring me word what he doth say to thee.
                              *Exeunt* ⌈*severally*⌉

**3.1**   *Enter* ⌈*at one door*⌉ *Artemidorus, the Soothsayer,*
          *and citizens. Flourish. Enter* ⌈*at another door*⌉
          *Caesar, Brutus, Cassius, Casca, Decius, Metellus,*
          *Trebonius, Cinna,* ⌈*Ligarius,*⌉ *Antony, Lepidus,*
          *Publius, Popillius,* ⌈*and other senators*⌉
CAESAR (*to the Soothsayer*) The ides of March are come.
SOOTHSAYER  Ay, Caesar, but not gone.
ARTEMIDORUS  Hail, Caesar! Read this schedule.
DECIUS (*to Caesar*)
Trebonius doth desire you to o'er-read
At your best leisure this his humble suit.           5
ARTEMIDORUS
O Caesar, read mine first, for mine's a suit
That touches Caesar nearer. Read it, great Caesar.
CAESAR
What touches us ourself shall be last served.
ARTEMIDORUS
Delay not, Caesar, read it instantly.
CAESAR
What, is the fellow mad?
PUBLIUS (*to Artemidorus*)       Sirrah, give place.    10
CASSIUS (*to Artemidorus*)
What, urge you your petitions in the street?
Come to the Capitol.
        ⌈*They walk about the stage*⌉
POPILLIUS (*aside to Cassius*)
I wish your enterprise today may thrive.
CASSIUS
What enterprise, Popillius?
POPILLIUS                    Fare you well.
        *He leaves Cassius, and makes to Caesar*
BRUTUS  What said Popillius Laena?                     15
CASSIUS
He wished today our enterprise might thrive.
I fear our purpose is discoverèd.
BRUTUS
Look how he makes to Caesar. Mark him.
CASSIUS
Casca, be sudden, for we fear prevention.—
Brutus, what shall be done? If this be known,          20
Cassius or Caesar never shall turn back,
For I will slay myself.
BRUTUS              Cassius, be constant.
Popillius Laena speaks not of our purposes,
For look, he smiles, and Caesar doth not change.
CASSIUS
Trebonius knows his time, for look you, Brutus,        25
He draws Mark Antony out of the way.
                        *Exeunt Trebonius and Antony*

DECIUS
  Where is Metellus Cimber? Let him go
  And presently prefer his suit to Caesar.
  ⌈Caesar sits⌉
BRUTUS
  He is addressed. Press near, and second him.
CINNA
  Casca, you are the first that rears your hand.                30
  ⌈The conspirators and the other senators take their
  places⌉
CAESAR
  Are we all ready? What is now amiss
  That Caesar and his Senate must redress?
METELLUS (coming forward and kneeling)
  Most high, most mighty, and most puissant Caesar,
  Metellus Cimber throws before thy seat
  An humble heart.
CAESAR              I must prevent thee, Cimber.           35
  These couchings and these lowly courtesies
  Might fire the blood of ordinary men,
  And turn preordinance and first decree
  Into the law of children. Be not fond
  To think that Caesar bears such rebel blood                 40
  That will be thawed from the true quality
  With that which melteth fools: I mean sweet words,
  Low-crookèd curtsies, and base spaniel fawning.
  Thy brother by decree is banishèd.
  If thou dost bend and pray and fawn for him,                45
  I spurn thee like a cur out of my way.
  Know Caesar doth not wrong but with just cause,
  Nor without cause will he be satisfied.
METELLUS
  Is there no voice more worthy than my own
  To sound more sweetly in great Caesar's ear                 50
  For the repealing of my banished brother?
BRUTUS (coming forward and kneeling)
  I kiss thy hand, but not in flattery, Caesar,
  Desiring thee that Publius Cimber may
  Have an immediate freedom of repeal.
CAESAR
  What, Brutus?
CASSIUS (coming forward and kneeling)
                 Pardon, Caesar; Caesar, pardon.           55
  As low as to thy foot doth Cassius fall
  To beg enfranchisement for Publius Cimber.
CAESAR
  I could be well moved if I were as you.
  If I could pray to move, prayers would move me.
  But I am constant as the Northern Star,                     60
  Of whose true fixed and resting quality
  There is no fellow in the firmament.
  The skies are painted with unnumbered sparks;
  They are all fire, and every one doth shine;
  But there's but one in all doth hold his place.             65
  So in the world: 'tis furnished well with men,
  And men are flesh and blood, and apprehensive;
  Yet in the number I do know but one
  That unassailable holds on his rank,
  Unshaked of motion; and that I am he                        70
  Let me a little show it even in this—
  That I was constant Cimber should be banished,
  And constant do remain to keep him so.
CINNA (coming forward and kneeling)
  O Caesar!
CAESAR       Hence! Wilt thou lift up Olympus?

DECIUS (coming forward ⌈with Ligarius⌉ and kneeling)
  Great Caesar!
CAESAR            Doth not Brutus bootless kneel?          75
CASCA (coming forward ⌈and kneeling⌉)
  Speak hands for me.
      They stab Caesar, ⌈Casca first, Brutus last⌉
CAESAR              Et tu, Brutè?—Then fall Caesar.
                                                  He dies
CINNA
  Liberty! Freedom! Tyranny is dead!
  Run hence, proclaim, cry it about the streets.
CASSIUS
  Some to the common pulpits, and cry out
  'Liberty, freedom, and enfranchisement!'                    80
BRUTUS
  People and senators, be not affrighted.
      ⌈Exeunt in a tumult Lepidus, Popillius, other
            senators, Artemidorus, Soothsayer, and
                                            citizens⌉
  Fly not! Stand still! Ambition's debt is paid.
CASCA Go to the pulpit, Brutus.
DECIUS And Cassius too.
BRUTUS Where's Publius?                                       85
CINNA
  Here, quite confounded with this mutiny.
METELLUS
  Stand fast together, lest some friend of Caesar's
  Should chance—
BRUTUS
  Talk not of standing.—Publius, good cheer!
  There is no harm intended to your person,                   90
  Nor to no Roman else—so tell them, Publius.
CASSIUS
  And leave us, Publius, lest that the people,
  Rushing on us, should do your age some mischief.
BRUTUS
  Do so; and let no man abide this deed
  But we the doers.                     ⌈Exit Publius⌉
      Enter Trebonius
CASSIUS Where is Antony?                                      96
TREBONIUS Fled to his house, amazed.
  Men, wives, and children stare, cry out, and run,
  As it were doomsday.
BRUTUS              Fates, we will know your pleasures.
  That we shall die, we know; 'tis but the time             100
  And drawing days out that men stand upon.
CASCA
  Why, he that cuts off twenty years of life
  Cuts off so many years of fearing death.
BRUTUS
  Grant that, and then is death a benefit.
  So are we Caesar's friends, that have abridged            105
  His time of fearing death. Stoop, Romans, stoop,
  And let us bathe our hands in Caesar's blood
  Up to the elbows, and besmear our swords;
  Then walk we forth even to the market-place,
  And, waving our red weapons o'er our heads,               110
  Let's all cry 'peace, freedom, and liberty!'
CASSIUS
  Stoop, then, and wash.
      They smear their hands with Caesar's blood
                     How many ages hence
  Shall this our lofty scene be acted over,
  In states unborn and accents yet unknown!
BRUTUS
  How many times shall Caesar bleed in sport,               115

That now on Pompey's basis lies along,
No worthier than the dust!
CASSIUS                    So oft as that shall be,
So often shall the knot of us be called
The men that gave their country liberty.
DECIUS
·What, shall we forth?
CASSIUS                    Ay, every man away.          120
Brutus shall lead, and we will grace his heels
With the most boldest and best hearts of Rome.
    *Enter Antony's Servant*
BRUTUS
Soft; who comes here? A friend of Antony's.
SERVANT (*kneeling and falling prostrate*)
Thus, Brutus, did my master bid me kneel.
Thus did Mark Antony bid me fall down,          125
And, being prostrate, thus he bade me say.
'Brutus is noble, wise, valiant, and honest.
Caesar was mighty, bold, royal, and loving.
Say I love Brutus, and I honour him.
Say I feared Caesar, honoured him, and loved him.
If Brutus will vouchsafe that Antony          131
May safely come to him and be resolved
How Caesar hath deserved to lie in death,
Mark Antony shall not love Caesar dead
So well as Brutus living, but will follow          135
The fortunes and affairs of noble Brutus
Thorough the hazards of this untrod state
With all true faith.' So says my master Antony.
BRUTUS
Thy master is a wise and valiant Roman.
I never thought him worse.          140
Tell him, so please him come unto this place,
He shall be satisfied, and, by my honour,
Depart untouched.
SERVANT ⌈*rising*⌉          I'll fetch him presently.          *Exit*
BRUTUS
I know that we shall have him well to friend.
CASSIUS ·
I wish we may. But yet have I a mind          145
That fears him much; and my misgiving still
Falls shrewdly to the purpose.
    *Enter Antony*
BRUTUS
But here comes Antony.—Welcome, Mark Antony.
ANTONY
O mighty Caesar! Dost thou lie so low?
Are all thy conquests, glories, triumphs, spoils,          150
Shrunk to this little measure? Fare thee well.—
I know not, gentlemen, what you intend—
Who else must be let blood, who else is rank.
If I myself, there is no hour so fit
As Caesar's death's hour, nor no instrument          155
Of half that worth as those your swords, made rich
With the most noble blood of all this world.
I do beseech ye, if you bear me hard,
Now, whilst your purpled hands do reek and smoke,
Fulfil your pleasure. Live a thousand years,          160
I shall not find myself so apt to die.
No place will please me so, no mean of death,
As here by Caesar, and by you cut off,
The choice and master spirits of this age.
BRUTUS
O Antony, beg not your death of us!          165
Though now we must appear bloody and cruel,
As by our hands and this our present act

You see we do, yet see you but our hands,
And this the bleeding business they have done.
Our hearts you see not; they are pitiful;          170
And pity to the general wrong of Rome—
As fire drives out fire, so pity pity—
Hath done this deed on Caesar. For your part,
To you our swords have leaden points, Mark Antony.
Our arms, unstrung of malice, and our hearts          175
Of brothers' temper, do receive you in
With all kind love, good thoughts, and reverence.
CASSIUS
Your voice shall be as strong as any man's
In the disposing of new dignities.
BRUTUS
Only be patient till we have appeased          180
The multitude, beside themselves with fear,
And then we will deliver you the cause
Why I, that did love Caesar when I struck him,
Have thus proceeded.
ANTONY                    I doubt not of your wisdom.
Let each man render me his bloody hand.          185
    *He shakes hands with the conspirators*
First, Marcus Brutus, will I shake with you.—
Next, Caius Cassius, do I take your hand.—
Now, Decius Brutus, yours;—now yours, Metellus;—
Yours, Cinna;—and my valiant Casca, yours;—
Though last, not least in love, yours, good Trebonius.
Gentlemen all—alas, what shall I say?          191
My credit now stands on such slippery ground
That one of two bad ways you must conceit me:
Either a coward or a flatterer.
That I did love thee, Caesar, O, 'tis true.          195
If then thy spirit look upon us now,
Shall it not grieve thee dearer than thy death
To see thy Antony making his peace,
Shaking the bloody fingers of thy foes—
Most noble!—in the presence of thy corpse?          200
Had I as many eyes as thou hast wounds,
Weeping as fast as they stream forth thy blood,
It would become me better than to close
In terms of friendship with thine enemies.
Pardon me, Julius. Here wast thou bayed, brave hart;
Here didst thou fall, and here thy hunters stand          206
Signed in thy spoil and crimsoned in thy lethe.
O world, thou wast the forest to this hart;
And this indeed, O world, the heart of thee.
How like a deer strucken by many princes          210
Dost thou here lie!
CASSIUS Mark Antony.
ANTONY Pardon me, Caius Cassius.
The enemies of Caesar shall say this;
Then in a friend it is cold modesty.          215
CASSIUS
I blame you not for praising Caesar so;
But what compact mean you to have with us?
Will you be pricked in number of our friends,
Or shall we on, and not depend on you?
ANTONY
Therefore I took your hands, but was indeed          220
Swayed from the point by looking down on Caesar.
Friends am I with you all, and love you all
Upon this hope: that you shall give me reasons
Why and wherein Caesar was dangerous.
BRUTUS
Or else were this a savage spectacle.          225
Our reasons are so full of good regard,

That were you, Antony, the son of Caesar,
You should be satisfied.
ANTONY            That's all I seek;
And am, moreover, suitor that I may
Produce his body to the market-place,     230
And in the pulpit, as becomes a friend,
Speak in the order of his funeral.
BRUTUS
You shall, Mark Antony.
CASSIUS          Brutus, a word with you.
(Aside to Brutus) You know not what you do. Do not
    consent
That Antony speak in his funeral.         235
Know you how much the people may be moved
By that which he will utter?
BRUTUS (aside to Cassius)        By your pardon,
I will myself into the pulpit first,
And show the reason of our Caesar's death.
What Antony shall speak I will protest     240
He speaks by leave and by permission;
And that we are contented Caesar shall
Have all true rites and lawful ceremonies,
It shall advantage more than do us wrong.
CASSIUS (aside to Brutus)
I know not what may fall. I like it not.      245
BRUTUS
Mark Antony, here, take you Caesar's body.
You shall not in your funeral speech blame us;
But speak all good you can devise of Caesar,
And say you do't by our permission;
Else shall you not have any hand at all      250
About his funeral. And you shall speak
In the same pulpit whereto I am going,
After my speech is ended.
ANTONY Be it so;
I do desire no more.                 255
BRUTUS
Prepare the body then, and follow us.
                *Exeunt all but Antony*
ANTONY
O pardon me, thou bleeding piece of earth,
That I am meek and gentle with these butchers.
Thou art the ruins of the noblest man
That ever lived in the tide of times.      260
Woe to the hand that shed this costly blood!
Over thy wounds now do I prophesy—
Which like dumb mouths do ope their ruby lips
To beg the voice and utterance of my tongue—
A curse shall light upon the limbs of men;    265
Domestic fury and fierce civil strife
Shall cumber all the parts of Italy;
Blood and destruction shall be so in use,
And dreadful objects so familiar,
That mothers shall but smile when they behold    270
Their infants quartered with the hands of war,
All pity choked with custom of fell deeds;
And Caesar's spirit, ranging for revenge,
With Ate by his side come hot from hell,
Shall in these confines with a monarch's voice    275
Cry 'havoc!' and let slip the dogs of war,
That this foul deed shall smell above the earth
With carrion men, groaning for burial.
        *Enter Octavius' Servant*
You serve Octavius Caesar, do you not?
SERVANT I do, Mark Antony.           280
ANTONY
Caesar did write for him to come to Rome.

SERVANT
He did receive his letters, and is coming,
And bid me say to you by word of mouth—
(*Seeing the body*) O Caesar!
ANTONY
Thy heart is big. Get thee apart and weep.    285
Passion, I see, is catching, for mine eyes,
Seeing those beads of sorrow stand in thine,
Began to water. Is thy master coming?
SERVANT
He lies tonight within seven leagues of Rome.
ANTONY
Post back with speed and tell him what hath
    chanced.                   290
Here is a mourning Rome, a dangerous Rome,
No Rome of safety for Octavius yet.
Hie hence and tell him so.—Yet stay awhile.
Thou shalt not back till I have borne this corpse
Into the market-place. There shall I try    295
In my oration how the people take
The cruel issue of these bloody men;
According to the which thou shalt discourse
To young Octavius of the state of things.
Lend me your hand.        *Exeunt with Caesar's body*

3.2     *Enter Brutus and Cassius, with the Plebeians*
ALL THE PLEBEIANS
We will be satisfied! Let us be satisfied!
BRUTUS
Then follow me, and give me audience, friends.
(*Aside to Cassius*) Cassius, go you into the other street,
And part the numbers.
(*To the Plebeians*)
Those that will hear me speak, let 'em stay here;    5
Those that will follow Cassius, go with him;
And public reasons shall be rendered
Of Caesar's death.
       *Brutus ascends to the pulpit*
FIRST PLEBEIAN       I will hear Brutus speak.
SECOND PLEBEIAN
I will hear Cassius, and compare their reasons
When severally we hear them rendered.     10
       *Exit Cassius, with some Plebeians*
     ⌈*Enter*⌉ *Brutus* ⌈*above*⌉ *in the pulpit*
THIRD PLEBEIAN
The noble Brutus is ascended. Silence.
BRUTUS Be patient till the last.
    Romans, countrymen, and lovers, hear me for my
cause, and be silent that you may hear. Believe me for
mine honour, and have respect to mine honour, that
you may believe. Censure me in your wisdom, and
awake your senses, that you may the better judge. If
there be any in this assembly, any dear friend of
Caesar's, to him I say that Brutus' love to Caesar was
no less than his. If then that friend demand why Brutus
rose against Caesar, this is my answer: not that I loved
Caesar less, but that I loved Rome more. Had you
rather Caesar were living, and die all slaves, than that
Caesar were dead, to live all free men? As Caesar loved
me, I weep for him. As he was fortunate, I rejoice at
it. As he was valiant, I honour him. But as he was
ambitious, I slew him. There is tears for his love, joy
for his fortune, honour for his valour, and death for
his ambition. Who is here so base that would be a
bondman? If any, speak, for him have I offended. Who
is here so rude that would not be a Roman? If any,

speak, for him have I offended. Who is here so vile that
will not love his country? If any, speak, for him have
I offended. I pause for a reply.
ALL THE PLEBEIANS None, Brutus, none.                    35
BRUTUS Then none have I offended. I have done no more
to Caesar than you shall do to Brutus. The question of
his death is enrolled in the Capitol, his glory not
extenuated wherein he was worthy, nor his offences
enforced for which he suffered death.                    40
    *Enter Mark Antony, with ⌈others bearing⌉ Caesar's*
    *body ⌈in a coffin⌉*
Here comes his body, mourned by Mark Antony, who,
though he had no hand in his death, shall receive the
benefit of his dying: a place in the commonwealth—as
which of you shall not? With this I depart: that as I
slew my best lover for the good of Rome, I have the
same dagger for myself when it shall please my country
to need my death.                                        47
ALL THE PLEBEIANS Live, Brutus, live, live!
FIRST PLEBEIAN
    Bring him with triumph home unto his house.
⌈FOURTH⌉ PLEBEIAN
    Give him a statue with his ancestors.                50
THIRD PLEBEIAN
    Let him be Caesar.
⌈FIFTH⌉ PLEBEIAN          Caesar's better parts
    Shall be crowned in Brutus.
FIRST PLEBEIAN
    We'll bring him to his house with shouts and
        clamours.
BRUTUS
    My countrymen.
⌈FOURTH⌉ PLEBEIAN Peace, silence. Brutus speaks.
FIRST PLEBEIAN Peace, ho!                                 55
BRUTUS
    Good countrymen, let me depart alone,
    And, for my sake, stay here with Antony.
    Do grace to Caesar's corpse, and grace his speech
    Tending to Caesar's glories, which Mark Antony,
    By our permission, is allowed to make.               60
    I do entreat you, not a man depart
    Save I alone till Antony have spoke.         *Exit*
FIRST PLEBEIAN
    Stay, ho, and let us hear Mark Antony.
THIRD PLEBEIAN
    Let him go up into the public chair.
    We'll hear him. Noble Antony, go up.                 65
ANTONY
    For Brutus' sake I am beholden to you.
    *Antony ascends to the pulpit*
⌈FIFTH⌉ PLEBEIAN
    What does he say of Brutus?
THIRD PLEBEIAN              He says, for Brutus' sake
    He finds himself beholden to us all.
⌈FIFTH⌉ PLEBEIAN
    'Twere best he speak no harm of Brutus here!
FIRST PLEBEIAN
    This Caesar was a tyrant.
THIRD PLEBEIAN              Nay, that's certain.        70
    We are blessed that Rome is rid of him.
    ⌈*Enter*⌉ *Antony in the pulpit*
⌈FOURTH⌉ PLEBEIAN
    Peace, let us hear what Antony can say.
ANTONY
    You gentle Romans.
ALL THE PLEBEIANS     Peace, ho! Let us hear him.

ANTONY
    Friends, Romans, countrymen, lend me your ears.
    I come to bury Caesar, not to praise him.            75
    The evil that men do lives after them;
    The good is oft interrèd with their bones.
    So let it be with Caesar. The noble Brutus
    Hath told you Caesar was ambitious.
    If it were so, it was a grievous fault,              80
    And grievously hath Caesar answered it.
    Here, under leave of Brutus and the rest—
    For Brutus is an honourable man,
    So are they all, all honourable men—
    Come I to speak in Caesar's funeral.                 85
    He was my friend, faithful and just to me.
    But Brutus says he was ambitious,
    And Brutus is an honourable man.
    He hath brought many captives home to Rome,
    Whose ransoms did the general coffers fill.          90
    Did this in Caesar seem ambitious?
    When that the poor have cried, Caesar hath wept.
    Ambition should be made of sterner stuff.
    Yet Brutus says he was ambitious,
    And Brutus is an honourable man.                     95
    You all did see that on the Lupercal
    I thrice presented him a kingly crown,
    Which he did thrice refuse. Was this ambition?
    Yet Brutus says he was ambitious,
    And sure he is an honourable man.                    100
    I speak not to disprove what Brutus spoke,
    But here I am to speak what I do know.
    You all did love him once, not without cause.
    What cause withholds you then to mourn for him?
    O judgement, thou art fled to brutish beasts,        105
    And men have lost their reason!
        *He weeps*
                             Bear with me.
    My heart is in the coffin there with Caesar,
    And I must pause till it come back to me.
FIRST PLEBEIAN
    Methinks there is much reason in his sayings.
⌈FOURTH⌉ PLEBEIAN
    If thou consider rightly of the matter,              110
    Caesar has had great wrong.
THIRD PLEBEIAN                Has he not, masters?
    I fear there will a worse come in his place.
⌈FIFTH⌉ PLEBEIAN
    Marked ye his words? He would not take the crown,
    Therefore 'tis certain he was not ambitious.
FIRST PLEBEIAN
    If it be found so, some will dear abide it.          115
⌈FOURTH⌉ PLEBEIAN
    Poor soul, his eyes are red as fire with weeping.
THIRD PLEBEIAN
    There's not a nobler man in Rome than Antony.
⌈FIFTH⌉ PLEBEIAN
    Now mark him; he begins again to speak.
ANTONY
    But yesterday the word of Caesar might
    Have stood against the world. Now lies he there,     120
    And none so poor to do him reverence.
    O masters, if I were disposed to stir
    Your hearts and minds to mutiny and rage,
    I should do Brutus wrong, and Cassius wrong,
    Who, you all know, are honourable men.               125
    I will not do them wrong. I rather choose
    To wrong the dead, to wrong myself and you,

Than I will wrong such honourable men.
But here's a parchment with the seal of Caesar.
I found it in his closet. 'Tis his will.                        130
Let but the commons hear this testament—
Which, pardon me, I do not mean to read—
And they would go and kiss dead Caesar's wounds,
And dip their napkins in his sacred blood,
Yea, beg a hair of him for memory,                             135
And, dying, mention it within their wills,
Bequeathing it as a rich legacy
Unto their issue.
⌜FIFTH⌝ PLEBEIAN
We'll hear the will. Read it, Mark Antony.
ALL THE PLEBEIANS
The will, the will! We will hear Caesar's will.                140
ANTONY
Have patience, gentle friends, I must not read it.
It is not meet you know how Caesar loved you.
You are not wood, you are not stones, but men;
And, being men, hearing the will of Caesar,
It will inflame you, it will make you mad.                     145
'Tis good you know not that you are his heirs,
For if you should, O what would come of it?
⌜FIFTH⌝ PLEBEIAN
Read the will. We'll hear it, Antony.
You shall read us the will, Caesar's will.
ANTONY
Will you be patient? Will you stay a while?                    150
I have o'ershot myself to tell you of it.
I fear I wrong the honourable men
Whose daggers have stabbed Caesar; I do fear it.
⌜FIFTH⌝ PLEBEIAN They were traitors. Honourable men?
ALL THE PLEBEIANS The will, the testament!                     155
⌜FOURTH⌝ PLEBEIAN They were villains, murderers. The
   will, read the will!
ANTONY
You will compel me then to read the will?
Then make a ring about the corpse of Caesar,
And let me show you him that made the will.                    160
Shall I descend? And will you give me leave?
ALL THE PLEBEIANS
Come down.
⌜FOURTH⌝ PLEBEIAN Descend.
THIRD PLEBEIAN              You shall have leave.
   *Antony descends from the pulpit*
⌜FIFTH⌝ PLEBEIAN                        A ring.
Stand round.
FIRST PLEBEIAN
            Stand from the hearse. Stand from the body.
⌜FOURTH⌝ PLEBEIAN
Room for Antony, most noble Antony!
   ⌜*Enter Antony below*⌝
ANTONY
Nay, press not so upon me. Stand farre off.                    165
ALL THE PLEBEIANS Stand back! Room! Bear back!
ANTONY
If you have tears, prepare to shed them now.
You all do know this mantle. I remember
The first time ever Caesar put it on.
'Twas on a summer's evening in his tent,                        170
That day he overcame the Nervii.
Look, in this place ran Cassius' dagger through.
See what a rent the envious Casca made.
Through this the well-belovèd Brutus stabbed;
And as he plucked his cursèd steel away,                        175

Mark how the blood of Caesar followed it,
As rushing out of doors to be resolved
If Brutus so unkindly knocked or no—
For Brutus, as you know, was Caesar's angel.
Judge, O you gods, how dearly Caesar loved him!               180
This was the most unkindest cut of all.
For when the noble Caesar saw him stab,
Ingratitude, more strong than traitors' arms,
Quite vanquished him. Then burst his mighty heart,
And in his mantle muffling up his face,                        185
Even at the base of Pompey's statue,
Which all the while ran blood, great Caesar fell.
O, what a fall was there, my countrymen!
Then I, and you, and all of us fell down,
Whilst bloody treason flourished over us.                      190
O now you weep, and I perceive you feel
The dint of pity. These are gracious drops.
Kind souls, what, weep you when you but behold
Our Caesar's vesture wounded? Look you here.
Here is himself, marred, as you see, with traitors.           195
   *He uncovers Caesar's body*
FIRST PLEBEIAN
O piteous spectacle!
⌜FOURTH⌝ PLEBEIAN      O noble Caesar!
THIRD PLEBEIAN O woeful day!
⌜FIFTH⌝ PLEBEIAN
O traitors, villains!
FIRST PLEBEIAN      O most bloody sight!
⌜FOURTH⌝ PLEBEIAN We will be revenged.
⌜ALL THE PLEBEIANS⌝
Revenge! About! Seek! Burn! Fire! Kill! Slay!                  200
Let not a traitor live!
ANTONY              Stay, countrymen.
FIRST PLEBEIAN Peace there, hear the noble Antony.
⌜FOURTH⌝ PLEBEIAN We'll hear him, we'll follow him, we'll
   die with him!
ANTONY
Good friends, sweet friends, let me not stir you up           205
To such a sudden flood of mutiny.
They that have done this deed are honourable.
What private griefs they have, alas, I know not,
That made them do it. They are wise and honourable,
And will no doubt with reasons answer you.                     210
I come not, friends, to steal away your hearts.
I am no orator as Brutus is,
But, as you know me all, a plain blunt man
That love my friend; and that they know full well
That gave me public leave to speak of him.                     215
For I have neither wit, nor words, nor worth,
Action, nor utterance, nor the power of speech,
To stir men's blood. I only speak right on.
I tell you that which you yourselves do know,
Show you sweet Caesar's wounds, poor poor dumb
   mouths,                                                      220
And bid them speak for me. But were I Brutus,
And Brutus Antony, there were an Antony
Would ruffle up your spirits, and put a tongue
In every wound of Caesar that should move
The stones of Rome to rise and mutiny.                         225
ALL THE PLEBEIANS
We'll mutiny.
FIRST PLEBEIAN We'll burn the house of Brutus.
THIRD PLEBEIAN
Away then! Come, seek the conspirators.
ANTONY
Yet hear me, countrymen, yet hear me speak.

ALL THE PLEBEIANS  -
    Peace, ho! Hear Antony, most noble Antony.
ANTONY
    Why, friends, you go to do you know not what.     230
    Wherein hath Caesar thus deserved your loves?
    Alas, you know not. I must tell you then.
    You have forgot the will I told you of.
ALL THE PLEBEIANS
    Most true. The will. Let's stay and hear the will.
ANTONY
    Here is the will, and under Caesar's seal.        235
    To every Roman citizen he gives—
    To every several man—seventy-five drachmas.
⌈FOURTH⌉ PLEBEIAN
    Most noble Caesar! We'll revenge his death.
THIRD PLEBEIAN
    O royal Caesar!
ANTONY                Hear me with patience.
ALL THE PLEBEIANS                        Peace, ho!
ANTONY
    Moreover he hath left you all his walks,          240
    His private arbours, and new-planted orchards,
    On this side Tiber. He hath left them you,
    And to your heirs for ever—common pleasures
    To walk abroad and recreate yourselves.
    Here was a Caesar. When comes such another?      245
FIRST PLEBEIAN
    Never, never! Come, away, away!
    We'll burn his body in the holy place,
    And with the brands fire the traitors' houses.
    Take up the body.
⌈FOURTH⌉ PLEBEIAN Go, fetch fire!                    250
THIRD PLEBEIAN Pluck down benches!
⌈FIFTH⌉ PLEBEIAN Pluck down forms, windows, anything!
            *Exeunt Plebeians ⌈with Caesar's body⌉*
ANTONY
    Now let it work. Mischief, thou art afoot.
    Take thou what course thou wilt.
            *Enter ⌈Octavius'⌉ Servant*
                                How now, fellow?
SERVANT
    Sir, Octavius is already come to Rome.            255
ANTONY Where is he?
SERVANT
    He and Lepidus are at Caesar's house.
ANTONY
    And thither will I straight to visit him.
    He comes upon a wish. Fortune is merry,
    And in this mood will give us anything.           260
SERVANT
    I heard him say Brutus and Cassius
    Are rid like madmen through the gates of Rome.
ANTONY
    Belike they had some notice of the people,
    How I had moved them. Bring me to Octavius.
                                          *Exeunt*

**3.3**  *Enter Cinna the poet*
CINNA
    I dreamt tonight that I did feast with Caesar,
    And things unlucky charge my fantasy.
    I have no will to wander forth of doors,
    Yet something leads me forth.
            *Enter the Plebeians*
FIRST PLEBEIAN What is your name?                       5

SECOND PLEBEIAN Whither are you going?
THIRD PLEBEIAN Where do you dwell?
FOURTH PLEBEIAN Are you a married man or a bachelor?
SECOND PLEBEIAN Answer every man directly.
FIRST PLEBEIAN Ay, and briefly.                        10
FOURTH PLEBEIAN Ay, and wisely.
THIRD PLEBEIAN Ay, and truly, you were best.
CINNA What is my name? Whither am I going? Where
    do I dwell? Am I a married man or a bachelor? Then
    to answer every man directly and briefly, wisely and
    truly: wisely, I say, I am a bachelor.             16
SECOND PLEBEIAN That's as much as to say they are fools
    that marry. You'll bear me a bang for that, I fear.
    Proceed directly.
CINNA Directly I am going to Caesar's funeral.         20
FIRST PLEBEIAN As a friend or an enemy?
CINNA As a friend.
SECOND PLEBEIAN That matter is answered directly.
FOURTH PLEBEIAN For your dwelling—briefly.
CINNA Briefly, I dwell by the Capitol.                 25
THIRD PLEBEIAN Your name, sir, truly.
CINNA Truly, my name is Cinna.
FIRST PLEBEIAN Tear him to pieces! He's a conspirator.
CINNA I am Cinna the poet, I am Cinna the poet.
FOURTH PLEBEIAN Tear him for his bad verses, tear him
    for his bad verses.                                31
CINNA I am not Cinna the conspirator.
FOURTH PLEBEIAN It is no matter, his name's Cinna. Pluck
    but his name out of his heart, and turn him going.
THIRD PLEBEIAN Tear him, tear him!                     35
        ⌈*They set upon Cinna*⌉
    Come, brands, ho! Firebrands! To Brutus', to Cassius'!
    Burn all! Some to Decius' house, and some to Casca's;
    some to Ligarius'. Away, go!
                    *Exeunt all the Plebeians, with Cinna*

**4.1**  *Enter Antony with papers, Octavius, and Lepidus*
ANTONY
    These many, then, shall die; their names are pricked.
OCTAVIUS (*to Lepidus*)
    Your brother too must die. Consent you, Lepidus?
LEPIDUS
    I do consent.
OCTAVIUS          Prick him down, Antony.
LEPIDUS
    Upon condition Publius shall not live,
    Who is your sister's son, Mark Antony.             5
ANTONY
    He shall not live. Look, with a spot I damn him.
    But Lepidus, go you to Caesar's house;
    Fetch the will hither, and we shall determine
    How to cut off some charge in legacies.
LEPIDUS What, shall I find you here?                   10
OCTAVIUS Or here or at the Capitol.          *Exit Lepidus*
ANTONY
    This is a slight, unmeritable man,
    Meet to be sent on errands. Is it fit,
    The three-fold world divided, he should stand
    One of the three to share it?
OCTAVIUS                        So you thought him,     15
    And took his voice who should be pricked to die
    In our black sentence and proscription.
ANTONY
    Octavius, I have seen more days than you,
    And though we lay these honours on this man
    To ease ourselves of divers sland'rous loads,      20

He shall but bear them as the ass bears gold,
To groan and sweat under the business,
Either led or driven as we point the way;
And having brought our treasure where we will,
Then take we down his load, and turn him off,     25
Like to the empty ass, to shake his ears
And graze in commons.

OCTAVIUS                        You may do your will;
But he's a tried and valiant soldier.

ANTONY
So is my horse, Octavius, and for that
I do appoint him store of provender.              30
It is a creature that I teach to fight,
To wind, to stop, to run directly on,
His corporal motion governed by my spirit;
And in some taste is Lepidus but so.
He must be taught, and trained, and bid go forth—  35
A barren-spirited fellow, one that feeds
On objects, arts, and imitations,
Which, out of use and staled by other men,
Begin his fashion. Do not talk of him
But as a property. And now, Octavius,             40
Listen great things. Brutus and Cassius
Are levying powers. We must straight make head.
Therefore let our alliance be combined,
Our best friends made, our meinies stretched,
And let us presently go sit in council,           45
How covert matters may be best disclosed,
And open perils surest answerèd.

OCTAVIUS
Let us do so, for we are at the stake
And bayed about with many enemies;
And some that smile have in their hearts, I fear,  50
Millions of mischiefs.                        Exeunt

4.2    *Drum. Enter Brutus, Lucius, and the army.*
       ⌐Lucillius,⌐ *Titinius, and Pindarus meet them*
BRUTUS Stand, ho!
⌐SOLDIER⌐ Give the word 'ho', and stand.
BRUTUS
What now, Lucillius: is Cassius near?
LUCILLIUS
He is at hand, and Pindarus is come
To do you salutation from his master.              5
BRUTUS
He greets me well. Your master, Pindarus,
In his own change or by ill officers,
Hath given me some worthy cause to wish
Things done undone. But if he be at hand,
I shall be satisfied.
PINDARUS              I do not doubt            10
But that my noble master will appear
Such as he is, full of regard and honour.
BRUTUS
He is not doubted.—A word, Lucillius.
       *Brutus and Lucillius speak apart*
How he received you let me be resolved.
LUCILLIUS
With courtesy and with respect enough,            15
But not with such familiar instances,
Nor with such free and friendly conference,
As he hath used of old.
BRUTUS                     Thou hast described
A hot friend cooling. Ever note, Lucillius:
When love begins to sicken and decay             20

It useth an enforcèd ceremony.
There are no tricks in plain and simple faith;
But hollow men, like horses hot at hand,
Make gallant show and promise of their mettle;
       *Low march within*
But when they should endure the bloody spur,      25
They fall their crests and, like deceitful jades,
Sink in the trial. Comes his army on?
LUCILLIUS
They mean this night in Sardis to be quartered.
The greater part, the horse in general,
Are come with Cassius.
       *Enter Cassius and his powers*
BRUTUS                      Hark, he is arrived.   30
March gently on to meet him.
       *The armies march*
CASSIUS Stand, ho!
BRUTUS Stand, ho! Speak the word along.
⌐FIRST SOLDIER⌐ Stand!
⌐SECOND SOLDIER⌐ Stand!                           35
⌐THIRD SOLDIER⌐ Stand!
CASSIUS
Most noble brother, you have done me wrong.
BRUTUS
Judge me, you gods: wrong I mine enemies?
And if not so, how should I wrong a brother?
CASSIUS
Brutus, this sober form of yours hides wrongs,    40
And when you do them—
BRUTUS                      Cassius, be content.
Speak your griefs softly. I do know you well.
Before the eyes of both our armies here,
Which should perceive nothing but love from us,
Let us not wrangle. Bid them move away,           45
Then in my tent, Cassius, enlarge your griefs,
And I will give you audience.
CASSIUS                          Pindarus,
Bid our commanders lead their charges off
A little from this ground.
BRUTUS
Lucillius, do you the like; and let no man        50
Come to our tent till we have done our conference.
Let Lucius and Titinius guard our door.
                          *Exeunt the armies
       Brutus and Cassius remain,⌐ ⌐with Titinius, and
       Lucius guarding the door⌐*
CASSIUS
That you have wronged me doth appear in this:
You have condemned and noted Lucius Pella
For taking bribes here of the Sardians,           55
Wherein my letters praying on his side,
Because I knew the man, was slighted off.
BRUTUS
You wronged yourself to write in such a case.
CASSIUS
In such a time as this it is not meet
That every nice offence should bear his comment.  60
BRUTUS
Let me tell you, Cassius, you yourself
Are much condemned to have an itching palm,
To sell and mart your offices for gold
To undeservers.
CASSIUS              I, an itching palm?
You know that you are Brutus that speaks this,    65
Or, by the gods, this speech were else your last.

BRUTUS
  The name of Cassius honours this corruption,
  And chastisement doth therefore hide his head.
CASSIUS Chastisement?
BRUTUS
  Remember March, the ides of March, remember.          70
  Did not great Julius bleed for justice' sake?
  What villain touched his body, that did stab,
  And not for justice? What, shall one of us,
  That struck the foremost man of all this world
  But for supporting robbers, shall we now              75
  Contaminate our fingers with base bribes,
  And sell the mighty space of our large honours
  For so much trash as may be graspèd thus?
  I had rather be a dog and bay the moon
  Than such a Roman.
CASSIUS                    Brutus, bay not me.          80
  I'll not endure it. You forget yourself
  To hedge me in. I am a soldier, I,
  Older in practice, abler than yourself
  To make conditions.
BRUTUS Go to, you are not, Cassius.                     85
CASSIUS I am.
BRUTUS I say you are not.
CASSIUS
  Urge me no more, I shall forget myself.
  Have mind upon your health. Tempt me no farther.
BRUTUS Away, slight man.                                90
CASSIUS Is't possible?
BRUTUS Hear me, for I will speak.
  Must I give way and room to your rash choler?
  Shall I be frighted when a madman stares?
CASSIUS
  O ye gods, ye gods! Must I endure all this?           95
BRUTUS
  All this? Ay, more. Fret till your proud heart break.
  Go show your slaves how choleric you are,
  And make your bondmen tremble. Must I budge?
  Must I observe you? Must I stand and crouch
  Under your testy humour? By the gods,                 100
  You shall digest the venom of your spleen,
  Though it do split you. For from this day forth
  I'll use you for my mirth, yea for my laughter,
  When you are waspish.
CASSIUS                    Is it come to this?
BRUTUS
  You say you are a better soldier.                     105
  Let it appear so, make your vaunting true,
  And it shall please me well. For mine own part,
  I shall be glad to learn of noble men.
CASSIUS
  You wrong me every way, you wrong me, Brutus.
  I said an elder soldier, not a better.                110
  Did I say better?
BRUTUS            If you did, I care not.
CASSIUS
  When Caesar lived he durst not thus have moved me.
BRUTUS
  Peace, peace; you durst not so have tempted him.
CASSIUS I durst not?
BRUTUS No.                                              115
CASSIUS What, durst not tempt him?
BRUTUS For your life you durst not.
CASSIUS
  Do not presume too much upon my love.
  I may do that I shall be sorry for.

BRUTUS
  You have done that you should be sorry for.           120
  There is no terror, Cassius, in your threats,
  For I am armed so strong in honesty
  That they pass by me as the idle wind,
  Which I respect not. I did send to you
  For certain sums of gold, which you denied me;        125
  For I can raise no money by vile means.
  By heaven, I had rather coin my heart
  And drop my blood for drachmas than to wring
  From the hard hands of peasants their vile trash
  By any indirection. I did send                        130
  To you for gold to pay my legions,
  Which you denied me. Was that done like Cassius?
  Should I have answered Caius Cassius so?
  When Marcus Brutus grows so covetous
  To lock such rascal counters from his friends,        135
  Be ready, gods, with all your thunderbolts;
  Dash him to pieces.
CASSIUS                    I denied you not.
BRUTUS
  You did.
CASSIUS     I did not. He was but a fool
  That brought my answer back. Brutus hath rived my
        heart.
  A friend should bear his friend's infirmities,        140
  But Brutus makes mine greater than they are.
BRUTUS
  I do not, till you practise them on me.
CASSIUS
  You love me not.
BRUTUS             I do not like your faults.
CASSIUS
  A friendly eye could never see such faults.
BRUTUS
  A flatterer's would not, though they do appear        145
  As huge as high Olympus.
CASSIUS
  Come, Antony and young Octavius, come,
  Revenge yourselves alone on Cassius;
  For Cassius is aweary of the world,
  Hated by one he loves, braved by his brother,         150
  Checked like a bondman; all his faults observed,
  Set in a notebook, learned and conned by rote,
  To cast into my teeth. O, I could weep
  My spirit from mine eyes! There is my dagger,
  And here my naked breast; within, a heart             155
  Dearer than Pluto's mine, richer than gold.
  If that thou beest a Roman, take it forth.
  I that denied thee gold will give my heart.
  Strike as thou didst at Caesar; for I know
  When thou didst hate him worst, thou loved'st him
        better                                          160
  Than ever thou loved'st Cassius.
BRUTUS                             Sheathe your dagger.
  Be angry when you will; it shall have scope.
  Do what you will; dishonour shall be humour.
  O Cassius, you are yokèd with a lamb
  That carries anger as the flint bears fire,           165
  Who, much enforcèd, shows a hasty spark
  And straight is cold again.
CASSIUS                       Hath Cassius lived
  To be but mirth and laughter to his Brutus
  When grief and blood ill-tempered vexeth him?
BRUTUS
  When I spoke that, I was ill-tempered too.            170

CASSIUS
　Do you confess so much? Give me your hand.
BRUTUS
　And my heart too.
　　　⌈*They embrace*⌉
CASSIUS　　　　　　O Brutus!
BRUTUS　　　　　　　What's the matter?
CASSIUS
　Have not you love enough to bear with me
　When that rash humour which my mother gave me
　Makes me forgetful?
BRUTUS　　　　　　Yes, Cassius, and from henceforth,
　When you are over-earnest with your Brutus,　176
　He'll think your mother chides, and leave you so.
　　　*Enter* ⌈*Lucillius and*⌉ *a Poet*
POET
　Let me go in to see the generals.
　There is some grudge between 'em; 'tis not meet
　They be alone.
LUCILLIUS　　　You shall not come to them.　180
POET
　Nothing but death shall stay me.
CASSIUS　　　　　　How now! What's the matter?
POET
　For shame, you generals, what do you mean?
　Love and be friends, as two such men should be,
　For I have seen more years, I'm sure, than ye.
CASSIUS
　Ha, ha! How vilely doth this cynic rhyme!　185
BRUTUS (*to the Poet*)
　Get you hence, sirrah; saucy fellow, hence!
CASSIUS
　Bear with him, Brutus, 'tis his fashion.
BRUTUS
　I'll know his humour when he knows his time.
　What should the wars do with these jigging fools?
　(*To the Poet*) Companion, hence!
CASSIUS (*to the Poet*)　　　　Away, away, be gone!
　　　　　　　　　　　　　*Exit Poet*
BRUTUS
　Lucillius and Titinius, bid the commanders　191
　Prepare to lodge their companies tonight.
CASSIUS
　And come yourselves, and bring Messala with you
　Immediately to us.　　*Exeunt Lucillius and Titinius*
BRUTUS　　　　　Lucius, a bowl of wine.
　　　　　　　　　　　　　*Exit Lucius*
CASSIUS
　I did not think you could have been so angry.　195
BRUTUS
　O Cassius, I am sick of many griefs.
CASSIUS
　Of your philosophy you make no use,
　If you give place to accidental evils.
BRUTUS
　No man bears sorrow better. Portia is dead.
CASSIUS Ha! Portia?　　　　　　　　200
BRUTUS She is dead.
CASSIUS
　How scaped I killing when I crossed you so?
　O insupportable and touching loss!
　Upon what sickness?
BRUTUS　　　　　Impatience of my absence,
　And grief that young Octavius with Mark Antony　205
　Have made themselves so strong—for with her death

That tidings came. With this, she fell distraught,
And, her attendants absent, swallowed fire.
CASSIUS
　And died so?
BRUTUS　　　　Even so.
CASSIUS　　　　　　O ye immortal gods!
　　　*Enter Lucius, with wine and tapers*
BRUTUS
　Speak no more of her. (*To Lucius*) Give me a bowl of
　　wine.　　　　　　　　　　　　　210
　(*To Cassius*) In this I bury all unkindness, Cassius.
　　　*He drinks*
CASSIUS
　My heart is thirsty for that noble pledge.
　Fill, Lucius, till the wine o'erswell the cup.
　I cannot drink too much of Brutus' love.
　　　*He drinks*　　　　　　⌈*Exit Lucius*⌉
　　　*Enter Titinius and Messala*
BRUTUS
　Come in, Titinius; welcome, good Messala.　215
　Now sit we close about this taper here,
　And call in question our necessities.
CASSIUS (*aside*)
　Portia, art thou gone?
BRUTUS　　　　　　No more, I pray you.
　　　⌈*They sit*⌉
　Messala, I have here receivèd letters
　That young Octavius and Mark Antony　220
　Come down upon us with a mighty power,
　Bending their expedition toward Philippi.
MESSALA
　Myself have letters of the selfsame tenor.
BRUTUS With what addition?
MESSALA
　That by proscription and bills of outlawry　225
　Octavius, Antony, and Lepidus
　Have put to death an hundred senators.
BRUTUS
　Therein our letters do not well agree.
　Mine speak of seventy senators that died
　By their proscriptions, Cicero being one.　230
CASSIUS
　Cicero one?
MESSALA　　　Ay, Cicero is dead,
　And by that order of proscription.
　(*To Brutus*)
　Had you your letters from your wife, my lord?
BRUTUS No, Messala.
MESSALA
　Nor nothing in your letters writ of her?　235
BRUTUS
　Nothing, Messala.
MESSALA　　　　　That methinks is strange.
BRUTUS
　Why ask you? Hear you aught of her in yours?
MESSALA No, my lord.
BRUTUS
　Now as you are a Roman, tell me true.
MESSALA
　Then like a Roman bear the truth I tell;　240
　For certain she is dead, and by strange manner.
BRUTUS
　Why, farewell, Portia. We must die, Messala.
　With meditating that she must die once,
　I have the patience to endure it now.

MESSALA
  Even so great men great losses should endure.    245
CASSIUS
  I have as much of this in art as you,
  But yet my nature could not bear it so.
BRUTUS
  Well, to our work alive. What do you think
  Of marching to Philippi presently?
CASSIUS
  I do not think it good.
BRUTUS               Your reason?
CASSIUS                This it is:    250
  'Tis better that the enemy seek us;
  So shall he waste his means, weary his soldiers,
  Doing himself offence; whilst we, lying still,
  Are full of rest, defence, and nimbleness.
BRUTUS
  Good reasons must of force give place to better.    255
  The people 'twixt Philippi and this ground
  Do stand but in a forced affection,
  For they have grudged us contribution.
  The enemy marching along by them
  By them shall make a fuller number up,    260
  Come on refreshed, new added, and encouraged;
  From which advantage shall we cut him off,
  If at Philippi we do face him there,
  These people at our back.
CASSIUS           Hear me, good brother.
BRUTUS
  Under your pardon. You must note beside    265
  That we have tried the utmost of our friends;
  Our legions are brim-full, our cause is ripe.
  The enemy increaseth every day;
  We at the height are ready to decline.
  There is a tide in the affairs of men    270
  Which, taken at the flood, leads on to fortune;
  Omitted, all the voyage of their life
  Is bound in shallows and in miseries.
  On such a full sea are we now afloat,
  And we must take the current when it serves,    275
  Or lose our ventures.
CASSIUS         Then, with your will, go on.
  We'll along ourselves, and meet them at Philippi.
BRUTUS
  The deep of night is crept upon our talk,
  And nature must obey necessity,
  Which we will niggard with a little rest.    280
  There is no more to say.
CASSIUS         No more. Good night.
  Early tomorrow will we rise and hence.
BRUTUS
  Lucius.
    *Enter Lucius*
      My gown.                 *Exit Lucius*
      Farewell, good Messala.
  Good night, Titinius. Noble, noble, Cassius,
  Good night and good repose.
CASSIUS         O my dear brother,    285
  This was an ill beginning of the night!
  Never come such division 'tween our souls.
  Let it not, Brutus.
    *Enter Lucius with the gown*
BRUTUS         Everything is well.
CASSIUS
  Good night, my lord.
BRUTUS         Good night, good brother.

TITINIUS *and* MESSALA
  Good night, Lord Brutus.
BRUTUS         Farewell, every one.    290
      *Exeunt Cassius, Titinius, and Messala*
  Give me the gown.
    ⌈*He puts on the gown*⌉
          Where is thy instrument?
LUCIUS
  Here in the tent.
BRUTUS        What, thou speak'st drowsily.
  Poor knave, I blame thee not; thou art o'erwatched.
  Call Claudio and some other of my men.
  I'll have them sleep on cushions in my tent.    295
LUCIUS
  Varrus and Claudio!
    *Enter Varrus and Claudio*
VARRUS         Calls my lord?
BRUTUS
  I pray you, sirs, lie in my tent and sleep.
  It may be I shall raise you by and by
  On business to my brother Cassius.
VARRUS
  So please you, we will stand and watch your pleasure.
BRUTUS
  I will not have it so. Lie down, good sirs.    301
  It may be I shall otherwise bethink me.
    *Varrus and Claudio lie down to sleep*
  Look, Lucius, here's the book I sought for so.
  I put it in the pocket of my gown.
LUCIUS
  I was sure your lordship did not give it me.    305
BRUTUS
  Bear with me, good boy, I am much forgetful.
  Canst thou hold up thy heavy eyes a while,
  And touch thy instrument a strain or two?
LUCIUS
  Ay, my lord, an't please you.
BRUTUS         It does, my boy.
  I trouble thee too much, but thou art willing.    310
LUCIUS It is my duty, sir.
BRUTUS
  I should not urge thy duty past thy might.
  I know young bloods look for a time of rest.
LUCIUS I have slept, my lord, already.
BRUTUS
  It was well done, and thou shalt sleep again.    315
  I will not hold thee long. If I do live,
  I will be good to thee.
    *Lucius plays music and sings a song, and so falls*
    *asleep*
  This is a sleepy tune. O murd'rous slumber,
  Lay'st thou thy leaden mace upon my boy
  That plays thee music?—Gentle knave, good night.
  I will not do thee so much wrong to wake thee.    321
  If thou dost nod thou break'st thy instrument;
  I'll take it from thee; and, good boy, good night.
    *He takes away Lucius' instrument, then opens*
    *the book*
  Let me see, let me see, is not the leaf turned down
  Where I left reading? Here it is, I think.    325
    *Enter the Ghost of Caesar*
  How ill this taper burns! Ha! Who comes here?
  I think it is the weakness of mine eyes
  That shapes this monstrous apparition.
  It comes upon me. Art thou any thing?
  Art thou some god, some angel, or some devil,    330

That mak'st my blood cold and my hair to stare?
Speak to me what thou art.
GHOST  Thy evil spirit, Brutus.
BRUTUS  Why com'st thou?
GHOST
To tell thee thou shalt see me at Philippi.          335
BRUTUS
Well; then I shall see thee again?
GHOST                              Ay, at Philippi.
BRUTUS
Why, I will see thee at Philippi then.        *Exit Ghost*
Now I have taken heart, thou vanishest.
Ill spirit, I would hold more talk with thee.—
Boy, Lucius, Varrus, Claudio, sirs, awake!          340
Claudio!
LUCIUS      The strings, my lord, are false.
BRUTUS
He thinks he still is at his instrument.—
Lucius, awake!
LUCIUS  My lord.
BRUTUS
Didst thou dream, Lucius, that thou so cried'st out?
LUCIUS
My lord, I do not know that I did cry.          346
BRUTUS
Yes, that thou didst. Didst thou see anything?
LUCIUS  Nothing, my lord.
BRUTUS
Sleep again, Lucius.—Sirrah Claudio!
(*To Varrus*)                        Fellow,
Thou, awake!                             350
VARRUS  My lord.
CLAUDIO  My lord.
BRUTUS
Why did you so cry out, sirs, in your sleep?
BOTH
Did we, my lord?
BRUTUS                  Ay. Saw you anything?
VARRUS
No, my lord, I saw nothing.
CLAUDIO                      Nor I, my lord.          355
BRUTUS
Go and commend me to my brother Cassius.
Bid him set on his powers betimes before,
And we will follow.
BOTH              It shall be done, my lord.
                *Exeunt ⌐Varrus and Claudio at one door, Brutus*
                          *and Lucius at another door⌐*

**5.1**   *Enter Octavius, Antony, and their army*
OCTAVIUS
Now, Antony, our hopes are answerèd.
You said the enemy would not come down,
But keep the hills and upper regions.
It proves not so; their battles are at hand.
They mean to warn us at Philippi here,          5
Answering before we do demand of them.
ANTONY
Tut, I am in their bosoms, and I know
Wherefore they do it. They could be content
To visit other places; and come down
With fearful bravery, thinking by this face          10
To fasten in our thoughts that they have courage;
But 'tis not so.
        *Enter a Messenger*
MESSENGER          Prepare you, generals.
The enemy comes on in gallant show.

Their bloody sign of battle is hung out,
And something to be done immediately.          15
ANTONY
Octavius, lead your battle softly on
Upon the left hand of the even field.
OCTAVIUS
Upon the right hand, I; keep thou the left.
ANTONY
Why do you cross me in this exigent?
OCTAVIUS
I do not cross you, but I will do so.          20
        ⌐*Drum. Antony and Octavius march with their*
        *army.*⌐
        *Drum within. Enter, marching, Brutus, Cassius,*
        *and their army, amongst them Titinius, Lucillius,*
        *and Messala.*
        *Octavius' and Antony's army makes a stand*
BRUTUS  They stand, and would have parley.
CASSIUS
Stand fast, Titinius. We must out and talk.
        *Brutus' and Cassius' army makes a stand*
OCTAVIUS
Mark Antony, shall we give sign of battle?
ANTONY
No, Caesar, we will answer on their charge.
Make forth, the generals would have some words.          25
OCTAVIUS (*to his army*)
Stir not until the signal.
        *Antony and Octavius meet Brutus and Cassius*
BRUTUS
Words before blows: is it so, countrymen?
OCTAVIUS
Not that we love words better, as you do.
BRUTUS
Good words are better than bad strokes, Octavius.
ANTONY
In your bad strokes, Brutus, you give good words.          30
Witness the hole you made in Caesar's heart,
Crying 'Long live, hail Caesar'.
CASSIUS                          Antony,
The posture of your blows are yet unknown;
But for your words, they rob the Hybla bees,
And leave them honeyless.          35
ANTONY  Not stingless too.
BRUTUS  O yes, and soundless too,
For you have stolen their buzzing, Antony,
And very wisely threat before you sting.
ANTONY
Villains, you did not so when your vile daggers          40
Hacked one another in the sides of Caesar.
You showed your teeth like apes, and fawned like
        hounds,
And bowed like bondmen, kissing Caesar's feet,
Whilst damnèd Casca, like a cur, behind,
Struck Caesar on the neck. O you flatterers!          45
CASSIUS
Flatterers? Now, Brutus, thank yourself.
This tongue had not offended so today
If Cassius might have ruled.
OCTAVIUS
Come, come, the cause. If arguing make us sweat,
The proof of it will turn to redder drops.          50
        *He draws*
Look, I draw a sword against conspirators.
When think you that the sword goes up again?
Never till Caesar's three and thirty wounds

Be well avengèd, or till another Caesar
Have added slaughter to the swords of traitors.          55
BRUTUS
Caesar, thou canst not die by traitors' hands,
Unless thou bring'st them with thee.
OCTAVIUS                              So I hope.
I was not born to die on Brutus' sword.
BRUTUS
O, if thou wert the noblest of thy strain,
Young man, thou couldst not die more honourable.
CASSIUS
A peevish schoolboy, worthless of such honour,          61
Joined with a masquer and a reveller!
ANTONY
Old Cassius still.
OCTAVIUS          Come, Antony, away.
Defiance, traitors, hurl we in your teeth.
If you dare fight today, come to the field.          65
If not, when you have stomachs.
              *Exeunt Octavius, Antony, and their army*
CASSIUS
Why, now blow wind, swell billow, and swim bark.
The storm is up, and all is on the hazard.
BRUTUS
Ho, Lucillius! Hark, a word with you.
LUCILLIUS                              My lord.
      *He stands forth, and speaks with Brutus*
CASSIUS
Messala.
MESSALA (*standing forth*)  What says my general?
CASSIUS                                        Messala,
This is my birthday; as this very day          71
Was Cassius born. Give me thy hand, Messala.
Be thou my witness that, against my will,
As Pompey was, am I compelled to set
Upon one battle all our liberties.          75
You know that I held Epicurus strong,
And his opinion. Now I change my mind,
And partly credit things that do presage.
Coming from Sardis, on our former ensigns
Two mighty eagles fell, and there they perched,          80
Gorging and feeding from our soldiers' hands,
Who to Philippi here consorted us.
This morning are they fled away and gone,
And in their steads do ravens, crows, and kites
Fly o'er our heads and downward look on us,          85
As we were sickly prey. Their shadows seem
A canopy most fatal, under which
Our army lies ready to give the ghost.
MESSALA
Believe not so.
CASSIUS          I but believe it partly,
For I am fresh of spirit, and resolved          90
To meet all perils very constantly.
BRUTUS
Even so, Lucillius.
CASSIUS (*joining Brutus*)  Now, most noble Brutus,
The gods today stand friendly, that we may,
Lovers in peace, lead on our days to age.
But since the affairs of men rest still incertain,          95
Let's reason with the worst that may befall.
If we do lose this battle, then is this
The very last time we shall speak together.
What are you then determinèd to do?
BRUTUS
Even by the rule of that philosophy          100
By which I did blame Cato for the death

Which he did give himself—I know not how,
But I do find it cowardly and vile
For fear of what might fall so to prevent
The time of life—arming myself with patience          105
To stay the providence of some high powers
That govern us below.
CASSIUS                    Then if we lose this battle,
You are contented to be led in triumph
Thorough the streets of Rome?
BRUTUS No, Cassius, no.          110
Think not, thou noble Roman,
That ever Brutus will go bound to Rome.
He bears too great a mind. But this same day
Must end that work the ides of March begun;
And whether we shall meet again I know not.          115
Therefore our everlasting farewell take.
For ever and for ever farewell, Cassius.
If we do meet again, why, we shall smile.
If not, why then, this parting was well made.
CASSIUS
For ever and for ever farewell, Brutus.          120
If we do meet again, we'll smile indeed.
If not, 'tis true this parting was well made.
BRUTUS
Why then, lead on. O that a man might know
The end of this day's business ere it come!
But it sufficeth that the day will end,          125
And then the end is known.—Come, ho, away!
                                        *Exeunt*

**5.2**    *Alarum. Enter Brutus and Messala*
BRUTUS
Ride, ride, Messala, ride, and give these bills
Unto the legions on the other side.
      *Loud alarum*
Let them set on at once, for I perceive
But cold demeanour in Octavio's wing,
And sudden push gives them the overthrow.          5
Ride, ride, Messala; let them all come down.
                              *Exeunt ⌐severally⌐*

**5.3**    *Alarums. Enter Cassius ⌐with an ensign⌐, and
          Titinius*
CASSIUS
O look, Titinius, look: the villains fly.
Myself have to mine own turned enemy:
This ensign here of mine was turning back;
I slew the coward, and did take it from him.
TITINIUS
O Cassius, Brutus gave the word too early,          5
Who, having some advantage on Octavius,
Took it too eagerly. His soldiers fell to spoil,
Whilst we by Antony are all enclosed.
      *Enter Pindarus*
PINDARUS
Fly further off, my lord, fly further off!
Mark Antony is in your tents, my lord;          10
Fly therefore, noble Cassius, fly farre off.
CASSIUS
This hill is far enough. Look, look, Titinius,
Are those my tents where I perceive the fire?
TITINIUS
They are, my lord.
CASSIUS                    Titinius, if thou lovest me,
Mount thou my horse, and hide thy spurs in him          15
Till he have brought thee up to yonder troops

And here again, that I may rest assured
Whether yon troops are friend or enemy.
TITINIUS
I will be here again even with a thought.          *Exit*
CASSIUS
Go, Pindarus, get higher on that hill.              20
My sight was ever thick. Regard, Titinius,
And tell me what thou not'st about the field.
                                    *Exit Pindarus*
This day I breathèd first. Time is come round,
And where I did begin, there shall I end.
My life is run his compass.
        *Enter Pindarus above*
                        Sirrah, what news?            25
PINDARUS  O my lord!
CASSIUS  What news?
PINDARUS
Titinius is enclosèd round about
With horsemen, that make to him on the spur.
Yet he spurs on. Now they are almost on him.       30
Now Titinius. Now some light. O, he lights too.
He's ta'en.
        *Shout within*
And hark, they shout for joy.
CASSIUS                 Come down; behold no more.
                                    *Exit Pindarus*
O coward that I am, to live so long
To see my best friend ta'en before my face!       35
        *Enter Pindarus below*
Come hither, sirrah. In Parthia did I take thee prisoner,
And then I swore thee, saving of thy life,
That whatsoever I did bid thee do
Thou shouldst attempt it. Come now, keep thine oath.
Now be a freeman, and, with this good sword        40
That ran through Caesar's bowels, search this bosom.
Stand not to answer. Here, take thou the hilts,
        *Pindarus takes the sword*
And when my face is covered, as 'tis now,
Guide thou the sword.
        *Pindarus stabs him*
                        Caesar, thou art revenged,
Even with the sword that killed thee.       *He dies*
PINDARUS
So, I am free, yet would not so have been          46
Durst I have done my will. O Cassius!
Far from this country Pindarus shall run,
Where never Roman shall take note of him.     *Exit*
        *Enter Titinius, wearing a wreath of victory, and
        Messala*
MESSALA
It is but change, Titinius, for Octavius            50
Is overthrown by noble Brutus' power,
As Cassius' legions are by Antony.
TITINIUS
These tidings will well comfort Cassius.
MESSALA
Where did you leave him?
TITINIUS                 All disconsolate,
With Pindarus his bondman, on this hill.           55
MESSALA
Is not that he that lies upon the ground?
TITINIUS
He lies not like the living.—O my heart!
MESSALA
Is not that he?
TITINIUS         No, this was he, Messala;
But Cassius is no more. O setting sun,

As in thy red rays thou dost sink tonight,         60
So in his red blood Cassius' day is set.
The sun of Rome is set. Our day is gone.
Clouds, dews, and dangers come. Our deeds are done.
Mistrust of my success hath done this deed.
MESSALA
Mistrust of good success hath done this deed.      65
O hateful Error, Melancholy's child,
Why dost thou show to the apt thoughts of men
The things that are not? O Error, soon conceived,
Thou never com'st unto a happy birth,
But kill'st the mother that engendered thee.       70
TITINIUS
What, Pindarus! Where art thou, Pindarus?
MESSALA
Seek him, Titinius, whilst I go to meet
The noble Brutus, thrusting this report
Into his ears. I may say 'thrusting' it,
For piercing steel and darts envenomèd            75
Shall be as welcome to the ears of Brutus
As tidings of this sight.
TITINIUS                 Hie you, Messala,
And I will seek for Pindarus the while.     *Exit Messala*
Why didst thou send me forth, brave Cassius?
Did I not meet thy friends, and did not they       80
Put on my brows this wreath of victory,
And bid me give it thee? Didst thou not hear their
        shouts?
Alas, thou hast misconstrued everything.
But hold thee, take this garland on thy brow.
Thy Brutus bid me give it thee, and I              85
Will do his bidding. Brutus, come apace,
And see how I regarded Caius Cassius.
By your leave, gods, this is a Roman's part:
Come Cassius' sword, and find Titinius' heart.
        *He stabs himself, and dies*
        *Alarum. Enter Brutus, Messala, young Cato,
        Strato, Volumnius, Lucillius,* ⌈*Labio, and Flavio*⌉
BRUTUS
Where, where, Messala, doth his body lie?          90
MESSALA
Lo yonder, and Titinius mourning it.
BRUTUS
Titinius' face is upward.
CATO                     He is slain.
BRUTUS
O Julius Caesar, thou art mighty yet.
Thy spirit walks abroad, and turns our swords
In our own proper entrails.
        *Low Alarums*
CATO                     Brave Titinius,             95
Look whe'er he have not crowned dead Cassius.
BRUTUS
Are yet two Romans living such as these?
The last of all the Romans, fare thee well.
It is impossible that ever Rome
Should breed thy fellow. Friends, I owe more tears  100
To this dead man than you shall see me pay.—
I shall find time, Cassius, I shall find time.
Come, therefore, and to Thasos send his body.
His funerals shall not be in our camp,
Lest it discomfort us. Lucillius, come;            105
And come, young Cato. Let us to the field.
Labio and Flavio, set our battles on.
'Tis three o'clock, and, Romans, yet ere night
We shall try fortune in a second fight.
                        *Exeunt* ⌈*with the bodies*⌉

**5.4** *Alarum. Enter Brutus, Messala, young Cato,*
*Lucillius, and Flavius*

BRUTUS
Yet, countrymen, O yet hold up your heads.
*⌈Exit with Messala and Flavius⌉*

CATO
What bastard doth not? Who will go with me?
I will proclaim my name about the field.
I am the son of Marcus Cato, ho!
A foe to tyrants, and my country's friend.            5
I am the son of Marcus Cato, ho!
*Enter Soldiers, and fight*

⌈LUCILLIUS⌉
And I am Brutus, Marcus Brutus, I,
Brutus, my country's friend. Know me for Brutus.
*Soldiers kill Cato*
O young and noble Cato, art thou down?
Why, now thou diest as bravely as Titinius,          10
And mayst be honoured, being Cato's son.

⌈FIRST⌉ SOLDIER
Yield, or thou diest.

LUCILLIUS                    Only I yield to die.
There is so much, that thou wilt kill me straight:
Kill Brutus, and be honoured in his death.

⌈FIRST⌉ SOLDIER
We must not.—A noble prisoner.                        15

SECOND SOLDIER
Room, ho! Tell Antony Brutus is ta'en.
*Enter Antony*

FIRST SOLDIER
I'll tell the news. Here comes the general.—
(*To Antony*) Brutus is ta'en, Brutus is ta'en, my lord.

ANTONY Where is he?

LUCILLIUS
Safe, Antony, Brutus is safe enough.                  20
I dare assure thee that no enemy
Shall ever take alive the noble Brutus.
The gods defend him from so great a shame.
When you do find him, or alive or dead,
He will be found like Brutus, like himself.           25

ANTONY (*to First Soldier*)
This is not Brutus, friend, but, I assure you,
A prize no less in worth. Keep this man safe.
Give him all kindness. I had rather have
Such men my friends than enemies.
⌈*To another Soldier*⌉                    Go on,
And see whe'er Brutus be alive or dead,               30
And bring us word unto Octavius' tent
How everything is chanced.
*Exeunt ⌈the Soldier at one door, Antony,*
*Lucillius and other Soldiers, some bearing*
*Cato's body, at another door⌉*

**5.5** *Enter Brutus, Dardanius, Clitus, Strato, and*
*Volumnius*

BRUTUS
Come, poor remains of friends, rest on this rock.
⌈*He sits. Strato rests and falls asleep*⌉

CLITUS
Statillius showed the torchlight, but, my lord,
He came not back. He is or ta'en or slain.

BRUTUS
Sit thee down, Clitus. Slaying is the word:
It is a deed in fashion. Hark thee, Clitus.           5
*He whispers*

CLITUS
What I, my lord? No, not for all the world.

BRUTUS
Peace, then, no words.

CLITUS                    I'll rather kill myself.
*He stands apart*

BRUTUS
Hark thee, Dardanius.
*He whispers*

DARDANIUS                    Shall I do such a deed?
*He joins Clitus*

CLITUS O Dardanius!

DARDANIUS O Clitus!                                   10

CLITUS
What ill request did Brutus make to thee?

DARDANIUS
To kill him, Clitus. Look, he meditates.

CLITUS
Now is that noble vessel full of grief,
That it runs over even at his eyes.

BRUTUS
Come hither, good Volumnius. List a word.            15

VOLUMNIUS
What says my lord?

BRUTUS                    Why this, Volumnius.
The ghost of Caesar hath appeared to me
Two several times by night—at Sardis once,
And this last night, here in Philippi fields.
I know my hour is come.

VOLUMNIUS                    Not so, my lord.         20

BRUTUS
Nay, I am sure it is, Volumnius.
Thou seest the world, Volumnius, how it goes.
Our enemies have beat us to the pit,
*Low alarums*
It is more worthy to leap in ourselves
Than tarry till they push us. Good Volumnius,         25
Thou know'st that we two went to school together.
Even for that, our love of old, I prithee,
Hold thou my sword hilts whilst I run on it.

VOLUMNIUS
That's not an office for a friend, my lord.
*Alarum still*

CLITUS
Fly, fly, my lord! There is no tarrying here.        30

BRUTUS
Farewell to you, and you, and you, Volumnius.—
Strato, thou hast been all this while asleep.
⌈*Strato wakes*⌉
Farewell to thee too, Strato. Countrymen,
My heart doth joy that yet in all my life
I found no man but he was true to me.                35
I shall have glory by this losing day,
More than Octavius and Mark Antony
By this vile conquest shall attain unto.
So fare you well at once, for Brutus' tongue
Hath almost ended his life's history.                40
Night hangs upon mine eyes. My bones would rest,
That have but laboured to attain this hour.
*Alarum. Cry within: 'Fly, fly, fly!'*

CLITUS
Fly, my lord, fly!

BRUTUS                    Hence; I will follow.
*Exeunt Clitus, Dardanius, and Volumnius*
I prithee, Strato, stay thou by thy lord.
Thou art a fellow of a good respect.                 45
Thy life hath had some smatch of honour in it.
Hold then my sword, and turn away thy face
While I do run upon it. Wilt thou, Strato?

STRATO
  Give me your hand first. Fare you well, my lord.
BRUTUS
  Farewell, good Strato.
        *Strato holds the sword, while Brutus runs on it*
                    Caesar, now be still.                    50
  I killed not thee with half so good a will.        *He dies*
        *Alarum. Retreat. Enter Antony, Octavius, Messala,*
        *Lucillius, and the army*
OCTAVIUS What man is that?
MESSALA
  My master's man. Strato, where is thy master?
STRATO
  Free from the bondage you are in, Messala.
  The conquerors can but make a fire of him,        55
  For Brutus only overcame himself,
  And no man else hath honour by his death.
LUCILLIUS
  So Brutus should be found. I thank thee, Brutus,
  That thou hast proved Lucillius' saying true.
OCTAVIUS
  All that served Brutus, I will entertain them.        60
  (*To Strato*)
  Fellow, wilt thou bestow thy time with me?
STRATO
  Ay, if Messala will prefer me to you.

OCTAVIUS
  Do so, good Messala.
MESSALA                    How died my master, Strato?
STRATO
  I held the sword, and he did run on it.
MESSALA
  Octavius, then take him to follow thee,        65
  That did the latest service to my master.
ANTONY
  This was the noblest Roman of them all.
  All the conspirators save only he
  Did that they did in envy of great Caesar.
  He only in a general honest thought        70
  And common good to all made one of them.
  His life was gentle, and the elements
  So mixed in him that nature might stand up
  And say to all the world 'This was a man'.
OCTAVIUS
  According to his virtue let us use him,        75
  With all respect and rites of burial.
  Within my tent his bones tonight shall lie,
  Most like a soldier, ordered honourably.
  So call the field to rest, and let's away
  To part the glories of this happy day.        80
                    *Exeunt ⌈with Brutus' body⌉*

# AS YOU LIKE IT

*As You Like It* is first heard of in the Stationers' Register on 4 August 1600, and was probably written not long before. In spite of its early entry for publication, it was not printed until 1623. This play, with its contrasts between court and country, its bucolic as well as its aristocratic characters, its inset songs and poems, its predominantly woodland setting, its conscious artifice and its romantic ending, is the one in which Shakespeare makes most use of the conventions of pastoral literature, though he does not wholly endorse them.

The story of the love between a high-born maiden—Rosalind—oppressed by the uncle—Duke Frederick—who has usurped his elder brother's dukedom, and Orlando, the third and youngest son of Duke Frederick's old enemy Sir Rowland de Bois, himself oppressed by his tyrannical eldest brother Oliver, derives from Thomas Lodge's *Rosalynde*, a prose romance interspersed with verses, which first appeared in 1590 and was several times reprinted. There are many indications that Shakespeare thought of the action as taking place in the Ardenne area of France, as in *Rosalynde*, even though there was also a Forest of Arden in Warwickshire. Like Lodge, Shakespeare counterpoints the developing love between Rosalind—who for much of the action is disguised as a boy, Ganymede—with the idealized pastoral romance of Silvius and Phoebe; he adds the down-to-earth, un-romantic affair between the jester Touchstone and Audrey. Once Rosalind and her cousin Celia (also disguised) reach the forest, plot is virtually suspended in favour of a series of scintillating conversations making much use of prose. The sudden flowering of love between Celia and Orlando's brother Oliver, newly converted to virtue, is based on *Rosalynde*, but Shakespeare alters the climax of the story, bringing Hymen, the god of marriage, on stage to resolve all complications. As well as Touchstone, Shakespeare added the melancholy courtier Jaques, both of whom act as commentators, though from very different standpoints.

The first performances of Shakespeare's text after his own time were given in 1740. It rapidly established itself in the theatrical repertoire, and has also been appreciated for its literary qualities. It has usually been played in picturesque settings, often since the late nineteenth century in the open air. Rosalind (written originally, of course, for a boy actor) is the dominant character, but other roles, especially Jaques, Touchstone, Audrey, Corin, and—in his single scene—William, have proved particularly effective when played by performers with a strong sense of their latent individuality.

# THE PERSONS OF THE PLAY

DUKE SENIOR, living in banishment

ROSALIND, his daughter, later disguised as Ganymede

AMIENS  
JAQUES } Lords attending on him

TWO PAGES

DUKE FREDERICK

CELIA, his daughter, later disguised as Aliena

LE BEAU, a courtier attending on him

CHARLES, Duke Frederick's wrestler

TOUCHSTONE, a jester

OLIVER, eldest son of Sir Rowland de Bois

JAQUES  
ORLANDO } his younger brothers

ADAM, a former servant of Sir Rowland

DENIS, Oliver's servant

SIR OLIVER MARTEXT, a country clergyman

CORIN, an old shepherd

SILVIUS, a young shepherd, in love with Phoebe

PHOEBE, a shepherdess

WILLIAM, a countryman, in love with Audrey

AUDREY, a goatherd, betrothed to Touchstone

HYMEN, god of marriage

Lords, pages, and other attendants

# As You Like It

**1.1** *Enter Orlando and Adam*

ORLANDO As I remember, Adam, it was upon this fashion bequeathed me by will but poor a thousand crowns, and, as thou sayst, charged my brother on his blessing to breed me well—and there begins my sadness. My brother Jaques he keeps at school, and report speaks goldenly of his profit. For my part, he keeps me rustically at home—or, to speak more properly, stays me here at home unkept; for call you that keeping for a gentleman of my birth, that differs not from the stalling of an ox? His horses are bred better, for besides that they are fair with their feeding, they are taught their manège, and to that end riders dearly hired. But I, his brother, gain nothing under him but growth, for the which his animals on his dunghills are as much bound to him as I. Besides this nothing that he so plentifully gives me, the something that nature gave me his countenance seems to take from me. He lets me feed with his hinds, bars me the place of a brother, and as much as in him lies, mines my gentility with my education. This is it, Adam, that grieves me; and the spirit of my father, which I think is within me, begins to mutiny against this servitude. I will no longer endure it, though yet I know no wise remedy how to avoid it.

*Enter Oliver*

ADAM Yonder comes my master, your brother. 24

ORLANDO Go apart, Adam, and thou shalt hear how he will shake me up.

*Adam stands aside*

OLIVER Now, sir, what make you here?

ORLANDO Nothing. I am not taught to make anything.

OLIVER What mar you then, sir? 29

ORLANDO Marry, sir, I am helping you to mar that which God made, a poor unworthy brother of yours, with idleness.

OLIVER Marry, sir, be better employed, and be nought awhile. 34

ORLANDO Shall I keep your hogs, and eat husks with them? What prodigal portion have I spent, that I should come to such penury?

OLIVER Know you where you are, sir?

ORLANDO O sir, very well; here in your orchard.

OLIVER Know you before whom, sir? 40

ORLANDO Ay, better than him I am before knows me. I know you are my eldest brother, and in the gentle condition of blood you should so know me. The courtesy of nations allows you my better, in that you are the first-born; but the same tradition takes not away my blood, were there twenty brothers betwixt us. I have as much of my father in me as you, albeit I confess your coming before me is nearer to his reverence.

OLIVER *(assailing him)* What, boy! 49

ORLANDO *(seizing him by the throat)* Come, come, elder brother, you are too young in this.

OLIVER Wilt thou lay hands on me, villain?

ORLANDO I am no villein. I am the youngest son of Sir Rowland de Bois. He was my father, and he is thrice a villain that says such a father begot villeins. Wert thou not my brother, I would not take this hand from thy throat till this other had pulled out thy tongue for saying so. Thou hast railed on thyself. 58

ADAM *(coming forward)* Sweet masters, be patient. For your father's remembrance, be at accord. 60

OLIVER *(to Orlando)* Let me go, I say.

ORLANDO I will not till I please. You shall hear me. My father charged you in his will to give me good education. You have trained me like a peasant, obscuring and hiding from me all gentleman-like qualities. The spirit of my father grows strong in me, and I will no longer endure it. Therefore allow me such exercises as may become a gentleman, or give me the poor allottery my father left me by testament. With that I will go buy my fortunes. 70

OLIVER And what wilt thou do—beg when that is spent? Well, sir, get you in. I will not long be troubled with you. You shall have some part of your will. I pray you, leave me.

ORLANDO I will no further offend you than becomes me for my good. 76

OLIVER *(to Adam)* Get you with him, you old dog.

ADAM Is 'old dog' my reward? Most true, I have lost my teeth in your service. God be with my old master, he would not have spoke such a word. 80

*Exeunt Orlando and Adam*

OLIVER Is it even so? Begin you to grow upon me? I will physic your rankness, and yet give no thousand crowns neither. Holla, Denis!

*Enter Denis*

DENIS Calls your worship?

OLIVER Was not Charles, the Duke's wrestler, here to speak with me? 86

DENIS So please you, he is here at the door, and importunes access to you.

OLIVER Call him in. *Exit Denis*

'Twill be a good way. And tomorrow the wrestling is.

*Enter Charles*

CHARLES Good morrow to your worship. 91

OLIVER Good Monsieur Charles—what's the new news at the new court?

CHARLES There's no news at the court, sir, but the old news: that is, the old Duke is banished by his younger brother, the new Duke, and three or four loving lords have put themselves into voluntary exile with him, whose lands and revenues enrich the new Duke; therefore he gives them good leave to wander. 99

OLIVER Can you tell if Rosalind, the Duke's daughter, be banished with her father?

CHARLES O no; for the Duke's daughter her cousin so loves her, being ever from their cradles bred together, that she would have followed her exile, or have died to stay behind her. She is at the court, and no less beloved of her uncle than his own daughter; and never two ladies loved as they do. 107

OLIVER Where will the old Duke live?

CHARLES They say he is already in the forest of Ardenne, and a many merry men with him; and there they live like the old Robin Hood of England. They say many young gentlemen flock to him every day, and fleet the time carelessly, as they did in the golden world.

OLIVER What, you wrestle tomorrow before the new Duke? 115

CHARLES Marry do I, sir, and I came to acquaint you with a matter. I am given, sir, secretly to understand that your younger brother, Orlando, hath a disposition to come in disguised against me to try a fall. Tomorrow, sir, I wrestle for my credit, and he that escapes me without some broken limb, shall acquit him well. Your brother is but young and tender, and for your love I would be loath to foil him, as I must for my own honour if he come in. Therefore out of my love to you I came hither to acquaint you withal, that either you might stay him from his intendment, or brook such disgrace well as he shall run into, in that it is a thing of his own search, and altogether against my will. 128

OLIVER Charles, I thank thee for thy love to me, which thou shalt find I will most kindly requite. I had myself notice of my brother's purpose herein, and have by underhand means laboured to dissuade him from it; but he is resolute. I'll tell thee, Charles, it is the stubbornest young fellow of France, full of ambition, an envious emulator of every man's good parts, a secret and villainous contriver against me his natural brother. Therefore use thy discretion. I had as lief thou didst break his neck as his finger. And thou wert best look to't; for if thou dost him any slight disgrace, or if he do not mightily grace himself on thee, he will practise against thee by poison, entrap thee by some treacherous device, and never leave thee till he hath ta'en thy life by some indirect means or other. For I assure thee—and almost with tears I speak it—there is not one so young and so villainous this day living. I speak but brotherly of him, but should I anatomize him to thee as he is, I must blush and weep, and thou must look pale and wonder. 148

CHARLES I am heartily glad I came hither to you. If he come tomorrow I'll give him his payment. If ever he go alone again, I'll never wrestle for prize more. And so God keep your worship.

OLIVER Farewell, good Charles.    *Exit Charles*
Now will I stir this gamester. I hope I shall see an end of him, for my soul—yet I know not why—hates nothing more than he. Yet he's gentle; never schooled, and yet learned; full of noble device; of all sorts enchantingly beloved; and, indeed, so much in the heart of the world, and especially of my own people, who best know him, that I am altogether misprized. But it shall not be so long. This wrestler shall clear all. Nothing remains but that I kindle the boy thither, which now I'll go about.    *Exit*

**1.2**    *Enter Rosalind and Celia*

CELIA I pray thee Rosalind, sweet my coz, be merry.

ROSALIND Dear Celia, I show more mirth than I am mistress of; and would you yet I were merrier? Unless you could teach me to forget a banished father you must not learn me how to remember any extraordinary pleasure. 6

CELIA Herein I see thou lovest me not with the full weight that I love thee. If my uncle, thy banished father, had banished thy uncle, the Duke my father, so thou hadst been still with me I could have taught my love to take thy father for mine. So wouldst thou, if the truth of thy love to me were so righteously tempered as mine is to thee.

ROSALIND Well, I will forget the condition of my estate to rejoice in yours. 15

CELIA You know my father hath no child but I, nor none is like to have. And truly, when he dies thou shalt be his heir; for what he hath taken away from thy father perforce, I will render thee again in affection. By mine honour I will, and when I break that oath, let me turn monster. Therefore, my sweet Rose, my dear Rose, be merry. 22

ROSALIND From henceforth I will, coz, and devise sports. Let me see, what think you of falling in love?

CELIA Marry, I prithee do, to make sport withal; but love no man in good earnest, nor no further in sport neither than with safety of a pure blush thou mayst in honour come off again.

ROSALIND What shall be our sport, then? 29

CELIA Let us sit and mock the good housewife Fortune from her wheel, that her gifts may henceforth be bestowed equally.

ROSALIND I would we could do so, for her benefits are mightily misplaced; and the bountiful blind woman doth most mistake in her gifts to women. 35

CELIA 'Tis true; for those that she makes fair she scarce makes honest, and those that she makes honest she makes very ill-favouredly.

ROSALIND Nay, now thou goest from Fortune's office to Nature's. Fortune reigns in gifts of the world, not in the lineaments of nature. 41

*Enter Touchstone the clown*

CELIA No. When Nature hath made a fair creature, may she not by Fortune fall into the fire? Though Nature hath given us wit to flout at Fortune, hath not Fortune sent in this fool to cut off the argument? 45

ROSALIND Indeed, there is Fortune too hard for Nature, when Fortune makes Nature's natural the cutter-off of Nature's wit.

CELIA Peradventure this is not Fortune's work, neither, but Nature's, who perceiveth our natural wits too dull to reason of such goddesses, and hath sent this natural for our whetstone; for always the dullness of the fool is the whetstone of the wits. How now, wit: whither wander you? 54

TOUCHSTONE Mistress, you must come away to your father.

CELIA Were you made the messenger?

TOUCHSTONE No, by mine honour, but I was bid to come for you.

ROSALIND Where learned you that oath, fool? 59

TOUCHSTONE Of a certain knight that swore 'by his honour' they were good pancakes, and swore 'by his honour' the mustard was naught. Now I'll stand to it the pancakes were naught and the mustard was good, and yet was not the knight forsworn. 64

CELIA How prove you that in the great heap of your knowledge?

ROSALIND Ay, marry, now unmuzzle your wisdom.

TOUCHSTONE Stand you both forth now. Stroke your chins, and swear by your beards that I am a knave.

CELIA By our beards—if we had them—thou art. 70

TOUCHSTONE By my knavery—if I had it—then I were; but if you swear by that that is not, you are not forsworn. No more was this knight, swearing by his honour, for he never had any; or if he had, he had sworn it away before ever he saw those pancakes or that mustard. 76

CELIA Prithee, who is't that thou meanest?

TOUCHSTONE One that old Frederick, your father, loves.

⌈CELIA⌉ My father's love is enough to honour him. Enough, speak no more of him; you'll be whipped for taxation one of these days. 81

TOUCHSTONE The more pity that fools may not speak wisely what wise men do foolishly.

CELIA By my troth, thou sayst true; for since the little wit that fools have was silenced, the little foolery that wise men have makes a great show. Here comes Monsieur Le Beau. 87

*Enter Le Beau*

ROSALIND With his mouth full of news.

CELIA Which he will put on us as pigeons feed their young. 90

ROSALIND Then shall we be news-crammed.

CELIA All the better: we shall be the more marketable. *Bonjour*, Monsieur Le Beau, what's the news?

LE BEAU Fair princess, you have lost much good sport.

CELIA Sport? Of what colour? 95

LE BEAU What colour, madam? How shall I answer you?

ROSALIND As wit and fortune will.

TOUCHSTONE Or as the destinies decrees.

CELIA Well said. That was laid on with a trowel.

TOUCHSTONE Nay, if I keep not my rank— 100

ROSALIND Thou losest thy old smell.

LE BEAU You amaze me, ladies. I would have told you of good wrestling, which you have lost the sight of.

ROSALIND Yet tell us the manner of the wrestling.

LE BEAU I will tell you the beginning, and if it please your ladyships you may see the end, for the best is yet to do, and here, where you are, they are coming to perform it. 108

CELIA Well, the beginning that is dead and buried.

LE BEAU There comes an old man and his three sons—

CELIA I could match this beginning with an old tale.

LE BEAU Three proper young men, of excellent growth and presence.

ROSALIND With bills on their necks: 'Be it known unto all men by these presents'— 115

LE BEAU The eldest of the three wrestled with Charles, the Duke's wrestler, which Charles in a moment threw him, and broke three of his ribs, that there is little hope of life in him. So he served the second, and so the third. Yonder they lie, the poor old man their father making such pitiful dole over them that all the beholders take his part with weeping. 122

ROSALIND Alas!

TOUCHSTONE But what is the sport, monsieur, that the ladies have lost? 125

LE BEAU Why, this that I speak of.

TOUCHSTONE Thus men may grow wiser every day. It is the first time that ever I heard breaking of ribs was sport for ladies.

CELIA Or I, I promise thee. 130

ROSALIND But is there any else longs to see this broken music in his sides? Is there yet another dotes upon rib-breaking? Shall we see this wrestling, cousin?

LE BEAU You must if you stay here, for here is the place appointed for the wrestling, and they are ready to perform it. 136

CELIA Yonder sure they are coming. Let us now stay and see it.

*Flourish. Enter Duke Frederick, Lords, Orlando, Charles, and attendants*

DUKE FREDERICK Come on. Since the youth will not be entreated, his own peril on his forwardness. 140

ROSALIND Is yonder the man?

LE BEAU Even he, madam.

CELIA Alas, he is too young. Yet he looks successfully.

DUKE FREDERICK How now, daughter and cousin; are you crept hither to see the wrestling? 145

ROSALIND Ay, my liege, so please you give us leave.

DUKE FREDERICK You will take little delight in it, I can tell you, there is such odds in the man. In pity of the challenger's youth I would fain dissuade him, but he will not be entreated. Speak to him, ladies; see if you can move him. 151

CELIA Call him hither, good Monsieur Le Beau.

DUKE FREDERICK Do so. I'll not be by.

*He stands aside*

LE BEAU (*to Orlando*) Monsieur the challenger, the Princess calls for you. 155

ORLANDO I attend them with all respect and duty.

ROSALIND Young man, have you challenged Charles the wrestler?

ORLANDO No, fair Princess. He is the general challenger; I come but in as others do, to try with him the strength of my youth. 161

CELIA Young gentleman, your spirits are too bold for your years. You have seen cruel proof of this man's strength. If you saw yourself with your eyes, or knew yourself with your judgement, the fear of your adventure would counsel you to a more equal enterprise. We pray you for your own sake to embrace your own safety and give over this attempt. 168

ROSALIND Do, young sir. Your reputation shall not therefore be misprized. We will make it our suit to the Duke that the wrestling might not go forward.

ORLANDO I beseech you, punish me not with your hard thoughts, wherein I confess me much guilty to deny so fair and excellent ladies anything. But let your fair eyes and gentle wishes go with me to my trial, wherein if I be foiled, there is but one shamed that was never gracious, if killed, but one dead that is willing to be so. I shall do my friends no wrong, for I have none to lament me; the world no injury, for in it I have nothing. Only in the world I fill up a place which may be better supplied when I have made it empty. 181

ROSALIND The little strength that I have, I would it were with you.

CELIA And mine, to eke out hers.

ROSALIND Fare you well. Pray heaven I be deceived in you. 186

CELIA Your heart's desires be with you.

CHARLES Come, where is this young gallant that is so desirous to lie with his mother earth?

ORLANDO Ready, sir; but his will hath in it a more modest working. 191

DUKE FREDERICK You shall try but one fall.

CHARLES No, I warrant your grace you shall not entreat him to a second that have so mightily persuaded him from a first. 195

ORLANDO You mean to mock me after; you should not have mocked me before. But come your ways.

ROSALIND (*to Orlando*) Now Hercules be thy speed, young man!

CELIA I would I were invisible, to catch the strong fellow by the leg. 201

*Charles and Orlando wrestle*

ROSALIND O excellent young man!

CELIA If I had a thunderbolt in mine eye, I can tell who should down.

*Orlando throws Charles. Shout*

DUKE FREDERICK
No more, no more.

ORLANDO         Yes, I beseech your grace. 205
I am not yet well breathed.

DUKE FREDERICK How dost thou, Charles?
LE BEAU He cannot speak, my lord.
DUKE FREDERICK Bear him away.
    *Attendants carry Charles off*
  What is thy name, young man?    210
ORLANDO Orlando, my liege, the youngest son of Sir
  Rowland de Bois.
DUKE FREDERICK
  I would thou hadst been son to some man else.
  The world esteemed thy father honourable,
  But I did find him still mine enemy.    215
  Thou shouldst have better pleased me with this deed
  Hadst thou descended from another house.
  But fare thee well, thou art a gallant youth.
  I would thou hadst told me of another father.
    *Exeunt Duke Frederick, Le Beau, ⌐Touchstone,⌐*
    *Lords, and attendants*
CELIA (*to Rosalind*)
  Were I my father, coz, would I do this?    220
ORLANDO
  I am more proud to be Sir Rowland's son,
  His youngest son, and would not change that calling
  To be adopted heir to Frederick.
ROSALIND
  My father loved Sir Rowland as his soul,
  And all the world was of my father's mind.    225
  Had I before known this young man his son
  I should have given him tears unto entreaties
  Ere he should thus have ventured.
CELIA               Gentle cousin,
  Let us go thank him, and encourage him.
  My father's rough and envious disposition    230
  Sticks me at heart.—Sir, you have well deserved.
  If you do keep your promises in love
  But justly, as you have exceeded all promise,
  Your mistress shall be happy.
ROSALIND (*giving him a chain from her neck*) Gentleman,
  Wear this for me—one out of suits with fortune,    235
  That could give more but that her hand lacks means.
  Shall we go, coz?
CELIA           Ay. Fare you well, fair gentleman.
    *Rosalind and Celia turn to go*
ORLANDO (*aside*)
  Can I not say 'I thank you'? My better parts
  Are all thrown down, and that which here stands up
  Is but a quintain, a mere lifeless block.    240
ROSALIND (*to Celia*)
  He calls us back. My pride fell with my fortunes,
  I'll ask him what he would.—Did you call, sir?
  Sir, you have wrestled well, and overthrown
  More than your enemies.
CELIA Will you go, coz?    245
ROSALIND Have with you. (*To Orlando*) Fare you well.
    *Exeunt Rosalind and Celia*
ORLANDO
  What passion hangs these weights upon my tongue?
  I cannot speak to her, yet she urged conference.
    *Enter Le Beau*
  O poor Orlando! Thou art overthrown.
  Or Charles or something weaker masters thee.    250
LE BEAU
  Good sir, I do in friendship counsel you
  To leave this place. Albeit you have deserved
  High commendation, true applause, and love,
  Yet such is now the Duke's condition

That he misconsters all that you have done.    255
  The Duke is humorous. What he is indeed
  More suits you to conceive than I to speak of.
ORLANDO
  I thank you, sir. And pray you tell me this,
  Which of the two was daughter of the Duke
  That here was at the wrestling?    260
LE BEAU
  Neither his daughter, if we judge by manners—
  But yet indeed the shorter is his daughter.
  The other is daughter to the banished Duke,
  And here detained by her usurping uncle
  To keep his daughter company, whose loves    265
  Are dearer than the natural bond of sisters.
  But I can tell you that of late this Duke
  Hath ta'en displeasure 'gainst his gentle niece,
  Grounded upon no other argument
  But that the people praise her for her virtues    270
  And pity her for her good father's sake.
  And, on my life, his malice 'gainst the lady
  Will suddenly break forth. Sir, fare you well.
  Hereafter, in a better world than this,
  I shall desire more love and knowledge of you.    275
ORLANDO
  I rest much bounden to you. Fare you well.
    *Exit Le Beau*
  Thus must I from the smoke into the smother,
  From tyrant Duke unto a tyrant brother.—
  But heavenly Rosalind!    *Exit*

**1.3**    *Enter Celia and Rosalind*
CELIA Why cousin, why Rosalind—Cupid have mercy,
  not a word?
ROSALIND Not one to throw at a dog.
CELIA No, thy words are too precious to be cast away
  upon curs. Throw some of them at me. Come, lame me
  with reasons.    6
ROSALIND Then there were two cousins laid up, when the
  one should be lamed with reasons and the other mad
  without any.
CELIA But is all this for your father?    10
ROSALIND No, some of it is for my child's father. O how
  full of briers is this working-day world!
CELIA They are but burs, cousin, thrown upon thee in
  holiday foolery. If we walk not in the trodden paths
  our very petticoats will catch them.    15
ROSALIND I could shake them off my coat. These burs are
  in my heart.
CELIA Hem them away.
ROSALIND I would try, if I could cry 'hem' and have him.
CELIA Come, come, wrestle with thy affections.    20
ROSALIND O, they take the part of a better wrestler than
  myself.
CELIA O, a good wish upon you! You will try in time, in
  despite of a fall. But turning these jests out of service,
  let us talk in good earnest. Is it possible on such a
  sudden you should fall into so strong a liking with old
  Sir Rowland's youngest son?    27
ROSALIND The Duke my father loved his father dearly.
CELIA Doth it therefore ensue that you should love his
  son dearly? By this kind of chase I should hate him,
  for my father hated his father dearly; yet I hate not
  Orlando.    32
ROSALIND No, faith, hate him not, for my sake.
CELIA Why should I not? Doth he not deserve well?

*Enter Duke Frederick, with Lords*

ROSALIND Let me love him for that, and do you love him
    because I do. Look, here comes the Duke.                    36

CELIA With his eyes full of anger.

DUKE FREDERICK (*to Rosalind*)
    Mistress, dispatch you with your safest haste,
    And get you from our court.

ROSALIND Me, uncle?                    40

DUKE FREDERICK You, cousin.
    Within these ten days if that thou beest found
    So near our public court as twenty miles,
    Thou diest for it.

ROSALIND            I do beseech your grace
    Let me the knowledge of my fault bear with me.     45
    If with myself I hold intelligence,
    Or have acquaintance with mine own desires,
    If that I do not dream, or be not frantic—
    As I do trust I am not—then, dear uncle,
    Never so much as in a thought unborn                50
    Did I offend your highness.

DUKE FREDERICK            Thus do all traitors.
    If their purgation did consist in words
    They are as innocent as grace itself.
    Let it suffice thee that I trust thee not.

ROSALIND
    Yet your mistrust cannot make me a traitor.        55
    Tell me whereon the likelihood depends?

DUKE FREDERICK
    Thou art thy father's daughter—there's enough.

ROSALIND
    So was I when your highness took his dukedom;
    So was I when your highness banished him.
    Treason is not inherited, my lord,                 60
    Or if we did derive it from our friends,
    What's that to me? My father was no traitor.
    Then, good my liege, mistake me not so much
    To think my poverty is treacherous.

CELIA Dear sovereign, hear me speak.                   65

DUKE FREDERICK
    Ay, Celia, we stayed her for your sake,
    Else had she with her father ranged along.

CELIA
    I did not then entreat to have her stay.
    It was your pleasure, and your own remorse.
    I was too young that time to value her,            70
    But now I know her. If she be a traitor,
    Why, so am I. We still have slept together,
    Rose at an instant, learned, played, eat together,
    And wheresoe'er we went, like Juno's swans
    Still we went coupled and inseparable.             75

DUKE FREDERICK
    She is too subtle for thee, and her smoothness,
    Her very silence, and her patience
    Speak to the people, and they pity her.
    Thou art a fool. She robs thee of thy name,
    And thou wilt show more bright and seem more
        virtuous                                       80
    When she is gone. Then open not thy lips.
    Firm and irrevocable is my doom
    Which I have passed upon her. She is banished.

CELIA
    Pronounce that sentence then on me, my liege.
    I cannot live out of her company.                  85

DUKE FREDERICK
    You are a fool.—You, niece, provide yourself.
    If you outstay the time, upon mine honour

    And in the greatness of my word, you die.
                *Exit Duke Frederick, with Lords*

CELIA
    O my poor Rosalind, whither wilt thou go?
    Wilt thou change fathers? I will give thee mine.   90
    I charge thee, be not thou more grieved than I am.

ROSALIND
    I have more cause.

CELIA            Thou hast not, cousin.
    Prithee, be cheerful. Know'st thou not the Duke
    Hath banished me, his daughter?

ROSALIND            That he hath not.

CELIA
    No, hath not? Rosalind, lack'st thou then the love  95
    Which teacheth thee that thou and I am one?
    Shall we be sundered? Shall we part, sweet girl?
    No. Let my father seek another heir.
    Therefore devise with me how we may fly,
    Whither to go, and what to bear with us,           100
    And do not seek to take your change upon you,
    To bear your griefs yourself, and leave me out.
    For by this heaven, now at our sorrows pale,
    Say what thou canst, I'll go along with thee.

ROSALIND Why, whither shall we go?                     105

CELIA
    To seek my uncle in the forest of Ardenne.

ROSALIND
    Alas, what danger will it be to us,
    Maids as we are, to travel forth so far!
    Beauty provoketh thieves sooner than gold.

CELIA
    I'll put myself in poor and mean attire,           110
    And with a kind of umber smirch my face.
    The like do you, so shall we pass along
    And never stir assailants.

ROSALIND            Were it not better,
    Because that I am more than common tall,
    That I did suit me all points like a man,          115
    A gallant curtal-axe upon my thigh,
    A boar-spear in my hand, and in my heart,
    Lie there what hidden woman's fear there will.
    We'll have a swashing and a martial outside,
    As many other mannish cowards have,                120
    That do outface it with their semblances.

CELIA
    What shall I call thee when thou art a man?

ROSALIND
    I'll have no worse a name than Jove's own page,
    And therefore look you call me Ganymede.
    But what will you be called?                       125

CELIA
    Something that hath a reference to my state.
    No longer Celia, but Aliena.

ROSALIND
    But cousin, what if we essayed to steal
    The clownish fool out of your father's court.
    Would he not be a comfort to our travel?           130

CELIA
    He'll go along o'er the wide world with me.
    Leave me alone to woo him. Let's away,
    And get our jewels and our wealth together,
    Devise the fittest time and safest way
    To hide us from pursuit that will be made          135
    After my flight. Now go we in content,
    To liberty, and not to banishment.        *Exeunt*

**2.1** *Enter Duke Senior, Amiens, and two or three Lords dressed as foresters*

DUKE SENIOR

Now, my co-mates and brothers in exile,
Hath not old custom made this life more sweet
Than that of painted pomp? Are not these woods
More free from peril than the envious court?
Here feel we not the penalty of Adam,                            5
The seasons' difference, as the icy fang
And churlish chiding of the winter's wind,
Which when it bites and blows upon my body
Even till I shrink with cold, I smile, and say
'This is no flattery. These are counsellors                      10
That feelingly persuade me what I am.'
Sweet are the uses of adversity
Which, like the toad, ugly and venomous,
Wears yet a precious jewel in his head;
And this our life, exempt from public haunt,                     15
Finds tongues in trees, books in the running brooks,
Sermons in stones, and good in everything.

AMIENS

I would not change it. Happy is your grace
That can translate the stubbornness of fortune
Into so quiet and so sweet a style.                              20

DUKE SENIOR

Come, shall we go and kill us venison?
And yet it irks me the poor dappled fools,
Being native burghers of this desert city,
Should in their own confines with forkèd heads
Have their round haunches gored.

FIRST LORD                          Indeed, my lord,          25
The melancholy Jaques grieves at that,
And in that kind swears you do more usurp
Than doth your brother that hath banished you.
Today my lord of Amiens and myself
Did steal behind him as he lay along                             30
Under an oak, whose antic root peeps out
Upon the brook that brawls along this wood,
To the which place a poor sequestered stag
That from the hunter's aim had ta'en a hurt
Did come to languish. And indeed, my lord,                       35
The wretched animal heaved forth such groans
That their discharge did stretch his leathern coat
Almost to bursting, and the big round tears
Coursed one another down his innocent nose
In piteous chase. And thus the hairy fool,                       40
Much markèd of the melancholy Jaques,
Stood on th'extremest verge of the swift brook,
Augmenting it with tears.

DUKE SENIOR                        But what said Jaques?
Did he not moralize this spectacle?

FIRST LORD

O yes, into a thousand similes.                                  45
First, for his weeping into the needless stream;
'Poor deer,' quoth he, 'thou mak'st a testament
As worldlings do, giving thy sum of more
To that which had too much.' Then being there
    alone,
Left and abandoned of his velvet friend,                         50
''Tis right,' quoth he, 'thus misery doth part
The flux of company.' Anon a careless herd
Full of the pasture jumps along by him
And never stays to greet him. 'Ay,' quoth Jaques,
'Sweep on, you fat and greasy citizens,                          55
'Tis just the fashion. Wherefore should you look
Upon that poor and broken bankrupt there?'

Thus most invectively he pierceth through
The body of the country, city, court,
Yea, and of this our life, swearing that we                      60
Are mere usurpers, tyrants, and what's worse,
To fright the animals and to kill them up
In their assigned and native dwelling place.

DUKE SENIOR

And did you leave him in this contemplation?

SECOND LORD

We did, my lord, weeping and commenting                          65
Upon the sobbing deer.

DUKE SENIOR                        Show me the place.
I love to cope him in these sullen fits,
For then he's full of matter.

FIRST LORD                        I'll bring you to him straight.
                                                        *Exeunt*

**2.2** *Enter Duke Frederick, with Lords*

DUKE FREDERICK

Can it be possible that no man saw them?
It cannot be. Some villains of my court
Are of consent and sufferance in this.

FIRST LORD

I cannot hear of any that did see her.
The ladies her attendants of her chamber                         5
Saw her abed, and in the morning early
They found the bed untreasured of their mistress.

SECOND LORD

My lord, the roynish clown at whom so oft
Your grace was wont to laugh is also missing.
Hisperia, the Princess' gentlewoman,                             10
Confesses that she secretly o'erheard
Your daughter and her cousin much commend
The parts and graces of the wrestler
That did but lately foil the sinewy Charles,
And she believes wherever they are gone                          15
That youth is surely in their company.

DUKE FREDERICK

Send to his brother; fetch that gallant hither.
If he be absent, bring his brother to me,
I'll make him find him. Do this suddenly,
And let not search and inquisition quail                         20
To bring again these foolish runaways.
                                            *Exeunt severally*

**2.3** *Enter Orlando and Adam, meeting*

ORLANDO Who's there?

ADAM

What, my young master, O my gentle master,
O my sweet master, O you memory
Of old Sir Rowland, why, what make you here!
Why are you virtuous? Why do people love you?                    5
And wherefore are you gentle, strong, and valiant?
Why would you be so fond to overcome
The bonny prizer of the humorous Duke?
Your praise is come too swiftly home before you.
Know you not, master, to some kind of men                        10
Their graces serve them but as enemies?
No more do yours. Your virtues, gentle master,
Are sanctified and holy traitors to you.
O, what a world is this, when what is comely
Envenoms him that bears it!                                      15

ORLANDO Why, what's the matter?

ADAM                              O, unhappy youth,
Come not within these doors. Within this roof

The enemy of all your graces lives,
Your brother—no, no brother—yet the son—                20
Yet not the son, I will not call him son—
Of him I was about to call his father,
Hath heard your praises, and this night he means
To burn the lodging where you use to lie,
And you within it. If he fail of that,                   25
He will have other means to cut you off.
I overheard him and his practices.
This is no place, this house is but a butchery.
Abhor it, fear it, do not enter it.

ORLANDO
Why, whither, Adam, wouldst thou have me go?             30

ADAM
No matter whither, so you come not here.

ORLANDO
What, wouldst thou have me go and beg my food,
Or with a base and boisterous sword enforce
A thievish living on the common road?
This I must do, or know not what to do.                  35
Yet this I will not do, do how I can.
I rather will subject me to the malice
Of a diverted blood and bloody brother.

ADAM
But do not so. I have five hundred crowns,
The thrifty hire I saved under your father,              40
Which I did store to be my foster-nurse
When service should in my old limbs lie lame,
And unregarded age in corners thrown.
Take that, and he that doth the ravens feed,
Yea providently caters for the sparrow,                  45
Be comfort to my age. Here is the gold.
All this I give you. Let me be your servant.
Though I look old, yet I am strong and lusty,
For in my youth I never did apply
Hot and rebellious liquors in my blood,                  50
Nor did not with unbashful forehead woo
The means of weakness and debility.
Therefore my age is as a lusty winter,
Frosty but kindly. Let me go with you,
I'll do the service of a younger man                     55
In all your business and necessities.

ORLANDO
O good old man, how well in thee appears
The constant service of the antique world,
When service sweat for duty, not for meed!
Thou art not for the fashion of these times,             60
Where none will sweat but for promotion,
And having that do choke their service up
Even with the having. It is not so with thee.
But, poor old man, thou prun'st a rotten tree,
That cannot so much as a blossom yield                   65
In lieu of all thy pains and husbandry.
But come thy ways. We'll go along together,
And ere we have thy youthful wages spent,
We'll light upon some settled low content.

ADAM
Master, go on, and I will follow thee                    70
To the last gasp with truth and loyalty.
From seventeen years till now almost fourscore
Here lived I, but now live here no more.
At seventeen years, many their fortunes seek,
But at fourscore, it is too late a week.                 75
Yet fortune cannot recompense me better
Than to die well, and not my master's debtor.    *Exeunt*

2.4   *Enter Rosalind in man's clothes as Ganymede; Celia*
      *as Aliena, a shepherdess; and Touchstone the*
      *clown*

ROSALIND O Jupiter, how weary are my spirits!

TOUCHSTONE I care not for my spirits, if my legs were not
weary.

ROSALIND I could find in my heart to disgrace my man's
apparel and to cry like a woman. But I must comfort
the weaker vessel, as doublet and hose ought to show
itself courageous to petticoat; therefore, courage, good
Aliena!                                                   8

CELIA I pray you, bear with me. I cannot go no further.

TOUCHSTONE For my part, I had rather bear with you than
bear you. Yet I should bear no cross if I did bear you,
for I think you have no money in your purse.

ROSALIND Well, this is the forest of Ardenne.           13

TOUCHSTONE Ay, now am I in Ardenne; the more fool I.
When I was at home I was in a better place; but
travellers must be content.
      *Enter Corin and Silvius*

ROSALIND Ay, be so, good Touchstone. Look you, who
comes here—a young man and an old in solemn talk.

CORIN (*to Silvius*)
That is the way to make her scorn you still.

SILVIUS
O Corin, that thou knew'st how I do love her!           20

CORIN
I partly guess; for I have loved ere now.

SILVIUS
No, Corin, being old thou canst not guess,
Though in thy youth thou wast as true a lover
As ever sighed upon a midnight pillow.
But if thy love were ever like to mine—                 25
As sure I think did never man love so—
How many actions most ridiculous
Hast thou been drawn to by thy fantasy?

CORIN
Into a thousand that I have forgotten.

SILVIUS
O, thou didst then never love so heartily.              30
If thou rememberest not the slightest folly
That ever love did make thee run into,
Thou hast not loved.
Or if thou hast not sat as I do now,
Wearing thy hearer in thy mistress' praise,             35
Thou hast not loved.
Or if thou hast not broke from company
Abruptly, as my passion now makes me,
Thou hast not loved.
O, Phoebe, Phoebe, Phoebe!                        *Exit*

ROSALIND
Alas, poor shepherd, searching of thy wound,            41
I have by hard adventure found mine own.

TOUCHSTONE And I mine. I remember when I was in love
I broke my sword upon a stone and bid him take that
for coming a-night to Jane Smile, and I remember the
kissing of her batlet, and the cow's dugs that her pretty
chapped hands had milked; and I remember the wooing
of a peascod instead of her, from whom I took two
cods, and giving her them again, said with weeping
tears, 'Wear these for my sake.' We that are true lovers
run into strange capers. But as all is mortal in nature,
so is all nature in love mortal in folly.               52

ROSALIND Thou speak'st wiser than thou art ware of.

TOUCHSTONE Nay, I shall ne'er be ware of mine own wit
till I break my shins against it.                        55

ROSALIND
  Jove, Jove, this shepherd's passion
  Is much upon my fashion.
TOUCHSTONE And mine, but it grows something stale with
  me.
CELIA
  I pray you, one of you question yon man      60
  If he for gold will give us any food.
  I faint almost to death.
TOUCHSTONE (to Corin) Holla, you clown!
ROSALIND Peace, fool, he's not thy kinsman.
CORIN Who calls?      65
TOUCHSTONE Your betters, sir.
CORIN Else are they very wretched.
ROSALIND (to Touchstone)
  Peace, I say. (To Corin) Good even to you, friend.
CORIN
  And to you, gentle sir, and to you all.
ROSALIND
  I prithee, shepherd, if that love or gold      70
  Can in this desert place buy entertainment,
  Bring us where we may rest ourselves, and feed.
  Here's a young maid with travel much oppressed,
  And faints for succour.
CORIN      Fair sir, I pity her,
  And wish, for her sake more than for mine own,      75
  My fortunes were more able to relieve her.
  But I am shepherd to another man,
  And do not shear the fleeces that I graze.
  My master is of churlish disposition,
  And little recks to find the way to heaven      80
  By doing deeds of hospitality.
  Besides, his cot, his flocks, and bounds of feed
  Are now on sale, and at our sheepcote now
  By reason of his absence there is nothing
  That you will feed on. But what is, come see,      85
  And in my voice most welcome shall you be.
ROSALIND
  What is he that shall buy his flock and pasture?
CORIN
  That young swain that you saw here but erewhile,
  That little cares for buying anything.
ROSALIND
  I pray thee, if it stand with honesty,      90
  Buy thou the cottage, pasture, and the flock,
  And thou shalt have to pay for it of us.
CELIA
  And we will mend thy wages. I like this place,
  And willingly could waste my time in it.
CORIN
  Assuredly the thing is to be sold.      95
  Go with me. If you like upon report
  The soil, the profit, and this kind of life,
  I will your very faithful feeder be,
  And buy it with your gold right suddenly.     *Exeunt*

**2.5**  *Enter Amiens, Jaques, and other Lords dressed as*
     *foresters*
⌜AMIENS⌝ (*sings*)
      Under the greenwood tree
      Who loves to lie with me,
      And turn his merry note
      Unto the sweet bird's throat,
   Come hither, come hither, come hither.     5
       Here shall he see
       No enemy
     But winter and rough weather.

JAQUES More, more, I prithee, more.
AMIENS It will make you melancholy, Monsieur Jaques.
JAQUES I thank it. More, I prithee, more. I can suck
  melancholy out of a song as a weasel sucks eggs. More,
  I prithee, more.     13
AMIENS My voice is ragged, I know I cannot please you.
JAQUES I do not desire you to please me, I do desire you
  to sing. Come, more; another stanza. Call you 'em
  stanzas?
AMIENS What you will, Monsieur Jaques.
JAQUES Nay, I care not for their names, they owe me
  nothing. Will you sing?     20
AMIENS More at your request than to please myself.
JAQUES Well then, if ever I thank any man, I'll thank
  you. But that they call compliment is like th'encounter
  of two dog-apes, and when a man thanks me heartily
  methinks I have given him a penny and he renders me
  the beggarly thanks. Come, sing; and you that will
  not, hold your tongues.     27
AMIENS Well, I'll end the song.—Sirs, cover the while.
     *Lords prepare food and drink*
  The Duke will drink under this tree. (*To Jaques*) He hath
  been all this day to look you.     30
JAQUES And I have been all this day to avoid him. He is
  too disputable for my company. I think of as many
  matters as he, but I give heaven thanks, and make no
  boast of them. Come, warble, come.

ALL (*sing*)  Who doth ambition shun,     35
      And loves to live i'th' sun,
      Seeking the food he eats
      And pleased with what he gets,
    Come hither, come hither, come hither.
       Here shall he see     40
       No enemy
     But winter and rough weather.

JAQUES I'll give you a verse to this note that I made
  yesterday in despite of my invention.
AMIENS And I'll sing it.     45
JAQUES Thus it goes:
      If it do come to pass
      That any man turn ass,
      Leaving his wealth and ease
      A stubborn will to please,     50
    Ducdame, ducdame, ducdame.
       Here shall he see
       Gross fools as he,
      An if he will come to me.
AMIENS What's that 'ducdame'?     55
JAQUES 'Tis a Greek invocation to call fools into a circle.
  I'll go sleep if I can. If I cannot, I'll rail against all the
  firstborn of Egypt.
AMIENS And I'll go seek the Duke; his banquet is prepared.
     *Exeunt*

**2.6**  *Enter Orlando and Adam*
ADAM Dear master, I can go no further. O, I die for food.
  Here lie I down and measure out my grave. Farewell,
  kind master.
ORLANDO Why, how now, Adam? No greater heart in
  thee? Live a little, comfort a little, cheer thyself a little.
  If this uncouth forest yield anything savage I will either
  be food for it or bring it for food to thee. Thy conceit
  is nearer death than thy powers. For my sake be
  comfortable. Hold death awhile at the arm's end. I will
  here be with thee presently, and if I bring thee not

something to eat, I will give thee leave to die. But if thou diest before I come, thou art a mocker of my labour. Well said. Thou lookest cheerly, and I'll be with thee quickly. Yet thou liest in the bleak air. Come, I will bear thee to some shelter, and thou shalt not die for lack of a dinner if there live anything in this desert. Cheerly, good Adam.                    *Orlando carries Adam off*

**2.7**    *Enter Duke Senior and Lords dressed as outlaws*
DUKE SENIOR
I think he be transformed into a beast,
For I can nowhere find him like a man.
FIRST LORD
My lord, he is but even now gone hence.
Here was he merry, hearing of a song.
DUKE SENIOR
If he, compact of jars, grow musical                    5
We shall have shortly discord in the spheres.
Go seek him. Tell him I would speak with him.
        *Enter Jaques*
FIRST LORD
He saves my labour by his own approach.
DUKE SENIOR
Why, how now, monsieur, what a life is this,
That your poor friends must woo your company!    10
What, you look merrily.
JAQUES
A fool, a fool, I met a fool i'th' forest,
A motley fool—a miserable world!—
As I do live by food, I met a fool,
Who laid him down and basked him in the sun,    15
And railed on Lady Fortune in good terms,
In good set terms, and yet a motley fool.
'Good morrow, fool,' quoth I. 'No, sir,' quoth he,
'Call me not fool till heaven hath sent me fortune.'
And then he drew a dial from his poke,                    20
And looking on it with lack-lustre eye
Says very wisely 'It is ten o'clock.
'Thus we may see', quoth he, 'how the world wags.
'Tis but an hour ago since it was nine,
And after one hour more 'twill be eleven.            25
And so from hour to hour we ripe and ripe,
And then from hour to hour we rot and rot;
And thereby hangs a tale.' When I did hear
The motley fool thus moral on the time
My lungs began to crow like chanticleer,            30
That fools should be so deep-contemplative,
And I did laugh sans intermission
An hour by his dial. O noble fool,
A worthy fool—motley's the only wear.
DUKE SENIOR    What fool is this?                    35
JAQUES
O worthy fool!—One that hath been a courtier,
And says 'If ladies be but young and fair
They have the gift to know it.' And in his brain,
Which is as dry as the remainder biscuit
After a voyage, he hath strange places crammed    40
With observation, the which he vents
In mangled forms. O that I were a fool,
I am ambitious for a motley coat.
DUKE SENIOR
Thou shalt have one.
JAQUES                    It is my only suit,
Provided that you weed your better judgements    45
Of all opinion that grows rank in them
That I am wise. I must have liberty

Withal, as large a charter as the wind,
To blow on whom I please, for so fools have;
And they that are most gallèd with my folly,        50
They most must laugh. And why, sir, must they so?
The why is plain as way to parish church:
He that a fool doth very wisely hit
Doth very foolishly, although he smart,
Seem aught but senseless of the bob. If not,        55
The wise man's folly is anatomized
Even by the squandering glances of the fool.
Invest me in my motley. Give me leave
To speak my mind, and I will through and through
Cleanse the foul body of th'infected world,        60
If they will patiently receive my medicine.
DUKE SENIOR
Fie on thee, I can tell what thou wouldst do.
JAQUES
What, for a counter, would I do but good?
DUKE SENIOR
Most mischievous foul sin, in chiding sin;
For thou thyself hast been a libertine,                65
As sensual as the brutish sting itself,
And all th'embossèd sores and headed evils
That thou with licence of free foot hast caught
Wouldst thou disgorge into the general world.
JAQUES Why, who cries out on pride                    70
That can therein tax any private party?
Doth it not flow as hugely as the sea,
Till that the weary very means do ebb?
What woman in the city do I name
When that I say the city-woman bears                75
The cost of princes on unworthy shoulders?
Who can come in and say that I mean her
When such a one as she, such is her neighbour?
Or what is he of basest function,
That says his bravery is not on my cost,            80
Thinking that I mean him, but therein suits
His folly to the mettle of my speech?
There then, how then, what then, let me see wherein
My tongue hath wronged him. If it do him right,
Then he hath wronged himself. If he be free,        85
Why then my taxing like a wild goose flies,
Unclaimed of any man. But who comes here?
        *Enter Orlando, with sword drawn*
ORLANDO
Forbear, and eat no more!
JAQUES                    Why, I have eat none yet.
ORLANDO
Nor shalt not till necessity be served.
JAQUES Of what kind should this cock come of?    90
DUKE SENIOR
Art thou thus boldened, man, by thy distress?
Or else a rude despiser of good manners,
That in civility thou seem'st so empty?
ORLANDO
You touched my vein at first. The thorny point
Of bare distress hath ta'en from me the show    95
Of smooth civility. Yet am I inland bred,
And know some nurture. But forbear, I say.
He dies that touches any of this fruit
Till I and my affairs are answerèd.
JAQUES An you will not be answered with reason, I must
die.                    101
DUKE SENIOR
What would you have? Your gentleness shall force
More than your force move us to gentleness.

ORLANDO
I almost die for food; and let me have it.
DUKE SENIOR
Sit down and feed, and welcome to our table.          105
ORLANDO
Speak you so gently? Pardon me, I pray you.
I thought that all things had been savage here,
And therefore put I on the countenance
Of stern commandment. But whate'er you are
That in this desert inaccessible,          110
Under the shade of melancholy boughs,
Lose and neglect the creeping hours of time,
If ever you have looked on better days,
If ever been where bells have knolled to church,
If ever sat at any good man's feast,          115
If ever from your eyelids wiped a tear,
And know what 'tis to pity, and be pitied,
Let gentleness my strong enforcement be.
In the which hope I blush, and hide my sword.
DUKE SENIOR
True is it that we have seen better days,          120
And have with holy bell been knolled to church,
And sat at good men's feasts, and wiped our eyes
Of drops that sacred pity hath engendered.
And therefore sit you down in gentleness,
And take upon command what help we have          125
That to your wanting may be ministered.
ORLANDO
Then but forbear your food a little while
Whiles, like a doe, I go to find my fawn
And give it food. There is an old poor man
Who after me hath many a weary step          130
Limped in pure love. Till he be first sufficed,
Oppressed with two weak evils, age and hunger,
I will not touch a bit.
DUKE SENIOR                    Go find him out,
And we will nothing waste till you return.
ORLANDO
I thank ye; and be blessed for your good comfort! *Exit*
DUKE SENIOR
Thou seest we are not all alone unhappy.
This wide and universal theatre
Presents more woeful pageants than the scene
Wherein we play in.
JAQUES                    All the world's a stage,
And all the men and women merely players.          140
They have their exits and their entrances,
And one man in his time plays many parts,
His acts being seven ages. At first the infant,
Mewling and puking in the nurse's arms.
Then the whining schoolboy with his satchel          145
And shining morning face, creeping like snail
Unwillingly to school. And then the lover,
Sighing like furnace, with a woeful ballad
Made to his mistress' eyebrow. Then, a soldier,
Full of strange oaths, and bearded like the pard,          150
Jealous in honour, sudden, and quick in quarrel,
Seeking the bubble reputation
Even in the cannon's mouth. And then the justice,
In fair round belly with good capon lined,
With eyes severe and beard of formal cut,          155
Full of wise saws and modern instances;
And so he plays his part. The sixth age shifts
Into the lean and slippered pantaloon,
With spectacles on nose and pouch on side,

His youthful hose, well saved, a world too wide          160
For his shrunk shank, and his big, manly voice,
Turning again toward childish treble, pipes
And whistles in his sound. Last scene of all,
That ends this strange, eventful history,
Is second childishness and mere oblivion,          165
Sans teeth, sans eyes, sans taste, sans everything.
          *Enter Orlando bearing Adam*
DUKE SENIOR
Welcome. Set down your venerable burden
And let him feed.
ORLANDO I thank you most for him.
ADAM So had you need;          170
I scarce can speak to thank you for myself.
DUKE SENIOR
Welcome. Fall to. I will not trouble you
As yet to question you about your fortunes.
Give us some music, and, good cousin, sing.

⌜AMIENS⌝ (*sings*)
     Blow, blow, thou winter wind,          175
     Thou art not so unkind
          As man's ingratitude.
     Thy tooth is not so keen,
     Because thou art not seen,
          Although thy breath be rude.          180
Hey-ho, sing hey-ho, unto the green holly.
Most friendship is feigning, most loving, mere folly.
     Then hey-ho, the holly;
     This life is most jolly.

     Freeze, freeze, thou bitter sky,          185
     That dost not bite so nigh
          As benefits forgot.
     Though thou the waters warp,
     Thy sting is not so sharp
          As friend remembered not.          190
Hey-ho, sing hey-ho, unto the green holly.
Most friendship is feigning, most loving, mere folly.
     Then hey-ho, the holly;
     This life is most jolly.

DUKE SENIOR (*to Orlando*)
If that you were the good Sir Rowland's son,          195
As you have whispered faithfully you were,
And as mine eye doth his effigies witness
Most truly limned and living in your face,
Be truly welcome hither. I am the Duke
That loved your father. The residue of your fortune,
Go to my cave and tell me. (*To Adam*) Good old man,
Thou art right welcome, as thy master is.—          202
(*To Lords*) Support him by the arm. (*To Orlando*) Give
     me your hand,
And let me all your fortunes understand.          *Exeunt*

**3.1**     *Enter Duke Frederick, Lords, and Oliver*
DUKE FREDERICK
Not see him since? Sir, sir, that cannot be.
But were I not the better part made mercy,
I should not seek an absent argument
Of my revenge, thou present. But look to it:
Find out thy brother wheresoe'er he is.          5
Seek him with candle. Bring him, dead or living,
Within this twelvemonth, or turn thou no more
To seek a living in our territory.
Thy lands, and all things that thou dost call thine

Worth seizure, do we seize into our hands          10
Till thou canst quit thee by thy brother's mouth
Of what we think against thee.
OLIVER
O that your highness knew my heart in this.
I never loved my brother in my life.
DUKE FREDERICK
More villain thou. (*To Lords*) Well, push him out of
     doors,                                         15
And let my officers of such a nature
Make an extent upon his house and lands.
Do this expediently, and turn him going.
                                    *Exeunt severally*

**3.2**   *Enter Orlando with a paper*
ORLANDO
Hang there, my verse, in witness of my love;
     And thou thrice-crownèd queen of night, survey
With thy chaste eye, from thy pale sphere above,
     Thy huntress' name that my full life doth sway.
O Rosalind, these trees shall be my books,         5
     And in their barks my thoughts I'll character
That every eye which in this forest looks
     Shall see thy virtue witnessed everywhere.
Run, run, Orlando; carve on every tree
The fair, the chaste, and unexpressive she.    *Exit*
     *Enter Corin and Touchstone the clown*
CORIN And how like you this shepherd's life, Master
Touchstone?                                        12
TOUCHSTONE Truly, shepherd, in respect of itself, it is a
good life; but in respect that it is a shepherd's life, it
is naught. In respect that it is solitary, I like it very
well; but in respect that it is private, it is a very vile
life. Now in respect it is in the fields, it pleaseth me
well; but in respect it is not in the court, it is tedious.
As it is a spare life, look you, it fits my humour well;
but as there is no more plenty in it, it goes much
against my stomach. Hast any philosophy in thee,
shepherd?                                          22
CORIN No more but that I know the more one sickens,
the worse at ease he is, and that he that wants money,
means, and content is without three good friends; that
the property of rain is to wet, and fire to burn; that
good pasture makes fat sheep; and that a great cause
of the night is lack of the sun; that he that hath learned
no wit by nature nor art may complain of good breeding
or comes of a very dull kindred.                   30
TOUCHSTONE Such a one is a natural philosopher. Wast
ever in court, shepherd?
CORIN No, truly.
TOUCHSTONE Then thou art damned.
CORIN Nay, I hope.                                 35
TOUCHSTONE Truly thou art damned, like an ill-roasted
egg, all on one side.
CORIN For not being at court? Your reason?
TOUCHSTONE Why, if thou never wast at court thou never
sawest good manners. If thou never sawest good
manners, then thy manners must be wicked, and
wickedness is sin, and sin is damnation. Thou art in a
parlous state, shepherd.                           43
CORIN Not a whit, Touchstone. Those that are good
manners at the court are as ridiculous in the country
as the behaviour of the country is most mockable at
the court. You told me you salute not at the court but
you kiss your hands. That courtesy would be uncleanly
if courtiers were shepherds.

TOUCHSTONE Instance, briefly; come, instance.      50
CORIN Why, we are still handling our ewes, and their
fells, you know, are greasy.
TOUCHSTONE Why, do not your courtier's hands sweat?
And is not the grease of a mutton as wholesome as the
sweat of a man? Shallow, shallow. A better instance,
I say. Come.                                       56
CORIN Besides, our hands are hard.
TOUCHSTONE Your lips will feel them the sooner. Shallow
again. A more sounder instance. Come.
CORIN And they are often tarred over with the surgery of
our sheep; and would you have us kiss tar? The
courtier's hands are perfumed with civet.          62
TOUCHSTONE Most shallow, man. Thou worms' meat in
respect of a good piece of flesh indeed, learn of the
wise, and perpend: civet is of a baser birth than tar,
the very uncleanly flux of a cat. Mend the instance,
shepherd.                                          67
CORIN You have too courtly a wit for me. I'll rest.
TOUCHSTONE Wilt thou rest damned? God help thee,
shallow man. God make incision in thee, thou art raw.
CORIN Sir, I am a true labourer. I earn that I eat, get that
I wear; owe no man hate, envy no man's happiness;
glad of other men's good, content with my harm; and
the greatest of my pride is to see my ewes graze and
my lambs suck.                                     75
TOUCHSTONE That is another simple sin in you, to bring
the ewes and the rams together, and to offer to get
your living by the copulation of cattle; to be bawd to
a bell-wether, and to betray a she-lamb of a twelvemonth to a crooked-pated old cuckoldly ram, out of all
reasonable match. If thou beest not damned for this,
the devil himself will have no shepherds. I cannot see
else how thou shouldst scape.
CORIN Here comes young Master Ganymede, my new
mistress's brother.                                85
     *Enter Rosalind as Ganymede*
ROSALIND (*reads*)
     'From the east to western Ind
     No jewel is like Rosalind.
     Her worth being mounted on the wind
     Through all the world bears Rosalind.
     All the pictures fairest lined                90
     Are but black to Rosalind.
     Let no face be kept in mind
     But the fair of Rosalind.'
TOUCHSTONE I'll rhyme you so eight years together,
dinners, and suppers, and sleeping-hours excepted. It
is the right butter-women's rank to market.        96
ROSALIND Out, fool.
TOUCHSTONE For a taste:
     If a hart do lack a hind,
     Let him seek out Rosalind.                    100
     If the cat will after kind,
     So, be sure, will Rosalind.
     Wintered garments must be lined,
     So must slender Rosalind.
     They that reap must sheaf and bind,           105
     Then to cart with Rosalind.
     'Sweetest nut hath sourest rind',
     Such a nut is Rosalind.
     He that sweetest rose will find
     Must find love's prick, and Rosalind.         110
This is the very false gallop of verses. Why do you
infect yourself with them?
ROSALIND Peace, you dull fool, I found them on a tree.

TOUCHSTONE Truly, the tree yields bad fruit.

ROSALIND I'll graft it with you, and then I shall graft it with a medlar; then it will be the earliest fruit i'th' country, for you'll be rotten ere you be half-ripe, and that's the right virtue of the medlar.

TOUCHSTONE You have said; but whether wisely or no, let the forest judge.                120

*Enter Celia, as Aliena, with a writing*

ROSALIND
Peace, here comes my sister, reading. Stand aside.

CELIA (*reads*)
'Why should this a desert be?
    For it is unpeopled? No.
Tongues I'll hang on every tree,
    That shall civil sayings show.          125
Some, how brief the life of man
    Runs his erring pilgrimage,
That the stretching of a span
    Buckles in his sum of age.
Some of violated vows                        130
    'Twixt the souls of friend and friend.
But upon the fairest boughs,
    Or at every sentence end,
Will I 'Rosalinda' write,
    Teaching all that read to know           135
The quintessence of every sprite
    Heaven would in little show.
Therefore heaven nature charged
    That one body should be filled
With all graces wide-enlarged.               140
    Nature presently distilled
Helen's cheek, but not her heart,
    Cleopatra's majesty,
Atalanta's better part,
    Sad Lucretia's modesty.                  145
Thus Rosalind of many parts
    By heavenly synod was devised
Of many faces, eyes, and hearts
    To have the touches dearest prized.
Heaven would that she these gifts should have  150
And I to live and die her slave.'

ROSALIND O most gentle Jupiter! What tedious homily of love have you wearied your parishioners withal, and never cried 'Have patience, good people.'

CELIA How now, back, friends. Shepherd, go off a little. Go with him, sirrah.                156

TOUCHSTONE Come, shepherd, let us make an honourable retreat, though not with bag and baggage, yet with scrip and scrippage.          *Exit with Corin*

CELIA Didst thou hear these verses?          160

ROSALIND O yes, I heard them all, and more, too, for some of them had in them more feet than the verses would bear.

CELIA That's no matter; the feet might bear the verses.

ROSALIND Ay, but the feet were lame, and could not bear themselves without the verse, and therefore stood lamely in the verse.                167

CELIA But didst thou hear without wondering how thy name should be hanged and carved upon these trees?

ROSALIND I was seven of the nine days out of the wonder before you came; for look here what I found on a palm-tree; (*showing Celia the verses*) I was never so berhymed since Pythagoras' time that I was an Irish rat, which I can hardly remember.

CELIA Trow you who hath done this?          175

ROSALIND Is it a man?

CELIA And a chain that you once wore about his neck. Change you colour?

ROSALIND I prithee, who?

CELIA O Lord, Lord, it is a hard matter for friends to meet. But mountains may be removed with earthquakes, and so encounter.                182

ROSALIND Nay, but who is it?

CELIA Is it possible?

ROSALIND Nay, I prithee now with most petitionary vehemence, tell me who it is.          186

CELIA O wonderful, wonderful, and most wonderful-wonderful, and yet again wonderful, and after that out of all whooping!

ROSALIND Good my complexion! Dost thou think, though I am caparisoned like a man, I have a doublet and hose in my disposition? One inch of delay more is a South Sea of discovery. I prithee tell me who is it quickly, and speak apace. I would thou couldst stammer, that thou mightst pour this concealed man out of thy mouth as wine comes out of a narrow-mouthed bottle—either too much at once, or none at all. I prithee, take the cork out of thy mouth, that I may drink thy tidings.

CELIA So you may put a man in your belly.          200

ROSALIND Is he of God's making? What manner of man? Is his head worth a hat? Or his chin worth a beard?

CELIA Nay, he hath but a little beard.

ROSALIND Why, God will send more, if the man will be thankful. Let me stay the growth of his beard, if thou delay me not the knowledge of his chin.          206

CELIA It is young Orlando, that tripped up the wrestler's heels and your heart both in an instant.

ROSALIND Nay, but the devil take mocking. Speak sad brow and true maid.                210

CELIA I'faith, coz, 'tis he.

ROSALIND Orlando?

CELIA Orlando.

ROSALIND Alas the day, what shall I do with my doublet and hose! What did he when thou sawest him? What said he? How looked he? Wherein went he? What makes he here? Did he ask for me? Where remains he? How parted he with thee? And when shalt thou see him again? Answer me in one word.          219

CELIA You must borrow me Gargantua's mouth first, 'tis a word too great for any mouth of this age's size. To say ay and no to these particulars is more than to answer in a catechism.

ROSALIND But doth he know that I am in this forest, and in man's apparel? Looks he as freshly as he did the day he wrestled?                226

CELIA It is as easy to count atomies as to resolve the propositions of a lover; but take a taste of my finding him, and relish it with good observance. I found him under a tree, like a dropped acorn—          230

ROSALIND It may well be called Jove's tree when it drops forth such fruit.

CELIA Give me audience, good madam.

ROSALIND Proceed.

CELIA There lay he, stretched along like a wounded knight—                236

ROSALIND Though it be pity to see such a sight, it well becomes the ground.

CELIA Cry 'holla' to thy tongue, I prithee: it curvets unseasonably.—He was furnished like a hunter—          240

ROSALIND O ominous—he comes to kill my heart.

CELIA I would sing my song without a burden; thou
  bringest me out of tune.

ROSALIND Do you not know I am a woman? When I
  think, I must speak.—Sweet, say on.                    245

  *Enter Orlando and Jaques*

CELIA You bring me out. Soft, comes he not here?

ROSALIND 'Tis he. Slink by, and note him.

  *Rosalind and Celia stand aside*

JAQUES (*to Orlando*) I thank you for your company, but,
  good faith, I had as lief have been myself alone.

ORLANDO And so had I. But yet for fashion' sake, I thank
  you too for your society.                    251

JAQUES God b'wi'you; let's meet as little as we can.

ORLANDO I do desire we may be better strangers.

JAQUES I pray you mar no more trees with writing love-
  songs in their barks.                    255

ORLANDO I pray you mar no more of my verses with
  reading them ill-favouredly.

JAQUES Rosalind is your love's name?

ORLANDO Yes, just.

JAQUES I do not like her name.                    260

ORLANDO There was no thought of pleasing you when
  she was christened.

JAQUES What stature is she of?

ORLANDO Just as high as my heart.

JAQUES You are full of pretty answers. Have you not been
  acquainted with goldsmiths' wives, and conned them
  out of rings?                    267

ORLANDO Not so; but I answer you right painted cloth,
  from whence you have studied your questions.

JAQUES You have a nimble wit; I think 'twas made of
  Atalanta's heels. Will you sit down with me, and we
  two will rail against our mistress the world, and all
  our misery?

ORLANDO I will chide no breather in the world but myself,
  against whom I know most faults.                    275

JAQUES The worst fault you have is to be in love.

ORLANDO 'Tis a fault I will not change for your best virtue.
  I am weary of you.

JAQUES By my troth, I was seeking for a fool when I found
  you.                    280

ORLANDO He is drowned in the brook. Look but in, and
  you shall see him.

JAQUES There I shall see mine own figure.

ORLANDO Which I take to be either a fool or a cipher.

JAQUES I'll tarry no longer with you. Farewell, good Signor
  Love.                    286

ORLANDO I am glad of your departure. Adieu, good
  Monsieur Melancholy.                    *Exit Jaques*

ROSALIND (*to Celia*) I will speak to him like a saucy lackey,
  and under that habit play the knave with him. (*To
  Orlando*) Do you hear, forester?                    291

ORLANDO Very well. What would you?

ROSALIND I pray you, what is't o'clock?

ORLANDO You should ask me what time o' day. There's
  no clock in the forest.                    295

ROSALIND Then there is no true lover in the forest, else
  sighing every minute and groaning every hour would
  detect the lazy foot of time as well as a clock.

ORLANDO And why not the swift foot of time? Had not
  that been as proper?                    300

ROSALIND By no means, sir. Time travels in divers paces
  with divers persons. I'll tell you who time ambles
  withal, who time trots withal, who time gallops withal,
  and who he stands still withal.

ORLANDO I prithee, who doth he trot withal?                    305

ROSALIND Marry, he trots hard with a young maid
  between the contract of her marriage and the day it is
  solemnized. If the interim be but a se'nnight, time's
  pace is so hard that it seems the length of seven year.

ORLANDO Who ambles time withal?                    310

ROSALIND With a priest that lacks Latin, and a rich man
  that hath not the gout; for the one sleeps easily because
  he cannot study, and the other lives merrily because
  he feels no pain, the one lacking the burden of lean
  and wasteful learning, the other knowing no burden
  of heavy tedious penury. These time ambles withal.

ORLANDO Who doth he gallop withal?                    317

ROSALIND With a thief to the gallows; for though he go
  as softly as foot can fall, he thinks himself too soon
  there.                    320

ORLANDO Who stays it still withal?

ROSALIND With lawyers in the vacation; for they sleep
  between term and term, and then they perceive not
  how time moves.

ORLANDO Where dwell you, pretty youth?                    325

ROSALIND With this shepherdess, my sister, here in the
  skirts of the forest, like fringe upon a petticoat.

ORLANDO Are you native of this place?

ROSALIND As the coney that you see dwell where she is
  kindled.                    330

ORLANDO Your accent is something finer than you could
  purchase in so removed a dwelling.

ROSALIND I have been told so of many; but indeed an old
  religious uncle of mine taught me to speak, who was
  in his youth an inland man; one that knew courtship
  too well, for there he fell in love. I have heard him
  read many lectures against it, and I thank God I am
  not a woman, to be touched with so many giddy
  offences as he hath generally taxed their whole sex
  withal.                    340

ORLANDO Can you remember any of the principal evils
  that he laid to the charge of women?

ROSALIND There were none principal; they were all like
  one another as halfpence are, every one fault seeming
  monstrous till his fellow-fault came to match it.                    345

ORLANDO I prithee, recount some of them.

ROSALIND No. I will not cast away my physic but on those
  that are sick. There is a man haunts the forest that
  abuses our young plants with carving Rosalind on their
  barks; hangs odes upon hawthorns and elegies on
  brambles; all, forsooth, deifying the name of Rosalind.
  If I could meet that fancy-monger, I would give him
  some good counsel, for he seems to have the quotidian
  of love upon him.                    354

ORLANDO I am he that is so love-shaked. I pray you, tell
  me your remedy.

ROSALIND There is none of my uncle's marks upon you.
  He taught me how to know a man in love, in which
  cage of rushes I am sure you are not prisoner.

ORLANDO What were his marks?                    360

ROSALIND A lean cheek, which you have not; a blue eye
  and sunken, which you have not; an unquestionable
  spirit, which you have not; a beard neglected, which
  you have not—but I pardon you for that, for simply
  your having in beard is a younger brother's revenue.
  Then your hose should be ungartered, your bonnet
  unbanded, your sleeve unbuttoned, your shoe untied,
  and everything about you demonstrating a careless
  desolation. But you are no such man. You are rather
  point-device in your accoutrements, as loving yourself
  than seeming the lover of any other.                    371

ORLANDO Fair youth, I would I could make thee believe I love.

ROSALIND Me believe it? You may as soon make her that you love believe it, which I warrant she is apter to do than to confess she does. That is one of the points in the which women still give the lie to their consciences. But in good sooth, are you he that hangs the verses on the trees wherein Rosalind is so admired?      379

ORLANDO I swear to thee, youth, by the white hand of Rosalind, I am that he, that unfortunate he.

ROSALIND But are you so much in love as your rhymes speak?

ORLANDO Neither rhyme nor reason can express how much.      385

ROSALIND Love is merely a madness, and I tell you, deserves as well a dark house and a whip as madmen do; and the reason why they are not so punished and cured is that the lunacy is so ordinary that the whippers are in love too. Yet I profess curing it by counsel.      390

ORLANDO Did you ever cure any so?

ROSALIND Yes, one; and in this manner. He was to imagine me his love, his mistress; and I set him every day to woo me. At which time would I, being but a moonish youth, grieve, be effeminate, changeable, longing and liking, proud, fantastical, apish, shallow, inconstant, full of tears, full of smiles; for every passion something, and for no passion truly anything, as boys and women are for the most part cattle of this colour—would now like him, now loathe him; then entertain him, then forswear him; now weep for him, then spit at him, that I drave my suitor from his mad humour of love to a living humour of madness, which was to forswear the full stream of the world and to live in a nook merely monastic. And thus I cured him, and this way will I take upon me to wash your liver as clean as a sound sheep's heart, that there shall not be one spot of love in't.

ORLANDO I would not be cured, youth.      409

ROSALIND I would cure you if you would but call me Rosalind and come every day to my cot, and woo me.

ORLANDO Now by the faith of my love, I will. Tell me where it is.      413

ROSALIND Go with me to it, and I'll show it you. And by the way you shall tell me where in the forest you live. Will you go?

ORLANDO With all my heart, good youth.

ROSALIND Nay, you must call me Rosalind.—Come, sister. Will you go?                                      *Exeunt*

**3.3** *Enter Touchstone the clown and Audrey, followed by Jaques*

TOUCHSTONE Come apace, good Audrey. I will fetch up your goats, Audrey. And how, Audrey, am I the man yet? Doth my simple feature content you?

AUDREY Your features, Lord warrant us—what features?

TOUCHSTONE I am here with thee and thy goats as the most capricious poet honest Ovid was among the Goths.

JAQUES (*aside*) O knowledge ill-inhabited; worse than Jove in a thatched house.      8

TOUCHSTONE When a man's verses cannot be understood, nor a man's good wit seconded with the forward child, understanding, it strikes a man more dead than a great reckoning in a little room. Truly, I would the gods had made thee poetical.

AUDREY I do not know what 'poetical' is. Is it honest in deed and word? Is it a true thing?      15

TOUCHSTONE No, truly; for the truest poetry is the most feigning, and lovers are given to poetry; and what they swear in poetry it may be said, as lovers, they do feign.

AUDREY Do you wish, then, that the gods had made me poetical?      20

TOUCHSTONE I do, truly; for thou swearest to me thou art honest. Now if thou wert a poet, I might have some hope thou didst feign.

AUDREY Would you not have me honest?

TOUCHSTONE No, truly, unless thou wert hard-favoured; for honesty coupled to beauty is to have honey a sauce to sugar.      27

JAQUES (*aside*) A material fool.

AUDREY Well, I am not fair, and therefore I pray the gods make me honest.      30

TOUCHSTONE Truly, and to cast away honesty upon a foul slut were to put good meat into an unclean dish.

AUDREY I am not a slut, though I thank the gods I am foul.

TOUCHSTONE Well, praised be the gods for thy foulness. Sluttishness may come hereafter. But be it as it may be, I will marry thee; and to that end I have been with Sir Oliver Martext, the vicar of the next village, who hath promised to meet me in this place of the forest, and to couple us.      40

JAQUES (*aside*) I would fain see this meeting.

AUDREY Well, the gods give us joy.

TOUCHSTONE Amen.—A man may, if he were of a fearful heart, stagger in this attempt; for here we have no temple but the wood, no assembly but horn-beasts. But what though? Courage. As horns are odious, they are necessary. It is said many a man knows no end of his goods. Right: many a man has good horns, and knows no end of them. Well, that is the dowry of his wife, 'tis none of his own getting. Horns? Even so. Poor men alone? No, no; the noblest deer hath them as huge as the rascal. Is the single man therefore blessed? No. As a walled town is more worthier than a village, so is the forehead of a married man more honourable than the bare brow of a bachelor. And by how much defence is better than no skill, by so much is a horn more precious than to want.      57

*Enter Sir Oliver Martext*

Here comes Sir Oliver.—Sir Oliver Martext, you are well met. Will you dispatch us here under this tree, or shall we go with you to your chapel?      60

SIR OLIVER MARTEXT Is there none here to give the woman?

TOUCHSTONE I will not take her on gift of any man.

SIR OLIVER MARTEXT Truly she must be given, or the marriage is not lawful.

JAQUES (*coming forward*) Proceed, proceed. I'll give her.

TOUCHSTONE Good even, good Monsieur What-ye-call't. How do you, sir? You are very well met. God'ield you for your last company. I am very glad to see you. Even a toy in hand here, sir.

*Jaques removes his hat*

Nay, pray be covered.      70

JAQUES Will you be married, motley?

TOUCHSTONE As the ox hath his bow, sir, the horse his curb, and the falcon her bells, so man hath his desires; and as pigeons bill, so wedlock would be nibbling.

JAQUES And will you, being a man of your breeding, be married under a bush, like a beggar? Get you to church, and have a good priest that can tell you what marriage is. This fellow will but join you together as they join wainscot; then one of you will prove a shrunk panel and, like green timber, warp, warp.      80

TOUCHSTONE I am not in the mind but I were better to be married of him than of another, for he is not like to marry me well, and not being well married, it will be a good excuse for me hereafter to leave my wife.

JAQUES Go thou with me, and let me counsel thee.    85

TOUCHSTONE

Come, sweet Audrey.

We must be married, or we must live in bawdry.

Farewell, good Master Oliver. Not

        O, sweet Oliver,

        O, brave Oliver,                    90

    Leave me not behind thee

but

        Wind away,

        Begone, I say,

    I will not to wedding with thee.        95

SIR OLIVER MARTEXT (aside) 'Tis no matter. Ne'er a fantastical knave of them all shall flout me out of my calling.                    Exeunt

3.4    Enter Rosalind as Ganymede and Celia as Aliena

ROSALIND Never talk to me. I will weep.

CELIA Do, I prithee, but yet have the grace to consider that tears do not become a man.

ROSALIND But have I not cause to weep?

CELIA As good cause as one would desire; therefore weep.

ROSALIND His very hair is of the dissembling colour.    6

CELIA Something browner than Judas's. Marry, his kisses are Judas's own children.

ROSALIND I'faith, his hair is of a good colour.

CELIA An excellent colour. Your chestnut was ever the only colour.                    11

ROSALIND And his kissing is as full of sanctity as the touch of holy bread.

CELIA He hath bought a pair of cast lips of Diana. A nun of winter's sisterhood kisses not more religiously. The very ice of chastity is in them.            16

ROSALIND But why did he swear he would come this morning, and comes not?

CELIA Nay, certainly, there is no truth in him.

ROSALIND Do you think so?                    20

CELIA Yes. I think he is not a pick-purse, nor a horse-stealer; but for his verity in love, I do think him as concave as a covered goblet, or a worm-eaten nut.

ROSALIND Not true in love?

CELIA Yes, when he is in. But I think he is not in.    25

ROSALIND You have heard him swear downright he was.

CELIA 'Was' is not 'is'. Besides, the oath of a lover is no stronger than the word of a tapster. They are both the confirmer of false reckonings. He attends here in the forest on the Duke your father.            30

ROSALIND I met the Duke yesterday, and had much question with him. He asked me of what parentage I was. I told him, of as good as he, so he laughed and let me go. But what talk we of fathers when there is such a man as Orlando?                    35

CELIA O that's a brave man. He writes brave verses, speaks brave words, swears brave oaths, and breaks them bravely, quite traverse, athwart the heart of his lover, as a puny tilter that spurs his horse but on one side breaks his staff, like a noble goose. But all's brave that youth mounts, and folly guides. Who comes here?

    Enter Corin

CORIN

Mistress and master, you have oft enquired    42

After the shepherd that complained of love

Who you saw sitting by me on the turf,

Praising the proud disdainful shepherdess    45

That was his mistress.

CELIA                    Well, and what of him?

CORIN

If you will see a pageant truly played

Between the pale complexion of true love

And the red glow of scorn and proud disdain,

Go hence a little, and I shall conduct you,    50

If you will mark it.

ROSALIND (to Celia)    O come, let us remove.

The sight of lovers feedeth those in love.

(To Corin) Bring us to this sight, and you shall say

I'll prove a busy actor in their play.        Exeunt

3.5    Enter Silvius and Phoebe

SILVIUS

Sweet Phoebe, do not scorn me, do not, Phoebe.

Say that you love me not, but say not so

In bitterness. The common executioner,

Whose heart th'accustomed sight of death makes hard,

Falls not the axe upon the humbled neck    5

But first begs pardon. Will you sterner be

Than he that dies and lives by bloody drops?

    Enter Rosalind as Ganymede, Celia as Aliena, and

    Corin, and stand aside

PHOEBE (to Silvius)

I would not be thy executioner.

I fly thee for I would not injure thee.

Thou tell'st me there is murder in mine eye.    10

'Tis pretty, sure, and very probable

That eyes, that are the frail'st and softest things,

Who shut their coward gates on atomies,

Should be called tyrants, butchers, murderers.

Now I do frown on thee with all my heart,    15

And if mine eyes can wound, now let them kill thee.

Now counterfeit to swoon, why now fall down;

Or if thou canst not, O, for shame, for shame,

Lie not, to say mine eyes are murderers.

Now show the wound mine eye hath made in thee.    20

Scratch thee but with a pin, and there remains

Some scar of it. Lean upon a rush,

The cicatrice and capable impressure

Thy palm some moment keeps. But now mine eyes,

Which I have darted at thee, hurt thee not;    25

Nor I am sure there is no force in eyes

That can do hurt.

SILVIUS O dear Phoebe,

If ever—as that ever may be near—

You meet in some fresh cheek the power of fancy,    30

Then shall you know the wounds invisible

That love's keen arrows make.

PHOEBE                    But till that time

Come not thou near me. And when that time comes,

Afflict me with thy mocks, pity me not,

As till that time I shall not pity thee.    35

ROSALIND (coming forward)

And why, I pray you? Who might be your mother,

That you insult, exult, and all at once,

Over the wretched? What though you have no beauty—

As, by my faith, I see no more in you

Than without candle may go dark to bed—    40

Must you be therefore proud and pitiless?

Why, what means this? Why do you look on me?

I see no more in you than in the ordinary
Of nature's sale-work.—'Od's my little life,
I think she means to tangle my eyes, too.                         45
No, faith, proud mistress, hope not after it.
'Tis not your inky brows, your black silk hair,
Your bugle eyeballs, nor your cheek of cream,
That can entame my spirits to your worship.
(*To Silvius*) You, foolish shepherd, wherefore do you
    follow her                                                    50
Like foggy south, puffing with wind and rain?
You are a thousand times a properer man
Than she a woman. 'Tis such fools as you
That makes the world full of ill-favoured children.
'Tis not her glass but you that flatters her,                     55
And out of you she sees herself more proper
Than any of her lineaments can show her.
(*To Phoebe*) But, mistress, know yourself; down on
    your knees
And thank heaven, fasting, for a good man's love;
For I must tell you friendly in your ear,                         60
Sell when you can. You are not for all markets.
Cry the man mercy, love him, take his offer;
Foul is most foul, being foul to be a scoffer.—
So, take her to thee, shepherd. Fare you well.

PHOEBE
Sweet youth, I pray you chide a year together.                    65
I had rather hear you chide than this man woo.

ROSALIND (*to Phoebe*) He's fallen in love with your foulness,
(*to Silvius*) and she'll fall in love with my anger. If it
be so, as fast as she answers thee with frowning looks,
I'll sauce her with bitter words.                                 70
(*To Phoebe*) Why look you so upon me?

PHOEBE
For no ill will I bear you.

ROSALIND
I pray you do not fall in love with me,
For I am falser than vows made in wine.
Besides, I like you not. If you will know my house,              75
'Tis at the tuft of olives, here hard by.
(*To Celia*) Will you go, sister? (*To Silvius*) Shepherd,
    ply her hard.—
Come, sister. (*To Phoebe*) Shepherdess, look on him
    better,
And be not proud. Though all the world could see,
None could be so abused in sight as he.—                         80
Come, to our flock.     *Exeunt Rosalind, Celia, and Corin*

PHOEBE (*aside*)
Dead shepherd, now I find thy saw of might:
'Who ever loved that loved not at first sight?'

SILVIUS
Sweet Phoebe—

PHOEBE              Ha, what sayst thou, Silvius?

SILVIUS Sweet Phoebe, pity me.                                    85

PHOEBE
Why, I am sorry for thee, gentle Silvius.

SILVIUS
Wherever sorrow is, relief would be.
If you do sorrow at my grief in love,
By giving love your sorrow and my grief
Were both extermined.                                             90

PHOEBE
Thou hast my love, is not that neighbourly?

SILVIUS
I would have you.

PHOEBE                Why, that were covetousness.
Silvius, the time was that I hated thee;

And yet it is not that I bear thee love.
But since that thou canst talk of love so well,                  95
Thy company, which erst was irksome to me,
I will endure; and I'll employ thee, too.
But do not look for further recompense
Than thine own gladness that thou art employed.

SILVIUS
So holy and so perfect is my love,                              100
And I in such a poverty of grace,
That I shall think it a most plenteous crop
To glean the broken ears after the man
That the main harvest reaps. Loose now and then
A scattered smile, and that I'll live upon.                     105

PHOEBE
Know'st thou the youth that spoke to me erewhile?

SILVIUS
Not very well, but I have met him oft,
And he hath bought the cottage and the bounds
That the old Carlot once was master of.

PHOEBE
Think not I love him, though I ask for him.                     110
'Tis but a peevish boy. Yet he talks well.
But what care I for words? Yet words do well
When he that speaks them pleases those that hear.
It is a pretty youth—not very pretty—
But sure he's proud; and yet his pride becomes him.
He'll make a proper man. The best thing in him                  116
Is his complexion; and faster than his tongue
Did make offence, his eye did heal it up.
He is not very tall; yet for his years he's tall.
His leg is but so-so; and yet 'tis well.                        120
There was a pretty redness in his lip,
A little riper and more lusty-red
Than that mixed in his cheek. 'Twas just the
    difference
Betwixt the constant red and mingled damask.
There be some women, Silvius, had they marked him
In parcels as I did, would have gone near               126
To fall in love with him; but for my part,
I love him not, nor hate him not. And yet
Have I more cause to hate him than to love him,
For what had he to do to chide at me?                   130
He said mine eyes were black, and my hair black,
And now I am remembered, scorned at me.
I marvel why I answered not again.
But that's all one. Omittance is no quittance.
I'll write to him a very taunting letter,               135
And thou shalt bear it. Wilt thou, Silvius?

SILVIUS
Phoebe, with all my heart.

PHOEBE                      I'll write it straight.
The matter's in my head and in my heart.
I will be bitter with him, and passing short.
Go with me, Silvius.                            *Exeunt*

4.1  *Enter Rosalind as Ganymede, Celia as Aliena, and
     Jaques*

JAQUES I prithee, pretty youth, let me be better acquainted
with thee.

ROSALIND They say you are a melancholy fellow.

JAQUES I am so. I do love it better than laughing.

ROSALIND Those that are in extremity of either are
abominable fellows, and betray themselves to every
modern censure worse than drunkards.

JAQUES Why, 'tis good to be sad and say nothing.

ROSALIND Why then, 'tis good to be a post.                       9

JAQUES I have neither the scholar's melancholy, which is emulation, nor the musician's, which is fantastical, nor the courtier's, which is proud, nor the soldier's, which is ambitious, nor the lawyer's, which is politic, nor the lady's, which is nice, nor the lover's, which is all these; but it is a melancholy of mine own, compounded of many simples, extracted from many objects, and indeed the sundry contemplation of my travels, in which my often rumination wraps me in a most humorous sadness.                                                   19

ROSALIND A traveller! By my faith, you have great reason to be sad. I fear you have sold your own lands to see other men's. Then to have seen much and to have nothing is to have rich eyes and poor hands.

JAQUES Yes, I have gained my experience.              24

*Enter Orlando*

ROSALIND And your experience makes you sad. I had rather have a fool to make me merry than experience to make me sad—and to travel for it too!

ORLANDO

Good day and happiness, dear Rosalind.

JAQUES Nay then, God b'wi'you an you talk in blank verse.                                                30

ROSALIND Farewell, Monsieur Traveller. Look you lisp, and wear strange suits; disable all the benefits of your own country; be out of love with your nativity, and almost chide God for making you that countenance you are, or I will scarce think you have swam in a gondola.                                      ⌜*Exit Jaques*⌝
Why, how now, Orlando? Where have you been all this while? You a lover? An you serve me such another trick, never come in my sight more.              39

ORLANDO My fair Rosalind, I come within an hour of my promise.

ROSALIND Break an hour's promise in love! He that will divide a minute into a thousand parts and break but a part of the thousand part of a minute in the affairs of love, it may be said of him that Cupid hath clapped him o'th' shoulder, but I'll warrant him heartwhole.

ORLANDO Pardon me, dear Rosalind.                     47

ROSALIND Nay, an you be so tardy, come no more in my sight. I had as lief be wooed of a snail.

ORLANDO Of a snail?                                    50

ROSALIND Ay, of a snail; for though he comes slowly, he carries his house on his head—a better jointure, I think, than you make a woman. Besides, he brings his destiny with him.

ORLANDO What's that?                                   55

ROSALIND Why, horns, which such as you are fain to be beholden to your wives for. But he comes armed in his fortune, and prevents the slander of his wife.

ORLANDO Virtue is no hornmaker, and my Rosalind is virtuous.                                             60

ROSALIND And I am your Rosalind.

CELIA It pleases him to call you so; but he hath a Rosalind of a better leer than you.

ROSALIND Come, woo me, woo me, for now I am in a holiday humour, and like enough to consent. What would you say to me now an I were your very, very Rosalind?                                              67

ORLANDO I would kiss before I spoke.

ROSALIND Nay, you were better speak first, and when you were gravelled for lack of matter you might take occasion to kiss. Very good orators, when they are out, they will spit; and for lovers, lacking—God warr'nt us—matter, the cleanliest shift is to kiss.

ORLANDO How if the kiss be denied?                     74

ROSALIND Then she puts you to entreaty, and there begins new matter.

ORLANDO Who could be out, being before his beloved mistress?

ROSALIND Marry, that should you if I were your mistress, or I should think my honesty ranker than my wit.   80

ORLANDO What, of my suit?

ROSALIND Not out of your apparel, and yet out of your suit. Am not I your Rosalind?

ORLANDO I take some joy to say you are because I would be talking of her.                                    85

ROSALIND Well, in her person I say I will not have you.

ORLANDO Then in mine own person I die.

ROSALIND No, faith; die by attorney. The poor world is almost six thousand years old, and in all this time there was not any man died in his own person, videlicet, in a love-cause. Troilus had his brains dashed out with a Grecian club, yet he did what he could to die before, and he is one of the patterns of love. Leander, he would have lived many a fair year though Hero had turned nun if it had not been for a hot midsummer night, for, good youth, he went but forth to wash him in the Hellespont and, being taken with the cramp, was drowned; and the foolish chroniclers of that age found it was Hero of Sestos. But these are all lies. Men have died from time to time, and worms have eaten them, but not for love.                                     101

ORLANDO I would not have my right Rosalind of this mind, for I protest her frown might kill me.

ROSALIND By this hand, it will not kill a fly. But come, now I will be your Rosalind in a more coming-on disposition; and ask me what you will, I will grant it.

ORLANDO Then love me, Rosalind.                       107

ROSALIND Yes, faith, will I, Fridays and Saturdays and all.

ORLANDO And wilt thou have me?                        110

ROSALIND Ay, and twenty such.

ORLANDO What sayst thou?

ROSALIND Are you not good?

ORLANDO I hope so.

ROSALIND Why then, can one desire too much of a good thing? (*To Celia*) Come, sister, you shall be the priest and marry us.—Give me your hand, Orlando.—What do you say, sister?

ORLANDO (*to Celia*) Pray thee, marry us.

CELIA I cannot say the words.                         120

ROSALIND You must begin, 'Will you, Orlando'—

CELIA Go to. Will you, Orlando, have to wife this Rosalind?

ORLANDO I will.

ROSALIND Ay, but when?                                125

ORLANDO Why now, as fast as she can marry us.

ROSALIND Then you must say, 'I take thee, Rosalind, for wife.'

ORLANDO I take thee, Rosalind, for wife.

ROSALIND I might ask you for your commission; but I do take thee, Orlando, for my husband. There's a girl goes before the priest; and certainly a woman's thought runs before her actions.                              133

ORLANDO So do all thoughts; they are winged.

ROSALIND Now tell me how long you would have her after you have possessed her?

ORLANDO For ever and a day.

ROSALIND Say a day without the ever. No, no, Orlando; men are April when they woo, December when they wed. Maids are May when they are maids, but the sky changes when they are wives. I will be more jealous of thee than a Barbary cock-pigeon over his hen, more clamorous than a parrot against rain, more new-fangled than an ape, more giddy in my desires than a monkey. I will weep for nothing, like Diana in the fountain, and I will do that when you are disposed to be merry. I will laugh like a hyena, and that when thou art inclined to sleep.

ORLANDO But will my Rosalind do so?

ROSALIND By my life, she will do as I do.          150

ORLANDO O, but she is wise.

ROSALIND Or else she could not have the wit to do this. The wiser, the waywarder. Make the doors upon a woman's wit, and it will out at the casement. Shut that, and 'twill out at the key-hole. Stop that, 'twill fly with the smoke out at the chimney.          156

ORLANDO A man that had a wife with such a wit, he might say 'Wit, whither wilt?'

ROSALIND Nay, you might keep that check for it till you met your wife's wit going to your neighbour's bed.

ORLANDO And what wit could wit have to excuse that?

ROSALIND Marry, to say she came to seek you there. You shall never take her without her answer unless you take her without her tongue. O, that woman that cannot make her fault her husband's occasion, let her never nurse her child herself, for she will breed it like a fool.          167

ORLANDO For these two hours, Rosalind, I will leave thee.

ROSALIND Alas, dear love, I cannot lack thee two hours.

ORLANDO I must attend the Duke at dinner. By two o'clock I will be with thee again.          171

ROSALIND Ay, go your ways, go your ways. I knew what you would prove; my friends told me as much, and I thought no less. That flattering tongue of yours won me. 'Tis but one cast away, and so, come, death! Two o'clock is your hour?          176

ORLANDO Ay, sweet Rosalind.

ROSALIND By my troth, and in good earnest, and so God mend me, and by all pretty oaths that are not dangerous, if you break one jot of your promise or come one minute behind your hour, I will think you the most pathetical break-promise, and the most hollow lover, and the most unworthy of her you call Rosalind that may be chosen out of the gross band of the unfaithful. Therefore beware my censure, and keep your promise.          186

ORLANDO With no less religion than if thou wert indeed my Rosalind. So, adieu.

ROSALIND Well, Time is the old justice that examines all such offenders; and let Time try. Adieu.     *Exit Orlando*

CELIA You have simply misused our sex in your love-prate. We must have your doublet and hose plucked over your head, and show the world what the bird hath done to her own nest.          194

ROSALIND O coz, coz, coz, my pretty little coz, that thou didst know how many fathom deep I am in love. But it cannot be sounded. My affection hath an unknown bottom, like the Bay of Portugal.

CELIA Or rather bottomless, that as fast as you pour affection in, it runs out.          200

ROSALIND No, that same wicked bastard of Venus, that was begot of thought, conceived of spleen, and born of madness, that blind rascally boy that abuses everyone's eyes because his own are out, let him be judge how deep I am in love. I'll tell thee, Aliena, I cannot be out

of the sight of Orlando. I'll go find a shadow and sigh till he come.          207

CELIA And I'll sleep.          *Exeunt*

**4.2**   *Enter Jaques and Lords dressed as foresters*

JAQUES Which is he that killed the deer?

FIRST LORD Sir, it was I.

JAQUES (*to the others*) Let's present him to the Duke like a Roman conqueror. And it would do well to set the deer's horns upon his head for a branch of victory. Have you no song, forester, for this purpose?          6

SECOND LORD Yes, sir.

JAQUES Sing it. 'Tis no matter how it be in tune, so it make noise enough.

LORDS (*sing*)
    What shall he have that killed the deer?          10
    His leather skin and horns to wear.
    Then sing him home; the rest shall bear
      This burden.
    Take thou no scorn to wear the horn;
    It was a crest ere thou wast born.          15
      Thy father's father wore it,
      And thy father bore it.
    The horn, the horn, the lusty horn
    Is not a thing to laugh to scorn.          *Exeunt*

**4.3**   *Enter Rosalind as Ganymede and Celia as Aliena*

ROSALIND How say you now? Is it not past two o'clock? And here much Orlando.

CELIA I warrant you, with pure love and troubled brain he hath ta'en his bow and arrows and is gone forth to sleep.          5
    ⌜*Enter Silvius*⌝
Look who comes here.

SILVIUS (*to Rosalind*)
My errand is to you, fair youth.
My gentle Phoebe did bid me give you this.
    *He offers Rosalind a letter, which she takes*
    *and reads*
I know not the contents, but as I guess
By the stern brow and waspish action          10
Which she did use as she was writing of it,
It bears an angry tenor. Pardon me;
I am but as a guiltless messenger.

ROSALIND
Patience herself would startle at this letter,
And play the swaggerer. Bear this, bear all.          15
She says I am not fair, that I lack manners;
She calls me proud, and that she could not love me
Were man as rare as Phoenix. 'Od's my will,
Her love is not the hare that I do hunt.
Why writes she so to me? Well, shepherd, well,          20
This is a letter of your own device.

SILVIUS
No, I protest; I know not the contents.
Phoebe did write it.

ROSALIND          Come, come, you are a fool,
And turned into the extremity of love.
I saw her hand. She has a leathern hand,          25
A free-stone coloured hand. I verily did think
That her old gloves were on; but 'twas her hands.
She has a housewife's hand—but that's no matter.
I say she never did invent this letter.
This is a man's invention, and his hand.          30

SILVIUS Sure, it is hers.

ROSALIND
Why, 'tis a boisterous and a cruel style,
A style for challengers. Why, she defies me,
Like Turk to Christian. Women's gentle brain
Could not drop forth such giant-rude invention,      35
Such Ethiop words, blacker in their effect
Than in their countenance. Will you hear the letter?
SILVIUS
So please you, for I never heard it yet,
Yet heard too much of Phoebe's cruelty.
ROSALIND
She Phoebes me. Mark how the tyrant writes:      40
(reads) 'Art thou god to shepherd turned,
        That a maiden's heart hath burned?'
Can a woman rail thus?
SILVIUS Call you this railing?
ROSALIND (reads)
        'Why, thy godhead laid apart,      45
        Warr'st thou with a woman's heart?'
Did you ever hear such railing?
        'Whiles the eye of man did woo me
        That could do no vengeance to me.'—
Meaning me a beast.      50
        'If the scorn of your bright eyne
        Have power to raise such love in mine,
        Alack, in me what strange effect
        Would they work in mild aspect?
        Whiles you chid me I did love;      55
        How then might your prayers move?
        He that brings this love to thee
        Little knows this love in me,
        And by him seal up thy mind
        Whether that thy youth and kind      60
        Will the faithful offer take
        Of me, and all that I can make,
        Or else by him my love deny,
        And then I'll study how to die.'
SILVIUS Call you this chiding?      65
CELIA Alas, poor shepherd.
ROSALIND Do you pity him? No, he deserves no pity. (To
Silvius) Wilt thou love such a woman? What, to make
thee an instrument, and play false strains upon thee?—
not to be endured. Well, go your way to her—for I see
love hath made thee a tame snake—and say this to
her: that if she love me, I charge her to love thee. If
she will not, I will never have her unless thou entreat
for her. If you be a true lover, hence, and not a word;
for here comes more company.      Exit Silvius
        Enter Oliver
OLIVER
Good morrow, fair ones. Pray you, if you know,      76
Where in the purlieus of this forest stands
A sheepcote fenced about with olive trees?
CELIA
West of this place, down in the neighbour bottom.
The rank of osiers by the murmuring stream      80
Left on your right hand brings you to the place.
But at this hour the house doth keep itself.
There's none within.
OLIVER
If that an eye may profit by a tongue,
Then should I know you by description.      85
Such garments, and such years. 'The boy is fair,
Of female favour, and bestows himself
Like a ripe sister. The woman low
And browner than her brother.' Are not you
The owner of the house I did enquire for?      90

CELIA
It is no boast, being asked, to say we are.
OLIVER
Orlando doth commend him to you both,
And to that youth he calls his Rosalind
He sends this bloody napkin. Are you he?
ROSALIND
I am. What must we understand by this?      95
OLIVER
Some of my shame, if you will know of me
What man I am, and how, and why, and where
This handkerchief was stained.
CELIA                          I pray you tell it.
OLIVER
When last the young Orlando parted from you,
He left a promise to return again      100
Within an hour, and pacing through the forest,
Chewing the food of sweet and bitter fancy,
Lo what befell. He threw his eye aside,
And mark what object did present itself.      104
Under an old oak, whose boughs were mossed with age
And high top bald with dry antiquity,
A wretched, ragged man, o'ergrown with hair,
Lay sleeping on his back. About his neck
A green and gilded snake had wreathed itself,
Who with her head, nimble in threats, approached
The opening of his mouth. But suddenly      111
Seeing Orlando, it unlinked itself,
And with indented glides did slip away
Into a bush, under which bush's shade
A lioness, with udders all drawn dry,      115
Lay couching, head on ground, with catlike watch
When that the sleeping man should stir. For 'tis
The royal disposition of that beast
To prey on nothing that doth seem as dead.
This seen, Orlando did approach the man      120
And found it was his brother, his elder brother.
CELIA
O, I have heard him speak of that same brother,
And he did render him the most unnatural
That lived amongst men.
OLIVER                    And well he might so do,
For well I know he was unnatural.      125
ROSALIND
But to Orlando. Did he leave him there,
Food to the sucked and hungry lioness?
OLIVER
Twice did he turn his back, and purposed so.
But kindness, nobler ever than revenge,
And nature, stronger than his just occasion,      130
Made him give battle to the lioness,
Who quickly fell before him; in which hurtling
From miserable slumber I awaked.
CELIA
Are you his brother?
ROSALIND                  Was't you he rescued?
CELIA
Was't you that did so oft contrive to kill him?      135
OLIVER
'Twas I, but 'tis not I. I do not shame
To tell you what I was, since my conversion
So sweetly tastes, being the thing I am.
ROSALIND
But for the bloody napkin?
OLIVER                          By and by.
When from the first to last betwixt us two      140
Tears our recountments had most kindly bathed—

647

As how I came into that desert place—
I' brief, he led me to the gentle Duke,
Who gave me fresh array, and entertainment,
Committing me unto my brother's love,　　　145
Who led me instantly unto his cave,
There stripped himself, and here upon his arm
The lioness had torn some flesh away,
Which all this while had bled. And now he fainted,
And cried in fainting upon Rosalind.　　　150
Brief, I recovered him, bound up his wound,
And after some small space, being strong at heart,
He sent me hither, stranger as I am,
To tell this story, that you might excuse
His broken promise, and to give this napkin,　　　155
Dyed in his blood, unto the shepherd youth
That he in sport doth call his Rosalind.

*Rosalind faints*

CELIA
Why, how now, Ganymede, sweet Ganymede!

OLIVER
Many will swoon when they do look on blood.

CELIA
There is more in it. Cousin Ganymede!　　　160

OLIVER Look, he recovers.

ROSALIND I would I were at home.

CELIA We'll lead you thither.

　(*To Oliver*) I pray you, will you take him by the arm?

OLIVER Be of good cheer, youth. You a man? You lack a
man's heart.　　　166

ROSALIND I do so, I confess it. Ah, sirrah, a body would
think this was well counterfeited. I pray you, tell your
brother how well I counterfeited. Heigh-ho!

OLIVER This was not counterfeit. There is too great
testimony in your complexion that it was a passion of
earnest.　　　172

ROSALIND Counterfeit, I assure you.

OLIVER Well then, take a good heart, and counterfeit to
be a man.　　　175

ROSALIND So I do; but, i'faith, I should have been a
woman by right.

CELIA Come, you look paler and paler. Pray you, draw
homewards. Good sir, go with us.

OLIVER
That will I, for I must bear answer back　　　180
How you excuse my brother, Rosalind.

ROSALIND I shall devise something. But I pray you
commend my counterfeiting to him. Will you go?

　　　　　　　　　　　　　　　　　　　*Exeunt*

**5.1**　*Enter Touchstone the clown and Audrey*

TOUCHSTONE We shall find a time, Audrey. Patience, gentle
Audrey.

AUDREY Faith, the priest was good enough, for all the old
gentleman's saying.

TOUCHSTONE A most wicked Sir Oliver, Audrey, a most
vile Martext. But, Audrey, there is a youth here in the
forest lays claim to you.　　　7

AUDREY Ay, I know who 'tis. He hath no interest in me
in the world. Here comes the man you mean.

　*Enter William*

TOUCHSTONE It is meat and drink to me to see a clown.
By my troth, we that have good wits have much to
answer for. We shall be flouting; we cannot hold.　　　12

WILLIAM Good ev'n, Audrey.

AUDREY God ye good ev'n, William.

WILLIAM (*to Touchstone*) And good ev'n to you, sir.　　　15

TOUCHSTONE Good ev'n, gentle friend. Cover thy head,
cover thy head. Nay, prithee, be covered. How old are
you, friend?

WILLIAM Five-and-twenty, sir.

TOUCHSTONE A ripe age. Is thy name William?　　　20

WILLIAM William, sir.

TOUCHSTONE A fair name. Wast born i'th' forest here?

WILLIAM Ay, sir, I thank God.

TOUCHSTONE Thank God—a good answer. Art rich?

WILLIAM Faith, sir, so-so.　　　25

TOUCHSTONE So-so is good, very good, very excellent good.
And yet it is not, it is but so-so. Art thou wise?

WILLIAM Ay, sir, I have a pretty wit.

TOUCHSTONE Why, thou sayst well. I do now remember a
saying: 'The fool doth think he is wise, but the wise
man knows himself to be a fool.' The heathen
philosopher, when he had a desire to eat a grape,
would open his lips when he put it into his mouth,
meaning thereby that grapes were made to eat, and
lips to open. You do love this maid?　　　35

WILLIAM I do, sir.

TOUCHSTONE Give me your hand. Art thou learned?

WILLIAM No, sir.

TOUCHSTONE Then learn this of me: to have is to have.
For it is a figure in rhetoric that drink, being poured
out of a cup into a glass, by filling the one doth empty
the other. For all your writers do consent that *ipse* is
he. Now you are not *ipse*, for I am he.　　　43

WILLIAM Which he, sir?

TOUCHSTONE He, sir, that must marry this woman.
Therefore, you clown, abandon—which is in the vulgar,
leave—the society—which in the boorish is company—
of this female—which in the common is woman; which
together is, abandon the society of this female, or,
clown, thou perishest; or, to thy better understanding,
diest; or, to wit, I kill thee, make thee away, translate
thy life into death, thy liberty into bondage. I will deal
in poison with thee, or in bastinado, or in steel. I will
bandy with thee in faction, I will o'errun thee with
policy. I will kill thee a hundred and fifty ways.
Therefore tremble, and depart.　　　56

AUDREY Do, good William.

WILLIAM God rest you merry, sir.　　　　　　　*Exit*

　*Enter Corin*

CORIN Our master and mistress seeks you. Come, away,
away.　　　60

TOUCHSTONE Trip, Audrey, trip, Audrey. (*To Corin*) I
attend, I attend.　　　　　　　　　　　　　*Exeunt*

**5.2**　*Enter Orlando and Oliver*

ORLANDO Is't possible that on so little acquaintance you
should like her? That but seeing, you should love her?
And loving, woo? And wooing, she should grant? And
will you persevere to enjoy her?

OLIVER Neither call the giddiness of it in question, the
poverty of her, the small acquaintance, my sudden
wooing, nor her sudden consenting; but say with me,
'I love Aliena'; say with her, that she loves me; consent
with both that we may enjoy each other. It shall be to
your good, for my father's house and all the revenue
that was old Sir Rowland's will I estate upon you, and
here live and die a shepherd.　　　12

　*Enter Rosalind as Ganymede*

ORLANDO You have my consent. Let your wedding be
tomorrow. Thither will I invite the Duke and all's

contented followers. Go you, and prepare Aliena; for look you, here comes my Rosalind.        16

ROSALIND God save you, brother.

OLIVER And you, fair sister.                    *Exit*

ROSALIND O, my dear Orlando, how it grieves me to see thee wear thy heart in a scarf.        20

ORLANDO It is my arm.

ROSALIND I thought thy heart had been wounded with the claws of a lion.

ORLANDO Wounded it is, but with the eyes of a lady.

ROSALIND Did your brother tell you how I counterfeited to swoon when he showed me your handkerchief?    26

ORLANDO Ay, and greater wonders than that.

ROSALIND O, I know where you are. Nay, 'tis true. There was never anything so sudden but the fight of two rams, and Caesar's thrasonical brag of 'I came, saw, and overcame', for your brother and my sister no sooner met but they looked; no sooner looked but they loved; no sooner loved but they sighed; no sooner sighed but they asked one another the reason; no sooner knew the reason but they sought the remedy; and in these degrees have they made a pair of stairs to marriage, which they will climb incontinent, or else be incontinent before marriage. They are in the very wrath of love, and they will together. Clubs cannot part them.

ORLANDO They shall be married tomorrow, and I will bid the Duke to the nuptial. But O, how bitter a thing it is to look into happiness through another man's eyes. By so much the more shall I tomorrow be at the height of heart-heaviness by how much I shall think my brother happy in having what he wishes for.        45

ROSALIND Why, then, tomorrow I cannot serve your turn for Rosalind?

ORLANDO I can live no longer by thinking.

ROSALIND I will weary you then no longer with idle talking. Know of me then—for now I speak to some purpose—that I know you are a gentleman of good conceit. I speak not this that you should bear a good opinion of my knowledge, insomuch I say I know you are; neither do I labour for a greater esteem than may in some little measure draw a belief from you to do yourself good, and not to grace me. Believe then, if you please, that I can do strange things. I have since I was three year old conversed with a magician, most profound in his art, and yet not damnable. If you do love Rosalind so near the heart as your gesture cries it out, when your brother marries Aliena shall you marry her. I know into what straits of fortune she is driven, and it is not impossible to me, if it appear not inconvenient to you, to set her before your eyes tomorrow, human as she is, and without any danger.

ORLANDO Speakest thou in sober meanings?        66

ROSALIND By my life, I do, which I tender dearly, though I say I am a magician. Therefore put you in your best array, bid your friends: for if you will be married tomorrow, you shall; and to Rosalind if you will.        70

*Enter Silvius and Phoebe*

Look, here comes a lover of mine and a lover of hers.

PHOEBE (*to Rosalind*)
Youth, you have done me much ungentleness,
To show the letter that I writ to you.

ROSALIND
I care not if I have. It is my study
To seem despiteful and ungentle to you.        75
You are there followed by a faithful shepherd.
Look upon him; love him. He worships you.

PHOEBE (*to Silvius*)
Good shepherd, tell this youth what 'tis to love.

SILVIUS
It is to be all made of sighs and tears,
And so am I for Phoebe.        80

PHOEBE And I for Ganymede.

ORLANDO And I for Rosalind.

ROSALIND And I for no woman.

SILVIUS
It is to be all made of faith and service,
And so am I for Phoebe.        85

PHOEBE And I for Ganymede.

ORLANDO And I for Rosalind.

ROSALIND And I for no woman.

SILVIUS
It is to be all made of fantasy,
All made of passion, and all made of wishes,        90
All adoration, duty, and observance,
All humbleness, all patience and impatience,
All purity, all trial, all obedience,
And so am I for Phoebe.

PHOEBE And so am I for Ganymede.        95

ORLANDO And so am I for Rosalind.

ROSALIND And so am I for no woman.

PHOEBE (*to Rosalind*)
If this be so, why blame you me to love you?

SILVIUS (*to Phoebe*)
If this be so, why blame you me to love you?

ORLANDO
If this be so, why blame you me to love you?        100

ROSALIND Why do you speak too, 'Why blame you me to love you?'

ORLANDO
To her that is not here nor doth not hear.

ROSALIND Pray you, no more of this, 'tis like the howling of Irish wolves against the moon. (*To Silvius*) I will help you if I can. (*To Phoebe*) I would love you if I could.— Tomorrow meet me all together. (*To Phoebe*) I will marry you if ever I marry woman, and I'll be married tomorrow. (*To Orlando*) I will satisfy you if ever I satisfy man, and you shall be married tomorrow. (*To Silvius*) I will content you if what pleases you contents you, and you shall be married tomorrow. (*To Orlando*) As you love Rosalind, meet. (*To Silvius*) As you love Phoebe, meet. And as I love no woman, I'll meet. So fare you well. I have left you commands.        115

SILVIUS I'll not fail, if I live.

PHOEBE Nor I.

ORLANDO Nor I.                    *Exeunt severally*

5.3    *Enter Touchstone the clown and Audrey*

TOUCHSTONE Tomorrow is the joyful day, Audrey, tomorrow will we be married.

AUDREY I do desire it with all my heart; and I hope it is no dishonest desire to desire to be a woman of the world. Here come two of the banished Duke's pages.

*Enter two Pages*

FIRST PAGE Well met, honest gentleman.        6

TOUCHSTONE By my troth, well met. Come, sit, sit, and a song.

SECOND PAGE We are for you. Sit i'th' middle.

FIRST PAGE Shall we clap into't roundly, without hawking, or spitting, or saying we are hoarse, which are the only prologues to a bad voice?        12

SECOND PAGE I'faith, i'faith, and both in a tune, like two gipsies on a horse.

BOTH PAGES (*sing*)
It was a lover and his lass,
With a hey, and a ho, and a hey-nonny-no,
That o'er the green cornfield did pass
In spring-time, the only pretty ring-time,
When birds do sing, hey ding-a-ding ding,
Sweet lovers love the spring.

Between the acres of the rye,
With a hey, and a ho, and a hey-nonny-no,
These pretty country folks would lie,
In spring-time, the only pretty ring-time,
When birds do sing, hey ding-a-ding ding,
Sweet lovers love the spring.

This carol they began that hour,
With a hey, and a ho, and a hey-nonny-no,
How that a life was but a flower,
In spring-time, the only pretty ring-time,
When birds do sing, hey ding-a-ding ding,
Sweet lovers love the spring.

And therefore take the present time,
With a hey, and a ho, and a hey-nonny-no,
For love is crownèd with the prime,
In spring time, the only pretty ring-time,
When birds do sing, hey ding-a-ding ding,
Sweet lovers love the spring.

TOUCHSTONE Truly, young gentlemen, though there was no great matter in the ditty, yet the note was very untunable.

FIRST PAGE You are deceived, sir, we kept time, we lost not our time.

TOUCHSTONE By my troth, yes, I count it but time lost to hear such a foolish song. God b'wi'you, and God mend your voices. Come, Audrey. *Exeunt severally*

**5.4** *Enter Duke Senior, Amiens, Jaques, Orlando, Oliver, and Celia as Aliena*

DUKE SENIOR
Dost thou believe, Orlando, that the boy
Can do all this that he hath promisèd?

ORLANDO
I sometimes do believe, and sometimes do not,
As those that fear they hope, and know they fear.
*Enter Rosalind as Ganymede, with Silvius and Phoebe*

ROSALIND
Patience once more, whiles our compact is urged.
(*To the Duke*) You say if I bring in your Rosalind
You will bestow her on Orlando here?

DUKE SENIOR
That would I, had I kingdoms to give with her.

ROSALIND (*to Orlando*)
And you say you will have her when I bring her?

ORLANDO
That would I, were I of all kingdoms king.

ROSALIND (*to Phoebe*)
You say you'll marry me if I be willing?

PHOEBE
That will I, should I die the hour after.

ROSALIND
But if you do refuse to marry me
You'll give yourself to this most faithful shepherd?

PHOEBE So is the bargain.

ROSALIND (*to Silvius*)
You say that you'll have Phoebe if she will.

SILVIUS
Though to have her and death were both one thing.

ROSALIND
I have promised to make all this matter even.
Keep you your word, O Duke, to give your daughter.
You yours, Orlando, to receive his daughter.
Keep your word, Phoebe, that you'll marry me,
Or else refusing me to wed this shepherd.
Keep your word, Silvius, that you'll marry her
If she refuse me; and from hence I go
To make these doubts all even.
*Exeunt Rosalind and Celia*

DUKE SENIOR
I do remember in this shepherd boy
Some lively touches of my daughter's favour.

ORLANDO
My lord, the first time that I ever saw him,
Methought he was a brother to your daughter.
But, my good lord, this boy is forest-born,
And hath been tutored in the rudiments
Of many desperate studies by his uncle,
Whom he reports to be a great magician
Obscurèd in the circle of this forest.
⌈*Enter Touchstone the clown and Audrey*⌉

JAQUES There is sure another flood toward, and these couples are coming to the ark. Here comes a pair of very strange beasts, which in all tongues are called fools.

TOUCHSTONE Salutation and greeting to you all.

JAQUES (*to the Duke*) Good my lord, bid him welcome. This is the motley-minded gentleman that I have so often met in the forest. He hath been a courtier, he swears.

TOUCHSTONE If any man doubt that, let him put me to my purgation. I have trod a measure, I have flattered a lady, I have been politic with my friend, smooth with mine enemy, I have undone three tailors, I have had four quarrels, and like to have fought one.

JAQUES And how was that ta'en up?

TOUCHSTONE Faith, we met, and found the quarrel was upon the seventh cause.

JAQUES How, seventh cause?—Good my lord, like this fellow.

DUKE SENIOR I like him very well.

TOUCHSTONE God'ield you, sir, I desire you of the like. I press in here, sir, amongst the rest of the country copulatives, to swear, and to forswear, according as marriage binds and blood breaks. A poor virgin, sir, an ill-favoured thing, sir, but mine own. A poor humour of mine, sir, to take that that no man else will. Rich honesty dwells like a miser, sir, in a poor house, as your pearl in your foul oyster.

DUKE SENIOR By my faith, he is very swift and sententious.

TOUCHSTONE According to the fool's bolt, sir, and such dulcet diseases.

JAQUES But for the seventh cause. How did you find the quarrel on the seventh cause?

TOUCHSTONE Upon a lie seven times removed.—Bear your body more seeming, Audrey.—As thus, sir: I did dislike the cut of a certain courtier's beard. He sent me word if I said his beard was not cut well, he was in the mind it was. This is called the Retort Courteous. If I sent him word again it was not well cut, he would send me word he cut it to please himself. This is called the Quip Modest. If again it was not well cut, he disabled my judgement. This is called the Reply Churlish. If again it was not well cut, he would answer I spake not true.

This is called the Reproof Valiant. If again it was not
well cut, he would say I lie. This is called the
Countercheck Quarrelsome. And so to the Lie Circumstantial, and the Lie Direct.                                  80

JAQUES And how oft did you say his beard was not well
cut?

TOUCHSTONE I durst go no further than the Lie
Circumstantial, nor he durst not give me the Lie Direct;
and so we measured swords, and parted.              85

JAQUES Can you nominate in order now the degrees of
the lie?

TOUCHSTONE O sir, we quarrel in print, by the book, as
you have books for good manners. I will name you the
degrees. The first, the Retort Courteous; the second,
the Quip Modest; the third, the Reply Churlish; the
fourth, the Reproof Valiant; the fifth, the Countercheck
Quarrelsome; the sixth, the Lie with Circumstance; the
seventh, the Lie Direct. All these you may avoid but
the Lie Direct; and you may avoid that, too, with an
'if'. I knew when seven justices could not take up a
quarrel, but when the parties were met themselves,
one of them thought but of an 'if', as 'If you said so,
then I said so', and they shook hands and swore
brothers. Your 'if' is the only peacemaker; much virtue
in 'if'.                                                              101

JAQUES (to the Duke) Is not this a rare fellow, my lord?
He's as good at anything, and yet a fool.

DUKE SENIOR He uses his folly like a stalking-horse, and
under the presentation of that he shoots his wit.   105
⌈Still music.⌉ Enter Hymen with Rosalind and Celia
    as themselves

HYMEN    Then is there mirth in heaven
    When earthly things made even
        Atone together.
    Good Duke, receive thy daughter;
    Hymen from heaven brought her,                110
        Yea, brought her hither,
    That thou mightst join her hand with his
    Whose heart within his bosom is.

ROSALIND (to the Duke)
    To you I give myself, for I am yours.
    (To Orlando) To you I give myself, for I am yours.   115

DUKE SENIOR
    If there be truth in sight, you are my daughter.

ORLANDO
    If there be truth in sight, you are my Rosalind.

PHOEBE
    If sight and shape be true,
    Why then, my love adieu!

ROSALIND (to the Duke)
    I'll have no father if you be not he.              120
    (To Orlando) I'll have no husband if you be not he,
    (To Phoebe) Nor ne'er wed woman if you be not she.

HYMEN    Peace, ho, I bar confusion.
    'Tis I must make conclusion
        Of these most strange events.              125
    Here's eight that must take hands
        To join in Hymen's bands,
        If truth holds true contents.
    (To Orlando and Rosalind)
    You and you no cross shall part.
    (To Oliver and Celia)
    You and you are heart in heart.                   130
    (To Phoebe)
    You to his love must accord,
    Or have a woman to your lord.

(To Touchstone and Audrey)
    You and you are sure together
    As the winter to foul weather.—
    Whiles a wedlock hymn we sing,                    135
    Feed yourselves with questioning,
    That reason wonder may diminish
    How thus we met, and these things finish.

                    Song
    Wedding is great Juno's crown,
        O blessèd bond of board and bed.          140
    'Tis Hymen peoples every town.
        High wedlock then be honourèd.
    Honour, high honour and renown
    To Hymen, god of every town.

DUKE SENIOR (to Celia)
    O my dear niece, welcome thou art to me,     145
    Even daughter; welcome in no less degree.

PHOEBE (to Silvius)
    I will not eat my word. Now thou art mine,
    Thy faith my fancy to thee doth combine.
        Enter Jaques de Bois, the second brother

JAQUES DE BOIS
    Let me have audience for a word or two.
    I am the second son of old Sir Rowland,        150
    That bring these tidings to this fair assembly.
    Duke Frederick, hearing how that every day
    Men of great worth resorted to this forest,
    Addressed a mighty power, which were on foot,
    In his own conduct, purposely to take           155
    His brother here, and put him to the sword.
    And to the skirts of this wild wood he came
    Where, meeting with an old religious man,
    After some question with him was converted
    Both from his enterprise and from the world,  160
    His crown bequeathing to his banished brother,
    And all their lands restored to them again
    That were with him exiled. This to be true
    I do engage my life.

DUKE SENIOR            Welcome, young man.
    Thou offer'st fairly to thy brothers' wedding:  165
    To one his lands withheld, and to the other
    A land itself at large, a potent dukedom.
    First, in this forest let us do those ends
    That here were well begun, and well begot.
    And after, every of this happy number            170
    That have endured shrewd days and nights with us
    Shall share the good of our returnèd fortune
    According to the measure of their states.
    Meantime, forget this new-fallen dignity
    And fall into our rustic revelry.                      175
    Play, music, and you brides and bridegrooms all,
    With measure heaped in joy to th' measures fall.

JAQUES
    Sir, by your patience. (To Jaques de Bois) If I heard you
        rightly
    The Duke hath put on a religious life
    And thrown into neglect the pompous court.   180

JAQUES DE BOIS He hath.

JAQUES
    To him will I. Out of these convertites
    There is much matter to be heard and learned.
    (To the Duke)
    You to your former honour I bequeath;
    Your patience and your virtue well deserves it.   185
    (To Orlando)
    You to a love that your true faith doth merit;

(*To Oliver*)
You to your land, and love, and great allies;
(*To Silvius*)
You to a long and well-deservèd bed;
(*To Touchstone*)
And you to wrangling, for thy loving voyage
Is but for two months victualled.—So, to your
    pleasures;                                                      190
I am for other than for dancing measures.
DUKE SENIOR  Stay, Jaques, stay.
JAQUES
To see no pastime, I. What you would have
I'll stay to know at your abandoned cave.                 *Exit*
DUKE SENIOR
Proceed, proceed. We'll so begin these rites              195
As we do trust they'll end, in true delights.
       ⌈*They dance; then*⌉ *exeunt all but Rosalind*

## Epilogue

ROSALIND (*to the audience*) It is not the fashion to see the
lady the epilogue; but it is no more unhandsome than
to see the lord the prologue. If it be true that good
wine needs no bush, 'tis true that a good play needs
no epilogue. Yet to good wine they do use good bushes,
and good plays prove the better by the help of good
epilogues. What a case am I in then, that am neither
a good epilogue nor cannot insinuate with you in the
behalf of a good play! I am not furnished like a beggar,
therefore to beg will not become me. My way is to
conjure you; and I'll begin with the women. I charge
you, O women, for the love you bear to men, to like
as much of this play as please you. And I charge you,
O men, for the love you bear to women—as I perceive
by your simpering none of you hates them—that
between you and the women the play may please. If I
were a woman I would kiss as many of you as had
beards that pleased me, complexions that liked me, and
breaths that I defied not. And I am sure, as many as
have good beards, or good faces, or sweet breaths will
for my kind offer, when I make curtsy, bid me farewell.
                                                                     *Exit*

# HAMLET

SEVERAL references from 1589 onwards witness the existence of a play about Hamlet, but Francis Meres did not attribute a play with this title to Shakespeare in 1598. The first clear reference to Shakespeare's play is its entry in the Stationers' Register on 26 July 1602 as *The Revenge of Hamlet Prince [of] Denmark*, when it was said to have been 'lately acted by the Lord Chamberlain his servants'. It survives in three versions; their relationship is a matter of dispute on which views about when Shakespeare wrote his play, and in what form, depend. In 1603 appeared an inferior text apparently assembled from actors' memories; it has only about 2,200 lines. In the following year, as if to put the record straight, James Roberts (to whom the play had been entered in 1602) published it as 'newly imprinted and enlarged to almost as much again as it was, according to the true and perfect copy'. At about 3,800 lines, this is the longest version. The 1623 Folio offers a still different text, some 230 lines shorter than the 1604 version, differing verbally from that at many points, and including about 70 additional lines. It is our belief that Shakespeare wrote *Hamlet* about 1600, and revised it later; that the 1604 edition was printed from his original papers; that the Folio represents the revised version; and that the 1603 edition represents a very imperfect report of an abridged version of the revision. So our text is based on the Folio; passages present in the 1604 quarto but absent from the Folio are printed as Additional Passages because we believe that, however fine they may be in themselves, Shakespeare decided that the play as a whole would be better without them.

The plot of *Hamlet* originates in a Scandinavian folk-tale told in the twelfth-century *Danish History* written in Latin by the Danish Saxo Grammaticus. François de Belleforest retold it in the fifth volume (1570) of his *Histoires Tragiques*, not translated into English until 1608. Saxo, through Belleforest, provided the basic story of a Prince of Denmark committed to revenge his father's murder by his own brother (Claudius) who has married the dead man's widow (Gertrude). As in Shakespeare, Hamlet pretends to be mad, kills his uncle's counsellor (Polonius) while he is eavesdropping, rebukes his mother, is sent to England under the escort of two retainers (Rosencrantz and Guildenstern) who bear orders that he be put to death on arrival, finds the letter containing the orders and alters it so that it is the retainers who are executed, returns to Denmark, and kills the King.

Belleforest's story differs at some points from Shakespeare's, and Shakespeare elaborates it, adding, for example, the Ghost of Hamlet's father, the coming of the actors to Elsinore, the performance of the play through which Hamlet tests his uncle's guilt, Ophelia's madness and death, Laertes' plot to revenge *his* father's death, the grave-digger, Ophelia's funeral, and the characters of Osric and Fortinbras. How much he owed to the lost Hamlet play we cannot tell; what is certain is that Shakespeare used his mastery of a wide range of diverse styles in both verse and prose, and his genius for dramatic effect, to create from these and other sources the most complex, varied, and exciting drama that had ever been seen on the English stage. Its popularity was instant and enduring. The play has had a profound influence on Western culture, and Shakespeare's Hamlet has himself entered the world of myth.

# THE PERSONS OF THE PLAY

GHOST of Hamlet, the late King of Denmark

KING CLAUDIUS, his brother

QUEEN GERTRUDE of Denmark, widow of King Hamlet, now wife of Claudius

Prince HAMLET, son of King Hamlet and Queen Gertrude

POLONIUS, a lord

LAERTES, son of Polonius

OPHELIA, daughter of Polonius

REYNALDO, servant of Polonius

HORATIO
ROSENCRANTZ } friends of Prince Hamlet
GUILDENSTERN

FRANCISCO
BARNARDO } soldiers
MARCELLUS

VALTEMAND
CORNELIUS
OSRIC } courtiers
GENTLEMEN

A SAILOR

Two CLOWNS, a gravedigger and his companion

A PRIEST

FORTINBRAS, Prince of Norway

A CAPTAIN in his army

AMBASSADORS from England

PLAYERS, who play the parts of the Prologue, Player King, Player Queen, and Lucianus, in 'The Mousetrap'

Lords, messengers, attendants, guards, soldiers, followers of Laertes, sailors

# The Tragedy of Hamlet, Prince of Denmark

**1.1** *Enter Barnardo and Francisco, two sentinels, at several doors*

BARNARDO Who's there?

FRANCISCO
Nay, answer me. Stand and unfold yourself.

BARNARDO
Long live the King!

FRANCISCO                Barnardo?

BARNARDO                        He.

FRANCISCO
You come most carefully upon your hour.

BARNARDO
'Tis now struck twelve. Get thee to bed, Francisco.    5

FRANCISCO
For this relief much thanks. 'Tis bitter cold,
And I am sick at heart.

BARNARDO                    Have you had quiet guard?

FRANCISCO
Not a mouse stirring.

BARNARDO                Well, good night.
If you do meet Horatio and Marcellus,
The rivals of my watch, bid them make haste.    10
*Enter Horatio and Marcellus*

FRANCISCO
I think I hear them.—Stand! Who's there?

HORATIO                    Friends to this ground.

MARCELLUS
And liegemen to the Dane.

FRANCISCO                    Give you good night.

MARCELLUS
O farewell, honest soldier. Who hath relieved you?

FRANCISCO
Barnardo has my place. Give you good night.    *Exit*

MARCELLUS Holla, Barnardo!    15

BARNARDO Say—what, is Horatio there?

HORATIO A piece of him.

BARNARDO
Welcome, Horatio. Welcome, good Marcellus.

MARCELLUS
What, has this thing appeared again tonight?

BARNARDO I have seen nothing.    20

MARCELLUS
Horatio says 'tis but our fantasy,
And will not let belief take hold of him
Touching this dreaded sight twice seen of us.
Therefore I have entreated him along
With us to watch the minutes of this night,    25
That if again this apparition come
He may approve our eyes and speak to it.

HORATIO
Tush, tush, 'twill not appear.

BARNARDO                    Sit down a while,
And let us once again assail your ears,
That are so fortified against our story,    30
What we two nights have seen.

HORATIO                    Well, sit we down,
And let us hear Barnardo speak of this.

BARNARDO Last night of all,
When yon same star that's westward from the pole
Had made his course t'illume that part of heaven    35

Where now it burns, Marcellus and myself,
The bell then beating one—
*Enter the Ghost in complete armour, holding a truncheon, with his beaver up*

MARCELLUS
Peace, break thee off. Look where it comes again.

BARNARDO
In the same figure like the King that's dead.

MARCELLUS *(to Horatio)*
Thou art a scholar—speak to it, Horatio.    40

BARNARDO
Looks it not like the King?—Mark it, Horatio.

HORATIO
Most like. It harrows me with fear and wonder.

BARNARDO
It would be spoke to.

MARCELLUS                Question it, Horatio.

HORATIO *(to the Ghost)*
What art thou that usurp'st this time of night,
Together with that fair and warlike form    45
In which the majesty of buried Denmark
Did sometimes march? By heaven, I charge thee speak.

MARCELLUS
It is offended.

BARNARDO        See, it stalks away.

HORATIO *(to the Ghost)*
Stay, speak, speak, I charge thee speak.    *Exit Ghost*

MARCELLUS 'Tis gone, and will not answer.    50

BARNARDO
How now, Horatio? You tremble and look pale.
Is not this something more than fantasy?
What think you on't?

HORATIO
Before my God, I might not this believe
Without the sensible and true avouch    55
Of mine own eyes.

MARCELLUS Is it not like the King?

HORATIO As thou art to thyself.
Such was the very armour he had on
When he th'ambitious Norway combated.    60
So frowned he once when in an angry parley
He smote the sledded Polacks on the ice.
'Tis strange.

MARCELLUS
Thus twice before, and just at this dead hour,
With martial stalk hath he gone by our watch.    65

HORATIO
In what particular thought to work I know not,
But in the gross and scope of my opinion
This bodes some strange eruption to our state.

MARCELLUS
Good now, sit down, and tell me, he that knows,
Why this same strict and most observant watch    70
So nightly toils the subject of the land,
And why such daily cast of brazen cannon,
And foreign mart for implements of war,
Why such impress of shipwrights, whose sore task
Does not divide the Sunday from the week:    75
What might be toward that this sweaty haste

Doth make the night joint-labourer with the day,
Who is't that can inform me?
HORATIO                              That can I—
At least the whisper goes so: our last king,
Whose image even but now appeared to us,          80
Was as you know by Fortinbras of Norway,
Thereto pricked on by a most emulate pride,
Dared to the combat; in which our valiant Hamlet—
For so this side of our known world esteemed him—
Did slay this Fortinbras, who by a sealed compact   85
Well ratified by law and heraldry
Did forfeit with his life all those his lands
Which he stood seized on to the conqueror;
Against the which a moiety competent
Was gagèd by our King, which had returned         90
To the inheritance of Fortinbras
Had he been vanquisher, as by the same cov'nant
And carriage of the article designed
His fell to Hamlet. Now sir, young Fortinbras,
Of unimprovèd mettle hot and full,                95
Hath in the skirts of Norway here and there
Sharked up a list of landless resolutes
For food and diet to some enterprise
That hath a stomach in't, which is no other—
And it doth well appear unto our state—          100
But to recover of us by strong hand
And terms compulsative those foresaid lands
So by his father lost. And this, I take it,
Is the main motive of our preparations,
The source of this our watch, and the chief head  105
Of this post-haste and rummage in the land.
        *Enter the Ghost, as before*
But soft, behold—lo where it comes again!
I'll cross it though it blast me.—Stay, illusion.
        *The Ghost spreads his arms*
If thou hast any sound or use of voice,
Speak to me.                                      110
If there be any good thing to be done
That may to thee do ease and grace to me,
Speak to me.
If thou art privy to thy country's fate
Which happily foreknowing may avoid,             115
O speak!
Or if thou hast uphoarded in thy life
Extorted treasure in the womb of earth—
For which, they say, you spirits oft walk in death—
        *The cock crows*
Speak of it, stay and speak.—Stop it, Marcellus.  120
MARCELLUS
Shall I strike at it with my partisan?
HORATIO
Do, if it will not stand.
BARNARDO                        'Tis here.
HORATIO                                    'Tis here.      *Exit Ghost*
MARCELLUS 'Tis gone.
We do it wrong, being so majestical,
To offer it the show of violence,               125
For it is as the air invulnerable,
And our vain blows malicious mockery.
BARNARDO
It was about to speak when the cock crew.
HORATIO
And then it started like a guilty thing
Upon a fearful summons. I have heard             130
The cock, that is the trumpet to the morn,
Doth with his lofty and shrill-sounding throat
Awake the god of day, and at his warning,

Whether in sea or fire, in earth or air,
Th'extravagant and erring spirit hies             135
To his confine; and of the truth herein
This present object made probation.
MARCELLUS
It faded on the crowing of the cock.
Some say that ever 'gainst that season comes
Wherein our saviour's birth is celebrated        140
The bird of dawning singeth all night long;
And then, they say, no spirit can walk abroad,
The nights are wholesome; then no planets strike,
No fairy takes, nor witch hath power to charm,
So hallowed and so gracious is the time.         145
HORATIO
So have I heard, and do in part believe it.
But look, the morn in russet mantle clad
Walks o'er the dew of yon high eastern hill.
Break we our watch up, and by my advice
Let us impart what we have seen tonight          150
Unto young Hamlet; for upon my life,
This spirit, dumb to us, will speak to him.
Do you consent we shall acquaint him with it,
As needful in our loves, fitting our duty?
MARCELLUS
Let's do't, I pray; and I this morning know      155
Where we shall find him most conveniently.      *Exeunt*

1.2     *Flourish. Enter Claudius, King of Denmark,*
        *Gertrude the Queen, members of the Council, such*
        *as Polonius, his son Laertes and daughter Ophelia,*
        *Prince Hamlet dressed in black, with others*
KING CLAUDIUS
Though yet of Hamlet our dear brother's death
The memory be green, and that it us befitted
To bear our hearts in grief and our whole kingdom
To be contracted in one brow of woe,
Yet so far hath discretion fought with nature      5
That we with wisest sorrow think on him
Together with remembrance of ourselves.
Therefore our sometime sister, now our queen,
Th'imperial jointress of this warlike state,
Have we as 'twere with a defeated joy,            10
With one auspicious and one dropping eye,
With mirth in funeral and with dirge in marriage,
In equal scale weighing delight and dole,
Taken to wife. Nor have we herein barred
Your better wisdoms, which have freely gone       15
With this affair along. For all, our thanks.
Now follows that you know young Fortinbras,
Holding a weak supposal of our worth,
Or thinking by our late dear brother's death
Our state to be disjoint and out of frame,        20
Co-leaguèd with the dream of his advantage,
He hath not failed to pester us with message
Importing the surrender of those lands
Lost by his father, with all bonds of law,
To our most valiant brother. So much for him.     25
        *Enter Valtemand and Cornelius*
Now for ourself, and for this time of meeting,
Thus much the business is: we have here writ
To Norway, uncle of young Fortinbras—
Who, impotent and bed-rid, scarcely hears
Of this his nephew's purpose—to suppress          30
His further gait herein, in that the levies,
The lists, and full proportions are all made
Out of his subject; and we here dispatch
You, good Cornelius, and you, Valtemand,

For bearers of this greeting to old Norway,              35
Giving to you no further personal power
To business with the King more than the scope
Of these dilated articles allow.
Farewell, and let your haste commend your duty.
VALTEMAND
In that and all things will we show our duty.            40
KING CLAUDIUS
We doubt it nothing, heartily farewell.
      *Exeunt Valtemand and Cornelius*
And now, Laertes, what's the news with you?
You told us of some suit. What is't, Laertes?
You cannot speak of reason to the Dane
And lose your voice. What wouldst thou beg, Laertes,
That shall not be my offer, not thy asking?              46
The head is not more native to the heart,
The hand more instrumental to the mouth,
Than is the throne of Denmark to thy father.
What wouldst thou have, Laertes?
LAERTES      Dread my lord,     50
Your leave and favour to return to France,
From whence though willingly I came to Denmark
To show my duty in your coronation,
Yet now I must confess, that duty done,
My thoughts and wishes bend again towards France
And bow them to your gracious leave and pardon.         56
KING CLAUDIUS
Have you your father's leave? What says Polonius?
POLONIUS
He hath, my lord, wrung from me my slow leave
By laboursome petition, and at last
Upon his will I sealed my hard consent.                 60
I do beseech you give him leave to go.
KING CLAUDIUS
Take thy fair hour, Laertes. Time be thine,
And thy best graces spend it at thy will.
But now, my cousin Hamlet, and my son—
HAMLET
A little more than kin and less than kind.              65
KING CLAUDIUS
How is it that the clouds still hang on you?
HAMLET
Not so, my lord, I am too much i'th' sun.
QUEEN GERTRUDE
Good Hamlet, cast thy nightly colour off,
And let thine eye look like a friend on Denmark.
Do not for ever with thy vailèd lids                    70
Seek for thy noble father in the dust.
Thou know'st 'tis common—all that lives must die,
Passing through nature to eternity.
HAMLET
Ay, madam, it is common.
QUEEN GERTRUDE    If it be,
Why seems it so particular with thee?                   75
HAMLET
Seems, madam? Nay, it *is*. I know not 'seems'.
'Tis not alone my inky cloak, good-mother,
Nor customary suits of solemn black,
Nor windy suspiration of forced breath,
No, nor the fruitful river in the eye,                  80
Nor the dejected haviour of the visage,
Together with all forms, moods, shows of grief
That can denote me truly. These indeed 'seem',
For they are actions that a man might play;
But I have that within which passeth show—              85
These but the trappings and the suits of woe.

KING CLAUDIUS
'Tis sweet and commendable in your nature, Hamlet,
To give these mourning duties to your father.
But you must know your father lost a father;
That father lost, lost his; and the survivor bound      90
In filial obligation for some term
To do obsequious sorrow. But to persever
In obstinate condolement is a course
Of impious stubbornness, 'tis unmanly grief,
It shows a will most incorrect to heaven,               95
A heart unfortified, a mind impatient,
An understanding simple and unschooled;
For what we know must be, and is as common
As any the most vulgar thing to sense,
Why should we in our peevish opposition                 100
Take it to heart? Fie, 'tis a fault to heaven,
A fault against the dead, a fault to nature,
To reason most absurd, whose common theme
Is death of fathers, and who still hath cried
From the first corpse till he that died today,          105
'This must be so'. We pray you throw to earth
This unprevailing woe, and think of us
As of a father; for let the world take note
You are the most immediate to our throne,
And with no less nobility of love                       110
Than that which dearest father bears his son
Do I impart towards you. For your intent
In going back to school in Wittenberg,
It is most retrograde to our desire,
And we beseech you bend you to remain                   115
Here in the cheer and comfort of our eye,
Our chiefest courtier, cousin, and our son.
QUEEN GERTRUDE
Let not thy mother lose her prayers, Hamlet.
I pray thee stay with us, go not to Wittenberg.
HAMLET
I shall in all my best obey you, madam.                 120
KING CLAUDIUS
Why, 'tis a loving and a fair reply.
Be as ourself in Denmark. (*To Gertrude*) Madam, come.
This gentle and unforced accord of Hamlet
Sits smiling to my heart; in grace whereof,
No jocund health that Denmark drinks today              125
But the great cannon to the clouds shall tell,
And the King's rouse the heavens shall bruit again,
Re-speaking earthly thunder. Come, away.
   ⌈*Flourish.*⌉ *Exeunt all but Hamlet*
HAMLET
O that this too too solid flesh would melt,
Thaw, and resolve itself into a dew,                    130
Or that the Everlasting had not fixed
His canon 'gainst self-slaughter! O God, O God,
How weary, stale, flat, and unprofitable
Seem to me all the uses of this world!
Fie on't, ah fie, fie! 'Tis an unweeded garden          135
That grows to seed; things rank and gross in nature
Possess it merely. That it should come to this—
But two months dead—nay, not so much, not two—
So excellent a king, that was to this
Hyperion to a satyr, so loving to my mother             140
That he might not beteem the winds of heaven
Visit her face too roughly! Heaven and earth,
Must I remember? Why, she would hang on him
As if increase of appetite had grown
By what it fed on, and yet within a month—              145

Let me not think on't; frailty, thy name is woman—
A little month, or ere those shoes were old
With which she followed my poor father's body,
Like Niobe, all tears, why she, even she—
O God, a beast that wants discourse of reason     150
Would have mourned longer!—married with mine
     uncle,
My father's brother, but no more like my father
Than I to Hercules; within a month,
Ere yet the salt of most unrighteous tears
Had left the flushing of her gallèd eyes,     155
She married. O most wicked speed, to post
With such dexterity to incestuous sheets!
It is not, nor it cannot come to good.
But break, my heart, for I must hold my tongue.
     *Enter Horatio, Marcellus, and Barnardo*
HORATIO
Hail to your lordship.
HAMLET                    I am glad to see you well.     160
Horatio—or I do forget myself.
HORATIO
The same, my lord, and your poor servant ever.
HAMLET
Sir, my good friend; I'll change that name with you.
And what make you from Wittenberg, Horatio?—
Marcellus.
MARCELLUS   My good lord.     165
HAMLET
I am very glad to see you. (*To Barnardo*) Good even,
     sir.—
But what in faith make you from Wittenberg?
HORATIO
A truant disposition, good my lord.
HAMLET
I would not have your enemy say so,
Nor shall you do mine ear that violence     170
To make it truster of your own report
Against yourself. I know you are no truant.
But what is your affair in Elsinore?
We'll teach you to drink deep ere you depart.
HORATIO
My lord, I came to see your father's funeral.     175
HAMLET
I prithee do not mock me, fellow-student;
I think it was to see my mother's wedding.
HORATIO
Indeed, my lord, it followed hard upon.
HAMLET
Thrift, thrift, Horatio. The funeral baked meats
Did coldly furnish forth the marriage tables.     180
Would I had met my dearest foe in heaven
Ere I had ever seen that day, Horatio.
My father—methinks I see my father.
HORATIO
O where, my lord?
HAMLET                    In my mind's eye, Horatio.
HORATIO
I saw him once. A was a goodly king.     185
HAMLET
A was a man. Take him for all in all,
I shall not look upon his like again.
HORATIO
My lord, I think I saw him yesternight.
HAMLET   Saw? Who?
HORATIO   My lord, the King your father.     190
HAMLET   The King my father?

HORATIO
Season your admiration for a while
With an attent ear till I may deliver,
Upon the witness of these gentlemen,
This marvel to you.
HAMLET                    For God's love let me hear!     195
HORATIO
Two nights together had these gentlemen,
Marcellus and Barnardo, on their watch,
In the dead waste and middle of the night,
Been thus encountered. A figure like your father,
Armed at all points exactly, cap-à-pie,     200
Appears before them, and with solemn march
Goes slow and stately by them. Thrice he walked
By their oppressed and fear-surprisèd eyes
Within his truncheon's length, whilst they distilled
Almost to jelly with the act of fear     205
Stand dumb and speak not to him. This to me
In dreadful secrecy impart they did,
And I with them the third night kept the watch,
Where, as they had delivered, both in time,
Form of the thing, each word made true and good,
The apparition comes. I knew your father;     211
These hands are not more like.
HAMLET                    But where was this?
MARCELLUS
My lord, upon the platform where we watched.
HAMLET
Did you not speak to it?
HORATIO                    My lord, I did,
But answer made it none; yet once methought     215
It lifted up it head and did address
Itself to motion like as it would speak,
But even then the morning cock crew loud,
And at the sound it shrunk in haste away
And vanished from our sight.
HAMLET                    'Tis very strange.     220
HORATIO
As I do live, my honoured lord, 'tis true,
And we did think it writ down in our duty
To let you know of it.
HAMLET
Indeed, indeed, sirs; but this troubles me.—
Hold you the watch tonight?
BARNARDO *and* MARCELLUS     We do, my lord.     225
HAMLET
Armed, say you?
BARNARDO *and* MARCELLUS   Armed, my lord.
HAMLET                    From top to toe?
BARNARDO *and* MARCELLUS
My lord, from head to foot.
HAMLET                    Then saw you not his face.
HORATIO
O yes, my lord, he wore his beaver up.
HAMLET
What looked he? Frowningly?
HORATIO                    A countenance more
In sorrow than in anger.
HAMLET                    Pale or red?     230
HORATIO
Nay, very pale.
HAMLET                    And fixed his eyes upon you?
HORATIO   Most constantly.
HAMLET   I would I had been there.
HORATIO   It would have much amazed you.
HAMLET
Very like, very like. Stayed it long?     235

HORATIO
While one with moderate haste might tell a hundred.
BARNARDO *and* MARCELLUS Longer, longer.
HORATIO Not when I saw't.
HAMLET His beard was grizzly, no?
HORATIO
It was as I have seen it in his life,                     240
A sable silvered.
HAMLET                    I'll watch tonight. Perchance
'Twill walk again.
HORATIO                    I warrant you it will.
HAMLET
If it assume my noble father's person
I'll speak to it though hell itself should gape
And bid me hold my peace. I pray you all,          245
If you have hitherto concealed this sight,
Let it be treble in your silence still,
And whatsoever else shall hap tonight,
Give it an understanding but no tongue.
I will requite your loves. So fare ye well.          250
Upon the platform 'twixt eleven and twelve
I'll visit you.
ALL THREE       Our duty to your honour.
HAMLET
Your love, as mine to you. Farewell.
                                *Exeunt all but Hamlet*
My father's spirit in arms! All is not well.
I doubt some foul play. Would the night were come.
Till then, sit still, my soul. Foul deeds will rise,   256
Though all the earth o'erwhelm them, to men's eyes.
                                                          *Exit*

**1.3**     *Enter Laertes and Ophelia, his sister*
LAERTES
My necessaries are inbarqued. Farewell.
And, sister, as the winds give benefit
And convoy is assistant, do not sleep
But let me hear from you.
OPHELIA                         Do you doubt that?
LAERTES
For Hamlet and the trifling of his favour,          5
Hold it a fashion and a toy in blood,
A violet in the youth of primy nature,
Forward not permanent, sweet not lasting,
The perfume and suppliance of a minute,
No more.
OPHELIA    No more but so?
LAERTES                         Think it no more.     10
For nature crescent does not grow alone
In thews and bulk, but as his temple waxes
The inward service of the mind and soul
Grows wide withal. Perhaps he loves you now,
And now no soil nor cautel doth besmirch          15
The virtue of his will; but you must fear,
His greatness weighed, his will is not his own,
For he himself is subject to his birth.
He may not, as unvalued persons do,
Carve for himself, for on his choice depends       20
The sanity and health of the whole state;
And therefore must his choice be circumscribed
Unto the voice and yielding of that body
Whereof he is the head. Then if he says he loves you,
It fits your wisdom so far to believe it              25
As he in his peculiar sect and force
May give his saying deed, which is no further
Than the main voice of Denmark goes withal.
Then weigh what loss your honour may sustain

If with too credent ear you list his songs,          30
Or lose your heart, or your chaste treasure open
To his unmastered importunity.
Fear it, Ophelia, fear it, my dear sister,
And keep within the rear of your affection,
Out of the shot and danger of desire.               35
The chariest maid is prodigal enough
If she unmask her beauty to the moon.
Virtue itself scapes not calumnious strokes.
The canker galls the infants of the spring
Too oft before their buttons be disclosed,          40
And in the morn and liquid dew of youth
Contagious blastments are most imminent.
Be wary then; best safety lies in fear;
Youth to itself rebels, though none else near.
OPHELIA
I shall th'effect of this good lesson keep          45
As watchman to my heart; but, good my brother,
Do not, as some ungracious pastors do,
Show me the steep and thorny way to heaven
Whilst like a puffed and reckless libertine
Himself the primrose path of dalliance treads      50
And recks not his own rede.
LAERTES                         O fear me not.
                    *Enter Polonius*
I stay too long—but here my father comes.
A double blessing is a double grace;
Occasion smiles upon a second leave.
POLONIUS
Yet here, Laertes? Aboard, aboard, for shame!    55
The wind sits in the shoulder of your sail,
And you are stayed for. There—my blessing with
      thee,
And these few precepts in thy memory
See thou character. Give thy thoughts no tongue,
Nor any unproportioned thought his act.           60
Be thou familiar but by no means vulgar.
The friends thou hast, and their adoption tried,
Grapple them to thy soul with hoops of steel,
But do not dull thy palm with entertainment
Of each new-hatched unfledged comrade. Beware   65
Of entrance to a quarrel, but being in,
Bear't that th'opposèd may beware of thee.
Give every man thine ear but few thy voice.
Take each man's censure, but reserve thy judgement.
Costly thy habit as thy purse can buy,             70
But not expressed in fancy; rich not gaudy;
For the apparel oft proclaims the man,
And they in France of the best rank and station
Are of all most select and generous chief in that.
Neither a borrower nor a lender be,               75
For loan oft loses both itself and friend,
And borrowing dulls the edge of husbandry.
This above all—to thine own self be true,
And it must follow, as the night the day,
Thou canst not then be false to any man.          80
Farewell—my blessing season this in thee.
LAERTES
Most humbly do I take my leave, my lord.
POLONIUS
The time invites you. Go; your servants tend.
LAERTES
Farewell, Ophelia, and remember well
What I have said to you.
OPHELIA                         'Tis in my memory locked,
And you yourself shall keep the key of it.          86
LAERTES Farewell.                              *Exit*

POLONIUS
  What is't, Ophelia, he hath said to you?
OPHELIA
  So please you, something touching the Lord Hamlet.
POLONIUS Marry, well bethought.            90
  'Tis told me he hath very oft of late
  Given private time to you, and you yourself
  Have of your audience been most free and bounteous.
  If it be so—as so 'tis put on me,
  And that in way of caution—I must tell you      95
  You do not understand yourself so clearly
  As it behoves my daughter and your honour.
  What is between you? Give me up the truth.
OPHELIA
  He hath, my lord, of late made many tenders
  Of his affection to me.              100
POLONIUS
  Affection, pooh! You speak like a green girl
  Unsifted in such perilous circumstance.
  Do you believe his 'tenders' as you call them?
OPHELIA
  I do not know, my lord, what I should think.
POLONIUS
  Marry, I'll teach you: think yourself a baby      105
  That you have ta'en his tenders for true pay,
  Which are not sterling. Tender yourself more dearly,
  Or—not to crack the wind of the poor phrase,
  Running it thus—you'll tender me a fool.
OPHELIA
  My lord, he hath importuned me with love      110
  In honourable fashion—
POLONIUS
  Ay, fashion you may call it. Go to, go to.
OPHELIA
  And hath given countenance to his speech, my lord,
  With all the vows of heaven.
POLONIUS
  Ay, springes to catch woodcocks. I do know      115
  When the blood burns how prodigal the soul
  Lends the tongue vows. These blazes, daughter,
  Giving more light than heat, extinct in both
  Even in their promise as it is a-making,
  You must not take for fire. From this time, daughter,
  Be somewhat scanter of your maiden presence.    121
  Set your entreatments at a higher rate
  Than a command to parley. For Lord Hamlet,
  Believe so much in him, that he is young,
  And with a larger tether may he walk          125
  Than may be given you. In few, Ophelia,
  Do not believe his vows, for they are brokers,
  Not of the dye which their investments show,
  But mere imploratators of unholy suits,
  Breathing like sanctified and pious bawds      130
  The better to beguile. This is for all—
  I would not, in plain terms, from this time forth
  Have you so slander any moment leisure
  As to give words or talk with the Lord Hamlet.
  Look to't, I charge you. Come your ways.      135
OPHELIA I shall obey, my lord.           *Exeunt*

**1.4**   *Enter Prince Hamlet, Horatio, and Marcellus*
HAMLET
  The air bites shrewdly, it is very cold.
HORATIO
  It is a nipping and an eager air.

HAMLET What hour now?
HORATIO I think it lacks of twelve.
MARCELLUS No, it is struck.            5
HORATIO
  Indeed? I heard it not. Then it draws near the season
  Wherein the spirit held his wont to walk.
    *A flourish of trumpets, and two pieces of ordnance*
    *goes off*
  What does this mean, my lord?
HAMLET
  The King doth wake tonight and takes his rouse,
  Keeps wassail, and the swagg'ring upspring reels,   10
  And as he drains his draughts of Rhenish down
  The kettle-drum and trumpet thus bray out
  The triumph of his pledge.
HORATIO Is it a custom?
HAMLET Ay, marry is't,             15
  And to my mind, though I am native here
  And to the manner born, it is a custom
  More honoured in the breach than the observance.
    *Enter the Ghost, as before*
HORATIO Look, my lord, it comes.
HAMLET
  Angels and ministers of grace defend us!      20
  Be thou a spirit of health or goblin damned,
  Bring with thee airs from heaven or blasts from hell,
  Be thy intents wicked or charitable,
  Thou com'st in such a questionable shape
  That I will speak to thee. I'll call thee Hamlet,    25
  King, father, royal Dane. O answer me!
  Let me not burst in ignorance, but tell
  Why thy canonized bones, hearsèd in death,
  Have burst their cerements, why the sepulchre
  Wherein we saw thee quietly enurned        30
  Hath oped his ponderous and marble jaws
  To cast thee up again. What may this mean,
  That thou, dead corpse, again in complete steel,
  Revisitst thus the glimpses of the moon,
  Making night hideous, and we fools of nature    35
  So horridly to shake our disposition
  With thoughts beyond the reaches of our souls?
  Say, why is this? Wherefore? What should we do?
    *The Ghost beckons Hamlet*
HORATIO
  It beckons you to go away with it
  As if it some impartment did desire          40
  To you alone.
MARCELLUS (*to Hamlet*) Look with what courteous action
  It wafts you to a more removèd ground.
  But do not go with it.
HORATIO (*to Hamlet*)    No, by no means.
HAMLET
  It will not speak. Then will I follow it.
HORATIO
  Do not, my lord.
HAMLET          Why, what should be the fear?    45
  I do not set my life at a pin's fee,
  And for my soul, what can it do to that,
  Being a thing immortal as itself?
    *The Ghost beckons Hamlet*
  It waves me forth again. I'll follow it.
HORATIO
  What if it tempt you toward the flood, my lord,    50
  Or to the dreadful summit of the cliff
  That beetles o'er his base into the sea,
  And there assume some other horrible form

Which might deprive your sovereignty of reason
And draw you into madness? Think of it. 55
*The Ghost beckons Hamlet*

HAMLET
It wafts me still. (*To the Ghost*) Go on, I'll follow thee.

MARCELLUS
You shall not go, my lord.

HAMLET                           Hold off your hand.

HORATIO
Be ruled. You shall not go.

HAMLET                           My fate cries out,
And makes each petty artere in this body
As hardy as the Nemean lion's nerve. 60
*The Ghost beckons Hamlet*
Still am I called. Unhand me, gentlemen.
By heav'n, I'll make a ghost of him that lets me.
I say, away! (*To the Ghost*) Go on, I'll follow thee.
*Exeunt the Ghost and Hamlet*

HORATIO
He waxes desperate with imagination.

MARCELLUS
Let's follow. 'Tis not fit thus to obey him. 65

HORATIO
Have after. To what issue will this come?

MARCELLUS
Something is rotten in the state of Denmark.

HORATIO
Heaven will direct it.

MARCELLUS                   Nay, let's follow him.    *Exeunt*

**1.5**    *Enter the Ghost, and Prince Hamlet following*

HAMLET
Whither wilt thou lead me? Speak. I'll go no further.

GHOST
Mark me.

HAMLET    I will.

GHOST              My hour is almost come
When I to sulph'rous and tormenting flames
Must render up myself.

HAMLET                   Alas, poor ghost!

GHOST
Pity me not, but lend thy serious hearing 5
To what I shall unfold.

HAMLET                   Speak, I am bound to hear.

GHOST
So art thou to revenge when thou shalt hear.

HAMLET What?

GHOST I am thy father's spirit,
Doomed for a certain term to walk the night, 10
And for the day confined to fast in fires
Till the foul crimes done in my days of nature
Are burnt and purged away. But that I am forbid
To tell the secrets of my prison-house
I could a tale unfold whose lightest word 15
Would harrow up thy soul, freeze thy young blood,
Make thy two eyes like stars start from their spheres,
Thy knotty and combinèd locks to part,
And each particular hair to stand on end
Like quills upon the fretful porcupine. 20
But this eternal blazon must not be
To ears of flesh and blood. List, Hamlet, list, O list!
If thou didst ever thy dear father love—

HAMLET O God!

GHOST
Revenge his foul and most unnatural murder. 25

HAMLET Murder?

GHOST
Murder most foul, as in the best it is,
But this most foul, strange, and unnatural.

HAMLET
Haste, haste me to know it, that with wings as swift
As meditation or the thoughts of love 30
May sweep to my revenge.

GHOST              I find thee apt,
And duller shouldst thou be than the fat weed
That rots itself in ease on Lethe wharf
Wouldst thou not stir in this. Now, Hamlet, hear.
'Tis given out that, sleeping in mine orchard, 35
A serpent stung me. So the whole ear of Denmark
Is by a forgèd process of my death
Rankly abused. But know, thou noble youth,
The serpent that did sting thy father's life
Now wears his crown. 40

HAMLET
O my prophetic soul! Mine uncle?

GHOST
Ay, that incestuous, that adulterate beast,
With witchcraft of his wit, with traitorous gifts—
O wicked wit and gifts, that have the power
So to seduce!—won to his shameful lust 45
The will of my most seeming-virtuous queen.
O Hamlet, what a falling off was there!—
From me, whose love was of that dignity
That it went hand-in-hand even with the vow
I made to her in marriage, and to decline 50
Upon a wretch whose natural gifts were poor
To those of mine.
But virtue, as it never will be moved,
Though lewdness court it in a shape of heaven,
So lust, though to a radiant angel linked, 55
Will sate itself in a celestial bed,
And prey on garbage.
But soft, methinks I scent the morning's air.
Brief let me be. Sleeping within mine orchard,
My custom always in the afternoon, 60
Upon my secure hour thy uncle stole
With juice of cursèd hebenon in a vial,
And in the porches of mine ears did pour
The leperous distilment, whose effect
Holds such an enmity with blood of man 65
That swift as quicksilver it courses through
The natural gates and alleys of the body,
And with a sudden vigour it doth posset
And curd, like eager droppings into milk,
The thin and wholesome blood. So did it mine; 70
And a most instant tetter barked about,
Most lazar-like, with vile and loathsome crust,
All my smooth body.
Thus was I, sleeping, by a brother's hand
Of life, of crown, of queen at once dispatched, 75
Cut off even in the blossoms of my sin,
Unhouseled, dis-appointed, unaneled,
No reck'ning made, but sent to my account
With all my imperfections on my head.
O horrible, O horrible, most horrible! 80
If thou hast nature in thee, bear it not.
Let not the royal bed of Denmark be
A couch for luxury and damnèd incest.
But howsoever thou pursuest this act,
Taint not thy mind, nor let thy soul contrive 85

Against thy mother aught. Leave her to heaven,
And to those thorns that in her bosom lodge
To prick and sting her. Fare thee well at once.
The glow-worm shows the matin to be near,
And gins to pale his uneffectual fire.                                    90
Adieu, adieu, Hamlet. Remember me.                           *Exit*
HAMLET
O all you host of heaven! O earth! What else?
And shall I couple hell? O fie! Hold, hold, my heart,
And you, my sinews, grow not instant old,
But bear me stiffly up. Remember thee?                           95
Ay, thou poor ghost, while memory holds a seat
In this distracted globe. Remember thee?
Yea, from the table of my memory
I'll wipe away all trivial fond records,
All saws of books, all forms, all pressures past,           100
That youth and observation copied there,
And thy commandment all alone shall live
Within the book and volume of my brain
Unmixed with baser matter. Yes, yes, by heaven.
O most pernicious woman!                                             105
O villain, villain, smiling, damnèd villain!
My tables,
My tables—meet it is I set it down
That one may smile and smile and be a villain.
At least I'm sure it may be so in Denmark.                  110
        *He writes*
So, uncle, there you are. Now to my word:
It is 'Adieu, adieu, remember me'.
I have sworn't.
HORATIO *and* MARCELLUS (*within*) My lord, my lord.
        *Enter Horatio and Marcellus*
MARCELLUS (*calling*) Lord Hamlet!                                  115
HORATIO Heaven secure him.
HAMLET So be it.
HORATIO (*calling*) Illo, ho, ho, my lord.
HAMLET
Hillo, ho, ho, boy; come, bird, come.
MARCELLUS How is't, my noble lord?                              120
HORATIO (*to Hamlet*) What news, my lord?
HAMLET O wonderful!
HORATIO
Good my lord, tell it.
HAMLET                    No, you'll reveal it.
HORATIO
Not I, my lord, by heaven.
MARCELLUS                    Nor I, my lord.
HAMLET
How say you then, would heart of man once think it?
But you'll be secret?
HORATIO *and* MARCELLUS Ay, by heav'n, my lord.      126
HAMLET
There's ne'er a villain dwelling in all Denmark
But he's an arrant knave.
HORATIO
There needs no ghost, my lord, come from the grave
To tell us this.
HAMLET                    Why, right, you are i'th' right,     130
And so without more circumstance at all
I hold it fit that we shake hands and part,
You as your business and desires shall point you—
For every man has business and desire,
Such as it is—and for mine own poor part,                     135
Look you, I'll go pray.
HORATIO
These are but wild and whirling words, my lord.

HAMLET
I'm sorry they offend you, heartily,
Yes, faith, heartily.
HORATIO                    There's no offence, my lord.
HAMLET
Yes, by Saint Patrick, but there is, Horatio,                 140
And much offence, too. Touching this vision here,
It is an honest ghost, that let me tell you.
For your desire to know what is between us,
O'ermaster't as you may. And now, good friends,
As you are friends, scholars, and soldiers,               145
Give me one poor request.
HORATIO                    What is't, my lord? We will.
HAMLET
Never make known what you have seen tonight.
HORATIO *and* MARCELLUS
My lord, we will not.
HAMLET                    Nay, but swear't.
HORATIO
In faith, my lord, not I.
MARCELLUS                    Nor I, my lord, in faith.
HAMLET
Upon my sword.
MARCELLUS            We have sworn, my lord, already.
HAMLET
Indeed, upon my sword, indeed.
        *The Ghost cries under the stage*
GHOST                                   Swear.                              151
HAMLET
Ah ha, boy, sayst thou so? Art thou there, truepenny?—
Come on. You hear this fellow in the cellarage.
Consent to swear.
HORATIO            Propose the oath, my lord.
HAMLET
Never to speak of this that you have seen,                  155
Swear by my sword.
GHOST (*under the stage*) Swear.
        ⌐They swear⌐
HAMLET
*Hic et ubique?* Then we'll shift our ground.—
Come hither, gentlemen,
And lay your hands again upon my sword.                   160
Never to speak of this that you have heard,
Swear by my sword.
GHOST (*under the stage*) Swear.
        ⌐They swear⌐
HAMLET
Well said, old mole. Canst work i'th' earth so fast?
A worthy pioneer.—Once more remove, good friends.
HORATIO
O day and night, but this is wondrous strange!            166
HAMLET
And therefore as a stranger give it welcome.
There are more things in heaven and earth, Horatio,
Than are dreamt of in our philosophy. But come,
Here as before, never, so help you mercy,                   170
How strange or odd soe'er I bear myself—
As I perchance hereafter shall think meet
To put an antic disposition on—
That you at such time seeing me never shall,
With arms encumbered thus, or this headshake,          175
Or by pronouncing of some doubtful phrase
As 'Well, we know' or 'We could an if we would',
Or 'If we list to speak', or 'There be, an if they might',
Or such ambiguous giving out, to note
That you know aught of me—this not to do,                 180
So grace and mercy at your most need help you, swear.

GHOST (*under the stage*) Swear.
⌈*They swear*⌉
HAMLET
  Rest, rest, perturbèd spirit.—So, gentlemen,
  With all my love I do commend me to you,
  And what so poor a man as Hamlet is          185
  May do t'express his love and friending to you,
  God willing, shall not lack. Let us go in together,
  And still your fingers on your lips, I pray.
  The time is out of joint. O cursèd spite
  That ever I was born to set it right!          190
  Nay, come, let's go together.          *Exeunt*

**2.1**  *Enter old Polonius with his man Reynaldo*
POLONIUS
  Give him this money and these notes, Reynaldo.
REYNALDO I will, my lord.
POLONIUS
  You shall do marv'lous wisely, good Reynaldo,
  Before you visit him to make enquire
  Of his behaviour.
REYNALDO          My lord, I did intend it.          5
POLONIUS
  Marry, well said, very well said. Look you, sir,
  Enquire me first what Danskers are in Paris,
  And how, and who, and what means, and where they keep,
  What company, at what expense; and finding
  By this encompassment and drift of question          10
  That they do know my son, come you more nearer
  Than your particular demands will touch it.
  Take you, as 'twere, some distant knowledge of him,
  As thus: 'I know his father and his friends,
  And in part him'—do you mark this, Reynaldo?          15
REYNALDO Ay, very well, my lord.
POLONIUS
  'And in part him, but', you may say, 'not well,
  But if't be he I mean, he's very wild,
  Addicted so and so'; and there put on him
  What forgeries you please—marry, none so rank          20
  As may dishonour him, take heed of that—
  But, sir, such wanton, wild, and usual slips
  As are companions noted and most known
  To youth and liberty.
REYNALDO As gaming, my lord?          25
POLONIUS Ay, or drinking, fencing, swearing,
  Quarrelling, drabbing—you may go so far.
REYNALDO
  My lord, that would dishonour him.
POLONIUS
  Faith, no, as you may season it in the charge.
  You must not put another scandal on him,          30
  That he is open to incontinency.
  That's not my meaning—but breathe his faults so
    quaintly
  That they may seem the taints of liberty,
  The flash and outbreak of a fiery mind,
  A savageness in unreclaimèd blood,          35
  Of general assault.
REYNALDO          But, my good lord—
POLONIUS
  Wherefore should you do this?
REYNALDO          Ay, my lord.
  I would know that.
POLONIUS          Marry, sir, here's my drift,
  And I believe it is a fetch of warrant:
  You laying these slight sullies on my son,          40

  As 'twere a thing a little soiled i'th' working,
  Mark you, your party in converse, him you would
    sound,
  Having ever seen in the prenominate crimes
  The youth you breathe of guilty, be assured
  He closes with you in this consequence:          45
  'Good sir', or so, or 'friend', or 'gentleman',
  According to the phrase and the addition
  Of man and country.
REYNALDO          Very good, my lord.
POLONIUS
  And then, sir, does a this—a does—
  what was I about to say? By the mass, I was about to
  say something. Where did I leave?          51
REYNALDO
  At 'closes in the consequence', at 'friend,
  Or so', and 'gentleman'.
POLONIUS
  At 'closes in the consequence'—ay, marry,
  He closes with you thus: 'I know the gentleman,          55
  I saw him yesterday'—or t'other day,
  Or then, or then—'with such and such, and, as you
    say,
  There was a gaming, there o'ertook in 's rouse,
  There falling out at tennis', or perchance
  'I saw him enter such a house of sale'—          60
  Videlicet, a brothel, or so forth. See you now,
  Your bait of falsehood takes this carp of truth;
  And thus do we of wisdom and of reach
  With windlasses and with assays of bias
  By indirections find directions out.          65
  So, by my former lecture and advice,
  Shall you my son. You have me, have you not?
REYNALDO My lord, I have.
POLONIUS God b'wi' ye. Fare ye well.
REYNALDO Good my lord.          70
POLONIUS
  Observe his inclination in yourself.
REYNALDO I shall, my lord.
POLONIUS And let him ply his music.
REYNALDO Well, my lord.
    *Enter Ophelia*
POLONIUS
  Farewell.          *Exit Reynaldo*
    How now, Ophelia, what's the matter?
OPHELIA
  Alas, my lord, I have been so affrighted.          76
POLONIUS With what, i'th' name of God?
OPHELIA
  My lord, as I was sewing in my chamber,
  Lord Hamlet, with his doublet all unbraced,
  No hat upon his head, his stockings fouled,          80
  Ungartered, and down-gyvèd to his ankle,
  Pale as his shirt, his knees knocking each other,
  And with a look so piteous in purport
  As if he had been loosèd out of hell
  To speak of horrors, he comes before me.          85
POLONIUS
  Mad for thy love?
OPHELIA          My lord, I do not know,
  But truly I do fear it.
POLONIUS          What said he?
OPHELIA
  He took me by the wrist and held me hard,
  Then goes he to the length of all his arm,
  And with his other hand thus o'er his brow          90

He falls to such perusal of my face
As a would draw it. Long stayed he so.
At last, a little shaking of mine arm,
And thrice his head thus waving up and down,
He raised a sigh so piteous and profound                    95
That it did seem to shatter all his bulk
And end his being. That done, he lets me go,
And, with his head over his shoulder turned,
He seemed to find his way without his eyes,
For out o' doors he went without their help,               100
And to the last bended their light on me.
POLONIUS
Come, go with me. I will go seek the King.
This is the very ecstasy of love,
Whose violent property fordoes itself
And leads the will to desperate undertakings               105
As oft as any passion under heaven
That does afflict our natures. I am sorry—
What, have you given him any hard words of late?
OPHELIA
No, my good lord, but as you did command
I did repel his letters and denied                         110
His access to me.
POLONIUS                    That hath made him mad.
I am sorry that with better speed and judgement
I had not quoted him. I feared he did but trifle
And meant to wreck thee. But beshrew my jealousy!
By heaven, it is as proper to our age                      115
To cast beyond ourselves in our opinions
As it is common for the younger sort
To lack discretion. Come, go we to the King.
This must be known, which, being kept close, might
       move
More grief to hide than hate to utter love.        *Exeunt*

**2.2**  ⌈*Flourish.*⌉ *Enter King Claudius and Queen*
       *Gertrude, Rosencrantz and Guildenstern, with others*
KING CLAUDIUS
Welcome, dear Rosencrantz and Guildenstern.
Moreover that we much did long to see you,
The need we have to use you did provoke
Our hasty sending. Something have you heard
Of Hamlet's transformation—so I call it,                     5
Since not th'exterior nor the inward man
Resembles that it was. What it should be,
More than his father's death, that thus hath put him
So much from th'understanding of himself,
I cannot deem of. I entreat you both                        10
That, being of so young days brought up with him,
And since so neighboured to his youth and humour,
That you vouchsafe your rest here in our court
Some little time, so by your companies
To draw him on to pleasures, and to gather,                 15
So much as from occasions you may glean,
Whether aught to us unknown afflicts him thus
That, opened, lies within our remedy.
QUEEN GERTRUDE
Good gentlemen, he hath much talked of you,
And sure I am two men there is not living                   20
To whom he more adheres. If it will please you
To show us so much gentry and good will
As to expend your time with us a while
For the supply and profit of our hope,
Your visitation shall receive such thanks                   25
As fits a king's remembrance.
ROSENCRANTZ                    Both your majesties
Might, by the sovereign power you have of us,

Put your dread pleasures more into command
Than to entreaty.
GUILDENSTERN            But we both obey,
And here give up ourselves in the full bent                 30
To lay our service freely at your feet
To be commanded.
KING CLAUDIUS
Thanks, Rosencrantz and gentle Guildenstern.
QUEEN GERTRUDE
Thanks, Guildenstern and gentle Rosencrantz.
And I beseech you instantly to visit                        35
My too-much changèd son.—Go, some of ye,
And bring the gentlemen where Hamlet is.
GUILDENSTERN
Heavens make our presence and our practices
Pleasant and helpful to him.
QUEEN GERTRUDE                    Ay, amen!
       *Exeunt Rosencrantz and Guildenstern* ⌈*with others*⌉
       *Enter Polonius*
POLONIUS
Th'ambassadors from Norway, my good lord,                   40
Are joyfully returned.
KING CLAUDIUS
Thou still hast been the father of good news.
POLONIUS
Have I, my lord? Assure you, my good liege,
I hold my duty, as I hold my soul,
Both to my God and to my gracious King.                     45
And I do think—or else this brain of mine
Hunts not the trail of policy so sure
As it hath used to do—that I have found
The very cause of Hamlet's lunacy.
KING CLAUDIUS
O speak of that, that I do long to hear!                    50
POLONIUS
Give first admittance to th'ambassadors.
My news shall be the fruit to that great feast.
KING CLAUDIUS
Thyself do grace to them, and bring them in.
                                        *Exit Polonius*
He tells me, my sweet queen, that he hath found
The head and source of all your son's distemper.            55
QUEEN GERTRUDE
I doubt it is no other but the main—
His father's death and our o'er-hasty marriage.
KING CLAUDIUS
Well, we shall sift him.
       *Enter Polonius, Valtemand, and Cornelius*
                    Welcome, my good friends.
Say, Valtemand, what from our brother Norway?
VALTEMAND
Most fair return of greetings and desires.                  60
Upon our first he sent out to suppress
His nephew's levies, which to him appeared
To be a preparation 'gainst the Polack;
But better looked into, he truly found
It was against your highness; whereat grieved               65
That so his sickness, age, and impotence
Was falsely borne in hand, sends out arrests
On Fortinbras, which he, in brief, obeys,
Receives rebuke from Norway, and, in fine,
Makes vow before his uncle never more                       70
To give th'essay of arms against your majesty;
Whereon old Norway, overcome with joy,
Gives him three thousand crowns in annual fee
And his commission to employ those soldiers

So levied as before, against the Polack,                     75
With an entreaty herein further shown,
*He gives a letter to Claudius*
That it might please you to give quiet pass
Through your dominions for his enterprise
On such regards of safety and allowance
As therein are set down.
KING CLAUDIUS              It likes us well,                  80
And at our more considered time we'll read,
Answer, and think upon this business.
Meantime we thank you for your well-took labour.
Go to your rest; at night we'll feast together.
Most welcome home.                                           85
                *Exeunt Valtemand and Cornelius*
POLONIUS
This business is very well ended.
My liege, and madam, to expostulate
What majesty should be, what duty is,
Why day is day, night night, and time is time,
Were nothing but to waste night, day, and time.              90
Therefore, since brevity is the soul of wit,
And tediousness the limbs and outward flourishes,
I will be brief. Your noble son is mad—
'Mad' call I it, for to define true madness,
What is't but to be nothing else but mad?                    95
But let that go.
QUEEN GERTRUDE   More matter with less art.
POLONIUS
Madam, I swear I use no art at all.
That he is mad, 'tis true; 'tis true 'tis pity,
And pity 'tis 'tis true—a foolish figure,
But farewell it, for I will use no art.                      100
Mad let us grant him, then; and now remains
That we find out the cause of this effect—
Or rather say 'the cause of this *defect*',
For this effect defective comes by cause.
Thus it remains, and the remainder thus.                     105
Perpend.
I have a daughter—have whilst she is mine—
Who in her duty and obedience, mark,
Hath given me this. Now gather and surmise.
    *He reads a letter*
'To the celestial and my soul's idol, the most beautified
Ophelia'—that's an ill phrase, a vile phrase, 'beautified'
is a vile phrase. But you shall hear—'these in her
excellent white bosom, these'.
QUEEN GERTRUDE  Came this from Hamlet to her?
POLONIUS
Good madam, stay a while. I will be faithful.                115
    'Doubt thou the stars are fire,
        Doubt that the sun doth move,
    Doubt truth to be a liar,
        But never doubt I love.                              119
O dear Ophelia, I am ill at these numbers. I have not
art to reckon my groans. But that I love thee best, O
most best, believe it. Adieu.
    Thine evermore, most dear lady, whilst this
    machine is to him,
                            Hamlet.'                         125
This in obedience hath my daughter showed me,
And more above hath his solicitings,
As they fell out by time, by means, and place,
All given to mine ear.
KING CLAUDIUS            But how hath she
    Received his love?
POLONIUS             What do you think of me?                130

KING CLAUDIUS
As of a man faithful and honourable.
POLONIUS
I would fain prove so. But what might you think,
When I had seen this hot love on the wing,
As I perceived it—I must tell you that—
Before my daughter told me, what might you,                  135
Or my dear majesty your queen here, think,
If I had played the desk or table-book,
Or given my heart a winking mute and dumb,
Or looked upon this love with idle sight—
What might you think? No, I went round to work,
And my young mistress thus I did bespeak:                    141
'Lord Hamlet is a prince out of thy star.
This must not be'. And then I precepts gave her,
That she should lock herself from his resort,
Admit no messengers, receive no tokens;                      145
Which done, she took the fruits of my advice,
And he, repulsèd—a short tale to make—
Fell into a sadness, then into a fast,
Thence to a watch, thence into a weakness,
Thence to a lightness, and, by this declension,             150
Into the madness wherein now he raves,
And all we wail for.
KING CLAUDIUS (*to Gertrude*) Do you think 'tis this?
QUEEN GERTRUDE It may be; very likely.
POLONIUS
Hath there been such a time—I'd fain know that—
That I have positively said ' 'Tis so'                       156
When it proved otherwise?
KING CLAUDIUS                     Not that I know.
POLONIUS (*touching his head, then his shoulder*)
Take this from this if this be otherwise.
If circumstances lead me I will find
Where truth is hid, though it were hid indeed                160
Within the centre.
KING CLAUDIUS       How may we try it further?
POLONIUS
You know sometimes he walks four hours together
Here in the lobby.
QUEEN GERTRUDE          So he does indeed.
POLONIUS
At such a time I'll loose my daughter to him.
(*To Claudius*) Be you and I behind an arras then.           165
Mark the encounter. If he love her not,
And be not from his reason fall'n thereon,
Let me be no assistant for a state,
But keep a farm and carters.
KING CLAUDIUS                    We will try it.
    *Enter Prince Hamlet, madly attired, reading on a book*
QUEEN GERTRUDE
But look where sadly the poor wretch comes reading.
POLONIUS
Away, I do beseech you both, away.                           171
I'll board him presently. O give me leave.
                        *Exeunt Claudius and Gertrude*
    How does my good Lord Hamlet?
HAMLET Well, God-'a'-mercy.
POLONIUS Do you know me, my lord?                            175
HAMLET Excellent, excellent well. You're a fishmonger.
POLONIUS Not I, my lord.
HAMLET Then I would you were so honest a man.
POLONIUS Honest, my lord?
HAMLET Ay, sir. To be honest, as this world goes, is to
    be one man picked out of ten thousand.                   181
POLONIUS That's very true, my lord.

HAMLET For if the sun breed maggots in a dead dog, being a good kissing carrion—have you a daughter?

POLONIUS I have, my lord. 185

HAMLET Let her not walk i'th' sun. Conception is a blessing, but not as your daughter may conceive. Friend, look to't.

POLONIUS (aside) How say you by that? Still harping on my daughter. Yet he knew me not at first—a said I was a fishmonger. A is far gone, far gone, and truly, in my youth I suffered much extremity for love, very near this. I'll speak to him again.—What do you read, my lord?

HAMLET Words, words, words. 195

POLONIUS What is the matter, my lord?

HAMLET Between who?

POLONIUS I mean the matter you read, my lord.

HAMLET Slanders, sir; for the satirical slave says here that old men have grey beards, that their faces are wrinkled, their eyes purging thick amber, or plum-tree gum, and that they have a plentiful lack of wit, together with most weak hams. All which, sir, though I most powerfully and potently believe, yet I hold it not honesty to have it thus set down; for you yourself, sir, should be old as I am—if, like a crab, you could go backward.

POLONIUS (aside) Though this be madness, yet there is method in't.—Will you walk out of the air, my lord?

HAMLET Into my grave. 209

POLONIUS Indeed, that is out o'th' air. (Aside) How pregnant sometimes his replies are! A happiness that often madness hits on, which reason and sanity could not so prosperously be delivered of. I will leave him, and suddenly contrive the means of meeting between him and my daughter.—My lord, I will take my leave of you. 216

HAMLET You cannot, sir, take from me anything that I will more willingly part withal—except my life, my life, my life.

POLONIUS (going) Fare you well, my lord. 220

HAMLET These tedious old fools!

⌈Enter Guildenstern and Rosencrantz⌉

POLONIUS You go to seek the Lord Hamlet. There he is.

ROSENCRANTZ God save you, sir.

GUILDENSTERN ⌈to Polonius⌉ Mine honoured lord.

⌈Exit Polonius⌉

ROSENCRANTZ (to Hamlet) My most dear lord. 225

HAMLET My ex'llent good friends. How dost thou, Guildenstern? Ah, Rosencrantz—good lads, how do ye both?

ROSENCRANTZ
As the indifferent children of the earth.

GUILDENSTERN
Happy in that we are not over-happy, 230
On Fortune's cap we are not the very button.

HAMLET Nor the soles of her shoe?

ROSENCRANTZ Neither, my lord.

HAMLET Then you live about her waist, or in the middle of her favour? 235

GUILDENSTERN Faith, her privates we.

HAMLET In the secret parts of Fortune? O, most true, she is a strumpet. What's the news?

ROSENCRANTZ None, my lord, but that the world's grown honest. 240

HAMLET Then is doomsday near. But your news is not true. Let me question more in particular. What have you, my good friends, deserved at the hands of Fortune that she sends you to prison hither?

GUILDENSTERN Prison, my lord? 245

HAMLET Denmark's a prison.

ROSENCRANTZ Then is the world one.

HAMLET A goodly one, in which there are many confines, wards, and dungeons, Denmark being one o'th' worst.

ROSENCRANTZ We think not so, my lord. 250

HAMLET Why, then 'tis none to you, for there is nothing either good or bad but thinking makes it so. To me it is a prison.

ROSENCRANTZ Why, then your ambition makes it one; 'tis too narrow for your mind. 255

HAMLET O God, I could be bounded in a nutshell and count myself a king of infinite space, were it not that I have bad dreams.

GUILDENSTERN Which dreams indeed are ambition; for the very substance of the ambitious is merely the shadow of a dream. 261

HAMLET A dream itself is but a shadow.

ROSENCRANTZ Truly, and I hold ambition of so airy and light a quality that it is but a shadow's shadow.

HAMLET Then are our beggars bodies, and our monarchs and outstretched heroes the beggars' shadows. Shall we to th' court? For, by my fay, I cannot reason. 267

ROSENCRANTZ and GUILDENSTERN We'll wait upon you.

HAMLET No such matter. I will not sort you with the rest of my servants, for, to speak to you like an honest man, I am most dreadfully attended. But in the beaten way of friendship, what make you at Elsinore? 272

ROSENCRANTZ To visit you, my lord, no other occasion.

HAMLET Beggar that I am, I am even poor in thanks, but I thank you; and sure, dear friends, my thanks are too dear a halfpenny. Were you not sent for? Is it your own inclining? Is it a free visitation? Come, deal justly with me. Come, come. Nay, speak. 278

GUILDENSTERN What should we say, my lord?

HAMLET Why, anything—but to th' purpose. You were sent for, and there is a kind of confession in your looks which your modesties have not craft enough to colour. I know the good King and Queen have sent for you.

ROSENCRANTZ To what end, my lord? 284

HAMLET That you must teach me. But let me conjure you by the rights of our fellowship, by the consonancy of our youth, by the obligation of our ever-preserved love, and by what more dear a better proposer could charge you withal, be even and direct with me whether you were sent for or no.

ROSENCRANTZ (to Guildenstern) What say you? 291

HAMLET Nay then, I have an eye of you—if you love me, hold not off.

GUILDENSTERN My lord, we were sent for.

HAMLET I will tell you why. So shall my anticipation prevent your discovery, and your secrecy to the King and Queen moult no feather. I have of late—but wherefore I know not—lost all my mirth, forgone all custom of exercise; and indeed it goes so heavily with my disposition that this goodly frame, the earth, seems to me a sterile promontory. This most excellent canopy the air, look you, this brave o'erhanging, this majestical roof fretted with golden fire—why, it appears no other thing to me than a foul and pestilent congregation of vapours. What a piece of work is a man! How noble in reason, how infinite in faculty, in form and moving how express and admirable, in action how like an angel, in apprehension how like a god—the beauty of the world, the paragon of animals! And yet to me what is this quintessence of dust? Man delights not me—no, nor woman neither, though by your smiling you seem to say so. 312

ROSENCRANTZ My lord, there was no such stuff in my thoughts.

HAMLET Why did you laugh, then, when I said 'Man delights not me'? 316

ROSENCRANTZ To think, my lord, if you delight not in man what lenten entertainment the players shall receive from you. We coted them on the way, and hither are they coming to offer you service. 320

HAMLET He that plays the King shall be welcome; his majesty shall have tribute of me. The adventurous Knight shall use his foil and target, the Lover shall not sigh gratis, the Humorous Man shall end his part in peace, the Clown shall make those laugh whose lungs are tickled o'th' sear, and the Lady shall say her mind freely, or the blank verse shall halt for't. What players are they?

ROSENCRANTZ Even those you were wont to take delight in, the tragedians of the city. 330

HAMLET How chances it they travel? Their residence both in reputation and profit was better both ways.

ROSENCRANTZ I think their inhibition comes by the means of the late innovation.

HAMLET Do they hold the same estimation they did when I was in the city? Are they so followed? 336

ROSENCRANTZ No, indeed, they are not.

HAMLET How comes it? Do they grow rusty?

ROSENCRANTZ Nay, their endeavour keeps in the wonted pace. But there is, sir, an eyrie of children, little eyases, that cry out on the top of question and are most tyrannically clapped for't. These are now the fashion, and so berattle the common stages—so they call them—that many wearing rapiers are afraid of goose-quills, and dare scarce come thither. 345

HAMLET What, are they children? Who maintains 'em? How are they escoted? Will they pursue the quality no longer than they can sing? Will they not say afterwards, if they should grow themselves to common players—as it is like most will, if their means are not better—their writers do them wrong to make them exclaim against their own succession? 352

ROSENCRANTZ Faith, there has been much to-do on both sides, and the nation holds it no sin to tarre them to controversy. There was for a while no money bid for argument unless the poet and the player went to cuffs in the question. 357

HAMLET Is't possible?

GUILDENSTERN O, there has been much throwing about of brains. 360

HAMLET Do the boys carry it away?

ROSENCRANTZ Ay, that they do, my lord, Hercules and his load too.

HAMLET It is not strange; for mine uncle is King of Denmark, and those that would make mows at him while my father lived give twenty, forty, an hundred ducats apiece for his picture in little. 'Sblood, there is something in this more than natural, if philosophy could find it out.

*A flourish for the Players*

GUILDENSTERN There are the players. 370

HAMLET Gentlemen, you are welcome to Elsinore. Your hands, come. Th'appurtenance of welcome is fashion and ceremony. Let me comply with you in the garb, lest my extent to the players—which, I tell you, must show fairly outward—should more appear like entertainment than yours. 376

*[He shakes hands with them]*

You are welcome. But my uncle-father and aunt-mother are deceived.

GUILDENSTERN In what, my dear lord?

HAMLET I am but mad north-north-west; when the wind is southerly, I know a hawk from a handsaw. 381

*Enter Polonius*

POLONIUS Well be with you, gentlemen.

HAMLET (*aside*) Hark you, Guildenstern, and you too—at each ear a hearer—that great baby you see there is not yet out of his swathing-clouts. 385

ROSENCRANTZ (*aside*) Haply he's the second time come to them, for they say an old man is twice a child.

HAMLET (*aside*) I will prophesy he comes to tell me of the players. Mark it.—You say right, sir, for o' Monday morning, 'twas so indeed. 390

POLONIUS My lord, I have news to tell you.

HAMLET My lord, I have news to tell you. When Roscius was an actor in Rome—

POLONIUS The actors are come hither, my lord.

HAMLET Buzz, buzz. 395

POLONIUS Upon mine honour—

HAMLET Then came each actor on his ass.

POLONIUS The best actors in the world, either for tragedy, comedy, history, pastoral, pastorical-comical, historical-pastoral, tragical-historical, tragical-comical-historical-pastoral, scene individable or poem unlimited. Seneca cannot be too heavy, nor Plautus too light. For the law of writ and the liberty, these are the only men.

HAMLET O Jephthah, judge of Israel, what a treasure hadst thou! 405

POLONIUS What a treasure had he, my lord?

HAMLET Why,
'One fair daughter and no more,
The which he lovèd passing well'.

POLONIUS (*aside*) Still on my daughter. 410

HAMLET Am I not i'th' right, old Jephthah?

POLONIUS If you call me Jephthah, my lord, I have a daughter that I love passing well.

HAMLET Nay, that follows not.

POLONIUS What follows then, my lord? 415

HAMLET Why
'As by lot
God wot',
and then you know
'It came to pass 420
As most like it was'—
the first row of the pious chanson will show you more, for look where my abridgements come.

*Enter four or five Players*

You're welcome, masters, welcome all.—I am glad to see thee well.—Welcome, good friends.—O, my old friend! Thy face is valanced since I saw thee last. Com'st thou to beard me in Denmark?—What, my young lady and mistress. By'r Lady, your ladyship is nearer heaven than when I saw you last by the altitude of a chopine. Pray God your voice, like a piece of uncurrent gold, be not cracked within the ring.—Masters, you are all welcome. We'll e'en to't like French falc'ners, fly at anything we see. We'll have a speech straight. Come, give us a taste of your quality. Come, a passionate speech. 435

FIRST PLAYER What speech, my good lord?

HAMLET I heard thee speak me a speech once, but it was
never acted, or, if it was, not above once; for the play,
I remember, pleased not the million. 'Twas caviare to
the general. But it was—as I received it, and others
whose judgements in such matters cried in the top of
mine—an excellent play, well digested in the scenes,
set down with as much modesty as cunning. I remember
one said there was no sallets in the lines to make the
matter savoury, nor no matter in the phrase that might
indict the author of affectation, but called it an honest
method, as wholesome as sweet, and by very much
more handsome than fine. One speech in it I chiefly
loved, 'twas Aeneas' tale to Dido, and thereabout of it
especially where he speaks of Priam's slaughter. If it
live in your memory, begin at this line—let me see, let
me see:     452
'The rugged Pyrrhus, like th'Hyrcanian beast'—
'tis not so. It begins with Pyrrhus—
'The rugged Pyrrhus, he whose sable arms,     455
Black as his purpose, did the night resemble
When he lay couchèd in the ominous horse,
Hath now this dread and black complexion smeared
With heraldry more dismal. Head to foot
Now is he total gules, horridly tricked     460
With blood of fathers, mothers, daughters, sons,
Baked and impasted with the parching streets,
That lend a tyrannous and damnèd light
To their vile murders. Roasted in wrath and fire,
And thus o'er-sizèd with coagulate gore,     465
With eyes like carbuncles the hellish Pyrrhus
Old grandsire Priam seeks.'
So, proceed you.
POLONIUS Fore God, my lord, well spoken, with good
accent and good discretion.     470
FIRST PLAYER               'Anon he finds him,
Striking too short at Greeks. His antique sword,
Rebellious to his arm, lies where it falls,
Repugnant to command. Unequal match,
Pyrrhus at Priam drives, in rage strikes wide;     475
But with the whiff and wind of his fell sword
Th'unnervèd father falls. Then senseless Ilium,
Seeming to feel his blow, with flaming top
Stoops to his base, and with a hideous crash
Takes prisoner Pyrrhus' ear. For lo, his sword,     480
Which was declining on the milky head
Of reverend Priam, seemed i'th' air to stick.
So, as a painted tyrant, Pyrrhus stood,
And, like a neutral to his will and matter,
Did nothing.     485
But as we often see against some storm
A silence in the heavens, the rack stand still,
The bold winds speechless, and the orb below
As hush as death, anon the dreadful thunder
Doth rend the region: so, after Pyrrhus' pause,     490
A rousèd vengeance sets him new a-work;
And never did the Cyclops' hammers fall
On Mars his armour, forged for proof eterne,
With less remorse than Pyrrhus' bleeding sword
Now falls on Priam.     495
Out, out, thou strumpet Fortune! All you gods,
In general synod, take away her power,
Break all the spokes and fellies from her wheel,
And bowl the round nave down the hill of heaven,
As low as to the fiends!'     500
POLONIUS This is too long.

HAMLET It shall to the barber's, with your beard. (To First
Player) Prithee, say on. He's for a jig or a tale of
bawdry, or he sleeps. Say on, come to Hecuba.
FIRST PLAYER
'But who, O who had seen the mobbled queen'—     505
HAMLET 'The mobbled queen'?
POLONIUS That's good; 'mobbled queen' is good.
FIRST PLAYER
'Run barefoot up and down, threat'ning the flames
With bisson rheum; a clout upon that head
Where late the diadem stood, and for a robe,     510
About her lank and all o'er-teemèd loins,
A blanket in th'alarm of fear caught up—
Who this had seen, with tongue in venom steeped,
'Gainst Fortune's state would treason have pronounced.
But if the gods themselves did see her then,     515
When she saw Pyrrhus make malicious sport
In mincing with his sword her husband's limbs,
The instant burst of clamour that she made—
Unless things mortal move them not at all—
Would have made milch the burning eyes of heaven,
And passion in the gods.'     521
POLONIUS Look whe'er he has not turned his colour, and
has tears in 's eyes. (To First Player) Prithee, no more.
HAMLET (to First Player) 'Tis well. I'll have thee speak out
the rest soon. (To Polonius) Good my lord, will you see
the players well bestowed? Do ye hear?—let them be
well used, for they are the abstracts and brief chronicles
of the time. After your death you were better have a
bad epitaph than their ill report while you live.     529
POLONIUS My lord, I will use them according to their
desert.
HAMLET God's bodykins, man, much better. Use every
man after his desert, and who should scape whipping?
Use them after your own honour and dignity—the less
they deserve, the more merit is in your bounty. Take
them in.     536
POLONIUS (to Players) Come, sirs.     Exit
HAMLET (to Players) Follow him, friends. We'll hear a play
tomorrow. Dost thou hear me, old friend? Can you play
the murder of Gonzago?     540
⌐PLAYERS⌐ Ay, my lord.
HAMLET We'll ha't tomorrow night. You could for a need
study a speech of some dozen or sixteen lines which I
would set down and insert in't, could ye not?
⌐PLAYERS⌐ Ay, my lord.     545
HAMLET Very well. Follow that lord, and look you mock
him not.                     ⌐Exeunt Players⌐
My good friends, I'll leave you till night. You are
welcome to Elsinore.
ROSENCRANTZ Good my lord.     550
HAMLET
Ay, so. God b'wi' ye.           Exeunt all but Hamlet
                Now I am alone.
O, what a rogue and peasant slave am I!
Is it not monstrous that this player here,
But in a fiction, in a dream of passion,
Could force his soul so to his whole conceit     555
That from her working all his visage wanned,
Tears in his eyes, distraction in 's aspect,
A broken voice, and his whole function suiting
With forms to his conceit? And all for nothing.
For Hecuba!     560
What's Hecuba to him, or he to Hecuba,
That he should weep for her? What would he do

Had he the motive and the cue for passion
That I have? He would drown the stage with tears,
And cleave the general ear with horrid speech,          565
Make mad the guilty and appal the free,
Confound the ignorant, and amaze indeed
The very faculty of eyes and ears. Yet I,
A dull and muddy-mettled rascal, peak
Like John-a-dreams, unpregnant of my cause,          570
And can say nothing—no, not for a king
Upon whose property and most dear life
A damned defeat was made. Am I a coward?
Who calls me villain, breaks my pate across,
Plucks off my beard and blows it in my face,          575
Tweaks me by th' nose, gives me the lie i'th' throat
As deep as to the lungs? Who does me this?
Ha? 'Swounds, I should take it; for it cannot be
But I am pigeon-livered and lack gall
To make oppression bitter, or ere this          580
I should 'a' fatted all the region kites
With this slave's offal. Bloody, bawdy villain!
Remorseless, treacherous, lecherous, kindless villain!
O, vengeance!—
Why, what an ass am I? Ay, sure, this is most brave,
That I, the son of the dear murderèd,          586
Prompted to my revenge by heaven and hell,
Must, like a whore, unpack my heart with words
And fall a-cursing like a very drab,
A scullion! Fie upon't, foh!—About, my brain.          590
I have heard that guilty creatures sitting at a play
Have by the very cunning of the scene
Been struck so to the soul that presently
They have proclaimed their malefactions;
For murder, though it have no tongue, will speak          595
With most miraculous organ. I'll have these players
Play something like the murder of my father
Before mine uncle. I'll observe his looks,
I'll tent him to the quick. If a but blench,
I know my course. The spirit that I have seen          600
May be the devil, and the devil hath power
T'assume a pleasing shape; yea, and perhaps,
Out of my weakness and my melancholy—
As he is very potent with such spirits—
Abuses me to damn me. I'll have grounds          605
More relative than this. The play's the thing
Wherein I'll catch the conscience of the King.          *Exit*

**3.1**    *Enter King Claudius, Queen Gertrude, Polonius,*
           *Ophelia, Rosencrantz, Guildenstern, and lords*
KING CLAUDIUS (*to Rosencrantz and Guildenstern*)
And can you by no drift of circumstance
Get from him why he puts on this confusion,
Grating so harshly all his days of quiet
With turbulent and dangerous lunacy?
ROSENCRANTZ
He does confess he feels himself distracted,          5
But from what cause a will by no means speak.
GUILDENSTERN
Nor do we find him forward to be sounded,
But with a crafty madness keeps aloof
When we would bring him on to some confession
Of his true state.          10
QUEEN GERTRUDE Did he receive you well?
ROSENCRANTZ Most like a gentleman.
GUILDENSTERN
But with much forcing of his disposition.

ROSENCRANTZ
Niggard of question, but of our demands
Most free in his reply.
QUEEN GERTRUDE          Did you assay him          15
To any pastime?
ROSENCRANTZ
Madam, it so fell out that certain players
We o'er-raught on the way. Of these we told him,
And there did seem in him a kind of joy
To hear of it. They are about the court,          20
And, as I think, they have already order
This night to play before him.
POLONIUS          'Tis most true,
And he beseeched me to entreat your majesties
To hear and see the matter.
KING CLAUDIUS
With all my heart; and it doth much content me          25
To hear him so inclined.—Good gentlemen,
Give him a further edge, and drive his purpose on
To these delights.
ROSENCRANTZ We shall, my lord.
          *Exeunt Rosencrantz and Guildenstern*
KING CLAUDIUS Sweet Gertrude, leave us too,          30
For we have closely sent for Hamlet hither,
That he, as 'twere by accident, may here
Affront Ophelia.
Her father and myself, lawful espials,
Will so bestow ourselves that, seeing unseen,          35
We may of their encounter frankly judge,
And gather by him, as he is behaved,
If't be th'affliction of his love or no
That thus he suffers for.
QUEEN GERTRUDE          I shall obey you.
And for your part, Ophelia, I do wish          40
That your good beauties be the happy cause
Of Hamlet's wildness; so shall I hope your virtues
Will bring him to his wonted way again,
To both your honours.
OPHELIA          Madam, I wish it may.
          *Exit Gertrude*
POLONIUS
Ophelia, walk you here.—Gracious, so please you,          45
We will bestow ourselves.—Read on this book,
That show of such an exercise may colour
Your loneliness. We are oft to blame in this:
'Tis too much proved that with devotion's visage
And pious action we do sugar o'er          50
The devil himself.
KING CLAUDIUS          O, 'tis too true.
(*Aside*) How smart a lash that speech doth give my
          conscience.
The harlot's cheek, beautied with plast'ring art,
Is not more ugly to the thing that helps it
Than is my deed to my most painted word.          55
O heavy burden!
POLONIUS
I hear him coming. Let's withdraw, my lord.
          *Exeunt Claudius and Polonius*
          *Enter Prince Hamlet*
HAMLET
To be, or not to be; that is the question:
Whether 'tis nobler in the mind to suffer
The slings and arrows of outrageous fortune,          60
Or to take arms against a sea of troubles,
And, by opposing, end them. To die, to sleep—
No more, and by a sleep to say we end

The heartache and the thousand natural shocks
That flesh is heir to—'tis a consummation     65
Devoutly to be wished. To die, to sleep.
To sleep, perchance to dream. Ay, there's the rub,
For in that sleep of death what dreams may come
When we have shuffled off this mortal coil
Must give us pause. There's the respect     70
That makes calamity of so long life,
For who would bear the whips and scorns of time,
Th'oppressor's wrong, the proud man's contumely,
The pangs of disprized love, the law's delay,
The insolence of office, and the spurns     75
That patient merit of th'unworthy takes,
When he himself might his quietus make
With a bare bodkin? Who would these fardels bear,
To grunt and sweat under a weary life,
But that the dread of something after death,     80
The undiscovered country from whose bourn
No traveller returns, puzzles the will,
And makes us rather bear those ills we have
Than fly to others that we know not of?
Thus conscience does make cowards of us all,     85
And thus the native hue of resolution
Is sicklied o'er with the pale cast of thought,
And enterprises of great pith and moment
With this regard their currents turn awry,
And lose the name of action. Soft you, now,     90
The fair Ophelia!—Nymph, in thy orisons
Be all my sins remembered.

OPHELIA                 Good my lord,
How does your honour for this many a day?

HAMLET
I humbly thank you, well, well, well.

OPHELIA
My lord, I have remembrances of yours     95
That I have longèd long to redeliver.
I pray you now receive them.

HAMLET
No, no, I never gave you aught.

OPHELIA
My honoured lord, you know right well you did,
And with them words of so sweet breath composed
As made the things more rich. Their perfume lost,     101
Take these again; for to the noble mind
Rich gifts wax poor when givers prove unkind.
There, my lord.

HAMLET Ha, ha? Are you honest?     105

OPHELIA My lord.

HAMLET Are you fair?

OPHELIA What means your lordship?

HAMLET That if you be honest and fair, your honesty
should admit no discourse to your beauty.     110

OPHELIA Could beauty, my lord, have better commerce
than with honesty?

HAMLET Ay, truly, for the power of beauty will sooner
transform honesty from what it is to a bawd than the
force of honesty can translate beauty into his likeness.
This was sometime a paradox, but now the time gives
it proof. I did love you once.     117

OPHELIA Indeed, my lord, you made me believe so.

HAMLET You should not have believed me, for virtue
cannot so inoculate our old stock but we shall relish
of it. I loved you not.     121

OPHELIA I was the more deceived.

HAMLET Get thee to a nunnery. Why wouldst thou be a
breeder of sinners? I am myself indifferent honest, but

yet I could accuse me of such things that it were better
my mother had not borne me. I am very proud,
revengeful, ambitious, with more offences at my beck
than I have thoughts to put them in, imagination to
give them shape, or time to act them in. What should
such fellows as I do crawling between heaven and
earth? We are arrant knaves, all. Believe none of us.
Go thy ways to a nunnery. Where's your father?     132

OPHELIA At home, my lord.

HAMLET Let the doors be shut upon him, that he may
play the fool nowhere but in 's own house. Farewell.

OPHELIA O help him, you sweet heavens!     136

HAMLET If thou dost marry, I'll give thee this plague for
thy dowry: be thou as chaste as ice, as pure as snow,
thou shalt not escape calumny. Get thee to a nunnery,
go, farewell. Or if thou wilt needs marry, marry a fool;
for wise men know well enough what monsters you
make of them. To a nunnery, go, and quickly, too.
Farewell.     143

OPHELIA O heavenly powers, restore him!

HAMLET I have heard of your paintings, too, well enough.
God hath given you one face, and you make yourselves
another. You jig, you amble, and you lisp, and
nickname God's creatures, and make your wantonness
your ignorance. Go to, I'll no more on't. It hath made
me mad. I say we will have no more marriages. Those
that are married already—all but one—shall live. The
rest shall keep as they are. To a nunnery, go.     *Exit*

OPHELIA
O what a noble mind is here o'erthrown!
The courtier's, soldier's, scholar's eye, tongue, sword,
Th'expectancy and rose of the fair state,     155
The glass of fashion and the mould of form,
Th'observed of all observers, quite, quite, down!
And I, of ladies most deject and wretched,
That sucked the honey of his music vows,
Now see that noble and most sovereign reason     160
Like sweet bells jangled out of tune and harsh;
That unmatched form and feature of blown youth
Blasted with ecstasy. O woe is me,
T'have seen what I have seen, see what I see!
*Enter King Claudius and Polonius*

KING CLAUDIUS
Love? His affections do not that way tend,     165
Nor what he spake, though it lacked form a little,
Was not like madness. There's something in his soul
O'er which his melancholy sits on brood,
And I do doubt the hatch and the disclose
Will be some danger; which to prevent     170
I have in quick determination
Thus set it down: he shall with speed to England
For the demand of our neglected tribute.
Haply the seas and countries different,
With variable objects, shall expel     175
This something-settled matter in his heart,
Whereon his brains still beating puts him thus
From fashion of himself. What think you on't?

POLONIUS
It shall do well. But yet do I believe
The origin and commencement of this grief     180
Sprung from neglected love.—How now, Ophelia?
You need not tell us what Lord Hamlet said;
We heard it all.—My lord, do as you please,
But, if you hold it fit, after the play
Let his queen mother all alone entreat him     185
To show his griefs. Let her be round with him,

And I'll be placed, so please you, in the ear
Of all their conference. If she find him not,
To England send him, or confine him where
Your wisdom best shall think.
KING CLAUDIUS                    It shall be so.    190
Madness in great ones must not unwatched go.
*Exeunt*

**3.2**    *Enter Prince Hamlet and two or three of the Players*
HAMLET Speak the speech, I pray you, as I pronounced
it to you—trippingly on the tongue; but if you mouth
it, as many of your players do, I had as lief the town-
crier had spoke my lines. Nor do not saw the air too
much with your hand, thus, but use all gently; for in
the very torrent, tempest, and as I may say the
whirlwind of your passion, you must acquire and beget
a temperance that may give it smoothness. O, it offends
me to the soul to hear a robustious, periwig-pated
fellow tear a passion to tatters, to very rags, to split
the ears of the groundlings, who for the most part are
capable of nothing but inexplicable dumb shows and
noise. I would have such a fellow whipped for o'erdoing
Termagant. It out-Herods Herod. Pray you avoid it.
A PLAYER I warrant your honour.                        15
HAMLET Be not too tame, neither; but let your own
discretion be your tutor. Suit the action to the word,
the word to the action, with this special observance:
that you o'erstep not the modesty of nature. For
anything so overdone is from the purpose of playing,
whose end, both at the first and now, was and is to
hold as 'twere the mirror up to nature, to show virtue
her own feature, scorn her own image, and the very
age and body of the time his form and pressure. Now
this overdone, or come tardy off, though it make the
unskilful laugh, cannot but make the judicious grieve;
the censure of the which one must in your allowance
o'erweigh a whole theatre of others. O, there be players
that I have seen play, and heard others praise, and
that highly, not to speak it profanely, that neither
having the accent of Christians nor the gait of Christian,
pagan, nor no man, have so strutted and bellowed that
I have thought some of nature's journeymen had made
men, and not made them well, they imitated humanity
so abominably.                                         35
A PLAYER I hope we have reformed that indifferently with
us, sir.
HAMLET O, reform it altogether. And let those that play
your clowns speak no more than is set down for them;
for there be of them that will themselves laugh to set
on some quantity of barren spectators to laugh too,
though in the mean time some necessary question of
the play be then to be considered. That's villainous,
and shows a most pitiful ambition in the fool that uses
it. Go make you ready.                    *Exeunt Players*
*Enter Polonius, Guildenstern, and Rosencrantz*
(*To Polonius*) How now, my lord? Will the King hear
this piece of work?                                    47
POLONIUS And the Queen too, and that presently.
HAMLET Bid the players make haste.          *Exit Polonius*
Will you two help to hasten them?
ROSENCRANTZ *and* GUILDENSTERN          We will, my lord.
*Exeunt*
HAMLET
What ho, Horatio!
*Enter Horatio*
HORATIO              Here, sweet lord, at your service.

HAMLET
Horatio, thou art e'en as just a man               52
As e'er my conversation coped withal.
HORATIO
O my dear lord—
HAMLET              Nay, do not think I flatter;
For what advancement may I hope from thee,         55
That no revenue hast but thy good spirits
To feed and clothe thee? Why should the poor be
    flattered?
No, let the candied tongue lick absurd pomp,
And crook the pregnant hinges of the knee
Where thrift may follow feigning. Dost thou hear?—
Since my dear soul was mistress of her choice      61
And could of men distinguish, her election
Hath sealed thee for herself; for thou hast been
As one in suff'ring all that suffers nothing,
A man that Fortune's buffets and rewards           65
Hath ta'en with equal thanks; and blest are those
Whose blood and judgement are so well commingled
That they are not a pipe for Fortune's finger
To sound what stop she please. Give me that man
That is not passion's slave, and I will wear him    70
In my heart's core, ay, in my heart of heart,
As I do thee. Something too much of this.
There is a play tonight before the King.
One scene of it comes near the circumstance
Which I have told thee of my father's death.        75
I prithee, when thou seest that act afoot,
Even with the very comment of thy soul
Observe mine uncle. If his occulted guilt
Do not itself unkennel in one speech,
It is a damnèd ghost that we have seen,             80
And my imaginations are as foul
As Vulcan's stithy. Give him heedful note,
For I mine eyes will rivet to his face,
And after, we will both our judgements join
To censure of his seeming.
HORATIO              Well, my lord.                  85
If a steal aught the whilst this play is playing
And scape detecting, I will pay the theft.
⌜*Sound a flourish*⌝
HAMLET
They are coming to the play. I must be idle.
Get you a place.
⌜*Danish march. Enter King Claudius, Queen
Gertrude, Polonius, Ophelia, Rosencrantz,
Guildenstern, and other lords attendant, with the
King's guard carrying torches*⌝
KING CLAUDIUS          How fares our cousin Hamlet?
HAMLET Excellent, i'faith, of the chameleon's dish. I eat
the air, promise-crammed. You cannot feed capons so.
KING CLAUDIUS I have nothing with this answer, Hamlet.
These words are not mine.
HAMLET No, nor mine now. (*To Polonius*) My lord, you
played once i'th' university, you say.              95
POLONIUS That I did, my lord, and was accounted a good
actor.
HAMLET And what did you enact?
POLONIUS I did enact Julius Caesar. I was killed i'th'
Capitol. Brutus killed me.                         100
HAMLET It was a brute part of him to kill so capital a calf
there.—Be the players ready?
ROSENCRANTZ Ay, my lord, they stay upon your patience.
QUEEN GERTRUDE
Come hither, my good Hamlet. Sit by me.

HAMLET No, good-mother, here's mettle more attractive.
    *He sits by Ophelia*
POLONIUS (*aside*) O ho, do you mark that?      106
HAMLET (*to Ophelia*) Lady, shall I lie in your lap?
OPHELIA No, my lord.
HAMLET I mean my head upon your lap?
OPHELIA Ay, my lord.      110
HAMLET Do you think I meant country matters?
OPHELIA I think nothing, my lord.
HAMLET That's a fair thought to lie between maids' legs.
OPHELIA What is, my lord?
HAMLET No thing.      115
OPHELIA You are merry, my lord.
HAMLET Who, I?
OPHELIA Ay, my lord.
HAMLET O God, your only jig-maker! What should a man do but be merry? For look you how cheerfully my mother looks, and my father died within 's two hours.
OPHELIA Nay, 'tis twice two months, my lord.      122
HAMLET So long? Nay then, let the devil wear black, for I'll have a suit of sables. O heavens, die two months ago and not forgotten yet! Then there's hope a great man's memory may outlive his life half a year. But, by'r Lady, a must build churches then, or else shall a suffer not thinking on, with the hobby-horse, whose epitaph is 'For O, for O, the hobby-horse is forgot.'
    *Hautboys play. The dumb show enters. Enter a King and a Queen very lovingly, the Queen embracing him. She kneels and makes show of protestation unto him. He takes her up and declines his head upon her neck. He lays him down upon a bank of flowers. She, seeing him asleep, leaves him. Anon comes in a fellow, takes off his crown, kisses it, and pours poison in the King's ears, and exits. The Queen returns, finds the King dead, and makes passionate action. The poisoner, with some two or three mutes, comes in again, seeming to lament with her. The dead body is carried away. The poisoner woos the Queen with gifts. She seems loath and unwilling a while, but in the end accepts his love. Exeunt the Players*
OPHELIA What means this, my lord?      130
HAMLET Marry, this is miching *malhecho*. That means mischief.
OPHELIA Belike this show imports the argument of the play.
    *Enter Prologue*
HAMLET We shall know by this fellow. The players cannot keep counsel, they'll tell all.      135
OPHELIA Will a tell us what this show meant?
HAMLET Ay, or any show that you'll show him. Be not you ashamed to show, he'll not shame to tell you what it means.
OPHELIA You are naught, you are naught. I'll mark the play.      141
PROLOGUE
    For us and for our tragedy
    Here stooping to your clemency,
    We beg your hearing patiently.      *Exit*
HAMLET Is this a prologue, or the posy of a ring?      145
OPHELIA 'Tis brief, my lord.
HAMLET As woman's love.
    *Enter the Player King and his Queen*
PLAYER KING
    Full thirty times hath Phoebus' cart gone round
    Neptune's salt wash and Tellus' orbèd ground,
    And thirty dozen moons with borrowed sheen      150
    About the world have times twelve thirties been

    Since love our hearts and Hymen did our hands
    Unite commutual in most sacred bands.
PLAYER QUEEN
    So many journeys may the sun and moon
    Make us again count o'er ere love be done.      155
    But woe is me, you are so sick of late,
    So far from cheer and from your former state,
    That I distrust you. Yet, though I distrust,
    Discomfort you my lord it nothing must.
    For women's fear and love holds quantity,      160
    In neither aught, or in extremity.
    Now what my love is, proof hath made you know,
    And as my love is sized, my fear is so.
PLAYER KING
    Faith, I must leave thee, love, and shortly too.
    My operant powers their functions leave to do,      165
    And thou shalt live in this fair world behind,
    Honoured, beloved; and haply one as kind
    For husband shalt thou—
PLAYER QUEEN           O, confound the rest!
    Such love must needs be treason in my breast.
    In second husband let me be accurst;      170
    None wed the second but who killed the first.
HAMLET Wormwood, wormwood.
PLAYER QUEEN
    The instances that second marriage move
    Are base respects of thrift, but none of love.
    A second time I kill my husband dead      175
    When second husband kisses me in bed.
PLAYER KING
    I do believe you think what now you speak;
    But what we do determine oft we break.
    Purpose is but the slave to memory,
    Of violent birth but poor validity,      180
    Which now like fruit unripe sticks on the tree,
    But fall unshaken when they mellow be.
    Most necessary 'tis that we forget
    To pay ourselves what to ourselves is debt.
    What to ourselves in passion we propose,      185
    The passion ending, doth the purpose lose.
    The violence of either grief or joy
    Their own enactures with themselves destroy.
    Where joy most revels, grief doth most lament;
    Grief joys, joy grieves, on slender accident.      190
    This world is not for aye, nor 'tis not strange
    That even our loves should with our fortunes change;
    For 'tis a question left us yet to prove
    Whether love lead fortune or else fortune love.
    The great man down, you mark his favourite flies;
    The poor advanced makes friends of enemies.      196
    And hitherto doth love on fortune tend,
    For who not needs shall never lack a friend,
    And who in want a hollow friend doth try
    Directly seasons him his enemy.      200
    But orderly to end where I begun,
    Our wills and fates do so contrary run
    That our devices still are overthrown;
    Our thoughts are ours, their ends none of our own.
    So think thou wilt no second husband wed;      205
    But die thy thoughts when thy first lord is dead.
PLAYER QUEEN
    Nor earth to me give food, nor heaven light,
    Sport and repose lock from me day and night,
    Each opposite that blanks the face of joy
    Meet what I would have well and it destroy,      210

Both here and hence pursue me lasting strife
If, once a widow, ever I be wife.
HAMLET If she should break it now!
PLAYER KING (to Player Queen)
  'Tis deeply sworn. Sweet, leave me here a while.
  My spirits grow dull, and fain I would beguile      215
  The tedious day with sleep.
PLAYER QUEEN                         Sleep rock thy brain,
  And never come mischance between us twain.
                         Player King sleeps. Player Queen exits
HAMLET (to Gertrude) Madam, how like you this play?
QUEEN GERTRUDE The lady protests too much, methinks.
HAMLET O, but she'll keep her word.                     220
KING CLAUDIUS Have you heard the argument? Is there
  no offence in't?
HAMLET No, no, they do but jest, poison in jest. No offence
  i'th' world.
KING CLAUDIUS What do you call the play?                 225
HAMLET The Mousetrap. Marry, how? Tropically. This play
  is the image of a murder done in Vienna. Gonzago is
  the Duke's name, his wife Baptista. You shall see anon.
  'Tis a knavish piece of work; but what o' that? Your
  majesty, and we that have free souls, it touches us not.
  Let the galled jade wince, our withers are unwrung.
      Enter Player Lucianus
  This is one Lucianus, nephew to the King.            232
OPHELIA You are as good as a chorus, my lord.
HAMLET I could interpret between you and your love if I
  could see the puppets dallying.                       235
OPHELIA You are keen, my lord, you are keen.
HAMLET It would cost you a groaning to take off mine edge.
OPHELIA Still better, and worse.
HAMLET So you mis-take your husbands. (To Lucianus)
  Begin, murderer. Pox, leave thy damnable faces and
  begin. Come: 'the croaking raven doth bellow for
  revenge'.                                             242
PLAYER LUCIANUS
  Thoughts black, hands apt, drugs fit, and time
      agreeing,
  Confederate season, else no creature seeing;
  Thou mixture rank of midnight weeds collected,        245
  With Hecate's ban thrice blasted, thrice infected,
  Thy natural magic and dire property
  On wholesome life usurp immediately.
      He pours the poison in the Player King's ear
HAMLET A poisons him i'th' garden for 's estate. His
  name's Gonzago. The story is extant, and writ in choice
  Italian. You shall see anon how the murderer gets the
  love of Gonzago's wife.                               252
OPHELIA The King rises.
HAMLET What, frighted with false fire?
QUEEN GERTRUDE (to Claudius) How fares my lord?         255
POLONIUS Give o'er the play.
KING CLAUDIUS Give me some light. Away.
⌈COURTIERS⌉ Lights, lights, lights!
                         Exeunt all but Hamlet and Horatio
HAMLET
  Why, let the stricken deer go weep,
      The hart ungallèd play,                           260
  For some must watch, while some must sleep,
      So runs the world away.
  Would not this, sir, and a forest of feathers, if the rest
  of my fortunes turn Turk with me, with two Provençal
  roses on my razed shoes, get me a fellowship in a cry
  of players, sir?                                      266
HORATIO Half a share.

HAMLET A whole one, I.
  For thou dost know, O Damon dear,
      This realm dismantled was                         270
  Of Jove himself, and now reigns here
      A very, very—pajock.
HORATIO You might have rhymed.
HAMLET O good Horatio, I'll take the Ghost's word for a
  thousand pound. Didst perceive?                       275
HORATIO Very well, my lord.
HAMLET Upon the talk of the pois'ning?
HORATIO I did very well note him.
      Enter Rosencrantz and Guildenstern
HAMLET Ah ha! Come, some music, come, the recorders,
  For if the King like not the comedy,                  280
  Why then, belike he likes it not, pardie.
  Come, some music.
GUILDENSTERN Good my lord, vouchsafe me a word with
  you.
HAMLET Sir, a whole history.                            285
GUILDENSTERN The King, sir—
HAMLET Ay, sir, what of him?
GUILDENSTERN Is in his retirement marvellous distempered.
HAMLET With drink, sir?
GUILDENSTERN No, my lord, rather with choler.           290
HAMLET Your wisdom should show itself more richer to
  signify this to his doctor, for for me to put him to his
  purgation would perhaps plunge him into far more
  choler.
GUILDENSTERN Good my lord, put your discourse into some
  frame, and start not so wildly from my affair.        296
HAMLET I am tame, sir. Pronounce.
GUILDENSTERN The Queen your mother, in most great
  affliction of spirit, hath sent me to you.
HAMLET You are welcome.                                 300
GUILDENSTERN Nay, good my lord, this courtesy is not of
  the right breed. If it shall please you to make me a
  wholesome answer, I will do your mother's
  commandment; if not, your pardon and my return
  shall be the end of my business.                      305
HAMLET Sir, I cannot.
GUILDENSTERN What, my lord?
HAMLET Make you a wholesome answer. My wit's
  diseased. But, sir, such answers as I can make, you
  shall command; or rather, as you say, my mother.
  Therefore no more, but to the matter. My mother, you
  say?                                                  312
ROSENCRANTZ Then thus she says: your behaviour hath
  struck her into amazement and admiration.
HAMLET O wonderful son, that can so astonish a mother!
  But is there no sequel at the heels of this mother's
  admiration?
ROSENCRANTZ She desires to speak with you in her closet
  ere you go to bed.                                    319
HAMLET We shall obey, were she ten times our mother.
  Have you any further trade with us?
ROSENCRANTZ My lord, you once did love me.
HAMLET So I do still, by these pickers and stealers.
ROSENCRANTZ Good my lord, what is your cause of
  distemper? You do freely bar the door of your own
  liberty if you deny your griefs to your friend.       326
HAMLET Sir, I lack advancement.
ROSENCRANTZ How can that be when you have the voice
  of the King himself for your succession in Denmark?
HAMLET Ay, but 'while the grass grows . . .'—the proverb
  is something musty.                                   331

*Enter one with a recorder*

O, the recorder. Let me see. (*To Rosencrantz and Guildenstern, taking them aside*) To withdraw with you, why do you go about to recover the wind of me as if you would drive me into a toil?    335

GUILDENSTERN O my lord, if my duty be too bold, my love is too unmannerly.

HAMLET I do not well understand that. Will you play upon this pipe?

GUILDENSTERN My lord, I cannot.    340

HAMLET I pray you.

GUILDENSTERN Believe me, I cannot.

HAMLET I do beseech you.

GUILDENSTERN I know no touch of it, my lord.

HAMLET 'Tis as easy as lying. Govern these ventages with your fingers and thumb, give it breath with your mouth, and it will discourse most excellent music. Look you, these are the stops.

GUILDENSTERN But these cannot I command to any utterance of harmony. I have not the skill.    350

HAMLET Why, look you now, how unworthy a thing you make of me! You would play upon me, you would seem to know my stops, you would pluck out the heart of my mystery, you would sound me from my lowest note to the top of my compass; and there is much music, excellent voice in this little organ, yet cannot you make it speak. 'Sblood, do you think I am easier to be played on than a pipe? Call me what instrument you will, though you can fret me, you cannot play upon me.    360

*Enter Polonius*

God bless you, sir.

POLONIUS My lord, the Queen would speak with you, and presently.

HAMLET Do you see yonder cloud that's almost in shape of a camel?    365

POLONIUS By th' mass, and 'tis: like a camel, indeed.

HAMLET Methinks it is like a weasel.

POLONIUS It is backed like a weasel.

HAMLET Or like a whale.

POLONIUS Very like a whale.    370

HAMLET Then will I come to my mother by and by. (*Aside*) They fool me to the top of my bent. (*To Polonius*) I will come by and by.

POLONIUS I will say so.

HAMLET 'By and by' is easily said.      *Exit Polonius*
Leave me, friends.    376

     *Exeunt Rosencrantz and Guildenstern*

'Tis now the very witching time of night,
When churchyards yawn, and hell itself breathes out
Contagion to this world. Now could I drink hot blood,
And do such bitter business as the day    380
Would quake to look on. Soft, now to my mother.
O heart, lose not thy nature! Let not ever
The soul of Nero enter this firm bosom.
Let me be cruel, not unnatural.
I will speak daggers to her, but use none.    385
My tongue and soul in this be hypocrites—
How in my words somever she be shent,
To give them seals never my soul consent.      *Exit*

**3.3**    *Enter King Claudius, Rosencrantz, and Guildenstern*

KING CLAUDIUS
I like him not, nor stands it safe with us
To let his madness range. Therefore prepare you.
I your commission will forthwith dispatch,
And he to England shall along with you.
The terms of our estate may not endure    5
Hazard so dangerous as doth hourly grow
Out of his lunacies.

GUILDENSTERN      We will ourselves provide.
Most holy and religious fear it is
To keep those many many bodies safe
That live and feed upon your majesty.    10

ROSENCRANTZ
The single and peculiar life is bound
With all the strength and armour of the mind
To keep itself from noyance; but much more
That spirit upon whose weal depends and rests
The lives of many. The cease of majesty    15
Dies not alone, but like a gulf doth draw
What's near it with it. It is a massy wheel
Fixed on the summit of the highest mount,
To whose huge spokes ten thousand lesser things
Are mortised and adjoined, which when it falls    20
Each small annexment, petty consequence,
Attends the boist'rous ruin. Never alone
Did the King sigh, but with a general groan.

KING CLAUDIUS
Arm you, I pray you, to this speedy voyage,
For we will fetters put upon this fear    25
Which now goes too free-footed.

ROSENCRANTZ *and* GUILDENSTERN      We will haste us.
     *Exeunt both*

*Enter Polonius*

POLONIUS
My lord, he's going to his mother's closet.
Behind the arras I'll convey myself
To hear the process. I'll warrant she'll tax him home.
And, as you said—and wisely was it said—    30
'Tis meet that some more audience than a mother,
Since nature makes them partial, should o'erhear
The speech of vantage. Fare you well, my liege.
I'll call upon you ere you go to bed,
And tell you what I know.

KING CLAUDIUS      Thanks, dear my lord.    35
     *Exit Polonius*

O, my offence is rank! It smells to heaven.
It hath the primal eldest curse upon't,
A brother's murder. Pray can I not.
Though inclination be as sharp as will,
My stronger guilt defeats my strong intent,    40
And like a man to double business bound
I stand in pause where I shall first begin,
And both neglect. What if this cursèd hand
Were thicker than itself with brother's blood,
Is there not rain enough in the sweet heavens    45
To wash it white as snow? Whereto serves mercy
But to confront the visage of offence?
And what's in prayer but this twofold force,
To be forestallèd ere we come to fall,
Or pardoned being down? Then I'll look up.    50
My fault is past—but O, what form of prayer
Can serve my turn? 'Forgive me my foul murder'?
That cannot be, since I am still possessed
Of those effects for which I did the murder—
My crown, mine own ambition, and my queen.    55
May one be pardoned and retain th'offence?
In the corrupted currents of this world
Offence's gilded hand may shove by justice,
And oft 'tis seen the wicked prize itself
Buys out the law. But 'tis not so above.    60

There is no shuffling, there the action lies
In his true nature, and we ourselves compelled
Even to the teeth and forehead of our faults
To give in evidence. What then? What rests?
Try what repentance can. What can it not?                    65
Yet what can it when one cannot repent?
O wretched state, O bosom black as death,
O limèd soul that, struggling to be free,
Art more engaged! Help, angels! Make assay.
Bow, stubborn knees; and heart with strings of steel,
Be soft as sinews of the new-born babe.                      71
All may be well.
    *He kneels.*
    *Enter Prince Hamlet behind him*
HAMLET
Now might I do it pat, now a is praying,
And now I'll do't,
    ⌈*He draws his sword*⌉
            and so a goes to heaven,
And so am I revenged. That would be scanned.                 75
A villain kills my father, and for that
I, his sole son, do this same villain send
To heaven.
O, this is hire and salary, not revenge!
A took my father grossly, full of bread,                     80
With all his crimes broad blown, as flush as May;
And how his audit stands, who knows save heaven?
But in our circumstance and course of thought
'Tis heavy with him. And am I then revenged
To take him in the purging of his soul,                      85
When he is fit and seasoned for his passage?
No.
    *He sheathes his sword*
Up, sword, and know thou a more horrid hint.
When he is drunk asleep, or in his rage,
Or in th'incestuous pleasure of his bed,                     90
At gaming, swearing, or about some act
That has no relish of salvation in't,
Then trip him that his heels may kick at heaven,
And that his soul may be as damned and black
As hell whereto it goes. My mother stays.                    95
This physic but prolongs thy sickly days.            *Exit*
KING CLAUDIUS
My words fly up, my thoughts remain below.
Words without thoughts never to heaven go.           *Exit*

**3.4**   *Enter Queen Gertrude and Polonius*
POLONIUS
A will come straight. Look you lay home to him.
Tell him his pranks have been too broad to bear with,
And that your grace hath screened and stood between
Much heat and him. I'll silence me e'en here.
Pray you be round with him.                                   5
HAMLET (*within*) Mother, mother, mother!
QUEEN GERTRUDE
I'll warr'nt you. Fear me not. Withdraw; I hear him
    coming.
    *Polonius hides behind the arras.*
    *Enter Prince Hamlet*
HAMLET Now, mother, what's the matter?
QUEEN GERTRUDE
Hamlet, thou hast thy father much offended.
HAMLET
Mother, you have my father much offended.                    10
QUEEN GERTRUDE
Come, come, you answer with an idle tongue.

HAMLET
Go, go, you question with a wicked tongue.
QUEEN GERTRUDE
Why, how now, Hamlet?
HAMLET                                    What's the matter now?
QUEEN GERTRUDE
Have you forgot me?
HAMLET                      No, by the rood, not so.
You are the Queen, your husband's brother's wife.    15
But—would you were not so—you are my mother.
QUEEN GERTRUDE
Nay, then, I'll set those to you that can speak.
HAMLET
Come, come, and sit you down. You shall not budge.
You go not till I set you up a glass
Where you may see the inmost part of you.            20
QUEEN GERTRUDE
What wilt thou do? Thou wilt not murder me?
Help, help, ho!
POLONIUS (*behind the arras*) What ho! Help, help, help!
HAMLET
How now, a rat? Dead for a ducat, dead.
    *He thrusts his sword through the arras*
POLONIUS
O, I am slain!
QUEEN GERTRUDE (*to Hamlet*) O me, what hast thou done?
HAMLET
Nay, I know not. Is it the King?                     25
QUEEN GERTRUDE
O, what a rash and bloody deed is this!
HAMLET
A bloody deed—almost as bad, good-mother,
As kill a king and marry with his brother.
QUEEN GERTRUDE
As kill a king?
HAMLET                Ay, lady, 'twas my word.
(*To Polonius*) Thou wretched, rash, intruding fool,
    farewell.                                         30
I took thee for thy better. Take thy fortune.
Thou find'st to be too busy is some danger.—
Leave wringing of your hands. Peace, sit you down,
And let me wring your heart; for so I shall
If it be made of penetrable stuff,                   35
If damnèd custom have not brassed it so
That it is proof and bulwark against sense.
QUEEN GERTRUDE
What have I done, that thou dar'st wag thy tongue
In noise so rude against me?
HAMLET                              Such an act
That blurs the grace and blush of modesty,           40
Calls virtue hypocrite, takes off the rose
From the fair forehead of an innocent love
And sets a blister there, makes marriage vows
As false as dicers' oaths—O, such a deed
As from the body of contraction plucks               45
The very soul, and sweet religion makes
A rhapsody of words. Heaven's face doth glow,
Yea, this solidity and compound mass
With tristful visage, as against the doom,
Is thought-sick at the act.
QUEEN GERTRUDE               Ay me, what act,        50
That roars so loud and thunders in the index?
HAMLET
Look here upon this picture, and on this,
The counterfeit presentment of two brothers.
See what a grace was seated on this brow—

Hyperion's curls, the front of Jove himself,                55
An eye like Mars, to threaten or command,
A station like the herald Mercury
New lighted on a heaven-kissing hill;
A combination and a form indeed
Where every god did seem to set his seal                    60
To give the world assurance of a man.
This *was* your husband. Look you now what follows.
Here *is* your husband, like a mildewed ear
Blasting his wholesome brother. Have you eyes?
Could you on this fair mountain leave to feed,              65
And batten on this moor? Ha, have you eyes?
You cannot call it love, for at your age
The heyday in the blood is tame, it's humble,
And waits upon the judgement; and what judgement
Would step from this to this? What devil was't            70
That thus hath cozened you at hood-man blind?
O shame, where is thy blush? Rebellious hell,
If thou canst mutine in a matron's bones,
To flaming youth let virtue be as wax
And melt in her own fire. Proclaim no shame                75
When the compulsive ardour gives the charge,
Since frost itself as actively doth burn,
And reason panders will.
QUEEN GERTRUDE                O Hamlet, speak no more!
Thou turn'st mine eyes into my very soul,
And there I see such black and grainèd spots              80
As will not leave their tinct.
HAMLET                        Nay, but to live
In the rank sweat of an enseamèd bed,
Stewed in corruption, honeying and making love
Over the nasty sty—
QUEEN GERTRUDE          O, speak to me no more!
These words like daggers enter in mine ears.              85
No more, sweet Hamlet.
HAMLET                    A murderer and a villain,
A slave that is not twenti'th part the tithe
Of your precedent lord, a vice of kings,
A cutpurse of the empire and the rule,
That from a shelf the precious diadem stole               90
And put it in his pocket—
QUEEN GERTRUDE No more.
HAMLET A king of shreds and patches—
          *Enter the Ghost in his nightgown*
Save me and hover o'er me with your wings,
You heavenly guards! (*To the Ghost*) What would
    you, gracious figure?                                    95
QUEEN GERTRUDE Alas, he's mad.
HAMLET (*to the Ghost*)
Do you not come your tardy son to chide,
That, lapsed in time and passion, lets go by
Th'important acting of your dread command?
O, say!
GHOST    Do not forget. This visitation                   100
Is but to whet thy almost blunted purpose.
But look, amazement on thy mother sits.
O, step between her and her fighting soul.
Conceit in weakest bodies strongest works.
Speak to her, Hamlet.                                     105
HAMLET How is it with you, lady?
QUEEN GERTRUDE Alas, how is't with you,
That you do bend your eye on vacancy,
And with th'incorporal air do hold discourse?
Forth at your eyes your spirits wildly peep,              110
And, as the sleeping soldiers in th'alarm,
Your bedded hair, like life in excrements,

Start up and stand on end. O gentle son,
Upon the heat and flame of thy distemper
Sprinkle cool patience! Whereon do you look?             115
HAMLET
On him, on him. Look you how pale he glares.
His form and cause conjoined, preaching to stones,
Would make them capable. (*To the Ghost*) Do not look
    upon me,
Lest with this piteous action you convert
My stern effects. Then what I have to do                  120
Will want true colour—tears perchance for blood.
QUEEN GERTRUDE
To whom do you speak this?
HAMLET                        Do you see nothing there?
QUEEN GERTRUDE
Nothing at all, yet all that is I see.
HAMLET
Nor did you nothing hear?
QUEEN GERTRUDE                No, nothing but ourselves.
HAMLET
Why, look you there. Look how it steals away.            125
My father, in his habit as he lived.
Look where he goes even now out at the portal.
                                       *Exit the Ghost*
QUEEN GERTRUDE
This is the very coinage of your brain.
This bodiless creation ecstasy
Is very cunning in.
HAMLET            Ecstasy?                                 130
My pulse as yours doth temperately keep time,
And makes as healthful music. It is not madness
That I have uttered. Bring me to the test,
And I the matter will reword, which madness
Would gambol from. Mother, for love of grace             135
Lay not a flattering unction to your soul
That not your trespass but my madness speaks.
It will but skin and film the ulcerous place
Whilst rank corruption, mining all within,
Infects unseen. Confess yourself to heaven;              140
Repent what's past, avoid what is to come,
And do not spread the compost o'er the weeds
To make them ranker. Forgive me this my virtue,
For in the fatness of these pursy times
Virtue itself of vice must pardon beg,                   145
Yea, curb and woo for leave to do him good.
QUEEN GERTRUDE
O Hamlet, thou hast cleft my heart in twain!
HAMLET
O, throw away the worser part of it,
And live the purer with the other half!
Good night—but go not to mine uncle's bed.               150
Assume a virtue if you have it not.
Refrain tonight,
And that shall lend a kind of easiness
To the next abstinence. Once more, good night;
And when you are desirous to be blest,                   155
I'll blessing beg of you. For this same lord,
I do repent. But heaven hath pleased it so
To punish me with this, and this with me,
That I must be their scourge and minister.
I will bestow him, and will answer well                  160
The death I gave him. So, again, good night.
I must be cruel only to be kind.
Thus bad begins, and worse remains behind.
QUEEN GERTRUDE What shall I do?

HAMLET
Not this, by no means, that I bid you do:    165
Let the bloat King tempt you again to bed,
Pinch wanton on your cheek, call you his mouse,
And let him for a pair of reechy kisses,
Or paddling in your neck with his damned fingers,
Make you to ravel all this matter out,    170
That I essentially am not in madness,
But mad in craft. 'Twere good you let him know,
For who that's but a queen, fair, sober, wise,
Would from a paddock, from a bat, a gib,
Such dear concernings hide? Who would do so?    175
No, in despite of sense and secrecy,
Unpeg the basket on the house's top,
Let the birds fly, and, like the famous ape,
To try conclusions in the basket creep,
And break your own neck down.    180
QUEEN GERTRUDE
Be thou assured, if words be made of breath,
And breath of life, I have no life to breathe
What thou hast said to me.
HAMLET                I must to England.
You know that?
QUEEN GERTRUDE     Alack, I had forgot.
'Tis so concluded on.
HAMLET           This man shall set me packing.
I'll lug the guts into the neighbour room.    186
Mother, good night indeed. This counsellor
Is now most still, most secret, and most grave,
Who was in life a foolish prating knave.—
Come, sir, to draw toward an end with you.—    190
Good night, mother.     *Exit, tugging in Polonius*

**4.1**    *Enter King Claudius to Queen Gertrude*
KING CLAUDIUS
There's matter in these sighs, these profound heaves;
You must translate. 'Tis fit we understand them.
Where is your son?
QUEEN GERTRUDE
Ah, my good lord, what have I seen tonight!
KING CLAUDIUS What, Gertrude? How does Hamlet?    5
QUEEN GERTRUDE
Mad as the sea and wind when both contend
Which is the mightier. In his lawless fit,
Behind the arras hearing something stir,
He whips his rapier out and cries 'A rat, a rat!',
And in his brainish apprehension kills    10
The unseen good old man.
KING CLAUDIUS          O heavy deed!
It had been so with us had we been there.
His liberty is full of threats to all—
To you yourself, to us, to everyone.
Alas, how shall this bloody deed be answered?    15
It will be laid to us, whose providence
Should have kept short, restrained, and out of haunt
This mad young man. But so much was our love,
We would not understand what was most fit,
But, like the owner of a foul disease,    20
To keep it from divulging, let it feed
Even on the pith of life. Where is he gone?
QUEEN GERTRUDE
To draw apart the body he hath killed,
O'er whom—his very madness, like some ore
Among a mineral of metals base,    25
Shows itself pure—a weeps for what is done.

KING CLAUDIUS O Gertrude, come away!
The sun no sooner shall the mountains touch
But we will ship him hence; and this vile deed
We must with all our majesty and skill    30
Both countenance and excuse.—Ho, Guildenstern!
*Enter Rosencrantz and Guildenstern*
Friends both, go join you with some further aid.
Hamlet in madness hath Polonius slain,
And from his mother's closet hath he dragged him.
Go seek him out, speak fair, and bring the body    35
Into the chapel. I pray you haste in this.
           *Exeunt Rosencrantz and Guildenstern*
Come, Gertrude, we'll call up our wisest friends
To let them know both what we mean to do
And what's untimely done. O, come away!
My soul is full of discord and dismay.     *Exeunt*

**4.2**    *Enter Prince Hamlet*
HAMLET Safely stowed.
ROSENCRANTZ *and* GUILDENSTERN (*within*)
Hamlet, Lord Hamlet!
HAMLET
What noise? Who calls on Hamlet?
     *Enter Rosencrantz and Guildenstern*
                 O, here they come.
ROSENCRANTZ
What have you done, my lord, with the dead body?
HAMLET
Compounded it with dust, whereto 'tis kin.    5
ROSENCRANTZ
Tell us where 'tis, that we may take it thence
And bear it to the chapel.
HAMLET Do not believe it.
ROSENCRANTZ Believe what?
HAMLET That I can keep your counsel and not mine own.
Besides, to be demanded of a sponge—what replication
should be made by the son of a king?    12
ROSENCRANTZ Take you me for a sponge, my lord?
HAMLET Ay, sir, that soaks up the King's countenance,
his rewards, his authorities. But such officers do the
King best service in the end. He keeps them, like an
ape an apple in the corner of his jaw, first mouthed to
be last swallowed. When he needs what you have
gleaned, it is but squeezing you, and, sponge, you shall
be dry again.    20
ROSENCRANTZ I understand you not, my lord.
HAMLET I am glad of it. A knavish speech sleeps in a
foolish ear.
ROSENCRANTZ My lord, you must tell us where the body
is, and go with us to the King.    25
HAMLET The body is with the King, but the King is not
with the body. The King is a thing—
GUILDENSTERN A thing, my lord?
HAMLET Of nothing. Bring me to him. Hide fox, and all
after.      *Exit running, pursued by the others*

**4.3**    *Enter King Claudius*
KING CLAUDIUS
I have sent to seek him, and to find the body.
How dangerous is it that this man goes loose!
Yet must not we put the strong law on him.
He's loved of the distracted multitude,
Who like not in their judgement but their eyes,    5
And where 'tis so, th'offender's scourge is weighed,
But never the offence. To bear all smooth and even,

This sudden sending him away must seem
Deliberate pause. Diseases desperate grown
By desperate appliance are relieved,                    10
Or not at all.
  *Enter Rosencrantz*
     How now, what hath befall'n?
ROSENCRANTZ
Where the dead body is bestowed, my lord,
We cannot get from him.
KING CLAUDIUS    But where is he?
ROSENCRANTZ
Without, my lord, guarded to know your pleasure.
KING CLAUDIUS Bring him before us.                    15
ROSENCRANTZ
Ho, Guildenstern! Bring in my lord.
  *Enter Prince Hamlet and Guildenstern*
KING CLAUDIUS
Now, Hamlet, where's Polonius?
HAMLET At supper.
KING CLAUDIUS At supper? Where?
HAMLET Not where he eats, but where a is eaten. A certain
convocation of politic worms are e'en at him. Your worm
is your only emperor for diet. We fat all creatures else to
fat us, and we fat ourselves for maggots. Your fat king
and your lean beggar is but variable service—two dishes,
but to one table. That's the end.                    25
KING CLAUDIUS Alas, alas!
HAMLET A man may fish with the worm that hath eat of
a king, and eat of the fish that hath fed of that worm.
KING CLAUDIUS What dost thou mean by this?
HAMLET Nothing but to show you how a king may go a
progress through the guts of a beggar.                    31
KING CLAUDIUS Where is Polonius?
HAMLET In heaven. Send thither to see. If your messenger
find him not there, seek him i'th' other place yourself.
But indeed, if you find him not this month, you shall
nose him as you go up the stairs into the lobby.                    36
KING CLAUDIUS ⌜to Rosencrantz⌝ Go seek him there.
HAMLET ⌜to Rosencrantz⌝ A will stay till ye come.
       *Exit ⌜Rosencrantz⌝*
KING CLAUDIUS
Hamlet, this deed of thine, for thine especial safety—
Which we do tender as we dearly grieve                    40
For that which thou hast done—must send thee hence
With fiery quickness. Therefore prepare thyself.
The barque is ready, and the wind at help,
Th'associates tend, and everything is bent
For England.                    45
HAMLET For England?
KING CLAUDIUS Ay, Hamlet.
HAMLET Good.
KING CLAUDIUS
So is it if thou knew'st our purposes.
HAMLET I see a cherub that sees them. But come, for
England. Farewell, dear mother.                    51
KING CLAUDIUS Thy loving father, Hamlet.
HAMLET My mother. Father and mother is man and wife,
man and wife is one flesh, and so my mother. Come,
for England.        *Exit*
KING CLAUDIUS ⌜to Guildenstern⌝
Follow him at foot. Tempt him with speed aboard.                    56
Delay it not. I'll have him hence tonight.
Away, for everything is sealed and done
That else leans on th'affair. Pray you, make haste.
       *Exit ⌜Guildenstern⌝*
And, England, if my love thou hold'st at aught—                    60

As my great power thereof may give thee sense,
Since yet thy cicatrice looks raw and red
After the Danish sword, and thy free awe
Pays homage to us—thou mayst not coldly set
Our sovereign process, which imports at full,                    65
By letters conjuring to that effect,
The present death of Hamlet. Do it, England,
For like the hectic in my blood he rages,
And thou must cure me. Till I know 'tis done,
Howe'er my haps, my joys were ne'er begun.  *Exit*

**4.4**  *Enter Fortinbras with an army over the stage*
FORTINBRAS
Go, captain, from me greet the Danish king.
Tell him that by his licence Fortinbras
Claims the conveyance of a promised march
Over his kingdom. You know the rendezvous.
If that his majesty would aught with us,                    5
We shall express our duty in his eye,
And let him know so.
CAPTAIN I will do't, my lord.    ⌜*Exit*⌝
FORTINBRAS Go safely on.    *Exeunt marching*

**4.5**  *Enter Queen Gertrude and Horatio*
QUEEN GERTRUDE
I will not speak with her.
HORATIO     She is importunate,
Indeed distraught. Her mood will needs be pitied.
QUEEN GERTRUDE What would she have?
HORATIO
She speaks much of her father, says she hears
There's tricks i'th' world, and hems, and beats her
  heart,                    5
Spurns enviously at straws, speaks things in doubt
That carry but half sense. Her speech is nothing,
Yet the unshapèd use of it doth move
The hearers to collection. They aim at it,
And botch the words up fit to their own thoughts,                    10
Which, as her winks and nods and gestures yield them,
Indeed would make one think there might be thought,
Though nothing sure, yet much unhappily.
QUEEN GERTRUDE
'Twere good she were spoken with, for she may strew
Dangerous conjectures in ill-breeding minds.                    15
Let her come in.
  ⌜*Horatio withdraws to admit Ophelia*⌝
To my sick soul, as sin's true nature is,
Each toy seems prologue to some great amiss.
So full of artless jealousy is guilt,
It spills itself in fearing to be spilt.                    20
  *Enter Ophelia mad,* ⌜*her hair down, with a lute*⌝
OPHELIA
Where is the beauteous majesty of Denmark?
QUEEN GERTRUDE How now, Ophelia?
OPHELIA (*sings*)
    How should I your true love know
      From another one?—
    By his cockle hat and staff,                    25
      And his sandal shoon.
QUEEN GERTRUDE
Alas, sweet lady, what imports this song?
OPHELIA Say you? Nay, pray you, mark.
 (*Sings*) He is dead and gone, lady,
      He is dead and gone.                    30
    At his head a grass-green turf,
      At his heels a stone.

QUEEN GERTRUDE Nay, but Ophelia—
OPHELIA Pray you, mark.
  *(Sings)*
    White his shroud as the mountain snow—  35
  *Enter King Claudius*
QUEEN GERTRUDE Alas, look here, my lord.
OPHELIA *(sings)*
    Larded with sweet flowers,
    Which bewept to the grave did—not—go
    With true-love showers.

KING CLAUDIUS How do ye, pretty lady?  40
OPHELIA Well, God'ield you. They say the owl was a
baker's daughter. Lord, we know what we are, but
know not what we may be. God be at your table!
KING CLAUDIUS *(to Gertrude)* Conceit upon her father.
OPHELIA Pray you, let's have no words of this, but when
they ask you what it means, say you this.  46
  *(Sings)*
    Tomorrow is Saint Valentine's day,
    All in the morning betime,
    And I a maid at your window
    To be your Valentine.  50

    Then up he rose, and donned his clothes,
    And dupped the chamber door;
    Let in the maid, that out a maid
    Never departed more.

KING CLAUDIUS Pretty Ophelia—  55
OPHELIA Indeed, la? Without an oath, I'll make an end
on't.
  *(Sings)*  By Gis, and by Saint Charity,
    Alack, and fie for shame!
    Young men will do't if they come to't,  60
    By Cock, they are to blame.

    Quoth she 'Before you tumbled me,
    You promised me to wed.'
    So would I 'a' done, by yonder sun,
    An thou hadst not come to my bed.  65

KING CLAUDIUS *(to Gertrude)* How long hath she been thus?
OPHELIA I hope all will be well. We must be patient. But
I cannot choose but weep to think they should lay him
i'th' cold ground. My brother shall know of it. And so
I thank you for your good counsel. Come, my coach!
Good night, ladies, good night, sweet ladies, good night,
good night.  *Exit*
KING CLAUDIUS *(to Horatio)*
Follow her close. Give her good watch, I pray you.  73
    *Exit Horatio*
O, this is the poison of deep grief! It springs
All from her father's death. O Gertrude, Gertrude,  75
When sorrows come they come not single spies,
But in battalions. First, her father slain;
Next, your son gone, and he most violent author
Of his own just remove; the people muddied,
Thick and unwholesome in their thoughts and whispers
For good Polonius' death; and we have done but
greenly  81
In hugger-mugger to inter him; poor Ophelia
Divided from herself and her fair judgement,
Without the which we are pictures or mere beasts;
Last, and as much containing as all these,  85
Her brother is in secret come from France,
Feeds on this wonder, keeps himself in clouds,
And wants not buzzers to infect his ear

With pestilent speeches of his father's death;
Wherein necessity, of matter beggared,  90
Will nothing stick our persons to arraign
In ear and ear. O my dear Gertrude, this,
Like to a murd'ring-piece, in many places
Gives me superfluous death.
  *A noise within*
QUEEN GERTRUDE  Alack, what noise is this?
KING CLAUDIUS
Where is my Switzers? Let them guard the door.  95
  *Enter a Messenger*
What is the matter?
MESSENGER  Save yourself, my lord.
The ocean, overpeering of his list,
Eats not the flats with more impetuous haste
Than young Laertes, in a riotous head,
O'erbears your officers. The rabble call him lord,  100
And, as the world were now but to begin,
Antiquity forgot, custom not known,
The ratifiers and props of every word,
They cry 'Choose we! Laertes shall be king.'
Caps, hands, and tongues applaud it to the clouds,
'Laertes shall be king, Laertes king.'  106
QUEEN GERTRUDE
How cheerfully on the false trail they cry!
  *A noise within*
O, this is counter, you false Danish dogs!
KING CLAUDIUS The doors are broke.
  *Enter Laertes ⌐with his followers at the door⌐*
LAERTES
Where is the King?—Sirs, stand you all without.  110
ALL HIS FOLLOWERS No, let's come in.
LAERTES I pray you, give me leave.
ALL HIS FOLLOWERS We will, we will.
LAERTES
I thank you. Keep the door.  ⌐*Exeunt followers*⌐
    O thou vile king,
Give me my father.
QUEEN GERTRUDE  Calmly, good Laertes.  115
LAERTES
That drop of blood that's calm proclaims me bastard,
Cries cuckold to my father, brands the harlot
Even here between the chaste unsmirchèd brow
Of my true mother.
KING CLAUDIUS  What is the cause, Laertes,
That thy rebellion looks so giant-like?—  120
Let him go, Gertrude. Do not fear our person.
There's such divinity doth hedge a king
That treason can but peep to what it would,
Acts little of his will.—Tell me, Laertes,
Why thou art thus incensed.—Let him go, Gertrude.—
Speak, man.
LAERTES  Where is my father?
KING CLAUDIUS  Dead.  126
QUEEN GERTRUDE *(to Laertes)*
But not by him.
KING CLAUDIUS  Let him demand his fill.
LAERTES
How came he dead? I'll not be juggled with.
To hell, allegiance! Vows to the blackest devil!
Conscience and grace to the profoundest pit!  130
I dare damnation. To this point I stand,
That both the worlds I give to negligence,
Let come what comes. Only I'll be revenged
Most throughly for my father.
KING CLAUDIUS Who shall stay you?  135

LAERTES My will, not all the world;
And for my means, I'll husband them so well
They shall go far with little.
KING CLAUDIUS                    Good Laertes,
If you desire to know the certainty                    139
Of your dear father's death, is't writ in your revenge
That, sweepstake, you will draw both friend and foe,
Winner and loser?
LAERTES None but his enemies.
KING CLAUDIUS Will you know them then?
LAERTES
To his good friends thus wide I'll ope my arms,    145
And, like the kind life-rend'ring pelican,
Repast them with my blood.
KING CLAUDIUS                    Why, now you speak
Like a good child and a true gentleman.
That I am guiltless of your father's death,
And am most sensibly in grief for it,                    150
It shall as level to your judgement pierce
As day does to your eye.
         A noise within
VOICES (within) Let her come in.
LAERTES How now, what noise is that?
         Enter Ophelia as before
O heat dry up my brains! Tears seven times salt    155
Burn out the sense and virtue of mine eye!
By heaven, thy madness shall be paid by weight
Till our scale turns the beam. O rose of May,
Dear maid, kind sister, sweet Ophelia!
O heavens, is't possible a young maid's wits    160
Should be as mortal as an old man's life?
Nature is fine in love, and where 'tis fine
It sends some precious instance of itself
After the thing it loves.
OPHELIA (sings)
         They bore him barefaced on the bier,    165
         Hey non nony, nony, hey nony,
         And on his grave rained many a tear—
Fare you well, my dove.
LAERTES
Hadst thou thy wits and didst persuade revenge,
It could not move thus.                    170
OPHELIA You must sing 'Down, a-down', and you, 'Call
him a-down-a'. O, how the wheel becomes it! It is the
false steward that stole his master's daughter.
LAERTES This nothing's more than matter.
OPHELIA There's rosemary, that's for remembrance. Pray,
love, remember. And there is pansies; that's for
thoughts.                    177
LAERTES
A document in madness—thoughts and remembrance
    fitted.
OPHELIA There's fennel for you, and columbines. There's
rue for you, and here's some for me. We may call it
herb-grace o' Sundays. O, you must wear your rue
with a difference. There's a daisy. I would give you
some violets, but they withered all when my father
died. They say a made a good end.
(Sings) For bonny sweet Robin is all my joy.    185
LAERTES
Thought and affliction, passion, hell itself
She turns to favour and to prettiness.
OPHELIA (sings)
         And will a not come again,
         And will a not come again?
         No, no, he is dead,                    190
         Go to thy death-bed,
         He never will come again.

His beard as white as snow,
All flaxen was his poll.
         He is gone, he is gone,                    195
         And we cast away moan.
God 'a' mercy on his soul.

And of all Christian souls, I pray God. God b'wi' ye.
         ⌐Exeunt Ophelia and Gertrude⌐
LAERTES Do you see this, O God?
KING CLAUDIUS
Laertes, I must commune with your grief,    200
Or you deny me right. Go but apart,
Make choice of whom your wisest friends you will,
And they shall hear and judge 'twixt you and me.
If by direct or by collateral hand
They find us touched, we will our kingdom give,    205
Our crown, our life, and all that we call ours,
To you in satisfaction. But if not,
Be you content to lend your patience to us,
And we shall jointly labour with your soul
To give it due content.
LAERTES                    Let this be so.    210
His means of death, his obscure burial—
No trophy, sword, nor hatchment o'er his bones,
No noble rite nor formal ostentation—
Cry to be heard, as 'twere from heaven to earth,
That I must call't in question.
KING CLAUDIUS                    So you shall;    215
And where th'offence is, let the great axe fall.
I pray you go with me.
                    Exeunt

4.6    Enter Horatio with a Servant
HORATIO
What are they that would speak with me?
SERVANT
Sailors, sir. They say they have letters for you.
HORATIO Let them come in.              Exit Servant
I do not know from what part of the world
I should be greeted if not from Lord Hamlet.    5
         Enter ⌐Sailors⌐
A SAILOR God bless you, sir.
HORATIO Let him bless thee too.
A SAILOR A shall, sir, an't please him. There's a letter for
you, sir. It comes from th'ambassador that was bound
for England—if your name be Horatio, as I am let to
know it is.                    11
HORATIO (reads) 'Horatio, when thou shalt have overlooked
this, give these fellows some means to the King. They
have letters for him. Ere we were two days old at sea,
a pirate of very warlike appointment gave us chase.
Finding ourselves too slow of sail, we put on a compelled
valour, and in the grapple I boarded them. On the
instant they got clear of our ship, so I alone became
their prisoner. They have dealt with me like thieves of
mercy; but they knew what they did: I am to do a
good turn for them. Let the King have the letters I have
sent, and repair thou to me with as much haste as
thou wouldst fly death. I have words to speak in thine
ear will make thee dumb, yet are they much too light
for the bore of the matter. These good fellows will bring
thee where I am. Rosencrantz and Guildenstern hold
their course for England. Of them I have much to tell
thee. Farewell.                    28
                    He that thou knowest thine,
                              Hamlet.'

Come, I will give you way for these your letters,
And do't the speedier that you may direct me
To him from whom you brought them.          *Exeunt*

**4.7**    *Enter King Claudius and Laertes*
KING CLAUDIUS
Now must your conscience my acquittance seal,
And you must put me in your heart for friend,
Sith you have heard, and with a knowing ear,
That he which hath your noble father slain
Pursued my life.
LAERTES              It well appears. But tell me       5
Why you proceeded not against these feats,
So crimeful and so capital in nature,
As by your safety, wisdom, all things else,
You mainly were stirred up.
KING CLAUDIUS              O, for two special reasons,
Which may to you perhaps seem much unsinewed,   10
And yet to me they're strong. The Queen his mother
Lives almost by his looks; and for myself—
My virtue or my plague, be it either which—
She's so conjunctive to my life and soul
That, as the star moves not but in his sphere,   15
I could not but by her. The other motive
Why to a public count I might not go
Is the great love the general gender bear him,
Who, dipping all his faults in their affection,
Would, like the spring that turneth wood to stone,  20
Convert his guilts to graces; so that my arrows,
Too slightly timbered for so loud a wind,
Would have reverted to my bow again,
And not where I had aimed them.
LAERTES
And so have I a noble father lost,                  25
A sister driven into desp'rate terms,
Who has, if praises may go back again,
Stood challenger, on mount, of all the age
For her perfections. But my revenge will come.
KING CLAUDIUS
Break not your sleeps for that. You must not think   30
That we are made of stuff so flat and dull
That we can let our beard be shook with danger,
And think it pastime. You shortly shall hear more.
I loved your father, and we love ourself.
And that, I hope, will teach you to imagine—         35
          *Enter a Messenger with letters*
How now? What news?
MESSENGER              Letters, my lord, from Hamlet.
This to your majesty; this to the Queen.
KING CLAUDIUS From Hamlet? Who brought them?
MESSENGER
Sailors, my lord, they say. I saw them not.
They were given me by Claudio. He received them.   40
KING CLAUDIUS
Laertes, you shall hear them.—Leave us.
                              *Exit Messenger*
(*Reads*) 'High and mighty, you shall know I am set
naked on your kingdom. Tomorrow shall I beg leave
to see your kingly eyes, when I shall, first asking your
pardon, thereunto recount th'occasions of my sudden
and more strange return.                            46
                              Hamlet.'
What should this mean? Are all the rest come back?
Or is it some abuse, and no such thing?
LAERTES
Know you the hand?
KING CLAUDIUS           'Tis Hamlet's character.      50
'Naked'—and in a postscript here he says

'Alone'. Can you advise me?
LAERTES
I'm lost in it, my lord. But let him come.
It warms the very sickness in my heart
That I shall live and tell him to his teeth,         55
'Thus diddest thou'.
KING CLAUDIUS              If it be so, Laertes—
As how should it be so, how otherwise?—
Will you be ruled by me?
LAERTES
If so you'll not o'errule me to a peace.
KING CLAUDIUS
To thine own peace. If he be now returned,           60
As checking at his voyage, and that he means
No more to undertake it, I will work him
To an exploit, now ripe in my device,
Under the which he shall not choose but fall;
And for his death no wind of blame shall breathe;    65
But even his mother shall uncharge the practice
And call it accident. Some two months since
Here was a gentleman of Normandy.
I've seen myself, and served against, the French,
And they can well on horseback; but this gallant     70
Had witchcraft in't. He grew into his seat,
And to such wondrous doing brought his horse
As had he been incorpsed and demi-natured
With the brave beast. So far he passed my thought
That I in forgery of shapes and tricks               75
Come short of what he did.
LAERTES                    A Norman was't?
KING CLAUDIUS A Norman.
LAERTES
Upon my life, Lamord.
KING CLAUDIUS              The very same.
LAERTES
I know him well. He is the brooch indeed,
And gem, of all the nation.
KING CLAUDIUS              He made confession of you,
And gave you such a masterly report                  81
For art and exercise in your defence,
And for your rapier most especially,
That he cried out 'twould be a sight indeed
If one could match you. Sir, this report of his      85
Did Hamlet so envenom with his envy
That he could nothing do but wish and beg
Your sudden coming o'er to play with him.
Now, out of this—
LAERTES              What out of this, my lord?
KING CLAUDIUS
Laertes, was your father dear to you?                90
Or are you like the painting of a sorrow,
A face without a heart?
LAERTES                    Why ask you this?
KING CLAUDIUS
Not that I think you did not love your father,
But that I know love is begun by time,
And that I see, in passages of proof,                95
Time qualifies the spark and fire of it.
Hamlet comes back. What would you undertake
To show yourself your father's son in deed
More than in words?
LAERTES              To cut his throat i'th' church.
KING CLAUDIUS
No place indeed should murder sanctuarize,          100
Revenge should have no bounds. But, good Laertes,
Will you do this?—keep close within your chamber.

Hamlet returned shall know you are come home.
We'll put on those shall praise your excellence,
And set a double varnish on the fame                    105
The Frenchman gave you; bring you, in fine, together,
And wager on your heads. He, being remiss,
Most generous, and free from all contriving,
Will not peruse the foils; so that with ease,
Or with a little shuffling, you may choose              110
A sword unbated, and, in a pass of practice,
Requite him for your father.

LAERTES                              I will do't,
And for that purpose I'll anoint my sword.
I bought an unction of a mountebank
So mortal that, but dip a knife in it,                  115
Where it draws blood no cataplasm so rare,
Collected from all simples that have virtue
Under the moon, can save the thing from death
That is but scratched withal. I'll touch my point
With this contagion, that if I gall him slightly,       120
It may be death.

KING CLAUDIUS          Let's further think of this;
Weigh what convenience both of time and means
May fit us to our shape. If this should fail,
And that our drift look through our bad performance,
'Twere better not essayed. Therefore this project      125
Should have a back or second that might hold
If this should blast in proof. Soft, let me see.
We'll make a solemn wager on your cunnings . . .
I ha't! When in your motion you are hot and dry—
As make your bouts more violent to that end—           130
And that he calls for drink, I'll have prepared him
A chalice for the nonce, whereon but sipping,
If he by chance escape your venomed stuck,
Our purpose may hold there.—
    *Enter Queen Gertrude*
                              How now, sweet Queen?

QUEEN GERTRUDE
One woe doth tread upon another's heel,                 135
So fast they follow. Your sister's drowned, Laertes.

LAERTES Drowned? O, where?

QUEEN GERTRUDE
There is a willow grows aslant a brook
That shows his hoar leaves in the glassy stream.
Therewith fantastic garlands did she make              140
Of crow-flowers, nettles, daisies, and long purples,
That liberal shepherds give a grosser name,
But our cold maids do dead men's fingers call them.
There on the pendent boughs her crownet weeds
Clamb'ring to hang, an envious sliver broke,           145
When down the weedy trophies and herself
Fell in the weeping brook. Her clothes spread wide,
And mermaid-like a while they bore her up;
Which time she chanted snatches of old tunes,
As one incapable of her own distress,                  150
Or like a creature native and endued
Unto that element. But long it could not be
Till that her garments, heavy with their drink,
Pulled the poor wretch from her melodious lay
To muddy death.                                         155

LAERTES Alas, then is she drowned.

QUEEN GERTRUDE Drowned, drowned.

LAERTES
Too much of water hast thou, poor Ophelia,
And therefore I forbid my tears. But yet
It is our trick; nature her custom holds,              160

Let shame say what it will.
    *He weeps*
                              When these are gone,
The woman will be out. Adieu, my lord.
I have a speech of fire that fain would blaze,
But that this folly douts it.                    *Exit*

KING CLAUDIUS              Let's follow, Gertrude.
How much I had to do to calm his rage!                  165
Now fear I this will give it start again;
Therefore let's follow.                       *Exeunt*

**5.1**    *Enter two Clowns ⌈carrying a spade and a pickaxe⌉*

FIRST CLOWN Is she to be buried in Christian burial that
wilfully seeks her own salvation?

SECOND CLOWN I tell thee she is, and therefore make her
grave straight. The coroner hath sat on her, and finds
it Christian burial.                                      5

FIRST CLOWN How can that be unless she drowned herself
in her own defence?

SECOND CLOWN Why, 'tis found so.

FIRST CLOWN It must be *se offendendo*, it cannot be else;
for here lies the point: if I drown myself wittingly, it
argues an act; and an act hath three branches: it is
to act, to do, and to perform. Argal she drowned herself
wittingly.                                               13

SECOND CLOWN Nay, but hear you, Goodman Delver.

FIRST CLOWN Give me leave. Here lies the water—good.
Here stands the man—good. If the man go to this water
and drown himself, it is, will he nill he, he goes. Mark
you that. But if the water come to him and drown him,
he drowns not himself; argal he that is not guilty of
his own death shortens not his own life.                 20

SECOND CLOWN But is this law?

FIRST CLOWN Ay, marry, is't: coroner's quest law.

SECOND CLOWN Will you ha' the truth on't? If this had
not been a gentlewoman, she should have been buried
out o' Christian burial.                                 25

FIRST CLOWN Why, there thou sayst, and the more pity
that great folk should have count'nance in this world
to drown or hang themselves more than their even
Christian. Come, my spade. There is no ancient
gentlemen but gardeners, ditchers, and gravemakers;
they hold up Adam's profession.                          31
    ⌈*First Clown digs*⌉

SECOND CLOWN Was he a gentleman?

FIRST CLOWN A was the first that ever bore arms.

SECOND CLOWN Why, he had none.

FIRST CLOWN What, art a heathen? How dost thou
understand the Scripture? The Scripture says Adam
digged. Could he dig without arms? I'll put another
question to thee. If thou answerest me not to the
purpose, confess thyself—

SECOND CLOWN Go to.                                      40

FIRST CLOWN What is he that builds stronger than either
the mason, the shipwright, or the carpenter?

SECOND CLOWN The gallows-maker; for that frame outlives
a thousand tenants.

FIRST CLOWN I like thy wit well, in good faith. The gallows
does well. But how does it well? It does well to those
that do ill. Now thou dost ill to say the gallows is built
stronger than the church, argal the gallows may do
well to thee. To't again, come.                          49

SECOND CLOWN 'Who builds stronger than a mason, a
shipwright, or a carpenter?'

FIRST CLOWN Ay, tell me that, and unyoke.

SECOND CLOWN Marry, now I can tell.

FIRST CLOWN To't.

SECOND CLOWN Mass, I cannot tell.                    55

*Enter Prince Hamlet and Horatio afar off*

FIRST CLOWN Cudgel thy brains no more about it, for your dull ass will not mend his pace with beating; and when you are asked this question next, say 'a grave-maker'; the houses that he makes lasts till doomsday. Go, get thee to Johan. Fetch me a stoup of liquor.    60

*Exit Second Clown*

(*Sings*)

> In youth when I did love, did love,
> Methought it was very sweet
> To contract-O-the time for-a-my behove,
> O methought there-a-was nothing-a-meet.

HAMLET Has this fellow no feeling of his business that a sings at grave-making?                           66

HORATIO Custom hath made it in him a property of easiness.

HAMLET 'Tis e'en so; the hand of little employment hath the daintier sense.                                70

FIRST CLOWN (*sings*)

> But age with his stealing steps
> Hath caught me in his clutch,
> And hath shipped me intil the land,
> As if I had never been such.

⌈*He throws up a skull*⌉

HAMLET That skull had a tongue in it and could sing once. How the knave jowls it to th' ground as if 'twere Cain's jawbone, that did the first murder! This might be the pate of a politician which this ass o'er-offices, one that would circumvent God, might it not?

HORATIO It might, my lord.                           80

HAMLET Or of a courtier, which could say 'Good morrow, sweet lord. How dost thou, good lord?' This might be my lord such a one, that praised my lord such a one's horse when a meant to beg it, might it not?

HORATIO Ay, my lord.                                 85

HAMLET Why, e'en so, and now my lady Worm's, chapless, and knocked about the mazard with a sexton's spade. Here's fine revolution, an we had the trick to see't. Did these bones cost no more the breeding but to play at loggats with 'em? Mine ache to think on't.  90

FIRST CLOWN (*sings*)

> A pickaxe and a spade, a spade,
> For and a shrouding-sheet;
> O, a pit of clay for to be made
> For such a guest is meet.

⌈*He throws up another skull*⌉

HAMLET There's another. Why might not that be the skull of a lawyer? Where be his quiddits now, his quillets, his cases, his tenures, and his tricks? Why does he suffer this rude knave now to knock him about the sconce with a dirty shovel, and will not tell him of his action of battery? H'm! This fellow might be in 's time a great buyer of land, with his statutes, his recognizances, his fines, his double vouchers, his recoveries. Is this the fine of his fines and the recovery of his recoveries, to have his fine pate full of fine dirt? Will his vouchers vouch him no more of his purchases, and double ones too, than the length and breadth of a pair of indentures? The very conveyances of his lands will hardly lie in this box; and must th'inheritor himself have no more, ha?

HORATIO Not a jot more, my lord.                    110

HAMLET Is not parchment made of sheepskins?

HORATIO Ay, my lord, and of calf-skins too.

HAMLET They are sheep and calves that seek out assurance in that. I will speak to this fellow. (*To the First Clown*) Whose grave's this, sirrah?    115

FIRST CLOWN Mine, sir.

(*Sings*)       O, a pit of clay for to be made
> For such a guest is meet.

HAMLET I think it be thine indeed, for thou liest in't.

FIRST CLOWN You lie out on't, sir, and therefore it is not yours. For my part, I do not lie in't, and yet it is mine.

HAMLET Thou dost lie in't, to be in't and say 'tis thine. 'Tis for the dead, not for the quick; therefore thou liest.

FIRST CLOWN 'Tis a quick lie, sir, 'twill away again from me to you.                                     125

HAMLET What man dost thou dig it for?

FIRST CLOWN For no man, sir.

HAMLET What woman, then?

FIRST CLOWN For none, neither.

HAMLET Who is to be buried in't?                    130

FIRST CLOWN One that was a woman, sir; but, rest her soul, she's dead.

HAMLET How absolute the knave is! We must speak by the card, or equivocation will undo us. By the Lord, Horatio, these three years I have taken note of it. The age is grown so picked that the toe of the peasant comes so near the heel of the courtier he galls his kibe. (*To the First Clown*) How long hast thou been a grave-maker?                                        139

FIRST CLOWN Of all the days i'th' year I came to't that day that our last King Hamlet o'ercame Fortinbras.

HAMLET How long is that since?

FIRST CLOWN Cannot you tell that? Every fool can tell that. It was the very day that young Hamlet was born—he that was mad and sent into England.   145

HAMLET Ay, marry, why was he sent into England?

FIRST CLOWN Why, because a was mad. A shall recover his wits there; or if a do not, 'tis no great matter there.

HAMLET Why?

FIRST CLOWN 'Twill not be seen in him there. There the men are as mad as he.                          151

HAMLET How came he mad?

FIRST CLOWN Very strangely, they say.

HAMLET How strangely?

FIRST CLOWN Faith, e'en with losing his wits.       155

HAMLET Upon what ground?

FIRST CLOWN Why, here in Denmark. I have been sexton here, man and boy, thirty years.

HAMLET How long will a man lie i'th' earth ere he rot?

FIRST CLOWN I'faith, if a be not rotten before a die—as we have many pocky corpses nowadays, that will scarce hold the laying in—a will last you some eight year or nine year. A tanner will last you nine year.  163

HAMLET Why he more than another?

FIRST CLOWN Why, sir, his hide is so tanned with his trade that a will keep out water a great while, and your water is a sore decayer of your whoreson dead body. Here's a skull, now. This skull has lain in the earth three-and-twenty years.

HAMLET Whose was it?                                 170

FIRST CLOWN A whoreson mad fellow's it was. Whose do you think it was?

HAMLET Nay, I know not.

FIRST CLOWN A pestilence on him for a mad rogue—a poured a flagon of Rhenish on my head once! This same skull, sir, was Yorick's skull, the King's jester.

HAMLET This?                                         177

FIRST CLOWN E'en that.

HAMLET Let me see. 179
*He takes the skull*
Alas, poor Yorick. I knew him, Horatio—a fellow of
infinite jest, of most excellent fancy. He hath borne me
on his back a thousand times; and now, how abhorred
my imagination is! My gorge rises at it. Here hung
those lips that I have kissed I know not how oft. Where
be your gibes now, your gambols, your songs, your
flashes of merriment that were wont to set the table
on a roar? Not one now to mock your own grinning?
Quite chop-fallen? Now get you to my lady's chamber
and tell her, let her paint an inch thick, to this favour
she must come. Make her laugh at that. Prithee,
Horatio, tell me one thing. 191
HORATIO What's that, my lord?
HAMLET Dost thou think Alexander looked o' this fashion
i'th' earth?
HORATIO E'en so. 195
HAMLET And smelt so? Pah!
⌈*He throws the skull down*⌉
HORATIO E'en so, my lord.
HAMLET To what base uses we may return, Horatio! Why
may not imagination trace the noble dust of Alexander
till a find it stopping a bung-hole? 200
HORATIO 'Twere to consider too curiously to consider so.
HAMLET No, faith, not a jot; but to follow him thither
with modesty enough, and likelihood to lead it, as thus:
Alexander died, Alexander was buried, Alexander
returneth into dust, the dust is earth, of earth we make
loam, and why of that loam whereto he was converted
might they not stop a beer-barrel? 207
Imperial Caesar, dead and turned to clay,
Might stop a hole to keep the wind away.
O, that that earth which kept the world in awe 210
Should patch a wall t'expel the winter's flaw!
But soft, but soft; aside.
　　　*Hamlet and Horatio stand aside. Enter King*
　　　*Claudius, Queen Gertrude, Laertes, and a coffin,*
　　　*with a Priest and lords attendant*
　　　　　　　Here comes the King,
The Queen, the courtiers—who is that they follow,
And with such maimèd rites? This doth betoken
The corpse they follow did with desp'rate hand 215
Fordo it own life. 'Twas of some estate.
Couch we a while, and mark.
LAERTES　　　　　　　What ceremony else?
HAMLET (*aside to Horatio*)
That is Laertes, a very noble youth. Mark.
LAERTES What ceremony else?
PRIEST
Her obsequies have been as far enlarged 220
As we have warrantise. Her death was doubtful,
And but that great command o'ersways the order
She should in ground unsanctified have lodged
Till the last trumpet. For charitable prayers,
Shards, flints, and pebbles should be thrown on her,
Yet here she is allowed her virgin rites, 226
Her maiden strewments, and the bringing home
Of bell and burial.
LAERTES Must there no more be done?
PRIEST No more be done. 230
We should profane the service of the dead
To sing sage requiem and such rest to her
As to peace-parted souls.
LAERTES　　　　　　　Lay her i'th' earth,
And from her fair and unpolluted flesh

May violets spring. I tell thee, churlish priest, 235
A minist'ring angel shall my sister be
When thou liest howling.
HAMLET (*aside*) What, the fair Ophelia!
QUEEN GERTRUDE (*scattering flowers*)
Sweets to the sweet. Farewell.
I hoped thou shouldst have been my Hamlet's wife.
I thought thy bride-bed to have decked, sweet maid,
And not t'have strewed thy grave.
LAERTES　　　　　　　O, treble woe 242
Fall ten times treble on that cursèd head
Whose wicked deed thy most ingenious sense
Deprived thee of!—Hold off the earth a while, 245
Till I have caught her once more in mine arms.
　　　*He leaps into the grave*
Now pile your dust upon the quick and dead
Till of this flat a mountain you have made
To o'ertop old Pelion, or the skyish head
Of blue Olympus.
HAMLET (*coming forward*) What is he whose grief 250
Bears such an emphasis, whose phrase of sorrow
Conjures the wand'ring stars and makes them stand
Like wonder-wounded hearers? This is I,
Hamlet the Dane.
　　　⌈*Hamlet leaps in after Laertes*⌉
LAERTES The devil take thy soul. 255
HAMLET Thou pray'st not well.
I prithee take thy fingers from my throat,
For though I am not splenative and rash,
Yet have I something in me dangerous,
Which let thy wiseness fear. Away thy hand. 260
KING CLAUDIUS (*to Lords*)
Pluck them asunder.
QUEEN GERTRUDE　　　　Hamlet, Hamlet!
ALL ⌈THE LORDS⌉
Gentlemen!
HORATIO (*to Hamlet*) Good my lord, be quiet.
HAMLET
Why, I will fight with him upon this theme
Until my eyelids will no longer wag.
QUEEN GERTRUDE O my son, what theme? 265
HAMLET
I loved Ophelia. Forty thousand brothers
Could not, with all their quantity of love,
Make up my sum.—What wilt thou do for her?
KING CLAUDIUS O, he is mad, Laertes.
QUEEN GERTRUDE (*to Laertes*) For love of God, forbear him.
HAMLET (*to Laertes*) 'Swounds, show me what thou'lt do.
Woot weep, woot fight, woot fast, woot tear thyself,
Woot drink up eisel, eat a crocodile?
I'll do't. Dost thou come here to whine,
To outface me with leaping in her grave? 275
Be buried quick with her, and so will I.
And if thou prate of mountains, let them throw
Millions of acres on us, till our ground,
Singeing his pate against the burning zone,
Make Ossa like a wart. Nay, an thou'lt mouth, 280
I'll rant as well as thou.
KING CLAUDIUS ⌈*to Laertes*⌉ This is mere madness,
And thus a while the fit will work on him.
Anon, as patient as the female dove
When that her golden couplets are disclosed,
His silence will sit drooping.
HAMLET (*to Laertes*)　　　　Hear you, sir, 285
What is the reason that you use me thus?
I loved you ever. But it is no matter.

Let Hercules himself do what he may,
The cat will mew, and dog will have his day.        *Exit*
KING CLAUDIUS
I pray you, good Horatio, wait upon him.    *Exit Horatio*
(*To Laertes*) Strengthen your patience in our last
      night's speech.                                   291
We'll put the matter to the present push.—
Good Gertrude, set some watch over your son.—
This grave shall have a living monument.                295
An hour of quiet shortly shall we see;
Till then, in patience our proceeding be.          *Exeunt*

**5.2**   *Enter Prince Hamlet and Horatio*
HAMLET
So much for this, sir. Now, let me see, the other.
You do remember all the circumstance?
HORATIO Remember it, my lord!
HAMLET
Sir, in my heart there was a kind of fighting
That would not let me sleep. Methought I lay       5
Worse than the mutines in the bilboes. Rashly—
And praised be rashness for it: let us know
Our indiscretion sometime serves us well
When our dear plots do pall, and that should teach us
There's a divinity that shapes our ends,           10
Rough-hew them how we will—
HORATIO That is most certain.
HAMLET Up from my cabin,
My sea-gown scarfed about me in the dark,
Groped I to find out them, had my desire,          15
Fingered their packet, and in fine withdrew
To mine own room again, making so bold,
My fears forgetting manners, to unseal
Their grand commission; where I found, Horatio—
O royal knavery!—an exact command,                 20
Larded with many several sorts of reasons
Importing Denmark's health, and England's, too,
With ho! such bugs and goblins in my life,
That on the supervise, no leisure bated,
No, not to stay the grinding of the axe,           25
My head should be struck off.
HORATIO                           Is't possible?
HAMLET (*giving it to him*)
Here's the commission. Read it at more leisure.
But wilt thou hear me how I did proceed?
HORATIO I beseech you.
HAMLET
Being thus benetted round with villainies—         30
Ere I could make a prologue to my brains,
They had begun the play—I sat me down,
Devised a new commission, wrote it fair.
I once did hold it, as our statists do,
A baseness to write fair, and laboured much        35
How to forget that learning; but, sir, now
It did me yeoman's service. Wilt thou know
Th'effect of what I wrote?
HORATIO                       Ay, good my lord.
HAMLET
An earnest conjuration from the King,
As England was his faithful tributary,             40
As love between them like the palm should flourish,
As peace should still her wheaten garland wear
And stand a comma 'tween their amities,
And many such like 'as'es of great charge,
That on the view and know of these contents,       45

Without debatement further more or less,
He should the bearers put to sudden death,
Not shriving-time allowed.
HORATIO                How was this sealed?
HAMLET
Why, even in that was heaven ordinant.
I had my father's signet in my purse,              50
Which was the model of that Danish seal;
Folded the writ up in the form of th'other,
Subscribed it, gave't th'impression, placed it safely,
The changeling never known. Now the next day
Was our sea-fight; and what to this was sequent    55
Thou know'st already.
HORATIO
So Guildenstern and Rosencrantz go to't.
HAMLET
Why, man, they did make love to this employment.
They are not near my conscience. Their defeat
Doth by their own insinuation grow.                60
'Tis dangerous when the baser nature comes
Between the pass and fell incensèd points
Of mighty opposites.
HORATIO              Why, what a king is this!
HAMLET
Does it not, think'st thee, stand me now upon—
He that hath killed my king and whored my mother,
Popped in between th'election and my hopes,        66
Thrown out his angle for my proper life,
And with such coz'nage—is't not perfect conscience
To quit him with this arm? And is't not to be damned
To let this canker of our nature come              70
In further evil?
HORATIO
It must be shortly known to him from England
What is the issue of the business there.
HAMLET
It will be short. The interim's mine,
And a man's life's no more than to say 'one'.      75
But I am very sorry, good Horatio,
That to Laertes I forgot myself;
For by the image of my cause I see
The portraiture of his. I'll court his favours.
But sure, the bravery of his grief did put me      80
Into a tow'ring passion.
HORATIO                   Peace, who comes here?
    *Enter young Osric, a courtier,* ⌈*taking off his hat*⌉
OSRIC
Your lordship is right welcome back to Denmark.
HAMLET I humbly thank you, sir. (*To Horatio*) Dost know
this water-fly?
HORATIO No, my good lord.                           85
HAMLET Thy state is the more gracious, for 'tis a vice to
know him. He hath much land, and fertile. Let a beast
be lord of beasts, and his crib shall stand at the king's
mess. 'Tis a chuff, but, as I say, spacious in the
possession of dirt.                                 90
OSRIC Sweet lord, if your friendship were at leisure I
should impart a thing to you from his majesty.
HAMLET I will receive it, sir, with all diligence of spirit.
Put your bonnet to his right use; 'tis for the head.
OSRIC I thank your lordship, 'tis very hot.         95
HAMLET No, believe me, 'tis very cold. The wind is
northerly.
OSRIC It is indifferent cold, my lord, indeed.
HAMLET Methinks it is very sultry and hot for my
complexion.                                         100

OSRIC Exceedingly, my lord. It is very sultry, as 'twere—
I cannot tell how. But, my lord, his majesty bade me
signify to you that a has laid a great wager on your
head. Sir, this is the matter.

HAMLET I beseech you, remember.                              105

OSRIC Nay, good my lord, for mine ease, in good faith.
Sir, you are not ignorant of what excellence Laertes is
at his weapon.

HAMLET What's his weapon?

OSRIC Rapier and dagger.                                    110

HAMLET That's two of his weapons. But well.

OSRIC The King, sir, hath wagered with him six Barbary
horses, against the which he imponed, as I take it, six
French rapiers and poniards, with their assigns as
girdle, hanger, or so. Three of the carriages, in faith,
are very dear to fancy, very responsive to the hilts,
most delicate carriages, and of very liberal conceit.

HAMLET What call you the carriages?                         118

OSRIC The carriages, sir, are the hangers.

HAMLET The phrase would be more germane to the matter
if we could carry cannon by our sides. I would it might
be hangers till then. But on: six Barbary horses against
six French swords, their assigns, and three liberal-
conceited carriages—that's the French bet against the
Danish. Why is this 'imponed', as you call it?            125

OSRIC The King, sir, hath laid, sir, that in a dozen passes
between you and him he shall not exceed you three
hits. He hath on't twelve for nine, and it would come
to immediate trial if your lordship would vouchsafe the
answer.                                                     130

HAMLET How if I answer no?

OSRIC I mean, my lord, the opposition of your person in
trial.

HAMLET Sir, I will walk here in the hall. If it please his
majesty, 'tis the breathing time of day with me. Let the
foils be brought; the gentleman willing, an the King
hold his purpose, I will win for him an I can. If not,
I'll gain nothing but my shame and the odd hits.          138

OSRIC Shall I re-deliver you e'en so?

HAMLET To this effect, sir; after what flourish your nature
will.

OSRIC I commend my duty to your lordship.

HAMLET Yours, yours.                              *Exit Osric*
He does well to commend it himself; there are no
tongues else for 's turn.                                   145

HORATIO This lapwing runs away with the shell on his
head.

HAMLET A did comply with his dug before a sucked it.
Thus has he—and many more of the same bevy that I
know the drossy age dotes on—only got the tune of
the time and outward habit of encounter, a kind of
yeasty collection which carries them through and
through the most fanned and winnowed opinions; and
do but blow them to their trial, the bubbles are out.

HORATIO You will lose this wager, my lord.                  155

HAMLET I do not think so. Since he went into France, I
have been in continual practice. I shall win at the odds.
But thou wouldst not think how all here about my
heart—but it is no matter.

HORATIO Nay, good my lord—                                  160

HAMLET It is but foolery, but it is such a kind of gain-
giving as would perhaps trouble a woman.

HORATIO If your mind dislike anything, obey it. I will
forestall their repair hither, and say you are not fit.

HAMLET Not a whit. We defy augury. There's a special
providence in the fall of a sparrow. If it be now, 'tis

not to come. If it be not to come, it will be now. If it
be not now, yet it will come. The readiness is all. Since
no man has aught of what he leaves, what is't to leave
betimes?                                                    170

*Enter King Claudius, Queen Gertrude, Laertes, and*
*lords, with Osric and other attendants with*
⌜*trumpets, drums, cushions*⌝, *foils, and gauntlets; a*
*table, and flagons of wine on it*

KING CLAUDIUS
Come, Hamlet, come, and take this hand from me.

HAMLET (*to Laertes*)
Give me your pardon, sir. I've done you wrong;
But pardon't as you are a gentleman.
This presence knows,
And you must needs have heard, how I am punished
With sore distraction. What I have done              176
That might your nature, honour, and exception
Roughly awake, I here proclaim was madness.
Was't Hamlet wronged Laertes? Never Hamlet.
If Hamlet from himself be ta'en away,                180
And when he's not himself does wrong Laertes,
Then Hamlet does it not, Hamlet denies it.
Who does it then? His madness. If't be so,
Hamlet is of the faction that is wronged.
His madness is poor Hamlet's enemy.                  185
Sir, in this audience
Let my disclaiming from a purposed evil
Free me so far in your most generous thoughts
That I have shot mine arrow o'er the house
And hurt my brother.

LAERTES                        I am satisfied in nature,    190
Whose motive in this case should stir me most
To my revenge. But in my terms of honour
I stand aloof, and will no reconcilement
Till by some elder masters of known honour
I have a voice and precedent of peace                195
To keep my name ungored; but till that time
I do receive your offered love like love,
And will not wrong it.

HAMLET                     I do embrace it freely,
And will this brothers' wager frankly play.—
(*To attendants*) Give us the foils. Come on.

LAERTES (*to attendants*)                 Come, one for me.

HAMLET
I'll be your foil, Laertes. In mine ignorance        201
Your skill shall, like a star i'th' darkest night,
Stick fiery off indeed.

LAERTES You mock me, sir.

HAMLET No, by this hand.                                    205

KING CLAUDIUS
Give them the foils, young Osric. Cousin Hamlet,
You know the wager?

HAMLET                    Very well, my lord.
Your grace hath laid the odds o'th' weaker side.

KING CLAUDIUS
I do not fear it; I have seen you both.
But since he is bettered, we have therefore odds.    210

LAERTES (*taking a foil*)
This is too heavy; let me see another.

HAMLET (*taking a foil*)
This likes me well. These foils have all a length?

OSRIC Ay, my good lord.
*Hamlet and Laertes prepare to play*

KING CLAUDIUS (*to attendants*)
Set me the stoups of wine upon that table.
If Hamlet give the first or second hit,              215

Or quit in answer of the third exchange,
Let all the battlements their ordnance fire.
The King shall drink to Hamlet's better breath,
And in the cup an union shall he throw
Richer than that which four successive kings          220
In Denmark's crown have worn. Give me the cups,
And let the kettle to the trumpet speak,
The trumpet to the cannoneer without,
The cannons to the heavens, the heaven to earth,
'Now the King drinks to Hamlet'.
  *Trumpets the while he drinks*
       Come, begin.          225
And you, the judges, bear a wary eye.
HAMLET (*to Laertes*) Come on, sir.
LAERTES Come, my lord.
  *They play*
HAMLET One.
LAERTES No.          230
HAMLET (*to Osric*) Judgement.
OSRIC A hit, a very palpable hit.
LAERTES Well, again.
KING CLAUDIUS
 Stay. Give me drink. Hamlet, this pearl is thine.
 Here's to thy health.—          235
  ⌈*Drum and*⌉ *trumpets sound, and shot goes off*
     Give him the cup.
HAMLET
 I'll play this bout first. Set it by a while.—
 Come.
  *They play again*
   Another hit. What say you?
LAERTES
 A touch, a touch, I do confess.
KING CLAUDIUS
 Our son shall win.
QUEEN GERTRUDE  He's fat and scant of breath.—          240
 Here, Hamlet, take my napkin. Rub thy brows.
 The Queen carouses to thy fortune, Hamlet.
HAMLET
 Good madam.
KING CLAUDIUS Gertrude, do not drink.
QUEEN GERTRUDE
 I will, my lord, I pray you pardon me.
  *She drinks, then offers the cup to Hamlet*
KING CLAUDIUS (*aside*)
 It is the poisoned cup; it is too late.          245
HAMLET
 I dare not drink yet, madam; by and by.
QUEEN GERTRUDE (*to Hamlet*) Come, let me wipe thy face.
LAERTES (*aside to Claudius*) My lord, I'll hit him now.
KING CLAUDIUS (*aside to Laertes*) I do not think't.
LAERTES (*aside*)
 And yet 'tis almost 'gainst my conscience.          250
HAMLET
 Come for the third, Laertes, you but dally.
 I pray you pass with your best violence.
 I am afeard you make a wanton of me.
LAERTES
 Say you so? Come on.
  *They play*
OSRIC     Nothing neither way.
LAERTES (*to Hamlet*)
 Have at you now!
  ⌈*Laertes wounds Hamlet.*⌉ *In scuffling, they change*
  *rapiers,* ⌈*and Hamlet wounds Laertes*⌉
KING CLAUDIUS (*to attendants*)
     Part them, they are incensed.          255

HAMLET (*to Laertes*)
 Nay, come again.
  ⌈*The Queen falls down*⌉
OSRIC    Look to the Queen there, ho!
HORATIO
 They bleed on both sides. (*To Hamlet*) How is't, my lord?
OSRIC How is't, Laertes?
LAERTES
 Why, as a woodcock to mine own springe, Osric.
 I am justly killed with mine own treachery.          260
HAMLET
 How does the Queen?
KING CLAUDIUS   She swoons to see them bleed.
QUEEN GERTRUDE
 No, no, the drink, the drink! O my dear Hamlet,
 The drink, the drink—I am poisoned. ⌈*She dies*⌉
HAMLET
 O villainy! Ho! Let the door be locked! ⌈*Exit Osric*⌉
 Treachery, seek it out.          265
LAERTES
 It is here, Hamlet. Hamlet, thou art slain.
 No med'cine in the world can do thee good.
 In thee there is not half an hour of life.
 The treacherous instrument is in thy hand,
 Unbated and envenomed. The foul practice          270
 Hath turned itself on me. Lo, here I lie,
 Never to rise again. Thy mother's poisoned.
 I can no more. The King, the King's to blame.
HAMLET
 The point envenomed too? Then, venom, to thy work.
  *He hurts King Claudius*
ALL THE COURTIERS Treason, treason!          275
KING CLAUDIUS
 O yet defend me, friends! I am but hurt.
HAMLET
 Here, thou incestuous, murd'rous, damnèd Dane,
 Drink off this potion. Is thy union here?
 Follow my mother.   *King Claudius dies*
LAERTES    He is justly served.
 It is a poison tempered by himself.          280
 Exchange forgiveness with me, noble Hamlet.
 Mine and my father's death come not upon thee,
 Nor thine on me.    *He dies*
HAMLET
 Heaven make thee free of it! I follow thee.
 I am dead, Horatio. Wretched Queen, adieu!          285
 You that look pale and tremble at this chance,
 That are but mutes or audience to this act,
 Had I but time—as this fell sergeant Death
 Is strict in his arrest—O, I could tell you—
 But let it be. Horatio, I am dead,          290
 Thou liv'st. Report me and my cause aright
 To the unsatisfied.
HORATIO    Never believe it.
 I am more an antique Roman than a Dane.
 Here's yet some liquor left.
HAMLET    As thou'rt a man,
 Give me the cup. Let go. By heaven, I'll ha't.          295
 O God, Horatio, what a wounded name,
 Things standing thus unknown, shall live behind me!
 If thou didst ever hold me in thy heart,
 Absent thee from felicity a while,
 And in this harsh world draw thy breath in pain          300
 To tell my story.
  *March afar off, and shout within*
    What warlike noise is this?

*Enter Osric*

OSRIC
Young Fortinbras, with conquest come from Poland,
To th'ambassadors of England gives
This warlike volley.
HAMLET                    O, I die, Horatio!
The potent poison quite o'ercrows my spirit.                    305
I cannot live to hear the news from England,
But I do prophesy th'election lights
On Fortinbras. He has my dying voice.
So tell him, with th'occurrents, more and less,
Which have solicited. The rest is silence.                    310
O, O, O, O!                                        *He dies*
HORATIO
Now cracks a noble heart. Good night, sweet prince,
And flights of angels sing thee to thy rest.—
Why does the drum come hither?
*Enter Fortinbras with the English ⌈Ambassadors⌉,*
*with a drummer, colours, and attendants*
FORTINBRAS Where is this sight?                    315
HORATIO What is it ye would see?
If aught of woe or wonder, cease your search.
FORTINBRAS
This quarry cries on havoc. O proud death,
What feast is toward in thine eternal cell
That thou so many princes at a shot                    320
So bloodily hast struck!
AMBASSADOR                    The sight is dismal,
And our affairs from England come too late.
The ears are senseless that should give us hearing
To tell him his commandment is fulfilled,
That Rosencrantz and Guildenstern are dead.                    325
Where should we have our thanks?
HORATIO                    Not from his mouth,
Had it th'ability of life to thank you.
He never gave commandment for their death.

But since so jump upon this bloody question
You from the Polack wars, and you from England,    330
Are here arrived, give order that these bodies
High on a stage be placèd to the view;
And let me speak to th' yet unknowing world
How these things came about. So shall you hear
Of carnal, bloody, and unnatural acts,                    335
Of accidental judgements, casual slaughters,
Of deaths put on by cunning and forced cause;
And, in this upshot, purposes mistook
Fall'n on th'inventors' heads. All this can I
Truly deliver.
FORTINBRAS                    Let us haste to hear it,    340
And call the noblest to the audience.
For me, with sorrow I embrace my fortune.
I have some rights of memory in this kingdom,
Which now to claim my vantage doth invite me.
HORATIO
Of that I shall have also cause to speak,                    345
And from his mouth whose voice will draw on more.
But let this same be presently performed,
Even whiles men's minds are wild, lest more
            mischance
On plots and errors happen.
FORTINBRAS                              Let four captains
Bear Hamlet like a soldier to the stage,                    350
For he was likely, had he been put on,
To have proved most royally; and for his passage,
The soldiers' music and the rites of war
Speak loudly for him.
Take up the body. Such a sight as this                    355
Becomes the field, but here shows much amiss.
Go, bid the soldiers shoot.
*Exeunt, marching, with the bodies; after the*
*which, a peal of ordnance are shot off*

## ADDITIONAL PASSAGES

A. Just before the second entrance of the Ghost in 1.1 (l. 106.1), Q2 has these additional lines:

BARNARDO
I think it be no other but e'en so.
Well may it sort that this portentous figure
Comes armèd through our watch so like the king
That was and is the question of these wars.
HORATIO
A mote it is to trouble the mind's eye.                    5
In the most high and palmy state of Rome,
A little ere the mightiest Julius fell,
The graves stood tenantless, and the sheeted dead
Did squeak and gibber in the Roman streets
At stars with trains of fire, and dews of blood,    10
Disasters in the sun; and the moist star,
Upon whose influence Neptune's empire stands,
Was sick almost to doomsday with eclipse.
And even the like precurse of feared events,
As harbingers preceding still the fates,                    15
And prologue to the omen coming on,
Have heaven and earth together demonstrated
Unto our climature and countrymen.

B. Just before the entrance of the Ghost in 1.4 (l. 18.1), Q2 has these additional lines continuing Hamlet's speech:

This heavy-headed revel east and west
Makes us traduced and taxed of other nations.

They clepe us drunkards, and with swinish phrase
Soil our addition; and indeed it takes
From our achievements, though performed at height,    5
The pith and marrow of our attribute.
So, oft it chances in particular men
That, for some vicious mole of nature in them—
As in their birth, wherein they are not guilty,
Since nature cannot choose his origin,                    10
By the o'ergrowth of some complexion,
Oft breaking down the pales and forts of reason,
Or by some habit that too much o'erleavens
The form of plausive manners—that these men,
Carrying, I say, the stamp of one defect,                    15
Being nature's livery or fortune's star,
His virtues else be they as pure as grace,
As infinite as man may undergo,
Shall in the general censure take corruption
From that particular fault. The dram of evil                    20
Doth all the noble substance over-daub
To his own scandal.

C. After 1.4.55, Q2 has these additional lines continuing Horatio's speech:

The very place puts toys of desperation,
Without more motive, into every brain
That looks so many fathoms to the sea
And hears it roar beneath.

D. After 3.2.163, Q2 has this additional couplet concluding the Player Queen's speech:

Where love is great, the littlest doubts are fear;
Where little fears grow great, great love grows there.

E. After 3.2.208, Q2 has this additional couplet in the middle of the Player Queen's speech:

To desperation turn my trust and hope;
An anchor's cheer in prison be my scope.

F. After 'this?' in 3.4.70, Q2 has this more expansive version of Hamlet's lines of which F retains only 'what devil . . . blind':

Sense sure you have,
Else could you not have motion; but sure that sense
Is apoplexed, for madness would not err,
Nor sense to ecstasy was ne'er so thralled
But it reserved some quantity of choice          5
To serve in such a difference. What devil was't
That thus hath cozened you at hoodman-blind?
Eyes without feeling, feeling without sight,
Ears without hands or eyes, smelling sans all,
Or but a sickly part of one true sense          10
Could not so mope.

G. After 3.4.151, Q2 has this more expansive version of Hamlet's lines of which F retains only 'refrain . . . abstinence':

That monster custom, who all sense doth eat,
Of habits devilish, is angel yet in this:
That to the use of actions fair and good
He likewise gives a frock or livery
That aptly is put on. Refrain tonight,          5
And that shall lend a kind of easiness
To the next abstinence, the next more easy—
For use almost can change the stamp of nature—
And either in the devil, or throw him out
With wondrous potency.          10

H. At 3.4.185, Q2 has these additional lines before 'This man . . .':

HAMLET
There's letters sealed, and my two schoolfellows—
Whom I will trust as I will adders fanged—
They bear the mandate, they must sweep my way
And marshal me to knavery. Let it work,
For 'tis the sport to have the engineer          5
Hoised with his own petard; and't shall go hard
But I will delve one yard below their mines
And blow them at the moon. O, 'tis most sweet
When in one line two crafts directly meet.

I. After 'done' in 4.1.39, Q2 has these additional lines continuing the King's speech (the first three words are an editorial conjecture):

So envious slander,
Whose whisper o'er the world's diameter,
As level as the cannon to his blank,
Transports his poisoned shot, may miss our name
And hit the woundless air.          5

J. Q2 has this more expansive version of the ending of 4.4:

CAPTAIN          I will do't, my lord.
FORTINBRAS
Go softly on.          *Exit with his army*

*Enter Prince Hamlet, Rosencrantz, Guildenstern, etc.*
HAMLET (*to the Captain*) Good sir, whose powers are these?
CAPTAIN
They are of Norway, sir.
HAMLET          How purposed, sir, I pray you?
CAPTAIN
Against some part of Poland.
HAMLET          Who commands them, sir?
CAPTAIN
The nephew to old Norway, Fortinbras.          5
HAMLET
Goes it against the main of Poland, sir,
Or for some frontier?
CAPTAIN
Truly to speak, and with no addition,
We go to gain a little patch of ground
That hath in it no profit but the name.          10
To pay five ducats, five, I would not farm it,
Nor will it yield to Norway or the Pole
A ranker rate, should it be sold in fee.
HAMLET
Why then, the Polack never will defend it.
CAPTAIN
Yes, it is already garrisoned.          15
HAMLET
Two thousand souls and twenty thousand ducats
Will now debate the question of this straw.
This is th'imposthume of much wealth and peace,
That inward breaks and shows no cause without
Why the man dies. I humbly thank you, sir.          20
CAPTAIN
God buy you, sir.          *Exit*
ROSENCRANTZ          Will't please you go, my lord?
HAMLET
I'll be with you straight. Go a little before.
                                        *Exeunt all but Hamlet*
How all occasions do inform against me
And spur my dull revenge! What is a man
If his chief good and market of his time          25
Be but to sleep and feed?—a beast, no more.
Sure, he that made us with such large discourse,
Looking before and after, gave us not
That capability and god-like reason
To fust in us unused. Now whether it be          30
Bestial oblivion, or some craven scruple
Of thinking too precisely on th'event—
A thought which, quartered, hath but one part wisdom
And ever three parts coward—I do not know
Why yet I live to say 'This thing's to do',          35
Sith I have cause, and will, and strength, and means,
To do't. Examples gross as earth exhort me,
Witness this army of such mass and charge,
Led by a delicate and tender prince,
Whose spirit with divine ambition puffed          40
Makes mouths at the invisible event,
Exposing what is mortal and unsure
To all that fortune, death, and danger dare,
Even for an eggshell. Rightly to be great
Is not to stir without great argument,          45

But greatly to find quarrel in a straw
When honour's at the stake. How stand I, then,
That have a father killed, a mother stained,
Excitements of my reason and my blood,
And let all sleep while, to my shame, I see          50
The imminent death of twenty thousand men
That, for a fantasy and trick of fame,
Go to their graves like beds, fight for a plot
Whereon the numbers cannot try the cause,
Which is not tomb enough and continent          55
To hide the slain. O, from this time forth
My thoughts be bloody or be nothing worth!          *Exit*

K. After 'accident' at 4.7.67, Q2 has these additional lines:

LAERTES          My lord, I will be ruled,
The rather if you could devise it so
That I might be the organ.
KING CLAUDIUS          It falls right.
You have been talked of, since your travel, much,
And that in Hamlet's hearing, for a quality          5
Wherein they say you shine. Your sum of parts
Did not together pluck such envy from him
As did that one, and that, in my regard,
Of the unworthiest siege.
LAERTES          What part is that, my lord?
KING CLAUDIUS
A very ribbon in the cap of youth,          10
Yet needful too, for youth no less becomes
The light and careless livery that it wears
Than settled age his sables and his weeds
Importing health and graveness.

L. After 'match you' at 4.7.85, Q2 has these additional lines continuing the King's speech:

                    Th'escrimers of their nation
He swore had neither motion, guard, nor eye
If you opposed them.

M. After 4.7.96, Q2 has these additional lines continuing the King's speech:

There lives within the very flame of love
A kind of wick or snuff that will abate it,
And nothing is at a like goodness still,
For goodness, growing to a plurisy,
Dies in his own too much. That we would do          5
We should do when we would, for this 'would' changes,
And hath abatements and delays as many
As there are tongues, are hands, are accidents;
And then this 'should' is like a spendthrift's sigh,
That hurts by easing. But to the quick of th'ulcer—          10

N. After 'Sir' at 5.2.107, Q2 has these lines (in place of F's 'you are not ignorant of what excellence Laertes is at his weapon'):

here is newly come to court Laertes, believe me, an absolute gentleman, full of most excellent differences, of very soft society and great showing. Indeed, to speak feelingly of him, he is the card or calendar of gentry, for you shall find in him the continent of what part a gentleman would see.          6
HAMLET Sir, his definement suffers no perdition in you, though I know to divide him inventorially would dizzy th'arithmetic of memory, and yet but yaw neither in respect of his quick sail. But in the verity of extolment, I take him to be a soul of great article, and his infusion of such dearth and rareness as, to make true diction of him, his semblable is his mirror, and who else would trace him his umbrage, nothing more.
OSRIC Your lordship speaks most infallibly of him.          15
HAMLET The concernancy, sir? Why do we wrap the gentleman in our more rawer breath?
OSRIC Sir?
HORATIO Is't not possible to understand in another tongue? You will to't, sir, rarely.          20
HAMLET What imports the nomination of this gentleman?
OSRIC Of Laertes?
HORATIO (*aside to Hamlet*) His purse is empty already; all 's golden words are spent.
HAMLET (*to Osric*) Of him, sir.          25
OSRIC I know you are not ignorant—
HAMLET I would you did, sir; yet, in faith, if you did it would not much approve me. Well, sir?
OSRIC You are not ignorant of what excellence Laertes is.
HAMLET I dare not confess that, lest I should compare with him in excellence. But to know a man well were to know himself.          32
OSRIC I mean, sir, for his weapon. But in the imputation laid on him by them, in his meed he's unfellowed.

O. After 5.2.118, Q2 has the following additional speech:

HORATIO (*aside to Hamlet*) I knew you must be edified by the margin ere you had done.

P. After 5.2.154, Q2 has the following (in place of F's 'HORATIO You will lose this wager, my lord'):

*Enter a Lord*
LORD (*to Hamlet*) My lord, his majesty commended him to you by young Osric, who brings back to him that you attend him in the hall. He sends to know if your pleasure hold to play with Laertes, or that you will take longer time.          5
HAMLET I am constant to my purposes; they follow the King's pleasure. If his fitness speaks, mine is ready, now or whensoever, provided I be so able as now.
LORD The King and Queen and all are coming down.
HAMLET In happy time.          10
LORD The Queen desires you to use some gentle entertainment to Laertes before you fall to play.
HAMLET She well instructs me.          *Exit Lord*
HORATIO You will lose, my lord.

# TWELFTH NIGHT

TWELFTH NIGHT, the end of the Christmas season, was traditionally a time of revelry and topsy-turvydom; Shakespeare's title for a play in which a servant aspires to his mistress's hand has no more specific reference. It was thought appropriate to the festive occasion of Candlemas (2 February) 1602 when, in the first known allusion to it, John Manningham, a law student of the Middle Temple in London, noted 'at our feast we had a play called *Twelfth Night, or What You Will*'. References to 'the Sophy'— the Shah of Persia (2.5.174; 3.4.271)—probably post-date Sir Robert Shirley's return from Persia, in a ship named *The Sophy*, in 1599; and 'the new map with the augmentation of the Indies' (3.2.75) appears to be one published in 1599 and reissued in 1600. Shakespeare may have picked up the name Orsino for his young duke from a Tuscan nobleman whom Queen Elizabeth entertained at Whitehall with a play performed by Shakespeare's company on Twelfth Night 1601. Probably he wrote *Twelfth Night* during that year.

*Twelfth Night*'s romantic setting is Illyria, the Greek and Roman name for Adriatic territory roughly corresponding to modern Yugoslavia. Manningham had noted that the play was 'much like *The Comedy of Errors* or *Menaechmi* in Plautus', thinking no doubt of the confusions created by identical twins. Shakespeare may also have known an anonymous Italian comedy, *Gl'Ingannati* (*The Deceived Ones*), acted in 1531 and first printed in 1537, which influenced a number of other plays and prose tales including Barnaby Riche's story of Apolonius and Silla printed as part of *Riche's Farewell to Military Profession* (1581). Riche gave Shakespeare his main plot of a shipwrecked girl (Viola) who, disguised as a boy (Cesario), serves a young Duke (Orsino) and undertakes love-errands on his behalf to a noble lady (Olivia) who falls in love with her but mistakenly marries her twin brother (Sebastian). Shakespeare idealizes Riche's characters and purges the story of some its explicit sexuality: Riche's Olivia, for example, is pregnant before marriage, and his Viola reveals her identity, in a manner impractical for a boy actor, by stripping to the waist. Shakespeare complicates the plot by giving Olivia a reprobate uncle, Sir Toby Belch, and two additional suitors, the asinine Sir Andrew Aguecheek and her steward, Malvolio, tricked by members of her household into believing that she loves him. More important to the play than to the plot is the entirely Shakespearian clown, Feste, a wry and oblique commentator whose wit in folly is opposed to Malvolio's folly in wit.

*Twelfth Night* is the consummation of Shakespeare's romantic comedy, a play of wide emotional range, extending from the robust, brilliantly orchestrated humour of the scene of midnight revelry (2.2) to the rapt wonder of the antiphon of recognition (5.1.224-56) between the reunited twins. In performance the balance shifts, favouring sometimes the exposure and celebration of folly, at other times the poignancy of unattained love and of unheeded wisdom; but few other plays have so consistently provided theatrical pleasure of so high an order.

# THE PERSONS OF THE PLAY

ORSINO, Duke of Illyria

VALENTINE

CURIO } attending on Orsino

FIRST OFFICER

SECOND OFFICER


VIOLA, a lady, later disguised as Cesario

A CAPTAIN

SEBASTIAN, her twin brother

ANTONIO, another sea-captain

OLIVIA, a Countess

MARIA, her waiting-gentlewoman

SIR TOBY Belch, Olivia's kinsman

SIR ANDREW Aguecheek, companion of Sir Toby

MALVOLIO, Olivia's steward

FABIAN, a member of Olivia's household

FESTE the Clown, her jester

A PRIEST

A SERVANT of Olivia

Musicians, sailors, lords, attendants

# Twelfth Night, or What You Will

**1.1** *Music. Enter Orsino Duke of Illyria, Curio, and
other lords*

ORSINO
If music be the food of love, play on,
Give me excess of it that, surfeiting,
The appetite may sicken and so die.
That strain again, it had a dying fall.
O, it came o'er my ear like the sweet sound     5
That breathes upon a bank of violets,
Stealing and giving odour. Enough, no more,
'Tis not so sweet now as it was before.
   ⌜*Music ceases*⌝
O spirit of love, how quick and fresh art thou
That, notwithstanding thy capacity     10
Receiveth as the sea, naught enters there,
Of what validity and pitch so e'er,
But falls into abatement and low price
Even in a minute! So full of shapes is fancy
That it alone is high fantastical.     15

CURIO
Will you go hunt, my lord?

ORSINO                 What, Curio?

CURIO                        The hart.

ORSINO
Why so I do, the noblest that I have.
O, when mine eyes did see Olivia first
Methought she purged the air of pestilence;
That instant was I turned into a hart,     20
And my desires, like fell and cruel hounds,
E'er since pursue me.
   *Enter Valentine*
                How now, what news from her?

VALENTINE
So please my lord, I might not be admitted,
But from her handmaid do return this answer:
The element itself till seven years' heat     25
Shall not behold her face at ample view,
But like a cloistress she will veilèd walk
And water once a day her chamber round
With eye-offending brine—all this to season
A brother's dead love, which she would keep fresh     30
And lasting in her sad remembrance.

ORSINO
O, she that hath a heart of that fine frame
To pay this debt of love but to a brother,
How will she love when the rich golden shaft
Hath killed the flock of all affections else     35
That live in her—when liver, brain, and heart,
These sovereign thrones, are all supplied, and filled
Her sweet perfections with one self king!
Away before me to sweet beds of flowers.
Love-thoughts lie rich when canopied with bowers.    40
                           *Exeunt*

**1.2** *Enter Viola, a Captain, and sailors*

VIOLA
  What country, friends, is this?

CAPTAIN             This is Illyria, lady.

VIOLA
  And what should I do in Illyria?

My brother, he is in Elysium.
Perchance he is not drowned. What think you sailors?

CAPTAIN
It is perchance that you yourself were saved.     5

VIOLA
O my poor brother!—and so perchance may he be.

CAPTAIN
True, madam, and to comfort you with chance,
Assure yourself, after our ship did split,
When you and those poor number savèd with you
Hung on our driving boat, I saw your brother,     10
Most provident in peril, bind himself—
Courage and hope both teaching him the practice—
To a strong mast that lived upon the sea,
Where, like Arion on the dolphin's back,
I saw him hold acquaintance with the waves     15
So long as I could see.

VIOLA (*giving money*)     For saying so, there's gold.
Mine own escape unfoldeth to my hope,
Whereto thy speech serves for authority,
The like of him. Know'st thou this country?

CAPTAIN
Ay, madam, well, for I was bred and born     20
Not three hours' travel from this very place.

VIOLA
Who governs here?

CAPTAIN            A noble duke, in nature
As in name.

VIOLA        What is his name?

CAPTAIN                  Orsino.

VIOLA
Orsino. I have heard my father name him.
He was a bachelor then.     25

CAPTAIN
And so is now, or was so very late,
For but a month ago I went from hence,
And then 'twas fresh in murmur—as, you know,
What great ones do the less will prattle of—
That he did seek the love of fair Olivia.     30

VIOLA What's she?

CAPTAIN
A virtuous maid, the daughter of a count
That died some twelvemonth since, then leaving her
In the protection of his son, her brother,
Who shortly also died, for whose dear love,     35
They say, she hath abjured the sight
And company of men.

VIOLA           O that I served that lady,
And might not be delivered to the world
Till I had made mine own occasion mellow,
What my estate is.

CAPTAIN          That were hard to compass,    40
Because she will admit no kind of suit,
No, not the Duke's.

VIOLA
There is a fair behaviour in thee, captain,
And though that nature with a beauteous wall
Doth oft close in pollution, yet of thee     45
I will believe thou hast a mind that suits
With this thy fair and outward character.

I pray thee—and I'll pay thee bounteously—
Conceal me what I am, and be my aid
For such disguise as haply shall become　　50
The form of my intent. I'll serve this duke.
Thou shalt present me as an eunuch to him.
It may be worth thy pains, for I can sing,
And speak to him in many sorts of music
That will allow me very worth his service.　　55
What else may hap, to time I will commit.
Only shape thou thy silence to my wit.

CAPTAIN
Be you his eunuch, and your mute I'll be.
When my tongue blabs, then let mine eyes not see.

VIOLA
I thank thee. Lead me on.　　　　　　*Exeunt*

1.3　*Enter Sir Toby Belch and Maria*

SIR TOBY What a plague means my niece to take the death
of her brother thus? I am sure care's an enemy to life.

MARIA By my troth, Sir Toby, you must come in earlier
o' nights. Your cousin, my lady, takes great exceptions
to your ill hours.　　5

SIR TOBY Why, let her except, before excepted.

MARIA Ay, but you must confine yourself within the
modest limits of order.

SIR TOBY Confine? I'll confine myself no finer than I am.
These clothes are good enough to drink in, and so be
these boots too; an they be not, let them hang
themselves in their own straps.　　12

MARIA That quaffing and drinking will undo you. I heard
my lady talk of it yesterday, and of a foolish knight
that you brought in one night here to be her wooer.

SIR TOBY Who, Sir Andrew Aguecheek?　　16

MARIA Ay, he.

SIR TOBY He's as tall a man as any's in Illyria.

MARIA What's that to th' purpose?

SIR TOBY Why, he has three thousand ducats a year.　　20

MARIA Ay, but he'll have but a year in all these ducats.
He's a very fool, and a prodigal.

SIR TOBY Fie that you'll say so! He plays o'th' viol-de-
gamboys, and speaks three or four languages word for
word without book, and hath all the good gifts of
nature.　　26

MARIA He hath indeed, almost natural, for besides that
he's a fool, he's a great quarreller, and but that he
hath the gift of a coward to allay the gust he hath in
quarrelling, 'tis thought among the prudent he would
quickly have the gift of a grave.　　31

SIR TOBY By this hand, they are scoundrels and sub-
stractors that say so of him. Who are they?

MARIA They that add, moreover, he's drunk nightly in
your company.　　35

SIR TOBY With drinking healths to my niece. I'll drink to
her as long as there is a passage in my throat and
drink in Illyria. He's a coward and a coistrel that will
not drink to my niece till his brains turn o'th' toe, like
a parish top. What wench, *Castiliano, vulgo*, for here
comes Sir Andrew Agueface.　　41

　　　*Enter Sir Andrew Aguecheek*

SIR ANDREW Sir Toby Belch! How now, Sir Toby Belch?

SIR TOBY Sweet Sir Andrew.

SIR ANDREW (*to Maria*) Bless you, fair shrew.

MARIA And you too, sir.　　45

SIR TOBY Accost, Sir Andrew, accost.

SIR ANDREW What's that?

SIR TOBY My niece's chambermaid.

SIR ANDREW Good Mistress Accost, I desire better
acquaintance.　　50

MARIA My name is Mary, sir.

SIR ANDREW Good Mistress Mary Accost.

SIR TOBY You mistake, knight. 'Accost' is front her, board
her, woo her, assail her.　　54

SIR ANDREW By my troth, I would not undertake her in
this company. Is that the meaning of 'accost'?

MARIA Fare you well, gentlemen.

SIR TOBY An thou let part so, Sir Andrew, would thou
mightst never draw sword again.　　59

SIR ANDREW An you part so, mistress, I would I might
never draw sword again. Fair lady, do you think you
have fools in hand?

MARIA Sir, I have not you by th' hand.

SIR ANDREW Marry, but you shall have, and here's my
hand.　　65

MARIA (*taking his hand*) Now sir, thought is free. I pray
you, bring your hand to th' buttery-bar, and let it
drink.

SIR ANDREW Wherefore, sweetheart? What's your metaphor?　　70

MARIA It's dry, sir.

SIR ANDREW Why, I think so. I am not such an ass but I
can keep my hand dry. But what's your jest?

MARIA A dry jest, sir.

SIR ANDREW Are you full of them?　　75

MARIA Ay, sir, I have them at my fingers' ends. Marry,
now I let go your hand I am barren.　　*Exit*

SIR TOBY O knight, thou lackest a cup of canary. When
did I see thee so put down?

SIR ANDREW Never in your life, I think, unless you see
canary put me down. Methinks sometimes I have no
more wit than a Christian or an ordinary man has;
but I am a great eater of beef, and I believe that does
harm to my wit.

SIR TOBY No question.　　85

SIR ANDREW An I thought that, I'd forswear it. I'll ride
home tomorrow, Sir Toby.

SIR TOBY *Pourquoi*, my dear knight?

SIR ANDREW What is 'Pourquoi'? Do, or not do? I would
I had bestowed that time in the tongues that I have
in fencing, dancing, and bear-baiting. O, had I but
followed the arts!　　92

SIR TOBY Then hadst thou had an excellent head of hair.

SIR ANDREW Why, would that have mended my hair?

SIR TOBY Past question, for thou seest it will not curl by
nature.　　96

SIR ANDREW But it becomes me well enough, does't not?

SIR TOBY Excellent, it hangs like flax on a distaff, and I
hope to see a housewife take thee between her legs and
spin it off.　　100

SIR ANDREW Faith, I'll home tomorrow, Sir Toby. Your
niece will not be seen, or if she be, it's four to one
she'll none of me. The Count himself here hard by woos
her.　　104

SIR TOBY She'll none o'th' Count. She'll not match above
her degree, neither in estate, years, nor wit, I have
heard her swear't. Tut, there's life in't, man.

SIR ANDREW I'll stay a month longer. I am a fellow o'th'
strangest mind i'th' world. I delight in masques and
revels sometimes altogether.　　110

SIR TOBY Art thou good at these kickshawses, knight?

SIR ANDREW As any man in Illyria, whatsoever he be,
under the degree of my betters; and yet I will not
compare with an old man.

SIR TOBY What is thy excellence in a galliard, knight?
SIR ANDREW Faith, I can cut a caper.
SIR TOBY And I can cut the mutton to't.
SIR ANDREW And I think I have the back-trick simply as
strong as any man in Illyria.                              119
SIR TOBY Wherefore are these things hid? Wherefore have
these gifts a curtain before 'em? Are they like to take
dust, like Mistress Mall's picture? Why dost thou not
go to church in a galliard, and come home in a coranto?
My very walk should be a jig. I would not so much as
make water but in a cinquepace. What dost thou mean?
Is it a world to hide virtues in? I did think by the
excellent constitution of thy leg it was formed under
the star of a galliard.                                    128
SIR ANDREW Ay, 'tis strong, and it does indifferent well
in a divers-coloured stock. Shall we set about some
revels?
SIR TOBY What shall we do else—were we not born under
Taurus?                                                    133
SIR ANDREW Taurus? That's sides and heart.
SIR TOBY No, sir, it is legs and thighs: let me see thee
caper.
          ⌈Sir Andrew capers⌉
Ha, higher! Ha ha, excellent.                   Exeunt

**1.4**    *Enter Valentine, and Viola (as Cesario) in man's
attire*

VALENTINE If the Duke continue these favours towards
you, Cesario, you are like to be much advanced. He
hath known you but three days, and already you are
no stranger.
VIOLA You either fear his humour or my negligence, that
you call in question the continuance of his love. Is he
inconstant, sir, in his favours?                             7
VALENTINE No, believe me.
          *Enter the Duke, Curio, and attendants*
VIOLA I thank you. Here comes the Count.
ORSINO Who saw Cesario, ho?                                 10
VIOLA On your attendance, my lord, here.
ORSINO (*to Curio and attendants*)
Stand you a while aloof. (*To Viola*) Cesario,
Thou know'st no less but all. I have unclasped
To thee the book even of my secret soul.
Therefore, good youth, address thy gait unto her,         15
Be not denied access, stand at her doors,
And tell them there thy fixèd foot shall grow
Till thou have audience.
VIOLA                                   Sure, my noble lord,
If she be so abandoned to her sorrow
As it is spoke, she never will admit me.                  20
ORSINO
Be clamorous, and leap all civil bounds,
Rather than make unprofited return.
VIOLA
Say I do speak with her, my lord, what then?
ORSINO
O then unfold the passion of my love,
Surprise her with discourse of my dear faith.             25
It shall become thee well to act my woes—
She will attend it better in thy youth
Than in a nuncio's of more grave aspect.
VIOLA
I think not so, my lord.
ORSINO                           Dear lad, believe it;
For they shall yet belie thy happy years                  30
That say thou art a man. Diana's lip

Is not more smooth and rubious; thy small pipe
Is as the maiden's organ, shrill and sound,
And all is semblative a woman's part.
I know thy constellation is right apt                     35
For this affair. (*To Curio and attendants*) Some four or
          five attend him.
All if you will, for I myself am best
When least in company. (*To Viola*) Prosper well in this
And thou shalt live as freely as thy lord,
To call his fortunes thine.
VIOLA                              I'll do my best         40
To woo your lady—⌈*aside*⌉ yet a barful strife—
Whoe'er I woo, myself would be his wife.        *Exeunt*

**1.5**    *Enter Maria, and Feste, the clown*

MARIA Nay, either tell me where thou hast been or I will
not open my lips so wide as a bristle may enter in way
of thy excuse. My lady will hang thee for thy absence.
FESTE Let her hang me. He that is well hanged in this
world needs to fear no colours.
MARIA Make that good.                                       6
FESTE He shall see none to fear.
MARIA A good lenten answer. I can tell thee where that
saying was born, of 'I fear no colours'.
FESTE Where, good Mistress Mary?                           10
MARIA In the wars, and that may you be bold to say in
your foolery.
FESTE Well, God give them wisdom that have it; and
those that are fools, let them use their talents.         14
MARIA Yet you will be hanged for being so long absent,
or to be turned away—is not that as good as a hanging
to you?
FESTE Many a good hanging prevents a bad marriage;
and for turning away, let summer bear it out.
MARIA You are resolute then?                               20
FESTE Not so neither, but I am resolved on two points.
MARIA That if one break, the other will hold; or if both
break, your gaskins fall.
FESTE Apt, in good faith, very apt. Well, go thy way. If
Sir Toby would leave drinking thou wert as witty a
piece of Eve's flesh as any in Illyria.                   26
MARIA Peace, you rogue, no more o' that. Here comes
my lady. Make your excuse wisely, you were best.
                                                        *Exit*
          *Enter Olivia, with Malvolio and attendants*
FESTE ⌈*aside*⌉ Wit, an't be thy will, put me into good
fooling! Those wits that think they have thee do very
oft prove fools, and I that am sure I lack thee may pass
for a wise man. For what says Quinapalus?—'Better a
witty fool than a foolish wit.' (*To Olivia*) God bless thee,
lady.
OLIVIA (*to attendants*) Take the fool away.              35
FESTE Do you not hear, fellows? Take away the lady.
OLIVIA Go to, you're a dry fool. I'll no more of you.
Besides, you grow dishonest.
FESTE Two faults, madonna, that drink and good counsel
will amend, for give the dry fool drink, then is the fool
not dry; bid the dishonest man mend himself: if he
mend, he is no longer dishonest; if he cannot, let the
botcher mend him. Anything that's mended is but
patched. Virtue that transgresses is but patched with
sin, and sin that amends is but patched with virtue. If
that this simple syllogism will serve, so. If it will not,
what remedy? As there is no true cuckold but calamity,
so beauty's a flower. The lady bade take away the fool,
therefore I say again, take her away.                     49

OLIVIA Sir, I bade them take away you.          50

FESTE Misprision in the highest degree! Lady, 'Cucullus non facit monachum'—that's as much to say as I wear not motley in my brain. Good madonna, give me leave to prove you a fool.

OLIVIA Can you do it?          55

FESTE Dexteriously, good madonna.

OLIVIA Make your proof.

FESTE I must catechize you for it, madonna. Good my mouse of virtue, answer me.

OLIVIA Well, sir, for want of other idleness I'll bide your proof.          61

FESTE Good madonna, why mournest thou?

OLIVIA Good fool, for my brother's death.

FESTE I think his soul is in hell, madonna.

OLIVIA I know his soul is in heaven, fool.          65

FESTE The more fool, madonna, to mourn for your brother's soul, being in heaven. Take away the fool, gentlemen.

OLIVIA What think you of this fool, Malvolio? Doth he not mend?          70

MALVOLIO Yes, and shall do till the pangs of death shake him. Infirmity, that decays the wise, doth ever make the better fool.

FESTE God send you, sir, a speedy infirmity for the better increasing your folly. Sir Toby will be sworn that I am no fox, but he will not pass his word for twopence that you are no fool.          77

OLIVIA How say you to that, Malvolio?

MALVOLIO I marvel your ladyship takes delight in such a barren rascal. I saw him put down the other day with an ordinary fool that has no more brain than a stone. Look you now, he's out of his guard already. Unless you laugh and minister occasion to him, he is gagged. I protest I take these wise men that crow so at these set kind of fools no better than the fools' zanies.          85

OLIVIA O, you are sick of self-love, Malvolio, and taste with a distempered appetite. To be generous, guiltless, and of free disposition is to take those things for birdbolts that you deem cannon bullets. There is no slander in an allowed fool, though he do nothing but rail; nor no railing in a known discreet man, though he do nothing but reprove.          92

FESTE Now Mercury indue thee with leasing, for thou speakest well of fools.

*Enter Maria*

MARIA Madam, there is at the gate a young gentleman much desires to speak with you.          96

OLIVIA From the Count Orsino, is it?

MARIA I know not, madam. 'Tis a fair young man, and well attended.

OLIVIA Who of my people hold him in delay?          100

MARIA Sir Toby, madam, your kinsman.

OLIVIA Fetch him off, I pray you, he speaks nothing but madman. Fie on him. Go you, Malvolio. If it be a suit from the Count, I am sick, or not at home—what you will to dismiss it.          *Exit Malvolio*
Now you see, sir, how your fooling grows old, and people dislike it.          107

FESTE Thou hast spoke for us, madonna, as if thy eldest son should be a fool, whose skull Jove cram with brains, for—here he comes—          110

*Enter Sir Toby*

one of thy kin has a most weak *pia mater.*

OLIVIA By mine honour, half-drunk. What is he at the gate, cousin?

SIR TOBY A gentleman.

OLIVIA A gentleman? What gentleman?          115

SIR TOBY 'Tis a gentleman here. (*He belches*) A plague o' these pickle herring! (*To Feste*) How now, sot?

FESTE Good Sir Toby.

OLIVIA Cousin, cousin, how have you come so early by this lethargy?          120

SIR TOBY Lechery? I defy lechery. There's one at the gate.

OLIVIA Ay, marry, what is he?

SIR TOBY Let him be the devil an he will, I care not. Give me faith, say I. Well, it's all one.          *Exit*

OLIVIA What's a drunken man like, fool?          125

FESTE Like a drowned man, a fool, and a madman—one draught above heat makes him a fool, the second mads him, and a third drowns him.

OLIVIA Go thou and seek the coroner, and let him sit o' my coz, for he's in the third degree of drink, he's drowned. Go look after him.          131

FESTE He is but mad yet, madonna, and the fool shall look to the madman.          *Exit*

*Enter Malvolio*

MALVOLIO Madam, yon young fellow swears he will speak with you. I told him you were sick—he takes on him to understand so much, and therefore comes to speak with you. I told him you were asleep—he seems to have a foreknowledge of that too, and therefore comes to speak with you. What is to be said to him, lady? He's fortified against any denial.          140

OLIVIA Tell him he shall not speak with me.

MALVOLIO He's been told so, and he says he'll stand at your door like a sheriff's post, and be the supporter to a bench, but he'll speak with you.

OLIVIA What kind o' man is he?          145

MALVOLIO Why, of mankind.

OLIVIA What manner of man?

MALVOLIO Of very ill manner: he'll speak with you, will you or no.

OLIVIA Of what personage and years is he?          150

MALVOLIO Not yet old enough for a man, nor young enough for a boy; as a squash is before 'tis a peascod, or a codling when 'tis almost an apple. 'Tis with him in standing water between boy and man. He is very well-favoured, and he speaks very shrewishly. One would think his mother's milk were scarce out of him.

OLIVIA
Let him approach. Call in my gentlewoman.          157

MALVOLIO Gentlewoman, my lady calls.          *Exit*

*Enter Maria*

OLIVIA
Give me my veil. Come, throw it o'er my face.
We'll once more hear Orsino's embassy.          160

*Enter Viola as Cesario*

VIOLA The honourable lady of the house, which is she?

OLIVIA Speak to me, I shall answer for her. Your will.

VIOLA Most radiant, exquisite, and unmatchable beauty. —I pray you, tell me if this be the lady of the house, for I never saw her. I would be loath to cast away my speech, for besides that it is excellently well penned, I have taken great pains to con it. Good beauties, let me sustain no scorn; I am very 'countable, even to the least sinister usage.

OLIVIA Whence came you, sir?          170

VIOLA I can say little more than I have studied, and that question's out of my part. Good gentle one, give me modest assurance if you be the lady of the house, that I may proceed in my speech.

OLIVIA Are you a comedian?     175

VIOLA No, my profound heart; and yet—by the very fangs of malice I swear—I am not that I play. Are you the lady of the house?

OLIVIA If I do not usurp myself, I am.

VIOLA Most certain if you are she you do usurp yourself, for what is yours to bestow is not yours to reserve. But this is from my commission. I will on with my speech in your praise, and then show you the heart of my message.     184

OLIVIA Come to what is important in't, I forgive you the praise.

VIOLA Alas, I took great pains to study it, and 'tis poetical.

OLIVIA It is the more like to be feigned, I pray you keep it in. I heard you were saucy at my gates, and allowed your approach rather to wonder at you than to hear you. If you be not mad, be gone. If you have reason, be brief. 'Tis not that time of moon with me to make one in so skipping a dialogue.

MARIA Will you hoist sail, sir? Here lies your way.     194

VIOLA No, good swabber, I am to hull here a little longer. (To Olivia) Some mollification for your giant, sweet lady. Tell me your mind, I am a messenger.

OLIVIA Sure, you have some hideous matter to deliver when the courtesy of it is so fearful. Speak your office.

VIOLA It alone concerns your ear. I bring no overture of war, no taxation of homage. I hold the olive in my hand. My words are as full of peace as matter.     202

OLIVIA Yet you began rudely. What are you? What would you?

VIOLA The rudeness that hath appeared in me have I learned from my entertainment. What I am and what I would are as secret as maidenhead; to your ears, divinity; to any others', profanation.

OLIVIA (to Maria ⌈and attendants⌉) Give us the place alone, we will hear this divinity.     210

*Exeunt Maria ⌈and attendants⌉*

Now sir, what is your text?

VIOLA Most sweet lady—

OLIVIA A comfortable doctrine, and much may be said of it. Where lies your text?

VIOLA In Orsino's bosom.     215

OLIVIA In his bosom? In what chapter of his bosom?

VIOLA To answer by the method, in the first of his heart.

OLIVIA O, I have read it. It is heresy. Have you no more to say?

VIOLA Good madam, let me see your face.     220

OLIVIA Have you any commission from your lord to negotiate with my face? You are now out of your text. But we will draw the curtain and show you the picture.

*She unveils*

Look you, sir, such a one I was this present. Is't not well done?     225

VIOLA Excellently done, if God did all.

OLIVIA 'Tis in grain, sir, 'twill endure wind and weather.

VIOLA

'Tis beauty truly blent, whose red and white
Nature's own sweet and cunning hand laid on.
Lady, you are the cruell'st she alive     230
If you will lead these graces to the grave
And leave the world no copy.

OLIVIA O sir, I will not be so hard-hearted. I will give out divers schedules of my beauty. It shall be inventoried and every particle and utensil labelled to my will, as, *item*, two lips, indifferent red; *item*, two grey eyes, with lids to them; *item*, one neck, one chin, and so forth. Were you sent hither to praise me?     238

VIOLA

I see you what you are, you are too proud,
But if you were the devil, you are fair.     240
My lord and master loves you. O, such love
Could be but recompensed though you were crowned
The nonpareil of beauty.

OLIVIA     How does he love me?

VIOLA

With adorations, fertile tears,
With groans that thunder love, with sighs of fire.     245

OLIVIA

Your lord does know my mind, I cannot love him.
Yet I suppose him virtuous, know him noble,
Of great estate, of fresh and stainless youth,
In voices well divulged, free, learned, and valiant,
And in dimension and the shape of nature     250
A gracious person; but yet I cannot love him.
He might have took his answer long ago.

VIOLA

If I did love you in my master's flame,
With such a suff'ring, such a deadly life,
In your denial I would find no sense,     255
I would not understand it.

OLIVIA     Why, what would you?

VIOLA

Make me a willow cabin at your gate
And call upon my soul within the house,
Write loyal cantons of contemnèd love,
And sing them loud even in the dead of night;     260
Halloo your name to the reverberate hills,
And make the babbling gossip of the air
Cry out 'Olivia!' O, you should not rest
Between the elements of air and earth
But you should pity me.     265

OLIVIA You might do much.
What is your parentage?

VIOLA

Above my fortunes, yet my state is well.
I am a gentleman.

OLIVIA     Get you to your lord.
I cannot love him. Let him send no more,     270
Unless, perchance, you come to me again
To tell me how he takes it. Fare you well.
I thank you for your pains. (*Offering a purse*) Spend this for me.

VIOLA

I am no fee'd post, lady. Keep your purse.
My master, not myself, lacks recompense.     275
Love make his heart of flint that you shall love,
And let your fervour, like my master's, be
Placed in contempt. Farewell, fair cruelty.     *Exit*

OLIVIA 'What is your parentage?'
'Above my fortunes, yet my state is well.     280
I am a gentleman.' I'll be sworn thou art.
Thy tongue, thy face, thy limbs, actions, and spirit
Do give thee five-fold blazon. Not too fast. Soft, soft—
Unless the master were the man. How now?
Even so quickly may one catch the plague?     285
Methinks I feel this youth's perfections
With an invisible and subtle stealth
To creep in at mine eyes. Well, let it be.
What ho, Malvolio.

*Enter Malvolio*

MALVOLIO     Here, madam, at your service.

OLIVIA

Run after that same peevish messenger     290
The County's man. He left this ring behind him,

Would I or not. Tell him I'll none of it.
Desire him not to flatter with his lord,
Nor hold him up with hopes. I am not for him.
If that the youth will come this way tomorrow,　295
I'll give him reasons for't. Hie thee, Malvolio.
MALVOLIO Madam, I will.　　　　　　　*Exit at one door*
OLIVIA
I do I know not what, and fear to find
Mine eye too great a flatterer for my mind.
Fate, show thy force. Ourselves we do not owe.　300
What is decreed must be; and be this so.
　　　　　　　　　　　　　*Exit at another door*

**2.1**　*Enter Antonio and Sebastian*

ANTONIO Will you stay no longer, nor will you not that
I go with you?
SEBASTIAN By your patience, no. My stars shine darkly
over me. The malignancy of my fate might perhaps
distemper yours, therefore I shall crave of you your
leave that I may bear my evils alone. It were a bad
recompense for your love to lay any of them on you.
ANTONIO Let me yet know of you whither you are bound.
SEBASTIAN No, sooth, sir. My determinate voyage is mere
extravagancy. But I perceive in you so excellent a touch
of modesty that you will not extort from me what I am
willing to keep in. Therefore it charges me in manners
the rather to express myself. You must know of me
then, Antonio, my name is Sebastian, which I called
Roderigo. My father was that Sebastian of Messaline
whom I know you have heard of. He left behind him
myself and a sister, both born in an hour. If the heavens
had been pleased, would we had so ended. But you,
sir, altered that, for some hour before you took me
from the breach of the sea was my sister drowned.　20
ANTONIO Alas the day!
SEBASTIAN A lady, sir, though it was said she much
resembled me, was yet of many accounted beautiful.
But though I could not with such estimable wonder
over-far believe that, yet thus far I will boldly publish
her: she bore a mind that envy could not but call fair.
She is drowned already, sir, with salt water, though I
seem to drown her remembrance again with more.
ANTONIO Pardon me, sir, your bad entertainment.
SEBASTIAN O good Antonio, forgive me your trouble.　30
ANTONIO If you will not murder me for my love, let me
be your servant.
SEBASTIAN If you will not undo what you have done—
that is, kill him whom you have recovered—desire it
not. Fare ye well at once. My bosom is full of kindness,
and I am yet so near the manners of my mother that
upon the least occasion more mine eyes will tell tales
of me. I am bound to the Count Orsino's court. Farewell.
　　　　　　　　　　　　　　　⌈*Exit*⌉
ANTONIO
The gentleness of all the gods go with thee!
I have many enemies in Orsino's court,　　　　40
Else would I very shortly see thee there.
But come what may, I do adore thee so
That danger shall seem sport, and I will go.　*Exit*

**2.2**　*Enter Viola as Cesario, and Malvolio, at several
doors*

MALVOLIO Were not you ev'n now with the Countess
Olivia?
VIOLA Even now, sir, on a moderate pace, I have since
arrived but hither.

MALVOLIO (*offering a ring*) She returns this ring to you,
sir. You might have saved me my pains to have taken
it away yourself. She adds, moreover, that you should
put your lord into a desperate assurance she will none
of him. And one thing more: that you be never so
hardy to come again in his affairs, unless it be to report
your lord's taking of this. Receive it so.　　　　11
VIOLA
She took the ring of me. I'll none of it.
MALVOLIO Come, sir, you peevishly threw it to her, and
her will is it should be so returned.
　　　*He throws the ring down*
If it be worth stooping for, there it lies, in your eye; if
not, be it his that finds it.　　　　　　　　*Exit*
VIOLA (*picking up the ring*)
I left no ring with her. What means this lady?　17
Fortune forbid my outside have not charmed her.
She made good view of me, indeed so much
That straight methought her eyes had lost her tongue,
For she did speak in starts, distractedly.　　　21
She loves me, sure. The cunning of her passion
Invites me in this churlish messenger.
None of my lord's ring! Why, he sent her none.
I am the man. If it be so—as 'tis—　　　　　25
Poor lady, she were better love a dream!
Disguise, I see thou art a wickedness
Wherein the pregnant enemy does much.
How easy is it for the proper false
In women's waxen hearts to set their forms!　　30
Alas, our frailty is the cause, not we,
For such as we are made of, such we be.
How will this fadge? My master loves her dearly,
And I, poor monster, fond as much on him,
And she, mistaken, seems to dote on me.　　　35
What will become of this? As I am man,
My state is desperate for my master's love.
As I am woman, now, alas the day,
What thriftless sighs shall poor Olivia breathe!
O time, thou must untangle this, not I.　　　40
It is too hard a knot for me t'untie.　　　　*Exit*

**2.3**　*Enter Sir Toby and Sir Andrew*

SIR TOBY Approach, Sir Andrew. Not to be abed after
midnight is to be up betimes, and *diliculo surgere*, thou
knowest.
SIR ANDREW Nay, by my troth, I know not; but I know
to be up late is to be up late.　　　　　　　5
SIR TOBY A false conclusion. I hate it as an unfilled can.
To be up after midnight and to go to bed then is early;
so that to go to bed after midnight is to go to bed
betimes. Does not our lives consist of the four elements?
SIR ANDREW Faith, so they say, but I think it rather
consists of eating and drinking.　　　　　　11
SIR TOBY Thou'rt a scholar; let us therefore eat and drink.
Marian, I say, a stoup of wine.
　　　*Enter Feste, the clown*
SIR ANDREW Here comes the fool, i'faith.
FESTE How now, my hearts. Did you never see the picture
of 'we three'?　　　　　　　　　　　　16
SIR TOBY Welcome, ass. Now let's have a catch.
SIR ANDREW By my troth, the fool has an excellent breast.
I had rather than forty shillings I had such a leg, and
so sweet a breath to sing, as the fool has. In sooth,
thou wast in very gracious fooling last night, when
thou spokest of Pigrogromitus, of the Vapians passing

the equinoctial of Queubus. 'Twas very good, i'faith. I sent thee sixpence for thy leman. Hadst it?    24

FESTE I did impeticos thy gratility; for Malvolio's nose is no whipstock. My lady has a white hand, and the Myrmidons are no bottle-ale houses.

SIR ANDREW Excellent! Why, this is the best fooling, when all is done. Now a song.    29

SIR TOBY (to Feste) Come on, there is sixpence for you. Let's have a song.

SIR ANDREW (to Feste) There's a testril of me, too. If one knight give a—

FESTE Would you have a love-song, or a song of good life?    35

SIR TOBY A love song, a love-song.

SIR ANDREW Ay, ay. I care not for good life.

FESTE (sings)
    O mistress mine, where are you roaming?
    O stay and hear, your true love's coming,
      That can sing both high and low.    40
    Trip no further, pretty sweeting.
    Journeys end in lovers meeting,
      Every wise man's son doth know.

SIR ANDREW Excellent good, i'faith.

SIR TOBY Good, good.    45

FESTE    What is love? 'Tis not hereafter,
    Present mirth hath present laughter.
      What's to come is still unsure.
    In delay there lies no plenty,
    Then come kiss me, sweet and twenty.    50
      Youth's a stuff will not endure.

SIR ANDREW A mellifluous voice, as I am true knight.

SIR TOBY A contagious breath.

SIR ANDREW Very sweet and contagious, i'faith.

SIR TOBY To hear by the nose, it is dulcet in contagion. But shall we make the welkin dance indeed? Shall we rouse the night-owl in a catch that will draw three souls out of one weaver? Shall we do that?

SIR ANDREW An you love me, let's do't. I am dog at a catch.    60

FESTE By'r Lady, sir, and some dogs will catch well.

SIR ANDREW Most certain. Let our catch be 'Thou knave'.

FESTE 'Hold thy peace, thou knave', knight. I shall be constrained in't to call thee knave, knight.

SIR ANDREW 'Tis not the first time I have constrained one to call me knave. Begin, fool. It begins 'Hold thy peace'.

FESTE I shall never begin if I hold my peace.    67

SIR ANDREW Good, i'faith. Come, begin.
    They sing the catch.
    Enter Maria

MARIA What a caterwauling do you keep here! If my lady have not called up her steward Malvolio and bid him turn you out of doors, never trust me.    71

SIR TOBY My lady's a Cathayan, we are politicians, Malvolio's a Peg-o'-Ramsey, and 'Three merry men be we'. Am not I consanguineous? Am I not of her blood? Tilly-vally—'lady'! 'There dwelt a man in Babylon, lady, lady.'    76

FESTE Beshrew me, the knight's in admirable fooling.

SIR ANDREW Ay, he does well enough if he be disposed, and so do I, too. He does it with a better grace, but I do it more natural.    80

SIR TOBY
    'O' the twelfth day of December'—

MARIA For the love o' God, peace.

*Enter Malvolio*

MALVOLIO My masters, are you mad? Or what are you? Have you no wit, manners, nor honesty, but to gabble like tinkers at this time of night? Do ye make an alehouse of my lady's house, that ye squeak out your coziers' catches without any mitigation or remorse of voice? Is there no respect of place, persons, nor time in you?    89

SIR TOBY We did keep time, sir, in our catches. Sneck up!

MALVOLIO Sir Toby, I must be round with you. My lady bade me tell you that though she harbours you as her kinsman she's nothing allied to your disorders. If you can separate yourself and your misdemeanours you are welcome to the house. If not, an it would please you to take leave of her she is very willing to bid you farewell.

SIR TOBY
    'Farewell, dear heart, since I must needs be gone.'

MARIA Nay, good Sir Toby.

FESTE
    'His eyes do show his days are almost done.'    100

MALVOLIO Is't even so?

SIR TOBY
    'But I will never die.'

FESTE
    'Sir Toby, there you lie.'

MALVOLIO This is much credit to you.

SIR TOBY
    'Shall I bid him go?'    105

FESTE
    'What an if you do?'

SIR TOBY
    'Shall I bid him go, and spare not?'

FESTE
    'O no, no, no, no, you dare not.'

SIR TOBY Out o' tune, sir, ye lie. (*To Malvolio*) Art any more than a steward? Dost thou think because thou art virtuous there shall be no more cakes and ale?   111

FESTE Yes, by Saint Anne, and ginger shall be hot i'th' mouth, too.

SIR TOBY Thou'rt i'th' right. (*To Malvolio*) Go, sir, rub your chain with crumbs. (*To Maria*) A stoup of wine, Maria.    116

MALVOLIO Mistress Mary, if you prized my lady's favour at anything more than contempt you would not give means for this uncivil rule. She shall know of it, by this hand.    *Exit*

MARIA Go shake your ears.    121

SIR ANDREW 'Twere as good a deed as to drink when a man's a-hungry to challenge him the field and then to break promise with him, and make a fool of him.

SIR TOBY Do't, knight. I'll write thee a challenge, or I'll deliver thy indignation to him by word of mouth.   126

MARIA Sweet Sir Toby, be patient for tonight. Since the youth of the Count's was today with my lady she is much out of quiet. For Monsieur Malvolio, let me alone with him. If I do not gull him into a nayword and make him a common recreation, do not think I have wit enough to lie straight in my bed. I know I can do it.    133

SIR TOBY Possess us, possess us, tell us something of him.

MARIA Marry, sir, sometimes he is a kind of puritan.

SIR ANDREW O, if I thought that I'd beat him like a dog.

SIR TOBY What, for being a puritan? Thy exquisite reason, dear knight.

SIR ANDREW I have no exquisite reason for't, but I have reason good enough.    140

MARIA The dev'l a puritan that he is, or anything
constantly but a time-pleaser, an affectioned ass that
cons state without book and utters it by great swathes;
the best persuaded of himself, so crammed, as he thinks,
with excellencies, that it is his grounds of faith that all
that look on him love him; and on that vice in him
will my revenge find notable cause to work.     147

SIR TOBY What wilt thou do?

MARIA I will drop in his way some obscure epistles of
love, wherein by the colour of his beard, the shape of
his leg, the manner of his gait, the expressure of his
eye, forehead, and complexion, he shall find himself
most feelingly personated. I can write very like my lady
your niece; on a forgotten matter we can hardly make
distinction of our hands.     155

SIR TOBY Excellent, I smell a device.

SIR ANDREW I have't in my nose too.

SIR TOBY He shall think by the letters that thou wilt drop
that they come from my niece, and that she's in love
with him.     160

MARIA My purpose is indeed a horse of that colour.

SIR ANDREW And your horse now would make him an
ass.

MARIA Ass I doubt not.

SIR ANDREW O, 'twill be admirable.     165

MARIA Sport royal, I warrant you. I know my physic will
work with him. I will plant you two—and let the fool
make a third—where he shall find the letter. Observe
his construction of it. For this night, to bed, and dream
on the event. Farewell.     *Exit*     171

SIR TOBY Good night, Penthesilea.

SIR ANDREW Before me, she's a good wench.

SIR TOBY She's a beagle true bred, and one that adores
me. What o' that?

SIR ANDREW I was adored once, too.     175

SIR TOBY Let's to bed, knight. Thou hadst need send for
more money.

SIR ANDREW If I cannot recover your niece, I am a foul
way out.

SIR TOBY Send for money, knight. If thou hast her not
i'th' end, call me cut.     181

SIR ANDREW If I do not, never trust me, take it how you
will.

SIR TOBY Come, come, I'll go burn some sack, 'tis too late
to go to bed now. Come knight, come knight.     *Exeunt*

**2.4**     *Enter the Duke, Viola as Cesario, Curio, and others*

ORSINO
Give me some music. Now good morrow, friends.
Now good Cesario, but that piece of song,
That old and antic song we heard last night.
Methought it did relieve my passion much,
More than light airs and recollected terms     5
Of these most brisk and giddy-pacèd times.
Come, but one verse.

CURIO He is not here, so please your lordship, that should
sing it.

ORSINO Who was it?     10

CURIO Feste the jester, my lord, a fool that the lady Olivia's
father took much delight in. He is about the house.

ORSINO
Seek him out, and play the tune the while.     *Exit Curio*
     *Music plays*
(*To Viola*) Come hither, boy. If ever thou shalt love,
In the sweet pangs of it remember me;     15

For such as I am, all true lovers are,
Unstaid and skittish in all motions else
Save in the constant image of the creature
That is beloved. How dost thou like this tune?

VIOLA
It gives a very echo to the seat     20
Where love is throned.

ORSINO                    Thou dost speak masterly.
My life upon't, young though thou art thine eye
Hath stayed upon some favour that it loves.
Hath it not, boy?

VIOLA                    A little, by your favour.

ORSINO
What kind of woman is't?

VIOLA                    Of your complexion.     25

ORSINO
She is not worth thee then. What years, i'faith?

VIOLA About your years, my lord.

ORSINO
Too old, by heaven. Let still the woman take
An elder than herself. So wears she to him;
So sways she level in her husband's heart.     30
For, boy, however we do praise ourselves,
Our fancies are more giddy and unfirm,
More longing, wavering, sooner lost and worn,
Than women's are.

VIOLA                    I think it well, my lord.

ORSINO
Then let thy love be younger than thyself,     35
Or thy affection cannot hold the bent;
For women are as roses, whose fair flower
Being once displayed, doth fall that very hour.

VIOLA
And so they are. Alas that they are so:
To die even when they to perfection grow.     40
     *Enter Curio and Feste the clown*

ORSINO (*to Feste*)
O fellow, come, the song we had last night.
Mark it, Cesario, it is old and plain.
The spinsters, and the knitters in the sun,
And the free maids that weave their thread with
     bones,
Do use to chant it. It is silly sooth,     45
And dallies with the innocence of love,
Like the old age.

FESTE Are you ready, sir?

ORSINO I prithee, sing.
     *Music*

FESTE (*sings*)
Come away, come away death,     50
     And in sad cypress let me be laid.
Fie away, fie away breath,
     I am slain by a fair cruel maid.
My shroud of white, stuck all with yew,
     O prepare it.     55
My part of death no one so true
     Did share it.
Not a flower, not a flower sweet
     On my black coffin let there be strewn.
Not a friend, not a friend greet     60
     My poor corpse, where my bones shall be thrown.
A thousand thousand sighs to save,
     Lay me O where
Sad true lover never find my grave,
     To weep there.     65

DUKE (*giving money*) There's for thy pains.
FESTE No pains, sir. I take pleasure in singing, sir.
ORSINO I'll pay thy pleasure then.
FESTE Truly, sir, and pleasure will be paid, one time or
  another.                                                    70
ORSINO Give me now leave to leave thee.
FESTE Now the melancholy god protect thee, and the
  tailor make thy doublet of changeable taffeta, for thy
  mind is a very opal. I would have men of such constancy
  put to sea, that their business might be everything,
  and their intent everywhere, for that's it that always
  makes a good voyage of nothing. Farewell.      *Exit*
ORSINO
  Let all the rest give place:      *Exeunt Curio and others*
                           Once more, Cesario,
  Get thee to yon same sovereign cruelty.
  Tell her my love, more noble than the world,        80
  Prizes not quantity of dirty lands.
  The parts that fortune hath bestowed upon her
  Tell her I hold as giddily as fortune;
  But 'tis that miracle and queen of gems
  That nature pranks her in attracts my soul.        85
VIOLA
  But if she cannot love you, sir?
ORSINO
  I cannot be so answered.
VIOLA                    Sooth, but you must.
  Say that some lady, as perhaps there is,
  Hath for your love as great a pang of heart
  As you have for Olivia. You cannot love her.        90
  You tell her so. Must she not then be answered?
ORSINO There is no woman's sides
  Can bide the beating of so strong a passion
  As love doth give my heart; no woman's heart
  So big, to hold so much. They lack retention.        95
  Alas, their love may be called appetite,
  No motion of the liver, but the palate,
  That suffer surfeit, cloyment, and revolt.
  But mine is all as hungry as the sea,
  And can digest as much. Make no compare        100
  Between that love a woman can bear me
  And that I owe Olivia.
VIOLA Ay, but I know—
ORSINO What dost thou know?
VIOLA
  Too well what love women to men may owe.        105
  In faith, they are as true of heart as we.
  My father had a daughter loved a man
  As it might be, perhaps, were I a woman
  I should your lordship.
ORSINO                  And what's her history?
VIOLA
  A blank, my lord. She never told her love,        110
  But let concealment, like a worm i'th' bud,
  Feed on her damask cheek. She pined in thought,
  And with a green and yellow melancholy
  She sat like patience on a monument,
  Smiling at grief. Was not this love indeed?        115
  We men may say more, swear more, but indeed
  Our shows are more than will; for still we prove
  Much in our vows, but little in our love.
ORSINO
  But died thy sister of her love, my boy?
VIOLA
  I am all the daughters of my father's house,        120

And all the brothers too; and yet I know not.
Sir, shall I to this lady?
ORSINO                Ay, that's the theme,
  To her in haste. Give her this jewel. Say
  My love can give no place, bide no denay.
                               *Exeunt severally*

**2.5**    *Enter Sir Toby, Sir Andrew, and Fabian*
SIR TOBY Come thy ways, Signor Fabian.
FABIAN Nay, I'll come. If I lose a scruple of this sport let
  me be boiled to death with melancholy.
SIR TOBY Wouldst thou not be glad to have the niggardly
  rascally sheep-biter come by some notable shame?    5
FABIAN I would exult, man. You know he brought me
  out o' favour with my lady about a bear-baiting here.
SIR TOBY To anger him we'll have the bear again, and
  we will fool him black and blue, shall we not, Sir
  Andrew?                                              10
SIR ANDREW An we do not, it is pity of our lives.
      *Enter Maria with a letter*
SIR TOBY Here comes the little villain. How now, my metal
  of India?
MARIA Get ye all three into the box-tree. Malvolio's
  coming down this walk. He has been yonder i' the sun
  practising behaviour to his own shadow this half-hour.
  Observe him, for the love of mockery, for I know this
  letter will make a contemplative idiot of him. Close, in
  the name of jesting!                                19
      *The men hide. Maria places the letter*
  Lie thou there, for here comes the trout that must be
  caught with tickling.                          *Exit*
      *Enter Malvolio*
MALVOLIO 'Tis but fortune, all is fortune. Maria once told
  me she did affect me, and I have heard herself come
  thus near, that should she fancy it should be one of
  my complexion. Besides, she uses me with a more
  exalted respect than anyone else that follows her. What
  should I think on't?                                27
SIR TOBY Here's an overweening rogue.
FABIAN O, peace! Contemplation makes a rare turkeycock
  of him—how he jets under his advanced plumes!    30
SIR ANDREW 'Slight, I could so beat the rogue.
SIR TOBY Peace, I say.
MALVOLIO To be Count Malvolio!
SIR TOBY Ah, rogue.
SIR ANDREW Pistol him, pistol him.                  35
SIR TOBY Peace, peace.
MALVOLIO There's example for't: the Lady of the Strachey
  married the yeoman of the wardrobe.
SIR ANDREW Fie on him, Jezebel.
FABIAN O peace, now he's deeply in. Look how imagina-
  tion blows him.                                    41
MALVOLIO Having been three months married to her,
  sitting in my state—
SIR TOBY O for a stone-bow to hit him in the eye!
MALVOLIO Calling my officers about me, in my branched
  velvet gown, having come from a day-bed where I have
  left Olivia sleeping—
SIR TOBY Fire and brimstone!
FABIAN O peace, now he's deeply in.                  49
MALVOLIO And then to have the humour of state and—
  after a demure travel of regard, telling them I know
  my place, as I would they should do theirs—to ask for
  my kinsman Toby.
SIR TOBY Bolts and shackles!

FABIAN O peace, peace, peace, now, now.                                55

MALVOLIO Seven of my people with an obedient start make out for him. I frown the while, and perchance wind up my watch, or play with my—(*touching his chain*) some rich jewel. Toby approaches; curtsies there to me.                                                                     60

SIR TOBY Shall this fellow live?

FABIAN Though our silence be drawn from us with cars, yet peace.

MALVOLIO I extend my hand to him thus, quenching my familiar smile with an austere regard of control—   65

SIR TOBY And does not Toby take you a blow o' the lips, then?

MALVOLIO Saying 'Cousin Toby, my fortunes, having cast me on your niece, give me this prerogative of speech'—

SIR TOBY What, what!                                                   70

MALVOLIO 'You must amend your drunkenness.'

SIR TOBY Out, scab.

FABIAN Nay, patience, or we break the sinews of our plot.

MALVOLIO 'Besides, you waste the treasure of your time with a foolish knight'—                                            75

SIR ANDREW That's me, I warrant you.

MALVOLIO 'One Sir Andrew.'

SIR ANDREW I knew 'twas I, for many do call me fool.

MALVOLIO (*seeing the letter*) What employment have we here?                                                                80

FABIAN Now is the woodcock near the gin.

SIR TOBY O peace, and the spirit of humours intimate reading aloud to him.

MALVOLIO (*taking up the letter*) By my life, this is my lady's hand. These be her very c's, her u's, and her t's, and thus makes she her great P's. It is in contempt of question her hand.                                                     87

SIR ANDREW Her c's, her u's, and her t's? Why that?

MALVOLIO (*reads*) 'To the unknown beloved, this, and my good wishes.' Her very phrases! (*Opening the letter*) By your leave, wax—soft, and the impressure her Lucrece, with which she uses to seal—'tis my lady. To whom should this be?

FABIAN This wins him, liver and all.

MALVOLIO          'Jove knows I love,                               95
                      But who?
                   Lips do not move,
                   No man must know.'

'No man must know.' What follows? The numbers altered. 'No man must know.' If this should be thee, Malvolio?                                                              101

SIR TOBY Marry, hang thee, brock.

MALVOLIO
          'I may command where I adore,
              But silence like a Lucrece knife
          With bloodless stroke my heart doth gore.   105
              M.O.A.I. doth sway my life.'

FABIAN A fustian riddle.

SIR TOBY Excellent wench, say I.

MALVOLIO 'M.O.A.I. doth sway my life.' Nay, but first let me see, let me see, let me see.                                    110

FABIAN What dish o' poison has she dressed him!

SIR TOBY And with what wing the staniel checks at it!

MALVOLIO 'I may command where I adore.' Why, she may command me. I serve her, she is my lady. Why, this is evident to any formal capacity. There is no obstruction in this. And the end—what should that alphabetical position portend? If I could make that resemble something in me. Softly—'M.O.A.I.'        118

SIR TOBY O ay, make up that, he is now at a cold scent.

FABIAN Sowter will cry upon't for all this, though it be as rank as a fox.

MALVOLIO 'M.' Malvolio—'M'—why, that begins my name.

FABIAN Did not I say he would work it out? The cur is excellent at faults.                                               125

MALVOLIO 'M.' But then there is no consonancy in the sequel. That suffers under probation. 'A' should follow, but 'O' does.

FABIAN And 'O' shall end, I hope.

SIR TOBY Ay, or I'll cudgel him, and make him cry 'O!'

MALVOLIO And then 'I' comes behind.                           131

FABIAN Ay, an you had any eye behind you you might see more detraction at your heels than fortunes before you.

MALVOLIO 'M.O.A.I.' This simulation is not as the former; and yet to crush this a little, it would bow to me, for every one of these letters are in my name. Soft, here follows prose: 'If this fall into thy hand, revolve. In my stars I am above thee, but be not afraid of greatness. Some are born great, some achieve greatness, and some have greatness thrust upon 'em. Thy fates open their hands, let thy blood and spirit embrace them, and to inure thyself to what thou art like to be, cast thy humble slough, and appear fresh. Be opposite with a kinsman, surly with servants. Let thy tongue tang arguments of state; put thyself into the trick of singularity. She thus advises thee that sighs for thee. Remember who commended thy yellow stockings, and wished to see thee ever cross-gartered. I say remember, go to, thou art made if thou desirest to be so; if not, let me see thee a steward still, the fellow of servants, and not worthy to touch Fortune's fingers. Farewell. She that would alter services with thee,                 153
                      The Fortunate-Unhappy.'
Daylight and champaign discovers not more. This is open. I will be proud, I will read politic authors, I will baffle Sir Toby, I will wash off gross acquaintance, I will be point-device the very man. I do not now fool myself, to let imagination jade me; for every reason excites to this, that my lady loves me. She did commend my yellow stockings of late, she did praise my leg, being cross-gartered, and in this she manifests herself to my love, and with a kind of injunction drives me to these habits of her liking. I thank my stars, I am happy. I will be strange, stout, in yellow stockings, and cross-gartered, even with the swiftness of putting on. Jove and my stars be praised. Here is yet a postscript. 'Thou canst not choose but know who I am. If thou entertainest my love, let it appear in thy smiling, thy smiles become thee well. Therefore in my presence still smile, dear my sweet, I prithee.' Jove, I thank thee. I will smile, I will do everything that thou wilt have me.
                                                              *Exit*

          *Sir Toby, Sir Andrew, and Fabian come from hiding*

FABIAN I will not give my part of this sport for a pension of thousands to be paid from the Sophy.

SIR TOBY I could marry this wench for this device.         175

SIR ANDREW So could I, too.

SIR TOBY And ask no other dowry with her but such another jest.

          *Enter Maria*

SIR ANDREW Nor I neither.

FABIAN Here comes my noble gull-catcher.                      180

SIR TOBY (*to Maria*) Wilt thou set thy foot o' my neck?

SIR ANDREW (*to Maria*) Or o' mine either?

SIR TOBY (*to Maria*) Shall I play my freedom at tray-trip, and become thy bondslave?

SIR ANDREW (*to Maria*) I'faith, or I either?        185

SIR TOBY (*to Maria*) Why, thou hast put him in such a dream that when the image of it leaves him, he must run mad.

MARIA Nay, but say true, does it work upon him?

SIR TOBY Like aqua vitae with a midwife.        190

MARIA If you will then see the fruits of the sport, mark his first approach before my lady. He will come to her in yellow stockings, and 'tis a colour she abhors, and cross-gartered, a fashion she detests; and he will smile upon her, which will now be so unsuitable to her disposition, being addicted to a melancholy as she is, that it cannot but turn him into a notable contempt. If you will see it, follow me.

SIR TOBY To the gates of Tartar, thou most excellent devil of wit.        200

SIR ANDREW I'll make one, too.        *Exeunt*

**3.1**   *Enter Viola as Cesario and Feste the clown, with* ⌈*pipe and*⌉ *tabor*

VIOLA Save thee, friend, and thy music. Dost thou live by thy tabor?

FESTE No, sir, I live by the church.

VIOLA Art thou a churchman?        4

FESTE No such matter, sir. I do live by the church for I do live at my house, and my house doth stand by the church.

VIOLA So thou mayst say the king lies by a beggar if a beggar dwell near him, or the church stands by thy tabor if thy tabor stand by the church.        10

FESTE You have said, sir. To see this age!—A sentence is but a cheveril glove to a good wit, how quickly the wrong side may be turned outward.

VIOLA Nay, that's certain. They that dally nicely with words may quickly make them wanton.        15

FESTE I would therefore my sister had had no name, sir.

VIOLA Why, man?

FESTE Why, sir, her name's a word, and to dally with that word might make my sister wanton. But indeed, words are very rascals since bonds disgraced them.

VIOLA Thy reason, man?        21

FESTE Troth, sir, I can yield you none without words, and words are grown so false I am loath to prove reason with them.

VIOLA I warrant thou art a merry fellow, and carest for nothing.        26

FESTE Not so, sir, I do care for something; but in my conscience, sir, I do not care for you. If that be to care for nothing, sir, I would it would make you invisible.

VIOLA Art not thou the Lady Olivia's fool?        30

FESTE No indeed, sir, the Lady Olivia has no folly, she will keep no fool, sir, till she be married, and fools are as like husbands as pilchards are to herrings—the husband's the bigger. I am indeed not her fool, but her corrupter of words.        35

VIOLA I saw thee late at the Count Orsino's.

FESTE Foolery, sir, does walk about the orb like the sun, it shines everywhere. I would be sorry, sir, but the fool should be as oft with your master as with my mistress. I think I saw your wisdom there.        40

VIOLA Nay, an thou pass upon me, I'll no more with thee. (*Giving money*) Hold, there's expenses for thee.

FESTE Now Jove in his next commodity of hair send thee a beard.        44

VIOLA By my troth I'll tell thee, I am almost sick for one, though I would not have it grow on *my* chin. Is thy lady within?

FESTE Would not a pair of these have bred, sir?

VIOLA Yes, being kept together and put to use.

FESTE I would play Lord Pandarus of Phrygia, sir, to bring a Cressida to this Troilus.        51

VIOLA (*giving money*) I understand you, sir, 'tis well begged.

FESTE The matter I hope is not great, sir; begging but a beggar—Cressida was a beggar. My lady is within, sir. I will conster to them whence you come. Who you are and what you would are out of my welkin—I might say 'element', but the word is over-worn.        *Exit*

VIOLA

This fellow is wise enough to play the fool,

And to do that well craves a kind of wit.        60

He must observe their mood on whom he jests,

The quality of persons, and the time,

And, like the haggard, check at every feather

That comes before his eye. This is a practice

As full of labour as a wise man's art,        65

For folly that he wisely shows is fit,

But wise men, folly-fall'n, quite taint their wit.

*Enter Sir Toby and Sir Andrew*

SIR TOBY Save you, gentleman.

VIOLA And you, sir.

SIR ANDREW *Dieu vous garde, monsieur.*        70

VIOLA *Et vous aussi, votre serviteur.*

SIR ANDREW I hope, sir, you are, and I am yours.

SIR TOBY Will you encounter the house? My niece is desirous you should enter if your trade be to her.

VIOLA I am bound to your niece, sir: I mean she is the list of my voyage.        76

SIR TOBY Taste your legs, sir, put them to motion.

VIOLA My legs do better understand me, sir, than I understand what you mean by bidding me taste my legs.        80

SIR TOBY I mean to go, sir, to enter.

VIOLA I will answer you with gait and entrance.

*Enter Olivia, and Maria, her gentlewoman*

But we are prevented. (*To Olivia*) Most excellent accomplished lady, the heavens rain odours on you.

SIR ANDREW (*to Sir Toby*) That youth's a rare courtier; 'rain odours'—well.        86

VIOLA My matter hath no voice, lady, but to your own most pregnant and vouchsafed ear.

SIR ANDREW (*to Sir Toby*) 'Odours', 'pregnant', and 'vouchsafed'—I'll get 'em all three all ready.        90

OLIVIA Let the garden door be shut, and leave me to my hearing.        *Exeunt Sir Toby, Sir Andrew, and Maria*

Give me your hand, sir.

VIOLA

My duty, madam, and most humble service.

OLIVIA What is your name?        95

VIOLA

Cesario is your servant's name, fair princess.

OLIVIA

My servant, sir? 'Twas never merry world

Since lowly feigning was called compliment.

You're servant to the Count Orsino, youth.

VIOLA

And he is yours, and his must needs be yours.        100

Your servant's servant is *your* servant, madam.

OLIVIA
For him, I think not on him. For his thoughts,
Would they were blanks rather than filled with me.
VIOLA
Madam, I come to whet your gentle thoughts
On his behalf.
OLIVIA          O by your leave, I pray you.          105
I bade you never speak again of him;
But would you undertake another suit,
I had rather hear you to solicit that
Than music from the spheres.
VIOLA                    Dear lady—
OLIVIA
Give me leave, beseech you. I did send,          110
After the last enchantment you did here,
A ring in chase of you. So did I abuse
Myself, my servant, and I fear me you.
Under your hard construction must I sit,
To force that on you in a shameful cunning          115
Which you knew none of yours. What might you
     think?
Have you not set mine honour at the stake
And baited it with all th'unmuzzled thoughts
That tyrannous heart can think? To one of your
     receiving
Enough is shown. A cypress, not a bosom,          120
Hides my heart. So let me hear you speak.
VIOLA
I pity you.
OLIVIA          That's a degree to love.
VIOLA
No, not a grece, for 'tis a vulgar proof
That very oft we pity enemies.
OLIVIA
Why then, methinks 'tis time to smile again.          125
O world, how apt the poor are to be proud!
If one should be a prey, how much the better
To fall before the lion than the wolf!
     *Clock strikes*
The clock upbraids me with the waste of time.
Be not afraid, good youth, I will not have you;          130
And yet when wit and youth is come to harvest
Your wife is like to reap a proper man.
There lies your way, due west.
VIOLA                    Then westward ho!
Grace and good disposition attend your ladyship.
You'll nothing, madam, to my lord by me?          135
OLIVIA
Stay. I prithee tell me what thou think'st of me.
VIOLA
That you do think you are not what you are.
OLIVIA
If I think so, I think the same of you.
VIOLA
Then think you right, I am not what I am.
OLIVIA
I would you were as I would have you be.          140
VIOLA
Would it be better, madam, than I am?
I wish it might, for now I am your fool.
OLIVIA (*aside*)
O, what a deal of scorn looks beautiful
In the contempt and anger of his lip!
A murd'rous guilt shows not itself more soon          145
Than love that would seem hid. Love's night is noon.
(*To Viola*) Cesario, by the roses of the spring,
By maidhood, honour, truth, and everything,

I love thee so that, maugre all thy pride,
Nor wit nor reason can my passion hide.          150
Do not extort thy reasons from this clause,
For that I woo, thou therefore hast no cause.
But rather reason thus with reason fetter:
Love sought is good, but given unsought, is better.
VIOLA
By innocence I swear, and by my youth,          155
I have one heart, one bosom, and one truth,
And that no woman has, nor never none
Shall mistress be of it save I alone.
And so adieu, good madam. Never more
Will I my master's tears to you deplore.          160
OLIVIA
Yet come again, for thou perhaps mayst move
That heart which now abhors, to like his love.
                              *Exeunt ⌐severally⌐*

**3.2**    *Enter Sir Toby, Sir Andrew, and Fabian*
SIR ANDREW No, faith, I'll not stay a jot longer.
SIR TOBY Thy reason, dear venom, give thy reason.
FABIAN You must needs yield your reason, Sir Andrew.
SIR ANDREW Marry, I saw your niece do more favours to
   the Count's servingman than ever she bestowed upon
   me. I saw't i'th' orchard.          6
SIR TOBY Did she see thee the while, old boy? Tell me
   that.
SIR ANDREW As plain as I see you now.
FABIAN This was a great argument of love in her toward
   you.          11
SIR ANDREW 'Slight, will you make an ass o' me?
FABIAN I will prove it legitimate, sir, upon the oaths of
   judgement and reason.
SIR TOBY And they have been grand-jurymen since before
   Noah was a sailor.          16
FABIAN She did show favour to the youth in your sight
   only to exasperate you, to awake your dormouse valour,
   to put fire in your heart and brimstone in your liver.
   You should then have accosted her, and with some
   excellent jests, fire-new from the mint, you should have
   banged the youth into dumbness. This was looked for
   at your hand, and this was balked. The double gilt of
   this opportunity you let time wash off, and you are
   now sailed into the north of my lady's opinion, where
   you will hang like an icicle on a Dutchman's beard
   unless you do redeem it by some laudable attempt
   either of valour or policy.          28
SIR ANDREW An't be any way, it must be with valour, for
   policy I hate. I had as lief be a Brownist as a politician.
SIR TOBY Why then, build me thy fortunes upon the basis
   of valour. Challenge me the Count's youth to fight with
   him, hurt him in eleven places. My niece shall take
   note of it; and assure thyself, there is no love-broker
   in the world can more prevail in man's commendation
   with woman than report of valour.          36
FABIAN There is no way but this, Sir Andrew.
SIR ANDREW Will either of you bear me a challenge to
   him?
SIR TOBY Go, write it in a martial hand, be curst and
   brief. It is no matter how witty so it be eloquent and
   full of invention. Taunt him with the licence of ink. If
   thou 'thou'st' him some thrice, it shall not be amiss,
   and as many lies as will lie in thy sheet of paper,
   although the sheet were big enough for the bed of
   Ware, in England, set 'em down, go about it. Let there
   be gall enough in thy ink; though thou write with a
   goose-pen, no matter. About it.          48
SIR ANDREW Where shall I find you?

SIR TOBY  We'll call thee at the cubiculo. Go.         50
*Exit Sir Andrew*

FABIAN  This is a dear manikin to you, Sir Toby.

SIR TOBY  I have been dear to him, lad, some two thousand strong or so.

FABIAN  We shall have a rare letter from him; but you'll not deliver't.         55

SIR TOBY  Never trust me then; and by all means stir on the youth to an answer. I think oxen and wain-ropes cannot hale them together. For Andrew, if he were opened and you find so much blood in his liver as will clog the foot of a flea, I'll eat the rest of th'anatomy.

FABIAN  And his opposite, the youth, bears in his visage no great presage of cruelty.         62

*Enter Maria*

SIR TOBY  Look where the youngest wren of nine comes.

MARIA  If you desire the spleen, and will laugh yourselves into stitches, follow me. Yon gull Malvolio is turned heathen, a very renegado, for there is no Christian that means to be saved by believing rightly can ever believe such impossible passages of grossness. He's in yellow stockings.

SIR TOBY  And cross-gartered?         70

MARIA  Most villainously, like a pedant that keeps a school i'th' church. I have dogged him like his murderer. He does obey every point of the letter that I dropped to betray him. He does smile his face into more lines than is in the new map with the augmentation of the Indies. You have not seen such a thing as 'tis. I can hardly forbear hurling things at him. I know my lady will strike him. If she do, he'll smile, and take't for a great favour.         79

SIR TOBY  Come bring us, bring us where he is.     *Exeunt*

**3.3**     *Enter Sebastian and Antonio*

SEBASTIAN
I would not by my will have troubled you,
But since you make your pleasure of your pains
I will no further chide you.

ANTONIO
I could not stay behind you. My desire,
More sharp than filèd steel, did spur me forth,         5
And not all love to see you—though so much
As might have drawn one to a longer voyage—
But jealousy what might befall your travel,
Being skilless in these parts, which to a stranger,
Unguided and unfriended, often prove         10
Rough and unhospitable. My willing love
The rather by these arguments of fear
Set forth in your pursuit.

SEBASTIAN              My kind Antonio,
I can no other answer make but thanks,
And thanks; and ever oft good turns         15
Are shuffled off with such uncurrent pay.
But were my worth as is my conscience firm,
You should find better dealing. What's to do?
Shall we go see the relics of this town?

ANTONIO
Tomorrow, sir. Best first go see your lodging.         20

SEBASTIAN
I am not weary, and 'tis long to night.
I pray you let us satisfy our eyes
With the memorials and the things of fame
That do renown this city.

ANTONIO                  Would you'd pardon me.
I do not without danger walk these streets.         25
Once in a sea-fight 'gainst the Count his galleys

I did some service, of such note indeed
That were I ta'en here it would scarce be answered.

SEBASTIAN
Belike you slew great number of his people.

ANTONIO
Th'offence is not of such a bloody nature,         30
Albeit the quality of the time and quarrel
Might well have given us bloody argument.
It might have since been answered in repaying
What we took from them, which for traffic's sake
Most of our city did. Only myself stood out,         35
For which if I be latchèd in this place
I shall pay dear.

SEBASTIAN              Do not then walk too open.

ANTONIO
It doth not fit me. Hold, sir, here's my purse.
In the south suburbs at the Elephant
Is best to lodge. I will bespeak our diet         40
Whiles you beguile the time and feed your knowledge
With viewing of the town. There shall you have me.

SEBASTIAN  Why I your purse?

ANTONIO
Haply your eye shall light upon some toy
You have desire to purchase; and your store         45
I think is not for idle markets, sir.

SEBASTIAN
I'll be your purse-bearer, and leave you
For an hour.

ANTONIO              To th' Elephant.

SEBASTIAN                          I do remember.
*Exeunt severally*

**3.4**     *Enter Olivia and Maria*

OLIVIA (*aside*)
I have sent after him, he says he'll come.
How shall I feast him? What bestow of him?
For youth is bought more oft than begged or
      borrowed.
I speak too loud.
(*To Maria*) Where's Malvolio? He is sad and civil,         5
And suits well for a servant with my fortunes.
Where is Malvolio?

MARIA  He's coming, madam, but in very strange manner.
He is sure possessed, madam.

OLIVIA
Why, what's the matter? Does he rave?         10

MARIA  No, madam, he does nothing but smile. Your ladyship were best to have some guard about you if he come, for sure the man is tainted in's wits.

OLIVIA
Go call him hither.                         *Exit Maria*
          I am as mad as he,
If sad and merry madness equal be.         15
*Enter Malvolio, cross-gartered and wearing yellow stockings, with Maria*
How now, Malvolio?

MALVOLIO  Sweet lady, ho, ho!

OLIVIA
Smil'st thou? I sent for thee upon a sad occasion.

MALVOLIO  Sad, lady? I could be sad. This does make some obstruction in the blood, this cross-gartering, but what of that? If it please the eye of one, it is with me as the very true sonnet is, 'Please one, and please all'.         22

⌈OLIVIA⌉
Why, how dost thou, man? What is the matter with thee?

MALVOLIO Not black in my mind, though yellow in my legs. It did come to his hands, and commands shall be executed. I think we do know the sweet roman hand.

OLIVIA Wilt thou go to bed, Malvolio?                                    27

MALVOLIO (*kissing his hand*) To bed? 'Ay, sweetheart, and I'll come to thee.'

OLIVIA God comfort thee. Why dost thou smile so, and kiss thy hand so oft?                                              31

MARIA How do you, Malvolio?

MALVOLIO At your request?—yes, nightingales answer daws.

MARIA Why appear you with this ridiculous boldness before my lady?                                                   36

MALVOLIO 'Be not afraid of greatness'—'twas well writ.

OLIVIA What meanest thou by that, Malvolio?

MALVOLIO 'Some are born great'—

OLIVIA Ha?                                                                    40

MALVOLIO 'Some achieve greatness'—

OLIVIA What sayst thou?

MALVOLIO 'And some have greatness thrust upon them.'

OLIVIA Heaven restore thee.

MALVOLIO 'Remember who commended thy yellow stockings'—                                                          46

OLIVIA 'Thy yellow stockings'?

MALVOLIO 'And wished to see thee cross-gartered.'

OLIVIA 'Cross-gartered'?

MALVOLIO 'Go to, thou art made, if thou desirest to be so.'                                                                     51

OLIVIA Am I made?

MALVOLIO 'If not, let me see thee a servant still.'

OLIVIA Why, this is very midsummer madness.

*Enter a Servant*

SERVANT Madam, the young gentleman of the Count Orsino's is returned. I could hardly entreat him back. He attends your ladyship's pleasure.                         57

OLIVIA I'll come to him.                              *Exit Servant*
Good Maria, let this fellow be looked to. Where's my cousin Toby? Let some of my people have a special care of him, I would not have him miscarry for the half of my dowry.       *Exeunt Olivia and Maria, severally*

MALVOLIO O ho, do you come near me now? No worse man than Sir Toby to look to me. This concurs directly with the letter, she sends him on purpose, that I may appear stubborn to him, for she incites me to that in the letter. 'Cast thy humble slough,' says she, 'be opposite with a kinsman, surly with servants, let thy tongue tang arguments of state, put thyself into the trick of singularity', and consequently sets down the manner how, as a sad face, a reverend carriage, a slow tongue, in the habit of some sir of note, and so forth. I have limed her, but it is Jove's doing, and Jove make me thankful. And when she went away now, 'let this fellow be looked to'. Fellow!—not 'Malvolio', nor after my degree, but 'fellow'. Why, everything adheres together that no dram of a scruple, no scruple of a scruple, no obstacle, no incredulous or unsafe circumstance—what can be said?—nothing that can be can come between me and the full prospect of my hopes. Well, Jove, not I, is the doer of this, and he is to be thanked.                                                         82

*Enter Sir Toby, Fabian, and Maria*

SIR TOBY Which way is he, in the name of sanctity? If all the devils of hell be drawn in little, and Legion himself possessed him, yet I'll speak to him.                       85

FABIAN Here he is, here he is. (*To Malvolio*) How is't with you, sir? How is't with you, man?

MALVOLIO Go off, I discard you. Let me enjoy my private. Go off.                                                              89

MARIA Lo, how hollow the fiend speaks within him. Did not I tell you? Sir Toby, my lady prays you to have a care of him.

MALVOLIO Aha, does she so?                                      93

SIR TOBY Go to, go to. Peace, peace, we must deal gently with him. Let me alone. How do you, Malvolio? How is't with you? What, man, defy the devil. Consider, he's an enemy to mankind.                                    97

MALVOLIO Do you know what you say?

MARIA La you, an you speak ill of the devil, how he takes it at heart. Pray God he be not bewitched.          100

FABIAN Carry his water to th' wise woman.

MARIA Marry, and it shall be done tomorrow morning, if I live. My lady would not lose him for more than I'll say.

MALVOLIO How now, mistress?                               105

MARIA O Lord!

SIR TOBY Prithee hold thy peace, this is not the way. Do you not see you move him? Let me alone with him.

FABIAN No way but gentleness, gently, gently. The fiend is rough, and will not be roughly used.             110

SIR TOBY Why how now, my bawcock? How dost thou, chuck?

MALVOLIO Sir!

SIR TOBY Ay, biddy, come with me. What man, 'tis not for gravity to play at cherry-pit with Satan. Hang him, foul collier.                                                      116

MARIA Get him to say his prayers. Good Sir Toby, get him to pray.

MALVOLIO My prayers, minx?

MARIA No, I warrant you, he will not hear of godliness.

MALVOLIO Go hang yourselves, all. You are idle shallow things, I am not of your element. You shall know more hereafter.                                                     *Exit*

SIR TOBY Is't possible?                                          124

FABIAN If this were played upon a stage, now, I could condemn it as an improbable fiction.

SIR TOBY His very genius hath taken the infection of the device, man.

MARIA Nay, pursue him now, lest the device take air and taint.                                                             130

FABIAN Why, we shall make him mad indeed.

MARIA The house will be the quieter.

SIR TOBY Come, we'll have him in a dark room and bound. My niece is already in the belief that he's mad. We may carry it thus for our pleasure and his penance till our very pastime, tired out of breath, prompt us to have mercy on him, at which time we will bring the device to the bar and crown thee for a finder of madmen. But see, but see.

*Enter Sir Andrew with a paper*

FABIAN More matter for a May morning.              140

SIR ANDREW Here's the challenge, read it. I warrant there's vinegar and pepper in't.

FABIAN Is't so saucy?

SIR ANDREW Ay—is't? I warrant him. Do but read.

SIR TOBY Give me.                                               145

(*Reads*) 'Youth, whatsoever thou art, thou art but a scurvy fellow.'

FABIAN Good, and valiant.

SIR TOBY 'Wonder not, nor admire not in thy mind why I do call thee so, for I will show thee no reason for't.'

FABIAN A good note, that keeps you from the blow of the law.                                                               152

SIR TOBY 'Thou comest to the Lady Olivia, and in my sight she uses thee kindly; but thou liest in thy throat, that is not the matter I challenge thee for.'     155

FABIAN Very brief, and to exceeding good sense (*aside*) -less.

SIR TOBY 'I will waylay thee going home, where if it be thy chance to kill me'—

FABIAN Good.     160

SIR TOBY 'Thou killest me like a rogue and a villain.'

FABIAN Still you keep o'th' windy side of the law—good.

SIR TOBY 'Fare thee well, and God have mercy upon one of our souls. He may have mercy upon mine, but my hope is better, and so look to thyself.     165
Thy friend as thou usest him, and thy sworn enemy,
                                    Andrew Aguecheek.'
If this letter move him not, his legs cannot. I'll give't him.

MARIA You may have very fit occasion for't. He is now in some commerce with my lady, and will by and by depart.     172

SIR TOBY Go, Sir Andrew. Scout me for him at the corner of the orchard like a bum-baily. So soon as ever thou seest him, draw, and as thou drawest, swear horrible, for it comes to pass oft that a terrible oath, with a swaggering accent sharply twanged off, gives manhood more approbation than ever proof itself would have earned him. Away.     179

SIR ANDREW Nay, let me alone for swearing.     *Exit*

SIR TOBY Now will not I deliver his letter, for the behaviour of the young gentleman gives him out to be of good capacity and breeding. His employment between his lord and my niece confirms no less. Therefore this letter, being so excellently ignorant, will breed no terror in the youth. He will find it comes from a clodpoll. But, sir, I will deliver his challenge by word of mouth, set upon Aguecheek a notable report of valour, and drive the gentleman—as I know his youth will aptly receive it—into a most hideous opinion of his rage, skill, fury, and impetuosity. This will so fright them both that they will kill one another by the look, like cockatrices.     192

*Enter Olivia, and Viola as Cesario*

FABIAN Here he comes with your niece. Give them way till he take leave, and presently after him.

SIR TOBY I will meditate the while upon some horrid message for a challenge.     196

*Exeunt Sir Toby, Fabian, and Maria*

OLIVIA
I have said too much unto a heart of stone,
And laid mine honour too unchary out.
There's something in me that reproves my fault,
But such a headstrong potent fault it is     200
That it but mocks reproof.

VIOLA                    With the same 'haviour
That your passion bears goes on my master's griefs.

OLIVIA (*giving a jewel*)
Here, wear this jewel for me, 'tis my picture—
Refuse it not, it hath no tongue to vex you—
And I beseech you come again tomorrow.     205
What shall you ask of me that I'll deny,
That honour, saved, may upon asking give?

VIOLA
Nothing but this: your true love for my master.

OLIVIA
How with mine honour may I give him that
Which I have given to you?

VIOLA                    I will acquit you.     210

OLIVIA
Well, come again tomorrow. Fare thee well.
A fiend like thee might bear my soul to hell.     *Exit*

*Enter Sir Toby and Fabian*

SIR TOBY Gentleman, God save thee.

VIOLA And you, sir.

SIR TOBY That defence thou hast, betake thee to't. Of what nature the wrongs are thou hast done him, I know not, but thy intercepter, full of despite, bloody as the hunter, attends thee at the orchard end. Dismount thy tuck, be yare in thy preparation, for thy assailant is quick, skilful, and deadly.     220

VIOLA You mistake, sir, I am sure no man hath any quarrel to me. My remembrance is very free and clear from any image of offence done to any man.

SIR TOBY You'll find it otherwise, I assure you. Therefore, if you hold your life at any price, betake you to your guard, for your opposite hath in him what youth, strength, skill, and wrath can furnish man withal.

VIOLA I pray you, sir, what is he?     228

SIR TOBY He is knight dubbed with unhatched rapier and on carpet consideration, but he is a devil in private brawl. Souls and bodies hath he divorced three, and his incensement at this moment is so implacable that satisfaction can be none but by pangs of death and sepulchre. Hob nob is his word, give't or take't.     234

VIOLA I will return again into the house and desire some conduct of the lady. I am no fighter. I have heard of some kind of men that put quarrels purposely on others, to taste their valour. Belike this is a man of that quirk.

SIR TOBY Sir, no. His indignation derives itself out of a very competent injury, therefore get you on, and give him his desire. Back you shall not to the house unless you undertake that with me which with as much safety you might answer him. Therefore on, or strip your sword stark naked, for meddle you must, that's certain, or forswear to wear iron about you.     245

VIOLA This is as uncivil as strange. I beseech you do me this courteous office, as to know of the knight what my offence to him is. It is something of my negligence, nothing of my purpose.     249

SIR TOBY I will do so. Signor Fabian, stay you by this gentleman till my return.     *Exit*

VIOLA Pray you, sir, do you know of this matter?

FABIAN I know the knight is incensed against you even to a mortal arbitrement, but nothing of the circumstance more.     255

VIOLA I beseech you, what manner of man is he?

FABIAN Nothing of that wonderful promise to read him by his form as you are like to find him in the proof of his valour. He is indeed, sir, the most skilful, bloody, and fatal opposite that you could possibly have found in any part of Illyria. Will you walk towards him, I will make your peace with him if I can.     262

VIOLA I shall be much bound to you for't. I am one that had rather go with Sir Priest than Sir Knight—I care not who knows so much of my mettle.     ⌈*Exeunt*⌉

*Enter Sir Toby and Sir Andrew*

SIR TOBY Why, man, he's a very devil, I have not seen such a virago. I had a pass with him, rapier, scabbard, and all, and he gives me the stuck-in with such a mortal motion that it is inevitable, and on the answer, he pays you as surely as your feet hits the ground they step on. They say he has been fencer to the Sophy.

SIR ANDREW Pox on't, I'll not meddle with him.     272

SIR TOBY Ay, but he will not now be pacified, Fabian can scarce hold him yonder.

SIR ANDREW Plague on't, an I thought he had been valiant and so cunning in fence I'd have seen him damned ere I'd have challenged him. Let him let the matter slip and I'll give him my horse, grey Capulet. 278

SIR TOBY I'll make the motion. Stand here, make a good show on't—this shall end without the perdition of souls. (*Aside*) Marry, I'll ride your horse as well as I ride you.
*Enter Fabian, and Viola as Cesario*
⌈*Aside to Fabian*⌉ I have his horse to take up the quarrel, I have persuaded him the youth's a devil.

FABIAN (*aside to Sir Toby*) He is as horribly conceited of him, and pants and looks pale as if a bear were at his heels. 286

SIR TOBY (*to Viola*) There's no remedy, sir, he will fight with you for's oath' sake. Marry, he hath better bethought him of his quarrel, and he finds that now scarce to be worth talking of. Therefore draw for the supportance of his vow, he protests he will not hurt you. 292

VIOLA (*aside*) Pray God defend me. A little thing would make me tell them how much I lack of a man.

FABIAN (*to Sir Andrew*) Give ground if you see him furious.

SIR TOBY Come, Sir Andrew, there's no remedy, the gentleman will for his honour's sake have one bout with you, he cannot by the duello avoid it, but he has promised me, as he is a gentleman and a soldier, he will not hurt you. Come on, to't. 300

SIR ANDREW Pray God he keep his oath.
*Enter Antonio*

VIOLA
I do assure you 'tis against my will.
*Sir Andrew and Viola draw their swords*

ANTONIO (*drawing his sword, to Sir Andrew*)
Put up your sword. If this young gentleman
Have done offence, I take the fault on me.
If you offend him, I for him defy you. 305

SIR TOBY You, sir? Why, what are you?

ANTONIO
One, sir, that for his love dares yet do more
Than you have heard him brag to you he will.

SIR TOBY (*drawing his sword*) Nay, if you be an undertaker, I am for you. 310
*Enter Officers*

FABIAN O, good Sir Toby, hold. Here come the officers.

SIR TOBY (*to Antonio*) I'll be with you anon.

VIOLA (*to Sir Andrew*) Pray, sir, put your sword up if you please. 314

SIR ANDREW Marry will I, sir, and for that I promised you I'll be as good as my word. He will bear you easily, and reins well.
*Sir Andrew and Viola put up their swords*

FIRST OFFICER This is the man, do thy office.

SECOND OFFICER Antonio, I arrest thee at the suit of Count Orsino. 320

ANTONIO You do mistake me, sir.

FIRST OFFICER
No, sir, no jot. I know your favour well,
Though now you have no seacap on your head.
(*To Second Officer*) Take him away, he knows I know him well.

ANTONIO
I must obey. (*To Viola*) This comes with seeking you.
But there's no remedy, I shall answer it. 326

What will you do now my necessity
Makes me to ask you for my purse? It grieves me
Much more for what I cannot do for you
Than what befalls myself. You stand amazed, 330
But be of comfort.

SECOND OFFICER          Come, sir, away.

ANTONIO (*to Viola*)
I must entreat of you some of that money.

VIOLA What money, sir?
For the fair kindness you have showed me here,
And part being prompted by your present trouble, 335
Out of my lean and low ability
I'll lend you something. My having is not much.
I'll make division of my present with you.
Hold, (*offering money*) there's half my coffer.

ANTONIO                    Will you deny me now?
Is't possible that my deserts to you 340
Can lack persuasion? Do not tempt my misery,
Lest that it make me so unsound a man
As to upbraid you with those kindnesses
That I have done for you.

VIOLA               I know of none,
Nor know I you by voice, or any feature. 345
I hate ingratitude more in a man
Than lying, vainness, babbling drunkenness,
Or any taint of vice whose strong corruption
Inhabits our frail blood.

ANTONIO             O heavens themselves!

SECOND OFFICER Come, sir, I pray you go. 350

ANTONIO
Let me speak a little. This youth that you see here
I snatched one half out of the jaws of death,
Relieved him with such sanctity of love,
And to his image, which methought did promise
Most venerable worth, did I devotion. 355

FIRST OFFICER
What's that to us? The time goes by, away.

ANTONIO
But O, how vile an idol proves this god!
Thou hast, Sebastian, done good feature shame.
In nature there's no blemish but the mind.
None can be called deformed but the unkind. 360
Virtue is beauty, but the beauteous evil
Are empty trunks o'er-flourished by the devil.

FIRST OFFICER
The man grows mad, away with him. Come, come, sir.

ANTONIO Lead me on.          *Exit with Officers*

VIOLA (*aside*)
Methinks his words do from such passion fly 365
That he believes himself. So do not I.
Prove true, imagination, O prove true,
That I, dear brother, be now ta'en for you!

SIR TOBY Come hither, knight. Come hither, Fabian. We'll whisper o'er a couplet or two of most sage saws. 370
*They stand aside*

VIOLA
He named Sebastian. I my brother know
Yet living in my glass. Even such and so
In favour was my brother, and he went
Still in this fashion, colour, ornament,
For him I imitate. O, if it prove, 375
Tempests are kind, and salt waves fresh in love! *Exit*

SIR TOBY (*to Sir Andrew*) A very dishonest, paltry boy, and more a coward than a hare. His dishonesty appears in leaving his friend here in necessity, and denying him; and for his cowardship, ask Fabian. 380

FABIAN A coward, a most devout coward, religious in it.

SIR ANDREW 'Slid, I'll after him again, and beat him.

SIR TOBY Do, cuff him soundly, but never draw thy sword.

SIR ANDREW An I do not—                                    *Exit*

FABIAN Come, let's see the event.                            385

SIR TOBY I dare lay any money 'twill be nothing yet.

*Exeunt*

**4.1**    *Enter Sebastian and Feste, the clown*

FESTE Will you make me believe that I am not sent for
you?

SEBASTIAN
Go to, go to, thou art a foolish fellow,
Let me be clear of thee.                                       4

FESTE Well held out, i'faith! No, I do not know you, nor
I am not sent to you by my lady to bid you come speak
with her, nor your name is not Master Cesario, nor
this is not my nose, neither. Nothing that is so, is so.

SEBASTIAN
I prithee vent thy folly somewhere else,
Thou know'st not me.                                          10

FESTE Vent my folly! He has heard that word of some
great man, and now applies it to a fool. Vent my folly—
I am afraid this great lubber the world will prove a
cockney. I prithee now ungird thy strangeness, and tell
me what I shall 'vent' to my lady? Shall I 'vent' to her
that thou art coming?                                         16

SEBASTIAN
I prithee, foolish Greek, depart from me.
There's money for thee. If you tarry longer
I shall give worse payment.

FESTE By my troth, thou hast an open hand. These wise
men that give fools money get themselves a good report,
after fourteen years' purchase.                               22

*Enter Sir Andrew, Sir Toby, and Fabian*

SIR ANDREW (*to Sebastian*) Now, sir, have I met you again?
(*Striking him*) There's for you.

SEBASTIAN ⌈*striking Sir Andrew with his dagger*⌉
Why, there's for thee, and there, and there.                  25
Are all the people mad?

SIR TOBY (*to Sebastian, holding him back*)  Hold, sir, or I'll
throw your dagger o'er the house.

FESTE This will I tell my lady straight, I would not be in
some of your coats for twopence.                          *Exit*

SIR TOBY Come on, sir, hold.                                   31

SIR ANDREW Nay, let him alone, I'll go another way to
work with him. I'll have an action of battery against
him if there be any law in Illyria. Though I struck him
first, yet it's no matter for that.                           35

SEBASTIAN Let go thy hand.

SIR TOBY Come, sir, I will not let you go. Come, my young
soldier, put up your iron. You are well fleshed. Come
on.

SEBASTIAN (*freeing himself*)
I will be free from thee. What wouldst thou now?           40
If thou dar'st tempt me further, draw thy sword.

SIR TOBY What, what? Nay then, I must have an ounce
or two of this malapert blood from you.

*Sir Toby and Sebastian draw their swords.*
*Enter Olivia*

OLIVIA
Hold, Toby, on thy life I charge thee hold.

SIR TOBY Madam.                                               45

OLIVIA
Will it be ever thus? Ungracious wretch,
Fit for the mountains and the barbarous caves,

Where manners ne'er were preached—out of my sight!
Be not offended, dear Cesario.
(*To Sir Toby*) Rudesby, be gone.

*Exeunt Sir Toby, Sir Andrew, and Fabian*
I prithee, gentle friend,
Let thy fair wisdom, not thy passion sway                   51
In this uncivil and unjust extent
Against thy peace. Go with me to my house,
And hear thou there how many fruitless pranks
This ruffian hath botched up, that thou thereby            55
Mayst smile at this. Thou shalt not choose but go.
Do not deny. Beshrew his soul for me,
He started one poor heart of mine in thee.

SEBASTIAN
What relish is in this? How runs the stream?
Or I am mad, or else this is a dream.                       60
Let fancy still my sense in Lethe steep.
If it be thus to dream, still let me sleep.

OLIVIA
Nay, come, I prithee, would thou'dst be ruled by me.

SEBASTIAN
Madam, I will.

OLIVIA                    O, say so, and so be.        *Exeunt*

**4.2**    *Enter Maria carrying a gown and false beard, and*
*Feste, the clown*

MARIA Nay, I prithee put on this gown and this beard,
make him believe thou art Sir Topas the curate. Do it
quickly. I'll call Sir Toby the whilst.                   *Exit*

FESTE Well, I'll put it on, and I will dissemble myself in't,
and I would I were the first that ever dissembled in
such a gown.                                                   6

*He disguises himself*
I am not tall enough to become the function well, nor
lean enough to be thought a good student, but to be
said 'an honest man and a good housekeeper' goes as
fairly as to say 'a careful man and a great scholar'.
The competitors enter.                                        11

*Enter Sir Toby and Maria*

SIR TOBY Jove bless thee, Master Parson.

FESTE *Bonos dies*, Sir Toby, for, as the old hermit of
Prague, that never saw pen and ink, very wittily said
to a niece of King Gorboduc, 'That that is, is.' So I,
being Master Parson, am Master Parson; for what is
'that' but 'that', and 'is' but 'is'?                         17

SIR TOBY To him, Sir Topas.

FESTE What ho, I say, peace in this prison.

SIR TOBY The knave counterfeits well—a good knave.   20

*Malvolio within*

MALVOLIO Who calls there?

FESTE Sir Topas the curate, who comes to visit Malvolio
the lunatic.

MALVOLIO Sir Topas, Sir Topas, good Sir Topas, go to my
lady.                                                         25

FESTE Out, hyperbolical fiend, how vexest thou this man!
Talkest thou nothing but of ladies?

SIR TOBY Well said, Master Parson.

MALVOLIO Sir Topas, never was man thus wronged. Good
Sir Topas, do not think I am mad. They have laid me
here in hideous darkness.                                     31

FESTE Fie, thou dishonest Satan—I call thee by the most
modest terms, for I am one of those gentle ones that
will use the devil himself with courtesy. Sayst thou that
house is dark?                                                35

MALVOLIO As hell, Sir Topas.

FESTE Why, it hath bay windows transparent as barricadoes, and the clerestories toward the south-north are as lustrous as ebony, and yet complainest thou of obstruction? 40

MALVOLIO I am not mad, Sir Topas; I say to you this house is dark.

FESTE Madman, thou errest. I say there is no darkness but ignorance, in which thou art more puzzled than the Egyptians in their fog. 45

MALVOLIO I say this house is as dark as ignorance, though ignorance were as dark as hell; and I say there was never man thus abused. I am no more mad than you are. Make the trial of it in any constant question.

FESTE What is the opinion of Pythagoras concerning wildfowl? 51

MALVOLIO That the soul of our grandam might haply inhabit a bird.

FESTE What thinkest thou of his opinion?

MALVOLIO I think nobly of the soul, and no way approve his opinion. 56

FESTE Fare thee well. Remain thou still in darkness. Thou shalt hold th'opinion of Pythagoras ere I will allow of thy wits, and fear to kill a woodcock lest thou dispossess the soul of thy grandam. Fare thee well. 60

MALVOLIO Sir Topas, Sir Topas!

SIR TOBY My most exquisite Sir Topas.

FESTE Nay, I am for all waters.

MARIA Thou mightst have done this without thy beard and gown, he sees thee not. 65

SIR TOBY (to Feste) To him in thine own voice, and bring me word how thou findest him. I would we were well rid of this knavery. If he may be conveniently delivered, I would he were, for I am now so far in offence with my niece that I cannot pursue with any safety this sport to the upshot. [To Maria] Come by and by to my chamber.    Exit [with Maria]

FESTE (sings) 'Hey Robin, jolly Robin,
                   Tell me how thy lady does.'

MALVOLIO Fool! 75

FESTE              'My lady is unkind, pardie.'

MALVOLIO Fool!

FESTE              'Alas, why is she so?'

MALVOLIO Fool, I say!

FESTE              'She loves another.' 80
Who calls, ha?

MALVOLIO Good fool, as ever thou wilt deserve well at my hand, help me to a candle and pen, ink, and paper. As I am a gentleman, I will live to be thankful to thee for't. 85

FESTE Master Malvolio?

MALVOLIO Ay, good fool.

FESTE Alas, sir, how fell you besides your five wits?

MALVOLIO Fool, there was never man so notoriously abused. I am as well in my wits, fool, as thou art. 90

FESTE But as well? Then you are mad indeed, if you be no better in your wits than a fool.

MALVOLIO They have here propertied me, keep me in darkness, send ministers to me, asses, and do all they can to face me out of my wits. 95

FESTE Advise you what you say, the minister is here. (As Sir Topas) Malvolio, Malvolio, thy wits the heavens restore. Endeavour thyself to sleep, and leave thy vain bibble-babble.

MALVOLIO Sir Topas. 100

FESTE (as Sir Topas) Maintain no words with him, good fellow. (As himself) Who I, sir? Not I, sir. God b'wi'

you, good Sir Topas. (As Sir Topas) Marry, amen. (As himself) I will, sir, I will.

MALVOLIO Fool, fool, fool, I say. 105

FESTE Alas, sir, be patient. What say you, sir? I am shent for speaking to you.

MALVOLIO Good fool, help me to some light and some paper. I tell thee I am as well in my wits as any man in Illyria. 110

FESTE Well-a-day that you were, sir.

MALVOLIO By this hand, I am. Good fool, some ink, paper, and light, and convey what I will set down to my lady. It shall advantage thee more than ever the bearing of letter did. 115

FESTE I will help you to't. But tell me true, are you not mad indeed, or do you but counterfeit?

MALVOLIO Believe me, I am not, I tell thee true.

FESTE Nay, I'll ne'er believe a madman till I see his brains. I will fetch you light, and paper, and ink. 120

MALVOLIO Fool, I'll requite it in the highest degree. I prithee, be gone.

FESTE              I am gone, sir,
                   And anon, sir,
                        I'll be with you again, 125
                   In a trice,
                   Like to the old Vice,
                        Your need to sustain,
                   Who with dagger of lath
                   In his rage and his wrath 130
                        Cries 'Aha,' to the devil,
                   Like a mad lad,
                   'Pare thy nails, dad,
                        Adieu, goodman devil.'    Exit

**4.3**　　*Enter Sebastian*

SEBASTIAN
This is the air, that is the glorious sun.
This pearl she gave me, I do feel't and see't,
And though 'tis wonder that enwraps me thus,
Yet 'tis not madness. Where's Antonio then?
I could not find him at the Elephant, 5
Yet there he was, and there I found this credit,
That he did range the town to seek me out.
His counsel now might do me golden service,
For though my soul disputes well with my sense
That this may be some error but no madness, 10
Yet doth this accident and flood of fortune
So far exceed all instance, all discourse,
That I am ready to distrust mine eyes
And wrangle with my reason that persuades me
To any other trust but that I am mad, 15
Or else the lady's mad. Yet if 'twere so
She could not sway her house, command her
                   followers,
Take and give back affairs and their dispatch
With such a smooth, discreet, and stable bearing
As I perceive she does. There's something in't 20
That is deceivable. But here the lady comes.
   *Enter Olivia and a Priest*
OLIVIA
Blame not this haste of mine. If you mean well
Now go with me, and with this holy man,
Into the chantry by. There before him,
And underneath that consecrated roof, 25
Plight me the full assurance of your faith,
That my most jealous and too doubtful soul

May live at peace. He shall conceal it
Whiles you are willing it shall come to note,
What time we will our celebration keep          30
According to my birth. What do you say?

SEBASTIAN
I'll follow this good man, and go with you,
And having sworn truth, ever will be true.

OLIVIA
Then lead the way, good father, and heavens so shine
That they may fairly note this act of mine.          *Exeunt*

**5.1**    *Enter Feste the clown and Fabian*
FABIAN Now, as thou lovest me, let me see his letter.
FESTE Good Master Fabian, grant me another request.
FABIAN Anything.
FESTE Do not desire to see this letter.
FABIAN This is to give a dog, and in recompense desire
my dog again.          6
    *Enter the Duke, Viola as Cesario, Curio, and lords*
ORSINO
Belong you to the Lady Olivia, friends?
FESTE Ay, sir, we are some of her trappings.
ORSINO
I know thee well. How dost thou, my good fellow?
FESTE Truly, sir, the better for my foes and the worse for
my friends.          11
ORSINO
Just the contrary—the better for thy friends.
FESTE No, sir, the worse.
ORSINO How can that be?
FESTE Marry, sir, they praise me, and make an ass of me.
Now my foes tell me plainly I am an ass, so that by
my foes, sir, I profit in the knowledge of myself, and
by my friends I am abused; so that, conclusions to be
as kisses, if your four negatives make your two
affirmatives, why then the worse for my friends and
the better for my foes.          21
ORSINO Why, this is excellent.
FESTE By my troth, sir, no, though it please you to be
one of my friends.
ORSINO (*giving money*)
Thou shalt not be the worse for me. There's gold.          25
FESTE But that it would be double-dealing, sir, I would
you could make it another.
ORSINO O, you give me ill counsel.
FESTE Put your grace in your pocket, sir, for this once,
and let your flesh and blood obey it.          30
ORSINO Well, I will be so much a sinner to be a double-
dealer. (*Giving money*) There's another.
FESTE *Primo, secundo, tertio* is a good play, and the old
saying is 'The third pays for all'. The triplex, sir, is a
good tripping measure, or the bells of Saint Bennet, sir,
may put you in mind—'one, two, three'.          36
ORSINO You can fool no more money out of me at this
throw. If you will let your lady know I am here to
speak with her, and bring her along with you, it may
awake my bounty further.          40
FESTE Marry, sir, lullaby to your bounty till I come again.
I go, sir, but I would not have you to think that my
desire of having is the sin of covetousness. But as you
say, sir, let your bounty take a nap, I will awake it
anon.          *Exit*
    *Enter Antonio and Officers*
VIOLA
Here comes the man, sir, that did rescue me.          46

ORSINO
That face of his I do remember well,
Yet when I saw it last it was besmeared
As black as Vulcan in the smoke of war.
A baubling vessel was he captain of,          50
For shallow draught and bulk unprizable,
With which such scatheful grapple did he make
With the most noble bottom of our fleet
That very envy and the tongue of loss
Cried fame and honour on him. What's the matter?

FIRST OFFICER
Orsino, this is that Antonio          56
That took the Phoenix and her freight from Candy,
And this is he that did the *Tiger* board
When your young nephew Titus lost his leg.
Here in the streets, desperate of shame and state,          60
In private brabble did we apprehend him.

VIOLA
He did me kindness, sir, drew on my side,
But in conclusion put strange speech upon me.
I know not what 'twas but distraction.

ORSINO (*to Antonio*)
Notable pirate, thou salt-water thief,          65
What foolish boldness brought thee to their mercies
Whom thou in terms so bloody and so dear
Hast made thine enemies?

ANTONIO                    Orsino, noble sir,
Be pleased that I shake off these names you give me.
Antonio never yet was thief or pirate,          70
Though, I confess, on base and ground enough
Orsino's enemy. A witchcraft drew me hither.
That most ingrateful boy there by your side
From the rude sea's enraged and foamy mouth
Did I redeem. A wreck past hope he was.          75
His life I gave him, and did thereto add
My love without retention or restraint,
All his in dedication. For his sake
Did I expose myself, pure for his love,
Into the danger of this adverse town,          80
Drew to defend him when he was beset,
Where being apprehended, his false cunning—
Not meaning to partake with me in danger—
Taught him to face me out of his acquaintance,
And grew a twenty years' removèd thing          85
While one would wink, denied me mine own purse,
Which I had recommended to his use
Not half an hour before.

VIOLA How can this be?
ORSINO When came he to this town?          90

ANTONIO
Today, my lord, and for three months before,
No int'rim, not a minute's vacancy,
Both day and night did we keep company.
    *Enter Olivia and attendants*

ORSINO
Here comes the Countess. Now heaven walks on
    earth.
But for thee, fellow—fellow, thy words are madness.
Three months this youth hath tended upon me.          96
But more of that anon. Take him aside.

OLIVIA
What would my lord, but that he may not have,
Wherein Olivia may seem serviceable?
Cesario, you do not keep promise with me.          100

VIOLA Madam—
ORSINO Gracious Olivia—

OLIVIA
What do you say, Cesario? Good my lord—
VIOLA
My lord would speak, my duty hushes me.
OLIVIA
If it be aught to the old tune, my lord,                    105
It is as fat and fulsome to mine ear
As howling after music.
ORSINO Still so cruel?
OLIVIA Still so constant, lord.
ORSINO
What, to perverseness? You uncivil lady,                    110
To whose ingrate and unauspicious altars
My soul the faithfull'st off'rings hath breathed out
That e'er devotion tendered—what shall I do?
OLIVIA
Even what it please my lord that shall become him.
ORSINO
Why should I not, had I the heart to do it,                 115
Like to th' Egyptian thief, at point of death
Kill what I love—a savage jealousy
That sometime savours nobly. But hear me this:
Since you to non-regardance cast my faith,
And that I partly know the instrument                       120
That screws me from my true place in your favour,
Live you the marble-breasted tyrant still.
But this your minion, whom I know you love,
And whom, by heaven I swear, I tender dearly,
Him will I tear out of that cruel eye                        125
Where he sits crownèd in his master's spite.
(To Viola) Come, boy, with me. My thoughts are ripe
    in mischief.
I'll sacrifice the lamb that I do love
To spite a raven's heart within a dove.
VIOLA
And I most jocund, apt, and willingly                       130
To do you rest a thousand deaths would die.
OLIVIA
Where goes Cesario?
VIOLA                    After him I love
More than I love these eyes, more than my life,
More by all mores than e'er I shall love wife.
If I do feign, you witnesses above,                         135
Punish my life for tainting of my love.
OLIVIA
Ay me detested, how am I beguiled!
VIOLA
Who does beguile you? Who does do you wrong?
OLIVIA
Hast thou forgot thyself? Is it so long?
Call forth the holy father.            Exit an attendant
ORSINO (to Viola)            Come, away.
OLIVIA
Whither, my lord? Cesario, husband, stay.                   141
ORSINO
Husband?
OLIVIA        Ay, husband. Can he that deny?
ORSINO (to Viola)
Her husband, sirrah?
VIOLA                    No, my lord, not I.
OLIVIA
Alas, it is the baseness of thy fear
That makes thee strangle thy propriety.                     145
Fear not, Cesario, take thy fortunes up,
Be that thou know'st thou art, and then thou art

As great as that thou fear'st.
    Enter the Priest
                        O welcome, father.
Father, I charge thee by thy reverence
Here to unfold—though lately we intended                    150
To keep in darkness what occasion now
Reveals before 'tis ripe—what thou dost know
Hath newly passed between this youth and me.
PRIEST
A contract of eternal bond of love,
Confirmed by mutual joinder of your hands,                  155
Attested by the holy close of lips,
Strengthened by interchangement of your rings,
And all the ceremony of this compact
Sealed in my function, by my testimony;
Since when, my watch hath told me, toward my grave
I have travelled but two hours.                             161
ORSINO (to Viola)
O thou dissembling cub, what wilt thou be
When time hath sowed a grizzle on thy case?
Or will not else thy craft so quickly grow
That thine own trip shall be thine overthrow?               165
Farewell, and take her, but direct thy feet
Where thou and I henceforth may never meet.
VIOLA
My lord, I do protest.
OLIVIA                O, do not swear!
Hold little faith, though thou hast too much fear.
    Enter Sir Andrew
SIR ANDREW For the love of God, a surgeon—send one
presently to Sir Toby.                                      171
OLIVIA What's the matter?
SIR ANDREW He's broke my head across, and has given
Sir Toby a bloody coxcomb, too. For the love of God,
your help! I had rather than forty pound I were at
home.                                                       176
OLIVIA Who has done this, Sir Andrew?
SIR ANDREW The Count's gentleman, one Cesario. We
took him for a coward, but he's the very devil
incarnate.                                                  180
ORSINO My gentleman, Cesario?
SIR ANDREW 'Od's lifelings, here he is. (To Viola) You broke
my head for nothing, and that that I did I was set on
to do't by Sir Toby.
VIOLA
Why do you speak to me? I never hurt you.                   185
You drew your sword upon me without cause,
But I bespake you fair, and hurt you not.
    Enter Sir Toby and Feste, the clown
SIR ANDREW If a bloody coxcomb be a hurt you have hurt
me. I think you set nothing by a bloody coxcomb. Here
comes Sir Toby, halting. You shall hear more; but if
he had not been in drink he would have tickled you
othergates than he did.                                     192
ORSINO (to Sir Toby)
How now, gentleman? How is't with you?
SIR TOBY That's all one, he's hurt me, and there's th'end
on't. (To Feste) Sot, didst see Dick Surgeon, sot?          195
FESTE O, he's drunk, Sir Toby, an hour agone. His eyes
were set at eight i'th' morning.
SIR TOBY Then he's a rogue, and a passy-measures pavan.
I hate a drunken rogue.
OLIVIA
Away with him! Who hath made this havoc with them?
SIR ANDREW I'll help you, Sir Toby, because we'll be
dressed together.                                           202

SIR TOBY Will *you* help—an ass-head, and a coxcomb,
and a knave; a thin-faced knave, a gull?
OLIVIA
Get him to bed, and let his hurt be looked to.          205
          *Exeunt Sir Toby, Sir Andrew, Feste, and Fabian*
          *Enter Sebastian*
SEBASTIAN (*to Olivia*)
I am sorry, madam, I have hurt your kinsman,
But had it been the brother of my blood
I must have done no less with wit and safety.
You throw a strange regard upon me, and by that
I do perceive it hath offended you.          210
Pardon me, sweet one, even for the vows
We made each other but so late ago.
ORSINO
One face, one voice, one habit, and two persons,
A natural perspective, that is and is not.
SEBASTIAN
Antonio! O, my dear Antonio,          215
How have the hours racked and tortured me
Since I have lost thee!
ANTONIO Sebastian are you?
SEBASTIAN Fear'st thou that, Antonio?
ANTONIO
How have you made division of yourself?          220
An apple cleft in two is not more twin
Than these two creatures. Which is Sebastian?
OLIVIA Most wonderful!
SEBASTIAN (*seeing Viola*)
Do I stand there? I never had a brother,
Nor can there be that deity in my nature          225
Of here and everywhere. I had a sister,
Whom the blind waves and surges have devoured.
Of charity, what kin are you to me?
What countryman? What name? What parentage?
VIOLA
Of Messaline. Sebastian was my father.          230
Such a Sebastian was my brother, too.
So went he suited to his watery tomb.
If spirits can assume both form and suit
You come to fright us.
SEBASTIAN          A spirit I am indeed,
But am in that dimension grossly clad          235
Which from the womb I did participate.
Were you a woman, as the rest goes even,
I should my tears let fall upon your cheek
And say 'Thrice welcome, drownèd Viola.'
VIOLA
My father had a mole upon his brow.          240
SEBASTIAN And so had mine.
VIOLA
And died that day when Viola from her birth
Had numbered thirteen years.
SEBASTIAN
O, that record is lively in my soul.
He finishèd indeed his mortal act          245
That day that made my sister thirteen years.
VIOLA
If nothing lets to make us happy both
But this my masculine usurped attire,
Do not embrace me till each circumstance
Of place, time, fortune do cohere and jump          250
That I am Viola, which to confirm
I'll bring you to a captain in this town
Where lie my maiden weeds, by whose gentle help
I was preserved to serve this noble count.

All the occurrence of my fortune since          255
Hath been between this lady and this lord.
SEBASTIAN (*to Olivia*)
So comes it, lady, you have been mistook.
But nature to her bias drew in that.
You would have been contracted to a maid,
Nor are you therein, by my life, deceived.          260
You are betrothed both to a maid and man.
ORSINO (*to Olivia*)
Be not amazed. Right noble is his blood.
If this be so, as yet the glass seems true,
I shall have share in this most happy wreck.
(*To Viola*) Boy, thou hast said to me a thousand times
Thou never shouldst love woman like to me.          266
VIOLA
And all those sayings will I overswear,
And all those swearings keep as true in soul
As doth that orbèd continent the fire
That severs day from night.
ORSINO          Give me thy hand,          270
And let me see thee in thy woman's weeds.
VIOLA
The captain that did bring me first on shore
Hath my maid's garments. He upon some action
Is now in durance, at Malvolio's suit,
A gentleman and follower of my lady's.          275
OLIVIA
He shall enlarge him. Fetch Malvolio hither—
And yet, alas, now I remember me,
They say, poor gentleman, he's much distraught.
          *Enter Feste the clown with a letter, and Fabian*
A most extracting frenzy of mine own
From my remembrance clearly banished his.          280
How does he, sirrah?
FESTE Truly, madam, he holds Beelzebub at the stave's
end as well as a man in his case may do. He's here
writ a letter to you. I should have given't you today
morning. But as a madman's epistles are no gospels,
so it skills not much when they are delivered.          286
OLIVIA Open't and read it.
FESTE Look then to be well edified when the fool delivers
the madman. (*Reads*) 'By the Lord, madam'—
OLIVIA How now, art thou mad?          290
FESTE No, madam, I do but read madness. An your
ladyship will have it as it ought to be you must allow
*vox.*
OLIVIA Prithee, read i'thy right wits.
FESTE So I do, madonna, but to read his right wits is to
read thus. Therefore perpend, my princess, and give
ear.          297
OLIVIA (*to Fabian*) Read it you, sirrah.
          *Feste gives the letter to Fabian*
FABIAN (*reads*) 'By the Lord, madam, you wrong me, and
the world shall know it. Though you have put me into
darkness and given your drunken cousin rule over me,
yet have I the benefit of my senses as well as your
ladyship. I have your own letter that induced me to
the semblance I put on, with the which I doubt not
but to do myself much right or you much shame. Think
of me as you please. I leave my duty a little unthought
of, and speak out of my injury.          307
          The madly-used Malvolio.'
OLIVIA Did he write this?
FESTE Ay, madam.          310
ORSINO
This savours not much of distraction.

713

**OLIVIA**
See him delivered, Fabian, bring him hither.
My lord, so please you—these things further thought
    on—
To think me as well a sister as a wife,
One day shall crown th'alliance on't, so please you,
Here at my house and at my proper cost.          316
**ORSINO**
Madam, I am most apt t'embrace your offer.
(*To Viola*) Your master quits you, and for your service
    done him
So much against the mettle of your sex,
So far beneath your soft and tender breeding,          320
And since you called me master for so long,
Here is my hand. You shall from this time be
Your master's mistress.
**OLIVIA** (*to Viola*)          A sister, you are she.
        *Enter Malvolio*
**ORSINO**
Is this the madman?
**OLIVIA**          Ay, my lord, this same.
How now, Malvolio?
**MALVOLIO**          Madam, you have done me wrong,
Notorious wrong.
**OLIVIA**          Have I, Malvolio? No.          326
**MALVOLIO** (*showing a letter*)
Lady, you have. Pray you peruse that letter.
You must not now deny it is your hand.
Write from it if you can, in hand or phrase,
Or say 'tis not your seal, not your invention.          330
You can say none of this. Well, grant it then,
And tell me in the modesty of honour
Why you have given me such clear lights of favour,
Bade me come smiling and cross-gartered to you,
To put on yellow stockings, and to frown          335
Upon Sir Toby and the lighter people,
And acting this in an obedient hope,
Why have you suffered me to be imprisoned,
Kept in a dark house, visited by the priest,
And made the most notorious geck and gull          340
That e'er invention played on? Tell me why?
**OLIVIA**
Alas, Malvolio, this is not my writing,
Though I confess much like the character,
But out of question, 'tis Maria's hand.
And now I do bethink me, it was she          345
First told me thou wast mad; then cam'st in smiling,
And in such forms which here were presupposed
Upon thee in the letter. Prithee be content;
This practice hath most shrewdly passed upon thee,
But when we know the grounds and authors of it          350
Thou shalt be both the plaintiff and the judge
Of thine own cause.
**FABIAN**          Good madam, hear me speak,
And let no quarrel nor no brawl to come
Taint the condition of this present hour,
Which I have wondered at. In hope it shall not,          355
Most freely I confess myself and Toby

Set this device against Malvolio here
Upon some stubborn and uncourteous parts
We had conceived against him. Maria writ
The letter, at Sir Toby's great importance,          360
In recompense whereof he hath married her.
How with a sportful malice it was followed
May rather pluck on laughter than revenge
If that the injuries be justly weighed
That have on both sides passed.          365
**OLIVIA** (*to Malvolio*)
Alas, poor fool, how have they baffled thee!
**FESTE** Why, 'Some are born great, some achieve great-
ness, and some have greatness thrown upon them.' I
was one, sir, in this interlude, one Sir Topas, sir; but
that's all one. 'By the Lord, fool, I am not mad'—but
do you remember, 'Madam, why laugh you at such a
barren rascal, an you smile not, he's gagged'—and
thus the whirligig of time brings in his revenges.
**MALVOLIO** I'll be revenged on the whole pack of you.
                                                        *Exit*

**OLIVIA**
He hath been most notoriously abused.          375
**ORSINO**
Pursue him, and entreat him to a peace.
He hath not told us of the captain yet.
                                ⌜*Exit one or more*⌝
When that is known, and golden time convents,
A solemn combination shall be made
Of our dear souls. Meantime, sweet sister,          380
We will not part from hence. Cesario, come—
For so you shall be while you are a man;
But when in other habits you are seen,
Orsino's mistress, and his fancy's queen.
                                *Exeunt all but Feste*
**FESTE** (*sings*)
When that I was and a little tiny boy,          385
    With hey, ho, the wind and the rain,
A foolish thing was but a toy,
    For the rain it raineth every day.

But when I came to man's estate,
    With hey, ho, the wind and the rain,          390
'Gainst knaves and thieves men shut their gate,
    For the rain it raineth every day.

But when I came, alas, to wive,
    With hey, ho, the wind and the rain,
By swaggering could I never thrive,          395
    For the rain it raineth every day.

But when I came unto my beds,
    With hey, ho, the wind and the rain,
With tosspots still had drunken heads,
    For the rain it raineth every day.          400

A great while ago the world begun,
    With hey ho, the wind and the rain,
But that's all one, our play is done,
    And we'll strive to please you every day.          *Exit*

# TROILUS AND CRESSIDA

*Troilus and Cressida*, first heard of in a Stationers' Register entry of 7 February 1603, was probably written within the previous eighteen months. This entry did not result in publication; the play was re-entered on 28 January 1609, and a quarto appeared during that year. The version printed in the 1623 Folio adds a Prologue, and has many variations in dialogue. It includes the epilogue spoken by Pandarus (which we print as an Additional Passage), but certain features of the text suggest that it does so by accident, and that the epilogue had been marked for omission. Our text is based in substance on the Folio in the belief that this represents the play in its later, revised form.

The story of the siege of Troy was the main subject of one of the greatest surviving works of classical literature, Homer's *Iliad*; probably Shakespeare read George Chapman's 1598 translation of Books 1-2 and 7-11. It also figures prominently in Virgil's *Aeneid* and Ovid's *Metamorphoses*, both of which Shakespeare knew well. The war between Greece and Troy had been provoked by the abduction of the Grecian Helen (better, if confusingly, known as Helen of Troy) by the Trojan hero Paris, son of King Priam. Shakespeare's play opens when the Greek forces, led by Menelaus' brother Agamemnon, have already been besieging Troy for seven years. Shakespeare concentrates on the opposition between the Greek hero Achilles and the Trojan Hector. In the Folio, *Troilus and Cressida* is printed among the tragedies; if there is a tragic hero, it is Hector.

Shakespeare also shows how the war caused by one love affair destroys another. The stories of the love between the Trojan Troilus and the Grecian Cressida, encouraged by her uncle Pandarus, and of Cressida's desertion of Troilus for the Greek Diomedes, are medieval additions to the heroic narrative. Chaucer's long poem *Troilus and Criseyde* was already a classic, and Shakespeare would also have known Robert Henryson's continuation, *The Testament of Cresseid*, in which Cressida, deserted by Diomedes, dwindles into a leprous beggar.

*Troilus and Cressida* is a demanding play, Shakespeare's third longest, highly philosophical in tone and with an exceptionally learned vocabulary. Possibly (as has often been conjectured) he wrote it for private performance; the 1603 Stationers' Register entry says it had been acted by the King's Men, and the original title-page of the 1609 quarto repeats this claim, but while the edition was being printed this title-page was replaced by one that does not mention performance, and an epistle was added claiming that it was 'a new play, never staled with the stage, never clapper-clawed with the palms of the vulgar'. An adaptation by John Dryden of 1679 was successfully acted from time to time for half a century, but the first verified performance of Shakespeare's play was in Germany in 1898, and that was heavily adapted. *Troilus and Cressida* has come into its own in the twentieth century, when its deflation of heroes, its radical questioning of human values (especially in relation to love and war), and its remorseless examination of the frailty of human aspirations in the face of the destructive powers of time have seemed particularly apposite to modern intellectual and ethical preoccupations.

# THE PERSONS OF THE PLAY

PROLOGUE

### Trojans

PRIAM, King of Troy

HECTOR
DEIPHOBUS
HELENUS, a priest
PARIS          } his sons
TROILUS
MARGARETON, a bastard

CASSANDRA, Priam's daughter, a prophetess

ANDROMACHE, wife of Hector

AENEAS     }
ANTENOR   } commanders

PANDARUS, a lord

CRESSIDA, his niece

CALCHAS, her father, who has joined the Greeks

HELEN, wife of Menelaus, now living with Paris

ALEXANDER, servant of Cressida

Servants of Troilus, musicians, soldiers, attendants

### Greeks

AGAMEMNON, Commander-in-Chief

MENELAUS, his brother

NESTOR

ULYSSES

ACHILLES

PATROCLUS, his companion

DIOMEDES

AJAX

THERSITES

MYRMIDONS, soldiers of Achilles

Servants of Diomedes, soldiers

# Troilus and Cressida

**Prologue** *Enter the Prologue armed*
PROLOGUE
In Troy there lies the scene. From isles of Greece
The princes orgulous, their high blood chafed,
Have to the port of Athens sent their ships,
Fraught with the ministers and instruments
Of cruel war. Sixty-and-nine, that wore                    5
Their crownets regal, from th'Athenian bay
Put forth toward Phrygia, and their vow is made
To ransack Troy, within whose strong immures
The ravished Helen, Menelaus' queen,
With wanton Paris sleeps—and that's the quarrel.    10
To Tenedos they come,
And the deep-drawing barques do there disgorge
Their warlike freightage; now on Dardan plains
The fresh and yet unbruisèd Greeks do pitch
Their brave pavilions. Priam's six-gated city—      15
Dardan and Timbria, Helias, Chetas, Troien,
And Antenorides—with massy staples
And corresponsive and full-filling bolts
Spar up the sons of Troy.
Now expectation, tickling skittish spirits          20
On one and other side, Trojan and Greek,
Sets all on hazard. And hither am I come,
A Prologue armed—but not in confidence
Of author's pen or actor's voice, but suited
In like conditions as our argument—                 25
To tell you, fair beholders, that our play
Leaps o'er the vaunt and firstlings of those broils,
Beginning in the middle, starting thence away
To what may be digested in a play.
Like or find fault; do as your pleasures are;       30
Now, good or bad, 'tis but the chance of war.    *Exit*

**1.1** *Enter Pandarus, and Troilus armed*
TROILUS
Call here my varlet. I'll unarm again.
Why should I war without the walls of Troy
That find such cruel battle here within?
Each Trojan that is master of his heart,
Let him to field—Troilus, alas, hath none.           5
PANDARUS Will this gear ne'er be mended?
TROILUS
The Greeks are strong, and skilful to their strength,
Fierce to their skill, and to their fierceness valiant.
But I am weaker than a woman's tear,
Tamer than sleep, fonder than ignorance,            10
Less valiant than the virgin in the night,
And skilless as unpractised infancy.
PANDARUS Well, I have told you enough of this. For my
part, I'll not meddle nor make no farther. He that will
have a cake out of the wheat must tarry the grinding.
TROILUS Have I not tarried?                          16
PANDARUS Ay, the grinding; but you must tarry the
boulting.
TROILUS Have I not tarried?
PANDARUS Ay, the boulting; but you must tarry the
leavening.                                           21
TROILUS Still have I tarried.

PANDARUS Ay, to the leavening; but here's yet in the
word 'hereafter' the kneading, the making of the cake,
the heating the oven, and the baking—nay, you must
stay the cooling too, or ye may chance burn your lips.
TROILUS
Patience herself, what goddess e'er she be,          27
Doth lesser blench at suff'rance than I do.
At Priam's royal table do I sit
And when fair Cressid comes into my thoughts—        30
So, traitor! 'When she comes'? When is she thence?
PANDARUS Well, she looked yesternight fairer than ever I
saw her look, or any woman else.
TROILUS
I was about to tell thee: when my heart,
As wedgèd with a sigh, would rive in twain,          35
Lest Hector or my father should perceive me
I have, as when the sun doth light askance,
Buried this sigh in wrinkle of a smile.
But sorrow that is couched in seeming gladness
Is like that mirth fate turns to sudden sadness.     40
PANDARUS An her hair were not somewhat darker than
Helen's—well, go to, there were no more comparison
between the women. But, for my part, she is my
kinswoman; I would not, as they term it, 'praise' her.
But I would somebody had heard her talk yesterday,
as I did. I will not dispraise your sister Cassandra's wit,
but—                                                 47
TROILUS
O Pandarus! I tell thee, Pandarus,
When I do tell thee 'There my hopes lie drowned',
Reply not in how many fathoms deep                   50
They lie endrenched. I tell thee I am mad
In Cressid's love; thou answer'st 'She is fair',
Pourest in the open ulcer of my heart
Her eyes, her hair, her cheek, her gait, her voice;
Handlest in thy discourse, O, that her hand,         55
In whose comparison all whites are ink
Writing their own reproach, to whose soft seizure
The cygnet's down is harsh, and spirit of sense
Hard as the palm of ploughman. This thou tell'st me—
As true thou tell'st me—when I say I love her.       60
But saying thus, instead of oil and balm
Thou lay'st in every gash that love hath given me
The knife that made it.
PANDARUS I speak no more than truth.
TROILUS Thou dost not speak so much.                 65
PANDARUS Faith, I'll not meddle in it. Let her be as she
is. If she be fair, 'tis the better for her; an she be not,
she has the mends in her own hands.
TROILUS Good Pandarus, how now, Pandarus!
PANDARUS I have had my labour for my travail. Ill thought
on of her and ill thought on of you. Gone between and
between, but small thanks for my labour.             72
TROILUS
What, art thou angry, Pandarus? What, with me?
PANDARUS Because she's kin to me, therefore she's not so
fair as Helen. An she were not kin to me, she would
be as fair o' Friday as Helen is on Sunday. But what
care I? I care not an she were a blackamoor. 'Tis all
one to me.                                           78

TROILUS Say I she is not fair?

PANDARUS I do not care whether you do or no. She's a fool to stay behind her father. Let her to the Greeks—and so I'll tell her the next time I see her. For my part, I'll meddle nor make no more i'th' matter.

TROILUS Pandarus—

PANDARUS Not I.                                                                    85

TROILUS Sweet Pandarus—

PANDARUS Pray you, speak no more to me. I will leave all as I found it. And there an end.                          *Exit*

  *Alarum*

TROILUS

Peace, you ungracious clamours! Peace, rude sounds!

Fools on both sides. Helen must needs be fair         90

When with your blood you daily paint her thus.

I cannot fight upon this argument.

It is too starved a subject for my sword.

But Pandarus—O gods, how do you plague me!

I cannot come to Cressid but by Pandar,                 95

And he's as tetchy to be wooed to woo

As she is stubborn-chaste against all suit.

Tell me, Apollo, for thy Daphne's love,

What Cressid is, what Pandar, and what we?

Her bed is India; there she lies, a pearl.              100

Between our Ilium and where she resides

Let it be called the wild and wand'ring flood,

Ourself the merchant, and this sailing Pandar

Our doubtful hope, our convoy, and our barque.

  *Alarum. Enter Aeneas*

AENEAS

How now, Prince Troilus? Wherefore not afield?    105

TROILUS

Because not there. This woman's answer sorts,

For womanish it is to be from thence.

What news, Aeneas, from the field today?

AENEAS

That Paris is returnèd home, and hurt.

TROILUS

By whom, Aeneas?

AENEAS                         Troilus, by Menelaus.          110

TROILUS

Let Paris bleed, 'tis but a scar to scorn:

Paris is gored with Menelaus' horn.

  *Alarum*

AENEAS

Hark what good sport is out of town today.

TROILUS

Better at home, if 'would I might' were 'may'.

But to the sport abroad—are you bound thither?    115

AENEAS

In all swift haste.

TROILUS                         Come, go we then together.    *Exeunt*

**1.2**    *Enter ⌜above⌝ Cressida and her servant Alexander*

CRESSIDA

Who were those went by?

ALEXANDER                         Queen Hecuba and Helen.

CRESSIDA

And whither go they?

ALEXANDER                         Up to the eastern tower,

Whose height commands as subject all the vale,

To see the battle. Hector, whose patience

Is as a virtue fixed, today was moved.                    5

He chid Andromache and struck his armourer

And, like as there were husbandry in war,

Before the sun rose he was harnessed light,

And to the field goes he, where every flower

Did as a prophet weep what it foresaw                   10

In Hector's wrath.

CRESSIDA                         What was his cause of anger?

ALEXANDER

The noise goes this: there is among the Greeks

A lord of Trojan blood, nephew to Hector;

They call him Ajax.

CRESSIDA                         Good, and what of him?

ALEXANDER

They say he is a very man *per se*,                        15

And stands alone.

CRESSIDA                         So do all men

Unless they are drunk, sick, or have no legs.

ALEXANDER This man, lady, hath robbed many beasts of their particular additions: he is as valiant as the lion, churlish as the bear, slow as the elephant—a man into whom nature hath so crowded humours that his valour is crushed into folly, his folly farced with discretion. There is no man hath a virtue that he hath not a glimpse of, nor any man an attaint but he carries some stain of it. He is melancholy without cause and merry against the hair; he hath the joints of everything, but everything so out of joint that he is a gouty Briareus, many hands and no use, or purblind Argus, all eyes and no sight.                                                                    29

CRESSIDA But how should this man that makes me smile make Hector angry?

ALEXANDER They say he yesterday coped Hector in the battle and struck him down, the disdain and shame whereof hath ever since kept Hector fasting and waking.

CRESSIDA Who comes here?                                        35

ALEXANDER Madam, your uncle Pandarus.

  ⌜*Enter Pandarus above*⌝

CRESSIDA Hector's a gallant man.

ALEXANDER As may be in the world, lady.

PANDARUS What's that? What's that?

CRESSIDA Good morrow, uncle Pandarus.                    40

PANDARUS Good morrow, cousin Cressid. What do you talk of?—Good morrow, Alexander.—How do you, cousin? When were you at Ilium?

CRESSIDA This morning, uncle.

PANDARUS What were you talking of when I came? Was Hector armed and gone ere ye came to Ilium? Helen was not up, was she?                                                         47

CRESSIDA

Hector was gone but Helen was not up?

PANDARUS E'en so. Hector was stirring early.

CRESSIDA

That were we talking of, and of his anger.            50

PANDARUS Was he angry?

CRESSIDA So he says here.

PANDARUS True, he was so. I know the cause too. He'll lay about him today, I can tell them that. And there's Troilus will not come far behind him. Let them take heed of Troilus, I can tell them that too.           56

CRESSIDA What, is he angry too?

PANDARUS Who, Troilus? Troilus is the better man of the two.

CRESSIDA

O Jupiter! There's no comparison.                          60

PANDARUS What, not between Troilus and Hector? Do you know a man if you see him?

CRESSIDA

Ay, if I ever saw him before and knew him.

PANDARUS Well, I say Troilus is Troilus.

CRESSIDA
Then you say as I say, for I am sure                    65
He is not Hector.
PANDARUS No, nor Hector is not Troilus, in some degrees.
CRESSIDA
'Tis just to each of them: he is himself.
PANDARUS Himself? Alas, poor Troilus, I would he were.
CRESSIDA So he is.                                     70
PANDARUS Condition I had gone barefoot to India.
CRESSIDA He is not Hector.
PANDARUS Himself? No, he's not himself. Would a were
himself! Well, the gods are above, time must friend or
end. Well, Troilus, well, I would my heart were in her
body. No, Hector is not a better man than Troilus.    76
CRESSIDA Excuse me.
PANDARUS He is elder.
CRESSIDA Pardon me, pardon me.
PANDARUS Th'other's not come to't. You shall tell me
another tale when th'other's come to't. Hector shall
not have his will this year.                           82
CRESSIDA
He shall not need it if he have his own.
PANDARUS Nor his qualities.
CRESSIDA No matter.                                    85
PANDARUS Nor his beauty.
CRESSIDA
'Twould not become him; his own's better.
PANDARUS You have no judgement, niece. Helen herself
swore th'other day that Troilus for a brown favour, for
so 'tis, I must confess—not brown neither—             90
CRESSIDA No, but brown.
PANDARUS Faith, to say truth, brown and not brown.
CRESSIDA To say the truth, true and not true.
PANDARUS She praised his complexion above Paris'.
CRESSIDA Why, Paris hath colour enough.                95
PANDARUS So he has.
CRESSIDA Then Troilus should have too much. If she
praised him above, his complexion is higher than his;
he having colour enough, and the other higher, is too
flaming a praise for a good complexion. I had as lief
Helen's golden tongue had commended Troilus for a
copper nose.                                          102
PANDARUS I swear to you, I think Helen loves him better
than Paris.
CRESSIDA Then she's a merry Greek indeed.             105
PANDARUS Nay, I am sure she does. She came to him
th'other day into the compassed window, and you
know he has not past three or four hairs on his chin—
CRESSIDA Indeed, a tapster's arithmetic may soon bring
his particulars therein to a total.                   110
PANDARUS Why, he is very young—and yet will he within
three pound lift as much as his brother Hector.
CRESSIDA Is he so young a man and so old a lifter?
PANDARUS But to prove to you that Helen loves him: she
came and puts me her white hand to his cloven chin.
CRESSIDA Juno have mercy! How came it cloven?         116
PANDARUS Why, you know, 'tis dimpled. I think his
smiling becomes him better than any man in all
Phrygia.
CRESSIDA O he smiles valiantly.                        120
PANDARUS Does he not?
CRESSIDA O yes, an't were a cloud in autumn.
PANDARUS Why, go to then. But to prove to you that
Helen loves Troilus—
CRESSIDA Troilus will stand to the proof if you'll prove it
so.                                                   126

PANDARUS Troilus? Why, he esteems her no more than I
esteem an addle egg.
CRESSIDA If you love an addle egg as well as you love an
idle head you would eat chickens i'th' shell.         130
PANDARUS I cannot choose but laugh to think how she
tickled his chin. Indeed, she has a marvellous white
hand, I must needs confess—
CRESSIDA Without the rack.
PANDARUS And she takes upon her to spy a white hair
on his chin.                                          136
CRESSIDA Alas, poor chin! Many a wart is richer.
PANDARUS But there was such laughing! Queen Hecuba
laughed that her eyes ran o'er.
CRESSIDA With millstones.                              140
PANDARUS And Cassandra laughed.
CRESSIDA But there was a more temperate fire under the
pot of her eyes—or did her eyes run o'er too?
PANDARUS And Hector laughed.
CRESSIDA At what was all this laughing?                145
PANDARUS Marry, at the white hair that Helen spied on
Troilus' chin.
CRESSIDA An't had been a green hair I should have
laughed too.
PANDARUS They laughed not so much at the hair as at
his pretty answer.                                    151
CRESSIDA What was his answer?
PANDARUS Quoth she, 'Here's but two-and-fifty hairs on
your chin, and one of them is white.'
CRESSIDA This is her question.                          155
PANDARUS That's true, make no question of that. 'Two-
and-fifty hairs,' quoth he, 'and one white? That white
hair is my father, and all the rest are his sons.' 'Jupiter!'
quoth she, 'which of these hairs is Paris my husband?'
'The forked one,' quoth he, 'pluck't out and give it
him.' But there was such laughing, and Helen so
blushed and Paris so chafed and all the rest so laughed,
that it passed.
CRESSIDA So let it now, for it has been a great while going
by.                                                   165
PANDARUS Well, cousin, I told you a thing yesterday.
Think on't.
CRESSIDA So I do.
PANDARUS I'll be sworn 'tis true. He will weep you an't
were a man born in April.                             170
CRESSIDA And I'll spring up in his tears an't were a nettle
against May.
  *A retreat is sounded*
PANDARUS Hark, they are coming from the field. Shall we
stand up here and see them as they pass toward Ilium?
Good niece, do, sweet niece Cressida.                 175
CRESSIDA At your pleasure.
PANDARUS Here, here, here's an excellent place, here we
may see most bravely. I'll tell you them all by their
names as they pass by, but mark Troilus above the
rest.                                                 180
  *Enter Aeneas passing by ⌜below⌝*
CRESSIDA Speak not so loud.
PANDARUS That's Aeneas. Is not that a brave man? He's
one of the flowers of Troy, I can tell you. But mark
Troilus; you shall see anon.
  *Enter Antenor passing by ⌜below⌝*
CRESSIDA Who's that?                                   185
PANDARUS That's Antenor. He has a shrewd wit, I can
tell you, and he's a man good enough. He's one o'th'
soundest judgements in Troy whosoever, and a proper

man of person. When comes Troilus? I'll show you
Troilus anon. If he see me you shall see him nod at
me.                                                              191
CRESSIDA Will he give you the nod?
PANDARUS You shall see.
CRESSIDA If he do, the rich shall have more.
            *Enter Hector passing by ⌜below⌝*
PANDARUS That's Hector, that, that, look you, that.
There's a fellow!—Go thy way, Hector!—There's a
brave man, niece. O brave Hector! Look how he looks.
There's a countenance. Is't not a brave man?
CRESSIDA O a brave man.                                          199
PANDARUS Is a not? It does a man's heart good. Look you
what hacks are on his helmet. Look you yonder, do
you see? Look you there. There's no jesting. There's
laying on, take't off who will, as they say. There be
hacks.
CRESSIDA Be those with swords?                                   205
            *Enter Paris passing by ⌜below⌝*
PANDARUS Swords, anything, he cares not. An the devil
come to him it's all one. By'God's lid it does one's heart
good. Yonder comes Paris, yonder comes Paris. Look
ye yonder, niece. Is't not a gallant man too? Is't not?
Why, this is brave now. Who said he came hurt home
today? He's not hurt. Why, this will do Helen's heart
good now, ha! Would I could see Troilus now. You
shall see Troilus anon.
            *Enter Helenus passing by ⌜below⌝*
CRESSIDA Who's that?                                             214
PANDARUS That's Helenus. I marvel where Troilus is.
That's Helenus. I think he went not forth today. That's
Helenus.
CRESSIDA Can Helenus fight, uncle?
PANDARUS Helenus? No—yes, he'll fight indifferent well.
I marvel where Troilus is.                                       220
            *⌜A Shout⌝*
Hark, do you not hear the people cry 'Troilus'? Helenus
is a priest.
            *Enter Troilus passing by ⌜below⌝*
CRESSIDA What sneaking fellow comes yonder?
PANDARUS Where? Yonder? That's Deiphobus.—'Tis
Troilus! There's a man, niece, h'm? Brave Troilus, the
prince of chivalry!                                              226
CRESSIDA Peace, for shame, peace.
PANDARUS Mark him, note him. O brave Troilus! Look
well upon him, niece. Look you how his sword is
bloodied and his helm more hacked than Hector's, and
how he looks and how he goes. O admirable youth!
He ne'er saw three-and-twenty. —Go thy way, Troilus,
go thy way!—Had I a sister were a grace, or a daughter
a goddess, he should take his choice. O admirable man!
Paris? Paris is dirt to him, and I warrant Helen to
change would give an eye to boot.                                236
            *Enter common soldiers passing by ⌜below⌝*
CRESSIDA Here comes more.
PANDARUS Asses, fools, dolts. Chaff and bran, chaff and
bran. Porridge after meat. I could live and die i'th' eyes
of Troilus. Ne'er look, ne'er look, the eagles are gone.
Crows and daws, crows and daws. I had rather be such
a man as Troilus than Agamemnon and all Greece.
CRESSIDA There is among the Greeks Achilles, a better
man than Troilus.                                                244
PANDARUS Achilles? A drayman, a porter, a very camel.
CRESSIDA Well, well.
PANDARUS Well, well? Why, have you any discretion?
Have you any eyes? Do you know what a man is? Is

not birth, beauty, good shape, discourse, manhood,
learning, gentleness, virtue, youth, liberality, and so
forth, the spice and salt that season a man?                     251
CRESSIDA Ay, a minced man—and then to be baked with
no date in the pie, for then the man's date is out.
PANDARUS You are such another woman! One knows not
at what ward you lie.                                            255
CRESSIDA Upon my back to defend my belly, upon my wit
to defend my wiles, upon my secrecy to defend mine
honesty, my mask to defend my beauty, and you to
defend all these—and at all these wards I lie at a
thousand watches.                                                260
PANDARUS Say one of your watches.
CRESSIDA 'Nay, I'll watch you for that'—and that's one
of the chiefest of them too. If I cannot ward what I
would not have hit, I can watch you for telling how I
took the blow—unless it swell past hiding, and then
it's past watching.                                              266
PANDARUS You are such another!
            *Enter Boy*
BOY Sir, my lord would instantly speak with you.
PANDARUS Where?
BOY At your own house.                                           270
PANDARUS Good boy, tell him I come.            *Exit Boy*
I doubt he be hurt. Fare ye well, good niece.
CRESSIDA Adieu, uncle.
PANDARUS I'll be with you, niece, by and by.
CRESSIDA To bring, uncle?                                        275
PANDARUS Ay, a token from Troilus.
CRESSIDA By the same token, you are a bawd.
            *Exeunt Pandarus ⌜and Alexander⌝*
Words, vows, gifts, tears, and love's full sacrifice
He offers in another's enterprise;
But more in Troilus thousandfold I see                           280
Than in the glass of Pandar's praise may be.
Yet hold I off. Women are angels, wooing;
Things won are done. Joy's soul lies in the doing.
That she beloved knows naught that knows not this:
Men price the thing ungained more than it is.                    285
That she was never yet that ever knew
Love got so sweet as when desire did sue.
Therefore this maxim out of love I teach:
Achievement is command; ungained, beseech.
Then though my heart's contents firm love doth bear,
Nothing of that shall from mine eyes appear.        *Exit*

1.3    *Sennet. Enter Agamemnon, Nestor, Ulysses,*
       *Diomedes, and Menelaus, with others*
AGAMEMNON
Princes, what grief hath set the jaundice on your
      cheeks?
The ample proposition that hope makes
In all designs begun on earth below
Fails in the promised largeness. Checks and disasters
Grow in the veins of actions highest reared,              5
As knots, by the conflux of meeting sap,
Infects the sound pine and diverts his grain
Tortive and errant from his course of growth.
Nor, princes, is it matter new to us
That we come short of our suppose so far           10
That after seven years' siege yet Troy walls stand,
Sith every action that hath gone before,
Whereof we have record, trial did draw
Bias and thwart, not answering the aim
And that unbodied figure of the thought           15
That gave't surmisèd shape. Why then, you princes,

Do you with cheeks abashed behold our works,
And think them shames, which are indeed naught else
But the protractive trials of great Jove
To find persistive constancy in men?                         20
The fineness of which mettle is not found
In fortune's love—for then the bold and coward,
The wise and fool, the artist and unread,
The hard and soft, seem all affined and kin.
But in the wind and tempest of her frown         25
Distinction with a loud and powerful fan,
Puffing at all, winnows the light away,
And what hath mass or matter by itself
Lies rich in virtue and unmingled.

NESTOR
With due observance of thy godly seat,           30
Great Agamemnon, Nestor shall apply
Thy latest words. In the reproof of chance
Lies the true proof of men. The sea being smooth,
How many shallow bauble-boats dare sail
Upon her patient breast, making their way        35
With those of nobler bulk!
But let the ruffian Boreas once enrage
The gentle Thetis, and anon behold
The strong-ribbed barque through liquid mountains
    cut,
Bounding between the two moist elements           40
Like Perseus' horse. Where's then the saucy boat
Whose weak untimbered sides but even now
Co-rivalled greatness? Either to harbour fled,
Or made a toast for Neptune. Even so
Doth valour's show and valour's worth divide      45
In storms of fortune. For in her ray and brightness
The herd hath more annoyance by the breese
Than by the tiger; but when the splitting wind
Makes flexible the knees of knotted oaks
And flies flee under shade, why then the thing of
    courage,                                       50
As roused with rage, with rage doth sympathize,
And with an accent tuned in selfsame key
Retorts to chiding fortune.

ULYSSES                     Agamemnon,
Thou great commander, nerve and bone of Greece,
Heart of our numbers, soul and only spirit         55
In whom the tempers and the minds of all
Should be shut up, hear what Ulysses speaks.
Besides th'applause and approbation
The which, (to Agamemnon) most mighty for thy place
    and sway,
And thou, (to Nestor) most reverend for thy stretched-
    out life,                                       60
I give to both your speeches—which were such
As, Agamemnon, every hand of Greece
Should hold up high in brass, and such again
As, venerable Nestor, hatched in silver,
Should with a bond of air, strong as the axle-tree   65
On which the heavens ride, knit all Greeks' ears
To his experienced tongue—yet let it please both,
Thou (to Agamemnon) great, and (to Nestor) wise, to
    hear Ulysses speak.

AGAMEMNON
Speak, Prince of Ithaca, and be't of less expect
That matter needless, of importless burden,          70
Divide thy lips, than we are confident
When rank Thersites opes his mastic jaws
We shall hear music, wit, and oracle.

ULYSSES
Troy, yet upon his basis, had been down
And the great Hector's sword had lacked a master    75
But for these instances:
The specialty of rule hath been neglected.
And look how many Grecian tents do stand
Hollow upon this plain: so many hollow factions.
When that the general is not like the hive           80
To whom the foragers shall all repair,
What honey is expected? Degree being vizarded,
Th'unworthiest shows as fairly in the masque
⌈                                               ⌉.
The heavens themselves, the planets, and this centre
Observe degree, priority, and place,                 86
Infixture, course, proportion, season, form,
Office and custom, in all line of order.
And therefore is the glorious planet Sol
In noble eminence enthroned and sphered             90
Amidst the other, whose med'cinable eye
Corrects the ill aspects of planets evil
And posts like the commandment of a king,
Sans check, to good and bad. But when the planets
In evil mixture to disorder wander,                  95
What plagues and what portents, what mutiny?
What raging of the sea, shaking of earth?
Commotion in the winds, frights, changes, horrors
Divert and crack, rend and deracinate
The unity and married calm of states                100
Quite from their fixture. O when degree is shaked,
Which is the ladder to all high designs,
The enterprise is sick. How could communities,
Degrees in schools, and brotherhoods in cities,
Peaceful commerce from dividable shores,            105
The primogenity and due of birth,
Prerogative of age, crowns, sceptres, laurels,
But by degree stand in authentic place?
Take but degree away, untune that string,
And hark what discord follows. Each thing meets     110
In mere oppugnancy. The bounded waters
Should lift their bosoms higher than the shores
And make a sop of all this solid globe;
Strength should be lord of imbecility,
And the rude son should strike his father dead.     115
Force should be right—or rather, right and wrong,
Between whose endless jar justice resides,
Should lose their names, and so should justice too.
Then everything includes itself in power,
Power into will, will into appetite;                 120
And appetite, an universal wolf,
So doubly seconded with will and power,
Must make perforce an universal prey,
And last eat up himself. Great Agamemnon,
This chaos, when degree is suffocate,                125
Follows the choking.
And this neglection of degree it is
That by a pace goes backward in a purpose
It hath to climb. The general's disdained
By him one step below; he, by the next;              130
That next, by him beneath. So every step,
Exampled by the first pace that is sick
Of his superior, grows to an envious fever
Of pale and bloodless emulation.
And 'tis this fever that keeps Troy on foot,          135
Not her own sinews. To end a tale of length:
Troy in our weakness lives, not in her strength.

**NESTOR**

Most wisely hath Ulysses here discovered
The fever whereof all our power is sick.

**AGAMEMNON**

The nature of the sickness found, Ulysses,     140
What is the remedy?

**ULYSSES**

The great Achilles, whom opinion crowns
The sinew and the forehand of our host,
Having his ear full of his airy fame
Grows dainty of his worth, and in his tent     145
Lies mocking our designs. With him Patroclus
Upon a lazy bed the livelong day
Breaks scurrile jests
And, with ridiculous and awkward action
Which, slanderer, he 'imitation' calls,     150
He pageants us. Sometime, great Agamemnon,
Thy topless deputation he puts on,
And like a strutting player, whose conceit
Lies in his hamstring and doth think it rich
To hear the wooden dialogue and sound     155
'Twixt his stretched footing and the scaffoldage,
Such to-be-pitied and o'er-wrested seeming
He acts thy greatness in. And when he speaks
'Tis like a chime a-mending, with terms unsquared
Which from the tongue of roaring Typhon dropped
Would seem hyperboles. At this fusty stuff     161
The large Achilles on his pressed bed lolling
From his deep chest laughs out a loud applause,
Cries 'Excellent! 'Tis Agamemnon just.
Now play me Nestor, hem and stroke thy beard,     165
As he being dressed to some oration.'
That's done as near as the extremest ends
Of parallels, as like as Vulcan and his wife.
Yet god Achilles still cries, 'Excellent!
'Tis Nestor right. Now play him me, Patroclus,     170
Arming to answer in a night alarm'.
And then forsooth the faint defects of age
Must be the scene of mirth: to cough and spit,
And with a palsy, fumbling on his gorget,
Shake in and out the rivet. And at this sport     175
Sir Valour dies, cries, 'O enough, Patroclus!
Or give me ribs of steel. I shall split all
In pleasure of my spleen.' And in this fashion
All our abilities, gifts, natures, shapes,
Severals and generals of grace exact,     180
Achievements, plots, orders, preventions,
Excitements to the field or speech for truce,
Success or loss, what is or is not, serves
As stuff for these two to make paradoxes.

**NESTOR**

And in the imitation of these twain     185
Who, as Ulysses says, opinion crowns
With an imperial voice, many are infect.
Ajax is grown self-willed and bears his head
In such a rein, in full as proud a place
As broad Achilles, and keeps his tent like him,     190
Makes factious feasts, rails on our state of war
Bold as an oracle, and sets Thersites,
A slave whose gall coins slanders like a mint,
To match us in comparisons with dirt,
To weaken and discredit our exposure,     195
How rank so ever rounded in with danger.

**ULYSSES**

They tax our policy and call it cowardice,
Count wisdom as no member of the war,

Forestall prescience and esteem no act
But that of hand. The still and mental parts     200
That do contrive how many hands shall strike
When fitness calls them on, and know by measure
Of their observant toil the enemy's weight,
Why, this hath not a finger's dignity.
They call this 'bed-work', 'mapp'ry', 'closet war'.     205
So that the ram that batters down the wall,
For the great swinge and rudeness of his poise
They place before his hand that made the engine,
Or those that with the finesse of their souls
By reason guide his execution.     210

**NESTOR**

Let this be granted, and Achilles' horse
Makes many Thetis' sons.
     *Tucket*

**AGAMEMNON**               What trumpet?
Look, Menelaus.

**MENELAUS**

From Troy.
     *Enter Aeneas ⌈and a trumpeter⌉*

**AGAMEMNON** What would you fore our tent?

**AENEAS**

Is this great Agamemnon's tent I pray you?     215

**AGAMEMNON** Even this.

**AENEAS**

May one that is a herald and a prince
Do a fair message to his kingly ears?

**AGAMEMNON**

With surety stronger than Achilles' arm,
Fore all the Greekish heads, which with one voice     220
Call Agamemnon heart and general.

**AENEAS**

Fair leave and large security. How may
A stranger to those most imperial looks
Know them from eyes of other mortals?

**AGAMEMNON**               How?

**AENEAS**

Ay, I ask that I might waken reverence     225
And on the cheek be ready with a blush
Modest as morning when she coldly eyes
The youthful Phoebus.
Which is that god in office, guiding men?
Which is the high and mighty Agamemnon?     230

**AGAMEMNON** (*to the Greeks*)

This Trojan scorns us, or the men of Troy
Are ceremonious courtiers.

**AENEAS**

Courtiers as free, as debonair, unarmed,
As bending angels—that's their fame in peace.
But when they would seem soldiers they have galls,
Good arms, strong joints, true swords—and great
     Jove's acorn     236
Nothing so full of heart. But peace, Aeneas,
Peace, Trojan; lay thy finger on thy lips.
The worthiness of praise distains his worth,
If that the praised himself bring the praise forth.     240
But what, repining, the enemy commends,
That breath fame blows; that praise, sole pure,
     transcends.

**AGAMEMNON**

Sir, you of Troy, call you yourself Aeneas?

**AENEAS**

Ay, Greek, that is my name.

**AGAMEMNON**             What's your affair, I pray you?

AENEAS

Sir, pardon, 'tis for Agamemnon's ears.  245

AGAMEMNON

He hears naught privately that comes from Troy.

AENEAS

Nor I from Troy come not to whisper him.
I bring a trumpet to awake his ear,
To set his sense on the attentive bent,
And then to speak.

AGAMEMNON                Speak frankly as the wind.  250
It is not Agamemnon's sleeping hour.
That thou shalt know, Trojan, he is awake,
He tells thee so himself.

AENEAS                        Trumpet, blow loud.
Send thy brass voice through all these lazy tents,
And every Greek of mettle let him know  255
What Troy means fairly shall be spoke aloud.

_The trumpet sounds_

We have, great Agamemnon, here in Troy
A prince called Hector—Priam is his father—
Who in this dull and long-continued truce
Is resty grown. He bade me take a trumpet  260
And to this purpose speak: 'Kings, princes, lords,
If there be one among the fair'st of Greece
That holds his honour higher than his ease,
That seeks his praise more than he fears his peril,
That knows his valour and knows not his fear,  265
That loves his mistress more than in confession
With truant vows to her own lips he loves,
And dare avow her beauty and her worth
In other arms than hers—to him this challenge.
Hector in view of Trojans and of Greeks  270
Shall make it good, or do his best to do it:
He hath a lady wiser, fairer, truer,
Than ever Greek did compass in his arms,
And will tomorrow with his trumpet call
Midway between your tents and walls of Troy  275
To rouse a Grecian that is true in love.
If any come, Hector shall honour him.
If none, he'll say in Troy when he retires
The Grecian dames are sunburnt and not worth
The splinter of a lance.' Even so much.  280

AGAMEMNON

This shall be told our lovers, Lord Aeneas.
If none of them have soul in such a kind,
We left them all at home. But we are soldiers,
And may that soldier a mere recreant prove
That means not, hath not, or is not in love.  285
If then one is, or hath, or means to be,
That one meets Hector. If none else, I'll be he.

NESTOR (_to Aeneas_)

Tell him of Nestor, one that was a man
When Hector's grandsire sucked. He is old now,
But if there be not in our Grecian mould  290
One noble man that hath one spark of fire
To answer for his love, tell him from me
I'll hide my silver beard in a gold beaver
And in my vambrace put this withered brawn,
And meeting him will tell him that my lady  295
Was fairer than his grandam, and as chaste
As may be in the world. His youth in flood,
I'll prove this truth with my three drops of blood.

AENEAS

Now heavens forbid such scarcity of youth.

ULYSSES Amen.  300

AGAMEMNON

Fair Lord Aeneas, let me touch your hand.
To our pavilion shall I lead you first.
Achilles shall have word of this intent;
So shall each lord of Greece, from tent to tent.
Yourself shall feast with us before you go,  305
And find the welcome of a noble foe.

_Exeunt all but Ulysses and Nestor_

ULYSSES

Nestor!

NESTOR  What says Ulysses?

ULYSSES                    I have a young
Conception in my brain; be you my time
To bring it to some shape.

NESTOR                      What is't?

ULYSSES                              This 'tis:
Blunt wedges rive hard knots. The seeded pride  310
That hath to this maturity blown up
In rank Achilles must or now be cropped
Or, shedding, breed a nursery of like evil
To overbulk us all.

NESTOR                  Well, and how?

ULYSSES

This challenge that the gallant Hector sends,  315
However it is spread in general name,
Relates in purpose only to Achilles.

NESTOR

The purpose is perspicuous, even as substance
Whose grossness little characters sum up.
And, in the publication, make no strain  320
But that Achilles, were his brain as barren
As banks of Libya—though, Apollo knows,
'Tis dry enough—will with great speed of judgement,
Ay with celerity, find Hector's purpose
Pointing on him.  325

ULYSSES

And wake him to the answer, think you?

NESTOR

Yes, 'tis most meet. Who may you else oppose,
That can from Hector bring his honour off,
If not Achilles? Though't be a sportful combat,
Yet in this trial much opinion dwells,  330
For here the Trojans taste our dear'st repute
With their fin'st palate. And trust to me, Ulysses,
Our imputation shall be oddly poised
In this wild action: for the success,
Although particular, shall give a scantling  335
Of good or bad unto the general—
And in such indices, although small pricks
To their subsequent volumes, there is seen
The baby figure of the giant mass
Of things to come at large. It is supposed  340
He that meets Hector issues from our choice,
And choice, being mutual act of all our souls,
Makes merit her election, and doth boil,
As 'twere, from forth us all a man distilled
Out of our virtues—who miscarrying,  345
What heart from hence receives the conqu'ring part
To steel a strong opinion to themselves?
Which entertained, limbs are e'en his instruments,
In no less working than are swords and bows
Directive by the limbs.

ULYSSES                    Give pardon to my speech:  350
Therefore 'tis meet Achilles meet not Hector.
Let us like merchants show our foulest wares
And think perchance they'll sell. If not,

The lustre of the better yet to show
Shall show the better. Do not consent      355
That ever Hector and Achilles meet,
For both our honour and our shame in this
Are dogged with two strange followers.
NESTOR
I see them not with my old eyes. What are they?
ULYSSES
What glory our Achilles shares from Hector,      360
Were he not proud we all should wear with him.
But he already is too insolent,
And we were better parch in Afric sun
Than in the pride and salt scorn of his eyes,
Should he scape Hector fair. If he were foiled,      365
Why then we did our main opinion crush
In taint of our best man. No, make a lott'ry,
And by device let blockish Ajax draw
The sort to fight with Hector. Among ourselves
Give him allowance as the worthier man—      370
For that will physic the great Myrmidon,
Who broils in loud applause, and make him fall
His crest, that prouder than blue Iris bends.
If the dull brainless Ajax come safe off,
We'll dress him up in voices; if he fail,      375
Yet go we under our opinion still
That we have better men. But hit or miss,
Our project's life this shape of sense assumes:
Ajax employed plucks down Achilles' plumes.
NESTOR
Now, Ulysses, I begin to relish thy advice,      380
And I will give a taste of it forthwith
To Agamemnon. Go we to him straight.
Two curs shall tame each other; pride alone
Must tarre the mastiffs on, as 'twere their bone.
           *Exeunt*

**2.1**     *Enter Ajax and Thersites*
AJAX Thersites.
THERSITES Agamemnon—how if he had boils, full, all
over, generally?
AJAX Thersites.
THERSITES And those boils did run? Say so, did not the
General run then? Were not that a botchy core?     6
AJAX Dog.
THERSITES Then there would come some matter from him.
I see none now.
AJAX Thou bitch-wolf's son, canst thou not hear? Feel
then.            11
     *He strikes Thersites*
THERSITES The plague of Greece upon thee, thou mongrel
beef-witted lord!
AJAX Speak then, thou unsifted leaven, speak! I will beat
thee into handsomeness.         15
THERSITES I shall sooner rail thee into wit and holiness.
But I think thy horse will sooner con an oration than
thou learn a prayer without book.
     ⌈*Ajax strikes him*⌉
Thou canst strike, canst thou? A red murrain o' thy
jade's tricks.         20
AJAX Toad's stool!
     ⌈*He strikes Thersites*⌉
Learn me the proclamation.
THERSITES Dost thou think I have no sense, thou strikest
me thus?
AJAX The proclamation.        25
THERSITES Thou art proclaimed a fool, I think.
AJAX Do not, porcupine, do not. My fingers itch.

THERSITES I would thou didst itch from head to foot. An
I had the scratching of thee, I would make thee the
loathsomest scab in Greece.        30
AJAX I say, the proclamation.
THERSITES Thou grumblest and railest every hour on
Achilles, and thou art as full of envy at his greatness
as Cerberus is at Proserpina's beauty, ay, that thou
barkest at him.        35
AJAX Mistress Thersites.
THERSITES Thou shouldst strike him.
AJAX Cobloaf.
THERSITES He would pun thee into shivers with his fist,
as a sailor breaks a biscuit.        40
AJAX You whoreson cur.
     ⌈*He strikes Thersites*⌉
THERSITES Do! Do!
AJAX Thou stool for a witch.
     ⌈*He strikes Thersites*⌉
THERSITES Ay, do, do! Thou sodden-witted lord, thou hast
in thy skull no more brain than I have in mine elbows.
An *asnico* may tutor thee. Thou scurvy valiant ass,
thou art here but to thrash Trojans, and thou art
bought and sold among those of any wit like a barbarian
slave. If thou use to beat me, I will begin at thy heel
and tell what thou art by inches, thou thing of no
bowels, thou.        51
AJAX You dog.
THERSITES You scurvy lord.
AJAX You cur.
     ⌈*He strikes Thersites*⌉
THERSITES Mars his idiot! Do, rudeness! Do, camel, do,
do!        56
     *Enter Achilles and Patroclus*
ACHILLES
Why, how now, Ajax? Wherefore do ye thus?
How now, Thersites? What's the matter, man?
THERSITES You see him there? Do you?
ACHILLES Ay. What's the matter?        60
THERSITES Nay, look upon him.
ACHILLES So I do. What's the matter?
THERSITES Nay, but regard him well.
ACHILLES 'Well'? Why, I do so.
THERSITES But yet you look not well upon him. For
whomsoever you take him to be, he is Ajax.      66
ACHILLES I know that, fool.
THERSITES Ay, but 'that fool' knows not himself.
AJAX Therefore I beat thee.
THERSITES Lo, lo, lo, lo, what modicums of wit he utters.
His evasions have ears thus long. I have bobbed his
brain more than he has beat my bones. I will buy nine
sparrows for a penny, and his *pia mater* is not worth
the ninth part of a sparrow. This lord, Achilles—Ajax,
who wears his wit in his belly and his guts in his
head—I'll tell you what I say of him.      76
ACHILLES What?
THERSITES I say, this Ajax—
     ⌈*Ajax threatens to strike him*⌉
ACHILLES Nay, good Ajax.
THERSITES Has not so much wit—        80
     ⌈*Ajax threatens to strike him*⌉
ACHILLES (*to Ajax*) Nay, I must hold you.
THERSITES As will stop the eye of Helen's needle, for whom
he comes to fight.
ACHILLES Peace, fool.
THERSITES I would have peace and quietness, but the fool
will not. He, there, that he, look you there.      86

AJAX O thou damned cur I shall—
ACHILLES (*to Ajax*) Will you set your wit to a fool's?
THERSITES No, I warrant you, for a fool's will shame it.
PATROCLUS Good words, Thersites.                                        90
ACHILLES (*to Ajax*) What's the quarrel?
AJAX I bade the vile owl go learn me the tenor of the
proclamation, and he rails upon me.
THERSITES I serve thee not.
AJAX Well, go to, go to.                                                95
THERSITES I serve here voluntary.
ACHILLES Your last service was sufferance. 'Twas not
voluntary: no man is beaten voluntary. Ajax was here
the voluntary, and you as under an impress.         99
THERSITES E'en so. A great deal of your wit, too, lies in
your sinews, or else there be liars. Hector shall have a
great catch an a knock out either of your brains. A
were as good crack a fusty nut with no kernel.
ACHILLES What, with me too, Thersites?            104
THERSITES There's Ulysses and old Nestor, whose wit was
mouldy ere your grandsires had nails on their toes,
yoke you like draught oxen and make you plough up
the war.
ACHILLES What? What?
THERSITES Yes, good sooth. To Achilles! To, Ajax, to—
AJAX I shall cut out your tongue.                        111
THERSITES 'Tis no matter. I shall speak as much wit as
thou afterwards.
PATROCLUS No more words, Thersites, peace.
THERSITES I will hold my peace when Achilles' brach bids
me, shall I?                                                          116
ACHILLES There's for you, Patroclus.
THERSITES I will see you hanged like clodpolls ere I come
any more to your tents. I will keep where there is wit
stirring, and leave the faction of fools.                *Exit*
PATROCLUS A good riddance.                              121
ACHILLES (*to Ajax*)
Marry, this, sir, is proclaimed through all our host:
That Hector, by the fifth hour of the sun,
Will with a trumpet 'twixt our tents and Troy
Tomorrow morning call some knight to arms       125
That hath a stomach, and such a one that dare
Maintain—I know not what. 'Tis trash. Farewell.
AJAX Farewell. Who shall answer him?
ACHILLES
I know not. 'Tis put to lott'ry. Otherwise,
He knew his man.              ⌈*Exeunt Achilles and Patroclus*⌉
AJAX O, meaning you? I will go learn more of it.   ⌈*Exit*⌉

**2.2**  ⌈*Sennet.*⌉ *Enter King Priam, Hector, Troilus, Paris,*
*and Helenus*
PRIAM
After so many hours, lives, speeches spent,
Thus once again says Nestor from the Greeks:
'Deliver Helen, and all damage else—
As honour, loss of time, travail, expense,
Wounds, friends, and what else dear that is consumed
In hot digestion of this cormorant war—              6
Shall be struck off.' Hector, what say you to't?
HECTOR
Though no man lesser fears the Greeks than I,
As far as toucheth my particular, yet, dread Priam,
There is no lady of more softer bowels,                10
More spongy to suck in the sense of fear,
More ready to cry out, 'Who knows what follows?'
Than Hector is. The wound of peace is surety,
Surety secure; but modest doubt is called

The beacon of the wise, the tent that searches       15
To th' bottom of the worst. Let Helen go.
Since the first sword was drawn about this question,
Every tithe-soul, 'mongst many thousand dimes,
Hath been as dear as Helen—I mean, of ours.
If we have lost so many tenths of ours                 20
To guard a thing not ours—nor worth to us,
Had it our name, the value of one ten—
What merit's in that reason which denies
The yielding of her up?
TROILUS                      Fie, fie, my brother!
Weigh you the worth and honour of a king            25
So great as our dread father in a scale
Of common ounces? Will you with counters sum
The past-proportion of his infinite,
And buckle in a waist most fathomless
With spans and inches so diminutive                    30
As fears and reasons? Fie, for godly shame!
HELENUS
No marvel though you bite so sharp at reasons,
You are so empty of them. Should not our father
Bear the great sway of his affairs with reason
Because your speech hath none that tells him so?   35
TROILUS
You are for dreams and slumbers, brother priest.
You fur your gloves with 'reason'. Here are your
reasons:
You know an enemy intends you harm,
You know a sword employed is perilous,
And reason flies the object of all harm.                40
Who marvels then, when Helenus beholds
A Grecian and his sword, if he do set
The very wings of reason to his heels
And fly like chidden Mercury from Jove,
Or like a star disorbed? Nay, if we talk of reason,   45
Let's shut our gates and sleep. Manhood and honour
Should have hare hearts, would they but fat their
thoughts
With this crammed reason. Reason and respect
Make livers pale and lustihood deject.
HECTOR
Brother, she is not worth what she doth cost        50
The holding.
TROILUS             What's aught but as 'tis valued?
HECTOR
But value dwells not in particular will.
It holds his estimate and dignity
As well wherein 'tis precious of itself
As in the prizer. 'Tis mad idolatry                        55
To make the service greater than the god;
And the will dotes that is inclinable
To what infectiously itself affects
Without some image of th'affected merit.
TROILUS
I take today a wife, and my election                     60
Is led on in the conduct of my will;
My will enkindled by mine eyes and ears,
Two traded pilots 'twixt the dangerous shores
Of will and judgement. How may I avoid—
Although my will distaste what it elected—           65
The wife I chose? There can be no evasion
To blench from this and to stand firm by honour.
We turn not back the silks upon the merchant
When we have spoiled them; nor the remainder viands
We do not throw in unrespective sewer                70
Because we now are full. It was thought meet

Paris should do some vengeance on the Greeks.
Your breath of full consent bellied his sails;
The seas and winds, old wranglers, took a truce
And did him service. He touched the ports desired, 75
And for an old aunt whom the Greeks held captive
He brought a Grecian queen, whose youth and freshness
Wrinkles Apollo's and makes stale the morning.
Why keep we her? The Grecians keep our aunt.
Is she worth keeping? Why, she is a pearl 80
Whose price hath launched above a thousand ships
And turned crowned kings to merchants.
If you'll avouch 'twas wisdom Paris went—
As you must needs, for you all cried, 'Go, go!';
If you'll confess he brought home noble prize— 85
As you must needs, for you all clapped your hands
And cried, 'Inestimable!'—why do you now
The issue of your proper wisdoms rate,
And do a deed that never fortune did:
Beggar the estimation which you prized 90
Richer than sea and land? O theft most base,
That we have stol'n what we do fear to keep!
But thieves unworthy of a thing so stol'n,
That in their country did them that disgrace
We fear to warrant in our native place. 95

CASSANDRA ⌈within⌉
Cry, Trojans, cry!
PRIAM             What noise? What shriek is this?
TROILUS
'Tis our mad sister. I do know her voice.
CASSANDRA ⌈within⌉ Cry, Trojans!
HECTOR It is Cassandra.
    ⌈Enter Cassandra raving, with her hair about her
    ears⌉
CASSANDRA
Cry, Trojans, cry! Lend me ten thousand eyes 100
And I will fill them with prophetic tears.
HECTOR Peace, sister, peace.
CASSANDRA
Virgins and boys, mid-age, and wrinkled old,
Soft infancy that nothing canst but cry,
Add to my clamours. Let us pay betimes 105
A moiety of that mass of moan to come.
Cry, Trojans, cry! Practise your eyes with tears.
Troy must not be, nor goodly Ilium stand.
Our firebrand brother, Paris, burns us all.
Cry, Trojans, cry! Ah Helen, and ah woe! 110
Cry, cry 'Troy burns!'—or else let Helen go.
                            Exit
HECTOR
Now, youthful Troilus, do not these high strains
Of divination in our sister work
Some touches of remorse? Or is your blood
So madly hot that no discourse of reason, 115
Nor fear of bad success in a bad cause,
Can qualify the same?
TROILUS             Why, brother Hector,
We may not think the justness of each act
Such and no other than the event doth form it,
Nor once deject the courage of our minds 120
Because Cassandra's mad. Her brainsick raptures
Cannot distaste the goodness of a quarrel
Which hath our several honours all engaged
To make it gracious. For my private part,
I am no more touched than all Priam's sons, 125
And Jove forbid there should be done amongst us
Such things as might offend the weakest spleen
To fight for and maintain.

PARIS
Else might the world convince of levity
As well my undertakings as your counsels. 130
But I attest the gods, your full consent
Gave wings to my propension and cut off
All fears attending on so dire a project.
For what, alas, can these my single arms?
What propugnation is in one man's valour 135
To stand the push and enmity of those
This quarrel would excite? Yet I protest,
Were I alone to pass the difficulties
And had as ample power as I have will,
Paris should ne'er retract what he hath done 140
Nor faint in the pursuit.
PRIAM            Paris, you speak
Like one besotted on your sweet delights.
You have the honey still, but these the gall.
So to be valiant is no praise at all.
PARIS
Sir, I propose not merely to myself 145
The pleasures such a beauty brings with it,
But I would have the soil of her fair rape
Wiped off in honourable keeping her.
What treason were it to the ransacked queen,
Disgrace to your great worths, and shame to me, 150
Now to deliver her possession up
On terms of base compulsion? Can it be
That so degenerate a strain as this
Should once set footing in your generous bosoms?
There's not the meanest spirit on our party 155
Without a heart to dare or sword to draw
When Helen is defended; nor none so noble
Whose life were ill bestowed or death unfamed
Where Helen is the subject. Then I say:
Well may we fight for her whom we know well 160
The world's large spaces cannot parallel.
HECTOR
Paris and Troilus, you have both said well,
But on the cause and question now in hand
Have glossed but superficially—not much
Unlike young men, whom Aristotle thought 165
Unfit to hear moral philosophy.
The reasons you allege do more conduce
To the hot passion of distempered blood
Than to make up a free determination
'Twixt right and wrong; for pleasure and revenge 170
Have ears more deaf than adders to the voice
Of any true decision. Nature craves
All dues be rendered to their owners. Now,
What nearer debt in all humanity
Than wife is to the husband? If this law 175
Of nature be corrupted through affection,
And that great minds, of partial indulgence
To their benumbèd wills, resist the same,
There is a law in each well-ordered nation
To curb those raging appetites that are 180
Most disobedient and refractory.
If Helen then be wife to Sparta's king,
As it is known she is, these moral laws
Of nature and of nations speak aloud
To have her back returned. Thus to persist 185
In doing wrong extenuates not wrong,
But makes it much more heavy. Hector's opinion
Is this in way of truth—yet ne'ertheless,
My sprightly brethren, I propend to you
In resolution to keep Helen still; 190

For 'tis a cause that hath no mean dependence
Upon our joint and several dignities.

TROILUS

Why, there you touched the life of our design.
Were it not glory that we more affected
Than the performance of our heaving spleens,      195
I would not wish a drop of Trojan blood
Spent more in her defence. But, worthy Hector,
She is a theme of honour and renown,
A spur to valiant and magnanimous deeds,
Whose present courage may beat down our foes,     200
And fame in time to come canonize us—
For I presume brave Hector would not lose
So rich advantage of a promised glory
As smiles upon the forehead of this action
For the wide world's revenue.

HECTOR                               I am yours,     205
You valiant offspring of great Priamus.
I have a roisting challenge sent amongst
The dull and factious nobles of the Greeks
Will shriek amazement to their drowsy spirits.
I was advertised their great general slept          210
Whilst emulation in the army crept;
This I presume will wake him.        ⌈*Flourish.*⌉ *Exeunt*

**2.3**  *Enter Thersites*

THERSITES How now, Thersites? What, lost in the
labyrinth of thy fury? Shall the elephant Ajax carry it
thus? He beats me and I rail at him. O worthy
satisfaction! Would it were otherwise: that I could beat
him whilst he railed at me. 'Sfoot, I'll learn to conjure
and raise devils but I'll see some issue of my spiteful
execrations. Then there's Achilles: a rare engineer. If
Troy be not taken till these two undermine it, the walls
will stand till they fall of themselves. O thou great
thunder-darter of Olympus, forget that thou art Jove,
the king of gods; and Mercury, lose all the serpentine
craft of thy caduceus, if ye take not that little, little,
less than little wit from them that they have—which
short-armed ignorance itself knows is so abundant-
scarce it will not in circumvention deliver a fly from a
spider without drawing their massy irons and cutting
the web. After this, the vengeance on the whole camp—
or rather, the Neapolitan bone-ache, for that methinks
is the curse dependent on those that war for a placket.
I have said my prayers, and devil Envy say 'Amen'.—
What ho! My lord Achilles!                             21

*Enter Patroclus* ⌈*at the door to the tent*⌉

PATROCLUS Who's there? Thersites? Good Thersites, come
in and rail.                                    ⌈*Exit*⌉

THERSITES If I could ha' remembered a gilt counterfeit,
thou wouldst not have slipped out of my contemplation;
but it is no matter. Thyself upon thyself! The common
curse of mankind, folly and ignorance, be thine in great
revenue! Heaven bless thee from a tutor, and discipline
come not near thee! Let thy blood be thy direction till
thy death! Then if she that lays thee out says thou art
a fair corpse, I'll be sworn and sworn upon't she never
shrouded any but lazars.                                32

⌈*Enter Patroclus*⌉
Amen.—Where's Achilles?

PATROCLUS What, art thou devout? Wast thou in prayer?
THERSITES Ay. The heavens hear me!                      35
PATROCLUS Amen.

*Enter Achilles*

ACHILLES Who's there?

PATROCLUS Thersites, my lord.                           38
ACHILLES Where? Where? O where?—Art thou come?
Why, my cheese, my digestion, why hast thou not
served thyself into my table so many meals? Come:
what's Agamemnon?
THERSITES Thy commander, Achilles.—Then tell me,
Patroclus, what's Achilles?                             44
PATROCLUS Thy lord, Thersites. Then tell me, I pray thee,
what's Thersites?
THERSITES Thy knower, Patroclus. Then tell me, Patroclus,
what art thou?
PATROCLUS Thou mayst tell, that knowest.
ACHILLES O tell, tell.                                  50
THERSITES I'll decline the whole question. Agamemnon
commands Achilles, Achilles is my lord, I am Patroclus'
knower, and Patroclus is a fool.
PATROCLUS You rascal.
THERSITES Peace, fool, I have not done.                 55
ACHILLES (*to Patroclus*) He is a privileged man.—Proceed,
Thersites.
THERSITES Agamemnon is a fool, Achilles is a fool,
Thersites is a fool, and as aforesaid Patroclus is a fool.
ACHILLES Derive this. Come.                             60
THERSITES Agamemnon is a fool to offer to command
Achilles; Achilles is a fool to be commanded of
Agamemnon; Thersites is a fool to serve such a fool;
and Patroclus is a fool positive.
PATROCLUS Why am I a fool?                              65
THERSITES Make that demand to the Creator. It suffices
me thou art. Look you, who comes here?

*Enter Agamemnon, Ulysses, Nestor, Diomedes,*
*Ajax, and Calchas*

ACHILLES Patroclus, I'll speak with nobody.—Come in
with me, Thersites.                                *Exit*
THERSITES Here is such patchery, such juggling and such
knavery. All the argument is a whore and a cuckold.
A good quarrel to draw emulous factions and bleed to
death upon. Now the dry serpigo on the subject, and
war and lechery confound all.                       *Exit*
AGAMEMNON (*to Patroclus*) Where is Achilles?           75
PATROCLUS
Within his tent; but ill-disposed, my lord.
AGAMEMNON
Let it be known to him that we are here.
He faced our messengers, and we lay by
Our appertainments, visiting of him.
Let him be told so, lest perchance he think       80
We dare not move the question of our place,
Or know not what we are.
PATROCLUS                    I shall so say to him.  ⌈*Exit*⌉
ULYSSES
We saw him at the opening of his tent.
He is not sick.                                     84
AJAX Yes, lion-sick: sick of proud heart. You may call it
'melancholy' if you will favour the man, but by my
head 'tis pride. But why? Why? Let him show us the
cause. ⌈*To Agamemnon*⌉ A word, my lord.
⌈*Ajax and Agamemnon talk apart*⌉
NESTOR What moves Ajax thus to bay at him?
ULYSSES Achilles hath inveigled his fool from him.     90
NESTOR Who? Thersites?
ULYSSES He.
NESTOR Then will Ajax lack matter, if he have lost his
argument.
ULYSSES No, you see, he *is* his argument that *has* his
argument: Achilles.                                 96

NESTOR All the better—their fraction is more our wish
  than their faction. But it was a strong council that a
  fool could disunite.
ULYSSES The amity that wisdom knits not, folly may easily
  untie.                                               101

      *Enter Patroclus*
Here comes Patroclus.
NESTOR No Achilles with him.
ULYSSES The elephant hath joints, but none for courtesy:
  his legs are legs for necessity, not for flexure.     105
PATROCLUS (*to Agamemnon*)
  Achilles bids me say he is much sorry
  If anything more than your sport and pleasure
  Did move your greatness and this noble state
  To call upon him. He hopes it is no other
  But for your health and your digestion's sake:     110
  An after-dinner's breath.
AGAMEMNON                   Hear you, Patroclus.
  We are too well acquainted with these answers.
  But his evasion, winged thus swift with scorn,
  Cannot outfly our apprehensions.
  Much attribute he hath, and much the reason     115
  Why we ascribe it to him. Yet all his virtues,
  Not virtuously on his own part beheld,
  Do in our eyes begin to lose their gloss,
  Yea, and like fair fruit in an unwholesome dish
  Are like to rot untasted. Go and tell him     120
  We come to speak with him—and you shall not sin
  If you do say we think him over-proud
  And under-honest, in self-assumption greater
  Than in the note of judgement. And worthier than
    himself
  Here tend the savage strangeness he puts on,     125
  Disguise the holy strength of their command,
  And underwrite in an observing kind
  His humorous predominance—yea, watch
  His pettish lunes, his ebbs, his flows, as if
  The passage and whole carriage of this action     130
  Rode on his tide. Go tell him this, and add
  That if he overhold his price so much
  We'll none of him, but let him, like an engine
  Not portable, lie under this report:
  'Bring action hither, this cannot go to war.'     135
  A stirring dwarf we do allowance give
  Before a sleeping giant. Tell him so.
PATROCLUS
  I shall, and bring his answer presently.
AGAMEMNON
  In second voice we'll not be satisfied;
  We come to speak with him.—Ulysses, enter you.   140
      *Exit Ulysses ⌈with Patroclus⌉*
AJAX What is he more than another?
AGAMEMNON No more than what he thinks he is.
AJAX Is he so much? Do you not think he thinks himself
  a better man than I am?
AGAMEMNON No question.                       145
AJAX Will you subscribe his thought, and say he is?
AGAMEMNON No, noble Ajax. You are as strong, as
  valiant, as wise, no less noble, much more gentle, and
  altogether more tractable.
AJAX Why should a man be proud? How doth pride
  grow? I know not what it is.                   151
AGAMEMNON Your mind is the clearer, Ajax, and your
  virtues the fairer. He that is proud eats up himself.
  Pride is his own glass, his own trumpet, his own
  chronicle—and whatever praises itself but in the deed
  devours the deed in the praise.              156

    *Enter Ulysses*
AJAX I do hate a proud man as I hate the engendering
  of toads.
NESTOR (*aside*) Yet he loves himself. Is't not strange?
ULYSSES
  Achilles will not to the field tomorrow.     160
AGAMEMNON
  What's his excuse?
ULYSSES              He doth rely on none,
  But carries on the stream of his dispose
  Without observance or respect of any,
  In will peculiar and in self-admission.
AGAMEMNON
  Why, will he not, upon our fair request,     165
  Untent his person and share the air with us?
ULYSSES
  Things small as nothing, for request's sake only,
  He makes important. Possessed he is with greatness,
  And speaks not to himself but with a pride
  That quarrels at self-breath. Imagined worth     170
  Holds in his blood such swoll'n and hot discourse
  That 'twixt his mental and his active parts
  Kingdomed Achilles in commotion rages
  And batters 'gainst himself. What should I say?
  He is so plaguy proud that the death tokens of it   175
  Cry 'No recovery'.
AGAMEMNON           Let Ajax go to him.
  (*To Ajax*) Dear lord, go you and greet him in his tent.
  'Tis said he holds you well and will be led,
  At your request, a little from himself.
ULYSSES
  O Agamemnon, let it not be so.     180
  We'll consecrate the steps that Ajax makes
  When they go from Achilles. Shall the proud lord
  That bastes his arrogance with his own seam
  And never suffers matter of the world
  Enter his thoughts, save such as do revolve     185
  And ruminate himself—shall he be worshipped
  Of that we hold an idol more than he?
  No, this thrice-worthy and right valiant lord
  Must not so stale his palm, nobly acquired,
  Nor by my will assubjugate his merit,     190
  As amply titled as Achilles' is,
  By going to Achilles—
  That were to enlard his fat-already pride
  And add more coals to Cancer when he burns
  With entertaining great Hyperion.     195
  This lord go to him? Jupiter forbid,
  And say in thunder 'Achilles, go to him'.
NESTOR (*aside to Diomedes*)
  O this is well. He rubs the vein of him.
DIOMEDES (*aside to Nestor*)
  And how his silence drinks up this applause.
AJAX
  If I go to him, with my armèd fist     200
  I'll pash him o'er the face.
AGAMEMNON           O no, you shall not go.
AJAX
  An a be proud with me, I'll feeze his pride.
  Let me go to him.
ULYSSES
  Not for the worth that hangs upon our quarrel.
AJAX A paltry insolent fellow.              205
NESTOR (*aside*) How he describes himself!
AJAX Can he not be sociable?
ULYSSES (*aside*) The raven chides blackness.

AJAX I'll let his humour's blood.                                   209
AGAMEMNON (aside) He will be the physician that should
  be the patient.
AJAX An all men were o' my mind—
ULYSSES (aside) Wit would be out of fashion.
AJAX A should not bear it so. A should eat swords first.
  Shall pride carry it?                                             215
NESTOR (aside) An't would, you'd carry half.
⌈AJAX⌉ A would have ten shares.
⌈ULYSSES⌉ (aside) I will knead him; I'll make him supple.
  He's not yet through warm.
NESTOR (aside) Farce him with praises. Pour in, pour in!
  His ambition is dry.                                              221
ULYSSES (to Agamemnon)
  My lord, you feed too much on this dislike.
NESTOR (to Agamemnon)
  Our noble general, do not do so.
DIOMEDES (to Agamemnon)
  You must prepare to fight without Achilles.
ULYSSES
  Why, 'tis this naming of him does him harm.                      225
  Here is a man—but 'tis before his face.
  I will be silent.
NESTOR                 Wherefore should you so?
  He is not emulous, as Achilles is.
ULYSSES
  Know the whole world he is as valiant—
AJAX A whoreson dog, that shall palter thus with us—
  would he were a Trojan!                                          231
NESTOR
  What a vice were it in Ajax now—
ULYSSES
  If he were proud—
DIOMEDES              Or covetous of praise—
ULYSSES
  Ay, or surly borne—
DIOMEDES                Or strange, or self-affected.
ULYSSES (to Ajax)
  Thank the heavens, lord, thou art of sweet composure.
  Praise him that got thee, she that gave thee suck.               236
  Famed be thy tutor, and thy parts of nature
  Thrice famed beyond, beyond all erudition.
  But he that disciplined thine arms to fight—
  Let Mars divide eternity in twain,                               240
  And give him half. And for thy vigour,
  Bull-bearing Milo his addition yield
  To sinewy Ajax. I will not praise thy wisdom,
  Which like a bourn, a pale, a shore confines
  Thy spacious and dilated parts. Here's Nestor,                   245
  Instructed by the antiquary times:
  He must, he is, he cannot but be, wise.
  But pardon, father Nestor: were your days
  As green as Ajax', and your brain so tempered,
  You should not have the eminence of him,                         250
  But be as Ajax.
AJAX              Shall I call you father?
ULYSSES
  Ay, my good son.
DIOMEDES            Be ruled by him, Lord Ajax.
ULYSSES (to Agamemnon)
  There is no tarrying here: the hart Achilles
  Keeps thicket. Please it our great general
  To call together all his state of war.                           255
  Fresh kings are come today to Troy; tomorrow
  We must with all our main of power stand fast.

  And here's a lord, come knights from east to west
  And cull their flower, Ajax shall cope the best.
AGAMEMNON
  Go we to counsel. Let Achilles sleep.                            260
  Light boats sail swift, though greater hulks draw
    deep.                                            Exeunt

3.1    Music sounds within. Enter Pandarus ⌈at one door⌉
       and a Servant ⌈at another door⌉
PANDARUS Friend? You. Pray you, a word. Do not you
  follow the young Lord Paris?
SERVANT Ay, sir, when he goes before me.
PANDARUS You depend upon him, I mean.
SERVANT Sir, I do depend upon the Lord.                              5
PANDARUS You depend upon a notable gentleman; I must
  needs praise him.
SERVANT The Lord be praised!
PANDARUS You know me—do you not?
SERVANT Faith, sir, superficially.                                  10
PANDARUS Friend, know me better. I am the Lord
  Pandarus.
SERVANT I hope I shall know your honour better.
PANDARUS I do desire it.
SERVANT You are in the state of grace?                              15
PANDARUS Grace? Not so, friend. 'Honour' and 'lordship'
  are my titles. What music is this?
SERVANT I do but partly know, sir. It is music in parts.
PANDARUS Know you the musicians?
SERVANT Wholly, sir.                                                20
PANDARUS Who play they to?
SERVANT To the hearers, sir.
PANDARUS At whose pleasure, friend?
SERVANT At mine, sir, and theirs that love music.
PANDARUS 'Command' I mean, friend.                                  25
SERVANT Who shall I command, sir?
PANDARUS Friend, we understand not one another. I am
  too courtly and thou too cunning. At whose request
  do these men play?
SERVANT That's to't indeed, sir. Marry, sir, at the request
  of Paris my lord, who's there in person; with him, the
  mortal Venus, the heart-blood of beauty, love's visible
  soul—
PANDARUS Who, my cousin Cressida?                                   34
SERVANT No, sir, Helen. Could not you find out that by
  her attributes?
PANDARUS It should seem, fellow, that thou hast not seen
  the Lady Cressid. I come to speak with Paris from the
  Prince Troilus. I will make a complimental assault upon
  him, for my business seethes.                                     40
SERVANT Sodden business! There's a stewed phrase,
  indeed.
       Enter Paris and Helen, attended ⌈by musicians⌉
PANDARUS Fair be to you, my lord, and to all this fair
  company. Fair desires in all fair measure fairly guide
  them—especially to you, fair Queen. Fair thoughts be
  your fair pillow.                                                 46
HELEN Dear lord, you are full of fair words.
PANDARUS You speak your fair pleasure, sweet Queen. (To
  Paris) Fair prince, here is good broken music.
PARIS You have broke it, cousin, and by my life you shall
  make it whole again. You shall piece it out with a piece
  of your performance.—Nell, he is full of harmony.                52
PANDARUS Truly, lady, no.
HELEN O sir.
       ⌈She tickles him⌉

PANDARUS Rude, in sooth, in good sooth very rude.    55
PARIS Well said, my lord. Will you say so in fits?
PANDARUS I have business to my lord, dear Queen.—My
  lord, will you vouchsafe me a word?
HELEN Nay, this shall not hedge us out. We'll hear you
  sing, certainly.    60
PANDARUS Well, sweet Queen, you are pleasant with
  me.—But marry, thus, my lord: my dear lord and most
  esteemed friend, your brother Troilus—
HELEN My lord Pandarus, honey-sweet lord.
PANDARUS Go to, sweet Queen, go to!—commends himself
  most affectionately to you.    66
HELEN You shall not bob us out of our melody. If you do,
  our melancholy upon your head.
PANDARUS Sweet Queen, sweet Queen, that's a sweet
  Queen. Ay, faith—    70
HELEN And to make a sweet lady sad is a sour offence.
PANDARUS Nay, that shall not serve your turn; that shall
  it not, in truth, la. Nay, I care not for such words. No,
  no.—And, my lord, he desires you that, if the King call
  for him at supper, you will make his excuse.    75
HELEN My lord Pandarus.
PANDARUS What says my sweet Queen, my very very
  sweet Queen?
PARIS What exploit's in hand? Where sups he tonight?
HELEN Nay, but my lord—    80
PANDARUS What says my sweet Queen? My cousin will
  fall out with you.
HELEN (to Paris) You must not know where he sups.
PARIS I'll lay my life, with my dispenser Cressida.
PANDARUS No, no! No such matter. You are wide. Come,
  your dispenser is sick.    86
PARIS Well, I'll make 's excuse.
PANDARUS Ay, good my lord. Why should you say
  Cressida? No, your poor dispenser's sick.
PARIS 'I spy.'    90
PANDARUS You spy? What do you spy?—⌈To a musician⌉
  Come, give me an instrument.—Now, sweet Queen.
HELEN Why, this is kindly done!
PANDARUS My niece is horrible in love with a thing you
  have, sweet Queen.    95
HELEN She shall have it, my lord—if it be not my lord
  Paris.
PANDARUS He? No, she'll none of him. They two are
  twain.    99
HELEN Falling in, after falling out, may make them three.
PANDARUS Come, come, I'll hear no more of this. I'll sing
  you a song now.
HELEN Ay, ay, prithee. Now by my troth, sweet lord, thou
  hast a fine forehead.
  ⌈She strokes his forehead⌉
PANDARUS Ay, you may, you may.    105
HELEN Let thy song be love. 'This love will undo us all.'
  O Cupid, Cupid, Cupid!
PANDARUS Love? Ay, that it shall, i'faith.
PARIS Ay, good now, 'Love, love, nothing but love'.
PANDARUS In good truth, it begins so.    110

(Sings)
  Love, love, nothing but love, still love, still more!
      For O love's bow
      Shoots buck and doe.
      The shaft confounds
      Not that it wounds,    115
  But tickles still the sore.

These lovers cry 'O! O!', they die.
    Yet that which seems the wound to kill
  Doth turn 'O! O!' to 'ha ha he!'
    So dying love lives still.    120
    'O! O!' a while, but 'ha ha ha!'
    'O! O!' groans out for 'ha ha ha!'—

Heigh-ho.
HELEN In love—ay, faith, to the very tip of the nose.
PARIS He eats nothing but doves, love, and that breeds
  hot blood, and hot blood begets hot thoughts, and hot
  thoughts beget hot deeds, and hot deeds is love.
PANDARUS Is this the generation of love: hot blood, hot
  thoughts, and hot deeds? Why, they are vipers. Is love
  a generation of vipers?    130
  ⌈Alarum⌉
  Sweet lord, who's afield today?
PARIS Hector, Deiphobus, Helenus, Antenor, and all the
  gallantry of Troy. I would fain have armed today, but
  my Nell would not have it so. How chance my brother
  Troilus went not?    135
HELEN He hangs the lip at something. You know all, Lord
  Pandarus.
PANDARUS Not I, honey-sweet Queen. I long to hear how
  they sped today.—You'll remember your brother's
  excuse?    140
PARIS To a hair.
PANDARUS Farewell, sweet Queen.
HELEN Commend me to your niece.
PANDARUS I will, sweet Queen.               Exit
    Sound a retreat
PARIS
  They're come from field. Let us to Priam's hall    145
  To greet the warriors. Sweet Helen, I must woo you
  To help unarm our Hector. His stubborn buckles,
  With these your white enchanting fingers touched,
  Shall more obey than to the edge of steel
  Or force of Greekish sinews. You shall do more    150
  Than all the island kings: disarm great Hector.
HELEN
  'Twill make us proud to be his servant, Paris;
  Yea, what he shall receive of us in duty
  Gives us more palm in beauty than we have—
  Yea, overshines ourself.
PARIS                Sweet above thought, I love thee!
                                    Exeunt

**3.2**   *Enter Pandarus ⌈at one door⌉ and Troilus' man ⌈at
    another door⌉*
PANDARUS How now, where's thy master? At my cousin
  Cressida's?
MAN No, sir, he stays for you to conduct him thither.
    *Enter Troilus*
PANDARUS O here he comes.—How now, how now?
TROILUS Sirrah, walk off.            *Exit Man*
PANDARUS Have you seen my cousin?    6
TROILUS
  No, Pandarus, I stalk about her door
  Like a strange soul upon the Stygian banks
  Staying for waftage. O be thou my Charon,
  And give me swift transportation to those fields    10
  Where I may wallow in the lily beds
  Proposed for the deserver. O gentle Pandar,
  From Cupid's shoulder pluck his painted wings
  And fly with me to Cressid.

PANDARUS Walk here i'th' orchard. I'll bring her straight.
*Exit*

TROILUS
I am giddy. Expectation whirls me round.                     16
Th'imaginary relish is so sweet
That it enchants my sense. What will it be
When that the wat'ry palates taste indeed
Love's thrice-repurèd nectar? Death, I fear me,              20
Swooning destruction, or some joy too fine,
Too subtle-potent, tuned too sharp in sweetness
For the capacity of my ruder powers.
I fear it much, and I do fear besides
That I shall lose distinction in my joys,                    25
As doth a battle when they charge on heaps
The enemy flying.
    *Enter Pandarus*
PANDARUS She's making her ready. She'll come straight.
You must be witty now. She does so blush, and fetches
her wind so short as if she were frayed with a spirit.
I'll fetch her. It is the prettiest villain! She fetches her
breath as short as a new-ta'en sparrow.         *Exit*

TROILUS
Even such a passion doth embrace my bosom.
My heart beats thicker than a feverous pulse,
And all my powers do their bestowing lose,                   35
Like vassalage at unawares encount'ring
The eye of majesty.
    *Enter Pandarus, with Cressida ⌈veiled⌉*
PANDARUS (*to Cressida*) Come, come, what need you blush?
Shame's a baby. (*To Troilus*) Here she is now. Swear
the oaths now to her that you have sworn to me. (*To
Cressida*) What, are you gone again? You must be
watched ere you be made tame, must you? Come your
ways, come your ways. An you draw backward, we'll
put you i'th' thills. (*To Troilus*) Why do you not speak
to her? (*To Cressida*) Come, draw this curtain, and let's
see your picture. ⌈*He unveils her*⌉ Alas the day! How
loath you are to offend daylight! An't were dark, you'd
close sooner. So, so. (*To Troilus*) Rub on, and kiss the
mistress. (*They kiss*) How now, a kiss in fee farm! Build
there, carpenter, the air is sweet. Nay, you shall fight
your hearts out ere I part you. The falcon as the tercel,
for all the ducks i'th' river. Go to, go to.              52
TROILUS You have bereft me of all words, lady.
PANDARUS Words pay no debts; give her deeds. But she'll
bereave you o'th' deeds too, if she call your activity in
question. (*They kiss*) What, billing again? Here's 'in
witness whereof the parties interchangeably'. Come in,
come in. I'll go get a fire.                        *Exit*
CRESSIDA Will you walk in, my lord?                       59
TROILUS O Cressida, how often have I wished me thus.
CRESSIDA Wished, my lord? The gods grant—O, my lord!
TROILUS What should they grant? What makes this pretty
abruption? What too-curious dreg espies my sweet lady
in the fountain of our love?                             64
CRESSIDA More dregs than water, if my fears have eyes.
TROILUS Fears make devils of cherubims; they never see
truly.
CRESSIDA Blind fear, that seeing reason leads, finds safer
footing than blind reason, stumbling without fear. To
fear the worst oft cures the worse.                      70
TROILUS O let my lady apprehend no fear. In all Cupid's
pageant there is presented no monster.
CRESSIDA Nor nothing monstrous neither?
TROILUS Nothing but our undertakings, when we vow to
weep seas, live in fire, eat rocks, tame tigers, thinking
it harder for our mistress to devise imposition enough

than for us to undergo any difficulty imposed. This is
the monstruosity in love, lady—that the will is infinite
and the execution confined; that the desire is boundless
and the act a slave to limit.                            80
CRESSIDA They say all lovers swear more performance
than they are able, and yet reserve an ability that they
never perform: vowing more than the perfection of ten,
and discharging less than the tenth part of one. They
that have the voice of lions and the act of hares, are
they not monsters?                                       86
TROILUS Are there such? Such are not we. Praise us as
we are tasted; allow us as we prove. Our head shall
go bare till merit crown it. No perfection in reversion
shall have a praise in present. We will not name desert
before his birth, and being born his addition shall be
humble. Few words to fair faith. Troilus shall be such
to Cressid as what envy can say worst shall be a mock
for his truth; and what truth can speak truest, not
truer than Troilus.                                      95
CRESSIDA Will you walk in, my lord?
    *Enter Pandarus*
PANDARUS What, blushing still? Have you not done
talking yet?
CRESSIDA Well, uncle, what folly I commit I dedicate to
you.                                                     100
PANDARUS I thank you for that. If my lord get a boy of
you, you'll give him me. Be true to my lord. If he
flinch, chide me for it.
TROILUS (*to Cressida*) You know now your hostages: your
uncle's word and my firm faith.                          105
PANDARUS Nay, I'll give my word for her too. Our kindred,
though they be long ere they are wooed, they are
constant being won. They are burrs, I can tell you:
they'll stick where they are thrown.
CRESSIDA
Boldness comes to me now, and brings me heart.           110
Prince Troilus, I have loved you night and day
For many weary months.
TROILUS
Why was my Cressid then so hard to win?
CRESSIDA
Hard to seem won; but I was won, my lord,
With the first glance that ever—pardon me:               115
If I confess much, you will play the tyrant.
I love you now, but till now not so much
But I might master it. In faith, I lie:
My thoughts were like unbridled children, grown
Too headstrong for their mother. See, we fools!          120
Why have I blabbed? Who shall be true to us,
When we are so unsecret to ourselves?
But though I loved you well, I wooed you not—
And yet, good faith, I wished myself a man,
Or that we women had men's privilege                     125
Of speaking first. Sweet, bid me hold my tongue,
For in this rapture I shall surely speak
The thing I shall repent. See, see, your silence,
Cunning in dumbness, in my weakness draws
My soul of counsel from me. Stop my mouth.               130
TROILUS
And shall, albeit sweet music issues thence.
    *He kisses her*
PANDARUS Pretty, i' faith.
CRESSIDA (*to Troilus*)
My lord, I do beseech you pardon me.
'Twas not my purpose thus to beg a kiss.
I am ashamed. O heavens, what have I done?               135
For this time I will take my leave, my lord.

TROILUS Your leave, sweet Cressid?

PANDARUS Leave? An you take leave till tomorrow morning—

CRESSIDA

Pray you, content you.

TROILUS                 What offends you, lady?   140

CRESSIDA Sir, mine own company.

TROILUS You cannot shun yourself.

CRESSIDA Let me go and try.

I have a kind of self resides with you—

But an unkind self, that itself will leave   145

To be another's fool. Where is my wit?

I would be gone. I speak I know not what.

TROILUS

Well know they what they speak that speak so wisely.

CRESSIDA

Perchance, my lord, I show more craft than love,

And fell so roundly to a large confession   150

To angle for your thoughts. But you are wise,

Or else you love not—for to be wise and love

Exceeds man's might: that dwells with gods above.

TROILUS

O that I thought it could be in a woman—

As, if it can, I will presume in you—   155

To feed for aye her lamp and flames of love,

To keep her constancy in plight and youth,

Outliving beauty's outward, with a mind

That doth renew swifter than blood decays;

Or that persuasion could but thus convince me   160

That my integrity and truth to you

Might be affronted with the match and weight

Of such a winnowed purity in love.

How were I then uplifted! But alas,

I am as true as truth's simplicity,   165

And simpler than the infancy of truth.

CRESSIDA

In that I'll war with you.

TROILUS              O virtuous fight,

When right with right wars who shall be most right.

True swains in love shall in the world to come

Approve their truth by Troilus. When their rhymes,

Full of protest, of oath and big compare,   171

Wants similes, truth tired with iteration—

'As true as steel, as plantage to the moon,

As sun to day, as turtle to her mate,

As iron to adamant, as earth to th' centre'—   175

Yet, after all comparisons of truth,

As truth's authentic author to be cited,

'As true as Troilus' shall crown up the verse

And sanctify the numbers.

CRESSIDA            Prophet may you be!

If I be false, or swerve a hair from truth,   180

When time is old and hath forgot itself,

When water drops have worn the stones of Troy

And blind oblivion swallowed cities up,

And mighty states characterless are grated

To dusty nothing, yet let memory   185

From false to false among false maids in love

Upbraid my falsehood. When they've said, 'as false

As air, as water, wind or sandy earth,

As fox to lamb, or wolf to heifer's calf,

Pard to the hind, or stepdame to her son',   190

Yea, let them say, to stick the heart of falsehood,

'As false as Cressid'.

PANDARUS Go to, a bargain made. Seal it, seal it. I'll be the witness. Here I hold your hand; here, my cousin's.

If ever you prove false one to another, since I have taken such pain to bring you together, let all pitiful goers-between be called to the world's end after my name: call them all panders. Let all constant men be Troiluses, all false women Cressids, and all brokers-between panders. Say 'Amen'.   200

TROILUS Amen.

CRESSIDA Amen.

PANDARUS Amen. Whereupon I will show you a chamber with a bed—which bed, because it shall not speak of your pretty encounters, press it to death. Away!   205

*Exeunt Troilus and Cressida*

And Cupid grant all tongue-tied maidens here

Bed, chamber, pander to provide this gear.     *Exit*

3.3   *Flourish. Enter Ulysses, Diomedes, Nestor,*
       *Agamemnon, Menelaus, Ajax, and Calchas*

CALCHAS

Now, princes, for the service I have done you,

Th'advantage of the time prompts me aloud

To call for recompense. Appear it to your mind

That through the sight I bear in things to come

I have abandoned Troy, left my profession,   5

Incurred a traitor's name, exposed myself

From certain and possessed conveniences

To doubtful fortunes, sequest'ring from me all

That time, acquaintance, custom, and condition

Made tame and most familiar to my nature,   10

And here to do you service am become

As new into the world, strange, unacquainted.

I do beseech you, as in way of taste,

To give me now a little benefit

Out of those many registered in promise   15

Which you say live to come in my behalf.

AGAMEMNON

What wouldst thou of us, Trojan? Make demand.

CALCHAS

You have a Trojan prisoner called Antenor,

Yesterday took. Troy holds him very dear.

Oft have you—often have you thanks therefor—   20

Desired my Cressid in right great exchange,

Whom Troy hath still denied. But this Antenor

I know is such a wrest in their affairs

That their negotiations all must slack,

Wanting his manage, and they will almost   25

Give us a prince of blood, a son of Priam,

In change of him. Let him be sent, great princes,

And he shall buy my daughter, and her presence

Shall quite strike off all service I have done

In most accepted pain.

AGAMEMNON        Let Diomedes bear him,   30

And bring us Cressid hither; Calchas shall have

What he requests of us. Good Diomed,

Furnish you fairly for this interchange;

Withal bring word if Hector will tomorrow

Be answered in his challenge. Ajax is ready.   35

DIOMEDES

This shall I undertake, and 'tis a burden

Which I am proud to bear.     *Exit with Calchas*

*Enter Achilles and Patroclus in their tent*

ULYSSES

Achilles stands i'th' entrance of his tent.

Please it our general pass strangely by him,

As if he were forgot; and, princes all,   40

Lay negligent and loose regard upon him.

I will come last. 'Tis like he'll question me

Why such unplausive eyes are bent, why turned on
    him.
If so, I have derision medicinable
To use between your strangeness and his pride,          45
Which his own will shall have desire to drink.
It may do good. Pride hath no other glass
To show itself but pride; for supple knees
Feed arrogance and are the proud man's fees.
AGAMEMNON
We'll execute your purpose and put on          50
A form of strangeness as we pass along.
So do each lord, and either greet him not
Or else disdainfully, which shall shake him more
Than if not looked on. I will lead the way.
    *They pass by the tent, in turn*
ACHILLES
What, comes the general to speak with me?          55
You know my mind: I'll fight no more 'gainst Troy.
AGAMEMNON (*to Nestor*)
What says Achilles? Would he aught with us?
NESTOR (*to Achilles*)
Would you, my lord, aught with the general?
ACHILLES                                    No.
NESTOR (*to Agamemnon*)
Nothing, my lord.
AGAMEMNON          The better.
          ⌐*Exeunt Agamemnon and Nestor*⌐
ACHILLES ⌐*to Menelaus*⌐          Good day, good day.
MENELAUS How do you? How do you?          ⌐*Exit*⌐
ACHILLES (*to Patroclus*)
What, does the cuckold scorn me?
AJAX                              How now, Patroclus?
ACHILLES
Good morrow, Ajax.
AJAX          Ha?
ACHILLES                    Good morrow.          62
AJAX Ay, and good next day too.          *Exit*
ACHILLES (*to Patroclus*)
What mean these fellows? Know they not Achilles?
PATROCLUS
They pass by strangely. They were used to bend,          65
To send their smiles before them to Achilles,
To come as humbly as they use to creep
To holy altars.
ACHILLES          What, am I poor of late?
'Tis certain, greatness once fall'n out with fortune
Must fall out with men too. What the declined is          70
He shall as soon read in the eyes of others
As feel in his own fall; for men, like butterflies,
Show not their mealy wings but to the summer,
And not a man, for being simply man,
Hath any honour, but honour for those honours          75
That are without him—as place, riches, and favour:
Prizes of accident as oft as merit;
Which, when they fall, as being slippery standers—
The love that leaned on them, as slippery too—
Doth one pluck down another, and together          80
Die in the fall. But 'tis not so with me.
Fortune and I are friends. I do enjoy
At ample point all that I did possess,
Save these men's looks—who do methinks find out
Something not worth in me such rich beholding          85
As they have often given. Here is Ulysses;
I'll interrupt his reading. How now, Ulysses?
ULYSSES Now, great Thetis' son.

ACHILLES What are you reading?
ULYSSES A strange fellow here          90
Writes me that man, how dearly ever parted,
How much in having, or without or in,
Cannot make boast to have that which he hath,
Nor feels not what he owes, but by reflection—
As when his virtues, shining upon others,          95
Heat them, and they retort that heat again
To the first givers.
ACHILLES                    This is not strange, Ulysses.
The beauty that is borne here in the face
The bearer knows not, but commends itself
To others' eyes. Nor doth the eye itself,          100
That most pure spirit of sense, behold itself,
Not going from itself; but eye to eye opposed
Salutes each other with each other's form.
For speculation turns not to itself
Till it hath travelled and is mirrored there          105
Where it may see itself. This is not strange at all.
ULYSSES
I do not strain at the position—
It is familiar—but at the author's drift;
Who in his circumstance expressly proves
That no man is the lord of anything,          110
Though in and of him there be much consisting,
Till he communicate his parts to others.
Nor doth he of himself know them for aught
Till he behold them formèd in th'applause          114
Where they're extended—who, like an arch, reverb'rate
The voice again; or, like a gate of steel
Fronting the sun, receives and renders back
His figure and his heat. I was much rapt in this,
And apprehended here immediately
The unknown Ajax.          120
Heavens, what a man is there! A very horse,
That has he knows not what. Nature, what things
    there are,
Most abject in regard and dear in use.
What things again, most dear in the esteem
And poor in worth. Now shall we see tomorrow          125
An act that very chance doth throw upon him.
Ajax renowned? O heavens, what some men do,
While some men leave to do.
How some men creep in skittish Fortune's hall
Whiles others play the idiots in her eyes;          130
How one man eats into another's pride
While pride is fasting in his wantonness.
To see these Grecian lords! Why, even already
They clap the lubber Ajax on the shoulder,
As if his foot were on brave Hector's breast          135
And great Troy shrinking.
ACHILLES                    I do believe it,
For they passed by me as misers do by beggars,
Neither gave to me good word nor look.
What, are my deeds forgot?
ULYSSES                    Time hath, my lord,
A wallet at his back, wherein he puts          140
Alms for oblivion, a great-sized monster
Of ingratitudes. Those scraps are good deeds past,
Which are devoured as fast as they are made,
Forgot as soon as done. Perseverance, dear my lord,
Keeps honour bright. To have done is to hang          145
Quite out of fashion, like a rusty mail
In monumental mock'ry. Take the instant way,
For honour travels in a strait so narrow,

Where one but goes abreast. Keep then the path,
For emulation hath a thousand sons      150
That one by one pursue: if you give way,
Or hedge aside from the direct forthright,
Like to an entered tide they all rush by
And leave you hindmost;
Or, like a gallant horse fall'n in first rank,    155
Lie there for pavement to the abject rear,
O'errun and trampled on. Then what they do in
   present,
Though less than yours in past, must o'ertop yours.
For Time is like a fashionable host,
That slightly shakes his parting guest by th' hand  160
And, with his arms outstretched as he would fly,
Grasps in the comer. Welcome ever smiles,
And Farewell goes out sighing. O let not virtue seek
Remuneration for the thing it was;
For beauty, wit,      165
High birth, vigour of bone, desert in service,
Love, friendship, charity, are subjects all
To envious and calumniating time.
One touch of nature makes the whole world kin—
That all with one consent praise new-born gauds,  170
Though they are made and moulded of things past,
And give to dust that is a little gilt
More laud than gilt o'er-dusted.
The present eye praises the present object.
Then marvel not, thou great and complete man,  175
That all the Greeks begin to worship Ajax,
Since things in motion sooner catch the eye
Than what not stirs. The cry went once on thee,
And still it might, and yet it may again,
If thou wouldst not entomb thyself alive    180
And case thy reputation in thy tent,
Whose glorious deeds but in these fields of late
Made emulous missions 'mongst the gods themselves,
And drove great Mars to faction.

ACHILLES          Of this my privacy
I have strong reasons.

ULYSSES        But 'gainst your privacy  185
The reasons are more potent and heroical.
'Tis known, Achilles, that you are in love
With one of Priam's daughters.

ACHILLES          Ha? Known?

ULYSSES           Is that a wonder?
The providence that's in a watchful state
Knows almost every grain of Pluto's gold,    190
Finds bottom in th'uncomprehensive deeps,
Keeps place with aught, and almost like the gods
Do infant thoughts unveil in their dumb cradles.
There is a mystery, with whom relation
Durst never meddle, in the soul of state,    195
Which hath an operation more divine
Than breath or pen can give expressure to.
All the commerce that you have had with Troy
As perfectly is ours as yours, my lord;
And better would it fit Achilles much    200
To throw down Hector than Polyxena.
But it must grieve young Pyrrhus now at home,
When fame shall in his island sound her trump
And all the Greekish girls shall tripping sing,
'Great Hector's sister did Achilles win,    205
But our great Ajax bravely beat down *him*'.
Farewell, my lord. I as your lover speak.
The fool slides o'er the ice that you should break. *Exit*

PATROCLUS
To this effect, Achilles, have I moved you.
A woman impudent and mannish grown    210
Is not more loathed than an effeminate man
In time of action. I stand condemned for this.
They think my little stomach to the war
And your great love to me restrains me thus.
Sweet, rouse yourself, and the weak wanton Cupid
Shall from your neck unloose his amorous fold  216
And like a dew-drop from the lion's mane
Be shook to air.

ACHILLES        Shall Ajax fight with Hector?

PATROCLUS
Ay, and perhaps receive much honour by him.

ACHILLES
I see my reputation is at stake.    220
My fame is shrewdly gored.

PATROCLUS         O then beware:
Those wounds heal ill that men do give themselves.
Omission to do what is necessary
Seals a commission to a blank of danger,
And danger like an ague subtly taints    225
Even then when we sit idly in the sun.

ACHILLES
Go call Thersites hither, sweet Patroclus.
I'll send the fool to Ajax, and desire him
T'invite the Trojan lords after the combat
To see us here unarmed. I have a woman's longing,
An appetite that I am sick withal,    231
To see great Hector in his weeds of peace,
   *Enter Thersites*
To talk with him and to behold his visage
Even to my full of view.—A labour saved.

THERSITES A wonder!    235

ACHILLES What?

THERSITES Ajax goes up and down the field, as asking for
himself.

ACHILLES How so?

THERSITES He must fight singly tomorrow with Hector,
and is so prophetically proud of an heroical cudgelling
that he raves in saying nothing.    242

ACHILLES How can that be?

THERSITES Why, a stalks up and down like a peacock—a
stride and a stand; ruminates like an hostess that hath
no arithmetic but her brain to set down her reckoning;
bites his lip with a politic regard, as who should say
'There were wit in this head, an't would out'—and so
there is; but it lies as coldly in him as fire in a flint,
which will not show without knocking. The man's
undone for ever, for if Hector break not his neck i'th'
combat he'll break't himself in vainglory. He knows
not me. I said, 'Good morrow, Ajax', and he replies,
'Thanks, Agamemnon'. What think you of this man
that takes me for the General? He's grown a very land-
fish, languageless, a monster. A plague of opinion! A
man may wear it on both sides like a leather jerkin.

ACHILLES Thou must be my ambassador to him, Thersites.

THERSITES Who, I? Why, he'll answer nobody. He
professes not answering. Speaking is for beggars. He
wears his tongue in's arms. I will put on his presence.
Let Patroclus make demands to me. You shall see the
pageant of Ajax.    263

ACHILLES To him, Patroclus. Tell him I humbly desire the
valiant Ajax to invite the most valorous Hector to come
unarmed to my tent, and to procure safe-conduct for

his person of the magnanimous and most illustrious six-or-seven-times-honoured captain-general of the Grecian army, Agamemnon; et cetera. Do this.

PATROCLUS (*to Thersites*) Jove bless great Ajax!          270

THERSITES H'm.

PATROCLUS I come from the worthy Achilles—

THERSITES Ha?

PATROCLUS Who most humbly desires you to invite Hector to his tent—          275

THERSITES H'm!

PATROCLUS And to procure safe-conduct from Agamemnon.

THERSITES Agamemnon?

PATROCLUS Ay, my lord.          280

THERSITES Ha!

PATROCLUS What say you to't?

THERSITES God b'wi' you, with all my heart.

PATROCLUS Your answer, sir?          284

THERSITES If tomorrow be a fair day, by eleven o'clock it will go one way or other. Howsoever, he shall pay for me ere he has me.

PATROCLUS Your answer, sir?

THERSITES Fare ye well, with all my heart.

ACHILLES Why, but he is not in this tune, is he?          290

THERSITES No, but he's out o' tune thus. What music will be in him when Hector has knocked out his brains, I know not. But I am feared none, unless the fiddler Apollo get his sinews to make catlings on.

ACHILLES Come, thou shalt bear a letter to him straight.          295

THERSITES Let me carry another to his horse, for that's the more capable creature.

ACHILLES
My mind is troubled like a fountain stirred,
And I myself see not the bottom of it.
                              *Exit with Patroclus*

THERSITES Would the fountain of your mind were clear again, that I might water an ass at it. I had rather be a tick in a sheep than such a valiant ignorance.   *Exit*

**4.1** *Enter at one door Aeneas with a torch; at another Paris, Deiphobus, Antenor, and Diomedes the Grecian, with torch-bearers*

PARIS See, ho! Who is that there?

DEIPHOBUS It is the Lord Aeneas.

AENEAS Is the Prince there in person?
Had I so good occasion to lie long
As you, Prince Paris, nothing but heavenly business  5
Should rob my bed-mate of my company.

DIOMEDES
That's my mind too. Good morrow, Lord Aeneas.

PARIS
A valiant Greek, Aeneas, take his hand.
Witness the process of your speech, wherein
You told how Diomed e'en a whole week by days  10
Did haunt you in the field.

AENEAS (*to Diomedes*)          Health to you, valiant sir,
During all question of the gentle truce.
But when I meet you armed, as black defiance
As heart can think or courage execute.

DIOMEDES
The one and other Diomed embraces.          15
Our bloods are now in calm; and so long, health.
But when contention and occasion meet,
By Jove I'll play the hunter for thy life
With all my force, pursuit, and policy.

AENEAS
And thou shalt hunt a lion that will fly          20
With his face backward. In humane gentleness,
Welcome to Troy. Now by Anchises' life,
Welcome indeed! By Venus' hand I swear
No man alive can love in such a sort
The thing he means to kill more excellently.          25

DIOMEDES
We sympathize. Jove, let Aeneas live—
If to my sword his fate be not the glory—
A thousand complete courses of the sun;
But, in mine emulous honour, let him die
With every joint a wound—and that, tomorrow.          30

AENEAS We know each other well.

DIOMEDES
We do, and long to know each other worse.

PARIS
This is the most despitefull'st gentle greeting,
The noblest hateful love, that e'er I heard of.
What business, lord, so early?          35

AENEAS
I was sent for to the King; but why, I know not.

PARIS
His purpose meets you: 'twas to bring this Greek
To Calchas' house, and there to render him,
For the enfreed Antenor, the fair Cressid.
Let's have your company, or if you please          40
Haste there before us. ⌈*Aside*⌉ I constantly do think—
Or rather, call my thought a certain knowledge—
My brother Troilus lodges there tonight.
Rouse him and give him note of our approach,
With the whole quality wherefore. I fear          45
We shall be much unwelcome.

AENEAS ⌈*aside*⌉          That I assure you.
Troilus had rather Troy were borne to Greece
Than Cressid borne from Troy.

PARIS ⌈*aside*⌉          There is no help.
The bitter disposition of the time
Will have it so.          50
⌈*Aloud*⌉ On, lord, we'll follow you.

AENEAS Good morrow all.          *Exit*

PARIS
And tell me, noble Diomed—faith, tell me true,
Even in the soul of sound good-fellowship—
Who in your thoughts merits fair Helen most,          55
Myself or Menelaus?

DIOMEDES          Both alike.
He merits well to have her that doth seek her,
Not making any scruple of her soilure,
With such a hell of pain and world of charge;
And you as well to keep her that defend her,          60
Not palating the taste of her dishonour,
With such a costly loss of wealth and friends.
He like a puling cuckold would drink up
The lees and dregs of a flat 'tamèd piece;
You like a lecher out of whorish loins          65
Are pleased to breed out your inheritors.
Both merits poised, each weighs nor less nor more,
But he as he: which heavier for a whore?

PARIS
You are too bitter to your countrywoman.

DIOMEDES
She's bitter to her country. Hear me, Paris.          70
For every false drop in her bawdy veins
A Grecian's life hath sunk; for every scruple
Of her contaminated carrion weight

A Trojan hath been slain. Since she could speak
She hath not given so many good words breath    75
As, for her, Greeks and Trojans suffered death.

PARIS
Fair Diomed, you do as chapmen do:
Dispraise the thing that you desire to buy.
But we in silence hold this virtue well:
We'll but commend what we intend to sell.—    80
Here lies our way.                          *Exeunt*

**4.2    *Enter Troilus and Cressida***

TROILUS
Dear, trouble not yourself. The morn is cold.

CRESSIDA
Then, sweet my lord, I'll call mine uncle down.
He shall unbolt the gates.

TROILUS                     Trouble him not.
To bed, to bed! Sleep lull those pretty eyes
And give as soft attachment to thy senses    5
As to infants empty of all thought.

CRESSIDA Good morrow, then.

TROILUS I prithee now, to bed.

CRESSIDA Are you aweary of me?

TROILUS
O Cressida! But that the busy day,           10
Waked by the lark, hath roused the ribald crows,
And dreaming night will hide our joys no longer,
I would not from thee.

CRESSIDA               Night hath been too brief.

TROILUS
Beshrew the witch! With venomous wights she stays
As hideously as hell, but flies the grasps of love    15
With wings more momentary-swift than thought.
You will catch cold and curse me.

CRESSIDA Prithee, tarry. You men will never tarry.
O foolish Cressid! I might have still held off,
And then you would have tarried.—Hark, there's one
up.                                          20
⌜*She veils herself*⌝

PANDARUS (*within*) What's all the doors open here?

TROILUS It is your uncle.

CRESSIDA
A pestilence on him! Now will he be mocking.
I shall have such a life.
⌜*Enter Pandarus*⌝

PANDARUS How now, how now, how go maidenheads?
(*To Cressida*) Here, you, maid! Where's my cousin
Cressid?                                     27

CRESSIDA ⌜*unveiling*⌝
Go hang yourself. You naughty, mocking uncle!
You bring me to do—and then you flout me too.

PANDARUS To do what? To do what?—Let her say what.—
What have I brought you to do?               31

CRESSIDA
Come, come, beshrew your heart. You'll ne'er be
good,
Nor suffer others.

PANDARUS Ha ha! Alas, poor wretch. Ah, poor *capocchia*,
hast not slept tonight? Would he not—a naughty
man—let it sleep? A bugbear take him.        36

CRESSIDA (*to Troilus*)
Did not I tell you? Would he were knocked i'th' head.
⌜*One knocks within*⌝
Who's that at door?—Good uncle, go and see.—
My lord, come you again into my chamber.
You smile and mock me, as if I meant naughtily.  40

TROILUS Ha ha!

CRESSIDA
Come, you are deceived, I think of no such thing.
*One knocks within*
How earnestly they knock! Pray you come in.
I would not for half Troy have you seen here.
*Exeunt* ⌜*Troilus and Cressida*⌝

PANDARUS Who's there? What's the matter? Will you
beat down the door?                          46
*He opens the door.* ⌜*Enter Aeneas*⌝
How now, what's the matter?

AENEAS Good morrow, lord, good morrow.

PANDARUS
Who's there? My Lord Aeneas? By my troth,
I knew you not. What news with you so early?    50

AENEAS
Is not Prince Troilus here?

PANDARUS                    Here? What should he do here?

AENEAS
Come, he is here, my lord. Do not deny him.
It doth import him much to speak with me.

PANDARUS Is he here, say you? It's more than I know,
I'll be sworn. For my own part, I came in late. What
should he do here?                           56

AENEAS
Whoa! Nay, then. Come, come, you'll do him wrong
Ere you are ware. You'll be so true to him
To be false to him. Do not you know of him,
But yet go fetch him hither. Go.    ⌜*Exit Pandarus*⌝
*Enter Troilus*

TROILUS How now, what's the matter?          61

AENEAS
My lord, I scarce have leisure to salute you,
My matter is so rash. There is at hand
Paris your brother and Deiphobus,
The Grecian Diomed, and our Antenor          65
Delivered to us—and for him forthwith,
Ere the first sacrifice, within this hour,
We must give up to Diomedes' hand
The Lady Cressida.

TROILUS           Is it so concluded?

AENEAS
By Priam and the general state of Troy.      70
They are at hand, and ready to effect it.

TROILUS How my achievements mock me.
I will go meet them—and, my Lord Aeneas,
We met by chance: you did not find me here.

AENEAS
Good, good, my lord: the secrecies of nature  75
Have not more gift in taciturnity.          *Exeunt*

**4.3    *Enter Pandarus and Cressida***

PANDARUS Is't possible? No sooner got but lost. The devil
take Antenor! The young prince will go mad. A plague
upon Antenor! I would they had broke 's neck.

CRESSIDA How now? What's the matter? Who was here?

PANDARUS Ah, ah!                             5

CRESSIDA Why sigh you so profoundly? Where's my lord?
Gone? Tell me, sweet uncle, what's the matter?

PANDARUS Would I were as deep under the earth as I am
above.

CRESSIDA O the gods! What's the matter?     10

PANDARUS Pray thee, get thee in. Would thou hadst ne'er
been born. I knew thou wouldst be his death. O poor
gentleman! A plague upon Antenor!

CRESSIDA Good uncle, I beseech you on my knees; I
  beseech you, what's the matter?                                 15
PANDARUS Thou must be gone, wench, thou must be
  gone. Thou art changed for Antenor. Thou must to thy
  father, and be gone from Troilus. 'Twill be his death.
  'Twill be his bane. He cannot bear it.
CRESSIDA
  O you immortal gods! I will not go.                             20
PANDARUS Thou must.
CRESSIDA
  I will not, uncle. I have forgot my father.
  I know no touch of consanguinity,
  No kin, no love, no blood, no soul, so near me
  As the sweet Troilus. O you gods divine,                        25
  Make Cressid's name the very crown of falsehood
  If ever she leave Troilus. Time, force, and death
  Do to this body what extremity you can,
  But the strong base and building of my love
  Is as the very centre of the earth,                             30
  Drawing all things to it. I'll go in and weep—
PANDARUS Do, do.
CRESSIDA
  Tear my bright hair, and scratch my praisèd cheeks,
  Crack my clear voice with sobs, and break my heart
  With sounding 'Troilus'. I will not go from Troy.               35
                                                        *Exeunt*

**4.4**  *Enter Paris, Troilus, Aeneas, Deiphobus, Antenor,*
        *and Diomedes*
PARIS
  It is great morning, and the hour prefixed
  Of her delivery to this valiant Greek
  Comes fast upon us. Good my brother Troilus,
  Tell you the lady what she is to do,
  And haste her to the purpose.
TROILUS                      Walk into her house.                  5
  I'll bring her to the Grecian presently—
  And to his hand when I deliver her,
  Think it an altar, and thy brother Troilus
  A priest, there off'ring to it his own heart.
PARIS I know what 'tis to love,                                   10
  And would, as I shall pity, I could help.—
  Please you walk in, my lords?                       ⌈*Exeunt*⌉

**4.5**  *Enter Pandarus and Cressida*
PANDARUS Be moderate, be moderate.
CRESSIDA
  Why tell you me of moderation?
  The grief is fine, full, perfect that I taste,
  And violenteth in a sense as strong
  As that which causeth it. How can I moderate it?                 5
  If I could temporize with my affection
  Or brew it to a weak and colder palate,
  The like allayment could I give my grief.
  My love admits no qualifying dross;
  No more my grief, in such a precious loss.                      10
      *Enter Troilus*
PANDARUS Here, here, here he comes. Ah, sweet ducks!
CRESSIDA (*embracing him*) O Troilus, Troilus!
PANDARUS What a pair of spectacles is here! Let me
  embrace you too. 'O heart', as the goodly saying is,
      'O heart, heavy heart,                                      15
          Why sigh'st thou without breaking?'
  where he answers again
      'Because thou canst not ease thy smart
          By friendship nor by speaking.'

There was never a truer rhyme. Let us cast away
nothing, for we may live to have need of such a verse.
We see it, we see it. How now, lambs?                             22
TROILUS
  Cressid, I love thee in so strained a purity
  That the blest gods, as angry with my fancy—
  More bright in zeal than the devotion which                     25
  Cold lips blow to their deities—take thee from me.
CRESSIDA Have the gods envy?
PANDARUS Ay, ay, ay, ay, 'tis too plain a case.
CRESSIDA
  And is it true that I must go from Troy?
TROILUS
  A hateful truth.
CRESSIDA                What, and from Troilus too?               30
TROILUS
  From Troy and Troilus.
CRESSIDA                  Is't possible?
TROILUS
  And suddenly—where injury of chance
  Puts back leave-taking, jostles roughly by
  All time of pause, rudely beguiles our lips
  Of all rejoindure, forcibly prevents                            35
  Our locked embrasures, strangles our dear vows
  Even in the birth of our own labouring breath.
  We two, that with so many thousand sighs
  Did buy each other, must poorly sell ourselves
  With the rude brevity and discharge of one.                     40
  Injurious Time now with a robber's haste
  Crams his rich thiev'ry up, he knows not how.
  As many farewells as be stars in heaven,
  With distinct breath and consigned kisses to them,
  He fumbles up into a loose adieu                                45
  And scants us with a single famished kiss,
  Distasted with the salt of broken tears.
      *Enter Aeneas*
AENEAS My lord, is the lady ready?
TROILUS (*to Cressida*)
  Hark, you are called. Some say the *genius* so
  Cries 'Come!' to him that instantly must die.                   50
  ⌈*To Pandarus*⌉ Bid them have patience. She shall come
      anon.
PANDARUS Where are my tears? Rain, to lay this wind,
  or my heart will be blown up by the root.
                                        ⌈*Exit with Aeneas*⌉
CRESSIDA
  I must then to the Grecians.
TROILUS                    No remedy.
CRESSIDA
  A woeful Cressid 'mongst the merry Greeks!                      55
  When shall we see again?
TROILUS
  Hear me, my love: be thou but true of heart—
CRESSIDA
  I true? How now! What wicked deem is this?
TROILUS
  Nay, we must use expostulation kindly,
  For it is parting from us.                                      60
  I speak not 'Be thou true' as fearing thee—
  For I will throw my glove to Death himself
  That there's no maculation in thy heart—
  But 'Be thou true' say I, to fashion in
  My sequent protestation: 'Be thou true,                         65
  And I will see thee'.
CRESSIDA
  O you shall be exposed, my lord, to dangers
  As infinite as imminent. But I'll be true.

TROILUS
And I'll grow friend with danger. Wear this sleeve.
CRESSIDA
And you this glove. When shall I see you?          70
TROILUS
I will corrupt the Grecian sentinels
To give thee nightly visitation.
But yet, be true.
CRESSIDA O heavens! 'Be true' again!
TROILUS Hear why I speak it, love.          75
The Grecian youths are full of quality,
Their loving well composed, with gifts of nature
    flowing,
And swelling o'er with arts and exercise.
How novelty may move, and parts with person,
Alas, a kind of godly jealousy—          80
Which I beseech you call a virtuous sin—
Makes me afeard.
CRESSIDA O heavens, you love me not!
TROILUS Die I a villain then!
In this I do not call your faith in question          85
So mainly as my merit. I cannot sing,
Nor heel the high lavolt, nor sweeten talk,
Nor play at subtle games—fair virtues all,
To which the Grecians are most prompt and
    pregnant.
But I can tell that in each grace of these          90
There lurks a still and dumb-discoursive devil
That tempts most cunningly. But be not tempted.
CRESSIDA Do you think I will?
TROILUS
No, but something may be done that we will not,
And sometimes we are devils to ourselves,          95
When we will tempt the frailty of our powers,
Presuming on their changeful potency.
AENEAS (within)
Nay, good my lord!
TROILUS                    Come, kiss, and let us part.
PARIS ⌈at the door⌉
Brother Troilus?
TROILUS                    Good brother, come you hither,
And bring Aeneas and the Grecian with you.          100
    ⌈Exit Paris⌉
CRESSIDA My lord, will you be true?
TROILUS
Who, I? Alas, it is my vice, my fault.
Whiles others fish with craft for great opinion,
I with great truth catch mere simplicity;
Whilst some with cunning gild their copper crowns,
With truth and plainness I do wear mine bare.          106
    Enter Paris, Aeneas, Antenor, Deiphobus, and
    Diomedes
Fear not my truth. The moral of my wit
Is 'plain and true!'; there's all the reach of it.—
Welcome, Sir Diomed. Here is the lady
Which for Antenor we deliver you.          110
At the port, lord, I'll give her to thy hand,
And by the way possess thee what she is.
Entreat her fair, and by my soul, fair Greek,
If e'er thou stand at mercy of my sword,
Name Cressid, and thy life shall be as safe          115
As Priam is in Ilium.
DIOMEDES                    Fair Lady Cressid,
So please you, save the thanks this prince expects.
The lustre in your eye, heaven in your cheek,

Pleads your fair usage; and to Diomed
You shall be mistress, and command him wholly.          120
TROILUS
Grecian, thou dost not use me courteously,
To shame the zeal of my petition towards thee
In praising her. I tell thee, lord of Greece,
She is as far high-soaring o'er thy praises
As thou unworthy to be called her servant.          125
I charge thee use her well, even for my charge;
For, by the dreadful Pluto, if thou dost not,
Though the great bulk Achilles be thy guard
I'll cut thy throat.
DIOMEDES                    O be not moved, Prince Troilus.
Let me be privileged by my place and message          130
To be a speaker free. When I am hence
I'll answer to my lust. And know you, lord,
I'll nothing do on charge. To her own worth
She shall be prized; but that you say 'Be't so',
I'll speak it in my spirit and honour 'No!'          135
TROILUS
Come, to the port.—I'll tell thee, Diomed,
This brave shall oft make thee to hide thy head.—
Lady, give me your hand, and as we walk
To our own selves bend we our needful talk.
    Exeunt Troilus, Cressida, and Diomedes
    A trumpet sounds
PARIS
Hark, Hector's trumpet.
AENEAS                    How have we spent this morning?
The Prince must think me tardy and remiss,          141
That swore to ride before him in the field.
PARIS
'Tis Troilus' fault. Come, come to field with him.
DEIPHOBUS Let us make ready straight.
AENEAS
Yea, with a bridegroom's fresh alacrity          145
Let us address to tend on Hector's heels.
The glory of our Troy doth this day lie
On his fair worth and single chivalry.          Exeunt

**4.6**  Enter Ajax armed, Achilles, Patroclus, Agamemnon,
    Menelaus, Ulysses, Nestor, a trumpeter, and others
AGAMEMNON
Here art thou in appointment fresh and fair,
Anticipating time with starting courage.
Give with thy trumpet a loud note to Troy,
Thou dreadful Ajax, that the appalled air
May pierce the head of the great combatant          5
And hale him hither.
AJAX                    Thou trumpet, there's my purse.
    He gives him money
Now crack thy lungs and split thy brazen pipe.
Blow, villain, till thy sphered bias cheek
Outswell the colic of puffed Aquilon.
Come, stretch thy chest and let thy eyes spout blood;
Thou blow'st for Hector.          11
    ⌈The trumpet sounds⌉
ULYSSES No trumpet answers.
ACHILLES 'Tis but early days.
AGAMEMNON
Is not yond Diomed with Calchas' daughter?
ULYSSES
'Tis he. I ken the manner of his gait.          15
He rises on the toe: that spirit of his
In aspiration lifts him from the earth.

*Enter Diomedes and Cressida*
AGAMEMNON (*to Diomedes*)
  Is this the Lady Cressid?
DIOMEDES                    Even she.
AGAMEMNON
  Most dearly welcome to the Greeks, sweet lady.
    *He kisses her*
NESTOR (*to Cressida*)
  Our General doth salute you with a kiss.          20
ULYSSES
  Yet is the kindness but particular;
  'Twere better she were kissed in general.
NESTOR
  And very courtly counsel. I'll begin.
    *He kisses her*
  So much for Nestor.
ACHILLES
  I'll take that winter from your lips, fair lady.   25
    *He kisses her*
  Achilles bids you welcome.
MENELAUS (*to Cressida*)
  I had good argument for kissing once.
PATROCLUS
  But that's no argument for kissing now;
  For thus ⌜*stepping between them*⌝ popped Paris in his
    hardiment,
  And parted thus you and your argument.            30
    *He kisses her*
ULYSSES ⌜*aside*⌝
  O deadly gall, and theme of all our scorns!
  For which we lose our heads to gild his horns.
PATROCLUS (*to Cressida*)
  The first was Menelaus' kiss; this, mine.
  Patroclus kisses you.
    *He kisses her again*
MENELAUS                    O this is trim.
PATROCLUS (*to Cressida*)
  Paris and I kiss evermore for him.                35
MENELAUS
  I'll have my kiss, sir.—Lady, by your leave.
CRESSIDA
  In kissing do you render or receive?
⌜MENELAUS⌝
  Both take and give.
CRESSIDA               I'll make my match to live,
  The kiss you take is better than you give.
  Therefore no kiss.                                40
MENELAUS
  I'll give you boot: I'll give you three for one.
CRESSIDA
  You are an odd man: give even or give none.
MENELAUS
  An odd man, lady? Every man is odd.
CRESSIDA
  No, Paris is not—for you know 'tis true
  That you are odd, and he is even with you.        45
MENELAUS
  You fillip me o'th' head.
CRESSIDA               No, I'll be sworn.
ULYSSES
  It were no match, your nail against his horn.
  May I, sweet lady, beg a kiss of you?
CRESSIDA
  You may.
ULYSSES     I do desire it.
CRESSIDA                    Why, beg too.

ULYSSES
  Why then, for Venus' sake, give me a kiss,        50
  When Helen is a maid again, and his—
CRESSIDA
  I am your debtor; claim it when 'tis due.
ULYSSES
  Never's my day, and then a kiss of you.
DIOMEDES
  Lady, a word. I'll bring you to your father.
    ⌜*They talk apart*⌝
NESTOR
  A woman of quick sense.
ULYSSES                    Fie, fie upon her!        55
  There's language in her eye, her cheek, her lip;
  Nay, her foot speaks. Her wanton spirits look out
  At every joint and motive of her body.
  O these encounterers so glib of tongue,
  That give accosting welcome ere it comes,         60
  And wide unclasp the tables of their thoughts
  To every ticklish reader, set them down
  For sluttish spoils of opportunity
  And daughters of the game.
                      ⌜*Exeunt Diomedes and Cressida*⌝
    *Flourish*
ALL  The Trojans' trumpet.                          65
    *Enter all of Troy: Hector* ⌜*armed*⌝, *Paris, Aeneas,*
    *Helenus, and attendants, among them Troilus*
AGAMEMNON  Yonder comes the troop.
AENEAS ⌜*coming forward*⌝
  Hail, all you state of Greece! What shall be done
  To him that victory commands? Or do you purpose
  A victor shall be known? Will you the knights
  Shall to the edge of all extremity                70
  Pursue each other, or shall they be divided
  By any voice or order of the field?
  Hector bade ask.
AGAMEMNON          Which way would Hector have it?
AENEAS
  He cares not; he'll obey conditions.
⌜ACHILLES⌝
  'Tis done like Hector—but securely done,          75
  A little proudly, and great deal disprising
  The knight opposed.
AENEAS               If not Achilles, sir,
  What is your name?
ACHILLES             If not Achilles, nothing.
AENEAS
  Therefore Achilles. But whate'er, know this:
  In the extremity of great and little,             80
  Valour and pride excel themselves in Hector,
  The one almost as infinite as all,
  The other blank as nothing. Weigh him well,
  And that which looks like pride is courtesy.
  This Ajax is half made of Hector's blood,         85
  In love whereof half Hector stays at home.
  Half heart, half hand, half Hector comes to seek
  This blended knight, half Trojan and half Greek.
ACHILLES
  A maiden battle, then? O I perceive you.
    *Enter Diomedes*
AGAMEMNON
  Here is Sir Diomed.—Go, gentle knight,            90
  Stand by our Ajax. As you and Lord Aeneas
  Consent upon the order of their fight,
  So be it: either to the uttermost

Or else a breath.
    ⌜*Exeunt Ajax, Diomedes, Hector, and Aeneas*⌝
    The combatants being kin
Half stints their strife before their strokes begin.  95
ULYSSES They are opposed already.
AGAMEMNON
 What Trojan is that same that looks so heavy?
ULYSSES
 The youngest son of Priam, a true knight:
 They call him Troilus.
 Not yet mature, yet matchless-firm of word,  100
 Speaking in deeds and deedless in his tongue;
 Not soon provoked, nor being provoked soon calmed;
 His heart and hand both open and both free.
 For what he has he gives; what thinks, he shows;
 Yet gives he not till judgement guide his bounty,  105
 Nor dignifies an impare thought with breath.
 Manly as Hector but more dangerous,
 For Hector in his blaze of wrath subscribes
 To tender objects, but he in heat of action
 Is more vindicative than jealous love.  110
 They call him Troilus, and on him erect
 A second hope as fairly built as Hector.
 Thus says Aeneas, one that knows the youth
 Even to his inches, and with private soul
 Did in great Ilium thus translate him to me.  115
    *Alarum*
AGAMEMNON They are in action.
NESTOR Now, Ajax, hold thine own!
TROILUS Hector, thou sleep'st! Awake thee!
AGAMEMNON
 His blows are well disposed. There, Ajax!  ⌜*Exeunt*⌝

**4.7** ⌜*Enter Hector and Ajax fighting, and Aeneas and*
  *Diomedes interposing.*⌝ *Trumpets cease*
DIOMEDES
 You must no more.
AENEAS    Princes, enough, so please you.
AJAX
 I am not warm yet. Let us fight again.
DIOMEDES
 As Hector pleases.
HECTOR    Why then will I no more.—
 Thou art, great lord, my father's sister's son,
 A cousin-german to great Priam's seed.  5
 The obligation of our blood forbids
 A gory emulation 'twixt us twain.
 Were thy commixtion Greek and Trojan so
 That thou couldst say 'This hand is Grecian all,
 And this is Trojan; the sinews of this leg  10
 All Greek, and this all Troy; my mother's blood
 Runs on the dexter cheek, and this sinister
 Bounds in my father's,' by Jove multipotent
 Thou shouldst not bear from me a Greekish member
 Wherein my sword had not impressure made  15
 Of our rank feud. But the just gods gainsay
 That any drop thou borrowed'st from thy mother,
 My sacred aunt, should by my mortal sword
 Be drained. Let me embrace thee, Ajax.
 By him that thunders, thou hast lusty arms.  20
 Hector would have them fall upon him thus.
 Cousin, all honour to thee.
AJAX    I thank thee, Hector.
 Thou art too gentle and too free a man.
 I came to kill thee, cousin, and bear hence
 A great addition earnèd in thy death.  25

HECTOR
 Not Neoptolemus so mirable,
 On whose bright crest Fame with her loud'st oyez
 Cries 'This is he!', could promise to himself
 A thought of added honour torn from Hector.
AENEAS
 There is expectance here from both the sides  30
 What further you will do.
HECTOR    We'll answer it:
 The issue is embracement.—Ajax, farewell.
AJAX
 If I might in entreaties find success,
 As seld I have the chance, I would desire
 My famous cousin to our Grecian tents.  35
DIOMEDES
 'Tis Agamemnon's wish—and great Achilles
 Doth long to see unarmed the valiant Hector.
HECTOR
 Aeneas, call my brother Troilus to me,
 And signify this loving interview
 To the expecters of our Trojan part.  40
 Desire them home.    ⌜*Exit Aeneas*⌝
    Give me thy hand, my cousin.
 I will go eat with thee, and see your knights.
    *Enter Agamemnon and the rest: Aeneas, Ulysses,*
    *Menelaus, Nestor, Achilles, Patroclus, Troilus, and*
    *others*
AJAX
 Great Agamemnon comes to meet us here.
HECTOR (*to Aeneas*)
 The worthiest of them, tell me name by name.
 But for Achilles, mine own searching eyes  45
 Shall find him by his large and portly size.
AGAMEMNON (*embracing him*)
 Worthy of arms, as welcome as to one
 That would be rid of such an enemy.
 But that's no welcome. Understand more clear:
 What's past and what's to come is strewed with husks
 And formless ruin of oblivion,  51
 But in this extant moment faith and troth,
 Strained purely from all hollow bias-drawing,
 Bids thee with most divine integrity
 From heart of very heart, 'Great Hector, welcome!' 55
HECTOR
 I thank thee, most imperious Agamemnon.
AGAMEMNON ⌜*to Troilus*⌝
 My well-famed lord of Troy, no less to you.
MENELAUS
 Let me confirm my princely brother's greeting.
 You brace of warlike brothers, welcome hither.
    ⌜*He embraces Hector and Troilus*⌝
HECTOR (*to Aeneas*)
 Who must we answer?
AENEAS    The noble Menelaus.  60
HECTOR
 O, you, my lord! By Mars his gauntlet, thanks.
 Mock not that I affect th'untraded oath.
 Your quondam wife swears still by Venus' glove.
 She's well, but bade me not commend her to you.
MENELAUS
 Name her not now, sir. She's a deadly theme.  65
HECTOR O, pardon. I offend.
NESTOR
 I have, thou gallant Trojan, seen thee oft,
 Labouring for destiny, make cruel way
 Through ranks of Greekish youth, and I have seen thee
 As hot as Perseus spur thy Phrygian steed,  70

And seen thee scorning forfeits and subduements,
When thou hast hung th'advancèd sword i'th' air,
Not letting it decline on the declined,
That I have said unto my standers-by,
'Lo, Jupiter is yonder, dealing life'. 75
And I have seen thee pause and take thy breath,
When that a ring of Greeks have hemmed thee in,
Like an Olympian, wrestling. This have I seen;
But this thy countenance, still locked in steel,
I never saw till now. I knew thy grandsire 80
And once fought with him. He was a soldier good,
But—by great Mars, the captain of us all—
Never like thee. Let an old man embrace thee;
And, worthy warrior, welcome to our tents.
    *He embraces Hector*
AENEAS (*to Hector*) 'Tis the old Nestor. 85
HECTOR
Let me embrace thee, good old chronicle,
That hast so long walked hand in hand with time.
Most reverend Nestor, I am glad to clasp thee.
NESTOR
I would my arms could match thee in contention
As they contend with thee in courtesy. 90
HECTOR   I would they could.
NESTOR
Ha! By this white beard I'd fight with thee tomorrow.
Well, welcome, welcome! I have seen the time.
ULYSSES
I wonder now how yonder city stands
When we have here her base and pillar by us? 95
HECTOR
I know your favour, Lord Ulysses, well.
Ah, sir, there's many a Greek and Trojan dead
Since first I saw yourself and Diomed
In Ilium on your Greekish embassy.
ULYSSES
Sir, I foretold you then what would ensue. 100
My prophecy is but half his journey yet;
For yonder walls that pertly front your town,
Yon towers whose wanton tops do buss the clouds,
Must kiss their own feet.
HECTOR       I must not believe you.
There they stand yet, and modestly I think 105
The fall of every Phrygian stone will cost
A drop of Grecian blood. The end crowns all,
And that old common arbitrator Time
Will one day end it.
ULYSSES       So to him we leave it.
Most gentle and most valiant Hector, welcome. 110
    ⌈*He embraces him*⌉
After the General, I beseech you next
To feast with me and see me at my tent.
ACHILLES
I shall forestall thee, Lord Ulysses. ⌈*To Hector*⌉ Thou!
Now, Hector, I have fed mine eyes on thee.
I have with exact view perused thee, Hector, 115
And quoted joint by joint.
HECTOR Is this Achilles?
ACHILLES I am Achilles.
HECTOR
Stand fair, I pray thee, let me look on thee.
ACHILLES
Behold thy fill.
HECTOR       Nay, I have done already. 120
ACHILLES
Thou art too brief. I will the second time,
As I would buy thee, view thee limb by limb.

HECTOR
O, like a book of sport thou'lt read me o'er.
But there's more in me than thou understand'st.
Why dost thou so oppress me with thine eye? 125
ACHILLES
Tell me, you heavens, in which part of his body
Shall I destroy him—whether there, or there, or
    there—
That I may give the local wound a name,
And make distinct the very breach whereout
Hector's great spirit flew? Answer me, heavens. 130
HECTOR
It would discredit the blest gods, proud man,
To answer such a question. Stand again.
Think'st thou to catch my life so pleasantly
As to prenominate in nice conjecture
Where thou wilt hit me dead?
ACHILLES       I tell thee, yea. 135
HECTOR
Wert thou the oracle to tell me so,
I'd not believe thee. Henceforth guard thee well.
For I'll not kill thee there, nor there, nor there,
But, by the forge that stithied Mars his helm,
I'll kill thee everywhere, yea, o'er and o'er.— 140
You wisest Grecians, pardon me this brag:
His insolence draws folly from my lips.
But I'll endeavour deeds to match these words,
Or may I never—
AJAX       Do not chafe thee, cousin.—
And you, Achilles, let these threats alone, 145
Till accident or purpose bring you to't.
You may have every day enough of Hector,
If you have stomach. The general state, I fear,
Can scarce entreat you to be odd with him.
HECTOR (*to Achilles*)
I pray you, let us see you in the field. 150
We have had pelting wars since you refused
The Grecians' cause.
ACHILLES       Dost thou entreat me, Hector?
Tomorrow do I meet thee, fell as death;
Tonight, all friends.
HECTOR       Thy hand upon that match.
AGAMEMNON
First, all you peers of Greece, go to my tent. 155
There in the full convive you. Afterwards,
As Hector's leisure and your bounties shall
Concur together, severally entreat him.
Beat loud the taborins, let the trumpets blow,
That this great soldier may his welcome know. 160
    *Flourish. Exeunt all but Troilus and Ulysses*
TROILUS
My Lord Ulysses, tell me, I beseech you,
In what place of the field doth Calchas keep?
ULYSSES
At Menelaus' tent, most princely Troilus.
There Diomed doth feast with him tonight—
Who neither looks on heaven nor on earth, 165
But gives all gaze and bent of amorous view
On the fair Cressid.
TROILUS
Shall I, sweet lord, be bound to you so much,
After we part from Agamemnon's tent,
To bring me thither?
ULYSSES       You shall command me, sir. 170
As gentle tell me, of what honour was
This Cressida in Troy? Had she no lover there
That wails her absence?

**TROILUS**
O sir, to such as boasting show their scars
A mock is due. Will you walk on, my lord? 175
She was beloved, she loved; she is, and doth.
But still sweet love is food for fortune's tooth. *Exeunt*

**5.1** *Enter Achilles and Patroclus*
**ACHILLES**
I'll heat his blood with Greekish wine tonight,
Which with my scimitar I'll cool tomorrow.
Patroclus, let us feast him to the height.
**PATROCLUS**
Here comes Thersites.
*Enter Thersites*
**ACHILLES** How now, thou core of envy,
Thou crusty botch of nature, what's the news? 5
**THERSITES** Why, thou picture of what thou seemest, and idol of idiot-worshippers, here's a letter for thee.
**ACHILLES** From whence, fragment?
**THERSITES** Why, thou full dish of fool, from Troy.
*Achilles reads the letter*
**PATROCLUS** Who keeps the tent now? 10
**THERSITES** The surgeon's box or the patient's wound.
**PATROCLUS** Well said, adversity. And what need these tricks?
**THERSITES** Prithee be silent, boy. I profit not by thy talk. Thou art thought to be Achilles' male varlet. 15
**PATROCLUS** 'Male varlet', you rogue? What's that?
**THERSITES** Why, his masculine whore. Now the rotten diseases of the south, guts-griping, ruptures, catarrhs, loads o' gravel i'th' back, lethargies, cold palsies, and the like, take and take again such preposterous discoveries! 21
**PATROCLUS** Why, thou damnable box of envy thou, what mean'st thou to curse thus?
**THERSITES** Do I curse thee?
**PATROCLUS** Why, no, you ruinous butt, you whoreson indistinguishable cur, no. 26
**THERSITES** No? Why art thou then exasperate? Thou idle immaterial skein of sleave-silk, thou green sarsenet flap for a sore eye, thou tassel of a prodigal's purse, thou! Ah, how the poor world is pestered with such waterflies! Diminutives of nature. 31
**PATROCLUS** Out, gall!
**THERSITES** Finch egg!
**ACHILLES**
My sweet Patroclus, I am thwarted quite
From my great purpose in tomorrow's battle. 35
Here is a letter from Queen Hecuba,
A token from her daughter, my fair love,
Both taxing me, and gaging me to keep
An oath that I have sworn. I will not break it.
Fall, Greeks; fail, fame; honour, or go or stay. 40
My major vow lies here; this I'll obey.—
Come, come, Thersites, help to trim my tent.
This night in banqueting must all be spent.—
Away, Patroclus. *Exeunt Achilles and Patroclus*
**THERSITES** With too much blood and too little brain these two may run mad, but if with too much brain and too little blood they do, I'll be a curer of madmen. Here's Agamemnon: an honest fellow enough, and one that loves quails, but he has not so much brain as ear-wax. And the goodly transformation of Jupiter there, his brother the bull, the primitive statue and oblique memorial of cuckolds, a thrifty shoeing-horn in a chain, hanging at his brother's leg: to what form but that he

is should wit larded with malice and malice farced with wit turn him to? To an ass were nothing: he is both ass and ox. To an ox were nothing: he is both ox and ass. To be a dog, a mule, a cat, a fitchew, a toad, a lizard, an owl, a puttock, or a herring without a roe, I would not care; but to be Menelaus!—I would conspire against destiny. Ask me not what I would be if I were not Thersites, for I care not to be the louse of a lazar, so I were not Menelaus.—Hey-day, sprites and fires. 63
*Enter Hector, Ajax, Agamemnon, Ulysses, Nestor, Menelaus, Troilus, and Diomedes, with lights*
**AGAMEMNON**
We go wrong, we go wrong.
**AJAX** No, yonder 'tis:
There, where we see the light.
**HECTOR** I trouble you. 65
**AJAX**
No, not a whit.
*Enter Achilles*
**ULYSSES** Here comes himself to guide you.
**ACHILLES**
Welcome, brave Hector. Welcome, princes all.
**AGAMEMNON** (*to Hector*)
So now, fair prince of Troy, I bid good night.
Ajax commands the guard to tend on you.
**HECTOR**
Thanks and good night to the Greeks' general. 70
**MENELAUS**
Good night, my lord.
**HECTOR** Good night, sweet Lord Menelaus.
**THERSITES** (*aside*) Sweet draught! 'Sweet', quoth a? Sweet sink, sweet sewer.
**ACHILLES**
Good night and welcome both at once, to those
That go or tarry. 75
**AGAMEMNON** Good night.
*Exeunt Agamemnon and Menelaus*
**ACHILLES**
Old Nestor tarries, and you too, Diomed.
Keep Hector company an hour or two.
**DIOMEDES**
I cannot, lord. I have important business
The tide whereof is now.—Good night, great Hector.
**HECTOR** Give me your hand. 81
**ULYSSES** (*aside to Troilus*)
Follow his torch, he goes to Calchas' tent.
I'll keep you company.
**TROILUS** (*aside*) Sweet sir, you honour me.
**HECTOR** (*to Diomedes*)
And so good night.
**ACHILLES** Come, come, enter my tent. 84
*Exeunt Diomedes, followed by Ulysses and Troilus, at one door; and Achilles, Hector, Ajax, and Nestor at another door*
**THERSITES** That same Diomed's a false-hearted rogue, a most unjust knave. I will no more trust him when he leers than I will a serpent when he hisses. He will spend his mouth and promise like Brabbler the hound, but when he performs astronomers foretell it: that is prodigious, there will come some change. The sun borrows of the moon when Diomed keeps his word. I will rather leave to see Hector than not to dog him. They say he keeps a Trojan drab, and uses the traitor Calchas his tent. I'll after.—Nothing but lechery! All incontinent varlets! *Exit*

**5.2** *Enter Diomedes*
DIOMEDES What, are you up here? Ho! Speak!
CALCHAS ⌈*at the door*⌉ Who calls?
DIOMEDES Diomed. Calchas, I think. Where's your
  daughter?
CALCHAS ⌈*at the door*⌉ She comes to you.    5
    *Enter Troilus and Ulysses, unseen*
ULYSSES (*aside*)
  Stand where the torch may not discover us.
TROILUS (*aside*)
  Cressid comes forth to him.
    *Enter Cressida*
DIOMEDES                How now, my charge?
CRESSIDA
  Now, my sweet guardian. Hark, a word with you.
    *She whispers to him.*
    ⌈*Enter Thersites, unseen*⌉
TROILUS (*aside*) Yea, so familiar?
ULYSSES (*aside*) She will sing any man at first sight.    10
THERSITES (*aside*) And any man may sing her, if he can
  take her clef. She's noted.
DIOMEDES Will you remember?
CRESSIDA Remember? Yes.
DIOMEDES Nay, but do then,    15
  And let your mind be coupled with your words.
TROILUS (*aside*) What should she remember?
ULYSSES (*aside*) List!
CRESSIDA
  Sweet honey Greek, tempt me no more to folly.
THERSITES (*aside*) Roguery.    20
DIOMEDES Nay, then!
CRESSIDA I'll tell you what—
DIOMEDES
  Fo, fo! Come, tell a pin. You are forsworn.
CRESSIDA
  In faith, I cannot. What would you have me do?
THERSITES (*aside*) A juggling trick: to be secretly open.    25
DIOMEDES
  What did you swear you would bestow on me?
CRESSIDA
  I prithee, do not hold me to mine oath.
  Bid me do anything but that, sweet Greek.
DIOMEDES Good night.
TROILUS (*aside*)
  Hold, patience!
ULYSSES (*aside*)    How now, Trojan?
CRESSIDA                Diomed.    30
DIOMEDES
  No, no, good night. I'll be your fool no more.
TROILUS (*aside*) Thy better must.
CRESSIDA Hark, one word in your ear.
    *She whispers to him*
TROILUS (*aside*) O plague and madness!
ULYSSES (*aside*)
  You are movèd, Prince. Let us depart, I pray you,    35
  Lest your displeasure should enlarge itself
  To wrathful terms. This place is dangerous,
  The time right deadly. I beseech you go.
TROILUS (*aside*)
  Behold, I pray you.
ULYSSES (*aside*)        Nay, good my lord, go off.
  You flow to great distraction. Come, my lord.    40
TROILUS (*aside*)
  I prithee, stay.
ULYSSES (*aside*)    You have not patience. Come.
TROILUS (*aside*)
  I pray you, stay. By hell and all hell's torments,

  I will not speak a word.
DIOMEDES            And so good night.
CRESSIDA
  Nay, but you part in anger.
TROILUS (*aside*)            Doth that grieve thee?
  O withered truth!
ULYSSES (*aside*)        Why, how now, lord?
TROILUS (*aside*)            By Jove,    45
  I will be patient.
    ⌈*Diomedes starts to go*⌉
CRESSIDA Guardian! Why, Greek!
DIOMEDES Fo, fo! Adieu. You palter.
CRESSIDA
  In faith, I do not. Come hither once again.
ULYSSES (*aside*)
  You shake, my lord, at something. Will you go?    50
  You will break out.
TROILUS (*aside*)        She strokes his cheek.
ULYSSES (*aside*)                Come, come.
TROILUS (*aside*)
  Nay, stay. By Jove, I will not speak a word.
  There is between my will and all offences
  A guard of patience. Stay a little while.
THERSITES (*aside*) How the devil Luxury with his fat rump
  and potato finger tickles these together! Fry, lechery,
  fry.    57
DIOMEDES But will you then?
CRESSIDA
  In faith, I will, la. Never trust me else.
DIOMEDES
  Give me some token for the surety of it.    60
CRESSIDA I'll fetch you one.            *Exit*
ULYSSES (*aside*) You have sworn patience.
TROILUS (*aside*) Fear me not, sweet lord.
  I will not be myself, nor have cognition
  Of what I feel. I am all patience.    65
    *Enter Cressida with Troilus' sleeve*
THERSITES (*aside*) Now the pledge! Now, now, now.
CRESSIDA Here Diomed, keep this sleeve.
TROILUS (*aside*) O beauty, where is thy faith?
ULYSSES (*aside*) My lord.
TROILUS (*aside*)
  I will be patient; outwardly I will.    70
CRESSIDA
  You look upon that sleeve. Behold it well.
  He loved me—O false wench!—give't me again.
    *She takes it back*
DIOMEDES Whose was't?
CRESSIDA
  It is no matter, now I ha't again.
  I will not meet with you tomorrow night.    75
  I prithee, Diomed, visit me no more.
THERSITES (*aside*) Now she sharpens. Well said, whetstone.
DIOMEDES I shall have it.
CRESSIDA What, this?
DIOMEDES Ay, that.    80
CRESSIDA
  O all you gods! O pretty pretty pledge!
  Thy master now lies thinking on his bed
  Of thee and me, and sighs, and takes my glove
  And gives memorial dainty kisses to it—
⌈DIOMEDES⌉
  As I kiss thee.
    ⌈*He snatches the sleeve*⌉
⌈CRESSIDA⌉        Nay, do not snatch it from me.    85
  He that takes that doth take my heart withal.

DIOMEDES
I had your heart before; this follows it.
TROILUS (*aside*) I did swear patience.
CRESSIDA
You shall not have it, Diomed. Faith, you shall not.
I'll give you something else.
DIOMEDES          I will have this. Whose was it?
CRESSIDA
It is no matter.
DIOMEDES      Come, tell me whose it was?     91
CRESSIDA
'Twas one's that loved me better than you will.
But now you have it, take it.
DIOMEDES          Whose was it?
CRESSIDA
By all Diana's waiting-women yond,
And by herself, I will not tell you whose.     95
DIOMEDES
Tomorrow will I wear it on my helm,
And grieve his spirit that dares not challenge it.
TROILUS (*aside*)
Wert thou the devil and wor'st it on thy horn,
It should be challenged.
CRESSIDA
Well, well, 'tis done, 'tis past—and yet it is not.     100
I will not keep my word.
DIOMEDES        Why then, farewell.
Thou never shalt mock Diomed again.
CRESSIDA
You shall not go. One cannot speak a word
But it straight starts you.
DIOMEDES      I do not like this fooling.
⌈TROILUS⌉ (*aside*)
Nor I, by Pluto—but that that likes not you     105
Pleases me best.
DIOMEDES       What, shall I come? The hour—
CRESSIDA
Ay, come. O Jove, do come. I shall be plagued.
DIOMEDES
Farewell till then.
CRESSIDA        Good night. I prithee, come.
                        *Exit Diomedes*
Troilus, farewell. One eye yet looks on thee,
But with my heart the other eye doth see.     110
Ah, poor our sex! This fault in us I find:
The error of our eye directs our mind.
What error leads must err. O then conclude:
Minds swayed by eyes are full of turpitude.     *Exit*
THERSITES (*aside*)
A proof of strength she could not publish more     115
Unless she said, 'My mind is now turned whore'.
ULYSSES
All's done, my lord.
TROILUS       It is.
ULYSSES         Why stay we then?
TROILUS
To make a recordation to my soul
Of every syllable that here was spoke.
But if I tell how these two did co-act,     120
Shall I not lie in publishing a truth?
Sith yet there is a credence in my heart,
An esperance so obstinately strong,
That doth invert th'attest of eyes and ears,
As if those organs had deceptious functions     125
Created only to calumniate.
Was Cressid here?
ULYSSES       I cannot conjure, Trojan.

TROILUS
She was not, sure.
ULYSSES        Most sure, she was.
TROILUS
Why, my negation hath no taste of madness.
ULYSSES
Nor mine, my lord. Cressid was here but now.     130
TROILUS
Let it not be believed, for womanhood.
Think: we had mothers. Do not give advantage
To stubborn critics, apt without a theme
For depravation to square the general sex
By Cressid's rule. Rather, think this not Cressid.     135
ULYSSES
What hath she done, Prince, that can soil our mothers?
TROILUS
Nothing at all, unless that this were she.
THERSITES (*aside*) Will a swagger himself out on's own
eyes?
TROILUS
This, she? No, this is Diomed's Cressida.     140
If beauty have a soul, this is not she.
If souls guide vows, if vows be sanctimonies,
If sanctimony be the gods' delight,
If there be rule in unity itself,
This is not she. O madness of discourse,     145
That cause sets up with and against thyself!
Bifold authority, where reason can revolt
Without perdition, and loss assume all reason
Without revolt! This is and is not Cressid.
Within my soul there doth conduce a fight     150
Of this strange nature, that a thing inseparate
Divides more wider than the sky and earth,
And yet the spacious breadth of this division
Admits no orifex for a point as subtle
As Ariachne's broken woof to enter.     155
Instance, O instance, strong as Pluto's gates:
Cressid is mine, tied with the bonds of heaven.
Instance, O instance, strong as heaven itself:
The bonds of heaven are slipped, dissolved, and loosed,
And with another knot, five-finger-tied,     160
The fractions of her faith, orts of her love,
The fragments, scraps, the bits and greasy relics
Of her o'er-eaten faith, are bound to Diomed.
ULYSSES
May worthy Troilus e'en be half attached
With that which here his passion doth express?     165
TROILUS
Ay, Greek, and that shall be divulgèd well
In characters as red as Mars his heart
Inflamed with Venus. Never did young man fancy
With so eternal and so fixed a soul.
Hark, Greek: as much as I do Cressid love,     170
So much by weight hate I her Diomed.
That sleeve is mine that he'll bear in his helm.
Were it a casque composed by Vulcan's skill,
My sword should bite it. Not the dreadful spout
Which shipmen do the hurricano call,     175
Constrainèd in mass by the almighty sun,
Shall dizzy with more clamour Neptune's ear
In his descent, than shall my prompted sword
Falling on Diomed.
THERSITES (*aside*) He'll tickle it for his concupy.     180
TROILUS
O Cressid, O false Cressid! False, false, false.
Let all untruths stand by thy stainèd name,

And they'll seem glorious.
ULYSSES                    O contain yourself.
  Your passion draws ears hither.
        *Enter Aeneas*
AENEAS (*to Troilus*)
  I have been seeking you this hour, my lord.         185
  Hector by this is arming him in Troy.
  Ajax your guard stays to conduct you home.
TROILUS
  Have with you, Prince.—My courteous lord, adieu.—
  Farewell, revolted fair; and Diomed,
  Stand fast and wear a castle on thy head.           190
ULYSSES
  I'll bring you to the gates.
TROILUS                    Accept distracted thanks.
        *Exeunt Troilus, Aeneas, and Ulysses*
THERSITES Would I could meet that rogue Diomed! I would
  croak like a raven. I would bode, I would bode.
  Patroclus will give me anything for the intelligence of
  this whore. The parrot will not do more for an almond
  than he for a commodious drab. Lechery, lechery, still
  wars and lechery! Nothing else holds fashion. A burning
  devil take them!                                    *Exit*

**5.3**    *Enter Hector armed, and Andromache*
ANDROMACHE
  When was my lord so much ungently tempered
  To stop his ears against admonishment?
  Unarm, unarm, and do not fight today.
HECTOR
  You train me to offend you. Get you in.
  By all the everlasting gods, I'll go.                 5
ANDROMACHE
  My dreams will sure prove ominous to the day.
HECTOR
  No more, I say.
        *Enter Cassandra*
CASSANDRA           Where is my brother Hector?
ANDROMACHE
  Here, sister, armed and bloody in intent.
  Consort with me in loud and dear petition,
  Pursue we him on knees—for I have dreamed          10
  Of bloody turbulence, and this whole night
  Hath nothing been but shapes and forms of slaughter.
CASSANDRA
  O 'tis true.
HECTOR        Ho! Bid my trumpet sound.
CASSANDRA
  No notes of sally, for the heavens, sweet brother.
HECTOR
  Begone, I say. The gods have heard me swear.        15
CASSANDRA
  The gods are deaf to hot and peevish vows.
  They are polluted off'rings, more abhorred
  Than spotted livers in the sacrifice.
ANDROMACHE (*to Hector*)
  O, be persuaded. Do not count it holy
  To hurt by being just. It is as lawful,              20
  For we would give much, to use violent thefts,
  And rob in the behalf of charity.
CASSANDRA
  It is the purpose that makes strong the vow,
  But vows to every purpose must not hold.
  Unarm, sweet Hector.
HECTOR                Hold you still, I say.           25
  Mine honour keeps the weather of my fate.
  Life every man holds dear, but the dear man

Holds honour far more precious-dear than life.
        *Enter Troilus, armed*
  How now, young man, mean'st thou to fight today?
ANDROMACHE ⌈*aside*⌉
  Cassandra, call my father to persuade.   *Exit Cassandra*
HECTOR
  No, faith, young Troilus. Doff thy harness, youth.   31
  I am today i'th' vein of chivalry.
  Let grow thy sinews till their knots be strong,
  And tempt not yet the brushes of the war.
  Unarm thee, go—and doubt thou not, brave boy,       35
  I'll stand today for thee and me and Troy.
TROILUS
  Brother, you have a vice of mercy in you,
  Which better fits a lion than a man.
HECTOR
  What vice is that? Good Troilus, chide me for it.
TROILUS
  When many times the captive Grecian falls            40
  Even in the fan and wind of your fair sword,
  You bid them rise and live.
HECTOR  O 'tis fair play.
TROILUS  Fool's play, by heaven, Hector.
HECTOR  How now! How now!                              45
TROILUS  For th' love of all the gods,
  Let's leave the hermit pity with our mother
  And, when we have our armours buckled on,
  The venomed vengeance ride upon our swords,
  Spur them to ruthful work, rein them from ruth.      50
HECTOR
  Fie, savage, fie!
TROILUS            Hector, then 'tis wars.
HECTOR
  Troilus, I would not have you fight today.
TROILUS  Who should withhold me?
  Not fate, obedience, nor the hand of Mars
  Beck'ning with fiery truncheon my retire,            55
  Not Priamus and Hecuba on knees,
  Their eyes o'er-gallèd with recourse of tears,
  Nor you, my brother, with your true sword drawn
  Opposed to hinder me, should stop my way
  But by my ruin.                                      60
        *Enter Priam and Cassandra*
CASSANDRA
  Lay hold upon him, Priam, hold him fast.
  He is thy crutch: now if thou loose thy stay,
  Thou on him leaning and all Troy on thee,
  Fall all together.
PRIAM              Come, Hector, come. Go back.
  Thy wife hath dreamt, thy mother hath had visions,
  Cassandra doth foresee, and I myself                 66
  Am like a prophet suddenly enrapt
  To tell thee that this day is ominous.
  Therefore come back.
HECTOR                Aeneas is afield,
  And I do stand engaged to many Greeks,               70
  Even in the faith of valour, to appear
  This morning to them.
PRIAM  Ay, but thou shalt not go.
HECTOR ⌈*kneeling*⌉ I must not break my faith.
  You know me dutiful; therefore, dear sire,           75
  Let me not shame respect, but give me leave
  To take that course, by your consent and voice,
  Which you do here forbid me, royal Priam.
CASSANDRA
  O Priam, yield not to him.
ANDROMACHE                Do not, dear father.

HECTOR
Andromache, I am offended with you.                    80
Upon the love you bear me, get you in.
*Exit Andromache*

TROILUS
This foolish, dreaming, superstitious girl
Makes all these bodements.

CASSANDRA                            O farewell, dear Hector.
Look how thou diest; look how thy eye turns pale;
Look how thy wounds do bleed at many vents.           85
Hark how Troy roars, how Hecuba cries out,
How poor Andromache shrills her dolours forth.
Behold: distraction, frenzy, and amazement
Like witless antics one another meet,
And all cry 'Hector, Hector's dead, O Hector!'         90

TROILUS Away, away!

CASSANDRA
Farewell. Yet soft: Hector, I take my leave.
Thou dost thyself and all our Troy deceive.    *Exit*

HECTOR (*to Priam*)
You are amazed, my liege, at her exclaim.
Go in and cheer the town. We'll forth and fight,      95
Do deeds of praise, and tell you them at night.

PRIAM
Farewell. The gods with safety stand about thee.
*Exeunt Priam and Hector severally. Alarum*

TROILUS
They are at it, hark! Proud Diomed, believe
I come to lose my arm or win my sleeve.
*Enter Pandarus*

PANDARUS Do you hear, my lord, do you hear?           100

TROILUS What now?

PANDARUS Here's a letter come from yon poor girl.

TROILUS Let me read.
*Troilus reads the letter*

PANDARUS A whoreson phthisic, a whoreson rascally
phthisic so troubles me, and the foolish fortune of this
girl, and what one thing, what another, that I shall
leave you one o' these days. And I have a rheum in
mine eyes too, and such an ache in my bones that
unless a man were cursed I cannot tell what to think
on't.—What says she there?                            110

TROILUS (*tearing the letter*)
Words, words, mere words, no matter from the heart.
Th'effect doth operate another way.
Go, wind, to wind: there turn and change together.
My love with words and errors still she feeds,
But edifies another with her deeds.                   115

PANDARUS Why, but hear you—

TROILUS
Hence, broker-lackey! Ignomy and shame
Pursue thy life, and live aye with thy name.
*Exeunt severally*

**5.4**    *Alarum. Enter Thersites [in] excursions*

THERSITES Now they are clapper-clawing one another. I'll
go look on. That dissembling abominable varlet Diomed
has got that same scurvy doting foolish young knave's
sleeve of Troy there in his helm. I would fain see them
meet, that that same young Trojan ass that loves the
whore there might send that Greekish whoremasterly
villain with the sleeve back to the dissembling luxurious
drab of a sleeveless errand. O'th' t'other side, the policy
of those crafty swearing rascals—that stale old mouse-
eaten dry cheese Nestor and that same dog-fox
Ulysses—is proved not worth a blackberry. They set
me up in policy that mongrel cur Ajax against that

dog of as bad a kind Achilles. And now is the cur Ajax
prouder than the cur Achilles, and will not arm today—
whereupon the Grecians began to proclaim barbarism,
and policy grows into an ill opinion.                  16
*Enter Diomedes, followed by Troilus*
Soft, here comes sleeve and t'other.

TROILUS (*to Diomedes*)
Fly not, for shouldst thou take the river Styx
I would swim after.

DIOMEDES                    Thou dost miscall retire.
I do not fly, but advantageous care                    20
Withdrew me from the odds of multitude. Have at
thee!
*They fight*

THERSITES Hold thy whore, Grecian! Now for thy whore,
Trojan! Now the sleeve, now the sleeve!
*Exit Diomedes [driving in] Troilus*
*Enter Hector [behind]*

HECTOR
What art thou, Greek? Art thou for Hector's match?
Art thou of blood and honour?                          25

THERSITES No, no, I am a rascal, a scurvy railing knave,
a very filthy rogue.

HECTOR I do believe thee: live.

THERSITES God-a-mercy, that thou wilt believe me—
*[Exit Hector]*
but a plague break thy neck for frighting me. What's
become of the wenching rogues? I think they have
swallowed one another. I would laugh at that miracle—
yet in a sort lechery eats itself. I'll seek them.   *Exit*

**5.5**    *Enter Diomedes and Servants*

DIOMEDES
Go, go, my servant, take thou Troilus' horse.
Present the fair steed to my Lady Cressid.
Fellow, commend my service to her beauty.
Tell her I have chastised the amorous Trojan,
And am her knight by proof.

SERVANT                        I go, my lord.    *Exit*
*Enter Agamemnon*

AGAMEMNON
Renew, renew! The fierce Polydamas                      6
Hath beat down Menon; bastard Margareton
Hath Doreus prisoner,
And stands colossus-wise waving his beam
Upon the pashèd corpses of the kings                   10
Epistropus and Cedius; Polixenes is slain,
Amphimacus and Thoas deadly hurt,
Patroclus ta'en or slain, and Palamedes
Sore hurt and bruised; the dreadful sagittary
Appals our numbers. Haste we, Diomed,                  15
To reinforcement, or we perish all.
*Enter Nestor [with Patroclus' body]*

NESTOR
Go, bear Patroclus' body to Achilles,
And bid the snail-paced Ajax arm for shame.
*[Exit one or more with the body]*
There is a thousand Hectors in the field.
Now here he fights on Galathe his horse,               20
And there lacks work; anon he's there afoot,
And there they fly or die, like scalèd schools
Before the belching whale. Then is he yonder,
And there the strawy Greeks, ripe for his edge,
Fall down before him like the mower's swath.           25
Here, there, and everywhere he leaves and takes,
Dexterity so obeying appetite
That what he will he does, and does so much

That proof is called impossibility.
*Enter Ulysses*

ULYSSES
O courage, courage, princes! Great Achilles                    30
Is arming, weeping, cursing, vowing vengeance.
Patroclus' wounds have roused his drowsy blood,
Together with his mangled Myrmidons,
That noseless, handless, hacked and chipped come to
   him
Crying on Hector. Ajax hath lost a friend                    35
And foams at mouth, and he is armed and at it,
Roaring for Troilus—who hath done today
Mad and fantastic execution,
Engaging and redeeming of himself
With such a careless force and forceless care                    40
As if that luck, in very spite of cunning,
Bade him win all.
*Enter Ajax*

AJAX  Troilus, thou coward Troilus!                    *Exit*
DIOMEDES  Ay, there, there!                    ⌈*Exit*⌉
NESTOR  So, so, we draw together.                    45
*Enter Achilles*

ACHILLES  Where is this Hector?
Come, come, thou brave boy-queller, show thy face.
Know what it is to meet Achilles angry.
Hector! Where's Hector? I will none but Hector.
                                   ⌈*Exeunt*⌉

**5.6**    *Enter Ajax*
AJAX
Troilus, thou coward Troilus! Show thy head!
*Enter Diomedes*

DIOMEDES
Troilus, I say! Where's Troilus?
AJAX                                   What wouldst thou?
DIOMEDES  I would correct him.
AJAX
Were I the general, thou shouldst have my office
Ere that correction.—Troilus, I say! What, Troilus!    5
*Enter Troilus*

TROILUS
O traitor Diomed! Turn thy false face, thou traitor,
And pay the life thou ow'st me for my horse.
DIOMEDES  Ha, art thou there?
AJAX
I'll fight with him alone. Stand, Diomed.
DIOMEDES
He is my prize; I will not look upon.                    10
TROILUS
Come, both you cogging Greeks, have at you both!
*They fight.*
*Enter Hector*

HECTOR
Yea, Troilus? O well fought, my youngest brother!
          *Exit Troilus* ⌈*driving Diomedes and Ajax in*⌉
*Enter Achilles* ⌈*behind*⌉

ACHILLES
Now do I see thee.—Ha! Have at thee, Hector.
*They fight.* ⌈*Achilles is bested*⌉

HECTOR  Pause, if thou wilt.
ACHILLES
I do disdain thy courtesy, proud Trojan.                    15
Be happy that my arms are out of use.
My rest and negligence befriends thee now;
But thou anon shalt here of me again.
Till when, go seek thy fortune.                    *Exit*
HECTOR                                   Fare thee well.

I would have been much more a fresher man                    20
Had I expected thee.
*Enter Troilus* ⌈*in haste*⌉
                              How now, my brother?
TROILUS
Ajax hath ta'en Aeneas. Shall it be?
No, by the flame of yonder glorious heaven,
He shall not carry him. I'll be ta'en too,
Or bring him off. Fate, hear me what I say:                    25
I reck not though thou end my life today.                    *Exit*
*Enter one in sumptuous armour*

HECTOR
Stand, stand, thou Greek! Thou art a goodly mark.
No? Wilt thou not? I like thy armour well.
I'll frush it and unlock the rivets all,
But I'll be master of it.                    ⌈*Exit one in armour*⌉
                    Wilt thou not, beast, abide?
Why then, fly on; I'll hunt thee for thy hide.                    *Exit*

**5.7**    *Enter Achilles with Myrmidons*
ACHILLES
Come here about me, you my Myrmidons.
Mark what I say. Attend me where I wheel;
Strike not a stroke, but keep yourselves in breath,
And when I have the bloody Hector found,
Empale him with your weapons round about.                    5
In fellest manner execute your arms.
Follow me, sirs, and my proceedings eye.
It is decreed Hector the great must die.                    *Exeunt*

**5.8**    *Enter Menelaus and Paris, fighting,* ⌈*then*⌉ *Thersites*
THERSITES  The cuckold and the cuckold-maker are at it.—
Now, bull! Now, dog! 'Loo, Paris, 'loo! Now, my
double-horned Spartan! 'Loo, Paris, 'loo! The bull has
the game. Ware horns, ho!
                    *Exit Menelaus* ⌈*driving in*⌉ *Paris*
*Enter Bastard* ⌈*behind*⌉

BASTARD  Turn, slave, and fight.                    5
THERSITES  What art thou?
BASTARD  A bastard son of Priam's.
THERSITES  I am a bastard, too. I love bastards. I am
bastard begot, bastard instructed, bastard in mind,
bastard in valour, in everything illegitimate. One bear
will not bite another, and wherefore should one
bastard? Take heed: the quarrel's most ominous to us.
If the son of a whore fight for a whore, he tempts
judgement. Farewell, bastard.                    ⌈*Exit*⌉
BASTARD  The devil take thee, coward.                    *Exit*

**5.9**    *Enter Hector* ⌈*dragging*⌉ *the one in sumptuous*
          *armour*
HECTOR ⌈*taking off the helmet*⌉
Most putrefièd core, so fair without,
Thy goodly armour thus hath cost thy life.
Now is my day's work done. I'll take good breath.
Rest, sword: thou hast thy fill of blood and death.
*He disarms.*
*Enter Achilles and his Myrmidons, surrounding*
*Hector*

ACHILLES
Look, Hector, how the sun begins to set,                    5
How ugly night comes breathing at his heels,
Even with the veil and dark'ning of the sun
To close the day up, Hector's life is done.
HECTOR
I am unarmed. Forgo this vantage, Greek.

747

ACHILLES
Strike, fellows, strike! This is the man I seek.      10
⌈The Myrmidons⌉ kill Hector
So, Ilium, fall thou. Now, Troy, sink down.
Here lies thy heart, thy sinews, and thy bone.—
On, Myrmidons, and cry you all amain,
'Achilles hath the mighty Hector slain!'
A retreat is sounded
Hark, a retire upon our Grecian part.      15
⌈Another retreat is sounded⌉
A MYRMIDON
The Trojan trumpets sound the like, my lord.
ACHILLES
The dragon wing of night o'erspreads the earth
And, stickler-like, the armies separates.
My half-supped sword, that frankly would have fed,
Pleased with this dainty bait, thus goes to bed.      20
He sheathes his sword
Come, tie his body to my horse's tail.
Along the field I will the Trojan trail.
*Exeunt, dragging the bodies*

**5.10**   *A retreat is sounded. Enter Agamemnon, Ajax,*
     *Menelaus, Nestor, Diomedes, and the rest,*
     *marching.* ⌈A shout within⌉
AGAMEMNON
Hark, hark! What shout is that?
NESTOR                 Peace, drums.
MYRMIDONS (*within*)                       Achilles!
Achilles! Hector's slain! Achilles!
DIOMEDES
The bruit is: Hector's slain, and by Achilles.
AJAX
If it be so, yet bragless let it be.
Great Hector was a man as good as he.      5
AGAMEMNON
March patiently along. Let one be sent
To pray Achilles see us at our tent.
If in his death the gods have us befriended,
Great Troy is ours, and our sharp wars are ended.
*Exeunt* ⌈*marching*⌉

**5.11**   *Enter Aeneas, Paris, Antenor, and Deiphobus*
AENEAS
Stand, ho! Yet are we masters of the field.
Never go home; here starve we out the night.
*Enter Troilus*
TROILUS
Hector is slain.
ALL THE OTHERS   Hector? The gods forbid.
TROILUS
He's dead, and at the murderer's horse's tail
In beastly sort dragged through the shameful field.      5
Frown on, you heavens; effect your rage with speed;
Sit, gods, upon your thrones, and smite at Troy.
I say, at once: let your brief plagues be mercy,
And linger not our sure destructions on.
AENEAS
My lord, you do discomfort all the host.      10
TROILUS
You understand me not that tell me so.
I do not speak of flight, of fear of death,
But dare all imminence that gods and men
Address their dangers in. Hector is gone.
Who shall tell Priam so, or Hecuba?      15
Let him that will a screech-owl aye be called
Go into Troy and say their Hector's dead.
There is a word will Priam turn to stone,
Make wells and Niobes of the maids and wives,
Cold statues of the youth, and in a word      20
Scare Troy out of itself. But march away.
Hector is dead; there is no more to say.
Stay yet.—You vile abominable tents
Thus proudly pitched upon our Phrygian plains,
Let Titan rise as early as he dare,      25
I'll through and through you! And thou great-sized
    coward,
No space of earth shall sunder our two hates.
I'll haunt thee like a wicked conscience still,
That mouldeth goblins swift as frenzy's thoughts.
Strike a free march! To Troy with comfort go:      30
Hope of revenge shall hide our inward woe.
⌈*Exeunt marching*⌉

## ADDITIONAL PASSAGES

A. The Quarto (below) gives a more elaborate version of Thersites' speech at 5.1.17–21.

THERSITES Why, his masculine whore. Now the rotten diseases of the south, the guts-griping, ruptures, loads o' gravel in the back, lethargies, cold palsies, raw eyes, dirt-rotten livers, wheezing lungs, bladders full of impostume, sciaticas, lime-kilns i'th' palm, incurable bone-ache, and the rivelled fee-simple of the tetter, take and take again such preposterous discoveries.

B. The Quarto gives a different ending to the play (which the Folio inadvertently repeats).

    *Enter Pandarus*
PANDARUS But hear you, hear you.
TROILUS
Hence, broker-lackey. ⌈*Strikes him*⌉ Ignomy and shame
Pursue thy life, and live aye with thy name.
*Exeunt all but Pandarus*
PANDARUS A goodly medicine for my aching bones. O world, world, world!—thus is the poor agent despised.

O traitors and bawds, how earnestly are you set a work, and how ill requited! Why should our endeavour be so desired and the performance so loathed? What verse for it? What instance for it? Let me see,
    Full merrily the humble-bee doth sing
    Till he hath lost his honey and his sting,
    And being once subdued in armèd tail,
    Sweet honey and sweet notes together fail.
Good traders in the flesh, set this in your painted cloths:
As many as be here of Pandar's hall,
Your eyes, half out, weep out at Pandar's fall.
Or if you cannot weep, yet give some groans,
Though not for me, yet for your aching bones.
Brethren and sisters of the hold-door trade,
Some two months hence my will shall here be made.
It should be now, but that my fear is this:
Some gallèd goose of Winchester would hiss.
Till then I'll sweat and seek about for eases,
And at that time bequeath you my diseases.     *Exit*

# SONNETS
# AND 'A LOVER'S COMPLAINT'

SHAKESPEARE'S Sonnets were published as a collection by Thomas Thorpe in 1609; the title-page declared that they were 'never before imprinted'. Versions of two of them—138 and 144—had appeared in 1599, in *The Passionate Pilgrim*, a collection ascribed to Shakespeare but including some poems certainly written by other authors; and in the previous year Francis Meres, in *Palladis Tamia*, had alluded to Shakespeare's 'sugared sonnets among his private friends'. The sonnet sequence had enjoyed a brief but intense vogue from the publication of Sir Philip Sidney's *Astrophil and Stella* in 1591 till about 1597. Some of Shakespeare's plays of this period reflect the fashion: in the comedy of *Love's Labour's Lost* the writing of sonnets is seen as a laughable symptom of love, and in the tragedy of *Romeo and Juliet* both speeches of the Chorus and the lovers' first conversation are in sonnet form. Later plays use it, too, but it seems likely that most, if not all, of Shakespeare's sonnets were first written during this period. But there are indications that some of them were revised; the two printed in *The Passionate Pilgrim* differ at certain points from Thorpe's version, and two other sonnets (2 and 106) exist in manuscript versions which also are not identical with those published in the sequence. We print these as 'Alternative Versions' of Sonnets 2, 106, 138, and 144.

The order in which Thorpe printed the Sonnets has often been questioned, but is not entirely haphazard: all the first seventeen, and no later ones, exhort a young man to marry; all those clearly addressed to a man are among the first 126, and all those clearly addressed to, or concerned with, a woman (the 'dark lady') follow. Some of the sonnets in the second group appear to refer to events that prompted sonnets in the first group; it seems likely that the poems were rearranged after composition. Moreover, the volume contains 'A Lover's Complaint', clearly ascribed to Shakespeare, which stylistic evidence suggests was written in the early seventeenth century and which may have been intended as a companion piece. So, printing the Sonnets in Thorpe's order, we place them according to the likely date of their revision.

Textual evidence suggests that Thorpe printed from a transcript by someone other than Shakespeare. His volume bears a dedication over his own initials to 'Mr W.H.'; we do not know whether this derives from the manuscript, and can only speculate about the dedicatee's identity. His initials are those of Shakespeare's only known dedicatee, Henry Wriothesley, Earl of Southampton, but in reverse order. We have even less clue as to the identity of the Sonnets' other personae, a rival poet and the dark woman.

Shakespeare's Sonnets may not be autobiographical, but they are certainly unconventional: the most idealistic poems celebrating love's mutuality are addressed by one man to another (Sonnet 20 implies that the relationship is not sexual), and the poems clearly addressed to a woman revile her morals, speak ill of her appearance, and explore the poet's self-disgust at his entanglement with her. The Sonnets include some of the finest love poems in the English language: the sequence itself presents an internal drama of great psychological complexity.

TO.THE.ONLY.BEGETTER.OF.
THESE.ENSUING.SONNETS.
M$^r$.W.H.  ALL.HAPPINESS.
AND.THAT.ETERNITY.
PROMISED.

BY.

OUR.EVER-LIVING.POET.

WISHETH.

THE.WELL-WISHING.
ADVENTURER.IN.
SETTING.
FORTH.

T.T.

# Sonnets

### 1

From fairest creatures we desire increase,
That thereby beauty's rose might never die,
But as the riper should by time decease,
His tender heir might bear his memory;
But thou, contracted to thine own bright eyes,      5
Feed'st thy light's flame with self-substantial fuel,
Making a famine where abundance lies,
Thyself thy foe, to thy sweet self too cruel.
Thou that art now the world's fresh ornament
And only herald to the gaudy spring      10
Within thine own bud buriest thy content,
And, tender churl, mak'st waste in niggarding.
    Pity the world, or else this glutton be:
    To eat the world's due, by the grave and thee.

### 2

When forty winters shall besiege thy brow
And dig deep trenches in thy beauty's field,
Thy youth's proud livery, so gazed on now,
Will be a tattered weed, of small worth held.
Then being asked where all thy beauty lies,      5
Where all the treasure of thy lusty days,
To say within thine own deep-sunken eyes
Were an all-eating shame and thriftless praise.
How much more praise deserved thy beauty's use
If thou couldst answer 'This fair child of mine      10
Shall sum my count, and make my old excuse',
Proving his beauty by succession thine.
    This were to be new made when thou art old,
    And see thy blood warm when thou feel'st it cold.

### 3

Look in thy glass, and tell the face thou viewest
Now is the time that face should form another,
Whose fresh repair if now thou not renewest
Thou dost beguile the world, unbless some mother.
For where is she so fair whose uneared womb      5
Disdains the tillage of thy husbandry?
Or who is he so fond will be the tomb
Of his self-love to stop posterity?
Thou art thy mother's glass, and she in thee
Calls back the lovely April of her prime;      10
So thou through windows of thine age shalt see,
Despite of wrinkles, this thy golden time.
    But if thou live remembered not to be,
    Die single, and thine image dies with thee.

### 4

Unthrifty loveliness, why dost thou spend
Upon thyself thy beauty's legacy?
Nature's bequest gives nothing, but doth lend,
And being frank, she lends to those are free.
Then, beauteous niggard, why dost thou abuse      5
The bounteous largess given thee to give?
Profitless usurer, why dost thou use
So great a sum of sums yet canst not live?

For having traffic with thyself alone,
Thou of thyself thy sweet self dost deceive.      10
Then how when nature calls thee to be gone:
What acceptable audit canst thou leave?
    Thy unused beauty must be tombed with thee,
    Which used, lives th'executor to be.

### 5

Those hours that with gentle work did frame
The lovely gaze where every eye doth dwell
Will play the tyrants to the very same,
And that unfair which fairly doth excel;
For never-resting time leads summer on      5
To hideous winter, and confounds him there,
Sap checked with frost, and lusty leaves quite gone,
Beauty o'er-snowed, and bareness everywhere.
Then were not summer's distillation left
A liquid prisoner pent in walls of glass,      10
Beauty's effect with beauty were bereft,
Nor it nor no remembrance what it was.
    But flowers distilled, though they with winter meet,
    Lose but their show; their substance still lives sweet.

### 6

Then let not winter's ragged hand deface
In thee thy summer ere thou be distilled.
Make sweet some vial, treasure thou some place
With beauty's treasure ere it be self-killed.
That use is not forbidden usury      5
Which happies those that pay the willing loan:
That's for thyself to breed another thee,
Or ten times happier, be it ten for one;
Ten times thyself were happier than thou art,
If ten of thine ten times refigured thee.      10
Then what could death do if thou shouldst depart,
Leaving thee living in posterity?
    Be not self-willed, for thou art much too fair
    To be death's conquest and make worms thine heir.

### 7

Lo, in the orient when the gracious light
Lifts up his burning head, each under eye
Doth homage to his new-appearing sight,
Serving with looks his sacred majesty,
And having climbed the steep-up heavenly hill,      5
Resembling strong youth in his middle age,
Yet mortal looks adore his beauty still,
Attending on his golden pilgrimage.
But when from highmost pitch, with weary car,
Like feeble age he reeleth from the day,      10
The eyes, 'fore duteous, now converted are
From his low tract, and look another way.
    So thou, thyself outgoing in thy noon,
    Unlooked on diest unless thou get a son.

### 8

Music to hear, why hear'st thou music sadly?
Sweets with sweets war not, joy delights in joy.
Why lov'st thou that which thou receiv'st not gladly,
Or else receiv'st with pleasure thine annoy?
If the true concord of well-tunèd sounds    5
By unions married do offend thine ear,
They do but sweetly chide thee, who confounds
In singleness the parts that thou shouldst bear.
Mark how one string, sweet husband to another,
Strikes each in each by mutual ordering,    10
Resembling sire and child and happy mother,
Who all in one one pleasing note do sing;
   Whose speechless song, being many, seeming one,
   Sings this to thee: 'Thou single wilt prove none.'

### 9

Is it for fear to wet a widow's eye
That thou consum'st thyself in single life?
Ah, if thou issueless shalt hap to die,
The world will wail thee like a makeless wife.
The world will be thy widow, and still weep    5
That thou no form of thee hast left behind,
When every private widow well may keep
By children's eyes her husband's shape in mind.
Look what an unthrift in the world doth spend
Shifts but his place, for still the world enjoys it;    10
But beauty's waste hath in the world an end,
And kept unused, the user so destroys it.
   No love toward others in that bosom sits
   That on himself such murd'rous shame commits.

### 10

For shame deny that thou bear'st love to any,
Who for thyself art so unprovident.
Grant, if thou wilt, thou art beloved of many,
But that thou none lov'st is most evident;
For thou art so possessed with murd'rous hate    5
That 'gainst thyself thou stick'st not to conspire,
Seeking that beauteous roof to ruinate
Which to repair should be thy chief desire.
O, change thy thought, that I may change my mind!
Shall hate be fairer lodged than gentle love?    10
Be as thy presence is, gracious and kind,
Or to thyself at least kind-hearted prove.
   Make thee another self for love of me,
   That beauty still may live in thine or thee.

### 11

As fast as thou shalt wane, so fast thou grow'st
In one of thine from that which thou departest,
And that fresh blood which youngly thou bestow'st
Thou mayst call thine when thou from youth
     convertest.
Herein lives wisdom, beauty, and increase;    5
Without this, folly, age, and cold decay.
If all were minded so, the times should cease,
And threescore year would make the world away.
Let those whom nature hath not made for store,
Harsh, featureless, and rude, barrenly perish.    10
Look whom she best endowed she gave the more,
Which bounteous gift thou shouldst in bounty cherish.
   She carved thee for her seal, and meant thereby
   Thou shouldst print more, not let that copy die.

### 12

When I do count the clock that tells the time,
And see the brave day sunk in hideous night;
When I behold the violet past prime,
And sable curls ensilvered o'er with white;
When lofty trees I see barren of leaves,    5
Which erst from heat did canopy the herd,
And summer's green all girded up in sheaves
Borne on the bier with white and bristly beard:
Then of thy beauty do I question make
That thou among the wastes of time must go,    10
Since sweets and beauties do themselves forsake,
And die as fast as they see others grow;
   And nothing 'gainst time's scythe can make defence
   Save breed to brave him when he takes thee hence.

### 13

O that you were yourself! But, love, you are
No longer yours than you yourself here live.
Against this coming end you should prepare,
And your sweet semblance to some other give.
So should that beauty which you hold in lease    5
Find no determination; then you were
Yourself again after your self's decease,
When your sweet issue your sweet form should bear.
Who lets so fair a house fall to decay,
Which husbandry in honour might uphold    10
Against the stormy gusts of winter's day,
And barren rage of death's eternal cold?
   O, none but unthrifts, dear my love, you know.
   You had a father; let your son say so.

### 14

Not from the stars do I my judgement pluck,
And yet methinks I have astronomy;
But not to tell of good or evil luck,
Of plagues, of dearths, or seasons' quality.
Nor can I fortune to brief minutes tell,    5
'Pointing to each his thunder, rain, and wind,
Or say with princes if it shall go well
By oft predict that I in heaven find;
But from thine eyes my knowledge I derive,
And, constant stars, in them I read such art    10
As truth and beauty shall together thrive
If from thyself to store thou wouldst convert.
   Or else of thee this I prognosticate:
   Thy end is truth's and beauty's doom and date.

### 15

When I consider every thing that grows
Holds in perfection but a little moment,
That this huge stage presenteth naught but shows
Whereon the stars in secret influence comment;
When I perceive that men as plants increase,    5
Cheerèd and checked even by the selfsame sky;
Vaunt in their youthful sap, at height decrease,
And wear their brave state out of memory:
Then the conceit of this inconstant stay
Sets you most rich in youth before my sight,    10
Where wasteful time debateth with decay
To change your day of youth to sullied night;
   And all in war with time for love of you,
   As he takes from you, I engraft you new.

### 16

But wherefore do not you a mightier way
Make war upon this bloody tyrant, time,
And fortify yourself in your decay
With means more blessèd than my barren rhyme?
Now stand you on the top of happy hours,　　5
And many maiden gardens yet unset
With virtuous wish would bear your living flowers,
Much liker than your painted counterfeit.
So should the lines of life that life repair
Which this time's pencil or my pupil pen　　10
Neither in inward worth nor outward fair
Can make you live yourself in eyes of men.
　　To give away yourself keeps yourself still,
　　And you must live drawn by your own sweet skill.

### 17

Who will believe my verse in time to come
If it were filled with your most high deserts?—
Though yet, heaven knows, it is but as a tomb
Which hides your life, and shows not half your parts.
If I could write the beauty of your eyes　　5
And in fresh numbers number all your graces,
The age to come would say 'This poet lies;
Such heavenly touches ne'er touched earthly faces.'
So should my papers, yellowed with their age,
Be scorned, like old men of less truth than tongue,　　10
And your true rights be termed a poet's rage
And stretchèd metre of an antique song.
　　But were some child of yours alive that time,
　　You should live twice: in it, and in my rhyme.

### 18

Shall I compare thee to a summer's day?
Thou art more lovely and more temperate.
Rough winds do shake the darling buds of May,
And summer's lease hath all too short a date.
Sometime too hot the eye of heaven shines,　　5
And often is his gold complexion dimmed,
And every fair from fair sometime declines,
By chance or nature's changing course untrimmed;
But thy eternal summer shall not fade
Nor lose possession of that fair thou ow'st,　　10
Nor shall death brag thou wander'st in his shade
When in eternal lines to time thou grow'st.
　　So long as men can breathe or eyes can see,
　　So long lives this, and this gives life to thee.

### 19

Devouring time, blunt thou the lion's paws,
And make the earth devour her own sweet brood;
Pluck the keen teeth from the fierce tiger's jaws,
And burn the long-lived phoenix in her blood.
Make glad and sorry seasons as thou fleet'st,　　5
And do whate'er thou wilt, swift-footed time,
To the wide world and all her fading sweets.
But I forbid thee one most heinous crime:
O, carve not with thy hours my love's fair brow,
Nor draw no lines there with thine antique pen.　　10
Him in thy course untainted do allow
For beauty's pattern to succeeding men.
　　Yet do thy worst, old time; despite thy wrong
　　My love shall in my verse ever live young.

### 20

A woman's face with nature's own hand painted
Hast thou, the master-mistress of my passion;
A woman's gentle heart, but not acquainted
With shifting change as is false women's fashion;
An eye more bright than theirs, less false in rolling,　　5
Gilding the object whereupon it gazeth;
A man in hue, all hues in his controlling,
Which steals men's eyes and women's souls amazeth.
And for a woman wert thou first created,
Till nature as she wrought thee fell a-doting,　　10
And by addition me of thee defeated
By adding one thing to my purpose nothing.
　　But since she pricked thee out for women's pleasure,
　　Mine be thy love and thy love's use their treasure.

### 21

So is it not with me as with that muse
Stirred by a painted beauty to his verse,
Who heaven itself for ornament doth use,
And every fair with his fair doth rehearse,
Making a couplement of proud compare　　5
With sun and moon, with earth, and sea's rich gems,
With April's first-born flowers, and all things rare
That heaven's air in this huge rondure hems.
O let me, true in love, but truly write,
And then believe me my love is as fair　　10
As any mother's child, though not so bright
As those gold candles fixed in heaven's air.
　　Let them say more that like of hearsay well;
　　I will not praise that purpose not to sell.

### 22

My glass shall not persuade me I am old
So long as youth and thou are of one date;
But when in thee time's furrows I behold,
Then look I death my days should expiate.
For all that beauty that doth cover thee　　5
Is but the seemly raiment of my heart,
Which in thy breast doth live, as thine in me;
How can I then be elder than thou art?
O therefore, love, be of thyself so wary
As I, not for myself, but for thee will,　　10
Bearing thy heart, which I will keep so chary
As tender nurse her babe from faring ill.
　　Presume not on thy heart when mine is slain:
　　Thou gav'st me thine not to give back again.

### 23

As an unperfect actor on the stage
Who with his fear is put besides his part,
Or some fierce thing replete with too much rage
Whose strength's abundance weakens his own heart,
So I, for fear of trust, forget to say　　5
The perfect ceremony of love's rite,
And in mine own love's strength seem to decay,
O'er-charged with burden of mine own might.
O let my books be then the eloquence
And dumb presagers of my speaking breast,　　10
Who plead for love, and look for recompense
More than that tongue that more hath more expressed.
　　O learn to read what silent love hath writ;
　　To hear with eyes belongs to love's fine wit.

## 24

Mine eye hath played the painter, and hath steeled
Thy beauty's form in table of my heart.
My body is the frame wherein 'tis held,
And perspective it is best painter's art;
For through the painter must you see his skill        5
To find where your true image pictured lies,
Which in my bosom's shop is hanging still,
That hath his windows glazèd with thine eyes.
Now see what good turns eyes for eyes have done:
Mine eyes have drawn thy shape, and thine for me    10
Are windows to my breast, wherethrough the sun
Delights to peep, to gaze therein on thee.
    Yet eyes this cunning want to grace their art:
    They draw but what they see, know not the heart.

## 25

Let those who are in favour with their stars
Of public honour and proud titles boast,
Whilst I, whom fortune of such triumph bars,
Unlooked-for joy in that I honour most.
Great princes' favourites their fair leaves spread     5
But as the marigold at the sun's eye,
And in themselves their pride lies burièd,
For at a frown they in their glory die.
The painful warrior famousèd for might,
After a thousand victories once foiled              10
Is from the book of honour razèd quite,
And all the rest forgot for which he toiled.
    Then happy I, that love and am beloved
    Where I may not remove nor be removed.

## 26

Lord of my love, to whom in vassalage
Thy merit hath my duty strongly knit,
To thee I send this written embassage
To witness duty, not to show my wit;
Duty so great which wit so poor as mine             5
May make seem bare in wanting words to show it,
But that I hope some good conceit of thine
In thy soul's thought, all naked, will bestow it,
Till whatsoever star that guides my moving
Points on me graciously with fair aspect,          10
And puts apparel on my tattered loving
To show me worthy of thy sweet respect.
    Then may I dare to boast how I do love thee;
    Till then, not show my head where thou mayst prove
      me.

## 27

Weary with toil I haste me to my bed,
The dear repose for limbs with travel tired;
But then begins a journey in my head
To work my mind when body's work's expired;
For then my thoughts, from far where I abide,       5
Intend a zealous pilgrimage to thee,
And keep my drooping eyelids open wide,
Looking on darkness which the blind do see:
Save that my soul's imaginary sight
Presents thy shadow to my sightless view,          10
Which like a jewel hung in ghastly night
Makes black night beauteous and her old face new.
    Lo, thus by day my limbs, by night my mind,
    For thee, and for myself, no quiet find.

## 28

How can I then return in happy plight,
That am debarred the benefit of rest,
When day's oppression is not eased by night,
But day by night and night by day oppressed,
And each, though enemies to either's reign,         5
Do in consent shake hands to torture me,
The one by toil, the other to complain
How far I toil, still farther off from thee?
I tell the day to please him thou art bright,
And do'st him grace when clouds do blot the heaven;
So flatter I the swart-complexioned night          11
When sparkling stars twire not thou gild'st the even.
    But day doth daily draw my sorrows longer,
    And night doth nightly make grief's strength seem
      stronger.

## 29

When, in disgrace with fortune and men's eyes,
I all alone beweep my outcast state,
And trouble deaf heaven with my bootless cries,
And look upon myself and curse my fate,
Wishing me like to one more rich in hope,           5
Featured like him, like him with friends possessed,
Desiring this man's art and that man's scope,
With what I most enjoy contented least:
Yet in these thoughts myself almost despising,
Haply I think on thee, and then my state,          10
Like to the lark at break of day arising
From sullen earth, sings hymns at heaven's gate;
    For thy sweet love remembered such wealth brings
    That then I scorn to change my state with kings'.

## 30

When to the sessions of sweet silent thought
I summon up remembrance of things past,
I sigh the lack of many a thing I sought,
And with old woes new wail my dear time's waste.
Then can I drown an eye unused to flow              5
For precious friends hid in death's dateless night,
And weep afresh love's long-since-cancelled woe,
And moan th'expense of many a vanished sight.
Then can I grieve at grievances foregone,
And heavily from woe to woe tell o'er              10
The sad account of fore-bemoanèd moan,
Which I new pay as if not paid before.
    But if the while I think on thee, dear friend,
    All losses are restored, and sorrows end.

## 31

Thy bosom is endearèd with all hearts
Which I by lacking have supposèd dead,
And there reigns love, and all love's loving parts,
And all those friends which I thought burièd.
How many a holy and obsequious tear                 5
Hath dear religious love stol'n from mine eye
As interest of the dead, which now appear
But things removed that hidden in thee lie!
Thou art the grave where buried love doth live,
Hung with the trophies of my lovers gone,          10
Who all their parts of me to thee did give:
That due of many now is thine alone.
    Their images I loved I view in thee,
    And thou, all they, hast all the all of me.

## 32

If thou survive my well-contented day
When that churl death my bones with dust shall cover,
And shalt by fortune once more resurvey
These poor rude lines of thy deceasèd lover,
Compare them with the bett'ring of the time,          5
And though they be outstripped by every pen,
Reserve them for my love, not for their rhyme
Exceeded by the height of happier men.
O then vouchsafe me but this loving thought:
'Had my friend's muse grown with this growing age,     10
A dearer birth than this his love had brought
To march in ranks of better equipage;
    But since he died, and poets better prove,
    Theirs for their style I'll read, his for his love.'

## 33

Full many a glorious morning have I seen
Flatter the mountain tops with sovereign eye,
Kissing with golden face the meadows green,
Gilding pale streams with heavenly alchemy;
Anon permit the basest clouds to ride             5
With ugly rack on his celestial face,
And from the forlorn world his visage hide,
Stealing unseen to west with this disgrace.
Even so my sun one early morn did shine
With all triumphant splendour on my brow;         10
But out, alack, he was but one hour mine;
The region cloud hath masked him from me now.
    Yet him for this my love no whit disdaineth:
    Suns of the world may stain when heaven's sun staineth.

## 34

Why didst thou promise such a beauteous day
And make me travel forth without my cloak,
To let base clouds o'ertake me in my way,
Hiding thy brav'ry in their rotten smoke?
'Tis not enough that through the cloud thou break     5
To dry the rain on my storm-beaten face,
For no man well of such a salve can speak
That heals the wound and cures not the disgrace.
Nor can thy shame give physic to my grief;
Though thou repent, yet I have still the loss.        10
Th'offender's sorrow lends but weak relief
To him that bears the strong offence's cross.
    Ah, but those tears are pearl which thy love sheds,
    And they are rich, and ransom all ill deeds.

## 35

No more be grieved at that which thou hast done:
Roses have thorns, and silver fountains mud.
Clouds and eclipses stain both moon and sun,
And loathsome canker lives in sweetest bud.
All men make faults, and even I in this,          5
Authorizing thy trespass with compare,
Myself corrupting salving thy amiss,
Excusing thy sins more than thy sins are;
For to thy sensual fault I bring in sense—
Thy adverse party is thy advocate—                10
And 'gainst myself a lawful plea commence.
Such civil war is in my love and hate
    That I an accessory needs must be
    To that sweet thief which sourly robs from me.

## 36

Let me confess that we two must be twain
Although our undivided loves are one;
So shall those blots that do with me remain
Without thy help by me be borne alone.
In our two loves there is but one respect,          5
Though in our lives a separable spite
Which, though it alter not love's sole effect,
Yet doth it steal sweet hours from love's delight.
I may not evermore acknowledge thee
Lest my bewailèd guilt should do thee shame,        10
Nor thou with public kindness honour me
Unless thou take that honour from thy name.
    But do not so. I love thee in such sort
    As, thou being mine, mine is thy good report.

## 37

As a decrepit father takes delight
To see his active child do deeds of youth,
So I, made lame by fortune's dearest spite,
Take all my comfort of thy worth and truth;
For whether beauty, birth, or wealth, or wit,       5
Or any of these all, or all, or more,
Entitled in thy parts do crownèd sit,
I make my love engrafted to this store.
So then I am not lame, poor, nor despised,
Whilst that this shadow doth such substance give    10
That I in thy abundance am sufficed
And by a part of all thy glory live.
    Look what is best, that best I wish in thee;
    This wish I have, then ten times happy me.

## 38

How can my muse want subject to invent
While thou dost breathe, that pour'st into my verse
Thine own sweet argument, too excellent
For every vulgar paper to rehearse?
O, give thyself the thanks if aught in me            5
Worthy perusal stand against thy sight;
For who's so dumb that cannot write to thee,
When thou thyself dost give invention light?
Be thou the tenth muse, ten times more in worth
Than those old nine which rhymers invocate,          10
And he that calls on thee, let him bring forth
Eternal numbers to outlive long date.
    If my slight muse do please these curious days,
    The pain be mine, but thine shall be the praise.

## 39

O, how thy worth with manners may I sing
When thou art all the better part of me?
What can mine own praise to mine own self bring,
And what is't but mine own when I praise thee?
Even for this let us divided live,                   5
And our dear love lose name of single one,
That by this separation I may give
That due to thee which thou deserv'st alone.
O absence, what a torment wouldst thou prove
Were it not thy sour leisure gave sweet leave        10
To entertain the time with thoughts of love,
Which time and thoughts so sweetly doth deceive,
    And that thou teachest how to make one twain
    By praising him here who doth hence remain!

### 40

Take all my loves, my love, yea, take them all:
What hast thou then more than thou hadst before?
No love, my love, that thou mayst true love call—
All mine was thine before thou hadst this more.
Then if for my love thou my love receivest,                    5
I cannot blame thee for my love thou usest;
But yet be blamed if thou this self deceivest
By wilful taste of what thyself refusest.
I do forgive thy robb'ry, gentle thief,
Although thou steal thee all my poverty;                      10
And yet love knows it is a greater grief
To bear love's wrong than hate's known injury.
　　Lascivious grace, in whom all ill well shows,
　　Kill me with spites, yet we must not be foes.

### 41

Those pretty wrongs that liberty commits
When I am sometime absent from thy heart
Thy beauty and thy years full well befits,
For still temptation follows where thou art.
Gentle thou art, and therefore to be won;                      5
Beauteous thou art, therefore to be assailed;
And when a woman woos, what woman's son
Will sourly leave her till he have prevailed?
Ay me, but yet thou mightst my seat forbear,
And chide thy beauty and thy straying youth                   10
Who lead thee in their riot even there
Where thou art forced to break a two-fold troth:
　　Hers, by thy beauty tempting her to thee,
　　Thine, by thy beauty being false to me.

### 42

That thou hast her, it is not all my grief,
And yet it may be said I loved her dearly;
That she hath thee is of my wailing chief,
A loss in love that touches me more nearly.
Loving offenders, thus I will excuse ye:                        5
Thou dost love her because thou know'st I love her,
And for my sake even so doth she abuse me,
Suff'ring my friend for my sake to approve her.
If I lose thee, my loss is my love's gain,
And losing her, my friend hath found that loss:               10
Both find each other, and I lose both twain,
And both for my sake lay on me this cross.
　　But here's the joy: my friend and I are one.
　　Sweet flattery! Then she loves but me alone.

### 43

When most I wink, then do mine eyes best see,
For all the day they view things unrespected;
But when I sleep, in dreams they look on thee,
And, darkly bright, are bright in dark directed.
Then thou, whose shadow shadows doth make bright,     5
How would thy shadow's form form happy show
To the clear day with thy much clearer light,
When to unseeing eyes thy shade shines so!
How would, I say, mine eyes be blessèd made
By looking on thee in the living day,                          10
When in dead night thy fair imperfect shade
Through heavy sleep on sightless eyes doth stay!
　　All days are nights to see till I see thee,
　　And nights bright days when dreams do show thee me.

### 44

If the dull substance of my flesh were thought,
Injurious distance should not stop my way;
For then, despite of space, I would be brought
From limits far remote where thou dost stay.
No matter then although my foot did stand                      5
Upon the farthest earth removed from thee;
For nimble thought can jump both sea and land
As soon as think the place where he would be.
But ah, thought kills me that I am not thought,
To leap large lengths of miles when thou art gone,            10
But that, so much of earth and water wrought,
I must attend time's leisure with my moan,
　　Receiving naught by elements so slow
　　But heavy tears, badges of either's woe.

### 45

The other two, slight air and purging fire,
Are both with thee wherever I abide;
The first my thought, the other my desire,
These present-absent with swift motion slide;
For when these quicker elements are gone                        5
In tender embassy of love to thee,
My life, being made of four, with two alone
Sinks down to death, oppressed with melancholy,
Until life's composition be recured
By those swift messengers returned from thee,                 10
Who even but now come back again assured
Of thy fair health, recounting it to me.
　　This told, I joy; but then no longer glad,
　　I send them back again and straight grow sad.

### 46

Mine eye and heart are at a mortal war
How to divide the conquest of thy sight.
Mine eye my heart thy picture's sight would bar,
My heart, mine eye the freedom of that right.
My heart doth plead that thou in him dost lie,                  5
A closet never pierced with crystal eyes;
But the defendant doth that plea deny,
And says in him thy fair appearance lies.
To 'cide this title is empanellèd
A quest of thoughts, all tenants to the heart,                10
And by their verdict is determinèd
The clear eye's moiety and the dear heart's part,
　　As thus: mine eye's due is thy outward part,
　　And my heart's right thy inward love of heart.

### 47

Betwixt mine eye and heart a league is took,
And each doth good turns now unto the other.
When that mine eye is famished for a look,
Or heart in love with sighs himself doth smother,
With my love's picture then my eye doth feast,                  5
And to the painted banquet bids my heart.
Another time mine eye is my heart's guest
And in his thoughts of love doth share a part.
So either by thy picture or my love,
Thyself away art present still with me;                       10
For thou no farther than my thoughts canst move,
And I am still with them, and they with thee;
　　Or if they sleep, thy picture in my sight
　　Awakes my heart to heart's and eye's delight.

### 48

How careful was I when I took my way
Each trifle under truest bars to thrust,
That to my use it might unusèd stay
From hands of falsehood, in sure wards of trust.
But thou, to whom my jewels trifles are,          5
Most worthy comfort, now my greatest grief,
Thou best of dearest and mine only care
Art left the prey of every vulgar thief.
Thee have I not locked up in any chest
Save where thou art not, though I feel thou art—          10
Within the gentle closure of my breast,
From whence at pleasure thou mayst come and part;
   And even thence thou wilt be stol'n, I fear,
   For truth proves thievish for a prize so dear.

### 49

Against that time—if ever that time come—
When I shall see thee frown on my defects,
Whenas thy love hath cast his utmost sum,
Called to that audit by advisèd respects;
Against that time when thou shalt strangely pass          5
And scarcely greet me with that sun, thine eye,
When love converted from the thing it was
Shall reasons find of settled gravity:
Against that time do I ensconce me here
Within the knowledge of mine own desert,          10
And this my hand against myself uprear
To guard the lawful reasons on thy part.
   To leave poor me thou hast the strength of laws,
   Since why to love I can allege no cause.

### 50

How heavy do I journey on the way,
When what I seek—my weary travel's end—
Doth teach that ease and that repose to say
'Thus far the miles are measured from thy friend.'
The beast that bears me, tired with my woe,          5
Plods dully on to bear that weight in me,
As if by some instinct the wretch did know
His rider loved not speed, being made from thee.
The bloody spur cannot provoke him on
That sometimes anger thrusts into his hide,          10
Which heavily he answers with a groan
More sharp to me than spurring to his side;
   For that same groan doth put this in my mind:
   My grief lies onward and my joy behind.

### 51

Thus can my love excuse the slow offence
Of my dull bearer when from thee I speed:
From where thou art why should I haste me thence?
Till I return, of posting is no need.
O what excuse will my poor beast then find          5
When swift extremity can seem but slow?
Then should I spur, though mounted on the wind;
In wingèd speed no motion shall I know.
Then can no horse with my desire keep pace;
Therefore desire, of perfect'st love being made,          10
Shall rein no dull flesh in his fiery race;
But love, for love, thus shall excuse my jade:
   Since from thee going he went wilful-slow,
   Towards thee I'll run and give him leave to go.

### 52

So am I as the rich whose blessèd key
Can bring him to his sweet up-lockèd treasure,
The which he will not ev'ry hour survey,
For blunting the fine point of seldom pleasure.
Therefore are feasts so solemn and so rare          5
Since, seldom coming, in the long year set
Like stones of worth they thinly placèd are,
Or captain jewels in the carcanet.
So is the time that keeps you as my chest,
Or as the wardrobe which the robe doth hide,          10
To make some special instant special blest
By new unfolding his imprisoned pride.
   Blessèd are you whose worthiness gives scope,
   Being had, to triumph; being lacked, to hope.

### 53

What is your substance, whereof are you made,
That millions of strange shadows on you tend?
Since every one hath, every one, one shade,
And you, but one, can every shadow lend.
Describe Adonis, and the counterfeit          5
Is poorly imitated after you.
On Helen's cheek all art of beauty set,
And you in Grecian tires are painted new.
Speak of the spring and foison of the year:
The one doth shadow of your beauty show,          10
The other as your bounty doth appear;
And you in every blessèd shape we know.
   In all external grace you have some part,
   But you like none, none you, for constant heart.

### 54

O how much more doth beauty beauteous seem
By that sweet ornament which truth doth give!
The rose looks fair, but fairer we it deem
For that sweet odour which doth in it live.
The canker blooms have full as deep a dye          5
As the perfumèd tincture of the roses,
Hang on such thorns, and play as wantonly
When summer's breath their maskèd buds discloses;
But for their virtue only is their show
They live unwooed and unrespected fade,          10
Die to themselves. Sweet roses do not so;
Of their sweet deaths are sweetest odours made:
   And so of you, beauteous and lovely youth,
   When that shall fade, by verse distils your truth.

### 55

Not marble nor the gilded monuments
Of princes shall outlive this powerful rhyme,
But you shall shine more bright in these contents
Than unswept stone besmeared with sluttish time.
When wasteful war shall statues overturn,          5
And broils root out the work of masonry,
Nor Mars his sword nor war's quick fire shall burn
The living record of your memory.
'Gainst death and all oblivious enmity
Shall you pace forth; your praise shall still find room          10
Even in the eyes of all posterity
That wear this world out to the ending doom.
   So, till the judgement that yourself arise,
   You live in this, and dwell in lovers' eyes.

### 56

Sweet love, renew thy force. Be it not said
Thy edge should blunter be than appetite,
Which but today by feeding is allayed,
Tomorrow sharpened in his former might.
So, love, be thou; although today thou fill          5
Thy hungry eyes even till they wink with fullness,
Tomorrow see again, and do not kill
The spirit of love with a perpetual dullness.
Let this sad int'rim like the ocean be
Which parts the shore where two contracted new       10
Come daily to the banks, that when they see
Return of love, more blessed may be the view;
   Or call it winter, which, being full of care,
   Makes summer's welcome, thrice more wished, more
     rare.

### 57

Being your slave, what should I do but tend
Upon the hours and times of your desire?
I have no precious time at all to spend,
Nor services to do, till you require;
Nor dare I chide the world-without-end hour          5
Whilst I, my sovereign, watch the clock for you,
Nor think the bitterness of absence sour
When you have bid your servant once adieu.
Nor dare I question with my jealous thought
Where you may be, or your affairs suppose,            10
But like a sad slave stay and think of naught
Save, where you are, how happy you make those.
   So true a fool is love that in your will,
   Though you do anything, he thinks no ill.

### 58

That god forbid, that made me first your slave,
I should in thought control your times of pleasure,
Or at your hand th'account of hours to crave,
Being your vassal bound to stay your leisure.
O let me suffer, being at your beck,                 5
Th' imprisoned absence of your liberty,
And patience, tame to sufferance, bide each check,
Without accusing you of injury.
Be where you list, your charter is so strong
That you yourself may privilege your time            10
To what you will; to you it doth belong
Yourself to pardon of self-doing crime.
   I am to wait, though waiting so be hell,
   Not blame your pleasure, be it ill or well.

### 59

If there be nothing new, but that which is
Hath been before, how are our brains beguiled,
Which, labouring for invention, bear amiss
The second burden of a former child!
O that record could with a backward look             5
Even of five hundred courses of the sun
Show me your image in some antique book
Since mind at first in character was done,
That I might see what the old world could say
To this composèd wonder of your frame;               10
Whether we are mended or whe'er better they,
Or whether revolution be the same.
   O, sure I am the wits of former days
   To subjects worse have given admiring praise.

### 60

Like as the waves make towards the pebbled shore,
So do our minutes hasten to their end,
Each changing place with that which goes before;
In sequent toil all forwards do contend.
Nativity, once in the main of light,                 5
Crawls to maturity, wherewith being crowned
Crookèd eclipses 'gainst his glory fight,
And time that gave doth now his gift confound.
Time doth transfix the flourish set on youth,
And delves the parallels in beauty's brow;           10
Feeds on the rarities of nature's truth,
And nothing stands but for his scythe to mow.
   And yet to times in hope my verse shall stand,
   Praising thy worth despite his cruel hand.

### 61

Is it thy will thy image should keep open
My heavy eyelids to the weary night?
Dost thou desire my slumbers should be broken
While shadows like to thee do mock my sight?
Is it thy spirit that thou send'st from thee         5
So far from home into my deeds to pry,
To find out shames and idle hours in me,
The scope and tenor of thy jealousy?
O no; thy love, though much, is not so great.
It is my love that keeps mine eye awake,             10
Mine own true love that doth my rest defeat,
To play the watchman ever for thy sake.
   For thee watch I whilst thou dost wake elsewhere,
   From me far off, with others all too near.

### 62

Sin of self-love possesseth all mine eye,
And all my soul, and all my every part;
And for this sin there is no remedy,
It is so grounded inward in my heart.
Methinks no face so gracious is as mine,             5
No shape so true, no truth of such account,
And for myself mine own worth do define
As I all other in all worths surmount.
But when my glass shows me myself indeed,
Beated and chapped with tanned antiquity,            10
Mine own self-love quite contrary I read;
Self so self-loving were iniquity.
   'Tis thee, my self, that for myself I praise,
   Painting my age with beauty of thy days.

### 63

Against my love shall be as I am now,
With time's injurious hand crushed and o'erworn;
When hours have drained his blood and filled his brow
With lines and wrinkles; when his youthful morn
Hath travelled on to age's steepy night,             5
And all those beauties whereof now he's king
Are vanishing, or vanished out of sight,
Stealing away the treasure of his spring:
For such a time do I now fortify
Against confounding age's cruel knife,               10
That he shall never cut from memory
My sweet love's beauty, though my lover's life.
   His beauty shall in these black lines be seen,
   And they shall live, and he in them still green.

### 64

When I have seen by time's fell hand defaced
The rich proud cost of outworn buried age;
When sometime-lofty towers I see down razed,
And brass eternal slave to mortal rage;
When I have seen the hungry ocean gain                                    5
Advantage on the kingdom of the shore,
And the firm soil win of the wat'ry main,
Increasing store with loss and loss with store;
When I have seen such interchange of state,
Or state itself confounded to decay,                                       10
Ruin hath taught me thus to ruminate:
That time will come and take my love away.
    This thought is as a death, which cannot choose
    But weep to have that which it fears to lose.

### 65

Since brass, nor stone, nor earth, nor boundless sea,
But sad mortality o'erswways their power,
How with this rage shall beauty hold a plea,
Whose action is no stronger than a flower?
O how shall summer's honey breath hold out                                 5
Against the wrackful siege of battering days
When rocks impregnable are not so stout,
Nor gates of steel so strong, but time decays?
O fearful meditation! Where, alack,
Shall time's best jewel from time's chest lie hid,                         10
Or what strong hand can hold his swift foot back,
Or who his spoil of beauty can forbid?
    O none, unless this miracle have might:
    That in black ink my love may still shine bright.

### 66

Tired with all these, for restful death I cry:
As, to behold desert a beggar born,
And needy nothing trimmed in jollity,
And purest faith unhappily forsworn,
And gilded honour shamefully misplaced,                                    5
And maiden virtue rudely strumpeted,
And right perfection wrongfully disgraced,
And strength by limping sway disablèd,
And art made tongue-tied by authority,
And folly, doctor-like, controlling skill,                                 10
And simple truth miscalled simplicity,
And captive good attending captain ill.
    Tired with all these, from these would I be gone,
    Save that to die I leave my love alone.

### 67

Ah, wherefore with infection should he live
And with his presence grace impiety,
That sin by him advantage should achieve
And lace itself with his society?
Why should false painting imitate his cheek,                               5
And steal dead seeming of his living hue?
Why should poor beauty indirectly seek
Roses of shadow, since his rose is true?
Why should he live now nature bankrupt is,
Beggared of blood to blush through lively veins,                          10
For she hath no exchequer now but his,
And proud of many, lives upon his gains?
    O, him she stores to show what wealth she had
    In days long since, before these last so bad.

### 68

Thus is his cheek the map of days outworn,
When beauty lived and died as flowers do now,
Before these bastard signs of fair were borne
Or durst inhabit on a living brow;
Before the golden tresses of the dead,                                     5
The right of sepulchres, were shorn away
To live a second life on second head;
Ere beauty's dead fleece made another gay.
In him those holy antique hours are seen
Without all ornament, itself and true,                                     10
Making no summer of another's green,
Robbing no old to dress his beauty new;
    And him as for a map doth nature store,
    To show false art what beauty was of yore.

### 69

Those parts of thee that the world's eye doth view
Want nothing that the thought of hearts can mend.
All tongues, the voice of souls, give thee that due,
Utt'ring bare truth even so as foes commend.
Thy outward thus with outward praise is crowned,                          5
But those same tongues that give thee so thine own
In other accents do this praise confound
By seeing farther than the eye hath shown.
They look into the beauty of thy mind,
And that in guess they measure by thy deeds.                              10
Then, churls, their thoughts—although their eyes were
        kind—
To thy fair flower add the rank smell of weeds.
    But why thy odour matcheth not thy show,
    The soil is this: that thou dost common grow.

### 70

That thou are blamed shall not be thy defect,
For slander's mark was ever yet the fair.
The ornament of beauty is suspect,
A crow that flies in heaven's sweetest air.
So thou be good, slander doth but approve                                 5
Thy worth the greater, being wooed of time;
For canker vice the sweetest buds doth love,
And thou present'st a pure unstainèd prime.
Thou hast passed by the ambush of young days
Either not assailed, or victor being charged;                             10
Yet this thy praise cannot be so thy praise
To tie up envy, evermore enlarged.
    If some suspect of ill masked not thy show,
    Then thou alone kingdoms of hearts shouldst owe.

### 71

No longer mourn for me when I am dead
Than you shall hear the surly sullen bell
Give warning to the world that I am fled
From this vile world with vilest worms to dwell.
Nay, if you read this line, remember not                                   5
The hand that writ it; for I love you so
That I in your sweet thoughts would be forgot
If thinking on me then should make you woe.
O, if, I say, you look upon this verse
When I perhaps compounded am with clay,                                   10
Do not so much as my poor name rehearse,
But let your love even with my life decay,
    Lest the wise world should look into your moan
    And mock you with me after I am gone.

### 72

O, lest the world should task you to recite
What merit lived in me that you should love,
After my death, dear love, forget me quite;
For you in me can nothing worthy prove—
Unless you would devise some virtuous lie    5
To do more for me than mine own desert,
And hang more praise upon deceasèd I
Than niggard truth would willingly impart.
O, lest your true love may seem false in this,
That you for love speak well of me untrue,    10
My name be buried where my body is,
And live no more to shame nor me nor you;
    For I am shamed by that which I bring forth,
    And so should you, to love things nothing worth.

### 73

That time of year thou mayst in me behold
When yellow leaves, or none, or few, do hang
Upon those boughs which shake against the cold,
Bare ruined choirs where late the sweet birds sang.
In me thou seest the twilight of such day    5
As after sunset fadeth in the west,
Which by and by black night doth take away,
Death's second self, that seals up all in rest.
In me thou seest the glowing of such fire
That on the ashes of his youth doth lie    10
As the death-bed whereon it must expire,
Consumed with that which it was nourished by.
    This thou perceiv'st, which makes thy love more strong,
    To love that well which thou must leave ere long.

### 74

But be contented when that fell arrest
Without all bail shall carry me away.
My life hath in this line some interest,
Which for memorial still with thee shall stay.
When thou reviewest this, thou dost review    5
The very part was consecrate to thee.
The earth can have but earth, which is his due;
My spirit is thine, the better part of me.
So then thou hast but lost the dregs of life,
The prey of worms, my body being dead,    10
The coward conquest of a wretch's knife,
Too base of thee to be rememberèd.
    The worth of that is that which it contains,
    And that is this, and this with thee remains.

### 75

So are you to my thoughts as food to life,
Or as sweet-seasoned showers are to the ground;
And for the peace of you I hold such strife
As 'twixt a miser and his wealth is found:
Now proud as an enjoyer, and anon    5
Doubting the filching age will steal his treasure;
Now counting best to be with you alone,
Then bettered that the world may see my pleasure;
Sometime all full with feasting on your sight,
And by and by clean starvèd for a look;    10
Possessing or pursuing no delight
Save what is had or must from you be took.
    Thus do I pine and surfeit day by day,
    Or gluttoning on all, or all away.

### 76

Why is my verse so barren of new pride,
So far from variation or quick change?
Why, with the time, do I not glance aside
To new-found methods and to compounds strange?
Why write I still all one, ever the same,    5
And keep invention in a noted weed,
That every word doth almost tell my name,
Showing their birth and where they did proceed?
O know, sweet love, I always write of you,
And you and love are still my argument;    10
So all my best is dressing old words new,
Spending again what is already spent;
    For as the sun is daily new and old,
    So is my love, still telling what is told.

### 77

Thy glass will show thee how thy beauties wear,
Thy dial how thy precious minutes waste,
The vacant leaves thy mind's imprint will bear,
And of this book this learning mayst thou taste:
The wrinkles which thy glass will truly show    5
Of mouthèd graves will give thee memory;
Thou by thy dial's shady stealth mayst know
Time's thievish progress to eternity;
Look what thy memory cannot contain
Commit to these waste blanks, and thou shalt find    10
Those children nursed, delivered from thy brain,
To take a new acquaintance of thy mind.
    These offices so oft as thou wilt look
    Shall profit thee and much enrich thy book.

### 78

So oft have I invoked thee for my muse
And found such fair assistance in my verse
As every alien pen hath got my use,
And under thee their poesy disperse.
Thine eyes, that taught the dumb on high to sing    5
And heavy ignorance aloft to fly,
Have added feathers to the learned's wing
And given grace a double majesty.
Yet be most proud of that which I compile,
Whose influence is thine and born of thee.    10
In others' works thou dost but mend the style,
And arts with thy sweet graces gracèd be;
    But thou art all my art, and dost advance
    As high as learning my rude ignorance.

### 79

Whilst I alone did call upon thy aid
My verse alone had all thy gentle grace;
But now my gracious numbers are decayed,
And my sick muse doth give another place.
I grant, sweet love, thy lovely argument    5
Deserves the travail of a worthier pen,
Yet what of thee thy poet doth invent
He robs thee of, and pays it thee again.
He lends thee virtue, and he stole that word
From thy behaviour; beauty doth he give,    10
And found it in thy cheek: he can afford
No praise to thee but what in thee doth live.
    Then thank him not for that which he doth say,
    Since what he owes thee thou thyself dost pay.

## 80

O, how I faint when I of you do write,
Knowing a better spirit doth use your name,
And in the praise thereof spends all his might,
To make me tongue-tied, speaking of your fame!
But since your worth, wide as the ocean is,          5
The humble as the proudest sail doth bear,
My saucy barque, inferior far to his,
On your broad main doth wilfully appear,
Your shallowest help will hold me up afloat
Whilst he upon your soundless deep doth ride;          10
Or, being wrecked, I am a worthless boat,
He of tall building and of goodly pride.
 Then if he thrive and I be cast away,
 The worst was this: my love was my decay.

## 81

Or I shall live your epitaph to make,
Or you survive when I in earth am rotten.
From hence your memory death cannot take,
Although in me each part will be forgotten.
Your name from hence immortal life shall have,          5
Though I, once gone, to all the world must die.
The earth can yield me but a common grave
When you entombèd in men's eyes shall lie.
Your monument shall be my gentle verse,
Which eyes not yet created shall o'er-read,          10
And tongues to be your being shall rehearse
When all the breathers of this world are dead.
 You still shall live—such virtue hath my pen—
 Where breath most breathes, even in the mouths of
  men.

## 82

I grant thou wert not married to my muse,
And therefore mayst without attaint o'erlook
The dedicated words which writers use
Of their fair subject, blessing every book.
Thou art as fair in knowledge as in hue,          5
Finding thy worth a limit past my praise,
And therefore art enforced to seek anew
Some fresher stamp of these time-bettering days.
And do so, love; yet when they have devised
What strainèd touches rhetoric can lend,          10
Thou, truly fair, wert truly sympathized
In true plain words by thy true-telling friend;
 And their gross painting might be better used
 Where cheeks need blood: in thee it is abused.

## 83

I never saw that you did painting need,
And therefore to your fair no painting set.
I found—or thought I found—you did exceed
The barren tender of a poet's debt;
And therefore have I slept in your report:          5
That you yourself, being extant, well might show
How far a modern quill doth come too short,
Speaking of worth, what worth in you doth grow.
This silence for my sin you did impute,
Which shall be most my glory, being dumb;          10
For I impair not beauty, being mute,
When others would give life, and bring a tomb.
 There lives more life in one of your fair eyes
 Than both your poets can in praise devise.

## 84

Who is it that says most which can say more
Than this rich praise: that you alone are you,
In whose confine immurèd is the store
Which should example where your equal grew?
Lean penury within that pen doth dwell          5
That to his subject lends not some small glory;
But he that writes of you, if he can tell
That you are you, so dignifies his story.
Let him but copy what in you is writ,
Not making worse what nature made so clear,          10
And such a counterpart shall fame his wit,
Making his style admirèd everywhere.
 You to your beauteous blessings add a curse,
 Being fond on praise, which makes your praises worse.

## 85

My tongue-tied muse in manners holds her still
While comments of your praise, richly compiled,
Reserve thy character with golden quill
And precious phrase by all the muses filed.
I think good thoughts whilst other write good words,          5
And like unlettered clerk still cry 'Amen'
To every hymn that able spirit affords
In polished form of well-refinèd pen.
Hearing you praised I say ''Tis so, 'tis true,'
And to the most of praise add something more;          10
But that is in my thought, whose love to you,
Though words come hindmost, holds his rank before.
 Then others for the breath of words respect,
 Me for my dumb thoughts, speaking in effect.

## 86

Was it the proud full sail of his great verse
Bound for the prize of all-too-precious you
That did my ripe thoughts in my brain inhearse,
Making their tomb the womb wherein they grew?
Was it his spirit, by spirits taught to write          5
Above a mortal pitch, that struck me dead?
No, neither he nor his compeers by night
Giving him aid my verse astonishèd.
He nor that affable familiar ghost
Which nightly gulls him with intelligence,          10
As victors, of my silence cannot boast;
I was not sick of any fear from thence.
 But when your countenance filled up his line,
 Then lacked I matter; that enfeebled mine.

## 87

Farewell—thou art too dear for my possessing,
And like enough thou know'st thy estimate.
The charter of thy worth gives thee releasing;
My bonds in thee are all determinate.
For how do I hold thee but by thy granting,          5
And for that riches where is my deserving?
The cause of this fair gift in me is wanting,
And so my patent back again is swerving.
Thyself thou gav'st, thy own worth then not knowing,
Or me to whom thou gav'st it else mistaking;          10
So thy great gift, upon misprision growing,
Comes home again, on better judgement making.
 Thus have I had thee as a dream doth flatter:
 In sleep a king, but waking no such matter.

### 88

When thou shalt be disposed to set me light
And place my merit in the eye of scorn,
Upon thy side against myself I'll fight,
And prove thee virtuous though thou art forsworn.
With mine own weakness being best acquainted,    5
Upon thy part I can set down a story
Of faults concealed wherein I am attainted,
That thou in losing me shall win much glory;
And I by this will be a gainer too;
For bending all my loving thoughts on thee,    10
The injuries that to myself I do,
Doing thee vantage, double vantage me.
    Such is my love, to thee I so belong,
    That for thy right myself will bear all wrong.

### 89

Say that thou didst forsake me for some fault,
And I will comment upon that offence;
Speak of my lameness, and I straight will halt,
Against thy reasons making no defence.
Thou canst not, love, disgrace me half so ill,    5
To set a form upon desirèd change,
As I'll myself disgrace, knowing thy will.
I will acquaintance strangle and look strange,
Be absent from thy walks, and in my tongue
Thy sweet belovèd name no more shall dwell,    10
Lest I, too much profane, should do it wrong,
And haply of our old acquaintance tell.
    For thee, against myself I'll vow debate;
    For I must ne'er love him whom thou dost hate.

### 90

Then hate me when thou wilt, if ever, now,
Now while the world is bent my deeds to cross,
Join with the spite of fortune, make me bow,
And do not drop in for an after-loss.
Ah do not, when my heart hath scaped this sorrow,    5
Come in the rearward of a conquered woe;
Give not a windy night a rainy morrow
To linger out a purposed overthrow.
If thou wilt leave me, do not leave me last,
When other petty griefs have done their spite,    10
But in the onset come; so shall I taste
At first the very worst of fortune's might,
    And other strains of woe, which now seem woe,
    Compared with loss of thee will not seem so.

### 91

Some glory in their birth, some in their skill,
Some in their wealth, some in their body's force,
Some in their garments (though new-fangled ill),
Some in their hawks and hounds, some in their horse,
And every humour hath his adjunct pleasure    5
Wherein it finds a joy above the rest.
But these particulars are not my measure;
All these I better in one general best.
Thy love is better than high birth to me,
Richer than wealth, prouder than garments' cost,    10
Of more delight than hawks or horses be,
And having thee of all men's pride I boast,
    Wretched in this alone: that thou mayst take
    All this away, and me most wretched make.

### 92

But do thy worst to steal thyself away,
For term of life thou art assurèd mine,
And life no longer than thy love will stay,
For it depends upon that love of thine.
Then need I not to fear the worst of wrongs    5
When in the least of them my life hath end.
I see a better state to me belongs
Than that which on thy humour doth depend.
Thou canst not vex me with inconstant mind,
Since that my life on thy revolt doth lie.    10
O, what a happy title do I find—
Happy to have thy love, happy to die!
    But what's so blessèd fair that fears no blot?
    Thou mayst be false, and yet I know it not.

### 93

So shall I live supposing thou art true
Like a deceivèd husband; so love's face
May still seem love to me, though altered new—
Thy looks with me, thy heart in other place.
For there can live no hatred in thine eye,    5
Therefore in that I cannot know thy change.
In many's looks the false heart's history
Is writ in moods and frowns and wrinkles strange;
But heaven in thy creation did decree
That in thy face sweet love should ever dwell;    10
Whate'er thy thoughts or thy heart's workings be,
Thy looks should nothing thence but sweetness tell.
    How like Eve's apple doth thy beauty grow
    If thy sweet virtue answer not thy show!

### 94

They that have power to hurt and will do none,
That do not do the thing they most do show,
Who moving others are themselves as stone,
Unmovèd, cold, and to temptation slow—
They rightly do inherit heaven's graces,    5
And husband nature's riches from expense;
They are the lords and owners of their faces,
Others but stewards of their excellence.
The summer's flower is to the summer sweet
Though to itself it only live and die,    10
But if that flower with base infection meet
The basest weed outbraves his dignity;
    For sweetest things turn sourest by their deeds:
    Lilies that fester smell far worse than weeds.

### 95

How sweet and lovely dost thou make the shame
Which, like a canker in the fragrant rose,
Doth spot the beauty of thy budding name!
O, in what sweets dost thou thy sins enclose!
That tongue that tells the story of thy days,    5
Making lascivious comments on thy sport,
Cannot dispraise, but in a kind of praise,
Naming thy name, blesses an ill report.
O, what a mansion have those vices got
Which for their habitation chose out thee,    10
Where beauty's veil doth cover every blot
And all things turns to fair that eyes can see!
    Take heed, dear heart, of this large privilege:
    The hardest knife ill used doth lose his edge.

## 96

Some say thy fault is youth, some wantonness;
Some say thy grace is youth and gentle sport.
Both grace and faults are loved of more and less;
Thou mak'st faults graces that to thee resort.
As on the finger of a thronèd queen     5
The basest jewel will be well esteemed,
So are those errors that in thee are seen
To truths translated and for true things deemed.
How many lambs might the stern wolf betray
If like a lamb he could his looks translate!     10
How many gazers mightst thou lead away
If thou wouldst use the strength of all thy state!
   But do not so: I love thee in such sort
   As, thou being mine, mine is thy good report.

## 97

How like a winter hath my absence been
From thee, the pleasure of the fleeting year!
What freezings have I felt, what dark days seen,
What old December's bareness everywhere!
And yet this time removed was summer's time,     5
The teeming autumn big with rich increase,
Bearing the wanton burden of the prime
Like widowed wombs after their lords' decease.
Yet this abundant issue seemed to me
But hope of orphans and unfathered fruit,     10
For summer and his pleasures wait on thee,
And thou away, the very birds are mute;
   Or if they sing, 'tis with so dull a cheer
   That leaves look pale, dreading the winter's near.

## 98

From you have I been absent in the spring
When proud-pied April, dressed in all his trim,
Hath put a spirit of youth in everything,
That heavy Saturn laughed and leapt with him.
Yet nor the lays of birds nor the sweet smell     5
Of different flowers in odour and in hue
Could make me any summer's story tell,
Or from their proud lap pluck them where they grew;
Nor did I wonder at the lily's white,
Nor praise the deep vermilion in the rose.     10
They were but sweet, but figures of delight
Drawn after you, you pattern of all those;
   Yet seemed it winter still, and, you away,
   As with your shadow I with these did play.

## 99

The forward violet thus did I chide:
Sweet thief, whence didst thou steal thy sweet that smells,
If not from my love's breath? The purple pride
Which on thy soft cheek for complexion dwells
In my love's veins thou hast too grossly dyed.     5
The lily I condemnèd for thy hand,
And buds of marjoram had stol'n thy hair;
The roses fearfully on thorns did stand,
One blushing shame, another white despair;
A third, nor red nor white, had stol'n of both,     10
And to his robb'ry had annexed thy breath;
But for his theft in pride of all his growth
A vengeful canker ate him up to death.
   More flowers I noted, yet I none could see
   But sweet or colour it had stol'n from thee.

## 100

Where art thou, muse, that thou forget'st so long
To speak of that which gives thee all thy might?
Spend'st thou thy fury on some worthless song,
Dark'ning thy power to lend base subjects light?
Return, forgetful muse, and straight redeem     5
In gentle numbers time so idly spent;
Sing to the ear that doth thy lays esteem
And gives thy pen both skill and argument.
Rise, resty muse, my love's sweet face survey
If time have any wrinkle graven there.     10
If any, be a satire to decay
And make time's spoils despisèd everywhere.
   Give my love fame faster than time wastes life;
   So, thou prevene'st his scythe and crookèd knife.

## 101

O truant muse, what shall be thy amends
For thy neglect of truth in beauty dyed?
Both truth and beauty on my love depends;
So dost thou too, and therein dignified.
Make answer, muse. Wilt thou not haply say     5
'Truth needs no colour with his colour fixed,
Beauty no pencil beauty's truth to lay,
But best is best if never intermixed'?
Because he needs no praise wilt thou be dumb?
Excuse not silence so, for't lies in thee     10
To make him much outlive a gilded tomb,
And to be praised of ages yet to be.
   Then do thy office, muse; I teach thee how
   To make him seem long hence as he shows now.

## 102

My love is strengthened, though more weak in seeming.
I love not less, though less the show appear.
That love is merchandized whose rich esteeming
The owner's tongue doth publish everywhere.
Our love was new and then but in the spring     5
When I was wont to greet it with my lays,
As Philomel in summer's front doth sing,
And stops her pipe in growth of riper days—
Not that the summer is less pleasant now
Than when her mournful hymns did hush the night,     10
But that wild music burdens every bough,
And sweets grown common lose their dear delight.
   Therefore like her I sometime hold my tongue,
   Because I would not dull you with my song.

## 103

Alack, what poverty my muse brings forth
That, having such a scope to show her pride,
The argument all bare is of more worth
Than when it hath my added praise beside!
O blame me not if I no more can write!     5
Look in your glass and there appears a face
That overgoes my blunt invention quite,
Dulling my lines and doing me disgrace.
Were it not sinful then, striving to mend,
To mar the subject that before was well?—     10
For to no other pass my verses tend
Than of your graces and your gifts to tell;
   And more, much more, than in my verse can sit
   Your own glass shows you when you look in it.

## 104

To me, fair friend, you never can be old;
For as you were when first your eye I eyed,
Such seems your beauty still. Three winters cold
Have from the forests shook three summers' pride;
Three beauteous springs to yellow autumn turned 5
In process of the seasons have I seen,
Three April perfumes in three hot Junes burned
Since first I saw you fresh, which yet are green.
Ah yet doth beauty, like a dial hand,
Steal from his figure and no pace perceived; 10
So your sweet hue, which methinks still doth stand,
Hath motion, and mine eye may be deceived.
   For fear of which, hear this, thou age unbred:
   Ere you were born was beauty's summer dead.

## 105

Let not my love be called idolatry,
Nor my belovèd as an idol show,
Since all alike my songs and praises be
To one, of one, still such, and ever so.
Kind is my love today, tomorrow kind, 5
Still constant in a wondrous excellence.
Therefore my verse, to constancy confined,
One thing expressing, leaves out difference.
'Fair, kind, and true' is all my argument,
'Fair, kind, and true' varying to other words, 10
And in this change is my invention spent,
Three themes in one, which wondrous scope affords.
   Fair, kind, and true have often lived alone,
   Which three till now never kept seat in one.

## 106

When in the chronicle of wasted time
I see descriptions of the fairest wights,
And beauty making beautiful old rhyme
In praise of ladies dead and lovely knights;
Then in the blazon of sweet beauty's best, 5
Of hand, of foot, of lip, of eye, of brow,
I see their antique pen would have expressed
Even such a beauty as you master now.
So all their praises are but prophecies
Of this our time, all you prefiguring, 10
And for they looked but with divining eyes
They had not skill enough your worth to sing;
   For we which now behold these present days
   Have eyes to wonder, but lack tongues to praise.

## 107

Not mine own fears nor the prophetic soul
Of the wide world dreaming on things to come
Can yet the lease of my true love control,
Supposed as forfeit to a confined doom.
The mortal moon hath her eclipse endured, 5
And the sad augurs mock their own presage;
Incertainties now crown themselves assured,
And peace proclaims olives of endless age.
Now with the drops of this most balmy time
My love looks fresh, and death to me subscribes, 10
Since spite of him I'll live in this poor rhyme
While he insults o'er dull and speechless tribes;
   And thou in this shalt find thy monument
   When tyrants' crests and tombs of brass are spent.

## 108

What's in the brain that ink may character
Which hath not figured to thee my true spirit?
What's new to speak, what now to register,
That may express my love or thy dear merit?
Nothing, sweet boy; but yet like prayers divine 5
I must each day say o'er the very same,
Counting no old thing old, thou mine, I thine,
Even as when first I hallowed thy fair name.
So that eternal love in love's fresh case
Weighs not the dust and injury of age, 10
Nor gives to necessary wrinkles place,
But makes antiquity for aye his page,
   Finding the first conceit of love there bred
   Where time and outward form would show it dead.

## 109

O never say that I was false of heart,
Though absence seemed my flame to qualify—
As easy might I from myself depart
As from my soul, which in thy breast doth lie.
That is my home of love. If I have ranged, 5
Like him that travels I return again,
Just to the time, not with the time exchanged,
So that myself bring water for my stain.
Never believe, though in my nature reigned
All frailties that besiege all kinds of blood, 10
That it could so preposterously be stained
To leave for nothing all thy sum of good;
   For nothing this wide universe I call
   Save thou my rose; in it thou art my all.

## 110

Alas, 'tis true, I have gone here and there
And made myself a motley to the view,
Gored mine own thoughts, sold cheap what is most dear,
Made old offences of affections new.
Most true it is that I have looked on truth 5
Askance and strangely. But, by all above,
These blenches gave my heart another youth,
And worse essays proved thee my best of love.
Now all is done, have what shall have no end;
Mine appetite I never more will grind 10
On newer proof to try an older friend,
A god in love, to whom I am confined.
   Then give me welcome, next my heaven the best,
   Even to thy pure and most most loving breast.

## 111

O, for my sake do you with fortune chide,
The guilty goddess of my harmful deeds,
That did not better for my life provide
Than public means which public manners breeds.
Thence comes it that my name receives a brand, 5
And almost thence my nature is subdued
To what it works in, like the dyer's hand.
Pity me then, and wish I were renewed,
Whilst like a willing patient I will drink
Potions of eisel 'gainst my strong infection; 10
No bitterness that I will bitter think,
Nor double penance to correct correction.
   Pity me then, dear friend, and I assure ye
   Even that your pity is enough to cure me.

## 112

Your love and pity doth th'impression fill
Which vulgar scandal stamped upon my brow;
For what care I who calls me well or ill,
So you o'er-green my bad, my good allow?
You are my all the world, and I must strive          5
To know my shames and praises from your tongue—
None else to me, nor I to none alive,
That my steeled sense or changes, right or wrong.
In so profound abyss I throw all care
Of others' voices that my adder's sense          10
To critic and to flatterer stoppèd are.
  Mark how with my neglect I do dispense:
    You are so strongly in my purpose bred
    That all the world besides, methinks, they're dead.

## 113

Since I left you mine eye is in my mind,
And that which governs me to go about
Doth part his function and is partly blind,
Seems seeing, but effectually is out;
For it no form delivers to the heart          5
Of bird, of flower, or shape which it doth latch.
Of his quick objects hath the mind no part,
Nor his own vision holds what it doth catch;
For if it see the rud'st or gentlest sight,
The most sweet favour or deformèd'st creature,          10
The mountain or the sea, the day or night,
The crow or dove, it shapes them to your feature.
  Incapable of more, replete with you,
    My most true mind thus makes mine eye untrue.

## 114

Or whether doth my mind, being crowned with you,
Drink up the monarch's plague, this flattery,
Or whether shall I say mine eye saith true,
And that your love taught it this alchemy,
To make of monsters and things indigest          5
Such cherubins as your sweet self resemble,
Creating every bad a perfect best
As fast as objects to his beams assemble?
O, 'tis the first, 'tis flatt'ry in my seeing,
And my great mind most kingly drinks it up.          10
Mine eye well knows what with his gust is 'greeing,
And to his palate doth prepare the cup.
  If it be poisoned, 'tis the lesser sin
    That mine eye loves it and doth first begin.

## 115

Those lines that I before have writ do lie,
Even those that said I could not love you dearer,
Yet then my judgement knew no reason why
My most full flame should afterwards burn clearer.
But reckoning time, whose millioned accidents          5
Creep in 'twixt vows and change decrees of kings,
Tan sacred beauty, blunt the sharp'st intents,
Divert strong minds to th' course of alt'ring things—
Alas, why, fearing of time's tyranny,
Might I not then say 'Now I love you best',          10
When I was certain o'er incertainty,
Crowning the present, doubting of the rest?
  Love is a babe; then might I not say so,
    To give full growth to that which still doth grow.

## 116

Let me not to the marriage of true minds
Admit impediments. Love is not love
Which alters when it alteration finds,
Or bends with the remover to remove.
O no, it is an ever fixèd mark          5
That looks on tempests and is never shaken;
It is the star to every wand'ring barque,
Whose worth's unknown although his height be taken.
Love's not time's fool, though rosy lips and cheeks          10
Within his bending sickle's compass come;
Love alters not with his brief hours and weeks,
But bears it out even to the edge of doom.
  If this be error and upon me proved,
    I never writ, nor no man ever loved.

## 117

Accuse me thus: that I have scanted all
Wherein I should your great deserts repay,
Forgot upon your dearest love to call,
Whereto all bonds do tie me day by day;
That I have frequent been with unknown minds,          5
And given to time your own dear-purchased right;
That I have hoisted sail to all the winds
Which should transport me farthest from your sight.
Book both my wilfulness and errors down,
And on just proof surmise accumulate;          10
Bring me within the level of your frown,
But shoot not at me in your wakened hate,
  Since my appeal says I did strive to prove
    The constancy and virtue of your love.

## 118

Like as, to make our appetites more keen,
With eager compounds we our palate urge;
As to prevent our maladies unseen
We sicken to shun sickness when we purge:
Even so, being full of your ne'er cloying sweetness,          5
To bitter sauces did I frame my feeding,
And, sick of welfare, found a kind of meetness
To be diseased ere that there was true needing.
Thus policy in love, t'anticipate
The ills that were not, grew to faults assured,          10
And brought to medicine a healthful state
Which, rank of goodness, would by ill be cured.
  But thence I learn, and find the lesson true:
    Drugs poison him that so fell sick of you.

## 119

What potions have I drunk of siren tears
Distilled from limbecks foul as hell within,
Applying fears to hopes and hopes to fears,
Still losing when I saw myself to win!
What wretched errors hath my heart committed          5
Whilst it hath thought itself so blessèd never!
How have mine eyes out of their spheres been fitted
In the distraction of this madding fever!
O benefit of ill! Now I find true
That better is by evil still made better,          10
And ruined love when it is built anew
Grows fairer than at first, more strong, far greater.
  So I return rebuked to my content,
    And gain by ills thrice more than I have spent.

## 120

That you were once unkind befriends me now,
And for that sorrow which I then did feel
Needs must I under my transgression bow,
Unless my nerves were brass or hammered steel.
For if you were by my unkindness shaken          5
As I by yours, you've past a hell of time,
And I, a tyrant, have no leisure taken
To weigh how once I suffered in your crime.
O that our night of woe might have remembered
My deepest sense how hard true sorrow hits,      10
And soon to you as you to me then tendered
The humble salve which wounded bosoms fits!
   But that your trespass now becomes a fee;
   Mine ransoms yours, and yours must ransom me.

## 121

'Tis better to be vile than vile esteemed
When not to be receives reproach of being,
And the just pleasure lost, which is so deemed
Not by our feeling but by others' seeing.
For why should others' false adulterate eyes      5
Give salutation to my sportive blood?
Or on my frailties why are frailer spies,
Which in their wills count bad what I think good?
No, I am that I am, and they that level
At my abuses reckon up their own.                 10
I may be straight, though they themselves be bevel;
By their rank thoughts my deeds must not be shown,
   Unless this general evil they maintain:
   All men are bad and in their badness reign.

## 122

Thy gift, thy tables, are within my brain
Full charactered with lasting memory,
Which shall above that idle rank remain
Beyond all date, even to eternity;
Or at the least so long as brain and heart        5
Have faculty by nature to subsist,
Till each to razed oblivion yield his part
Of thee, thy record never can be missed.
That poor retention could not so much hold,
Nor need I tallies thy dear love to score;        10
Therefore to give them from me was I bold,
To trust those tables that receive thee more.
   To keep an adjunct to remember thee
   Were to import forgetfulness in me.

## 123

No, time, thou shalt not boast that I do change!
Thy pyramids built up with newer might
To me are nothing novel, nothing strange,
They are but dressings of a former sight.
Our dates are brief, and therefore we admire      5
What thou dost foist upon us that is old,
And rather make them born to our desire
Than think that we before have heard them told.
Thy registers and thee I both defy,
Not wond'ring at the present nor the past;         10
For thy records and what we see doth lie,
Made more or less by thy continual haste.
   This I do vow, and this shall ever be:
   I will be true despite thy scythe and thee.

## 124

If my dear love were but the child of state
It might for fortune's bastard be unfathered,
As subject to time's love or to time's hate,
Weeds among weeds or flowers with flowers gathered.
No, it was builded far from accident;              5
It suffers not in smiling pomp, nor falls
Under the blow of thrallèd discontent
Whereto th'inviting time our fashion calls.
It fears not policy, that heretic
Which works on leases of short-numbered hours,     10
But all alone stands hugely politic,
That it nor grows with heat nor drowns with showers.
   To this I witness call the fools of time,
   Which die for goodness, who have lived for crime.

## 125

Were't aught to me I bore the canopy,
With my extern the outward honouring,
Or laid great bases for eternity
Which proves more short than waste or ruining?
Have I not seen dwellers on form and favour        5
Lose all and more by paying too much rent,
For compound sweet forgoing simple savour,
Pitiful thrivers in their gazing spent?
No, let me be obsequious in thy heart,
And take thou my oblation, poor but free,          10
Which is not mixed with seconds, knows no art
But mutual render, only me for thee.
   Hence, thou suborned informer! A true soul
   When most impeached stands least in thy control.

## 126

O thou my lovely boy, who in thy power
Dost hold time's fickle glass, his sickle-hour;
Who hast by waning grown, and therein show'st
Thy lovers withering as thy sweet self grow'st—
If nature, sovereign mistress over wrack,          5
As thou goest onwards still will pluck thee back,
She keeps thee to this purpose: that her skill
May time disgrace, and wretched minutes kill!
Yet fear her, O thou minion of her pleasure!
She may detain but not still keep her treasure.    10
   Her audit, though delayed, answered must be,
   And her quietus is to render thee.

## 127

In the old age black was not counted fair,
Or if it were, it bore not beauty's name;
But now is black beauty's successive heir,
And beauty slandered with a bastard shame:
For since each hand hath put on nature's power,    5
Fairing the foul with art's false borrowed face,
Sweet beauty hath no name, no holy bower,
But is profaned, if not lives in disgrace.
Therefore my mistress' eyes are raven-black,
Her brow so suited, and they mourners seem         10
At such who, not born fair, no beauty lack,
Sland'ring creation with a false esteem.
   Yet so they mourn, becoming of their woe,
   That every tongue says beauty should look so.

## 128

How oft, when thou, my music, music play'st
Upon that blessèd wood whose motion sounds
With thy sweet fingers when thou gently sway'st
The wiry concord that mine ear confounds,
Do I envy those jacks that nimble leap                    5
To kiss the tender inward of thy hand
Whilst my poor lips, which should that harvest reap,
At the wood's boldness by thee blushing stand!
To be so tickled they would change their state
And situation with those dancing chips                    10
O'er whom thy fingers walk with gentle gait,
Making dead wood more blessed than living lips.
    Since saucy jacks so happy are in this,
    Give them thy fingers, me thy lips to kiss.

## 129

Th'expense of spirit in a waste of shame
Is lust in action; and till action, lust
Is perjured, murd'rous, bloody, full of blame,
Savage, extreme, rude, cruel, not to trust,
Enjoyed no sooner but despisèd straight,                  5
Past reason hunted, and no sooner had
Past reason hated as a swallowed bait
On purpose laid to make the taker mad;
Mad in pursuit and in possession so,
Had, having, and in quest to have, extreme;               10
A bliss in proof and proved, a very woe;
Before, a joy proposed; behind, a dream.
    All this the world well knows, yet none knows well
    To shun the heaven that leads men to this hell.

## 130

My mistress' eyes are nothing like the sun;
Coral is far more red than her lips' red.
If snow be white, why then her breasts are dun;
If hairs be wires, black wires grow on her head.
I have seen roses damasked, red and white,                5
But no such roses see I in her cheeks;
And in some perfumes is there more delight
Than in the breath that from my mistress reeks.
I love to hear her speak, yet well I know
That music hath a far more pleasing sound.                10
I grant I never saw a goddess go:
My mistress when she walks treads on the ground.
    And yet, by heaven, I think my love as rare
    As any she belied with false compare.

## 131

Thou art as tyrannous so as thou art
As those whose beauties proudly make them cruel,
For well thou know'st to my dear doting heart
Thou art the fairest and most precious jewel.
Yet, in good faith, some say that thee behold             5
Thy face hath not the power to make love groan.
To say they err I dare not be so bold,
Although I swear it to myself alone,
And, to be sure that is not false I swear,
A thousand groans but thinking on thy face                10
One on another's neck do witness bear
Thy black is fairest in my judgement's place.
    In nothing art thou black save in thy deeds,
    And thence this slander, as I think, proceeds.

## 132

Thine eyes I love, and they, as pitying me—
Knowing thy heart torment me with disdain—
Have put on black, and loving mourners be,
Looking with pretty ruth upon my pain;
And truly, not the morning sun of heaven                  5
Better becomes the gray cheeks of the east,
Nor that full star that ushers in the even
Doth half that glory to the sober west,
As those two mourning eyes become thy face.
O, let it then as well beseem thy heart                   10
To mourn for me, since mourning doth thee grace,
And suit thy pity like in every part.
    Then will I swear beauty herself is black,
    And all they foul that thy complexion lack.

## 133

Beshrew that heart that makes my heart to groan
For that deep wound it gives my friend and me!
Is't not enough to torture me alone,
But slave to slavery my sweet'st friend must be?
Me from myself thy cruel eye hath taken,                  5
And my next self thou harder hast engrossed.
Of him, myself, and thee I am forsaken—
A torment thrice threefold thus to be crossed.
Prison my heart in thy steel bosom's ward,
But then my friend's heart let my poor heart bail;        10
Whoe'er keeps me, let my heart be his guard;
Thou canst not then use rigour in my jail.
    And yet thou wilt; for I, being pent in thee,
    Perforce am thine, and all that is in me.

## 134

So, now I have confessed that he is thine,
And I myself am mortgaged to thy will,
Myself I'll forfeit, so that other mine
Thou wilt restore to be my comfort still.
But thou wilt not, nor he will not be free,              5
For thou art covetous, and he is kind.
He learned but surety-like to write for me
Under that bond that him as fast doth bind.
The statute of thy beauty thou wilt take,
Thou usurer that putt'st forth all to use,               10
And sue a friend came debtor for my sake;
So him I lose through my unkind abuse.
    Him have I lost; thou hast both him and me;
    He pays the whole, and yet am I not free.

## 135

Whoever hath her wish, thou hast thy Will,
And Will to boot, and Will in overplus.
More than enough am I that vex thee still,
To thy sweet will making addition thus.
Wilt thou, whose will is large and spacious,             5
Not once vouchsafe to hide my will in thine?
Shall will in others seem right gracious,
And in my will no fair acceptance shine?
The sea, all water, yet receives rain still,
And in abundance addeth to his store;                    10
So thou, being rich in Will, add to thy Will
One will of mine to make thy large Will more.
    Let no unkind no fair beseechers kill;
    Think all but one, and me in that one Will.

## 136

If thy soul check thee that I come so near,
Swear to thy blind soul that I was thy Will,
And will, thy soul knows, is admitted there;
Thus far for love my love-suit, sweet, fulfil.
Will will fulfil the treasure of thy love,                    5
Ay, fill it full with wills, and my will one.
In things of great receipt with ease we prove
Among a number one is reckoned none.
Then in the number let me pass untold,
Though in thy store's account I one must be;                 10
For nothing hold me, so it please thee hold
That nothing me a something, sweet, to thee.
    Make but my name thy love, and love that still,
    And then thou lov'st me for my name is Will.

## 137

Thou blind fool love, what dost thou to mine eyes
That they behold and see not what they see?
They know what beauty is, see where it lies,
Yet what the best is take the worst to be.
If eyes corrupt by over-partial looks                        5
Be anchored in the bay where all men ride,
Why of eyes' falsehood hast thou forgèd hooks
Whereto the judgement of my heart is tied?
Why should my heart think that a several plot
Which my heart knows the wide world's common place?—
Or mine eyes, seeing this, say this is not,                  11
To put fair truth upon so foul a face?
    In things right true my heart and eyes have erred,
    And to this false plague are they now transferred.

## 138

When my love swears that she is made of truth
I do believe her though I know she lies,
That she might think me some untutored youth
Unlearnèd in the world's false subtleties.
Thus vainly thinking that she thinks me young,              5
Although she knows my days are past the best,
Simply I credit her false-speaking tongue;
On both sides thus is simple truth suppressed.
But wherefore says she not she is unjust,
And wherefore say not I that I am old?                      10
O, love's best habit is in seeming trust,
And age in love loves not to have years told.
    Therefore I lie with her, and she with me,
    And in our faults by lies we flattered be.

## 139

O, call not me to justify the wrong
That thy unkindness lays upon my heart.
Wound me not with thine eye but with thy tongue;
Use power with power, and slay me not by art.
Tell me thou lov'st elsewhere, but in my sight,            5
Dear heart, forbear to glance thine eye aside.
What need'st thou wound with cunning when thy
      might
Is more than my o'erpressed defence can bide?
Let me excuse thee: 'Ah, my love well knows
Her pretty looks have been mine enemies,                   10
And therefore from my face she turns my foes
That they elsewhere might dart their injuries.'
    Yet do not so; but since I am near slain,
    Kill me outright with looks, and rid my pain.

## 140

Be wise as thou art cruel; do not press
My tongue-tied patience with too much disdain,
Lest sorrow lend me words, and words express
The manner of my pity-wanting pain.
If I might teach thee wit, better it were,                  5
Though not to love, yet, love, to tell me so—
As testy sick men when their deaths be near
No news but health from their physicians know.
For if I should despair I should grow mad,
And in my madness might speak ill of thee.                  10
Now this ill-wresting world is grown so bad
Mad slanderers by mad ears believèd be.
    That I may not be so, nor thou belied,
    Bear thine eyes straight, though thy proud heart go
      wide.

## 141

In faith, I do not love thee with mine eyes,
For they in thee a thousand errors note;
But 'tis my heart that loves what they despise,
Who in despite of view is pleased to dote.
Nor are mine ears with thy tongue's tune delighted,       5
Nor tender feeling to base touches prone;
Nor taste nor smell desire to be invited
To any sensual feast with thee alone;
But my five wits nor my five senses can
Dissuade one foolish heart from serving thee,              10
Who leaves unswayed the likeness of a man,
Thy proud heart's slave and vassal-wretch to be.
    Only my plague thus far I count my gain:
    That she that makes me sin awards me pain.

## 142

Love is my sin, and thy dear virtue hate,
Hate of my sin grounded on sinful loving.
O, but with mine compare thou thine own state,
And thou shalt find it merits not reproving;
Or if it do, not from those lips of thine                  5
That have profaned their scarlet ornaments
And sealed false bonds of love as oft as mine,
Robbed others' beds' revenues of their rents.
Be it lawful I love thee as thou lov'st those
Whom thine eyes woo as mine importune thee.                10
Root pity in thy heart, that when it grows
Thy pity may deserve to pitied be.
    If thou dost seek to have what thou dost hide,
    By self example mayst thou be denied!

## 143

Lo, as a care-full housewife runs to catch
One of her feathered creatures broke away,
Sets down her babe and makes all swift dispatch
In pursuit of the thing she would have stay,
Whilst her neglected child holds her in chase,             5
Cries to catch her whose busy care is bent
To follow that which flies before her face,
Not prizing her poor infant's discontent:
So runn'st thou after that which flies from thee,
Whilst I, thy babe, chase thee afar behind;                10
But if thou catch thy hope, turn back to me
And play the mother's part: kiss me, be kind.
    So will I pray that thou mayst have thy Will
    If thou turn back and my loud crying still.

## 144

Two loves I have, of comfort and despair,
Which like two spirits do suggest me still.
The better angel is a man right fair,
The worser spirit a woman coloured ill.
To win me soon to hell my female evil      5
Tempteth my better angel from my side,
And would corrupt my saint to be a devil,
Wooing his purity with her foul pride;
And whether that my angel be turned fiend
Suspect I may, yet not directly tell;      10
But being both from me, both to each friend,
I guess one angel in another's hell.
     Yet this shall I ne'er know, but live in doubt
     Till my bad angel fire my good one out.

## 145

Those lips that love's own hand did make
Breathed forth the sound that said 'I hate'
To me that languished for her sake;
But when she saw my woeful state,
Straight in her heart did mercy come,      5
Chiding that tongue that ever sweet
Was used in giving gentle doom,
And taught it thus anew to greet:
'I hate' she altered with an end
That followed it as gentle day      10
Doth follow night who, like a fiend,
From heaven to hell is flown away.
     'I hate' from hate away she threw,
     And saved my life, saying 'not you.'

## 146

Poor soul, the centre of my sinful earth,
⌈        ⌉ these rebel powers that thee array;
Why dost thou pine within and suffer dearth,
Painting thy outward walls so costly gay?
Why so large cost, having so short a lease,      5
Dost thou upon thy fading mansion spend?
Shall worms, inheritors of this excess,
Eat up thy charge? Is this thy body's end?
Then, soul, live thou upon thy servant's loss,
And let that pine to aggravate thy store.      10
Buy terms divine in selling hours of dross;
Within be fed, without be rich no more.
     So shalt thou feed on death, that feeds on men,
     And death once dead, there's no more dying then.

## 147

My love is as a fever, longing still
For that which longer nurseth the disease,
Feeding on that which doth preserve the ill,
Th'uncertain sickly appetite to please.
My reason, the physician to my love,      5
Angry that his prescriptions are not kept,
Hath left me, and I desperate now approve
Desire is death, which physic did except.
Past cure I am, now reason is past care,
And frantic mad with evermore unrest.      10
My thoughts and my discourse as madmen's are,
At random from the truth vainly expressed;
     For I have sworn thee fair, and thought thee bright,
     Who art as black as hell, as dark as night.

## 148

O me, what eyes hath love put in my head,
Which have no correspondence with true sight!
Or if they have, where is my judgement fled,
That censures falsely what they see aright?
If that be fair whereon my false eyes dote,      5
What means the world to say it is not so?
If it be not, then love doth well denote
Love's eye is not so true as all men's. No,
How can it, O, how can love's eye be true,
That is so vexed with watching and with tears?      10
No marvel then though I mistake my view:
The sun itself sees not till heaven clears.
     O cunning love, with tears thou keep'st me blind
     Lest eyes, well seeing, thy foul faults should find!

## 149

Canst thou, O cruel, say I love thee not
When I against myself with thee partake?
Do I not think on thee when I forgot
Am of myself, all-tyrant, for thy sake?
Who hateth thee that I do call my friend?      5
On whom frown'st thou that I do fawn upon?
Nay, if thou lour'st on me, do I not spend
Revenge upon myself with present moan?
What merit do I in myself respect
That is so proud thy service to despise,      10
When all my best doth worship thy defect,
Commanded by the motion of thine eyes?
     But, love, hate on; for now I know thy mind.
     Those that can see thou lov'st, and I am blind.

## 150

O, from what power hast thou this powerful might
With insufficiency my heart to sway,
To make me give the lie to my true sight
And swear that brightness doth not grace the day?
Whence hast thou this becoming of things ill,      5
That in the very refuse of thy deeds
There is such strength and warrantise of skill
That in my mind thy worst all best exceeds?
Who taught thee how to make me love thee more
The more I hear and see just cause of hate?      10
O, though I love what others do abhor,
With others thou shouldst not abhor my state.
     If thy unworthiness raised love in me,
     More worthy I to be beloved of thee.

## 151

Love is too young to know what conscience is,
Yet who knows not conscience is born of love?
Then, gentle cheater, urge not my amiss,
Lest guilty of my faults thy sweet self prove.
For, thou betraying me, I do betray      5
My nobler part to my gross body's treason.
My soul doth tell my body that he may
Triumph in love; flesh stays no farther reason,
But rising at thy name doth point out thee
As his triumphant prize. Proud of this pride,      10
He is contented thy poor drudge to be,
To stand in thy affairs, fall by thy side.
     No want of conscience hold it that I call
     Her 'love' for whose dear love I rise and fall.

### 152

In loving thee thou know'st I am forsworn,
But thou art twice forsworn to me love swearing:
In act thy bed-vow broke, and new faith torn
In vowing new hate after new love bearing.
But why of two oaths' breach do I accuse thee          5
When I break twenty? I am perjured most,
For all my vows are oaths but to misuse thee,
And all my honest faith in thee is lost.
For I have sworn deep oaths of thy deep kindness,
Oaths of thy love, thy truth, thy constancy,          10
And to enlighten thee gave eyes to blindness,
Or made them swear against the thing they see.
    For I have sworn thee fair—more perjured eye
    To swear against the truth so foul a lie.

### 153

Cupid laid by his brand and fell asleep.
A maid of Dian's this advantage found,
And his love-kindling fire did quickly steep
In a cold valley-fountain of that ground,
Which borrowed from this holy fire of love          5
A dateless lively heat, still to endure,
And grew a seething bath which yet men prove
Against strange maladies a sovereign cure.
But at my mistress' eye love's brand new fired,
The boy for trial needs would touch my breast.          10
I, sick withal, the help of bath desired,
And thither hied, a sad distempered guest,
    But found no cure; the bath for my help lies
    Where Cupid got new fire: my mistress' eyes.

### 154

The little love-god lying once asleep
Laid by his side his heart-inflaming brand,
Whilst many nymphs that vowed chaste life to keep
Came tripping by; but in her maiden hand
The fairest votary took up that fire          5
Which many legions of true hearts had warmed,
And so the general of hot desire
Was sleeping by a virgin hand disarmed.
This brand she quenchèd in a cool well by,
Which from love's fire took heat perpetual,          10
Growing a bath and healthful remedy
For men diseased; but I, my mistress' thrall,
    Came there for cure; and this by that I prove:
    Love's fire heats water, water cools not love.

# A Lover's Complaint

From off a hill whose concave womb re-worded
A plaintful story from a sist'ring vale,
My spirits t'attend this double voice accorded,
And down I laid to list the sad-tuned tale;
Ere long espied a fickle maid full pale, 5
Tearing of papers, breaking rings a-twain,
Storming her world with sorrow's wind and rain.

Upon her head a plaited hive of straw
Which fortified her visage from the sun,
Whereon the thought might think sometime it saw 10
The carcass of a beauty spent and done.
Time had not scythèd all that youth begun,
Nor youth all quit; but spite of heaven's fell rage,
Some beauty peeped through lattice of seared age.

Oft did she heave her napkin to her eyne, 15
Which on it had conceited characters,
Laund'ring the silken figures in the brine
That seasoned woe had pelleted in tears,
And often reading what contents it bears;
As often shrieking undistinguished woe 20
In clamours of all size, both high and low.

Sometimes her levelled eyes their carriage ride
As they did batt'ry to the spheres intend;
Sometime diverted their poor balls are tied
To th'orbèd earth; sometimes they do extend 25
Their view right on; anon their gazes lend
To every place at once, and nowhere fixed,
The mind and sight distractedly commixed.

Her hair, nor loose nor tied in formal plait,
Proclaimed in her a careless hand of pride; 30
For some, untucked, descended her sheaved hat,
Hanging her pale and pinèd cheek beside.
Some in her threaden fillet still did bide,
And, true to bondage, would not break from thence,
Though slackly braided in loose negligence. 35

A thousand favours from a maund she drew
Of amber, crystal, and of beaded jet,
Which one by one she in a river threw
Upon whose weeping margin she was set;
Like usury applying wet to wet, 40
Or monarch's hands that lets not bounty fall
Where want cries some, but where excess begs all.

Of folded schedules had she many a one
Which she perused, sighed, tore, and gave the flood;
Cracked many a ring of posied gold and bone, 45
Bidding them find their sepulchres in mud;
Found yet more letters sadly penned in blood,
With sleided silk feat and affectedly
Enswathed and sealed to curious secrecy.

These often bathed she in her fluxive eyes, 50
And often kissed, and often 'gan to tear;
Cried 'O false blood, thou register of lies,
What unapprovèd witness dost thou bear!

Ink would have seemed more black and damnèd
    here!'
This said, in top of rage the lines she rents, 55
Big discontent so breaking their contents.

A reverend man that grazed his cattle nigh,
Sometime a blusterer that the ruffle knew
Of court, of city, and had let go by
The swiftest hours observèd as they flew, 60
Towards this afflicted fancy fastly drew,
And, privileged by age, desires to know
In brief the grounds and motives of her woe.

So slides he down upon his grainèd bat,
And comely distant sits he by her side, 65
When he again desires her, being sat,
Her grievance with his hearing to divide.
If that from him there may be aught applied
Which may her suffering ecstasy assuage,
'Tis promised in the charity of age. 70

'Father,' she says, 'though in me you behold
The injury of many a blasting hour,
Let it not tell your judgement I am old;
Not age, but sorrow over me hath power.
I might as yet have been a spreading flower, 75
Fresh to myself, if I had self-applied
Love to myself, and to no love beside.

'But, woe is me, too early I attended
A youthful suit—it was to gain my grace—
O, one by nature's outwards so commended 80
That maidens' eyes stuck over all his face.
Love lacked a dwelling and made him her place,
And when in his fair parts she did abide
She was new-lodged and newly deified.

'His browny locks did hang in crookèd curls, 85
And every light occasion of the wind
Upon his lips their silken parcels hurls.
What's sweet to do, to do will aptly find.
Each eye that saw him did enchant the mind,
For on his visage was in little drawn 90
What largeness thinks in paradise was sawn.

'Small show of man was yet upon his chin;
His phoenix down began but to appear,
Like unshorn velvet, on that termless skin
Whose bare outbragged the web it seemed to wear; 95
Yet showed his visage by that cost more dear,
And nice affections wavering stood in doubt
If best were as it was, or best without.

'His qualities were beauteous as his form,
For maiden-tongued he was, and thereof free. 100
Yet if men moved him, was he such a storm
As oft twixt May and April is to see
When winds breathe sweet, unruly though they be.
His rudeness so with his authorized youth
Did livery falseness in a pride of truth. 105

771

'Well could he ride, and often men would say
"That horse his mettle from his rider takes;
Proud of subjection, noble by the sway,
What rounds, what bounds, what course, what stop
    he makes!"
And controversy hence a question takes,    110
Whether the horse by him became his deed,
Or he his manège by th' well-doing steed.

'But quickly on this side the verdict went:
His real habitude gave life and grace
To appertainings and to ornament,    115
Accomplished in himself, not in his case.
All aids, themselves made fairer by their place,
Came for additions; yet their purposed trim
Pieced not his grace, but were all graced by him.

'So on the tip of his subduing tongue    120
All kind of arguments and question deep,
All replication prompt, and reason strong,
For his advantage still did wake and sleep.
To make the weeper laugh, the laugher weep,
He had the dialect and different skill,    125
Catching all passions in his craft of will,

'That he did in the general bosom reign
Of young, of old, and sexes both enchanted,
To dwell with him in thoughts, or to remain
In personal duty, following where he haunted.    130
Consents bewitched, ere he desire, have granted,
And dialogued for him what he would say,
Asked their own wills, and made their wills obey.

'Many there were that did his picture get
To serve their eyes, and in it put their mind,    135
Like fools that in th'imagination set
The goodly objects which abroad they find
Of lands and mansions, theirs in thought assigned,
And labour in more pleasures to bestow them
Than the true gouty landlord which doth owe them.

'So many have, that never touched his hand,    141
Sweetly supposed them mistress of his heart.
My woeful self, that did in freedom stand,
And was my own fee-simple, not in part,
What with his art in youth, and youth in art,    145
Threw my affections in his charmèd power,
Reserved the stalk and gave him all my flower.

'Yet did I not, as some my equals did,
Demand of him, nor being desirèd yielded.
Finding myself in honour so forbid,    150
With safest distance I mine honour shielded.
Experience for me many bulwarks builded
Of proofs new bleeding, which remained the foil
Of this false jewel and his amorous spoil.

'But ah, who ever shunned by precedent    155
The destined ill she must herself assay,
Or forced examples 'gainst her own content
To put the by-past perils in her way?
Counsel may stop a while what will not stay,
For when we rage, advice is often seen,    160
By blunting us, to make our wills more keen.

'Nor gives it satisfaction to our blood
That we must curb it upon others' proof,
To be forbod the sweets that seems so good
For fear of harms that preach in our behoof.    165
O appetite, from judgement stand aloof!
The one a palate hath that needs will taste,
Though reason weep, and cry it is thy last.

'For further I could say this man's untrue,
And knew the patterns of his foul beguiling;    170
Heard where his plants in others' orchards grew,
Saw how deceits were gilded in his smiling,
Knew vows were ever brokers to defiling,
Thought characters and words merely but art,
And bastards of his foul adulterate heart.    175

'And long upon these terms I held my city
Till thus he gan besiege me: "Gentle maid,
Have of my suffering youth some feeling pity,
And be not of my holy vows afraid.
That's to ye sworn to none was ever said;    180
For feasts of love I have been called unto,
Till now did ne'er invite nor never woo.

' "All my offences that abroad you see
Are errors of the blood, none of the mind.
Love made them not; with acture they may be,    185
Where neither party is nor true nor kind.
They sought their shame that so their shame did find,
And so much less of shame in me remains
By how much of me their reproach contains.

' "Among the many that mine eyes have seen,    190
Not one whose flame my heart so much as warmèd
Or my affection put to th' smallest teen,
Or any of my leisures ever charmèd.
Harm have I done to them, but ne'er was harmèd;
Kept hearts in liveries, but mine own was free,    195
And reigned commanding in his monarchy.

' "Look here what tributes wounded fancies sent me
Of pallid pearls and rubies red as blood,
Figuring that they their passions likewise lent me
Of grief and blushes, aptly understood    200
In bloodless white and the encrimsoned mood—
Effects of terror and dear modesty,
Encamped in hearts, but fighting outwardly.

' "And lo, behold, these talents of their hair,
With twisted mettle amorously impleached,    205
I have received from many a several fair,
Their kind acceptance weepingly beseeched,
With th'annexations of fair gems enriched,
And deep-brained sonnets that did amplify
Each stone's dear nature, worth, and quality.    210

' "The diamond?—why, 'twas beautiful and hard,
Whereto his invised properties did tend;
The deep-green em'rald, in whose fresh regard
Weak sights their sickly radiance do amend;
The heaven-hued sapphire and the opal blend    215
With objects manifold; each several stone,
With wit well blazoned, smiled or made some moan.

' "Lo, all these trophies of affections hot,
Of pensived and subdued desires the tender,
Nature hath charged me that I hoard them not,        220
But yield them up where I myself must render—
That is to you, my origin and ender;
For these of force must your oblations be,
Since I their altar, you enpatron me.

' "O then advance of yours that phraseless hand        225
Whose white weighs down the airy scale of praise.
Take all these similes to your own command,
Hallowed with sighs that burning lungs did raise.
What me, your minister for you, obeys,
Works under you, and to your audit comes        230
Their distract parcels in combinèd sums.

' "Lo, this device was sent me from a nun,
A sister sanctified of holiest note,
Which late her noble suit in court did shun,
Whose rarest havings made the blossoms dote;        235
For she was sought by spirits of richest coat,
But kept cold distance, and did thence remove
To spend her living in eternal love.

' "But O, my sweet, what labour is't to leave
The thing we have not, mast'ring what not strives,
Planing the place which did no form receive,        241
Playing patient sports in unconstrainèd gyves!
She that her fame so to herself contrives
The scars of battle scapeth by the flight,
And makes her absence valiant, not her might.        245

' "O, pardon me, in that my boast is true!
The accident which brought me to her eye
Upon the moment did her force subdue,
And now she would the cagèd cloister fly.
Religious love put out religion's eye.        250
Not to be tempted would she be immured,
And now, to tempt, all liberty procured.

' "How mighty then you are, O hear me tell!
The broken bosoms that to me belong
Have emptied all their fountains in my well,        255
And mine I pour your ocean all among.
I strong o'er them, and you o'er me being strong,
Must for your victory us all congest,
As compound love to physic your cold breast.

' "My parts had power to charm a sacred nun,        260
Who disciplined, ay dieted in grace,
Believed her eyes when they t' assail begun,
All vows and consecrations giving place.
O most potential love: vow, bond, nor space
In thee hath neither sting, knot, nor confine,        265
For thou art all, and all things else are thine.

' "When thou impressest, what are precepts worth
Of stale example? When thou wilt inflame,
How coldly those impediments stand forth
Of wealth, of filial fear, law, kindred, fame.        270
Love's arms are peace, 'gainst rule, 'gainst sense,
        'gainst shame;
And sweetens in the suff'ring pangs it bears
The aloes of all forces, shocks, and fears.

' "Now all these hearts that do on mine depend,
Feeling it break, with bleeding groans they pine,        275
And supplicant their sighs to you extend
To leave the batt'ry that you make 'gainst mine,
Lending soft audience to my sweet design,
And credent soul to that strong-bonded oath
That shall prefer and undertake my troth."        280

'This said, his wat'ry eyes he did dismount,
Whose sights till then were levelled on my face.
Each cheek a river running from a fount
With brinish current downward flowed apace.
O, how the channel to the stream gave grace,        285
Who glazed with crystal gate the glowing roses
That flame through water which their hue encloses.

'O father, what a hell of witchcraft lies
In the small orb of one particular tear!
But with the inundation of the eyes        290
What rocky heart to water will not wear?
What breast so cold that is not warmèd here?
O cleft effect! Cold modesty, hot wrath,
Both fire from hence and chill extincture hath.

'For lo, his passion, but an art of craft,        295
Even there resolved my reason into tears.
There my white stole of chastity I daffed,
Shook off my sober guards and civil fears;
Appear to him as he to me appears,
All melting, though our drops this diff'rence bore:        300
His poisoned me, and mine did him restore.

'In him a plenitude of subtle matter,
Applied to cautels, all strange forms receives,
Of burning blushes or of weeping water,
Or swooning paleness; and he takes and leaves,        305
In either's aptness, as it best deceives,
To blush at speeches rank, to weep at woes,
Or to turn white and swoon at tragic shows,

'That not a heart which in his level came
Could scape the hail of his all-hurting aim,        310
Showing fair nature is both kind and tame,
And, veiled in them, did win whom he would maim.
Against the thing he sought he would exclaim;
When he most burned in heart-wished luxury,
He preached pure maid and praised cold chastity.        315

'Thus merely with the garment of a grace
The naked and concealèd fiend he covered,
That th'unexperient gave the tempter place,
Which like a cherubin above them hovered,
Who, young and simple, would not be so lovered?        320
Ay me, I fell, and yet do question make
What I should do again for such a sake.

'O that infected moisture of his eye,
O that false fire which in his cheek so glowed,
O that forced thunder from his heart did fly,        325
O that sad breath his spongy lungs bestowed,
O all that borrowed motion seeming owed
Would yet again betray the fore-betrayed,
And new pervert a reconcilèd maid.'

ALTERNATIVE VERSIONS OF SONNETS 2, 106, 138, AND 144

Each of the four sonnets printed below exists in an alternative version. To the left, we give the text as it appeared in the volume of Shakespeare's sonnets printed in 1609. 'Spes Altera' and 'On his Mistress' Beauty' derive from seventeenth-century manuscripts. The alternative versions of Sonnets 138 and 144 are from *The Passionate Pilgrim* (1599).

## 2

When forty winters shall besiege thy brow
And dig deep trenches in thy beauty's field,
Thy youth's proud livery, so gazed on now,
Will be a tattered weed, of small worth held.
Then being asked where all thy beauty lies,                    5
Where all the treasure of thy lusty days,
To say within thine own deep-sunken eyes
Were an all-eating shame and thriftless praise.
How much more praise deserved thy beauty's use
If thou couldst answer 'This fair child of mine            10
Shall sum my count, and make my old excuse',
Proving his beauty by succession thine.
     This were to be new made when thou art old,
     And see thy blood warm when thou feel'st it cold.

## *Spes Altera*

When forty winters shall besiege thy brow
And trench deep furrows in that lovely field,
Thy youth's fair liv'ry, so accounted now,
Shall be like rotten weeds of no worth held.
Then being asked where all thy beauty lies,                    5
Where all the lustre of thy youthful days,
To say 'Within these hollow sunken eyes'
Were an all-eaten truth and worthless praise.
O how much better were thy beauty's use
If thou couldst say 'This pretty child of mine             10
Saves my account and makes my old excuse',
Making his beauty by succession thine.
     This were to be new born when thou art old,
     And see thy blood warm when thou feel'st it cold.

## 106

When in the chronicle of wasted time
I see descriptions of the fairest wights,
And beauty making beautiful old rhyme
In praise of ladies dead and lovely knights;
Then in the blazon of sweet beauty's best,                     5
Of hand, of foot, of lip, of eye, of brow,
I see their antique pen would have expressed
Even such a beauty as you master now.
So all their praises are but prophecies
Of this our time, all you prefiguring,                     10
And for they looked but with divining eyes
They had not skill enough your worth to sing;
     For we which now behold these present days
     Have eyes to wonder, but lack tongues to praise.

## On his Mistress' Beauty

When in the annals of all-wasting time
I see descriptions of the fairest wights,
And beauty making beautiful old rhyme
In praise of ladies dead and lovely knights;
Then in the blazon of sweet beauty's best,                     5
Of face, of hand, of lip, of eye, or brow,
I see their antique pen would have expressed
E'en such a beauty as you master now.
So all their praises were but prophecies
Of these our days, all you prefiguring,                    10
And for they saw but with divining eyes
They had not skill enough your worth to sing;
     For we which now behold these present days
     Have eyes to wonder, but no tongues to praise.

## 138

When my love swears that she is made of truth
I do believe her though I know she lies,
That she might think me some untutored youth
Unlearnèd in the world's false subtleties.
Thus vainly thinking that she thinks me young,                 5
Although she knows my days are past the best,
Simply I credit her false-speaking tongue;
On both sides thus is simple truth suppressed.
But wherefore says she not she is unjust,
And wherefore say not I that I am old?                     10
O, love's best habit is in seeming trust,
And age in love loves not to have years told.
     Therefore I lie with her, and she with me,
     And in our faults by lies we flattered be.

## 138

When my love swears that she is made of truth
I do believe her though I know she lies,
That she might think me some untutored youth
Unskilful in the world's false forgeries.
Thus vainly thinking that she thinks me young,                 5
Although I know my years be past the best,
I, smiling, credit her false-speaking tongue,
Outfacing faults in love with love's ill rest.
But wherefore says my love that she is young,
And wherefore say not I that I am old?                     10
O, love's best habit's in a soothing tongue,
And age in love loves not to have years told.
     Therefore I'll lie with love, and love with me,
     Since that our faults in love thus smothered be.

## 144

Two loves I have, of comfort and despair,
Which like two spirits do suggest me still.
The better angel is a man right fair,
The worser spirit a woman coloured ill.
To win me soon to hell my female evil                     5
Tempteth my better angel from my side,
And would corrupt my saint to be a devil,
Wooing his purity with her foul pride;
And whether that my angel be turned fiend
Suspect I may, yet not directly tell;                    10
But being both from me, both to each friend,
I guess one angel in another's hell.
  Yet this shall I ne'er know, but live in doubt
  Till my bad angel fire my good one out.

## 144

Two loves I have, of comfort and despair,
That like two spirits do suggest me still.
My better angel is a man right fair,
My worser spirit a woman coloured ill.
To win me soon to hell my female evil                     5
Tempteth my better angel from my side,
And would corrupt my saint to be a devil,
Wooing his purity with her fair pride;
And whether that my angel be turned fiend,
Suspect I may, yet not directly tell;                    10
For being both to me, both to each friend,
I guess one angel in another's hell.
  The truth I shall not know, but live in doubt
  Till my bad angel fire my good one out.

# VARIOUS POEMS

A POET like Shakespeare may frequently have been asked to write verses for a variety of occasions, and it is entirely possible that he is the author of song lyrics and other short poems published without attribution or attributed only to 'W.S.' The poems in this section (arranged in an approximate chronological order) were all explicitly ascribed to him either in his lifetime or not long afterwards. Because they are short it is impossible to be sure, on stylistic grounds alone, of Shakespeare's authorship; but none of the poems is ever attributed to anyone else.

'Shall I die?' is transcribed, with Shakespeare's name appended, in a manuscript collection of poems, dating probably from the late 1630s, which is now in the Bodleian Library, Oxford; another, unascribed version is in the Beinecke Library, Yale University. The poem exhibits many parallels with plays and poems that Shakespeare wrote about 1593-5. Its stanza form has not been found elsewhere in the period, but most closely resembles Robin Goodfellow's lines spoken over the sleeping Lysander (*A Midsummer Night's Dream*, 3.2.36-46). Extended over nine stanzas it becomes a virtuoso exercise: every third word rhymes. The strain shows in a number of ellipses, but there is no strong reason to doubt the ascription: the Oxford manuscript is generally reliable, and if the poem is of no great consequence, that might explain why it did not reach print.

Perhaps the most trivial verse ever ascribed to a great poet is the 'posy' said to have accompanied a pair of gloves given by a Stratford schoolmaster, Alexander Aspinall, to his second wife, whom he married in 1594. The ascription is found in a manuscript compiled by Sir Francis Fane of Bulbeck (1611-80).

In 1599 William Jaggard published a collection of poems, which he ascribed to Shakespeare, under the title *The Passionate Pilgrim*. It includes versions of two of Shakespeare's Sonnets (which we print as Alternative Versions), three extracts from *Love's Labour's Lost*, which had already appeared in print, several poems known to be by other poets, and eleven poems of unknown authorship. A reprint of 1612 added nine poems by Thomas Heywood, who promptly protested against the 'manifest injury' done to him by printing his poems 'in a less volume, under the name of another, which may put the world in opinion I might steal them from him . . . But as I must acknowledge my lines not worthy his patronage under whom he hath published them, so the author I know much offended with Master Jaggard that, altogether unknown to him, presumed to make so bold with his name.' Probably as a result, the original title-page of the 1612 edition was replaced with one that did not mention Shakespeare's name. We print below the poems of unknown authorship since the attribution to Shakespeare has not been disproved.

The finest poem in this section, 'The Phoenix and Turtle', was ascribed to Shakespeare in 1601 when it appeared, without title, as one of the 'Poetical Essays' appended to Robert Chester's *Love's Martyr: or Rosalind's Complaint*, which is described as 'allegorically shadowing the truth of love in the constant fate of the phoenix and turtle'. Chester's poem appears to have been composed as a compliment to Sir John and Lady Salusbury, his patron. We know of no link between Shakespeare and the Salusbury family; possibly his poem was not written specifically for the volume in which it appeared. Since the early nineteenth century it has been known as 'The Phoenix and the Turtle' or (following the title-page) 'The Phoenix and Turtle'. An incantatory elegy, it may well have irrecoverable allegorical significance.

It is not clear whether the two stanzas engraved at opposite ends of the Stanley tomb in the parish church of Tong, in Shropshire, constitute one epitaph or two, or which member (or members) of the family they commemorate. They are ascribed to Shakespeare in two manuscript miscellanies of the 1630s and by the antiquary Sir William Dugdale in a manuscript appended to his Visitation of Shropshire in 1664. Shakespeare had professional connections with the Stanleys early in his career: *Titus Andronicus* and *1 Henry VI* were performed by a theatrical company patronized by the family.

The satirical completion of an epitaph on Ben Jonson (written during his lifetime) is ascribed to Shakespeare in two different seventeenth-century manuscripts.

Shakespeare probably knew Elias James (c.1578–1610), who managed a brewery in the Blackfriars district of London. His epitaph is ascribed to Shakespeare in the same Oxford manuscript as 'Shall I die?'

The Combe family of Stratford-upon-Avon were friends of Shakespeare. He bequeathed his sword to one of them, and John Combe, who died in 1614, left Shakespeare £5. Several mock epitaphs similar to the first epitaph on John Combe have survived, one (on an unnamed usurer) printed as early as 1608; later versions mention three other men as the usurer. Shakespeare may have adapted some existing lines; or some existing lines may have been adapted anonymously in Stratford, and later attributed to Stratford's most famous poet. The ascription to him dates from 1634, and is supported by four other seventeenth-century manuscripts. The second Combe epitaph is found in only one manuscript; it seems entirely original, and alludes to a bequest to the poor made in Combe's will.

The lines on King James first appear, unattributed, beneath an engraving of the King printed as the frontispiece to the 1616 edition of his works. They are attributed to Shakespeare—the leading writer of the theatre company of which King James was patron—in at least two seventeenth-century manuscripts; the same attribution was recorded in a printed broadside now apparently lost.

Shakespeare's own epitaph is written in the first person; the tradition that he composed it himself is recorded in several manuscripts from the middle to the late seventeenth century.

# Various Poems

## A Song

### 1

Shall I die? Shall I fly
Lovers' baits and deceits,
  sorrow breeding?
Shall I tend? Shall I send?
Shall I sue, and not rue           5
  my proceeding?
In all duty her beauty
Binds me her servant for ever.
If she scorn, I mourn,
I retire to despair, joining never.       10

### 2

Yet I must vent my lust
And explain inward pain
  by my love conceiving.
If she smiles, she exiles
All my moan; if she frown,       15
  all my hopes deceiving—
Suspicious doubt, O keep out,
For thou art my tormentor.
Fie away, pack away;
I will love, for hope bids me venture.     20

### 3

'Twere abuse to accuse
My fair love, ere I prove
  her affection.
Therefore try! Her reply
Gives thee joy—or annoy,       25
  or affliction.
Yet howe'er, I will bear
Her pleasure with patience, for beauty
Sure will not seem to blot
Her deserts, wronging him doth her duty.   30

### 4

In a dream it did seem—
But alas, dreams do pass
  as do shadows—
I did walk, I did talk
With my love, with my dove,      35
  through fair meadows.
Still we passed till at last
We sat to repose us for pleasure.
Being set, lips met,
Arms twined, and did bind my heart's treasure.  40

### 5

Gentle wind sport did find
Wantonly to make fly
  her gold tresses.

As they shook I did look,
But her fair did impair       45
  all my senses.
As amazed, I gazed
On more than a mortal complexion.
You that love can prove
Such force in beauty's inflection.     50

### 6

Next her hair, forehead fair,
Smooth and high; neat doth lie,
  without wrinkle,
Her fair brows; under those,
Star-like eyes win love's prize     55
  when they twinkle.
In her cheeks who seeks
Shall find there displayed beauty's banner;
  O admiring desiring
Breeds, as I look still upon her.     60

### 7

Thin lips red, fancy's fed
With all sweets when he meets,
  and is granted
There to trade, and is made
Happy, sure, to endure      65
  still undaunted.
Pretty chin doth win
Of all their culled commendations;
  Fairest neck, no speck;
All her parts merit high admirations.   70

### 8

Pretty bare, past compare,
Parts those plots which besots
  still asunder.
It is meet naught but sweet
Should come near that so rare    75
  'tis a wonder.
No mis-shape, no scape
Inferior to nature's perfection;
  No blot, no spot:
She's beauty's queen in election.     80

### 9

Whilst I dreamt, I, exempt
From all care, seemed to share
  pleasure's plenty;
But awake, care take—
For I find to my mind      85
  pleasures scanty.
Therefore I will try
To compass my heart's chief contenting.
  To delay, some say,
In such a case causeth repenting.    90

'Upon a pair of gloves that master sent to his mistress'

The gift is small,
The will is all:
Alexander Aspinall

---

Poems from *The Passionate Pilgrim*

4

Sweet Cytherea, sitting by a brook
With young Adonis, lovely, fresh, and green,
Did court the lad with many a lovely look,
Such looks as none could look but beauty's queen.
She told him stories to delight his ear, 5
She showed him favours to allure his eye;
To win his heart she touched him here and there—
Touches so soft still conquer chastity.
But whether unripe years did want conceit,
Or he refused to take her figured proffer, 10
The tender nibbler would not touch the bait,
But smile and jest at every gentle offer.
 Then fell she on her back, fair queen and toward:
 He rose and ran away—ah, fool too froward!

6

Scarce had the sun dried up the dewy morn,
And scarce the herd gone to the hedge for shade,
When Cytherea, all in love forlorn,
A longing tarriance for Adonis made
Under an osier growing by a brook, 5
A brook where Adon used to cool his spleen.
Hot was the day, she hotter, that did look
For his approach that often there had been.
Anon he comes and throws his mantle by,
And stood stark naked on the brook's green brim. 10
The sun looked on the world with glorious eye,
Yet not so wistly as this queen on him.
 He, spying her, bounced in whereas he stood.
 'O Jove,' quoth she, 'why was not I a flood?'

7

Fair is my love, but not so fair as fickle,
Mild as a dove, but neither true nor trusty,
Brighter than glass, and yet, as glass is, brittle;
Softer than wax, and yet as iron rusty;
 A lily pale, with damask dye to grace her, 5
 None fairer, nor none falser to deface her.
Her lips to mine how often hath she joined,
Between each kiss her oaths of true love swearing.
How many tales to please me hath she coined,
Dreading my love, the loss whereof still fearing. 10
 Yet in the midst of all her pure protestings
 Her faith, her oaths, her tears, and all were jestings.
She burnt with love as straw with fire flameth,
She burnt out love as soon as straw out burneth.
She framed the love, and yet she foiled the framing, 15
She bade love last, and yet she fell a-turning.
 Was this a lover or a lecher whether,
 Bad in the best, though excellent in neither?

9

Fair was the morn when the fair queen of love,
⌈ ⌉
Paler for sorrow than her milk-white dove,
For Adon's sake, a youngster proud and wild,
Her stand she takes upon a steep-up hill. 5
Anon Adonis comes with horn and hounds.
She, seely queen, with more than love's good will
Forbade the boy he should not pass those grounds.
'Once,' quoth she, 'did I see a fair sweet youth
Here in these brakes deep-wounded with a boar, 10
Deep in the thigh, a spectacle of ruth.
See in my thigh,' quoth she, 'here was the sore.'
 She showèd hers; he saw more wounds than one,
 And blushing fled, and left her all alone.

10

Sweet rose, fair flower, untimely plucked, soon faded—
Plucked in the bud and faded in the spring;
Bright orient pearl, alack, too timely shaded;
Fair creature, killed too soon by death's sharp sting,
 Like a green plum that hangs upon a tree 5
 And falls through wind before the fall should be.
I weep for thee, and yet no cause I have,
For why: thou left'st me nothing in thy will,
And yet thou left'st me more than I did crave,
For why: I cravèd nothing of thee still. 10
 O yes, dear friend, I pardon crave of thee:
 Thy discontent thou didst bequeath to me.

12

Crabbèd age and youth cannot live together:
Youth is full of pleasance, age is full of care;
Youth like summer morn, age like winter weather;
Youth like summer brave, age like winter bare.
Youth is full of sport, age's breath is short. 5
Youth is nimble, age is lame,
Youth is hot and bold, age is weak and cold.
Youth is wild and age is tame.
 Age, I do abhor thee; youth, I do adore thee.
 O my love, my love is young. 10
 Age, I do defy thee. O sweet shepherd, hie thee,
 For methinks thou stay'st too long.

13

Beauty is but a vain and doubtful good,
A shining gloss that fadeth suddenly,
A flower that dies when first it 'gins to bud,
A brittle glass that's broken presently.
 A doubtful good, a gloss, a glass, a flower, 5
 Lost, faded, broken, dead within an hour.
And as goods lost are seld or never found,
As faded gloss no rubbing will refresh,
As flowers dead lie withered on the ground,
As broken glass no cement can redress, 10
 So beauty blemished once, for ever lost,
 In spite of physic, painting, pain, and cost.

14

Good night, good rest—ah, neither be my share.
She bade good night that kept my rest away,

And daffed me to a cabin hanged with care
To descant on the doubts of my decay.
   'Farewell,' quoth she, 'and come again tomorrow.'  5
   Fare well I could not, for I supped with sorrow.

Yet at my parting sweetly did she smile,
In scorn or friendship nill I conster whether.
'Tmay be she joyed to jest at my exile,
'Tmay be, again to make me wander thither.  10
   'Wander'—a word for shadows like myself,
   As take the pain but cannot pluck the pelf.

Lord, how mine eyes throw gazes to the east!
My heart doth charge the watch, the morning rise
Doth cite each moving sense from idle rest,  15
Not daring trust the office of mine eyes.
   While Philomela sings I sit and mark,
   And wish her lays were tunèd like the lark.

For she doth welcome daylight with her dite,
And daylight drives away dark dreaming night.  20
The night so packed, I post unto my pretty;
Heart hath his hope, and eyes their wishèd sight,
   Sorrow changed to solace, and solace mixed with
     sorrow,
   Forwhy she sighed and bade me come tomorrow.

Were I with her, the night would post too soon,  25
But now are minutes added to the hours.
To spite me now each minute seems a moon,
Yet not for me, shine sun to succour flowers!
   Pack night, peep day; good day, of night now
     borrow;
   Short night tonight, and length thyself tomorrow.  30

## Sonnets
### to Sundry Notes of Music

### 15

It was a lording's daughter, the fairest one of three,
That likèd of her master as well as well might be,
Till looking on an Englishman, the fairest that eye
   could see,
   Her fancy fell a-turning.

Long was the combat doubtful that love with love did
   fight:  5
To leave the master loveless, or kill the gallant knight.
To put in practice either, alas, it was a spite
   Unto the seely damsel.

But one must be refusèd, more mickle was the pain
That nothing could be usèd to turn them both to gain.
For of the two the trusty knight was wounded with
   disdain—
   Alas, she could not help it.

Thus art with arms contending was victor of the day,
Which by a gift of learning did bear the maid away.
Then lullaby, the learned man hath got the lady gay;
   For now my song is ended.

### 17

My flocks feed not, my ewes breed not,
   My rams speed not, all is amiss.
Love is dying, faith's defying,
   Heart's denying causer of this.
All my merry jigs are quite forgot,  5
All my lady's love is lost, God wot.
Where her faith was firmly fixed in love,
There a nay is placed without remove.
   One seely cross wrought all my loss—
   O frowning fortune, cursèd fickle dame!  10
   For now I see inconstancy
   More in women than in men remain.

In black mourn I, all fears scorn I,
   Love hath forlorn me, living in thrall.
Heart is bleeding, all help needing—  15
   O cruel speeding, freighted with gall.
My shepherd's pipe can sound no deal,
My wether's bell rings doleful knell,
My curtal dog that wont to have played
Plays not at all, but seems afraid,  20
   With sighs so deep procures to weep
   In howling wise to see my doleful plight.
   How sighs resound through heartless ground,
   Like a thousand vanquished men in bloody fight!

Clear wells spring not, sweet birds sing not,  25
   Green plants bring not forth their dye.
Herd stands weeping, flocks all sleeping,
   Nymphs back peeping fearfully.
All our pleasure known to us poor swains,
All our merry meetings on the plains,  30
All our evening sport from us is fled,
All our love is lost, for love is dead.
   Farewell, sweet lass, thy like ne'er was
   For a sweet content, the cause of all my moan.
   Poor Corydon must live alone,  35
   Other help for him I see that there is none.

### 18

Whenas thine eye hath chose the dame
And stalled the deer that thou shouldst strike,
Let reason rule things worthy blame
As well as fancy, partial might.
   Take counsel of some wiser head,  5
   Neither too young nor yet unwed,

And when thou com'st thy tale to tell,
Smooth not thy tongue with filèd talk
Lest she some subtle practice smell:
A cripple soon can find a halt.  10
   But plainly say thou lov'st her well,
   And set her person forth to sale,

And to her will frame all thy ways.
Spare not to spend, and chiefly there
Where thy desert may merit praise  15
By ringing in thy lady's ear.
   The strongest castle, tower, and town,
   The golden bullet beats it down.

Serve always with assurèd trust,
And in thy suit be humble-true;                          20
Unless thy lady prove unjust,
Press never thou to choose anew.
  When time shall serve, be thou not slack
  To proffer, though she put thee back.

What though her frowning brows be bent,     25
Her cloudy looks will calm ere night,
And then too late she will repent
That thus dissembled her delight,
  And twice desire, ere it be day,
  That which with scorn she put away.          30

What though she strive to try her strength,
And ban, and brawl, and say thee nay,
Her feeble force will yield at length
When craft hath taught her thus to say:
  'Had women been so strong as men,            35
  In faith you had not had it then.'

The wiles and guiles that women work,
Dissembled with an outward show,
The tricks and toys that in them lurk
The cock that treads them shall not know.      40
  Have you not heard it said full oft
  A woman's nay doth stand for nought?

Think women still to strive with men,
To sin and never for to saint.
There is no heaven; be holy then                   45
When time with age shall them attaint.
  Were kisses all the joys in bed,
  One woman would another wed.

But soft, enough—too much, I fear,
Lest that my mistress hear my song               50
She will not stick to round me on th'ear
To teach my tongue to be so long.
  Yet will she blush (here be it said)
  To hear her secrets so bewrayed.

## The Phoenix and Turtle

Let the bird of loudest lay
On the sole Arabian tree
Herald sad and trumpet be,
To whose sound chaste wings obey.

But thou shrieking harbinger,                        5
Foul precurrer of the fiend,
Augur of the fever's end—
To this troupe come thou not near.

From this session interdict
Every fowl of tyrant wing                             10
Save the eagle, feathered king.
Keep the obsequy so strict.

Let the priest in surplice white
That defunctive music can,
Be the death-divining swan,                          15
Lest the requiem lack his right.

And thou treble-dated crow,
That thy sable gender mak'st
With the breath thou giv'st and tak'st,
'Mongst our mourners shalt thou go.           20

Here the anthem doth commence:
Love and constancy is dead,
Phoenix and the turtle fled
In a mutual flame from hence.

So they loved as love in twain                      25
Had the essence but in one,
Two distincts, division none.
Number there in love was slain.

Hearts remote yet not asunder,
Distance and no space was seen                   30
'Twixt this turtle and his queen.
But in them it were a wonder.

So between them love did shine
That the turtle saw his right
Flaming in the Phoenix' sight.                       35
Either was the other's mine.

Property was thus appalled
That the self was not the same.
Single nature's double name
Neither two nor one was called.                     40

Reason, in itself confounded,
Saw division grow together
To themselves, yet either neither,
Simple were so well compounded

That it cried 'How true a twain                      45
Seemeth this concordant one!
Love hath reason, reason none,
If what parts can so remain.'

Whereupon it made this threne
To the phoenix and the dove,                         50
Co-supremes and stars of love,
As chorus to their tragic scene.

*Threnos*

Beauty, truth, and rarity,
Grace in all simplicity,
Here enclosed in cinders lie.                          55

Death is now the phoenix' nest,
And the turtle's loyal breast
To eternity doth rest.

Leaving no posterity
'Twas not their infirmity,                                 60
It was married chastity.

Truth may seem but cannot be,
Beauty brag, but 'tis not she.
Truth and beauty buried be.

To this urn let those repair                              65
That are either true or fair.
For these dead birds sigh a prayer.

## Verses upon the Stanley Tomb at Tong

### Written upon the east end of the tomb

Ask who lies here, but do not weep.
He is not dead; he doth but sleep.
This stony register is for his bones;
His fame is more perpetual than these stones,
And his own goodness, with himself being gone,     5
Shall live when earthly monument is none.

### Written upon the West end thereof

Not monumental stone preserves our fame,
Nor sky-aspiring pyramids our name.
The memory of him for whom this stands
Shall outlive marble and defacers' hands.
When all to time's consumption shall be given,     5
Stanley for whom this stands shall stand in heaven.

---

## On Ben Jonson

Master Ben Jonson and Master William Shakespeare
being merry at a tavern, Master Jonson having begun
this for his epitaph:

Here lies Ben Jonson
That was once one,

he gives it to Master Shakespeare to make up who
presently writes:

Who while he lived was a slow thing,
And now, being dead, is nothing.

---

## An Epitaph on Elias James

When God was pleased, the world unwilling yet,
Elias James to nature paid his debt,
And here reposeth. As he lived, he died,
The saying strongly in him verified:
'Such life, such death'. Then, a known truth to tell,   5
He lived a godly life, and died as well.

## An extemporary epitaph on John Combe, a noted usurer

Ten in the hundred here lies engraved;
A hundred to ten his soul is not saved.
If anyone ask who lies in this tomb,
'O ho!' quoth the devil, ''tis my John-a-Combe.'

---

## Another Epitaph on John Combe

He being dead, and making the poor his heirs, William
Shakespeare after writes this for his epitaph:

Howe'er he livèd judge not,
John Combe shall never be forgot
While poor hath memory, for he did gather
To make the poor his issue; he, their father,
As record of his tilth and seed                          5
Did crown him in his latter deed.

---

## Upon the King

At the foot of the effigy of King James I, before his
*Works* (1616)

Crowns have their compass; length of days, their date;
Triumphs, their tombs; felicity, her fate.
Of more than earth can earth make none partaker,
But knowledge makes the king most like his maker.

---

## Epitaph on Himself

Good friend, for Jesus' sake forbear
To dig the dust enclosèd here.
Blessed be the man that spares these stones,
And cursed be he that moves my bones.

# SIR THOMAS MORE

## PASSAGES ATTRIBUTED TO SHAKESPEARE

IN the British Library is a manuscript play described on its first leaf as 'The Booke'—
that is, the theatre manuscript—'of Sir Thomas Moore'. It is a heavily revised text with
contributions in six different hands as well as annotations by the Master of the Revels. The
basic manuscript appears to have been a fair copy by the dramatist Anthony Munday of a
play that he wrote in collaboration with Henry Chettle and, perhaps, another writer.
Alterations and additions were made by Chettle, Thomas Dekker, possibly Thomas
Heywood, and the author of the pages in Addition II ascribed to 'Hand D', whom many
scholars believe to be William Shakespeare.

The theory that Shakespeare was a contributor, first mooted in 1871, has led to intensive
study of the manuscript. Our view is that the original play, dating from the early 1590s,
was submitted in the normal way to the Master of the Revels, Sir Edmund Tilney, for a
licence. But Tilney, disturbed by the play's political implications, called for substantial
revisions which, if they had been carried out, would have required that about half the play
be scrapped.

What happened next is not clear. The alterations and additions to the basic play do not
meet Tilney's objections. Perhaps they had been made before the play was submitted for a
licence. More probably (in our view) the original play was laid aside after Tilney had
objected to it, and taken up again soon after Queen Elizabeth's death, in 1603, when the
political objections would no longer be felt.

*Sir Thomas More*, based mainly on Holinshed's *Chronicles* and on William Roper's manuscript *Life of More*, is an episodic treatment of its hero's rise and fall, ending with his death
on the scaffold. The principal episode attributed to Shakespeare comes towards the end of
the scenes early in the play portraying events leading up to the riots of Londoners against
resident foreigners on the 'ill May Day' of 1517. The leaders are John Lincoln, Williamson
and his wife Doll, George and Ralph Betts, and Sherwin. Outraged by the illegal activities
of foreign groups in London, they have planned that 'on May Day next in the morning
we'll go forth a-maying, but make it the worst May Day for the strangers'—i.e. foreigners—
'that ever they saw'. The authorities, dismayed by the violence, have sent More as a
peacemaker. Shakespeare—if indeed he wrote the scene—seems not to have known the
rest of the play; he was probably revising an original scene, now lost, with no other
sources. The ascription of this scene to Shakespeare is based partly on comparison between
the few surviving specimens of Shakespeare's handwriting (almost entirely in signatures)
with that of Hand D; partly on spelling links with printed texts apparently deriving directly
from Shakespeare's own papers; and partly on considerations of style and imagery.

Also attributed to Shakespeare is a soliloquy by More apparently intended to show his
state of mind after having been appointed Lord Chancellor. It is written in the hand of a
professional scribe (Addition III).

# Sir Thomas More

**Add.II.D** *John Lincoln (a broker), Doll, Betts, ⌐Sherwin (a goldsmith),⌐ and prentices armed; ⌐Thomas More (sheriff of the City of London), the other sheriff, Sir Thomas Palmer, Sir Roger Cholmeley, and a serjeant-at-arms stand aloof ⌐*

LINCOLN (*to the prentices*) Peace, hear me! He that will not see a red herring at a Harry groat, butter at eleven pence a pound, meal at nine shillings a bushel, and beef at four nobles a stone, list to me.

OTHER It will come to that pass if strangers be suffered. Mark him. 6

LINCOLN Our country is a great eating country; argo, they eat more in our country than they do in their own.

OTHER By a halfpenny loaf a day, troy weight. 10

LINCOLN They bring in strange roots, which is merely to the undoing of poor prentices, for what's a sorry parsnip to a good heart?

OTHER Trash, trash. They breed sore eyes, and 'tis enough to infect the city with the palsy. 15

LINCOLN Nay, it has infected it with the palsy, for these bastards of dung—as you know, they grow in dung— have infected us, and it is our infection will make the city shake, which partly comes through the eating of parsnips. 20

OTHER True, and pumpions together.

SERJEANT ⌐*coming forward*⌐
What say you to the mercy of the King? Do you refuse it?

LINCOLN You would have us upon th'hip, would you? No, marry, do we not. We accept of the King's mercy; but we will show no mercy upon the strangers. 26

SERJEANT         You are the simplest things That ever stood in such a question.

LINCOLN How say you now? Prentices 'simple'? (*To the prentices*) Down with him! 30

ALL Prentices simple! Prentices simple!
*Enter the Lord Mayor, the Earl of Surrey, and the Earl of Shrewsbury*
⌐SHERIFF⌐ (*to the prentices*)
Hold in the King's name! Hold!

SURREY (*to the prentices*) Friends, masters, countrymen—

MAYOR (*to the prentices*)
Peace ho, peace! I charge you, keep the peace!

SHREWSBURY (*to the prentices*) My masters, countrymen—

⌐SHERWIN⌐ The noble Earl of Shrewsbury, let's hear him.

BETTS We'll hear the Earl of Surrey. 36

LINCOLN The Earl of Shrewsbury.

BETTS We'll hear both.

ALL Both, both, both, both!

LINCOLN Peace, I say peace! Are you men of wisdom, or what are you? 41

SURREY
What you will have them, but not men of wisdom.

⌐SOME⌐ We'll not hear my Lord of Surrey.

⌐OTHERS⌐ No, no, no, no, no! Shrewsbury, Shrewsbury!

MORE (*to the nobles and officers*)
Whiles they are o'er the bank of their obedience, 45
Thus will they bear down all things.

LINCOLN (*to the prentices*) Sheriff More speaks. Shall we hear Sheriff More speak?

DOLL Let's hear him. A keeps a plentiful shrievaltry, and a made my brother Arthur Watchins Sergeant Safe's yeoman. Let's hear Sheriff More. 51

ALL Sheriff More, More, More, Sheriff More!

MORE
Even by the rule you have among yourselves, Command still audience.

SOME Surrey, Surrey! 55

OTHERS More, More!

LINCOLN and BETTS Peace, peace, silence, peace!

MORE
You that have voice and credit with the number, Command them to a stillness.

LINCOLN A plague on them! They will not hold their peace. The devil cannot rule them. 61

MORE
Then what a rough and riotous charge have you, To lead those that the devil cannot rule.
(*To the prentices*) Good masters, hear me speak.

DOLL Ay, by th' mass, will we. More, thou'rt a good housekeeper, and I thank thy good worship for my brother Arthur Watchins. 67

ALL Peace, peace!

MORE
Look, what you do offend you cry upon, That is the peace. Not one of you here present, 70
Had there such fellows lived when you were babes That could have topped the peace as now you would, The peace wherein you have till now grown up Had been ta'en from you, and the bloody times Could not have brought you to the state of men. 75
Alas, poor things, what is it you have got, Although we grant you get the thing you seek?

BETTS Marry, the removing of the strangers, which cannot choose but much advantage the poor handicrafts of the city. 80

MORE
Grant them removed, and grant that this your noise Hath chid down all the majesty of England. Imagine that you see the wretched strangers, Their babies at their backs, with their poor luggage Plodding to th' ports and coasts for transportation, 85
And that you sit as kings in your desires, Authority quite silenced by your brawl And you in ruff of your opinions clothed: What had you got? I'll tell you. You had taught How insolence and strong hand should prevail, 90
How order should be quelled—and by this pattern Not one of you should live an agèd man, For other ruffians as their fancies wrought With selfsame hand, self reasons, and self right Would shark on you, and men like ravenous fishes 95
Would feed on one another.

DOLL Before God, that's as true as the gospel.

BETTS Nay, this' a sound fellow, I tell you. Let's mark him.

MORE
Let me set up before your thoughts, good friends, 100
One supposition, which if you will mark You shall perceive how horrible a shape Your innovation bears. First, 'tis a sin

Which oft th'apostle did forewarn us of,
Urging obedience to authority;                          105
And 'twere no error if I told you all
You were in arms 'gainst God.
ALL Marry, God forbid that!
MORE Nay, certainly you are.
For to the King God hath his office lent          110
Of dread, of justice, power and command,
Hath bid him rule and willed you to obey;
And to add ampler majesty to this,
He hath not only lent the King his figure,
His throne and sword, but given him his own name,
Calls him a god on earth. What do you then,    116
Rising 'gainst him that God himself installs,
But rise 'gainst God? What do you to your souls
In doing this? O desperate as you are,
Wash your foul minds with tears, and those same
    hands                                                       120
That you like rebels lift against the peace
Lift up for peace; and your unreverent knees,
Make them your feet. To kneel to be forgiven
Is safer wars than ever you can make,
Whose discipline is riot.                               125
In, in, to your obedience! Why, even your hurly
Cannot proceed but by obedience.
What rebel captain,
As mut'nies are incident, by his name
Can still the rout? Who will obey a traitor?    130
Or how can well that proclamation sound,
When there is no addition but 'a rebel'
To qualify a rebel? You'll put down strangers,
Kill them, cut their throats, possess their houses,
And lead the majesty of law in lyam              135
To slip him like a hound—alas, alas!
Say now the King,
As he is clement if th'offender mourn,
Should so much come too short of your great trespass
As but to banish you: whither would you go?    140
What country, by the nature of your error,
Should give you harbour? Go you to France or
    Flanders,
To any German province, Spain or Portugal,
Nay, anywhere that not adheres to England—
Why, you must needs be strangers. Would you be
    pleased                                                     145
To find a nation of such barbarous temper

That breaking out in hideous violence
Would not afford you an abode on earth,
Whet their detested knives against your throats,
Spurn you like dogs, and like as if that God    150
Owed not nor made not you, nor that the elements
Were not all appropriate to your comforts
But chartered unto them, what would you think
To be thus used? This is the strangers' case,
And this your mountainish inhumanity.          155
⌜ONE⌝ (to the others) Faith, a says true. Let's do as we may
    be done by.
⌜ANOTHER⌝ (to More) We'll be ruled by you, Master More,
    if you'll stand our friend to procure our pardon.
MORE
Submit you to these noble gentlemen,            160
Entreat their mediation to the King,
Give up yourself to form, obey the magistrate,
And there's no doubt but mercy may be found,
If you so seek it.

---

**Add.III**    *Enter Sir Thomas More*
MORE
It is in heaven that I am thus and thus,
And that which we profanely term our fortunes
Is the provision of the power above,
Fitted and shaped just to that strength of nature
Which we are born withal. Good God, good God,    5
That I from such an humble bench of birth
Should step as 'twere up to my country's head
And give the law out there; ay, in my father's life
To take prerogative and tithe of knees
From elder kinsmen, and him bind by my place    10
To give the smooth and dexter way to me
That owe it him by nature! Sure these things,
Not physicked by respect, might turn our blood
To much corruption. But More, the more thou hast
Either of honour, office, wealth and calling,    15
Which might accite thee to embrace and hug them,
The more do thou e'en serpents' natures think them:
Fear their gay skins, with thought of their sharp
    stings,
And let this be thy maxim: to be great
Is, when the thread of hazard is once spun,     20
A bottom great wound up, greatly undone.

# MEASURE FOR MEASURE

*Measure for Measure*, first printed in the 1623 Folio, was performed at court on 26 December 1604. Plague had caused London's theatres to be closed from May 1603 to April 1604; the play was probably written and first acted during 1604. Dislocations and other features of the text as printed suggest that it may have undergone adaptation after Shakespeare's death. Someone—perhaps Thomas Middleton, to judge by the style—seems to have supplied a new, seedy opening to Act 1, Scene 2; and an adapter seems also to have altered 3.1.517–4.1.63 by transposing the Duke's two soliloquies, by introducing a stanza from a popular song, and by supplying dialogue to follow it. We print the text in what we believe to be its adapted form; a conjectured reconstruction of Shakespeare's original version of the adapted sections is given in the Additional Passages.

The story of a woman who, in seeking to save the life of a male relative, arouses the lust of a man in authority was an ancient one that reached literary form in the mid sixteenth century. Shakespeare may have known the prose version in Giambattista Cinzio Giraldi's *Gli Ecatommiti* (1565, translated into French in 1583) and the same author's play *Epitia* (1573, published in 1583), but his main source was George Whetstone's unsuccessful, unperformed two-part tragicomedy *Promos and Cassandra*, published in 1578.

Shakespeare's title comes from Saint Matthew's account of Christ's Sermon on the Mount: 'with what measure ye mete, it shall be measured to you again'. The title is not expressive of the play's morality, but it alerts the spectator to Shakespeare's exploration of moral issues. His heroine, Isabella, is not merely, as in Whetstone, a virtuous young maiden: she is about to enter a nunnery. Her brother, Claudio, has not, as in Whetstone, been accused (however unjustly) of rape: his union with the girl (Juliet) he has made pregnant has been ratified by a betrothal ceremony, and lacks only the church's formal blessing. So Angelo, deputizing for the absent Duke of Vienna, seems peculiarly harsh in attempting to enforce the city's laws against fornication by insisting on Claudio's execution; and Angelo's hypocrisy in demanding Isabella's chastity in return for her brother's life seems correspondingly greater. By adding the character of Mariana, to whom Angelo himself had once been betrothed, and by employing the traditional motif of the 'bed-trick', by which Mariana substitutes for Isabella in Angelo's bed, Shakespeare permits Isabella both to retain her virtue and to forgive Angelo without marrying him.

Although *Measure for Measure*, like *The Merchant of Venice*, is much concerned with justice and mercy, its more explicit concern with sex and death along with the intense emotional reality, at least in the earlier part of the play, of its portrayal of Angelo, Isabella, and Claudio, creates a deeper seriousness of tone which takes it out of the world of romantic comedy into that of tragicomedy or, as the twentieth-century label has it, 'problem play'. Its low-life characters inhabit a diseased world of brothels and prisons, but there is a life-enhancing quality in their frank acknowledgement of sexuality; and the Duke's manipulation of events casts a tinge of romance over the play's later scenes.

*Measure for Measure*'s subtle and passionate exploration of issues of sexual morality, of the uses and abuses of power, has given it a special appeal in the later part of the twentieth century. Each of the 'good' characters fails in some respect; none of the 'bad' ones lacks some redeeming quality; all are, in the last analysis, 'desperately mortal' (4.2.148).

# THE PERSONS OF THE PLAY

Vincentio, the DUKE of Vienna

ANGELO, appointed his deputy

ESCALUS, an old lord, appointed Angelo's secondary

CLAUDIO, a young gentleman

JULIET, betrothed to Claudio

ISABELLA, Claudio's sister, novice to a sisterhood of nuns

LUCIO, 'a fantastic'

Two other such GENTLEMEN

FROTH, a foolish gentleman

MISTRESS OVERDONE, a bawd

POMPEY, her clownish servant

A PROVOST

ELBOW, a simple constable

A JUSTICE

ABHORSON, an executioner

BARNARDINE, a dissolute condemned prisoner

MARIANA, betrothed to Angelo

A BOY, attendant on Mariana

FRIAR PETER

FRANCESCA, a nun

VARRIUS, a lord, friend to the Duke

Lords, officers, citizens, servants

# Measure for Measure

**1.1** *Enter the Duke, Escalus, and other lords*

DUKE Escalus.

ESCALUS My lord.

DUKE
Of government the properties to unfold
Would seem in me t'affect speech and discourse,
Since I am put to know that your own science     5
Exceeds in that the lists of all advice
My strength can give you. Then no more remains
But this: to your sufficiency, as your worth is able,
And let them work. The nature of our people,
Our city's institutions and the terms     10
For common justice, you're as pregnant in
As art and practice hath enrichèd any
That we remember.
      *He gives Escalus papers*
                  There is our commission,
From which we would not have you warp.
*(To a lord)*               Call hither,
I say bid come before us, Angelo.        *Exit lord*
*(To Escalus)* What figure of us think you he will
    bear?—     16
For you must know we have with special soul
Elected him our absence to supply,
Lent him our terror, dressed him with our love,
And given his deputation all the organs     20
Of our own power. What think you of it?

ESCALUS
If any in Vienna be of worth
To undergo such ample grace and honour,
It is Lord Angelo.
      *Enter Angelo*

DUKE             Look where he comes.

ANGELO
Always obedient to your grace's will,     25
I come to know your pleasure.

DUKE                 Angelo,
There is a kind of character in thy life
That to th'observer doth thy history
Fully unfold. Thyself and thy belongings
Are not thine own so proper as to waste     30
Thyself upon thy virtues, they on thee.
Heaven doth with us as we with torches do,
Not light them for themselves; for if our virtues
Did not go forth of us, 'twere all alike
As if we had them not. Spirits are not finely touched
But to fine issues; nor nature never lends     36
The smallest scruple of her excellence
But, like a thrifty goddess, she determines
Herself the glory of a creditor,
Both thanks and use. But I do bend my speech     40
To one that can my part in him advertise.
Hold therefore, Angelo.
In our remove be thou at full ourself.
Mortality and mercy in Vienna
Live in thy tongue and heart. Old Escalus,     45
Though first in question, is thy secondary.
Take thy commission.

ANGELO           Now good my lord,
Let there be some more test made of my metal
Before so noble and so great a figure
Be stamped upon it.

DUKE           No more evasion.     50
We have with leavened and preparèd choice
Proceeded to you; therefore take your honours.
      ⌈*Angelo takes his commission*⌉
Our haste from hence is of so quick condition
That it prefers itself, and leaves unquestioned
Matters of needful value. We shall write to you     55
As time and our concernings shall importune,
How it goes with us; and do look to know
What doth befall you here. So fare you well.
To th' hopeful execution do I leave you
Of your commissions.

ANGELO          Yet give leave, my lord,     60
That we may bring you something on the way.

DUKE My haste may not admit it;
Nor need you, on mine honour, have to do
With any scruple. Your scope is as mine own,
So to enforce or qualify the laws     65
As to your soul seems good. Give me your hand.
I'll privily away. I love the people,
But do not like to stage me to their eyes.
Though it do well, I do not relish well
Their loud applause and *aves* vehement;     70
Nor do I think the man of safe discretion
That does affect it. Once more, fare you well.

ANGELO
The heavens give safety to your purposes!

ESCALUS
Lead forth and bring you back in happiness!

DUKE I thank you. Fare you well.         *Exit*

ESCALUS
I shall desire you, sir, to give me leave     76
To have free speech with you; and it concerns me
To look into the bottom of my place.
A power I have, but of what strength and nature
I am not yet instructed.     80

ANGELO
'Tis so with me. Let us withdraw together,
And we may soon our satisfaction have
Touching that point.

ESCALUS           I'll wait upon your honour. *Exeunt*

**1.2** *Enter Lucio, and two other Gentlemen*

LUCIO If the Duke with the other dukes come not to
composition with the King of Hungary, why then, all
the dukes fall upon the King.

FIRST GENTLEMAN Heaven grant us its peace, but not the
King of Hungary's!     5

SECOND GENTLEMAN Amen.

LUCIO Thou concludest like the sanctimonious pirate, that
went to sea with the Ten Commandments, but scraped
one out of the table.

SECOND GENTLEMAN 'Thou shalt not steal'?     10

LUCIO Ay, that he razed.

FIRST GENTLEMAN Why, 'twas a commandment to command the captain and all the rest from their functions:
they put forth to steal. There's not a soldier of us all
that in the thanksgiving before meat do relish the
petition well that prays for peace.     16

SECOND GENTLEMAN I never heard any soldier dislike it.

791

LUCIO I believe thee, for I think thou never wast where grace was said.

SECOND GENTLEMAN No? A dozen times at least.          20

FIRST GENTLEMAN What, in metre?

LUCIO In any proportion, or in any language.

FIRST GENTLEMAN I think, or in any religion.

LUCIO Ay, why not? Grace is grace despite of all controversy; as for example, thou thyself art a wicked villain despite of all grace.          26

FIRST GENTLEMAN Well, there went but a pair of shears between us.

LUCIO I grant—as there may between the lists and the velvet. Thou art the list.          30

FIRST GENTLEMAN And thou the velvet. Thou art good velvet, thou'rt a three-piled piece, I warrant thee. I had as lief be a list of an English kersey as be piled as thou art pilled, for a French velvet. Do I speak feelingly now?

LUCIO I think thou dost, and indeed with most painful feeling of thy speech. I will out of thine own confession learn to begin thy health, but whilst I live forget to drink after thee.

FIRST GENTLEMAN I think I have done myself wrong, have I not?          40

SECOND GENTLEMAN Yes, that thou hast, whether thou art tainted or free.

*Enter Mistress Overdone*

LUCIO Behold, behold, where Madam Mitigation comes! I have purchased as many diseases under her roof as come to—          45

SECOND GENTLEMAN To what, I pray?

LUCIO Judge.

SECOND GENTLEMAN To three thousand dolours a year?

FIRST GENTLEMAN Ay, and more.

LUCIO A French crown more.          50

FIRST GENTLEMAN Thou art always figuring diseases in me, but thou art full of error—I am sound.

LUCIO Nay not, as one would say, healthy, but so sound as things that are hollow—thy bones are hollow, impiety has made a feast of thee.          55

FIRST GENTLEMAN (*to Mistress Overdone*) How now, which of your hips has the most profound sciatica?

MISTRESS OVERDONE Well, well! There's one yonder arrested and carried to prison was worth five thousand of you all.          60

SECOND GENTLEMAN Who's that, I pray thee?

MISTRESS OVERDONE Marry sir, that's Claudio, Signor Claudio.

FIRST GENTLEMAN Claudio to prison? 'Tis not so.          64

MISTRESS OVERDONE Nay, but I know 'tis so. I saw him arrested, saw him carried away; and, which is more, within these three days his head to be chopped off.

LUCIO But after all this fooling, I would not have it so. Art thou sure of this?          69

MISTRESS OVERDONE I am too sure of it, and it is for getting Madame Julietta with child.

LUCIO Believe me, this may be. He promised to meet me two hours since and he was ever precise in promise-keeping.          74

SECOND GENTLEMAN Besides, you know, it draws something near to the speech we had to such a purpose.

FIRST GENTLEMAN But most of all agreeing with the proclamation.

LUCIO Away; let's go learn the truth of it.          79

*Exeunt Lucio and Gentlemen*

MISTRESS OVERDONE Thus, what with the war, what with the sweat, what with the gallows, and what with poverty, I am custom-shrunk.

*Enter Pompey*

How now, what's the news with you?

POMPEY You have not heard of the proclamation, have you?          85

MISTRESS OVERDONE What proclamation, man?

POMPEY All houses in the suburbs of Vienna must be plucked down.

MISTRESS OVERDONE And what shall become of those in the city?          90

POMPEY They shall stand for seed. They had gone down too, but that a wise burgher put in for them.

MISTRESS OVERDONE But shall all our houses of resort in the suburbs be pulled down?

POMPEY To the ground, mistress.          95

MISTRESS OVERDONE Why, here's a change indeed in the commonwealth. What shall become of me?

POMPEY Come, fear not you. Good counsellors lack no clients. Though you change your place, you need not change your trade. I'll be your tapster still. Courage, there will be pity taken on you. You that have worn your eyes almost out in the service, you will be considered.

⌈*A noise within*⌉

MISTRESS OVERDONE What's to do here, Thomas Tapster? Let's withdraw!          105

*Enter the Provost, Claudio, Juliet, and officers;
Lucio and the two Gentlemen*

POMPEY Here comes Signor Claudio, led by the Provost to prison; and there's Madame Juliet.

*Exeunt Mistress Overdone and Pompey*

CLAUDIO (*to the Provost*)
Fellow, why dost thou show me thus to th' world?
Bear me to prison, where I am committed.

PROVOST
I do it not in evil disposition,          110
But from Lord Angelo by special charge.

CLAUDIO
Thus can the demigod Authority
Make us pay down for our offence, by weight,
The bonds of heaven. On whom it will, it will;
On whom it will not, so; yet still 'tis just.          115

LUCIO
Why, how now, Claudio? Whence comes this
     restraint?

CLAUDIO
From too much liberty, my Lucio, liberty.
As surfeit is the father of much fast,
So every scope, by the immoderate use,
Turns to restraint. Our natures do pursue,          120
Like rats that raven down their proper bane,
A thirsty evil; and when we drink, we die.

LUCIO If I could speak so wisely under an arrest, I would send for certain of my creditors. And yet, to say the truth, I had as lief have the foppery of freedom as the morality of imprisonment. What's thy offence, Claudio?

CLAUDIO
What but to speak of would offend again.          127

LUCIO
What, is't murder?

CLAUDIO                              No.

LUCIO                                      Lechery?

CLAUDIO                                            Call it so.

PROVOST Away, sir; you must go.

CLAUDIO
One word, good friend.
⌈*The Provost shows assent*⌉
                              Lucio, a word with you.          130

LUCIO  A hundred, if they'll do you any good.

⌈*Claudio and Lucio speak apart*⌉

Is lechery so looked after?

CLAUDIO

Thus stands it with me. Upon a true contract,
I got possession of Julietta's bed.
You know the lady; she is fast my wife,          135
Save that we do the denunciation lack
Of outward order. This we came not to
Only for propagation of a dower
Remaining in the coffer of her friends,
From whom we thought it meet to hide our love    140
Till time had made them for us. But it chances
The stealth of our most mutual entertainment
With character too gross is writ on Juliet.

LUCIO

With child, perhaps?

CLAUDIO                    Unhapp'ly even so.
And the new deputy now for the Duke—             145
Whether it be the fault and glimpse of newness,
Or whether that the body public be
A horse whereon the governor doth ride,
Who, newly in the seat, that it may know
He can command, lets it straight feel the spur—  150
Whether the tyranny be in his place,
Or in his eminence that fills it up—
I stagger in. But this new governor
Awakes me all the enrollèd penalties
Which have, like unscoured armour, hung by th' wall
So long that fourteen zodiacs have gone round,   156
And none of them been worn; and, for a name,
Now puts the drowsy and neglected act
Freshly on me. 'Tis surely for a name.

LUCIO  I warrant it is; and thy head stands so tickle on
thy shoulders that a milkmaid, if she be in love, may
sigh it off. Send after the Duke, and appeal to him.

CLAUDIO

I have done so, but he's not to be found.
I prithee, Lucio, do me this kind service.
This day my sister should the cloister enter,    165
And there receive her approbation.
Acquaint her with the danger of my state.
Implore her in my voice that she make friends
To the strict deputy. Bid herself assay him.
I have great hope in that, for in her youth      170
There is a prone and speechless dialect
Such as move men; beside, she hath prosperous art
When she will play with reason and discourse,
And well she can persuade.

LUCIO  I pray she may—as well for the encouragement of
thy like, which else would stand under grievous
imposition, as for the enjoying of thy life, who I would
be sorry should be thus foolishly lost at a game of tick-
tack. I'll to her.

CLAUDIO  I thank you, good friend Lucio.         180

LUCIO  Within two hours.

CLAUDIO  Come, officer; away.

*Exeunt* ⌈*Lucio and gentlemen at one door;*
*Claudio, Juliet, Provost, and officers at another*⌉

**1.3**  *Enter the Duke and a Friar*

DUKE

No, holy father, throw away that thought.
Believe not that the dribbling dart of love
Can pierce a complete bosom. Why I desire thee
To give me secret harbour hath a purpose
More grave and wrinkled than the aims and ends   5
Of burning youth.

FRIAR                    May your grace speak of it?

DUKE

My holy sir, none better knows than you
How I have ever loved the life removed,
And held in idle price to haunt assemblies
Where youth and cost a witless bravery keeps.    10
I have delivered to Lord Angelo—
A man of stricture and firm abstinence—
My absolute power and place here in Vienna;
And he supposes me travelled to Poland—
For so I have strewed it in the common ear,      15
And so it is received. Now, pious sir,
You will demand of me why I do this.

FRIAR  Gladly, my lord.

DUKE

We have strict statutes and most biting laws,
The needful bits and curbs to headstrong weeds,  20
Which for this fourteen years we have let slip;
Even like an o'ergrown lion in a cave
That goes not out to prey. Now, as fond fathers,
Having bound up the threat'ning twigs of birch
Only to stick it in their children's sight       25
For terror, not to use, in time the rod
More mocked becomes than feared: so our decrees,
Dead to infliction, to themselves are dead;
And Liberty plucks Justice by the nose,
The baby beats the nurse, and quite athwart      30
Goes all decorum.

FRIAR                It rested in your grace
To unloose this tied-up Justice when you pleased,
And it in you more dreadful would have seemed
Than in Lord Angelo.

DUKE                    I do fear, too dreadful.
Sith 'twas my fault to give the people scope,    35
'Twould be my tyranny to strike and gall them
For what I bid them do—for we bid this be done
When evil deeds have their permissive pass,
And not the punishment. Therefore indeed, my father,
I have on Angelo imposed the office,             40
Who may in th'ambush of my name strike home,
And yet my nature never in the fight
T'allow in slander. And to behold his sway,
I will as 'twere a brother of your order
Visit both prince and people. Therefore, I prithee, 45
Supply me with the habit, and instruct me
How I may formally in person bear
Like a true friar. More reasons for this action
At our more leisure shall I render you.
Only this one: Lord Angelo is precise,           50
Stands at a guard with envy, scarce confesses
That his blood flows, or that his appetite
Is more to bread than stone. Hence shall we see
If power change purpose, what our seemers be. *Exeunt*

**1.4**  *Enter Isabella, and Francesca, a nun*

ISABELLA

And have you nuns no farther privileges?

FRANCESCA  Are not these large enough?

ISABELLA

Yes, truly. I speak not as desiring more,
But rather wishing a more strict restraint
Upon the sisterhood, the votarists of Saint Clare. 5

LUCIO (*within*)
  Ho, peace be in this place!
ISABELLA ⌈*to Francesca*⌉          Who's that which calls?
FRANCESCA
  It is a man's voice. Gentle Isabella.
  Turn you the key, and know his business of him.
  You may, I may not; you are yet unsworn.
  When you have vowed, you must not speak with men
  But in the presence of the prioress.          11
  Then if you speak, you must not show your face;
  Or if you show your face, you must not speak.
          *Lucio calls within*
  He calls again. I pray you answer him.
          ⌈*She stands aside*⌉
ISABELLA
  Peace and prosperity! Who is't that calls?          15
          *She opens the door.*
          *Enter Lucio*
LUCIO
  Hail, virgin, if you be—as those cheek-roses
  Proclaim you are no less. Can you so stead me
  As bring me to the sight of Isabella,
  A novice of this place, and the fair sister
  To her unhappy brother Claudio?          20
ISABELLA
  Why her unhappy brother? Let me ask,
  The rather for I now must make you know
  I am that Isabella, and his sister.
LUCIO
  Gentle and fair, your brother kindly greets you.
  Not to be weary with you, he's in prison.          25
ISABELLA Woe me! For what?
LUCIO
  For that which, if myself might be his judge,
  He should receive his punishment in thanks.
  He hath got his friend with child.
ISABELLA                    Sir, make me not your story.
LUCIO
  'Tis true. I would not—though 'tis my familiar sin          30
  With maids to seem the lapwing, and to jest
  Tongue far from heart—play with all virgins so.
  I hold you as a thing enskied and sainted
  By your renouncement, an immortal spirit,
  And to be talked with in sincerity          35
  As with a saint.
ISABELLA
  You do blaspheme the good in mocking me.
LUCIO
  Do not believe it. Fewness and truth, 'tis thus:
  Your brother and his lover have embraced.
  As those that feed grow full, as blossoming time          40
  That from the seedness the bare fallow brings
  To teeming foison, even so her plenteous womb
  Expresseth his full tilth and husbandry.
ISABELLA
  Someone with child by him? My cousin Juliet?
LUCIO Is she your cousin?          45
ISABELLA
  Adoptedly, as schoolmaids change their names
  By vain though apt affection.
LUCIO                    She it is.
ISABELLA
  O, let him marry her!
LUCIO                    This is the point.
  The Duke is very strangely gone from hence;
  Bore many gentlemen—myself being one—          50
  In hand and hope of action; but we do learn,

By those that know the very nerves of state,
His giving out were of an infinite distance
From his true-meant design. Upon his place,
And with full line of his authority,          55
Governs Lord Angelo—a man whose blood
Is very snow-broth; one who never feels
The wanton stings and motions of the sense,
But doth rebate and blunt his natural edge
With profits of the mind, study, and fast.          60
He, to give fear to use and liberty,
Which have for long run by the hideous law
As mice by lions, hath picked out an act
Under whose heavy sense your brother's life
Falls into forfeit. He arrests him on it,          65
And follows close the rigour of the statute
To make him an example. All hope is gone,
Unless you have the grace by your fair prayer
To soften Angelo. And that's my pith
Of business 'twixt you and your poor brother.          70
ISABELLA
  Doth he so seek his life?
LUCIO                    Has censured him already,
  And, as I hear, the Provost hath a warrant
  For's execution.
ISABELLA          Alas, what poor
  Ability's in me to do him good?
LUCIO Assay the power you have.          75
ISABELLA My power? Alas, I doubt.
LUCIO Our doubts are traitors,
  And makes us lose the good we oft might win,
  By fearing to attempt. Go to Lord Angelo;
  And let him learn to know, when maidens sue,          80
  Men give like gods, but when they weep and kneel,
  All their petitions are as freely theirs
  As they themselves would owe them.
ISABELLA                    I'll see what I can do.
LUCIO
  But speedily.
ISABELLA          I will about it straight,
  No longer staying but to give the Mother          85
  Notice of my affair. I humbly thank you.
  Commend me to my brother. Soon at night
  I'll send him certain word of my success.
LUCIO
  I take my leave of you.
ISABELLA                    Good sir, adieu.
          *Exeunt ⌈Isabella and Francesca at one door,*
          *Lucio at another door⌉*

          ❦

2.1          *Enter Angelo, Escalus, and servants; a Justice*
ANGELO
  We must not make a scarecrow of the law,
  Setting it up to fear the birds of prey,
  And let it keep one shape till custom make it
  Their perch, and not their terror.
ESCALUS                    Ay, but yet
  Let us be keen, and rather cut a little          5
  Than fall and bruise to death. Alas, this gentleman
  Whom I would save had a most noble father.
  Let but your honour know—
  Whom I believe to be most strait in virtue—
  That in the working of your own affections,          10
  Had time cohered with place, or place with wishing,
  Or that the resolute acting of your blood
  Could have attained th'effect of your own purpose—
  Whether you had not sometime in your life

Erred in this point which now you censure him,        15
And pulled the law upon you.

ANGELO
'Tis one thing to be tempted, Escalus,
Another thing to fall. I not deny
The jury passing on the prisoner's life
May in the sworn twelve have a thief or two        20
Guiltier than him they try. What knows the law
That thieves do pass on thieves? What's open made to
        justice,
That justice seizes. 'Tis very pregnant:
The jewel that we find, we stoop and take't
Because we see it, but what we do not see        25
We tread upon and never think of it.
You may not so extenuate his offence
For I have had such faults; but rather tell me,
When I that censure him do so offend,
Let mine own judgement pattern out my death,        30
And nothing come in partial. Sir, he must die.

ESCALUS
Be it as your wisdom will.

ANGELO                                   Where is the Provost?
        *Enter Provost*

PROVOST
Here, if it like your honour.

ANGELO                                   See that Claudio
Be execute by nine tomorrow morning.
Bring him his confessor, let him be prepared,        35
For that's the utmost of his pilgrimage.        *Exit Provost*

ESCALUS
Well, heaven forgive him, and forgive us all!
Some rise by sin, and some by virtue fall.
Some run from brakes of vice, and answer none;
And some condemnèd for a fault alone.        40
        *Enter Elbow, Froth, Pompey, and officers*

ELBOW Come, bring them away. If these be good people
in a commonweal, that do nothing but use their abuses
in common houses, I know no law. Bring them away.

ANGELO
How now, sir? What's your name? And what's the
matter?

ELBOW If it please your honour, I am the poor Duke's
constable, and my name is Elbow. I do lean upon
justice, sir; and do bring in here before your good
honour two notorious benefactors.

ANGELO
Benefactors? Well! What benefactors are they?
Are they not malefactors?        50

ELBOW If it please your honour, I know not well what
they are; but precise villains they are, that I am sure
of, and void of all profanation in the world that good
Christians ought to have.

ESCALUS (*to Angelo*)
This comes off well; here's a wise officer!        55

ANGELO Go to, what quality are they of? Elbow is your
name? Why dost thou not speak, Elbow?

POMPEY He cannot, sir; he's out at elbow.

ANGELO What are you, sir?        59

ELBOW He, sir? A tapster, sir, parcel bawd; one that
serves a bad woman whose house, sir, was, as they
say, plucked down in the suburbs; and now she
professes a hot-house, which I think is a very ill house
too.

ESCALUS How know you that?        65

ELBOW My wife, sir, whom I detest before heaven and
your honour—

ESCALUS How, thy wife?

ELBOW Ay, sir, whom I thank heaven is an honest
woman—        70

ESCALUS Dost thou detest her therefor?

ELBOW I say, sir, I will detest myself also, as well as she,
that this house, if it be not a bawd's house, it is pity
of her life, for it is a naughty house.

ESCALUS How dost thou know that, constable?        75

ELBOW Marry, sir, by my wife, who, if she had been a
woman cardinally given, might have been accused in
fornication, adultery, and all uncleanliness there.

ESCALUS By the woman's means?

ELBOW Ay, sir, by Mistress Overdone's means. But as she
spit in his face, so she defied him.        81

POMPEY (*to Escalus*) Sir, if it please your honour, this is
not so.

ELBOW Prove it before these varlets here, thou honourable
man, prove it.        85

ESCALUS (*to Angelo*) Do you hear how he misplaces?

POMPEY Sir, she came in great with child, and longing—
saving your honour's reverence—for stewed prunes.
Sir, we had but two in the house, which at that very
distant time stood, as it were, in a fruit dish—a dish
of some threepence; your honours have seen such
dishes; they are not china dishes, but very good dishes.

ESCALUS Go to, go to, no matter for the dish, sir.        93

POMPEY No, indeed, sir, not of a pin; you are therein in
the right. But to the point. As I say, this Mistress Elbow,
being, as I say, with child, and being great-bellied, and
longing, as I said, for prunes; and having but two in
the dish, as I said, Master Froth here, this very man,
having eaten the rest, as I said, and, as I say, paying
for them very honestly; for, as you know, Master Froth,
I could not give you threepence again.        101

FROTH No, indeed.

POMPEY Very well. You being, then, if you be remembered,
cracking the stones of the foresaid prunes—

FROTH Ay, so I did indeed.        105

POMPEY Why, very well.—I telling you then, if you be
remembered, that such a one and such a one were past
cure of the thing you wot of, unless they kept very
good diet, as I told you—

FROTH All this is true.        110

POMPEY Why, very well then—

ESCALUS Come, you are a tedious fool. To the purpose.
What was done to Elbow's wife that he hath cause to
complain of? Come me to what was done to her.

POMPEY Sir, your honour cannot come to that yet.        115

ESCALUS No, sir, nor I mean it not.

POMPEY Sir, but you shall come to it, by your honour's
leave. And I beseech you, look into Master Froth here,
sir, a man of fourscore pound a year, whose father died
at Hallowmas—was't not at Hallowmas, Master Froth?

FROTH All Hallow Eve.        121

POMPEY Why, very well. I hope here be truths. He, sir,
sitting, as I say, in a lower chair, sir—'twas in the
Bunch of Grapes, where indeed you have a delight to
sit, have you not?        125

FROTH I have so, because it is an open room, and good
for winter.

POMPEY Why, very well then. I hope here be truths.

ANGELO
This will last out a night in Russia,
When nights are longest there. (*To Escalus*) I'll take
        my leave,        130
And leave you to the hearing of the cause,
Hoping you'll find good cause to whip them all.

ESCALUS
I think no less. Good morrow to your lordship.

*Exit Angelo*

Now, sir, come on, what was done to Elbow's wife,
once more?                                                    135

POMPEY Once, sir? There was nothing done to her once.

ELBOW I beseech you, sir, ask him what this man did to
my wife.

POMPEY I beseech your honour, ask me.

ESCALUS Well, sir, what did this gentleman to her?    140

POMPEY I beseech you, sir, look in this gentleman's face.
Good Master Froth, look upon his honour. 'Tis for a
good purpose. Doth your honour mark his face?

ESCALUS Ay, sir, very well.

POMPEY Nay, I beseech you, mark it well.             145

ESCALUS Well, I do so.

POMPEY Doth your honour see any harm in his face?

ESCALUS Why, no.

POMPEY I'll be supposed upon a book his face is the worst
thing about him. Good, then—if his face be the worst
thing about him, how could Master Froth do the
constable's wife any harm? I would know that of your
honour.                                                          153

ESCALUS He's in the right, constable; what say you to it?

ELBOW First, an it like you, the house is a respected
house; next, this is a respected fellow; and his mistress
is a respected woman.

POMPEY (*to Escalus*) By this hand, sir, his wife is a more
respected person than any of us all.                    159

ELBOW Varlet, thou liest; thou liest, wicked varlet. The
time is yet to come that she was ever respected with
man, woman, or child.

POMPEY Sir, she was respected with him before he married
with her.                                                          164

ESCALUS Which is the wiser here, justice or iniquity? (*To
Elbow*) Is this true?

ELBOW (*to Pompey*) O thou caitiff, O thou varlet, O thou
wicked Hannibal! I respected with her before I was
married to her? (*To Escalus*) If ever I was respected
with her, or she with me, let not your worship think
me the poor Duke's officer. (*To Pompey*) Prove this,
thou wicked Hannibal, or I'll have mine action of
battery on thee.

ESCALUS If he took you a box o'th' ear you might have
your action of slander too.                               175

ELBOW Marry, I thank your good worship for it. What
is't your worship's pleasure I shall do with this wicked
caitiff?

ESCALUS Truly, officer, because he hath some offences in
him that thou wouldst discover if thou couldst, let him
continue in his courses till thou knowest what they
are.                                                                182

ELBOW Marry, I thank your worship for it.—Thou seest,
thou wicked varlet now, what's come upon thee. Thou
art to continue now, thou varlet, thou art to continue.

ESCALUS (*to Froth*) Where were you born, friend?    186

FROTH Here in Vienna, sir.

ESCALUS Are you of fourscore pounds a year?

FROTH Yes, an't please you, sir.

ESCALUS So. (*To Pompey*) What trade are you of, sir?    190

POMPEY A tapster, a poor widow's tapster.

ESCALUS Your mistress's name?

POMPEY Mistress Overdone.

ESCALUS Hath she had any more than one husband?

POMPEY Nine, sir—Overdone by the last.                195

ESCALUS Nine?—Come hither to me, Master Froth. Master
Froth, I would not have you acquainted with tapsters.

They will draw you, Master Froth, and you will hang
them. Get you gone, and let me hear no more of you.

FROTH I thank your worship. For mine own part, I never
come into any room in a tap-house but I am drawn in.

ESCALUS Well, no more of it, Master Froth. Farewell.    202

*Exit Froth*

Come you hither to me, Master Tapster. What's your
name, Master Tapster?

POMPEY Pompey.                                               205

ESCALUS What else?

POMPEY Bum, sir.

ESCALUS Troth, and your bum is the greatest thing about
you; so that, in the beastliest sense, you are Pompey
the Great. Pompey, you are partly a bawd, Pompey,
howsoever you colour it in being a tapster, are you
not? Come, tell me true; it shall be the better for you.

POMPEY Truly, sir, I am a poor fellow that would live.

ESCALUS How would you live, Pompey? By being a bawd?
What do you think of the trade, Pompey? Is it a lawful
trade?                                                              216

POMPEY If the law would allow it, sir.

ESCALUS But the law will not allow it, Pompey; nor it
shall not be allowed in Vienna.

POMPEY Does your worship mean to geld and spay all the
youth of the city?                                            221

ESCALUS No, Pompey.

POMPEY Truly, sir, in my poor opinion they will to't then.
If your worship will take order for the drabs and the
knaves, you need not to fear the bawds.             225

ESCALUS There is pretty orders beginning, I can tell you.
It is but heading and hanging.

POMPEY If you head and hang all that offend that way
but for ten year together, you'll be glad to give out a
commission for more heads. If this law hold in Vienna
ten year, I'll rent the fairest house in it after threepence
a bay. If you live to see this come to pass, say Pompey
told you so.                                                       233

ESCALUS Thank you, good Pompey; and in requital of
your prophecy, hark you. I advise you, let me not find
you before me again upon any complaint whatsoever;
no, not for dwelling where you do. If I do, Pompey, I
shall beat you to your tent, and prove a shrewd Caesar
to you; in plain dealing, Pompey, I shall have you
whipped. So for this time, Pompey, fare you well.    240

POMPEY I thank your worship for your good counsel;
⌈*aside*⌉ but I shall follow it as the flesh and fortune shall
better determine.

Whip me? No, no; let carman whip his jade.

The valiant heart's not whipped out of his trade.    *Exit*

ESCALUS Come hither to me, Master Elbow; come hither,
Master Constable. How long have you been in this
place of constable?

ELBOW Seven year and a half, sir.                       249

ESCALUS I thought, by the readiness in the office, you had
continued in it some time. You say seven years
together?

ELBOW And a half, sir.

ESCALUS Alas, it hath been great pains to you. They do
you wrong to put you so oft upon't. Are there not men
in your ward sufficient to serve it?                      256

ELBOW Faith, sir, few of any wit in such matters. As they
are chosen, they are glad to choose me for them. I do
it for some piece of money, and go through with all.

ESCALUS Look you bring me in the names of some six or
seven, the most sufficient of your parish.            261

ELBOW To your worship's house, sir?

ESCALUS  To my house. Fare you well.
                                        *Exit Elbow with officers*
   What's o'clock, think you?
JUSTICE  Eleven, sir.                                         265
ESCALUS  I pray you home to dinner with me.
JUSTICE  I humbly thank you.
ESCALUS
   It grieves me for the death of Claudio,
   But there's no remedy.
JUSTICE  Lord Angelo is severe.                               270
ESCALUS  It is but needful.
   Mercy is not itself that oft looks so.
   Pardon is still the nurse of second woe.
   But yet, poor Claudio! There is no remedy.
   Come, sir.                                        *Exeunt*

**2.2**  *Enter the Provost and a Servant*
SERVANT
   He's hearing of a cause; he will come straight.
   I'll tell him of you.
PROVOST               Pray you do.          *Exit Servant*
                        I'll know
   His pleasure; maybe he will relent. Alas,
   He hath but as offended in a dream.
   All sects, all ages, smack of this vice; and he       5
   To die for't!
        *Enter Angelo*
ANGELO       Now, what's the matter, Provost?
PROVOST
   Is it your will Claudio shall die tomorrow?
ANGELO
   Did not I tell thee yea? Hadst thou not order?
   Why dost thou ask again?
PROVOST               Lest I might be too rash.
   Under your good correction, I have seen        10
   When after execution judgement hath
   Repented o'er his doom.
ANGELO               Go to; let that be mine.
   Do you your office, or give up your place,
   And you shall well be spared.
PROVOST               I crave your honour's pardon.
   What shall be done, sir, with the groaning Juliet?  15
   She's very near her hour.
ANGELO               Dispose of her
   To some more fitter place, and that with speed.
        *Enter Servant*
SERVANT
   Here is the sister of the man condemned
   Desires access to you.
ANGELO               Hath he a sister?
PROVOST
   Ay, my good lord; a very virtuous maid,        20
   And to be shortly of a sisterhood,
   If not already.
ANGELO       Well, let her be admitted.   *Exit Servant*
   See you the fornicatress be removed.
   Let her have needful but not lavish means.
   There shall be order for't.
        *Enter Lucio and Isabella*
PROVOST               God save your honour.        25
ANGELO
   Stay a little while. (*To Isabella*) You're welcome.
   What's your will?
ISABELLA
   I am a woeful suitor to your honour.
   Please but your honour hear me.
ANGELO               Well, what's your suit?

ISABELLA
   There is a vice that most I do abhor,
   And most desire should meet the blow of justice,   30
   For which I would not plead, but that I must;
   For which I must not plead, but that I am
   At war 'twixt will and will not.
ANGELO               Well, the matter?
ISABELLA
   I have a brother is condemned to die.
   I do beseech you, let it be his fault,        35
   And not my brother.
PROVOST (*aside*)       Heaven give thee moving graces!
ANGELO
   Condemn the fault, and not the actor of it?
   Why, every fault's condemned ere it be done.
   Mine were the very cipher of a function,
   To fine the faults whose fine stands in record,   40
   And let go by the actor.
ISABELLA               O just but severe law!
   I had a brother, then. Heaven keep your honour.
LUCIO (*aside to Isabella*)
   Give't not o'er so. To him again; entreat him.
   Kneel down before him; hang upon his gown.
   You are too cold. If you should need a pin,     45
   You could not with more tame a tongue desire it.
   To him, I say!
ISABELLA (*to Angelo*) Must he needs die?
ANGELO  Maiden, no remedy.
ISABELLA
   Yes, I do think that you might pardon him,        50
   And neither heaven nor man grieve at the mercy.
ANGELO
   I will not do't.
ISABELLA               But can you if you would?
ANGELO
   Look what I will not, that I cannot do.
ISABELLA
   But might you do't, and do the world no wrong,
   If so your heart were touched with that remorse   55
   As mine is to him?
ANGELO  He's sentenced; 'tis too late.
LUCIO (*aside to Isabella*) You are too cold.
ISABELLA
   Too late? Why, no; I that do speak a word
   May call it again. Well, believe this,        60
   No ceremony that to great ones 'longs,
   Not the king's crown, nor the deputed sword,
   The marshal's truncheon, nor the judge's robe,
   Become them with one half so good a grace
   As mercy does.                                   65
   If he had been as you and you as he,
   You would have slipped like him, but he, like you,
   Would not have been so stern.
ANGELO               Pray you be gone.
ISABELLA
   I would to heaven I had your potency,
   And you were Isabel! Should it then be thus?     70
   No; I would tell what 'twere to be a judge,
   And what a prisoner.
LUCIO (*aside to Isabella*)  Ay, touch him; there's the vein.
ANGELO
   Your brother is a forfeit of the law,
   And you but waste your words.
ISABELLA               Alas, alas!
   Why, all the souls that were were forfeit once,   75
   And He that might the vantage best have took
   Found out the remedy. How would you be

If He which is the top of judgement should
But judge you as you are? O, think on that,
And mercy then will breathe within your lips,　80
Like man new made.
ANGELO　　　　　　　Be you content, fair maid.
It is the law, not I, condemn your brother.
Were he my kinsman, brother, or my son,
It should be thus with him. He must die tomorrow.

ISABELLA
Tomorrow? O, that's sudden! Spare him, spare him!
He's not prepared for death. Even for our kitchens　86
We kill the fowl of season. Shall we serve heaven
With less respect than we do minister
To our gross selves? Good good my lord, bethink you:
Who is it that hath died for this offence?　90
There's many have committed it.

LUCIO (aside)　　　　　　Ay, well said.

ANGELO
The law hath not been dead, though it hath slept.
Those many had not dared to do that evil
If the first that did th'edict infringe
Had answered for his deed. Now 'tis awake,　95
Takes note of what is done, and, like a prophet,
Looks in a glass that shows what future evils,
Either raw, or by remissness new conceived
And so in progress to be hatched and born,
Are now to have no successive degrees,　100
But ere they live, to end.

ISABELLA　　　　　　Yet show some pity.

ANGELO
I show it most of all when I show justice,
For then I pity those I do not know
Which a dismissed offence would after gall,
And do him right that, answering one foul wrong,　105
Lives not to act another. Be satisfied.
Your brother dies tomorrow. Be content.

ISABELLA
So you must be the first that gives this sentence,
And he that suffers. O, it is excellent
To have a giant's strength, but it is tyrannous　110
To use it like a giant.

LUCIO (aside to Isabella) That's well said.

ISABELLA Could great men thunder
As Jove himself does, Jove would never be quiet,
For every pelting petty officer　115
Would use his heaven for thunder, nothing but
　　thunder.
Merciful heaven,
Thou rather with thy sharp and sulphurous bolt
Split'st the unwedgeable and gnarlèd oak
Than the soft myrtle. But man, proud man,　120
Dressed in a little brief authority,
Most ignorant of what he's most assured,
His glassy essence, like an angry ape
Plays such fantastic tricks before high heaven
As makes the angels weep, who, with our spleens,　125
Would all themselves laugh mortal.

LUCIO (aside to Isabella)
O, to him, to him, wench! He will relent.
He's coming; I perceive't.

PROVOST (aside)　　　　　　Pray heaven she win him!

ISABELLA
We cannot weigh our brother with ourself.
Great men may jest with saints; 'tis wit in them,　130
But in the less, foul profanation.

LUCIO (aside to Isabella) Thou'rt i'th' right, girl. More o'
　that.

ISABELLA
That in the captain's but a choleric word,
Which in the soldier is flat blasphemy.　135

LUCIO (aside to Isabella) Art advised o' that? More on't.

ANGELO
Why do you put these sayings upon me?

ISABELLA
Because authority, though it err like others,
Hath yet a kind of medicine in itself
That skins the vice o'th' top. Go to your bosom;　140
Knock there, and ask your heart what it doth know
That's like my brother's fault. If it confess
A natural guiltiness, such as is his,
Let it not sound a thought upon your tongue
Against my brother's life.

ANGELO (aside)　　　　　She speaks, and 'tis such sense
That my sense breeds with it. (To Isabella) Fare you
　well.　146

ISABELLA Gentle my lord, turn back.

ANGELO
I will bethink me. Come again tomorrow.

ISABELLA
Hark how I'll bribe you; good my lord, turn back.

ANGELO How, bribe me?　150

ISABELLA
Ay, with such gifts that heaven shall share with you.

LUCIO (aside to Isabella) You had marred all else.

ISABELLA
Not with fond shekels of the tested gold,
Or stones, whose rate are either rich or poor
As fancy values them; but with true prayers,　155
That shall be up at heaven and enter there
Ere sunrise, prayers from preservèd souls,
From fasting maids whose minds are dedicate
To nothing temporal.

ANGELO Well, come to me tomorrow.　160

LUCIO (aside to Isabella) Go to; 'tis well; away.

ISABELLA Heaven keep your honour safe.

ANGELO (aside) Amen;
For I am that way going to temptation,
Where prayer is crossed.

ISABELLA　　　　　　At what hour tomorrow　165
Shall I attend your lordship?

ANGELO　　　　　　At any time fore noon.

ISABELLA
God save your honour.

ANGELO (aside)　　　　From thee; even from thy virtue.
　　　Exeunt Isabella, Lucio, and Provost
What's this? What's this? Is this her fault or mine?
The tempter or the tempted, who sins most, ha?
Not she; nor doth she tempt; but it is I　170
That, lying by the violet in the sun,
Do, as the carrion does, not as the flower,
Corrupt with virtuous season. Can it be
That modesty may more betray our sense
Than woman's lightness? Having waste ground enough,
Shall we desire to raze the sanctuary,　176
And pitch our evils there? O, fie, fie, fie!
What dost thou, or what art thou, Angelo?
Dost thou desire her foully for those things
That make her good? O, let her brother live!　180
Thieves for their robbery have authority,
When judges steal themselves. What, do I love her,
That I desire to hear her speak again,
And feast upon her eyes? What is't I dream on?
O cunning enemy, that, to catch a saint,　185
With saints dost bait thy hook! Most dangerous

Is that temptation that doth goad us on
To sin in loving virtue. Never could the strumpet,
With all her double vigour—art and nature—
Once stir my temper; but this virtuous maid          190
Subdues me quite. Ever till now
When men were fond, I smiled, and wondered how.
                                                      Exit

**2.3**   *Enter ⌜at one door⌝ the Duke, disguised as a friar,*
        *and ⌜at another door⌝ the Provost*

DUKE
Hail to you, Provost!—so I think you are.
PROVOST
I am the Provost. What's your will, good friar?
DUKE
Bound by my charity and my blest order,
I come to visit the afflicted spirits
Here in the prison. Do me the common right             5
To let me see them, and to make me know
The nature of their crimes, that I may minister
To them accordingly.
PROVOST
I would do more than that, if more were needful.
        *Enter Juliet*
Look, here comes one, a gentlewoman of mine,          10
Who, falling in the flaws of her own youth,
Hath blistered her report. She is with child,
And he that got it, sentenced—a young man
More fit to do another such offence
Than die for this.                                     15
DUKE When must he die?
PROVOST As I do think, tomorrow.
    *(To Juliet)* I have provided for you. Stay a while,
And you shall be conducted.
DUKE
Repent you, fair one, of the sin you carry?           20
JULIET
I do, and bear the shame most patiently.
DUKE
I'll teach you how you shall arraign your conscience,
And try your penitence if it be sound
Or hollowly put on.
JULIET I'll gladly learn.                              25
DUKE Love you the man that wronged you?
JULIET
Yes, as I love the woman that wronged him.
DUKE
So then it seems your most offenceful act
Was mutually committed?
JULIET                    Mutually.
DUKE
Then was your sin of heavier kind than his.           30
JULIET
I do confess it and repent it, father.
DUKE
'Tis meet so, daughter. But lest you do repent
As that the sin hath brought you to this shame—
Which sorrow is always toward ourselves, not heaven,
Showing we would not spare heaven as we love it,      35
But as we stand in fear—
JULIET
I do repent me as it is an evil,
And take the shame with joy.
DUKE                    There rest.
Your partner, as I hear, must die tomorrow,
And I am going with instruction to him.               40
Grace go with you. *Benedicite!*                    *Exit*

JULIET
Must die tomorrow? O injurious law,
That respites me a life whose very comfort
Is still a dying horror!
PROVOST                    'Tis pity of him.    *Exeunt*

**2.4**   *Enter Angelo*

ANGELO
When I would pray and think, I think and pray
To several subjects: heaven hath my empty words,
Whilst my invention, hearing not my tongue,
Anchors on Isabel; God in my mouth,
As if I did but only chew his name,                    5
And in my heart the strong and swelling evil
Of my conception. The state whereon I studied
Is like a good thing, being often read,
Grown seared and tedious. Yea, my gravity,
Wherein—let no man hear me—I take pride,             10
Could I with boot change for an idle plume
Which the air beats in vain. O place, O form,
How often dost thou with thy case, thy habit,
Wrench awe from fools, and tie the wiser souls
To thy false seeming! Blood, thou art blood.           15
Let's write 'good angel' on the devil's horn—
'Tis now the devil's crest.
        *Enter Servant*
                        How now? Who's there?
SERVANT One Isabel, a sister, desires access to you.
ANGELO
Teach her the way.                        *Exit Servant*
            O heavens,
Why does my blood thus muster to my heart,             20
Making both it unable for itself,
And dispossessing all my other parts
Of necessary fitness?
So play the foolish throngs with one that swoons—
Come all to help him, and so stop the air              25
By which he should revive—and even so
The general subject to a well-wished king
Quit their own part and, in obsequious fondness,
Crowd to his presence, where their untaught love
Must needs appear offence.
        *Enter Isabella*
                        How now, fair maid?           30
ISABELLA I am come to know your pleasure.
ANGELO *(aside)*
That you might know it would much better please me
Than to demand what 'tis. *(To Isabella)* Your brother
            cannot live.
ISABELLA Even so. Heaven keep your honour.
ANGELO
Yet may he live a while, and it may be                  35
As long as you or I. Yet he must die.
ISABELLA Under your sentence?
ANGELO Yea.
ISABELLA
When, I beseech you?—that in his reprieve,
Longer or shorter, he may be so fitted                 40
That his soul sicken not.
ANGELO
Ha, fie, these filthy vices! It were as good
To pardon him that hath from nature stolen
A man already made, as to remit
Their saucy sweetness that do coin God's image         45
In stamps that are forbid. 'Tis all as easy

Falsely to take away a life true made
As to put metal in restrainèd moulds,
To make a false one.

ISABELLA
'Tis set down so in heaven, but not in earth.    50

ANGELO
Say you so? Then I shall pose you quickly.
Which had you rather: that the most just law
Now took your brother's life, or, to redeem him,
Give up your body to such sweet uncleanness
As she that he hath stained?

ISABELLA             Sir, believe this.    55
I had rather give my body than my soul.

ANGELO
I talk not of your soul. Our compelled sins
Stand more for number than for account.

ISABELLA             How say you?

ANGELO
Nay, I'll not warrant that, for I can speak
Against the thing I say. Answer to this.    60
I now, the voice of the recorded law,
Pronounce a sentence on your brother's life.
Might there not be a charity in sin
To save this brother's life?

ISABELLA         Please you to do't,
I'll take it as a peril to my soul    65
It is no sin at all, but charity.

ANGELO
Pleased you to do't at peril of your soul
Were equal poise of sin and charity.

ISABELLA
That I do beg his life, if it be sin,
Heaven let me bear it. You granting of my suit,    70
If that be sin, I'll make it my morn prayer
To have it added to the faults of mine,
And nothing of your answer.

ANGELO          Nay, but hear me.
Your sense pursues not mine. Either you are ignorant,
Or seem so craftily, and that's not good.    75

ISABELLA
Let me be ignorant, and in nothing good
But graciously to know I am no better.

ANGELO
Thus wisdom wishes to appear most bright
When it doth tax itself: as these black masks
Proclaim an enshield beauty ten times louder    80
Than beauty could, displayed. But mark me.
To be receivèd plain, I'll speak more gross.
Your brother is to die.

ISABELLA So.

ANGELO
And his offence is so, as it appears,    85
Accountant to the law upon that pain.

ISABELLA True.

ANGELO
Admit no other way to save his life—
As I subscribe not that nor any other—
But, in the loss of question, that you his sister,    90
Finding yourself desired of such a person
Whose credit with the judge, or own great place,
Could fetch your brother from the manacles
Of the all-binding law, and that there were
No earthly mean to save him, but that either    95
You must lay down the treasures of your body
To this supposed, or else to let him suffer—
What would you do?

ISABELLA
As much for my poor brother as myself.
That is, were I under the terms of death,    100
Th'impression of keen whips I'd wear as rubies,
And strip myself to death as to a bed
That longing have been sick for, ere I'd yield
My body up to shame.

ANGELO Then must your brother die.    105

ISABELLA And 'twere the cheaper way.
Better it were a brother died at once
Than that a sister, by redeeming him,
Should die for ever.

ANGELO
Were not you then as cruel as the sentence    110
That you have slandered so?

ISABELLA
Ignominy in ransom and free pardon
Are of two houses; lawful mercy
Is nothing kin to foul redemption.

ANGELO
You seemed of late to make the law a tyrant,    115
And rather proved the sliding of your brother
A merriment than a vice.

ISABELLA
O pardon me, my lord. It oft falls out
To have what we would have, we speak not what we
     mean.
I something do excuse the thing I hate    120
For his advantage that I dearly love.

ANGELO
We are all frail.

ISABELLA        Else let my brother die—
If not a federy, but only he,
Owe and succeed thy weakness.

ANGELO          Nay, women are frail too.

ISABELLA
Ay, as the glasses where they view themselves,    125
Which are as easy broke as they make forms.
Women? Help, heaven! Men their creation mar
In profiting by them. Nay, call us ten times frail,
For we are soft as our complexions are,
And credulous to false prints.

ANGELO         I think it well,    130
And from this testimony of your own sex,
Since I suppose we are made to be no stronger
Than faults may shake our frames, let me be bold.
I do arrest your words. Be that you are;
That is, a woman. If you be more, you're none.    135
If you be one, as you are well expressed
By all external warrants, show it now,
By putting on the destined livery.

ISABELLA
I have no tongue but one. Gentle my lord,
Let me entreat you speak the former language.    140

ANGELO Plainly conceive, I love you.

ISABELLA
My brother did love Juliet,
And you tell me that he shall die for it.

ANGELO
He shall not, Isabel, if you give me love.

ISABELLA
I know your virtue hath a licence in't,    145
Which seems a little fouler than it is,
To pluck on others.

ANGELO        Believe me, on mine honour,
My words express my purpose.

ISABELLA
  Ha, little honour to be much believed,
  And most pernicious purpose! Seeming, seeming!  150
  I will proclaim thee, Angelo; look for't.
  Sign me a present pardon for my brother,
  Or with an outstretched throat I'll tell the world aloud
  What man thou art.
ANGELO                    Who will believe thee, Isabel?
  My unsoiled name, th'austereness of my life,  155
  My vouch against you, and my place i'th' state,
  Will so your accusation overweigh
  That you shall stifle in your own report,
  And smell of calumny. I have begun,
  And now I give my sensual race the rein.  160
  Fit thy consent to my sharp appetite,
  Lay by all nicety and prolixious blushes
  That banish what they sue for. Redeem thy brother
  By yielding up thy body to my will,
  Or else he must not only die the death,  165
  But thy unkindness shall his death draw out
  To ling'ring sufferance. Answer me tomorrow,
  Or by the affection that now guides me most,
  I'll prove a tyrant to him. As for you,
  Say what you can, my false o'erweighs your true.  170
                              Exit

ISABELLA
  To whom should I complain? Did I tell this,
  Who would believe me? O perilous mouths,
  That bear in them one and the selfsame tongue
  Either of condemnation or approof,
  Bidding the law make curtsy to their will,  175
  Hooking both right and wrong to th'appetite,
  To follow as it draws! I'll to my brother.
  Though he hath fall'n by prompture of the blood,
  Yet hath he in him such a mind of honour
  That had he twenty heads to tender down  180
  On twenty bloody blocks, he'd yield them up
  Before his sister should her body stoop
  To such abhorred pollution.
  Then Isabel live chaste, and brother die:
  More than our brother is our chastity.  185
  I'll tell him yet of Angelo's request,
  And fit his mind to death, for his soul's rest.  Exit

3.1   *Enter the Duke, disguised as a friar, Claudio, and
       the Provost*
DUKE
  So then you hope of pardon from Lord Angelo?
CLAUDIO
  The miserable have no other medicine
  But only hope.
  I've hope to live, and am prepared to die.
DUKE
  Be absolute for death. Either death or life  5
  Shall thereby be the sweeter. Reason thus with life.
  If I do lose thee, I do lose a thing
  That none but fools would keep. A breath thou art,
  Servile to all the skyey influences
  That dost this habitation where thou keep'st  10
  Hourly afflict. Merely thou art death's fool,
  For him thou labour'st by thy flight to shun,
  And yet runn'st toward him still. Thou art not noble,
  For all th'accommodations that thou bear'st
  Are nursed by baseness. Thou'rt by no means valiant,
  For thou dost fear the soft and tender fork  16

  Of a poor worm. Thy best of rest is sleep,
  And that thou oft provok'st, yet grossly fear'st
  Thy death, which is no more. Thou art not thyself,
  For thou exist'st on many a thousand grains  20
  That issue out of dust. Happy thou art not,
  For what thou hast not, still thou striv'st to get,
  And what thou hast, forget'st. Thou art not certain,
  For thy complexion shifts to strange effects
  After the moon. If thou art rich, thou'rt poor,  25
  For like an ass whose back with ingots bows,
  Thou bear'st thy heavy riches but a journey,
  And death unloads thee. Friend hast thou none,
  For thine own bowels, which do call thee sire,
  The mere effusion of thy proper loins,  30
  Do curse the gout, serpigo, and the rheum,
  For ending thee no sooner. Thou hast nor youth nor
      age,
  But as it were an after-dinner's sleep
  Dreaming on both; for all thy blessèd youth
  Becomes as agèd, and doth beg the alms  35
  Of palsied eld; and when thou art old and rich,
  Thou hast neither heat, affection, limb, nor beauty,
  To make thy riches pleasant. What's in this
  That bears the name of life? Yet in this life
  Lie hid more thousand deaths; yet death we fear  40
  That makes these odds all even.
CLAUDIO               I humbly thank you.
  To sue to live, I find I seek to die,
  And seeking death, find life. Let it come on.
ISABELLA (*within*)
  What ho! Peace here, grace, and good company!
PROVOST
  Who's there? Come in; the wish deserves a welcome.
DUKE (*to Claudio*)
  Dear sir, ere long I'll visit you again.  46
CLAUDIO Most holy sir, I thank you.
       *Enter Isabella*
ISABELLA
  My business is a word or two with Claudio.
PROVOST
  And very welcome. Look, signor, here's your sister.
DUKE
  Provost, a word with you.
PROVOST               As many as you please.  50
       *The Duke and Provost draw aside*
DUKE
  Bring me to hear them speak where I may be
      concealed.
       *They conceal themselves*
CLAUDIO Now sister, what's the comfort?
ISABELLA
  Why, as all comforts are: most good, most good
      indeed.
  Lord Angelo, having affairs to heaven,
  Intends you for his swift ambassador,  55
  Where you shall be an everlasting leiger.
  Therefore your best appointment make with speed.
  Tomorrow you set on.
CLAUDIO            Is there no remedy?
ISABELLA
  None but such remedy as, to save a head,
  To cleave a heart in twain.  60
CLAUDIO But is there any?
ISABELLA Yes, brother, you may live.
  There is a devilish mercy in the judge,

If you'll implore it, that will free your life,
But fetter you till death.
CLAUDIO                    Perpetual durance?    65
ISABELLA
Ay, just, perpetual durance; a restraint,
Though all the world's vastidity you had,
To a determined scope.
CLAUDIO                    But in what nature?
ISABELLA
In such a one as you consenting to't
Would bark your honour from that trunk you bear,
And leave you naked.
CLAUDIO                    Let me know the point.    71
ISABELLA
O, I do fear thee, Claudio, and I quake
Lest thou a feverous life shouldst entertain,
And six or seven winters more respect
Than a perpetual honour. Dar'st thou die?    75
The sense of death is most in apprehension,
And the poor beetle that we tread upon
In corporal sufferance finds a pang as great
As when a giant dies.
CLAUDIO                    Why give you me this shame?
Think you I can a resolution fetch    80
From flow'ry tenderness? If I must die,
I will encounter darkness as a bride,
And hug it in mine arms.
ISABELLA
There spake my brother; there my father's grave
Did utter forth a voice. Yes, thou must die.    85
Thou art too noble to conserve a life
In base appliances. This outward-sainted deputy,
Whose settled visage and deliberate word
Nips youth i'th' head and follies doth enew
As falcon doth the fowl, is yet a devil.    90
His filth within being cast, he would appear
A pond as deep as hell.
CLAUDIO                    The precise Angelo?
ISABELLA
O, 'tis the cunning livery of hell
The damnedest body to invest and cover
In precise guards! Dost thou think, Claudio:    95
If I would yield him my virginity,
Thou might'st be freed!
CLAUDIO                    O heavens, it cannot be!
ISABELLA
Yes, he would give't thee, from this rank offence,
So to offend him still. This night's the time
That I should do what I abhor to name,    100
Or else thou diest tomorrow.
CLAUDIO Thou shalt not do't.
ISABELLA O, were it but my life,
I'd throw it down for your deliverance
As frankly as a pin.
CLAUDIO                    Thanks, dear Isabel.    105
ISABELLA
Be ready, Claudio, for your death tomorrow.
CLAUDIO
Yes. Has he affections in him
That thus can make him bite the law by th' nose
When he would force it? Sure it is no sin,
Or of the deadly seven it is the least.    110
ISABELLA Which is the least?
CLAUDIO
If it were damnable, he being so wise,
Why would he for the momentary trick

Be perdurably fined? O Isabel!
ISABELLA What says my brother?    115
CLAUDIO Death is a fearful thing.
ISABELLA And shamèd life a hateful.
CLAUDIO
Ay, but to die, and go we know not where;
To lie in cold obstruction, and to rot;
This sensible warm motion to become    120
A kneaded clod, and the dilated spirit
To bathe in fiery floods, or to reside
In thrilling region of thick-ribbèd ice;
To be imprisoned in the viewless winds,
And blown with restless violence round about    125
The pendent world; or to be worse than worst
Of those that lawless and incertain thought
Imagine howling—'tis too horrible!
The weariest and most loathèd worldly life
That age, ache, penury, and imprisonment    130
Can lay on nature is a paradise
To what we fear of death.
ISABELLA Alas, alas!
CLAUDIO Sweet sister, let me live.
What sin you do to save a brother's life,    135
Nature dispenses with the deed so far
That it becomes a virtue.
ISABELLA                    O, you beast!
O faithless coward, O dishonest wretch,
Wilt thou be made a man out of my vice?
Is't not a kind of incest to take life    140
From thine own sister's shame? What should I think?
Heaven shield my mother played my father fair,
For such a warpèd slip of wilderness
Ne'er issued from his blood. Take my defiance,
Die, perish! Might but my bending down    145
Reprieve thee from thy fate, it should proceed.
I'll pray a thousand prayers for thy death,
No word to save thee.
CLAUDIO Nay, hear me, Isabel.
ISABELLA O fie, fie, fie!    150
Thy sin's not accidental, but a trade.
Mercy to thee would prove itself a bawd.
'Tis best that thou diest quickly.
              ⌜She parts from Claudio⌝
CLAUDIO O hear me, Isabella.
DUKE (coming forward to Isabella)
Vouchsafe a word, young sister, but one word.    155
ISABELLA What is your will?
DUKE Might you dispense with your leisure, I would by
and by have some speech with you. The satisfaction I
would require is likewise your own benefit.
ISABELLA I have no superfluous leisure; my stay must be
stolen out of other affairs; but I will attend you a while.
DUKE ⌜standing aside with Claudio⌝ Son, I have overheard
what hath passed between you and your sister. Angelo
had never the purpose to corrupt her; only he hath
made an assay of her virtue, to practise his judgement
with the disposition of natures. She, having the truth
of honour in her, hath made him that gracious denial
which he is most glad to receive. I am confessor to
Angelo, and I know this to be true. Therefore prepare
yourself to death. Do not falsify your resolution with
hopes that are fallible. Tomorrow you must die. Go to
your knees and make ready.    172
CLAUDIO Let me ask my sister pardon. I am so out of love
with life that I will sue to be rid of it.

DUKE Hold you there. Farewell.                    175
  [*Claudio joins Isabella*]
  Provost, a word with you.
PROVOST (*coming forward*) What's your will, father?
DUKE That now you are come, you will be gone. Leave
  me a while with the maid. My mind promises with my
  habit no loss shall touch her by my company.    180
PROVOST In good time.                  *Exit* [*with Claudio*]
DUKE The hand that hath made you fair hath made you
  good. The goodness that is cheap in beauty makes
  beauty brief in goodness; but grace, being the soul of
  your complexion, shall keep the body of it ever fair.
  The assault that Angelo hath made to you fortune hath
  conveyed to my understanding; and but that frailty
  hath examples for his falling, I should wonder at
  Angelo. How will you do to content this substitute, and
  to save your brother?                            190
ISABELLA I am now going to resolve him. I had rather
  my brother die by the law than my son should be
  unlawfully born. But O, how much is the good Duke
  deceived in Angelo! If ever he return and I can speak
  to him, I will open my lips in vain, or discover his
  government.                                      196
DUKE That shall not be much amiss. Yet as the matter
  now stands, he will avoid your accusation: he made
  trial of you only. Therefore fasten your ear on my
  advisings. To the love I have in doing good, a remedy
  presents itself. I do make myself believe that you may
  most uprighteously do a poor wronged lady a merited
  benefit, redeem your brother from the angry law, do
  no stain to your own gracious person, and much please
  the absent Duke, if peradventure he shall ever return
  to have hearing of this business.                206
ISABELLA Let me hear you speak farther. I have spirit to
  do anything that appears not foul in the truth of my
  spirit.                                          209
DUKE Virtue is bold, and goodness never fearful. Have you
  not heard speak of Mariana, the sister of Frederick, the
  great soldier who miscarried at sea?
ISABELLA I have heard of the lady, and good words went
  with her name.                                   214
DUKE She should this Angelo have married, was affianced
  to her oath, and the nuptial appointed; between which
  time of the contract and limit of the solemnity, her
  brother Frederick was wrecked at sea, having in that
  perished vessel the dowry of his sister. But mark how
  heavily this befell to the poor gentlewoman. There she
  lost a noble and renowned brother, in his love toward
  her ever most kind and natural; with him, the portion
  and sinew of her fortune, her marriage dowry; with
  both, her combinate husband, this well-seeming
  Angelo.                                          225
ISABELLA Can this be so? Did Angelo so leave her?
DUKE Left her in her tears, and dried not one of them
  with his comfort; swallowed his vows whole, pretending in her discoveries of dishonour; in few,
  bestowed her on her own lamentation, which she yet
  wears for his sake; and he, a marble to her tears, is
  washed with them, but relents not.               232
ISABELLA What a merit were it in death to take this poor
  maid from the world! What corruption in this life, that
  it will let this man live! But how out of this can she
  avail?                                           236
DUKE It is a rupture that you may easily heal, and the
  cure of it not only saves your brother, but keeps you
  from dishonour in doing it.

ISABELLA Show me how, good father.                240
DUKE This forenamed maid hath yet in her the
  continuance of her first affection. His unjust unkindness, that in all reason should have quenched her love,
  hath, like an impediment in the current, made it more
  violent and unruly. Go you to Angelo, answer his
  requiring with a plausible obedience, agree with his
  demands to the point; only refer yourself to this
  advantage: first, that your stay with him may not be
  long; that the time may have all shadow and silence
  in it; and the place answer to convenience. This being
  granted in course, and now follows all. We shall advise
  this wronged maid to stead up your appointment, go
  in your place. If the encounter acknowledge itself
  hereafter, it may compel him to her recompense; and
  hear, by this is your brother saved, your honour
  untainted, the poor Mariana advantaged, and the
  corrupt deputy scaled. The maid will I frame and make
  fit for his attempt. If you think well to carry this, as
  you may, the doubleness of the benefit defends the
  deceit from reproof. What think you of it?       260
ISABELLA The image of it gives me content already, and
  I trust it will grow to a most prosperous perfection.
DUKE It lies much in your holding up. Haste you speedily
  to Angelo. If for this night he entreat you to his bed,
  give him promise of satisfaction. I will presently to
  Saint Luke's; there at the moated grange resides this
  dejected Mariana. At that place call upon me; and
  dispatch with Angelo, that it may be quickly.    268
ISABELLA I thank you for this comfort. Fare you well,
  good father.                                     *Exit*
    *Enter Elbow, Clown, and officers*
ELBOW Nay, if there be no remedy for it but that you will
  needs buy and sell men and women like beasts, we
  shall have all the world drink brown and white bastard.
DUKE O heavens, what stuff is here?               274
POMPEY 'Twas never merry world since, of two usuries,
  the merriest was put down, and the worser allowed by
  order of law, a furred gown to keep him warm—and
  furred with fox on lambskins too, to signify that craft,
  being richer than innocency, stands for the facing.
ELBOW Come your way, sir.—Bless you, good father friar.
DUKE And you, good brother father. What offence hath
  this man made you, sir?                          282
ELBOW Marry, sir, he hath offended the law; and, sir, we
  take him to be a thief, too, sir, for we have found upon
  him, sir, a strange picklock, which we have sent to the
  deputy.                                          286
DUKE (*to Pompey*)
  Fie, sirrah, a bawd, a wicked bawd!
  The evil that thou causest to be done,
  That is thy means to live. Do thou but think
  What 'tis to cram a maw or clothe a back          290
  From such a filthy vice. Say to thyself,
  'From their abominable and beastly touches
  I drink, I eat, array myself, and live'.
  Canst thou believe thy living is a life,
  So stinkingly depending? Go mend, go mend.        295
POMPEY Indeed it does stink in some sort, sir. But yet, sir,
  I would prove—
DUKE
  Nay, if the devil have given thee proofs for sin,
  Thou wilt prove his.—Take him to prison, officer.
  Correction and instruction must both work          300
  Ere this rude beast will profit.

ELBOW He must before the deputy, sir; he has given him
warning. The deputy cannot abide a whoremaster. If
he be a whoremonger and comes before him, he were
as good go a mile on his errand.                    305

DUKE
That we were all as some would seem to be—
Free from our faults, or faults from seeming free.

ELBOW His neck will come to your waist: a cord, sir.

*Enter Lucio*

POMPEY I spy comfort, I cry bail. Here's a gentleman, and
a friend of mine.                                   310

LUCIO How now, noble Pompey? What, at the wheels of
Caesar? Art thou led in triumph? What, is there none
of Pygmalion's images newly made woman to be had
now, for putting the hand in the pocket and extracting
clutched? What reply, ha? What sayst thou to this
tune, matter, and method? Is't not drowned i'th' last
rain, ha? What sayst thou, trot? Is the world as it was,
man? Which is the way? Is it sad and few words? Or
how? The trick of it?

DUKE Still thus and thus; still worse!                320

LUCIO How doth my dear morsel thy mistress? Procures
she still, ha?

POMPEY Troth, sir, she hath eaten up all her beef, and
she is herself in the tub.                          324

LUCIO Why, 'tis good, it is the right of it, it must be so.
Ever your fresh whore and your powdered bawd; an
unshunned consequence, it must be so. Art going to
prison, Pompey?

POMPEY Yes, faith, sir.                               329

LUCIO Why 'tis not amiss, Pompey. Farewell. Go; say I
sent thee thither. For debt, Pompey, or how?

ELBOW For being a bawd, for being a bawd.

LUCIO Well then, imprison him. If imprisonment be the
due of a bawd, why, 'tis his right. Bawd is he doubtless,
and of antiquity too—bawd born. Farewell, good
Pompey. Commend me to the prison, Pompey. You will
turn good husband now, Pompey; you will keep the
house.                                              338

POMPEY I hope, sir, your good worship will be my bail?

LUCIO No, indeed, will I not, Pompey; it is not the wear.
I will pray, Pompey, to increase your bondage. If you
take it not patiently, why, your mettle is the more.
Adieu, trusty Pompey.—Bless you, friar.

DUKE And you.

LUCIO Does Bridget paint still, Pompey, ha?          345

ELBOW (*to Pompey*) Come your ways, sir, come.

POMPEY (*to Lucio*) You will not bail me then, sir?

LUCIO Then, Pompey, nor now.—What news abroad,
friar, what news?

ELBOW (*to Pompey*) Come your ways, sir, come.       350

LUCIO Go to kennel, Pompey, go.

*Exeunt Elbow, Pompey, and officers*

What news, friar, of the Duke?

DUKE I know none. Can you tell me of any?

LUCIO Some say he is with the Emperor of Russia; other
some, he is in Rome. But where is he, think you?    355

DUKE I know not where; but wheresoever, I wish him
well.

LUCIO It was a mad, fantastical trick of him to steal from
the state, and usurp the beggary he was never born
to. Lord Angelo dukes it well in his absence; he puts
transgression to't.                                 361

DUKE He does well in't.

LUCIO A little more lenity to lechery would do no harm
in him. Something too crabbed that way, friar.

DUKE It is too general a vice, and severity must cure it.

LUCIO Yes, in good sooth, the vice is of a great kindred,
it is well allied. But it is impossible to extirp it quite,
friar, till eating and drinking be put down. They say
this Angelo was not made by man and woman, after
this downright way of creation. Is it true, think you?

DUKE How should he be made, then?                    371

LUCIO Some report a sea-maid spawned him, some that
he was begot between two stockfishes. But it is certain
that when he makes water his urine is congealed ice;
that I know to be true. And he is a motion ungenerative;
that's infallible.                                  376

DUKE You are pleasant, sir, and speak apace.

LUCIO Why, what a ruthless thing is this in him, for the
rebellion of a codpiece to take away the life of a man!
Would the Duke that is absent have done this? Ere he
would have hanged a man for the getting a hundred
bastards, he would have paid for the nursing a
thousand. He had some feeling of the sport, he knew
the service, and that instructed him to mercy.      384

DUKE I never heard the absent Duke much detected for
women; he was not inclined that way.

LUCIO O sir, you are deceived.

DUKE 'Tis not possible.

LUCIO Who, not the Duke? Yes, your beggar of fifty; and
his use was to put a ducat in her clack-dish. The Duke
had crochets in him. He would be drunk too, that let
me inform you.                                      392

DUKE You do him wrong, surely.

LUCIO Sir, I was an inward of his. A shy fellow was the
Duke, and I believe I know the cause of his withdrawing.

DUKE What, I prithee, might be the cause?            396

LUCIO No, pardon, 'tis a secret must be locked within the
teeth and the lips. But this I can let you understand.
The greater file of the subject held the Duke to be wise.

DUKE Wise? Why, no question but he was.             400

LUCIO A very superficial, ignorant, unweighing fellow.

DUKE Either this is envy in you, folly, or mistaking. The
very stream of his life, and the business he hath helmed,
must, upon a warranted need, give him a better
proclamation. Let him be but testimonied in his own
bringings-forth, and he shall appear to the envious a
scholar, a statesman, and a soldier. Therefore you speak
unskilfully, or, if your knowledge be more, it is much
darkened in your malice.

LUCIO Sir, I know him and I love him.                410

DUKE Love talks with better knowledge, and knowledge
with dearer love.

LUCIO Come, sir, I know what I know.

DUKE I can hardly believe that, since you know not what
you speak. But if ever the Duke return, as our prayers
are he may, let me desire you to make your answer
before him. If it be honest you have spoke, you have
courage to maintain it. I am bound to call upon you;
and I pray you, your name?                          419

LUCIO Sir, my name is Lucio, well known to the Duke.

DUKE He shall know you better, sir, if I may live to report
you.

LUCIO I fear you not.

DUKE O, you hope the Duke will return no more, or you
imagine me too unhurtful an opposite. But indeed I
can do you little harm; you'll forswear this again.  426

LUCIO I'll be hanged first. Thou art deceived in me, friar.
But no more of this. Canst thou tell if Claudio die
tomorrow or no?

DUKE Why should he die, sir?                                    430

LUCIO Why? For filling a bottle with a tundish. I would the Duke we talk of were returned again; this ungenitured agent will unpeople the province with continency. Sparrows must not build in his house-eaves, because they are lecherous. The Duke yet would have dark deeds darkly answered: he would never bring them to light. Would he were returned. Marry, this Claudio is condemned for untrussing. Farewell, good friar. I prithee pray for me. The Duke, I say to thee again, would eat mutton on Fridays. He's not past it yet, and, I say to thee, he would mouth with a beggar, though she smelt brown bread and garlic. Say that I said so. Farewell.                                    *Exit*

DUKE
No might nor greatness in mortality
Can censure scape; back-wounding calumny              445
The whitest virtue strikes. What king so strong
Can tie the gall up in the slanderous tongue?
    *Enter Escalus, the Provost, and Mistress Overdone*
But who comes here?

ESCALUS (*to the Provost*) Go, away with her to prison.

MISTRESS OVERDONE Good my lord, be good to me. Your honour is accounted a merciful man, good my lord.

ESCALUS Double and treble admonition, and still forfeit in the same kind! This would make mercy swear and play the tyrant.                                          454

PROVOST A bawd of eleven years' continuance, may it please your honour.

MISTRESS OVERDONE My lord, this is one Lucio's information against me. Mistress Kate Keepdown was with child by him in the Duke's time; he promised her marriage. His child is a year and a quarter old come Philip and Jacob. I have kept it myself; and see how he goes about to abuse me.                            462

ESCALUS That fellow is a fellow of much licence. Let him be called before us. Away with her to prison. Go to, no more words. Provost, my brother Angelo will not be altered; Claudio must die tomorrow. Let him be furnished with divines, and have all charitable preparation. If my brother wrought by my pity, it should not be so with him.                          469

PROVOST So please you, this friar hath been with him and advised him for th'entertainment of death.
    ⌈*Exeunt Provost and Mistress Overdone*⌉

ESCALUS Good even, good father.

DUKE Bliss and goodness on you.

ESCALUS Of whence are you?

DUKE
Not of this country, though my chance is now       475
To use it for my time. I am a brother
Of gracious order, late come from the See
In special business from his Holiness.

ESCALUS What news abroad i'th' world?

DUKE None, but that there is so great a fever on goodness that the dissolution of it must cure it. Novelty is only in request, and it is as dangerous to be aged in any kind of course as it is virtuous to be inconstant in any undertaking. There is scarce truth enough alive to make societies secure, but security enough to make fellowships accursed. Much upon this riddle runs the wisdom of the world. This news is old enough, yet it is every day's news. I pray you, sir, of what disposition was the Duke?                                          489

ESCALUS One that, above all other strifes, contended especially to know himself.

DUKE What pleasure was he given to?

ESCALUS Rather rejoicing to see another merry than merry at anything which professed to make him rejoice; a gentleman of all temperance. But leave we him to his events, with a prayer they may prove prosperous, and let me desire to know how you find Claudio prepared. I am made to understand that you have lent him visitation.                                          499

DUKE He professes to have received no sinister measure from his judge, but most willingly humbles himself to the determination of justice. Yet had he framed to himself, by the instruction of his frailty, many deceiving promises of life, which I, by my good leisure, have discredited to him; and now is he resolved to die.   505

ESCALUS You have paid the heavens your function, and the prisoner the very debt of your calling. I have laboured for the poor gentleman to the extremest shore of my modesty, but my brother-justice have I found so severe that he hath forced me to tell him he is indeed Justice.                                           511

DUKE If his own life answer the straitness of his proceeding, it shall become him well; wherein if he chance to fail, he hath sentenced himself.

ESCALUS I am going to visit the prisoner. Fare you well.

DUKE Peace be with you.                          *Exit Escalus*
He who the sword of heaven will bear          517
Should be as holy as severe,
Pattern in himself to know,
Grace to stand, and virtue go,                 520
More nor less to others paying
Than by self-offences weighing.
Shame to him whose cruel striking
Kills for faults of his own liking!
Twice treble shame on Angelo,                  525
To weed my vice, and let his grow!
O, what may man within him hide,
Though angel on the outward side!
How may likeness made in crimes
Make my practice on the times                  530
To draw with idle spiders' strings
Most ponderous and substantial things?
Craft against vice I must apply.
With Angelo tonight shall lie
His old betrothèd but despisèd.                535
So disguise shall, by th' disguisèd,
Pay with falsehood false exacting,
And perform an old contracting.               *Exit*

**4.1**    *Mariana ⌈discovered⌉ with a Boy singing*

BOY
Take, O take those lips away
    That so sweetly were forsworn,
And those eyes, the break of day
    Lights that do mislead the morn;
But my kisses bring again, bring again,        5
Seals of love, though sealed in vain, sealed in vain.
    *Enter the Duke, disguised as a friar*

MARIANA
Break off thy song, and haste thee quick away.
Here comes a man of comfort, whose advice
Hath often stilled my brawling discontent.    *Exit Boy*
I cry you mercy, sir, and well could wish      10
You had not found me here so musical.
Let me excuse me, and believe me so:
My mirth it much displeased, but pleased my woe.

DUKE
'Tis good; though music oft hath such a charm
To make bad good, and good provoke to harm.          15
I pray you tell me, hath anybody enquired for me here
today? Much upon this time have I promised here to
meet.

MARIANA You have not been enquired after; I have sat
here all day.          20
*Enter Isabella*

DUKE I do constantly believe you; the time is come even
now. I shall crave your forbearance a little. Maybe I
will call upon you anon, for some advantage to yourself.

MARIANA I am always bound to you.          *Exit*

DUKE Very well met, and welcome.          25
What is the news from this good deputy?

ISABELLA
He hath a garden circummured with brick,
Whose western side is with a vineyard backed;
And to that vineyard is a plankèd gate,
That makes his opening with this bigger key.          30
This other doth command a little door
Which from the vineyard to the garden leads.
There have I made my promise
Upon the heavy middle of the night
To call upon him.          35

DUKE
But shall you on your knowledge find this way?

ISABELLA
I have ta'en a due and wary note upon't.
With whispering and most guilty diligence,
In action all of precept, he did show me
The way twice o'er.

DUKE          Are there no other tokens          40
Between you 'greed concerning her observance?

ISABELLA
No, none, but only a repair i'th' dark,
And that I have possessed him my most stay
Can be but brief, for I have made him know
I have a servant comes with me along          45
That stays upon me, whose persuasion is
I come about my brother.

DUKE          'Tis well borne up.
I have not yet made known to Mariana
A word of this.—What ho, within! Come forth!
*Enter Mariana*

(*To Mariana*) I pray you be acquainted with this maid.
She comes to do you good.

ISABELLA          I do desire the like.          51

DUKE (*to Mariana*)
Do you persuade yourself that I respect you?

MARIANA
Good friar, I know you do, and so have found it.

DUKE
Take then this your companion by the hand,
Who hath a story ready for your ear.          55
I shall attend your leisure; but make haste,
The vaporous night approaches.

MARIANA (*to Isabella*)          Will't please you walk aside?
⌜*Exeunt Mariana and Isabella*⌝

DUKE
O place and greatness, millions of false eyes
Are stuck upon thee; volumes of report
Run with their false and most contrarious quest          60
Upon thy doings; thousand escapes of wit
Make thee the father of their idle dream,

And rack thee in their fancies.
⌜*Enter Mariana and Isabella*⌝
Welcome. How agreed?

ISABELLA
She'll take the enterprise upon her, father,
If you advise it.

DUKE          It is not my consent,          65
But my entreaty too.

ISABELLA (*to Mariana*) Little have you to say
When you depart from him but, soft and low,
'Remember now my brother'.

MARIANA          Fear me not.

DUKE
Nor, gentle daughter, fear you not at all.
He is your husband on a pre-contract.          70
To bring you thus together 'tis no sin,
Sith that the justice of your title to him
Doth flourish the deceit. Come, let us go.
Our corn's to reap, for yet our tilth's to sow.          *Exeunt*

**4.2**     *Enter the Provost and Pompey*

PROVOST Come hither, sirrah. Can you cut off a man's
head?

POMPEY If the man be a bachelor, sir, I can; but if he be
a married man, he's his wife's head, and I can never
cut off a woman's head.          5

PROVOST Come, sir, leave me your snatches, and yield me
a direct answer. Tomorrow morning are to die Claudio
and Barnardine. Here is in our prison a common
executioner, who in his office lacks a helper. If you will
take it on you to assist him, it shall redeem you from
your gyves; if not, you shall have your full time of
imprisonment, and your deliverance with an unpitied
whipping; for you have been a notorious bawd.          13

POMPEY Sir, I have been an unlawful bawd time out of
mind, but yet I will be content to be a lawful hangman.
I would be glad to receive some instruction from my
fellow partner.

PROVOST What ho, Abhorson! Where's Abhorson there?
*Enter Abhorson*

ABHORSON Do you call, sir?          19

PROVOST Sirrah, here's a fellow will help you tomorrow
in your execution. If you think it meet, compound with
him by the year, and let him abide here with you; if
not, use him for the present, and dismiss him. He
cannot plead his estimation with you; he hath been a
bawd.          25

ABHORSON A bawd, sir? Fie upon him, he will discredit
our mystery.

PROVOST Go to, sir, you weigh equally; a feather will turn
the scale.          *Exit*

POMPEY Pray, sir, by your good favour—for surely, sir, a
good favour you have, but that you have a hanging
look—do you call, sir, your occupation a mystery?          32

ABHORSON Ay, sir, a mystery.

POMPEY Painting, sir, I have heard say is a mystery; and
your whores, sir, being members of my occupation,
using painting, do prove my occupation a mystery. But
what mystery there should be in hanging, if I should
be hanged I cannot imagine.

ABHORSON Sir, it is a mystery.

POMPEY Proof.          40

ABHORSON Every true man's apparel fits your thief—

POMPEY If it be too little for your thief, your true man
thinks it big enough. If it be too big for your thief, your

thief thinks it little enough. So every true man's apparel
fits your thief.                                                    45
*Enter Provost*
PROVOST Are you agreed?
POMPEY Sir, I will serve him, for I do find your hangman
is a more penitent trade than your bawd—he doth
oftener ask forgiveness.
PROVOST (*to Abhorson*) You, sirrah, provide your block and
your axe tomorrow, four o'clock.                                   51
ABHORSON (*to Pompey*) Come on, bawd, I will instruct thee
in my trade. Follow.
POMPEY I do desire to learn, sir, and I hope, if you have
occasion to use me for your own turn, you shall find
me yare. For truly, sir, for your kindness I owe you a
good turn.                                                         57
PROVOST
Call hither Barnardine and Claudio.
                              *Exeunt Abhorson and Pompey*
Th'one has my pity; not a jot the other,
Being a murderer, though he were my brother.                       60
*Enter Claudio*
Look, here's the warrant, Claudio, for thy death.
'Tis now dead midnight, and by eight tomorrow
Thou must be made immortal. Where's Barnardine?
CLAUDIO
As fast locked up in sleep as guiltless labour
When it lies starkly in the travailer's bones.                     65
He will not wake.
PROVOST                Who can do good on him?
Well, go prepare yourself.
*Knocking within*
                              But hark, what noise?
Heaven give your spirits comfort!          *Exit Claudio*
⌜*Knocking again*⌝
                                          By and by!
I hope it is some pardon or reprieve
For the most gentle Claudio.
*Enter the Duke, disguised as a friar*
                              Welcome, father.                     70
DUKE
The best and wholesom'st spirits of the night
Envelop you, good Provost! Who called here of late?
PROVOST None since the curfew rung.
DUKE Not Isabel?
PROVOST No.                                                        75
DUKE They will then, ere't be long.
PROVOST What comfort is for Claudio?
DUKE There's some in hope.
PROVOST It is a bitter deputy.
DUKE
Not so, not so; his life is paralleled                             80
Even with the stroke and line of his great justice.
He doth with holy abstinence subdue
That in himself which he spurs on his power
To qualify in others. Were he mealed with that
Which he corrects, then were he tyrannous;                         85
But this being so, he's just.
*Knocking within*
                              Now are they come.
⌜*The Provost goes to a door*⌝
This is a gentle Provost. Seldom when
The steelèd jailer is the friend of men.
*Knocking within*
(*To Provost*) How now, what noise? That spirit's
    possessed with haste
That wounds th'unlisting postern with these strokes.

PROVOST
There he must stay until the officer                               91
Arise to let him in. He is called up.
DUKE
Have you no countermand for Claudio yet,
But he must die tomorrow?
PROVOST                        None, sir, none.
DUKE
As near the dawning, Provost, as it is,                            95
You shall hear more ere morning.
PROVOST                            Happily
You something know, yet I believe there comes
No countermand. No such example have we;
Besides, upon the very siege of justice
Lord Angelo hath to the public ear                                 100
Professed the contrary.
*Enter a Messenger*
This is his lordship's man.
⌜DUKE⌝ And here comes Claudio's pardon.
MESSENGER (*giving a paper to Provost*) My lord hath sent
you this note, and by me this further charge: that you
swerve not from the smallest article of it, neither in
time, matter, or other circumstance. Good morrow;
for, as I take it, it is almost day.
PROVOST I shall obey him.          *Exit Messenger*
DUKE (*aside*)
This is his pardon, purchased by such sin                          110
For which the pardoner himself is in.
Hence hath offence his quick celerity,
When it is borne in high authority.
When vice makes mercy, mercy's so extended
That for the fault's love is th'offender friended.—                115
Now sir, what news?
PROVOST I told you: Lord Angelo, belike thinking me
remiss in mine office, awakens me with this unwonted
putting-on; methinks strangely, for he hath not used
it before.                                                         120
DUKE Pray you let's hear.
⌜PROVOST⌝ (*reading the letter*) 'Whatsoever you may hear
to the contrary, let Claudio be executed by four of the
clock, and in the afternoon Barnardine. For my better
satisfaction, let me have Claudio's head sent me by five.
Let this be duly performed, with a thought that more
depends on it than we must yet deliver. Thus fail not
to do your office, as you will answer it at your peril.'
What say you to this, sir?                                         129
DUKE What is that Barnardine, who is to be executed in
th'afternoon?
PROVOST A Bohemian born, but here nursed up and bred;
one that is a prisoner nine years old.
DUKE How came it that the absent Duke had not either
delivered him to his liberty or executed him? I have
heard it was ever his manner to do so.                             136
PROVOST His friends still wrought reprieves for him; and
indeed his fact, till now in the government of Lord
Angelo, came not to an undoubtful proof.
DUKE It is now apparent?                                           140
PROVOST Most manifest, and not denied by himself.
DUKE Hath he borne himself penitently in prison? How
seems he to be touched?
PROVOST A man that apprehends death no more dreadfully
but as a drunken sleep; careless, reckless, and fearless
of what's past, present, or to come; insensible of
mortality, and desperately mortal.                                 147
DUKE He wants advice.

PROVOST He will hear none. He hath evermore had the liberty of the prison. Give him leave to escape hence, he would not. Drunk many times a day, if not many days entirely drunk. We have very oft awaked him as if to carry him to execution, and showed him a seeming warrant for it; it hath not moved him at all.          154

DUKE More of him anon. There is written in your brow, Provost, honesty and constancy. If I read it not truly, my ancient skill beguiles me. But in the boldness of my cunning, I will lay myself in hazard. Claudio, whom here you have warrant to execute, is no greater forfeit to the law than Angelo who hath sentenced him. To make you understand this in a manifested effect, I crave but four days' respite, for the which you are to do me both a present and a dangerous courtesy.

PROVOST Pray sir, in what?

DUKE In the delaying death.          165

PROVOST Alack, how may I do it, having the hour limited, and an express command under penalty to deliver his head in the view of Angelo? I may make my case as Claudio's to cross this in the smallest.

DUKE By the vow of mine order, I warrant you, if my instructions may be your guide, let this Barnardine be this morning executed, and his head borne to Angelo.

PROVOST Angelo hath seen them both, and will discover the favour.          174

DUKE O, death's a great disguiser, and you may add to it. Shave the head and tie the beard, and say it was the desire of the penitent to be so bared before his death; you know the course is common. If anything fall to you upon this more than thanks and good fortune, by the saint whom I profess, I will plead against it with my life.          181

PROVOST Pardon me, good father, it is against my oath.

DUKE Were you sworn to the Duke or to the deputy?

PROVOST To him and to his substitutes.

DUKE You will think you have made no offence if the Duke avouch the justice of your dealing?          186

PROVOST But what likelihood is in that?

DUKE Not a resemblance, but a certainty. Yet since I see you fearful, that neither my coat, integrity, nor persuasion can with ease attempt you, I will go further than I meant, to pluck all fears out of you. (*Showing a letter*) Look you, sir, here is the hand and seal of the Duke. You know the character, I doubt not, and the signet is not strange to you?

PROVOST I know them both.          195

DUKE The contents of this is the return of the Duke. You shall anon over-read it at your pleasure, where you shall find within these two days he will be here. This is a thing that Angelo knows not, for he this very day receives letters of strange tenor, perchance of the Duke's death, perchance entering into some monastery; but by chance nothing of what is writ. Look, th'unfolding star calls up the shepherd. Put not yourself into amazement how these things should be. All difficulties are but easy when they are known. Call your executioner, and off with Barnardine's head. I will give him a present shrift, and advise him for a better place. Yet you are amazed; but this shall absolutely resolve you. Come away, it is almost clear dawn.          *Exeunt*

**4.3**    *Enter Pompey*

POMPEY I am as well acquainted here as I was in our house of profession. One would think it were Mistress Overdone's own house, for here be many of her old customers. First, here's young Master Rash; he's in for a commodity of brown paper and old ginger, nine score and seventeen pounds, of which he made five marks ready money. Marry, then ginger was not much in request, for the old women were all dead. Then is there here one Master Caper, at the suit of Master Threepile the mercer, for some four suits of peach-coloured satin, which now peaches him a beggar. Then have we here young Dizzy, and young Master Deepvow, and Master Copperspur and Master Starve-lackey the rapier and dagger man, and young Drop-hair that killed lusty Pudding, and Master Forthright the tilter, and brave Master Shoe-tie the great traveller, and wild Half-can that stabbed Pots, and I think forty more, all great doers in our trade, and are now 'for the Lord's sake'.

*Enter Abhorson*

ABHORSON Sirrah, bring Barnardine hither.          19

POMPEY Master Barnardine! You must rise and be hanged, Master Barnardine!

ABHORSON What ho, Barnardine!

BARNARDINE (*within*) A pox o' your throats! Who makes that noise there? What are you?          24

POMPEY Your friends, sir; the hangman. You must be so good, sir, to rise and be put to death.

BARNARDINE Away, you rogue, away! I am sleepy.

ABHORSON Tell him he must awake, and that quickly too.

POMPEY Pray, Master Barnardine, awake till you are executed, and sleep afterwards.          30

ABHORSON Go in to him and fetch him out.

POMPEY He is coming, sir, he is coming. I hear his straw rustle.

ABHORSON Is the axe upon the block, sirrah?

POMPEY Very ready, sir.          35

*Enter Barnardine*

BARNARDINE How now, Abhorson, what's the news with you?

ABHORSON Truly, sir, I would desire you to clap into your prayers, for, look you, the warrant's come.

BARNARDINE You rogue, I have been drinking all night. I am not fitted for't.          41

POMPEY O, the better, sir; for he that drinks all night, and is hanged betimes in the morning, may sleep the sounder all the next day.

*Enter the Duke, disguised as a friar*

ABHORSON (*to Barnardine*) Look you, sir, here comes your ghostly father. Do we jest now, think you?          46

DUKE (*to Barnardine*) Sir, induced by my charity, and hearing how hastily you are to depart, I am come to advise you, comfort you, and pray with you.

BARNARDINE Friar, not I. I have been drinking hard all night, and I will have more time to prepare me, or they shall beat out my brains with billets. I will not consent to die this day, that's certain.

DUKE
O sir, you must; and therefore, I beseech you,
Look forward on the journey you shall go.          55

BARNARDINE I swear I will not die today, for any man's persuasion.

DUKE But hear you—

BARNARDINE Not a word. If you have anything to say to me, come to my ward, for thence will not I today.          60
*Exit*

DUKE
Unfit to live or die. O gravel heart!
After him, fellows; bring him to the block.
*Exeunt Abhorson and Pompey*

*Enter Provost*

PROVOST
    Now, sir, how do you find the prisoner?
DUKE
    A creature unprepared, unmeet for death;
    And to transport him in the mind he is                          65
    Were damnable.
PROVOST                Here in the prison, father,
    There died this morning of a cruel fever
    One Ragusine, a most notorious pirate,
    A man of Claudio's years, his beard and head
    Just of his colour. What if we do omit                          70
    This reprobate till he were well inclined,
    And satisfy the deputy with the visage
    Of Ragusine, more like to Claudio?
DUKE
    O, 'tis an accident that heaven provides.
    Dispatch it presently; the hour draws on                        75
    Prefixed by Angelo. See this be done,
    And sent according to command, whiles I
    Persuade this rude wretch willingly to die.
PROVOST
    This shall be done, good father, presently.
    But Barnardine must die this afternoon;                         80
    And how shall we continue Claudio,
    To save me from the danger that might come
    If he were known alive?
DUKE                Let this be done:
    Put them in secret holds, both Barnardine and Claudio.
    Ere twice the sun hath made his journal greeting                85
    To yonder generation, you shall find
    Your safety manifested.
PROVOST                I am your free dependant.
DUKE
    Quick, dispatch, and send the head to Angelo.
                                                    *Exit Provost*
    Now will I write letters to Angelo—
    The Provost, he shall bear them—whose contents                 90
    Shall witness to him I am near at home,
    And that by great injunctions I am bound
    To enter publicly. Him I'll desire
    To meet me at the consecrated fount
    A league below the city, and from thence,                       95
    By cold gradation and well-balanced form,
    We shall proceed with Angelo.
        *Enter the Provost, with Ragusine's head*
PROVOST
    Here is the head; I'll carry it myself.
DUKE
    Convenient is it. Make a swift return,
    For I would commune with you of such things                    100
    That want no ear but yours.
PROVOST I'll make all speed.                              *Exit*
ISABELLA (*within*) Peace, ho, be here!
DUKE
    The tongue of Isabel. She's come to know
    If yet her brother's pardon be come hither;                    105
    But I will keep her ignorant of her good,
    To make her heavenly comforts of despair
    When it is least expected.
ISABELLA ⌈*within*⌉                Ho, by your leave!
        ⌈*Enter Isabella*⌉
DUKE
    Good morning to you, fair and gracious daughter.
ISABELLA
    The better, given me by so holy a man.                         110
    Hath yet the deputy sent my brother's pardon?

DUKE
    He hath released him, Isabel, from the world.
    His head is off and sent to Angelo.
ISABELLA
    Nay, but it is not so.
DUKE                It is no other.
    Show your wisdom, daughter, in your close patience.
ISABELLA
    O, I will to him and pluck out his eyes!                       116
DUKE
    You shall not be admitted to his sight.
ISABELLA (*weeping*)
    Unhappy Claudio! Wretched Isabel!
    Injurious world! Most damnèd Angelo!
DUKE
    This nor hurts him, nor profits you a jot.                     120
    Forbear it, therefore; give your cause to heaven.
    Mark what I say, which you shall find
    By every syllable a faithful verity.
    The Duke comes home tomorrow—nay, dry your
        eyes—
    One of our convent, and his confessor,                         125
    Gives me this instance. Already he hath carried
    Notice to Escalus and Angelo,
    Who do prepare to meet him at the gates,
    There to give up their power. If you can pace your
        wisdom
    In that good path that I would wish it go,                     130
    And you shall have your bosom on this wretch,
    Grace of the Duke, revenges to your heart,
    And general honour.
ISABELLA                I am directed by you.
DUKE
    This letter, then, to Friar Peter give.
    'Tis that he sent me of the Duke's return.                     135
    Say by this token I desire his company
    At Mariana's house tonight. Her cause and yours
    I'll perfect him withal, and he shall bring you
    Before the Duke, and to the head of Angelo
    Accuse him home and home. For my poor self,         140
    I am combinèd by a sacred vow,
    And shall be absent. (*Giving the letter*) Wend you with
        this letter.
    Command these fretting waters from your eyes
    With a light heart. Trust not my holy order
    If I pervert your course.
        *Enter Lucio*
                        Who's here?
LUCIO                        Good even.             145
    Friar, where's the Provost?
DUKE                        Not within, sir.
LUCIO O pretty Isabella, I am pale at mine heart to see
    thine eyes so red. Thou must be patient. I am fain to
    dine and sup with water and bran; I dare not for my
    head fill my belly; one fruitful meal would set me to't.
    But they say the Duke will be here tomorrow. By my
    troth, Isabel, I loved thy brother. If the old fantastical
    Duke of dark corners had been at home, he had lived.
                                        ⌈*Exit Isabella*⌉
DUKE Sir, the Duke is marvellous little beholden to your
    reports; but the best is, he lives not in them.         155
LUCIO Friar, thou knowest not the Duke so well as I do.
    He's a better woodman than thou tak'st him for.
DUKE Well, you'll answer this one day. Fare ye well.
LUCIO Nay, tarry, I'll go along with thee. I can tell thee
    pretty tales of the Duke.                               160

DUKE You have told me too many of him already, sir, if
they be true; if not true, none were enough.

LUCIO I was once before him for getting a wench with
child.

DUKE Did you such a thing?             165

LUCIO Yes, marry, did I; but I was fain to forswear it.
They would else have married me to the rotten medlar.

DUKE Sir, your company is fairer than honest. Rest you
well.

LUCIO By my troth, I'll go with thee to the lane's end. If
bawdy talk offend you, we'll have very little of it. Nay,
friar, I am a kind of burr; I shall stick.      *Exeunt*

**4.4**    *Enter Angelo and Escalus*

ESCALUS Every letter he hath writ hath disvouched other.

ANGELO In most uneven and distracted manner. His
actions show much like to madness. Pray heaven his
wisdom be not tainted. And why meet him at the gates,
and redeliver our authorities there?       5

ESCALUS I guess not.

ANGELO And why should we proclaim it in an hour before
his entering, that if any crave redress of injustice, they
should exhibit their petitions in the street?     9

ESCALUS He shows his reason for that—to have a dispatch
of complaints, and to deliver us from devices hereafter,
which shall then have no power to stand against us.

ANGELO
Well, I beseech you let it be proclaimed.
Betimes i'th' morn I'll call you at your house.
Give notice to such men of sort and suit       15
As are to meet him.

ESCALUS I shall, sir. Fare you well.

ANGELO Good night.             *Exit Escalus*
This deed unshapes me quite, makes me unpregnant
And dull to all proceedings. A deflowered maid,   20
And by an eminent body that enforced
The law against it! But that her tender shame
Will not proclaim against her maiden loss,
How might she tongue me! Yet reason dares her no,
For my authority bears off a credent bulk,     25
That no particular scandal once can touch
But it confounds the breather. He should have lived,
Save that his riotous youth, with dangerous sense,
Might in the times to come have ta'en revenge
By so receiving a dishonoured life       30
With ransom of such shame. Would yet he had lived.
Alack, when once our grace we have forgot,
Nothing goes right; we would, and we would not. *Exit*

**4.5**    *Enter the Duke, in his own habit, and Friar Peter*

DUKE
These letters at fit time deliver me.
The Provost knows our purpose and our plot.
The matter being afoot, keep your instruction,
And hold you ever to our special drift,
Though sometimes you do blench from this to that   5
As cause doth minister. Go call at Flavio's house,
And tell him where I stay. Give the like notice
To Valentinus, Rowland, and to Crassus,
And bid them bring the trumpets to the gate.
But send me Flavius first.

FRIAR             It shall be speeded well. *Exit*
    *Enter Varrius*

DUKE
I thank thee, Varrius; thou hast made good haste.
Come, we will walk. There's other of our friends
Will greet us here anon. My gentle Varrius!      *Exeunt*

**4.6**    *Enter Isabella and Mariana*

ISABELLA
To speak so indirectly I am loath—
I would say the truth, but to accuse him so,
That is your part—yet I am advised to do it,
He says, to veil full purpose.

MARIANA             Be ruled by him.

ISABELLA
Besides, he tells me that if peradventure     5
He speak against me on the adverse side,
I should not think it strange, for 'tis a physic
That's bitter to sweet end.
    *Enter Friar Peter*

MARIANA I would Friar Peter—

ISABELLA O, peace; the friar is come.     10

FRIAR PETER
Come, I have found you out a stand most fit,
Where you may have such vantage on the Duke
He shall not pass you. Twice have the trumpets
       sounded.
The generous and gravest citizens
Have hent the gates, and very near upon     15
The Duke is ent'ring; therefore hence, away.    *Exeunt*

**5.1**    *Enter ⌈at one door⌉ the Duke, Varrius, and lords, ⌈at
     another door⌉ Angelo, Escalus, Lucio, citizens, ⌈and
     officers⌉*

DUKE *(to Angelo)*
My very worthy cousin, fairly met.
*(To Escalus)* Our old and faithful friend, we are glad to
       see you.

ANGELO *and* ESCALUS
Happy return be to your royal grace.

DUKE
Many and hearty thankings to you both.
We have made enquiry of you, and we hear     5
Such goodness of your justice that our soul
Cannot but yield you forth to public thanks,
Forerunning more requital.

ANGELO          You make my bonds still greater.

DUKE
O, your desert speaks loud, and I should wrong it
To lock it in the wards of covert bosom,     10
When it deserves with characters of brass
A forted residence 'gainst the tooth of time
And razure of oblivion. Give me your hand,
And let the subject see, to make them know
That outward courtesies would fain proclaim     15
Favours that keep within. Come, Escalus,
You must walk by us on our other hand,
And good supporters are you.
    ⌈*They walk forward.*⌉
    *Enter Friar Peter and Isabella*

FRIAR PETER
Now is your time. Speak loud, and kneel before him.

ISABELLA *(kneeling)*
Justice, O royal Duke! Vail your regard     20
Upon a wronged—I would fain have said, a maid.
O worthy prince, dishonour not your eye
By throwing it on any other object,
Till you have heard me in my true complaint,
And given me justice, justice, justice, justice!     25

DUKE
Relate your wrongs. In what? By whom? Be brief.
Here is Lord Angelo shall give you justice.

Reveal yourself to him.

ISABELLA                        O worthy Duke,
You bid me seek redemption of the devil.
Hear me yourself, for that which I must speak          30
Must either punish me, not being believed,
Or wring redress from you. Hear me, O hear me, hear!

ANGELO
My lord, her wits, I fear me, are not firm.
She hath been a suitor to me for her brother,
Cut off by course of justice.

ISABELLA ⌈standing⌉            By course of justice!      35

ANGELO
And she will speak most bitterly and strange.

ISABELLA
Most strange, but yet most truly, will I speak.
That Angelo's forsworn, is it not strange?
That Angelo's a murderer, is't not strange?
That Angelo is an adulterous thief,                    40
An hypocrite, a virgin-violator,
Is it not strange, and strange?

DUKE                          Nay, it is ten times strange!

ISABELLA
It is not truer he is Angelo
Than this is all as true as it is strange.
Nay, it is ten times true, for truth is truth          45
To th'end of reck'ning.

DUKE                    Away with her. Poor soul,
She speaks this in th'infirmity of sense.

ISABELLA
O prince, I conjure thee, as thou believ'st
There is another comfort than this world,
That thou neglect me not with that opinion            50
That I am touched with madness. Make not
      impossible
That which but seems unlike. 'Tis not impossible
But one, the wicked'st caitiff on the ground,
May seem as shy, as grave, as just, as absolute,
As Angelo; even so may Angelo,                         55
In all his dressings, characts, titles, forms,
Be an arch-villain. Believe it, royal prince,
If he be less, he's nothing; but he's more,
Had I more name for badness.

DUKE                          By mine honesty,
If she be mad, as I believe no other,                  60
Her madness hath the oddest frame of sense,
Such a dependency of thing on thing
As e'er I heard in madness.

ISABELLA                      O gracious Duke,
Harp not on that, nor do not banish reason
For inequality; but let your reason serve             65
To make the truth appear where it seems hid,
And hide the false seems true.

DUKE                          Many that are not mad
Have sure more lack of reason. What would you say?

ISABELLA
I am the sister of one Claudio,
Condemned upon the act of fornication                 70
To lose his head, condemned by Angelo.
I, in probation of a sisterhood,
Was sent to by my brother, one Lucio
As then the messenger.

LUCIO                    That's I, an't like your grace.
I came to her from Claudio, and desired her           75
To try her gracious fortune with Lord Angelo
For her poor brother's pardon.

ISABELLA                        That's he indeed.

DUKE (to Lucio)
You were not bid to speak.

LUCIO                      No, my good lord,
Nor wished to hold my peace.

DUKE
I wish you now, then. Pray you take note of it;       80
And when you have a business for yourself,
Pray heaven you then be perfect.

LUCIO                          I warrant your honour.

DUKE
The warrant's for yourself; take heed to't.

ISABELLA
This gentleman told somewhat of my tale—

LUCIO Right.                                           85

DUKE
It may be right, but you are i'the wrong
To speak before your time. (To Isabella) Proceed.

ISABELLA                                      I went
To this pernicious caitiff deputy—

DUKE
That's somewhat madly spoken.

ISABELLA                        Pardon it;
The phrase is to the matter.

DUKE                        Mended again.             90
The matter; proceed.

ISABELLA
In brief, to set the needless process by,
How I persuaded, how I prayed and kneeled,
How he refelled me, and how I replied—
For this was of much length—the vile conclusion      95
I now begin with grief and shame to utter.
He would not, but by gift of my chaste body
To his concupiscible intemperate lust,
Release my brother; and after much debatement,
My sisterly remorse confutes mine honour,            100
And I did yield to him. But the next morn betimes,
His purpose surfeiting, he sends a warrant
For my poor brother's head.

DUKE                          This is most likely!

ISABELLA
O, that it were as like as it is true!

DUKE
By heaven, fond wretch, thou know'st not what thou
      speak'st,                                       105
Or else thou art suborned against his honour
In hateful practice. First, his integrity
Stands without blemish. Next, it imports no reason
That with such vehemency he should pursue
Faults proper to himself. If he had so offended,     110
He would have weighed thy brother by himself,
And not have cut him off. Someone hath set you on.
Confess the truth, and say by whose advice
Thou cam'st here to complain.

ISABELLA                      And is this all?
Then, O you blessèd ministers above,                 115
Keep me in patience, and with ripened time
Unfold the evil which is here wrapped up
In countenance! Heaven shield your grace from woe,
As I, thus wronged, hence unbelievèd go.

DUKE
I know you'd fain be gone. An officer!                120
To prison with her.
      An officer guards Isabella
                        Shall we thus permit
A blasting and a scandalous breath to fall

On him so near us? This needs must be a practice.
Who knew of your intent and coming hither?

ISABELLA
One that I would were here, Friar Lodowick.          125
                              ⌜Exit, guarded⌝

DUKE
A ghostly father, belike. Who knows that Lodowick?

LUCIO
My lord, I know him. 'Tis a meddling friar;
I do not like the man. Had he been lay, my lord,
For certain words he spake against your grace
In your retirement, I had swinged him soundly.          130

DUKE
Words against me? This' a good friar, belike!
And to set on this wretched woman here
Against our substitute! Let this friar be found.
                              ⌜Exit one or more⌝

LUCIO
But yesternight, my lord, she and that friar,
I saw them at the prison. A saucy friar,          135
A very scurvy fellow.

FRIAR PETER                    Blessed be your royal grace!
I have stood by, my lord, and I have heard
Your royal ear abused. First hath this woman
Most wrongfully accused your substitute,
Who is as free from touch or soil with her          140
As she from one ungot.

DUKE                    We did believe no less.
Know you that Friar Lodowick that she speaks of?

FRIAR PETER
I know him for a man divine and holy,
Not scurvy, nor a temporary meddler,
As he's reported by this gentleman;          145
And, on my trust, a man that never yet
Did, as he vouches, misreport your grace.

LUCIO My lord, most villainously; believe it.

FRIAR PETER
Well, he in time may come to clear himself;
But at this instant he is sick, my lord,          150
Of a strange fever. Upon his mere request,
Being come to knowledge that there was complaint
Intended 'gainst Lord Angelo, came I hither
To speak, as from his mouth, what he doth know
Is true and false, and what he with his oath          155
And all probation will make up full clear
Whensoever he's convented. First, for this woman:
To justify this worthy nobleman,
So vulgarly and personally accused,
Her shall you hear disprovèd to her eyes,          160
Till she herself confess it.

DUKE                    Good friar, let's hear it.
                              ⌜Exit Friar Peter⌝
Do you not smile at this, Lord Angelo?
O heaven, the vanity of wretched fools!
Give us some seats.
        ⌜Seats are brought in⌝
                    Come, cousin Angelo,
In this I'll be impartial; be you judge          165
Of your own cause.
        The Duke and Angelo sit.
        Enter ⌜Friar Peter, and⌝ Mariana, veiled
                    Is this the witness, friar?
First let her show her face, and after speak.

MARIANA
Pardon, my lord, I will not show my face
Until my husband bid me.

DUKE What, are you married?          170

MARIANA No, my lord.

DUKE Are you a maid?

MARIANA No, my lord.

DUKE A widow then?

MARIANA Neither, my lord.          175

DUKE Why, you are nothing then; neither maid, widow,
nor wife!

LUCIO My lord, she may be a punk, for many of them are
neither maid, widow, nor wife.

DUKE Silence that fellow. I would he had some cause to
prattle for himself.          181

LUCIO Well, my lord.

MARIANA
My lord, I do confess I ne'er was married,
And I confess besides, I am no maid.
I have known my husband, yet my husband          185
Knows not that ever he knew me.

LUCIO He was drunk then, my lord, it can be no better.

DUKE For the benefit of silence, would thou wert so too.

LUCIO Well, my lord.

DUKE
This is no witness for Lord Angelo.          190

MARIANA Now I come to't, my lord.
She that accuses him of fornication
In self-same manner doth accuse my husband,
And charges him, my lord, with such a time
When I'll depose I had him in mine arms          195
With all th'effect of love.

ANGELO                    Charges she more than me?

MARIANA
Not that I know.

DUKE                    No? You say your husband.

MARIANA
Why just, my lord, and that is Angelo,
Who thinks he knows that he ne'er knew my body,
But knows, he thinks, that he knows Isabel's.          200

ANGELO
This is a strange abuse. Let's see thy face.

MARIANA (unveiling)
My husband bids me; now I will unmask.
This is that face, thou cruel Angelo,
Which once thou swor'st was worth the looking on.
This is the hand which, with a vowed contract,          205
Was fast belocked in thine. This is the body
That took away the match from Isabel,
And did supply thee at thy garden-house
In her imagined person.

DUKE (to Angelo) Know you this woman?          210

LUCIO Carnally, she says.

DUKE Sirrah, no more!

LUCIO Enough, my lord.

ANGELO
My lord, I must confess I know this woman;
And five years since there was some speech of
                    marriage          215
Betwixt myself and her, which was broke off,
Partly for that her promisèd proportions
Came short of composition, but in chief
For that her reputation was disvalued
In levity; since which time of five years          220
I never spake with her, saw her, nor heard from her,
Upon my faith and honour.

MARIANA ⌜kneeling before the Duke⌝ Noble prince,
As there comes light from heaven, and words from
                    breath,

As there is sense in truth, and truth in virtue,
I am affianced this man's wife, as strongly          225
As words could make up vows. And, my good lord,
But Tuesday night last gone, in's garden-house,
He knew me as a wife. As this is true,
Let me in safety raise me from my knees,
Or else forever be confixèd here,          230
A marble monument.

ANGELO                    I did but smile till now.
Now, good my lord, give me the scope of justice.
My patience here is touched. I do perceive
These poor informal women are no more
But instruments of some more mightier member          235
That sets them on. Let me have way, my lord,
To find this practice out.

DUKE (standing)                    Ay, with my heart,
And punish them even to your height of pleasure.—
Thou foolish friar, and thou pernicious woman          239
Compact with her that's gone, think'st thou thy oaths,
Though they would swear down each particular saint,
Were testimonies against his worth and credit
That's sealed in approbation? You, Lord Escalus,
Sit with my cousin; lend him your kind pains
To find out this abuse, whence 'tis derived.          245
There is another friar that set them on.
Let him be sent for.

*Escalus sits*

FRIAR PETER
Would he were here, my lord, for he indeed
Hath set the women on to this complaint.
Your Provost knows the place where he abides,          250
And he may fetch him.

DUKE (to one or more)          Go, do it instantly.

*Exit one or more*

(To Angelo) And you, my noble and well-warranted
          cousin,
Whom it concerns to hear this matter forth,
Do with your injuries as seems you best
In any chastisement. I for a while will leave you,          255
But stir not you till you have well determined
Upon these slanderers.

ESCALUS                    My lord, we'll do it throughly.

*Exit Duke*

Signor Lucio, did not you say you knew that Friar
Lodowick to be a dishonest person?          259

LUCIO *Cucullus non facit monachum*: honest in nothing but
in his clothes; and one that hath spoke most villainous
speeches of the Duke.

ESCALUS We shall entreat you to abide here till he come,
and enforce them against him. We shall find this friar
a notable fellow.          265

LUCIO As any in Vienna, on my word.

ESCALUS Call that same Isabel here once again; I would
speak with her.          *Exit one or more*
(To Angelo) Pray you, my lord, give me leave to question.
You shall see how I'll handle her.          270

LUCIO Not better than he, by her own report.

ESCALUS Say you?

LUCIO Marry, sir, I think if you handled her privately, she
would sooner confess; perchance publicly she'll be
ashamed.          275

ESCALUS I will go darkly to work with her.

LUCIO That's the way, for women are light at midnight.

*Enter Isabella, guarded*

ESCALUS (to Isabella) Come on, mistress, here's a gentlewoman denies all that you have said.

*Enter the Duke, disguised as a friar, hooded, and the
Provost*

LUCIO My lord, here comes the rascal I spoke of, here
with the Provost.          281

ESCALUS In very good time. Speak not you to him till we
call upon you.

LUCIO Mum.

ESCALUS (to the Duke) Come, sir, did you set these women
on to slander Lord Angelo? They have confessed you
did.          287

DUKE 'Tis false.

ESCALUS How! Know you where you are?

DUKE
Respect to your great place, and let the devil          290
Be sometime honoured fore his burning throne.
Where is the Duke? 'Tis he should hear me speak.

ESCALUS
The Duke's in us, and we will hear you speak.
Look you speak justly.

DUKE                    Boldly at least.
(To Isabella and Mariana)          But O, poor souls,
Come you to seek the lamb here of the fox,          295
Good night to your redress! Is the Duke gone?
Then is your cause gone too. The Duke's unjust
Thus to retort your manifest appeal,
And put your trial in the villain's mouth
Which here you come to accuse.          300

LUCIO
This is the rascal, this is he I spoke of.

ESCALUS
Why, thou unreverend and unhallowed friar,
Is't not enough thou hast suborned these women
To accuse this worthy man but, in foul mouth,
And in the witness of his proper ear,          305
To call him villain, and then to glance from him
To th' Duke himself, to tax him with injustice?
Take him hence; to th' rack with him. We'll touse you
Joint by joint—but we will know his purpose.
What, 'unjust'?

DUKE                    Be not so hot. The Duke          310
Dare no more stretch this finger of mine than he
Dare rack his own. His subject am I not,
Nor here provincial. My business in this state
Made me a looker-on here in Vienna,
Where I have seen corruption boil and bubble          315
Till it o'errun the stew; laws for all faults,
But faults so countenanced that the strong statutes
Stand like the forfeits in a barber's shop,
As much in mock as mark.

ESCALUS Slander to th' state!          320
Away with him to prison.

ANGELO
What can you vouch against him, Signor Lucio?
Is this the man that you did tell us of?

LUCIO 'Tis he, my lord.—Come hither, goodman Bald-
pate. Do you know me?          325

DUKE I remember you, sir, by the sound of your voice. I
met you at the prison, in the absence of the Duke.

LUCIO O, did you so? And do you remember what you
said of the Duke?

DUKE Most notedly, sir.          330

LUCIO Do you so, sir? And was the Duke a fleshmonger,
a fool, and a coward, as you then reported him to be?

DUKE You must, sir, change persons with me ere you
make that my report. You indeed spoke so of him, and
much more, much worse.          335

LUCIO O, thou damnable fellow! Did not I pluck thee by
the nose for thy speeches?

DUKE I protest I love the Duke as I love myself.

ANGELO Hark how the villain would close now, after his
treasonable abuses.                                    340

ESCALUS Such a fellow is not to be talked withal. Away
with him to prison. Where is the Provost? Away with
him to prison. Lay bolts enough upon him. Let him
speak no more. Away with those giglets too, and with
the other confederate companion.                       345
⌜Mariana is raised to her feet, and is guarded.⌝
*The Provost makes to seize the Duke*

DUKE Stay, sir, stay a while.

ANGELO What, resists he? Help him, Lucio.

LUCIO (*to the Duke*) Come, sir; come, sir; come, sir! Foh,
sir! Why, you bald-pated lying rascal, you must be
hooded, must you? Show your knave's visage, with a
pox to you! Show your sheep-biting face, and be hanged
an hour! Will't not off?                               352
*He pulls off the friar's hood, and discovers the Duke.*
⌜*Angelo and Escalus rise*⌝

DUKE
Thou art the first knave that e'er madest a duke.
First, Provost, let me bail these gentle three.
(*To Lucio*) Sneak not away, sir, for the friar and you
Must have a word anon. (*To one or more*) Lay hold on
him.                                                   356

LUCIO This may prove worse than hanging.

DUKE (*to Escalus*)
What you have spoke, I pardon. Sit you down.
We'll borrow place of him.
⌜*Escalus sits*⌝
(*To Angelo*)                Sir, by your leave.
⌜*He takes Angelo's seat*⌝
Hast thou or word or wit or impudence          360
That yet can do thee office? If thou hast,
Rely upon it till my tale be heard,
And hold no longer out.

ANGELO                    O my dread lord,
I should be guiltier than my guiltiness
To think I can be undiscernible,               365
When I perceive your grace, like power divine,
Hath looked upon my passes. Then, good prince,
No longer session hold upon my shame,
But let my trial be mine own confession.
Immediate sentence then, and sequent death,    370
Is all the grace I beg.

DUKE                    Come hither, Mariana.
(*To Angelo*) Say, wast thou e'er contracted to this
woman?

ANGELO I was, my lord.

DUKE
Go, take her hence and marry her instantly.
Do you the office, friar; which consummate,    375
Return him here again. Go with him, Provost.
*Exeunt Angelo, Mariana, Friar Peter, and the
Provost*

ESCALUS
My lord, I am more amazed at his dishonour
Than at the strangeness of it.

DUKE                    Come hither, Isabel.
Your friar is now your prince. As I was then
Advertising and holy to your business,         380
Not changing heart with habit I am still
Attorneyed at your service.

ISABELLA                   O, give me pardon,
That I, your vassal, have employed and pained

Your unknown sovereignty.

DUKE                    You are pardoned, Isabel.
And now, dear maid, be you as free to us.       385
Your brother's death I know sits at your heart,
And you may marvel why I obscured myself,
Labouring to save his life, and would not rather
Make rash remonstrance of my hidden power
Than let him so be lost. O most kind maid,      390
It was the swift celerity of his death,
Which I did think with slower foot came on,
That brained my purpose. But peace be with him!
That life is better life, past fearing death,
Than that which lives to fear. Make it your comfort,
So happy is your brother.

ISABELLA                 I do, my lord.         396
*Enter Angelo, Mariana, Friar Peter, and the Provost*

DUKE
For this new-married man approaching here,
Whose salt imagination yet hath wronged
Your well-defended honour, you must pardon
For Mariana's sake; but as he adjudged your
brother—                                        400
Being criminal in double violation
Of sacred chastity and of promise-breach,
Thereon dependent, for your brother's life—
The very mercy of the law cries out
Most audible, even from his proper tongue,      405
'An Angelo for Claudio, death for death'.
Haste still pays haste, and leisure answers leisure;
Like doth quit like, and measure still for measure.
Then, Angelo, thy fault's thus manifested,
Which, though thou wouldst deny, denies thee
vantage.                                        410
We do condemn thee to the very block
Where Claudio stooped to death, and with like haste.
Away with him.

MARIANA          O my most gracious lord,
I hope you will not mock me with a husband!

DUKE
It is your husband mocked you with a husband.   415
Consenting to the safeguard of your honour,
I thought your marriage fit; else imputation,
For that he knew you, might reproach your life,
And choke your good to come. For his possessions,
Although by confiscation they are ours,         420
We do enstate and widow you withal,
To buy you a better husband.

MARIANA                    O my dear lord,
I crave no other, nor no better man.

DUKE
Never crave him; we are definitive.

MARIANA
Gentle my liege—

DUKE              You do but lose your labour.—   425
Away with him to death. (*To Lucio*) Now, sir, to you.

MARIANA (*kneeling*)
O my good lord!—Sweet Isabel, take my part;
Lend me your knees, and all my life to come
I'll lend you all my life to do you service.

DUKE
Against all sense you do importune her.         430
Should she kneel down in mercy of this fact,
Her brother's ghost his pavèd bed would break,
And take her hence in horror.

MARIANA                    Isabel,
Sweet Isabel, do yet but kneel by me.
Hold up your hands; say nothing; I'll speak all.  435

They say best men are moulded out of faults,
And, for the most, become much more the better
For being a little bad. So may my husband.
O Isabel, will you not lend a knee?

DUKE
He dies for Claudio's death.

ISABELLA (*kneeling*)          Most bounteous sir,          440
Look, if it please you, on this man condemned
As if my brother lived. I partly think
A due sincerity governed his deeds,
Till he did look on me. Since it is so,
Let him not die. My brother had but justice,          445
In that he did the thing for which he died.
For Angelo,
His act did not o'ertake his bad intent,
And must be buried but as an intent
That perished by the way. Thoughts are no subjects,
Intents but merely thoughts.

MARIANA          Merely, my lord.          451

DUKE
Your suit's unprofitable. Stand up, I say.
⌈*Mariana and Isabella stand*⌉
I have bethought me of another fault.
Provost, how came it Claudio was beheaded
At an unusual hour?

PROVOST          It was commanded so.          455

DUKE
Had you a special warrant for the deed?

PROVOST
No, my good lord, it was by private message.

DUKE
For which I do discharge you of your office.
Give up your keys.

PROVOST          Pardon me, noble lord.
I thought it was a fault, but knew it not,          460
Yet did repent me after more advice;
For testimony whereof one in the prison
That should by private order else have died
I have reserved alive.

DUKE What's he?          465

PROVOST His name is Barnardine.

DUKE
I would thou hadst done so by Claudio.
Go fetch him hither. Let me look upon him.
          *Exit Provost*

ESCALUS
I am sorry one so learned and so wise
As you, Lord Angelo, have still appeared,          470
Should slip so grossly, both in the heat of blood
And lack of tempered judgement afterward.

ANGELO
I am sorry that such sorrow I procure,
And so deep sticks it in my penitent heart
That I crave death more willingly than mercy.          475
'Tis my deserving, and I do entreat it.
          *Enter Barnardine and the Provost; Claudio, muffled,*
          *and Juliet*

DUKE
Which is that Barnardine?

PROVOST          This, my lord.

DUKE
There was a friar told me of this man.
(*To Barnardine*) Sirrah, thou art said to have a
          stubborn soul
That apprehends no further than this world,          480

And squar'st thy life according. Thou'rt condemned;
But, for those earthly faults, I quit them all,
And pray thee take this mercy to provide
For better times to come.—Friar, advise him.
I leave him to your hand. (*To Provost*) What muffled
          fellow's that?          485

PROVOST
This is another prisoner that I saved,
Who should have died when Claudio lost his head,
As like almost to Claudio as himself.
          *He unmuffles Claudio*

DUKE (*to Isabella*)
If he be like your brother, for his sake
Is he pardoned; and for your lovely sake          490
Give me your hand, and say you will be mine.
He is my brother too. But fitter time for that.
By this Lord Angelo perceives he's safe.
Methinks I see a quick'ning in his eye.
Well, Angelo, your evil quits you well.          495
Look that you love your wife, her worth worth yours.
I find an apt remission in myself;
And yet here's one in place I cannot pardon.
(*To Lucio*) You, sirrah, that knew me for a fool, a
          coward,
One all of luxury, an ass, a madman,          500
Wherein have I so deserved of you
That you extol me thus?

LUCIO Faith, my lord, I spoke it but according to the trick.
If you will hang me for it, you may; but I had rather
it would please you I might be whipped.          505

DUKE Whipped first, sir, and hanged after.
Proclaim it, Provost, round about the city,
If any woman wronged by this lewd fellow,
As I have heard him swear himself there's one
Whom he begot with child, let her appear,          510
And he shall marry her. The nuptial finished,
Let him be whipped and hanged.

LUCIO I beseech your highness, do not marry me to a
whore. Your highness said even now I made you a
duke; good my lord, do not recompense me in making
me a cuckold.          516

DUKE
Upon mine honour, thou shalt marry her.
Thy slanders I forgive, and therewithal
Remit thy other forfeits.—Take him to prison,
And see our pleasure herein executed.          520

LUCIO Marrying a punk, my lord, is pressing to death,
whipping, and hanging.

DUKE Slandering a prince deserves it.
          ⌈*Exit Lucio guarded*⌉
She, Claudio, that you wronged, look you restore.
Joy to you, Mariana. Love her, Angelo.          525
I have confessed her, and I know her virtue.
Thanks, good friend Escalus, for thy much goodness.
There's more behind that is more gratulate.
Thanks, Provost, for thy care and secrecy.
We shall employ thee in a worthier place.          530
Forgive him, Angelo, that brought you home
The head of Ragusine for Claudio's.
Th'offence pardons itself. Dear Isabel,
I have a motion much imports your good,
Whereto, if you'll a willing ear incline,          535
What's mine is yours, and what is yours is mine.
(*To all*) So bring us to our palace, where we'll show
What's yet behind that's meet you all should know.
          *Exeunt*

## ADDITIONAL PASSAGES

The text of *Measure for Measure* given in this edition is probably that of an adapted version made for Shakespeare's company after his death. Adaptation seems to have affected two passages, printed below as we believe Shakespeare to have written them.

### A. 1.2.0.1–116

A.2–9 ('. . . by him') are lines which the adapter (whom we believe to be Thomas Middleton) evidently intended to be replaced by 1.2.56–79 of the play as we print it. The adapter must have contributed all of 1.2.0.1–83, which in the earliest and subsequent printed texts precede the discussion between the Clown (Pompey) and the Bawd (Mistress Overdone) about Claudio's arrest. Lucio's entry alone at l. 40.1 below, some eleven lines after his re-entry with the two Gentlemen and the Provost's party in the adapted text, probably represents Shakespeare's original intention. In his version, Juliet, present but silent in the adapted text both in 1.2 and 5.1, probably did not appear in either scene; accordingly, the words 'and there's Madam Juliet' (1.2.107) must also be the reviser's work, and do not appear below.

> *Enter Pompey and Mistress Overdone, ⌈meeting⌉*
> MISTRESS OVERDONE How now, what's the news with you?
> POMPEY Yonder man is carried to prison.
> MISTRESS OVERDONE Well! What has he done?
> POMPEY A woman.
> MISTRESS OVERDONE But what's his offence? 5
> POMPEY Groping for trouts in a peculiar river.
> MISTRESS OVERDONE What, is there a maid with child by him?
> POMPEY No, but there's a woman with maid by him: you have not heard of the proclamation, have you? 10
> MISTRESS OVERDONE What proclamation, man?
> POMPEY All houses in the suburbs of Vienna must be plucked down.
> MISTRESS OVERDONE And what shall become of those in the city? 15
> POMPEY They shall stand for seed. They had gone down too, but that a wise burgher put in for them.
> MISTRESS OVERDONE But shall all our houses of resort in the suburbs be pulled down?
> POMPEY To the ground, mistress. 20
> MISTRESS OVERDONE Why, here's a change indeed in the commonwealth. What shall become of me?
> POMPEY Come, fear not you. Good counsellors lack no clients. Though you change your place, you need not change your trade. I'll be your tapster still. Courage, there will be pity taken on you. You that have worn your eyes almost out in the service, you will be considered.
> ⌈*A noise within*⌉
> MISTRESS OVERDONE What's to do here, Thomas Tapster? Let's withdraw! 30
> *Enter the Provost and Claudio*
> POMPEY Here comes Signor Claudio, led by the Provost to prison. *Exeunt Mistress Overdone and Pompey*
> CLAUDIO
> Fellow, why dost thou show me thus to th' world?
> Bear me to prison, where I am committed.

PROVOST
I do it not in evil disposition, 35
But from Lord Angelo by special charge.
CLAUDIO
Thus can the demigod Authority
Make us pay down for our offence, by weight,
The bonds of heaven. On whom it will, it will;
On whom it will not, so; yet still 'tis just. 40
⌈*Enter Lucio*⌉
LUCIO
Why, how now, Claudio? Whence comes this restraint?

### B. 3.1.515–4.1.65

Before revision there would have been no act-break and no song; the lines immediately following the song would also have been absent. The Duke's soliloquies 'He who the sword of heaven will bear' and 'O place and greatness' have evidently been transposed in revision; in the original, the end of 'O place and greatness' would have led straight on to the Duke's meeting with Isabella and then Mariana.

ESCALUS I am going to visit the prisoner. Fare you well.
DUKE Peace be with you. *Exit Escalus*
O place and greatness, millions of false eyes
Are stuck upon thee; volumes of report
Run with their false and most contrarious quest 5
Upon thy doings; thousand escapes of wit
Make thee the father of their idle dream,
And rack thee in their fancies.
*Enter Isabella*
Very well met.
What is the news from this good deputy?
ISABELLA
He hath a garden circummured with brick, 10
Whose western side is with a vineyard backed;
And to that vineyard is a planckèd gate,
That makes his opening with this bigger key.
This other doth command a little door
Which from the vineyard to the garden leads. 15
There have I made my promise
Upon the heavy middle of the night
To call upon him.
DUKE
But shall you on your knowledge find this way?
ISABELLA
I have ta'en a due and wary note upon't. 20
With whispering and most guilty diligence,
In action all of precept, he did show me
The way twice o'er.
DUKE Are there no other tokens
Between you 'greed concerning her observance?
ISABELLA
No, none, but only a repair i'th' dark, 25
And that I have possessed him my most stay
Can be but brief, for I have made him know
I have a servant comes with me along
That stays upon me, whose persuasion is
I come about my brother.
DUKE 'Tis well borne up. 30
I have not yet made known to Mariana
A word of this.—What ho, within! Come forth!

*Enter Mariana*
(*To Mariana*) I pray you be acquainted with this maid.
She comes to do you good.
ISABELLA                    I do desire the like.
DUKE (*to Mariana*)
Do you persuade yourself that I respect you?          35
MARIANA
Good friar, I know you do, and so have found it.
DUKE
Take then this your companion by the hand,
Who hath a story ready for your ear.
I shall attend your leisure; but make haste,
The vaporous night approaches.
MARIANA                      Will't please you walk aside.
                ⌈*Exeunt Mariana and Isabella*⌉
DUKE
He who the sword of heaven will bear          41
Should be as holy as severe,
Pattern in himself to know,
Grace to stand, and virtue go,
More nor less to others paying          45
Than by self-offences weighing.

Shame to him whose cruel striking
Kills for faults of his own liking!
Twice treble shame on Angelo,
To weed my vice, and let his grow!          50
O, what may man within him hide,
Though angel on the outward side!
How may likeness made in crimes
Make my practice on the times
To draw with idle spiders' strings          55
Most ponderous and substantial things?
Craft against vice I must apply.
With Angelo tonight shall lie
His old betrothed but despisèd.
So disguise shall, by th' disguisèd,          60
Pay with falsehood false exacting,
And perform an old contracting.
                ⌈*Enter Mariana and Isabella*⌉
Welcome. How agreed?
ISABELLA
She'll take the enterprise upon her, father,
If you advise it.          65

# OTHELLO

*Othello* was given before James I in the Banqueting House at Whitehall on 1 November 1604. Information about the Turkish invasion of Cyprus appears to derive from Richard Knolles's *History of the Turks*, published no earlier than 30 September 1603, so Shakespeare probably completed his play some time between that date and the summer of 1604. It first appeared in print in a quarto of 1622; the version printed in the 1623 Folio is about 160 lines longer, and has over a thousand differences in wording. It seems that Shakespeare partially revised his play, adding, for example, Desdemona's willow song (4.3) and building up Emilia's role in the closing scenes. We base our text on the Folio as that seems to represent Shakespeare's second thoughts.

Shakespeare's decision to make a black man a tragic hero was bold and original: by an ancient tradition, blackness was associated with sin and death; and blackamoors in plays before Shakespeare are generally villainous (like Aaron in *Titus Andronicus*). The story of a Moorish commander deluded by his ensign (standard-bearer) into believing that his young wife has been unfaithful to him with another soldier derives from a prose tale by the Italian Giambattista Cinzio Giraldi first published in 1565 in a collection of linked tales, *Gli Ecatommiti* (*The Hundred Tales*). Shakespeare must have read it either in Italian or in a French translation of 1584; he may have looked at both. Giraldi tells the tale in a few pages of compressed, matter-of-fact narrative interspersed with brief conversations. His main characters are a Moor of Venice (Othello), his Venetian wife (Desdemona), his ensign (Iago), his ensign's wife (Emilia), and a corporal (Cassio) 'who was very dear to the Moor'. Only Desdemona is named. Shakespeare's invented characters include Roderigo, a young, disappointed suitor of Desdemona, and Brabanzio, Desdemona's father, who opposes her marriage to Othello. Bianca, Cassio's mistress, is developed from a few hints in the source. Shakespeare also introduces the military action between Turkey and Venice—infidels and Christians—which gives especial importance to Othello's posting to Cyprus, a Venetian protectorate which the Turks attacked in 1570 and conquered in the following year. In the source, Othello and Desdemona are already happily settled into married life when they go to Cyprus; Shakespeare compresses the time-scheme and makes many changes to the narrative.

*Othello*, a great success in Shakespeare's time, was one of the first plays to be acted after the reopening of the theatres in 1660, and since that time has remained one of the most popular plays on the English stage.

# THE PERSONS OF THE PLAY

OTHELLO, the Moor of Venice
DESDEMONA, his wife
Michael CASSIO, his lieutenant
BIANCA, a courtesan, in love with Cassio
IAGO, the Moor's ensign
EMILIA, Iago's wife
A CLOWN, a servant of Othello

The DUKE of Venice
BRABANZIO, Desdemona's father, a Senator of Venice

GRAZIANO, Brabanzio's brother
LODOVICO, kinsman of Brabanzio
SENATORS of Venice
RODERIGO, a Venetian gentleman, in love with Desdemona

MONTANO, Governor of Cyprus
A HERALD

A MESSENGER
Attendants, officers, sailors, gentlemen of Cyprus, musicians

# The Tragedy of Othello the Moor of Venice

**1.1** *Enter Iago and Roderigo*

RODERIGO
Tush, never tell me! I take it much unkindly
That thou, Iago, who hast had my purse
As if the strings were thine, shouldst know of this.

IAGO 'Sblood, but you'll not hear me!
If ever I did dream of such a matter, abhor me. 5

RODERIGO
Thou told'st me thou didst hold him in thy hate.

IAGO Despise me
If I do not. Three great ones of the city,
In personal suit to make me his lieutenant,
Off-capped to him; and by the faith of man 10
I know my price, I am worth no worse a place.
But he, as loving his own pride and purposes,
Evades them with a bombast circumstance
Horribly stuffed with epithets of war,
Nonsuits my mediators; for 'Certes,' says he, 15
'I have already chose my officer.'
And what was he?
Forsooth, a great arithmetician,
One Michael Cassio, a Florentine,
A fellow almost damned in a fair wife, 20
That never set a squadron in the field
Nor the division of a battle knows
More than a spinster—unless the bookish theoric,
Wherein the togaed consuls can propose
As masterly as he. Mere prattle without practice 25
Is all his soldiership; but he, sir, had th'election,
And I—of whom his eyes had seen the proof
At Rhodes, at Cyprus, and on other grounds
Christened and heathen—must be beleed and calmed
By debitor and creditor. This counter-caster, 30
He in good time must his lieutenant be,
And I—God bless the mark!—his Moorship's ensign.

RODERIGO
By heaven, I rather would have been his hangman.

IAGO
Why, there's no remedy. 'Tis the curse of service.
Preferment goes by letter and affection, 35
And not by old gradation, where each second
Stood heir to th' first. Now, sir, be judge yourself
Whether I in any just term am affined
To love the Moor.

RODERIGO I would not follow him then. 40

IAGO O sir, content you.
I follow him to serve my turn upon him.
We cannot all be masters, nor all masters
Cannot be truly followed. You shall mark
Many a duteous and knee-crooking knave 45
That, doting on his own obsequious bondage,
Wears out his time much like his master's ass
For naught but provender, and when he's old,
  cashiered.
Whip me such honest knaves. Others there are
Who, trimmed in forms and visages of duty, 50
Keep yet their hearts attending on themselves,
And, throwing but shows of service on their lords,
Do well thrive by 'em, and when they have lined their
  coats,

Do themselves homage. These fellows have some soul,
And such a one do I profess myself—for, sir, 55
It is as sure as you are Roderigo,
Were I the Moor I would not be Iago.
In following him I follow but myself.
Heaven is my judge, not I for love and duty,
But seeming so for my peculiar end. 60
For when my outward action doth demonstrate
The native act and figure of my heart
In compliment extern, 'tis not long after
But I will wear my heart upon my sleeve
For daws to peck at. I am not what I am. 65

RODERIGO
What a full fortune does the thick-lips owe
If he can carry't thus!

IAGO Call up her father,
Rouse him, make after him, poison his delight,
Proclaim him in the streets; incense her kinsmen,
And, though he in a fertile climate dwell, 70
Plague him with flies. Though that his joy be joy,
Yet throw such chances of vexation on't
As it may lose some colour.

RODERIGO
Here is her father's house. I'll call aloud.

IAGO
Do, with like timorous accent and dire yell 75
As when, by night and negligence, the fire
Is spied in populous cities.

RODERIGO *(calling)*
What ho, Brabanzio, Signor Brabanzio, ho!

IAGO *(calling)*
Awake, what ho, Brabanzio, thieves, thieves, thieves!
Look to your house, your daughter, and your bags. 80
Thieves, thieves!

*Enter Brabanzio in his nightgown at a window
above*

BRABANZIO
What is the reason of this terrible summons?
What is the matter there?

RODERIGO
Signor, is all your family within?

IAGO
Are your doors locked?

BRABANZIO Why, wherefore ask you this?

IAGO
'Swounds, sir, you're robbed. For shame, put on your
  gown. 86
Your heart is burst, you have lost half your soul.
Even now, now, very now, an old black ram
Is tupping your white ewe. Arise, arise!
Awake the snorting citizens with the bell, 90
Or else the devil will make a grandsire of you.
Arise, I say.

BRABANZIO What, have you lost your wits?

RODERIGO
Most reverend signor, do you know my voice?

BRABANZIO Not I. What are you?

RODERIGO My name is Roderigo. 95

BRABANZIO The worser welcome.
I have charged thee not to haunt about my doors.

In honest plainness thou hast heard me say
My daughter is not for thee, and now in madness,
Being full of supper and distempering draughts,      100
Upon malicious bravery dost thou come
To start my quiet.

RODERIGO Sir, sir, sir.

BRABANZIO But thou must needs be sure
My spirits and my place have in their power      105
To make this bitter to thee.

RODERIGO                         Patience, good sir.

BRABANZIO
What tell'st thou me of robbing? This is Venice.
My house is not a grange.

RODERIGO                         Most grave Brabanzio,
In simple and pure soul I come to you.

IAGO (to Brabanzio) 'Swounds, sir, you are one of those
that will not serve God if the devil bid you. Because we
come to do you service and you think we are ruffians,
you'll have your daughter covered with a Barbary
horse, you'll have your nephews neigh to you, you'll
have coursers for cousins and jennets for germans.

BRABANZIO What profane wretch art thou?      116

IAGO I am one, sir, that comes to tell you your daughter
and the Moor are now making the beast with two
backs.

BRABANZIO
Thou art a villain.

IAGO                         You are a senator.      120

BRABANZIO
This thou shalt answer. I know thee, Roderigo.

RODERIGO
Sir, I will answer anything. But I beseech you,
If't be your pleasure and most wise consent—
As partly I find it is—that your fair daughter,
At this odd-even and dull watch o'th' night,      125
Transported with no worse nor better guard
But with a knave of common hire, a gondolier,
To the gross clasps of a lascivious Moor—
If this be known to you, and your allowance,
We then have done you bold and saucy wrongs.      130
But if you know not this, my manners tell me
We have your wrong rebuke. Do not believe
That, from the sense of all civility,
I thus would play and trifle with your reverence.
Your daughter, if you have not given her leave,      135
I say again hath made a gross revolt,
Tying her duty, beauty, wit, and fortunes
In an extravagant and wheeling stranger
Of here and everywhere. Straight satisfy yourself.
If she be in her chamber or your house,      140
Let loose on me the justice of the state
For thus deluding you.

BRABANZIO (calling)         Strike on the tinder, ho!
Give me a taper, call up all my people.
This accident is not unlike my dream;
Belief of it oppresses me already.      145
Light, I say, light!                         Exit

IAGO                         Farewell, for I must leave you.
It seems not meet nor wholesome to my place
To be produced—as, if I stay, I shall—
Against the Moor, for I do know the state,
However this may gall him with some check,      150
Cannot with safety cast him, for he's embarked
With such loud reason to the Cyprus wars,
Which even now stands in act, that, for their souls,
Another of his fathom they have none

To lead their business, in which regard—      155
Though I do hate him as I do hell pains—
Yet for necessity of present life
I must show out a flag and sign of love,
Which is indeed but sign. That you shall surely find
him,
Lead to the Sagittary the raisèd search,      160
And there will I be with him. So farewell.      Exit
Enter below Brabanzio in his nightgown, and
servants with torches

BRABANZIO
It is too true an evil. Gone she is,
And what's to come of my despisèd time
Is naught but bitterness. Now, Roderigo,
Where didst thou see her?—O unhappy girl!—      165
With the Moor, sayst thou?—Who would be a
father?—
How didst thou know 'twas she?—O, she deceives me
Past thought!—What said she to you? (To servants)
Get more tapers,
Raise all my kindred.                         [Exit one or more]
(To Roderigo)                         Are they married, think you?

RODERIGO Truly, I think they are.      170

BRABANZIO
O heaven, how got she out? O, treason of the blood!
Fathers, from hence trust not your daughters' minds
By what you see them act. Is there not charms
By which the property of youth and maidhood
May be abused? Have you not read, Roderigo,      175
Of some such thing?

RODERIGO                         Yes, sir, I have indeed.

BRABANZIO (to servants)
Call up my brother. (To Roderigo) O, would you had
had her.
(To servants) Some one way, some another.
[Exit one or more]
(To Roderigo)                         Do you know
Where we may apprehend her and the Moor?

RODERIGO
I think I can discover him, if you please      180
To get good guard and go along with me.

BRABANZIO
Pray you lead on. At every house I'll call;
I may command at most. (Calling) Get weapons, ho,
And raise some special officers of night.
On, good Roderigo. I will deserve your pains.      Exeunt

1.2     Enter Othello, Iago, and attendants with torches

IAGO
Though in the trade of war I have slain men,
Yet do I hold it very stuff o'th' conscience
To do no contrived murder. I lack iniquity,
Sometime, to do me service. Nine or ten times
I had thought to've yerked him here, under the ribs.      5

OTHELLO
'Tis better as it is.

IAGO                         Nay, but he prated,
And spoke such scurvy and provoking terms
Against your honour
That, with the little godliness I have,
I did full hard forbear him. But I pray you, sir,      10
Are you fast married? Be assured of this:
That the magnifico is much beloved,
And hath in his effect a voice potential
As double as the Duke's. He will divorce you,
Or put upon you what restraint or grievance      15

The law, with all his might to enforce it on,
Will give him cable.
OTHELLO        Let him do his spite.
My services which I have done the signory
Shall out-tongue his complaints. 'Tis yet to know—
Which, when I know that boasting is an honour,    20
I shall promulgate—I fetch my life and being
From men of royal siege, and my demerits
May speak unbonneted to as proud a fortune
As this that I have reached. For know, Iago,
But that I love the gentle Desdemona      25
I would not my unhousèd free condition
Put into circumscription and confine
For the seas' worth.
       *Enter Cassio and officers, with torches*
            But look, what lights come yond?
IAGO
Those are the raisèd father and his friends.
You were best go in.
OTHELLO          Not I. I must be found.    30
My parts, my title, and my perfect soul
Shall manifest me rightly. Is it they?
IAGO By Janus, I think no.
OTHELLO
The servants of the Duke, and my lieutenant!
The goodness of the night upon you, friends.    35
What is the news?
CASSIO        The Duke does greet you, general,
And he requires your haste-post-haste appearance
Even on the instant.
OTHELLO        What is the matter, think you?
CASSIO
Something from Cyprus, as I may divine;
It is a business of some heat. The galleys      40
Have sent a dozen sequent messengers
This very night at one another's heels,
And many of the consuls, raised and met,
Are at the Duke's already. You have been hotly called
    for,
When, being not at your lodging to be found,    45
The senate sent about three several quests
To search you out.
OTHELLO        'Tis well I am found by you.
I will but spend a word here in the house
And go with you.                 *Exit*
CASSIO        Ensign, what makes he here?
IAGO
Faith, he tonight hath boarded a land-carrack.    50
If it prove lawful prize, he's made for ever.
CASSIO
I do not understand.
IAGO        He's married.
CASSIO           To who?
       *Enter Brabanzio, Roderigo, and officers, with lights*
       *and weapons*
IAGO
Marry, to—
       *Enter Othello*
(*To Othello*) Come, captain, will you go?
OTHELLO              Have with you.
CASSIO
Here comes another troop to seek for you.    55
IAGO
It is Brabanzio. General, be advised.
He comes to bad intent.
OTHELLO           Holla, stand, there!

RODERIGO (*to Brabanzio*)
Signor, it is the Moor.
BRABANZIO          Down with him, thief!
IAGO (*drawing his sword*)
You, Roderigo? Come, sir, I am for you.
OTHELLO
Keep up your bright swords, for the dew will rust 'em.
(*To Brabanzio*) Good signor, you shall more command
    with years             61
Than with your weapons.
BRABANZIO
O thou foul thief, where hast thou stowed my
    daughter?
Damned as thou art, thou hast enchanted her,
For I'll refer me to all things of sense,      65
If she in chains of magic were not bound,
Whether a maid so tender, fair, and happy,
So opposite to marriage that she shunned
The wealthy curlèd darlings of our nation,
Would ever have, t'incur a general mock,    70
Run from her guardage to the sooty bosom
Of such a thing as thou—to fear, not to delight.
Judge me the world if 'tis not gross in sense
That thou hast practised on her with foul charms,
Abused her delicate youth with drugs or minerals    75
That weakens motion. I'll have't disputed on.
'Tis probable, and palpable to thinking.
I therefore apprehend and do attach thee
For an abuser of the world, a practiser
Of arts inhibited and out of warrant.      80
(*To officers*) Lay hold upon him. If he do resist,
Subdue him at his peril.
OTHELLO          Hold your hands,
Both you of my inclining and the rest.
Were it my cue to fight, I should have known it
Without a prompter. Whither will you that I go    85
To answer this your charge?
BRABANZIO          To prison, till fit time
Of law and course of direct session
Call thee to answer.
OTHELLO        What if I do obey?
How may the Duke be therewith satisfied,
Whose messengers are here about my side    90
Upon some present business of the state
To bring me to him?
OFFICER (*to Brabanzio*) 'Tis true, most worthy signor.
The Duke's in council, and your noble self,
I am sure, is sent for.
BRABANZIO        How, the Duke in council?
In this time of the night? Bring him away.    95
Mine's not an idle cause. The Duke himself,
Or any of my brothers of the state,
Cannot but feel this wrong as 'twere their own;
For if such actions may have passage free,
Bondslaves and pagans shall our statesmen be.    *Exeunt*

**1.3**    *Enter the Duke and Senators set at a table, with*
        *lights and officers*
DUKE
There is no composition in these news
That gives them credit.
FIRST SENATOR       Indeed, they are disproportioned.
My letters say a hundred and seven galleys.
DUKE
And mine a hundred-forty.
SECOND SENATOR        And mine two hundred.
But though they jump not on a just account—    5

As, in these cases, where the aim reports
'Tis oft with difference—yet do they all confirm
A Turkish fleet, and bearing up to Cyprus.

DUKE
Nay, it is possible enough to judgement.
I do not so secure me in the error,                        10
But the main article I do approve
In fearful sense.

SAILOR (*within*)     What ho, what ho, what ho!
        *Enter a Sailor*

OFFICER
A messenger from the galleys.

DUKE                          Now, what's the business?

SAILOR
The Turkish preparation makes for Rhodes.
So was I bid report here to the state                       15
By Signor Angelo.

DUKE (*to Senators*) How say you by this change?

FIRST SENATOR This cannot be,
By no assay of reason—'tis a pageant
To keep us in false gaze. When we consider                  20
The importancy of Cyprus to the Turk,
And let ourselves again but understand
That, as it more concerns the Turk than Rhodes,
So may he with more facile question bear it,
For that it stands not in such warlike brace,               25
But altogether lacks th'abilities
That Rhodes is dressed in—if we make thought of this,
We must not think the Turk is so unskilful
To leave that latest which concerns him first,
Neglecting an attempt of ease and gain                      30
To wake and wage a danger profitless.

DUKE
Nay, in all confidence, he's not for Rhodes.

OFFICER Here is more news.
        *Enter a Messenger*

MESSENGER
The Ottomites, reverend and gracious,
Steering with due course toward the Isle of Rhodes,         35
Have there injointed them with an after fleet.

FIRST SENATOR
Ay, so I thought. How many, as you guess?

MESSENGER
Of thirty sail, and now they do restem
Their backward course, bearing with frank appearance
Their purposes toward Cyprus. Signor Montano,               40
Your trusty and most valiant servitor,
With his free duty recommends you thus,
And prays you to believe him.

DUKE                          'Tis certain then for Cyprus.
Marcus Luccicos, is not he in town?

FIRST SENATOR He's now in Florence.                         45

DUKE
Write from us to him post-post-haste. Dispatch.
        *Enter Brabanzio, Othello, Roderigo, Iago, Cassio,*
        *and officers*

FIRST SENATOR
Here comes Brabanzio and the valiant Moor.

DUKE
Valiant Othello, we must straight employ you
Against the general enemy Ottoman.
(*To Brabanzio*) I did not see you. Welcome, gentle
        signor.                                             50
We lacked your counsel and your help tonight.

BRABANZIO
So did I yours. Good your grace, pardon me.
Neither my place, nor aught I heard of business,

Hath raised me from my bed, nor doth the general
        care
Take hold on me; for my particular grief                    55
Is of so floodgate and o'erbearing nature
That it engluts and swallows other sorrows,
And it is still itself.

DUKE                          Why, what's the matter?

BRABANZIO
My daughter, O, my daughter!

⌈SENATORS⌉                          Dead?

BRABANZIO                          Ay, to me.
She is abused, stol'n from me, and corrupted                60
By spells and medicines bought of mountebanks.
For nature so preposterously to err,
Being not deficient, blind, or lame of sense,
Sans witchcraft could not.

DUKE
Whoe'er he be that in this foul proceeding                  65
Hath thus beguiled your daughter of herself
And you of her, the bloody book of law
You shall yourself read in the bitter letter
After your own sense, yea, though our proper son
Stood in your action.

BRABANZIO                 Humbly I thank your grace.         70
Here is the man, this Moor, whom now it seems
Your special mandate for the state affairs
Hath hither brought.

SENATORS                 We are very sorry for't.

DUKE (*to Othello*)
What in your own part can you say to this?

BRABANZIO Nothing but this is so.                           75

OTHELLO
Most potent, grave, and reverend signors,
My very noble and approved good masters,
That I have ta'en away this old man's daughter,
It is most true, true I have married her.
The very head and front of my offending                     80
Hath this extent, no more. Rude am I in my speech,
And little blessed with the soft phrase of peace,
For since these arms of mine had seven years' pith
Till now some nine moons wasted, they have used
Their dearest action in the tented field,                   85
And little of this great world can I speak
More than pertains to feats of broils and battle.
And therefore little shall I grace my cause
In speaking for myself. Yet, by your gracious patience,
I will a round unvarnished tale deliver                      90
Of my whole course of love, what drugs, what charms,
What conjuration and what mighty magic—
For such proceeding I am charged withal—
I won his daughter.

BRABANZIO                 A maiden never bold,
Of spirit so still and quiet that her motion                95
Blushed at herself—and she in spite of nature,
Of years, of country, credit, everything,
To fall in love with what she feared to look on!
It is a judgement maimed and most imperfect
That will confess perfection so could err                   100
Against all rules of nature, and must be driven
To find out practices of cunning hell
Why this should be. I therefore vouch again
That with some mixtures powerful o'er the blood,
Or with some dram conjured to this effect,                  105
He wrought upon her.

DUKE                 To vouch this is no proof
Without more wider and more overt test

Than these thin habits and poor likelihoods
Of modern seeming do prefer against him.
A SENATOR  But Othello, speak.                                    110
  Did you by indirect and forcèd courses
  Subdue and poison this young maid's affections,
  Or came it by request and such fair question
  As soul to soul affordeth?
OTHELLO                        I do beseech you,
  Send for the lady to the Sagittary,                             115
  And let her speak of me before her father.
  If you do find me foul in her report,
  The trust, the office I do hold of you
  Not only take away, but let your sentence
  Even fall upon my life.
DUKE (to officers)          Fetch Desdemona hither.               120
OTHELLO
  Ensign, conduct them. You best know the place.
              Exit Iago with two or three officers
  And till she come, as truly as to heaven
  I do confess the vices of my blood,
  So justly to your grave ears I'll present
  How I did thrive in this fair lady's love,                      125
  And she in mine.
DUKE              Say it, Othello.
OTHELLO
  Her father loved me, oft invited me,
  Still questioned me the story of my life
  From year to year, the battles, sieges, fortunes
  That I have passed.                                             130
  I ran it through even from my boyish days
  To th' very moment that he bade me tell it,
  Wherein I spoke of most disastrous chances,
  Of moving accidents by flood and field,
  Of hair-breadth scapes i'th' imminent deadly breach,
  Of being taken by the insolent foe                              136
  And sold to slavery, of my redemption thence,
  And portance in my traveller's history,
  Wherein of antres vast and deserts idle,
  Rough quarries, rocks, and hills whose heads touch
        heaven,                                                   140
  It was my hint to speak. Such was my process,
  And of the cannibals that each other eat,
  The Anthropophagi, and men whose heads
  Do grow beneath their shoulders. These things to hear
  Would Desdemona seriously incline,                              145
  But still the house affairs would draw her thence,
  Which ever as she could with haste dispatch
  She'd come again, and with a greedy ear
  Devour up my discourse; which I observing,
  Took once a pliant hour, and found good means      150
  To draw from her a prayer of earnest heart
  That I would all my pilgrimage dilate,
  Whereof by parcels she had something heard,
  But not intentively. I did consent,
  And often did beguile her of her tears                          155
  When I did speak of some distressful stroke
  That my youth suffered. My story being done,
  She gave me for my pains a world of kisses.
  She swore in faith 'twas strange, 'twas passing strange,
  'Twas pitiful, 'twas wondrous pitiful.                          160
  She wished she had not heard it, yet she wished
  That heaven had made her such a man. She thankèd
        me,
  And bade me, if I had a friend that loved her,
  I should but teach him how to tell my story,
  And that would woo her. Upon this hint I spake.    165

  She loved me for the dangers I had passed,
  And I loved her that she did pity them.
  This only is the witchcraft I have used.
              Enter Desdemona, Iago, and attendants
  Here comes the lady. Let her witness it.
DUKE
  I think this tale would win my daughter, too.—                  170
  Good Brabanzio,
  Take up this mangled matter at the best.
  Men do their broken weapons rather use
  Than their bare hands.
BRABANZIO              I pray you hear her speak.
  If she confess that she was half the wooer,                     175
  Destruction on my head if my bad blame
  Light on the man! Come hither, gentle mistress.
  Do you perceive in all this noble company
  Where most you owe obedience?
DESDEMONA                      My noble father,
  I do perceive here a divided duty.                              180
  To you I am bound for life and education.
  My life and education both do learn me
  How to respect you. You are the lord of duty,
  I am hitherto your daughter. But here's my husband,
  And so much duty as my mother showed                            185
  To you, preferring you before her father,
  So much I challenge that I may profess
  Due to the Moor my lord.
BRABANZIO              God b'wi'you, I ha' done.
  Please it your grace, on to the state affairs.
  I had rather to adopt a child than get it.                      190
  Come hither, Moor.
  I here do give thee that with all my heart
  Which, but thou hast already, with all my heart
  I would keep from thee. (To Desdemona) For your sake,
        jewel,
  I am glad at soul I have no other child,                        195
  For thy escape would teach me tyranny,
  To hang clogs on 'em. I have done, my lord.
DUKE
  Let me speak like yourself, and lay a sentence
  Which, as a grece or step, may help these lovers
  Into your favour.                                               200
  When remedies are past, the griefs are ended
  By seeing the worst which late on hopes depended.
  To mourn a mischief that is past and gone
  Is the next way to draw new mischief on.
  What cannot be preserved when fortune takes,                    205
  Patience her injury a mockery makes.
  The robbed that smiles steals something from the thief;
  He robs himself that spends a bootless grief.
BRABANZIO
  So let the Turk of Cyprus us beguile,
  We lose it not so long as we can smile.                         210
  He bears the sentence well that nothing bears
  But the free comfort which from thence he hears,
  But he bears both the sentence and the sorrow
  That, to pay grief, must of poor patience borrow.
  These sentences, to sugar or to gall,                           215
  Being strong on both sides, are equivocal.
  But words are words. I never yet did hear
  That the bruisèd heart was piercèd through the ear.
  I humbly beseech you proceed to th'affairs of state.
DUKE  The Turk with a most mighty preparation makes
  for Cyprus. Othello, the fortitude of the place is best
  known to you, and though we have there a substitute
  of most allowed sufficiency, yet opinion, a more

sovereign mistress of effects, throws a more safer voice
on you. You must therefore be content to slubber the
gloss of your new fortunes with this more stubborn
and boisterous expedition.                                227

OTHELLO
The tyrant custom, most grave senators,
Hath made the flinty and steel couch of war
My thrice-driven bed of down. I do agnize           230
A natural and prompt alacrity
I find in hardness, and do undertake
This present wars against the Ottomites.
Most humbly therefore bending to your state,
I crave fit disposition for my wife,                235
Due reference of place and exhibition,
With such accommodation and besort
As levels with her breeding.

DUKE Why, at her father's!
BRABANZIO I will not have it so.                    240
OTHELLO Nor I.
DESDEMONA Nor would I there reside,
To put my father in impatient thoughts
By being in his eye. Most gracious Duke,
To my unfolding lend your prosperous ear,           245
And let me find a charter in your voice
T'assist my simpleness.

DUKE                       What would you, Desdemona?
DESDEMONA
That I did love the Moor to live with him,
My downright violence and storm of fortunes
May trumpet to the world. My heart's subdued        250
Even to the very quality of my lord.
I saw Othello's visage in his mind,
And to his honours and his valiant parts
Did I my soul and fortunes consecrate;
So that, dear lords, if I be left behind,           255
A moth of peace, and he go to the war,
The rites for why I love him are bereft me,
And I a heavy interim shall support
By his dear absence. Let me go with him.

OTHELLO (to the Duke) Let her have your voice.      260
Vouch with me heaven, I therefor beg it not
To please the palate of my appetite,
Nor to comply with heat—the young affects
In me defunct—and proper satisfaction,
But to be free and bounteous to her mind;           265
And heaven defend your good souls that you think
I will your serious and great business scant
When she is with me. No, when light-winged toys
Of feathered Cupid seel with wanton dullness
My speculative and officed instruments,             270
That my disports corrupt and taint my business,
Let housewives make a skillet of my helm,
And all indign and base adversities
Make head against my estimation.

DUKE
Be it as you shall privately determine,             275
Either for her stay or going. Th'affair cries haste,
And speed must answer it.

A SENATOR (to Othello)        You must away tonight.
DESDEMONA
Tonight, my lord?
DUKE                This night.
OTHELLO                              With all my heart.
DUKE
At nine i'th' morning here we'll meet again.
Othello, leave some officer behind,                 280

And he shall our commission bring to you,
And such things else of quality and respect
As doth import you.

OTHELLO            So please your grace, my ensign.
A man he is of honesty and trust.
To his conveyance I assign my wife,                 285
With what else needful your good grace shall think
To be sent after me.

DUKE            Let it be so.
Good night to everyone. (To Brabanzio) And, noble
  signor,
If virtue no delighted beauty lack,
Your son-in-law is far more fair than black.        290

A SENATOR
Adieu, brave Moor. Use Desdemona well.
BRABANZIO
Look to her, Moor, if thou hast eyes to see.
She has deceived her father, and may thee.
        ⌜Exeunt Duke, Brabanzio, Cassio, Senators, and
                                                officers⌝
OTHELLO
My life upon her faith. Honest Iago,
My Desdemona must I leave to thee.                  295
I prithee let thy wife attend on her,
And bring them after in the best advantage.
Come, Desdemona. I have but an hour
Of love, of worldly matter and direction
To spend with thee. We must obey the time.         300
                    Exeunt Othello and Desdemona

RODERIGO Iago.
IAGO What sayst thou, noble heart?
RODERIGO What will I do, think'st thou?
IAGO Why, go to bed and sleep.
RODERIGO I will incontinently drown myself.        305
IAGO If thou dost, I shall never love thee after. Why, thou
  silly gentleman!
RODERIGO It is silliness to live when to live is torment;
  and then have we a prescription to die when death is
  our physician.                                   310
IAGO O, villainous! I ha' looked upon the world for four
  times seven years, and since I could distinguish betwixt
  a benefit and an injury I never found man that knew
  how to love himself. Ere I would say I would drown
  myself for the love of a guinea-hen, I would change
  my humanity with a baboon.                        316
RODERIGO What should I do? I confess it is my shame to
  be so fond, but it is not in my virtue to amend it.
IAGO Virtue? A fig! 'Tis in ourselves that we are thus or
  thus. Our bodies are our gardens, to the which our
  wills are gardeners; so that if we will plant nettles or
  sow lettuce, set hyssop and weed up thyme, supply it
  with one gender of herbs or distract it with many,
  either to have it sterile with idleness or manured with
  industry, why, the power and corrigible authority of
  this lies in our wills. If the beam of our lives had not
  one scale of reason to peise another of sensuality, the
  blood and baseness of our natures would conduct us
  to most preposterous conclusions. But we have reason
  to cool our raging motions, our carnal stings, our
  unbitted lusts; whereof I take this that you call love to
  be a sect or scion.                              332
RODERIGO It cannot be.
IAGO It is merely a lust of the blood and a permission of
  the will. Come, be a man. Drown thyself? Drown cats
  and blind puppies. I have professed me thy friend, and
  I confess me knit to thy deserving with cables of

perdurable toughness. I could never better stead thee than now. Put money in thy purse. Follow thou the wars, defeat thy favour with an usurped beard. I say, put money in thy purse. It cannot be long that Desdemona should continue her love to the Moor—put money in thy purse—nor he his to her. It was a violent commencement in her, and thou shalt see an answerable sequestration—put but money in thy purse. These Moors are changeable in their wills—fill thy purse with money. The food that to him now is as luscious as locusts shall be to him shortly as bitter as coloquintida. She must change for youth. When she is sated with his body, she will find the error of her choice. Therefore put money in thy purse. If thou wilt needs damn thyself, do it a more delicate way than drowning. Make all the money thou canst. If sanctimony and a frail vow betwixt an erring barbarian and a super-subtle Venetian be not too hard for my wits and all the tribe of hell, thou shalt enjoy her; therefore make money. A pox o' drowning thyself—it is clean out of the way. Seek thou rather to be hanged in compassing thy joy than to be drowned and go without her.                                                    360

RODERIGO Wilt thou be fast to my hopes if I depend on the issue?

IAGO Thou art sure of me. Go, make money. I have told thee often, and I re-tell thee again and again, I hate the Moor. My cause is hearted, thine hath no less reason. Let us be conjunctive in our revenge against him. If thou canst cuckold him, thou dost thyself a pleasure, me a sport. There are many events in the womb of time, which will be delivered. Traverse, go, provide thy money. We will have more of this tomorrow. Adieu.                                                         371

RODERIGO
Where shall we meet i'th' morning?

IAGO                                              At my lodging.

RODERIGO
I'll be with thee betimes.

IAGO                                     Go to, farewell—
Do you hear, Roderigo?

RODERIGO                          I'll sell all my land.       *Exit*

IAGO
Thus do I ever make my fool my purse—                        375
For I mine own gained knowledge should profane
If I would time expend with such a snipe
But for my sport and profit. I hate the Moor,
And it is thought abroad that 'twixt my sheets
He has done my office. I know not if't be true,            380
But I, for mere suspicion in that kind,
Will do as if for surety. He holds me well:
The better shall my purpose work on him.
Cassio's a proper man. Let me see now,
To get his place, and to plume up my will                    385
In double knavery—how, how? Let's see.
After some time to abuse Othello's ears
That he is too familiar with his wife;
He hath a person and a smooth dispose
To be suspected, framed to make women false.                 390
The Moor is of a free and open nature,
That thinks men honest that but seem to be so,
And will as tenderly be led by th' nose
As asses are.
I ha't. It is ingendered. Hell and night                     395
Must bring this monstrous birth to the world's light.
                                                            *Exit*

**2.1**    *Enter below Montano, Governor of Cyprus; two
           other gentlemen ⌈above⌉*

MONTANO
What from the cape can you discern at sea?

FIRST GENTLEMAN
Nothing at all. It is a high-wrought flood.
I cannot 'twixt the heaven and the main
Descry a sail.

MONTANO
Methinks the wind hath spoke aloud at land.                    5
A fuller blast ne'er shook our battlements.
If it ha' ruffianed so upon the sea,
What ribs of oak, when mountains melt on them,
Can hold the mortise? What shall we hear of this?

SECOND GENTLEMAN
A segregation of the Turkish fleet;                           10
For do but stand upon the foaming shore,
The chidden billow seems to pelt the clouds,
The wind-shaked surge with high and monstrous mane
Seems to cast water on the burning Bear
And quench the guards of th'ever-fixèd Pole.                   15
I never did like molestation view
On the enchafèd flood.

MONTANO                    If that the Turkish fleet
Be not ensheltered and embayed, they are drowned.
It is impossible to bear it out.
                *Enter a third Gentleman*

THIRD GENTLEMAN News, lads! Our wars are done.                 20
The desperate tempest hath so banged the Turks
That their designment halts. A noble ship of Venice
Hath seen a grievous wrack and sufferance
On most part of their fleet.

MONTANO How, is this true?                                     25

THIRD GENTLEMAN The ship is here put in,
A Veronessa. Michael Cassio,
Lieutenant to the warlike Moor Othello,
Is come on shore; the Moor himself at sea,
And is in full commission here for Cyprus.                     30

MONTANO
I am glad on't; 'tis a worthy governor.

THIRD GENTLEMAN
But this same Cassio, though he speak of comfort
Touching the Turkish loss, yet he looks sadly,
And prays the Moor be safe, for they were parted
With foul and violent tempest.

MONTANO                      Pray heavens he be,               35
For I have served him, and the man commands
Like a full soldier. Let's to the sea-side, ho!—
As well to see the vessel that's come in
As to throw out our eyes for brave Othello,
Even till we make the main and th'aerial blue                  40
An indistinct regard.

THIRD GENTLEMAN           Come, let's do so,
For every minute is expectancy
Of more arrivance.
                *Enter Cassio*

CASSIO
Thanks, you the valiant of this warlike isle
That so approve the Moor! O, let the heavens                   45
Give him defence against the elements,
For I have lost him on a dangerous sea.

MONTANO Is he well shipped?

CASSIO
His barque is stoutly timbered, and his pilot
Of very expert and approved allowance.                         50

Therefore my hopes, not surfeited to death,
Stand in bold cure.
VOICES (*within*)          A sail, a sail, a sail!
CASSIO What noise?
A GENTLEMAN
The town is empty. On the brow o'th' sea
Stand ranks of people, and they cry 'A sail!'                    55
CASSIO
My hopes do shape him for the governor.
    *A shot*
A GENTLEMAN
They do discharge their shot of courtesy—
Our friends, at least.
CASSIO                    I pray you, sir, go forth,
And give us truth who 'tis that is arrived.
A GENTLEMAN I shall.                              *Exit*
MONTANO
But, good lieutenant, is your general wived?          61
CASSIO
Most fortunately. He hath achieved a maid
That paragons description and wild fame,
One that excels the quirks of blazoning pens,
And in th'essential vesture of creation                    65
Does tire the engineer.
    *Enter Gentleman*
                    How now, who has put in?
GENTLEMAN
'Tis one Iago, ensign to the general.
CASSIO
He's had most favourable and happy speed.
Tempests themselves, high seas, and howling winds,
The guttered rocks and congregated sands,          70
Traitors ensteeped to enclog the guiltless keel,
As having sense of beauty do omit
Their mortal natures, letting go safely by
The divine Desdemona.
MONTANO                    What is she?
CASSIO
She that I spake of, our great captain's captain,          75
Left in the conduct of the bold Iago,
Whose footing here anticipates our thoughts
A sennight's speed. Great Jove, Othello guard,
And swell his sail with thine own powerful breath,
That he may bless this bay with his tall ship,          80
Make love's quick pants in Desdemona's arms,
Give renewed fire to our extincted spirits,
And bring all Cyprus comfort.
    *Enter Desdemona, Iago, Emilia, and Roderigo*
                    O, behold,
The riches of the ship is come on shore!
You men of Cyprus, let her have your knees.          85
    *Montano and the Gentlemen make curtsy to Desdemona*
Hail to thee, lady, and the grace of heaven
Before, behind thee, and on every hand
Enwheel thee round!
DESDEMONA                    I thank you, valiant Cassio.
What tidings can you tell me of my lord?
CASSIO
He is not yet arrived, nor know I aught          90
But that he's well and will be shortly here.
DESDEMONA
O, but I fear—how lost you company?
CASSIO
The great contention of the sea and skies
Parted our fellowship.

VOICES (*within*) A sail, a sail!          95
CASSIO But hark, a sail.
    *A shot*
A GENTLEMAN
They give their greeting to the citadel.
This likewise is a friend.
CASSIO                    See for the news.
                    *Exit Gentleman*
Good ensign, you are welcome. (*Kissing Emilia*)
    Welcome, mistress.
Let it not gall your patience, good Iago,          100
That I extend my manners. 'Tis my breeding
That gives me this bold show of courtesy.
IAGO
Sir, would she give you so much of her lips
As of her tongue she oft bestows on me,
You would have enough.          105
DESDEMONA Alas, she has no speech!
IAGO In faith, too much.
I find it still when I ha' leave to sleep.
Marry, before your ladyship, I grant,
She puts her tongue a little in her heart,          110
And chides with thinking.
EMILIA                    You ha' little cause to say so.
IAGO
Come on, come on. You are pictures out of door,
Bells in your parlours; wildcats in your kitchens,
Saints in your injuries; devils being offended,
Players in your housewifery, and hussies in your beds.
DESDEMONA
O, fie upon thee, slanderer!          116
IAGO
Nay, it is true, or else I am a Turk.
You rise to play and go to bed to work.
EMILIA
You shall not write my praise.
IAGO                    No, let me not.
DESDEMONA
What wouldst write of me, if thou shouldst praise me?
IAGO
O, gentle lady, do not put me to't,          121
For I am nothing if not critical.
DESDEMONA
Come on, essay—there's one gone to the harbour?
IAGO Ay, madam.
DESDEMONA
I am not merry, but I do beguile          125
The thing I am by seeming otherwise.
Come, how wouldst thou praise me?
IAGO
I am about it, but indeed my invention
Comes from my pate as birdlime does from frieze—
It plucks out brains and all. But my muse labours, 130
And thus she is delivered:
If she be fair and wise, fairness and wit,
The one's for use, the other useth it.
DESDEMONA Well praised! How if she be black and witty?
IAGO
If she be black and thereto have a wit,          135
She'll find a white that shall her blackness fit.
DESDEMONA
Worse and worse.
EMILIA                    How if fair and foolish?
IAGO
She never yet was foolish that was fair,
For even her folly helped her to an heir.

DESDEMONA These are old fond paradoxes, to make fools
    laugh i'th' alehouse. 141
    What miserable praise hast thou for her
    That's foul and foolish?

IAGO
    There's none so foul and foolish thereunto,
    But does foul pranks which fair and wise ones do. 145

DESDEMONA O heavy ignorance! Thou praisest the worst
    best. But what praise couldst thou bestow on a
    deserving woman indeed—one that, in the authority
    of her merit, did justly put on the vouch of very malice
    itself? 150

IAGO
    She that was ever fair and never proud,
    Had tongue at will and yet was never loud,
    Never lacked gold and yet went never gay,
    Fled from her wish, and yet said 'Now I may';
    She that, being angered, her revenge being nigh, 155
    Bade her wrong stay and her displeasure fly;
    She that in wisdom never was so frail
    To change the cod's head for the salmon's tail;
    She that could think and ne'er disclose her mind,
    See suitors following, and not look behind— 160
    She was a wight, if ever such wights were—

DESDEMONA To do what?

IAGO
    To suckle fools, and chronicle small beer.

DESDEMONA O most lame and impotent conclusion! Do
    not learn of him, Emilia, though he be thy husband.
    How say you, Cassio, is he not a most profane and
    liberal counsellor?

CASSIO He speaks home, madam. You may relish him
    more in the soldier than in the scholar. 169

    *Cassio and Desdemona talk apart*

IAGO (*aside*) He takes her by the palm. Ay, well said—
    whisper. With as little a web as this will I ensnare as
    great a fly as Cassio. Ay, smile upon her, do. I will
    gyve thee in thine own courtship. You say true, 'tis so
    indeed. If such tricks as these strip you out of your
    lieutenantry, it had been better you had not kissed
    your three fingers so oft, which you now again you are
    most apt to play the sir in. Very good, well kissed, an
    excellent curtsy, 'tis so indeed; yet again your fingers
    to your lips? Would they were clyster-pipes for your
    sake. 180

    *Trumpets within*

(*Aloud*) The Moor—I know his trumpet.

CASSIO                'Tis truly so.

DESDEMONA
    Let's meet him and receive him.

CASSIO              Lo where he comes!

    *Enter Othello and attendants*

OTHELLO (*to Desdemona*)
    O my fair warrior!

DESDEMONA       My dear Othello.

OTHELLO
    It gives me wonder great as my content
    To see you here before me. O my soul's joy, 185
    If after every tempest come such calms,
    May the winds blow till they have wakened death,
    And let the labouring barque climb hills of seas
    Olympus-high, and duck again as low
    As hell's from heaven. If it were now to die 190
    'Twere now to be most happy, for I fear
    My soul hath her content so absolute

    That not another comfort like to this
    Succeeds in unknown fate.

DESDEMONA         The heavens forbid
    But that our loves and comforts should increase 195
    Even as our days do grow.

OTHELLO         Amen to that, sweet powers!
    I cannot speak enough of this content.
    It stops me here, it is too much of joy.
    And this, (*they kiss*) and this, the greatest discords be
    That e'er our hearts shall make.

IAGO (*aside*)        O, you are well tuned now,
    But I'll set down the pegs that make this music, 201
    As honest as I am.

OTHELLO        Come, let us to the castle.
    News, friends: our wars are done, the Turks are
      drowned.
    How does my old acquaintance of this isle?—
    Honey, you shall be well desired in Cyprus, 205
    I have found great love amongst them. O my sweet,
    I prattle out of fashion, and I dote
    In mine own comforts. I prithee, good Iago,
    Go to the bay and disembark my coffers.
    Bring thou the master to the citadel. 210
    He is a good one, and his worthiness
    Does challenge much respect. Come, Desdemona.—
    Once more, well met at Cyprus!

    *Exeunt Othello and Desdemona with all but Iago*
                              *and Roderigo*

IAGO (*to an attendant as he goes out*) Do thou meet me
    presently at the harbour. (*To Roderigo*) Come hither. If
    thou beest valiant—as they say base men being in love
    have then a nobility in their natures more than is
    native to them—list me. The lieutenant tonight watches
    on the court of guard. First, I must tell thee this:
    Desdemona is directly in love with him. 220

RODERIGO With him? Why, 'tis not possible!

IAGO Lay thy finger thus, and let thy soul be instructed.
    Mark me with what violence she first loved the Moor,
    but for bragging and telling her fantastical lies. To love
    him still for prating?—let not thy discreet heart think
    it. Her eye must be fed, and what delight shall she
    have to look on the devil? When the blood is made
    dull with the act of sport, there should be again to
    inflame it, and to give satiety a fresh appetite, loveliness
    in favour, sympathy in years, manners, and beauties,
    all which the Moor is defective in. Now, for want of
    these required conveniences, her delicate tenderness
    will find itself abused, begin to heave the gorge, disrelish
    and abhor the Moor. Very nature will instruct her in
    it and compel her to some second choice. Now, sir, this
    granted—as it is a most pregnant and unforced
    position—who stands so eminent in the degree of this
    fortune as Cassio does?—a knave very voluble, no
    further conscionable than in putting on the mere form
    of civil and humane seeming for the better compass of
    his salt and most hidden loose affection. Why, none;
    why, none—a slipper and subtle knave, a finder of
    occasion, that has an eye can stamp and counterfeit
    advantages, though true advantage never present itself,
    a devilish knave! Besides, the knave is handsome,
    young, and hath all those requisites in him that folly
    and green minds look after. A pestilent complete knave,
    and the woman hath found him already.

RODERIGO I cannot believe that in her. She's full of most
    blessed condition. 250

IAGO Blessed fig's end! The wine she drinks is made of grapes. If she had been blessed, she would never have loved the Moor. Blessed pudding! Didst thou not see her paddle with the palm of his hand? Didst not mark that?                                                                    255

RODERIGO Yes, that I did, but that was but courtesy.

IAGO Lechery, by this hand; an index and obscure prologue to the history of lust and foul thoughts. They met so near with their lips that their breaths embraced together. Villainous thoughts, Roderigo! When these mutualities so marshal the way, hard at hand comes the master and main exercise, th'incorporate conclusion. Pish! But, sir, be you ruled by me. I have brought you from Venice. Watch you tonight. For the command, I'll lay't upon you. Cassio knows you not; I'll be far from you. Do you find some occasion to anger Cassio, either by speaking too loud, or tainting his discipline, or from what other course you please, which the time shall more favourably minister.

RODERIGO Well.                                                            270

IAGO Sir, he's rash and very sudden in choler, and haply may strike at you. Provoke him that he may, for even out of that will I cause these of Cyprus to mutiny, whose qualification shall come into no true taste again but by the displanting of Cassio. So shall you have a shorter journey to your desires by the means I shall then have to prefer them, and the impediment most profitably removed, without the which there were no expectation of our prosperity.                                 279

RODERIGO I will do this, if you can bring it to any opportunity.

IAGO I warrant thee. Meet me by and by at the citadel. I must fetch his necessaries ashore. Farewell.

RODERIGO Adieu.                                                     *Exit*

IAGO
That Cassio loves her, I do well believe it.                        285
That she loves him, 'tis apt and of great credit.
The Moor—howbe't that I endure him not—
Is of a constant, loving, noble nature,
And I dare think he'll prove to Desdemona
A most dear husband. Now I do love her too,                          290
Not out of absolute lust—though peradventure
I stand accountant for as great a sin—
But partly led to diet my revenge
For that I do suspect the lusty Moor
Hath leapt into my seat, the thought whereof                         295
Doth, like a poisonous mineral, gnaw my inwards;
And nothing can or shall content my soul
Till I am evened with him, wife for wife—
Or failing so, yet that I put the Moor
At least into a jealousy so strong                                   300
That judgement cannot cure, which thing to do,
If this poor trash of Venice whom I trace
For his quick hunting stand the putting on,
I'll have our Michael Cassio on the hip,
Abuse him to the Moor in the rank garb—                              305
For I fear Cassio with my nightcap, too—
Make the Moor thank me, love me, and reward me
For making him egregiously an ass,
And practising upon his peace and quiet
Even to madness. 'Tis here, but yet confused.                        310
Knavery's plain face is never seen till used.            *Exit*

**2.2**    *Enter Othello's Herald reading a proclamation*

HERALD It is Othello's pleasure—our noble and valiant general—that, upon certain tidings now arrived importing the mere perdition of the Turkish fleet, every man put himself into triumph: some to dance, some to make bonfires, each man to what sport and revels his addiction leads him; for besides these beneficial news, it is the celebration of his nuptial. So much was his pleasure should be proclaimed. All offices are open, and there is full liberty of feasting from this present hour of five till the bell have told eleven. Heaven bless the isle of Cyprus and our noble general, Othello!     *Exit*

**2.3**    *Enter Othello, Desdemona, Cassio, and attendants*

OTHELLO
Good Michael, look you to the guard tonight.
Let's teach ourselves that honourable stop
Not to outsport discretion.

CASSIO
Iago hath direction what to do,
But notwithstanding, with my personal eye                            5
Will I look to't.

OTHELLO                Iago is most honest.
Michael, good night. Tomorrow with your earliest
Let me have speech with you. (*To Desdemona*) Come,
    my dear love,
The purchase made, the fruits are to ensue.
That profit's yet to come 'tween me and you.                         10
(*To Cassio*) Good night.
            *Exeunt Othello, Desdemona, and attendants*
        *Enter Iago*

CASSIO
Welcome, Iago. We must to the watch.

IAGO Not this hour, lieutenant; 'tis not yet ten o'th' clock. Our general cast us thus early for the love of his Desdemona, who let us not therefore blame. He hath not yet made wanton the night with her, and she is sport for Jove.

CASSIO She's a most exquisite lady.

IAGO And I'll warrant her full of game.                              19

CASSIO Indeed, she's a most fresh and delicate creature.

IAGO What an eye she has! Methinks it sounds a parley to provocation.

CASSIO An inviting eye, and yet, methinks, right modest.

IAGO And when she speaks, is it not an alarum to love?

CASSIO She is indeed perfection.                                     25

IAGO Well, happiness to their sheets. Come, lieutenant. I have a stoup of wine, and here without are a brace of Cyprus gallants that would fain have a measure to the health of black Othello.

CASSIO Not tonight, good Iago. I have very poor and unhappy brains for drinking. I could well wish courtesy would invent some other custom of entertainment.   32

IAGO O, they are our friends! But one cup. I'll drink for you.

CASSIO I ha' drunk but one cup tonight, and that was craftily qualified, too, and behold what innovation it makes here! I am infortunate in the infirmity, and dare not task my weakness with any more.

IAGO What, man, 'tis a night of revels, the gallants desire it!                                                                       40

CASSIO Where are they?

IAGO
Here at the door. I pray you call them in.

CASSIO I'll do't, but it dislikes me.                           *Exit*

IAGO
If I can fasten but one cup upon him,
With that which he hath drunk tonight already             45
He'll be as full of quarrel and offence

As my young mistress' dog. Now my sick fool Roderigo,
Whom love hath turned almost the wrong side out,
To Desdemona hath tonight caroused
Potations pottle-deep, and he's to watch.                    50
Three else of Cyprus—noble swelling spirits
That hold their honours in a wary distance,
The very elements of this warlike isle—
Have I tonight flustered with flowing cups,
And they watch too. Now 'mongst this flock of
    drunkards                                                55
Am I to put our Cassio in some action
That may offend the isle.
        *Enter Montano, Cassio, Gentlemen, ⌈and servants⌉*
        *with wine*
                        But here they come.
If consequence do but approve my dream,
My boat sails freely both with wind and stream.

CASSIO
Fore God, they have given me a rouse already.            60

MONTANO
Good faith, a little one; not past a pint,
As I am a soldier.

IAGO                    Some wine, ho!
    *(Sings)* And let me the cannikin clink, clink,
        And let me the cannikin clink.
        A soldier's a man,                                 65
        O, man's life's but a span,
        Why then, let a soldier drink.
Some wine, boys!

CASSIO Fore God, an excellent song.

IAGO I learned it in England, where indeed they are most
potent in potting. Your Dane, your German, and your
swag-bellied Hollander—drink, ho!—are nothing to
your English.                                              73

CASSIO Is your Englishman so exquisite in his drinking?

IAGO Why, he drinks you with facility your Dane dead
drunk. He sweats not to overthrow your Almain. He
gives your Hollander a vomit ere the next pottle can
be filled.

CASSIO To the health of our general!                       79

MONTANO I am for it, lieutenant, and I'll do you justice.

IAGO O sweet England!
    *(Sings)* King Stephen was and a worthy peer,
        His breeches cost him but a crown;
        He held them sixpence all too dear,
        With that he called the tailor lown.              85
        He was a wight of high renown,
        And thou art but of low degree.
        'Tis pride that pulls the country down,
        Then take thy auld cloak about thee.
Some wine, ho!                                             90

CASSIO Fore God, this is a more exquisite song than the
other.

IAGO Will you hear't again?

CASSIO No, for I hold him to be unworthy of his place
that does these things. Well, God's above all, and there
be souls must be saved, and there be souls must not
be saved.                                                 97

IAGO It's true, good lieutenant.

CASSIO For mine own part—no offence to the general, nor
any man of quality—I hope to be saved.                   100

IAGO And so do I too, lieutenant.

CASSIO Ay, but, by your leave, not before me. The
lieutenant is to be saved before the ensign. Let's ha'
no more of this. Let's to our affairs. God forgive us our
sins. Gentlemen, let's look to our business. Do not

think, gentlemen, I am drunk. This is my ensign, this
is my right hand, and this is my left. I am not drunk
now. I can stand well enough, and I speak well enough.

GENTLEMEN Excellent well.                                 109

CASSIO Why, very well then. You must not think then
that I am drunk.                                     *Exit*

MONTANO
To th' platform, masters. Come, let's set the watch.
                                        *Exeunt Gentlemen*

IAGO
You see this fellow that is gone before—
He's a soldier fit to stand by Caesar
And give direction; and do but see his vice.             115
'Tis to his virtue a just equinox,
The one as long as th'other. 'Tis pity of him.
I fear the trust Othello puts him in,
On some odd time of his infirmity,
Will shake this island.

MONTANO                    But is he often thus?          120

IAGO
'Tis evermore his prologue to his sleep.
He'll watch the horologe a double set
If drink rock not his cradle.

MONTANO                        It were well
The general were put in mind of it.
Perhaps he sees it not, or his good nature                125
Prizes the virtue that appears in Cassio,
And looks not on his evils. Is not this true?
        *Enter Roderigo*

IAGO ⌈*aside*⌉ How now, Roderigo!
I pray you after the lieutenant, go.         *Exit Roderigo*

MONTANO
And 'tis great pity that the noble Moor                   130
Should hazard such a place as his own second
With one of an engraffed infirmity.
It were an honest action to say so
To the Moor.

IAGO            Not I, for this fair island!
I do love Cassio well, and would do much                  135
To cure him of this evil.

VOICES *(within)* Help, help!

IAGO                    But hark, what noise?
        *Enter Cassio, driving in Roderigo*

CASSIO 'Swounds, you rogue, you rascal!

MONTANO What's the matter, lieutenant?                    140

CASSIO A knave teach me my duty?—I'll beat the knave
into a twiggen bottle.

RODERIGO Beat me?

CASSIO Dost thou prate, rogue?

MONTANO Nay, good lieutenant, I pray you, sir, hold your
hand.                                                     146

CASSIO Let me go, sir, or I'll knock you o'er the mazard.

MONTANO Come, come, you're drunk.

CASSIO Drunk?
        *They fight*

IAGO *(to Roderigo)*
Away, I say. Go out and cry a mutiny.        *Exit Roderigo*
Nay, good lieutenant. God's will, gentlemen!              151
Help, ho! Lieutenant! Sir! Montano! Sir!
Help, masters. Here's a goodly watch indeed.
        *A bell rung*
Who's that which rings the bell? Diablo, ho!
The town will rise. God's will, lieutenant, hold.         155
You'll be ashamed for ever.
        *Enter Othello and attendants, with weapons*

OTHELLO                            What is the matter here?

MONTANO
'Swounds, I bleed still. I am hurt to th' death.
    (*Attacking Cassio*) He dies.
OTHELLO Hold, for your lives!
IAGO
Hold, ho, lieutenant, sir, Montano, gentlemen!
Have you forgot all place of sense and duty?          160
Hold, the general speaks to you. Hold, hold, for shame.
OTHELLO
Why, how now, ho? From whence ariseth this?
Are we turned Turks, and to ourselves do that
Which heaven hath forbid the Ottomites?
For Christian shame, put by this barbarous brawl.     165
He that stirs next to carve for his own rage
Holds his soul light. He dies upon his motion.
Silence that dreadful bell—it frights the isle
From her propriety.
    ⌈*Bell stops*⌉
                    What is the matter, masters?
Honest Iago, that looks dead with grieving,            170
Speak. Who began this? On thy love I charge thee.
IAGO
I do not know. Friends all but now, even now,
In quarter and in terms like bride and groom
Devesting them for bed; and then but now—
As if some planet had unwitted men—                    175
Swords out, and tilting one at others' breasts
In opposition bloody. I cannot speak
Any beginning to this peevish odds,
And would in action glorious I had lost
Those legs that brought me to a part of it.            180
OTHELLO
How comes it, Michael, you are thus forgot?
CASSIO
I pray you pardon me. I cannot speak.
OTHELLO
Worthy Montano, you were wont be civil.
The gravity and stillness of your youth
The world hath noted, and your name is great          185
In mouths of wisest censure. What's the matter,
That you unlace your reputation thus,
And spend your rich opinion for the name
Of a night-brawler? Give me answer to it.
MONTANO
Worthy Othello, I am hurt to danger.                   190
Your officer Iago can inform you,
While I spare speech—which something now offends
    me—
Of all that I do know; nor know I aught
By me that's said or done amiss this night,
Unless self-charity be sometimes a vice,               195
And to defend ourselves it be a sin
When violence assails us.
OTHELLO                      Now, by heaven,
My blood begins my safer guides to rule,
And passion, having my best judgement collied,
Essays to lead the way. 'Swounds, if I stir,          200
Or do but lift this arm, the best of you
Shall sink in my rebuke. Give me to know
How this foul rout began, who set it on,
And he that is approved in this offence,
Though he had twinned with me, both at a birth,       205
Shall lose me. What, in a town of war
Yet wild, the people's hearts brimful of fear,
To manage private and domestic quarrel

In night, and on the court and guard of safety!
'Tis monstrous. Iago, who began't?                     210
MONTANO (*to Iago*)
If partially affined or leagued in office
Thou dost deliver more or less than truth,
Thou art no soldier.
IAGO                      Touch me not so near.
I had rather ha' this tongue cut from my mouth
Than it should do offence to Michael Cassio.          215
Yet I persuade myself to speak the truth
Shall nothing wrong him. This it is, general.
Montano and myself being in speech,
There comes a fellow crying out for help,
And Cassio following him with determined sword        220
To execute upon him. Sir, this gentleman
Steps in to Cassio, and entreats his pause.
Myself the crying fellow did pursue,
Lest by his clamour, as it so fell out,
The town might fall in fright. He, swift of foot,     225
Outran my purpose, and I returned, the rather
For that I heard the clink and fall of swords
And Cassio high in oath, which till tonight
I ne'er might say before. When I came back—
For this was brief—I found them close together        230
At blow and thrust, even as again they were
When you yourself did part them.
More of this matter cannot I report,
But men are men. The best sometimes forget.
Though Cassio did some little wrong to him,            235
As men in rage strike those that wish them best,
Yet surely Cassio, I believe, received
From him that fled some strange indignity
Which patience could not pass.
OTHELLO                      I know, Iago,
Thy honesty and love doth mince this matter,          240
Making it light to Cassio. Cassio, I love thee,
But never more be officer of mine.
    *Enter Desdemona, attended*
Look if my gentle love be not raised up.
I'll make thee an example.
DESDEMONA What is the matter, dear?                    245
OTHELLO All's well now, sweeting.
Come away to bed. (*To Montano*) Sir, for your hurts
Myself will be your surgeon. (*To attendants*) Lead him
    off.                    *Exeunt attendants with Montano*
Iago, look with care about the town,
And silence those whom this vile brawl distracted.    250
Come, Desdemona. 'Tis the soldier's life
To have their balmy slumbers waked with strife.
                    *Exeunt all but Iago and Cassio*
IAGO What, are you hurt, lieutenant?
CASSIO Ay, past all surgery.
IAGO Marry, God forbid.                                255
CASSIO Reputation, reputation, reputation—O, I ha' lost
    my reputation, I ha' lost the immortal part of myself,
    and what remains is bestial! My reputation, Iago, my
    reputation.                                        259
IAGO As I am an honest man, I thought you had received
    some bodily wound. There is more sense in that than
    in reputation. Reputation is an idle and most false
    imposition, oft got without merit and lost without
    deserving. You have lost no reputation at all unless
    you repute yourself such a loser. What, man, there are
    more ways to recover the general again. You are but
    now cast in his mood—a punishment more in policy

than in malice, even so as one would beat his offenceless
dog to affright an imperious lion. Sue to him again,
and he's yours.                                              270

CASSIO I will rather sue to be despised than to deceive so
good a commander with so slight, so drunken, and so
indiscreet an officer. Drunk, and speak parrot, and
squabble? Swagger, swear, and discourse fustian with
one's own shadow? O thou invisible spirit of wine, if
thou hast no name to be known by, let us call thee
devil.                                                       277

IAGO What was he that you followed with your sword?
What had he done to you?

CASSIO I know not.                                           280

IAGO Is't possible?

CASSIO I remember a mass of things, but nothing distinctly; a quarrel, but nothing wherefore. O God, that
men should put an enemy in their mouths to steal
away their brains! That we should with joy, pleasance,
revel, and applause transform ourselves into beasts!

IAGO Why, but you are now well enough. How came you
thus recovered?                                              288

CASSIO It hath pleased the devil drunkenness to give place
to the devil wrath. One unperfectness shows me
another, to make me frankly despise myself.

IAGO Come, you are too severe a moraller. As the time,
the place, and the condition of this country stands, I
could heartily wish this had not befallen; but since it
is as it is, mend it for your own good.                      295

CASSIO I will ask him for my place again. He shall tell me
I am a drunkard. Had I as many mouths as Hydra,
such an answer would stop them all. To be now a
sensible man, by and by a fool, and presently a beast!
O, strange! Every inordinate cup is unblessed, and the
ingredient is a devil.                                       301

IAGO Come, come. Good wine is a good familiar creature,
if it be well used. Exclaim no more against it. And,
good lieutenant, I think you think I love you.

CASSIO I have well approved it, sir—I drunk?             305

IAGO You or any man living may be drunk at a time,
man. I'll tell you what you shall do. Our general's wife
is now the general. I may say so in this respect, for
that he hath devoted and given up himself to the
contemplation, mark, and denotement of her parts and
graces. Confess yourself freely to her. Importune her
help to put you in your place again. She is of so free,
so kind, so apt, so blessed a disposition, she holds it a
vice in her goodness not to do more than she is
requested. This broken joint between you and her
husband entreat her to splinter, and, my fortunes
against any lay worth naming, this crack of your love
shall grow stronger than it was before.

CASSIO You advise me well.                                   319

IAGO I protest, in the sincerity of love and honest kindness.

CASSIO I think it freely, and betimes in the morning I will
beseech the virtuous Desdemona to undertake for me.
I am desperate of my fortunes if they check me here.

IAGO You are in the right. Good night, lieutenant. I must
to the watch.                                                325

CASSIO Good night, honest Iago.                         Exit

IAGO
And what's he then that says I play the villain,
When this advice is free I give, and honest,
Probal to thinking, and indeed the course
To win the Moor again? For 'tis most easy              330
Th'inclining Desdemona to subdue
In any honest suit. She's framed as fruitful
As the free elements; and then for her

To win the Moor, were't to renounce his baptism,
All seals and symbols of redeemèd sin,                 335
His soul is so enfettered to her love
That she may make, unmake, do what she list,
Even as her appetite shall play the god
With his weak function. How am I then a villain,
To counsel Cassio to this parallel course              340
Directly to his good? Divinity of hell:
When devils will the blackest sins put on,
They do suggest at first with heavenly shows,
As I do now; for whiles this honest fool
Plies Desdemona to repair his fortune,                 345
And she for him pleads strongly to the Moor,
I'll pour this pestilence into his ear:
That she repeals him for her body's lust,
And by how much she strives to do him good
She shall undo her credit with the Moor.               350
So will I turn her virtue into pitch,
And out of her own goodness make the net
That shall enmesh them all.
      *Enter Roderigo*
                              How now, Roderigo?

RODERIGO I do follow here in the chase, not like a hound
that hunts, but one that fills up the cry. My money is
almost spent, I ha' been tonight exceedingly well
cudgelled, and I think the issue will be I shall have so
much experience for my pains: and so, with no money
at all and a little more wit, return again to Venice.

IAGO
How poor are they that ha' not patience!               360
What wound did ever heal but by degrees?
Thou know'st we work by wit and not by witchcraft,
And wit depends on dilatory time.
Does't not go well? Cassio hath beaten thee,
And thou by that small hurt hast cashiered Cassio.
Though other things grow fair against the sun,         366
Yet fruits that blossom first will first be ripe.
Content thyself a while. By the mass, 'tis morning.
Pleasure and action make the hours seem short.
Retire thee. Go where thou art billeted.               370
Away, I say. Thou shalt know more hereafter.
Nay, get thee gone.                       *Exit Roderigo*
                    Two things are to be done.
My wife must move for Cassio to her mistress.
I'll set her on.
Myself a while to draw the Moor apart,                 375
And bring him jump when he may Cassio find
Soliciting his wife. Ay, that's the way.
Dull not device by coldness and delay.          *Exit*

**3.1**  *Enter Cassio with Musicians*
CASSIO
Masters, play here—I will content your pains—
Something that's brief, and bid 'Good morrow, general'.
      *Music. Enter Clown*
CLOWN Why, masters, ha' your instruments been in
Naples, that they speak i'th' nose thus?
MUSICIAN How, sir, how?                                       5
CLOWN Are these, I pray you, wind instruments?
MUSICIAN Ay, marry are they, sir.
CLOWN O, thereby hangs a tail.
MUSICIAN Whereby hangs a tale, sir?
CLOWN Marry, sir, by many a wind instrument that I
know. But masters, here's money for you, and the
general so likes your music that he desires you, for
love's sake, to make no more noise with it.            13

MUSICIAN Well, sir, we will not.

CLOWN If you have any music that may not be heard, to't again; but, as they say, to hear music the general does not greatly care.

MUSICIAN We ha' none such, sir.        18

CLOWN Then put up your pipes in your bag, for I'll away. Go, vanish into air, away.      *Exeunt Musicians*

CASSIO Dost thou hear, my honest friend?

CLOWN No, I hear not your honest friend, I hear you.

CASSIO Prithee, keep up thy quillets. There's a poor piece of gold for thee. If the gentlewoman that attends the general's wife be stirring, tell her there's one Cassio entreats her a little favour of speech. Wilt thou do this?

CLOWN She is stirring, sir. If she will stir hither, I shall seem to notify unto her.      28

CASSIO

Do, good my friend.            *Exit Clown*

     *Enter Iago*

             In happy time, Iago.

IAGO

You ha' not been abed, then.

CASSIO           Why, no. The day had broke

Before we parted. I ha' made bold, Iago,     31

To send in to your wife. My suit to her

Is that she will to virtuous Desdemona

Procure me some access.

IAGO

I'll send her to you presently,          35

And I'll devise a mean to draw the Moor

Out of the way, that your converse and business

May be more free.

CASSIO       I humbly thank you for't.     *Exit Iago*

I never knew a Florentine more kind and honest.

     *Enter Emilia*

EMILIA

Good morrow, good lieutenant. I am sorry    40

For your displeasure, but all will sure be well.

The general and his wife are talking of it,

And she speaks for you stoutly. The Moor replies

That he you hurt is of great fame in Cyprus,

And great affinity, and that in wholesome wisdom   45

He might not but refuse you. But he protests he loves you,

And needs no other suitor but his likings

To take the saf'st occasion by the front

To bring you in again.

CASSIO          Yet I beseech you,

If you think fit, or that it may be done,     50

Give me advantage of some brief discourse

With Desdemon alone.

EMILIA           Pray you come in.

I will bestow you where you shall have time

To speak your bosom freely.

CASSIO          I am much bound to you.

                          *Exeunt*

**3.2**    *Enter Othello, Iago, and Gentlemen*

OTHELLO

These letters give, Iago, to the pilot,

And by him do my duties to the senate.

That done, I will be walking on the works.

Repair there to me.

IAGO       Well, my good lord, I'll do't.    *Exit*

OTHELLO

This fortification, gentlemen—shall we see't?    5

A GENTLEMAN We'll wait upon your lordship.    *Exeunt*

**3.3**    *Enter Desdemona, Cassio, and Emilia*

DESDEMONA

Be thou assured, good Cassio, I will do

All my abilities in thy behalf.

EMILIA

Good madam, do. I warrant it grieves my husband

As if the cause were his.

DESDEMONA

O, that's an honest fellow. Do not doubt, Cassio,   5

But I will have my lord and you again

As friendly as you were.

CASSIO          Bounteous madam,

Whatever shall become of Michael Cassio

He's never anything but your true servant.

DESDEMONA

I know't. I thank you. You do love my lord.    10

You have known him long, and be you well assured

He shall in strangeness stand no farther off

Than in a politic distance.

CASSIO          Ay, but, lady,

That policy may either last so long,

Or feed upon such nice and wat'rish diet,     15

Or breed itself so out of circumstance,

That, I being absent and my place supplied,

My general will forget my love and service.

DESDEMONA

Do not doubt that. Before Emilia here

I give thee warrant of thy place. Assure thee,    20

If I do vow a friendship I'll perform it

To the last article. My lord shall never rest.

I'll watch him tame, and talk him out of patience.

His bed shall seem a school, his board a shrift.

I'll intermingle everything he does        25

With Cassio's suit. Therefore be merry, Cassio,

For thy solicitor shall rather die

Than give thy cause away.

     *Enter Othello and Iago*

EMILIA          Madam, here comes my lord.

CASSIO

Madam, I'll take my leave.

DESDEMONA        Why, stay, and hear me speak.

CASSIO

Madam, not now. I am very ill at ease,     30

Unfit for mine own purposes.

DESDEMONA Well, do your discretion.    *Exit Cassio*

IAGO Ha! I like not that.

OTHELLO What dost thou say?

IAGO

Nothing, my lord. Or if, I know not what.    35

OTHELLO

Was not that Cassio parted from my wife?

IAGO

Cassio, my lord? No, sure, I cannot think it,

That he would steal away so guilty-like

Seeing your coming.

OTHELLO I do believe 'twas he.      40

DESDEMONA How now, my lord?

I have been talking with a suitor here,

A man that languishes in your displeasure.

OTHELLO Who is't you mean?

DESDEMONA

Why, your lieutenant, Cassio; good my lord,    45

If I have any grace or power to move you,

His present reconciliation take;

For if he be not one that truly loves you,

That errs in ignorance and not in cunning,

I have no judgement in an honest face.            50
I prithee call him back.
OTHELLO Went he hence now?
DESDEMONA Yes, faith, so humbled
  That he hath left part of his grief with me
  To suffer with him. Good love, call him back.    55
OTHELLO
  Not now, sweet Desdemon. Some other time.
DESDEMONA
  But shall't be shortly?
OTHELLO                    The sooner, sweet, for you.
DESDEMONA
  Shall't be tonight at supper?
OTHELLO                    No, not tonight.
DESDEMONA
  Tomorrow dinner, then?
OTHELLO                    I shall not dine at home.
  I meet the captains at the citadel.               60
DESDEMONA
  Why then, tomorrow night, or Tuesday morn,
  On Tuesday noon, or night, on Wednesday morn—
  I prithee name the time, but let it not
  Exceed three days. In faith, he's penitent,
  And yet his trespass, in our common reason—       65
  Save that, they say, the wars must make example
  Out of her best—is not almost a fault
  T'incur a private check. When shall he come?
  Tell me, Othello. I wonder in my soul
  What you would ask me that I should deny,          70
  Or stand so mamm'ring on? What, Michael Cassio,
  That came a-wooing with you, and so many a time
  When I have spoke of you dispraisingly
  Hath ta'en your part—to have so much to-do
  To bring him in? By'r Lady, I could do much.      75
OTHELLO
  Prithee, no more. Let him come when he will.
  I will deny thee nothing.
DESDEMONA                    Why, this is not a boon.
  'Tis as I should entreat you wear your gloves,
  Or feed on nourishing dishes, or keep you warm,
  Or sue to you to do a peculiar profit              80
  To your own person. Nay, when I have a suit
  Wherein I mean to touch your love indeed,
  It shall be full of poise and difficult weight,
  And fearful to be granted.
OTHELLO                    I will deny thee nothing,
  Whereon I do beseech thee grant me this:           85
  To leave me but a little to myself.
DESDEMONA
  Shall I deny you? No. Farewell, my lord.
OTHELLO
  Farewell, my Desdemona. I'll come to thee straight.
DESDEMONA
  Emilia, come. (To Othello) Be as your fancies teach you.
  Whate'er you be, I am obedient.                    90
                    Exeunt Desdemona and Emilia
OTHELLO
  Excellent wretch! Perdition catch my soul
  But I do love thee, and when I love thee not,
  Chaos is come again.
IAGO My noble lord.
OTHELLO What dost thou say, Iago?                    95
IAGO
  Did Michael Cassio, when you wooed my lady,
  Know of your love?

OTHELLO
  He did, from first to last. Why dost thou ask?
IAGO
  But for a satisfaction of my thought,
  No further harm.
OTHELLO          Why of thy thought, Iago?          100
IAGO
  I did not think he had been acquainted with her.
OTHELLO
  O yes, and went between us very oft.
IAGO Indeed?
OTHELLO
  Indeed? Ay, indeed. Discern'st thou aught in that?
  Is he not honest?                                  105
IAGO Honest, my lord?
OTHELLO Honest? Ay, honest.
IAGO
  My lord, for aught I know.
OTHELLO                    What dost thou think?
IAGO Think, my lord?
OTHELLO
  'Think, my lord?' By heaven, thou echo'st me      110
  As if there were some monster in thy thought
  Too hideous to be shown! Thou dost mean something.
  I heard thee say even now thou liked'st not that,
  When Cassio left my wife. What didst not like?
  And when I told thee he was of my counsel          115
  In my whole course of wooing, thou cried'st 'Indeed?'
  And didst contract and purse thy brow together
  As if thou then hadst shut up in thy brain
  Some horrible conceit. If thou dost love me,
  Show me thy thought.                               120
IAGO My lord, you know I love you.
OTHELLO I think thou dost,
  And for I know thou'rt full of love and honesty,
  And weigh'st thy words before thou giv'st them
    breath,
  Therefore these stops of thine fright me the more; 125
  For such things in a false disloyal knave
  Are tricks of custom, but in a man that's just,
  They're close dilations, working from the heart
  That passion cannot rule.
IAGO                    For Michael Cassio,
  I dare be sworn I think that he is honest.          130
OTHELLO
  I think so too.
IAGO          Men should be what they seem,
  Or those that be not, would they might seem none.
OTHELLO
  Certain, men should be what they seem.
IAGO
  Why then, I think Cassio's an honest man.
OTHELLO Nay, yet there's more in this.               135
  I prithee speak to me as to thy thinkings,
  As thou dost ruminate, and give thy worst of thoughts
  The worst of words.
IAGO                    Good my lord, pardon me.
  Though I am bound to every act of duty,
  I am not bound to that all slaves are free to.      140
  Utter my thoughts? Why, say they are vile and false,
  As where's that palace whereinto foul things
  Sometimes intrude not? Who has that breast so pure
  But some uncleanly apprehensions
  Keep leets and law-days, and in sessions sit        145
  With meditations lawful?

OTHELLO
Thou dost conspire against thy friend, Iago,
If thou but think'st him wronged and mak'st his ear
A stranger to thy thoughts.
IAGO                 I do beseech you,
Though I perchance am vicious in my guess—    150
As I confess it is my nature's plague
To spy into abuses, and oft my jealousy
Shapes faults that are not—that your wisdom then,
From one that so imperfectly conceits,
Would take no notice, nor build yourself a trouble   155
Out of his scattering and unsure observance.
It were not for your quiet nor your good,
Nor for my manhood, honesty, and wisdom,
To let you know my thoughts.
OTHELLO            What dost thou mean?
IAGO
Good name in man and woman, dear my lord,    160
Is the immediate jewel of their souls.
Who steals my purse steals trash; 'tis something, nothing;
'Twas mine, 'tis his, and has been slave to thousands.
But he that filches from me my good name
Robs me of that which not enriches him       165
And makes me poor indeed.
OTHELLO       By heaven, I'll know thy thoughts.
IAGO
You cannot, if my heart were in your hand;
Nor shall not whilst 'tis in my custody.
OTHELLO
Ha!
IAGO   O, beware, my lord, of jealousy.
It is the green-eyed monster which doth mock    170
The meat it feeds on. That cuckold lives in bliss
Who, certain of his fate, loves not his wronger.
But O, what damnèd minutes tells he o'er
Who dotes yet doubts, suspects yet fondly loves!
OTHELLO   O misery!                175
IAGO
Poor and content is rich, and rich enough,
But riches fineless is as poor as winter
To him that ever fears he shall be poor.
Good God the souls of all my tribe defend
From jealousy!
OTHELLO       Why, why is this?       180
Think'st thou I'd make a life of jealousy,
To follow still the changes of the moon
With fresh suspicions? No, to be once in doubt
Is once to be resolved. Exchange me for a goat
When I shall turn the business of my soul     185
To such exsufflicate and blowed surmises
Matching thy inference. 'Tis not to make me jealous
To say my wife is fair, feeds well, loves company,
Is free of speech, sings, plays, and dances well.
Where virtue is, these are more virtuous,     190
Nor from mine own weak merits will I draw
The smallest fear or doubt of her revolt,
For she had eyes and chose me. No, Iago,
I'll see before I doubt; when I doubt, prove;
And on the proof, there is no more but this:    195
Away at once with love or jealousy.
IAGO
I am glad of this, for now I shall have reason
To show the love and duty that I bear you
With franker spirit. Therefore, as I am bound,
Receive it from me. I speak not yet of proof.    200

Look to your wife. Observe her well with Cassio.
Wear your eyes thus: not jealous, nor secure.
I would not have your free and noble nature
Out of self-bounty be abused. Look to't.
I know our country disposition well.       205
In Venice they do let God see the pranks
They dare not show their husbands; their best conscience
Is not to leave't undone, but keep't unknown.
OTHELLO   Dost thou say so?
IAGO
She did deceive her father, marrying you,     210
And when she seemed to shake and fear your looks
She loved them most.
OTHELLO       And so she did.
IAGO                Why, go to, then.
She that so young could give out such a seeming,
To seel her father's eyes up close as oak,
He thought 'twas witchcraft! But I am much to blame.
I humbly do beseech you of your pardon     216
For too much loving you.
OTHELLO       I am bound to thee for ever.
IAGO
I see this hath a little dashed your spirits.
OTHELLO
Not a jot, not a jot.
IAGO           I'faith, I fear it has.
I hope you will consider what is spoke      220
Comes from my love. But I do see you're moved.
I am to pray you not to strain my speech
To grosser issues, nor to larger reach
Than to suspicion.
OTHELLO   I will not.              225
IAGO   Should you do so, my lord,
My speech should fall into such vile success
Which my thoughts aimed not. Cassio's my worthy friend.
My lord, I see you're moved.
OTHELLO          No, not much moved.
I do not think but Desdemona's honest.     230
IAGO
Long live she so, and long live you to think so!
OTHELLO
And yet how nature, erring from itself—
IAGO
Ay, there's the point; as, to be bold with you,
Not to affect many proposèd matches
Of her own clime, complexion, and degree,    235
Whereto we see in all things nature tends.
Foh, one may smell in such a will most rank,
Foul disproportions, thoughts unnatural!
But pardon me. I do not in position
Distinctly speak of her, though I may fear     240
Her will, recoiling to her better judgement,
May fall to match you with her country forms
And happily repent.
OTHELLO       Farewell, farewell.
If more thou dost perceive, let me know more.
Set on thy wife to observe. Leave me, Iago.    245
IAGO (going) My lord, I take my leave.
OTHELLO
Why did I marry? This honest creature doubtless
Sees and knows more, much more, than he unfolds.
IAGO (returning)
My lord, I would I might entreat your honour
To scan this thing no farther. Leave it to time.   250

Although 'tis fit that Cassio have his place—
For sure he fills it up with great ability—
Yet, if you please to hold him off a while,
You shall by that perceive him and his means.
Note if your lady strain his entertainment                          255
With any strong or vehement importunity.
Much will be seen in that. In the mean time,
Let me be thought too busy in my fears—
As worthy cause I have to fear I am—
And hold her free, I do beseech your honour.                        260

OTHELLO
Fear not my government.

IAGO                          I once more take my leave.
                                                                    *Exit*

OTHELLO
This fellow's of exceeding honesty,
And knows all qualities with a learned spirit
Of human dealings. If I do prove her haggard,
Though that her jesses were my dear heart-strings 265
I'd whistle her off and let her down the wind
To prey at fortune. Haply for I am black,
And have not those soft parts of conversation
That chamberers have; or for I am declined
Into the vale of years—yet that's not much—                         270
She's gone. I am abused, and my relief
Must be to loathe her. O curse of marriage,
That we can call these delicate creatures ours
And not their appetites! I had rather be a toad
And live upon the vapour of a dungeon                               275
Than keep a corner in the thing I love
For others' uses. Yet 'tis the plague of great ones;
Prerogatived are they less than the base.
'Tis destiny unshunnable, like death.
Even then this forkèd plague is fated to us                         280
When we do quicken.
        *Enter Desdemona and Emilia*
                        Look where she comes.
If she be false, O then heaven mocks itself!
I'll not believe't.

DESDEMONA          How now, my dear Othello?
Your dinner, and the generous islanders
By you invited, do attend your presence.                            285

OTHELLO I am to blame.

DESDEMONA
Why do you speak so faintly? Are you not well?

OTHELLO
I have a pain upon my forehead here.

DESDEMONA
Faith, that's with watching. 'Twill away again.
Let me but bind it hard, within this hour                           290
It will be well.

OTHELLO          Your napkin is too little.
        *He puts the napkin from him. It drops.*
Let it alone. Come, I'll go in with you.

DESDEMONA
I am very sorry that you are not well.
                *Exeunt Othello and Desdemona*

EMILIA (*taking up the napkin*)
I am glad I have found this napkin.
This was her first remembrance from the Moor.                       295
My wayward husband hath a hundred times
Wooed me to steal it, but she so loves the token—
For he conjured her she should ever keep it—
That she reserves it evermore about her
To kiss and talk to. I'll ha' the work ta'en out,                   300
And give't Iago. What he will do with it,

Heaven knows, not I.
I nothing, but to please his fantasy.
        *Enter Iago*

IAGO
How now, what do you here alone?

EMILIA
Do not you chide. I have a thing for you.                           305

IAGO
You have a thing for me? It is a common thing.

EMILIA Ha?

IAGO To have a foolish wife.

EMILIA
O, is that all? What will you give me now
For that same handkerchief?                                         310

IAGO What handkerchief?

EMILIA What handkerchief?
Why, that the Moor first gave to Desdemona,
That which so often you did bid me steal.

IAGO Hast stol'n it from her?                                       315

EMILIA
No, faith, she let it drop by negligence,
And to th'advantage I, being here, took't up.
Look, here 'tis.

IAGO                A good wench! Give it me.

EMILIA
What will you do with it, that you have been so earnest
To have me filch it?

IAGO                Why, what is that to you?        320
        *He takes the napkin*

EMILIA
If it be not for some purpose of import,
Give't me again. Poor lady, she'll run mad
When she shall lack it.

IAGO
Be not acknown on't. I have use for it. Go, leave me.
                                                *Exit Emilia*
I will in Cassio's lodging lose this napkin,                        325
And let him find it. Trifles light as air
Are to the jealous confirmations strong
As proofs of holy writ. This may do something.
The Moor already changes with my poison.
Dangerous conceits are in their natures poisons,                   330
Which at the first are scarce found to distaste,
But, with a little act upon the blood,
Burn like the mines of sulphur.
        *Enter Othello*
                                I did say so.
Look where he comes. Not poppy nor mandragora
Nor all the drowsy syrups of the world                             335
Shall ever medicine thee to that sweet sleep
Which thou owedst yesterday.

OTHELLO Ha, ha, false to me?

IAGO
Why, how now, general? No more of that.

OTHELLO
Avaunt, be gone. Thou hast set me on the rack.                     340
I swear 'tis better to be much abused
Than but to know't a little.

IAGO                        How now, my lord?

OTHELLO
What sense had I of her stol'n hours of lust?
I saw't not, thought it not; it harmed not me.
I slept the next night well, fed well, was free and
        merry.                                                     345
I found not Cassio's kisses on her lips.
He that is robbed, not wanting what is stol'n,
Let him not know't and he's not robbed at all.

IAGO I am sorry to hear this.

OTHELLO
I had been happy if the general camp,                    350
Pioneers and all, had tasted her sweet body,
So I had nothing known. O, now for ever
Farewell the tranquil mind, farewell content,
Farewell the plumèd troops and the big wars
That makes ambition virtue! O, farewell,                 355
Farewell the neighing steed and the shrill trump,
The spirit-stirring drum, th'ear-piercing fife,
The royal banner, and all quality,
Pride, pomp, and circumstance of glorious war!
And O, you mortal engines whose rude throats            360
Th'immortal Jove's dread clamours counterfeit,
Farewell! Othello's occupation's gone.

IAGO Is't possible, my lord?

OTHELLO ⌈taking Iago by the throat⌉
Villain, be sure thou prove my love a whore.
Be sure of it. Give me the ocular proof,                 365
Or, by the worth of mine eternal soul,
Thou hadst been better have been born a dog
Than answer my waked wrath.

IAGO                                Is't come to this?

OTHELLO
Make me to see't, or at the least so prove it
That the probation bear no hinge nor loop                370
To hang a doubt on, or woe upon thy life.

IAGO My noble lord.

OTHELLO
If thou dost slander her and torture me,
Never pray more; abandon all remorse,
On horror's head horrors accumulate,                     375
Do deeds to make heaven weep, all earth amazed,
For nothing canst thou to damnation add
Greater than that.

IAGO                        O grace, O heaven forgive me!
Are you a man? Have you a soul or sense?
God buy you, take mine office. O wretched fool,          380
That lov'st to make thine honesty a vice!
O monstrous world, take note, take note, O world,
To be direct and honest is not safe!
I thank you for this profit, and from hence
I'll love no friend, sith love breeds such offence.      385

OTHELLO Nay, stay. Thou shouldst be honest.

IAGO
I should be wise, for honesty's a fool,
And loses that it works for.

OTHELLO                        By the world,
I think my wife be honest, and think she is not.
I think that thou art just, and think thou art not.      390
I'll have some proof. My name, that was as fresh
As Dian's visage, is now begrimed and black
As mine own face. If there be cords, or knives,
Poison, or fire, or suffocating streams,
I'll not endure it. Would I were satisfied!              395

IAGO
I see, sir, you are eaten up with passion.
I do repent me that I put it to you.
You would be satisfied?

OTHELLO                        Would? Nay, and I will.

IAGO
And may. But how, how satisfied, my lord?
Would you, the supervisor, grossly gape on,              400
Behold her topped?

OTHELLO                        Death and damnation! O!

IAGO
It were a tedious difficulty, I think,
To bring them to that prospect. Damn them then
If ever mortal eyes do see them bolster
More than their own! What then, how then?                405
What shall I say? Where's satisfaction?
It is impossible you should see this,
Were they as prime as goats, as hot as monkeys,
As salt as wolves in pride, and fools as gross
As ignorance made drunk. But yet I say,                  410
If imputation, and strong circumstances
Which lead directly to the door of truth,
Will give you satisfaction, you might ha't.

OTHELLO
Give me a living reason she's disloyal.

IAGO I do not like the office,                           415
But sith I am entered in this cause so far,
Pricked to't by foolish honesty and love,
I will go on. I lay with Cassio lately,
And being troubled with a raging tooth,
I could not sleep. There are a kind of men              420
So loose of soul that in their sleeps
Will mutter their affairs. One of this kind is Cassio.
In sleep I heard him say 'Sweet Desdemona,
Let us be wary, let us hide our loves',
And then, sir, would he grip and wring my hand,          425
Cry 'O, sweet creature!', then kiss me hard,
As if he plucked up kisses by the roots,
That grew upon my lips, lay his leg o'er my thigh,
And sigh, and kiss, and then cry 'Cursèd fate,
That gave thee to the Moor!'                             430

OTHELLO O, monstrous, monstrous!

IAGO Nay, this was but his dream.

OTHELLO
But this denoted a foregone conclusion.

IAGO
'Tis a shrewd doubt, though it be but a dream,
And this may help to thicken other proofs                435
That do demonstrate thinly.

OTHELLO                              I'll tear her all to pieces.

IAGO
Nay, yet be wise; yet we see nothing done.
She may be honest yet. Tell me but this:
Have you not sometimes seen a handkerchief
Spotted with strawberries in your wife's hand?           440

OTHELLO
I gave her such a one. 'Twas my first gift.

IAGO
I know not that, but such a handkerchief—
I am sure it was your wife's—did I today
See Cassio wipe his beard with.

OTHELLO                              If it be that—

IAGO
If it be that, or any that was hers,                     445
It speaks against her with the other proofs.

OTHELLO
O that the slave had forty thousand lives!
One is too poor, too weak for my revenge.
Now do I see 'tis true. Look here, Iago.
All my fond love thus do I blow to heaven—'tis gone.
Arise, black vengeance, from the hollow hell.            451
Yield up, O love, thy crown and hearted throne
To tyrannous hate! Swell, bosom, with thy freight,
For 'tis of aspics' tongues.

IAGO                              Yet be content.

OTHELLO
O, blood, blood, blood!
IAGO                    Patience, I say. Your mind may change.
OTHELLO
Never, Iago. Like to the Pontic Sea,                                    456
Whose icy current and compulsive course
Ne'er knows retiring ebb, but keeps due on
To the Propontic and the Hellespont,
Even so my bloody thoughts with violent pace            460
Shall ne'er look back, ne'er ebb to humble love,
Till that a capable and wide revenge
Swallow them up.
        ⌈He kneels⌉
                    Now, by yon marble heaven,
In the due reverence of a sacred vow
I here engage my words.
IAGO                    Do not rise yet.            465
        Iago kneels
Witness you ever-burning lights above,
You elements that clip us round about,
Witness that here Iago doth give up
The execution of his wit, hands, heart
To wronged Othello's service. Let him command,       470
And to obey shall be in me remorse,
What bloody business ever.
        ⌈They rise⌉
OTHELLO                    I greet thy love,
Not with vain thanks, but with acceptance bounteous,
And will upon the instant put thee to't.
Within these three days let me hear thee say            475
That Cassio's not alive.
IAGO                    My friend is dead.
'Tis done at your request; but let her live.
OTHELLO
Damn her, lewd minx! O, damn her, damn her!
Come, go with me apart. I will withdraw
To furnish me with some swift means of death           480
For the fair devil. Now art thou my lieutenant.
IAGO I am your own for ever.                    Exeunt

3.4    Enter Desdemona, Emilia, and the Clown
DESDEMONA Do you know, sirrah, where Lieutenant Cassio
    lies?
CLOWN I dare not say he lies anywhere.
DESDEMONA Why, man?
CLOWN He's a soldier, and for me to say a soldier lies, 'tis
    stabbing.                                             6
DESDEMONA Go to. Where lodges he?
CLOWN To tell you where he lodges is to tell you where I
    lie.
DESDEMONA Can anything be made of this?            10
CLOWN I know not where he lodges, and for me to devise
    a lodging and say he lies here, or he lies there, were
    to lie in mine own throat.
DESDEMONA Can you enquire him out, and be edified by
    report?                                             15
CLOWN I will catechize the world for him; that is, make
    questions, and by them answer.
DESDEMONA Seek him, bid him come hither, tell him I
    have moved my lord on his behalf, and hope all will
    be well.                                             20
CLOWN To do this is within the compass of man's wit,
    and therefore I will attempt the doing it.        Exit
DESDEMONA
    Where should I lose the handkerchief, Emilia?
EMILIA I know not, madam.

DESDEMONA
Believe me, I had rather have lost my purse          25
Full of crusadoes, and but my noble Moor
Is true of mind, and made of no such baseness
As jealous creatures are, it were enough
To put him to ill thinking.
EMILIA                    Is he not jealous?
DESDEMONA
Who, he? I think the sun where he was born          30
Drew all such humours from him.
        Enter Othello
EMILIA                    Look where he comes.
DESDEMONA
I will not leave him now till Cassio
Be called to him. How is't with you, my lord?
OTHELLO
Well, my good lady. (Aside) O hardness to dissemble!—
How do you, Desdemona?
DESDEMONA                    Well, my good lord.       35
OTHELLO
Give me your hand. This hand is moist, my lady.
DESDEMONA
It hath felt no age, nor known no sorrow.
OTHELLO
This argues fruitfulness and liberal heart.
Hot, hot and moist—this hand of yours requires
A sequester from liberty; fasting, and prayer,       40
Much castigation, exercise devout,
For here's a young and sweating devil here
That commonly rebels. 'Tis a good hand,
A frank one.
DESDEMONA        You may indeed say so,
For 'twas that hand that gave away my heart.         45
OTHELLO
A liberal hand. The hearts of old gave hands,
But our new heraldry is hands, not hearts.
DESDEMONA
I cannot speak of this. Come now, your promise.
OTHELLO What promise, chuck?
DESDEMONA
I have sent to bid Cassio come speak with you.       50
OTHELLO
I have a salt and sorry rheum offends me.
Lend me thy handkerchief.
DESDEMONA (offering a handkerchief) Here, my lord.
OTHELLO
That which I gave you.
DESDEMONA                I have it not about me.
OTHELLO Not?
DESDEMONA
No, faith, my lord.
OTHELLO                That's a fault. That handkerchief
Did an Egyptian to my mother give.                    56
She was a charmer, and could almost read
The thoughts of people. She told her, while she kept it
'Twould make her amiable, and subdue my father
Entirely to her love; but if she lost it,            60
Or made a gift of it, my father's eye
Should hold her loathèd, and his spirits should hunt
After new fancies. She, dying, gave it me,
And bid me, when my fate would have me wived,
To give it her. I did so, and take heed on't.        65
Make it a darling, like your precious eye.
To lose't or give't away were such perdition
As nothing else could match.
DESDEMONA                Is't possible?

OTHELLO
'Tis true. There's magic in the web of it.
A sibyl that had numbered in the world                    70
The sun to course two hundred compasses
In her prophetic fury sewed the work.
The worms were hallowed that did breed the silk,
And it was dyed in mummy, which the skilful
Conserved of maidens' hearts.
DESDEMONA                    I'faith, is't true?    75
OTHELLO
Most veritable. Therefore look to't well.
DESDEMONA
Then would to God that I had never seen it!
OTHELLO Ha, wherefore?
DESDEMONA
Why do you speak so startingly and rash?
OTHELLO
Is't lost? Is't gone? Speak, is't out o'th' way?    80
DESDEMONA Heaven bless us!
OTHELLO Say you?
DESDEMONA
It is not lost, but what an if it were?
OTHELLO How?
DESDEMONA
I say it is not lost.
OTHELLO                    Fetch't, let me see't.    85
DESDEMONA
Why, so I can, sir, but I will not now.
This is a trick to put me from my suit.
Pray you let Cassio be received again.
OTHELLO
Fetch me the handkerchief. My mind misgives.
DESDEMONA
Come, come, you'll never meet a more sufficient man.
OTHELLO
The handkerchief.
DESDEMONA                    I pray, talk me of Cassio.    91
OTHELLO
The handkerchief.
DESDEMONA                    A man that all his time
Hath founded his good fortunes on your love,
Shared dangers with you—
OTHELLO The handkerchief.    95
DESDEMONA I'faith, you are to blame.
OTHELLO 'Swounds!                    Exit
EMILIA
Is not this man jealous?
DESDEMONA                    I ne'er saw this before.
Sure there's some wonder in this handkerchief.
I am most unhappy in the loss of it.    100
EMILIA
'Tis not a year or two shows us a man.
They are all but stomachs, and we all but food.
They eat us hungrily, and when they are full,
They belch us.
        Enter Iago and Cassio
                Look you, Cassio and my husband.
IAGO (to Cassio)
There is no other way. 'Tis she must do't,    105
And lo, the happiness! Go and importune her.
DESDEMONA
How now, good Cassio? What's the news with you?
CASSIO
Madam, my former suit. I do beseech you
That by your virtuous means I may again

Exist and be a member of his love    110
Whom I, with all the office of my heart,
Entirely honour. I would not be delayed.
If my offence be of such mortal kind
That nor my service past, nor present sorrows,
Nor purposed merit in futurity    115
Can ransom me into his love again,
But to know so must be my benefit.
So shall I clothe me in a forced content,
And shut myself up in some other course
To fortune's alms.
DESDEMONA                    Alas, thrice-gentle Cassio!    120
My advocation is not now in tune.
My lord is not my lord, nor should I know him
Were he in favour as in humour altered.
So help me every spirit sanctified
As I have spoken for you all my best,    125
And stood within the blank of his displeasure
For my free speech! You must a while be patient.
What I can do I will, and more I will
Than for myself I dare. Let that suffice you.
IAGO
Is my lord angry?
EMILIA                    He went hence but now,    130
And certainly in strange unquietness.
IAGO
Can he be angry? I have seen the cannon
When it hath blown his ranks into the air,
And, like the devil, from his very arm
Puffed his own brother; and is he angry?    135
Something of moment then. I will go meet him.
There's matter in't indeed, if he be angry.
DESDEMONA
I prithee do so.                    Exit Iago
                Something sure of state,
Either from Venice or some unhatched practice
Made demonstrable here in Cyprus to him,    140
Hath puddled his clear spirit; and in such cases
Men's natures wrangle with inferior things,
Though great ones are their object. 'Tis even so;
For let our finger ache and it indues
Our other, healthful members even to a sense    145
Of pain. Nay, we must think men are not gods,
Nor of them look for such observancy
As fits the bridal. Beshrew me much, Emilia,
I was—unhandsome warrior as I am—
Arraigning his unkindness with my soul;    150
But now I find I had suborned the witness,
And he's indicted falsely.
EMILIA                    Pray heaven it be
State matters, as you think, and no conception
Nor no jealous toy concerning you.
DESDEMONA
Alas the day, I never gave him cause.    155
EMILIA
But jealous souls will not be answered so.
They are not ever jealous for the cause,
But jealous for they're jealous. It is a monster
Begot upon itself, born on itself.
DESDEMONA
Heaven keep the monster from Othello's mind.    160
EMILIA Lady, amen.
DESDEMONA
I will go seek him. Cassio, walk here about.
If I do find him fit I'll move your suit,

And seek to effect it to my uttermost.

CASSIO
I humbly thank your ladyship.                           165

*Exeunt Desdemona and Emilia*
*Enter Bianca*

BIANCA
Save you, friend Cassio.

CASSIO                    What make you from home?
How is't with you, my most fair Bianca?
I'faith, sweet love, I was coming to your house.

BIANCA
And I was going to your lodging, Cassio.
What, keep a week away? Seven days and nights, 170
Eightscore-eight hours, and lovers' absent hours
More tedious than the dial eightscore times!
O weary reckoning!

CASSIO                          Pardon me, Bianca,
I have this while with leaden thoughts been pressed,
But I shall in a more continuate time                    175
Strike off this score of absence. Sweet Bianca,
Take me this work out.

*He gives her Desdemona's napkin*

BIANCA                          O Cassio, whence came this?
This is some token from a newer friend.
To the felt absence now I feel a cause.
Is't come to this? Well, well.

CASSIO                              Go to, woman.           180
Throw your vile guesses in the devil's teeth,
From whence you have them. You are jealous now
That this is from some mistress, some remembrance.
No, by my faith, Bianca.

BIANCA                          Why, whose is it?

CASSIO
I know not, neither. I found it in my chamber.        185
I like the work well. Ere it be demanded—
As like enough it will—I would have it copied.
Take it, and do't, and leave me for this time.

BIANCA Leave you? Wherefore?

CASSIO
I do attend here on the general,                        190
And think it no addition, nor my wish,
To have him see me womaned.

BIANCA                          Why, I pray you?

CASSIO
Not that I love you not.

BIANCA                          But that you do not love me.
I pray you bring me on the way a little,
And say if I shall see you soon at night.               195

CASSIO
'Tis but a little way that I can bring you,
For I attend here; but I'll see you soon.

BIANCA
'Tis very good. I must be circumstanced.        *Exeunt*

**4.1**   *Enter Iago and Othello*

IAGO
Will you think so?

OTHELLO                  Think so, Iago?

IAGO
What, to kiss in private?

OTHELLO                  An unauthorized kiss.

IAGO
Or to be naked with her friend in bed
An hour or more, not meaning any harm?

OTHELLO
Naked in bed, Iago, and not mean harm?                  5

It is hypocrisy against the devil.
They that mean virtuously and yet do so,
The devil their virtue tempts, and they tempt heaven.

IAGO
If they do nothing, 'tis a venial slip.
But if I give my wife a handkerchief—                    10

OTHELLO What then?

IAGO
Why then, 'tis hers, my lord, and being hers,
She may, I think, bestow't on any man.

OTHELLO
She is protectress of her honour, too.
May she give that?                                       15

IAGO
Her honour is an essence that's not seen.
They have it very oft that have it not.
But for the handkerchief—

OTHELLO
By heaven, I would most gladly have forgot it.
Thou said'st—O, it comes o'er my memory                  20
As doth the raven o'er the infectious house,
Boding to all!—he had my handkerchief.

IAGO
Ay, what of that?

OTHELLO                  That's not so good now.

IAGO
What if I had said I had seen him do you wrong,
Or heard him say—as knaves be such abroad,               25
Who having by their own importunate suit
Or voluntary dotage of some mistress
Convincèd or supplied them, cannot choose
But they must blab—

OTHELLO                  Hath he said anything?

IAGO
He hath, my lord. But, be you well assured,              30
No more than he'll unswear.

OTHELLO                  What hath he said?

IAGO
Faith, that he did—I know not what he did.

OTHELLO
What, what?

IAGO                  Lie—

OTHELLO          With her?

IAGO                  With her, on her, what you will.

OTHELLO Lie with her? Lie on her? We say 'lie on her'
when they belie her. Lie with her? 'Swounds, that's
fulsome! Handkerchief—confessions—hankerchief. To
confess and be hanged for his labour. First to be hanged
and then to confess! I tremble at it. Nature would not
invest herself in such shadowing passion without some
instruction. It is not words that shakes me thus. Pish!
Noses, ears, and lips! Is't possible? Confess? Handkerchief? O devil!                                       42

*He falls down in a trance*

IAGO
Work on; my medicine works. Thus credulous fools
    are caught,
And many worthy and chaste dames even thus,
All guiltless, meet reproach. What ho, my lord!          45
My lord, I say. Othello!

*Enter Cassio*

How now, Cassio?

CASSIO What's the matter?

IAGO
My lord is fall'n into an epilepsy.
This is his second fit. He had one yesterday.

CASSIO
Rub him about the temples.

IAGO                          No, forbear.                        50
The lethargy must have his quiet course.
If not, he foams at mouth, and by and by
Breaks out to savage madness. Look, he stirs.
Do you withdraw yourself a little while,
He will recover straight. When he is gone          55
I would on great occasion speak with you.
                                        *Exit Cassio*
How is it, general? Have you not hurt your head?

OTHELLO
Dost thou mock me?

IAGO                        I mock you not, by heaven.
Would you would bear your fortune like a man.

OTHELLO
A hornèd man's a monster and a beast.              60

IAGO
There's many a beast then in a populous city,
And many a civil monster.

OTHELLO Did he confess it?

IAGO Good sir, be a man.
Think every bearded fellow that's but yoked        65
May draw with you. There's millions now alive
That nightly lie in those unproper beds
Which they dare swear peculiar. Your case is better.
O, 'tis the spite of hell, the fiend's arch-mock,
To lip a wanton in a secure couch                  70
And to suppose her chaste! No, let me know,
And knowing what I am, I know what she shall be.

OTHELLO
O, thou art wise, 'tis certain.

IAGO                              Stand you a while apart.
Confine yourself but in a patient list.
Whilst you were here, o'erwhelmèd with your grief—
A passion most unsuiting such a man—               76
Cassio came hither. I shifted him away,
And laid good 'scuse upon your ecstasy,
Bade him anon return and here speak with me,
The which he promised. Do but encave yourself,     80
And mark the fleers, the gibes and notable scorns
That dwell in every region of his face.
For I will make him tell the tale anew,
Where, how, how oft, how long ago, and when
He hath and is again to cope your wife.            85
I say, but mark his gesture. Marry, patience,
Or I shall say you're all-in-all in spleen,
And nothing of a man.

OTHELLO               Dost thou hear, Iago?
I will be found most cunning in my patience,
But—dost thou hear?—most bloody.

IAGO                              That's not amiss,
But yet keep time in all. Will you withdraw?       91
                              *Othello stands apart*
Now will I question Cassio of Bianca,
A hussy that by selling her desires
Buys herself bread and cloth. It is a creature
That dotes on Cassio—as 'tis the strumpet's plague  95
To beguile many and be beguiled by one.
He, when he hears of her, cannot restrain
From the excess of laughter.
                  *Enter Cassio*
                              Here he comes.
As he shall smile, Othello shall go mad;
And his unbookish jealousy must conster            100
Poor Cassio's smiles, gestures, and light behaviours
Quite in the wrong. How do you now, lieutenant?

CASSIO
The worser that you give me the addition
Whose want even kills me.

IAGO
Ply Desdemona well and you are sure on't.          105
Now, if this suit lay in Bianca's power,
How quickly should you speed!

CASSIO (*laughing*) Alas, poor caitiff!

OTHELLO (*aside*) Look how he laughs already.

IAGO
I never knew a woman love man so.                  110

CASSIO
Alas, poor rogue! I think i'faith she loves me.

OTHELLO (*aside*)
Now he denies it faintly, and laughs it out.

IAGO
Do you hear, Cassio?

OTHELLO (*aside*)           Now he importunes him
To tell it o'er. Go to, well said, well said.

IAGO
She gives it out that you shall marry her.         115
Do you intend it?

CASSIO                Ha, ha, ha!

OTHELLO (*aside*)
Do ye triumph, Roman, do you triumph?

CASSIO I marry! What, a customer? Prithee, bear some
charity to my wit—do not think it so unwholesome.
Ha, ha, ha!                                        120

OTHELLO (*aside*) So, so, so, so. They laugh that wins.

IAGO Faith, the cry goes that you marry her.

CASSIO Prithee, say true.

IAGO I am a very villain else.

OTHELLO (*aside*) Ha' you scored me? Well.         125

CASSIO This is the monkey's own giving out. She is
persuaded I will marry her out of her own love and
flattery, not out of my promise.

OTHELLO (*aside*) Iago beckons me. Now he begins the story.
                  *Othello draws closer*

CASSIO She was here even now. She haunts me in every
place. I was the other day talking on the sea-bank with
certain Venetians, and thither comes the bauble, and
falls me thus about my neck.

OTHELLO (*aside*) Crying 'O dear Cassio!' as it were. His
gesture imports it.                                135

CASSIO So hangs and lolls and weeps upon me, so shakes
and pulls me—ha, ha, ha!

OTHELLO (*aside*) Now he tells how she plucked him to my
chamber. O, I see that nose of yours, but not that dog
I shall throw it to!                               140

CASSIO Well, I must leave her company.
                  *Enter Bianca*

IAGO Before me, look where she comes.

CASSIO 'Tis such another fitchew! Marry, a perfumed one.
(*To Bianca*) What do you mean by this haunting of me?

BIANCA Let the devil and his dam haunt you. What did
you mean by that same handkerchief you gave me
even now? I was a fine fool to take it. I must take out
the whole work—a likely piece of work, that you should
find it in your chamber and know not who left it there.
This is some minx's token, and I must take out the
work. There, give it your hobby-horse. (*Giving Cassio
the napkin*) Wheresoever you had it, I'll take out no
work on't.

CASSIO How now, my sweet Bianca, how now, how now?

OTHELLO (*aside*)
By heaven, that should be my handkerchief.         155

BIANCA An you'll come to supper tonight, you may. An you will not, come when you are next prepared for.

*Exit*

IAGO After her, after her.

CASSIO Faith, I must, she'll rail in the streets else.

IAGO Will you sup there?                                    160

CASSIO Faith, I intend so.

IAGO Well, I may chance to see you, for I would very fain speak with you.

CASSIO Prithee, come, will you?

IAGO Go to, say no more.                          *Exit Cassio*

OTHELLO How shall I murder him, Iago?           166

IAGO Did you perceive how he laughed at his vice?

OTHELLO O Iago!

IAGO And did you see the handkerchief?

OTHELLO Was that mine?                                    170

IAGO Yours, by this hand. And to see how he prizes the foolish woman your wife. She gave it him, and he hath given it his whore.

OTHELLO I would have him nine years a-killing. A fine woman, a fair woman, a sweet woman.      175

IAGO Nay, you must forget that.

OTHELLO Ay, let her rot and perish, and be damned tonight, for she shall not live. No, my heart is turned to stone; I strike it, and it hurts my hand. O, the world hath not a sweeter creature! She might lie by an emperor's side, and command him tasks.      181

IAGO Nay, that's not your way.

OTHELLO Hang her, I do but say what she is—so delicate with her needle, an admirable musician. O, she will sing the savageness out of a bear! Of so high and plenteous wit and invention.               186

IAGO She's the worse for all this.

OTHELLO O, a thousand, a thousand times! And then of so gentle a condition.

IAGO Ay, too gentle.                                        190

OTHELLO Nay, that's certain. But yet the pity of it, Iago. O, Iago, the pity of it, Iago!

IAGO If you are so fond over her iniquity, give her patent to offend; for if it touch not you, it comes near nobody.

OTHELLO I will chop her into messes. Cuckold me!   195

IAGO O, 'tis foul in her.

OTHELLO With mine officer.

IAGO That's fouler.

OTHELLO Get me some poison, Iago, this night. I'll not expostulate with her, lest her body and beauty unprovide my mind again. This night, Iago.      201

IAGO Do it not with poison. Strangle her in her bed, even the bed she hath contaminated.

OTHELLO Good, good, the justice of it pleases, very good.

IAGO And for Cassio, let me be his undertaker. You shall hear more by midnight.                           206

OTHELLO Excellent good.

*A trumpet*

What trumpet is that same?

IAGO I warrant, something from Venice.

*Enter Lodovico, Desdemona, and attendants*

'Tis Lodovico. This comes from the Duke. See, your wife's with him.                                        211

LODOVICO God save the worthy general.

OTHELLO With all my heart, sir.

LODOVICO (*giving Othello a letter*) The Duke and the senators of Venice greet you.                               215

OTHELLO I kiss the instrument of their pleasures.

*He reads the letter*

DESDEMONA
   And what's the news, good cousin Lodovico?

IAGO (*to Lodovico*) I am very glad to see you, signor. Welcome to Cyprus.

LODOVICO I thank you. How does Lieutenant Cassio?   220

IAGO Lives, sir.

DESDEMONA
   Cousin, there's fall'n between him and my lord
   An unkind breach. But you shall make all well.

OTHELLO Are you sure of that?

DESDEMONA My lord.                                        225

OTHELLO (*reads*) 'This fail you not to do as you will'—

LODOVICO
   He did not call, he's busy in the paper.
   Is there division 'twixt my lord and Cassio?

DESDEMONA
   A most unhappy one. I would do much
   T'atone them, for the love I bear to Cassio.      230

OTHELLO
   Fire and brimstone!

DESDEMONA                My lord?

OTHELLO                          Are you wise?

DESDEMONA
   What, is he angry?

LODOVICO               Maybe the letter moved him,
   For, as I think, they do command him home,
   Deputing Cassio in his government.

DESDEMONA By my troth, I am glad on't.            235

OTHELLO Indeed!

DESDEMONA My lord?

OTHELLO (*to Desdemona*) I am glad to see you mad.

DESDEMONA Why, sweet Othello!

OTHELLO Devil!                                            240

*He strikes her*

DESDEMONA I have not deserved this.

LODOVICO
   My lord, this would not be believed in Venice,
   Though I should swear I saw't. 'Tis very much.
   Make her amends, she weeps.

OTHELLO                      O, devil, devil!
   If that the earth could teem with woman's tears,   245
   Each drop she falls would prove a crocodile.
   Out of my sight!

DESDEMONA (*going*) I will not stay to offend you.

LODOVICO
   Truly, an obedient lady.
   I do beseech your lordship call her back.

OTHELLO Mistress!                                         250

DESDEMONA (*returning*) My lord?

OTHELLO (*to Lodovico*) What would you with her, sir?

LODOVICO Who, I, my lord?

OTHELLO
   Ay, you did wish that I would make her turn.
   Sir, she can turn and turn, and yet go on           255
   And turn again, and she can weep, sir, weep,
   And she's obedient, as you say, obedient,
   Very obedient. (*To Desdemona*) Proceed you in your tears.
   (*To Lodovico*) Concerning this, sir—(*To Desdemona*) O well painted passion!
   (*To Lodovico*) I am commanded home. (*To Desdemona*) Get you away.                                           260
   I'll send for you anon. (*To Lodovico*) Sir, I obey the mandate,
   And will return to Venice. (*To Desdemona*) Hence, avaunt!                          *Exit Desdemona*

(*To Lodovico*) Cassio shall have my place, and, sir,
    tonight
I do entreat that we may sup together.
You are welcome, sir, to Cyprus. Goats and monkeys!
                                                    *Exit*

LODOVICO
Is this the noble Moor whom our full senate          266
Call all-in-all sufficient? Is this the nature
Whom passion could not shake, whose solid virtue
The shot of accident nor dart of chance
Could neither graze nor pierce?
IAGO                           He is much changed.
LODOVICO
Are his wits safe? Is he not light of brain?         271
IAGO
He's that he is. I may not breathe my censure
What he might be. If what he might he is not,
I would to heaven he were.
LODOVICO            What, strike his wife!
IAGO
Faith, that was not so well. Yet would I knew         275
That stroke would prove the worst.
LODOVICO                        Is it his use,
Or did the letters work upon his blood
And new-create his fault?
IAGO                Alas, alas.
It is not honesty in me to speak
What I have seen and known. You shall observe him,
And his own courses will denote him so               281
That I may save my speech. Do but go after,
And mark how he continues.
LODOVICO
I am sorry that I am deceived in him.       *Exeunt*

**4.2**    *Enter Othello and Emilia*
OTHELLO You have seen nothing then?
EMILIA
Nor ever heard, nor ever did suspect.
OTHELLO
Yes, you have seen Cassio and she together.
EMILIA
But then I saw no harm, and then I heard
Each syllable that breath made up between 'em.        5
OTHELLO What, did they never whisper?
EMILIA Never, my lord.
OTHELLO Nor send you out o'th' way?
EMILIA Never.
OTHELLO
To fetch her fan, her gloves, her mask, nor nothing?
EMILIA Never, my lord.                                11
OTHELLO That's strange.
EMILIA
I durst, my lord, to wager she is honest,
Lay down my soul at stake. If you think other,
Remove your thought; it doth abuse your bosom.       15
If any wretch ha' put this in your head,
Let heaven requite it with the serpent's curse,
For if she be not honest, chaste, and true,
There's no man happy; the purest of their wives
Is foul as slander.
OTHELLO        Bid her come hither. Go.               20
                                       *Exit Emilia*
She says enough, yet she's a simple bawd
That cannot say as much. This is a subtle whore,
A closet lock and key of villainous secrets,
And yet she'll kneel and pray—I ha' seen her do't.

*Enter Desdemona and Emilia*
DESDEMONA
My lord, what is your will?
OTHELLO            Pray you, chuck, come hither.
DESDEMONA
What is your pleasure?
OTHELLO            Let me see your eyes.              26
Look in my face.
DESDEMONA What horrible fancy's this?
OTHELLO (*to Emilia*) Some of your function, mistress.
Leave procreants alone, and shut the door,            30
Cough or cry 'Hem' if anybody come.
Your mystery, your mystery—nay, dispatch.
                                        *Exit Emilia*
DESDEMONA
Upon my knees, what doth your speech import?
I understand a fury in your words,
But not the words.
OTHELLO        Why, what art thou?                    35
DESDEMONA
Your wife, my lord, your true and loyal wife.
OTHELLO Come, swear it, damn thyself,
Lest, being like one of heaven, the devils themselves
Should fear to seize thee. Therefore be double-damned:
Swear thou art honest.
DESDEMONA        Heaven doth truly know it.           40
OTHELLO
Heaven truly knows that thou art false as hell.
DESDEMONA
To whom, my lord? With whom? How am I false?
OTHELLO (*weeping*)
Ah, Desdemon, away, away, away!
DESDEMONA
Alas, the heavy day! Why do you weep?
Am I the motive of these tears, my lord?              45
If haply you my father do suspect
An instrument of this your calling back,
Lay not your blame on me. If you have lost him,
I have lost him too.
OTHELLO        Had it pleased God
To try me with affliction; had He rained              50
All kind of sores and shames on my bare head,
Steeped me in poverty to the very lips,
Given to captivity me and my utmost hopes,
I should have found in some place of my soul
A drop of patience. But, alas, to make me             55
The fixèd figure for the time of scorn
To point his slow and moving finger at—
Yet could I bear that too, well, very well.
But there where I have garnered up my heart,
Where either I must live or bear no life,             60
The fountain from the which my current runs
Or else dries up—to be discarded thence,
Or keep it as a cistern for foul toads
To knot and gender in! Turn thy complexion there,
Patience, thou young and rose-lipped cherubin,        65
Ay, here look grim as hell.
DESDEMONA
I hope my noble lord esteems me honest.
OTHELLO
O, ay—as summer flies are in the shambles,
That quicken even with blowing. O thou weed,
Who art so lovely fair, and smell'st so sweet,        70
That the sense aches at thee—would thou hadst ne'er
    been born!
DESDEMONA
Alas, what ignorant sin have I committed?

OTHELLO
Was this fair paper, this most goodly book,
Made to write 'whore' upon? What committed?
Committed? O thou public commoner,                    75
I should make very forges of my cheeks,
That would to cinders burn up modesty,
Did I but speak thy deeds. What committed?
Heaven stops the nose at it, and the moon winks;
The bawdy wind, that kisses all it meets,              80
Is hushed within the hollow mine of earth
And will not hear't. What committed?
DESDEMONA  By heaven, you do me wrong.
OTHELLO  Are not you a strumpet?
DESDEMONA  No, as I am a Christian.                    85
If to preserve this vessel for my lord
From any other foul unlawful touch
Be not to be a strumpet, I am none.
OTHELLO
What, not a whore?
DESDEMONA                No, as I shall be saved.
OTHELLO  Is't possible?                                90
DESDEMONA  O heaven forgive us!
OTHELLO  I cry you mercy then.
I took you for that cunning whore of Venice
That married with Othello. (Calling) You, mistress,
That have the office opposite to Saint Peter          95
And keeps the gate of hell,
          Enter Emilia
                          you, you, ay, you.
We ha' done our course. (Giving money) There's
          money for your pains.
I pray you, turn the key and keep our counsel.    Exit
EMILIA
Alas, what does this gentleman conceive?
How do you, madam? How do you, my good lady?
DESDEMONA  Faith, half asleep.                        101
EMILIA
Good madam, what's the matter with my lord?
DESDEMONA
With who?
EMILIA          Why, with my lord, madam.
DESDEMONA
Who is thy lord?
EMILIA                He that is yours, sweet lady.
DESDEMONA
I ha' none. Do not talk to me, Emilia.                105
I cannot weep, nor answers have I none
But what should go by water. Prithee tonight
Lay on my bed my wedding sheets, remember.
And call thy husband hither.
EMILIA                    Here's a change indeed.
                                              Exit
DESDEMONA
'Tis meet I should be used so, very meet.             110
How have I been behaved, that he might stick
The small'st opinion on my least misuse?
          Enter Iago and Emilia
IAGO
What is your pleasure, madam? How is't with you?
DESDEMONA
I cannot tell. Those that do teach young babes
Do it with gentle means and easy tasks.              115
He might ha' chid me so, for, in good faith,
I am a child to chiding.
IAGO                    What is the matter, lady?

EMILIA
Alas, Iago, my lord hath so bewhored her,
Thrown such despite and heavy terms upon her,
That true hearts cannot bear it.                      120
DESDEMONA  Am I that name, Iago?
IAGO  What name, fair lady?
DESDEMONA
Such as she said my lord did say I was.
EMILIA
He called her whore. A beggar in his drink
Could not have laid such terms upon his callet.       125
IAGO  Why did he so?
DESDEMONA
I do not know. I am sure I am none such.
IAGO
Do not weep, do not weep. Alas the day!
EMILIA
Hath she forsook so many noble matches,
Her father and her country and her friends,           130
To be called whore? Would it not make one weep?
DESDEMONA
It is my wretched fortune.
IAGO                        Beshrew him for't.
How comes this trick upon him?
DESDEMONA                        Nay, heaven doth know.
EMILIA
I will be hanged if some eternal villain,
Some busy and insinuating rogue,                      135
Some cogging, cozening slave, to get some office,
Have not devised this slander. I will be hanged else.
IAGO
Fie, there is no such man. It is impossible.
DESDEMONA
If any such there be, heaven pardon him.
EMILIA
A halter pardon him, and hell gnaw his bones!         140
Why should he call her whore? Who keeps her
          company?
What place, what time, what form, what likelihood?
The Moor's abused by some most villainous knave,
Some base, notorious knave, some scurvy fellow.
O heaven, that such companions thou'dst unfold,       145
And put in every honest hand a whip
To lash the rascals naked through the world,
Even from the east to th' west!
IAGO                            Speak within door.
EMILIA
O, fie upon them. Some such squire he was
That turned your wit the seamy side without,          150
And made you to suspect me with the Moor.
IAGO
You are a fool. Go to.
DESDEMONA                O God, Iago,
What shall I do to win my lord again?
Good friend, go to him; for by this light of heaven,
I know not how I lost him.
          She kneels
                          Here I kneel.                155
If e'er my will did trespass 'gainst his love,
Either in discourse of thought or actual deed,
Or that mine eyes, mine ears, or any sense
Delighted them in any other form,
Or that I do not yet, and ever did,                   160
And ever will—though he do shake me off
To beggarly divorcement—love him dearly,
Comfort forswear me. Unkindness may do much,

And his unkindness may defeat my life,
But never taint my love.
　　　　⌈*She rises*⌉
　　　　　　　　　I cannot say 'whore'.　165
It does abhor me now I speak the word.
To do the act that might the addition earn,
Not the world's mass of vanity could make me.
IAGO
I pray you, be content. 'Tis but his humour.
The business of the state does him offence,　170
And he does chide with you.
DESDEMONA If 'twere no other!
IAGO It is but so, I warrant.
　　　　*Flourish within*
Hark how these instruments summon you to supper.
The messengers of Venice stays the meat.　175
Go in, and weep not. All things shall be well.
　　　　　　　*Exeunt Desdemona and Emilia*
　　　　*Enter Roderigo*
How now, Roderigo?
RODERIGO
I do not find that thou deal'st justly with me.
IAGO What in the contrary?
RODERIGO Every day thou daff'st me with some device,
Iago, and rather, as it seems to me now, keep'st from
me all conveniency than suppliest me with the least
advantage of hope. I will indeed no longer endure it,
nor am I yet persuaded to put up in peace what already
I have foolishly suffered.　185
IAGO Will you hear me, Roderigo?
RODERIGO Faith, I have heard too much, for your words
and performances are no kin together.
IAGO You charge me most unjustly.
RODERIGO With naught but truth. I have wasted myself
out of my means. The jewels you have had from me
to deliver Desdemona would half have corrupted a
votarist. You have told me she hath received 'em, and
returned me expectations and comforts of sudden
respect and acquaintance, but I find none.　195
IAGO Well, go to, very well.
RODERIGO 'Very well', 'go to'! I cannot go to, man, nor
'tis not very well. Nay, I think it is scurvy, and begin
to find myself fopped in it.
IAGO Very well.　200
RODERIGO I tell you 'tis not very well. I will make myself
known to Desdemona. If she will return me my jewels,
I will give over my suit and repent my unlawful
solicitation. If not, assure yourself I will seek satisfaction
of you.　205
IAGO You have said now.
RODERIGO Ay, and said nothing but what I protest
intendment of doing.
IAGO Why, now I see there's mettle in thee, and even
from this instant do build on thee a better opinion than
ever before. Give me thy hand, Roderigo. Thou hast
taken against me a most just exception, but yet I protest
I have dealt most directly in thy affair.
RODERIGO It hath not appeared.　214
IAGO I grant, indeed, it hath not appeared, and your
suspicion is not without wit and judgement. But,
Roderigo, if thou hast that in thee indeed which I have
greater reason to believe now than ever—I mean
purpose, courage, and valour—this night show it. If
thou the next night following enjoy not Desdemona,
take me from this world with treachery, and devise
engines for my life.　222

RODERIGO Well, what is it? Is it within reason and
compass?
IAGO Sir, there is especial commission come from Venice
to depute Cassio in Othello's place.　226
RODERIGO Is that true? Why then, Othello and Desdemona
return again to Venice.
IAGO O no, he goes into Mauritania, and takes away with
him the fair Desdemona, unless his abode be lingered
here by some accident, wherein none can be so
determinate as the removing of Cassio.　232
RODERIGO How do you mean 'removing' of him?
IAGO Why, by making him uncapable of Othello's place—
knocking out his brains.　235
RODERIGO And that you would have me to do.
IAGO Ay, if you dare do yourself a profit and a right. He
sups tonight with a harlotry, and thither will I go to
him. He knows not yet of his honourable fortune. If
you will watch his going thence, which I will fashion
to fall out between twelve and one, you may take him
at your pleasure. I will be near, to second your attempt,
and he shall fall between us. Come, stand not amazed
at it, but go along with me. I will show you such a
necessity in his death that you shall think yourself
bound to put it on him. It is now high supper-time,
and the night grows to waste. About it.　247
RODERIGO I will hear further reason for this.
IAGO And you shall be satisfied.　　　　　*Exeunt*

**4.3**　*Enter Othello, Desdemona, Lodovico, Emilia, and
attendants*
LODOVICO
I do beseech you, sir, trouble yourself no further.
OTHELLO
O, pardon me, 'twill do me good to walk.
LODOVICO (*to Desdemona*)
Madam, good night. I humbly thank your ladyship.
DESDEMONA
Your honour is most welcome.
OTHELLO　　　　　　　　Will you walk, sir?
O, Desdemona!　　　　　　　　　　　　　5
DESDEMONA My lord?
OTHELLO Get you to bed on th'instant. I will be returned
forthwith. Dismiss your attendant there. Look't be done.
DESDEMONA I will, my lord.
　　　　　　*Exeunt Othello, Lodovico, and attendants*
EMILIA How goes it now? He looks gentler than he did.　10
DESDEMONA
He says he will return incontinent.
He hath commanded me to go to bed,
And bid me to dismiss you.
EMILIA　　　　　　　　Dismiss me?
DESDEMONA
It was his bidding. Therefore, good Emilia,
Give me my nightly wearing, and adieu.　15
We must not now displease him.
EMILIA I would you had never seen him.
DESDEMONA
So would not I. My love doth so approve him
That even his stubbornness, his checks, his frowns—
Prithee unpin me—have grace and favour in them.　20
　　　　*Emilia helps Desdemona to undress*
EMILIA
I have laid those sheets you bade me on the bed.
DESDEMONA
All's one. Good faith, how foolish are our minds!

If I do die before thee, prithee shroud me
In one of these same sheets.
EMILIA                              Come, come, you talk.
DESDEMONA
My mother had a maid called Barbary.                  25
She was in love, and he she loved proved mad
And did forsake her. She had a song of willow.
An old thing 'twas, but it expressed her fortune,
And she died singing it. That song tonight
Will not go from my mind. I have much to do          30
But to go hang my head all at one side
And sing it, like poor Barbary. Prithee, dispatch.
EMILIA
Shall I go fetch your nightgown?
DESDEMONA                              No. Unpin me here.
This Lodovico is a proper man.
EMILIA
A very handsome man.
DESDEMONA                              He speaks well.        35
EMILIA I know a lady in Venice would have walked
barefoot to Palestine for a touch of his nether lip.
DESDEMONA (sings)
'The poor soul sat sighing by a sycamore tree,
     Sing all a green willow.
Her hand on her bosom, her head on her knee,      40
     Sing willow, willow, willow.
The fresh streams ran by her and murmured her
     moans,
     Sing willow, willow, willow.
Her salt tears fell from her and softened the stones,
     Sing willow'—                                               45
Lay by these.—
                    'willow, willow.'
Prithee, hie thee. He'll come anon.
'Sing all a green willow must be my garland.

'Let nobody blame him, his scorn I approve'—      50
Nay, that's not next. Hark, who is't that knocks?
EMILIA It's the wind.
DESDEMONA (sings)
'I called my love false love, but what said he then?
     Sing willow, willow, willow.
If I court more women, you'll couch with more men.'
So, get thee gone. Good night. Mine eyes do itch.   56
Doth that bode weeping?
EMILIA                              'Tis neither here nor there.
DESDEMONA
I have heard it said so. O, these men, these men!
Dost thou in conscience think—tell me, Emilia—
That there be women do abuse their husbands        60
In such gross kind?
EMILIA                    There be some such, no question.
DESDEMONA
Wouldst thou do such a deed for all the world?
EMILIA
Why, would not you?
DESDEMONA                    No, by this heavenly light.
EMILIA Nor I neither, by this heavenly light. I might do't
as well i'th' dark.                                              65
DESDEMONA
Wouldst thou do such a deed for all the world?
EMILIA The world's a huge thing. It is a great price for
a small vice.
DESDEMONA In truth, I think thou wouldst not.      69
EMILIA In truth, I think I should, and undo't when I had

done. Marry, I would not do such a thing for a joint
ring, nor for measures of lawn, nor for gowns,
petticoats, nor caps, nor any petty exhibition; but for
all the whole world? Ud's pity, who would not make
her husband a cuckold to make him a monarch? I
should venture purgatory for't.                            76
DESDEMONA
Beshrew me if I would do such a wrong
For the whole world.
EMILIA Why, the wrong is but a wrong i'th' world, and
having the world for your labour, 'tis a wrong in your
own world, and you might quickly make it right.    81
DESDEMONA
I do not think there is any such woman.
EMILIA
Yes, a dozen, and as many
To th' vantage as would store the world they played
     for.
But I do think it is their husbands' faults             85
If wives do fall. Say that they slack their duties,
And pour our treasures into foreign laps,
Or else break out in peevish jealousies,
Throwing restraint upon us; or say they strike us,
Or scant our former having in despite:                  90
Why, we have galls; and though we have some grace,
Yet have we some revenge. Let husbands know
Their wives have sense like them. They see, and smell,
And have their palates both for sweet and sour,
As husbands have. What is it that they do            95
When they change us for others? Is it sport?
I think it is. And doth affection breed it?
I think it doth. Is't frailty that thus errs?
It is so, too. And have not we affections,
Desires for sport, and frailty, as men have?          100
Then let them use us well, else let them know
The ills we do, their ills instruct us so.
DESDEMONA
Good night, good night. God me such uses send
Not to pick bad from bad, but by bad mend!    Exeunt

5.1    *Enter Iago and Roderigo*
IAGO
Here, stand behind this bulk. Straight will he come.
Wear thy good rapier bare, and put it home.
Quick, quick, fear nothing. I'll be at thy elbow.
It makes us or it mars us. Think on that,
And fix most firm thy resolution.                         5
RODERIGO
Be near at hand. I may miscarry in't.
IAGO
Here at thy hand. Be bold, and take thy stand.
RODERIGO (aside)
I have no great devotion to the deed,
And yet he hath given me satisfying reasons.
'Tis but a man gone. Forth my sword—he dies!     10
IAGO (aside)
I have rubbed this young quat almost to the sense,
And he grows angry. Now, whether he kill Cassio
Or Cassio him, or each do kill the other,
Every way makes my gain. Live Roderigo,
He calls me to a restitution large                          15
Of gold and jewels that I bobbed from him
As gifts to Desdemona.
It must not be. If Cassio do remain,
He hath a daily beauty in his life

That makes me ugly; and besides, the Moor    20
May unfold me to him—there stand I in much peril.
No, he must die. But so, I hear him coming.
            *Enter Cassio*

RODERIGO
I know his gait, 'tis he. (*Attacking Cassio*) Villain, thou
    diest.

CASSIO
That thrust had been mine enemy indeed,
But that my coat is better than thou know'st.    25
I will make proof of thine.
            *He stabs Roderigo, who falls*

RODERIGO                    O, I am slain!
            *Iago wounds Cassio in the leg from behind. Exit Iago*

CASSIO (*falling*)
I am maimed for ever. Help, ho, murder, murder!
            *Enter Othello ⌈above⌉*

OTHELLO
The voice of Cassio. Iago keeps his word.

RODERIGO O, villain that I am!

OTHELLO It is even so.    30

CASSIO O, help, ho! Light, a surgeon!

OTHELLO
'Tis he. O brave Iago, honest and just,
That hast such noble sense of thy friend's wrong—
Thou teachest me. Minion, your dear lies dead,
And your unblessed fate hies. Strumpet, I come.    35
Forth of my heart those charms, thine eyes, are blotted.
Thy bed, lust-stained, shall with lust's blood be spotted.
                                    *Exit*

            *Enter Lodovico and Graziano*

CASSIO
What ho, no watch, no passage? Murder, murder!

GRAZIANO
'Tis some mischance. The voice is very direful.

CASSIO O, help!    40

LODOVICO Hark.

RODERIGO O wretched villain!

LODOVICO
Two or three groan. 'Tis heavy night.
These may be counterfeits. Let's think't unsafe
To come into the cry without more help.    45

RODERIGO
Nobody come? Then shall I bleed to death.
            *Enter Iago with a light*

LODOVICO Hark.

GRAZIANO
Here's one comes in his shirt, with light and weapons.

IAGO
Who's there? Whose noise is this that cries on murder?

LODOVICO
We do not know.

IAGO                    Do not you hear a cry?    50

CASSIO
Here, here. For heaven's sake, help me.

IAGO                            What's the matter?

GRAZIANO (*to Lodovico*)
This is Othello's ensign, as I take it.

LODOVICO
The same indeed, a very valiant fellow.

IAGO (*to Cassio*)
What are you here that cry so grievously?

CASSIO
Iago—O, I am spoiled, undone by villains.    55
Give me some help.

IAGO
O me, lieutenant, what villains have done this?

CASSIO
I think that one of them is hereabout
And cannot make away.

IAGO                    O treacherous villains!
(*To Lodovico and Graziano*)
What are you there? Come in and give some help.    60

RODERIGO O, help me there!

CASSIO That's one of 'em.

IAGO (*stabbing Roderigo*) O murderous slave! O villain!

RODERIGO
O damned Iago! O inhuman dog!

IAGO
Kill men i'th' dark? Where be these bloody thieves?
How silent is this town! Ho, murder, murder!    66
(*To Lodovico and Graziano*)
What may you be? Are you of good or evil?

LODOVICO
As you shall prove us, praise us.

IAGO                            Signor Lodovico.

LODOVICO He, sir.

IAGO
I cry you mercy. Here's Cassio hurt by villains.    70

GRAZIANO Cassio?

IAGO How is't, brother?

CASSIO My leg is cut in two.

IAGO Marry, heaven forbid!
Light, gentlemen. I'll bind it with my shirt.    75
            *Enter Bianca*

BIANCA
What is the matter, ho? Who is't that cried?

IAGO
Who is't that cried?

BIANCA                    O my dear Cassio,
My sweet Cassio, O, Cassio, Cassio!

IAGO
O notable strumpet! Cassio, may you suspect
Who they should be that have thus mangled you?    80

CASSIO No.

GRAZIANO
I am sorry to find you thus. I have been to seek you.

IAGO
Lend me a garter. So. O for a chair,
To bear him easily hence!

BIANCA
Alas, he faints. O, Cassio, Cassio, Cassio!    85

IAGO
Gentlemen all, I do suspect this trash
To be a party in this injury.
Patience a while, good Cassio. Come, come,
Lend me a light. (*Going to Roderigo*) Know we this face
    or no?
Alas, my friend, and my dear countryman.    90
Roderigo? No—yes, sure—O heaven, Roderigo!

GRAZIANO What, of Venice?

IAGO Even he, sir. Did you know him?

GRAZIANO Know him? Ay.

IAGO
Signor Graziano, I cry your gentle pardon.    95
These bloody accidents must excuse my manners
That so neglected you.

GRAZIANO                    I am glad to see you.

IAGO
How do you, Cassio? O, a chair, a chair!

GRAZIANO Roderigo.

848

IAGO
He, he, 'tis he.
              *Enter attendants with a chair*
              O, that's well said, the chair!          100
Some good man bear him carefully from hence.
I'll fetch the general's surgeon. (*To Bianca*) For you,
    mistress,
Save you your labour. He that lies slain here, Cassio,
Was my dear friend. What malice was between you?
CASSIO
None in the world, nor do I know the man.          105
IAGO (*to Bianca*)
What, look you pale? (*To attendants*) O, bear him out
    o'th' air.
(*To Lodovico and Graziano*)
Stay you, good gentlemen.
              *Exeunt attendants with Cassio in the chair*
              ⌈*and with Roderigo's body*⌉
(*To Bianca*) Look you pale, mistress?
(*To Lodovico and Graziano*)
Do you perceive the ghastness of her eye?
(*To Bianca*) Nay, an you stare we shall hear more
    anon.
(*To Lodovico and Graziano*)
Behold her well; I pray you look upon her.          110
Do you see, gentlemen? Nay, guiltiness
Will speak, though tongues were out of use.
              *Enter Emilia*
EMILIA
Alas, what is the matter? What is the matter,
    husband?
IAGO
Cassio hath here been set on in the dark
By Roderigo and fellows that are scaped.          115
He's almost slain, and Roderigo dead.
EMILIA
Alas, good gentleman! Alas, good Cassio!
IAGO
This is the fruits of whoring. Prithee, Emilia,
Go know of Cassio where he supped tonight.
(*To Bianca*) What, do you shake at that?          120
BIANCA
He supped at my house, but I therefore shake not.
IAGO
O, did he so? I charge you go with me.
EMILIA (*to Bianca*) O, fie upon thee, strumpet!
BIANCA
I am no strumpet, but of life as honest
As you that thus abuse me.
EMILIA                    As I? Fough, fie upon thee!
IAGO
Kind gentlemen, let's go see poor Cassio dressed.          126
(*To Bianca*) Come, mistress, you must tell's another tale.
Emilia, run you to the citadel
And tell my lord and lady what hath happed.
Will you go on afore?                    *Exit Emilia*
(*Aside*)          This is the night
That either makes me or fordoes me quite.          *Exeunt*

5.2     *Enter Othello with a light.* ⌈*He draws back a*
        *curtain, revealing*⌉ *Desdemona asleep in her bed*
OTHELLO
It is the cause, it is the cause, my soul.
Let me not name it to you, you chaste stars.
It is the cause. Yet I'll not shed her blood,
Nor scar that whiter skin of hers than snow,

And smooth as monumental alabaster.          5
Yet she must die, else she'll betray more men.
Put out the light, and then put out the light.
If I quench thee, thou flaming minister,
I can again thy former light restore
Should I repent me; but once put out thy light,          10
Thou cunning'st pattern of excelling nature,
I know not where is that Promethean heat
That can thy light relume. When I have plucked thy
    rose
I cannot give it vital growth again.
It needs must wither. I'll smell thee on the tree.          15
              ⌈*He kisses her*⌉
O balmy breath, that dost almost persuade
Justice to break her sword! One more, one more.
Be thus when thou art dead, and I will kill thee
And love thee after. One more, and that's the last.
              *He kisses her*
So sweet was ne'er so fatal. I must weep,          20
But they are cruel tears. This sorrow's heavenly,
It strikes where it doth love. She wakes.
DESDEMONA Who's there? Othello?
OTHELLO Ay, Desdemona.
DESDEMONA Will you come to bed, my lord?          25
OTHELLO Have you prayed tonight, Desdemon?
DESDEMONA Ay, my lord.
OTHELLO
If you bethink yourself of any crime
Unreconciled as yet to heaven and grace,
Solicit for it straight.          30
DESDEMONA
Alack, my lord, what may you mean by that?
OTHELLO
Well, do it, and be brief. I will walk by.
I would not kill thy unpreparèd spirit.
No, heavens forfend! I would not kill thy soul.
DESDEMONA
Talk you of killing?
OTHELLO                    Ay, I do.
DESDEMONA                              Then heaven          35
Have mercy on me.
OTHELLO                    Amen, with all my heart.
DESDEMONA
If you say so, I hope you will not kill me.
OTHELLO Hum!
DESDEMONA
And yet I fear you, for you're fatal then
When your eyes roll so. Why I should fear I know not,
Since guiltiness I know not, but yet I feel I fear.          41
OTHELLO Think on thy sins.
DESDEMONA They are loves I bear to you.
OTHELLO Ay, and for that thou diest.
DESDEMONA
That death's unnatural that kills for loving.          45
Alas, why gnaw you so your nether lip?
Some bloody passion shakes your very frame.
These are portents, but yet I hope, I hope
They do not point on me.
OTHELLO                    Peace, and be still.
DESDEMONA
I will so. What's the matter?
OTHELLO                    That handkerchief          50
Which I so loved and gave thee, thou gav'st to Cassio.
DESDEMONA
No, by my life and soul. Send for the man
And ask him.

OTHELLO
  Sweet soul, take heed, take heed of perjury.
  Thou art on thy deathbed.                                    55
DESDEMONA  Ay, but not yet to die.
OTHELLO  Presently,
  Therefore confess thee freely of thy sin,
  For to deny each article with oath
  Cannot remove nor choke the strong conception     60
  That I do groan withal. Thou art to die.
DESDEMONA
  Then Lord have mercy on me.
OTHELLO                                     I say amen.
DESDEMONA
  And have you mercy, too. I never did
  Offend you in my life, never loved Cassio
  But with such general warranty of heaven           65
  As I might love. I never gave him token.
OTHELLO
  By heaven, I saw my handkerchief in's hand.
  O perjured woman! Thou dost stone my heart,
  And makes me call what I intend to do
  A murder, which I thought a sacrifice.              70
  I saw the handkerchief.
DESDEMONA                    He found it, then.
  I never gave it him. Send for him hither.
  Let him confess a truth.
OTHELLO                       He hath confessed.
DESDEMONA
  What, my lord?
OTHELLO              That he hath usèd thee.
DESDEMONA  How, unlawfully?                             75
OTHELLO  Ay.
DESDEMONA  He will not say so.
OTHELLO  No, his mouth is stopped.
  Honest Iago hath ta'en order for't.
DESDEMONA
  O, my fear interprets! What, is he dead?           80
OTHELLO
  Had all his hairs been lives, my great revenge
  Had stomach for 'em all.
DESDEMONA (weeping)
  Alas, he is betrayed, and I undone.
OTHELLO
  Out, strumpet! Weep'st thou for him to my face?
DESDEMONA
  O, banish me, my lord, but kill me not.            85
OTHELLO (attacking her)  Down, strumpet.
DESDEMONA
  Kill me tomorrow; let me live tonight.
OTHELLO  Nay, an you strive—
DESDEMONA  But half an hour.
OTHELLO  Being done, there is no pause.                 90
DESDEMONA  But while I say one prayer.
OTHELLO  It is too late.
       He smothers her
DESDEMONA  O Lord, Lord, Lord!
       Emilia at the door
EMILIA (within)
  My lord, my lord, what ho, my lord, my lord!
OTHELLO
  What noise is this? Not dead? Not yet quite dead?  95
  I that am cruel am yet merciful.
  I would not have thee linger in thy pain.
  So, so.
EMILIA (within)  What ho, my lord, my lord!
OTHELLO                                    Who's there?

EMILIA (within)
  O, good my lord, I would speak a word with you.
OTHELLO (calling)
  Yes.—'Tis Emilia. (Calling) By and by!—She's dead.
  'Tis like she comes to speak of Cassio's death.    101
  The noise was high. Ha! No more moving.
  Still as the grave. Shall she come in? Were't good?
  I think she stirs again. No. What's best to do?
  If she come in, she'll sure speak to my wife.      105
  My wife, my wife! What wife? I ha' no wife.
  O insupportable, O heavy hour!
  Methinks it should be now a huge eclipse
  Of sun and moon, and that th'affrighted globe
  Should yawn at alteration.
EMILIA (within)                  I do beseech you      110
  That I may speak with you. O, good my lord!
OTHELLO
  I had forgot thee.—O, come in, Emilia.—
  Soft, by and by. Let me the curtains draw.
       He closes the bed-curtains
  Where art thou?
       He opens the door. Enter Emilia
           What's the matter with thee now?
EMILIA
  O my good lord, yonder's foul murder's done.       115
OTHELLO  What, now?
EMILIA  But now, my lord.
OTHELLO
  It is the very error of the moon,
  She comes more nearer earth than she was wont,
  And makes men mad.                                 120
EMILIA
  Cassio, my lord, hath killed a young Venetian
  Called Roderigo.
OTHELLO              Roderigo killed
  And Cassio killed?
EMILIA                 No, Cassio is not killed.
OTHELLO
  Not Cassio killed? Then murder's out of tune,
  And sweet revenge grows harsh.                     125
DESDEMONA  O, falsely, falsely murdered!
EMILIA  O Lord, what cry is that?
OTHELLO  That? What?
EMILIA
  Out and alas, that was my lady's voice!
       ⌈She opens the bed-curtains⌉
  Help, help, ho, help! O lady, speak again!         130
  Sweet Desdemona, O sweet mistress, speak!
DESDEMONA
  A guiltless death I die.
EMILIA                       O, who hath done this deed?
DESDEMONA
  Nobody, I myself. Farewell.
  Commend me to my kind lord. O, farewell!    She dies
OTHELLO
  Why, how should she be murdered?
EMILIA                              Alas, who knows?
OTHELLO
  You heard her say herself it was not I.            136
EMILIA
  She said so, I must needs report the truth.
OTHELLO
  She's like a liar gone to burning hell.
  'Twas I that killed her.
EMILIA
  O, the more angel she, and you the blacker devil!  140

OTHELLO
She turned to folly, and she was a whore.
EMILIA
Thou dost belie her, and thou art a devil.
OTHELLO
She was false as water.
EMILIA                                    Thou art rash as fire
To say that she was false. O, she was heavenly true!
OTHELLO
Cassio did top her. Ask thy husband else.          145
O, I were damned beneath all depth in hell
But that I did proceed upon just grounds
To this extremity! Thy husband knew it all.
EMILIA
My husband?
OTHELLO          Thy husband.
EMILIA                              That she was false to wedlock?
OTHELLO
Ay, with Cassio. Had she been true,                150
If heaven would make me such another world
Of one entire and perfect chrysolite
I'd not have sold her for it.
EMILIA                              My husband?
OTHELLO
Ay, 'twas he that told me on her first.
An honest man he is, and hates the slime           155
That sticks on filthy deeds.
EMILIA                              My husband?
OTHELLO
What needs this iterance? Woman, I say thy husband.
EMILIA
O mistress, villainy hath made mocks with love.
My husband say she was false?
OTHELLO                              He, woman.
I say thy husband. Dost understand the word?       160
My friend, thy husband, honest, honest Iago.
EMILIA
If he say so, may his pernicious soul
Rot half a grain a day. He lies to th' heart.
She was too fond of her most filthy bargain.
OTHELLO Ha?                                         165
EMILIA Do thy worst.
This deed of thine is no more worthy heaven
Than thou wast worthy her.
OTHELLO                              Peace, you were best.
EMILIA
Thou hast not half that power to do me harm
As I have to be hurt. O gull, O dolt,              170
As ignorant as dirt! Thou hast done a deed—
I care not for thy sword, I'll make thee known
Though I lost twenty lives. Help, help, ho! Help!
The Moor hath killed my mistress. Murder, murder!
        Enter Montano, Graziano, and Iago
MONTANO
What is the matter? How now, general?              175
EMILIA
O, are you come, Iago? You have done well,
That men must lay their murders on your neck.
GRAZIANO What is the matter?
EMILIA (to Iago)
Disprove this villain if thou beest a man.
He says thou told'st him that his wife was false.  180
I know thou didst not. Thou'rt not such a villain.
Speak, for my heart is full.
IAGO
I told him what I thought, and told no more
Than what he found himself was apt and true.

EMILIA
But did you ever tell him she was false?           185
IAGO I did.
EMILIA
You told a lie, an odious, damnèd lie,
Upon my soul a lie, a wicked lie.
She false with Cassio? Did you say with Cassio?
IAGO
With Cassio, mistress. Go to, charm your tongue.   190
EMILIA
I will not charm my tongue. I am bound to speak.
My mistress here lies murdered in her bed.
⌈MONTANO and GRAZIANO⌉ O heavens forfend!
EMILIA (to Iago)
And your reports have set the murder on.
OTHELLO
Nay, stare not, masters. It is true indeed.        195
GRAZIANO
'Tis a strange truth.
MONTANO                         O monstrous act!
EMILIA
Villainy, villainy, villainy!
I think upon't, I think. I smell't. O villainy!
I thought so then. I'll kill myself for grief.
O villainy, villainy!                              200
IAGO
What, are you mad? I charge you get you home.
EMILIA
Good gentlemen, let me have leave to speak.
'Tis proper I obey him, but not now.
Perchance, Iago, I will ne'er go home.
OTHELLO
O, O, O!
        Othello falls on the bed
EMILIA          Nay, lay thee down and roar,         205
For thou hast killed the sweetest innocent
That e'er did lift up eye.
OTHELLO ⌈rising⌉              O, she was foul!
(To Graziano) I scarce did know you, uncle. There lies
        your niece,
Whose breath indeed these hands have newly
        stopped.
I know this act shows horrible and grim.           210
GRAZIANO
Poor Desdemon, I am glad thy father's dead.
Thy match was mortal to him, and pure grief
Shore his old thread in twain. Did he live now
This sight would make him do a desperate turn,
Yea, curse his better angel from his side,         215
And fall to reprobance.
OTHELLO
'Tis pitiful. But yet Iago knows
That she with Cassio hath the act of shame
A thousand times committed. Cassio confessed it,
And she did gratify his amorous works              220
With that recognizance and pledge of love
Which I first gave her. I saw it in his hand.
It was a handkerchief, an antique token
My father gave my mother.
EMILIA                              O God! O heavenly God!
IAGO
'Swounds, hold your peace!
EMILIA                              'Twill out, 'twill out. I peace?
No, I will speak as liberal as the north.          226
Let heaven, and men, and devils, let 'em all,
All, all cry shame against me, yet I'll speak.

IAGO  Be wise and get you home.
EMILIA  I will not.                                                    230
    *Iago draws his sword*
GRAZIANO (*to Iago*) Fie, your sword upon a woman?
EMILIA
O thou dull Moor, that handkerchief thou speak'st of
I found by fortune and did give my husband,
For often, with a solemn earnestness—
More than indeed belonged to such a trifle—          235
He begged of me to steal't.
IAGO                 Villainous whore!
EMILIA
She give it Cassio? No, alas, I found it,
And I did give't my husband.
IAGO                 Filth, thou liest.
EMILIA
By heaven I do not, I do not, gentlemen.
O murderous coxcomb! What should such a fool     240
Do with so good a wife?
OTHELLO           Are there no stones in heaven
But what serves for the thunder? Precious villain!
    *The Moor runs at Iago. Montano disarms Othello.*
    *Iago wounds his wife*
GRAZIANO
The woman falls. Sure he hath killed his wife.
EMILIA
Ay, ay. O, lay me by my mistress' side!     *Exit Iago*
GRAZIANO  He's gone, but his wife's killed.     245
MONTANO
'Tis a notorious villain. Take you this weapon
Which I have here recovered from the Moor.
Come, guard the door without. Let him not pass,
But kill him rather. I'll after that same villain,
For 'tis a damnèd slave.   *Exeunt Montano and Graziano*
OTHELLO        I am not valiant neither,   250
But every puny whipster gets my sword.
But why should honour outlive honesty?
Let it go all.
EMILIA      What did thy song bode, lady?
Hark, canst thou hear me? I will play the swan,
And die in music. (*Sings*) 'Willow, willow, willow.'—
Moor, she was chaste. She loved thee, cruel Moor.  256
So come my soul to bliss as I speak true.
So, speaking as I think, alas, I die.     *She dies*
OTHELLO
I have another weapon in this chamber.
It was a sword of Spain, the ice-brook's temper.   260
O, here it is. (*Calling*) Uncle, I must come forth.
GRAZIANO (*within*)
If thou attempt it, it will cost thee dear.
Thou hast no weapon, and perforce must suffer.
OTHELLO
Look in upon me then, and speak with me,
Or, naked as I am, I will assault thee.     265
    *Enter Graziano*
GRAZIANO
What is the matter?
OTHELLO      Behold, I have a weapon;
A better never did itself sustain
Upon a soldier's thigh. I have seen the day
That, with this little arm and this good sword,
I have made my way through more impediments   270
Than twenty times your stop. But O, vain boast!
Who can control his fate? 'Tis not so now.
Be not afraid, though you do see me weaponed.
Here is my journey's end, here is my butt

And very sea-mark of my utmost sail.     275
Do you go back dismayed? 'Tis a lost fear.
Man but a rush against Othello's breast
And he retires. Where should Othello go?
(*To Desdemona*) Now, how dost thou look now? O ill-
    starred wench,
Pale as thy smock! When we shall meet at count   280
This look of thine will hurl my soul from heaven,
And fiends will snatch at it.
    ⌜*He touches her*⌝
                Cold, cold, my girl,
Even like thy chastity. O cursèd, cursèd slave!
Whip me, ye devils,
From the possession of this heavenly sight.     285
Blow me about in winds, roast me in sulphur,
Wash me in steep-down gulfs of liquid fire!
O Desdemon! Dead Desdemon! Dead! O! O!
    *Enter Lodovico, Montano, Iago guarded by officers,*
    *with Cassio in a chair*
LODOVICO
Where is this rash and most unfortunate man?
OTHELLO
That's he that was Othello. Here I am.     290
LODOVICO
Where is that viper? Bring the villain forth.
OTHELLO
I look down towards his feet, but that's a fable.
(*To Iago*) If that thou beest a devil I cannot kill thee.
    *He wounds Iago*
LODOVICO
Wrench his sword from him.
    *Officers disarm Othello*
IAGO          I bleed, sir, but not killed.
OTHELLO
I am not sorry, neither. I'd have thee live,     295
For in my sense 'tis happiness to die.
LODOVICO
O thou Othello, that was once so good,
Fall'n in the practice of a cursèd slave,
What shall be said to thee?
OTHELLO        Why, anything;
An honourable murderer, if you will,     300
For naught I did in hate, but all in honour.
LODOVICO
This wretch hath part confessed his villainy.
Did you and he consent in Cassio's death?
OTHELLO  Ay.
CASSIO
Dear general, I never gave you cause.     305
OTHELLO
I do believe it, and I ask your pardon.
Will you, I pray, demand that demi-devil
Why he hath thus ensnared my soul and body?
IAGO
Demand me nothing. What you know, you know.
From this time forth I never will speak word.     310
LODOVICO (*to Iago*) What, not to pray?
GRAZIANO (*to Iago*) Torments will ope your lips.
OTHELLO  Well, thou dost best.
LODOVICO
Sir, you shall understand what hath befall'n,
Which as, I think, you know not. Here is a letter,   315
Found in the pocket of the slain Roderigo,
And here another. The one of them imports
The death of Cassio, to be undertook
By Roderigo.

OTHELLO O villain!                                                    320
CASSIO Most heathenish and most gross.
LODOVICO
  Now here's another discontented paper
  Found in his pocket too, and this it seems
  Roderigo meant to've sent this damnèd villain,
  But that, belike, Iago in the interim                              325
  Came in and satisfied him.
OTHELLO (to Iago) O thou pernicious caitiff!
  How came you, Cassio, by that handkerchief
  That was my wife's?
CASSIO                    I found it in my chamber,
  And he himself confessed it, but even now,                         330
  That there he dropped it for a special purpose
  Which wrought to his desire.
OTHELLO                          O fool, fool, fool!
CASSIO
  There is besides in Roderigo's letter
  How he upbraids Iago, that he made him
  Brave me upon the watch, whereon it came                           335
  That I was cast; and even but now he spake
  After long seeming dead, Iago hurt him,
  Iago set him on.
LODOVICO (to Othello)
  You must forsake this room and go with us.
  Your power and your command is taken off,                          340
  And Cassio rules in Cyprus. For this slave,
  If there be any cunning cruelty
  That can torment him much and hold him long,
  It shall be his. You shall close prisoner rest
  Till that the nature of your fault be known                        345
  To the Venetian state. (To officers) Come, bring away.
OTHELLO
  Soft you, a word or two before you go.
  I have done the state some service, and they know't.
  No more of that. I pray you, in your letters,
  When you shall these unlucky deeds relate,                         350
  Speak of me as I am. Nothing extenuate,
  Nor set down aught in malice. Then must you speak

Of one that loved not wisely but too well,
Of one not easily jealous but, being wrought,
Perplexed in the extreme; of one whose hand,              355
Like the base Indian, threw a pearl away
Richer than all his tribe; of one whose subdued eyes,
Albeit unusèd to the melting mood,
Drops tears as fast as the Arabian trees
Their medicinable gum. Set you down this,                 360
And say besides that in Aleppo once,
Where a malignant and a turbaned Turk
Beat a Venetian and traduced the state,
I took by th' throat the circumcisèd dog
And smote him thus.                                       365
      *He stabs himself*
LODOVICO O bloody period!
GRAZIANO All that is spoke is marred.
OTHELLO (to Desdemona)
  I kissed thee ere I killed thee. No way but this:
  Killing myself, to die upon a kiss.
                  *He kisses Desdemona and dies*
CASSIO
  This did I fear, but thought he had no weapon,          370
  For he was great of heart.
LODOVICO (to Iago)              O Spartan dog,
  More fell than anguish, hunger, or the sea,
  Look on the tragic loading of this bed.
  This is thy work. The object poisons sight.
  Let it be hid.
      ⌜*They close the bed-curtains*⌝
                  Graziano, keep the house,               375
  And seize upon the fortunes of the Moor,
  For they succeed on you. (To Cassio) To you, Lord
     Governor,
  Remains the censure of this hellish villain.
  The time, the place, the torture, O, enforce it!
  Myself will straight aboard, and to the state           380
  This heavy act with heavy heart relate.
                  *Exeunt ⌜with Emilia's body⌝*

# ALL'S WELL THAT ENDS WELL

*All's Well That Ends Well*, first printed in the 1623 Folio, is often paired with *Measure for Measure*. Though we lack external evidence as to its date of composition, internal evidence suggests that it, too, is an early Jacobean play. Like *Measure for Measure*, it places its central characters in more painful situations than those in which the heroes and heroines of the earlier, more romantic comedies usually find themselves. The touching ardour with which Helen, 'a poor physician's daughter', pursues the young Bertram, son of her guardian the Countess of Rousillon, creates embarrassments for both of them. When the King whose illness she cures by her semi-magical skills brings about their marriage as a reward, Bertram's flight to the wars seems to destroy all her chances of happiness. She achieves consummation of the marriage only by the ruse (resembling Isabella's 'bed-trick' in *Measure for Measure*) of substituting herself for the Florentine maiden Diana whom Bertram believes himself to be seducing. The play's conclusion, in which the deception is exposed and Bertram is shamed into acknowledging Helen as his wife, offers only a tentatively happy ending.

Shakespeare based the story of Bertram and Helen on a tale from Boccaccio's *Decameron* either in the original or in the version included in William Painter's *Palace of Pleasure* (1566–7, revised 1575). But he created several important characters, including the Countess and the old Lord, Lafeu. He also invented the accompanying action exposing the roguery of Bertram's flashy friend Paroles, a man of words (as his name indicates) descending from the braggart soldier of Roman comedy.

Versions of the play performed in the eighteenth and nineteenth centuries, mostly emphasizing either the comedy of Paroles or the sentimental appeal of Helen, had little success; but some twentieth-century productions have shown it in a more favourable light, demonstrating, for example, that the role of the Countess is (in Bernard Shaw's words) 'the most beautiful old woman's part ever written', that the discomfiture of Paroles provides comedy that is subtle as well as highly laughable, and that the relationship of Bertram and Helen is profoundly convincing in its emotional reality.

# THE PERSONS OF THE PLAY

The Dowager COUNTESS of Roussillon

BERTRAM, Count of Roussillon, her son

HELEN, an orphan, attending on the Countess

LAVATCH, a Clown, the Countess's servant

REYNALDO, the Countess's steward

PAROLES, Bertram's companion


The KING of France

LAFEU, an old lord

FIRST LORD DUMAINE ⎫
SECOND LORD DUMAINE ⎭ brothers

INTERPRETER, a French soldier

An AUSTRINGER

The DUKE of Florence

WIDOW Capilet

DIANA, her daughter

MARIANA, a friend of the Widow

Lords, attendants, soldiers, citizens

# All's Well That Ends Well

**1.1** *Enter young Bertram Count of Roussillon, his mother the Countess, Helen, and Lord Lafeu, all in black*

COUNTESS In delivering my son from me I bury a second husband.

BERTRAM And I in going, madam, weep o'er my father's death anew; but I must attend his majesty's command, to whom I am now in ward, evermore in subjection.

LAFEU You shall find of the King a husband, madam; you, sir, a father. He that so generally is at all times good must of necessity hold his virtue to you, whose worthiness would stir it up where it wanted rather than lack it where there is such abundance. 10

COUNTESS What hope is there of his majesty's amendment?

LAFEU He hath abandoned his physicians, madam, under whose practices he hath persecuted time with hope, and finds no other advantage in the process but only the losing of hope by time. 15

COUNTESS This young gentlewoman had a father—O that 'had': how sad a passage 'tis!—whose skill was almost as great as his honesty; had it stretched so far, would have made nature immortal, and death should have play for lack of work. Would for the King's sake he were living. I think it would be the death of the King's disease. 22

LAFEU How called you the man you speak of, madam?

COUNTESS He was famous, sir, in his profession, and it was his great right to be so: Gérard de Narbonne. 25

LAFEU He was excellent indeed, madam. The King very lately spoke of him, admiringly and mournfully. He was skilful enough to have lived still, if knowledge could be set up against mortality.

BERTRAM What is it, my good lord, the King languishes of? 31

LAFEU A fistula, my lord.

BERTRAM I heard not of it before.

LAFEU I would it were not notorious.—Was this gentlewoman the daughter of Gérard de Narbonne? 35

COUNTESS His sole child, my lord, and bequeathed to my overlooking. I have those hopes of her good that her education promises; her dispositions she inherits, which makes fair gifts fairer—for where an unclean mind carries virtuous qualities, there commendations go with pity: they are virtues and traitors too. In her they are the better for their simpleness. She derives her honesty and achieves her goodness. 43

LAFEU Your commendations, madam, get from her tears.

COUNTESS 'Tis the best brine a maiden can season her praise in. The remembrance of her father never approaches her heart but the tyranny of her sorrows takes all livelihood from her cheek.—No more of this, Helen. Go to, no more, lest it be rather thought you affect a sorrow than to have— 50

HELEN I do affect a sorrow indeed, but I have it too.

LAFEU Moderate lamentation is the right of the dead, excessive grief the enemy to the living.

COUNTESS If the living be not enemy to the grief, the excess makes it soon mortal. 55

BERTRAM (*kneeling*) Madam, I desire your holy wishes.

LAFEU How understand we that?

COUNTESS
Be thou blessed, Bertram, and succeed thy father
In manners as in shape. Thy blood and virtue
Contend for empire in thee, and thy goodness 60
Share with thy birthright. Love all, trust a few,
Do wrong to none. Be able for thine enemy
Rather in power than use, and keep thy friend
Under thy own life's key. Be checked for silence
But never taxed for speech. What heaven more will 65
That thee may furnish and my prayers pluck down,
Fall on thy head. Farewell. (*To Lafeu*) My lord,
'Tis an unseasoned courtier. Good my lord,
Advise him.

LAFEU                    He cannot want the best
That shall attend his love. 70

COUNTESS Heaven bless him!—Farewell, Bertram.

BERTRAM (*rising*) The best wishes that can be forged in your thoughts be servants to you.          ⌜*Exit Countess*⌝
(*To Helen*) Be comfortable to my mother, your mistress, and make much of her. 75

LAFEU Farewell, pretty lady. You must hold the credit of your father.          *Exeunt Bertram and Lafeu*

HELEN
O were that all! I think not on my father,
And these great tears grace his remembrance more
Than those I shed for him. What was he like? 80
I have forgot him. My imagination
Carries no favour in't but Bertram's.
I am undone. There is no living, none,
If Bertram be away. 'Twere all one
That I should love a bright particular star 85
And think to wed it, he is so above me.
In his bright radiance and collateral light
Must I be comforted, not in his sphere.
Th'ambition in my love thus plagues itself.
The hind that would be mated by the lion 90
Must die for love. 'Twas pretty, though a plague,
To see him every hour, to sit and draw
His archèd brows, his hawking eye, his curls,
In our heart's table—heart too capable
Of every line and trick of his sweet favour. 95
But now he's gone, and my idolatrous fancy
Must sanctify his relics. Who comes here?
          *Enter Paroles*
One that goes with him. I love him for his sake—
And yet I know him a notorious liar,
Think him a great way fool, solely a coward. 100
Yet these fixed evils sit so fit in him
That they take place when virtue's steely bones
Looks bleak i'th' cold wind. Withal, full oft we see
Cold wisdom waiting on superfluous folly.

PAROLES Save you, fair queen. 105

HELEN And you, monarch.

PAROLES No.

HELEN And no.

PAROLES Are you meditating on virginity?

HELEN Ay. You have some stain of soldier in you, let me ask you a question. Man is enemy to virginity: how may we barricado it against him? 112

PAROLES Keep him out.

HELEN But he assails, and our virginity, though valiant in the defence, yet is weak. Unfold to us some warlike resistance.                                                   116

PAROLES There is none. Man, setting down before you, will undermine you and blow you up.

HELEN Bless our poor virginity from underminers and blowers-up. Is there no military policy how virgins might blow up men?                                           121

PAROLES Virginity being blown down, man will quicklier be blown up. Marry, in blowing him down again, with the breach yourselves made you lose your city. It is not politic in the commonwealth of nature to preserve virginity. Loss of virginity is rational increase, and there was never virgin got till virginity was first lost. That you were made of is mettle to make virgins. Virginity by being once lost may be ten times found; by being ever kept it is ever lost. 'Tis too cold a companion, away with't.                                                        131

HELEN I will stand for't a little, though therefore I die a virgin.

PAROLES There's little can be said in't. 'Tis against the rule of nature. To speak on the part of virginity is to accuse your mothers, which is most infallible disobedience. He that hangs himself is a virgin: virginity murders itself, and should be buried in highways, out of all sanctified limit, as a desperate offendress against nature. Virginity breeds mites, much like a cheese; consumes itself to the very paring, and so dies with feeding his own stomach. Besides, virginity is peevish, proud, idle, made of self-love—which is the most inhibited sin in the canon. Keep it not, you cannot choose but lose by't. Out with't! Within t'one year it will make itself two, which is a goodly increase, and the principal itself not much the worse. Away with't.

HELEN How might one do, sir, to lose it to her own liking?

PAROLES Let me see. Marry, ill, to like him that ne'er it likes. 'Tis a commodity will lose the gloss with lying: the longer kept, the less worth. Off with't while 'tis vendible. Answer the time of request. Virginity like an old courtier wears her cap out of fashion, richly suited but unsuitable, just like the brooch and the toothpick, which wear not now. Your date is better in your pie and your porridge than in your cheek, and your virginity, your old virginity, is like one of our French withered pears: it looks ill, it eats drily, marry, 'tis a withered pear—it was formerly better, marry, yet 'tis a withered pear. Will you anything with it?        160

HELEN Not my virginity, yet . . .

There shall your master have a thousand loves,
A mother and a mistress and a friend,
A phoenix, captain, and an enemy,
A guide, a goddess, and a sovereign,                      165
A counsellor, a traitress, and a dear:
His humble ambition, proud humility,
His jarring concord and his discord dulcet,
His faith, his sweet disaster, with a world
Of pretty fond adoptious christendoms                    170
That blinking Cupid gossips. Now shall he—
I know not what he shall. God send him well.
The court's a learning place, and he is one—

PAROLES What one, i'faith?

HELEN That I wish well. 'Tis pity.                        175

PAROLES What's pity?

HELEN

That wishing well had not a body in't

Which might be felt, that we, the poorer born,
Whose baser stars do shut us up in wishes,
Might with effects of them follow our friends            180
And show what we alone must think, which never
Returns us thanks.

*Enter a Page*

PAGE Monsieur Paroles, my lord calls for you.    [*Exit*]

PAROLES Little Helen, farewell. If I can remember thee I will think of thee at court.                                185

HELEN Monsieur Paroles, you were born under a charitable star.

PAROLES Under Mars, I.

HELEN I especially think *under* Mars.

PAROLES Why '*under* Mars'?                               190

HELEN The wars hath so kept you under that you must needs be born under Mars.

PAROLES When he was predominant.

HELEN When he was retrograde, I think rather.

PAROLES Why think you so?                                 195

HELEN You go so much backward when you fight.

PAROLES That's for advantage.

HELEN So is running away, when fear proposes the safety. But the composition that your valour and fear makes in you is a virtue of a good wing, and I like the wear well.                                                        201

PAROLES I am so full of businesses I cannot answer thee acutely. I will return perfect courtier, in the which my instruction shall serve to naturalize thee, so thou wilt be capable of a courtier's counsel and understand what advice shall thrust upon thee; else thou diest in thine unthankfulness, and thine ignorance makes thee away. Farewell. When thou hast leisure say thy prayers; when thou hast none remember thy friends. Get thee a good husband and use him as he uses thee. So farewell.                                                    *Exit*

HELEN

Our remedies oft in ourselves do lie                      212
Which we ascribe to heaven. The fated sky
Gives us free scope, only doth backward pull
Our slow designs when we ourselves are dull.              215
What power is it which mounts my love so high,
That makes me see and cannot feed mine eye?
The mightiest space in fortune nature brings
To join like likes and kiss like native things.
Impossible be strange attempts to those                   220
That weigh their pains in sense and do suppose
What hath been cannot be. Who ever strove
To show her merit that did miss her love?
The King's disease—my project may deceive me,
But my intents are fixed and will not leave me.    *Exit*

**1.2**    *A flourish of cornetts. Enter the King of France*
        *with letters, the two Lords Dumaine, ⌈and divers*
        *attendants⌉*

KING

The Florentines and Sienese are by th'ears,
Have fought with equal fortune, and continue
A braving war.

FIRST LORD DUMAINE So 'tis reported, sir.

KING

Nay, 'tis most credible: we here receive it
A certainty vouched from our cousin Austria,              5
With caution that the Florentine will move us
For speedy aid—wherein our dearest friend

Prejudicates the business, and would seem
To have us make denial.
FIRST LORD DUMAINE          His love and wisdom
Approved so to your majesty may plead          10
For amplest credence.
KING          He hath armed our answer,
And Florence is denied before he comes.
Yet for our gentlemen that mean to see
The Tuscan service, freely have they leave
To stand on either part.
SECOND LORD DUMAINE          It well may serve          15
A nursery to our gentry, who are sick
For breathing and exploit.
KING          What's he comes here?
     *Enter Bertram, Lafeu, and Paroles*
FIRST LORD DUMAINE
It is the Count Roussillon, my good lord,
Young Bertram.
KING (*to Bertram*)  Youth, thou bear'st thy father's face.
Frank nature, rather curious than in haste,          20
Hath well composed thee. Thy father's moral parts
Mayst thou inherit, too. Welcome to Paris.
BERTRAM
My thanks and duty are your majesty's.
KING
I would I had that corporal soundness now
As when thy father and myself in friendship          25
First tried our soldiership. He did look far
Into the service of the time, and was
Discipled of the bravest. He lasted long,
But on us both did haggish age steal on,
And wore us out of act. It much repairs me          30
To talk of your good father. In his youth
He had the wit which I can well observe
Today in our young lords, but they may jest
Till their own scorn return to them unnoted
Ere they can hide their levity in honour.          35
So like a courtier, contempt nor bitterness
Were in his pride or sharpness; if they were
His equal had awaked them, and his honour—
Clock to itself—knew the true minute when
Exception bid him speak, and at this time          40
His tongue obeyed his hand. Who were below him
He used as creatures of another place,
And bowed his eminent top to their low ranks,
Making them proud of his humility.
In their poor praise he humbled. Such a man          45
Might be a copy to these younger times,
Which followed well would demonstrate them now
But goers-backward.
BERTRAM          His good remembrance, sir,
Lies richer in your thoughts than on his tomb.
So in approof lives not his epitaph          50
As in your royal speech.
KING
Would I were with him! He would always say—
Methinks I hear him now; his plausive words
He scattered not in ears, but grafted them
To grow there and to bear. 'Let me not live'—          55
This his good melancholy oft began
On the catastrophe and heel of pastime,
When it was out—'Let me not live', quoth he,
'After my flame lacks oil, to be the snuff
Of younger spirits, whose apprehensive senses          60
All but new things disdain, whose judgements are
Mere fathers of their garments, whose constancies

Expire before their fashions.' This he wished.
I after him do after him wish too,
Since I nor wax nor honey can bring home,          65
I quickly were dissolvèd from my hive
To give some labourers room.
SECOND LORD DUMAINE          You're lovèd, sir.
They that least lend it you shall lack you first.
KING
I fill a place, I know't.—How long is't, Count,
Since the physician at your father's died?          70
He was much famed.
BERTRAM          Some six months since, my lord.
KING
If he were living I would try him yet.—
Lend me an arm.—The rest have worn me out
With several applications. Nature and sickness
Debate it at their leisure. Welcome, Count.          75
My son's no dearer.
BERTRAM          Thank your majesty.
          ⌐Flourish.⌐ *Exeunt*

**1.3**     *Enter the Countess, Reynaldo her steward, and*
          ⌐*behind*⌐ *Lavatch her clown*
COUNTESS  I will now hear. What say you of this gentle-
     woman?
REYNALDO  Madam, the care I have had to even your
     content I wish might be found in the calendar of my
     past endeavours, for then we wound our modesty and
     make foul the clearness of our deservings, when of
     ourselves we publish them.          7
COUNTESS  What does this knave here? (*To Lavatch*) Get
     you gone, sirrah. The complaints I have heard of you
     I do not all believe. 'Tis my slowness that I do not, for
     I know you lack not folly to commit them and have
     ability enough to make such knaveries yours.          12
LAVATCH  'Tis not unknown to you, madam, I am a poor
     fellow.
COUNTESS  Well, sir?          15
LAVATCH  No, madam, 'tis not so well that I am poor,
     though many of the rich are damned. But if I may
     have your ladyship's good will to go to the world, Isbel
     the woman and I will do as we may.
COUNTESS  Wilt thou needs be a beggar?          20
LAVATCH  I do beg your good will in this case.
COUNTESS  In what case?
LAVATCH  In Isbel's case and mine own. Service is no
     heritage, and I think I shall never have the blessing of
     God till I have issue o' my body, for they say bairns
     are blessings.          26
COUNTESS  Tell me thy reason why thou wilt marry.
LAVATCH  My poor body, madam, requires it. I am driven
     on by the flesh, and he must needs go that the devil
     drives.          30
COUNTESS  Is this all your worship's reason?
LAVATCH  Faith, madam, I have other holy reasons, such
     as they are.
COUNTESS  May the world know them?          34
LAVATCH  I have been, madam, a wicked creature, as
     you—and all flesh and blood—are, and indeed I do
     marry that I may repent.
COUNTESS  Thy marriage sooner than thy wickedness.
LAVATCH  I am out o' friends, madam, and I hope to have
     friends for my wife's sake.          40
COUNTESS  Such friends are thine enemies, knave.
LAVATCH  You're shallow, madam—in great friends, for
     the knaves come to do that for me which I am aweary

of. He that ears my land spares my team, and gives
me leave to in the crop. If I be his cuckold, he's my
drudge. He that comforts my wife is the cherisher of
my flesh and blood; he that cherishes my flesh and
blood loves my flesh and blood; he that loves my flesh
and blood is my friend; *ergo*, he that kisses my wife is
my friend. If men could be contented to be what they
are, there were no fear in marriage. For young
Chairbonne the puritan and old Poisson the papist,
howsome'er their hearts are severed in religion, their
heads are both one: they may jowl horns together like
any deer i'th' herd.                                              55

COUNTESS Wilt thou ever be a foul-mouthed and
calumnious knave?

LAVATCH A prophet? Ay, madam, and I speak the truth
the next way.

⌈*He sings*⌉

    For I the ballad will repeat,                    60
    Which men full true shall find:
    Your marriage comes by destiny,
    Your cuckoo sings by kind.

COUNTESS Get you gone, sir. I'll talk with you more anon.

REYNALDO May it please you, madam, that he bid Helen
come to you? Of her I am to speak.                     66

COUNTESS (*to Lavatch*) Sirrah, tell my gentlewoman I would
speak with her. Helen, I mean.

LAVATCH ⌈*sings*⌉

'Was this fair face the cause', quoth she,
'Why the Grecians sackèd Troy?                          70
Fond done, done fond. Was this King Priam's joy?'
With that she sighèd as she stood,
With that she sighèd as she stood,
    And gave this sentence then:
'Among nine bad if one be good,                          75
Among nine bad if one be good,
    There's yet one good in ten.'

COUNTESS What, 'one good in ten'? You corrupt the song,
sirrah.                                                        79

LAVATCH One good *woman* in ten, madam, which is a
purifying o'th' song. Would God would serve the world
so all the year! We'd find no fault with the tithe-
woman if I were the parson. One in ten, quoth a? An
we might have a good woman born but ere every
blazing star, or at an earthquake, 'twould mend the
lottery well. A man may draw his heart out ere a pluck
one.

COUNTESS You'll be gone, sir knave, and do as I command
you.                                                           89

LAVATCH That man should be at woman's command, and
yet no hurt done! Though honesty be no puritan, yet
it will do no hurt; it will wear the surplice of humility
over the black gown of a big heart. I am going, forsooth.
The business is for Helen to come hither.        *Exit*

COUNTESS Well now.                                             95

REYNALDO I know, madam, you love your gentlewoman
entirely.

COUNTESS Faith, I do. Her father bequeathed her to me,
and she herself without other advantage may lawfully
make title to as much love as she finds. There is more
owing her than is paid, and more shall be paid her
than she'll demand.                                          102

REYNALDO Madam, I was very late more near her than I
think she wished me. Alone she was, and did
communicate to herself, her own words to her own
ears; she thought, I dare vow for her, they touched
not any stranger sense. Her matter was, she loved your

son. Fortune, she said, was no goddess, that had put
such difference betwixt their two estates; Love no god,
that would not extend his might only where qualities
were level; Dian no queen of virgins, that would suffer
her poor knight surprised without rescue in the first
assault or ransom afterward. This she delivered in the
most bitter touch of sorrow that e'er I heard virgin
exclaim in; which I held my duty speedily to acquaint
you withal, sithence in the loss that may happen it
concerns you something to know it.                   117

COUNTESS You have discharged this honestly. Keep it to
yourself. Many likelihoods informed me of this before,
which hung so tott'ring in the balance that I could
neither believe nor misdoubt. Pray you, leave me. Stall
this in your bosom, and I thank you for your honest
care. I will speak with you further anon.           123

                        *Exit Steward*

*Enter Helen*

COUNTESS (*aside*)
Even so it was with me when I was young.
If ever we are nature's, these are ours: this thorn
Doth to our rose of youth rightly belong.           126
  Our blood to us, this to our blood is born;
It is the show and seal of nature's truth,
Where love's strong passion is impressed in youth.
By our remembrances of days foregone,             130
Such were our faults—or then we thought them
    none.
Her eye is sick on't. I observe her now.

HELEN
What is your pleasure, madam?

COUNTESS              You know, Helen,
I am a mother to you.

HELEN
Mine honourable mistress.

COUNTESS             Nay, a mother.       135
Why not a mother? When I said 'a mother',
Methought you saw a serpent. What's in 'mother'
That you start at it? I say I am your mother,
And put you in the catalogue of those
That were enwombèd mine. 'Tis often seen        140
Adoption strives with nature, and choice breeds
A native slip to us from foreign seeds.
You ne'er oppressed me with a mother's groan,
Yet I express to you a mother's care.
God's mercy, maiden! Does it curd thy blood     145
To say I am thy mother? What's the matter,
That this distempered messenger of wet,
The many-coloured Iris, rounds thine eye?
Why, that you are my daughter?

HELEN               That I am not.

COUNTESS
I say I am your mother.

HELEN             Pardon, madam.        150
The Count Roussillon cannot be my brother.
I am from humble, he from honoured name;
No note upon my parents, his all noble.
My master, my dear lord he is, and I
His servant live and will his vassal die.              155
He must not be my brother.

COUNTESS            Nor I your mother?

HELEN
You are my mother, madam. Would you were—
So that my lord your son were not my brother—
Indeed my mother! Or were you both our mothers
I care no more for than I do for heaven,             160

So I were not his sister. Can 't no other
But, I your daughter, he must be my brother?
COUNTESS
Yes, Helen, you might be my daughter-in-law.
God shield you mean it not! 'Daughter' and 'mother'
So strive upon your pulse. What, pale again?          165
My fear hath catched your fondness. Now I see
The myst'ry of your loneliness, and find
Your salt tears' head. Now to all sense 'tis gross:
You love my son. Invention is ashamed
Against the proclamation of thy passion          170
To say thou dost not. Therefore tell me true,
But tell me then 'tis so—for look, thy cheeks
Confess it t'one to th'other, and thine eyes
See it so grossly shown in thy behaviours
That in their kind they speak it. Only sin          175
And hellish obstinacy tie thy tongue,
That truth should be suspected. Speak, is't so?
If it be so you have wound a goodly clew;
If it be not, forswear't. Howe'er, I charge thee,
As heaven shall work in me for thine avail,          180
To tell me truly.
HELEN          Good madam, pardon me.
COUNTESS
Do you love my son?
HELEN          Your pardon, noble mistress.
COUNTESS
Love you my son?
HELEN          Do not you love him, madam?
COUNTESS
Go not about. My love hath in't a bond
Whereof the world takes note. Come, come, disclose
The state of your affection, for your passions          186
Have to the full appeached.
HELEN          Then I confess,
Here on my knee, before high heaven and you,
That before you and next unto high heaven
I love your son.          190
My friends were poor but honest; so's my love.
Be not offended, for it hurts not him
That he is loved of me. I follow him not
By any token of presumptuous suit,
Nor would I have him till I do deserve him,          195
Yet never know how that desert should be.
I know I love in vain, strive against hope;
Yet in this captious and intenable sieve
I still pour in the waters of my love
And lack not to lose still. Thus, Indian-like,          200
Religious in mine error, I adore
The sun that looks upon his worshipper
But knows of him no more. My dearest madam,
Let not your hate encounter with my love
For loving where you do; but if yourself,          205
Whose agèd honour cites a virtuous youth,
Did ever in so true a flame of liking
Wish chastely and love dearly, that your Dian
Was both herself and Love, O then give pity
To her whose state is such that cannot choose          210
But lend and give where she is sure to lose,
That seeks to find not that her search implies,
But riddle-like lives sweetly where she dies.
COUNTESS
Had you not lately an intent—speak truly—
To go to Paris?          215
HELEN Madam, I had.
COUNTESS Wherefore? Tell true.

HELEN
I will tell truth, by grace itself I swear.
You know my father left me some prescriptions
Of rare and proved effects, such as his reading          220
And manifest experience had collected
For general sovereignty, and that he willed me
In heedfull'st reservation to bestow them,
As notes whose faculties inclusive were
More than they were in note. Amongst the rest          225
There is a remedy, approved, set down,
To cure the desperate languishings whereof
The King is rendered lost.
COUNTESS          This was your motive
For Paris, was it? Speak.
HELEN
My lord your son made me to think of this,          230
Else Paris and the medicine and the King
Had from the conversation of my thoughts
Haply been absent then.
COUNTESS          But think you, Helen,
If you should tender your supposèd aid,
He would receive it? He and his physicians          235
Are of a mind: he, that they cannot help him;
They, that they cannot help. How shall they credit
A poor unlearnèd virgin, when the schools,
Embowelled of their doctrine, have left off
The danger to itself?
HELEN          There's something in't          240
More than my father's skill, which was the great'st
Of his profession, that his good receipt
Shall for my legacy be sanctified
By th' luckiest stars in heaven, and would your
          honour
But give me leave to try success, I'd venture          245
The well-lost life of mine on his grace's cure
By such a day, an hour.
COUNTESS Dost thou believe't?
HELEN Ay, madam, knowingly.
COUNTESS
Why, Helen, thou shalt have my leave and love,          250
Means and attendants, and my loving greetings
To those of mine in court. I'll stay at home
And pray God's blessing into thy attempt.
Be gone tomorrow, and be sure of this:
What I can help thee to, thou shalt not miss.          *Exeunt*

**2.1**     *Flourish of cornetts. Enter the King ⌈carried in a*
          *chair⌉, with the two Lords Dumaine, divers young*
          *lords taking leave for the Florentine war, and*
          *Bertram and Paroles*
KING
Farewell, young lords. These warlike principles
Do not throw from you. And you, my lords, farewell.
Share the advice betwixt you; if both gain all,
The gift doth stretch itself as 'tis received,
And is enough for both.
FIRST LORD DUMAINE          'Tis our hope, sir,          5
After well-entered soldiers, to return
And find your grace in health.
KING
No, no, it cannot be—and yet my heart
Will not confess he owes the malady
That doth my life besiege. Farewell, young lords.          10
Whether I live or die, be you the sons
Of worthy Frenchmen; let higher Italy—
Those bated that inherit but the fall

Of the last monarchy—see that you come
Not to woo honour but to wed it. When                    15
The bravest questant shrinks, find what you seek,
That fame may cry you loud. I say farewell.

FIRST LORD DUMAINE
Health at your bidding serve your majesty.

KING
Those girls of Italy, take heed of them.
They say our French lack language to deny             20
If they demand. Beware of being captives
Before you serve.

BOTH LORDS DUMAINE  Our hearts receive your warnings.

KING  Farewell.—Come hither to me.
          ⌐Some lords stand aside with the King⌐

FIRST LORD DUMAINE (to Bertram)
O my sweet lord, that you will stay behind us.

PAROLES
'Tis not his fault, the spark.

SECOND LORD DUMAINE          O 'tis brave wars.       25

PAROLES
Most admirable! I have seen those wars.

BERTRAM
I am commanded here, and kept a coil with
'Too young' and 'the next year' and ''tis too early'.

PAROLES
An thy mind stand to't, boy, steal away bravely.

BERTRAM
I shall stay here the forehorse to a smock,           30
Creaking my shoes on the plain masonry,
Till honour be bought up, and no sword worn
But one to dance with. By heaven, I'll steal away.

FIRST LORD DUMAINE
There's honour in the theft.

PAROLES                    Commit it, Count.

SECOND LORD DUMAINE
I am your accessary. And so, farewell.                35

BERTRAM  I grow to you,
And our parting is a tortured body.

FIRST LORD DUMAINE
Farewell, captain.

SECOND LORD DUMAINE  Sweet Monsieur Paroles.

PAROLES  Noble heroes, my sword and yours are kin. Good
sparks and lustrous, a word, good mettles. You shall
find in the regiment of the Spinii one Captain Spurio,
with his cicatrice, an emblem of war, here on his
sinister cheek. It was this very sword entrenched it.
Say to him I live, and observe his reports for me.

FIRST LORD DUMAINE  We shall, noble captain.         45

PAROLES  Mars dote on you for his novices.
                        Exeunt both Lords Dumaine
(To Bertram) What will ye do?

BERTRAM  Stay the King.

PAROLES  Use a more spacious ceremony to the noble lords.
You have restrained yourself within the list of too cold
an adieu. Be more expressive to them, for they wear
themselves in the cap of the time, there do muster true
gait; eat, speak, and move under the influence of the
most received star—and though the devil lead the
measure, such are to be followed. After them, and take
a more dilated farewell.                              56

BERTRAM  And I will do so.

PAROLES  Worthy fellows, and like to prove most sinewy
sword-men.            Exeunt ⌐Bertram and Paroles⌐
          Enter Lafeu to the King

LAFEU (kneeling)
Pardon, my lord, for me and for my tidings.          60

KING  I'll fee thee to stand up.

LAFEU (rising)
Then here's a man stands that has bought his pardon.
I would you had kneeled, my lord, to ask me mercy,
And that at my bidding you could so stand up.

KING
I would I had, so I had broke thy pate              65
And asked thee mercy for't.

LAFEU              Good faith, across!
But my good lord, 'tis thus: will you be cured
Of your infirmity?

KING                No.

LAFEU                    O will you eat
No grapes, my royal fox? Yes, but you will,
My noble grapes, an if my royal fox                 70
Could reach them. I have seen a medicine
That's able to breathe life into a stone,
Quicken a rock, and make you dance canary
With sprightly fire and motion; whose simple touch
Is powerful to araise King Pépin, nay,              75
To give great Charlemagne a pen in's hand,
And write to her a love-line.

KING                    What 'her' is this?

LAFEU
Why, Doctor She. My lord, there's one arrived,
If you will see her. Now by my faith and honour,
If seriously I may convey my thoughts               80
In this my light deliverance, I have spoke
With one that in her sex, her years, profession,
Wisdom and constancy, hath amazed me more
Than I dare blame my weakness. Will you see her—
For that is her demand—and know her business?  85
That done, laugh well at me.

KING                Now, good Lafeu,
Bring in the admiration, that we with thee
May spend our wonder too, or take off thine
By wond'ring how thou took'st it.

LAFEU                    Nay, I'll fit you,
And not be all day neither.                          90
          ⌐He goes to the door⌐

KING
Thus he his special nothing ever prologues.

LAFEU (to Helen, within) Nay, come your ways.
          Enter Helen ⌐disguised⌐

KING  This haste hath wings indeed.

LAFEU (to Helen) Nay, come your ways.
This is his majesty. Say your mind to him.          95
A traitor you do look like, but such traitors
His majesty seldom fears. I am Cressid's uncle,
That dare leave two together. Fare you well.
          Exeunt ⌐all but the King and Helen⌐

KING
Now, fair one, does your business follow us?

HELEN
Ay, my good lord. Gérard de Narbonne was my father;
In what he did profess, well found.

KING                    I knew him.      101

HELEN
The rather will I spare my praises towards him;
Knowing him is enough. On's bed of death
Many receipts he gave me, chiefly one
Which, as the dearest issue of his practice,        105
And of his old experience th'only darling,
He bade me store up as a triple eye
Safer than mine own two, more dear. I have so,
And hearing your high majesty is touched
With that malignant cause wherein the honour       110

Of my dear father's gift stands chief in power,
I come to tender it and my appliance
With all bound humbleness.
KING                                  We thank you, maiden,
But may not be so credulous of cure,
When our most learnèd doctors leave us, and          115
The congregated College have concluded
That labouring art can never ransom nature
From her inaidable estate. I say we must not
So stain our judgement or corrupt our hope,
To prostitute our past-cure malady          120
To empirics, or to dissever so
Our great self and our credit, to esteem
A senseless help, when help past sense we deem.
HELEN
My duty then shall pay me for my pains.
I will no more enforce mine office on you,          125
Humbly entreating from your royal thoughts
A modest one to bear me back again.
KING
I cannot give thee less, to be called grateful.
Thou thought'st to help me, and such thanks I give
As one near death to those that wish him live.          130
But what at full I know, thou know'st no part;
I knowing all my peril, thou no art.
HELEN
What I can do can do no hurt to try,
Since you set up your rest 'gainst remedy.
He that of greatest works is finisher          135
Oft does them by the weakest minister.
So holy writ in babes hath judgement shown
When judges have been babes; great floods have
          flow'n
From simple sources, and great seas have dried.
When miracles have by th' great'st been denied          140
⌈                                                              ⌉
Oft expectation fails, and most oft there
Where most it promises, and oft it hits
Where hope is coldest and despair most fits.
KING
I must not hear thee. Fare thee well, kind maid.          145
Thy pains, not used, must by thyself be paid:
Proffers not took reap thanks for their reward.
HELEN
Inspirèd merit so by breath is barred.
It is not so with him that all things knows
As 'tis with us that square our guess by shows;          150
But most it is presumption in us when
The help of heaven we count the act of men.
Dear sir, to my endeavours give consent.
Of heaven, not me, make an experiment.
I am not an impostor, that proclaim          155
Myself against the level of mine aim,
But know I think, and think I know most sure,
My art is not past power, nor you past cure.
KING
Art thou so confident? Within what space
Hop'st thou my cure?
HELEN                        The great'st grace lending grace,
Ere twice the horses of the sun shall bring          161
Their fiery coacher his diurnal ring,
Ere twice in murk and occidental damp
Moist Hesperus hath quenched her sleepy lamp,
Or four-and-twenty times the pilot's glass          165
Hath told the thievish minutes how they pass,
What is infirm from your sound parts shall fly,
Health shall live free, and sickness freely die.

KING
Upon thy certainty and confidence
What dar'st thou venture?
HELEN                        Tax of impudence,          170
A strumpet's boldness, a divulgèd shame;
Traduced by odious ballads, my maiden's name
Seared otherwise, nay—worse of worst—extended
With vilest torture, let my life be ended.
KING
Methinks in thee some blessèd spirit doth speak,          175
His powerful sound within an organ weak;
And what impossibility would slay
In common sense, sense saves another way.
Thy life is dear, for all that life can rate
Worth name of life in thee hath estimate:          180
Youth, beauty, wisdom, courage, all
That happiness and prime can happy call.
Thou this to hazard needs must intimate
Skill infinite, or monstrous desperate.
Sweet practiser, thy physic I will try,          185
That ministers thine own death if I die.
HELEN
If I break time, or flinch in property
Of what I spoke, unpitied let me die,
And well deserved. Not helping, death's my fee.
But if I help, what do you promise me?          190
KING
Make thy demand.
HELEN                        But will you make it even?
KING
Ay, by my sceptre and my hopes of heaven.
HELEN
Then shalt thou give me with thy kingly hand
What husband in thy power I will command.
Exempted be from me the arrogance          195
To choose from forth the royal blood of France,
My low and humble name to propagate
With any branch or image of thy state;
But such a one, thy vassal, whom I know
Is free for me to ask, thee to bestow.          200
KING
Here is my hand. The premises observed,
Thy will by my performance shall be served.
So make the choice of thy own time, for I,
Thy resolved patient, on thee still rely.
More should I question thee, and more I must,          205
Though more to know could not be more to trust:
From whence thou cam'st, how tended on—but rest
Unquestioned welcome, and undoubted blessed.—
Give me some help here, ho! If thou proceed
As high as word, my deed shall match thy deed.          210
          *Flourish. Exeunt the King, ⌈carried⌉, and Helen*

**2.2**          *Enter the Countess and Lavatch the clown*
COUNTESS Come on, sir. I shall now put you to the height
of your breeding.
LAVATCH I will show myself highly fed and lowly taught.
I know my business is but to the court.
COUNTESS 'To the court'? Why, what place make you
special, when you put off that with such contempt?
'But to the court'!          7
LAVATCH Truly, madam, if God have lent a man any
manners he may easily put it off at court. He that
cannot make a leg, put off's cap, kiss his hand, and
say nothing, has neither leg, hands, lip, nor cap, and
indeed such a fellow, to say precisely, were not for the
court. But for me, I have an answer will serve all men.

COUNTESS Marry, that's a bountiful answer that fits all questions. 15

LAVATCH It is like a barber's chair that fits all buttocks: the pin-buttock, the quatch-buttock, the brawn-buttock, or any buttock.

COUNTESS Will your answer serve fit to all questions?

LAVATCH As fit as ten groats is for the hand of an attorney, as your French crown for your taffeta punk, as Tib's rush for Tom's forefinger, as a pancake for Shrove Tuesday, a morris for May Day, as the nail to his hole, the cuckold to his horn, as a scolding quean to a wrangling knave, as the nun's lip to the friar's mouth, nay as the pudding to his skin. 26

COUNTESS Have you, I say, an answer of such fitness for all questions?

LAVATCH From beyond your duke to beneath your constable, it will fit any question. 30

COUNTESS It must be an answer of most monstrous size that must fit all demands.

LAVATCH But a trifle neither, in good faith, if the learned should speak truth of it. Here it is, and all that belongs to't. Ask me if I am a courtier. It shall do you no harm to learn. 36

COUNTESS To be young again, if we could! I will be a fool in question, hoping to be the wiser by your answer. I pray you, sir, are you a courtier?

LAVATCH O Lord, sir!—There's a simple putting off. More, more, a hundred of them. 41

COUNTESS Sir, I am a poor friend of yours that loves you.

LAVATCH O Lord, sir!—Thick, thick, spare not me.

COUNTESS I think, sir, you can eat none of this homely meat. 45

LAVATCH O Lord, sir!—Nay, put me to't, I warrant you.

COUNTESS You were lately whipped, sir, as I think.

LAVATCH O Lord, sir!—Spare not me.

COUNTESS Do you cry 'O Lord, sir!' at your whipping, and 'spare not me'? Indeed, your 'O Lord, sir!' is very sequent to your whipping. You would answer very well to a whipping, if you were but bound to't. 52

LAVATCH I ne'er had worse luck in my life in my 'O Lord, sir!' I see things may serve long, but not serve ever.

COUNTESS I play the noble housewife with the time, to entertain it so merrily with a fool. 56

LAVATCH O Lord, sir!—Why, there't serves well again.

COUNTESS
An end, sir! To your business: give Helen this,
    *She gives him a letter*
And urge her to a present answer back.
Commend me to my kinsmen and my son. 60
This is not much.

LAVATCH Not much commendation to them?

COUNTESS Not much employment for you. You understand me.

LAVATCH Most fruitfully. I am there before my legs. 65

COUNTESS Haste you again.     *Exeunt severally*

**2.3**   *Enter Bertram, Lafeu ⌈with a ballad⌉, and Paroles*

LAFEU They say miracles are past, and we have our philosophical persons to make modern and familiar things supernatural and causeless. Hence is it that we make trifles of terrors, ensconcing ourselves into seeming knowledge when we should submit ourselves to an unknown fear. 6

PAROLES Why, 'tis the rarest argument of wonder that hath shot out in our latter times.

BERTRAM And so 'tis.

LAFEU To be relinquished of the artists— 10

PAROLES So I say—both of Galen and Paracelsus.

LAFEU Of all the learned and authentic Fellows—

PAROLES Right, so I say.

LAFEU That gave him out incurable—

PAROLES Why, there 'tis, so say I too. 15

LAFEU Not to be helped.

PAROLES Right, as 'twere a man assured of a—

LAFEU Uncertain life and sure death.

PAROLES Just, you say well, so would I have said.

LAFEU I may truly say it is a novelty to the world. 20

PAROLES It is indeed. If you will have it in showing, you shall read it in ⌈*pointing to the ballad*⌉ what-do-ye-call there.

LAFEU ⌈*reads*⌉ 'A showing of a heavenly effect in an earthly actor.' 25

PAROLES That's it, I would have said the very same.

LAFEU Why, your dolphin is not lustier. Fore me, I speak in respect—

PAROLES Nay, 'tis strange, 'tis very strange, that is the brief and the tedious of it, and he's of a most facinorous spirit that will not acknowledge it to be the— 31

LAFEU Very hand of heaven.

PAROLES Ay, so I say.

LAFEU In a most weak—

PAROLES And debile minister great power, great transcendence, which should indeed give us a further use to be made than alone the recov'ry of the king, as to be— 38

LAFEU Generally thankful.
    *Enter the King, Helen, and attendants*

PAROLES I would have said it, you say well. Here comes the King.

LAFEU *Lustig*, as the Dutchman says. I'll like a maid the better whilst I have a tooth in my head.
    ⌈*The King and Helen dance*⌉
Why, he's able to lead her a coranto.

PAROLES *Mort du vinaigre*, is not this Helen? 45

LAFEU Fore God, I think so.

KING
Go call before me all the lords in court.
                  *Exit one or more*
Sit, my preserver, by thy patient's side,
    ⌈*The King and Helen sit*⌉
And with this healthful hand whose banished sense
Thou hast repealed, a second time receive 50
The confirmation of my promised gift,
Which but attends thy naming.
    *Enter four Lords*
Fair maid, send forth thine eye. This youthful parcel
Of noble bachelors stand at my bestowing,
O'er whom both sovereign power and father's voice
I have to use. Thy frank election make. 56
Thou hast power to choose, and they none to forsake.

HELEN
To each of you one fair and virtuous mistress
Fall when love please. Marry, to each but one.

LAFEU (*aside*)
I'd give bay Curtal and his furniture 60
My mouth no more were broken than these boys',
And writ as little beard.

KING (*to Helen*)     Peruse them well.
Not one of these but had a noble father.

HELEN Gentlemen,
Heaven hath through me restored the King to health.

⌜ALL BUT HELEN⌝
We understand it, and thank heaven for you.          66
HELEN
I am a simple maid, and therein wealthiest
That I protest I simply am a maid.—
Please it your majesty, I have done already.
The blushes in my cheeks thus whisper me:          70
'We blush that thou shouldst choose; but, be refused,
Let the white death sit on thy cheek for ever,
We'll ne'er come there again.'
KING                              Make choice and see.
Who shuns thy love shuns all his love in me.
HELEN (rising)
Now, Dian, from thy altar do I fly,          75
And to imperial Love, that god most high,
Do my sighs stream.
          ⌜She addresses her to a Lord⌝
                    Sir, will you hear my suit?
FIRST LORD
And grant it.
HELEN                    Thanks, sir. All the rest is mute.
LAFEU (aside) I had rather be in this choice than throw
ambs-ace for my life.          80
HELEN (to another Lord)
The honour, sir, that flames in your fair eyes,
Before I speak, too threat'ningly replies.
Love make your fortunes twenty times above
Her that so wishes, and her humble love.
SECOND LORD
No better, if you please.
HELEN                    My wish receive,          85
Which great Love grant. And so I take my leave.
LAFEU (aside) Do all they deny her? An they were sons of
mine I'd have them whipped, or I would send them to
th' Turk to make eunuchs of.
HELEN (to another Lord)
Be not afraid that I your hand should take;          90
I'll never do you wrong for your own sake.
Blessing upon your vows, and in your bed
Find fairer fortune, if you ever wed.
LAFEU (aside) These boys are boys of ice, they'll none have
her. Sure they are bastards to the English, the French
ne'er got 'em.          96
HELEN (to another Lord)
You are too young, too happy, and too good
To make yourself a son out of my blood.
FOURTH LORD Fair one, I think not so.
LAFEU (aside) There's one grape yet. I am sure thy father
drunk wine, but if thou beest not an ass I am a youth
of fourteen. I have known thee already.          102
HELEN (to Bertram)
I dare not say I take you, but I give
Me and my service ever whilst I live
Into your guiding power.—This is the man.          105
KING
Why then, young Bertram, take her, she's thy wife.
BERTRAM
My wife, my liege? I shall beseech your highness,
In such a business give me leave to use
The help of mine own eyes.
KING                    Know'st thou not, Bertram,
What she has done for me?
BERTRAM                    Yes, my good lord,          110
But never hope to know why I should marry her.
KING
Thou know'st she has raised me from my sickly bed.

BERTRAM
But follows it, my lord, to bring me down
Must answer for your raising? I know her well:
She had her breeding at my father's charge.          115
A poor physician's daughter, my wife? Disdain
Rather corrupt me ever.
KING
'Tis only title thou disdain'st in her, the which
I can build up. Strange is it that our bloods,
Of colour, weight, and heat, poured all together,          120
Would quite confound distinction, yet stands off
In differences so mighty. If she be
All that is virtuous, save what thou dislik'st—
'A poor physician's daughter'—thou dislik'st
Of virtue for the name. But do not so.          125
From lowest place when virtuous things proceed,
The place is dignified by th' doer's deed.
Where great additions swell's, and virtue none,
It is a dropsied honour. Good alone
Is good without a name, vileness is so:          130
The property by what it is should go,
Not by the title. She is young, wise, fair.
In these to nature she's immediate heir,
And these breed honour. That is honour's scorn
Which challenges itself as honour's born          135
And is not like the sire; honours thrive
When rather from our acts we them derive
Than our foregoers. The mere word's a slave,
Debauched on every tomb, on every grave
A lying trophy, and as oft is dumb          140
Where dust and dammed oblivion is the tomb
Of honoured bones indeed. What should be said?
If thou canst like this creature as a maid,
I can create the rest. Virtue and she
Is her own dower; honour and wealth from me.          145
BERTRAM
I cannot love her, nor will strive to do't.
KING
Thou wrong'st thyself. If thou shouldst strive to
          choose—
HELEN
That you are well restored, my lord, I'm glad.
Let the rest go.
KING
My honour's at the stake, which to defeat          150
I must produce my power. Here, take her hand,
Proud, scornful boy, unworthy this good gift,
That dost in vile misprision shackle up
My love and her desert; that canst not dream
We, poising us in her defective scale,          155
Shall weigh thee to the beam; that wilt not know
It is in us to plant thine honour where
We please to have it grow. Check thy contempt;
Obey our will, which travails in thy good;
Believe not thy disdain, but presently          160
Do thine own fortunes that obedient right
Which both thy duty owes and our power claims,
Or I will throw thee from my care for ever
Into the staggers and the careless lapse
Of youth and ignorance, both my revenge and hate
Loosing upon thee in the name of justice          166
Without all terms of pity. Speak. Thine answer.
BERTRAM (kneeling)
Pardon, my gracious lord, for I submit
My fancy to your eyes. When I consider
What great creation and what dole of honour          170

Flies where you bid it, I find that she, which late
Was in my nobler thoughts most base, is now
The praisèd of the King; who, so ennobled,
Is as 'twere born so.

KING               Take her by the hand
And tell her she is thine; to whom I promise    175
A counterpoise, if not to thy estate
A balance more replete.

BERTRAM (rising)         I take her hand.

KING
Good fortune and the favour of the King
Smile upon this contract, whose ceremony
Shall seem expedient on the now-born brief,    180
And be performed tonight. The solemn feast
Shall more attend upon the coming space,
Expecting absent friends. As thou lov'st her
Thy love's to me religious; else, does err.

      ⌈Flourish.⌉ Exeunt all but Paroles and Lafeu,
      who stay behind, commenting on this wedding

LAFEU Do you hear, monsieur? A word with you.    185

PAROLES Your pleasure, sir.

LAFEU Your lord and master did well to make his
recantation.

PAROLES Recantation? My lord? My master?

LAFEU Ay. Is it not a language I speak?    190

PAROLES A most harsh one, and not to be understood
without bloody succeeding. My master?

LAFEU Are you companion to the Count Roussillon?

PAROLES To any count, to all counts, to what is man.

LAFEU To what is count's man; count's master is of
another style.    196

PAROLES You are too old, sir. Let it satisfy you, you are
too old.

LAFEU I must tell thee, sirrah, I write 'Man', to which
title age cannot bring thee.    200

PAROLES What I dare too well do I dare not do.

LAFEU I did think thee for two ordinaries to be a pretty
wise fellow. Thou didst make tolerable vent of thy
travel; it might pass. Yet the scarves and the bannerets
about thee did manifoldly dissuade me from believing
thee a vessel of too great a burden. I have now found
thee; when I lose thee again I care not. Yet art thou
good for nothing but taking up, and that thou'rt scarce
worth.    209

PAROLES Hadst thou not the privilege of antiquity upon
thee—

LAFEU Do not plunge thyself too far in anger, lest thou
hasten thy trial, which if—Lord have mercy on thee
for a hen! So, my good window of lattice, fare thee
well. Thy casement I need not open, for I look through
thee. Give me thy hand.    216

PAROLES My lord, you give me most egregious indignity.

LAFEU Ay, with all my heart, and thou art worthy of it.

PAROLES I have not, my lord, deserved it.

LAFEU Yes, good faith, every dram of it, and I will not
bate thee a scruple.    221

PAROLES Well, I shall be wiser.

LAFEU E'en as soon as thou canst, for thou hast to pull
at a smack o' th' contrary. If ever thou beest bound in
thy scarf and beaten thou shall find what it is to be
proud of thy bondage. I have a desire to hold my
acquaintance with thee, or rather my knowledge, that
I may say in the default, 'He is a man I know'.    228

PAROLES My lord, you do me most insupportable vexation.

LAFEU I would it were hell-pains for thy sake, and my

poor doing eternal; for doing I am past, as I will by
thee, in what motion age will give me leave.      Exit

PAROLES Well, thou hast a son shall take this disgrace off
me. Scurvy, old, filthy, scurvy lord. Well, I must be
patient. There is no fettering of authority. I'll beat him,
by my life, if I can meet him with any convenience,
an he were double and double a lord. I'll have no more
pity of his age than I would have of—I'll beat him, an
if I could but meet him again.    239

      Enter Lafeu

LAFEU Sirrah, your lord and master's married. There's
news for you: you have a new mistress.

PAROLES I most unfeignedly beseech your lordship to make
some reservation of your wrongs. He is my good lord;
whom I serve above is my master.

LAFEU Who? God?    245

PAROLES Ay, sir.

LAFEU The devil it is that's thy master. Why dost thou
garter up thy arms o' this fashion? Dost make hose of
thy sleeves? Do other servants so? Thou wert best set
thy lower part where thy nose stands. By mine honour,
if I were but two hours younger I'd beat thee. Methink'st
thou art a general offence and every man should beat
thee. I think thou wast created for men to breathe
themselves upon thee.    254

PAROLES This is hard and undeserved measure, my lord.

LAFEU Go to, sir. You were beaten in Italy for picking a
kernel out of a pomegranate, you are a vagabond and
no true traveller, you are more saucy with lords and
honourable personages than the commission of your
birth and virtue gives you heraldry. You are not worth
another word, else I'd call you knave. I leave you.    261
                                       Exit

PAROLES Good, very good, it is so then. Good, very good,
let it be concealed awhile.

      ⌈Enter Bertram⌉

BERTRAM
Undone and forfeited to cares for ever.

PAROLES What's the matter, sweetheart?    265

BERTRAM
Although before the solemn priest I have sworn,
I will not bed her.

PAROLES What, what, sweetheart?

BERTRAM
O my Paroles, they have married me.
I'll to the Tuscan wars and never bed her.    270

PAROLES
France is a dog-hole, and it no more merits
The tread of a man's foot. To th' wars!

BERTRAM
There's letters from my mother. What th'import is
I know not yet.

PAROLES
Ay, that would be known. To th' wars, my boy, to th'
    wars!    275
He wears his honour in a box unseen
That hugs his kicky-wicky here at home,
Spending his manly marrow in her arms,
Which should sustain the bound and high curvet
Of Mars's fiery steed. To other regions!    280
France is a stable, we that dwell in't jades.
Therefore to th' war.

BERTRAM
It shall be so. I'll send her to my house,
Acquaint my mother with my hate to her,

And wherefore I am fled, write to the King     285
That which I durst not speak. His present gift
Shall furnish me to those Italian fields
Where noble fellows strike. Wars is no strife
To the dark house and the detested wife.

PAROLES
Will this *capriccio* hold in thee? Art sure?     290

BERTRAM
Go with me to my chamber and advise me.
I'll send her straight away. Tomorrow
I'll to the wars, she to her single sorrow.

PAROLES
Why, these balls bound, there's noise in it. 'Tis hard:
A young man married is a man that's marred.     295
Therefore away, and leave her bravely. Go.
The King has done you wrong, but hush 'tis so.
                                        *Exeunt*

**2.4**   *Enter Helen reading a letter, and Lavatch the clown*

HELEN
My mother greets me kindly. Is she well?

LAVATCH She is not well, but yet she has her health. She's
very merry, but yet she is not well. But thanks be given
she's very well and wants nothing i'th' world. But yet
she is not well.     5

HELEN
If she be very well, what does she ail
That she's not very well?

LAVATCH Truly, she's very well indeed, but for two things.

HELEN What two things?

LAVATCH One, that she's not in heaven, whither God send
her quickly. The other, that she's in earth, from whence
God send her quickly.     12

       *Enter Paroles*

PAROLES Bless you, my fortunate lady.

HELEN
I hope, sir, I have your good will to have
Mine own good fortunes.     15

PAROLES You had my prayers to lead them on, and to
keep them on have them still.—O my knave, how does
my old lady?

LAVATCH So that you had her wrinkles and I her money,
I would she did as you say.     20

PAROLES Why, I say nothing.

LAVATCH Marry, you are the wiser man, for many a
man's tongue shakes out his master's undoing. To say
nothing, to do nothing, to know nothing, and to have
nothing, is to be a great part of your title, which is
within a very little of nothing.     26

PAROLES Away, thou'rt a knave.

LAVATCH You should have said, sir, 'Before a knave,
thou'rt a knave'—that's 'Before me, thou'rt a knave'.
This had been truth, sir.     30

PAROLES Go to, thou art a witty fool. I have found thee.

LAVATCH Did you find me in yourself, sir, or were you
taught to find me?

⌜PAROLES⌝ In myself, knave.

LAVATCH The search, sir, was profitable, and much fool
may you find in you, even to the world's pleasure and
the increase of laughter.     37

PAROLES (*to Helen*) A good knave, i'faith, and well fed.
Madam, my lord will go away tonight.
A very serious business calls on him.     40
The great prerogative and rite of love,
Which as your due time claims, he does acknowledge,
But puts it off to a compelled restraint:

Whose want and whose delay is strewed with sweets,
Which they distil now in the curbèd time,     45
To make the coming hour o'erflow with joy,
And pleasure drown the brim.

HELEN                           What's his will else?

PAROLES
That you will take your instant leave o'th' King,
And make this haste as your own good proceeding,
Strengthened with what apology you think     50
May make it probable need.

HELEN                         What more commands he?

PAROLES
That having this obtained, you presently
Attend his further pleasure.

HELEN              In everything
I wait upon his will.

PAROLES                I shall report it so.

HELEN I pray you.                    ⌜*Exit Paroles at one door*⌝
Come, sirrah.                   *Exeunt* ⌜*at another door*⌝

**2.5**   *Enter Lafeu and Bertram*

LAFEU But I hope your lordship thinks not him a soldier.

BERTRAM Yes, my lord, and of very valiant approof.

LAFEU You have it from his own deliverance.

BERTRAM And by other warranted testimony.

LAFEU Then my dial goes not true. I took this lark for a
bunting.     6

BERTRAM I do assure you, my lord, he is very great in
knowledge, and accordingly valiant.

LAFEU I have then sinned against his experience and
transgressed against his valour—and my state that way
is dangerous, since I cannot yet find in my heart to
repent. Here he comes. I pray you make us friends. I
will pursue the amity.

       *Enter Paroles*

PAROLES (*to Bertram*) These things shall be done, sir.

LAFEU (*to Bertram*) Pray you, sir, who's his tailor?     15

PAROLES Sir!

LAFEU O, I know him well. Ay, 'Sir', he; 'Sir' 's a good
workman, a very good tailor.

BERTRAM (*aside to Paroles*) Is she gone to the King?

PAROLES She is.     20

BERTRAM Will she away tonight?

PAROLES As you'll have her.

BERTRAM
I have writ my letters, casketed my treasure,
Given order for our horses, and tonight,
When I should take possession of the bride,     25
End ere I do begin.

LAFEU (*aside*) A good traveller is something at the latter
end of a dinner, but one that lies three-thirds and uses
a known truth to pass a thousand nothings with,
should be once heard and thrice beaten. (*To Paroles*)
God save you, captain.     31

BERTRAM (*to Paroles*) Is there any unkindness between my
lord and you, monsieur?

PAROLES I know not how I have deserved to run into my
lord's displeasure.     35

LAFEU You have made shift to run into't, boots and spurs
and all, like him that leaped into the custard, and out
of it you'll run again, rather than suffer question for
your residence.

BERTRAM It may be you have mistaken him, my lord.     40

LAFEU And shall do so ever, though I took him at's
prayers. Fare you well, my lord, and believe this of me:

there can be no kernel in this light nut. The soul of
this man is his clothes. Trust him not in matter of
heavy consequence. I have kept of them tame, and
know their natures.—Farewell, monsieur. I have
spoken better of you than you have wit or will to
deserve at my hand, but we must do good against evil.

                                   *Exit*

PAROLES An idle lord, I swear.
BERTRAM I think not so.                      50
PAROLES Why, do you not know him?
BERTRAM
  Yes, I do know him well, and common speech
  Gives him a worthy pass. Here comes my clog.
      *Enter Helen, ⌐attended⌐*
HELEN
  I have, sir, as I was commanded from you,
  Spoke with the King, and have procured his leave    55
  For present parting; only he desires
  Some private speech with you.
BERTRAM                 I shall obey his will.
  You must not marvel, Helen, at my course,
  Which holds not colour with the time, nor does
  The ministration and requirèd office           60
  On my particular. Prepared I was not
  For such a business, therefore am I found
  So much unsettled. This drives me to entreat you
  That presently you take your way for home,
  And rather muse than ask why I entreat you,    65
  For my respects are better than they seem,
  And my appointments have in them a need
  Greater than shows itself at the first view
  To you that know them not. This to my mother.
      *He gives her a letter*
  'Twill be two days ere I shall see you, so      70
  I leave you to your wisdom.
HELEN                Sir, I can nothing say
  But that I am your most obedient servant.
BERTRAM
  Come, come, no more of that.
HELEN               And ever shall
  With true observance seek to eke out that
  Wherein toward me my homely stars have failed    75
  To equal my great fortune.
BERTRAM              Let that go.
  My haste is very great. Farewell. Hie home.
HELEN
  Pray sir, your pardon.
BERTRAM              Well, what would you say?
HELEN
  I am not worthy of the wealth I owe,
  Nor dare I say 'tis mine—and yet it is—      80
  But like a timorous thief most fain would steal
  What law does vouch mine own.
BERTRAM              What would you have?
HELEN
  Something, and scarce so much: nothing indeed.
  I would not tell you what I would, my lord. Faith,
    yes:
  Strangers and foes do sunder and not kiss.    85
BERTRAM
  I pray you, stay not, but in haste to horse.
HELEN
  I shall not break your bidding, good my lord.—
  Where are my other men?—Monsieur, farewell.
      *Exeunt Helen ⌐and attendants at one door⌐*

BERTRAM
  Go thou toward home, where I will never come
  Whilst I can shake my sword or hear the drum.—    90
  Away, and for our flight.
PAROLES            Bravely. *Coraggio!*
                            *Exeunt ⌐at another door⌐*

**3.1**   *Flourish of trumpets. Enter the Duke of Florence*
        *and the two Lords Dumaine, with a troop of*
        *soldiers*
DUKE
  So that from point to point now have you heard
  The fundamental reasons of this war,
  Whose great decision hath much blood let forth,
  And more thirsts after.
FIRST LORD DUMAINE       Holy seems the quarrel
  Upon your grace's part; black and fearful       5
  On the opposer.
DUKE
  Therefore we marvel much our cousin France
  Would in so just a business shut his bosom
  Against our borrowing prayers.
SECOND LORD DUMAINE       Good my lord,
  The reasons of our state I cannot yield       10
  But like a common and an outward man
  That the great figure of a council frames
  By self-unable motion; therefore dare not
  Say what I think of it, since I have found
  Myself in my incertain grounds to fail       15
  As often as I guessed.
DUKE             Be it his pleasure.
FIRST LORD DUMAINE
  But I am sure the younger of our nation,
  That surfeit on their ease, will day by day
  Come here for physic.
DUKE            Welcome shall they be,
  And all the honours that can fly from us      20
  Shall on them settle. You know your places well;
  When better fall, for your avails they fell.
  Tomorrow to the field.        *Flourish. Exeunt*

**3.2**   *Enter the Countess with a letter, and Lavatch*
COUNTESS It hath happened all as I would have had it,
  save that he comes not along with her.
LAVATCH By my troth, I take my young lord to be a very
  melancholy man.
COUNTESS By what observance, I pray you?      5
LAVATCH Why, he will look upon his boot and sing, mend
  the ruff and sing, ask questions and sing, pick his teeth
  and sing. I know a man that had this trick of
  melancholy sold a goodly manor for a song.
COUNTESS Let me see what he writes, and when he means
  to come.                                           11
      *She opens the letter and reads*
LAVATCH (*aside*) I have no mind to Isbel since I was at
  court. Our old lings and our Isbels o'th' country are
  nothing like your old ling and your Isbels o'th' court.
  The brains of my Cupid's knocked out, and I begin to
  love as an old man loves money: with no stomach.
COUNTESS What have we here?                   17
LAVATCH E'en that you have there.             *Exit*
COUNTESS (*reads the letter aloud*) 'I have sent you a
  daughter-in-law. She hath recovered the King and
  undone me. I have wedded her, not bedded her, and
  sworn to make the "not" eternal. You shall hear I am
  run away; know it before the report come. If there be

breadth enough in the world I will hold a long distance.
My duty to you.                                                  25
                              Your unfortunate son,
                                   Bertram.'
This is not well, rash and unbridled boy,
To fly the favours of so good a King,
To pluck his indignation on thy head                             30
By the misprizing of a maid too virtuous
For the contempt of empire.
     *Enter Lavatch*
LAVATCH O madam, yonder is heavy news within,
     between two soldiers and my young lady.
COUNTESS What is the matter?                                     35
LAVATCH Nay, there is some comfort in the news, some
     comfort. Your son will not be killed so soon as I thought
     he would.
COUNTESS Why should he be killed?                                39
LAVATCH So say I, madam—if he run away, as I hear he
     does. The danger is in standing to't; that's the loss of
     men, though it be the getting of children. Here they
     come will tell you more. For my part, I only heard
     your son was run away.                          *[Exit]*
     *Enter Helen with a letter, and the two Lords*
     *Dumaine*
SECOND LORD DUMAINE *(to the Countess)*
Save you, good madam.                                            45
HELEN
Madam, my lord is gone, for ever gone.
FIRST LORD DUMAINE Do not say so.
COUNTESS *(to Helen)*
Think upon patience.—Pray you, gentlemen,
I have felt so many quirks of joy and grief
That the first face of neither on the start                      50
Can woman me unto't. Where is my son, I pray you?
FIRST LORD DUMAINE
Madam, he's gone to serve the Duke of Florence.
We met him thitherward, for thence we came,
And, after some dispatch in hand at court,
Thither we bend again.                                           55
HELEN
Look on his letter, madam: here's my passport.
     *[She] reads aloud*
'When thou canst get the ring upon my finger, which
never shall come off, and show me a child begotten of
thy body that I am father to, then call me husband;
but in such a "then" I write a "never".'                         60
This is a dreadful sentence.
COUNTESS
Brought you this letter, gentlemen?
FIRST LORD DUMAINE                              Ay, madam,
And for the contents' sake are sorry for our pains.
COUNTESS
I prithee, lady, have a better cheer.
If thou engrossest all the griefs are thine                      65
Thou robb'st me of a moiety. He was my son,
But I do wash his name out of my blood,
And thou art all my child.—Towards Florence is he?
FIRST LORD DUMAINE
Ay, madam.
COUNTESS        And to be a soldier?
FIRST LORD DUMAINE
Such is his noble purpose, and—believe't—                        70
The Duke will lay upon him all the honour
That good convenience claims.
COUNTESS                        Return you thither?
SECOND LORD DUMAINE
Ay, madam, with the swiftest wing of speed.

HELEN 'Till I have no wife, I have nothing in France.'
'Tis bitter.                                                     75
COUNTESS Find you that there?
HELEN Ay, madam.
SECOND LORD DUMAINE
'Tis but the boldness of his hand,
Haply, which his heart was not consenting to.
COUNTESS
Nothing in France until he have no wife?                         80
There's nothing here that is too good for him
But only she, and she deserves a lord
That twenty such rude boys might tend upon
And call her, hourly, mistress. Who was with him?
SECOND LORD DUMAINE
A servant only, and a gentleman                                  85
Which I have sometime known.
COUNTESS Paroles, was it not?
SECOND LORD DUMAINE Ay, my good lady, he.
COUNTESS
A very tainted fellow, and full of wickedness.
My son corrupts a well-derivèd nature                            90
With his inducement.
SECOND LORD DUMAINE   Indeed, good lady,
The fellow has a deal of that too much,
Which holds him much to have.
COUNTESS                      You're welcome, gentlemen.
I will entreat you when you see my son
To tell him that his sword can never win                         95
The honour that he loses. More I'll entreat you
Written to bear along.
FIRST LORD DUMAINE       We serve you, madam,
In that and all your worthiest affairs.
COUNTESS
Not so, but as we change our courtesies.                         99
Will you draw near?              *Exeunt all but Helen*
HELEN 'Till I have no wife I have nothing in France.'
Nothing in France until he has no wife.
Thou shalt have none, Roussillon, none in France;
Then hast thou all again. Poor lord, is't I
That chase thee from thy country and expose                      105
Those tender limbs of thine to the event
Of the none-sparing war? And is it I
That drive thee from the sportive court, where thou
Wast shot at with fair eyes, to be the mark
Of smoky muskets? O you leaden messengers                        110
That ride upon the violent speed of fire,
Fly with false aim, cleave the still-piecing air
That sings with piercing, do not touch my lord.
Whoever shoots at him, I set him there.
Whoever charges on his forward breast,                           115
I am the caitiff that do hold him to't,
And though I kill him not, I am the cause
His death was so effected. Better 'twere
I met the ravin lion when he roared
With sharp constraint of hunger; better 'twere                   120
That all the miseries which nature owes
Were mine at once. No, come thou home, Roussillon,
Whence honour but of danger wins a scar,
As oft it loses all. I will be gone;
My being here it is that holds thee hence.                       125
Shall I stay here to do't? No, no, although
The air of paradise did fan the house
And angels officed all. I will be gone,
That pitiful rumour may report my flight
To consolate thine ear. Come night, end day;                     130
For with the dark, poor thief, I'll steal away.      *Exit*

**3.3**  *Flourish of trumpets. Enter the Duke of Florence,*
*Bertram, a drummer and trumpeters, soldiers, and*
*Paroles*

DUKE (*to Bertram*)
  The general of our horse thou art, and we,
  Great in our hope, lay our best love and credence
  Upon thy promising fortune.

BERTRAM                              Sir, it is
  A charge too heavy for my strength, but yet
  We'll strive to bear it for your worthy sake    5
  To th'extreme edge of hazard.

DUKE                              Then go thou forth,
  And Fortune play upon thy prosperous helm
  As thy auspicious mistress.

BERTRAM                              This very day,
  Great Mars, I put myself into thy file.
  Make me but like my thoughts, and I shall prove   10
  A lover of thy drum, hater of love.          *Exeunt*

**3.4**  *Enter the Countess and Reynaldo her steward, with*
*a letter*

COUNTESS
  Alas! And would you take the letter of her?
  Might you not know she would do as she has done,
  By sending me a letter? Read it again.

REYNALDO (*reads the letter*)
  'I am Saint Jaques' pilgrim, thither gone.
    Ambitious love hath so in me offended       5
  That barefoot plod I the cold ground upon
    With sainted vow my faults to have amended.
  Write, write, that from the bloody course of war
    My dearest master, your dear son, may hie.
  Bless him at home in peace, whilst I from far   10
    His name with zealous fervour sanctify.
  His taken labours bid him me forgive;
    I, his despiteful Juno, sent him forth
  From courtly friends, with camping foes to live,
    Where death and danger dogs the heels of worth.
  He is too good and fair for death and me;       16
    Whom I myself embrace to set him free.'

COUNTESS
  Ah, what sharp stings are in her mildest words!
  Reynaldo, you did never lack advice so much
  As letting her pass so. Had I spoke with her,   20
  I could have well diverted her intents,
  Which thus she hath prevented.

REYNALDO                           Pardon me, madam.
  If I had given you this at over-night
  She might have been o'erta'en—and yet she writes
  Pursuit would be but vain.

COUNTESS                    What angel shall       25
  Bless this unworthy husband? He cannot thrive
  Unless her prayers, whom heaven delights to hear
  And loves to grant, reprieve him from the wrath
  Of greatest justice. Write, write, Reynaldo,
  To this unworthy husband of his wife,            30
  Let every word weigh heavy of her worth,
  That he does weigh too light; my greatest grief,
  Though little he do feel it, set down sharply.
  Dispatch the most convenient messenger.
  When haply he shall hear that she is gone,       35
  He will return, and hope I may that she,
  Hearing so much, will speed her foot again,
  Led hither by pure love. Which of them both
  Is dearest to me I have no skill in sense

  To make distinction. Provide this messenger.    40
  My heart is heavy and mine age is weak;
  Grief would have tears, and sorrow bids me speak.
                                        *Exeunt*

**3.5**  *A tucket afar off. Enter an old Widow, her daughter*
*Diana, and Mariana, with other Florentine citizens*

WIDOW Nay, come, for if they do approach the city we
  shall lose all the sight.

DIANA They say the French Count has done most
  honourable service.

WIDOW It is reported that he has taken their greatest
  commander, and that with his own hand he slew the
  Duke's brother. (*Tucket*) We have lost our labour; they
  are gone a contrary way. Hark. You may know by
  their trumpets.                                 9

MARIANA Come, let's return again, and suffice ourselves
  with the report of it.—Well, Diana, take heed of this
  French earl. The honour of a maid is her name, and
  no legacy is so rich as honesty.

WIDOW (*to Diana*) I have told my neighbour how you have
  been solicited by a gentleman, his companion.    15

MARIANA I know that knave, hang him! One Paroles. A
  filthy officer he is in those suggestions for the young
  earl. Beware of them, Diana; their promises, entice-
  ments, oaths, tokens, and all their engines of lust,
  are not the things they go under. Many a maid hath
  been seduced by them; and the misery is, example,
  that so terrible shows in the wreck of maidenhood,
  cannot for all that dissuade succession, but that they
  are limed with the twigs that threatens them. I hope I
  need not to advise you further, but I hope your own
  grace will keep you where you are, though there were
  no further danger known but the modesty which is so
  lost.                                           28

DIANA You shall not need to fear me.
    *Enter Helen dressed as a pilgrim*

WIDOW I hope so. Look, here comes a pilgrim. I know she
  will lie at my house; thither they send one another.
  I'll question her.
  God save you, pilgrim. Whither are you bound?

HELEN To Saint Jaques le Grand.
  Where do the palmers lodge, I do beseech you?    35

WIDOW
  At the 'Saint Francis' here beside the port.

HELEN
  Is this the way?

WIDOW                     Ay, marry, is't.
    *Sound of a march, far off*
  Hark you, they come this way. If you will tarry,
  Holy pilgrim, but till the troops come by,
  I will conduct you where you shall be lodged,    40
  The rather for I think I know your hostess
  As ample as myself.

HELEN Is it yourself?

WIDOW If you shall please so, pilgrim.

HELEN
  I thank you, and will stay upon your leisure.    45

WIDOW
  You came, I think, from France?

HELEN                              I did so.

WIDOW
  Here you shall see a countryman of yours
  That has done worthy service.

HELEN                          His name, I pray you?

DIANA
The Count Roussillon. Know you such a one?

HELEN
But by the ear, that hears most nobly of him;     50
His face I know not.

DIANA                    Whatsome'er he is,
He's bravely taken here. He stole from France,
As 'tis reported; for the King had married him
Against his liking. Think you it is so?

HELEN
Ay, surely, mere the truth. I know his lady.     55

DIANA
There is a gentleman that serves the Count
Reports but coarsely of her.

HELEN                    What's his name?

DIANA
Monsieur Paroles.

HELEN                    O, I believe with him:
In argument of praise, or to the worth
Of the great Count himself, she is too mean     60
To have her name repeated. All her deserving
Is a reservèd honesty, and that
I have not heard examined.

DIANA                    Alas, poor lady.
'Tis a hard bondage to become the wife
Of a detesting lord.     65

WIDOW
I warr'nt, good creature, wheresoe'er she is
Her heart weighs sadly. This young maid might do her
A shrewd turn if she pleased.

HELEN                    How do you mean?
Maybe the amorous Count solicits her
In the unlawful purpose.

WIDOW                    He does indeed,     70
And brokes with all that can in such a suit
Corrupt the tender honour of a maid.
But she is armed for him, and keeps her guard
In honestest defence.

MARIANA The gods forbid else.     75
⌜Enter, with drummer and colours, Bertram,
Paroles, and the whole army⌝

WIDOW So, now they come.
That is Antonio, the Duke's eldest son;
That, Escalus.

HELEN                    Which is the Frenchman?

DIANA                    He—
That with the plume. 'Tis a most gallant fellow.
I would he loved his wife. If he were honester     80
He were much goodlier. Is't not
A handsome gentleman?

HELEN I like him well.

DIANA 'Tis pity he is not honest.
Yond's that same knave that leads him to those
places.     85
Were I his lady, I would poison
That vile rascal.

HELEN                    Which is he?

DIANA                    That jackanapes
With scarves. Why is he melancholy?

HELEN Perchance he's hurt i'th' battle.

PAROLES (aside) Lose our drum? Well.     90

MARIANA He's shrewdly vexed at something.
Look, he has spied us.

WIDOW (to Paroles)     Marry, hang you!

MARIANA (to Paroles)
And your courtesy, for a ring-carrier.
Exeunt Bertram, Paroles, and the army

WIDOW
The troop is past. Come, pilgrim, I will bring you
Where you shall host. Of enjoined penitents     95
There's four or five to great Saint Jaques bound
Already at my house.

HELEN                    I humbly thank you.
Please it this matron and this gentle maid
To eat with us tonight, the charge and thanking
Shall be for me. And to requite you further,     100
I will bestow some precepts of this virgin
Worthy the note.

WIDOW and MARIANA We'll take your offer kindly. Exeunt

3.6     Enter Bertram and the two Captains Dumaine

SECOND LORD DUMAINE (to Bertram) Nay, good my lord,
put him to't. Let him have his way.

FIRST LORD DUMAINE (to Bertram) If your lordship find him
not a hilding, hold me no more in your respect.

SECOND LORD DUMAINE (to Bertram) On my life, my lord, a
bubble.     6

BERTRAM Do you think I am so far deceived in him?

SECOND LORD DUMAINE Believe it, my lord. In mine own
direct knowledge—without any malice, but to speak of
him as my kinsman—he's a most notable coward, an
infinite and endless liar, an hourly promise-breaker,
the owner of no one good quality worthy your lordship's
entertainment.     13

FIRST LORD DUMAINE (to Bertram) It were fit you knew
him, lest reposing too far in his virtue, which he hath
not, he might at some great and trusty business, in a
main danger, fail you.

BERTRAM I would I knew in what particular action to try
him.     19

FIRST LORD DUMAINE None better than to let him fetch off
his drum, which you hear him so confidently undertake
to do.

SECOND LORD DUMAINE (to Bertram) I, with a troop of
Florentines, will suddenly surprise him. Such I will
have whom I am sure he knows not from the enemy;
we will bind and hoodwink him so, that he shall
suppose no other but that he is carried into the laager
of the adversary's when we bring him to our own tents.
Be but your lordship present at his examination: if he
do not, for the promise of his life and in the highest
compulsion of base fear, offer to betray you, and deliver
all the intelligence in his power against you, and that
with the divine forfeit of his soul upon oath, never trust
my judgement in anything.     34

FIRST LORD DUMAINE (to Bertram) O, for the love of laughter,
let him fetch his drum. He says he has a stratagem
for't. When your lordship sees the bottom of his success
in't, and to what metal this counterfeit lump of ore
will be melted, if you give him not John Drum's
entertainment, your inclining cannot be removed. Here
he comes.     41

Enter Paroles

SECOND LORD DUMAINE O ⌜aside⌝ for the love of laughter
⌜aloud⌝ hinder not the honour of his design; let him
fetch off his drum in any hand.

BERTRAM (to Paroles) How now, monsieur? This drum
sticks sorely in your disposition.     46

FIRST LORD DUMAINE A pox on't, let it go. 'Tis but a drum.

PAROLES But a drum? Is't but a drum? A drum so lost!
There was excellent command: to charge in with our
horse upon our own wings and to rend our own
soldiers!     51

FIRST LORD DUMAINE That was not to be blamed in the command of the service. It was a disaster of war that Caesar himself could not have prevented, if he had been there to command.                                              55

BERTRAM Well, we cannot greatly condemn our success. Some dishonour we had in the loss of that drum, but it is not to be recovered.

PAROLES It might have been recovered.

BERTRAM It might, but it is not now.                              60

PAROLES It *is* to be recovered. But that the merit of service is seldom attributed to the true and exact performer, I would have that drum or another, or 'hic iacet'.

BERTRAM Why, if you have a stomach, to't, monsieur. If you think your mystery in stratagem can bring this instrument of honour again into his native quarter, be magnanimous in the enterprise and go on. I will grace the attempt for a worthy exploit. If you speed well in it, the Duke shall both speak of it and extend to you what further becomes his greatness, even to the utmost syllable of your worthiness.                              71

PAROLES By the hand of a soldier, I will undertake it.

BERTRAM But you must not now slumber in it.

PAROLES I'll about it this evening, and I will presently pen down my dilemmas, encourage myself in my certainty, put myself into my mortal preparation; and by midnight look to hear further from me.          77

BERTRAM May I be bold to acquaint his grace you are gone about it?

PAROLES I know not what the success will be, my lord, but the attempt I vow.                              81

BERTRAM I know thou'rt valiant, and to the possibility of thy soldiership will subscribe for thee. Farewell.

PAROLES I love not many words.                    *Exit*

SECOND LORD DUMAINE No more than a fish loves water. (*To Bertram*) Is not this a strange fellow, my lord, that so confidently seems to undertake this business, which he knows is not to be done? Damns himself to do, and dares better be damned than to do't.            89

FIRST LORD DUMAINE (*to Bertram*) You do not know him, my lord, as we do. Certain it is that he will steal himself into a man's favour, and for a week escape a great deal of discoveries, but when you find him out, you have him ever after.                              94

BERTRAM Why, do you think he will make no deed at all of this that so seriously he does address himself unto?

SECOND LORD DUMAINE None in the world, but return with an invention, and clap upon you two or three probable lies. But we have almost embosked him. You shall see his fall tonight; for indeed he is not for your lordship's respect.                                          101

FIRST LORD DUMAINE (*to Bertram*) We'll make you some sport with the fox ere we case him. He was first smoked by the old Lord Lafeu. When his disguise and he is parted, tell me what a sprat you shall find him, which you shall see this very night.                  106

SECOND LORD DUMAINE
I must go look my twigs. He shall be caught.

BERTRAM
Your brother, he shall go along with me.

⌜SECOND⌝ LORD DUMAINE As't please your lordship. I'll leave you.                                          *Exit*

BERTRAM
Now will I lead you to the house, and show you   111
The lass I spoke of.

⌜FIRST⌝ LORD DUMAINE But you say she's honest.

BERTRAM
That's all the fault. I spoke with her but once
And found her wondrous cold, but I sent to her
By this same coxcomb that we have i'th' wind   115
Tokens and letters, which she did re-send,
And this is all I have done. She's a fair creature.
Will you go see her?

⌜FIRST⌝ LORD DUMAINE   With all my heart, my lord.

*Exeunt*

**3.7**   *Enter Helen and the Widow*

HELEN
If you misdoubt me that I am not she,
I know not how I shall assure you further
But I shall lose the grounds I work upon.

WIDOW
Though my estate be fall'n, I was well born,
Nothing acquainted with these businesses,          5
And would not put my reputation now
In any staining act.

HELEN                    Nor would I wish you.
First give me trust the Count he is my husband,
And what to your sworn counsel I have spoken
Is so from word to word, and then you cannot,   10
By the good aid that I of you shall borrow,
Err in bestowing it.

WIDOW                    I should believe you,
For you have showed me that which well approves
You're great in fortune.

HELEN                    Take this purse of gold,
And let me buy your friendly help thus far,       15
Which I will over-pay, and pay again
When I have found it. The Count he woos your
         daughter,
Lays down his wanton siege before her beauty,
Resolved to carry her. Let her in fine consent,
As we'll direct her how 'tis best to bear it.       20
Now his important blood will naught deny
That she'll demand. A ring the County wears,
That downward hath succeeded in his house
From son to son some four or five descents
Since the first father wore it. This ring he holds   25
In most rich choice; yet in his idle fire
To buy his will it would not seem too dear,
Howe'er repented after.

WIDOW
Now I see the bottom of your purpose.

HELEN
You see it lawful then. It is no more               30
But that your daughter ere she seems as won
Desires this ring; appoints him an encounter;
In fine, delivers me to fill the time,
Herself most chastely absent. After,
To marry her I'll add three thousand crowns        35
To what is passed already.

WIDOW                    I have yielded.
Instruct my daughter how she shall persever,
That time and place with this deceit so lawful
May prove coherent. Every night he comes
With musics of all sorts, and songs composed      40
To her unworthiness. It nothing steads us
To chide him from our eaves, for he persists
As if his life lay on't.

HELEN                    Why then tonight
Let us essay our plot, which if it speed
Is wicked meaning in a lawful deed                 45

And lawful meaning in a wicked act,
Where both not sin, and yet a sinful fact.
But let's about it.                                          *Exeunt*

**4.1**   *Enter ⌐Second Lord Dumaine¬, with five or six other*
          *soldiers, in ambush*
⌐SECOND¬ LORD DUMAINE He can come no other way but
by this hedge corner. When you sally upon him, speak
what terrible language you will. Though you under-
stand it not yourselves, no matter, for we must not
seem to understand him, unless some one among us,
whom we must produce for an interpreter.          6
INTERPRETER Good captain, let me be th'interpreter.
⌐SECOND¬ LORD DUMAINE Art not acquainted with him?
Knows he not thy voice?
INTERPRETER No, sir, I warrant you.                  10
⌐SECOND¬ LORD DUMAINE But what linsey-woolsey hast
thou to speak to us again?
INTERPRETER E'en such as you speak to me.
⌐SECOND¬ LORD DUMAINE He must think us some band of
strangers i'th' adversary's entertainment. Now he hath
a smack of all neighbouring languages, therefore we
must every one be a man of his own fancy. Not to
know what we speak one to another, so we seem to
know, is to know straight our purpose: choughs'
language, gabble enough and good enough. As for you,
interpreter, you must seem very politic. But couch, ho!
Here he comes, to beguile two hours in a sleep, and
then to return and swear the lies he forges.          23
          *They hide. Enter Paroles. ⌐Clock strikes¬*
PAROLES Ten o'clock. Within these three hours 'twill be
time enough to go home. What shall I say I have done?
It must be a very plausive invention that carries it.
They begin to smoke me, and disgraces have of late
knocked too often at my door. I find my tongue is too
foolhardy, but my heart hath the fear of Mars before
it, and of his creatures, not daring the reports of my
tongue.                                                          31
⌐SECOND¬ LORD DUMAINE (*aside*) This is the first truth that
e'er thine own tongue was guilty of.
PAROLES What the devil should move me to undertake
the recovery of this drum, being not ignorant of the
impossibility, and knowing I had no such purpose? I
must give myself some hurts, and say I got them in
exploit. Yet slight ones will not carry it. They will say,
'Came you off with so little?' And great ones I dare
not give. Wherefore, what's the instance? Tongue, I
must put you into a butter-woman's mouth, and buy
myself another of Bajazet's mute, if you prattle me into
these perils.
⌐SECOND¬ LORD DUMAINE (*aside*) Is it possible he should
know what he is, and be that he is?              45
PAROLES I would the cutting of my garments would serve
the turn, or the breaking of my Spanish sword.
⌐SECOND¬ LORD DUMAINE (*aside*) We cannot afford you so.
PAROLES Or the baring of my beard, and to say it was in
stratagem.
⌐SECOND¬ LORD DUMAINE (*aside*) 'Twould not do.      51
PAROLES Or to drown my clothes, and say I was stripped.
⌐SECOND¬ LORD DUMAINE (*aside*) Hardly serve.
PAROLES Though I swore I leapt from the window of the
citadel?                                                          55
⌐SECOND¬ LORD DUMAINE (*aside*) How deep?
PAROLES Thirty fathom.
⌐SECOND¬ LORD DUMAINE (*aside*) Three great oaths would
scarce make that be believed.                        59

PAROLES I would I had any drum of the enemy's. I would
swear I recovered it.
⌐SECOND¬ LORD DUMAINE (*aside*) You shall hear one anon.
PAROLES A drum now of the enemy's—
          *Alarum within. ⌐The ambush rushes forth¬*
⌐SECOND¬ LORD DUMAINE *Throca movousus, cargo, cargo,*
*cargo.*                                                          65
⌐SOLDIERS¬ (*severally*) *Cargo, cargo, cargo, villianda par corbo,*
*cargo.*
          *⌐They seize and blindfold him¬*
PAROLES
O ransom, ransom, do not hide mine eyes.
INTERPRETER *Boskos thromuldo boskos.*
PAROLES
I know you are the Moscows regiment,             70
And I shall lose my life for want of language.
If there be here German or Dane, Low Dutch,
Italian, or French, let him speak to me,
I'll discover that which shall undo the Florentine.
INTERPRETER *Boskos vauvado.*—                          75
I understand thee, and can speak thy tongue.—
*Kerelybonto.*—Sir,
Betake thee to thy faith, for seventeen poniards
Are at thy bosom.
PAROLES                          O!
INTERPRETER                              O pray, pray, pray!—
*Manka revania dulche?*                              80
⌐SECOND¬ LORD DUMAINE
*Oscorbidulchos volivorco.*
INTERPRETER
The general is content to spare thee yet,
And, hoodwinked as thou art, will lead thee on
To gather from thee. Haply thou mayst inform
Something to save thy life.
PAROLES                          O let me live,          85
And all the secrets of our camp I'll show,
Their force, their purposes; nay, I'll speak that
Which you will wonder at.
INTERPRETER                              But wilt thou faithfully?
PAROLES
If I do not, damn me.
INTERPRETER                    *Acordo linta.*—
Come on, thou art granted space.               90
                          *Exeunt all but ⌐Second¬ Lord Dumaine*
                                                *and a Soldier*
          *A short alarum within*
⌐SECOND¬ LORD DUMAINE
Go tell the Count Roussillon and my brother
We have caught the woodcock, and will keep him
          muffled
Till we do hear from them.
SOLDIER                          Captain, I will.
⌐SECOND¬ LORD DUMAINE
A will betray us all unto ourselves.
Inform on that.
SOLDIER                    So I will, sir.                    95
⌐SECOND¬ LORD DUMAINE
Till then I'll keep him dark and safely locked.
                                          *Exeunt severally*

**4.2**   *Enter Bertram and the maid called Diana*
BERTRAM
They told me that your name was Fontibel.
DIANA
No, my good lord, Diana.
BERTRAM                              Titled goddess,
And worth it, with addition. But, fair soul,

In your fine frame hath love no quality?
If the quick fire of youth light not your mind,                    5
You are no maiden but a monument,
When you are dead you should be such a one
As you are now, for you are cold and stern,
And now you should be as your mother was
When your sweet self was got.                                      10
DIANA  She then was honest.
BERTRAM  So should you be.
DIANA                           No.
  My mother did but duty; such, my lord,
  As you owe to your wife.
BERTRAM                     No more o' that.
  I prithee do not strive against my vows.                         15
  I was compelled to her, but I love thee
  By love's own sweet constraint, and will for ever
  Do thee all rights of service.
DIANA                          Ay, so you serve us
  Till we serve you. But when you have our roses,
  You barely leave our thorns to prick ourselves,                  20
  And mock us with our bareness.
BERTRAM                         How have I sworn!
DIANA
  'Tis not the many oaths that makes the truth,
  But the plain single vow that is vowed true.
  What is not holy, that we swear not by,
  But take the high'st to witness; then pray you, tell me,
  If I should swear by Jove's great attributes                     26
  I loved you dearly, would you believe my oaths
  When I did love you ill? This has no holding,
  To swear by him whom I protest to love
  That I will work against him. Therefore your oaths               30
  Are words and poor conditions but unsealed,
  At least in my opinion.
BERTRAM                  Change it, change it.
  Be not so holy-cruel. Love is holy,
  And my integrity ne'er knew the crafts
  That you do charge men with. Stand no more off,                  35
  But give thyself unto my sick desires,
  Who then recovers. Say thou art mine, and ever
  My love as it begins shall so persever.
DIANA
  I see that men make toys e'en such a surance
  That we'll forsake ourselves. Give me that ring.                 40
BERTRAM
  I'll lend it thee, my dear, but have no power
  To give it from me.
DIANA               Will you not, my lord?
BERTRAM
  It is an honour 'longing to our house,
  Bequeathèd down from many ancestors,
  Which were the greatest obloquy i'th' world                      45
  In me to lose.
DIANA          Mine honour's such a ring.
  My chastity's the jewel of our house,
  Bequeathèd down from many ancestors,
  Which were the greatest obloquy i'th' world
  In me to lose. Thus your own proper wisdom                       50
  Brings in the champion Honour on my part
  Against your vain assault.
BERTRAM                     Here, take my ring.
  My house, mine honour, yea my life be thine,
  And I'll be bid by thee.
DIANA
  When midnight comes, knock at my chamber window.
  I'll order take my mother shall not hear.                        56
  Now will I charge you in the bond of truth,

When you have conquered my yet maiden bed,
  Remain there but an hour, nor speak to me—
  My reasons are most strong, and you shall know them
  When back again this ring shall be delivered—                    61
  And on your finger in the night I'll put
  Another ring that, what in time proceeds,
  May token to the future our past deeds.
  Adieu till then; then, fail not. You have won                    65
  A wife of me, though there my hope be done.
BERTRAM
  A heaven on earth I have won by wooing thee.
DIANA
  For which live long to thank both heaven and me.
  You may so in the end.                    [Exit Bertram]
  My mother told me just how he would woo,                         70
  As if she sat in's heart. She says all men
  Have the like oaths. He had sworn to marry me
  When his wife's dead; therefore I'll lie with him
  When I am buried. Since Frenchmen are so braid,
  Marry that will; I live and die a maid.                          75
  Only, in this disguise I think't no sin
  To cozen him that would unjustly win.                  Exit

**4.3**  *Enter the two Captains Dumaine and some two or*
       *three soldiers*
FIRST LORD DUMAINE  You have not given him his mother's
  letter?
SECOND LORD DUMAINE  I have delivered it an hour since.
  There is something in't that stings his nature, for on
  the reading it he changed almost into another man.
FIRST LORD DUMAINE  He has much worthy blame laid
  upon him for shaking off so good a wife and so sweet
  a lady.                                                           8
SECOND LORD DUMAINE  Especially he hath incurred the
  everlasting displeasure of the King, who had even tuned
  his bounty to sing happiness to him. I will tell you a
  thing, but you shall let it dwell darkly with you.
FIRST LORD DUMAINE  When you have spoken it 'tis dead,
  and I am the grave of it.                                        14
SECOND LORD DUMAINE  He hath perverted a young gentlewoman here in Florence of a most chaste renown, and
  this night he fleshes his will in the spoil of her honour.
  He hath given her his monumental ring, and thinks
  himself made in the unchaste composition.                        19
FIRST LORD DUMAINE  Now God delay our rebellion! As we
  are ourselves, what things are we.
SECOND LORD DUMAINE  Merely our own traitors. And as in
  the common course of all treasons we still see them
  reveal themselves till they attain to their abhorred ends,
  so he that in this action contrives against his own
  nobility, in his proper stream o'erflows himself.                26
FIRST LORD DUMAINE  Is it not meant damnable in us to be
  trumpeters of our unlawful intents? We shall not then
  have his company tonight?
SECOND LORD DUMAINE  Not till after midnight, for he is
  dieted to his hour.                                              31
FIRST LORD DUMAINE  That approaches apace. I would
  gladly have him see his company anatomized, that he
  might take a measure of his own judgements, wherein
  so curiously he had set this counterfeit.                        35
SECOND LORD DUMAINE  We will not meddle with him till
  he come, for his presence must be the whip of the
  other.
FIRST LORD DUMAINE  In the mean time, what hear you of
  these wars?                                                      40
SECOND LORD DUMAINE  I hear there is an overture of peace.
FIRST LORD DUMAINE  Nay, I assure you, a peace concluded.

SECOND LORD DUMAINE What will Count Roussillon do then? Will he travel higher, or return again into France?                                                              45
FIRST LORD DUMAINE I perceive by this demand you are not altogether of his council.
SECOND LORD DUMAINE Let it be forbid, sir; so should I be a great deal of his act.
FIRST LORD DUMAINE Sir, his wife some two months since fled from his house. Her pretence is a pilgrimage to Saint Jaques le Grand, which holy undertaking with most austere sanctimony she accomplished, and there residing, the tenderness of her nature became as a prey to her grief: in fine, made a groan of her last breath, and now she sings in heaven.                              56
SECOND LORD DUMAINE How is this justified?
FIRST LORD DUMAINE The stronger part of it by her own letters, which makes her story true even to the point of her death. Her death itself, which could not be her office to say is come, was faithfully confirmed by the rector of the place.                                         62
SECOND LORD DUMAINE Hath the Count all this intelligence?
FIRST LORD DUMAINE Ay, and the particular confirmations, point from point, to the full arming of the verity.     65
SECOND LORD DUMAINE I am heartily sorry that he'll be glad of this.
FIRST LORD DUMAINE How mightily sometimes we make us comforts of our losses.
SECOND LORD DUMAINE And how mightily some other times we drown our gain in tears. The great dignity that his valour hath here acquired for him shall at home be encountered with a shame as ample.                       73
FIRST LORD DUMAINE The web of our life is of a mingled yarn, good and ill together. Our virtues would be proud if our faults whipped them not, and our crimes would despair if they were not cherished by our virtues.
        *Enter a Servant*
How now? Where's your master?                           78
SERVANT He met the Duke in the street, sir, of whom he hath taken a solemn leave. His lordship will next morning for France. The Duke hath offered him letters of commendations to the King.
SECOND LORD DUMAINE They shall be no more than needful there, if they were more than they can commend.     84
        *Enter Bertram*
⌈FIRST LORD DUMAINE⌉ They cannot be too sweet for the King's tartness. Here's his lordship now. How now, my lord, is't not after midnight?
BERTRAM I have tonight dispatched sixteen businesses, a month's length apiece. By an abstract of success: I have *congéd* with the Duke, done my adieu with his nearest, buried a wife, mourned for her, writ to my lady mother I am returning, entertained my convoy, and between these main parcels of dispatch affected many nicer needs. The last was the greatest, but that I have not ended yet.                                   95
SECOND LORD DUMAINE If the business be of any difficulty, and this morning your departure hence, it requires haste of your lordship.
BERTRAM I mean the business is not ended, as fearing to hear of it hereafter. But shall we have this dialogue between the Fool and the Soldier? Come, bring forth this counterfeit model, has deceived me like a double-meaning prophesier.
SECOND LORD DUMAINE Bring him forth. *Exit one or more*
He's sat i'th' stocks all night, poor gallant knave.  105

BERTRAM No matter, his heels have deserved it in usurping his spurs so long. How does he carry himself?
SECOND LORD DUMAINE I have told your lordship already, the stocks carry him. But to answer you as you would be understood, he weeps like a wench that had shed her milk. He hath confessed himself to Morgan, whom he supposes to be a friar, from the time of his remembrance to this very instant disaster of his setting i'th' stocks. And what think you he hath confessed?
BERTRAM Nothing of me, has a?                         115
SECOND LORD DUMAINE His confession is taken, and it shall be read to his face. If your lordship be in't, as I believe you are, you must have the patience to hear it.
        *Enter Paroles ⌈guarded and⌉ blindfolded, with the Interpreter*
BERTRAM A plague upon him! Muffled! He can say nothing of me.                                          120
⌈FIRST LORD DUMAINE⌉ (*aside to Bertram*) Hush, hush.
⌈SECOND⌉ LORD DUMAINE (*aside to Bertram*) Hoodman comes. (*Aloud*) *Porto tartarossa.*
INTERPRETER (*to Paroles*) He calls for the tortures. What will you say without 'em?                           125
PAROLES I will confess what I know without constraint. If ye pinch me like a pasty I can say no more.
INTERPRETER *Bosko chimurcho.*
⌈SECOND⌉ LORD DUMAINE *Boblibindo chicurmurco.*
INTERPRETER You are a merciful general.—Our general bids you answer to what I shall ask you out of a note.
PAROLES And truly, as I hope to live.                 132
INTERPRETER ⌈*reads*⌉ 'First demand of him how many horse the Duke is strong.'—What say you to that?
PAROLES Five or six thousand, but very weak and unserviceable. The troops are all scattered and the commanders very poor rogues, upon my reputation and credit, and as I hope to live.
INTERPRETER Shall I set down your answer so?         139
PAROLES Do. I'll take the sacrament on't, how and which way you will.
⌈FIRST LORD DUMAINE⌉ (*aside*) All's one to him.
BERTRAM (*aside*) What a past-saving slave is this!    143
FIRST LORD DUMAINE (*aside*) You're deceived, my lord. This is Monsieur Paroles, the 'gallant militarist'—that was his own phrase—that had the whole theoric of war in the knot of his scarf, and the practice in the chape of his dagger.                                           148
SECOND LORD DUMAINE (*aside*) I will never trust a man again for keeping his sword clean, nor believe he can have everything in him by wearing his apparel neatly.
INTERPRETER (*to Paroles*) Well, that's set down.
PAROLES 'Five or six thousand horse,' I said—I will say true—'or thereabouts' set down, for I'll speak truth.
FIRST LORD DUMAINE (*aside*) He's very near the truth in this.                                                156
BERTRAM (*aside*) But I con him no thanks for't in the nature he delivers it.
PAROLES 'Poor rogues', I pray you say.
INTERPRETER Well, that's set down.                     160
PAROLES I humbly thank you, sir. A truth's a truth. The rogues are marvellous poor.
INTERPRETER ⌈*reads*⌉ 'Demand of him of what strength they are a-foot.'—What say you to that?
PAROLES By my troth, sir, if I were to die this present hour, I will tell true. Let me see, Spurio a hundred and fifty; Sebastian so many; Corambus so many; Jaques so many; Guillaume, Cosmo, Lodowick, and Gratii, two

hundred fifty each; mine own company, Chitopher, Vaumond, Bentii, two hundred fifty each. So that the muster file, rotten and sound, upon my life amounts not to fifteen thousand poll, half of the which dare not shake the snow from off their cassocks lest they shake themselves to pieces.

BERTRAM (*aside*) What shall be done to him?          175

FIRST LORD DUMAINE (*aside*) Nothing, but let him have thanks. (*To Interpreter*) Demand of him my condition, and what credit I have with the Duke.

INTERPRETER (*to Paroles*) Well, that's set down. ⌜*Reads*⌝ 'You shall demand of him, whether one Captain Dumaine be i'th' camp, a Frenchman; what his reputation is with the Duke; what his valour, honesty, and expertness in wars; or whether he thinks it were not possible with well-weighing sums of gold to corrupt him to a revolt.'—What say you to this? What do you know of it?          186

PAROLES I beseech you let me answer to the particular of the inter'gatories. Demand them singly.

INTERPRETER Do you know this Captain Dumaine?

PAROLES I know him. A was a botcher's prentice in Paris, from whence he was whipped for getting the sheriff's fool with child—a dumb innocent that could not say him nay.          193

BERTRAM (*aside to First Lord Dumaine*) Nay, by your leave, hold your hands, though I know his brains are forfeit to the next tile that falls.

INTERPRETER Well, is this captain in the Duke of Florence's camp?

PAROLES Upon my knowledge he is, and lousy.          199

FIRST LORD DUMAINE (*aside*) Nay, look not so upon me: we shall hear of your lordship anon.

INTERPRETER What is his reputation with the Duke?

PAROLES The Duke knows him for no other but a poor officer of mine, and writ to me this other day to turn him out o'th' band. I think I have his letter in my pocket.          206

INTERPRETER Marry, we'll search.

PAROLES In good sadness, I do not know. Either it is there, or it is upon a file with the Duke's other letters in my tent.          210

INTERPRETER Here 'tis, here's a paper. Shall I read it to you?

PAROLES I do not know if it be it or no.

BERTRAM (*aside*) Our interpreter does it well.

FIRST LORD DUMAINE (*aside*) Excellently.          215

INTERPRETER (*reads the letter*)
'Dian, the Count's a fool, and full of gold.'

PAROLES That is not the Duke's letter, sir. That is an advertisement to a proper maid in Florence, one Diana, to take heed of the allurement of one Count Roussillon, a foolish idle boy, but for all that very ruttish. I pray you, sir, put it up again.          221

INTERPRETER Nay, I'll read it first, by your favour.

PAROLES My meaning in't, I protest, was very honest in the behalf of the maid, for I knew the young Count to be a dangerous and lascivious boy, who is a whale to virginity, and devours up all the fry it finds.          226

BERTRAM (*aside*) Damnable both-sides rogue.

INTERPRETER (*reads*)
'When he swears oaths, bid him drop gold, and take it.
After he scores he never pays the score.
Half-won is match well made; match, and well make it.
He ne'er pays after-debts, take it before.          231

And say a soldier, Dian, told thee this:
Men are to mell with, boys are not to kiss.
For count of this, the Count's a fool, I know it,
Who pays before, but not when he does owe it.          235
          Thine, as he vowed to thee in thine ear,
                                        Paroles.'

BERTRAM (*aside*) He shall be whipped through the army with this rhyme in's forehead.

SECOND LORD DUMAINE (*aside*) This is your devoted friend, sir, the manifold linguist and the armipotent soldier.

BERTRAM (*aside*) I could endure anything before but a cat, and now he's a cat to me.

INTERPRETER I perceive, sir, by the general's looks, we shall be fain to hang you.          245

PAROLES My life, sir, in any case! Not that I am afraid to die, but that, my offences being many, I would repent out the remainder of nature. Let me live, sir, in a dungeon, i'th' stocks, or anywhere, so I may live.

INTERPRETER We'll see what may be done, so you confess freely. Therefore once more to this Captain Dumaine. You have answered to his reputation with the Duke, and to his valour. What is his honesty?          253

PAROLES He will steal, sir, an egg out of a cloister. For rapes and ravishments he parallels Nessus. He professes not keeping of oaths; in breaking 'em he is stronger than Hercules. He will lie, sir, with such volubility that you would think truth were a fool. Drunkenness is his best virtue, for he will be swine-drunk, and in his sleep he does little harm, save to his bedclothes; but they about him know his conditions, and lay him in straw. I have but little more to say, sir, of his honesty. He has everything that an honest man should not have; what an honest man should have, he has nothing.          264

FIRST LORD DUMAINE (*aside*) I begin to love him for this.

BERTRAM (*aside*) For this description of thine honesty? A pox upon him! For me, he's more and more a cat.

INTERPRETER What say you to his expertness in war?

PAROLES Faith, sir, he's led the drum before the English tragedians. To belie him I will not, and more of his soldiership I know not, except in that country he had the honour to be the officer at a place there called Mile End, to instruct for the doubling of files. I would do the man what honour I can, but of this I am not certain.

FIRST LORD DUMAINE (*aside*) He hath out-villained villainy so far that the rarity redeems him.          276

BERTRAM (*aside*) A pox on him! He's a cat still.

INTERPRETER His qualities being at this poor price, I need not to ask you if gold will corrupt him to revolt.

PAROLES Sir, for a *quart d'écu* he will sell the fee-simple of his salvation, the inheritance of it, and cut th'entail from all remainders, and a perpetual succession for it perpetually.

INTERPRETER What's his brother, the other Captain Dumaine?          285

SECOND LORD DUMAINE (*aside*) Why does he ask him of me?

INTERPRETER What's he?

PAROLES E'en a crow o'th' same nest. Not altogether so great as the first in goodness, but greater a great deal in evil. He excels his brother for a coward, yet his brother is reputed one of the best that is. In a retreat he outruns any lackey; marry, in coming on he has the cramp.

INTERPRETER If your life be saved will you undertake to betray the Florentine?          295

PAROLES Ay, and the captain of his horse, Count Roussillon.

INTERPRETER I'll whisper with the general and know his
pleasure.                                                    299
PAROLES I'll no more drumming. A plague of all drums!
Only to seem to deserve well, and to beguile the
supposition of that lascivious young boy, the Count,
have I run into this danger. Yet who would have
suspected an ambush where I was taken?            304
INTERPRETER There is no remedy, sir, but you must die.
The general says you that have so traitorously
discovered the secrets of your army, and made such
pestiferous reports of men very nobly held, can serve
the world for no honest use; therefore you must die.—
Come, headsman, off with his head.                 310
PAROLES O Lord, sir!—Let me live, or let me see my death!
INTERPRETER That shall you, and take your leave of all
your friends.
          *He unmuffles Paroles*
So, look about you. Know you any here?
BERTRAM Good morrow, noble captain.                 315
SECOND LORD DUMAINE God bless you, Captain Paroles.
FIRST LORD DUMAINE God save you, noble captain.
SECOND LORD DUMAINE Captain, what greeting will you to
my Lord Lafeu? I am for France.
FIRST LORD DUMAINE Good captain, will you give me a
copy of the sonnet you writ to Diana in behalf of the
Count Roussillon? An I were not a very coward I'd
compel it of you. But fare you well.
          *Exeunt all but Paroles and Interpreter*
INTERPRETER You are undone, captain—all but your scarf;
that has a knot on't yet.                            325
PAROLES Who cannot be crushed with a plot?
INTERPRETER If you could find out a country where but
women were that had received so much shame, you
might begin an impudent nation. Fare ye well, sir. I
am for France too. We shall speak of you there.   *Exit*
PAROLES
Yet am I thankful. If my heart were great         331
'Twould burst at this. Captain I'll be no more,
But I will eat and drink and sleep as soft
As captain shall. Simply the thing I am
Shall make me live. Who knows himself a braggart,
Let him fear this, for it will come to pass         336
That every braggart shall be found an ass.
Rust, sword; cool, blushes; and Paroles live
Safest in shame; being fooled, by fool'ry thrive.
There's place and means for every man alive.      340
I'll after them.                                       *Exit*

**4.4**  *Enter Helen, the Widow, and Diana*
HELEN
That you may well perceive I have not wronged you,
One of the greatest in the Christian world
Shall be my surety; fore whose throne 'tis needful,
Ere I can perfect mine intents, to kneel.
Time was, I did him a desirèd office                  5
Dear almost as his life; which gratitude
Through flinty Tartar's bosom would peep forth
And answer 'Thanks'. I duly am informed
His grace is at Marseilles, to which place
We have convenient convoy. You must know         10
I am supposèd dead. The army breaking,
My husband hies him home, where, heaven aiding,
And by the leave of my good lord the King,
We'll be before our welcome.
WIDOW                          Gentle madam,
You never had a servant to whose trust            15

Your business was more welcome.
HELEN                          Nor you, mistress,
Ever a friend whose thoughts more truly labour
To recompense your love. Doubt not but heaven
Hath brought me up to be your daughter's dower,
As it hath fated her to be my motive                20
And helper to a husband. But O, strange men,
That can such sweet use make of what they hate,
When saucy trusting of the cozened thoughts
Defiles the pitchy night; so lust doth play
With what it loathes, for that which is away.        25
But more of this hereafter. You, Diana,
Under my poor instructions yet must suffer
Something in my behalf.
DIANA                          Let death and honesty
Go with your impositions, I am yours,
Upon your will to suffer.
HELEN                          Yet, I pray you.—          30
But with that word the time will bring on summer,
When briers shall have leaves as well as thorns
And be as sweet as sharp. We must away,
Our wagon is prepared, and time revives us.
All's well that ends well; still the fine's the crown.  35
Whate'er the course, the end is the renown.    *Exeunt*

**4.5**  *Enter Lavatch, the old Countess, and Lafeu*
LAFEU No, no, no, your son was misled with a snipped-
taffeta fellow there, whose villainous saffron would
have made all the unbaked and doughy youth of a
nation in his colour. Else, your daughter-in-law had
been alive at this hour, and your son here at home,
more advanced by the King than by that red-tailed
humble-bee I speak of.                                7
COUNTESS I would a had not known him. It was the death
of the most virtuous gentlewoman that ever nature
had praise for creating. If she had partaken of my flesh
and cost me the dearest groans of a mother I could not
have owed her a more rooted love.                    12
LAFEU 'Twas a good lady, 'twas a good lady. We may
pick a thousand salads ere we light on such another
herb.                                                   15
LAVATCH Indeed, sir, she was the sweet marjoram of the
salad, or rather the herb of grace.
LAFEU They are not grass, you knave, they are nose-
herbs.
LAVATCH I am no great Nebuchadnezzar, sir, I have not
much skill in grace.                                   21
LAFEU Whether dost thou profess thyself, a knave or a
fool?
LAVATCH A fool, sir, at a woman's service, and a knave
at a man's.                                             25
LAFEU Your distinction?
LAVATCH I would cozen the man of his wife and do his
service.
LAFEU So you were a knave at his service indeed.
LAVATCH And I would give his wife my bauble, sir, to do
her service.                                            31
LAFEU I will subscribe for thee, thou art both knave and
fool.
LAVATCH At your service.
LAFEU No, no, no.                                       35
LAVATCH Why, sir, if I cannot serve you I can serve as
great a prince as you are.
LAFEU Who's that? A Frenchman?
LAVATCH Faith, sir, a has an English name, but his
phys'namy is more hotter in France than there.     40

LAFEU What prince is that?

LAVATCH The Black Prince, sir, alias the prince of darkness, alias the devil.

LAFEU Hold thee, there's my purse. I give thee not this to suggest thee from thy master thou talk'st of; serve him still.                                                          46

LAVATCH I am a woodland fellow, sir, that always loved a great fire, and the master I speak of ever keeps a good fire. But since he is the prince of the world, let the nobility remain in's court; I am for the house with the narrow gate, which I take to be too little for pomp to enter. Some that humble themselves may, but the many will be too chill and tender, and they'll be for the flow'ry way that leads to the broad gate and the great fire.                                                          55

LAFEU Go thy ways. I begin to be aweary of thee, and I tell thee so before, because I would not fall out with thee. Go thy ways. Let my horses be well looked to, without any tricks.                                                          59

LAVATCH If I put any tricks upon 'em, sir, they shall be jades' tricks, which are their own right by the law of nature.                                                    *Exit*

LAFEU A shrewd knave and an unhappy.                        63

COUNTESS So a is. My lord that's gone made himself much sport out of him; by his authority he remains here, which he thinks is a patent for his sauciness, and indeed he has no pace, but runs where he will.

LAFEU I like him well, 'tis not amiss. And I was about to tell you, since I heard of the good lady's death and that my lord your son was upon his return home, I moved the King my master to speak in the behalf of my daughter; which, in the minority of them both, his majesty out of a self-gracious remembrance did first propose. His highness hath promised me to do it; and to stop up the displeasure he hath conceived against your son, there is no fitter matter. How does your ladyship like it?                                                          77

COUNTESS With very much content, my lord, and I wish it happily effected.

LAFEU His highness comes post from Marseilles, of as able body as when he numbered thirty. A will be here tomorrow, or I am deceived by him that in such intelligence hath seldom failed.                                  83

COUNTESS It rejoices me that I hope I shall see him ere I die. I have letters that my son will be here tonight. I shall beseech your lordship to remain with me till they meet together.

LAFEU Madam, I was thinking with what manners I might safely be admitted.                                                          89

COUNTESS You need but plead your honourable privilege.

LAFEU Lady, of that I have made a bold charter, but, I thank my God, it holds yet.

*Enter Lavatch*

LAVATCH O madam, yonder's my lord your son with a patch of velvet on's face. Whether there be a scar under't or no, the velvet knows; but 'tis a goodly patch of velvet. His left cheek is a cheek of two pile and a half, but his right cheek is worn bare.                        97

LAFEU A scar nobly got, or a noble scar, is a good liv'ry of honour. So belike is that.

LAVATCH But it is your carbonadoed face.                   100

LAFEU (*to the Countess*) Let us go see your son, I pray you. I long to talk with the young noble soldier.

LAVATCH Faith, there's a dozen of 'em, with delicate fine hats, and most courteous feathers, which bow the head and nod at every man.                                          *Exeunt*

**5.1**    *Enter Helen, the Widow, and Diana, with two attendants*

HELEN
But this exceeding posting day and night
Must wear your spirits low. We cannot help it.
But since you have made the days and nights as one
To wear your gentle limbs in my affairs,
Be bold you do so grow in my requital                         5
As nothing can unroot you.
    *Enter a Gentleman Austringer*
                                        In happy time!
This man may help me to his majesty's ear,
If he would spend his power.—God save you, sir.

GENTLEMAN And you.

HELEN
Sir, I have seen you in the court of France.               10

GENTLEMAN I have been sometimes there.

HELEN
I do presume, sir, that you are not fall'n
From the report that goes upon your goodness,
And therefore, goaded with most sharp occasions
Which lay nice manners by, I put you to                       15
The use of your own virtues, for the which
I shall continue thankful.

GENTLEMAN                              What's your will?

HELEN That it will please you
To give this poor petition to the King,
And aid me with that store of power you have                  20
To come into his presence.

GENTLEMAN The King's not here.

HELEN Not here, sir?

GENTLEMAN                         Not indeed.
He hence removed last night, and with more haste
Than is his use.                                             25

WIDOW Lord, how we lose our pains.

HELEN All's well that ends well yet,
Though time seem so adverse, and means unfit.—
I do beseech you, whither is he gone?

GENTLEMAN
Marry, as I take it, to Roussillon,                         30
Whither I am going.

HELEN                       I do beseech you, sir,
Since you are like to see the King before me,
Commend the paper to his gracious hand,
Which I presume shall render you no blame,
But rather make you thank your pains for it.               35
I will come after you with what good speed
Our means will make us means.

GENTLEMAN (*taking the paper*)       This I'll do for you.

HELEN
And you shall find yourself to be well thanked,
Whate'er falls more. We must to horse again.—
Go, go, provide.                              *Exeunt severally*

**5.2**    *Enter Lavatch and Paroles, with a letter*

PAROLES Good Master Lavatch, give my Lord Lafeu this letter. I have ere now, sir, been better known to you, when I have held familiarity with fresher clothes. But I am now, sir, muddied in Fortune's mood, and smell somewhat strong of her strong displeasure.                        5

LAVATCH Truly, Fortune's displeasure is but sluttish if it smell so strongly as thou speakest of. I will henceforth eat no fish of Fortune's butt'ring. Prithee allow the wind.

PAROLES Nay, you need not to stop your nose, sir, I spake but by a metaphor.                                          11

LAVATCH Indeed, sir, if your metaphor stink I will stop my nose, or against any man's metaphor. Prithee get thee further.

PAROLES Pray you, sir, deliver me this paper.   15

LAVATCH Foh, prithee stand away. A paper from Fortune's close-stool to give to a nobleman! Look, here he comes himself.

    *Enter Lafeu*

    Here is a pur of Fortune's, sir, or of Fortune's cat—but not a musk-cat—that has fallen into the unclean fish-pond of her displeasure and, as he says, is muddied withal. Pray you, sir, use the carp as you may, for he looks like a poor, decayed, ingenious, foolish, rascally knave. I do pity his distress in my similes of comfort, and leave him to your lordship.    *Exit*

PAROLES My lord, I am a man whom Fortune hath cruelly scratched.   27

LAFEU And what would you have me to do? 'Tis too late to pare her nails now. Wherein have you played the knave with Fortune that she should scratch you, who of herself is a good lady and would not have knaves thrive long under her? There's a *quart d'écu* for you. Let the justices make you and Fortune friends; I am for other business.   34

PAROLES I beseech your honour to hear me one single word—

LAFEU You beg a single penny more. Come, you shall ha't. Save your word.

PAROLES My name, my good lord, is Paroles.   39

LAFEU You beg more than one word then. Cox my passion! Give me your hand. How does your drum?

PAROLES O my good lord, you were the first that found me.

LAFEU Was I, in sooth? And I was the first that lost thee.

PAROLES It lies in you, my lord, to bring me in some grace, for you did bring me out.   46

LAFEU Out upon thee, knave! Dost thou put upon me at once both the office of God and the devil? One brings thee in grace, and the other brings thee out.

    *Trumpets sound*

    The King's coming; I know by his trumpets. Sirrah, enquire further after me. I had talk of you last night. Though you are a fool and a knave, you shall eat. Go to, follow.   53

PAROLES I praise God for you.    ⌐*Exeunt*⌐

**5.3**   *Flourish of trumpets. Enter the King, the old Countess, Lafeu, and attendants*

KING
We lost a jewel of her, and our esteem
Was made much poorer by it. But your son,
As mad in folly, lacked the sense to know
Her estimation home.

COUNTESS         'Tis past, my liege,
And I beseech your majesty to make it   5
Natural rebellion done i'th' blade of youth,
When oil and fire, too strong for reason's force,
O'erbears it and burns on.

KING         My honoured lady,
I have forgiven and forgotten all,
Though my revenges were high bent upon him   10
And watched the time to shoot.

LAFEU         This I must say—
But first I beg my pardon—the young lord
Did to his majesty, his mother, and his lady
Offence of mighty note, but to himself
The greatest wrong of all. He lost a wife   15

Whose beauty did astonish the survey
Of richest eyes, whose words all ears took captive,
Whose dear perfection hearts that scorned to serve
Humbly called mistress.

KING         Praising what is lost
Makes the remembrance dear. Well, call him hither.
We are reconciled, and the first view shall kill   21
All repetition. Let him not ask our pardon.
The nature of his great offence is dead,
And deeper than oblivion we do bury
Th'incensing relics of it. Let him approach   25
A stranger, no offender; and inform him
So 'tis our will he should.

ATTENDANT         I shall, my liege.    *Exit*

KING (*to Lafeu*)
What says he to your daughter? Have you spoke?

LAFEU
All that he is hath reference to your highness.

KING
Then shall we have a match. I have letters sent me   30
That sets him high in fame.

    *Enter Bertram* ⌐*with a patch of velvet on his left cheek, and kneels*⌐

LAFEU He looks well on't.

KING (*to Bertram*) I am not a day of season,
For thou mayst see a sunshine and a hail
In me at once. But to the brightest beams   35
Distracted clouds give way; so stand thou forth.
The time is fair again.

BERTRAM         My high-repented blames,
Dear sovereign, pardon to me.

KING         All is whole.
Not one word more of the consumèd time.
Let's take the instant by the forward top,   40
For we are old, and on our quick'st decrees
Th'inaudible and noiseless foot of time
Steals ere we can effect them. You remember
The daughter of this lord?

BERTRAM
Admiringly, my liege. At first   45
I stuck my choice upon her, ere my heart
Durst make too bold a herald of my tongue;
Where, the impression of mine eye enfixing,
Contempt his scornful perspective did lend me,
Which warped the line of every other favour,   50
Stained a fair colour or expressed it stolen,
Extended or contracted all proportions
To a most hideous object. Thence it came
That she whom all men praised and whom myself,
Since I have lost, have loved, was in mine eye   55
The dust that did offend it.

KING         Well excused.
That thou didst love her strikes some scores away
From the great count. But love that comes too late,
Like a remorseful pardon slowly carried,
To the grace-sender turns a sour offence,   60
Crying, 'That's good that's gone.' Our rash faults
Make trivial price of serious things we have,
Not knowing them until we know their grave.
Oft our displeasures, to ourselves unjust,
Destroy our friends and after weep their dust.   65
Our own love waking cries to see what's done,
While shameful hate sleeps out the afternoon.
Be this sweet Helen's knell, and now forget her.
Send forth your amorous token for fair Maudlin.
The main consents are had, and here we'll stay   70
To see our widower's second marriage day.

⌈COUNTESS⌉
Which better than the first, O dear heaven, bless!
Or ere they meet, in me, O nature, cease.
LAFEU (to Bertram)
Come on, my son, in whom my house's name
Must be digested, give a favour from you                    75
To sparkle in the spirits of my daughter,
That she may quickly come.
          Bertram gives Lafeu a ring
                              By my old beard
And ev'ry hair that's on't, Helen that's dead
Was a sweet creature. Such a ring as this,
The last that ere I took her leave at court,                80
I saw upon her finger.
BERTRAM                         Hers it was not.
KING
Now pray you let me see it; for mine eye,
While I was speaking, oft was fastened to't.
          Lafeu gives him the ring
This ring was mine, and when I gave it Helen
I bade her, if her fortunes ever stood                      85
Necessitied to help, that by this token
I would relieve her. Had you that craft to reave her
Of what should stead her most?
BERTRAM                         My gracious sovereign,
Howe'er it pleases you to take it so,
The ring was never hers.
COUNTESS                         Son, on my life                90
I have seen her wear it, and she reckoned it
At her life's rate.
LAFEU                   I am sure I saw her wear it.
BERTRAM
You are deceived, my lord, she never saw it.
In Florence was it from a casement thrown me,
Wrapped in a paper which contained the name                 95
Of her that threw it. Noble she was, and thought
I stood ingaged. But when I had subscribed
To mine own fortune, and informed her fully
I could not answer in that course of honour
As she had made the overture, she ceased                    100
In heavy satisfaction, and would never
Receive the ring again.
KING                         Plutus himself,
That knows the tinct and multiplying med'cine,
Hath not in nature's mystery more science
Than I have in this ring. 'Twas mine, 'twas Helen's,
Whoever gave it you. Then if you know                       106
That you are well acquainted with yourself,
Confess 'twas hers, and by what rough enforcement
You got it from her. She called the saints to surety
That she would never put it from her finger                 110
Unless she gave it to yourself in bed,
Where you have never come, or sent it us
Upon her great disaster.
BERTRAM                         She never saw it.
KING
Thou speak'st it falsely, as I love mine honour,
And mak'st conjectural fears to come into me                115
Which I would fain shut out. If it should prove
That thou art so inhuman—'twill not prove so.
And yet I know not. Thou didst hate her deadly,
And she is dead, which nothing but to close
Her eyes myself could win me to believe,                    120
More than to see this ring.—Take him away.
My fore-past proofs, howe'er the matter fall,
Shall tax my fears of little vanity,

Having vainly feared too little. Away with him.
We'll sift this matter further.
BERTRAM               If you shall prove            125
This ring was ever hers, you shall as easy
Prove that I husbanded her bed in Florence,
Where yet she never was.            Exit guarded
          Enter the Gentleman Austringer with a paper
KING I am wrapped in dismal thinkings.
GENTLEMAN Gracious sovereign,                      130
Whether I have been to blame or no, I know not.
Here's a petition from a Florentine
Who hath for four or five removes come short
To tender it herself. I undertook it,
Vanquished thereto by the fair grace and speech   135
Of the poor suppliant, who by this I know
Is here attending. Her business looks in her
With an importing visage, and she told me
In a sweet verbal brief it did concern
Your highness with herself.                        140
⌈KING⌉ (reads a letter) 'Upon his many protestations to
marry me when his wife was dead, I blush to say it,
he won me. Now is the Count Roussillon a widower,
his vows are forfeited to me, and my honour's paid to
him. He stole from Florence, taking no leave, and I
follow him to his country for justice. Grant it me, O
King! In you it best lies; otherwise a seducer flourishes
and a poor maid is undone.                         148
                              Diana Capilet.'
LAFEU I will buy me a son-in-law in a fair, and toll for
this. I'll none of him.
KING
The heavens have thought well on thee, Lafeu,
To bring forth this discov'ry.—Seek these suitors.
Go speedily and bring again the Count.
                              Exit one or more
I am afeard the life of Helen, lady,               155
Was foully snatched.
          ⌈Enter Bertram guarded⌉
COUNTESS               Now justice on the doers!
KING (to Bertram)
I wonder, sir, since wives are monsters to you,
And that you fly them as you swear them lordship,
Yet you desire to marry.
          Enter the Widow and Diana
                         What woman's that?
DIANA
I am, my lord, a wretched Florentine,             160
Derivèd from the ancient Capilet.
My suit, as I do understand, you know,
And therefore know how far I may be pitied.
WIDOW (to the King)
I am her mother, sir, whose age and honour
Both suffer under this complaint we bring,         165
And both shall cease without your remedy.
KING
Come hither, Count. Do you know these women?
BERTRAM
My lord, I neither can nor will deny
But that I know them. Do they charge me further?
DIANA
Why do you look so strange upon your wife?          170
BERTRAM (to the King)
She's none of mine, my lord.
DIANA                    If you shall marry
You give away this hand, and that is mine;
You give away heaven's vows, and those are mine;

You give away myself, which is known mine,
For I by vow am so embodied yours                     175
That she which marries you must marry me,
Either both or none.
LAFEU (*to Bertram*) Your reputation comes too short for
my daughter, you are no husband for her.
BERTRAM (*to the King*)
My lord, this is a fond and desp'rate creature       180
Whom sometime I have laughed with. Let your
highness
Lay a more noble thought upon mine honour
Than for to think that I would sink it here.
KING
Sir, for my thoughts, you have them ill to friend
Till your deeds gain them. Fairer prove your honour
Than in my thought it lies.
DIANA                         Good my lord,          186
Ask him upon his oath if he does think
He had not my virginity.
KING What sayst thou to her?
BERTRAM She's impudent, my lord,                     190
And was a common gamester to the camp.
DIANA (*to the King*)
He does me wrong, my lord. If I were so
He might have bought me at a common price.
Do not believe him. O behold this ring,
Whose high respect and rich validity                 195
Did lack a parallel; yet for all that
He gave it to a commoner o'th' camp,
If I be one.
COUNTESS        He blushes and 'tis hit.
Of six preceding ancestors, that gem;
Conferred by testament to th' sequent issue          200
Hath it been owed and worn. This is his wife.
That ring's a thousand proofs.
KING (*to Diana*)                      Methought you said
You saw one here in court could witness it.
DIANA
I did, my lord, but loath am to produce
So bad an instrument. His name's Paroles.            205
LAFEU
I saw the man today, if man he be.
KING
Find him and bring him hither.          *Exit one*
BERTRAM                      What of him?
He's quoted for a most perfidious slave
With all the spots o'th' world taxed and debauched,
Whose nature sickens but to speak a truth.           210
Am I or that or this for what he'll utter,
That will speak anything?
KING                       She hath that ring of yours.
BERTRAM
I think she has. Certain it is I liked her
And boarded her i'th' wanton way of youth.
She knew her distance and did angle for me,          215
Madding my eagerness with her restraint,
As all impediments in fancy's course
Are motives of more fancy; and in fine
Her inf'nite cunning with her modern grace
Subdued me to her rate. She got the ring,            220
And I had that which my inferior might
At market price have bought.
DIANA                        I must be patient.
You that have turned off a first so noble wife
May justly diet me. I pray you yet—
Since you lack virtue I will lose a husband—         225

Send for your ring, I will return it home,
And give me mine again.
BERTRAM I have it not.
KING (*to Diana*) What ring was yours, I pray you?
DIANA
Sir, much like the same upon your finger.            230
KING
Know you this ring? This ring was his of late.
DIANA
And this was it I gave him being abed.
KING
The story then goes false you threw it him
Out of a casement?
DIANA                I have spoke the truth.
*Enter Paroles*
BERTRAM (*to the King*)
My lord, I do confess the ring was hers.             235
KING
You boggle shrewdly; every feather starts you.—
Is this the man you speak of?
DIANA                         Ay, my lord.
KING (*to Paroles*)
Tell me, sirrah—but tell me true, I charge you,
Not fearing the displeasure of your master,
Which on your just proceeding I'll keep off—          240
By him and by this woman here what know you?
PAROLES So please your majesty, my master hath been
an honourable gentleman. Tricks he hath had in him
which gentlemen have.
KING
Come, come, to th' purpose. Did he love this woman?
PAROLES Faith, sir, he did love her, but how?        246
KING How, I pray you?
PAROLES He did love her, sir, as a gentleman loves a
woman.
KING How is that?                                    250
PAROLES He loved her, sir, and loved her not.
KING As thou art a knave and no knave. What an
equivocal companion is this!
PAROLES I am a poor man, and at your majesty's
command.                                             255
LAFEU (*to the King*) He's a good drum, my lord, but a
naughty orator.
DIANA (*to Paroles*) Do you know he promised me marriage?
PAROLES Faith, I know more than I'll speak.
KING But wilt thou not speak all thou know'st?       260
PAROLES Yes, so please your majesty. I did go between
them, as I said; but more than that, he loved her, for
indeed he was mad for her and talked of Satan and of
limbo and of Furies and I know not what. Yet I was
in that credit with them at that time that I knew of
their going to bed and of other motions, as promising
her marriage and things which would derive me ill will
to speak of. Therefore I will not speak what I know.
KING Thou hast spoken all already, unless thou canst say
they are married. But thou art too fine in thy evidence,
therefore stand aside.—                              271
This ring you say was yours.
DIANA                        Ay, my good lord.
KING
Where did you buy it? Or who gave it you?
DIANA
It was not given me, nor I did not buy it.
KING
Who lent it you?
DIANA             It was not lent me neither.         275

KING
  Where did you find it then?
DIANA                              I found it not.
KING
  If it were yours by none of all these ways,
  How could you give it him?
DIANA                        I never gave it him.
LAFEU (to the King) This woman's an easy glove, my lord,
  she goes off and on at pleasure.                    280
KING (to Diana)
  This ring was mine. I gave it his first wife.
DIANA
  It might be yours or hers for aught I know.
KING (to attendants)
  Take her away, I do not like her now.
  To prison with her. And away with him.—
  Unless thou tell'st me where thou hadst this ring   285
  Thou diest within this hour.
DIANA                        I'll never tell you.
KING (to attendants)
  Take her away.
DIANA            I'll put in bail, my liege.
KING
  I think thee now some common customer.
DIANA
  By Jove, if ever I knew man 'twas you.
KING
  Wherefore hast thou accused him all this while?   290
DIANA
  Because he's guilty, and he is not guilty.
  He knows I am no maid, and he'll swear to't;
  I'll swear I am a maid, and he knows not.
  Great King, I am no strumpet; by my life,
  I am either maid or else this old man's wife.      295
KING (to attendants)
  She does abuse our ears. To prison with her.
DIANA
  Good mother, fetch my bail.           Exit Widow
                           Stay, royal sir.
  The jeweller that owes the ring is sent for,
  And he shall surety me. But for this lord,
  Who hath abused me as he knows himself,            300
  Though yet he never harmed me, here I quit him.
  He knows himself my bed he hath defiled,
  And at that time he got his wife with child.
  Dead though she be she feels her young one kick.
  So there's my riddle; one that's dead is quick.    305
  And now behold the meaning.

*Enter Helen and the Widow*

KING                        Is there no exorcist
  Beguiles the truer office of mine eyes?
  Is't real that I see?
HELEN              No, my good lord,
  'Tis but the shadow of a wife you see,
  The name and not the thing.
BERTRAM                    Both, both. O, pardon!
HELEN
  O, my good lord, when I was like this maid         311
  I found you wondrous kind. There is your ring.
  And, look you, here's your letter. This it says:
  'When from my finger you can get this ring,
  And are by me with child,' et cetera. This is done.  315
  Will you be mine now you are doubly won?
BERTRAM (to the King)
  If she, my liege, can make me know this clearly
  I'll love her dearly, ever ever dearly.
HELEN
  If it appear not plain and prove untrue,
  Deadly divorce step between me and you.—          320
  O my dear mother, do I see you living?
LAFEU
  Mine eyes smell onions, I shall weep anon.
  (To Paroles) Good Tom Drum, lend me a handkerchief.
  So, I thank thee. Wait on me home, I'll make sport
  with thee. Let thy curtsies alone, they are scurvy ones.
KING (to Helen)
  Let us from point to point this story know          326
  To make the even truth in pleasure flow.
  (To Diana) If thou be'st yet a fresh uncroppèd flower,
  Choose thou thy husband and I'll pay thy dower.
  For I can guess that by thy honest aid              330
  Thou kept'st a wife herself, thyself a maid.
  Of that and all the progress more and less
  Resolvèdly more leisure shall express.
  All yet seems well; and if it end so meet,
  The bitter past, more welcome is the sweet.         335
  *Flourish of trumpets*

## Epilogue

  The King's a beggar now the play is done.
  All is well ended if this suit be won:
  That you express content, which we will pay
  With strife to please you, day exceeding day.
  Ours be your patience then, and yours our parts:    5
  Your gentle hands lend us, and take our hearts.
                                            *Exeunt*

# TIMON OF ATHENS

## BY WILLIAM SHAKESPEARE AND THOMAS MIDDLETON

WE know no more of *Timon of Athens* than we can deduce from the text printed in the 1623 Folio. Some episodes, such as the emblematic opening dialogue featuring a Poet and a Painter, are elegantly finished, but the play has more unpolished dialogue and loose ends of plot than usual: for example, the episode (3.6) in which Alcibiades pleads for a soldier's life is only tenuously related to the main structure; and the final stretch of action seems imperfectly worked out. Various theories of collaboration and revision have been advanced to explain the play's peculiarities. During the 1970s and 1980s strong linguistic and other evidence has been adduced in support of the belief that it is a product of collaboration between Shakespeare and Thomas Middleton, a dramatist born in 1580 and educated at Queen's College, Oxford, who was writing for the stage by 1602 and was to develop into a great playwright. The major passages for which Middleton seems to have taken prime responsibility are Act 1, Scene 2; all of Act 3 except for parts of Scene 7; and the closing episode (4.3.460–537) of Act 4. The theory of collaboration explains some features of the text—Middleton's verse, for example, was less regular than Shakespeare's. There is no record of early performance; the play is conjecturally assigned to 1604.

The story of Timon was well known and had been told in an anonymous play which seems to have been acted at one of the Inns of Court in 1602 or 1603; Middleton has even been suggested as its author. The classical sources of Timon's story are a brief, anecdotal passage in Plutarch's Life of Mark Antony, and a Greek dialogue by Lucian, who wrote during the second century AD; the former was certainly known to the authors of *Timon of Athens*; the latter influences them directly or indirectly. Plutarch records two epitaphs, one written by Timon himself, which recur, conflated as one epitaph, almost word for word in the play. In Lucian, as in the play, Timon is a misanthrope because his friends flattered and sponged on him in prosperity but abandoned him in poverty. The first part of the play dramatizes this process; in the second part, as in Lucian, Timon finds gold and suddenly becomes attractive again to his old friends.

*Timon of Athens* is an exceptionally schematic play falling into two sharply contrasting parts, the second a kind of mirror image of the first. Many of the characters are presented two-dimensionally, as if the dramatists were more concerned with the play's pattern of ideas than with psychological realism. The overall tone is harsh and bitter; there are passages of magnificent invective along with some brilliant satire, but there is also tenderness in the portrayal of Timon's servants, especially his 'one honest man', Flavius. In the play's comparatively rare performances some adaptation has usually been found necessary; but the exceptionally long role of Timon offers great opportunities to an actor who can convey his vulnerability as well as his virulence, especially in the strange music of the closing scenes which suggests in him a vision beyond the ordinary.

# THE PERSONS OF THE PLAY

TIMON of Athens

A POET
A PAINTER
A JEWELLER
A MERCHANT
A mercer
LUCILIUS, one of Timon's servants
An OLD ATHENIAN

LORDS and SENATORS of Athens
VENTIDIUS, one of Timon's false friends
ALCIBIADES, an Athenian captain
APEMANTUS, a churlish philosopher

One dressed as CUPID in the masque
LADIES dressed as Amazons in the masque

FLAVIUS, Timon's steward
FLAMINIUS ⎱
SERVILIUS ⎰ Timon's servants
Other SERVANTS of Timon

A FOOL
A PAGE

CAPHIS ⎫
ISIDORE'S SERVANT ⎬ servants to Timon's creditors
Two of VARRO'S SERVANTS ⎭

LUCULLUS ⎫
LUCIUS ⎬ flattering lords
SEMPRONIUS ⎭
LUCULLUS' SERVANT
LUCIUS' SERVANT
Three STRANGERS, one called Hostilius

TITUS' SERVANT ⎫
HORTENSIUS' SERVANT ⎬ other servants to Timon's creditors
PHILOTUS' SERVANT ⎭

PHRYNIA ⎫
TIMANDRA ⎬ whores with Alcibiades
The banditti, THIEVES
SOLDIER of Alcibiades' army

Messengers, attendants, soldiers

# The Life of Timon of Athens

**1.1** *Enter Poet ⌜at one door⌝, Painter carrying a picture ⌜at another door⌝, ⌜followed by⌝ Jeweller, Merchant, and Mercer, at several doors*

POET
Good day, sir.

PAINTER        I am glad you're well.

POET
I have not seen you long. How goes the world?

PAINTER
It wears, sir, as it grows.

POET        Ay, that's well known.
But what particular rarity, what strange,
Which manifold record not matches?—See,     5
Magic of bounty, all these spirits thy power
Hath conjured to attend.

       ⌜*Merchant and Jeweller meet. Mercer passes over the stage, and exits*⌝

       I know the merchant.

PAINTER
I know them both. Th'other's a jeweller.

MERCHANT (*to Jeweller*)
O, 'tis a worthy lord!

JEWELLER        Nay, that's most fixed.

MERCHANT
A most incomparable man, breathed, as it were,   10
To an untirable and continuate goodness.
He passes.

JEWELLER (*showing a jewel*) I have a jewel here.

MERCHANT
O, pray, let's see't. For the Lord Timon, sir?

JEWELLER
If he will touch the estimate. But for that—

POET (*to himself*)
'When we for recompense have praised the vile,   15
It stains the glory in that happy verse
Which aptly sings the good.'

MERCHANT (*to Jeweller*)        'Tis a good form.

JEWELLER
And rich. Here is a water, look ye.

PAINTER (*to Poet*)
You are rapt, sir, in some work, some dedication
To the great lord.

POET        A thing slipped idly from me.   20
Our poesy is as a gum which oozes
From whence 'tis nourished. The fire i'th' flint
Shows not till it be struck; our gentle flame
Provokes itself, and like the current flies
Each bound it chafes. What have you there?   25

PAINTER
A picture, sir. When comes your book forth?

POET
Upon the heels of my presentment, sir.
Let's see your piece.

PAINTER (*showing the picture*) 'Tis a good piece.

POET
So 'tis. This comes off well and excellent.

PAINTER
Indifferent.

POET        Admirable. How this grace   30
Speaks his own standing! What a mental power
This eye shoots forth! How big imagination

Moves in this lip! To th' dumbness of the gesture
One might interpret.

PAINTER
It is a pretty mocking of the life.   35
Here is a touch; is't good?

POET        I will say of it,
It tutors nature. Artificial strife
Lives in these touches livelier than life.

*Enter certain Senators*

PAINTER How this lord is followed!

POET
The senators of Athens. Happy man!   40

PAINTER Look, more.

       ⌜*The Senators pass over the stage, and exeunt*⌝

POET
You see this confluence, this great flood of visitors.
I have in this rough work shaped out a man
Whom this beneath world doth embrace and hug
With amplest entertainment. My free drift   45
Halts not particularly, but moves itself
In a wide sea of tax. No levelled malice
Infects one comma in the course I hold,
But flies an eagle flight, bold and forth on,
Leaving no tract behind.   50

PAINTER How shall I understand you?

POET I will unbolt to you.
You see how all conditions, how all minds,
As well of glib and slipp'ry creatures as
Of grave and austere quality, tender down   55
Their service to Lord Timon. His large fortune,
Upon his good and gracious nature hanging,
Subdues and properties to his love and tendance
All sorts of hearts; yea, from the glass-faced flatterer
To Apemantus, that few things loves better   60
Than to abhor himself; even he drops down
The knee before him, and returns in peace,
Most rich in Timon's nod.

PAINTER        I saw them speak together.

POET
Sir, I have upon a high and pleasant hill
Feigned Fortune to be throned. The base o'th' mount
Is ranked with all deserts, all kind of natures   66
That labour on the bosom of this sphere
To propagate their states. Amongst them all
Whose eyes are on this sovereign lady fixed
One do I personate of Lord Timon's frame,   70
Whom Fortune with her ivory hand wafts to her,
Whose present grace to present slaves and servants
Translates his rivals.

PAINTER        'Tis conceived to scope.
This throne, this Fortune, and this hill, methinks,
With one man beckoned from the rest below,   75
Bowing his head against the steepy mount
To climb his happiness, would be well expressed
In our condition.

POET        Nay, sir, but hear me on.
All those which were his fellows but of late,
Some better than his value, on the moment   80
Follow his strides, his lobbies fill with tendance,
Rain sacrificial whisperings in his ear,

Make sacred even his stirrup, and through him
Drink the free air.
PAINTER                    Ay, marry, what of these?
POET
When Fortune in her shift and change of mood          85
Spurns down her late belovèd, all his dependants,
Which laboured after him to the mountain's top
Even on their knees and hands, let him fall down,
Not one accompanying his declining foot.
PAINTER 'Tis common.                                  90
A thousand moral paintings I can show
That shall demonstrate these quick blows of Fortune's
More pregnantly than words. Yet you do well
To show Lord Timon that mean eyes have seen
The foot above the head.                              95
    *Trumpets sound. Enter Timon ⌐wearing a rich*
    *jewel⌐, with a Messenger from Ventidius; Lucilius*
    *⌐and other Servants⌐ attending. Timon addresses*
    *himself courteously to every suitor, then speaks to*
    *the Messenger*
TIMON Imprisoned is he, say you?
MESSENGER
Ay, my good lord. Five talents is his debt,
His means most short, his creditors most strait.
Your honourable letter he desires
To those have shut him up, which failing,            100
Periods his comfort.
TIMON                    Noble Ventidius! Well,
I am not of that feather to shake off
My friend when he must need me. I do know him
A gentleman that well deserves a help,
Which he shall have. I'll pay the debt and free him.
MESSENGER Your lordship ever binds him.              106
TIMON
Commend me to him. I will send his ransom;
And, being enfranchised, bid him come to me.
'Tis not enough to help the feeble up,
But to support him after. Fare you well.             110
MESSENGER All happiness to your honour.        *Exit*
    *Enter an Old Athenian*
OLD ATHENIAN
Lord Timon, hear me speak.
TIMON                    Freely, good father.
OLD ATHENIAN
Thou hast a servant named Lucilius.
TIMON I have so. What of him?
OLD ATHENIAN
Most noble Timon, call the man before thee.          115
TIMON
Attends he here or no? Lucilius!
LUCILIUS (*coming forward*) Here at your lordship's service.
OLD ATHENIAN
This fellow here, Lord Timon, this thy creature,
By night frequents my house. I am a man
That from my first have been inclined to thrift,     120
And my estate deserves an heir more raised
Than one which holds a trencher.
TIMON                    Well, what further?
OLD ATHENIAN
One only daughter have I, no kin else
On whom I may confer what I have got.
The maid is fair, o'th' youngest for a bride,        125
And I have bred her at my dearest cost
In qualities of the best. This man of thine
Attempts her love. I prithee, noble lord,

Join with me to forbid him her resort.
Myself have spoke in vain.                           130
TIMON The man is honest.
OLD ATHENIAN Therefore he will be, Timon.
His honesty rewards him in itself;
It must not bear my daughter.
TIMON Does she love him?                             135
OLD ATHENIAN She is young and apt.
Our own precedent passions do instruct us
What levity's in youth.
TIMON (*to Lucilius*)        Love you the maid?
LUCILIUS
Ay, my good lord, and she accepts of it.
OLD ATHENIAN
If in her marriage my consent be missing,            140
I call the gods to witness, I will choose
Mine heir from forth the beggars of the world,
And dispossess her all.
TIMON                    How shall she be endowed
If she be mated with an equal husband?
OLD ATHENIAN
Three talents on the present; in future, all.        145
TIMON
This gentleman of mine hath served me long.
To build his fortune I will strain a little,
For 'tis a bond in men. Give him thy daughter.
What you bestow in him I'll counterpoise,
And make him weigh with her.
OLD ATHENIAN                    Most noble lord,      150
Pawn me to this your honour, she is his.
TIMON
My hand to thee; mine honour on my promise.
LUCILIUS
Humbly I thank your lordship. Never may
That state or fortune fall into my keeping
Which is not owed to you.                            155
    *Exeunt Lucilius and Old Athenian*
POET (*presenting a poem to Timon*)
Vouchsafe my labour, and long live your lordship!
TIMON
I thank you. You shall hear from me anon.
Go not away. (*To Painter*) What have you there, my
    friend?
PAINTER
A piece of painting, which I do beseech
Your lordship to accept.
TIMON                    Painting is welcome.         160
The painting is almost the natural man;
For since dishonour traffics with man's nature,
He is but outside; these pencilled figures are
Even such as they give out. I like your work,
And you shall find I like it. Wait attendance        165
Till you hear further from me.
PAINTER                    The gods preserve ye!
TIMON
Well fare you, gentleman. Give me your hand.
We must needs dine together. (*To Jeweller*) Sir, your jewel
Hath suffered under praise.
JEWELLER                    What, my lord, dispraise?
TIMON
A mere satiety of commendations.                     170
If I should pay you for't as 'tis extolled
It would unclew me quite.
JEWELLER                    My lord, 'tis rated
As those which sell would give; but you well know
Things of like value differing in the owners

Are prizèd by their masters. Believe't, dear lord,    175
You mend the jewel by the wearing it.
TIMON  Well mocked.
MERCHANT
No, my good lord, he speaks the common tongue
Which all men speak with him.
        *Enter Apemantus*
TIMON                                      Look who comes here.
Will you be chid?                                          180
JEWELLER  We will bear, with your lordship.
MERCHANT  He'll spare none.
TIMON
Good morrow to thee, gentle Apemantus.
APEMANTUS
Till I be gentle, stay thou for thy good morrow—
When thou art Timon's dog, and these knaves honest.
TIMON
Why dost thou call them knaves? Thou know'st them
    not.                                                  186
APEMANTUS  Are they not Athenians?
TIMON  Yes.
APEMANTUS  Then I repent not.
JEWELLER  You know me, Apemantus?                          190
APEMANTUS
Thou know'st I do. I called thee by thy name.
TIMON  Thou art proud, Apemantus!
APEMANTUS  Of nothing so much as that I am not like
Timon.
TIMON  Whither art going?                                  195
APEMANTUS  To knock out an honest Athenian's brains.
TIMON  That's a deed thou'lt die for.
APEMANTUS  Right, if doing nothing be death by th' law.
TIMON
How likest thou this picture, Apemantus?
APEMANTUS  The best for the innocence.                     200
TIMON
Wrought he not well that painted it?
APEMANTUS  He wrought better that made the painter, and
yet he's but a filthy piece of work.
PAINTER  You're a dog.
APEMANTUS  Thy mother's of my generation. What's she,
if I be a dog?                                             206
TIMON  Wilt dine with me, Apemantus?
APEMANTUS  No, I eat not lords.
TIMON  An thou shouldst, thou'dst anger ladies.
APEMANTUS  O, they eat lords. So they come by great bellies.
TIMON
That's a lascivious apprehension.                          211
APEMANTUS
So thou apprehend'st it; take it for thy labour.
TIMON
How dost thou like this jewel, Apemantus?
APEMANTUS  Not so well as plain dealing, which will not
cost a man a doit.                                         215
TIMON
What dost thou think 'tis worth?
APEMANTUS                          Not worth my thinking.
How now, poet?
POET  How now, philosopher?
APEMANTUS  Thou liest.
POET  Art not one?                                         220
APEMANTUS  Yes.
POET  Then I lie not.
APEMANTUS  Art not a poet?
POET  Yes.
APEMANTUS  Then thou liest. Look in thy last work, where
thou hast feigned him a worthy fellow.                     226

POET  That's not feigned, he is so.
APEMANTUS  Yes, he is worthy of thee, and to pay thee for
thy labour. He that loves to be flattered is worthy o'th'
flatterer. Heavens, that I were a lord!                    230
TIMON  What wouldst do then, Apemantus?
APEMANTUS  E'en as Apemantus does now: hate a lord
with my heart.
TIMON  What, thyself?
APEMANTUS  Ay.                                             235
TIMON  Wherefore?
APEMANTUS  That I had no augury but to be a lord.—Art
not thou a merchant?
MERCHANT  Ay, Apemantus.
APEMANTUS
Traffic confound thee, if the gods will not!               240
MERCHANT  If traffic do it, the gods do it.
APEMANTUS
Traffic's thy god, and thy god confound thee!
        *Trumpet sounds. Enter a Messenger*
TIMON  What trumpet's that?
MESSENGER
'Tis Alcibiades, and some twenty horse
All of companionship.                                      245
TIMON (*to Servants*)
Pray entertain them. Give them guide to us.
                    ⌐*Exit one or more Servants*⌐
⌐*To Jeweller*⌐ You must needs dine with me.
⌐*To Poet*⌐                        Go not you hence
Till I have thanked you. ⌐*To Painter*⌐ When dinner's done
Show me this piece. ⌐*To all*⌐ I am joyful of your sights.
        *Enter Alcibiades with* ⌐*his horsemen*⌐
Most welcome, sir!                                         250
APEMANTUS ⌐*aside*⌐ So, so, there.
Aches contract and starve your supple joints!
That there should be small love 'mongst these sweet
    knaves,
And all this courtesy! The strain of man's bred out
Into baboon and monkey.                                    255
ALCIBIADES (*to Timon*)
Sir, you have saved my longing, and I feed
Most hungrily on your sight.
TIMON                              Right welcome, sir!
Ere we depart, we'll share a bounteous time
In different pleasures. Pray you, let us in.
                    *Exeunt all but Apemantus*
        *Enter two Lords*
FIRST LORD
What time o' day is't, Apemantus?                          260
APEMANTUS
Time to be honest.
FIRST LORD                  That time serves still.
APEMANTUS
The most accursèd thou, that still omitt'st it.
SECOND LORD
Thou art going to Lord Timon's feast?
APEMANTUS
Ay, to see meat fill knaves, and wine heat fools.
SECOND LORD  Fare thee well, fare thee well.               265
APEMANTUS
Thou art a fool to bid me farewell twice.
SECOND LORD  Why, Apemantus?
APEMANTUS  Shouldst have kept one to thyself, for I mean
to give thee none.
FIRST LORD  Hang thyself!                                  270
APEMANTUS  No, I will do nothing at thy bidding. Make
thy requests to thy friend.

SECOND LORD Away, unpeaceable dog, or I'll spurn thee
hence.
APEMANTUS I will fly, like a dog, the heels o'th' ass.     *Exit*
FIRST LORD
He's opposite to humanity. Come, shall we in,       276
And taste Lord Timon's bounty? He outgoes
The very heart of kindness.
SECOND LORD
He pours it out. Plutus the god of gold
Is but his steward; no meed but he repays          280
Sevenfold above itself; no gift to him
But breeds the giver a return exceeding
All use of quittance.
FIRST LORD                The noblest mind he carries
That ever governed man.
SECOND LORD
Long may he live in fortunes! Shall we in?         285
⌜FIRST LORD⌝ I'll keep you company.          *Exeunt*

1.2   *Hautboys playing loud music. A great banquet*
*served in, ⌜Flavius and Servants attending⌝; and*
*then enter Timon, Alcibiades, the Senators, the*
*Athenian Lords, and Ventidius which Timon*
*redeemed from prison. Then comes, dropping after*
*all, Apemantus, discontentedly, like himself*
VENTIDIUS
Most honoured Timon, it hath pleased the gods to
remember
My father's age and call him to long peace.
He is gone happy, and has left me rich.
Then, as in grateful virtue I am bound
To your free heart, I do return those talents,      5
Doubled with thanks and service, from whose help
I derived liberty.
TIMON                O, by no means,
Honest Ventidius. You mistake my love.
I gave it freely ever, and there's none
Can truly say he gives if he receives.             10
If our betters play at that game, we must not dare
To imitate them. Faults that are rich are fair.
VENTIDIUS
A noble spirit!
⌜*The Lords stand with ceremony*⌝
TIMON              Nay, my lords,
Ceremony was but devised at first
To set a gloss on faint deeds, hollow welcomes,     15
Recanting goodness, sorry ere 'tis shown;
But where there is true friendship, there needs none.
Pray sit. More welcome are ye to my fortunes
Than my fortunes to me.
⌜*They sit*⌝
FIRST LORD
My lord, we always have confessed it.               20
APEMANTUS
Ho, ho, confessed it? Hanged it, have you not?
TIMON
O, Apemantus! You are welcome.
APEMANTUS                        No,
You shall not make me welcome.
I come to have thee thrust me out of doors.
TIMON
Fie, thou'rt a churl. Ye've got a humour there      25
Does not become a man; 'tis much to blame.
They say, my lords, *Ira furor brevis est*,
But yon man is ever angry.

Go, let him have a table by himself,
For he does neither affect company                  30
Nor is he fit for't, indeed.
APEMANTUS
Let me stay at thine apperil, Timon.
I come to observe, I give thee warning on't.
TIMON
I take no heed of thee; thou'rt an Athenian,
Therefore welcome. I myself would have no power: 35
Prithee, let my meat make thee silent.
APEMANTUS I scorn thy meat. 'Twould choke me, for I
should ne'er flatter thee. O you gods, what a number
of men eats Timon, and he sees 'em not! It grieves me
to see so many dip their meat in one man's blood; and
all the madness is, he cheers them up, too.         41
I wonder men dare trust themselves with men.
Methinks they should invite them without knives:
Good for their meat, and safer for their lives.
There's much example for't. The fellow that sits next
him, now parts bread with him, pledges the breath of
him in a divided draught, is the readiest man to kill
him. 'T'as been proved. If I were a huge man, I should
fear to drink at meals,                             49
Lest they should spy my windpipe's dangerous notes.
Great men should drink with harness on their throats.
TIMON (*drinking to a Lord*)
My lord, in heart; and let the health go round.
SECOND LORD
Let it flow this way, my good lord.
APEMANTUS 'Flow this way'? A brave fellow; he keeps his
tides well. Those healths will make thee and thy state
look ill, Timon.                                    56
Here's that which is too weak to be a sinner:
Honest water, which ne'er left man i'th' mire.
This and my food are equals; there's no odds.
Feasts are too proud to give thanks to the gods.    60
    *Apemantus' grace*
    Immortal gods, I crave no pelf.
    I pray for no man but myself.
    Grant I may never prove so fond
    To trust man on his oath or bond,
    Or a harlot for her weeping,                     65
    Or a dog that seems a-sleeping,
    Or a keeper with my freedom,
    Or my friends if I should need 'em.
    Amen. So fall to't.
    Rich men sin, and I eat root.                    70
    ⌜*He eats*⌝
Much good dich thy good heart, Apemantus.
TIMON Captain Alcibiades, your heart's in the field now.
ALCIBIADES My heart is ever at your service, my lord.
TIMON You had rather be at a breakfast of enemies than
a dinner of friends.                                75
ALCIBIADES So they were bleeding new, my lord; there's
no meat like 'em. I could wish my best friend at such
a feast.
APEMANTUS
Would all those flatterers were thine enemies then,
That thou mightst kill 'em and bid me to 'em.       80
FIRST LORD (*to Timon*) Might we but have that happiness,
my lord, that you would once use our hearts, whereby
we might express some part of our zeals, we should
think ourselves for ever perfect.                   84
TIMON O, no doubt, my good friends, but the gods
themselves have provided that I shall have much help
from you. How had you been my friends else? Why

have you that charitable title from thousands, did not you chiefly belong to my heart? I have told more of you to myself than you can with modesty speak in your own behalf; and thus far I confirm you. 'O you gods,' think I, 'what need we have any friends if we should ne'er have need of 'em? They were the most needless creatures living, should we ne'er have use for 'em, and would most resemble sweet instruments hung up in cases, that keeps their sounds to themselves.' Why, I have often wished myself poorer, that I might come nearer to you. We are born to do benefits; and what better or properer can we call our own than the riches of our friends? O, what a precious comfort 'tis to have so many like brothers commanding one another's fortunes! O, joy's e'en made away ere't can be born: mine eyes cannot hold out water, methinks. To forget their faults, I drink to you.          104

APEMANTUS Thou weep'st to make them drink, Timon.

SECOND LORD (*to Timon*)
Joy had the like conception in our eyes,
And at that instant like a babe sprung up.

APEMANTUS
Ho, ho, I laugh to think that babe a bastard.

THIRD LORD (*to Timon*)
I promise you, my lord, you moved me much.

APEMANTUS Much!          110

*A tucket sounds within*

TIMON What means that trump?

*Enter a Servant*

How now?

SERVANT Please you, my lord, there are certain ladies most desirous of admittance.

TIMON Ladies? What are their wills?          115

SERVANT There comes with them a forerunner, my lord, which bears that office to signify their pleasures.

TIMON I pray let them be admitted.

*Enter one as Cupid*

CUPID
Hail to thee, worthy Timon, and to all
That of his bounties taste! The five best senses          120
Acknowledge thee their patron, and come freely
To gratulate thy plenteous bosom. Th'ear,
Taste, touch, smell, all, pleased from thy table rise.
They only now come but to feast thine eyes.          124

TIMON
They're welcome all. Let 'em have kind admittance.
Music make their welcome!          *Exit Cupid*

⌈FIRST LORD⌉
You see, my lord, how ample you're beloved.

*Music. Enter a masque of Ladies as Amazons, with lutes in their hands, dancing and playing*

APEMANTUS
Hey-day, what a sweep of vanity comes this way!
They dance? They are madwomen.
Like madness is the glory of this life          130
As this pomp shows to a little oil and root.
We make ourselves fools to disport ourselves,
And spend our flatteries to drink those men
Upon whose age we void it up again
With poisonous spite and envy.          135
Who lives that's not depravèd or depraves?
Who dies that bears not one spurn to their graves
Of their friends' gift?
I should fear those that dance before me now
Would one day stamp upon me. 'T'as been done.          140
Men shut their doors against a setting sun.

*The Lords rise from table with much adoring of Timon; and to show their loves each singles out an Amazon, and all dance, men with women, a lofty strain or two to the hautboys; and cease*

TIMON
You have done our pleasures much grace, fair ladies,
Set a fair fashion on our entertainment,
Which was not half so beautiful and kind.
You have added worth unto't and lustre,          145
And entertained me with mine own device.
I am to thank you for't.

FIRST ⌈LADY⌉
My lord, you take us even at the best.

APEMANTUS Faith; for the worst is filthy, and would not hold taking, I doubt me.          150

TIMON
Ladies, there is an idle banquet 'tends you.
Please you to dispose yourselves.

ALL LADIES Most thankfully, my lord.          *Exeunt Ladies*

TIMON Flavius.

FLAVIUS My lord.          155

TIMON The little casket bring me hither.

FLAVIUS Yes, my lord. (*Aside*) More jewels yet?
There is no crossing him in's humour,
Else I should tell him well, i'faith I should.
When all's spent, he'd be crossed then, an he could.
'Tis pity bounty had not eyes behind,          161
That man might ne'er be wretched for his mind.          *Exit*

FIRST LORD Where be our men?

SERVANT Here, my lord, in readiness.

SECOND LORD Our horses.          ⌈*Exit Servant*⌉

*Enter Flavius with the casket. He gives it to Timon, ⌈and exits⌉*

TIMON
O my friends, I have one word to say to you.          166
Look you, my good lord,
I must entreat you honour me so much
As to advance this jewel. Accept and wear it,
Kind my lord.          170

FIRST LORD
I am so far already in your gifts.

ALL LORDS So are we all.

⌈*Timon gives them jewels.*⌉

*Enter a Servant*

FIRST SERVANT My lord, there are certain nobles of the senate newly alighted and come to visit you.          174

TIMON They are fairly welcome.          *Exit Servant*

*Enter Flavius*

FLAVIUS I beseech your honour, vouchsafe me a word; it does concern you near.

TIMON
Near? Why then, another time I'll hear thee.
I prithee, let's be provided to show them entertainment.

FLAVIUS I scarce know how.          180

*Enter a Second Servant*

SECOND SERVANT
May it please your honour, Lord Lucius
Out of his free love hath presented to you
Four milk-white horses trapped in silver.

TIMON
I shall accept them fairly. Let the presents
Be worthily entertained.          *Exit Servant*

*Enter a Third Servant*

How now, what news?          185

THIRD SERVANT Please you, my lord, that honourable gentleman Lord Lucullus entreats your company

tomorrow to hunt with him, and has sent your honour
two brace of greyhounds.
TIMON
I'll hunt with him, and let them be received          190
Not without fair reward.                      *Exit Servant*
FLAVIUS (*aside*)                What will this come to?
He commands us to provide and give great gifts,
And all out of an empty coffer;
Nor will he know his purse, or yield me this:
To show him what a beggar his heart is,              195
Being of no power to make his wishes good.
His promises fly so beyond his state
That what he speaks is all in debt, he owes
For every word. He is so kind that he now
Pays interest for't. His land's put to their books.    200
Well, would I were gently put out of office
Before I were forced out.
Happier is he that has no friend to feed
Than such that do e'en enemies exceed.
I bleed inwardly for my lord.                      *Exit*
TIMON (*to the Lords*)          You do yourselves   205
Much wrong, you bate too much of your own merits.
(*To Second Lord*) Here, my lord, a trifle of our love.
SECOND LORD
With more than common thanks I will receive it.
THIRD LORD
O, he's the very soul of bounty!                    209
TIMON (*to First Lord*) And now I remember, my lord, you
gave good words the other day of a bay courser I rode
on. 'Tis yours, because you liked it.
FIRST LORD
O I beseech you pardon me, my lord, in that.
TIMON
You may take my word, my lord, I know no man
Can justly praise but what he does affect.           215
I weigh my friends' affection with mine own.
I'll tell you true, I'll call to you.
ALL LORDS                    O, none so welcome.
TIMON
I take all and your several visitations
So kind to heart, 'tis not enough to give.
Methinks I could deal kingdoms to my friends,      220
And ne'er be weary. Alcibiades,
Thou art a soldier, therefore seldom rich.
⌜*Giving a present*⌝ It comes in charity to thee, for all
       thy living
Is 'mongst the dead, and all the lands thou hast
Lie in a pitched field.
ALCIBIADES                    Ay, defiled land, my lord.   225
FIRST LORD We are so virtuously bound—
TIMON And so am I to you.
SECOND LORD So infinitely endeared—
TIMON All to you. Lights, more lights!
FIRST LORD
The best of happiness, honour, and fortunes         230
Keep with you, Lord Timon.
TIMON Ready for his friends.
                    *Exeunt all but Timon and Apemantus*
APEMANTUS What a coil's here,
Serving of becks and jutting-out of bums!
I doubt whether their legs be worth the sums        235
That are given for 'em. Friendship's full of dregs.
Methinks false hearts should never have sound legs.
Thus honest fools lay out their wealth on curtseys.
TIMON
Now, Apemantus, if thou wert not sullen
I would be good to thee.                            240

APEMANTUS No, I'll nothing; for if I should be bribed too,
there would be none left to rail upon thee, and then
thou wouldst sin the faster. Thou giv'st so long, Timon,
I fear me thou wilt give away thyself in paper shortly.
What needs these feasts, pomps, and vainglories?    245
TIMON Nay, an you begin to rail on society once, I am
sworn not to give regard to you.
Farewell, and come with better music.             *Exit*
APEMANTUS                                So.
Thou wilt not hear me now, thou shalt not then.
I'll lock thy heaven from thee. O, that men's ears
       should be                                    250
To counsel deaf, but not to flattery!             *Exit*

2.1     *Enter a Senator* ⌜*with bonds*⌝
SENATOR
And late five thousand. To Varro and to Isidore
He owes nine thousand, besides my former sum,
Which makes it five-and-twenty. Still in motion
Of raging waste! It cannot hold, it will not.
If I want gold, steal but a beggar's dog               5
And give it Timon, why, the dog coins gold.
If I would sell my horse and buy twenty more
Better than he, why, give my horse to Timon—
Ask nothing, give it him—it foals me straight,
And able horses. No porter at his gate,              10
But rather one that smiles and still invites
All that pass by. It cannot hold. No reason
Can sound his state in safety. Caphis ho!
Caphis, I say!
       *Enter Caphis*
CAPHIS          Here, sir. What is your pleasure?
SENATOR
Get on your cloak and haste you to Lord Timon.     15
Importune him for my moneys. Be not ceased
With slight denial, nor then silenced when
'Commend me to your master', and the cap
Plays in the right hand, thus; but tell him
My uses cry to me, I must serve my turn             20
Out of mine own, his days and times are past,
And my reliances on his fracted dates
Have smit my credit. I love and honour him,
But must not break my back to heal his finger.
Immediate are my needs, and my relief               25
Must not be tossed and turned to me in words,
But find supply immediate. Get you gone.
Put on a most importunate aspect,
A visage of demand, for I do fear
When every feather sticks in his own wing           30
Lord Timon will be left a naked gull,
Which flashes now a phoenix. Get you gone.
CAPHIS
I go, sir.
SENATOR ⌜*giving him bonds*⌝
                    Take the bonds along with you,
And have the dates in count.
CAPHIS                          I will, sir.
SENATOR                                Go.
                              *Exeunt* ⌜*severally*⌝

2.2     *Enter Flavius, with many bills in his hand*
FLAVIUS
No care, no stop; so senseless of expense
That he will neither know how to maintain it
Nor cease his flow of riot, takes no account
How things go from him, nor resumes no care

Of what is to continue. Never mind                              5
Was to be so unwise to be so kind.
What shall be done? He will not hear till feel.
⌈*A sound of horns within*⌉
I must be round with him, now he comes from hunting.
Fie, fie, fie, fie!
    *Enter Caphis* ⌈*at one door*⌉ *and Servants of Isidore*
    *and Varro* ⌈*at another door*⌉
CAPHIS
Good even, Varro. What, you come for money?       10
VARRO'S SERVANT Is't not your business too?
CAPHIS
It is; and yours too, Isidore?
ISIDORE'S SERVANT                    It is so.
CAPHIS
Would we were all discharged.
VARRO'S SERVANT                    I fear it.
CAPHIS                              Here comes the lord.
    *Enter Timon and his train, amongst them*
    *Alcibiades,* ⌈*as from hunting*⌉
TIMON
So soon as dinner's done we'll forth again,
My Alcibiades.
    *Caphis meets Timon*
                    With me? What is your will?       15
CAPHIS
My lord, here is a note of certain dues.
TIMON Dues? Whence are you?
CAPHIS Of Athens here, my lord.
TIMON Go to my steward.
CAPHIS
Please it your lordship, he hath put me off,       20
To the succession of new days, this month.
My master is awaked by great occasion
To call upon his own, and humbly prays you
That with your other noble parts you'll suit
In giving him his right.
TIMON                    Mine honest friend,       25
I prithee but repair to me next morning.
CAPHIS
Nay, good my lord.
TIMON                    Contain thyself, good friend.
VARRO'S SERVANT
One Varro's servant, my good lord.
ISIDORE'S SERVANT (*to Timon*)
From Isidore. He humbly prays your speedy payment.
CAPHIS (*to Timon*)
If you did know, my lord, my master's wants—       30
VARRO'S SERVANT (*to Timon*)
'Twas due on forfeiture, my lord, six weeks and past.
ISIDORE'S SERVANT (*to Timon*)
Your steward puts me off, my lord, and I
Am sent expressly to your lordship.
TIMON                    Give me breath.—
I do beseech you, good my lords, keep on.
I'll wait upon you instantly.
    *Exeunt Alcibiades and Timon's train*
(*To Flavius*)                    Come hither. Pray you,
How goes the world, that I am thus encountered       36
With clamorous demands of broken bonds
And the detention of long-since-due debts,
Against my honour?
FLAVIUS (*to Servants*)    Please you, gentlemen,
The time is unagreeable to this business;       40
Your importunacy cease till after dinner,

That I may make his lordship understand
Wherefore you are not paid.
TIMON (*to Servants*)                    Do so, my friends.
(*To Flavius*) See them well entertained.       *Exit*
FLAVIUS                    Pray draw near.
                                   *Exit*
    *Enter Apemantus and Fool*
CAPHIS
Stay, stay, here comes the fool with Apemantus.       45
Let's ha' some sport with 'em.
VARRO'S SERVANT Hang him, he'll abuse us.
ISIDORE'S SERVANT A plague upon him, dog!
VARRO'S SERVANT How dost, fool?
APEMANTUS Dost dialogue with thy shadow?       50
VARRO'S SERVANT I speak not to thee.
APEMANTUS No, 'tis to thyself. (*To Fool*) Come away.
ISIDORE'S SERVANT (*to Varro's Servant*) There's the fool
hangs on your back already.
APEMANTUS No, thou stand'st single: thou'rt not on him
yet.                              56
CAPHIS (*to Isidore's Servant*) Where's the fool now?
APEMANTUS He last asked the question. Poor rogues' and
usurers' men, bawds between gold and want.
ALL SERVANTS What are we, Apemantus?       60
APEMANTUS Asses.
ALL SERVANTS Why?
APEMANTUS That you ask me what you are, and do not
know yourselves. Speak to 'em, fool.
FOOL How do you, gentlemen?       65
ALL SERVANTS Gramercies, good fool. How does your
mistress?
FOOL She's e'en setting on water to scald such chickens
as you are. Would we could see you at Corinth.
APEMANTUS Good; gramercy.       70
    *Enter Page with two letters*
FOOL Look you, here comes my mistress' page.
PAGE Why, how now, captain? What do you in this wise
company? How dost thou, Apemantus?
APEMANTUS Would I had a rod in my mouth, that I might
answer thee profitably.       75
PAGE Prithee, Apemantus, read me the superscription of
these letters. I know not which is which.
APEMANTUS Canst not read?
PAGE No.
APEMANTUS There will little learning die then that day
thou art hanged. This is to Lord Timon, this to
Alcibiades. Go, thou wast born a bastard, and thou'lt
die a bawd.       83
PAGE Thou wast whelped a dog, and thou shalt famish a
dog's death. Answer not; I am gone.       *Exit*
APEMANTUS E'en so thou outrunn'st grace. Fool, I will go
with you to Lord Timon's.
FOOL Will you leave me there?
APEMANTUS If Timon stay at home. (*To Servants*) You
three serve three usurers?       90
ALL SERVANTS Ay. Would they served us.
APEMANTUS So would I: as good a trick as ever hangman
served thief.
FOOL Are you three usurers' men?
ALL SERVANTS Ay, fool.       95
FOOL I think no usurer but has a fool to his servant. My
mistress is one, and I am her fool. When men come to
borrow of your masters they approach sadly and go
away merry, but they enter my mistress's house merrily
and go away sadly. The reason of this?       100
VARRO'S SERVANT I could render one.

APEMANTUS Do it then, that we may account thee a
   whoremaster and a knave, which notwithstanding thou
   shalt be no less esteemed.
VARRO'S SERVANT What is a whoremaster, fool?        105
FOOL A fool in good clothes, and something like thee. 'Tis
   a spirit; sometime 't appears like a lord, sometime like
   a lawyer, sometime like a philosopher with two stones
   more than's artificial one. He is very often like a knight;
   and generally in all shapes that man goes up and down
   in from fourscore to thirteen, this spirit walks in.     111
VARRO'S SERVANT Thou art not altogether a fool.
FOOL Nor thou altogether a wise man. As much foolery
   as I have, so much wit thou lack'st.
APEMANTUS That answer might have become Apemantus.
     *Enter Timon and Flavius*
ALL SERVANTS Aside, aside, here comes Lord Timon.    116
APEMANTUS Come with me, fool, come.
FOOL I do not always follow lover, elder brother, and
   woman: sometime the philosopher.
                    *Exeunt Apemantus and Fool*
FLAVIUS (*to Servants*)
   Pray you, walk near. I'll speak with you anon.    120
                           *Exeunt Servants*
TIMON
   You make me marvel wherefore ere this time
   Had you not fully laid my state before me,
   That I might so have rated my expense
   As I had leave of means.
FLAVIUS               You would not hear me.
   At many leisures I proposed—
TIMON              Go to.    125
   Perchance some single vantages you took,
   When my indisposition put you back,
   And that unaptness made your minister
   Thus to excuse yourself.
FLAVIUS           O my good lord,
   At many times I brought in my accounts,    130
   Laid them before you; you would throw them off
   And say you summed them in mine honesty.
   When for some trifling present you have bid me
   Return so much, I have shook my head and wept,
   Yea, 'gainst th'authority of manners prayed you    135
   To hold your hand more close. I did endure
   Not seldom nor no slight checks when I have
   Prompted you in the ebb of your estate
   And your great flow of debts. My lovèd lord—
   Though you hear now too late, yet now's a time—
   The greatest of your having lacks a half    141
   To pay your present debts.
TIMON             Let all my land be sold.
FLAVIUS
   'Tis all engaged, some forfeited and gone,
   And what remains will hardly stop the mouth
   Of present dues. The future comes apace.    145
   What shall defend the interim, and at length
   How goes our reck'ning?
TIMON
   To Lacedaemon did my land extend.
FLAVIUS
   O my good lord, the world is but a word.
   Were it all yours to give it in a breath,    150
   How quickly were it gone.
TIMON            You tell me true.
FLAVIUS
   If you suspect my husbandry or falsehood,
   Call me before th'exactest auditors

And set me on the proof. So the gods bless me,
When all our offices have been oppressed    155
With riotous feeders, when our vaults have wept
With drunken spilth of wine, when every room
Hath blazed with lights and brayed with minstrelsy,
I have retired me to a wasteful cock,
And set mine eyes at flow.
TIMON          Prithee, no more.    160
FLAVIUS
   'Heavens,' have I said, 'the bounty of this lord!
   How many prodigal bits have slaves and peasants
   This night englutted! Who is not Timon's?
   What heart, head, sword, force, means, but is Lord
      Timon's?
   Great Timon, noble, worthy, royal Timon!    165
   Ah, when the means are gone that buy this praise,
   The breath is gone whereof this praise is made.
   Feast won, fast lost; one cloud of winter show'rs,
   These flies are couched.'
TIMON        Come, sermon me no further.
   No villainous bounty yet hath passed my heart.    170
   Unwisely, not ignobly, have I given.
   Why dost thou weep? Canst thou the conscience lack
   To think I shall lack friends? Secure thy heart.
   If I would broach the vessels of my love
   And try the argument of hearts by borrowing,    175
   Men and men's fortunes could I frankly use
   As I can bid thee speak.
FLAVIUS        Assurance bless your thoughts!
TIMON
   And in some sort these wants of mine are crowned
   That I account them blessings, for by these
   Shall I try friends. You shall perceive how you    180
   Mistake my fortunes. I am wealthy in my friends.—
   Within there, Flaminius, Servilius!
     *Enter Flaminius, Servilius, and a Third Servant*
ALL SERVANTS
   My lord, my lord.
TIMON          I will dispatch you severally,
   (*To Servilius*) You to Lord Lucius,
   (*To Flaminius*)          to Lord Lucullus you—
   I hunted with his honour today—    185
   (*To Third Servant*) You to Sempronius. Commend me
     to their loves,
   And I am proud, say, that my occasions have
   Found time to use 'em toward a supply of money.
   Let the request be fifty talents.
FLAMINIUS As you have said, my lord.    *Exeunt Servants*
FLAVIUS
   Lord Lucius and Lucullus? Hmh!    191
TIMON
   Go you, sir, to the senators,
   Of whom, even to the state's best health, I have
   Deserved this hearing. Bid 'em send o'th' instant
   A thousand talents to me.
FLAVIUS        I have been bold,    195
   For that I knew it the most general way,
   To them, to use your signet and your name;
   But they do shake their heads, and I am here
   No richer in return.
TIMON         Is't true? Can't be?
FLAVIUS
   They answer in a joint and corporate voice    200
   That now they are at fall, want treasure, cannot
   Do what they would, are sorry, you are honourable,

But yet they could have wished—they know not—
Something hath been amiss—a noble nature
May catch a wrench—would all were well—'tis pity;
And so, intending other serious matters,                                    206
After distasteful looks and these hard fractions,
With certain half-caps and cold moving nods
They froze me into silence.

TIMON                                                You gods reward them!
Prithee, man, look cheerly. These old fellows      210
Have their ingratitude in them hereditary.
Their blood is caked, 'tis cold, it seldom flows.
'Tis lack of kindly warmth they are not kind;
And nature as it grows again toward earth
Is fashioned for the journey dull and heavy.      215
Go to Ventidius. Prithee, be not sad.
Thou art true and honest—ingenuously I speak—
No blame belongs to thee. Ventidius lately
Buried his father, by whose death he's stepped
Into a great estate. When he was poor,                   220
Imprisoned, and in scarcity of friends,
I cleared him with five talents. Greet him from me.
Bid him suppose some good necessity
Touches his friend, which craves to be remembered
With those five talents. That had, give't these fellows
To whom 'tis instant due. Ne'er speak or think      226
That Timon's fortunes 'mong his friends can sink.

FLAVIUS
I would I could not think it. That thought is bounty's
    foe:
Being free itself, it thinks all others so.
                                        *Exeunt ⌈severally⌉*

**3.1**   *Enter Flaminius, with a box under his cloak,*
         *waiting to speak with Lucullus. From his master,*
         *enters a Servant to him*

LUCULLUS' SERVANT I have told my lord of you. He is
coming down to you.
FLAMINIUS I thank you, sir.
    *Enter Lucullus*
LUCULLUS' SERVANT Here's my lord.                            4
LUCULLUS (*aside*) One of Lord Timon's men? A gift, I
warrant. Why, this hits right; I dreamt of a silver basin
and ewer tonight.—Flaminius, honest Flaminius, you
are very respectively welcome, sir. (*To his Servant*) Fill
me some wine.                                    *Exit Servant*
And how does that honourable, complete, free-hearted
gentleman of Athens, thy very bountiful good lord and
master?                                                            12
FLAMINIUS His health is well, sir.
LUCULLUS I am right glad that his health is well, sir. And
what hast thou there under thy cloak, pretty Flaminius?
FLAMINIUS Faith, nothing but an empty box, sir, which
in my lord's behalf I come to entreat your honour to
supply, who, having great and instant occasion to use
fifty talents, hath sent to your lordship to furnish him,
nothing doubting your present assistance therein.
LUCULLUS La, la, la, la, 'nothing doubting' says he? Alas,
good lord! A noble gentleman 'tis, if he would not keep
so good a house. Many a time and often I ha' dined
with him and told him on't, and come again to supper
to him of purpose to have him spend less; and yet he
would embrace no counsel, take no warning by my
coming. Every man has his fault, and honesty is his. I
ha' told him on't, but I could ne'er get him from't.
    *Enter Servant, with wine*
SERVANT Please your lordship, here is the wine.       29

LUCULLUS Flaminius, I have noted thee always wise.
    (*Drinking*) Here's to thee!
FLAMINIUS Your lordship speaks your pleasure.       32
LUCULLUS I have observed thee always for a towardly
prompt spirit, give thee thy due, and one that knows
what belongs to reason; and canst use the time well if
the time use thee well. (*Drinking*) Good parts in thee!
(*To his Servant*) Get you gone, sirrah.       *Exit Servant*
Draw nearer, honest Flaminius. Thy lord's a bountiful
gentleman; but thou art wise, and thou know'st well
enough, although thou com'st to me, that this is no
time to lend money, especially upon bare friendship
without security. (*Giving coins*) Here's three solidares
for thee. Good boy, wink at me, and say thou saw'st
me not. Fare thee well.
FLAMINIUS
Is't possible the world should so much differ,        45
And we alive that lived?
    *He throws the coins at Lucullus*
                                        Fly, damnèd baseness,
To him that worships thee.
LUCULLUS Ha! Now I see thou art a fool, and fit for thy
master.                                                         *Exit*
FLAMINIUS
May these add to the number that may scald thee.   50
Let molten coin be thy damnation,
Thou disease of a friend, and not himself.
Has friendship such a faint and milky heart
It turns in less than two nights? O you gods,
I feel my master's passion! This slave                     55
Unto this hour has my lord's meat in him.
Why should it thrive and turn to nutriment,
When he is turned to poison?
O, may diseases only work upon't;
And when he's sick to death, let not that part of nature
Which my lord paid for be of any power                61
To expel sickness, but prolong his hour.        *Exit*

**3.2**   *Enter Lucius, with three Strangers*
LUCIUS Who, the Lord Timon? He is my very good friend,
and an honourable gentleman.
FIRST STRANGER We know him for no less, though we are
but strangers to him. But I can tell you one thing, my
lord, and which I hear from common rumours: now
Lord Timon's happy hours are done and past, and his
estate shrinks from him.
LUCIUS Fie, no, do not believe it. He cannot want for
money.                                                             9
SECOND STRANGER But believe you this, my lord, that not
long ago one of his men was with the Lord Lucullus
to borrow so many talents—nay, urged extremely for't,
and showed what necessity belonged to't, and yet was
denied.
LUCIUS How?                                                     15
SECOND STRANGER I tell you, denied, my lord.
LUCIUS What a strange case was that! Now before the
gods, I am ashamed on't. Denied that honourable man?
There was very little honour showed in't. For my own
part, I must needs confess I have received some small
kindnesses from him, as money, plate, jewels, and
suchlike trifles—nothing comparing to his; yet had he
not mistook him and sent to me, I should ne'er have
denied his occasion so many talents.                   24
    *Enter Servilius*
SERVILIUS (*aside*) See, by good hap yonder's my lord. I
have sweat to see his honour. (*To Lucius*) My honoured
lord!

⌈LUCIUS⌉ Servilius! You are kindly met, sir. Fare thee well.
  Commend me to thy honourable virtuous lord, my very
  exquisite friend.                                              30
SERVILIUS May it please your honour, my lord hath sent—
LUCIUS Ha! What has he sent? I am so much endeared
  to that lord, he's ever sending. How shall I thank him,
  think'st thou? And what has he sent now?
SERVILIUS He's only sent his present occasion now, my
  lord, requesting your lordship to supply his instant use
  with so many talents.                                         37
⌈LUCIUS⌉
  I know his lordship is but merry with me.
  He cannot want fifty-five hundred talents.
SERVILIUS
  But in the mean time he wants less, my lord.                  40
  If his occasion were not virtuous
  I should not urge it half so faithfully.
LUCIUS
  Dost thou speak seriously, Servilius?
SERVILIUS Upon my soul, 'tis true, sir.                         44
LUCIUS What a wicked beast was I to disfurnish myself
  against such a good time when I might ha' shown
  myself honourable! How unluckily it happened that I
  should purchase the day before a little part, and undo
  a great deal of honour! Servilius, now before the gods
  I am not able to do, the more beast I, I say. I was
  sending to use Lord Timon myself—these gentlemen
  can witness—but I would not for the wealth of Athens
  I had done't now. Commend me bountifully to his good
  lordship; and I hope his honour will conceive the fairest
  of me because I have no power to be kind. And tell
  him this from me: I count it one of my greatest
  afflictions, say, that I cannot pleasure such an
  honourable gentleman. Good Servilius, will you befriend
  me so far as to use mine own words to him?
SERVILIUS Yes, sir, I shall.                                    60
⌈LUCIUS⌉
  I'll look you out a good turn, Servilius.   *Exit Servilius*
  True as you said: Timon is shrunk indeed;
  And he that's once denied will hardly speed.   *Exit*
FIRST STRANGER
  Do you observe this, Hostilius?
SECOND STRANGER                       Ay, too well.             64
FIRST STRANGER
  Why, this is the world's soul, and just of the same piece
  Is every flatterer's spirit. Who can call him his friend
  That dips in the same dish? For, in my knowing,
  Timon has been this lord's father
  And kept his credit with his purse,
  Supported his estate; nay, Timon's money                      70
  Has paid his men their wages. He ne'er drinks,
  But Timon's silver treads upon his lip;
  And yet—O see the monstrousness of man
  When he looks out in an ungrateful shape!—
  He does deny him, in respect of his,                          75
  What charitable men afford to beggars.
THIRD STRANGER
  Religion groans at it.
FIRST STRANGER              For mine own part,
  I never tasted Timon in my life,
  Nor came any of his bounties over me
  To mark me for his friend; yet I protest,                     80
  For his right noble mind, illustrious virtue,
  And honourable carriage,
  Had his necessity made use of me
  I would have put my wealth into donation

  And the best half should have returned to him,                85
  So much I love his heart. But I perceive
  Men must learn now with pity to dispense,
  For policy sits above conscience.                  *Exeunt*

3.3   *Enter Timon's Third Servant, with Sempronius,*
      *another of Timon's friends*
SEMPRONIUS
  Must he needs trouble me in't? Hmh! 'Bove all others?
  He might have tried Lord Lucius or Lucullus;
  And now Ventidius is wealthy too,
  Whom he redeemed from prison. All these
  Owes their estates unto him.
SERVANT               My lord,                                   5
  They have all been touched and found base metal,
  For they have all denied him.
SEMPRONIUS                   How, have they denied him?
  Has Ventidius and Lucullus denied him,
  And does he send to me? Three? Hmh!
  It shows but little love or judgement in him.                 10
  Must I be his last refuge? His friends, like physicians,
  Thrive, give him over; must I take th' cure upon me?
  He's much disgraced me in't. I'm angry at him,
  That might have known my place. I see no sense for't
  But his occasions might have wooed me first,                  15
  For, in my conscience, I was the first man
  That e'er received gift from him.
  And does he think so backwardly of me now
  That I'll requite it last? No.
  So it may prove an argument of laughter                       20
  To th' rest, and I 'mongst lords be thought a fool.
  I'd rather than the worth of thrice the sum
  He'd sent to me first, but for my mind's sake.
  I'd such a courage to do him good. But now return,
  And with their faint reply this answer join:                  25
  Who bates mine honour shall not know my coin.   *Exit*
SERVANT Excellent. Your lordship's a goodly villain. The
  devil knew not what he did when he made man
  politic—he crossed himself by't, and I cannot think but
  in the end the villainies of man will set him clear. How
  fairly this lord strives to appear foul! Takes virtuous
  copies to be wicked, like those that under hot ardent
  zeal would set whole realms on fire; of such a nature
  is his politic love.
  This was my lord's best hope. Now all are fled                35
  Save only the gods. Now his friends are dead.
  Doors that were ne'er acquainted with their wards
  Many a bounteous year must be employed
  Now to guard sure their master;
  And this is all a liberal course allows:                      40
  Who cannot keep his wealth must keep his house.
                                                  *Exit*

3.4   *Enter Varro's two Servants, meeting others, all*
      *Servants of Timon's creditors, to wait for his coming*
      *out. Then enter ⌈Servants of⌉ Lucius, Titus, and*
      *Hortensius*
VARRO'S ⌈FIRST⌉ SERVANT
  Well met; good morrow, Titus and Hortensius.
TITUS' SERVANT The like to you, kind Varro.
HORTENSIUS' SERVANT
  Lucius, what, do we meet together?
LUCIUS' SERVANT
  Ay, and I think one business does command us all,
  For mine is money.
TITUS' SERVANT           So is theirs and ours.                 5

*Enter ⌈a Servant of ⌉ Philotus*

LUCIUS' SERVANT
And Sir Philotus too!

PHILOTUS' SERVANT          Good day at once.

LUCIUS' SERVANT
Welcome, good brother. What do you think the hour?

PHILOTUS' SERVANT Labouring for nine.

LUCIUS' SERVANT So much?

PHILOTUS' SERVANT Is not my lord seen yet?          10

LUCIUS' SERVANT Not yet.

PHILOTUS' SERVANT
I wonder on't; he was wont to shine at seven.

LUCIUS' SERVANT
Ay, but the days are waxed shorter with him.
You must consider that a prodigal course
Is like the sun's,                                   15
But not, like his, recoverable. I fear
'Tis deepest winter in Lord Timon's purse; that is,
One may reach deep enough, and yet find little.

PHILOTUS' SERVANT I am of your fear for that.

TITUS' SERVANT
I'll show you how t'observe a strange event.       20
Your lord sends now for money?

HORTENSIUS' SERVANT          Most true, he does.

TITUS' SERVANT
And he wears jewels now of Timon's gift,
For which I wait for money.

HORTENSIUS' SERVANT It is against my heart.

LUCIUS' SERVANT Mark how strange it shows.          25
Timon in this should pay more than he owes,
And e'en as if your lord should wear rich jewels
And send for money for 'em.

HORTENSIUS' SERVANT
I'm weary of this charge, the gods can witness.
I know my lord hath spent of Timon's wealth,       30
And now ingratitude makes it worse than stealth.

VARRO'S FIRST SERVANT
Yes; mine's three thousand crowns. What's yours?

LUCIUS' SERVANT          Five thousand, mine.

VARRO'S FIRST SERVANT
'Tis much deep, and it should seem by th' sum
Your master's confidence was above mine,
Else surely his had equalled.

*Enter Flaminius*

TITUS' SERVANT          One of Lord Timon's men.

LUCIUS' SERVANT
Flaminius! Sir, a word. Pray, is my lord          36
Ready to come forth?

FLAMINIUS No, indeed he is not.

TITUS' SERVANT We attend his lordship.
Pray signify so much.

FLAMINIUS          I need not tell          40
Him that; he knows you are too diligent.

*Enter Flavius, muffled in a cloak*

LUCIUS' SERVANT
Ha, is not that his steward muffled so?
He goes away in a cloud. Call him, call him.

TITUS' SERVANT (*to Flavius*) Do you hear, sir?

VARRO'S SECOND SERVANT (*to Flavius*) By your leave, sir.

FLAVIUS What do ye ask of me, my friend?          46

TITUS' SERVANT
We wait for certain money here, sir.

FLAVIUS          Ay,
If money were as certain as your waiting,
'Twere sure enough.
Why then preferred you not your sums and bills     50

When your false masters ate of my lord's meat?
Then they could smile and fawn upon his debts,
And take down th'int'rest into their glutt'nous maws.
You do yourselves but wrong to stir me up.
Let me pass quietly.                                55
Believe't, my lord and I have made an end.
I have no more to reckon, he to spend.

LUCIUS' SERVANT
Ay, but this answer will not serve.

FLAVIUS
If 'twill not serve 'tis not so base as you,       59
For you serve knaves.                          *Exit*

VARRO'S FIRST SERVANT How? What does his cashiered
worship mutter?

VARRO'S SECOND SERVANT No matter what; he's poor, and
that's revenge enough. Who can speak broader than
he that has no house to put his head in? Such may
rail against great buildings.                      66

*Enter Servilius*

TITUS' SERVANT O, here's Servilius. Now we shall know
some answer.

SERVILIUS If I might beseech you, gentlemen, to repair
some other hour, I should derive much from't; for,
take't of my soul, my lord leans wondrously to
discontent. His comfortable temper has forsook him.
He's much out of health, and keeps his chamber.

LUCIUS' SERVANT
Many do keep their chambers are not sick,
And if it be so far beyond his health             75
Methinks he should the sooner pay his debts
And make a clear way to the gods.

SERVILIUS          Good gods!

TITUS' SERVANT
We cannot take this for an answer, sir.

FLAMINIUS (*within*)
Servilius, help! My lord, my lord!

*Enter Timon in a rage*

TIMON
What, are my doors opposed against my passage?     80
Have I been ever free, and must my house
Be my retentive enemy, my jail?
The place which I have feasted, does it now,
Like all mankind, show me an iron heart?

LUCIUS' SERVANT
Put in now, Titus.

TITUS' SERVANT          My lord, here is my bill.   85

LUCIUS' SERVANT
Here's mine.

⌈HORTENSIUS' SERVANT⌉ And mine, my lord.

VARRO'S ⌈FIRST and⌉ SECOND SERVANTS  And ours, my lord.

PHILOTUS' SERVANT All our bills.

TIMON
Knock me down with 'em, cleave me to the girdle.

LUCIUS' SERVANT Alas, my lord.

TIMON Cut my heart in sums.                        90

TITUS' SERVANT Mine fifty talents.

TIMON
Tell out my blood.

LUCIUS' SERVANT          Five thousand crowns, my lord.

TIMON
Five thousand drops pays that. What yours? And
yours?

VARRO'S FIRST SERVANT My lord—

VARRO'S SECOND SERVANT My lord—                     95

TIMON
Tear me, take me, and the gods fall upon you.   *Exit*

HORTENSIUS' SERVANT Faith, I perceive our masters may
  throw their caps at their money. These debts may well
  be called desperate ones, for a madman owes 'em.    99
                                              *Exeunt*

**3.5**   *Enter Timon and Flavius*
TIMON
  They have e'en put my breath from me, the slaves.
  Creditors? Devils!
FLAVIUS My dear lord—
TIMON What if it should be so?
FLAVIUS My lord—                                        5
TIMON
  I'll have it so. My steward!
FLAVIUS                         Here, my lord.
TIMON
  So fitly? Go bid all my friends again:
  Lucius, Lucullus, and Sempronius—all luxors, all.
  I'll once more feast the rascals.
FLAVIUS                            O my lord,
  You only speak from your distracted soul.           10
  There is not so much left to furnish out
  A moderate table.
TIMON            Be it not in thy care.
  Go, I charge thee, invite them all. Let in the tide
  Of knaves once more. My cook and I'll provide.
                                 *Exeunt ⌈severally⌉*

**3.6**   *Enter three Senators at one door*
FIRST SENATOR
  My lords, you have my voice to't. The fault's bloody.
  'Tis necessary he should die.
  Nothing emboldens sin so much as mercy.
SECOND SENATOR Most true; the law shall bruise 'im.
    ⌈*Enter Alcibiades at another door, with attendants*⌉
ALCIBIADES
  Honour, health, and compassion to the senate!       5
FIRST SENATOR Now, captain.
ALCIBIADES
  I am an humble suitor to your virtues;
  For pity is the virtue of the law,
  And none but tyrants use it cruelly.
  It pleases time and fortune to lie heavy            10
  Upon a friend of mine, who in hot blood
  Hath stepped into the law, which is past depth
  To those that without heed do plunge into't.
  He is a man, setting his feat aside,
  Of comely virtues;                                  15
  Nor did he soil the fact with cowardice—
  An honour in him which buys out his fault—
  But with a noble fury and fair spirit,
  Seeing his reputation touched to death,
  He did oppose his foe;                              20
  And with such sober and unnoted passion
  He did behave his anger, ere 'twas spent,
  As if he had but proved an argument.
FIRST SENATOR
  You undergo too strict a paradox,
  Striving to make an ugly deed look fair.            25
  Your words have took such pains as if they laboured
  To bring manslaughter into form, and set quarrelling
  Upon the head of valour—which indeed
  Is valour misbegot, and came into the world
  When sects and factions were newly born.            30
  He's truly valiant that can wisely suffer
  The worst that man can breathe, and make his
      wrongs his outsides

To wear them like his raiment carelessly,
And ne'er prefer his injuries to his heart
To bring it into danger.                              35
If wrongs be evils and enforce us kill,
What folly 'tis to hazard life for ill!
ALCIBIADES
  My lord—
FIRST SENATOR You cannot make gross sins look clear.
  To revenge is no valour, but to bear.
ALCIBIADES
  My lords, then, under favour, pardon me             40
  If I speak like a captain.
  Why do fond men expose themselves to battle,
  And not endure all threats, sleep upon't,
  And let the foes quietly cut their throats
  Without repugnancy? If there be                     45
  Such valour in the bearing, what make we
  Abroad? Why then, women are more valiant
  That stay at home if bearing carry it,
  And the ass more captain than the lion, the felon
  Loaden with irons wiser than the judge,             50
  If wisdom be in suffering. O my lords,
  As you are great, be pitifully good.
  Who cannot condemn rashness in cold blood?
  To kill, I grant, is sin's extremest gust,
  But in defence, by mercy, 'tis most just.           55
  To be in anger is impiety,
  But who is man that is not angry?
  Weigh but the crime with this.
SECOND SENATOR           You breathe in vain.
ALCIBIADES                                In vain?
  His service done at Lacedaemon and Byzantium
  Were a sufficient briber for his life.              60
FIRST SENATOR
  What's that?
ALCIBIADES Why, I say, my lords, he's done fair service,
  And slain in fight many of your enemies.
  How full of valour did he bear himself
  In the last conflict, and made plenteous wounds!
SECOND SENATOR
  He has made too much plenty with 'em.               65
  He's a sworn rioter; he has a sin
  That often drowns him and takes his valour prisoner.
  If there were no foes, that were enough
  To overcome him. In that beastly fury
  He has been known to commit outrages                70
  And cherish factions. 'Tis inferred to us
  His days are foul and his drink dangerous.
FIRST SENATOR
  He dies.
ALCIBIADES Hard fate! He might have died in war.
  My lords, if not for any parts in him—
  Though his right arm might purchase his own time
  And be in debt to none—yet more to move you,        76
  Take my deserts to his and join 'em both.
  And for I know
  Your reverend ages love security,
  I'll pawn my victories, all my honour to you        80
  Upon his good returns.
  If by this crime he owes the law his life,
  Why, let the war receive't in valiant gore,
  For law is strict, and war is nothing more.
FIRST SENATOR
  We are for law; he dies. Urge it no more,           85
  On height of our displeasure. Friend or brother,
  He forfeits his own blood that spills another.

ALCIBIADES
  Must it be so? It must not be.
  My lords, I do beseech you know me.
SECOND SENATOR                    How?
ALCIBIADES
  Call me to your remembrances.
THIRD SENATOR                    What?                    90
ALCIBIADES
  I cannot think but your age has forgot me.
  It could not else be I should prove so base
  To sue and be denied such common grace.
  My wounds ache at you.
FIRST SENATOR                    Do you dare our anger?
  'Tis in few words, but spacious in effect:                    95
  We banish thee for ever.
ALCIBIADES                    Banish me?
  Banish your dotage, banish usury
  That makes the senate ugly.
FIRST SENATOR                    If after two days' shine
  Athens contain thee, attend our weightier judgement;
  And, not to swell your spirit, he shall be                    100
  Executed presently.                    *Exeunt Senators ⌈and attendants⌉*
ALCIBIADES
  Now the gods keep you old enough that you may live
  Only in bone, that none may look on you!
  I'm worse than mad. I have kept back their foes
  While they have told their money and let out                    105
  Their coin upon large interest—I myself,
  Rich only in large hurts. All those for this?
  Is this the balsam that the usuring senate
  Pours into captains' wounds? Banishment!
  It comes not ill; I hate not to be banished.                    110
  It is a cause worthy my spleen and fury,
  That I may strike at Athens. I'll cheer up
  My discontented troops, and lay for hearts.
  'Tis honour with most lands to be at odds.
  Soldiers should brook as little wrongs as gods.                    *Exit*

**3.7**    *Enter divers of Timon's friends, ⌈amongst them*
    *Lucullus, Lucius, Sempronius, and other Lords and*
    *Senators,⌉ at several doors*
FIRST LORD The good time of day to you, sir.
SECOND LORD I also wish it to you. I think this honourable
  lord did but try us this other day.
FIRST LORD Upon that were my thoughts tiring when we
  encountered. I hope it is not so low with him as he
  made it seem in the trial of his several friends.                    6
SECOND LORD It should not be, by the persuasion of his
  new feasting.
FIRST LORD I should think so. He hath sent me an earnest
  inviting, which many my near occasions did urge me
  to put off, but he hath conjured me beyond them, and
  I must needs appear.                    12
SECOND LORD In like manner was I in debt to my
  importunate business, but he would not hear my
  excuse. I am sorry when he sent to borrow of me that
  my provision was out.                    16
FIRST LORD I am sick of that grief too, as I understand
  how all things go.
SECOND LORD Every man hears so. What would he have
  borrowed of you?                    20
FIRST LORD A thousand pieces.
SECOND LORD A thousand pieces?
FIRST LORD What of you?
SECOND LORD He sent to me, sir—

⌈*Loud music.*⌉ *Enter Timon and attendants*
Here he comes.                    25
TIMON With all my heart, gentlemen both; and how fare
  you?
FIRST LORD Ever at the best, hearing well of your lordship.
SECOND LORD The swallow follows not summer more
  willing than we your lordship.                    30
TIMON (*aside*) Nor more willingly leaves winter, such
  summer birds are men.—Gentlemen, our dinner will
  not recompense this long stay. Feast your ears with
  the music a while, if they will fare so harshly o'th'
  trumpets' sound; we shall to't presently.                    35
FIRST LORD I hope it remains not unkindly with your
  lordship that I returned you an empty messenger.
TIMON O sir, let it not trouble you.
SECOND LORD My noble lord—
TIMON Ah, my good friend, what cheer?                    40
    ⌈*A table and stools are⌉ brought in*
SECOND LORD My most honourable lord, I am e'en sick of
  shame that when your lordship this other day sent to
  me I was so unfortunate a beggar.
TIMON Think not on't, sir.
SECOND LORD If you had sent but two hours before—                    45
TIMON Let it not cumber your better remembrance.—
  Come, bring in all together.
    ⌈*Enter Servants with covered dishes⌉*
SECOND LORD All covered dishes.
FIRST LORD Royal cheer, I warrant you.
THIRD LORD Doubt not that, if money and the season can
  yield it.                    51
FIRST LORD How do you? What's the news?
THIRD LORD Alcibiades is banished. Hear you of it?
FIRST *and* SECOND LORDS Alcibiades banished?
THIRD LORD 'Tis so, be sure of it.                    55
FIRST LORD How, how?
SECOND LORD I pray you, upon what?
TIMON My worthy friends, will you draw near?
THIRD LORD I'll tell you more anon. Here's a noble feast
  toward.                    60
SECOND LORD This is the old man still.
THIRD LORD Will't hold, will't hold?
SECOND LORD It does; but time will—and so—
THIRD LORD I do conceive.                    64
TIMON Each man to his stool with that spur as he would
  to the lip of his mistress. Your diet shall be in all places
  alike. Make not a city feast of it, to let the meat cool
  ere we can agree upon the first place. Sit, sit. The gods
  require our thanks.                    69
    *They sit*
You great benefactors, sprinkle our society with
thankfulness. For your own gifts make yourselves
praised; but reserve still to give, lest your deities be
despised. Lend to each man enough that one need not
lend to another; for were your godheads to borrow of
men, men would forsake the gods. Make the meat be
beloved more than the man that gives it. Let no
assembly of twenty be without a score of villains. If
there sit twelve women at the table, let a dozen of them
be as they are. The rest of your foes, O gods—the
senators of Athens, together with the common tag of
people—what is amiss in them, you gods, make suitable
for destruction. For these my present friends, as they
are to me nothing, so in nothing bless them; and to
nothing are they welcome.—Uncover, dogs, and lap.
    *The dishes are uncovered, and seen to be full of*
    *steaming water ⌈and stones⌉*

SOME LORDS  What does his lordship mean?          85
OTHER LORDS  I know not.
TIMON
  May you a better feast never behold,
  You knot of mouth-friends. Smoke and lukewarm water
  Is your perfection. This is Timon's last,
  Who, stuck and spangled with your flattery,          90
  Washes it off, and sprinkles in your faces
  Your reeking villainy.
      ⌈He throws water in their faces⌉
                        Live loathed and long,
  Most smiling, smooth, detested parasites,
  Courteous destroyers, affable wolves, meek bears,
  You fools of fortune, trencher-friends, time's flies,          95
  Cap-and-knee slaves, vapours, and minute-jacks!
  Of man and beast the infinite malady
  Crust you quite o'er.
      ⌈A Lord is going⌉
                        What, dost thou go?
  Soft, take thy physic first. Thou too, and thou.
      ⌈He beats them⌉
  Stay, I will lend thee money, borrow none.          100
      Exeunt Lords, leaving caps and gowns
  What, all in motion? Henceforth be no feast
  Whereat a villain's not a welcome guest.
  Burn house! Sink Athens! Henceforth hated be
  Of Timon man and all humanity!          Exit
      Enter the Senators and other Lords
FIRST LORD  How now, my lords?          105
SECOND LORD
  Know you the quality of Lord Timon's fury?
THIRD LORD
  Push! Did you see my cap?
FOURTH LORD                    I have lost my gown.
FIRST LORD  He's but a mad lord, and naught but humours
  sways him. He gave me a jewel th'other day, and now
  he has beat it out of my hat.          110
  Did you see my jewel?
⌈THIRD⌉ LORD          Did you see my cap?
⌈SECOND⌉ LORD
  Here 'tis.
FOURTH LORD  Here lies my gown.
FIRST LORD                    Let's make no stay.
SECOND LORD
  Lord Timon's mad.
THIRD LORD          I feel't upon my bones.
FOURTH LORD
  One day he gives us diamonds, next day stones.
                                        Exeunt

4.1  *Enter Timon*
TIMON
  Let me look back upon thee. O thou wall
  That girdles in those wolves, dive in the earth,
  And fence not Athens! Matrons, turn incontinent!
  Obedience fail in children! Slaves and fools,
  Pluck the grave wrinkled senate from the bench          5
  And minister in their steads! To general filths
  Convert o'th' instant, green virginity!
  Do't in your parents' eyes. Bankrupts, hold fast!
  Rather than render back, out with your knives,
  And cut your trusters' throats. Bound servants, steal!
  Large-handed robbers your grave masters are,          11
  And pill by law. Maid, to thy master's bed!
  Thy mistress is o'th' brothel. Son of sixteen,
  Pluck the lined crutch from thy old limping sire;

With it beat out his brains! Piety and fear,          15
  Religion to the gods, peace, justice, truth,
  Domestic awe, night rest, and neighbourhood,
  Instruction, manners, mysteries, and trades,
  Degrees, observances, customs, and laws,
  Decline to your confounding contraries,          20
  And let confusion live! Plagues incident to men,
  Your potent and infectious fevers heap
  On Athens, ripe for stroke! Thou cold sciatica,
  Cripple our senators, that their limbs may halt
  As lamely as their manners! Lust and liberty,          25
  Creep in the minds and marrows of our youth,
  That 'gainst the stream of virtue they may strive
  And drown themselves in riot! Itches, blains,
  Sow all th'Athenian bosoms, and their crop
  Be general leprosy! Breath infect breath,          30
  That their society, as their friendship, may
  Be merely poison!
      ⌈He tears off his clothes⌉
                  Nothing I'll bear from thee
  But nakedness, thou detestable town;
  Take thou that too, with multiplying bans.
  Timon will to the woods, where he shall find          35
  Th'unkindest beast more kinder than mankind.
  The gods confound—hear me you good gods all—
  Th'Athenians, both within and out that wall;
  And grant, as Timon grows, his hate may grow
  To the whole race of mankind, high and low.          40
  Amen.          Exit

4.2    *Enter Flavius, with two or three Servants*
FIRST SERVANT
  Hear you, master steward, where's our master?
  Are we undone, cast off, nothing remaining?
FLAVIUS
  Alack, my fellows, what should I say to you?
  Let me be recorded: by the righteous gods,
  I am as poor as you.
FIRST SERVANT          Such a house broke,          5
  So noble a master fall'n? All gone, and not
  One friend to take his fortune by the arm
  And go along with him?
SECOND SERVANT          As we do turn our backs
  From our companion thrown into his grave,
  So his familiars to his buried fortunes          10
  Slink all away, leave their false vows with him
  Like empty purses picked; and his poor self,
  A dedicated beggar to the air,
  With his disease of all-shunned poverty,
  Walks like contempt alone.
      *Enter other Servants*
                        More of our fellows.          15
FLAVIUS
  All broken implements of a ruined house.
THIRD SERVANT
  Yet do our hearts wear Timon's livery.
  That see I by our faces. We are fellows still,
  Serving alike in sorrow. Leaked is our barque,
  And we, poor mates, stand on the dying deck          20
  Hearing the surges' threat. We must all part
  Into this sea of air.
FLAVIUS          Good fellows all,
  The latest of my wealth I'll share amongst you.
  Wherever we shall meet, for Timon's sake
  Let's yet be fellows. Let's shake our heads and say,          25

As 'twere a knell unto our master's fortunes,
'We have seen better days.'
   *He gives them money*
                   Let each take some.
Nay, put out all your hands. Not one word more.
Thus part we rich in sorrow, parting poor.
   *They embrace, and the Servants part several ways*
O, the fierce wretchedness that glory brings us!   30
Who would not wish to be from wealth exempt,
Since riches point to misery and contempt?
Who would be so mocked with glory, or to live
But in a dream of friendship,
To have his pomp and all what state compounds   35
But only painted like his varnished friends?
Poor honest lord, brought low by his own heart,
Undone by goodness! Strange, unusual blood
When man's worst sin is he does too much good!
Who then dares to be half so kind again?   40
For bounty, that makes gods, does still mar men.
My dearest lord, blessed to be most accursed,
Rich only to be wretched, thy great fortunes
Are made thy chief afflictions. Alas, kind lord!
He's flung in rage from this ingrateful seat   45
Of monstrous friends;
Nor has he with him to supply his life,
Or that which can command it.
I'll follow and enquire him out.
I'll ever serve his mind with my best will.   50
Whilst I have gold I'll be his steward still.   *Exit*

**4.3**   *Enter Timon ⌈from his cave⌉ in the woods, ⌈half*
       *naked, and with a spade⌉*

TIMON
O blessèd breeding sun, draw from the earth
Rotten humidity; below thy sister's orb
Infect the air. Twinned brothers of one womb,
Whose procreation, residence, and birth
Scarce is dividant, touch them with several fortunes,
The greater scorns the lesser. Not nature,   6
To whom all sores lay siege, can bear great fortune
But by contempt of nature.
It is the pasture lards the brother's sides,
The want that makes him lean.   10
Raise me this beggar and demit that lord,
The senator shall bear contempt hereditary,
The beggar native honour. Who dares, who dares
In purity of manhood stand upright
And say 'This man's a flatterer'? If one be,   15
So are they all, for every grece of fortune
Is smoothed by that below. The learnèd pate
Ducks to the golden fool. All's obliquy;
There's nothing level in our cursèd natures
But direct villainy. Therefore be abhorred   20
All feasts, societies, and throngs of men.
His semblable, yea, himself, Timon disdains.
Destruction fang mankind. Earth, yield me roots.
   *He digs*
Who seeks for better of thee, sauce his palate
With thy most operant poison.
   *He finds gold*
                  What is here?   25
Gold? Yellow, glittering, precious gold?
No, gods, I am no idle votarist:
Roots, you clear heavens. Thus much of this will
   make
Black white, foul fair, wrong right,

Base noble, old young, coward valiant.   30
Ha, you gods! Why this, what, this, you gods? Why,
this
Will lug your priests and servants from your sides,
Pluck stout men's pillows from below their heads.
This yellow slave
Will knit and break religions, bless th'accursed,   35
Make the hoar leprosy adored, place thieves,
And give them title, knee, and approbation
With senators on the bench. This is it
That makes the wappered widow wed again.
She whom the spittle house and ulcerous sores   40
Would cast the gorge at, this embalms and spices
To th' April day again. Come, damnèd earth,
Thou common whore of mankind, that puts odds
Among the rout of nations; I will make thee
Do thy right nature.
   *March afar off*
             Ha, a drum! Thou'rt quick;   45
But yet I'll bury thee.
   *He buries gold*
              Thou'lt go, strong thief,
When gouty keepers of thee cannot stand.
   *He keeps some gold*
Nay, stay thou out for earnest.
   *Enter Alcibiades, with soldiers playing drum and*
   *fife, in warlike manner; and Phrynia and Timandra*
ALCIBIADES           What art thou there? Speak.
TIMON
A beast, as thou art. The canker gnaw thy heart
For showing me again the eyes of man.   50
ALCIBIADES
What is thy name? Is man so hateful to thee
That art thyself a man?
TIMON
I am Misanthropos, and hate mankind.
For thy part, I do wish thou wert a dog,
That I might love thee something.
ALCIBIADES           I know thee well,
But in thy fortunes am unlearned and strange.   56
TIMON
I know thee too, and more than that I know thee
I not desire to know. Follow thy drum.
With man's blood paint the ground gules, gules.
Religious canons, civil laws, are cruel;   60
Then what should war be? This fell whore of thine
Hath in her more destruction than thy sword,
For all her cherubin look.
PHRYNIA            Thy lips rot off!
TIMON
I will not kiss thee; then the rot returns
To thine own lips again.   65
ALCIBIADES
How came the noble Timon to this change?
TIMON
As the moon does, by wanting light to give.
But then renew I could not like the moon;
There were no suns to borrow of.
ALCIBIADES
Noble Timon, what friendship may I do thee?   70
TIMON
None but to maintain my opinion.
ALCIBIADES What is it, Timon?
TIMON Promise me friendship, but perform none. If thou
wilt promise, the gods plague thee, for thou art a man. If
thou dost not perform, confound thee, for thou art a
man.   76

ALCIBIADES
  I have heard in some sort of thy miseries.
TIMON
  Thou saw'st them when I had prosperity.
ALCIBIADES
  I see them now; then was a blessèd time.
TIMON
  As thine is now, held with a brace of harlots.    80
TIMANDRA
  Is this th'Athenian minion, whom the world
  Voiced so regardfully?
TIMON               Art thou Timandra?
TIMANDRA   Yes.
TIMON
  Be a whore still. They love thee not that use thee.
  Give them diseases, leaving with thee their lust.    85
  Make use of thy salt hours: season the slaves
  For tubs and baths, bring down rose-cheeked youth
  To the tub-fast and the diet.
TIMANDRA           Hang thee, monster!
ALCIBIADES
  Pardon him, sweet Timandra, for his wits
  Are drowned and lost in his calamities.    90
  I have but little gold of late, brave Timon,
  The want whereof doth daily make revolt
  In my penurious band. I have heard and grieved
  How cursèd Athens, mindless of thy worth,
  Forgetting thy great deeds, when neighbour states    95
  But for thy sword and fortune trod upon them—
TIMON
  I prithee, beat thy drum and get thee gone.
ALCIBIADES
  I am thy friend, and pity thee, dear Timon.
TIMON
  How dost thou pity him whom thou dost trouble?
  I had rather be alone.
ALCIBIADES         Why, fare thee well.    100
  Here is some gold for thee.
TIMON            Keep it. I cannot eat it.
ALCIBIADES
  When I have laid proud Athens on a heap—
TIMON
  Warr'st thou 'gainst Athens?
ALCIBIADES         Ay, Timon, and have cause.
TIMON
  The gods confound them all in thy conquest,
  And thee after, when thou hast conquerèd.    105
ALCIBIADES
  Why me, Timon?
TIMON           That by killing of villains
  Thou wast born to conquer my country.
  Put up thy gold.
       *He gives Alcibiades gold*
           Go on; here's gold; go on.
  Be as a planetary plague when Jove
  Will o'er some high-viced city hang his poison    110
  In the sick air. Let not thy sword skip one.
  Pity not honoured age for his white beard;
  He is an usurer. Strike me the counterfeit matron;
  It is her habit only that is honest,
  Herself's a bawd. Let not the virgin's cheek    115
  Make soft thy trenchant sword; for those milk paps
  That through the window-bars bore at men's eyes
  Are not within the leaf of pity writ;
  But set them down horrible traitors. Spare not the
    babe

  Whose dimpled smiles from fools exhaust their mercy.
  Think it a bastard whom the oracle    121
  Hath doubtfully pronounced thy throat shall cut,
  And mince it sans remorse. Swear against objects.
  Put armour on thine ears and on thine eyes
  Whose proof nor yells of mothers, maids, nor babes,
  Nor sight of priests in holy vestments bleeding,    126
  Shall pierce a jot. There's gold to pay thy soldiers.
  Make large confusion, and, thy fury spent,
  Confounded be thyself. Speak not. Be gone.
ALCIBIADES
  Hast thou gold yet? I'll take the gold thou giv'st me,
  Not all thy counsel.    131
TIMON
  Dost thou or dost thou not, heaven's curse upon thee!
PHRYNIA *and* TIMANDRA
  Give us some gold, good Timon. Hast thou more?
TIMON
  Enough to make a whore forswear her trade,
  And to make wholesomeness a bawd. Hold up, you
    sluts,    135
  Your aprons mountant.
       ⌐*He throws gold into their aprons*⌐
              You are not oathable,
  Although I know you'll swear, terribly swear,
  Into strong shudders and to heavenly agues
  Th'immortal gods that hear you. Spare your oaths;
  I'll trust to your conditions. Be whores still,    140
  And he whose pious breath seeks to convert you,
  Be strong in whore, allure him, burn him up.
  Let your close fire predominate his smoke;
  And be no turncoats. Yet may your pain-sick months
  Be quite contrary, and thatch your poor thin roofs    145
  With burdens of the dead—some that were hanged,
  No matter. Wear them, betray with them; whore still;
  Paint till a horse may mire upon your face.
  A pox of wrinkles!
PHRYNIA *and* TIMANDRA   Well, more gold; what then?
  Believe't that we'll do anything for gold.    150
TIMON   Consumptions sow
  In hollow bones of man, strike their sharp shins,
  And mar men's spurring. Crack the lawyer's voice,
  That he may never more false title plead
  Nor sound his quillets shrilly. Hoar the flamen    155
  That scolds against the quality of flesh
  And not believes himself. Down with the nose,
  Down with it flat; take the bridge quite away
  Of him that his particular to foresee
  Smells from the general weal. Make curled-pate
    ruffians bald,    160
  And let the unscarred braggarts of the war
  Derive some pain from you. Plague all,
  That your activity may defeat and quell
  The source of all erection. There's more gold.
  Do you damn others, and let this damn you;    165
  And ditches grave you all!
PHRYNIA *and* TIMANDRA
  More counsel with more money, bounteous Timon.
TIMON
  More whore, more mischief first; I have given you
    earnest.
ALCIBIADES
  Strike up the drum towards Athens. Farewell, Timon.
  If I thrive well, I'll visit thee again.    170
TIMON
  If I hope well, I'll never see thee more.

ALCIBIADES I never did thee harm.
TIMON Yes, thou spok'st well of me.
ALCIBIADES Call'st thou that harm?
TIMON
Men daily find it. Get thee away,                                         175
And take thy beagles with thee.
ALCIBIADES                           We but offend him. Strike!
          *Exeunt ⌐to drum and fife⌐ all but Timon*
TIMON
That nature, being sick of man's unkindness,
Should yet be hungry!
          *He digs the earth*
                              Common mother—thou
Whose womb unmeasurable and infinite breast
Teems and feeds all, whose selfsame mettle             180
Whereof thy proud child, arrogant man, is puffed
Engenders the black toad and adder blue,
The gilded newt and eyeless venomed worm,
With all th'abhorrèd births below crisp heaven
Whereon Hyperion's quick'ning fire doth shine—        185
Yield him who all thy human sons do hate
From forth thy plenteous bosom, one poor root.
Ensear thy fertile and conceptious womb;
Let it no more bring out ingrateful man.
Go great with tigers, dragons, wolves, and bears;     190
Teem with new monsters whom thy upward face
Hath to the marbled mansion all above
Never presented.
          *He finds a root*
                              O, a root! Dear thanks.
Dry up thy marrows, vines, and plough-torn leas,
Whereof ingrateful man with liquorish draughts        195
And morsels unctuous greases his pure mind,
That from it all consideration slips!—
          *Enter Apemantus*
More man? Plague, plague!
APEMANTUS
I was directed hither. Men report
Thou dost affect my manners, and dost use them.       200
TIMON
'Tis then because thou dost not keep a dog
Whom I would imitate. Consumption catch thee!
APEMANTUS
This is in thee a nature but infected,
A poor unmanly melancholy, sprung
From change of fortune. Why this spade, this place,
This slave-like habit, and these looks of care?       206
Thy flatterers yet wear silk, drink wine, lie soft,
Hug their diseased perfumes, and have forgot
That ever Timon was. Shame not these woods
By putting on the cunning of a carper.                210
Be thou a flatterer now, and seek to thrive
By that which has undone thee. Hinge thy knee,
And let his very breath whom thou'lt observe
Blow off thy cap. Praise his most vicious strain,
And call it excellent. Thou wast told thus.           215
Thou gav'st thine ears like tapsters that bade welcome
To knaves and all approachers. 'Tis most just
That thou turn rascal. Hadst thou wealth again,
Rascals should have't. Do not assume my likeness.
TIMON
Were I like thee, I'd throw away myself.              220
APEMANTUS
Thou hast cast away thyself being like thyself—
A madman so long, now a fool. What, think'st
That the bleak air, thy boisterous chamberlain,
Will put thy shirt on warm? Will these mossed trees

That have outlived the eagle page thy heels           225
And skip when thou point'st out? Will the cold brook,
Candied with ice, caudle thy morning taste
To cure thy o'ernight's surfeit? Call the creatures
Whose naked natures live in all the spite
Of wreakful heaven, whose bare unhousèd trunks        230
To the conflicting elements exposed
Answer mere nature; bid them flatter thee.
O, thou shalt find—
TIMON                           A fool of thee! Depart.
APEMANTUS
I love thee better now than e'er I did.
TIMON
I hate thee worse.
APEMANTUS                  Why?
TIMON                                  Thou flatter'st misery.       235
APEMANTUS
I flatter not, but say thou art a caitiff.
TIMON
Why dost thou seek me out?
APEMANTUS                          To vex thee.
TIMON
Always a villain's office, or a fool's.
Dost please thyself in't?
APEMANTUS                      Ay.
TIMON                                  What, a knave too?
APEMANTUS
If thou didst put this sour cold habit on             240
To castigate thy pride, 'twere well; but thou
Dost it enforcèdly. Thou'dst courtier be again
Wert thou not beggar. Willing misery
Outlives incertain pomp, is crowned before.
The one is filling still, never complete;             245
The other at high wish. Best state, contentless,
Hath a distracted and most wretched being,
Worse than the worst, content.
Thou shouldst desire to die, being miserable.
TIMON
Not by his breath that is more miserable.             250
Thou art a slave whom fortune's tender arm
With favour never clasped, but bred a dog.
Hadst thou like us from our first swathe proceeded
The sweet degrees that this brief world affords
To such as may the passive drudges of it              255
Freely command, thou wouldst have plunged thyself
In general riot, melted down thy youth
In different beds of lust, and never learned
The icy precepts of respect, but followed
The sugared game before thee. But myself,             260
Who had the world as my confectionary,
The mouths, the tongues, the eyes and hearts of men
At duty, more than I could frame employment,
That numberless upon me stuck, as leaves
Do on the oak, have with one winter's brush           265
Fell from their boughs, and left me open, bare
For every storm that blows—I to bear this,
That never knew but better, is some burden.
Thy nature did commence in sufferance, time
Hath made thee hard in't. Why shouldst thou hate men?
They never flattered thee. What hast thou given?      271
If thou wilt curse, thy father, that poor rag,
Must be thy subject, who in spite put stuff
To some she-beggar and compounded thee
Poor rogue hereditary. Hence, be gone.                275
If thou hadst not been born the worst of men
Thou hadst been a knave and flatterer.

901

APEMANTUS Art thou proud yet?

TIMON Ay, that I am not thee.

APEMANTUS I that I was                                    280
No prodigal.

TIMON          I that I am one now.
Were all the wealth I have shut up in thee
I'd give thee leave to hang it. Get thee gone.
That the whole life of Athens were in this!
Thus would I eat it.
        *He bites the root*

APEMANTUS ⌈*offering food*⌉ Here, I will mend thy feast.

TIMON
First mend my company: take away thyself.        286

APEMANTUS
So I shall mend mine own by th' lack of thine.

TIMON
'Tis not well mended so, it is but botched;
If not, I would it were.

APEMANTUS          What wouldst thou have to Athens?

TIMON
Thee thither in a whirlwind. If thou wilt,        290
Tell them there I have gold. Look, so I have.

APEMANTUS
Here is no use for gold.

TIMON                    The best and truest,
For here it sleeps and does no hirèd harm.

APEMANTUS Where liest a-nights, Timon?

TIMON Under that's above me. Where feed'st thou a-days,
Apemantus?                                        296

APEMANTUS Where my stomach finds meat; or rather,
where I eat it.

TIMON Would poison were obedient, and knew my mind!

APEMANTUS Where wouldst thou send it?             300

TIMON To sauce thy dishes.

APEMANTUS The middle of humanity thou never knewest,
but the extremity of both ends. When thou wast in thy
gilt and thy perfume, they mocked thee for too much
curiosity; in thy rags thou know'st none, but art
despised for the contrary. There's a medlar for thee;
eat it.                                           307

TIMON On what I hate I feed not.

APEMANTUS Dost hate a medlar?

TIMON Ay, though it look like thee.              310

APEMANTUS An thou'dst hated meddlers sooner, thou
shouldst have loved thyself better now. What man didst
thou ever know unthrift that was beloved after his
means?

TIMON Who, without those means thou talk'st of, didst
thou ever know beloved?                           316

APEMANTUS Myself.

TIMON I understand thee: thou hadst some means to keep
a dog.

APEMANTUS What things in the world canst thou nearest
compare to thy flatterers?                        321

TIMON Women nearest; but men, men are the things
themselves. What wouldst thou do with the world,
Apemantus, if it lay in thy power?

APEMANTUS Give it the beasts, to be rid of the men.   325

TIMON Wouldst thou have thyself fall in the confusion of
men, and remain a beast with the beasts?

APEMANTUS Ay, Timon.

TIMON A beastly ambition, which the gods grant thee
t'attain to. If thou wert the lion, the fox would beguile
thee. If thou wert the lamb, the fox would eat thee. If
thou wert the fox, the lion would suspect thee when

peradventure thou wert accused by the ass. If thou
wert the ass, thy dullness would torment thee, and still
thou lived'st but as a breakfast to the wolf. If thou wert
the wolf, thy greediness would afflict thee, and oft thou
shouldst hazard thy life for thy dinner. Wert thou the
unicorn, pride and wrath would confound thee, and
make thine own self the conquest of thy fury. Wert
thou a bear, thou wouldst be killed by the horse. Wert
thou a horse, thou wouldst be seized by the leopard.
Wert thou a leopard, thou wert german to the lion,
and the spots of thy kindred were jurors on thy life;
all thy safety were remotion, and thy defence absence.
What beast couldst thou be that were not subject to a
beast? And what a beast art thou already, that seest
not thy loss in transformation!                   347

APEMANTUS If thou couldst please me with speaking to
me, thou mightst have hit upon it here. The commonwealth of Athens is become a forest of beasts.    350

TIMON How, has the ass broke the wall, that thou art out
of the city?

APEMANTUS Yonder comes a poet and a painter. The
plague of company light upon thee! I will fear to catch
it, and give way. When I know not what else to do,
I'll see thee again.                              356

TIMON When there is nothing living but thee, thou shalt
be welcome. I had rather be a beggar's dog than
Apemantus.

APEMANTUS
Thou art the cap of all the fools alive.          360

TIMON
Would thou wert clean enough to spit upon.

APEMANTUS
A plague on thee! Thou art too bad to curse.

TIMON
All villains that do stand by thee are pure.

APEMANTUS
There is no leprosy but what thou speak'st.

TIMON If I name thee.                             365
I'd beat thee, but I should infect my hands.

APEMANTUS
I would my tongue could rot them off.

TIMON
Away, thou issue of a mangy dog!
Choler does kill me that thou art alive.
I swoon to see thee.                              370

APEMANTUS Would thou wouldst burst!

TIMON Away, thou tedious rogue!
        ⌈*He throws a stone at Apemantus*⌉
I am sorry I shall lose a stone by thee.

APEMANTUS Beast!

TIMON Slave!                                      375

APEMANTUS Toad!

TIMON Rogue, rogue, rogue!
I am sick of this false world, and will love naught
But even the mere necessities upon't.
Then, Timon, presently prepare thy grave.         380
Lie where the light foam of the sea may beat
Thy gravestone daily. Make thine epitaph,
That death in me at others' lives may laugh.
        *He looks on the gold*
O, thou sweet king-killer, and dear divorce
'Twixt natural son and sire; thou bright defiler  385
Of Hymen's purest bed; thou valiant Mars;
Thou ever young, fresh, loved, and delicate wooer,
Whose blush doth thaw the consecrated snow
That lies on Dian's lap; thou visible god,

That sold'rest close impossibilities                    390
And mak'st them kiss, that speak'st with every tongue
To every purpose; O thou touch of hearts:
Think thy slave man rebels, and by thy virtue
Set them into confounding odds, that beasts
May have the world in empire.
APEMANTUS                        Would 'twere so,    395
But not till I am dead. I'll say thou'st gold.
Thou wilt be thronged to shortly.
TIMON                        Thronged to?
APEMANTUS                                Ay.
TIMON
Thy back, I prithee.
APEMANTUS            Live, and love thy misery.
TIMON
Long live so, and so die. I am quit.
        *Enter the Banditti, thieves*
APEMANTUS
More things like men. Eat, Timon, and abhor them.
                                    *Exit*
FIRST THIEF Where should he have this gold? It is some
poor fragment, some slender ort of his remainder. The
mere want of gold and the falling-from of his friends
drove him into this melancholy.            404
SECOND THIEF It is noised he hath a mass of treasure.
THIRD THIEF Let us make the assay upon him. If he care
not for't, he will supply us easily. If he covetously
reserve it, how shall 's get it?
SECOND THIEF True, for he bears it not about him; 'tis hid.
FIRST THIEF Is not this he?                410
OTHER THIEVES Where?
SECOND THIEF 'Tis his description.
THIRD THIEF He, I know him.
ALL THIEVES (*coming forward*) Save thee, Timon.
TIMON Now, thieves.                        415
ALL THIEVES
Soldiers, not thieves.
TIMON                Both, too, and women's sons.
ALL THIEVES
We are not thieves, but men that much do want.
TIMON
Your greatest want is, you want much of meat.
Why should you want? Behold, the earth hath roots.
Within this mile break forth a hundred springs.    420
The oaks bear mast, the briars scarlet hips.
The bounteous housewife nature on each bush
Lays her full mess before you. Want? Why want?
FIRST THIEF
We cannot live on grass, on berries, water,
As beasts and birds and fishes.            425
TIMON
Nor on the beasts themselves, the birds and fishes;
You must eat men. Yet thanks I must you con
That you are thieves professed, that you work not
In holier shapes; for there is boundless theft
In limited professions. (*Giving gold*) Rascal thieves,    430
Here's gold. Go suck the subtle blood o'th' grape
Till the high fever seethe your blood to froth,
And so scape hanging. Trust not the physician:
His antidotes are poison, and he slays
More than you rob. Take wealth and lives together.
Do villainy; do, since you protest to do't,    436
Like workmen. I'll example you with thievery.
The sun's a thief, and with his great attraction
Robs the vast sea. The moon's an arrant thief,
And her pale fire she snatches from the sun.    440

The sea's a thief, whose liquid surge resolves
The moon into salt tears. The earth's a thief,
That feeds and breeds by a composture stol'n
From gen'ral excrement. Each thing's a thief.
The laws, your curb and whip, in their rough power
Has unchecked theft. Love not yourselves. Away,    446
Rob one another. There's more gold. Cut throats;
All that you meet are thieves. To Athens go,
Break open shops; nothing can you steal
But thieves do lose it. Steal no less for this I give you,
And gold confound you howsoe'er. Amen.    451
THIRD THIEF He's almost charmed me from my profession
by persuading me to it.
FIRST THIEF 'Tis in the malice of mankind that he thus
advises us, not to have us thrive in our mystery.    455
SECOND THIEF I'll believe him as an enemy, and give over
my trade.
FIRST THIEF Let us first see peace in Athens. There is no
time so miserable but a man may be true.
                            *Exeunt Thieves*
        *Enter Flavius to Timon*
FLAVIUS O you gods!                        460
Is yon despised and ruinous man my lord,
Full of decay and failing? O monument
And wonder of good deeds evilly bestowed!
What an alteration of honour has desp'rate want made!
What viler thing upon the earth than friends,    465
Who can bring noblest minds to basest ends!
How rarely does it meet with this time's guise,
When man was wished to love his enemies!
Grant I may ever love and rather woo
Those that would mischief me than those that do!    470
        *Timon sees him*
He's caught me in his eye. I will present
My honest grief unto him, and as my lord
Still serve him with my life.—My dearest master.
TIMON
Away! What art thou?
FLAVIUS                Have you forgot me, sir?
TIMON
Why dost ask that? I have forgot all men;    475
Then if thou grant'st thou'rt man, I have forgot thee.
FLAVIUS An honest poor servant of yours.
TIMON
Then I know thee not. I never had
Honest man about me; ay, all I kept were knaves,
To serve in meat to villains.
FLAVIUS                The gods are witness,    480
Ne'er did poor steward wear a truer grief
For his undone lord than mine eyes for you.
TIMON
What, dost thou weep? Come nearer then; I love thee
Because thou art a woman, and disclaim'st
Flinty mankind whose eyes do never give    485
But thorough lust and laughter. Pity's sleeping.
Strange times, that weep with laughing, not with
weeping!
FLAVIUS
I beg of you to know me, good my lord,
T'accept my grief,
        ⌈*He offers his money*⌉
                    and whilst this poor wealth lasts
To entertain me as your steward still.    490
TIMON Had I a steward
So true, so just, and now so comfortable?
It almost turns my dangerous nature mild.

Let me behold thy face. Surely this man
Was born of woman. 495
Forgive my general and exceptless rashness,
You perpetual sober gods! I do proclaim
One honest man—mistake me not, but one,
No more, I pray—and he's a steward.
How fain would I have hated all mankind, 500
And thou redeem'st thyself! But all save thee
I fell with curses.
Methinks thou art more honest now than wise,
For by oppressing and betraying me
Thou mightst have sooner got another service; 505
For many so arrive at second masters
Upon their first lord's neck. But tell me true—
For I must ever doubt, though ne'er so sure—
Is not thy kindness subtle, covetous,
A usuring kindness, and, as rich men deal gifts, 510
Expecting in return twenty for one?

FLAVIUS
No, my most worthy master, in whose breast
Doubt and suspect, alas, are placed too late.
You should have feared false times when you did feast.
Suspect still comes where an estate is least. 515
That which I show, heaven knows, is merely love,
Duty and zeal to your unmatchèd mind,
Care of your food and living; and, believe it,
My most honoured lord,
For any benefit that points to me, 520
Either in hope or present, I'd exchange
For this one wish: that you had power and wealth
To requite me by making rich yourself.

TIMON
Look thee, 'tis so. Thou singly honest man,
⌈He gives Flavius gold⌉
Here, take. The gods, out of my misery, 525
Has sent thee treasure. Go, live rich and happy,
But thus conditioned: thou shalt build from men,
Hate all, curse all, show charity to none,
But let the famished flesh slide from the bone
Ere thou relieve the beggar. Give to dogs 530
What thou deniest to men. Let prisons swallow 'em,
Debts wither 'em to nothing; be men like blasted woods,
And may diseases lick up their false bloods.
And so farewell, and thrive.

FLAVIUS            O, let me stay
And comfort you, my master.

TIMON            If thou hat'st curses, 535
Stay not. Fly whilst thou art blest and free.
Ne'er see thou man, and let me ne'er see thee.
       Exeunt ⌈Timon into his cave, Flavius another way⌉

5.1    Enter Poet and Painter

PAINTER As I took note of the place, it cannot be far where
he abides.

POET What's to be thought of him? Does the rumour hold
for true that he's so full of gold? 4

PAINTER Certain. Alcibiades reports it. Phrynia and
Timandra had gold of him. He likewise enriched poor
straggling soldiers with great quantity. 'Tis said he gave
unto his steward a mighty sum.

POET Then this breaking of his has been but a try for his
friends? 10

PAINTER Nothing else. You shall see him a palm in Athens
again, and flourish with the highest. Therefore 'tis not
amiss we tender our loves to him in this supposed distress
of his. It will show honestly in us, and is very likely to
load our purposes with what they travail for, if it be a
just and true report that goes of his having. 16

POET What have you now to present unto him?

PAINTER Nothing at this time, but my visitation; only I
will promise him an excellent piece.

POET I must serve him so too, tell him of an intent that's
coming toward him. 21

PAINTER Good as the best.
       ⌈Enter Timon from his cave, unobserved⌉
Promising is the very air o'th' time; it opens the eyes of
expectation. Performance is ever the duller for his act,
and but in the plainer and simpler kind of people the
deed of saying is quite out of use. To promise is most
courtly and fashionable. Performance is a kind of will or
testament which argues a great sickness in his judgement
that makes it.

TIMON (aside) Excellent workman, thou canst not paint a
man so bad as is thyself. 31

POET (to Painter) I am thinking what I shall say I have
provided for him. It must be a personating of himself, a
satire against the softness of prosperity, with a discovery
of the infinite flatteries that follow youth and opulency.

TIMON (aside) Must thou needs stand for a villain in thine
own work? Wilt thou whip thine own faults in other
men? Do so; I have gold for thee.

POET (to Painter) Nay, let's seek him.
Then do we sin against our own estate 40
When we may profit meet and come too late.

PAINTER True.
When the day serves, before black-cornered night,
Find what thou want'st by free and offered light.
Come. 45

TIMON (aside)
I'll meet you at the turn. What a god's gold,
That he is worshipped in a baser temple
Than where swine feed!
'Tis thou that rigg'st the barque and plough'st the foam,
Settlest admirèd reverence in a slave. 50
To thee be worship, and thy saints for aye
Be crowned with plagues, that thee alone obey.
Fit I meet them.
       He comes forward to them

POET
Hail, worthy Timon!

PAINTER            Our late noble master!

TIMON
Have I once lived to see two honest men? 55

POET
Sir, having often of your open bounty tasted,
Hearing you were retired, your friends fall'n off,
Whose thankless natures, O abhorrèd spirits,
Not all the whips of heaven are large enough—
What, to you, 60
Whose star-like nobleness gave life and influence
To their whole being! I am rapt, and cannot cover
The monstrous bulk of this ingratitude
With any size of words.

TIMON
Let it go naked; men may see't the better. 65
You that are honest, by being what you are
Make them best seen and known.

PAINTER            He and myself
Have travelled in the great show'r of your gifts,
And sweetly felt it.

TIMON            Ay, you are honest men.

PAINTER
We are hither come to offer you our service. 70

TIMON
　Most honest men. Why, how shall I requite you?
　Can you eat roots and drink cold water? No.
POET *and* PAINTER
　What we can do we'll do to do you service.
TIMON
　You're honest men. You've heard that I have gold,
　I am sure you have. Speak truth; you're honest men.
PAINTER
　So it is said, my noble lord, but therefor　　　　76
　Came not my friend nor I.
TIMON
　Good honest men. (*To Painter*) Thou draw'st a
　　counterfeit
　Best in all Athens; thou'rt indeed the best;
　Thou counterfeit'st most lively.
PAINTER　　　　　　　　　　So so, my lord.　　80
TIMON
　E'en so, sir, as I say. (*To Poet*) And for thy fiction,
　Why, thy verse swells with stuff so fine and smooth
　That thou art even natural in thine art.
　But for all this, my honest-natured friends,
　I must needs say you have a little fault.　　　85
　Marry, 'tis not monstrous in you, neither wish I
　You take much pains to mend.
POET *and* PAINTER　　　　　　　Beseech your honour
　To make it known to us.
TIMON　　　　　　　　　You'll take it ill.
POET *and* PAINTER Most thankfully, my lord.
TIMON Will you indeed?　　　　　　　　　90
POET *and* PAINTER Doubt it not, worthy lord.
TIMON
　There's never a one of you but trusts a knave
　That mightily deceives you.
POET *and* PAINTER　　　　　　Do we, my lord?
TIMON
　Ay, and you hear him cog, see him dissemble,
　Know his gross patchery, love him, feed him,　　95
　Keep in your bosom; yet remain assured
　That he's a made-up villain.
PAINTER I know none such, my lord.
POET Nor I.
TIMON
　Look you, I love you well. I'll give you gold,　　100
　Rid me these villains from your companies.
　Hang them or stab them, drown them in a draught,
　Confound them by some course, and come to me,
　I'll give you gold enough.
POET *and* PAINTER
　　　　　　　Name them, my lord, let's know them.
TIMON
　You that way and you this—but two in company—
　Each man apart, all single and alone,　　　106
　Yet an arch-villain keeps him company.
　⌈*To Painter*⌉ If where thou art two villains shall not be,
　Come not near him. ⌈*To Poet*⌉ If thou wouldst not
　　reside
　But where one villain is, then him abandon.　　110
　Hence; pack! ⌈*Striking him*⌉ There's gold. You came
　　for gold, ye slaves.
　⌈*Striking Painter*⌉ You have work for me; there's
　　payment. Hence!
　⌈*Striking Poet*⌉ You are an alchemist; make gold of that.
　Out, rascal dogs!　　*Exeunt* ⌈*Poet and Painter one way,*
　　　　　　　　　　　　　*Timon into his cave*⌉

**5.2**　　*Enter Flavius and two Senators*
FLAVIUS
　It is in vain that you would speak with Timon,
　For he is set so only to himself
　That nothing but himself which looks like man
　Is friendly with him.
FIRST SENATOR　　　　　Bring us to his cave.
　It is our part and promise to th' Athenians　　5
　To speak with Timon.
SECOND SENATOR　　　At all times alike
　Men are not still the same. 'Twas time and griefs
　That framed him thus. Time with his fairer hand
　Offering the fortunes of his former days,
　The former man may make him. Bring us to him,　10
　And chance it as it may.
FLAVIUS　　　　　　　Here is his cave.
　(*Calling*) Peace and content be here! Lord Timon,
　　Timon,
　Look out and speak to friends. Th'Athenians
　By two of their most reverend senate greet thee.
　Speak to them, noble Timon.　　　　　　15
　　　　　*Enter Timon out of his cave*
TIMON
　Thou sun that comforts, burn! Speak and be hanged.
　For each true word a blister, and each false
　Be as a cantherizing to the root o'th' tongue,
　Consuming it with speaking.
FIRST SENATOR　　　　　　Worthy Timon—
TIMON
　Of none but such as you, and you of Timon.　　20
FIRST SENATOR
　The senators of Athens greet thee, Timon.
TIMON
　I thank them, and would send them back the plague
　Could I but catch it for them.
FIRST SENATOR　　　　　　O, forget
　What we are sorry for, ourselves in thee.
　The senators with one consent of love　　　25
　Entreat thee back to Athens, who have thought
　On special dignities which vacant lie
　For thy best use and wearing.
SECOND SENATOR　　　　　　They confess
　Toward thee forgetfulness too general-gross,
　Which now the public body, which doth seldom　30
　Play the recanter, feeling in itself
　A lack of Timon's aid, hath sense withal
　Of it own fail, restraining aid to Timon;
　And send forth us to make their sorrowed render,
　Together with a recompense more fruitful　　35
　Than their offence can weigh down by the dram;
　Ay, even such heaps and sums of love and wealth
　As shall to thee blot out what wrongs were theirs,
　And write in thee the figures of their love,
　Ever to read them thine.
TIMON　　　　　　　　You witch me in it,　　40
　Surprise me to the very brink of tears.
　Lend me a fool's heart and a woman's eyes,
　And I'll beweep these comforts, worthy senators.
FIRST SENATOR
　Therefore so please thee to return with us,
　And of our Athens, thine and ours, to take　　45
　The captainship, thou shalt be met with thanks,
　Allowed with absolute power, and thy good name
　Live with authority. So soon we shall drive back
　Of Alcibiades th'approaches wild,

Who, like a boar too savage, doth root up       50
  His country's peace.
SECOND SENATOR         And shakes his threat'ning sword
  Against the walls of Athens.
FIRST SENATOR            Therefore, Timon—
TIMON
  Well, sir, I will; therefore I will, sir, thus.
  If Alcibiades kill my countrymen,
  Let Alcibiades know this of Timon:         55
  That Timon cares not. But if he sack fair Athens,
  And take our goodly agèd men by th' beards,
  Giving our holy virgins to the stain
  Of contumelious, beastly, mad-brained war,
  Then let him know, and tell him Timon speaks it   60
  In pity of our agèd and our youth,
  I cannot choose but tell him that I care not;
  And—let him take't at worst—for their knives care
    not
  While you have throats to answer. For myself,
  There's not a whittle in th' unruly camp      65
  But I do prize it at my love before
  The reverend'st throat in Athens. So I leave you
  To the protection of the prosperous gods,
  As thieves to keepers.
FLAVIUS (to Senators)      Stay not; all's in vain.
TIMON
  Why, I was writing of my epitaph.        70
  It will be seen tomorrow. My long sickness
  Of health and living now begins to mend,
  And nothing brings me all things. Go; live still.
  Be Alcibiades your plague, you his,
  And last so long enough.
FIRST SENATOR         We speak in vain.    75
TIMON
  But yet I love my country, and am not
  One that rejoices in the common wrack
  As common bruit doth put it.
FIRST SENATOR           That's well spoke.
TIMON
  Commend me to my loving countrymen—
FIRST SENATOR
  These words become your lips as they pass through
    them.              80
SECOND SENATOR
  And enter in our ears like great triumphers
  In their applauding gates.
TIMON              Commend me to them,
  And tell them that to ease them of their griefs,
  Their fears of hostile strokes, their aches, losses,
  Their pangs of love, with other incident throes   85
  That nature's fragile vessel doth sustain
  In life's uncertain voyage, I will some kindness do them.
  I'll teach them to prevent wild Alcibiades' wrath.
FIRST SENATOR (aside)
  I like this well; he will return again.
TIMON
  I have a tree which grows here in my close   90
  That mine own use invites me to cut down,
  And shortly must I fell it. Tell my friends,
  Tell Athens, in the sequence of degree
  From high to low throughout, that whoso please
  To stop affliction, let him take his haste,     95
  Come hither ere my tree hath felt the axe,
  And hang himself. I pray you do my greeting.
FLAVIUS (to Senators)
  Trouble him no further. Thus you still shall find him.

TIMON
  Come not to me again, but say to Athens,
  Timon hath made his everlasting mansion    100
  Upon the beachèd verge of the salt flood,
  Who once a day with his embossèd froth
  The turbulent surge shall cover. Thither come,
  And let my gravestone be your oracle.
  Lips, let four words go by, and language end.   105
  What is amiss, plague and infection mend.
  Graves only be men's works, and death their gain.
  Sun, hide thy beams. Timon hath done his reign.
                          Exit ⌈into his cave⌉
FIRST SENATOR
  His discontents are unremovably
  Coupled to nature.                   110
SECOND SENATOR
  Our hope in him is dead. Let us return,
  And strain what other means is left unto us
  In our dear peril.
FIRST SENATOR        It requires swift foot.    Exeunt

**5.3**    *Enter two other Senators, with a Messenger*
⌈THIRD⌉ SENATOR
  Thou hast painfully discovered. Are his files
  As full as thy report?
MESSENGER          I have spoke the least.
  Besides, his expedition promises
  Present approach.
⌈FOURTH⌉ SENATOR
  We stand much hazard if they bring not Timon.   5
MESSENGER
  I met a courier, one mine ancient friend,
  Whom, though in general part we were opposed,
  Yet our old love made a particular force
  And made us speak like friends. This man was riding
  From Alcibiades to Timon's cave       10
  With letters of entreaty which imported
  His fellowship i'th' cause against your city,
  In part for his sake moved.
              *Enter the other Senators*
⌈THIRD⌉ SENATOR         Here come our brothers.
⌈FIRST⌉ SENATOR
  No talk of Timon; nothing of him expect.
  The enemy's drum is heard, and fearful scouring   15
  Doth choke the air with dust. In, and prepare.
  Ours is the fall, I fear, our foe's the snare.   Exeunt

**5.4**    *Enter a Soldier, in the woods, seeking Timon*
SOLDIER
  By all description, this should be the place.
  Who's here? Speak, ho! No answer?
        ⌈He discovers a gravestone⌉
                        What is this?
  Dead, sure, and this his grave. What's on this tomb
  I cannot read. The character I'll take with wax.
  Our captain hath in every figure skill,      5
  An aged interpreter, though young in days.
  Before proud Athens he's set down by this,
  Whose fall the mark of his ambition is.    Exit

**5.5**    *Trumpets sound. Enter Alcibiades with his powers,*
      *before Athens*
ALCIBIADES
  Sound to this coward and lascivious town
  Our terrible approach.

*A parley sounds. The Senators appear upon the walls*
Till now you have gone on and filled the time
With all licentious measure, making your wills
The scope of justice. Till now myself and such                5
As slept within the shadow of your power
Have wandered with our traversèd arms, and breathed
Our sufferance vainly. Now the time is flush
When crouching marrow, in the bearer strong,
Cries of itself 'No more'; now breathless wrong          10
Shall sit and pant in your great chairs of ease,
And pursy insolence shall break his wind
With fear and horrid flight.
FIRST SENATOR                        Noble and young,
When thy first griefs were but a mere conceit,
Ere thou hadst power or we had cause of fear,            15
We sent to thee to give thy rages balm,
To wipe out our ingratitude with loves
Above their quantity.
SECOND SENATOR              So did we woo
Transformèd Timon to our city's love
By humble message and by promised means.           20
We were not all unkind, nor all deserve
The common stroke of war.
FIRST SENATOR                        These walls of ours
Were not erected by their hands from whom
You have received your grief; nor are they such
That these great tow'rs, trophies, and schools should fall
For private faults in them.
SECOND SENATOR                    Nor are they living       26
Who were the motives that you first went out.
Shame that they wanted cunning, in excess,
Hath broke their hearts. March, noble lord,
Into our city with thy banners spread.                          30
By decimation and a tithèd death,
If thy revenges hunger for that food
Which nature loathes, take thou the destined tenth,
And by the hazard of the spotted die
Let die the spotted.
FIRST SENATOR                All have not offended.         35
For those that were, it is not square to take,
On those that are, revenges. Crimes like lands
Are not inherited. Then, dear countryman,
Bring in thy ranks, but leave without thy rage.
Spare thy Athenian cradle and those kin                     40
Which, in the bluster of thy wrath, must fall
With those that have offended. Like a shepherd
Approach the fold and cull th'infected forth,
But kill not all together.
SECOND SENATOR                What thou wilt,
Thou rather shalt enforce it with thy smile                  45
Than hew to't with thy sword.
FIRST SENATOR                        Set but thy foot
Against our rampired gates and they shall ope,

So thou wilt send thy gentle heart before
To say thou'lt enter friendly.
SECOND SENATOR              Throw thy glove,
Or any token of thine honour else,                                 50
That thou wilt use the wars as thy redress,
And not as our confusion. All thy powers
Shall make their harbour in our town till we
Have sealed thy full desire.
ALCIBIADES ⌈*throwing up a glove*⌉ Then there's my glove.
Descend, and open your unchargèd ports.                   55
Those enemies of Timon's and mine own
Whom you yourselves shall set out for reproof
Fall, and no more; and to atone your fears
With my more noble meaning, not a man
Shall pass his quarter or offend the stream               60
Of regular justice in your city's bounds
But shall be remedied to your public laws
At heaviest answer.
BOTH SENATORS 'Tis most nobly spoken.
ALCIBIADES Descend, and keep your words.              65
        ⌈*Trumpets sound. Exeunt Senators from the walls.*⌉
        *Enter Soldier, with a tablet of wax*
SOLDIER
My noble general, Timon is dead,
Entombed upon the very hem o'th' sea;
And on his gravestone this insculpture, which
With wax I brought away, whose soft impression
Interprets for my poor ignorance.                                  70
        *Alcibiades reads the epitaph*
ALCIBIADES
'Here lies a wretched corpse,
    Of wretched soul bereft.
Seek not my name. A plague consume
    You wicked caitiffs left!
Here lie I, Timon, who alive                                           75
    All living men did hate.
Pass by and curse thy fill, but pass
    And stay not here thy gait.'
These well express in thee thy latter spirits.
Though thou abhorred'st in us our human griefs,     80
Scorned'st our brains' flow and those our droplets which
From niggard nature fall, yet rich conceit
Taught thee to make vast Neptune weep for aye
On thy low grave, on faults forgiven. Dead
Is noble Timon, of whose memory                                85
Hereafter more.
        ⌈*Enter Senators through the gates*⌉
                Bring me into your city,
And I will use the olive with my sword,
Make war breed peace, make peace stint war, make each
Prescribe to other as each other's leech.
Let our drums strike.                                                    90
        ⌈*Drums.*⌉ *Exeunt* ⌈*through the gates*⌉

# THE HISTORY OF KING LEAR
## THE QUARTO TEXT

*King Lear* first appeared in print in a quarto of 1608. A substantially different text appeared in the 1623 Folio. Until now, editors, assuming that each of these early texts imperfectly represented a single play, have conflated them. But research conducted mainly during the 1970s and 1980s confirms an earlier view that the 1608 quarto represents the play as Shakespeare originally wrote it, and the 1623 Folio as he substantially revised it. He revised other plays, too, but usually by making many small changes in the dialogue and adding or omitting passages, as in *Hamlet, Troilus and Cressida*, and *Othello*. For these plays we print the revised text in so far as it can be ascertained. But in *King Lear* revisions are not simply local but structural, too; conflation, as Harley Granville-Barker wrote, 'may make for redundancy or confusion', so we print an edited version of each text. The first, printed in the following pages, represents the play as Shakespeare first conceived it, probably before it was performed.

The story of a king who, angry with the failure of his virtuous youngest daughter (Cordelia) to respond as he desires in a love-test, divides his kingdom between her two malevolent sisters (Gonoril and Regan), had been often told; Shakespeare would have come upon it in Holinshed's *Chronicles* and in *A Mirror for Magistrates* while reading for his plays on English history. It is told also (though briefly) in Edmund Spenser's *Faerie Queene* (Book 2, canto 10), and had been dramatized in a play of unknown authorship— *The True Chronicle History of King Leir and his three daughters*—published in 1605, but probably written some fifteen years earlier. This play particularly gave Shakespeare much, including suggestions for the characters of Lear's loyal servant, Kent, and of Gonoril's husband, Albany, and her steward, Oswald; for the storm; for Lear's kneeling to Cordelia; and for many details of language. Nevertheless, his play is a highly original creation. Lear's madness and the harrowing series of disasters in *King Lear*'s final stages are of Shakespeare's invention, and he complicates the plot by adding the story (based on an episode of Sir Philip Sidney's *Arcadia*) of Gloucester and his two sons, Edmund and Edgar. Edgar's love and loyalty to the father who, failing to see the truth, has rejected him in favour of the villainous Edmund makes him a counterpart to Cordelia; and the horrific blinding of Gloucester brought about by Edmund creates a physical parallel to Lear's madness which reaches its consummation in the scene (Sc. 20) at Dover Cliff when the mad and the blind old men commune together.

The clear-eyed intensity of Shakespeare's tragic vision in *King Lear* has been too much for some audiences, and Nahum Tate's adaptation, which gave the play a happy ending, held the stage from 1681 to 1843; since then, increased understanding of Shakespeare's stagecraft along with a greater seriousness in theatre audiences has assisted in the rehabilitation of a play that is now recognized as one of the profoundest of all artistic explorations of the human condition.

# THE PERSONS OF THE PLAY

LEAR, King of Britain

GONORIL, Lear's eldest daughter

Duke of ALBANY, her husband

REGAN, Lear's second daughter

Duke of CORNWALL, her husband

CORDELIA, Lear's youngest daughter

King of FRANCE ⎱
Duke of BURGUNDY ⎰ suitors of Cordelia

Earl of KENT, later disguised as Caius

Earl of GLOUCESTER

EDGAR, elder son of Gloucester, later disguised as Tom o' Bedlam

EDMUND, bastard son of Gloucester

OLD MAN, a tenant of Gloucester

CURAN, Gloucester's retainer

Lear's FOOL

OSWALD, Gonoril's steward

Three SERVANTS of Cornwall

DOCTOR, attendant on Cordelia

Three CAPTAINS

A HERALD

A KNIGHT

A MESSENGER

Gentlemen, servants, soldiers, followers, trumpeters, others

# The History of King Lear

**Sc. 1** *Enter the Earl of Kent, the Duke of Gloucester, and Edmund the bastard*

KENT I thought the King had more affected the Duke of Albany than Cornwall.

GLOUCESTER It did always seem so to us, but now in the division of the kingdoms it appears not which of the Dukes he values most; for equalities are so weighed that curiosity in neither can make choice of either's moiety. 7

KENT Is not this your son, my lord?

GLOUCESTER His breeding, sir, hath been at my charge. I have so often blushed to acknowledge him that now I am brazed to it. 11

KENT I cannot conceive you.

GLOUCESTER Sir, this young fellow's mother could, whereupon she grew round-wombed and had indeed, sir, a son for her cradle ere she had a husband for her bed. Do you smell a fault? 16

KENT I cannot wish the fault undone, the issue of it being so proper.

GLOUCESTER But I have, sir, a son by order of law, some year elder than this, who yet is no dearer in my account. Though this knave came something saucily into the world before he was sent for, yet was his mother fair, there was good sport at his making, and the whoreson must be acknowledged. (*To Edmund*) Do you know this noble gentleman, Edmund? 25

EDMUND No, my lord.

GLOUCESTER (*to Edmund*) My lord of Kent. Remember him hereafter as my honourable friend.

EDMUND (*to Kent*) My services to your lordship.

KENT I must love you, and sue to know you better. 30

EDMUND Sir, I shall study deserving.

GLOUCESTER (*to Kent*) He hath been out nine years, and away he shall again.

*Sound a sennet*

The King is coming. 34

*Enter one bearing a coronet, then King Lear, then the Dukes of Albany and Cornwall; next Gonoril, Regan, Cordelia, with followers*

LEAR

Attend my lords of France and Burgundy, Gloucester.

GLOUCESTER I shall, my liege. ⌐*Exit*⌐

LEAR

Meantime we will express our darker purposes.
The map there. Know we have divided
In three our kingdom, and 'tis our first intent
To shake all cares and business off our state, 40
Confirming them on younger years.
The two great princes, France and Burgundy—
Great rivals in our youngest daughter's love—
Long in our court have made their amorous sojourn,
And here are to be answered. Tell me, my daughters,
Which of you shall we say doth love us most, 46
That we our largest bounty may extend
Where merit doth most challenge it?
Gonoril, our eldest born, speak first.

GONORIL

Sir, I do love you more than words can wield the
matter;                                          50
Dearer than eyesight, space, or liberty;
Beyond what can be valued, rich or rare;
No less than life; with grace, health, beauty, honour;
As much as child e'er loved, or father, friend;
A love that makes breath poor and speech unable. 55
Beyond all manner of so much I love you.

CORDELIA (*aside*)

What shall Cordelia do? Love and be silent.

LEAR (*to Gonoril*)

Of all these bounds even from this line to this,
With shady forests and wide skirted meads,
We make thee lady. To thine and Albany's issue 60
Be this perpetual.—What says our second daughter?
Our dearest Regan, wife to Cornwall, speak.

REGAN Sir, I am made
Of the self-same mettle that my sister is,
And prize me at her worth. In my true heart 65
I find she names my very deed of love—
Only she came short, that I profess
Myself an enemy to all other joys
Which the most precious square of sense possesses,
And find I am alone felicitate 70
In your dear highness' love.

CORDELIA (*aside*)                       Then poor Cordelia—
And yet not so, since I am sure my love's
More richer than my tongue.

LEAR (*to Regan*)

To thee and thine hereditary ever
Remain this ample third of our fair kingdom, 75
No less in space, validity, and pleasure
Than that confirmed on Gonoril. (*To Cordelia*) But
now our joy,
Although the last, not least in our dear love:
What can you say to win a third more opulent
Than your sisters? 80

CORDELIA Nothing, my lord.

LEAR

How? Nothing can come of nothing. Speak again.

CORDELIA

Unhappy that I am, I cannot heave
My heart into my mouth. I love your majesty
According to my bond, nor more nor less. 85

LEAR

Go to, go to, mend your speech a little
Lest it may mar your fortunes.

CORDELIA                       Good my lord,
You have begot me, bred me, loved me.
I return those duties back as are right fit—
Obey you, love you, and most honour you. 90
Why have my sisters husbands if they say
They love you all? Haply when I shall wed
That lord whose hand must take my plight shall carry
Half my love with him, half my care and duty.
Sure, I shall never marry like my sisters, 95
To love my father all.

LEAR But goes this with thy heart?

CORDELIA Ay, good my lord.

LEAR So young and so untender?

CORDELIA So young, my lord, and true. 100

LEAR

Well, let it be so. Thy truth then be thy dower;

For by the sacred radiance of the sun,
The mysteries of Hecate and the night,
By all the operation of the orbs
From whom we do exist and cease to be,    105
Here I disclaim all my paternal care,
Propinquity, and property of blood,
And as a stranger to my heart and me
Hold thee from this for ever. The barbarous Scythian,
Or he that makes his generation    110
Messes to gorge his appetite,
Shall be as well neighboured, pitied, and relieved
As thou, my sometime daughter.

KENT                    Good my liege—

LEAR
  Peace, Kent. Come not between the dragon and his
      wrath.
I loved her most, and thought to set my rest    115
On her kind nursery. [*To Cordelia*] Hence, and avoid
    my sight!—
So be my grave my peace as here I give
Her father's heart from her. Call France. Who stirs?
Call Burgundy.             [*Exit one or more*]
            Cornwall and Albany,
With my two daughters' dowers digest this third.    120
Let pride, which she calls plainness, marry her.
I do invest you jointly in my power,
Pre-eminence, and all the large effects
That troop with majesty. Ourself by monthly course,
With reservation of an hundred knights    125
By you to be sustained, shall our abode
Make with you by due turns. Only we still retain
The name and all the additions to a king.
The sway, revenue, execution of the rest,
Belovèd sons, be yours; which to confirm,    130
This crownet part betwixt you.

KENT                 Royal Lear,
Whom I have ever honoured as my king,
Loved as my father, as my master followed,
As my great patron thought on in my prayers—

LEAR
  The bow is bent and drawn; make from the shaft.  135

KENT
  Let it fall rather, though the fork invade
The region of my heart. Be Kent unmannerly
When Lear is mad. What wilt thou do, old man?
Think'st thou that duty shall have dread to speak
When power to flattery bows? To plainness honour's
    bound    140
When majesty stoops to folly. Reverse thy doom,
And in thy best consideration check
This hideous rashness. Answer my life my judgement,
Thy youngest daughter does not love thee least,
Nor are those empty-hearted whose low sound    145
Reverbs no hollowness.

LEAR            Kent, on thy life, no more!

KENT
  My life I never held but as a pawn
To wage against thy enemies, nor fear to lose it,
Thy safety being the motive.

LEAR              Out of my sight!

KENT
  See better, Lear, and let me still remain    150
The true blank of thine eye.

LEAR            Now, by Apollo—

KENT
  Now, by Apollo, King, thou swear'st thy gods in vain.

LEAR [*making to strike him*]
  Vassal, recreant!

KENT          Do, kill thy physician,
And the fee bestow upon the foul disease.
Revoke thy doom, or whilst I can vent clamour    155
From my throat I'll tell thee thou dost evil.

LEAR
  Hear me; on thy allegiance hear me!
Since thou hast sought to make us break our vow,
Which we durst never yet, and with strayed pride
To come between our sentence and our power,    160
Which nor our nature nor our place can bear,
Our potency made good take thy reward:
Four days we do allot thee for provision
To shield thee from dis-eases of the world,
And on the fifth to turn thy hated back    165
Upon our kingdom. If on the next day following
Thy banished trunk be found in our dominions,
The moment is thy death. Away! By Jupiter,
This shall not be revoked.

KENT
  Why, fare thee well, King; since thus thou wilt
    appear,    170
Friendship lives hence, and banishment is here.
(*To Cordelia*) The gods to their protection take thee,
    maid,
That rightly thinks, and hast most justly said.
(*To Gonoril and Regan*)
And your large speeches may your deeds approve,
That good effects may spring from words of love.    175
Thus Kent, O princes, bids you all adieu;
He'll shape his old course in a country new.     *Exit*
    *Enter the King of France and the Duke of*
    *Burgundy, with the Duke of Gloucester*

GLOUCESTER
  Here's France and Burgundy, my noble lord.

LEAR  My lord of Burgundy,
We first address towards you, who with a king    180
Hath rivalled for our daughter: what in the least
Will you require in present dower with her
Or cease your quest of love?

BURGUNDY            Royal majesty,
I crave no more than what your highness offered;
Nor will you tender less.

LEAR             Right noble Burgundy,    185
When she was dear to us we did hold her so;
But now her price is fallen. Sir, there she stands.
If aught within that little seeming substance,
Or all of it, with our displeasure pieced,
And nothing else, may fitly like your grace,    190
She's there, and she is yours.

BURGUNDY             I know no answer.

LEAR
  Sir, will you with those infirmities she owes,
Unfriended, new-adopted to our hate,
Covered with our curse and strangered with our oath,
Take her or leave her?

BURGUNDY           Pardon me, royal sir.    195
Election makes not up on such conditions.

LEAR
  Then leave her, sir; for by the power that made me,
I tell you all her wealth. (*To France*) For you, great
    King,
I would not from your love make such a stray
To match you where I hate, therefore beseech you    200

To avert your liking a more worthier way
Than on a wretch whom nature is ashamed
Almost to acknowledge hers.
FRANCE
    This is most strange, that she that even but now
Was your best object, the argument of your praise,
Balm of your age, most best, most dearest,                    206
Should in this trice of time commit a thing
So monstrous to dismantle
So many folds of favour. Sure, her offence
Must be of such unnatural degree                              210
That monsters it, or your fore-vouched affections
Fall'n into taint; which to believe of her
Must be a faith that reason without miracle
Could never plant in me.
CORDELIA (to Lear)
    I yet beseech your majesty,                               215
If for I want that glib and oily art
To speak and purpose not—since what I well intend,
I'll do't before I speak—that you acknow
It is no vicious blot, murder, or foulness,
No unclean action or dishonoured step                        220
That hath deprived me of your grace and favour,
But even the want of that for which I am rich—
A still-soliciting eye, and such a tongue
As I am glad I have not, though not to have it
Hath lost me in your liking.
LEAR                              Go to, go to.               225
    Better thou hadst not been born than not to have
        pleased me better.
FRANCE
    Is it no more but this—a tardiness in nature,
That often leaves the history unspoke
That it intends to do?—My lord of Burgundy,
What say you to the lady? Love is not love                    230
When it is mingled with respects that stands
Aloof from the entire point. Will you have her?
She is herself a dower.
BURGUNDY                          Royal Lear,
    Give but that portion which yourself proposed,
And here I take Cordelia by the hand,                         235
Duchess of Burgundy—
LEAR                              Nothing. I have sworn.
BURGUNDY (to Cordelia)
    I am sorry, then, you have so lost a father
That you must lose a husband.
CORDELIA
    Peace be with Burgundy; since that respects
Of fortune are his love, I shall not be his wife.            240
FRANCE
    Fairest Cordelia, that art most rich, being poor;
Most choice, forsaken; and most loved, despised:
Thee and thy virtues here I seize upon.
Be it lawful, I take up what's cast away.
Gods, gods! 'Tis strange that from their cold'st neglect
My love should kindle to inflamed respect.—                  246
Thy dowerless daughter, King, thrown to my chance,
Is queen of us, of ours, and our fair France.
Not all the dukes in wat'rish Burgundy
Shall buy this unprized precious maid of me.—                250
Bid them farewell, Cordelia, though unkind.
Thou losest here, a better where to find.
LEAR
    Thou hast her, France. Let her be thine, for we
Have no such daughter, nor shall ever see

That face of hers again. Therefore be gone,                  255
Without our grace, our love, our benison.—
Come, noble Burgundy.
            ⌐Flourish.⌐ Exeunt Lear and Burgundy, then
                Albany, Cornwall, Gloucester, ⌐Edmund,⌐
                                        and followers
FRANCE (to Cordelia)          Bid farewell to your sisters.
CORDELIA
    Ye jewels of our father, with washed eyes
Cordelia leaves you. I know you what you are,
And like a sister am most loath to call                      260
Your faults as they are named. Use well our father.
To your professèd bosoms I commit him.
But yet, alas, stood I within his grace
I would prefer him to a better place.
So farewell to you both.                                     265
GONORIL  Prescribe not us our duties.
REGAN  Let your study
Be to content your lord, who hath received you
At fortune's alms. You have obedience scanted,
And well are worth the worst that you have wanted.
CORDELIA
    Time shall unfold what pleated cunning hides.            271
Who covers faults, at last shame them derides.
Well may you prosper.
FRANCE                        Come, fair Cordelia.
                        Exeunt France and Cordelia
GONORIL  Sister, it is not a little I have to say of what
most nearly appertains to us both. I think our father
will hence tonight.                                          276
REGAN  That's most certain, and with you. Next month
with us.
GONORIL  You see how full of changes his age is. The
observation we have made of it hath not been little.
He always loved our sister most, and with what poor
judgement he hath now cast her off appears too gross.
REGAN  'Tis the infirmity of his age; yet he hath ever but
slenderly known himself.                                     284
GONORIL  The best and soundest of his time hath been but
rash; then must we look to receive from his age not
alone the imperfection of long-engrafted condition, but
therewithal unruly waywardness that infirm and
choleric years bring with them.                              289
REGAN  Such unconstant starts are we like to have from
him as this of Kent's banishment.
GONORIL  There is further compliment of leave-taking
between France and him. Pray, let's hit together. If our
father carry authority with such dispositions as he
bears, this last surrender of his will but offend us.  295
REGAN  We shall further think on't.
GONORIL  We must do something, and i'th' heat.  Exeunt

Sc. 2  Enter Edmund the bastard
EDMUND
    Thou, nature, art my goddess. To thy law
My services are bound. Wherefore should I
Stand in the plague of custom and permit
The curiosity of nations to deprive me
For that I am some twelve or fourteen moonshines    5
Lag of a brother? Why 'bastard'? Wherefore 'base',
When my dimensions are as well compact,
My mind as generous, and my shape as true
As honest madam's issue?
Why brand they us with 'base, base bastardy',       10
Who in the lusty stealth of nature take
More composition and fierce quality

Than doth within a stale, dull-eyed bed go
To the creating a whole tribe of fops
Got 'tween a sleep and wake? Well then,     15
Legitimate Edgar, I must have your land.
Our father's love is to the bastard Edmund
As to the legitimate. Well, my legitimate, if
This letter speed and my invention thrive,
Edmund the base shall to th' legitimate.     20
I grow, I prosper. Now gods, stand up for bastards!

*Enter the Duke of Gloucester. Edmund reads*
*a letter*

GLOUCESTER
Kent banished thus, and France in choler parted,
And the King gone tonight, subscribed his power,
Confined to exhibition—all this done
Upon the gad?—Edmund, how now? What news?   25

EDMUND So please your lordship, none.

GLOUCESTER Why so earnestly seek you to put up that
letter?

EDMUND I know no news, my lord.

GLOUCESTER What paper were you reading?     30

EDMUND Nothing, my lord.

GLOUCESTER No? What needs then that terrible dispatch
of it into your pocket? The quality of nothing hath not
such need to hide itself. Let's see. Come, if it be nothing
I shall not need spectacles.     35

EDMUND I beseech you, sir, pardon me. It is a letter from
my brother that I have not all o'er-read; for so much
as I have perused, I find it not fit for your liking.

GLOUCESTER Give me the letter, sir.     39

EDMUND I shall offend either to detain or give it. The
contents, as in part I understand them, are to blame.

GLOUCESTER Let's see, let's see.

EDMUND I hope for my brother's justification he wrote
this but as an assay or taste of my virtue.     44

*He gives Gloucester a letter*

GLOUCESTER (*reads*) 'This policy of age makes the world
bitter to the best of our times, keeps our fortunes from
us till our oldness cannot relish them. I begin to find
an idle and fond bondage in the oppression of aged
tyranny, who sways not as it hath power but as it is
suffered. Come to me, that of this I may speak more.
If our father would sleep till I waked him, you should
enjoy half his revenue for ever and live the beloved of
your brother,     53
                              Edgar.'
Hum, conspiracy! 'Slept till I waked him, you should
enjoy half his revenue'—my son Edgar! Had he a hand
to write this, a heart and brain to breed it in? When
came this to you? Who brought it?     58

EDMUND It was not brought me, my lord, there's the
cunning of it. I found it thrown in at the casement of
my closet.

GLOUCESTER You know the character to be your brother's?

EDMUND If the matter were good, my lord, I durst swear
it were his; but in respect of that, I would fain think
it were not.     65

GLOUCESTER It is his.

EDMUND It is his hand, my lord, but I hope his heart is
not in the contents.

GLOUCESTER Hath he never heretofore sounded you in this
business?     70

EDMUND Never, my lord; but I have often heard him
maintain it to be fit that, sons at perfect age and fathers
declining, his father should be as ward to the son, and
the son manage the revenue.

GLOUCESTER O villain, villain—his very opinion in the
letter! Abhorred villain, unnatural, detested, brutish
villain—worse than brutish! Go, sir, seek him, ay,
apprehend him. Abominable villain! Where is he?   78

EDMUND I do not well know, my lord. If it shall please
you to suspend your indignation against my brother
till you can derive from him better testimony of this
intent, you should run a certain course; where if you
violently proceed against him, mistaking his purpose,
it would make a great gap in your own honour and
shake in pieces the heart of his obedience. I dare pawn
down my life for him he hath wrote this to feel my
affection to your honour, and to no further pretence of
danger.

GLOUCESTER Think you so?     89

EDMUND If your honour judge it meet, I will place you
where you shall hear us confer of this, and by an
auricular assurance have your satisfaction, and that
without any further delay than this very evening.

GLOUCESTER He cannot be such a monster.

EDMUND Nor is not, sure.     95

GLOUCESTER To his father, that so tenderly and entirely
loves him—heaven and earth! Edmund seek him out,
wind me into him. I pray you, frame your business
after your own wisdom. I would unstate myself to be
in a due resolution.     100

EDMUND I shall seek him, sir, presently, convey the
business as I shall see means, and acquaint you withal.

GLOUCESTER These late eclipses in the sun and moon
portend no good to us. Though the wisdom of nature
can reason thus and thus, yet nature finds itself
scourged by the sequent effects. Love cools, friendship
falls off, brothers divide; in cities mutinies, in countries
discords, palaces treason, the bond cracked between
son and father. Find out this villain, Edmund; it shall
lose thee nothing. Do it carefully. And the noble and
true-hearted Kent banished, his offence honesty!
Strange, strange!                       *Exit*

EDMUND This is the excellent foppery of the world: that
when we are sick in fortune—often the surfeit of our
own behaviour—we make guilty of our disasters the
sun, the moon, and the stars, as if we were villains by
necessity, fools by heavenly compulsion, knaves,
thieves, and treacherers by spherical predominance,
drunkards, liars, and adulterers by an enforced
obedience of planetary influence, and all that we are
evil in by a divine thrusting on. An admirable evasion
of whoremaster man, to lay his goatish disposition to
the charge of stars! My father compounded with my
mother under the Dragon's tail and my nativity was
under Ursa Major, so that it follows I am rough and
lecherous. Fut! I should have been that I am had the
maidenliest star of the firmament twinkled on my
bastardy. Edgar . . .     128

*Enter Edgar*

and on's cue out he comes, like the catastrophe of the
old comedy; mine is villainous melancholy, with a sigh
like them of Bedlam.—O, these eclipses do portend these
divisions.

EDGAR How now, brother Edmund, what serious contemplation are you in?     134

EDMUND I am thinking, brother, of a prediction I read this
other day, what should follow these eclipses.

EDGAR Do you busy yourself about that?

EDMUND I promise you, the effects he writ of succeed
unhappily, as of unnaturalness between the child and

the parent, death, dearth, dissolutions of ancient amities, divisions in state, menaces and maledictions against king and nobles, needless diffidences, banishment of friends, dissipation of cohorts, nuptial breaches, and I know not what.                                    144

EDGAR How long have you been a sectary astronomical?

EDMUND Come, come, when saw you my father last?

EDGAR Why, the night gone by.

EDMUND Spake you with him?

EDGAR Two hours together.                                    149

EDMUND Parted you in good terms? Found you no displeasure in him by word or countenance?

EDGAR None at all.

EDMUND Bethink yourself wherein you may have offended him, and at my entreaty forbear his presence till some little time hath qualified the heat of his displeasure, which at this instant so rageth in him that with the mischief of your person it would scarce allay.    157

EDGAR Some villain hath done me wrong.

EDMUND That's my fear, brother. I advise you to the best. Go armed. I am no honest man if there be any good meaning towards you. I have told you what I have seen and heard but faintly, nothing like the image and horror of it. Pray you, away.                            163

EDGAR Shall I hear from you anon?

EDMUND I do serve you in this business.        *Exit Edgar*
A credulous father, and a brother noble,
Whose nature is so far from doing harms
That he suspects none; on whose foolish honesty
My practices ride easy. I see the business.
Let me, if not by birth, have lands by wit.        170
All with me's meet that I can fashion fit.        *Exit*

**Sc. 3**    *Enter Gonoril and Oswald, her gentleman*

GONORIL
Did my father strike my gentleman
For chiding of his fool?

OSWALD                          Yes, madam.

GONORIL
By day and night he wrongs me. Every hour
He flashes into one gross crime or other
That sets us all at odds. I'll not endure it.        5
His knights grow riotous, and himself upbraids us
On every trifle. When he returns from hunting
I will not speak with him. Say I am sick.
If you come slack of former services
You shall do well; the fault of it I'll answer.        10
        ⌈*Hunting horns within*⌉

OSWALD He's coming, madam. I hear him.

GONORIL
Put on what weary negligence you please,
You and your fellow servants. I'd have it come in
        question.
If he dislike it, let him to our sister,
Whose mind and mine I know in that are one,        15
Not to be overruled. Idle old man,
That still would manage those authorities
That he hath given away! Now, by my life,
Old fools are babes again, and must be used
With checks as flatteries, when they are seen abused.
Remember what I tell you.

OSWALD                          Very well, madam.        21

GONORIL
And let his knights have colder looks among you.
What grows of it, no matter. Advise your fellows so.

I would breed from hence occasions, and I shall,
That I may speak. I'll write straight to my sister        25
To hold my very course. Go prepare for dinner.
        *Exeunt severally*

**Sc. 4**    *Enter the Earl of Kent, disguised*

KENT
If but as well I other accents borrow
That can my speech diffuse, my good intent
May carry through itself to that full issue
For which I razed my likeness. Now, banished Kent,
If thou canst serve where thou dost stand condemned,
Thy master, whom thou lov'st, shall find thee full of
        labour.                                    6
        *Enter King Lear and servants from hunting*

LEAR Let me not stay a jot for dinner. Go get it ready.
        ⌈*Exit one*⌉

(*To Kent*) How now, what art thou?

KENT A man, sir.

LEAR What dost thou profess? What wouldst thou with us?                                    11

KENT I do profess to be no less than I seem, to serve him truly that will put me in trust, to love him that is honest, to converse with him that is wise and says little, to fear judgement, to fight when I cannot choose, and to eat no fish.                                    16

LEAR What art thou?

KENT A very honest-hearted fellow, and as poor as the King.

LEAR If thou be as poor for a subject as he is for a king, thou'rt poor enough. What wouldst thou?        21

KENT Service.

LEAR Who wouldst thou serve?

KENT You.

LEAR Dost thou know me, fellow?                            25

KENT No, sir, but you have that in your countenance which I would fain call master.

LEAR What's that?

KENT Authority.

LEAR What services canst do?                            30

KENT I can keep honest counsel, ride, run, mar a curious tale in telling it, and deliver a plain message bluntly. That which ordinary men are fit for I am qualified in; and the best of me is diligence.

LEAR How old art thou?                                    35

KENT Not so young to love a woman for singing, nor so old to dote on her for anything. I have years on my back forty-eight.

LEAR Follow me. Thou shalt serve me, if I like thee no worse after dinner. I will not part from thee yet.— Dinner, ho, dinner! Where's my knave, my fool? Go you and call my fool hither.        ⌈*Exit one*⌉
        *Enter Oswald the steward*
You, sirrah, where's my daughter?                        43

OSWALD So please you—                                *Exit*

LEAR What says the fellow there? Call the clotpoll back.
        *Exeunt Servant* ⌈*and Kent*⌉
Where's my fool? Ho, I think the world's asleep.
        *Enter the Earl of Kent* ⌈*and a Servant*⌉
How now, where's that mongrel?

KENT He says, my lord, your daughter is not well.

LEAR Why came not the slave back to me when I called him?                                    50

SERVANT Sir, he answered me in the roundest manner he would not.

LEAR A would not?

SERVANT My lord, I know not what the matter is, but to my judgement your highness is not entertained with that ceremonious affection as you were wont. There's a great abatement appears as well in the general dependants as in the Duke himself also, and your daughter.

LEAR Ha, sayst thou so?                                         60

SERVANT I beseech you pardon me, my lord, if I be mistaken, for my duty cannot be silent when I think your highness wronged.

LEAR Thou but rememberest me of mine own conception. I have perceived a most faint neglect of late, which I have rather blamed as mine own jealous curiosity than as a very pretence and purport of unkindness. I will look further into't. But where's this fool? I have not seen him these two days.                                        69

SERVANT Since my young lady's going into France, sir, the fool hath much pined away.

LEAR No more of that, I have noted it. Go you and tell my daughter I would speak with her.          ⌈Exit one⌉
Go you, call hither my fool.                        ⌈Exit one⌉
         Enter Oswald the steward ⌈crossing the stage⌉
O you, sir, you, sir, come you hither. Who am I, sir?

OSWALD My lady's father.                                        76

LEAR My lady's father? My lord's knave, you whoreson dog, you slave, you cur!

OSWALD I am none of this, my lord, I beseech you pardon me.                                                        80

LEAR Do you bandy looks with me, you rascal?
         ⌈Lear strikes him⌉

OSWALD I'll not be struck, my lord—

KENT (tripping him) Nor tripped neither, you base football player.

LEAR (to Kent) I thank thee, fellow. Thou serv'st me, and I'll love thee.                                              86

KENT (to Oswald) Come, sir, I'll teach you differences. Away, away. If you will measure your lubber's length again, tarry; but away if you have wisdom.
                                                    Exit Oswald

LEAR Now, friendly knave, I thank thee.                        90
         Enter Lear's Fool
There's earnest of thy service.
         He gives Kent money

FOOL Let me hire him, too. (To Kent) Here's my coxcomb.

LEAR How now, my pretty knave, how dost thou?

FOOL (to Kent) Sirrah, you were best take my coxcomb.

KENT Why, fool?                                                 95

FOOL Why, for taking one's part that's out of favour. Nay, an thou canst not smile as the wind sits, thou'lt catch cold shortly. There, take my coxcomb. Why, this fellow hath banished two on's daughters and done the third a blessing against his will. If thou follow him, thou must needs wear my coxcomb. (To Lear) How now, nuncle? Would I had two coxcombs and two daughters.

LEAR Why, my boy?                                              103

FOOL If I gave them my living I'd keep my coxcombs myself. There's mine; beg another off thy daughters.

LEAR Take heed, sirrah—the whip.

FOOL Truth is a dog that must to kennel. He must be whipped out when Lady the brach may stand by the fire and stink.

LEAR A pestilent gall to me!                                   110

FOOL ⌈to Kent⌉ Sirrah, I'll teach thee a speech.

LEAR Do.

FOOL Mark it, uncle.
         Have more than thou showest,
         Speak less than thou knowest,                          115

         Lend less than thou owest,
         Ride more than thou goest,
         Learn more than thou trowest,
         Set less than thou throwest,
         Leave thy drink and thy whore,                        120
         And keep in-a-door,
         And thou shalt have more
         Than two tens to a score.

LEAR This is nothing, fool.                                    124

FOOL Then, like the breath of an unfee'd lawyer, you gave me nothing for't. Can you make no use of nothing, uncle?

LEAR Why no, boy. Nothing can be made out of nothing.

FOOL (to Kent) Prithee, tell him so much the rent of his land comes to. He will not believe a fool.            130

LEAR A bitter fool.

FOOL Dost know the difference, my boy, between a bitter fool and a sweet fool?

LEAR No, lad. Teach me.

FOOL ⌈sings⌉    That lord that counselled thee                  135
                  To give away thy land,
              Come, place him here by me;
                  Do thou for him stand.
              The sweet and bitter fool
                  Will presently appear,                       140
              The one in motley here,
                  The other found out there.

LEAR Dost thou call me fool, boy?

FOOL All thy other titles thou hast given away. That thou wast born with.                                         145

KENT (to Lear) This is not altogether fool, my lord.

FOOL No, faith; lords and great men will not let me. If I had a monopoly out, they would have part on't, and ladies too, they will not let me have all the fool to myself—they'll be snatching. Give me an egg, nuncle, and I'll give thee two crowns.                            151

LEAR What two crowns shall they be?

FOOL Why, after I have cut the egg in the middle and eat up the meat, the two crowns of the egg. When thou clovest thy crown i'th' middle and gavest away both parts, thou borest thy ass o'th' back o'er the dirt. Thou hadst little wit in thy bald crown when thou gavest thy golden one away. If I speak like myself in this, let him be whipped that first finds it so.
⌈Sings⌉
         Fools had ne'er less wit in a year,                    160
              For wise men are grown foppish.
         They know not how their wits do wear,
              Their manners are so apish.

LEAR When were you wont to be so full of songs, sirrah?

FOOL I have used it, nuncle, ever since thou madest thy daughters thy mother; for when thou gavest them the rod and puttest down thine own breeches,               167
⌈Sings⌉    Then they for sudden joy did weep,
              And I for sorrow sung,
         That such a king should play bo-peep                   170
              And go the fools among.
Prithee, nuncle, keep a schoolmaster that can teach thy fool to lie. I would fain learn to lie.

LEAR An you lie, we'll have you whipped.                       174

FOOL I marvel what kin thou and thy daughters are. They'll have me whipped for speaking true, thou wilt have me whipped for lying, and sometime I am whipped for holding my peace. I had rather be any kind of thing than a fool; and yet I would not be thee, nuncle. Thou hast pared thy wit o' both sides and left nothing in the middle.                                                181

*Enter Gonoril*

Here comes one of the parings.

LEAR

How now, daughter, what makes that frontlet on?
Methinks you are too much o' late i'th' frown.

FOOL  Thou wast a pretty fellow when thou hadst no need
to care for her frown. Now thou art an O without a
figure. I am better than thou art, now. I am a fool;
thou art nothing. ⌈*To Gonoril*⌉ Yes, forsooth, I will hold
my tongue; so your face bids me, though you say
nothing.                                                   190

⌈*Sings*⌉ Mum, mum.
              He that keeps neither crust nor crumb,
              Weary of all, shall want some.

That's a shelled peascod.

GONORIL (*to Lear*)

Not only, sir, this your all-licensed fool,          195
But other of your insolent retinue
Do hourly carp and quarrel, breaking forth
In rank and not-to-be-endurèd riots.
Sir, I had thought by making this well known unto
    you
To have found a safe redress, but now grow fearful,
By what yourself too late have spoke and done,      201
That you protect this course, and put it on
By your allowance; which if you should, the fault
Would not scape censure, nor the redress sleep
Which in the tender of a wholesome weal            205
Might in their working do you that offence,
That else were shame, that then necessity
Must call discreet proceedings.

FOOL (*to Lear*)  For, you trow, nuncle,
⌈*Sings*⌉
              The hedge-sparrow fed the cuckoo so long  210
              That it had it head bit off by it young;
so out went the candle, and we were left darkling.

LEAR (*to Gonoril*)  Are you our daughter?

GONORIL

Come, sir, I would you would make use of that good
    wisdom
Whereof I know you are fraught, and put away      215
These dispositions that of late transform you
From what you rightly are.

FOOL  May not an ass know when the cart draws the
horse? ⌈*Sings*⌉ 'Whoop, jug, I love thee!'

LEAR

Doth any here know me? Why, this is not Lear.     220
Doth Lear walk thus, speak thus? Where are his eyes?
Either his notion weakens, or his discernings
Are lethargied. Sleeping or waking, ha?
Sure, 'tis not so.
Who is it that can tell me who I am?               225
Lear's shadow? I would learn that, for by the marks
Of sovereignty, knowledge, and reason
I should be false persuaded I had daughters.

FOOL  Which they will make an obedient father.

LEAR (*to Gonoril*)
Your name, fair gentlewoman?

GONORIL                          Come, sir,          230
This admiration is much of the savour
Of other your new pranks. I do beseech you
Understand my purposes aright,
As you are old and reverend, should be wise.
Here do you keep a hundred knights and squires,   235
Men so disordered, so debauched and bold
That this our court, infected with their manners,

Shows like a riotous inn, epicurism
And lust make more like to a tavern, or brothel,
Than a great palace. The shame itself doth speak   240
For instant remedy. Be thou desired,
By her that else will take the thing she begs,
A little to disquantity your train,
And the remainder that shall still depend
To be such men as may besort your age,            245
That know themselves and you.

LEAR                          Darkness and devils!
Saddle my horses, call my train together!—
                              ⌈*Exit one or more*⌉
Degenerate bastard, I'll not trouble thee.
Yet have I left a daughter.

GONORIL

You strike my people, and your disordered rabble   250
Make servants of their betters.
          *Enter the Duke of Albany*

LEAR

We that too late repent's—O sir, are you come?
Is it your will that we—prepare my horses.
                              ⌈*Exit one or more*⌉
Ingratitude, thou marble-hearted fiend,
More hideous when thou show'st thee in a child    255
Than the sea-monster—(*to Gonoril*) detested kite, thou
    liest.
My train are men of choice and rarest parts,
That all particulars of duty know,
And in the most exact regard support
The worships of their name. O most small fault,    260
How ugly didst thou in Cordelia show,
That, like an engine, wrenched my frame of nature
From the fixed place, drew from my heart all love,
And added to the gall! O Lear, Lear!
Beat at this gate that let thy folly in            265
And thy dear judgement out.—Go, go, my people!

ALBANY

My lord, I am guiltless as I am ignorant.

LEAR

It may be so, my lord. Hark, nature, hear:
Dear goddess, suspend thy purpose if
Thou didst intend to make this creature fruitful.  270
Into her womb convey sterility.
Dry up in her the organs of increase,
And from her derogate body never spring
A babe to honour her. If she must teem,
Create her child of spleen, that it may live       275
And be a thwart disnatured torment to her.
Let it stamp wrinkles in her brow of youth,
With cadent tears fret channels in her cheeks,
Turn all her mother's pains and benefits
To laughter and contempt, that she may feel—      280
That she may feel
How sharper than a serpent's tooth it is
To have a thankless child.—Go, go, my people!
          *Exeunt Lear, ⌈Kent, Fool, and servants⌉*

ALBANY

Now, gods that we adore, whereof comes this?

GONORIL

Never afflict yourself to know the cause,          285
But let his disposition have that scope
That dotage gives it.
          *Enter King Lear ⌈and his Fool⌉*

LEAR

What, fifty of my followers at a clap?
Within a fortnight?

ALBANY                    What is the matter, sir?

917

LEAR
I'll tell thee. (*To Gonoril*) Life and death! I am
    ashamed                                            290
That thou hast power to shake my manhood thus,
That these hot tears, that break from me perforce
And should make thee—worst blasts and fogs upon
    thee!
Untented woundings of a father's curse
Pierce every sense about thee! Old fond eyes,      295
Beweep this cause again I'll pluck you out
And cast you, with the waters that you make,
To temper clay. Yea,
Is't come to this? Yet have I left a daughter
Whom, I am sure, is kind and comfortable.           300
When she shall hear this of thee, with her nails
She'll flay thy wolvish visage. Thou shalt find
That I'll resume the shape which thou dost think
I have cast off for ever; thou shalt, I warrant thee.
                                                *Exit*

GONORIL Do you mark that, my lord?                 305
ALBANY
I cannot be so partial, Gonoril,
To the great love I bear you—
GONORIL                        Come, sir, no more.—
You, more knave than fool, after your master!
FOOL Nuncle Lear, nuncle Lear, tarry, and take the fool
    with thee.                                      310
        A fox when one has caught her,
        And such a daughter,
        Should sure to the slaughter,
        If my cap would buy a halter.
        So, the fool follows after.           *Exit*
GONORIL What, Oswald, ho!                           316
    *Enter Oswald*
OSWALD Here, madam.
GONORIL
What, have you writ this letter to my sister?
OSWALD Yes, madam.
GONORIL
Take you some company, and away to horse.          320
Inform her full of my particular fears,
And thereto add such reasons of your own
As may compact it more. Get you gone,
And after, your retinue.                *Exit Oswald*
                        Now, my lord,
This milky gentleness and course of yours,          325
Though I dislike not, yet under pardon
You're much more ataxed for want of wisdom
Than praised for harmful mildness.
ALBANY
How far your eyes may pierce I cannot tell.
Striving to better aught, we mar what's well.      330
GONORIL Nay, then—
ALBANY Well, well, the event.               *Exeunt*

**Sc. 5**  *Enter King Lear, the Earl of Kent disguised, and
    Lear's Fool*
LEAR ⌈*to Kent*⌉ Go you before to Gloucester with these
    letters. Acquaint my daughter no further with anything
    you know than comes from her demand out of the
    letter. If your diligence be not speedy, I shall be there
    before you.                                       5
KENT I will not sleep, my lord, till I have delivered your
    letter.                                        *Exit*
FOOL If a man's brains were in his heels, were't not in
    danger of kibes?

LEAR Ay, boy.                                        10
FOOL Then, I prithee, be merry: thy wit shall ne'er go
    slipshod.
LEAR Ha, ha, ha!
FOOL Shalt see thy other daughter will use thee kindly,
    for though she's as like this as a crab is like an apple,
    yet I con what I can tell.                        16
LEAR Why, what canst thou tell, my boy?
FOOL She'll taste as like this as a crab doth to a crab.
    Thou canst not tell why one's nose stands in the middle
    of his face?                                      20
LEAR No.
FOOL Why, to keep his eyes on either side 's nose, that
    what a man cannot smell out, a may spy into.
LEAR I did her wrong.
FOOL Canst tell how an oyster makes his shell?       25
LEAR No.
FOOL Nor I neither; but I can tell why a snail has a
    house.
LEAR Why?
FOOL Why, to put his head in, not to give it away to his
    daughter and leave his horns without a case.       31
LEAR
I will forget my nature. So kind a father!
Be my horses ready?
FOOL Thy asses are gone about them. The reason why
    the seven stars are no more than seven is a pretty
    reason.                                          36
LEAR Because they are not eight.
FOOL Yes. Thou wouldst make a good fool.
LEAR
To take't again perforce—monster ingratitude!
FOOL If thou wert my fool, nuncle, I'd have thee beaten
    for being old before thy time.                    41
LEAR How's that?
FOOL Thou shouldst not have been old before thou hadst
    been wise.
LEAR
O, let me not be mad, sweet heaven!                  45
I would not be mad.
Keep me in temper. I would not be mad.
    *Enter a Servant*
Are the horses ready?
SERVANT                        Ready, my lord.
LEAR (*to Fool*) Come, boy.        *Exeunt Lear and Servant*
FOOL
She that is maid now, and laughs at my departure,  50
Shall not be a maid long, except things be cut shorter.
                                                *Exit*

**Sc. 6**  *Enter Edmund the bastard, and Curan, meeting*
EDMUND Save thee, Curan.
CURAN And you, sir. I have been with your father, and
    given him notice that the Duke of Cornwall and his
    duchess will be here with him tonight.
EDMUND How comes that?                               5
CURAN Nay, I know not. You have heard of the news
    abroad?—I mean the whispered ones, for there are yet
    but ear-bussing arguments.
EDMUND Not. I pray you, what are they?
CURAN Have you heard of no likely wars towards twixt
    the two Dukes of Cornwall and Albany?             11
EDMUND Not a word.
CURAN You may then in time. Fare you well, sir.
                                                *Exit*

EDMUND
 The Duke be here tonight! The better, best.
 This weaves itself perforce into my business.          15
  ⌜Enter Edgar at a window above⌝
 My father hath set guard to take my brother,
 And I have one thing of a queasy question
 Which must ask briefness. Wit and fortune help!—
 Brother, a word. Descend, brother, I say.
  ⌜Edgar climbs down⌝
 My father watches. O, fly this place.                   20
 Intelligence is given where you are hid.
 You have now the good advantage of the night.
 Have you not spoken 'gainst the Duke of Cornwall
  aught?
 He's coming hither now, in the night, i'th' haste,
 And Regan with him. Have you nothing said            25
 Upon his party against the Duke of Albany?
 Advise you—
EDGAR                  I am sure on't, not a word.
EDMUND
 I hear my father coming. Pardon me.
 In cunning I must draw my sword upon you.
 Seem to defend yourself. Now, quit you well.         30
 (Calling) Yield, come before my father. Light here,
  here!
 (To Edgar) Fly, brother, fly! (Calling) Torches, torches!
  (To Edgar) So, farewell.                        Exit Edgar
 Some blood drawn on me would beget opinion
 Of my more fierce endeavour.
   He wounds his arm
                        I have seen
 Drunkards do more than this in sport. (Calling) Father,
  father!                                             35
 Stop, stop! Ho, help!
   Enter the Duke of Gloucester ⌜and others⌝
GLOUCESTER            Now, Edmund, where is the villain?
EDMUND
 Here stood he in the dark, his sharp sword out,
 Warbling of wicked charms, conjuring the moon
 To stand 's auspicious mistress.
GLOUCESTER                     But where is he?
EDMUND
 Look, sir, I bleed.
GLOUCESTER            Where is the villain, Edmund?    40
EDMUND
 Fled this way, sir, when by no means he could—
GLOUCESTER
 Pursue him, go after.                       Exeunt others
                By no means what?
EDMUND
 Persuade me to the murder of your lordship,
 But that I told him the revengive gods
 'Gainst parricides did all their thunders bend,         45
 Spoke with how manifold and strong a bond
 The child was bound to the father. Sir, in fine,
 Seeing how loathly opposite I stood
 To his unnatural purpose, with fell motion,
 With his preparèd sword he charges home            50
 My unprovided body, lanced mine arm;
 But when he saw my best alarumed spirits
 Bold in the quarrel's rights, roused to the encounter,
 Or whether ghasted by the noise I made
 Or ⌜              ⌝ I know not,            55
 But suddenly he fled.
GLOUCESTER            Let him fly far,
 Not in this land shall he remain uncaught,

And found, dispatch. The noble Duke my master,
 My worthy arch and patron, comes tonight.
 By his authority I will proclaim it                    60
 That he which finds him shall deserve our thanks,
 Bringing the murderous caitiff to the stake;
 He that conceals him, death.
EDMUND
 When I dissuaded him from his intent
 And found him pitched to do it, with curst speech    65
 I threatened to discover him. He replied,
 'Thou unpossessing bastard, dost thou think
 If I would stand against thee, could the reposure
 Of any trust, virtue, or worth in thee
 Make thy words faithed? No, what I should deny—   70
 As this I would, ay, though thou didst produce
 My very character—I'd turn it all
 To thy suggestion, plot, and damnèd pretence,
 And thou must make a dullard of the world
 If they not thought the profits of my death          75
 Were very pregnant and potential spurs
 To make thee seek it.'
GLOUCESTER            Strong and fastened villain!
 Would he deny his letter? I never got him.
   Trumpets within
 Hark, the Duke's trumpets. I know not why he comes.
 All ports I'll bar. The villain shall not scape.        80
 The Duke must grant me that; besides, his picture
 I will send far and near, that all the kingdom
 May have note of him—and of my land,
 Loyal and natural boy, I'll work the means
 To make thee capable.                             85
   Enter the Duke of Cornwall and Regan
CORNWALL
 How now, my noble friend? Since I came hither,
 Which I can call but now, I have heard strange news.
REGAN
 If it be true, all vengeance comes too short
 Which can pursue the offender. How dost, my lord?
GLOUCESTER
 Madam, my old heart is cracked, is cracked.       90
REGAN
 What, did my father's godson seek your life?
 He whom my father named, your Edgar?
GLOUCESTER
 Ay, lady, lady; shame would have it hid.
REGAN
 Was he not companion with the riotous knights
 That tend upon my father?                         95
GLOUCESTER
 I know not, madam. 'Tis too bad, too bad.
EDMUND Yes, madam, he was.
REGAN
 No marvel, then, though he were ill affected.
 'Tis they have put him on the old man's death,
 To have the spoil and waste of his revenues.        100
 I have this present evening from my sister
 Been well informed of them, and with such cautions
 That if they come to sojourn at my house
 I'll not be there.
CORNWALL            Nor I, assure thee, Regan.
 Edmund, I heard that you have shown your father
 A childlike office.
EDMUND            'Twas my duty, sir.             106
GLOUCESTER (to Cornwall)
 He did betray his practice, and received
 This hurt you see striving to apprehend him.

CORNWALL
Is he pursued?

GLOUCESTER      Ay, my good lord.

CORNWALL
If he be taken, he shall never more     110
Be feared of doing harm. Make your own purpose
How in my strength you please. For you, Edmund,
Whose virtue and obedience doth this instant
So much commend itself, you shall be ours.
Natures of such deep trust we shall much need.    115
You we first seize on.

EDMUND          I shall serve you truly,
However else.

GLOUCESTER (to Cornwall) For him I thank your grace.

CORNWALL
You know not why we came to visit you—

REGAN
This out-of-season threat'ning dark-eyed night—
Occasions, noble Gloucester, of some poise,    120
Wherein we must have use of your advice.
Our father he hath writ, so hath our sister,
Of differences which I least thought it fit
To answer from our home. The several messengers
From hence attend dispatch. Our good old friend,   125
Lay comforts to your bosom, and bestow
Your needful counsel to our business,
Which craves the instant use.

GLOUCESTER I serve you, madam.
Your graces are right welcome.       *Exeunt*

**Sc. 7**   *Enter the Earl of Kent, disguised, at one door, and*
      *Oswald the steward, at another door*

OSWALD Good even to thee, friend. Art of the house?

KENT Ay.

OSWALD Where may we set our horses?

KENT I'th' mire.

OSWALD Prithee, if thou love me, tell me.      5

KENT I love thee not.

OSWALD Why then, I care not for thee.

KENT If I had thee in Lipsbury pinfold I would make thee
care for me.

OSWALD Why dost thou use me thus? I know thee not.

KENT Fellow, I know thee.      11

OSWALD What dost thou know me for?

KENT A knave, a rascal, an eater of broken meats, a base,
proud, shallow, beggarly, three-suited, hundred-pound,
filthy worsted-stocking knave; a lily-livered, action-
taking knave; a whoreson, glass-gazing, superfinical
rogue; one-trunk-inheriting slave; one that wouldst be
a bawd in way of good service, and art nothing but
the composition of a knave, beggar, coward, pander,
and the son and heir of a mongrel bitch, whom I will
beat into clamorous whining if thou deny the least
syllable of the addition.      22

OSWALD What a monstrous fellow art thou, thus to rail
on one that's neither known of thee nor knows thee!

KENT What a brazen-faced varlet art thou, to deny thou
knowest me! Is it two days ago since I beat thee and
tripped up thy heels before the King? Draw, you rogue;
for though it be night, the moon shines.
     ⌈*He draws his sword*⌉
I'll make a sop of the moonshine o' you. Draw, you
whoreson, cullionly barber-monger, draw!    30

OSWALD Away. I have nothing to do with thee.

KENT Draw, you rascal. You bring letters against the
King, and take Vanity the puppet's part against the

royalty of her father. Draw, you rogue, or I'll so
carbonado your shanks—draw, you rascal, come your
ways!      36

OSWALD Help, ho, murder, help!

KENT Strike, you slave! Stand, rogue! Stand, you neat
slave, strike!

OSWALD Help, ho, murder, help!      40
     *Enter Edmund the bastard with his rapier drawn,*
     ⌈*then*⌉ *the Duke of Gloucester,* ⌈*then*⌉ *the Duke of*
     *Cornwall and Regan the Duchess*

EDMUND ⌈*parting them*⌉ How now, what's the matter?

KENT With you, goodman boy. An you please come, I'll
flesh you. Come on, young master.

GLOUCESTER Weapons? Arms? What's the matter here?

CORNWALL Keep peace, upon your lives. He dies that
strikes again. What's the matter?      46

REGAN The messengers from our sister and the King.

CORNWALL (to Kent and Oswald) What's your difference?
Speak.

OSWALD I am scarce in breath, my lord.      50

KENT No marvel, you have so bestirred your valour, you
cowardly rascal. Nature disclaims in thee; a tailor made
thee.

CORNWALL Thou art a strange fellow—a tailor make a
man?      55

KENT Ay, a tailor, sir. A stone-cutter or a painter could
not have made him so ill though he had been but two
hours at the trade.

GLOUCESTER Speak yet; how grew your quarrel?

OSWALD This ancient ruffian, sir, whose life I have spared
at suit of his grey beard—      61

KENT Thou whoreson Z, thou unnecessary letter—(to
Cornwall) my lord, if you'll give me leave I will tread
this unboulted villain into mortar and daub the walls
of a jakes with him. (To Oswald) Spare my grey beard,
you wagtail?      66

CORNWALL
Peace, sir. You beastly knave, have you no reverence?

KENT
Yes, sir, but anger has a privilege.

CORNWALL Why art thou angry?

KENT
That such a slave as this should wear a sword,    70
That wears no honesty. Such smiling rogues
As these, like rats, oft bite those cords in twain
Which are too entrenched to unloose, smooth every
     passion
That in the natures of their lords rebel,
Bring oil to fire, snow to their colder moods,    75
Renege, affirm, and turn their halcyon beaks
With every gale and vary of their masters,
Knowing naught, like dogs, but following.
(To Oswald) A plague upon your epileptic visage!
Smile you my speeches as I were a fool?    80
Goose, an I had you upon Sarum Plain
I'd send you cackling home to Camelot.

CORNWALL
What, art thou mad, old fellow?

GLOUCESTER ⌈*to Kent*⌉      How fell you out? Say that.

KENT
No contraries hold more antipathy
Than I and such a knave.

CORNWALL      Why dost thou call him knave?
What's his offence?

KENT      His countenance likes me not.    86

CORNWALL
No more perchance does mine, or his, or hers.
KENT
Sir, 'tis my occupation to be plain:
I have seen better faces in my time
Than stands on any shoulder that I see          90
Before me at this instant.
CORNWALL                          This is a fellow
Who, having been praised for bluntness, doth affect
A saucy roughness, and constrains the garb
Quite from his nature. He cannot flatter, he.
He must be plain, he must speak truth.          95
An they will take't, so; if not, he's plain.
These kind of knaves I know, which in this plainness
Harbour more craft and more corrupter ends
Than twenty silly-ducking observants
That stretch their duties nicely.               100
KENT
Sir, in good sooth, or in sincere verity,
Under the allowance of your grand aspect,
Whose influence, like the wreath of radiant fire
In flickering Phoebus' front—
CORNWALL                What mean'st thou by this?
KENT To go out of my dialect, which you discommend so
much. I know, sir, I am no flatterer. He that beguiled
you in a plain accent was a plain knave, which for my
part I will not be, though I should win your displeasure
to entreat me to't.                             109
CORNWALL (to Oswald)
What's the offence you gave him?
OSWALD                          I never gave him any.
It pleased the King his master very late
To strike at me upon his misconstruction,
When he, conjunct, and flattering his displeasure,
Tripped me behind; being down, insulted, railed,
And put upon him such a deal of man that        115
That worthied him, got praises of the King
For him attempting who was self-subdued,
And in the fleshment of this dread exploit
Drew on me here again.
KENT                    None of these rogues and cowards
But Ajax is their fool.
CORNWALL ⌈calling⌉     Bring forth the stocks, ho!—
You stubborn, ancient knave, you reverend braggart,
We'll teach you.
KENT            I am too old to learn.           122
Call not your stocks for me. I serve the King,
On whose employments I was sent to you.
You should do small respect, show too bold malice
Against the grace and person of my master,      126
Stocking his messenger.
CORNWALL ⌈calling⌉     Fetch forth the stocks!—
As I have life and honour, there shall he sit till noon.
REGAN
Till noon?—till night, my lord, and all night too.
KENT
Why, madam, if I were your father's dog         130
You could not use me so.
REGAN                    Sir, being his knave, I will.
⌈Stocks brought out⌉
CORNWALL
This is a fellow of the selfsame nature
Our sister speaks of.—Come, bring away the stocks.
GLOUCESTER
Let me beseech your grace not to do so.

His fault is much, and the good King his master    135
Will check him for't. Your purposed low correction
Is such as basest and contemnèd wretches
For pilf'rings and most common trespasses
Are punished with. The King must take it ill
That he's so slightly valued in his messenger,    140
Should have him thus restrained.
CORNWALL                          I'll answer that.
REGAN
My sister may receive it much more worse
To have her gentlemen abused, assaulted,
For following her affairs. Put in his legs.
They put Kent in the stocks
Come, my good lord, away!                         145
Exeunt all but Gloucester and Kent
GLOUCESTER
I am sorry for thee, friend. 'Tis the Duke's pleasure,
Whose disposition, all the world well knows,
Will not be rubbed nor stopped. I'll entreat for thee.
KENT
Pray you, do not, sir. I have watched and travelled
hard.
Some time I shall sleep out; the rest I'll whistle.    150
A good man's fortune may grow out at heels.
Give you good morrow.
GLOUCESTER
The Duke's to blame in this; 'twill be ill took.    Exit
KENT
Good King, that must approve the common say:
Thou out of heaven's benediction com'st           155
To the warm sun.
⌈He takes out a letter⌉
Approach, thou beacon to this under globe,
That by thy comfortable beams I may
Peruse this letter. Nothing almost sees miracles
But misery. I know 'tis from Cordelia,            160
Who hath now fortunately been informed
Of my obscurèd course, and shall find time
For this enormous state, seeking to give
Losses their remedies. All weary and overwatched,
Take vantage, heavy eyes, not to behold           165
This shameful lodging. Fortune, good night;
Smile; once more turn thy wheel.                  He sleeps
Enter Edgar
EDGAR                          I heard myself proclaimed,
And by the happy hollow of a tree
Escaped the hunt. No port is free, no place
That guard and most unusual vigilance             170
Does not attend my taking. While I may scape
I will preserve myself, and am bethought
To take the basest and most poorest shape
That ever penury in contempt of man
Brought near to beast. My face I'll grime with filth,
Blanket my loins, elf all my hair with knots,     176
And with presented nakedness outface
The wind and persecution of the sky.
The country gives me proof and precedent
Of Bedlam beggars who with roaring voices         180
Strike in their numbed and mortified bare arms
Pins, wooden pricks, nails, sprigs of rosemary,
And with this horrible object from low farms,
Poor pelting villages, sheep-cotes and mills
Sometime with lunatic bans, sometime with prayers
Enforce their charity. 'Poor Tuelygod, Poor Tom!'    186
That's something yet. Edgar I nothing am.         Exit

*Enter King Lear, his Fool, and a Knight*

LEAR
'Tis strange that they should so depart from home
And not send back my messenger.

KNIGHT                                        As I learned,
The night before there was no purpose          190
Of his remove.

KENT (*waking*)    Hail to thee, noble master.

LEAR
How! Mak'st thou this shame thy pastime?

FOOL Ha, ha, look, he wears cruel garters! Horses are
tied by the heads, dogs and bears by th' neck, monkeys
by th' loins, and men by th' legs. When a man's overlusty at legs, then he wears wooden nether-stocks.

LEAR (*to Kent*)
What's he that hath so much thy place mistook    197
To set thee here?

KENT                    It is both he and she:
Your son and daughter.

LEAR                        No.

KENT                            Yes.

LEAR                                No, I say.

KENT
I say yea.

LEAR        No, no, they would not.

KENT                                    Yes, they have.    200

LEAR
By Jupiter, I swear no. They durst not do't,
They would not, could not do't. 'Tis worse than murder,
To do upon respect such violent outrage.
Resolve me with all modest haste which way
Thou mayst deserve or they propose this usage,    205
Coming from us.

KENT                    My lord, when at their home
I did commend your highness' letters to them,
Ere I was risen from the place that showed
My duty kneeling, came there a reeking post
Stewed in his haste, half breathless, panting forth    210
From Gonoril, his mistress, salutations,
Delivered letters spite of intermission,
Which presently they read, on whose contents
They summoned up their meiny, straight took horse,
Commanded me to follow and attend    215
The leisure of their answer, gave me cold looks;
And meeting here the other messenger,
Whose welcome I perceived had poisoned mine—
Being the very fellow that of late
Displayed so saucily against your highness—    220
Having more man than wit about me, drew.
He raised the house with loud and coward cries.
Your son and daughter found this trespass worth
This shame which here it suffers.

LEAR
O, how this mother swells up toward my heart!    225
*Histerica passio*, down, thou climbing sorrow;
Thy element's below.—Where is this daughter?

KENT
With the Earl, sir, within.

LEAR                            Follow me not; stay there.
*Exit*

KNIGHT (*to Kent*)
Made you no more offence than what you speak of?

KENT
No. How chance the King comes with so small a train?

FOOL An thou hadst been set in the stocks for that
question, thou hadst well deserved it.    232

KENT Why, fool?

FOOL We'll set thee to school to an ant, to teach thee
there's no labouring in the winter. All that follow their
noses are led by their eyes but blind men, and there's
not a nose among a hundred but can smell him that's
stinking. Let go thy hold when a great wheel runs
down a hill, lest it break thy neck with following it;
but the great one that goes up the hill, let him draw
thee after. When a wise man gives thee better counsel,
give me mine again. I would have none but knaves
follow it, since a fool gives it.

⌐*Sings*⌐    That sir that serves for gain
                And follows but for form,          245
            Will pack when it begin to rain,
                And leave thee in the storm.

            But I will tarry, the fool will stay,
                And let the wise man fly.
            The knave turns fool that runs away,    250
                The fool no knave, pardie.

KENT Where learnt you this, fool?

FOOL Not in the stocks.

*Enter King Lear and the Duke of Gloucester*

LEAR
Deny to speak with me? They're sick, they're weary?
They travelled hard tonight?—mere insolence,    255
Ay, the images of revolt and flying off.
Fetch me a better answer.

GLOUCESTER                    My dear lord,
You know the fiery quality of the Duke,
How unremovable and fixed he is    259
In his own course.

LEAR                    Vengeance, death, plague, confusion!
What 'fiery quality'? Why, Gloucester, Gloucester, I'd
Speak with the Duke of Cornwall and his wife.

GLOUCESTER Ay, my good lord.

LEAR
The King would speak with Cornwall; the dear father
Would with his daughter speak, commands, tends
            service.                              265
'Fiery'? The Duke?—tell the hot Duke that Lear—
No, but not yet. Maybe he is not well.
Infirmity doth still neglect all office
Whereto our health is bound. We are not ourselves
When nature, being oppressed, commands the mind
To suffer with the body. I'll forbear,    271
And am fallen out with my more headier will,
To take the indisposed and sickly fit
For the sound man.—Death on my state,
Wherefore should he sit here? This act persuades me
That this remotion of the Duke and her    276
Is practice only. Give me my servant forth.
Tell the Duke and 's wife I'll speak with them,
Now, presently. Bid them come forth and hear me,
Or at their chamber door I'll beat the drum    280
Till it cry sleep to death.

GLOUCESTER                    I would have all well
Betwixt you.                                    *Exit*

LEAR            O, my heart, my heart!

FOOL Cry to it, nuncle, as the cockney did to the eels
when she put 'em i'th' paste alive. She rapped 'em
o'th' coxcombs with a stick, and cried 'Down, wantons,
down!' 'Twas her brother that, in pure kindness to his
horse, buttered his hay.    287

*Enter the Duke of Cornwall and Regan, the Duke of
Gloucester, and others*

LEAR Good morrow to you both.

CORNWALL  Hail to your grace.
　　　　⌈*Kent here set at liberty*⌉
REGAN  I am glad to see your highness.                    290
LEAR
　　Regan, I think you are. I know what reason
　　I have to think so. If thou shouldst not be glad
　　I would divorce me from thy mother's shrine,
　　Sepulchring an adultress. (*To Kent*) Yea, are you free?
　　Some other time for that.—Belovèd Regan,        295
　　Thy sister is naught. O, Regan, she hath tied
　　Sharp-toothed unkindness like a vulture here.
　　I can scarce speak to thee. Thou'lt not believe
　　Of how deplored a quality—O, Regan!
REGAN
　　I pray you, sir, take patience. I have hope           300
　　You less know how to value her desert
　　Than she to slack her duty.
LEAR  My curses on her.
REGAN  O sir, you are old.
　　Nature in you stands on thè very verge               305
　　Of her confine. You should be ruled and led
　　By some discretion that discerns your state
　　Better than you yourself. Therefore I pray
　　That to our sister you do make return;
　　Say you have wronged her, sir.
LEAR　　　　　　　　　　　Ask her forgiveness?
　　Do you mark how this becomes the house?       311
　　⌈*Kneeling*⌉ 'Dear daughter, I confess that I am old.
　　Age is unnecessary. On my knees I beg
　　That you'll vouchsafe me raiment, bed, and food.'
REGAN
　　Good sir, no more. These are unsightly tricks.    315
　　Return you to my sister.
LEAR ⌈*rising*⌉　　　　　No, Regan.
　　She hath abated me of half my train,
　　Looked black upon me, struck me with her tongue
　　Most serpent-like upon the very heart.
　　All the stored vengeances of heaven fall             320
　　On her ungrateful top! Strike her young bones,
　　You taking airs, with lameness!
CORNWALL　　　　　　　　　　Fie, fie, sir.
LEAR
　　You nimble lightnings, dart your blinding flames
　　Into her scornful eyes. Infect her beauty,
　　You fen-sucked fogs drawn by the pow'rful sun   325
　　To fall and blast her pride.
REGAN　　　　　　　　　　O, the blest gods!
　　So will you wish on me when the rash mood—
LEAR
　　No, Regan. Thou shalt never have my curse.
　　Thy tender-hested nature shall not give
　　Thee o'er to harshness. Her eyes are fierce, but thine
　　Do comfort and not burn. 'Tis not in thee          331
　　To grudge my pleasures, to cut off my train,
　　To bandy hasty words, to scant my sizes,
　　And, in conclusion, to oppose the bolt
　　Against my coming in. Thou better know'st       335
　　The offices of nature, bond of childhood,
　　Effects of courtesy, dues of gratitude.
　　Thy half of the kingdom hast thou not forgot,
　　Wherein I thee endowed.
REGAN　　　　　　　　Good sir, to th' purpose.
LEAR
　　Who put my man i'th' stocks?
　　⌈*Trumpets within*⌉
CORNWALL　　　　　　　　What trumpet's that?

*Enter Oswald the steward*
REGAN
　　I know't, my sister's. This approves her letters    341
　　That she would soon be here. (*To Oswald*) Is your lady
　　　come?
LEAR
　　This is a slave whose easy-borrowed pride
　　Dwells in the fickle grace of her a follows.
　　⌈*He strikes Oswald*⌉
　　Out, varlet, from my sight!
CORNWALL　　　　　　　What means your grace?
　　*Enter Gonoril*
GONORIL
　　Who struck my servant? Regan, I have good hope  346
　　Thou didst not know on't.
LEAR　　　　　　　　Who comes here? O heavens,
　　If you do love old men, if your sweet sway
　　Allow obedience, if yourselves are old,
　　Make it your cause! Send down and take my part.  350
　　(*To Gonoril*) Art not ashamed to look upon this
　　　beard?
　　O Regan, wilt thou take her by the hand?
GONORIL
　　Why not by the hand, sir? How have I offended?
　　All's not offence that indiscretion finds
　　And dotage terms so.
LEAR　　　　　　　O sides, you are too tough!    355
　　Will you yet hold?—How came my man i'th' stocks?
CORNWALL
　　I set him there, sir; but his own disorders
　　Deserved much less advancement.
LEAR　　　　　　　　　　You? Did you?
REGAN
　　I pray you, father, being weak, seem so.
　　If till the expiration of your month                     360
　　You will return and sojourn with my sister,
　　Dismissing half your train, come then to me.
　　I am now from home, and out of that provision
　　Which shall be needful for your entertainment.
LEAR
　　Return to her, and fifty men dismissed?           365
　　No, rather I abjure all roofs, and choose
　　To be a comrade with the wolf and owl,
　　To wage against the enmity of the air
　　Necessity's sharp pinch. Return with her?
　　Why, the hot-blood in France that dowerless took  370
　　Our youngest born—I could as well be brought
　　To knee his throne and, squire-like, pension beg
　　To keep base life afoot. Return with her?
　　Persuade me rather to be slave and sumpter
　　To this detested groom.
GONORIL　　　　　　　At your choice, sir.          375
LEAR
　　Now I prithee, daughter, do not make me mad.
　　I will not trouble thee, my child. Farewell.
　　We'll no more meet, no more see one another.
　　But yet thou art my flesh, my blood, my daughter—
　　Or rather a disease that lies within my flesh,       380
　　Which I must needs call mine. Thou art a boil,
　　A plague-sore, an embossèd carbuncle
　　In my corrupted blood. But I'll not chide thee.
　　Let shame come when it will, I do not call it.
　　I do not bid the thunder-bearer shoot,              385
　　Nor tell tales of thee to high-judging Jove.
　　Mend when thou canst; be better at thy leisure.

I can be patient, I can stay with Regan,
I and my hundred knights.

REGAN                              Not altogether so, sir.
I look not for you yet, nor am provided          390
For your fit welcome. Give ear, sir, to my sister;
For those that mingle reason with your passion
Must be content to think you are old, and so—
But she knows what she does.

LEAR                              Is this well spoken now?

REGAN
I dare avouch it, sir. What, fifty followers?     395
Is it not well? What should you need of more,
Yea, or so many, sith that both charge and danger
Speaks 'gainst so great a number? How in a house
Should many people under two commands
Hold amity? 'Tis hard, almost impossible.          400

GONORIL
Why might not you, my lord, receive attendance
From those that she calls servants, or from mine?

REGAN
Why not, my lord? If then they chanced to slack you,
We could control them. If you will come to me—
For now I spy a danger—I entreat you              405
To bring but five-and-twenty; to no more
Will I give place or notice.

LEAR I gave you all.

REGAN And in good time you gave it.

LEAR
Made you my guardians, my depositaries,           410
But kept a reservation to be followed
With such a number. What, must I come to you
With five-and-twenty, Regan? Said you so?

REGAN
And speak't again, my lord. No more with me.

LEAR
Those wicked creatures yet do seem well favoured  415
When others are more wicked. Not being the worst
Stands in some rank of praise. (*To Gonoril*) I'll go with
   thee.
Thy fifty yet doth double five-and-twenty,
And thou art twice her love.

GONORIL                     Hear me, my lord.
What need you five-and-twenty, ten, or five,       420
To follow in a house where twice so many
Have a command to tend you?

REGAN                       What needs one?

LEAR
O, reason not the need! Our basest beggars
Are in the poorest thing superfluous.
Allow not nature more than nature needs,           425
Man's life is cheap as beast's. Thou art a lady.
If only to go warm were gorgeous,
Why, nature needs not what thou, gorgeous, wearest,
Which scarcely keeps thee warm. But for true need—
You heavens, give me that patience, patience I need.
You see me here, you gods, a poor old fellow,      431
As full of grief as age, wretchèd in both.
If it be you that stirs these daughters' hearts
Against their father, fool me not so much
To bear it tamely. Touch me with noble anger.      435
O, let not women's weapons, water-drops,
Stain my man's cheeks! No, you unnatural hags,
I will have such revenges on you both
That all the world shall—I will do such things—
What they are, yet I know not; but they shall be   440
The terrors of the earth. You think I'll weep.

No, I'll not weep.
   ⌈*Storm within*⌉
I have full cause of weeping, but this heart
Shall break into a hundred thousand flaws
Or ere I'll weep.—O fool, I shall go mad!          445
   *Exeunt Lear, Gloucester, Kent,* ⌈*Knight,*⌉
                                          *and Fool*

CORNWALL
Let us withdraw. 'Twill be a storm.

REGAN
This house is little. The old man and his people
Cannot be well bestowed.

GONORIL                  'Tis his own blame;
Hath put himself from rest, and must needs taste his
   folly.

REGAN
For his particular I'll receive him gladly,        450
But not one follower.

CORNWALL
So am I purposed. Where is my lord of Gloucester?

REGAN
Followed the old man forth.
   *Enter the Duke of Gloucester*
                              He is returned.

GLOUCESTER
The King is in high rage, and will I know not whither.

REGAN
'Tis good to give him way. He leads himself.       455

GONORIL (*to Gloucester*)
My lord, entreat him by no means to stay.

GLOUCESTER
Alack, the night comes on, and the bleak winds
Do sorely rustle. For many miles about
There's not a bush.

REGAN              O sir, to wilful men
The injuries that they themselves procure          460
Must be their schoolmasters. Shut up your doors.
He is attended with a desperate train,
And what they may incense him to, being apt
To have his ear abused, wisdom bids fear.

CORNWALL
Shut up your doors, my lord. 'Tis a wild night.    465
My Regan counsels well. Come out o'th' storm. *Exeunt*

Sc. 8  *Storm. Enter the Earl of Kent disguised, and First*
       *Gentleman, at several doors*

KENT
What's here, beside foul weather?

FIRST GENTLEMAN            One minded like the weather,
Most unquietly.

KENT           I know you. Where's the King?

FIRST GENTLEMAN
Contending with the fretful element;
Bids the wind blow the earth into the sea
Or swell the curlèd waters 'bove the main,          5
That things might change or cease; tears his white
   hair,
Which the impetuous blasts, with eyeless rage,
Catch in their fury and make nothing of;
Strives in his little world of man to outstorm
The to-and-fro-conflicting wind and rain.          10
This night, wherein the cub-drawn bear would couch,
The lion and the belly-pinchèd wolf
Keep their fur dry, unbonneted he runs,
And bids what will take all.

KENT                        But who is with him?

FIRST GENTLEMAN
   None but the fool, who labours to outjest          15
   His heart-struck injuries.
KENT                         Sir, I do know you,
   And dare upon the warrant of my art
   Commend a dear thing to you. There is division,
   Although as yet the face of it be covered
   With mutual cunning, 'twixt Albany and Cornwall;
   But true it is. From France there comes a power    21
   Into this scattered kingdom, who already,
   Wise in our negligence, have secret feet
   In some of our best ports, and are at point
   To show their open banner. Now to you:             25
   If on my credit you dare build so far
   To make your speed to Dover, you shall find
   Some that will thank you, making just report
   Of how unnatural and bemadding sorrow
   The King hath cause to plain.                      30
   I am a gentleman of blood and breeding,
   And from some knowledge and assurance offer
   This office to you.
FIRST GENTLEMAN I will talk farther with you.
KENT No, do not.                                      35
   For confirmation that I am much more
   Than my out-wall, open this purse, and take
   What it contains. If you shall see Cordelia—
   As fear not but you shall—show her this ring
   And she will tell you who your fellow is,           40
   That yet you do not know. Fie on this storm!
   I will go seek the King.
FIRST GENTLEMAN            Give me your hand.
   Have you no more to say?
KENT                         Few words, but to effect
   More than all yet: that when we have found the King—
   In which endeavour I'll this way, you that—        45
   He that first lights on him holla the other.
                                    *Exeunt severally*

Sc. 9   *Storm. Enter King Lear and his Fool*
LEAR
   Blow, wind, and crack your cheeks! Rage, blow,
   You cataracts and hurricanoes, spout
   Till you have drenched the steeples, drowned the
      cocks!
   You sulphurous and thought-executing fires,
   Vaunt-couriers to oak-cleaving thunderbolts,        5
   Singe my white head; and thou all-shaking thunder,
   Smite flat the thick rotundity of the world,
   Crack nature's mould, all germens spill at once
   That make ingrateful man.                           9
FOOL O nuncle, court holy water in a dry house is better
   than this rain-water out o' door. Good nuncle, in, and
   ask thy daughters blessing. Here's a night pities neither
   wise man nor fool.
LEAR
   Rumble thy bellyful; spit, fire; spout, rain.
   Nor rain, wind, thunder, fire are my daughters.    15
   I tax not you, you elements, with unkindness.
   I never gave you kingdom, called you children.
   You owe me no subscription. Why then, let fall
   Your horrible pleasure. Here I stand your slave,
   A poor, infirm, weak and despised old man,          20
   But yet I call you servile ministers,
   That have with two pernicious daughters joined
   Your high engendered battle 'gainst a head
   So old and white as this. O, 'tis foul!

FOOL He that has a house to put his head in has a good
   headpiece.                                          26
   ⌜*Sings*⌝      The codpiece that will house
                     Before the head has any,
                  The head and he shall louse,
                     So beggars marry many.            30

                  The man that makes his toe
                     What he his heart should make
                  Shall have a corn cry woe,
                     And turn his sleep to wake—
   for there was never yet fair woman but she made
   mouths in a glass.                                  36
LEAR
   No, I will be the pattern of all patience.
      ⌜*He sits.*⌝ *Enter the Earl of Kent disguised*
   I will say nothing.
KENT Who's there?
FOOL Marry, here's grace and a codpiece—that's a wise
   man and a fool.                                     41
KENT (*to Lear*)
   Alas, sir, sit you here? Things that love night
   Love not such nights as these. The wrathful skies
   Gallow the very wanderers of the dark
   And makes them keep their caves. Since I was man   45
   Such sheets of fire, such bursts of horrid thunder,
   Such groans of roaring wind and rain I ne'er
   Remember to have heard. Man's nature cannot carry
   The affliction nor the force.
LEAR                         Let the great gods,
   That keep this dreadful pother o'er our heads,      50
   Find out their enemies now. Tremble, thou wretch
   That hast within thee undivulgèd crimes
   Unwhipped of justice; hide thee, thou bloody hand,
   Thou perjured and thou simular man of virtue
   That art incestuous; caitiff, in pieces shake,      55
   That under covert and convenient seeming
   Hast practisèd on man's life;
   Close pent-up guilts, rive your concealèd centres
   And cry these dreadful summoners grace.
   I am a man more sinned against than sinning.        60
KENT Alack, bare-headed?
   Gracious my lord, hard by here is a hovel.
   Some friendship will it lend you 'gainst the tempest.
   Repose you there whilst I to this hard house—
   More hard than is the stone whereof 'tis raised,    65
   Which even but now, demanding after you,
   Denied me to come in—return and force
   Their scanted courtesy.
LEAR                       My wit begins to turn.
   (*To Fool*) Come on, my boy. How dost, my boy? Art
      cold?
   I am cold myself.—Where is this straw, my fellow?   70
   The art of our necessities is strange,
   That can make vile things precious. Come, your
      hovel.
   Poor fool and knave, I have one part of my heart
   That sorrows yet for thee.
FOOL ⌜*sings*⌝
                  He that has a little tiny wit,       75
                     With heigh-ho, the wind and the rain,
                  Must make content with his fortunes fit,
                     For the rain it raineth every day.
LEAR
   True, my good boy. (*To Kent*) Come, bring us to this
      hovel.                                  *Exeunt*

**Sc. 10** *Enter the Duke of Gloucester and Edmund the bastard, with lights*

GLOUCESTER
Alack, alack, Edmund, I like not this
Unnatural dealing. When I desired their leave
That I might pity him, they took from me
The use of mine own house, charged me on pain
Of their displeasure neither to speak of him,          5
Entreat for him, nor any way sustain him.

EDMUND Most savage and unnatural!

GLOUCESTER Go to, say you nothing. There's a division betwixt the Dukes, and a worse matter than that. I have received a letter this night—'tis dangerous to be spoken—I have locked the letter in my closet. These injuries the King now bears will be revenged home. There's part of a power already landed. We must incline to the King. I will seek him and privily relieve him. Go you and maintain talk with the Duke, that my charity be not of him perceived. If he ask for me, I am ill and gone to bed. Though I die for't—as no less is threatened me—the King my old master must be relieved. There is some strange thing toward. Edmund, pray you be careful.                              *Exit*

EDMUND
This courtesy, forbid thee, shall the Duke          21
Instantly know, and of that letter too.
This seems a fair deserving, and must draw me
That which my father loses: no less than all.
The younger rises when the old do fall.          *Exit*

**Sc. 11** *Storm. Enter King Lear, the Earl of Kent disguised, and Lear's Fool*

KENT
Here is the place, my lord. Good my lord, enter.
The tyranny of the open night's too rough
For nature to endure.

LEAR                    Let me alone.

KENT
Good my lord, enter here.

LEAR                              Wilt break my heart?

KENT
I had rather break mine own. Good my lord, enter.     5

LEAR
Thou think'st 'tis much that this contentious storm
Invades us to the skin. So 'tis to thee;
But where the greater malady is fixed,
The lesser is scarce felt. Thou'dst shun a bear,
But if thy flight lay toward the roaring sea          10
Thou'dst meet the bear i'th' mouth. When the mind's free,
The body's delicate. This tempest in my mind
Doth from my senses take all feeling else
Save what beats there: filial ingratitude.
Is it not as this mouth should tear this hand          15
For lifting food to't? But I will punish sure.
No, I will weep no more.—
In such a night as this! O Regan, Gonoril,
Your old kind father, whose frank heart gave you all—
O, that way madness lies. Let me shun that.          20
No more of that.

KENT                    Good my lord, enter.

LEAR
Prithee, go in thyself. Seek thy own ease.
This tempest will not give me leave to ponder
On things would hurt me more; but I'll go in.
                              *Exit Fool*
Poor naked wretches, wheresoe'er you are,          25

That bide the pelting of this pitiless night,
How shall your houseless heads and unfed sides,
Your looped and windowed raggedness, defend you
From seasons such as these? O, I have ta'en
Too little care of this. Take physic, pomp,          30
Expose thyself to feel what wretches feel,
That thou mayst shake the superflux to them
And show the heavens more just.
          *Enter Lear's Fool*

FOOL Come not in here, nuncle; here's a spirit. Help me, help me!          35

KENT Give me thy hand. Who's there?

FOOL A spirit. He says his name's Poor Tom.

KENT
What art thou that dost grumble there in the straw?
Come forth.
          *Enter Edgar as a Bedlam beggar*

EDGAR Away, the foul fiend follows me. Through the sharp hawthorn blows the cold wind. Go to thy cold bed and warm thee.          42

LEAR
Hast thou given all to thy two daughters,
And art thou come to this?

EDGAR Who gives anything to Poor Tom, whom the foul fiend hath led through fire and through ford and whirlpool, o'er bog and quagmire; that has laid knives under his pillow and halters in his pew, set ratsbane by his potage, made him proud of heart to ride on a bay trotting-horse over four-inched bridges, to course his own shadow for a traitor. Bless thy five wits, Tom's a-cold! Bless thee from whirlwinds, star-blasting, and taking. Do Poor Tom some charity, whom the foul fiend vexes. There could I have him, now, and there, and there again.          55

LEAR
What, has his daughters brought him to this pass?
(*To Edgar*) Couldst thou save nothing? Didst thou give them all?

FOOL Nay, he reserved a blanket, else we had been all shamed.

LEAR (*to Edgar*)
Now all the plagues that in the pendulous air          60
Hang fated o'er men's faults fall on thy daughters!

KENT He hath no daughters, sir.

LEAR
Death, traitor! Nothing could have subdued nature
To such a lowness but his unkind daughters.
(*To Edgar*) Is it the fashion that discarded fathers          65
Should have thus little mercy on their flesh?
Judicious punishment: 'twas this flesh begot
Those pelican daughters.

EDGAR Pillicock sat on pillicock's hill; a lo, lo, lo.          69

FOOL This cold night will turn us all to fools and madmen.

EDGAR Take heed o'th' foul fiend; obey thy parents; keep thy word justly; swear not; commit not with man's sworn spouse: set not thy sweet heart on proud array. Tom's a-cold.

LEAR What hast thou been?          75

EDGAR A servingman, proud in heart and mind, that curled my hair, wore gloves in my cap, served the lust of my mistress' heart, and did the act of darkness with her; swore as many oaths as I spake words, and broke them in the sweet face of heaven; one that slept in the contriving of lust, and waked to do it. Wine loved I deeply, dice dearly, and in woman out-paramoured the Turk. False of heart, light of ear, bloody of hand; hog in sloth, fox in stealth, wolf in greediness, dog in madness, lion in prey. Let not the creaking of shoes nor the rustlings of silks betray thy poor heart to women. Keep thy foot out of brothel, thy hand out of placket, thy pen from lender's book, and defy the foul fiend. Still through the hawthorn blows the cold wind. Heigh no nonny. Dolphin, my boy, my boy! Cease, let him trot by.      91

LEAR Why, thou wert better in thy grave than to answer with thy uncovered body this extremity of the skies. Is man no more but this? Consider him well. Thou owest the worm no silk, the beast no hide, the sheep no wool, the cat no perfume. Here's three on 's are sophisticated; thou art the thing itself. Unaccommodated man is no more but such a poor, bare, forked animal as thou art. Off, off, you lendings! Come on, be true.      99

FOOL Prithee, nuncle, be content. This is a naughty night to swim in. Now a little fire in a wild field were like an old lecher's heart—a small spark, all the rest on 's body cold. Look, here comes a walking fire.

*Enter the Duke of Gloucester with a ⌐torch⌐*

EDGAR This is the foul fiend Flibbertigibbet. He begins at curfew and walks till the first cock. He gives the web and the pin, squinies the eye, and makes the harelip; mildews the white wheat, and hurts the poor creature of earth.

⌐*Sings*⌐

> Swithin footed thrice the wold,
> A met the night mare and her nine foal;   110
>   Bid her alight
>   And her troth plight,
> And aroint thee, witch, aroint thee!

KENT (*to Lear*)
  How fares your grace?

LEAR               What's he?

KENT (*to Gloucester*) Who's there? What is't you seek?

GLOUCESTER What are you there? Your names?   116

EDGAR Poor Tom, that eats the swimming frog, the toad, the tadpole, the wall-newt and the water; that in the fury of his heart, when the foul fiend rages, eats cowdung for salads, swallows the old rat and the ditch-dog, drinks the green mantle of the standing pool; who is whipped from tithing to tithing, and stock-punished, and imprisoned; who hath had three suits to his back, six shirts to his body,

> Horse to ride, and weapon to wear.   125
> But mice and rats and such small deer
> Hath been Tom's food for seven long year—

Beware my follower. Peace, Smolking; peace, thou fiend!

GLOUCESTER (*to Lear*)
  What, hath your grace no better company?   130

EDGAR
  The Prince of Darkness is a gentleman;
  Modo he's called, and Mahu—

GLOUCESTER (*to Lear*)
  Our flesh and blood is grown so vile, my lord,
  That it doth hate what gets it.

EDGAR            Poor Tom's a-cold.

GLOUCESTER (*to Lear*)
  Go in with me. My duty cannot suffer   135
  To obey in all your daughters' hard commands.
  Though their injunction be to bar my doors
  And let this tyrannous night take hold upon you,
  Yet have I ventured to come seek you out
  And bring you where both food and fire is ready.  140

LEAR
  First let me talk with this philosopher.
  (*To Edgar*) What is the cause of thunder?

KENT               My good lord,
  Take his offer; go into the house.

LEAR
  I'll talk a word with this most learnèd Theban.
  (*To Edgar*) What is your study?   145

EDGAR
  How to prevent the fiend, and to kill vermin.

LEAR
  Let me ask you one word in private.
      *They converse apart*

KENT (*to Gloucester*)
  Importune him to go, my lord.
  His wits begin to unsettle.

GLOUCESTER         Canst thou blame him?
  His daughters seek his death. O, that good Kent,   150
  He said it would be thus, poor banished man!
  Thou sayst the King grows mad; I'll tell thee, friend,
  I am almost mad myself. I had a son,
  Now outlawed from my blood; a sought my life
  But lately, very late. I loved him, friend;   155
  No father his son dearer. True to tell thee,
  The grief hath crazed my wits. What a night's this!
  (*To Lear*) I do beseech your grace—

LEAR               O, cry you mercy.
  (*To Edgar*) Noble philosopher, your company.

EDGAR              Tom's a-cold.

GLOUCESTER
  In, fellow, there in t'hovel; keep thee warm.   160

LEAR
  Come, let's in all.

KENT        This way, my lord.

LEAR               With him!
  I will keep still with my philosopher.

KENT (*to Gloucester*)
  Good my lord, soothe him; let him take the fellow.

GLOUCESTER Take him you on.

KENT ⌐*to Edgar*⌐
  Sirrah, come on. Go along with us.   165

LEAR (*to Edgar*)
  Come, good Athenian.

GLOUCESTER      No words, no words. Hush.

EDGAR    Child Roland to the dark tower come,
    His word was still 'Fie, fo, and fum;
    I smell the blood of a British man.'   *Exeunt*

**Sc. 12** *Enter the Duke of Cornwall and Edmund the bastard*

CORNWALL I will have my revenge ere I depart the house.

EDMUND How, my lord, I may be censured, that nature thus gives way to loyalty, something fears me to think of.

CORNWALL I now perceive it was not altogether your brother's evil disposition made him seek his death, but a provoking merit set a-work by a reprovable badness in himself.   8

EDMUND How malicious is my fortune, that I must repent to be just! This is the letter he spoke of, which approves him an intelligent party to the advantages of France. O heavens, that his treason were not, or not I the detector!

CORNWALL Go with me to the Duchess.   14

EDMUND If the matter of this paper be certain, you have mighty business in hand.

CORNWALL True or false, it hath made thee Earl of
Gloucester. Seek out where thy father is, that he may
be ready for our apprehension.                        19
EDMUND ⌈*aside*⌉ If I find him comforting the King, it will
stuff his suspicion more fully. (*To Cornwall*) I will
persever in my course of loyalty, though the conflict
be sore between that and my blood.
CORNWALL I will lay trust upon thee, and thou shalt find
a dearer father in my love.                    *Exeunt*

**Sc. 13** *Enter the Duke of Gloucester and King Lear, the
Earl of Kent disguised, Lear's Fool, and Edgar as a
Bedlam beggar*
GLOUCESTER Here is better than the open air; take it
thankfully. I will piece out the comfort with what
addition I can. I will not be long from you.
KENT All the power of his wits have given way to
impatience; the gods discern your kindness!          5
                              ⌈*Exit Gloucester*⌉
EDGAR Frateretto calls me, and tells me Nero is an angler
in the lake of darkness. Pray, innocent; beware the
foul fiend.
FOOL (*to Lear*) Prithee, nuncle, tell me whether a madman
be a gentleman or a yeoman.                           10
LEAR
A king, a king! To have a thousand
With red burning spits come hissing in upon them!
EDGAR The foul fiend bites my back.
FOOL (*to Lear*) He's mad that trusts in the tameness of a
wolf, a horse's health, a boy's love, or a whore's oath.
LEAR
It shall be done. I will arraign them straight.      16
⌈*To Edgar*⌉ Come, sit thou here, most learnèd justicer.
⌈*To Fool*⌉ Thou sapient sir, sit here.—No, you she-
foxes—
EDGAR Look where he stands and glares. Want'st thou
eyes at troll-madam?                                 20
⌈*Sings*⌉ Come o'er the burn, Bessy, to me.
FOOL ⌈*sings*⌉
          Her boat hath a leak,
          And she must not speak
          Why she dares not come over to thee.       24
EDGAR The foul fiend haunts Poor Tom in the voice of a
nightingale. Hoppedance cries in Tom's belly for two
white herring. Croak not, black angel: I have no food
for thee.
KENT (*to Lear*)
How do you, sir? Stand you not so amazed?
Will you lie down and rest upon the cushions?        30
LEAR
I'll see their trial first. Bring in the evidence.
⌈*To Edgar*⌉ Thou robèd man of justice, take thy place;
⌈*To Fool*⌉ And thou, his yokefellow of equity,
Bench by his side. ⌈*To Kent*⌉ You are o'th'
          commission,
Sit you, too.                                        35
EDGAR Let us deal justly.
     ⌈*Sings*⌉
          Sleepest or wakest thou, jolly shepherd?
          Thy sheep be in the corn,
          And for one blast of thy minikin mouth
          Thy sheep shall take no harm.              40
Purr, the cat is grey.
LEAR Arraign her first. 'Tis Gonoril. I here take my oath
before this honourable assembly she kicked the poor
King her father.
FOOL Come hither, mistress. Is your name Gonoril?    45

LEAR She cannot deny it.
FOOL Cry you mercy, I took you for a join-stool.
LEAR
And here's another, whose warped looks proclaim
What store her heart is made on. Stop her there.
Arms, arms, sword, fire, corruption in the place!    50
False justicer, why hast thou let her scape?
EDGAR Bless thy five wits.
KENT (*to Lear*)
O pity! Sir, where is the patience now
That you so oft have boasted to retain?
EDGAR (*aside*)
My tears begin to take his part so much              55
They'll mar my counterfeiting.
LEAR                          The little dogs and all,
Tray, Blanch, and Sweetheart—see, they bark at me.
EDGAR Tom will throw his head at them.—Avaunt, you
curs!
          Be thy mouth or black or white,            60
          Tooth that poisons if it bite,
          Mastiff, greyhound, mongrel grim,
          Hound or spaniel, brach or him,
          Bobtail tyke or trundle-tail,
          Tom will make them weep and wail;          65
          For with throwing thus my head,
          Dogs leap the hatch, and all are fled.
Loudla, doodla! Come, march to wakes and fairs
And market towns. Poor Tom, thy horn is dry.         69
LEAR Then let them anatomize Regan; see what breeds
about her heart. Is there any cause in nature that
makes this hardness? (*To Edgar*) You, sir, I entertain
you for one of my hundred, only I do not like the
fashion of your garments. You'll say they are Persian
attire; but let them be changed.                     75
KENT
Now, good my lord, lie here a while.
LEAR Make no noise, make no noise. Draw the curtains.
So, so, so. We'll go to supper i'th' morning. So, so, so.
          *He sleeps. Enter the Duke of Gloucester*
GLOUCESTER (*to Kent*)
Come hither, friend. Where is the King my master?
KENT
Here, sir, but trouble him not; his wits are gone.   80
GLOUCESTER
Good friend, I prithee take him in thy arms.
I have o'erheard a plot of death upon him.
There is a litter ready. Lay him in't
And drive towards Dover, friend, where thou shalt meet
Both welcome and protection. Take up thy master.     85
If thou shouldst dally half an hour, his life,
With thine and all that offer to defend him,
Stand in assurèd loss. Take up, take up,
And follow me, that will to some provision
Give thee quick conduct.
KENT (*to Lear*)              Oppressèd nature sleeps.   90
This rest might yet have balmed thy broken sinews
Which, if convenience will not allow,
Stand in hard cure. (*To Fool*) Come, help to bear thy
          master.
Thou must not stay behind.
GLOUCESTER                   Come, come away.
                              *Exeunt all but Edgar*
EDGAR
When we our betters see bearing our woes,            95
We scarcely think our miseries our foes.
Who alone suffers, suffers most i'th' mind,
Leaving free things and happy shows behind.

But then the mind much sufferance doth o'erskip
When grief hath mates, and bearing fellowship.     100
How light and portable my pain seems now,
When that which makes me bend, makes the King
    bow.
He childed as I fathered. Tom, away.
Mark the high noises, and thyself bewray
When false opinion, whose wrong thoughts defile thee,
In thy just proof repeals and reconciles thee.     106
What will hap more tonight, safe scape the King!
Lurk, lurk.                                    *Exit*

**Sc. 14**  *Enter the Duke of Cornwall and Regan, and*
           *Gonoril and Edmund the bastard, and Servants*
CORNWALL (*to Gonoril*)
    Post speedily to my lord your husband.
    Show him this letter. The army of France is landed.
    (*To Servants*) Seek out the villain Gloucester.
                                      *Exeunt some*
REGAN                        Hang him instantly.
GONORIL
    Pluck out his eyes.
CORNWALL                 Leave him to my displeasure.—
    Edmund, keep you our sister company.          5
    The revenges we are bound to take upon your traitorous
    father are not fit for your beholding. Advise the Duke
    where you are going, to a most festinate preparation;
    we are bound to the like. Our posts shall be swift, and
    intelligence betwixt us.—                     10
    Farewell, dear sister. Farewell, my lord of Gloucester.
              *Enter Oswald the steward*
    How now, where's the King?
OSWALD
    My lord of Gloucester hath conveyed him hence.
    Some five- or six-and-thirty of his knights,
    Hot questants after him, met him at gate,     15
    Who, with some other of the lord's dependants,
    Are gone with him towards Dover, where they boast
    To have well-armèd friends.
CORNWALL  Get horses for your mistress.     *Exit Oswald*
GONORIL  Farewell, sweet lord, and sister.         20
CORNWALL
    Edmund, farewell.         *Exeunt Gonoril and Edmund*
    (*To Servants*)      Go seek the traitor Gloucester.
    Pinion him like a thief; bring him before us.
                              *Exeunt other Servants*
    Though we may not pass upon his life
    Without the form of justice, yet our power
    Shall do a curtsy to our wrath, which men     25
    May blame but not control. Who's there—the traitor?
         *Enter the Duke of Gloucester brought in by two or*
         *three*
REGAN
    Ingrateful fox, 'tis he.
CORNWALL (*to Servants*) Bind fast his corky arms.
GLOUCESTER
    What means your graces? Good my friends, consider
    You are my guests. Do me no foul play, friends.
CORNWALL (*to Servants*)
    Bind him, I say—
REGAN                Hard, hard! O filthy traitor!     30
GLOUCESTER
    Unmerciful lady as you are, I am true.
CORNWALL (*to Servants*)
    To this chair bind him. (*To Gloucester*) Villain, thou
        shalt find—
         *Regan plucks Gloucester's beard*

GLOUCESTER
    By the kind gods, 'tis most ignobly done,
    To pluck me by the beard.
REGAN  So white, and such a traitor!               35
GLOUCESTER  Naughty lady,
    These hairs which thou dost ravish from my chin
    Will quicken and accuse thee. I am your host.
    With robbers' hands my hospitable favours
    You should not ruffle thus. What will you do?      40
CORNWALL
    Come, sir, what letters had you late from France?
REGAN
    Be simple, answerer, for we know the truth.
CORNWALL
    And what confederacy have you with the traitors
    Late footed in the kingdom?
REGAN                       To whose hands
    You have sent the lunatic King. Speak.         45
GLOUCESTER
    I have a letter guessingly set down,
    Which came from one that's of a neutral heart,
    And not from one opposed.
CORNWALL                     Cunning.
REGAN                                    And false.
CORNWALL
    Where hast thou sent the King?
GLOUCESTER                         To Dover.
REGAN
    Wherefore to Dover? Wast thou not charged at peril—
CORNWALL
    Wherefore to Dover? Let him first answer that.     51
GLOUCESTER
    I am tied to th' stake, and I must stand the course.
REGAN  Wherefore to Dover, sir?
GLOUCESTER
    Because I would not see thy cruel nails
    Pluck out his poor old eyes, nor thy fierce sister     55
    In his anointed flesh rash boarish fangs.
    The sea, with such a storm as his bowed head
    In hell-black night endured, would have buoyed up
    And quenched the stellèd fires. Yet, poor old heart,
    He holped the heavens to rage. Yet, poor old heart,    60
    If wolves had at thy gate howled that dern time,
    Thou shouldst have said 'Good porter, turn the key;
    All cruels I'll subscribe.' But I shall see
    The wingèd vengeance overtake such children.
CORNWALL
    See't shalt thou never.—Fellows, hold the chair.—     65
    Upon those eyes of thine I'll set my foot.
GLOUCESTER
    He that will think to live till he be old
    Give me some help!—O cruel! O ye gods!
         ⌜*Cornwall pulls out one of Gloucester's eyes and*
         *stamps on it*⌝
REGAN (*to Cornwall*)
    One side will mock another; t'other, too.
CORNWALL (*to Gloucester*)
    If you see vengeance—
SERVANT                  Hold your hand, my lord.      70
    I have served you ever since I was a child,
    But better service have I never done you
    Than now to bid you hold.
REGAN                        How now, you dog!
SERVANT
    If you did wear a beard upon your chin
    I'd shake it on this quarrel. ⌜*To Cornwall*⌝ What do
        you mean?                                       75

CORNWALL  My villein!
SERVANT
Why then, come on, and take the chance of anger.
   *They draw and fight*
REGAN ⌈*to another Servant*⌉
Give me thy sword. A peasant stand up thus!
   *She takes a sword and runs at him behind*
SERVANT (*to Gloucester*)
O, I am slain, my lord! Yet have you one eye left
To see some mischief on him.
   ⌈*Regan stabs him again*⌉
                   O!         *He dies*
CORNWALL
Lest it see more, prevent it. Out, vile jelly!    81
   *He* ⌈*pulls out*⌉ *Gloucester's other eye*
Where is thy lustre now?
GLOUCESTER
All dark and comfortless. Where's my son Edmund?
Edmund, enkindle all the sparks of nature
To quite this horrid act.
REGAN            Out, villain!    85
Thou call'st on him that hates thee. It was he
That made the overture of thy treasons to us,
Who is too good to pity thee.
GLOUCESTER
O, my follies! Then Edgar was abused.
Kind gods, forgive me that, and prosper him!    90
REGAN (*to Servants*)
Go thrust him out at gates, and let him smell
His way to Dover. (*To Cornwall*) How is't, my lord?
How look you?
CORNWALL
I have received a hurt. Follow me, lady.
(*To Servants*) Turn out that eyeless villain. Throw this
   slave
Upon the dunghill.     *Exit one or more with Gloucester*
                      ⌈*and the body*⌉
           Regan, I bleed apace.    95
Untimely comes this hurt. Give me your arm.
         *Exeunt Cornwall and Regan*
SECOND SERVANT
I'll never care what wickedness I do
If this man come to good.
THIRD SERVANT       If she live long
And in the end meet the old course of death,
Women will all turn monsters.    100
SECOND SERVANT
Let's follow the old Earl and get the bedlam
To lead him where he would. His roguish madness
Allows itself to anything.
THIRD SERVANT
Go thou. I'll fetch some flax and whites of eggs
To apply to his bleeding face. Now heaven help him!
           *Exeunt severally*

**Sc. 15**    *Enter Edgar as a Bedlam beggar*
EDGAR
Yet better thus and known to be contemned
Than still contemned and flattered. To be worst,
The low'st and most dejected thing of fortune,
Stands still in esperance, lives not in fear.
The lamentable change is from the best;    5
The worst returns to laughter.
   *Enter the Duke of Gloucester led by an Old Man*
Who's here? My father, parti-eyed? World, world, O
world!

But that thy strange mutations make us hate thee,
Life would not yield to age.
   ⌈*Edgar stands aside*⌉
OLD MAN (*to Gloucester*)      O my good lord,
I have been your tenant and your father's tenant    10
This fourscore—
GLOUCESTER
Away, get thee away, good friend, be gone.
Thy comforts can do me no good at all;
Thee they may hurt.
OLD MAN
Alack, sir, you cannot see your way.    15
GLOUCESTER
I have no way, and therefore want no eyes.
I stumbled when I saw. Full oft 'tis seen
Our means secure us, and our mere defects
Prove our commodities. Ah dear son Edgar,
The food of thy abusèd father's wrath—    20
Might I but live to see thee in my touch
I'd say I had eyes again.
OLD MAN          How now? Who's there?
EDGAR (*aside*)
O gods! Who is't can say 'I am at the worst'?
I am worse than e'er I was.
OLD MAN        'Tis poor mad Tom.
EDGAR (*aside*)
And worse I may be yet. The worst is not    25
As long as we can say 'This is the worst.'
OLD MAN (*to Edgar*) Fellow, where goest?
GLOUCESTER  Is it a beggarman?
OLD MAN  Madman and beggar too.
GLOUCESTER
A has some reason, else he could not beg.    30
In the last night's storm I such a fellow saw,
Which made me think a man a worm. My son
Came then into my mind, and yet my mind
Was then scarce friends with him. I have heard more
   since.
As flies to wanton boys are we to th' gods;    35
They kill us for their sport.
EDGAR (*aside*)       How should this be?
Bad is the trade that must play fool to sorrow,
Ang'ring itself and others.
   ⌈*He comes forward*⌉
           Bless thee, master.
GLOUCESTER
Is that the naked fellow?
OLD MAN       Ay, my lord.
GLOUCESTER
Then prithee, get thee gone. If for my sake    40
Thou wilt o'ertake us hence a mile or twain
I'th' way toward Dover, do it for ancient love,
And bring some covering for this naked soul,
Who I'll entreat to lead me.
OLD MAN        Alack, sir, he is mad.
GLOUCESTER
'Tis the time's plague when madmen lead the blind.
Do as I bid thee; or rather do thy pleasure.    46
Above the rest, be gone.
OLD MAN
I'll bring him the best 'parel that I have,
Come on't what will.           *Exit*
GLOUCESTER      Sirrah, naked fellow!
EDGAR
Poor Tom's a-cold. I cannot dance it farther.    50

GLOUCESTER  Come hither, fellow.

EDGAR  Bless thy sweet eyes, they bleed.

GLOUCESTER  Know'st thou the way to Dover?

EDGAR  Both stile and gate, horseway and footpath. Poor
Tom hath been scared out of his good wits. Bless thee,
goodman, from the foul fiend. Five fiends have been in
Poor Tom at once, as Obidicut of lust, Hobbididence
prince of dumbness, Mahu of stealing, Modo of murder,
Flibbertigibbet of mocking and mowing, who since
possesses chambermaids and waiting-women. So bless
thee, master.                                                                      61

GLOUCESTER
Here, take this purse, thou whom the heavens' plagues
Have humbled to all strokes. That I am wretched
Makes thee the happier. Heavens deal so still.
Let the superfluous and lust-dieted man                          65
That stands your ordinance, that will not see
Because he does not feel, feel your power quickly.
So distribution should undo excess,
And each man have enough. Dost thou know Dover?

EDGAR  Ay, master.                                                          70

GLOUCESTER
There is a cliff whose high and bending head
Looks saucily in the confinèd deep.
Bring me but to the very brim of it
And I'll repair the misery thou dost bear
With something rich about me. From that place          75
I shall no leading need.

EDGAR  Give me thy arm.
Poor Tom shall lead thee.

                              *Exit Edgar guiding Gloucester*

Sc. 16   *Enter ⌈at one door⌉ Gonoril and Edmund the*
              *bastard*

GONORIL
Welcome, my lord. I marvel our mild husband
Not met us on the way.
              *Enter ⌈at another door⌉ Oswald the steward*
                              Now, where's your master?

OSWALD
Madam, within; but never man so changed.
I told him of the army that was landed;
He smiled at it. I told him you were coming;              5
His answer was 'The worse.' Of Gloucester's treachery
And of the loyal service of his son
When I informed him, then he called me sot,
And told me I had turned the wrong side out.
What he should most defy seems pleasant to him;     10
What like, offensive.

GONORIL (*to Edmund*)  Then shall you go no further.
It is the cowish terror of his spirit
That dares not undertake. He'll not feel wrongs
Which tie him to an answer. Our wishes on the way
May prove effects. Back, Edmund, to my brother.    15
Hasten his musters and conduct his powers.
I must change arms at home, and give the distaff
Into my husband's hands. This trusty servant
Shall pass between us. Ere long you are like to hear,
If you dare venture in your own behalf,                     20
A mistress's command. Wear this. Spare speech.
Decline your head. This kiss, if it durst speak,
Would stretch thy spirits up into the air.
              ⌈*She kisses him*⌉
Conceive, and fare you well.

EDMUND  Yours in the ranks of death.                         25

GONORIL  My most dear Gloucester.           ⌈*Exit Edmund*⌉
To thee a woman's services are due;
My foot usurps my body.

OSWALD                      Madam, here comes my lord.
                                                                                  *Exit*
              *Enter the Duke of Albany*

GONORIL
I have been worth the whistling.

ALBANY                      O Gonoril,
You are not worth the dust which the rude wind     30
Blows in your face. I fear your disposition.
That nature which contemns it origin
Cannot be bordered certain in itself.
She that herself will sliver and disbranch
From her material sap perforce must wither,             35
And come to deadly use.

GONORIL              No more. The text is foolish.

ALBANY
Wisdom and goodness to the vile seem vile;
Filths savour but themselves. What have you done?
Tigers, not daughters, what have you performed?
A father, and a gracious, agèd man,                             40
Whose reverence even the head-lugged bear would
      lick,
Most barbarous, most degenerate, have you madded.
Could my good-brother suffer you to do it—
A man, a prince by him so benefacted?
If that the heavens do not their visible spirits            45
Send quickly down to tame these vile offences,
It will come,
Humanity must perforce prey on itself,
Like monsters of the deep.

GONORIL                      Milk-livered man,
That bear'st a cheek for blows, a head for wrongs;   50
Who hast not in thy brows an eye discerning
Thine honour from thy suffering; that not know'st
Fools do those villains pity who are punished
Ere they have done their mischief: where's thy drum?
France spreads his banners in our noiseless land,    55
With plumèd helm thy flaxen biggin threats,
Whiles thou, a moral fool, sits still and cries
'Alack, why does he so?'

ALBANY                      See thyself, devil.
Proper deformity shows not in the fiend
So horrid as in woman.

GONORIL                      O vain fool!                               60

ALBANY
Thou changèd and self-covered thing, for shame
Bemonster not thy feature. Were't my fitness
To let these hands obey my blood,
They are apt enough to dislocate and tear
Thy flesh and bones. Howe'er thou art a fiend,         65
A woman's shape doth shield thee.

GONORIL  Marry your manhood, mew—
              *Enter ⌈Second⌉ Gentleman*

ALBANY  What news?

⌈SECOND⌉ GENTLEMAN
O my good lord, the Duke of Cornwall's dead,
Slain by his servant going to put out                           70
The other eye of Gloucester.

ALBANY                      Gloucester's eyes?

⌈SECOND⌉ GENTLEMAN
A servant that he bred, thralled with remorse,
Opposed against the act, bending his sword
To his great master, who thereat enraged
Flew on him, and amongst them felled him dead,    75

But not without that harmful stroke which since
Hath plucked him after.
ALBANY                       This shows you are above,
You justicers, that these our nether crimes
So speedily can venge. But O, poor Gloucester!
Lost he his other eye?
[SECOND] GENTLEMAN       Both, both, my lord.                80
(*To Gonoril*) This letter, madam, craves a speedy
    answer.
'Tis from your sister.
GONORIL (*aside*)         One way I like this well;
But being widow, and my Gloucester with her,
May all the building on my fancy pluck
Upon my hateful life. Another way                              85
The news is not so took.—I'll read and answer.     *Exit*
ALBANY
Where was his son when they did take his eyes?
[SECOND] GENTLEMAN
Come with my lady hither.
ALBANY                            He is not here.
[SECOND] GENTLEMAN
No, my good lord; I met him back again.
ALBANY Knows he the wickedness?                               90
[SECOND] GENTLEMAN
Ay, my good lord; 'twas he informed against him,
And quit the house on purpose that their punishment
Might have the freer course.
ALBANY                          Gloucester, I live
To thank thee for the love thou showed'st the King,
And to revenge thy eyes.—Come hither, friend.    95
Tell me what more thou knowest.                  *Exeunt*

**Sc. 17**  *Enter the Earl of Kent disguised, and* [First]
            *Gentleman*
KENT Why the King of France is so suddenly gone back
    know you no reason?
[FIRST] GENTLEMAN
Something he left imperfect in the state
Which, since his coming forth, is thought of; which
Imports to the kingdom so much fear and danger      5
That his personal return was most required
And necessary.
KENT
Who hath he left behind him general?
[FIRST] GENTLEMAN
The Maréchal of France, Monsieur la Far.
KENT Did your letters pierce the Queen to any
    demonstration of grief?                             11
[FIRST] GENTLEMAN
Ay, sir. She took them, read them in my presence,
And now and then an ample tear trilled down
Her delicate cheek. It seemed she was a queen
Over her passion who, most rebel-like,                 15
Sought to be king o'er her.
KENT                         O, then it moved her.
[FIRST] GENTLEMAN
Not to a rage. Patience and sorrow strove
Who should express her goodliest. You have seen
Sunshine and rain at once; her smiles and tears
Were like, a better way. Those happy smilets           20
That played on her ripe lip seemed not to know
What guests were in her eyes, which parted thence
As pearls from diamonds dropped. In brief,
Sorrow would be a rarity most beloved                  24
If all could so become it.
KENT                         Made she no verbal question?

[FIRST] GENTLEMAN
Faith, once or twice she heaved the name of 'father'
Pantingly forth as if it pressed her heart,
Cried 'Sisters, sisters, shame of ladies, sisters,
Kent, father, sisters, what, i'th' storm, i'th' night,
Let piety not be believed!' There she shook            30
The holy water from her heavenly eyes
And clamour mastered, then away she started
To deal with grief alone.
KENT                      It is the stars,
The stars above us govern our conditions,
Else one self mate and make could not beget            35
Such different issues. You spoke not with her since?
[FIRST] GENTLEMAN No.
KENT
Was this before the King returned?
[FIRST] GENTLEMAN                       No, since.
KENT
Well, sir, the poor distressèd Lear's i'th' town,
Who sometime in his better tune remembers             40
What we are come about, and by no means
Will yield to see his daughter.
[FIRST] GENTLEMAN              Why, good sir?
KENT
A sovereign shame so elbows him: his own unkindness,
That stripped her from his benediction, turned her
To foreign casualties, gave her dear rights           45
To his dog-hearted daughters—these things sting
His mind so venomously that burning shame
Detains him from Cordelia.
[FIRST] GENTLEMAN          Alack, poor gentleman!
KENT
Of Albany's and Cornwall's powers you heard not?
[FIRST] GENTLEMAN 'Tis so; they are afoot.            50
KENT
Well, sir, I'll bring you to our master Lear,
And leave you to attend him. Some dear cause
Will in concealment wrap me up a while.
When I am known aright you shall not grieve
Lending me this acquaintance. I pray you go           55
Along with me.                              *Exeunt*

**Sc. 18**  *Enter Queen Cordelia, a Doctor, and others*
CORDELIA
Alack, 'tis he! Why, he was met even now,
As mad as the racked sea, singing aloud,
Crowned with rank fumitor and furrow-weeds,
With burdocks, hemlock, nettles, cuckoo-flowers,
Darnel, and all the idle weeds that grow               5
In our sustaining corn. The centuries send forth,
Search every acre in the high-grown field,
And bring him to our eye.        [*Exit one or more*]
                            What can man's wisdom
In the restoring his bereavèd sense,
He that can help him                                  10
Take all my outward worth.
DOCTOR There is means, madam.
Our foster-nurse of nature is repose,
The which he lacks. That to provoke in him
Are many simples operative, whose power               15
Will close the eye of anguish.
CORDELIA               All blest secrets,
All you unpublished virtues of the earth,
Spring with my tears, be aidant and remediate
In the good man's distress!—Seek, seek for him,

Lest his ungoverned rage dissolve the life          20
That wants the means to lead it.
   *Enter a Messenger*
MESSENGER       News, madam.
 The British powers are marching hitherward.
CORDELIA
 'Tis known before; our preparation stands
 In expectation of them.—O dear father,
 It is thy business that I go about;          25
 Therefore great France
 My mourning and important tears hath pitied.
 No blown ambition doth our arms incite,
 But love, dear love, and our aged father's right.
 Soon may I hear and see him!   *Exeunt*


Sc. 19 *Enter Regan and Oswald, Gonoril's steward*
REGAN
 But are my brother's powers set forth?
OSWALD         Ay, madam.
REGAN
 Himself in person?
OSWALD     Madam, with much ado.
 Your sister is the better soldier.
REGAN
 Lord Edmund spake not with your lord at home?
OSWALD No, madam.         5
REGAN
 What might import my sister's letters to him?
OSWALD I know not, lady.
REGAN
 Faith, he is posted hence on serious matter.
 It was great ignorance, Gloucester's eyes being out,
 To let him live. Where he arrives he moves          10
 All hearts against us. Edmund, I think, is gone,
 In pity of his misery, to dispatch
 His 'nighted life, moreover to descry
 The strength o'th' army.
OSWALD
 I must needs after with my letters, madam.          15
REGAN
 Our troop sets forth tomorrow. Stay with us.
 The ways are dangerous.
OSWALD      I may not, madam.
 My lady charged my duty in this business.
REGAN
 Why should she write to Edmund? Might not you
 Transport her purposes by word? Belike—          20
 Something, I know not what. I'll love thee much:
 Let me unseal the letter.
OSWALD     Madam, I'd rather—
REGAN
 I know your lady does not love her husband.
 I am sure of that, and at her late being here
 She gave strange oeillades and most speaking looks 25
 To noble Edmund. I know you are of her bosom.
OSWALD I, madam?
REGAN
 I speak in understanding, for I know't.
 Therefore I do advise you take this note.
 My lord is dead. Edmund and I have talked,          30
 And more convenient is he for my hand
 Than for your lady's. You may gather more.
 If you do find him, pray you give him this,
 And when your mistress hears thus much from you,
 I pray desire her call her wisdom to her.          35

So, farewell.
 If you do chance to hear of that blind traitor,
 Preferment falls on him that cuts him off.
OSWALD
 Would I could meet him, madam. I would show
 What lady I do follow.
REGAN      Fare thee well. *Exeunt severally*


Sc. 20 *Enter Edgar disguised as a peasant, with a staff,*
    *guiding the blind Duke of Gloucester*
GLOUCESTER
 When shall we come to th' top of that same hill?
EDGAR
 You do climb up it now. Look how we labour.
GLOUCESTER
 Methinks the ground is even.
EDGAR       Horrible steep.
 Hark, do you hear the sea?
GLOUCESTER     No, truly.
EDGAR
 Why, then your other senses grow imperfect          5
 By your eyes' anguish.
GLOUCESTER    So may it be indeed.
 Methinks thy voice is altered, and thou speak'st
 With better phrase and matter than thou didst.
EDGAR
 You're much deceived. In nothing am I changed          9
 But in my garments.
GLOUCESTER   Methinks you're better spoken.
EDGAR
 Come on, sir, here's the place. Stand still. How fearful
 And dizzy 'tis to cast one's eyes so low!
 The crows and choughs that wing the midway air
 Show scarce so gross as beetles. Halfway down
 Hangs one that gathers samphire, dreadful trade!          15
 Methinks he seems no bigger than his head.
 The fishermen that walk upon the beach
 Appear like mice, and yon tall anchoring barque
 Diminished to her cock, her cock a buoy
 Almost too small for sight. The murmuring surge          20
 That on the unnumbered idle pebble chafes
 Cannot be heard, it's so high. I'll look no more,
 Lest my brain turn and the deficient sight
 Topple down headlong.
GLOUCESTER    Set me where you stand.
EDGAR
 Give me your hand. You are now within a foot          25
 Of th'extreme verge. For all beneath the moon
 Would I not leap upright.
GLOUCESTER     Let go my hand.
 Here, friend, 's another purse; in it a jewel
 Well worth a poor man's taking. Fairies and gods
 Prosper it with thee! Go thou farther off.          30
 Bid me farewell, and let me hear thee going.
EDGAR
 Now fare you well, good sir.
   *He stands aside*
GLOUCESTER      With all my heart.
EDGAR *(aside)*
 Why I do trifle thus with his despair
 Is done to cure it.
GLOUCESTER    O you mighty gods,
   *He kneels*
 This world I do renounce, and in your sights          35
 Shake patiently my great affliction off!
 If I could bear it longer, and not fall

To quarrel with your great opposeless wills,
My snuff and loathèd part of nature should
Burn itself out. If Edgar live, O bless him!—　　40
Now, fellow, fare thee well.

EDGAR　　　　　　　　　　　Gone, sir. Farewell.
　　　*Gloucester falls forward*
(*Aside*) And yet I know not how conceit may rob
The treasury of life, when life itself
Yields to the theft. Had he been where he thought,
By this had thought been past.—Alive or dead?　　45
(*To Gloucester*) Ho you, sir; hear you, sir? Speak.
(*Aside*) Thus might he pass indeed. Yet he revives.
(*To Gloucester*) What are you, sir?

GLOUCESTER　　　　　　　　　Away, and let me die.

EDGAR
Hadst thou been aught but goss'mer, feathers, air,
So many fathom down precipitating　　50
Thou hadst shivered like an egg. But thou dost breathe,
Hast heavy substance, bleed'st not, speak'st, art sound.
Ten masts a-length make not the altitude
Which thou hast perpendicularly fell.
Thy life's a miracle. Speak yet again.　　55

GLOUCESTER But have I fallen, or no?

EDGAR
From the dread summit of this chalky bourn.
Look up a-height. The shrill-gorged lark so far
Cannot be seen or heard. Do but look up.

GLOUCESTER Alack, I have no eyes.　　60
Is wretchedness deprived that benefit
To end itself by death? 'Twas yet some comfort
When misery could beguile the tyrant's rage
And frustrate his proud will.

EDGAR　　　　　　　　　　Give me your arm.
Up. So, how now? Feel you your legs? You stand.　　65

GLOUCESTER
Too well, too well.

EDGAR　　　　　This is above all strangeness.
Upon the crown of the cliff what thing was that
Which parted from you?

GLOUCESTER　　　　　A poor unfortunate beggar.

EDGAR
As I stood here below, methoughts his eyes
Were two full moons. A had a thousand noses,　　70
Horns whelked and wavèd like the enridgèd sea.
It was some fiend. Therefore, thou happy father,
Think that the clearest gods, who made their honours
Of men's impossibilities, have preserved thee.

GLOUCESTER
I do remember now. Henceforth I'll bear　　75
Affliction till it do cry out itself
'Enough, enough,' and die. That thing you speak of,
I took it for a man. Often would it say
'The fiend, the fiend!' He led me to that place.

EDGAR
Bear free and patient thoughts.
　　*Enter King Lear mad, ⌈crowned with weeds and
　　flowers⌉*
　　　　　　　　　　　But who comes here?
The safer sense will ne'er accommodate　　81
His master thus.

LEAR No, they cannot touch me for coining. I am the
King himself.

EDGAR O thou side-piercing sight!　　85

LEAR Nature is above art in that respect. There's your
press-money. That fellow handles his bow like a crow-

keeper. Draw me a clothier's yard. Look, look, a mouse!
Peace, peace, this toasted cheese will do it. There's my
gauntlet. I'll prove it on a giant. Bring up the brown
bills. O, well flown, bird, in the air. Ha! Give the word.

EDGAR Sweet marjoram.

LEAR Pass.

GLOUCESTER I know that voice.　　94

LEAR Ha, Gonoril! Ha, Regan! They flattered me like a
dog, and told me I had white hairs in my beard ere
the black ones were there. To say 'ay' and 'no' to
everything I said 'ay' and 'no' to was no good divinity.
When the rain came to wet me once, and the wind to
make me chatter, when the thunder would not peace
at my bidding, there I found them, there I smelt them
out. Go to, they are not men of their words. They told
me I was everything; 'tis a lie, I am not ague-proof.

GLOUCESTER
The trick of that voice I do well remember.
Is't not the King?

LEAR　　　　　Ay, every inch a king.　　105
　⌈*Gloucester kneels*⌉
When I do stare, see how the subject quakes!
I pardon that man's life. What was thy cause?
Adultery? Thou shalt not die for adultery.
No, the wren goes to't, and the small gilded fly
Does lecher in my sight.　　110
Let copulation thrive, for Gloucester's bastard son
Was kinder to his father than my daughters
Got 'tween the lawful sheets. To't, luxury, pell-mell,
For I lack soldiers. Behold yon simp'ring dame,
Whose face between her forks presageth snow,　　115
That minces virtue, and does shake the head
To hear of pleasure's name:
The fitchew nor the soilèd horse goes to't
With a more riotous appetite. Down from the waist
They're centaurs, though women all above.　　120
But to the girdle do the gods inherit;
Beneath is all the fiend's. There's hell, there's
　　darkness,
There's the sulphury pit, burning, scalding,
Stench, consummation. Fie, fie, fie; pah, pah!
Give me an ounce of civet, good apothecary,　　125
To sweeten my imagination.
There's money for thee.

GLOUCESTER　　　　　O, let me kiss that hand!

LEAR Here, wipe it first; it smells of mortality.

GLOUCESTER
O ruined piece of nature! This great world
Shall so wear out to naught. Do you know me?　　130

LEAR I remember thy eyes well enough. Dost thou squiny
on me?
No, do thy worst, blind Cupid, I'll not love.
Read thou that challenge. Mark the penning of't.

GLOUCESTER
Were all the letters suns, I could not see one.　　135

EDGAR (*aside*)
I would not take this from report; it is,
And my heart breaks at it.

LEAR (*to Gloucester*) Read.

GLOUCESTER What—with the case of eyes?　　139

LEAR O ho, are you there with me? No eyes in your head,
nor no money in your purse? Your eyes are in a heavy
case, your purse in a light; yet you see how this world
goes.

GLOUCESTER I see it feelingly.　　144

LEAR What, art mad? A man may see how the world
goes with no eyes; look with thy ears. See how yon
justice rails upon yon simple thief. Hark in thy ear:
handy-dandy, which is the thief, which is the justice?
Thou hast seen a farmer's dog bark at a beggar?
GLOUCESTER Ay, sir.                                             150
LEAR An the creature run from the cur, there thou
mightst behold the great image of authority. A dog's
obeyed in office.
Thou rascal beadle, hold thy bloody hand.
Why dost thou lash that whore? Strip thine own back.
Thy blood as hotly lusts to use her in that kind      156
For which thou whip'st her. The usurer hangs the
cozener.
Through tattered rags small vices do appear;
Robes and furred gowns hides all. Get thee glass eyes,
And, like a scurvy politician, seem                    160
To see the things thou dost not. No tears, now.
Pull off my boots. Harder, harder! So.
EDGAR (aside)
O, matter and impertinency mixed—
Reason in madness!
LEAR
If thou wilt weep my fortune, take my eyes.           165
I know thee well enough: thy name is Gloucester.
Thou must be patient. We came crying hither.
Thou know'st the first time that we smell the air
We wail and cry. I will preach to thee. Mark me.
GLOUCESTER Alack, alack, the day!                      170
LEAR ⌈removing his crown of weeds⌉
When we are born, we cry that we are come
To this great stage of fools. This' a good block.
It were a delicate stratagem to shoe
A troop of horse with felt; and when I have stole
upon
These son-in-laws, then kill, kill, kill, kill, kill, kill! 175
    Enter three Gentlemen
⌈FIRST⌉ GENTLEMAN
O, here he is. Lay hands upon him, sirs.
(To Lear) Your most dear—
LEAR
No rescue? What, a prisoner? I am e'en
The natural fool of fortune. Use me well.
You shall have ransom. Let me have a surgeon;      180
I am cut to the brains.
⌈FIRST⌉ GENTLEMAN You shall have anything.
LEAR No seconds? All myself?
Why, this would make a man a man of salt,
To use his eyes for garden water-pots,               185
Ay, and laying autumn's dust.
⌈FIRST⌉ GENTLEMAN                      Good sir—
LEAR
I will die bravely, like a bridegroom.
What, I will be jovial. Come, come,
I am a king, my masters, know you that?
⌈FIRST⌉ GENTLEMAN
You are a royal one, and we obey you.               190
LEAR Then there's life in't. Nay, an you get it, you shall
get it with running.
          Exit running, pursued by two Gentlemen
⌈FIRST⌉ GENTLEMAN
A sight most pitiful in the meanest wretch,
Past speaking in a king. Thou hast one daughter
Who redeems nature from the general curse            195
Which twain hath brought her to.
EDGAR Hail, gentle sir.
⌈FIRST⌉ GENTLEMAN Sir, speed you. What's your will?

EDGAR
Do you hear aught of a battle toward?
⌈FIRST⌉ GENTLEMAN
Most sure and vulgar, everyone hears that            200
That can distinguish sense.
EDGAR But, by your favour,
How near's the other army?
⌈FIRST⌉ GENTLEMAN
Near and on speedy foot, the main; descriers
Stands on the hourly thoughts.
EDGAR                              I thank you, sir. That's all.
⌈FIRST⌉ GENTLEMAN
Though that the Queen on special cause is here,      206
Her army is moved on.
EDGAR                    I thank you, sir.   Exit Gentleman
GLOUCESTER
You ever gentle gods, take my breath from me.
Let not my worser spirit tempt me again
To die before you please.                            210
EDGAR Well pray you, father.
GLOUCESTER Now, good sir, what are you?
EDGAR
A most poor man, made lame by fortune's blows,
Who by the art of known and feeling sorrows
Am pregnant to good pity. Give me your hand,         215
I'll lead you to some biding.
GLOUCESTER ⌈rising⌉                      Hearty thanks.
The bounty and the benison of heaven
To send thee boot to boot.
        Enter Oswald the steward
OSWALD                      A proclaimed prize! Most happy!
That eyeless head of thine was first framed flesh
To raise my fortunes. Thou most unhappy traitor,    220
Briefly thyself remember. The sword is out
That must destroy thee.
GLOUCESTER                  Now let thy friendly hand
Put strength enough to't.
OSWALD (to Edgar)                Wherefore, bold peasant,
Durst thou support a published traitor? Hence,
Lest the infection of his fortune take               225
Like hold on thee. Let go his arm.
EDGAR 'Chill not let go, sir, without 'cagion.
OSWALD Let go, slave, or thou diest.
EDGAR Good gentleman, go your gate. Let poor volk pass.
An 'chud have been swaggered out of my life, it would
not have been so long by a vortnight. Nay, come not
near the old man. Keep out, 'che vor' ye, or I'll try
whether your costard or my baton be the harder; I'll
be plain with you.
OSWALD Out, dunghill!                                235
        They fight
EDGAR 'Chill pick your teeth, sir. Come, no matter for
your foins.
        ⌈Edgar knocks him down⌉
OSWALD
Slave, thou hast slain me. Villain, take my purse.
If ever thou wilt thrive, bury my body,
And give the letters which thou find'st about me     240
To Edmund, Earl of Gloucester. Seek him out
Upon the British party. O untimely death! Death!
                                          He dies
EDGAR
I know thee well—a serviceable villain,
As duteous to the vices of thy mistress
As badness would desire.                             245
GLOUCESTER What, is he dead?

EDGAR  Sit you down, father. Rest you.
> *Gloucester sits*
Let's see his pockets. These letters that he speaks of
May be my friends. He's dead; I am only sorrow
He had no other deathsman. Let us see.                    250
Leave, gentle wax; and manners, blame us not.
To know our enemies' minds we'd rip their hearts;
Their papers is more lawful.
> *He reads a letter*
'Let your reciprocal vows be remembered. You have
many opportunities to cut him off. If your will want
not, time and place will be fruitfully offered. There is
nothing done if he return the conqueror; then am I
the prisoner, and his bed my jail, from the loathed
warmth whereof, deliver me, and supply the place for
your labour.                    260
> Your—wife, so I would say—your affectionate
> servant, and for you her own for venture,
>                                   Gonoril.'
O indistinguished space of woman's wit—
A plot upon her virtuous husband's life,          265
And the exchange my brother!—Here in the sands
Thee I'll rake up, the post unsanctified
Of murderous lechers, and in the mature time
With this ungracious paper strike the sight
Of the death-practisèd Duke. For him 'tis well          270
That of thy death and business I can tell.
>                          ⌈*Exit with the body*⌉

GLOUCESTER
The King is mad. How stiff is my vile sense,
That I stand up and have ingenious feeling
Of my huge sorrows! Better I were distraught;
So should my thoughts be fencèd from my griefs,          275
And woes by wrong imaginations lose
The knowledge of themselves.
> *A drum afar off.* ⌈*Enter Edgar*⌉
EDGAR                                   Give me your hand.
Far off methinks I hear the beaten drum.
Come, father, I'll bestow you with a friend.
>                  *Exit Edgar guiding Gloucester*

Sc. 21  ⌈*Soft music.*⌉ *Enter Queen Cordelia, and the Earl*
> *of Kent, disguised*
CORDELIA  O thou good Kent,
How shall I live and work to match thy goodness?
My life will be too short, and every measure fail me.
KENT
To be acknowledged, madam, is o'erpaid.
All my reports go with the modest truth,          5
Nor more, nor clipped, but so.
CORDELIA                          Be better suited.
These weeds are memories of those worser hours.
I prithee put them off.
KENT                          Pardon me, dear madam.
Yet to be known shortens my made intent.
My boon I make it that you know me not          10
Till time and I think meet.
CORDELIA                          Then be't so, my good lord.
> ⌈*Enter the Doctor and First Gentleman*⌉
How does the King?
DOCTOR                          Madam, sleeps still.
CORDELIA                          O you kind gods,
Cure this great breach in his abusèd nature;
The untuned and hurrying senses O wind up
Of this child-changèd father!
DOCTOR                          So please your majesty
That we may wake the King? He hath slept long.          16

CORDELIA
Be governed by your knowledge, and proceed
I'th' sway of your own will. Is he arrayed?
⌈FIRST GENTLEMAN⌉
Ay, madam. In the heaviness of his sleep
We put fresh garments on him.          20
⌈DOCTOR⌉
Good madam, be by when we do awake him.
I doubt not of his temperance.
CORDELIA                          Very well.
DOCTOR
Please you draw near. Louder the music there!
> *King Lear is* ⌈*discovered*⌉ *asleep*
CORDELIA
O my dear father, restoration hang
Thy medicine on my lips, and let this kiss          25
Repair those violent harms that my two sisters
Have in thy reverence made!
KENT                          Kind and dear princess!
CORDELIA
Had you not been their father, these white flakes
Had challenged pity of them. Was this a face
To be exposed against the warring winds,          30
To stand against the deep dread-bolted thunder
In the most terrible and nimble stroke
Of quick cross-lightning, to watch—poor *perdu*—
With this thin helm? Mine injurer's mean'st dog,
Though he had bit me, should have stood that night
Against my fire. And wast thou fain, poor father,          36
To hovel thee with swine and rogues forlorn
In short and musty straw? Alack, alack,
'Tis wonder that thy life and wits at once
Had not concluded all! (*To the Doctor*) He wakes.
> Speak to him.          40
DOCTOR  Madam, do you; 'tis fittest.
CORDELIA (*to Lear*)
How does my royal lord? How fares your majesty?
LEAR
You do me wrong to take me out o'th' grave.
Thou art a soul in bliss, but I am bound
Upon a wheel of fire, that mine own tears          45
Do scald like molten lead.
CORDELIA                          Sir, know me.
LEAR
You're a spirit, I know. Where did you die?
CORDELIA (*to the Doctor*) Still, still far wide!
DOCTOR
He's scarce awake. Let him alone a while.
LEAR
Where have I been? Where am I? Fair daylight?          50
I am mightily abused. I should e'en die with pity
To see another thus. I know not what to say.
I will not swear these are my hands. Let's see:
I feel this pin prick. Would I were assured
Of my condition.
CORDELIA (*kneeling*) O look upon me, sir,          55
And hold your hands in benediction o'er me.
No, sir, you must not kneel.
LEAR                          Pray do not mock.
I am a very foolish, fond old man,
Fourscore and upward, and to deal plainly,
I fear I am not in my perfect mind.          60
Methinks I should know you, and know this man;
Yet I am doubtful, for I am mainly ignorant
What place this is; and all the skill I have
Remembers not these garments; nor I know not
Where I did lodge last night. Do not laugh at me,          65

For as I am a man, I think this lady
To be my child, Cordelia.
CORDELIA                    And so I am.
LEAR
  Be your tears wet? Yes, faith. I pray, weep not.
  If you have poison for me, I will drink it.
  I know you do not love me; for your sisters     70
  Have, as I do remember, done me wrong.
  You have some cause; they have not.
CORDELIA                    No cause, no cause.
LEAR  Am I in France?
KENT  In your own kingdom, sir.
LEAR  Do not abuse me.                            75
DOCTOR
  Be comforted, good madam. The great rage
  You see is cured in him, and yet it is danger
  To make him even o'er the time he has lost.
  Desire him to go in; trouble him no more
  Till further settling.                          80
CORDELIA (to Lear)  Will't please your highness walk?
LEAR  You must bear with me.
  Pray now, forget and forgive. I am old
  And foolish.     Exeunt all but Kent and ⌈First⌉ Gentleman
⌈FIRST⌉ GENTLEMAN  Holds it true, sir, that the Duke
  Of Cornwall was so slain?
KENT                    Most certain, sir.       85
⌈FIRST⌉ GENTLEMAN
  Who is conductor of his people?
KENT                    As 'tis said,
  The bastard son of Gloucester.
⌈FIRST⌉ GENTLEMAN          They say Edgar,
  His banished son, is with the Earl of Kent
  In Germany.
KENT          Report is changeable.
  'Tis time to look about. The powers of the kingdom  90
  Approach apace.
⌈FIRST⌉ GENTLEMAN  The arbitrement is
  Like to be bloody. Fare you well, sir.          Exit
KENT
  My point and period will be throughly wrought,
  Or well or ill, as this day's battle's fought.  Exit

Sc. 22  Enter Edmund, Regan, and their powers
EDMUND
  Know of the Duke if his last purpose hold,
  Or whether since he is advised by aught
  To change the course. He's full of abdication
  And self-reproving. Bring his constant pleasure.
                              Exit one or more
REGAN
  Our sister's man is certainly miscarried.        5
EDMUND
  'Tis to be doubted, madam.
REGAN                    Now, sweet lord,
  You know the goodness I intend upon you.
  Tell me but truly—but then speak the truth—
  Do you not love my sister?
EDMUND                    Ay: honoured love.
REGAN
  But have you never found my brother's way        10
  To the forfended place?
EDMUND  That thought abuses you.
REGAN  I am doubtful
  That you have been conjunct and bosomed with her,
  As far as we call hers.
EDMUND                    No, by mine honour, madam.

REGAN
  I never shall endure her. Dear my lord,          16
  Be not familiar with her.
EDMUND  Fear me not.
  She and the Duke her husband—
              Enter the Duke of Albany and Gonoril with troops
GONORIL (aside)
  I had rather lose the battle than that sister    20
  Should loosen him and me.
ALBANY (to Regan)
  Our very loving sister, well bemet,
  For this I hear: the King is come to his daughter,
  With others whom the rigour of our state
  Forced to cry out. Where I could not be honest   25
  I never yet was valiant. For this business,
  It touches us as France invades our land;
  Yet bold's the King, with others whom I fear.
  Most just and heavy causes make oppose.
EDMUND
  Sir, you speak nobly.
REGAN                    Why is this reasoned?     30
GONORIL
  Combine together 'gainst the enemy;
  For these domestic poor particulars
  Are not to question here.
ALBANY
  Let us then determine with the ensign of war
  On our proceedings.
EDMUND              I shall attend you            35
  Presently at your tent.          ⌈Exit with his powers⌉
REGAN                    Sister, you'll go with us?
GONORIL  No.
REGAN
  'Tis most convenient. Pray you go with us.
GONORIL ⌈aside⌉
  O ho, I know the riddle! (To Regan) I will go.
              Enter Edgar disguised as a peasant
EDGAR (to Albany)
  If e'er your grace had speech with man so poor,  40
  Hear me one word.
ALBANY (to the others)  I'll overtake you.
                    Exeunt all but Albany and Edgar
                                        Speak.
EDGAR
  Before you fight the battle, ope this letter.
  If you have victory, let the trumpet sound
  For him that brought it. Wretched though I seem,
  I can produce a champion that will prove         45
  What is avouchèd there. If you miscarry,
  Your business of the world hath so an end.
  Fortune love you—
ALBANY  Stay till I have read the letter.
EDGAR  I was forbid it.                            50
  When time shall serve, let but the herald cry,
  And I'll appear again.
ALBANY  Why, fare thee well.
  I will o'erlook the paper.              Exit Edgar
              Enter Edmund
EDMUND
  The enemy's in view; draw up your powers.        55
          He ⌈offers⌉ Albany a paper
  Here is the guess of their great strength and forces
  By diligent discovery; but your haste
  Is now urged on you.
ALBANY                    We will greet the time.   Exit
EDMUND
  To both these sisters have I sworn my love,

Each jealous of the other as the stung                    60
Are of the adder. Which of them shall I take?—
Both?—one?—or neither? Neither can be enjoyed
If both remain alive. To take the widow
Exasperates, makes mad, her sister Gonoril,
And hardly shall I carry out my side,                    65
Her husband being alive. Now then, we'll use
His countenance for the battle, which being done,
Let her that would be rid of him devise
His speedy taking off. As for his mercy
Which he intends to Lear and to Cordelia,                    70
The battle done, and they within our power,
Shall never see his pardon; for my state
Stands on me to defend, not to debate.                    *Exit*

Sc. 23    *Alarum. The powers of France pass over the stage*
*⌈led by⌉ Queen Cordelia with her father in her hand.*
*Then enter Edgar disguised as a peasant, guiding the*
*blind Duke of Gloucester*

EDGAR
Here, father, take the shadow of this bush
For your good host; pray that the right may thrive.
If ever I return to you again
I'll bring you comfort.                    *Exit*
GLOUCESTER                    Grace go with you, sir.
*Alarum and retreat. Enter Edgar*
EDGAR
Away, old man. Give me thy hand. Away.                    5
King Lear hath lost, he and his daughter ta'en.
Give me thy hand. Come on.
GLOUCESTER
No farther, sir. A man may rot even here.
EDGAR
What, in ill thoughts again? Men must endure
Their going hence even as their coming hither.                    10
Ripeness is all. Come on.    *Exit Edgar guiding Gloucester*

Sc. 24    *Enter Edmund with King Lear and Queen Cordelia*
*prisoners, a Captain, and soldiers*
EDMUND
Some officers take them away. Good guard
Until their greater pleasures best be known
That are to censure them.
CORDELIA (*to Lear*)                    We are not the first
Who with best meaning have incurred the worst.
For thee, oppressèd King, am I cast down,                    5
Myself could else outfrown false fortune's frown.
Shall we not see these daughters and these sisters?
LEAR
No, no. Come, let's away to prison.
We two alone will sing like birds i'th' cage.
When thou dost ask me blessing, I'll kneel down                    10
And ask of thee forgiveness; so we'll live,
And pray, and sing, and tell old tales, and laugh
At gilded butterflies, and hear poor rogues
Talk of court news, and we'll talk with them too—
Who loses and who wins, who's in, who's out,                    15
And take upon 's the mystery of things
As if we were God's spies; and we'll wear out
In a walled prison packs and sects of great ones
That ebb and flow by th' moon.
EDMUND (*to soldiers*)                    Take them away.
LEAR (*to Cordelia*)
Upon such sacrifices, my Cordelia,                    20
The gods themselves throw incense. Have I caught
    thee?
He that parts us shall bring a brand from heaven

And fire us hence like foxes. Wipe thine eyes.
The goodyear shall devour 'em, flesh and fell,
Ere they shall make us weep. We'll see 'em starve
    first. Come.    *Exeunt all but Edmund and the Captain*
EDMUND Come hither, captain. Hark.                    26
Take thou this note. Go follow them to prison.
One step I have advanced thee; if thou dost
As this instructs thee, thou dost make thy way
To noble fortunes. Know thou this: that men                    30
Are as the time is. To be tender-minded
Does not become a sword. Thy great employment
Will not bear question. Either say thou'lt do't,
Or thrive by other means.
CAPTAIN                    I'll do't, my lord.
EDMUND
About it, and write 'happy' when thou hast done.                    35
Mark, I say, instantly, and carry it so
As I have set it down.
CAPTAIN                    I cannot draw a cart,
Nor eat dried oats. If it be man's work, I'll do't.    *Exit*
*Enter the Duke of Albany, the two ladies Gonoril*
*and Regan, ⌈another Captain,⌉ and others*
ALBANY (*to Edmund*)
Sir, you have showed today your valiant strain,
And fortune led you well. You have the captives                    40
That were the opposites of this day's strife.
We do require then of you, so to use them
As we shall find their merits and our safety
May equally determine.
EDMUND                    Sir, I thought it fit
To send the old and miserable King                    45
To some retention and appointed guard,
Whose age has charms in it, whose title more,
To pluck the common bosom on his side
And turn our impressed lances in our eyes
Which do command them. With him I sent the Queen,
My reason all the same, and they are ready                    51
Tomorrow, or at further space, to appear
Where you shall hold your session. At this time
We sweat and bleed. The friend hath lost his friend,
And the best quarrels in the heat are cursed                    55
By those that feel their sharpness.
The question of Cordelia and her father
Requires a fitter place.
ALBANY                    Sir, by your patience,
I hold you but a subject of this war,
Not as a brother.
REGAN                    That's as we list to grace him.                    60
Methinks our pleasure should have been demanded
Ere you had spoke so far. He led our powers,
Bore the commission of my place and person,
The which immediate may well stand up
And call itself your brother.
GONORIL                    Not so hot.                    65
In his own grace he doth exalt himself
More than in your advancement.
REGAN                    In my right
By me invested, he compeers the best.
GONORIL
That were the most if he should husband you.
REGAN
Jesters do oft prove prophets.
GONORIL                    Holla, holla—                    70
That eye that told you so looked but asquint.
REGAN
Lady, I am not well, else I should answer
From a full-flowing stomach. (*To Edmund*) General,

Take thou my soldiers, prisoners, patrimony.
Witness the world that I create thee here          75
My lord and master.
GONORIL                    Mean you to enjoy him, then?
ALBANY
The let-alone lies not in your good will.
EDMUND
Nor in thine, lord.
ALBANY                    Half-blooded fellow, yes.
EDMUND
Let the drum strike and prove my title good.
ALBANY
Stay yet, hear reason. Edmund, I arrest thee      80
On capital treason, and in thine attaint
This gilded serpent. (*To Regan*) For your claim, fair
    sister,
I bar it in the interest of my wife.
'Tis she is subcontracted to this lord,
And I, her husband, contradict the banns.          85
If you will marry, make your love to me.
My lady is bespoke.—Thou art armed, Gloucester.
If none appear to prove upon thy head
Thy heinous, manifest, and many treasons,
    ⌜*He throws down a glove*⌝
There is my pledge. I'll prove it on thy heart,    90
Ere I taste bread, thou art in nothing less
Than I have here proclaimed thee.
REGAN Sick, O sick!
GONORIL (*aside*) If not, I'll ne'er trust poison.
EDMUND (*to Albany,* ⌜*throwing down a glove*⌝)
There's my exchange. What in the world he is       95
That names me traitor, villain-like he lies.
Call by thy trumpet. He that dares, approach;
On him, on you—who not?—I will maintain
My truth and honour firmly.
ALBANY A herald, ho!                                100
EDMUND A herald, ho, a herald!
ALBANY
Trust to thy single virtue, for thy soldiers,
All levied in my name, have in my name
Took their discharge.
REGAN                    This sickness grows upon me.
ALBANY
She is not well. Convey her to my tent.            105
                    *Exit one or more with Regan*
    ⌜*Enter a Herald and a trumpeter*⌝
Come hither, herald. Let the trumpet sound,
And read out this.
SECOND CAPTAIN Sound, trumpet!
    *Trumpeter sounds*
HERALD (*reads*) 'If any man of quality or degree in the
host of the army will maintain upon Edmund, supposed
Earl of Gloucester, that he's a manifold traitor, let him
appear at the third sound of the trumpet. He is bold in
his defence.'
EDMUND Sound! (*Trumpeter sounds*) Again!
    *Enter Edgar, armed, at the third sound, a trumpeter*
    *before him*
ALBANY (*to the Herald*)
Ask him his purposes, why he appears               115
Upon this call o'th' trumpet.
HERALD (*to Edgar*)            What are you?
Your name and quality, and why you answer
This present summons?
EDGAR                    O, know my name is lost,
By treason's tooth bare-gnawn and canker-bit.

Yet ere I move't, where is the adversary            120
I come to cope withal?
ALBANY                    Which is that adversary?
EDGAR
What's he that speaks for Edmund, Earl of Gloucester?
EDMUND
Himself. What sayst thou to him?
EDGAR                    Draw thy sword,
That if my speech offend a noble heart
Thy arm may do thee justice. Here is mine.          125
    *He draws his sword*
Behold, it is the privilege of my tongue,
My oath, and my profession. I protest,
Maugre thy strength, youth, place, and eminence,
Despite thy victor-sword and fire-new fortune,
Thy valour and thy heart, thou art a traitor,       130
False to thy gods, thy brother, and thy father,
Conspirant 'gainst this high illustrious prince,
And from th'extremest upward of thy head
To the descent and dust beneath thy feet
A most toad-spotted traitor. Say thou no,           135
This sword, this arm, and my best spirits are bent
To prove upon thy heart, whereto I speak,
Thou liest.
EDMUND    In wisdom I should ask thy name,
But since thy outside looks so fair and warlike,
And that thy tongue some say of breeding breathes,
My right of knighthood I disdain and spurn.         141
Here do I toss those treasons to thy head,
With the hell-hated lie o'erturn thy heart,
Which, for they yet glance by and scarcely bruise,
This sword of mine shall give them instant way      145
Where they shall rest for ever. Trumpets, speak!
    ⌜*Flourish.*⌝ *They fight. Edmund is vanquished*
⌜ALL⌝
Save him, save him!
GONORIL                    This is mere practice, Gloucester.
By the law of arms thou art not bound to answer
An unknown opposite. Thou art not vanquished,
But cozened and beguiled.
ALBANY                    Stop your mouth, dame,
Or with this paper shall I stopple it.              151
Thou worse than anything, read thine own evil.
Nay, no tearing, lady. I perceive you know't.
GONORIL
Say if I do, the laws are mine, not thine.
Who shall arraign me for't?
ALBANY                    Most monstrous!       155
Know'st thou this paper?
GONORIL                    Ask me not what I know.
                                        *Exit*
ALBANY
Go after her. She's desperate. Govern her.
                                *Exit one or more*
EDMUND
What you have charged me with, that have I done,
And more, much more. The time will bring it out.
'Tis past, and so am I. (*To Edgar*) But what art thou,
That hast this fortune on me? If thou beest noble,  161
I do forgive thee.
EDGAR                    Let's exchange charity.
I am no less in blood than thou art, Edmund.
If more, the more ignobly thou hast wronged me.
    ⌜*He takes off his helmet*⌝
My name is Edgar, and thy father's son.             165
The gods are just, and of our pleasant vices

Make instruments to scourge us.
The dark and vicious place where thee he got
Cost him his eyes.
EDMUND                    Thou hast spoken truth.
The wheel is come full circled. I am here.        170
ALBANY (to Edgar)
Methought thy very gait did prophesy
A royal nobleness. I must embrace thee.
Let sorrow split my heart if I did ever hate
Thee or thy father.
EDGAR  Worthy prince, I know't.                  175
ALBANY  Where have you hid yourself?
How have you known the miseries of your father?
EDGAR
By nursing them, my lord. List a brief tale,
And when 'tis told, O that my heart would burst!
The bloody proclamation to escape               180
That followed me so near—O, our lives' sweetness,
That with the pain of death would hourly die
Rather than die at once!—taught me to shift
Into a madman's rags, to assume a semblance
That very dogs disdained; and in this habit      185
Met I my father with his bleeding rings,
The precious stones new-lost; became his guide,
Led him, begged for him, saved him from despair;
Never—O father!—revealed myself unto him
Until some half hour past, when I was armed.      190
Not sure, though hoping, of this good success,
I asked his blessing, and from first to last
Told him my pilgrimage; but his flawed heart—
Alack, too weak the conflict to support—
'Twixt two extremes of passion, joy and grief,    195
Burst smilingly.
EDMUND                This speech of yours hath moved me,
And shall perchance do good. But speak you on—
You look as you had something more to say.
ALBANY
If there be more, more woeful, hold it in,
For I am almost ready to dissolve,              200
Hearing of this.
EDGAR                This would have seemed a period
To such as love not sorrow; but another
To amplify, too much would make much more,
And top extremity.
Whilst I was big in clamour came there in a man   205
Who, having seen me in my worst estate,
Shunned my abhorred society; but then, finding
Who 'twas that so endured, with his strong arms
He fastened on my neck and bellowed out
As he'd burst heaven; threw him on my father,     210
Told the most piteous tale of Lear and him
That ever ear received, which in recounting
His grief grew puissant and the strings of life
Began to crack. Twice then the trumpets sounded,
And there I left him tranced.                    
ALBANY                    But who was this?       215
EDGAR
Kent, sir, the banished Kent, who in disguise
Followed his enemy king, and did him service
Improper for a slave.
        Enter ⌈Second⌉ Gentleman with a bloody knife
⌈SECOND⌉ GENTLEMAN    Help, help!
ALBANY                    What kind of help?
What means that bloody knife?
⌈SECOND⌉ GENTLEMAN                It's hot, it smokes.
It came even from the heart of—
ALBANY                    Who, man? Speak.

⌈SECOND⌉ GENTLEMAN
Your lady, sir, your lady; and her sister         221
By her is poisonèd—she hath confessed it.
EDMUND
I was contracted to them both; all three
Now marry in an instant.
ALBANY
Produce their bodies, be they alive or dead.      225
This justice of the heavens, that makes us tremble,
Touches us not with pity.
        Enter Kent as himself
EDGAR                    Here comes Kent, sir.
ALBANY
O, 'tis he; the time will not allow
The compliment that very manners urges.
KENT  I am come                                  230
To bid my king and master aye good night.
Is he not here?
ALBANY                Great thing of us forgot!—
Speak, Edmund; where's the King, and where's
            Cordelia?
        The bodies of Gonoril and Regan are brought in
Seest thou this object, Kent?
KENT  Alack, why thus?                           235
EDMUND  Yet Edmund was beloved.
The one the other poisoned for my sake,
And after slew herself.
ALBANY                Even so.—Cover their faces.
EDMUND
I pant for life. Some good I mean to do,
Despite of my own nature. Quickly send,          240
Be brief in't, to th' castle; for my writ
Is on the life of Lear and on Cordelia.
Nay, send in time.
ALBANY                Run, run, O run!
EDGAR
To who, my lord? Who hath the office? Send
Thy token of reprieve.                           245
EDMUND
Well thought on! Take my sword. The captain,
Give it the captain.
ALBANY                Haste thee for thy life.
                    Exit ⌈Second Captain⌉
EDMUND
He hath commission from thy wife and me
To hang Cordelia in the prison, and
To lay the blame upon her own despair,           250
That she fordid herself.
ALBANY
The gods defend her!—Bear him hence a while.
                    Exeunt some with Edmund
        Enter King Lear with Queen Cordelia in his arms,
        ⌈followed by the Second Captain⌉
LEAR
Howl, howl, howl, howl! O, you are men of stones.
Had I your tongues and eyes, I would use them so
That heaven's vault should crack. She's gone for ever.
I know when one is dead and when one lives.       256
She's dead as earth.
        ⌈He lays her down⌉
                    Lend me a looking-glass.
If that her breath will mist or stain the stone,
Why, then she lives.
KENT                Is this the promised end?
EDGAR
Or image of that horror?
ALBANY                Fall and cease.            260

LEAR
    This feather stirs. She lives. If it be so,
    It is a chance which does redeem all sorrows
    That ever I have felt.
KENT ⌜kneeling⌝          Ah, my good master!
LEAR
    Prithee, away.
EDGAR          'Tis noble Kent, your friend.
LEAR
    A plague upon you, murderous traitors all.          265
    I might have saved her; now she's gone for ever.—
    Cordelia, Cordelia: stay a little. Ha?
    What is't thou sayst?—Her voice was ever soft,
    Gentle, and low, an excellent thing in women.—
    I killed the slave that was a-hanging thee.          270
⌜SECOND⌝ CAPTAIN
    'Tis true, my lords, he did.
LEAR                     Did I not, fellow?
    I have seen the day with my good biting falchion
    I would have made them skip. I am old now,
    And these same crosses spoil me. (To Kent) Who are you?
    Mine eyes are not o' the best, I'll tell you straight.  275
KENT
    If fortune bragged of two she loved or hated,
    One of them we behold.
LEAR                     Are not you Kent?
KENT
    The same, your servant Kent. Where is your servant
        Caius?
LEAR
    He's a good fellow, I can tell you that.
    He'll strike, and quickly too. He's dead and rotten.  280
KENT
    No, my good lord, I am the very man—
LEAR I'll see that straight.
KENT
    That from your first of difference and decay
    Have followed your sad steps.
LEAR                     You're welcome hither.
KENT
    Nor no man else. All's cheerless, dark, and deadly.
    Your eldest daughters have fordone themselves,     286
    And desperately are dead.
LEAR                     So think I, too.
ALBANY
    He knows not what he sees; and vain it is
    That we present us to him.
EDGAR                     Very bootless.

                     Enter another Captain
⌜THIRD⌝ CAPTAIN (to Albany)
    Edmund is dead, my lord.
ALBANY                     That's but a trifle here.—
    You lords and noble friends, know our intent.     291
    What comfort to this great decay may come
    Shall be applied; for us, we will resign
    During the life of this old majesty
    To him our absolute power; (to Edgar and Kent) you
        to your rights,                              295
    With boot and such addition as your honours
    Have more than merited. All friends shall taste
    The wages of their virtue, and all foes
    The cup of their deservings.—O see, see!
LEAR
    And my poor fool is hanged. No, no life.          300
    Why should a dog, a horse, a rat have life,
    And thou no breath at all? O, thou wilt come no more.
    Never, never, never.—Pray you, undo
    This button. Thank you, sir. O, O, O, O!
EDGAR He faints. (To Lear) My lord, my lord!        305
LEAR Break, heart, I prithee break.
EDGAR Look up, my lord.
KENT
    Vex not his ghost. O, let him pass. He hates him
    That would upon the rack of this tough world
    Stretch him out longer.
                     ⌜Lear dies⌝
EDGAR                     O, he is gone indeed.       310
KENT
    The wonder is he hath endured so long.
    He but usurped his life.
ALBANY (to attendants)
    Bear them from hence. Our present business
    Is to general woe. (To Kent and Edgar) Friends of my
        soul, you twain
    Rule in this kingdom, and the gored state sustain.  315
KENT
    I have a journey, sir, shortly to go:
    My master calls, and I must not say no.
ALBANY
    The weight of this sad time we must obey,
    Speak what we feel, not what we ought to say.
    The oldest have borne most. We that are young    320
    Shall never see so much, nor live so long.
                     Exeunt carrying the bodies

# THE TRAGEDY OF KING LEAR
## THE FOLIO TEXT

THE text of *King Lear* given here represents the revision made probably two or three years after the first version had been written and performed; it is based on the text printed in the 1623 Folio. This is a more obviously theatrical text. It makes a number of significant cuts, amounting to some 300 lines. The most conspicuous ones are the dialogue in which Lear's Fool implicitly calls his master a fool (Quarto Sc. 4, 136-51); Kent's account of the French invasion of England (Quarto Sc. 8, 21-33); Lear's mock-trial, in his madness, of his daughters (Quarto Sc. 13, 13-52); Edgar's generalizing couplets at the end of that scene (Quarto Sc. 13, 97-110); the brief, compassionate dialogue of two of Gloucester's servants after his blinding (Quarto Sc. 14, 97-106); parts of Albany's protest to Goneril about the sisters' treatment of Lear (in Quarto Sc. 16); the entire scene (Quarto Sc. 17) in which a Gentleman tells Kent of Cordelia's grief on hearing of her father's condition; the presence of the Doctor and the musical accompaniment to the reunion of Lear and Cordelia (Quarto Sc. 21); and Edgar's account of his meeting with Kent in which Kent's 'strings of life | Began to crack' (Quarto Sc. 24, 201-18). The Folio also adds about 100 lines that are not in the Quarto—mostly in short passages, including Kent's statement that Albany and Cornwall have servants who are in the pay of France (3.1.13-20), Merlin's prophecy spoken by the Fool at the end of 3.2, and the last lines of both the Fool and Lear. In addition, several speeches are differently assigned, and there are many variations in wording.

The reasons for these variations, and their effect on the play, are to some extent matters of speculation and of individual interpretation. Certainly they streamline the play's action, removing some reflective passages, particularly at the ends of scenes. They affect the characterization of, especially, Edgar, Albany, and Kent, and there are significant differences in the play's closing passages. Structurally the principal differences lie in the presentation of the military actions in the later part of the play; in the Folio-based text Cordelia is more clearly in charge of the forces that come to Lear's assistance, and they are less clearly a French invasion force. The absence from this text of passages that appeared in the 1608 text implies no criticism of them in themselves. The play's revision may have been dictated in whole or in part by theatrical exigencies, or it may have emerged from Shakespeare's own dissatisfaction with what he had first written. Each version has its own integrity, which is distorted by the practice, traditional since the early eighteenth century, of conflation.

# THE PERSONS OF THE PLAY

LEAR, King of Britain

GONERIL, Lear's eldest daughter

Duke of ALBANY, her husband

REGAN, Lear's second daughter

Duke of CORNWALL, her husband

CORDELIA, Lear's youngest daughter

King of FRANCE

Duke of BURGUNDY } suitors of Cordelia

Earl of KENT, later disguised as Caius

Earl of GLOUCESTER

EDGAR, elder son of Gloucester, later disguised as Tom o' Bedlam

EDMOND, bastard son of Gloucester

OLD MAN, Gloucester's tenant

CURAN, Gloucester's retainer

Lear's FOOL

OSWALD, Goneril's steward

A SERVANT of Cornwall

A KNIGHT

A HERALD

A CAPTAIN

Gentlemen, servants, soldiers, attendants, messengers

# The Tragedy of King Lear

**1.1**  *Enter the Earl of Kent, the Duke of Gloucester, and Edmond*

KENT  I thought the King had more affected the Duke of Albany than Cornwall.

GLOUCESTER  It did always seem so to us, but now in the division of the kingdom it appears not which of the Dukes he values most; for qualities are so weighed that curiosity in neither can make choice of either's moiety.

KENT  Is not this your son, my lord?  7

GLOUCESTER  His breeding, sir, hath been at my charge. I have so often blushed to acknowledge him that now I am brazed to't.  10

KENT  I cannot conceive you.

GLOUCESTER  Sir, this young fellow's mother could, whereupon she grew round-wombed and had indeed, sir, a son for her cradle ere she had a husband for her bed. Do you smell a fault?  15

KENT  I cannot wish the fault undone, the issue of it being so proper.

GLOUCESTER  But I have a son, sir, by order of law, some year older than this, who yet is no dearer in my account. Though this knave came something saucily to the world before he was sent for, yet was his mother fair, there was good sport at his making, and the whoreson must be acknowledged. (*To Edmond*) Do you know this noble gentleman, Edmond?

EDMOND  No, my lord.  25

GLOUCESTER  (*to Edmond*) My lord of Kent. Remember him hereafter as my honourable friend.

EDMOND  (*to Kent*) My services to your lordship.

KENT  I must love you, and sue to know you better.

EDMOND  Sir, I shall study deserving.  30

GLOUCESTER  (*to Kent*) He hath been out nine years, and away he shall again.

*Sennet*

The King is coming.

*Enter King Lear, the Dukes of Cornwall and Albany, Goneril, Regan, Cordelia, and attendants*

LEAR

Attend the lords of France and Burgundy, Gloucester.

GLOUCESTER  I shall, my lord.  *Exit*

LEAR

Meantime we shall express our darker purpose.  36
Give me the map there. Know that we have divided
In three our kingdom, and 'tis our fast intent
To shake all cares and business from our age,
Conferring them on younger strengths while we  40
Unburdened crawl toward death. Our son of Cornwall,
And you, our no less loving son of Albany,
We have this hour a constant will to publish
Our daughters' several dowers, that future strife
May be prevented now. The princes France and
    Burgundy—  45
Great rivals in our youngest daughter's love—
Long in our court have made their amorous sojourn,
And here are to be answered. Tell me, my daughters—
Since now we will divest us both of rule,
Interest of territory, cares of state—  50
Which of you shall we say doth love us most,
That we our largest bounty may extend

Where nature doth with merit challenge? Goneril,
Our eldest born, speak first.

GONERIL

Sir, I love you more than words can wield the matter;
Dearer than eyesight, space, and liberty;  56
Beyond what can be valued, rich or rare,
No less than life; with grace, health, beauty, honour;
As much as child e'er loved or father found;
A love that makes breath poor and speech unable.  60
Beyond all manner of so much I love you.

CORDELIA  (*aside*)

What shall Cordelia speak? Love and be silent.

LEAR  (*to Goneril*)

Of all these bounds even from this line to this,
With shadowy forests and with champaigns riched,
With plenteous rivers and wide-skirted meads,  65
We make thee lady. To thine and Albany's issues
Be this perpetual.—What says our second daughter?
Our dearest Regan, wife of Cornwall?

REGAN

I am made of that self mettle as my sister,
And prize me at her worth. In my true heart  70
I find she names my very deed of love—
Only she comes too short, that I profess
Myself an enemy to all other joys
Which the most precious square of sense possesses,
And find I am alone felicitate  75
In your dear highness' love.

CORDELIA  (*aside*)                Then poor Cordelia—
And yet not so, since I am sure my love's
More ponderous than my tongue.

LEAR  (*to Regan*)

To thee and thine hereditary ever
Remain this ample third of our fair kingdom,  80
No less in space, validity, and pleasure
Than that conferred on Goneril. (*To Cordelia*) Now our
    joy,
Although our last and least, to whose young love
The vines of France and milk of Burgundy
Strive to be interested: what can you say to draw  85
A third more opulent than your sisters? Speak.

CORDELIA  Nothing, my lord.

LEAR  Nothing?

CORDELIA  Nothing.

LEAR

Nothing will come of nothing. Speak again.  90

CORDELIA

Unhappy that I am, I cannot heave
My heart into my mouth. I love your majesty
According to my bond, no more nor less.

LEAR

How, how, Cordelia? Mend your speech a little
Lest you may mar your fortunes.

CORDELIA                Good my lord,  95
You have begot me, bred me, loved me.
I return those duties back as are right fit—
Obey you, love you, and most honour you.
Why have my sisters husbands if they say
They love you all? Haply when I shall wed  100
That lord whose hand must take my plight shall carry

Half my love with him, half my care and duty.
Sure, I shall never marry like my sisters.
LEAR But goes thy heart with this?
CORDELIA Ay, my good lord.                                    105
LEAR So young and so untender?
CORDELIA So young, my lord, and true.
LEAR
Let it be so. Thy truth then be thy dower;
For by the sacred radiance of the sun,
The mysteries of Hecate and the night,                        110
By all the operation of the orbs
From whom we do exist and cease to be,
Here I disclaim all my paternal care,
Propinquity, and property of blood,
And as a stranger to my heart and me                          115
Hold thee from this for ever. The barbarous Scythian,
Or he that makes his generation messes
To gorge his appetite, shall to my bosom
Be as well neighboured, pitied, and relieved
As thou, my sometime daughter.
KENT                              Good my liege—             120
LEAR Peace, Kent.
Come not between the dragon and his wrath.
I loved her most, and thought to set my rest
On her kind nursery. ⌜To Cordelia⌝ Hence, and avoid
    my sight!—
So be my grave my peace as here I give                        125
Her father's heart from her. Call France. Who stirs?
Call Burgundy.                        ⌜Exit one or more⌝
                    Cornwall and Albany,
With my two daughters' dowers digest the third.
Let pride, which she calls plainness, marry her.
I do invest you jointly with my power,                        130
Pre-eminence, and all the large effects
That troop with majesty. Ourself by monthly course,
With reservation of an hundred knights
By you to be sustained, shall our abode
Make with you by due turn. Only we shall retain              135
The name and all th'addition to a king. The sway,
Revenue, execution of the rest,
Belovèd sons, be yours; which to confirm,
This crownet part between you.
KENT                              Royal Lear,
Whom I have ever honoured as my king,                         140
Loved as my father, as my master followed,
As my great patron thought on in my prayers—
LEAR
The bow is bent and drawn; make from the shaft.
KENT
Let it fall rather, though the fork invade
The region of my heart. Be Kent unmannerly                    145
When Lear is mad. What wouldst thou do, old man?
Think'st thou that duty shall have dread to speak
When power to flattery bows? To plainness honour's
    bound
When majesty falls to folly. Reserve thy state,
And in thy best consideration check                           150
This hideous rashness. Answer my life my judgement,
Thy youngest daughter does not love thee least,
Nor are those empty-hearted whose low sounds
Reverb no hollowness.
LEAR                      Kent, on thy life, no more!
KENT
My life I never held but as a pawn                            155
To wage against thine enemies, ne'er feared to lose it,
Thy safety being motive.
LEAR                      Out of my sight!

KENT
See better, Lear, and let me still remain
The true blank of thine eye.
LEAR                              Now, by Apollo—
KENT
Now, by Apollo, King, thou swear'st thy gods in vain.
LEAR ⌜making to strike him⌝
O vassal! Miscreant!
ALBANY and ⌜CORDELIA⌝ Dear sir, forbear.                     161
KENT (to Lear)
Kill thy physician, and thy fee bestow
Upon the foul disease. Revoke thy gift,
Or whilst I can vent clamour from my throat
I'll tell thee thou dost evil.                                165
LEAR
Hear me, recreant; on thine allegiance hear me!
That thou hast sought to make us break our vows,
Which we durst never yet, and with strained pride
To come betwixt our sentence and our power,
Which nor our nature nor our place can bear,                  170
Our potency made good take thy reward:
Five days we do allot thee for provision
To shield thee from disasters of the world,
And on the sixth to turn thy hated back
Upon our kingdom. If on the seventh day following
Thy banished trunk be found in our dominions,                176
The moment is thy death. Away! By Jupiter,
This shall not be revoked.
KENT
Fare thee well, King; sith thus thou wilt appear,
Freedom lives hence, and banishment is here.                 180
(To Cordelia) The gods to their dear shelter take thee,
    maid,
That justly think'st, and hast most rightly said.
(To Goneril and Regan) And your large speeches may
    your deeds approve,
That good effects may spring from words of love.
Thus Kent, O princes, bids you all adieu;                    185
He'll shape his old course in a country new.        Exit
        Flourish. Enter the Duke of Gloucester with the
        King of France, the Duke of Burgundy, and attendants
⌜CORDELIA⌝
Here's France and Burgundy, my noble lord.
LEAR My lord of Burgundy,
We first address toward you, who with this King
Hath rivalled for our daughter: what in the least            190
Will you require in present dower with her
Or cease your quest of love?
BURGUNDY                      Most royal majesty,
I crave no more than hath your highness offered;
Nor will you tender less.
LEAR                      Right noble Burgundy,
When she was dear to us we did hold her so;                  195
But now her price is fallen. Sir, there she stands.
If aught within that little seeming substance,
Or all of it, with our displeasure pieced,
And nothing more, may fitly like your grace,
She's there, and she is yours.
BURGUNDY                      I know no answer.              200
LEAR
Will you with those infirmities she owes,
Unfriended, new adopted to our hate,
Dowered with our curse and strangered with our oath,
Take her or leave her?
BURGUNDY                      Pardon me, royal sir.
Election makes not up in such conditions.                    205

LEAR
  Then leave her, sir; for by the power that made me,
  I tell you all her wealth. (*To France*) For you, great King,
  I would not from your love make such a stray
  To match you where I hate, therefore beseech you
  T'avert your liking a more worthier way                210
  Than on a wretch whom nature is ashamed
  Almost t'acknowledge hers.
FRANCE                          This is most strange,
  That she whom even but now was your best object,
  The argument of your praise, balm of your age,
  The best, the dear'st, should in this trice of time    215
  Commit a thing so monstrous to dismantle
  So many folds of favour. Sure, her offence
  Must be of such unnatural degree
  That monsters it, or your fore-vouched affection
  Fall into taint; which to believe of her               220
  Must be a faith that reason without miracle
  Should never plant in me.
CORDELIA (*to Lear*)
  I yet beseech your majesty,
  If for I want that glib and oily art
  To speak and purpose not—since what I well intend,
  I'll do't before I speak—that you make known           226
  It is no vicious blot, murder, or foulness,
  No unchaste action or dishonoured step
  That hath deprived me of your grace and favour,
  But even the want of that for which I am richer—       230
  A still-soliciting eye, and such a tongue
  That I am glad I have not, though not to have it
  Hath lost me in your liking.
LEAR                          Better thou
  Hadst not been born than not t'have pleased me better.
FRANCE
  Is it but this—a tardiness in nature,                  235
  Which often leaves the history unspoke
  That it intends to do?—My lord of Burgundy,
  What say you to the lady? Love's not love
  When it is mingled with regards that stands
  Aloof from th'entire point. Will you have her?         240
  She is herself a dowry.
BURGUNDY (*to Lear*)          Royal King,
  Give but that portion which yourself proposed,
  And here I take Cordelia by the hand,
  Duchess of Burgundy.
LEAR Nothing. I have sworn. I am firm.                   245
BURGUNDY (*to Cordelia*)
  I am sorry, then, you have so lost a father
  That you must lose a husband.
CORDELIA                      Peace be with Burgundy;
  Since that respect and fortunes are his love,
  I shall not be his wife.
FRANCE
  Fairest Cordelia, that art most rich, being poor;      250
  Most choice, forsaken; and most loved, despised:
  Thee and thy virtues here I seize upon.
  Be it lawful, I take up what's cast away.
  Gods, gods! 'Tis strange that from their cold'st neglect
  My love should kindle to inflamed respect.—            255
  Thy dowerless daughter, King, thrown to my chance,
  Is queen of us, of ours, and our fair France.
  Not all the dukes of wat'rish Burgundy
  Can buy this unprized precious maid of me.—
  Bid them farewell, Cordelia, though unkind.            260
  Thou losest here, a better where to find.

LEAR
  Thou hast her, France. Let her be thine, for we
  Have no such daughter, nor shall ever see
  That face of hers again. Therefore be gone,
  Without our grace, our love, our benison.—             265
  Come, noble Burgundy.  *Flourish. Exeunt all but France
                                          and the sisters*
FRANCE    Bid farewell to your sisters.
CORDELIA
  Ye jewels of our father, with washed eyes
  Cordelia leaves you. I know you what you are,
  And like a sister am most loath to call
  Your faults as they are named. Love well our father.
  To your professèd bosoms I commit him.                 271
  But yet, alas, stood I within his grace
  I would prefer him to a better place.
  So farewell to you both.
REGAN Prescribe not us our duty.                         275
GONERIL Let your study
  Be to content your lord, who hath received you
  At fortune's alms. You have obedience scanted,
  And well are worth the want that you have wanted.
CORDELIA
  Time shall unfold what pleated cunning hides,          280
  Who covert faults at last with shame derides.
  Well may you prosper.
FRANCE                   Come, my fair Cordelia.
                              *Exeunt France and Cordelia*
GONERIL Sister, it is not little I have to say of what most
  nearly appertains to us both. I think our father will
  hence tonight.                                         285
REGAN That's most certain, and with you. Next month
  with us.
GONERIL You see how full of changes his age is. The
  observation we have made of it hath been little. He
  always loved our sister most, and with what poor
  judgement he hath now cast her off appears too grossly.
REGAN 'Tis the infirmity of his age; yet he hath ever but
  slenderly known himself.                               293
GONERIL The best and soundest of his time hath been but
  rash; then must we look from his age to receive not
  alone the imperfections of long-engrafted condition, but
  therewithal the unruly waywardness that infirm and
  choleric years bring with them.
REGAN Such unconstant starts are we like to have from
  him as this of Kent's banishment.                      300
GONERIL There is further compliment of leave-taking
  between France and him. Pray you, let us sit together.
  If our father carry authority with such disposition as
  he bears, this last surrender of his will but offend us.
REGAN We shall further think of it.                      305
GONERIL We must do something, and i'th' heat.  *Exeunt*

1.2    *Enter Edmond the bastard*
EDMOND
  Thou, nature, art my goddess. To thy law
  My services are bound. Wherefore should I
  Stand in the plague of custom and permit
  The curiosity of nations to deprive me
  For that I am some twelve or fourteen moonshines       5
  Lag of a brother? Why 'bastard'? Wherefore 'base',
  When my dimensions are as well compact,
  My mind as generous, and my shape as true
  As honest madam's issue? Why brand they us
  With 'base', with 'baseness, bastardy—base, base'—

Who in the lusty stealth of nature take 11
More composition and fierce quality
Than doth within a dull, stale, tirèd bed
Go to th' creating a whole tribe of fops
Got 'tween a sleep and wake? Well then, 15
Legitimate Edgar, I must have your land.
Our father's love is to the bastard Edmond
As to th' legitimate. Fine word, 'legitimate'.
Well, my legitimate, if this letter speed
And my invention thrive, Edmond the base 20
Shall to th' legitimate. I grow, I prosper.
Now gods, stand up for bastards!
          *Enter the Duke of Gloucester. Edmond reads a*
          *letter*
GLOUCESTER
Kent banished thus, and France in choler parted,
And the King gone tonight, prescribed his power,
Confined to exhibition—all this done 25
Upon the gad?—Edmond, how now? What news?
EDMOND So please your lordship, none.
GLOUCESTER Why so earnestly seek you to put up that
letter?
EDMOND I know no news, my lord. 30
GLOUCESTER What paper were you reading?
EDMOND Nothing, my lord.
GLOUCESTER No? What needed then that terrible dispatch
of it into your pocket? The quality of nothing hath not
such need to hide itself. Let's see. Come, if it be nothing
I shall not need spectacles. 36
EDMOND I beseech you, sir, pardon me. It is a letter from
my brother that I have not all o'er-read; and for so
much as I have perused, I find it not fit for your
o'erlooking. 40
GLOUCESTER Give me the letter, sir.
EDMOND I shall offend either to detain or give it. The
contents, as in part I understand them, are to blame.
GLOUCESTER Let's see, let's see.
EDMOND I hope for my brother's justification he wrote
this but as an assay or taste of my virtue. 46
          *He gives Gloucester a letter*
GLOUCESTER (*reads*) 'This policy and reverence of age makes
the world bitter to the best of our times, keeps our
fortunes from us till our oldness cannot relish them. I
begin to find an idle and fond bondage in the oppression
of aged tyranny, who sways not as it hath power but
as it is suffered. Come to me, that of this I may speak
more. If our father would sleep till I waked him, you
should enjoy half his revenue for ever and live the
beloved of your brother, 55
                              Edgar.'
Hum, conspiracy! 'Sleep till I wake him, you should
enjoy half his revenue'—my son Edgar! Had he a hand
to write this, a heart and brain to breed it in? When
came you to this? Who brought it? 60
EDMOND It was not brought me, my lord, there's the
cunning of it. I found it thrown in at the casement of
my closet.
GLOUCESTER You know the character to be your brother's?
EDMOND If the matter were good, my lord, I durst swear
it were his; but in respect of that, I would fain think
it were not. 67
GLOUCESTER It is his.
EDMOND It is his hand, my lord, but I hope his heart is
not in the contents. 70
GLOUCESTER Has he never before sounded you in this
business?

EDMOND Never, my lord; but I have heard him oft
maintain it to be fit that, sons at perfect age and fathers
declined, the father should be as ward to the son, and
the son manage his revenue. 76
GLOUCESTER O villain, villain—his very opinion in the
letter! Abhorred villain, unnatural, detested, brutish
villain—worse than brutish! Go, sirrah, seek him. I'll
apprehend him. Abominable villain! Where is he? 80
EDMOND I do not well know, my lord. If it shall please you
to suspend your indignation against my brother till you
can derive from him better testimony of his intent, you
should run a certain course; where if you violently
proceed against him, mistaking his purpose, it would
make a great gap in your own honour and shake in
pieces the heart of his obedience. I dare pawn down my
life for him that he hath writ this to feel my affection to
your honour, and to no other pretence of danger.
GLOUCESTER Think you so? 90
EDMOND If your honour judge it meet, I will place you
where you shall hear us confer of this, and by an
auricular assurance have your satisfaction, and that
without any further delay than this very evening. 94
GLOUCESTER He cannot be such a monster. Edmond, seek
him out, wind me into him, I pray you. Frame the
business after your own wisdom. I would unstate myself
to be in a due resolution.
EDMOND I will seek him, sir, presently, convey the business
as I shall find means, and acquaint you withal. 100
GLOUCESTER These late eclipses in the sun and moon
portend no good to us. Though the wisdom of nature
can reason it thus and thus, yet nature finds itself
scourged by the sequent effects. Love cools, friendship
falls off, brothers divide; in cities, mutinies; in countries,
discord; in palaces, treason; and the bond cracked
'twixt son and father. This villain of mine comes under
the prediction: there's son against father. The King
falls from bias of nature: there's father against child.
We have seen the best of our time. Machinations,
hollowness, treachery, and all ruinous disorders follow
us disquietly to our graves. Find out this villain,
Edmond; it shall lose thee nothing. Do it carefully. And
the noble and true-hearted Kent banished, his offence
honesty! 'Tis strange.                              *Exit*
EDMOND This is the excellent foppery of the world: that
when we are sick in fortune—often the surfeits of our
own behaviour—we make guilty of our disasters the
sun, the moon, and stars, as if we were villains on
necessity, fools by heavenly compulsion, knaves, thieves,
and treachers by spherical predominance, drunkards,
liars, and adulterers by an enforced obedience of
planetary influence, and all that we are evil in by a
divine thrusting on. An admirable evasion of whoremaster man, to lay his goatish disposition on the charge
of a star! My father compounded with my mother under
the Dragon's tail and my nativity was under Ursa Major,
so that it follows I am rough and lecherous. Fut! I should
have been that I am had the maidenliest star in the
firmament twinkled on my bastardizing. 130
          *Enter Edgar*
Pat he comes, like the catastrophe of the old comedy.
My cue is villainous melancholy, with a sigh like Tom
o' Bedlam.
          ⌈*He reads a book*⌉
—O, these eclipses do portend these divisions. Fa, so,
la, mi. 135
EDGAR How now, brother Edmond, what serious contemplation are you in?

EDMOND I am thinking, brother, of a prediction I read this
  other day, what should follow these eclipses.
EDGAR Do you busy yourself with that?                    140
EDMOND I promise you, the effects he writes of succeed
  unhappily. When saw you my father last?
EDGAR The night gone by.
EDMOND Spake you with him?
EDGAR Ay, two hours together.                            145
EDMOND Parted you in good terms? Found you no
  displeasure in him by word nor countenance?
EDGAR None at all.
EDMOND Bethink yourself wherein you may have offended
  him, and at my entreaty forbear his presence until
  some little time hath qualified the heat of his
  displeasure, which at this instant so rageth in him that
  with the mischief of your person it would scarcely allay.
EDGAR Some villain hath done me wrong.                   154
EDMOND That's my fear. I pray you have a continent
  forbearance till the speed of his rage goes slower; and,
  as I say, retire with me to my lodging, from whence I
  will fitly bring you to hear my lord speak. Pray ye, go.
  There's my key. If you do stir abroad, go armed.
EDGAR Armed, brother?                                    160
EDMOND Brother, I advise you to the best. I am no honest
  man if there be any good meaning toward you. I have
  told you what I have seen and heard but faintly,
  nothing like the image and horror of it. Pray you,
  away.                                                  165
EDGAR Shall I hear from you anon?
EDMOND I do serve you in this business.      *Exit Edgar*
  A credulous father, and a brother noble,
  Whose nature is so far from doing harms
  That he suspects none; on whose foolish honesty   170
  My practices ride easy. I see the business.
  Let me, if not by birth, have lands by wit.
  All with me's meet that I can fashion fit.           *Exit*

**1.3**   *Enter Goneril and Oswald, her steward*
GONERIL
  Did my father strike my gentleman
  For chiding of his fool?
OSWALD                        Ay, madam.
GONERIL
  By day and night he wrongs me. Every hour
  He flashes into one gross crime or other
  That sets us all at odds. I'll not endure it.          5
  His knights grow riotous, and himself upbraids us
  On every trifle. When he returns from hunting
  I will not speak with him. Say I am sick.
  If you come slack of former services
  You shall do well; the fault of it I'll answer.        10
        ⌈*Horns within*⌉
OSWALD He's coming, madam. I hear him.
GONERIL
  Put on what weary negligence you please,
  You and your fellows. I'd have it come to question.
  If he distaste it, let him to my sister,
  Whose mind and mine I know in that are one.            15
  Remember what I have said.
OSWALD                        Well, madam.
GONERIL
  And let his knights have colder looks among you.
  What grows of it, no matter. Advise your fellows so.
  I'll write straight to my sister to hold my course.
  Prepare for dinner.              *Exeunt severally*

**1.4**   *Enter the Earl of Kent, disguised*
KENT
  If but as well I other accents borrow
  That can my speech diffuse, my good intent
  May carry through itself to that full issue
  For which I razed my likeness. Now, banished Kent,
  If thou canst serve where thou dost stand condemned,
  So may it come thy master, whom thou lov'st,           6
  Shall find thee full of labours.
        *Horns within. Enter King Lear and attendants from
        hunting*
LEAR Let me not stay a jot for dinner. Go get it ready.
        ⌈*Exit one*⌉
  (*To Kent*) How now, what art thou?
KENT A man, sir.                                          10
LEAR What dost thou profess? What wouldst thou with
  us?
KENT I do profess to be no less than I seem, to serve him
  truly that will put me in trust, to love him that is
  honest, to converse with him that is wise and says
  little, to fear judgement, to fight when I cannot choose,
  and to eat no fish.                                    17
LEAR What art thou?
KENT A very honest-hearted fellow, and as poor as the
  King.                                                  20
LEAR If thou be'st as poor for a subject as he's for a king,
  thou'rt poor enough. What wouldst thou?
KENT Service.
LEAR Who wouldst thou serve?
KENT You.                                                25
LEAR Dost thou know me, fellow?
KENT No, sir, but you have that in your countenance
  which I would fain call master.
LEAR What's that?
KENT Authority.                                          30
LEAR What services canst do?
KENT I can keep honest counsel, ride, run, mar a curious
  tale in telling it, and deliver a plain message bluntly.
  That which ordinary men are fit for I am qualified in;
  and the best of me is diligence.                       35
LEAR How old art thou?
KENT Not so young, sir, to love a woman for singing, nor
  so old to dote on her for anything. I have years on my
  back forty-eight.                                      39
LEAR Follow me. Thou shalt serve me, if I like thee no
  worse after dinner. I will not part from thee yet. Dinner,
  ho, dinner! Where's my knave, my fool? Go you and
  call my fool hither.                        ⌈*Exit one*⌉
        *Enter Oswald the steward*
  You, you, sirrah, where's my daughter?                 44
OSWALD So please you—                             *Exit*
LEAR What says the fellow there? Call the clotpoll back.
        *Exit a knight*
  Where's my fool? Ho, I think the world's asleep.
        *Enter a Knight*
  How now? Where's that mongrel?
KNIGHT He says, my lord, your daughter is not well.      49
LEAR Why came not the slave back to me when I called
  him?
KNIGHT Sir, he answered me in the roundest manner he
  would not.
LEAR A would not?                                        54
KNIGHT My lord, I know not what the matter is, but to
  my judgement your highness is not entertained with
  that ceremonious affection as you were wont. There's
  a great abatement of kindness appears as well in the

general dependants as in the Duke himself also, and
your daughter.   60

LEAR Ha, sayst thou so?

KNIGHT I beseech you pardon me, my lord, if I be
mistaken, for my duty cannot be silent when I think
your highness wronged.   64

LEAR Thou but rememberest me of mine own conception.
I have perceived a most faint neglect of late, which I
have rather blamed as mine own jealous curiosity than
as a very pretence and purpose of unkindness. I will
look further into't. But where's my fool? I have not
seen him these two days.   70

KNIGHT Since my young lady's going into France, sir, the
fool hath much pined away.

LEAR No more of that, I have noted it well. Go you and
tell my daughter I would speak with her.   ⌜Exit one⌝
Go you, call hither my fool.   ⌜Exit one⌝
    *Enter Oswald the steward ⌜crossing the stage⌝*
O you, sir, you, come you hither, sir, who am I, sir?

OSWALD My lady's father?   77

LEAR My lady's father? My lord's knave, you whoreson
dog, you slave, you cur!

OSWALD I am none of these, my lord, I beseech your
pardon.   81

LEAR Do you bandy looks with me, you rascal?
    ⌜*Lear strikes him*⌝

OSWALD I'll not be strucken, my lord.

KENT ⌜*tripping him*⌝ Nor tripped neither, you base football
player.   85

LEAR (*to Kent*) I thank thee, fellow. Thou serv'st me, and
I'll love thee.

KENT (*to Oswald*) Come, sir, arise, away. I'll teach you
differences. Away, away. If you will measure your
lubber's length again, tarry; but away, go to. Have
you wisdom? So.   *Exit Oswald*

LEAR Now, my friendly knave, I thank thee.   92
    *Enter Lear's Fool*
There's earnest of thy service.
    *He gives Kent money*

FOOL Let me hire him, too. (*To Kent*) Here's my coxcomb.

LEAR How now, my pretty knave, how dost thou?   95

FOOL (*to Kent*) Sirrah, you were best take my coxcomb.

LEAR Why, my boy?

FOOL Why? For taking one's part that's out of favour. (*To
Kent*) Nay, an thou canst not smile as the wind sits,
thou'lt catch cold shortly. There, take my coxcomb.
Why, this fellow has banished two on's daughters and
did the third a blessing against his will. If thou follow
him, thou must needs wear my coxcomb. (*To Lear*)
How now, nuncle? Would I had two coxcombs and
two daughters.   105

LEAR Why, my boy?

FOOL If I gave them all my living I'd keep my coxcombs
myself. There's mine; beg another off thy daughters.

LEAR Take heed, sirrah—the whip.   109

FOOL Truth's a dog must to kennel. He must be whipped
out when the Lady Brach may stand by th' fire and
stink.

LEAR A pestilent gall to me!

FOOL ⌜*to Kent*⌝ Sirrah, I'll teach thee a speech.

LEAR Do.   115

FOOL Mark it, nuncle:
    Have more than thou showest,
    Speak less than thou knowest,
    Lend less than thou owest,
    Ride more than thou goest,   120

    Learn more than thou trowest,
    Set less than thou throwest,
    Leave thy drink and thy whore,
    And keep in-a-door,
    And thou shalt have more   125
    Than two tens to a score.

KENT This is nothing, fool.

FOOL Then 'tis like the breath of an unfee'd lawyer: you
gave me nothing for't. (*To Lear*) Can you make no use
of nothing, nuncle?   130

LEAR Why no, boy. Nothing can be made out of nothing.

FOOL (*to Kent*) Prithee, tell him so much the rent of his
land comes to. He will not believe a fool.

LEAR A bitter fool.

FOOL Dost know the difference, my boy, between a bitter
fool and a sweet one?   136

LEAR No, lad. Teach me.

FOOL Nuncle, give me an egg, and I'll give thee two
crowns.

LEAR What two crowns shall they be?   140

FOOL Why, after I have cut the egg i'th' middle and eat
up the meat, the two crowns of the egg. When thou
clovest thy crown i'th' middle and gavest away both
parts, thou borest thine ass o'th' back o'er the dirt.
Thou hadst little wit in thy bald crown when thou
gavest thy golden one away. If I speak like myself in
this, let him be whipped that first finds it so.   147
    ⌜*Sings*⌝ Fools had ne'er less grace in a year,
        For wise men are grown foppish,
      And know not how their wits to wear,   150
        Their manners are so apish.

LEAR When were you wont to be so full of songs, sirrah?

FOOL I have used it, nuncle, e'er since thou madest thy
daughters thy mothers; for when thou gavest them the
rod and puttest down thine own breeches,   155
    ⌜*Sings*⌝ Then they for sudden joy did weep,
        And I for sorrow sung,
      That such a king should play bo-peep
        And go the fools among.
Prithee, nuncle, keep a schoolmaster that can teach
thy fool to lie. I would fain learn to lie.   161

LEAR An you lie, sirrah, we'll have you whipped.

FOOL I marvel what kin thou and thy daughters are.
They'll have me whipped for speaking true, thou'lt
have me whipped for lying, and sometimes I am
whipped for holding my peace. I had rather be any
kind o' thing than a fool; and yet I would not be thee,
nuncle. Thou hast pared thy wit o' both sides and left
nothing i'th' middle.
    *Enter Goneril*
Here comes one o' the parings.   170

LEAR
How now, daughter? What makes that frontlet on?
You are too much of late i'th' frown.

FOOL Thou wast a pretty fellow when thou hadst no need
to care for her frowning. Now thou art an O without
a figure. I am better than thou art, now. I am a fool;
thou art nothing. ⌜*To Goneril*⌝ Yes, forsooth, I will hold
my tongue; so your face bids me, though you say
nothing.
    ⌜*Sings*⌝ Mum, mum.
        He that keeps nor crust nor crumb,   180
        Weary of all, shall want some.
That's a shelled peascod.

GONERIL (*to Lear*)
Not only, sir, this your all-licensed fool,

But other of your insolent retinue
Do hourly carp and quarrel, breaking forth    185
In rank and not-to-be-endurèd riots. Sir,
I had thought by making this well known unto you
To have found a safe redress, but now grow fearful,
By what yourself too late have spoke and done,
That you protect this course, and put it on    190
By your allowance; which if you should, the fault
Would not scape censure, nor the redresses sleep
Which in the tender of a wholesome weal
Might in their working do you that offence,
Which else were shame, that then necessity    195
Will call discreet proceeding.

FOOL (to Lear) For, you know, nuncle,
⌈Sings⌉    The hedge-sparrow fed the cuckoo so long
        That it's had it head bit off by it young;
so out went the candle, and we were left darkling.   200

LEAR (to Goneril) Are you our daughter?

GONERIL
I would you would make use of your good wisdom,
Whereof I know you are fraught, and put away
These dispositions which of late transport you
From what you rightly are.    205

FOOL May not an ass know when the cart draws the
horse? ⌈Sings⌉ 'Whoop, jug, I love thee!'

LEAR
Does any here know me? This is not Lear.
Does Lear walk thus, speak thus? Where are his eyes?
Either his notion weakens, his discernings    210
Are lethargied—ha, waking? 'Tis not so.
Who is it that can tell me who I am?

FOOL Lear's shadow.

LEAR (to Goneril) Your name, fair gentlewoman?

GONERIL
This admiration, sir, is much o'th' savour    215
Of other your new pranks. I do beseech you
To understand my purposes aright,
As you are old and reverend, should be wise.
Here do you keep a hundred knights and squires,
Men so disordered, so debauched and bold    220
That this our court, infected with their manners,
Shows like a riotous inn. Epicurism and lust
Makes it more like a tavern or a brothel
Than a graced palace. The shame itself doth speak
For instant remedy. Be then desired,    225
By her that else will take the thing she begs,
A little to disquantity your train,
And the remainders that shall still depend
To be such men as may besort your age,
Which know themselves and you.

LEAR            Darkness and devils!
Saddle my horses, call my train together!—    231
                    ⌈Exit one or more⌉
Degenerate bastard, I'll not trouble thee.
Yet have I left a daughter.

GONERIL
You strike my people, and your disordered rabble
Make servants of their betters.
      Enter the Duke of Albany

LEAR           Woe that too late repents!
Is it your will? Speak, sir.—Prepare my horses.   236
                    ⌈Exit one or more⌉
Ingratitude, thou marble-hearted fiend,
More hideous when thou show'st thee in a child
Than the sea-monster—

ALBANY Pray sir, be patient.    240

LEAR (to Goneril) Detested kite, thou liest.
My train are men of choice and rarest parts,
That all particulars of duty know,
And in the most exact regard support
The worships of their name. O most small fault,   245
How ugly didst thou in Cordelia show,
Which, like an engine, wrenched my frame of nature
From the fixed place, drew from my heart all love,
And added to the gall! O Lear, Lear, Lear!
Beat at this gate that let thy folly in    250
And thy dear judgement out.—Go, go, my people!

ALBANY
My lord, I am guiltless, as I am ignorant
Of what hath moved you.

LEAR            It may be so, my lord.
Hear, nature; hear, dear goddess, hear:
Suspend thy purpose if thou didst intend    255
To make this creature fruitful.
Into her womb convey sterility.
Dry up in her the organs of increase,
And from her derogate body never spring
A babe to honour her. If she must teem,    260
Create her child of spleen, that it may live
And be a thwart disnatured torment to her.
Let it stamp wrinkles in her brow of youth,
With cadent tears fret channels in her cheeks,
Turn all her mother's pains and benefits    265
To laughter and contempt, that she may feel—
That she may feel
How sharper than a serpent's tooth it is
To have a thankless child. Away, away!
       Exeunt Lear, ⌈Kent, and attendants⌉

ALBANY
Now, gods that we adore, whereof comes this?   270

GONERIL
Never afflict yourself to know more of it,
But let his disposition have that scope
As dotage gives it.
      Enter King Lear

LEAR
What, fifty of my followers at a clap?
Within a fortnight?

ALBANY           What's the matter, sir?    275

LEAR
I'll tell thee. (To Goneril) Life and death! I am ashamed
That thou hast power to shake my manhood thus,
That these hot tears, which break from me perforce,
Should make thee worth them. Blasts and fogs upon
     thee!
Th'untented woundings of a father's curse    280
Pierce every sense about thee! Old fond eyes,
Beweep this cause again I'll pluck ye out
And cast you, with the waters that you loose,
To temper clay. Ha! Let it be so.
I have another daughter    285
Who, I am sure, is kind and comfortable.
When she shall hear this of thee, with her nails
She'll flay thy wolvish visage. Thou shalt find
That I'll resume the shape which thou dost think
I have cast off for ever.           Exit

GONERIL        Do you mark that?    290

ALBANY
I cannot be so partial, Goneril,
To the great love I bear you—

GONERIL
Pray you, content. What, Oswald, ho!—
You, sir, more knave than fool, after your master.

FOOL
Nuncle Lear, nuncle Lear,                               295
Tarry, take the fool with thee.
A fox when one has caught her,
And such a daughter,
Should sure to the slaughter,
If my cap would buy a halter.                          300
So, the fool follows after.                      *Exit*

GONERIL
This man hath had good counsel—a hundred
    knights?
'Tis politic and safe to let him keep
At point a hundred knights, yes, that on every dream,
Each buzz, each fancy, each complaint, dislike,   305
He may enguard his dotage with their powers
And hold our lives in mercy.—Oswald, I say!

ALBANY
Well, you may fear too far.

GONERIL                          Safer than trust too far.
Let me still take away the harms I fear,
Not fear still to be taken. I know his heart.       310
What he hath uttered I have writ my sister.
If she sustain him and his hundred knights
When I have showed th'unfitness—
    *Enter Oswald the steward*
                                    How now, Oswald?
What, have you writ that letter to my sister?

OSWALD Ay, madam.                                       315

GONERIL
Take you some company, and away to horse.
Inform her full of my particular fear,
And thereto add such reasons of your own
As may compact it more. Get you gone,
And hasten your return.                   *Exit Oswald*
                        No, no, my lord,                320
This milky gentleness and course of yours,
Though I condemn not, yet under pardon
You are much more attasked for want of wisdom
Than praised for harmful mildness.

ALBANY
How far your eyes may pierce I cannot tell.         325
Striving to better, oft we mar what's well.

GONERIL Nay, then—

ALBANY Well, well, th'event.                    *Exeunt*

**1.5**   *Enter King Lear, the Earl of Kent disguised, the*
          *First Gentleman, and Lear's Fool*

LEAR ⌈*to the Gentleman, giving him a letter*⌉ Go you before
to Gloucester with these letters.      ⌈*Exit Gentleman*⌉
⌈*To Kent, giving him a letter*⌉ Acquaint my daughter no
further with anything you know than comes from her
demand out of the letter. If your diligence be not speedy,
I shall be there afore you.                               6

KENT I will not sleep, my lord, till I have delivered your
letter.                                              *Exit*

FOOL If a man's brains were in's heels, were't not in
danger of kibes?                                        10

LEAR Ay, boy.

FOOL Then, I prithee, be merry: thy wit shall not go
slipshod.

LEAR Ha, ha, ha!

FOOL Shalt see thy other daughter will use thee kindly,
for though she's as like this as a crab's like an apple,
yet I can tell what I can tell.                         17

LEAR What canst tell, boy?

FOOL She will taste as like this as a crab does to a crab.
Thou canst tell why one's nose stands i'th' middle
on 's face?                                             21

LEAR No.

FOOL Why, to keep one's eyes of either side 's nose, that
what a man cannot smell out, a may spy into.

LEAR I did her wrong.                                   25

FOOL Canst tell how an oyster makes his shell?

LEAR No.

FOOL Nor I neither; but I can tell why a snail has a
house.

LEAR Why?                                               30

FOOL Why, to put 's head in, not to give it away to his
daughters and leave his horns without a case.

LEAR
I will forget my nature. So kind a father!
Be my horses ready?                                     34

FOOL Thy asses are gone about 'em. The reason why the
seven stars are no more than seven is a pretty reason.

LEAR Because they are not eight.

FOOL Yes, indeed, thou wouldst make a good fool.

LEAR
To take't again perforce—monster ingratitude!

FOOL If thou wert my fool, nuncle, I'd have thee beaten
for being old before thy time.                          41

LEAR How's that?

FOOL Thou shouldst not have been old till thou hadst
been wise.

LEAR
O, let me not be mad, not mad, sweet heaven!        45
Keep me in temper. I would not be mad.
    ⌈*Enter the First Gentleman*⌉
How now, are the horses ready?

⌈FIRST⌉ GENTLEMAN                         Ready, my lord.

LEAR (*to Fool*) Come, boy.   ⌈*Exeunt Lear and Gentleman*⌉

FOOL
She that's a maid now, and laughs at my departure,
Shall not be a maid long, unless things be cut shorter.
                                                  ⌈*Exit*⌉

**2.1**   *Enter Edmond the bastard, and Curan, severally*

EDMOND Save thee, Curan.

CURAN And you, sir. I have been with your father, and
given him notice that the Duke of Cornwall and Regan
his duchess will be here with him this night.

EDMOND How comes that?                                   5

CURAN Nay, I know not. You have heard of the news
abroad?—I mean the whispered ones, for they are yet
but ear-kissing arguments.

EDMOND Not I. Pray you, what are they?

CURAN Have you heard of no likely wars toward twixt
the Dukes of Cornwall and Albany?                       11

EDMOND Not a word.

CURAN You may do then in time. Fare you well, sir.
                                                    *Exit*

EDMOND
The Duke be here tonight! The better, best.
This weaves itself perforce into my business.       15
    ⌈*Enter Edgar at a window above*⌉
My father hath set guard to take my brother,
And I have one thing of a queasy question
Which I must act. Briefness and fortune work!—
Brother, a word, descend. Brother, I say.
    ⌈*Edgar climbs down*⌉
My father watches. O sir, fly this place.           20
Intelligence is given where you are hid.
You have now the good advantage of the night.

Have you not spoken 'gainst the Duke of Cornwall?
He's coming hither, now, i'th' night, i'th' haste,
And Regan with him. Have you nothing said          25
Upon his party 'gainst the Duke of Albany?
Advise yourself.
EDGAR          I am sure on't, not a word.
EDMOND
I hear my father coming. Pardon me.
In cunning I must draw my sword upon you.
Draw. Seem to defend yourself. Now, quit you well.  30
(*Calling*) Yield, come before my father. Light ho, here!
(*To Edgar*) Fly, brother! (*Calling*) Torches, torches!
(*To Edgar*)                              So, farewell.
                              *Exit Edgar*
Some blood drawn on me would beget opinion
Of my more fierce endeavour.
          *He wounds his arm*
                              I have seen drunkards
Do more than this in sport. (*Calling*) Father, father!  35
Stop, stop! Ho, help!
          *Enter the Duke of Gloucester, and servants with*
          *torches*
GLOUCESTER          Now, Edmond, where's the villain?
EDMOND
Here stood he in the dark, his sharp sword out,
Mumbling of wicked charms, conjuring the moon
To stand 's auspicious mistress.
GLOUCESTER                              But where is he?
EDMOND
Look, sir, I bleed.
GLOUCESTER          Where is the villain, Edmond?    40
EDMOND
Fled this way, sir, when by no means he could—
GLOUCESTER
Pursue him, ho! Go after.          *Exeunt servants*
                              By no means what?
EDMOND
Persuade me to the murder of your lordship,
But that I told him the revenging gods
'Gainst parricides did all the thunder bend,          45
Spoke with how manifold and strong a bond
The child was bound to th' father. Sir, in fine,
Seeing how loathly opposite I stood
To his unnatural purpose, in fell motion
With his preparèd sword he charges home          50
My unprovided body, latched mine arm;
And when he saw my best alarumed spirits
Bold in the quarrel's right, roused to th'encounter,
Or whether ghasted by the noise I made,
Full suddenly he fled.
GLOUCESTER          Let him fly far,          55
Not in this land shall he remain uncaught,
And found, dispatch. The noble Duke my master,
My worthy arch and patron, comes tonight.
By his authority I will proclaim it
That he which finds him shall deserve our thanks,  60
Bringing the murderous coward to the stake;
He that conceals him, death.
EDMOND
When I dissuaded him from his intent
And found him pitched to do it, with curst speech
I threatened to discover him. He replied,          65
'Thou unpossessing bastard, dost thou think
If I would stand against thee, would the reposal
Of any trust, virtue, or worth in thee
Make thy words faithed? No, what I should deny—
As this I would, ay, though thou didst produce  70

My very character—I'd turn it all
To thy suggestion, plot, and damnèd practice,
And thou must make a dullard of the world
If they not thought the profits of my death
Were very pregnant and potential spirits          75
To make thee seek it.'
GLOUCESTER          O strange and fastened villain!
Would he deny his letter, said he?
          *Tucket within*
Hark, the Duke's trumpets. I know not why he comes.
All ports I'll bar. The villain shall not scape.
The Duke must grant me that; besides, his picture  80
I will send far and near, that all the kingdom
May have due note of him—and of my land,
Loyal and natural boy, I'll work the means
To make thee capable.
          *Enter the Duke of Cornwall, Regan, and attendants*
CORNWALL
How now, my noble friend? Since I came hither,  85
Which I can call but now, I have heard strange news.
REGAN
If it be true, all vengeance comes too short
Which can pursue th'offender. How dost, my lord?
GLOUCESTER
O madam, my old heart is cracked, it's cracked.
REGAN
What, did my father's godson seek your life?          90
He whom my father named, your Edgar?
GLOUCESTER
O lady, lady, shame would have it hid!
REGAN
Was he not companion with the riotous knights
That tend upon my father?
GLOUCESTER
I know not, madam. 'Tis too bad, too bad.          95
EDMOND
Yes, madam, he was of that consort.
REGAN
No marvel, then, though he were ill affected.
'Tis they have put him on the old man's death,
To have th'expense and spoil of his revenues.
I have this present evening from my sister          100
Been well informed of them, and with such cautions
That if they come to sojourn at my house
I'll not be there.
CORNWALL          Nor I, assure thee, Regan.
Edmond, I hear that you have shown your father
A childlike office.
EDMOND          It was my duty, sir,          105
GLOUCESTER (*to Cornwall*)
He did bewray his practice, and received
This hurt you see striving to apprehend him.
CORNWALL
Is he pursued?
GLOUCESTER          Ay, my good lord.
CORNWALL
If he be taken, he shall never more
Be feared of doing harm. Make your own purpose  110
How in my strength you please. For you, Edmond,
Whose virtue and obedience doth this instant
So much commend itself, you shall be ours.
Natures of such deep trust we shall much need.
You we first seize on.
EDMOND          I shall serve you, sir,          115
Truly, however else.
GLOUCESTER (*to Cornwall*) For him I thank your grace.

CORNWALL
You know not why we came to visit you—
REGAN
Thus out of season, threading dark-eyed night—
Occasions, noble Gloucester, of some poise,
Wherein we must have use of your advice.     120
Our father he hath writ, so hath our sister,
Of differences which I least thought it fit
To answer from our home. The several messengers
From hence attend dispatch. Our good old friend,
Lay comforts to your bosom, and bestow     125
Your needful counsel to our businesses,
Which craves the instant use.
GLOUCESTER   I serve you, madam.
Your graces are right welcome.      *Flourish. Exeunt*

**2.2**    *Enter the Earl of Kent, disguised, and Oswald the*
     *steward, severally*
OSWALD Good dawning to thee, friend. Art of this house?
KENT Ay.
OSWALD Where may we set our horses?
KENT I'th' mire.
OSWALD Prithee, if thou lov'st me, tell me.      5
KENT I love thee not.
OSWALD Why then, I care not for thee.
KENT If I had thee in Lipsbury pinfold I would make thee
care for me.
OSWALD Why dost thou use me thus? I know thee not.
KENT Fellow, I know thee.      11
OSWALD What dost thou know me for?
KENT A knave, a rascal, an eater of broken meats, a base,
proud, shallow, beggarly, three-suited, hundred-pound,
filthy worsted-stocking knave; a lily-livered, action-
taking, whoreson, glass-gazing, super-serviceable,
finical rogue; one-trunk-inheriting slave; one that
wouldst be a bawd in way of good service, and art
nothing but the composition of a knave, beggar,
coward, pander, and the son and heir of a mongrel
bitch, one whom I will beat into clamorous whining if
thou deniest the least syllable of thy addition.     22
OSWALD Why, what a monstrous fellow art thou, thus to
rail on one that is neither known of thee nor knows
thee!      25
KENT What a brazen-faced varlet art thou, to deny thou
knowest me! Is it two days since I tripped up thy heels
and beat thee before the King? Draw, you rogue; for
though it be night, yet the moon shines.
⌈*He draws his sword*⌉
I'll make a sop o'th' moonshine of you, you whoreson,
cullionly barber-monger, draw!      31
OSWALD Away. I have nothing to do with thee.
KENT Draw, you rascal. You come with letters against
the King, and take Vanity the puppet's part against the
royalty of her father. Draw, you rogue, or I'll so
carbonado your shanks—draw, you rascal, come your
ways!      37
OSWALD Help, ho, murder, help!
KENT Strike, you slave! Stand, rogue! Stand, you neat
slave, strike!      40
OSWALD Help, ho, murder, murder!
     *Enter Edmond the bastard, ⌈then⌉ the Duke of*
     *Cornwall, Regan, the Duke of Gloucester, and*
     *servants*
EDMOND How now, what's the matter? Part.
KENT With you, goodman boy. If you please, come, I'll
flesh ye. Come on, young master.      44

GLOUCESTER Weapons? Arms? What's the matter here?
CORNWALL
Keep peace, upon your lives. He dies that strikes again.
What is the matter?
REGAN The messengers from our sister and the King.
CORNWALL (*to Kent and Oswald*) What is your difference?
Speak.      50
OSWALD I am scarce in breath, my lord.
KENT No marvel, you have so bestirred your valour, you
cowardly rascal. Nature disclaims in thee; a tailor made
thee.
CORNWALL Thou art a strange fellow—a tailor make a
man?      56
KENT A tailor, sir. A stone-cutter or a painter could not
have made him so ill though they had been but two
years o'th' trade.
CORNWALL Speak yet; how grew your quarrel?     60
OSWALD This ancient ruffian, sir, whose life I have spared
at suit of his grey beard—
KENT Thou whoreson Z, thou unnecessary letter— (*to
Cornwall*) my lord, if you'll give me leave I will tread
this unbolted villain into mortar and daub the wall of
a jakes with him. (*To Oswald*) Spare my grey beard,
you wagtail?      67
CORNWALL Peace, sirrah.
You beastly knave, know you no reverence?
KENT
Yes, sir, but anger hath a privilege.      70
CORNWALL Why art thou angry?
KENT
That such a slave as this should wear a sword,
Who wears no honesty. Such smiling rogues as these,
Like rats, oft bite the holy cords a-twain
Which are too intrince t'unloose, smooth every
    passion      75
That in the natures of their lords rebel;
Being oil to fire, snow to the colder moods,
Renege, affirm, and turn their halcyon beaks
With every gall and vary of their masters,
Knowing naught, like dogs, but following.      80
⌈*To Oswald*⌉ A plague upon your epileptic visage!
Smile you my speeches as I were a fool?
Goose, an I had you upon Sarum Plain
I'd drive ye cackling home to Camelot.
CORNWALL
What, art thou mad, old fellow?
GLOUCESTER ⌈*to Kent*⌉      How fell you out? Say that.
KENT
No contraries hold more antipathy      86
Than I and such a knave.
CORNWALL      Why dost thou call him knave?
What is his fault?
KENT      His countenance likes me not.
CORNWALL
No more perchance does mine, nor his, nor hers.
KENT
Sir, 'tis my occupation to be plain:      90
I have seen better faces in my time
Than stands on any shoulder that I see
Before me at this instant.
CORNWALL      This is some fellow
Who, having been praised for bluntness, doth affect
A saucy roughness, and constrains the garb      95
Quite from his nature. He cannot flatter, he;

An honest mind and plain, he must speak truth.
An they will take't, so; if not, he's plain.
These kind of knaves I know, which in this plainness
Harbour more craft and more corrupter ends          100
Than twenty silly-ducking observants
That stretch their duties nicely.

KENT
Sir, in good faith, in sincere verity,
Under th'allowance of your great aspect,
Whose influence, like the wreath of radiant fire          105
On flick'ring Phoebus' front—

CORNWALL                              What mean'st by this?

KENT  To go out of my dialect, which you discommend so
much. I know, sir, I am no flatterer. He that beguiled
you in a plain accent was a plain knave, which for my
part I will not be, though I should win your displeasure
to entreat me to't.          111

CORNWALL (to Oswald)
What was th'offence you gave him?

OSWALD                              I never gave him any.
It pleased the King his master very late
To strike at me upon his misconstruction,
When he, compact, and flattering his displeasure,          115
Tripped me behind; being down, insulted, railed,
And put upon him such a deal of man
That worthied him, got praises of the King
For him attempting who was self-subdued,
And in the fleshment of this dread exploit          120
Drew on me here again.

KENT                              None of these rogues and cowards
But Ajax is their fool.

CORNWALL                              Fetch forth the stocks!
                              ⌜Exeunt some servants⌝
You stubborn, ancient knave, you reverend braggart,
We'll teach you.

KENT                              Sir, I am too old to learn.
Call not your stocks for me. I serve the King,          125
On whose employment I was sent to you.
You shall do small respect, show too bold malice
Against the grace and person of my master,
Stocking his messenger.

CORNWALL ⌜calling⌝          Fetch forth the stocks!—
As I have life and honour, there shall he sit till noon.

REGAN
Till noon?—till night, my lord, and all night too.          131

KENT
Why, madam, if I were your father's dog
You should not use me so.

REGAN                              Sir, being his knave, I will.
                              Stocks brought out

CORNWALL
This is a fellow of the selfsame colour
Our sister speaks of.—Come, bring away the stocks.

GLOUCESTER
Let me beseech your grace not to do so.          136
The King his master needs must take it ill
That he, so slightly valued in his messenger,
Should have him thus restrained.

CORNWALL                              I'll answer that.
                              ⌜They put Kent in the stocks⌝

REGAN
My sister may receive it much more worse          140
To have her gentlemen abused, assaulted.

CORNWALL  Come, my good lord, away!
                              Exeunt all but Gloucester and Kent

GLOUCESTER
I am sorry for thee, friend. 'Tis the Duke's pleasure,
Whose disposition, all the world well knows,
Will not be rubbed nor stopped. I'll entreat for thee.

KENT
Pray do not, sir. I have watched and travelled hard.
Some time I shall sleep out; the rest I'll whistle.          147
A good man's fortune may grow out at heels.
Give you good morrow.

GLOUCESTER
The Duke's to blame in this; 'twill be ill taken.          Exit

KENT
Good King, that must approve the common say:          151
Thou out of heaven's benediction com'st
To the warm sun.
                              ⌜He takes out a letter⌝
Approach, thou beacon to this under globe,
That by thy comfortable beams I may          155
Peruse this letter. Nothing almost sees miracles
But misery. I know 'tis from Cordelia,
Who hath now fortunately been informed
Of my obscurèd course, and shall find time
For this enormous state, seeking to give          160
Losses their remedies. All weary and o'erwatched,
Take vantage, heavy eyes, not to behold
This shameful lodging. Fortune, good night;
Smile once more; turn thy wheel.          He sleeps
                              Enter Edgar

EDGAR                              I heard myself proclaimed,
And by the happy hollow of a tree          165
Escaped the hunt. No port is free, no place
That guard and most unusual vigilance
Does not attend my taking. Whiles I may scape
I will preserve myself, and am bethought
To take the basest and most poorest shape          170
That ever penury in contempt of man
Brought near to beast. My face I'll grime with filth,
Blanket my loins, elf all my hairs in knots,
And with presented nakedness outface
The winds and persecutions of the sky.          175
The country gives me proof and precedent
Of Bedlam beggars who with roaring voices
Strike in their numbed and mortifièd arms
Pins, wooden pricks, nails, sprigs of rosemary,
And with this horrible object from low farms,          180
Poor pelting villages, sheep-cotes and mills
Sometime with lunatic bans, sometime with prayers
Enforce their charity. 'Poor Tuelygod, Poor Tom.'
That's something yet. Edgar I nothing am.
                              Exit
                    Enter King Lear, his Fool, and ⌜the First⌝ Gentleman

LEAR
'Tis strange that they should so depart from home          185
And not send back my messenger.

⌜FIRST⌝ GENTLEMAN                              As I learned,
The night before there was no purpose in them
Of this remove.

KENT (waking)          Hail to thee, noble master.

LEAR
Ha! Mak'st thou this shame thy pastime?

KENT                              No, my lord.

FOOL  Ha, ha, he wears cruel garters! Horses are tied by
the heads, dogs and bears by th' neck, monkeys by th'
loins, and men by th' legs. When a man's overlusty at
legs, then he wears wooden nether-stocks.          193

LEAR (*to Kent*)
What's he that hath so much thy place mistook
To set thee here?
KENT                          It is both he and she:        195
Your son and daughter.
LEAR                          No.
KENT                          Yes.
LEAR                          No, I say.
KENT
I say yea.
LEAR          By Jupiter, I swear no.
KENT
By Juno, I swear ay.
LEAR                          They durst not do't,
They could not, would not do't. 'Tis worse than
    murder,
To do upon respect such violent outrage.        200
Resolve me with all modest haste which way
Thou mightst deserve or they impose this usage,
Coming from us.
KENT                          My lord, when at their home
I did commend your highness' letters to them,
Ere I was risen from the place that showed        205
My duty kneeling, came there a reeking post
Stewed in his haste, half breathless, painting forth
From Goneril, his mistress, salutations,
Delivered letters spite of intermission,
Which presently they read, on whose contents        210
They summoned up their meiny, straight took horse,
Commanded me to follow and attend
The leisure of their answer, gave me cold looks;
And meeting here the other messenger,
Whose welcome I perceived had poisoned mine—        215
Being the very fellow which of late
Displayed so saucily against your highness—
Having more man than wit about me, drew.
He raised the house with loud and coward cries.
Your son and daughter found this trespass worth        220
The shame which here it suffers.
FOOL Winter's not gone yet if the wild geese fly that way.
    ⌜*Sings*⌝          Fathers that wear rags
                    Do make their children blind,
                But fathers that bear bags        225
                    Shall see their children kind.
                Fortune, that arrant whore,
                    Ne'er turns the key to th' poor.
But for all this thou shalt have as many dolours for
thy daughters as thou canst tell in a year.        230
LEAR
O, how this mother swells up toward my heart!
*Historica passio* down, thou climbing sorrow;
Thy element's below.—Where is this daughter?
KENT
With the Earl, sir, here within.
LEAR                          Follow me not; stay here.
                                                *Exit*
⌜FIRST⌝ GENTLEMAN (*to Kent*)
Made you no more offence but what you speak of?
KENT None.        236
How chance the King comes with so small a number?
FOOL An thou hadst been set i'th' stocks for that question,
    thou'dst well deserved it.
KENT Why, Fool?        240
FOOL We'll set thee to school to an ant, to teach thee
    there's no labouring i'th' winter. All that follow their
    noses are led by their eyes but blind men, and there's
    not a nose among twenty but can smell him that's

stinking. Let go thy hold when a great wheel runs
down a hill, lest it break thy neck with following; but
the great one that goes upward, let him draw thee
after. When a wise man gives thee better counsel, give
me mine again. I would have none but knaves follow
it, since a fool gives it.        250
⌜*Sings*⌝
        That sir which serves and seeks for gain
            And follows but for form,
        Will pack when it begin to rain,
            And leave thee in the storm.

        But I will tarry, the fool will stay,        255
            And let the wise man fly.
        The knave turns fool that runs away,
            The fool no knave, pardie.

KENT Where learned you this, Fool?
FOOL Not i'th' stocks, fool.        260
            *Enter King Lear and the Duke of Gloucester*
LEAR
Deny to speak with me? They are sick, they are weary,
They have travelled all the night?—mere fetches,
The images of revolt and flying off.
Fetch me a better answer.
GLOUCESTER                          My dear lord,
You know the fiery quality of the Duke,        265
How unremovable and fixed he is
In his own course.
LEAR                          Vengeance, plague, death, confusion!
'Fiery'? What 'quality'? Why, Gloucester, Gloucester,
I'd speak with the Duke of Cornwall and his wife.
GLOUCESTER
Well, my good lord, I have informed them so.        270
LEAR
'Informed them'? Dost thou understand me, man?
GLOUCESTER Ay, my good lord.
LEAR
The King would speak with Cornwall; the dear father
Would with his daughter speak, commands, tends
    service.
Are they 'informed' of this? My breath and blood—
'Fiery'? The 'fiery' Duke—tell the hot Duke that—        276
No, but not yet. Maybe he is not well.
Infirmity doth still neglect all office
Whereto our health is bound. We are not ourselves
When nature, being oppressed, commands the mind
To suffer with the body. I'll forbear,        281
And am fallen out with my more headier will,
To take the indisposed and sickly fit
For the sound man.—Death on my state, wherefore
Should he sit here? This act persuades me        285
That this remotion of the Duke and her
Is practice only. Give me my servant forth.
Go tell the Duke and 's wife I'd speak with them,
Now, presently. Bid them come forth and hear me,
Or at their chamber door I'll beat the drum        290
Till it cry sleep to death.
GLOUCESTER                    I would have all well betwixt you.
                                                *Exit*
LEAR
O me, my heart! My rising heart! But down.
FOOL Cry to it, nuncle, as the cockney did to the eels
when she put 'em i'th' paste alive. She knapped 'em
o'th' coxcombs with a stick, and cried 'Down, wantons,
down!' 'Twas her brother that, in pure kindness to his
horse, buttered his hay.        297

*Enter the Duke of Cornwall, Regan, the Duke of*
*Gloucester, and servants*
LEAR  Good morrow to you both.
CORNWALL  Hail to your grace.
          *Kent here set at liberty*
REGAN  I am glad to see your highness.              300
LEAR
  Regan, I think you are. I know what reason
  I have to think so. If thou shouldst not be glad
  I would divorce me from thy mother's shrine,
  Sepulchring an adultress. (*To Kent*) O, are you free?
  Some other time for that.          ⌜*Exit Kent*⌝
                    Belovèd Regan,            305
  Thy sister's naught. O, Regan, she hath tied
  Sharp-toothed unkindness like a vulture here.
  I can scarce speak to thee. Thou'lt not believe
  With how depraved a quality—O, Regan!
REGAN
  I pray you, sir, take patience. I have hope      310
  You less know how to value her desert
  Than she to scant her duty.
LEAR                    Say, how is that?
REGAN
  I cannot think my sister in the least
  Would fail her obligation. If, sir, perchance
  She have restrained the riots of your followers,  315
  'Tis on such ground and to such wholesome end
  As clears her from all blame.
LEAR  My curses on her.
REGAN  O sir, you are old.
  Nature in you stands on the very verge        320
  Of his confine. You should be ruled and led
  By some discretion that discerns your state
  Better than you yourself. Therefore I pray you
  That to our sister you do make return;
  Say you have wronged her.
LEAR                    Ask her forgiveness?    325
  Do you but mark how this becomes the house?
  ⌜*Kneeling*⌝ 'Dear daughter, I confess that I am old.
  Age is unnecessary. On my knees I beg
  That you'll vouchsafe me raiment, bed, and food.'
REGAN
  Good sir, no more. These are unsightly tricks.  330
  Return you to my sister.
LEAR ⌜*rising*⌝          Never, Regan.
  She hath abated me of half my train,
  Looked black upon me, struck me with her tongue
  Most serpent-like upon the very heart.
  All the stored vengeances of heaven fall      335
  On her ingrateful top! Strike her young bones,
  You taking airs, with lameness!
CORNWALL                    Fie, sir, fie.
LEAR
  You nimble lightnings, dart your blinding flames
  Into her scornful eyes. Infect her beauty,
  You fen-sucked fogs drawn by the pow'rful sun   340
  To fall and blister.
REGAN              O, the blest gods!
  So will you wish on me when the rash mood is on.
LEAR
  No, Regan. Thou shalt never have my curse.
  Thy tender-hafted nature shall not give
  Thee o'er to harshness. Her eyes are fierce, but thine
  Do comfort and not burn. 'Tis not in thee      346
  To grudge my pleasures, to cut off my train,
  To bandy hasty words, to scant my sizes,

And, in conclusion, to oppose the bolt
Against my coming in. Thou better know'st      350
The offices of nature, bond of childhood,
Effects of courtesy, dues of gratitude.
Thy half o'th' kingdom hast thou not forgot,
Wherein I thee endowed.
REGAN                    Good sir, to th' purpose.
LEAR
  Who put my man i'th' stocks?
        *Tucket within*
CORNWALL                    What trumpet's that?
        *Enter Oswald the steward*
REGAN
  I know't, my sister's. This approves her letter   356
  That she would soon be here. (*To Oswald*) Is your lady
    come?
LEAR
  This is a slave whose easy-borrowed pride
  Dwells in the sickly grace of her a follows.
  (*To Oswald*) Out, varlet, from my sight!
CORNWALL                    What means your grace?
        *Enter Goneril*
LEAR
  Who stocked my servant? Regan, I have good hope
  Thou didst not know on't. Who comes here? O heavens,
  If you do love old men, if your sweet sway
  Allow obedience, if you yourselves are old,
  Make it your cause! Send down and take my part.  365
  (*To Goneril*) Art not ashamed to look upon this beard?
  O Regan, will you take her by the hand?
GONERIL
  Why not by th' hand, sir? How have I offended?
  All's not offence that indiscretion finds
  And dotage terms so.
LEAR              O sides, you are too tough!     370
  Will you yet hold?—How came my man i'th' stocks?
CORNWALL
  I set him there, sir; but his own disorders
  Deserved much less advancement.
LEAR                          You? Did you?
REGAN
  I pray you, father, being weak, seem so.
  If till the expiration of your month            375
  You will return and sojourn with my sister,
  Dismissing half your train, come then to me.
  I am now from home, and out of that provision
  Which shall be needful for your entertainment.
LEAR
  Return to her, and fifty men dismissed?        380
  No, rather I abjure all roofs, and choose
  To be a comrade with the wolf and owl,
  To wage against the enmity o'th' air
  Necessity's sharp pinch. Return with her?
  Why, the hot-blooded France, that dowerless took  385
  Our youngest born—I could as well be brought
  To knee his throne and, squire-like, pension beg
  To keep base life afoot. Return with her?
  Persuade me rather to be slave and sumpter
  To this detested groom.
GONERIL                    At your choice, sir.    390
LEAR
  I prithee, daughter, do not make me mad.
  I will not trouble thee, my child. Farewell.
  We'll no more meet, no more see one another.
  But yet thou art my flesh, my blood, my daughter—

Or rather a disease that's in my flesh,    395
Which I must needs call mine. Thou art a boil,
A plague-sore or embossèd carbuncle
In my corrupted blood. But I'll not chide thee.
Let shame come when it will, I do not call it.
I do not bid the thunder-bearer shoot,    400
Nor tell tales of thee to high-judging Jove.
Mend when thou canst; be better at thy leisure.
I can be patient, I can stay with Regan,
I and my hundred knights.

REGAN            Not altogether so.
I looked not for you yet, nor am provided    405
For your fit welcome. Give ear, sir, to my sister;
For those that mingle reason with your passion
Must be content to think you old, and so—
But she knows what she does.

LEAR            Is this well spoken?

REGAN
I dare avouch it, sir. What, fifty followers?    410
Is it not well? What should you need of more,
Yea, or so many, sith that both charge and danger
Speak 'gainst so great a number? How in one house
Should many people under two commands
Hold amity? 'Tis hard, almost impossible.    415

GONERIL
Why might not you, my lord, receive attendance
From those that she calls servants, or from mine?

REGAN
Why not, my lord? If then they chanced to slack ye,
We could control them. If you will come to me—
For now I spy a danger—I entreat you    420
To bring but five-and-twenty; to no more
Will I give place or notice.

LEAR I gave you all.

REGAN And in good time you gave it.

LEAR
Made you my guardians, my depositaries,    425
But kept a reservation to be followed
With such a number. What, must I come to you
With five-and-twenty? Regan, said you so?

REGAN
And speak't again, my lord. No more with me.

LEAR
Those wicked creatures yet do look well favoured    430
When others are more wicked. Not being the worst
Stands in some rank of praise. (To Goneril) I'll go with
   thee.
Thy fifty yet doth double five-and-twenty,
And thou art twice her love.

GONERIL            Hear me, my lord.
What need you five-and-twenty, ten, or five,    435
To follow in a house where twice so many
Have a command to tend you?

REGAN            What need one?

LEAR
O, reason not the need! Our basest beggars
Are in the poorest thing superfluous.
Allow not nature more than nature needs,    440
Man's life is cheap as beast's. Thou art a lady.
If only to go warm were gorgeous,
Why, nature needs not what thou, gorgeous, wear'st,
Which scarcely keeps thee warm. But for true need—
You heavens, give me that patience, patience I need.
You see me here, you gods, a poor old man,    446
As full of grief as age, wretchèd in both.

If it be you that stirs these daughters' hearts
Against their father, fool me not so much
To bear it tamely. Touch me with noble anger,    450
And let not women's weapons, water-drops,
Stain my man's cheeks. No, you unnatural hags,
I will have such revenges on you both
That all the world shall—I will do such things—
What they are, yet I know not; but they shall be    455
The terrors of the earth. You think I'll weep.
No, I'll not weep. I have full cause of weeping,
     *Storm and tempest*
But this heart shall break into a hundred thousand
   flaws
Or ere I'll weep.—O Fool, I shall go mad!
     *Exeunt Lear, Fool, Gentleman, and Gloucester*

CORNWALL
Let us withdraw. 'Twill be a storm.    460

REGAN
This house is little. The old man and 's people
Cannot be well bestowed.

GONERIL            'Tis his own blame;
Hath put himself from rest, and must needs taste his
   folly.

REGAN
For his particular I'll receive him gladly,
But not one follower.

GONERIL            So am I purposed.    465
Where is my lord of Gloucester?

CORNWALL
Followed the old man forth.
     ⌈*Enter the Duke of Gloucester*⌉
           He is returned.

GLOUCESTER
The King is in high rage.

CORNWALL            Whither is he going?

GLOUCESTER
He calls to horse, but will I know not whither.

CORNWALL
'Tis best to give him way. He leads himself.    470

GONERIL (*to Gloucester*)
My lord, entreat him by no means to stay.

GLOUCESTER
Alack, the night comes on, and the high winds
Do sorely ruffle. For many miles about
There's scarce a bush.

REGAN            O sir, to wilful men
The injuries that they themselves procure    475
Must be their schoolmasters. Shut up your doors.
He is attended with a desperate train,
And what they may incense him to, being apt
To have his ear abused, wisdom bids fear.

CORNWALL
Shut up your doors, my lord. 'Tis a wild night.    480
My Regan counsels well. Come out o'th' storm.    *Exeunt*

3.1    *Storm still. Enter the Earl of Kent disguised and*
     ⌈*the First*⌉ *Gentleman, severally*

KENT
Who's there, besides foul weather?

⌈FIRST⌉ GENTLEMAN        One minded like the weather,
Most unquietly.

KENT            I know you. Where's the King?

⌈FIRST⌉ GENTLEMAN
Contending with the fretful elements;
Bids the wind blow the earth into the sea

Or swell the curlèd waters 'bove the main,          5
That things might change or cease.
KENT                              But who is with him?
⌈FIRST⌉ GENTLEMAN
None but the Fool, who labours to outjest
His heart-struck injuries.
KENT                              Sir, I do know you,
And dare upon the warrant of my note
Commend a dear thing to you. There is division,          10
Although as yet the face of it is covered
With mutual cunning, 'twixt Albany and Cornwall,
Who have—as who have not that their great stars
Throned and set high—servants, who seem no less,
Which are to France the spies and speculations          15
Intelligent of our state. What hath been seen,
Either in snuffs and packings of the Dukes,
Or the hard rein which both of them hath borne
Against the old kind King; or something deeper,
Whereof perchance these are but furnishings—          20
⌈FIRST⌉ GENTLEMAN
I will talk further with you.
KENT                              No, do not.
For confirmation that I am much more
Than my out-wall, open this purse, and take
What it contains. If you shall see Cordelia—
As fear not but you shall—show her this ring          25
And she will tell you who that fellow is
That yet you do not know. Fie on this storm!
I will go seek the King.
⌈FIRST⌉ GENTLEMAN
Give me your hand. Have you no more to say?
KENT
Few words, but to effect more than all yet:          30
That when we have found the King—in which your
    pain
That way, I'll this—he that first lights on him
Holla the other.                    *Exeunt severally*

**3.2**   *Storm still. Enter King Lear and his Fool*
LEAR
Blow, winds, and crack your cheeks! Rage, blow,
You cataracts and hurricanoes, spout
Till you have drenched our steeples, drowned the
    cocks!
You sulph'rous and thought-executing fires,
Vaunt-couriers of oak-cleaving thunderbolts,          5
Singe my white head; and thou all-shaking thunder,
Strike flat the thick rotundity o'th' world,
Crack nature's moulds, all germens spill at once
That makes ingrateful man.          9
FOOL O nuncle, court holy water in a dry house is better
than this rain-water out o' door. Good nuncle, in, ask
thy daughters blessing. Here's a night pities neither
wise men nor fools.
LEAR
Rumble thy bellyful; spit, fire; spout, rain.
Nor rain, wind, thunder, fire are my daughters.          15
I tax not you, you elements, with unkindness.
I never gave you kingdom, called you children.
You owe me no subscription. Then let fall
Your horrible pleasure. Here I stand your slave,
A poor, infirm, weak and despised old man,          20
But yet I call you servile ministers,
That will with two pernicious daughters join
Your high-engendered battles 'gainst a head
So old and white as this. O, ho, 'tis foul!

FOOL He that has a house to put 's head in has a good
head-piece.
⌈*Sings*⌉     The codpiece that will house          27
                    Before the head has any,
              The head and he shall louse,
                    So beggars marry many.          30

              The man that makes his toe
                    What he his heart should make
              Shall of a corn cry woe,
                    And turn his sleep to wake—
for there was never yet fair woman but she made
mouths in a glass.          36
        *Enter the Earl of Kent disguised*
LEAR
No, I will be the pattern of all patience.
I will say nothing.
KENT Who's there?
FOOL Marry, here's grace and a codpiece—that's a wise
man and a fool.          41
KENT (*to Lear*)
Alas, sir, are you here? Things that love night
Love not such nights as these. The wrathful skies
Gallow the very wanderers of the dark
And make them keep their caves. Since I was man          45
Such sheets of fire, such bursts of horrid thunder,
Such groans of roaring wind and rain I never
Remember to have heard. Man's nature cannot carry
Th'affliction nor the fear.
LEAR                    Let the great gods,
That keep this dreadful pother o'er our heads,          50
Find out their enemies now. Tremble, thou wretch
That hast within thee undivulgèd crimes
Unwhipped of justice; hide thee, thou bloody hand,
Thou perjured and thou simular of virtue
That art incestuous; caitiff, to pieces shake,          55
That under covert and convenient seeming
Has practised on man's life; close pent-up guilts,
Rive your concealing continents and cry
These dreadful summoners grace. I am a man
More sinned against than sinning.
KENT                              Alack, bare-headed?
Gracious my lord, hard by here is a hovel.          61
Some friendship will it lend you 'gainst the tempest.
Repose you there while I to this hard house—
More harder than the stones whereof 'tis raised,
Which even but now, demanding after you,          65
Denied me to come in—return and force
Their scanted courtesy.
LEAR                    My wits begin to turn.
(*To Fool*) Come on, my boy. How dost, my boy? Art
    cold?
I am cold myself.—Where is this straw, my fellow?
The art of our necessities is strange,          70
And can make vile things precious. Come, your hovel.—
Poor fool and knave, I have one part in my heart
That's sorry yet for thee.
FOOL ⌈*Sings*⌉
              He that has and a little tiny wit,
                    With heigh-ho, the wind and the rain,          75
              Must make content with his fortunes fit,
                    Though the rain it raineth every day.
LEAR
True, boy. (*To Kent*) Come, bring us to this hovel.
                              *Exeunt Lear and Kent*

FOOL  This is a brave night to cool a courtesan. I'll speak
a prophecy ere I go:　　　　　　　　　　　　　　　80
　When priests are more in word than matter;
　When brewers mar their malt with water;
　When nobles are their tailors' tutors,
　No heretics burned, but wenches' suitors,
　Then shall the realm of Albion　　　　　　　　85
　Come to great confusion.

　When every case in law is right;
　No squire in debt nor no poor knight;
　When slanders do not live in tongues,
　Nor cutpurses come not to throngs;　　　　　　90
　When usurers tell their gold i'th' field,
　And bawds and whores do churches build,
　Then comes the time, who lives to see't,
　That going shall be used with feet.
This prophecy Merlin shall make; for I live before his
time.　　　　　　　　　　　　　　　　　　　*Exit*

**3.3**　*Enter the Duke of Gloucester and Edmond*

GLOUCESTER  Alack, alack, Edmond, I like not this
unnatural dealing. When I desired their leave that I
might pity him, they took from me the use of mine
own house, charged me on pain of perpetual displeasure
neither to speak of him, entreat for him, or any way
sustain him.　　　　　　　　　　　　　　　　6
EDMOND  Most savage and unnatural!
GLOUCESTER  Go to, say you nothing. There is division
between the Dukes, and a worse matter than that. I
have received a letter this night—'tis dangerous to be
spoken—I have locked the letter in my closet. These
injuries the King now bears will be revenged home.
There is part of a power already footed. We must incline
to the King. I will look him and privily relieve him. Go
you and maintain talk with the Duke, that my charity
be not of him perceived. If he ask for me, I am ill and
gone to bed. If I die for't—as no less is threatened me—
the King my old master must be relieved. There is strange
things toward, Edmond; pray you be careful.　　*Exit*
EDMOND
　This courtesy, forbid thee, shall the Duke　　　20
　Instantly know, and of that letter too.
　This seems a fair deserving, and must draw me
　That which my father loses: no less than all.
　The younger rises when the old doth fall.　　*Exit*

**3.4**　*Enter King Lear, the Earl of Kent disguised, and
　　　Lear's Fool*

KENT
　Here is the place, my lord. Good my lord, enter.
　The tyranny of the open night's too rough
　For nature to endure.
　　　*Storm still*
LEAR　　　　　　　　Let me alone.
KENT
　Good my lord, enter here.
LEAR　　　　　　　　Wilt break my heart?
KENT
　I had rather break mine own. Good my lord, enter.　5
LEAR
　Thou think'st 'tis much that this contentious storm
　Invades us to the skin. So 'tis to thee;
　But where the greater malady is fixed,
　The lesser is scarce felt. Thou'dst shun a bear,
　But if thy flight lay toward the roaring sea　　10

Thou'dst meet the bear i'th' mouth. When the mind's
　free,
The body's delicate. This tempest in my mind
Doth from my senses take all feeling else
Save what beats there: filial ingratitude.
Is it not as this mouth should tear this hand　　15
For lifting food to't? But I will punish home.
No, I will weep no more.—In such a night
To shut me out? Pour on, I will endure.
In such a night as this! O Regan, Goneril,
Your old kind father, whose frank heart gave all—　20
O, that way madness lies. Let me shun that.
No more of that.
KENT　　　　　　　Good my lord, enter here.
LEAR
　Prithee, go in thyself. Seek thine own ease.
　This tempest will not give me leave to ponder
　On things would hurt me more; but I'll go in.　25
　(*To Fool*) In, boy; go first. ⌜*Kneeling*⌝ You houseless
　　poverty—
　Nay, get thee in. I'll pray, and then I'll sleep.　*Exit Fool*
　Poor naked wretches, wheresoe'er you are,
　That bide the pelting of this pitiless storm,
　How shall your houseless heads and unfed sides,　30
　Your looped and windowed raggedness, defend you
　From seasons such as these? O, I have ta'en
　Too little care of this. Take physic, pomp,
　Expose thyself to feel what wretches feel,
　That thou mayst shake the superflux to them　35
　And show the heavens more just.
　　　*Enter Lear's Fool, ⌜and Edgar as a Bedlam beggar
　　　in the hovel⌝*
EDGAR
Fathom and half! Fathom and half! Poor Tom!
FOOL  Come not in here, nuncle. Here's a spirit. Help me,
help me!
KENT  Give me thy hand. Who's there?　　　　　40
FOOL  A spirit, a spirit. He says his name's Poor Tom.
KENT
　What art thou that dost grumble there i'th' straw?
　Come forth.
　　　⌜*Edgar comes forth*⌝
EDGAR　　　　　Away, the foul fiend follows me.
　Thorough the sharp hawthorn blow the winds. Hm!
　Go to thy cold bed and warm thee.　　　　　45
LEAR
　Didst thou give all to thy two daughters,
　And art thou come to this?
EDGAR  Who gives anything to Poor Tom, whom the foul
fiend hath led through fire and through flame, through
ford and whirlpool, o'er bog and quagmire; that hath
laid knives under his pillow and halters in his pew, set
ratsbane by his porridge, made him proud of heart to
ride on a bay trotting-horse over four-inched bridges, to
course his own shadow for a traitor. Bless thy five wits,
Tom's a-cold! O, do, de, do, de, do de. Bless thee from
whirlwinds, star-blasting, and taking. Do Poor Tom some
charity, whom the foul fiend vexes. There could I have
him now, and there, and there again, and there.
　　　*Storm still*
LEAR
　Has his daughters brought him to this pass?
　(*To Edgar*) Couldst thou save nothing? Wouldst thou
　　give 'em all?　　　　　　　　　　　　60
FOOL  Nay, he reserved a blanket, else we had been all
shamed.

LEAR (*to Edgar*)
  Now all the plagues that in the pendulous air
  Hang fated o'er men's faults light on thy daughters!
KENT He hath no daughters, sir.     65
LEAR
  Death, traitor! Nothing could have subdued nature
  To such a lowness but his unkind daughters.
  (*To Edgar*) Is it the fashion that discarded fathers
  Should have thus little mercy on their flesh?
  Judicious punishment: 'twas this flesh begot     70
  Those pelican daughters.
EDGAR Pillicock sat on Pillicock Hill; alow, alow, loo, loo.
FOOL This cold night will turn us all to fools and madmen.
EDGAR Take heed o'th' foul fiend; obey thy parents; keep
  thy words' justice; swear not; commit not with man's
  sworn spouse; set not thy sweet heart on proud array.
  Tom's a-cold.     77
LEAR What hast thou been?
EDGAR A servingman, proud in heart and mind, that
  curled my hair, wore gloves in my cap, served the lust
  of my mistress' heart, and did the act of darkness with
  her; swore as many oaths as I spake words, and broke
  them in the sweet face of heaven; one that slept in the
  contriving of lust, and waked to do it. Wine loved I
  deeply, dice dearly, and in woman out-paramoured the
  Turk. False of heart, light of ear, bloody of hand; hog
  in sloth, fox in stealth, wolf in greediness, dog in
  madness, lion in prey. Let not the creaking of shoes
  nor the rustling of silks betray thy poor heart to woman.
  Keep thy foot out of brothels, thy hand out of plackets,
  thy pen from lenders' books, and defy the foul fiend.
  Still through the hawthorn blows the cold wind, says
  suum, mun, nonny. Dauphin, my boy! Boy, *cessez*; let
  him trot by.     94
  *Storm still*
LEAR Thou wert better in a grave than to answer with
  thy uncovered body this extremity of the skies. Is man
  no more than this? Consider him well. Thou owest the
  worm no silk, the beast no hide, the sheep no wool,
  the cat no perfume. Ha, here's three on 's are
  sophisticated; thou art the thing itself. Unaccommodated man is no more but such a poor, bare, forked
  animal as thou art. Off, off, you lendings! Come,
  unbutton here.     103
  *Enter the Duke of Gloucester with a torch*
FOOL Prithee, nuncle, be contented. 'Tis a naughty night
  to swim in. Now a little fire in a wild field were like
  an old lecher's heart—a small spark, all the rest on 's
  body cold. Look, here comes a walking fire.
EDGAR This is the foul fiend Flibbertigibbet. He begins at
  curfew and walks till the first cock. He gives the web
  and the pin, squints the eye, and makes the harelip;
  mildews the white wheat, and hurts the poor creature
  of earth.     112
  ⌜*Sings*⌝
    Swithin footed thrice the wold,
    A met the night mare and her nine foal,
      Bid her alight     115
      And her troth plight,
    And aroint thee, witch, aroint thee!
KENT (*to Lear*)
  How fares your grace?
LEAR          What's he?
KENT (*to Gloucester*) Who's there? What is't you seek?
GLOUCESTER What are you there? Your names?     120

EDGAR Poor Tom, that eats the swimming frog, the toad,
  the tadpole, the wall-newt and the water; that in the
  fury of his heart, when the foul fiend rages, eats
  cowdung for salads, swallows the old rat and the ditch-
  dog, drinks the green mantle of the standing pool; who
  is whipped from tithing to tithing, and stocked,
  punished, and imprisoned; who hath had three suits
  to his back, six shirts to his body,
    Horse to ride, and weapon to wear;
    But mice and rats and such small deer     130
    Have been Tom's food for seven long year.
  Beware my follower. Peace, Smulkin; peace, thou fiend!
GLOUCESTER (*to Lear*)
  What, hath your grace no better company?
EDGAR
  The Prince of Darkness is a gentleman.
  Modo he's called, and Mahu.     135
GLOUCESTER (*to Lear*)
  Our flesh and blood, my lord, is grown so vile
  That it doth hate what gets it.
EDGAR          Poor Tom's a-cold.
GLOUCESTER (*to Lear*)
  Go in with me. My duty cannot suffer
  T'obey in all your daughters' hard commands.
  Though their injunction be to bar my doors     140
  And let this tyrannous night take hold upon you,
  Yet have I ventured to come seek you out
  And bring you where both fire and food is ready.
LEAR
  First let me talk with this philosopher.
  (*To Edgar*) What is the cause of thunder?     145
KENT
  Good my lord, take his offer; go into th' house.
LEAR
  I'll talk a word with this same learnèd Theban.
  (*To Edgar*) What is your study?
EDGAR
  How to prevent the fiend, and to kill vermin.
LEAR
  Let me ask you one word in private.     150
  *They converse apart*
KENT (*to Gloucester*)
  Importune him once more to go, my lord.
  His wits begin t'unsettle.
GLOUCESTER          Canst thou blame him?
  *Storm still*
  His daughters seek his death. Ah, that good Kent,
  He said it would be thus, poor banished man!     154
  Thou sayst the King grows mad; I'll tell thee, friend,
  I am almost mad myself. I had a son,
  Now outlawed from my blood; a sought my life
  But lately, very late. I loved him, friend,
  No father his son dearer. True to tell thee,     159
  The grief hath crazed my wits. What a night's this!
  (*To Lear*) I do beseech your grace—
LEAR          O, cry you mercy, sir!
  (*To Edgar*) Noble philosopher, your company.
EDGAR          Tom's a-cold.
GLOUCESTER
  In, fellow, there in t'hovel; keep thee warm.
LEAR
  Come, let's in all.
KENT          This way, my lord.
LEAR          With him!
  I will keep still with my philosopher.     165

KENT (*to Gloucester*)
Good my lord, soothe him; let him take the fellow.
GLOUCESTER Take him you on.
KENT ⌈*to Edgar*⌉
Sirrah, come on. Go along with us.
LEAR (*to Edgar*)
Come, good Athenian.
GLOUCESTER          No words, no words. Hush.
EDGAR
    Child Roland to the dark tower came,     170
    His word was still 'Fie, fo, and fum;
    I smell the blood of a British man.'     *Exeunt*

**3.5**   *Enter the Duke of Cornwall and Edmond*
CORNWALL I will have my revenge ere I depart his house.
EDMOND How, my lord, I may be censured, that nature
thus gives way to loyalty, something fears me to think
of.
CORNWALL I now perceive it was not altogether your
brother's evil disposition made him seek his death, but
a provoking merit set a-work by a reprovable badness
in himself.     8
EDMOND How malicious is my fortune, that I must repent
to be just! This is the letter which he spoke of, which
approves him an intelligent party to the advantages of
France. O heavens, that this treason were not, or not
I the detector!
CORNWALL Go with me to the Duchess.     14
EDMOND If the matter of this paper be certain, you have
mighty business in hand.
CORNWALL True or false, it hath made thee Earl of
Gloucester. Seek out where thy father is, that he may
be ready for our apprehension.     19
EDMOND ⌈*aside*⌉ If I find him comforting the King, it will
stuff his suspicion more fully. (*To Cornwall*) I will
persever in my course of loyalty, though the conflict
be sore between that and my blood.
CORNWALL I will lay trust upon thee, and thou shalt find
a dearer father in my love.     *Exeunt*

**3.6**   *Enter the Earl of Kent disguised, and the Duke of
       Gloucester*
GLOUCESTER Here is better than the open air; take it
thankfully. I will piece out the comfort with what
addition I can. I will not be long from you.
KENT All the power of his wits have given way to his
impatience; the gods reward your kindness!     5
                       *Exit Gloucester*
    *Enter King Lear, Edgar as a Bedlam beggar, and
       Lear's Fool*
EDGAR Fraretetto calls me, and tells me Nero is an angler
in the lake of darkness. Pray, innocent, and beware
the foul fiend.
FOOL Prithee, nuncle, tell me whether a madman be a
gentleman or a yeoman.     10
LEAR A king, a king!
FOOL No, he's a yeoman that has a gentleman to his son;
for he's a mad yeoman that sees his son a gentleman
before him.
LEAR
    To have a thousand with red burning spits     15
    Come hissing in upon 'em!
EDGAR                Bless thy five wits.
KENT (*to Lear*)
    O, pity! Sir, where is the patience now
    That you so oft have boasted to retain?

EDGAR (*aside*)
    My tears begin to take his part so much     19
    They mar my counterfeiting.
LEAR                The little dogs and all,
    Tray, Blanch, and Sweetheart—see, they bark at me.
EDGAR Tom will throw his head at them.—Avaunt, you
curs!
    Be thy mouth or black or white,
    Tooth that poisons if it bite,     25
    Mastiff, greyhound, mongrel grim,
    Hound or spaniel, brach or him,
    Bobtail tyke or trundle-tail,
    Tom will make him weep and wail;
    For with throwing thus my head,     30
    Dogs leapt the hatch, and all are fled.
    Do, de, de, de. Sese! Come, march to wakes and fairs
    And market towns. Poor Tom, thy horn is dry.
LEAR Then let them anatomize Regan; see what breeds
about her heart. Is there any cause in nature that
makes these hard-hearts? (*To Edgar*) You, sir, I entertain
for one of my hundred, only I do not like the fashion
of your garments. You will say they are Persian; but
let them be changed.
KENT
    Now, good my lord, lie here and rest a while.     40
LEAR Make no noise, make no noise. Draw the curtains.
    So, so. We'll go to supper i'th' morning.
       ⌈*He sleeps*⌉
FOOL And I'll go to bed at noon.
    *Enter the Duke of Gloucester*
GLOUCESTER (*to Kent*)
    Come hither, friend. Where is the King my master?
KENT
    Here, sir, but trouble him not; his wits are gone.     45
GLOUCESTER
    Good friend, I prithee take him in thy arms.
    I have o'erheard a plot of death upon him.
    There is a litter ready. Lay him in't
    And drive toward Dover, friend, where thou shalt meet
    Both welcome and protection. Take up thy master.     50
    If thou shouldst dally half an hour, his life,
    With thine and all that offer to defend him,
    Stand in assurèd loss. Take up, take up,
    And follow me, that will to some provision
    Give thee quick conduct. Come, come away.     55
       *Exeunt,* ⌈*Kent carrying Lear in his arms*⌉

**3.7**   *Enter the Duke of Cornwall, Regan, Goneril,
       Edmond the bastard, and Servants*
CORNWALL (*to Goneril*)
    Post speedily to my lord your husband.
    Show him this letter. The army of France is landed.
    (*To Servants*) Seek out the traitor Gloucester.
                            *Exeunt some*
REGAN                  Hang him instantly.
GONERIL
    Pluck out his eyes.
CORNWALL         Leave him to my displeasure.
    Edmond, keep you our sister company.     5
    The revenges we are bound to take upon your traitorous
    father are not fit for your beholding. Advise the Duke
    where you are going, to a most festinate preparation;
    we are bound to the like. Our posts shall be swift and
    intelligent betwixt us. (*To Goneril*) Farewell, dear sister.
    (*To Edmond*) Farewell, my lord of Gloucester.     11
    *Enter Oswald the steward*
    How now, where's the King?

OSWALD
My lord of Gloucester hath conveyed him hence.
Some five- or six-and-thirty of his knights,
Hot questrists after him, met him at gate,                    15
Who, with some other of the lord's dependants,
Are gone with him toward Dover, where they boast
To have well-armèd friends.
CORNWALL  Get horses for your mistress.       *Exit Oswald*
GONERIL  Farewell, sweet lord, and sister.                    20
CORNWALL
Edmond, farewell.              *Exeunt Goneril and Edmond*
(*To Servants*)       Go seek the traitor Gloucester.
Pinion him like a thief; bring him before us.
                              *Exeunt other Servants*
Though well we may not pass upon his life
Without the form of justice, yet our power
Shall do a curtsy to our wrath, which men          25
May blame but not control.
           *Enter the Duke of Gloucester and Servants*
                    Who's there—the traitor?
REGAN
Ingrateful fox, 'tis he.
CORNWALL (*to Servants*) Bind fast his corky arms.
GLOUCESTER
What means your graces? Good my friends, consider
You are my guests. Do me no foul play, friends.
CORNWALL (*to Servants*)
Bind him, I say.
REGAN            Hard, hard! O filthy traitor!          30
GLOUCESTER
Unmerciful lady as you are, I'm none.
CORNWALL (*to Servants*)
To this chair bind him. (*To Gloucester*) Villain, thou
   shalt find—
        *Regan plucks Gloucester's beard*
GLOUCESTER
By the kind gods, 'tis most ignobly done,
To pluck me by the beard.
REGAN So white, and such a traitor?                    35
GLOUCESTER Naughty lady,
These hairs which thou dost ravish from my chin
Will quicken and accuse thee. I am your host.
With robbers' hands my hospitable favours
You should not ruffle thus. What will you do?          40
CORNWALL
Come, sir, what letters had you late from France?
REGAN
Be simple-answered, for we know the truth.
CORNWALL
And what confederacy have you with the traitors
Late footed in the kingdom?
REGAN            To whose hands
You have sent the lunatic King. Speak.                    45
GLOUCESTER
I have a letter guessingly set down,
Which came from one that's of a neutral heart,
And not from one opposed.
CORNWALL            Cunning.
REGAN                      And false.
CORNWALL
Where hast thou sent the King?
GLOUCESTER            To Dover.
REGAN
Wherefore to Dover? Wast thou not charged at peril—
CORNWALL
Wherefore to Dover?—Let him answer that.          51

GLOUCESTER
I am tied to th' stake, and I must stand the course.
REGAN Wherefore to Dover?
GLOUCESTER
Because I would not see thy cruel nails
Pluck out his poor old eyes, nor thy fierce sister          55
In his anointed flesh stick boarish fangs.
The sea, with such a storm as his bare head
In hell-black night endured, would have buoyed up
And quenched the stellèd fires.
Yet, poor old heart, he holp the heavens to rain.          60
If wolves had at thy gate howled that stern time,
Thou shouldst have said 'Good porter, turn the key;
All cruels I'll subscribe.' But I shall see
The wingèd vengeance overtake such children.
CORNWALL
See't shalt thou never.—Fellows, hold the chair.—          65
Upon these eyes of thine I'll set my foot.
GLOUCESTER
He that will think to live till he be old
Give me some help!—O cruel! O you gods!
        ⌐*Cornwall pulls out one of Gloucester's eyes and*
        *stamps on it*⌐
REGAN (*to Cornwall*)
One side will mock another; th'other, too.
CORNWALL (*to Gloucester*)
If you see vengeance—
SERVANT            Hold your hand, my lord.          70
I have served you ever since I was a child,
But better service have I never done you
Than now to bid you hold.
REGAN            How now, you dog!
SERVANT
If you did wear a beard upon your chin
I'd shake it on this quarrel. ⌐*To Cornwall*⌐ What do
   you mean?                                          75
CORNWALL My villein!
SERVANT
Nay then, come on, and take the chance of anger.
        *They draw and fight*
REGAN (*to another Servant*)
Give me thy sword. A peasant stand up thus!
        ⌐*She takes a sword and runs at him behind*⌐
SERVANT (*to Gloucester*)
O, I am slain. My lord, you have one eye left
To see some mischief on him.
        ⌐*Regan stabs him again*⌐
                    O!            *He dies*
CORNWALL
Lest it see more, prevent it. Out, vile jelly!          81
        *He* ⌐*pulls out*⌐ *Gloucester's other eye*
Where is thy lustre now?
GLOUCESTER
All dark and comfortless. Where's my son Edmond?
Edmond, enkindle all the sparks of nature
To quite this horrid act.
REGAN            Out, treacherous villain!          85
Thou call'st on him that hates thee. It was he
That made the overture of thy treasons to us,
Who is too good to pity thee.
GLOUCESTER
O, my follies! Then Edgar was abused.
Kind gods, forgive me that, and prosper him!          90
REGAN (*to Servants*)
Go thrust him out at gates, and let him smell
His way to Dover.       *Exit one or more with Gloucester*
                    How is't, my lord? How look you?

**CORNWALL**
I have received a hurt. Follow me, lady.
(*To Servants*) Turn out that eyeless villain. Throw this
   slave
Upon the dunghill. Regan, I bleed apace.       95
Untimely comes this hurt. Give me your arm.
                          *Exeunt ⌐with the body⌐*

**4.1**    *Enter Edgar as a Bedlam beggar*
**EDGAR**
Yet better thus and known to be contemned
Than still contemned and flattered. To be worst,
The low'st and most dejected thing of fortune,
Stands still in esperance, lives not in fear.
The lamentable change is from the best;       5
The worst returns to laughter. Welcome, then,
Thou unsubstantial air that I embrace.
The wretch that thou hast blown unto the worst
Owes nothing to thy blasts.
      *Enter the Duke of Gloucester led by an Old Man*
                    But who comes here?
My father, parti-eyed? World, world, O world!       10
But that thy strange mutations make us hate thee,
Life would not yield to age.
      ⌐*Edgar stands aside*⌐
**OLD MAN** (*to Gloucester*)       O my good lord,
I have been your tenant and your father's tenant
These fourscore years.
**GLOUCESTER**
Away, get thee away, good friend, be gone.       15
Thy comforts can do me no good at all;
Thee they may hurt.
**OLD MAN**            You cannot see your way.
**GLOUCESTER**
I have no way, and therefore want no eyes.
I stumbled when I saw. Full oft 'tis seen
Our means secure us, and our mere defects       20
Prove our commodities. O dear son Edgar,
The food of thy abusèd father's wrath—
Might I but live to see thee in my touch
I'd say I had eyes again.
**OLD MAN**            How now? Who's there?
**EDGAR** (*aside*)
O gods! Who is't can say 'I am at the worst'?
I am worse than e'er I was.       25
**OLD MAN** (*to Gloucester*)       'Tis poor mad Tom.
**EDGAR** (*aside*)
And worse I may be yet. The worst is not
So long as we can say 'This is the worst.'
**OLD MAN** (*to Edgar*) Fellow, where goest?
**GLOUCESTER** Is it a beggarman?
**OLD MAN** Madman and beggar too.       30
**GLOUCESTER**
A has some reason, else he could not beg.
I'th' last night's storm I such a fellow saw,
Which made me think a man a worm. My son
Came then into my mind, and yet my mind       35
Was then scarce friends with him. I have heard more
   since.
As flies to wanton boys are we to th' gods;
They kill us for their sport.
**EDGAR** (*aside*)           How should this be?
Bad is the trade that must play fool to sorrow,
Ang'ring itself and others.
      ⌐*He comes forward*⌐
                  Bless thee, master.       40

**GLOUCESTER**
Is that the naked fellow?
**OLD MAN**           Ay, my lord.
**GLOUCESTER**
Get thee away. If for my sake
Thou wilt o'ertake us hence a mile or twain
I'th' way toward Dover, do it for ancient love,
And bring some covering for this naked soul,       45
Which I'll entreat to lead me.
**OLD MAN**           Alack, sir, he is mad.
**GLOUCESTER**
'Tis the time's plague when madmen lead the blind.
Do as I bid thee; or rather do thy pleasure.
Above the rest, be gone.
**OLD MAN**
I'll bring him the best 'parel that I have,       50
Come on't what will.               *Exit*
**GLOUCESTER**       Sirrah, naked fellow!
**EDGAR**
Poor Tom's a-cold. (*Aside*) I cannot daub it further.
**GLOUCESTER**
Come hither, fellow.
**EDGAR** (*aside*)           And yet I must.
(*To Gloucester*) Bless thy sweet eyes, they bleed.
**GLOUCESTER**           Know'st thou the way to Dover?
**EDGAR** Both stile and gate, horseway and footpath. Poor
Tom hath been scared out of his good wits. Bless thee,
goodman's son, from the foul fiend.       57
**GLOUCESTER**
Here, take this purse, thou whom the heavens'
   plagues
Have humbled to all strokes. That I am wretched
Makes thee the happier. Heavens deal so still.       60
Let the superfluous and lust-dieted man
That slaves your ordinance, that will not see
Because he does not feel, feel your power quickly.
So distribution should undo excess,
And each man have enough. Dost thou know Dover?
**EDGAR** Ay, master.       66
**GLOUCESTER**
There is a cliff whose high and bending head
Looks fearfully in the confinèd deep.
Bring me but to the very brim of it
And I'll repair the misery thou dost bear       70
With something rich about me. From that place
I shall no leading need.
**EDGAR** Give me thy arm.
Poor Tom shall lead thee. *Exit Edgar guiding Gloucester*

**4.2**    *Enter Goneril and Edmond the bastard ⌐at one door⌐*
         *and Oswald the steward ⌐at another⌐*
**GONERIL**
Welcome, my lord. I marvel our mild husband
Not met us on the way. (*To Oswald*) Now, where's
   your master?
**OSWALD**
Madam, within; but never man so changed.
I told him of the army that was landed;
He smiled at it. I told him you were coming;       5
His answer was 'The worse'. Of Gloucester's treachery
And of the loyal service of his son
When I informed him, then he called me sot,
And told me I had turned the wrong side out.
What most he should dislike seems pleasant to him;
What like, offensive.
**GONERIL** (*to Edmond*)       Then shall you go no further.       11
It is the cowish terror of his spirit

That dares not undertake. He'll not feel wrongs
Which tie him to an answer. Our wishes on the way
May prove effects. Back, Edmond, to my brother.    15
Hasten his musters and conduct his powers.
I must change names at home, and give the distaff
Into my husband's hands. This trusty servant
Shall pass between us. Ere long you are like to hear,
If you dare venture in your own behalf,    20
A mistress's command. Wear this. Spare speech.
Decline your head. This kiss, if it durst speak,
Would stretch thy spirits up into the air.
        ⌈*She kisses him*⌉
Conceive, and fare thee well.

EDMOND Yours in the ranks of death.    25
GONERIL My most dear Gloucester.        *Exit Edmond*
O, the difference of man and man!
To thee a woman's services are due;
My fool usurps my body.
OSWALD                    Madam, here comes my lord.
        *Enter the Duke of Albany*
GONERIL
I have been worth the whistling.
ALBANY                    O Goneril,    30
You are not worth the dust which the rude wind
Blows in your face.
GONERIL            Milk-livered man,
That bear'st a cheek for blows, a head for wrongs;
Who hast not in thy brows an eye discerning
Thine honour from thy suffering—
ALBANY                    See thyself, devil.
Proper deformity shows not in the fiend    36
So horrid as in woman.
GONERIL            O vain fool!
        *Enter a Messenger*
MESSENGER
O my good lord, the Duke of Cornwall's dead,
Slain by his servant going to put out
The other eye of Gloucester.
ALBANY                    Gloucester's eyes?    40
MESSENGER
A servant that he bred, thrilled with remorse,
Opposed against the act, bending his sword
To his great master, who thereat enraged
Flew on him, and amongst them felled him dead,
But not without that harmful stroke which since    45
Hath plucked him after.
ALBANY                    This shows you are above,
You justicers, that these our nether crimes
So speedily can venge. But O, poor Gloucester!
Lost he his other eye?
MESSENGER            Both, both, my lord.—
This letter, madam, craves a speedy answer.    50
'Tis from your sister.
GONERIL (*aside*)        One way I like this well;
But being widow, and my Gloucester with her,
May all the building in my fancy pluck
Upon my hateful life. Another way
The news is not so tart.—I'll read and answer.    55
        ⌈*Exit with Oswald*⌉
ALBANY
Where was his son when they did take his eyes?
MESSENGER
Come with my lady hither.
ALBANY                    He is not here.
MESSENGER
No, my good lord; I met him back again.

ALBANY Knows he the wickedness?
MESSENGER
Ay, my good lord; 'twas he informed against him,    60
And quit the house on purpose that their punishment
Might have the freer course.
ALBANY                    Gloucester, I live
To thank thee for the love thou showed'st the King,
And to revenge thine eyes.—Come hither, friend.
Tell me what more thou know'st.        *Exeunt*

**4.3**    *Enter with a drummer and colours, Queen Cordelia,*
        *Gentlemen, and soldiers*
CORDELIA
Alack, 'tis he! Why, he was met even now,
As mad as the vexed sea, singing aloud,
Crowned with rank fumitor and furrow-weeds,
With burdocks, hemlock, nettles, cuckoo-flowers,
Darnel, and all the idle weeds that grow    5
In our sustaining corn. A century send forth.
Search every acre in the high-grown field,
And bring him to our eye.        ⌈*Exit one or more*⌉
                    What can man's wisdom
In the restoring his bereavèd sense,
He that helps him take all my outward worth.    10
⌈FIRST⌉ GENTLEMAN There is means, madam.
Our foster-nurse of nature is repose,
The which he lacks. That to provoke in him
Are many simples operative, whose power
Will close the eye of anguish.
CORDELIA                    All blest secrets,    15
All you unpublished virtues of the earth,
Spring with my tears, be aidant and remediate
In the good man's distress!—Seek, seek for him,
Lest his ungoverned rage dissolve the life
That wants the means to lead it.
        *Enter a Messenger*
MESSENGER                    News, madam.    20
The British powers are marching hitherward.
CORDELIA
'Tis known before; our preparation stands
In expectation of them.—O dear father,
It is thy business that I go about;
Therefore great France    25
My mourning and importuned tears hath pitied.
No blown ambition doth our arms incite,
But love, dear love, and our aged father's right.
Soon may I hear and see him!        *Exeunt*

**4.4**    *Enter Regan and Oswald the steward*
REGAN
But are my brother's powers set forth?
OSWALD                    Ay, madam.
REGAN
Himself in person there?
OSWALD                    Madam, with much ado.
Your sister is the better soldier.
REGAN
Lord Edmond spake not with your lord at home?
OSWALD No, madam.    5
REGAN
What might import my sister's letters to him?
OSWALD I know not, lady.
REGAN
Faith, he is posted hence on serious matter.
It was great ignorance, Gloucester's eyes being out,
To let him live. Where he arrives he moves    10
All hearts against us. Edmond, I think, is gone,

In pity of his misery, to dispatch
His 'nighted life, moreover to descry
The strength o'th' enemy.

OSWALD
I must needs after, madam, with my letter.     15

REGAN
Our troops set forth tomorrow. Stay with us.
The ways are dangerous.

OSWALD                          I may not, madam.
My lady charged my duty in this business.

REGAN
Why should she write to Edmond? Might not you
Transport her purposes by word? Belike—     20
Some things—I know not what. I'll love thee much:
Let me unseal the letter.

OSWALD              Madam, I had rather—

REGAN
I know your lady does not love her husband.
I am sure of that, and at her late being here
She gave strange oeillades and most speaking looks     25
To noble Edmond. I know you are of her bosom.

OSWALD I, madam?

REGAN
I speak in understanding. Y'are, I know't.
Therefore I do advise you take this note.
My lord is dead. Edmond and I have talked,     30
And more convenient is he for my hand
Than for your lady's. You may gather more.
If you do find him, pray you give him this,
And when your mistress hears thus much from you,
I pray desire her call her wisdom to her.     35
So, fare you well.
If you do chance to hear of that blind traitor,
Preferment falls on him that cuts him off.

OSWALD
Would I could meet him, madam. I should show
What party I do follow.

REGAN                    Fare thee well.     40
                    *Exeunt severally*

**4.5**   *Enter Edgar disguised as a peasant, with a staff,*
        *guiding the blind Duke of Gloucester*

GLOUCESTER
When shall I come to th' top of that same hill?

EDGAR
You do climb up it now. Look how we labour.

GLOUCESTER
Methinks the ground is even.

EDGAR                      Horrible steep.
Hark, do you hear the sea?

GLOUCESTER              No, truly.

EDGAR
Why, then your other senses grow imperfect     5
By your eyes' anguish.

GLOUCESTER        So may it be indeed.
Methinks thy voice is altered, and thou speak'st
In better phrase and matter than thou didst.

EDGAR
You're much deceived. In nothing am I changed
But in my garments.

GLOUCESTER      Methinks you're better spoken.

EDGAR
Come on, sir, here's the place. Stand still. How fearful
And dizzy 'tis to cast one's eyes so low!     12
The crows and choughs that wing the midway air
Show scarce so gross as beetles. Halfway down

Hangs one that gathers samphire, dreadful trade!     15
Methinks he seems no bigger than his head.
The fishermen that walk upon the beach
Appear like mice, and yon tall anchoring barque
Diminished to her cock, her cock a buoy
Almost too small for sight. The murmuring surge     20
That on th'unnumbered idle pebble chafes
Cannot be heard so high. I'll look no more,
Lest my brain turn and the deficient sight
Topple down headlong.

GLOUCESTER          Set me where you stand.

EDGAR
Give me your hand. You are now within a foot     25
Of th'extreme verge. For all beneath the moon
Would I not leap upright.

GLOUCESTER          Let go my hand.
Here, friend, 's another purse; in it a jewel
Well worth a poor man's taking. Fairies and gods
Prosper it with thee! Go thou further off.     30
Bid me farewell, and let me hear thee going.

EDGAR
Now fare ye well, good sir.
        *He stands aside*

GLOUCESTER              With all my heart.

EDGAR (*aside*)
Why I do trifle thus with his despair
Is done to cure it.

GLOUCESTER (*kneeling*) O you mighty gods,
This world I do renounce, and in your sights     35
Shake patiently my great affliction off!
If I could bear it longer, and not fall
To quarrel with your great opposeless wills,
My snuff and loathèd part of nature should
Burn itself out. If Edgar live, O bless him!—     40
Now, fellow, fare thee well.

EDGAR                Gone, sir. Farewell.
        *Gloucester falls forward*
(*Aside*) And yet I know not how conceit may rob
The treasury of life, when life itself
Yields to the theft. Had he been where he thought,
By this had thought been past.—Alive or dead?     45
(*To Gloucester*) Ho, you, sir, friend; hear you, sir?
        Speak.
(*Aside*) Thus might he pass indeed. Yet he revives.
(*To Gloucester*) What are you, sir?

GLOUCESTER              Away, and let me die.

EDGAR
Hadst thou been aught but gossamer, feathers, air,
So many fathom down precipitating     50
Thou'dst shivered like an egg. But thou dost breathe,
Hast heavy substance, bleed'st not, speak'st, art sound.
Ten masts a-length make not the altitude
Which thou hast perpendicularly fell.
Thy life's a miracle. Speak yet again.     55

GLOUCESTER But have I fall'n, or no?

EDGAR
From the dread summit of this chalky bourn.
Look up a-height. The shrill-gorged lark so far
Cannot be seen or heard. Do but look up.

GLOUCESTER Alack, I have no eyes.     60
Is wretchedness deprived that benefit
To end itself by death? 'Twas yet some comfort
When misery could beguile the tyrant's rage
And frustrate his proud will.

EDGAR                Give me your arm.
Up, so. How is't? Feel you your legs? You stand.     65

**GLOUCESTER**
Too well, too well.

**EDGAR**                         This is above all strangeness.
Upon the crown o'th' cliff what thing was that
Which parted from you?

**GLOUCESTER**                         A poor unfortunate beggar.

**EDGAR**
As I stood here below, methoughts his eyes
Were two full moons. He had a thousand noses,          70
Horns whelked and wavèd like the enragèd sea.
It was some fiend. Therefore, thou happy father,
Think that the clearest gods, who make them honours
Of men's impossibilities, have preserved thee.

**GLOUCESTER**
I do remember now. Henceforth I'll bear          75
Affliction till it do cry out itself
'Enough, enough,' and die. That thing you speak of,
I took it for a man. Often 'twould say
'The fiend, the fiend!' He led me to that place.

**EDGAR**
Bear free and patient thoughts.
*Enter King Lear mad, ⌈crowned with weeds and
flowers⌉*
                         But who comes here?
The safer sense will ne'er accommodate          81
His master thus.

**LEAR** No, they cannot touch me for crying. I am the King
himself.

**EDGAR** O thou side-piercing sight!          85

**LEAR** Nature's above art in that respect. There's your
press-money. That fellow handles his bow like a crow-
keeper. Draw me a clothier's yard. Look, look, a mouse!
Peace, peace, this piece of toasted cheese will do't.
There's my gauntlet. I'll prove it on a giant. Bring up
the brown bills. O, well flown, bird, i'th' clout, i'th'
clout! Whew! Give the word.

**EDGAR** Sweet marjoram.

**LEAR** Pass.

**GLOUCESTER** I know that voice.          95

**LEAR** Ha! Goneril with a white beard? They flattered me
like a dog, and told me I had the white hairs in my
beard ere the black ones were there. To say 'ay' and
'no' to everything that I said 'ay' and 'no' to was no
good divinity. When the rain came to wet me once,
and the wind to make me chatter; when the thunder
would not peace at my bidding, there I found 'em,
there I smelt 'em out. Go to, they are not men o' their
words. They told me I was everything; 'tis a lie, I am
not ague-proof.          105

**GLOUCESTER**
The trick of that voice I do well remember.
Is't not the King?

**LEAR**                         Ay, every inch a king.
⌈*Gloucester kneels*⌉
When I do stare, see how the subject quakes!
I pardon that man's life. What was thy cause?
Adultery? Thou shalt not die. Die for adultery!          110
No, the wren goes to't, and the small gilded fly
Does lecher in my sight. Let copulation thrive,
For Gloucester's bastard son
Was kinder to his father than my daughters
Got 'tween the lawful sheets. To't, luxury, pell-mell,
For I lack soldiers. Behold yon simp'ring dame,          116
Whose face between her forks presages snow,
That minces virtue, and does shake the head
To hear of pleasure's name.

The fitchew nor the soilèd horse goes to't          120
With a more riotous appetite. Down from the waist
They're centaurs, though women all above.
But to the girdle do the gods inherit;
Beneath is all the fiend's. There's hell, there's darkness,
there is the sulphurous pit, burning, scalding, stench,
consumption. Fie, fie, fie; pah, pah! Give me an ounce
of civet, good apothecary, sweeten my imagination.
There's money for thee.

**GLOUCESTER**                         O, let me kiss that hand!

**LEAR** Let me wipe it first; it smells of mortality.

**GLOUCESTER**
O ruined piece of nature! This great world          130
Shall so wear out to naught. Dost thou know me?

**LEAR** I remember thine eyes well enough. Dost thou
squiny at me?
No, do thy worst, blind Cupid, I'll not love.
Read thou this challenge. Mark but the penning of it.

**GLOUCESTER**
Were all thy letters suns, I could not see.          136

**EDGAR** (*aside*)
I would not take this from report; it is,
And my heart breaks at it.

**LEAR** (*to Gloucester*) Read.

**GLOUCESTER** What—with the case of eyes?          140

**LEAR** O ho, are you there with me? No eyes in your head,
nor no money in your purse? Your eyes are in a heavy
case, your purse in a light; yet you see how this world
goes.

**GLOUCESTER** I see it feelingly.          145

**LEAR** What, art mad? A man may see how this world
goes with no eyes; look with thine ears. See how yon
justice rails upon yon simple thief. Hark in thine ear:
change places, and handy-dandy, which is the justice,
which is the thief? Thou hast seen a farmer's dog bark
at a beggar?          151

**GLOUCESTER** Ay, sir.

**LEAR** An the creature run from the cur, there thou
mightst behold the great image of authority. A dog's
obeyed in office.          155
Thou rascal beadle, hold thy bloody hand.
Why dost thou lash that whore? Strip thy own back.
Thou hotly lusts to use her in that kind
For which thou whip'st her. The usurer hangs the
cozener.
Through tattered clothes great vices do appear;          160
Robes and furred gowns hide all. Plate sin with gold,
And the strong lance of justice hurtless breaks;
Arm it in rags, a pygmy's straw does pierce it.
None does offend, none, I say none. I'll able 'em.
Take that of me, my friend, who have the power          165
To seal th'accuser's lips. Get thee glass eyes,
And, like a scurvy politician, seem
To see the things thou dost not. Now, now, now, now!
Pull off my boots. Harder, harder! So.

**EDGAR** (*aside*)
O, matter and impertinency mixed—          170
Reason in madness!

**LEAR**
If thou wilt weep my fortunes, take my eyes.
I know thee well enough: thy name is Gloucester.
Thou must be patient. We came crying hither.
Thou know'st the first time that we smell the air          175
We waul and cry. I will preach to thee. Mark.

**GLOUCESTER** Alack, alack the day!

LEAR ⌈removing his crown of weeds⌉
When we are born, we cry that we are come
To this great stage of fools. This' a good block.
It were a delicate stratagem to shoe                          180
A troop of horse with felt. I'll put't in proof,
And when I have stol'n upon these son-in-laws,
Then kill, kill, kill, kill, kill, kill!
        Enter ⌈two⌉ Gentlemen
⌈FIRST⌉ GENTLEMAN
O, here he is. Lay hand upon him. ⌈To Lear⌉ Sir,
Your most dear daughter—                                      185
LEAR
No rescue? What, a prisoner? I am even
The natural fool of fortune. Use me well.
You shall have ransom. Let me have surgeons;
I am cut to th' brains.
⌈FIRST⌉ GENTLEMAN You shall have anything.                    190
LEAR No seconds? All myself?
Why, this would make a man a man of salt,
To use his eyes for garden water-pots.
I will die bravely, like a smug bridegroom. What,
I will be jovial. Come, come, I am a king.                    195
Masters, know you that?
⌈FIRST⌉ GENTLEMAN
You are a royal one, and we obey you.
LEAR Then there's life in't. Come, an you get it, you shall
get it by running. Sa, sa, sa, sa!
        Exit running ⌈pursued by a Gentleman⌉
⌈FIRST⌉ GENTLEMAN
A sight most pitiful in the meanest wretch,                   200
Past speaking in a king. Thou hast a daughter
Who redeems nature from the general curse
Which twain have brought her to.
EDGAR Hail, gentle sir.
⌈FIRST⌉ GENTLEMAN Sir, speed you. What's your will?
EDGAR
Do you hear aught, sir, of a battle toward?                   206
⌈FIRST⌉ GENTLEMAN
Most sure and vulgar, everyone hears that
That can distinguish sound.
EDGAR But, by your favour,
How near's the other army?                                    210
⌈FIRST⌉ GENTLEMAN
Near and on speedy foot. The main descry
Stands in the hourly thought.
EDGAR                         I thank you, sir. That's all.
⌈FIRST⌉ GENTLEMAN
Though that the Queen on special cause is here,
Her army is moved on.
EDGAR                    I thank you, sir. Exit Gentleman
GLOUCESTER
You ever gentle gods, take my breath from me.                 215
Let not my worser spirit tempt me again
To die before you please.
EDGAR Well pray you, father.
GLOUCESTER Now, good sir, what are you?
EDGAR
A most poor man, made tame to fortune's blows,               220
Who by the art of known and feeling sorrows
Am pregnant to good pity. Give me your hand,
I'll lead you to some biding.
GLOUCESTER ⌈rising⌉           Hearty thanks.
The bounty and the benison of heaven
To boot and boot.
        Enter Oswald the steward
OSWALD            A proclaimed prize! Most happy!
That eyeless head of thine was first framed flesh             226

To raise my fortunes. Thou old unhappy traitor,
Briefly thyself remember. The sword is out
That must destroy thee.
GLOUCESTER               Now let thy friendly hand
Put strength enough to't.
OSWALD (to Edgar)          Wherefore, bold peasant,           230
Durst thou support a published traitor? Hence,
Lest that th'infection of his fortune take
Like hold on thee. Let go his arm.
EDGAR 'Chill not let go, sir, without vurther 'cagion.
OSWALD Let go, slave, or thou diest.                          235
EDGAR Good gentleman, go your gate, and let poor volk
    pass. An 'chud ha' been swaggered out of my life,
    'twould not ha' been so long as 'tis by a vortnight.
    Nay, come not near th'old man. Keep out, 'che vor'
    ye, or I's' try whether your costard or my baton be the
    harder; I'll be plain with you.                           241
OSWALD Out, dunghill!
EDGAR 'Chill pick your teeth, sir. Come, no matter vor
    your foins.
        ⌈Edgar knocks him down⌉
OSWALD
Slave, thou hast slain me. Villain, take my purse.            245
If ever thou wilt thrive, bury my body,
And give the letters which thou find'st about me
To Edmond, Earl of Gloucester. Seek him out
Upon the English party. O untimely death! Death!
                                        He dies
EDGAR
I know thee well—a serviceable villain,                       250
As duteous to the vices of thy mistress
As badness would desire.
GLOUCESTER What, is he dead?
EDGAR Sit you down, father. Rest you.
        Gloucester sits
Let's see these pockets. The letters that he speaks of
May be my friends. He's dead; I am only sorrow               256
He had no other deathsman. Let us see.
Leave, gentle wax, and manners; blame us not.
To know our enemies' minds we rip their hearts;
Their papers is more lawful.                                  260
        He reads the letter
'Let our reciprocal vows be remembered. You have
many opportunities to cut him off. If your will want
not, time and place will be fruitfully offered. There is
nothing done if he return the conqueror; then am I
the prisoner, and his bed my jail, from the loathed
warmth whereof, deliver me, and supply the place for
your labour.                                                  267
        Your—wife, so I would say,—affectionate
    servant, and for you her own for venture,
                                        Goneril.'
O indistinguished space of woman's will—                     271
A plot upon her virtuous husband's life,
And the exchange my brother!—Here in the sands
Thee I'll rake up, the post unsanctified
Of murderous lechers, and in the mature time                 275
With this ungracious paper strike the sight
Of the death-practised Duke. For him 'tis well
That of thy death and business I can tell.
                                ⌈Exit with the body⌉
GLOUCESTER
The King is mad. How stiff is my vile sense,
That I stand up and have ingenious feeling                    280
Of my huge sorrows! Better I were distraught,
So should my thoughts be severed from my griefs,

*Drum afar off*
And woes by wrong imaginations lose
The knowledge of themselves.
    ⌐Enter Edgar⌐

EDGAR                Give me your hand.
Far off methinks I hear the beaten drum.    285
Come, father, I'll bestow you with a friend.
         *Exit Edgar guiding Gloucester*

**4.6**    *Enter Queen Cordelia, the Earl of Kent disguised,*
         *and ⌐the First⌐ Gentleman*

CORDELIA
O thou good Kent, how shall I live and work
To match thy goodness? My life will be too short,
And every measure fail me.

KENT
To be acknowledged, madam, is o'erpaid.
All my reports go with the modest truth,       5
Nor more, nor clipped, but so.

CORDELIA           Be better suited.
These weeds are memories of those worser hours.
I prithee put them off.

KENT              Pardon, dear madam.
Yet to be known shortens my made intent.
My boon I make it that you know me not     10
Till time and I think meet.

CORDELIA         Then be't so, my good lord.—
How does the King?

⌐FIRST⌐ GENTLEMAN    Madam, sleeps still.

CORDELIA            O you kind gods,
Cure this great breach in his abusèd nature;
Th'untuned and jarring senses O wind up
Of this child-changèd father!

⌐FIRST⌐ GENTLEMAN       So please your majesty
That we may wake the King? He hath slept long.   16

CORDELIA
Be governed by your knowledge, and proceed
I'th' sway of your own will. Is he arrayed?

⌐FIRST⌐ GENTLEMAN
Ay, madam. In the heaviness of sleep
We put fresh garments on him.           20
     *Enter King Lear asleep, in a chair carried by servants*
Be by, good madam, when we do awake him.
I doubt not of his temperance.

CORDELIA
O my dear father, restoration hang
Thy medicine on my lips, and let this kiss
Repair those violent harms that my two sisters   25
Have in thy reverence made!

KENT         Kind and dear princess!

CORDELIA
Had you not been their father, these white flakes
Did challenge pity of them. Was this a face
To be opposed against the warring winds?
Mine enemy's dog, though he had bit me, should
    have stood                       30
That night against my fire. And wast thou fain, poor
    father,
To hovel thee with swine and rogues forlorn
In short and musty straw? Alack, alack,
'Tis wonder that thy life and wits at once
Had not concluded all! (*To the Gentleman*) He wakes.
    Speak to him.                     35

⌐FIRST⌐ GENTLEMAN Madam, do you; 'tis fittest.

CORDELIA (*to Lear*)
How does my royal lord? How fares your majesty?

LEAR
You do me wrong to take me out o'th' grave.
Thou art a soul in bliss, but I am bound
Upon a wheel of fire, that mine own tears      40
Do scald like molten lead.

CORDELIA          Sir, do you know me?

LEAR
You are a spirit, I know. Where did you die?

CORDELIA (*to the Gentleman*) Still, still far wide!

⌐FIRST⌐ GENTLEMAN
He's scarce awake. Let him alone a while.

LEAR
Where have I been? Where am I? Fair daylight?    45
I am mightily abused. I should ev'n die with pity
To see another thus. I know not what to say.
I will not swear these are my hands. Let's see:
I feel this pin prick. Would I were assured
Of my condition.

CORDELIA (*kneeling*) O look upon me, sir,      50
And hold your hands in benediction o'er me.
You must not kneel.

LEAR             Pray do not mock.
I am a very foolish, fond old man,
Fourscore and upward,
Not an hour more nor less; and to deal plainly,   55
I fear I am not in my perfect mind.
Methinks I should know you, and know this man;
Yet I am doubtful, for I am mainly ignorant
What place this is; and all the skill I have
Remembers not these garments; nor I know not   60
Where I did lodge last night. Do not laugh at me,
For as I am a man, I think this lady
To be my child, Cordelia.

CORDELIA         And so I am, I am.

LEAR
Be your tears wet? Yes, faith. I pray, weep not.
If you have poison for me, I will drink it.      65
I know you do not love me; for your sisters
Have, as I do remember, done me wrong.
You have some cause; they have not.

CORDELIA          No cause, no cause.

LEAR Am I in France?

KENT In your own kingdom, sir.           70

LEAR Do not abuse me.

⌐FIRST⌐ GENTLEMAN
Be comforted, good madam. The great rage
You see is killed in him. Desire him to go in.
Trouble him no more till further settling.

CORDELIA (*to Lear*) Will't please your highness walk?   75

LEAR
You must bear with me. Pray you now, forget
And forgive. I am old and foolish.        *Exeunt*

**5.1**    *Enter with a drummer and colours Edmond, Regan,*
         *Gentlemen, and soldiers*

EDMOND
Know of the Duke if his last purpose hold,
Or whether since he is advised by aught
To change the course. He's full of abdication
And self-reproving. Bring his constant pleasure.
                     *Exit one or more*

REGAN
Our sister's man is certainly miscarried.       5

EDMOND
'Tis to be doubted, madam.

REGAN         Now, sweet lord,
You know the goodness I intend upon you.

Tell me but truly—but then speak the truth—
Do you not love my sister?

EDMOND                    In honoured love.

REGAN
But have you never found my brother's way          10
To the forfended place?

EDMOND                    No, by mine honour, madam.

REGAN
I never shall endure her. Dear my lord,
Be not familiar with her.

EDMOND Fear me not.
She and the Duke her husband—          15
*Enter with a drummer and colours the Duke of*
*Albany, Goneril, and soldiers*

ALBANY (*to Regan*)
Our very loving sister, well bemet.
(*To Edmond*) Sir, this I heard: the King is come to his
    daughter,
With others whom the rigour of our state
Forced to cry out.

REGAN                    Why is this reasoned?

GONERIL
Combine together 'gainst the enemy;          20
For these domestic and particular broils
Are not the question here.

ALBANY
Let's then determine with th'ensign of war
On our proceeding.

REGAN                    Sister, you'll go with us?

GONERIL No.          25

REGAN
'Tis most convenient. Pray go with us.

GONERIL (*aside*)
O ho, I know the riddle! (*To Regan*) I will go.
*Enter Edgar disguised as a peasant*

EDGAR (*to Albany*)
If e'er your grace had speech with man so poor,
Hear me one word.

ALBANY (*to the others*) I'll overtake you.
*Exeunt both the armies*
Speak.

EDGAR
Before you fight the battle, ope this letter.          30
If you have victory, let the trumpet sound
For him that brought it. Wretched though I seem,
I can produce a champion that will prove
What is avouchèd there. If you miscarry,
Your business of the world hath so an end,          35
And machination ceases. Fortune love you.

ALBANY
Stay till I have read the letter.

EDGAR                    I was forbid it.
When time shall serve, let but the herald cry,
And I'll appear again.

ALBANY Why, fare thee well.          40
I will o'erlook thy paper.          *Exit Edgar*
*Enter Edmond*

EDMOND
The enemy's in view; draw up your powers.
*He ⌐offers⌐ Albany a paper*
Here is the guess of their true strength and forces
By diligent discovery; but your haste
Is now urged on you.

ALBANY                    We will greet the time.          45
*Exit*

EDMOND
To both these sisters have I sworn my love,
Each jealous of the other as the stung
Are of the adder. Which of them shall I take?—
Both?—one?—or neither? Neither can be enjoyed
If both remain alive. To take the widow          50
Exasperates, makes mad, her sister Goneril,
And hardly shall I carry out my side,
Her husband being alive. Now then, we'll use
His countenance for the battle, which being done,
Let her who would be rid of him devise          55
His speedy taking off. As for the mercy
Which he intends to Lear and to Cordelia,
The battle done, and they within our power,
Shall never see his pardon; for my state
Stands on me to defend, not to debate.          *Exit*

5.2     *Alarum within. Enter with a drummer and colours*
        *King Lear, Queen Cordelia, and soldiers over the*
        *stage; and exeunt. Enter Edgar disguised as a*
        *peasant, guiding the blind Duke of Gloucester*

EDGAR
Here, father, take the shadow of this tree
For your good host; pray that the right may thrive.
If ever I return to you again
I'll bring you comfort.

GLOUCESTER                    Grace go with you, sir.
*Exit Edgar*

*Alarum and retreat within. Enter Edgar*

EDGAR
Away, old man. Give me thy hand. Away.          5
King Lear hath lost, he and his daughter ta'en.
Give me thy hand. Come on.

GLOUCESTER
No further, sir. A man may rot even here.

EDGAR
What, in ill thoughts again? Men must endure
Their going hence even as their coming hither.          10
Ripeness is all. Come on.

GLOUCESTER                    And that's true, too.
*Exit Edgar guiding Gloucester*

5.3     *Enter in conquest with a drummer and colours*
        *Edmond; King Lear and Queen Cordelia as*
        *prisoners; soldiers; a Captain*

EDMOND
Some officers take them away. Good guard
Until their greater pleasures first be known
That are to censure them.

CORDELIA (*to Lear*)          We are not the first
Who with best meaning have incurred the worst.
For thee, oppressèd King, I am cast down,          5
Myself could else outfrown false fortune's frown.
Shall we not see these daughters and these sisters?

LEAR
No, no, no, no. Come, let's away to prison.
We two alone will sing like birds i'th' cage.
When thou dost ask me blessing, I'll kneel down          10
And ask of thee forgiveness; so we'll live,
And pray, and sing, and tell old tales, and laugh
At gilded butterflies, and hear poor rogues
Talk of court news, and we'll talk with them too—
Who loses and who wins, who's in, who's out,          15
And take upon 's the mystery of things
As if we were God's spies; and we'll wear out

In a walled prison packs and sects of great ones
That ebb and flow by th' moon.
EDMOND (*to soldiers*)                    Take them away.
LEAR
Upon such sacrifices, my Cordelia,                             20
The gods themselves throw incense. Have I caught
      thee?
He that parts us shall bring a brand from heaven
And fire us hence like foxes. Wipe thine eyes.
The goodyear shall devour them, flesh and fell,
Ere they shall make us weep. We'll see 'em starved
      first. Come.  *Exeunt all but Edmond and the Captain*
EDMOND  Come hither, captain. Hark.                      26
Take thou this note. Go follow them to prison.
One step I have advanced thee; if thou dost
As this instructs thee, thou dost make thy way
To noble fortunes. Know thou this: that men       30
Are as the time is. To be tender-minded
Does not become a sword. Thy great employment
Will not bear question. Either say thou'lt do't,
Or thrive by other means.
CAPTAIN                        I'll do't, my lord.
EDMOND
About it, and write 'happy' when thou'st done.   35
Mark, I say, instantly, and carry it so
As I have set it down.                *Exit the Captain*
      *Flourish. Enter the Duke of Albany, Goneril, Regan,*
      ⌐*drummer, trumpeter*⌐ *and soldiers*
ALBANY
Sir, you have showed today your valiant strain,
And fortune led you well. You have the captives
Who were the opposites of this day's strife.        40
I do require them of you, so to use them
As we shall find their merits and our safety
May equally determine.
EDMOND                        Sir, I thought it fit
To send the old and miserable King
To some retention and appointed guard,              45
Whose age had charms in it, whose title more,
To pluck the common bosom on his side
And turn our impressed lances in our eyes
Which do command them. With him I sent the Queen,
My reason all the same, and they are ready          50
Tomorrow, or at further space, t'appear
Where you shall hold your session.
ALBANY                              Sir, by your patience,
I hold you but a subject of this war,
Not as a brother.
REGAN                That's as we list to grace him.
Methinks our pleasure might have been demanded  55
Ere you had spoke so far. He led our powers,
Bore the commission of my place and person,
The which immediacy may well stand up
And call itself your brother.
GONERIL                        Not so hot.
In his own grace he doth exalt himself              60
More than in your addition.
REGAN                        In my rights
By me invested, he compeers the best.
ALBANY
That were the most if he should husband you.
REGAN
Jesters do oft prove prophets.
GONERIL                        Holla, holla—
That eye that told you so looked but asquint.     65

REGAN
Lady, I am not well, else I should answer
From a full-flowing stomach. (*To Edmond*) General,
Take thou my soldiers, prisoners, patrimony.
Dispose of them, of me. The walls is thine.
Witness the world that I create thee here          70
My lord and master.
GONERIL                        Mean you to enjoy him?
ALBANY
The let-alone lies not in your good will.
EDMOND
Nor in thine, lord.
ALBANY                Half-blooded fellow, yes.
REGAN (*to Edmond*)
Let the drum strike and prove my title thine.
ALBANY
Stay yet, hear reason. Edmond, I arrest thee      75
On capital treason, and in thy attaint
This gilded serpent. (*To Regan*) For your claim, fair
      sister,
I bar it in the interest of my wife.
'Tis she is subcontracted to this lord,
And I, her husband, contradict your banns.        80
If you will marry, make your loves to me.
My lady is bespoke.
GONERIL                        An interlude!
ALBANY
Thou art armed, Gloucester. Let the trumpet sound.
If none appear to prove upon thy person
Thy heinous, manifest, and many treasons,         85
There is my pledge.
      ⌐*He throws down a glove*⌐
                              I'll make it on thy heart,
Ere I taste bread, thou art in nothing less
Than I have here proclaimed thee.
REGAN Sick, O sick!
GONERIL (*aside*) If not, I'll ne'er trust medicine.   90
EDMOND (*to Albany,* ⌐*throwing down a glove*⌐)
There's my exchange. What in the world he is
That names me traitor, villain-like he lies.
Call by the trumpet. He that dares, approach;
On him, on you,—who not?—I will maintain
My truth and honour firmly.
ALBANY                        A herald, ho!                 95
      *Enter a Herald*
(*To Edmond*) Trust to thy single virtue, for thy soldiers,
All levied in my name, have in my name
Took their discharge.
REGAN                        My sickness grows upon me.
ALBANY
She is not well. Convey her to my tent.
                              *Exit one or more with Regan*
Come hither, herald. Let the trumpet sound,    100
And read out this.
      *A trumpet sounds*
HERALD (*reads*) 'If any man of quality or degree within
      the lists of the army will maintain upon Edmond,
      supposed Earl of Gloucester, that he is a manifold
      traitor, let him appear by the third sound of the trumpet.
      He is bold in his defence.'                        106
      *First trumpet*
Again.
      *Second trumpet*
Again.

*Third trumpet.*
*Trumpet answers within. Enter Edgar, armed*
ALBANY (*to the Herald*)
    Ask him his purposes, why he appears
    Upon this call o'th' trumpet.
HERALD (*to Edgar*)                    What are you?        110
    Your name, your quality, and why you answer
    This present summons?
EDGAR                    Know, my name is lost,
    By treason's tooth bare-gnawn and canker-bit.
    Yet am I noble as the adversary
    I come to cope.
ALBANY                Which is that adversary?        115
EDGAR
    What's he that speaks for Edmond, Earl of Gloucester?
EDMOND
    Himself. What sayst thou to him?
EDGAR                            Draw thy sword,
    That if my speech offend a noble heart
    Thy arm may do thee justice. Here is mine.
    *He draws his sword*
    Behold, it is the privilege of mine honour,        120
    My oath, and my profession. I protest,
    Maugre thy strength, place, youth, and eminence,
    Despite thy victor-sword and fire-new fortune,
    Thy valour and thy heart, thou art a traitor,
    False to thy gods, thy brother, and thy father,        125
    Conspirant 'gainst this high illustrious prince,
    And from th'extremest upward of thy head
    To the descent and dust below thy foot
    A most toad-spotted traitor. Say thou no,
    This sword, this arm, and my best spirits are bent        130
    To prove upon thy heart, whereto I speak,
    Thou liest.
EDMOND        In wisdom I should ask thy name,
    But since thy outside looks so fair and warlike,
    And that thy tongue some say of breeding breathes,
    What safe and nicely I might well demand        135
    By rule of knighthood I disdain and spurn.
    Back do I toss those treasons to thy head,
    With the hell-hated lie o'erwhelm thy heart,
    Which, for they yet glance by and scarcely bruise,
    This sword of mine shall give them instant way        140
    Where they shall rest for ever. Trumpets, speak!
    *Alarums. They fight. Edmond is vanquished*
⌈ALL⌉
    Save him, save him!
GONERIL                This is practice, Gloucester.
    By th' law of arms thou wast not bound to answer
    An unknown opposite. Thou art not vanquished,
    But cozened and beguiled.
ALBANY                Shut your mouth, dame,
    Or with this paper shall I stopple it.        146
    ⌈*To Edmond*⌉ Hold, sir, thou worse than any name:
        read thine own evil.
    (*To Goneril*) No tearing, lady. I perceive you know it.
GONERIL
    Say if I do, the laws are mine, not thine.
    Who can arraign me for't?                Exit
ALBANY                    Most monstrous!—
    O, know'st thou this paper?
EDMOND                Ask me not what I know.
ALBANY
    Go after her. She's desperate. Govern her.        152
                    *Exit one or more*

EDMOND
    What you have charged me with, that have I done,
    And more, much more. The time will bring it out.
    'Tis past, and so am I. (*To Edgar*) But what art thou,
    That hast this fortune on me? If thou'rt noble,        156
    I do forgive thee.
EDGAR                Let's exchange charity.
    I am no less in blood than thou art, Edmond.
    If more, the more thou'st wronged me.
        ⌈*He takes off his helmet*⌉
    My name is Edgar, and thy father's son.        160
    The gods are just, and of our pleasant vices
    Make instruments to plague us.
    The dark and vicious place where thee he got
    Cost him his eyes.
EDMOND                Thou'st spoken right. 'Tis true.
    The wheel is come full circle. I am here.        165
ALBANY (*to Edgar*)
    Methought thy very gait did prophesy
    A royal nobleness. I must embrace thee.
    Let sorrow split my heart if ever I
    Did hate thee or thy father.
EDGAR Worthy prince, I know't.        170
ALBANY Where have you hid yourself?
    How have you known the miseries of your father?
EDGAR
    By nursing them, my lord. List a brief tale,
    And when 'tis told, O that my heart would burst!
    The bloody proclamation to escape        175
    That followed me so near—O, our lives' sweetness,
    That we the pain of death would hourly die
    Rather than die at once!—taught me to shift
    Into a madman's rags, t'assume a semblance
    That very dogs disdained; and in this habit        180
    Met I my father with his bleeding rings,
    Their precious stones new-lost; became his guide,
    Led him, begged for him, saved him from despair;
    Never—O fault!—revealed myself unto him
    Until some half hour past, when I was armed.        185
    Not sure, though hoping, of this good success,
    I asked his blessing, and from first to last
    Told him our pilgrimage; but his flawed heart—
    Alack, too weak the conflict to support—
    'Twixt two extremes of passion, joy and grief,        190
    Burst smilingly.
EDMOND                This speech of yours hath moved me,
    And shall perchance do good. But speak you on—
    You look as you had something more to say.
ALBANY
    If there be more, more woeful, hold it in,
    For I am almost ready to dissolve,        195
    Hearing of this.
        *Enter a Gentleman with a bloody knife*
GENTLEMAN
    Help, help, O help!
EDGAR                What kind of help?
ALBANY                                Speak, man.
EDGAR
    What means this bloody knife?
GENTLEMAN                        'Tis hot, it smokes.
    It came even from the heart of—O, she's dead!
ALBANY Who dead? Speak, man.        200
GENTLEMAN
    Your lady, sir, your lady; and her sister
    By her is poisoned. She confesses it.

972

EDMOND
I was contracted to them both; all three
Now marry in an instant.
EDGAR                              Here comes Kent.
    *Enter the Earl of Kent as himself*
ALBANY
Produce the bodies, be they alive or dead.            205
    *Goneril's and Regan's bodies brought out*
This judgement of the heavens, that makes us tremble,
Touches us not with pity.—O, is this he?
(*To Kent*) The time will not allow the compliment
Which very manners urges.
KENT                            I am come
To bid my king and master aye good night.            210
Is he not here?
ALBANY            Great thing of us forgot!—
Speak, Edmond; where's the King, and where's
    Cordelia?—
Seest thou this object, Kent?
KENT  Alack, why thus?
EDMOND  Yet Edmond was beloved.                      215
The one the other poisoned for my sake,
And after slew herself.
ALBANY                     Even so.—Cover their faces.
EDMOND
I pant for life. Some good I mean to do,
Despite of mine own nature. Quickly send,
Be brief in it, to th' castle; for my writ          220
Is on the life of Lear and on Cordelia.
Nay, send in time.
ALBANY                Run, run, O run!
EDGAR
To who, my lord?—Who has the office? Send
Thy token of reprieve.
EDMOND
Well thought on! Take my sword. The captain,        225
Give it the captain.
EDGAR              Haste thee for thy life.
                              *Exit ⌐the Gentleman⌐*
EDMOND (*to Albany*)
He hath commission from thy wife and me
To hang Cordelia in the prison, and
To lay the blame upon her own despair,
That she fordid herself.                             230
ALBANY
The gods defend her!—Bear him hence a while.
                      *Exeunt some with Edmond*
    *Enter King Lear with Queen Cordelia in his arms,*
    ⌐*followed by the Gentleman*⌐
LEAR
Howl, howl, howl, howl! O, you are men of stones.
Had I your tongues and eyes, I'd use them so
That heaven's vault should crack. She's gone for ever.
I know when one is dead and when one lives.          235
She's dead as earth.
    ⌐*He lays her down*⌐
                    Lend me a looking-glass.
If that her breath will mist or stain the stone,
Why, then she lives.
KENT                 Is this the promised end?
EDGAR
Or image of that horror?
ALBANY                   Fall and cease.
LEAR
This feather stirs. She lives. If it be so,         240
It is a chance which does redeem all sorrows

That ever I have felt.
KENT ⌐*kneeling*⌐          O, my good master!
LEAR
Prithee, away.
EDGAR            'Tis noble Kent, your friend.
LEAR
A plague upon you, murderers, traitors all.
I might have saved her; now she's gone for ever.—
Cordelia, Cordelia: stay a little. Ha?              246
What is't thou sayst?—Her voice was ever soft,
Gentle, and low, an excellent thing in woman.—
I killed the slave that was a-hanging thee.
GENTLEMAN
'Tis true, my lords, he did.
LEAR                    Did I not, fellow?          250
I have seen the day with my good biting falchion
I would have made them skip. I am old now,
And these same crosses spoil me. (*To Kent*) Who are
    you?
Mine eyes are not o'th' best, I'll tell you straight.
KENT
If fortune brag of two she loved and hated,         255
One of them we behold.
LEAR                 This' a dull sight.
Are you not Kent?
KENT  The same, your servant Kent.
Where is your servant Caius?
LEAR
He's a good fellow, I can tell you that.            260
He'll strike, and quickly too. He's dead and rotten.
KENT
No, my good lord, I am the very man—
LEAR  I'll see that straight.
KENT
That from your first of difference and decay
Have followed your sad steps.
LEAR                 You're welcome hither.
KENT
Nor no man else. All's cheerless, dark, and deadly.
Your eldest daughters have fordone themselves,      267
And desperately are dead.
LEAR                  Ay, so think I.
ALBANY
He knows not what he says; and vain is it
That we present us to him.
    *Enter a Messenger*
EDGAR                  Very bootless.                270
MESSENGER (*to Albany*)
Edmond is dead, my lord.
ALBANY              That's but a trifle here.—
You lords and noble friends, know our intent.
What comfort to this great decay may come
Shall be applied; for us, we will resign
During the life of this old majesty                 275
To him our absolute power;
(*To Edgar and Kent*)          you to your rights,
With boot and such addition as your honours
Have more than merited. All friends shall taste
The wages of their virtue, and all foes
The cup of their deservings.—O see, see!            280
LEAR
And my poor fool is hanged. No, no, no life?
Why should a dog, a horse, a rat have life,
And thou no breath at all? Thou'lt come no more.
Never, never, never, never, never.
⌐*To Kent*⌐ Pray you, undo this button. Thank you, sir.

Do you see this? Look on her. Look, her lips.        286
Look there, look there.                        *He dies*
EDGAR                    He faints. (*To Lear*) My lord, my lord!
KENT ⌈*to Lear*⌉
  Break, heart, I prithee break.
EDGAR (*to Lear*)                    Look up, my lord.
KENT
  Vex not his ghost. O, let him pass. He hates him
  That would upon the rack of this tough world        290
  Stretch him out longer.
EDGAR                    He is gone indeed.
KENT
  The wonder is he hath endured so long.
  He but usurped his life.

ALBANY
  Bear them from hence. Our present business
  Is general woe. (*To Edgar and Kent*) Friends of my
      soul, you twain                            295
  Rule in this realm, and the gored state sustain.
KENT
  I have a journey, sir, shortly to go:
  My master calls me; I must not say no.
EDGAR
  The weight of this sad time we must obey,
  Speak what we feel, not what we ought to say.        300
  The oldest hath borne most. We that are young
  Shall never see so much, nor live so long.
                    *Exeunt with a dead march, carrying the bodies*

# MACBETH

## *BY WILLIAM SHAKESPEARE*

## *(ADAPTED BY THOMAS MIDDLETON)*

---

SHORTLY after James VI of Scotland succeeded to the English throne, in 1603, he gave his patronage to Shakespeare's company; the Lord Chamberlain's Men became the King's Men, entering into a special relationship with their sovereign. *Macbeth* is the play of Shakespeare's that most clearly reflects this relationship. James regarded the virtuous and noble Banquo, Macbeth's comrade at the start of the action, as his direct ancestor; eight Stuart kings were said to have preceded James, just as, in the play, Banquo points to 'a show of eight kings' as his descendants (4.1.127.1–140); and in the play the English king (historically Edward the Confessor) is praised for the capacity, on which James also prided himself, to cure 'the king's evil' (scrofula). *Macbeth* is obviously a Jacobean play, composed probably in 1606.

But the first printed text, in the 1623 Folio, shows signs of having been adapted at a later date. It is exceptionally short by comparison with Shakespeare's other tragedies; and it includes episodes which there is good reason to believe are not by Shakespeare. These are Act 3, Scene 5 and parts of Act 4, Scene 1: 38.1–60 and 141–8.1. These episodes feature Hecate, who does not appear elsewhere in the play; they are composed largely in octosyllabic couplets in a style conspicuously different from the rest of the play; and they call for the performance of two songs that are found in *The Witch*, a play of uncertain date by Thomas Middleton. Probably Middleton himself adapted Shakespeare's play some years after its first performance. We do not attempt to excise passages probably not written by Shakespeare, because the adapter's hand may have affected the text at other, indeterminable points. The Folio text of *Macbeth* cites only the opening words of the songs; drawing on *The Witch*, we attempt a reconstruction of their staging in *Macbeth*.

Shakespeare took materials for his story from the account in Raphael Holinshed's *Chronicle* of the reigns of Duncan and Macbeth (AD 1034–57). Occasionally (especially in the English episodes of Act 4, Scene 2) he closely followed Holinshed's wording; but essentially the play's structure is his own. He invented the framework of the three witches who tempted both Macbeth and Banquo with prophecies of greatness. His Macbeth is both more introspective and more intensely evil than the competent warrior-king portrayed by Holinshed; conversely, Shakespeare made Duncan, the king whom Macbeth murders, far more venerable and saintly. Some of the play's features, notably the character of Lady Macbeth, originate in Holinshed's account of the murder of an earlier Scottish king, Duff; he was killed in his castle at Forres by Donwald, who had been 'set on' by his wife.

*Macbeth* can be enjoyed at many levels. It is an exciting story of witchcraft, murder, and retribution that can also be seen as a study in the philosophy and psychology of evil. The witches are not easily made credible in modern performances, and Shakespeare seems deliberately to have drained colour away from some parts of his composition in order to concentrate attention on Macbeth and his Lady. It is Macbeth's neurotic self-absorption, his fear, his anger, and his despair, along with his wife's steely determination, her invoking of the powers of evil, and her eventual revelation in sleep of her repressed humanity, that have given the play its long-proven power to fascinate readers and to challenge performers.

# THE PERSONS OF THE PLAY

KING DUNCAN of Scotland

MALCOLM ⎫
         ⎬ his sons
DONALBAIN ⎭

A CAPTAIN in Duncan's army

MACBETH, Thane of Glamis, later Thane of Cawdor, then King of Scotland

A PORTER at Macbeth's castle

Three MURDERERS attending on Macbeth

SEYTON, servant of Macbeth

LADY MACBETH, Macbeth's wife

A DOCTOR of Physic ⎫
                   ⎬ attending on Lady Macbeth
A Waiting-GENTLEWOMAN ⎭

BANQUO, a Scottish thane

FLEANCE, his son

MACDUFF, Thane of Fife

LADY MACDUFF, his wife

MACDUFF'S SON

LENNOX ⎫
ROSS     ⎪
ANGUS   ⎬ Scottish Thanes
CAITHNESS ⎪
MENTEITH ⎭

SIWARD, Earl of Northumberland

YOUNG SIWARD, his son

An English DOCTOR

HECATE, Queen of the Witches

Six WITCHES

Three APPARITIONS, one an armed head, one a bloody child, one a child crowned

A SPIRIT LIKE A CAT

Other SPIRITS

An OLD MAN

A MESSENGER

MURDERERS

SERVANTS

A show of eight kings; Lords and Thanes, attendants, soldiers, drummers

# The Tragedy of Macbeth

**1.1** *Thunder and lightning. Enter three Witches*

FIRST WITCH
When shall we three meet again?
In thunder, lightning, or in rain?

SECOND WITCH
When the hurly-burly's done,
When the battle's lost and won.

THIRD WITCH
That will be ere the set of sun.                    5

FIRST WITCH
Where the place?

SECOND WITCH                    Upon the heath.

THIRD WITCH
There to meet with Macbeth.

FIRST WITCH
I come, Grimalkin.

SECOND WITCH
Paddock calls.

THIRD WITCH                    Anon.

ALL
Fair is foul, and foul is fair,                    10
Hover through the fog and filthy air.      *Exeunt*


**1.2** *Alarum within. Enter King Duncan, Malcolm,*
*Donalbain, Lennox, with attendants, meeting a*
*bleeding Captain*

KING DUNCAN
What bloody man is that? He can report,
As seemeth by his plight, of the revolt
The newest state.

MALCOLM                    This is the sergeant
Who like a good and hardy soldier fought
'Gainst my captivity. Hail, brave friend.                    5
Say to the King the knowledge of the broil
As thou didst leave it.

CAPTAIN                    Doubtful it stood,
As two spent swimmers that do cling together
And choke their art. The merciless Macdonald—
Worthy to be a rebel, for to that                    10
The multiplying villainies of nature
Do swarm upon him—from the Western Isles
Of kerns and galloglasses is supplied,
And fortune on his damnèd quarry smiling
Showed like a rebel's whore. But all's too weak,      15
For brave Macbeth—well he deserves that name!—
Disdaining fortune, with his brandished steel
Which smoked with bloody execution,
Like valour's minion
Carved out his passage till he faced the slave,      20
Which ne'er shook hands nor bade farewell to him
Till he unseamed him from the nave to th' chops,
And fixed his head upon our battlements.

KING DUNCAN
O valiant cousin, worthy gentleman!

CAPTAIN
As whence the sun 'gins his reflection                    25
Shipwrecking storms and direful thunders break,
So from that spring whence comfort seemed to come
Discomfort swells. Mark, King of Scotland, mark.
No sooner justice had, with valour armed,
Compelled these skipping kerns to trust their heels  30
But the Norwegian lord, surveying vantage,
With furbished arms and new supplies of men
Began a fresh assault.

KING DUNCAN
Dismayed not this our captains, Macbeth and
Banquo?

CAPTAIN
Yes, as sparrows eagles, or the hare the lion!      35
If I say sooth I must report they were
As cannons overcharged with double cracks,
So they doubly redoubled strokes upon the foe.
Except they meant to bathe in reeking wounds
Or memorize another Golgotha,                    40
I cannot tell—
But I am faint. My gashes cry for help.

KING DUNCAN
So well thy words become thee as thy wounds:
They smack of honour both.—Go get him surgeons.
                    *Exit Captain with attendants*
                    *Enter Ross and Angus*
Who comes here?

MALCOLM                    The worthy Thane of Ross.      45

LENNOX
What haste looks through his eyes! So should he look
That seems to speak things strange.

ROSS                    God save the King.

KING DUNCAN
Whence cam'st thou, worthy thane?

ROSS                    From Fife, great King,
Where the Norwegian banners flout the sky
And fan our people cold.                    50
Norway himself, with terrible numbers,
Assisted by that most disloyal traitor
The Thane of Cawdor, began a dismal conflict,
Till that Bellona's bridegroom, lapped in proof,
Confronted him with self-comparisons,                    55
Point against point, rebellious arm 'gainst arm,
Curbing his lavish spirit; and to conclude,
The victory fell on us—

KING DUNCAN                    Great happiness.

ROSS                    That now
Sweno, the Norways' king, craves composition;
Nor would we deign him burial of his men      60
Till he disbursèd at Saint Colum's inch
Ten thousand dollars to our general use.

KING DUNCAN
No more that Thane of Cawdor shall deceive
Our bosom interest. Go pronounce his present death,
And with his former title greet Macbeth.                    65

ROSS I'll see it done.

KING DUNCAN
What he hath lost, noble Macbeth hath won.
                    *Exeunt severally*


**1.3** *Thunder. Enter the three Witches*

FIRST WITCH
Where hast thou been, sister?

SECOND WITCH
Killing swine.

THIRD WITCH                    Sister, where thou?

FIRST WITCH
A sailor's wife had chestnuts in her lap,
And munched, and munched, and munched. 'Give
  me,' quoth I.
'Aroint thee, witch,' the rump-fed runnion cries.   5
Her husband's to Aleppo gone, master o'th' Tiger.
  But in a sieve I'll thither sail,
  And like a rat without a tail
  I'll do, I'll do, and I'll do.
SECOND WITCH
I'll give thee a wind.   10
FIRST WITCH
Thou'rt kind.
THIRD WITCH
And I another.
FIRST WITCH
I myself have all the other,
And the very ports they blow,
All the quarters that they know   15
I'th' shipman's card.
I'll drain him dry as hay.
Sleep shall neither night nor day
Hang upon his penthouse lid.
He shall live a man forbid.   20
Weary sennights nine times nine
Shall he dwindle, peak, and pine.
Though his barque cannot be lost,
Yet it shall be tempest-tossed.
Look what I have.
SECOND WITCH          Show me, show me.   25
FIRST WITCH
Here I have a pilot's thumb,
Wrecked as homeward he did come.
  *Drum within*
THIRD WITCH
A drum, a drum—
Macbeth doth come.
ALL (*dancing in a ring*)
The weird sisters hand in hand,
Posters of the sea and land,   30
Thus do go about, about,
Thrice to thine, and thrice to mine,
And thrice again to make up nine.
Peace! The charm's wound up.   35
  *Enter Macbeth and Banquo*
MACBETH
So foul and fair a day I have not seen.
BANQUO
How far is't called to Forres?—What are these,
So withered, and so wild in their attire,
That look not like th'inhabitants o'th' earth
And yet are on't?—Live you, or are you aught   40
That man may question? You seem to understand me
By each at once her choppy finger laying
Upon her skinny lips. You should be women,
And yet your beards forbid me to interpret
That you are so.
MACBETH (*to the Witches*)
        Speak, if you can. What are you?   45
FIRST WITCH
All hail, Macbeth! Hail to thee, Thane of Glamis.
SECOND WITCH
All hail, Macbeth! Hail to thee, Thane of Cawdor.
THIRD WITCH
All hail, Macbeth, that shalt be king hereafter!

BANQUO
Good sir, why do you start and seem to fear
Things that do sound so fair? (*To the Witches*) I'th'
  name of truth,   50
Are ye fantastical or that indeed
Which outwardly ye show? My noble partner
You greet with present grace and great prediction
Of noble having and of royal hope,
That he seems rapt withal. To me you speak not.   55
If you can look into the seeds of time
And say which grain will grow and which will not,
Speak then to me, who neither beg nor fear
Your favours nor your hate.
FIRST WITCH Hail!   60
SECOND WITCH Hail!
THIRD WITCH Hail!
FIRST WITCH
Lesser than Macbeth, and greater.
SECOND WITCH
Not so happy, yet much happier.
THIRD WITCH
Thou shalt get kings, though thou be none.   65
So all hail, Macbeth and Banquo!
FIRST WITCH
Banquo and Macbeth, all hail!
MACBETH
Stay, you imperfect speakers, tell me more.
By Sinel's death I know I am Thane of Glamis,
But how of Cawdor? The Thane of Cawdor lives,   70
A prosperous gentleman, and to be king
Stands not within the prospect of belief,
No more than to be Cawdor. Say from whence
You owe this strange intelligence, or why
Upon this blasted heath you stop our way   75
With such prophetic greeting. Speak, I charge you.
  *The Witches vanish*
BANQUO
The earth hath bubbles, as the water has,
And these are of them. Whither are they vanished?
MACBETH
Into the air, and what seemed corporal
Melted as breath into the wind. Would they had stayed.
BANQUO
Were such things here as we do speak about,   81
Or have we eaten on the insane root
That takes the reason prisoner?
MACBETH
Your children shall be kings.
BANQUO         You shall be king.
MACBETH
And Thane of Cawdor too. Went it not so?   85
BANQUO
To th' self-same tune and words. Who's here?
  *Enter Ross and Angus*
ROSS
The King hath happily received, Macbeth,
The news of thy success, and when he reads
Thy personal venture in the rebels' sight
His wonders and his praises do contend   90
Which should be thine or his; silenced with that,
In viewing o'er the rest o'th' self-same day
He finds thee in the stout Norwegian ranks,
Nothing afeard of what thyself didst make,
Strange images of death. As thick as hail   95
Came post with post, and every one did bear

Thy praises in his kingdom's great defence,
And poured them down before him.
ANGUS (*to Macbeth*)                        We are sent
To give thee from our royal master thanks;
Only to herald thee into his sight,                    100
Not pay thee.
ROSS
And, for an earnest of a greater honour,
He bade me from him call thee Thane of Cawdor,
In which addition, hail, most worthy thane,
For it is thine.
BANQUO               What, can the devil speak true?    105
MACBETH
The Thane of Cawdor lives. Why do you dress me
In borrowed robes?
ANGUS                    Who was the thane lives yet,
But under heavy judgement bears that life
Which he deserves to lose. Whether he was combined
With those of Norway, or did line the rebel       110
With hidden help and vantage, or that with both
He laboured in his country's wrack, I know not;
But treasons capital, confessed, and proved
Have overthrown him.
MACBETH (*aside*)        Glamis, and Thane of Cawdor.
The greatest is behind. (*To Ross and Angus*) Thanks for
    your pains.                                        115
(*To Banquo*) Do you not hope your children shall be kings
When those that gave the thane of Cawdor to me
Promised no less to them?
BANQUO                       That, trusted home,
Might yet enkindle you unto the crown,
Besides the thane of Cawdor. But 'tis strange,    120
And oftentimes to win us to our harm
The instruments of darkness tell us truths,
Win us with honest trifles to betray's
In deepest consequence.
(*To Ross and Angus*) Cousins, a word, I pray you.   125
MACBETH (*aside*) Two truths are told
As happy prologues to the swelling act
Of the imperial theme. (*To Ross and Angus*) I thank
    you, gentlemen.
(*Aside*) This supernatural soliciting
Cannot be ill, cannot be good. If ill,                130
Why hath it given me earnest of success
Commencing in a truth? I am Thane of Cawdor.
If good, why do I yield to that suggestion
Whose horrid image doth unfix my hair
And make my seated heart knock at my ribs        135
Against the use of nature? Present fears
Are less than horrible imaginings.
My thought, whose murder yet is but fantastical,
Shakes so my single state of man that function
Is smothered in surmise, and nothing is              140
But what is not.
BANQUO (*to Ross and Angus*)
                    Look how our partner's rapt.
MACBETH (*aside*)
If chance will have me king, why, chance may crown
    me
Without my stir.
BANQUO (*to Ross and Angus*)
                    New honours come upon him,
Like our strange garments, cleave not to their mould
But with the aid of use.
MACBETH (*aside*)        Come what come may,      145
Time and the hour runs through the roughest day.

BANQUO
Worthy Macbeth, we stay upon your leisure.
MACBETH
Give me your favour. My dull brain was wrought
With things forgotten. (*To Ross and Angus*) Kind
    gentlemen, your pains
Are registered where every day I turn                150
The leaf to read them. Let us toward the King.
(*Aside to Banquo*) Think upon what hath chanced, and
    at more time,
The interim having weighed it, let us speak
Our free hearts each to other.
BANQUO  Very gladly.                                  155
MACBETH
Till then, enough. (*To Ross and Angus*) Come, friends.
                                            *Exeunt*

**1.4**    *Flourish. Enter King Duncan, Lennox, Malcolm,*
          *Donalbain, and attendants*
KING DUNCAN
Is execution done on Cawdor? Are not
Those in commission yet returned?
MALCOLM                                My liege,
They are not yet come back. But I have spoke
With one that saw him die, who did report
That very frankly he confessed his treasons,           5
Implored your highness' pardon, and set forth
A deep repentance. Nothing in his life
Became him like the leaving it. He died
As one that had been studied in his death
To throw away the dearest thing he owed             10
As 'twere a careless trifle.
KING DUNCAN                  There's no art
To find the mind's construction in the face.
He was a gentleman on whom I built
An absolute trust.
      *Enter Macbeth, Banquo, Ross, and Angus*
(*To Macbeth*)        O worthiest cousin,
The sin of my ingratitude even now                    15
Was heavy on me! Thou art so far before
That swiftest wing of recompense is slow
To overtake thee. Would thou hadst less deserved,
That the proportion both of thanks and payment
Might have been mine. Only I have left to say,     20
'More is thy due than more than all can pay'.
MACBETH
The service and the loyalty I owe,
In doing it, pays itself. Your highness' part
Is to receive our duties, and our duties
Are to your throne and state children and servants 25
Which do but what they should by doing everything
Safe toward your love and honour.
KING DUNCAN                        Welcome hither.
I have begun to plant thee, and will labour
To make thee full of growing.—Noble Banquo,
That hast no less deserved, nor must be known     30
No less to have done so, let me enfold thee
And hold thee to my heart.
BANQUO                      There if I grow
The harvest is your own.
KING DUNCAN               My plenteous joys,
Wanton in fullness, seek to hide themselves
In drops of sorrow. Sons, kinsmen, thanes,         35
And you whose places are the nearest, know
We will establish our estate upon
Our eldest, Malcolm, whom we name hereafter

The Prince of Cumberland; which honour must
Not unaccompanied invest him only,      40
But signs of nobleness, like stars, shall shine
On all deservers. (*To Macbeth*) From hence to Inverness,
And bind us further to you.

MACBETH
The rest is labour which is not used for you.
I'll be myself the harbinger, and make joyful      45
The hearing of my wife with your approach;
So humbly take my leave.

KING DUNCAN          My worthy Cawdor.

MACBETH (*aside*)
The Prince of Cumberland—that is a step
On which I must fall down or else o'erleap,
For in my way it lies. Stars, hide your fires,      50
Let not light see my black and deep desires;
The eye wink at the hand; yet let that be
Which the eye fears, when it is done, to see.      *Exit*

KING DUNCAN
True, worthy Banquo, he is full so valiant,
And in his commendations I am fed.      55
It is a banquet to me. Let's after him,
Whose care is gone before to bid us welcome.
It is a peerless kinsman.      *Flourish. Exeunt*

**1.5**    *Enter Lady Macbeth, with a letter*

LADY MACBETH (*reading*) 'They met me in the day of success,
and I have learned by the perfect'st report they have
more in them than mortal knowledge. When I burned in
desire to question them further, they made themselves
air, into which they vanished. Whiles I stood rapt in the
wonder of it came missives from the King, who all-hailed
me "Thane of Cawdor", by which title before these weird
sisters saluted me, and referred me to the coming on of
time with "Hail, King that shalt be!" This have I thought
good to deliver thee, my dearest partner of greatness,
that thou mightst not lose the dues of rejoicing by being
ignorant of what greatness is promised thee. Lay it to
thy heart, and farewell.'
Glamis thou art, and Cawdor, and shalt be
What thou art promised. Yet do I fear thy nature.      15
It is too full o'th' milk of human kindness
To catch the nearest way. Thou wouldst be great,
Art not without ambition, but without
The illness should attend it. What thou wouldst highly,
That wouldst thou holily; wouldst not play false,      20
And yet wouldst wrongly win. Thou'dst have, great
     Glamis,
That which cries 'Thus thou must do' if thou have it,
And that which rather thou dost fear to do
Than wishest should be undone. Hie thee hither,
That I may pour my spirits in thine ear      25
And chastise with the valour of my tongue
All that impedes thee from the golden round
Which fate and metaphysical aid doth seem
To have thee crowned withal.
     *Enter a Servant*
                 What is your tidings?

SERVANT
The King comes here tonight.

LADY MACBETH         Thou'rt mad to say it.
Is not thy master with him, who, were't so,      31
Would have informed for preparation?

SERVANT
So please you, it is true. Our thane is coming,

One of my fellows had the speed of him,
Who, almost dead for breath, had scarcely more      35
Than would make up his message.

LADY MACBETH           Give him tending;
He brings great news.               *Exit Servant*
              The raven himself is hoarse
That croaks the fatal entrance of Duncan
Under my battlements. Come, you spirits
That tend on mortal thoughts, unsex me here,      40
And fill me from the crown to the toe top-full
Of direst cruelty. Make thick my blood,
Stop up th'access and passage to remorse,
That no compunctious visitings of nature
Shake my fell purpose, nor keep peace between      45
Th'effect and it. Come to my woman's breasts,
And take my milk for gall, you murd'ring ministers,
Wherever in your sightless substances
You wait on nature's mischief. Come, thick night,
And pall thee in the dunnest smoke of hell,      50
That my keen knife see not the wound it makes,
Nor heaven peep through the blanket of the dark
To cry 'Hold, hold!'
     *Enter Macbeth*
            Great Glamis, worthy Cawdor,
Greater than both by the all-hail hereafter,
Thy letters have transported me beyond      55
This ignorant present, and I feel now
The future in the instant.

MACBETH           My dearest love,
Duncan comes here tonight.

LADY MACBETH         And when goes hence?

MACBETH
Tomorrow, as he purposes.

LADY MACBETH          O never
Shall sun that morrow see.      60
Your face, my thane, is as a book where men
May read strange matters. To beguile the time,
Look like the time; bear welcome in your eye,
Your hand, your tongue; look like the innocent flower,
But be the serpent under't. He that's coming      65
Must be provided for; and you shall put
This night's great business into my dispatch,
Which shall to all our nights and days to come
Give solely sovereign sway and masterdom.

MACBETH
We will speak further.

LADY MACBETH        Only look up clear.      70
To alter favour ever is to fear.
Leave all the rest to me.               *Exeunt*

**1.6**    ⌜*Hautboys and torches.*⌝ *Enter King Duncan,*
     *Malcolm, Donalbain, Banquo, Lennox, Macduff,*
     *Ross, Angus, and attendants*

KING DUNCAN
This castle hath a pleasant seat. The air
Nimbly and sweetly recommends itself
Unto our gentle senses.

BANQUO           This guest of summer,
The temple-haunting martlet, does approve
By his loved mansionry that the heavens' breath      5
Smells wooingly here. No jutty, frieze,
Buttress, nor coign of vantage but this bird
Hath made his pendant bed and procreant cradle;
Where they most breed and haunt I have observed
The air is delicate.

*Enter Lady Macbeth*

KING DUNCAN  See, see, our honoured hostess!                    10
    The love that follows us sometime is our trouble,
    Which still we thank as love. Herein I teach you
    How you shall bid God 'ield us for your pains,
    And thank us for your trouble.

LADY MACBETH                        All our service
    In every point twice done, and then done double,     15
    Were poor and single business to contend
    Against those honours deep and broad wherewith
    Your majesty loads our house. For those of old,
    And the late dignities heaped up to them,
    We rest your hermits.

KING DUNCAN             Where's the Thane of Cawdor?
    We coursed him at the heels, and had a purpose       21
    To be his purveyor; but he rides well,
    And his great love, sharp as his spur, hath holp him
    To his home before us. Fair and noble hostess,
    We are your guest tonight.

LADY MACBETH                    Your servants ever        25
    Have theirs, themselves, and what is theirs in count
    To make their audit at your highness' pleasure,
    Still to return your own.

KING DUNCAN                 Give me your hand.
    Conduct me to mine host. We love him highly,
    And shall continue our graces towards him.            30
    By your leave, hostess.                        *Exeunt*

1.7    *Hautboys. Torches. Enter a sewer and divers*
       *servants with dishes and service over the stage.*
       *Then enter Macbeth*

MACBETH
    If it were done when 'tis done, then 'twere well
    It were done quickly. If th'assassination
    Could trammel up the consequence, and catch
    With his surcease success: that but this blow
    Might be the be-all and the end-all, here,            5
    But here upon this bank and shoal of time,
    We'd jump the life to come. But in these cases
    We still have judgement here, that we but teach
    Bloody instructions which, being taught, return
    To plague th'inventor. This even-handed justice      10
    Commends th'ingredience of our poisoned chalice
    To our own lips. He's here in double trust:
    First, as I am his kinsman and his subject,
    Strong both against the deed; then, as his host,
    Who should against his murderer shut the door,       15
    Not bear the knife myself. Besides, this Duncan
    Hath borne his faculties so meek, hath been
    So clear in his great office, that his virtues
    Will plead like angels, trumpet-tongued against
    The deep damnation of his taking-off,                20
    And pity, like a naked new-born babe,
    Striding the blast, or heaven's cherubin, horsed
    Upon the sightless couriers of the air,
    Shall blow the horrid deed in every eye
    That tears shall drown the wind. I have no spur       25
    To prick the sides of my intent, but only
    Vaulting ambition which o'erleaps itself
    And falls on th'other.
            *Enter Lady Macbeth*
                        How now? What news?

LADY MACBETH
    He has almost supped. Why have you left the
    chamber?

MACBETH
    Hath he asked for me?

LADY MACBETH                Know you not he has?         30

MACBETH
    We will proceed no further in this business.
    He hath honoured me of late, and I have bought
    Golden opinions from all sorts of people,
    Which would be worn now in their newest gloss,
    Not cast aside so soon.

LADY MACBETH            Was the hope drunk              35
    Wherein you dressed yourself? Hath it slept since?
    And wakes it now to look so green and pale
    At what it did so freely? From this time
    Such I account thy love. Art thou afeard
    To be the same in thine own act and valour           40
    As thou art in desire? Wouldst thou have that
    Which thou esteem'st the ornament of life,
    And live a coward in thine own esteem,
    Letting 'I dare not' wait upon 'I would',
    Like the poor cat i'th' adage?

MACBETH                    Prithee, peace.              45
    I dare do all that may become a man;
    Who dares do more is none.

LADY MACBETH                What beast was't then
    That made you break this enterprise to me?
    When you durst do it, then you were a man;
    And to be more than what you were, you would         50
    Be so much more the man. Nor time nor place
    Did then adhere, and yet you would make both.
    They have made themselves, and that their fitness now
    Does unmake you. I have given suck, and know
    How tender 'tis to love the babe that milks me.      55
    I would, while it was smiling in my face,
    Have plucked my nipple from his boneless gums
    And dashed the brains out, had I so sworn
    As you have done to this.

MACBETH                    If we should fail?

LADY MACBETH                            We fail!
    But screw your courage to the sticking-place         60
    And we'll not fail. When Duncan is asleep—
    Whereto the rather shall his day's hard journey
    Soundly invite him—his two chamberlains
    Will I with wine and wassail so convince
    That memory, the warder of the brain,                65
    Shall be a fume, and the receipt of reason
    A limbeck only. When in swinish sleep
    Their drenchèd natures lies as in a death,
    What cannot you and I perform upon
    Th'unguarded Duncan? What not put upon                70
    His spongy officers, who shall bear the guilt
    Of our great quell?

MACBETH            Bring forth men-children only,
    For thy undaunted mettle should compose
    Nothing but males. Will it not be received,
    When we have marked with blood those sleepy two
    Of his own chamber and used their very daggers,      76
    That they have done't?

LADY MACBETH            Who dares receive it other,
    As we shall make our griefs and clamour roar
    Upon his death?

MACBETH            I am settled, and bend up
    Each corporal agent to this terrible feat.           80
    Away, and mock the time with fairest show.
    False face must hide what the false heart doth know.
                                            *Exeunt*

**2.1** *Enter Banquo and Fleance, with a torch before him*
BANQUO How goes the night, boy?
FLEANCE
　The moon is down. I have not heard the clock.
BANQUO
　And she goes down at twelve.
FLEANCE　　　　　　　　　I take't 'tis later, sir.
BANQUO *(giving Fleance his sword)*
　Hold, take my sword. There's husbandry in heaven,
　Their candles are all out. Take thee that, too.　　5
　A heavy summons lies like lead upon me,
　And yet I would not sleep. Merciful powers,
　Restrain in me the cursèd thoughts that nature
　Gives way to in repose.
　　　　*Enter Macbeth, and a servant with a torch*
　　　　　　Give me my sword. Who's there?
MACBETH　A friend.　　　　　　　　　　　　10
BANQUO
　What, sir, not yet at rest? The King's a-bed.
　He hath been in unusual pleasure, and
　Sent forth great largesse to your offices.
　This diamond he greets your wife withal
　By the name of most kind hostess, and shut up　　15
　In measureless content.
MACBETH　　　　　　　Being unprepared
　Our will became the servant to defect,
　Which else should free have wrought.
BANQUO　　　　　　　　　　　　All's well.
　I dreamt last night of the three weird sisters.
　To you they have showed some truth.
MACBETH　　　　　　　　　I think not of them;
　Yet, when we can entreat an hour to serve,　　21
　We would spend it in some words upon that business
　If you would grant the time.
BANQUO　　　　　　　　At your kind'st leisure.
MACBETH
　If you shall cleave to my consent when 'tis,
　It shall make honour for you.
BANQUO　　　　　　　　So I lose none　　25
　In seeking to augment it, but still keep
　My bosom franchised and allegiance clear,
　I shall be counselled.
MACBETH　Good repose the while.
BANQUO　Thanks, sir. The like to you.　　　　30
　　　　　　　*Exeunt Banquo and Fleance*
MACBETH *(to the Servant)*
　Go bid thy mistress, when my drink is ready,
　She strike upon the bell. Get thee to bed.　*Exit Servant*
　Is this a dagger which I see before me,
　The handle toward my hand? Come, let me clutch thee.
　I have thee not, and yet I see thee still.　　35
　Art thou not, fatal vision, sensible
　To feeling as to sight? Or art thou but
　A dagger of the mind, a false creation
　Proceeding from the heat-oppressèd brain?
　I see thee yet, in form as palpable　　40
　As this which now I draw.
　Thou marshall'st me the way that I was going,
　And such an instrument I was to use.
　Mine eyes are made the fools o'th' other senses,
　Or else worth all the rest. I see thee still,　　45
　And on thy blade and dudgeon gouts of blood,
　Which was not so before. There's no such thing.
　It is the bloody business which informs
　Thus to mine eyes. Now o'er the one half-world
　Nature seems dead, and wicked dreams abuse　　50

　The curtained sleep. Witchcraft celebrates
　Pale Hecate's offerings, and withered murder,
　Alarumed by his sentinel the wolf,
　Whose howl's his watch, thus with his stealthy pace,
　With Tarquin's ravishing strides, towards his design
　Moves like a ghost. Thou sure and firm-set earth,　　56
　Hear not my steps which way they walk, for fear
　Thy very stones prate of my whereabout,
　And take the present horror from the time,
　Which now suits with it. Whiles I threat, he lives.　　60
　Words to the heat of deeds too cold breath gives.
　　　　*A bell rings*
　I go, and it is done. The bell invites me.
　Hear it not, Duncan; for it is a knell
　That summons thee to heaven or to hell.　　　*Exit*

**2.2**　*Enter Lady Macbeth*
LADY MACBETH
　That which hath made them drunk hath made me bold.
　What hath quenched them hath given me fire. Hark,
　　peace!—
　It was the owl that shrieked, the fatal bellman
　Which gives the stern'st good-night. He is about it.
　The doors are open, and the surfeited grooms　　5
　Do mock their charge with snores. I have drugged
　　their possets
　That death and nature do contend about them
　Whether they live or die.
　　　　*Enter Macbeth ⌜above⌝*
MACBETH　　　　　　Who's there? What ho?　*Exit*
LADY MACBETH
　Alack, I am afraid they have awaked,
　And 'tis not done. Th'attempt and not the deed　　10
　Confounds us. Hark!—I laid their daggers ready;
　He could not miss 'em. Had he not resembled
　My father as he slept, I had done't.
　　　⌜*Enter Macbeth below*⌝
　　　　　　　　　My husband!
MACBETH
　I have done the deed. Didst thou not hear a noise?
LADY MACBETH
　I heard the owl scream and the crickets cry.　　15
　Did not you speak?
MACBETH　　　　　When?
LADY MACBETH　　　　　　　Now.
MACBETH　　　　　　　　As I descended?
LADY MACBETH
　Ay.
MACBETH　Hark!—Who lies i'th' second chamber?
LADY MACBETH
　Donalbain.
MACBETH *(looking at his hands)* This is a sorry sight.
LADY MACBETH
　A foolish thought, to say a sorry sight.
MACBETH
　There's one did laugh in's sleep, and one cried 'Murder!'
　That they did wake each other. I stood and heard
　　them.　　　　　　　　　　　　　　21
　But they did say their prayers and addressed them
　Again to sleep.
LADY MACBETH　　There are two lodged together.
MACBETH
　One cried 'God bless us' and 'Amen' the other,
　As they had seen me with these hangman's hands.　　25
　List'ning their fear I could not say 'Amen'
　When they did say 'God bless us.'

LADY MACBETH
  Consider it not so deeply.
MACBETH
  But wherefore could not I pronounce 'Amen'?
  I had most need of blessing, and 'Amen'                    30
  Stuck in my throat.
LADY MACBETH              These deeds must not be thought
  After these ways. So, it will make us mad.
MACBETH
  Methought I heard a voice cry 'Sleep no more,
  Macbeth does murder sleep'—the innocent sleep,
  Sleep that knits up the ravelled sleave of care,          35
  The death of each day's life, sore labour's bath,
  Balm of hurt minds, great nature's second course,
  Chief nourisher in life's feast—
LADY MACBETH                          What do you mean?
MACBETH
  Still it cried 'Sleep no more' to all the house,
  'Glamis hath murdered sleep, and therefore Cawdor         40
  Shall sleep no more, Macbeth shall sleep no more.'
LADY MACBETH
  Who was it that thus cried? Why, worthy thane,
  You do unbend your noble strength to think
  So brain-sickly of things. Go get some water
  And wash this filthy witness from your hand.              45
  Why did you bring these daggers from the place?
  They must lie there. Go, carry them, and smear
  The sleepy grooms with blood.
MACBETH                          I'll go no more.
  I am afraid to think what I have done,
  Look on't again I dare not.
LADY MACBETH                    Infirm of purpose!          50
  Give me the daggers. The sleeping and the dead
  Are but as pictures. 'Tis the eye of childhood
  That fears a painted devil. If he do bleed
  I'll gild the faces of the grooms withal,
  For it must seem their guilt.                       Exit
      Knock within
MACBETH                        Whence is that knocking?—
  How is't with me when every noise appals me?             56
  What hands are here! Ha, they pluck out mine eyes.
  Will all great Neptune's ocean wash this blood
  Clean from my hand? No, this my hand will rather
  The multitudinous seas incarnadine,                      60
  Making the green one red.
      Enter Lady Macbeth
LADY MACBETH
  My hands are of your colour, but I shame
  To wear a heart so white.
      Knock within
                              I hear a knocking
  At the south entry. Retire we to our chamber.
  A little water clears us of this deed.                   65
  How easy is it then! Your constancy
  Hath left you unattended.
      Knock within
                        Hark, more knocking.
  Get on your nightgown, lest occasion call us
  And show us to be watchers. Be not lost
  So poorly in your thoughts.                              70
MACBETH
  To know my deed 'twere best not know myself.
      Knock within
  Wake Duncan with thy knocking. I would thou
  couldst.                                         Exeunt

2.3    Enter a Porter. Knocking within
PORTER Here's a knocking indeed! If a man were porter
  of hell-gate he should have old turning the key.
      Knock within
  Knock, knock, knock. Who's there, i'th' name of
  Beelzebub? Here's a farmer that hanged himself on
  th'expectation of plenty. Come in time! Have napkins
  enough about you; here you'll sweat for't.                6
      Knock within
  Knock, knock. Who's there, in th'other devil's name?
  Faith, here's an equivocator that could swear in both
  the scales against either scale, who committed treason
  enough for God's sake, yet could not equivocate to
  heaven. O, come in, equivocator.                         11
      Knock within
  Knock, knock, knock. Who's there? 'Faith, here's an
  English tailor come hither for stealing out of a French
  hose. Come in, tailor. Here you may roast your goose.
      Knock within
  Knock, knock. Never at quiet. What are you?—But this
  place is too cold for hell. I'll devil-porter it no further.
  I had thought to have let in some of all professions
  that go the primrose way to th'everlasting bonfire.
      Knock within
  Anon, anon!
      He opens the gate
  I pray you remember the porter.                          20
      Enter Macduff and Lennox
MACDUFF
  Was it so late, friend, ere you went to bed
  That you do lie so late?
PORTER Faith, sir, we were carousing till the second cock,
  and drink, sir, is a great provoker of three things.
MACDUFF What three things does drink especially
  provoke?                                                 26
PORTER Marry, sir, nose-painting, sleep, and urine.
  Lechery, sir, it provokes and unprovokes: it provokes
  the desire but it takes away the performance. Therefore
  much drink may be said to be an equivocator with
  lechery: it makes him and it mars him; it sets him on
  and it takes him off; it persuades him and disheartens
  him, makes him stand to and not stand to; in
  conclusion, equivocates him in a sleep, and, giving him
  the lie, leaves him.                                     35
MACDUFF I believe drink gave thee the lie last night.
PORTER That it did, sir, i'the very throat on me; but I
  requited him for his lie, and, I think, being too strong
  for him, though he took up my legs sometime, yet I
  made a shift to cast him.                                40
MACDUFF Is thy master stirring?
      Enter Macbeth
  Our knocking has awaked him: here he comes.
                                        ⌜Exit Porter⌝
LENNOX (to Macbeth)
  Good morrow, noble sir.
MACBETH                        Good morrow, both.
MACDUFF
  Is the King stirring, worthy thane?
MACBETH                                      Not yet.
MACDUFF
  He did command me to call timely on him.                 45
  I have almost slipped the hour.
MACBETH                            I'll bring you to him.
MACDUFF
  I know this is a joyful trouble to you,
  But yet 'tis one.

**MACBETH**
The labour we delight in physics pain.
This is the door.
**MACDUFF**                    I'll make so bold to call,                    50
For 'tis my limited service.                    *Exit Macduff*
**LENNOX**
Goes the King hence today?
**MACBETH**                    He does; he did appoint so.
**LENNOX**
The night has been unruly. Where we lay
Our chimneys were blown down, and, as they say,
Lamentings heard i'th' air, strange screams of death,
And prophesying with accents terrible                    56
Of dire combustion and confused events
New-hatched to th' woeful time. The obscure bird
Clamoured the livelong night. Some say the earth
Was feverous and did shake.
**MACBETH**                    'Twas a rough night.                    60
**LENNOX**
My young remembrance cannot parallel
A fellow to it.
          *Enter Macduff*
**MACDUFF**          O horror, horror, horror!
Tongue nor heart cannot conceive nor name thee.
**MACBETH** *and* **LENNOX** What's the matter?
**MACDUFF**
Confusion now hath made his masterpiece.                    65
Most sacrilegious murder hath broke ope
The Lord's anointed temple and stole thence
The life o'th' building.
**MACBETH** What is't you say—the life?
**LENNOX** Mean you his majesty?                    70
**MACDUFF**
Approach the chamber and destroy your sight
With a new Gorgon. Do not bid me speak.
See, and then speak yourselves.
                    *Exeunt Macbeth and Lennox*
                    Awake, awake!
Ring the alarum bell. Murder and treason!
Banquo and Donalbain, Malcolm, awake!                    75
Shake off this downy sleep, death's counterfeit,
And look on death itself. Up, up, and see
The great doom's image. Malcolm, Banquo,
As from your graves rise up, and walk like sprites
To countenance this horror.
          *Bell rings. Enter Lady Macbeth*
**LADY MACBETH**                    What's the business,                    80
That such a hideous trumpet calls to parley
The sleepers of the house? Speak, speak.
**MACDUFF**                    O gentle lady,
'Tis not for you to hear what I can speak.
The repetition in a woman's ear
Would murder as it fell.
          *Enter Banquo*
                    O Banquo, Banquo,                    85
Our royal master's murdered!
**LADY MACBETH**                    Woe, alas—
What, in our house?
**BANQUO**                    Too cruel anywhere.
Dear Duff, I prithee contradict thyself,
And say it is not so.
          *Enter Macbeth, Lennox, ⌈and Ross⌉*
**MACBETH**
Had I but died an hour before this chance                    90
I had lived a blessèd time, for from this instant
There's nothing serious in mortality.

All is but toys. Renown and grace is dead.
The wine of life is drawn, and the mere lees
Is left this vault to brag of.                    95
          *Enter Malcolm and Donalbain*
**DONALBAIN** What is amiss?
**MACBETH** You are, and do not know't.
The spring, the head, the fountain of your blood
Is stopped, the very source of it is stopped.
**MACDUFF**
Your royal father's murdered.
**MALCOLM**                    O, by whom?                    100
**LENNOX**
Those of his chamber, as it seemed, had done't.
Their hands and faces were all badged with blood,
So were their daggers, which, unwiped, we found
Upon their pillows. They stared and were distracted.
No man's life was to be trusted with them.                    105
**MACBETH**
O, yet I do repent me of my fury
That I did kill them.
**MACDUFF**                    Wherefore did you so?
**MACBETH**
Who can be wise, amazed, temp'rate and furious,
Loyal and neutral in a moment? No man.
Th'expedition of my violent love                    110
Outran the pauser, reason. Here lay Duncan,
His silver skin laced with his golden blood,
And his gashed stabs looked like a breach in nature
For ruin's wasteful entrance; there the murderers,
Steeped in the colours of their trade, their daggers                    115
Unmannerly breeched with gore. Who could refrain,
That had a heart to love, and in that heart
Courage to make 's love known?
**LADY MACBETH**                    Help me hence, ho!
**MACDUFF**
Look to the lady.
**MALCOLM** (*aside to Donalbain*)
                    Why do we hold our tongues,
That most may claim this argument for ours?                    120
**DONALBAIN** (*aside to Malcolm*)
What should be spoken here, where our fate,
Hid in an auger-hole, may rush and seize us?
Let's away. Our tears are not yet brewed.
**MALCOLM** (*aside to Donalbain*)                    Nor our strong sorrow
Upon the foot of motion.
**BANQUO**                    Look to the lady;
                    *Exit Lady Macbeth, attended*
And when we have our naked frailties hid,                    125
That suffer in exposure, let us meet
And question this most bloody piece of work,
To know it further. Fears and scruples shake us.
In the great hand of God I stand, and thence
Against the undivulged pretence I fight                    130
Of treasonous malice.
**MACDUFF**                    And so do I.
**ALL**                    So all.
**MACBETH**
Let's briefly put on manly readiness,
And meet i'th' hall together.
**ALL**                    Well contented.
          *Exeunt all but Malcolm and Donalbain*
**MALCOLM**
What will you do? Let's not consort with them.
To show an unfelt sorrow is an office                    135
Which the false man does easy. I'll to England.

DONALBAIN
    To Ireland, I. Our separated fortune
    Shall keep us both the safer. Where we are
    There's daggers in men's smiles. The nea'er in blood,
    The nearer bloody.
MALCOLM           This murderous shaft that's shot
    Hath not yet lighted, and our safest way    141
    Is to avoid the aim. Therefore to horse,
    And let us not be dainty of leave-taking,
    But shift away. There's warrant in that theft
    Which steals itself when there's no mercy left.    *Exeunt*

**2.4**   *Enter Ross with an Old Man*
OLD MAN
    Threescore and ten I can remember well,
    Within the volume of which time I have seen
    Hours dreadful and things strange, but this sore night
    Hath trifled former knowings.
ROSS               Ha, good father,
    Thou seest the heavens, as troubled with man's act,  5
    Threatens his bloody stage. By th' clock 'tis day,
    And yet dark night strangles the travelling lamp.
    Is't night's predominance or the day's shame
    That darkness does the face of earth entomb
    When living light should kiss it?
OLD MAN            'Tis unnatural,   10
    Even like the deed that's done. On Tuesday last
    A falcon, tow'ring in her pride of place,
    Was by a mousing owl hawked at and killed.
ROSS
    And Duncan's horses—a thing most strange and
        certain—
    Beauteous and swift, the minions of their race,   15
    Turned wild in nature, broke their stalls, flung out,
    Contending 'gainst obedience, as they would
    Make war with mankind.
OLD MAN          'Tis said they ate each other.
ROSS
    They did so, to th'amazement of mine eyes
    That looked upon't.
    *Enter Macduff*
           Here comes the good Macduff.  20
    How goes the world, sir, now?
MACDUFF          Why, see you not?
ROSS
    Is't known who did this more than bloody deed?
MACDUFF
    Those that Macbeth hath slain.
ROSS             Alas the day,
    What good could they pretend?
MACDUFF          They were suborned.
    Malcolm and Donalbain, the King's two sons,  25
    Are stol'n away and fled, which puts upon them
    Suspicion of the deed.
ROSS           'Gainst nature still.
    Thriftless ambition, that will raven up
    Thine own life's means! Then 'tis most like
    The sovereignty will fall upon Macbeth.  30
MACDUFF
    He is already named and gone to Scone
    To be invested.
ROSS    Where is Duncan's body?
MACDUFF    Carried to Colmekill,
    The sacred storehouse of his predecessors,  35
    And guardian of their bones.
ROSS              Will you to Scone?

MACDUFF
    No, cousin, I'll to Fife.
ROSS           Well, I will thither.
MACDUFF
    Well, may you see things well done there. Adieu,
    Lest our old robes sit easier than our new.
ROSS   Farewell, father.    40
OLD MAN
    God's benison go with you, and with those
    That would make good of bad, and friends of foes.
                     *Exeunt severally*

**3.1**   *Enter Banquo*
BANQUO
    Thou hast it now: King, Cawdor, Glamis, all
    As the weird women promised; and I fear
    Thou played'st most foully for't. Yet it was said
    It should not stand in thy posterity,
    But that myself should be the root and father  5
    Of many kings. If there come truth from them—
    As upon thee, Macbeth, their speeches shine—
    Why by the verities on thee made good
    May they not be my oracles as well,
    And set me up in hope? But hush, no more.   10
    *Sennet sounded. Enter Macbeth as King, Lady Macbeth*
    *as Queen, Lennox, Ross, lords, and attendants*
MACBETH
    Here's our chief guest.
LADY MACBETH      If he had been forgotten
    It had been as a gap in our great feast,
    And all-thing unbecoming.
MACBETH (*to Banquo*)
    Tonight we hold a solemn supper, sir,
    And I'll request your presence.
BANQUO           Let your highness  15
    Command upon me, to the which my duties
    Are with a most indissoluble tie
    For ever knit.
MACBETH    Ride you this afternoon?
BANQUO    Ay, my good lord.    20
MACBETH
    We should have else desired your good advice,
    Which still hath been both grave and prosperous,
    In this day's council; but we'll talk tomorrow.
    Is't far you ride?
BANQUO
    As far, my lord, as will fill up the time   25
    'Twixt this and supper. Go not my horse the better,
    I must become a borrower of the night
    For a dark hour or twain.
MACBETH    Fail not our feast.
BANQUO    My lord, I will not.    30
MACBETH
    We hear our bloody cousins are bestowed
    In England and in Ireland, not confessing
    Their cruel parricide, filling their hearers
    With strange invention. But of that tomorrow,
    When therewithal we shall have cause of state  35
    Craving us jointly. Hie you to horse. Adieu,
    Till you return at night. Goes Fleance with you?
BANQUO
    Ay, my good lord. Our time does call upon 's.
MACBETH
    I wish your horses swift and sure of foot,
    And so I do commend you to their backs.  40
    Farewell.                      *Exit Banquo*

Let every man be master of his time
Till seven at night. To make society
The sweeter welcome, we will keep ourself
Till supper-time alone. While then, God be with you.
                      *Exeunt all but Macbeth and a Servant*
Sirrah, a word with you. Attend those men     46
Our pleasure?
SERVANT
They are, my lord, without the palace gate.
MACBETH
Bring them before us.               *Exit Servant*
           To be thus is nothing
But to be safely thus. Our fears in Banquo     50
Stick deep, and in his royalty of nature
Reigns that which would be feared. 'Tis much he dares,
And to that dauntless temper of his mind
He hath a wisdom that doth guide his valour
To act in safety. There is none but he     55
Whose being I do fear, and under him
My genius is rebuked as, it is said,
Mark Antony's was by Caesar. He chid the sisters
When first they put the name of king upon me,
And bade them speak to him. Then, prophet-like,     60
They hailed him father to a line of kings.
Upon my head they placed a fruitless crown,
And put a barren sceptre in my grip,
Thence to be wrenched with an unlineal hand,
No son of mine succeeding. If't be so,     65
For Banquo's issue have I filed my mind,
For them the gracious Duncan have I murdered,
Put rancours in the vessel of my peace
Only for them, and mine eternal jewel
Given to the common enemy of man     70
To make them kings, the seeds of Banquo kings.
Rather than so, come fate into the list
And champion me to th'utterance. Who's there?
    *Enter Servant and two Murderers*
(*To the Servant*) Now go to the door, and stay there till
    we call.              *Exit Servant*
Was it not yesterday we spoke together?     75
MURDERERS
It was, so please your highness.
MACBETH             Well then, now
Have you considered of my speeches? Know
That it was he in the times past which held you
So under fortune, which you thought had been
Our innocent self. This I made good to you     80
In our last conference, passed in probation with you
How you were borne in hand, how crossed, the
    instruments,
Who wrought with them, and all things else that
    might
To half a soul, and to a notion crazed,
Say 'Thus did Banquo'.
FIRST MURDERER        You made it known to us.   85
MACBETH
I did so, and went further, which is now
Our point of second meeting. Do you find
Your patience so predominant in your nature
That you can let this go? Are you so gospelled
To pray for this good man and for his issue,     90
Whose heavy hand hath bowed you to the grave
And beggared yours for ever?
FIRST MURDERER         We are men, my liege.
MACBETH
Ay, in the catalogue ye go for men,

As hounds and greyhounds, mongrels, spaniels, curs,
Shoughs, water-rugs, and demi-wolves are clept     95
All by the name of dogs. The valued file
Distinguishes the swift, the slow, the subtle,
The housekeeper, the hunter, every one
According to the gift which bounteous nature
Hath in him closed; whereby he does receive     100
Particular addition from the bill
That writes them all alike. And so of men.
Now, if you have a station in the file,
Not i'th' worst rank of manhood, say't,
And I will put that business in your bosoms     105
Whose execution takes your enemy off,
Grapples you to the heart and love of us,
Who wear our health but sickly in his life,
Which in his death were perfect.
SECOND MURDERER         I am one, my liege,
Whom the vile blows and buffets of the world     110
Hath so incensed that I am reckless what
I do to spite the world.
FIRST MURDERER        And I another,
So weary with disasters, tugged with fortune,
That I would set my life on any chance
To mend it or be rid on't.
MACBETH            Both of you     115
Know Banquo was your enemy.
MURDERERS             True, my lord.
MACBETH
So is he mine, and in such bloody distance
That every minute of his being thrusts
Against my near'st of life; and though I could
With barefaced power sweep him from my sight     120
And bid my will avouch it, yet I must not,
For certain friends that are both his and mine,
Whose loves I may not drop, but wail his fall
Who I myself struck down. And thence it is
That I to your assistance do make love,     125
Masking the business from the common eye
For sundry weighty reasons.
SECOND MURDERER        We shall, my lord,
Perform what you command us.
FIRST MURDERER            Though our lives—
MACBETH
Your spirits shine through you. Within this hour at most
I will advise you where to plant yourselves,     130
Acquaint you with the perfect spy o'th' time,
The moment on't; for't must be done tonight,
And something from the palace; always thought
That I require a clearness; and with him,
To leave no rubs nor botches in the work,     135
Fleance, his son, that keeps him company—
Whose absence is no less material to me
Than is his father's—must embrace the fate
Of that dark hour. Resolve yourselves apart.
I'll come to you anon.
MURDERERS           We are resolved, my lord.   140
MACBETH
I'll call upon you straight. Abide within.
                    *Exeunt Murderers*
It is concluded. Banquo, thy soul's flight,
If it find heaven, must find it out tonight.     *Exit*

**3.2**    *Enter Lady Macbeth and a Servant*
LADY MACBETH Is Banquo gone from court?
SERVANT
    Ay, madam, but returns again tonight.

LADY MACBETH
  Say to the King I would attend his leisure
  For a few words.
SERVANT  Madam, I will.                                          *Exit*
LADY MACBETH  Naught's had, all's spent,                              6
  Where our desire is got without content.
  'Tis safer to be that which we destroy
  Than by destruction dwell in doubtful joy.
      *Enter Macbeth*
  How now, my lord, why do you keep alone,                         10
  Of sorriest fancies your companions making,
  Using those thoughts which should indeed have died
  With them they think on? Things without all remedy
  Should be without regard. What's done is done.
MACBETH
  We have scorched the snake, not killed it.                       15
  She'll close and be herself, whilst our poor malice
  Remains in danger of her former tooth.
  But let the frame of things disjoint, both the worlds
        suffer,
  Ere we will eat our meal in fear, and sleep
  In the affliction of these terrible dreams                       20
  That shake us nightly. Better be with the dead,
  Whom we to gain our peace have sent to peace,
  Than on the torture of the mind to lie
  In restless ecstasy. Duncan is in his grave.
  After life's fitful fever he sleeps well.                        25
  Treason has done his worst. Nor steel nor poison,
  Malice domestic, foreign levy, nothing
  Can touch him further.
LADY MACBETH                  Come on, gentle my lord,
  Sleek o'er your rugged looks, be bright and jovial
  Among your guests tonight.
MACBETH                       So shall I, love,                    30
  And so I pray be you. Let your remembrance
  Apply to Banquo. Present him eminence
  Both with eye and tongue; unsafe the while that we
  Must lave our honours in these flattering streams
  And make our faces visors to our hearts,                         35
  Disguising what they are.
LADY MACBETH                 You must leave this.
MACBETH
  O, full of scorpions is my mind, dear wife!
  Thou know'st that Banquo and his Fleance lives.
LADY MACBETH
  But in them nature's copy's not eterne.
MACBETH
  There's comfort yet, they are assailable.                        40
  Then be thou jocund. Ere the bat hath flown
  His cloistered flight, ere to black Hecate's summons
  The shard-borne beetle with his drowsy hums
  Hath rung night's yawning peal, there shall be done
  A deed of dreadful note.
LADY MACBETH               What's to be done?                      45
MACBETH
  Be innocent of the knowledge, dearest chuck,
  Till thou applaud the deed.—Come, seeling night,
  Scarf up the tender eye of pitiful day,
  And with thy bloody and invisible hand
  Cancel and tear to pieces that great bond                        50
  Which keeps me pale. Light thickens, and the crow
  Makes wing to th' rooky wood.
  Good things of day begin to droop and drowse,
  Whiles night's black agents to their preys do rouse.
  Thou marvell'st at my words; but hold thee still.                55
  Things bad begun make strong themselves by ill.
  So prithee go with me.                                    *Exeunt*

**3.3**  *Enter three Murderers*
FIRST MURDERER (*to Third Murderer*)
  But who did bid thee join with us?
THIRD MURDERER                         Macbeth.
SECOND MURDERER (*to First Murderer*)
  He needs not our mistrust, since he delivers
  Our offices and what we have to do
  To the direction just.
FIRST MURDERER (*to Third Murderer*)  Then stand with us.
  The west yet glimmers with some streaks of day.                  5
  Now spurs the lated traveller apace
  To gain the timely inn, and near approaches
  The subject of our watch.
THIRD MURDERER                 Hark, I hear horses.
BANQUO (*within*)
  Give us a light there, ho!
SECOND MURDERER                Then 'tis he. The rest
  That are within the note of expectation                          10
  Already are i'th' court.
FIRST MURDERER              His horses go about.
THIRD MURDERER
  Almost a mile; but he does usually,
  So all men do, from hence to th' palace gate
  Make it their walk.
      *Enter Banquo and Fleance with a torch*
SECOND MURDERER (*aside*)  A light, a light.
THIRD MURDERER (*aside*)                        'Tis he.
FIRST MURDERER (*aside*)  Stand to't.                              15
BANQUO
  It will be rain tonight.
FIRST MURDERER              Let it come down.
      *First Murderer strikes out the torch. The others
        attack Banquo*
BANQUO
  O, treachery! Fly, good Fleance, fly, fly, fly!
  Thou mayst revenge.—O slave!   *He dies. Exit Fleance*
THIRD MURDERER  Who did strike out the light?
FIRST MURDERER  Was't not the way?                                 20
THIRD MURDERER
  There's but one down. The son is fled.
SECOND MURDERER
  We have lost best half of our affair.
FIRST MURDERER
  Well, let's away and say how much is done.
                              *Exeunt with Banquo's body*

**3.4**  *Banquet prepared. Enter Macbeth as King, Lady
        Macbeth as Queen, Ross, Lennox, Lords, and
        attendants. ⌈Lady Macbeth sits⌉*
MACBETH
  You know your own degrees; sit down. At first and last
  The hearty welcome.
LORDS                  Thanks to your majesty.
      *They sit*
MACBETH
  Ourself will mingle with society
  And play the humble host. Our hostess keeps her
        state,
  But in best time we will require her welcome.                    5
LADY MACBETH
  Pronounce it for me, sir, to all our friends,
  For my heart speaks they are welcome.
      *Enter First Murderer ⌈to the door⌉*
MACBETH
  See, they encounter thee with their hearts' thanks.
  Both sides are even. Here I'll sit, i'th' midst.

Be large in mirth. Anon we'll drink a measure       10
The table round. (*To First Murderer*) There's blood
      upon thy face.
FIRST MURDERER (*aside to Macbeth*) 'Tis Banquo's, then.
MACBETH
'Tis better thee without than he within.
Is he dispatched?
FIRST MURDERER
My lord, his throat is cut. That I did for him.       15
MACBETH
Thou art the best o'th' cut-throats. Yet he's good
That did the like for Fleance. If thou didst it,
Thou art the nonpareil.
FIRST MURDERER                    Most royal sir,
Fleance is scaped.
MACBETH
Then comes my fit again; I had else been perfect,    20
Whole as the marble, founded as the rock,
As broad and general as the casing air,
But now I am cabined, cribbed, confined, bound in
To saucy doubts and fears. But Banquo's safe?
FIRST MURDERER
Ay, my good lord. Safe in a ditch he bides,       25
With twenty trenchèd gashes on his head,
The least a death to nature.
MACBETH                       Thanks for that.
There the grown serpent lies. The worm that's fled
Hath nature that in time will venom breed,
No teeth for th' present. Get thee gone. Tomorrow   30
We'll hear ourselves again.       *Exit First Murderer*
LADY MACBETH                     My royal lord,
You do not give the cheer. The feast is sold
That is not often vouched, while 'tis a-making,
'Tis given with welcome. To feed were best at home.
From thence the sauce to meat is ceremony,       35
Meeting were bare without it.
      *Enter the Ghost of Banquo, and sits in Macbeth's
      place*
MACBETH                     Sweet remembrancer.
Now good digestion wait on appetite,
And health on both.
LENNOX              May't please your highness sit?
MACBETH
Here had we now our country's honour roofed
Were the graced person of our Banquo present,    40
Who may I rather challenge for unkindness
Than pity for mischance.
ROSS                   His absence, sir,
Lays blame upon his promise. Please't your highness
To grace us with your royal company?
MACBETH
The table's full.
LENNOX           Here is a place reserved, sir.       45
MACBETH  Where?
LENNOX
Here, my good lord. What is't that moves your
      highness?
MACBETH
Which of you have done this?
LORDS                     What, my good lord?
MACBETH (*to the Ghost*)
Thou canst not say I did it. Never shake
Thy gory locks at me.       50
ROSS (*rising*)
Gentlemen, rise. His highness is not well.

LADY MACBETH (*rising*)
Sit, worthy friends. My lord is often thus,
And hath been from his youth. Pray you, keep seat.
The fit is momentary. Upon a thought
He will again be well. If much you note him       55
You shall offend him, and extend his passion.
Feed, and regard him not.
      *She speaks apart with Macbeth*
                     Are you a man?
MACBETH
Ay, and a bold one, that dare look on that
Which might appal the devil.
LADY MACBETH                    O proper stuff!
This is the very painting of your fear;       60
This is the air-drawn dagger which you said
Led you to Duncan. O, these flaws and starts,
Impostors to true fear, would well become
A woman's story at a winter's fire
Authorized by her grandam. Shame itself,       65
Why do you make such faces? When all's done
You look but on a stool.
MACBETH
Prithee see there. Behold, look, lo—how say you?
Why, what care I? If thou canst nod, speak, too!
If charnel-houses and our graves must send       70
Those that we bury back, our monuments
Shall be the maws of kites.                *Exit Ghost*
LADY MACBETH              What, quite unmanned in folly?
MACBETH
If I stand here, I saw him.
LADY MACBETH                    Fie, for shame!
MACBETH
Blood hath been shed ere now, i'th' olden time,
Ere human statute purged the gentle weal;       75
Ay, and since, too, murders have been performed
Too terrible for the ear. The time has been
That, when the brains were out, the man would die,
And there an end. But now they rise again
With twenty mortal murders on their crowns,       80
And push us from our stools. This is more strange
Than such a murder is.
LADY MACBETH (*aloud*)      My worthy lord,
Your noble friends do lack you.
MACBETH                     I do forget.
Do not muse at me, my most worthy friends.
I have a strange infirmity which is nothing       85
To those that know me. Come, love and health to all,
Then I'll sit down.
(*To an attendant*)   Give me some wine. Fill full.
      *Enter Ghost*
I drink to th' general joy of th' whole table,
And to our dear friend Banquo, whom we miss.
Would he were here. To all and him we thirst,       90
And all to all.
LORDS              Our duties, and the pledge.
      *They drink*
MACBETH (*seeing the Ghost*)
Avaunt, and quit my sight! Let the earth hide thee.
Thy bones are marrowless, thy blood is cold.
Thou hast no speculation in those eyes
Which thou dost glare with.
LADY MACBETH              Think of this, good peers,
But as a thing of custom. 'Tis no other;       96
Only it spoils the pleasure of the time.
MACBETH  What man dare, I dare.
Approach thou like the ruggèd Russian bear,

The armed rhinoceros, or th'Hyrcan tiger;          100
Take any shape but that, and my firm nerves
Shall never tremble. Or be alive again,
And dare me to the desert with thy sword.
If trembling I inhabit then, protest me
The baby of a girl. Hence, horrible shadow,          105
Unreal mock'ry, hence!                    *Exit Ghost*
                    Why so, being gone,
I am a man again. Pray you sit still.

LADY MACBETH
You have displaced the mirth, broke the good meeting
With most admired disorder.

MACBETH                              Can such things be
And overcome us like a summer's cloud,          110
Without our special wonder? You make me strange
Even to the disposition that I owe,
When now I think you can behold such sights
And keep the natural ruby of your cheeks
When mine is blanched with fear.

ROSS                                    What sights, my lord?

LADY MACBETH
I pray you, speak not. He grows worse and worse.     116
Question enrages him. At once, good night.
Stand not upon the order of your going,
But go at once.

LENNOX              Good night, and better health
Attend his majesty.

LADY MACBETH              A kind good-night to all.     120
                              *Exeunt Lords*

MACBETH
It will have blood, they say. Blood will have blood.
Stones have been known to move, and trees to speak,
Augurs and understood relations have
By maggot-pies and choughs and rooks brought forth
The secret'st man of blood. What is the night?     125

LADY MACBETH
Almost at odds with morning, which is which.

MACBETH
How sayst thou that Macduff denies his person
At our great bidding?

LADY MACBETH              Did you send to him, sir?

MACBETH
I hear it by the way, but I will send.
There's not a one of them but in his house          130
I keep a servant fee'd. I will tomorrow,
And betimes I will, to the weird sisters.
More shall they speak, for now I am bent to know
By the worst means the worst. For mine own good
All causes shall give way. I am in blood          135
Stepped in so far that, should I wade no more,
Returning were as tedious as go o'er.
Strange things I have in head that will to hand,
Which must be acted ere they may be scanned.

LADY MACBETH
You lack the season of all natures, sleep.          140

MACBETH
Come, we'll to sleep. My strange and self-abuse
Is the initiate fear that wants hard use.
We are yet but young in deed.                    *Exeunt*

**3.5**    *Thunder. Enter the three Witches meeting Hecate*
FIRST WITCH
Why, how now, Hecate? You look angerly.

HECATE
Have I not reason, beldams as you are?
Saucy and over-bold, how did you dare

To trade and traffic with Macbeth
In riddles and affairs of death,                    5
And I, the mistress of your charms,
The close contriver of all harms,
Was never called to bear my part
Or show the glory of our art?—
And, which is worse, all you have done          10
Hath been but for a wayward son,
Spiteful and wrathful, who, as others do,
Loves for his own ends, not for you.
But make amends now. Get you gone,
And at the pit of Acheron                    15
Meet me i'th' morning. Thither he
Will come to know his destiny.
Your vessels and your spells provide,
Your charms and everything beside.
I am for th'air. This night I'll spend          20
Unto a dismal and a fatal end.
Great business must be wrought ere noon.
Upon the corner of the moon
There hangs a vap'rous drop profound.
I'll catch it ere it come to ground,          25
And that, distilled by magic sleights,
Shall raise such artificial sprites
As by the strength of their illusion
Shall draw him on to his confusion.
He shall spurn fate, scorn death, and bear          30
His hopes 'bove wisdom, grace, and fear;
And you all know security
Is mortals' chiefest enemy.

SPIRITS (*singing dispersedly within*)
Come away, come away.
Hecate, Hecate, come away.                    35

HECATE
Hark, I am called! My little spirit, see,
Sits in a foggy cloud and stays for me.

                              *The Song*

SPIRITS ⌈*within*⌉
Come away, come away,
Hecate, Hecate, come away.

HECATE
I come, I come, I come, I come,                    40
With all the speed I may,
With all the speed I may.
Where's Stadlin?

SPIRIT ⌈*within*⌉                    Here.

HECATE                              Where's Puckle?

ANOTHER SPIRIT ⌈*within*⌉                         Here.

OTHER SPIRITS ⌈*within*⌉
And Hoppo, too, and Hellwain, too,
We lack but you, we lack but you.                    45
Come away, make up the count.

HECATE
I will but 'noint, and then I mount.
⌈*Spirits appear above.*⌉ *A Spirit like a Cat descends*

SPIRITS ⌈*above*⌉
There's one comes down to fetch his dues,
A kiss, a coll, a sip of blood,
And why thou stay'st so long I muse, I muse,     50
Since the air's so sweet and good.

HECATE
O, art thou come? What news, what news?

SPIRIT LIKE A CAT
All goes still to our delight.
Either come, or else refuse, refuse.

HECATE    Now I am furnished for the flight.          55

*She ascends with the spirit and sings*
Now I go, now I fly,
Malkin my sweet spirit and I.
⌈SPIRITS *and* HECATE⌉
O what a dainty pleasure 'tis
    To ride in the air
    When the moon shines fair,                    60
And sing, and dance, and toy, and kiss.
Over woods, high rocks and mountains,
Over seas and misty fountains,
Over steeples, towers and turrets,
We fly by night 'mongst troops of spirits.       65
No ring of bells to our ears sounds,
No howls of wolves, no yelps of hounds.
No, not the noise of waters-breach
Or cannons' throat our height can reach.
SPIRITS ⌈*above*⌉
No ring of bells to our ears sounds,
No howls of wolves, no yelps of hounds.          70
No, not the noise of waters-breach
Or cannons' throat our height can reach.
                    *Exeunt into the heavens the*
                    *Spirit like a Cat and Hecate*
FIRST WITCH
Come, let's make haste. She'll soon be back again.
                                        *Exeunt*

**3.6**    *Enter Lennox and another Lord*
LENNOX
My former speeches have but hit your thoughts,
Which can interpret farther. Only I say
Things have been strangely borne. The gracious
    Duncan
Was pitied of Macbeth: marry, he was dead;
And the right valiant Banquo walked too late,    5
Whom you may say, if't please you, Fleance killed,
For Fleance fled: men must not walk too late.
Who cannot want the thought how monstrous
It was for Malcolm and for Donalbain
To kill their gracious father? Damnèd fact,       10
How it did grieve Macbeth! Did he not straight
In pious rage the two delinquents tear,
That were the slaves of drink, and thralls of sleep?
Was not that nobly done? Ay, and wisely too,
For 'twould have angered any heart alive          15
To hear the men deny't. So that I say
He has borne all things well, and I do think
That had he Duncan's sons under his key—
As, an't please heaven, he shall not—they should find
What 'twere to kill a father. So should Fleance.  20
But peace, for from broad words, and 'cause he failed
His presence at the tyrant's feast, I hear
Macduff lives in disgrace. Sir, can you tell
Where he bestows himself?
LORD                    The son of Duncan
From whom this tyrant holds the due of birth     25
Lives in the English court, and is received
Of the most pious Edward with such grace
That the malevolence of fortune nothing
Takes from his high respect. Thither Macduff
Is gone to pray the holy King upon his aid        30
To wake Northumberland and warlike Siward,
That by the help of these—with Him above
To ratify the work—we may again
Give to our tables meat, sleep to our nights,

Free from our feasts and banquets bloody knives,  35
Do faithful homage, and receive free honours,
All which we pine for now. And this report
Hath so exasperate their king that he
Prepares for some attempt of war.
LENNOX Sent he to Macduff?                         40
LORD
He did, and with an absolute 'Sir, not I,'
The cloudy messenger turns me his back
And hums, as who should say 'You'll rue the time
That clogs me with this answer.'
LENNOX                    And that well might
Advise him to a caution t'hold what distance      45
His wisdom can provide. Some holy angel
Fly to the court of England and unfold
His message ere he come, that a swift blessing
May soon return to this our suffering country
Under a hand accursed.
LORD                    I'll send my prayers with him.
                                        *Exeunt*

                    ❀

**4.1**    *A Cauldron. Thunder. Enter the three Witches*
FIRST WITCH
Thrice the brinded cat hath mewed.
SECOND WITCH
Thrice, and once the hedge-pig whined.
THIRD WITCH
Harpier cries ''Tis time, 'tis time.'
FIRST WITCH
Round about the cauldron go,
In the poisoned entrails throw.                    5
Toad that under cold stone
Days and nights has thirty-one
Sweltered venom sleeping got,
Boil thou first i'th' charmèd pot.
ALL
Double, double, toil and trouble,                  10
Fire burn, and cauldron bubble.
SECOND WITCH
Fillet of a fenny snake,
In the cauldron boil and bake.
Eye of newt and toe of frog,
Wool of bat and tongue of dog,                     15
Adder's fork and blind-worm's sting,
Lizard's leg and owlet's wing,
For a charm of powerful trouble,
Like a hell-broth boil and bubble.
ALL
Double, double, toil and trouble,                  20
Fire burn, and cauldron bubble.
THIRD WITCH
Scale of dragon, tooth of wolf,
Witches' mummy, maw and gulf
Of the ravined salt-sea shark,
Root of hemlock digged i'th' dark,                 25
Liver of blaspheming Jew,
Gall of goat, and slips of yew
Slivered in the moon's eclipse,
Nose of Turk, and Tartar's lips,
Finger of birth-strangled babe                     30
Ditch-delivered by a drab,
Make the gruel thick and slab.
Add thereto a tiger's chaudron
For th'ingredience of our cauldron.

ALL
      Double, double, toil and trouble,                    35
      Fire burn, and cauldron bubble.
SECOND WITCH
      Cool it with a baboon's blood,
      Then the charm is firm and good.
            *Enter Hecate and the other three Witches*
HECATE
      O, well done! I commend your pains,
      And everyone shall share i'th' gains.               40
      And now about the cauldron sing
      Like elves and fairies in a ring,
      Enchanting all that you put in.
            *Music and a song*
HECATE
      Black spirits and white, red spirits and grey,
      Mingle, mingle, mingle, you that mingle may.        45
FOURTH WITCH
      Titty, Tiffin, keep it stiff in;
      Firedrake, Puckey, make it lucky;
      Liard, Robin, you must bob in.
ALL   Round, around, around, about, about,
      All ill come running in, all good keep out.         50
FOURTH WITCH
      Here's the blood of a bat.
HECATE
      Put in that, O put in that!
FIFTH WITCH
      Here's leopard's bane.
HECATE
      Put in a grain.
FOURTH WITCH
      The juice of toad, the oil of adder.                55
FIFTH WITCH
      Those will make the younker madder.
HECATE
      Put in, there's all, and rid the stench.
A WITCH
      Nay, here's three ounces of a red-haired wench.
ALL   Round, around, around, about, about,
      All ill come running in, all good keep out.         60
SECOND WITCH
      By the pricking of my thumbs,
      Something wicked this way comes.
            ⌈*Knock within*⌉
      Open, locks, whoever knocks.
            *Enter Macbeth*
MACBETH
      How now, you secret, black, and midnight hags,
      What is't you do?
ALL THE WITCHES      A deed without a name.               65
MACBETH
      I conjure you by that which you profess,
      Howe'er you come to know it, answer me.
      Though you untie the winds and let them fight
      Against the churches, though the yeasty waves
      Confound and swallow navigation up,                 70
      Though bladed corn be lodged and trees blown down,
      Though castles topple on their warders' heads,
      Though palaces and pyramids do slope
      Their heads to their foundations, though the treasure
      Of nature's germens tumble all together             75
      Even till destruction sicken, answer me
      To what I ask you.
FIRST WITCH          Speak.
SECOND WITCH               Demand.
THIRD WITCH                         We'll answer.

FIRST WITCH
      Say if thou'dst rather hear it from our mouths
      Or from our masters.
MACBETH                   Call 'em, let me see 'em.
FIRST WITCH
      Pour in sow's blood that hath eaten                 80
      Her nine farrow; grease that's sweaten
      From the murderer's gibbet throw
      Into the flame.
ALL THE WITCHES      Come high or low,
      Thyself and office deftly show.
            *Thunder. First Apparition: an armed head*
MACBETH
      Tell me, thou unknown power—
FIRST WITCH                        He knows thy thought.
      Hear his speech, but say thou naught.               86
FIRST APPARITION
      Macbeth, Macbeth, Macbeth, beware Macduff,
      Beware the Thane of Fife. Dismiss me. Enough.
            *Apparition descends*
MACBETH
      Whate'er thou art, for thy good caution thanks.
      Thou hast harped my fear aright. But one word
            more—                                         90
FIRST WITCH
      He will not be commanded. Here's another,
      More potent than the first.
            *Thunder. Second Apparition: a bloody child*
SECOND APPARITION   Macbeth, Macbeth, Macbeth.
MACBETH   Had I three ears I'd hear thee.
SECOND APPARITION
      Be bloody, bold, and resolute. Laugh to scorn       95
      The power of man, for none of woman born
      Shall harm Macbeth.
            *Apparition descends*
MACBETH
      Then live, Macduff—what need I fear of thee?
      But yet I'll make assurance double sure,
      And take a bond of fate thou shalt not live,        100
      That I may tell pale-hearted fear it lies,
      And sleep in spite of thunder.
            *Thunder. Third Apparition: a child crowned, with a
            tree in his hand*
                              What is this
      That rises like the issue of a king,
      And wears upon his baby-brow the round
      And top of sovereignty?
ALL THE WITCHES               Listen, but speak not to't.  105
THIRD APPARITION
      Be lion-mettled, proud, and take no care
      Who chafes, who frets, or where conspirers are.
      Macbeth shall never vanquished be until
      Great Birnam Wood to high Dunsinane Hill
      Shall come against him.
            *Apparition descends*
MACBETH               That will never be.                 110
      Who can impress the forest, bid the tree
      Unfix his earth-bound root? Sweet bodements, good!
      Rebellious dead, rise never till the wood
      Of Birnam rise, and on's high place Macbeth
      Shall live the lease of nature, pay his breath      115
      To time and mortal custom. Yet my heart
      Throbs to know one thing. Tell me, if your art
      Can tell so much, shall Banquo's issue ever
      Reign in this kingdom?
ALL THE WITCHES               Seek to know no more.

MACBETH
    I will be satisfied. Deny me this,                              120
    And an eternal curse fall on you! Let me know.
        *The cauldron sinks. Hautboys*
    Why sinks that cauldron? And what noise is this?
FIRST WITCH  Show.
SECOND WITCH  Show.
THIRD WITCH  Show.                                                 125
ALL THE WITCHES
        Show his eyes and grieve his heart,
        Come like shadows, so depart.
            *A show of eight kings, the last with a glass in his
            hand; and Banquo*
MACBETH
    Thou art too like the spirit of Banquo. Down!
    Thy crown does sear mine eyeballs. And thy hair,
    Thou other gold-bound brow, is like the first.                 130
    A third is like the former. Filthy hags,
    Why do you show me this?—A fourth? Start, eyes!
    What, will the line stretch out to th' crack of doom?
    Another yet? A seventh? I'll see no more—
    And yet the eighth appears, who bears a glass                  135
    Which shows me many more; and some I see
    That twofold balls and treble sceptres carry.
    Horrible sight! Now I see 'tis true,
    For the blood-baltered Banquo smiles upon me,
    And points at them for his.
                            *Exeunt kings and Banquo*
                    What, is this so?                              140
⌐HECATE⌐
        Ay, sir, all this is so. But why
        Stands Macbeth thus amazedly?
        Come, sisters, cheer we up his sprites,
        And show the best of our delights.
        I'll charm the air to give a sound                         145
        While you perform your antic round,
        That this great king may kindly say
        Our duties did his welcome pay.
            *Music. The Witches dance, and vanish*
MACBETH
    Where are they? Gone? Let this pernicious hour
    Stand aye accursèd in the calendar.                            150
    Come in, without there.
        *Enter Lennox*
LENNOX                      What's your grace's will?
MACBETH
    Saw you the weird sisters?
LENNOX                          No, my lord.
MACBETH
    Came they not by you?
LENNOX                      No, indeed, my lord.
MACBETH
    Infected be the air whereon they ride,
    And damned all those that trust them. I did hear              155
    The galloping of horse. Who was't came by?
LENNOX
    'Tis two or three, my lord, that bring you word
    Macduff is fled to England.
MACBETH                         Fled to England?
LENNOX  Ay, my good lord.
MACBETH  (*aside*)
    Time, thou anticipat'st my dread exploits.                    160
    The flighty purpose never is o'ertook
    Unless the deed go with it. From this moment
    The very firstlings of my heart shall be
    The firstlings of my hand. And even now,

To crown my thoughts with acts, be it thought and
    done:                                                          165
    The castle of Macduff I will surprise,
    Seize upon Fife, give to th'edge o'th' sword
    His wife, his babes, and all unfortunate souls
    That trace him in his line. No boasting like a fool;
    This deed I'll do before this purpose cool.                   170
    But no more sights! (*To Lennox*) Where are these
        gentlemen?
    Come bring me where they are.              *Exeunt*

**4.2**   *Enter Macduff's Wife, her Son, and Ross*
LADY MACDUFF
    What had he done to make him fly the land?
ROSS
    You must have patience, madam.
LADY MACDUFF                          He had none.
    His flight was madness. When our actions do not,
    Our fears do make us traitors.
ROSS                          You know not
    Whether it was his wisdom or his fear.                          5
LADY MACDUFF
    Wisdom—to leave his wife, to leave his babes,
    His mansion, and his titles in a place
    From whence himself does fly? He loves us not,
    He wants the natural touch, for the poor wren,
    The most diminutive of birds, will fight,                      10
    Her young ones in her nest, against the owl.
    All is the fear and nothing is the love;
    As little is the wisdom, where the flight
    So runs against all reason.
ROSS                          My dearest coz,
    I pray you school yourself. But for your husband,             15
    He is noble, wise, judicious, and best knows
    The fits o'th' season. I dare not speak much further,
    But cruel are the times when we are traitors
    And do not know ourselves; when we hold rumour
    From what we fear, yet know not what we fear,                 20
    But float upon a wild and violent sea
    Each way and none. I take my leave of you;
    Shall not be long but I'll be here again.
    Things at the worst will cease, or else climb upward
    To what they were before. My pretty cousin,                   25
    Blessing upon you!
LADY MACDUFF
    Fathered he is, and yet he's fatherless.
ROSS
    I am so much a fool, should I stay longer
    It would be my disgrace and your discomfort.
    I take my leave at once.                          *Exit*
LADY MACDUFF              Sirrah, your father's dead,             30
    And what will you do now? How will you live?
MACDUFF'S SON
    As birds do, mother.
LADY MACDUFF              What, with worms and flies?
MACDUFF'S SON
    With what I get, I mean, and so do they.
LADY MACDUFF
    Poor bird, thou'dst never fear the net nor lime,
    The pitfall nor the gin.                                       35
MACDUFF'S SON
    Why should I, mother? Poor birds they are not set for.
    My father is not dead, for all your saying.
LADY MACDUFF  Yes, he is dead. How wilt thou do for a
    father?
MACDUFF'S SON  Nay, how will you do for a husband?  40

LADY MACDUFF  Why, I can buy me twenty at any market.

MACDUFF'S SON  Then you'll buy 'em to sell again.

LADY MACDUFF  Thou speak'st with all thy wit, and yet, i'faith, with wit enough for thee.

MACDUFF'S SON  Was my father a traitor, mother?  45

LADY MACDUFF  Ay, that he was.

MACDUFF'S SON  What is a traitor?

LADY MACDUFF  Why, one that swears and lies.

MACDUFF'S SON  And be all traitors that do so?

LADY MACDUFF  Everyone that does so is a traitor, and must be hanged.  51

MACDUFF'S SON  And must they all be hanged that swear and lie?

LADY MACDUFF  Every one.

MACDUFF'S SON  Who must hang them?  55

LADY MACDUFF  Why, the honest men.

MACDUFF'S SON  Then the liars and swearers are fools, for there are liars and swearers enough to beat the honest men and hang up them.

LADY MACDUFF  Now God help thee, poor monkey! But how wilt thou do for a father?  61

MACDUFF'S SON  If he were dead you'd weep for him. If you would not, it were a good sign that I should quickly have a new father.

LADY MACDUFF  Poor prattler, how thou talk'st!  65

*Enter a Messenger*

MESSENGER
Bless you, fair dame. I am not to you known,
Though in your state of honour I am perfect.
I doubt some danger does approach you nearly.
If you will take a homely man's advice,
Be not found here. Hence with your little ones!  70
To fright you thus methinks I am too savage,
To do worse to you were fell cruelty,
Which is too nigh your person. Heaven preserve you.
I dare abide no longer.  *Exit Messenger*

LADY MACDUFF                           Whither should I fly?
I have done no harm. But I remember now  75
I am in this earthly world, where to do harm
Is often laudable, to do good sometime
Accounted dangerous folly. Why then, alas,
Do I put up that womanly defence
To say I have done no harm?

*Enter Murderers*

                                    What are these faces?

A MURDERER  Where is your husband?  81

LADY MACDUFF
I hope in no place so unsanctified
Where such as thou mayst find him.

A MURDERER                           He's a traitor.

MACDUFF'S SON
Thou liest, thou shag-haired villain.

A MURDERER  (*stabbing him*)          What, you egg!
Young fry of treachery!

MACDUFF'S SON          He has killed me, mother.  85
Run away, I pray you.

⌐*He dies.*⌐ *Exit Macduff's Wife crying 'Murder!'*
*followed by Murderers* ⌐*with the Son's body*⌐

**4.3**    *Enter Malcolm and Macduff*

MALCOLM
Let us seek out some desolate shade, and there
Weep our sad bosoms empty.

MACDUFF                           Let us rather
Hold fast the mortal sword, and like good men
Bestride our downfall birthdom. Each new morn

New widows howl, new orphans cry, new sorrows  5
Strike heaven on the face that it resounds
As if it felt with Scotland and yelled out
Like syllable of dolour.

MALCOLM                    What I believe I'll wail,
What know believe; and what I can redress,
As I shall find the time to friend, I will.  10
What you have spoke it may be so, perchance.
This tyrant, whose sole name blisters our tongues,
Was once thought honest. You have loved him well.
He hath not touched you yet. I am young, but something
You may discern of him through me: and wisdom  15
To offer up a weak poor innocent lamb
T'appease an angry god.

MACDUFF  I am not treacherous.

MALCOLM  But Macbeth is.
A good and virtuous nature may recoil  20
In an imperial charge. But I shall crave your pardon.
That which you are my thoughts cannot transpose.
Angels are bright still, though the brightest fell.
Though all things foul would wear the brows of grace,
Yet grace must still look so.

MACDUFF                    I have lost my hopes.  25

MALCOLM
Perchance even there where I did find my doubts.
Why in that rawness left you wife and child,
Those precious motives, those strong knots of love,
Without leave-taking? I pray you,
Let not my jealousies be your dishonours,  30
But mine own safeties. You may be rightly just,
Whatever I shall think.

MACDUFF                    Bleed, bleed, poor country!
Great tyranny, lay thou thy basis sure,
For goodness dare not check thee. Wear thou thy wrongs;
The title is affeered. Fare thee well, lord.  35
I would not be the villain that thou think'st
For the whole space that's in the tyrant's grasp,
And the rich east to boot.

MALCOLM                    Be not offended.
I speak not as in absolute fear of you.
I think our country sinks beneath the yoke.  40
It weeps, it bleeds, and each new day a gash
Is added to her wounds. I think withal
There would be hands uplifted in my right,
And here from gracious England have I offer
Of goodly thousands. But for all this,  45
When I shall tread upon the tyrant's head,
Or wear it on my sword, yet my poor country
Shall have more vices than it had before,
More suffer, and more sundry ways, than ever,
By him that shall succeed.

MACDUFF                    What should he be?  50

MALCOLM
It is myself I mean, in whom I know
All the particulars of vice so grafted
That when they shall be opened black Macbeth
Will seem as pure as snow, and the poor state
Esteem him as a lamb, being compared  55
With my confineless harms.

MACDUFF                    Not in the legions
Of horrid hell can come a devil more damned
In evils to top Macbeth.

MALCOLM                    I grant him bloody,
Luxurious, avaricious, false, deceitful,

Sudden, malicious, smacking of every sin                    60
That has a name. But there's no bottom, none,
In my voluptuousness. Your wives, your daughters,
Your matrons, and your maids could not fill up
The cistern of my lust, and my desire
All continent impediments would o'erbear                    65
That did oppose my will. Better Macbeth
Than such an one to reign.

MACDUFF                              Boundless intemperance
In nature is a tyranny. It hath been
Th'untimely emptying of the happy throne,
And fall of many kings. But fear not yet                    70
To take upon you what is yours. You may
Convey your pleasures in a spacious plenty
And yet seem cold. The time you may so hoodwink.
We have willing dames enough. There cannot be
That vulture in you to devour so many                       75
As will to greatness dedicate themselves,
Finding it so inclined.

MALCOLM                              With this there grows
In my most ill-composed affection such
A staunchless avarice that were I king
I should cut off the nobles for their lands,                80
Desire his jewels and this other's house,
And my more having would be as a sauce
To make me hunger more, that I should forge
Quarrels unjust against the good and loyal,
Destroying them for wealth.

MACDUFF                              This avarice            85
Sticks deeper, grows with more pernicious root
Than summer-seeming lust, and it hath been
The sword of our slain kings. Yet do not fear.
Scotland hath foisons to fill up your will
Of your mere own. All these are portable,                   90
With other graces weighed.

MALCOLM
But I have none. The king-becoming graces,
As justice, verity, temp'rance, stableness,
Bounty, perseverance, mercy, lowliness,
Devotion, patience, courage, fortitude,                     95
I have no relish of them, but abound
In the division of each several crime,
Acting it many ways. Nay, had I power I should
Pour the sweet milk of concord into hell,
Uproar the universal peace, confound                        100
All unity on earth.

MACDUFF                              O Scotland, Scotland!

MALCOLM
If such a one be fit to govern, speak.
I am as I have spoken.

MACDUFF                              Fit to govern?
No, not to live. O nation miserable,
With an untitled tyrant bloody-sceptered,                   105
When shalt thou see thy wholesome days again,
Since that the truest issue of thy throne
By his own interdiction stands accursed
And does blaspheme his breed? Thy royal father
Was a most sainted king. The Queen that bore thee,
Oft'ner upon her knees than on her feet,                    111
Died every day she lived. Fare thee well.
These evils thou repeat'st upon thyself
Hath banished me from Scotland. O, my breast—
Thy hope ends here!

MALCOLM                              Macduff, this noble passion,   115
Child of integrity, hath from my soul
Wiped the black scruples, reconciled my thoughts

To thy good truth and honour. Devilish Macbeth
By many of these trains hath sought to win me
Into his power, and modest wisdom plucks me                 120
From over-credulous haste; but God above
Deal between thee and me, for even now
I put myself to thy direction and
Unspeak mine own detraction, here abjure
The taints and blames I laid upon myself                    125
For strangers to my nature. I am yet
Unknown to woman, never was forsworn,
Scarcely have coveted what was mine own,
At no time broke my faith, would not betray
The devil to his fellow, and delight                        130
No less in truth than life. My first false-speaking
Was this upon myself. What I am truly
Is thine and my poor country's to command,
Whither indeed, before thy here-approach,
Old Siward with ten thousand warlike men,                   135
Already at a point, was setting forth.
Now we'll together; and the chance of goodness
Be like our warranted quarrel!—Why are you silent?

MACDUFF
Such welcome and unwelcome things at once
'Tis hard to reconcile.                                     140
                    *Enter a Doctor*

MALCOLM
Well, more anon. (*To the Doctor*) Comes the King
          forth, I pray you?

DOCTOR
Ay, sir. There are a crew of wretched souls
That stay his cure. Their malady convinces
The great essay of art, but at his touch,
Such sanctity hath Heaven given his hand,                   145
They presently amend.

MALCOLM                              I thank you, doctor. *Exit Doctor*

MACDUFF
What's the disease he means?

MALCOLM                              'Tis called the evil—
A most miraculous work in this good King,
Which often since my here-remain in England
I have seen him do. How he solicits heaven                  150
Himself best knows, but strangely visited people,
All swoll'n and ulcerous, pitiful to the eye,
The mere despair of surgery, he cures,
Hanging a golden stamp about their necks,
Put on with holy prayers; and 'tis spoken,                  155
To the succeeding royalty he leaves
The healing benediction. With this strange virtue
He hath a heavenly gift of prophecy,
And sundry blessings hang about his throne
That speak him full of grace.
                    *Enter Ross*

MACDUFF                              See who comes here.     160

MALCOLM
My countryman, but yet I know him not.

MACDUFF
My ever gentle cousin, welcome hither.

MALCOLM
I know him now. Good God betimes remove
The means that makes us strangers!

ROSS                              Sir, amen.

MACDUFF
Stands Scotland where it did?

ROSS                              Alas, poor country,          165
Almost afraid to know itself. It cannot
Be called our mother, but our grave, where nothing

But who knows nothing is once seen to smile;
Where sighs and groans and shrieks that rend the air
Are made, not marked; where violent sorrow seems
A modern ecstasy. The dead man's knell            171
Is there scarce asked for who, and good men's lives
Expire before the flowers in their caps,
Dying or ere they sicken.

MACDUFF                    O relation
Too nice and yet too true!

MALCOLM                    What's the newest grief?

ROSS
That of an hour's age doth hiss the speaker;        176
Each minute teems a new one.

MACDUFF                    How does my wife?

ROSS
Why, well.

MACDUFF    And all my children?

ROSS                         Well, too.

MACDUFF
The tyrant has not battered at their peace?

ROSS
No, they were well at peace when I did leave 'em.  180

MACDUFF
Be not a niggard of your speech. How goes't?

ROSS
When I came hither to transport the tidings
Which I have heavily borne, there ran a rumour
Of many worthy fellows that were out,
Which was to my belief witnessed the rather         185
For that I saw the tyrant's power afoot.
Now is the time of help. (To Malcolm) Your eye in
        Scotland
Would create soldiers, make our women fight
To doff their dire distresses.

MALCOLM                       Be't their comfort
We are coming thither. Gracious England hath       190
Lent us good Siward and ten thousand men;
An older and a better soldier none
That Christendom gives out.

ROSS                         Would I could answer
This comfort with the like. But I have words
That would be howled out in the desert air          195
Where hearing should not latch them.

MACDUFF                       What concern they—
The general cause, or is it a fee-grief
Due to some single breast?

ROSS                         No mind that's honest
But in it shares some woe, though the main part
Pertains to you alone.

MACDUFF                    If it be mine,            200
Keep it not from me; quickly let me have it.

ROSS
Let not your ears despise my tongue for ever,
Which shall possess them with the heaviest sound
That ever yet they heard.

MACDUFF                    H'm, I guess at it.

ROSS
Your castle is surprised, your wife and babes       205
Savagely slaughtered. To relate the manner
Were on the quarry of these murdered deer
To add the death of you.

MALCOLM                    Merciful heaven!
(To Macduff) What, man, ne'er pull your hat upon
        your brows.
Give sorrow words. The grief that does not speak    210
Whispers the o'erfraught heart and bids it break.

MACDUFF
My children too?

ROSS                Wife, children, servants, all
That could be found.

MACDUFF               And I must be from thence!
My wife killed too?

ROSS                I have said.

MALCOLM               Be comforted.
Let's make us medicines of our great revenge        215
To cure this deadly grief.

MACDUFF
He has no children. All my pretty ones?
Did you say all? O hell-kite! All?
What, all my pretty chickens and their dam
At one fell swoop?                                  220

MALCOLM  Dispute it like a man.

MACDUFF  I shall do so,
But I must also feel it as a man.
I cannot but remember such things were
That were most precious to me. Did heaven look on
And would not take their part? Sinful Macduff,      226
They were all struck for thee. Naught that I am,
Not for their own demerits but for mine
Fell slaughter on their souls. Heaven rest them now.

MALCOLM
Be this the whetstone of your sword. Let grief      230
Convert to anger: blunt not the heart, enrage it.

MACDUFF
O, I could play the woman with mine eyes
And braggart with my tongue! But gentle heavens
Cut short all intermission. Front to front
Bring thou this fiend of Scotland and myself.       235
Within my sword's length set him. If he scape,
Heaven forgive him too.

MALCOLM               This tune goes manly.
Come, go we to the King. Our power is ready;
Our lack is nothing but our leave. Macbeth
Is ripe for shaking, and the powers above           240
Put on their instruments. Receive what cheer you may:
The night is long that never finds the day.    Exeunt

❦

5.1    Enter a Doctor of Physic and a Waiting-
       Gentlewoman

DOCTOR  I have two nights watched with you, but can
perceive no truth in your report. When was it she last
walked?

GENTLEWOMAN  Since his majesty went into the field I have
seen her rise from her bed, throw her nightgown upon
her, unlock her closet, take forth paper, fold it, write
upon't, read it, afterwards seal it, and again return to
bed, yet all this while in a most fast sleep.        8

DOCTOR  A great perturbation in nature, to receive at once
the benefit of sleep and do the effects of watching. In
this slumbery agitation besides her walking and other
actual performances, what at any time have you heard
her say?                                             13

GENTLEWOMAN  That, sir, which I will not report after her.

DOCTOR  You may to me; and 'tis most meet you should.

GENTLEWOMAN  Neither to you nor anyone, having no
witness to confirm my speech.

       Enter Lady Macbeth with a taper

Lo you, here she comes. This is her very guise, and,
upon my life, fast asleep. Observe her. Stand close.

DOCTOR  How came she by that light?                  20

GENTLEWOMAN Why, it stood by her. She has light by her
continually. 'Tis her command.

DOCTOR You see her eyes are open.

GENTLEWOMAN Ay, but their sense are shut.                    24

DOCTOR What is it she does now? Look how she rubs her
hands.

GENTLEWOMAN It is an accustomed action with her, to
seem thus washing her hands. I have known her
continue in this a quarter of an hour.

LADY MACBETH Yet here's a spot.                              30

DOCTOR Hark, she speaks. I will set down what comes
from her to satisfy my remembrance the more strongly.

LADY MACBETH Out, damned spot; out, I say. One, two,—
why, then 'tis time to do't. Hell is murky. Fie, my lord,
fie, a soldier and afeard? What need we fear who knows
it when none can call our power to account? Yet who
would have thought the old man to have had so much
blood in him?

DOCTOR Do you mark that?                                     39

LADY MACBETH The Thane of Fife had a wife. Where is
she now? What, will these hands ne'er be clean? No
more o' that, my lord, no more o' that. You mar all
with this starting.

DOCTOR Go to, go to. You have known what you should
not.                                                         45

GENTLEWOMAN She has spoke what she should not, I am
sure of that. Heaven knows what she has known.

LADY MACBETH Here's the smell of the blood still. All the
perfumes of Arabia will not sweeten this little hand. O,
O, O!                                                        50

DOCTOR What a sigh is there! The heart is sorely charged.

GENTLEWOMAN I would not have such a heart in my
bosom for the dignity of the whole body.

DOCTOR Well, well, well.

GENTLEWOMAN Pray God it be, sir.                             55

DOCTOR This disease is beyond my practice. Yet I have
known those which have walked in their sleep who
have died holily in their beds.

LADY MACBETH Wash your hands, put on your nightgown,
look not so pale. I tell you yet again, Banquo's buried.
He cannot come out on's grave.                               61

DOCTOR Even so?

LADY MACBETH To bed, to bed. There's knocking at the
gate. Come, come, come, come, give me your hand.
What's done cannot be undone. To bed, to bed, to bed.

*Exit*

DOCTOR Will she go now to bed?                               66

GENTLEWOMAN Directly.

DOCTOR
Foul whisp'rings are abroad. Unnatural deeds
Do breed unnatural troubles; infected minds
To their deaf pillows will discharge their secrets.         70
More needs she the divine than the physician.
God, God forgive us all! Look after her.
Remove from her the means of all annoyance,
And still keep eyes upon her. So, good night.
My mind she has mated, and amazed my sight.                 75
I think, but dare not speak.

GENTLEWOMAN                    Good night, good doctor.

*Exeunt*

**5.2** *Enter Menteith, Caithness, Angus, Lennox, soldiers,*
*with a drummer and colours*

MENTEITH
The English power is near, led on by Malcolm,

His uncle Siward, and the good Macduff.
Revenges burn in them, for their dear causes
Would to the bleeding and the grim alarm
Excite the mortified man.

ANGUS                         Near Birnam Wood              5
Shall we well meet them. That way are they coming.

CAITHNESS
Who knows if Donalbain be with his brother?

LENNOX
For certain, sir, he is not. I have a file
Of all the gentry. There is Siward's son,
And many unrough youths that even now                      10
Protest their first of manhood.

MENTEITH                      What does the tyrant?

CAITHNESS
Great Dunsinane he strongly fortifies.
Some say he's mad, others that lesser hate him
Do call it valiant fury; but for certain
He cannot buckle his distempered cause                     15
Within the belt of rule.

ANGUS                         Now does he feel
His secret murders sticking on his hands.
Now minutely revolts upbraid his faith-breach.
Those he commands move only in command,
Nothing in love. Now does he feel his title                20
Hang loose about him, like a giant's robe
Upon a dwarfish thief.

MENTEITH                      Who then shall blame
His pestered senses to recoil and start
When all that is within him does condemn
Itself for being there?

CAITHNESS                     Well, march we on             25
To give obedience where 'tis truly owed.
Meet we the medicine of the sickly weal,
And with him pour we in our country's purge,
Each drop of us.

LENNOX                        Or so much as it needs
To dew the sovereign flower and drown the weeds.           30
Make we our march towards Birnam.

*Exeunt, marching*

**5.3** *Enter Macbeth, the Doctor of Physic, and attendants*

MACBETH
Bring me no more reports. Let them fly all.
Till Birnam Wood remove to Dunsinane
I cannot taint with fear. What's the boy Malcolm?
Was he not born of woman? The spirits that know
All mortal consequences have pronounced me thus:          5
'Fear not, Macbeth. No man that's born of woman
Shall e'er have power upon thee.' Then fly, false
thanes,
And mingle with the English epicures.
The mind I sway by and the heart I bear
Shall never sag with doubt nor shake with fear.           10

*Enter Servant*

The devil damn thee black, thou cream-faced loon!
Where gott'st thou that goose look?

SERVANT There is ten thousand—

MACBETH Geese, villain?

SERVANT                       Soldiers, sir.               15

MACBETH
Go prick thy face and over-red thy fear,
Thou lily-livered boy. What soldiers, patch?
Death of thy soul, those linen cheeks of thine
Are counsellors to fear. What soldiers, whey-face?

SERVANT The English force, so please you.                          20
MACBETH
Take thy face hence.                              *Exit Servant*
                    Seyton!—I am sick at heart
When I behold—Seyton, I say!—This push
Will cheer me ever or disseat me now.
I have lived long enough. My way of life
Is fall'n into the sere, the yellow leaf,                          25
And that which should accompany old age,
As honour, love, obedience, troops of friends,
I must not look to have, but in their stead
Curses, not loud but deep, mouth-honour, breath                    29
Which the poor heart would fain deny and dare not.
Seyton!
       *Enter Seyton*
SEYTON  What's your gracious pleasure?
MACBETH                                   What news more?
SEYTON
All is confirmed, my lord, which was reported.
MACBETH
I'll fight till from my bones my flesh be hacked.
Give me my armour.
SEYTON  'Tis not needed yet.                                       35
MACBETH  I'll put it on.
Send out more horses. Skirr the country round.
Hang those that talk of fear. Give me mine armour.
How does your patient, doctor?
DOCTOR                             Not so sick, my lord,
As she is troubled with thick-coming fancies                       40
That keep her from her rest.
MACBETH                      Cure her of that.
Canst thou not minister to a mind diseased,
Pluck from the memory a rooted sorrow,
Raze out the written troubles of the brain,
And with some sweet oblivious antidote                             45
Cleanse the fraught bosom of that perilous stuff
Which weighs upon the heart?
DOCTOR                        Therein the patient
Must minister to himself.
MACBETH
Throw physic to the dogs; I'll none of it.
(*To an attendant*) Come, put mine armour on. Give me
    my staff.                                                      50
Seyton, send out. Doctor, the thanes fly from me.
(*To an attendant*) Come, sir, dispatch.—If thou couldst,
    doctor, cast
The water of my land, find her disease,
And purge it to a sound and pristine health,
I would applaud thee to the very echo,                             55
That should applaud again. (*To an attendant*) Pull't off,
    I say.
(*To the Doctor*) What rhubarb, cyme, or what
    purgative drug
Would scour these English hence? Hear'st thou of
    them?
DOCTOR
Ay, my good lord. Your royal preparation
Makes us hear something.
MACBETH (*To an attendant*)   Bring it after me.                   60
I will not be afraid of death and bane
Till Birnam Forest come to Dunsinane.
DOCTOR (*aside*)
Were I from Dunsinane away and clear,
Profit again should hardly draw me here.
                                              *Exeunt*

5.4    *Enter Malcolm, Siward, Macduff, Siward's Son,*
       *Menteith, Caithness, Angus, and soldiers,*
       *marching, with a drummer and colours*
MALCOLM
Cousins, I hope the days are near at hand
That chambers will be safe.
MENTEITH                    We doubt it nothing.
SIWARD
What wood is this before us?
MENTEITH                     The wood of Birnam.
MALCOLM
Let every soldier hew him down a bough
And bear't before him. Thereby shall we shadow              5
The numbers of our host, and make discovery
Err in report of us.
A SOLDIER            It shall be done.
SIWARD
We learn no other but the confident tyrant
Keeps still in Dunsinane, and will endure
Our setting down before't.
MALCOLM                    'Tis his main hope,             10
For where there is advantage to be gone,
Both more and less have given him the revolt,
And none serve with him but constrainèd things,
Whose hearts are absent too.
MACDUFF                      Let our just censures
Attend the true event, and put we on                       15
Industrious soldiership.
SIWARD                   The time approaches
That will with due decision make us know
What we shall say we have, and what we owe.
Thoughts speculative their unsure hopes relate,
But certain issue strokes must arbitrate;                  20
Towards which, advance the war.    *Exeunt, marching*

5.5    *Enter Macbeth, Seyton, and soldiers, with a*
       *drummer and colours*
MACBETH
Hang out our banners on the outward walls.
The cry is still 'They come.' Our castle's strength
Will laugh a siege to scorn. Here let them lie
Till famine and the ague eat them up.
Were they not forced with those that should be ours   5
We might have met them dareful, beard to beard,
And beat them backward home.
       *A cry within of women*
                                       What is that noise?
SEYTON
It is the cry of women, my good lord.              ⌈*Exit*⌉
MACBETH
I have almost forgot the taste of fears.
The time has been my senses would have cooled        10
To hear a night-shriek, and my fell of hair
Would at a dismal treatise rouse and stir
As life were in't. I have supped full with horrors.
Direness, familiar to my slaughterous thoughts,
Cannot once start me.
       ⌈*Enter Seyton*⌉
                       Wherefore was that cry?       15
SEYTON
The Queen, my lord, is dead.
MACBETH                      She should have died hereafter.
There would have been a time for such a word.
Tomorrow, and tomorrow, and tomorrow
Creeps in this petty pace from day to day
To the last syllable of recorded time,               20

And all our yesterdays have lighted fools
The way to dusty death. Out, out, brief candle.
Life's but a walking shadow, a poor player
That struts and frets his hour upon the stage,
And then is heard no more. It is a tale                    25
Told by an idiot, full of sound and fury,
Signifying nothing.
      *Enter a Messenger*
                Thou com'st to use
Thy tongue: thy story quickly.
MESSENGER              Gracious my lord,
I should report that which I say I saw,
But know not how to do't.
MACBETH              Well, say, sir.                    30
MESSENGER
As I did stand my watch upon the hill
I looked toward Birnam, and anon methought
The wood began to move.
MACBETH              Liar and slave!
MESSENGER
Let me endure your wrath if't be not so.
Within this three mile may you see it coming.                    35
I say, a moving grove.
MACBETH            If thou speak'st false
Upon the next tree shall thou hang alive
Till famine cling thee. If thy speech be sooth,
I care not if thou dost for me as much.
I pall in resolution, and begin                    40
To doubt th'equivocation of the fiend,
That lies like truth. 'Fear not till Birnam Wood
Do come to Dunsinane'—and now a wood
Comes toward Dunsinane. Arm, arm, and out.
If this which he avouches does appear                    45
There is nor flying hence nor tarrying here.
I 'gin to be aweary of the sun,
And wish th'estate o'th' world were now undone.
Ring the alarum bell. ⌐Alarums⌐ Blow wind, come wrack,
At least we'll die with harness on our back.     *Exeunt*

**5.6**   *Enter Malcolm, Siward, Macduff, and their army*
      *with boughs, with a drummer and colours*
MALCOLM
Now near enough. Your leafy screens throw down,
And show like those you are.
      ⌐*They throw down the boughs*⌐
                You, worthy uncle,
Shall with my cousin, your right noble son,
Lead our first battle. Worthy Macduff and we
Shall take upon's what else remains to do                    5
According to our order.
SIWARD             Fare you well.
Do we but find the tyrant's power tonight,
Let us be beaten if we cannot fight.
MACDUFF
Make all our trumpets speak, give them all breath,
Those clamorous harbingers of blood and death.                    10
      *Exeunt. Alarums continued*

**5.7**   *Enter Macbeth*
MACBETH
They have tied me to a stake. I cannot fly,
But bear-like I must fight the course. What's he
That was not born of woman? Such a one
Am I to fear, or none.
      *Enter Young Siward*
YOUNG SIWARD What is thy name?                    5
MACBETH Thou'lt be afraid to hear it.

YOUNG SIWARD
No, though thou call'st thyself a hotter name
Than any is in hell.
MACBETH          My name's Macbeth.
YOUNG SIWARD
The devil himself could not pronounce a title
More hateful to mine ear.
MACBETH         No, nor more fearful.                    10
YOUNG SIWARD
Thou liest, abhorrèd tyrant. With my sword
I'll prove the lie thou speak'st.
      *They fight, and Young Siward is slain*
MACBETH        Thou wast born of woman,
But swords I smile at, weapons laugh to scorn,
Brandished by man that's of a woman born.
      *Exit* ⌐*with the body*⌐

**5.8**   *Alarums. Enter Macduff*
MACDUFF
That way the noise is. Tyrant, show thy face!
If thou beest slain and with no stroke of mine,
My wife and children's ghosts will haunt me still.
I cannot strike at wretched kerns, whose arms
Are hired to bear their staves. Either thou, Macbeth,                    5
Or else my sword with an unbattered edge
I sheathe again undeeded. There thou shouldst be;
By this great clatter one of greatest note
Seems bruited. Let me find him, fortune,
And more I beg not.        *Exit. Alarums*

**5.9**   *Enter Malcolm and Siward*
SIWARD
This way, my lord. The castle's gently rendered.
The tyrant's people on both sides do fight.
The noble thanes do bravely in the war.
The day almost itself professes yours,
And little is to do.
MALCOLM       We have met with foes                    5
That strike beside us.
SIWARD         Enter, sir, the castle.
      *Exeunt. Alarum*

**5.10**   *Enter Macbeth*
MACBETH
Why should I play the Roman fool, and die
On mine own sword? Whiles I see lives, the gashes
Do better upon them.
      *Enter Macduff*
MACDUFF        Turn, hell-hound, turn.
MACBETH
Of all men else I have avoided thee.
But get thee back. My soul is too much charged                    5
With blood of thine already.
MACDUFF         I have no words;
My voice is in my sword, thou bloodier villain
Than terms can give thee out.
      *They fight; alarum*
MACBETH         Thou losest labour.
As easy mayst thou the intrenchant air
With thy keen sword impress as make me bleed.                    10
Let fall thy blade on vulnerable crests;
I bear a charmèd life, which must not yield
To one of woman born.
MACDUFF        Despair thy charm,
And let the angel whom thou still hast served
Tell thee Macduff was from his mother's womb                    15
Untimely ripped.

MACBETH
  Accursèd be that tongue that tells me so,
  For it hath cowed my better part of man;
  And be these juggling fiends no more believed,
  That palter with us in a double sense,      20
  That keep the word of promise to our ear
  And break it to our hope. I'll not fight with thee.
MACDUFF  Then yield thee, coward,
  And live to be the show and gaze o'th' time.
  We'll have thee as our rarer monsters are,  25
  Painted upon a pole, and underwrit
  'Here may you see the tyrant.'
MACBETH             I will not yield
  To kiss the ground before young Malcolm's feet,
  And to be baited with the rabble's curse.
  Though Birnam Wood be come to Dunsinane,  30
  And thou opposed being of no woman born,
  Yet I will try the last. Before my body
  I throw my warlike shield. Lay on, Macduff,
  And damned be him that first cries 'Hold, enough!'
           *Exeunt fighting. Alarums*

    *They enter fighting, and Macbeth is slain. ⌈Exit*
    *Macduff with Macbeth's body⌉*

**5.11**  *Retreat and flourish. Enter with a drummer and colours*
    *Malcolm, Siward, Ross, thanes, and soldiers*
MALCOLM
  I would the friends we miss were safe arrived.
SIWARD
  Some must go off; and yet by these I see
  So great a day as this is cheaply bought.
MALCOLM
  Macduff is missing, and your noble son.
ROSS (*to Siward*)
  Your son, my lord, has paid a soldier's debt.  5
  He only lived but till he was a man,
  The which no sooner had his prowess confirmed
  In the unshrinking station where he fought,
  But like a man he died.
SIWARD          Then he is dead?
ROSS
  Ay, and brought off the field. Your cause of sorrow  10

  Must not be measured by his worth, for then
  It hath no end.
SIWARD       Had he his hurts before?
ROSS
  Ay, on the front.
SIWARD      Why then, God's soldier be he.
  Had I as many sons as I have hairs
  I would not wish them to a fairer death;  15
  And so his knell is knolled.
MALCOLM        He's worth more sorrow,
  And that I'll spend for him.
SIWARD          He's worth no more.
  They say he parted well and paid his score,
  And so God be with him. Here comes newer comfort.
    *Enter Macduff with Macbeth's head*
MACDUFF (*to Malcolm*)
  Hail, King, for so thou art. Behold where stands  20
  Th'usurper's cursèd head. The time is free.
  I see thee compassed with thy kingdom's pearl,
  That speak my salutation in their minds,
  Whose voices I desire aloud with mine:
  Hail, King of Scotland!
ALL BUT MALCOLM      Hail, King of Scotland!  25
    *Flourish*
MALCOLM
  We shall not spend a large expense of time
  Before we reckon with your several loves
  And make us even with you. My thanes and kinsmen,
  Henceforth be earls, the first that ever Scotland
  In such an honour named. What's more to do  30
  Which would be planted newly with the time,
  As calling home our exiled friends abroad,
  That fled the snares of watchful tyranny,
  Producing forth the cruel ministers
  Of this dead butcher and his fiend-like queen—  35
  Who, as 'tis thought, by self and violent hands
  Took off her life—this and what needful else
  That calls upon us, by the grace of grace
  We will perform in measure, time, and place.
  So thanks to all at once, and to each one,  40
  Whom we invite to see us crowned at Scone.
    *Flourish. Exeunt Omnes*

# ANTONY AND CLEOPATRA

FIRST printed in the 1623 Folio, *Antony and Cleopatra* had been entered on the Stationers' Register on 20 May 1608. Echoes of it in Barnabe Barnes's tragedy *The Devil's Charter*, acted by Shakespeare's company in February 1607, suggest that Shakespeare wrote his play no later than 1606, and stylistic evidence supports that date.

The Life of Marcus Antonius in Sir Thomas North's translation of Plutarch's *Lives of the Noble Grecians and Romans* (1579) was one of the sources for *Julius Caesar*; it also provided Shakespeare with most of his material for *Antony and Cleopatra*, in which he draws upon its language to a remarkable extent even in some of the play's most poetic passages. For example, Enobarbus' famous description of Cleopatra in her barge (2.2.197-224) incorporates phrase after phrase of North's prose. And the play's action stays close to North's account, though with significant adjustments, particularly compressions of the time-scheme. It opens in 40 BC, two years after the end of *Julius Caesar*, and portrays events that took place over a period of ten years. Mark Antony has become an older man, though Octavius is still 'scarce-bearded'. Plutarch, who was a connoisseur of human behaviour, also afforded many hints for the characterization; but some characters, particularly Antony's comrade Domitius Enobarbus and Cleopatra's women, Charmian and Iras, are largely created by Shakespeare.

In the earlier play, Mark Antony had formed a triumvirate with Octavius Caesar and Lepidus. In *Antony and Cleopatra* the triumvirate is in a state of disintegration, partly because Mark Antony—married at the play's opening to Fulvia, who is rebelling against Octavius Caesar—is infatuated with Cleopatra, Queen of Egypt (and the former mistress of Julius Caesar). The play's action swings between Rome and Alexandria as Antony is torn between the claims of Rome—strengthened for a while by his marriage, after Fulvia's death, to Octavius Caesar's sister Octavia—and the temptations of Egypt. Gradually opposition between Antony and Octavius increases, until they engage in a sea-fight near Actium (in Greece), in which Antony follows Cleopatra's navy in ignominious retreat. The closing stages of the double tragedy portray Antony's shame, humiliation, and suicide after Cleopatra falsely causes him to believe that she has killed herself; faced with the threat that Caesar will take her captive to Rome, Cleopatra too commits suicide. According to Plutarch, she was thirty-eight years old; as for Antony, 'some say that he lived three-and-fifty years, and others say, six-and-fifty'.

In *Antony and Cleopatra* the classical restraint of *Julius Caesar* gives way to a fine excess of language, of dramatic action, and of individual behaviour. The style is hyperbolical, overflowing the measure of the iambic pentameter. The action is amazingly fluid, shifting with an ease and rapidity that caused bewilderment to ages unfamiliar with the conventions of Shakespeare's theatre. And the characterization is correspondingly extravagant, delighting in the quirks of individual behaviour, above all in the paradoxes and inconsistencies of the Egyptian queen who contains within herself the capacity for every extreme of feminine behaviour, from vanity, meanness, and frivolity to the sublime self-transcendence with which she faces and embraces death.

# THE PERSONS OF THE PLAY

Mark ANTONY (Marcus Antonius), triumvir of Rome

DEMETRIUS

PHILO

Domitius ENOBARBUS

VENTIDIUS

SILIUS — friends and followers of Antony

EROS

CAMIDIUS

SCARUS

DECRETAS

Octavius CAESAR, triumvir of Rome

OCTAVIA, his sister

MAECENAS

AGRIPPA

TAURUS

DOLABELLA — friends and followers of Caesar

THIDIAS

GALLUS

PROCULEIUS

LEPIDUS, triumvir of Rome

Sextus POMPEY (Pompeius)

MENECRATES

MENAS — friends of Pompey

VARRIUS

CLEOPATRA, Queen of Egypt

CHARMIAN

IRAS

ALEXAS

MARDIAN, a eunuch — attending on Cleopatra

DIOMED

SELEUCUS

A SOOTHSAYER

An AMBASSADOR

MESSENGERS

A BOY who sings

A SENTRY and men of his WATCH

Men of the GUARD

An EGYPTIAN

A CLOWN

SERVANTS

SOLDIERS

Eunuchs, attendants, captains, soldiers, servants

# The Tragedy of Antony and Cleopatra

**1.1** *Enter Demetrius and Philo*

PHILO

Nay, but this dotage of our General's
O'erflows the measure. Those his goodly eyes,
That o'er the files and musters of the war
Have glowed like plated Mars, now bend, now turn
The office and devotion of their view                     5
Upon a tawny front. His captain's heart,
Which in the scuffles of great fights hath burst
The buckles on his breast, reneges all temper,
And is become the bellows and the fan
To cool a gipsy's lust.

    *Flourish. Enter Antony, Cleopatra, her ladies, the*
    *train, with eunuchs fanning her*
                Look where they come.          10
Take but good note, and you shall see in him
The triple pillar of the world transformed
Into a strumpet's fool. Behold and see.

CLEOPATRA *(to Antony)*

If it be love indeed, tell me how much.

ANTONY

There's beggary in the love that can be reckoned.    15

CLEOPATRA

I'll set a bourn how far to be beloved.

ANTONY

Then must thou needs find out new heaven, new earth.
    *Enter a Messenger*

MESSENGER News, my good lord, from Rome.

ANTONY Grates me: the sum.

CLEOPATRA Nay, hear them, Antony.                         20
Fulvia perchance is angry; or who knows
If the scarce-bearded Caesar have not sent
His powerful mandate to you: 'Do this, or this,
Take in that kingdom and enfranchise that.
Perform't, or else we damn thee.'                         25

ANTONY How, my love?

CLEOPATRA Perchance? Nay, and most like.
You must not stay here longer. Your dismission
Is come from Caesar, therefore hear it, Antony.
Where's Fulvia's process—Caesar's, I would say—
    both?                                                30
Call in the messengers. As I am Egypt's queen,
Thou blushest, Antony, and that blood of thine
Is Caesar's homager; else so thy cheek pays shame
When shrill-tongued Fulvia scolds. The messengers!

ANTONY

Let Rome in Tiber melt, and the wide arch               35
Of the ranged empire fall. Here is my space.
Kingdoms are clay. Our dungy earth alike
Feeds beast as man. The nobleness of life
Is to do thus; when such a mutual pair
And such a twain can do't—in which I bind              40
On pain of punishment the world to weet—
We stand up peerless.

CLEOPATRA ⌈*aside*⌉     Excellent falsehood!
Why did he marry Fulvia and not love her?
I'll seem the fool I am not. *(To Antony)* Antony
Will be himself.

ANTONY       But stirred by Cleopatra.          45
Now, for the love of Love and her soft hours
Let's not confound the time with conference harsh.

There's not a minute of our lives should stretch
Without some pleasure now. What sport tonight?

CLEOPATRA

Hear the ambassadors.

ANTONY         Fie, wrangling queen,          50
Whom everything becomes—to chide, to laugh,
To weep; how every passion fully strives
To make itself, in thee, fair and admired!
No messenger but thine; and all alone
Tonight we'll wander through the streets and note   55
The qualities of people. Come, my queen.
Last night you did desire it. *(To the Messenger)* Speak
    not to us.
    *Exeunt Antony and Cleopatra with the train,*
    ⌈*and by another door the Messenger*⌉

DEMETRIUS

Is Caesar with Antonius prized so slight?

PHILO

Sir, sometimes when he is not Antony
He comes too short of that great property               60
Which still should go with Antony.

DEMETRIUS       I am full sorry
That he approves the common liar who
Thus speaks of him at Rome; but I will hope
Of better deeds tomorrow. Rest you happy.    *Exeunt*

**1.2** *Enter Enobarbus, a Soothsayer, Charmian, Iras,*
    *Mardian the eunuch, Alexas, ⌈and attendants⌉*

CHARMIAN Lord Alexas, sweet Alexas, most anything
Alexas, almost most absolute Alexas, where's the
soothsayer that you praised so to th' Queen?
O that I knew this husband, which you say
Must charge his horns with garlands!

ALEXAS      Soothsayer!          5

SOOTHSAYER Your will?

CHARMIAN

Is this the man? Is't you, sir, that know things?

SOOTHSAYER

In nature's infinite book of secrecy
A little I can read.

ALEXAS *(to Charmian)* Show him your hand.          10

ENOBARBUS *(calling)* Bring in the banquet quickly,
Wine enough Cleopatra's health to drink.
    ⌈*Enter servants with food and wine, and exeunt*⌉

CHARMIAN *(to Soothsayer)* Good sir, give me good fortune.

SOOTHSAYER I make not, but foresee.

CHARMIAN

Pray then, foresee me one.

SOOTHSAYER      You shall be yet          15
Far fairer than you are.

CHARMIAN      He means in flesh.

IRAS

No, you shall paint when you are old.

CHARMIAN         Wrinkles forbid!

ALEXAS

Vex not his prescience. Be attentive.

CHARMIAN         Hush!

SOOTHSAYER

You shall be more beloving than beloved.

CHARMIAN I had rather heat my liver with drinking.    20

ALEXAS Nay, hear him.

CHARMIAN Good now, some excellent fortune! Let me be
married to three kings in a forenoon and widow them
all. Let me have a child at fifty to whom Herod of Jewry
may do homage. Find me to marry me with Octavius
Caesar, and companion me with my mistress.    26
SOOTHSAYER
You shall outlive the lady whom you serve.
CHARMIAN O, excellent! I love long life better than figs.
SOOTHSAYER
You have seen and proved a fairer former fortune
Than that which is to approach.    30
CHARMIAN Then belike my children shall have no names.
Prithee, how many boys and wenches must I have?
SOOTHSAYER
If every of your wishes had a womb,
And fertile every wish, a million.
CHARMIAN Out, fool—I forgive thee for a witch.    35
ALEXAS You think none but your sheets are privy to your
wishes.
CHARMIAN (to the Soothsayer) Nay, come, tell Iras hers.
ALEXAS We'll know all our fortunes.
ENOBARBUS Mine, and most of our fortunes, tonight shall
be drunk to bed.    41
IRAS (showing her hand to the Soothsayer) There's a palm
presages chastity, if nothing else.
CHARMIAN E'en as the o'erflowing Nilus presageth famine.
IRAS Go, you wild bedfellow, you cannot soothsay.    45
CHARMIAN Nay, if an oily palm be not a fruitful prognos-
tication, I cannot scratch mine ear. (To the Soothsayer)
Prithee, tell her but a workaday fortune.
SOOTHSAYER Your fortunes are alike.
IRAS
But how, but how? Give me particulars.    50
SOOTHSAYER I have said.
IRAS Am I not an inch of fortune better than she?
CHARMIAN Well, if you were but an inch of fortune better
than I, where would you choose it?
IRAS Not in my husband's nose.    55
CHARMIAN Our worser thoughts heavens mend! Alexas—
come, his fortune, his fortune. O, let him marry a
woman that cannot go, sweet Isis, I beseech thee, and
let her die too, and give him a worse, and let worse
follow worse till the worst of all follow him laughing
to his grave, fiftyfold a cuckold. Good Isis, hear me this
prayer, though thou deny me a matter of more weight;
good Isis, I beseech thee.    63
IRAS Amen, dear goddess, hear that prayer of the people.
For as it is a heart-breaking to see a handsome man
loose-wived, so it is a deadly sorrow to behold a foul
knave uncuckolded. Therefore, dear Isis, keep decorum,
and fortune him accordingly.
CHARMIAN Amen.    69
ALEXAS Lo now, if it lay in their hands to make me a
cuckold, they would make themselves whores but
they'd do't.
     Enter Cleopatra
ENOBARBUS
Hush, here comes Antony.
CHARMIAN             Not he, the Queen.
CLEOPATRA
Saw you my lord?
ENOBARBUS        No, lady.
CLEOPATRA           Was he not here?
CHARMIAN No, madam.    75
CLEOPATRA
He was disposed to mirth, but on the sudden
A Roman thought hath struck him. Enobarbus!

ENOBARBUS Madam?
CLEOPATRA
Seek him, and bring him hither. Where's Alexas?
ALEXAS
Here at your service. My lord approaches.    80
     Enter Antony with a Messenger
CLEOPATRA
We will not look upon him. Go with us.
     Exeunt all but Antony and the Messenger
MESSENGER
Fulvia thy wife first came into the field.
ANTONY Against my brother Lucius?
MESSENGER
Ay, but soon that war had end, and the time's state
Made friends of them, jointing their force 'gainst
     Caesar,    85
Whose better issue in the war from Italy
Upon the first encounter drave them.
ANTONY             Well, what worst?
MESSENGER
The nature of bad news infects the teller.
ANTONY
When it concerns the fool or coward. On.
Things that are past are done. With me 'tis thus:    90
Who tells me true, though in his tale lie death,
I hear him as he flattered.
MESSENGER            Labienus—
This is stiff news—hath with his Parthian force
Extended Asia; from Euphrates
His conquering banner shook, from Syria    95
To Lydia and to Ionia,
Whilst—
ANTONY     Antony, thou wouldst say—
MESSENGER             O, my lord!
ANTONY
Speak to me home. Mince not the general tongue.
Name Cleopatra as she is called in Rome.
Rail thou in Fulvia's phrase, and taunt my faults    100
With such full licence as both truth and malice
Have power to utter. O, then we bring forth weeds
When our quick winds lie still, and our ills told us
Is as our earing. Fare thee well a while.
MESSENGER At your noble pleasure.      Exit Messenger
     Enter another Messenger
ANTONY
From Sicyon, ho, the news? Speak there.    106
⌈SECOND MESSENGER⌉
The man from Sicyon—
⌈ANTONY⌉            Is there such a one?
⌈SECOND MESSENGER⌉
He stays upon your will.
ANTONY           Let him appear.
         Exit Second Messenger
These strong Egyptian fetters I must break,
Or lose myself in dotage.
     Enter another Messenger with a letter
              What are you?    110
⌈THIRD MESSENGER⌉
Fulvia thy wife is dead.
ANTONY           Where died she?
THIRD MESSENGER In Sicyon.
Her length of sickness, with what else more serious
Importeth thee to know, this bears.
     He gives Antony the letter
ANTONY           Forbear me.
         ⌈Exit Third Messenger⌉
There's a great spirit gone. Thus did I desire it.    115

What our contempts doth often hurl from us
We wish it ours again. The present pleasure,
By revolution low'ring, does become
The opposite of itself. She's good being gone;
The hand could pluck her back that shoved her on.
I must from this enchanting queen break off.         121
Ten thousand harms more than the ills I know
My idleness doth hatch. How now, Enobarbus!
⌐Enter Enobarbus⌐
ENOBARBUS
What's your pleasure, sir?
ANTONY                         I must with haste from hence.
ENOBARBUS Why, then we kill all our women. We see
how mortal an unkindness is to them; if they suffer
our departure, death's the word.                    127
ANTONY I must be gone.
ENOBARBUS Under a compelling occasion let women die.
It were pity to cast them away for nothing, though
between them and a great cause they should be
esteemed nothing. Cleopatra catching but the least
noise of this dies instantly. I have seen her die twenty
times upon far poorer moment. I do think there is
mettle in death, which commits some loving act upon
her, she hath such a celerity in dying.             136
ANTONY She is cunning past man's thought.
ENOBARBUS Alack, sir, no. Her passions are made of
nothing but the finest part of pure love. We cannot call
her winds and waters sighs and tears; they are greater
storms and tempests than almanacs can report. This
cannot be cunning in her; if it be, she makes a shower
of rain as well as Jove.
ANTONY Would I had never seen her!                  144
ENOBARBUS O, sir, you had then left unseen a wonderful
piece of work, which not to have been blessed withal
would have discredited your travel.
ANTONY Fulvia is dead.
ENOBARBUS Sir.
ANTONY Fulvia is dead.                              150
ENOBARBUS Fulvia?
ANTONY Dead.
ENOBARBUS Why, sir, give the gods a thankful sacrifice.
When it pleaseth their deities to take the wife of a man
from him, it shows to man the tailors of the earth;
comforting therein that when old robes are worn out
there are members to make new. If there were no more
women but Fulvia, then had you indeed a cut, and the
case to be lamented. This grief is crowned with consola-
tion; your old smock brings forth a new petticoat, and
indeed the tears live in an onion that should water this
sorrow.
ANTONY
The business she hath broachèd in the state
Cannot endure my absence.                           164
ENOBARBUS And the business you have broached here
cannot be without you, especially that of Cleopatra's,
which wholly depends on your abode.
ANTONY
No more light answers. Let our officers
Have notice what we purpose. I shall break
The cause of our expedience to the Queen,           170
And get her leave to part; for not alone
The death of Fulvia, with more urgent touches,
Do strongly speak to us, but the letters too
Of many our contriving friends in Rome
Petition us at home. Sextus Pompeius                175

Hath given the dare to Caesar and commands
The empire of the sea. Our slippery people,
Whose love is never linked to the deserver
Till his deserts are past, begin to throw
Pompey the Great and all his dignities              180
Upon his son, who—high in name and power,
Higher than both in blood and life—stands up
For the main soldier; whose quality, going on,
The sides o'th' world may danger. Much is breeding
Which, like the courser's hair, hath yet but life,  185
And not a serpent's poison. Say our pleasure,
To such whose place is under us, requires
Our quick remove from hence.
ENOBARBUS                         I shall do't.
                                   Exeunt severally

1.3   Enter Cleopatra, Charmian, Alexas, and Iras
CLEOPATRA
Where is he?
CHARMIAN       I did not see him since.
CLEOPATRA ⌐to Alexas⌐
See where he is, who's with him, what he does.
I did not send you. If you find him sad,
Say I am dancing; if in mirth, report
That I am sudden sick. Quick, and return.           5
                                   Exit ⌐Alexas⌐
CHARMIAN
Madam, methinks, if you did love him dearly,
You do not hold the method to enforce
The like from him.
CLEOPATRA               What should I do I do not?
CHARMIAN
In each thing give him way; cross him in nothing.
CLEOPATRA
Thou teachest like a fool, the way to lose him.     10
CHARMIAN
Tempt him not so too far. Iwis, forbear.
In time we hate that which we often fear.
      Enter Antony
But here comes Antony.
CLEOPATRA                    I am sick and sullen.
ANTONY
I am sorry to give breathing to my purpose.
CLEOPATRA
Help me away, dear Charmian, I shall fall.          15
It cannot be thus long—the sides of nature
Will not sustain it.
ANTONY               Now, my dearest queen.
CLEOPATRA
Pray you, stand farther from me.
ANTONY                             What's the matter?
CLEOPATRA
I know by that same eye there's some good news.
What says the married woman—you may go?            20
Would she had never given you leave to come.
Let her not say 'tis I that keep you here.
I have no power upon you; hers you are.
ANTONY
The gods best know—
CLEOPATRA              O, never was there queen
So mightily betrayed! Yet at the first              25
I saw the treasons planted.
ANTONY                        Cleopatra—
CLEOPATRA
Why should I think you can be mine and true—
Though you in swearing shake the thronèd gods—

Who have been false to Fulvia? Riotous madness,
To be entangled with those mouth-made vows    30
Which break themselves in swearing.
ANTONY                  Most sweet queen—
CLEOPATRA
Nay, pray you, seek no colour for your going,
But bid farewell and go. When you sued staying,
Then was the time for words; no going then.
Eternity was in our lips and eyes,    35
Bliss in our brow's bent; none our parts so poor
But was a race of heaven. They are so still,
Or thou, the greatest soldier of the world,
Art turned the greatest liar.
ANTONY                  How now, lady!
CLEOPATRA
I would I had thy inches. Thou shouldst know    40
There were a heart in Egypt.
ANTONY                  Hear me, Queen.
The strong necessity of time commands
Our services a while, but my full heart
Remains in use with you. Our Italy
Shines o'er with civil swords. Sextus Pompeius    45
Makes his approaches to the port of Rome.
Equality of two domestic powers
Breed scrupulous faction. The hated, grown to
     strength,
Are newly grown to love. The condemned Pompey,
Rich in his father's honour, creeps apace    50
Into the hearts of such as have not thrived
Upon the present state, whose numbers threaten;
And quietness, grown sick of rest, would purge
By any desperate change. My more particular,
And that which most with you should safe my going,
Is Fulvia's death.    56
CLEOPATRA
Though age from folly could not give me freedom,
It does from childishness. Can Fulvia die?
ANTONY She's dead, my queen.
     *He offers letters*
Look here, and at thy sovereign leisure read    60
The garboils she awaked. At the last, best,
See when and where she died.
CLEOPATRA            O most false love!
Where be the sacred vials thou shouldst fill
With sorrowful water? Now I see, I see,
In Fulvia's death how mine received shall be.    65
ANTONY
Quarrel no more, but be prepared to know
The purposes I bear, which are or cease
As you shall give th'advice. By the fire
That quickens Nilus' slime, I go from hence
Thy soldier-servant, making peace or war    70
As thou affects.
CLEOPATRA       Cut my lace, Charmian, come.
But let it be. I am quickly ill and well;
So Antony loves.
ANTONY       My precious queen, forbear,
And give true evidence to his love, which stands
An honourable trial.
CLEOPATRA        So Fulvia told me.    75
I prithee turn aside and weep for her,
Then bid adieu to me, and say the tears
Belong to Egypt. Good now, play one scene
Of excellent dissembling, and let it look
Like perfect honour.
ANTONY        You'll heat my blood. No more.

CLEOPATRA
You can do better yet; but this is meetly.    81
ANTONY
Now by my sword—
CLEOPATRA        And target. Still he mends.
But this is not the best. Look, prithee, Charmian,
How this Herculean Roman does become
The carriage of his chafe.    85
ANTONY I'll leave you, lady.
CLEOPATRA Courteous lord, one word.
Sir, you and I must part; but that's not it.
Sir, you and I have loved; but there's not it;
That you know well. Something it is I would—    90
O, my oblivion is a very Antony,
And I am all forgotten.
ANTONY        But that your royalty
Holds idleness your subject, I should take you
For idleness itself.
CLEOPATRA       'Tis sweating labour
To bear such idleness so near the heart    95
As Cleopatra this. But sir, forgive me,
Since my becomings kill me when they do not
Eye well to you. Your honour calls you hence,
Therefore be deaf to my unpitied folly,
And all the gods go with you. Upon your sword    100
Sit laurel victory, and smooth success
Be strewed before your feet.
ANTONY            Let us go.
Come. Our separation so abides and flies
That thou residing here goes yet with me,
And I hence fleeting, here remain with thee.    105
Away.                   *Exeunt severally*

**1.4**     *Enter Octavius reading a letter, Lepidus, and their*
         *train*
CAESAR
You may see, Lepidus, and henceforth know,
It is not Caesar's natural vice to hate
Our great competitor. From Alexandria
This is the news: he fishes, drinks, and wastes
The lamps of night in revel; is not more manlike    5
Than Cleopatra, nor the queen of Ptolemy
More womanly than he; hardly gave audience
Or vouchsafed to think he had partners. You shall find
     there
A man who is the abstract of all faults
That all men follow.
LEPIDUS       I must not think there are    10
Evils enough to darken all his goodness.
His faults in him seem as the spots of heaven,
More fiery by night's blackness; hereditary
Rather than purchased; what he cannot change
Than what he chooses.    15
CAESAR
You are too indulgent. Let's grant it is not
Amiss to tumble on the bed of Ptolemy,
To give a kingdom for a mirth, to sit
And keep the turn of tippling with a slave,
To reel the streets at noon, and stand the buffet    20
With knaves that smells of sweat. Say this becomes
     him—
As his composure must be rare indeed
Whom these things cannot blemish—yet must Antony
No way excuse his foils when we do bear
So great weight in his lightness. If he filled    25

His vacancy with his voluptuousness,
Full surfeits and the dryness of his bones
Call on him for't. But to confound such time
That drums him from his sport, and speaks as loud
As his own state and ours—'tis to be chid          30
As we rate boys who, being mature in knowledge,
Pawn their experience to their present pleasure,
And so rebel to judgement.
                        *Enter a Messenger*
LEPIDUS                              Here's more news.
MESSENGER
Thy biddings have been done, and every hour,
Most noble Caesar, shalt thou have report          35
How 'tis abroad. Pompey is strong at sea,
And it appears he is beloved of those
That only have feared Caesar. To the ports
The discontents repair, and men's reports
Give him much wronged.                    ⌈*Exit*⌉
CAESAR                    I should have known no less.
It hath been taught us from the primal state          41
That he which is was wished until he were,
And the ebbed man, ne'er loved till ne'er worth love,
Comes deared by being lacked. This common body,
Like to a vagabond flag upon the stream,          45
Goes to, and back, lackeying the varying tide,
To rot itself with motion.
              ⌈*Enter a second Messenger*⌉
SECOND MESSENGER          Caesar, I bring thee word
Menecrates and Menas, famous pirates,
Makes the sea serve them, which they ear and wound
With keels of every kind. Many hot inroads          50
They make in Italy. The borders maritime
Lack blood to think on't, and flush youth revolt.
No vessel can peep forth but 'tis as soon
Taken as seen; for Pompey's name strikes more
Than could his war resisted.                    ⌈*Exit*⌉
CAESAR                              Antony,          55
Leave thy lascivious wassails. When thou once
Was beaten from Modena, where thou slew'st
Hirtius and Pansa, consuls, at thy heel
Did famine follow, whom thou fought'st against—
Though daintily brought up—with patience more          60
Than savages could suffer. Thou didst drink
The stale of horses, and the gilded puddle
Which beasts would cough at. Thy palate then did
        deign
The roughest berry on the rudest hedge.
Yea, like the stag when snow the pasture sheets,          65
The barks of trees thou browsed. On the Alps
It is reported thou didst eat strange flesh,
Which some did die to look on; and all this—
It wounds thine honour that I speak it now—
Was borne so like a soldier that thy cheek          70
So much as lanked not.
LEPIDUS 'Tis pity of him.
CAESAR Let his shames quickly
Drive him to Rome. 'Tis time we twain
Did show ourselves i'th' field; and to that end          75
Assemble we immediate council. Pompey
Thrives in our idleness.
LEPIDUS                    Tomorrow, Caesar,
I shall be furnished to inform you rightly
Both what by sea and land I can be able
To front this present time.
CAESAR                    Till which encounter          80
It is my business, too. Farewell.

LEPIDUS
Farewell, my lord. What you shall know meantime
Of stirs abroad I shall beseech you, sir,
To let me be partaker.
CAESAR
Doubt not, sir. I knew it for my bond.          *Exeunt*

1.5          *Enter Cleopatra, Charmian, Iras, and Mardian*
CLEOPATRA Charmian!
CHARMIAN Madam?
CLEOPATRA (*yawning*)
Ha, ha. Give me to drink mandragora.
CHARMIAN Why, madam?
CLEOPATRA
That I might sleep out this great gap of time          5
My Antony is away.
CHARMIAN                    You think of him too much.
CLEOPATRA
O, 'tis treason!
CHARMIAN          Madam, I trust not so.
CLEOPATRA
Thou, eunuch Mardian!
MARDIAN                    What's your highness' pleasure?
CLEOPATRA
Not now to hear thee sing. I take no pleasure
In aught an eunuch has. 'Tis well for thee          10
That, being unseminared, thy freer thoughts
May not fly forth of Egypt. Hast thou affections?
MARDIAN Yes, gracious madam.
CLEOPATRA Indeed?
MARDIAN
Not in deed, madam, for I can do nothing          15
But what indeed is honest to be done.
Yet have I fierce affections, and think
What Venus did with Mars.
CLEOPATRA                    O, Charmian,
Where think'st thou he is now? Stands he or sits he?
Or does he walk? Or is he on his horse?          20
O happy horse, to bear the weight of Antony!
Do bravely, horse, for wot'st thou whom thou
        mov'st?—
The demi-Atlas of this earth, the arm
And burgonet of men. He's speaking now,
Or murmuring 'Where's my serpent of old Nile?'—          25
For so he calls me. Now I feed myself
With most delicious poison. Think on me,
That am with Phoebus' amorous pinches black,
And wrinkled deep in time. Broad-fronted Caesar,
When thou wast here above the ground I was          30
A morsel for a monarch, and great Pompey
Would stand and make his eyes grow in my brow.
There would he anchor his aspect, and die
With looking on his life.
              *Enter Alexas*
ALEXAS                    Sovereign of Egypt, hail!
CLEOPATRA
How much unlike art thou Mark Antony!          35
Yet, coming from him, that great medicine hath
With his tinct gilded thee. How goes it
With my brave Mark Antony?
ALEXAS                    Last thing he did, dear Queen,
He kissed—the last of many doubled kisses—
This orient pearl. His speech sticks in my heart.          40
CLEOPATRA
Mine ear must pluck it thence.
ALEXAS                    'Good friend,' quoth he,
'Say the firm Roman to great Egypt sends

This treasure of an oyster; at whose foot,
To mend the petty present, I will piece
Her opulent throne with kingdoms. All the East,    45
Say thou, shall call her mistress.' So he nodded,
And soberly did mount an arm-jaunced steed,
Who neighed so high that what I would have spoke
Was beastly dumbed by him.

CLEOPATRA                          What, was he sad or merry?

ALEXAS
Like to the time o'th' year between the extremes    50
Of hot and cold, he was nor sad nor merry.

CLEOPATRA
O well divided disposition! Note him,
Note him, good Charmian, 'tis the man; but note him.
He was not sad, for he would shine on those
That make their looks by his; he was not merry,    55
Which seemed to tell them his remembrance lay
In Egypt with his joy; but between both.
O heavenly mingle! Be'st thou sad or merry,
The violence of either thee becomes;
So does it no man else. Met'st thou my posts?    60

ALEXAS
Ay, madam, twenty several messengers.
Why do you send so thick?

CLEOPATRA                          Who's born that day
When I forget to send to Antony
Shall die a beggar. Ink and paper, Charmian!
Welcome, my good Alexas. Did I, Charmian,    65
Ever love Caesar so?

CHARMIAN                      O, that brave Caesar!

CLEOPATRA
Be choked with such another emphasis!
Say 'the brave Antony'.

CHARMIAN                      The valiant Caesar.

CLEOPATRA
By Isis, I will give thee bloody teeth
If thou with Caesar paragon again    70
My man of men.

CHARMIAN          By your most gracious pardon,
I sing but after you.

CLEOPATRA          My salad days,
When I was green in judgement, cold in blood,
To say as I said then. But come, away,
Get me ink and paper.    75
He shall have every day a several greeting,
Or I'll unpeople Egypt.                          Exeunt

**2.1**    Enter Pompey, Menecrates, and Menas, in warlike
           manner

POMPEY
If the great gods be just, they shall assist
The deeds of justest men.

⌜MENECRATES⌝                      Know, worthy Pompey,
That what they do delay they not deny.

POMPEY
Whiles we are suitors to their throne, decays
The thing we sue for.

⌜MENECRATES⌝          We, ignorant of ourselves,    5
Beg often our own harms, which the wise powers
Deny us for our good; so find we profit
By losing of our prayers.

POMPEY                      I shall do well.
The people love me, and the sea is mine.
My powers are crescent, and my auguring hope    10
Says it will come to th' full. Mark Antony
In Egypt sits at dinner, and will make

No wars without doors. Caesar gets money where
He loses hearts. Lepidus flatters both,
Of both is flattered; but he neither loves,    15
Nor either cares for him.

⌜MENAS⌝                          Caesar and Lepidus
Are in the field; a mighty strength they carry.

POMPEY
Where have you this? 'Tis false.

⌜MENAS⌝                          From Silvius, sir.

POMPEY
He dreams. I know they are in Rome together,
Looking for Antony. But all the charms of love,    20
Salt Cleopatra, soften thy waned lip.
Let witchcraft join with beauty, lust with both
Tie up the libertine, in a field of feasts
Keep his brain fuming; Epicurean cooks
Sharpen with cloyless sauce his appetite,    25
That sleep and feeding may prorogue his honour
Even till a Lethe'd dullness—
           Enter Varrius
                          How now, Varrius?

VARRIUS
This is most certain that I shall deliver:
Mark Antony is every hour in Rome
Expected. Since he went from Egypt, 'tis    30
A space for farther travel.

POMPEY                      I could have given less matter
A better ear. Menas, I did not think
This amorous surfeiter would have donned his helm
For such a petty war. His soldiership
Is twice the other twain. But let us rear    35
The higher our opinion, that our stirring
Can from the lap of Egypt's widow pluck
The ne'er lust-wearied Antony.

MENAS                          I cannot hope
Caesar and Antony shall well greet together.
His wife that's dead did trespasses to Caesar,    40
His brother warred upon him, although, I think,
Not moved by Antony.

POMPEY                      I know not, Menas,
How lesser enmities may give way to greater.
Were't not that we stand up against them all,
'Twere pregnant they should square between
           themselves,    45
For they have entertainèd cause enough
To draw their swords. But how the fear of us
May cement their divisions, and bind up
The petty difference, we yet not know.
Be't as our gods will have't; it only stands    50
Our lives upon to use our strongest hands.
Come, Menas.                          Exeunt

**2.2**    Enter Enobarbus and Lepidus

LEPIDUS
Good Enobarbus, 'tis a worthy deed,
And shall become you well, to entreat your captain
To soft and gentle speech.

ENOBARBUS                      I shall entreat him
To answer like himself. If Caesar move him,
Let Antony look over Caesar's head    5
And speak as loud as Mars. By Jupiter,
Were I the wearer of Antonio's beard
I would not shave't today.

LEPIDUS                      'Tis not a time
For private stomaching.

ENOBARBUS                      Every time
Serves for the matter that is then born in't.    10

LEPIDUS
But small to greater matters must give way.

ENOBARBUS
Not if the small come first.

LEPIDUS                              Your speech is passion.
But pray you, stir no embers up. Here comes
The noble Antony.

*Enter at one door Antony and Ventidius*

ENOBARBUS              And yonder Caesar.

*Enter at another door Caesar, Maecenas, and*
*Agrippa*

ANTONY (*to Ventidius*)
If we compose well here, to Parthia.                    15
Hark, Ventidius.

CAESAR              I do not know,
Maecenas; ask Agrippa.

LEPIDUS (*to Caesar and Antony*) Noble friends,
That which combined us was most great; and let not
A leaner action rend us. What's amiss,
May it be gently heard. When we debate            20
Our trivial difference loud, we do commit
Murder in healing wounds. Then, noble partners,
The rather for I earnestly beseech,
Touch you the sourest points with sweetest terms,
Nor curstness grow to th' matter.

ANTONY                              'Tis spoken well.    25
Were we before our armies, and to fight,
I should do thus.

⌈*Antony and Caesar embrace.*⌉ *Flourish*

CAESAR  Welcome to Rome.

ANTONY  Thank you.

CAESAR  Sit.                                            30

ANTONY  Sit, sir.

CAESAR  Nay then.

*They sit*

ANTONY
I learn you take things ill which are not so,
Or being, concern you not.

CAESAR                        I must be laughed at
If or for nothing or a little I                         35
Should say myself offended, and with you
Chiefly i'th' world; more laughed at that I should
Once name you derogately, when to sound your name
It not concernèd me.

ANTONY
My being in Egypt, Caesar, what was't to you?          40

CAESAR
No more than my residing here at Rome
Might be to you in Egypt. Yet if you there
Did practise on my state, your being in Egypt
Might be my question.

ANTONY                    How intend you 'practised'?

CAESAR
You may be pleased to catch at mine intent            45
By what did here befall me. Your wife and brother
Made wars upon me, and their contestation
Was theme for you. You were the word of war.

ANTONY
You do mistake the business. My brother never
Did urge me in his act. I did enquire it,             50
And have my learning from some true reports
That drew their swords with you. Did he not rather
Discredit my authority with yours,
And make the wars alike against my stomach,
Having alike your cause? Of this, my letters         55
Before did satisfy you. If you'll patch a quarrel,

As matter whole you have to make it with,
It must not be with this.

CAESAR              You praise yourself
By laying defects of judgement to me, but
You patched up your excuses.

ANTONY                        Not so, not so.          60
I know you could not lack, I am certain on't,
Very necessity of this thought, that I,
Your partner in the cause 'gainst which he fought,
Could not with graceful eyes attend those wars
Which fronted mine own peace. As for my wife,        65
I would you had her spirit in such another.
The third o'th' world is yours, which with a snaffle
You may pace easy, but not such a wife.

ENOBARBUS  Would we had all such wives, that the men
might go to wars with the women.                      70

ANTONY
So much uncurable, her garboils, Caesar,
Made out of her impatience—which not wanted
Shrewdness of policy too—I grieving grant
Did you too much disquiet, for that you must
But say I could not help it.

CAESAR              I wrote to you                      75
When, rioting in Alexandria, you
Did pocket up my letters, and with taunts
Did gibe my missive out of audience.

ANTONY
Sir, he fell upon me ere admitted, then.
Three kings I had newly feasted, and did want        80
Of what I was i'th' morning; but next day
I told him of myself, which was as much
As to have asked him pardon. Let this fellow
Be nothing of our strife. If we contend,
Out of our question wipe him.

CAESAR              You have broken                    85
The article of your oath, which you shall never
Have tongue to charge me with.

LEPIDUS  Soft, Caesar.

ANTONY  No, Lepidus, let him speak.
The honour is sacred which he talks on now,          90
Supposing that I lacked it. But on, Caesar:
The article of my oath—

CAESAR
To lend me arms and aid when I required them,
The which you both denied.

ANTONY                    Neglected, rather,
And then when poisoned hours had bound me up         95
From mine own knowledge. As nearly as I may
I'll play the penitent to you, but mine honesty
Shall not make poor my greatness, nor my power
Work without it. Truth is that Fulvia,
To have me out of Egypt, made wars here,            100
For which myself, the ignorant motive, do
So far ask pardon as befits mine honour
To stoop in such a case.

LEPIDUS                    'Tis noble spoken.

MAECENAS
If it might please you to enforce no further
The griefs between ye; to forget them quite         105
Were to remember that the present need
Speaks to atone you.

LEPIDUS              Worthily spoken, Maecenas.

ENOBARBUS  Or if you borrow one another's love for the
instant, you may, when you hear no more words of
Pompey, return it again. You shall have time to wrangle
in when you have nothing else to do.                111

ANTONY
  Thou art a soldier only. Speak no more.
ENOBARBUS That truth should be silent I had almost forgot.
ANTONY
  You wrong this presence, therefore speak no more.
ENOBARBUS Go to, then; your considerate stone.   115
CAESAR
  I do not much dislike the matter, but
  The manner of his speech, for't cannot be
  We shall remain in friendship, our conditions
  So diff'ring in their acts. Yet if I knew
  What hoop should hold us staunch, from edge to edge
  O'th' world I would pursue it.   121
AGRIPPA Give me leave, Caesar.
CAESAR Speak, Agrippa.
AGRIPPA
  Thou hast a sister by the mother's side,
  Admired Octavia. Great Mark Antony   125
  Is now a widower.
CAESAR           Say not so, Agrippa.
  If Cleopatra heard you, your reproof
  Were well deserved of rashness.
ANTONY
  I am not married, Caesar. Let me hear
  Agrippa further speak.   130
AGRIPPA
  To hold you in perpetual amity,
  To make you brothers, and to knit your hearts
  With an unslipping knot, take Antony
  Octavia to his wife; whose beauty claims
  No worse a husband than the best of men;   135
  Whose virtue and whose general graces speak
  That which none else can utter. By this marriage
  All little jealousies which now seem great,
  And all great fears which now import their dangers,
  Would then be nothing. Truths would be tales   140
  Where now half-tales be truths. Her love to both
  Would each to other and all loves to both
  Draw after her. Pardon what I have spoke,
  For 'tis a studied, not a present thought,
  By duty ruminated.
ANTONY          Will Caesar speak?   145
CAESAR
  Not till he hears how Antony is touched
  With what is spoke already.
ANTONY What power is in Agrippa,
  If I would say 'Agrippa, be it so',
  To make this good?
CAESAR          The power of Caesar,   150
  And his power unto Octavia.
ANTONY          May I never
  To this good purpose, that so fairly shows,
  Dream of impediment! Let me have thy hand.
  Further this act of grace, and from this hour
  The heart of brothers govern in our loves   155
  And sway our great designs.
CAESAR          There's my hand.
    *Antony and Caesar clasp hands*
  A sister I bequeath you whom no brother
  Did ever love so dearly. Let her live
  To join our kingdoms and our hearts; and never
  Fly off our loves again.
LEPIDUS          Happily, amen.   160
ANTONY
  I did not think to draw my sword 'gainst Pompey,
  For he hath laid strange courtesies and great

Of late upon me. I must thank him only,
Lest my remembrance suffer ill report;
At heel of that, defy him.
LEPIDUS          Time calls upon's.   165
  Of us must Pompey presently be sought,
  Or else he seeks out us.
ANTONY          Where lies he?
CAESAR
  About the Mount Misena.
ANTONY          What is his strength
  By land?
CAESAR   Great and increasing, but by sea
  He is an absolute master.
ANTONY          So is the fame.   170
  Would we had spoke together. Haste we for it;
  Yet ere we put ourselves in arms, dispatch we
  The business we have talked of.
CAESAR          With most gladness,
  And do invite you to my sister's view,
  Whither straight I'll lead you.
ANTONY          Let us, Lepidus,   175
  Not lack your company.
LEPIDUS          Noble Antony,
  Not sickness should detain me.
      *Flourish. Exeunt all but Enobarbus, Agrippa,*
                *and Maecenas*
MAECENAS (*to Enobarbus*) Welcome from Egypt, sir.
ENOBARBUS Half the heart of Caesar, worthy Maecenas!
  My honourable friend, Agrippa!   180
AGRIPPA Good Enobarbus!
MAECENAS We have cause to be glad that matters are so
  well digested. You stayed well by't in Egypt.
ENOBARBUS Ay, sir, we did sleep day out of countenance,
  and made the night light with drinking.   185
MAECENAS Eight wild boars roasted whole at a breakfast
  and but twelve persons there—is this true?
ENOBARBUS This was but as a fly by an eagle. We had
  much more monstrous matter of feast, which worthily
  deserved noting.   190
MAECENAS She's a most triumphant lady, if report be
  square to her.
ENOBARBUS When she first met Mark Antony, she pursed
  up his heart upon the river of Cydnus.
AGRIPPA There she appeared indeed, or my reporter
  devised well for her.   196
ENOBARBUS I will tell you.
  The barge she sat in, like a burnished throne
  Burned on the water. The poop was beaten gold;
  Purple the sails, and so perfumèd that   200
  The winds were love-sick with them. The oars were silver,
  Which to the tune of flutes kept stroke, and made
  The water which they beat to follow faster,
  As amorous of their strokes. For her own person,
  It beggared all description. She did lie   205
  In her pavilion—cloth of gold, of tissue—
  O'er-picturing that Venus where we see
  The fancy outwork nature. On each side her
  Stood pretty dimpled boys, like smiling Cupids,
  With divers-coloured fans whose wind did seem   210
  To glow the delicate cheeks which they did cool,
  And what they undid did.
AGRIPPA          O, rare for Antony!
ENOBARBUS
  Her gentlewomen, like the Nereides,

So many mermaids, tended her i'th' eyes,
And made their bends adornings. At the helm          215
A seeming mermaid steers. The silken tackle
Swell with the touches of those flower-soft hands
That yarely frame the office. From the barge
A strange invisible perfume hits the sense
Of the adjacent wharfs. The city cast          220
Her people out upon her, and Antony,
Enthroned i'th' market-place, did sit alone,
Whistling to th'air, which but for vacancy
Had gone to gaze on Cleopatra too,
And made a gap in nature.

AGRIPPA                    Rare Egyptian!          225

ENOBARBUS
Upon her landing Antony sent to her,
Invited her to supper. She replied
It should be better he became her guest,
Which she entreated. Our courteous Antony,
Whom ne'er the word of 'No' woman heard speak,
Being barbered ten times o'er, goes to the feast,          231
And for his ordinary pays his heart
For what his eyes eat only.

AGRIPPA                    Royal wench!
She made great Caesar lay his sword to bed.
He ploughed her, and she cropped.

ENOBARBUS                    I saw her once          235
Hop forty paces through the public street,
And having lost her breath, she spoke and panted,
That she did make defect perfection,
And breathless, pour breath forth.

MAECENAS                    Now Antony
Must leave her utterly.

ENOBARBUS                    Never. He will not.          240
Age cannot wither her, nor custom stale
Her infinite variety. Other women cloy
The appetites they feed, but she makes hungry
Where most she satisfies. For vilest things
Become themselves in her, that the holy priests          245
Bless her when she is riggish.

MAECENAS
If beauty, wisdom, modesty can settle
The heart of Antony, Octavia is
A blessèd lottery to him.

AGRIPPA                    Let us go.
Good Enobarbus, make yourself my guest          250
Whilst you abide here.

ENOBARBUS                    Humbly, sir, I thank you.
                                   Exeunt

**2.3**  *Enter Antony and Caesar; Octavia between them*

ANTONY
The world and my great office will sometimes
Divide me from your bosom.

OCTAVIA                    All which time,
Before the gods my knee shall bow my prayers
To them for you.

ANTONY                    Good night, sir. My Octavia,
Read not my blemishes in the world's report.          5
I have not kept my square, but that to come
Shall all be done by th' rule. Good night, dear lady.
Good night, sir.

CAESAR Good night.                    *Exeunt Caesar and Octavia*
          *Enter Soothsayer*

ANTONY
Now, sirrah. You do wish yourself in Egypt?          10

SOOTHSAYER
Would I had never come from thence, nor you
Gone thither.

ANTONY                    If you can, your reason?

SOOTHSAYER
I see it in my motion, have it not in my tongue.
But yet hie you to Egypt again.

ANTONY                    Say to me
Whose fortunes shall rise higher: Caesar's or mine?          15

SOOTHSAYER
Caesar's. Therefore, O Antony, stay not by his side.
Thy daemon, that thy spirit which keeps thee, is
Noble, courageous, high, unmatchable,
Where Caesar's is not. But near him thy angel
Becomes afeard, as being o'erpowered. Therefore          20
Make space enough between you.

ANTONY                    Speak this no more.

SOOTHSAYER
To none but thee; no more but when to thee.
If thou dost play with him at any game
Thou art sure to lose; and of that natural luck
He beats thee 'gainst the odds. Thy lustre thickens          25
When he shines by. I say again, thy spirit
Is all afraid to govern thee near him;
But he away, 'tis noble.

ANTONY                    Get thee gone.
Say to Ventidius I would speak with him.
                                   *Exit Soothsayer*
He shall to Parthia; be it art or hap,          30
He hath spoken true. The very dice obey him,
And in our sports my better cunning faints
Under his chance. If we draw lots, he speeds.
His cocks do win the battle still of mine
When it is all to nought, and his quails ever          35
Beat mine, inhooped, at odds. I will to Egypt;
And though I make this marriage for my peace,
I'th' East my pleasure lies.
          *Enter Ventidius*
                                   O, come, Ventidius.
You must to Parthia, your commission's ready.
Follow me, and receive't.                    *Exeunt*

**2.4**  *Enter Lepidus, Maecenas, and Agrippa*

LEPIDUS
Trouble yourselves no further. Pray you, hasten
Your generals after.

AGRIPPA                    Sir, Mark Antony
Will e'en but kiss Octavia, and we'll follow.

LEPIDUS
Till I shall see you in your soldier's dress,
Which will become you both, farewell.

MAECENAS                    We shall,          5
As I conceive the journey, be at the Mount
Before you, Lepidus.

LEPIDUS                    Your way is shorter.
My purposes do draw me much about.
You'll win two days upon me.

MAECENAS *and* AGRIPPA                    Sir, good success.

LEPIDUS Farewell.          *Exeunt Maecenas and Agrippa at one*
                              *door, Lepidus at another*

**2.5**  *Enter Cleopatra, Charmian, Iras, and Alexas*

CLEOPATRA
Give me some music—music, moody food
Of us that trade in love.

CHARMIAN, IRAS, *and* ALEXAS The music, ho!

*Enter Mardian, the eunuch*

CLEOPATRA

Let it alone. Let's to billiards. Come, Charmian.

CHARMIAN

My arm is sore. Best play with Mardian.

CLEOPATRA

As well a woman with an eunuch played                              5
As with a woman. Come, you'll play with me, sir?

MARDIAN As well as I can, madam.

CLEOPATRA

And when good will is showed, though't come too
     short
The actor may plead pardon. I'll none now.
Give me mine angle. We'll to th' river. There,         10
My music playing far off, I will betray
Tawny-finned fishes. My bended hook shall pierce
Their slimy jaws, and as I draw them up
I'll think them every one an Antony,
And say 'Ah ha, you're caught!'

CHARMIAN                              'Twas merry when
You wagered on your angling, when your diver      16
Did hang a salt fish on his hook, which he
With fervency drew up.

CLEOPATRA                    That time—O times!—
I laughed him out of patience, and that night
I laughed him into patience, and next morn,         20
Ere the ninth hour, I drunk him to his bed,
Then put my tires and mantles on him whilst
I wore his sword Philippan.

     *Enter a Messenger*

                                        O, from Italy.
Ram thou thy fruitful tidings in mine ears,
That long time have been barren.

MESSENGER                              Madam, madam!  25

CLEOPATRA

Antonio's dead. If thou say so, villain,
Thou kill'st thy mistress; but well and free,
If thou so yield him, there is gold, and here
My bluest veins to kiss—a hand that kings
Have lipped, and trembled kissing.

MESSENGER                              First, madam, he is well.

CLEOPATRA

Why, there's more gold. But, sirrah, mark: we use  31
To say the dead are well. Bring it to that,
The gold I give thee will I melt and pour
Down thy ill-uttering throat.

MESSENGER Good madam, hear me.                         35

CLEOPATRA Well, go to, I will.

But there's no goodness in thy face. If Antony
Be free and healthful, so tart a favour
To trumpet such good tidings! If not well,
Thou shouldst come like a Fury crowned with snakes,
Not like a formal man.

MESSENGER                    Will't please you hear me?  41

CLEOPATRA

I have a mind to strike thee ere thou speak'st.
Yet if thou say Antony lives, is well,
Or friends with Caesar, or not captive to him,
I'll set thee in a shower of gold, and hail         45
Rich pearls upon thee.

MESSENGER                    Madam, he's well.

CLEOPATRA                              Well said.

MESSENGER

And friends with Caesar.

CLEOPATRA                              Thou'rt an honest man.

MESSENGER

Caesar and he are greater friends than ever.

CLEOPATRA

Make thee a fortune from me.

MESSENGER                              But yet, madam—

CLEOPATRA

I do not like 'But yet'; it does allay              50
The good precedence. Fie upon 'But yet'.
'But yet' is as a jailer to bring forth
Some monstrous malefactor. Prithee, friend,
Pour out the pack of matter to mine ear,
The good and bad together. He's friends with Caesar,
In state of health, thou sayst; and, thou sayst, free. 56

MESSENGER

Free, madam? No, I made no such report.
He's bound unto Octavia.

CLEOPATRA                              For what good turn?

MESSENGER

For the best turn i'th' bed.

CLEOPATRA                              I am pale, Charmian.

MESSENGER

Madam, he's married to Octavia.                      60

CLEOPATRA

The most infectious pestilence upon thee!
     *She strikes him down*

MESSENGER

Good madam, patience!

CLEOPATRA                              What say you?
     *She strikes him*
Hence, horrible villain, or I'll spurn thine eyes
Like balls before me. I'll unhair thy head,
     *She hales him up and down*
Thou shalt be whipped with wire and stewed in brine,
Smarting in ling'ring pickle.

MESSENGER                              Gracious madam,      66
I that do bring the news made not the match.

CLEOPATRA

Say 'tis not so, a province I will give thee,
And make thy fortunes proud. The blow thou hadst
Shall make thy peace for moving me to rage,         70
And I will boot thee with what gift beside
Thy modesty can beg.

MESSENGER                    He's married, madam.

CLEOPATRA

Rogue, thou hast lived too long.
     *She draws a knife*

MESSENGER                              Nay then, I'll run.
What mean you, madam? I have made no fault.   *Exit*

CHARMIAN

Good madam, keep yourself within yourself.          75
The man is innocent.

CLEOPATRA

Some innocents 'scape not the thunderbolt.
Melt Egypt into Nile, and kindly creatures
Turn all to serpents! Call the slave again.
Though I am mad I will not bite him. Call!          80

CHARMIAN

He is afeard to come.

CLEOPATRA                              I will not hurt him.
     *[Exit Charmian]*
These hands do lack nobility that they strike
A meaner than myself, since I myself
Have given myself the cause.
     *Enter the Messenger again [with Charmian]*
                                   Come hither, sir.
Though it be honest, it is never good               85

To bring bad news. Give to a gracious message
An host of tongues, but let ill tidings tell
Themselves when they be felt.
MESSENGER  I have done my duty.
CLEOPATRA  Is he married?                                          90
I cannot hate thee worser than I do
If thou again say 'Yes'.
MESSENGER                    He's married, madam.
CLEOPATRA
The gods confound thee! Dost thou hold there still?
MESSENGER
Should I lie, madam?
CLEOPATRA                    O, I would thou didst,
So half my Egypt were submerged and made     95
A cistern for scaled snakes. Go, get thee hence.
Hadst thou Narcissus in thy face, to me
Thou wouldst appear most ugly. He is married?
MESSENGER
I crave your highness' pardon.
CLEOPATRA                    He is married?
MESSENGER
Take no offence that I would not offend you.    100
To punish me for what you make me do
Seems much unequal. He's married to Octavia.
CLEOPATRA
O that his fault should make a knave of thee,
That act not what thou'rt sure of! Get thee hence.
The merchandise which thou hast brought from Rome
Are all too dear for me. Lie they upon thy hand,   106
And be undone by 'em.            Exit Messenger
CHARMIAN            Good your highness, patience.
CLEOPATRA
In praising Antony I have dispraised Caesar.
CHARMIAN  Many times, madam.
CLEOPATRA
I am paid for't now. Lead me from hence.        110
I faint. O Iras, Charmian—'tis no matter.
Go to the fellow, good Alexas, bid him
Report the feature of Octavia: her years,
Her inclination; let him not leave out
The colour of her hair. Bring me word quickly.    115
                                        Exit Alexas

Let him for ever go—let him not, Charmian;
Though he be painted one way like a Gorgon,
The other way's a Mars. ⌜To Mardian⌝ Bid you Alexas
Bring me word how tall she is. Pity me, Charmian,
But do not speak to me. Lead me to my chamber.    120
                                        Exeunt


2.6        Flourish. Enter Pompey and Menas at one door,
           with a drummer and a trumpeter; at another,
           Caesar, Lepidus, Antony, Enobarbus, Maecenas,
           Agrippa, with soldiers marching

POMPEY
Your hostages I have, so have you mine,
And we shall talk before we fight.
CAESAR                    Most meet
That first we come to words, and therefore have we
Our written purposes before us sent,
Which if thou hast considered, let us know      5
If 'twill tie up thy discontented sword
And carry back to Sicily much tall youth
That else must perish here.
POMPEY                    To you all three,
The senators alone of this great world,
Chief factors for the gods: I do not know      10

Wherefore my father should revengers want,
Having a son and friends, since Julius Caesar,
Who at Philippi the good Brutus ghosted,
There saw you labouring for him. What was't
That moved pale Cassius to conspire? And what    15
Made the all-honoured, honest Roman Brutus,
With the armed rest, courtiers of beauteous freedom,
To drench the Capitol but that they would
Have one man but a man? And that is it
Hath made me rig my navy, at whose burden     20
The angered ocean foams; with which I meant
To scourge th'ingratitude that despiteful Rome
Cast on my noble father.
CAESAR                    Take your time.
ANTONY
Thou canst not fear us, Pompey, with thy sails.
We'll speak with thee at sea. At land thou know'st  25
How much we do o'ercount thee.
POMPEY                    At land indeed
Thou dost o'ercount me of my father's house,
But since the cuckoo builds not for himself,
Remain in't as thou mayst.
LEPIDUS                    Be pleased to tell us—
For this is from the present—how you take     30
The offers we have sent you.
CAESAR                    There's the point.
ANTONY
Which do not be entreated to, but weigh
What it is worth, embraced.
CAESAR                    And what may follow,
To try a larger fortune?
POMPEY                    You have made me offer
Of Sicily, Sardinia; and I must               35
Rid all the sea of pirates; then to send
Measures of wheat to Rome; this 'greed upon,
To part with unhacked edges, and bear back
Our targes undinted.
CAESAR, ANTONY, and LEPIDUS  That's our offer.
POMPEY                    Know, then,
I came before you here a man prepared         40
To take this offer. But Mark Antony
Put me to some impatience. Though I lose
The praise of it by telling, you must know,
When Caesar and your brother were at blows,
Your mother came to Sicily, and did find      45
Her welcome friendly.
ANTONY                    I have heard it, Pompey,
And am well studied for a liberal thanks
Which I do owe you.
POMPEY                    Let me have your hand.
           Pompey and Antony shake hands
I did not think, sir, to have met you here.
ANTONY
The beds i'th' East are soft; and thanks to you,   50
That called me timelier than my purpose hither;
For I have gained by't.
CAESAR (to Pompey)            Since I saw you last
There is a change upon you.
POMPEY                    Well, I know not
What counts harsh fortune casts upon my face,
But in my bosom shall she never come          55
To make my heart her vassal.
LEPIDUS                    Well met here.
POMPEY
I hope so, Lepidus. Thus we are agreed.
I crave our composition may be written

And sealed between us.

CAESAR                    That's the next to do.

POMPEY
We'll feast each other ere we part, and let's          60
Draw lots who shall begin.

ANTONY That will I, Pompey.

POMPEY No, Antony, take the lot.
But, first or last, your fine Egyptian cookery
Shall have the fame. I have heard that Julius Caesar
Grew fat with feasting there.

ANTONY                    You have heard much.

POMPEY I have fair meanings, sir.          67

ANTONY And fair words to them.

POMPEY Then so much have I heard,
And I have heard Apollodorus carried—          70

ENOBARBUS
No more o' that, he did so.

POMPEY                    What, I pray you?

ENOBARBUS
A certain queen to Caesar in a mattress.

POMPEY
I know thee now. How far'st thou, soldier?

ENOBARBUS
Well, and well am like to do, for I perceive
Four feasts are toward.

POMPEY                    Let me shake thy hand.          75
          *Pompey and Enobarbus shake hands*
I never hated thee. I have seen thee fight
When I have envied thy behaviour.

ENOBARBUS
Sir, I never loved you much, but I ha' praised ye
When you have well deserved ten times as much
As I have said you did.          80

POMPEY
Enjoy thy plainness. It nothing ill becomes thee.
Aboard my galley I invite you all.
Will you lead, lords?

CAESAR, ANTONY, *and* LEPIDUS Show's the way, sir.

POMPEY                              Come.
          *Exeunt all but Enobarbus and Menas*

MENAS (*aside*)
Thy father, Pompey, would ne'er have made this
          treaty.
(*To Enobarbus*) You and I have known, sir.          85

ENOBARBUS At sea, I think.

MENAS We have, sir.

ENOBARBUS You have done well by water.

MENAS And you by land.          89

ENOBARBUS I will praise any man that will praise me,
though it cannot be denied what I have done by land.

MENAS Nor what I have done by water.

ENOBARBUS Yes, something you can deny for your own
safety. You have been a great thief by sea.

MENAS And you by land.          95

ENOBARBUS There I deny my land service; but give me
your hand, Menas. If our eyes had authority, here they
might take two thieves kissing.
          *They shake hands*

MENAS All men's faces are true, whatsome'er their hands
are.          100

ENOBARBUS But there is never a fair woman has a true
face.

MENAS No slander; they steal hearts.

ENOBARBUS We came hither to fight with you.

MENAS For my part, I am sorry it is turned to a drinking.
Pompey doth this day laugh away his fortune.          106

ENOBARBUS If he do, sure he cannot weep't back again.

MENAS You've said, sir. We looked not for Mark Antony
here. Pray you, is he married to Cleopatra?

ENOBARBUS Caesar's sister is called Octavia.          110

MENAS True, sir. She was the wife of Caius Marcellus.

ENOBARBUS But she is now the wife of Marcus Antonius.

MENAS Pray ye, sir?

ENOBARBUS 'Tis true.

MENAS Then is Caesar and he for ever knit together.          115

ENOBARBUS If I were bound to divine of this unity I would
not prophesy so.

MENAS I think the policy of that purpose made more in
the marriage than the love of the parties.

ENOBARBUS I think so, too. But you shall find the band
that seems to tie their friendship together will be the
very strangler of their amity. Octavia is of a holy, cold,
and still conversation.

MENAS Who would not have his wife so?          124

ENOBARBUS Not he that himself is not so, which is Mark
Antony. He will to his Egyptian dish again; then shall
the sighs of Octavia blow the fire up in Caesar, and, as
I said before, that which is the strength of their amity
shall prove the immediate author of their variance.
Antony will use his affection where it is. He married
but his occasion here.          131

MENAS And thus it may be. Come, sir, will you aboard?
I have a health for you.

ENOBARBUS I shall take it, sir. We have used our throats
in Egypt.          135

MENAS Come, let's away.                    *Exeunt*

2.7     *Music plays. Enter two or three Servants with a*
          *banquet*

FIRST SERVANT Here they'll be, man. Some o' their plants
are ill rooted already; the least wind i'th' world will
blow them down.

SECOND SERVANT Lepidus is high-coloured.

FIRST SERVANT They have made him drink alms-drink.          5

SECOND SERVANT As they pinch one another by the
disposition, he cries out 'No more!'—reconciles them
to his entreaty and himself to th' drink.

FIRST SERVANT But it raises the greater war between him
and his discretion.          10

SECOND SERVANT Why, this it is to have a name in great
men's fellowship. I had as lief have a reed that will do
me no service as a partisan I could not heave.

FIRST SERVANT To be called into a huge sphere and not
to be seen to move in't, are the holes where eyes should
be which pitifully disaster the cheeks.          16
          *A sennet sounded. Enter Caesar, Antony, Pompey,*
          *Lepidus, Agrippa, Maecenas, Enobarbus, and*
          *Menas, with other captains ⌐and a boy⌐*

ANTONY (*to Caesar*)
Thus do they, sir: they take the flow o'th' Nile
By certain scales i'th' pyramid. They know
By th' height, the lowness, or the mean, if dearth
Or foison follow. The higher Nilus swells          20
The more it promises; as it ebbs, the seedsman
Upon the slime and ooze scatters his grain,
And shortly comes to harvest.

LEPIDUS You've strange serpents there?

ANTONY Ay, Lepidus.          25

LEPIDUS Your serpent of Egypt is bred now of your mud
by the operation of your sun; so is your crocodile.

ANTONY They are so.

POMPEY
  Sit, and some wine. A health to Lepidus!       29
  ⌐Antony, Pompey, and Lepidus sit⌐
LEPIDUS I am not so well as I should be, but I'll ne'er out.
ENOBARBUS Not till you have slept—I fear me you'll be in
  till then.
LEPIDUS Nay, certainly, I have heard the Ptolemies'
  pyramises are very goodly things: without contra-
  diction I have heard that.                     35
MENAS (aside to Pompey)
  Pompey, a word.
POMPEY (aside to Menas) Say in mine ear; what is't?
MENAS (aside to Pompey)
  Forsake thy seat, I do beseech thee, captain,
  And hear me speak a word.
POMPEY (aside to Menas)       Forbear me till anon.
  (Aloud) This wine for Lepidus!
      Menas whispers in Pompey's ear
LEPIDUS What manner o' thing is your crocodile?   40
ANTONY It is shaped, sir, like itself, and it is as broad as
  it hath breadth. It is just so high as it is, and moves
  with it own organs. It lives by that which nourisheth
  it, and the elements once out of it, it transmigrates.
LEPIDUS What colour is it of?                 45
ANTONY Of it own colour, too.
LEPIDUS 'Tis a strange serpent.
ANTONY 'Tis so, and the tears of it are wet.
CAESAR (to Antony)
  Will this description satisfy him?
ANTONY With the health that Pompey gives him; else he
  is a very epicure.                         51
POMPEY (aside to Menas)
  Go hang, sir, hang! Tell me of that? Away,
  Do as I bid you. (Aloud) Where's this cup I called for?
MENAS (aside to Pompey)
  If for the sake of merit thou wilt hear me,
  Rise from thy stool.
POMPEY ⌐rising⌐     I think thou'rt mad. The matter?
  ⌐Menas and Pompey stand apart⌐
MENAS
  I have ever held my cap off to thy fortunes.    56
POMPEY
  Thou hast served me with much faith. What's else to
    say?
  Be jolly, lords.
ANTONY       These quicksands, Lepidus,
  Keep off them, for you sink.
MENAS
  Wilt thou be lord of all the world?
POMPEY               What sayst thou?
MENAS
  Wilt thou be lord of the whole world? That's twice. 61
POMPEY
  How should that be?
MENAS          But entertain it
  And, though thou think me poor, I am the man
  Will give thee all the world.
POMPEY          Hast thou drunk well?
MENAS
  No, Pompey, I have kept me from the cup.     65
  Thou art, if thou dar'st be, the earthly Jove.
  Whate'er the ocean pales or sky inclips
  Is thine, if thou wilt ha't.
POMPEY         Show me which way!
MENAS
  These three world-sharers, these competitors,
  Are in thy vessel. Let me cut the cable;      70

  And when we are put off, fall to their throats.
  All there is thine.
POMPEY       Ah, this thou shouldst have done
  And not have spoke on't. In me 'tis villainy,
  In thee 't had been good service. Thou must know
  'Tis not my profit that does lead mine honour;    75
  Mine honour, it. Repent that e'er thy tongue
  Hath so betrayed thine act. Being done unknown,
  I should have found it afterwards well done,
  But must condemn it now. Desist, and drink.
      He returns to the others
MENAS (aside)
  For this, I'll never follow thy palled fortunes more.  80
  Who seeks and will not take when once 'tis offered,
  Shall never find it more.
POMPEY         This health to Lepidus!
ANTONY
  Bear him ashore.—I'll pledge it for him, Pompey.
ENOBARBUS
  Here's to thee, Menas!
MENAS          Enobarbus, welcome.
POMPEY
  Fill till the cup be hid.
      One lifts Lepidus, drunk, and carries him off
ENOBARBUS         There's a strong fellow, Menas.
MENAS Why?                       86
ENOBARBUS
  A bears the third part of the world, man; seest not?
MENAS
  The third part then is drunk. Would it were all,
  That it might go on wheels.
ENOBARBUS      Drink thou, increase the reels.
MENAS Come.                     90
POMPEY
  This is not yet an Alexandrian feast.
ANTONY
  It ripens towards it. Strike the vessels, ho!
  Here's to Caesar!
CAESAR       I could well forbear't.
  It's monstrous labour when I wash my brain,
  An it grow fouler.                     95
ANTONY Be a child o'th' time.
CAESAR Possess it, I'll make answer.
  But I had rather fast from all, four days,
  Than drink so much in one.
ENOBARBUS (to Antony)     Ha, my brave Emperor,
  Shall we dance now the Egyptian bacchanals,   100
  And celebrate our drink?
POMPEY Let's ha't, good soldier.
ANTONY Come, let's all take hands
  Till that the conquering wine hath steeped our sense
  In soft and delicate Lethe.
ENOBARBUS         All take hands.   105
  Make battery to our ears with the loud music.
  The while I'll place you, then the boy shall sing.
  The holding every man shall beat as loud
  As his strong sides can volley.
      Music plays. Enobarbus places them hand in hand
⌐BOY⌐ (sings)
      Come, thou monarch of the vine,   110
      Plumpy Bacchus, with pink eyne!
      In thy vats our cares be drowned,
      With thy grapes our hairs be crowned!
      Cup us till the world go round,
      Cup us till the world go round!   115

CAESAR
What would you more? Pompey, good night.
(*To Antony*) Good-brother,
Let me request you off. Our graver business
Frowns at this levity. Gentle lords, let's part.
You see we have burnt our cheeks. Strong Enobarb
Is weaker than the wine, and mine own tongue 120
Splits what it speaks. The wild disguise hath almost
Anticked us all. What needs more words? Good night.
Good Antony, your hand.

POMPEY I'll try you on the shore.

ANTONY
And shall, sir. Give's your hand.

POMPEY O Antony,
You have my father's house. But what, we are friends!
Come down into the boat. 126

*Exeunt all but Enobarbus and Menas*

ENOBARBUS
Take heed you fall not, Menas.

MENAS I'll not on shore.
No, to my cabin. These drums, these trumpets, flutes,
what!
Let Neptune hear we bid a loud farewell
To these great fellows. Sound and be hanged, sound out!

*Sound a flourish, with drums*

ENOBARBUS (*throwing his cap in the air*)
Hoo, says a! There's my cap.

MENAS Ho, noble captain, come!

*Exeunt*

**3.1** *Enter Ventidius, with Silius and other Roman*
*soldiers, as it were in triumph; the dead body of*
*Pacorus borne before him*

VENTIDIUS
Now, darting Parthia, art thou struck; and now
Pleased fortune does of Marcus Crassus' death
Make me revenger. Bear the King's son's body
Before our army. Thy Pacorus, Orodes,
Pays this for Marcus Crassus.

SILIUS Noble Ventidius, 5
Whilst yet with Parthian blood thy sword is warm,
The fugitive Parthians follow. Spur through Media,
Mesopotamia, and the shelters whither
The routed fly. So thy grand captain, Antony,
Shall set thee on triumphant chariots and 10
Put garlands on thy head.

VENTIDIUS O Silius, Silius,
I have done enough. A lower place, note well,
May make too great an act. For learn this, Silius:
Better to leave undone than by our deed
Acquire too high a fame when him we serve's away.
Caesar and Antony have ever won 16
More in their officer than person. Sossius,
One of my place in Syria, his lieutenant,
For quick accumulation of renown,
Which he achieved by th' minute, lost his favour. 20
Who does i'th' wars more than his captain can
Becomes his captain's captain; and ambition,
The soldier's virtue, rather makes choice of loss
Than gain which darkens him.
I could do more to do Antonius good, 25
But 'twould offend him, and in his offence
Should my performance perish.

SILIUS Thou hast, Ventidius, that
Without the which a soldier and his sword
Grants scarce distinction. Thou wilt write to Antony?

VENTIDIUS
I'll humbly signify what in his name, 30
That magical word of war, we have effected;
How, with his banners and his well-paid ranks,
The ne'er-yet-beaten horse of Parthia
We have jaded out o'th' field.

SILIUS Where is he now?

VENTIDIUS
He purposeth to Athens; whither, with what haste 35
The weight we must convey with's will permit,
We shall appear before him.—On there; pass along.

*Exeunt*

**3.2** *Enter Agrippa at one door, Enobarbus at another*

AGRIPPA What, are the brothers parted?

ENOBARBUS
They have dispatched with Pompey; he is gone.
The other three are sealing. Octavia weeps
To part from Rome, Caesar is sad, and Lepidus
Since Pompey's feast, as Menas says, is troubled 5
With the green-sickness.

AGRIPPA 'Tis a noble Lepidus.

ENOBARBUS
A very fine one. O, how he loves Caesar!

AGRIPPA
Nay, but how dearly he adores Mark Antony!

ENOBARBUS
Caesar? Why, he's the Jupiter of men.

AGRIPPA
What's Antony—the god of Jupiter? 10

ENOBARBUS
Spake you of Caesar? How, the nonpareil?

AGRIPPA
O Antony, O thou Arabian bird!

ENOBARBUS
Would you praise Caesar, say 'Caesar'; go no further.

AGRIPPA
Indeed, he plied them both with excellent praises.

ENOBARBUS
But he loves Caesar best; yet he loves Antony— 15
Hoo! Hearts, tongues, figures, scribes, bards, poets,
cannot
Think, speak, cast, write, sing, number—hoo!—
His love to Antony. But as for Caesar—
Kneel down, kneel down, and wonder.

AGRIPPA Both he loves.

ENOBARBUS
They are his shards, and he their beetle.

⌈*Trumpet within*⌉
So, 20
This is to horse. Adieu, noble Agrippa.

AGRIPPA
Good fortune, worthy soldier, and farewell.

*Enter Caesar, Antony, Lepidus, and Octavia*

ANTONY (*to Caesar*) No further, sir.

CAESAR
You take from me a great part of myself.
Use me well in't. Sister, prove such a wife 25
As my thoughts make thee, and as my farthest bond
Shall pass on thy approof. Most noble Antony,
Let not the piece of virtue which is set
Betwixt us as the cement of our love
To keep it builded, be the ram to batter 30
The fortress of it; for better might we
Have loved without this mean if on both parts

This be not cherished.

ANTONY                    Make me not offended
  In your distrust.
CAESAR            I have said.
ANTONY                        You shall not find,
  Though you be therein curious, the least cause          35
  For what you seem to fear. So, the gods keep you,
  And make the hearts of Romans serve your ends.
  We will here part.
CAESAR
  Farewell, my dearest sister, fare thee well.
  The elements be kind to thee, and make          40
  Thy spirits all of comfort. Fare thee well.
OCTAVIA (*weeping*) My noble brother!
ANTONY
  The April's in her eyes; it is love's spring,
  And these the showers to bring it on. Be cheerful.
OCTAVIA
  Sir, look well to my husband's house, and—          45
CAESAR
  What, Octavia?
OCTAVIA            I'll tell you in your ear.
        *She whispers to Caesar*
ANTONY
  Her tongue will not obey her heart, nor can
  Her heart inform her tongue—the swan's-down
        feather,
  That stands upon the swell at full of tide,
  And neither way inclines.          50
ENOBARBUS (*aside to Agrippa*) Will Caesar weep?
AGRIPPA (*aside to Enobarbus*) He has a cloud in's face.
ENOBARBUS (*aside to Agrippa*)
  He were the worse for that were he a horse;
  So is he, being a man.
AGRIPPA (*aside to Enobarbus*) Why, Enobarbus,
  When Antony found Julius Caesar dead          55
  He cried almost to roaring, and he wept
  When at Philippi he found Brutus slain.
ENOBARBUS (*aside to Agrippa*)
  That year indeed he was troubled with a rheum.
  What willingly he did confound he wailed,
  Believe't, till I wept too.
CAESAR                No, sweet Octavia,          60
  You shall hear from me still. The time shall not
  Outgo my thinking on you.
ANTONY                Come, sir, come,
  I'll wrestle with you in my strength of love.
  Look, here I have you (*embracing Caesar*); thus I let
        you go,
  And give you to the gods.
CAESAR                Adieu, be happy.          65
LEPIDUS
  Let all the number of the stars give light
  To thy fair way.
CAESAR            Farewell, farewell.
        *He kisses Octavia*
ANTONY                    Farewell.
        *Trumpets sound. Exeunt Antony, Octavia, and
            Enobarbus at one door, Caesar, Lepidus, and
                    Agrippa at another*

**3.3**  *Enter Cleopatra, Charmian, Iras, and Alexas*
CLEOPATRA
  Where is the fellow?
ALEXAS                Half afeard to come.

CLEOPATRA
  Go to, go to.
        *Enter the Messenger as before*
                    Come hither, sir.
ALEXAS                Good majesty,
  Herod of Jewry dare not look upon you
  But when you are well pleased.
CLEOPATRA                That Herod's head
  I'll have; but how, when Antony is gone,          5
  Through whom I might command it?
  (*To the Messenger*)            Come thou near.
MESSENGER
  Most gracious majesty!
CLEOPATRA                Didst thou behold
  Octavia?
MESSENGER  Ay, dread Queen.
CLEOPATRA                Where?
MESSENGER                Madam, in Rome.
  I looked her in the face, and saw her led
  Between her brother and Mark Antony.          10
CLEOPATRA
  Is she as tall as me?
MESSENGER            She is not, madam.
CLEOPATRA
  Didst hear her speak? Is she shrill-tongued or low?
MESSENGER
  Madam, I heard her speak. She is low-voiced.
CLEOPATRA
  That's not so good. He cannot like her long.
CHARMIAN
  Like her? O Isis, 'tis impossible!          15
CLEOPATRA
  I think so, Charmian. Dull of tongue, and dwarfish.
  What majesty is in her gait? Remember
  If e'er thou looked'st on majesty.
MESSENGER                She creeps.
  Her motion and her station are as one.
  She shows a body rather than a life,          20
  A statue than a breather.
CLEOPATRA                Is this certain?
MESSENGER
  Or I have no observance.
CHARMIAN                Three in Egypt
  Cannot make better note.
CLEOPATRA                He's very knowing,
  I do perceive't. There's nothing in her yet.
  The fellow has good judgement.
CHARMIAN                Excellent.          25
CLEOPATRA (*to the Messenger*)
  Guess at her years, I prithee.
MESSENGER                Madam,
  She was a widow—
CLEOPATRA                Widow? Charmian, hark.
MESSENGER  And I do think she's thirty.
CLEOPATRA
  Bear'st thou her face in mind? Is't long or round?
MESSENGER  Round, even to faultiness.          30
CLEOPATRA
  For the most part, too, they are foolish that are so.
  Her hair—what colour?
MESSENGER                Brown, madam; and her forehead
  As low as she would wish it.
CLEOPATRA (*giving money*)        There's gold for thee.
  Thou must not take my former sharpness ill.
  I will employ thee back again. I find thee          35

Most fit for business. Go, make thee ready.
Our letters are prepared.         *Exit Messenger*
CHARMIAN         A proper man.
CLEOPATRA
Indeed he is so. I repent me much
That so I harried him. Why, methinks, by him,
This creature's no such thing.
CHARMIAN         Nothing, madam.    40
CLEOPATRA
The man hath seen some majesty, and should know.
CHARMIAN
Hath he seen majesty? Isis else defend,
And serving you so long!
CLEOPATRA
I have one thing more to ask him yet, good
   Charmian.
But 'tis no matter. Thou shalt bring him to me    45
Where I will write. All may be well enough.
CHARMIAN   I warrant you, madam.         *Exeunt*

**3.4**    *Enter Antony and Octavia*
ANTONY
Nay, nay, Octavia, not only that,
That were excusable, that and thousands more
Of semblable import; but he hath waged
New wars 'gainst Pompey, made his will and read it
To public ear, spoke scantly of me;    5
When perforce he could not
But pay me terms of honour, cold and sickly
He vented them, most narrow measure lent me.
When the best hint was given him, he not took't,
Or did it from his teeth.
OCTAVIA         O my good lord,    10
Believe not all, or if you must believe,
Stomach not all. A more unhappy lady,
If this division chance, ne'er stood between,
Praying for both parts.
The good gods will mock me presently,    15
When I shall pray 'O, bless my lord and husband!',
Undo that prayer by crying out as loud
'O, bless my brother!' Husband win, win brother
Prays and destroys the prayer; no midway
'Twixt these extremes at all.
ANTONY         Gentle Octavia,    20
Let your best love draw to that point which seeks
Best to preserve it. If I lose mine honour,
I lose myself. Better I were not yours
Than yours so branchless. But, as you requested,
Yourself shall go between's. The meantime, lady,    25
I'll raise the preparation of a war
Shall stain your brother. Make your soonest haste;
So your desires are yours.
OCTAVIA         Thanks to my lord.
The Jove of power make me most weak, most weak,
Your reconciler! Wars 'twixt you twain would be    30
As if the world should cleave, and that slain men
Should solder up the rift.
ANTONY
When it appears to you where this begins,
Turn your displeasure that way, for our faults
Can never be so equal that your love    35
Can equally move with them. Provide your going,
Choose your own company, and command what cost
Your heart has mind to.
                        *Exeunt*

**3.5**    *Enter Enobarbus and Eros, meeting*
ENOBARBUS   How now, friend Eros?
EROS   There's strange news come, sir.
ENOBARBUS   What, man?
EROS   Caesar and Lepidus have made wars upon Pompey.
ENOBARBUS   This is old. What is the success?    5
EROS   Caesar, having made use of him in the wars 'gainst
   Pompey, presently denied him rivality, would not let
   him partake in the glory of the action, and, not resting
   here, accuses him of letters he had formerly wrote to
   Pompey; upon his own appeal seizes him; so the poor
   third is up, till death enlarge his confine.    11
ENOBARBUS
Then, world, thou hast a pair of chops, no more,
And throw between them all the food thou hast,
They'll grind the one the other. Where's Antony?
EROS
He's walking in the garden, thus, and spurns    15
The rush that lies before him, cries 'Fool Lepidus!'
And threats the throat of that his officer
That murdered Pompey.
ENOBARBUS         Our great navy's rigged.
EROS
For Italy and Caesar. More, Domitius:
My lord desires you presently. My news    20
I might have told hereafter.
ENOBARBUS         'Twill be naught.
But let it be; bring me to Antony.
EROS         Come, sir.        *Exeunt*

**3.6**    *Enter Agrippa, Maecenas, and Caesar*
CAESAR
Contemning Rome, he has done all this and more
In Alexandria. Here's the manner of't:
I'th' market place on a tribunal silvered,
Cleopatra and himself in chairs of gold
Were publicly enthroned. At the feet sat    5
Caesarion, whom they call my father's son,
And all the unlawful issue that their lust
Since then hath made between them. Unto her
He gave the stablishment of Egypt; made her
Of lower Syria, Cyprus, Lydia,    10
Absolute queen.
MAECENAS         This in the public eye?
CAESAR
I'th' common showplace, where they exercise.
His sons he there proclaimed the kings of kings;
Great Media, Parthia, and Armenia
He gave to Alexander. To Ptolemy he assigned    15
Syria, Cilicia, and Phoenicia. She
In th'habiliments of the goddess Isis
That day appeared, and oft before gave audience,
As 'tis reported, so.
MAECENAS         Let Rome be thus informed.
AGRIPPA
Who, queasy with his insolence already,
Will their good thoughts call from him.    20
CAESAR         The people knows it,
And have now received his accusations.
AGRIPPA   Who does he accuse?
CAESAR
Caesar, and that having in Sicily
Sextus Pompeius spoiled, we had not rated him    25
His part o'th' isle. Then does he say he lent me
Some shipping, unrestored. Lastly, he frets
That Lepidus of the triumvirate

Should be deposed; and being, that we detain
All his revenue."
AGRIPPA                  Sir, this should be answered.      30
CAESAR
'Tis done already, and the messenger gone.
I have told him Lepidus was grown too cruel,
That he his high authority abused
And did deserve his change. For what I have
          conquered,
I grant him part; but then in his Armenia,      35
And other of his conquered kingdoms,
I demand the like.
MAECENAS            He'll never yield to that.
CAESAR
Nor must not then be yielded to in this.
          Enter Octavia with her train
OCTAVIA
Hail, Caesar, and my lord; hail, most dear Caesar!
CAESAR
That ever I should call thee castaway!      40
OCTAVIA
You have not called me so, nor have you cause.
CAESAR
Why have you stol'n upon us thus? You come not
Like Caesar's sister. The wife of Antony
Should have an army for an usher, and
The neighs of horse to tell of her approach      45
Long ere she did appear. The trees by th' way
Should have borne men, and expectation fainted,
Longing for what it had not. Nay, the dust
Should have ascended to the roof of heaven,
Raised by your populous troops. But you are come      50
A market maid to Rome, and have prevented
The ostentation of our love; which, left unshown,
Is often left unloved. We should have met you
By sea and land, supplying every stage
With an augmented greeting.
OCTAVIA                  Good my lord,      55
To come thus was I not constrained, but did it
On my free will. My lord, Mark Antony,
Hearing that you prepared for war, acquainted
My grievèd ear withal, whereon I begged
His pardon for return.
CAESAR            Which soon he granted,      60
Being an obstruct 'tween his lust and him.
OCTAVIA
Do not say so, my lord.
CAESAR                  I have eyes upon him,
And his affairs come to me on the wind.
Where is he now?
OCTAVIA            My lord, in Athens.
CAESAR
No, my most wrongèd sister. Cleopatra      65
Hath nodded him to her. He hath given his empire
Up to a whore; who now are levying
The kings o'th' earth for war. He hath assembled
Bocchus, the King of Libya; Archelaus
Of Cappadocia; Philadelphos, King      70
Of Paphlagonia; the Thracian King Adallas;
King Malchus of Arabia; King of Pont;
Herod of Jewry; Mithridates, King
Of Comagene; Polemon and Amyntas,
The Kings of Mede and Lycaonia;      75
With a more larger list of sceptres.
OCTAVIA                  Ay me most wretched,
That have my heart parted betwixt two friends

That does afflict each other!
CAESAR            Welcome hither.
Your letters did withhold our breaking forth
Till we perceived both how you were wrong led      80
And we in negligent danger. Cheer your heart.
Be you not troubled with the time, which drives
O'er your content these strong necessities;
But let determined things to destiny
Hold unbewailed their way. Welcome to Rome;      85
Nothing more dear to me. You are abused
Beyond the mark of thought, and the high gods,
To do you justice, makes their ministers
Of us and those that love you. Best of comfort,
And ever welcome to us.
AGRIPPA                  Welcome, lady.      90
MAECENAS Welcome, dear madam.
Each heart in Rome does love and pity you.
Only th'adulterous Antony, most large
In his abominations, turns you off,
And gives his potent regiment to a trull      95
That noises it against us.
OCTAVIA            Is it so, sir?
CAESAR
Most certain. Sister, welcome. Pray you
Be ever known to patience. My dear'st sister!      Exeunt

3.7      Enter Cleopatra and Enobarbus
CLEOPATRA
I will be even with thee, doubt it not.
ENOBARBUS But why, why, why?
CLEOPATRA
Thou hast forspoke my being in these wars,
And sayst it is not fit.
ENOBARBUS            Well, is it, is it?
CLEOPATRA
Is't not denounced against us? Why should not we      5
Be there in person?
ENOBARBUS ⌜aside⌝      Well, I could reply
If we should serve with horse and mares together,
The horse were merely lost; the mares would bear
A soldier and his horse.
CLEOPATRA            What is't you say?
ENOBARBUS
Your presence needs must puzzle Antony,      10
Take from his heart, take from his brain, from's time
What should not then be spared. He is already
Traduced for levity; and 'tis said in Rome
That Photinus, an eunuch, and your maids
Manage this war.
CLEOPATRA            Sink Rome, and their tongues rot      15
That speak against us! A charge we bear i'th' war,
And as the president of my kingdom will
Appear there for a man. Speak not against it.
I will not stay behind.
          Enter Antony and Camidius
ENOBARBUS            Nay, I have done.
Here comes the Emperor.
ANTONY                  Is it not strange, Camidius,
That from Tarentum and Brundisium      21
He could so quickly cut the Ionian Sea
And take in Toryne?—You have heard on't, sweet?
CLEOPATRA
Celerity is never more admired
Than by the negligent.
ANTONY                  A good rebuke,      25
Which might have well becomed the best of men

To taunt at slackness. Camidius, we
Will fight with him by sea.
CLEOPATRA                    By sea—what else?
CAMIDIUS
Why will my lord do so?
ANTONY                    For that he dares us to't.
ENOBARBUS
So hath my lord dared him to single fight.                    30
CAMIDIUS
Ay, and to wage this battle at Pharsalia,
Where Caesar fought with Pompey. But these offers
Which serve not for his vantage, he shakes off,
And so should you.
ENOBARBUS                    Your ships are not well manned,
Your mariners are muleters, reapers, people                    35
Engrossed by swift impress. In Caesar's fleet
Are those that often have 'gainst Pompey fought.
Their ships are yare, yours heavy. No disgrace
Shall fall you for refusing him at sea,
Being prepared for land.
ANTONY                    By sea, by sea.                    40
ENOBARBUS
Most worthy sir, you therein throw away
The absolute soldiership you have by land;
Distract your army, which doth most consist
Of war-marked footmen; leave unexecuted
Your own renownèd knowledge; quite forgo                    45
The way which promises assurance, and
Give up yourself merely to chance and hazard
From firm security.
ANTONY                    I'll fight at sea.
CLEOPATRA
I have sixty sails, Caesar none better.
ANTONY
Our overplus of shipping will we burn,                    50
And with the rest full-manned, from th'head of
   Actium
Beat th'approaching Caesar. But if we fail,
We then can do't at land.
      Enter a Messenger
                    Thy business?
MESSENGER
The news is true, my lord. He is descried.
Caesar has taken Toryne.                    55
ANTONY
Can he be there in person? 'Tis impossible;
Strange that his power should be. Camidius,
Our nineteen legions thou shalt hold by land,
And our twelve thousand horse. We'll to our ship.
Away, my Thetis!
      Enter a Soldier
                    How now, worthy soldier?                    60
SOLDIER
O noble Emperor, do not fight by sea.
Trust not to rotten planks. Do you misdoubt
This sword and these my wounds? Let th'Egyptians
And the Phoenicians go a-ducking; we
Have used to conquer standing on the earth,                    65
And fighting foot to foot.
ANTONY                    Well, well; away!
      Exeunt Antony, Cleopatra, and Enobarbus
SOLDIER
By Hercules, I think I am i'th' right.
CAMIDIUS
Soldier, thou art; but his whole action grows

Not in the power on't. So our leader's led,
And we are women's men.
SOLDIER                    You keep by land                    70
The legions and the horse whole, do you not?
CAMIDIUS
Marcus Octavius, Marcus Justeius,
Publicola and Caelius are for sea,
But we keep whole by land. This speed of Caesar's
Carries beyond belief.
SOLDIER                    While he was yet in Rome                    75
His power went out in such distractions
As beguiled all spies.
CAMIDIUS                    Who's his lieutenant, hear you?
SOLDIER
They say, one Taurus.
CAMIDIUS                    Well I know the man.
      Enter a Messenger
MESSENGER
The Emperor calls Camidius.
CAMIDIUS
With news the time's in labour, and throws forth                    80
Each minute some.                    Exeunt

**3.8**    *Enter Caesar with his army, marching, and Taurus*
CAESAR Taurus!
TAURUS My lord?
CAESAR
Strike not by land. Keep whole. Provoke not battle
Till we have done at sea. (*Giving a scroll*) Do not
   exceed
The prescript of this scroll. Our fortune lies                    5
Upon this jump.
      *Exit Caesar and his army at one door, Taurus at
                                                another*

**3.9**    *Enter Antony and Enobarbus*
ANTONY
Set we our squadrons on yon side o'th' hill
In eye of Caesar's battle, from which place
We may the number of the ships behold,
And so proceed accordingly.                    *Exeunt*

**3.10**    *Camidius marcheth with his land army one way
         over the stage, and Taurus, the lieutenant of
         Caesar, with his army the other way. After their
         going in is heard the noise of a sea-fight. Alarum.
         Enter Enobarbus*
ENOBARBUS
Naught, naught, all naught! I can behold no longer.
Th'*Antoniad*, the Egyptian admiral,
With all their sixty, fly and turn the rudder.
To see't mine eyes are blasted.
      *Enter Scarus*
SCARUS                    Gods and goddesses—
All the whole synod of them!
ENOBARBUS                    What's thy passion?                    5
SCARUS
The greater cantle of the world is lost
With very ignorance; we have kissed away
Kingdoms and provinces.
ENOBARBUS                    How appears the fight?
SCARUS
On our side like the tokened pestilence,
Where death is sure. Yon riband-red nag of Egypt—
Whom leprosy o'ertake!—i'th' midst o'th' fight—                    11
When vantage like a pair of twins appeared,

Both as the same, or rather ours the elder—
The breese upon her, like a cow in June,
Hoists sails and flies.

ENOBARBUS           That I beheld.    15
Mine eyes did sicken at the sight, and could not
Endure a further view.

SCARUS           She once being luffed,
The noble ruin of her magic, Antony,
Claps on his sea-wing and, like a doting mallard,
Leaving the fight in height, flies after her.    20
I never saw an action of such shame.
Experience, manhood, honour, ne'er before
Did violate so itself.

ENOBARBUS        Alack, alack!

*Enter Camidius*

CAMIDIUS
Our fortune on the sea is out of breath,
And sinks most lamentably. Had our general    25
Been what he knew himself, it had gone well.
O, he has given example for our flight
Most grossly by his own.

ENOBARBUS
Ay, are you thereabouts? Why then, good night
indeed!

CAMIDIUS
Toward Peloponnesus are they fled.    30

SCARUS
'Tis easy to't, and there I will attend
What further comes.

CAMIDIUS        To Caesar will I render
My legions and my horse. Six kings already
Show me the way of yielding.

ENOBARBUS         I'll yet follow
The wounded chance of Antony, though my reason
Sits in the wind against me.    ⌜*Exeunt severally*⌝

**3.11** *Enter Antony with Attendants*

ANTONY
Hark, the land bids me tread no more upon't,
It is ashamed to bear me. Friends, come hither.
I am so lated in the world that I
Have lost my way for ever. I have a ship
Laden with gold. Take that; divide it, fly,    5
And make your peace with Caesar.

ATTENDANTS          Fly? Not we.

ANTONY
I have fled myself, and have instructed cowards
To run and show their shoulders. Friends, be gone.
I have myself resolved upon a course
Which has no need of you. Be gone.    10
My treasure's in the harbour. Take it. O,
I followed that I blush to look upon.
My very hairs do mutiny, for the white
Reprove the brown for rashness, and they them
For fear and doting. Friends, be gone. You shall    15
Have letters from me to some friends that will
Sweep your way for you. Pray you, look not sad,
Nor make replies of loathness. Take the hint
Which my despair proclaims. Let that be left
Which leaves itself. To the seaside straightway!    20
I will possess you of that ship and treasure.
Leave me, I pray, a little. Pray you now,
Nay, do so; for indeed I have lost command.
Therefore I pray you; I'll see you by and by.

*Exeunt attendants*

*He sits down.*
*Enter Cleopatra led by Charmian, Iras, and Eros*

EROS
Nay, gentle madam, to him. Comfort him.    25

IRAS Do, most dear Queen.

CHARMIAN Do. Why, what else?

CLEOPATRA Let me sit down. O Juno!

*She sits down*

ANTONY No, no, no, no, no.

EROS (*to Antony*) See you here, sir?    30

ANTONY O fie, fie, fie!

CHARMIAN Madam.

IRAS Madam. O good Empress!

EROS Sir, sir.

ANTONY
Yes, my lord, yes. He at Philippi kept    35
His sword e'en like a dancer, while I struck
The lean and wrinkled Cassius; and 'twas I
That the mad Brutus ended. He alone
Dealt on lieutenantry, and no practice had
In the brave squares of war. Yet now—no matter.    40

CLEOPATRA (⌜*rising,*⌝ *to Charmian and Iras*) Ah, stand by.

EROS The Queen, my lord, the Queen.

IRAS Go to him, madam.
Speak to him. He's unqualitied
With very shame.

CLEOPATRA        Well then, sustain me. O!    45

EROS
Most noble sir, arise. The Queen approaches.
Her head's declined, and death will seize her but
Your comfort makes the rescue.

ANTONY
I have offended reputation;
A most unnoble swerving.

EROS           Sir, the Queen.    50

ANTONY ⌜*rising*⌝
O, whither hast thou led me, Egypt? See
How I convey my shame out of thine eyes
By looking back what I have left behind
'Stroyed in dishonour.

CLEOPATRA        O, my lord, my lord,
Forgive my fearful sails! I little thought    55
You would have followed.

ANTONY        Egypt, thou knew'st too well
My heart was to thy rudder tied by th' strings,
And thou shouldst tow me after. O'er my spirit
Thy full supremacy thou knew'st, and that
Thy beck might from the bidding of the gods    60
Command me.

CLEOPATRA        O, my pardon!

ANTONY           Now I must
To the young man send humble treaties, dodge
And palter in the shifts of lowness, who
With half the bulk o'th' world played as I pleased,
Making and marring fortunes. You did know    65
How much you were my conqueror, and that
My sword, made weak by my affection, would
Obey it on all cause.

CLEOPATRA        Pardon, pardon!

ANTONY
Fall not a tear, I say. One of them rates
All that is won and lost. Give me a kiss.    70

*He kisses her*

Even this repays me. (*To an Attendant*) We sent our
schoolmaster;

Is a come back? (*To Cleopatra*) Love, I am full of lead.
(*Calling*) Some wine
Within there, and our viands! Fortune knows
We scorn her most when most she offers blows.

             *Exeunt*

**3.12** *Enter Caesar, ⌈Agrippa,⌉ Thidias, and Dolabella,*
   *with others*

CAESAR
Let him appear that's come from Antony.
Know you him?

DOLABELLA   Caesar, 'tis his schoolmaster;
An argument that he is plucked, when hither
He sends so poor a pinion of his wing,
Which had superfluous kings for messengers   5
Not many moons gone by.

   *Enter Ambassador from Antony*

CAESAR       Approach and speak.

AMBASSADOR
Such as I am, I come from Antony.
I was of late as petty to his ends
As is the morn-dew on the myrtle leaf
To his grand sea.

CAESAR     Be't so. Declare thine office.  10

AMBASSADOR
Lord of his fortunes he salutes thee, and
Requires to live in Egypt; which not granted,
He lessens his requests, and to thee sues
To let him breathe between the heavens and earth,
A private man in Athens. This for him.   15
Next, Cleopatra does confess thy greatness,
Submits her to thy might, and of thee craves
The circle of the Ptolemies for her heirs,
Now hazarded to thy grace.

CAESAR      For Antony,
I have no ears to his request. The Queen  20
Of audience nor desire shall fail, so she
From Egypt drive her all-disgracèd friend,
Or take his life there. This if she perform
She shall not sue unheard. So to them both.

AMBASSADOR
Fortune pursue thee!

CAESAR     Bring him through the bands. 25

       *Exit Ambassador, attended*

(*To Thidias*) To try thy eloquence now 'tis time.
Dispatch.
From Antony win Cleopatra. Promise,
And in our name, what she requires. Add more
As thine invention offers. Women are not
In their best fortunes strong, but want will perjure 30
The ne'er-touched vestal. Try thy cunning, Thidias.
Make thine own edict for thy pains, which we
Will answer as a law.

THIDIAS    Caesar, I go.

CAESAR
Observe how Antony becomes his flaw,
And what thou think'st his very action speaks  35
In every power that moves.

THIDIAS     Caesar, I shall.

   *Exeunt Caesar and his train at one door, and*
         *Thidias at another*

**3.13** *Enter Cleopatra, Enobarbus, Charmian, and Iras*

CLEOPATRA
What shall we do, Enobarbus?

ENOBARBUS     Think, and die.

CLEOPATRA
Is Antony or we in fault for this?

ENOBARBUS
Antony only, that would make his will
Lord of his reason. What though you fled
From that great face of war, whose several ranges 5
Frighted each other? Why should he follow?
The itch of his affection should not then
Have nicked his captainship, at such a point,
When half to half the world opposed, he being
The mooted question. 'Twas a shame no less  10
Than was his loss, to course your flying flags
And leave his navy gazing.

CLEOPATRA     Prithee, peace.

   *Enter the Ambassador with Antony*

ANTONY
Is that his answer?

AMBASSADOR     Ay, my lord.

ANTONY
The Queen shall then have courtesy, so she
Will yield us up.

AMBASSADOR   He says so.

ANTONY       Let her know't.  15
(*To Cleopatra*) To the boy Caesar send this grizzled head,
And he will fill thy wishes to the brim
With principalities.

CLEOPATRA    That head, my lord?

ANTONY (*to the Ambassador*)
To him again. Tell him he wears the rose
Of youth upon him, from which the world should note
Something particular. His coin, ships, legions,  21
May be a coward's, whose ministers would prevail
Under the service of a child as soon
As i'th' command of Caesar. I dare him therefore
To lay his gay caparisons apart   25
And answer me declined, sword against sword,
Ourselves alone. I'll write it. Follow me.

     *Exeunt Antony and Ambassador*

ENOBARBUS (*aside*)
Yes, like enough, high-battled Caesar will
Unstate his happiness and be staged to th' show
Against a sworder! I see men's judgements are  30
A parcel of their fortunes, and things outward
Do draw the inward quality after them
To suffer all alike. That he should dream,
Knowing all measures, the full Caesar will
Answer his emptiness! Caesar, thou hast subdued 35
His judgement, too.

   *Enter a Servant*

SERVANT     A messenger from Caesar.

CLEOPATRA
What, no more ceremony? See, my women:
Against the blown rose may they stop their nose,
That kneeled unto the buds. Admit him, sir.

          *Exit Servant*

ENOBARBUS (*aside*)
Mine honesty and I begin to square.  40
The loyalty well held to fools does make
Our faith mere folly; yet he that can endure
To follow with allegiance a fall'n lord
Does conquer him that did his master conquer,
And earns a place i'th' story.

   *Enter Thidias*

CLEOPATRA     Caesar's will?  45

THIDIAS
Hear it apart.

CLEOPATRA   None but friends; say boldly.

THIDIAS
  So haply are they friends to Antony.
ENOBARBUS
  He needs as many, sir, as Caesar has,
  Or needs not us. If Caesar please, our master
  Will leap to be his friend. For us, you know,    50
  Whose he is, we are: and that is Caesar's.
THIDIAS
  So. (*To Cleopatra*) Thus, then, thou most renowned:
    Caesar entreats
  Not to consider in what case thou stand'st
  Further than he is Caesar.
CLEOPATRA           Go on; right royal.
THIDIAS
  He knows that you embraced not Antony    55
  As you did love, but as you feared him.
CLEOPATRA O.
THIDIAS
  The scars upon your honour therefore he
  Does pity as constrainèd blemishes,
  Not as deserved.
CLEOPATRA       He is a god, and knows    60
  What is most right. Mine honour was not yielded,
  But conquered merely.
ENOBARBUS (*aside*)      To be sure of that
  I will ask Antony. Sir, sir, thou art so leaky
  That we must leave thee to thy sinking, for
  Thy dearest quit thee.                *Exit*
THIDIAS        Shall I say to Caesar    65
  What you require of him?—For he partly begs
  To be desired to give. It much would please him
  That of his fortunes you should make a staff
  To lean upon. But it would warm his spirits
  To hear from me you had left Antony,    70
  And put your self under his shroud,
  The universal landlord.
CLEOPATRA        What's your name?
THIDIAS
  My name is Thidias.
CLEOPATRA       Most kind messenger,
  Say to great Caesar this in deputation:
  I kiss his conqu'ring hand. Tell him I am prompt    75
  To lay my crown at's feet, and there to kneel
  Till from his all-obeying breath I hear
  The doom of Egypt.
THIDIAS        'Tis your noblest course.
  Wisdom and fortune combating together,
  If that the former dare but what it can,    80
  No chance may shake it. Give me grace to lay
  My duty on your hand.
     *He kisses Cleopatra's hand*
CLEOPATRA       Your Caesar's father oft,
  When he hath mused of taking kingdoms in,
  Bestowed his lips on that unworthy place,
  As it rained kisses.
     *Enter Antony and Enobarbus*
ANTONY       Favours, by Jove that thunders!    85
  What art thou, fellow?
THIDIAS        One that but performs
  The bidding of the fullest man, and worthiest
  To have command obeyed.
ENOBARBUS       You will be whipped.
ANTONY (*calling*)
  Approach, there!—Ah, you kite! Now, gods and
     devils,
  Authority melts from me of late. When I cried 'Ho!',

Like boys unto a muss kings would start forth,    91
And cry 'Your will?'—Have you no ears? I am
Antony yet.
     *Enter servants*
         Take hence this jack, and whip him.
ENOBARBUS ⌈*aside to Thidias*⌉
  'Tis better playing with a lion's whelp
  Than with an old one dying.
ANTONY           Moon and stars!    95
  Whip him! Were't twenty of the greatest tributaries
  That do acknowledge Caesar, should I find them
  So saucy with the hand of she here—what's her name
  Since she was Cleopatra? Whip him, fellows,
  Till like a boy you see him cringe his face,    100
  And whine aloud for mercy. Take him hence.
THIDIAS
  Mark Antony—
ANTONY       Tug him away. Being whipped,
  Bring him again. This jack of Caesar's shall
  Bear us an errand to him.
     *Exeunt servants with Thidias*
  You were half blasted ere I knew you. Ha,    105
  Have I my pillow left unpressed in Rome,
  Forborne the getting of a lawful race,
  And by a gem of women, to be abused
  By one that looks on feeders?
CLEOPATRA Good my lord—    110
ANTONY You have been a boggler ever.
  But when we in our viciousness grow hard—
  O misery on't!—the wise gods seel our eyes,
  In our own filth drop our clear judgements, make us
  Adore our errors, laugh at's while we strut    115
  To our confusion.
CLEOPATRA       O, is't come to this?
ANTONY
  I found you as a morsel cold upon
  Dead Caesar's trencher; nay, you were a fragment
  Of Gnaeus Pompey's, besides what hotter hours
  Unregistered in vulgar fame you have    120
  Luxuriously picked out. For I am sure,
  Though you can guess what temperance should be,
  You know not what it is.
CLEOPATRA       Wherefore is this?
ANTONY
  To let a fellow that will take rewards
  And say 'God quit you' be familiar with    125
  My playfellow your hand, this kingly seal
  And plighter of high hearts! O that I were
  Upon the hill of Basan to outroar
  The hornèd herd! For I have savage cause,
  And to proclaim it civilly were like    130
  A haltered neck which does the hangman thank
  For being yare about him.
     *Enter a Servant with Thidias*
          Is he whipped?
SERVANT Soundly, my lord.
ANTONY Cried he, and begged a pardon?
SERVANT He did ask favour.    135
ANTONY (*to Thidias*)
  If that thy father live, let him repent
  Thou wast not made his daughter; and be thou sorry
  To follow Caesar in his triumph, since
  Thou hast been whipped for following him. Henceforth
  The white hand of a lady fever thee,    140
  Shake thou to look on't. Get thee back to Caesar;
  Tell him thy entertainment. Look thou say
  He makes me angry with him, for he seems

Proud and disdainful, harping on what I am,
Not what he knew I was. He makes me angry,     145
And at this time most easy 'tis to do 't,
When my good stars that were my former guides
Have empty left their orbs, and shot their fires
Into th'abyss of hell. If he mislike
My speech and what is done, tell him he has     150
Hipparchus, my enfranchèd bondman, whom
He may at pleasure whip, or hang, or torture,
As he shall like, to quit me. Urge it thou.
Hence, with thy stripes, be gone!

                          *Exit ⌜Servant with⌝ Thidias*
CLEOPATRA  Have you done yet?     155
ANTONY  Alack, our terrene moon
  Is now eclipsed, and it portends alone
  The fall of Antony.
CLEOPATRA (*aside*)     I must stay his time.
ANTONY
  To flatter Caesar would you mingle eyes
  With one that ties his points?
CLEOPATRA                  Not know me yet?     160
ANTONY
  Cold-hearted toward me?
CLEOPATRA            Ah, dear, if I be so,
  From my cold heart let heaven engender hail,
  And poison it in the source, and the first stone
  Drop in my neck: as it determines, so
  Dissolve my life! The next Caesarion smite,     165
  Till by degrees the memory of my womb,
  Together with my brave Egyptians all,
  By the discandying of this pelleted storm
  Lie graveless till the flies and gnats of Nile
  Have buried them for prey!
ANTONY            I am satisfied.     170
  Caesar sits down in Alexandria, where
  I will oppose his fate. Our force by land
  Hath nobly held; our severed navy too
  Have knit again, and fleet, threat'ning most sea-like.
  Where hast thou been, my heart? Dost thou hear,
      lady?     175
  If from the field I shall return once more
  To kiss these lips, I will appear in blood.
  I and my sword will earn our chronicle.
  There's hope in 't yet.
CLEOPATRA            That's my brave lord.
ANTONY
  I will be treble-sinewed, hearted, breathed,     180
  And fight maliciously; for when mine hours
  Were nice and lucky, men did ransom lives
  Of me for jests; but now I'll set my teeth,
  And send to darkness all that stop me. Come,
  Let's have one other gaudy night. Call to me     185
  All my sad captains. Fill our bowls once more.
  Let's mock the midnight bell.
CLEOPATRA                  It is my birthday.
  I had thought to've held it poor, but since my lord
  Is Antony again, I will be Cleopatra.
ANTONY  We will yet do well.     190
CLEOPATRA
  Call all his noble captains to my lord!
ANTONY
  Do so. We'll speak to them, and tonight I'll force
  The wine peep through their scars. Come on, my queen,
  There's sap in 't yet. The next time I do fight
  I'll make death love me, for I will contend     195
  Even with his pestilent scythe.

                          *Exeunt all but Enobarbus*

ENOBARBUS
  Now he'll outstare the lightning. To be furious
  Is to be frighted out of fear, and in that mood
  The dove will peck the estridge; and I see still
  A diminution in our captain's brain     200
  Restores his heart. When valour preys on reason,
  It eats the sword it fights with. I will seek
  Some way to leave him.                  *Exit*

**4.1**   *Enter Caesar, reading a letter, with Agrippa,*
          *Maecenas, and his army*
CAESAR
  He calls me boy, and chides as he had power
  To beat me out of Egypt. My messenger
  He hath whipped with rods, dares me to personal
      combat,
  Caesar to Antony. Let the old ruffian know
  I have many other ways to die; meantime,     5
  Laugh at his challenge.
MAECENAS            Caesar must think,
  When one so great begins to rage, he's hunted
  Even to falling. Give him no breath, but now
  Make boot of his distraction. Never anger
  Made good guard for itself.
CAESAR            Let our best heads     10
  Know that tomorrow the last of many battles
  We mean to fight. Within our files there are,
  Of those that served Mark Antony but late,
  Enough to fetch him in. See it done,
  And feast the army. We have store to do 't,     15
  And they have earned the waste. Poor Antony!

                                      *Exeunt*

**4.2**   *Enter Antony, Cleopatra, Enobarbus, Charmian,*
          *Iras, Alexas, with others*
ANTONY
  He will not fight with me, Domitius?
ENOBARBUS                  No.
ANTONY  Why should he not?
ENOBARBUS
  He thinks, being twenty times of better fortune,
  He is twenty men to one.
ANTONY            Tomorrow, soldier,
  By sea and land I'll fight. Or I will live     5
  Or bathe my dying honour in the blood
  Shall make it live again. Woot thou fight well?
ENOBARBUS
  I'll strike, and cry 'Take all!'
ANTONY            Well said. Come on!
  Call forth my household servants. Let's tonight
  Be bounteous at our meal.
      *Enter Servitors*
                          Give me thy hand.     10
  Thou hast been rightly honest; so hast thou,
  Thou, and thou, and thou; you have served me well,
  And kings have been your fellows.
CLEOPATRA (*to Enobarbus*)     What means this?
ENOBARBUS (*to Cleopatra*)
  'Tis one of those odd tricks which sorrow shoots
  Out of the mind.
ANTONY (*to a Servitor*)  And thou art honest too.     15
  I wish I could be made so many men,
  And all of you clapped up together in
  An Antony, that I might do you service
  So good as you have done.
SERVITORS            The gods forbid!

ANTONY
  Well, my good fellows, wait on me tonight.                    20
  Scant not my cups, and make as much of me
  As when mine empire was your fellow too,
  And suffered my command.
CLEOPATRA (aside to Enobarbus)  What does he mean?
ENOBARBUS (aside to Cleopatra)
  To make his followers weep.
ANTONY                              Tend me tonight.
  Maybe it is the period of your duty.                          25
  Haply you shall not see me more; or if,
  A mangled shadow. Perchance tomorrow
  You'll serve another master. I look on you
  As one that takes his leave. Mine honest friends,
  I turn you not away, but, like a master                       30
  Married to your good service, stay till death.
  Tend me tonight two hours. I ask no more;
  And the gods yield you for't!
ENOBARBUS                    What mean you, sir,
  To give them this discomfort? Look, they weep,
  And I, an ass, am onion-eyed. For shame,                      35
  Transform us not to women.
ANTONY                          Ho, ho, ho,
  Now the witch take me if I meant it thus!
  Grace grow where those drops fall. My hearty friends,
  You take me in too dolorous a sense;
  For I spake to you for your comfort, did desire you           40
  To burn this night with torches. Know, my hearts,
  I hope well of tomorrow, and will lead you
  Where rather I'll expect victorious life
  Than death and honour. Let's to supper, come,
  And drown consideration.                    Exeunt

4.3    Enter a company of Soldiers
FIRST SOLDIER
  Brother, good night. Tomorrow is the day.
SECOND SOLDIER
  It will determine one way. Fare you well.
  Heard you of nothing strange about the streets?
FIRST SOLDIER Nothing. What news?
SECOND SOLDIER
  Belike 'tis but a rumour. Good night to you.                  5
FIRST SOLDIER
  Well, sir, good night.
  Enter other Soldiers, meeting them
SECOND SOLDIER          Soldiers, have careful watch.
THIRD SOLDIER
  And you. Good night, good night.
  They place themselves in every corner of the stage
SECOND SOLDIER               Here we; an if tomorrow
  Our navy thrive, I have an absolute hope
  Our landmen will stand up.
FIRST SOLDIER               'Tis a brave army,
  And full of purpose.
  Music of the hautboys is under the stage
SECOND SOLDIER          Peace, what noise?
FIRST SOLDIER                        List, list!    10
SECOND SOLDIER
  Hark!
FIRST SOLDIER Music i'th' air.
THIRD SOLDIER               Under the earth.
FOURTH SOLDIER
  It signs well, does it not?
THIRD SOLDIER          No.
FIRST SOLDIER          Peace, I say!
  What should this mean?

SECOND SOLDIER
  'Tis the god Hercules, whom Antony loved,
  Now leaves him.
FIRST SOLDIER          Walk. Let's see if other watchmen    15
  Do hear what we do.
SECOND SOLDIER          How now, masters?
ALL (speaking together)                        How now?
  How now? Do you hear this?
FIRST SOLDIER               Ay. Is't not strange?
THIRD SOLDIER
  Do you hear, masters? Do you hear?
FIRST SOLDIER
  Follow the noise so far as we have quarter.
  Let's see how it will give off.
ALL                          Content. 'Tis strange.    20
                                        Exeunt

4.4    Enter Antony and Cleopatra, with Charmian and
       others
ANTONY (calling)
  Eros, mine armour, Eros!
CLEOPATRA               Sleep a little.
ANTONY
  No, my chuck. Eros, come, mine armour, Eros!
  Enter Eros with armour
  Come, good fellow, put thine iron on.
  If fortune be not ours today, it is
  Because we brave her. Come.
CLEOPATRA               Nay, I'll help, too.    5
  What's this for?
ANTONY          Ah, let be, let be! Thou art
  The armourer of my heart. False, false! This, this!
CLEOPATRA
  Sooth, la, I'll help. Thus it must be.
  She helps Antony to arm
ANTONY                          Well, well,
  We shall thrive now. Seest thou, my good fellow?
  Go put on thy defences.
EROS                    Briefly, sir.    10
CLEOPATRA
  Is not this buckled well?
ANTONY               Rarely, rarely.
  He that unbuckles this, till we do please
  To doff't for our repose, shall hear a storm.
  Thou fumblest, Eros, and my queen's a squire
  More tight at this than thou. Dispatch. O love,    15
  That thou couldst see my wars today, and knew'st
  The royal occupation! Thou shouldst see
  A workman in't.
  Enter an armed Soldier
                Good morrow to thee. Welcome.
  Thou look'st like him that knows a warlike charge.
  To business that we love we rise betime,    20
  And go to't with delight.
SOLDIER                    A thousand, sir,
  Early though't be, have on their riveted trim,
  And at the port expect you.
  Shout within. Trumpets flourish. Enter ⌈Captains⌉
  and Soldiers
CAPTAIN
  The morn is fair. Good morrow, General.
SOLDIERS
  Good morrow, General.
ANTONY                    'Tis well blown, lads.    25
  This morning, like the spirit of a youth
  That means to be of note, begins betimes.
  So, so. Come, give me that. This way. Well said.

Fare thee well, dame. Whate'er becomes of me,
This is a soldier's kiss.
    *He kisses Cleopatra*
                    Rebukable           30
And worthy shameful check it were to stand
On more mechanic compliment. I'll leave thee
Now like a man of steel. You that will fight,
Follow me close. I'll bring you to't. Adieu.
    *Exeunt all but Cleopatra and Charmian*

CHARMIAN
Please you retire to your chamber?

CLEOPATRA                   Lead me.     35
He goes forth gallantly. That he and Caesar might
Determine this great war in single fight!
Then, Antony—but now! Well, on.        *Exeunt*

**4.5**   *Trumpets sound. Enter Antony and Eros, meeting a*
       *Soldier*

SOLDIER
The gods make this a happy day to Antony!

ANTONY
Would thou and those thy scars had once prevailed
To make me fight at land!

SOLDIER              Hadst thou done so,
The kings that have revolted, and the soldier
That has this morning left thee, would have still    5
Followed thy heels.

ANTONY             Who's gone this morning?

SOLDIER
Who? One ever near thee. Call for Enobarbus,
He shall not hear thee, or from Caesar's camp
Say 'I am none of thine'.

ANTONY             What sayest thou?

SOLDIER
Sir, he is with Caesar.

EROS (*to Antony*)       Sir, his chests and treasure    10
He has not with him.

ANTONY          Is he gone?

SOLDIER               Most certain.

ANTONY
Go, Eros, send his treasure after. Do it.
Detain no jot, I charge thee. Write to him—
I will subscribe—gentle adieus and greetings.
Say that I wish he never find more cause          15
To change a master. O, my fortunes have
Corrupted honest men! Dispatch. Enobarbus!    *Exeunt*

**4.6**   *Flourish. Enter Agrippa, Caesar, with Enobarbus*
       *and Dolabella*

CAESAR
Go forth, Agrippa, and begin the fight.
Our will is Antony be took alive.
Make it so known.

AGRIPPA        Caesar, I shall.         *Exit*

CAESAR
The time of universal peace is near.
Prove this a prosp'rous day, the three-nooked world   5
Shall bear the olive freely.
    *Enter a Messenger*

MESSENGER          Antony
Is come into the field.

CAESAR           Go charge Agrippa
Plant those that have revolted in the van,
That Antony may seem to spend his fury
Upon himself.                             10
    *Exeunt Messenger* ⌈*at one door*⌉, *Caesar and*
            *Dolabella* ⌈*at another*⌉

ENOBARBUS
Alexas did revolt, and went to Jewry on
Affairs of Antony; there did dissuade
Great Herod to incline himself to Caesar
And leave his master, Antony. For this pains,
Caesar hath hanged him. Camidius and the rest    15
That fell away have entertainment but
No honourable trust. I have done ill,
Of which I do accuse myself so sorely
That I will joy no more.
    *Enter a Soldier of Caesar's*

SOLDIER            Enobarbus, Antony
Hath after thee sent all thy treasure, with       20
His bounty overplus. The messenger
Came on my guard, and at thy tent is now
Unloading of his mules.

ENOBARBUS I give it you.

SOLDIER Mock not, Enobarbus,            25
I tell you true. Best you safed the bringer
Out of the host. I must attend mine office,
Or would have done't myself. Your Emperor
Continues still a Jove.                  *Exit*

ENOBARBUS
I am alone the villain of the earth,           30
And feel I am so most. O Antony,
Thou mine of bounty, how wouldst thou have paid
My better service, when my turpitude
Thou dost so crown with gold! This blows my heart.
If swift thought break it not, a swifter mean    35
Shall outstrike thought; but thought will do't, I feel.
I fight against thee? No, I will go seek
Some ditch wherein to die. The foul'st best fits
My latter part of life.                  *Exit*

**4.7**   *Alarum. Enter Agrippa* ⌈*with drummers and*
       *trumpeters*⌉

ACRIPPA
Retire! We have engaged our selves too far.
Caesar himself has work, and our oppression
Exceeds what we expected.              *Exeunt*

**4.8**   *Alarums. Enter Antony, and Scarus wounded*

SCARUS
O my brave Emperor, this is fought indeed!
Had we done so at first, we had droven them home
With clouts about their heads.

ANTONY             Thou bleed'st apace.

SCARUS
I had a wound here that was like a T,
But now 'tis made an H.
    *Retreat sounded far off*

ANTONY              They do retire.      5

SCARUS
We'll beat 'em into bench-holes. I have yet
Room for six scotches more.
    *Enter Eros*

EROS
They are beaten, sir, and our advantage serves
For a fair victory.

SCARUS          Let us score their backs
And snatch 'em up as we take hares, behind.     10
'Tis sport to maul a runner.

ANTONY (*to Eros*)        I will reward thee
Once for thy sprightly comfort, and tenfold
For thy good valour. Come thee on.

SCARUS             I'll halt after.
                                 *Exeunt*

**4.9**   *Alarum. Enter Antony again in a march; drummers and trumpeters; Scarus, with others*

ANTONY
We have beat him to his camp. Run one before,
And let the Queen know of our gests.    ⌜*Exit a soldier*⌝
                           Tomorrow,
Before the sun shall see's, we'll spill the blood
That has today escaped. I thank you all,
For doughty-handed are you, and have fought     5
Not as you served the cause, but as't had been
Each man's like mine. You have shown all Hectors.
Enter the city, clip your wives, your friends,
Tell them your feats whilst they with joyful tears
Wash the congealment from your wounds, and kiss
The honoured gashes whole.
     *Enter Cleopatra*
(*To Scarus*)             Give me thy hand.    11
To this great fairy I'll commend thy acts,
Make her thanks bless thee.
(*To Cleopatra, embracing her*) O thou day o'th' world,
Chain mine armed neck; leap thou, attire and all,
Through proof of harness to my heart, and there    15
Ride on the pants triumphing.
CLEOPATRA            Lord of lords!
O infinite virtue, com'st thou smiling from
The world's great snare uncaught?
ANTONY                 My nightingale,
We have beat them to their beds. What, girl, though
    grey
Do something mingle with our younger brown, yet
    ha' we                                  20
A brain that nourishes our nerves, and can
Get goal for goal of youth. Behold this man.
Commend unto his lips thy favouring hand;
Kiss it, my warrior.
     *Scarus kisses Cleopatra's hand*
                He hath fought today
As if a god, in hate of mankind, had         25
Destroyed in such a shape.
CLEOPATRA           I'll give thee, friend,
An armour all of gold. It was a king's.
ANTONY
He has deserved it, were it carbuncled
Like holy Phoebus' car. Give me thy hand.
Through Alexandria make a jolly march.      30
Bear our hacked targets like the men that owe them.
Had our great palace the capacity
To camp this host, we all would sup together
And drink carouses to the next day's fate,
Which promises royal peril. Trumpeters,      35
With brazen din blast you the city's ear;
Make mingle with our rattling taborins,
That heaven and earth may strike their sounds
    together,
Applauding our approach.      *Trumpets sound. Exeunt*

**4.10**   *Enter a Sentry and his company; Enobarbus follows*
SENTRY
If we be not relieved within this hour
We must return to th' court of guard. The night
Is shiny, and they say we shall embattle
By th' second hour i'th' morn.
FIRST WATCH            This last day was
   A shrewd one to's.
ENOBARBUS        O bear me witness, night—    5

SECOND WATCH
What man is this?
FIRST WATCH         Stand close, and list him.
ENOBARBUS
Be witness to me, O thou blessèd moon,
When men revolted shall upon record
Bear hateful memory, poor Enobarbus did
Before thy face repent.
SENTRY            Enobarbus?
SECOND WATCH            Peace; hark further.
ENOBARBUS
O sovereign mistress of true melancholy,      11
The poisonous damp of night disponge upon me,
That life, a very rebel to my will,
May hang no longer on me. Throw my heart
Against the flint and hardness of my fault,      15
Which, being dried with grief, will break to powder,
And finish all foul thoughts. O Antony,
Nobler than my revolt is infamous,
Forgive me in thine own particular,
But let the world rank me in register      20
A master-leaver and a fugitive.
O Antony! O Antony!            *He dies*
FIRST WATCH Let's speak to him.
SENTRY
Let's hear him, for the things he speaks
May concern Caesar.
SECOND WATCH       Let's do so. But he sleeps.    25
SENTRY
Swoons, rather; for so bad a prayer as his
Was never yet for sleep.
FIRST WATCH            Go we to him.
SECOND WATCH
Awake, sir, awake; speak to us.
FIRST WATCH            Hear you, sir?
SENTRY
The hand of death hath raught him.
     *Drums afar off*
                         Hark, the drums
Demurely wake the sleepers. Let us bear him     30
To th' court of guard; he is of note. Our hour
Is fully out.
SECOND WATCH
Come on, then. He may recover yet.
                       *Exeunt with the body*

**4.11**   *Enter Antony and Scarus with their army*
ANTONY
Their preparation is today by sea;
We please them not by land.
SCARUS            For both, my lord.
ANTONY
I would they'd fight i'th' fire or i'th' air;
We'd fight there too. But this it is: our foot
Upon the hills adjoining to the city         5
Shall stay with us. Order for sea is given.
They have put forth the haven—
Where their appointment we may best discover,
And look on their endeavour.        *Exeunt*

**4.12**   *Enter Caesar and his army*
CAESAR
But being charged, we will be still by land—
Which, as I take't, we shall, for his best force
Is forth to man his galleys. To the vales,
And hold our best advantage.        *Exeunt*

**4.13** ⌈*Alarum afar off, as at a sea fight.*⌉
    *Enter Antony and Scarus*
ANTONY
  Yet they are not joined. Where yon pine does stand
  I shall discover all. I'll bring thee word
  Straight how 'tis like to go.         *Exit*
SCARUS           Swallows have built
  In Cleopatra's sails their nests. The augurs
  Say they know not, they cannot tell, look grimly,   5
  And dare not speak their knowledge. Antony
  Is valiant, and dejected, and by starts
  His fretted fortunes give him hope and fear
  Of what he has and has not.
    *Enter Antony*
ANTONY           All is lost.
  This foul Egyptian hath betrayèd me.     10
  My fleet hath yielded to the foe, and yonder
  They cast their caps up, and carouse together
  Like friends long lost. Triple-turned whore! 'Tis thou
  Hast sold me to this novice, and my heart
  Makes only wars on thee. Bid them all fly;   15
  For when I am revengèd upon my charm,
  I have done all. Bid them all fly. Be gone. ⌈*Exit Scarus*⌉
  O sun, thy uprise shall I see no more.
  Fortune and Antony part here; even here
  Do we shake hands. All come to this? The hearts  20
  That spanieled me at heels, to whom I gave
  Their wishes, do discandy, melt their sweets
  On blossoming Caesar; and this pine is barked
  That overtopped them all. Betrayed I am.
  O this false soul of Egypt! This grave charm,  25
  Whose eye becked forth my wars and called them home,
  Whose bosom was my crownet, my chief end,
  Like a right gipsy hath at fast and loose
  Beguiled me to the very heart of loss.
  What, Eros, Eros!
    *Enter Cleopatra*
           Ah, thou spell! Avaunt.    30
CLEOPATRA
  Why is my lord enraged against his love?
ANTONY
  Vanish, or I shall give thee thy deserving
  And blemish Caesar's triumph. Let him take thee
  And hoist thee up to the shouting plebeians;
  Follow his chariot, like the greatest spot   35
  Of all thy sex; most monster-like be shown
  For poor'st diminutives, for dolts, and let
  Patient Octavia plough thy visage up
  With her preparèd nails.     *Exit Cleopatra*
           'Tis well thou'rt gone,
  If it be well to live. But better 'twere   40
  Thou fell'st into my fury, for one death
  Might have prevented many. Eros, ho!
  The shirt of Nessus is upon me. Teach me,
  Alcides, thou mine ancestor, thy rage.
  Let me lodge Lichas on the horns o'th' moon,   45
  And with those hands that grasped the heaviest club
  Subdue my worthiest self. The witch shall die.
  To the young Roman boy she hath sold me, and I fall
  Under this plot. She dies for't. Eros, ho!    *Exit*

**4.14** *Enter Cleopatra, Charmian, Iras, Mardian*
CLEOPATRA
  Help me, my women! O, he's more mad
  Than Telamon for his shield; the boar of Thessaly

Was never so embossed.
CHARMIAN          To th' monument!
  There lock yourself, and send him word you are dead.
  The soul and body rive not more in parting   5
  Than greatness going off.
CLEOPATRA         To th' monument!
  Mardian, go tell him I have slain myself.
  Say that the last I spoke was 'Antony',
  And word it, prithee, piteously. Hence, Mardian,
  And bring me how he takes my death. To th'
    monument!          *Exeunt*

**4.15** *Enter Antony and Eros*
ANTONY
  Eros, thou yet behold'st me?
EROS          Ay, noble lord.
ANTONY
  Sometime we see a cloud that's dragonish,
  A vapour sometime like a bear or lion,
  A towered citadel, a pendent rock,
  A forkèd mountain, or blue promontory   5
  With trees upon't that nod unto the world
  And mock our eyes with air. Thou hast seen these signs;
  They are black vesper's pageants.
EROS          Ay, my lord.
ANTONY
  That which is now a horse even with a thought
  The rack distains, and makes it indistinct   10
  As water is in water.
EROS        It does, my lord.
ANTONY
  My good knave Eros, now thy captain is
  Even such a body. Here I am Antony,
  Yet cannot hold this visible shape, my knave.
  I made these wars for Egypt, and the Queen—   15
  Whose heart I thought I had, for she had mine,
  Which whilst it was mine had annexed unto't
  A million more, now lost—she, Eros, has
  Packed cards with Caesar, and false-played my glory
  Unto an enemy's triumph.   20
  Nay, weep not, gentle Eros. There is left us
  Ourselves to end ourselves.
    *Enter Mardian*
           O thy vile lady,
  She has robbed me of my sword!
MARDIAN         No, Antony,
  My mistress loved thee, and her fortunes mingled
  With thine entirely.
ANTONY       Hence, saucy eunuch, peace!   25
  She hath betrayed me, and shall die the death.
MARDIAN
  Death of one person can be paid but once,
  And that she has discharged. What thou wouldst do
  Is done unto thy hand. The last she spake
  Was 'Antony, most noble Antony!'   30
  Then in the midst a tearing groan did break
  The name of Antony. It was divided
  Between her heart and lips. She rendered life,
  Thy name so buried in her.
ANTONY         Dead, then?
MARDIAN          Dead.
ANTONY
  Unarm, Eros. The long day's task is done,   35
  And we must sleep. (*To Mardian*) That thou depart'st
    hence safe

Does pay thy labour richly. Go.                    *Exit Mardian*
                              Off, pluck off.
        *Eros helps Antony to unarm*
The seven-fold shield of Ajax cannot keep
The battery from my heart. O, cleave, my sides!
Heart, once be stronger than thy continent;                40
Crack thy frail case. Apace, Eros, apace.
No more a soldier. Bruisèd pieces, go;
You have been nobly borne.—From me a while.
                                        *Exit Eros*
I will o'ertake thee, Cleopatra, and
Weep for my pardon. So it must be, for now         45
All length is torture. Since the torch is out,
Lie down, and stray no farther. Now all labour
Mars what it does; yea, very force entangles
Itself with strength. Seal, then, and all is done.
Eros!—I come, my queen.—Eros!—Stay for me.         50
Where souls do couch on flowers we'll hand in hand,
And with our sprightly port make the ghosts gaze.
Dido and her Aeneas shall want troops,
And all the haunt be ours. Come, Eros, Eros!
        *Enter Eros*
EROS
  What would my lord?
ANTONY                    Since Cleopatra died        55
I have lived in such dishonour that the gods
Detest my baseness. I, that with my sword
Quartered the world, and o'er green Neptune's back
With ships made cities, condemn myself to lack
The courage of a woman; less noble mind             60
Than she which by her death our Caesar tells
'I am conqueror of myself.' Thou art sworn, Eros,
That when the exigent should come, which now
Is come indeed—when I should see behind me
Th'inevitable prosecution of                         65
Disgrace and horror—that on my command
Thou then wouldst kill me. Do't. The time is come.
Thou strik'st not me; 'tis Caesar thou defeat'st.
Put colour in thy cheek.
EROS                    The gods withhold me!
Shall I do that which all the Parthian darts,        70
Though enemy, lost aim and could not?
ANTONY                              Eros,
Wouldst thou be windowed in great Rome and see
Thy master thus with pleached arms, bending down
His corrigible neck, his face subdued
To penetrative shame, whilst the wheeled seat       75
Of fortunate Caesar, drawn before him, branded
His baseness that ensued?
EROS                    I would not see't.
ANTONY
  Come then; for with a wound I must be cured.
Draw that thy honest sword, which thou hast worn
Most useful for thy country.
EROS                    O sir, pardon me!            80
ANTONY
  When I did make thee free, swor'st thou not then
To do this when I bade thee? Do it at once,
Or thy precedent services are all
But accidents unpurposed. Draw, and come.
EROS
  Turn from me then that noble countenance          85
Wherein the worship of the whole world lies.
ANTONY (*turning away*) Lo thee!

EROS
  My sword is drawn.
ANTONY              Then let it do at once
The thing why thou hast drawn it.
EROS                          My dear master,
My captain, and my Emperor: let me say,             90
Before I strike this bloody stroke, farewell.
ANTONY 'Tis said, man; and farewell.
EROS
  Farewell, great chief. Shall I strike now?
ANTONY                              Now, Eros.
        *Eros stabs himself*
EROS
  Why, there then, thus I do escape the sorrow
Of Antony's death.                    *He dies*
ANTONY              Thrice nobler than myself,        95
Thou teachest me, O valiant Eros, what
I should and thou couldst not. My queen and Eros
Have by their brave instruction got upon me
A nobleness in record. But I will be
A bridegroom in my death, and run into't            100
As to a lover's bed. Come then; and, Eros,
Thy master dies thy scholar. To do thus
I learned of thee.
        *He stabs himself*
                    How, not dead? Not dead?
The guard, ho! O, dispatch me!
        *Enter a guard [and Decretas]*
FIRST GUARD                      What's the noise?
ANTONY
  I have done my work ill, friends. O, make an end   105
Of what I have begun!
SECOND GUARD          The star is fall'n.
FIRST GUARD
  And time is at his period.
ALL THE GUARDS              Alas
  And woe!
ANTONY      Let him that loves me strike me dead.
FIRST GUARD
  Not I.
SECOND GUARD Nor I.
THIRD GUARD          Nor anyone.      *Exeunt the guard*
DECRETAS
  Thy death and fortunes bid thy followers fly.      110
        *He takes Antony's sword*
This sword but shown to Caesar, with this tidings,
Shall enter me with him.
        *Enter Diomedes*
DIOMEDES                Where's Antony?
DECRETAS
  There, Diomed, there.
DIOMEDES              Lives he? Wilt thou not answer, man?
                              *Exit Decretas*
ANTONY
  Art thou there, Diomed? Draw thy sword, and give me
Sufficing strokes for death.
DIOMEDES              Most absolute lord,            115
My mistress Cleopatra sent me to thee.
ANTONY
  When did she send thee?
DIOMEDES              Now, my lord.
ANTONY                          Where is she?
DIOMEDES
  Locked in her monument. She had a prophesying fear
Of what hath come to pass; for when she saw—
Which never shall be found—you did suspect          120

She had disposed with Caesar, and that your rage
Would not be purged, she sent word she was dead;
But fearing since how it might work, hath sent
Me to proclaim the truth; and I am come,
I dread, too late.                                                    125
ANTONY
Too late, good Diomed. Call my guard, I prithee.
DIOMEDES
What ho, the Emperor's guard! The guard, what ho!
Come, your lord calls.
          *Enter four or five of the guard of Antony*
ANTONY
Bear me, good friends, where Cleopatra bides.
'Tis the last service that I shall command you.          130
FIRST GUARD
Woe, woe are we, sir, you may not live to wear
All your true followers out.
ALL THE GUARDS                    Most heavy day!
ANTONY
Nay, good my fellows, do not please sharp fate
To grace it with your sorrows. Bid that welcome
Which comes to punish us, and we punish it,          135
Seeming to bear it lightly. Take me up.
I have led you oft; carry me now, good friends,
And have my thanks for all.
          *Exeunt bearing Antony ⌈and Eros⌉*

**4.16** *Enter Cleopatra ⌈and her maids aloft⌉, with*
          *Charmian and Iras*
CLEOPATRA
O Charmian, I will never go from hence.
CHARMIAN
Be comforted, dear madam.
CLEOPATRA                              No, I will not.
All strange and terrible events are welcome,
But comforts we despise. Our size of sorrow,
Proportioned to our cause, must be as great          5
As that which makes it.
          *Enter Diomedes ⌈below⌉*
                              How now? Is he dead?
DIOMEDES
His death's upon him, but not dead.
Look out o'th' other side your monument.
His guard have brought him thither.
          *Enter below Antony, borne by the guard*
CLEOPATRA                              O sun,
Burn the great sphere thou mov'st in; darkling stand
The varying shore o'th' world! O Antony,          11
Antony, Antony! Help, Charmian,
Help, Iras, help, help, friends below!
Let's draw him hither.
ANTONY                         Peace. Not Caesar's valour
Hath o'erthrown Antony, but Antony's          15
Hath triumphed on itself.
CLEOPATRA                    So it should be,
That none but Antony should conquer Antony.
But woe 'tis so!
ANTONY
I am dying, Egypt, dying. Only
I here importune death awhile until          20
Of many thousand kisses the poor last
I lay upon thy lips.
CLEOPATRA                    I dare not, dear,
Dear, my lord, pardon. I dare not,
Lest I be taken. Nor th'imperious show
Of the full-fortuned Caesar ever shall          25

Be brooched with me, if knife, drugs, serpents, have
Edge, sting, or operation. I am safe.
Your wife, Octavia, with her modest eyes
And still conclusion, shall acquire no honour
Demuring upon me. But come, come, Antony.—          30
Help me, my women.—We must draw thee up.
Assist, good friends.
ANTONY                    O quick, or I am gone!
CLEOPATRA
Here's sport indeed. How heavy weighs my lord!
Our strength is all gone into heaviness,
That makes the weight. Had I great Juno's power          35
The strong-winged Mercury should fetch thee up
And set thee by Jove's side. Yet come a little.
Wishers were ever fools. O come, come, come!
          *They heave Antony aloft to Cleopatra*
And welcome, welcome! Die when thou hast lived,
Quicken with kissing. Had my lips that power,          40
Thus would I wear them out.
          *They kiss*
ALL THE LOOKERS-ON A heavy sight.
ANTONY I am dying, Egypt, dying.
Give me some wine, and let me speak a little.
CLEOPATRA
No, let me speak, and let me rail so high          45
That the false hussy Fortune break her wheel,
Provoked by my offence.
ANTONY                    One word, sweet queen.
Of Caesar seek your honour, with your safety. O!
CLEOPATRA
They do not go together.
ANTONY                    Gentle, hear me.
None about Caesar trust but Proculeius.          50
CLEOPATRA
My resolution and my hands I'll trust,
None about Caesar.
ANTONY
The miserable change now at my end
Lament nor sorrow at, but please your thoughts
In feeding them with those my former fortunes,          55
Wherein I lived the greatest prince o'th' world,
The noblest; and do now not basely die,
Not cowardly put off my helmet to
My countryman; a Roman by a Roman
Valiantly vanquished. Now my spirit is going;          60
I can no more.
CLEOPATRA          Noblest of men, woot die?
Hast thou no care of me? Shall I abide
In this dull world, which in thy absence is
No better than a sty?
          *Antony dies*
                    O see, my women,
The crown o'th' earth doth melt. My lord!          65
O, withered is the garland of the war.
The soldier's pole is fall'n. Young boys and girls
Are level now with men. The odds is gone,
And there is nothing left remarkable
Beneath the visiting moon.          70
          *She falls*
CHARMIAN O, quietness, lady!
IRAS She's dead, too, our sovereign.
CHARMIAN
Lady!
IRAS     Madam!
CHARMIAN          O, madam, madam, madam!

IRAS
　Royal Egypt, Empress!
CHARMIAN　　　　　　　Peace, peace, Iras!
CLEOPATRA (*recovering*)
　No more but e'en a woman, and commanded　　75
　By such poor passion as the maid that milks
　And does the meanest chores. It were for me
　To throw my sceptre at the injurious gods,
　To tell them that this world did equal theirs
　Till they had stol'n our jewel. All's but naught.　　80
　Patience is sottish, and impatience does
　Become a dog that's mad. Then is it sin
　To rush into the secret house of death
　Ere death dare come to us? How do you, women?
　What, what, good cheer! Why, how now, Charmian?
　My noble girls! Ah, women, women! Look,　　86
　Our lamp is spent, it's out. Good sirs, take heart;
　We'll bury him, and then what's brave, what's noble,
　Let's do it after the high Roman fashion,
　And make death proud to take us. Come, away.　　90
　This case of that huge spirit now is cold.
　Ah, women, women! Come. We have no friend
　But resolution, and the briefest end.
　　　　　*Exeunt, those above bearing off Antony's body*

**5.1**　*Enter Caesar with his council of war: Agrippa,*
　　　*Dolabella, Maecenas, Gallus, Proculeius*
CAESAR
　Go to him, Dolabella, bid him yield.
　Being so frustrate, tell him, he but mocks
　The pauses that he makes.
DOLABELLA　　　　　Caesar, I shall.　　　　*Exit*
　　　*Enter Decretas with the sword of Antony*
CAESAR
　Wherefore is that? And what art thou that dar'st
　Appear thus to us?
DECRETAS　　　　I am called Decretas.　　5
　Mark Antony I served, who best was worthy
　Best to be served. Whilst he stood up and spoke
　He was my master, and I wore my life
　To spend upon his haters. If thou please
　To take me to thee, as I was to him　　10
　I'll be to Caesar; if thou pleasest not,
　I yield thee up my life.
CAESAR　　　　　　What is't thou sayst?
DECRETAS
　I say, O Caesar, Antony is dead.
CAESAR
　The breaking of so great a thing should make
　A greater crack. The rivèd world　　15
　Should have shook lions into civil streets,
　And citizens to their dens. The death of Antony
　Is not a single doom; in that name lay
　A moiety of the world.
DECRETAS　　　　　He is dead, Caesar,
　Not by a public minister of justice,　　20
　Nor by a hirèd knife; but that self hand
　Which writ his honour in the acts it did
　Hath, with the courage which the heart did lend it,
　Splitted the heart. This is his sword;
　I robbed his wound of it. Behold it stained　　25
　With his most noble blood.
CAESAR (*weeping*)　　　　Look you, sad friends,
　The gods rebuke me; but it is a tidings
　To wash the eyes of kings.
AGRIPPA　　　　　　And strange it is
　That nature must compel us to lament

　Our most persisted deeds.
MAECENAS　　　　　His taints and honours　　30
　Waged equal with him.
AGRIPPA　　　　　　A rarer spirit never
　Did steer humanity; but you gods will give us
　Some faults to make us men. Caesar is touched.
MAECENAS
　When such a spacious mirror's set before him
　He needs must see himself.
CAESAR　　　　　O Antony,　　35
　I have followed thee to this. But we do lance
　Diseases in our bodies. I must perforce
　Have shown to thee such a declining day,
　Or look on thine. We could not stall together
　In the whole world. But yet let me lament　　40
　With tears as sovereign as the blood of hearts,
　That thou, my brother, my competitor
　In top of all design, my mate in empire,
　Friend and companion in the front of war,
　The arm of mine own body, and the heart　　45
　Where mine his thoughts did kindle—that our stars,
　Unreconciliable, should divide
　Our equalness to this. Hear me, good friends—
　　　　*Enter an Egyptian*
　But I will tell you at some meeter season.
　The business of this man looks out of him;　　50
　We'll hear him what he says.—Whence are you?
EGYPTIAN
　A poor Egyptian, yet the Queen my mistress,
　Confined in all she has, her monument,
　Of thy intents desires instruction,
　That she preparèdly may frame herself　　55
　To th' way she's forced to.
CAESAR　　　　　Bid her have good heart.
　She soon shall know of us, by some of ours,
　How honourable and how kindly we
　Determine for her. For Caesar cannot live
　To be ungentle.
EGYPTIAN　　　So; the gods preserve thee!　　*Exit*
CAESAR
　Come hither, Proculeius. Go, and say　　61
　We purpose her no shame. Give her what comforts
　The quality of her passion shall require,
　Lest in her greatness, by some mortal stroke,
　She do defeat us; for her life in Rome　　65
　Would be eternal in our triumph. Go,
　And with your speediest bring us what she says
　And how you find of her.
PROCULEIUS　　　　Caesar, I shall.　　*Exit*
CAESAR
　Gallus, go you along.　　　　*Exit Gallus*
　　　　　　　　Where's Dolabella,
　To second Proculeius?
ALL BUT CAESAR　　　Dolabella!　　70
CAESAR
　Let him alone; for I remember now
　How he's employed. He shall in time be ready.
　Go with me to my tent, where you shall see
　How hardly I was drawn into this war,
　How calm and gentle I proceeded still　　75
　In all my writings. Go with me, and see
　What I can show in this.　　　　*Exeunt*

**5.2**　*Enter Cleopatra, Charmian, Iras, and Mardian*
CLEOPATRA
　My desolation does begin to make
　A better life. 'Tis paltry to be Caesar.

Not being Fortune, he's but Fortune's knave,
A minister of her will. And it is great
To do that thing that ends all other deeds,    5
Which shackles accidents and bolts up change,
Which sleeps and never palates more the dung,
The beggar's nurse, and Caesar's.

*Enter Proculeius*

PROCULEIUS
Caesar sends greeting to the Queen of Egypt,
And bids thee study on what fair demands    10
Thou mean'st to have him grant thee.

CLEOPATRA             What's thy name?

PROCULEIUS
My name is Proculeius.

CLEOPATRA             Antony
Did tell me of you, bade me trust you; but
I do not greatly care to be deceived,
That have no use for trusting. If your master    15
Would have a queen his beggar, you must tell him
That majesty, to keep decorum, must
No less beg than a kingdom. If he please
To give me conquered Egypt for my son,
He gives me so much of mine own as I    20
Will kneel to him with thanks.

PROCULEIUS             Be of good cheer.
You're fall'n into a princely hand; fear nothing.
Make your full reference freely to my lord,
Who is so full of grace that it flows over
On all that need. Let me report to him    25
Your sweet dependency, and you shall find
A conqueror that will pray in aid for kindness,
Where he for grace is kneeled to.

CLEOPATRA             Pray you, tell him
I am his fortune's vassal, and I send him
The greatness he has got. I hourly learn    30
A doctrine of obedience, and would gladly
Look him i'th' face.

PROCULEIUS             This I'll report, dear lady;
Have comfort, for I know your plight is pitied
Of him that caused it.

⌈*Enter Roman soldiers from behind*⌉

PROCULEIUS (*to the soldiers*)
You see how easily she may be surprised.    35
Guard her till Caesar come.

IRAS             Royal Queen—

CHARMIAN
O Cleopatra, thou art taken, Queen!

CLEOPATRA (*drawing a dagger*)
Quick, quick, good hands!

PROCULEIUS (*disarming Cleopatra*)
            Hold, worthy lady, hold!
Do not yourself such wrong, who are in this
Relieved but not betrayed.

CLEOPATRA          What, of death too,    40
That rids our dogs of languish?

PROCULEIUS             Cleopatra,
Do not abuse my master's bounty by
Th'undoing of yourself. Let the world see
His nobleness well acted, which your death
Will never let come forth.

CLEOPATRA          Where art thou, death?    45
Come hither, come. Come, come, and take a queen
Worth many babes and beggars.

PROCULEIUS             O temperance, lady!

CLEOPATRA
Sir, I will eat no meat. I'll not drink, sir.

If idle talk will once be necessary,
I'll not sleep, neither. This mortal house I'll ruin,    50
Do Caesar what he can. Know, sir, that I
Will not wait pinioned at your master's court,
Nor once be chastised with the sober eye
Of dull Octavia. Shall they hoist me up
And show me to the shouting varletry    55
Of censuring Rome? Rather a ditch in Egypt
Be gentle grave unto me; rather on Nilus' mud
Lay me stark naked, and let the waterflies
Blow me into abhorring; rather make
My country's high pyramides my gibbet,    60
And hang me up in chains.

PROCULEIUS          You do extend
These thoughts of horror further than you shall
Find cause in Caesar.

*Enter Dolabella*

DOLABELLA          Proculeius,
What thou hast done thy master Caesar knows,
And he hath sent for thee. For the Queen,    65
I'll take her to my guard.

PROCULEIUS          So, Dolabella,
It shall content me best. Be gentle to her.
(*To Cleopatra*) To Caesar I will speak what you shall
         please,
If you'll employ me to him.

CLEOPATRA          Say I would die.

*Exit Proculeius*

DOLABELLA
Most noble Empress, you have heard of me.    70

CLEOPATRA
I cannot tell.

DOLABELLA      Assurèdly you know me.

CLEOPATRA
No matter, sir, what I have heard or known.
You laugh when boys or women tell their dreams;
Is't not your trick?

DOLABELLA          I understand not, madam.

CLEOPATRA
I dreamt there was an Emperor Antony.    75
O, such another sleep, that I might see
But such another man!

DOLABELLA          If it might please ye—

CLEOPATRA
His face was as the heav'ns, and therein stuck
A sun and moon, which kept their course and lighted
The little O o'th' earth.

DOLABELLA          Most sovereign creature—    80

CLEOPATRA
His legs bestrid the ocean; his reared arm
Crested the world. His voice was propertied
As all the tunèd spheres, and that to friends;
But when he meant to quail and shake the orb,
He was as rattling thunder. For his bounty,    85
There was no winter in't; an autumn 'twas,
That grew the more by reaping. His delights
Were dolphin-like; they showed his back above
The element they lived in. In his livery
Walked crowns and crownets. Realms and islands were
As plates dropped from his pocket.

DOLABELLA          Cleopatra—    91

CLEOPATRA
Think you there was, or might be, such a man
As this I dreamt of?

DOLABELLA          Gentle madam, no.

CLEOPATRA
You lie, up to the hearing of the gods.
But if there be, or ever were one such,                    95
It's past the size of dreaming. Nature wants stuff
To vie strange forms with fancy; yet t'imagine
An Antony were nature's piece 'gainst fancy,
Condemning shadows quite.
DOLABELLA                          Hear me, good madam:
Your loss is as yourself, great, and you bear it          100
As answering to the weight. Would I might never
O'ertake pursued success but I do feel,
By the rebound of yours, a grief that smites
My very heart at root.
CLEOPATRA                     I thank you, sir.
Know you what Caesar means to do with me?          105
DOLABELLA
I am loath to tell you what I would you knew.
CLEOPATRA
Nay, pray you, sir.
DOLABELLA                     Though he be honourable—
CLEOPATRA
He'll lead me then in triumph.
DOLABELLA                          Madam, he will, I know't.
          *Flourish. Enter Caesar, with Proculeius, Gallus,*
          *Maecenas, and others of his train*
ALL
Make way, there! Caesar!
CAESAR                          Which is the Queen of Egypt?
DOLABELLA (*to Cleopatra*)
It is the Emperor, madam.
          *Cleopatra kneels*
CAESAR                          Arise! You shall not kneel.
I pray you rise, rise, Egypt.
CLEOPATRA (*rising*)               Sir, the gods          111
Will have it thus. My master and my lord
I must obey.
CAESAR          Take to you no hard thoughts.
The record of what injuries you did us,
Though written in our flesh, we shall remember          115
As things but done by chance.
CLEOPATRA                          Sole sir o'th' world,
I cannot project mine own cause so well
To make it clear, but do confess I have
Been laden with like frailties which before
Have often shamed our sex.
CAESAR                          Cleopatra, know          120
We will extenuate rather than enforce.
If you apply yourself to our intents,
Which towards you are most gentle, you shall find
A benefit in this change; but if you seek
To lay on me a cruelty by taking          125
Antony's course, you shall bereave yourself
Of my good purposes and put your children
To that destruction which I'll guard them from,
If thereon you rely. I'll take my leave.
CLEOPATRA
And may through all the world! 'Tis yours, and we,
Your scutcheons and your signs of conquest, shall  131
Hang in what place you please. (*Giving a paper*) Here,
          my good lord.
CAESAR
You shall advise me in all for Cleopatra.
CLEOPATRA
This is the brief of money, plate, and jewels
I am possessed of. 'Tis exactly valued,          135
Not petty things admitted. Where's Seleucus?

          *Enter Seleucus*
SELEUCUS Here, madam.
CLEOPATRA (*to Caesar*)
This is my treasurer. Let him speak, my lord,
Upon his peril, that I have reserved
To myself nothing. Speak the truth, Seleucus.          140
SELEUCUS
Madam, I had rather seal my lips
Than to my peril speak that which is not.
CLEOPATRA What have I kept back?
SELEUCUS
Enough to purchase what you have made known.
CAESAR
Nay, blush not, Cleopatra. I approve          145
Your wisdom in the deed.
CLEOPATRA                          See, Caesar! O, behold
How pomp is followed! Mine will now be yours,
And should we shift estates, yours would be mine.
The ingratitude of this Seleucus does
Even make me wild.—O slave, of no more trust          150
Than love that's hired! What, goest thou back? Thou
          shalt
Go back, I warrant thee; but I'll catch thine eyes
Though they had wings. Slave, soulless villain, dog!
O rarely base!
CAESAR          Good Queen, let us entreat you.
CLEOPATRA
O Caesar, what a wounding shame is this,          155
That thou vouchsafing here to visit me,
Doing the honour of thy lordliness
To one so meek—that mine own servant should
Parcel the sum of my disgraces by
Addition of his envy. Say, good Caesar,          160
That I some lady trifles have reserved,
Immoment toys, things of such dignity
As we greet modern friends withal; and say
Some nobler token I have kept apart
For Livia and Octavia, to induce          165
Their mediation—must I be unfolded
With one that I have bred? The gods! It smites me
Beneath the fall I have. (*To Seleucus*) Prithee, go hence,
Or I shall show the cinders of my spirits
Through th'ashes of my chance. Wert thou a man  170
Thou wouldst have mercy on me.
CAESAR                          Forbear, Seleucus.
          *Exit Seleucus*
CLEOPATRA
Be it known that we, the greatest, are misthought
For things that others do; and when we fall
We answer others' merits in our name,
Are therefore to be pitied.
CAESAR                          Cleopatra,          175
Not what you have reserved nor what acknowledged
Put we i'th' roll of conquest. Still be't yours.
Bestow it at your pleasure, and believe
Caesar's no merchant, to make prize with you
Of things that merchants sold. Therefore be cheered.
Make not your thoughts your prisons. No, dear
          Queen;          181
For we intend so to dispose you as
Yourself shall give us counsel. Feed and sleep.
Our care and pity is so much upon you
That we remain your friend; and so adieu.          185
CLEOPATRA
My master and my lord!
CAESAR                     Not so. Adieu.
          *Flourish. Exeunt Caesar and his train*

CLEOPATRA
He words me, girls, he words me, that I should not
Be noble to myself. But hark thee, Charmian.
*She whispers to Charmian*
IRAS
Finish, good lady. The bright day is done,
And we are for the dark.
CLEOPATRA (*to Charmian*)    Hie thee again.    190
I have spoke already, and it is provided.
Go put it to the haste.
CHARMIAN            Madam, I will.
*Enter Dolabella*
DOLABELLA
Where's the Queen?
CHARMIAN            Behold, sir.            *Exit*
CLEOPATRA            Dolabella!
DOLABELLA
Madam, as thereto sworn by your command—
Which my love makes religion to obey—    195
I tell you this: Caesar through Syria
Intends his journey, and within three days
You with your children will he send before.
Make your best use of this. I have performed
Your pleasure, and my promise.
CLEOPATRA            Dolabella,    200
I shall remain your debtor.
DOLABELLA            I your servant.
Adieu, good Queen. I must attend on Caesar.
CLEOPATRA
Farewell, and thanks.            *Exit Dolabella*
                Now, Iras, what think'st thou?
Thou, an Egyptian puppet shall be shown
In Rome, as well as I. Mechanic slaves    205
With greasy aprons, rules, and hammers shall
Uplift us to the view. In their thick breaths,
Rank of gross diet, shall we be enclouded,
And forced to drink their vapour.
IRAS                The gods forbid!
CLEOPATRA
Nay, 'tis most certain, Iras. Saucy lictors    210
Will catch at us like strumpets, and scald rhymers
Ballad us out o' tune. The quick comedians
Extemporally will stage us, and present
Our Alexandrian revels. Antony
Shall be brought drunken forth, and I shall see    215
Some squeaking Cleopatra boy my greatness
I'th' posture of a whore.
IRAS                O, the good gods!
CLEOPATRA Nay, that's certain.
IRAS
I'll never see't! For I am sure my nails
Are stronger than mine eyes.
CLEOPATRA            Why, that's the way    220
To fool their preparation and to conquer
Their most absurd intents.
*Enter Charmian*
                Now, Charmian!
Show me, my women, like a queen. Go fetch
My best attires. I am again for Cydnus
To meet Mark Antony. Sirrah Iras, go.    225
Now, noble Charmian, we'll dispatch indeed,
And when thou hast done this chore I'll give thee
    leave
To play till doomsday.—Bring our crown and all.
                ⌜*Exit Iras*⌝

*A noise within*
Wherefore's this noise?
*Enter a Guardsman*
GUARDSMAN            Here is a rural fellow
That will not be denied your highness' presence.    230
He brings you figs.
CLEOPATRA
Let him come in.            *Exit Guardsman*
                What poor an instrument
May do a noble deed! He brings me liberty.
My resolution's placed, and I have nothing
Of woman in me. Now from head to foot    235
I am marble-constant. Now the fleeting moon
No planet is of mine.
*Enter Guardsman, and Clown with a basket*
GUARDSMAN            This is the man.
CLEOPATRA
Avoid, and leave him.            *Exit Guardsman*
                Hast thou the pretty worm
Of Nilus there, that kills and pains not?    239
CLOWN Truly, I have him; but I would not be the party
    that should desire you to touch him, for his biting is
    immortal; those that do die of it do seldom or never
    recover.            243
CLEOPATRA Remember'st thou any that have died on't?
CLOWN Very many, men, and women too. I heard of one
    of them no longer than yesterday, a very honest
    woman, but something given to lie, as a woman should
    not do but in the way of honesty, how she died of
    the biting of it, what pain she felt. Truly, she makes a
    very good report o'th' worm; but he that will believe
    all that they say shall never be saved by half that
    they do; but this is most falliable: the worm's an odd
    worm.
CLEOPATRA Get thee hence, farewell.
CLOWN I wish you all joy of the worm.    255
CLEOPATRA Farewell.
CLOWN You must think this, look you, that the worm will
    do his kind.
CLEOPATRA Ay, ay; farewell.    259
CLOWN Look you, the worm is not to be trusted but in
    the keeping of wise people; for indeed there is no
    goodness in the worm.
CLEOPATRA Take thou no care; it shall be heeded.
CLOWN Very good. Give it nothing, I pray you, for it is
    not worth the feeding.    265
CLEOPATRA Will it eat me?
CLOWN You must not think I am so simple but I know
    the devil himself will not eat a woman; I know that a
    woman is a dish for the gods, if the devil dress her not.
    But truly, these same whoreson devils do the gods great
    harm in their women; for in every ten that they make,
    the devils mar five.    272
CLEOPATRA Well, get thee gone, farewell.
CLOWN Yes, forsooth. I wish you joy o'th' worm.
                *Exit, leaving the basket*
*Enter* ⌜*Iras*⌝ *with a robe, crown, and other jewels*
CLEOPATRA
Give me my robe. Put on my crown. I have    275
Immortal longings in me. Now no more
The juice of Egypt's grape shall moist this lip.
*Charmian and Iras help her to dress*
Yare, yare, good Iras, quick—methinks I hear
Antony call. I see him rouse himself
To praise my noble act. I hear him mock    280

The luck of Caesar, which the gods give men
To excuse their after wrath. Husband, I come.
Now to that name my courage prove my title.
I am fire and air; my other elements
I give to baser life. So, have you done?                    285
Come then, and take the last warmth of my lips.
    *She kisses them*
Farewell, kind Charmian. Iras, long farewell.
    *Iras falls and dies*
Have I the aspic in my lips? Dost fall?
If thou and nature can so gently part,
The stroke of death is as a lover's pinch,                    290
Which hurts and is desired. Dost thou lie still?
If thus thou vanishest, thou tell'st the world
It is not worth leave-taking.

CHARMIAN
Dissolve, thick cloud, and rain, that I may say
The gods themselves do weep.

CLEOPATRA                                     This proves me base.
If she first meet the curlèd Antony                    296
He'll make demand of her, and spend that kiss
Which is my heaven to have.
    *She takes an aspic from the basket and puts it to her*
    *breast*
                      Come, thou mortal wretch,
With thy sharp teeth this knot intrinsicate
Of life at once untie. Poor venomous fool,                    300
Be angry, and dispatch. O, couldst thou speak,
That I might hear thee call great Caesar ass
Unpolicied!

CHARMIAN     O eastern star!

CLEOPATRA                                     Peace, peace.
Dost thou not see my baby at my breast,
That sucks the nurse asleep?

CHARMIAN                              O, break! O, break!     305

CLEOPATRA
As sweet as balm, as soft as air, as gentle.
O Antony!
    *She puts another aspic to her arm*
             Nay, I will take thee too.
What should I stay—                              *She dies*

CHARMIAN                    In this vile world? So, fare thee well.
Now boast thee, death, in thy possession lies
A lass unparalleled. Downy windows, close,                    310
And golden Phoebus never be beheld
Of eyes again so royal. Your crown's awry.
I'll mend it, and then play—
    *Enter the Guard, rustling in*

FIRST GUARD  Where's the Queen?

CHARMIAN  Speak softly. Wake her not.                    315

FIRST GUARD
Caesar hath sent—

CHARMIAN                    Too slow a messenger.
    *She applies an aspic*
O come apace, dispatch! I partly feel thee.

FIRST GUARD
Approach, ho! All's not well. Caesar's beguiled.

SECOND GUARD
There's Dolabella sent from Caesar. Call him.
    ⌈*Exit a Guardsman*⌉

FIRST GUARD
What work is here, Charmian? Is this well done?     320

CHARMIAN
It is well done, and fitting for a princess
Descended of so many royal kings.
Ah, soldier!                                         *She dies*
    *Enter Dolabella*

DOLABELLA
How goes it here?

SECOND GUARD          All dead.

DOLABELLA                              Caesar, thy thoughts
Touch their effects in this. Thyself art coming     325
To see performed the dreaded act which thou
So sought'st to hinder.

ALL                              A way there, a way for Caesar!
    *Enter Caesar and all his train, marching*

DOLABELLA (*to Caesar*)
O sir, you are too sure an augurer.
That you did fear is done.

CAESAR                              Bravest at the last,
She levelled at our purposes, and, being royal,     330
Took her own way. The manner of their deaths?
I do not see them bleed.

DOLABELLA (*to a Guardsman*) Who was last with them?

FIRST GUARD
A simple countryman that brought her figs.
This was his basket.

CAESAR                    Poisoned, then.

FIRST GUARD                              O Caesar,
This Charmian lived but now; she stood and spake.
I found her trimming up the diadem                    336
On her dead mistress; tremblingly she stood,
And on the sudden dropped.

CAESAR                              O, noble weakness!
If they had swallowed poison, 'twould appear
By external swelling; but she looks like sleep,     340
As she would catch another Antony
In her strong toil of grace.

DOLABELLA                    Here on her breast
There is a vent of blood, and something blown.
The like is on her arm.

FIRST GUARD                    This is an aspic's trail,
And these fig-leaves have slime upon them such     345
As th'aspic leaves upon the caves of Nile.

CAESAR Most probable
That so she died; for her physician tells me
She hath pursued conclusions infinite
Of easy ways to die. Take up her bed,                    350
And bear her women from the monument.
She shall be buried by her Antony.
No grave upon the earth shall clip in it
A pair so famous. High events as these
Strike those that make them, and their story is     355
No less in pity than his glory which
Brought them to be lamented. Our army shall
In solemn show attend this funeral,
And then to Rome. Come, Dolabella, see
High order in this great solemnity.                    360
    *Exeunt all, soldiers bearing Cleopatra* ⌈*on her*
    *bed*⌉*, Charmian, and Iras*

# PERICLES

## BY WILLIAM SHAKESPEARE AND GEORGE WILKINS
## A RECONSTRUCTED TEXT

ON 20 May 1608 *Pericles* was entered on the Stationers' Register to Edward Blount; but he did not publish it. Probably the players allowed him to license it in the hope of preventing its publication by anyone else, for it was one of the most popular plays of the period. Its success was exploited, also in 1608, by the publication of a novel, by George Wilkins, 'The *Painful Adventures of Pericles Prince of Tyre*, Being the True History of the Play of *Pericles*, as it was lately presented by the worthy and ancient poet John Gower'. The play itself appeared in print in the following year, with an ascription to Shakespeare, but in a manifestly corrupt text that gives every sign of having been put together from memory. This quarto was several times reprinted; but the play was not included in the 1623 Folio (perhaps because Heminges and Condell knew that Shakespeare was responsible for only part of it).

In putting together *The Painful Adventures*, Wilkins drew on an earlier version of the tale, *The Pattern of Painful Adventures*, by Laurence Twine, written in the mid-1570s and re-printed in 1607. Twine's book is also a source of the play, which draws too on the story of Apollonius of Tyre as told by John Gower in his *Confessio Amantis*, and, to a lesser extent, on Sir Philip Sidney's *Arcadia*. Wilkins not only incorporated verbatim passages from Twine's book, he also drew heavily on *Pericles* itself. Since the play text is so corrupt, it is quite likely that Wilkins reports parts of it both more accurately and more fully than the quarto. And he may have had special qualifications for doing so. He was a dramatist whose popular play *The Miseries of Enforced Marriage* had been performed by Shakespeare's company. *Pericles* has usually been regarded as either a collaborative play or one in which Shakespeare revised a pre-existing script. Our edition is based on the hypothesis (not new) that Wilkins was its joint author. Our attempt to reconstruct the play draws more heavily than is usual on Wilkins's novel, especially in the first nine scenes (which he probably wrote); in general, because of its obvious corruption, the original text is more freely emended than usual. So that readers may experience the play as originally printed, an unemended reprint of the 1609 quarto is given in our original-spelling edition. The deficiencies of the text are in part compensated for by the survival of an unusual amount of relevant visual material, reproduced overleaf.

The complex textual background of *Pericles* should not be allowed to draw attention away from the merits of this dramatic romance, which we hope will be more apparent as the result of our treatment of the text. If the original play had survived, it might well have been as highly valued as *The Winter's Tale* or *The Tempest*; as it is, it contains some hauntingly beautiful episodes, above all that in Scene 21 in which Marina, Pericles' long-lost daughter, draws him out of the comatose state to which his sufferings have reduced him.

The true History of the Play of *Pericles*, as it was lately presented by the worthy and ancient Poet *John Gower*.

John Gower

## The Description of John Gower

Large he was; his height was long;
Broad of breast; his limbs were strong,
But colour pale, and wan his look—
Such have they that plyen their book.
His head was grey and quaintly shorn.
Neatly was his beard worn.
His visage grave, stern and grim.
Cato was most like to him.
His bonnet was a hat of blue;
His sleeves straight of that same hue.
A surcoat of a tawny dye
Hung in pleats over his thigh;
A breech, close unto his dock,
Handsomed with a long stock.
Pricked before were his shoon;
He wore such as others doone.
A bag of red by his side,
And by that his napkin tied.
Thus John Gower did appear,
Quaint attired, as you hear.

10. From the title-page of *The Painful Adventures of Pericles Prince of Tyre* (1608), by George Wilkins; artist unknown. Since Gower is not a character in Wilkins's novel, the choice of woodcut undoubtedly reflects both the play's popularity and Gower's own impact in early performances, and it is as likely to reflect the visual detail of performance as any early title-page. The sprig of laurel (or posy) in Gower's left hand is symbolic of his poetic status.

11. From *Greene's Vision* (1592), sig. C1r–C1v; probably by Robert Greene. The description here fits reasonably well the *Painful Adventures* title-page, though the woodcut does not contain the 'bag of red', 'napkin', or tight-fitting 'breech'.

Bridge Gate

Me pompæ prouexit apex.
The desire of renowne hath promoted me, or set me forward.

Qui me alit, me extinguit.
He that nourisheth me, killeth me.

13. From *The Heroical Devices of M. Claudius Paradin*, translated by P.S. (1591), sig. V3. This is the source for the impresa of the Third Knight, in Sc. 6.

14. From *The Heroical Devices of M. Claudius Paradin*, translated by P.S. (1591), sig. Z3. This is the source for the impresa of the Fourth Knight, in Sc. 6.

12. Severed heads displayed on the gate of London Bridge, from an etching by Claes Jan Visscher (1616). In the play's sources, and *Painful Adventures*, the heads of previous suitors (Sc. 1) are placed on the 'gate' of Antioch. In performance they could have been thrust out on poles from the upper stage; but the timing and method of their display is not clear.

16. A miniature of Diana by Isaac Oliver (1615): the dress is yellow, the scarf a gauzy pink-white, the cloak over her right shoulder blue; the leaf-shaped brooch topped by the crescent moon, gold. In Samuel Daniel's masque *The Vision of the Twelve Goddesses* (1604), 'Diana, in a green mantle embroidered with silver half moons, and a crescent of pearl on her head, presents a bow and quiver' (sig. A5). The 'crescent of pearl'—an ornamental crescent moon, also detectable in Jones's sketch—can be seen in many emblematic representations of the goddess.

15. An Inigo Jones sketch of Diana, probably for Ben Jonson's masque *Time Vindicated* (1623). The goddess of chastity appeared as a character in court entertainments, masques, and plays, and her representation was governed by iconographic convention. As goddess of hunting, she was most often identified by her 'silver bow' (21.234). In Thomas Heywood's *The Golden Age* (1611), stage directions refer to '*Diana's bow*' (sig. E1ᵛ) and her '*buskins*' (sig. E3ᵛ); her '*nymphs*' explicitly, and by inference she, have '*garlands on their heads, and javelins in their hands . . . bows and quivers*' (sig. D3ᵛ). The bow, quiver, and javelin, all visible in Jones's sketch, were commonplace in emblematic representations. As a huntress, Diana could naturally be envisaged in a chariot: in Aurelian Townshend's masque *Albion's Triumph* (1631), she descends 'in her chariot' (pp. 2, 12); in *Time Vindicated*, 'Diana descends' (l. 446). Such descents for deities were used in the public theatres, too, usually in a chair or chariot (21.224.2).

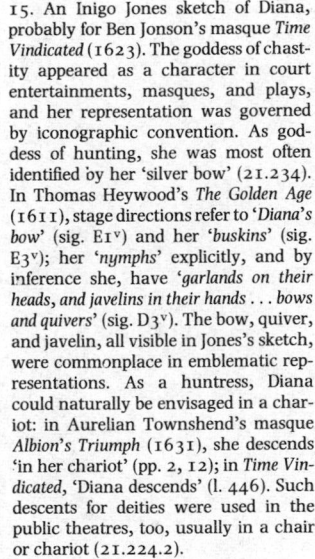

17. For the pastoral *Florimène* (1635), Inigo Jones designed two scenic views of 'The Temple of Diana' (see l. 22.17.1). Though such scenes were not used in the public theatres in Shakespeare's time, the columns supporting the overhanging roof of the public stage (see General Introduction, pp. xxvii–xxix) could have created a scenic effect roughly similar to Jones's recessed classical temple. Statues were also available as props in the public theatre; in *Pericles*, as in *The Winter's Tale*, the statue could have been impersonated by an actor on a pedestal. Whether or not a statue was visible, the temple could be identified by an altar (as in *The Two Noble Kinsmen*).

# THE PERSONS OF THE PLAY

John GOWER, the Presenter

ANTIOCHUS, King of Antioch
His DAUGHTER
THALIART, a villain

PERICLES, Prince of Tyre
HELICANUS ⎱
AESCHINES ⎰ two grave counsellors of Tyre
MARINA, Pericles' daughter

CLEON, Governor of Tarsus
DIONIZA, his wife
LEONINE, a murderer

KING SIMONIDES, of Pentapolis
THAISA, his daughter
Three FISHERMEN, his subjects

Five PRINCES, suitors of Thaisa
A MARSHAL
LYCHORIDA, Thaisa's nurse

CERIMON, a physician of Ephesus
PHILEMON, his servant

LYSIMACHUS, Governor of Mytilene
A BAWD
A PANDER
BOULT, a leno

DIANA, goddess of chastity

Lords, ladies, pages, messengers, sailors, gentlemen

# A Reconstructed Text of
# Pericles, Prince of Tyre

**Sc. 1** *Enter Gower as Prologue*

GOWER
To sing a song that old was sung
From ashes ancient Gower is come,
Assuming man's infirmities
To glad your ear and please your eyes.
It hath been sung at festivals, 5
On ember-eves and holy-ales,
And lords and ladies in their lives
Have read it for restoratives.
The purchase is to make men glorious,
*Et bonum quo antiquius eo melius.* 10
If you, born in these latter times
When wit's more ripe, accept my rhymes,
And that to hear an old man sing
May to your wishes pleasure bring,
I life would wish, and that I might 15
Waste it for you like taper-light.
This' Antioch, then; Antiochus the Great
Built up this city for his chiefest seat,
The fairest in all Syria.
I tell you what mine authors say. 20
This king unto him took a fere
Who died, and left a female heir
So buxom, blithe, and full of face
As heav'n had lent her all his grace,
With whom the father liking took, 25
And her to incest did provoke.
Bad child, worse father, to entice his own
To evil should be done by none.
By custom what they did begin
Was with long use account' no sin. 30
The beauty of this sinful dame
Made many princes thither frame
To seek her as a bedfellow,
In marriage pleasures playfellow,
Which to prevent he made a law 35
To keep her still, and men in awe,
That whoso asked her for his wife,
His riddle told not, lost his life.
So for her many a wight did die,
⌐*A row of heads is revealed*⌐
As yon grim looks do testify. 40
What now ensues, to th' judgement of your eye
I give, my cause who best can justify. *Exit*
⌐*Sennet.*⌐ *Enter King Antiochus, Prince Pericles, and*
⌐*lords and peers in their richest ornaments*⌐

ANTIOCHUS
Young Prince of Tyre, you have at large received
The danger of the task you undertake.

PERICLES
I have, Antiochus, and with a soul 45
Emboldened with the glory of her praise
Think death no hazard in this enterprise.

ANTIOCHUS Music!
*Music sounds*
Bring in our daughter, clothèd like a bride

Fit for th'embracements ev'n of Jove himself, 50
At whose conception, till Lucina reigned,
Nature this dowry gave to glad her presence:
The senate-house of planets all did sit,
In her their best perfections to knit.
*Enter Antiochus' Daughter*

PERICLES
See where she comes, apparelled like the spring, 55
Graces her subjects, and her thoughts the king
Of ev'ry virtue gives renown to men;
Her face the book of praises, where is read
Nothing but curious pleasures, as from thence
Sorrow were ever razed and testy wrath 60
Could never be her mild companion.
You gods that made me man, and sway in love,
That have inflamed desire in my breast
To taste the fruit of yon celestial tree
Or die in the adventure, be my helps, 65
As I am son and servant to your will,
To compass such a boundless happiness.

ANTIOCHUS Prince Pericles—
PERICLES
That would be son to great Antiochus.

ANTIOCHUS
Before thee stands this fair Hesperides, 70
With golden fruit, but dang'rous to be touched,
⌐*He gestures towards the heads*⌐
For death-like dragons here affright thee hard.
⌐*He gestures towards his daughter*⌐
Her heav'n-like face enticeth thee to view
Her countless glory, which desert must gain;
And which without desert, because thine eye 75
Presumes to reach, all the whole heap must die.
Yon sometimes famous princes, like thyself
Drawn by report, advent'rous by desire,
Tell thee with speechless tongues and semblants
   bloodless
That without covering save yon field of stars 80
Here they stand, martyrs slain in Cupid's wars,
And with dead cheeks advise thee to desist
From going on death's net, whom none resist.

PERICLES
Antiochus, I thank thee, who hath taught
My frail mortality to know itself, 85
And by those fearful objects to prepare
This body, like to them, to what I must;
For death remembered should be like a mirror
Who tells us life's but breath, to trust it error.
I'll make my will then, and, as sick men do, 90
Who know the world, see heav'n, but feeling woe
Grip not at earthly joys as erst they did,
So I bequeath a happy peace to you
And all good men, as ev'ry prince should do;
My riches to the earth from whence they came, 95
(*To the Daughter*) But my unspotted fire of love to you.
(*To Antiochus*) Thus ready for the way of life or death,
I wait the sharpest blow, Antiochus.

ANTIOCHUS
Scorning advice, read the conclusion then,
⌜*He angrily throws down the riddle*⌝
Which read and not expounded, 'tis decreed, 100
As these before thee, thou thyself shalt bleed.
DAUGHTER (*to Pericles*)
Of all 'sayed yet, mayst thou prove prosperous;
Of all 'sayed yet, I wish thee happiness.
PERICLES
Like a bold champion I assume the lists,
Nor ask advice of any other thought 105
But faithfulness and courage.
⌜*He takes up and*⌝ *reads aloud the riddle*
I am no viper, yet I feed
On mother's flesh which did me breed.
I sought a husband, in which labour
I found that kindness in a father. 110
He's father, son, and husband mild;
I mother, wife, and yet his child.
How this may be and yet in two,
As you will live resolve it you.
Sharp physic is the last. ⌜*Aside*⌝ But O, you powers 115
That gives heav'n countless eyes to view men's acts,
Why cloud they not their sights perpetually
If this be true which makes me pale to read it?
⌜*He gazes on the Daughter*⌝
Fair glass of light, I loved you, and could still,
Were not this glorious casket stored with ill. 120
But I must tell you now my thoughts revolt,
For he's no man on whom perfections wait
That, knowing sin within, will touch the gate.
You're a fair viol, and your sense the strings
Who, fingered to make man his lawful music, 125
Would draw heav'n down and all the gods to hearken,
But, being played upon before your time,
Hell only danceth at so harsh a chime.
Good sooth, I care not for you.
ANTIOCHUS
Prince Pericles, touch not, upon thy life, 130
For that's an article within our law
As dang'rous as the rest. Your time's expired.
Either expound now, or receive your sentence.
PERICLES Great King,
Few love to hear the sins they love to act. 135
'Twould braid yourself too near for me to tell it.
Who has a book of all that monarchs do,
He's more secure to keep it shut than shown,
For vice repeated, like the wand'ring wind,
Blows dust in others' eyes to spread itself; 140
And yet the end of all is bought thus dear,
The breath is gone, and the sore eyes see clear
To stop the air would hurt them. The blind mole casts
Copped hills towards heav'n to tell the earth is thronged
By man's oppression, and the poor worm doth die for't.
Kings are earth's gods; in vice their law's their will,
And if Jove stray, who dares say Jove doth ill? 147
It is enough you know, and it is fit,
What being more known grows worse, to smother it.
All love the womb that their first being bred; 150
Then give my tongue like leave to love my head.
ANTIOCHUS (*aside*)
Heav'n, that I had thy head! He's found the meaning.
But I will gloze with him. —Young Prince of Tyre,
Though by the tenor of our strict edict,
Your exposition misinterpreting, 155
We might proceed to cancel of your days,

Yet hope, succeeding from so fair a tree
As your fair self, doth tune us otherwise.
Forty days longer we do respite you,
If by which time our secret be undone, 160
This mercy shows we'll joy in such a son.
And until then your entertain shall be
As doth befit your worth and our degree.
⌜*Flourish.*⌝ *Exeunt all but Pericles*
PERICLES
How courtesy would seem to cover sin
When what is done is like an hypocrite, 165
The which is good in nothing but in sight.
If it be true that I interpret false,
Then were it certain you were not so bad
As with foul incest to abuse your soul,
Where now you're both a father and a son 170
By your uncomely claspings with your child—
Which pleasures fits a husband, not a father—
And she, an eater of her mother's flesh,
By the defiling of her parents' bed,
And both like serpents are, who though they feed 175
On sweetest flowers, yet they poison breed.
Antioch, farewell, for wisdom sees those men
Blush not in actions blacker than the night
Will 'schew no course to keep them from the light.
One sin, I know, another doth provoke. 180
Murder's as near to lust as flame to smoke.
Poison and treason are the hands of sin,
Ay, and the targets to put off the shame.
Then, lest my life be cropped to keep you clear, 184
By flight I'll shun the danger which I fear. *Exit*
*Enter Antiochus*
ANTIOCHUS
He hath found the meaning, for the which we mean
To have his head. He must not live
To trumpet forth my infamy, nor tell the world
Antiochus doth sin in such a loathèd manner,
And therefore instantly this prince must die, 190
For by his fall my honour must keep high.
Who attends us there?
*Enter Thaliart*
THALIART                    Doth your highness call?
ANTIOCHUS
Thaliart, you are of our chamber, Thaliart,
And to your secrecy our mind partakes
Her private actions. For your faithfulness 195
We will advance you, Thaliart. Behold,
Here's poison, and here's gold.
We hate the Prince of Tyre, and thou must kill him.
It fits thee not to ask the reason. Why?
Because we bid it. Say, is it done? 200
THALIART My lord, 'tis done.
ANTIOCHUS Enough.
*Enter a Messenger hastily*
Let your breath cool yourself, telling your haste.
MESSENGER
Your majesty, Prince Pericles is fled. ⌜*Exit*⌝
ANTIOCHUS (*to Thaliart*)
As thou wilt live, fly after; like an arrow 205
Shot from a well-experienced archer hits
The mark his eye doth level at, so thou
Never return unless it be to say
'Your majesty, Prince Pericles is dead.'
THALIART
If I can get him in my pistol's length 210
I'll make him sure enough. Farewell, your highness.

ANTIOCHUS
  Thaliart, adieu.                              ⌈*Exit Thaliart*⌉
          Till Pericles be dead
  My heart can lend no succour to my head.
                              *Exit.* ⌈*The heads are concealed*⌉

Sc. 2   *Enter Pericles, distempered, with his lords*
PERICLES
  Let none disturb us.                          *Exeunt lords*
            Why should this change of thoughts,
  The sad companion, dull-eyed melancholy,
  Be my so used a guest as not an hour
  In the day's glorious walk or peaceful night,
  The tomb where grief should sleep, can breed me
      quiet?                                              5
  Here pleasures court mine eyes, and mine eyes shun
      them,
  And danger, which I feared, 's at Antioch,
  Whose arm seems far too short to hit me here.
  Yet neither pleasure's art can joy my spirits,
  Nor yet care's author's distance comfort me.           10
  Then it is thus: the passions of the mind,
  That have their first conception by misdread,
  Have after-nourishment and life by care,
  And what was first but fear what might be done
  Grows elder now, and cares it be not done.             15
  And so with me. The great Antiochus,
  'Gainst whom I am too little to contend,
  Since he's so great can make his will his act,
  Will think me speaking though I swear to silence,
  Nor boots it me to say I honour him                    20
  If he suspect I may dishonour him.
  And what may make him blush in being known,
  He'll stop the course by which it might be known.
  With hostile forces he'll o'erspread the land,
  And with th'ost of war will look so huge               25
  Amazement shall drive courage from the state,
  Our men be vanquished ere they do resist,
  And subjects punished that ne'er thought offence,
  Which care of them, not pity of myself,
  Who am no more but as the tops of trees                30
  Which fence the roots they grow by and defend them,
  Makes both my body pine and soul to languish,
  And punish that before that he would punish.
            *Enter all the Lords, among them old Helicanus, to*
            *Pericles*
FIRST LORD
  Joy and all comfort in your sacred breast!
SECOND LORD
  And keep your mind peaceful and comfortable.           35
HELICANUS
  Peace, peace, and give experience tongue.
  (*To Pericles*) You do not well so to abuse yourself,
  To waste your body here with pining sorrow,
  Upon whose safety doth depend the lives
  And the prosperity of a whole kingdom.                 40
  'Tis ill in you to do it, and no less
  Ill in your council not to contradict it.
  They do abuse the King that flatter him,
  For flatt'ry is the bellows blows up sin;
  The thing the which is flattered, but a spark,         45
  To which that wind gives heat and stronger glowing;
  Whereas reproof, obedient and in order,
  Fits kings as they are men, for they may err.
  When Signor Sooth here does proclaim a peace
  He flatters you, makes war upon your life.             50
  ⌈*He kneels*⌉

  Prince, pardon me, or strike me if you please.
  I cannot be much lower than my knees.
PERICLES
  All leave us else; but let your cares o'erlook
  What shipping and what lading's in our haven,
  And then return to us.                     *Exeunt Lords*
            Helicane, thou                              55
  Hast movèd us. What seest thou in our looks?
HELICANUS  An angry brow, dread lord.
PERICLES
  If there be such a dart in princes' frowns,
  How durst thy tongue move anger to our brows?
HELICANUS
  How dares the plants look up to heav'n from whence
  They have their nourishment?                           61
PERICLES
  Thou knowest I have pow'r to take thy life from thee.
HELICANUS
  I have ground the axe myself; do you but strike the
      blow.
PERICLES ⌈*lifting him up*⌉
  Rise, prithee, rise. Sit down. Thou art no flatterer,
  I thank thee for it, and the heav'ns forbid           65
  That kings should let their ears hear their faults hid.
  Fit counsellor and servant for a prince,
  Who by thy wisdom mak'st a prince thy servant,
  What wouldst thou have me do?
HELICANUS                          To bear with patience
  Such griefs as you do lay upon yourself.               70
PERICLES
  Thou speak'st like a physician, Helicanus,
  That ministers a potion unto me
  That thou wouldst tremble to receive thyself.
  Attend me, then. I went to Antioch,
  Where, as thou know'st, against the face of death      75
  I sought the purchase of a glorious beauty
  From whence an issue I might propagate,
  As children are heav'n's blessings: to parents,
      objects;
  Are arms to princes, and bring joys to subjects.
  Her face was to mine eye beyond all wonder,            80
  The rest—hark in thine ear—as black as incest,
  Which by my knowledge found, the sinful father
  Seemed not to strike, but smooth. But thou know'st
      this,
  'Tis time to fear when tyrants seems to kiss;
  Which fear so grew in me I hither fled                 85
  Under the covering of careful night,
  Who seemed my good protector, and being here
  Bethought me what was past, what might succeed.
  I knew him tyrannous, and tyrants' fears
  Decrease not, but grow faster than the years.          90
  And should he doubt—as doubt no doubt he doth—
  That I should open to the list'ning air
  How many worthy princes' bloods were shed
  To keep his bed of blackness unlaid ope,
  To lop that doubt he'll fill this land with arms,      95
  And make pretence of wrong that I have done him,
  When all for mine—if I may call—offence
  Must feel war's blow, who spares not innocence;
  Which love to all, of which thyself art one,
  Who now reproved'st me for't—
HELICANUS                            Alas, sir.          100
PERICLES
  Drew sleep out of mine eyes, blood from my cheeks,
  Musings into my mind, with thousand doubts,

How I might stop this tempest ere it came,
And, finding little comfort to relieve them,
I thought it princely charity to grieve them.     105

HELICANUS
Well, my lord, since you have giv'n me leave to speak,
Freely will I speak. Antiochus you fear,
And justly too, I think, you fear the tyrant,
Who either by public war or private treason
Will take away your life.     110
Therefore, my lord, go travel for a while,
Till that his rage and anger be forgot,
Or destinies do cut his thread of life.
Your rule direct to any; if to me,
Day serves not light more faithful than I'll be.     115

PERICLES  I do not doubt thy faith,
But should he in my absence wrong thy liberties?

HELICANUS
We'll mingle our bloods together in the earth
From whence we had our being and our birth.

PERICLES
Tyre, I now look from thee then, and to Tarsus     120
Intend my travel, where I'll hear from thee,
And by whose letters I'll dispose myself.
The care I had and have of subjects' good
On thee I lay, whose wisdom's strength can bear it.
I'll take thy word for faith, not ask thine oath;     125
Who shuns not to break one will sure crack both.
But in our orbs we'll live so round and safe
That time of both this truth shall ne'er convince:
Thou showed'st a subject's shine, I a true prince.
                                     *Exeunt*

**Sc. 3**    *Enter Thaliart*

THALIART  So this is Tyre, and this the court. Here must
I kill King Pericles, and if I do it and am caught I am
like to be hanged abroad, but if I do it not, I am sure
to be hanged at home. 'Tis dangerous. Well, I perceive
he was a wise fellow and had good discretion that,
being bid to ask what he would of the King, desired he
might know none of his secrets. Now do I see he had
some reason for't, for if a king bid a man be a villain,
he's bound by the indenture of his oath to be one.
Hush, here comes the lords of Tyre.     10

     *Enter Helicanus and Aeschines, with other lords*

HELICANUS
You shall not need, my fellow peers of Tyre,
Further to question of your King's departure.
His sealed commission left in trust with me
Does speak sufficiently he's gone to travel.

THALIART (*aside*)  How? The King gone?     15

HELICANUS
If further yet you will be satisfied
Why, as it were unlicensed of your loves,
He would depart, I'll give some light unto you.
Being at Antioch—

THALIART (*aside*)     What from Antioch?

HELICANUS
Royal Antiochus, on what cause I know not,     20
Took some displeasure at him—at least he judged so—
And doubting lest that he had erred or sinned,
To show his sorrow he'd correct himself;
So puts himself unto the ship-man's toil,
With whom each minute threatens life or death.     25

THALIART (*aside*)
Well, I perceive I shall not be hanged now,
Although I would.
But since he's gone, the King's ears it must please

He scaped the land to perish on the seas.
I'll present myself.—Peace to the lords of Tyre.     30
Lord Thaliart am I, of Antioch.

⌈HELICANUS⌉
Lord Thaliart of Antioch is welcome.

THALIART
From King Antiochus I come
With message unto princely Pericles,
But since my landing I have understood     35
Your lord's betook himself to unknown travels.
Now my message must return from whence it came.

HELICANUS
We have no reason to enquire it,
Commended to our master, not to us.
Yet ere you shall depart, this we desire:     40
As friends to Antioch, we may feast in Tyre.     *Exeunt*

**Sc. 4**    *Enter Cleon, the Governor of Tarsus, with Dionyza*
           *his wife, and others*

CLEON
My Dionyza, shall we rest us here
And, by relating tales of others' griefs,
See if 'twill teach us to forget our own?

DIONYZA
That were to blow at fire in hope to quench it,
For who digs hills because they do aspire     5
Throws down one mountain to cast up a higher.
O my distressèd lord, e'en such our griefs are;
Here they're but felt and seen with midges' eyes,
But like to groves, being topped they higher rise.

CLEON  O Dionyza,     10
Who wanteth food and will not say he wants it,
Or can conceal his hunger till he famish?
Our tongues our sorrows dictate to sound deep
Our woes into the air, our eyes to weep
Till lungs fetch breath that may proclaim them louder,
That, if heav'n slumber while their creatures want,     16
They may awake their helps to comfort them.
I'll then discourse our woes, felt sev'ral years,
And, wanting breath to speak, help me with tears.

DIONYZA  As you think best, sir.     20

CLEON
This Tarsus o'er which I have the government,
A city o'er whom plenty held full hand,
For riches strewed herself ev'n in the streets,
Whose tow'rs bore heads so high they kissed the clouds,
And strangers ne'er beheld but wondered at,     25
Whose men and dames so jetted and adorned
Like one another's glass to trim them by;
Their tables were stored full to glad the sight,
And not so much to feed on as delight;
All poverty was scorned, and pride so great     30
The name of help grew odious to repeat.

DIONYZA  O, 'tis too true.

CLEON
But see what heav'n can do by this our change.
Those mouths who but of late earth, sea, and air
Were all too little to content and please,     35
Although they gave their creatures in abundance,
As houses are defiled for want of use,
They are now starved for want of exercise.
Those palates who, not yet two summers younger,
Must have inventions to delight the taste     40
Would now be glad of bread and beg for it.
Those mothers who to nuzzle up their babes
Thought naught too curious are ready now

To eat those little darlings whom they loved.
So sharp are hunger's teeth that man and wife          45
Draw lots who first shall die to lengthen life.
Here weeping stands a lord, there lies a lady dying,
Here many sink, yet those which see them fall
Have scarce strength left to give them burial.
Is not this true?                                                      50

DIONYZA
Our cheeks and hollow eyes do witness it.

CLEON
O, let those cities that of plenty's cup
And her prosperities so largely taste
With their superfluous riots, heed these tears!
The misery of Tarsus may be theirs.                    55

*Enter a ⌈fainting⌉ Lord of Tarsus ⌈slowly⌉*

LORD Where's the Lord Governor?

CLEON
Here. Speak out thy sorrows which thou bring'st in
     haste,
For comfort is too far for us t'expect.

LORD
We have descried upon our neighbouring shore
A portly sail of ships make hitherward.                  60

CLEON I thought as much.
One sorrow never comes but brings an heir
That may succeed as his inheritor,
And so in ours. Some neighbour nation,
Taking advantage of our misery,                          65
Hath stuffed these hollow vessels with their power
To beat us down, the which are down already,
And make a conquest of unhappy men,
Whereas no glory's got to overcome.

LORD
That's the least fear, for by the semblance         70
Of their white flags displayed they bring us peace,
And come to us as favourers, not foes.

CLEON
Thou speak'st like him's untutored to repeat;
Who makes the fairest show means most deceit.
But bring they what they will and what they can,   75
What need we fear?
Our grave's the low'st, and we are half-way there.
Go tell their gen'ral we attend him here
To know for what he comes, and whence he comes.

LORD I go, my lord.                                          *Exit*

CLEON
Welcome is peace, if he on peace consist;           81
If wars, we are unable to resist.

*Enter ⌈the Lord again conducting⌉ Pericles with
attendants*

PERICLES (*to Cleon*)
Lord Governor, for so we hear you are,
Let not our ships and number of our men
Be like a beacon fixed t'amaze your eyes.           85
We have heard your miseries as far as Tyre,
Since entering your unshut gates have witnessed
The widowed desolation of your streets;
Nor come we to add sorrow to your hearts,
But to relieve them of their heavy load;              90
And these our ships, you happily may think
Are like the Trojan horse was fraught within
With bloody veins importing overthrow,
Are stored with corn to make your needy bread,
And give them life whom hunger starved half dead. 95

ALL OF TARSUS ⌈*falling on their knees and weeping*⌉
The gods of Greece protect you, and we'll pray for you!

PERICLES Arise, I pray you, rise.

We do not look for reverence but for love,
And harbourage for me, my ships and men.

CLEON
The which when any shall not gratify,                 100
Or pay you with unthankfulness in thought,
Be it our wives, our children, or ourselves,
The curse of heav'n and men succeed their evils!
Till when—the which I hope shall ne'er be seen—
Your grace is welcome to our town and us.          105

PERICLES
Which welcome we'll accept, feast here a while,
Until our stars that frown lend us a smile.          *Exeunt*

**Sc. 5** *Enter Gower*

GOWER
Here have you seen a mighty king
His child, iwis, to incest bring;
A better prince and benign lord
Prove awe-full both in deed and word.
Be quiet then, as men should be,                         5
Till he hath passed necessity.
I'll show you those in trouble's reign,
Losing a mite, a mountain gain.
The good in conversation,
To whom I give my benison,                                10
Is still at Tarsus where each man
Thinks all is writ he speken can,
And to remember what he does
His statue build to make him glorious.
But tidings to the contrary                                 15
Are brought your eyes. What need speak I?

*Dumb show.*
*Enter at one door Pericles talking with Cleon, all
the train with them. Enter at another door a
gentleman with a letter to Pericles. Pericles shows
the letter to Cleon. Pericles gives the messenger a
reward, and knights him. Exeunt with their trains
Pericles at one door and Cleon at another*

Good Helicane that stayed at home,
Not to eat honey like a drone
From others' labours, for that he strive
To killen bad, keep good alive,                           20
And to fulfil his prince' desire
Sent word of all that haps in Tyre;
How Thaliart came full bent with sin
And hid intent to murdren him,
And that in Tarsus was not best                          25
Longer for him to make his rest.
He deeming so put forth to seas,
Where when men been there's seldom ease,
For now the wind begins to blow;
Thunder above and deeps below                         30
Makes such unquiet that the ship
Should house him safe is wrecked and split,
And he, good prince, having all lost,
By waves from coast to coast is tossed.
All perishen of man, of pelf,                              35
Ne aught escapend but himself,
Till fortune, tired with doing bad,
Threw him ashore to give him glad.

⌈*Enter Pericles wet and half-naked*⌉
And here he comes. What shall be next
Pardon old Gower; this 'longs the text.      *Exit*
⌈*Thunder and lightning*⌉

PERICLES
Yet cease your ire, you angry stars of heaven!     41
Wind, rain, and thunder, remember earthly man

Is but a substance that must yield to you,
And I, as fits my nature, do obey you.
Alas, the seas hath cast me on the rocks,    45
Washed me from shore to shore, and left my breath
Nothing to think on but ensuing death.
Let it suffice the greatness of your powers
To have bereft a prince of all his fortunes,
And, having thrown him from your wat'ry grave,    50
Here to have death in peace is all he'll crave.
⌜*He sits.*⌝
*Enter two poor Fishermen: one the Master, the*
*other his man*

MASTER ⌜*calling*⌝ What ho, Pilch!

SECOND FISHERMAN ⌜*calling*⌝ Ha, come and bring away the
nets.

MASTER ⌜*calling*⌝ What, Patchbreech, I say!    55
⌜*Enter a Third rough Fisherman with a hood upon*
*his head and a filthy leathern pelt upon his back,*
*unseemly clad, and homely to behold. He brings nets*
*to dry and repair*⌝

THIRD FISHERMAN What say you, master?

MASTER Look how thou stirrest now. Come away, or I'll
fetch th' with a wanion.

THIRD FISHERMAN Faith, master, I am thinking of the poor
men that were cast away before us even now.    60

MASTER Alas, poor souls, it grieved my heart to hear what
pitiful cries they made to us to help them when, well-
a-day, we could scarce help ourselves.

THIRD FISHERMAN Nay, master, said not I as much when
I saw the porpoise how he bounced and tumbled? They
say they're half fish, half flesh. A plague on them, they
ne'er come but I look to be washed. Master, I marvel
how the fishes live in the sea.    68

MASTER Why, as men do a-land—the great ones eat up
the little ones. I can compare our rich misers to nothing
so fitly as to a whale: a plays and tumbles, driving the
poor fry before him, and at last devours them all at a
mouthful. Such whales have I heard on o'th' land, who
never leave gaping till they swallowed the whole parish:
church, steeple, bells, and all.    75

PERICLES (*aside*) A pretty moral.

THIRD FISHERMAN But, master, if I had been the sexton,
I would have been that day in the belfry.

SECOND FISHERMAN Why, man?

THIRD FISHERMAN Because he should have swallowed me,
too, and when I had been in his belly I would have
kept such a jangling of the bells that he should never
have left till he cast bells, steeple, church, and parish
up again. But if the good King Simonides were of my
mind—    85

PERICLES (*aside*) Simonides?

THIRD FISHERMAN We would purge the land of these
drones that rob the bee of her honey.

PERICLES (*aside*)
How from the finny subject of the sea
These fishers tell th'infirmities of men,    90
And from their wat'ry empire recollect
All that may men approve or men detect!
⌜*Coming forward*⌝ Peace be at your labour, honest
fishermen.

SECOND FISHERMAN Honest, good fellow? What's that? If
it be a day fits you, scratch't out of the calendar, and
nobody look after it.    96

PERICLES
May see the sea hath cast upon your coast—

SECOND FISHERMAN What a drunken knave was the sea
to cast thee in our way!

PERICLES
A man, whom both the waters and the wind    100
In that vast tennis-court hath made the ball
For them to play upon, entreats you pity him.
He asks of you that never used to beg.

MASTER No, friend, cannot you beg? Here's them in our
country of Greece gets more with begging than we can
do with working.    106

SECOND FISHERMAN Canst thou catch any fishes, then?

PERICLES I never practised it.

SECOND FISHERMAN Nay, then thou wilt starve, sure; for
here's nothing to be got nowadays unless thou canst
fish for't.    111

PERICLES
What I have been, I have forgot to know,
But what I am, want teaches me to think on:
A man thronged up with cold; my veins are chill,
And have no more of life than may suffice    115
To give my tongue that heat to crave your help,
Which if you shall refuse, when I am dead,
For that I am a man, pray see me burièd.
⌜*He falls down*⌝

MASTER Die, quotha? Now, gods forbid't an I have a gown
here! ⌜*To Pericles, lifting him up from the ground*⌝ Come,
put it on, keep thee warm. Now, afore me, a handsome
fellow! Come, thou shalt go home, and we'll have flesh
for holidays, fish for fasting-days, and moreo'er
puddings and flapjacks, and thou shalt be welcome.

PERICLES I thank you, sir.    125

SECOND FISHERMAN Hark you, my friend, you said you
could not beg?

PERICLES I did but crave.

SECOND FISHERMAN But crave? Then I'll turn craver too,
an so I shall scape whipping.    130

PERICLES Why, are all your beggars whipped, then?

SECOND FISHERMAN O, not all, my friend, not all; for if all
your beggars were whipped I would wish no better
office than to be beadle.

MASTER Thine office, knave—    135

SECOND FISHERMAN Is to draw up the other nets. I'll go.
*Exit with Third Fisherman*

PERICLES (*aside*)
How well this honest mirth becomes their labour!

MASTER ⌜*seating himself by Pericles*⌝ Hark you, sir, do you
know where ye are?

PERICLES Not well.    140

MASTER Why, I'll tell you. This is called Pentapolis, and
our king the good Simonides.

PERICLES
'The good Simonides' do you call him?

MASTER Ay, sir, and he deserves so to be called for his
peaceable reign and good government.    145

PERICLES
He is a happy king, since from his subjects
He gains the name of good by his government.
How far is his court distant from this shore?

MASTER Marry, sir, some half a day's journey. And I'll
tell you, he hath a fair daughter, and tomorrow is her
birthday, and there are princes and knights come from
all parts of the world to joust and tourney for her love.

PERICLES
Were but my fortunes answerable
To my desires I could wish to make one there.    154

MASTER O, sir, things must be as they may, and what a
man cannot get himself, he may lawfully deal for with
his wife's soul.

*Enter the other two Fishermen drawing up a net*

SECOND FISHERMAN Help, master, help! Here's a fish hangs
in the net like a poor man's right in the law; 'twill
hardly come out.                                                    160
        ⌜*Before help comes, up comes their prize*⌝
Ha, bots on't, 'tis come at last, and 'tis turned to a
rusty armour.

PERICLES
An armour, friends? I pray you let me see it.
(*Aside*) Thanks, fortune, yet that after all thy crosses
Thou giv'st me somewhat to repair my losses,         165
And though it was mine own, part of my heritage
Which my dead father did bequeath to me
With this strict charge ev'n as he left his life:
'Keep it, my Pericles; it hath been a shield
'Twixt me and death,' and pointed to this brace,     170
'For that it saved me, keep it. In like necessity,
The which the Gods forfend, the same may defend thee.'
It kept where I kept, I so dearly loved it,
Till the rough seas that spares not any man
Took it in rage, though calmed have giv'n't again.
I thank thee for't. My shipwreck now's no ill,        176
Since I have here my father gave in 's will.

MASTER What mean you, sir?

PERICLES
To beg of you, kind friends, this coat of worth,
For it was sometime target to a king.                 180
I know it by this mark. He loved me dearly,
And for his sake I wish the having of it,
And that you'd guide me to your sov'reign's court,
Where with't I may appear a gentleman.
And if that ever my low fortune's better,             185
I'll pay your bounties, till then rest your debtor.

MASTER Why, wilt thou tourney for the lady?

PERICLES
I'll show the virtue I have learned in arms.

MASTER Why, d'ye take it, and the gods give thee good
on't!                                                 190

SECOND FISHERMAN Ay, but hark you, my friend, 'twas
we that made up this garment through the rough
seams of the waters. There are certain condolements,
certain vails. I hope, sir, if you thrive, you'll remember
from whence you had this.                             195

PERICLES Believe't, I will.
By your furtherance I'm clothed in steel,
And spite of all the rapture of the sea
This jewel holds his building on my arm.
Unto thy value I will mount myself                    200
Upon a courser whose delightsome steps
Shall make the gazer joy to see him tread.
Only, my friends, I yet am unprovided
Of a pair of bases.                                   204

SECOND FISHERMAN We'll sure provide. Thou shalt have
my best gown to make thee a pair, and I'll bring thee
to the court myself.

PERICLES
Then honour be but equal to my will,
This day I'll rise, or else add ill to ill.           209
                              *Exeunt with nets and armour*

Sc. 6  ⌜*Sennet.*⌝ *Enter King Simonides and Thaisa, with
        Lords in attendance,* ⌜*and sit on two thrones*⌝

KING SIMONIDES
Are the knights ready to begin the triumph?

FIRST LORD They are, my liege,
And stay your coming to present themselves.

KING SIMONIDES
Return them we are ready; and our daughter,
In honour of whose birth these triumphs are,          5
Sits here like beauty's child, whom nature gat
For men to see and, seeing, wonder at.    ⌜*Exit one*⌝

THAISA
It pleaseth you, my father, to express
My commendations great, whose merit's less.

KING SIMONIDES
It's fit it should be so, for princes are            10
A model which heav'n makes like to itself.
As jewels lose their glory if neglected,
So princes their renown, if not respected.
'Tis now your office, daughter, to entertain
The labour of each knight in his device.             15

THAISA
Which, to preserve mine honour, I'll perform.
        ⌜*Flourish.*⌝ *The first knight passes by* ⌜*richly armed,
        and his page before him, bearing his device on his
        shield, delivers it to the Lady Thaisa*⌝

KING SIMONIDES
Who is the first that doth prefer himself?

THAISA
A knight of Sparta, my renownèd father,
And the device he bears upon his shield
Is a black Ethiop reaching at the sun.               20
The word, *Lux tua vita mihi.*
        ⌜*She presents it to the King*⌝

KING SIMONIDES
He loves you well that holds his life of you.
        ⌜*He returns it to the page, who exits with the first
        knight.*⌝
        ⌜*Flourish.*⌝ *The second knight passes by* ⌜*richly
        armed, and his page before him, bearing his device
        on his shield, delivers it to the Lady Thaisa*⌝
Who is the second that presents himself?

THAISA
A prince of Macedon, my royal father,
And the device he bears upon his shield              25
An armèd knight that's conquered by a lady.
The motto thus: *Piùe per dolcezza che per forza.*
        ⌜*She presents it to the King*⌝

KING SIMONIDES
You win him more by lenity than force.
        ⌜*He returns it to the page, who exits with the
        second knight.*⌝
        ⌜*Flourish.*⌝ *The third knight passes by* ⌜*richly armed,
        and his page before him, bearing his device on his
        shield, delivers it to the Lady Thaisa*⌝
And what's the third?

THAISA                              The third of Antioch,
And his device a wreath of chivalry.                 30
The word, *Me pompae provexit apex.*
        ⌜*She presents it to the King*⌝

KING SIMONIDES
Desire of renown he doth devise,
The which hath drawn him to this enterprise.
        ⌜*He returns it to the page, who exits with the third
        knight.*⌝
        ⌜*Flourish.*⌝ *The fourth knight passes by* ⌜*richly
        armed, and his page before him, bearing his device
        on his shield, delivers it to the Lady Thaisa*⌝
What is the fourth?

THAISA                    A knight of Athens bearing
A burning torch that's turnèd upside down.           35
The word, *Qui me alit me extinguit.*
        ⌜*She presents it to the King*⌝

KING SIMONIDES
　Which shows that beauty hath this power and will,
　Which can as well inflame as it can kill.
　　⌈He returns it to the page, who exits with the fourth
　　knight.⌉
　　⌈Flourish.⌉ The fifth Knight passes by ⌈richly armed,
　　and his page before him, bearing his device on his
　　shield, delivers it to the Lady Thaisa⌉
　And who the fifth?
THAISA　　　　　　　The fifth, a prince of Corinth,
　Presents an hand environèd with clouds,　　　　40
　Holding out gold that's by the touchstone tried.
　The motto thus: Sic spectanda fides.
　　⌈She presents it to the King⌉
KING SIMONIDES
　So faith is to be looked into.
　　⌈He returns it to the page, who exits with the fifth
　　knight.⌉
　　⌈Flourish.⌉ The sixth knight, Pericles, in a rusty
　　armour, who, having neither page to deliver his
　　shield nor shield to deliver, presents his device unto
　　the Lady Thaisa
　And what's the sixth and last, the which the knight
　　himself
　With such a graceful courtesy delivereth?　　　45
THAISA
　He seems to be a stranger, but his present is
　A withered branch that's only green at top.
　The motto, In hac spe vivo.
KING SIMONIDES
　From the dejected state wherein he is
　He hopes by you his fortunes yet may flourish.　　50
FIRST LORD
　He had need mean better than his outward show
　Can any way speak in his just commend,
　For by his rusty outside he appears
　T'have practised more the whipstock than the lance.
SECOND LORD
　He well may be a stranger, for he comes　　　　55
　Unto an honoured triumph strangely furnished.
THIRD LORD
　And on set purpose let his armour rust
　Until this day, to scour it in the dust.
KING SIMONIDES
　Opinion's but a fool, that makes us scan
　The outward habit for the inward man.　　　　60
　　⌈Cornetts⌉
　But stay, the knights are coming. We will withdraw
　Into the gallery.　　　　　　　　　⌈Exeunt⌉
　　⌈Cornetts and⌉ great shouts ⌈within⌉, and all cry
　　'The mean knight!'

Sc. 7　⌈A stately banquet is brought in.⌉ Enter King
　　Simonides, Thaisa ⌈and their train at one door⌉,
　　and ⌈at another door⌉ a Marshal ⌈conducting⌉
　　Pericles and the other knights from tilting
KING SIMONIDES (to the knights)
　To say you're welcome were superfluous.
　To place upon the volume of your deeds
　As in a title page your worth in arms
　Were more than you expect, or more than's fit,
　Since every worth in show commends itself.　　5
　Prepare for mirth, for mirth becomes a feast.
　You're princes, and my guests.
THAISA (to Pericles)　　　　But you, my knight and guest;
　To whom this wreath of victory I give,
　And crown you king of this day's happiness.

PERICLES
　'Tis more by fortune, lady, than my merit.　　10
KING SIMONIDES
　Call it by what you will, the day is yours,
　And here I hope is none that envies it.
　In framing artists art hath thus decreed,
　To make some good, but others to exceed.
　You are her laboured scholar. (To Thaisa) Come,
　　queen o'th' feast—　　　　　　　　15
　For, daughter, so you are—here take your place.
　(To Marshal) Marshal the rest as they deserve their
　　grace.
KNIGHTS
　We are honoured much by good Simonides.
KING SIMONIDES
　Your presence glads our days; honour we love,
　For who hates honour hates the gods above.　　20
MARSHAL (to Pericles)
　Sir, yonder is your place.
PERICLES　　　　　　Some other is more fit.
FIRST KNIGHT
　Contend not, sir, for we are gentlemen
　Have neither in our hearts nor outward eyes
　Envied the great, nor shall the low despise.
PERICLES
　You are right courteous knights.
KING SIMONIDES　　　　　　Sit, sir, sit.　　25
　　⌈Pericles sits directly over against the King and
　　Thaisa. The guests feed apace. Pericles sits still and
　　eats nothing⌉
　⌈Aside⌉ By Jove I wonder, that is king of thoughts,
　These cates distaste me, he but thought upon.
THAISA ⌈aside⌉
　By Juno, that is queen of marriage,
　I am amazed all viands that I eat
　Do seem unsavoury, wishing him my meat.　　30
　⌈To the King⌉ Sure he's a gallant gentleman.
KING SIMONIDES
　He's but a country gentleman.
　He's done no more than other knights have done.
　He's broke a staff or so, so let it pass.
THAISA ⌈aside⌉
　To me he seems like diamond to glass.　　　35
PERICLES ⌈aside⌉
　Yon king's to me like to my father's picture,
　Which tells me in what glory once he was—
　Had princes sit like stars about his throne,
　And he the sun for them to reverence.
　None that beheld him but like lesser lights　　40
　Did vail their crowns to his supremacy;
　Where now his son's a glow-worm in the night,
　The which hath fire in darkness, none in light;
　Whereby I see that time's the king of men;
　He's both their parent and he is their grave,　　45
　And gives them what he will, not what they crave.
KING SIMONIDES What, are you merry, knights?
⌈THE OTHER KNIGHTS⌉
　Who can be other in this royal presence?
KING SIMONIDES
　Here with a cup that's stored unto the brim,
　As you do love, full to your mistress' lips,　　50
　We drink this health to you.
⌈THE OTHER KNIGHTS⌉　　　　We thank your grace.
KING SIMONIDES
　Yet pause a while. Yon knight doth sit too
　　melancholy,

As if the entertainment in our court
Had not a show might countervail his worth.
Note it not you, Thaisa?

THAISA                  What is't to me, my father?

KING SIMONIDES
O, attend, my daughter. Princes in this    56
Should live like gods above, who freely give
To everyone that come to honour them.
And princes not so doing are like gnats
Which make a sound but, killed, are wondered at.  60
Therefore to make his entertain more sweet,
Here bear this standing-bowl of wine to him.

THAISA
Alas, my father, it befits not me
Unto a stranger knight to be so bold.
He may my proffer take for an offence,    65
Since men take women's gifts for impudence.

KING SIMONIDES
How? Do as I bid you, or you'll move me else.

THAISA (aside)
Now, by the gods, he could not please me better.

KING SIMONIDES
Furthermore, tell him we desire to know
Of whence he is, his name and parentage.    70
⌜Thaisa bears the cup to Pericles⌝

THAISA
The King my father, sir, has drunk to you,
Wishing it so much blood unto your life.

PERICLES
I thank both him and you, and pledge him freely.
He pledges the King

THAISA
And further he desires to know of you
Of whence you are, your name and parentage.  75

PERICLES
A gentleman of Tyre, my name Pericles,
My education been in arts and arms,
Who, looking for adventures in the world,
Was by the rough unconstant seas bereft
Unfortunately both of ships and men,    80
And after shipwreck driven upon this shore.
⌜Thaisa returns to the King⌝

THAISA
He thanks your grace, names himself Pericles,
A gentleman of Tyre, who, seeking adventures,
Was solely by misfortune of the seas
Bereft of ships and men, cast on this shore.  85

KING SIMONIDES
Now by the gods I pity his mishaps,
And will awake him from his melancholy.
⌜Simonides, rising from his state, goes forthwith and
    embraces Pericles⌝
Be cheered, for what misfortune hath impaired you of,
Fortune by my help can repair to you.
My self and country both shall be your friends,  90
And presently a goodly milk-white steed
And golden spurs I first bestow upon you,
The prizes due your merit, and ordained
For this day's enterprise.

PERICLES
Your kingly courtesy I thankfully accept.  95

KING SIMONIDES
Come, gentlemen, we sit too long on trifles,
And waste the time which looks for other revels.
Ev'n in your armours, as you are addressed,
Your limbs will well become a soldier's dance.
I will not have excuse with saying this,  100
'Loud music is too harsh for ladies' heads',
Since they love men in arms as well as beds.
The knights dance
So this was well asked, 'twas so well performed.
Come, here's a lady that wants breathing too.
(To Pericles) And I have heard, sir, that the knights of
Tyre    105
Are excellent in making ladies trip,
And that their measures are as excellent.

PERICLES
In those that practise them they are, my lord.

KING SIMONIDES
O, that's as much as you would be denied
Of your fair courtesy. Unclasp, unclasp.  110
They dance
Thanks, gentlemen, to all. All have done well,
(To Pericles) But you the best.—Lights, pages, to
conduct
These knights unto their sev'ral lodgings.—Yours, sir,
We have giv'n order should be next our own.

PERICLES I am at your grace's pleasure.  115

KING SIMONIDES
Princes, it is too late to talk of love,
And that's the mark I know you level at.
Therefore each one betake him to his rest;
Tomorrow all for speeding do their best.
Exeunt ⌜severally⌝

Sc. 8    Enter Helicanus and Aeschines

HELICANUS
No, Aeschines, know this of me:
Antiochus from incest lived not free,
For which the most high gods, not minding longer
To hold the vengeance that they had in store
Due to this heinous capital offence,    5
Even in the height and pride of all his glory,
When he was seated in a chariot
Of an inestimable value, and
His daughter with him, both apparelled all in jewels,
A fire from heaven came and shrivelled up  10
Their bodies e'en to loathing, for they so stunk
That all those eyes adored them ere their fall
Scorn now their hands should give them burial.

AESCHINES
'Twas very strange.

HELICANUS          And yet but justice, for though
This king were great, his greatness was no guard  15
To bar heav'n's shaft, but sin had his reward.

AESCHINES 'Tis very true.
Enter three Lords, and stand aside

FIRST LORD
See, not a man in private conference
Or council has respect with him but he.

SECOND LORD
It shall no longer grieve without reproof.  20

THIRD LORD
And cursed be he that will not second it.

FIRST LORD
Follow me, then.—Lord Helicane, a word.

HELICANUS
With me? And welcome. Happy day, my lords.

FIRST LORD
Know that our griefs are risen to the top,
And now at length they overflow their banks.  25

HELICANUS
Your griefs? For what? Wrong not your prince you
    love.

FIRST LORD
Wrong not yourself, then, noble Helicane,
But if the prince do live, let us salute him
Or know what ground's made happy by his step,
And be resolved he lives to govern us,      30
Or dead, give 's cause to mourn his funeral
And leave us to our free election.

SECOND LORD
Whose death indeed's the strongest in our censure,
And knowing this—kingdoms without a head,
Like goodly buildings left without a roof,      35
Soon fall to utter ruin—your noble self,
That best know how to rule and how to reign,
We thus submit unto as sovereign.

ALL ⌈kneeling⌉ Live, noble Helicane!

HELICANUS
By honour's cause, forbear your suffrages.      40
If that you love Prince Pericles, forbear.
    ⌈The lords rise⌉
Take I your wish I leap into the seas
Where's hourly trouble for a minute's ease,
But if I cannot win you to this love,
A twelvemonth longer then let me entreat you      45
Further to bear the absence of your king;
If in which time expired he not return,
I shall with agèd patience bear your yoke.
Go, seek your noble prince like noble subjects,
And in your search spend your adventurous worth,
Whom if you find and win unto return,      51
You shall like diamonds sit about his crown.

FIRST LORD
To wisdom he's a fool that will not yield,
And since Lord Helicane enjoineth us,
We with our travels will endeavour us.      55
If in the world he live we'll seek him out;
If in his grave he rest, we'll find him there.

HELICANUS
Then you love us, we you, and we'll clasp hands.
When peers thus knit, a kingdom ever stands.    *Exeunt*

**Sc. 8a**   *Enter Pericles with Gentlemen with lights*

FIRST GENTLEMAN
Here is your lodging, sir.

PERICLES               Pray leave me private.
Only for instant solace pleasure me
With some delightful instrument, with which,
And with my former practice, I intend
To pass away the tediousness of night,      5
Though slumbers were more fitting.

FIRST GENTLEMAN            Presently.
                 *Exit First Gentleman*

SECOND GENTLEMAN
Your will's obeyed in all things, for our master
Commanded you be disobeyed in nothing.
     *Enter First Gentleman with a stringed instrument*

PERICLES
I thank you. Now betake you to your pillows,
And to the nourishment of quiet sleep.      10
                 *Exeunt Gentlemen*
     *Pericles plays and sings*
Day—that hath still that sovereignty to draw back
The empire of the night, though for a while

In darkness she usurp—brings morning on.
I will go give his grace that salutation
Morning requires of me.      *Exit with instrument*

**Sc. 9**   *Enter King Simonides at one door reading of a
    letter, the Knights enter ⌈at another door⌉ and meet
    him*

FIRST KNIGHT
Good morrow to the good Simonides.

KING SIMONIDES
Knights, from my daughter this I let you know:
That for this twelvemonth she'll not undertake
A married life. Her reason to herself
Is only known, which from her none can get.      5

SECOND KNIGHT
May we not have access to her, my lord?

KING SIMONIDES
Faith, by no means. It is impossible,
She hath so strictly tied her to her chamber.
One twelve moons more she'll wear Diana's liv'ry.
This by the eye of Cynthia hath she vowed,      10
And on her virgin honour will not break it.

THIRD KNIGHT
Loath to bid farewell, we take our leaves.
                 *Exeunt Knights*

KING SIMONIDES
So, they are well dispatched. Now to my daughter's
    letter.
She tells me here she'll wed the stranger knight,
Or never more to view nor day nor light.      15
I like that well. Nay, how absolute she's in't,
Not minding whether I dislike or no!
Mistress, 'tis well, I do commend your choice,
And will no longer have it be delayed.
     *Enter Pericles*
Soft, here he comes. I must dissemble that      20
In show, I have determined on in heart.

PERICLES
All fortune to the good Simonides.

KING SIMONIDES
To you as much, sir. I am beholden to you
For your sweet music this last night. My ears,
I do protest, were never better fed      25
With such delightful pleasing harmony.

PERICLES
It is your grace's pleasure to commend,
Not my desert.

KING SIMONIDES    Sir, you are music's master.

PERICLES
The worst of all her scholars, my good lord.

KING SIMONIDES
Let me ask you one thing. What think you of my
    daughter?      30

PERICLES
A most virtuous princess.

KING SIMONIDES          And fair, too, is she not?

PERICLES
As a fair day in summer; wondrous fair.

KING SIMONIDES
My daughter, sir, thinks very well of you;
So well indeed that you must be her master
And she will be your scholar; therefore look to it.      35

PERICLES
I am unworthy for her schoolmaster.

KING SIMONIDES
She thinks not so. Peruse this writing else.

*He gives the letter to Pericles, who reads*

PERICLES *(aside)*

What's here?—a letter that she loves the knight of Tyre?
'Tis the King's subtlety to have my life.

⌜*He prostrates himself at the King's feet*⌝

O, seek not to entrap me, gracious lord,                    40
A stranger and distressèd gentleman
That never aimed so high to love your daughter,
But bent all offices to honour her.
Never did thought of mine levy offence,
Nor never did my actions yet commence              45
A deed might gain her love or your displeasure.

KING SIMONIDES

Thou liest like a traitor.

PERICLES                          Traitor?

KING SIMONIDES                                Ay, traitor,
That thus disguised art stol'n into my court
With witchcraft of thy actions to bewitch
The yielding spirit of my tender child.                50

PERICLES ⌜*rising*⌝

Who calls me traitor, unless it be the King,
Ev'n in his bosom I will write the lie.

KING SIMONIDES *(aside)*

Now, by the gods, I do applaud his courage.

PERICLES

My actions are as noble as my blood,
That never relished of a base descent.               55
I came unto your court in search of honour,
And not to be a rebel to your state;
And he that otherwise accounts of me,
This sword shall prove he's honour's enemy.

KING SIMONIDES

I shall prove otherwise, since both your practice      60
And her consent therein is evident
There, by my daughter's hand, as she can witness.

*Enter Thaisa*

PERICLES *(to Thaisa)*

Then as you are as virtuous as fair,
By what you hope of heaven or desire
By your best wishes here i'th' world fulfilled,        65
Resolve your angry father if my tongue
Did e'er solicit, or my hand subscribe
To any syllable made love to you.

THAISA Why, sir, say if you had,
Who takes offence at that would make me glad?           70

KING SIMONIDES

How, minion, are you so peremptory?
*(Aside)* I am glad on't.—Is this a fit match for you?
A straggling Theseus, born we know not where,
One that hath neither blood nor merit
For thee to hope for, or himself to challenge           75
Of thy perfections e'en the least allowance.

THAISA *(kneeling)*

Suppose his birth were base, when that his life
Shows that he is not so, yet he hath virtue,
The very ground of all nobility,
Enough to make him noble. I entreat you                80
To remember that I am in love,
The power of which love cannot be confined
By th' power of your will. Most royal father,
What with my pen I have in secret written
With my tongue now I openly confirm,                   85
Which is I have no life but in his love,
Nor any being but in joying of his worth.

KING SIMONIDES

Equals to equals, good to good is joined.
This not being so, the bavin of your mind

In rashness kindled must again be quenched,            90
Or purchase our displeasure.—And for you, sir,
First learn to know I banish from my court,
And yet I scorn our rage should stoop so low.
For your ambition, sir, I'll have your life.

THAISA *(to Pericles)*

For every drop of blood he sheds of yours              95
He'll draw another from his only child.

KING SIMONIDES

I'll tame you, yea, I'll bring you in subjection.
Will you not having my consent
Bestow your love and your affections
Upon a stranger?—*(aside)* who for aught I know      100
May be, nor can I think the contrary,
As great in blood as I myself.

⌜*He catches Thaisa rashly by the hand*⌝

Therefore hear you, mistress: either frame your will to
mine—

⌜*He catches Pericles rashly by the hand*⌝

And you, sir, hear you: either be ruled by me—
Or I shall make you

⌜*He claps their hands together*⌝

                                  man and wife.       105
Nay, come, your hands and lips must seal it too,
      *Pericles and Thaisa kiss*
And being joined, I'll thus your hopes destroy,
      ⌜*He parts them*⌝
And for your further grief, God give you joy.
What, are you pleased?

THAISA                    Yes, *(to Pericles)* if you love me, sir.

PERICLES

Ev'n as my life my blood that fosters it.              110

KING SIMONIDES

What, are you both agreed?

PERICLES *and* THAISA            Yes, if't please your majesty.

KING SIMONIDES

It pleaseth me so well that I will see you wed,
Then with what haste you can, get you to bed.   *Exeunt*

Sc. 10  *Enter Gower*

GOWER

Now sleep y-slackèd hath the rout,
No din but snores the house about,
Made louder by the o'erfed breast
Of this most pompous marriage feast.
The cat with eyne of burning coal                      5
Now couches fore the mouse's hole,
And crickets sing at th'oven's mouth
As the blither for their drouth.
Hymen hath brought the bride to bed,
Where by the loss of maidenhead                        10
A babe is moulded. Be attent,
And time that is so briefly spent
With your fine fancies quaintly eche.
What's dumb in show, I'll plain with speech.
          *Dumb show.*
*Enter Pericles and Simonides at one door with
attendants. A messenger comes ⌜hastily⌝ in to them,
kneels, and gives Pericles a letter. Pericles shows it
Simonides; the lords kneel to him. Then enter
Thaisa with child, with Lychorida, a nurse. The
King shows her the letter. She rejoices. She and
Pericles take leave of her father and depart with
Lychorida at one door; Simonides ⌜and attendants⌝
depart at another*
By many a dern and painful perch                       15
Of Pericles the care-full search,

By the four opposing coigns
Which the world together joins,
Is made with all due diligence
That horse and sail and high expense          20
Can stead the quest. At last from Tyre
Fame answering the most strange enquire,
To th' court of King Simonides
Are letters brought, the tenor these:
Antiochus and his daughter dead,               25
The men of Tyrus on the head
Of Helicanus would set on
The crown of Tyre, but he will none.
The mutiny there he hastes t'appease,
Says to 'em if King Pericles                        30
Come not home in twice six moons
He, obedient to their dooms,
Will take the crown. The sum of this
Brought hither to Pentapolis
Y-ravishèd the regions round,                     35
And everyone with claps can sound
'Our heir-apparent is a king!
Who dreamt, who thought of such a thing?'
Brief he must hence depart to Tyre;
His queen with child makes her desire—          40
Which who shall cross?—along to go.
Omit we all their dole and woe.
Lychorida her nurse she takes,
And so to sea. Their vessel shakes
On Neptune's billow. Half the flood              45
Hath their keel cut, but fortune's mood
Varies again. The grizzled north
Disgorges such a tempest forth
That as a duck for life that dives,
So up and down the poor ship drives.             50
The lady shrieks, and well-a-near
Does fall in travail with her fear,
And what ensues in this fell storm
Shall for itself itself perform;
I nill relate; action may                               55
Conveniently the rest convey,
Which might not what by me is told.
In your imagination hold
This stage the ship, upon whose deck
The sea-tossed Pericles appears to speke.       *Exit*

**Sc. 11** ⌐*Thunder and lightning.*⌐ *Enter Pericles a-shipboard*
PERICLES
The god of this great vast rebuke these surges
Which wash both heav'n and hell; and thou that hast
Upon the winds command, bind them in brass,
Having called them from the deep. O still
Thy deaf'ning dreadful thunders, gently quench    5
Thy nimble sulph'rous flashes.—O, ho, Lychorida!
How does my queen?—Thou stormest venomously.
Wilt thou spit all thyself? The seaman's whistle
Is as a whisper in the ears of death,
Unheard.—Lychorida!—Lucina, O!                    10
Divinest patroness, and midwife gentle
To those that cry by night, convey thy deity
Aboard our dancing boat, make swift the pangs
Of my queen's travails!—Now, Lychorida.
       *Enter Lychorida with an infant*
LYCHORIDA
Here is a thing too young for such a place,        15
Who, if it had conceit, would die, as I

Am like to do. Take in your arms this piece
Of your dead queen.
PERICLES                         How, how, Lychorida?
LYCHORIDA
Patience, good sir, do not assist the storm.
Here's all that is left living of your queen,      20
A little daughter. For the sake of it
Be manly, and take comfort.
PERICLES                              O you gods!
Why do you make us love your goodly gifts,
And snatch them straight away? We here below
Recall not what we give, and therein may          25
Use honour with you.
LYCHORIDA                         Patience, good sir,
E'en for this charge.
       *She gives him the infant.* ⌐*Pericles, looking
       mournfully upon it, shakes his head, and weeps*⌐
PERICLES                         Now mild may be thy life,
For a more blust'rous birth had never babe;
Quiet and gentle thy conditions, for
Thou art the rudeliest welcome to this world      30
That e'er was prince's child; happy what follows.
Thou hast as chiding a nativity
As fire, air, water, earth, and heav'n can make
To herald thee from th' womb. Poor inch of nature,
Ev'n at the first thy loss is more than can        35
Thy partage quit with all thou canst find here.
Now the good gods throw their best eyes upon't.
       *Enter* ⌐*the Master*⌐ *and a Sailor*
⌐MASTER⌐ What, courage, sir! God save you.
PERICLES
Courage enough, I do not fear the flaw;
It hath done to me its worst. Yet for the love     40
Of this poor infant, this fresh new seafarer,
I would it would be quiet.
⌐MASTER⌐ (*calling*) Slack the bow-lines, there.—Thou wilt
not, wilt thou? Blow, and split thyself.
SAILOR But searoom, an the brine and cloudy billow kiss
the moon, I care not.                               46
⌐MASTER⌐ (*to Pericles*) Sir, your queen must overboard.
The sea works high, the wind is loud, and will not lie
till the ship be cleared of the dead.
PERICLES
That's but your superstition.                       50
⌐MASTER⌐ Pardon us, sir; with us at sea it hath been still
observed, and we are strong in custom. Therefore briefly
yield 'er, for she must overboard straight.
PERICLES
As you think meet. Most wretched queen!
LYCHORIDA                              Here she lies, sir.
       *She* ⌐*draws the curtains and discovers*⌐ *the body of
       Thaisa in a* ⌐*bed. Pericles gives Lychorida the
       infant*⌐
PERICLES (*to Thaisa*)
A terrible childbed hast thou had, my dear,        55
No light, no fire. Th'unfriendly elements
Forgot thee utterly, nor have I time
To give thee hallowed to thy grave, but straight
Must cast thee, scarcely coffined, in the ooze,
Where, for a monument upon thy bones              60
And aye-remaining lamps, the belching whale
And humming water must o'erwhelm thy corpse,
Lying with simple shells.—O Lychorida,
Bid Nestor bring me spices, ink, and paper,
My casket and my jewels, and bid Nicander         65

Bring me the satin coffer. Lay the babe
Upon the pillow. Hie thee whiles I say
A priestly farewell to her. Suddenly, woman.
                              *Exit Lychorida*
⌈SAILOR⌉ Sir, we have a chest beneath the hatches caulked
    and bitumed ready.                              70
PERICLES
    I thank thee. ⌈*To the Master*⌉ Mariner, say, what coast
    is this?
⌈MASTER⌉
    We are near Tarsus.
PERICLES                    Thither, gentle mariner,
    Alter thy course from Tyre. When canst thou reach it?
⌈MASTER⌉
    By break of day, if the wind cease.
PERICLES                            Make for Tarsus.
    There will I visit Cleon, for the babe          75
    Cannot hold out to Tyrus. There I'll leave it
    At careful nursing. Go thy ways, good mariner.
    I'll bring the body presently.
            ⌈*Exit Master at one door and Sailor beneath*
            *the hatches. Exit Pericles to Thaisa,*
            *closing the curtains*⌉

**Sc. 12**  *Enter Lord Cerimon with a ⌈poor man and a⌉*
            *servant*
CERIMON
    Philemon, ho!
    *Enter Philemon*
PHILEMON        Doth my lord call?
CERIMON
    Get fire and meat for those poor men.
                              ⌈*Exit Philemon*⌉
    'T'as been a turbulent and stormy night.
SERVANT
    I have seen many, but such a night as this
    Till now I ne'er endured.                        5
CERIMON
    Your master will be dead ere you return.
    There's nothing can be ministered in nature
    That can recover him. ⌈*To poor man*⌉ Give this to th'
        pothecary
    And tell me how it works.
                    ⌈*Exeunt poor man and servant*⌉
    *Enter two Gentlemen*
FIRST GENTLEMAN           Good morrow.
SECOND GENTLEMAN
    Good morrow to your lordship.
CERIMON                       Gentlemen,       10
    Why do you stir so early?
FIRST GENTLEMAN           Sir,
    Our lodgings, standing bleak upon the sea,
    Shook as the earth did quake.
    The very principals did seem to rend
    And all to topple. Pure surprise and fear      15
    Made me to quit the house.
SECOND GENTLEMAN
    That is the cause we trouble you so early;
    'Tis not our husbandry.
CERIMON                 O, you say well.
FIRST GENTLEMAN
    But I much marvel that your lordship should,
    Having rich tire about you, at this hour        20
    Shake off the golden slumber of repose. 'Tis most
        strange,

Nature to be so conversant with pain,
Being thereto not compelled.
CERIMON               I held it ever
Virtue and cunning were endowments greater
Than nobleness and riches. Careless heirs           25
May the two latter darken and dispend,
But immortality attends the former,
Making a man a god. 'Tis known I ever
Have studied physic, through which secret art,
By turning o'er authorities, I have,                30
Together with my practice, made familiar
To me and to my aid the blest infusions
That dwells in vegetives, in metals, stones,
And so can speak of the disturbances
That nature works, and of her cures, which doth     35
    give me
A more content and cause of true delight
Than to be thirsty after tott'ring honour,
Or tie my pleasure up in silken bags
To glad the fool and death.
SECOND GENTLEMAN           Your honour has
Through Ephesus poured forth your charity,          40
And hundreds call themselves your creatures who by
    you
Have been restored. And not alone your knowledge,
Your personal pain, but e'en your purse still open
Hath built Lord Cerimon such strong renown
As time shall never—                                45
        *Enter ⌈Philemon and one or⌉ two with a chest*
⌈PHILEMON⌉ So, lift there.
CERIMON What's that?
⌈PHILEMON⌉ Sir, even now
    The sea tossed up upon our shore this chest.
    'Tis off some wreck.
CERIMON           Set't down. Let's look upon't.     50
SECOND GENTLEMAN
    'Tis like a coffin, sir.
CERIMON               Whate'er it be,
    'Tis wondrous heavy.—Did the sea cast it up?
⌈PHILEMON⌉
    I never saw so huge a billow, sir,
    Or a more eager.
CERIMON           Wrench it open straight.
    *The others start to work*
    If the sea's stomach be o'ercharged with gold   55
    'Tis by a good constraint of queasy fortune
    It belches upon us.
SECOND GENTLEMAN     'Tis so, my lord.
CERIMON
    How close 'tis caulked and bitumed!
        ⌈*They force the lid*⌉
                              Soft, it smells
    Most sweetly in my sense.
SECOND GENTLEMAN       A delicate odour.
CERIMON
    As ever hit my nostril. So, up with it.         60
        *They take the lid off*
    O you most potent gods! What's here—a corpse?
SECOND GENTLEMAN
    Most strange.
CERIMON       Shrouded in cloth of state, and crowned,
    Balmed and entreasured with full bags of spices.
    A passport, too!
        *He takes a paper from the chest*
    Apollo perfect me i'th' characters.             65

'Here I give to understand,
If e'er this coffin drives a-land,
I, King Pericles, have lost
This queen worth all our mundane cost.
Who finds her, give her burying;　　　　70
She was the daughter of a king.
Besides this treasure for a fee,
The gods requite his charity.'
If thou liv'st, Pericles, thou hast a heart
That even cracks for woe. This chanced tonight.　　75
SECOND GENTLEMAN
Most likely, sir.
CERIMON　　　　Nay, certainly tonight,
For look how fresh she looks. They were too rash
That threw her in the sea. Make a fire within.
Fetch hither all my boxes in my closet. ⌈Exit Philemon⌉
Death may usurp on nature many hours,　　　　80
And yet the fire of life kindle again
The o'erpressed spirits. I have heard
Of an Egyptian nine hours dead
Who was by good appliances recovered.
　　　Enter ⌈Philemon⌉ with napkins and fire
Well said, well said, the fire and cloths.　　　　85
The still and woeful music that we have,
Cause it to sound, beseech you.
　　　Music
　　　　　　　　The vial once more.
How thou stirr'st, thou block! The music there!
I pray you give her air. Gentlemen,
This queen will live. Nature awakes, a warmth　　90
Breathes out of her. She hath not been entranced
Above five hours. See how she 'gins to blow
Into life's flow'r again.
FIRST GENTLEMAN　　　　The heavens
Through you increase our wonder, and set up
Your fame for ever.
CERIMON　　　　She is alive. Behold,　　　　95
Her eyelids, cases to those heav'nly jewels
Which Pericles hath lost,
Begin to part their fringes of bright gold.
The diamonds of a most praisèd water
Doth appear to make the world twice rich.—Live,　100
And make us weep to hear your fate, fair creature,
Rare as you seem to be.
　　　She moves
THAISA　　　　　　O dear Diana,
Where am I? Where's my lord? What world is this?
SECOND GENTLEMAN
Is not this strange?
FIRST GENTLEMAN　　　Most rare.
CERIMON　　　　　　Hush, gentle neighbours.
Lend me your hands. To the next chamber bear her.
Get linen. Now this matter must be looked to,　106
For her relapse is mortal. Come, come,
And Aesculapius guide us. They carry her away. Exeunt

Sc. 13　Enter Pericles at Tarsus, with Cleon and Dionyza,
　　　and Lychorida with a babe
PERICLES
Most honoured Cleon, I must needs be gone.
My twelve months are expired, and Tyrus stands
In a litigious peace. You and your lady
Take from my heart all thankfulness. The gods
Make up the rest upon you!
CLEON　　　　　　Your strokes of fortune,　5
Though they hurt you mortally, yet glance

Full woundingly on us.
DIONYZA　　　　O your sweet queen!
That the strict fates had pleased you'd brought her
　hither
T'have blessed mine eyes with her!
PERICLES　　　　　　We cannot but obey　10
The pow'rs above us. Should I rage and roar
As doth the sea she lies in, yet the end
Must be as 'tis. My gentle babe Marina,
Whom for she was born at sea I have named so,
Here I charge your charity withal, and leave her
The infant of your care, beseeching you　　　15
To give her princely training, that she may be
Mannered as she is born.
CLEON　　　　　　Fear not, my lord, but think
Your grace, that fed my country with your corn—
For which the people's pray'rs still fall upon you—
Must in your child be thought on. If neglection　20
Should therein make me vile, the common body
By you relieved would force me to my duty.
But if to that my nature need a spur,
The gods revenge it upon me and mine
To th' end of generation.
PERICLES　　　　I believe you.　　25
Your honour and your goodness teach me to't
Without your vows.—Till she be married, madam,
By bright Diana, whom we honour all,
Unscissored shall this hair of mine remain,
Though I show ill in't. So I take my leave.　　30
Good madam, make me blessèd in your care
In bringing up my child.
DIONYZA　　　　I have one myself,
Who shall not be more dear to my respect
Than yours, my lord.
PERICLES　　　　Madam, my thanks and prayers.
CLEON
We'll bring your grace e'en to the edge o'th' shore,　35
Then give you up to th' masted Neptune and
The gentlest winds of heaven.
PERICLES
I will embrace your offer.—Come, dear'st madam.—
O, no tears, Lychorida, no tears.
Look to your little mistress, on whose grace　　40
You may depend hereafter.—Come, my lord.　Exeunt

Sc. 14　Enter Cerimon and Thaisa
CERIMON
Madam, this letter and some certain jewels
Lay with you in your coffer, which are all
At your command. Know you the character?
THAISA
It is my lord's. That I was shipped at sea
I well remember, ev'n on my eaning time,　　5
But whether there delivered, by th' holy gods
I cannot rightly say. But since King Pericles,
My wedded lord, I ne'er shall see again,
A vestal liv'ry will I take me to,
And never more have joy.　　　　10
CERIMON
Madam, if this you purpose as ye speak,
Diana's temple is not distant far,
Where till your date expire you may abide.
Moreover, if you please a niece of mine
Shall there attend you.　　　　15
THAISA
My recompense is thanks, that's all,
Yet my good will is great, though the gift small. Exeunt

**Sc. 15** *Enter Gower*

GOWER
Imagine Pericles arrived at Tyre,
Welcomed and settled to his own desire.
His woeful queen we leave at Ephesus,
Unto Diana there 's a votaress.
  Now to Marina bend your mind,                          5
  Whom our fast-growing scene must find
  At Tarsus, and by Cleon trained
  In music, letters; who hath gained
  Of education all the grace,
  Which makes her both the heart and place    10
  Of gen'ral wonder. But, alack,
  That monster envy, oft the wrack
  Of earnèd praise, Marina's life
  Seeks to take off by treason's knife,
  And in this kind our Cleon has            15
  One daughter, and a full-grown lass
  E'en ripe for marriage-rite. This maid
  Hight Philoten, and it is said
  For certain in our story she
  Would ever with Marina be,              20
  Be't when they weaved the sleided silk
  With fingers long, small, white as milk;
  Or when she would with sharp nee'le wound
  The cambric which she made more sound
  By hurting it, or when to th' lute        25
  She sung, and made the night bird mute,
  That still records with moan; or when
  She would with rich and constant pen
  Vail to her mistress Dian. Still
  This Philoten contends in skill          30
  With absolute Marina; so
  With dove of Paphos might the crow
  Vie feathers white. Marina gets
  All praises which are paid as debts,
  And not as given. This so darks         35
  In Philoten all graceful marks
  That Cleon's wife with envy rare
  A present murder does prepare
  For good Marina, that her daughter
  Might stand peerless by this slaughter.   40
  The sooner her vile thoughts to stead
  Lychorida, our nurse, is dead,
    ⌈*A tomb is revealed*⌉
  And cursèd Dionyza hath
  The pregnant instrument of wrath
  Pressed for this blow. Th'unborn event   45
  I do commend to your content,
  Only I carry wingèd Time
  Post on the lame feet of my rhyme,
  Which never could I so convey
  Unless your thoughts went on my way.    50
    ⌈*Enter Dionyza with Leonine*⌉
  Dionyza does appear,
  With Leonine, a murderer.                          *Exit*

DIONYZA
Thy oath remember. Thou hast sworn to do't.
'Tis but a blow, which never shall be known.
Thou canst not do a thing i'th' world so soon    55
To yield thee so much profit. Let not conscience,
Which is but cold, or fanning love thy bosom
Unflame too nicely, nor let pity, which
E'en women have cast off, melt thee; but be
A soldier to thy purpose.
LEONINE                         I will do't;      60
But yet she is a goodly creature.

DIONYZA
The fitter then the gods should have her.
    *Enter Marina* ⌈*to the tomb*⌉ *with a basket of flowers*
Here she comes, weeping her only nurse's death.
Thou art resolved.
LEONINE                    I am resolved.
MARINA
No, I will rob Tellus of her weed               65
To strew thy grave with flow'rs. The yellows, blues,
The purple violets and marigolds
Shall as a carpet hang upon thy tomb
While summer days doth last. Ay me, poor maid,
Born in a tempest when my mother died,       70
This world to me is but a ceaseless storm
Whirring me from my friends.
DIONYZA
How now, Marina, why do you keep alone?
How chance my daughter is not with you?
Do not consume your blood with sorrowing.     75
Have you a nurse of me. Lord, how your favour
Is changed with this unprofitable woe!
Give me your flowers. Come, o'er the sea margin
Walk with Leonine. The air is piercing there,
And quick; it sharps the stomach. Come, Leonine,   80
Take her by th' arm. Walk with her.
MARINA                        No, I pray you,
I'll not bereave you of your servant.
DIONYZA                           Come, come,
I love the King your father and yourself
With more than foreign heart. We ev'ry day
Expect him here. When he shall come and find   85
Our paragon to all reports thus blasted,
He will repent the breadth of his great voyage,
Blame both my lord and me, that we have taken
No care to your best courses. Go, I pray you,
Walk and be cheerful once again; resume      90
That excellent complexion which did steal
The eyes of young and old. Care not for me.
I can go home alone.
MARINA                 Well, I will go,
But truly I have no desire to it.
DIONYZA
Nay, I know 'tis good for you. Walk half an hour,   95
Leonine, at the least; remember
What I have said.
LEONINE             I warr'nt you, madam.
DIONYZA (*to Marina*)
I'll leave you, my sweet lady, for a while.
Pray you walk softly, do not heat your blood.
What, I must have care of you!
MARINA                  My thanks, sweet madam.
                                    *Exit Dionyza*
Is this wind westerly that blows?
LEONINE                        South-west.      101
MARINA
When I was born the wind was north.
LEONINE                          Was't so?
MARINA
My father, as nurse says, did never fear,
But cried 'Good seamen' to the mariners,
Galling his kingly hands with haling ropes,     105
And, clasping to the mast, endured a sea
That almost burst the deck.
LEONINE When was this?
MARINA When I was born.
Never was waves nor wind more violent.          110

Once from the ladder tackle washes off
A canvas-climber. 'Ha!' says one, 'wolt out?'
And with a dropping industry they skip
From stem to stern. The boatswain whistles, and
The master calls and trebles their confusion.                115
LEONINE Come, say your prayers.
MARINA What mean you?
LEONINE
If you require a little space for prayer
I grant it. Pray, but be not tedious.
The gods are quick of ear, and I am sworn                    120
To do my work with haste.
MARINA                          Why would you kill me?
LEONINE
To satisfy my lady.
MARINA                          Why would she have me killed?
Now, as I can remember, by my troth
I never did her hurt in all my life.
I never spake bad word, nor did ill turn                     125
To any living creature. Believe me, la.
I never killed a mouse nor hurt a fly.
I trod once on a worm against my will,
But I wept for it. How have I offended
Wherein my death might yield her any profit                  130
Or my life imply her danger?
LEONINE                          My commission
Is not to reason of the deed, but do't.
MARINA
You will not do't for all the world, I hope.
You are well favoured, and your looks foreshow
You have a gentle heart. I saw you lately                    135
When you caught hurt in parting two that fought.
Good sooth, it showed well in you. Do so now.
Your lady seeks my life. Come you between,
And save poor me, the weaker.
LEONINE ⌈drawing out his sword⌉   I am sworn,
And will dispatch.                                           140
          Enter Pirates ⌈running⌉
FIRST PIRATE Hold, villain.
          Leonine runs away ⌈and hides behind the tomb⌉
SECOND PIRATE A prize, a prize.
THIRD PIRATE Half-part, mates, half-part. Come, let's have
her aboard suddenly.
               Exeunt Pirates ⌈carrying⌉ Marina
     Leonine ⌈steals back⌉
LEONINE
These roguing thieves serve the great pirate Valdes.
An they have seized Marina, let her go.                      146
There's no hope she'll return. I'll swear she's dead
And thrown into the sea; but I'll see further.
Perhaps they will but please themselves upon her,
Not carry her aboard. If she remain,                         150
Whom they have ravished must by me be slain.
               Exit. ⌈The tomb is concealed⌉

Sc. 16 ⌈A brothel sign.⌉ Enter the Pander, his wife the
     Bawd, and their man Boult
PANDER Boult.
BOULT Sir.
PANDER Search the market narrowly. Mytilene is full of
gallants. We lose too much money this mart by being
wenchless.                                                     5
BAWD We were never so much out of creatures. We have
but poor three, and they can do no more than they
can do, and they with continual action are even as
good as rotten.                                                9

PANDER Therefore let's have fresh ones, whate'er we pay
for them. If there be not a conscience to be used in
every trade, we shall never prosper.
BAWD Thou sayst true. 'Tis not our bringing up of poor
bastards—as I think I have brought up some eleven—
BOULT Ay, to eleven, and brought them down again. But
shall I search the market?                                    16
BAWD What else, man? The stuff we have, a strong wind
will blow it to pieces, they are so pitifully sodden.
PANDER Thou sayst true. They're too unwholesome, o'
conscience. The poor Transylvanian is dead that lay
with the little baggage.                                      21
BOULT Ay, she quickly pooped him, she made him roast
meat for worms. But I'll go search the market.      Exit
PANDER Three or four thousand chequins were as pretty
a proportion to live quietly, and so give over.             25
BAWD Why to give over, I pray you? Is it a shame to get
when we are old?
PANDER O, our credit comes not in like the commodity,
nor the commodity wages not with the danger. Therefore if in our youths we could pick up some pretty
estate, 'twere not amiss to keep our door hatched.
Besides, the sore terms we stand upon with the gods
will be strong with us for giving o'er.
BAWD Come, other sorts offend as well as we.                 34
PANDER As well as we? Ay, and better too; we offend
worse. Neither is our profession any mystery, it's no
calling. But here comes Boult.
          Enter Boult with the Pirates and Marina
BOULT ⌈to the Pirates⌉ Come your ways, my masters, you
say she's a virgin?
A PIRATE O sir, we doubt it not.                             40
BOULT (to Pander) Master, I have gone through for this
piece you see. If you like her, so; if not, I have lost my
earnest.
BAWD Boult, has she any qualities?
BOULT She has a good face, speaks well, and has excellent
good clothes. There's no farther necessity of qualities
can make her be refused.                                     47
BAWD What's her price, Boult?
BOULT I cannot be bated one doit of a hundred sesterces.
PANDER (to Pirates) Well, follow me, my masters. You
shall have your money presently. (To Bawd) Wife, take
her in, instruct her what she has to do, that she may
not be raw in her entertainment.                             53
               Exeunt Pander and Pirates
BAWD Boult, take you the marks of her, the colour of her
hair, complexion, height, her age, with warrant of her
virginity, and cry 'He that will give most shall have
her first.' Such a maidenhead were no cheap thing if
men were as they have been. Get this done as I
command you.                                                 59
BOULT Performance shall follow.                        Exit
MARINA
Alack that Leonine was so slack, so slow.
He should have struck, not spoke; or that these pirates,
Not enough barbarous, had but o'erboard thrown me
To seek my mother.
BAWD Why lament you, pretty one?                             65
MARINA That I am pretty.
BAWD Come, the gods have done their part in you.
MARINA I accuse them not.
BAWD You are light into my hands, where you are like
to live.                                                     70
MARINA The more my fault
To scape his hands where I was like to die.

BAWD Ay, and you shall live in pleasure.

MARINA No.                                                        74

BAWD Yes, indeed shall you, and taste gentlemen of all fashions. You shall fare well. You shall have the difference of all complexions. What, do you stop your ears?

MARINA Are you a woman?                                           79

BAWD What would you have me be an I be not a woman?

MARINA An honest woman, or not a woman.

BAWD Marry, whip the gosling! I think I shall have something to do with you. Come, you're a young foolish sapling, and must be bowed as I would have you.

MARINA The gods defend me!                                        85

BAWD If it please the gods to defend you by men, then men must comfort you, men must feed you, men must stir you up.

   *Enter Boult*

Now, sir, hast thou cried her through the market?

BOULT I have cried her almost to the number of her hairs. I have drawn her picture with my voice.             91

BAWD And I prithee tell me, how dost thou find the inclination of the people, especially of the younger sort?

BOULT Faith, they listened to me as they would have hearkened to their fathers' testament. There was a Spaniard's mouth watered as he went to bed to her very description.                                          97

BAWD We shall have him here tomorrow with his best ruff on.

BOULT Tonight, tonight. But mistress, do you know the French knight that cowers i' the hams?          101

BAWD Who, Monsieur Veroles?

BOULT Ay, he. He offered to cut a caper at the proclamation, but he made a groan at it, and swore he would see her tomorrow.                                 105

BAWD Well, well, as for him, he brought his disease hither. Here he does but repair it. I know he will come in our shadow to scatter his crowns of the sun.

BOULT Well, if we had of every nation a traveller, we should lodge them all with this sign.           110

BAWD (*to Marina*) Pray you, come hither a while. You have fortunes coming upon you. Mark me, you must seem to do that fearfully which you commit willingly, to despise profit where you have most gain. To weep that you live as ye do makes pity in your lovers. Seldom but that pity begets you a good opinion, and that opinion a mere profit.                             117

MARINA I understand you not.

BOULT (*to Bawd*) O, take her home, mistress, take her home. These blushes of hers must be quenched with some present practice.                              121

BAWD Thou sayst true, i'faith, so they must, for your bride goes to that with shame which is her way to go with warrant.

BOULT Faith, some do and some do not. But mistress, if I have bargained for the joint—                    126

BAWD Thou mayst cut a morsel off the spit.

BOULT I may so.

BAWD Who should deny it? (*To Marina*) Come, young one, I like the manner of your garments well.      130

BOULT Ay, by my faith, they shall not be changed yet.

BAWD (*giving him money*) Boult, spend thou that in the town. Report what a sojourner we have. You'll lose nothing by custom. When nature framed this piece she meant thee a good turn. Therefore say what a paragon she is, and thou reapest the harvest out of thine own setting forth.                                      137

BOULT I warrant you, mistress, thunder shall not so awake the beds of eels as my giving out her beauty stirs up the lewdly inclined. I'll bring home some tonight.   140

   ⌈*Exit*⌉

BAWD Come your ways, follow me.

MARINA

If fires be hot, knives sharp, or waters deep,
Untied I still my virgin knot will keep.
Diana aid my purpose.

BAWD What have we to do with Diana? Pray you, will you go with me?          *Exeunt.* ⌈*The sign is removed*⌉

**Sc. 17** *Enter* ⌈*in mourning garments*⌉ *Cleon and Dionyza*

DIONYZA

Why, are you foolish? Can it be undone?

CLEON

O Dionyza, such a piece of slaughter
The sun and moon ne'er looked upon.

DIONYZA

I think you'll turn a child again.

CLEON

Were I chief lord of all this spacious world            5
I'd give it to undo the deed. A lady
Much less in blood than virtue, yet a princess
To equal any single crown o'th' earth
I'th' justice of compare. O villain Leonine,
Whom thou hast poisoned too,                            10
If thou hadst drunk to him 't'ad been a kindness
Becoming well thy fact. What canst thou say
When noble Pericles demands his child?

DIONYZA

That she is dead. Nurses are not the fates.
To foster is not ever to preserve.                      15
She died at night. I'll say so. Who can cross it,
Unless you play the pious innocent
And, for an honest attribute, cry out
'She died by foul play.'

CLEON      O, go to. Well, well,
Of all the faults beneath the heav'ns the gods          20
Do like this worst.

DIONYZA     Be one of those that thinks
The petty wrens of Tarsus will fly hence
And open this to Pericles. I do shame
To think of what a noble strain you are,
And of how cowed a spirit.

CLEON      To such proceeding       25
Whoever but his approbation added,
Though not his prime consent, he did not flow
From honourable sources.

DIONYZA     Be it so, then.
Yet none does know but you how she came dead,
Nor none can know, Leonine being gone.                  30
She did distain my child, and stood between
Her and her fortunes. None would look on her,
But cast their gazes on Marina's face
Whilst ours was blurted at, and held a malkin
Not worth the time of day. It pierced me through,       35
And though you call my course unnatural,
You not your child well loving, yet I find
It greets me as an enterprise of kindness
Performed to your sole daughter.

CLEON Heavens forgive it.                               40

DIONYZA And as for Pericles,
What should he say? We wept after her hearse,
And yet we mourn. Her monument
Is almost finished, and her epitaphs

In glitt'ring golden characters express 45
A gen'ral praise to her and care in us,
At whose expense 'tis done.
CLEON               Thou art like the harpy,
Which, to betray, dost, with thine angel face,
Seize in thine eagle talons.
DIONYZA
Ye're like one that superstitiously 50
Do swear to th' gods that winter kills the flies,
But yet I know you'll do as I advise.       *Exeunt*

Sc. 18   *Enter Gower*
GOWER
Thus time we waste, and long leagues make we short,
Sail seas in cockles, have and wish but for't,
Making to take imagination
From bourn to bourn, region to region.
By you being pardoned, we commit no crime 5
To use one language in each sev'ral clime
Where our scene seems to live. I do beseech you
To learn of me, who stand i'th' gaps to teach you
The stages of our story: Pericles
Is now again thwarting the wayward seas, 10
Attended on by many a lord and knight,
To see his daughter, all his life's delight.
Old Helicanus goes along. Behind
Is left to govern, if you bear in mind,
Old Aeschines, whom Helicanus late 15
Advanced in Tyre to great and high estate.
Well sailing ships and bounteous winds have brought
This king to Tarsus—think his pilot thought;
So with his steerage shall your thoughts go on—
To fetch his daughter home, who first is gone. 20
Like motes and shadows see them move a while;
Your ears unto your eyes I'll reconcile.
       *Dumb show.*
   *Enter Pericles at one door with all his train, Cleon*
   *and Dionyza ⌜in mourning garments⌝ at the other.*
   *Cleon ⌜draws the curtain and⌝ shows Pericles the*
   *tomb, whereat Pericles makes lamentation, puts on*
   *sack-cloth, and in a mighty passion departs,*
   *followed by his train. Cleon and Dionyza depart at*
   *the other door*
See how belief may suffer by foul show.
This borrowed passion stands for true-owed woe,
And Pericles, in sorrow all devoured, 25
With sighs shot through, and biggest tears o'ershow'red,
Leaves Tarsus, and again embarks. He swears
Never to wash his face nor cut his hairs.
He puts on sack-cloth, and to sea. He bears
A tempest which his mortal vessel tears, 30
And yet he rides it out. Now please you wit
The epitaph is for Marina writ
By wicked Dionyza.
     *He reads Marina's epitaph on the tomb*
'The fairest, sweetest, best lies here,
Who withered in her spring of year. 35
In nature's garden, though by growth a bud,
She was the chiefest flower: she was good.'
No visor does become black villainy
So well as soft and tender flattery.
Let Pericles believe his daughter's dead 40
And bear his courses to be orderèd
By Lady Fortune, while our scene must play
His daughter's woe and heavy well-a-day
In her unholy service. Patience then,
And think you now are all in Mytilene.       *Exit*

Sc. 19   ⌜*A brothel sign.*⌝ *Enter two Gentlemen*
FIRST GENTLEMAN  Did you ever hear the like?
SECOND GENTLEMAN  No, nor never shall do in such a place
as this, she being once gone.
FIRST GENTLEMAN  But to have divinity preached there—
did you ever dream of such a thing? 5
SECOND GENTLEMAN  No, no. Come, I am for no more
bawdy houses. Shall 's go hear the vestals sing?
FIRST GENTLEMAN  I'll do anything now that is virtuous,
but I am out of the road of rutting for ever.    *Exeunt*
     *Enter Pander, Bawd, and Boult*
PANDER  Well, I had rather than twice the worth of her
she had ne'er come here. 11
BAWD  Fie, fie upon her, she's able to freeze the god
Priapus and undo the whole of generation. We must
either get her ravished or be rid of her. When she
should do for clients her fitment and do me the kindness
of our profession, she has me her quirks, her reasons,
her master reasons, her prayers, her knees, that she
would make a puritan of the devil if he should cheapen
a kiss of her. 19
BOULT  Faith, I must ravish her, or she'll disfurnish us of
all our cavalleria and make our swearers priests.
PANDER  Now, the pox upon her green-sickness for me.
BAWD  Faith, there's no way to be rid on't but by the way
to the pox.
     *Enter Lysimachus, disguised*
Here comes the Lord Lysimachus, disguised. 25
BOULT  We should have both lord and loon if the peevish
baggage would but give way to custom.
LYSIMACHUS  How now, how a dozen of virginities?
BAWD  Now, the gods to-bless your honour!
BOULT  I am glad to see your honour in good health. 30
LYSIMACHUS  You may so. 'Tis the better for you that your
resorters stand upon sound legs. How now, wholesome
iniquity have you, that a man may deal withal and
defy the surgeon? 34
BAWD  We have here one, sir, if she would—but there
never came her like in Mytilene.
LYSIMACHUS  If she'd do the deed of darkness, thou wouldst
say. 38
BAWD  Your honour knows what 'tis to say well enough.
LYSIMACHUS  Well, call forth, call forth.      ⌜*Exit Pander*⌝
BOULT  For flesh and blood, sir, white and red, you shall
see a rose. And she were a rose indeed, if she had but—
LYSIMACHUS  What, prithee?
BOULT  O sir, I can be modest. 44
LYSIMACHUS  That dignifies the renown of a bawd no less
than it gives a good report to a noble to be chaste.
     ⌜*Enter Pander with Marina*⌝
BAWD  Here comes that which grows to the stalk, never
plucked yet, I can assure you. Is she not a fair creature?
LYSIMACHUS  Faith, she would serve after a long voyage
at sea. Well, there's for you. Leave us. 50
     ⌜*He pays the Bawd*⌝
BAWD  I beseech your honour give me leave: a word, and
I'll have done presently.
LYSIMACHUS  I beseech you, do.
BAWD  (*aside to Marina*) First, I would have you note this
is an honourable man. 55
MARINA  I desire to find him so, that I may honourably
know him.
BAWD  Next, he's the governor of this country, and a man
whom I am bound to. 59
MARINA  If he govern the country you are bound to him
indeed, but how honourable he is in that, I know not.

BAWD Pray you, without any more virginal fencing, will
you use him kindly? He will line your apron with gold.

MARINA What he will do graciously I will thankfully
receive.                                                        65

LYSIMACHUS (*to Bawd*) Ha' you done?

BAWD My lord, she's not paced yet. You must take some
pains to work her to your manège. (*To Boult and Pander*)
Come, we will leave his honour and hers together. Go
thy ways.            *Exeunt Pander, Bawd, and Boult*

LYSIMACHUS
Fair one, how long have you been at this trade?    71

MARINA What trade, sir?

LYSIMACHUS
I cannot name it but I shall offend.

MARINA
I cannot be offended with my trade.
Please you to name it.

LYSIMACHUS              How long have you been    75
Of this profession?

MARINA                 E'er since I can remember.

LYSIMACHUS
Did you go to't so young? Were you a gamester
At five, or seven?

MARINA              Earlier too, sir,
If now I be one.

LYSIMACHUS         Why, the house you dwell in
Proclaimeth you a creature of sale.                80

MARINA
And do you know this house to be a place
Of such resort and will come into it?
I hear say you're of honourable blood,
And are the governor of this whole province.

LYSIMACHUS
What, hath your principal informed you who I am?

MARINA
Who is my principal?

LYSIMACHUS            Why, your herb-woman;    86
She that sets seeds of shame, roots of iniquity.
⌈*Marina weeps*⌉
O, you've heard something of my pow'r, and so
Stand off aloof for a more serious wooing.
But I protest to thee,                            90
Pretty one, my authority can wink
At blemishes, or can on faults look friendly;
Or my displeasure punish at my pleasure,
From which displeasure, not thy beauty shall
Privilege thee, nor my affection, which         95
Hath drawn me here, abate with further ling'ring.
Come bring me to some private place. Come, come.

MARINA
Let not authority, which teaches you
To govern others, be the means to make you
Misgovern much yourself.                         100
If you were born to honour, show it now;
If put upon you, make the judgement good
That thought you worthy of it. What reason's in
Your justice, who hath power over all,
To undo any? If you take from me               105
Mine honour, you're like him that makes a gap
Into forbidden ground, whom after
Too many enter, and of all their evils
Yourself are guilty. My life is yet unspotted;
My chastity unstainèd ev'n in thought.          110
Then if your violence deface this building,
The workmanship of heav'n, you do kill your honour,

Abuse your justice, and impoverish me.
My yet good lord, if there be fire before me,
Must I straight fly and burn myself? Suppose this
house—                                           115
Which too too many feel such houses are—
Should be the doctor's patrimony, and
The surgeon's feeding; follows it, that I
Must needs infect myself to give them maint'nance?

LYSIMACHUS
How's this, how's this? Some more. Be sage.

MARINA ⌈*kneeling*⌉                    For me
That am a maid, though most ungentle fortune    121
Have franked me in this sty, where since I came
Diseases have been sold dearer than physic—
That the gods would set me free from this unhallowed
place,
Though they did change me to the meanest bird   125
That flies i'th' purer air!

LYSIMACHUS ⌈*moved*⌉      I did not think
Thou couldst have spoke so well, ne'er dreamt thou
couldst.
⌈*He lifts her up with his hands*⌉
Though I brought hither a corrupted mind,
Thy speech hath altered it,
⌈*He wipes the wet from her eyes*⌉
                        and my foul thoughts
Thy tears so well hath laved that they're now white.
I came here meaning but to pay the price,       131
A piece of gold for thy virginity;
Here's twenty to relieve thine honesty.
Persever still in that clear way thou goest,
And the gods strengthen thee.

MARINA                   The good gods preserve you!

LYSIMACHUS
The very doors and windows savour vilely.       136
Fare thee well. Thou art a piece of virtue,
The best wrought up that ever nature made,
And I doubt not thy training hath been noble.
A curse upon him, die he like a thief,          140
That robs thee of thy honour. Hold, here's more gold.
If thou dost hear from me, it shall be for thy good.
⌈*Enter Boult standing ready at the door, making his
obeisance unto him as Lysimachus should go out*⌉

BOULT I beseech your honour, one piece for me.

LYSIMACHUS
Avaunt, thou damnèd door-keeper!
Your house, but for this virgin that doth prop it,  145
Would sink and overwhelm you. Away.        *Exit*

BOULT How's this? We must take another course with
you. If your peevish chastity, which is not worth a
breakfast in the cheapest country under the cope, shall
undo a whole household, let me be gelded like a spaniel.
Come your ways.                                  151

MARINA Whither would you have me?

BOULT I must have your maidenhead taken off, or the
common executioner shall do it. We'll have no more
gentlemen driven away. Come your ways, I say.   155
*Enter Bawd and Pander*

BAWD How now, what's the matter?

BOULT Worse and worse, mistress, she has here spoken
holy words to the Lord Lysimachus.

BAWD O, abominable!

BOULT She makes our profession as it were to stink afore
the face of the gods.                            161

BAWD Marry hang her up for ever!

BOULT The nobleman would have dealt with her like a
  nobleman, and she sent him away as cold as a snowball,
  saying his prayers, too.                              165
⌈PANDER⌉ Boult, take her away. Use her at thy pleasure.
  Crack the ice of her virginity, and make the rest
  malleable.
BOULT An if she were a thornier piece of ground than she
  is, she shall be ploughed.                            170
MARINA Hark, hark, you gods!
BAWD She conjures. Away with her! Would she had never
  come within my doors.—Marry, hang you!—She's born
  to undo us.—Will you not go the way of womenkind?
  Marry, come up, my dish of chastity with rosemary
  and bays.                         Exeunt Bawd and Pander
BOULT ⌈catching her rashly by the hand⌉ Come, mistress,
  come your way with me.
MARINA Whither wilt thou have me?
BOULT To take from you the jewel you hold so dear.   180
MARINA Prithee, tell me one thing first.
BOULT Come, now, your one thing.
MARINA
  What canst thou wish thine enemy to be?
BOULT Why, I could wish him to be my master, or rather
  my mistress.                                          185
MARINA
  Neither of these can be so bad as thou art,
  Since they do better thee in their command.
  Thou hold'st a place the painèd'st fiend of hell
  Would not in reputation change with thee,
  Thou damnèd doorkeeper to ev'ry coistrel           190
  That comes enquiring for his Tib.
  To th' choleric fisting of ev'ry rogue
  Thy ear is liable. Thy food is such
  As hath been belched on by infected lungs.         194
BOULT What would you have me do? Go to the wars,
  would you, where a man may serve seven years for
  the loss of a leg, and have not money enough in the
  end to buy him a wooden one?
MARINA
  Do anything but this thou dost. Empty
  Old receptacles or common sew'rs of filth,          200
  Serve by indenture to the public hangman—
  Any of these are yet better than this.
  For what thou professest a baboon, could he speak,
  Would own a name too dear. Here's gold for thee.
  If that thy master would make gain by me,           205
  Proclaim that I can sing, weave, sew, and dance,
  With other virtues which I'll keep from boast,
  And I will undertake all these to teach.
  I doubt not but this populous city will
  Yield many scholars.                                 210
BOULT But can you teach all this you speak of?
MARINA
  Prove that I cannot, take me home again
  And prostitute me to the basest groom
  That doth frequent your house.
BOULT Well, I will see what I can do for thee. If I can
  place thee, I will.                                   216
MARINA But amongst honest women.
BOULT Faith, my acquaintance lies little amongst them;
  but since my master and mistress hath bought you,
  there's no going but by their consent. Therefore I will
  make them acquainted with your purpose, and I doubt
  not but I shall find them tractable enough. Come, I'll
  do for thee what I can. Come your ways.               223
                      Exeunt. ⌈The sign is removed⌉

Sc. 20  *Enter Gower*
GOWER
  Marina thus the brothel scapes, and chances
    Into an honest house, our story says.
  She sings like one immortal, and she dances
    As goddess-like to her admirèd lays.
  Deep clerks she dumbs, and with her nee'le composes
    Nature's own shape, of bud, bird, branch, or berry,
  That e'en her art sisters the natural roses.           7
  Her inkle, silk, twin with the rubied cherry;
    That pupils lacks she none of noble race,
    Who pour their bounty on her, and her gain          10
  She gives the cursèd Bawd. Here we her place,
    And to her father turn our thoughts again.
  We left him on the sea. Waves there him tossed,
    Whence, driven tofore the winds, he is arrived
  Here where his daughter dwells, and on this coast     15
    Suppose him now at anchor. The city strived
  God Neptune's annual feast to keep, from whence
    Lysimachus our Tyrian ship espies,
  His banners sable, trimmed with rich expense;
    And to him in his barge with fervour hies.          20
  In your supposing once more put your sight;
    Of heavy Pericles think this the barque,
  Where what is done in action, more if might,
    Shall be discovered. Please you sit and hark.   *Exit*

Sc. 21  *Enter Helicanus ⌈above; below, enter⌉ to him at
        the first door two Sailors, ⌈one of Tyre, the other of
        Mytilene⌉*
SAILOR OF TYRE (*to Sailor of Mytilene*)
  Lord Helicanus can resolve you, sir.
  (*To Helicanus*) There is a barge put off from Mytilene.
  In it, Lysimachus, the governor,
  Who craves to come aboard. What is your will?
HELICANUS
  That he have his.      ⌈*Exit Sailor of Mytilene at first door*⌉
                  Call up some gentlemen.               5
                              ⌈*Exit Helicanus above*⌉
⌈SAILOR OF TYRE⌉
  Ho, my lord calls!
      *Enter ⌈from below the stage⌉ two or three
      Gentlemen; ⌈to them, enter Helicanus⌉*
FIRST GENTLEMAN       What is your lordship's pleasure?
HELICANUS
  Gentlemen, some of worth would come aboard.
  I pray you, greet him fairly.
      *Enter Lysimachus ⌈at first door, with the Sailor and
      Lords of Mytilene⌉*
⌈SAILOR OF MYTILENE⌉ (*to Lysimachus*)
  This is the man that can in aught resolve you.
LYSIMACHUS (*to Helicanus*)
  Hail, reverend sir; the gods preserve you!           10
HELICANUS
  And you, sir, to outlive the age I am,
  And die as I would do.
LYSIMACHUS             You wish me well.
  I am the governor of Mytilene;
  Being on shore, honouring of Neptune's triumphs,
  Seeing this goodly vessel ride before us,            15
  I made to it to know of whence you are.
HELICANUS
  Our vessel is of Tyre, in it our king,
  A man who for this three months hath not spoken
  To anyone, nor taken sustenance
  But to prorogue his grief.                           20

LYSIMACHUS
Upon what ground grew his distemp'rature?
HELICANUS
'Twould be too tedious to tell it over,
But the main grief springs from the precious loss
Of a belovèd daughter and a wife.
LYSIMACHUS
May we not see him?
HELICANUS                    See him, sir, you may,          25
But bootless is your sight. He will not speak
To any.
LYSIMACHUS  Let me yet obtain my wish.
HELICANUS
Behold him.
      ⌈*Helicanus draws a curtain, revealing Pericles lying
      upon a couch with a long overgrown beard, diffused
      hair, undecent nails on his fingers, and attired in
      sack-cloth*⌉
                    This was a goodly person
Till the disaster of one mortal night
Drove him to this.                                           30
LYSIMACHUS (*to Pericles*)
Sir, King, all hail. Hail, royal sir.
      ⌈*Pericles shrinks himself down upon his pillow*⌉
HELICANUS
It is in vain. He will not speak to you.
LORD OF MYTILENE
Sir, we have a maid in Mytilene I durst wager
Would win some words of him.
LYSIMACHUS                    'Tis well bethought.
She questionless, with her sweet harmony          35
And other choice attractions, would alarum
And make a batt'ry through his deafened ports,
Which now are midway stopped. She in all happy,
As the fair'st of all, among her fellow maids
Dwells now i'th' leafy shelter that abuts          40
Against the island's side. Go fetch her hither.
                                        ⌈*Exit Lord*⌉
HELICANUS
Sure, all effectless; yet nothing we'll omit
That bears recov'ry's name. But since your kindness
We have stretched thus far, let us beseech you
That for our gold we may provision have,          45
Wherein we are not destitute for want,
But weary for the staleness.
LYSIMACHUS                    O sir, a courtesy
Which if we should deny, the most just gods
For every graft would send a caterpillar,
And so inflict our province. Yet once more          50
Let me entreat to know at large the cause
Of your king's sorrow.
HELICANUS            Sit, sir. I will recount it.
      ⌈*Enter Lord with Marina and another maid*⌉
But see, I am prevented.
LYSIMACHUS
O, here's the lady that I sent for.—
Welcome, fair one.—Is't not a goodly presence?          55
HELICANUS  She's a gallant lady.
LYSIMACHUS
She's such a one that, were I well assured
Came of gentle kind or noble stock, I'd wish
No better choice to think me rarely wed.—
Fair one, all goodness that consists in bounty          60
Expect e'en here, where is a kingly patient;
If that thy prosperous and artificial feat
Can draw him but to answer thee in aught,

Thy sacred physic shall receive such pay
As thy desires can wish.
MARINA            Sir, I will use          65
My utmost skill in his recure, provided
That none but I and my companion maid
Be suffered to come near him.
LYSIMACHUS (*to the others*)      Let us leave her,
And the gods prosper her.      ⌈*The men stand aside*⌉
      *The Song*
LYSIMACHUS ⌈*coming forward*⌉ Marked he your music?
⌈MAID⌉
No, nor looked on us.
LYSIMACHUS (*to the others*) See, she will speak to him.  70
MARINA (*to Pericles*)
Hail, sir; my lord, lend ear.
PERICLES  Hmh, ha!
      ⌈*He roughly repulses her*⌉
MARINA                    I am a maid,
My lord, that ne'er before invited eyes,
But have been gazed on like a comet. She speaks,  75
My lord, that maybe hath endured a grief
Might equal yours, if both were justly weighed.
Though wayward fortune did malign my state,
My derivation was from ancestors
Who stood equivalent with mighty kings,          80
But time hath rooted out my parentage,
And to the world and awkward casualties
Bound me in servitude. (*Aside*) I will desist.
But there is something glows upon my cheek,
And whispers in mine ear 'Stay till he speak.'          85
PERICLES
My fortunes, parentage, good parentage,
To equal mine? Was it not thus? What say you?
MARINA
I said if you did know my parentage,
My lord, you would not do me violence.
PERICLES
I do think so. Pray you, turn your eyes upon me.          90
You're like something that—what countrywoman?
Here of these shores?
MARINA            No, nor of any shores,
Yet I was mortally brought forth, and am
No other than I seem.
PERICLES ⌈*aside*⌉
I am great with woe, and shall deliver weeping.          95
My dearest wife was like this maid, and such
My daughter might have been. My queen's square
      brows,
Her stature to an inch, as wand-like straight,
As silver-voiced, her eyes as jewel-like,
And cased as richly, in pace another Juno,          100
Who starves the ears she feeds, and makes them
      hungry
The more she gives them speech.—Where do you live?
MARINA
Where I am but a stranger. From the deck
You may discern the place.
PERICLES            Where were you bred,
And how achieved you these endowments which          105
You make more rich to owe?
MARINA                    If I should tell
My history, it would seem like lies
Disdained in the reporting.
PERICLES            Prithee speak.
Falseness cannot come from thee, for thou look'st
Modest as justice, and thou seem'st a palace          110

For the crownèd truth to dwell in. I will believe thee,
And make my senses credit thy relation
To points that seem impossible. Thou show'st
Like one I loved indeed. What were thy friends?
Didst thou not say, when I did push thee back—          115
Which was when I perceived thee—that thou cam'st
From good descending?

MARINA                    So indeed I did.

PERICLES
Report thy parentage. I think thou said'st
Thou hadst been tossed from wrong to injury,
And that thou thought'st thy griefs might equal mine,
If both were opened.

MARINA                    Some such thing I said,          121
And said no more but what my circumstance
Did warrant me was likely.

PERICLES                    Tell thy story.
If thine considered prove the thousandth part
Of my endurance, thou art a man, and I          125
Have suffered like a girl. Yet thou dost look
Like patience gazing on kings' graves, and smiling
Extremity out of act. What were thy friends?
How lost thou them? Thy name, my most kind virgin?
Recount, I do beseech thee. Come, sit by me.          130
          *She sits*

MARINA
My name, sir, is Marina.

PERICLES                    O, I am mocked,
And thou by some incensèd god sent hither
To make the world to laugh at me.

MARINA                    Patience, good sir,
Or here I'll cease.

PERICLES                    Nay, I'll be patient.
Thou little know'st how thou dost startle me          135
To call thyself Marina.

MARINA                    The name
Was given me by one that had some power:
My father, and a king.

PERICLES                    How, a king's daughter,
And called Marina?

MARINA                    You said you would believe me,
But not to be a troubler of your peace          140
I will end here.

PERICLES          But are you flesh and blood?
Have you a working pulse and are no fairy?
Motion as well? Speak on. Where were you born,
And wherefore called Marina?

MARINA                    Called Marina
For I was born at sea.

PERICLES                    At sea? What mother?          145

MARINA
My mother was the daughter of a king,
Who died when I was born, as my good nurse
Lychorida hath oft recounted weeping.

PERICLES
O, stop there a little! ⌜*Aside*⌝ This is the rarest dream
That e'er dulled sleep did mock sad fools withal.          150
This cannot be my daughter, buried. Well.
(*To Marina*) Where were you bred? I'll hear you more
          to th' bottom
Of your story, and never interrupt you.

MARINA
You will scarce believe me. 'Twere best I did give o'er.

PERICLES
I will believe you by the syllable          155
Of what you shall deliver. Yet give me leave.
How came you in these parts? Where were you bred?

MARINA
The King my father did in Tarsus leave me,
Till cruel Cleon, with his wicked wife,
Did seek to murder me, and wooed a villain          160
To attempt the deed; who having drawn to do't,
A crew of pirates came and rescued me.
To Mytilene they brought me. But, good sir,
What will you of me? Why do you weep? It may be
You think me an impostor. No, good faith,          165
I am the daughter to King Pericles,
If good King Pericles be.

PERICLES ⌜*rising*⌝ Ho, Helicanus!

HELICANUS (*coming forward*) Calls my lord?

PERICLES
Thou art a grave and noble counsellor,          170
Most wise in gen'ral. Tell me if thou canst
What this maid is, or what is like to be,
That thus hath made me weep.

HELICANUS                    I know not.
But here's the regent, sir, of Mytilene
Speaks nobly of her.

LYSIMACHUS                    She would never tell          175
Her parentage. Being demanded that,
She would sit still and weep.

PERICLES
O Helicanus, strike me, honoured sir,
Give me a gash, put me to present pain,
Lest this great sea of joys rushing upon me          180
O'erbear the shores of my mortality
And drown me with their sweetness! (*To Marina*) O,
          come hither,
⌜*Marina stands*⌝
Thou that begett'st him that did thee beget,
Thou that wast born at sea, buried at Tarsus,
And found at sea again!—O Helicanus,          185
Down on thy knees, thank the holy gods as loud
As thunder threatens us, this is Marina!
(*To Marina*) What was thy mother's name? Tell me
          but that,
For truth can never be confirmed enough,
Though doubts did ever sleep.

MARINA                    First, sir, I pray,          190
What is your title?

PERICLES                    I am Pericles
Of Tyre. But tell me now my drowned queen's name.
As in the rest thou hast been godlike perfect,
So prove but true in that, thou art my daughter,
The heir of kingdoms, and another life          195
To Pericles thy father.

MARINA ⌜*kneeling*⌝          Is it no more
To be your daughter than to say my mother's name?
Thaisa was my mother, who did end
The minute I began.

PERICLES
Now blessing on thee! Rise. Thou art my child.          200
          ⌜*Marina stands. He kisses her*⌝
⌜*To attendants*⌝ Give me fresh garments.—Mine own,
          Helicanus!
Not dead at Tarsus, as she should have been
By savage Cleon. She shall tell thee all,
When thou shalt kneel and justify in knowledge
She is thy very princess. Who is this?          205

HELICANUS
Sir, 'tis the governor of Mytilene,
Who, hearing of your melancholy state,

Did come to see you.

PERICLES (*to Lysimachus*) I embrace you, sir.—
Give me my robes.
⌐*He is attired in fresh robes*⌐
           I am wild in my beholding.
O heavens, bless my girl!
⌐*Celestial music*⌐
           But hark, what music?    210
Tell Helicanus, my Marina, tell him
O'er point by point, for yet he seems to doubt,
How sure you are my daughter. But what music?

HELICANUS My lord, I hear none.

PERICLES
None? The music of the spheres! List, my Marina.  215

LYSIMACHUS (*aside to the others*)
It is not good to cross him. Give him way.

PERICLES Rar'st sounds. Do ye not hear?

LYSIMACHUS Music, my lord?

PERICLES I hear most heav'nly music.
It raps me unto list'ning, and thick slumber    220
Hangs upon mine eyelids. Let me rest.
    *He sleeps*

LYSIMACHUS
A pillow for his head.
⌐*To Marina and others*⌐ Companion friends,
If this but answer to my just belief
I'll well remember you. So leave him all.
           *Exeunt all but Pericles*
    *Diana* ⌐*descends from the heavens*⌐

DIANA
My temple stands in Ephesus. Hie thee thither,  225
    And do upon mine altar sacrifice.
There when my maiden priests are met together,
    At large discourse thy fortunes in this wise:
With a full voice before the people all,
    Reveal how thou at sea didst lose thy wife.  230
To mourn thy crosses, with thy daughter's, call
    And give them repetition to the life.
Perform my bidding, or thou liv'st in woe;
    Do't, and rest happy, by my silver bow.
Awake, and tell thy dream.    235
        ⌐*Diana ascends into the heavens*⌐

PERICLES
Celestial Dian, goddess argentine,
I will obey thee. (*Calling*) Helicanus!
    *Enter Helicanus, Lysimachus, and Marina*

HELICANUS          Sir?

PERICLES
My purpose was for Tarsus, there to strike
Th'inhospitable Cleon, but I am
For other service first. Toward Ephesus  240
Turn our blown sails. Eftsoons I'll tell thee why.
        ⌐*Exit Helicanus*⌐
Shall we refresh us, sir, upon your shore,
And give you gold for such provision
As our intents will need?

LYSIMACHUS      With all my heart, sir,
And when you come ashore I have a suit.  245

PERICLES
You shall prevail, were it to woo my daughter,
For it seems you have been noble towards her.

LYSIMACHUS
Sir, lend me your arm.

PERICLES        Come, my Marina.
    ⌐*Exit Pericles with Lysimachus at one arm,*
           *Marina at the other*⌐

**Sc. 22**  *Enter Gower*

GOWER
Now our sands are almost run;
More a little, and then dumb.
This my last boon give me,
For such kindness must relieve me,
That you aptly will suppose  5
What pageantry, what feats, what shows,
What minstrelsy and pretty din
The regent made in Mytilene
To greet the King. So well he thrived
That he is promised to be wived  10
To fair Marina, but in no wise
Till he had done his sacrifice
As Dian bade, whereto being bound
The int'rim, pray you, all confound.
In feathered briefness sails are filled,  15
And wishes fall out as they're willed.
At Ephesus the temple see:
⌐*An altar, Thaisa and other vestals are revealed*⌐
Our king, and all his company.
⌐*Enter Pericles, Marina, Lysimachus, Helicanus,*
    *Cerimon, with attendants*⌐
That he can hither come so soon
Is by your fancies' thankful doom.  20
⌐*Gower stands aside*⌐

PERICLES
Hail, Dian. To perform thy just command
I here confess myself the King of Tyre,
Who, frighted from my country, did espouse
The fair Thaisa
    ⌐*Thaisa starts*⌐
          at Pentapolis.
At sea in childbed died she, but brought forth  25
A maid child called Marina, who, O goddess,
Wears yet thy silver liv'ry. She at Tarsus
Was nursed with Cleon, whom at fourteen years
He sought to murder, but her better stars
Bore her to Mytilene, 'gainst whose shore riding  30
Her fortunes brought the maid aboard our barque,
Where, by her own most clear remembrance, she
Made known herself my daughter.

THAISA          Voice and favour—
You are, you are—O royal Pericles!
    *She falls*

PERICLES
What means the nun? She dies. Help, gentlemen!  35

CERIMON Noble sir,
If you have told Diana's altar true,
This is your wife.

PERICLES      Reverend appearer, no.
I threw her overboard with these same arms.

CERIMON
Upon this coast, I warr'nt you.

PERICLES         'Tis most certain.  40

CERIMON
Look to the lady. O, she's but o'erjoyed.
Early one blustering morn this lady
Was thrown upon this shore. I oped the coffin,
Found there rich jewels, recovered her, and placed her
Here in Diana's temple.

PERICLES      May we see them?  45

CERIMON
Great sir, they shall be brought you to my house,
Whither I invite you. Look, Thaisa is
Recoverèd.

THAISA    O, let me look upon him!
If he be none of mine, my sanctity

Will to my sense bend no licentious ear,                    50
But curb it, spite of seeing. O, my lord,
Are you not Pericles? Like him you spake,
Like him you are. Did you not name a tempest,
A birth and death?
PERICLES The voice of dead Thaisa!                          55
THAISA That Thaisa
Am I, supposèd dead and drowned.
PERICLES ⌈taking Thaisa's hand⌉ Immortal Dian!
THAISA Now I know you better.
When we with tears parted Pentapolis,                       60
The King my father gave you such a ring.
PERICLES
This, this! No more, you gods. Your present kindness
Makes my past miseries sports; you shall do well
That on the touching of her lips I may
Melt, and no more be seen.—O come, be buried               65
A second time within these arms.
⌈They embrace and kiss⌉
MARINA (kneeling to Thaisa)          My heart
Leaps to be gone into my mother's bosom.
PERICLES
Look who kneels here: flesh of thy flesh, Thaisa,
Thy burden at the sea, and called Marina
For she was yielded there.
THAISA ⌈embracing Marina⌉ Blessed, and mine own!  70
HELICANUS ⌈kneeling to Thaisa⌉
Hail, madam, and my queen.
THAISA                              I know you not.
PERICLES
You have heard me say, when I did fly from Tyre,
I left behind an ancient substitute.
Can you remember what I called the man?
I have named him oft.                                       75
THAISA 'Twas Helicanus then.
PERICLES Still confirmation.
Embrace him, dear Thaisa; this is he.
Now do I long to hear how you were found,
How possibly preserved, and who to thank—                   80
Besides the gods—for this great miracle.
THAISA
Lord Cerimon, my lord. This is the man
Through whom the gods have shown their pow'r,
    that can
From first to last resolve you.
PERICLES (to Cerimon)           Reverend sir,
The gods can have no mortal officer                         85

More like a god than you. Will you deliver
How this dead queen re-lives?
CERIMON                    I will, my lord.
Beseech you, first go with me to my house,
Where shall be shown you all was found with her,
And told how in this temple she came placed,                90
No needful thing omitted.
PERICLES                    Pure Diana,
I bless thee for thy vision, and will offer
Nightly oblations to thee.—Beloved Thaisa,
This prince, the fair betrothèd of your daughter,
At Pentapolis shall marry her.                              95
(To Marina) And now this ornament
Makes me look dismal will I clip to form,
And what this fourteen years no razor touched,
To grace thy marriage day I'll beautify.
THAISA
Lord Cerimon hath letters of good credit,                  100
Sir, from Pentapolis: my father's dead.
PERICLES
Heav'n make a star of him! Yet there, my queen,
We'll celebrate their nuptials, and ourselves
Will in that kingdom spend our following days.
Our son and daughter shall in Tyrus reign.—                105
Lord Cerimon, we do our longing stay
To hear the rest untold. Sir, lead 's the way.
                              Exeunt ⌈all but Gower⌉
GOWER
In Antiochus and his daughter you have heard
Of monstrous lust the due and just reward;
In Pericles, his queen, and daughter seen,                 110
Although assailed with fortune fierce and keen,
Virtue preserved from fell destruction's blast,
Led on by heav'n, and crowned with joy at last.
In Helicanus may you well descry
A figure of truth, of faith, of loyalty.                   115
In reverend Cerimon there well appears
The worth that learnèd charity aye wears.
For wicked Cleon and his wife, when fame
Had spread their cursèd deed to th' honoured name
Of Pericles, to rage the city turn,                        120
That him and his they in his palace burn.
The gods for murder seemèd so content
To punish that, although not done, but meant.
So on your patience evermore attending,
New joy wait on you. Here our play has ending.
                                              Exit

## ADDITIONAL PASSAGES

Q gives this more expansive version of Marina's Epitaph
(18.34–7):

'The fairest, sweetest, best lies here,
Who withered in her spring of year.
She was of Tyrus the King's daughter,
On whom foul death hath made this slaughter.

Marina was she called, and at her birth                     5
Thetis, being proud, swallowed some part o'th' earth;
Therefore the earth, fearing to be o'erflowed,
Hath Thetis' birth-child on the heav'ns bestowed,
Wherefore she does, and swears she'll never stint,
Make raging batt'ry upon shores of flint.'                  10

# CORIOLANUS

FOR *Coriolanus*, Shakespeare turned once more to Roman history as told by Plutarch and translated by Sir Thomas North in the *Lives of the Noble Grecians and Romans* published in 1579. This time he dramatized early events, not much subsequent to those he had written about many years previously in *The Rape of Lucrece*. Plutarch gave him most of his material, but he also drew on other writings, including William Camden's *Remains of a Greater Work Concerning Britain*, published in 1605, for Menenius' fable of the belly (1.1). Though he needed no source other than Plutarch for the insurrections and corn riots of ancient Rome, similar happenings in England during 1607 and 1608 may have stimulated his interest in the story. The cumulative evidence suggests that *Coriolanus*, first printed in the 1623 Folio, is Shakespeare's last Roman play, written around 1608.

In the fifth century BC, following the expulsion of the Tarquins, Rome was an aristocratically controlled republic in which power was invested primarily in two annually elected magistrates, or consuls. For many years the main issues confronting the republic were the internal class struggle between patricians and plebeians, and the external struggle for domination over neighbouring peoples. Among the republic's early enemies were the Volsci (or Volscians), who inhabited an area to the south and south-east of Rome; their towns included Antium and Corioli. According to ancient historians, Rome's greatest leader in campaigns against the Volsci was the patrician Gnaeus (or Caius) Marcius, who, at a time of famine which caused the plebeians to rebel against the patricians, led an army against the Volsci and captured Corioli; as a reward he was granted the cognomen, or surname, of Coriolanus. After this he is said to have been charged with behaving tyrannically in opposing the distribution of corn to starving plebeians, and as a result to have abandoned Rome, joined the Volsci, and led a Volscian army against his native city.

This is the story of conflict between public and private issues that Shakespeare dramatizes, concentrating on the later part of Plutarch's Life and speeding up its time-scheme, while also alluding retrospectively to earlier incidents. He increases the responsibility of the Tribunes, Sicinius Velutus and Junius Brutus, for Coriolanus' banishment, and greatly develops certain characters, such as the Volscian leader Tullus Aufidius and the patrician Menenius Agrippa. The roles of the womenfolk are almost entirely of Shakespeare's devising up to the scene (5.3) of their embassy; here, as in certain other set speeches, Shakespeare draws heavily on the language of North's translation.

*Coriolanus* is an austere play, gritty in style, deeply serious in its concern with the relationship between personal characteristics and national destiny, but relieved by flashes of comedy (especially in the scenes in which Coriolanus begs for the plebeians' votes in his election campaign for the consulship) which are more apparent on the stage than on the page. Though Coriolanus is arrogant, choleric, and self-centered, he is also a blazingly successful warrior, conspicuous for integrity, who ultimately yields to a tenderness which, he knows, will destroy him. *Coriolanus* is a deeply human as well as a profoundly political play.

# THE PERSONS OF THE PLAY

Caius MARTIUS, later surnamed CORIOLANUS

MENENIUS Agrippa — patricians of Rome

Titus LARTIUS ⎱ generals
COMINIUS ⎰

VOLUMNIA, Coriolanus' mother
VIRGILIA, his wife
YOUNG MARTIUS, his son
VALERIA, a chaste lady of Rome

SICINIUS Velutus ⎱ tribunes of the Roman people
Junius BRUTUS ⎰

CITIZENS of Rome
SOLDIERS in the Roman army

Tullus AUFIDIUS, general of the Volscian army
His LIEUTENANT
His SERVINGMEN
CONSPIRATORS with Aufidius
Volscian LORDS
Volscian CITIZENS
SOLDIERS in the Volscian army

ADRIAN, a Roman
NICANOR, a Volscian
A Roman HERALD
MESSENGERS
AEDILES

A gentlewoman, an usher, Roman and Volscian senators and nobles, captains in the Roman army, officers, lictors

# The Tragedy of Coriolanus

**1.1** *Enter a company of mutinous Citizens with staves, clubs, and other weapons*

FIRST CITIZEN Before we proceed any further, hear me speak.

ALL Speak, speak.

FIRST CITIZEN You are all resolved rather to die than to famish?     5

ALL Resolved, resolved.

FIRST CITIZEN First, you know Caius Martius is chief enemy to the people.

ALL We know't, we know't.

FIRST CITIZEN Let us kill him, and we'll have corn at our own price. Is't a verdict?     11

ALL No more talking on't, let it be done. Away, away.

SECOND CITIZEN One word, good citizens.

FIRST CITIZEN We are accounted poor citizens, the patricians good. What authority surfeits on would relieve us. If they would yield us but the superfluity while it were wholesome we might guess they relieved us humanely, but they think we are too dear. The leanness that afflicts us, the object of our misery, is as an inventory to particularize their abundance; our sufferance is a gain to them. Let us revenge this with our pikes ere we become rakes; for the gods know I speak this in hunger for bread, not in thirst for revenge.

SECOND CITIZEN Would you proceed especially against Caius Martius?     25

⌈THIRD CITIZEN⌉ Against him first.

⌈FOURTH CITIZEN⌉ He's a very dog to the commonalty.

SECOND CITIZEN Consider you what services he has done for his country?

FIRST CITIZEN Very well, and could be content to give him good report for't, but that he pays himself with being proud.     32

⌈FIFTH CITIZEN⌉ Nay, but speak not maliciously.

FIRST CITIZEN I say unto you, what he hath done famously, he did it to that end—though soft-conscienced men can be content to say 'it was for his country', 'he did it to please his mother, and to be partly proud'—which he is even to the altitude of his virtue.     38

SECOND CITIZEN What he cannot help in his nature you account a vice in him. You must in no way say he is covetous.

FIRST CITIZEN If I must not, I need not be barren of accusations. He hath faults, with surplus, to tire in repetition.     44

*Shouts within*

What shouts are these? The other side o'th' city is risen. Why stay we prating here? To th' Capitol!

ALL Come, come.

*Enter Menenius*

FIRST CITIZEN Soft, who comes here?

SECOND CITIZEN Worthy Menenius Agrippa, one that hath always loved the people.     50

FIRST CITIZEN He's one honest enough. Would all the rest were so!

MENENIUS
What work's, my countrymen, in hand? Where go you
With bats and clubs? The matter. Speak, I pray you.

⌈FIRST⌉ CITIZEN Our business is not unknown to th' senate. They have had inkling this fortnight what we intend to do, which now we'll show 'em in deeds. They say poor suitors have strong breaths; they shall know we have strong arms, too.

MENENIUS
Why, masters, my good friends, mine honest neighbours,     60
Will you undo yourselves?

⌈FIRST⌉ CITIZEN
We cannot, sir. We are undone already.

MENENIUS
I tell you, friends, most charitable care
Have the patricians of you. For your wants,
Your suffering in this dearth, you may as well     65
Strike at the heaven with your staves as lift them
Against the Roman state, whose course will on
The way it takes, cracking ten thousand curbs
Of more strong link asunder than can ever
Appear in your impediment. For the dearth,     70
The gods, not the patricians, make it, and
Your knees to them, not arms, must help. Alack,
You are transported by calamity
Thither where more attends you, and you slander
The helms o'th' state, who care for you like fathers,
When you curse them as enemies.     76

⌈FIRST⌉ CITIZEN Care for us? True, indeed! They ne'er cared for us yet: suffer us to famish, and their storehouses crammed with grain; make edicts for usury to support usurers; repeal daily any wholesome act established against the rich; and provide more piercing statutes daily to chain up and restrain the poor. If the wars eat us not up, they will; and there's all the love they bear us.

MENENIUS Either you must     85
Confess yourselves wondrous malicious
Or be accused of folly. I shall tell you
A pretty tale. It may be you have heard it,
But since it serves my purpose, I will venture
To stale't a little more.     90

⌈FIRST⌉ CITIZEN Well, I'll hear it, sir. Yet you must not think to fob off our disgrace with a tale. But an't please you, deliver.

MENENIUS
There was a time when all the body's members,
Rebelled against the belly, thus accused it:     95
That only like a gulf it did remain
I'th' midst o'th' body, idle and unactive,
Still cupboarding the viand, never bearing
Like labour with the rest; where th'other instruments
Did see and hear, devise, instruct, walk, feel,     100
And, mutually participate, did minister
Unto the appetite and affection common
Of the whole body. The belly answered—

⌈FIRST⌉ CITIZEN
Well, sir, what answer made the belly?

MENENIUS
Sir, I shall tell you. With a kind of smile,     105
Which ne'er came from the lungs, but even thus—
For look you, I may make the belly smile

As well as speak—it tauntingly replied
To th' discontented members, the mutinous parts
That envied his receipt; even so most fitly          110
As you malign our senators for that
They are not such as you.
⌐FIRST⌐ CITIZEN                    Your belly's answer—what?
The kingly crownèd head, the vigilant eye,
The counsellor heart, the arm our soldier,
Our steed the leg, the tongue our trumpeter,          115
With other muniments and petty helps
In this our fabric, if that they—
MENENIUS                    What then?
Fore me, this fellow speaks! What then? What then?
⌐FIRST⌐ CITIZEN
Should by the cormorant belly be restrained,
Who is the sink o'th' body—
MENENIUS                    Well, what then?     120
⌐FIRST⌐ CITIZEN
The former agents, if they did complain,
What could the belly answer?
MENENIUS                    I will tell you,
If you'll bestow a small of what you have little—
Patience—a while, you'st hear the belly's answer.
⌐FIRST⌐ CITIZEN
You're long about it.
MENENIUS                    Note me this, good friend:     125
Your most grave belly was deliberate,
Not rash like his accusers, and thus answered:
'True is it, my incorporate friends,' quoth he,
'That I receive the general food at first
Which you do live upon, and fit it is,          130
Because I am the storehouse and the shop
Of the whole body. But, if you do remember,
I send it through the rivers of your blood
Even to the court, the heart, to th' seat o'th' brain;
And through the cranks and offices of man          135
The strongest nerves and small inferior veins
From me receive that natural competency
Whereby they live. And though that all at once'—
You my good friends, this says the belly, mark me—
⌐FIRST⌐ CITIZEN
Ay, sir, well, well.
MENENIUS                    'Though all at once cannot     140
See what I do deliver out to each,
Yet I can make my audit up that all
From me do back receive the flour of all
And leave me but the bran.' What say you to't?
⌐FIRST⌐ CITIZEN
It was an answer. How apply you this?          145
MENENIUS
The senators of Rome are this good belly,
And you the mutinous members. For examine
Their counsels and their cares, digest things rightly
Touching the weal o'th' common, you shall find
No public benefit which you receive          150
But it proceeds or comes from them to you,
And no way from yourselves. What do you think,
You, the great toe of this assembly?
⌐FIRST⌐ CITIZEN
I the great toe? Why the great toe?
MENENIUS
For that, being one o'th' lowest, basest, poorest     155
Of this most wise rebellion, thou goest foremost.
Thou rascal, that art worst in blood to run,
Lead'st first to win some vantage.
But make you ready your stiff bats and clubs.

Rome and her rats are at the point of battle.     160
The one side must have bale.
    *Enter Martius*
                        Hail, noble Martius!
MARTIUS
Thanks.—What's the matter, you dissentious rogues,
That, rubbing the poor itch of your opinion,
Make yourselves scabs?
⌐FIRST⌐ CITIZEN          We have ever your good word.
MARTIUS
He that will give good words to thee will flatter     165
Beneath abhorring. What would you have, you curs
That like nor peace nor war? The one affrights you,
The other makes you proud. He that trusts to you,
Where he should find you lions finds you hares,
Where foxes, geese. You are no surer, no,          170
Than is the coal of fire upon the ice,
Or hailstone in the sun. Your virtue is
To make him worthy whose offence subdues him,
And curse that justice did it. Who deserves greatness
Deserves your hate, and your affections are          175
A sick man's appetite, who desires most that
Which would increase his evil. He that depends
Upon your favours swims with fins of lead,
And hews down oaks with rushes. Hang ye! Trust
ye?
With every minute you do change a mind,          180
And call him noble that was now your hate,
Him vile that was your garland. What's the matter,
That in these several places of the city
You cry against the noble senate, who,
Under the gods, keep you in awe, which else          185
Would feed on one another?
*(To Menenius)*          What's their seeking?
MENENIUS
For corn at their own rates, whereof they say
The city is well stored.
MARTIUS                    Hang 'em! They say?
They'll sit by th' fire and presume to know
What's done i'th' Capitol, who's like to rise,          190
Who thrives and who declines; side factions and give
out
Conjectural marriages, making parties strong
And feebling such as stand not in their liking
Below their cobbled shoes. They say there's grain
enough!
Would the nobility lay aside their ruth          195
And let me use my sword, I'd make a quarry
With thousands of these quartered slaves as high
As I could pitch my lance.
MENENIUS
Nay, these are all most thoroughly persuaded,
For though abundantly they lack discretion,          200
Yet are they passing cowardly. But I beseech you,
What says the other troop?
MARTIUS                    They are dissolved. Hang 'em.
They said they were an-hungry, sighed forth
proverbs—
That hunger broke stone walls, that dogs must eat,
That meat was made for mouths, that the gods sent
not          205
Corn for the rich men only. With these shreds
They vented their complainings, which being
answered,
And a petition granted them—a strange one,
To break the heart of generosity

And make bold power look pale—they threw their caps
As they would hang them on the horns o'th' moon,
Shouting their emulation.
MENENIUS                         What is granted them?    212
MARTIUS
Five tribunes to defend their vulgar wisdoms,
Of their own choice. One's Junius Brutus,
Sicinius Velutus, and I know not. 'Sdeath,    215
The rabble should have first unroofed the city
Ere so prevailed with me! It will in time
Win upon power and throw forth greater themes
For insurrection's arguing.
MENENIUS  This is strange.    220
MARTIUS (to the Citizens) Go get you home, you fragments.
          Enter a Messenger hastily
MESSENGER  Where's Caius Martius?
MARTIUS  Here. What's the matter?
MESSENGER
The news is, sir, the Volsces are in arms.
MARTIUS
I am glad on't. Then we shall ha' means to vent    225
Our musty superfluity.
          Enter Sicinius, Brutus, Cominius, Lartius, with
          other Senators
                         See, our best elders.
FIRST SENATOR
Martius, 'tis true that you have lately told us.
The Volsces are in arms.
MARTIUS                   They have a leader,
Tullus Aufidius, that will put you to't.
I sin in envying his nobility,    230
And were I anything but what I am,
I would wish me only he.
COMINIUS              You have fought together!
MARTIUS
Were half to half the world by th' ears and he
Upon my party, I'd revolt to make
Only my wars with him. He is a lion    235
That I am proud to hunt.
FIRST SENATOR            Then, worthy Martius,
Attend upon Cominius to these wars.
COMINIUS (to Martius)
It is your former promise.
MARTIUS                   Sir, it is,
And I am constant. Titus Lartius, thou
Shalt see me once more strike at Tullus' face.    240
What, art thou stiff? Stand'st out?
LARTIUS                   No, Caius Martius.
I'll lean upon one crutch and fight with th'other
Ere stay behind this business.
MENENIUS                  O true bred!
⌈FIRST⌉ SENATOR
Your company to th' Capitol, where I know
Our greatest friends attend us.
LARTIUS (to Cominius)            Lead you on.    245
(To Martius) Follow Cominius. We must follow you,
Right worthy your priority.
COMINIUS                   Noble Martius.
⌈FIRST⌉ SENATOR (to the Citizens)
Hence to your homes, be gone.
MARTIUS                   Nay, let them follow.
The Volsces have much corn. Take these rats thither
To gnaw their garners.    Citizens steal away
                         Worshipful mutineers,    250
Your valour puts well forth. (To the Senators) Pray
    follow.    Exeunt all but Sicinius and Brutus

SICINIUS
Was ever man so proud as is this Martius?
BRUTUS  He has no equal.
SICINIUS
When we were chosen tribunes for the people—
BRUTUS
Marked you his lip and eyes?
SICINIUS                   Nay, but his taunts.    255
BRUTUS
Being moved, he will not spare to gird the gods.
SICINIUS  Bemock the modest moon.
BRUTUS
The present wars devour him. He is grown
Too proud to be so valiant.
SICINIUS                   Such a nature,
Tickled with good success, disdains the shadow    260
Which he treads on at noon. But I do wonder
His insolence can brook to be commanded
Under Cominius.
BRUTUS          Fame, at the which he aims—
In whom already he's well graced—cannot
Better be held nor more attained than by    265
A place below the first; for what miscarries
Shall be the general's fault, though he perform
To th' utmost of a man, and giddy censure
Will then cry out of Martius 'O, if he
Had borne the business!'
SICINIUS              Besides, if things go well,
Opinion, that so sticks on Martius, shall    271
Of his demerits rob Cominius.
BRUTUS                   Come,
Half all Cominius' honours are to Martius,
Though Martius earned them not; and all his faults
To Martius shall be honours, though indeed    275
In aught he merit not.
SICINIUS              Let's hence and hear
How the dispatch is made, and in what fashion,
More than his singularity, he goes
Upon this present action.
BRUTUS                   Let's along.    Exeunt

1.2    Enter Aufidius, with Senators of Corioles
FIRST SENATOR
So, your opinion is, Aufidius,
That they of Rome are entered in our counsels
And know how we proceed.
AUFIDIUS              Is it not yours?
What ever have been thought on in this state
That could be brought to bodily act ere Rome    5
Had circumvention? 'Tis not four days gone
Since I heard thence. These are the words. I think
I have the letter here—yes, here it is.
          ⌈He reads the letter⌉
'They have pressed a power, but it is not known
Whether for east or west. The dearth is great,    10
The people mutinous, and it is rumoured
Cominius, Martius your old enemy,
Who is of Rome worse hated than of you,
And Titus Lartius, a most valiant Roman,
These three lead on this preparation    15
Whither 'tis bent. Most likely 'tis for you.
Consider of it.'
FIRST SENATOR  Our army's in the field.
We never yet made doubt but Rome was ready
To answer us.
AUFIDIUS      Nor did you think it folly
To keep your great pretences veiled till when    20

They needs must show themselves, which in the
    hatching,
It seemed, appeared to Rome. By the discovery
We shall be shortened in our aim, which was
To take in many towns ere, almost, Rome
Should know we were afoot.

SECOND SENATOR            Noble Aufidius,    25
Take your commission, hie you to your bands.
Let us alone to guard Corioles.
If they set down before's, for the remove
Bring up your army, but I think you'll find
They've not prepared for us.

AUFIDIUS            O, doubt not that.    30
I speak from certainties. Nay, more,
Some parcels of their power are forth already,
And only hitherward. I leave your honours.
If we and Caius Martius chance to meet,
'Tis sworn between us we shall ever strike    35
Till one can do no more.

ALL THE SENATORS        The gods assist you!

AUFIDIUS
And keep your honours safe.

FIRST SENATOR          Farewell.

SECOND SENATOR             Farewell.

ALL                             Farewell.
            *Exeunt, ⌈Aufidius at one door,*
                 *Senators at another door⌉*

**1.3**    *Enter Volumnia and Virgilia, mother and wife to*
       *Martius. They set them down on two low stools*
       *and sew*

VOLUMNIA I pray you, daughter, sing, or express yourself
in a more comfortable sort. If my son were my husband,
I should freelier rejoice in that absence wherein he won
honour than in the embracements of his bed where he
would show most love. When yet he was but tender-
bodied and the only son of my womb, when youth
with comeliness plucked all gaze his way, when for a
day of kings' entreaties a mother should not sell him
an hour from her beholding, I, considering how honour
would become such a person—that it was no better
than, picture-like, to hang by th' wall if renown made
it not stir—was pleased to let him seek danger where
he was like to find fame. To a cruel war I sent him,
from whence he returned his brows bound with oak. I
tell thee, daughter, I sprang not more in joy at first
hearing he was a man-child than now in first seeing
he had proved himself a man.    17

VIRGILIA But had he died in the business, madam, how
then?

VOLUMNIA Then his good report should have been my
son. I therein would have found issue. Hear me profess
sincerely: had I a dozen sons, each in my love alike,
and none less dear than thine and my good Martius',
I had rather had eleven die nobly for their country
than one voluptuously surfeit out of action.    25
       *Enter a Gentlewoman*

GENTLEWOMAN Madam, the Lady Valeria is come to visit
you.

VIRGILIA (*to Volumnia*) Beseech you give me leave to retire
myself.

VOLUMNIA Indeed you shall not.    30
Methinks I hear hither your husband's drum,
See him pluck Aufidius down by th' hair;
As children from a bear, the Volsces shunning him.
Methinks I see him stamp thus, and call thus:

'Come on, you cowards, you were got in fear    35
Though you were born in Rome!' His bloody brow
With his mailed hand then wiping, forth he goes,
Like to a harvest-man that's tasked to mow
Or all or lose his hire.

VIRGILIA
His bloody brow? O Jupiter, no blood!    40

VOLUMNIA
Away, you fool! It more becomes a man
Than gilt his trophy. The breasts of Hecuba
When she did suckle Hector looked not lovelier
Than Hector's forehead when it spit forth blood
At Grecian sword, contemning.
(*To the Gentlewoman*)           Tell Valeria    45
We are fit to bid her welcome.      *Exit Gentlewoman*

VIRGILIA
Heavens bless my lord from fell Aufidius!

VOLUMNIA
He'll beat Aufidius' head below his knee
And tread upon his neck.
       *Enter Valeria, with an usher and the Gentlewoman*

VALERIA My ladies both, good day to you.    50

VOLUMNIA Sweet madam.

VIRGILIA I am glad to see your ladyship.

VALERIA How do you both? You are manifest
housekeepers. What are you sewing here? A fine spot,
in good faith. How does your little son?    55

VIRGILIA
I thank your ladyship; well, good madam.

VOLUMNIA He had rather see the swords and hear a drum
than look upon his schoolmaster.

VALERIA O' my word, the father's son! I'll swear 'tis a
very pretty boy. O' my troth, I looked upon him o'
Wednesday half an hour together. He's such a
confirmed countenance! I saw him run after a gilded
butterfly, and when he caught it he let it go again,
and after it again, and over and over he comes, and
up again, catched it again. Or whether his fall enraged
him, or how 'twas, he did so set his teeth and tear it!
O, I warrant, how he mammocked it!    67

VOLUMNIA One on's father's moods.

VALERIA Indeed, la, 'tis a noble child.

VIRGILIA A crack, madam.    70

VALERIA Come, lay aside your stitchery. I must have you
play the idle housewife with me this afternoon.

VIRGILIA No, good madam, I will not out of doors.

VALERIA Not out of doors?

VOLUMNIA She shall, she shall.    75

VIRGILIA Indeed, no, by your patience. I'll not over the
threshold till my lord return from the wars.

VALERIA Fie, you confine yourself most unreasonably.
Come, you must go visit the good lady that lies in.

VIRGILIA I will wish her speedy strength, and visit her
with my prayers, but I cannot go thither.    81

VOLUMNIA Why, I pray you?

VIRGILIA 'Tis not to save labour, nor that I want love.

VALERIA You would be another Penelope. Yet they say
all the yarn she spun in Ulysses' absence did but fill
Ithaca full of moths. Come, I would your cambric were
sensible as your finger, that you might leave pricking
it for pity. Come, you shall go with us.

VIRGILIA No, good madam, pardon me, indeed I will not
forth.    90

VALERIA In truth, la, go with me, and I'll tell you excellent
news of your husband.

VIRGILIA O, good madam, there can be none yet.

VALERIA Verily, I do not jest with you: there came news
from him last night.                                         95
VIRGILIA Indeed, madam?
VALERIA In earnest, it's true. I heard a senator speak it.
Thus it is: the Volsces have an army forth, against
whom Cominius the general is gone with one part of
our Roman power. Your lord and Titus Lartius are set
down before their city Corioles. They nothing doubt
prevailing, and to make it brief wars. This is true, on
mine honour; and so, I pray, go with us.
VIRGILIA Give me excuse, good madam, I will obey you
in everything hereafter.                                     105
VOLUMNIA (*to Valeria*) Let her alone, lady. As she is now
she will but disease our better mirth.
VALERIA In truth, I think she would. Fare you well, then.
Come, good sweet lady. Prithee, Virgilia, turn thy
solemness out o' door and go along with us.       110
VIRGILIA No, at a word, madam. Indeed, I must not. I
wish you much mirth.
VALERIA Well then, farewell.
                *Exeunt ⌈Valeria, Volumnia, and usher at one
                door, Virgilia and Gentlewoman at another door⌉*

**1.4**    *Enter Martius, Lartius with a drummer, ⌈a
          trumpeter,⌉ and colours, with captains and Soldiers
          ⌈carrying scaling ladders⌉, as before the city
          Corioles; to them a Messenger*
MARTIUS
Yonder comes news. A wager they have met.
LARTIUS
My horse to yours, no.
MARTIUS                          'Tis done.
LARTIUS                                        Agreed.
MARTIUS (*to the Messenger*)
Say, has our general met the enemy?
MESSENGER
They lie in view, but have not spoke as yet.
LARTIUS
So, the good horse is mine.
MARTIUS                          I'll buy him of you.    5
LARTIUS
No, I'll nor sell nor give him. Lend you him I will,
For half a hundred years.
(*To the trumpeter*)          Summon the town.
MARTIUS (*to the Messenger*)
How far off lie these armies?
MESSENGER                        Within this mile and half.
MARTIUS
Then shall we hear their 'larum, and they ours.
Now Mars, I prithee, make us quick in work,       10
That we with smoking swords may march from hence
To help our fielded friends.
(*To the trumpeter*)          Come, blow thy blast.
          *They sound a parley. Enter two Senators, with
          others, on the walls of Corioles*
(*To the Senators*) Tullus Aufidius, is he within your walls?
FIRST SENATOR
No, nor a man that fears you less than he:
That's lesser than a little.
          *Drum afar off*
⌈*To the Volscians*⌉          Hark, our drums    15
Are bringing forth our youth. We'll break our walls
Rather than they shall pound us up. Our gates,
Which yet seem shut, we have but pinned with rushes.
They'll open of themselves.
          *Alarum far off*
(*To the Romans*)          Hark you, far off

There is Aufidius. List what work he makes    20
Amongst your cloven army.
                    ⌈*Exeunt Volscians from the walls*⌉
MARTIUS                    O, they are at it!
LARTIUS
Their noise be our instruction. Ladders, ho!
          ⌈*They prepare to assault the walls.*⌉
          *Enter the army of the Volsces from the gates*
MARTIUS
They fear us not, but issue forth their city.
Now put your shields before your hearts, and fight
With hearts more proof than shields. Advance, brave
Titus.                                                        25
They do disdain us much beyond our thoughts,
Which makes me sweat with wrath. Come on, my
fellows.
He that retires, I'll take him for a Volsce,
And he shall feel mine edge.
          *Alarum. The Romans are beat back ⌈and exeunt⌉ to
          their trenches, ⌈the Volsces following⌉*

**1.5**    *Enter ⌈Roman Soldiers, in retreat, followed by⌉
          Martius, cursing*
MARTIUS
All the contagion of the south light on you,
You shames of Rome! You herd of—boils and plagues
Plaster you o'er, that you may be abhorred
Farther than seen, and one infect another
Against the wind a mile! You souls of geese    5
That bear the shapes of men, how have you run
From slaves that apes would beat! Pluto and hell:
All hurt behind! Backs red, and faces pale
With flight and agued fear! Mend and charge home,
Or by the fires of heaven I'll leave the foe    10
And make my wars on you. Look to't. Come on.
If you'll stand fast, we'll beat them to their wives,
As they us to our trenches. Follow.
          ⌈*The Romans come forward towards the walls.*⌉
          *Another alarum, and ⌈enter the army of the Volsces.⌉
          Martius beats them back ⌈through⌉ the gates*
So, now the gates are ope. Now prove good seconds.
'Tis for the followers fortune widens them,    15
Not for the fliers. Mark me, and do the like.
          *He enters the gates*
FIRST SOLDIER
Foolhardiness! Not I.
SECOND SOLDIER          Nor I.
          *Alarum continues. The gates close, and Martius is
          shut in*
FIRST SOLDIER
See, they have shut him in.
⌈THIRD SOLDIER⌉                    To th' pot, I warrant him.
          *Enter Lartius*
LARTIUS
What is become of Martius?
⌈FOURTH SOLDIER⌉              Slain, sir, doubtless.
FIRST SOLDIER
Following the fliers at the very heels,    20
With them he enters, who upon the sudden
Clapped-to their gates. He is himself alone
To answer all the city.
LARTIUS              O noble fellow,
Who sensibly outdares his senseless sword    24
And, when it bows, stand'st up! Thou art lost, Martius.
A carbuncle entire, as big as thou art,
Were not so rich a jewel. Thou wast a soldier

Even to Cato's wish, not fierce and terrible
Only in strokes, but with thy grim looks and
The thunder-like percussion of thy sounds          30
Thou mad'st thine enemies shake as if the world
Were feverous and did tremble.
    *Enter Martius, bleeding, assaulted by the enemy*
FIRST SOLDIER               Look, sir.
LARTIUS               O, 'tis Martius!
Let's fetch him off, or make remain alike.
    *They fight, and all exeunt into the city*

**1.6**   *Enter certain Romans with spoils*
FIRST ROMAN  This will I carry to Rome.
SECOND ROMAN  And I this.
THIRD ROMAN  A murrain on't, I took this for silver.
    ⌈*He throws it away.*⌉
    *Alarum continues still afar off. Enter Martius,*
    *bleeding, and Lartius with a trumpeter. Exeunt*
    *Romans with spoils*
MARTIUS
See here these movers that do prize their honours
At a cracked drachma! Cushions, leaden spoons,          5
Irons of a doit, doublets that hangmen would
Bury with those that wore them, these base slaves,
Ere yet the fight be done, pack up. Down with them!
And hark what noise the general makes. To him.
There is the man of my soul's hate, Aufidius,          10
Piercing our Romans. Then, valiant Titus, take
Convenient numbers to make good the city,
Whilst I, with those that have the spirit, will haste
To help Cominius.
LARTIUS         Worthy sir, thou bleed'st.
Thy exercise hath been too violent          15
For a second course of fight.
MARTIUS         Sir, praise me not.
My work hath yet not warmed me. Fare you well.
The blood I drop is rather physical
Than dangerous to me. To Aufidius thus
I will appear and fight.
LARTIUS         Now the fair goddess fortune
Fall deep in love with thee, and her great charms          21
Misguide thy opposers' swords! Bold gentleman,
Prosperity be thy page.
MARTIUS         Thy friend no less
Than those she placeth highest. So farewell.
LARTIUS  Thou worthiest Martius!     *Exit Martius*
Go sound thy trumpet in the market-place.          26
Call thither all the officers o'th' town,
Where they shall know our mind. Away.
    *Exeunt* ⌈*severally*⌉

**1.7**   *Enter Cominius, as it were in retire, with soldiers*
COMINIUS
Breathe you, my friends. Well fought. We are come off
Like Romans, neither foolish in our stands
Nor cowardly in retire. Believe me, sirs,
We shall be charged again. Whiles we have struck,
By interims and conveying gusts we have heard          5
The charges of our friends. The Roman gods
Lead their successes as we wish our own,
That both our powers, with smiling fronts
    encount'ring,
May give you thankful sacrifice!
    *Enter a Messenger*
                      Thy news?

MESSENGER
The citizens of Corioles have issued,          10
And given to Lartius and to Martius battle.
I saw our party to their trenches driven,
And then I came away.
COMINIUS        Though thou speak'st truth,
Methinks thou speak'st not well. How long is't since?
MESSENGER  Above an hour, my lord.          15
COMINIUS
'Tis not a mile; briefly we heard their drums.
How couldst thou in a mile confound an hour,
And bring thy news so late?
MESSENGER         Spies of the Volsces
Held me in chase, that I was forced to wheel
Three or four miles about; else had I, sir,          20
Half an hour since brought my report.     ⌈*Exit*⌉
    *Enter Martius, bloody*
COMINIUS              Who's yonder,
That does appear as he were flayed? O gods!
He has the stamp of Martius, and I have
Before-time seen him thus.
MARTIUS         Come I too late?
COMINIUS
The shepherd knows not thunder from a tabor          25
More than I know the sound of Martius' tongue
From every meaner man.
MARTIUS         Come I too late?
COMINIUS
Ay, if you come not in the blood of others,
But mantled in your own.
MARTIUS         O, let me clip ye
In arms as sound as when I wooed, in heart          30
As merry as when our nuptial day was done,
And tapers burnt to bedward!
    ⌈*They embrace*⌉
COMINIUS
Flower of warriors! How is't with Titus Lartius?
MARTIUS
As with a man busied about decrees,
Condemning some to death and some to exile,          35
Ransoming him or pitying, threat'ning th'other;
Holding Corioles in the name of Rome
Even like a fawning greyhound in the leash,
To let him slip at will.
COMINIUS         Where is that slave
Which told me they had beat you to your trenches?
Where is he? Call him hither.
MARTIUS         Let him alone.          41
He did inform the truth. But for our gentlemen,
The common file—a plague—tribunes for them?—
The mouse ne'er shunned the cat as they did budge
From rascals worse than they.
COMINIUS         But how prevailed you?
MARTIUS
Will the time serve to tell? I do not think.          46
Where is the enemy? Are you lords o'th' field?
If not, why cease you till you are so?
COMINIUS
Martius, we have at disadvantage fought,
And did retire to win our purpose.          50
MARTIUS
How lies their battle? Know you on which side
They have placed their men of trust?
COMINIUS         As I guess, Martius,
Their bands i'th' vanguard are the Antiates,
Of their best trust; o'er them Aufidius,

Their very heart of hope.
MARTIUS                        I do beseech you          55
  By all the battles wherein we have fought,
  By th' blood we have shed together, by th' vows we
    have made
  To endure friends, that you directly set me
  Against Aufidius and his Antiates,
  And that you not delay the present, but,          60
  Filling the air with swords advanced and darts,
  We prove this very hour.
COMINIUS                        Though I could wish
  You were conducted to a gentle bath
  And balms applied to you, yet dare I never
  Deny your asking. Take your choice of those          65
  That best can aid your action.
MARTIUS                        Those are they
  That most are willing. If any such be here—
  As it were sin to doubt—that love this painting
  Wherein you see me smeared; if any fear
  Lesser his person than an ill report;          70
  If any think brave death outweighs bad life,
  And that his country's dearer than himself,
  Let him alone, or so many so minded,
    *He waves his sword*
  Wave thus to express his disposition,
  And follow Martius.          75
    *They all shout and wave their swords, ⌈then some⌉*
    *take him up in their arms and they cast up their caps*
  O' me alone, make you a sword of me?
  If these shows be not outward, which of you
  But is four Volsces? None of you but is
  Able to bear against the great Aufidius
  A shield as hard as his. A certain number—          80
  Though thanks to all—must I select from all.
  The rest shall bear the business in some other fight
  As cause will be obeyed. Please you to march,
  And I shall quickly draw out my command,
  Which men are best inclined.
COMINIUS                        March on, my fellows.
  Make good this ostentation, and you shall          86
  Divide in all with us.          *Exeunt marching*

**1.8**   *Enter Lartius ⌈through the gates of Corioles⌉, with*
    *a drummer and a trumpeter, a Lieutenant, other*
    *soldiers, and a scout*
LARTIUS (*to the Lieutenant*)
  So, let the ports be guarded. Keep your duties
  As I have set them down. If I do send, dispatch
  Those centuries to our aid. The rest will serve
  For a short holding. If we lose the field
  We cannot keep the town.          5
LIEUTENANT Fear not our care, sir.
LARTIUS Hence, and shut your gates upon's.
                        ⌈*Exit Lieutenant*⌉
  (*To the scout*) Our guider, come; to th' Roman camp
    conduct us.
          *Exeunt towards Cominius and Caius Martius*

**1.9**   *Alarum, as in battle. Enter Martius, bloody, and*
    *Aufidius, at several doors*
MARTIUS
  I'll fight with none but thee, for I do hate thee
  Worse than a promise-breaker.
AUFIDIUS                        We hate alike.
  Not Afric owns a serpent I abhor
  More than thy fame and envy. Fix thy foot.

MARTIUS
  Let the first budger die the other's slave,          5
  And the gods doom him after.
AUFIDIUS                        If I fly, Martius,
  Holla me like a hare.
MARTIUS                        Within these three hours, Tullus,
  Alone I fought in your Corioles' walls,
  And made what work I pleased. 'Tis not my blood
  Wherein thou seest me masked. For thy revenge,          10
  Wrench up thy power to th' highest.
AUFIDIUS                        Wert thou the Hector
  That was the whip of your bragged progeny,
  Thou shouldst not scape me here.
    *Here they fight, and certain Volsces come in the aid*
    *of Aufidius. Martius fights till the Volsces be driven*
    *in breathless, ⌈Martius following⌉*
  Officious and not valiant, you have shamed me
  In your condemnèd seconds.
                        *Exit*

**1.10**   *Alarum. A retreat is sounded. ⌈Flourish.⌉ Enter at*
    *one door Cominius with the Romans, at another*
    *door Martius with his arm in a scarf*
COMINIUS (*to Martius*)
  If I should tell thee o'er this thy day's work
  Thou'lt not believe thy deeds. But I'll report it
  Where senators shall mingle tears with smiles,
  Where great patricians shall attend and shrug,
  I'th' end admire; where ladies shall be frighted          5
  And, gladly quaked, hear more; where the dull
    tribunes,
  That with the fusty plebeians hate thine honours,
  Shall say against their hearts 'We thank the gods
  Our Rome hath such a soldier.'
  Yet cam'st thou to a morsel of this feast,          10
  Having fully dined before.
    *Enter Lartius, with his power, from the pursuit*
LARTIUS                        O general,
  Here is the steed, we the caparison.
  Hadst thou beheld—
MARTIUS                        Pray now, no more. My mother,
  Who has a charter to extol her blood,
  When she does praise me grieves me. I have done          15
  As you have done, that's what I can; induced
  As you have been, that's for my country.
  He that has but effected his good will
  Hath overta'en mine act.
COMINIUS                        You shall not be
  The grave of your deserving. Rome must know          20
  The value of her own. 'Twere a concealment
  Worse than a theft, no less than a traducement,
  To hide your doings and to silence that
  Which, to the spire and top of praises vouched,
  Would seem but modest. Therefore, I beseech you—
  In sign of what you are, not to reward          26
  What you have done—before our army hear me.
MARTIUS
  I have some wounds upon me, and they smart
  To hear themselves remembered.
COMINIUS                        Should they not,
  Well might they fester 'gainst ingratitude,          30
  And tent themselves with death. Of all the horses—
  Whereof we have ta'en good, and good store—of all
  The treasure in this field achieved and city,
  We render you the tenth, to be ta'en forth

Before the common distribution
At your only choice.                                                 35
MARTIUS                     I thank you, general,
But cannot make my heart consent to take
A bribe to pay my sword. I do refuse it,
And stand upon my common part with those
That have upheld the doing.                                          40
    *A long flourish. They all cry 'Martius, Martius!',*
    *casting up their caps and lances. Cominius and*
    *Lartius stand bare*
May these same instruments which you profane
Never sound more. When drums and trumpets shall
I'th' field prove flatterers, let courts and cities be
Made all of false-faced soothing. When steel grows
Soft as the parasite's silk, let him be made                        45
An overture for th' wars. No more, I say.
For that I have not washed my nose that bled,
Or foiled some debile wretch, which without note
Here's many else have done, you shout me forth
In acclamations hyperbolical,                                       50
As if I loved my little should be dieted
In praises sauced with lies.
COMINIUS                  Too modest are you,
More cruel to your good report than grateful
To us that give you truly. By your patience,
If 'gainst yourself you be incensed we'll put you,                  55
Like one that means his proper harm, in manacles,
Then reason safely with you. Therefore be it known,
As to us, to all the world, that Caius Martius
Wears this war's garland, in token of the which
My noble steed, known to the camp, I give him,                      60
With all his trim belonging; and from this time,
For what he did before Corioles, call him,
With all th'applause and clamour of the host,
Martius Caius Coriolanus. Bear th'addition
Nobly ever!                                                         65
    *Flourish. Trumpets sound, and drums*
ALL Martius Caius Coriolanus!
CORIOLANUS (*to Cominius*) I will go wash,
And when my face is fair you shall perceive
Whether I blush or no. Howbeit, I thank you.
I mean to stride your steed, and at all times                       70
To undercrest your good addition
To th' fairness of my power.
COMINIUS                  So, to our tent,
Where, ere we do repose us, we will write
To Rome of our success. You, Titus Lartius,
Must to Corioles back. Send us to Rome                              75
The best, with whom we may articulate
For their own good and ours.
LARTIUS                  I shall, my lord.
CORIOLANUS The gods begin to mock me. I, that now
Refused most princely gifts, am bound to beg
Of my lord general.
COMINIUS            Take't, 'tis yours. What is't?                  80
CORIOLANUS
I sometime lay here in Corioles,
And at a poor man's house. He used me kindly.
He cried to me; I saw him prisoner;
But then Aufidius was within my view,
And wrath o'erwhelmed my pity. I request you                        85
To give my poor host freedom.
COMINIUS              O, well begged!
Were he the butcher of my son he should
Be free as is the wind. Deliver him, Titus.

LARTIUS
Martius, his name?
CORIOLANUS          By Jupiter, forgot!
I am weary, yea, my memory is tired.                                90
Have we no wine here?
COMINIUS          Go we to our tent.
The blood upon your visage dries; 'tis time
It should be looked to. Come.
                    ⌜*A flourish of cornetts.*⌝ *Exeunt*

**1.11** *Enter Aufidius, bloody, with two or three Soldiers*
AUFIDIUS The town is ta'en.
A SOLDIER
'Twill be delivered back on good condition.
AUFIDIUS Condition?
I would I were a Roman, for I cannot,
Being a Volsce, be that I am. Condition?                            5
What good condition can a treaty find
I'th' part that is at mercy? Five times, Martius,
I have fought with thee; so often hast thou beat me,
And wouldst do so, I think, should we encounter
As often as we eat. By th' elements,                                10
If e'er again I meet him beard to beard,
He's mine, or I am his! Mine emulation
Hath not that honour in't it had, for where
I thought to crush him in an equal force,
True sword to sword, I'll potch at him some way                     15
Or wrath or craft may get him.
A SOLDIER                He's the devil.
AUFIDIUS
Bolder, though not so subtle. My valour, poisoned
With only suff'ring stain by him, for him
Shall fly out of itself. Nor sleep nor sanctuary,
Being naked, sick, nor fane nor Capitol,                            20
The prayers of priests nor times of sacrifice—
Embargements all of fury—shall lift up
Their rotten privilege and custom 'gainst
My hate to Martius. Where I find him, were it
At home upon my brother's guard, even there,                        25
Against the hospitable canon, would I
Wash my fierce hand in's heart. Go you to th' city.
Learn how 'tis held, and what they are that must
Be hostages for Rome.
A SOLDIER            Will not you go?
AUFIDIUS
I am attended at the cypress grove. I pray you—                     30
'Tis south the city mills—bring me word thither
How the world goes, that to the pace of it
I may spur on my journey.
A SOLDIER            I shall, sir.
            *Exeunt* ⌜*Aufidius at one door, Soldiers at*
                                        *another door*⌝

                        ✿

**2.1** *Enter Menenius with the two tribunes of the people,*
        *Sicinius and Brutus*
MENENIUS The augurer tells me we shall have news
tonight.
BRUTUS Good or bad?
MENENIUS Not according to the prayer of the people, for
they love not Martius.                                              5
SICINIUS Nature teaches beasts to know their friends.
MENENIUS Pray you, who does the wolf love?
SICINIUS The lamb.
MENENIUS Ay, to devour him, as the hungry plebeians
would the noble Martius.                                            10

BRUTUS He's a lamb indeed that baas like a bear.

MENENIUS He's a bear indeed that lives like a lamb. You two are old men. Tell me one thing that I shall ask you.

SICINIUS *and* BRUTUS Well, sir?                                    15

MENENIUS In what enormity is Martius poor in that you two have not in abundance?

BRUTUS He's poor in no one fault, but stored with all.

SICINIUS Especially in pride.

BRUTUS And topping all others in boasting.                          20

MENENIUS This is strange now. Do you two know how you are censured here in the city—I mean of us o'th' right-hand file. Do you?

SICINIUS *and* BRUTUS Why, how are we censured?

MENENIUS Because—you talk of pride now—will you not be angry?                                                   26

SICINIUS *and* BRUTUS Well, well, sir, well?

MENENIUS Why, 'tis no great matter, for a very little thief of occasion will rob you of a great deal of patience. Give your dispositions the reins, and be angry at your pleasures—at the least, if you take it as a pleasure to you in being so. You blame Martius for being proud?

BRUTUS We do it not alone, sir.                                     33

MENENIUS I know you can do very little alone, for your helps are many, or else your actions would grow wondrous single. Your abilities are too infant-like for doing much alone. You talk of pride. O that you could turn your eyes toward the napes of your necks, and make but an interior survey of your good selves! O that you could!                                              40

SICINIUS *and* BRUTUS What then, sir?

MENENIUS Why, then you should discover a brace of unmeriting, proud, violent, testy magistrates, alias fools, as any in Rome.

SICINIUS Menenius, you are known well enough too.    45

MENENIUS I am known to be a humorous patrician, and one that loves a cup of hot wine with not a drop of allaying Tiber in't; said to be something imperfect in favouring the first complaint, hasty and tinder-like upon too trivial motion; one that converses more with the buttock of the night than with the forehead of the morning. What I think, I utter, and spend my malice in my breath. Meeting two such wealsmen as you are— I cannot call you Lycurguses—if the drink you give me touch my palate adversely, I make a crooked face at it. I cannot say your worships have delivered the matter well, when I find the ass in compound with the major part of your syllables. And though I must be content to bear with those that say you are reverend grave men, yet they lie deadly that tell you have good faces. If you see this in the map of my microcosm, follows it that I am known well enough too? What harm can your bisson conspectuities glean out of this character, if I be known well enough too?

BRUTUS Come, sir, come, we know you well enough.     65

MENENIUS You know neither me, yourselves, nor anything. You are ambitious for poor knaves' caps and legs. You wear out a good wholesome forenoon in hearing a cause between an orange-wife and a faucet-seller, and then rejourn the controversy of threepence to a second day of audience. When you are hearing a matter between party and party, if you chance to be pinched with the colic, you make faces like mummers, set up the bloody flag against all patience, and in roaring for a chamber-pot, dismiss the controversy bleeding, the more entangled by your hearing. All the peace you make in their cause is calling both the parties knaves. You are a pair of strange ones.                              78

BRUTUS Come, come, you are well understood to be a perfecter giber for the table than a necessary bencher in the Capitol.

MENENIUS Our very priests must become mockers if they shall encounter such ridiculous subjects as you are. When you speak best unto the purpose it is not worth the wagging of your beards, and your beards deserve not so honourable a grave as to stuff a botcher's cushion or to be entombed in an ass's pack-saddle. Yet you must be saying 'Martius is proud', who, in a cheap estimation, is worth all your predecessors since Deucalion, though peradventure some of the best of 'em were hereditary hangmen. Good e'en to your worships. More oˈ your conversation would infect my brain, being the herdsmen of the beastly plebeians. I will be bold to take my leave of you.                            94

*He leaves Brutus and Sicinius, who stand aside.*

*Enter in haste Volumnia, Virgilia, and Valeria*

How now, my as fair as noble ladies—and the moon, were she earthly, no nobler—whither do you follow your eyes so fast?

VOLUMNIA Honourable Menenius, my boy Martius approaches. For the love of Juno, let's go.

MENENIUS Ha, Martius coming home?                                  100

VOLUMNIA Ay, worthy Menenius, and with most prosperous approbation.

MENENIUS ⌈*throwing up his cap*⌉ Take my cap, Jupiter, and I thank thee! Hoo, Martius coming home?

VIRGILIA *and* VALERIA Nay, 'tis true.                             105

VOLUMNIA Look, here's a letter from him. The state hath another, his wife another, and I think there's one at home for you.

MENENIUS I will make my very house reel tonight. A letter for me?                                                       110

VIRGILIA Yes, certain, there's a letter for you; I saw't.

MENENIUS A letter for me? It gives me an estate of seven years' health, in which time I will make a lip at the physician. The most sovereign prescription in Galen is but empiricutic and, to this preservative, of no better report than a horse-drench. Is he not wounded? He was wont to come home wounded.

VIRGILIA O, no, no, no!

VOLUMNIA O, he is wounded, I thank the gods for't!    119

MENENIUS So do I, too, if it be not too much. Brings a victory in his pocket, the wounds become him.

VOLUMNIA On's brows, Menenius. He comes the third time home with the oaken garland.

MENENIUS Has he disciplined Aufidius soundly?         124

VOLUMNIA Titus Lartius writes they fought together, but Aufidius got off.

MENENIUS And 'twas time for him too, I'll warrant him that. An he had stayed by him, I would not have been so fidiussed for all the chests in Corioles and the gold that's in them. Is the senate possessed of this?       130

VOLUMNIA Good ladies, let's go. Yes, yes, yes. The senate has letters from the general, wherein he gives my son the whole name of the war. He hath in this action outdone his former deeds doubly.                          134

VALERIA In truth, there's wondrous things spoke of him.

MENENIUS Wondrous, ay, I warrant you; and not without his true purchasing.

VIRGILIA The gods grant them true.

VOLUMNIA True? Pooh-whoo!                                          139

MENENIUS True? I'll be sworn they are true. Where is he
  wounded? (*To the tribunes*) God save your good
  worships. Martius is coming home. He has more cause
  to be proud. (*To Volumnia*) Where is he wounded?
VOLUMNIA I'th' shoulder and i'th' left arm. There will be
  large cicatrices to show the people when he shall stand
  for his place. He received in the repulse of Tarquin
  seven hurts i'th' body.                                    147
MENENIUS One i'th' neck and two i'th' thigh—there's nine
  that I know.
VOLUMNIA He had before this last expedition twenty-five
  wounds upon him.                                           151
MENENIUS Now it's twenty-seven. Every gash was an
  enemy's grave.
      *A shout and flourish*
  Hark, the trumpets.
VOLUMNIA These are the ushers of Martius. Before him
  he carries noise, and behind him he leaves tears.   156
  Death, that dark spirit, in's nervy arm doth lie,
  Which being advanced, declines; and then men die.
      *Trumpets sound a sennet. Enter ⌈in state⌉ Cominius
      the general and Lartius, between them Coriolanus,
      crowned with an oaken garland, with captains and
      soldiers and a Herald*
HERALD
  Know, Rome, that all alone Martius did fight
  Within Corioles' gates, where he hath won            160
  With fame a name to 'Martius Caius'; these
  In honour follows 'Coriolanus'.
  Welcome to Rome, renownèd Coriolanus!
      *A flourish sounds*
ALL
  Welcome to Rome, renownèd Coriolanus!
CORIOLANUS
  No more of this, it does offend my heart.            165
  Pray now, no more.
COMINIUS                Look, sir, your mother.
CORIOLANUS (*to Volumnia*)                      O,
  You have, I know, petitioned all the gods
  For my prosperity!
      *He kneels*
VOLUMNIA              Nay, my good soldier, up,
  My gentle Martius, worthy Caius,
      ⌈He rises⌉
  And, by deed-achieving honour newly named—      170
  What is it?—'Coriolanus' must I call thee?
  But O, thy wife!
CORIOLANUS (*to Virgilia*) My gracious silence, hail.
  Wouldst thou have laughed had I come coffined
      home,
  That weep'st to see me triumph? Ah, my dear,
  Such eyes the widows in Corioles wear,               175
  And mothers that lack sons.
MENENIUS                    Now the gods crown thee!
⌈CORIOLANUS⌉ (*to Valeria*)
  And live you yet? O my sweet lady, pardon.
VOLUMNIA
  I know not where to turn. O, welcome home!
  And welcome, general, and you're welcome all!
MENENIUS
  A hundred thousand welcomes! I could weep     180
  And I could laugh, I am light and heavy. Welcome!
  A curse begnaw at very root on's heart
  That is not glad to see thee. You are three
  That Rome should dote on. Yet, by the faith of men,

We have some old crab-trees here at home that will not
Be grafted to your relish. Yet welcome, warriors!   186
We call a nettle but a nettle, and
The faults of fools but folly.
COMINIUS Ever right.
CORIOLANUS Menenius, ever, ever.                    190
HERALD
  Give way there, and go on.
CORIOLANUS ⌈to Volumnia and Virgilia⌉
                                Your hand, and yours.
  Ere in our own house I do shade my head
  The good patricians must be visited,
  From whom I have received not only greetings,
  But with them change of honours.
VOLUMNIA                        I have lived    195
  To see inherited my very wishes,
  And the buildings of my fancy. Only
  There's one thing wanting, which I doubt not but
  Our Rome will cast upon thee.
CORIOLANUS                    Know, good mother,
  I had rather be their servant in my way          200
  Than sway with them in theirs.
COMINIUS                On, to the Capitol.
      *A flourish of cornetts. Exeunt in state, as before, all
      but Brutus and Sicinius, who come forward*
BRUTUS
  All tongues speak of him, and the bleared sights
  Are spectacled to see him. Your prattling nurse
  Into a rapture lets her baby cry
  While she chats him; the kitchen malkin pins   205
  Her richest lockram 'bout her reechy neck,
  Clamb'ring the walls to eye him. Stalls, bulks, windows
  Are smothered up, leads filled and ridges horsed
  With variable complexions, all agreeing
  In earnestness to see him. Seld-shown flamens   210
  Do press among the popular throngs, and puff
  To win a vulgar station. Our veiled dames
  Commit the war of white and damask in
  Their nicely guarded cheeks to th' wanton spoil
  Of Phoebus' burning kisses. Such a pother        215
  As if that whatsoever god who leads him
  Were slily crept into his human powers
  And gave him graceful posture.
SICINIUS                      On the sudden
  I warrant him consul.
BRUTUS                Then our office may
  During his power go sleep.                        220
SICINIUS
  He cannot temp'rately transport his honours
  From where he should begin and end, but will
  Lose those he hath won.
BRUTUS                In that there's comfort.
SICINIUS                                Doubt not
  The commoners, for whom we stand, but they
  Upon their ancient malice will forget             225
  With the least cause these his new honours, which
  That he will give them make I as little question
  As he is proud to do't.
BRUTUS                I heard him swear,
  Were he to stand for consul, never would he
  Appear i'th' market-place nor on him put          230
  The napless vesture of humility,
  Nor, showing, as the manner is, his wounds
  To th' people, beg their stinking breaths.
SICINIUS                                'Tis right.

BRUTUS
It was his word. O, he would miss it rather
Than carry it, but by the suit of the gentry to him,
And the desire of the nobles.
SICINIUS                              I wish no better      236
Than have him hold that purpose, and to put it
In execution.
BRUTUS                  'Tis most like he will.
SICINIUS
It shall be to him then, as our good wills,
A sure destruction.
BRUTUS                  So it must fall out      240
To him, or our authority's for an end.
We must suggest the people in what hatred
He still hath held them; that to's power he would
Have made them mules, silenced their pleaders,
And dispropertied their freedoms, holding them      245
In human action and capacity
Of no more soul nor fitness for the world
Than camels in their war, who have their provand
Only for bearing burdens, and sore blows
For sinking under them.
SICINIUS                  This, as you say, suggested
At some time when his soaring insolence      251
Shall touch the people—which time shall not want
If he be put upon't, and that's as easy
As to set dogs on sheep—will be his fire
To kindle their dry stubble, and their blaze      255
Shall darken him for ever.
        *Enter a Messenger*
BRUTUS                  What's the matter?
MESSENGER
You are sent for to the Capitol. 'Tis thought
That Martius shall be consul. I have seen
The dumb men throng to see him, and the blind
To hear him speak. Matrons flung gloves,      260
Ladies and maids their scarves and handkerchiefs,
Upon him as he passed. The nobles bended
As to Jove's statue, and the commons made
A shower and thunder with their caps and shouts.
I never saw the like.
BRUTUS                  Let's to the Capitol,      265
And carry with us ears and eyes for th' time,
But hearts for the event.
SICINIUS                  Have with you.      *Exeunt*

2.2  *Enter two Officers, to lay cushions, as it were in the
    Capitol*
FIRST OFFICER Come, come, they are almost here. How
    many stand for consulships?
SECOND OFFICER Three, they say, but 'tis thought of every-
    one Coriolanus will carry it.
FIRST OFFICER That's a brave fellow, but he's vengeance
    proud and loves not the common people.      6
SECOND OFFICER Faith, there hath been many great men
    that have flattered the people who ne'er loved them;
    and there be many that they have loved they know
    not wherefore, so that if they love they know not why,
    they hate upon no better a ground. Therefore for
    Coriolanus neither to care whether they love or hate
    him manifests the true knowledge he has in their
    disposition, and out of his noble carelessness lets them
    plainly see't.      15
FIRST OFFICER If he did not care whether he had their
    love or no he waved indifferently 'twixt doing them

neither good nor harm; but he seeks their hate with
greater devotion than they can render it him, and
leaves nothing undone that may fully discover him
their opposite. Now to seem to affect the malice and
displeasure of the people is as bad as that which he
dislikes, to flatter them for their love.      23
SECOND OFFICER He hath deserved worthily of his country,
    and his assent is not by such easy degrees as those
    who, having been supple and courteous to the people,
    bonneted, without any further deed to have them at
    all into their estimation and report. But he hath so
    planted his honours in their eyes and his actions in
    their hearts that for their tongues to be silent and not
    confess so much were a kind of ingrateful injury. To
    report otherwise were a malice that, giving itself the
    lie, would pluck reproof and rebuke from every ear that
    heard it.      34
FIRST OFFICER No more of him. He's a worthy man. Make
    way, they are coming.
        *A sennet. Enter the Patricians, and Sicinius and
        Brutus, the tribunes of the people, lictors before
        them; Coriolanus, Menenius, Cominius the consul.
        ⌈The Patricians take their places and sit.⌉ Sicinius
        and Brutus take their places by themselves.
        Coriolanus stands*
MENENIUS
Having determined of the Volsces, and
To send for Titus Lartius, it remains
As the main point of this our after-meeting
To gratify his noble service that      40
Hath thus stood for his country. Therefore please
    you,
Most reverend and grave elders, to desire
The present consul and last general
In our well-found successes to report
A little of that worthy work performed      45
By Martius Caius Coriolanus, whom
We met here both to thank and to remember
With honours like himself.
        *⌈Coriolanus sits⌉*
FIRST SENATOR                  Speak, good Cominius.
Leave nothing out for length, and make us think
Rather our state's defective for requital      50
Than we to stretch it out.
*(To the tribunes)*                  Masters o'th' people,
We do request your kindest ears and, after,
Your loving motion toward the common body
To yield what passes here.
SICINIUS                  We are convented
Upon a pleasing treaty, and have hearts      55
Inclinable to honour and advance
The theme of our assembly.
BRUTUS                  Which the rather
We shall be blessed to do if he remember
A kinder value of the people than
He hath hereto prized them at.
MENENIUS                  That's off, that's off.      60
I would you rather had been silent. Please you
To hear Cominius speak?
BRUTUS                  Most willingly,
But yet my caution was more pertinent
Than the rebuke you give it.
MENENIUS                  He loves your people,
But tie him not to be their bedfellow.      65
Worthy Cominius, speak.

*Coriolanus rises and offers to go away*
(*To Coriolanus*) Nay, keep your place.
⌈FIRST⌉ SENATOR Sit, Coriolanus. Never shame to hear
  What you have nobly done.
CORIOLANUS            Your honours' pardon,
  I had rather have my wounds to heal again
  Than hear say how I got them.
BRUTUS            Sir, I hope     70
  My words disbenched you not?
CORIOLANUS         No, sir, yet oft
  When blows have made me stay I fled from words.
  You soothed not, therefore hurt not; but your people,
  I love them as they weigh—
MENENIUS           Pray now, sit down.
CORIOLANUS
  I had rather have one scratch my head i'th' sun    75
  When the alarum were struck than idly sit
  To hear my nothings monstered.        *Exit*
MENENIUS        Masters of the people,
  Your multiplying spawn how can he flatter—
  That's thousand to one good one—when you now see
  He had rather venture all his limbs for honour    80
  Than one on's ears to hear it? Proceed, Cominius.
COMINIUS
  I shall lack voice; the deeds of Coriolanus
  Should not be uttered feebly. It is held
  That valour is the chiefest virtue, and
  Most dignifies the haver. If it be,        85
  The man I speak of cannot in the world
  Be singly counterpoised. At sixteen years,
  When Tarquin made a head for Rome, he fought
  Beyond the mark of others. Our then dictator,
  Whom with all praise I point at, saw him fight    90
  When with his Amazonian chin he drove
  The bristled lips before him. He bestrid
  An o'erpressed Roman, and, i'th' consul's view,
  Slew three opposers. Tarquin's self he met,
  And struck him on his knee. In that day's feats,    95
  When he might act the woman in the scene,
  He proved best man i'th' field, and for his meed
  Was brow-bound with the oak. His pupil age
  Man-entered thus, he waxèd like a sea,
  And in the brunt of seventeen battles since     100
  He lurched all swords of the garland. For this last
  Before and in Corioles, let me say
  I cannot speak him home. He stopped the fliers,
  And by his rare example made the coward
  Turn terror into sport. As weeds before       105
  A vessel under sail, so men obeyed
  And fell below his stem. His sword, death's stamp,
  Where it did mark, it took. From face to foot
  He was a thing of blood, whose every motion
  Was timed with dying cries. Alone he entered    110
  The mortal gate of th' city, which he, painted
  With shunless destiny, aidless came off,
  And with a sudden reinforcement struck
  Corioles like a planet. Now all's his.
  When by and by the din of war gan pierce     115
  His ready sense, then straight his doubled spirit
  Requickened what in flesh was fatigate,
  And to the battle came he, where he did
  Run reeking o'er the lives of men as if
  'Twere a perpetual spoil; and till we called    120
  Both field and city ours he never stood
  To ease his breast with panting.
MENENIUS            Worthy man.

⌈FIRST⌉ SENATOR
  He cannot but with measure fit the honours
  Which we devise him.
COMINIUS          Our spoils he kicked at,
  And looked upon things precious as they were    125
  The common muck of the world. He covets less
  Than misery itself would give, rewards
  His deeds with doing them, and is content
  To spend the time to end it.
MENENIUS           He's right noble.
  Let him be called for.                130
⌈FIRST⌉ SENATOR Call Coriolanus.
OFFICER He doth appear.
        *Enter Coriolanus*
MENENIUS
  The senate, Coriolanus, are well pleased
  To make thee consul.
CORIOLANUS         I do owe them still
  My life and services.
MENENIUS          It then remains     135
  That you do speak to the people.
CORIOLANUS         I do beseech you,
  Let me o'erleap that custom, for I cannot
  Put on the gown, stand naked, and entreat them
  For my wounds' sake to give their suffrage.
  Please you that I may pass this doing.
SICINIUS          Sir, the people
  Must have their voices, neither will they bate    141
  One jot of ceremony.
MENENIUS (*to Coriolanus*) Put them not to't.
  Pray you, go fit you to the custom and
  Take to you, as your predecessors have,
  Your honour with your form.
CORIOLANUS        It is a part    145
  That I shall blush in acting, and might well
  Be taken from the people.
BRUTUS (*to Sicinius*)      Mark you that?
CORIOLANUS
  To brag unto them 'Thus I did, and thus',
  Show them th'unaching scars, which I should hide,
  As if I had received them for the hire       150
  Of their breath only!
MENENIUS         Do not stand upon't.—
  We recommend to you, tribunes of the people,
  Our purpose to them; and to our noble consul
  Wish we all joy and honour.
SENATORS
  To Coriolanus come all joy and honour!      155
         *A flourish of cornetts, then exeunt all but*
                     *Sicinius and Brutus*
BRUTUS
  You see how he intends to use the people.
SICINIUS
  May they perceive's intent! He will require them
  As if he did contemn what he requested
  Should be in them to give.
BRUTUS        Come, we'll inform them
  Of our proceedings here. On th' market-place    160
  I know they do attend us.           *Exeunt*

**2.3**    *Enter seven or eight Citizens*
FIRST CITIZEN Once, if he do require our voices we ought
  not to deny him.
SECOND CITIZEN We may, sir, if we will.
THIRD CITIZEN We have power in ourselves to do it, but
  it is a power that we have no power to do. For if he

show us his wounds and tell us his deeds, we are to
put our tongues into those wounds and speak for them;
so if he tell us his noble deeds we must also tell him
our noble acceptance of them. Ingratitude is monstrous,
and for the multitude to be ingrateful were to make a
monster of the multitude, of the which we, being
members, should bring ourselves to be monstrous
members.                                                          13

FIRST CITIZEN And to make us no better thought of, a
little help will serve; for once we stood up about the
corn, he himself stuck not to call us the many-headed
multitude.                                                        17

THIRD CITIZEN We have been called so of many, not that
our heads are some brown, some black, some abram,
some bald, but that our wits are so diversely coloured;
and truly I think if all our wits were to issue out of
one skull, they would fly east, west, north, south, and
their consent of one direct way should be at once to
all the points o'th' compass.                                     24

SECOND CITIZEN Think you so? Which way do you judge
my wit would fly?

THIRD CITIZEN Nay, your wit will not so soon out as
another man's will, 'tis strongly wedged up in a
blockhead. But if it were at liberty, 'twould sure
southward.                                                        30

SECOND CITIZEN Why that way?

THIRD CITIZEN To lose itself in a fog where, being three
parts melted away with rotten dews, the fourth would
return for conscience' sake, to help to get thee a wife.

SECOND CITIZEN You are never without your tricks. You
may, you may.                                                     36

THIRD CITIZEN Are you all resolved to give your voices?
But that's no matter, the greater part carries it. I say,
if he would incline to the people there was never a
worthier man.                                                     40

*Enter Coriolanus in a gown of humility, with
Menenius*

Here he comes, and in the gown of humility. Mark his
behaviour. We are not to stay all together, but to come
by him where he stands by ones, by twos, and by
threes. He's to make his requests by particulars, wherein
every one of us has a single honour in giving him our
own voices with our own tongues. Therefore follow
me, and I'll direct you how you shall go by him.        47

ALL THE CITIZENS Content, content.        *Exeunt Citizens*

MENENIUS
O sir, you are not right. Have you not known
The worthiest men have done't?

CORIOLANUS                         What must I say?      50
'I pray, sir'? Plague upon't, I cannot bring
My tongue to such a pace. 'Look, sir, my wounds.
I got them in my country's service, when
Some certain of your brethren roared and ran
From th' noise of our own drums'?

MENENIUS                         O me, the gods!    55
You must not speak of that, you must desire them
To think upon you.

CORIOLANUS               Think upon me? Hang 'em.
I would they would forget me like the virtues
Which our divines lose by 'em.

MENENIUS                    You'll mar all.
I'll leave you. Pray you, speak to 'em, I pray you,     60
In wholesome manner.

CORIOLANUS                 Bid them wash their faces
And keep their teeth clean.            *Exit Menenius*

*Enter three of the Citizens*
                                    So, here comes a brace.
You know the cause, sir, of my standing here.

THIRD CITIZEN
We do, sir. Tell us what hath brought you to't.

CORIOLANUS Mine own desert.                              65

SECOND CITIZEN Your own desert?

CORIOLANUS Ay, but not mine own desire.

THIRD CITIZEN How not your own desire?

CORIOLANUS No, sir, 'twas never my desire yet to trouble
the poor with begging.                                   70

THIRD CITIZEN You must think if we give you anything
we hope to gain by you.

CORIOLANUS Well then, I pray, your price o'th' consulship?

FIRST CITIZEN The price is to ask it kindly.             74

CORIOLANUS Kindly, sir, I pray let me ha't. I have wounds
to show you which shall be yours in private. (*To Second
Citizen*) Your good voice, sir. What say you?

SECOND CITIZEN You shall ha't, worthy sir.

CORIOLANUS A match, sir. There's in all two worthy voices
begged. I have your alms. Adieu.                         80

THIRD CITIZEN (*to the other Citizens*) But this is something
odd.

SECOND CITIZEN An 'twere to give again—but 'tis no
matter.                                     *Exeunt Citizens*

*Enter two other Citizens*

CORIOLANUS Pray you now, if it may stand with the tune
of your voices that I may be consul, I have here the
customary gown.                                          87

⌈FOURTH⌉ CITIZEN You have deserved nobly of your
country, and you have not deserved nobly.

CORIOLANUS Your enigma?                                  90

⌈FOURTH⌉ CITIZEN You have been a scourge to her enemies,
you have been a rod to her friends. You have not,
indeed, loved the common people.

CORIOLANUS You should account me the more virtuous
that I have not been common in my love. I will, sir,
flatter my sworn brother the people to earn a dearer
estimation of them. 'Tis a condition they account gentle.
And since the wisdom of their choice is rather to have
my hat than my heart, I will practise the insinuating
nod and be off to them most counterfeitly; that is, sir,
I will counterfeit the bewitchment of some popular
man, and give it bountiful to the desirers. Therefore,
beseech you I may be consul.

⌈FIFTH⌉ CITIZEN We hope to find you our friend, and
therefore give you our voices heartily.                 105

⌈FOURTH⌉ CITIZEN You have received many wounds for
your country.

CORIOLANUS I will not seal your knowledge with showing
them. I will make much of your voices, and so trouble
you no farther.                                          110

BOTH CITIZENS The gods give you joy, sir, heartily.

CORIOLANUS Most sweet voices.            *Exeunt Citizens*
Better it is to die, better to starve,
Than crave the hire which first we do deserve.
Why in this womanish toge should I stand here     115
To beg of Hob and Dick that does appear
Their needless vouches? Custom calls me to't.
What custom wills, in all things should we do't,
The dust on antique time would lie unswept,
And mountainous error be too highly heaped         120
For truth to o'erpeer. Rather than fool it so,
Let the high office and the honour go
To one that would do thus. I am half through.
The one part suffered, the other will I do.

*Enter three Citizens more*
Here come more voices.                                    125
Your voices! For your voices I have fought,
Watched for your voices, for your voices bear
Of wounds two dozen odd; battles thrice six
I have seen and heard of for your voices, have
Done many things, some less, some more. Your
voices!                                                   130
Indeed I would be consul.
⌈SIXTH⌉ CITIZEN He has done nobly, and cannot go without
any honest man's voice.
⌈SEVENTH⌉ CITIZEN Therefore let him be consul. The gods
give him joy and make him good friend to the people!
ALL THE CITIZENS Amen, amen. God save thee, noble
consul!                                                   137
CORIOLANUS Worthy voices.                *Exeunt Citizens*
*Enter Menenius with Brutus and Sicinius*
MENENIUS
You have stood your limitation, and the tribunes
Endue you with the people's voice. Remains               140
That in th' official marks invested, you
Anon do meet the senate.
CORIOLANUS                    Is this done?
SICINIUS
The custom of request you have discharged.
The people do admit you, and are summoned
To meet anon upon your approbation.                      145
CORIOLANUS
Where, at the senate-house?
SICINIUS                    There, Coriolanus.
CORIOLANUS
May I change these garments?
SICINIUS                    You may, sir.
CORIOLANUS
That I'll straight do, and, knowing myself again,
Repair to th' senate-house.
MENENIUS
I'll keep you company. (*To the tribunes*) Will you
along?                                                    150
BRUTUS
We stay here for the people.
SICINIUS                    Fare you well.
*Exeunt Coriolanus and Menenius*
He has it now, and by his looks methinks
'Tis warm at's heart.
BRUTUS                    With a proud heart he wore
His humble weeds. Will you dismiss the people?
*Enter the Plebeians*
SICINIUS
How now, my masters, have you chose this man? 155
FIRST CITIZEN He has our voices, sir.
BRUTUS
We pray the gods he may deserve your loves.
SECOND CITIZEN
Amen, sir. To my poor unworthy notice
He mocked us when he begged our voices.
THIRD CITIZEN
Certainly. He flouted us downright.                      160
FIRST CITIZEN
No, 'tis his kind of speech. He did not mock us.
SECOND CITIZEN
Not one amongst us save yourself but says
He used us scornfully. He should have showed us
His marks of merit, wounds received for's country.
SICINIUS
Why, so he did, I am sure.
ALL THE CITIZENS              No, no; no man saw 'em.

THIRD CITIZEN
He said he had wounds which he could show in
private,                                                  166
And with his hat, thus waving it in scorn,
'I would be consul,' says he. 'Agèd custom
But by your voices will not so permit me.
Your voices therefore.' When we granted that,            170
Here was 'I thank you for your voices, thank you.
Your most sweet voices. Now you have left your voices
I have no further with you.' Was not this mockery?
SICINIUS
Why either were you ignorant to see't,
Or, seeing it, of such childish friendliness             175
To yield your voices?
BRUTUS (*to the Citizens*) Could you not have told him
As you were lessoned: when he had no power
But was a petty servant to the state,
He was your enemy, ever spake against
Your liberties and the charters that you bear            180
I'th' body of the weal; and now arriving
A place of potency and sway o'th' state,
If he should still malignantly remain
Fast foe to th' plebeii, your voices might
Be curses to yourselves. You should have said            185
That as his worthy deeds did claim no less
Than what he stood for, so his gracious nature
Would think upon you for your voices and
Translate his malice towards you into love,
Standing your friendly lord.
SICINIUS (*to the Citizens*)          Thus to have said   190
As you were fore-advised had touched his spirit
And tried his inclination, from him plucked
Either his gracious promise which you might,
As cause had called you up, have held him to,
Or else it would have galled his surly nature,           195
Which easily endures not article
Tying him to aught. So putting him to rage,
You should have ta'en th'advantage of his choler
And passed him unelected.
BRUTUS (*to the Citizens*)          Did you perceive
He did solicit you in free contempt                      200
When he did need your loves, and do you think
That his contempt shall not be bruising to you
When he hath power to crush? Why, had your bodies
No heart among you? Or had you tongues to cry
Against the rectorship of judgement?
SICINIUS (*to the Citizens*)          Have you          205
Ere now denied the asker, and now again,
Of him that did not ask but mock, bestow
Your sued-for tongues?
THIRD CITIZEN
He's not confirmed, we may deny him yet.
SECOND CITIZEN And will deny him.                         210
I'll have five hundred voices of that sound.
FIRST CITIZEN
I twice five hundred, and their friends to piece 'em.
BRUTUS
Get you hence instantly, and tell those friends
They have chose a consul that will from them take
Their liberties, make them of no more voice              215
Than dogs that are as often beat for barking,
As therefor kept to do so.
SICINIUS (*to the Citizens*)          Let them assemble,
And on a safer judgement all revoke
Your ignorant election. Enforce his pride
And his old hate unto you. Besides, forget not           220

With what contempt he wore the humble weed,
How in his suit he scorned you; but your loves,
Thinking upon his services, took from you
Th'apprehension of his present portance,
Which most gibingly, ungravely he did fashion          225
After the inveterate hate he bears you.
BRUTUS (*to the Citizens*)                    Lay
A fault on us your tribunes, that we laboured
No impediment between, but that you must
Cast your election on him.
SICINIUS (*to the Citizens*)      Say you chose him
More after our commandment than as guided          230
By your own true affections, and that your minds,
Preoccupied with what you rather must do
Than what you should, made you against the grain
To voice him consul. Lay the fault on us.
BRUTUS (*to the Citizens*)
Ay, spare us not. Say we read lectures to you,          235
How youngly he began to serve his country,
How long continued, and what stock he springs of,
The noble house o'th' Martians, from whence came
That Ancus Martius, Numa's daughter's son,
Who after great Hostilius here was king;          240
Of the same house Publius and Quintus were,
That our best water brought by conduits hither;
And Censorinus that was so surnamed,
And nobly named so, twice being censor,
Was his great ancestor.
SICINIUS (*to the Citizens*)  One thus descended,          245
That hath beside well in his person wrought
To be set high in place, we did commend
To your remembrances, but you have found,
Scaling his present bearing with his past,
That he's your fixèd enemy, and revoke          250
Your sudden approbation.
BRUTUS (*to the Citizens*)        Say you ne'er had done't—
Harp on that still—but by our putting on;
And presently when you have drawn your number,
Repair to th' Capitol.
⌈A CITIZEN⌉              We will so.
⌈ANOTHER CITIZEN⌉                    Almost all
Repent in their election.              *Exeunt Citizens*
BRUTUS            Let them go on.          255
This mutiny were better put in hazard
Than stay, past doubt, for greater.
If, as his nature is, he fall in rage
With their refusal, both observe and answer
The vantage of his anger.
SICINIUS            To th' Capitol, come.          260
We will be there before the stream o'th' people,
And this shall seem, as partly 'tis, their own,
Which we have goaded onward.              *Exeunt*

❦

3.1    *Cornetts. Enter Coriolanus, Menenius, all the*
       *gentry; Cominius, Lartius, and other Senators*
CORIOLANUS
Tullus Aufidius then had made new head?
LARTIUS
He had, my lord, and that it was which caused
Our swifter composition.
CORIOLANUS
So then the Volsces stand but as at first,
Ready when time shall prompt them to make raid          5
Upon's again.
COMINIUS        They are worn, lord consul, so
That we shall hardly in our ages see

Their banners wave again.
CORIOLANUS (*to Lartius*)        Saw you Aufidius?
LARTIUS
On safeguard he came to me, and did curse
Against the Volsces for they had so vilely          10
Yielded the town. He is retired to Antium.
CORIOLANUS
Spoke he of me?
LARTIUS            He did, my lord.
CORIOLANUS                    How? What?
LARTIUS
How often he had met you sword to sword;
That of all things upon the earth he hated
Your person most; that he would pawn his fortunes
To hopeless restitution, so he might          16
Be called your vanquisher.
CORIOLANUS At Antium lives he?
LARTIUS At Antium.
CORIOLANUS
I wish I had a cause to seek him there,          20
To oppose his hatred fully. Welcome home.
         *Enter Sicinius and Brutus*
Behold, these are the tribunes of the people,
The tongues o'th' common mouth. I do despise them,
For they do prank them in authority
Against all noble sufferance.          25
SICINIUS Pass no further.
CORIOLANUS Ha, what is that?
BRUTUS
It will be dangerous to go on. No further.
CORIOLANUS What makes this change?
MENENIUS The matter?          30
COMINIUS
Hath he not passed the noble and the common?
BRUTUS
Cominius, no.
CORIOLANUS        Have I had children's voices?
⌈FIRST⌉ SENATOR
Tribunes, give way. He shall to th' market-place.
BRUTUS
The people are incensed against him.
SICINIUS                    Stop,
Or all will fall in broil.
CORIOLANUS            Are these your herd?          35
Must these have voices, that can yield them now
And straight disclaim their tongues? What are your
         offices?
You being their mouths, why rule you not their
         teeth?
Have you not set them on?
MENENIUS            Be calm, be calm.
CORIOLANUS
It is a purposed thing, and grows by plot          40
To curb the will of the nobility.
Suffer't, and live with such as cannot rule
Nor ever will be ruled.
BRUTUS            Call't not a plot.
The people cry you mocked them, and of late
When corn was given them gratis, you repined,          45
Scandalled the suppliants for the people, called them
Time-pleasers, flatterers, foes to nobleness.
CORIOLANUS
Why, this was known before.
BRUTUS                Not to them all.
CORIOLANUS
Have you informed them sithence?
BRUTUS                How, I inform them?

⌈CORIOLANUS⌉
You are like to do such business.                                    50
BRUTUS Not unlike
Each way to better yours.
CORIOLANUS
Why then should I be consul? By yon clouds,
Let me deserve so ill as you, and make me
Your fellow tribune.
SICINIUS                    You show too much of that    55
For which the people stir. If you will pass
To where you are bound, you must enquire your way,
Which you are out of, with a gentler spirit,
Or never be so noble as a consul,
Nor yoke with him for tribune.
MENENIUS                          Let's be calm.          60
COMINIUS
The people are abused, set on. This palt'ring
Becomes not Rome, nor has Coriolanus
Deserved this so dishonoured rub, laid falsely
I'th' plain way of his merit.
CORIOLANUS                        Tell me of corn?
This was my speech, and I will speak't again.           65
MENENIUS Not now, not now.
⌈FIRST⌉ SENATOR Not in this heat, sir, now.
CORIOLANUS Now as I live,
I will. My nobler friends, I crave their pardons.
For the mutable rank-scented meinie,                     70
Let them regard me, as I do not flatter,
And therein behold themselves. I say again,
In soothing them we nourish 'gainst our Senate
The cockle of rebellion, insolence, sedition,
Which we ourselves have ploughed for, sowed, and
    scattered                                           75
By mingling them with us, the honoured number
Who lack not virtue, no, nor power, but that
Which they have given to beggars.
MENENIUS                          Well, no more.
⌈FIRST⌉ SENATOR
No more words, we beseech you.
CORIOLANUS                        How, no more?
As for my country I have shed my blood,                  80
Not fearing outward force, so shall my lungs
Coin words till their decay against those measles
Which we disdain should tetter us, yet sought
The very way to catch them.
BRUTUS
You speak o'th' people as if you were a god              85
To punish, not a man of their infirmity.
SICINIUS
'Twere well we let the people know't.
MENENIUS                          What, what, his choler?
CORIOLANUS
Choler? Were I as patient as the midnight sleep,
By Jove, 'twould be my mind.
SICINIUS                          It is a mind
That shall remain a poison where it is,                  90
Not poison any further.
CORIOLANUS                        'Shall remain'?
Hear you this Triton of the minnows? Mark you
His absolute 'shall'?
COMINIUS                        'Twas from the canon.
CORIOLANUS                                    'Shall'?
O good but most unwise patricians, why,
You grave but reckless senators, have you thus          95
Given Hydra here to choose an officer
That, with his peremptory 'shall', being but

The horn and noise o'th' monster's, wants not spirit
To say he'll turn your current in a ditch
And make your channel his? If he have power,            100
Then vail your impotence; if none, awake
Your dangerous lenity. If you are learned,
Be not as common fools; if you are not,
Let them have cushions by you. You are plebeians
If they be senators, and they are no less               105
When, both your voices blended, the great'st taste
Most palates theirs. They choose their magistrate,
And such a one as he, who puts his 'shall',
His popular 'shall', against a graver bench
Than ever frowned in Greece. By Jove himself,          110
It makes the consuls base, and my soul aches
To know, when two authorities are up,
Neither supreme, how soon confusion
May enter 'twixt the gap of both and take
The one by th' other.
COMINIUS                    Well, on to th' market-place.
CORIOLANUS
Whoever gave that counsel to give forth                 116
The corn o'th' storehouse gratis, as 'twas used
Sometime in Greece—
MENENIUS                    Well, well, no more of that.
CORIOLANUS
Though there the people had more absolute power—
I say they nourished disobedience, fed                  120
The ruin of the state.
BRUTUS                    Why shall the people give
One that speaks thus their voice?
CORIOLANUS                              I'll give my reasons,
More worthier than their voices. They know the corn
Was not our recompense, resting well assured
They ne'er did service for't. Being pressed to th' war,
Even when the navel of the state was touched,           126
They would not thread the gates. This kind of service
Did not deserve corn gratis. Being i'th' war,
Their mutinies and revolts, wherein they showed
Most valour, spoke not for them. Th'accusation          130
Which they have often made against the senate,
All cause unborn, could never be the native
Of our so frank donation. Well, what then?
How shall this bosom multiplied digest
The senate's courtesy? Let deeds express               135
What's like to be their words: 'We did request it,
We are the greater poll, and in true fear
They gave us our demands.' Thus we debase
The nature of our seats, and make the rabble
Call our cares fears, which will in time                140
Break ope the locks o'th' senate and bring in
The crows to peck the eagles.
MENENIUS                          Come, enough.
BRUTUS
Enough with over-measure.
CORIOLANUS                        No, take more.
What may be sworn by, both divine and human,
Seal what I end withal! This double worship,            145
Where one part does disdain with cause, the other
Insult without all reason, where gentry, title, wisdom
Cannot conclude but by the yea and no
Of general ignorance, it must omit
Real necessities, and give way the while                150
To unstable slightness. Purpose so barred, it follows
Nothing is done to purpose. Therefore beseech you—
You that will be less fearful than discreet,
That love the fundamental part of state

More than you doubt the change on't, that prefer 155
A noble life before a long, and wish
To jump a body with a dangerous physic
That's sure of death without it—at once pluck out
The multitudinous tongue; let them not lick
The sweet which is their poison. Your dishonour 160
Mangles true judgement, and bereaves the state
Of that integrity which should become't,
Not having the power to do the good it would
For th'ill which doth control't.

BRUTUS                                    He's said enough.

SICINIUS
He's spoken like a traitor, and shall answer 165
As traitors do.

CORIOLANUS          Thou wretch, despite o'erwhelm thee!
What should the people do with these bald tribunes,
On whom depending, their obedience fails
To th' greater bench? In a rebellion,
When what's not meet but what must be was law,
Then were they chosen. In a better hour 171
Let what is meet be said it must be meet,
And throw their power i'th' dust.

BRUTUS
Manifest treason.

SICINIUS                    This a consul? No.

BRUTUS
The aediles, ho!
          *Enter an Aedile*
                    Let him be apprehended. 175

SICINIUS
Go call the people,                    ⌈*Exit Aedile*⌉
(*To Coriolanus*)    in whose name myself
Attach thee as a traitorous innovator,
A foe to th' public weal. Obey, I charge thee,
And follow to thine answer.

CORIOLANUS                    Hence, old goat!

ALL ⌈THE PATRICIANS⌉
We'll surety him.

COMINIUS (*to Sicinius*) Aged sir, hands off. 180

CORIOLANUS (*to Sicinius*)
Hence, rotten thing, or I shall shake thy bones
Out of thy garments.

SICINIUS              Help, ye citizens!
          *Enter a rabble of Plebeians, with the Aediles*

MENENIUS
On both sides more respect.

SICINIUS                    Here's he
That would take from you all your power.

BRUTUS                          Seize him, aediles.

ALL ⌈THE CITIZENS⌉
Down with him, down with him!

SECOND SENATOR        Weapons, weapons, weapons!
          *They all bustle about Coriolanus*
⌈CITIZENS *and* PATRICIANS⌉ ⌈*in dispersed cries*⌉
Tribunes! Patricians! Citizens! What ho! 186
Sicinius! Brutus! Coriolanus! Citizens!
⌈SOME CITIZENS *and* PATRICIANS⌉
Peace, peace, peace! Stay! Hold! Peace!

MENENIUS
What is about to be? I am out of breath.
Confusion's near; I cannot speak. You tribunes 190
To th' people, Coriolanus, patience!
Speak, good Sicinius.

SICINIUS              Hear me, people, peace.

ALL ⌈THE CITIZENS⌉
Let's hear our tribune! Peace! Speak, speak, speak!

SICINIUS
You are at point to lose your liberties.
Martius would have all from you—Martius 195
Whom late you have named for consul.

MENENIUS                          Fie, fie, fie,
This is the way to kindle, not to quench.

⌈FIRST⌉ SENATOR
To unbuild the city, and to lay all flat.

SICINIUS
What is the city but the people?

ALL ⌈THE CITIZENS⌉                    True,
The people are the city.

BRUTUS                    By the consent of all 200
We were established the people's magistrates.

ALL ⌈THE CITIZENS⌉
You so remain.

MENENIUS        And so are like to do.

⌈CORIOLANUS⌉
That is the way to lay the city flat,
To bring the roof to the foundation,
And bury all which yet distinctly ranges 205
In heaps and piles of ruin.

SICINIUS                    This deserves death.

BRUTUS
Or let us stand to our authority,
Or let us lose it. We do here pronounce,
Upon the part o'th' people in whose power
We were elected theirs, Martius is worthy 210
Of present death.

SICINIUS              Therefore lay hold of him,
Bear him to th' rock Tarpeian; and from thence
Into destruction cast him.

BRUTUS              Aediles, seize him.

ALL THE CITIZENS
Yield, Martius, yield.

MENENIUS          Hear me one word.
Beseech you, tribunes, hear me but a word. 215

AEDILES Peace, peace!

MENENIUS (*to the tribunes*)
Be that you seem, truly your country's friend,
And temp'rately proceed to what you would
Thus violently redress.

BRUTUS              Sir, those cold ways
That seem like prudent helps are very poisons 220
Where the disease is violent. Lay hands upon him,
And bear him to the rock.
          *Coriolanus draws his sword*

CORIOLANUS          No, I'll die here.
There's some among you have beheld me fighting.
Come, try upon yourselves what you have seen me.

MENENIUS
Down with that sword. Tribunes, withdraw a while.

BRUTUS
Lay hands upon him.

MENENIUS              Help Martius, help! 226
You that be noble, help him, young and old.

ALL ⌈THE CITIZENS⌉ Down with him, down with him!
          *In this mutiny the tribunes, the Aediles, and the
          people are beat in*

MENENIUS (*to Coriolanus*)
Go get you to your house. Be gone, away!
All will be naught else.

SECOND SENATOR (*to Coriolanus*) Get you gone. 230

⌈CORIOLANUS⌉
Stand fast; we have as many friends as enemies.

MENENIUS
 Shall it be put to that?
⌈FIRST⌉ SENATOR          The gods forbid!
 (*To Coriolanus*) I prithee, noble friend, home to thy
     house.
 Leave us to cure this cause.
MENENIUS                    For 'tis a sore upon us
 You cannot tent yourself. Be gone, beseech you.   235
⌈COMINIUS⌉ Come, sir, along with us.
⌈CORIOLANUS⌉
 I would they were barbarians, as they are,
 Though in Rome littered; not Romans, as they are
     not,
 Though calved i'th' porch o'th' Capitol.
⌈MENENIUS⌉                    Be gone.
 Put not your worthy rage into your tongue.   240
 One time will owe another.
CORIOLANUS               On fair ground
 I could beat forty of them.
MENENIUS               I could myself
 Take up a brace o'th' best of them, yea, the two
     tribunes.
COMINIUS
 But now 'tis odds beyond arithmetic,
 And manhood is called foolery when it stands   245
 Against a falling fabric.
 (*To Coriolanus*)          Will you hence
 Before the tag return, whose rage doth rend
 Like interrupted waters, and o'erbear
 What they are used to bear?
MENENIUS (*to Coriolanus*)          Pray you be gone.
 I'll try whether my old wit be in request   250
 With those that have but little. This must be patched
 With cloth of any colour.
COMINIUS Nay, come away.
                    *Exeunt Coriolanus and Cominius*
A PATRICIAN This man has marred his fortune.
MENENIUS
 His nature is too noble for the world.   255
 He would not flatter Neptune for his trident
 Or Jove for's power to thunder. His heart's his mouth.
 What his breast forges, that his tongue must vent,
 And, being angry, does forget that ever
 He heard the name of death.
     *A noise within*
                    Here's goodly work.   260
A PATRICIAN
 I would they were abed.
MENENIUS               I would they were in Tiber.
 What the vengeance, could he not speak 'em fair?
     *Enter Brutus and Sicinius, with the rabble again*
SICINIUS Where is this viper
 That would depopulate the city and
 Be every man himself?
MENENIUS               You worthy tribunes—   265
SICINIUS
 He shall be thrown down the Tarpeian rock
 With rigorous hands. He hath resisted law,
 And therefore law shall scorn him further trial
 Than the severity of the public power,
 Which he so sets at naught.
FIRST CITIZEN               He shall well know   270
 The noble tribunes are the people's mouths,
 And we their hands.
ALL ⌈THE CITIZENS⌉     He shall, sure on't.
MENENIUS                    Sir, sir.

SICINIUS Peace!
MENENIUS
 Do not cry havoc where you should but hunt
 With modest warrant.
SICINIUS          Sir, how comes't that you   275
 Have holp to make this rescue?
MENENIUS               Hear me speak.
 As I do know the consul's worthiness,
 So can I name his faults.
SICINIUS Consul? What consul?
MENENIUS The consul Coriolanus.   280
BRUTUS He consul?
ALL ⌈THE CITIZENS⌉ No, no, no, no, no!
MENENIUS
 If, by the tribunes' leave and yours, good people,
 I may be heard, I would crave a word or two,
 The which shall turn you to no further harm   285
 Than so much loss of time.
SICINIUS               Speak briefly, then,
 For we are peremptory to dispatch
 This viperous traitor. To eject him hence
 Were but our danger, and to keep him here
 Our certain death. Therefore it is decreed   290
 He dies tonight.
MENENIUS          Now the good gods forbid
 That our renownèd Rome, whose gratitude
 Towards her deservèd children is enrolled
 In Jove's own book, like an unnatural dam
 Should now eat up her own!   295
SICINIUS
 He's a disease that must be cut away.
MENENIUS
 O, he's a limb that has but a disease—
 Mortal to cut it off, to cure it easy.
 What has he done to Rome that's worthy death?
 Killing our enemies, the blood he hath lost—   300
 Which I dare vouch is more than that he hath
 By many an ounce—he dropped it for his country;
 And what is left, to lose it by his country
 Were to us all that do't and suffer it
 A brand to th' end o'th' world.
SICINIUS               This is clean cam.   305
BRUTUS
 Merely awry. When he did love his country
 It honoured him.
⌈SICINIUS⌉          The service of the foot,
 Being once gangrened, is not then respected
 For what before it was.
BRUTUS               We'll hear no more.
 Pursue him to his house and pluck him thence,   310
 Lest his infection, being of catching nature,
 Spread further.
MENENIUS          One word more, one word!
 This tiger-footed rage, when it shall find
 The harm of unscanned swiftness, will too late
 Tie leaden pounds to's heels. Proceed by process,   315
 Lest parties—as he is beloved—break out
 And sack great Rome with Romans.
BRUTUS If it were so?
SICINIUS (*to Menenius*) What do ye talk?
 Have we not had a taste of his obedience:   320
 Our aediles smote, ourselves resisted? Come.
MENENIUS
 Consider this: he has been bred i'th' wars
 Since a could draw a sword, and is ill-schooled
 In bolted language. Meal and bran together

He throws without distinction. Give me leave,        325
I'll go to him and undertake to bring him
Where he shall answer by a lawful form,
In peace, to his utmost peril.
FIRST SENATOR                    Noble tribunes,
It is the humane way. The other course
Will prove too bloody, and the end of it        330
Unknown to the beginning.
SICINIUS                    Noble Menenius,
Be you then as the people's officer.
(*To the Citizens*) Masters, lay down your weapons.
BRUTUS                                    Go not home.
SICINIUS
Meet on the market-place. (*To Menenius*) We'll attend
    you there,
Where if you bring not Martius, we'll proceed        335
In our first way.
MENENIUS            I'll bring him to you.
(*To the Senators*) Let me desire your company. He must
    come,
Or what is worst will follow.
⌈FIRST⌉ SENATOR            Pray you, let's to him.
        *Exeunt* ⌈*tribunes and Citizens at one door,*
                    *Patricians at another door*⌉

3.2    *Enter Coriolanus, with Nobles*
CORIOLANUS
Let them pull all about mine ears, present me
Death on the wheel or at wild horses' heels,
Or pile ten hills on the Tarpeian rock,
That the precipitation might down stretch
Below the beam of sight, yet will I still        5
Be thus to them.
        *Enter Volumnia*
A PATRICIAN            You do the nobler.
CORIOLANUS                    I muse my mother
Does not approve me further, who was wont
To call them woollen vassals, things created
To buy and sell with groats, to show bare heads
In congregations, to yawn, be still, and wonder,        10
When one but of my ordinance stood up
To speak of peace or war. (*To Volumnia*) I talk of you.
Why did you wish me milder? Would you have me
False to my nature? Rather say I play
The man I am.
VOLUMNIA            O, sir, sir, sir,        15
I would have had you put your power well on
Before you had worn it out.
CORIOLANUS                    Let go.
VOLUMNIA
You might have been enough the man you are
With striving less to be so. Lesser had been
The taxings of your dispositions if        20
You had not showed them how ye were disposed
Ere they lacked power to cross you.
CORIOLANUS                    Let them hang.
VOLUMNIA  Ay, and burn too.
        *Enter Menenius, with the Senators*
MENENIUS (*to Coriolanus*)
Come, come, you have been too rough, something too
    rough.
You must return and mend it.
⌈FIRST⌉ SENATOR            There's no remedy        25
Unless, by not so doing, our good city
Cleave in the midst and perish.
VOLUMNIA (*to Coriolanus*)            Pray be counselled.
I have a heart as little apt as yours,

But yet a brain that leads my use of anger
To better vantage.
MENENIUS            Well said, noble woman.        30
Before he should thus stoop to th' herd, but that
The violent fit o'th' time craves it as physic
For the whole state, I would put mine armour on,
Which I can scarcely bear.
CORIOLANUS  What must I do?        35
MENENIUS  Return to th' tribunes.
CORIOLANUS  Well, what then, what then?
MENENIUS  Repent what you have spoke.
CORIOLANUS
For them? I cannot do it to the gods.
Must I then do't to them?
VOLUMNIA            You are too absolute,        40
Though therein you can never be too noble,
But when extremities speak. I have heard you say,
Honour and policy, like unsevered friends,
I'th' war do grow together. Grant that, and tell me
In peace what each of them by th' other lose        45
That they combine not there.
CORIOLANUS            Tush, tush!
MENENIUS                    A good demand.
VOLUMNIA
If it be honour in your wars to seem
The same you are not, which for your best ends
You adopt your policy, how is it less or worse
That it shall hold companionship in peace        50
With honour, as in war, since that to both
It stands in like request?
CORIOLANUS            Why force you this?
VOLUMNIA
Because that now it lies you on to speak to th' people,
Not by your own instruction, nor by th' matter
Which your heart prompts you, but with such words
That are but roted in your tongue, though but        56
Bastards and syllables of no allowance
To your bosom's truth. Now this no more
Dishonours you at all than to take in
A town with gentle words, which else would put you
To your fortune and the hazard of much blood.        61
I would dissemble with my nature where
My fortunes and my friends at stake required
I should do so in honour. I am in this
Your wife, your son, these senators, the nobles;        65
And you will rather show our general louts
How you can frown than spend a fawn upon 'em
For the inheritance of their loves and safeguard
Of what that want might ruin.
MENENIUS                    Noble lady!
(*To Coriolanus*) Come, go with us, speak fair. You may
    salve so,        70
Not what is dangerous present, but the loss
Of what is past.
VOLUMNIA            I prithee now, my son,
    ⌈*She takes his bonnet*⌉
Go to them with this bonnet in thy hand,
And thus far having stretched it—here be with
    them—
Thy knee bussing the stones—for in such business        75
Action is eloquence, and the eyes of th' ignorant
More learnèd than the ears—waving thy head,
With often, thus, correcting thy stout heart,
Now humble as the ripest mulberry
That will not hold the handling; or say to them        80
Thou art their soldier and, being bred in broils,

Hast not the soft way which, thou dost confess,
Were fit for thee to use as they to claim,
In asking their good loves; but thou wilt frame
Thyself, forsooth, hereafter theirs so far          85
As thou hast power and person.
MENENIUS (*to Coriolanus*)            This but done
Even as she speaks, why, their hearts were yours;
For they have pardons, being asked, as free
As words to little purpose.
VOLUMNIA (*to Coriolanus*)            Prithee now,
Go, and be ruled, although I know thou hadst rather
Follow thine enemy in a fiery gulf                  91
Than flatter him in a bower.
        *Enter Cominius*
                            Here is Cominius.
COMINIUS
I have been i'th' market-place; and, sir, 'tis fit
You make strong party, or defend yourself
By calmness or by absence. All's in anger.          95
MENENIUS
Only fair speech.
COMINIUS            I think 'twill serve, if he
Can thereto frame his spirit.
VOLUMNIA                        He must, and will.
Prithee now, say you will, and go about it.
CORIOLANUS
Must I go show them my unbarbèd sconce?
Must I with my base tongue give to my noble heart
A lie that it must bear? Well, I will do't.         101
Yet were there but this single plot to lose,
This mould of Martius they to dust should grind it
And throw't against the wind. To th' market-place.
You have put me now to such a part which never     105
I shall discharge to th' life.
COMINIUS                    Come, come, we'll prompt you.
VOLUMNIA
I prithee now, sweet son, as thou hast said
My praises made thee first a soldier, so,
To have my praise for this, perform a part
Thou hast not done before.
CORIOLANUS                      Well, I must do't.   110
Away, my disposition; and possess me
Some harlot's spirit! My throat of war be turned,
Which choired with my drum, into a pipe
Small as an eunuch or the virgin voice
That babies lull asleep! The smiles of knaves       115
Tent in my cheeks, and schoolboys' tears take up
The glasses of my sight! A beggar's tongue
Make motion through my lips, and my armed knees,
Who bowed but in my stirrup, bend like his
That hath received an alms! I will not do't,        120
Lest I surcease to honour mine own truth,
And by my body's action teach my mind
A most inherent baseness.
VOLUMNIA                    At thy choice, then.
To beg of thee it is my more dishonour
Than thou of them. Come all to ruin. Let            125
Thy mother rather feel thy pride than fear
Thy dangerous stoutness, for I mock at death
With as big heart as thou. Do as thou list.
Thy valiantness was mine, thou suck'st it from me,
But owe thy pride thyself.
CORIOLANUS                      Pray be content.     130
Mother, I am going to the market-place.
Chide me no more. I'll mountebank their loves,
Cog their hearts from them, and come home beloved

Of all the trades in Rome. Look, I am going.
Commend me to my wife. I'll return consul,          135
Or never trust to what my tongue can do
I'th' way of flattery further.
VOLUMNIA                        Do your will.
                                    *Exit Volumnia*
COMINIUS
Away! The tribunes do attend you. Arm yourself
To answer mildly, for they are prepared
With accusations, as I hear, more strong            140
Than are upon you yet.
CORIOLANUS
The word is 'mildly'. Pray you let us go.
Let them accuse me by invention, I
Will answer in mine honour.
MENENIUS Ay, but mildly.                             145
CORIOLANUS Well, mildly be it, then—mildly.   *Exeunt*

**3.3**    *Enter Sicinius and Brutus*
BRUTUS
In this point charge him home: that he affects
Tyrannical power. If he evade us there,
Enforce him with his envy to the people,
And that the spoil got on the Antiats
Was ne'er distributed.
        *Enter an Aedile*
                        What, will he come?          5
AEDILE
He's coming.
BRUTUS            How accompanied?
AEDILE
With old Menenius, and those senators
That always favoured him.
SICINIUS                    Have you a catalogue
Of all the voices that we have procured,
Set down by th' poll?
AEDILE                    I have, 'tis ready.         10
SICINIUS
Have you collected them by tribes?
AEDILE                                I have.
SICINIUS
Assemble presently the people hither,
And when they hear me say 'It shall be so
I'th' right and strength o'th' commons', be it either
For death, for fine, or banishment, then let them,  15
If I say 'Fine', cry 'Fine!', if 'Death', cry 'Death!',
Insisting on the old prerogative
And power i'th' truth o'th' cause.
AEDILE                                I shall inform them.
BRUTUS
And when such time they have begun to cry,
Let them not cease, but with a din confused         20
Enforce the present execution
Of what we chance to sentence.
AEDILE                            Very well.
SICINIUS
Make them be strong, and ready for this hint
When we shall hap to give't them.
BRUTUS ⌜*to the Aedile*⌝            Go about it.
                                ⌜*Exit Aedile*⌝
Put him to choler straight. He hath been used       25
Ever to conquer and to have his worth
Of contradiction. Being once chafed, he cannot
Be reined again to temperance. Then he speaks
What's in his heart, and that is there which looks
With us to break his neck.                          30

*Enter Coriolanus, Menenius, and Cominius, with*
    *other ⌈Senators and Patricians⌉*
SICINIUS Well, here he comes.
MENENIUS (*to Coriolanus*) Calmly, I do beseech you.
CORIOLANUS
    Ay, as an hostler that for th' poorest piece
    Will bear the knave by th' volume.—Th'honoured
        gods
    Keep Rome in safety and the chairs of justice      35
    Supplied with worthy men, plant love among's,
    Throng our large temples with the shows of peace,
    And not our streets with war!
FIRST SENATOR Amen, amen.
MENENIUS A noble wish.                                 40
    *Enter the Aedile with the Citizens*
SICINIUS
    Draw near, ye people.
AEDILE                    List to your tribunes. Audience!
    Peace, I say.
CORIOLANUS     First, hear me speak.
SICINIUS *and* BRUTUS            Well, say.—Peace ho!
CORIOLANUS
    Shall I be charged no further than this present?
    Must all determine here?
SICINIUS                    I do demand
    If you submit you to the people's voices,          45
    Allow their officers, and are content
    To suffer lawful censure for such faults
    As shall be proved upon you.
CORIOLANUS                    I am content.
MENENIUS
    Lo, citizens, he says he is content.
    The warlike service he has done, consider. Think   50
    Upon the wounds his body bears, which show
    Like graves i'th' holy churchyard.
CORIOLANUS                    Scratches with briers,
    Scars to move laughter only.
MENENIUS                    Consider further
    That when he speaks not like a citizen,
    You find him like a soldier. Do not take           55
    His rougher accents for malicious sounds,
    But, as I say, such as become a soldier
    Rather than envy you.
COMINIUS Well, well, no more.
CORIOLANUS What is the matter                          60
    That, being passed for consul with full voice,
    I am so dishonoured that the very hour
    You take it off again?
SICINIUS Answer to us.
CORIOLANUS Say, then. 'Tis true I ought so.            65
SICINIUS
    We charge you that you have contrived to take
    From Rome all seasoned office, and to wind
    Yourself into a power tyrannical,
    For which you are a traitor to the people.
CORIOLANUS
    How, traitor?
MENENIUS        Nay, temperately—your promise.         70
CORIOLANUS
    The fires i'th' lowest hell fold in the people!
    Call me their traitor, thou injurious tribune?
    Within thine eyes sat twenty thousand deaths,
    In thy hands clutched as many millions, in
    Thy lying tongue both numbers, I would say         75
    'Thou liest' unto thee with a voice as free
    As I do pray the gods.

SICINIUS Mark you this, people?
ALL ⌈THE CITIZENS⌉ To th' rock, to th' rock with him!
SICINIUS Peace!                                        80
    We need not put new matter to his charge.
    What you have seen him do and heard him speak,
    Beating your officers, cursing yourselves,
    Opposing laws with strokes, and here defying
    Those whose great power must try him—              85
    Even this, so criminal and in such capital kind,
    Deserves th'extremest death.
BRUTUS                    But since he hath
    Served well for Rome—
CORIOLANUS            What do you prate of service?
BRUTUS
    I talk of that that know it.
CORIOLANUS            You?
MENENIUS
    Is this the promise that you made your mother?     90
COMINIUS
    Know, I pray you—
CORIOLANUS            I'll know no further.
    Let them pronounce the steep Tarpeian death,
    Vagabond exile, flaying, pent to linger
    But with a grain a day, I would not buy
    Their mercy at the price of one fair word,         95
    Nor check my courage for what they can give
    To have't with saying 'Good morrow'.
SICINIUS                    For that he has,
    As much as in him lies, from time to time
    Inveighed against the people, seeking means
    To pluck away their power, as now at last          100
    Given hostile strokes, and that not in the presence
    Of dreaded justice, but on the ministers
    That doth distribute it, in the name o'th' people,
    And in the power of us the tribunes, we
    E'en from this instant banish him our city         105
    In peril of precipitation
    From off the rock Tarpeian, never more
    To enter our Rome gates. I'th' people's name
    I say it shall be so.
ALL ⌈THE CITIZENS⌉    It shall be so,
    It shall be so. Let him away. He's banished,       110
    And it shall be so.
COMINIUS
    Hear me, my masters and my common friends.
SICINIUS
    He's sentenced. No more hearing.
COMINIUS                    Let me speak.
    I have been consul, and can show for Rome
    Her enemies' marks upon me. I do love             115
    My country's good with a respect more tender,
    More holy and profound, than mine own life,
    My dear wife's estimate, her womb's increase,
    And treasure of my loins. Then if I would
    Speak that—
SICINIUS        We know your drift. Speak what?        120
BRUTUS
    There's no more to be said, but he is banished,
    As enemy to the people and his country.
    It shall be so.
ALL ⌈THE CITIZENS⌉ It shall be so, it shall be so.
CORIOLANUS
    You common cry of curs, whose breath I hate
    As reek o'th' rotten fens, whose loves I prize     125
    As the dead carcasses of unburied men
    That do corrupt my air: I banish you.

And here remain with your uncertainty.
Let every feeble rumour shake your hearts;
Your enemies, with nodding of their plumes,          130
Fan you into despair! Have the power still
To banish your defenders, till at length
Your ignorance—which finds not till it feels—
Making but reservation of yourselves,
Still your own foes, deliver you          135
As most abated captives to some nation
That won you without blows! Despising
For you the city, thus I turn my back.
There is a world elsewhere.

          *Exeunt Coriolanus, Cominius, and Menenius,*
          *with the rest of the Patricians. The Citizens*
          *all shout, and throw up their caps*

AEDILE
The people's enemy is gone, is gone.          140
ALL THE CITIZENS
Our enemy is banished, he is gone. Hoo-oo!
SICINIUS
Go see him out at gates, and follow him
As he hath followed you, with all despite.
Give him deserved vexation. Let a guard
Attend us through the city.          145
ALL THE CITIZENS
Come, come, let's see him out at gates. Come.
The gods preserve our noble tribunes! Come.          *Exeunt*

                    ❧

**4.1**     *Enter Coriolanus, Volumnia, Virgilia, Menenius,*
          *and Cominius, with the young nobility of Rome*
CORIOLANUS
Come, leave your tears. A brief farewell. The beast
With many heads butts me away. Nay, mother,
Where is your ancient courage? You were used
To say extremities was the trier of spirits,
That common chances common men could bear,          5
That when the sea was calm all boats alike
Showed mastership in floating; fortune's blows
When most struck home, being gentle wounded craves
A noble cunning. You were used to load me
With precepts that would make invincible          10
The heart that conned them.
VIRGILIA O heavens, O heavens!
CORIOLANUS Nay, I prithee, woman—
VOLUMNIA
Now the red pestilence strike all trades in Rome,
And occupations perish!
CORIOLANUS          What, what, what?          15
I shall be loved when I am lacked. Nay, mother,
Resume that spirit when you were wont to say,
If you had been the wife of Hercules
Six of his labours you'd have done, and saved
Your husband so much sweat. Cominius,          20
Droop not. Adieu. Farewell, my wife, my mother.
I'll do well yet. Thou old and true Menenius,
Thy tears are salter than a younger man's,
And venomous to thine eyes. My sometime general,
I have seen thee stern, and thou hast oft beheld          25
Heart-hard'ning spectacles. Tell these sad women
'Tis fond to wail inevitable strokes
As 'tis to laugh at 'em. My mother, you wot well
My hazards still have been your solace, and—
Believe't not lightly—though I go alone,          30
Like to a lonely dragon that his fen
Makes feared and talked of more than seen, your son

Will or exceed the common or be caught
With cautelous baits and practice.
VOLUMNIA                    My first son,
Whither will thou go? Take good Cominius          35
With thee a while. Determine on some course
More than a wild exposure to each chance
That starts i'th' way before thee.
⌈VIRGILIA⌉                    O the gods!
COMINIUS
I'll follow thee a month, devise with thee
Where thou shalt rest, that thou mayst hear of us          40
And we of thee. So, if the time thrust forth
A cause for thy repeal, we shall not send
O'er the vast world to seek a single man,
And lose advantage, which doth ever cool
I'th' absence of the needer.
CORIOLANUS                    Fare ye well.          45
Thou hast years upon thee, and thou art too full
Of the wars' surfeits to go rove with one
That's yet unbruised. Bring me but out at gate.
Come, my sweet wife, my dearest mother, and
My friends of noble touch. When I am forth,          50
Bid me farewell, and smile. I pray you come.
While I remain above the ground you shall
Hear from me still, and never of me aught
But what is like me formerly.
MENENIUS                    That's worthily
As any ear can hear. Come, let's not weep.          55
If I could shake off but one seven years
From these old arms and legs, by the good gods,
I'd with thee every foot.
CORIOLANUS                    Give me thy hand. Come.
                              *Exeunt*

**4.2**     *Enter the two tribunes, Sicinius and Brutus, with*
          *the Aedile*
SICINIUS (*to the Aedile*)
Bid them all home. He's gone, and we'll no further.
The nobility are vexed, whom we see have sided
In his behalf.
BRUTUS                    Now we have shown our power,
Let us seem humbler after it is done
Than when it was a-doing.
SICINIUS (*to the Aedile*)          Bid them home.          5
Say their great enemy is gone, and they
Stand in their ancient strength.
BRUTUS                    Dismiss them home.
                              *Exit Aedile*
          *Enter Volumnia, Virgilia, ⌈weeping,⌉ and Menenius*
Here comes his mother.
SICINIUS Let's not meet her.
BRUTUS Why?          10
SICINIUS They say she's mad.
BRUTUS
They have ta'en note of us. Keep on your way.
VOLUMNIA
O, you're well met! Th'hoarded plague o'th' gods
Requite your love!
MENENIUS          Peace, peace, be not so loud.
VOLUMNIA (*to the tribunes*)
If that I could for weeping, you should hear—          15
Nay, and you shall hear some. Will you be gone?
VIRGILIA (*to the tribunes*)
You shall stay, too. I would I had the power
To say so to my husband.
SICINIUS (*to Volumnia*)          Are you mankind?

VOLUMNIA
Ay, fool. Is that a shame? Note but this, fool:
Was not a man my father? Hadst thou foxship     20
To banish him that struck more blows for Rome
Than thou hast spoken words?
SICINIUS                                O blessèd heavens!
VOLUMNIA
More noble blows than ever thou wise words,
And for Rome's good. I'll tell thee what—yet go.
Nay, but thou shalt stay too. I would my son     25
Were in Arabia, and thy tribe before him,
His good sword in his hand.
SICINIUS                             What then?
VIRGILIA                                        What then?
He'd make an end of thy posterity.
VOLUMNIA Bastards and all.
Good man, the wounds that he does bear for Rome!
MENENIUS Come, come, peace.                      31
SICINIUS
I would he had continued to his country
As he began, and not unknit himself
The noble knot he made.
BRUTUS                          I would he had.
VOLUMNIA
'I would he had'! 'Twas you incensed the rabble—  35
Cats that can judge as fitly of his worth
As I can of those mysteries which heaven
Will not have earth to know.
BRUTUS (to Sicinius) Pray, let's go.
VOLUMNIA Now pray, sir, get you gone.             40
You have done a brave deed. Ere you go, hear this:
As far as doth the Capitol exceed
The meanest house in Rome, so far my son—
This lady's husband here, this, do you see?—
Whom you have banished does exceed you all.       45
BRUTUS
Well, well, we'll leave you.
SICINIUS                         Why stay we to be baited
With one that wants her wits?        Exeunt tribunes
VOLUMNIA                          Take my prayers with you.
I would the gods had nothing else to do
But to confirm my curses. Could I meet 'em
But once a day, it would unclog my heart           50
Of what lies heavy to't.
MENENIUS                    You have told them home
And, by my troth, you have cause. You'll sup with me?
VOLUMNIA
Anger's my meat, I sup upon myself,
And so shall starve with feeding.
(To Virgilia)                    Come, let's go.
Leave this faint puling and lament as I do,        55
In anger, Juno-like. Come, come, come.
                        Exeunt Volumnia and Virgilia
MENENIUS                            Fie, fie, fie.
                                            Exit

4.3   Enter Nicanor, a Roman, and Adrian, a Volscian
NICANOR I know you well, sir, and you know me. Your
   name, I think, is Adrian.
ADRIAN It is so, sir. Truly, I have forgot you.
NICANOR I am a Roman, and my services are, as you are,
   against 'em. Know you me yet?                    5
ADRIAN Nicanor, no?
NICANOR The same, sir.
ADRIAN You had more beard when I last saw you, but
   your favour is well approved by your tongue. What's

the news in Rome? I have a note from the Volscian
state to find you out there. You have well saved me a
day's journey.                                      12
NICANOR There hath been in Rome strange insurrections,
   the people against the senators, patricians, and nobles.
ADRIAN Hath been?—is it ended then? Our state thinks
   not so. They are in a most warlike preparation, and
   hope to come upon them in the heat of their division.
NICANOR The main blaze of it is past, but a small thing
   would make it flame again, for the nobles receive so to
   heart the banishment of that worthy Coriolanus that
   they are in a ripe aptness to take all power from the
   people, and to pluck from them their tribunes for ever.
   This lies glowing, I can tell you, and is almost mature
   for the violent breaking out.
ADRIAN Coriolanus banished?                         25
NICANOR Banished, sir.
ADRIAN You will be welcome with this intelligence,
   Nicanor.
NICANOR The day serves well for them now. I have heard
   it said the fittest time to corrupt a man's wife is when
   she's fallen out with her husband. Your noble Tullus
   Aufidius will appear well in these wars, his great
   opposer Coriolanus being now in no request of his
   country.                                          34
ADRIAN He cannot choose. I am most fortunate thus
   accidentally to encounter you. You have ended my
   business, and I will merrily accompany you home.
NICANOR I shall between this and supper tell you most
   strange things from Rome, all tending to the good of
   their adversaries. Have you an army ready, say you?
ADRIAN A most royal one—the centurions and their
   charges distinctly billeted already in th'entertainment,
   and to be on foot at an hour's warning.           43
NICANOR I am joyful to hear of their readiness, and am
   the man, I think, that shall set them in present action.
   So, sir, heartily well met, and most glad of your
   company.
ADRIAN You take my part from me, sir. I have the most
   cause to be glad of yours.                        49
NICANOR Well, let us go together.              Exeunt

4.4   Enter Coriolanus in mean apparel, disguised and
       muffled
CORIOLANUS
A goodly city is this Antium. City,
'Tis I that made thy widows. Many an heir
Of these fair edifices fore my wars
Have I heard groan and drop. Then know me not,
Lest that thy wives with spits and boys with stones  5
In puny battle slay me.
       Enter a Citizen
                         Save you, sir.
CITIZEN
And you.
CORIOLANUS Direct me, if it be your will,
Where great Aufidius lies. Is he in Antium?
CITIZEN
He is, and feasts the nobles of the state
At his house this night.
CORIOLANUS              Which is his house, beseech you?
CITIZEN
This here before you.
CORIOLANUS              Thank you, sir. Farewell.    11
                                      Exit Citizen

O world, thy slippery turns! Friends now fast sworn,
Whose double bosoms seem to wear one heart,
Whose hours, whose bed, whose meal and exercise
Are still together, who twin as 'twere in love     15
Unseparable, shall within this hour,
On a dissension of a doit, break out
To bitterest enmity. So fellest foes,
Whose passions and whose plots have broke their
    sleep
To take the one the other, by some chance,     20
Some trick not worth an egg, shall grow dear friends
And interjoin their issues. So with me.
My birthplace hate I, and my love's upon
This enemy town. I'll enter. If he slay me,
He does fair justice; if he give me way,     25
I'll do his country service.                 *Exit*

**4.5**     *Music plays. Enter a Servingman*
FIRST SERVINGMAN Wine, wine, wine! What service is
here? I think our fellows are asleep.       ⌈*Exit*⌉
    *Enter a Second Servingman*
SECOND SERVINGMAN Where's Cotus? My master calls for
him. Cotus!                             *Exit*
    *Enter Coriolanus, as before*
CORIOLANUS A goodly house. The feast     5
Smells well, but I appear not like a guest.
    *Enter the First Servingman*
FIRST SERVINGMAN What would you have, friend? Whence
are you? Here's no place for you. Pray go to the door.
                                   *Exit*
CORIOLANUS
I have deserved no better entertainment
In being Coriolanus.                 10
    *Enter Second Servingman*
SECOND SERVINGMAN Whence are you, sir? Has the porter
his eyes in his head, that he gives entrance to such
companions? Pray get you out.
CORIOLANUS Away!
SECOND SERVINGMAN Away? Get you away.     15
CORIOLANUS Now thou'rt troublesome.
SECOND SERVINGMAN Are you so brave? I'll have you
talked with anon.
    *Enter Third Servingman. The First meets him*
THIRD SERVINGMAN What fellow's this?     19
FIRST SERVINGMAN A strange one as ever I looked on. I
cannot get him out o'th' house. Prithee, call my master
to him.
THIRD SERVINGMAN (*to Coriolanus*) What have you to do
here, fellow? Pray you, avoid the house.
CORIOLANUS
Let me but stand. I will not hurt your hearth.     25
THIRD SERVINGMAN What are you?
CORIOLANUS A gentleman.
THIRD SERVINGMAN A marvellous poor one.
CORIOLANUS True, so I am.     29
THIRD SERVINGMAN Pray you, poor gentleman, take up
some other station. Here's no place for you. Pray you,
avoid. Come.
CORIOLANUS
Follow your function. Go and batten on cold bits.
    *He pushes him away from him*
THIRD SERVINGMAN What, you will not?—Prithee tell my
master what a strange guest he has here.     35
SECOND SERVINGMAN And I shall.
                    *Exit Second Servingman*
THIRD SERVINGMAN Where dwell'st thou?

CORIOLANUS Under the canopy.
THIRD SERVINGMAN Under the canopy?
CORIOLANUS Ay.     40
THIRD SERVINGMAN Where's that?
CORIOLANUS I'th' city of kites and crows.
THIRD SERVINGMAN I'th' city of kites and crows? What an
ass it is! Then thou dwell'st with daws, too?
CORIOLANUS No, I serve not thy master.     45
THIRD SERVINGMAN How, sir? Do you meddle with my
master?
CORIOLANUS Ay, 'tis an honester service than to meddle
with thy mistress. Thou prat'st and prat'st. Serve with
thy trencher. Hence!     50
    *He beats him away.*
    *Enter Aufidius, with the Second Servingman*
AUFIDIUS Where is this fellow?
SECOND SERVINGMAN Here, sir. I'd have beaten him like a
dog but for disturbing the lords within.
    ⌈*The Servingmen stand aside*⌉
AUFIDIUS
Whence com'st thou? What wouldst thou? Thy name?
Why speak'st not? Speak, man. What's thy name?
CORIOLANUS ⌈*unmuffling his head*⌉          If, Tullus,
Not yet thou know'st me, and seeing me dost not     56
Think me for the man I am, necessity
Commands me name myself.
AUFIDIUS                 What is thy name?
CORIOLANUS
A name unmusical to the Volscians' ears
And harsh in sound to thine.
AUFIDIUS               Say, what's thy name?
Thou hast a grim appearance, and thy face     61
Bears a command in't. Though thy tackle's torn,
Thou show'st a noble vessel. What's thy name?
CORIOLANUS
Prepare thy brow to frown. Know'st thou me yet?
AUFIDIUS I know thee not. Thy name?     65
CORIOLANUS
My name is Caius Martius, who hath done
To thee particularly, and to all the Volsces,
Great hurt and mischief. Thereto witness may
My surname Coriolanus. The painful service,
The extreme dangers, and the drops of blood     70
Shed for my thankless country, are requited
But with that surname—a good memory
And witness of the malice and displeasure
Which thou shouldst bear me. Only that name
    remains.
The cruelty and envy of the people,     75
Permitted by our dastard nobles, who
Have all forsook me, hath devoured the rest,
And suffered me by th' voice of slaves to be
Whooped out of Rome. Now this extremity
Hath brought me to thy hearth. Not out of hope—     80
Mistake me not—to save my life, for if
I had feared death, of all the men i'th' world
I would have 'voided thee, but in mere spite
To be full quit of those my banishers
Stand I before thee here. Then if thou hast     85
A heart of wreak in thee, that wilt revenge
Thine own particular wrongs and stop those maims
Of shame seen through thy country, speed thee
    straight,
And make my misery serve thy turn. So use it
That my revengeful services may prove     90
As benefits to thee; for I will fight

Against my cankered country with the spleen
Of all the under-fiends. But if so be
Thou dar'st not this, and that to prove more fortunes
Thou'rt tired, then, in a word, I also am                    95
Longer to live most weary, and present
My throat to thee and to thy ancient malice,
Which not to cut would show thee but a fool,
Since I have ever followed thee with hate,
Drawn tuns of blood out of thy country's breast,            100
And cannot live but to thy shame unless
It be to do thee service.
AUFIDIUS                    O Martius, Martius!
Each word thou hast spoke hath weeded from my heart
A root of ancient envy. If Jupiter
Should from yon cloud speak divine things                   105
And say ' 'Tis true', I'd not believe them more
Than thee, all-noble Martius. Let me twine
Mine arms about that body whereagainst
My grainèd ash an hundred times hath broke,
And scarred the moon with splinters.
    (He embraces Coriolanus)
                             Here I clip               110
The anvil of my sword, and do contest
As hotly and as nobly with thy love
As ever in ambitious strength I did
Contend against thy valour. Know thou first,
I loved the maid I married; never man                  115
Sighed truer breath. But that I see thee here,
Thou noble thing, more dances my rapt heart
Than when I first my wedded mistress saw
Bestride my threshold. Why, thou Mars, I tell thee
We have a power on foot, and I had purpose             120
Once more to hew thy target from thy brawn,
Or lose mine arm for't. Thou hast beat me out
Twelve several times, and I have nightly since
Dreamt of encounters 'twixt thyself and me—
We have been down together in my sleep,                125
Unbuckling helms, fisting each other's throat—
And waked half dead with nothing. Worthy Martius,
Had we no other quarrel else to Rome but that
Thou art thence banished, we would muster all
From twelve to seventy, and, pouring war               130
Into the bowels of ungrateful Rome,
Like a bold flood o'erbear't. O, come, go in,
And take our friendly senators by th' hands
Who now are here taking their leaves of me,
Who am prepared against your territories,              135
Though not for Rome itself.
CORIOLANUS                    You bless me, gods.
AUFIDIUS
Therefore, most absolute sir, if thou wilt have
The leading of thine own revenges, take
Th'one half of my commission and set down—
As best thou art experienced, since thou know'st        140
Thy country's strength and weakness—thine own ways:
Whether to knock against the gates of Rome,
Or rudely visit them in parts remote
To fright them ere destroy. But come in.
Let me commend thee first to those that shall           145
Say yea to thy desires. A thousand welcomes!
And more a friend than ere an enemy;
Yet, Martius, that was much. Your hand. Most
    welcome!                                    Exeunt
    [The two Servingmen come forward]
FIRST SERVINGMAN Here's a strange alteration!           149

SECOND SERVINGMAN By my hand, I had thought to have
strucken him with a cudgel, and yet my mind gave me
his clothes made a false report of him.
FIRST SERVINGMAN What an arm he has! He turned me
about with his finger and his thumb as one would set
up a top.                                            155
SECOND SERVINGMAN Nay, I knew by his face that there
was something in him. He had, sir, a kind of face,
methought—I cannot tell how to term it.
FIRST SERVINGMAN He had so, looking, as it were—would
I were hanged but I thought there was more in him
than I could think.                                   161
SECOND SERVINGMAN So did I, I'll be sworn. He is simply
the rarest man i'th' world.
FIRST SERVINGMAN I think he is yet a greater soldier than
he you wot on.                                        165
SECOND SERVINGMAN Who, my master?
FIRST SERVINGMAN Nay, it's no matter for that.
SECOND SERVINGMAN Worth six on him.
FIRST SERVINGMAN Nay, not so, neither; but I take him
to be the greater soldier.                            170
SECOND SERVINGMAN Faith, look you, one cannot tell how
to say that. For the defence of a town our general is
excellent.
FIRST SERVINGMAN Ay, and for an assault too.
    Enter the Third Servingman
THIRD SERVINGMAN O, slaves, I can tell you news—news,
you rascals!                                          176
FIRST and SECOND SERVINGMEN What, what, what? Let's
partake.
THIRD SERVINGMAN I would not be a Roman of all nations.
I had as lief be a condemned man.                     180
FIRST and SECOND SERVINGMEN Wherefore? Wherefore?
THIRD SERVINGMAN Why, here's he that was wont to
thwack our general, Caius Martius.
FIRST SERVINGMAN Why do you say 'thwack our general'?
THIRD SERVINGMAN I do not say 'thwack our general'; but
he was always good enough for him.                    186
SECOND SERVINGMAN Come, we are fellows and friends. He
was ever too hard for him. I have heard him say so
himself.
FIRST SERVINGMAN He was too hard for him directly. To
say the truth on't, before Corioles he scotched him and
notched him like a carbonado.                         192
SECOND SERVINGMAN An he had been cannibally given, he
might have broiled and eaten him too.
FIRST SERVINGMAN But more of thy news!                195
THIRD SERVINGMAN Why, he is so made on here within
as if he were son and heir to Mars; set at upper end
o'th' table, no question asked him by any of the senators
but they stand bald before him. Our general himself
makes a mistress of him, sanctifies himself with's hand,
and turns up the white o'th' eye to his discourse. But
the bottom of the news is, our general is cut i'th'
middle, and but one half of what he was yesterday, for
the other has half by the entreaty and grant of the
whole table. He'll go, he says, and sowl the porter of
Rome gates by th' ears. He will mow all down before
him, and leave his passage polled.                    207
SECOND SERVINGMAN And he's as like to do't as any man
I can imagine.
THIRD SERVINGMAN Do't? He will do't; for look you, sir,
he has as many friends as enemies; which friends, sir,
as it were durst not—look you, sir—show themselves,
as we term it, his friends whilst he's in dejectitude.
FIRST SERVINGMAN Dejectitude? What's that?           214

THIRD SERVINGMAN But when they shall see, sir, his crest
up again and the man in blood, they will out of their
burrows like conies after rain, and revel all with him.

FIRST SERVINGMAN But when goes this forward?

THIRD SERVINGMAN Tomorrow, today, presently. You shall
have the drum struck up this afternoon. 'Tis as it were
a parcel of their feast, and to be executed ere they wipe
their lips.     222

SECOND SERVINGMAN Why, then we shall have a stirring
world again. This peace is nothing but to rust iron,
increase tailors, and breed ballad-makers.     225

FIRST SERVINGMAN Let me have war, say I. It exceeds
peace as far as day does night. It's sprightly walking,
audible and full of vent. Peace is a very apoplexy,
lethargy; mulled, deaf, sleepy, insensible; a getter of
more bastard children than war's a destroyer of men.

SECOND SERVINGMAN 'Tis so, and as war in some sort may
be said to be a ravisher, so it cannot be denied but
peace is a great maker of cuckolds.

FIRST SERVINGMAN Ay, and it makes men hate one
another.     235

THIRD SERVINGMAN Reason; because they then less need
one another. The wars for my money. I hope to see
Romans as cheap as Volscians.
⌈*A sound within*⌉
They are rising, they are rising.

FIRST *and* SECOND SERVINGMEN In, in, in, in.     *Exeunt*

**4.6**   *Enter the two tribunes, Sicinius and Brutus*

SICINIUS
We hear not of him, neither need we fear him.
His remedies are tame—the present peace
And quietness of the people, which before
Were in wild hurry. Here do we make his friends
Blush that the world goes well, who rather had,     5
Though they themselves did suffer by't, behold
Dissentious numbers pest'ring streets than see
Our tradesmen singing in their shops and going
About their functions friendly.
    *Enter Menenius*

BRUTUS
We stood to't in good time. Is this Menenius?     10

SICINIUS
'Tis he, 'tis he. O, he is grown most kind of late.
Hail, sir.

MENENIUS Hail to you both.

SICINIUS
Your Coriolanus is not much missed
But with his friends. The commonwealth doth stand,
And so would do were he more angry at it.     16

MENENIUS
All's well, and might have been much better if
He could have temporized.

SICINIUS Where is he, hear you?

MENENIUS Nay, I hear nothing.     20
His mother and his wife hear nothing from him.
    *Enter three or four Citizens*

ALL THE CITIZENS (*to the tribunes*)
The gods preserve you both.

SICINIUS     Good e'en, our neighbours.

BRUTUS
Good e'en to you all, good e'en to you all.

FIRST CITIZEN
Ourselves, our wives and children, on our knees
Are bound to pray for you both.

SICINIUS     Live and thrive.     25

BRUTUS Farewell, kind neighbours.
We wished Coriolanus had loved you as we did.

ALL THE CITIZENS
Now the gods keep you!

SICINIUS *and* BRUTUS     Farewell, farewell.
    *Exeunt Citizens*

SICINIUS
This is a happier and more comely time
Than when these fellows ran about the streets     30
Crying confusion.

BRUTUS     Caius Martius was
A worthy officer i'th' war, but insolent,
O'ercome with pride, ambitious past all thinking,
Self-loving—

SICINIUS     And affecting one sole throne
Without assistance.

MENENIUS     I think not so.     35

SICINIUS
We should by this, to all our lamentation,
If he had gone forth consul found it so.

BRUTUS
The gods have well prevented it, and Rome
Sits safe and still without him.
    *Enter an Aedile*

AEDILE     Worthy tribunes,
There is a slave whom we have put in prison     40
Reports the Volsces, with two several powers,
Are entered in the Roman territories,
And with the deepest malice of the war
Destroy what lies before 'em.

MENENIUS     'Tis Aufidius,
Who, hearing of our Martius' banishment,     45
Thrusts forth his horns again into the world,
Which were inshelled when Martius stood for Rome,
And durst not once peep out.

SICINIUS     Come, what talk you of Martius?

BRUTUS (*to the Aedile*)
Go see this rumourer whipped. It cannot be
The Volsces dare break with us.

MENENIUS     Cannot be?     50
We have record that very well it can,
And three examples of the like hath been
Within my age. But reason with the fellow,
Before you punish him, where he heard this,
Lest you shall chance to whip your information     55
And beat the messenger who bids beware
Of what is to be dreaded.

SICINIUS     Tell not me.
I know this cannot be.

BRUTUS     Not possible.
    *Enter a Messenger*

MESSENGER
The nobles in great earnestness are going
All to the senate-house. Some news is come     60
That turns their countenances.

SICINIUS     'Tis this slave.
(*To the Aedile*) Go whip him fore the people's eyes.—
His raising,
Nothing but his report.     *Exit Aedile*

MESSENGER     Yes, worthy sir,
The slave's report is seconded, and more,
More fearful, is delivered.

SICINIUS     What more fearful?     65

MESSENGER
It is spoke freely out of many mouths—
How probable I do not know—that Martius,

Joined with Aufidius, leads a power 'gainst Rome,
And vows revenge as spacious as between
The young'st and oldest thing.
SICINIUS                              This is most likely!    70
BRUTUS
  Raised only that the weaker sort may wish
  Good Martius home again.
SICINIUS The very trick on't.
MENENIUS This is unlikely.
  He and Aufidius can no more atone                         75
  Than violent'st contrariety.
          *Enter another Messenger*
SECOND MESSENGER
  You are sent for to the senate.
  A fearful army, led by Caius Martius
  Associated with Aufidius, rages
  Upon our territories, and have already                    80
  O'erborne their way, consumed with fire and took
  What lay before them.
          *Enter Cominius*
COMINIUS O, you have made good work!
MENENIUS What news? What news?
COMINIUS
  You have holp to ravish your own daughters and            85
  To melt the city leads upon your pates,
  To see your wives dishonoured to your noses.
MENENIUS What's the news? What's the news?
COMINIUS
  Your temples burnèd in their cement, and
  Your franchises, whereon you stood, confined              90
  Into an auger's bore.
MENENIUS                        Pray now, your news?
  (*To the tribunes*) You have made fair work, I fear me.
  (*To Cominius*)                        Pray, your news.
  If Martius should be joined wi'th' Volscians—
COMINIUS
  If? He is their god. He leads them like a thing
  Made by some other deity than nature,                     95
  That shapes man better, and they follow him
  Against us brats with no less confidence
  Than boys pursuing summer butterflies,
  Or butchers killing flies.
MENENIUS (*to the tribunes*) You have made good work,
  You and your apron-men, you that stood so much   100
  Upon the voice of occupation and
  The breath of garlic-eaters!
COMINIUS (*to the tribunes*)
  He'll shake your Rome about your ears.
MENENIUS
  As Hercules did shake down mellow fruit.
  (*To the tribunes*) You have made fair work.              105
BRUTUS But is this true, sir?
COMINIUS Ay, and you'll look pale
  Before you find it other. All the regions
  Do smilingly revolt, and who resists
  Are mocked for valiant ignorance,                         110
  And perish constant fools. Who is't can blame him?
  Your enemies and his find something in him.
MENENIUS We are all undone unless
  The noble man have mercy.
COMINIUS                        Who shall ask it?
  The tribunes cannot do't, for shame; the people    115
  Deserve such pity of him as the wolf
  Does of the shepherds. For his best friends, if they
  Should say 'Be good to Rome', they charged him even

As those should do that had deserved his hate,
And therein showed like enemies.
MENENIUS                        'Tis true.            120
  If he were putting to my house the brand
  That should consume it, I have not the face
  To say 'Beseech you, cease.'
  (*To the tribunes*)          You have made fair hands,
  You and your crafts! You have crafted fair!
COMINIUS (*to the tribunes*)            You have brought
  A trembling upon Rome such as was never              125
  S'incapable of help.
SICINIUS *and* BRUTUS Say not we brought it.
MENENIUS How? Was't we?
  We loved him, but like beasts and cowardly nobles
  Gave way unto your clusters, who did hoot           130
  Him out o'th' city.
COMINIUS            But I fear
  They'll roar him in again. Tullus Aufidius,
  The second name of men, obeys his points
  As if he were his officer. Desperation
  Is all the policy, strength, and defence            135
  That Rome can make against them.
          *Enter a troop of Citizens*
MENENIUS                        Here come the clusters.
  (*To the Citizens*) And is Aufidius with him? You are they
  That made the air unwholesome when you cast
  Your stinking greasy caps in hooting at
  Coriolanus' exile. Now he's coming,                 140
  And not a hair upon a soldier's head
  Which will not prove a whip. As many coxcombs
  As you threw caps up will he tumble down,
  And pay you for your voices. 'Tis no matter.
  If he could burn us all into one coal,              145
  We have deserved it.
ALL THE CITIZENS Faith, we hear fearful news.
FIRST CITIZEN For mine own part,
  When I said 'banish him' I said 'twas pity.
SECOND CITIZEN And so did I.                           150
THIRD CITIZEN And so did I, and to say the truth so did
  very many of us. That we did, we did for the best, and
  though we willingly consented to his banishment, yet
  it was against our will.
COMINIUS
  You're goodly things, you voices.
MENENIUS                        You have made good work,
  You and your cry. Shall's to the Capitol?           156
COMINIUS O, ay, what else?
          *Exeunt Menenius and Cominius*
SICINIUS
  Go, masters, get you home. Be not dismayed.
  These are a side that would be glad to have
  This true which they so seem to fear. Go home,      160
  And show no sign of fear.
FIRST CITIZEN The gods be good to us! Come, masters,
  let's home. I ever said we were i'th' wrong when we
  banished him.
SECOND CITIZEN So did we all. But come, let's home.  165
          *Exeunt Citizens*
BRUTUS
  I do not like this news.
SICINIUS            Nor I.
BRUTUS
  Let's to the Capitol. Would half my wealth
  Would buy this for a lie.
SICINIUS                  Pray let's go.        *Exeunt*

**4.7** *Enter Aufidius with his Lieutenant*

AUFIDIUS Do they still fly to th' Roman?

LIEUTENANT
I do not know what witchcraft's in him, but
Your soldiers use him as the grace fore meat,
Their talk at table, and their thanks at end,
And you are darkened in this action, sir,                    5
Even by your own.

AUFIDIUS                    I cannot help it now,
Unless by using means I lame the foot
Of our design. He bears himself more proudlier,
Even to my person, than I thought he would
When first I did embrace him. Yet his nature           10
In that's no changeling, and I must excuse
What cannot be amended.

LIEUTENANT                    Yet I wish, sir—
I mean for your particular—you had not
Joined in commission with him, but either
Have borne the action of yourself or else              15
To him had left it solely.

AUFIDIUS
I understand thee well, and be thou sure,
When he shall come to his account, he knows not
What I can urge against him. Although it seems—
And so he thinks, and is no less apparent              20
To th' vulgar eye—that he bears all things fairly
And shows good husbandry for the Volscian state,
Fights dragon-like, and does achieve as soon
As draw his sword, yet he hath left undone
That which shall break his neck or hazard mine         25
Whene'er we come to our account.

LIEUTENANT
Sir, I beseech you, think you he'll carry Rome?

AUFIDIUS
All places yields to him ere he sits down,
And the nobility of Rome are his.
The senators and patricians love him too.              30
The tribunes are no soldiers, and their people
Will be as rash in the repeal as hasty
To expel him thence. I think he'll be to Rome
As is the osprey to the fish, who takes it
By sovereignty of nature. First he was                 35
A noble servant to them, but he could not
Carry his honours even. Whether 'twas pride,
Which out of daily fortune ever taints
The happy man; whether defect of judgement,
To fail in the disposing of those chances              40
Which he was lord of; or whether nature,
Not to be other than one thing, not moving
From th' casque to th' cushion, but commanding peace
Even with the same austerity and garb
As he controlled the war: but one of these—            45
As he hath spices of them all—not all,
For I dare so far free him—made him feared,
So hated, and so banished. But he has a merit
To choke it in the utt'rance. So our virtues
Lie in th'interpretation of the time,                  50
And power, unto itself most commendable,
Hath not a tomb so evident as a chair
T'extol what it hath done.
One fire drives out one fire, one nail one nail;
Rights by rights falter, strengths by strengths do fail.
Come, let's away. When, Caius, Rome is thine,          56
Thou art poor'st of all; then shortly art thou mine.

*Exeunt*

**5.1** *Enter Menenius, Cominius, Sicinius and Brutus, the
two tribunes, with others*

MENENIUS
No, I'll not go. You hear what he hath said
Which was sometime his general, who loved him
In a most dear particular. He called me father,
But what o' that? (*To the tribunes*) Go, you that
     banished him.
A mile before his tent fall down, and knee             5
The way into his mercy. Nay, if he coyed
To hear Cominius speak, I'll keep at home.

COMINIUS
He would not seem to know me.

MENENIUS (*to the tribunes*)            Do you hear?

COMINIUS
Yet one time he did call me by my name.
I urged our old acquaintance and the drops             10
That we have bled together. 'Coriolanus'
He would not answer to, forbade all names.
He was a kind of nothing, titleless,
Till he had forged himself a name o'th' fire
Of burning Rome.

MENENIUS (*to the tribunes*)
                    Why, so! You have made good work.
A pair of tribunes that have wracked fair Rome         16
To make coals cheap—a noble memory!

COMINIUS
I minded him how royal 'twas to pardon
When it was less expected. He replied
It was a bare petition of a state                      20
To one whom they had punished.

MENENIUS                    Very well.
Could he say less?

COMINIUS
I offered to awaken his regard
For's private friends. His answer to me was
He could not stay to pick them in a pile               25
Of noisome, musty chaff. He said 'twas folly,
For one poor grain or two, to leave unburnt
And still to nose th'offence.

MENENIUS                    For one poor grain or two?
I am one of those. His mother, wife, his child,
And this brave fellow too—we are the grains.           30
(*To the tribunes*) You are the musty chaff, and you are
     smelt
Above the moon. We must be burnt for you.

SICINIUS
Nay, pray be patient. If you refuse your aid
In this so never-needed help, yet do not
Upbraid's with our distress. But sure, if you          35
Would be your country's pleader, your good tongue,
More than the instant army we can make,
Might stop our countryman.

MENENIUS                    No, I'll not meddle.

SICINIUS
Pray you go to him.

MENENIUS                    What should I do?

BRUTUS
Only make trial what your love can do                  40
For Rome towards Martius.

MENENIUS
Well, and say that Martius return me,
As Cominius is returned, unheard—what then?
But as a discontented friend, grief-shot               44
With his unkindness? Say't be so?

SICINIUS                    Yet your good will
Must have that thanks from Rome after the measure

As you intended well.
MENENIUS                    I'll undertake't.
I think he'll hear me. Yet to bite his lip
And 'hmh' at good Cominius much unhearts me.
He was not taken well, he had not dined.                    50
The veins unfilled, our blood is cold, and then
We pout upon the morning, are unapt
To give or to forgive; but when we have stuffed
These pipes and these conveyances of our blood
With wine and feeding, we have suppler souls                    55
Than in our priest-like fasts. Therefore I'll watch him
Till he be dieted to my request,
And then I'll set upon him.
BRUTUS
You know the very road into his kindness,
And cannot lose your way.
MENENIUS                    Good faith, I'll prove him.
Speed how it will, I shall ere long have knowledge                    61
Of my success.                    Exit
COMINIUS                    He'll never hear him.
SICINIUS                    Not?
COMINIUS
I tell you, he does sit in gold, his eye
Red as 'twould burn Rome, and his injury
The jailer to his pity. I kneeled before him;                    65
'Twas very faintly he said 'Rise', dismissed me
Thus with his speechless hand. What he would do
He sent in writing after me, what he would not,
Bound with an oath to hold to his conditions.
So that all hope is vain unless his noble mother                    70
And his wife, who as I hear mean to solicit him
For mercy to his country. Therefore let's hence,
And with our fair entreaties haste them on.                    Exeunt

**5.2**    *Enter Menenius to the Watch or guard*
FIRST WATCHMAN Stay. Whence are you?
SECOND WATCHMAN Stand, and go back.
MENENIUS You guard like men; 'tis well.
But, by your leave, I am an officer
Of state, and come to speak with Coriolanus.                    5
FIRST WATCHMAN From whence?
MENENIUS
From Rome.
FIRST WATCHMAN You may not pass, you must return.
Our general will no more hear from thence.
SECOND WATCHMAN
You'll see your Rome embraced with fire before
You'll speak with Coriolanus.
MENENIUS                    Good my friends,                    10
If you have heard your general talk of Rome
And of his friends there, it is lots to blanks
My name hath touched your ears. It is Menenius.
FIRST WATCHMAN
Be it so; go back. The virtue of your name
Is not here passable.
MENENIUS                    I tell thee, fellow,                    15
Thy general is my lover. I have been
The book of his good acts, whence men have read
His fame unparalleled happily amplified;
For I have ever verified my friends,
Of whom he's chief, with all the size that verity                    20
Would without lapsing suffer. Nay, sometimes,
Like to a bowl upon a subtle ground,
I have tumbled past the throw, and in his praise
Have almost stamped the leasing. Therefore, fellow,
I must have leave to pass.                    25

FIRST WATCHMAN Faith, sir, if you had told as many lies
in his behalf as you have uttered words in your own,
you should not pass here, no, though it were as virtuous
to lie as to live chastely. Therefore go back.                    29
MENENIUS Prithee, fellow, remember my name is
Menenius, always factionary on the party of your
general.
SECOND WATCHMAN Howsoever you have been his liar, as
you say you have, I am one that, telling true under
him, must say you cannot pass. Therefore go back.                    35
MENENIUS Has he dined, canst thou tell? For I would not
speak with him till after dinner.
FIRST WATCHMAN You are a Roman, are you?
MENENIUS I am as thy general is.                    39
FIRST WATCHMAN Then you should hate Rome as he does.
Can you, when you have pushed out your gates the
very defender of them, and in a violent popular
ignorance given your enemy your shield, think to front
his revenges with the easy groans of old women, the
virginal palms of your daughters, or with the palsied
intercession of such a decayed dotant as you seem to
be? Can you think to blow out the intended fire your
city is ready to flame in with such weak breath as this?
No, you are deceived, therefore back to Rome, and
prepare for your execution. You are condemned, our
general has sworn you out of reprieve and pardon.                    51
MENENIUS Sirrah, if thy captain knew I were here, he
would use me with estimation.
FIRST WATCHMAN Come, my captain knows you not.
MENENIUS I mean thy general.                    55
FIRST WATCHMAN My general cares not for you. Back, I
say, go, lest I let forth your half pint of blood. Back.
That's the utmost of your having. Back.
MENENIUS Nay, but fellow, fellow—
                    *Enter Coriolanus with Aufidius*
CORIOLANUS What's the matter?                    60
MENENIUS (*to First Watchman*) Now, you companion, I'll
say an errand for you. You shall know now that I am
in estimation. You shall perceive that a jack guardant
cannot office me from my son Coriolanus. Guess but
by my entertainment with him if thou stand'st not i'th'
state of hanging, or of some death more long in
spectatorship and crueller in suffering. Behold now
presently, and swoon for what's to come upon thee.
(*To Coriolanus*) The glorious gods sit in hourly synod
about thy particular prosperity, and love thee no worse
than thy old father Menenius does! (*Weeping*) O, my
son, my son, thou art preparing fire for us. Look thee,
here's water to quench it. I was hardly moved to come
to thee, but being assured none but myself could move
thee, I have been blown out of our gates with sighs,
and conjure thee to pardon Rome and thy petitionary
countrymen. The good gods assuage thy wrath and
turn the dregs of it upon this varlet here, this, who
like a block hath denied my access to thee!
CORIOLANUS Away!                    80
MENENIUS How? Away?
CORIOLANUS
Wife, mother, child, I know not. My affairs
Are servanted to others. Though I owe
My revenge properly, my remission lies
In Volscian breasts. That we have been familiar,                    85
Ingrate forgetfulness shall poison rather
Than pity note how much. Therefore be gone.
Mine ears against your suits are stronger than
Your gates against my force. Yet, for I loved thee,

*He gives him a letter*
Take this along. I writ it for thy sake,　　　　90
And would have sent it. Another word, Menenius,
I will not hear thee speak.—This man, Aufidius,
Was my beloved in Rome; yet thou behold'st.
AUFIDIUS You keep a constant temper.
　　　　　　　　　*Exeunt Coriolanus and Aufidius*
FIRST WATCHMAN Now, sir, is your name Menenius?　95
SECOND WATCHMAN 'Tis a spell, you see, of much power.
　You know the way home again.
FIRST WATCHMAN Do you hear how we are shent for
　keeping your greatness back?
SECOND WATCHMAN What cause do you think I have to
　swoon?　　　　　　　　　　　　　　　101
MENENIUS I neither care for th' world nor your general.
　For such things as you, I can scarce think there's any,
　you're so slight. He that hath a will to die by himself
　fears it not from another. Let your general do his worst.
　For you, be that you are long, and your misery increase
　with your age. I say to you as I was said to, 'Away!'
　　　　　　　　　　　　　　　　　　　*Exit*
FIRST WATCHMAN A noble fellow, I warrant him.　108
SECOND WATCHMAN The worthy fellow is our general. He's
　the rock, the oak, not to be wind-shaken.　*Exeunt*

**5.3**　*Enter Coriolanus and Aufidius, with Volscian*
　　　*soldiers.* ⌈*Coriolanus and Aufidius sit*⌉
CORIOLANUS
　We will before the walls of Rome tomorrow
　Set down our host. My partner in this action,
　You must report to th' Volscian lords how plainly
　I have borne this business.
AUFIDIUS　　　　　　　Only their ends
　You have respected, stopped your ears against　　5
　The general suit of Rome, never admitted
　A private whisper, no, not with such friends
　That thought them sure of you.
CORIOLANUS　　　　　　This last old man,
　Whom with a cracked heart I have sent to Rome,
　Loved me above the measure of a father,　　10
　Nay, godded me indeed. Their latest refuge
　Was to send him, for whose old love I have—
　Though I showed sourly to him—once more offered
　The first conditions, which they did refuse
　And cannot now accept, to grace him only　　15
　That thought he could do more. A very little
　I have yielded to. Fresh embassies and suits,
　Nor from the state nor private friends, hereafter
　Will I lend ear to.
　　　　*Shout within*
　　　　　　Ha, what shout is this?
　Shall I be tempted to infringe my vow　　20
　In the same time 'tis made? I will not.
　　　*Enter Virgilia, Volumnia, Valeria, Young Martius,*
　　　*with attendants*
　My wife comes foremost, then the honoured mould
　Wherein this trunk was framed, and in her hand
　The grandchild to her blood. But out, affection!
　All bond and privilege of nature break;　　25
　Let it be virtuous to be obstinate.
　　　⌈*Virgilia*⌉ *curtsies*
　What is that curtsy worth? Or those dove's eyes
　Which can make gods forsworn? I melt, and am not
　Of stronger earth than others.
　　　*Volumnia bows*
　　　　　　　My mother bows,

As if Olympus to a molehill should　　30
In supplication nod; and my young boy
Hath an aspect of intercession which
Great nature cries 'Deny not'.—Let the Volsces
Plough Rome and harrow Italy! I'll never
Be such a gosling to obey instinct, but stand　35
As if a man were author of himself
And knew no other kin.
VIRGILIA　　　　My lord and husband.
CORIOLANUS
These eyes are not the same I wore in Rome.
VIRGILIA
The sorrow that delivers us thus changed
Makes you think so.
CORIOLANUS　　　Like a dull actor now　　40
I have forgot my part, and I am out
Even to a full disgrace. ⌈*Rising*⌉ Best of my flesh,
Forgive my tyranny, but do not say
For that 'Forgive our Romans'.
　　⌈*Virgilia kisses him*⌉
　　　　　　O, a kiss
Long as my exile, sweet as my revenge!　　45
Now, by the jealous queen of heaven, that kiss
I carried from thee, dear, and my true lip
Hath virgined it e'er since. You gods, I prate,
And the most noble mother of the world
Leave unsaluted! Sink, my knee, i'th' earth.　50
　*He kneels*
Of thy deep duty more impression show
Than that of common sons.
VOLUMNIA　　　　O, stand up blest,
　⌈*Coriolanus rises*⌉
Whilst with no softer cushion than the flint
I kneel before thee, and unproperly
Show duty as mistaken all this while　　55
Between the child and parent.
　*She kneels*
CORIOLANUS　　　　　What's this?
Your knees to me? To your corrected son?
　⌈*He raises her*⌉
Then let the pebbles on the hungry beach
Fillip the stars; then let the mutinous winds
Strike the proud cedars 'gainst the fiery sun,　60
Murd'ring impossibility to make
What cannot be slight work.
VOLUMNIA　　　　Thou art my warrior.
I holp to frame thee. Do you know this lady?
CORIOLANUS
The noble sister of Publicola,
The moon of Rome, chaste as the icicle　　65
That's candied by the frost from purest snow
And hangs on Dian's temple—dear Valeria!
VOLUMNIA (*showing Coriolanus his son*)
This is a poor epitome of yours,
Which by th' interpretation of full time
May show like all yourself.
CORIOLANUS (*to Young Martius*) The god of soldiers,　70
With the consent of supreme Jove, inform
Thy thoughts with nobleness, that thou mayst prove
To shame unvulnerable, and stick i'th' wars
Like a great sea-mark standing every flaw
And saving those that eye thee!　　75
VOLUMNIA (*to Young Martius*) Your knee, sirrah.
　⌈*Young Martius kneels*⌉
CORIOLANUS That's my brave boy.

VOLUMNIA
  Even he, your wife, this lady, and myself
  Are suitors to you.
CORIOLANUS          I beseech you, peace.
  Or if you'd ask, remember this before:                    80
  The things I have forsworn to grant may never
  Be held by you denials. Do not bid me
  Dismiss my soldiers, or capitulate
  Again with Rome's mechanics. Tell me not
  Wherein I seem unnatural. Desire not t'allay             85
  My rages and revenges with your colder reasons.
VOLUMNIA O, no more, no more!
  You have said you will not grant us anything—
  For we have nothing else to ask but that
  Which you deny already. Yet we will ask,                 90
  That, if you fail in our request, the blame
  May hang upon your hardness. Therefore hear us.
CORIOLANUS
  Aufidius and you Volsces, mark, for we'll
  Hear naught from Rome in private.
    ⌜He sits⌝
                                Your request?
VOLUMNIA
  Should we be silent and not speak, our raiment           95
  And state of bodies would bewray what life
  We have led since thy exile. Think with thyself
  How more unfortunate than all living women
  Are we come hither, since that thy sight, which should
  Make our eyes flow with joy, hearts dance with
    comforts,                                              100
  Constrains them weep and shake with fear and
    sorrow,
  Making the mother, wife, and child to see
  The son, the husband, and the father tearing
  His country's bowels out; and to poor we
  Thine enmity's most capital. Thou barr'st us             105
  Our prayers to the gods, which is a comfort
  That all but we enjoy. For how can we,
  Alas, how can we for our country pray,
  Whereto we are bound, together with thy victory,
  Whereto we are bound? Alack, or we must lose             110
  The country, our dear nurse, or else thy person,
  Our comfort in the country. We must find
  An evident calamity, though we had
  Our wish which side should win. For either thou
  Must as a foreign recreant be led                        115
  With manacles thorough our streets, or else
  Triumphantly tread on thy country's ruin,
  And bear the palm for having bravely shed
  Thy wife and children's blood. For myself, son,
  I purpose not to wait on fortune till                    120
  These wars determine. If I cannot persuade thee
  Rather to show a noble grace to both parts
  Than seek the end of one, thou shalt no sooner
  March to assault thy country than to tread—
  Trust to't, thou shalt not—on thy mother's womb  125
  That brought thee to this world.
VIRGILIA                      Ay, and mine,
  That brought you forth this boy to keep your name
  Living to time.
YOUNG MARTIUS  A shall not tread on me.
  I'll run away till I am bigger, but then I'll fight.
CORIOLANUS
  Not of a woman's tenderness to be                        130
  Requires nor child nor woman's face to see.

  I have sat too long.
    ⌜He rises and turns away⌝
VOLUMNIA                Nay, go not from us thus.
  If it were so that our request did tend
  To save the Romans, thereby to destroy
  The Volsces whom you serve, you might condemn us
  As poisonous of your honour. No, our suit               136
  Is that you reconcile them: while the Volsces
  May say 'This mercy we have showed', the Romans
  'This we received', and each in either side
  Give the all-hail to thee and cry 'Be blest             140
  For making up this peace!' Thou know'st, great son,
  The end of war's uncertain; but this certain,
  That if thou conquer Rome, the benefit
  Which thou shalt thereby reap is such a name
  Whose repetition will be dogged with curses,            145
  Whose chronicle thus writ: 'The man was noble,
  But with his last attempt he wiped it out,
  Destroyed his country, and his name remains
  To th' ensuing age abhorred.' Speak to me, son,
  Thou hast affected the fine strains of honour,          150
  To imitate the graces of the gods,
  To tear with thunder the wide cheeks o'th' air,
  And yet to charge thy sulphur with a bolt
  That should but rive an oak. Why dost not speak?
  Think'st thou it honourable for a noble man             155
  Still to remember wrongs? Daughter, speak you,
  He cares not for your weeping. Speak thou, boy.
  Perhaps thy childishness will move him more
  Than can our reasons. There's no man in the world
  More bound to's mother, yet here he lets me prate 160
  Like one i'th' stocks. Thou hast never in thy life
  Showed thy dear mother any courtesy,
  When she, poor hen, fond of no second brood,
  Has clucked thee to the wars and safely home,
  Loaden with honour. Say my request's unjust,           165
  And spurn me back. But if it be not so,
  Thou art not honest, and the gods will plague thee
  That thou restrain'st from me the duty which
  To a mother's part belongs.—He turns away.
  Down, ladies. Let us shame him with our knees.         170
  To his surname 'Coriolanus' 'longs more pride
  Than pity to our prayers. Down! An end.
  This is the last.
    The ladies and Young Martius kneel
                    So we will home to Rome,
  And die among our neighbours.—Nay, behold's.
  This boy, that cannot tell what he would have,         175
  But kneels and holds up hands for fellowship,
  Does reason our petition with more strength
  Than thou hast to deny't.—Come, let us go.
  This fellow had a Volscian to his mother.
  His wife is in Corioles, and this child                180
  Like him by chance.—Yet give us our dispatch.
  I am hushed until our city be afire,
  And then I'll speak a little.
    He holds her by the hand, silent
CORIOLANUS               O mother, mother!
  What have you done? Behold, the heavens do ope,
  The gods look down, and this unnatural scene           185
  They laugh at. O my mother, mother, O!
  You have won a happy victory to Rome;
  But for your son, believe it, O believe it,
  Most dangerously you have with him prevailed,
  If not most mortal to him. But let it come.            190
    ⌜The ladies and Young Martius rise⌝

Aufidius, though I cannot make true wars,
I'll frame convenient peace. Now, good Aufidius,
Were you in my stead would you have heard
A mother less, or granted less, Aufidius?

AUFIDIUS
I was moved withal.

CORIOLANUS                    I dare be sworn you were.          195
And, sir, it is no little thing to make
Mine eyes to sweat compassion. But, good sir,
What peace you'll make, advise me. For my part,
I'll not to Rome; I'll back with you, and pray you
Stand to me in this cause.—O mother! Wife!          200

AUFIDIUS (aside)
I am glad thou hast set thy mercy and thy honour
At difference in thee. Out of that I'll work
Myself a former fortune.

CORIOLANUS (to Volumnia and Virgilia) Ay, by and by.
But we will drink together, and you shall bear
A better witness back than words, which we          205
On like conditions will have counter-sealed.
Come, enter with us. Ladies, you deserve
To have a temple built you. All the swords
In Italy, and her confederate arms,
Could not have made this peace.          Exeunt

**5.4**     *Enter Menenius and Sicinius*

MENENIUS See you yon coign o'th' Capitol, yon corner-
stone?

SICINIUS Why, what of that?

MENENIUS If it be possible for you to displace it with your
little finger, there is some hope the ladies of Rome,
especially his mother, may prevail with him. But I say
there is no hope in't, our throats are sentenced and
stay upon execution.

SICINIUS Is't possible that so short a time can alter the
condition of a man?          10

MENENIUS There is differency between a grub and a
butterfly, yet your butterfly was a grub. This Martius
is grown from man to dragon. He has wings, he's more
than a creeping thing.

SICINIUS He loved his mother dearly.          15

MENENIUS So did he me, and he no more remembers his
mother now than an eight-year old horse. The tartness
of his face sours ripe grapes. When he walks, he moves
like an engine, and the ground shrinks before his
treading. He is able to pierce a corslet with his eye,
talks like a knell, and his 'hmh!' is a battery. He sits
in his state as a thing made for Alexander. What he
bids be done is finished with his bidding. He wants
nothing of a god but eternity and a heaven to throne
in.          25

SICINIUS Yes: mercy, if you report him truly.

MENENIUS I paint him in the character. Mark what mercy
his mother shall bring from him. There is no more
mercy in him than there is milk in a male tiger. That
shall our poor city find; and all this is 'long of you.          30

SICINIUS The gods be good unto us!

MENENIUS No, in such a case the gods will not be good
unto us. When we banished him we respected not
them, and, he returning to break our necks, they
respect not us.          35

*Enter a Messenger*

MESSENGER (to Sicinius)
Sir, if you'd save your life, fly to your house.
The plebeians have got your fellow tribune

And hale him up and down, all swearing if
The Roman ladies bring not comfort home
They'll give him death by inches.

*Enter another Messenger*

SICINIUS                              What's the news?  40

SECOND MESSENGER
Good news, good news. The ladies have prevailed,
The Volscians are dislodged, and Martius gone.
A merrier day did never yet greet Rome,
No, not th'expulsion of the Tarquins.

SICINIUS                              Friend,
Art thou certain this is true? Is't most certain?          45

SECOND MESSENGER
As certain as I know the sun is fire.
Where have you lurked that you make doubt of it?
Ne'er through an arch so hurried the blown tide
As the recomforted through th' gates.

*Trumpets, hautboys, drums, beat all together*
                                   Why, hark you,
The trumpets, sackbuts, psalteries, and fifes,          50
Tabors and cymbals and the shouting Romans
Make the sun dance.

*A shout within*
                         Hark you!

MENENIUS                         This is good news.
I will go meet the ladies. This Volumnia
Is worth of consuls, senators, patricians,
A city full; of tribunes such as you,          55
A sea and land full. You have prayed well today.
This morning for ten thousand of your throats
I'd not have given a doit.

*Music sounds still with the shouts*
                         Hark how they joy!

SICINIUS (to the Messenger)
First, the gods bless you for your tidings. Next,
⌐Giving money⌐ Accept my thankfulness.          60

SECOND MESSENGER
Sir, we have all great cause to give great thanks.

SICINIUS
They are near the city.

SECOND MESSENGER          Almost at point to enter.

SICINIUS We'll meet them, and help the joy.          Exeunt

**5.5**     *Enter ⌐at one door⌐ Lords ⌐and Citizens⌐, ⌐at another
            door⌐ two Senators with the ladies Volumnia,
            Virgilia, and Valeria, passing over the stage*

A SENATOR
Behold our patroness, the life of Rome!
Call all your tribes together, praise the gods,
And make triumphant fires. Strew flowers before them.
Unshout the noise that banished Martius,
Repeal him with the welcome of his mother.          5
Cry 'Welcome, ladies, welcome!'

ALL                              Welcome, ladies, welcome!
                    *A flourish with drums and trumpets. Exeunt*

**5.6**     *Enter Tullus Aufidius with attendants*

AUFIDIUS
Go tell the lords o'th' city I am here.
Deliver them this paper. Having read it,
Bid them repair to th' market-place, where I,
Even in theirs and in the commons' ears,
Will vouch the truth of it. Him I accuse          5
The city ports by this hath entered, and
Intends t'appear before the people, hoping

To purge himself with words. Dispatch.

*Exeunt attendants*

*Enter three or four Conspirators of Aufidius' faction*

Most welcome.

FIRST CONSPIRATOR
How is it with our general?

AUFIDIUS                                Even so
As with a man by his own alms impoisoned,                10
And with his charity slain.

SECOND CONSPIRATOR                Most noble sir,
If you do hold the same intent wherein
You wished us parties, we'll deliver you
Of your great danger.

AUFIDIUS                        Sir, I cannot tell.
We must proceed as we do find the people.                15

THIRD CONSPIRATOR
The people will remain uncertain whilst
'Twixt you there's difference, but the fall of either
Makes the survivor heir of all.

AUFIDIUS                                I know it,
And my pretext to strike at him admits
A good construction. I raised him, and I pawned          20
Mine honour for his truth; who being so heightened,
He watered his new plants with dews of flattery,
Seducing so my friends; and to this end
He bowed his nature, never known before
But to be rough, unswayable, and free.                   25

THIRD CONSPIRATOR Sir, his stoutness
When he did stand for consul, which he lost
By lack of stooping—

AUFIDIUS                        That I would have spoke of.
Being banished for't, he came unto my hearth,
Presented to my knife his throat. I took him,           30
Made him joint-servant with me, gave him way
In all his own desires; nay, let him choose
Out of my files, his projects to accomplish,
My best and freshest men; served his designments
In mine own person, holp to reap the fame              35
Which he did end all his, and took some pride
To do myself this wrong, till at the last
I seemed his follower, not partner, and
He waged me with his countenance as if
I had been mercenary.

FIRST CONSPIRATOR        So he did, my lord.            40
The army marvelled at it, and in the last,
When he had carried Rome and that we looked
For no less spoil than glory—

AUFIDIUS                        There was it,
For which my sinews shall be stretched upon him.
At a few drops of women's rheum, which are             45
As cheap as lies, he sold the blood and labour
Of our great action; therefore shall he die,
And I'll renew me in his fall.

*Drums and trumpets sound, with great shouts of*
*the people*

But hark.

FIRST CONSPIRATOR
Your native town you entered like a post,
And had no welcomes home; but he returns              50
Splitting the air with noise.

SECOND CONSPIRATOR        And patient fools,
Whose children he hath slain, their base throats tear
With giving him glory.

THIRD CONSPIRATOR        Therefore, at your vantage,
Ere he express himself or move the people
With what he would say, let him feel your sword,      55

Which we will second. When he lies along,
After your way his tale pronounced shall bury
His reasons with his body.

*Enter the Lords of the city*

AUFIDIUS                        Say no more.
Here come the lords.

ALL THE LORDS You are most welcome home.              60

AUFIDIUS I have not deserved it.
But, worthy lords, have you with heed perused
What I have written to you?

ALL THE LORDS                We have.

FIRST LORD                        And grieve to hear't.
What faults he made before the last, I think
Might have found easy fines. But there to end         65
Where he was to begin, and give away
The benefit of our levies, answering us
With our own charge, making a treaty where
There was a yielding—this admits no excuse.

AUFIDIUS He approaches. You shall hear him.           70

*Enter Coriolanus marching with drum and colours,*
*the Commoners being with him*

CORIOLANUS
Hail, lords! I am returned your soldier,
No more infected with my country's love
Than when I parted hence, but still subsisting
Under your great command. You are to know
That prosperously I have attempted, and               75
With bloody passage led your wars even to
The gates of Rome. Our spoils we have brought home
Doth more than counterpoise a full third part
The charges of the action. We have made peace
With no less honour to the Antiates                   80
Than shame to th' Romans. And we here deliver,
Subscribed by th' consuls and patricians,
Together with the seal o'th' senate, what
We have compounded on.

*He gives the Lords a paper*

AUFIDIUS                        Read it not, noble lords,
But tell the traitor in the highest degree            85
He hath abused your powers.

CORIOLANUS Traitor? How now?

AUFIDIUS Ay, traitor, Martius.

CORIOLANUS Martius?

AUFIDIUS
Ay, Martius, Caius Martius. Dost thou think           90
I'll grace thee with that robbery, thy stol'n name,
'Coriolanus', in Corioles?
You lords and heads o'th' state, perfidiously
He has betrayed your business, and given up,
For certain drops of salt, your city, Rome—           95
I say your city—to his wife and mother,
Breaking his oath and resolution like
A twist of rotten silk, never admitting
Counsel o'th' war. But at his nurse's tears
He whined and roared away your victory,              100
That pages blushed at him, and men of heart
Looked wond'ring each at others.

CORIOLANUS                        Hear'st thou, Mars?

AUFIDIUS
Name not the god, thou boy of tears.

CORIOLANUS                                Ha?

AUFIDIUS                                No more.

CORIOLANUS
Measureless liar, thou hast made my heart
Too great for what contains it. 'Boy'? O slave!—     105

Pardon me, lords, 'tis the first time that ever
I was forced to scold. Your judgements, my grave lords,
Must give this cur the lie, and his own notion—
Who wears my stripes impressed upon him, that
Must bear my beating to his grave—shall join          110
To thrust the lie unto him.
FIRST LORD                         Peace both, and hear me speak.
CORIOLANUS
Cut me to pieces, Volsces. Men and lads,
Stain all your edges on me. 'Boy'! False hound,
If you have writ your annals true, 'tis there
That, like an eagle in a dove-cote, I                 115
Fluttered your Volscians in Corioles.
Alone I did it. 'Boy'!
AUFIDIUS                      Why, noble lords,
Will you be put in mind of his blind fortune,
Which was your shame, by this unholy braggart,
Fore your own eyes and ears?
ALL THE CONSPIRATORS              Let him die for't.      120
ALL THE PEOPLE ⌈shouting dispersedly⌉
Tear him to pieces! Do it presently!
He killed my son! My daughter! He killed my cousin
Marcus! He killed my father!
SECOND LORD                       Peace, ho! No outrage, peace.
The man is noble, and his fame folds in
This orb o'th' earth. His last offences to us          125
Shall have judicious hearing. Stand, Aufidius,
And trouble not the peace.
CORIOLANUS ⌈drawing his sword⌉
O that I had him with six Aufidiuses,
Or more, his tribe, to use my lawful sword!
AUFIDIUS ⌈drawing his sword⌉
Insolent villain!
ALL THE CONSPIRATORS  Kill, kill, kill, kill, kill him!   130
    Two Conspirators draw and kill Martius, who falls.
    Aufidius ⌈and Conspirators⌉ stand on him

LORDS
Hold, hold, hold, hold!
AUFIDIUS                  My noble masters, hear me speak.
FIRST LORD
O Tullus!
SECOND LORD (to Aufidius)
              Thou hast done a deed whereat
Valour will weep.
THIRD LORD ⌈to Aufidius and the Conspirators⌉
                 Tread not upon him, masters.
All be quiet. Put up your swords.
AUFIDIUS My lords,                                       135
When you shall know—as in this rage
Provoked by him you cannot—the great danger
Which this man's life did owe you, you'll rejoice
That he is thus cut off. Please it your honours
To call me to your senate, I'll deliver                 140
Myself your loyal servant, or endure
Your heaviest censure.
FIRST LORD                Bear from hence his body,
And mourn you for him. Let him be regarded
As the most noble corpse that ever herald
Did follow to his urn.
SECOND LORD               His own impatience            145
Takes from Aufidius a great part of blame.
Let's make the best of it.
AUFIDIUS                  My rage is gone,
And I am struck with sorrow. Take him up.
Help three o'th' chiefest soldiers; I'll be one.
Beat thou the drum, that it speak mournfully.          150
Trail your steel pikes. Though in this city he
Hath widowed and unchilded many a one,
Which to this hour bewail the injury,
Yet he shall have a noble memory. Assist.
                       *A dead march sounded. Exeunt*
                       *bearing the body of Martius*

# THE WINTER'S TALE

THE astrologer Simon Forman saw *The Winter's Tale* at the Globe on 15 May 1611. Just how much earlier the play was written is not certainly known. During the sheep-shearing feast in Act 4, twelve countrymen perform a satyrs' dance that three of them are said to have already 'danced before the King'. This is not necessarily a topical reference, but satyrs danced in Ben Jonson's *Masque of Oberon*, performed before King James on 1 January 1611. It seems likely that this dance was incorporated in *The Winter's Tale* (just as, later, another masque dance seems to have been transferred to *The Two Noble Kinsmen*). But it occurs in a self-contained passage that may well have been added after Shakespeare wrote the play itself. *The Winter's Tale*, first printed in the 1623 Folio, is usually thought to have been written after *Cymbeline*, but stylistic evidence places it before that play, perhaps in 1609–10.

A mid sixteenth-century book classes 'winter tales' along with 'old wives' tales'; Shakespeare's title prepared his audiences for a tale of romantic improbability, one to be wondered at rather than believed; and within the play itself characters compare its events to 'an old tale' (5.2.61; 5.3.118). The comparison is just: Shakespeare is dramatizing a story by his old rival Robert Greene, published as *Pandosto: The Triumph of Time* in or before 1588. This gave Shakespeare his plot outline, of a king (Leontes) who believes his wife (Hermione) to have committed adultery with another king (Polixenes), his boyhood friend, and who casts off his new-born daughter (Perdita—the lost one) in the belief that she is his friend's bastard. In both versions the baby is brought up as a shepherdess, falls in love with her supposed father's son (Florizel in the play), and returns to her real father's court where she is at last recognized as his daughter. In both versions, too, the wife's innocence is demonstrated by the pronouncement of the Delphic oracle, and her husband passes the period of his daughter's absence in penitence; but Shakespeare alters the ending of his source story, bringing it into line with the conventions of romance. He adopts Greene's tripartite structure, but greatly develops it, adding for instance Leontes' steward Antigonus and his redoubtable wife Paulina, along with the comic rogue Autolycus, 'snapper-up of unconsidered trifles'.

The intensity of poetic suffering with which Leontes expresses his irrational jealousy is matched by the lyrical rapture of the love episodes between Florizel and Perdita. In both verse and prose *The Winter's Tale* shows Shakespeare's verbal powers at their greatest, and his theatrical mastery is apparent in, for example, Hermione's trial (3.1) and the daring final scene in which time brings about its triumph.

# THE PERSONS OF THE PLAY

LEONTES, King of Sicily

HERMIONE, his wife

MAMILLIUS, his son

PERDITA, his daughter

CAMILLO ⎫
ANTIGONUS ⎪
CLEOMENES ⎬ Lords at Leontes's court
DION ⎭

PAULINA, Antigonus's wife

EMILIA, a lady attending on Hermione

A JAILER

A MARINER

Other Lords and Gentlemen, Ladies, Officers, and Servants at
  Leontes's court

POLIXENES, King of Bohemia

FLORIZEL, his son, in love with Perdita; known as Doricles

ARCHIDAMUS, a Bohemian lord

AUTOLYCUS, a rogue, once in the service of Florizel

CLOWN, his son

MOPSA ⎫
      ⎬ shepherdesses
DORCAS ⎭

SERVANT of the Old Shepherd

Other Shepherds and Shepherdesses

Twelve countrymen disguised as satyrs

TIME, as chorus

# The Winter's Tale

**1.1** *Enter Camillo and Archidamus*

ARCHIDAMUS If you shall chance, Camillo, to visit Bohemia
on the like occasion whereon my services are now on
foot, you shall see, as I have said, great difference
betwixt our Bohemia and your Sicilia.                                              4

CAMILLO I think this coming summer the King of Sicilia
means to pay Bohemia the visitation which he justly
owes him.

ARCHIDAMUS Wherein our entertainment shall shame us,
we will be justified in our loves; for indeed—

CAMILLO Beseech you—                                                                    10

ARCHIDAMUS Verily, I speak it in the freedom of my
knowledge. We cannot with such magnificence—in so
rare—I know not what to say.—We will give you sleepy
drinks, that your senses, unintelligent of our
insufficience, may, though they cannot praise us, as
little accuse us.                                                                              16

CAMILLO You pay a great deal too dear for what's given
freely.

ARCHIDAMUS Believe me, I speak as my understanding
instructs me, and as mine honesty puts it to utterance.

CAMILLO Sicilia cannot show himself over-kind to
Bohemia. They were trained together in their
childhoods, and there rooted betwixt them then such
an affection which cannot choose but branch now.
Since their more mature dignities and royal necessities
made separation of their society, their encounters—
though not personal—hath been royally attorneyed
with interchange of gifts, letters, loving embassies, that
they have seemed to be together, though absent; shook
hands as over a vast; and embraced as it were from
the ends of opposed winds. The heavens continue their
loves.                                                                                           32

ARCHIDAMUS I think there is not in the world either malice
or matter to alter it. You have an unspeakable comfort
of your young prince, Mamillius. It is a gentleman of
the greatest promise that ever came into my note.      36

CAMILLO I very well agree with you in the hopes of him.
It is a gallant child; one that, indeed, physics the
subject, makes old hearts fresh. They that went on
crutches ere he was born desire yet their life to see him
a man.                                                                                           41

ARCHIDAMUS Would they else be content to die?

CAMILLO Yes—if there were no other excuse why they
should desire to live.

ARCHIDAMUS If the King had no son they would desire to
live on crutches till he had one.                          *Exeunt*

**1.2** *Enter Leontes, Hermione, Mamillius, Polixenes, and*
⌜*Camillo*⌝

POLIXENES
Nine changes of the wat'ry star hath been
The shepherd's note since we have left our throne
Without a burden. Time as long again
Would be filled up, my brother, with our thanks,
And yet we should for perpetuity                                                      5
Go hence in debt. And therefore, like a cipher,
Yet standing in rich place, I multiply
With one 'We thank you' many thousands more

That go before it.

LEONTES                          Stay your thanks a while,
And pay them when you part.

POLIXENES                                      Sir, that's tomorrow.  10
I am questioned by my fears of what may chance
Or breed upon our absence, that may blow
No sneaping winds at home to make us say
'This is put forth too truly.' Besides, I have stayed
To tire your royalty.

LEONTES                          We are tougher, brother,              15
Than you can put us to't.

POLIXENES                                    No longer stay.

LEONTES
One sennight longer.

POLIXENES                              Very sooth, tomorrow.

LEONTES
We'll part the time between's, then; and in that
I'll no gainsaying.

POLIXENES                          Press me not, beseech you, so.
There is no tongue that moves, none, none i'th' world
So soon as yours, could win me. So it should now,  21
Were there necessity in your request, although
'Twere needful I denied it. My affairs
Do even drag me homeward; which to hinder
Were, in your love, a whip to me; my stay              25
To you a charge and trouble. To save both,
Farewell, our brother.

LEONTES                          Tongue-tied, our queen? Speak you.

HERMIONE
I had thought, sir, to have held my peace until
You had drawn oaths from him not to stay. You, sir,
Charge him too coldly. Tell him you are sure        30
All in Bohemia's well. This satisfaction
The bygone day proclaimed. Say this to him,
He's beat from his best ward.

LEONTES                                            Well said, Hermione!

HERMIONE
To tell he longs to see his son were strong.
But let him say so then, and let him go.                  35
But let him swear so and he shall not stay,
We'll thwack him hence with distaffs.
(*To Polixenes*) Yet of your royal presence I'll adventure
The borrow of a week. When at Bohemia
You take my lord, I'll give him my commission        40
To let him there a month behind the gest
Prefixed for's parting.—Yet, good deed, Leontes,
I love thee not a jar o'th' clock behind
What lady she her lord.—You'll stay?

POLIXENES                                                      No, madam.

HERMIONE Nay, but you will?                                                  45

POLIXENES I may not, verily.

HERMIONE Verily?
You put me off with limber vows. But I,
Though you would seek t'unsphere the stars with
    oaths,
Should yet say 'Sir, no going.' Verily                      50
You shall not go. A lady's 'verily' 's
As potent as a lord's. Will you go yet?
Force me to keep you as a prisoner,
Not like a guest: so you shall pay your fees

1103

When you depart, and save your thanks. How say
    you?                                         55
My prisoner? or my guest? By your dread 'verily',
One of them you shall be.
POLIXENES               Your guest then, madam.
To be your prisoner should import offending,
Which is for me less easy to commit
Than you to punish.
HERMIONE          Not your jailer then,    60
But your kind hostess. Come, I'll question you
Of my lord's tricks and yours when you were boys.
You were pretty lordings then?
POLIXENES            We were, fair Queen,
Two lads that thought there was no more behind
But such a day tomorrow as today,    65
And to be boy eternal.
HERMIONE Was not my lord
The verier wag o'th' two?
POLIXENES
We were as twinned lambs that did frisk i'th' sun,
And bleat the one at th'other. What we changed    70
Was innocence for innocence. We knew not
The doctrine of ill-doing, nor dreamed
That any did. Had we pursued that life,
And our weak spirits ne'er been higher reared
With stronger blood, we should have answered
    heaven                                     75
Boldly, 'Not guilty', the imposition cleared
Hereditary ours.
HERMIONE        By this we gather
You have tripped since.
POLIXENES        O my most sacred lady,
Temptations have since then been born to's; for
In those unfledged days was my wife a girl.    80
Your precious self had then not crossed the eyes
Of my young playfellow.
HERMIONE        Grace to boot!
Of this make no conclusion, lest you say
Your queen and I are devils. Yet go on.
Th'offences we have made you do we'll answer,    85
If you first sinned with us, and that with us
You did continue fault, and that you slipped not
With any but with us.
LEONTES        Is he won yet?
HERMIONE
He'll stay, my lord.
LEONTES        At my request he would not.
Hermione, my dearest, thou never spok'st    90
To better purpose.
HERMIONE        Never?
LEONTES        Never but once.
HERMIONE
What, have I twice said well? When was't before?
I prithee tell me. Cram's with praise, and make's
As fat as tame things. One good deed dying tongueless
Slaughters a thousand waiting upon that.    95
Our praises are our wages. You may ride's
With one soft kiss a thousand furlongs ere
With spur we heat an acre. But to th' goal.
My last good deed was to entreat his stay.
What was my first? It has an elder sister,    100
Or I mistake you. O, would her name were Grace!
But once before I spoke to th' purpose? When?
Nay, let me have't. I long.
LEONTES        Why, that was when
Three crabbèd months had soured themselves to death
Ere I could make thee open thy white hand    105

And clap thyself my love. Then didst thou utter,
'I am yours for ever.'
HERMIONE        'Tis grace indeed.
Why lo you now; I have spoke to th' purpose twice.
The one for ever earned a royal husband;
Th'other, for some while a friend.
    ⌈*She gives her hand to Polixenes.*⌉
    *They stand aside*
LEONTES (*aside*)        Too hot, too hot:
To mingle friendship farre is mingling bloods.    111
I have *tremor cordis* on me. My heart dances,
But not for joy, not joy. This entertainment
May a free face put on, derive a liberty
From heartiness, from bounty, fertile bosom,    115
And well become the agent. 'T may, I grant.
But to be paddling palms and pinching fingers,
As now they are, and making practised smiles
As in a looking-glass; and then to sigh, as 'twere
The mort o'th' deer—O, that is entertainment    120
My bosom likes not, nor my brows.—Mamillius,
Art thou my boy?
MAMILLIUS        Ay, my good lord.
LEONTES        I'fecks,
Why, that's my bawcock. What? Hast smutched thy
    nose?
They say it is a copy out of mine. Come, captain,
We must be neat—not neat, but cleanly, captain.    125
And yet the steer, the heifer, and the calf
Are all called neat.—Still virginalling
Upon his palm?—How now, you wanton calf—
Art thou my calf?
MAMILLIUS        Yes, if you will, my lord.
LEONTES
Thou want'st a rough pash and the shoots that I have,
To be full like me. Yet they say we are    131
Almost as like as eggs. Women say so,
That will say anything. But were they false
As o'er-dyed blacks, as wind, as waters, false
As dice are to be wished by one that fixes    135
No bourn 'twixt his and mine, yet were it true
To say this boy were like me. Come, sir page,
Look on me with your welkin eye. Sweet villain,
Most dear'st, my collop! Can thy dam—may't be?—
Affection, thy intention stabs the centre.    140
Thou dost make possible things not so held,
Communicat'st with dreams—how can this be?—
With what's unreal thou coactive art,
And fellow'st nothing. Then 'tis very credent
Thou mayst co-join with something, and thou dost—
And that beyond commission; and I find it—    146
And that to the infection of my brains
And hard'ning of my brows.
POLIXENES        What means Sicilia?
HERMIONE
He something seems unsettled.
POLIXENES        How, my lord!
LEONTES
What cheer? How is't with you, best brother?
HERMIONE        You look
As if you held a brow of much distraction.    151
Are you moved, my lord?
LEONTES        No, in good earnest.
How sometimes nature will betray its folly,
Its tenderness, and make itself a pastime
To harder bosoms! Looking on the lines    155
Of my boy's face, methoughts I did recoil

Twenty-three years, and saw myself unbreeched,
In my green velvet coat; my dagger muzzled,
Lest it should bite its master, and so prove,
As ornament oft does, too dangerous.                     160
How like, methought, I then was to this kernel,
This squash, this gentleman.—Mine honest friend,
Will you take eggs for money?

MAMILLIUS                        No, my lord, I'll fight.

LEONTES
You will? Why, happy man be's dole!—My brother,
Are you so fond of your young prince as we        165
Do seem to be of ours?

POLIXENES                 If at home, sir,
He's all my exercise, my mirth, my matter;
Now my sworn friend, and then mine enemy;
My parasite, my soldier, statesman, all.
He makes a July's day short as December,         170
And with his varying childness cures in me
Thoughts that would thick my blood.

LEONTES                           So stands this squire
Officed with me. We two will walk, my lord,
And leave you to your graver steps. Hermione,
How thou lov'st us show in our brother's welcome.
Let what is dear in Sicily be cheap.             176
Next to thyself and my young rover, he's
Apparent to my heart.

HERMIONE              If you would seek us,
We are yours i'th' garden. Shall's attend you there?

LEONTES
To your own bents dispose you. You'll be found,    180
Be you beneath the sky. (Aside) I am angling now,
Though you perceive me not how I give line.
Go to, go to!
How she holds up the neb, the bill to him,
And arms her with the boldness of a wife           185
To her allowing husband!

                    Exeunt Polixenes and Hermione
                    Gone already.
Inch-thick, knee-deep, o'er head and ears a forked
    one!—
Go play, boy, play. Thy mother plays, and I
Play too; but so disgraced a part, whose issue
Will hiss me to my grave. Contempt and clamour   190
Will be my knell. Go play, boy, play. There have been,
Or I am much deceived, cuckolds ere now,
And many a man there is, even at this present,
Now, while I speak this, holds his wife by th'arm,
That little thinks she has been sluiced in's absence,
And his pond fished by his next neighbour, by     196
Sir Smile, his neighbour. Nay, there's comfort in't,
Whiles other men have gates, and those gates opened,
As mine, against their will. Should all despair
That have revolted wives, the tenth of mankind    200
Would hang themselves. Physic for't there's none.
It is a bawdy planet, that will strike
Where 'tis predominant; and 'tis powerful. Think it:
From east, west, north, and south, be it concluded,
No barricado for a belly. Know't,                  205
It will let in and out the enemy
With bag and baggage. Many thousand on's
Have the disease and feel't not.—How now, boy?

MAMILLIUS
I am like you, they say.

LEONTES               Why, that's some comfort.
What, Camillo there!

CAMILLO ⌈coming forward⌉ Ay, my good lord.        210

LEONTES
Go play, Mamillius, thou'rt an honest man.
                              Exit Mamillius
Camillo, this great sir will yet stay longer.

CAMILLO
You had much ado to make his anchor hold.
When you cast out, it still came home.

LEONTES                              Didst note it?

CAMILLO
He would not stay at your petitions, made          215
His business more material.

LEONTES             Didst perceive it?
(Aside) They're here with me already, whisp'ring,
    rounding,
'Sicilia is a so-forth'. 'Tis far gone
When I shall gust it last.—How came't, Camillo,
That he did stay?

CAMILLO          At the good Queen's entreaty.      220

LEONTES
'At the Queen's' be't. 'Good' should be pertinent,
But so it is, it is not. Was this taken
By any understanding pate but thine?
For thy conceit is soaking, will draw in
More than the common blocks. Not noted, is't,     225
But of the finer natures? By some severals
Of head-piece extraordinary? Lower messes
Perchance are to this business purblind? Say.

CAMILLO
Business, my lord? I think most understand
Bohemia stays here longer.                          230

LEONTES Ha?

CAMILLO Stays here longer.

LEONTES Ay, but why?

CAMILLO
To satisfy your highness, and the entreaties
Of our most gracious mistress.

LEONTES                       Satisfy?               235
Th'entreaties of your mistress? Satisfy?
Let that suffice. I have trusted thee, Camillo,
With all the near'st things to my heart, as well
My chamber-counsels, wherein, priest-like, thou
Hast cleansed my bosom, I from thee departed       240
Thy penitent reformed. But we have been
Deceived in thy integrity, deceived
In that which seems so.

CAMILLO                Be it forbid, my lord.

LEONTES
To bide upon't: thou art not honest; or
If thou inclin'st that way, thou art a coward,     245
Which hoxes honesty behind, restraining
From course required. Or else thou must be counted
A servant grafted in my serious trust
And therein negligent, or else a fool
That seest a game played home, the rich stake drawn,
And tak'st it all for jest.

CAMILLO          My gracious lord,                  251
I may be negligent, foolish, and fearful.
In every one of these no man is free,
But that his negligence, his folly, fear,
Among the infinite doings of the world             255
Sometime puts forth. In your affairs, my lord,
If ever I were wilful-negligent,
It was my folly. If industriously
I played the fool, it was my negligence,
Not weighing well the end. If ever fearful          260
To do a thing where I the issue doubted,

Whereof the execution did cry out
Against the non-performance, 'twas a fear
Which oft infects the wisest. These, my lord,
Are such allowed infirmities that honesty 265
Is never free of. But beseech your grace
Be plainer with me, let me know my trespass
By its own visage. If I then deny it,
'Tis none of mine.
LEONTES         Ha' not you seen, Camillo—
But that's past doubt; you have, or your eye-glass 270
Is thicker than a cuckold's horn—or heard—
For, to a vision so apparent, rumour
Cannot be mute—or thought—for cogitation
Resides not in that man that does not think—
My wife is slippery? If thou wilt confess— 275
Or else be impudently negative
To have nor eyes, nor ears, nor thought—then say
My wife's a hobby-horse, deserves a name
As rank as any flax-wench that puts to
Before her troth-plight. Say't, and justify't. 280
CAMILLO
I would not be a stander-by to hear
My sovereign mistress clouded so without
My present vengeance taken. 'Shrew my heart,
You never spoke what did become you less
Than this, which to reiterate were sin 285
As deep as that, though true.
LEONTES          Is whispering nothing?
Is leaning cheek to cheek? Is meeting noses?
Kissing with inside lip? Stopping the career
Of laughter with a sigh?—a note infallible
Of breaking honesty. Horsing foot on foot? 290
Skulking in corners? Wishing clocks more swift,
Hours minutes, noon midnight? And all eyes
Blind with the pin and web but theirs, theirs only,
That would unseen be wicked? Is this nothing?
Why then the world and all that's in't is nothing, 295
The covering sky is nothing, Bohemia nothing,
My wife is nothing, nor nothing have these nothings
If this be nothing.
CAMILLO       Good my lord, be cured
Of this diseased opinion, and betimes,
For 'tis most dangerous.
LEONTES       Say it be, 'tis true. 300
CAMILLO
No, no, my lord.
LEONTES       It is. You lie, you lie.
I say thou liest, Camillo, and I hate thee,
Pronounce thee a gross lout, a mindless slave,
Or else a hovering temporizer, that
Canst with thine eyes at once see good and evil, 305
Inclining to them both. Were my wife's liver
Infected as her life, she would not live
The running of one glass.
CAMILLO       Who does infect her?
LEONTES
Why, he that wears her like her medal, hanging
About his neck, Bohemia, who, if I 310
Had servants true about me, that bare eyes
To see alike mine honour as their profits,
Their own particular thrifts, they would do that
Which should undo more doing. Ay, and thou
His cupbearer, whom I from meaner form 315
Have benched, and reared to worship, who mayst see
Plainly as heaven sees earth and earth sees heaven,
How I am galled, mightst bespice a cup
To give mine enemy a lasting wink,

Which draught to me were cordial.
CAMILLO         Sir, my lord, 320
I could do this, and that with no rash potion,
But with a ling'ring dram, that should not work
Maliciously, like poison. But I cannot
Believe this crack to be in my dread mistress,
So sovereignly being honourable. 325
I have loved thee—
LEONTES       Make that thy question, and go rot!
Dost think I am so muddy, so unsettled,
To appoint myself in this vexation?
Sully the purity and whiteness of my sheets—
Which to preserve is sleep, which being spotted 330
Is goads, thorns, nettles, tails of wasps—
Give scandal to the blood o'th' prince, my son—
Who I do think is mine, and love as mine—
Without ripe moving to't? Would I do this?
Could man so blench?
CAMILLO       I must believe you, sir. 335
I do, and will fetch off Bohemia for't,
Provided that when he's removed your highness
Will take again your queen as yours at first,
Even for your son's sake, and thereby for sealing
The injury of tongues in courts and kingdoms 340
Known and allied to yours.
LEONTES       Thou dost advise me
Even so as I mine own course have set down.
I'll give no blemish to her honour, none.
CAMILLO
My lord, go then, and with a countenance as clear
As friendship wears at feasts, keep with Bohemia 345
And with your queen. I am his cupbearer,
If from me he have wholesome beverage,
Account me not your servant.
LEONTES         This is all.
Do't, and thou hast the one half of my heart; 349
Do't not, thou splitt'st thine own.
CAMILLO       I'll do't, my lord.
LEONTES
I will seem friendly, as thou hast advised me.     *Exit*
CAMILLO
O miserable lady. But for me,
What case stand I in? I must be the poisoner
Of good Polixenes, and my ground to do't
Is the obedience to a master—one 355
Who in rebellion with himself, will have
All that are his so too. To do this deed,
Promotion follows. If I could find example
Of thousands that had struck anointed kings
And flourished after, I'd not do't. But since 360
Nor brass, nor stone, nor parchment bears not one,
Let villainy itself forswear't. I must
Forsake the court. To do't, or no, is certain
To me a break-neck.
    *Enter Polixenes*
            Happy star reign now!
Here comes Bohemia.
POLIXENES (*aside*)     This is strange. Methinks 365
My favour here begins to warp. Not speak?—
Good day, Camillo.
CAMILLO       Hail, most royal sir.
POLIXENES
What is the news i'th' court?
CAMILLO       None rare, my lord.
POLIXENES
The King hath on him such a countenance
As he had lost some province, and a region 370

Loved as he loves himself. Even now I met him
With customary compliment, when he,
Wafting his eyes to th' contrary, and falling
A lip of much contempt, speeds from me, and
So leaves me to consider what is breeding          375
That changes thus his manners.
CAMILLO                      I dare not know, my lord.
POLIXENES
How, 'dare not'? Do not? Do you know, and dare not?
Be intelligent to me. 'Tis thereabouts.
For to yourself what you do know you must,
And cannot say you 'dare not'. Good Camillo,        380
Your changed complexions are to me a mirror
Which shows me mine changed, too; for I must be
A party in this alteration, finding
Myself thus altered with't.
CAMILLO                        There is a sickness
Which puts some of us in distemper, but            385
I cannot name th' disease, and it is caught
Of you that yet are well.
POLIXENES                        How caught of me?
Make me not sighted like the basilisk.
I have looked on thousands who have sped the better
By my regard, but killed none so. Camillo,          390
As you are certainly a gentleman, thereto
Clerk-like experienced, which no less adorns
Our gentry than our parents' noble names,
In whose success we are gentle: I beseech you,
If you know aught which does behove my knowledge
Thereof to be informed, imprison't not             396
In ignorant concealment.
CAMILLO                        I may not answer.
POLIXENES
A sickness caught of me, and yet I well?
I must be answered. Dost thou hear, Camillo,
I conjure thee, by all the parts of man            400
Which honour does acknowledge, whereof the least
Is not this suit of mine, that thou declare
What incidency thou dost guess of harm
Is creeping toward me; how far off, how near,
Which way to be prevented, if to be;               405
If not, how best to bear it.
CAMILLO                        Sir, I will tell you,
Since I am charged in honour, and by him
That I think honourable. Therefore mark my counsel,
Which must be e'en as swiftly followed as
I mean to utter it; or both yourself and me        410
Cry lost, and so good night!
POLIXENES                        On, good Camillo.
CAMILLO
I am appointed him to murder you.
POLIXENES
By whom, Camillo?
CAMILLO                      By the King.
POLIXENES                                 For what?
CAMILLO
He thinks, nay, with all confidence he swears
As he had seen't, or been an instrument            415
To vice you to't, that you have touched his queen
Forbiddenly.
POLIXENES        O, then my best blood turn
To an infected jelly, and my name
Be yoked with his that did betray the Best!
Turn then my freshest reputation to               420
A savour that may strike the dullest nostril
Where I arrive, and my approach be shunned,

Nay hated, too, worse than the great'st infection
That e'er was heard or read.
CAMILLO                        Swear his thought over
By each particular star in heaven, and             425
By all their influences, you may as well
Forbid the sea for to obey the moon
As or by oath remove or counsel shake
The fabric of his folly, whose foundation
Is piled upon his faith, and will continue          430
The standing of his body.
POLIXENES                        How should this grow?
CAMILLO
I know not, but I am sure 'tis safer to
Avoid what's grown than question how 'tis born.
If therefore you dare trust my honesty,
That lies enclosèd in this trunk which you         435
Shall bear along impawned, away tonight!
Your followers I will whisper to the business,
And will by twos and threes at several posterns
Clear them o'th' city. For myself, I'll put
My fortunes to your service, which are here        440
By this discovery lost. Be not uncertain,
For by the honour of my parents, I
Have uttered truth; which if you seek to prove,
I dare not stand by; nor shall you be safer
Than one condemnèd by the King's own mouth,       445
Thereon his execution sworn.
POLIXENES                        I do believe thee,
I saw his heart in's face. Give me thy hand.
Be pilot to me, and thy places shall
Still neighbour mine. My ships are ready, and
My people did expect my hence departure            450
Two days ago. This jealousy
Is for a precious creature. As she's rare
Must it be great; and as his person's mighty
Must it be violent; and as he does conceive
He is dishonoured by a man which ever              455
Professed to him, why, his revenges must
In that be made more bitter. Fear o'ershades me.
Good expedition be my friend and comfort
The gracious Queen, part of his theme, but nothing
Of his ill-ta'en suspicion. Come, Camillo,         460
I will respect thee as a father if
Thou bear'st my life off hence. Let us avoid.
CAMILLO
It is in mine authority to command
The keys of all the posterns. Please your highness
To take the urgent hour. Come, sir, away.    Exeunt

2.1   *Enter Hermione, Mamillius, and Ladies*
HERMIONE
Take the boy to you. He so troubles me
'Tis past enduring.
FIRST LADY                Come, my gracious lord,
Shall I be your play-fellow?
MAMILLIUS No, I'll none of you.
FIRST LADY Why, my sweet lord?                        5
MAMILLIUS
You'll kiss me hard, and speak to me as if
I were a baby still. (*To Second Lady*) I love you better.
SECOND LADY
And why so, my lord?
MAMILLIUS                        Not for because
Your brows are blacker—yet black brows they say
Become some women best, so that there be not       10
Too much hair there, but in a semicircle,

Or a half-moon made with a pen.
SECOND LADY                          Who taught 'this?
MAMILLIUS
I learned it out of women's faces. Pray now,
What colour are your eyebrows?
FIRST LADY                           Blue, my lord.
MAMILLIUS
Nay, that's a mock. I have seen a lady's nose       15
That has been blue, but not her eyebrows.
FIRST LADY                           Hark ye,
The Queen your mother rounds apace. We shall
Present our services to a fine new prince
One of these days, and then you'd wanton with us,
If we would have you.
SECOND LADY          She is spread of late          20
Into a goodly bulk, good time encounter her.
HERMIONE
What wisdom stirs amongst you? Come sir, now
I am for you again. Pray you sit by us,
And tell's a tale.
MAMILLIUS Merry or sad shall't be?                  25
HERMIONE As merry as you will.
MAMILLIUS
A sad tale's best for winter. I have one
Of sprites and goblins.
HERMIONE          Let's have that, good sir.
Come on, sit down, come on, and do your best
To fright me with your sprites. You're powerful at it.
MAMILLIUS
There was a man—
HERMIONE          Nay, come sit down, then on.      31
MAMILLIUS (sitting)
Dwelt by a churchyard.—I will tell it softly,
Yon crickets shall not hear it.
HERMIONE
Come on then, and give't me in mine ear.
    Enter apart Leontes, Antigonus, and Lords
LEONTES
Was he met there? His train? Camillo with him?     35
A LORD
Behind the tuft of pines I met them. Never
Saw I men scour so on their way. I eyed them
Even to their ships.
LEONTES          How blest am I
In my just censure, in my true opinion!
Alack, for lesser knowledge—how accursed           40
In being so blest! There may be in the cup
A spider steeped, and one may drink, depart,
And yet partake no venom, for his knowledge
Is not infected; but if one present
Th'abhorred ingredient to his eye, make known      45
How he hath drunk, he cracks his gorge, his sides,
With violent hefts. I have drunk, and seen the spider.
Camillo was his help in this, his pander.
There is a plot against my life, my crown.
All's true that is mistrusted. That false villain   50
Whom I employed was pre-employed by him.
He has discovered my design, and I
Remain a pinched thing, yea, a very trick
For them to play at will. How came the posterns
So easily open?
A LORD          By his great authority,              55
Which often hath no less prevailed than so
On your command.
LEONTES          I know't too well.
(To Hermione) Give me the boy. I am glad you did not
    nurse him.

Though he does bear some signs of me, yet you
Have too much blood in him.
HERMIONE          What is this? Sport?               60
LEONTES (to a Lord)
Bear the boy hence. He shall not come about her.
Away with him, and let her sport herself
With that she's big with, (to Hermione) for 'tis
    Polixenes
Has made thee swell thus.      Exit one with Mamillius
HERMIONE          But I'd say he had not,
And I'll be sworn you would believe my saying,      65
Howe'er you lean to th' nayward.
LEONTES                    You, my lords,
Look on her, mark her well. Be but about
To say she is a goodly lady, and
The justice of your hearts will thereto add
''Tis pity she's not honest, honourable.'           70
Praise her but for this her without-door form—
Which on my faith deserves high speech—and
    straight
The shrug, the 'hum' or 'ha', these petty brands
That calumny doth use—O, I am out,
That mercy does, for calumny will sear              75
Virtue itself—these shrugs, these 'hum's and 'ha's',
When you have said she's goodly, come between
Ere you can say she's honest. But be't known
From him that has most cause to grieve it should be,
She's an adultress.
HERMIONE          Should a villain say so,           80
The most replenished villain in the world,
He were as much more villain. You, my lord,
Do but mistake.
LEONTES          You have mistook, my lady—
Polixenes for Leontes. O, thou thing,
Which I'll not call a creature of thy place         85
Lest barbarism, making me the precedent,
Should a like language use to all degrees,
And mannerly distinguishment leave out
Betwixt the prince and beggar. I have said
She's an adultress, I have said with whom.          90
More, she's a traitor, and Camillo is
A federary with her, and one that knows
What she should shame to know herself
But with her most vile principal: that she's
A bed-swerver, even as bad as those                 95
That vulgars give bold'st titles; ay, and privy
To this their late escape.
HERMIONE          No, by my life,
Privy to none of this. How will this grieve you
When you shall come to clearer knowledge, that
You thus have published me? Gentle my lord,         100
You scarce can right me throughly then to say
You did mistake.
LEONTES          No. If I mistake
In those foundations which I build upon,
The centre is not big enough to bear
A schoolboy's top.—Away with her to prison!         105
He who shall speak for her is afar-off guilty,
But that he speaks.
HERMIONE          There's some ill planet reigns.
I must be patient till the heavens look
With an aspect more favourable. Good my lords,
I am not prone to weeping, as our sex               110
Commonly are; the want of which vain dew
Perchance shall dry your pities. But I have
That honourable grief lodged here which burns
Worse than tears drown. Beseech you all, my lords,

With thoughts so qualified as your charities        115
Shall best instruct you, measure me; and so
The King's will be performed.
LEONTES                                    Shall I be heard?
HERMIONE
Who is't that goes with me? Beseech your highness
My women may be with me, for you see
My plight requires it.—Do not weep, good fools,        120
There is no cause. When you shall know your
    mistress
Has deserved prison, then abound in tears
As I come out. This action I now go on
Is for my better grace.—Adieu, my lord.
I never wished to see you sorry; now        125
I trust I shall. My women, come, you have leave.
LEONTES Go, do our bidding. Hence!
                    *Exit Hermione, guarded, with Ladies*
A LORD
Beseech your highness, call the Queen again.
ANTIGONUS (*to Leontes*)
Be certain what you do, sir, lest your justice
Prove violence, in the which three great ones suffer—
Yourself, your queen, your son.
A LORD (*to Leontes*)                    For her, my lord,        131
I dare my life lay down, and will do't, sir,
Please you t'accept it, that the Queen is spotless
I'th' eyes of heaven and to you—I mean
In this which you accuse her.
ANTIGONUS (*to Leontes*)                    If it prove        135
She's otherwise, I'll keep my stables where
I lodge my wife, I'll go in couples with her;
Than when I feel and see her, no farther trust her.
For every inch of woman in the world,
Ay, every dram of woman's flesh is false        140
If she be.
LEONTES        Hold your peaces.
A LORD                                    Good my lord—
ANTIGONUS (*to Leontes*)
It is for you we speak, not for ourselves.
You are abused, and by some putter-on
That will be damned for't. Would I knew the villain—
I would land-damn him. Be she honour-flawed—        145
I have three daughters: the eldest is eleven;
The second and the third nine and some five;
If this prove true, they'll pay for't. By mine honour,
I'll geld 'em all. Fourteen they shall not see,
To bring false generations. They are co-heirs,        150
And I had rather glib myself than they
Should not produce fair issue.
LEONTES                                    Cease, no more!
You smell this business with a sense as cold
As is a dead man's nose. But I do see't and feel't
As you feel doing thus; and see withal        155
The instruments that feel.
ANTIGONUS                    If it be so,
We need no grave to bury honesty;
There's not a grain of it the face to sweeten
Of the whole dungy earth.
LEONTES                    What? Lack I credit?
A LORD
I had rather you did lack than I, my lord,        160
Upon this ground; and more it would content me
To have her honour true than your suspicion,
Be blamed for't how you might.
LEONTES                    Why, what need we
Commune with you of this, but rather follow
Our forceful instigation? Our prerogative        165

Calls not your counsels, but our natural goodness
Imparts this; which, if you—or stupefied
Or seeming so in skill—cannot or will not
Relish a truth like us, inform yourselves
We need no more of your advice. The matter,        170
The loss, the gain, the ord'ring on't, is all
Properly ours.
ANTIGONUS        And I wish, my liege,
You had only in your silent judgement tried it
Without more overture.
LEONTES                    How could that be?
Either thou art most ignorant by age        175
Or thou wert born a fool. Camillo's flight
Added to their familiarity,
Which was as gross as ever touched conjecture
That lacked sight only, naught for approbation
But only seeing, all other circumstances        180
Made up to th' deed—doth push on this proceeding.
Yet for a greater confirmation—
For in an act of this importance 'twere
Most piteous to be wild—I have dispatched in post
To sacred Delphos, to Apollo's temple,        185
Cleomenes and Dion, whom you know
Of stuffed sufficiency. Now from the oracle
They will bring all, whose spiritual counsel had
Shall stop or spur me. Have I done well?
A LORD Well done, my lord.        190
LEONTES
Though I am satisfied, and need no more
Than what I know, yet shall the oracle
Give rest to th' minds of others such as he,
Whose ignorant credulity will not
Come up to th' truth. So have we thought it good        195
From our free person she should be confined,
Lest that the treachery of the two fled hence
Be left her to perform. Come, follow us.
We are to speak in public; for this business
Will raise us all.
ANTIGONUS (*aside*) To laughter, as I take it,        200
If the good truth were known.        *Exeunt*

**2.2**        *Enter Paulina, a Gentleman, and attendants*
PAULINA
The keeper of the prison, call to him.
Let him have knowledge who I am.        *Exit Gentleman*
                                    Good lady,
No court in Europe is too good for thee.
What dost thou then in prison?
                    *Enter Jailer and Gentleman*
                                    Now, good sir,
You know me, do you not?
JAILER                    For a worthy lady,        5
And one who much I honour.
PAULINA Pray you then,
Conduct me to the Queen.
JAILER
I may not, madam. To the contrary
I have express commandment.
PAULINA                    Here's ado,        10
To lock up honesty and honour from
Th'access of gentle visitors. Is't lawful, pray you,
To see her women? Any of them? Emilia?
JAILER So please you, madam,
To put apart these your attendants, I        15
Shall bring Emilia forth.
PAULINA I pray now call her.—
Withdraw yourselves. *Exeunt Gentleman and attendants*

JAILER And, madam,
  I must be present at your conference.                    20
PAULINA Well, be't so, prithee.                    *Exit Jailer*
  Here's such ado, to make no stain a stain
  As passes colouring.
     *Enter Jailer and Emilia*
              Dear gentlewoman,
  How fares our gracious lady?
EMILIA
  As well as one so great and so forlorn                    25
  May hold together. On her frights and griefs,
  Which never tender lady hath borne greater,
  She is, something before her time, delivered.
PAULINA
  A boy?
EMILIA                    A daughter, and a goodly babe,
  Lusty, and like to live. The Queen receives                    30
  Much comfort in't; says, 'My poor prisoner,
  I am innocent as you.'
PAULINA                    I dare be sworn.
  These dangerous, unsafe lunes i'th' King, beshrew
    them!
  He must be told on't, and he shall. The office
  Becomes a woman best. I'll take't upon me.                    35
  If I prove honey-mouthed, let my tongue blister,
  And never to my red-looked anger be
  The trumpet any more. Pray you, Emilia,
  Commend my best obedience to the Queen.
  If she dares trust me with her little babe                    40
  I'll show't the King, and undertake to be
  Her advocate to th' loud'st. We do not know
  How he may soften at the sight o'th' child.
  The silence often of pure innocence
  Persuades when speaking fails.
EMILIA                    Most worthy madam,
  Your honour and your goodness is so evident                    46
  That your free undertaking cannot miss
  A thriving issue. There is no lady living
  So meet for this great errand. Please your ladyship
  To visit the next room, I'll presently                    50
  Acquaint the Queen of your most noble offer,
  Who but today hammered of this design
  But durst not tempt a minister of honour
  Lest she should be denied.
PAULINA                    Tell her, Emilia,
  I'll use that tongue I have. If wit flow from't                    55
  As boldness from my bosom, let't not be doubted
  I shall do good.
EMILIA                    Now be you blest for it!
  I'll to the Queen. Please you come something nearer.
JAILER
  Madam, if't please the Queen to send the babe
  I know not what I shall incur to pass it,                    60
  Having no warrant.
PAULINA                    You need not fear it, sir.
  This child was prisoner to the womb, and is
  By law and process of great nature thence
  Freed and enfranchised, not a party to
  The anger of the King, nor guilty of—                    65
  If any be—the trespass of the Queen.
JAILER I do believe it.
PAULINA
  Do not you fear. Upon mine honour,
  I will stand twixt you and danger.                    69
                          *Exeunt*

**2.3**    *Enter Leontes*
LEONTES
  Nor night nor day, no rest! It is but weakness
  To bear the matter thus, mere weakness. If
  The cause were not in being—part o'th' cause,
  She, th'adultress; for the harlot King
  Is quite beyond mine arm, out of the blank                    5
  And level of my brain, plot-proof; but she
  I can hook to me. Say that she were gone,
  Given to the fire, a moiety of my rest
  Might come to me again. Who's there?
     *Enter a Servant*
SERVANT                    My lord.
LEONTES
  How does the boy?
SERVANT                    He took good rest tonight.                    10
  'Tis hoped his sickness is discharged.
LEONTES To see his nobleness!
  Conceiving the dishonour of his mother
  He straight declined, drooped, took it deeply,
  Fastened and fixed the shame on't in himself;                    15
  Threw off his spirit, his appetite, his sleep,
  And downright languished. Leave me solely. Go,
  See how he fares.                    *Exit Servant*
              Fie, fie, no thought of him.
  The very thought of my revenges that way
  Recoil upon me. In himself too mighty,                    20
  And in his parties, his alliance. Let him be
  Until a time may serve. For present vengeance,
  Take it on her. Camillo and Polixenes
  Laugh at me, make their pastime at my sorrow.
  They should not laugh if I could reach them, nor                    25
  Shall she, within my power.
     *Enter Paulina, carrying a babe, with Antigonus,*
      *Lords, and the Servant, trying to restrain her*
A LORD                    You must not enter.
PAULINA
  Nay rather, good my lords, be second to me.
  Fear you his tyrannous passion more, alas,
  Than the Queen's life?—a gracious, innocent soul,
  More free than he is jealous.
ANTIGONUS                    That's enough.                    30
SERVANT
  Madam, he hath not slept tonight, commanded
  None should come at him.
PAULINA                    Not so hot, good sir.
  I come to bring him sleep. 'Tis such as you,
  That creep like shadows by him, and do sigh
  At each his needless heavings, such as you                    35
  Nourish the cause of his awaking. I
  Do come with words as medicinal as true,
  Honest as either, to purge him of that humour
  That presses him from sleep.
LEONTES                    What noise there, ho?
PAULINA
  No noise, my lord, but needful conference                    40
  About some gossips for your highness.
LEONTES                    How?
  Away with that audacious lady! Antigonus,
  I charged thee that she should not come about me.
  I knew she would.
ANTIGONUS                    I told her so, my lord,
  On your displeasure's peril and on mine,                    45
  She should not visit you.
LEONTES                    What, canst not rule her?

PAULINA
From all dishonesty he can. In this,
Unless he take the course that you have done—
Commit me for committing honour—trust it,
He shall not rule me.
ANTIGONUS                   La you now, you hear.        50
When she will take the rein I let her run,
But she'll not stumble.
PAULINA (to Leontes)        Good my liege, I come—
And I beseech you hear me, who professes
Myself your loyal servant, your physician,
Your most obedient counsellor; yet that dares      55
Less appear so in comforting your evils
Than such as most seem yours—I say, I come
From your good queen.
LEONTES  Good queen?
PAULINA
Good queen, my lord, good queen, I say good queen,
And would by combat make her good, so were I      61
A man, the worst about you.
LEONTES (to Lords)            Force her hence.
PAULINA
Let him that makes but trifles of his eyes
First hand me. On mine own accord, I'll off.
But first I'll do my errand. The good Queen—       65
For she is good—hath brought you forth a daughter—
Here 'tis—commends it to your blessing.
            *She lays down the babe*
LEONTES                            Out!
A mankind witch! Hence with her, out o'door—
A most intelligencing bawd.
PAULINA                 Not so.
I am as ignorant in that as you                    70
In so entitling me, and no less honest
Than you are mad, which is enough, I'll warrant,
As this world goes, to pass for honest.
LEONTES (to Lords)                Traitors,
Will you not push her out?
(To Antigonus)            Give her the bastard.
Thou dotard, thou art woman-tired, unroosted       75
By thy Dame Partlet here. Take up the bastard,
Take't up, I say. Give't to thy crone.
PAULINA (to Antigonus)           For ever
Unvenerable be thy hands if thou
Tak'st up the princess by that forcèd baseness
Which he has put upon't.
LEONTES             He dreads his wife.           80
PAULINA
So I would you did. Then 'twere past all doubt
You'd call your children yours.
LEONTES             A nest of traitors.
ANTIGONUS
I am none, by this good light.
PAULINA               Nor I, nor any
But one that's here, and that's himself, for he
The sacred honour of himself, his queen's,         85
His hopeful son's, his babe's, betrays to slander,
Whose sting is sharper than the sword's; and will
    not—
For as the case now stands, it is a curse
He cannot be compelled to't—once remove
The root of his opinion, which is rotten           90
As ever oak or stone was sound.
LEONTES (to Lords)            A callat
Of boundless tongue, who late hath beat her husband,
And now baits me! This brat is none of mine.

It is the issue of Polixenes.
Hence with it, and together with the dam           95
Commit them to the fire.
PAULINA              It is yours,
And might we lay th'old proverb to your charge,
So like you 'tis the worse. Behold, my lords,
Although the print be little, the whole matter
And copy of the father: eye, nose, lip,           100
The trick of's frown, his forehead, nay, the valley,
The pretty dimples of his chin and cheek, his smiles,
The very mould and frame of hand, nail, finger.
And thou good goddess Nature, which hast made it
So like to him that got it, if thou hast          105
The ordering of the mind too, 'mongst all colours
No yellow in't, lest she suspect, as he does,
Her children not her husband's.
LEONTES (to Antigonus)           A gross hag!—
And lozel, thou art worthy to be hanged,
That wilt not stay her tongue.
ANTIGONUS            Hang all the husbands
That cannot do that feat, you'll leave yourself   111
Hardly one subject.
LEONTES         Once more, take her hence.
PAULINA
A most unworthy and unnatural lord
Can do no more.
LEONTES       I'll ha' thee burnt.
PAULINA                    I care not.
It is an heretic that makes the fire,             115
Not she which burns in't. I'll not call you tyrant;
But this most cruel usage of your queen—
Not able to produce more accusation
Than your own weak-hinged fancy—something
    savours
Of tyranny, and will ignoble make you,            120
Yea, scandalous to the world.
LEONTES (to Antigonus)           On your allegiance,
Out of the chamber with her! Were I a tyrant,
Where were her life? She durst not call me so
If she did know me one. Away with her!
PAULINA
I pray you do not push me, I'll be gone.          125
Look to your babe, my lord; 'tis yours. Jove send her
A better guiding spirit. What needs these hands?
You that are thus so tender o'er his follies
Will never do him good, not one of you.
So, so. Farewell, we are gone.            *Exit*
LEONTES (to Antigonus)
Thou, traitor, hast set on thy wife to this.      131
My child? Away with't! Even thou, that hast
A heart so tender o'er it, take it hence
And see it instantly consumed with fire.
Even thou, and none but thou. Take it up straight.
Within this hour bring me word 'tis done,         136
And by good testimony, or I'll seize thy life,
With what thou else call'st thine. If thou refuse
And wilt encounter with my wrath, say so.
The bastard brains with these my proper hands     140
Shall I dash out. Go, take it to the fire;
For thou set'st on thy wife.
ANTIGONUS             I did not, sir.
These lords, my noble fellows, if they please
Can clear me in't.
LORDS           We can. My royal liege,
He is not guilty of her coming hither.            145
LEONTES  You're liars all.

A LORD
Beseech your highness, give us better credit.
We have always truly served you, and beseech
So to esteem of us. And on our knees we beg,
As recompense of our dear services                     150
Past and to come, that you do change this purpose
Which, being so horrible, so bloody, must
Lead on to some foul issue. We all kneel.

LEONTES
I am a feather for each wind that blows.
Shall I live on, to see this bastard kneel             155
And call me father? Better burn it now
Than curse it then. But be it. Let it live.
It shall not neither.
(*To Antigonus*)        You, sir, come you hither,
You that have been so tenderly officious
With Lady Margery your midwife there,                  160
To save this bastard's life—for 'tis a bastard,
So sure as this beard's grey. What will you adventure
To save this brat's life?

ANTIGONUS                 Anything, my lord,
That my ability may undergo,
And nobleness impose. At least thus much,              165
I'll pawn the little blood which I have left
To save the innocent; anything possible.

LEONTES
It shall be possible. Swear by this sword
Thou wilt perform my bidding.

ANTIGONUS                          I will, my lord.

LEONTES
Mark, and perform it. Seest thou? For the fail        170
Of any point in't shall not only be
Death to thyself but to thy lewd-tongued wife,
Whom for this time we pardon. We enjoin thee,
As thou art liegeman to us, that thou carry
This female bastard hence, and that thou bear it      175
To some remote and desert place, quite out
Of our dominions; and that there thou leave it,
Without more mercy, to it own protection
And favour of the climate. As by strange fortune
It came to us, I do in justice charge thee,           180
On thy soul's peril and thy body's torture,
That thou commend it strangely to some place
Where chance may nurse or end it. Take it up.

ANTIGONUS
I swear to do this, though a present death
Had been more merciful. Come on, poor babe,           185
Some powerful spirit instruct the kites and ravens
To be thy nurses. Wolves and bears, they say,
Casting their savageness aside, have done
Like offices of pity. Sir, be prosperous
In more than this deed does require; (*to the babe*) and
blessing                                              190
Against this cruelty, fight on thy side,
Poor thing, condemned to loss.    *Exit with the babe*

LEONTES                            No, I'll not rear
Another's issue.
              *Enter a Servant*

SERVANT        Please your highness, posts
From those you sent to th'oracle are come
An hour since. Cleomenes and Dion,                    195
Being well arrived from Delphos, are both landed,
Hasting to th' court.

A LORD (*to Leontes*)    So please you, sir, their speed
Hath been beyond account.

LEONTES                    Twenty-three days
They have been absent. 'Tis good speed, foretells

The great Apollo suddenly will have                   200
The truth of this appear. Prepare you, lords.
Summon a session, that we may arraign
Our most disloyal lady; for as she hath
Been publicly accused, so shall she have
A just and open trial. While she lives                205
My heart will be a burden to me. Leave me,
And think upon my bidding.
                           *Exeunt severally*

3.1    *Enter Cleomenes and Dion*

CLEOMENES
The climate's delicate, the air most sweet;
Fertile the isle, the temple much surpassing
The common praise it bears.

DION                        I shall report,
For most it caught me, the celestial habits—
Methinks I so should term them—and the reverence    5
Of the grave wearers. O, the sacrifice—
How ceremonious, solemn, and unearthly
It was i'th' off'ring!

CLEOMENES            But of all, the burst
And the ear-deaf'ning voice o'th' oracle,
Kin to Jove's thunder, so surprised my sense         10
That I was nothing.

DION                If th'event o'th' journey
Prove as successful to the Queen—O, be't so!—
As it hath been to us rare, pleasant, speedy,
The time is worth the use on't.

CLEOMENES                       Great Apollo
Turn all to th' best! These proclamations,           15
So forcing faults upon Hermione,
I little like.

DION           The violent carriage of it
Will clear or end the business. When the oracle,
Thus by Apollo's great divine sealed up,
Shall the contents discover, something rare          20
Even then will rush to knowledge. Go. Fresh horses!
And gracious be the issue.
                                    *Exeunt*

3.2    *Enter Leontes, Lords, and Officers*

LEONTES
This sessions, to our great grief we pronounce,
Even pushes 'gainst our heart: the party tried
The daughter of a king, our wife, and one
Of us too much beloved. Let us be cleared
Of being tyrannous since we so openly                5
Proceed in justice, which shall have due course
Even to the guilt or the purgation.
Produce the prisoner.

OFFICER              It is his highness' pleasure
That the Queen appear in person here in court.
    *Enter Hermione guarded, with Paulina and Ladies*
Silence.                                             10

LEONTES  Read the indictment.

OFFICER (*reads*) Hermione, queen to the worthy Leontes,
King of Sicilia, thou art here accused and arraigned of
high treason in committing adultery with Polixenes,
King of Bohemia, and conspiring with Camillo to take
away the life of our sovereign lord the King, thy royal
husband; the pretence whereof being by circumstances
partly laid open, thou, Hermione, contrary to the faith
and allegiance of a true subject, didst counsel and aid
them for their better safety to fly away by night.   20

HERMIONE
Since what I am to say must be but that
Which contradicts my accusation, and
The testimony on my part no other
But what comes from myself, it shall scarce boot me
To say 'Not guilty'. Mine integrity                              25
Being counted falsehood shall, as I express it,
Be so received. But thus: if powers divine
Behold our human actions—as they do—
I doubt not then but innocence shall make
False accusation blush, and tyranny                            30
Tremble at patience. You, my lord, best know—
Who least will seem to do so—my past life
Hath been as continent, as chaste, as true
As I am now unhappy; which is more
Than history can pattern, though devised                       35
And played to take spectators. For behold me,
A fellow of the royal bed, which owe
A moiety of the throne; a great king's daughter,
The mother to a hopeful prince, here standing
To prate and talk for life and honour, fore                    40
Who please to come and hear. For life, I prize it
As I weigh grief, which I would spare. For honour,
'Tis a derivative from me to mine,
And only that I stand for. I appeal
To your own conscience, sir, before Polixenes                  45
Came to your court how I was in your grace,
How merited to be so; since he came,
With what encounter so uncurrent I
Have strained t'appear thus. If one jot beyond
The bound of honour, or in act or will                         50
That way inclining, hardened be the hearts
Of all that hear me, and my near'st of kin
Cry 'Fie' upon my grave.
LEONTES                             I ne'er heard yet
That any of these bolder vices wanted
Less impudence to gainsay what they did                        55
Than to perform it first.
HERMIONE                        That's true enough,
Though 'tis a saying, sir, not due to me.
LEONTES
You will not own it.
HERMIONE                    More than mistress of
Which comes to me in name of fault, I must not
At all acknowledge. For Polixenes,                             60
With whom I am accused, I do confess
I loved him as in honour he required,
With such a kind of love as might become
A lady like me; with a love, even such,
So, and no other, as yourself commanded;                       65
Which not to have done I think had been in me
Both disobedience and ingratitude
To you and toward your friend, whose love had spoke
Even since it could speak, from an infant, freely
That it was yours. Now for conspiracy,                         70
I know not how it tastes, though it be dished
For me to try how. All I know of it
Is that Camillo was an honest man;
And why he left your court, the gods themselves,
Wotting no more than I, are ignorant.                          75
LEONTES
You knew of his departure, as you know
What you have underta'en to do in's absence.
HERMIONE Sir,
You speak a language that I understand not.

My life stands in the level of your dreams,                    80
Which I'll lay down.
LEONTES                     Your actions are my 'dreams'.
You had a bastard by Polixenes,
And I but dreamed it. As you were past all shame—
Those of your fact are so—so past all truth;
Which to deny concerns more than avails; for as                85
Thy brat hath been cast out, like to itself,
No father owning it—which is indeed
More criminal in thee than it—so thou
Shalt feel our justice, in whose easiest passage
Look for no less than death.
HERMIONE                              Sir, spare your threats.   90
The bug which you would fright me with, I seek.
To me can life be no commodity.
The crown and comfort of my life, your favour,
I do give lost, for I do feel it gone
But know not how it went. My second joy,                       95
And first fruits of my body, from his presence
I am barred, like one infectious. My third comfort,
Starred most unluckily, is from my breast,
The innocent milk in it most innocent mouth,
Haled out to murder; myself on every post                      100
Proclaimed a strumpet, with immodest hatred
The childbed privilege denied, which 'longs
To women of all fashion; lastly, hurried
Here, to this place, i'th' open air, before
I have got strength of limit. Now, my liege,                   105
Tell me what blessings I have here alive,
That I should fear to die. Therefore proceed.
But yet hear this—mistake me not—no life,
I prize it not a straw; but for mine honour,
Which I would free: if I shall be condemned                    110
Upon surmises, all proofs sleeping else
But what your jealousies awake, I tell you
'Tis rigour, and not law. Your honours all,
I do refer me to the oracle.
Apollo be my judge.
A LORD                      This your request                   115
Is altogether just. Therefore bring forth,
And in Apollo's name, his oracle.
                        ⌈Exeunt certain Officers⌉
HERMIONE
The Emperor of Russia was my father.
O that he were alive, and here beholding
His daughter's trial; that he did but see                      120
The flatness of my misery—yet with eyes
Of pity, not revenge.
              ⌈Enter Officers with Cleomenes and Dion⌉
OFFICER
You here shall swear upon this sword of justice
That you, Cleomenes and Dion, have
Been both at Delphos, and from thence have brought
This sealed-up oracle, by the hand delivered                   126
Of great Apollo's priest; and that since then
You have not dared to break the holy seal,
Nor read the secrets in't.
CLEOMENES and DION All this we swear.                           130
LEONTES Break up the seals, and read.
OFFICER (reads) Hermione is chaste, Polixenes blameless,
   Camillo a true subject, Leontes a jealous tyrant, his
   innocent babe truly begotten, and the King shall live
   without an heir if that which is lost be not found.         135
LORDS
Now blessèd be the great Apollo!
HERMIONE                                    Praised!

LEONTES Hast thou read truth?
OFFICER
   Ay, my lord, even so as it is here set down.
LEONTES
   There is no truth at all i'th' oracle.
   The sessions shall proceed. This is mere falsehood.  140
        *Enter a Servant*
SERVANT
   My lord the King! The King!
LEONTES                    What is the business?
SERVANT
   O sir, I shall be hated to report it.
   The prince your son, with mere conceit and fear
   Of the Queen's speed, is gone.
LEONTES                    How, 'gone'?
SERVANT                               Is dead.
LEONTES
   Apollo's angry, and the heavens themselves  145
   Do strike at my injustice.
        *Hermione falls to the ground*
                         How now there?
PAULINA
   This news is mortal to the Queen. Look down
   And see what death is doing.
LEONTES                    Take her hence.
   Her heart is but o'ercharged. She will recover.
   I have too much believed mine own suspicion.  150
   Beseech you, tenderly apply to her
   Some remedies for life.
        *Exeunt Paulina and Ladies, carrying Hermione*
                         Apollo, pardon
   My great profaneness 'gainst thine oracle.
   I'll reconcile me to Polixenes,
   New woo my queen, recall the good Camillo,  155
   Whom I proclaim a man of truth, of mercy;
   For being transported by my jealousies
   To bloody thoughts and to revenge, I chose
   Camillo for the minister to poison
   My friend Polixenes, which had been done,  160
   But that the good mind of Camillo tardied
   My swift command. Though I with death and with
   Reward did threaten and encourage him,
   Not doing it, and being done, he, most humane
   And filled with honour, to my kingly guest  165
   Unclasped my practice, quit his fortunes here—
   Which you knew great—and to the certain hazard
   Of all incertainties himself commended,
   No richer than his honour. How he glisters
   Through my rust! And how his piety  170
   Does my deeds make the blacker!
        *Enter Paulina*
PAULINA                    Woe the while!
   O cut my lace, lest my heart, cracking it,
   Break too.
A LORD    What fit is this, good lady?
PAULINA (*to Leontes*)
   What studied torments, tyrant, hast for me?
   What wheels, racks, fires? What flaying, boiling  175
   In leads or oils? What old or newer torture
   Must I receive, whose every word deserves
   To taste of thy most worst? Thy tyranny,
   Together working with thy jealousies—
   Fancies too weak for boys, too green and idle  180
   For girls of nine—O think what they have done,
   And then run mad indeed, stark mad, for all
   Thy bygone fooleries were but spices of it.

   That thou betrayed'st Polixenes, 'twas nothing.
   That did but show thee, of a fool, inconstant,  185
   And damnable ingrateful. Nor was't much
   Thou wouldst have poisoned good Camillo's honour
   To have him kill a king—poor trespasses,
   More monstrous standing by, whereof I reckon
   The casting forth to crows thy baby daughter  190
   To be or none or little, though a devil
   Would have shed water out of fire ere done't.
   Nor is't directly laid to thee the death
   Of the young prince, whose honourable thoughts—
   Thoughts high for one so tender—cleft the heart  195
   That could conceive a gross and foolish sire
   Blemished his gracious dam. This is not, no,
   Laid to thy answer. But the last—O lords,
   When I have said, cry woe! The Queen, the Queen,
   The sweet'st, dear'st creature's dead, and vengeance
        for't                                          200
   Not dropped down yet.
A LORD                    The higher powers forbid!
PAULINA
   I say she's dead. I'll swear't. If word nor oath
   Prevail not, go and see. If you can bring
   Tincture or lustre in her lip, her eye,
   Heat outwardly or breath within, I'll serve you  205
   As I would do the gods. But O thou tyrant,
   Do not repent these things, for they are heavier
   Than all thy woes can stir. Therefore betake thee
   To nothing but despair. A thousand knees,
   Ten thousand years together, naked, fasting,  210
   Upon a barren mountain, and still winter
   In storm perpetual, could not move the gods
   To look that way thou wert.
LEONTES                    Go on, go on.
   Thou canst not speak too much. I have deserved
   All tongues to talk their bitt'rest.
A LORD (*to Paulina*)                    Say no more.  215
   Howe'er the business goes, you have made fault
   I'th' boldness of your speech.
PAULINA                    I am sorry for't.
   All faults I make, when I shall come to know them
   I do repent. Alas, I have showed too much
   The rashness of a woman. He is touched  220
   To th' noble heart. What's gone and what's past help
   Should be past grief.
   (*To Leontes*)         Do not receive affliction
   At my petition. I beseech you, rather
   Let me be punished, that have minded you
   Of what you should forget. Now, good my liege,  225
   Sir, royal sir, forgive a foolish woman.
   The love I bore your queen—lo, fool again!
   I'll speak of her no more, nor of your children.
   I'll not remember you of my own lord,
   Who is lost too. Take your patience to you,  230
   And I'll say nothing.
LEONTES                    Thou didst speak but well
   When most the truth, which I receive much better
   Than to be pitied of thee. Prithee bring me
   To the dead bodies of my queen and son.
   One grave shall be for both. Upon them shall  235
   The causes of their death appear, unto
   Our shame perpetual. Once a day I'll visit
   The chapel where they lie, and tears shed there
   Shall be my recreation. So long as nature
   Will bear up with this exercise, so long  240
   I daily vow to use it. Come, and lead me
   To these sorrows.                         *Exeunt*

**3.3**    *Enter Antigonus, carrying the babe, with a Mariner*

ANTIGONUS
Thou art perfect then our ship hath touched upon
The deserts of Bohemia?

MARINER                            Ay, my lord, and fear
We have landed in ill time. The skies look grimly
And threaten present blusters. In my conscience,
The heavens with that we have in hand are angry,    5
And frown upon's.

ANTIGONUS
Their sacred wills be done. Go get aboard.
Look to thy barque. I'll not be long before
I call upon thee.

MARINER                Make your best haste, and go not
Too far i'th' land. 'Tis like to be loud weather.    10
Besides, this place is famous for the creatures
Of prey that keep upon't.

ANTIGONUS                          Go thou away.
I'll follow instantly.

MARINER                      I am glad at heart
To be so rid o'th' business.                    *Exit*

ANTIGONUS                        Come, poor babe.
I have heard, but not believed, the spirits o'th' dead
May walk again. If such thing be, thy mother    16
Appeared to me last night, for ne'er was dream
So like a waking. To me comes a creature,
Sometimes her head on one side, some another.
I never saw a vessel of like sorrow,              20
So filled and so becoming. In pure white robes
Like very sanctity she did approach
My cabin where I lay, thrice bowed before me,
And, gasping to begin some speech, her eyes
Became two spouts. The fury spent, anon        25
Did this break from her: 'Good Antigonus,
Since fate, against thy better disposition,
Hath made thy person for the thrower-out
Of my poor babe according to thine oath,
Places remote enough are in Bohemia.            30
There weep, and leave it crying; and for the babe
Is counted lost for ever, Perdita
I prithee call't. For this ungentle business
Put on thee by my lord, thou ne'er shalt see
Thy wife Paulina more.' And so with shrieks      35
She melted into air. Affrighted much,
I did in time collect myself, and thought
This was so, and no slumber. Dreams are toys,
Yet for this once, yea superstitiously,
I will be squared by this. I do believe          40
Hermione hath suffered death, and that
Apollo would—this being indeed the issue
Of King Polixenes—it should here be laid,
Either for life or death, upon the earth
Of its right father. Blossom, speed thee well!    45
    *He lays down the babe and a scroll*
There lie, and there thy character.
    *He lays down a box*
                            There these,
Which may, if fortune please, both breed thee, pretty,
And still rest thine.
    ⌈*Thunder*⌉
                    The storm begins. Poor wretch,
That for thy mother's fault art thus exposed
To loss and what may follow! Weep I cannot,      50
But my heart bleeds, and most accursed am I
To be by oath enjoined to this. Farewell.
The day frowns more and more. Thou'rt like to have

A lullaby too rough. I never saw
The heavens so dim by day. A savage clamour!     55
Well may I get aboard. This is the chase.
I am gone for ever!          *Exit, pursued by a bear*
    *Enter an Old Shepherd*

OLD SHEPHERD I would there were no age between ten
and three-and-twenty, or that youth would sleep out
the rest; for there is nothing in the between but getting
wenches with child, wronging the ancientry, stealing,
fighting—hark you now, would any but these boiled-
brains of nineteen and two-and-twenty hunt this
weather? They have scared away two of my best sheep,
which I fear the wolf will sooner find than the master.
If anywhere I have them, 'tis by the seaside, browsing
of ivy. Good luck, an't be thy will!              67
    *He sees the babe*
What have we here? Mercy on's, a bairn! A very pretty
bairn. A boy or a child, I wonder? A pretty one, a very
pretty one. Sure some scape. Though I am not bookish,
yet I can read 'waiting-gentlewoman' in the scape. This
has been some stair-work, some trunk-work, some
behind-door-work. They were warmer that got this
than the poor thing is here. I'll take it up for pity; yet
I'll tarry till my son come. He hallooed but even now.
Whoa-ho-hoa!                                      76
    *Enter Clown*

CLOWN Hilloa, loa!
OLD SHEPHERD What, art so near? If thou'lt see a thing
to talk on when thou art dead and rotten, come hither.
What ail'st thou, man?                            80
CLOWN I have seen two such sights, by sea and by land!
But I am not to say it is a sea, for it is now the sky.
Betwixt the firmament and it you cannot thrust a
bodkin's point.
OLD SHEPHERD Why, boy, how is it?                   85
CLOWN I would you did but see how it chafes, how it
rages, how it takes up the shore. But that's not to the
point. O, the most piteous cry of the poor souls!
Sometimes to see 'em, and not to see 'em; now the
ship boring the moon with her mainmast, and anon
swallowed with yeast and froth, as you'd thrust a cork
into a hogshead. And then for the land-service, to see
how the bear tore out his shoulder-bone, how he cried
to me for help, and said his name was Antigonus, a
nobleman! But to make an end of the ship—to see how
the sea flap-dragoned it! But first, how the poor souls
roared, and the sea mocked them, and how the poor
gentleman roared, and the bear mocked him, both
roaring louder than the sea or weather.
OLD SHEPHERD Name of mercy, when was this, boy?    100
CLOWN Now, now. I have not winked since I saw these
sights. The men are not yet cold under water, nor the
bear half dined on the gentleman. He's at it now.
OLD SHEPHERD Would I had been by to have helped the
old man!                                          105
CLOWN I would you had been by the ship side, to have
helped her. There your charity would have lacked
footing.
OLD SHEPHERD Heavy matters, heavy matters. But look
thee here, boy. Now bless thyself. Thou metst with
things dying, I with things new-born. Here's a sight
for thee. Look thee, a bearing-cloth for a squire's child.
    *He points to the box*
Look thee here, take up, take up, boy. Open't. So, let's
see. It was told me I should be rich by the fairies. This
is some changeling. Open't. What's within, boy?    115

CLOWN (*opening the box*) You're a made old man. If the sins of your youth are forgiven you, you're well to live. Gold, all gold!

OLD SHEPHERD This is fairy gold, boy, and 'twill prove so. Up with't, keep it close. Home, home, the next way. We are lucky, boy, and to be so still requires nothing but secrecy. Let my sheep go. Come, good boy, the next way home.                                                123

CLOWN Go you the next way with your findings. I'll go see if the bear be gone from the gentleman, and how much he hath eaten. They are never curst but when they are hungry. If there be any of him left, I'll bury it.                                                          128

OLD SHEPHERD That's a good deed. If thou mayst discern by that which is left of him what he is, fetch me to th' sight of him.

CLOWN Marry will I; and you shall help to put him i'th' ground.                                                      133

OLD SHEPHERD 'Tis a lucky day, boy, and we'll do good deeds on't.                                              *Exeunt*

❧

**4.1**    *Enter Time, the Chorus*

TIME
I that please some, try all; both joy and terror
Of good and bad; that makes and unfolds error,
Now take upon me in the name of Time
To use my wings. Impute it not a crime
To me or my swift passage that I slide                     5
O'er sixteen years and leave the growth untried
Of that wide gap, since it is in my power
To o'erthrow law, and in one self-born hour
To plant and o'erwhelm custom. Let me pass
The same I am ere ancient'st order was              10
Or what is now received. I witness to
The times that brought them in; so shall I do
To th' freshest things now reigning, and make stale
The glistering of this present as my tale
Now seems to it. Your patience this allowing,      15
I turn my glass, and give my scene such growing
As you had slept between. Leontes leaving
Th'effects of his fond jealousies, so grieving
That he shuts up himself, imagine me,
Gentle spectators, that I now may be                    20
In fair Bohemia, and remember well
I mentionèd a son o'th' King's, which Florizel
I now name to you; and with speed so pace
To speak of Perdita, now grown in grace
Equal with wond'ring. What of her ensues              25
I list not prophesy, but let Time's news
Be known when 'tis brought forth. A shepherd's daughter
And what to her adheres, which follows after,
Is th'argument of Time. Of this allow,
If ever you have spent time worse ere now.             30
If never, yet that Time himself doth say
He wishes earnestly you never may.              *Exit*

**4.2**    *Enter Polixenes and Camillo*

POLIXENES I pray thee, good Camillo, be no more importunate. 'Tis a sickness denying thee anything, a death to grant this.

CAMILLO It is sixteen years since I saw my country. Though I have for the most part been aired abroad, I desire to lay my bones there. Besides, the penitent King,

my master, hath sent for me, to whose feeling sorrows I might be some allay—or I o'erween to think so— which is another spur to my departure.                    9

POLIXENES As thou lov'st me, Camillo, wipe not out the rest of thy services by leaving me now. The need I have of thee thine own goodness hath made. Better not to have had thee than thus to want thee. Thou, having made me businesses which none without thee can sufficiently manage, must either stay to execute thyself or take away with thee the very services thou hast done; which if I have not enough considered—as too much I cannot—to be more thankful to thee shall be my study, and my profit therein, the heaping friendships. Of that fatal country Sicilia, prithee speak no more, whose very naming punishes me with the remembrance of that penitent—as thou callest him— and reconciled King my brother, whose loss of his most precious queen and children are even now to be afresh lamented. Say to me, when sawest thou the Prince Florizel, my son? Kings are no less unhappy, their issue not being gracious, than they are in losing them when they have approved their virtues.                       28

CAMILLO Sir, it is three days since I saw the Prince. What his happier affairs may be are to me unknown; but I have missingly noted he is of late much retired from court, and is less frequent to his princely exercises than formerly he hath appeared.                                    33

POLIXENES I have considered so much, Camillo, and with some care, so far that I have eyes under my service which look upon his removedness, from whom I have this intelligence: that he is seldom from the house of a most homely shepherd, a man, they say, that from very nothing, and beyond the imagination of his neighbours, is grown into an unspeakable estate.       40

CAMILLO I have heard, sir, of such a man, who hath a daughter of most rare note. The report of her is extended more than can be thought to begin from such a cottage.

POLIXENES That's likewise part of my intelligence; but, I fear, the angle that plucks our son thither. Thou shalt accompany us to the place, where we will, not appearing what we are, have some question with the shepherd; from whose simplicity I think it not uneasy to get the cause of my son's resort thither. Prithee, be my present partner in this business, and lay aside the thoughts of Sicilia.                                          51

CAMILLO I willingly obey your command.

POLIXENES My best Camillo! We must disguise ourselves.
                                                *Exeunt*

**4.3**    *Enter Autolycus singing*

AUTOLYCUS
When daffodils begin to peer,
   With heigh, the doxy over the dale,
Why then comes in the sweet o'the year,
   For the red blood reigns in the winter's pale.

The white sheet bleaching on the hedge,          5
   With heigh, the sweet birds, O how they sing!
Doth set my pugging tooth on edge,
   For a quart of ale is a dish for a king.

The lark, that tirra-lirra chants,
   With heigh, with heigh, the thrush and the jay,
Are summer songs for me and my aunts            11
   While we lie tumbling in the hay.

I have served Prince Florizel, and in my time wore three-pile, but now I am out of service.

But shall I go mourn for that, my dear?          15
  The pale moon shines by night,
And when I wander here and there
  I then do most go right.

If tinkers may have leave to live,
  And bear the sow-skin budget,          20
Then my account I well may give,
  And in the stocks avouch it.

My traffic is sheets. When the kite builds, look to lesser linen. My father named me Autolycus, who being, as I am, littered under Mercury, was likewise a snapper-up of unconsidered trifles. With die and drab I purchased this caparison, and my revenue is the silly cheat. Gallows and knock are too powerful on the highway. Beating and hanging are terrors to me. For the life to come, I sleep out the thought of it. A prize, a prize!
  *Enter Clown*

CLOWN Let me see. Every 'leven wether tods, every tod yields pound and odd shilling. Fifteen hundred shorn, what comes the wool to?

AUTOLYCUS (*aside*) If the springe hold, the cock's mine.   34

CLOWN I cannot do't without counters. Let me see, what am I to buy for our sheep-shearing feast? Three pound of sugar, five pound of currants, rice—what will this sister of mine do with rice? But my father hath made her mistress of the feast, and she lays it on. She hath made me four-and-twenty nosegays for the shearers—three-man-song-men, all, and very good ones—but they are most of them means and basses, but one Puritan amongst them, and he sings psalms to hornpipes. I must have saffron to colour the warden pies; mace; dates, none—that's out of my note; nutmegs, seven; a race or two of ginger—but that I may beg; four pound of prunes, and as many of raisins o'th' sun.

AUTOLYCUS (*grovelling on the ground*) O, that ever I was born!          50

CLOWN I'th' name of me!

AUTOLYCUS O help me, help me! Pluck but off these rags, and then death, death!

CLOWN Alack, poor soul, thou hast need of more rags to lay on thee rather than have these off.          55

AUTOLYCUS O sir, the loathsomeness of them offend me more than the stripes I have received, which are mighty ones and millions.

CLOWN Alas, poor man, a million of beating may come to a great matter.          60

AUTOLYCUS I am robbed, sir, and beaten; my money and apparel ta'en from me, and these detestable things put upon me.

CLOWN What, by a horseman, or a footman?

AUTOLYCUS A footman, sweet sir, a footman.          65

CLOWN Indeed, he should be a footman, by the garments he has left with thee. If this be a horseman's coat it hath seen very hot service. Lend me thy hand, I'll help thee. Come, lend me thy hand.
  *He helps Autolycus up*

AUTOLYCUS O, good sir, tenderly. O!          70

CLOWN Alas, poor soul!

AUTOLYCUS O, good sir, softly, good sir! I fear, sir, my shoulder-blade is out.

CLOWN How now? Canst stand?

AUTOLYCUS Softly, dear sir. Good sir, softly.          75
  ⌈*He picks the Clown's pocket*⌉
You ha' done me a charitable office.

CLOWN (*reaching for his purse*) Dost lack any money? I have a little money for thee.

AUTOLYCUS No, good sweet sir, no, I beseech you, sir. I have a kinsman not past three-quarters of a mile hence, unto whom I was going. I shall there have money, or anything I want. Offer me no money, I pray you. That kills my heart.          83

CLOWN What manner of fellow was he that robbed you?

AUTOLYCUS A fellow, sir, that I have known to go about with troll-madams. I knew him once a servant of the Prince. I cannot tell, good sir, for which of his virtues it was, but he was certainly whipped out of the court.          89

CLOWN His vices, you would say. There's no virtue whipped out of the court. They cherish it to make it stay there; and yet it will no more but abide.

AUTOLYCUS Vices, I would say, sir. I know this man well. He hath been since an ape-bearer, then a process-server—a bailiff—then he compassed a motion of the Prodigal Son, and married a tinker's wife within a mile where my land and living lies, and having flown over many knavish professions, he settled only in rogue. Some call him Autolycus.          99

CLOWN Out upon him! Prig, for my life, prig! He haunts wakes, fairs, and bear-baitings.

AUTOLYCUS Very true, sir. He, sir, he. That's the rogue that put me into this apparel.

CLOWN Not a more cowardly rogue in all Bohemia. If you had but looked big and spit at him, he'd have run.

AUTOLYCUS I must confess to you, sir, I am no fighter. I am false of heart that way, and that he knew, I warrant him.          108

CLOWN How do you now?

AUTOLYCUS Sweet sir, much better than I was. I can stand, and walk. I will even take my leave of you, and pace softly towards my kinsman's.

CLOWN Shall I bring thee on the way?

AUTOLYCUS No, good-faced sir, no, sweet sir.          114

CLOWN Then fare thee well. I must go buy spices for our sheep-shearing.

AUTOLYCUS Prosper you, sweet sir.          *Exit the Clown*
Your purse is not hot enough to purchase your spice. I'll be with you at your sheep-shearing, too. If I make not this cheat bring out another, and the shearers prove sheep, let me be unrolled and my name put in the book of virtue.          122

(*Sings*)  Jog on, jog on, the footpath way,
    And merrily hent the stile-a.
  A merry heart goes all the day,          125
    Your sad tires in a mile-a.          *Exit*

**4.4**  *Enter Florizel dressed as Doricles a countryman,*
    *and Perdita as Queen of the Feast*

FLORIZEL
These your unusual weeds to each part of you
Does give a life; no shepherdess, but Flora
Peering in April's front. This your sheep-shearing
Is as a meeting of the petty gods,
And you the queen on't.

PERDITA          Sir, my gracious lord,          5
To chide at your extremes it not becomes me—
O, pardon that I name them! Your high self,
The gracious mark o'th' land, you have obscured
With a swain's wearing, and me, poor lowly maid,
Most goddess-like pranked up. But that our feasts          10
In every mess have folly, and the feeders

Digest it with a custom, I should blush
To see you so attired; swoon, I think,
To show myself a glass.

FLORIZEL             I bless the time
When my good falcon made her flight across     15
Thy father's ground.

PERDITA           Now Jove afford you cause!
To me the difference forges dread; your greatness
Hath not been used to fear. Even now I tremble
To think your father by some accident
Should pass this way, as you did. O, the fates!     20
How would he look to see his work, so noble,
Vilely bound up? What would he say? Or how
Should I, in these my borrowed flaunts, behold
The sternness of his presence?

FLORIZEL           Apprehend
Nothing but jollity. The gods themselves,     25
Humbling their deities to love, have taken
The shapes of beasts upon them. Jupiter
Became a bull, and bellowed; the green Neptune
A ram, and bleated; and the fire-robed god,
Golden Apollo, a poor humble swain,     30
As I seem now. Their transformations
Were never for a piece of beauty rarer,
Nor in a way so chaste, since my desires
Run not before mine honour, nor my lusts
Burn hotter than my faith.

PERDITA           O, but sir,     35
Your resolution cannot hold when 'tis
Opposed, as it must be, by th' power of the King.
One of these two must be necessities,
Which then will speak that you must change this
       purpose,
Or I my life.

FLORIZEL     Thou dearest Perdita,     40
With these forced thoughts I prithee darken not
The mirth o'th' feast. Or I'll be thine, my fair,
Or not my father's. For I cannot be
Mine own, nor anything to any, if
I be not thine. To this I am most constant,     45
Though destiny say no. Be merry, gentle;
Strangle such thoughts as these with anything
That you behold the while. Your guests are coming.
Lift up your countenance as it were the day
Of celebration of that nuptial which     50
We two have sworn shall come.

PERDITA           O Lady Fortune,
Stand you auspicious!

FLORIZEL        See, your guests approach.
Address yourself to entertain them sprightly,
And let's be red with mirth.

*Enter the Old Shepherd, with Polixenes and Camillo,*
*disguised, the Clown, Mopsa, Dorcas, and others*

OLD SHEPHERD (*to Perdita*)
Fie, daughter, when my old wife lived, upon     55
This day she was both pantler, butler, cook,
Both dame and servant, welcomed all, served all,
Would sing her song and dance her turn, now here
At upper end o'th' table, now i'th' middle,
On his shoulder, and his, her face afire     60
With labour, and the thing she took to quench it
She would to each one sip. You are retired
As if you were a feasted one and not
The hostess of the meeting. Pray you bid
These unknown friends to's welcome, for it is     65
A way to make us better friends, more known.

Come, quench your blushes, and present yourself
That which you are, mistress o'th' feast. Come on,
And bid us welcome to your sheep-shearing,
As your good flock shall prosper.

PERDITA (*to Polixenes*)        Sir, welcome.     70
It is my father's will I should take on me
The hostess-ship o'th' day.
(*To Camillo*)        You're welcome, sir.
Give me those flowers there, Dorcas. Reverend sirs,
For you there's rosemary and rue. These keep
Seeming and savour all the winter long.     75
Grace and remembrance be to you both,
And welcome to our shearing.

POLIXENES        Shepherdess,
A fair one are you. Well you fit our ages
With flowers of winter.

PERDITA        Sir, the year growing ancient,
Not yet on summer's death, nor on the birth     80
Of trembling winter, the fairest flowers o'th' season
Are our carnations and streaked gillyvors,
Which some call nature's bastards. Of that kind
Our rustic garden's barren, and I care not
To get slips of them.

POLIXENES        Wherefore, gentle maiden,     85
Do you neglect them?

PERDITA        For I have heard it said
There is an art which in their piedness shares
With great creating nature.

POLIXENES        Say there be,
Yet nature is made better by no mean
But nature makes that mean. So over that art     90
Which you say adds to nature is an art
That nature makes. You see, sweet maid, we marry
A gentler scion to the wildest stock,
And make conceive a bark of baser kind
By bud of nobler race. This is an art     95
Which does mend nature—change it rather; but
The art itself is nature.

PERDITA        So it is.

POLIXENES
Then make your garden rich in gillyvors,
And do not call them bastards.

PERDITA        I'll not put
The dibble in earth to set one slip of them,     100
No more than, were I painted, I would wish
This youth should say 'twere well, and only therefore
Desire to breed by me. Here's flowers for you:
Hot lavender, mints, savory, marjoram,
The marigold, that goes to bed wi'th' sun,     105
And with him rises, weeping. These are flowers
Of middle summer, and I think they are given
To men of middle age. You're very welcome.

*She gives them flowers*

CAMILLO
I should leave grazing were I of your flock,
And only live by gazing.

PERDITA        Out, alas,     110
You'd be so lean that blasts of January
Would blow you through and through.
(*To Florizel*)        Now, my fair'st friend,
I would I had some flowers o'th' spring that might
Become your time of day; (*to Mopsa and Dorcas*) and
yours, and yours,
That wear upon your virgin branches yet     115
Your maidenheads growing. O Proserpina,
For the flowers now that, frighted, thou letst fall

From Dis's wagon!—daffodils,
That come before the swallow dares, and take
The winds of March with beauty; violets, dim,   120
But sweeter than the lids of Juno's eyes
Or Cytherea's breath; pale primroses,
That die unmarried ere they can behold
Bright Phoebus in his strength—a malady
Most incident to maids; bold oxlips, and   125
The crown imperial; lilies of all kinds,
The flower-de-luce being one. O, these I lack,
To make you garlands of, and my sweet friend,
To strew him o'er and o'er.
FLORIZEL               What, like a corpse?
PERDITA
No, like a bank, for love to lie and play on,   130
Not like a corpse—or if, not to be buried,
But quick and in mine arms. Come, take your flowers.
Methinks I play as I have seen them do
In Whitsun pastorals. Sure this robe of mine
Does change my disposition.
FLORIZEL            What you do   135
Still betters what is done. When you speak, sweet,
I'd have you do it ever; when you sing,
I'd have you buy and sell so, so give alms,
Pray so; and for the ord'ring your affairs,
To sing them too. When you do dance, I wish you   140
A wave o'th' sea, that you might ever do
Nothing but that, move still, still so,
And own no other function. Each your doing,
So singular in each particular,
Crowns what you are doing in the present deeds,   145
That all your acts are queens.
PERDITA             O Doricles,
Your praises are too large. But that your youth
And the true blood which peeps so fairly through't
Do plainly give you out an unstained shepherd,
With wisdom I might fear, my Doricles,   150
You wooed me the false way.
FLORIZEL            I think you have
As little skill to fear as I have purpose
To put you to't. But come, our dance, I pray;
Your hand, my Perdita. So turtles pair,
That never mean to part.
PERDITA          I'll swear for 'em.   155
POLIXENES (to Camillo)
This is the prettiest low-born lass that ever
Ran on the greensward. Nothing she does or seems
But smacks of something greater than herself,
Too noble for this place.
CAMILLO         He tells her something
That makes her blood look out. Good sooth, she is   160
The queen of curds and cream.
CLOWN            Come on, strike up!
DORCAS Mopsa must be your mistress. Marry, garlic to
mend her kissing with!
MOPSA Now, in good time!
CLOWN Not a word, a word, we stand upon our manners.
Come, strike up!   166
   *Music. Here a dance of shepherds and shepherdesses*
POLIXENES
Pray, good shepherd, what fair swain is this
Which dances with your daughter?
OLD SHEPHERD
They call him Doricles, and boasts himself
To have a worthy feeding; but I have it   170
Upon his own report, and I believe it.

He looks like sooth. He says he loves my daughter.
I think so, too, for never gazed the moon
Upon the water as he'll stand and read,
As 'twere, my daughter's eyes; and to be plain,   175
I think there is not half a kiss to choose
Who loves another best.
POLIXENES          She dances featly.
OLD SHEPHERD
So she does anything, though I report it
That should be silent. If young Doricles
Do light upon her, she shall bring him that   180
Which he not dreams of.
   *Enter a Servant*
SERVANT O, master, if you did but hear the pedlar at the
door, you would never dance again after a tabor and
pipe. No, the bagpipe could not move you. He sings
several tunes faster than you'll tell money. He utters
them as he had eaten ballads, and all men's ears grew
to his tunes.   187
CLOWN He could never come better. He shall come in. I
love a ballad but even too well, if it be doleful matter
merrily set down, or a very pleasant thing indeed, and
sung lamentally.   191
SERVANT He hath songs for man or woman, of all sizes.
No milliner can so fit his customers with gloves. He
has the prettiest love songs for maids, so without
bawdry, which is strange, with such delicate burdens
of dildos and fadings, 'Jump her, and thump her'; and
where some stretch-mouthed rascal would, as it were,
mean mischief and break a foul gap into the matter,
he makes the maid to answer, 'Whoop, do me no harm,
good man'; puts him off, slights him, with 'Whoop, do
me no harm, good man!'   201
POLIXENES This is a brave fellow.
CLOWN Believe me, thou talkest of an admirable conceited
fellow. Has he any unbraided wares?
SERVANT He hath ribbons of all the colours i'th' rainbow;
points more than all the lawyers in Bohemia can
learnedly handle, though they come to him by th'
gross; inkles, caddises, cambrics, lawns—why, he sings
'em over as they were gods or goddesses. You would
think a smock were a she-angel, he so chants to the
sleeve-hand and the work about the square on't.   211
CLOWN Prithee bring him in, and let him approach
singing.
PERDITA Forewarn him that he use no scurrilous words
in's tunes.               *Exit Servant*
CLOWN You have of these pedlars that have more in them
than you'd think, sister.   217
PERDITA Ay, good brother, or go about to think.
   *Enter Autolycus, wearing a false beard, carrying his
   pack, and singing*
AUTOLYCUS
Lawn as white as driven snow,
Cypress black as e'er was crow,   220
Gloves as sweet as damask roses,
Masks for faces, and for noses;
Bugle-bracelet, necklace amber,
Perfume for a lady's chamber;
Golden coifs, and stomachers   225
For my lads to give their dears;
Pins and poking-sticks of steel,
What maids lack from head to heel
Come buy of me, come, come buy, come buy,
Buy, lads, or else your lasses cry. Come buy!   230

CLOWN If I were not in love with Mopsa thou shouldst take no money of me, but being enthralled as I am, it will also be the bondage of certain ribbons and gloves.

MOPSA I was promised them against the feast, but they come not too late now.                               235

DORCAS He hath promised you more than that, or there be liars.

MOPSA He hath paid you all he promised you. Maybe he has paid you more, which will shame you to give him again.                                                240

CLOWN Is there no manners left among maids? Will they wear their plackets where they should bear their faces? Is there not milking-time, when you are going to bed, or kiln-hole, to whistle of these secrets, but you must be tittle-tattling before all our guests? 'Tis well they are whispering. Clammer your tongues, and not a word more.                                                247

MOPSA I have done. Come, you promised me a tawdry-lace and a pair of sweet gloves.

CLOWN Have I not told thee how I was cozened by the way, and lost all my money?                              251

AUTOLYCUS And indeed, sir, there are cozeners abroad, therefore it behoves men to be wary.

CLOWN Fear not thou, man, thou shalt lose nothing here.

AUTOLYCUS I hope so, sir, for I have about me many parcels of charge.                                       256

CLOWN What hast here? Ballads?

MOPSA Pray now, buy some. I love a ballad in print, alife, for then we are sure they are true.

AUTOLYCUS Here's one to a very doleful tune, how a usurer's wife was brought to bed of twenty money-bags at a burden, and how she longed to eat adders' heads and toads carbonadoed.

MOPSA Is it true, think you?

AUTOLYCUS Very true, and but a month old.           265

DORCAS Bless me from marrying a usurer!

AUTOLYCUS Here's the midwife's name to't, one Mistress Tail-Porter, and five or six honest wives' that were present. Why should I carry lies abroad?

MOPSA (to Clown) Pray you now, buy it.                270

CLOWN Come on, lay it by, and let's first see more ballads. We'll buy the other things anon.

AUTOLYCUS Here's another ballad, of a fish that appeared upon the coast on Wednesday the fourscore of April, forty thousand fathom above water, and sung this ballad against the hard hearts of maids. It was thought she was a woman, and was turned into a cold fish for she would not exchange flesh with one that loved her. The ballad is very pitiful, and as true.

DORCAS Is it true too, think you?                     280

AUTOLYCUS Five justices' hands at it, and witnesses more than my pack will hold.

CLOWN Lay it by, too. Another.

AUTOLYCUS This is a merry ballad, but a very pretty one.

MOPSA Let's have some merry ones.                     285

AUTOLYCUS Why, this is a passing merry one, and goes to the tune of 'Two Maids Wooing a Man'. There's scarce a maid westward but she sings it. 'Tis in request, I can tell you.

MOPSA We can both sing it. If thou'lt bear a part thou shalt hear; 'tis in three parts.                        291

DORCAS We had the tune on't a month ago.

AUTOLYCUS I can bear my part, you must know, 'tis my occupation. Have at it with you.

*They sing*

AUTOLYCUS
    Get you hence, for I must go                    295
    Where it fits not you to know.
DORCAS    Whither?
MOPSA           O whither?
DORCAS                Whither?
MOPSA    It becomes thy oath full well
    Thou to me thy secrets tell.
DORCAS    Me too. Let me go thither.                   300
MOPSA    Or thou go'st to th' grange or mill,
DORCAS    If to either, thou dost ill.
AUTOLYCUS    Neither.
DORCAS        What neither?
AUTOLYCUS             Neither.
DORCAS    Thou hast sworn my love to be.
MOPSA    Thou hast sworn it more to me.                305
    Then whither goest? Say, whither?

CLOWN We'll have this song out anon by ourselves. My father and the gentlemen are in sad talk, and we'll not trouble them. Come, bring away thy pack after me. Wenches, I'll buy for you both. Pedlar, let's have the first choice. Follow me, girls.                         311
          *Exit with Dorcas and Mopsa*
AUTOLYCUS And you shall pay well for 'em.

(*Sings*)    Will you buy any tape,
    Or lace for your cape,
    My dainty duck, my dear-a?                        315
    Any silk, any thread,
    Any toys for your head,
    Of the new'st and fin'st, fin'st wear-a?
    Come to the pedlar,
    Money's a meddler,                               320
    That doth utter all men's ware-a.        *Exit*

*Enter Servant*

SERVANT Master, there is three carters, three shepherds, three neatherds, three swineherds that have made themselves all men of hair. They call themselves saultiers, and they have a dance which the wenches say is a gallimaufry of gambols, because they are not in't. But they themselves are o' th' mind, if it be not too rough for some that know little but bowling, it will please plentifully.                                     329

OLD SHEPHERD Away. We'll none on't. Here has been too much homely foolery already. (*To Polixenes*) I know, sir, we weary you.

POLIXENES You weary those that refresh us. Pray, let's see these four threes of herdsmen.                    334

SERVANT One three of them, by their own report, sir, hath danced before the King, and not the worst of the three but jumps twelve foot and a half by th' square.

OLD SHEPHERD Leave your prating. Since these good men are pleased, let them come in—but quickly, now.

SERVANT Why, they stay at door, sir.                   340
    *Here a dance of twelve satyrs*
POLIXENES (*to the Old Shepherd*)
    O, father, you'll know more of that hereafter.
    (*To Camillo*) Is it not too far gone? 'Tis time to part them.
    He's simple, and tells much.
    (*To Florizel*)        How now, fair shepherd,
    Your heart is full of something that does take
    Your mind from feasting. Sooth, when I was young
    And handed love as you do, I was wont          346

To load my she with knacks. I would have ransacked
The pedlar's silken treasury, and have poured it
To her acceptance. You have let him go,
And nothing marted with him. If your lass          350
Interpretation should abuse, and call this
Your lack of love or bounty, you were straited
For a reply, at least if you make a care
Of happy holding her.
FLORIZEL                    Old sir, I know
She prizes not such trifles as these are.          355
The gifts she looks from me are packed and locked
Up in my heart, which I have given already,
But not delivered.
(*To Perdita*)          O, hear me breathe my life
Before this ancient sir, who, it should seem,
Hath sometime loved. I take thy hand, this hand          360
As soft as dove's down, and as white as it,
Or Ethiopian's tooth, or the fanned snow that's bolted
By th' northern blasts twice o'er.
POLIXENES                    What follows this?
How prettily the young swain seems to wash
The hand was fair before! I have put you out.          365
But to your protestation. Let me hear
What you profess.
FLORIZEL          Do, and be witness to't.
POLIXENES
And this my neighbour too?
FLORIZEL                    And he, and more
Than he; and men, the earth, the heavens, and all,
That were I crowned the most imperial monarch,          370
Thereof most worthy, were I the fairest youth
That ever made eye swerve, had force and knowledge
More than was ever man's, I would not prize them
Without her love; for her employ them all,
Commend them and condemn them to her service          375
Or to their own perdition.
POLIXENES                    Fairly offered.
CAMILLO
This shows a sound affection.
OLD SHEPHERD                    But, my daughter,
Say you the like to him?
PERDITA                    I cannot speak
So well, nothing so well, no, nor mean better.
By th' pattern of mine own thoughts I cut out          380
The purity of his.
OLD SHEPHERD          Take hands, a bargain;
And, friends unknown, you shall bear witness to't.
I give my daughter to him, and will make
Her portion equal his.
FLORIZEL                    O, that must be
I'th' virtue of your daughter. One being dead,          385
I shall have more than you can dream of yet,
Enough then for your wonder. But come on,
Contract us fore these witnesses.
OLD SHEPHERD                    Come, your hand;
And, daughter, yours.
POLIXENES          Soft, swain, a while, beseech you.
Have you a father?          390
FLORIZEL I have. But what of him?
POLIXENES Knows he of this?
FLORIZEL He neither does nor shall.
POLIXENES Methinks a father
Is at the nuptial of his son a guest          395
That best becomes the table. Pray you once more,
Is not your father grown incapable
Of reasonable affairs? Is he not stupid
With age and alt'ring rheums? Can he speak, hear,

Know man from man? Dispute his own estate?          400
Lies he not bed-rid, and again does nothing
But what he did being childish?
FLORIZEL                    No, good sir.
He has his health, and ampler strength indeed
Than most have of his age.
POLIXENES                    By my white beard,
You offer him, if this be so, a wrong          405
Something unfilial. Reason my son
Should choose himself a wife, but as good reason
The father, all whose joy is nothing else
But fair posterity, should hold some counsel
In such a business.
FLORIZEL                    I yield all this;          410
But for some other reasons, my grave sir,
Which 'tis not fit you know, I not acquaint
My father of this business.
POLIXENES                    Let him know't.
FLORIZEL
He shall not.
POLIXENES                    Prithee let him.
FLORIZEL                    No, he must not.
OLD SHEPHERD
Let him, my son. He shall not need to grieve          415
At knowing of thy choice.
FLORIZEL                    Come, come, he must not.
Mark our contract.
POLIXENES (*removing his disguise*)
                    Mark your divorce, young sir,
Whom son I dare not call. Thou art too base
To be acknowledged. Thou a sceptre's heir,
That thus affects a sheep-hook?
(*To the Old Shepherd*)          Thou, old traitor,          420
I am sorry that by hanging thee I can but
Shorten thy life one week.
(*To Perdita*)          And thou, fresh piece
Of excellent witchcraft, who of force must know
The royal fool thou cop'st with—
OLD SHEPHERD                    O, my heart!
POLIXENES
I'll have thy beauty scratched with briers and made
More homely than thy state.
(*To Florizel*)          For thee, fond boy,          426
If I may ever know thou dost but sigh
That thou no more shalt see this knack, as never
I mean thou shalt, we'll bar thee from succession,
Not hold thee of our blood, no, not our kin,          430
Farre than Deucalion off. Mark thou my words.
Follow us to the court.
(*To the Old Shepherd*)          Thou churl, for this time,
Though full of our displeasure, yet we free thee
From the dead blow of it.
(*To Perdita*)          And you, enchantment,
Worthy enough a herdsman—yea, him too,          435
That makes himself, but for our honour therein,
Unworthy thee—if ever henceforth thou
These rural latches to his entrance open,
Or hoop his body more with thy embraces,
I will devise a death as cruel for thee          440
As thou art tender to't.          *Exit*
PERDITA                    Even here undone.
I was not much afeard, for once or twice
I was about to speak, and tell him plainly
The selfsame sun that shines upon his court
Hides not his visage from our cottage, but          445
Looks on alike. Will't please you, sir, be gone?
I told you what would come of this. Beseech you,

Of your own state take care. This dream of mine
Being now awake, I'll queen it no inch farther,
But milk my ewes and weep.

CAMILLO (*to the Old Shepherd*)　　Why, how now, father?
Speak ere thou diest.

OLD SHEPHERD　　　　I cannot speak, nor think,　451
Nor dare to know that which I know.
(*To Florizel*)　　　　　　　　O sir,
You have undone a man of fourscore-three,
That thought to fill his grave in quiet, yea,
To die upon the bed my father died,　455
To lie close by his honest bones. But now
Some hangman must put on my shroud, and lay me
Where no priest shovels in dust.
(*To Perdita*)　　　　　　　O cursed wretch,
That knew'st this was the Prince, and wouldst
　adventure
To mingle faith with him. Undone, undone!　460
If I might die within this hour, I have lived
To die when I desire.　　　　　　　*Exit*

FLORIZEL (*to Perdita*)　　Why look you so upon me?
I am but sorry, not afeard; delayed,
But nothing altered. What I was, I am,
More straining on for plucking back, not following　465
My leash unwillingly.

CAMILLO　　　　　Gracious my lord,
You know your father's temper. At this time
He will allow no speech—which I do guess
You do not purpose to him; and as hardly
Will he endure your sight as yet, I fear.　470
Then till the fury of his highness settle,
Come not before him.

FLORIZEL　　　　　I not purpose it.
I think, Camillo?

CAMILLO　　　　Even he, my lord.

PERDITA (*to Florizel*)
How often have I told you 'twould be thus?
How often said my dignity would last　475
But till 'twere known?

FLORIZEL　　　　It cannot fail but by
The violation of my faith, and then
Let nature crush the sides o'th' earth together
And mar the seeds within. Lift up thy looks.
From my succession wipe me, father! I　480
Am heir to my affection.

CAMILLO　　　　　Be advised.

FLORIZEL
I am, and by my fancy. If my reason
Will thereto be obedient, I have reason.
If not, my senses, better pleased with madness,
Do bid it welcome.

CAMILLO　　　This is desperate, sir.　485

FLORIZEL
So call it. But it does fulfil my vow.
I needs must think it honesty. Camillo,
Not for Bohemia, nor the pomp that may
Be thereat gleaned; for all the sun sees, or
The close earth wombs, or the profound seas hides
In unknown fathoms, will I break my oath　491
To this my fair beloved. Therefore, I pray you,
As you have ever been my father's honoured friend,
When he shall miss me—as, in faith, I mean not
To see him any more—cast your good counsels　495
Upon his passion. Let myself and fortune
Tug for the time to come. This you may know,
And so deliver: I am put to sea

With her who here I cannot hold on shore;
And most opportune to her need, I have　500
A vessel rides fast by, but not prepared
For this design. What course I mean to hold
Shall nothing benefit your knowledge, nor
Concern me the reporting.

CAMILLO　　　　　O my lord,
I would your spirit were easier for advice,　505
Or stronger for your need.

FLORIZEL　　　　　Hark, Perdita—
(*To Camillo*) I'll hear you by and by.

CAMILLO (*aside*)　　　　　He's irremovable,
Resolved for flight. Now were I happy if
His going I could frame to serve my turn,
Save him from danger, do him love and honour,　510
Purchase the sight again of dear Sicilia
And that unhappy king, my master, whom
I so much thirst to see.

FLORIZEL　　　Now, good Camillo,
I am so fraught with curious business that
I leave out ceremony.

CAMILLO　　　　Sir, I think　515
You have heard of my poor services i'th' love
That I have borne your father?

FLORIZEL　　　　　Very nobly
Have you deserved. It is my father's music
To speak your deeds, not little of his care
To have them recompensed as thought on.

CAMILLO　　　　　　Well, my lord,
If you may please to think I love the King,　521
And through him what's nearest to him, which is
Your gracious self, embrace but my direction,
If your more ponderous and settled project
May suffer alteration. On mine honour,　525
I'll point you where you shall have such receiving
As shall become your highness, where you may
Enjoy your mistress—from the whom I see
There's no disjunction to be made but by,
As heavens forfend, your ruin—marry her,　530
And with my best endeavours in your absence
Your discontenting father strive to qualify
And bring him up to liking.

FLORIZEL　　　　　How, Camillo,
May this, almost a miracle, be done?—
That I may call thee something more than man,　535
And after that trust to thee.

CAMILLO　　　　　Have you thought on
A place whereto you'll go?

FLORIZEL　　　　Not any yet.
But as th'unthought-on accident is guilty
To what we wildly do, so we profess
Ourselves to be the slaves of chance, and flies　540
Of every wind that blows.

CAMILLO　　　　Then list to me.
This follows, if you will not change your purpose
But undergo this flight: make for Sicilia,
And there present yourself and your fair princess,
For so I see she must be, fore Leontes.　545
She shall be habited as it becomes
The partner of your bed. Methinks I see
Leontes opening his free arms and weeping
His welcomes forth; asks thee there 'Son, forgiveness!'
As 'twere i'th' father's person, kisses the hands　550
Of your fresh princess; o'er and o'er divides him
'Twixt his unkindness and his kindness. Th'one
He chides to hell, and bids the other grow

Faster than thought or time.

**FLORIZEL**                          Worthy Camillo,
What colour for my visitation shall I                                    555
Hold up before him?

**CAMILLO**                          Sent by the King your father
To greet him, and to give him comforts. Sir,
The manner of your bearing towards him, with
What you, as from your father, shall deliver—
Things known betwixt us three—I'll write you down,
The which shall point you forth at every sitting   561
What you must say, that he shall not perceive
But that you have your father's bosom there,
And speak his very heart.

**FLORIZEL**                          I am bound to you.
There is some sap in this.

**CAMILLO**                          A course more promising
Than a wild dedication of yourselves                          566
To unpathed waters, undreamed shores; most certain,
To miseries enough—no hope to help you,
But as you shake off one, to take another;
Nothing so certain as your anchors, who            570
Do their best office if they can but stay you
Where you'll be loath to be. Besides, you know,
Prosperity's the very bond of love,
Whose fresh complexion and whose heart together
Affliction alters.

**PERDITA**                          One of these is true.                          575
I think affliction may subdue the cheek
But not take in the mind.

**CAMILLO**                          Yea, say you so?
There shall not at your father's house these seven
                                                              years
Be born another such.

**FLORIZEL**                          My good Camillo,
She's as forward of her breeding as                          580
She is i'th' rear our birth.

**CAMILLO**                          I cannot say 'tis pity
She lacks instructions, for she seems a mistress
To most that teach.

**PERDITA**                          Your pardon, sir. For this
I'll blush you thanks.

**FLORIZEL**                          My prettiest Perdita!
But O, the thorns we stand upon! Camillo,           585
Preserver of my father, now of me,
The medicine of our house, how shall we do?
We are not furnished like Bohemia's son,
Nor shall appear so in Sicilia.

**CAMILLO** My lord,                                                       590
Fear none of this. I think you know my fortunes
Do all lie there. It shall be so my care
To have you royally appointed as if
The scene you play were mine. For instance, sir,   594
That you may know you shall not want—one word.
            *They speak apart.*
            *Enter Autolycus*

**AUTOLYCUS** Ha, ha! What a fool honesty is, and trust—
his sworn brother—a very simple gentleman! I have
sold all my trumpery; not a counterfeit stone, not a
ribbon, glass, pomander, brooch, table-book, ballad,
knife, tape, glove, shoe-tie, bracelet, horn-ring to keep
my pack from fasting. They throng who should buy
first, as if my trinkets had been hallowed, and brought
a benediction to the buyer; by which means I saw
whose purse was best in picture; and what I saw, to
my good use I remembered. My clown, who wants but
something to be a reasonable man, grew so in love
with the wenches' song that he would not stir his

pettitoes till he had both tune and words, which so
drew the rest of the herd to me that all their other
senses stuck in ears. You might have pinched a placket,
it was senseless. 'Twas nothing to geld a codpiece of a
purse. I could have filed keys off that hung in chains.
No hearing, no feeling but my sir's song, and admiring
the nothing of it. So that in this time of lethargy I
picked and cut most of their festival purses, and had
not the old man come in with a hubbub against his
daughter and the King's son, and scared my choughs
from the chaff, I had not left a purse alive in the whole
army.
            *Camillo, Florizel, and Perdita come forward*

**CAMILLO**
Nay, but my letters by this means being there       620
So soon as you arrive shall clear that doubt.

**FLORIZEL**
And those that you'll procure from King Leontes—

**CAMILLO**
Shall satisfy your father.

**PERDITA**                          Happy be you!
All that you speak shows fair.

**CAMILLO** (*seeing Autolycus*)       Who have we here?
We'll make an instrument of this, omit             625
Nothing may give us aid.

**AUTOLYCUS** (*aside*) If they have overheard me now—why,
hanging!

**CAMILLO** How now, good fellow? Why shakest thou so?
Fear not, man. Here's no harm intended to thee.   630

**AUTOLYCUS** I am a poor fellow, sir.

**CAMILLO** Why, be so still. Here's nobody will steal that
from thee. Yet for the outside of thy poverty, we must
make an exchange. Therefore discase thee instantly—
thou must think there's a necessity in't—and change
garments with this gentleman. Though the pennyworth
on his side be the worst, yet hold thee, (*giving him
money*) there's some boot.

**AUTOLYCUS** I am a poor fellow, sir. (*Aside*) I know ye well
enough.                                                              640

**CAMILLO** Nay prithee, dispatch—the gentleman is half
flayed already.

**AUTOLYCUS** Are you in earnest, sir? (*Aside*) I smell the
trick on't.

**FLORIZEL** Dispatch, I prithee.                                   645

**AUTOLYCUS** Indeed, I have had earnest, but I cannot with
conscience take it.

**CAMILLO** Unbuckle, unbuckle.
            *Florizel and Autolycus exchange clothes*
(*To Perdita*) Fortunate mistress—let my prophecy
Come home to ye!—you must retire yourself        650
Into some covert, take your sweetheart's hat
And pluck it o'er your brows, muffle your face,
Dismantle you, and, as you can, disliken
The truth of your own seeming, that you may—
For I do fear eyes—over to shipboard                655
Get undescried.

**PERDITA**                          I see the play so lies
That I must bear a part.

**CAMILLO**                          No remedy.
(*To Florizel*) Have you done there?

**FLORIZEL**                          Should I now meet my father
He would not call me son.

**CAMILLO**                          Nay, you shall have no hat.
            *He gives the hat to Perdita*
Come, lady, come. Farewell, my friend.

**AUTOLYCUS**                          Adieu, sir.   660

FLORIZEL
O Perdita, what have we twain forgot!
Pray you, a word.
*They speak aside*
CAMILLO (*aside*)
What I do next shall be to tell the King
Of this escape, and whither they are bound;
Wherein my hope is I shall so prevail                    665
To force him after, in whose company
I shall re-view Sicilia, for whose sight
I have a woman's longing.
FLORIZEL                          Fortune speed us!
Thus we set on, Camillo, to th' seaside.
CAMILLO  The swifter speed the better.                   670
          *Exeunt Florizel, Perdita, and Camillo*
AUTOLYCUS  I understand the business, I hear it. To have
an open ear, a quick eye, and a nimble hand is
necessary for a cutpurse. A good nose is requisite also,
to smell out work for th'other senses. I see this is the
time that the unjust man doth thrive. What an
exchange had this been without boot! What a boot is
here with this exchange! Sure the gods do this year
connive at us, and we may do anything extempore.
The Prince himself is about a piece of iniquity, stealing
away from his father with his clog at his heels. If I
thought it were a piece of honesty to acquaint the King
withal, I would not do't. I hold it the more knavery
to conceal it, and therein am I constant to my profession.
          *Enter the Clown and the Old Shepherd, carrying a*
          *fardel and a box*
Aside, aside! Here is more matter for a hot brain. Every
lane's end, every shop, church, session, hanging, yields
a careful man work.                                      686
CLOWN  See, see, what a man you are now! There is no
other way but to tell the King she's a changeling, and
none of your flesh and blood.
OLD SHEPHERD  Nay, but hear me.                          690
CLOWN  Nay, but hear *me*.
OLD SHEPHERD  Go to, then.
CLOWN  She being none of your flesh and blood, your flesh
and blood has not offended the King, and so your flesh
and blood is not to be punished by him. Show those
things you found about her, those secret things, all but
what she has with her. This being done, let the law go
whistle, I warrant you.                                  698
OLD SHEPHERD  I will tell the King all, every word, yea,
and his son's pranks, too, who, I may say, is no honest
man, neither to his father nor to me, to go about to
make me the King's brother-in-law.
CLOWN  Indeed, brother-in-law was the farthest off you
could have been to him, and then your blood had been
the dearer by I know not how much an ounce.             705
AUTOLYCUS (*aside*)  Very wisely, puppies.
OLD SHEPHERD  Well, let us to the King. There is that in
this fardel will make him scratch his beard.
AUTOLYCUS (*aside*)  I know not what impediment this
complaint may be to the flight of my master.            710
CLOWN  Pray heartily he be at' palace.
AUTOLYCUS (*aside*)  Though I am not naturally honest, I
am so sometimes by chance. Let me pocket up my
pedlar's excrement.
          *He removes his false beard*
—How now, rustics, whither are you bound?               715
OLD SHEPHERD  To th' palace, an it like your worship.
AUTOLYCUS  Your affairs there? What? With whom? The
condition of that fardel? The place of your dwelling?

Your names? Your ages? Of what having, breeding,
and anything that is fitting to be known, discover.     720
CLOWN  We are but plain fellows, sir.
AUTOLYCUS  A lie, you are rough and hairy. Let me have
no lying. It becomes none but tradesmen, and they
often give us soldiers the lie, but we pay them for it
with stamped coin, not stabbing steel, therefore they
do not *give* us the lie.                               726
CLOWN  Your worship had like to have given us one if you
had not taken yourself with the manner.
OLD SHEPHERD  Are you a courtier, an't like you, sir?
AUTOLYCUS  Whether it like me or no, I am a courtier.
Seest thou not the air of the court in these enfoldings?
Hath not my gait in it the measure of the court?
Receives not thy nose court-odour from me? Reflect I
not on thy baseness court-contempt? Thinkest thou,
for that I insinuate to toze from thee thy business, I
am therefore no courtier? I am courtier cap-à-pie, and
one that will either push on or pluck back thy business
there. Whereupon I command thee to open thy affair.
OLD SHEPHERD  My business, sir, is to the King.
AUTOLYCUS  What advocate hast thou to him?              740
OLD SHEPHERD  I know not, an't like you.
CLOWN (*aside to the Old Shepherd*)  'Advocate' 's the court
word for a pheasant. Say you have none.
OLD SHEPHERD
None, sir. I have no pheasant, cock nor hen.
AUTOLYCUS (*aside*)
How blessed are we that are not simple men!             745
Yet nature might have made me as these are,
Therefore I will not disdain.
CLOWN  This cannot be but a great courtier.
OLD SHEPHERD  His garments are rich, but he wears them
not handsomely.                                         750
CLOWN  He seems to be the more noble in being fantastical.
A great man, I'll warrant. I know by the picking on's
teeth.
AUTOLYCUS  The fardel there, what's i'th' fardel? Wherefore
that box?                                               755
OLD SHEPHERD  Sir, there lies such secrets in this fardel
and box which none must know but the King, and
which he shall know within this hour, if I may come
to th' speech of him.
AUTOLYCUS  Age, thou hast lost thy labour.              760
OLD SHEPHERD  Why, sir?
AUTOLYCUS  The King is not at the palace, he is gone
aboard a new ship to purge melancholy and air himself;
for if thou beest capable of things serious, thou must
know the King is full of grief.                         765
OLD SHEPHERD  So 'tis said, sir; about his son, that should
have married a shepherd's daughter.
AUTOLYCUS  If that shepherd be not in handfast, let him
fly. The curses he shall have, the tortures he shall feel,
will break the back of man, the heart of monster.       770
CLOWN  Think you so, sir?
AUTOLYCUS  Not he alone shall suffer what wit can make
heavy and vengeance bitter, but those that are germane
to him, though removed fifty times, shall all come
under the hangman, which, though it be great pity,
yet it is necessary. An old sheep-whistling rogue, a
ram-tender, to offer to have his daughter come into
grace! Some say he shall be stoned; but that death is
too soft for him, say I. Draw our throne into a
sheepcote? All deaths are too few, the sharpest too
easy.                                                   781
CLOWN  Has the old man e'er a son, sir, do you hear, an't
like you, sir?

AUTOLYCUS He has a son, who shall be flayed alive, then 'nointed over with honey, set on the head of a wasps' nest, then stand till he be three-quarters-and-a-dram dead, then recovered again with aqua-vitae, or some other hot infusion, then, raw as he is, and in the hottest day prognostication proclaims, shall he be set against a brick wall, the sun looking with a southward eye upon him, where he is to behold him with flies blown to death. But what talk we of these traitorly rascals, whose miseries are to be smiled at, their offences being so capital? Tell me, for you seem to be honest plain men, what you have to the King. Being something gently considered, I'll bring you where he is aboard, tender your persons to his presence, whisper him in your behalfs, and if it be in man, besides the King, to effect your suits, here is man shall do it.                         799

CLOWN (to the Old Shepherd) He seems to be of great authority. Close with him, give him gold; and though authority be a stubborn bear, yet he is oft led by the nose with gold. Show the inside of your purse to the outside of his hand, and no more ado. Remember— 'stoned', and 'flayed alive'.                         805

OLD SHEPHERD An't please you, sir, to undertake the business for us, here is that gold I have. I'll make it as much more, and leave this young man in pawn till I bring it you.

AUTOLYCUS After I have done what I promised?        810

OLD SHEPHERD Ay, sir.

AUTOLYCUS Well, give me the moiety. (To the Clown) Are you a party in this business?

CLOWN In some sort, sir. But though my case be a pitiful one, I hope I shall not be flayed out of it.        815

AUTOLYCUS O, that's the case of the shepherd's son. Hang him, he'll be made an example.

CLOWN (to the Old Shepherd) Comfort, good comfort. We must to the King, and show our strange sights. He must know 'tis none of your daughter, nor my sister. We are gone else. (To Autolycus) Sir, I will give you as much as this old man does when the business is performed, and remain, as he says, your pawn till it be brought you.                         824

AUTOLYCUS I will trust you. Walk before toward the seaside. Go on the right hand. I will but look upon the hedge, and follow you.

CLOWN (to the Old Shepherd) We are blessed in this man, as I may say, even blessed.                         829

OLD SHEPHERD Let's before, as he bids us. He was provided to do us good.                         Exit with the Clown

AUTOLYCUS If I had a mind to be honest, I see fortune would not suffer me. She drops booties in my mouth. I am courted now with a double occasion: gold, and a means to do the Prince my master good, which who knows how that may turn back to my advancement? I will bring these two moles, these blind ones, aboard him. If he think it fit to shore them again, and that the complaint they have to the King concerns him nothing, let him call me rogue for being so far officious, for I am proof against that title, and what shame else belongs to't. To him will I present them. There may be matter in it.                         Exit

☙

**5.1**  *Enter Leontes, Cleomenes, Dion, and Paulina*

CLEOMENES (to Leontes)
    Sir, you have done enough, and have performed
    A saint-like sorrow. No fault could you make

Which you have not redeemed, indeed, paid down
More penitence than done trespass. At the last
Do as the heavens have done, forget your evil.        5
With them, forgive yourself.

LEONTES                    Whilst I remember
Her and her virtues I cannot forget
My blemishes in them, and so still think of
The wrong I did myself, which was so much
That heirless it hath made my kingdom, and        10
Destroyed the sweet'st companion that e'er man
Bred his hopes out of. True?

PAULINA                    Too true, my lord.
If one by one you wedded all the world,
Or from the all that are took something good
To make a perfect woman, she you killed        15
Would be unparalleled.

LEONTES                    I think so. Killed?
She I killed? I did so. But thou strik'st me
Sorely to say I did; it is as bitter
Upon thy tongue as in my thought. Now, good now,
Say so but seldom.

CLEOMENES            Not at all, good lady.        20
You might have spoke a thousand things that would
Have done the time more benefit, and graced
Your kindness better.

PAULINA                You are one of those
Would have him wed again.

DION                    If you would not so
You pity not the state, nor the remembrance        25
Of his most sovereign name, consider little
What dangers, by his highness' fail of issue,
May drop upon his kingdom and devour
Incertain lookers-on. What were more holy
Than to rejoice the former queen is well?        30
What holier, than for royalty's repair,
For present comfort and for future good,
To bless the bed of majesty again
With a sweet fellow to't?

PAULINA            There is none worthy
Respecting her that's gone. Besides, the gods        35
Will have fulfilled their secret purposes.
For has not the divine Apollo said?
Is't not the tenor of his oracle
That King Leontes shall not have an heir
Till his lost child be found? Which that it shall        40
Is all as monstrous to our human reason
As my Antigonus to break his grave
And come again to me, who, on my life,
Did perish with the infant. 'Tis your counsel
My lord should to the heavens be contrary,        45
Oppose against their wills.
(To Leontes)            Care not for issue.
The crown will find an heir. Great Alexander
Left his to th' worthiest, so his successor
Was like to be the best.

LEONTES            Good Paulina,
Who hast the memory of Hermione,        50
I know, in honour—O, that ever I
Had squared me to thy counsel! Then even now
I might have looked upon my queen's full eyes,
Have taken treasure from her lips.

PAULINA                    And left them
More rich for what they yielded.

LEONTES                Thou speak'st truth.
No more such wives, therefore no wife. One worse,        56
And better used, would make her sainted spirit

Again possess her corpse, and on this stage,
Where we offenders mourn, appear soul-vexed,
And begin, 'Why to me?'
PAULINA                        Had she such power       60
She had just cause.
LEONTES                   She had, and would incense me
To murder her I married.
PAULINA                          I should so.
Were I the ghost that walked I'd bid you mark
Her eye, and tell me for what dull part in't
You chose her. Then I'd shriek that even your ears  65
Should rift to hear me, and the words that followed
Should be, 'Remember mine'.
LEONTES                               Stars, stars,
And all eyes else, dead coals! Fear thou no wife.
I'll have no wife, Paulina.
PAULINA                           Will you swear
Never to marry but by my free leave?                  70
LEONTES
Never, Paulina, so be blest, my spirit.
PAULINA
Then, good my lords, bear witness to his oath.
CLEOMENES
You tempt him over-much.
PAULINA                           Unless another
As like Hermione as is her picture
Affront his eye—
CLEOMENES        Good madam, I have done.           75
PAULINA
Yet if my lord will marry—if you will, sir;
No remedy but you will—give me the office
To choose your queen. She shall not be so young
As was your former, but she shall be such
As, walked your first queen's ghost, it should take joy
To see her in your arms.
LEONTES                       My true Paulina,          81
We shall not marry till thou bidd'st us.
PAULINA                                          That
Shall be when your first queen's again in breath.
Never till then.
       *Enter a Servant*
SERVANT
One that gives out himself Prince Florizel,            85
Son of Polixenes, with his princess—she
The fairest I have yet beheld—desires access
To your high presence.
LEONTES                     What with him? He comes not
Like to his father's greatness. His approach,
So out of circumstance and sudden, tells us            90
'Tis not a visitation framed, but forced
By need and accident. What train?
SERVANT                             But few,
And those but mean.
LEONTES                    His princess, say you, with him?
SERVANT
Ay, the most peerless piece of earth, I think,
That e'er the sun shone bright on.
PAULINA                              O, Hermione,       95
As every present time doth boast itself
Above a better, gone, so must thy grave
Give way to what's seen now!
(*To the Servant*)              Sir, you yourself
Have said and writ so; but your writing now
Is colder than that theme. She had not been          100
Nor was not to be equalled—thus your verse
Flowed with her beauty once. 'Tis shrewdly ebbed

To say you have seen a better.
SERVANT                       Pardon, madam.
The one I have almost forgot—your pardon!
The other, when she has obtained your eye,           105
Will have your tongue too. This is a creature,
Would she begin a sect, might quench the zeal
Of all professors else; make proselytes
Of who she but bid follow.
PAULINA                        How? Not women!
SERVANT
Women will love her that she is a woman              110
More worth than any man; men, that she is
The rarest of all women.
LEONTES                     Go, Cleomenes.
Yourself, assisted with your honoured friends,
Bring them to our embracement.        *Exit Cleomenes*
                                  Still 'tis strange
He thus should steal upon us.
PAULINA                        Had our prince,        115
Jewel of children, seen this hour, he had paired
Well with this lord. There was not full a month
Between their births.
LEONTES                Prithee no more, cease. Thou know'st
He dies to me again when talked of. Sure,
When I shall see this gentleman thy speeches         120
Will bring me to consider that which may
Unfurnish me of reason. They are come.
       *Enter Florizel, Perdita, Cleomenes, and others*
Your mother was most true to wedlock, Prince,
For she did print your royal father off,
Conceiving you. Were I but twenty-one,               125
Your father's image is so hit in you,
His very air, that I should call you brother,
As I did him, and speak of something wildly
By us performed before. Most dearly welcome,
And your fair princess—goddess! O, alas,             130
I lost a couple that 'twixt heaven and earth
Might thus have stood, begetting wonder, as
You, gracious couple, do; and then I lost—
All mine own folly—the society,
Amity too, of your brave father, whom,               135
Though bearing misery, I desire my life
Once more to look on him.
FLORIZEL                    By his command
Have I here touched Sicilia, and from him
Give you all greetings that a king at friend
Can send his brother; and but infirmity,             140
Which waits upon worn times, hath something seized
His wished ability, he had himself
The lands and waters 'twixt your throne and his
Measured to look upon you, whom he loves—
He bade me say so—more than all the sceptres,        145
And those that bear them, living.
LEONTES                            O, my brother!
Good gentleman, the wrongs I have done thee stir
Afresh within me, and these thy offices,
So rarely kind, are as interpreters
Of my behindhand slackness. Welcome hither,          150
As is the spring to th'earth! And hath he too
Exposed this paragon to th' fearful usage—
At least ungentle—of the dreadful Neptune
To greet a man not worth her pains, much less
Th'adventure of her person?
FLORIZEL                      Good my lord,            155
She came from Libya.
LEONTES                  Where the warlike Smalus,
That noble honoured lord, is feared and loved?

FLORIZEL
  Most royal sir, from thence; from him whose daughter
  His tears proclaimed his, parting with her. Thence,
  A prosperous south wind friendly, we have crossed,
  To execute the charge my father gave me          161
  For visiting your highness. My best train
  I have from your Sicilian shores dismissed;
  Who for Bohemia bend, to signify
  Not only my success in Libya, sir,               165
  But my arrival, and my wife's, in safety
  Here where we are.
LEONTES            The blessèd gods
  Purge all infection from our air whilst you
  Do climate here! You have a holy father,
  A graceful gentleman, against whose person,      170
  So sacred as it is, I have done sin,
  For which the heavens, taking angry note,
  Have left me issueless; and your father's blessed,
  As he from heaven merits it, with you,
  Worthy his goodness. What might I have been,     175
  Might I a son and daughter now have looked on,
  Such goodly things as you?
*Enter a Lord*
LORD               Most noble sir,
  That which I shall report will bear no credit
  Were not the proof so nigh. Please you, great sir,
  Bohemia greets you from himself by me;           180
  Desires you to attach his son, who has,
  His dignity and duty both cast off,
  Fled from his father, from his hopes, and with
  A shepherd's daughter.
LEONTES         Where's Bohemia? Speak.
LORD
  Here in your city. I now came from him.          185
  I speak amazedly, and it becomes
  My marvel and my message. To your court
  Whiles he was hast'ning—in the chase, it seems,
  Of this fair couple—meets he on the way
  The father of this seeming lady and              190
  Her brother, having both their country quitted
  With this young prince.
FLORIZEL       Camillo has betrayed me,
  Whose honour and whose honesty till now
  Endured all weathers.
LORD          Lay't so to his charge.
  He's with the King your father.
LEONTES         Who, Camillo?     195
LORD
  Camillo, sir. I spake with him, who now
  Has these poor men in question. Never saw I
  Wretches so quake. They kneel, they kiss the earth,
  Forswear themselves as often as they speak.
  Bohemia stops his ears, and threatens them       200
  With divers deaths in death.
PERDITA        O, my poor father!
  The heaven sets spies upon us, will not have
  Our contract celebrated.
LEONTES        You are married?
FLORIZEL
  We are not, sir, nor are we like to be.
  The stars, I see, will kiss the valleys first.   205
  The odds for high and low's alike.
LEONTES         My lord,
  Is this the daughter of a king?
FLORIZEL        She is,
  When once she is my wife.

LEONTES
  That 'once', I see, by your good father's speed
  Will come on very slowly. I am sorry,            210
  Most sorry, you have broken from his liking
  Where you were tied in duty; and as sorry
  Your choice is not so rich in worth as beauty,
  That you might well enjoy her.
FLORIZEL *(to Perdita)*      Dear, look up.
  Though fortune, visible an enemy,                215
  Should chase us with my father, power no jot
  Hath she to change our loves.—Beseech you, sir,
  Remember since you owed no more to time
  Than I do now. With thought of such affections,
  Step forth mine advocate. At your request        220
  My father will grant precious things as trifles.
LEONTES
  Would he do so, I'd beg your precious mistress,
  Which he counts but a trifle.
PAULINA         Sir, my liege,
  Your eye hath too much youth in't. Not a month
  Fore your queen died she was more worth such gazes
  Than what you look on now.
LEONTES        I thought of her     226
  Even in these looks I made.
*(To Florizel)*      But your petition
  Is yet unanswered. I will to your father.
  Your honour not o'erthrown by your desires,
  I am friend to them and you. Upon which errand   230
  I now go toward him. Therefore follow me,
  And mark what way I make. Come, good my lord.
                        *Exeunt*

**5.2**    *Enter Autolycus and a Gentleman*
AUTOLYCUS Beseech you, sir, were you present at this
  relation?
FIRST GENTLEMAN I was by at the opening of the fardel,
  heard the old shepherd deliver the manner how he
  found it; whereupon, after a little amazedness, we were
  all commanded out of the chamber. Only this,
  methought I heard the shepherd say he found the child.
AUTOLYCUS I would most gladly know the issue of it.    8
FIRST GENTLEMAN I make a broken delivery of the business,
  but the changes I perceived in the King and Camillo
  were very notes of admiration. They seemed almost,
  with staring on one another, to tear the cases of their
  eyes. There was speech in their dumbness, language in
  their very gesture. They looked as they had heard of a
  world ransomed, or one destroyed. A notable passion
  of wonder appeared in them, but the wisest beholder,
  that knew no more but seeing, could not say if
  th'importance were joy or sorrow. But in the extremity
  of the one, it must needs be.                          19
*Enter another Gentleman*
  Here comes a gentleman that happily knows more. The
  news, Ruggiero!
SECOND GENTLEMAN Nothing but bonfires. The oracle is
  fulfilled. The King's daughter is found. Such a deal of
  wonder is broken out within this hour, that ballad-
  makers cannot be able to express it.                   25
*Enter another Gentleman*
  Here comes the Lady Paulina's steward. He can deliver
  you more.—How goes it now, sir? This news which is
  called true is so like an old tale that the verity of it is
  in strong suspicion. Has the King found his heir?      29
THIRD GENTLEMAN Most true, if ever truth were pregnant
  by circumstance. That which you hear you'll swear

you see, there is such unity in the proofs. The mantle of Queen Hermione's, her jewel about the neck of it, the letters of Antigonus found with it, which they know to be his character; the majesty of the creature, in resemblance of the mother; the affection of nobleness which nature shows above her breeding, and many other evidences proclaim her with all certainty to be the King's daughter. Did you see the meeting of the two kings? 40

SECOND GENTLEMAN No.

THIRD GENTLEMAN Then have you lost a sight which was to be seen, cannot be spoken of. There might you have beheld one joy crown another, so and in such manner that it seemed sorrow wept to take leave of them, for their joy waded in tears. There was casting up of eyes, holding up of hands, with countenance of such distraction that they were to be known by garment, not by favour. Our king being ready to leap out of himself for joy of his found daughter, as if that joy were now become a loss cries, 'O, thy mother, thy mother!', then asks Bohemia forgiveness, then embraces his son-in-law, then again worries he his daughter with clipping her. Now he thanks the old shepherd, which stands by like a weather-bitten conduit of many kings' reigns. I never heard of such another encounter, which lames report to follow it, and undoes description to do it.

SECOND GENTLEMAN What, pray you, became of Antigonus, that carried hence the child? 60

THIRD GENTLEMAN Like an old tale still, which will have matter to rehearse though credit be asleep and not an ear open. He was torn to pieces with a bear. This avouches the shepherd's son, who has not only his innocence, which seems much, to justify him, but a handkerchief and rings of his, that Paulina knows. 66

FIRST GENTLEMAN What became of his barque and his followers?

THIRD GENTLEMAN Wrecked the same instant of their master's death, and in the view of the shepherd; so that all the instruments which aided to expose the child were even then lost when it was found. But O, the noble combat that 'twixt joy and sorrow was fought in Paulina! She had one eye declined for the loss of her husband, another elevated that the oracle was fulfilled. She lifted the Princess from the earth, and so locks her in embracing as if she would pin her to her heart, that she might no more be in danger of losing. 78

FIRST GENTLEMAN The dignity of this act was worth the audience of kings and princes, for by such was it acted.

THIRD GENTLEMAN One of the prettiest touches of all, and that which angled for mine eyes—caught the water, though not the fish—was when at the relation of the Queen's death, with the manner how she came to't bravely confessed and lamented by the King, how attentiveness wounded his daughter till from one sign of dolour to another she did, with an 'Alas', I would fain say bleed tears; for I am sure my heart wept blood. Who was most marble there changed colour. Some swooned, all sorrowed. If all the world could have seen't, the woe had been universal. 91

FIRST GENTLEMAN Are they returned to the court?

THIRD GENTLEMAN No. The Princess, hearing of her mother's statue, which is in the keeping of Paulina, a piece many years in doing, and now newly performed by that rare Italian master Giulio Romano, who, had he himself eternity and could put breath into his work, would beguile nature of her custom, so perfectly he is her ape. He so near to Hermione hath done Hermione that they say one would speak to her and stand in hope of answer. Thither with all greediness of affection are they gone, and there they intend to sup. 102

SECOND GENTLEMAN I thought she had some great matter there in hand, for she hath privately twice or thrice a day, ever since the death of Hermione, visited that removed house. Shall we thither, and with our company piece the rejoicing? 107

FIRST GENTLEMAN Who would be thence, that has the benefit of access? Every wink of an eye some new grace will be born. Our absence makes us unthrifty to our knowledge. Let's along. *Exeunt Gentlemen*

AUTOLYCUS Now, had I not the dash of my former life in me, would preferment drop on my head. I brought the old man and his son aboard the Prince; told him I heard them talk of a fardel, and I know not what. But he at that time over-fond of the shepherd's daughter—so he then took her to be—who began to be much seasick, and himself little better, extremity of weather continuing, this mystery remained undiscovered. But 'tis all one to me, for had I been the finder-out of this secret it would not have relished among my other discredits. 122

*Enter the Old Shepherd and the Clown, dressed as gentlemen*

Here come those I have done good to against my will, and already appearing in the blossoms of their fortune.

OLD SHEPHERD Come, boy; I am past more children, but thy sons and daughters will be all gentlemen born.

CLOWN (*to Autolycus*) You are well met, sir. You denied to fight with me this other day because I was no gentleman born. See you these clothes? Say you see them not, and think me still no gentleman born. You were best say these robes are not gentlemen born. Give me the lie, do, and try whether I am not now a gentleman born. 133

AUTOLYCUS I know you are now, sir, a gentleman born.

CLOWN Ay, and have been so any time these four hours.

OLD SHEPHERD And so have I, boy.

CLOWN So you have; but I was a gentleman born before my father, for the King's son took me by the hand and called me brother; and then the two kings called my father brother; and then the Prince my brother and the Princess my sister called my father father; and so we wept; and there was the first gentleman-like tears that ever we shed. 143

OLD SHEPHERD We may live, son, to shed many more.

CLOWN Ay, or else 'twere hard luck, being in so preposterous estate as we are.

AUTOLYCUS I humbly beseech you, sir, to pardon me all the faults I have committed to your worship, and to give me your good report to the Prince my master.

OLD SHEPHERD Prithee, son, do, for we must be gentle now we are gentlemen. 151

CLOWN Thou wilt amend thy life?

AUTOLYCUS Ay, an it like your good worship.

CLOWN Give me thy hand. I will swear to the Prince thou art as honest a true fellow as any is in Bohemia. 155

OLD SHEPHERD You may say it, but not swear it.

CLOWN Not swear it now I am a gentleman? Let boors and franklins say it; I'll swear it.

OLD SHEPHERD How if it be false, son? 159

CLOWN If it be ne'er so false, a true gentleman may swear it in the behalf of his friend, (*to Autolycus*) and I'll

swear to the Prince thou art a tall fellow of thy hands
and that thou wilt not be drunk; but I know thou art
no tall fellow of thy hands and that thou wilt be drunk;
but I'll swear it, and I would thou wouldst be a tall
fellow of thy hands.                                     166

AUTOLYCUS I will prove so, sir, to my power.

CLOWN Ay, by any means prove a tall fellow. If I do not
wonder how thou dar'st venture to be drunk, not being
a tall fellow, trust me not.                             170

⌜*Flourish within*⌝

Hark, the kings and princes, our kindred, are going to
see the Queen's picture. Come, follow us. We'll be thy
good masters.                                           *Exeunt*

**5.3**    *Enter Leontes, Polixenes, Florizel, Perdita, Camillo,*
         *Paulina, Lords, and attendants*

LEONTES
O grave and good Paulina, the great comfort
That I have had of thee!

PAULINA                        What, sovereign sir,
I did not well, I meant well. All my services
You have paid home, but that you have vouchsafed
With your crowned brother and these young
       contracted                                          5
Heirs of your kingdoms my poor house to visit,
It is a surplus of your grace which never
My life may last to answer.

LEONTES                                O Paulina,
We honour you with trouble. But we came
To see the statue of our queen. Your gallery              10
Have we passed through, not without much content
In many singularities; but we saw not
That which my daughter came to look upon,
The statue of her mother.

PAULINA                        As she lived peerless,
So her dead likeness I do well believe                    15
Excels what ever yet you looked upon,
Or hand of man hath done. Therefore I keep it
Lonely, apart. But here it is. Prepare
To see the life as lively mocked as ever
Still sleep mocked death. Behold, and say 'tis well.      20

*She draws a curtain and reveals the figure of*
*Hermione, standing like a statue*

I like your silence; it the more shows off
Your wonder. But yet speak; first you, my liege.
Comes it not something near?

LEONTES                        Her natural posture.
Chide me, dear stone, that I may say indeed
Thou art Hermione; or rather, thou art she               25
In thy not chiding, for she was as tender
As infancy and grace. But yet, Paulina,
Hermione was not so much wrinkled, nothing
So agèd as this seems.

POLIXENES                    O, not by much.

PAULINA
So much the more our carver's excellence,                30
Which lets go by some sixteen years, and makes her
As she lived now.

LEONTES               As now she might have done,
So much to my good comfort as it is
Now piercing to my soul. O, thus she stood,
Even with such life of majesty—warm life,                35
As now it coldly stands—when first I wooed her.
I am ashamed. Does not the stone rebuke me
For being more stone than it? O royal piece!
There's magic in thy majesty, which has

My evils conjured to remembrance, and                    40
From thy admiring daughter took the spirits,
Standing like stone with thee.

PERDITA                        And give me leave,
And do not say 'tis superstition, that
I kneel and then implore her blessing. Lady,
Dear Queen, that ended when I but began,                 45
Give me that hand of yours to kiss.

PAULINA                        O, patience!
The statue is but newly fixed; the colour's
Not dry.

CAMILLO (*to Leontes*)
My lord, your sorrow was too sore laid on,
Which sixteen winters cannot blow away,                  50
So many summers dry. Scarce any joy
Did ever so long live; no sorrow
But killed itself much sooner.

POLIXENES (*to Leontes*)        Dear my brother,
Let him that was the cause of this have power
To take off so much grief from you as he                 55
Will piece up in himself.

PAULINA (*to Leontes*)          Indeed, my lord,
If I had thought the sight of my poor image
Would thus have wrought you—for the stone is mine—
I'd not have showed it.

*She makes to draw the curtain*

LEONTES                        Do not draw the curtain.

PAULINA
No longer shall you gaze on't, lest your fancy           60
May think anon it moves.

LEONTES                    Let be, let be!
Would I were dead but that methinks already.
What was he that did make it? See, my lord,
Would you not deem it breathed, and that those veins
Did verily bear blood?

POLIXENES                    Masterly done.               65
The very life seems warm upon her lip.

LEONTES
The fixture of her eye has motion in't,
As we are mocked with art.

PAULINA                    I'll draw the curtain.
My lord's almost so far transported that
He'll think anon it lives.

LEONTES                    O sweet Paulina,               70
Make me to think so twenty years together.
No settled senses of the world can match
The pleasure of that madness. Let't alone.

PAULINA
I am sorry, sir, I have thus far stirred you; but
I could afflict you farther.

LEONTES                    Do, Paulina,                   75
For this affliction has a taste as sweet
As any cordial comfort. Still methinks
There is an air comes from her. What fine chisel
Could ever yet cut breath? Let no man mock me,
For I will kiss her.

PAULINA               Good my lord, forbear.              80
The ruddiness upon her lip is wet.
You'll mar it if you kiss it, stain your own
With oily painting. Shall I draw the curtain?

LEONTES
No, not these twenty years.

PERDITA                    So long could I
Stand by, a looker-on.

PAULINA                    Either forbear,                85
Quit presently the chapel, or resolve you

For more amazement. If you can behold it,
I'll make the statue move indeed, descend,
And take you by the hand. But then you'll think—
Which I protest against—I am assisted                    90
By wicked powers.
LEONTES                    What you can make her do
I am content to look on; what to speak,
I am content to hear; for 'tis as easy
To make her speak as move.
PAULINA                    It is required
You do awake your faith. Then, all stand still.          95
Or those that think it is unlawful business
I am about, let them depart.
LEONTES                    Proceed.
No foot shall stir.
PAULINA                    Music; awake her; strike!
    *Music*
(*To Hermione*) 'Tis time. Descend. Be stone no more.
    Approach.
Strike all that look upon with marvel. Come,            100
I'll fill your grave up. Stir. Nay, come away.
Bequeath to death your numbness, for from him
Dear life redeems you.
(*To Leontes*)            You perceive she stirs.
    *Hermione slowly descends*
Start not. Her actions shall be holy as
You hear my spell is lawful. Do not shun her            105
Until you see her die again, for then
You kill her double. Nay, present your hand.
When she was young, you wooed her. Now, in age,
Is she become the suitor?
LEONTES                    O, she's warm!
If this be magic, let it be an art                      110
Lawful as eating.
POLIXENES    She embraces him.
CAMILLO  She hangs about his neck.
If she pertain to life, let her speak too.
POLIXENES
Ay, and make it manifest where she has lived,           115
Or how stol'n from the dead.
PAULINA                    That she is living,
Were it but told you, should be hooted at
Like an old tale. But it appears she lives,
Though yet she speak not. Mark a little while.

(*To Perdita*) Please you to interpose, fair madam.
    Kneel,                                               120
And pray your mother's blessing.—Turn, good lady,
Our Perdita is found.
HERMIONE                    You gods, look down,
And from your sacred vials pour your graces
Upon my daughter's head.—Tell me, mine own,
Where hast thou been preserved? Where lived? How
    found                                               125
Thy father's court? For thou shalt hear that I,
Knowing by Paulina that the oracle
Gave hope thou wast in being, have preserved
Myself to see the issue.
PAULINA                    There's time enough for that,
Lest they desire upon this push to trouble             130
Your joys with like relation. Go together,
You precious winners all; your exultation
Partake to everyone. I, an old turtle,
Will wing me to some withered bough, and there
My mate, that's never to be found again,               135
Lament till I am lost.
LEONTES                    O peace, Paulina!
Thou shouldst a husband take by my consent,
As I by thine a wife. This is a match,
And made between's by vows. Thou hast found mine,
But how is to be questioned, for I saw her,            140
As I thought, dead, and have in vain said many
A prayer upon her grave. I'll not seek far—
For him, I partly know his mind—to find thee
An honourable husband. Come, Camillo,
And take her by the hand, whose worth and honesty
Is richly noted, and here justified                    146
By us, a pair of kings. Let's from this place.
(*To Hermione*) What, look upon my brother. Both your
    pardons,
That e'er I put between your holy looks
My ill suspicion. This' your son-in-law                150
And son unto the King, whom heavens directing
Is troth-plight to your daughter. Good Paulina,
Lead us from hence, where we may leisurely
Each one demand and answer to his part
Performed in this wide gap of time since first         155
We were dissevered. Hastily lead away.        *Exeunt*

# CYMBELINE

OUR first reference to *Cymbeline* is a note by the astrologer Simon Forman that he saw the play, probably not long before his death on 8 September 1611. He refers to the heroine as 'Innogen', and this name occurs in the sources; the form 'Imogen', found only in the Folio, appears to be a misprint. The play's courtly tone, and the masque-like quality of, particularly, the episode (5.5.186.1–2) in which Jupiter 'descends in thunder and lightning, sitting upon an eagle', and 'throws a thunderbolt', suggest that as Shakespeare wrote he may have had in mind the audiences and the stage equipment of the Blackfriars theatre, which his company used from the autumn of 1609; and stylistic evidence places the play about 1610–11. It was first printed in the 1623 Folio, as the last of the tragedies. In fact it is a tragicomedy, or a romance, telling a complex and implausible tale of events which cause the deaths of certain subsidiary characters (Cloten, and the Queen) and bring major characters (including the heroine, Innogen) close to death, but which are miraculously resolved in the reunions and reconciliations of the closing scene.

Shakespeare's plot reflects a wide range of reading. He took his title and setting from the name and reign of the legendary British king Cymbeline, or Cunobelinus, said to have reigned from 33 BC till shortly after the birth of Christ. *Cymbeline* is no chronicle history, but Shakespeare derived some ideas, and many of his characters' names, from accounts of early British history in Holinshed's *Chronicles* and elsewhere. Drawing partially, it seems, on an old play, *The Rare Triumphs of Love and Fortune* (acted 1582, printed 1589), he gives Cymbeline a daughter, Innogen, and a wicked second Queen with a loutish, vicious son, Cloten, whom she wishes to see on the throne in her husband's place. Cymbeline, disapproving of his daughter's marriage to 'a poor but worthy gentleman', Posthumus Leonatus, banishes him. The strand of plot showing the outcome of a wager that Posthumus, in Rome, lays on his wife's chastity is indebted, directly or indirectly, to Boccaccio's *Decameron*. Another old play, *Sir Clyomon and Clamydes* (printed in 1599), may have suggested the bizarre scene (4.2) in which Innogen mistakes Cloten's headless body for that of Posthumus; and Holinshed's *History of Scotland* supplied the episode in which Cymbeline's two sons, Guiderius and Arviragus, helped only by the old man (Belarius) who has brought them up in the wilds of Wales, defeat the entire Roman army.

The tone of *Cymbeline* has puzzled commentators. Its prose and verse style is frequently ornate, sometimes grotesque. Its characterization often seems deliberately artificial. Extremes are violently juxtaposed, most daringly when Innogen, supposed dead, is laid beside Cloten's headless body: the beauty of the verse in which she is mourned, and of the flowers strewn over the bodies, contrasts with the hideous spectacle of the headless corpse; her waking speech is one of Shakespeare's most thrillingly difficult challenges to his performers. The appearance of Jupiter lifts the action to a new level of even greater implausibility, preparing us for the extraordinary series of revelations by which the play advances to its impossibly happy ending. *Cymbeline* has been valued mostly for its portrayal of Innogen, ideal of womanhood to, especially, Victorian readers and theatre-goers. The play as a whole is a fantasy, an experimental exercise in virtuosity.

# THE PERSONS OF THE PLAY

CYMBELINE, King of Britain

Princess INNOGEN, his daughter, later disguised as a man named Fidele

GUIDERIUS, known as Polydore ⎱ Cymbeline's sons, stolen by
ARVIRAGUS, known as Cadwal ⎰ Belarius

QUEEN, Cymbeline's wife, Innogen's stepmother

Lord CLOTEN, her son

BELARIUS, a banished lord, calling himself Morgan

CORNELIUS, a physician

HELEN, a lady attending on Innogen

Two LORDS attending on Cloten

Two GENTLEMEN

Two British CAPTAINS

Two JAILERS

POSTHUMUS Leonatus, a poor gentleman, Innogen's husband

PISANIO, his servant

FILARIO, a friend of Posthumus

GIACOMO, an Italian ⎫
A FRENCHMAN ⎪
                ⎬ Filario's friends
A DUTCHMAN ⎪
A SPANIARD ⎭

Caius LUCIUS, ambassador from Rome, later General of the Roman forces

Two Roman SENATORS

Roman TRIBUNES

A Roman CAPTAIN

Philharmonus, a SOOTHSAYER

JUPITER

Ghost of SICILIUS Leonatus, father of Posthumus

Ghost of the MOTHER of Posthumus

Ghosts of the BROTHERS of Posthumus

Lords attending on Cymbeline, ladies attending on the Queen, musicians attending on Cloten, messengers, soldiers

# Cymbeline, King of Britain

1.1 *Enter two Gentlemen*

FIRST GENTLEMAN
You do not meet a man but frowns. Our bloods
No more obey the heavens than our courtiers
Still seem as does the King.
SECOND GENTLEMAN                     But what's the matter?
FIRST GENTLEMAN
His daughter, and the heir of 's kingdom, whom          5
He purposed to his wife's sole son—a widow
That late he married—hath referred herself
Unto a poor but worthy gentleman. She's wedded,
Her husband banished, she imprisoned. All
Is outward sorrow, though I think the King
Be touched at very heart.
SECOND GENTLEMAN          None but the King?          10
FIRST GENTLEMAN
He that hath lost her, too. So is the Queen,
That most desired the match. But not a courtier—
Although they wear their faces to the bent
Of the King's looks—hath a heart that is not
Glad of the thing they scowl at.
SECOND GENTLEMAN                     And why so?          15
FIRST GENTLEMAN
He that hath missed the Princess is a thing
Too bad for bad report, and he that hath her—
I mean that married her—alack, good man,
And therefore banished!—is a creature such
As, to seek through the regions of the earth          20
For one his like, there would be something failing
In him that should compare. I do not think
So fair an outward and such stuff within
Endows a man but he.
SECOND GENTLEMAN          You speak him far.
FIRST GENTLEMAN
I do extend him, sir, within himself;          25
Crush him together rather than unfold
His measure duly.
SECOND GENTLEMAN  What's his name and birth?
FIRST GENTLEMAN
I cannot delve him to the root. His father
Was called Sicilius, who did join his honour
Against the Romans with Cassibelan          30
But had his titles by Tenantius, whom
He served with glory and admired success,
So gained the sur-addition 'Leonatus';
And had, besides this gentleman in question,
Two other sons who in the wars o'th' time          35
Died with their swords in hand; for which their father,
Then old and fond of issue, took such sorrow
That he quit being, and his gentle lady,
Big of this gentleman, our theme, deceased
As he was born. The King, he takes the babe          40
To his protection, calls him Posthumus Leonatus,
Breeds him, and makes him of his bedchamber;
Puts to him all the learnings that his time
Could make him the receiver of, which he took
As we do air, fast as 'twas ministered,          45
And in 's spring became a harvest; lived in court—
Which rare it is to do—most praised, most loved;

A sample to the youngest, to th' more mature
A glass that feated them, and to the graver
A child that guided dotards. To his mistress,          50
For whom he now is banished, her own price
Proclaims how she esteemed him and his virtue.
By her election may be truly read
What kind of man he is.
SECOND GENTLEMAN          I honour him
Even out of your report. But pray you tell me,          55
Is she sole child to th' King?
FIRST GENTLEMAN          His only child.
He had two sons—if this be worth your hearing,
Mark it: the eld'st of them at three years old,
I'th' swathing clothes the other, from their nursery
Were stol'n, and to this hour no guess in knowledge
Which way they went.          61
SECOND GENTLEMAN How long is this ago?
FIRST GENTLEMAN Some twenty years.
SECOND GENTLEMAN
That a king's children should be so conveyed,
So slackly guarded, and the search so slow          65
That could not trace them!
FIRST GENTLEMAN          Howsoe'er 'tis strange,
Or that the negligence may well be laughed at,
Yet is it true, sir.
SECOND GENTLEMAN I do well believe you.
          *Enter the Queen, Posthumus, and Innogen*
FIRST GENTLEMAN
We must forbear. Here comes the gentleman,
The Queen and Princess.     *Exeunt the two Gentlemen*
QUEEN
No, be assured you shall not find me, daughter,          71
After the slander of most stepmothers,
Evil-eyed unto you. You're my prisoner, but
Your jailer shall deliver you the keys
That lock up your restraint. For you, Posthumus,          75
So soon as I can win th'offended King
I will be known your advocate. Marry, yet
The fire of rage is in him, and 'twere good
You leaned unto his sentence with what patience
Your wisdom may inform you.
POSTHUMUS          Please your highness,
I will from hence today.
QUEEN          You know the peril.          81
I'll fetch a turn about the garden, pitying
The pangs of barred affections, though the King
Hath charged you should not speak together.     *Exit*
INNOGEN
O dissembling courtesy! How fine this tyrant          85
Can tickle where she wounds! My dearest husband,
I something fear my father's wrath, but nothing—
Always reserved my holy duty—what
His rage can do on me. You must be gone,
And I shall here abide the hourly shot          90
Of angry eyes, not comforted to live
But that there is this jewel in the world
That I may see again.
POSTHUMUS          My queen, my mistress!
O lady, weep no more, lest I give cause

To be suspected of more tenderness     95
Than doth become a man. I will remain
The loyal'st husband that did e'er plight troth;
My residence in Rome at one Filario's,
Who to my father was a friend, to me
Known but by letter; thither write, my queen,     100
And with mine eyes I'll drink the words you send
Though ink be made of gall.
       *Enter Queen*
QUEEN                  Be brief, I pray you.
If the King come, I shall incur I know not
How much of his displeasure. (*Aside*) Yet I'll move him
To walk this way. I never do him wrong     105
But he does buy my injuries, to be friends,
Pays dear for my offences.              *Exit*
POSTHUMUS        Should we be taking leave
As long a term as yet we have to live,
The loathness to depart would grow. Adieu.
INNOGEN Nay, stay a little.     110
Were you but riding forth to air yourself
Such parting were too petty. Look here, love:
This diamond was my mother's. Take it, heart;
       *She gives him a ring*
But keep it till you woo another wife
When Innogen is dead.
POSTHUMUS           How, how? Another?     115
You gentle gods, give me but this I have,
And cere up my embracements from a next
With bonds of death! Remain, remain thou here
       *He puts on the ring*
While sense can keep it on; and, sweetest, fairest,
As I my poor self did exchange for you     120
To your so infinite loss, so in our trifles
I still win of you. For my sake wear this.
       *He gives her a bracelet*
It is a manacle of love. I'll place it
Upon this fairest prisoner.
INNOGEN            O the gods!
When shall we see again?
       *Enter Cymbeline and lords*
POSTHUMUS          Alack, the King!     125
CYMBELINE
Thou basest thing, avoid hence, from my sight!
If after this command thou fraught the court
With thy unworthiness, thou diest. Away.
Thou'rt poison to my blood.
POSTHUMUS          The gods protect you,
And bless the good remainders of the court!     130
I am gone.                       *Exit*
INNOGEN      There cannot be a pinch in death
More sharp than this is.
CYMBELINE          O disloyal thing,
That shouldst repair my youth, thou heap'st
A year's age on me.
INNOGEN          I beseech you, sir,
Harm not yourself with your vexation.     135
I am senseless of your wrath. A touch more rare
Subdues all pangs, all fears.
CYMBELINE          Past grace, obedience—
INNOGEN
Past hope and in despair: that way past grace.
CYMBELINE
That mightst have had the sole son of my queen!
INNOGEN
O blessèd that I might not! I chose an eagle     140
And did avoid a puttock.

CYMBELINE
Thou took'st a beggar, wouldst have made my throne
A seat for baseness.
INNOGEN          No, I rather added
A lustre to it.
CYMBELINE        O thou vile one!
INNOGEN              Sir,
It is your fault that I have loved Posthumus.     145
You bred him as my playfellow, and he is
A man worth any woman, over-buys me
Almost the sum he pays.
CYMBELINE          What, art thou mad?
INNOGEN
Almost, sir. Heaven restore me! Would I were
A neatherd's daughter, and my Leonatus     150
Our neighbour shepherd's son.
       *Enter Queen*
CYMBELINE          Thou foolish thing.
(*To Queen*) They were again together; you have done
Not after our command. (*To lords*) Away with her,
And pen her up.
QUEEN        Beseech your patience, peace,
Dear lady daughter, peace. Sweet sovereign,     155
Leave us to ourselves, and make yourself some comfort
Out of your best advice.
CYMBELINE        Nay, let her languish
A drop of blood a day, and, being aged,
Die of this folly.            *Exit with lords*
QUEEN        Fie, you must give way.
       *Enter Pisanio*
Here is your servant. How now, sir? What news?     160
PISANIO
My lord your son drew on my master.
QUEEN                Ha!
No harm, I trust, is done?
PISANIO           There might have been,
But that my master rather played than fought,
And had no help of anger. They were parted
By gentlemen at hand.
QUEEN          I am very glad on't.     165
INNOGEN
Your son's my father's friend; he takes his part
To draw upon an exile—O brave sir!
I would they were in Afric both together,
Myself by with a needle, that I might prick
The goer-back. (*To Pisanio*) Why came you from your
     master?     170
PISANIO
On his command. He would not suffer me
To bring him to the haven, left these notes
Of what commands I should be subject to
When't pleased you to employ me.
QUEEN          This hath been
Your faithful servant. I dare lay mine honour     175
He will remain so.
PISANIO I humbly thank your highness.
QUEEN Pray walk a while.            ⌜*Exit*⌝
INNOGEN
About some half hour hence, pray you speak with me.
You shall at least go see my lord aboard.     180
For this time leave me.          *Exeunt severally*

**1.2**    *Enter Cloten and two Lords*
FIRST LORD Sir, I would advise you to shift a shirt. The
violence of action hath made you reek as a sacrifice.

Where air comes out, air comes in. There's none abroad
so wholesome as that you vent.

CLOTEN If my shirt were bloody, then to shift it. Have I
hurt him?                                                          6

SECOND LORD (*aside*) No, faith, not so much as his patience.

FIRST LORD Hurt him? His body's a passable carcass if he
be not hurt. It is a thoroughfare for steel if he be not
hurt.                                                             10

SECOND LORD (*aside*) His steel was in debt—it went o'th'
backside the town.

CLOTEN The villain would not stand me.

SECOND LORD (*aside*) No, but he fled forward still, toward
your face.                                                        15

FIRST LORD Stand you? You have land enough of your
own, but he added to your having, gave you some
ground.

SECOND LORD (*aside*) As many inches as you have oceans.
Puppies!                                                          20

CLOTEN I would they had not come between us.

SECOND LORD (*aside*) So would I, till you had measured
how long a fool you were upon the ground.

CLOTEN And that she should love this fellow and refuse
me!                                                               25

SECOND LORD (*aside*) If it be a sin to make a true election,
she is damned.

FIRST LORD Sir, as I told you always, her beauty and her
brain go not together. She's a good sign, but I have
seen small reflection of her wit.                                 30

SECOND LORD (*aside*) She shines not upon fools lest the
reflection should hurt her.

CLOTEN Come, I'll to my chamber. Would there had been
some hurt done.

SECOND LORD (*aside*) I wish not so, unless it had been the
fall of an ass, which is no great hurt.                           36

CLOTEN (*to Second Lord*) You'll go with us?

FIRST LORD I'll attend your lordship.

CLOTEN Nay, come, let's go together.

SECOND LORD Well, my lord.

                                                        *Exeunt*

**1.3    *Enter Innogen and Pisanio***

INNOGEN
I would thou grew'st unto the shores o'th' haven
And questionedst every sail. If he should write
And I not have it, 'twere a paper lost
As offered mercy is. What was the last
That he spake to thee?

PISANIO                    It was his queen, his queen.     5

INNOGEN
Then waved his handkerchief?

PISANIO                          And kissed it, madam.

INNOGEN
Senseless linen, happier therein than I!
And that was all?

PISANIO              No, madam. For so long
As he could make me with this eye or ear
Distinguish him from others he did keep              10
The deck, with glove or hat or handkerchief
Still waving, as the fits and stirs of 's mind
Could best express how slow his soul sailed on,
How swift his ship.

INNOGEN                Thou shouldst have made him
As little as a crow, or less, ere left                15
To after-eye him.

PISANIO            Madam, so I did.

INNOGEN
I would have broke mine eye-strings, cracked them,
    but
To look upon him till the diminution
Of space had pointed him sharp as my needle;
Nay, followed him till he had melted from            20
The smallness of a gnat to air, and then
Have turned mine eye and wept. But, good Pisanio,
When shall we hear from him?

PISANIO Be assured, madam,
With his next vantage.                                25

INNOGEN
I did not take my leave of him, but had
Most pretty things to say. Ere I could tell him
How I would think on him at certain hours,
Such thoughts and such, or I could make him swear
The shes of Italy should not betray                   30
Mine interest and his honour, or have charged him
At the sixth hour of morn, at noon, at midnight
T'encounter me with orisons—for then
I am in heaven for him—or ere I could
Give him that parting kiss which I had set            35
Betwixt two charming words, comes in my father,
And, like the tyrannous breathing of the north,
Shakes all our buds from growing.
          *Enter a Lady*

LADY                          The Queen, madam,
Desires your highness' company.

INNOGEN (*to Pisanio*)
Those things I bid you do, get them dispatched.       40
I will attend the Queen.

PISANIO                    Madam, I shall.
          *Exeunt Innogen and Lady at one door, Pisanio
                    at another*

**1.4** ⌈*A table brought out, with a banquet upon it.*⌉ *Enter
      Filario, Giacomo, a Frenchman, a Dutchman, and a
      Spaniard*

GIACOMO Believe it, sir, I have seen him in Britain. He
was then of a crescent note, expected to prove so worthy
as since he hath been allowed the name of. But I could
then have looked on him without the help of admiration,
though the catalogue of his endowments had been
tabled by his side and I to peruse him by items.

FILARIO You speak of him when he was less furnished
than now he is with that which makes him both
without and within.                                    9

FRENCHMAN I have seen him in France. We had very
many there could behold the sun with as firm eyes as
he.

GIACOMO This matter of marrying his king's daughter,
wherein he must be weighed rather by her value than
his own, words him, I doubt not, a great deal from the
matter.                                               16

FRENCHMAN And then his banishment.

GIACOMO Ay, and the approbation of those that weep this
lamentable divorce under her colours are wonderfully
to extend him, be it but to fortify her judgement, which
else an easy battery might lay flat for taking a beggar
without less quality. But how comes it he is to sojourn
with you? How creeps acquaintance?

FILARIO His father and I were soldiers together, to whom
I have been often bound for no less than my life.      25
          *Enter Posthumus*
Here comes the Briton. Let him be so entertained

amongst you as suits with gentlemen of your knowing to a stranger of his quality. I beseech you all, be better known to this gentleman, whom I commend to you as a noble friend of mine. How worthy he is I will leave to appear hereafter rather than story him in his own hearing.

FRENCHMAN (*to Posthumus*) Sir, we have known together in Orléans.                                                                           34

POSTHUMUS Since when I have been debtor to you for courtesies which I will be ever to pay, and yet pay still.

FRENCHMAN Sir, you o'er-rate my poor kindness. I was glad I did atone my countryman and you. It had been pity you should have been put together with so mortal a purpose as then each bore, upon importance of so slight and trivial a nature.                                             41

POSTHUMUS By your pardon, sir, I was then a young traveller, rather shunned to go even with what I heard than in my every action to be guided by others' experiences; but upon my mended judgement—if I offend not to say it is mended—my quarrel was not altogether slight.                                                         47

FRENCHMAN Faith, yes, to be put to the arbitrement of swords, and by such two that would by all likelihood have confounded one the other, or have fallen both.

GIACOMO Can we with manners ask what was the difference?                                                                              52

FRENCHMAN Safely, I think. 'Twas a contention in public, which may without contradiction suffer the report. It was much like an argument that fell out last night, where each of us fell in praise of our country mistresses, this gentleman at that time vouching—and upon warrant of bloody affirmation—his to be more fair, virtuous, wise, chaste, constant, qualified, and less attemptable than any the rarest of our ladies in France.

GIACOMO That lady is not now living, or this gentleman's opinion by this worn out.                                               62

POSTHUMUS She holds her virtue still, and I my mind.

GIACOMO You must not so far prefer her fore ours of Italy.

POSTHUMUS Being so far provoked as I was in France I would abate her nothing, though I profess myself her adorer, not her friend.                                                         67

GIACOMO As fair and as good—a kind of hand-in-hand comparison—had been something too fair and too good for any lady in Britain. If she went before others I have seen—as that diamond of yours outlustres many I have beheld—I could not but believe she excelled many; but I have not seen the most precious diamond that is, nor you the lady.                                                                    74

POSTHUMUS I praised her as I rated her; so do I my stone.

GIACOMO What do you esteem it at?

POSTHUMUS More than the world enjoys.

GIACOMO Either your unparagoned mistress is dead, or she's outprized by a trifle.                                               79

POSTHUMUS You are mistaken. The one may be sold or given, or if there were wealth enough for the purchase or merit for the gift. The other is not a thing for sale, and only the gift of the gods.

GIACOMO Which the gods have given you?

POSTHUMUS Which, by their graces, I will keep.                         85

GIACOMO You may wear her in title yours; but, you know, strange fowl light upon neighbouring ponds. Your ring may be stolen too; so your brace of unprizable estimations, the one is but frail, and the other casual. A cunning thief or a that-way accomplished courtier would hazard the winning both of first and last.       91

POSTHUMUS Your Italy contains none so accomplished a courtier to convince the honour of my mistress if in the holding or loss of that you term her frail. I do nothing doubt you have store of thieves; notwithstanding, I fear not my ring.                                      96

FILARIO Let us leave here, gentlemen.

POSTHUMUS Sir, with all my heart. This worthy signor, I thank him, makes no stranger of me. We are familiar at first.                                                                              100

GIACOMO With five times so much conversation I should get ground of your fair mistress, make her go back even to the yielding, had I admittance and opportunity to friend.

POSTHUMUS No, no.                                                          105

GIACOMO I dare thereupon pawn the moiety of my estate to your ring, which in my opinion o'ervalues it something. But I make my wager rather against your confidence than her reputation, and, to bar your offence herein too, I durst attempt it against any lady in the world.                                                                          111

POSTHUMUS You are a great deal abused in too bold a persuasion, and I doubt not you sustain what you're worthy of by your attempt.

GIACOMO What's that?                                                       115

POSTHUMUS A repulse; though your attempt, as you call it, deserve more—a punishment, too.

FILARIO Gentlemen, enough of this. It came in too suddenly. Let it die as it was born; and, I pray you, be better acquainted.                                                          120

GIACOMO Would I had put my estate and my neighbour's on th'approbation of what I have spoke.

POSTHUMUS What lady would you choose to assail?

GIACOMO Yours, whom in constancy you think stands so safe. I will lay you ten thousand ducats to your ring that, commend me to the court where your lady is, with no more advantage than the opportunity of a second conference, and I will bring from thence that honour of hers which you imagine so reserved.      129

POSTHUMUS I will wage against your gold, gold to it; my ring I hold dear as my finger, 'tis part of it.

GIACOMO You are a friend, and therein the wiser. If you buy ladies' flesh at a million a dram, you cannot preserve it from tainting. But I see you have some religion in you, that you fear.                                    135

POSTHUMUS This is but a custom in your tongue. You bear a graver purpose, I hope.

GIACOMO I am the master of my speeches, and would undergo what's spoken, I swear.

POSTHUMUS Will you? I shall but lend my diamond till your return. Let there be covenants drawn between 's. My mistress exceeds in goodness the hugeness of your unworthy thinking. I dare you to this match. Here's my ring.

FILARIO I will have it no lay.                                           145

GIACOMO By the gods, it is one. If I bring you no sufficient testimony that I have enjoyed the dearest bodily part of your mistress, my ten thousand ducats are yours; so is your diamond too. If I come off and leave her in such honour as you have trust in, she your jewel, this your jewel, and my gold are yours, provided I have your commendation for my more free entertainment.

POSTHUMUS I embrace these conditions; let us have articles betwixt us. Only thus far you shall answer: if you make your voyage upon her and give me directly to understand you have prevailed, I am no further your enemy; she is not worth our debate. If she remain

unseduced, you not making it appear otherwise, for
your ill opinion and th'assault you have made to her
chastity you shall answer me with your sword.    160
GIACOMO Your hand, a covenant. We will have these
things set down by lawful counsel, and straight away
for Britain, lest the bargain should catch cold and
starve. I will fetch my gold and have our two wagers
recorded.                                        165
POSTHUMUS Agreed.                    ⌐Exit with Giacomo⌐
FRENCHMAN Will this hold, think you?
FILARIO Signor Giacomo will not from it. Pray let us
follow 'em.                    Exeunt. ⌐Table is removed⌐

**1.5**    *Enter Queen, Ladies, and Cornelius, a doctor*
QUEEN
    Whiles yet the dew's on ground, gather those flowers.
    Make haste. Who has the note of them?
A LADY                                    I, madam.
QUEEN Dispatch.                        *Exeunt Ladies*
    Now, Master Doctor, have you brought those drugs?
CORNELIUS
    Pleaseth your highness, ay. Here they are, madam.  5
        *He gives her a box*
    But I beseech your grace, without offence—
    My conscience bids me ask—wherefore you have
    Commanded of me these most poisonous compounds,
    Which are the movers of a languishing death,
    But though slow, deadly.
QUEEN                    I wonder, doctor,        10
    Thou ask'st me such a question. Have I not been
    Thy pupil long? Hast thou not learned me how
    To make perfumes, distil, preserve—yea, so
    That our great King himself doth woo me oft
    For my confections? Having thus far proceeded,  15
    Unless thou think'st me devilish, is't not meet
    That I did amplify my judgement in
    Other conclusions? I will try the forces
    Of these thy compounds on such creatures as
    We count not worth the hanging, but none human,
    To try the vigour of them, and apply          21
    Allayments to their act, and by them gather
    Their several virtues and effects.
CORNELIUS                    Your highness
    Shall from this practice but make hard your heart.
    Besides, the seeing these effects will be      25
    Both noisome and infectious.
QUEEN                        O, content thee.
        *Enter Pisanio*
    (*Aside*) Here comes a flattering rascal; upon him
    Will I first work. He's factor for his master,
    And enemy to my son. (*Aloud*) How now, Pisanio?—
    Doctor, your service for this time is ended.    30
    Take your own way.
CORNELIUS (*aside*)        I do suspect you, madam.
    But you shall do no harm.
QUEEN (*to Pisanio*)            Hark thee, a word.
CORNELIUS (*aside*)
    I do not like her. She doth think she has
    Strange ling'ring poisons. I do know her spirit,
    And will not trust one of her malice with      35
    A drug of such damned nature. Those she has
    Will stupefy and dull the sense a while,
    Which first, perchance, she'll prove on cats and dogs,
    Then afterward up higher; but there is
    No danger in what show of death it makes        40
    More than the locking up the spirits a time,

To be more fresh, reviving. She is fooled
    With a most false effect, and I the truer
    So to be false with her.
QUEEN                No further service, doctor,   44
    Until I send for thee.
CORNELIUS            I humbly take my leave.    *Exit*
QUEEN (*to Pisanio*)
    Weeps she still, sayst thou? Dost thou think in time
    She will not quench, and let instructions enter
    Where folly now possesses? Do thou work.
    When thou shalt bring me word she loves my son
    I'll tell thee on the instant thou art then    50
    As great as is thy master—greater, for
    His fortunes all lie speechless, and his name
    Is at last gasp. Return he cannot, nor
    Continue where he is. To shift his being
    Is to exchange one misery with another,        55
    And every day that comes comes to decay
    A day's work in him. What shalt thou expect
    To be depender on a thing that leans,
    Who cannot be new built nor has no friends
    So much as but to prop him?
        ⌐*She drops her box. He takes it up*⌐
                        Thou tak'st up        60
    Thou know'st not what; but take it for thy labour.
    It is a thing I made which hath the King
    Five times redeemed from death. I do not know
    What is more cordial. Nay, I prithee take it.
    It is an earnest of a farther good            65
    That I mean to thee. Tell thy mistress how
    The case stands with her; do't as from thyself.
    Think what a chance thou changest on, but think
    Thou hast thy mistress still; to boot, my son,
    Who shall take notice of thee. I'll move the King  70
    To any shape of thy preferment, such
    As thou'lt desire; and then myself, I chiefly,
    That set thee on to this desert, am bound
    To load thy merit richly. Call my women.
    Think on my words.                *Exit Pisanio*
                    A sly and constant knave,
    Not to be shaked; the agent for his master,    76
    And the remembrancer of her to hold
    The hand-fast to her lord. I have given him that
    Which, if he take, shall quite unpeople her
    Of liegers for her sweet, and which she after,  80
    Except she bend her humour, shall be assured
    To taste of too.
        *Enter Pisanio and Ladies*
                So, so; well done, well done.
    The violets, cowslips, and the primroses
    Bear to my closet. Fare thee well, Pisanio.
    Think on my words, Pisanio.
PISANIO                    And shall do.        85
                    *Exeunt Queen and Ladies*
    But when to my good lord I prove untrue,
    I'll choke myself—there's all I'll do for you.    *Exit*

**1.6**    *Enter Innogen*
INNOGEN
    A father cruel and a stepdame false,
    A foolish suitor to a wedded lady
    That hath her husband banished. O, that husband,
    My supreme crown of grief, and those repeated
    Vexations of it! Had I been thief-stol'n,        5
    As my two brothers, happy; but most miserable
    Is the desire that's glorious. Blest be those,

How mean soe'er, that have their honest wills,
Which seasons comfort.
*Enter Pisanio and Giacomo*
                              Who may this be? Fie!
PISANIO
Madam, a noble gentleman of Rome                    10
Comes from my lord with letters.
GIACOMO                              Change you, madam?
The worthy Leonatus is in safety,
And greets your highness dearly.
*He gives her the letters*
INNOGEN                              Thanks, good sir.
You're kindly welcome.
*She reads the letters*
GIACOMO (*aside*)
All of her that is out of door most rich!            15
If she be furnished with a mind so rare
She is alone, th'Arabian bird, and I
Have lost the wager. Boldness be my friend;
Arm me audacity from head to foot,
Or, like the Parthian, I shall flying fight;         20
Rather, directly fly.
INNOGEN (*reads aloud*) 'He is one of the noblest note, to
whose kindnesses I am most infinitely tied. Reflect upon
him accordingly, as you value                        24
                              Your truest
                                        Leonatus.'
(*To Giacomo*) So far I read aloud,
But even the very middle of my heart
Is warmed by th' rest, and takes it thankfully.
You are as welcome, worthy sir, as I                 30
Have words to bid you, and shall find it so
In all that I can do.
GIACOMO              Thanks, fairest lady.
What, are men mad? Hath nature given them eyes
To see this vaulted arch and the rich crop
Of sea and land, which can distinguish 'twixt        35
The fiery orbs above and the twinned stones
Upon th'unnumbered beach, and can we not
Partition make with spectacles so precious
'Twixt fair and foul?
INNOGEN              What makes your admiration?
GIACOMO
It cannot be i'th' eye—for apes and monkeys,         40
'Twixt two such shes, would chatter this way and
Contemn with mows the other; nor i'th' judgement,
For idiots in this case of favour would
Be wisely definite; nor i'th' appetite—
Sluttery, to such neat excellence opposed,           45
Should make desire vomit emptiness,
Not so allured to feed.
INNOGEN What is the matter, trow?
GIACOMO The cloyèd will,
That satiate yet unsatisfied desire, that tub        50
Both filled and running, ravening first the lamb,
Longs after for the garbage.
INNOGEN              What, dear sir,
Thus raps you? Are you well?
GIACOMO
Thanks, madam, well. (*To Pisanio*) Beseech you, sir,
Desire my man's abode where I did leave him.         55
He's strange and peevish.
PISANIO              I was going, sir,
To give him welcome.                              *Exit*
INNOGEN              Continues well my lord?
His health, beseech you?
GIACOMO              Well, madam.

INNOGEN
Is he disposed to mirth? I hope he is.
GIACOMO
Exceeding pleasant, none a stranger there            60
So merry and so gamesome. He is called
The Briton Reveller.
INNOGEN              When he was here
He did incline to sadness, and oft-times
Not knowing why.
GIACOMO              I never saw him sad.
There is a Frenchman his companion, one              65
An eminent monsieur that, it seems, much loves
A Gallian girl at home. He furnaces
The thick sighs from him, whiles the jolly Briton—
Your lord, I mean—laughs from 's free lungs,
          cries 'O,
Can my sides hold, to think that man, who knows      70
By history, report or his own proof
What woman is, yea, what she cannot choose
But must be, will 's free hours languish
For assurèd bondage?'
INNOGEN              Will my lord say so?
GIACOMO
Ay, madam, with his eyes in flood with laughter.     75
It is a recreation to be by
And hear him mock the Frenchman. But heavens
          know
Some men are much to blame.
INNOGEN              Not he, I hope.
GIACOMO
Not he; but yet heaven's bounty towards him might
Be used more thankfully. In himself 'tis much;      80
In you, which I count his, beyond all talents.
Whilst I am bound to wonder, I am bound
To pity too.
INNOGEN      What do you pity, sir?
GIACOMO
Two creatures heartily.
INNOGEN              Am I one, sir?
You look on me; what wreck discern you in me         85
Deserves your pity?
GIACOMO              Lamentable! What,
To hide me from the radiant sun, and solace
I'th' dungeon by a snuff?
INNOGEN              I pray you, sir,
Deliver with more openness your answers
To my demands. Why do you pity me?                   90
GIACOMO That others do—
I was about to say enjoy your—but
It is an office of the gods to venge it,
Not mine to speak on't.
INNOGEN              You do seem to know
Something of me, or what concerns me. Pray you,      95
Since doubting things go ill often hurts more
Than to be sure they do—for certainties
Either are past remedies, or, timely knowing,
The remedy then born—discover to me
What both you spur and stop.
GIACOMO              Had I this cheek         100
To bathe my lips upon; this hand whose touch,
Whose every touch, would force the feeler's soul
To th'oath of loyalty; this object which
Takes prisoner the wild motion of mine eye,
Firing it only here: should I, damned then,          105
Slaver with lips as common as the stairs
That mount the Capitol; join grips with hands

Made hard with hourly falsehood—falsehood as
With labour; then by-peeping in an eye
Base and illustrous as the smoky light                        110
That's fed with stinking tallow—it were fit
That all the plagues of hell should at one time
Encounter such revolt.

INNOGEN                          My lord, I fear,
Has forgot Britain.

GIACOMO                          And himself. Not I
Inclined to this intelligence pronounce                       115
The beggary of his change, but 'tis your graces
That from my mutest conscience to my tongue
Charms this report out.

INNOGEN                          Let me hear no more.

GIACOMO
O dearest soul, your cause doth strike my heart
With pity that doth make me sick. A lady                      120
So fair, and fastened to an empery
Would make the great'st king double, to be partnered
With tomboys hired with that self exhibition
Which your own coffers yield; with diseased ventures
That play with all infirmities for gold                       125
Which rottenness can lend to nature; such boiled stuff
As well might poison poison! Be revenged,
Or she that bore you was no queen, and you
Recoil from your great stock.

INNOGEN                          Revenged?
How should I be revenged? If this be true—                    130
As I have such a heart that both mine ears
Must not in haste abuse—if it be true,
How should I be revenged?

GIACOMO                          Should he make me
Live like Diana's priest betwixt cold sheets
Whiles he is vaulting variable ramps,                         135
In your despite, upon your purse—revenge it.
I dedicate myself to your sweet pleasure,
More noble than that runagate to your bed,
And will continue fast to your affection,
Still close as sure.

INNOGEN                          What ho, Pisanio!             140

GIACOMO
Let me my service tender on your lips.

INNOGEN
Away, I do condemn mine ears that have
So long attended thee. If thou wert honourable
Thou wouldst have told this tale for virtue, not
For such an end thou seek'st, as base as strange.             145
Thou wrong'st a gentleman who is as far
From thy report as thou from honour, and
Solicit'st here a lady that disdains
Thee and the devil alike. What ho, Pisanio!
The King my father shall be made acquainted                   150
Of thy assault. If he shall think it fit
A saucy stranger in his court to mart
As in a Romish stew, and to expound
His beastly mind to us, he hath a court
He little cares for, and a daughter who                       155
He not respects at all. What ho, Pisanio!

GIACOMO
O happy Leonatus! I may say
The credit that thy lady hath of thee
Deserves thy trust, and thy most perfect goodness
Her assured credit. Blessèd live you long,                    160
A lady to the worthiest sir that ever
Country called his; and you his mistress, only

For the most worthiest fit. Give me your pardon.
I have spoke this to know if your affiance
Were deeply rooted, and shall make your lord                  165
That which he is new o'er; and he is one
The truest mannered, such a holy witch
That he enchants societies into him;
Half all men's hearts are his.

INNOGEN                          You make amends.

GIACOMO
He sits 'mongst men like a descended god.                     170
He hath a kind of honour sets him off
More than a mortal seeming. Be not angry,
Most mighty princess, that I have adventured
To try your taking of a false report, which hath
Honoured with confirmation your great judgement
In the election of a sir so rare                              176
Which you know cannot err. The love I bear him
Made me to fan you thus, but the gods made you,
Unlike all others, chaffless. Pray, your pardon.

INNOGEN
All's well, sir. Take my power i'th' court for yours.

GIACOMO
My humble thanks. I had almost forgot                         181
T'entreat your grace but in a small request,
And yet of moment too, for it concerns
Your lord; myself and other noble friends
Are partners in the business.

INNOGEN                          Pray what is't?              185

GIACOMO
Some dozen Romans of us, and your lord—
Best feather of our wing—have mingled sums
To buy a present for the Emperor,
Which I, the factor for the rest, have done
In France. 'Tis plate of rare device, and jewels             190
Of rich and exquisite form; their value's great,
And I am something curious, being strange,
To have them in safe stowage. May it please you
To take them in protection?

INNOGEN                          Willingly,
And pawn mine honour for their safety; since                  195
My lord hath interest in them, I will keep them
In my bedchamber.

GIACOMO                          They are in a trunk
Attended by my men. I will make bold
To send them to you, only for this night.
I must aboard tomorrow.

INNOGEN                          O, no, no!                   200

GIACOMO
Yes, I beseech, or I shall short my word
By length'ning my return. From Gallia
I crossed the seas on purpose and on promise
To see your grace.

INNOGEN                          I thank you for your pains;
But not away tomorrow!

GIACOMO                          O, I must, madam.            205
Therefore I shall beseech you, if you please
To greet your lord with writing, do't tonight.
I have outstood my time, which is material
To th' tender of our present.

INNOGEN                          I will write.
Send your trunk to me, it shall safe be kept,                 210
And truly yielded you. You're very welcome.
                                        *Exeunt severally*

                          1139

**2.1**    *Enter Cloten and the two Lords*

CLOTEN Was there ever man had such luck? When I
kissed the jack upon an upcast, to be hit away! I had
a hundred pound on 't, and then a whoreson jackanapes
must take me up for swearing, as if I borrowed mine
oaths of him, and might not spend them at my pleasure.

FIRST LORD What got he by that? You have broke his
pate with your bowl.

SECOND LORD (*aside*) If his wit had been like him that broke
it, it would have run all out.                                           9

CLOTEN When a gentleman is disposed to swear it is not
for any standers-by to curtail his oaths, ha?

SECOND LORD No, my lord (*aside*)—nor crop the ears of
them.

CLOTEN Whoreson dog! I give him satisfaction? Would he
had been one of my rank.                                                15

SECOND LORD (*aside*) To have smelled like a fool.

CLOTEN I am not vexed more at anything in th'earth. A
pox on 't, I had rather not be so noble as I am. They
dare not fight with me because of the Queen, my
mother. Every jack-slave hath his bellyful of fighting,
and I must go up and down like a cock that nobody
can match.

SECOND LORD (*aside*) You are cock and capon too an you
crow cock with your comb on.

CLOTEN Sayst thou?                                                      25

SECOND LORD It is not fit your lordship should undertake
every companion that you give offence to.

CLOTEN No, I know that, but it is fit I should commit
offence to my inferiors.

SECOND LORD Ay, it is fit for your lordship only.                       30

CLOTEN Why, so I say.

FIRST LORD Did you hear of a stranger that's come to
court tonight?

CLOTEN A stranger, and I not know on 't?

SECOND LORD (*aside*) He's a strange fellow himself and
knows it not.                                                           36

FIRST LORD There's an Italian come, and, 'tis thought,
one of Leonatus' friends.

CLOTEN Leonatus? A banished rascal; and he's another,
whatsoever he be. Who told you of this stranger?        40

FIRST LORD One of your lordship's pages.

CLOTEN Is it fit I went to look upon him? Is there no
derogation in 't?

SECOND LORD You cannot derogate, my lord.

CLOTEN Not easily, I think.                                             45

SECOND LORD (*aside*) You are a fool granted, therefore your
issues, being foolish, do not derogate.

CLOTEN Come, I'll go see this Italian. What I have lost
today at bowls I'll win tonight of him. Come, go.

SECOND LORD I'll attend your lordship.                                  50

*Exeunt Cloten and First Lord*

That such a crafty devil as is his mother
Should yield the world this ass!—a woman that
Bears all down with her brain, and this her son
Cannot take two from twenty, for his heart,
And leave eighteen. Alas, poor princess,                                55
Thou divine Innogen, what thou endur'st,
Betwixt a father by thy stepdame governed,
A mother hourly coining plots, a wooer
More hateful than the foul expulsion is
Of thy dear husband, than that horrid act                               60
Of the divorce he'd make! The heavens hold firm
The walls of thy dear honour, keep unshaked
That temple, thy fair mind, that thou mayst stand
T'enjoy thy banished lord and this great land!        *Exit*

**2.2**    *A trunk ⌈and arras⌉. A bed is ⌈thrust forth⌉ with
Innogen in it, reading a book. Enter to her Helen, a
lady*

INNOGEN
Who's there? My woman Helen?

HELEN                                   Please you, madam.

INNOGEN
What hour is it?

HELEN                        Almost midnight, madam.

INNOGEN
I have read three hours then. Mine eyes are weak.
Fold down the leaf where I have left. To bed.
Take not away the taper; leave it burning,                              5
And if thou canst awake by four o' th' clock,
I prithee call me. Sleep hath seized me wholly.

⌈*Exit Helen*⌉

To your protection I commend me, gods.
From fairies and the tempters of the night
Guard me, beseech ye.                                                   10
*She sleeps.*
*Giacomo comes from the trunk*

GIACOMO
The crickets sing, and man's o'er-laboured sense
Repairs itself by rest. Our Tarquin thus
Did softly press the rushes ere he wakened
The chastity he wounded. Cytherea,
How bravely thou becom'st thy bed! Fresh lily,                          15
And whiter than the sheets! That I might touch,
But kiss, one kiss! Rubies unparagoned,
How dearly they do 't! 'Tis her breathing that
Perfumes the chamber thus. The flame o' th' taper
Bows toward her, and would underpeep her lids,                          20
To see th'enclosèd lights, now canopied
Under these windows, white and azure-laced
With blue of heaven's own tinct. But my design—
To note the chamber. I will write all down.
*He writes in his tables*
Such and such pictures, there the window, such          25
Th'adornment of her bed, the arras, figures,
Why, such and such; and the contents o' th' story.
Ah, but some natural notes about her body
Above ten thousand meaner movables
Would testify t'enrich mine inventory.                                  30
O sleep, thou ape of death, lie dull upon her,
And be her sense but as a monument
Thus in a chapel lying. Come off, come off;
As slippery as the Gordian knot was hard.
*He takes the bracelet from her arm*
'Tis mine, and this will witness outwardly,                             35
As strongly as the conscience does within,
To th' madding of her lord. On her left breast
A mole, cinque-spotted, like the crimson drops
I' th' bottom of a cowslip. Here's a voucher
Stronger than ever law could make. This secret          40
Will force him think I have picked the lock and
ta'en
The treasure of her honour. No more. To what end?
Why should I write this down that's riveted,
Screwed to my memory? She hath been reading late,
The tale of Tereus. Here the leaf's turned down         45
Where Philomel gave up. I have enough.
To th' trunk again, and shut the spring of it.
Swift, swift, you dragons of the night, that dawning
May bare the raven's eye! I lodge in fear.
Though this' a heavenly angel, hell is here.                            50

*Clock strikes*
One, two, three. Time, time!
                *Exit into the trunk.* ⌜*The bed and trunk are*
                                                    *removed*⌝

**2.3**    *Enter Cloten and the two Lords*
FIRST LORD  Your lordship is the most patient man in loss,
    the most coldest that ever turned up ace.
CLOTEN  It would make any man cold to lose.
FIRST LORD  But not every man patient after the noble
    temper of your lordship. You are most hot and furious
    when you win.                                               6
CLOTEN  Winning will put any man into courage. If I could
    get this foolish Innogen I should have gold enough. It's
    almost morning, is't not?
FIRST LORD  Day, my lord.                                      10
CLOTEN  I would this music would come. I am advised to
    give her music o' mornings; they say it will penetrate.
        *Enter Musicians*
    Come on, tune. If you can penetrate her with your
    fingering, so; we'll try with tongue too. If none will
    do, let her remain; but I'll never give o'er. First, a very
    excellent good-conceited thing; after, a wonderful sweet
    air with admirable rich words to it; and then let her
    consider.
        ⌜*Music*⌝
    ⌜MUSICIAN⌝ (*sings*)
        Hark, hark, the lark at heaven gate sings,
            And Phoebus gins arise,                            20
        His steeds to water at those springs
            On chaliced flowers that lies,
        And winking Mary-buds begin to ope their golden eyes;
        With everything that pretty is, my lady sweet, arise,
            Arise, arise!                                      25
CLOTEN  So, get you gone. If this penetrate I will consider
    your music the better; if it do not, it is a vice in her
    ears which horse hairs and calves' guts nor the voice
    of unpaved eunuch to boot can never amend.
                            *Exeunt Musicians*
        *Enter Cymbeline and the Queen*
SECOND LORD  Here comes the King.                              30
CLOTEN  I am glad I was up so late, for that's the reason
    I was up so early. He cannot choose but take this
    service I have done fatherly. Good morrow to your
    majesty, and to my gracious mother.
CYMBELINE
    Attend you here the door of our stern daughter?           35
    Will she not forth?
CLOTEN  I have assailed her with musics, but she
    vouchsafes no notice.
CYMBELINE
    The exile of her minion is too new.
    She hath not yet forgot him. Some more time               40
    Must wear the print of his remembrance out,
    And then she's yours.
QUEEN (*to Cloten*)        You are most bound to th' King,
    Who lets go by no vantages that may
    Prefer you to his daughter. Frame yourself
    To orderly solicits, and be friended                      45
    With aptness of the season. Make denials
    Increase your services; so seem as if
    You were inspired to do those duties which
    You tender to her; that you in all obey her,
    Save when command to your dismission tends,               50
    And therein you are senseless.
CLOTEN                              Senseless? Not so.

        *Enter a Messenger*
MESSENGER (*to Cymbeline*)
    So like you, sir, ambassadors from Rome;
    The one is Caius Lucius.
CYMBELINE                        A worthy fellow,
    Albeit he comes on angry purpose now:
    But that's no fault of his. We must receive him            55
    According to the honour of his sender,
    And towards himself, his goodness forespent on us,
    We must extend our notice. Our dear son,
    When you have given good morning to your mistress,
    Attend the Queen and us. We shall have need               60
    T'employ you towards this Roman. Come, our queen.
                        *Exeunt all but Cloten*
CLOTEN
    If she be up, I'll speak with her; if not,
    Let her lie still and dream.
        ⌜*He knocks*⌝
                            By your leave, ho!—
    I know her women are about her; what
    If I do line one of their hands? 'Tis gold                 65
    Which buys admittance—oft it doth—yea, and makes
    Diana's rangers false themselves, yield up
    Their deer to th' stand o'th' stealer; and 'tis gold
    Which makes the true man killed and saves the thief,
    Nay, sometime hangs both thief and true man. What
    Can it not do and undo? I will make                       71
    One of her women lawyer to me, for
    I yet not understand the case myself.—
    By your leave.
        *Knocks. Enter a Lady*
LADY
    Who's there that knocks?
CLOTEN                        A gentleman.
LADY                                    No more?            75
CLOTEN
    Yes, and a gentlewoman's son.
LADY                                That's more
    ⌜*Aside*⌝ Than some whose tailors are as dear as
        yours
    Can justly boast of. (*To him*) What's your lordship's
        pleasure?
CLOTEN
    Your lady's person. Is she ready?
LADY                                Ay.
    ⌜*Aside*⌝ To keep her chamber.
CLOTEN                              There is gold for you.   80
    Sell me your good report.
LADY
    How, my good name?—or to report of you
    What I shall think is good?
        *Enter Innogen*
                            The Princess.        ⌜*Exit*⌝
CLOTEN
    Good morrow, fairest. Sister, your sweet hand.
INNOGEN
    Good morrow, sir. You lay out too much pains            85
    For purchasing but trouble. The thanks I give
    Is telling you that I am poor of thanks,
    And scarce can spare them.
CLOTEN                          Still I swear I love you.
INNOGEN
    If you but said so, 'twere as deep with me.
    If you swear still, your recompense is still            90
    That I regard it not.
CLOTEN                    This is no answer.

INNOGEN
But that you shall not say I yield being silent,
I would not speak. I pray you, spare me. Faith,
I shall unfold equal discourtesy
To your best kindness. One of your great knowing    95
Should learn, being taught, forbearance.

CLOTEN
To leave you in your madness, 'twere my sin.
I will not.

INNOGEN          Fools cure not mad folks.

CLOTEN
Do you call me fool?

INNOGEN                    As I am mad, I do.
If you'll be patient, I'll no more be mad;          100
That cures us both. I am much sorry, sir,
You put me to forget a lady's manners
By being so verbal; and learn now for all
That I, which know my heart, do here pronounce
By th' very truth of it: I care not for you,        105
And am so near the lack of charity
To accuse myself I hate you, which I had rather
You felt than make't my boast.

CLOTEN                            You sin against
Obedience which you owe your father. For
The contract you pretend with that base wretch,     110
One bred of alms and fostered with cold dishes,
With scraps o'th' court, it is no contract, none.
And though it be allowed in meaner parties—
Yet who than he more mean?—to knit their souls,
On whom there is no more dependency                 115
But brats and beggary, in self-figured knot,
Yet you are curbed from that enlargement by
The consequence o'th' crown, and must not foil
The precious note of it with a base slave,
A hilding for a livery, a squire's cloth,           120
A pantler—not so eminent.

INNOGEN                    Profane fellow,
Wert thou the son of Jupiter, and no more
But what thou art besides, thou wert too base
To be his groom; thou wert dignified enough,
Even to the point of envy, if 'twere made           125
Comparative for your virtues to be styled
The under-hangman of his kingdom, and hated
For being preferred so well.

CLOTEN                    The south-fog rot him!

INNOGEN
He never can meet more mischance than come
To be but named of thee. His meanest garment        130
That ever hath but clipped his body is dearer
In my respect than all the hairs above thee,
Were they all made such men. How now, Pisanio!
    Enter Pisanio

CLOTEN His garment? Now the devil—

INNOGEN (to Pisanio)
To Dorothy, my woman, hie thee presently.           135

CLOTEN
His garment?

INNOGEN (to Pisanio) I am sprited with a fool,
Frighted, and angered worse. Go bid my woman
Search for a jewel that too casually
Hath left mine arm. It was thy master's. 'Shrew me
If I would lose it for a revenue                     140
Of any king's in Europe! I do think
I saw't this morning; confident I am
Last night 'twas on mine arm; I kissed it.
I hope it be not gone to tell my lord

That I kiss aught but he.

PISANIO                    'Twill not be lost.       145

INNOGEN
I hope so. Go and search.                   Exit Pisanio

CLOTEN                    You have abused me.
'His meanest garment'?

INNOGEN                    Ay, I said so, sir.
If you will make't an action, call witness to't.

CLOTEN
I will inform your father.

INNOGEN                    Your mother too.
She's my good lady, and will conceive, I hope,      150
But the worst of me. So I leave you, sir,
To th' worst of discontent.                         Exit

CLOTEN                    I'll be revenged.
'His meanest garment'? Well!                        Exit

**2.4**    *Enter Posthumus and Filario*

POSTHUMUS
Fear it not, sir. I would I were so sure
To win the King as I am bold her honour
Will remain hers.

FILARIO                    What means do you make to him?

POSTHUMUS
Not any; but abide the change of time,
Quake in the present winter's state, and wish       5
That warmer days would come. In these seared hopes
I barely gratify your love; they failing,
I must die much your debtor.

FILARIO
Your very goodness and your company
O'erpays all I can do. By this, your king           10
Hath heard of great Augustus. Caius Lucius
Will do 's commission throughly. And I think
He'll grant the tribute, send th'arrearages,
Ere look upon our Romans, whose remembrance
Is yet fresh in their grief.

POSTHUMUS                    I do believe,           15
Statist though I am none, nor like to be,
That this will prove a war, and you shall hear
The legions now in Gallia sooner landed
In our not-fearing Britain than have tidings
Of any penny tribute paid. Our countrymen           20
Are men more ordered than when Julius Caesar
Smiled at their lack of skill but found their courage
Worthy his frowning at. Their discipline,
Now wing-led with their courage, will make known
To their approvers they are people such             25
That mend upon the world.
    *Enter Giacomo*

FILARIO                    See, Giacomo.

POSTHUMUS (to Giacomo)
The swiftest harts have posted you by land,
And winds of all the corners kissed your sails
To make your vessel nimble.

FILARIO (to Giacomo)                    Welcome, sir.

POSTHUMUS (to Giacomo)
I hope the briefness of your answer made            30
The speediness of your return.

GIACOMO                    Your lady is
One of the fair'st that I have looked upon—

POSTHUMUS
And therewithal the best, or let her beauty
Look through a casement to allure false hearts,
And be false with them.

GIACOMO                    Here are letters for you.    35

POSTHUMUS
Their tenor good, I trust.
GIACOMO            'Tis very like.
    *Posthumus reads the letters*
⌈FILARIO⌉
Was Caius Lucius in the Briton court
When you were there?
GIACOMO          He was expected then,
But not approached.
POSTHUMUS         All is well yet.
Sparkles this stone as it was wont, or is't not    40
Too dull for your good wearing?
GIACOMO           If I had lost it
I should have lost the worth of it in gold.
I'll make a journey twice as far t'enjoy
A second night of such sweet shortness which
Was mine in Britain; for the ring is won.    45
POSTHUMUS
The stone's too hard to come by.
GIACOMO            Not a whit,
Your lady being so easy.
POSTHUMUS        Make not, sir,
Your loss your sport. I hope you know that we
Must not continue friends.
GIACOMO         Good sir, we must,
If you keep covenant. Had I not brought    50
The knowledge of your mistress home I grant
We were to question farther, but I now
Profess myself the winner of her honour,
Together with your ring, and not the wronger
Of her or you, having proceeded but    55
By both your wills.
POSTHUMUS      If you can make't apparent
That you have tasted her in bed, my hand
And ring is yours. If not, the foul opinion
You had of her pure honour gains or loses
Your sword or mine, or masterless leaves both    60
To who shall find them.
GIACOMO        Sir, my circumstances,
Being so near the truth as I will make them,
Must first induce you to believe; whose strength
I will confirm with oath, which I doubt not
You'll give me leave to spare when you shall find    65
You need it not.
POSTHUMUS      Proceed.
GIACOMO        First, her bedchamber—
Where I confess I slept not, but profess
Had that was well worth watching—it was hanged
With tapestry of silk and silver; the story
Proud Cleopatra when she met her Roman,    70
And Cydnus swelled above the banks, or for
The press of boats or pride: a piece of work
So bravely done, so rich, that it did strive
In workmanship and value; which I wondered
Could be so rarely and exactly wrought,    75
Such the true life on't was.
POSTHUMUS        This is true,
And this you might have heard of here, by me
Or by some other.
GIACOMO     More particulars
Must justify my knowledge.
POSTHUMUS        So they must,
Or do your honour injury.
GIACOMO         The chimney    80
Is south the chamber, and the chimney-piece

Chaste Dian bathing. Never saw I figures
So likely to report themselves; the cutter
Was as another nature; dumb, outwent her,
Motion and breath left out.
POSTHUMUS       This is a thing    85
Which you might from relation likewise reap,
Being, as it is, much spoke of.
GIACOMO        The roof o'th' chamber
With golden cherubins is fretted. Her andirons—
I had forgot them—were two winking Cupids
Of silver, each on one foot standing, nicely    90
Depending on their brands.
POSTHUMUS       This is her honour!
Let it be granted you have seen all this—and praise
Be given to your remembrance—the description
Of what is in her chamber nothing saves
The wager you have laid.
GIACOMO        Then, if you can    95
Be pale, I beg but leave to air this jewel. See!
    *He shows the bracelet*
And now 'tis up again; it must be married
To that your diamond. I'll keep them.
POSTHUMUS        Jove!
Once more let me behold it. Is it that
Which I left with her?
GIACOMO       Sir, I thank her, that.    100
She stripped it from her arm. I see her yet.
Her pretty action did outsell her gift,
And yet enriched it too. She gave it me,
And said she prized it once.
POSTHUMUS       Maybe she plucked it off
To send it me.
GIACOMO      She writes so to you, doth she?    105
POSTHUMUS
O, no, no, no—'tis true! Here, take this too.
    *He gives Giacomo his ring*
It is a basilisk unto mine eye,
Kills me to look on't. Let there be no honour
Where there is beauty, truth where semblance, love
Where there's another man. The vows of women    110
Of no more bondage be to where they are made
Than they are to their virtues, which is nothing!
O, above measure false!
FILARIO        Have patience, sir,
And take your ring again; 'tis not yet won.
It may be probable she lost it, or    115
Who knows if one her woman, being corrupted,
Hath stol'n it from her?
POSTHUMUS       Very true,
And so I hope he came by't. Back my ring.
    *He takes his ring again*
Render to me some corporal sign about her
More evident than this; for this was stol'n.    120
GIACOMO
By Jupiter, I had it from her arm.
POSTHUMUS
Hark you, he swears, by Jupiter he swears.
'Tis true, nay, keep the ring, 'tis true. I am sure
She would not lose it. Her attendants are
All sworn and honourable. They induced to steal it?
And by a stranger? No, he hath enjoyed her.    126
The cognizance of her incontinency
Is this. She hath bought the name of whore thus
  dearly.
    *He gives Giacomo his ring*

There, take thy hire, and all the fiends of hell
Divide themselves between you!
FILARIO                    Sir, be patient.                    130
This is not strong enough to be believed
Of one persuaded well of.
POSTHUMUS                    Never talk on't.
She hath been colted by him.
GIACOMO                    If you seek
For further satisfying, under her breast—
Worthy the pressing—lies a mole, right proud    135
Of that most delicate lodging. By my life,
I kissed it, and it gave me present hunger
To feed again, though full. You do remember
This stain upon her?
POSTHUMUS                    Ay, and it doth confirm
Another stain as big as hell can hold,          140
Were there no more but it.
GIACOMO                    Will you hear more?
POSTHUMUS
Spare your arithmetic, never count the turns.
Once, and a million!
GIACOMO                    I'll be sworn.
POSTHUMUS                    No swearing.
If you will swear you have not done't, you lie,
And I will kill thee if thou dost deny          145
Thou'st made me cuckold.
GIACOMO                    I'll deny nothing.
POSTHUMUS
O that I had her here to tear her limb-meal!
I will go there and do't i'th' court, before
Her father. I'll do something.                  Exit
FILARIO                    Quite besides
The government of patience! You have won.        150
Let's follow and pervert the present wrath
He hath against himself.
GIACOMO                    With all my heart.    Exeunt

**2.5**    *Enter Posthumus*
POSTHUMUS
Is there no way for men to be, but women
Must be half-workers? We are bastards all,
And that most venerable man which I
Did call my father was I know not where
When I was stamped. Some coiner with his tools    5
Made me a counterfeit; yet my mother seemed
The Dian of that time: so doth my wife
The nonpareil of this. O vengeance, vengeance!
Me of my lawful pleasure she restrained,
And prayed me oft forbearance; did it with        10
A pudency so rosy the sweet view on't
Might well have warmed old Saturn; that I thought
    her
As chaste as unsunned snow. O all the devils!
This yellow Giacomo in an hour—was't not?—
Or less—at first? Perchance he spoke not, but     15
Like a full-acorned boar, a German one,
Cried 'O!' and mounted; found no opposition
But what he looked for should oppose and she
Should from encounter guard. Could I find out
The woman's part in me—for there's no motion      20
That tends to vice in man but I affirm
It is the woman's part; be it lying, note it,
The woman's; flattering, hers; deceiving, hers;
Lust and rank thoughts, hers, hers; revenges, hers;
Ambitions, covetings, change of prides, disdain,  25
Nice longing, slanders, mutability,

All faults that man can name, nay, that hell knows,
Why, hers in part or all, but rather all—
For even to vice
They are not constant, but are changing still     30
One vice but of a minute old for one
Not half so old as that. I'll write against them,
Detest them, curse them, yet 'tis greater skill
In a true hate to pray they have their will.
The very devils cannot plague them better.    Exit

**3.1**    ⌈*Flourish.*⌉ *Enter in state Cymbeline, the Queen,*
*Cloten, and lords at one door, and at another,*
*Caius Lucius and attendants*
CYMBELINE
Now say, what would Augustus Caesar with us?
LUCIUS
When Julius Caesar—whose remembrance yet
Lives in men's eyes, and will to ears and tongues
Be theme and hearing ever—was in this Britain
And conquered it, Cassibelan, thine uncle,         5
Famous in Caesar's praises no whit less
Than in his feats deserving it, for him
And his succession granted Rome a tribute,
Yearly three thousand pounds, which by thee lately
Is left untendered.
QUEEN                    And, to kill the marvel,   10
Shall be so ever.
CLOTEN                    There will be many Caesars
Ere such another Julius. Britain's a world
By itself, and we will nothing pay
For wearing our own noses.
QUEEN                    That opportunity
Which then they had to take from 's, to resume     15
We have again. Remember, sir, my liege,
The kings your ancestors, together with
The natural bravery of your isle, which stands
As Neptune's park, ribbed and paled in
With banks unscalable and roaring waters,          20
With sands that will not bear your enemies' boats,
But suck them up to th' topmast. A kind of conquest
Caesar made here, but made not here his brag
Of 'came and saw and overcame'. With shame—
The first that ever touched him—he was carried     25
From off our coast, twice beaten; and his shipping,
Poor ignorant baubles, on our terrible seas
Like eggshells moved upon their surges, cracked
As easily 'gainst our rocks; for joy whereof
The famed Cassibelan, who was once at point—       30
O giglot fortune!—to master Caesar's sword,
Made Lud's town with rejoicing fires bright,
And Britons strut with courage.
CLOTEN Come, there's no more tribute to be paid. Our
kingdom is stronger than it was at that time, and, as
I said, there is no more such Caesars. Other of them
may have crooked noses, but to owe such straight
arms, none.
CYMBELINE Son, let your mother end.                39
CLOTEN We have yet many among us can grip as hard
as Cassibelan. I do not say I am one, but I have a
hand. Why tribute? Why should we pay tribute? If
Caesar can hide the sun from us with a blanket, or put
the moon in his pocket, we will pay him tribute for
light; else, sir, no more tribute, pray you now.   45
CYMBELINE (*to Lucius*) You must know,
Till the injurious Romans did extort

This tribute from us we were free. Caesar's ambition,
Which swelled so much that it did almost stretch
The sides o'th' world, against all colour here          50
Did put the yoke upon 's, which to shake off
Becomes a warlike people, whom we reckon
Ourselves to be. We do say then to Caesar,
Our ancestor was that Mulmutius which
Ordained our laws, whose use the sword of Caesar  55
Hath too much mangled, whose repair and franchise
Shall by the power we hold be our good deed,
Though Rome be therefore angry. Mulmutius made
  our laws,
Who was the first of Britain which did put
His brows within a golden crown and called          60
Himself a king.
LUCIUS                    I am sorry, Cymbeline,
That I am to pronounce Augustus Caesar—
Caesar, that hath more kings his servants than
Thyself domestic officers—thine enemy.
Receive it from me, then: war and confusion          65
In Caesar's name pronounce I 'gainst thee. Look
For fury not to be resisted. Thus defied,
I thank thee for myself.
CYMBELINE              Thou art welcome, Caius.
Thy Caesar knighted me; my youth I spent
Much under him; of him I gathered honour,          70
Which he to seek of me again perforce
Behoves me keep at utterance. I am perfect
That the Pannonians and Dalmatians for
Their liberties are now in arms, a precedent
Which not to read would show the Britons cold;     75
So Caesar shall not find them.
LUCIUS                              Let proof speak.
CLOTEN His majesty bids you welcome. Make pastime with
us a day or two or longer. If you seek us afterwards in
other terms, you shall find us in our salt-water girdle.
If you beat us out of it, it is yours; if you fall in the
adventure, our crows shall fare the better for you; and
there's an end.                                                      82
LUCIUS So, sir.
CYMBELINE
I know your master's pleasure, and he mine.
All the remain is 'Welcome'.          ⌜Flourish.⌝ Exeunt

3.2    Enter Pisanio, reading of a letter
PISANIO
How? Of adultery? Wherefore write you not
What monster's her accuser? Leonatus,
O master, what a strange infection
Is fall'n into thy ear! What false Italian,
As poisonous tongued as handed, hath prevailed     5
On thy too ready hearing? Disloyal? No.
She's punished for her truth, and undergoes,
More goddess-like than wife-like, such assaults
As would take in some virtue. O my master,
Thy mind to hers is now as low as were          10
Thy fortunes. How? That I should murder her,
Upon the love and truth and vows which I
Have made to thy command? I her? Her blood?
If it be so to do good service, never
Let me be counted serviceable. How look I,          15
That I should seem to lack humanity
So much as this fact comes to? (Reads) 'Do't. The letter
That I have sent her, by her own command
Shall give thee opportunity.' O damned paper,
Black as the ink that's on thee! Senseless bauble,   20

Art thou a fedary for this act, and look'st
So virgin-like without?
    Enter Innogen
                              Lo, here she comes.
I am ignorant in what I am commanded.
INNOGEN How now, Pisanio?
PISANIO
Madam, here is a letter from my lord.          25
INNOGEN
Who, thy lord that is my lord, Leonatus?
O learned indeed were that astronomer
That knew the stars as I his characters—
He'd lay the future open. You good gods,
Let what is here contained relish of love,          30
Of my lord's health, of his content—yet not
That we two are asunder; let that grieve him.
Some griefs are med'cinable; that is one of them,
For it doth physic love—of his content
All but in that. Good wax, thy leave. Blest be     35
You bees that make these locks of counsel! Lovers
And men in dangerous bonds pray not alike;
Though forfeiters you cast in prison, yet
You clasp young Cupid's tables. Good news, gods!
    She opens and reads the letter
'Justice and your father's wrath, should he take me in
his dominion, could not be so cruel to me as you, O
the dearest of creatures, would even renew me with
your eyes. Take notice that I am in Cambria, at Milford
Haven. What your own love will out of this advise you,
follow. So he wishes you all happiness, that remains
loyal to his vow, and your increasing in love,     46
                              Leonatus Posthumus.'
O for a horse with wings! Hear'st thou, Pisanio?
He is at Milford Haven. Read, and tell me
How far 'tis thither. If one of mean affairs          50
May plod it in a week, why may not I
Glide thither in a day? Then, true Pisanio,
Who long'st like me to see thy lord, who long'st—
O let me bate—but not like me—yet long'st
But in a fainter kind—O, not like me,          55
For mine's beyond beyond; say, and speak thick—
Love's counsellor should fill the bores of hearing,
To th' smothering of the sense—how far it is
To this same blessèd Milford. And by th' way
Tell me how Wales was made so happy as          60
T'inherit such a haven. But first of all,
How we may steal from hence; and for the gap
That we shall make in time from our hence-going
Till our return, to excuse; but first, how get hence.
Why should excuse be born or ere begot?          65
We'll talk of that hereafter. Prithee speak,
How many score of miles may we well ride
'Twixt hour and hour?
PISANIO                    One score 'twixt sun and sun,
Madam, 's enough for you, and too much too.
INNOGEN
Why, one that rode to 's execution, man,          70
Could never go so slow. I have heard of riding wagers
Where horses have been nimbler than the sands
That run i'th' clock's behalf. But this is fool'ry.
Go bid my woman feign a sickness, say
She'll home to her father; and provide me presently
A riding-suit no costlier than would fit          76
A franklin's housewife.
PISANIO                    Madam, you're best consider.

INNOGEN
I see before me, man. Nor here, nor here,
Nor what ensues, but have a fog in them
That I cannot look through. Away, I prithee,      80
Do as I bid thee. There's no more to say:
Accessible is none but Milford way.          *Exeunt*

**3.3**    *Enter Belarius, followed by Guiderius and*
       *Arviragus, ⌈from a cave in the woods⌉*
BELARIUS
A goodly day not to keep house with such
Whose roof's as low as ours. Stoop, boys; this gate
Instructs you how t'adore the heavens, and bows you
To a morning's holy office. The gates of monarchs
Are arched so high that giants may jet through     5
And keep their impious turbans on without
Good morrow to the sun. Hail, thou fair heaven!
We house i'th' rock, yet use thee not so hardly
As prouder livers do.
GUIDERIUS          Hail, heaven!
ARVIRAGUS                Hail, heaven!
BELARIUS
Now for our mountain sport. Up to yon hill,     10
Your legs are young; I'll tread these flats. Consider,
When you above perceive me like a crow,
That it is place which lessens and sets off,
And you may then revolve what tales I have told you
Of courts, of princes, of the tricks in war;      15
That service is not service, so being done,
But being so allowed. To apprehend thus
Draws us a profit from all things we see,
And often to our comfort shall we find
The sharded beetle in a safer hold         20
Than is the full-winged eagle. O, this life
Is nobler than attending for a check,
Richer than doing nothing for a bauble,
Prouder than rustling in unpaid-for silk;
Such gain the cap of him that makes 'em fine,    25
Yet keeps his book uncrossed. No life to ours.
GUIDERIUS
Out of your proof you speak. We, poor unfledged,
Have never winged from view o'th' nest, nor know
     not
What air's from home. Haply this life is best,
If quiet life be best; sweeter to you        30
That have a sharper known; well corresponding
With your stiff age, but unto us it is
A cell of ignorance, travelling abed,
A prison for a debtor, that not dares
To stride a limit.
ARVIRAGUS (*to Belarius*) What should we speak of   35
When we are old as you? When we shall hear
The rain and wind beat dark December, how,
In this our pinching cave, shall we discourse
The freezing hours away? We have seen nothing.
We are beastly: subtle as the fox for prey,     40
Like warlike as the wolf for what we eat.
Our valour is to chase what flies; our cage
We make a choir, as doth the prisoned bird,
And sing our bondage freely.
BELARIUS             How you speak!
Did you but know the city's usuries,        45
And felt them knowingly; the art o'th' court,
As hard to leave as keep, whose top to climb
Is certain falling, or so slipp'ry that
The fear's as bad as falling; the toil o'th' war,
A pain that only seems to seek out danger     50

I'th' name of fame and honour, which dies i'th' search
And hath as oft a sland'rous epitaph
As record of fair act; nay, many times
Doth ill deserve by doing well; what's worse,
Must curtsy at the censure. O boys, this story    55
The world may read in me. My body's marked
With Roman swords, and my report was once
First with the best of note. Cymbeline loved me,
And when a soldier was the theme my name
Was not far off. Then was I as a tree       60
Whose boughs did bend with fruit; but in one night
A storm or robbery, call it what you will,
Shook down my mellow hangings, nay, my leaves,
And left me bare to weather.
GUIDERIUS         Uncertain favour!
BELARIUS
My fault being nothing, as I have told you oft,    65
But that two villains, whose false oaths prevailed
Before my perfect honour, swore to Cymbeline
I was confederate with the Romans. So
Followed my banishment, and this twenty years
This rock and these demesnes have been my world, 70
Where I have lived at honest freedom, paid
More pious debts to heaven than in all
The fore-end of my time. But up to th' mountains!
This is not hunter's language. He that strikes
The venison first shall be the lord o'th' feast,    75
To him the other two shall minister,
And we will fear no poison which attends
In place of greater state. I'll meet you in the valleys.
             *Exeunt Guiderius and Arviragus*
How hard it is to hide the sparks of nature!
These boys know little they are sons to th' King,   80
Nor Cymbeline dreams that they are alive.
They think they are mine, and though trained up
     thus meanly
I'th' cave wherein they bow, their thoughts do hit
The roofs of palaces, and nature prompts them
In simple and low things to prince it much      85
Beyond the trick of others. This Polydore,
The heir of Cymbeline and Britain, who
The King his father called Guiderius—Jove,
When on my three-foot stool I sit and tell
The warlike feats I have done, his spirits fly out   90
Into my story: say 'Thus mine enemy fell,
And thus I set my foot on 's neck', even then
The princely blood flows in his cheek, he sweats,
Strains his young nerves, and puts himself in posture
That acts my words. The younger brother, Cadwal, 95
Once Arviragus, in as like a figure
Strikes life into my speech, and shows much more
His own conceiving.
     ⌈*A hunting-horn sounds*⌉
            Hark, the game is roused!
O Cymbeline, heaven and my conscience knows
Thou didst unjustly banish me, whereon      100
At three and two years old I stole these babes,
Thinking to bar thee of succession as
Thou reft'st me of my lands. Euriphile,
Thou wast their nurse; they took thee for their
     mother,
And every day do honour to her grave.        105
Myself, Belarius, that am Morgan called,
They take for natural father.
     ⌈*A hunting-horn sounds*⌉
                The game is up.     *Exit*

**3.4**    *Enter Pisanio, and Innogen in a riding-suit*

INNOGEN
Thou told'st me when we came from horse the place
Was near at hand. Ne'er longed my mother so
To see me first as I have now. Pisanio, man,
Where is Posthumus? What is in thy mind
That makes thee stare thus? Wherefore breaks that
     sigh      5
From th'inward of thee? One but painted thus
Would be interpreted a thing perplexed
Beyond self-explication. Put thyself
Into a haviour of less fear, ere wildness
Vanquish my staider senses. What's the matter?    10
     *Pisanio gives her a letter*
Why tender'st thou that paper to me with
A look untender? If't be summer news,
Smile to't before; if winterly, thou need'st
But keep that count'nance still. My husband's hand?
That drug-damned Italy hath out-craftied him,    15
And he's at some hard point. Speak, man. Thy tongue
May take off some extremity which to read
Would be even mortal to me.

PISANIO           Please you read,
And you shall find me, wretched man, a thing
The most disdained of fortune.    20

INNOGEN (*reads*) 'Thy mistress, Pisanio, hath played the
strumpet in my bed, the testimonies whereof lies
bleeding in me. I speak not out of weak surmises but
from proof as strong as my grief and as certain as I
expect my revenge. That part thou, Pisanio, must act
for me, if thy faith be not tainted with the breach of
hers. Let thine own hands take away her life. I shall
give thee opportunity at Milford Haven. She hath my
letter for the purpose, where if thou fear to strike and
to make me certain it is done, thou art the pander to
her dishonour and equally to me disloyal.'    31

PISANIO (*aside*)
What shall I need to draw my sword? The paper
Hath cut her throat already. No, 'tis slander,
Whose edge is sharper than the sword, whose tongue
Outvenoms all the worms of Nile, whose breath    35
Rides on the posting winds and doth belie
All corners of the world. Kings, queens, and states,
Maids, matrons, nay, the secrets of the grave
This viperous slander enters. (*To Innogen*) What cheer,
     madam?

INNOGEN
False to his bed? What is it to be false?    40
To lie in watch there and to think on him?
To weep 'twixt clock and clock? If sleep charge nature,
To break it with a fearful dream of him
And cry myself awake? That's false to 's bed, is it?

PISANIO Alas, good lady.    45

INNOGEN
I false? Thy conscience witness, Giacomo,
Thou didst accuse him of incontinency.
Thou then lookedst like a villain; now, methinks,
Thy favour's good enough. Some jay of Italy,
Whose mother was her painting, hath betrayed him.
Poor I am stale, a garment out of fashion,    51
And for I am richer than to hang by th' walls
I must be ripped. To pieces with me! O,
Men's vows are women's traitors. All good seeming,
By thy revolt, O husband, shall be thought    55
Put on for villainy; not born where't grows,
But worn a bait for ladies.

PISANIO           Good madam, hear me.

INNOGEN
True honest men being heard like false Aeneas
Were in his time thought false, and Sinon's weeping
Did scandal many a holy tear, took pity    60
From most true wretchedness. So thou, Posthumus,
Wilt lay the leaven on all proper men.
Goodly and gallant shall be false and perjured
From thy great fail. (*To Pisanio*) Come, fellow, be thou
     honest,
Do thou thy master's bidding. When thou seest
     him,    65
A little witness my obedience. Look,
I draw the sword myself. Take it, and hit
The innocent mansion of my love, my heart.
Fear not, 'tis empty of all things but grief.
Thy master is not there, who was indeed    70
The riches of it. Do his bidding; strike.
Thou mayst be valiant in a better cause,
But now thou seem'st a coward.

PISANIO           Hence, vile instrument,
Thou shalt not damn my hand!

INNOGEN           Why, I must die,
And if I do not by thy hand thou art    75
No servant of thy master's. Against self-slaughter
There is a prohibition so divine
That cravens my weak hand. Come, here's my heart.
Something's afore't. Soft, soft, we'll no defence;
Obedient as the scabbard. What is here?    80
     *She takes letters from her bosom*
The scriptures of the loyal Leonatus,
All turned to heresy? Away, away,
Corrupters of my faith, you shall no more
Be stomachers to my heart. Thus may poor fools
Believe false teachers. Though those that are betrayed
Do feel the treason sharply, yet the traitor    86
Stands in worse case of woe. And thou, Posthumus,
That didst set up my disobedience 'gainst the King
My father, and make me put into contempt the suits
Of princely fellows, shalt hereafter find    90
It is no act of common passage but
A strain of rareness; and I grieve myself
To think, when thou shalt be disedged by her
That now thou tirest on, how thy memory
Will then be panged by me. (*To Pisanio*) Prithee,
     dispatch.    95
The lamb entreats the butcher. Where's thy knife?
Thou art too slow to do thy master's bidding
When I desire it too.

PISANIO           O gracious lady,
Since I received command to do this business
I have not slept one wink.

INNOGEN           Do't, and to bed, then.    100

PISANIO
I'll wake mine eyeballs out first.

INNOGEN           Wherefore then
Didst undertake it? Why hast thou abused
So many miles with a pretence?—this place,
Mine action, and thine own? Our horses' labour,
The time inviting thee? The perturbed court,    105
For my being absent, whereunto I never
Purpose return? Why hast thou gone so far
To be unbent when thou hast ta'en thy stand,
Th'elected deer before thee?

PISANIO           But to win time
To lose so bad employment, in the which    110

I have considered of a course. Good lady,
Hear me with patience.

INNOGEN                    Talk thy tongue weary. Speak.
I have heard I am a strumpet, and mine ear,
Therein false struck, can take no greater wound,
Nor tent to bottom that. But speak.

PISANIO                    Then, madam,    115
I thought you would not back again.

INNOGEN                    Most like,
Bringing me here to kill me.

PISANIO                    Not so, neither.
But if I were as wise as honest, then
My purpose would prove well. It cannot be
But that my master is abused. Some villain,    120
Ay, and singular in his art, hath done you both
This cursèd injury.

INNOGEN  Some Roman courtesan.

PISANIO  No, on my life.
I'll give but notice you are dead, and send him    125
Some bloody sign of it, for 'tis commanded
I should do so. You shall be missed at court,
And that will well confirm it.

INNOGEN                    Why, good fellow,
What shall I do the while, where bide, how live,
Or in my life what comfort when I am    130
Dead to my husband?

PISANIO                    If you'll back to th' court—

INNOGEN
No court, no father, nor no more ado
With that harsh, churlish, noble, simple nothing,
That Cloten, whose love suit hath been to me
As fearful as a siege.

PISANIO                    If not at court,    135
Then not in Britain must you bide.

INNOGEN                    Where then?
Hath Britain all the sun that shines? Day, night,
Are they not but in Britain? I'th' world's volume
Our Britain seems as of it but not in't,
In a great pool a swan's nest. Prithee, think    140
There's livers out of Britain.

PISANIO                    I am most glad
You think of other place. Th'ambassador,
Lucius the Roman, comes to Milford Haven
Tomorrow. Now if you could wear a mind
Dark as your fortune is, and but disguise    145
That which t'appear itself must not yet be
But by self-danger, you should tread a course
Pretty and full of view; yea, haply near
The residence of Posthumus; so nigh, at least,
That though his actions were not visible, yet    150
Report should render him hourly to your ear
As truly as he moves.

INNOGEN                    O, for such means,
Though peril to my modesty, not death on't,
I would adventure.

PISANIO                    Well then, here's the point:
You must forget to be a woman; change    155
Command into obedience, fear and niceness—
The handmaids of all women, or more truly
Woman it pretty self—into a waggish courage,
Ready in gibes, quick-answered, saucy and
As quarrelous as the weasel. Nay, you must    160
Forget that rarest treasure of your cheek,
Exposing it—but O, the harder heart!—
Alack, no remedy—to the greedy touch
Of common-kissing Titan, and forget

Your laboursome and dainty trims wherein    165
You made great Juno angry.

INNOGEN                    Nay, be brief.
I see into thy end, and am almost
A man already.

PISANIO                    First, make yourself but like one.
Forethinking this, I have already fit—
'Tis in my cloak-bag—doublet, hat, hose, all    170
That answer to them. Would you in their serving,
And with what imitation you can borrow
From youth of such a season, fore noble Lucius
Present yourself, desire his service, tell him
Wherein you're happy—which will make him know
If that his head have ear in music—doubtless    176
With joy he will embrace you, for he's honourable,
And, doubling that, most holy. Your means abroad—
You have me, rich, and I will never fail
Beginning nor supplyment.

INNOGEN                    Thou art all the comfort
The gods will diet me with. Prithee away.    181
There's more to be considered, but we'll even
All that good time will give us. This attempt
I am soldier to, and will abide it with
A prince's courage. Away, I prithee.    185

PISANIO
Well, madam, we must take a short farewell
Lest, being missed, I be suspected of
Your carriage from the court. My noble mistress,
Here is a box. I had it from the Queen.
What's in't is precious. If you are sick at sea    190
Or stomach-qualmed at land, a dram of this
Will drive away distemper. To some shade,
And fit you to your manhood. May the gods
Direct you to the best.

INNOGEN                    Amen. I thank thee.

*Exeunt severally*

3.5    ⌜*Flourish.*⌝ *Enter Cymbeline, the Queen, Cloten,*
*Lucius, and lords*

CYMBELINE (*to Lucius*)
Thus far, and so farewell.

LUCIUS                    Thanks, royal sir.
My emperor hath wrote I must from hence;
And am right sorry that I must report ye
My master's enemy.

CYMBELINE                    Our subjects, sir,
Will not endure his yoke, and for ourself    5
To show less sovereignty than they must needs
Appear unkinglike.

LUCIUS                    So, sir, I desire of you
A conduct over land to Milford Haven.
(*To the Queen*) Madam, all joy befall your grace,
⌜*to Cloten*⌝ and you.

CYMBELINE
My lords, you are appointed for that office.    10
The due of honour in no point omit.
So farewell, noble Lucius.

LUCIUS                    Your hand, my lord.

CLOTEN
Receive it friendly, but from this time forth
I wear it as your enemy.

LUCIUS                    Sir, the event
Is yet to name the winner. Fare you well.    15

CYMBELINE
Leave not the worthy Lucius, good my lords,
Till he have crossed the Severn. Happiness.

*Exeunt Lucius and lords*

QUEEN
He goes hence frowning, but it honours us
That we have given him cause.
CLOTEN                                    'Tis all the better.
Your valiant Britons have their wishes in it.        20
CYMBELINE
Lucius hath wrote already to the Emperor
How it goes here. It fits us therefore ripely
Our chariots and our horsemen be in readiness.
The powers that he already hath in Gallia
Will soon be drawn to head, from whence he moves
His war for Britain.
QUEEN                        'Tis not sleepy business,        26
But must be looked to speedily and strongly.
CYMBELINE
Our expectation that it would be thus
Hath made us forward. But, my gentle queen,
Where is our daughter? She hath not appeared        30
Before the Roman, nor to us hath tendered
The duty of the day. She looks us like
A thing more made of malice than of duty.
We have noted it. Call her before us, for
We have been too slight in sufferance.
                              *Exit one or more*
QUEEN                              Royal sir,        35
Since the exile of Posthumus most retired
Hath her life been, the cure whereof, my lord,
'Tis time must do. Beseech your majesty
Forbear sharp speeches to her. She's a lady
So tender of rebukes that words are strokes,        40
And strokes death to her.
                              *Enter a Messenger*
CYMBELINE                        Where is she, sir? How
Can her contempt be answered?
MESSENGER                        Please you, sir,
Her chambers are all locked, and there's no answer
That will be given to th' loud'st of noise we make.
QUEEN
My lord, when last I went to visit her        45
She prayed me to excuse her keeping close,
Whereto constrained by her infirmity,
She should that duty leave unpaid to you
Which daily she was bound to proffer. This
She wished me to make known, but our great
          court        50
Made me to blame in memory.
CYMBELINE                        Her doors locked?
Not seen of late? Grant heavens that which I
Fear prove false.                              *Exit*
QUEEN          Son, I say, follow the King.
CLOTEN
That man of hers, Pisanio, her old servant,
I have not seen these two days.
QUEEN          Go, look after.        55
                              *Exit Cloten*
Pisanio, thou that stand'st so for Posthumus!
He hath a drug of mine. I pray his absence
Proceed by swallowing that, for he believes
It is a thing most precious. But for her,
Where is she gone? Haply despair hath seized her,        60
Or, winged with fervour of her love, she's flown
To her desired Posthumus. Gone she is
To death or to dishonour, and my end
Can make good use of either. She being down,
I have the placing of the British crown.        65

                              *Enter Cloten*
How now, my son?
CLOTEN                        'Tis certain she is fled.
Go in and cheer the King. He rages, none
Dare come about him.
QUEEN                        All the better. May
This night forestall him of the coming day.        *Exit*
CLOTEN
I love and hate her. For she's fair and royal,        70
And that she hath all courtly parts more exquisite
Than lady, ladies, woman—from every one
The best she hath, and she, of all compounded,
Outsells them all—I love her therefore; but
Disdaining me, and throwing favours on        75
The low Posthumus, slanders so her judgement
That what's else rare is choked; and in that point
I will conclude to hate her, nay, indeed,
To be revenged upon her. For when fools
Shall—
                              *Enter Pisanio*
          Who is here? What, are you packing, sirrah?
Come hither. Ah, you precious pander! Villain,        81
Where is thy lady? In a word, or else
Thou art straightway with the fiends.
PISANIO                        O good my lord!
CLOTEN
Where is thy lady?—or, by Jupiter,
I will not ask again. Close villain,        85
I'll have this secret from thy tongue or rip
Thy heart to find it. Is she with Posthumus,
From whose so many weights of baseness cannot
A dram of worth be drawn?
PISANIO                        Alas, my lord,
How can she be with him? When was she missed?        90
He is in Rome.
CLOTEN                        Where is she, sir? Come nearer.
No farther halting. Satisfy me home
What is become of her.
PISANIO          O my all-worthy lord!
CLOTEN All-worthy villain,        95
Discover where thy mistress is at once,
At the next word. No more of 'worthy lord'.
Speak, or thy silence on the instant is
Thy condemnation and thy death.
PISANIO                        Then, sir,
This paper is the history of my knowledge        100
Touching her flight.
                              *He gives Cloten a letter*
CLOTEN                        Let's see't. I will pursue her
Even to Augustus' throne.
PISANIO [*aside*]          Or this or perish.
She's far enough, and what he learns by this
May prove his travel, not her danger.
CLOTEN                        Hum!
PISANIO (*aside*)
I'll write to my lord she's dead. O Innogen,        105
Safe mayst thou wander, safe return again!
CLOTEN
Sirrah, is this letter true?
PISANIO                        Sir, as I think.
CLOTEN It is Posthumus' hand; I know't. Sirrah, if thou
wouldst not be a villain but do me true service, undergo
those employments wherein I should have cause to use
thee with a serious industry—that is, what villainy
soe'er I bid thee do, to perform it directly and truly—I
would think thee an honest man. Thou shouldst neither

want my means for thy relief nor my voice for thy
preferment.     115

PISANIO Well, my good lord.

CLOTEN Wilt thou serve me? For since patiently and
constantly thou hast stuck to the bare fortune of that
beggar Posthumus, thou canst not in the course of
gratitude but be a diligent follower of mine. Wilt thou
serve me?     121

PISANIO Sir, I will.

CLOTEN Give me thy hand. Here's my purse. Hast any of
thy late master's garments in thy possession?

PISANIO I have, my lord, at my lodging the same suit he
wore when he took leave of my lady and mistress.

CLOTEN The first service thou dost me, fetch that suit
hither. Let it be thy first service. Go.     128

PISANIO I shall, my lord.      *Exit*

CLOTEN Meet thee at Milford Haven! I forgot to ask him
one thing; I'll remember't anon. Even there, thou
villain Posthumus, will I kill thee. I would these
garments were come. She said upon a time—the
bitterness of it I now belch from my heart—that she
held the very garment of Posthumus in more respect
than my noble and natural person, together with the
adornment of my qualities. With that suit upon my
back will I ravish her—first kill him, and in her eyes;
there shall she see my valour, which will then be a
torment to her contempt. He on the ground, my speech
of insultment ended on his dead body, and when my
lust hath dined—which, as I say, to vex her I will
execute in the clothes that she so praised—to the court
I'll knock her back, foot her home again. She hath
despised me rejoicingly, and I'll be merry in my revenge.

     *Enter Pisanio with Posthumus' suit*

Be those the garments?

PISANIO          Ay, my noble lord.     146

CLOTEN

How long is't since she went to Milford Haven?

PISANIO She can scarce be there yet.

CLOTEN Bring this apparel to my chamber. That is the
second thing that I have commanded thee. The third
is that thou wilt be a voluntary mute to my design. Be
but duteous, and true preferment shall tender itself to
thee. My revenge is now at Milford. Would I had wings
to follow it. Come, and be true.      *Exit*

PISANIO

Thou bidd'st me to my loss, for true to thee    155
Were to prove false, which I will never be
To him that is most true. To Milford go,
And find not her whom thou pursuest. Flow, flow,
You heavenly blessings, on her. This fool's speed
Be crossed with slowness; labour be his meed.    *Exit*

**3.6**    *Enter Innogen, dressed as a man, before the cave*

INNOGEN

I see a man's life is a tedious one.
I have tired myself, and for two nights together
Have made the ground my bed. I should be sick,
But that my resolution helps me. Milford,
When from the mountain-top Pisanio showed thee,   5
Thou wast within a ken. O Jove, I think
Foundations fly the wretched—such, I mean,
Where they should be relieved. Two beggars told me
I could not miss my way. Will poor folks lie,
That have afflictions on them, knowing 'tis     10
A punishment or trial? Yes. No wonder,
When rich ones scarce tell true. To lapse in fullness

Is sorer than to lie for need, and falsehood
Is worse in kings than beggars. My dear lord,
Thou art one o'th' false ones. Now I think on thee   15
My hunger's gone, but even before I was
At point to sink for food. But what is this?
Here is a path to't. 'Tis some savage hold.
I were best not call; I dare not call; yet famine,
Ere clean it o'erthrow nature, makes it valiant.    20
Plenty and peace breeds cowards, hardness ever
Of hardiness is mother. Ho! Who's here?
If anything that's civil, speak; if savage,
Take or lend. Ho! No answer? Then I'll enter.
Best draw my sword, and if mine enemy     25
But fear the sword like me he'll scarcely look on't.
Such a foe, good heavens!      *Exit into the cave*

     *Enter Belarius, Guiderius, and Arviragus*

BELARIUS

You, Polydore, have proved best woodman and
Are master of the feast. Cadwal and I
Will play the cook and servant; 'tis our match.    30
The sweat of industry would dry and die
But for the end it works to. Come, our stomachs
Will make what's homely savoury. Weariness
Can snore upon the flint when resty sloth
Finds the down pillow hard. Now peace be here,   35
Poor house, that keep'st thyself.

GUIDERIUS          I am throughly weary.

ARVIRAGUS

I am weak with toil yet strong in appetite.

GUIDERIUS

There is cold meat i'th' cave. We'll browse on that
Whilst what we have killed be cooked.

BELARIUS (*looking into the cave*)      Stay, come not in.
But that it eats our victuals I should think     40
Here were a fairy.

GUIDERIUS        What's the matter, sir?

BELARIUS

By Jupiter, an angel—or, if not,
An earthly paragon. Behold divineness
No elder than a boy.

     *Enter Innogen from the cave, dressed as a man*

INNOGEN        Good masters, harm me not.
Before I entered here I called, and thought     45
To have begged or bought what I have took. Good
truth,
I have stol'n naught, nor would not, though I had
found
Gold strewed i'th' floor. Here's money for my meat.
I would have left it on the board so soon
As I had made my meal, and parted      50
With prayers for the provider.

GUIDERIUS          Money, youth?

ARVIRAGUS

All gold and silver rather turn to dirt,
As 'tis no better reckoned but of those
Who worship dirty gods.

INNOGEN        I see you're angry.
Know, if you kill me for my fault, I should     55
Have died had I not made it.

BELARIUS          Whither bound?

INNOGEN

To Milford Haven.

BELARIUS      What's your name?

INNOGEN

Fidele, sir. I have a kinsman who

Is bound for Italy. He embarked at Milford,
To whom being going, almost spent with hunger,        60
I am fall'n in this offence.
BELARIUS                           Prithee, fair youth,
Think us no churls, nor measure our good minds
By this rude place we live in. Well encountered.
'Tis almost night. You shall have better cheer
Ere you depart, and thanks to stay and eat it.        65
Boys, bid him welcome.
GUIDERIUS                        Were you a woman, youth,
I should woo hard but be your groom in honesty,
Ay, bid for you as I'd buy.
ARVIRAGUS                        I'll make't my comfort
He is a man, I'll love him as my brother.
(To Innogen) And such a welcome as I'd give to him
After long absence, such is yours. Most welcome.      71
Be sprightly, for you fall 'mongst friends.
INNOGEN                                   'Mongst friends
If brothers. (Aside) Would it had been so that they
Had been my father's sons. Then had my price
Been less, and so more equal ballasting               75
To thee, Posthumus.
        The three men speak apart
BELARIUS                        He wrings at some distress.
GUIDERIUS
Would I could free't.
ARVIRAGUS                    Or I, whate'er it be,
What pain it cost, what danger. Gods!
BELARIUS                                  Hark, boys.
        They whisper
INNOGEN (aside)                          Great men
That had a court no bigger than this cave,            80
That did attend themselves and had the virtue
Which their own conscience sealed them, laying by
That nothing-gift of differing multitudes,
Could not outpeer these twain. Pardon me, gods,
I'd change my sex to be companion with them,          85
Since Leonatus' false.
BELARIUS                     It shall be so.
Boys, we'll go dress our hunt. Fair youth, come in.
Discourse is heavy, fasting. When we have supped
We'll mannerly demand thee of thy story,
So far as thou wilt speak it.
GUIDERIUS                        Pray draw near.        90
ARVIRAGUS
The night to th' owl and morn to th' lark less
        welcome.
INNOGEN Thanks, sir.
ARVIRAGUS I pray draw near.        Exeunt into the cave

3.7    Enter two Roman Senators, and Tribunes
FIRST SENATOR
This is the tenor of the Emperor's writ:
That since the common men are now in action
'Gainst the Pannonians and Dalmatians,
And that the legions now in Gallia are
Full weak to undertake our wars against                5
The fall'n-off Britons, that we do incite
The gentry to this business. He creates
Lucius pro-consul, and to you the tribunes,
For this immediate levy, he commends
His absolute commission. Long live Caesar!            10
A TRIBUNE
Is Lucius general of the forces?
SECOND SENATOR                          Ay.

A TRIBUNE
Remaining now in Gallia?
FIRST SENATOR                With those legions
Which I have spoke of, whereunto your levy
Must be supplyant. The words of your commission
Will tie you to the numbers and the time              15
Of their dispatch.
A TRIBUNE                We will discharge our duty.
                                            Exeunt

4.1    Enter Cloten, in Posthumus' suit
CLOTEN I am near to th' place where they should meet,
if Pisanio have mapped it truly. How fit his garments
serve me! Why should his mistress, who was made by
him that made the tailor, not be fit too?—the rather—
saving reverence of the word—for 'tis said a woman's
fitness comes by fits. Therein I must play the workman.
I dare speak it to myself, for it is not vainglory for a
man and his glass to confer in his own chamber. I
mean the lines of my body are as well drawn as his:
no less young, more strong, not beneath him in
fortunes, beyond him in the advantage of the time,
above him in birth, alike conversant in general services,
and more remarkable in single oppositions. Yet this
imperceiverant thing loves him in my despite. What
mortality is! Posthumus, thy head which now is
growing upon thy shoulders shall within this hour be
off, thy mistress enforced, thy garments cut to pieces
before thy face; and all this done, spurn her home to
her father, who may haply be a little angry for my so
rough usage; but my mother, having power of his
testiness, shall turn all into my commendations. My
horse is tied up safe. Out, sword, and to a sore purpose!
Fortune, put them into my hand. This is the very
description of their meeting-place, and the fellow dares
not deceive me.                                      Exit

4.2    Enter Belarius, Guiderius, Arviragus, and Innogen
        dressed as a man, from the cave
BELARIUS (to Innogen)
You are not well. Remain here in the cave.
We'll come to you from hunting.
ARVIRAGUS (to Innogen)                Brother, stay here.
Are we not brothers?
INNOGEN                    So man and man should be,
But clay and clay differs in dignity,
Whose dust is both alike. I am very sick.             5
GUIDERIUS (to Belarius and Arviragus)
Go you to hunting. I'll abide with him.
INNOGEN
So sick I am not, yet I am not well;
But not so citizen a wanton as
To seem to die ere sick. So please you, leave me.
Stick to your journal course. The breach of custom   10
Is breach of all. I am ill, but your being by me
Cannot amend me. Society is no comfort
To one not sociable. I am not very sick,
Since I can reason of it. Pray you, trust me here.
I'll rob none but myself; and let me die,             15
Stealing so poorly.
GUIDERIUS            I love thee: I have spoke it;
How much the quantity, the weight as much,
As I do love my father.
BELARIUS                    What, how, how?

ARVIRAGUS
If it be sin to say so, sir, I yoke me
In my good brother's fault. I know not why      20
I love this youth, and I have heard you say
Love's reason's without reason. The bier at door
And a demand who is't shall die, I'd say
'My father, not this youth'.
BELARIUS (aside)           O noble strain!
O worthiness of nature, breed of greatness!    25
Cowards father cowards, and base things sire base.
Nature hath meal and bran, contempt and grace.
I'm not their father, yet who this should be
Doth miracle itself, loved before me.
  (Aloud) 'Tis the ninth hour o'th' morn.
ARVIRAGUS (to Innogen)        Brother, farewell.
INNOGEN
I wish ye sport.
ARVIRAGUS       You health.—So please you, sir.    31
INNOGEN (aside)
These are kind creatures. Gods, what lies I have heard!
Our courtiers say all's savage but at court.
Experience, O thou disprov'st report!
Th'imperious seas breeds monsters; for the dish    35
Poor tributary rivers as sweet fish.
I am sick still, heart-sick. Pisanio,
I'll now taste of thy drug.
  ⌈She swallows the drug.⌉ The men speak apart
GUIDERIUS         I could not stir him.
He said he was gentle but unfortunate,
Dishonestly afflicted but yet honest.       40
ARVIRAGUS
Thus did he answer me, yet said hereafter
I might know more.
BELARIUS       To th' field, to th' field!
  (To Innogen) We'll leave you for this time. Go in and
    rest.
ARVIRAGUS (to Innogen)
We'll not be long away.
BELARIUS (to Innogen)      Pray be not sick,
For you must be our housewife.
INNOGEN           Well or ill,    45
I am bound to you.                        Exit
BELARIUS        And shalt be ever.
This youth, howe'er distressed, appears hath had
Good ancestors.
ARVIRAGUS How angel-like he sings!
GUIDERIUS But his neat cookery!           50
⌈BELARIUS⌉
He cut our roots in characters,
And sauced our broths as Juno had been sick
And he her dieter.
ARVIRAGUS       Nobly he yokes
A smiling with a sigh, as if the sigh
Was that it was for not being such a smile;    55
The smile mocking the sigh that it would fly
From so divine a temple to commix
With winds that sailors rail at.
GUIDERIUS         I do note
That grief and patience, rooted in him both,
Mingle their spurs together.
ARVIRAGUS       Grow patience,    60
And let the stinking elder, grief, untwine
His perishing root with the increasing vine.
BELARIUS
It is great morning. Come away. Who's there?

                  *Enter Cloten in Posthumus' suit*
CLOTEN
I cannot find those runagates. That villain
Hath mocked me. I am faint.
BELARIUS (aside to Arviragus and Guiderius)
                   'Those runagates'?    65
Means he not us? I partly know him; 'tis
Cloten, the son o'th' Queen. I fear some ambush.
I saw him not these many years, and yet
I know 'tis he. We are held as outlaws. Hence!
GUIDERIUS (aside to Arviragus and Belarius)
He is but one. You and my brother search    70
What companies are near. Pray you, away.
Let me alone with him.
              *Exeunt Arviragus and Belarius*
CLOTEN           Soft, what are you
That fly me thus? Some villain mountaineers?
I have heard of such. What slave art thou?
GUIDERIUS                  A thing
More slavish did I ne'er than answering    75
A slave without a knock.
CLOTEN           Thou art a robber,
A law-breaker, a villain. Yield thee, thief.
GUIDERIUS
To who? To thee? What art thou? Have not I
An arm as big as thine, a heart as big?
Thy words, I grant, are bigger, for I wear not    80
My dagger in my mouth. Say what thou art,
Why I should yield to thee.
CLOTEN           Thou villain base,
Know'st me not by my clothes?
GUIDERIUS        No, nor thy tailor, rascal,
Who is thy grandfather. He made those clothes,
Which, as it seems, make thee.
CLOTEN           Thou precious varlet,
My tailor made them not.
GUIDERIUS        Hence, then, and thank    86
The man that gave them thee. Thou art some fool.
I am loath to beat thee.
CLOTEN           Thou injurious thief,
Hear but my name and tremble.
GUIDERIUS           What's thy name?
CLOTEN Cloten, thou villain.           90
GUIDERIUS
Cloten, thou double villain, be thy name,
I cannot tremble at it. Were it toad or adder, spider,
'Twould move me sooner.
CLOTEN         To thy further fear,
Nay, to thy mere confusion, thou shalt know
I am son to th' Queen.
GUIDERIUS       I am sorry for't, not seeming
So worthy as thy birth.
CLOTEN         Art not afeard?    96
GUIDERIUS
Those that I reverence, those I fear, the wise.
At fools I laugh, not fear them.
CLOTEN           Die the death.
When I have slain thee with my proper hand
I'll follow those that even now fled hence,    100
And on the gates of Lud's town set your heads.
Yield, rustic mountaineer.       *Fight and exeunt*
        *Enter Belarius and Arviragus*
BELARIUS           No company's abroad?
ARVIRAGUS
None in the world. You did mistake him, sure.

BELARIUS
I cannot tell. Long is it since I saw him,
But time hath nothing blurred those lines of favour
Which then he wore. The snatches in his voice      106
And burst of speaking were as his. I am absolute
'Twas very Cloten.
ARVIRAGUS            In this place we left them.
I wish my brother make good time with him,
You say he is so fell.
BELARIUS            Being scarce made up,      110
I mean to man, he had not apprehension
Of roaring terrors; for defect of judgement
Is oft the cause of fear.
        *Enter Guiderius with Cloten's head*
                    But see, thy brother.
GUIDERIUS
This Cloten was a fool, an empty purse,
There was no money in't. Not Hercules      115
Could have knocked out his brains, for he had none.
Yet I not doing this, the fool had borne
My head as I do his.
BELARIUS            What hast thou done?
GUIDERIUS
I am perfect what: cut off one Cloten's head,
Son to the Queen after his own report,      120
Who called me traitor, mountaineer, and swore
With his own single hand he'd take us in,
Displace our heads where—thanks, ye gods—they
        grow,
And set them on Lud's town.
BELARIUS                    We are all undone.
GUIDERIUS
Why, worthy father, what have we to lose      125
But that he swore to take, our lives? The law
Protects not us: then why should we be tender
To let an arrogant piece of flesh threat us,
Play judge and executioner all himself,
For we do fear the law? What company      130
Discover you abroad?
BELARIUS            No single soul
Can we set eye on, but in all safe reason
He must have some attendants. Though his humour
Was nothing but mutation, ay, and that
From one bad thing to worse, not frenzy,      135
Not absolute madness, could so far have raved
To bring him here alone. Although perhaps
It may be heard at court that such as we
Cave here, hunt here, are outlaws, and in time
May make some stronger head, the which he
        hearing—      140
As it is like him—might break out, and swear
He'd fetch us in, yet is't not probable
To come alone, either he so undertaking,
Or they so suffering. Then on good ground we fear
If we do fear this body hath a tail      145
More perilous than the head.
ARVIRAGUS            Let ord'nance
Come as the gods foresay it; howsoe'er,
My brother hath done well.
BELARIUS            I had no mind
To hunt this day. The boy Fidele's sickness      149
Did make my way long forth.
GUIDERIUS            With his own sword,
Which he did wave against my throat, I have ta'en
His head from him. I'll throw't into the creek

Behind our rock, and let it to the sea
And tell the fishes he's the Queen's son, Cloten.
That's all I reck.            *Exit with Cloten's head*
BELARIUS            I fear 'twill be revenged.      155
Would, Polydore, thou hadst not done't, though
        valour
Becomes thee well enough.
ARVIRAGUS            Would I had done't,
So the revenge alone pursued me. Polydore,
I love thee brotherly, but envy much
Thou hast robbed me of this deed. I would revenges
That possible strength might meet would seek us
        through      161
And put us to our answer.
BELARIUS            Well, 'tis done.
We'll hunt no more today, nor seek for danger
Where there's no profit. I prithee, to our rock.
You and Fidele play the cooks. I'll stay      165
Till hasty Polydore return, and bring him
To dinner presently.
ARVIRAGUS            Poor sick Fidele!
I'll willingly to him. To gain his colour
I'd let a parish of such Clotens blood,
And praise myself for charity.      *Exit into the cave*
BELARIUS            O thou goddess,      170
Thou divine Nature, how thyself thou blazon'st
In these two princely boys! They are as gentle
As zephyrs blowing below the violet,
Not wagging his sweet head; and yet as rough,
Their royal blood enchafed, as the rud'st wind      175
That by the top doth take the mountain pine
And make him stoop to th' vale. 'Tis wonder
That an invisible instinct should frame them
To royalty unlearned, honour untaught,
Civility not seen from other, valour      180
That wildly grows in them, but yields a crop
As if it had been sowed. Yet still it's strange
What Cloten's being here to us portends,
Or what his death will bring us.
        *Enter Guiderius*
GUIDERIUS            Where's my brother?
I have sent Cloten's clotpoll down the stream      185
In embassy to his mother. His body's hostage
For his return.
        *Solemn music*
BELARIUS            My ingenious instrument!—
Hark, Polydore, it sounds. But what occasion
Hath Cadwal now to give it motion? Hark!
GUIDERIUS
Is he at home?
BELARIUS            He went hence even now.      190
GUIDERIUS
What does he mean? Since death of my dear'st mother
It did not speak before. All solemn things
Should answer solemn accidents. The matter?
Triumphs for nothing and lamenting toys
Is jollity for apes and grief for boys.      195
Is Cadwal mad?
        *Enter from the cave Arviragus with Innogen, dead,*
        *bearing her in his arms*
BELARIUS            Look, here he comes,
And brings the dire occasion in his arms
Of what we blame him for.
ARVIRAGUS            The bird is dead
That we have made so much on. I had rather
Have skipped from sixteen years of age to sixty,      200

To have turned my leaping time into a crutch,
Than have seen this.
GUIDERIUS (*to Innogen*)  O sweetest, fairest lily!
My brother wears thee not one half so well
As when thou grew'st thyself.
BELARIUS                          O melancholy,
Who ever yet could sound thy bottom, find          205
The ooze to show what coast thy sluggish crare
Might easiliest harbour in? Thou blessèd thing,
Jove knows what man thou mightst have made;
     but I,
Thou diedst a most rare boy, of melancholy.
(*To Arviragus*) How found you him?
ARVIRAGUS                          Stark, as you see,
Thus smiling as some fly had tickled slumber,      211
Not as death's dart being laughed at; his right cheek
Reposing on a cushion.
GUIDERIUS                  Where?
ARVIRAGUS                          O'th' floor,
His arms thus leagued. I thought he slept, and put
My clouted brogues from off my feet, whose rudeness
Answered my steps too loud.
GUIDERIUS                  Why, he but sleeps.  216
If he be gone he'll make his grave a bed.
With female fairies will his tomb be haunted,
(*To Innogen*) And worms will not come to thee.
ARVIRAGUS (*to Innogen*)          With fairest flowers
Whilst summer lasts and I live here, Fidele,        220
I'll sweeten thy sad grave. Thou shalt not lack
The flower that's like thy face, pale primrose, nor
The azured harebell, like thy veins; no, nor
The leaf of eglantine, whom not to slander
Outsweetened not thy breath. The ruddock would     225
With charitable bill—O bill sore shaming
Those rich-left heirs that let their fathers lie
Without a monument!—bring thee all this,
Yea, and furred moss besides, when flowers are none,
To winter-gown thy corpse.
GUIDERIUS                  Prithee, have done,       230
And do not play in wench-like words with that
Which is so serious. Let us bury him,
And not protract with admiration what
Is now due debt. To th' grave.
ARVIRAGUS                  Say, where shall 's lay him?
GUIDERIUS
By good Euriphile, our mother.
ARVIRAGUS                  Be't so,                  235
And let us, Polydore, though now our voices
Have got the mannish crack, sing him to th' ground
As once our mother; use like note and words,
Save that 'Euriphile' must be 'Fidele'.
GUIDERIUS Cadwal,                                    240
I cannot sing. I'll weep, and word it with thee,
For notes of sorrow out of tune are worse
Than priests and fanes that lie.
ARVIRAGUS                  We'll speak it then.
BELARIUS
Great griefs, I see, medicine the less, for Cloten
Is quite forgot. He was a queen's son, boys,        245
And though he came our enemy, remember
He was paid for that. Though mean and mighty
     rotting
Together have one dust, yet reverence,
That angel of the world, doth make distinction
Of place 'tween high and low. Our foe was princely,

And though you took his life as being our foe,      251
Yet bury him as a prince.
GUIDERIUS                  Pray you, fetch him hither.
Thersites' body is as good as Ajax'
When neither are alive.
ARVIRAGUS (*to Belarius*)  If you'll go fetch him,
We'll say our song the whilst.          *Exit Belarius*
                          Brother, begin.             255
GUIDERIUS
Nay, Cadwal, we must lay his head to th'east.
My father hath a reason for't.
ARVIRAGUS                          'Tis true.
GUIDERIUS
Come on, then, and remove him.
ARVIRAGUS                          So, begin.
GUIDERIUS

Fear no more the heat o'th' sun,
     Nor the furious winter's rages.                  260
Thou thy worldly task hast done,
     Home art gone and ta'en thy wages.
Golden lads and girls all must,
     As chimney-sweepers, come to dust.
ARVIRAGUS

Fear no more the frown o'th' great,                  265
     Thou art past the tyrant's stroke.
Care no more to clothe and eat,
     To thee the reed is as the oak.
The sceptre, learning, physic, must
All follow this and come to dust.                    270
GUIDERIUS

Fear no more the lightning flash,
ARVIRAGUS  Nor th'all-dreaded thunder-stone.
GUIDERIUS

Fear not slander, censure rash.
ARVIRAGUS  Thou hast finished joy and moan.
GUIDERIUS *and* ARVIRAGUS
All lovers young, all lovers must                    275
Consign to thee and come to dust.
GUIDERIUS

No exorcisor harm thee,
ARVIRAGUS
Nor no witchcraft charm thee.
GUIDERIUS

Ghost unlaid forbear thee.
ARVIRAGUS
Nothing ill come near thee.                          280
GUIDERIUS *and* ARVIRAGUS
Quiet consummation have,
And renownèd be thy grave.

*Enter Belarius with the body of Cloten in*
*Posthumus' suit*
GUIDERIUS
We have done our obsequies. Come, lay him down.
BELARIUS
Here's a few flowers, but 'bout midnight more;
The herbs that have on them cold dew o'th' night   285
Are strewings fitt'st for graves upon th'earth's face.
You were as flowers, now withered; even so
These herblets shall, which we upon you strow.
Come on, away; apart upon our knees
⌐                                          ⌐          290
The ground that gave them first has them again.
Their pleasures here are past, so is their pain.
                *Exeunt Belarius, Arviragus, and Guiderius*

INNOGEN (*awakes*)
Yes, sir, to Milford Haven. Which is the way?
I thank you. By yon bush? Pray, how far thither?
'Od's pitykins, can it be six mile yet?                              295
I have gone all night. 'Faith, I'll lie down and sleep.
    *She sees Cloten*
But soft, no bedfellow! O gods and goddesses!
These flowers are like the pleasures of the world,
This bloody man the care on't. I hope I dream,
For so I thought I was a cavekeeper,                                 300
And cook to honest creatures. But 'tis not so.
'Twas but a bolt of nothing, shot of nothing,
Which the brain makes of fumes. Our very eyes
Are sometimes like our judgements, blind. Good faith,
I tremble still with fear; but if there be                           305
Yet left in heaven as small a drop of pity
As a wren's eye, feared gods, a part of it!
The dream's here still. Even when I wake it is
Without me as within me; not imagined, felt.
A headless man? The garments of Posthumus?                           310
I know the shape of 's leg; this is his hand,
His foot Mercurial, his Martial thigh,
The brawns of Hercules; but his Jovial face—
Murder in heaven! How? 'Tis gone. Pisanio,
All curses madded Hecuba gave the Greeks,                            315
And mine to boot, be darted on thee! Thou,
Conspired with that irregulous devil Cloten,
Hath here cut off my lord. To write and read
Be henceforth treacherous! Damned Pisanio
Hath with his forgèd letters—damned Pisanio—                         320
From this most bravest vessel of the world
Struck the main-top! O Posthumus, alas,
Where is thy head? Where's that? Ay me, where's
    that?
Pisanio might have killed thee at the heart
And left thy head on. How should this be? Pisanio?
'Tis he and Cloten. Malice and lucre in them                         326
Have laid this woe here. O, 'tis pregnant, pregnant!
The drug he gave me, which he said was precious
And cordial to me, have I not found it
Murd'rous to th' senses? That confirms it home.                      330
This is Pisanio's deed, and Cloten—O,
Give colour to my pale cheek with thy blood,
That we the horrider may seem to those
Which chance to find us!
    ⌜*She smears her face with blood*⌝
               O my lord, my lord!
    ⌜*She faints.*⌝
    *Enter Lucius, Roman Captains, and a Soothsayer*
A ROMAN CAPTAIN (*to Lucius*)
To them the legions garrisoned in Gallia                             335
After your will have crossed the sea, attending
You here at Milford Haven with your ships.
They are hence in readiness.
LUCIUS              But what from Rome?
A ROMAN CAPTAIN
The senate hath stirred up the confiners
And gentlemen of Italy, most willing spirits                         340
That promise noble service, and they come
Under the conduct of bold Giacomo,
Siena's brother.
LUCIUS         When expect you them?
A ROMAN CAPTAIN
With the next benefit o'th' wind.
LUCIUS             This forwardness
Makes our hopes fair. Command our present numbers

Be mustered; bid the captains look to't.                            345
    ⌜*Exit one or more*⌝
(*To Soothsayer*)              Now, sir,
What have you dreamed of late of this war's purpose?
SOOTHSAYER
Last night the very gods showed me a vision—
I fast, and prayed for their intelligence—thus:
I saw Jove's bird, the Roman eagle, winged                           350
From the spongy south to this part of the west,
There vanished in the sunbeams; which portends,
Unless my sins abuse my divination,
Success to th' Roman host.
LUCIUS             Dream often so,
And never false.
    *He sees Cloten's body*
           Soft, ho, what trunk is here                355
Without his top? The ruin speaks that sometime
It was a worthy building. How, a page?
Or dead or sleeping on him? But dead rather,
For nature doth abhor to make his bed
With the defunct, or sleep upon the dead.                            360
Let's see the boy's face.
A ROMAN CAPTAIN        He's alive, my lord.
LUCIUS
He'll then instruct us of this body. Young one,
Inform us of thy fortunes, for it seems
They crave to be demanded. Who is this
Thou mak'st thy bloody pillow? Or who was he                         365
That, otherwise than noble nature did,
Hath altered that good picture? What's thy interest
In this sad wreck? How came't? Who is't?
What art thou?
INNOGEN         I am nothing; or if not,
Nothing to be were better. This was my master,                      370
A very valiant Briton, and a good,
That here by mountaineers lies slain. Alas,
There is no more such masters. I may wander
From east to occident, cry out for service,
Try many, all good; serve truly, never                              375
Find such another master.
LUCIUS           'Lack, good youth,
Thou mov'st no less with thy complaining than
Thy master in bleeding. Say his name, good friend.
INNOGEN
Richard du Champ. (*Aside*) If I do lie and do
No harm by it, though the gods hear I hope                          380
They'll pardon it. (*Aloud*) Say you, sir?
LUCIUS            Thy name?
INNOGEN              Fidele, sir.
LUCIUS
Thou dost approve thyself the very same.
Thy name well fits thy faith, thy faith thy name.
Wilt take thy chance with me? I will not say
Thou shalt be so well mastered, but be sure,                        385
No less beloved. The Roman Emperor's letters
Sent by a consul to me should not sooner
Than thine own worth prefer thee. Go with me.
INNOGEN
I'll follow, sir. But first, an't please the gods,
I'll hide my master from the flies as deep                          390
As these poor pickaxes can dig; and when
With wild-wood leaves and weeds I ha' strewed his
    grave
And on it said a century of prayers,
Such as I can, twice o'er I'll weep and sigh,

And leaving so his service, follow you,      395
So please you entertain me.
LUCIUS               Ay, good youth,
And rather father thee than master thee. My friends,
The boy hath taught us manly duties. Let us
Find out the prettiest daisied plot we can,
And make him with our pikes and partisans      400
A grave. Come, arm him. Boy, he is preferred
By thee to us, and he shall be interred
As soldiers can. Be cheerful. Wipe thine eyes.
Some falls are means the happier to arise.
                *Exeunt with Cloten's body*

**4.3**    *Enter Cymbeline, Lords, and Pisanio*
CYMBELINE
Again, and bring me word how 'tis with her.
                *Exit one or more*
A fever with the absence of her son,
A madness of which her life's in danger—heavens,
How deeply you at once do touch me! Innogen,
The great part of my comfort, gone; my queen     5
Upon a desperate bed, and in a time
When fearful wars point at me; her son gone,
So needful for this present! It strikes me past
The hope of comfort. (*To Pisanio*) But for thee, fellow,
Who needs must know of her departure and     10
Dost seem so ignorant, we'll enforce it from thee
By a sharp torture.
PISANIO             Sir, my life is yours.
I humbly set it at your will. But for my mistress,
I nothing know where she remains, why gone,
Nor when she purposes return. Beseech your
     highness,                      15
Hold me your loyal servant.
A LORD             Good my liege,
The day that she was missing he was here.
I dare be bound he's true, and shall perform
All parts of his subjection loyally. For Cloten,
There wants no diligence in seeking him,      20
And will no doubt be found.
CYMBELINE           The time is troublesome.
(*To Pisanio*) We'll slip you for a season, but our jealousy
Does yet depend.
A LORD        So please your majesty,
The Roman legions, all from Gallia drawn,
Are landed on your coast with a supply      25
Of Roman gentlemen by the senate sent.
CYMBELINE
Now for the counsel of my son and queen!
I am amazed with matter.
A LORD         Good my liege,
Your preparation can affront no less
Than what you hear of. Come more, for more you're
     ready.                        30
The want is but to put those powers in motion
That long to move.
CYMBELINE     I thank you. Let's withdraw,
And meet the time as it seeks us. We fear not
What can from Italy annoy us, but
We grieve at chances here. Away.      35
             *Exeunt Cymbeline and Lords*
PISANIO
I heard no letter from my master since
I wrote him Innogen was slain. 'Tis strange.
Nor hear I from my mistress, who did promise
To yield me often tidings. Neither know I

What is betid to Cloten, but remain      40
Perplexed in all. The heavens still must work.
Wherein I am false I am honest; not true, to be true.
These present wars shall find I love my country
Even to the note o'th' King, or I'll fall in them.
All other doubts, by time let them be cleared:     45
Fortune brings in some boats that are not steered.
                       *Exit*

**4.4**    *Enter Belarius, Guiderius, and Arviragus*
GUIDERIUS
The noise is round about us.
BELARIUS              Let us from it.
ARVIRAGUS
What pleasure, sir, find we in life to lock it
From action and adventure?
GUIDERIUS          Nay, what hope
Have we in hiding us? This way the Romans
Must or for Britains slay us, or receive us      5
For barbarous and unnatural revolts
During their use, and slay us after.
BELARIUS             Sons,
We'll higher to the mountains; there secure us.
To the King's party there's no going. Newness
Of Cloten's death—we being not known, not mustered
Among the bands—may drive us to a render     11
Where we have lived, and so extort from 's that
Which we have done, whose answer would be death
Drawn on with torture.
GUIDERIUS         This is, sir, a doubt
In such a time nothing becoming you      15
Nor satisfying us.
ARVIRAGUS         It is not likely
That when they hear the Roman horses neigh,
Behold their quartered files, have both their eyes
And ears so cloyed importantly as now,
That they will waste their time upon our note,    20
To know from whence we are.
BELARIUS           O, I am known
Of many in the army. Many years,
Though Cloten then but young, you see, not wore him
From my remembrance. And besides, the King
Hath not deserved my service nor your loves,     25
Who find in my exile the want of breeding,
The certainty of this hard life; aye hopeless
To have the courtesy your cradle promised,
But to be still hot summer's tanlings, and
The shrinking slaves of winter.
GUIDERIUS         Than be so,     30
Better to cease to be. Pray, sir, to th'army.
I and my brother are not known; yourself
So out of thought, and thereto so o'ergrown,
Cannot be questioned.
ARVIRAGUS        By this sun that shines,
I'll thither. What thing is't that I never      35
Did see man die, scarce ever looked on blood
But that of coward hares, hot goats, and venison,
Never bestrid a horse save one that had
A rider like myself, who ne'er wore rowel
Nor iron on his heel! I am ashamed      40
To look upon the holy sun, to have
The benefit of his blest beams, remaining
So long a poor unknown.
GUIDERIUS        By heavens, I'll go.
If you will bless me, sir, and give me leave,
I'll take the better care; but if you will not,    45

The hazard therefore due fall on me by
The hands of Romans.
ARVIRAGUS                    So say I, amen.
BELARIUS
No reason I, since of your lives you set
So slight a valuation, should reserve
My cracked one to more care. Have with you, boys!
If in your country wars you chance to die,            51
That is my bed, too, lads, and there I'll lie.
Lead, lead. (*Aside*) The time seems long. Their blood
    thinks scorn
Till it fly out and show them princes born.    *Exeunt*

✤

5.1    *Enter Posthumus, dressed as an Italian gentleman,*
       *carrying a bloody cloth*
POSTHUMUS
Yea, bloody cloth, I'll keep thee, for I once wished
Thou shouldst be coloured thus. You married ones,
If each of you should take this course, how many
Must murder wives much better than themselves
For wrying but a little! O Pisanio,                    5
Every good servant does not all commands,
No bond but to do just ones. Gods, if you
Should have ta'en vengeance on my faults, I never
Had lived to put on this; so had you saved
The noble Innogen to repent, and struck              10
Me, wretch, more worth your vengeance. But alack,
You snatch some hence for little faults; that's love,
To have them fall no more. You some permit
To second ills with ills, each elder worse,
And make them dread ill, to the doer's thrift.       15
But Innogen is your own. Do your blest wills,
And make me blest to obey. I am brought hither
Among th'Italian gentry, and to fight
Against my lady's kingdom. 'Tis enough
That, Britain, I have killed thy mistress-piece;     20
I'll give no wound to thee. Therefore, good heavens,
Hear patiently my purpose. I'll disrobe me
Of these Italian weeds, and suit myself
As does a Briton peasant.
       ⌐He disrobes himself ⌐
                           So I'll fight
Against the part I come with; so I'll die            25
For thee, O Innogen, even for whom my life
Is every breath a death; and, thus unknown,
Pitied nor hated, to the face of peril
Myself I'll dedicate. Let me make men know
More valour in me than my habits show.               30
Gods, put the strength o'th' Leonati in me.
To shame the guise o'th' world, I will begin
The fashion—less without and more within.    *Exit*

5.2    ⌐A march.⌐ *Enter Lucius, Giacomo, and the Roman*
       *army at one door, and the Briton army at another,*
       *Leonatus Posthumus following like a poor soldier.*
       *They march over and go out.* ⌐Alarums.⌐
       *Then enter again in skirmish Giacomo and*
       *Posthumus: he vanquisheth and disarmeth*
       *Giacomo, and then leaves him*
GIACOMO
The heaviness and guilt within my bosom
Takes off my manhood. I have belied a lady,
The princess of this country, and the air on't
Revengingly enfeebles me; or could this carl,
A very drudge of nature's, have subdued me            5

In my profession? Knighthoods and honours borne
As I wear mine are titles but of scorn.
If that thy gentry, Britain, go before
This lout as he exceeds our lords, the odds
Is that we scarce are men and you are gods.    *Exit*

5.3    *The battle continues.* ⌐Alarums. Excursions. The
       *trumpets sound a retreat.⌐ The Britons fly,*
       *Cymbeline is taken. Then enter to his rescue*
       *Belarius, Guiderius, and Arviragus*
BELARIUS
Stand, stand, we have th'advantage of the ground.
The lane is guarded. Nothing routs us but
The villainy of our fears.
GUIDERIUS *and* ARVIRAGUS    Stand, stand, and fight.
       *Enter Posthumus like a poor soldier, and seconds*
       *the Britons. They rescue Cymbeline and exeunt*

5.4    ⌐The trumpets sound a retreat,⌐ *then enter Lucius,*
       *Giacomo, and Innogen*
LUCIUS (*to Innogen*)
Away, boy, from the troops, and save thyself;
For friends kill friends, and the disorder's such
As war were hoodwinked.
GIACOMO                     'Tis their fresh supplies.
LUCIUS
It is a day turned strangely. Or betimes            4
Let's reinforce, or fly.                       *Exeunt*

5.5    *Enter Posthumus like a poor soldier, and a Briton*
       *Lord*
LORD
Cam'st thou from where they made the stand?
POSTHUMUS                                 I did,
Though you, it seems, come from the fliers.
LORD                                      Ay.
POSTHUMUS
No blame be to you, sir, for all was lost,
But that the heavens fought. The King himself
Of his wings destitute, the army broken,           5
And but the backs of Britons seen, all flying
Through a strait lane; the enemy full-hearted,
Lolling the tongue with slaught'ring, having work
More plentiful than tools to do't, struck down
Some mortally, some slightly touched, some falling  10
Merely through fear, that the strait pass was dammed
With dead men hurt behind, and cowards living
To die with lengthened shame.
LORD                          Where was this lane?
POSTHUMUS
Close by the battle, ditched, and walled with turf;
Which gave advantage to an ancient soldier,        15
An honest one, I warrant, who deserved
So long a breeding as his white beard came to,
In doing this for 's country. Athwart the lane
He with two striplings—lads more like to run
The country base than to commit such slaughter;    20
With faces fit for masks, or rather fairer
Than those for preservation cased, or shame—
Made good the passage, cried to those that fled
'Our Britain's harts die flying, not her men.
To darkness fleet souls that fly backwards. Stand,  25
Or we are Romans, and will give you that
Like beasts which you shun beastly, and may save
But to look back in frown. Stand, stand.' These three,
Three thousand confident, in act as many—

For three performers are the file when all　　30
The rest do nothing—with this word 'Stand, stand',
Accommodated by the place, more charming
With their own nobleness, which could have turned
A distaff to a lance, gilded pale looks;
Part shame, part spirit renewed, that some, turned
　　coward　　35
But by example—O, a sin in war,
Damned in the first beginners!—gan to look
The way that they did and to grin like lions
Upon the pikes o'th' hunters. Then began
A stop i'th' chaser, a retire. Anon　　40
A rout, confusion thick; forthwith they fly
Chickens the way which they stooped eagles; slaves,
The strides they victors made; and now our cowards,
Like fragments in hard voyages, became　　44
The life o'th' need. Having found the back door open
Of the unguarded hearts, heavens, how they wound!
Some slain before, some dying, some their friends
O'erborne i'th' former wave, ten chased by one,
Are now each one the slaughterman of twenty.
Those that would die or ere resist are grown　　50
The mortal bugs o'th' field.

LORD　　　　　　　　　This was strange chance:
A narrow lane, an old man, and two boys.

POSTHUMUS
Nay, do not wonder at it. Yet you are made
Rather to wonder at the things you hear
Than to work any. Will you rhyme upon't,　　55
And vent it for a mock'ry? Here is one:
'Two boys, an old man twice a boy, a lane,
Preserved the Britons, was the Romans' bane.'

LORD
Nay, be not angry, sir.

POSTHUMUS　　　　　　　'Lack, to what end?
Who dares not stand his foe, I'll be his friend,　　60
For if he'll do as he is made to do,
I know he'll quickly fly my friendship too.
You have put me into rhyme.

LORD　　　　　　　　　Farewell; you're angry.
　　　　　　　　　　　　　　　Exit

POSTHUMUS
Still going? This a lord? O noble misery,
To be i'th' field and ask 'What news?' of me!　　65
Today how many would have given their honours
To have saved their carcasses—took heel to do't,
And yet died too! I, in mine own woe charmed,
Could not find death where I did hear him groan,
Nor feel him where he struck. Being an ugly monster,
'Tis strange he hides him in fresh cups, soft beds,　　71
Sweet words, or hath more ministers than we
That draw his knives i'th' war. Well, I will find him;
For being now a favourer to the Briton,
No more a Briton, I have resumed again　　75
The part I came in. Fight I will no more,
But yield me to the veriest hind that shall
Once touch my shoulder. Great the slaughter is
Here made by th' Roman; great the answer be
Britons must take. For me, my ransom's death,　　80
On either side I come to spend my breath,
Which neither here I'll keep nor bear again,
But end it by some means for Innogen.
　　　Enter two Briton Captains, and soldiers

FIRST CAPTAIN
Great Jupiter be praised, Lucius is taken.
'Tis thought the old man and his sons were angels. 85

SECOND CAPTAIN
There was a fourth man, in a seely habit,
That gave th'affront with them.

FIRST CAPTAIN　　　　　　　So 'tis reported,
But none of 'em can be found. Stand, who's there?

POSTHUMUS A Roman,
Who had not now been drooping here if seconds　　90
Had answered him.

SECOND CAPTAIN (to soldiers) Lay hands on him, a dog!
A leg of Rome shall not return to tell
What crows have pecked them here. He brags his
　　service
As if he were of note. Bring him to th' King.
　　⌈Flourish.⌉ Enter Cymbeline ⌈and his train⌉,
　　Belarius, Guiderius, Arviragus, Pisanio, and
　　Roman captives. The Captains present Posthumus to
　　Cymbeline, who delivers him over to a Jailer.
　　Exeunt all but Posthumus and two Jailers, ⌈who
　　lock gyves on his legs⌉

FIRST JAILER
You shall not now be stol'n. You have locks upon you,
So graze as you find pasture.

SECOND JAILER　　　　　　Ay, or a stomach.　　96
　　　　　　　　　　　　　　　Exeunt Jailers

POSTHUMUS
Most welcome, bondage, for thou art a way,
I think, to liberty. Yet am I better
Than one that's sick o'th' gout, since he had rather
Groan so in perpetuity than be cured　　100
By th' sure physician, death, who is the key
T'unbar these locks. My conscience, thou art fettered
More than my shanks and wrists. You good gods give
　　me
The penitent instrument to pick that bolt,
Then free for ever. Is't enough I am sorry?　　105
So children temporal fathers do appease;
Gods are more full of mercy. Must I repent,
I cannot do it better than in gyves
Desired more than constrained. To satisfy,
If of my freedom 'tis the main part, take　　110
No stricter render of me than my all.
I know you are more clement than vile men
Who of their broken debtors take a third,
A sixth, a tenth, letting them thrive again
On their abatement. That's not my desire.　　115
For Innogen's dear life take mine, and though
'Tis not so dear, yet 'tis a life; you coined it.
'Tween man and man they weigh not every stamp;
Though light, take pieces for the figure's sake;
You rather mine, being yours. And so, great powers,
If you will make this audit, take this life,　　121
And cancel these cold bonds. O Innogen,
I'll speak to thee in silence!
　　　He sleeps. Solemn music. Enter, as in an apparition,
　　　Sicilius Leonatus (father to Posthumus, an old
　　　man), attired like a warrior, leading in his hand an
　　　ancient matron, his wife, and mother to
　　　Posthumus, with music before them.
　　　Then, after other music, follows the two young
　　　Leonati, brothers to Posthumus, with wounds as
　　　they died in the wars. They circle Posthumus round
　　　as he lies sleeping

SICILIUS
No more, thou thunder-master, show
　　Thy spite on mortal flies.　　125
With Mars fall out, with Juno chide,
　　That thy adulteries

Rates and revenges.
Hath my poor boy done aught but well,
  Whose face I never saw?          130
I died whilst in the womb he stayed,
  Attending nature's law,
Whose father then—as men report
  Thou orphans' father art—
Thou shouldst have been, and shielded him   135
  From this earth-vexing smart.

MOTHER
Lucina lent not me her aid,
  But took me in my throes,
That from me was Posthumus ripped,
  Came crying 'mongst his foes,       140
A thing of pity.

SICILIUS
Great nature like his ancestry
  Moulded the stuff so fair
That he deserved the praise o'th' world
  As great Sicilius' heir.        145

FIRST BROTHER
When once he was mature for man,
  In Britain where was he
That could stand up his parallel,
  Or fruitful object be
In eye of Innogen, that best       150
  Could deem his dignity?

MOTHER
With marriage wherefore was he mocked,
  To be exiled, and thrown
From Leonati seat and cast
  From her his dearest one,      155
Sweet Innogen?

SICILIUS
Why did you suffer Giacomo,
  Slight thing of Italy,
To taint his nobler heart and brain
  With needless jealousy,       160
And to become the geck and scorn
  O'th' other's villainy?

SECOND BROTHER
For this from stiller seats we come,
  Our parents and us twain,
That striking in our country's cause   165
  Fell bravely and were slain,
Our fealty and Tenantius' right
  With honour to maintain.

FIRST BROTHER
Like hardiment Posthumus hath
  To Cymbeline performed.      170
Then, Jupiter, thou king of gods,
  Why hast thou thus adjourned
The graces for his merits due,
  Being all to dolours turned?

SICILIUS
Thy crystal window ope; look out;   175
  No longer exercise
Upon a valiant race thy harsh
  And potent injuries.

MOTHER
Since, Jupiter, our son is good,
  Take off his miseries.      180

SICILIUS
Peep through thy marble mansion. Help,
  Or we poor ghosts will cry
To th' shining synod of the rest
  Against thy deity.

BROTHERS
Help, Jupiter, or we appeal,      185
  And from thy justice fly.
*Jupiter descends in thunder and lightning, sitting*
*upon an eagle. He throws a thunderbolt. The ghosts*
*fall on their knees*

JUPITER
No more, you petty spirits of region low,
  Offend our hearing. Hush! How dare you ghosts
Accuse the thunderer, whose bolt, you know,
  Sky-planted, batters all rebelling coasts?   190
Poor shadows of Elysium, hence, and rest
  Upon your never-withering banks of flowers.
Be not with mortal accidents oppressed;
  No care of yours it is; you know 'tis ours.
Whom best I love, I cross, to make my gift,   195
  The more delayed, delighted. Be content.
Your low-laid son our godhead will uplift.
  His comforts thrive, his trials well are spent.
Our Jovial star reigned at his birth, and in
  Our temple was he married. Rise, and fade.  200
He shall be lord of Lady Innogen,
  And happier much by his affliction made.
This tablet lay upon his breast, wherein
  Our pleasure his full fortune doth confine.
   *He gives the ghosts a tablet which they lay upon*
   *Posthumus' breast*
And so away. No farther with your din   205
  Express impatience, lest you stir up mine.
Mount, eagle, to my palace crystalline.
        *He ascends into the heavens*

SICILIUS
He came in thunder. His celestial breath
Was sulphurous to smell. The holy eagle
Stooped, as to foot us. His ascension is   210
More sweet than our blest fields. His royal bird
Preens the immortal wing and claws his beak
As when his god is pleased.

ALL THE GHOSTS        Thanks, Jupiter.

SICILIUS
The marble pavement closes, he is entered
His radiant roof. Away, and, to be blest,   215
Let us with care perform his great behest.
         *The ghosts vanish*
   *Posthumus awakes*

POSTHUMUS
Sleep, thou hast been a grandsire, and begot
A father to me; and thou hast created
A mother and two brothers. But, O scorn,
Gone! They went hence so soon as they were born,
And so I am awake. Poor wretches that depend  221
On greatness' favour dream as I have done,
Wake and find nothing. But, alas, I swerve.
Many dream not to find, neither deserve,
And yet are steeped in favours; so am I,   225
That have this golden chance and know not why.
What fairies haunt this ground? A book? O rare one,
Be not, as is our fangled world, a garment
Nobler than that it covers. Let thy effects
So follow to be most unlike our courtiers,  230
As good as promise.
   *He reads*
'Whenas a lion's whelp shall, to himself unknown,
without seeking find, and be embraced by a piece of
tender air; and when from a stately cedar shall be
lopped branches which, being dead many years, shall
after revive, be jointed to the old stock, and freshly

grow; then shall Posthumus end his miseries, Britain
be fortunate and flourish in peace and plenty.'
'Tis still a dream, or else such stuff as madmen
Tongue, and brain not; either both, or nothing,     240
Or senseless speaking, or a speaking such
As sense cannot untie. Be what it is,
The action of my life is like it, which I'll keep,
If but for sympathy.
    *Enter Jailer*

JAILER Come, sir, are you ready for death?     245
POSTHUMUS Over-roasted rather; ready long ago.
JAILER Hanging is the word, sir. If you be ready for that,
you are well cooked.
POSTHUMUS So, if I prove a good repast to the spectators,
the dish pays the shot.     250
JAILER A heavy reckoning for you, sir. But the comfort
is, you shall be called to no more payments, fear no
more tavern bills, which are as often the sadness of
parting as the procuring of mirth. You come in faint
for want of meat, depart reeling with too much drink,
sorry that you have paid too much and sorry that you
are paid too much; purse and brain both empty: the
brain the heavier for being too light, the purse too
light, being drawn of heaviness. Of this contradiction
you shall now be quit. O, the charity of a penny cord!
It sums up thousands in a trice. You have no true
debitor and creditor but it: of what's past, is, and to
come the discharge. Your neck, sir, is pen, book, and
counters; so the acquittance follows.
POSTHUMUS I am merrier to die than thou art to live.  265
JAILER Indeed, sir, he that sleeps feels not the toothache;
but a man that were to sleep your sleep, and a hangman
to help him to bed, I think he would change places
with his officer; for look you, sir, you know not which
way you shall go.     270
POSTHUMUS Yes, indeed do I, fellow.
JAILER Your death has eyes in 's head, then. I have not
seen him so pictured. You must either be directed by
some that take upon them to know, or take upon
yourself that which I am sure you do not know, or
jump the after-enquiry on your own peril; and how
you shall speed in your journey's end I think you'll
never return to tell on.     278
POSTHUMUS I tell thee, fellow, there are none want eyes
to direct them the way I am going but such as wink
and will not use them.
JAILER What an infinite mock is this, that a man should
have the best use of eyes to see the way of blindness!
I am sure hanging's the way of winking.     284
    *Enter a Messenger*
MESSENGER Knock off his manacles, bring your prisoner
to the King.
POSTHUMUS Thou bring'st good news, I am called to be
made free.
JAILER I'll be hanged then.     289
POSTHUMUS Thou shalt be then freer than a jailer; no
bolts for the dead.
JAILER (*aside*) Unless a man would marry a gallows and
beget young gibbets, I never saw one so prone. Yet, on
my conscience, there are verier knaves desire to live,
for all he be a Roman; and there be some of them, too,
that die against their wills; so should I if I were one. I
would we were all of one mind, and one mind good.
O, there were desolation of jailers and gallowses! I
speak against my present profit, but my wish hath a
preferment in't.     *Exeunt*

**5.6**   ⌜*Flourish.*⌝ *Enter Cymbeline, Belarius, Guiderius,*
    *Arviragus, Pisanio, and lords*
CYMBELINE (*to Belarius, Guiderius, and Arviragus*)
Stand by my side, you whom the gods have made
Preservers of my throne. Woe is my heart
That the poor soldier that so richly fought,
Whose rags shamed gilded arms, whose naked breast
Stepped before targs of proof, cannot be found.     5
He shall be happy that can find him, if
Our grace can make him so.
BELARIUS                   I never saw
Such noble fury in so poor a thing,
Such precious deeds in one that promised naught
But beggary and poor looks.
CYMBELINE             No tidings of him?     10
PISANIO
He hath been searched among the dead and living,
But no trace of him.
CYMBELINE         To my grief I am
The heir of his reward, which I will add
(*To Belarius, Guiderius, and Arviragus*)
To you, the liver, heart, and brain of Britain,
By whom I grant she lives. 'Tis now the time     15
To ask of whence you are. Report it.
BELARIUS                  Sir,
In Cambria are we born, and gentlemen.
Further to boast were neither true nor modest,
Unless I add we are honest.
CYMBELINE             Bow your knees.
    *They kneel. He knights them*
Arise, my knights o'th' battle. I create you     20
Companions to our person, and will fit you
With dignities becoming your estates.
    *Belarius, Guiderius, and Arviragus rise.*
    *Enter Cornelius and Ladies*
There's business in these faces. Why so sadly
Greet you our victory? You look like Romans,
And not o'th' court of Britain.
CORNELIUS             Hail, great King!     25
To sour your happiness I must report
The Queen is dead.
CYMBELINE         Who worse than a physician
Would this report become? But I consider
By medicine life may be prolonged, yet death
Will seize the doctor too. How ended she?     30
CORNELIUS
With horror, madly dying, like her life,
Which being cruel to the world, concluded
Most cruel to herself. What she confessed
I will report, so please you. These her women
Can trip me if I err, who with wet cheeks     35
Were present when she finished.
CYMBELINE             Prithee, say.
CORNELIUS
First, she confessed she never loved you, only
Affected greatness got by you, not you;
Married your royalty, was wife to your place,
Abhorred your person.
CYMBELINE           She alone knew this,     40
And but she spoke it dying, I would not
Believe her lips in opening it. Proceed.
CORNELIUS
Your daughter, whom she bore in hand to love
With such integrity, she did confess
Was as a scorpion to her sight, whose life,     45

But that her flight prevented it, she had
Ta'en off by poison.
CYMBELINE                O most delicate fiend!
Who is't can read a woman? Is there more?
CORNELIUS
More, sir, and worse. She did confess she had
For you a mortal mineral which, being took,                    50
Should by the minute feed on life, and, ling'ring,
By inches waste you. In which time she purposed
By watching, weeping, tendance, kissing, to
O'ercome you with her show; and in fine,
When she had fit you with her craft, to work                    55
Her son into th'adoption of the crown;
But failing of her end by his strange absence,
Grew shameless-desperate, opened in despite
Of heaven and men her purposes, repented
The evils she hatched were not effected; so                    60
Despairing died.
CYMBELINE            Heard you all this, her women?
⌈LADIES⌉
We did, so please your highness.
CYMBELINE                Mine eyes
Were not in fault, for she was beautiful;
Mine ears that heard her flattery, nor my heart
That thought her like her seeming. It had been vicious
To have mistrusted her. Yet, O my daughter,                    66
That it was folly in me thou mayst say,
And prove it in thy feeling. Heaven mend all!
        *Enter Lucius, Giacomo, Soothsayer, and other*
        *Roman prisoners, Posthumus behind, and Innogen*
        *dressed as a man, all guarded by Briton soldiers*
Thou com'st not, Caius, now for tribute. That
The Britons have razed out, though with the loss                    70
Of many a bold one; whose kinsmen have made suit
That their good souls may be appeased with slaughter
Of you, their captives, which ourself have granted.
So think of your estate.
LUCIUS
Consider, sir, the chance of war. The day                    75
Was yours by accident. Had it gone with us,
We should not, when the blood was cool, have
        threatened
Our prisoners with the sword. But since the gods
Will have it thus, that nothing but our lives
May be called ransom, let it come. Sufficeth                    80
A Roman with a Roman's heart can suffer.
Augustus lives to think on't; and so much
For my peculiar care. This one thing only
I will entreat:
        *He presents Innogen to Cymbeline*
                my boy, a Briton born,
Let him be ransomed. Never master had                    85
A page so kind, so duteous, diligent,
So tender over his occasions, true,
So feat, so nurse-like; let his virtue join
With my request, which I'll make bold your highness
Cannot deny. He hath done no Briton harm,                    90
Though he have served a Roman. Save him, sir,
And spare no blood beside.
CYMBELINE                I have surely seen him.
His favour is familiar to me. Boy,
Thou hast looked thyself into my grace,
And art mine own. I know not why, wherefore,                    95
To say 'Live, boy'. Ne'er thank thy master. Live,
And ask of Cymbeline what boon thou wilt
Fitting my bounty and thy state, I'll give it,

Yea, though thou do demand a prisoner
The noblest ta'en.
INNOGEN            I humbly thank your highness.                    100
LUCIUS
I do not bid thee beg my life, good lad,
And yet I know thou wilt.
INNOGEN                No, no. Alack,
There's other work in hand. I see a thing
Bitter to me as death. Your life, good master,
Must shuffle for itself.
LUCIUS            The boy disdains me.                    105
He leaves me, scorns me. Briefly die their joys
That place them on the truth of girls and boys.
Why stands he so perplexed?
CYMBELINE (*to Innogen*)            What wouldst thou, boy?
I love thee more and more; think more and more
What's best to ask. Know'st him thou look'st on?
        Speak,                    110
Wilt have him live? Is he thy kin, thy friend?
INNOGEN
He is a Roman, no more kin to me
Than I to your highness, who, being born your vassal,
Am something nearer.
CYMBELINE            Wherefore ey'st him so?
INNOGEN
I'll tell you, sir, in private, if you please                    115
To give me hearing.
CYMBELINE            Ay, with all my heart,
And lend my best attention. What's thy name?
INNOGEN
Fidele, sir.
CYMBELINE    Thou'rt my good youth, my page.
I'll be thy master. Walk with me, speak freely.
        *Cymbeline and Innogen speak apart*
BELARIUS (*aside to Guiderius and Arviragus*)
Is not this boy revived from death?
ARVIRAGUS                One sand another
Not more resembles that sweet rosy lad                    121
Who died, and was Fidele. What think you?
GUIDERIUS The same dead thing alive.
BELARIUS
Peace, peace, see further. He eyes us not. Forbear.
Creatures may be alike. Were't he, I am sure                    125
He would have spoke to us.
GUIDERIUS                But we see him dead.
BELARIUS
Be silent; let's see further.
PISANIO (*aside*)            It is my mistress.
Since she is living, let the time run on
To good or bad.
CYMBELINE (*to Innogen*) Come, stand thou by our side,
Make thy demand aloud. (*To Giacomo*) Sir, step you
        forth.                    130
Give answer to this boy, and do it freely,
Or, by our greatness and the grace of it,
Which is our honour, bitter torture shall
Winnow the truth from falsehood.
(*To Innogen*)                On, speak to him.
INNOGEN
My boon is that this gentleman may render                    135
Of whom he had this ring.
POSTHUMUS (*aside*)            What's that to him?
CYMBELINE (*to Giacomo*)
That diamond upon your finger, say,
How came it yours?

GIACOMO
Thou'lt torture me to leave unspoken that
Which to be spoke would torture thee.

CYMBELINE                                            How, me?    140

GIACOMO
I am glad to be constrained to utter that
Torments me to conceal. By villainy
I got this ring; 'twas Leonatus' jewel,
Whom thou didst banish; and, which more may
        grieve thee,
As it doth me, a nobler sir ne'er lived          145
'Twixt sky and ground. Wilt thou hear more, my lord?

CYMBELINE
All that belongs to this.

GIACOMO                            That paragon thy daughter,
For whom my heart drops blood, and my false spirits
Quail to remember—give me leave, I faint.

CYMBELINE
My daughter? What of her? Renew thy strength.    150
I had rather thou shouldst live while nature will
Than die ere I hear more. Strive, man, and speak.

GIACOMO
Upon a time—unhappy was the clock
That struck the hour—it was in Rome—accursed
The mansion where—'twas at a feast—O, would    155
Our viands had been poisoned, or at least
Those which I heaved to head!—the good Posthumus—
What should I say?—he was too good to be
Where ill men were, and was the best of all
Amongst the rar'st of good ones—sitting sadly,    160
Hearing us praise our loves of Italy
For beauty that made barren the swelled boast
Of him that best could speak; for feature laming
The shrine of Venus or straight-pitched Minerva,
Postures beyond brief nature; for condition,    165
A shop of all the qualities that man
Loves woman for; besides that hook of wiving,
Fairness which strikes the eye—

CYMBELINE                                    I stand on fire.
Come to the matter.

GIACOMO                    All too soon I shall,
Unless thou wouldst grieve quickly. This Posthumus,
Most like a noble lord in love and one          171
That had a royal lover, took his hint,
And not dispraising whom we praised—therein
He was as calm as virtue—he began
His mistress' picture, which by his tongue being made,
And then a mind put in't, either our brags    176
Were cracked of kitchen-trulls, or his description
Proved us unspeaking sots.

CYMBELINE                        Nay, nay, to th' purpose.

GIACOMO
Your daughter's chastity—there it begins.
He spake of her as Dian had hot dreams          180
And she alone were cold, whereat I, wretch,
Made scruple of his praise, and wagered with him
Pieces of gold 'gainst this which then he wore
Upon his honoured finger, to attain
In suit the place of 's bed and win this ring    185
By hers and mine adultery. He, true knight,
No lesser of her honour confident
Than I did truly find her, stakes this ring—
And would so had it been a carbuncle
Of Phoebus' wheel, and might so safely had it    190
Been all the worth of 's car. Away to Britain
Post I in this design. Well may you, sir,
Remember me at court, where I was taught

Of your chaste daughter the wide difference
'Twixt amorous and villainous. Being thus quenched
Of hope, not longing, mine Italian brain          196
Gan in your duller Britain operate
Most vilely; for my vantage, excellent.
And, to be brief, my practice so prevailed
That I returned with simular proof enough          200
To make the noble Leonatus mad
By wounding his belief in her renown
With tokens thus and thus; averring notes
Of chamber-hanging, pictures, this her bracelet—
O cunning, how I got it!—nay, some marks          205
Of secret on her person, that he could not
But think her bond of chastity quite cracked,
I having ta'en the forfeit. Whereupon—
Methinks I see him now—

POSTHUMUS (coming forward) Ay, so thou dost,
Italian fiend! Ay me, most credulous fool,          210
Egregious murderer, thief, anything
That's due to all the villains past, in being,
To come! O, give me cord, or knife, or poison,
Some upright justicer! Thou, King, send out
For torturers ingenious. It is I                    215
That all th' abhorrèd things o'th' earth amend
By being worse than they. I am Posthumus,
That killed thy daughter—villain-like, I lie;
That caused a lesser villain than myself,
A sacrilegious thief, to do't. The temple          220
Of virtue was she; yea, and she herself.
Spit and throw stones, cast mire upon me, set
The dogs o'th' street to bay me. Every villain
Be called Posthumus Leonatus, and
Be 'villain' less than 'twas! O Innogen!            225
My queen, my life, my wife, O Innogen,
Innogen, Innogen!

INNOGEN (approaching him) Peace, my lord. Hear, hear.

POSTHUMUS
Shall 's have a play of this? Thou scornful page,
There lie thy part.
        He strikes her down

PISANIO (coming forward) O gentlemen, help!
Mine and your mistress! O my lord Posthumus,    230
You ne'er killed Innogen till now. Help, help!
(To Innogen) Mine honoured lady.

CYMBELINE                                Does the world go round?

POSTHUMUS
How comes these staggers on me?

PISANIO (to Innogen)                    Wake, my mistress.

CYMBELINE
If this be so, the gods do mean to strike me
To death with mortal joy.                          235

PISANIO (to Innogen) How fares my mistress?

INNOGEN O, get thee from my sight!
Thou gav'st me poison. Dangerous fellow, hence.
Breathe not where princes are.

CYMBELINE                        The tune of Innogen.

PISANIO
Lady, the gods throw stones of sulphur on me if    240
That box I gave you was not thought by me
A precious thing. I had it from the Queen.

CYMBELINE
New matter still.

INNOGEN                It poisoned me.

CORNELIUS                            O gods!
I left out one thing which the Queen confessed
(To Pisanio) Which must approve thee honest. 'If
        Pisanio                                        245

Have', said she, 'given his mistress that confection
Which I gave him for cordial, she is served
As I would serve a rat.'
CYMBELINE                    What's this, Cornelius?
CORNELIUS
The Queen, sir, very oft importuned me
To temper poisons for her, still pretending            250
The satisfaction of her knowledge only
In killing creatures vile, as cats and dogs
Of no esteem. I, dreading that her purpose
Was of more danger, did compound for her
A certain stuff which, being ta'en, would cease       255
The present power of life, but in short time
All offices of nature should again
Do their due functions. (To Innogen) Have you ta'en
   of it?
INNOGEN
Most like I did, for I was dead.
BELARIUS (aside to Guiderius and Arviragus)  My boys,
There was our error.
GUIDERIUS             This is sure Fidele.           260
INNOGEN (to Posthumus)
Why did you throw your wedded lady from you?
Think that you are upon a lock, and now
Throw me again.
   She throws her arms about his neck
POSTHUMUS            Hang there like fruit, my soul,
Till the tree die.
CYMBELINE (to Innogen)  How now, my flesh, my child?
What, mak'st thou me a dullard in this act?          265
Wilt thou not speak to me?
INNOGEN (kneeling)            Your blessing, sir.
BELARIUS (aside to Guiderius and Arviragus)
Though you did love this youth, I blame ye not.
You had a motive for't.
CYMBELINE               My tears that fall
Prove holy water on thee!
   ⌜He raises her⌝
                        Innogen,
Thy mother's dead.
INNOGEN            I am sorry for't, my lord.         270
CYMBELINE
O, she was naught, and 'long of her it was
That we meet here so strangely. But her son
Is gone, we know not how nor where.
PISANIO                            My lord,
Now fear is from me I'll speak truth. Lord Cloten,
Upon my lady's missing, came to me                   275
With his sword drawn, foamed at the mouth, and
   swore
If I discovered not which way she was gone
It was my instant death. By accident
I had a feignèd letter of my master's
Then in my pocket, which directed him                280
To seek her on the mountains near to Milford,
Where in a frenzy, in my master's garments,
Which he enforced from me, away he posts
With unchaste purpose, and with oath to violate
My lady's honour. What became of him                 285
I further know not.
GUIDERIUS          Let me end the story.
I slew him there.
CYMBELINE        Marry, the gods forfend!
I would not thy good deeds should from my lips
Pluck a hard sentence. Prithee, valiant youth,
Deny't again.                                        290

GUIDERIUS  I have spoke it, and I did it.
CYMBELINE  He was a prince.
GUIDERIUS
A most incivil one. The wrongs he did me
Were nothing prince-like, for he did provoke me
With language that would make me spurn the sea   295
If it could so roar to me. I cut off 's head,
And am right glad he is not standing here
To tell this tale of mine.
CYMBELINE                I am sorrow for thee.
By thine own tongue thou art condemned, and must
Endure our law. Thou'rt dead.
INNOGEN                      That headless man   300
I thought had been my lord.
CYMBELINE (to soldiers)        Bind the offender,
And take him from our presence.
BELARIUS                        Stay, sir King.
This boy is better than the man he slew,
As well descended as thyself, and hath
More of thee merited than a band of Clotens      305
Had ever scar for. Let his arms alone;
They were not born for bondage.
CYMBELINE                       Why, old soldier,
Wilt thou undo the worth thou art unpaid for
By tasting of our wrath? How of descent
As good as we?
ARVIRAGUS       In that he spake too far.         310
CYMBELINE ⌜to Belarius⌝
And thou shalt die for't.
BELARIUS                 We will die all three
But I will prove that two on 's are as good
As I have given out him. My sons, I must
For mine own part unfold a dangerous speech,
Though haply well for you.
ARVIRAGUS                 Your danger's ours.     315
GUIDERIUS
And our good his.
BELARIUS          Have at it then. By leave,
Thou hadst, great King, a subject who
Was called Belarius.
CYMBELINE            What of him? He is
A banished traitor.
BELARIUS           He it is that hath
Assumed this age. Indeed, a banished man;        320
I know not how a traitor.
CYMBELINE (to soldiers)     Take him hence.
The whole world shall not save him.
BELARIUS                            Not too hot.
First pay me for the nursing of thy sons,
And let it be confiscate all so soon
As I have received it.
CYMBELINE            Nursing of my sons?          325
BELARIUS
I am too blunt and saucy. (Kneeling) Here's my knee.
Ere I arise I will prefer my sons,
Then spare not the old father. Mighty sir,
These two young gentlemen that call me father
And think they are my sons are none of mine.     330
They are the issue of your loins, my liege,
And blood of your begetting.
CYMBELINE                   How, my issue?
BELARIUS
So sure as you your father's. I, old Morgan,
Am that Belarius whom you sometime banished.
Your pleasure was my mere offence, my punishment
Itself, and all my treason. That I suffered       336

Was all the harm I did. These gentle princes—
For such and so they are—these twenty years
Have I trained up. Those arts they have as I
Could put into them. My breeding was, sir,          340
As your highness knows. Their nurse Euriphile,
Whom for the theft I wedded, stole these children
Upon my banishment. I moved her to't,
Having received the punishment before
For that which I did then. Beaten for loyalty        345
Excited me to treason. Their dear loss,
The more of you 'twas felt, the more it shaped
Unto my end of stealing them. But, gracious sir,
Here are your sons again, and I must lose
Two of the sweet'st companions in the world.        350
The benediction of these covering heavens
Fall on their heads like dew, for they are worthy
To inlay heaven with stars.

CYMBELINE                              Thou weep'st, and speak'st.
The service that you three have done is more
Unlike than this thou tell'st. I lost my children.   355
If these be they, I know not how to wish
A pair of worthier sons.

BELARIUS ⌈rising⌉          Be pleased a while.
This gentleman, whom I call Polydore,
Most worthy prince, as yours, is true Guiderius.
      ⌈Guiderius kneels⌉
This gentleman, my Cadwal, Arviragus,               360
Your younger princely son.
      ⌈Arviragus kneels⌉
                              He, sir, was lapped
In a most curious mantle wrought by th' hand
Of his queen mother, which for more probation
I can with ease produce.

CYMBELINE                    Guiderius had
Upon his neck a mole, a sanguine star.              365
It was a mark of wonder.

BELARIUS                    This is he,
Who hath upon him still that natural stamp.
It was wise nature's end in the donation
To be his evidence now.

CYMBELINE              O, what am I?
A mother to the birth of three? Ne'er mother       370
Rejoiced deliverance more. Blest pray you be,
That, after this strange starting from your orbs,
You may reign in them now!
      ⌈Guiderius and Arviragus rise⌉
                              O Innogen,
Thou hast lost by this a kingdom.

INNOGEN                              No, my lord,
I have got two worlds by't. O my gentle brothers,   375
Have we thus met? O, never say hereafter
But I am truest speaker. You called me brother
When I was but your sister; I you brothers
When ye were so indeed.

CYMBELINE              Did you e'er meet?

ARVIRAGUS
Ay, my good lord.

GUIDERIUS          And at first meeting loved,        380
Continued so until we thought he died.

CORNELIUS
By the Queen's dram she swallowed.

CYMBELINE                              O rare instinct!
When shall I hear all through? This fierce abridgement
Hath to it circumstantial branches which
Distinction should be rich in. Where? How lived you?

And when came you to serve our Roman captive?       386
How parted with your brothers? How first met them?
Why fled you from the court? And whither? These,
And your three motives to the battle, with
I know not how much more, should be demanded,
And all the other by-dependences,                   391
From chance to chance. But nor the time nor place
Will serve our long inter'gatories. See,
Posthumus anchors upon Innogen,
And she, like harmless lightning, throws her eye    395
On him, her brothers, me, her master, hitting
Each object with a joy. The counterchange
Is severally in all. Let's quit this ground,
And smoke the temple with our sacrifices.
      (To Belarius) Thou art my brother; so we'll hold thee
            ever.                                    400

INNOGEN (to Belarius)
You are my father too, and did relieve me
To see this gracious season.

CYMBELINE                    All o'erjoyed,
Save these in bonds. Let them be joyful too,
For they shall taste our comfort.

INNOGEN (to Lucius)              My good master,
I will yet do you service.

LUCIUS                    Happy be you!            405

CYMBELINE
The forlorn soldier that so nobly fought,
He would have well becomed this place, and graced
The thankings of a king.

POSTHUMUS                    I am, sir,
The soldier that did company these three
In poor beseeming. 'Twas a fitment for              410
The purpose I then followed. That I was he,
Speak, Giacomo; I had you down, and might
Have made you finish.

GIACOMO (kneeling)          I am down again,
But now my heavy conscience sinks my knee
As then your force did. Take that life, beseech you,
Which I so often owe; but your ring first,          416
And here the bracelet of the truest princess
That ever swore her faith.

POSTHUMUS (raising him)          Kneel not to me.
The power that I have on you is to spare you,
The malice towards you to forgive you. Live,        420
And deal with others better.

CYMBELINE                    Nobly doomed!
We'll learn our freeness of a son-in-law.
Pardon's the word to all.

ARVIRAGUS (to Posthumus)    You holp us, sir,
As you did mean indeed to be our brother.
Joyed are we that you are.                          425

POSTHUMUS
Your servant, princes. (To Lucius) Good my lord of
            Rome,
Call forth your soothsayer. As I slept, methought
Great Jupiter, upon his eagle backed,
Appeared to me with other spritely shows
Of mine own kindred. When I waked I found           430
This label on my bosom, whose containing
Is so from sense in hardness that I can
Make no collection of it. Let him show
His skill in the construction.

LUCIUS                    Philharmonus.            434

SOOTHSAYER
Here, my good lord.

LUCIUS                    Read, and declare the meaning.

SOOTHSAYER (*reads the tablet*) 'Whenas a lion's whelp shall, to himself unknown, without seeking find, and be embraced by a piece of tender air; and when from a stately cedar shall be lopped branches which, being dead many years, shall after revive, be jointed to the old stock, and freshly grow: then shall Posthumus end his miseries, Britain be fortunate and flourish in peace and plenty.'
Thou, Leonatus, art the lion's whelp.
The fit and apt construction of thy name,                        445
Being *leo-natus*, doth import so much.
(*To Cymbeline*) The piece of tender air thy virtuous
    daughter,
Which we call '*mollis aer*'; and '*mollis aer*'
We term it '*mulier*', (*to Posthumus*) which '*mulier*' I
    divine
Is this most constant wife, who even now,                        450
Answering the letter of the oracle,
Unknown to you, unsought, were clipped about
With this most tender air.
CYMBELINE                                    This hath some seeming.
SOOTHSAYER
The lofty cedar, royal Cymbeline,
Personates thee, and thy lopped branches point         455
Thy two sons forth, who, by Belarius stol'n,
For many years thought dead, are now revived,
To the majestic cedar joined, whose issue
Promises Britain peace and plenty.
CYMBELINE                                              Well,
My peace we will begin; and, Caius Lucius,              460

Although the victor, we submit to Caesar
And to the Roman empire, promising
To pay our wonted tribute, from the which
We were dissuaded by our wicked queen,
Whom heavens in justice both on her and hers     465
Have laid most heavy hand.
SOOTHSAYER
The fingers of the powers above do tune
The harmony of this peace. The vision,
Which I made known to Lucius ere the stroke
Of this yet scarce-cold battle, at this instant         470
Is full accomplished. For the Roman eagle,
From south to west on wing soaring aloft,
Lessened herself, and in the beams o'th' sun
So vanished; which foreshowed our princely eagle
Th'imperial Caesar should again unite                     475
His favour with the radiant Cymbeline,
Which shines here in the west.
CYMBELINE                                      Laud we the gods,
And let our crookèd smokes climb to their nostrils
From our blest altars. Publish we this peace
To all our subjects. Set we forward, let                    480
A Roman and a British ensign wave
Friendly together. So through Lud's town march,
And in the temple of great Jupiter
Our peace we'll ratify, seal it with feasts.
Set on there. Never was a war did cease,                   485
Ere bloody hands were washed, with such a peace.
                     ⌈*Flourish.*⌉ *Exeunt* ⌈*in triumph*⌉

# THE TEMPEST

THE King's Men acted *The Tempest* before their patron, James I, at Whitehall on 1 November 1611. (It was also chosen for performance during the festivities for the marriage of James's daughter, Princess Elizabeth, to the Elector Palatine during the winter of 1612-13.) Shakespeare's play takes place on a desert island somewhere between Tunis and Naples; he derived some details of it from his reading of travel literature, including accounts of an expedition of nine ships taking five hundred colonists from Plymouth to Virginia, which set sail in May 1609. On 29 July the flagship, the *Sea-Adventure*, was wrecked by a storm on the coast of the Bermudas. She was presumed lost, but on 23 May 1610 those aboard her arrived safely in Jamestown, Virginia, having found shelter on the island of Bermuda, where they were able to build the pinnaces in which they completed their journey. Accounts of the voyage soon reached England; the last-written that Shakespeare seems to have known is a letter by William Strachey, who was on the *Sea-Adventure*, dated 15 July 1610; though it was not published until 1625, it circulated in manuscript. So it seems clear that Shakespeare wrote *The Tempest* during the later part of 1610 or in 1611. It was first printed in the 1623 Folio, where it is the opening play.

Though other items of Shakespeare's reading—including both Arthur Golding's translation and Ovid's original *Metamorphoses* (closely echoed in Prospero's farewell to his magic), John Florio's translation of essays by Michel de Montaigne, and (less locally but no less pervasively) Virgil's *Aeneid*—certainly fed Shakespeare's imagination as he wrote *The Tempest*, he appears to have devised the main plot himself. Many of its elements are based on the familiar stuff of romance literature: the long-past shipwreck after a perilous voyage of Prospero and his daughter Miranda; the shipwreck, depicted in the opening scene, of Prospero's brother, Antonio, with Alonso, King of Naples, and others; the separation and estrangement of relatives—Antonio usurped Prospero's dukedom, and Alonso believes his son, Ferdinand, is drowned; the chaste love, subjected to trials, of the handsome Ferdinand and the beautiful Miranda; the influence of the supernatural exercised through Prospero's magic powers; and the final reunions and reconciliations along with the happy conclusion of the love affair. Shakespeare had employed such conventions from the beginning of his career in his comedies, and with especial concentration, shortly before he wrote *The Tempest*, in *Pericles*, *The Winter's Tale*, and *Cymbeline*. But whereas those plays unfold the events as they happen, taking us on a journey through time and space, in *The Tempest* (as elsewhere only in *The Comedy of Errors*) Shakespeare gives us only the end of the story, concentrating the action into a few hours and locating it in a single place, but informing us about the past, as in the long, romance-type narrative (1.2) in which Prospero tells Miranda of her childhood. The supernatural, a strong presence in all Shakespeare's late plays, is particularly pervasive in *The Tempest*; Prospero is a 'white' magician—a beneficent one—attended by the spirit Ariel and the sub-human Caliban, two of Shakespeare's most obviously symbolic characters; and a climax of the play is the supernaturally induced wedding masque that Prospero conjures up for the entertainment and edification of the young lovers, and which vanishes as he remembers Caliban's plot against his life.

# THE PERSONS OF THE PLAY

PROSPERO, the rightful Duke of Milan

MIRANDA, his daughter

ANTONIO, his brother, the usurping Duke of Milan

ALONSO, King of Naples

SEBASTIAN, his brother

FERDINAND, Alonso's son

GONZALO, an honest old counsellor of Naples

ADRIAN
FRANCISCO } lords

ARIEL, an airy spirit attendant upon Prospero

CALIBAN, a savage and deformed native of the island, Prospero's slave

TRINCULO, Alonso's jester

STEFANO, Alonso's drunken butler

The MASTER of a ship

BOATSWAIN

MARINERS

SPIRITS

*The Masque*

Spirits appearing as:

IRIS

CERES

JUNO

Nymphs, reapers

# The Tempest

**1.1** *A tempestuous noise of thunder and lightning heard.*
*Enter ⌈severally⌉ a Shipmaster and a Boatswain*

MASTER Boatswain!

BOATSWAIN Here, Master. What cheer?

MASTER Good, speak to th' mariners. Fall to't yarely, or
we run ourselves aground. Bestir, bestir!                    *Exit*
*Enter Mariners*

BOATSWAIN Heigh, my hearts! Cheerly, cheerly, my hearts!
Yare, yare! Take in the topsail! Tend to th' Master's
whistle!—Blow till thou burst thy wind, if room enough.
*Enter Alonso, Sebastian, Antonio, Ferdinand,*
*Gonzalo, and others*

ALONSO Good Boatswain, have care. Where's the Master?
(*To the Mariners*) Play the men!

BOATSWAIN I pray now, keep below.                             10

ANTONIO Where is the Master, Boatswain?

BOATSWAIN Do you not hear him? You mar our labour.
Keep your cabins; you do assist the storm.

GONZALO Nay, good, be patient.

BOATSWAIN When the sea is. Hence! What cares these
roarers for the name of king? To cabin! Silence; trouble
us not.                                                       17

GONZALO Good, yet remember whom thou hast aboard.

BOATSWAIN None that I more love than myself. You are
a councillor; if you can command these elements to
silence and work peace of the present, we will not hand
a rope more. Use your authority. If you cannot, give
thanks you have lived so long and make yourself ready
in your cabin for the mischance of the hour, if it so
hap. (*To the Mariners*) Cheerly, good hearts! (*To Gonzalo*)
Out of our way, I say!                                        *Exit*

GONZALO I have great comfort from this fellow. Methinks
he hath no drowning mark upon him; his complexion
is perfect gallows. Stand fast, good Fate, to his hanging.
Make the rope of his destiny our cable, for our own
doth little advantage. If he be not born to be hanged,
our case is miserable.                            *Exeunt ⌈Courtiers⌉*
*Enter Boatswain*

BOATSWAIN Down with the topmast! Yare! Lower, lower!
Bring her to try wi'th' main-course!                          34
*A cry within*
A plague upon this howling! They are louder than the
weather, or our office.
*Enter Sebastian, Antonio, and Gonzalo*
Yet again? What do you here? Shall we give o'er and
drown? Have you a mind to sink?

SEBASTIAN A pox o'your throat, you bawling, blasphemous,
incharitable dog!                                             40

BOATSWAIN Work you, then.

ANTONIO Hang, cur, hang, you whoreson insolent noise-
maker. We are less afraid to be drowned than thou art.
⌈*Exeunt Mariners*⌉

GONZALO I'll warrant him for drowning, though the ship
were no stronger than a nutshell and as leaky as an
unstanched wench.                                             46

BOATSWAIN Lay her a-hold, a-hold! Set her two courses!
Off to sea again! Lay her off!
*Enter Mariners, wet*

MARINERS All lost! To prayers, to prayers! All lost!
⌈*Exeunt Mariners*⌉

BOATSWAIN What, must our mouths be cold?                      50

GONZALO
The King and Prince at prayers! Let's assist them,
For our case is as theirs.

SEBASTIAN                    I'm out of patience.

ANTONIO
We are merely cheated of our lives by drunkards.
This wide-chopped rascal—would thou mightst lie
drowning
The washing of ten tides.

GONZALO                    He'll be hanged yet,                55
Though every drop of water swear against it
And gape at wid'st to glut him.
*A confused noise within*

MARINERS (*within*)                    Mercy on us!
We split, we split! Farewell, my wife and children!
Farewell, brother! We split, we split, we split!
⌈*Exit Boatswain*⌉

ANTONIO
Let's all sink wi'th' King.

SEBASTIAN                    Let's take leave of him.          60
*Exeunt Antonio and Sebastian*

GONZALO Now would I give a thousand furlongs of sea
for an acre of barren ground: long heath, broom, furze,
anything. The wills above be done, but I would fain
die a dry death.                                              *Exit*

**1.2** *Enter Prospero ⌈in his magic cloak, with a staff⌉,*
*and Miranda*

MIRANDA
If by your art, my dearest father, you have
Put the wild waters in this roar, allay them.
The sky, it seems, would pour down stinking pitch,
But that the sea, mounting to th' welkin's cheek,
Dashes the fire out. O, I have suffered                       5
With those that I saw suffer! A brave vessel,
Who had, no doubt, some noble creature in her,
Dashed all to pieces! O, the cry did knock
Against my very heart! Poor souls, they perished.
Had I been any god of power, I would                          10
Have sunk the sea within the earth, or ere
It should the good ship so have swallowed and
The fraughting souls within her.

PROSPERO                    Be collected.
No more amazement. Tell your piteous heart
There's no harm done.

MIRANDA                    O woe the day!

PROSPERO                    No harm.                           15
I have done nothing but in care of thee,
Of thee, my dear one, thee, my daughter, who
Art ignorant of what thou art, naught knowing
Of whence I am, nor that I am more better
Than Prospero, master of a full poor cell                     20
And thy no greater father.

MIRANDA                    More to know
Did never meddle with my thoughts.

PROSPERO                    'Tis time
I should inform thee farther. Lend thy hand,
And pluck my magic garment from me.
*Miranda removes Prospero's cloak, ⌈and he lays it*
*on the ground⌉*
                                   So.

1169

Lie there, my art.—Wipe thou thine eyes; have comfort.
The direful spectacle of the wreck, which touched 26
The very virtue of compassion in thee,
I have with such provision in mine art
So safely ordered that there is no soul—
No, not so much perdition as an hair 30
Betid to any creature in the vessel,
Which thou heard'st cry, which thou saw'st sink. Sit down,
For thou must now know farther.
    *Miranda sits*

MIRANDA                                   You have often
Begun to tell me what I am, but stopped
And left me to a bootless inquisition, 35
Concluding 'Stay; not yet'.

PROSPERO                         The hour's now come.
The very minute bids thee ope thine ear,
Obey, and be attentive. Canst thou remember
A time before we came unto this cell?
I do not think thou canst, for then thou wast not 40
Out three years old.

MIRANDA                         Certainly, sir, I can.

PROSPERO
By what? By any other house or person?
Of anything the image tell me that
Hath kept with thy remembrance.

MIRANDA                                   'Tis far off,
And rather like a dream than an assurance 45
That my remembrance warrants. Had I not
Four or five women once that tended me?

PROSPERO
Thou hadst, and more, Miranda. But how is it
That this lives in thy mind? What seest thou else
In the dark backward and abyss of time? 50
If thou rememb'rest aught ere thou cam'st here,
How thou cam'st here thou mayst.

MIRANDA                                   But that I do not.

PROSPERO
Twelve year since, Miranda, twelve year since,
Thy father was the Duke of Milan, and
A prince of power—

MIRANDA                         Sir, are not you my father? 55

PROSPERO
Thy mother was a piece of virtue, and
She said thou wast my daughter; and thy father
Was Duke of Milan, and his only heir
And princess no worse issued.

MIRANDA                                   O the heavens!
What foul play had we that we came from thence? 60
Or blessèd was't we did?

PROSPERO                         Both, both, my girl.
By foul play, as thou sayst, were we heaved thence,
But blessedly holp hither.

MIRANDA                         O, my heart bleeds
To think o'th' teen that I have turned you to,
Which is from my remembrance. Please you, farther.

PROSPERO
My brother and thy uncle called Antonio— 66
I pray thee mark me, that a brother should
Be so perfidious—he whom next thyself
Of all the world I loved, and to him put
The manage of my state—as at that time 70
Through all the signories it was the first,
And Prospero the prime duke—being so reputed
In dignity, and for the liberal arts
Without a parallel—those being all my study,

The government I cast upon my brother, 75
And to my state grew stranger, being transported
And rapt in secret studies. Thy false uncle—
Dost thou attend me?

MIRANDA                         Sir, most heedfully.

PROSPERO
Being once perfected how to grant suits,
How to deny them, who t'advance and who 80
To trash for over-topping, new created
The creatures that were mine, I say—or changed 'em
Or else new formed 'em; having both the key
Of officer and office, set all hearts i'th' state
To what tune pleased his ear, that now he was 85
The ivy which had hid my princely trunk
And sucked my verdure out on't. Thou attend'st not!

MIRANDA
O good sir, I do.

PROSPERO                         I pray thee mark me.
I, thus neglecting worldly ends, all dedicated
To closeness and the bettering of my mind 90
With that which but by being so retired
O'er-priced all popular rate, in my false brother
Awaked an evil nature; and my trust,
Like a good parent, did beget of him
A falsehood, in its contrary as great 95
As my trust was, which had indeed no limit,
A confidence sans bound. He being thus lorded
Not only with what my revenue yielded
But what my power might else exact, like one
Who having into truth, by telling oft, 100
Made such a sinner of his memory
To credit his own lie, he did believe
He was indeed the Duke. Out o'th' substitution,
And executing th'outward face of royalty 104
With all prerogative, hence his ambition growing—
Dost thou hear?

MIRANDA                         Your tale, sir, would cure deafness.

PROSPERO
To have no screen between this part he played
And him he played it for, he needs will be
Absolute Milan. Me, poor man—my library
Was dukedom large enough—of temporal royalties
He thinks me now incapable; confederates, 111
So dry he was for sway, wi'th' King of Naples
To give him annual tribute, do him homage,
Subject his coronet to his crown, and bend
The dukedom, yet unbowed—alas, poor Milan— 115
To most ignoble stooping.

MIRANDA                         O the heavens!

PROSPERO
Mark his condition and th'event, then tell me
If this might be a brother.

MIRANDA                         I should sin
To think but nobly of my grandmother.
Good wombs have borne bad sons.

PROSPERO                                   Now the condition.
This King of Naples, being an enemy 121
To me inveterate, hearkens my brother's suit;
Which was that he, in lieu o'th' premises
Of homage and I know not how much tribute,
Should presently extirpate me and mine 125
Out of the dukedom, and confer fair Milan,
With all the honours, on my brother. Whereon,
A treacherous army levied, one midnight
Fated to th' purpose did Antonio open
The gates of Milan; and, i'th' dead of darkness, 130
The ministers for th' purpose hurried thence

Me and thy crying self.

MIRANDA            Alack, for pity!
I, not rememb'ring how I cried out then,
Will cry it o'er again; it is a hint
That wrings mine eyes to't.

PROSPERO ⌈sitting⌉       Hear a little further,  135
And then I'll bring thee to the present business
Which now's upon's, without the which this story
Were most impertinent.

MIRANDA         Wherefore did they not
That hour destroy us?

PROSPERO        Well demanded, wench;
My tale provokes that question. Dear, they durst not,
So dear the love my people bore me; nor set  141
A mark so bloody on the business, but
With colours fairer painted their foul ends.
In few, they hurried us aboard a barque,
Bore us some leagues to sea, where they prepared  145
A rotten carcass of a butt, not rigged,
Nor tackle, sail, nor mast—the very rats
Instinctively have quit it. There they hoist us,
To cry to th' sea that roared to us, to sigh
To th'winds, whose pity, sighing back again,  150
Did us but loving wrong.

MIRANDA        Alack, what trouble
Was I then to you!

PROSPERO      O, a cherubin
Thou wast that did preserve me. Thou didst smile,
Infusèd with a fortitude from heaven,
When I have decked the sea with drops full salt,  155
Under my burden groaned; which raised in me
An undergoing stomach, to bear up
Against what should ensue.

MIRANDA How came we ashore?

PROSPERO By providence divine.  160
Some food we had, and some fresh water, that
A noble Neapolitan, Gonzalo,
Out of his charity—who being then appointed
Master of this design—did give us; with
Rich garments, linens, stuffs, and necessaries  165
Which since have steaded much. So, of his gentleness,
Knowing I loved my books, he furnished me
From mine own library with volumes that
I prize above my dukedom.

MIRANDA        Would I might
But ever see that man!

PROSPERO      Now I arise.  170
⌈He stands and puts on his cloak⌉
Sit still, and hear the last of our sea-sorrow.
Here in this island we arrived, and here
Have I thy schoolmaster made thee more profit
Than other princes can, that have more time
For vainer hours and tutors not so careful.  175

MIRANDA
Heavens thank you for't. And now I pray you, sir—
For still 'tis beating in my mind—your reason
For raising this sea-storm.

PROSPERO      Know thus far forth.
By accident most strange, bountiful Fortune,
Now my dear lady, hath mine enemies  180
Brought to this shore; and by my prescience
I find my zenith doth depend upon
A most auspicious star, whose influence
If now I court not, but omit, my fortunes
Will ever after droop. Here cease more questions.  185
Thou art inclined to sleep; 'tis a good dullness,
And give it way. I know thou canst not choose.

*Miranda sleeps*
Come away, servant, come! I am ready now.
Approach, my Ariel, come!
*Enter Ariel*

ARIEL
All hail, great master, grave sir, hail. I come  190
To answer thy best pleasure. Be't to fly,
To swim, to dive into the fire, to ride
On the curled clouds, to thy strong bidding task
Ariel and all his quality.

PROSPERO      Hast thou, spirit,
Performed to point the tempest that I bade thee?  195

ARIEL To every article.
I boarded the King's ship. Now on the beak,
Now in the waste, the deck, in every cabin,
I flamed amazement. Sometime I'd divide,
And burn in many places; on the top-mast,  200
The yards, and bowsprit, would I flame distinctly;
Then meet and join. Jove's lightning, the precursors
O'th' dreadful thunderclaps, more momentary
And sight-outrunning were not. The fire and cracks
Of sulphurous roaring the most mighty Neptune  205
Seem to besiege, and make his bold waves tremble,
Yea, his dread trident shake.

PROSPERO       My brave spirit!
Who was so firm, so constant, that this coil
Would not infect his reason?

ARIEL        Not a soul
But felt a fever of the mad, and played  210
Some tricks of desperation. All but mariners
Plunged in the foaming brine and quit the vessel,
Then all afire with me. The King's son Ferdinand,
With hair upstaring—then like reeds, not hair—
Was the first man that leaped; cried 'Hell is empty,
And all the devils are here'.

PROSPERO      Why, that's my spirit!
But was not this nigh shore?

ARIEL       Close by, my master.  217

PROSPERO
But are they, Ariel, safe?

ARIEL       Not a hair perished.
On their sustaining garments not a blemish,
But fresher than before. And, as thou bad'st me,  220
In troops I have dispersed them 'bout the isle.
The King's son have I landed by himself,
Whom I left cooling of the air with sighs
In an odd angle of the isle, and sitting,
His arms in this sad knot.

PROSPERO      Of the King's ship,  225
The mariners, say how thou hast disposed,
And all the rest o'th' fleet.

ARIEL      Safely in harbour
Is the King's ship, in the deep nook where once
Thou called'st me up at midnight to fetch dew
From the still-vexed Bermudas, there she's hid;  230
The mariners all under hatches stowed,
Who, with a charm joined to their suffered labour,
I have left asleep. And for the rest o'th' fleet,
Which I dispersed, they all have met again,
And are upon the Mediterranean float  235
Bound sadly home for Naples,
Supposing that they saw the King's ship wrecked,
And his great person perish.

PROSPERO      Ariel, thy charge
Exactly is performed; but there's more work.
What is the time o'th' day?

ARIEL       Past the mid season.  240

PROSPERO
  At least two glasses. The time 'twixt six and now
  Must by us both be spent most preciously.
ARIEL
  Is there more toil? Since thou dost give me pains,
  Let me remember thee what thou hast promised
  Which is not yet performed me.
PROSPERO              How now? Moody?
  What is't thou canst demand?
ARIEL                My liberty.    246
PROSPERO
  Before the time be out? No more!
ARIEL                I prithee,
  Remember I have done thee worthy service,
  Told thee no lies, made thee no mistakings, served
  Without or grudge or grumblings. Thou did promise
  To bate me a full year.
PROSPERO           Dost thou forget    251
  From what a torment I did free thee?
ARIEL                No.
PROSPERO
  Thou dost, and think'st it much to tread the ooze
  Of the salt deep,
  To run upon the sharp wind of the north,    255
  To do me business in the veins o'th' earth
  When it is baked with frost.
ARIEL             I do not, sir.
PROSPERO
  Thou liest, malignant thing. Hast thou forgot
  The foul witch Sycorax, who with age and envy
  Was grown into a hoop? Hast thou forgot her?    260
ARIEL
  No, sir.
PROSPERO  Thou hast. Where was she born? Speak, tell me!
ARIEL
  Sir, in Algiers.
PROSPERO         O, was she so! I must
  Once in a month recount what thou hast been,
  Which thou forget'st. This damned witch Sycorax,
  For mischiefs manifold and sorceries terrible    265
  To enter human hearing, from Algiers
  Thou know'st was banished. For one thing she did
  They would not take her life. Is not this true?
ARIEL  Ay, sir.
PROSPERO
  This blue-eyed hag was hither brought with child,  270
  And here was left by th' sailors. Thou, my slave,
  As thou report'st thyself, was then her servant;
  And for thou wast a spirit too delicate
  To act her earthy and abhorred commands,
  Refusing her grand hests, she did confine thee    275
  By help of her more potent ministers,
  And in her most unmitigable rage,
  Into a cloven pine; within which rift
  Imprisoned thou didst painfully remain
  A dozen years, within which space she died    280
  And left thee there, where thou didst vent thy groans
  As fast as mill-wheels strike. Then was this island—
  Save for the son that she did litter here,
  A freckled whelp, hag-born—not honoured with
  A human shape.
ARIEL         Yes, Caliban her son.    285
PROSPERO
  Dull thing, I say so: he, that Caliban
  Whom now I keep in service. Thou best know'st

What torment I did find thee in. Thy groans
Did make wolves howl, and penetrate the breasts
Of ever-angry bears; it was a torment    290
To lay upon the damned, which Sycorax
Could not again undo. It was mine art,
When I arrived and heard thee, that made gape
The pine and let thee out.
ARIEL         I thank thee, master.
PROSPERO
If thou more murmur'st, I will rend an oak,    295
And peg thee in his knotty entrails till
Thou hast howled away twelve winters.
ARIEL           Pardon, master.
I will be correspondent to command,
And do my spriting gently.
PROSPERO  Do so, and after two days    300
I will discharge thee.
ARIEL         That's my noble master!
What shall I do? Say what, what shall I do?
PROSPERO
Go make thyself like to a nymph o'th' sea. Be subject
To no sight but thine and mine, invisible
To every eyeball else. Go take this shape,    305
And hither come in't. Go; hence with diligence!
                        *Exit Ariel*
Awake, dear heart, awake! Thou hast slept well;
Awake.
MIRANDA (*awaking*) The strangeness of your story put
Heaviness in me.
PROSPERO        Shake it off. Come on;
We'll visit Caliban my slave, who never    310
Yields us kind answer.
MIRANDA          'Tis a villain, sir,
I do not love to look on.
PROSPERO          But as 'tis,
We cannot miss him. He does make our fire,
Fetch in our wood, and serves in offices
That profit us.—What ho! Slave, Caliban!    315
Thou earth, thou, speak!
CALIBAN (*within*)        There's wood enough within.
PROSPERO
Come forth, I say! There's other business for thee.
Come, thou tortoise! When?
     *Enter Ariel, like a water-nymph*
Fine apparition! My quaint Ariel,
Hark in thine ear.
     *He whispers*
ARIEL         My lord, it shall be done.    *Exit*
PROSPERO
Thou poisonous slave, got by the devil himself    321
Upon thy wicked dam, come forth!
     *Enter Caliban*
CALIBAN
As wicked dew as e'er my mother brushed
With raven's feather from unwholesome fen
Drop on you both! A southwest blow on ye,    325
And blister you all o'er!
PROSPERO
For this be sure tonight thou shalt have cramps,
Side-stitches that shall pen thy breath up. Urchins
Shall forth at vast of night, that they may work
All exercise on thee. Thou shalt be pinched    330
As thick as honeycomb, each pinch more stinging
Than bees that made 'em.
CALIBAN         I must eat my dinner.
This island's mine, by Sycorax my mother,
Which thou tak'st from me. When thou cam'st first,

Thou strok'st me and made much of me, wouldst give me
Water with berries in't, and teach me how          336
To name the bigger light, and how the less,
That burn by day and night; and then I loved thee,
And showed thee all the qualities o'th' isle,
The fresh springs, brine-pits, barren place and fertile—
Cursed be I that did so! All the charms            341
Of Sycorax, toads, beetles, bats, light on you;
For I am all the subjects that you have,
Which first was mine own king, and here you sty me
In this hard rock, whiles you do keep from me      345
The rest o'th' island.
PROSPERO                    Thou most lying slave,
Whom stripes may move, not kindness! I have used
    thee,
Filth as thou art, with human care, and lodged thee
In mine own cell, till thou didst seek to violate
The honour of my child.                            350
CALIBAN
O ho, O ho! Would't had been done!
Thou didst prevent me; I had peopled else
This isle with Calibans.
MIRANDA                    Abhorrèd slave,
Which any print of goodness wilt not take,
Being capable of all ill! I pitied thee,           355
Took pains to make thee speak, taught thee each hour
One thing or other. When thou didst not, savage,
Know thine own meaning, but wouldst gabble like
A thing most brutish, I endowed thy purposes
With words that made them known. But thy vile race,
Though thou didst learn, had that in't which good
    natures                                        361
Could not abide to be with; therefore wast thou
Deservedly confined into this rock,
Who hadst deserved more than a prison.
CALIBAN
You taught me language, and my profit on't         365
Is I know how to curse. The red plague rid you
For learning me your language!
PROSPERO                    Hag-seed, hence!
Fetch us in fuel. And be quick, thou'rt best,
To answer other business.—Shrug'st thou, malice?
If thou neglect'st or dost unwillingly              370
What I command, I'll rack thee with old cramps,
Fill all thy bones with aches, make thee roar,
That beasts shall tremble at thy din.
CALIBAN                          No, pray thee.
(Aside) I must obey. His art is of such power
It would control my dam's god Setebos,             375
And make a vassal of him.
PROSPERO              So, slave, hence!
                                    Exit Caliban
Enter Ariel ⌈like a water-nymph⌉, playing and
singing, invisible to Ferdinand, who follows.
⌈Prospero and Miranda stand aside⌉
                    Song
ARIEL     Come unto these yellow sands,
              And then take hands;
          Curtsied when you have and kissed—
              The wild waves whist—           380
          Foot it featly here and there,
              And, sweet sprites, bear
          The burden. Hark, hark.
⌈SPIRITS⌉ (dispersedly within)
                    Bow-wow!
⌈ARIEL⌉   The watch-dogs bark.                 385

⌈SPIRITS⌉ (within) Bow-wow!
ARIEL        Hark, hark, I hear
             The strain of strutting Chanticleer
                Cry 'cock-a-diddle-dow'.

FERDINAND
Where should this music be? I'th' air or th'earth?  390
It sounds no more; and sure it waits upon
Some god o'th' island. Sitting on a bank,
Weeping again the King my father's wreck,
This music crept by me upon the waters,
Allaying both their fury and my passion            395
With its sweet air. Thence I have followed it—
Or it hath drawn me rather. But 'tis gone.
No, it begins again.

                    Song

ARIEL      Full fathom five thy father lies.
               Of his bones are coral made;    400
           Those are pearls that were his eyes;
               Nothing of him that doth fade
           But doth suffer a sea-change
           Into something rich and strange.
               Sea-nymphs hourly ring his knell:  405
⌈SPIRITS⌉ (within) Ding dong.
ARIEL      Hark, now I hear them.
⌈SPIRITS⌉ (within)              Ding-dong bell. ⌈etc.⌉
FERDINAND
The ditty does remember my drowned father.
This is no mortal business, nor no sound
That the earth owes.
          ⌈Music⌉
                         I hear it now above me.   410
PROSPERO (to Miranda)
The fringèd curtains of thine eye advance,
And say what thou seest yon.
MIRANDA                    What is't? A spirit?
Lord, how it looks about! Believe me, sir,
It carries a brave form. But 'tis a spirit.
PROSPERO
No, wench, it eats and sleeps, and hath such senses
As we have, such. This gallant which thou seest    416
Was in the wreck, and but he's something stained
With grief, that's beauty's canker, thou mightst call
    him
A goodly person. He hath lost his fellows,
And strays about to find 'em.
MIRANDA                    I might call him        420
A thing divine, for nothing natural
I ever saw so noble.
PROSPERO (aside)      It goes on, I see,
As my soul prompts it. (To Ariel) Spirit, fine spirit, I'll
free thee
Within two days for this.
FERDINAND ⌈aside⌉          Most sure the goddess
On whom these airs attend. (To Miranda) Vouchsafe
    my prayer                                      425
May know if you remain upon this island,
And that you will some good instruction give
How I may bear me here. My prime request,
Which I do last pronounce, is—O you wonder—
If you be maid or no?
MIRANDA              No wonder, sir,               430
But certainly a maid.
FERDINAND            My language! Heavens!
I am the best of them that speak this speech,

Were I but where 'tis spoken.
PROSPERO                          How, the best?
What wert thou if the King of Naples heard thee?
FERDINAND
A single thing, as I am now that wonders          435
To hear thee speak of Naples. He does hear me,
And that he does I weep. Myself am Naples,
Who with mine eyes, never since at ebb, beheld
The King my father wrecked.
MIRANDA                          Alack, for mercy!
FERDINAND
Yes, faith, and all his lords, the Duke of Milan   440
And his brave son being twain.
PROSPERO (aside)                 The Duke of Milan
And his more braver daughter could control thee,
If now 'twere fit to do't. At the first sight
They have changed eyes.—Delicate Ariel,
I'll set thee free for this. (To Ferdinand) A word, good
      sir.                                         445
I fear you have done yourself some wrong. A word.
MIRANDA (aside)
Why speaks my father so ungently? This
Is the third man that e'er I saw, the first
That e'er I sighed for. Pity move my father
To be inclined my way.
FERDINAND              O, if a virgin,             450
And your affection not gone forth, I'll make you
The Queen of Naples.
PROSPERO              Soft, sir! One word more.
(Aside) They are both in either's powers. But this swift
      business
I must uneasy make, lest too light winning
Make the prize light. (To Ferdinand) One word more. I
      charge thee                                  455
That thou attend me. Thou dost here usurp
The name thou ow'st not; and hast put thyself
Upon this island as a spy, to win it
From me the lord on't.
FERDINAND              No, as I am a man.
MIRANDA
There's nothing ill can dwell in such a temple.    460
If the ill spirit have so fair a house,
Good things will strive to dwell with't.
PROSPERO (to Ferdinand)              Follow me.
(To Miranda) Speak not you for him; he's a traitor.
      (To Ferdinand) Come!
I'll manacle thy neck and feet together.
Sea-water shalt thou drink; thy food shall be     465
The fresh-brook mussels, withered roots, and husks
Wherein the acorn cradled. Follow!
FERDINAND              No.
I will resist such entertainment till
Mine enemy has more power.
      He draws, and is charmed from moving
MIRANDA              O dear father,
Make not too rash a trial of him, for              470
He's gentle, and not fearful.
PROSPERO              What, I say,
My foot my tutor? Put thy sword up, traitor,
Who mak'st a show but dar'st not strike, thy
      conscience
Is so possessed with guilt. Come from thy ward,
For I can here disarm thee with this stick         475
And make thy weapon drop.
MIRANDA              Beseech you, father!

PROSPERO
Hence! Hang not on my garments.
MIRANDA              Sir, have pity.
I'll be his surety.
PROSPERO              Silence! One word more
Shall make me chide thee, if not hate thee. What,
An advocate for an impostor? Hush!                 480
Thou think'st there is no more such shapes as he,
Having seen but him and Caliban. Foolish wench!
To th' most of men this is a Caliban,
And they to him are angels.
MIRANDA              My affections
Are then most humble. I have no ambition           485
To see a goodlier man.
PROSPERO (to Ferdinand) Come on; obey.
Thy nerves are in their infancy again,
And have no vigour in them.
FERDINAND              So they are.
My spirits, as in a dream, are all bound up.
My father's loss, the weakness which I feel,       490
The wreck of all my friends, nor this man's threats
To whom I am subdued, are but light to me,
Might I but through my prison once a day
Behold this maid. All corners else o'th' earth
Let liberty make use of; space enough              495
Have I in such a prison.
PROSPERO (aside)        It works. (To Ferdinand) Come on.—
Thou hast done well, fine Ariel. (To Ferdinand) Follow
      me.
(To Ariel) Hark what thou else shalt do me.
MIRANDA (to Ferdinand)              Be of comfort.
My father's of a better nature, sir,
Than he appears by speech. This is unwonted        500
Which now came from him.
PROSPERO (to Ariel)        Thou shalt be as free
As mountain winds; but then exactly do
All points of my command.
ARIEL To th' syllable.
PROSPERO (to Ferdinand)
Come, follow. (To Miranda) Speak not for him.   Exeunt

2.1    Enter Alonso, Sebastian, Antonio, Gonzalo, Adrian,
       and Francisco
GONZALO (to Alonso)
Beseech you, sir, be merry. You have cause,
So have we all, of joy; for our escape
Is much beyond our loss. Our hint of woe
Is common; every day some sailor's wife,
The masters of some merchant, and the merchant,  5
Have just our theme of woe. But for the miracle,
I mean our preservation, few in millions
Can speak like us. Then wisely, good sir, weigh
Our sorrow with our comfort.
ALONSO              Prithee, peace.
SEBASTIAN (to Antonio) He receives comfort like cold
      porridge.                                    11
ANTONIO The visitor will not give him o'er so.
SEBASTIAN Look, he's winding up the watch of his wit.
By and by it will strike.
GONZALO (to Alonso) Sir—                           15
SEBASTIAN (to Antonio) One: tell.
GONZALO (to Alonso)
When every grief is entertained that's offered,
Comes to th'entertainer—
SEBASTIAN A dollar.

GONZALO Dolour comes to him indeed. You have spoken
truer than you purposed.                                    21
SEBASTIAN You have taken it wiselier than I meant you
should.
GONZALO (to Alonso) Therefore my lord—
ANTONIO (to Sebastian) Fie, what a spendthrift is he of his
tongue!                                                     26
ALONSO (to Gonzalo) I prithee, spare.
GONZALO Well, I have done. But yet—
SEBASTIAN (to Antonio) He will be talking.
ANTONIO Which of he or Adrian, for a good wager, first
begins to crow?                                             31
SEBASTIAN The old cock.
ANTONIO The cockerel.
SEBASTIAN Done. The wager?
ANTONIO A laughter.                                         35
SEBASTIAN A match!
ADRIAN (to Gonzalo) Though this island seem to be desert—
⌜ANTONIO⌝ (to Sebastian) Ha, ha, ha!
⌜SEBASTIAN⌝ So, you're paid.
ADRIAN Uninhabitable, and almost inaccessible—             40
SEBASTIAN (to Antonio) Yet—
ADRIAN Yet—
ANTONIO (to Sebastian) He could not miss't.
ADRIAN It must needs be of subtle, tender, and delicate
temperance.                                                 45
ANTONIO (to Sebastian) Temperance was a delicate wench.
SEBASTIAN Ay, and a subtle, as he most learnedly
delivered.
ADRIAN (to Gonzalo) The air breathes upon us here most
sweetly.                                                    50
SEBASTIAN (to Antonio) As if it had lungs, and rotten ones.
ANTONIO Or as 'twere perfumed by a fen.
GONZALO (to Adrian) Here is everything advantageous to
life.
ANTONIO (to Sebastian) True, save means to live.           55
SEBASTIAN Of that there's none, or little.
GONZALO (to Adrian) How lush and lusty the grass looks!
How green!
ANTONIO The ground indeed is tawny.
SEBASTIAN With an eye of green in't.                        60
ANTONIO He misses not much.
SEBASTIAN No, he doth but mistake the truth totally.
GONZALO (to Adrian) But the rarity of it is, which is indeed
almost beyond credit—
SEBASTIAN (to Antonio) As many vouched rarities are.       65
GONZALO (to Adrian) That our garments being, as they
were, drenched in the sea, hold notwithstanding their
freshness and glosses, being rather new-dyed than
stained with salt water.
ANTONIO (to Sebastian) If but one of his pockets could
speak, would it not say he lies?                            71
SEBASTIAN Ay, or very falsely pocket up his report.
GONZALO (to Adrian) Methinks our garments are now as
fresh as when we put them on first in Afric, at the
marriage of the King's fair daughter Claribel to the
King of Tunis.                                              76
SEBASTIAN 'Twas a sweet marriage, and we prosper well
in our return.
ADRIAN Tunis was never graced before with such a
paragon to their queen.                                     80
GONZALO Not since widow Dido's time.
ANTONIO (to Sebastian) Widow? A pox o'that! How came
that 'widow' in? Widow Dido!
SEBASTIAN What if he had said 'widower Aeneas' too?
Good Lord, how you take it!                                 85

ADRIAN (to Gonzalo) 'Widow Dido' said you? You make
me study of that: she was of Carthage, not of Tunis.
GONZALO This Tunis, sir, was Carthage.
ADRIAN Carthage?
GONZALO I assure you, Carthage.                            90
ANTONIO (to Sebastian) His word is more than the
miraculous harp.
SEBASTIAN He hath raised the wall, and houses too.
ANTONIO What impossible matter will he make easy next?
SEBASTIAN I think he will carry this island home in his
pocket, and give it his son for an apple.                  96
ANTONIO And sowing the kernels of it in the sea, bring
forth more islands.
GONZALO (to Adrian) Ay.
ANTONIO (to Sebastian) Why, in good time.                  100
GONZALO (to Alonso) Sir, we were talking that our garments
seem now as fresh as when we were at Tunis, at the
marriage of your daughter, who is now queen.
ANTONIO And the rarest that e'er came there.
SEBASTIAN Bate, I beseech you, widow Dido.                 105
ANTONIO O, widow Dido? Ay, widow Dido.
GONZALO (to Alonso) Is not, sir, my doublet as fresh as the
first day I wore it? I mean in a sort.
ANTONIO (to Sebastian) That 'sort' was well fished for.
GONZALO (to Alonso) When I wore it at your daughter's
marriage.                                                   111
ALONSO
You cram these words into mine ears against
The stomach of my sense. Would I had never
Married my daughter there! For, coming thence,
My son is lost; and, in my rate, she too,                  115
Who is so far from Italy removed
I ne'er again shall see her. O thou mine heir
Of Naples and of Milan, what strange fish
Hath made his meal on thee?
FRANCISCO                     Sir, he may live.
I saw him beat the surges under him                        120
And ride upon their backs. He trod the water,
Whose enmity he flung aside, and breasted
The surge, most swoll'n, that met him. His bold head
'Bove the contentious waves he kept, and oared
Himself with his good arms in lusty stroke                 125
To th' shore, that o'er his wave-worn basis bowed,
As stooping to relieve him. I not doubt
He came alive to land.
ALONSO                     No, no; he's gone.
SEBASTIAN (to Alonso)
Sir, you may thank yourself for this great loss,
That would not bless our Europe with your daughter,
But rather loose her to an African,                        131
Where she, at least, is banished from your eye,
Who hath cause to wet the grief on't.
ALONSO                     Prithee, peace.
SEBASTIAN
You were kneeled to and importuned otherwise
By all of us, and the fair soul herself                    135
Weighed between loathness and obedience at
Which end o'th' beam should bow. We have lost your
son,
I fear, for ever. Milan and Naples have
More widows in them of this business' making
Than we bring men to comfort them. The fault's your
own.                                                        140
ALONSO
So is the dear'st o'th' loss.
GONZALO                     My lord Sebastian,
The truth you speak doth lack some gentleness

And time to speak it in. You rub the sore
When you should bring the plaster.
SEBASTIAN (*to Antonio*) Very well.                              145
ANTONIO And most chirurgeonly.
GONZALO (*to Alonso*)
  It is foul weather in us all, good sir,
  When you are cloudy.
SEBASTIAN (*to Antonio*)    Fowl weather?
ANTONIO                               Very foul.
GONZALO (*to Alonso*)
  Had I plantation of this isle, my lord—
ANTONIO (*to Sebastian*)
  He'd sow't with nettle-seed.
SEBASTIAN                Or docks, or mallows.    150
GONZALO
  And were the king on't, what would I do?
SEBASTIAN (*to Antonio*)
  Scape being drunk, for want of wine.
GONZALO
  I'th' commonwealth I would by contraries
  Execute all things. For no kind of traffic
  Would I admit, no name of magistrate;           155
  Letters should not be known; riches, poverty,
  And use of service, none; contract, succession,
  Bourn, bound of land, tilth, vineyard, none;
  No use of metal, corn, or wine, or oil;
  No occupation, all men idle, all;               160
  And women too—but innocent and pure;
  No sovereignty—
SEBASTIAN (*to Antonio*) Yet he would be king on't.
ANTONIO The latter end of his commonwealth forgets the
  beginning.
GONZALO (*to Alonso*)
  All things in common nature should produce      165
  Without sweat or endeavour. Treason, felony,
  Sword, pike, knife, gun, or need of any engine,
  Would I not have; but nature should bring forth
  Of it own kind all foison, all abundance,
  To feed my innocent people.                     170
SEBASTIAN (*to Antonio*) No marrying 'mong his subjects?
ANTONIO None, man, all idle: whores and knaves.
GONZALO (*to Alonso*)
  I would with such perfection govern, sir,
  T'excel the Golden Age.
SEBASTIAN                Save his majesty!
ANTONIO
  Long live Gonzalo!
GONZALO (*to Alonso*) And—do you mark me, sir?    175
ALONSO
  Prithee, no more. Thou dost talk nothing to me.
GONZALO I do well believe your highness, and did it to
  minister occasion to these gentlemen, who are of such
  sensible and nimble lungs that they always use to laugh
  at nothing.                                     180
ANTONIO 'Twas you we laughed at.
GONZALO Who, in this kind of merry fooling, am nothing
  to you. So you may continue, and laugh at nothing
  still.
ANTONIO What a blow was there given!              185
SEBASTIAN An it had not fallen flat-long.
GONZALO You are gentlemen of brave mettle. You would
  lift the moon out of her sphere, if she would continue
  in it five weeks without changing.
    *Enter Ariel, invisible, playing solemn music*
SEBASTIAN We would so, and then go a-bat-fowling.   190
ANTONIO (*to Gonzalo*) Nay, good my lord, be not angry.

GONZALO No, I warrant you, I will not adventure my
  discretion so weakly. Will you laugh me asleep? For I
  am very heavy.
ANTONIO Go sleep, and hear us.                    195
    *Gonzalo, Adrian, and Francisco sleep*
ALONSO
  What, all so soon asleep? I wish mine eyes
  Would, with themselves, shut up my thoughts.—I find
  They are inclined to do so.
SEBASTIAN                Please you, sir,
  Do not omit the heavy offer of it.
  It seldom visits sorrow; when it doth,          200
  It is a comforter.
ANTONIO            We two, my lord,
  Will guard your person while you take your rest,
  And watch your safety.
ALONSO        Thank you. Wondrous heavy.
    *He sleeps. ⌜Exit Ariel⌝*
SEBASTIAN
  What a strange drowsiness possesses them!
ANTONIO
  It is the quality o'th' climate.
SEBASTIAN                    Why                   205
  Doth it not then our eyelids sink? I find
  Not myself disposed to sleep.
ANTONIO            Nor I; my spirits are nimble.
  They fell together all, as by consent;
  They dropped as by a thunderstroke. What might,
  Worthy Sebastian, O, what might—? No more!—     210
  And yet methinks I see it in thy face.
  What thou shouldst be th'occasion speaks thee, and
  My strong imagination sees a crown
  Dropping upon thy head.
SEBASTIAN            What, art thou waking?
ANTONIO
  Do you not hear me speak?
SEBASTIAN            I do, and surely             215
  It is a sleepy language, and thou speak'st
  Out of thy sleep. What is it thou didst say?
  This is a strange repose, to be asleep
  With eyes wide open; standing, speaking, moving,
  And yet so fast asleep.
ANTONIO            Noble Sebastian,               220
  Thou letst thy fortune sleep, die rather; wink'st
  Whiles thou art waking.
SEBASTIAN            Thou dost snore distinctly;
  There's meaning in thy snores.
ANTONIO
  I am more serious than my custom. You
  Must be so too if heed me, which to do           225
  Trebles thee o'er.
SEBASTIAN            Well, I am standing water.
ANTONIO
  I'll teach you how to flow.
SEBASTIAN            Do so; to ebb
  Hereditary sloth instructs me.
ANTONIO                O,
  If you but knew how you the purpose cherish
  Whiles thus you mock it; how in stripping it     230
  You more invest it! Ebbing men, indeed,
  Most often do so near the bottom run
  By their own fear or sloth.
SEBASTIAN            Prithee, say on.
  The setting of thine eye and cheek proclaim
  A matter from thee, and a birth, indeed,         235

Which throes thee much to yield.
ANTONIO                                    Thus, sir.
Although this lord of weak remembrance, this,
Who shall be of as little memory
When he is earthed, hath here almost persuaded—
For he's a spirit of persuasion, only          240
Professes to persuade—the King his son's alive,
'Tis as impossible that he's undrowned
As he that sleeps here swims.
SEBASTIAN                          I have no hope
That he's undrowned.
ANTONIO                    O, out of that 'no hope'
What great hope have you! No hope that way is   245
Another way so high a hope that even
Ambition cannot pierce a wink beyond,
But doubt discovery there. Will you grant with me
That Ferdinand is drowned?
SEBASTIAN                          He's gone.
ANTONIO                              Then tell me,
Who's the next heir of Naples?
SEBASTIAN                          Claribel.          250
ANTONIO
She that is Queen of Tunis; she that dwells
Ten leagues beyond man's life; she that from Naples
Can have no note—unless the sun were post—
The man i'th' moon's too slow—till new-born chins
Be rough and razorable; she that from whom    255
We all were sea-swallowed, though some cast again—
And by that destiny, to perform an act
Whereof what's past is prologue, what to come
In yours and my discharge.
SEBASTIAN            What stuff is this? How say you?
'Tis true my brother's daughter's Queen of Tunis;  260
So is she heir of Naples; 'twixt which regions
There is some space.
ANTONIO            A space whose every cubit
Seems to cry out 'How shall that Claribel
Measure us back to Naples? Keep in Tunis,
And let Sebastian wake.' Say this were death    265
That now hath seized them; why, they were no worse
Than now they are. There be that can rule Naples
As well as he that sleeps, lords that can prate
As amply and unnecessarily
As this Gonzalo; I myself could make            270
A chough of as deep chat. O, that you bore
The mind that I do, what a sleep were this
For your advancement! Do you understand me?
SEBASTIAN
Methinks I do.
ANTONIO          And how does your content
Tender your own good fortune?
SEBASTIAN                    I remember          275
You did supplant your brother Prospero.
ANTONIO                              True;
And look how well my garments sit upon me,
Much feater than before. My brother's servants
Were then my fellows; now they are my men.
SEBASTIAN But for your conscience.              280
ANTONIO
Ay, sir, where lies that? If 'twere a kibe
'Twould put me to my slipper; but I feel not
This deity in my bosom. Twenty consciences
That stand 'twixt me and Milan, candied be they,
And melt ere they molest. Here lies your brother,  285
No better than the earth he lies upon
If he were that which now he's like—that's dead;
Whom I with this obedient steel, three inches of it,

Can lay to bed for ever; whiles you, doing thus,
To the perpetual wink for aye might put         290
This ancient morsel, this Sir Prudence, who
Should not upbraid our course. For all the rest,
They'll take suggestion as a cat laps milk;
They'll tell the clock to any business that
We say befits the hour.
SEBASTIAN            Thy case, dear friend,      295
Shall be my precedent. As thou got'st Milan,
I'll come by Naples. Draw thy sword. One stroke
Shall free thee from the tribute which thou payest,
And I the King shall love thee.
ANTONIO                    Draw together,
And when I rear my hand, do you the like         300
To fall it on Gonzalo.
        They draw
SEBASTIAN            O, but one word.
        Enter Ariel, invisible, with music
ARIEL (to Gonzalo)
My master through his art foresees the danger
That you his friend are in—and sends me forth,
For else his project dies, to keep them living.
        He sings in Gonzalo's ear
            While you here do snoring lie,       305
        Open-eyed conspiracy
            His time doth take.
        If of life you keep a care,
        Shake off slumber, and beware.
            Awake, awake!                         310
ANTONIO (to Sebastian)
Then let us both be sudden.
GONZALO (awaking)              Now good angels
Preserve the King!
ALONSO (awaking)
Why, how now? Ho, awake!
        The others awake
(To Antonio and Sebastian)    Why are you drawn?
(To Gonzalo) Wherefore this ghastly looking?
GONZALO                          What's the matter?
SEBASTIAN
Whiles we stood here securing your repose,       315
Even now we heard a hollow burst of bellowing,
Like bulls, or rather lions. Did't not wake you?
It struck mine ear most terribly.
ALONSO                        I heard nothing.
ANTONIO
O, 'twas a din to fright a monster's ear,
To make an earthquake! Sure it was the roar      320
Of a whole herd of lions.
ALONSO                Heard you this, Gonzalo?
GONZALO
Upon mine honour, sir, I heard a humming,
And that a strange one too, which did awake me.
I shaked you, sir, and cried. As mine eyes opened
I saw their weapons drawn. There was a noise,    325
That's verily. 'Tis best we stand upon our guard,
Or that we quit this place. Let's draw our weapons.
ALONSO
Lead off this ground, and let's make further search
For my poor son.
GONZALO            Heavens keep him from these beasts!
For he is sure i'th' island.
ALONSO                    Lead away.             330
        Exeunt all but Ariel
ARIEL
Prospero my lord shall know what I have done.
So, King, go safely on to seek thy son.          Exit

1177

**2.2**    *Enter Caliban, wearing a gaberdine, and with a*
       *burden of wood*
CALIBAN ⌈*throwing down his burden*⌉
   All the infections that the sun sucks up
   From bogs, fens, flats, on Prosper fall, and make him
   By inch-meal a disease!
       ⌈*A noise of thunder heard*⌉
                 His spirits hear me,
   And yet I needs must curse. But they'll nor pinch,
   Fright me with urchin-shows, pitch me i'th' mire,    5
   Nor lead me like a fire-brand in the dark
   Out of my way, unless he bid 'em. But
   For every trifle are they set upon me;
   Sometime like apes, that mow and chatter at me
   And after bite me; then like hedgehogs, which    10
   Lie tumbling in my barefoot way and mount
   Their pricks at my footfall; sometime am I
   All wound with adders, who with cloven tongues
   Do hiss me into madness.
       *Enter Trinculo*
                 Lo now, lo!
   Here comes a spirit of his, and to torment me    15
   For bringing wood in slowly. I'll fall flat.
   Perchance he will not mind me.
       *He lies down*
TRINCULO Here's neither bush nor shrub to bear off any
   weather at all, and another storm brewing. I hear it
   sing i'th' wind. Yon same black cloud, yon huge one,
   looks like a foul bombard that would shed his liquor.
   If it should thunder as it did before, I know not where
   to hide my head. Yon same cloud cannot choose but
   fall by pailfuls. (*Seeing Caliban*) What have we here, a
   man or a fish? Dead or alive?—A fish, he smells like
   a fish; a very ancient and fish-like smell; a kind of not-
   of-the-newest poor-john. A strange fish! Were I in
   England now, as once I was, and had but this fish
   painted, not a holiday-fool there but would give a piece
   of silver. There would this monster make a man. Any
   strange beast there makes a man. When they will not
   give a doit to relieve a lame beggar, they will lay out
   ten to see a dead Indian. Legged like a man, and his
   fins like arms! Warm, o'my troth! I do now let loose
   my opinion, hold it no longer. This is no fish, but an
   islander that hath lately suffered by a thunderbolt.   36
       ⌈*Thunder*⌉
   Alas, the storm is come again. My best way is to creep
   under his gaberdine; there is no other shelter hereabout.
   Misery acquaints a man with strange bedfellows. I will
   here shroud till the dregs of the storm be past.    40
       *He hides under Caliban's gaberdine.*
       *Enter Stefano, singing, with a wooden bottle in his*
       *hand*
STEFANO      I shall no more to sea, to sea,
            Here shall I die ashore—
   This is a very scurvy tune to sing at a man's funeral.
   Well, here's my comfort.
       *He drinks, then sings*
   The master, the swabber, the boatswain, and I,    45
      The gunner and his mate,
   Loved Mall, Meg, and Marian, and Margery,
     But none of us cared for Kate.
      For she had a tongue with a tang,
      Would cry to a sailor 'Go hang!'    50
   She loved not the savour of tar nor of pitch,
   Yet a tailor might scratch her where'er she did itch.
     Then to sea, boys, and let her go hang!
     Then to sea, *etc.*

This is a scurvy tune, too. But here's my comfort.   55
       *He drinks*
CALIBAN (*to* Stefano) Do not torment me! O!
STEFANO What's the matter? Have we devils here? Do
   you put tricks upon's with savages and men of Ind,
   ha? I have not scaped drowning to be afeard now of
   your four legs. For it hath been said: 'As proper a man
   as ever went on four legs cannot make him give
   ground.' And it shall be said so again, while Stefano
   breathes at' nostrils.
CALIBAN The spirit torments me. O!    64
STEFANO This is some monster of the isle with four legs,
   who hath got, as I take it, an ague. Where the devil
   should he learn our language? I will give him some
   relief, if it be but for that. If I can recover him and
   keep him tame and get to Naples with him, he's a
   present for any emperor that ever trod on neat's leather.
CALIBAN (*to* Trinculo) Do not torment me, prithee! I'll
   bring my wood home faster.    72
STEFANO He's in his fit now, and does not talk after the
   wisest. He shall taste of my bottle. If he have never
   drunk wine afore, it will go near to remove his fit. If I
   can recover him and keep him tame, I will not take
   too much for him. He shall pay for him that hath him,
   and that soundly.    78
CALIBAN (*to* Trinculo) Thou dost me yet but little hurt.
   Thou wilt anon, I know it by thy trembling. Now
   Prosper works upon thee.
STEFANO Come on your ways. Open your mouth. Here is
   that which will give language to you, cat. Open your
   mouth. This will shake your shaking, I can tell you,
   and that soundly. You cannot tell who's your friend.
   Open your chaps again.    86
       *Caliban drinks*
TRINCULO I should know that voice. It should be—but he
   is drowned, and these are devils. O, defend me!
STEFANO Four legs and two voices—a most delicate
   monster! His forward voice now is to speak well of his
   friend; his backward voice is to utter foul speeches and
   to detract. If all the wine in my bottle will recover him,
   I will help his ague. Come.
       *Caliban drinks*
   Amen. I will pour some in thy other mouth.
TRINCULO Stefano!    95
STEFANO Doth thy other mouth call me? Mercy, mercy!
   This is a devil, and no monster. I will leave him. I have
   no long spoon.
TRINCULO Stefano! If thou beest Stefano, touch me and
   speak to me, for I am Trinculo. Be not afeard. Thy good
   friend Trinculo.    101
STEFANO If thou beest Trinculo, come forth. I'll pull thee by
   the lesser legs. If any be Trinculo's legs, these are they.
       *He pulls out Trinculo by the legs*
   Thou art very Trinculo indeed! How cam'st thou to be
   the siege of this moon-calf? Can he vent Trinculos?
TRINCULO (*rising*) I took him to be killed with a
   thunderstroke. But art thou not drowned, Stefano? I
   hope now thou art not drowned. Is the storm
   overblown? I hid me under the dead moon-calf's
   gaberdine for fear of the storm. And art thou living,
   Stefano? O Stefano, two Neapolitans scaped!    111
       ⌈*He dances Stefano round*⌉
STEFANO Prithee, do not turn me about. My stomach is
   not constant.
CALIBAN
   These be fine things, an if they be not spirits.

That's a brave god, and bears celestial liquor.    115
I will kneel to him.
⌈*He kneels*⌉
STEFANO (*to Trinculo*) How didst thou scape? How cam'st
thou hither? Swear by this bottle how thou cam'st
hither. I escaped upon a butt of sack which the sailors
heaved o'erboard, by this bottle—which I made of the
bark of a tree with mine own hands since I was cast
ashore.                                            122
CALIBAN I'll swear upon that bottle to be thy true subject,
for the liquor is not earthly.
STEFANO (*offering Trinculo the bottle*) Here. Swear then how
thou escapedst.                                    126
TRINCULO Swum ashore, man, like a duck. I can swim
like a duck, I'll be sworn.
STEFANO Here, kiss the book.
       *Trinculo drinks*
Though thou canst swim like a duck, thou art made
like a goose.                                      131
TRINCULO O Stefano, hast any more of this?
STEFANO The whole butt, man. My cellar is in a rock by
th' seaside, where my wine is hid.
       ⌈*Caliban rises*⌉
How now, moon-calf? How does thine ague?           135
CALIBAN Hast thou not dropped from heaven?
STEFANO Out o'th' moon, I do assure thee. I was the man
i'th' moon when time was.
CALIBAN
I have seen thee in her, and I do adore thee.      139
My mistress showed me thee, and thy dog and thy bush.
STEFANO Come, swear to that. Kiss the book. I will furnish
it anon with new contents. Swear.
       *Caliban drinks*
TRINCULO By this good light, this is a very shallow
monster! I afeard of him? A very weak monster! The
man i'th' moon? A most poor, credulous monster! Well
drawn, monster, in good sooth!                     146
CALIBAN (*to Stefano*)
I'll show thee every fertile inch o'th' island,
And I will kiss thy foot. I prithee, be my god.
TRINCULO By this light, a most perfidious and drunken
monster! When's god's asleep, he'll rob his bottle.
CALIBAN (*to Stefano*)
I'll kiss thy foot. I'll swear myself thy subject. 151
STEFANO Come on then; down, and swear.
       ⌈*Caliban kneels*⌉
TRINCULO I shall laugh myself to death at this puppy-
headed monster. A most scurvy monster! I could find
in my heart to beat him—                           155
STEFANO (*to Caliban*) Come, kiss.
       ⌈*Caliban kisses his foot*⌉
TRINCULO But that the poor monster's in drink. An
abominable monster!
CALIBAN
I'll show thee the best springs; I'll pluck thee berries;
I'll fish for thee, and get thee wood enough.      160
A plague upon the tyrant that I serve!
I'll bear him no more sticks, but follow thee,
Thou wondrous man.
TRINCULO A most ridiculous monster, to make a wonder
of a poor drunkard!                                165
CALIBAN (*to Stefano*)
I prithee, let me bring thee where crabs grow,
And I with my long nails will dig thee pig-nuts,
Show thee a jay's nest, and instruct thee how
To snare the nimble marmoset. I'll bring thee

To clust'ring filberts, and sometimes I'll get thee  170
Young seamews from the rock. Wilt thou go with
me?
STEFANO I prithee now, lead the way without any more
talking.—Trinculo, the King and all our company else
being drowned, we will inherit here.—Here, bear my
bottle.—Fellow Trinculo, we'll fill him by and by again.
CALIBAN (*sings drunkenly*) Farewell, master, farewell,
farewell!                                          177
TRINCULO A howling monster, a drunken monster!
CALIBAN (*sings*)
       No more dams I'll make for fish,
              Nor fetch in firing                   180
          At requiring,
       Nor scrape trenchering, nor wash dish.
          'Ban, 'ban, Cacaliban
              Has a new master.—Get a new man!
Freedom, high-day! High-day, freedom! Freedom, high-
day, freedom!                                      186
STEFANO O brave monster! Lead the way.    *Exeunt*

❀

**3.1**    *Enter Ferdinand, bearing a log*
FERDINAND
There be some sports are painful, and their labour
Delight in them sets off. Some kinds of baseness
Are nobly undergone, and most poor matters
Point to rich ends. This my mean task
Would be as heavy to me as odious, but                5
The mistress which I serve quickens what's dead,
And makes my labours pleasures. O, she is
Ten times more gentle than her father's crabbed,
And he's composed of harshness. I must remove
Some thousands of these logs and pile them up,       10
Upon a sore injunction. My sweet mistress
Weeps when she sees me work, and says such
       baseness
Had never like executor. I forget,
But these sweet thoughts do even refresh my labours,
Most busil'est when I do it.
       *Enter Miranda, and Prospero following at a distance*
MIRANDA                    Alas now, pray you        15
Work not so hard. I would the lightning had
Burnt up those logs that you are enjoined to pile.
Pray set it down, and rest you. When this burns
'Twill weep for having wearied you. My father
Is hard at study. Pray now, rest yourself.           20
He's safe for these three hours.
FERDINAND                    O most dear mistress,
The sun will set before I shall discharge
What I must strive to do.
MIRANDA                    If you'll sit down
I'll bear your logs the while. Pray give me that;
I'll carry it to the pile.
FERDINAND                    No, precious creature.  25
I had rather crack my sinews, break my back,
Than you should such dishonour undergo
While I sit lazy by.
MIRANDA              It would become me
As well as it does you; and I should do it
With much more ease, for my good will is to it,      30
And yours it is against.
PROSPERO (*aside*)        Poor worm, thou art infected.
This visitation shows it.
MIRANDA (*to Ferdinand*)   You look wearily.

**FERDINAND**
No, noble mistress, 'tis fresh morning with me
When you are by at night. I do beseech you,
Chiefly that I might set it in my prayers,          35
What is your name?

**MIRANDA**                    Miranda. O my father,
I have broke your hest to say so!

**FERDINAND**                    Admired Miranda!
Indeed the top of admiration, worth
What's dearest to the world. Full many a lady
I have eyed with best regard, and many a time      40
Th'harmony of their tongues hath into bondage
Brought my too diligent ear. For several virtues
Have I liked several women; never any
With so full soul but some defect in her
Did quarrel with the noblest grace she owed        45
And put it to the foil. But you, O you,
So perfect and so peerless, are created
Of every creature's best.

**MIRANDA**                    I do not know
One of my sex, no woman's face remember
Save from my glass mine own; nor have I seen       50
More that I may call men than you, good friend,
And my dear father. How features are abroad
I am skilless of; but, by my modesty,
The jewel in my dower, I would not wish
Any companion in the world but you;                55
Nor can imagination form a shape
Besides yourself to like of. But I prattle
Something too wildly, and my father's precepts
I therein do forget.

**FERDINAND**                    I am in my condition
A prince, Miranda, I do think a king—              60
I would not so—and would no more endure
This wooden slavery than to suffer
The flesh-fly blow my mouth. Hear my soul speak.
The very instant that I saw you did
My heart fly to your service; there resides        65
To make me slave to it. And for your sake
Am I this patient log-man.

**MIRANDA**                    Do you love me?

**FERDINAND**
O heaven, O earth, bear witness to this sound,
And crown what I profess with kind event
If I speak true! If hollowly, invert               70
What best is boded me to mischief! I,
Beyond all limit of what else i'th' world,
Do love, prize, honour you.

**MIRANDA** (*weeping*)          I am a fool
To weep at what I am glad of.

**PROSPERO** (*aside*)           Fair encounter
Of two most rare affections! Heavens rain grace    75
On that which breeds between 'em.

**FERDINAND** (*to Miranda*)          Wherefore weep you?

**MIRANDA**
At mine unworthiness, that dare not offer
What I desire to give, and much less take
What I shall die to want. But this is trifling,
And all the more it seeks to hide itself           80
The bigger bulk it shows. Hence, bashful cunning,
And prompt me, plain and holy innocence.
I am your wife, if you will marry me.
If not, I'll die your maid. To be your fellow
You may deny me, but I'll be your servant          85
Whether you will or no.

**FERDINAND** ⌜*kneeling*⌝          My mistress, dearest;
And I thus humble ever.

**MIRANDA** My husband then?

**FERDINAND** Ay, with a heart as willing
As bondage e'er of freedom. Here's my hand.        90

**MIRANDA**
And mine, with my heart in't. And now farewell
Till half an hour hence.

**FERDINAND**                    A thousand thousand.
                    *Exeunt severally Miranda and Ferdinand*

**PROSPERO**
So glad of this as they I cannot be,
Who are surprised with all; but my rejoicing
At nothing can be more. I'll to my book,           95
For yet ere supper-time must I perform
Much business appertaining.
                                        *Exit*

**3.2**    *Enter Caliban, Stefano, and Trinculo*

**STEFANO** (*to Caliban*) Tell not me. When the butt is out
we will drink water, not a drop before. Therefore bear
up and board 'em. Servant monster, drink to me.

**TRINCULO** Servant monster? The folly of this island! They
say there's but five upon this isle. We are three of
them; if th'other two be brained like us, the state
totters.                                            7

**STEFANO** Drink, servant monster, when I bid thee. Thy
eyes are almost set in thy head.

**TRINCULO** Where should they be set else? He were a brave
monster indeed if they were set in his tail.        11

**STEFANO** My man-monster hath drowned his tongue in
sack. For my part, the sea cannot drown me. I swam,
ere I could recover the shore, five and thirty leagues,
off and on. By this light, thou shalt be my lieutenant,
monster, or my standard.                            16

**TRINCULO** Your lieutenant if you list; he's no standard.

**STEFANO** We'll not run, Monsieur Monster.

**TRINCULO** Nor go neither; but you'll lie like dogs, and yet
say nothing neither.                                20

**STEFANO** Moon-calf, speak once in thy life, if thou beest
a good moon-calf.

**CALIBAN**
How does thy honour? Let me lick thy shoe.
I'll not serve him; he is not valiant.

**TRINCULO** Thou liest, most ignorant monster! I am in case
to jostle a constable. Why, thou debauched fish, thou,
was there ever man a coward that hath drunk so much
sack as I today? Wilt thou tell a monstrous lie, being
but half a fish and half a monster?                 29

**CALIBAN** (*to Stefano*) Lo, how he mocks me! Wilt thou let
him, my lord?

**TRINCULO** 'Lord' quoth he? That a monster should be
such a natural!

**CALIBAN** (*to Stefano*)
Lo, lo, again! Bite him to death, I prithee.        34

**STEFANO** Trinculo, keep a good tongue in your head. If
you prove a mutineer, the next tree. The poor monster's
my subject, and he shall not suffer indignity.

**CALIBAN**
I thank my noble lord. Wilt thou be pleased
To hearken once again to the suit I made to thee?   39

**STEFANO** Marry, will I. Kneel and repeat it. I will stand,
and so shall Trinculo.
                    ⌜*Caliban kneels.*⌝
                    *Enter Ariel, invisible*

**CALIBAN** As I told thee before, I am subject to a tyrant,
a sorcerer, that by his cunning hath cheated me of the
island.

ARIEL  Thou liest.                                                              45
CALIBAN (*to Trinculo*)
  Thou liest, thou jesting monkey, thou.
  I would my valiant master would destroy thee.
  I do not lie.
STEFANO  Trinculo, if you trouble him any more in's tale,
  by this hand, I will supplant some of your teeth.          50
TRINCULO  Why, I said nothing.
STEFANO  Mum, then, and no more. (*To Caliban*) Proceed.
CALIBAN
  I say by sorcery he got this isle;
  From me he got it. If thy greatness will
  Revenge it on him—for I know thou dar'st,                  55
  But this thing dare not—
STEFANO  That's most certain.
CALIBAN
  Thou shalt be lord of it, and I'll serve thee.
STEFANO  How now shall this be compassed? Canst thou
  bring me to the party?                                     60
CALIBAN
  Yea, yea, my lord. I'll yield him thee asleep
  Where thou mayst knock a nail into his head.
ARIEL  Thou liest, thou canst not.
CALIBAN
  What a pied ninny's this! (*To Trinculo*) Thou scurvy
    patch!
  (*To Stefano*) I do beseech thy greatness give him blows,
  And take his bottle from him. When that's gone             66
  He shall drink naught but brine, for I'll not show him
  Where the quick freshes are.
STEFANO  Trinculo, run into no further danger. Interrupt
  the monster one word further, and, by this hand, I'll
  turn my mercy out o'doors and make a stockfish of
  thee.                                                      72
TRINCULO  Why, what did I? I did nothing. I'll go farther
  off.
STEFANO  Didst thou not say he lied?                         75
ARIEL  Thou liest.
STEFANO  Do I so? (*Striking Trinculo*) Take thou that. As
  you like this, give me the lie another time.
TRINCULO  I did not give the lie. Out o'your wits and
  hearing too? A pox o'your bottle! This can sack and
  drinking do. A murrain on your monster, and the devil
  take your fingers.                                         82
CALIBAN  Ha, ha, ha!
STEFANO  Now forward with your tale. (*To Trinculo*)
  Prithee, stand further off.                                85
CALIBAN
  Beat him enough; after a little time
  I'll beat him too.
STEFANO (*to Trinculo*)
              Stand farther. (*To Caliban*) Come, proceed.
CALIBAN
  Why, as I told thee, 'tis a custom with him
  I'th' afternoon to sleep. There thou mayst brain him,
  Having first seized his books; or with a log              90
  Batter his skull, or paunch him with a stake,
  Or cut his weasand with thy knife. Remember
  First to possess his books, for without them
  He's but a sot as I am, nor hath not
  One spirit to command—they all do hate him                95
  As rootedly as I. Burn but his books.
  He has brave utensils, for so he calls them,
  Which when he has a house he'll deck withal.
  And that most deeply to consider is

The beauty of his daughter. He himself                      100
Calls her a nonpareil. I never saw a woman
But only Sycorax my dam and she,
But she as far surpasseth Sycorax
As great'st does least.
STEFANO              Is it so brave a lass?
CALIBAN
  Ay, lord. She will become thy bed, I warrant,             105
  And bring thee forth brave brood.
STEFANO  Monster, I will kill this man. His daughter and
  I will be king and queen—save our graces!—and
  Trinculo and thyself shall be viceroys. Dost thou like
  the plot, Trinculo?                                        110
TRINCULO  Excellent.
STEFANO  Give me thy hand. I am sorry I beat thee. But
  while thou liv'st, keep a good tongue in thy head.
CALIBAN
  Within this half hour will he be asleep.
  Wilt thou destroy him then?                                115
STEFANO  Ay, on mine honour.
ARIEL (*aside*) This will I tell my master.
CALIBAN
  Thou mak'st me merry; I am full of pleasure.
  Let us be jocund. Will you troll the catch
  You taught me but while-ere?                               120
STEFANO  At thy request, monster, I will do reason, any
  reason.—Come on, Trinculo, let us sing.
  (*Sings*)        Flout 'em and cout 'em,
                   And scout 'em and flout 'em.
                        Thought is free.                     125
CALIBAN  That's not the tune.
       *Ariel plays the tune on a tabor and pipe*
STEFANO  What is this same?
TRINCULO  This is the tune of our catch, played by the
  picture of Nobody.                                         129
STEFANO (*calls towards Ariel*) If thou beest a man, show
  thyself in thy likeness. If thou beest a devil, take't as
  thou list.
TRINCULO  O, forgive me my sins!
STEFANO  He that dies pays all debts. (*Calls*) I defy thee.—
  Mercy upon us!                                             135
CALIBAN  Art thou afeard?
STEFANO  No, monster, not I.
CALIBAN
  Be not afeard. The isle is full of noises,
  Sounds, and sweet airs, that give delight and hurt
    not.
  Sometimes a thousand twangling instruments               140
  Will hum about mine ears, and sometime voices
  That if I then had waked after long sleep
  Will make me sleep again; and then in dreaming
  The clouds methought would open and show riches
  Ready to drop upon me, that when I waked                  145
  I cried to dream again.
STEFANO  This will prove a brave kingdom to me, where
  I shall have my music for nothing.
CALIBAN  When Prospero is destroyed.                         149
STEFANO  That shall be by and by. I remember the story.
                        *Exit Ariel, playing music*
TRINCULO  The sound is going away. Let's follow it, and
  after do our work.
STEFANO  Lead, monster; we'll follow.—I would I could
  see this taborer. He lays it on.
TRINCULO (*to Caliban*) Wilt come? I'll follow Stefano.     155
                                        *Exeunt*

**3.3**    *Enter Alonso, Sebastian, Antonio, Gonzalo, Adrian,*
       *and Francisco*

GONZALO (*to Alonso*)
By'r la'kin, I can go no further, sir.
My old bones ache. Here's a maze trod indeed
Through forthrights and meanders. By your patience,
I needs must rest me.

ALONSO            Old lord, I cannot blame thee,
Who am myself attached with weariness      5
To th' dulling of my spirits. Sit down and rest.
Even here I will put off my hope, and keep it
No longer for my flatterer. He is drowned
Whom thus we stray to find, and the sea mocks
Our frustrate search on land. Well, let him go.    10
     ⌐They sit⌐

ANTONIO (*aside to Sebastian*)
I am right glad that he's so out of hope.
Do not for one repulse forgo the purpose
That you resolved t'effect.

SEBASTIAN (*aside to Antonio*) The next advantage
Will we take throughly.

ANTONIO (*aside to Sebastian*) Let it be tonight,
For now they are oppressed with travel. They    15
Will not nor cannot use such vigilance
As when they are fresh.

SEBASTIAN (*aside to Antonio*) I say tonight. No more.
     *Solemn and strange music. Enter Prospero on the*
       *top, invisible*

ALONSO
What harmony is this? My good friends, hark.

GONZALO Marvellous sweet music.
     *Enter spirits, in several strange shapes, bringing in*
     *a table and a banquet, and dance about it with*
     *gentle actions of salutations, and, inviting the King*
     *and his companions to eat, they depart*

ALONSO
Give us kind keepers, heavens! What were these?    20

SEBASTIAN
A living drollery. Now I will believe
That there are unicorns; that in Arabia
There is one tree, the phoenix' throne, one phoenix
At this hour reigning there.

ANTONIO          I'll believe both;
And what does else want credit come to me,    25
And I'll be sworn 'tis true. Travellers ne'er did lie,
Though fools at home condemn 'em.

GONZALO          If in Naples
I should report this now, would they believe me—
If I should say I saw such islanders?
For certes these are people of the island,    30
Who though they are of monstrous shape, yet note
Their manners are more gentle-kind than of
Our human generation you shall find
Many, nay, almost any.

PROSPERO (*aside*)        Honest lord,
Thou hast said well, for some of you there present    35
Are worse than devils.

ALONSO       I cannot too much muse.
Such shapes, such gesture, and such sound,
     expressing—
Although they want the use of tongue—a kind
Of excellent dumb discourse.

PROSPERO (*aside*)       Praise in departing.

FRANCISCO
They vanished strangely.

SEBASTIAN        No matter, since    40
They have left their viands behind, for we have
     stomachs.
Will't please you taste of what is here?

ALONSO                   Not I.

GONZALO
Faith, sir, you need not fear. When we were boys,
Who would believe that there were mountaineers
Dewlapped like bulls, whose throats had hanging at 'em
Wallets of flesh? Or that there were such men    46
Whose heads stood in their breasts? Which now we
     find
Each putter-out of five for one will bring us
Good warrant of.

ALONSO ⌐*rising*⌐      I will stand to and feed,
Although my last—no matter, since I feel    50
The best is past. Brother, my lord the Duke,
Stand to, and do as we.
     ⌐*Alonso, Sebastian, and Antonio approach the table.*⌐
     *Thunder and lightning. Ariel* ⌐*descends*⌐ *like a*
     *harpy, claps his wings upon the table, and, with a*
     *quaint device, the banquet vanishes*

ARIEL
You are three men of sin, whom destiny—
That hath to instrument this lower world
And what is in't—the never-surfeited sea    55
Hath caused to belch up you, and on this island
Where man doth not inhabit, you 'mongst men
Being most unfit to live. I have made you mad,
And even with suchlike valour men hang and drown
Their proper selves.
     *Alonso, Sebastian, and Antonio draw*
            You fools! I and my fellows    60
Are ministers of fate. The elements
Of whom your swords are tempered may as well
Wound the loud winds, or with bemocked-at stabs
Kill the still-closing waters, as diminish
One dowl that's in my plume. My fellow ministers    65
Are like invulnerable. If you could hurt,
Your swords are now too massy for your strengths
And will not be uplifted.
     *Alonso, Sebastian, and Antonio stand amazed*
            But remember,
For that's my business to you, that you three
From Milan did supplant good Prospero;    70
Exposed unto the sea, which hath requit it,
Him and his innocent child; for which foul deed,
The powers, delaying not forgetting, have
Incensed the seas and shores, yea, all the creatures,
Against your peace. Thee of thy son, Alonso,    75
They have bereft, and do pronounce by me
Ling'ring perdition—worse than any death
Can be at once—shall step by step attend
You and your ways; whose wraths to guard you
     from—
Which here in this most desolate isle else falls    80
Upon your heads—is nothing but heart's sorrow
And a clear life ensuing.
     *He* ⌐*ascends and*⌐ *vanishes in thunder. Then, to soft*
     *music, enter the spirits again, and dance with mocks*
     *and mows, and they depart, carrying out the table*

PROSPERO
Bravely the figure of this harpy hast thou
Performed, my Ariel; a grace it had devouring.
Of my instruction hast thou nothing bated    85
In what thou hadst to say. So with good life
And observation strange my meaner ministers

Their several kinds have done. My high charms work,
And these mine enemies are all knit up
In their distractions. They now are in my power;      90
And in these fits I leave them, while I visit
Young Ferdinand, whom they suppose is drowned,
And his and mine loved darling.                    *Exit*
⌈*Gonzalo, Adrian, and Francisco go towards the others*⌉
GONZALO
I'th' name of something holy, sir, why stand you
In this strange stare?
ALONSO                    O, it is monstrous, monstrous!   95
Methought the billows spoke and told me of it,
The winds did sing it to me, and the thunder,
That deep and dreadful organ-pipe, pronounced
The name of Prosper. It did bass my trespass.
Therefor my son i'th' ooze is bedded, and          100
I'll seek him deeper than e'er plummet sounded,
And with him there lie mudded.                     *Exit*
SEBASTIAN                    But one fiend at a time,
I'll fight their legions o'er.
ANTONIO                    I'll be thy second.
                              *Exeunt Sebastian and Antonio*
GONZALO
All three of them are desperate. Their great guilt,
Like poison given to work a great time after,      105
Now 'gins to bite the spirits. I do beseech you
That are of suppler joints, follow them swiftly,
And hinder them from what this ecstasy
May now provoke them to.
ADRIAN                    Follow, I pray you.    *Exeunt*

**4.1**    *Enter Prospero, Ferdinand, and Miranda*
PROSPERO (*to Ferdinand*)
If I have too austerely punished you,
Your compensation makes amends, for I
Have given you here a third of mine own life—
Or that for which I live—who once again
I tender to thy hand. All thy vexations              5
Were but my trials of thy love, and thou
Hast strangely stood the test. Here, afore heaven,
I ratify this my rich gift. O Ferdinand,
Do not smile at me that I boast of her,
For thou shalt find she will outstrip all praise,    10
And make it halt behind her.
FERDINAND                    I do believe it
Against an oracle.
PROSPERO
Then, as my gift and thine own acquisition
Worthily purchased, take my daughter. But
If thou dost break her virgin-knot before            15
All sanctimonious ceremonies may
With full and holy rite be ministered,
No sweet aspersion shall the heavens let fall
To make this contract grow; but barren hate,
Sour-eyed disdain, and discord, shall bestrew        20
The union of your bed with weeds so loathly
That you shall hate it both. Therefore take heed,
As Hymen's lamps shall light you.
FERDINAND                    As I hope
For quiet days, fair issue, and long life
With such love as 'tis now, the murkiest den,        25
The most opportune place, the strong'st suggestion
Our worser genius can, shall never melt
Mine honour into lust to take away
The edge of that day's celebration;

When I shall think or Phoebus' steeds are foundered  30
Or night kept chained below.
PROSPERO                    Fairly spoke.
Sit, then, and talk with her. She is thine own.
        *Ferdinand and Miranda sit and talk together*
What, Ariel, my industrious servant Ariel!
        *Enter Ariel*
ARIEL
What would my potent master? Here I am.
PROSPERO
Thou and thy meaner fellows your last service        35
Did worthily perform, and I must use you
In such another trick. Go bring the rabble,
O'er whom I give thee power, here to this place.
Incite them to quick motion, for I must
Bestow upon the eyes of this young couple            40
Some vanity of mine art. It is my promise,
And they expect it from me.
ARIEL                    Presently?
PROSPERO Ay, with a twink.
ARIEL                    Before you can say 'Come' and 'Go',
        And breathe twice, and cry 'So, so',         45
        Each one tripping on his toe
        Will be here with mop and mow.
        Do you love me, master? No?
PROSPERO
Dearly, my delicate Ariel. Do not approach
Till thou dost hear me call.
ARIEL                    Well; I conceive.    *Exit*
PROSPERO (*to Ferdinand*)
Look thou be true. Do not give dalliance            51
Too much the rein. The strongest oaths are straw
To th' fire i'th' blood. Be more abstemious,
Or else, good night your vow.
FERDINAND                    I warrant you, sir,
The white cold virgin snow upon my heart            55
Abates the ardour of my liver.
PROSPERO                    Well.—
Now come, my Ariel! Bring a corollary
Rather than want a spirit. Appear, and pertly.
        *Soft music*
(*To Ferdinand and Miranda*) No tongue, all eyes! Be silent.
        *Enter Iris*
IRIS
Ceres, most bounteous lady, thy rich leas           60
Of wheat, rye, barley, vetches, oats, and peas;
Thy turfy mountains where live nibbling sheep,
And flat meads thatched with stover, them to keep;
Thy banks with peonied and twillèd brims
Which spongy April at thy hest betrims              65
To make cold nymphs chaste crowns; and thy broom-
        groves,
Whose shadow the dismissèd bachelor loves,
Being lass-lorn; thy pole-clipped vineyard,
And thy sea-marge, sterile and rocky-hard,
Where thou thyself dost air: the Queen o'th' Sky,    70
Whose wat'ry arch and messenger am I,
Bids thee leave these, and with her sovereign grace
        *Juno* ⌈*appears in the air*⌉
Here on this grass-plot, in this very place,
To come and sport.—Her peacocks fly amain.
Approach, rich Ceres, her to entertain.              75
        *Enter* ⌈*Ariel as*⌉ *Ceres*
CERES
Hail, many-coloured messenger, that ne'er
Dost disobey the wife of Jupiter;

Who with thy saffron wings upon my flowers
Diffusest honey-drops, refreshing showers,
And with each end of thy blue bow dost crown      80
My bosky acres and my unshrubbed down,
Rich scarf to my proud earth. Why hath thy queen
Summoned me hither to this short-grassed green?

IRIS
A contract of true love to celebrate,
And some donation freely to estate      85
On the blest lovers.

CERES                          Tell me, heavenly bow,
If Venus or her son, as thou dost know,
Do now attend the Queen. Since they did plot
The means that dusky Dis my daughter got,
Her and her blind boy's scandalled company      90
I have forsworn.

IRIS                    Of her society
Be not afraid. I met her deity
Cutting the clouds towards Paphos, and her son
Dove-drawn with her. Here thought they to have
        done
Some wanton charm upon this man and maid,      95
Whose vows are that no bed-right shall be paid
Till Hymen's torch be lighted—but in vain.
Mars's hot minion is returned again.
Her waspish-headed son has broke his arrows,
Swears he will shoot no more, but play with
        sparrows,      100
And be a boy right out.
        *Music. Juno descends to the stage*
CERES                          Highest queen of state,
Great Juno, comes; I know her by her gait.

JUNO
How does my bounteous sister? Go with me
To bless this twain, that they may prosperous be,
And honoured in their issue.      105
        *Ceres joins Juno, and they sing*

JUNO        Honour, riches, marriage-blessing,
        Long continuance and increasing,
        Hourly joys be still upon you!
        Juno sings her blessings on you.
*CERES*     Earth's increase, and foison plenty,      110
        Barns and garners never empty,
        Vines with clust'ring bunches growing,
        Plants with goodly burden bowing;
        Spring come to you at the farthest,
        In the very end of harvest.      115
        Scarcity and want shall shun you,
        Ceres' blessing so is on you.

FERDINAND
This is a most majestic vision, and
Harmonious charmingly. May I be bold
To think these spirits?

PROSPERO                          Spirits, which by mine art      120
I have from their confines called to enact
My present fancies.

FERDINAND                    Let me live here ever!
So rare a wondered father and a wise
Makes this place paradise.
        *Juno and Ceres whisper, and send Iris on
        employment*

PROSPERO                    Sweet now, silence.
Juno and Ceres whisper seriously.      125
There's something else to do. Hush, and be mute,
Or else our spell is marred.

IRIS
You nymphs called naiads of the wind'ring brooks,
With your sedged crowns and ever-harmless looks,
Leave your crisp channels, and on this green land      130
Answer your summons; Juno does command.
Come, temperate nymphs, and help to celebrate
A contract of true love. Be not too late.
        *Enter certain nymphs*
You sunburned sicklemen, of August weary,
Come hither from the furrow and be merry;      135
Make holiday, your rye-straw hats put on,
And these fresh nymphs encounter every one
In country footing.
        *Enter certain reapers, properly habited. They join
        with the nymphs in a graceful dance; towards the
        end whereof Prospero starts suddenly, and speaks*

PROSPERO *(aside)*
I had forgot that foul conspiracy
Of the beast Caliban and his confederates      140
Against my life. The minute of their plot
Is almost come. *(To the spirits)* Well done! Avoid; no
        more!
        *To a strange, hollow, and confused noise, the spirits
        in the pageant heavily vanish.*
        *⌈Ferdinand and Miranda rise⌉*
FERDINAND *(to Miranda)*
This is strange. Your father's in some passion
That works him strongly.

MIRANDA                          Never till this day
Saw I him touched with anger so distempered.      145

PROSPERO
You do look, my son, in a moved sort,
As if you were dismayed. Be cheerful, sir.
Our revels now are ended. These our actors,
As I foretold you, were all spirits, and
Are melted into air, into thin air;      150
And like the baseless fabric of this vision,
The cloud-capped towers, the gorgeous palaces,
The solemn temples, the great globe itself,
Yea, all which it inherit, shall dissolve;
And, like this insubstantial pageant faded,      155
Leave not a rack behind. We are such stuff
As dreams are made on, and our little life
Is rounded with a sleep. Sir, I am vexed.
Bear with my weakness. My old brain is troubled.
Be not disturbed with my infirmity.      160
If you be pleased, retire into my cell,
And there repose. A turn or two I'll walk
To still my beating mind.

FERDINAND *and* MIRANDA          We wish your peace.
        *Exeunt Ferdinand and Miranda*

PROSPERO
Come with a thought! I thank thee, Ariel. Come!
        *Enter Ariel*
ARIEL
Thy thoughts I cleave to. What's thy pleasure?
PROSPERO                                                        Spirit,
We must prepare to meet with Caliban.      166
ARIEL
Ay, my commander. When I presented Ceres
I thought to have told thee of it, but I feared
Lest I might anger thee.
PROSPERO
Say again: where didst thou leave these varlets?      170
ARIEL
I told you, sir, they were red-hot with drinking;

So full of valour that they smote the air
For breathing in their faces, beat the ground
For kissing of their feet; yet always bending
Towards their project. Then I beat my tabor,                175
At which like unbacked colts they pricked their ears,
Advanced their eyelids, lifted up their noses
As they smelt music. So I charmed their ears
That calf-like they my lowing followed, through
Toothed briars, sharp furzes, pricking gorse, and
    thorns,                                                 180
Which entered their frail shins. At last I left them
I'th' filthy-mantled pool beyond your cell,
There dancing up to th' chins, that the foul lake
O'er-stunk their feet.

PROSPERO                    This was well done, my bird.
Thy shape invisible retain thou still.                      185
The trumpery in my house, go bring it hither
For stale to catch these thieves.

ARIEL                                      I go, I go.      *Exit*

PROSPERO
A devil, a born devil, on whose nature
Nurture can never stick; on whom my pains,
Humanely taken, all, all lost, quite lost,                  190
And, as with age his body uglier grows,
So his mind cankers. I will plague them all,
Even to roaring.
        *Enter Ariel, laden with glistening apparel, etc.*
                    Come, hang them on this lime.
        *Ariel hangs up the apparel. ⌈Exeunt Prospero and
        Ariel.⌉*
        *Enter Caliban, Stefano, and Trinculo, all wet*

CALIBAN
Pray you, tread softly, that the blind mole may
Not hear a foot fall. We now are near his cell.             195

STEFANO Monster, your fairy, which you say is a harmless
fairy, has done little better than played the Jack with
us.

TRINCULO Monster, I do smell all horse-piss, at which my
nose is in great indignation.                               200

STEFANO So is mine. Do you hear, monster? If I should
take a displeasure against you, look you—

TRINCULO Thou wert but a lost monster.

CALIBAN
Good my lord, give me thy favour still.
Be patient, for the prize I'll bring thee to                205
Shall hoodwink this mischance. Therefore speak softly.
All's hushed as midnight yet.

TRINCULO Ay, but to lose our bottles in the pool!

STEFANO There is not only disgrace and dishonour in that,
monster, but an infinite loss.                              210

TRINCULO That's more to me than my wetting. Yet this is
your harmless fairy, monster.

STEFANO I will fetch off my bottle, though I be o'er ears
for my labour.

CALIBAN
Prithee, my king, be quiet. Seest thou here;               215
This is the mouth o'th' cell. No noise, and enter.
Do that good mischief which may make this island
Thine own for ever, and I thy Caliban
For aye thy foot-licker.

STEFANO                       Give me thy hand.
I do begin to have bloody thoughts.                         220

TRINCULO (*seeing the apparel*) O King Stefano, O peer! O
worthy Stefano, look what a wardrobe here is for thee!

CALIBAN
Let it alone, thou fool, it is but trash.

TRINCULO (*putting on a gown*) O ho, monster, we know
what belongs to a frippery! O King Stefano!                 225

STEFANO Put off that gown, Trinculo. By this hand, I'll
have that gown.

TRINCULO Thy grace shall have it.

CALIBAN
The dropsy drown this fool! What do you mean
To dote thus on such luggage? Let't alone,                  230
And do the murder first. If he awake,
From toe to crown he'll fill our skins with pinches,
Make us strange stuff.

STEFANO Be you quiet, monster.—Mistress lime, is not
this my jerkin? Now is the jerkin under the line. Now,
jerkin, you are like to lose your hair and prove a bald
jerkin.
        *Stefano and Trinculo take garments*

TRINCULO Do, do! We steal by line and level, an't like
your grace.                                                 239

STEFANO I thank thee for that jest. Here's a garment for't.
Wit shall not go unrewarded while I am king of this
country. 'Steal by line and level' is an excellent pass
of pate. There's another garment for't.

TRINCULO Monster, come, put some lime upon your
fingers, and away with the rest.                            245

CALIBAN
I will have none on't. We shall lose our time,
And all be turned to barnacles, or to apes
With foreheads villainous low.

STEFANO Monster, lay to your fingers. Help to bear this
away where my hogshead of wine is, or I'll turn you
out of my kingdom. Go to, carry this.                       251

TRINCULO And this.

STEFANO Ay, and this.
        *They load Caliban with apparel.
        A noise of hunters heard. Enter divers spirits in
        shape of dogs and hounds, hunting them about;
        Prospero and Ariel setting them on*

PROSPERO
Hey, Mountain, hey!

ARIEL                      Silver! There it goes, Silver!

PROSPERO
Fury, Fury! There, Tyrant, there! Hark, hark!              255
        *Exeunt Stefano, Trinculo, and Caliban, pursued
                                                by spirits*
(*To Ariel*) Go, charge my goblins that they grind their
    joints
With dry convulsions, shorten up their sinews
With agèd cramps, and more pinch-spotted make
    them
Than pard or cat o'mountain.
        *Cries within*

ARIEL                                Hark, they roar!

PROSPERO
Let them be hunted soundly. At this hour                   260
Lies at my mercy all mine enemies.
Shortly shall all my labours end, and thou
Shalt have the air at freedom. For a little,
Follow, and do me service.                          *Exeunt*

                            ❀

5.1    *Enter Prospero, in his magic robes, and Ariel*

PROSPERO
Now does my project gather to a head.
My charms crack not, my spirits obey, and time
Goes upright with his carriage. How's the day?

ARIEL

On the sixth hour; at which time, my lord,
You said our work should cease.

PROSPERO　　　　　　　　I did say so　　5
When first I raised the tempest. Say, my spirit,
How fares the King and's followers?

ARIEL　　　　　　　　Confined together
In the same fashion as you gave in charge,
Just as you left them; all prisoners, sir,
In the lime-grove which weather-fends your cell.　　10
They cannot budge till your release. The King,
His brother, and yours, abide all three distracted,
And the remainder mourning over them,
Brimful of sorrow and dismay; but chiefly
Him that you termed, sir, the good old lord Gonzalo:
His tears run down his beard like winter's drops　　16
From eaves of reeds. Your charm so strongly works 'em
That if you now beheld them your affections
Would become tender.

PROSPERO　　　　　　　　Dost thou think so, spirit?

ARIEL

Mine would, sir, were I human.

PROSPERO　　　　　　　　And mine shall.　　20
Hast thou, which art but air, a touch, a feeling
Of their afflictions, and shall not myself,
One of their kind, that relish all as sharply
Passion as they, be kindlier moved than thou art?
Though with their high wrongs I am struck to th'
　　quick,　　25
Yet with my nobler reason 'gainst my fury
Do I take part. The rarer action is
In virtue than in vengeance. They being penitent,
The sole drift of my purpose doth extend
Not a frown further. Go release them, Ariel,　　30
My charms I'll break, their senses I'll restore,
And they shall be themselves.

ARIEL　　　　　　　　I'll fetch them, sir.　　Exit
⌈Prospero draws a circle with his staff⌉

PROSPERO

Ye elves of hills, brooks, standing lakes and groves,
And ye that on the sands with printless foot
Do chase the ebbing Neptune, and do fly him　　35
When he comes back; you demi-puppets that
By moonshine do the green sour ringlets make
Whereof the ewe not bites; and you whose pastime
Is to make midnight mushrooms, that rejoice
To hear the solemn curfew; by whose aid,　　40
Weak masters though ye be, I have bedimmed
The noontide sun, called forth the mutinous winds,
And 'twixt the green sea and the azured vault
Set roaring war—to the dread rattling thunder
Have I given fire, and rifted Jove's stout oak　　45
With his own bolt; the strong-based promontory
Have I made shake, and by the spurs plucked up
The pine and cedar; graves at my command
Have waked their sleepers, oped, and let 'em forth
By my so potent art. But this rough magic　　50
I here abjure. And when I have required
Some heavenly music—which even now I do—
To work mine end upon their senses that
This airy charm is for, I'll break my staff,
Bury it certain fathoms in the earth,　　55
And deeper than did ever plummet sound
I'll drown my book.

　　　Solemn music. Here enters first Ariel, invisible;
　　　then Alonso, with a frantic gesture, attended by

Gonzalo; Sebastian and Antonio, in like manner,
attended by Adrian and Francisco. They all enter
the circle which Prospero had made, and there stand
charmed; which Prospero observing, speaks
(To Alonso) A solemn air, and the best comforter
To an unsettled fancy, cure thy brains,
Now useless, boiled within thy skull.
(To Sebastian and Antonio)　　　　　　　There stand,　　60
For you are spell-stopped.—
Holy Gonzalo, honourable man,
Mine eyes, ev'n sociable to the show of thine,
Fall fellowly drops. (Aside) The charm dissolves apace,
And as the morning steals upon the night,　　65
Melting the darkness, so their rising senses
Begin to chase the ignorant fumes that mantle
Their clearer reason.—O good Gonzalo,
My true preserver, and a loyal sir
To him thou follow'st, I will pay thy graces　　70
Home both in word and deed.—Most cruelly
Didst thou, Alonso, use me and my daughter.
Thy brother was a furtherer in the act.—
Thou art pinched for't now, Sebastian.
(To Antonio)　　　　　　　Flesh and blood,
You, brother mine, that entertained ambition,　　75
Expelled remorse and nature, whom, with Sebastian—
Whose inward pinches therefore are most strong,—
Would here have killed your king, I do forgive thee,
Unnatural though thou art. (Aside) Their understanding
Begins to swell, and the approaching tide　　80
Will shortly fill the reasonable shores
That now lie foul and muddy. Not one of them
That yet looks on me, or would know me.—Ariel,
Fetch me the hat and rapier in my cell.
I will discase me, and myself present　　85
As I was sometime Milan. Quickly, spirit!
Thou shalt ere long be free.

　　　Ariel sings and helps to attire him as Duke of Milan

ARIEL　　Where the bee sucks, there suck I:
In a cowslip's bell I lie;
There I couch when owls do cry.　　90
On the bat's back I do fly
After summer merrily.
Merrily, merrily shall I live now
Under the blossom that hangs on the bough.
Merrily, merrily shall I live now　　95
Under the blossom that hangs on the bough.

PROSPERO

Why, that's my dainty Ariel! I shall miss thee,
But yet thou shalt have freedom.—So, so, so.—
To the King's ship, invisible as thou art!
There shalt thou find the mariners asleep　　100
Under the hatches. The Master and the Boatswain
Being awake, enforce them to this place,
And presently, I prithee.

ARIEL

I drink the air before me, and return
Or ere your pulse twice beat.　　　　　　Exit

GONZALO

All torment, trouble, wonder, and amazement　　106
Inhabits here. Some heavenly power guide us
Out of this fearful country!

PROSPERO　　　　　　　　Behold, sir King,
The wrongèd Duke of Milan, Prospero.
For more assurance that a living prince　　110
Does now speak to thee, I embrace thy body;

And to thee and thy company I bid
A hearty welcome.
    *He embraces Alonso*
ALONSO           Whe'er thou beest he or no,
Or some enchanted trifle to abuse me,
As late I have been, I not know. Thy pulse    115
Beats as of flesh and blood; and since I saw thee
Th'affliction of my mind amends, with which
I fear a madness held me. This must crave—
An if this be at all—a most strange story.
Thy dukedom I resign, and do entreat    120
Thou pardon me my wrongs. But how should
    Prospero
Be living and be here?
PROSPERO (*to Gonzalo*)    First, noble friend,
Let me embrace thine age, whose honour cannot
Be measured or confined.
    *He embraces Gonzalo*
GONZALO           Whether this be
Or be not, I'll not swear.
PROSPERO         You do yet taste    125
Some subtleties o'th' isle that will not let you
Believe things certain.—Welcome, my friends all.
(*Aside to Sebastian and Antonio*)
But you, my brace of lords, were I so minded,
I here could pluck his highness' frown upon you
And justify you traitors. At this time    130
I will tell no tales.
SEBASTIAN (*to Antonio*)    The devil speaks in him.
PROSPERO           No.
(*To Antonio*) For you, most wicked sir, whom to call
    brother
Would even infect my mouth, I do forgive
Thy rankest fault, all of them, and require
My dukedom of thee, which perforce I know    135
Thou must restore.
ALONSO        If thou beest Prospero,
Give us particulars of thy preservation,
How thou hast met us here, whom three hours since
Were wrecked upon this shore, where I have lost—
How sharp the point of this remembrance is!—    140
My dear son Ferdinand.
PROSPERO        I am woe for't, sir.
ALONSO
Irreparable is the loss, and patience
Says it is past her cure.
PROSPERO        I rather think
You have not sought her help, of whose soft grace
For the like loss I have her sovereign aid,    145
And rest myself content.
ALONSO        You the like loss?
PROSPERO
As great to me as late; and supportable
To make the dear loss have I means much weaker
Than you may call to comfort you, for I
Have lost my daughter.
ALONSO        A daughter?    150
O heavens, that they were living both in Naples,
The king and queen there! That they were, I wish
Myself were mudded in that oozy bed
Where my son lies. When did you lose your daughter?
PROSPERO
In this last tempest. I perceive these lords    155
At this encounter do so much admire
That they devour their reason, and scarce think

Their eyes do offices of truth, these words
Are natural breath. But howsoe'er you have
Been jostled from your senses, know for certain    160
That I am Prospero, and that very Duke
Which was thrust forth of Milan, who most strangely,
Upon this shore where you were wrecked, was landed
To be the lord on't. No more yet of this,
For 'tis a chronicle of day by day,    165
Not a relation for a breakfast, nor
Befitting this first meeting. Welcome, sir.
This cell's my court. Here have I few attendants,
And subjects none abroad. Pray you, look in.
My dukedom since you have given me again,    170
I will requite you with as good a thing;
At least bring forth a wonder to content ye
As much as me my dukedom.
    *Here Prospero discovers Ferdinand and Miranda,*
    *playing at chess*
MIRANDA
Sweet lord, you play me false.
FERDINAND    No, my dearest love,    175
I would not for the world.
MIRANDA
Yes, for a score of kingdoms you should wrangle,
An I would call it fair play.
ALONSO        If this prove
A vision of the island, one dear son
Shall I twice lose.
SEBASTIAN        A most high miracle.    180
FERDINAND (*coming forward*)
Though the seas threaten, they are merciful.
I have cursed them without cause.
    *He kneels*
ALONSO        Now all the blessings
Of a glad father compass thee about.
Arise and say how thou cam'st here.
    *Ferdinand rises*
MIRANDA (*coming forward*)    O wonder!
How many goodly creatures are there here!    185
How beauteous mankind is! O brave new world
That has such people in't!
PROSPERO        'Tis new to thee.
ALONSO (*to Ferdinand*)
What is this maid with whom thou wast at play?
Your eld'st acquaintance cannot be three hours.
Is she the goddess that hath severed us,    190
And brought us thus together?
FERDINAND        Sir, she is mortal;
But by immortal providence she's mine.
I chose her when I could not ask my father
For his advice, nor thought I had one. She
Is daughter to this famous Duke of Milan,    195
Of whom so often I have heard renown,
But never saw before; of whom I have
Received a second life; and second father
This lady makes him to me.
ALONSO        I am hers.
But O, how oddly will it sound, that I    200
Must ask my child forgiveness!
PROSPERO        There, sir, stop.
Let us not burden our remembrance with
A heaviness that's gone.
GONZALO        I have inly wept,
Or should have spoke ere this. Look down, you gods,
And on this couple drop a blessèd crown,    205

For it is you that have chalked forth the way
Which brought us hither.
ALONSO                          I say amen, Gonzalo.
GONZALO
Was Milan thrust from Milan, that his issue
Should become kings of Naples? O rejoice
Beyond a common joy! And set it down        210
With gold on lasting pillars: in one voyage
Did Claribel her husband find at Tunis,
And Ferdinand her brother found a wife
Where he himself was lost; Prospero his dukedom
In a poor isle; and all of us ourselves,      215
When no man was his own.
ALONSO (*to Ferdinand and Miranda*) Give me your hands.
Let grief and sorrow still embrace his heart
That doth not wish you joy.
GONZALO                          Be it so! Amen!
  *Enter Ariel, with the Master and Boatswain*
  *amazedly following*
O look, sir, look, sir, here is more of us!
I prophesied if a gallows were on land       220
This fellow could not drown. (*To the Boatswain*) Now,
  blasphemy,
That swear'st grace o'erboard: not an oath on shore?
Hast thou no mouth by land? What is the news?
BOATSWAIN
The best news is that we have safely found
Our King and company. The next, our ship,     225
Which but three glasses since we gave out split,
Is tight and yare and bravely rigged, as when
We first put out to sea.
ARIEL (*aside to Prospero*)   Sir, all this service
Have I done since I went.
PROSPERO (*aside to Ariel*)   My tricksy spirit!
ALONSO
These are not natural events; they strengthen   230
From strange to stranger. Say, how came you hither?
BOATSWAIN
If I did think, sir, I were well awake
I'd strive to tell you. We were dead of sleep,
And—how we know not—all clapped under hatches,
Where but even now, with strange and several noises
Of roaring, shrieking, howling, jingling chains,  236
And more diversity of sounds, all horrible,
We were awaked; straightway at liberty;
Where we in all her trim freshly beheld
Our royal, good, and gallant ship, our Master   240
Cap'ring to eye her. On a trice, so please you,
Even in a dream, were we divided from them,
And were brought moping hither.
ARIEL (*aside to Prospero*)        Was't well done?
PROSPERO (*aside to Ariel*)
Bravely, my diligence. Thou shalt be free.
ALONSO
This is as strange a maze as e'er men trod,     245
And there is in this business more than nature
Was ever conduct of. Some oracle
Must rectify our knowledge.
PROSPERO                          Sir, my liege,
Do not infest your mind with beating on
The strangeness of this business. At picked leisure, 250
Which shall be shortly, single I'll resolve you,
Which to you shall seem probable, of every
These happened accidents; till when be cheerful,
And think of each thing well. (*Aside to Ariel*) Come
  hither, spirit.

Set Caliban and his companions free.           255
Untie the spell.                          *Exit Ariel*
(*To Alonso*)    How fares my gracious sir?
There are yet missing of your company
Some few odd lads that you remember not.
  *Enter Ariel, driving in Caliban, Stefano, and*
  *Trinculo, in their stolen apparel*
STEFANO Every man shift for all the rest, and let no man
take care for himself, for all is but fortune. Coragio,
bully-monster, coragio!                      261
TRINCULO If these be true spies which I wear in my head,
here's a goodly sight.
CALIBAN
O Setebos, these be brave spirits indeed!
How fine my master is! I am afraid           265
He will chastise me.
SEBASTIAN
Ha, ha! What things are these, my lord Antonio?
Will money buy 'em?
ANTONIO                          Very like; one of them
Is a plain fish, and no doubt marketable.
PROSPERO
Mark but the badges of these men, my lords,     270
Then say if they be true. This misshapen knave,
His mother was a witch, and one so strong
That could control the moon, make flows and ebbs,
And deal in her command without her power.
These three have robbed me, and this demi-devil, 275
For he's a bastard one, had plotted with them
To take my life. Two of these fellows you
Must know and own. This thing of darkness I
Acknowledge mine.
CALIBAN                   I shall be pinched to death.
ALONSO
Is not this Stefano, my drunken butler?         280
SEBASTIAN
He is drunk now. Where had he wine?
ALONSO
And Trinculo is reeling ripe. Where should they
Find this grand liquor that hath gilded 'em?
(*To Trinculo*) How cam'st thou in this pickle?  284
TRINCULO I have been in such a pickle since I saw you
last that, I fear me, will never out of my bones. I shall
not fear fly-blowing.
SEBASTIAN Why, how now, Stefano?
STEFANO O, touch me not! I am not Stefano, but a cramp.
PROSPERO You'd be king o'the isle, sirrah?       290
STEFANO I should have been a sore one, then.
ALONSO (*pointing to Caliban*) This is a strange thing as e'er
I looked on.
PROSPERO
He is as disproportioned in his manners
As in his shape. (*To Caliban*) Go, sirrah, to my cell. 295
Take with you your companions. As you look
To have my pardon, trim it handsomely.
CALIBAN
Ay, that I will; and I'll be wise hereafter,
And seek for grace. What a thrice-double ass
Was I to take this drunkard for a god,          300
And worship this dull fool!
PROSPERO                          Go to, away!  *Exit Caliban*
ALONSO (*to Stefano and Trinculo*)
Hence, and bestow your luggage where you found it.
SEBASTIAN Or stole it, rather.
                          *Exeunt Stefano and Trinculo*

PROSPERO (*to Alonso*)
  Sir, I invite your highness and your train
  To my poor cell, where you shall take your rest       305
  For this one night; which part of it I'll waste
  With such discourse as I not doubt shall make it
  Go quick away: the story of my life,
  And the particular accidents gone by
  Since I came to this isle. And in the morn             310
  I'll bring you to your ship, and so to Naples,
  Where I have hope to see the nuptial
  Of these our dear-belovèd solemnized;
  And thence retire me to my Milan, where
  Every third thought shall be my grave.
ALONSO                                    I long          315
  To hear the story of your life, which must
  Take the ear strangely.
PROSPERO                      I'll deliver all,
  And promise you calm seas, auspicious gales,
  And sail so expeditious that shall catch
  Your royal fleet far off. (*Aside to Ariel*) My Ariel, chick,
  That is thy charge. Then to the elements               321
  Be free, and fare thou well.                *Exit Ariel*
                    Please you, draw near.
                    *Exeunt ⌈all but Prospero⌉*

**Epilogue**
PROSPERO
  Now my charms are all o'erthrown,
  And what strength I have's mine own,
  Which is most faint. Now 'tis true
  I must be here confined by you
  Or sent to Naples. Let me not,                          5
  Since I have my dukedom got,
  And pardoned the deceiver, dwell
  In this bare island by your spell;
  But release me from my bands
  With the help of your good hands.                       10
  Gentle breath of yours my sails
  Must fill, or else my project fails,
  Which was to please. Now I want
  Spirits to enforce, art to enchant;
  And my ending is despair                                15
  Unless I be relieved by prayer,
  Which pierces so, that it assaults
  Mercy itself, and frees all faults.
  As you from crimes would pardoned be,
  Let your indulgence set me free.                        20
                    *He awaits applause, then exit*

# CARDENIO

## *A BRIEF ACCOUNT*

MANY plays acted in Shakespeare's time have failed to survive; they may easily include some that he wrote. The mystery of *Love's Labour's Won* is discussed elsewhere (p. 349). Certain manuscript records of the seventeenth century suggest that at least one other play in which he had a hand may have disappeared. On 9 September 1653 the London publisher Humphrey Moseley entered in the Stationers' Register a batch of plays including 'The History of Cardenio, by Mr Fletcher and Shakespeare'. Cardenio is a character in Part One of Cervantes' *Don Quixote*, published in English translation in 1612. Two earlier allusions suggest that the King's Men owned a play on this subject at the time that Shakespeare was collaborating with John Fletcher (1579-1625). On 20 May 1613 the Privy Council authorized payment of £20 to John Heminges, as leader of the King's Men, for the presentation at court of six plays, one listed as 'Cardenno'. On 9 July of the same year Heminges received £6 13s. 4d. for his company's performance of a play 'called Cardenna' before the ambassador of the Duke of Savoy.

No more information about this play survives from the seventeenth century, but in 1728 Lewis Theobald published a play based on the story of Cardenio and called *Double Falsehood, or The Distrest Lovers*, which he claimed to have 'revised and adapted' from one 'written originally by W. Shakespeare'. It had been successfully produced at Drury Lane on 13 December 1727, and was given thirteen times up to 1 May 1728. Other performances are recorded in 1740, 1741, 1767 (when it was reprinted), 1770, and 1847. In 1770 a newspaper stated that 'the original manuscript' was 'treasured up in the Museum of Covent Garden Playhouse'; fire destroyed the theatre, including its library, in 1808.

Theobald claimed to own several manuscripts of an original play by Shakespeare, and remarked that some of his contemporaries thought the style was Fletcher's, not Shakespeare's. When he himself came to edit Shakespeare's plays he did not include either *Double Falsehood* or the play on which he claimed to have based it; he simply edited the plays of the First Folio, not adding either *Pericles* or *The Two Noble Kinsmen*, though he believed they were partly by Shakespeare. It is quite possible that *Double Falsehood* is based (however distantly) on a play of Shakespeare's time; if so, the play is likely to have been the one performed by the King's Men and ascribed by Moseley in 1653 to Fletcher and Shakespeare.

*Double Falsehood* is a tragicomedy; the characters' names differ from those in *Don Quixote*, and the story is varied. Henriquez rapes Violante, then falls in love with Leonora, loved by his friend Julio. Her parents agree to the marriage, but Julio interrupts the ceremony. Leonora (who had intended to kill herself) swoons and later takes sanctuary in a nunnery. Julio goes mad with desire for vengeance on his false friend; and the wronged Violante, disguised as a boy, joins a group of shepherds, and is almost raped by one of them. Henriquez's virtuous brother, Roderick, ignorant of his villainy, helps him to abduct Leonora. Leonora and Violante both denounce Henriquez to Roderick. Finally Henriquez repents and marries Violante, while Julio (now sane) marries Leonora.

Some of the motifs of *Double Falsehood*, such as the disguised heroine wronged by her lover and, particularly, the reuniting and reconciliation of parents with children, recall Shakespeare's late plays. But most of the dialogue seems un-Shakespearian. Though the play deserved its limited success, it is now no more than an interesting curiosity.

# ALL IS TRUE

## (HENRY VIII)

### BY WILLIAM SHAKESPEARE AND JOHN FLETCHER

---

ON 29 June 1613 the firing of cannon at the Globe Theatre ignited its thatch and burned it to the ground. According to a letter of 4 July the house was full of spectators who had come to see 'a new play called *All is True*, which had been acted not passing two or three times before'. No one was hurt 'except one man who was scalded with the fire by adventuring in to save a child which otherwise had been burnt'. This establishes the play's date with unusual precision. Though two other accounts of the fire refer to a play 'of'—which may mean simply 'about'—Henry VIII, yet another two unequivocally call it *All is True*; and these words also end the refrain of a ballad about the fire. When the play came to be printed as the last of the English history plays—all named after kings—in the 1623 Folio it was as *The Famous History of the Life of King Henry the Eighth*. We restore the title by which it was known to its first audiences.

No surviving account of the fire says who wrote the play that caused it. In 1850, James Spedding (prompted by Tennyson) suggested that Shakespeare collaborated on it with John Fletcher (1579-1625). We have external evidence that the two dramatists worked together in or around 1613 on the lost *Cardenio* and on *The Two Noble Kinsmen*. For their collaboration in *All is True* the evidence is wholly internal, stemming from the initial perception of two distinct verse styles within the play; later, more rigorous examination of evidence provided by both the play's language and its dramatic technique has convinced most scholars of Fletcher's hand in it. The passages most confidently attributed to Shakespeare are Act 1, Scenes 1 and 2; Act 2, Scenes 3 and 4; Act 3, Scene 2 to line 204; and Act 5, Scene 1.

The historical material derives, often closely, from the chronicles of Raphael Holinshed and Edward Hall, supplemented by John Foxe's *Book of Martyrs* (1563, etc.) for the Cranmer episodes in Act 5. It covers only part of Henry's reign, from the opening description of the Field of the Cloth of Gold, of 1520, to the christening of Princess Elizabeth, in 1533. It depicts the increasing abuse of power by Cardinal Wolsey; the execution, brought about by Wolsey's machinations, of the Duke of Buckingham; the King's abandonment of his Queen, Katherine of Aragon; the rise to the King's favour of Anne Boleyn; Wolsey's disgrace; and the birth to Henry and Anne of a daughter instead of the hoped-for son.

Sir Henry Wotton, writing of the fire, said that the play represented 'some principal pieces of the reign of Henry 8, which was set forth with many extraordinary circumstances of pomp and majesty'. It has continued popular in performance for the opportunities that it affords for spectacle and for the dramatic power of certain episodes such as Buckingham's speeches before execution (2.1), Queen Katherine's defence of the validity of her marriage (2.4), Wolsey's farewell to his greatness (3.2), and Katherine's dying scene (4.2). Though the play depicts a series of falls from greatness, it works towards the birth of the future Elizabeth I, fulsomely celebrated in the last scene (not attributed to Shakespeare) along with her successor, the patron of the King's Men.

# THE PERSONS OF THE PLAY

PROLOGUE

KING HENRY the Eighth

Duke of BUCKINGHAM

Lord ABERGAVENNY  ⎫
Earl of SURREY  ⎬ his sons-in-law
Duke of NORFOLK

Duke of SUFFOLK

LORD CHAMBERLAIN

LORD CHANCELLOR

Lord SANDS (also called Sir William Sands)

Sir Thomas LOVELL

Sir Anthony DENNY

Sir Henry GUILDFORD

CARDINAL WOLSEY

Two SECRETARIES

Buckingham's SURVEYOR

CARDINAL CAMPEIUS

GARDINER, the King's new secretary, later Bishop of Winchester

His PAGE

Thomas CROMWELL

CRANMER, Archbishop of Canterbury

QUEEN KATHERINE, later KATHERINE, Princess Dowager

GRIFFITH, her gentleman usher

PATIENCE, her waiting-woman

Other WOMEN

Six spirits, who dance before Katherine in a vision

A MESSENGER

Lord CAPUTIUS

ANNE Boleyn

An OLD LADY

BRANDON  ⎫
SERJEANT-at-arms  ⎬ who arrest Buckingham and Abergavenny

Sir Nicholas VAUX  ⎫
Tipstaves  ⎪
Halberdiers  ⎬ after Buckingham's arraignment
Common people  ⎭

Two vergers  ⎫
Two SCRIBES  ⎪
Archbishop of Canterbury  ⎪
Bishop of LINCOLN  ⎪
Bishop of Ely  ⎪
Bishop of Rochester  ⎬ appearing at the Legatine Court
Bishop of Saint Asaph  ⎪
Two priests  ⎪
Serjeant-at-arms  ⎪
Two noblemen  ⎪
A CRIER  ⎭

Three GENTLEMEN  ⎫
Two judges  ⎪
Choristers  ⎪
Lord Mayor of London  ⎪
Garter King of Arms  ⎬ appearing in the Coronation
Marquis Dorset  ⎪
Four Barons of the Cinque Ports  ⎪
Stokesley, Bishop of London  ⎪
Old Duchess of Norfolk  ⎪
Countesses  ⎭

A DOOR-KEEPER  ⎫
Doctor BUTTS, the King's physician  ⎬ at Cranmer's trial
Pursuivants, pages, footboys, grooms  ⎭

A PORTER  ⎫
His MAN  ⎪
Two aldermen  ⎪
Lord Mayor of London  ⎪
GARTER King of Arms  ⎬ at the Christening
Six noblemen  ⎪
Old Duchess of Norfolk, godmother  ⎪
The child, Princess Elizabeth  ⎪
Marchioness Dorset, godmother  ⎭

EPILOGUE

Ladies, gentlemen, a SERVANT, guards, attendants, trumpeters

# All Is True

**Prologue** *Enter Prologue*

PROLOGUE
I come no more to make you laugh. Things now
That bear a weighty and a serious brow,
Sad, high, and working, full of state and woe—
Such noble scenes as draw the eye to flow
We now present. Those that can pity here 5
May, if they think it well, let fall a tear.
The subject will deserve it. Such as give
Their money out of hope they may believe,
May here find truth, too. Those that come to see
Only a show or two, and so agree 10
The play may pass, if they be still, and willing,
I'll undertake may see away their shilling
Richly in two short hours. Only they
That come to hear a merry bawdy play,
A noise of targets, or to see a fellow 15
In a long motley coat guarded with yellow,
Will be deceived. For, gentle hearers, know
To rank our chosen truth with such a show
As fool and fight is, beside forfeiting
Our own brains, and the opinion that we bring 20
To make that only true we now intend,
Will leave us never an understanding friend.
Therefore, for goodness' sake, and as you are known
The first and happiest hearers of the town,
Be sad as we would make ye. Think ye see 25
The very persons of our noble story
As they were living; think you see them great,
And followed with the general throng and sweat
Of thousand friends; then, in a moment, see
How soon this mightiness meets misery. 30
And if you can be merry then, I'll say
A man may weep upon his wedding day. *Exit*

**1.1** ⌈*A cloth of state throughout the play.*⌉ *Enter the Duke
of Norfolk at one door; at the other door enter the
Duke of Buckingham and the Lord Abergavenny*

BUCKINGHAM (*to Norfolk*)
Good morrow, and well met. How have ye done
Since last we saw in France?

NORFOLK                               I thank your grace,
Healthful, and ever since a fresh admirer
Of what I saw there.

BUCKINGHAM                An untimely ague
Stayed me a prisoner in my chamber when 5
Those suns of glory, those two lights of men,
Met in the vale of Ardres.

NORFOLK                         'Twixt Guisnes and Ardres.
I was then present, saw them salute on horseback,
Beheld them when they lighted, how they clung
In their embracement as they grew together, 10
Which had they, what four throned ones could have
      weighed
Such a compounded one?

BUCKINGHAM                All the whole time
I was my chamber's prisoner.

NORFOLK                         Then you lost
The view of earthly glory. Men might say
Till this time pomp was single, but now married 15
To one above itself. Each following day

Became the next day's master, till the last
Made former wonders its. Today the French,
All clinquant all in gold, like heathen gods
Shone down the English; and tomorrow they 20
Made Britain India. Every man that stood
Showed like a mine. Their dwarfish pages were
As cherubim, all gilt; the *mesdames*, too,
Not used to toil, did almost sweat to bear
The pride upon them, that their very labour 25
Was to them as a painting. Now this masque
Was cried incomparable, and th'ensuing night
Made it a fool and beggar. The two kings
Equal in lustre, were now best, now worst,
As presence did present them. Him in eye 30
Still him in praise, and being present both,
'Twas said they saw but one, and no discerner
Durst wag his tongue in censure. When these suns—
For so they phrase 'em—by their heralds challenged
The noble spirits to arms, they did perform 35
Beyond thought's compass, that former fabulous story
Being now seen possible enough, got credit
That *Bevis* was believed.

BUCKINGHAM                O, you go far!

NORFOLK
As I belong to worship, and affect
In honour honesty, the tract of ev'rything 40
Would by a good discourser lose some life
Which action's self was tongue to. All was royal.
To the disposing of it naught rebelled.
Order gave each thing view. The office did
Distinctly his full function.

BUCKINGHAM                Who did guide— 45
I mean, who set the body and the limbs
Of this great sport together, as you guess?

NORFOLK
One, certes, that promises no element
In such a business.

BUCKINGHAM          I pray you who, my lord?

NORFOLK
All this was ordered by the good discretion 50
Of the right reverend Cardinal of York.

BUCKINGHAM
The devil speed him! No man's pie is freed
From his ambitious finger. What had he
To do in these fierce vanities? I wonder
That such a keech can, with his very bulk, 55
Take up the rays o'th' beneficial sun,
And keep it from the earth.

NORFOLK                         Surely, sir,
There's in him stuff that puts him to these ends.
For being not propped by ancestry, whose grace
Chalks successors their way, nor called upon 60
For high feats done to th' crown, neither allied
To eminent assistants, but spider-like,
Out of his self-drawing web, a gives us note
The force of his own merit makes his way—
A gift that heaven gives for him which buys 65
A place next to the King.

ABERGAVENNY                I cannot tell
What heaven hath given him—let some graver eye
Pierce into that; but I can see his pride

Peep through each part of him. Whence has he that?
If not from hell, the devil is a niggard     70
Or has given all before, and he begins
A new hell in himself.
BUCKINGHAM          Why the devil,
Upon this French going out, took he upon him
Without the privity o'th' King t'appoint
Who should attend on him? He makes up the file    75
Of all the gentry, for the most part such
To whom as great a charge as little honour
He meant to lay upon; and his own letter,
The honourable board of council out,
Must fetch him in, he papers.
ABERGAVENNY         I do know    80
Kinsmen of mine—three at the least—that have
By this so sickened their estates that never
They shall abound as formerly.
BUCKINGHAM          O, many
Have broke their backs with laying manors on 'em
For this great journey. What did this vanity    85
But minister communication of
A most poor issue?
NORFOLK         Grievingly I think
The peace between the French and us not values
The cost that did conclude it.
BUCKINGHAM         Every man,
After the hideous storm that followed, was    90
A thing inspired, and, not consulting, broke
Into a general prophecy—that this tempest,
Dashing the garment of this peace, aboded
The sudden breach on't.
NORFOLK        Which is budded out—
For France hath flawed the league, and hath attached
Our merchants' goods at Bordeaux.
ABERGAVENNY        Is it therefore    96
Th'ambassador is silenced?
NORFOLK        Marry is't.
ABERGAVENNY
A proper title of a peace, and purchased
At a superfluous rate.
BUCKINGHAM        Why, all this business
Our reverend Cardinal carried.
NORFOLK        Like it your grace,    100
The state takes notice of the private difference
Betwixt you and the Cardinal. I advise you—
And take it from a heart that wishes towards you
Honour and plenteous safety—that you read
The Cardinal's malice and his potency    105
Together; to consider further that
What his high hatred would effect wants not
A minister in his power. You know his nature,
That he's revengeful; and I know his sword
Hath a sharp edge—it's long, and't may be said    110
It reaches far; and where 'twill not extend
Thither he darts it. Bosom up my counsel,
You'll find it wholesome. Lo, where comes that rock
That I advise your shunning.
     *Enter Cardinal Wolsey, the purse containing the*
     *great seal borne before him. Enter with him certain*
     *of the guard, and two secretaries with papers. The*
     *Cardinal in his passage fixeth his eye on Buckingham*
     *and Buckingham on him, both full of disdain*
CARDINAL WOLSEY (*to a secretary*)
The Duke of Buckingham's surveyor, ha?    115
Where's his examination?
SECRETARY        Here, so please you.

CARDINAL WOLSEY
Is he in person ready?
SECRETARY        Ay, please your grace.
CARDINAL WOLSEY
Well, we shall then know more, and Buckingham
Shall lessen this big look.    *Exeunt Wolsey and his train*
BUCKINGHAM
This butcher's cur is venom-mouthed, and I    120
Have not the power to muzzle him; therefore best
Not wake him in his slumber. A beggar's book
Outworths a noble's blood.
NORFOLK        What, are you chafed?
Ask God for temp'rance; that's th'appliance only
Which your disease requires.
BUCKINGHAM        I read in's looks    125
Matter against me, and his eye reviled
Me as his abject object. At this instant
He bores me with some trick. He's gone to th' King—
I'll follow, and outstare him.
NORFOLK        Stay, my lord,
And let your reason with your choler question    130
What 'tis you go about. To climb steep hills
Requires slow pace at first. Anger is like
A full hot horse who, being allowed his way,
Self-mettle tires him. Not a man in England
Can advise me like you. Be to yourself    135
As you would to your friend.
BUCKINGHAM        I'll to the King,
And from a mouth of honour quite cry down
This Ipswich fellow's insolence, or proclaim
There's difference in no persons.
NORFOLK        Be advised.
Heat not a furnace for your foe so hot    140
That it do singe yourself. We may outrun
By violent swiftness that which we run at,
And lose by over-running. Know you not
The fire that mounts the liquor till't run o'er
In seeming to augment it wastes it? Be advised.    145
I say again there is no English soul
More stronger to direct you than yourself,
If with the sap of reason you would quench
Or but allay the fire of passion.
BUCKINGHAM        Sir,
I am thankful to you, and I'll go along    150
By your prescription; but this top-proud fellow—
Whom from the flow of gall I name not, but
From sincere motions—by intelligence,
And proofs as clear as founts in July when
We see each grain of gravel, I do know    155
To be corrupt and treasonous.
NORFOLK        Say not 'treasonous'.
BUCKINGHAM
To th' King I'll say't, and make my vouch as strong
As shore of rock. Attend: this holy fox,
Or wolf, or both—for he is equal rav'nous
As he is subtle, and as prone to mischief    160
As able to perform't, his mind and place
Infecting one another, yea, reciprocally—
Only to show his pomp as well in France
As here at home, suggests the King our master
To this last costly treaty, th'interview    165
That swallowed so much treasure and, like a glass,
Did break i'th' rinsing.
NORFOLK        Faith, and so it did.
BUCKINGHAM
Pray give me favour, sir. This cunning Cardinal,
The articles o'th' combination drew

As himself pleased, and they were ratified     170
As he cried 'Thus let be', to as much end
As give a crutch to th' dead. But our count-Cardinal
Has done this, and 'tis well for worthy Wolsey,
Who cannot err, he did it. Now this follows—
Which, as I take it, is a kind of puppy          175
To th' old dam, treason—Charles the Emperor,
Under pretence to see the Queen his aunt—
For 'twas indeed his colour, but he came
To whisper Wolsey—here makes visitation.
His fears were that the interview betwixt         180
England and France might through their amity
Breed him some prejudice, for from this league
Peeped harms that menaced him. Privily he
Deals with our Cardinal and, as I trow—
Which I do well, for I am sure the Emperor        185
Paid ere he promised, whereby his suit was granted
Ere it was asked—but when the way was made,
And paved with gold, the Emperor thus desired
That he would please to alter the King's course
And break the foresaid peace. Let the King know,  190
As soon he shall by me, that thus the Cardinal
Does buy and sell his honour as he pleases,
And for his own advantage.
NORFOLK                          I am sorry
To hear this of him, and could wish he were
Something mistaken in't.
BUCKINGHAM                    No, not a syllable.    195
I do pronounce him in that very shape
He shall appear in proof.
        *Enter Brandon, a serjeant-at-arms before him, and*
        *two or three of the guard*
BRANDON
Your office, serjeant, execute it.
SERJEANT                              Sir.
(*To Buckingham*) My lord the Duke of Buckingham and
        Earl
Of Hereford, Stafford, and Northampton, I          200
Arrest thee of high treason in the name
Of our most sovereign King.
BUCKINGHAM [*to Norfolk*]        Lo you, my lord,
The net has fall'n upon me. I shall perish
Under device and practice.
BRANDON                     I am sorry
To see you ta'en from liberty to look on           205
The business present. 'Tis his highness' pleasure
You shall to th' Tower.
BUCKINGHAM            It will help me nothing
To plead mine innocence, for that dye is on me
Which makes my whit'st part black. The will of
        heav'n
Be done in this and all things. I obey.            210
O, my lord Abergavenny, fare you well.
BRANDON
Nay, he must bear you company.
(*To Abergavenny*)                The King
Is pleased you shall to th' Tower till you know
How he determines further.
ABERGAVENNY              As the Duke said,
The will of heaven be done and the King's pleasure
By me obeyed.
BRANDON          Here is a warrant from           216
The King t'attach Lord Montague and the bodies
Of the duke's confessor, John de la Car,
One Gilbert Perk, his chancellor—
BUCKINGHAM                So, so;
These are the limbs o' th' plot. No more, I hope.  220

BRANDON
A monk o' th' Chartreux.
BUCKINGHAM              O, Nicholas Hopkins?
BRANDON He.
BUCKINGHAM
My surveyor is false. The o'er-great Cardinal
Hath showed him gold. My life is spanned already.
I am the shadow of poor Buckingham,                225
Whose figure even this instant cloud puts on
By dark'ning my clear sun. (*To Norfolk*) My lord,
    farewell.
        *Exeunt* ⌈*Norfolk at one door, Buckingham and*
                *Abergavenny under guard at another*⌉

1.2    *Cornetts. Enter King Henry leaning on Cardinal*
       *Wolsey's shoulder. Enter with them Wolsey's two*
       *secretaries, the nobles, and Sir Thomas Lovell. The*
       *King ascends to his seat under the cloth of state;*
       *Wolsey places himself under the King's feet on his*
       *right side*
KING HENRY ⌈*to Wolsey*⌉
My life itself and the best heart of it
Thanks you for this great care. I stood i' th' level
Of a full-charged confederacy, and give thanks
To you that choked it. Let be called before us
That gentleman of Buckingham's. In person          5
I'll hear him his confessions justify,
And point by point the treasons of his master
He shall again relate.
⌈CRIER⌉ (*within*)
Room for the Queen, ushered by the Duke of Norfolk.
        *Enter Queen Katherine, the Duke of Norfolk, and*
        *the Duke of Suffolk. She kneels. King Henry riseth*
        *from his state, takes her up, and kisses her*
QUEEN KATHERINE
Nay, we must longer kneel. I am a suitor.           10
KING HENRY
Arise, and take place by us.
        *He placeth her by him*
                              Half your suit
Never name to us. You have half our power,
The other moiety ere you ask is given.
Repeat your will and take it.
QUEEN KATHERINE              Thank your majesty.
That you would love yourself, and in that love      15
Not unconsidered leave your honour nor
The dignity of your office, is the point
Of my petition.
KING HENRY      Lady mine, proceed.
QUEEN KATHERINE
I am solicited, not by a few,
And those of true condition, that your subjects     20
Are in great grievance. There have been commissions
Sent down among 'em which hath flawed the heart
Of all their loyalties; wherein, although,
My good lord Cardinal, they vent reproaches
Most bitterly on you, as putter-on                  25
Of these exactions, yet the King our master—
Whose honour heaven shield from soil—even he
        escapes not
Language unmannerly, yea, such which breaks
The sides of loyalty, and almost appears
In loud rebellion.
NORFOLK          Not 'almost appears'—              30
It doth appear; for upon these taxations
The clothiers all, not able to maintain
The many to them 'longing, have put off

The spinsters, carders, fullers, weavers, who,
Unfit for other life, compelled by hunger                    35
And lack of other means, in desperate manner
Daring th'event to th' teeth, are all in uproar,
And danger serves among them.
KING HENRY                                        Taxation?
Wherein, and what taxation? My lord Cardinal,
You that are blamed for it alike with us,                    40
Know you of this taxation?
CARDINAL WOLSEY                        Please you, sir,
I know but of a single part in aught
Pertains to th' state, and front but in that file
Where others tell steps with me.
QUEEN KATHERINE                          No, my lord?
You know no more than others? But you frame          45
Things that are known alike, which are not wholesome
To those which would not know them, and yet must
Perforce be their acquaintance. These exactions
Whereof my sovereign would have note, they are
Most pestilent to th' hearing, and to bear 'em            50
The back is sacrifice to th' load. They say
They are devised by you, or else you suffer
Too hard an exclamation.
KING HENRY                              Still exaction!
The nature of it? In what kind, let's know,
Is this exaction?
QUEEN KATHERINE   I am much too venturous            55
In tempting of your patience, but am boldened
Under your promised pardon. The subjects' grief
Comes through commissions which compels from each
The sixth part of his substance to be levied
Without delay, and the pretence for this                    60
Is named your wars in France. This makes bold mouths.
Tongues spit their duties out, and cold hearts freeze
Allegiance in them. Their curses now
Live where their prayers did, and it's come to pass
This tractable obedience is a slave                            65
To each incensèd will. I would your highness
Would give it quick consideration, for
There is no primer business.
KING HENRY                              By my life,
This is against our pleasure.
CARDINAL WOLSEY                        And for me,
I have no further gone in this than by                      70
A single voice, and that not passed me but
By learnèd approbation of the judges. If I am
Traduced by ignorant tongues, which neither know
My faculties nor person yet will be
The chronicles of my doing, let me say                      75
'Tis but the fate of place, and the rough brake
That virtue must go through. We must not stint
Our necessary actions in the fear
To cope malicious censurers, which ever,
As rav'nous fishes, do a vessel follow                        80
That is new trimmed, but benefit no further
Than vainly longing. What we oft do best,
By sick interpreters, once weak ones, is
Not ours or not allowed; what worst, as oft,
Hitting a grosser quality, is cried up                        85
For our best act. If we shall stand still,
In fear our motion will be mocked or carped at,
We should take root here where we sit,
Or sit state-statues only.
KING HENRY                              Things done well,
And with a care, exempt themselves from fear;        90
Things done without example, in their issue
Are to be feared. Have you a precedent

Of this commission? I believe not any.
We must not rend our subjects from our laws
And stick them in our will. Sixth part of each?        95
A trembling contribution! Why, we take
From every tree lop, bark, and part o'th' timber,
And though we leave it with a root, thus hacked
The air will drink the sap. To every county
Where this is questioned send our letters with        100
Free pardon to each man that has denied
The force of this commission. Pray look to't—
I put it to your care.
CARDINAL WOLSEY (to a secretary) A word with you.
Let there be letters writ to every shire
Of the King's grace and pardon.
(Aside to the secretary)              The grievèd commons
Hardly conceive of me. Let it be noised                  106
That through our intercession this revokement
And pardon comes. I shall anon advise you
Further in the proceeding.                  Exit secretary
          Enter Buckingham's Surveyor
QUEEN KATHERINE (to the King)
I am sorry that the Duke of Buckingham              110
Is run in your displeasure.
KING HENRY                              It grieves many.
The gentleman is learnèd, and a most rare speaker,
To nature none more bound; his training such
That he may furnish and instruct great teachers
And never seek for aid out of himself. Yet see,       115
When these so noble benefits shall prove
Not well disposed, the mind growing once corrupt,
They turn to vicious forms ten times more ugly
Than ever they were fair. This man so complete,
Who was enrolled 'mongst wonders—and when we
Almost with ravished list'ning could not find         121
His hour of speech a minute—he, my lady,
Hath into monstrous habits put the graces
That once were his, and is become as black
As if besmeared in hell. Sit by us. You shall hear—
This was his gentleman in trust of him—               126
Things to strike honour sad.
(To Wolsey)                        Bid him recount
The fore-recited practices whereof
We cannot feel too little, hear too much.
CARDINAL WOLSEY (to the Surveyor)
Stand forth, and with bold spirit relate what you   130
Most like a careful subject have collected
Out of the Duke of Buckingham.
KING HENRY (to the Surveyor)              Speak freely.
BUCKINGHAM'S SURVEYOR
First, it was usual with him, every day
It would infect his speech, that if the King
Should without issue die, he'll carry it so              135
To make the sceptre his. These very words
I've heard him utter to his son-in-law,
Lord Abergavenny, to whom by oath he menaced
Revenge upon the Cardinal.
CARDINAL WOLSEY (to the King)
                                        Please your highness note
His dangerous conception in this point,                 140
Not friended by his wish to your high person.
His will is most malignant, and it stretches
Beyond you to your friends.
QUEEN KATHERINE            My learned Lord Cardinal,
Deliver all with charity.
KING HENRY (to the Surveyor) Speak on.
How grounded he his title to the crown              145

Upon our fail? To this point hast thou heard him
At any time speak aught?
BUCKINGHAM'S SURVEYOR        He was brought to this
By a vain prophecy of Nicholas Hopkins.
KING HENRY
What was that Hopkins?
BUCKINGHAM'S SURVEYOR        Sir, a Chartreux friar,
His confessor, who fed him every minute                      150
With words of sovereignty.
KING HENRY                          How know'st thou this?
BUCKINGHAM'S SURVEYOR
Not long before your highness sped to France,
The Duke being at the Rose, within the parish
Saint Lawrence Poutney, did of me demand
What was the speech among the Londoners                      155
Concerning the French journey. I replied
Men feared the French would prove perfidious,
To the King's danger; presently the Duke
Said 'twas the fear indeed, and that he doubted
'Twould prove the verity of certain words                    160
Spoke by a holy monk that oft, says he,
'Hath sent to me, wishing me to permit
John de la Car, my chaplain, a choice hour
To hear from him a matter of some moment;
Whom after under the confession's seal                       165
He solemnly had sworn, that what he spoke
My chaplain to no creature living but
To me should utter, with demure confidence
This pausingly ensued: "neither the King nor's heirs",
Tell you the Duke, "shall prosper. Bid him strive            170
To win the love o'th' commonalty. The Duke
Shall govern England." '
QUEEN KATHERINE        If I know you well,
You were the Duke's surveyor, and lost your office
On the complaint o'th' tenants. Take good heed
You charge not in your spleen a noble person                 175
And spoil your nobler soul. I say, take heed;
Yes, heartily beseech you.
KING HENRY                          Let him on.
(To the Surveyor) Go forward.
BUCKINGHAM'S SURVEYOR  On my soul I'll speak but truth.
I told my lord the Duke, by th' devil's illusions
The monk might be deceived, and that 'twas
     dangerous                                               180
To ruminate on this so far until
It forged him some design which, being believed,
It was much like to do. He answered, 'Tush,
It can do me no damage', adding further
That had the King in his last sickness failed,              185
The Cardinal's and Sir Thomas Lovell's heads
Should have gone off.
KING HENRY                  Ha? What, so rank? Ah, ha!
There's mischief in this man. Canst thou say further?
BUCKINGHAM'S SURVEYOR
I can, my liege.
KING HENRY              Proceed.
BUCKINGHAM'S SURVEYOR      Being at Greenwich,
After your highness had reproved the Duke                    190
About Sir William Bulmer—
KING HENRY                        I remember
Such a time, being my sworn servant,
The Duke retained him his. But on—what hence?
BUCKINGHAM'S SURVEYOR
'If', quoth he, 'I for this had been committed'—
As to the Tower, I thought—'I would have played 195
The part my father meant to act upon

Th'usurper Richard who, being at Salisbury,
Made suit to come in's presence; which if granted,
As he made semblance of his duty, would
Have put his knife into him.'
KING HENRY                          A giant traitor!        200
CARDINAL WOLSEY (to the Queen)
Now, madam, may his highness live in freedom,
And this man out of prison?
QUEEN KATHERINE            God mend all.
KING HENRY (to the Surveyor)
There's something more would out of thee—what
     sayst?
BUCKINGHAM'S SURVEYOR
After 'the Duke his father', with 'the knife',
He stretched him, and with one hand on his dagger,
Another spread on's breast, mounting his eyes,              206
He did discharge a horrible oath whose tenor
Was, were he evil used, he would outgo
His father by as much as a performance
Does an irresolute purpose.
KING HENRY                    There's his period—          210
To sheathe his knife in us. He is attached.
Call him to present trial. If he may
Find mercy in the law, 'tis his; if none,
Let him not seek't of us. By day and night,
He's traitor to th' height.            ⌈Flourish.⌉ Exeunt

1.3    Enter the Lord Chamberlain and Lord Sands
LORD CHAMBERLAIN
Is't possible the spells of France should juggle
Men into such strange mysteries?
SANDS                              New customs,
Though they be never so ridiculous—
Nay, let 'em be unmanly—yet are followed.
LORD CHAMBERLAIN
As far as I see, all the good our English                   5
Have got by the late voyage is but merely
A fit or two o'th' face. But they are shrewd ones,
For when they hold 'em you would swear directly
Their very noses had been counsellors
To Pépin or Clotharius, they keep state so.                 10
SANDS
They have all new legs, and lame ones; one would
     take it,
That never see 'em pace before, the spavin
Or spring-halt reigned among 'em.
LORD CHAMBERLAIN                    Death, my lord,
Their clothes are after such a pagan cut to't
That sure they've worn out Christendom.
     Enter Sir Thomas Lovell
                                        How now—
What news, Sir Thomas Lovell?
LOVELL                          Faith, my lord,            16
I hear of none but the new proclamation
That's clapped upon the court gate.
LORD CHAMBERLAIN                    What is't for?
LOVELL
The reformation of our travelled gallants
That fill the court with quarrels, talk, and tailors.      20
LORD CHAMBERLAIN
I'm glad 'tis there. Now I would pray our 'messieurs'
To think an English courtier may be wise
And never see the Louvre.
LOVELL                      They must either,
For so run the conditions, leave those remnants
Of fool and feather that they got in France,               25

With all their honourable points of ignorance
Pertaining thereunto—as fights and fireworks,
Abusing better men than they can be
Out of a foreign wisdom, renouncing clean
The faith they have in tennis and tall stockings,      30
Short blistered breeches, and those types of travel—
And understand again like honest men,
Or pack to their old playfellows. There, I take it,
They may, *cum privilegio*, 'oui' away
The lag end of their lewdness and be laughed at.      35

SANDS
'Tis time to give 'em physic, their diseases
Are grown so catching.

LORD CHAMBERLAIN          What a loss our ladies
Will have of these trim vanities!

LOVELL                         Ay, marry,
There will be woe indeed, lords. The sly whoresons
Have got a speeding trick to lay down ladies.      40
A French song and a fiddle has no fellow.

SANDS
The devil fiddle 'em! I am glad they are going,
For sure there's no converting of 'em. Now
An honest country lord, as I am, beaten
A long time out of play, may bring his plainsong      45
And have an hour of hearing, and; by'r Lady,
Held current music, too.

LORD CHAMBERLAIN          Well said, Lord Sands.
Your colt's tooth is not cast yet?

SANDS                         No, my lord,
Nor shall not while I have a stump.

LORD CHAMBERLAIN (*to Lovell*)          Sir Thomas,
Whither were you a-going?

LOVELL                    To the Cardinal's.      50
Your lordship is a guest too.

LORD CHAMBERLAIN          O, 'tis true.
This night he makes a supper, and a great one,
To many lords and ladies. There will be
The beauty of this kingdom, I'll assure you.

LOVELL
That churchman bears a bounteous mind indeed,      55
A hand as fruitful as the land that feeds us.
His dews fall everywhere.

LORD CHAMBERLAIN          No doubt he's noble.
He had a black mouth that said other of him.

SANDS
He may, my lord; he's wherewithal. In him
Sparing would show a worse sin than ill doctrine.      60
Men of his way should be most liberal.
They are set here for examples.

LORD CHAMBERLAIN          True, they are so,
But few now give so great ones. My barge stays.
Your lordship shall along. (*To Lovell*) Come, good Sir
   Thomas,
We shall be late else, which I would not be,      65
For I was spoke to, with Sir Henry Guildford,
This night to be comptrollers.

SANDS                    I am your lordship's.
                                        *Exeunt*

**1.4**   *Hautboys. ⌈Enter servants with⌉ a small table for
        Cardinal Wolsey ⌈which they place⌉ under the cloth
        of state, and a longer table for the guests. Then
        enter at one door Anne Boleyn and divers other
        ladies and gentlemen as guests, and at another door
        enter Sir Henry Guildford*

GUILDFORD
Ladies, a general welcome from his grace

Salutes ye all. This night he dedicates
To fair content and you. None here, he hopes,
In all this noble bevy, has brought with her
One care abroad. He would have all as merry      5
As feast, good company, good wine, good welcome
Can make good people.
        *Enter the Lord Chamberlain, Lord Sands, and Sir
        Thomas Lovell*
(*To the Lord Chamberlain*) O, my lord, you're tardy.
The very thought of this fair company
Clapped wings to me.

LORD CHAMBERLAIN    You are young, Sir Harry Guildford.

SANDS
Sir Thomas Lovell, had the Cardinal      10
But half my lay thoughts in him, some of these
Should find a running banquet, ere they rested,
I think would better please 'em. By my life,
They are a sweet society of fair ones.

LOVELL
O, that your lordship were but now confessor      15
To one or two of these.

SANDS                    I would I were.
They should find easy penance.

LOVELL                         Faith, how easy?

SANDS
As easy as a down bed would afford it.

LORD CHAMBERLAIN
Sweet ladies, will it please you sit?
(*To Guildford*)                    Sir Harry,
Place you that side, I'll take the charge of this.      20
        *They sit about the longer table. ⌈A noise within⌉*
His grace is ent'ring. Nay, you must not freeze—
Two women placed together makes cold weather.
My lord Sands, you are one will keep 'em waking.
Pray sit between these ladies.

SANDS                    By my faith,
And thank your lordship.
        *He sits between Anne and another*
                         By your leave, sweet ladies.
If I chance to talk a little wild, forgive me.      26
I had it from my father.

ANNE                    Was he mad, sir?

SANDS
O, very mad; exceeding mad—in love, too.
But he would bite none. Just as I do now,
He would kiss you twenty with a breath.
        *He kisses her*

LORD CHAMBERLAIN               Well said, my lord.
So now you're fairly seated. Gentlemen,      31
The penance lies on you if these fair ladies
Pass away frowning.

SANDS For my little cure,
Let me alone.      35
        *Hautboys. Enter Cardinal Wolsey who takes his
        seat at the small table under the state*

CARDINAL WOLSEY
You're welcome, my fair guests. That noble lady
Or gentleman that is not freely merry
Is not my friend. This, to confirm my welcome,
And to you all, good health!
        *He drinks*

SANDS                    Your grace is noble.
Let me have such a bowl may hold my thanks,      40
And save me so much talking.

CARDINAL WOLSEY          My lord Sands,
I am beholden to you. Cheer your neighbours.
Ladies, you are not merry! Gentlemen,

Whose fault is this?
SANDS                     The red wine first must rise
  In their fair cheeks, my lord, then we shall have 'em
  Talk us to silence.
ANNE                 You are a merry gamester,        46
  My lord Sands.
SANDS               Yes, if I make my play.
  Here's to your ladyship; and pledge it, madam,
  For 'tis to such a thing—
ANNE                    You cannot show me.
SANDS (to Wolsey)
  I told your grace they would talk anon.
        Drum and trumpet. Chambers discharged
CARDINAL WOLSEY                   What's that?
LORD CHAMBERLAIN (to the servants)
  Look out there, some of ye.          Exit a servant
CARDINAL WOLSEY            What warlike voice,     51
  And to what end is this? Nay, ladies, fear not.
  By all the laws of war you're privileged.
        Enter the servant
LORD CHAMBERLAIN
  How now—what is't?
SERVANT             A noble troop of strangers,
  For so they seem. They've left their barge and landed,
  And hither make as great ambassadors          56
  From foreign princes.
CARDINAL WOLSEY      Good Lord Chamberlain,
  Go give 'em welcome—you can speak the French
      tongue.
  And pray receive 'em nobly, and conduct 'em
  Into our presence where this heaven of beauty    60
  Shall shine at full upon them. Some attend him.
        Exit Chamberlain, attended
        All rise, and some servants remove the tables
  You have now a broken banquet, but we'll mend it.
  A good digestion to you all, and once more
  I shower a welcome on ye—welcome all.
        Hautboys. Enter, ushered by the Lord Chamberlain,
        King Henry and others as masquers habited like
        shepherds. They pass directly before Cardinal
        Wolsey and gracefully salute him
  A noble company. What are their pleasures?     65
LORD CHAMBERLAIN
  Because they speak no English, thus they prayed
  To tell your grace, that, having heard by fame
  Of this so noble and so fair assembly
  This night to meet here, they could do no less,
  Out of the great respect they bear to beauty,    70
  But leave their flocks, and, under your fair conduct,
  Crave leave to view these ladies, and entreat
  An hour of revels with 'em.
CARDINAL WOLSEY         Say, Lord Chamberlain,
  They have done my poor house grace, for which I pay
      'em
  A thousand thanks, and pray 'em take their pleasures.
        The masquers choose ladies. The King chooses Anne
        Boleyn
KING HENRY (to Anne)
  The fairest hand I ever touched. O beauty,       76
  Till now I never knew thee.
        Music. They dance
CARDINAL WOLSEY (to the Lord Chamberlain) My lord.
LORD CHAMBERLAIN Your grace.
CARDINAL WOLSEY Pray tell 'em thus much from me.  80
  There should be one amongst 'em by his person
  More worthy this place than myself, to whom,
  If I but knew him, with my love and duty

I would surrender it.
LORD CHAMBERLAIN       I will, my lord.
        He whispers with the masquers
CARDINAL WOLSEY
  What say they?
LORD CHAMBERLAIN Such a one they all confess   85
  There is indeed, which they would have your grace
  Find out, and he will take it.
CARDINAL WOLSEY (standing)     Let me see then.
  By all your good leaves, gentlemen, here I'll make
  My royal choice.
        He bows before the King
KING HENRY (unmasking) Ye have found him, Cardinal.
  You hold a fair assembly. You do well, lord.    90
  You are a churchman, or I'll tell you, Cardinal,
  I should judge now unhappily.
CARDINAL WOLSEY             I am glad
  Your grace is grown so pleasant.
KING HENRY              My Lord Chamberlain,
  Prithee come hither.
  (Gesturing towards Anne) What fair lady's that?   94
LORD CHAMBERLAIN
  An't please your grace, Sir Thomas Boleyn's daughter—
  The Viscount Rochford—one of her highness' women.
KING HENRY
  By heaven, she is a dainty one. (To Anne) Sweetheart,
  I were unmannerly to take you out
  And not to kiss you (kisses her). A health, gentlemen;
        He drinks
  Let it go round.                    100
CARDINAL WOLSEY
  Sir Thomas Lovell, is the banquet ready
  I'th' privy chamber?
LOVELL              Yes, my lord.
CARDINAL WOLSEY (to the King)      Your grace
  I fear with dancing is a little heated.
KING HENRY I fear too much.
CARDINAL WOLSEY There's fresher air, my lord,    105
  In the next chamber.
KING HENRY
  Lead in your ladies, every one. (To Anne) Sweet partner,
  I must not yet forsake you. (To Wolsey) Let's be merry,
  Good my lord Cardinal. I have half a dozen healths
  To drink to these fair ladies, and a measure      110
  To lead 'em once again, and then let's dream
  Who's best in favour. Let the music knock it.
        Exeunt with trumpets

                    ✿

2.1   Enter two Gentlemen, at several doors
FIRST GENTLEMAN
  Whither away so fast?
SECOND GENTLEMAN     O, God save ye.
  Ev'n to the hall to hear what shall become
  Of the great Duke of Buckingham.
FIRST GENTLEMAN                I'll save you
  That labour, sir. All's now done but the ceremony
  Of bringing back the prisoner.
SECOND GENTLEMAN          Were you there?    5
FIRST GENTLEMAN
  Yes, indeed was I.
SECOND GENTLEMAN Pray speak what has happened.
FIRST GENTLEMAN
  You may guess quickly what.
SECOND GENTLEMAN       Is he found guilty?
FIRST GENTLEMAN
  Yes, truly is he, and condemned upon't.

SECOND GENTLEMAN I am sorry for't.

FIRST GENTLEMAN So are a number more.                    10

SECOND GENTLEMAN But pray, how passed it?

FIRST GENTLEMAN
I'll tell you in a little. The great Duke
Came to the bar, where to his accusations
He pleaded still not guilty, and allegèd
Many sharp reasons to defeat the law.                    15
The King's attorney, on the contrary,
Urged on the examinations, proofs, confessions,
Of divers witnesses, which the Duke desired
To him brought *viva voce* to his face—
At which appeared against him his surveyor,                    20
Sir Gilbert Perk his chancellor, and John Car,
Confessor to him, with that devil-monk,
Hopkins, that made this mischief.

SECOND GENTLEMAN                    That was he
That fed him with his prophecies.

FIRST GENTLEMAN                    The same.
All these accused him strongly, which he fain                    25
Would have flung from him, but indeed he could not.
And so his peers, upon this evidence,
Have found him guilty of high treason. Much
He spoke, and learnèdly, for life, but all
Was either pitied in him or forgotten.                    30

SECOND GENTLEMAN
After all this, how did he bear himself?

FIRST GENTLEMAN
When he was brought again to th' bar to hear
His knell rung out, his judgement, he was stirred
With such an agony he sweat extremely,
And something spoke in choler, ill and hasty;                    35
But he fell to himself again, and sweetly
In all the rest showed a most noble patience.

SECOND GENTLEMAN
I do not think he fears death.

FIRST GENTLEMAN                    Sure he does not.
He never was so womanish. The cause
He may a little grieve at.

SECOND GENTLEMAN                    Certainly                    40
The Cardinal is the end of this.

FIRST GENTLEMAN                    'Tis likely
By all conjectures: first, Kildare's attainder,
Then deputy of Ireland, who, removed,
Earl Surrey was sent thither—and in haste, too,
Lest he should help his father.

SECOND GENTLEMAN                    That trick of state                    45
Was a deep envious one.

FIRST GENTLEMAN                    At his return
No doubt he will requite it. This is noted,
And generally: whoever the King favours,
The Card'nal instantly will find employment—
And far enough from court, too.

SECOND GENTLEMAN                    All the commons                    50
Hate him perniciously and, o' my conscience,
Wish him ten fathom deep. This Duke as much
They love and dote on, call him 'bounteous
Buckingham,
The mirror of all courtesy'—

*Enter the Duke of Buckingham from his arraignment,*
*tipstaves before him, the axe with the edge towards*
*him, halberdiers on each side, accompanied with Sir*
*Thomas Lovell, Sir Nicholas Vaux, Sir William*
*Sands, and common people*

FIRST GENTLEMAN                    Stay there, sir,
And see the noble ruined man you speak of.                    55

SECOND GENTLEMAN
Let's stand close and behold him.
*They stand apart*

BUCKINGHAM (*to the common people*) All good people,
You that thus far have come to pity me,
Hear what I say, and then go home and lose me.
I have this day received a traitor's judgement,
And by that name must die. Yet, heaven bear witness,
And if I have a conscience let it sink me,                    61
Even as the axe falls, if I be not faithful.
The law I bear no malice for my death.
'T has done, upon the premises, but justice.
But those that sought it I could wish more Christians.
Be what they will, I heartily forgive 'em.                    66
Yet let 'em look they glory not in mischief,
Nor build their evils on the graves of great men,
For then my guiltless blood must cry against 'em.
For further life in this world I ne'er hope,                    70
Nor will I sue, although the King have mercies
More than I dare make faults. You few that loved me,
And dare be bold to weep for Buckingham,
His noble friends and fellows, whom to leave
Is only bitter to him, only dying,                    75
Go with me like good angels to my end,
And, as the long divorce of steel falls on me,
Make of your prayers one sweet sacrifice,
And lift my soul to heaven. (*To the guard*) Lead on, i'
God's name.

LOVELL
I do beseech your grace, for charity,                    80
If ever any malice in your heart
Were hid against me, now to forgive me frankly.

BUCKINGHAM
Sir Thomas Lovell, I as free forgive you
As I would be forgiven. I forgive all.
There cannot be those numberless offences                    85
'Gainst me that I cannot take peace with. No black envy
Shall mark my grave. Commend me to his grace,
And if he speak of Buckingham, pray tell him
You met him half in heaven. My vows and prayers
Yet are the King's, and, till my soul forsake,                    90
Shall cry for blessings on him. May he live
Longer than I have time to tell his years;
Ever beloved and loving may his rule be;
And, when old time shall lead him to his end,
Goodness and he fill up one monument.                    95

LOVELL
To th' waterside I must conduct your grace,
Then give my charge up to Sir Nicholas Vaux,
Who undertakes you to your end.

VAUX (*to an attendant*)                    Prepare there—
The Duke is coming. See the barge be ready,
And fit it with such furniture as suits                    100
The greatness of his person.

BUCKINGHAM                    Nay, Sir Nicholas,
Let it alone. My state now will but mock me.
When I came hither I was Lord High Constable
And Duke of Buckingham; now, poor Edward Bohun.
Yet I am richer than my base accusers,                    105
That never knew what truth meant. I now seal it,
And with that blood will make 'em one day groan for't.
My noble father, Henry of Buckingham,
Who first raised head against usurping Richard,
Flying for succour to his servant Banister,                    110
Being distressed, was by that wretch betrayed,

And without trial fell. God's peace be with him.
Henry the Seventh succeeding, truly pitying
My father's loss, like a most royal prince,
Restored me to my honours, and out of ruins      115
Made my name once more noble. Now his son,
Henry the Eighth, life, honour, name, and all
That made me happy, at one stroke has taken
For ever from the world. I had my trial,
And must needs say a noble one; which makes me
A little happier than my wretched father.        121
Yet thus far we are one in fortunes: both
Fell by our servants, by those men we loved most—
A most unnatural and faithless service.
Heaven has an end in all. Yet, you that hear me,  125
This from a dying man receive as certain—
Where you are liberal of your loves and counsels,
Be sure you be not loose; for those you make friends
And give your hearts to, when they once perceive
The least rub in your fortunes, fall away        130
Like water from ye, never found again
But where they mean to sink ye. All good people
Pray for me. I must now forsake ye. The last hour
Of my long weary life is come upon me.
Farewell, and when you would say something that is
    sad,                                         135
Speak how I fell. I have done, and God forgive me.
                    *Exeunt Buckingham and train*
    *The two Gentlemen come forward*
FIRST GENTLEMAN
O, this is full of pity, sir; it calls,
I fear, too many curses on their heads
That were the authors.
SECOND GENTLEMAN         If the Duke be guiltless,
'Tis full of woe. Yet I can give you inkling     140
Of an ensuing evil, if it fall,
Greater than this.
FIRST GENTLEMAN     Good angels keep it from us.
What may it be? You do not doubt my faith, sir?
SECOND GENTLEMAN
This secret is so weighty, 'twill require
A strong faith to conceal it.
FIRST GENTLEMAN         Let me have it—         145
I do not talk much.
SECOND GENTLEMAN     I am confident;
You shall, sir. Did you not of late days hear
A buzzing of a separation
Between the King and Katherine?
FIRST GENTLEMAN             Yes, but it held not.
For when the King once heard it, out of anger    150
He sent command to the Lord Mayor straight
To stop the rumour and allay those tongues
That durst disperse it.
SECOND GENTLEMAN     But that slander, sir,
Is found a truth now, for it grows again
Fresher than e'er it was, and held for certain   155
The King will venture at it. Either the Cardinal
Or some about him near have, out of malice
To the good Queen, possessed him with a scruple
That will undo her. To confirm this, too,
Cardinal Campeius is arrived, and lately,         160
As all think, for this business.
FIRST GENTLEMAN             'Tis the Cardinal;
And merely to revenge him on the Emperor
For not bestowing on him at his asking
The Archbishopric of Toledo this is purposed.
SECOND GENTLEMAN
I think you have hit the mark. But is't not cruel 165

That she should feel the smart of this? The Cardinal
Will have his will, and she must fall.
FIRST GENTLEMAN                        'Tis woeful.
We are too open here to argue this.
Let's think in private more.         *Exeunt*

**2.2**    *Enter the Lord Chamberlain with a letter*
LORD CHAMBERLAIN (*reads*) 'My lord, the horses your
lordship sent for, with all the care I had, I saw well
chosen, ridden, and furnished. They were young and
handsome, and of the best breed in the north. When
they were ready to set out for London, a man of my
lord Cardinal's, by commission and main power, took
'em from me with this reason—his master would be
served before a subject, if not before the King; which
stopped our mouths, sir.'
I fear he will indeed. Well, let him have them.     10
He will have all, I think.
    *Enter to the Lord Chamberlain the Dukes of Norfolk
    and Suffolk*
NORFOLK Well met, my Lord Chamberlain.
LORD CHAMBERLAIN Good day to both your graces.
SUFFOLK
How is the King employed?
LORD CHAMBERLAIN             I left him private,
Full of sad thoughts and troubles.
NORFOLK                        What's the cause?
LORD CHAMBERLAIN
It seems the marriage with his brother's wife       16
Has crept too near his conscience.
SUFFOLK                        No, his conscience
Has crept too near another lady.
NORFOLK                        'Tis so.
This is the Cardinal's doing. The King-Cardinal,
That blind priest, like the eldest son of fortune,  20
Turns what he list. The King will know him one day.
SUFFOLK
Pray God he do. He'll never know himself else.
NORFOLK
How holily he works in all his business,
And with what zeal! For now he has cracked the
    league
Between us and the Emperor, the Queen's great-
    nephew,                                        25
He dives into the King's soul and there scatters
Dangers, doubts, wringing of the conscience,
Fears, and despairs—and all these for his marriage.
And out of all these, to restore the King,
He counsels a divorce—a loss of her               30
That like a jewel has hung twenty years
About his neck, yet never lost her lustre;
Of her that loves him with that excellence
That angels love good men with; even of her
That, when the greatest stroke of fortune falls,    35
Will bless the King—and is not this course pious?
LORD CHAMBERLAIN
Heaven keep me from such counsel! 'Tis most true—
These news are everywhere, every tongue speaks 'em,
And every true heart weeps for't. All that dare
Look into these affairs see this main end—         40
The French king's sister. Heaven will one day open
The King's eyes, that so long have slept, upon
This bold bad man.
SUFFOLK     And free us from his slavery.
NORFOLK We had need pray,
And heartily, for our deliverance,                  45

Or this imperious man will work us all
From princes into pages. All men's honours
Lie like one lump before him, to be fashioned
Into what pitch he please.

SUFFOLK            For me, my lords,      50
I love him not, nor fear him—there's my creed.
As I am made without him, so I'll stand,
If the King please. His curses and his blessings
Touch me alike; they're breath I not believe in.
I knew him, and I know him; so I leave him      55
To him that made him proud—the Pope.

NORFOLK                        Let's in,
And with some other business put the King
From these sad thoughts that work too much upon him.
(*To the Lord Chamberlain*)
My lord, you'll bear us company?

LORD CHAMBERLAIN           Excuse me,
The King has sent me otherwise. Besides,      60
You'll find a most unfit time to disturb him.
Health to your lordships.

NORFOLK           Thanks, my good Lord Chamberlain.
                 *Exit the Lord Chamberlain*
       *King Henry draws the curtain, and sits reading*
       *pensively*

SUFFOLK
How sad he looks! Sure he is much afflicted.

KING HENRY
Who's there? Ha?

NORFOLK          Pray God he be not angry.

KING HENRY
Who's there, I say? How dare you thrust yourselves
Into my private meditations!        66
Who am I? Ha?

NORFOLK
A gracious king that pardons all offences
Malice ne'er meant. Our breach of duty this way
Is business of estate, in which we come      70
To know your royal pleasure.

KING HENRY        Ye are too bold.
Go to, I'll make ye know your times of business.
Is this an hour for temporal affairs? Ha?
       *Enter Cardinal Wolsey and Cardinal Campeius, the*
       *latter with a commission*
Who's there? My good lord Cardinal? O, my Wolsey,
The quiet of my wounded conscience,      75
Thou art a cure fit for a king.
(*To Campeius*)           You're welcome,
Most learnèd reverend sir, into our kingdom.
Use us, and it. (*To Wolsey*) My good lord, have great
     care
I be not found a talker.

CARDINAL WOLSEY        Sir, you cannot.
I would your grace would give us but an hour      80
Of private conference.

KING HENRY (*to Norfolk and Suffolk*) We are busy; go.
       *Norfolk and Suffolk speak privately to one another*
       *as they depart*

NORFOLK
This priest has no pride in him!

SUFFOLK           Not to speak of.
I would not be so sick, though, for his place—
But this cannot continue.

NORFOLK        If it do
I'll venture one have-at-him.

SUFFOLK          I another.      85
               *Exeunt Norfolk and Suffolk*

CARDINAL WOLSEY (*to the King*)
Your grace has given a precedent of wisdom
Above all princes in committing freely
Your scruple to the voice of Christendom.
Who can be angry now? What envy reach you?
The Spaniard, tied by blood and favour to her,      90
Must now confess, if they have any goodness,
The trial just and noble. All the clerks—
I mean the learnèd ones in Christian kingdoms—
Have their free voices. Rome, the nurse of judgement,
Invited by your noble self, hath sent      95
One general tongue unto us: this good man,
This just and learnèd priest, Card'nal Campeius,
Whom once more I present unto your highness.

KING HENRY (*embracing Campeius*)
And once more in mine arms I bid him welcome,
And thank the holy conclave for their loves.      100
They have sent me such a man I would have wished for.

CARDINAL CAMPEIUS
Your grace must needs deserve all strangers' loves,
You are so noble. To your highness' hand
I tender my commission,
       *He gives the commission to the King*
(*To Wolsey*)           by whose virtue,
The Court of Rome commanding, you, my lord      105
Cardinal of York, are joined with me their servant
In the unpartial judging of this business.

KING HENRY
Two equal men. The Queen shall be acquainted
Forthwith for what you come. Where's Gardiner?

CARDINAL WOLSEY
I know your majesty has always loved her      110
So dear in heart not to deny her that
A woman of less place might ask by law—
Scholars allowed freely to argue for her.

KING HENRY
Ay, and the best she shall have, and my favour
To him that does her best, God forbid else. Cardinal,      115
Prithee call Gardiner to me, my new secretary.
       *Cardinal Wolsey goes to the door and calls Gardiner*
I find him a fit fellow.
       *Enter Gardiner*

CARDINAL WOLSEY (*aside to Gardiner*)
Give me your hand. Much joy and favour to you.
You are the King's now.

GARDINER (*aside to Wolsey*) But to be commanded
For ever by your grace, whose hand has raised me.

KING HENRY Come hither, Gardiner.      121
       *The King walks with Gardiner and whispers with him*

CARDINAL CAMPEIUS (*to Wolsey*)
My lord of York, was not one Doctor Pace
In this man's place before him?

CARDINAL WOLSEY        Yes, he was.

CARDINAL CAMPEIUS
Was he not held a learnèd man?

CARDINAL WOLSEY        Yes, surely.

CARDINAL CAMPEIUS
Believe me, there's an ill opinion spread then,      125
Even of yourself, lord Cardinal.

CARDINAL WOLSEY        How? Of me?

CARDINAL CAMPEIUS
They will not stick to say you envied him,
And fearing he would rise, he was so virtuous,
Kept him a foreign man still, which so grieved him
That he ran mad and died.

CARDINAL WOLSEY        Heav'n's peace be with him—
That's Christian care enough. For living murmurers

There's places of rebuke. He was a fool,                    132
For he would needs be virtuous.
(*Gesturing towards Gardiner*)         That good fellow,
If I command him, follows my appointment.
I will have none so near else. Learn this, brother:  135
We live not to be griped by meaner persons.
KING HENRY (*to Gardiner*)
Deliver this with modesty to th' Queen.     *Exit Gardiner*
The most convenient place that I can think of
For such receipt of learning is Blackfriars;
There ye shall meet about this weighty business.    140
My Wolsey, see it furnished. O, my lord,
Would it not grieve an able man to leave
So sweet a bedfellow? But conscience, conscience—
O, 'tis a tender place, and I must leave her.     *Exeunt*

**2.3**     *Enter Anne Boleyn and an Old Lady*
ANNE
Not for that neither. Here's the pang that pinches—
His highness having lived so long with her, and she
So good a lady that no tongue could ever
Pronounce dishonour of her—by my life,
She never knew harm-doing—O now, after               5
So many courses of the sun enthronèd,
Still growing in a majesty and pomp the which
To leave a thousandfold more bitter than
'Tis sweet at first t'acquire—after this process,
To give her the avaunt, it is a pity                  10
Would move a monster.
OLD LADY                        Hearts of most hard temper
Melt and lament for her.
ANNE                    O, God's will! Much better
She ne'er had known pomp; though't be temporal,
Yet if that quarrel, fortune, do divorce
It from the bearer, 'tis a sufferance panging         15
As soul and bodies severing.
OLD LADY                        Alas, poor lady!
She's a stranger now again.
ANNE                          So much the more
Must pity drop upon her. Verily,
I swear, 'tis better to be lowly born
And range with humble livers in content              20
Than to be perked up in a glist'ring grief
And wear a golden sorrow.
OLD LADY                   Our content
Is our best having.
ANNE                    By my troth and maidenhead,
I would not be a queen.
OLD LADY                  Beshrew me, I would—
And venture maidenhead for't; and so would you,    25
For all this spice of your hypocrisy.
You, that have so fair parts of woman on you,
Have, too, a woman's heart which ever yet
Affected eminence, wealth, sovereignty;
Which, to say sooth, are blessings; and which gifts,
Saving your mincing, the capacity                     31
Of your soft cheveril conscience would receive
If you might please to stretch it.
ANNE                      Nay, good troth.
OLD LADY
Yes, troth and troth. You would not be a queen?
ANNE
No, not for all the riches under heaven.              35
OLD LADY
'Tis strange. A threepence bowed would hire me,
Old as I am, to queen it. But I pray you,
What think you of a duchess? Have you limbs

To bear that load of title?
ANNE                          No, in truth.
OLD LADY
Then you are weakly made. Pluck off a little;        40
I would not be a young count in your way
For more than blushing comes to. If your back
Cannot vouchsafe this burden, 'tis too weak
Ever to get a boy.
ANNE                    How you do talk!
I swear again, I would not be a queen                 45
For all the world.
OLD LADY                In faith, for little England
You'd venture an emballing; I myself
Would for Caernarfonshire, although there 'longed
No more to th' crown but that. Lo, who comes here?
*Enter the Lord Chamberlain*
LORD CHAMBERLAIN
Good morrow, ladies. What were't worth to know       50
The secret of your conference?
ANNE                          My good lord,
Not your demand; it values not your asking.
Our mistress' sorrows we were pitying.
LORD CHAMBERLAIN
It was a gentle business, and becoming
The action of good women. There is hope              55
All will be well.
ANNE                    Now I pray God, amen.
LORD CHAMBERLAIN
You bear a gentle mind, and heav'nly blessings
Follow such creatures. That you may, fair lady,
Perceive I speak sincerely, and high note's
Ta'en of your many virtues, the King's majesty      60
Commends his good opinion of you, and
Does purpose honour to you no less flowing
Than Marchioness of Pembroke; to which title
A thousand pound a year annual support
Out of his grace he adds.
ANNE                        I do not know             65
What kind of my obedience I should tender.
More than my all is nothing; nor my prayers
Are not words duly hallowed, nor my wishes
More worth than empty vanities; yet prayers and wishes
Are all I can return. Beseech your lordship,         70
Vouchsafe to speak my thanks and my obedience,
As from a blushing handmaid to his highness,
Whose health and royalty I pray for.
LORD CHAMBERLAIN                          Lady,
I shall not fail t'approve the fair conceit
The King hath of you. (*Aside*) I have perused her well.
Beauty and honour in her are so mingled              76
That they have caught the King, and who knows yet
But from this lady may proceed a gem
To lighten all this isle. (*To Anne*) I'll to the King
And say I spoke with you.                             80
ANNE My honoured lord.          *Exit the Lord Chamberlain*
OLD LADY Why, this it is—see, see!
I have been begging sixteen years in court,
Am yet a courtier beggarly, nor could
Come pat betwixt too early and too late              85
For any suit of pounds; and you—O, fate!—
A very fresh fish here—fie, fie upon
This compelled fortune!—have your mouth filled up
Before you open it.
ANNE                    This is strange to me.
OLD LADY
How tastes it? Is it bitter? Forty pence, no.        90
There was a lady once—'tis an old story—

That would not be a queen, that would she not,
For all the mud in Egypt. Have you heard it?
ANNE
Come, you are pleasant.
OLD LADY                          With your theme I could
O'ermount the lark. The Marchioness of Pembroke?
A thousand pounds a year, for pure respect?          96
No other obligation? By my life,
That promises more thousands. Honour's train
Is longer than his foreskirt. By this time
I know your back will bear a duchess. Say,          100
Are you not stronger than you were?
ANNE                                    Good lady,
Make yourself mirth with your particular fancy,
And leave me out on't. Would I had no being,
If this salute my blood a jot. It faints me
To think what follows.                              105
The Queen is comfortless, and we forgetful
In our long absence. Pray do not deliver
What here you've heard to her.
OLD LADY                          What do you think me—
                                          *Exeunt*

2.4    *Trumpets: sennet. Then cornetts. Enter two vergers*
       *with short silver wands; next them two Scribes in the*
       *habit of doctors; after them the Archbishop of*
       *Canterbury alone; after him the Bishops of Lincoln,*
       *Ely, Rochester, and Saint Asaph; next them, with*
       *some small distance, follows a gentleman bearing*
       *both the purse containing the great seal and a*
       *cardinal's hat; then two priests bearing each a*
       *silver cross; then a gentleman usher, bare-headed,*
       *accompanied with a serjeant-at-arms bearing a silver*
       *mace; then two gentlemen bearing two great silver*
       *pillars; after them, side by side, the two cardinals,*
       *Wolsey and Campeius; then two noblemen with*
       *the sword and mace. The King ⌈ascends⌉ to his seat*
       *under the cloth of state; the two cardinals sit under*
       *him as judges; the Queen, attended by Griffith her*
       *gentleman usher, takes place some distance from the*
       *King; the Bishops place themselves on each side*
       *the court in the manner of a consistory; below*
       *them, the Scribes. The lords sit next the Bishops.*
       *The rest of the attendants stand in convenient order*
       *about the stage*
CARDINAL WOLSEY
Whilst our commission from Rome is read
Let silence be commanded.
KING HENRY                          What's the need?
It hath already publicly been read,
And on all sides th'authority allowed.
You may then spare that time.
CARDINAL WOLSEY                Be't so. Proceed.          5
SCRIBE (*to the Crier*)
Say, 'Henry, King of England, come into the court'.
CRIER
Henry, King of England, come into the court.
KING HENRY Here.
SCRIBE (*to the Crier*)
Say, 'Katherine, Queen of England, come into the court'.
CRIER
Katherine, Queen of England, come into the court.    10
       *The Queen makes no answer, but rises out of her*
       *chair, goes about the court, comes to the King, and*
       *kneels at his feet. Then she speaks*
QUEEN KATHERINE
Sir, I desire you do me right and justice,

And to bestow your pity on me; for
I am a most poor woman, and a stranger,
Born out of your dominions, having here
No judge indifferent, nor no more assurance         15
Of equal friendship and proceeding. Alas, sir,
In what have I offended you? What cause
Hath my behaviour given to your displeasure
That thus you should proceed to put me off,
And take your good grace from me? Heaven witness
I have been to you a true and humble wife,           21
At all times to your will conformable,
Ever in fear to kindle your dislike,
Yea, subject to your countenance, glad or sorry
As I saw it inclined. When was the hour             25
I ever contradicted your desire,
Or made it not mine too? Or which of your friends
Have I not strove to love, although I knew
He were mine enemy? What friend of mine
That had to him derived your anger did I            30
Continue in my liking? Nay, gave notice
He was from thence discharged? Sir, call to mind
That I have been your wife in this obedience
Upward of twenty years, and have been blessed
With many children by you. If, in the course        35
And process of this time, you can report—
And prove it, too—against mine honour aught,
My bond to wedlock, or my love and duty
Against your sacred person, in God's name
Turn me away, and let the foul'st contempt          40
Shut door upon me, and so give me up
To the sharp'st kind of justice. Please you, sir,
The King your father was reputed for
A prince most prudent, of an excellent
And unmatched wit and judgement. Ferdinand          45
My father, King of Spain, was reckoned one
The wisest prince that there had reigned by many
A year before. It is not to be questioned
That they had gathered a wise council to them
Of every realm, that did debate this business,      50
Who deemed our marriage lawful. Wherefore I humbly
Beseech you, sir, to spare me till I may
Be by my friends in Spain advised, whose counsel
I will implore. If not, i'th' name of God,
Your pleasure be fulfilled.
CARDINAL WOLSEY                You have here, lady,     55
And of your choice, these reverend fathers, men
Of singular integrity and learning,
Yea, the elect o'th' land, who are assembled
To plead your cause. It shall be therefore bootless
That longer you desire the court, as well            60
For your own quiet, as to rectify
What is unsettled in the King.
CARDINAL CAMPEIUS                His grace
Hath spoken well and justly. Therefore, madam,
It's fit this royal session do proceed,
And that without delay their arguments               65
Be now produced and heard.
QUEEN KATHERINE (*to Wolsey*)    Lord Cardinal,
To you I speak.
CARDINAL WOLSEY Your pleasure, madam.
QUEEN KATHERINE                          Sir,
I am about to weep, but thinking that
We are a queen, or long have dreamed so, certain
The daughter of a king, my drops of tears            70
I'll turn to sparks of fire.
CARDINAL WOLSEY              Be patient yet.

QUEEN KATHERINE
I will when you are humble! Nay, before,
Or God will punish me. I do believe,
Induced by potent circumstances, that
You are mine enemy, and make my challenge          75
You shall not be my judge. For it is you
Have blown this coal betwixt my lord and me,
Which God's dew quench. Therefore I say again,
I utterly abhor, yea, from my soul,
Refuse you for my judge, whom yet once more         80
I hold my most malicious foe, and think not
At all a friend to truth.
CARDINAL WOLSEY            I do profess
You speak not like yourself, who ever yet
Have stood to charity, and displayed th'effects
Of disposition gentle and of wisdom                85
O'er-topping woman's power. Madam, you do me wrong.
I have no spleen against you, nor injustice
For you or any. How far I have proceeded,
Or how far further shall, is warranted
By a commission from the consistory,               90
Yea, the whole consistory of Rome. You charge me
That I 'have blown this coal'. I do deny it.
The King is present. If it be known to him
That I gainsay my deed, how may he wound,
And worthily, my falsehood—yea, as much            95
As you have done my truth. If he know
That I am free of your report, he knows
I am not of your wrong. Therefore in him
It lies to cure me, and the cure is to
Remove these thoughts from you. The which before
His highness shall speak in, I do beseech          101
You, gracious madam, to unthink your speaking,
And to say so no more.
QUEEN KATHERINE            My lord, my lord—
I am a simple woman, much too weak
T'oppose your cunning. You're meek and humble-
                                        mouthed;  105
You sign your place and calling, in full seeming,
With meekness and humility—but your heart
Is crammed with arrogancy, spleen, and pride.
You have by fortune and his highness' favours      109
Gone slightly o'er low steps, and now are mounted
Where powers are your retainers, and your words,
Domestics to you, serve your will as't please
Yourself pronounce their office. I must tell you,
You tender more your person's honour than
Your high profession spiritual, that again         115
I do refuse you for my judge, and here,
Before you all, appeal unto the Pope,
To bring my whole cause 'fore his holiness,
And to be judged by him.
            *She curtsies to the King and begins to depart*
CARDINAL CAMPEIUS          The Queen is obstinate,
Stubborn to justice, apt to accuse it, and         120
Disdainful to be tried by't. 'Tis not well.
She's going away.
KING HENRY (*to the Crier*) Call her again.
CRIER
Katherine, Queen of England, come into the court.
GRIFFITH (*to the Queen*) Madam, you are called back.
QUEEN KATHERINE
What need you note it? Pray you keep your way.     125
When *you* are called, return. Now the Lord help.
They vex me past my patience. Pray you, pass on.
I will not tarry; no, nor ever more

Upon this business my appearance make
In any of their courts.
            *Exeunt Queen Katherine and her attendants*
KING HENRY          Go thy ways, Kate.             130
That man i'th' world who shall report he has
A better wife, let him in naught be trusted
For speaking false in that. Thou art alone—
If thy rare qualities, sweet gentleness,
Thy meekness saint-like, wife-like government,     135
Obeying in commanding, and thy parts
Sovereign and pious else could speak thee out—
The queen of earthly queens. She's noble born,
And like her true nobility she has
Carried herself towards me.
CARDINAL WOLSEY          Most gracious sir,         140
In humblest manner I require your highness
That it shall please you to declare in hearing
Of all these ears—for where I am robbed and bound,
There must I be unloosed, although not there
At once and fully satisfied—whether ever I         145
Did broach this business to your highness, or
Laid any scruple in your way which might
Induce you to the question on't, or ever
Have to you, but with thanks to God for such
A royal lady, spake one the least word that might  150
Be to the prejudice of her present state,
Or touch of her good person?
KING HENRY          My lord Cardinal,
I do excuse you; yea, upon mine honour,
I free you from't. You are not to be taught
That you have many enemies that know not           155
Why they are so, but, like to village curs,
Bark when their fellows do. By some of these
The Queen is put in anger. You're excused.
But will you be more justified? You ever
Have wished the sleeping of this business, never desired
It to be stirred, but oft have hindered, oft,      161
The passages made toward it. On my honour
I speak my good lord Card'nal to this point,
And thus far clear him. Now, what moved me to't,
I will be bold with time and your attention.       165
Then mark th'inducement. Thus it came—give heed to't.
My conscience first received a tenderness,
Scruple, and prick, on certain speeches uttered
By th' Bishop of Bayonne, then French Ambassador,
Who had been hither sent on the debating           170
A marriage 'twixt the Duke of Orléans and
Our daughter Mary. I'th' progress of this business,
Ere a determinate resolution, he—
I mean the Bishop—did require a respite
Wherein he might the King his lord advertise       175
Whether our daughter were legitimate,
Respecting this our marriage with the dowager,
Sometimes our brother's wife. This respite shook
The bosom of my conscience, entered me,
Yea, with a spitting power, and made to tremble    180
The region of my breast; which forced such way
That many mazed considerings did throng
And prest in with this caution. First, methought
I stood not in the smile of heaven, who had
Commanded nature that my lady's womb,              185
If it conceived a male child by me, should
Do no more offices of life to't than
The grave does yield to th' dead. For her male issue
Or died where they were made, or shortly after
This world had aired them. Hence I took a thought
This was a judgement on me that my kingdom,        191

Well worthy the best heir o'th' world, should not
Be gladded in't by me. Then follows that
I weighed the danger which my realms stood in
By this my issue's fail, and that gave to me          195
Many a groaning throe. Thus hulling in
The wild sea of my conscience, I did steer
Toward this remedy, whereupon we are
Now present here together—that's to say
I meant to rectify my conscience, which               200
I then did feel full sick, and yet not well,
By all the reverend fathers of the land
And doctors learned. First I began in private
With you, my lord of Lincoln. You remember
How under my oppression I did reek                    205
When I first moved you.

LINCOLN                        Very well, my liege.

KING HENRY
I have spoke long. Be pleased yourself to say
How far you satisfied me.

LINCOLN                        So please your highness,
The question did at first so stagger me,
Bearing a state of mighty moment in't                 210
And consequence of dread, that I committed
The daring'st counsel which I had to doubt,
And did entreat your highness to this course
Which you are running here.

KING HENRY (to Canterbury)        I then moved you,
My lord of Canterbury, and got your leave             215
To make this present summons. Unsolicited
I left no reverend person in this court,
But by particular consent proceeded
Under your hands and seals. Therefore, go on,
For no dislike i'th' world against the person         220
Of the good Queen, but the sharp thorny points
Of my allegèd reasons, drives this forward.
Prove but our marriage lawful, by my life
And kingly dignity, we are contented
To wear our mortal state to come with her,            225
Katherine, our queen, before the primest creature
That's paragoned o'th' world.

CARDINAL CAMPEIUS              So please your highness,
The Queen being absent, 'tis a needful fitness
That we adjourn this court till further day.
Meanwhile must be an earnest motion                   230
Made to the Queen to call back her appeal
She intends unto his holiness.

KING HENRY (aside)              I may perceive
These cardinals trifle with me. I abhor
This dilatory sloth and tricks of Rome.
My learned and well-belovèd servant, Cranmer,         235
Prithee return. With thy approach I know
My comfort comes along. (Aloud) Break up the court.
I say, set on.            Exeunt in manner as they entered

                          ⬡

**3.1**    Enter Queen Katherine and her women, as at work

QUEEN KATHERINE
Take thy lute, wench. My soul grows sad with troubles.
Sing, and disperse 'em if thou canst. Leave working.

GENTLEWOMAN (sings)
     Orpheus with his lute made trees,
     And the mountain tops that freeze,
         Bow themselves when he did sing.          5
     To his music plants and flowers
     Ever sprung, as sun and showers
         There had made a lasting spring.

     Everything that heard him play,
     Even the billows of the sea,                  10
         Hung their heads, and then lay by.
     In sweet music is such art,
     Killing care and grief of heart
         Fall asleep, or hearing, die.

     Enter Griffith, a gentleman

QUEEN KATHERINE How now?                            15

GRIFFITH
An't please your grace, the two great cardinals
Wait in the presence.

QUEEN KATHERINE        Would they speak with me?

GRIFFITH
They willed me say so, madam.

QUEEN KATHERINE        Pray their graces
To come near.                    Exit Griffith
               What can be their business
With me, a poor weak woman, fall'n from favour?    20
I do not like their coming, now I think on't;
They should be good men, their affairs as righteous—
But all hoods make not monks.

     Enter the two cardinals, Wolsey and Campeius,
     ushered by Griffith

CARDINAL WOLSEY                Peace to your highness.

QUEEN KATHERINE
Your graces find me here part of a housewife—
I would be all, against the worst may happen.      25
What are your pleasures with me, reverend lords?

CARDINAL WOLSEY
May it please you, noble madam, to withdraw
Into your private chamber, we shall give you
The full cause of our coming.

QUEEN KATHERINE              Speak it here.
There's nothing I have done yet, o' my conscience, 30
Deserves a corner. Would all other women
Could speak this with as free a soul as I do.
My lords, I care not—so much I am happy
Above a number—if my actions
Were tried by ev'ry tongue, ev'ry eye saw 'em,     35
Envy and base opinion set against 'em,
I know my life so even. If your business
Seek me out and that way I am wife in,
Out with it boldly. Truth loves open dealing.

CARDINAL WOLSEY
Tanta est erga te mentis integritas, Regina serenissima—

QUEEN KATHERINE O, good my lord, no Latin.         41
I am not such a truant since my coming
As not to know the language I have lived in.
A strange tongue makes my cause more strange
     suspicious—
Pray, speak in English. Here are some will thank you,
If you speak truth, for their poor mistress' sake.  46
Believe me, she has had much wrong. Lord Cardinal,
The willing'st sin I ever yet committed
May be absolved in English.

CARDINAL WOLSEY              Noble lady,
I am sorry my integrity should breed—              50
And service to his majesty and you—
So deep suspicion, where all faith was meant.
We come not by the way of accusation,
To taint that honour every good tongue blesses,
Nor to betray you any way to sorrow—               55
You have too much, good lady—but to know
How you stand minded in the weighty difference
Between the King and you, and to deliver,

Like free and honest men, our just opinions
And comforts to your cause.
CARDINAL CAMPEIUS                Most honoured madam,
My lord of York, out of his noble nature,                61
Zeal, and obedience he still bore your grace,
Forgetting, like a good man, your late censure
Both of his truth and him—which was too far—
Offers, as I do, in a sign of peace,                     65
His service and his counsel.
QUEEN KATHERINE (*aside*)          To betray me.
(*Aloud*) My lords, I thank you both for your good
     wills.
Ye speak like honest men—pray God ye prove so.
But how to make ye suddenly an answer
In such a point of weight, so near mine honour—          70
More near my life, I fear—with my weak wit,
And to such men of gravity and learning,
In truth I know not. I was set at work
Among my maids, full little—God knows—looking
Either for such men or such business.                    75
For her sake that I have been—for I feel
The last fit of my greatness—good your graces,
Let me have time and counsel for my cause.
Alas, I am a woman friendless, hopeless.
CARDINAL WOLSEY
Madam, you wrong the King's love with these fears.
Your hopes and friends are infinite.
QUEEN KATHERINE                    In England      81
But little for my profit. Can you think, lords,
That any Englishman dare give me counsel,
Or be a known friend 'gainst his highness' pleasure—
Though he be grown so desperate to be honest—            85
And live a subject? Nay, forsooth, my friends,
They that must weigh out my afflictions,
They that my trust must grow to, live not here.
They are, as all my other comforts, far hence,
In mine own country, lords.
CARDINAL CAMPEIUS          I would your grace       90
Would leave your griefs and take my counsel.
QUEEN KATHERINE                         How, sir?
CARDINAL CAMPEIUS
Put your main cause into the King's protection.
He's loving and most gracious. 'Twill be much
Both for your honour better and your cause,
For if the trial of the law o'ertake ye                  95
You'll part away disgraced.
CARDINAL WOLSEY (*to the Queen*) He tells you rightly.
QUEEN KATHERINE
Ye tell me what ye wish for both—my ruin.
Is this your Christian counsel? Out upon ye!
Heaven is above all yet—there sits a judge
That no king can corrupt.
CARDINAL CAMPEIUS          Your rage mistakes us.   100
QUEEN KATHERINE
The more shame for ye! Holy men I thought ye,
Upon my soul, two reverend cardinal virtues—
But cardinal sins and hollow hearts I fear ye.
Mend 'em, for shame, my lords! Is this your comfort?
The cordial that ye bring a wretched lady,               105
A woman lost among ye, laughed at, scorned?
I will not wish ye half my miseries—
I have more charity. But say I warned ye.
Take heed, for heaven's sake take heed, lest at once
The burden of my sorrows fall upon ye.                   110
CARDINAL WOLSEY
Madam, this is a mere distraction.
You turn the good we offer into envy.

QUEEN KATHERINE
Ye turn me into nothing. Woe upon ye,
And all such false professors. Would you have me—
If you have any justice, any pity,                       115
If ye be anything but churchmen's habits—
Put my sick cause into his hands that hates me?
Alas, he's banished me his bed already—
His love, too, long ago. I am old, my lords,
And all the fellowship I hold now with him               120
Is only my obedience. What can happen
To me above this wretchedness? All your studies
Make me accursed like this.
CARDINAL CAMPEIUS               Your fears are worse.
QUEEN KATHERINE
Have I lived thus long—let me speak myself,
Since virtue finds no friends—a wife, a true one?       125
A woman, I dare say, without vainglory,
Never yet branded with suspicion?
Have I with all my full affections
Still met the King, loved him next heav'n, obeyed him,
Been out of fondness superstitious to him,               130
Almost forgot my prayers to content him?
And am I thus rewarded? 'Tis not well, lords.
Bring me a constant woman to her husband,
One that ne'er dreamed a joy beyond his pleasure,
And to that woman when she has done most,                135
Yet will I add an honour, a great patience.
CARDINAL WOLSEY
Madam, you wander from the good we aim at.
QUEEN KATHERINE
My lord, I dare not make myself so guilty
To give up willingly that noble title
Your master wed me to. Nothing but death                140
Shall e'er divorce my dignities.
CARDINAL WOLSEY                    Pray, hear me.
QUEEN KATHERINE
Would I had never trod this English earth,
Or felt the flatteries that grow upon it.
Ye have angels' faces, but heaven knows your hearts.
What will become of me now, wretched lady?               145
I am the most unhappy woman living.
(*To her women*) Alas, poor wenches, where are now
     your fortunes?
Shipwrecked upon a kingdom where no pity,
No friends, no hope, no kindred weep for me?
Almost no grave allowed me? Like the lily,               150
That once was mistress of the field and flourished,
I'll hang my head and perish.
CARDINAL WOLSEY               If your grace
Could but be brought to know our ends are honest,
You'd feel more comfort. Why should we, good lady,
Upon what cause, wrong you? Alas, our places,            155
The way of our profession, is against it.
We are to cure such sorrows, not to sow 'em.
For goodness' sake, consider what you do,
How you may hurt yourself, ay, utterly
Grow from the King's acquaintance by this carriage.
The hearts of princes kiss obedience,                    161
So much they love it, but to stubborn spirits
They swell and grow as terrible as storms.
I know you have a gentle noble temper,
A soul as even as a calm. Pray, think us                 165
Those we profess—peacemakers, friends, and servants.
CARDINAL CAMPEIUS
Madam, you'll find it so. You wrong your virtues
With these weak women's fears. A noble spirit,

As yours was put into you, ever casts
Such doubts as false coin from it. The King loves you.
Beware you lose it not. For us, if you please　　171
To trust us in your business, we are ready
To use our utmost studies in your service.

QUEEN KATHERINE
Do what ye will, my lords, and pray forgive me.
If I have used myself unmannerly,　　175
You know I am a woman, lacking wit
To make a seemly answer to such persons.
Pray do my service to his majesty.
He has my heart yet, and shall have my prayers
While I shall have my life. Come, reverend fathers,
Bestow your counsels on me. She now begs　　181
That little thought, when she set footing here,
She should have bought her dignities so dear. *Exeunt*

**3.2**　*Enter the Duke of Norfolk, the Duke of Suffolk, Lord
Surrey, and the Lord Chamberlain*

NORFOLK
If you will now unite in your complaints,
And force them with a constancy, the Cardinal
Cannot stand under them. If you omit
The offer of this time, I cannot promise
But that you shall sustain more new disgraces　　5
With these you bear already.

SURREY　　　　　　I am joyful
To meet the least occasion that may give me
Remembrance of my father-in-law the Duke,
To be revenged on him.

SUFFOLK　　　　　　Which of the peers
Have uncontemned gone by him, or at least　　10
Strangely neglected? When did he regard
The stamp of nobleness in any person
Out of himself?

LORD CHAMBERLAIN My lords, you speak your pleasures.
What he deserves of you and me I know;
What we can do to him—though now the time　　15
Gives way to us—I much fear. If you cannot
Bar his access to th' King, never attempt
Anything on him, for he hath a witchcraft
Over the King in's tongue.

NORFOLK　　　　　　O, fear him not.
His spell in that is out. The King hath found　　20
Matter against him that for ever mars
The honey of his language. No, he's settled,
Not to come off, in his displeasure.

SURREY　　　　　　Sir,
I should be glad to hear such news as this
Once every hour.

NORFOLK　　　　　Believe it, this is true.　　25
In the divorce his contrary proceedings
Are all unfolded, wherein he appears
As I would wish mine enemy.

SURREY　　　　　　How came
His practices to light?

SUFFOLK　　　　　Most strangely.

SURREY　　　　　　　　O, how, how?

SUFFOLK
The Cardinal's letters to the Pope miscarried,　　30
And came to th'eye o'th' King, wherein was read
How that the Cardinal did entreat his holiness
To stay the judgement o'th' divorce, for if
It did take place, 'I do', quoth he, 'perceive
My king is tangled in affection to　　35

A creature of the Queen's, Lady Anne Boleyn'.

SURREY
Has the King this?

SUFFOLK　　　　　Believe it.

SURREY　　　　　　　　Will this work?

LORD CHAMBERLAIN
The King in this perceives him how he coasts
And hedges his own way. But in this point
All his tricks founder, and he brings his physic　　40
After his patient's death. The King already
Hath married the fair lady.

SURREY　　　　　　Would he had.

SUFFOLK
May you be happy in your wish, my lord,
For I profess you have it.

SURREY　　　　　Now all my joy
Trace the conjunction.

SUFFOLK　　　　　My amen to't.

NORFOLK　　　　　　　　All men's.　　45

SUFFOLK
There's order given for her coronation.
Marry, this is yet but young, and may be left
To some ears unrecounted. But, my lords,
She is a gallant creature, and complete
In mind and feature. I persuade me, from her　　50
Will fall some blessing to this land which shall
In it be memorized.

SURREY　　　　　But will the King
Digest this letter of the Cardinal's?
The Lord forbid!

NORFOLK　　　　　Marry, amen.

SUFFOLK　　　　　　　　No, no—
There be more wasps that buzz about his nose　　55
Will make this sting the sooner. Cardinal Campeius
Is stol'n away to Rome; hath ta'en no leave;
Has left the cause o'th' King unhandled, and
Is posted as the agent of our Cardinal
To second all his plot. I do assure you　　60
The King cried 'Ha!' at this.

LORD CHAMBERLAIN　　　Now God incense him,
And let him cry 'Ha!' louder.

NORFOLK　　　　　　But, my lord,
When returns Cranmer?

SUFFOLK
He is returned in his opinions, which
Have satisfied the King for his divorce,　　65
Together with all famous colleges,
Almost, in Christendom. Shortly, I believe,
His second marriage shall be published, and
Her coronation. Katherine no more
Shall be called 'Queen', but 'Princess Dowager',　　70
And 'widow to Prince Arthur'.

NORFOLK　　　　　　This same Cranmer's
A worthy fellow, and hath ta'en much pain
In the King's business.

SUFFOLK　　　　　He has, and we shall see him
For it an archbishop.

NORFOLK　　　　　So I hear.

SUFFOLK　　　　　　　　'Tis so.
*Enter Cardinal Wolsey and Cromwell*
The Cardinal.

NORFOLK　　　　Observe, observe—he's moody.　　75
*They stand apart and observe Wolsey and Cromwell*

CARDINAL WOLSEY (*to Cromwell*)
The packet, Cromwell—gave't you the King?

CROMWELL
    To his own hand, in's bedchamber.
CARDINAL WOLSEY               Looked he
    O'th' inside of the paper?
CROMWELL               Presently
    He did unseal them, and the first he viewed
    He did it with a serious mind; a heed          80
    Was in his countenance. You he bade
    Attend him here this morning.
CARDINAL WOLSEY           Is he ready
    To come abroad?
CROMWELL I think by this he is.
CARDINAL WOLSEY Leave me a while.      *Exit Cromwell*
    (*Aside*) It shall be to the Duchess of Alençon,   86
    The French King's sister—he shall marry her.
    Anne Boleyn? No, I'll no Anne Boleyns for him.
    There's more in't than fair visage. Boleyn?
    No, we'll no Boleyns. Speedily I wish        90
    To hear from Rome. The Marchioness of Pembroke?
    *The nobles speak among themselves*
NORFOLK
    He's discontented.
SUFFOLK            Maybe he hears the King
    Does whet his anger to him.
SURREY              Sharp enough,
    Lord, for thy justice.                  94
CARDINAL WOLSEY (*aside*)
    The late Queen's gentlewoman? A knight's daughter
    To be her mistress' mistress? The Queen's queen?
    This candle burns not clear; 'tis I must snuff it,
    Then out it goes. What though I know her virtuous
    And well deserving? Yet I know her for
    A spleeny Lutheran, and not wholesome to    100
    Our cause, that she should lie i'th' bosom of
    Our hard-ruled King. Again, there is sprung up
    An heretic, an arch-one, Cranmer, one
    Hath crawled into the favour of the King
    And is his oracle.
    *The nobles speak among themselves*
NORFOLK          He is vexed at something.   105
    *Enter King Henry reading a schedule, and Lovell*
    *with him*
SURREY
    I would 'twere something that would fret the string,
    The master-cord on's heart!
SUFFOLK            The King, the King!
KING HENRY ⌐*aside*⌐
    What piles of wealth hath he accumulated
    To his own portion? And what expense by th' hour
    Seems to flow from him? How i'th' name of thrift  110
    Does he rake this together? (*To the nobles*) Now, my lords,
    Saw you the Cardinal?
NORFOLK           My lord, we have
    Stood here observing him. Some strange commotion
    Is in his brain. He bites his lip, and starts,
    Stops on a sudden, looks upon the ground,    115
    Then lays his finger on his temple, straight
    Springs out into fast gait, then stops again,
    Strikes his breast hard, and anon he casts
    His eye against the moon. In most strange postures
    We have seen him set himself.
KING HENRY           It may well be   120
    There is a mutiny in's mind. This morning
    Papers of state he sent me to peruse
    As I required, and wot you what I found

    There, on my conscience put unwittingly?
    Forsooth, an inventory thus importing       125
    The several parcels of his plate, his treasure,
    Rich stuffs, and ornaments of household which
    I find at such proud rate that it outspeaks
    Possession of a subject.
NORFOLK            It's heaven's will.
    Some spirit put this paper in the packet    130
    To bless your eye withal.
KING HENRY          If we did think
    His contemplation were above the earth
    And fixed on spiritual object, he should still
    Dwell in his musings. But I am afraid
    His thinkings are below the moon, not worth   135
    His serious considering.
    *The King takes his seat and whispers with Lovell,*
    *who then goes to the Cardinal*
CARDINAL WOLSEY        Heaven forgive me!
    ⌐*To the King*⌐ Ever God bless your highness!
KING HENRY               Good my lord,
    You are full of heavenly stuff, and bear the inventory
    Of your best graces in your mind, the which
    You were now running o'er. You have scarce time
    To steal from spiritual leisure a brief span    141
    To keep your earthly audit. Sure, in that,
    I deem you an ill husband, and am glad
    To have you therein my companion.
CARDINAL WOLSEY            Sir,
    For holy offices I have a time; a time       145
    To think upon the part of business which
    I bear i'th' state; and nature does require
    Her times of preservation which, perforce,
    I, her frail son, amongst my brethren mortal,
    Must give my tendance to.
KING HENRY          You have said well.   150
CARDINAL WOLSEY
    And ever may your highness yoke together,
    As I will lend you cause, my doing well
    With my well-saying.
KING HENRY          'Tis well said again,
    And 'tis a kind of good deed to say well—
    And yet words are no deeds. My father loved you. 155
    He said he did, and with his deed did crown
    His word upon you. Since I had my office,
    I have kept you next my heart, have not alone
    Employed you where high profits might come home,
    But pared my present havings to bestow    160
    My bounties upon you.
CARDINAL WOLSEY (*aside*)  What should this mean?
SURREY ⌐*aside*⌐
    The Lord increase this business!
KING HENRY             Have I not made you
    The prime man of the state? I pray you tell me
    If what I now pronounce you have found true,
    And, if you may confess it, say withal    165
    If you are bound to us or no. What say you?
CARDINAL WOLSEY
    My sovereign, I confess your royal graces
    Showered on me daily have been more than could
    My studied purposes requite, which went
    Beyond all man's endeavours. My endeavours  170
    Have ever come too short of my desires,
    Yet filed with my abilities. Mine own ends
    Have been mine so that evermore they pointed
    To th' good of your most sacred person and

The profit of the state. For your great graces          175
Heaped upon me, poor undeserver, I
Can nothing render but allegiant thanks,
My prayers to heaven for you, my loyalty,
Which ever has and ever shall be growing,
Till death, that winter, kill it.
KING HENRY                              Fairly answered.          180
A loyal and obedient subject is
Therein illustrated. The honour of it
Does pay the act of it, as, i'th' contrary,
The foulness is the punishment. I presume
That as my hand has opened bounty to you,          185
My heart dropped love, my power rained honour, more
On you than any, so your hand and heart,
Your brain, and every function of your power,
Should, notwithstanding that your bond of duty,
As 'twere in love's particular, be more          190
To me, your friend, than any.
CARDINAL WOLSEY          . I do profess
That for your highness' good I ever laboured
More than mine own; that am, have, and will be—
Though all the world should crack their duty to you,
And throw it from their soul, though perils did          195
Abound, as thick as thought could make 'em, and
Appear in forms more horrid—yet, my duty,
As doth a rock against the chiding flood,
Should the approach of this wild river break,
And stand unshaken yours.
KING HENRY                              'Tis nobly spoken.          200
Take notice, lords, he has a loyal breast,
For you have seen him open't. (*To Wolsey*) Read o'er this,
          *He gives him a paper*
And after this (*giving him another paper*), and then to
          breakfast with
What appetite you have.
                              *Exit King Henry, frowning upon the*
                              *Cardinal. The nobles throng after*
                              *the King, smiling and whispering*
CARDINAL WOLSEY What should this mean?
What sudden anger's this? How have I reaped it?          205
He parted frowning from me, as if ruin
Leaped from his eyes. So looks the chafèd lion
Upon the daring huntsman that has galled him,
Then makes him nothing. I must read this paper—
I fear, the story of his anger.
          *He reads one of the papers*
                              'Tis so.          210
This paper has undone me. 'Tis th'account
Of all that world of wealth I have drawn together
For mine own ends—indeed, to gain the popedom,
And fee my friends in Rome. O negligence,
Fit for a fool to fall by! What cross devil          215
Made me put this main secret in the packet
I sent the King? Is there no way to cure this?
No new device to beat this from his brains?
I know 'twill stir him strongly. Yet I know
A way, if it take right, in spite of fortune          220
Will bring me off again. What's this?
          *He reads the other paper*
                              'To th' Pope'?
The letter, as I live, with all the business
I writ to's holiness. Nay then, farewell.
I have touched the highest point of all my greatness,
And from that full meridian of my glory          225
I haste now to my setting. I shall fall
Like a bright exhalation in the evening,
And no man see me more.

          *Enter to Cardinal Wolsey the Dukes of Norfolk and*
          *Suffolk, the Earl of Surrey, and the Lord Chamberlain*
NORFOLK
Hear the King's pleasure, Cardinal, who commands you
To render up the great seal presently          230
Into our hands, and to confine yourself
To Asher House, my lord of Winchester's,
Till you hear further from his highness.
CARDINAL WOLSEY                              Stay—
Where's your commission, lords? Words cannot carry
Authority so weighty.
SUFFOLK                              Who dare cross 'em          235
Bearing the King's will from his mouth expressly?
CARDINAL WOLSEY
Till I find more than will or words to do it—
I mean your malice—know, officious lords,
I dare and must deny it. Now I feel
Of what coarse metal ye are moulded—envy.          240
How eagerly ye follow my disgraces
As if it fed ye, and how sleek and wanton
Ye appear in everything may bring my ruin!
Follow your envious courses, men of malice.
You have Christian warrant for 'em, and no doubt
In time will find their fit rewards. That seal          246
You ask with such a violence, the King,
Mine and your master, with his own hand gave me,
Bade me enjoy it, with the place and honours,
During my life; and, to confirm his goodness,          250
Tied it by letters patents. Now, who'll take it?
SURREY
The King that gave it.
CARDINAL WOLSEY          It must be himself then.
SURREY
Thou art a proud traitor, priest.
CARDINAL WOLSEY                              Proud lord, thou liest.
Within these forty hours Surrey durst better
Have burnt that tongue than said so.
SURREY                              Thy ambition,
Thou scarlet sin, robbed this bewailing land          256
Of noble Buckingham, my father-in-law.
The heads of all thy brother cardinals
With thee and all thy best parts bound together
Weighed not a hair of his. Plague of your policy,          260
You sent me deputy for Ireland,
Far from his succour, from the King, from all
That might have mercy on the fault thou gav'st him;
Whilst your great goodness, out of holy pity,
Absolved him with an axe.
CARDINAL WOLSEY                    This, and all else          265
This talking lord can lay upon my credit,
I answer is most false. The Duke by law
Found his deserts. How innocent I was
From any private malice in his end,
His noble jury and foul cause can witness.          270
If I loved many words, lord, I should tell you
You have as little honesty as honour,
That in the way of loyalty and truth
Toward the King, my ever royal master,
Dare mate a sounder man than Surrey can be,          275
And all that love his follies.
SURREY                              By my soul,
Your long coat, priest, protects you; thou shouldst feel
My sword i'th' life-blood of thee else. My lords,
Can ye endure to hear this arrogance,
And from this fellow? If we live thus tamely,          280
To be thus jaded by a piece of scarlet,
Farewell nobility. Let his grace go forward

And dare us with his cap, like larks.
CARDINAL WOLSEY                              All goodness
   Is poison to thy stomach.
SURREY                     Yes, that goodness
   Of gleaning all the land's wealth into one,         285
   Into your own hands, Card'nal, by extortion;
   The goodness of your intercepted packets
   You writ to th' Pope against the King; your
      goodness—
   Since you provoke me—shall be most notorious.
   My lord of Norfolk, as you are truly noble,         290
   As you respect the common good, the state
   Of our despised nobility, our issues—
   Whom if he live will scarce be gentlemen—
   Produce the grand sum of his sins, the articles
   Collected from his life. (*To Wolsey*) I'll startle you   295
   Worse than the sacring-bell when the brown wench
   Lay kissing in your arms, lord Cardinal.
CARDINAL WOLSEY ⌈*aside*⌉
   How much, methinks, I could despise this man,
   But that I am bound in charity against it.
NORFOLK (*to Surrey*)
   Those articles, my lord, are in the King's hand;    300
   But thus much—they are foul ones.
CARDINAL WOLSEY                         So much fairer
   And spotless shall mine innocence arise
   When the King knows my truth.
SURREY                          This cannot save you.
   I thank my memory I yet remember
   Some of these articles, and out they shall.         305
   Now, if you can blush and cry 'Guilty', Cardinal,
   You'll show a little honesty.
CARDINAL WOLSEY                    Speak on, sir;
   I dare your worst objections. If I blush,
   It is to see a nobleman want manners.
SURREY
   I had rather want those than my head. Have at you!
   First, that without the King's assent or knowledge  311
   You wrought to be a legate, by which power
   You maimed the jurisdiction of all bishops.
NORFOLK (*to Wolsey*)
   Then, that in all you writ to Rome, or else
   To foreign princes, '*Ego et Rex meus*'            315
   Was still inscribed—in which you brought the King
   To be your servant.
SUFFOLK (*to Wolsey*)     Then, that without the knowledge
   Either of King or Council, when you went
   Ambassador to the Emperor, you made bold
   To carry into Flanders the great seal.             320
SURREY (*to Wolsey*)
   Item, you sent a large commission
   To Gregory de Cassado, to conclude,
   Without the King's will or the state's allowance,
   A league between his highness and Ferrara.
SUFFOLK (*to Wolsey*)
   That out of mere ambition you have caused          325
   Your holy hat to be stamped on the King's coin.
SURREY (*to Wolsey*)
   Then, that you have sent innumerable substance—
   By what means got, I leave to your own conscience—
   To furnish Rome, and to prepare the ways
   You have for dignities to the mere undoing         330
   Of all the kingdom. Many more there are,
   Which since they are of you, and odious,
   I will not taint my mouth with.
LORD CHAMBERLAIN                       O, my lord,
   Press not a falling man too far. 'Tis virtue.

His faults lie open to the laws. Let them,            335
   Not you, correct him. My heart weeps to see him
   So little of his great self.
SURREY                     I forgive him.
SUFFOLK
   Lord Cardinal, the King's further pleasure is—
   Because all those things you have done of late,
   By your power legantine within this kingdom,       340
   Fall into th' compass of a praemunire—
   That therefore such a writ be sued against you,
   To forfeit all your goods, lands, tenements,
   Chattels, and whatsoever, and to be
   Out of the King's protection. This is my charge.   345
NORFOLK (*to Wolsey*)
   And so we'll leave you to your meditations
   How to live better. For your stubborn answer
   About the giving back the great seal to us,
   The King shall know it and, no doubt, shall thank you.
   So fare you well, my little good lord Cardinal.    350
                    *Exeunt all but Wolsey*
CARDINAL WOLSEY
   So farewell—to the little good you bear me.
   Farewell, a long farewell, to all my greatness!
   This is the state of man. Today he puts forth
   The tender leaves of hopes; tomorrow blossoms,
   And bears his blushing honours thick upon him;     355
   The third day comes a frost, a killing frost,
   And when he thinks, good easy man, full surely
   His greatness is a-ripening, nips his root,
   And then he falls, as I do. I have ventured,
   Like little wanton boys that swim on bladders,     360
   This many summers in a sea of glory,
   But far beyond my depth; my high-blown pride
   At length broke under me, and now has left me
   Weary, and old with service, to the mercy
   Of a rude stream that must for ever hide me.       365
   Vain pomp and glory of this world, I hate ye!
   I feel my heart new opened. O, how wretched
   Is that poor man that hangs on princes' favours!
   There is betwixt that smile we would aspire to,
   That sweet aspect of princes, and their ruin,      370
   More pangs and fears than wars or women have,
   And when he falls, he falls like Lucifer,
   Never to hope again.
          *Enter Cromwell, who then stands amazed*
                          Why, how now, Cromwell?
CROMWELL
   I have no power to speak, sir.
CARDINAL WOLSEY              What, amazed
   At my misfortunes? Can thy spirit wonder           375
   A great man should decline?
          ⌈*Cromwell begins to weep*⌉
                          Nay, an you weep
   I am fall'n indeed.
CROMWELL              How does your grace?
CARDINAL WOLSEY                              Why, well—
   Never so truly happy, my good Cromwell.
   I know myself now, and I feel within me
   A peace above all earthly dignities,               380
   A still and quiet conscience. The King has cured me.
   I humbly thank his grace, and from these shoulders,
   These ruined pillars, out of pity, taken
   A load would sink a navy—too much honour.
   O, 'tis a burden, Cromwell, 'tis a burden          385
   Too heavy for a man that hopes for heaven.
CROMWELL
   I am glad your grace has made that right use of it.

CARDINAL WOLSEY
  I hope I have. I am able now, methinks,
  Out of a fortitude of soul I feel,
  To endure more miseries and greater far          390
  Than my weak-hearted enemies dare offer.
  What news abroad?
CROMWELL                    The heaviest and the worst
  Is your displeasure with the King.
CARDINAL WOLSEY                    God bless him.
CROMWELL
  The next is that Sir Thomas More is chosen       394
  Lord Chancellor in your place.
CARDINAL WOLSEY                    That's somewhat sudden.
  But he's a learnèd man. May he continue
  Long in his highness' favour, and do justice
  For truth's sake and his conscience, that his bones,
  When he has run his course and sleeps in blessings,
  May have a tomb of orphans' tears wept on him.   400
  What more?
CROMWELL    That Cranmer is returned with welcome,
  Installed lord Archbishop of Canterbury.
CARDINAL WOLSEY
  That's news indeed.
CROMWELL                    Last, that the Lady Anne,
  Whom the King hath in secrecy long married,
  This day was viewed in open as his queen,        405
  Going to chapel, and the voice is now
  Only about her coronation.
CARDINAL WOLSEY
  There was the weight that pulled me down. O,
      Cromwell,
  The King has gone beyond me. All my glories
  In that one woman I have lost for ever.          410
  No sun shall ever usher forth mine honours,
  Or gild again the noble troops that waited
  Upon my smiles. Go, get thee from me, Cromwell.
  I am a poor fall'n man, unworthy now
  To be thy lord and master. Seek the King—        415
  That sun I pray may never set—I have told him
  What and how true thou art. He will advance thee.
  Some little memory of me will stir him.
  I know his noble nature not to let
  Thy hopeful service perish too. Good Cromwell,   420
  Neglect him not. Make use now, and provide
  For thine own future safety.
CROMWELL ⌈weeping⌉            O, my lord,
  Must I then leave you? Must I needs forgo
  So good, so noble, and so true a master?
  Bear witness, all that have not hearts of iron,  425
  With what a sorrow Cromwell leaves his lord.
  The King shall have my service, but my prayers
  For ever and for ever shall be yours.
CARDINAL WOLSEY (weeping)
  Cromwell, I did not think to shed a tear
  In all my miseries, but thou hast forced me,     430
  Out of thy honest truth, to play the woman.
  Let's dry our eyes, and thus far hear me, Cromwell,
  And when I am forgotten, as I shall be,
  And sleep in dull cold marble, where no mention
  Of me more must be heard of, say I taught thee—  435
  Say Wolsey, that once trod the ways of glory,
  And sounded all the depths and shoals of honour,
  Found thee a way, out of his wreck, to rise in,
  A sure and safe one, though thy master missed it.
  Mark but my fall, and that that ruined me.       440
  Cromwell, I charge thee, fling away ambition.

  By that sin fell the angels. How can man, then,
  The image of his maker, hope to win by it?
  Love thyself last. Cherish those hearts that hate thee.
  Corruption wins not more than honesty.           445
  Still in thy right hand carry gentle peace
  To silence envious tongues. Be just, and fear not.
  Let all the ends thou aim'st at be thy country's,
  Thy God's, and truth's. Then if thou fall'st, O
      Cromwell,
  Thou fall'st a blessèd martyr.                   450
  Serve the King. And prithee, lead me in—
  There take an inventory of all I have:
  To the last penny 'tis the King's. My robe,
  And my integrity to heaven, is all
  I dare now call mine own. O Cromwell, Cromwell, 455
  Had I but served my God with half the zeal
  I served my King, He would not in mine age
  Have left me naked to mine enemies.
CROMWELL
  Good sir, have patience.
CARDINAL WOLSEY            So I have. Farewell
  The hopes of court; my hopes in heaven do dwell. 460
                                           Exeunt

                        ❀

4.1   *Enter the two Gentlemen meeting one another. The*
      *first holds a paper*
FIRST GENTLEMAN
  You're well met once again.
SECOND GENTLEMAN            So are you.
FIRST GENTLEMAN
  You come to take your stand here and behold
  The Lady Anne pass from her coronation?
SECOND GENTLEMAN
  'Tis all my business. At our last encounter
  The Duke of Buckingham came from his trial.       5
FIRST GENTLEMAN
  'Tis very true. But that time offered sorrow,
  This, general joy.
SECOND GENTLEMAN  'Tis well. The citizens,
  I am sure, have shown at full their royal minds—
  As, let 'em have their rights, they are ever forward—
  In celebration of this day with shows,           10
  Pageants, and sights of honour.
FIRST GENTLEMAN                    Never greater,
  Nor, I'll assure you, better taken, sir.
SECOND GENTLEMAN
  May I be bold to ask what that contains,
  That paper in your hand?
FIRST GENTLEMAN            Yes, 'tis the list
  Of those that claim their offices this day       15
  By custom of the coronation.
  The Duke of Suffolk is the first, and claims
  To be High Steward; next, the Duke of Norfolk,
  He to be Earl Marshal. You may read the rest.
                *He gives him the paper*
SECOND GENTLEMAN
  I thank you, sir. Had I not known those customs, 20
  I should have been beholden to your paper.
  But I beseech you, what's become of Katherine,
  The Princess Dowager? How goes her business?
FIRST GENTLEMAN
  That I can tell you too. The Archbishop
  Of Canterbury, accompanied with other            25
  Learnèd and reverend fathers of his order,
  Held a late court at Dunstable, six miles off

From Ampthill, where the Princess lay; to which
She was often cited by them, but appeared not.
And, to be short, for not appearance, and            30
The King's late scruple, by the main assent
Of all these learnèd men, she was divorced,
And the late marriage made of none effect,
Since which she was removed to Kimbolton,
Where she remains now sick.

SECOND GENTLEMAN                   Alas, good lady!     35
*Flourish of trumpets within*
The trumpets sound. Stand close. The Queen is coming.
*Enter the coronation procession, which passes over
the stage in order and state. Hautboys, within,
⌈play during the procession⌉*

THE ORDER OF THE CORONATION

*1. First, ⌈enter⌉ trumpeters, who play a lively
flourish.
2. Then, enter two judges.
3. Then, enter the Lord Chancellor, with both the
purse containing the great seal and the mace borne
before him.
4. Then, enter choristers singing; ⌈with them,
musicians playing.⌉
5. Then, enter the Lord Mayor of London bearing
the mace, followed by Garter King-of-Arms wearing
his coat of arms and a gilt copper crown.
6. Then, enter Marquis Dorset bearing a sceptre of
gold, and wearing, on his head, a demi-coronal of
gold and, about his neck, a collar of esses. With him
enter the Earl of Surrey bearing the rod of silver
with the dove, crowned with an earl's coronet, and
also wearing a collar of esses.
7. Next, enter the Duke of Suffolk as High Steward,
in his robe of estate, with his coronet on his head,
and bearing a long white wand. With him, enter the
Duke of Norfolk with the rod of marshalship and
a coronet on his head. Each wears a collar of esses.
8. Then, under a canopy borne by four barons of the
Cinque Ports, enter Anne, the new Queen, in her robe.
Her hair, which hangs loose, is richly adorned with
pearl. She wears a crown. Accompanying her on either
side are the Bishops of London and Winchester.
9. Next, enter the old Duchess of Norfolk, in a
coronal of gold wrought with flowers, bearing the
Queen's train.
10. Finally, enter certain ladies or countesses, with
plain circlets of gold without flowers.

The two Gentlemen comment on the procession as it
passes over the stage*

SECOND GENTLEMAN
A royal train, believe me. These I know.
Who's that that bears the sceptre?

FIRST GENTLEMAN                      Marquis Dorset. ·
And that, the Earl of Surrey with the rod.

SECOND GENTLEMAN
A bold brave gentleman. That should be          40
The Duke of Suffolk?

FIRST GENTLEMAN          'Tis the same: High Steward.

SECOND GENTLEMAN
And that, my lord of Norfolk?

FIRST GENTLEMAN          Yes.

SECOND GENTLEMAN (*seeing Anne*)   Heaven bless thee!
Thou hast the sweetest face I ever looked on.
Sir, as I have a soul, she is an angel.
Our King has all the Indies in his arms,          45

And more, and richer, when he strains that lady.
I cannot blame his conscience.

FIRST GENTLEMAN          They that bear
The cloth of honour over her are four barons
Of the Cinque Ports.

SECOND GENTLEMAN Those men are happy,          50
And so are all are near her.
I take it she that carries up the train
Is that old noble lady, Duchess of Norfolk.

FIRST GENTLEMAN
It is. And all the rest are countesses.

SECOND GENTLEMAN
Their coronets say so. These are stars indeed—     55
⌈FIRST GENTLEMAN⌉
And sometimes falling ones.

SECOND GENTLEMAN          No more of that.
*Exit the last of the procession, and then
a great flourish of trumpets within
Enter a third Gentleman ⌈in a sweat⌉*

FIRST GENTLEMAN
God save you, sir. Where have you been broiling?

THIRD GENTLEMAN
Among the crowd i'th' Abbey, where a finger
Could not be wedged in more. I am stifled
With the mere rankness of their joy.          60

SECOND GENTLEMAN
You saw the ceremony?

THIRD GENTLEMAN          That I did.

FIRST GENTLEMAN How was it?

THIRD GENTLEMAN
Well worth the seeing.

SECOND GENTLEMAN          Good sir, speak it to us.

THIRD GENTLEMAN
As well as I am able. The rich stream
Of lords and ladies, having brought the Queen     65
To a prepared place in the choir, fell off
A distance from her, while her grace sat down
To rest a while—some half an hour or so—
In a rich chair of state, opposing freely
The beauty of her person to the people.          70
Believe me, sir, she is the goodliest woman
That ever lay by man; which when the people
Had the full view of, such a noise arose
As the shrouds make at·sea in a stiff tempest,
As loud and to as many tunes. Hats, cloaks—     75
Doublets, I think—flew up, and had their faces
Been loose, this day they had been lost. Such joy
I never saw before. Great-bellied women,
That had not half a week to go, like rams
In the old time of war, would shake the press,     80
And make 'em reel before 'em. No man living
Could say 'This is my wife' there, all were woven
So strangely in one piece.

SECOND GENTLEMAN          But what followed?

THIRD GENTLEMAN
At length her grace rose, and with modest paces
Came to the altar, where she kneeled, and saint-like
Cast her fair eyes to heaven, and prayed devoutly,     86
Then rose again, and bowed her to the people,
When by the Archbishop of Canterbury
She had all the royal makings of a queen,
As holy oil, Edward Confessor's crown,          90
The rod and bird of peace, and all such emblems
Laid nobly on her. Which performed, the choir,
With all the choicest music of the kingdom,
Together sung *Te Deum*. So she parted,
And with the same full state paced back again     95

To York Place, where the feast is held.
FIRST GENTLEMAN                  Sir,
You must no more call it York Place—that's past,
For since the Cardinal fell, that title's lost.
'Tis now the King's, and called Whitehall.
THIRD GENTLEMAN              ·I know it,
But 'tis so lately altered that the old name     100
Is fresh about me.
SECOND GENTLEMAN   What two reverend bishops
Were those that went on each side of the Queen?
THIRD GENTLEMAN
Stokesley and Gardiner, the one of Winchester—
Newly preferred from the King's secretary—
The other London.
SECOND GENTLEMAN   He of Winchester     105
Is held no great good lover of the Archbishop's,
The virtuous Cranmer.
THIRD GENTLEMAN         All the land knows that.
However, yet there is no great breach. When it
    comes,
Cranmer will find a friend will not shrink from him.
SECOND GENTLEMAN
Who may that be, I pray you?
THIRD GENTLEMAN           Thomas Cromwell,    110
A man in much esteem with th' King, and truly
A worthy friend. The King has made him
Master o'th' Jewel House,
And one already of the Privy Council.
SECOND GENTLEMAN
He will deserve more.
THIRD GENTLEMAN       Yes, without all doubt.     115
Come, gentlemen, ye shall go my way,
Which is to th' court, and there ye shall be my
    guests.
Something I can command. As I walk thither
I'll tell ye more.
FIRST and SECOND GENTLEMEN   You may command us, sir.
                                     *Exeunt*

**4.2**   ⌜*Three chairs.*⌝ *Enter Katherine Dowager, sick, led*
      *between Griffith her gentleman usher, and Patience*
      *her woman*
GRIFFITH
How does your grace?
KATHERINE            O Griffith, sick to death.
My legs, like loaden branches, bow to th' earth,
Willing to leave their burden. Reach a chair.
    *A chair is brought to her. She sits*
So now, methinks, I feel a little ease.
Didst thou not tell me, Griffith, as thou led'st me,    5
That the great child of honour, Cardinal Wolsey,
Was dead?
GRIFFITH       Yes, madam, but I think your grace,
Out of the pain you suffered, gave no ear to't.
KATHERINE
Prithee, good Griffith, tell me how he died.
If well, he stepped before me happily           10
For my example.
GRIFFITH       Well, the voice goes, madam.
For after the stout Earl Northumberland
Arrested him at York, and brought him forward,
As a man sorely tainted, to his answer,
He fell sick, suddenly, and grew so ill           15
He could not sit his mule.
KATHERINE          Alas, poor man.
GRIFFITH
At last, with easy roads, he came to Leicester,

Lodged in the abbey, where the reverend abbot,
With all his convent, honourably received him,
To whom he gave these words: 'O father abbot,    20
An old man broken with the storms of state
Is come to lay his weary bones among ye.
Give him a little earth, for charity.'
So went to bed, where eagerly his sickness
Pursued him still, and three nights after this,    25
About the hour of eight, which he himself
Foretold should be his last, full of repentance,
Continual meditations, tears, and sorrows,
He gave his honours to the world again,
His blessèd part to heaven, and slept in peace.    30
KATHERINE
So may he rest, his faults lie gently on him.
Yet thus far, Griffith, give me leave to speak him,
And yet with charity. He was a man
Of an unbounded stomach, ever ranking
Himself with princes; one that by suggestion    35
Tied all the kingdom. Simony was fair play.
His own opinion was his law. I'th' presence
He would say untruths, and be ever double
Both in his words and meaning. He was never,
But where he meant to ruin, pitiful.           40
His promises were, as he then was, mighty;
But his performance, as he is now, nothing.
Of his own body he was ill, and gave
The clergy ill example.
GRIFFITH            Noble madam,
Men's evil manners live in brass, their virtues    45
We write in water. May it please your highness
To hear me speak his good now?
KATHERINE            Yes, good Griffith,
I were malicious else.
GRIFFITH            This cardinal,
Though from an humble stock, undoubtedly
Was fashioned to much honour. From his cradle    50
He was a scholar, and a ripe and good one,
Exceeding wise, fair-spoken, and persuading;
Lofty and sour to them that loved him not,
But to those men that sought him, sweet as summer.
And though he were unsatisfied in getting—    55
Which was a sin—yet in bestowing, madam,
He was most princely: ever witness for him
Those twins of learning that he raised in you,
Ipswich and Oxford—one of which fell with him,
Unwilling to outlive the good that did it;       60
The other, though unfinished, yet so famous,
So excellent in art, and still so rising,
That Christendom shall ever speak his virtue.
His overthrow heaped happiness upon him,
For then, and not till then, he felt himself,     65
And found the blessèdness of being little.
And to add greater honours to his age
Than man could give him, he died fearing God.
KATHERINE
After my death I wish no other herald,
No other speaker of my living actions         70
To keep mine honour from corruption
But such an honest chronicler as Griffith.
Whom I most hated living, thou hast made me,
With thy religious truth and modesty,
Now in his ashes honour. Peace be with him.    75
(*To her woman*) Patience, be near me still, and set me
    lower.
I have not long to trouble thee. Good Griffith,
Cause the musicians play me that sad note

I named my knell, whilst I sit meditating
On that celestial harmony I go to.                    80
    *Sad and solemn music. Katherine sleeps*
GRIFFITH (*to the woman*)
    She is asleep. Good wench, let's sit down quiet
For fear we wake her. Softly, gentle Patience.
    *They sit*

### THE VISION

*Enter, solemnly tripping one after another, six
personages clad in white robes, wearing on their
heads garlands of bays, and golden visors on their
faces. They carry branches of bays or palm in their
hands. They first congé unto Katherine, then
dance; and, at certain changes, the first two hold a
spare garland over her head at which the other four
make reverent curtsies. Then the two that held the
garland deliver the same to the other next two,
who observe the same order in their changes and
holding the garland over her head. Which done, they
deliver the same garland to the last two who
likewise observe the same order. At which, as it
were by inspiration, she makes in her sleep signs of
rejoicing, and holdeth up her hands to heaven. And
so in their dancing vanish, carrying the garland
with them. The music continues*

KATHERINE (*waking*)
    Spirits of peace, where are ye? Are ye all gone,
And leave me here in wretchedness behind ye?
    *Griffith and Patience rise and come forward*
GRIFFITH
    Madam, we are here.
KATHERINE          It is not you I call for.    85
    Saw ye none enter since I slept?
GRIFFITH           None, madam.
KATHERINE
    No? Saw you not even now a blessèd troop
Invite me to a banquet, whose bright faces
Cast thousand beams upon me, like the sun?
They promised me eternal happiness,                    90
And brought me garlands, Griffith, which I feel
I am not worthy yet to wear. I shall,
Assuredly.
GRIFFITH
    I am most joyful, madam, such good dreams
Possess your fancy.
KATHERINE        Bid the music leave.    95
    They are harsh and heavy to me.
    *Music ceases*
PATIENCE (*to Griffith*)      Do you note
How much her grace is altered on the sudden?
How long her face is drawn? How pale she looks,
And of an earthy colour? Mark her eyes?
GRIFFITH
    She is going, wench. Pray, pray.
PATIENCE        Heaven comfort her.
    *Enter a Messenger*
MESSENGER (*to Katherine*)
    An't like your grace—
KATHERINE      You are a saucy fellow—    101
    Deserve we no more reverence?
GRIFFITH (*to the Messenger*)    You are to blame,
    Knowing she will not lose her wonted greatness,
To use so rude behaviour. Go to, kneel.
MESSENGER (*kneeling before Katherine*)
    I humbly do entreat your highness' pardon.    105

My haste made me unmannerly. There is staying
A gentleman sent from the King to see you.
KATHERINE
    Admit him entrance, Griffith. But this fellow
Let me ne'er see again.      *Exit Messenger*
    *Enter Lord Caputius ⌈ushered by Griffith⌉*
            If my sight fail not,
You should be lord ambassador from the Emperor,    110
My royal nephew, and your name Caputius.
CAPUTIUS
    Madam, the same, ⌈*bowing*⌉ your servant.
KATHERINE          O, my lord,
The times and titles now are altered strangely
With me since first you knew me. But I pray you,
What is your pleasure with me?
CAPUTIUS         Noble lady,    115
First mine own service to your grace; the next,
The King's request that I would visit you,
Who grieves much for your weakness, and by me
Sends you his princely commendations,
And heartily entreats you take good comfort.    120
KATHERINE
    O, my good lord, that comfort comes too late,
'Tis like a pardon after execution.
That gentle physic, given in time, had cured me;
But now I am past all comforts here but prayers.
How does his highness?
CAPUTIUS        Madam, in good health.    125
KATHERINE
    So may he ever do, and ever flourish
When I shall dwell with worms, and my poor name
Banished the kingdom. (*To her woman*) Patience, is
    that letter
I caused you write yet sent away?
PATIENCE        No, madam.
KATHERINE (*to Caputius*)
    Sir, I most humbly pray you to deliver    130
This to my lord the King.
    *The letter is given to Caputius*
CAPUTIUS        Most willing, madam.
KATHERINE
    In which I have commended to his goodness
The model of our chaste loves, his young daughter—
The dews of heaven fall thick in blessings on her—
Beseeching him to give her virtuous breeding.    135
She is young, and of a noble modest nature.
I hope she will deserve well—and a little
To love her for her mother's sake, that loved him,
Heaven knows how dearly. My next poor petition
Is that his noble grace would have some pity    140
Upon my wretched women, that so long
Have followed both my fortunes faithfully;
Of which there is not one, I dare avow—
And now I should not lie—but will deserve,
For virtue and true beauty of the soul,    145
For honesty and decent carriage,
A right good husband. Let him be a noble,
And sure those men are happy that shall have 'em.
The last is for my men—they are the poorest,
But poverty could never draw 'em from me—    150
That they may have their wages duly paid 'em,
And something over to remember me by.
If heaven had pleased to have given me longer life,
And able means, we had not parted thus.
These are the whole contents; and, good my lord,    155
By that you love the dearest in this world,
As you wish Christian peace to souls departed,

Stand these poor people's friend and urge the King
To do me this last rite.
CAPUTIUS                    By heaven I will,
Or let me lose the fashion of a man.                    160
KATHERINE
I thank you, honest lord. Remember me
In all humility unto his highness.
Say his long trouble now is passing
Out of this world. Tell him, in death I blessed him,
For so I will. Mine eyes grow dim. Farewell,      165
My lord. Griffith, farewell.
(*To her woman*)                    Nay, Patience,
You must not leave me yet. I must to bed.
Call in more women. When I am dead, good wench,
Let me be used with honour. Strew me over
With maiden flowers, that all the world may know
I was a chaste wife to my grave. Embalm me,      171
Then lay me forth. Although unqueened, yet like
A queen and daughter to a king inter me.
I can no more.
                    *Exeunt ⌈Caputius and Griffith at one door;*
                    *Patience⌉ leading Katherine ⌈at another⌉*

                    ❧

5.1    *Enter ⌈at one door⌉ Gardiner, Bishop of*
            *Winchester; before him, a Page with a torch*
GARDINER
It's one o'clock, boy, is't not?
PAGE                    It hath struck.
GARDINER
These should be hours for necessities,
Not for delights; times to repair our nature
With comforting repose, and not for us
To waste these times.
            *Enter ⌈at another door⌉ Sir Thomas Lovell, meeting*
            *them*
                    Good hour of night, Sir Thomas!
Whither so late?
LOVELL                    Came you from the King, my lord?    6
GARDINER
I did, Sir Thomas, and left him at primero
With the Duke of Suffolk.
LOVELL                    I must to him too,
Before he go to bed. I'll take my leave.
GARDINER
Not yet, Sir Thomas Lovell—what's the matter?      10
It seems you are in haste. An if there be
No great offence belongs to't, give your friend
Some touch of your late business. Affairs that walk,
As they say spirits do, at midnight, have
In them a wilder nature than the business              15
That seeks dispatch by day.
LOVELL                    My lord, I love you,
And durst commend a secret to your ear
Much weightier than this work. The Queen's in labour—
They say in great extremity—and feared
She'll with the labour end.
GARDINER                    The fruit she goes with      20
I pray for heartily, that it may find
Good time, and live. But, for the stock, Sir Thomas,
I wish it grubbed up now.
LOVELL                    Methinks I could
Cry the amen, and yet my conscience says
She's a good creature and, sweet lady, does          25
Deserve our better wishes.
GARDINER                    But sir, sir,
Hear me, Sir Thomas. You're a gentleman

Of mine own way. I know you wise, religious.
And let me tell you, it will ne'er be well—
'Twill not, Sir Thomas Lovell, take't of me—      30
Till Cranmer, Cromwell—her two hands—and she,
Sleep in their graves.
LOVELL                    Now, sir, you speak of two
The most remarked i'th' kingdom. As for Cromwell,
Beside that of the Jewel House is made Master
O'th' Rolls and the King's secretary. Further, sir,    35
Stands in the gap and trade of more preferments
With which the time will load him. Th'Archbishop
Is the King's hand and tongue, and who dare speak
One syllable against him?
GARDINER                    Yes, yes, Sir Thomas—
There are that dare, and I myself have ventured      40
To speak my mind of him, and, indeed, this day,
Sir—I may tell it you, I think—I have
Incensed the lords o'th' Council that he is—
For so I know he is, they know he is—
A most arch heretic, a pestilence                    45
That does infect the land; with which they, moved,
Have broken with the King, who hath so far
Given ear to our complaint, of his great grace
And princely care, foreseeing those fell mischiefs
Our reasons laid before him, hath commanded      50
Tomorrow morning to the Council board
He be convented. He's a rank weed, Sir Thomas,
And we must root him out. From your affairs
I hinder you too long. Good night, Sir Thomas.
LOVELL
Many good nights, my lord; I rest your servant.      55
            *Exeunt Gardiner and Page at one door*
            *Enter King Henry and Suffolk at another door*
KING HENRY (*to Suffolk*)
Charles, I will play no more tonight.
My mind's not on't. You are too hard for me.
SUFFOLK
Sir, I did never win of you before.
KING HENRY But little, Charles,
Nor shall not when my fancy's on my play.          60
Now, Lovell, from the Queen what is the news?
LOVELL
I could not personally deliver to her
What you commanded me, but by her woman
I sent your message, who returned her thanks        64
In the great'st humbleness, and desired your highness
Most heartily to pray for her.
KING HENRY                    What sayst thou? Ha?
To pray for her? What, is she crying out?
LOVELL
So said her woman, and that her suffrance made
Almost each pang a death.
KING HENRY                    Alas, good lady.
SUFFOLK
God safely quit her of her burden, and                70
With gentle travail, to the gladding of
Your highness with an heir.
KING HENRY                    'Tis midnight, Charles.
Prithee to bed, and in thy prayers remember
Th'estate of my poor queen. Leave me alone,
For I must think of that which company              75
Would not be friendly to.
SUFFOLK                    I wish your highness
A quiet night, and my good mistress will
Remember in my prayers.
KING HENRY                    Charles, good night.
                    *Exit Suffolk*

*Enter Sir Anthony Denny*

Well, sir, what follows?

DENNY
Sir, I have brought my lord the Archbishop,          80
As you commanded me.

KING HENRY                    Ha, Canterbury?

DENNY
Ay, my good lord.

KING HENRY                    'Tis true—where is he, Denny?

DENNY
He attends your highness' pleasure.

KING HENRY                              Bring him to us.
                                        *Exit Denny*

LOVELL (*aside*)
This is about that which the Bishop spake.
I am happily come hither.                            85
          *Enter Cranmer the Archbishop, ushered by Denny*

KING HENRY (*to Lovell and Denny*) Avoid the gallery.
          ⌈*Denny begins to depart.*⌉ *Lovell seems to stay*
Ha? I have said. Be gone.
What?                         *Exeunt Lovell and Denny*

CRANMER (*aside*)
          I am fearful. Wherefore frowns he thus?
'Tis his aspect of terror. All's not well.

KING HENRY
How now, my lord? You do desire to know              90
Wherefore I sent for you.

CRANMER (*kneeling*)          It is my duty
T'attend your highness' pleasure.

KING HENRY                    Pray you, arise,
My good and gracious Lord of Canterbury.
Come, you and I must walk a turn together.
I have news to tell you. Come, come—give me your
          hand.                                      95
          ⌈*Cranmer rises. They walk*⌉
Ah, my good lord, I grieve at what I speak,
And am right sorry to repeat what follows.
I have, and most unwillingly, of late
Heard many grievous—I do say, my lord,
Grievous—complaints of you, which, being considered,
Have moved us and our Council that you shall        101
This morning come before us, where I know
You cannot with such freedom purge yourself
But that, till further trial in those charges
Which will require your answer, you must take       105
Your patience to you, and be well contented
To make your house our Tower. You a brother of us,
It fits we thus proceed, or else no witness
Would come against you.

CRANMER (*kneeling*)          I humbly thank your highness,
And am right glad to catch this good occasion       110
Most throughly to be winnowed, where my chaff
And corn shall fly asunder. For I know
There's none stands under more calumnious tongues
Than I myself, poor man.

KING HENRY                    Stand up, good Canterbury.
Thy truth and thy integrity is rooted               115
In us, thy friend. Give me thy hand. Stand up.
Prithee, let's walk.
          *Cranmer rises. They walk*
                    Now, by my halidom,
What manner of man are you? My lord, I looked
You would have given me your petition that
I should have ta'en some pains to bring together    120
Yourself and your accusers, and to have heard you
Without indurance further.

CRANMER                    Most dread liege,
The good I stand on is my truth and honesty.

If they shall fail, I with mine enemies
Will triumph o'er my person, which I weigh not,     125
Being of those virtues vacant. I fear nothing
What can be said against me.

KING HENRY                    Know you not
How your state stands i'th' world, with the whole
          world?
Your enemies are many, and not small; their practices
Must bear the same proportion, and not ever         130
The justice and the truth o'th' question carries
The dew o'th' verdict with it. At what ease
Might corrupt minds procure knaves as corrupt
To swear against you? Such things have been
          done.
You are potently opposed, and with a malice         135
Of as great size. Ween you of better luck,
I mean in perjured witness, than your master,
Whose minister you are, whiles here he lived
Upon this naughty earth? Go to, go to.
You take a precipice for no leap of danger,         140
And woo your own destruction.

CRANMER                    God and your majesty
Protect mine innocence, or I fall into
The trap is laid for me.

KING HENRY                    Be of good cheer.
They shall no more prevail than we give way to.
Keep comfort to you, and this morning see           145
You do appear before them. If they shall chance,
In charging you with matters, to commit you,
The best persuasions to the contrary
Fail not to use, and with what vehemency
Th'occasion shall instruct you. If entreaties       150
Will render you no remedy, ⌈*giving his ring*⌉ this ring
Deliver them, and your appeal to us
There make before them.
          *Cranmer weeps*
                    Look, the good man weeps.
He's honest, on mine honour. God's blest mother,
I swear he is true-hearted, and a soul              155
None better in my kingdom. Get you gone,
And do as I have bid you.               *Exit Cranmer*
                    He has strangled
His language in his tears.
          *Enter the Old Lady*
⌈LOVELL⌉ (*within*)          Come back! What mean you?
          ⌈*Enter Lovell, following her*⌉

OLD LADY
I'll not come back. The tidings that I bring
Will make my boldness manners. (*To the King*) Now
          good angels                               160
Fly o'er thy royal head, and shade thy person
Under their blessèd wings.

KING HENRY                    Now by thy looks
I guess thy message. Is the Queen delivered?
Say, 'Ay, and of a boy.'

OLD LADY                    Ay, ay, my liege,
And of a lovely boy. The God of heaven              165
Both now and ever bless her! 'Tis a girl
Promises boys hereafter. Sir, your queen
Desires your visitation, and to be
Acquainted with this stranger. 'Tis as like you
As cherry is to cherry.

KING HENRY                    Lovell—

LOVELL                         Sir?                 170

KING HENRY
Give her an hundred marks. I'll to the Queen.  *Exit*

**OLD LADY**
An hundred marks? By this light, I'll ha' more.
An ordinary groom is for such payment.
I will have more, or scold it out of him.
Said I for this the girl was like to him? I'll    175
Have more, or else unsay't; and now, while 'tis hot,
I'll put it to the issue.         *Exeunt*

**5.2**    *Enter ⌐pursuivants, pages, footboys, and grooms.*
    *Then enter⌐ Cranmer, Archbishop of Canterbury*
**CRANMER**
I hope I am not too late, and yet the gentleman
That was sent to me from the council prayed me
To make great haste. All fast? What means this?
    (*Calling at the door*) Ho!
Who waits there?
    *Enter a Doorkeeper*
         Sure you know me?
**DOORKEEPER**         Yes, my lord,
But yet I cannot help you.
**CRANMER**         Why?    5
    ⌐*Enter Doctor Butts, passing over the stage*⌐
**DOORKEEPER**
Your grace must wait till you be called for.
**CRANMER**         So.
**BUTTS** (*aside*)
This is a piece of malice. I am glad
I came this way so happily. The King
Shall understand it presently.         *Exit*
**CRANMER** (*aside*)         'Tis Butts,
The King's physician. As he passed along    10
How earnestly he cast his eyes upon me!
Pray heaven he found not my disgrace. For certain
This is of purpose laid by some that hate me—
God turn their hearts, I never sought their malice—
To quench mine honour. They would shame to make me
Wait else at door, a fellow Councillor,    16
'Mong boys, grooms, and lackeys. But their pleasures
Must be fulfilled, and I attend with patience.
    . *Enter King Henry and Doctor Butts at a window,*
    *above*
**BUTTS**
I'll show your grace the strangest sight—
**KING HENRY**         What's that, Butts?
**BUTTS**
I think your highness saw this many a day.    20
**KING HENRY**
Body o'me, where is it?
**BUTTS** (*pointing at Cranmer, below*) There, my lord.
The high promotion of his grace of Canterbury,
Who holds his state at door, 'mongst pursuivants,
Pages, and footboys.
**KING HENRY**         Ha? 'Tis he indeed.
Is this the honour they do one another?    25
'Tis well there's one above 'em yet. I had thought
They had parted so much honesty among 'em—
At least good manners—as not thus to suffer
A man of his place and so near our favour
To dance attendance on their lordships' pleasures,    30
And at the door, too, like a post with packets!
By holy Mary, Butts, there's knavery!
Let 'em alone, and draw the curtain close.
We shall hear more anon.
    ⌐*Cranmer and the doorkeeper stand to one side.*
         *Exeunt the lackeys*⌐

*Above, Butts ⌐partly⌐ draws the curtain close.*
*Below, a council table is brought in along with*
*chairs and stools, and placed under the cloth of state.*
*Enter the Lord Chancellor, who places himself at the*
*upper end of the table, on the left hand, leaving a*
*seat void above him at the table's head as for*
*Canterbury's seat. The Duke of Suffolk, the Duke of*
*Norfolk, the Earl of Surrey, the Lord Chamberlain,*
*and Gardiner, the Bishop of Winchester, seat themselves*
*in order on each side of the table. Cromwell sits at the*
*lower end, and acts as secretary*
**LORD CHANCELLOR** (*to Cromwell*)
Speak to the business, master secretary.    35
Why are we met in council?
**CROMWELL**         Please your honours,
The chief cause concerns his grace of Canterbury.
**GARDINER**
Has he had knowledge of it?
**CROMWELL**         Yes.
**NORFOLK** (*to the Doorkeeper*)    Who waits there?
**DOORKEEPER** ⌐*coming forward*⌐
Without, my noble lords?
**GARDINER**         Yes.
**DOORKEEPER**         My lord Archbishop;
And has done half an hour, to know your pleasures.
**LORD CHANCELLOR**
Let him come in.
**DOORKEEPER** (*to Cranmer*) Your grace may enter now.    41
    *Cranmer approaches the Council table*
**LORD CHANCELLOR**
My good lord Archbishop, I'm very sorry
To sit here at this present and behold
That chair stand empty, but we all are men
In our own natures frail, and capable    45
Of our flesh; few are angels; out of which frailty
And want of wisdom, you, that best should teach us,
Have misdemeaned yourself, and not a little,
Toward the King first, then his laws, in filling
The whole realm, by your teaching and your chaplains'—
For so we are informed—with new opinions,    51
Diverse and dangerous, which are heresies,
And, not reformed, may prove pernicious.
**GARDINER**
Which reformation must be sudden too,
My noble lords; for those that tame wild horses    55
Pace 'em not in their hands to make 'em gentle,
But stop their mouths with stubborn bits and spur 'em
Till they obey the manège. If we suffer,
Out of our easiness and childish pity
To one man's honour, this contagious sickness,    60
Farewell all physic—and what follows then?
Commotions, uproars—with a general taint
Of the whole state, as of late days our neighbours,
The upper Germany, can dearly witness,
Yet freshly pitied in our memories.    65
**CRANMER**
My good lords, hitherto in all the progress
Both of my life and office, I have laboured,
And with no little study, that my teaching
And the strong course of my authority
Might go one way, and safely; and the end    70
Was ever to do well. Nor is there living—
I speak it with a single heart, my lords—
A man that more detests, more stirs against,
Both in his private conscience and his place,

Defacers of a public peace than I do.                              75
Pray heaven the King may never find a heart
With less allegiance in it. Men that make
Envy and crooked malice nourishment
Dare bite the best. I do beseech your lordships
That, in this case of justice, my accusers,                        80
Be what they will, may stand forth face to face,
And freely urge against me.

SUFFOLK                           Nay, my lord,
That cannot be. You are a Councillor,
And by that virtue no man dare accuse you.                         84

GARDINER (to Cranmer)
My lord, because we have business of more moment,
We will be short with you. 'Tis his highness' pleasure
And our consent, for better trial of you,
From hence you be committed to the Tower
Where, being but a private man again,
You shall know many dare accuse you boldly,                        90
More than, I fear, you are provided for.

CRANMER
Ah, my good lord of Winchester, I thank you.
You are always my good friend. If your will pass,
I shall both find your lordship judge and juror,
You are so merciful. I see your end—                               95
'Tis my undoing. Love and meekness, lord,
Become a churchman better than ambition.
Win straying souls with modesty again;
Cast none away. That I shall clear myself,
Lay all the weight ye can upon my patience,                        100
I make as little doubt as you do conscience
In doing daily wrongs. I could say more,
But reverence to your calling makes me modest.

GARDINER
My lord, my lord—you are a sectary,                                104
That's the plain truth. Your painted gloss discovers,
To men that understand you, words and weakness.

CROMWELL (to Gardiner)
My lord of Winchester, you're a little,
By your good favour, too sharp. Men so noble,
However faulty, yet should find respect
For what they have been. 'Tis a cruelty                            110
To load a falling man.

GARDINER                      Good master secretary,
I cry your honour mercy. You may worst
Of all this table say so.

CROMWELL                       Why, my lord?

GARDINER
Do not I know you for a favourer
Of this new sect? Ye are not sound.

CROMWELL                              Not sound?                    115

GARDINER
Not sound, I say.

CROMWELL         Would you were half so honest!
Men's prayers then would seek you, not their fears.

GARDINER
I shall remember this bold language.

CROMWELL                      Do.
Remember your bold life, too.

LORD CHANCELLOR              This is too much.
Forbear, for shame, my lords.

GARDINER                      I have done.

CROMWELL                      And I.                               120

LORD CHANCELLOR (to Cranmer)
Then thus for you, my lord. It stands agreed,
I take it, by all voices, that forthwith
You be conveyed to th' Tower a prisoner,

There to remain till the King's further pleasure
Be known unto us. Are you all agreed, lords?                       125

ALL THE COUNCIL
We are.

CRANMER  Is there no other way of mercy,
But I must needs to th' Tower, my lords?

GARDINER                              What other
Would you expect? You are strangely troublesome.
Let some o'th' guard be ready there.
          Enter the guard

CRANMER                              For me?
Must I go like a traitor thither?

GARDINER (to the guard)          Receive him,                      130
And see him safe i'th' Tower.

CRANMER                      Stay, good my lords.
I have a little yet to say. Look there, my lords—
          He shows the King's ring
By virtue of that ring I take my cause
Out of the grips of cruel men, and give it
To a most noble judge, the King my master.                        135

LORD CHAMBERLAIN
This is the King's ring.

SURREY                      'Tis no counterfeit.

SUFFOLK
'Tis the right ring, by heav'n. I told ye all
When we first put this dangerous stone a-rolling
'Twould fall upon ourselves.

NORFOLK                      Do you think, my lords,
The King will suffer but the little finger                         140
Of this man to be vexed?

LORD CHAMBERLAIN          'Tis now too certain.
How much more is his life in value with him!
Would I were fairly out on't.
          ⌈Exit King with Butts above⌉

CROMWELL                      My mind gave me,
In seeking tales and informations
Against this man, whose honesty the devil                          145
And his disciples only envy at,
Ye blew the fire that burns ye. Now have at ye!
          Enter, below, King Henry frowning on them. He
          takes his seat

GARDINER
Dread sovereign, how much are we bound to heaven
In daily thanks, that gave us such a prince,
Not only good and wise, but most religious.                        150
One that in all obedience makes the church
The chief aim of his honour, and, to strengthen
That holy duty, out of dear respect,
His royal self in judgement comes to hear
The cause betwixt her and this great offender.                     155

KING HENRY
You were ever good at sudden commendations,
Bishop of Winchester. But know I come not
To hear such flattery now; and in my presence
They are too thin and base to hide offences.
To me you cannot reach. You play the spaniel,                      160
And think with wagging of your tongue to win me.
But whatsoe'er thou tak'st me for, I'm sure
Thou hast a cruel nature and a bloody.
(To Cranmer) Good man, sit down.
          Cranmer takes his seat at the head of the Council table
                              Now let me see the proudest,
He that dares most, but wag his finger at thee.                    165
By all that's holy, he had better starve
Than but once think this place becomes thee not.

SURREY
May it please your grace—
KING HENRY                    No, sir, it does not please me!
I had thought I had had men of some understanding
And wisdom of my Council, but I find none.        170
Was it discretion, lords, to let this man,
This good man—few of you deserve that title—
This honest man, wait like a lousy footboy
At chamber door? And one as great as you are?
Why, what a shame was this! Did my commission 175
Bid ye so far forget yourselves? I gave ye
Power as he was a Councillor to try him,
Not as a groom. There's some of ye, I see,
More out of malice than integrity,
Would try him to the utmost, had ye mean;        180
Which ye shall never have while I live.
LORD CHANCELLOR                    Thus far,
My most dread sovereign, may it like your grace
To let my tongue excuse all. What was purposed
Concerning his imprisonment was rather—
If there be faith in men—meant for his trial     185
And fair purgation to the world than malice,
I'm sure, in me.
KING HENRY          Well, well, my lords—respect him.
Take him and use him well, he's worthy of it.
I will say thus much for him—if a prince
May be beholden to a subject, I                   190
Am for his love and service so to him.
Make me no more ado, but all embrace him.
Be friends, for shame, my lords. (To Cranmer) My lord
   of Canterbury,
I have a suit which you must not deny me:         194
That is a fair young maid that yet wants baptism—
You must be godfather, and answer for her.
CRANMER
The greatest monarch now alive may glory
In such an honour; how may I deserve it,
That am a poor and humble subject to you?        199
KING HENRY Come, come, my lord—you'd spare your
   spoons. You shall have two noble partners with you—
   the old Duchess of Norfolk and Lady Marquis Dorset.
   Will these please you?
   (To Gardiner) Once more, my lord of Winchester, I
      charge you
   Embrace and love this man.
GARDINER                    With a true heart      205
And brother-love I do it.
      ⌈Gardiner and Cranmer embrace⌉
CRANMER (weeping)          And let heaven
Witness how dear I hold this confirmation.
KING HENRY
Good man, those joyful tears show thy true heart.
The common voice, I see, is verified
Of thee which says thus, 'Do my lord of Canterbury
A shrewd turn, and he's your friend for ever.'    211
Come, lords, we trifle time away. I long
To have this young one made a Christian.
As I have made ye one, lords, one remain—
So I grow stronger, you more honour gain.        Exeunt


5.3     Noise and tumult within. Enter Porter ⌈with rushes⌉
        and his man ⌈with a broken cudgel⌉
PORTER (to those within)
   You'll leave your noise anon, ye rascals. Do you take
   The court for Paris Garden, ye rude slaves?
   Leave your gaping.

ONE (within)
   Good master porter, I belong to th' larder.
PORTER
   Belong to th' gallows, and be hanged, ye rogue!      5
   Is this a place to roar in?
   (To his man)
   Fetch me a dozen crab-tree staves, and strong ones,
   ⌈Raising his rushes⌉ These are but switches to 'em.
   (To those within)          I'll scratch your heads.
   You must be seeing christenings? Do you look
   For ale and cakes here, you rude rascals?            10
MAN
   Pray, sir, be patient. 'Tis as much impossible,
   Unless we sweep 'em from the door with cannons,
   To scatter 'em as 'tis to make 'em sleep
   On May-day morning—which will never be.
   We may as well push against Paul's as stir 'em.      15
PORTER How got they in, and be hanged?
MAN
   Alas, I know not. How gets the tide in?
   As much as one sound cudgel of four foot—
        He raises his cudgel
   You see the poor remainder—could distribute,
   I made no spare, sir.
PORTER                    You did nothing, sir.          20
MAN
   I am not Samson, nor Sir Guy, nor Colbrand,
   To mow 'em down before me; but if I spared any
   That had a head to hit, either young or old,
   He or she, cuckold or cuckold-maker,
   Let me ne'er hope to see a chine again—              25
   And that I would not for a cow, God save her!
ONE (within) Do you hear, master porter?
PORTER
   I shall be with you presently,
   Good master puppy. (To his man) Keep the door close,
      sirrah.                                           29
MAN
   What would you have me do?
PORTER                    What should you do,
   but knock 'em down by th' dozens? Is this Moorfields
   to muster in? Or have we some strange Indian with
   the great tool come to court, the women so besiege us?
   Bless me, what a fry of fornication is at door! On my
   Christian conscience, this one christening will beget a
   thousand. Here will be father, godfather, and all
   together.                                            37
MAN The spoons will be the bigger, sir. There is a fellow
   somewhat near the door, he should be a brazier by his
   face, for o' my conscience twenty of the dog-days now
   reign in's nose. All that stand about him are under the
   line—they need no other penance. That fire-drake did
   I hit three times on the head, and three times was his
   nose discharged against me. He stands there like a
   mortar-piece, to blow us. There was a haberdasher's
   wife of small wit near him, that railed upon me till her
   pinked porringer fell off her head, for kindling such a
   combustion in the state. I missed the meteor once, and
   hit that woman, who cried out 'Clubs!', when I might
   see from far some forty truncheoners draw to her
   succour, which were the hope o'th' Strand, where she
   was quartered. They fell on. I made good my place. At
   length they came to th' broomstaff to me. I defied 'em
   still, when suddenly a file of boys behind 'em, loose
   shot, delivered such a shower of pebbles that I was fain

# ALL IS TRUE

Act 5 Scene 4

to draw mine honour in and let 'em win the work. The
devil was amongst 'em, I think, surely.                          57
PORTER  These are the youths that thunder at a playhouse,
and fight for bitten apples, that no audience but the
tribulation of Tower Hill or the limbs of Limehouse,
their dear brothers, are able to endure. I have some of
'em in *limbo patrum*, and there they are like to dance
these three days, besides the running banquet of two
beadles that is to come.
    *Enter the Lord Chamberlain*
LORD CHAMBERLAIN
Mercy o' me, what a multitude are here!                          65
They grow still, too—from all parts they are coming,
As if we kept a fair here! Where are these porters,
These lazy knaves? (*To the Porter and his man*) You've
    made a fine hand, fellows!
There's a trim rabble let in—are all these
Your faithful friends o'th' suburbs? We shall have        70
Great store of room, no doubt, left for the ladies
When they pass back from the christening!
PORTER                                          An't please your honour,
We are but men, and what so many may do,
Not being torn a-pieces, we have done.
An army cannot rule 'em.
LORD CHAMBERLAIN                        As I live,         75
If the King blame me for't, I'll lay ye all
By th' heels, and suddenly—and on your heads
Clap round fines for neglect. You're lazy knaves,
And here ye lie baiting of bombards when
Ye should do service.
    *Flourish of trumpets within*
                Hark, the trumpets sound.        80
They're come, already, from the christening.
Go break among the press, and find a way out
To let the troop pass fairly, or I'll find
A Marshalsea shall hold ye play these two months.
    ⌈*As they leave, the Porter and his man call within*⌉
PORTER
Make way there for the Princess!
MAN                                          You great fellow,        85
Stand close up, or I'll make your head ache.
PORTER
You i'th' camlet, get up o'th' rail—
I'll peck you o'er the pales else.                      *Exeunt*

5.4   *Enter trumpeters, sounding. Then enter two aldermen,*
    *the Lord Mayor of London, Garter King-of-Arms,*
    *Cranmer the Archbishop of Canterbury, the Duke of*
    *Norfolk with his marshal's staff, the Duke of*
    *Suffolk, two noblemen bearing great standing*
    *bowls for the christening gifts; then enter four*
    *noblemen bearing a canopy, under which is the*
    *Duchess of Norfolk, godmother, bearing the child*
    *Elizabeth richly habited in a mantle, whose train is*
    *borne by a lady. Then follows the Marchioness*
    *Dorset, the other godmother, and ladies. The troop*
    *pass once about the stage and Garter speaks*
GARTER  Heaven, from thy endless goodness send
prosperous life, long, and ever happy, to the high and
mighty Princess of England, Elizabeth.
    *Flourish. Enter King Henry and guard*
CRANMER (*kneeling*)
And to your royal grace, and the good Queen!
My noble partners and myself thus pray
All comfort, joy, in this most gracious lady,              5

Heaven ever laid up to make parents happy,
May hourly fall upon ye.
KING HENRY                        Thank you, good lord Archbishop.
What is her name?
CRANMER                        Elizabeth.
KING HENRY                        Stand up, lord.
    *Cranmer rises*
    (*To the child*) With this kiss take my blessing—
    *He kisses the child*
                  God protect thee,
Into whose hand I give thy life.
CRANMER                                          Amen.        11
KING HENRY (*to Cranmer, old Duchess, and Marchioness*)
My noble gossips, you've been too prodigal.
I thank ye heartily. So shall this lady,
When she has so much English.
CRANMER                        Let me speak, sir,
For heaven now bids me, and the words I utter        15
Let none think flattery, for they'll find 'em truth.
This royal infant—heaven still move about her—
Though in her cradle, yet now promises
Upon this land a thousand thousand blessings
Which time shall bring to ripeness. She shall be—        20
But few now living can behold that goodness—
A pattern to all princes living with her,
And all that shall succeed. Saba was never
More covetous of wisdom and fair virtue
Than this pure soul shall be. All princely graces        25
That mould up such a mighty piece as this is,
With all the virtues that attend the good,
Shall still be doubled on her. Truth shall nurse her,
Holy and heavenly thoughts still counsel her.
She shall be loved and feared. Her own shall bless her;
Her foes shake like a field of beaten corn,              31
And hang their heads with sorrow. Good grows with
    her.
In her days every man shall eat in safety
Under his own vine what he plants, and sing
The merry songs of peace to all his neighbours.        35
God shall be truly known, and those about her
From her shall read the perfect ways of honour,
And by those claim their greatness, not by blood.
Nor shall this peace sleep with her, but, as when
The bird of wonder dies—the maiden phoenix—        40
Her ashes new create another heir
As great in admiration as herself,
So shall she leave her blessèdness to one,
When heaven shall call her from this cloud of darkness,
Who from the sacred ashes of her honour              45
Shall star-like rise as great in fame as she was,
And so stand fixed. Peace, plenty, love, truth, terror,
That were the servants to this chosen infant,
Shall then be his, and, like a vine, grow to him.
Wherever the bright sun of heaven shall shine,        50
His honour and the greatness of his name
Shall be, and make new nations. He shall flourish,
And like a mountain cedar reach his branches
To all the plains about him. Our children's children
Shall see this, and bless heaven.
KING HENRY                        Thou speakest wonders.
CRANMER
She shall be, to the happiness of England,              56
An agèd princess. Many days shall see her,
And yet no day without a deed to crown it.
Would I had known no more. But she must die—

She must, the saints must have her—yet a virgin,   60
A most unspotted lily shall she pass
To th' ground, and all the world shall mourn her.
KING HENRY O lord Archbishop,
Thou hast made me now a man. Never before
This happy child did I get anything.   65
This oracle of comfort has so pleased me
That when I am in heaven I shall desire
To see what this child does, and praise my maker.
I thank ye all. To you, my good Lord Mayor,
And your good brethren, I am much beholden.   70
I have received much honour by your presence,
And ye shall find me thankful. Lead the way, lords.
Ye must all see the Queen, and she must thank ye.
She will be sick else. This day, no man think
He's business at his house, for all shall stay—   75
This little one shall make it holiday. ⌈*Flourish.*⌉ *Exeunt*

**Epilogue**   *Enter Epilogue*
EPILOGUE
'Tis ten to one this play can never please
All that are here. Some come to take their ease,
And sleep an act or two; but those, we fear,
We've frighted with our trumpets; so, 'tis clear,
They'll say 'tis naught. Others to hear the city   5
Abused extremely, and to cry 'That's witty!'—
Which we have not done neither; that, I fear,
All the expected good we're like to hear
For this play at this time is only in
The merciful construction of good women,   10
For such a one we showed 'em. If they smile,
And say ''Twill do', I know within a while
  All the best men are ours—for 'tis ill hap
  If they hold when their ladies bid 'em clap.   *Exit*

# THE TWO NOBLE KINSMEN

## BY JOHN FLETCHER AND WILLIAM SHAKESPEARE

WHEN it first appeared in print, in 1634, *The Two Noble Kinsmen* was stated to be 'by the memorable worthies of their time, Mr John Fletcher, and Mr William Shakespeare'. There is no reason to disbelieve this ascription: many plays of the period were not printed till long after they were acted, and there is other evidence that Shakespeare collaborated with Fletcher (1579-1625). The morris dance in Act 3, Scene 5, contains characters who also appear in Francis Beaumont's *Masque of the Inner Temple and Gray's Inn* performed before James I on 20 February 1613. Their dance was a great success with the King; probably the King's Men—some of whom may have taken part in the masque—decided to exploit its success by incorporating it in a play written soon afterwards, in the last year of Shakespeare's playwriting life.

*The Two Noble Kinsmen*, a tragicomedy of the kind that became popular during the last years of the first decade of the seventeenth century, is based on Chaucer's *Knight's Tale*, on which Shakespeare had already drawn for episodes of *A Midsummer Night's Dream*. It tells a romantic tale of the conflicting claims of love and friendship: the 'two noble kinsmen', Palamon and Arcite, are the closest of friends until each falls in love with Emilia, sister-in-law of Theseus, Duke of Athens. Their conflict is finally resolved by a formal combat with Emilia as the prize, in which the loser is to be executed. Arcite wins, and Palamon's head is on the block as news arrives that Arcite has been thrown from his horse. Dying, Arcite commends Emilia to his friend, and Theseus rounds off the play with a meditation on the paradoxes of fortune.

Studies of style suggest that Shakespeare was primarily responsible for the rhetorically and ritualistically impressive Act 1; for Act 2, Scene 1; Act 3, Scenes 1 and 2; and for most of Act 5 (Scene 4 excepted), which includes emblematically spectacular episodes related to his other late plays. Fletcher appears mainly to have written the scenes showing the rivalry of Palamon and Arcite along with the sub-plots concerned with the Jailer's daughter's love for Palamon and the rustics' entertainment for Theseus.

Though the play was adapted by William Davenant as *The Rivals* (1664), its first known performances since the seventeenth century were at the Old Vic in 1928; it has been played only occasionally since then, but was chosen to open the Swan Theatre in Stratford-upon-Avon in 1986. Critical interest, too, has been slight; but Shakespeare's contributions are entirely characteristic of his late style, and Fletcher's scenes are both touching and funny.

# THE PERSONS OF THE PLAY

PROLOGUE

THESEUS, Duke of Athens

HIPPOLYTA, Queen of the Amazons, later wife of Theseus

EMILIA, her sister

PIRITHOUS, friend of Theseus

PALAMON }
ARCITE  } the two noble kinsmen, cousins, nephews of Creon, the King of Thebes

Hymen, god of marriage

A BOY, who sings

ARTESIUS, an Athenian soldier

Three QUEENS, widows of kings killed in the siege of Thebes

VALERIUS, a Theban

A HERALD

WOMAN, attending Emilia

An Athenian GENTLEMAN

MESSENGERS

Six KNIGHTS, three attending Arcite and three Palamon

A SERVANT

A JAILER in charge of Theseus' prison

The JAILER'S DAUGHTER

The JAILER'S BROTHER

The WOOER of the Jailer's daughter

Two FRIENDS of the Jailer

A DOCTOR

Six COUNTRYMEN, one dressed as a babion, or baboon

Gerald, a SCHOOLMASTER

NELL, a country wench

Four other country wenches: Friz, Madeline, Luce, and Barbara

Timothy, a TABORER

EPILOGUE

Nymphs, attendants, maids, executioner, guard

# The Two Noble Kinsmen

**Prologue**  *Flourish. Enter Prologue*
PROLOGUE
New plays and maidenheads are near akin:
Much followed both, for both much money giv'n
If they stand sound and well. And a good play,
Whose modest scenes blush on his marriage day
And shake to lose his honour, is like her                    5
That after holy tie and first night's stir
Yet still is modesty, and still retains
More of the maid to sight than husband's pains.
We pray our play may be so, for I am sure
It has a noble breeder and a pure,                           10
A learnèd, and a poet never went
More famous yet 'twixt Po and silver Trent.
Chaucer, of all admired, the story gives:
There constant to eternity it lives.
If we let fall the nobleness of this                         15
And the first sound this child hear be a hiss,
How will it shake the bones of that good man,
And make him cry from under ground, 'O fan
From me the witless chaff of such a writer,
That blasts my bays and my famed works makes
      lighter                                                20
Than Robin Hood'? This is the fear we bring,
For to say truth, it were an endless thing
And too ambitious to aspire to him,
Weak as we are, and almost breathless swim
In this deep water. Do but you hold out                     25
Your helping hands and we shall tack about
And something do to save us. You shall hear
Scenes, though below his art, may yet appear
Worth two hours' travail. To his bones, sweet sleep;
Content to you. If this play do not keep                    30
A little dull time from us, we perceive
Our losses fall so thick we must needs leave.
                                          *Flourish. Exit*

**1.1**  *Music. Enter Hymen with a torch burning, a Boy*
      *in a white robe before, singing and strewing*
      *flowers. After Hymen, a nymph encompassed in her*
      *tresses, bearing a wheaten garland. Then Theseus*
      *between two other nymphs with wheaten chaplets*
      *on their heads. Then Hippolyta, the bride, led by*
      *Pirithous and another holding a garland over her*
      *head, her tresses likewise hanging. After her, Emilia*
      *holding up her train. Then Artesius ⌈and other*
      *attendants⌉*

BOY (*sings during procession*)
      Roses, their sharp spines being gone,
      Not royal in their smells alone,
            But in their hue;
      Maiden pinks, of odour faint,
      Daisies smell-less, yet most quaint,            5
            And sweet thyme true;

      Primrose, first-born child of Ver,
      Merry springtime's harbinger,
            With harebells dim;
      Oxlips, in their cradles growing,               10
      Marigolds, on deathbeds blowing,
            Lark's-heels trim;

      All dear nature's children sweet,
      Lie fore bride and bridegroom's feet,

*He strews flowers*
            Blessing their sense.                     15
      Not an angel of the air,
      Bird melodious, or bird fair,
            Is absent hence.

      The crow, the sland'rous cuckoo, nor
      The boding raven, nor chough hoar,              20
            Nor chatt'ring pie,
      May on our bridehouse perch or sing,
      Or with them any discord bring,
            But from it fly.

*Enter three Queens in black, with veils stained, with*
*imperial crowns. The First Queen falls down at the*
*foot of Theseus; the Second falls down at the foot of*
*Hippolyta; the Third, before Emilia*
FIRST QUEEN (*to Theseus*)
For pity's sake and true gentility's,                        25
Hear and respect me.
SECOND QUEEN (*to Hippolyta*) For your mother's sake,
And as you wish your womb may thrive with fair ones,
Hear and respect me.
THIRD QUEEN (*to Emilia*)
Now for the love of him whom Jove hath marked
The honour of your bed, and for the sake                     30
Of clear virginity, be advocate
For us and our distresses. This good deed
Shall raze you out o'th' Book of Trespasses
All you are set down there.
THESEUS (*to First Queen*)
Sad lady, rise.
HIPPOLYTA (*to Second Queen*) Stand up.
EMILIA (*to Third Queen*)                  No knees to me.
What woman I may stead that is distressed                    36
Does bind me to her.
THESEUS (*to First Queen*)
What's your request? Deliver you for all.
FIRST QUEEN ⌈*kneeling still*⌉
We are three queens whose sovereigns fell before
The wrath of cruel Creon; who endured                        40
The beaks of ravens, talons of the kites,
And pecks of crows in the foul fields of Thebes.
He will not suffer us to burn their bones,
To urn their ashes, nor to take th'offence
Of mortal loathsomeness from the blest eye                   45
Of holy Phoebus, but infects the winds
With stench of our slain lords. O pity, Duke!
Thou purger of the earth, draw thy feared sword
That does good turns to'th' world; give us the bones
Of our dead kings that we may chapel them;                   50
And of thy boundless goodness take some note
That for our crownèd heads we have no roof,
Save this, which is the lion's and the bear's,
And vault to everything.
THESEUS                    Pray you, kneel not:
I was transported with your speech, and suffered             55
Your knees to wrong themselves. I have heard the
      fortunes

Of your dead lords, which gives me such lamenting
As wakes my vengeance and revenge for 'em.
King Capaneus was your lord: the day
That he should marry you—at such a season     60
As now it is with me—I met your groom
By Mars's altar. You were that time fair,
Not Juno's mantle fairer than your tresses,
Nor in more bounty spread her. Your wheaten wreath
Was then nor threshed nor blasted; fortune at you   65
Dimpled her cheek with smiles; Hercules our
    kinsman—
Then weaker than your eyes—laid by his club.
He tumbled down upon his Nemean hide
And swore his sinews thawed. O grief and time,
Fearful consumers, you will all devour.     70
FIRST QUEEN ⌐*kneeling still*⌐ O, I hope some god,
Some god hath put his mercy in your manhood,
Whereto he'll infuse power and press you forth
Our undertaker.
THESEUS          O no knees, none, widow:
    ⌐*The First Queen rises*⌐
Unto the helmeted Bellona use them     75
And pray for me, your soldier. Troubled I am.
    *He turns away*
SECOND QUEEN ⌐*kneeling still*⌐ Honoured Hippolyta,
Most dreaded Amazonian, that hast slain
The scythe-tusked boar, that with thy arm, as strong
As it is white, wast near to make the male     80
To thy sex captive, but that this, thy lord—
Born to uphold creation in that honour
First nature styled it in—shrunk thee into
The bound thou wast o'erflowing, at once subduing
Thy force and thy affection; soldieress,     85
That equally canst poise sternness with pity,
Whom now I know hast much more power on him
Than ever he had on thee, who ow'st his strength,
And his love too, who is a servant for
The tenor of thy speech; dear glass of ladies,     90
Bid him that we, whom flaming war doth scorch,
Under the shadow of his sword may cool us.
Require him he advance it o'er our heads.
Speak't in a woman's key, like such a woman
As any of us three. Weep ere you fail.     95
Lend us a knee:
But touch the ground for us no longer time
Than a dove's motion when the head's plucked off.
Tell him, if he i'th' blood-sized field lay swoll'n,
Showing the sun his teeth, grinning at the moon,   100
What you would do.
HIPPOLYTA         Poor lady, say no more.
I had as lief trace this good action with you
As that whereto I am going, and never yet
Went I so willing way. My lord is taken
Heart-deep with your distress. Let him consider.   105
I'll speak anon.
    ⌐*The Second Queen rises*⌐
THIRD QUEEN (*kneeling* ⌐*still*⌐ *to Emilia*)
         O, my petition was
Set down in ice, which by hot grief uncandied
Melts into drops; so sorrow, wanting form,
Is pressed with deeper matter.
EMILIA             Pray stand up:
Your grief is written in your cheek.
THIRD QUEEN           O woe,    110
You cannot read it there; there, through my tears,

Like wrinkled pebbles in a glassy stream,
You may behold 'em.
    ⌐*The Third Queen rises*⌐
         Lady, lady, alack—
He that will all the treasure know o'th' earth
Must know the centre too; he that will fish     115
For my least minnow, let him lead his line
To catch one at my heart. O, pardon me:
Extremity, that sharpens sundry wits,
Makes me a fool.
EMILIA         Pray you, say nothing, pray you.
Who cannot feel nor see the rain, being in't,     120
Knows neither wet nor dry. If that you were
The ground-piece of some painter, I would buy you
T'instruct me 'gainst a capital grief, indeed
Such heart-pierced demonstration; but, alas,
Being a natural sister of our sex,     125
Your sorrow beats so ardently upon me
That it shall make a counter-reflect 'gainst
My brother's heart, and warm it to some pity,
Though it were made of stone. Pray have good
    comfort.
THESEUS
Forward to th' temple. Leave not out a jot     130
O'th' sacred ceremony.
FIRST QUEEN         O, this celebration
Will longer last and be more costly than
Your suppliants' war. Remember that your fame
Knolls in the ear o'th' world: what you do quickly
Is not done rashly; your first thought is more     135
Than others' laboured meditance; your premeditating
More than their actions. But, O Jove, your actions,
Soon as they move, as ospreys do the fish,
Subdue before they touch. Think, dear Duke, think
What beds our slain kings have.
SECOND QUEEN        What griefs our beds,
That our dear lords have none.
THIRD QUEEN        None fit for th' dead.
Those that with cords, knives, drams, precipitance,
Weary of this world's light, have to themselves
Been death's most horrid agents, human grace
Affords them dust and shadow.
FIRST QUEEN         But our lords    145
Lie blist'ring fore the visiting sun,
And were good kings, when living.
THESEUS          It is true,
And I will give you comfort to give your dead lords
    graves,
The which to do must make some work with Creon.
FIRST QUEEN
And that work presents itself to th' doing.     150
Now 'twill take form, the heats are gone tomorrow.
Then, bootless toil must recompense itself
With its own sweat; now he's secure,
Not dreams we stand before your puissance
Rinsing our holy begging in our eyes     155
To make petition clear.
SECOND QUEEN       Now you may take him,
Drunk with his victory.
THIRD QUEEN        And his army full
Of bread and sloth.
THESEUS       Artesius, that best knowest
How to draw out, fit to this enterprise
The prim'st for this proceeding and the number   160
To carry such a business: forth and levy
Our worthiest instruments, whilst we dispatch

This grand act of our life, this daring deed
Of fate in wedlock.
FIRST QUEEN (*to the other two Queens*)
                    Dowagers, take hands;
Let us be widows to our woes; delay                        165
Commends us to a famishing hope.
ALL THREE QUEENS                    Farewell.
SECOND QUEEN
We come unseasonably, but when could grief
Cull forth, as unpanged judgement can, fitt'st time
For best solicitation?
THESEUS                    Why, good ladies,
This is a service whereto I am going                        170
Greater than any war—it more imports me
Than all the actions that I have foregone,
Or futurely can cope.
FIRST QUEEN                    The more proclaiming
Our suit shall be neglected when her arms,
Able to lock Jove from a synod, shall                        175
By warranting moonlight corslet thee! O when
Her twinning cherries shall their sweetness fall
Upon thy tasteful lips, what wilt thou think
Of rotten kings or blubbered queens? What care
For what thou feel'st not, what thou feel'st being able
To make Mars spurn his drum? O, if thou couch                181
But one night with her, every hour in't will
Take hostage of thee for a hundred, and
Thou shalt remember nothing more than what
That banquet bids thee to.
HIPPOLYTA (*to Theseus*)                    Though much unlike    185
You should be so transported, as much sorry
I should be such a suitor—yet I think
Did I not by th'abstaining of my joy,
Which breeds a deeper longing, cure their surfeit
That craves a present medicine, I should pluck                190
All ladies' scandal on me. ⌜*Kneels*⌝ Therefore, sir,
As I shall here make trial of my prayers,
Either presuming them to have some force,
Or sentencing for aye their vigour dumb,
Prorogue this business we are going about, and hang
Your shield afore your heart—about that neck                196
Which is my fee, and which I freely lend
To do these poor queens service.
ALL THREE QUEENS (*to Emilia*)                    O, help now,
Our cause cries for your knee.
EMILIA (*kneels to Theseus*)                    If you grant not
My sister her petition in that force                        200
With that celerity and nature which
She makes it in, from henceforth I'll not dare
To ask you anything, nor be so hardy
Ever to take a husband.
THESEUS                    Pray stand up.
            ⌜*They rise*⌝
I am entreating of myself to do                        205
That which you kneel to have me.—Pirithous,
Lead on the bride: get you and pray the gods
For success and return; omit not anything
In the pretended celebration.—Queens,
Follow your soldier. (*To Artesius*) As before, hence you,
And at the banks of Aulis meet us with                211
The forces you can raise, where we shall find
The moiety of a number for a business
More bigger looked.                    *Exit Artesius*
(*To Hippolyta*)                    Since that our theme is haste,
I stamp this kiss upon thy current lip—                215
Sweet, keep it as my token. (*To the wedding party*) Set
    you forward,

For I will see you gone.
(*To Emilia*) Farewell, my beauteous sister.—Pirithous,
Keep the feast full: bate not an hour on't.
PIRITHOUS                    Sir,
I'll follow you at heels. The feast's solemnity                220
Shall want till your return.
THESEUS                    Cousin, I charge you
Budge not from Athens. We shall be returning
Ere you can end this feast, of which, I pray you,
Make no abatement.—Once more, farewell all.
            *Exeunt Hippolyta, Emilia, Pirithous, and train*
                    *towards the temple*
FIRST QUEEN
Thus dost thou still make good the tongue o'th' world.
SECOND QUEEN
And earn'st a deity equal with Mars—                226
THIRD QUEEN If not above him, for
Thou being but mortal mak'st affections bend
To godlike honours; they themselves, some say,
Groan under such a mast'ry.
THESEUS                    As we are men,                230
Thus should we do; being sensually subdued
We lose our human title. Good cheer, ladies.
Now turn we towards your comforts. ⌜*Flourish.*⌝ *Exeunt*

**1.2**    *Enter Palamon and Arcite*
ARCITE
Dear Palamon, dearer in love than blood,
And our prime cousin, yet unhardened in
The crimes of nature, let us leave the city,
Thebes, and the temptings in't, before we further
Sully our gloss of youth.                        5
And here to keep in abstinence we shame
As in incontinence; for not to swim
I'th' aid o'th' current were almost to sink—
At least to frustrate striving; and to follow
The common stream 'twould bring us to an eddy        10
Where we should turn or drown; if labour through,
Our gain but life and weakness.
PALAMON                    Your advice
Is cried up with example. What strange ruins
Since first we went to school may we perceive
Walking in Thebes? Scars and bare weeds                15
The gain o'th' martialist who did propound
To his bold ends honour and golden ingots,
Which though he won, he had not; and now flirted
By peace for whom he fought. Who then shall offer
To Mars's so-scorned altar? I do bleed                20
When such I meet, and wish great Juno would
Resume her ancient fit of jealousy
To get the soldier work, that peace might purge
For her repletion and retain anew
Her charitable heart, now hard and harsher                25
Than strife or war could be.
ARCITE                    Are you not out?
Meet you no ruin but the soldier in
The cranks and turns of Thebes? You did begin
As if you met decays of many kinds.
Perceive you none that do arouse your pity                30
But th'unconsidered soldier?
PALAMON                    Yes, I pity
Decays where'er I find them, but such most
That, sweating in an honourable toil,
Are paid with ice to cool 'em.
ARCITE                    'Tis not this
I did begin to speak of. This is virtue,                35
Of no respect in Thebes. I spake of Thebes,

How dangerous, if we will keep our honours,
It is for our residing where every evil
Hath a good colour, where every seeming good's
A certain evil, where not to be ev'n jump        40
As they are here were to be strangers, and
Such things to be; mere monsters.

PALAMON                              'Tis in our power,
Unless we fear that apes can tutor's, to
Be masters of our manners. What need I
Affect another's gait, which is not catching      45
Where there is faith? Or to be fond upon
Another's way of speech, when by mine own
I may be reasonably conceived—saved, too—
Speaking it truly? Why am I bound
By any generous bond to follow him               50
Follows his tailor, haply so long until
The followed make pursuit? Or let me know
Why mine own barber is unblest—with him
My poor chin, too—for 'tis not scissored just
To such a favourite's glass? What canon is there   55
That does command my rapier from my hip
To dangle't in my hand? Or to go tiptoe
Before the street be foul? Either I am
The fore-horse in the team or I am none
That draw i'th' sequent trace. These poor slight
    sores                                         60
Need not a plantain. That which rips my bosom
Almost to th' heart's—

ARCITE                    Our uncle Creon.

PALAMON                                    He,
A most unbounded tyrant, whose successes
Makes heaven unfeared and villainy assured
Beyond its power there's nothing; almost puts     65
Faith in a fever, and deifies alone
Voluble chance; who only attributes
The faculties of other instruments
To his own nerves and act; commands men's service,
And what they win in't, boot and glory; one       70
That fears not to do harm, good dares not. Let
The blood of mine that's sib to him be sucked
From me with leeches. Let them break and fall
Off me with that corruption.

ARCITE                        Clear-spirited cousin,
Let's leave his court that we may nothing share   75
Of his loud infamy: for our milk
Will relish of the pasture, and we must
Be vile or disobedient; not his kinsmen
In blood unless in quality.

PALAMON                    Nothing truer.
I think the echoes of his shames have deafed      80
The ears of heav'nly justice. Widows' cries
Descend again into their throats and have not
    Enter Valerius
Due audience of the gods—Valerius.

VALERIUS
The King calls for you; yet be leaden-footed
Till his great rage be off him. Phoebus, when     85
He broke his whipstock and exclaimed against
The horses of the sun, but whispered to
The loudness of his fury.

PALAMON              Small winds shake him.
But what's the matter?

VALERIUS
Theseus, who where he threats, appals, hath sent   90
Deadly defiance to him and pronounces
Ruin to Thebes, who is at hand to seal

The promise of his wrath.

ARCITE              Let him approach.
But that we fear the gods in him, he brings not
A jot of terror to us. Yet what man               95
Thirds his own worth—the case is each of ours—
When that his action's dregged with mind assured
'Tis bad he goes about.

PALAMON              Leave that unreasoned.
Our services stand now for Thebes, not Creon,
Yet to be neutral to him were dishonour,         100
Rebellious to oppose. Therefore we must
With him stand to the mercy of our fate,
Who hath bounded our last minute.

ARCITE                          So we must.
Is't said this war's afoot? Or it shall be
On fail of some condition?

VALERIUS                    'Tis in motion,        105
The intelligence of state came in the instant
With the defier.

PALAMON          Let's to the King, who, were he
A quarter carrier of that honour which
His enemy come in, the blood we venture
Should be as for our health, which were not spent,
Rather laid out for purchase. But, alas,          111
Our hands advanced before our hearts, what will
The fall o'th' stroke do damage?

ARCITE                      Let th'event—
That never-erring arbitrator—tell us
When we know all ourselves, and let us follow     115
The becking of our chance.              Exeunt

1.3    Enter Pirithous, Hippolyta, and Emilia
PIRITHOUS
No further.

HIPPOLYTA    Sir, farewell. Repeat my wishes
To our great lord, of whose success I dare not
Make any timorous question; yet I wish him
Excess and overflow of power, an't might be,
To dure ill-dealing fortune. Speed to him;         5
Store never hurts good governors.

PIRITHOUS                        Though I know
His ocean needs not my poor drops, yet they
Must yield their tribute there. (To Emilia) My precious
    maid,
Those best affections that the heavens infuse
In their best-tempered pieces keep enthroned      10
In your dear heart.

EMILIA                Thanks, sir. Remember me
To our all-royal brother, for whose speed
The great Bellona I'll solicit; and
Since in our terrene state petitions are not
Without gifts understood, I'll offer to her        15
What I shall be advised she likes. Our hearts
Are in his army, in his tent.

HIPPOLYTA                    In's bosom.
We have been soldiers, and we cannot weep
When our friends don their helms, or put to sea,
Or tell of babes broached on the lance, or women   20
That have sod their infants in—and after eat them—
The brine they wept at killing 'em: then if
You stay to see of us such spinsters, we
Should hold you here forever.

PIRITHOUS                    Peace be to you
As I pursue this war, which shall be then          25
Beyond further requiring.              Exit Pirithous

EMILIA                    How his longing
Follows his friend! Since his depart, his sports,

Though craving seriousness and skill, passed slightly
His careless execution, where nor gain
Made him regard or loss consider, but                    30
Playing one business in his hand, another
Directing in his head, his mind nurse equal
To these so diff'ring twins. Have you observed him
Since our great lord departed?
HIPPOLYTA                          With much labour;
And I did love him for't. They two have cabined          35
In many as dangerous as poor a corner,
Peril and want contending; they have skiffed
Torrents whose roaring tyranny and power
I'th' least of these was dreadful, and they have
Fought out together where death's self was lodged;      40
Yet fate hath brought them off. Their knot of love,
Tied, weaved, entangled with so true, so long,
And with a finger of so deep a cunning,
May be outworn, never undone. I think
Theseus cannot be umpire to himself,                     45
Cleaving his conscience into twain and doing
Each side like justice, which he loves best.
EMILIA                                  Doubtless
There is a best, and reason has no manners
To say it is not you. I was acquainted
Once with a time when I enjoyed a playfellow;            50
You were at wars when she the grave enriched,
Who made too proud the bed; took leave o'th'
    moon—
Which then looked pale at parting—when our count
Was each eleven.
HIPPOLYTA            'Twas Flavina.
EMILIA                            Yes.
You talk of Pirithous' and Theseus' love:                55
Theirs has more ground, is more maturely seasoned,
More buckled with strong judgement, and their needs
The one of th'other may be said to water
Their intertangled roots of love; but I
And she I sigh and spoke of were things innocent,        60
Loved for we did, and like the elements,
That know not what, nor why, yet do effect
Rare issues by their operance, our souls
Did so to one another. What she liked
Was then of me approved; what not, condemned—            65
No more arraignment. The flower that I would pluck
And put between my breasts—O then but beginning
To swell about the blossom—she would long
Till she had such another, and commit it
To the like innocent cradle, where, phoenix-like,        70
They died in perfume. On my head no toy
But was her pattern. Her affections—pretty,
Though happily her careless wear—I followed
For my most serious decking. Had mine ear
Stol'n some new air, or at adventure hummed one,         75
From musical coinage, why, it was a note
Whereon her spirits would sojourn—rather dwell on—
And sing it in her slumbers. This rehearsal—
Which, seely innocence wots well, comes in
Like old emportment's bastard—has this end:             80
That the true love 'tween maid and maid may be
More than in sex dividual.
HIPPOLYTA                You're out of breath,
And this high-speeded pace is but to say
That you shall never, like the maid Flavina,
Love any that's called man.                             85
EMILIA  I am sure I shall not.

HIPPOLYTA  Now alack, weak sister,
I must no more believe thee in this point—
Though in't I know thou dost believe thyself—
Than I will trust a sickly appetite                      90
That loathes even as it longs. But sure, my sister,
If I were ripe for your persuasion, you
Have said enough to shake me from the arm
Of the all-noble Theseus, for whose fortunes
I will now in and kneel, with great assurance            95
That we more than his Pirithous possess
The high throne in his heart.
EMILIA                          I am not
Against your faith, yet I continue mine.        *Exeunt*

1.4       *Cornetts. A battle struck within. Then a retreat.*
          *Flourish. Then enter Theseus, victor. The three*
          *Queens meet him and fall on their faces before him.*
          ⌐*Also enter a Herald, and attendants bearing*
          *Palamon and Arcite on two hearses*⌐
FIRST QUEEN (*to Theseus*)
To thee no star be dark.
SECOND QUEEN (*to Theseus*)  Both heaven and earth
Friend thee for ever.
THIRD QUEEN (*to Theseus*)  All the good that may
Be wished upon thy head, I cry 'Amen' to't.
THESEUS
Th'impartial gods, who from the mounted heavens
View us their mortal herd, behold who err         5
And in their time chastise. Go and find out
The bones of your dead lords and honour them
With treble ceremony: rather than a gap
Should be in their dear rites we would supply't.
But those we will depute which shall invest        10
You in your dignities, and even each thing
Our haste does leave imperfect. So adieu,
And heaven's good eyes look on you.
                          *Exeunt the Queens*
                          What are those?
HERALD
Men of great quality, as may be judged
By their appointment. Some of Thebes have told's   15
They are sisters' children, nephews to the King.
THESEUS
By th' helm of Mars I saw them in the war,
Like to a pair of lions smeared with prey,
Make lanes in troops aghast. I fixed my note
Constantly on them, for they were a mark           20
Worth a god's view. What prisoner was't that told me
When I enquired their names?
HERALD                      Wi' leave, they're called
Arcite and Palamon.
THESEUS              'Tis right: those, those.
They are not dead?
HERALD
Nor in a state of life. Had they been taken        25
When their last hurts were given, 'twas possible
They might have been recovered. Yet they breathe,
And have the name of men.
THESEUS                    Then like men use 'em.
The very lees of such, millions of rates
Exceed the wine of others. All our surgeons        30
Convent in their behoof; our richest balms,
Rather than niggard, waste. Their lives concern us
Much more than Thebes is worth. Rather than have
    'em
Freed of this plight and in their morning state—

Sound and at liberty—I would 'em dead;                              35
But forty-thousandfold we had rather have 'em
Prisoners to us, than death. Bear 'em speedily
From our kind air, to them unkind, and minister
What man to man may do—for our sake, more,
Since I have known frights, fury, friends' behests,      40
Love's provocations, zeal, a mistress' task,
Desire of liberty, a fever, madness,
Hath set a mark which nature could not reach to
Without some imposition, sickness in will
O'er-wrestling strength in reason. For our love          45
And great Apollo's mercy, all our best
Their best skill tender.—Lead into the city
Where, having bound things scattered, we will post
To Athens fore our army.                    *Flourish. Exeunt*

**1.5**  *Music. Enter the three Queens with the hearses of*
        *their lords in a funeral solemnity, with attendants*

                          *Song*

Urns and odours, bring away,
Vapours, sighs, darken the day;
    Our dole more deadly looks than dying.
Balms and gums and heavy cheers,
Sacred vials filled with tears,                              5
    And clamours through the wild air flying:

Come all sad and solemn shows,
That are quick-eyed pleasure's foes.
We convent naught else but woes,
We convent naught else but woes.                           10

THIRD QUEEN
  This funeral path brings to your household's grave—
  Joy seize on you again, peace sleep with him.
SECOND QUEEN
  And this to yours.
FIRST QUEEN                Yours this way. Heavens lend
  A thousand differing ways to one sure end.
THIRD QUEEN
  This world's a city full of straying streets,             15
  And death's the market-place where each one meets.
                                    *Exeunt severally*

✿

**2.1**  *Enter the Jailer and the Wooer*
JAILER I may depart with little, while I live; something I
  may cast to you, not much. Alas, the prison I keep,
  though it be for great ones, yet they seldom come;
  before one salmon you shall take a number of minnows.
  I am given out to be better lined than it can appear to
  me report is a true speaker. I would I were really that
  I am delivered to be. Marry, what I have—be it what
  it will—I will assure upon my daughter at the day of
  my death.                                                  9
WOOER Sir, I demand no more than your own offer, and
  I will estate your daughter in what I have promised.
JAILER Well, we will talk more of this when the solemnity
  is past. But have you a full promise of her?
    *Enter the Jailer's Daughter with rushes*
  When that shall be seen, I tender my consent.
WOOER I have, sir. Here she comes.                          15
JAILER (*to Daughter*) Your friend and I have chanced to
  name you here, upon the old business—but no more
  of that now. So soon as the court hurry is over we will
  have an end of it. I'th' mean time, look tenderly to the
  two prisoners. I can tell you they are princes.           20

JAILER'S DAUGHTER These strewings are for their chamber.
  'Tis pity they are in prison, and 'twere pity they should
  be out. I do think they have patience to make any
  adversity ashamed; the prison itself is proud of 'em,
  and they have all the world in their chamber.             25
JAILER They are famed to be a pair of absolute men.
JAILER'S DAUGHTER By my troth, I think fame but stam-
  mers 'em—they stand a grece above the reach of report.
JAILER I heard them reported in the battle to be the only
  doers.                                                     30
JAILER'S DAUGHTER Nay, most likely, for they are noble
  sufferers. I marvel how they would have looked had
  they been victors, that with such a constant nobility
  enforce a freedom out of bondage, making misery their
  mirth, and affliction a toy to jest at.                   35
JAILER Do they so?
JAILER'S DAUGHTER It seems to me they have no more
  sense of their captivity than I of ruling Athens. They
  eat well, look merrily, discourse of many things, but
  nothing of their own restraint and disasters. Yet
  sometime a divided sigh—martyred as 'twere i'th'
  deliverance—will break from one of them, when the
  other presently gives it so sweet a rebuke that I could
  wish myself a sigh to be so chid, or at least a sigher
  to be comforted.                                           45
WOOER I never saw 'em.
JAILER The Duke himself came privately in the night,
        *Palamon and Arcite appear ⌐at a window⌐ above*
  and so did they. What the reason of it is I know not.
  Look, yonder they are. That's Arcite looks out.           49
JAILER'S DAUGHTER No, sir, no—that's Palamon. Arcite is
  the lower of the twain—(*pointing at Arcite*) you may
  perceive a part of him.
JAILER Go to, leave your pointing. They would not make
  us their object. Out of their sight.                      54
JAILER'S DAUGHTER It is a holiday to look on them. Lord,
  the difference of men!                          *Exeunt*

**2.2**  *Enter Palamon and Arcite in prison, ⌐in shackles,*
        *above⌐*
PALAMON
  How do you, noble cousin?
ARCITE                     How do you, sir?
PALAMON
  Why, strong enough to laugh at misery
  And bear the chance of war. Yet we are prisoners,
  I fear, for ever, cousin.
ARCITE                  I believe it,
  And to that destiny have patiently                         5
  Laid up my hour to come.
PALAMON              O, cousin Arcite,
  Where is Thebes now? Where is our noble country?
  Where are our friends and kindreds? Never more
  Must we behold those comforts, never see
  The hardy youths strive for the games of honour,          10
  Hung with the painted favours of their ladies,
  Like tall ships under sail; then start amongst 'em
  And, as an east wind, leave 'em all behind us,
  Like lazy clouds, whilst Palamon and Arcite,
  Even in the wagging of a wanton leg,                       15
  Outstripped the people's praises, won the garlands
  Ere they have time to wish 'em ours. O never
  Shall we two exercise, like twins of honour,
  Our arms again and feel our fiery horses
  Like proud seas under us. Our good swords, now—           20
  Better the red-eyed god of war ne'er wore—

Ravished our sides, like age must run to rust
And deck the temples of those gods that hate us.
These hands shall never draw 'em out like lightning
To blast whole armies more.
ARCITE            No, Palamon,      25
Those hopes are prisoners with us. Here we are,
And here the graces of our youths must wither,
Like a too-timely spring. Here age must find us
And, which is heaviest, Palamon, unmarried—
The sweet embraces of a loving wife      30
Loaden with kisses, armed with thousand Cupids,
Shall never clasp our necks; no issue know us;
No figures of ourselves shall we e'er see
To glad our age, and, like young eagles, teach 'em
Boldly to gaze against bright arms and say,      35
'Remember what your fathers were, and conquer.'
The fair-eyed maids shall weep our banishments,
And in their songs curse ever-blinded fortune,
Till she for shame see what a wrong she has done
To youth and nature. This is all our world.      40
We shall know nothing here but one another,
Hear nothing but the clock that tells our woes.
The vine shall grow, but we shall never see it;
Summer shall come, and with her all delights,
But dead-cold winter must inhabit here still.      45
PALAMON
'Tis too true, Arcite. To our Theban hounds
That shook the agèd forest with their echoes,
No more now must we holler; no more shake
Our pointed javelins whilst the angry swine
Flies like a Parthian quiver from our rages,      50
Struck with our well-steeled darts. All valiant uses—
The food and nourishment of noble minds—
In us two here shall perish; we shall die—
Which is the curse of honour—lastly,
Children of grief and ignorance.
ARCITE            Yet, cousin,      55
Even from the bottom of these miseries,
From all that fortune can inflict upon us,
I see two comforts rising—two mere blessings,
If the gods please, to hold here a brave patience
And the enjoying of our griefs together.      60
Whilst Palamon is with me, let me perish
If I think this our prison.
PALAMON           Certainly
'Tis a main goodness, cousin, that our fortunes
Were twined together. 'Tis most true, two souls
Put in two noble bodies, let 'em suffer      65
The gall of hazard, so they grow together,
Will never sink; they must not, say they could.
A willing man dies sleeping and all's done.
ARCITE
Shall we make worthy uses of this place
That all men hate so much?
PALAMON          How, gentle cousin?      70
ARCITE
Let's think this prison holy sanctuary,
To keep us from corruption of worse men.
We are young, and yet desire the ways of honour
That liberty and common conversation,
The poison of pure spirits, might, like women,      75
Woo us to wander from. What worthy blessing
Can be, but our imaginations
May make it ours? And here being thus together,
We are an endless mine to one another:

We are one another's wife, ever begetting      80
New births of love; we are father, friends,
    acquaintance;
We are in one another, families—
I am your heir, and you are mine; this place
Is our inheritance: no hard oppressor
Dare take this from us. Here, with a little patience,      85
We shall live long and loving. No surfeits seek us—
The hand of war hurts none here, nor the seas
Swallow their youth. Were we at liberty
A wife might part us lawfully, or business;
Quarrels consume us; envy of ill men      90
Crave our acquaintance. I might sicken, cousin,
Where you should never know it, and so perish
Without your noble hand to close mine eyes,
Or prayers to the gods. A thousand chances,
Were we from hence, would sever us.
PALAMON           You have made me—
I thank you, cousin Arcite—almost wanton      96
With my captivity. What a misery
It is to live abroad, and everywhere!
'Tis like a beast, methinks. I find the court here;
I am sure, a more content; and all those pleasures      100
That woo the wills of men to vanity
I see through now, and am sufficient
To tell the world 'tis but a gaudy shadow,
That old Time, as he passes by, takes with him.
What had we been, old in the court of Creon,      105
Where sin is justice, lust and ignorance
The virtues of the great ones? Cousin Arcite,
Had not the loving gods found this place for us,
We had died as they do, ill old men, unwept,
And had their epitaphs, the people's curses.      110
Shall I say more?
ARCITE          I would hear you still.
PALAMON            Ye shall.
Is there record of any two that loved
Better than we do, Arcite?
ARCITE          Sure there cannot.
PALAMON
I do not think it possible our friendship
Should ever leave us.
ARCITE          Till our deaths it cannot,      115
    *Enter Emilia and her Woman ⌈below⌉. Palamon sees*
    *Emilia and is silent*
And after death our spirits shall be led
To those that love eternally. Speak on, sir.
EMILIA (*to her Woman*)
This garden has a world of pleasure in't.
What flower is this?
WOMAN          'Tis called narcissus, madam.
EMILIA
That was a fair boy, certain, but a fool      120
To love himself. Were there not maids enough?
ARCITE (*to Palamon*)
Pray forward.
PALAMON      Yes.
EMILIA (*to her Woman*) Or were they all hard-hearted?
WOMAN
They could not be to one so fair.
EMILIA          Thou wouldst not.
WOMAN
I think I should not, madam.
EMILIA          That's a good wench—
But take heed to your kindness, though.
WOMAN          Why, madam?

EMILIA
Men are mad things.

ARCITE (*to Palamon*)     Will ye go forward, cousin?     126

EMILIA (*to her Woman*)
Canst not thou work such flowers in silk, wench?

WOMAN                                                          Yes.

EMILIA
I'll have a gown full of 'em, and of these.
This is a pretty colour—will't not do
Rarely upon a skirt, wench?

WOMAN                         Dainty, madam.     130

ARCITE (*to Palamon*)
Cousin, cousin, how do you, sir? Why, Palamon!

PALAMON
Never till now was I in prison, Arcite.

ARCITE
Why, what's the matter, man?

PALAMON                         Behold and wonder!

    *Arcite sees Emilia*
By heaven, she is a goddess!

ARCITE                       Ha!

PALAMON                          Do reverence.
She is a goddess, Arcite.

EMILIA (*to her Woman*)     Of all flowers     135
Methinks a rose is best.

WOMAN                     Why, gentle madam?

EMILIA
It is the very emblem of a maid—
For when the west wind courts her gently,
How modestly she blows, and paints the sun
With her chaste blushes! When the north comes near
    her,                                                        140
Rude and impatient, then, like chastity,
She locks her beauties in her bud again,
And leaves him to base briers.

WOMAN                   Yet, good madam,
Sometimes her modesty will blow so far
She falls for't—a maid,                                        145
If she have any honour, would be loath
To take example by her.

EMILIA                   Thou art wanton.

ARCITE (*to Palamon*)
She is wondrous fair.

PALAMON              She is all the beauty extant.     148

EMILIA (*to her Woman*)
The sun grows high—let's walk in. Keep these flowers.
We'll see how close art can come near their colours.
I am wondrous merry-hearted—I could laugh now.

WOMAN
I could lie down, I am sure.

EMILIA                       And take one with you?

WOMAN
That's as we bargain, madam.

EMILIA                       Well, agree then.
    *Exeunt Emilia and her Woman*

PALAMON
What think you of this beauty?

ARCITE                         'Tis a rare one.

PALAMON
Is't but a rare one?

ARCITE              Yes, a matchless beauty.     155

PALAMON
Might not a man well lose himself and love her?

ARCITE
I cannot tell what you have done; I have,
Beshrew mine eyes for't. Now I feel my shackles.

PALAMON You love her then?

ARCITE Who would not?                                          160

PALAMON And desire her?

ARCITE Before my liberty.

PALAMON
I saw her first.

ARCITE           That's nothing.

PALAMON                          But it shall be.

ARCITE
I saw her too.

PALAMON        Yes, but you must not love her.

ARCITE
I will not, as you do, to worship her                          165
As she is heavenly and a blessèd goddess!
I love her as a woman, to enjoy her—
So both may love.

PALAMON           You shall not love at all.

ARCITE
Not love at all—who shall deny me?

PALAMON
I that first saw her, I that took possession                   170
First with mine eye of all those beauties
In her revealed to mankind. If thou lov'st her,
Or entertain'st a hope to blast my wishes,
Thou art a traitor, Arcite, and a fellow
False as thy title to her. Friendship, blood,                  175
And all the ties between us I disclaim,
If thou once think upon her.

ARCITE                       Yes, I love her—
And if the lives of all my name lay on it,
I must do so. I love her with my soul—
If that will lose ye, farewell, Palamon!                       180
I say again,
I love her, and in loving her maintain
I am as worthy and as free a lover,
And have as just a title to her beauty,
As any Palamon, or any living                                  185
That is a man's son.

PALAMON              Have I called thee friend?

ARCITE
Yes, and have found me so. Why are you moved
    thus?
Let me deal coldly with you. Am not I
Part of your blood, part of your soul? You have
    told me
That I was Palamon and you were Arcite.

PALAMON                                  Yes.     190

ARCITE
Am not I liable to those affections,
Those joys, griefs, angers, fears, my friend shall
    suffer?

PALAMON
Ye may be.

ARCITE     Why then would you deal so cunningly,
So strangely, so unlike a noble kinsman,
To love alone? Speak truly. Do you think me              195
Unworthy of her sight?

PALAMON                 No, but unjust
If thou pursue that sight.

ARCITE                     Because another
First sees the enemy, shall I stand still,
And let mine honour down, and never charge?

PALAMON
Yes, if he be but one.

ARCITE                 But say that one          200
Had rather combat me?

PALAMON                Let that one say so,
And use thy freedom; else, if thou pursuest her,

Be as that cursèd man that hates his country,
A branded villain.
ARCITE                    You are mad.
PALAMON                              I must be.
Till thou art worthy, Arcite, it concerns me;            205
And in this madness if I hazard thee
And take thy life, I deal but truly.
ARCITE                                  Fie, sir.
You play the child extremely. I will love her,
I must, I ought to do so, and I dare—
And all this justly.
PALAMON            O, that now, that now            210
Thy false self and thy friend had but this fortune—
To be one hour at liberty and grasp
Our good swords in our hands! I would quickly teach
    thee
What 'twere to filch affection from another.
Thou art baser in it than a cutpurse.            215
Put but thy head out of this window more
And, as I have a soul, I'll nail thy life to't.
ARCITE
Thou dar'st not, fool; thou canst not; thou art feeble.
Put my head out? I'll throw my body out
And leap the garden when I see her next,            220
    Enter the Jailer [above]
And pitch between her arms to anger thee.
PALAMON
No more—the keeper's coming. I shall live
To knock thy brains out with my shackles.
ARCITE                                  Do.
JAILER
By your leave, gentlemen.
PALAMON            Now, honest keeper?
JAILER
Lord Arcite, you must presently to th' Duke.            225
The cause I know not yet.
ARCITE            I am ready, keeper.
JAILER
Prince Palamon, I must a while bereave you
Of your fair cousin's company.
                    Exeunt Arcite and the Jailer
PALAMON                    And me, too,
Even when you please, of life. Why is he sent for?
It may be he shall marry her—he's goodly,            230
And like enough the Duke hath taken notice
Both of his blood and body. But his falsehood!
Why should a friend be treacherous? If that
Get him a wife so noble and so fair,
Let honest men ne'er love again. Once more            235
I would but see this fair one. Blessèd garden,
And fruit and flowers more blessèd, that still blossom
As her bright eyes shine on ye! Would I were,
For all the fortune of my life hereafter,
Yon little tree, yon blooming apricot—            240
How I would spread and fling my wanton arms
In at her window! I would bring her fruit
Fit for the gods to feed on; youth and pleasure
Still as she tasted should be doubled on her;
And if she be not heavenly, I would make her            245
So near the gods in nature they should fear her—
    Enter the Jailer [above]
And then I am sure she would love me. How now,
    keeper,
Where's Arcite?
JAILER            Banished—Prince Pirithous
Obtained his liberty; but never more,

Upon his oath and life, must he set foot            250
Upon this kingdom.
PALAMON [aside]            He's a blessèd man.
He shall see Thebes again, and call to arms
The bold young men that, when he bids 'em charge,
Fall on like fire. Arcite shall have a fortune,
If he dare make himself a worthy lover,            255
Yet in the field to strike a battle for her;
And if he lose her then, he's a cold coward.
How bravely may he bear himself to win her
If he be noble Arcite; thousand ways!
Were I at liberty I would do things            260
Of such a virtuous greatness that this lady,
This blushing virgin, should take manhood to her
And seek to ravish me.
JAILER            My lord, for you
I have this charge to—
PALAMON            To discharge my life.
JAILER
No, but from this place to remove your lordship—            265
The windows are too open.
PALAMON            Devils take 'em
That are so envious to me—prithee kill me.
JAILER
And hang for't afterward?
PALAMON            By this good light,
Had I a sword I would kill thee.
JAILER            Why, my lord?
PALAMON
Thou bring'st such pelting scurvy news continually,
Thou art not worthy life. I will not go.            271
JAILER
Indeed you must, my lord.
PALAMON            May I see the garden?
JAILER
No.
PALAMON Then I am resolved—I will not go.
JAILER
I must constrain you, then; and for you are dangerous,
I'll clap more irons on you.
PALAMON            Do, good keeper.            275
I'll shake 'em so ye shall not sleep:
I'll make ye a new morris. Must I go?
JAILER
There is no remedy.
PALAMON            Farewell, kind window.
May rude wind never hurt thee. O, my lady,
If ever thou hast felt what sorrow was,            280
Dream how I suffer. Come, now bury me.
                    Exeunt Palamon and the Jailer

2.3    Enter Arcite
ARCITE
Banished the kingdom? 'Tis a benefit,
A mercy I must thank 'em for; but banished
The free enjoying of that face I die for—
O, 'twas a studied punishment, a death
Beyond imagination; such a vengeance            5
That, were I old and wicked, all my sins
Could never pluck upon me. Palamon,
Thou hast the start now—thou shalt stay and see
Her bright eyes break each morning 'gainst thy
    window,
And let in life into thee. Thou shalt feed            10
Upon the sweetness of a noble beauty
That nature ne'er exceeded, nor ne'er shall.

Good gods! What happiness has Palamon!
Twenty to one he'll come to speak to her,
And if she be as gentle as she's fair,      15
I know she's his—he has a tongue will tame
Tempests and make the wild rocks wanton.
Come what can come,
The worst is death. I will not leave the kingdom.
I know mine own is but a heap of ruins,      20
And no redress there. If I go he has her.
I am resolved another shape shall make me,
Or end my fortunes. Either way I am happy—
I'll see her and be near her, or no more.

*Enter four Country People, one of whom carries a*
*garland before them. Arcite stands apart*

FIRST COUNTRYMAN
My masters, I'll be there—that's certain.      25
SECOND COUNTRYMAN And I'll be there.
THIRD COUNTRYMAN And I.
FOURTH COUNTRYMAN
Why then, have with ye, boys! 'Tis but a chiding—
Let the plough play today, I'll tickle't out
Of the jades' tails tomorrow.
FIRST COUNTRYMAN      I am sure      30
To have my wife as jealous as a turkey—
But that's all one. I'll go through, let her mumble.
SECOND COUNTRYMAN
Clap her aboard tomorrow night and stow her,
And all's made up again.
THIRD COUNTRYMAN      Ay, do but put
A fescue in her fist and you shall see her      35
Take a new lesson out and be a good wench.
Do we all hold against the maying?
FOURTH COUNTRYMAN
Hold? What should ail us?
THIRD COUNTRYMAN      Arcas will be there.
SECOND COUNTRYMAN And Sennois, and Rycas, and three
better lads ne'er danced under green tree; and ye know
what wenches, ha? But will the dainty dominie, the
schoolmaster, keep touch, do you think? For he does
all, ye know.      43
THIRD COUNTRYMAN He'll eat a hornbook ere he fail. Go
to, the matter's too far driven between him and the
tanner's daughter to let slip now, and she must see the
Duke, and she must dance too.
FOURTH COUNTRYMAN Shall we be lusty?      48
SECOND COUNTRYMAN All the boys in Athens blow wind
i'th' breech on's! And here I'll be and there I'll be, for
our town, and here again and there again—ha, boys,
hey for the weavers!
FIRST COUNTRYMAN This must be done i'th' woods.
FOURTH COUNTRYMAN O, pardon me.      54
SECOND COUNTRYMAN By any means, our thing of learning
said so; where he himself will edify the Duke most
parlously in our behalfs—he's excellent i'th' woods,
bring him to th' plains, his learning makes no cry.    58
THIRD COUNTRYMAN We'll see the sports, then every man
to's tackle—and, sweet companions, let's rehearse, by
any means, before the ladies see us, and do sweetly,
and God knows what may come on't.
FOURTH COUNTRYMAN Content—the sports once ended,
we'll perform. Away boys, and hold.
ARCITE (*coming forward*)
By your leaves, honest friends, pray you whither go
you?      65
FOURTH COUNTRYMAN
Whither? Why, what a question's that?

ARCITE Yet 'tis a question
To me that know not.
THIRD COUNTRYMAN      To the games, my friend.
SECOND COUNTRYMAN
Where were you bred, you know it not?
ARCITE      Not far, sir—
Are there such games today?
FIRST COUNTRYMAN      Yes, marry, are there,    70
And such as you never saw. The Duke himself
Will be in person there.
ARCITE      What pastimes are they?
SECOND COUNTRYMAN
Wrestling and running. (*To the others*) 'Tis a pretty
fellow.
THIRD COUNTRYMAN (*to Arcite*)
Thou wilt not go along?
ARCITE      Not yet, sir.
FOURTH COUNTRYMAN      Well, sir,
Take your own time. (*To the others*) Come, boys.
FIRST COUNTRYMAN      My mind misgives me—
This fellow has a vengeance trick o'th' hip:      76
Mark how his body's made for't.
SECOND COUNTRYMAN      I'll be hanged though
If he dare venture; hang him, plum porridge!
He wrestle? He roast eggs! Come, let's be gone, lads.
*Exeunt the four Countrymen*
ARCITE
This is an offered opportunity      80
I durst not wish for. Well I could have wrestled—
The best men called it excellent—and run
Swifter than wind upon a field of corn,
Curling the wealthy ears, never flew. I'll venture,
And in some poor disguise be there. Who knows    85
Whether my brows may not be girt with garlands,
And happiness prefer me to a place
Where I may ever dwell in sight of her?      *Exit*

**2.4**    *Enter the Jailer's Daughter*
JAILER'S DAUGHTER
Why should I love this gentleman? 'Tis odds
He never will affect me. I am base,
My father the mean keeper of his prison,
And he a prince. To marry him is hopeless,
To be his whore is witless. Out upon't,      5
What pushes are we wenches driven to
When fifteen once has found us? First, I saw him;
I, seeing, thought he was a goodly man;
He has as much to please a woman in him—
If he please to bestow it so—as ever      10
These eyes yet looked on. Next, I pitied him,
And so would any young wench, o'my conscience,
That ever dreamed or vowed her maidenhead
To a young handsome man. Then, I loved him,
Extremely loved him, infinitely loved him—      15
And yet he had a cousin fair as he, too.
But in my heart was Palamon, and there,
Lord, what a coil he keeps! To hear him
Sing in an evening, what a heaven it is!
And yet his songs are sad ones. Fairer spoken    20
Was never gentleman. When I come in
To bring him water in a morning, first
He bows his noble body, then salutes me, thus:
'Fair, gentle maid, good morrow. May thy goodness
Get thee a happy husband.' Once he kissed me—    25
I loved my lips the better ten days after.

Would he would do so every day! He grieves much,
And me as much to see his misery.
What should I do to make him know I love him?
For I would fain enjoy him. Say I ventured          30
To set him free? What says the law then? Thus much
For law or kindred! I will do it,
And this night; ere tomorrow he shall love me.     *Exit*

**2.5**  *Short flourish of cornetts and shouts within. Enter*
*Theseus, Hippolyta, Pirithous, Emilia, Arcite*
*disguised, with a garland, and attendants*

THESEUS
You have done worthily. I have not seen
Since Hercules a man of tougher sinews.
Whate'er you are, you run the best and wrestle
That these times can allow.
ARCITE                                    I am proud to please you.
THESEUS
What country bred you?
ARCITE                      This—but far off, prince.      5
THESEUS
Are you a gentleman?
ARCITE                              My father said so,
And to those gentle uses gave me life.
THESEUS
Are you his heir?
ARCITE                        His youngest, sir.
THESEUS                                        Your father
Sure is a happy sire, then. What proves you?
ARCITE
A little of all noble qualities.                          10
I could have kept a hawk and well have hollered
To a deep cry of dogs; I dare not praise
My feat in horsemanship, yet they that knew me
Would say it was my best piece; last and greatest,
I would be thought a soldier.
THESEUS                                  You are perfect.    15
PIRITHOUS
Upon my soul, a proper man.
EMILIA                                      He is so.
PIRITHOUS (*to Hippolyta*)
How do you like him, lady?
HIPPOLYTA                            I admire him.
I have not seen so young a man so noble—
If he say true—of his sort.
EMILIA                                Believe
His mother was a wondrous handsome woman—      20
His face methinks goes that way.
HIPPOLYTA                                  But his body
And fiery mind illustrate a brave father.
PIRITHOUS
Mark how his virtue, like a hidden sun,
Breaks through his baser garments.
HIPPOLYTA                                He's well got, sure.
THESEUS (*to Arcite*)
What made you seek this place, sir?
ARCITE                                      Noble Theseus,    25
To purchase name and do my ablest service
To such a well-found wonder as thy worth,
For only in thy court of all the world
Dwells fair-eyed honour.
PIRITHOUS                        All his words are worthy.
THESEUS (*to Arcite*)
Sir, we are much indebted to your travel,              30
Nor shall you lose your wish.—Pirithous,

Dispose of this fair gentleman.
PIRITHOUS                            Thanks, Theseus.
(*To Arcite*) Whate'er you are, you're mine, and I shall
    give you
To a most noble service, to this lady,
This bright young virgin; pray observe her goodness.
You have honoured her fair birthday with your
    virtues,                                              36
And as your due you're hers. Kiss her fair hand, sir.
ARCITE
Sir, you're a noble giver. (*To Emilia*) Dearest beauty,
Thus let me seal my vowed faith.
    *He kisses her hand*
                                    When your servant,
Your most unworthy creature, but offends you,      40
Command him die, he shall.
EMILIA                              That were too cruel.
If you deserve well, sir, I shall soon see't.
You're mine, and somewhat better than your rank I'll
use you.
PIRITHOUS (*to Arcite*)
I'll see you furnished, and, because you say
You are a horseman, I must needs entreat you       45
This afternoon to ride—but 'tis a rough one.
ARCITE
I like him better, prince—I shall not then
Freeze in my saddle.
THESEUS (*to Hippolyta*) Sweet, you must be ready—
And you, Emilia, ⌐to Pirithous⌐ and you, friend—and
    all,
Tomorrow by the sun, to do observance              50
To flow'ry May in Dian's wood. (*To Arcite*) Wait well,
    sir,
Upon your mistress.—Emily, I hope
He shall not go afoot.
EMILIA                        That were a shame, sir,
While I have horses. (*To Arcite*) Take your choice, and
    what
You want, at any time, let me but know it.         55
If you serve faithfully, I dare assure you,
You'll find a loving mistress.
ARCITE                              If I do not,
Let me find that my father ever hated—
Disgrace and blows.
THESEUS                    Go, lead the way—you have won it.
It shall be so: you shall receive all dues          60
Fit for the honour you have won. 'Twere wrong else.
(*To Emilia*) Sister, beshrew my heart, you have a
    servant
That, if I were a woman, would be master.
But you are wise.
EMILIA                  I hope too wise for that, sir.
                                    *Flourish. Exeunt*

**2.6**  *Enter the Jailer's Daughter*
JAILER'S DAUGHTER
Let all the dukes and all the devils roar—
He is at liberty! I have ventured for him,
And out I have brought him. To a little wood
A mile hence I have sent him, where a cedar
Higher than all the rest spreads like a plane,      5
Fast by a brook—and there he shall keep close
Till I provide him files and food, for yet
His iron bracelets are not off. O Love,
What a stout-hearted child thou art! My father
Durst better have endured cold iron than done it.  10

I love him beyond love and beyond reason
Or wit or safety. I have made him know it—
I care not, I am desperate. If the law
Find me and then condemn me fo'rt, some wenches,
Some honest-hearted maids, will sing my dirge          15
And tell to memory my death was noble,
Dying almost a martyr. That way he takes,
I purpose, is my way too. Sure, he cannot
Be so unmanly as to leave me here.
If he do, maids will not so easily          20
Trust men again. And yet, he has not thanked me
For what I have done—no, not so much as kissed me—
And that, methinks, is not so well. Nor scarcely
Could I persuade him to become a free man,
He made such scruples of the wrong he did          25
To me and to my father. Yet, I hope
When he considers more, this love of mine
Will take more root within him. Let him do
What he will with me—so he use me kindly.
For use me, so he shall, or I'll proclaim him,          30
And to his face, no man. I'll presently
Provide him necessaries and pack my clothes up,
And where there is a patch of ground I'll venture,
So he be with me. By him, like a shadow,
I'll ever dwell. Within this hour the hubbub          35
Will be all o'er the prison—I am then
Kissing the man they look for. Farewell, father:
Get many more such prisoners and such daughters,
And shortly you may keep yourself. Now to him.   *Exit*

❧

3.1   ⌈*A bush in place.*⌉ *Cornetts in sundry places. Noise*
      *and hollering as of people a-Maying.*
      *Enter Arcite*

ARCITE
The Duke has lost Hippolyta—each took
A several laund. This is a solemn rite
They owe bloomed May, and the Athenians pay it
To th' heart of ceremony. O, Queen Emilia,
Fresher than May, sweeter          5
Than her gold buttons on the boughs, or all
Th'enamelled knacks o'th' mead or garden—yea,
We challenge too the bank of any nymph
That makes the stream seem flowers; thou, O jewel
O'th' wood, o'th' world, hast likewise blessed a pace
With thy sole presence in thy ⌈          11
                              ⌉ rumination
That I, poor man, might eftsoons come between
And chop on some cold thought. Thrice blessèd
      chance
To drop on such a mistress, expectation          15
Most guiltless on't! Tell me, O Lady Fortune,
Next after Emily my sovereign, how far
I may be proud. She takes strong note of me,
Hath made me near her, and this beauteous morn,
The prim'st of all the year, presents me with          20
A brace of horses—two such steeds might well
Be by a pair of kings backed, in a field
That their crowns' titles tried. Alas, alas,
Poor cousin Palamon, poor prisoner—thou
So little dream'st upon my fortune that          25
Thou think'st thyself the happier thing to be
So near Emilia. Me thou deem'st at Thebes,
And therein wretched, although free. But if
Thou knew'st my mistress breathed on me, and that

I eared her language, lived in her eye—O, coz,          30
What passion would enclose thee!
      *Enter Palamon as out of a bush with his shackles.*
      *He bends his fist at Arcite*
PALAMON                                      Traitor kinsman,
Thou shouldst perceive my passion if these signs
Of prisonment were off me, and this hand
But owner of a sword. By all oaths in one,
I and the justice of my love would make thee          35
A confessed traitor. O thou most perfidious
That ever gently looked, the void'st of honour
That e'er bore gentle token, falsest cousin
That ever blood made kin—call'st thou her thine?
I'll prove it in my shackles, with these hands,          40
Void of appointment, that thou liest and art
A very thief in love, a chaffy lord
Not worth the name of villain. Had I a sword
And these house-clogs away—
ARCITE                              Dear cousin Palamon—
PALAMON
Cozener Arcite, give me language such          45
As thou hast showed me feat.
ARCITE                              Not finding in
The circuit of my breast any gross stuff
To form me like your blazon holds me to
This gentleness of answer—'tis your passion
That thus mistakes, the which, to you being enemy,
Cannot to me be kind. Honour and honesty          51
I cherish and depend on, howsoe'er
You skip them in me, and with them, fair coz,
I'll maintain my proceedings. Pray be pleased
To show in generous terms your griefs, since that          55
Your question's with your equal, who professes
To clear his own way with the mind and sword
Of a true gentleman.
PALAMON                  That thou durst, Arcite!
ARCITE
My coz, my coz, you have been well advertised
How much I dare; you've seen me use my sword          60
Against th'advice of fear. Sure, of another
You would not hear me doubted, but your silence
Should break out, though i'th' sanctuary.
PALAMON                                      Sir,
I have seen you move in such a place which well
Might justify your manhood; you were called          65
A good knight and a bold. But the whole week's not
      fair
If any day it rain: their valiant temper
Men lose when they incline to treachery,
And then they fight like compelled bears—would fly
Were they not tied.
ARCITE              Kinsman, you might as well          70
Speak this and act it in your glass as to
His ear which now disdains you.
PALAMON                        Come up to me,
Quit me of these cold gyves, give me a sword,
Though it be rusty, and the charity
Of one meal lend me. Come before me then,          75
A good sword in thy hand, and do but say
That Emily is thine—I will forgive
The trespass thou hast done me, yea, my life,
If then thou carry't; and brave souls in shades
That have died manly, which will seek of me          80
Some news from earth, they shall get none but this—

That thou art brave and noble.

ARCITE　　　　　　　　　　　Be content,
Again betake you to your hawthorn house.
With counsel of the night I will be here
With wholesome viands. These impediments　　85
Will I file off. You shall have garments and
Perfumes to kill the smell o'th' prison. After,
When you shall stretch yourself and say but 'Arcite,
I am in plight', there shall be at your choice
Both sword and armour.

PALAMON　　　　　　　O, you heavens, dares any　90
So noble bear a guilty business! None
But only Arcite, therefore none but Arcite
In this kind is so bold.

ARCITE　　　　　　　　　Sweet Palamon.

PALAMON
I do embrace you and your offer—for
Your offer do't I only, sir; your person,　　95
Without hypocrisy, I may not wish
　　　*Wind horns within*
More than my sword's edge on't.

ARCITE　　　　　　　　　You hear the horns—
Enter your muset lest this match between's
Be crossed ere met. Give me your hand, farewell.
I'll bring you every needful thing—I pray you,　100
Take comfort and be strong.

PALAMON　　　　　　　Pray hold your promise,
And do the deed with a bent brow. Most certain
You love me not—be rough with me and pour
This oil out of your language. By this air,
I could for each word give a cuff, my stomach　105
Not reconciled by reason.

ARCITE　　　　　　　　Plainly spoken,
Yet—pardon me—hard language: when I spur
　　　*Wind horns within*
My horse I chide him not. Content and anger
In me have but one face. Hark, sir, they call
The scattered to the banquet. You must guess　110
I have an office there.

PALAMON　　　　　　　Sir, your attendance
Cannot please heaven, and I know your office
Unjustly is achieved.

ARCITE　　　　　　　'Tis a good title.
I am persuaded this question, sick between's,
By bleeding must be cured. I am a suitor　115
That to your sword you will bequeath this plea
And talk of it no more.

PALAMON　　　　　　　But this one word:
You are going now to gaze upon my mistress—
For note you, mine she is—

ARCITE　　　　　　　　Nay then—

PALAMON　　　　　　　　　Nay, pray you—
You talk of feeding me to breed me strength—　120
You are going now to look upon a sun
That strengthens what it looks on. There you have
A vantage o'er me, but enjoy it till
I may enforce my remedy. Farewell.
　　　*Exeunt severally,* ⌈*Palamon as into the bush*⌉

**3.2**　　*Enter the Jailer's Daughter, with a file*

JAILER'S DAUGHTER
He has mistook the brake I meant, is gone
After his fancy. 'Tis now wellnigh morning.
No matter—would it were perpetual night,
And darkness lord o'th' world. Hark, 'tis a wolf!

In me hath grief slain fear, and, but for one thing,　5
I care for nothing—and that's Palamon.
I reck not if the wolves would jaw me, so
He had this file. What if I hollered for him?
I cannot holler. If I whooped, what then?
If he not answered, I should call a wolf　　10
And do him but that service. I have heard
Strange howls this livelong night—why may't not be
They have made prey of him? He has no weapons;
He cannot run; the jangling of his gyves
Might call fell things to listen, who have in them　15
A sense to know a man unarmed, and can
Smell where resistance is. I'll set it down
He's torn to pieces: they howled many together
And then they fed on him. So much for that.
Be bold to ring the bell. How stand I then?　　20
All's chared when he is gone. No, no, I lie:
My father's to be hanged for his escape,
Myself to beg, if I prized life so much
As to deny my act—but that I would not,
Should I try death by dozens. I am moped—　　25
Food took I none these two days,
Sipped some water. I have not closed mine eyes
Save when my lids scoured off their brine. Alas,
Dissolve, my life; let not my sense unsettle,
Lest I should drown or stab or hang myself.　　30
O state of nature, fail together in me,
Since thy best props are warped. So which way now?
The best way is the next way to a grave,
Each errant step beside is torment. Lo,
The moon is down, the crickets chirp, the screech-owl
Calls in the dawn. All offices are done　　36
Save what I fail in: but the point is this,
An end, and that is all.　　　　　　*Exit*

**3.3**　　*Enter Arcite with a bundle containing meat, wine,*
　　　　*and files*

ARCITE
I should be near the place. Ho, cousin Palamon!
　　　*Enter Palamon* ⌈*as from the bush*⌉

PALAMON
Arcite.

ARCITE　　The same. I have brought you food and files.
Come forth and fear not, here's no Theseus.

PALAMON
Nor none so honest, Arcite.

ARCITE　　　　　　　　That's no matter—
We'll argue that hereafter. Come, take courage—　5
You shall not die thus beastly. Here, sir, drink;
I know you are faint. Then I'll talk further with you.

PALAMON
Arcite, thou mightst now poison me.

ARCITE　　　　　　　　　　I might—
But I must fear you first. Sit down and, good now,
No more of these vain parleys. Let us not,　　10
Having our ancient reputation with us,
Make talk for fools and cowards. To your health, sir.

PALAMON
Do.
　　　⌈*Arcite drinks*⌉

ARCITE　Pray sit down, then, and let me entreat you,
By all the honesty and honour in you,
No mention of this woman—'twill disturb us.　15
We shall have time enough.

PALAMON　　　　　　　Well, sir, I'll pledge you.

*Palamon drinks*

ARCITE
Drink a good hearty draught; it breeds good blood,
    man.
Do not you feel it thaw you?

PALAMON                           Stay, I'll tell you
After a draught or two more.
            *Palamon drinks*

ARCITE                           Spare it not—
The Duke has more, coz. Eat now.

PALAMON                           Yes.
            *Palamon eats*

ARCITE                                    I am glad    20
You have so good a stomach.

PALAMON                           I am gladder
I have so good meat to't.

ARCITE                           Is't not mad, lodging
Here in the wild woods, cousin?

PALAMON                           Yes, for them
That have wild consciences.

ARCITE                           How tastes your victuals?
Your hunger needs no sauce, I see.

PALAMON                                    Not much.    25
But if it did, yours is too tart, sweet cousin.
What is this?

ARCITE          Venison.

PALAMON                           'Tis a lusty meat—
Give me more wine. Here, Arcite, to the wenches
We have known in our days. ⌈*Drinking*⌉ The lord
    steward's daughter.
Do you remember her?

ARCITE                           After you, coz.    30

PALAMON
She loved a black-haired man.

ARCITE                           She did so; well, sir.

PALAMON
And I have heard some call him Arcite, and—

ARCITE
Out with't, faith.

PALAMON                 She met him in an arbour—
What did she there, coz? Play o'th' virginals?

ARCITE
Something she did, sir—

PALAMON                 Made her groan a month for't—
Or two, or three, or ten.

ARCITE                 The marshal's sister    36
Had her share too, as I remember, cousin,
Else there be tales abroad. You'll pledge her?

PALAMON                           Yes.
            ⌈*They drink*⌉

ARCITE
A pretty brown wench 'tis. There was a time
When young men went a-hunting, and a wood,    40
And a broad beech, and thereby hangs a tale—
Heigh-ho!

PALAMON     For Emily, upon my life! Fool,
Away with this strained mirth. I say again,
That sigh was breathed for Emily. Base cousin,
Dar'st thou break first?

ARCITE                           You are wide.

PALAMON                           By heaven and earth,
There's nothing in thee honest.

ARCITE                           Then I'll leave you—
You are a beast now.

PALAMON                 As thou mak'st me, traitor.    47

ARCITE (*pointing to the bundle*)
There's all things needful: files and shirts and
    perfumes—
I'll come again some two hours hence and bring
That that shall quiet all.

PALAMON                 A sword and armour.    50

ARCITE
Fear me not. You are now too foul. Farewell.
Get off your trinkets: you shall want naught.

PALAMON                           Sirrah—

ARCITE
I'll hear no more.                                *Exit*

PALAMON                 If he keep touch, he dies for't.
            *Exit* ⌈*as into the bush*⌉

**3.4**   *Enter the Jailer's Daughter*

JAILER'S DAUGHTER
I am very cold, and all the stars are out too,
The little stars and all, that look like aglets—
The sun has seen my folly. Palamon!
Alas, no, he's in heaven. Where am I now?
Yonder's the sea and there's a ship—how't tumbles!    5
And there's a rock lies watching under water—
Now, now, it beats upon it—now, now, now,
There's a leak sprung, a sound one—how they cry!
Open her before the wind—you'll lose all else.
Up with a course or two and tack about, boys.    10
Good night, good night, you're gone. I am very
    hungry.
Would I could find a fine frog—he would tell me
News from all parts o'th' world, then would I make
A carrack of a cockle-shell, and sail
By east and north-east to the King of Pygmies,    15
For he tells fortunes rarely. Now my father,
Twenty to one, is trussed up in a trice
Tomorrow morning. I'll say never a word.

(*She sings*)
For I'll cut my green coat, a foot above my knee,
And I'll clip my yellow locks, an inch below mine eye,
    Hey nonny, nonny, nonny,                        21
He s'buy me a white cut, forth for to ride,
And I'll go seek him, through the world that is so wide,
    Hey nonny, nonny, nonny.

O for a prick now, like a nightingale,    25
To put my breast against. I shall sleep like a top else.
                                            *Exit*

**3.5**   *Enter Gerald (a schoolmaster), five Countrymen, one
    of whom is dressed as a Babion, five Wenches, and
    Timothy, a taborer. All are attired as morris dancers*

SCHOOLMASTER  Fie, fie,
What tediosity and disinsanity
Is here among ye! Have my rudiments
Been laboured so long with ye, milked unto ye,
And, by a figure, even the very plum-broth    5
And marrow of my understanding laid upon ye?
And do you still cry 'where?' and 'how?' and
    'wherefore?'
You most coarse frieze capacities, ye jean judgements,
Have I said, 'thus let be', and 'there let be',
And 'then let be', and no man understand me?    10
*Proh deum, medius fidius*—ye are all dunces.
Forwhy, here stand I. Here the Duke comes. There are
    you,
Close in the thicket. The Duke appears. I meet him,

And unto him I utter learnèd things
And many figures. He hears, and nods, and hums,    15
And then cries, 'Rare!', and I go forward. At length
I fling my cap up—mark there—then do you,
As once did Meleager and the boar,
Break comely out before him, like true lovers,
Cast yourselves in a body decently,    20
And sweetly, by a figure, trace and turn, boys.

FIRST COUNTRYMAN
And sweetly we will do it, master Gerald.

SECOND COUNTRYMAN
Draw up the company. Where's the taborer?

THIRD COUNTRYMAN
Why, Timothy!

TABORER            Here, my mad boys, have at ye!

SCHOOLMASTER
But I say, where's these women?

FOURTH COUNTRYMAN            Here's Friz and Madeline.

SECOND COUNTRYMAN
And little Luce with the white legs, and bouncing
Barbara.    26

FIRST COUNTRYMAN
And freckled Nell, that never failed her master.

SCHOOLMASTER
Where be your ribbons, maids? Swim with your bodies
And carry it sweetly and deliverly,
And now and then a favour and a frisk.    30

NELL
Let us alone, sir.

SCHOOLMASTER            Where's the rest o'th' music?

THIRD COUNTRYMAN
Dispersed as you commanded.

SCHOOLMASTER            Couple, then,
And see what's wanting. Where's the babion?
(To the Babion) My friend, carry your tail without
offence
Or scandal to the ladies; and be sure    35
You tumble with audacity and manhood,
And when you bark, do it with judgement.

BABION            Yes, sir.

SCHOOLMASTER
Quousque tandem? Here is a woman wanting!

FOURTH COUNTRYMAN
We may go whistle—all the fat's i'th' fire.

SCHOOLMASTER  We have,    40
As learnèd authors utter, washed a tile;
We have been fatuus, and laboured vainly.

SECOND COUNTRYMAN
This is that scornful piece, that scurvy hilding
That gave her promise faithfully she would be here—
Cicely, the seamstress' daughter.    45
The next gloves that I give her shall be dogskin.
Nay, an she fail me once—you can tell, Arcas,
She swore by wine and bread she would not break.

SCHOOLMASTER  An eel and woman,
A learnèd poet says, unless by th' tail    50
And with thy teeth thou hold, will either fail—
In manners this was false position.

FIRST COUNTRYMAN
A fire-ill take her! Does she flinch now?

THIRD COUNTRYMAN            What
Shall we determine, sir?

SCHOOLMASTER            Nothing;
Our business is become a nullity,    55
Yea, and a woeful and a piteous nullity.

FOURTH COUNTRYMAN
Now, when the credit of our town lay on it,
Now to be frampold, now to piss o'th' nettle!
Go thy ways—I'll remember thee, I'll fit thee!
Enter the Jailer's Daughter

JAILER'S DAUGHTER (sings)
The George Alow came from the south,    60
From the coast of Barbary-a;
And there he met with brave gallants of war,
By one, by two, by three-a.
'Well hailed, well hailed, you jolly gallants,
And whither now are you bound-a?    65
O let me have your company
Till I come to the sound-a.'
There was three fools fell out about an owlet—
The one he said it was an owl,
The other he said nay,    70
The third he said it was a hawk,
And her bells were cut away.

THIRD COUNTRYMAN
There's a dainty madwoman, master,
Comes i'th' nick, as mad as a March hare.
If we can get her dance, we are made again.    75
I warrant her, she'll do the rarest gambols.

FIRST COUNTRYMAN
A madwoman? We are made, boys.

SCHOOLMASTER (to the Jailer's Daughter)
And are you mad, good woman?

JAILER'S DAUGHTER            I would be sorry else.
Give me your hand.

SCHOOLMASTER            Why?

JAILER'S DAUGHTER            I can tell your fortune.
⌜She examines his hand⌝
You are a fool. Tell ten—I have posed him. Buzz!    80
Friend, you must eat no white bread—if you do,
Your teeth will bleed extremely. Shall we dance, ho?
I know you—you're a tinker. Sirrah tinker,
Stop no more holes but what you should.

SCHOOLMASTER            Dii boni—
A tinker, damsel?

JAILER'S DAUGHTER  Or a conjurer—    85
Raise me a devil now and let him play
Qui passa o'th' bells and bones.

SCHOOLMASTER            Go, take her,
And fluently persuade her to a peace.
Et opus exegi, quod nec Iovis ira, nec ignis—
Strike up, and lead her in.

SECOND COUNTRYMAN            Come, lass, let's trip it.    90

JAILER'S DAUGHTER  I'll lead.

THIRD COUNTRYMAN  Do, do.

SCHOOLMASTER
Persuasively and cunningly—
Wind horns within
            away, boys,
I hear the horns. Give me some meditation,
And mark your cue.
Exeunt all but Gerald the Schoolmaster
            Pallas inspire me.    95
Enter Theseus, Pirithous, Hippolyta, Emilia, Arcite,
and train

THESEUS  This way the stag took.

SCHOOLMASTER  Stay and edify.

THESEUS  What have we here?

PIRITHOUS
Some country sport, upon my life, sir.

THESEUS (*to the Schoolmaster*)
　Well, sir, go forward—we will edify.　　　　100
　Ladies, sit down—we'll stay it.
　　　*They sit:* ⌜*Theseus*⌝ *in a chair, the others on stools*
SCHOOLMASTER
　Thou doughty Duke, all hail! All hail, sweet ladies.
THESEUS This is a cold beginning.
SCHOOLMASTER
　If you but favour, our country pastime made is.
　We are a few of those collected here,　　　105
　That ruder tongues distinguish 'villager';
　And to say verity, and not to fable,
　We are a merry rout, or else a rabble,
　Or company, or, by a figure, chorus,
　That fore thy dignity will dance a morris.　　110
　And I, that am the rectifier of all,
　By title *pedagogus*, that let fall
　The birch upon the breeches of the small ones,
　And humble with a ferula the tall ones,
　Do here present this machine, or this frame;　115
　And dainty Duke, whose doughty dismal fame
　From Dis to Daedalus, from post to pillar,
　Is blown abroad, help me, thy poor well-willer,
　And with thy twinkling eyes, look right and straight
　Upon this mighty 'Moor'—of mickle weight—　120
　'Ice' now comes in, which, being glued together,
　Makes 'morris', and the cause that we came hither.
　The body of our sport, of no small study,
　I first appear, though rude, and raw, and muddy,
　To speak, before thy noble grace, this tenor　125
　At whose great feet I offer up my penner.
　The next, the Lord of May and Lady bright;
　The Chambermaid and Servingman, by night
　That seek out silent hanging; then mine Host
　And his fat Spouse, that welcomes, to their cost,　130
　The gallèd traveller, and with a beck'ning
　Informs the tapster to inflame the reck'ning;
　Then the beest-eating Clown; and next, the Fool;
　The babion with long tail and eke long tool,
　*Cum multis aliis* that make a dance—　　135
　Say 'ay', and all shall presently advance.
THESEUS
　Ay, ay, by any means, dear dominie.
PIRITHOUS　　　　　　　　　　Produce.
SCHOOLMASTER (*knocks for the dance*)
　*Intrate filii*, come forth and foot it.
　　⌜*He flings up his cap.*⌝ *Music.*
　　⌜*The Schoolmaster ushers in*
　*May Lord,　　　May Lady.*
　*Servingman,　　Chambermaid.*
　*A Country Clown,*
　　*or Shepherd,　Country Wench.*
　*An Host,　　　Hostess.*
　*A He-babion,　She-babion.*
　*A He-fool,　　The Jailer's Daughter as*
　　　　　　　　*She-fool.*
　*All these persons apparelled to the life, the men issuing out of one door and the wenches from the other. They dance a morris*⌝
　Ladies, if we have been merry,
　And have pleased ye with a derry,　　140
　And a derry, and a down,
　Say the schoolmaster's no clown.
　Duke, if we have pleased thee too,
　And have done as good boys should do,

　Give us but a tree or twain　　　　145
　For a maypole, and again,
　Ere another year run out,
　We'll make thee laugh, and all this rout.
THESEUS
　Take twenty, dominie. (*To Hippolyta*) How does my sweetheart?
HIPPOLYTA
　Never so pleased, sir.
EMILIA　　　　　　'Twas an excellent dance,　150
　And for a preface, I never heard a better.
THESEUS
　Schoolmaster, I thank you. One see 'em all rewarded.
PIRITHOUS
　And here's something to paint your pole withal.
　　*He gives them money*
THESEUS Now to our sports again.
SCHOOLMASTER
　　May the stag thou hunt'st stand long,　155
　　And thy dogs be swift and strong;
　　May they kill him without lets,
　　And the ladies eat his dowsets.
　　　*Exeunt Theseus and train. Wind horns within*
　Come, we are all made. *Dii deaeque omnes,*
　Ye have danced rarely, wenches.　　*Exeunt*

**3.6**　*Enter Palamon from the bush*
PALAMON
　About this hour my cousin gave his faith
　To visit me again, and with him bring
　Two swords and two good armours; if he fail,
　He's neither man nor soldier. When he left me,
　I did not think a week could have restored　5
　My lost strength to me, I was grown so low
　And crest-fall'n with my wants. I thank thee, Arcite,
　Thou art yet a fair foe, and I feel myself,
　With this refreshing, able once again
　To out-dure danger. To delay it longer　10
　Would make the world think, when it comes to hearing,
　That I lay fatting, like a swine, to fight,
　And not a soldier. Therefore this blest morning
　Shall be the last; and that sword he refuses,
　If it but hold, I kill him with; 'tis justice.　15
　So, love and fortune for me!
　　*Enter Arcite with two armours and two swords*
　　　　　　　　　　O, good morrow.
ARCITE
　Good morrow, noble kinsman.
PALAMON　　　　　　　I have put you
　To too much pains, sir.
ARCITE　　　　　　That too much, fair cousin,
　Is but a debt to honour, and my duty.
PALAMON
　Would you were so in all, sir—I could wish ye　20
　As kind a kinsman, as you force me find
　A beneficial foe, that my embraces
　Might thank ye, not my blows.
ARCITE　　　　　　I shall think either,
　Well done, a noble recompense.
PALAMON　　　　　　Then I shall quit you.
ARCITE
　Defy me in these fair terms, and you show　25
　More than a mistress to me—no more anger,
　As you love anything that's honourable.
　We were not bred to talk, man. When we are armed

And both upon our guards, then let our fury,
Like meeting of two tides, fly strongly from us;　　30
And then to whom the birthright of this beauty
Truly pertains—without upbraidings, scorns,
Despisings of our persons, and such poutings
Fitter for girls and schoolboys—will be seen,
And quickly, yours or mine. Will't please you arm,
　　sir?　　35
Or, if you feel yourself not fitting yet,
And furnished with your old strength, I'll stay,
　　cousin,
And every day discourse you into health,
As I am spared. Your person I am friends with,
And I could wish I had not said I loved her,　　40
Though I had died; but loving such a lady,
And justifying my love, I must not fly from't.

PALAMON
Arcite, thou art so brave an enemy
That no man but thy cousin's fit to kill thee.
I am well and lusty—choose your arms.

ARCITE　　　　　　　　　　　　Choose you, sir.

PALAMON
Wilt thou exceed in all, or dost thou do it　　46
To make me spare thee?

ARCITE　　　　　　　If you think so, cousin,
You are deceived, for as I am a soldier,
I will not spare you.

PALAMON　　　　　　That's well said.

ARCITE　　　　　　　　　　　You'll find it.

PALAMON
Then as I am an honest man, and love　　50
With all the justice of affection,
I'll pay thee soundly.
　　*He chooses one armour*
　　　　　　　This I'll take.

ARCITE (*indicating the remaining armour*)
　　　　　　　　　　That's mine, then.
I'll arm you first.

PALAMON　　　Do.
　　*Arcite arms Palamon*
　　　　　　　Pray thee tell me, cousin,
Where gott'st thou this good armour?

ARCITE　　　　　　　　　　'Tis the Duke's,
And to say true, I stole it. Do I pinch you?

PALAMON　　　　　　　　　　No.　　55

ARCITE
Is't not too heavy?

PALAMON　　　　　I have worn a lighter—
But I shall make it serve.

ARCITE　　　　　　I'll buckle't close.

PALAMON
By any means.

ARCITE　　　You care not for a grand guard?

PALAMON
No, no, we'll use no horses. I perceive
You would fain be at that fight.

ARCITE　　　　　　　　　I am indifferent.　　60

PALAMON
Faith, so am I. Good cousin, thrust the buckle
Through far enough.

ARCITE　　　　I warrant you.

PALAMON　　　　　　　　My casque now.

ARCITE
Will you fight bare-armed?

PALAMON　　　　　　　We shall be the nimbler.

ARCITE
But use your gauntlets, though—those are o'th' least.
Prithee take mine, good cousin.

PALAMON　　　　　　　　Thank you, Arcite.　　65
How do I look? Am I fall'n much away?

ARCITE
Faith, very little—love has used you kindly.

PALAMON
I'll warrant thee, I'll strike home.

ARCITE　　　　　　　　Do, and spare not—
I'll give you cause, sweet cousin.

PALAMON　　　　　　　　Now to you, sir.
　　*Palamon arms Arcite*
Methinks this armour's very like that, Arcite,　　70
Thou wor'st that day the three kings fell, but lighter.

ARCITE
That was a very good one, and that day,
I well remember, you outdid me, cousin.
I never saw such valour. When you charged
Upon the left wing of the enemy,　　75
I spurred hard to come up, and under me
I had a right good horse.

PALAMON　　　　　　You had indeed—
A bright bay, I remember.

ARCITE　　　　　　　Yes. But all
Was vainly laboured in me—you outwent me,
Nor could my wishes reach you. Yet a little　　80
I did by imitation.

PALAMON　　　　More by virtue—
You are modest, cousin.

ARCITE　　　　　　When I saw you charge first,
Methought I heard a dreadful clap of thunder
Break from the troop.

PALAMON　　　　　But still before that flew
The lightning of your valour. Stay a little,　　85
Is not this piece too strait?

ARCITE　　　　　　　No, no, 'tis well.

PALAMON
I would have nothing hurt thee but my sword—
A bruise would be dishonour.

ARCITE　　　　　　　Now I am perfect.

PALAMON
Stand off, then.

ARCITE　　　Take my sword; I hold it better.

PALAMON
I thank ye. No, keep it—your life lies on it.　　90
Here's one—if it but hold, I ask no more
For all my hopes. My cause and honour guard me.

ARCITE
And me, my love.
　　*They bow several ways, then advance and stand*
　　　　　　Is there aught else to say?

PALAMON
This only, and no more. Thou art mine aunt's son,
And that blood we desire to shed is mutual:　　95
In me, thine, and in thee, mine. My sword
Is in my hand, and if thou kill'st me,
The gods and I forgive thee. If there be
A place prepared for those that sleep in honour,
I wish his weary soul that falls may win it.　　100
Fight bravely, cousin. Give me thy noble hand.

ARCITE
Here, Palamon. This hand shall never more
Come near thee with such friendship.

PALAMON　　　　　　　　I commend thee.

ARCITE
If I fall, curse me, and say I was a coward—
For none but such dare die in these just trials.
Once more farewell, my cousin.                              105

PALAMON                               Farewell, Arcite.
 *Fight. Horns within; they stand*

ARCITE
Lo, cousin, lo, our folly has undone us.

PALAMON                               Why?

ARCITE
This is the Duke a-hunting, as I told you.
If we be found, we are wretched. O, retire,
For honour's sake, and safely, presently,        110
Into your bush again. Sir, we shall find
Too many hours to die. In, gentle cousin—
If you be seen, you perish instantly
For breaking prison, and I, if you reveal me,
For my contempt. Then all the world will scorn us,
And say we had a noble difference,                116
But base disposers of it.

PALAMON                         No, no, cousin,
I will no more be hidden, nor put off
This great adventure to a second trial.
I know your cunning and I know your cause—   120
He that faints now, shame take him! Put thyself
Upon thy present guard—

ARCITE                         You are not mad?

PALAMON
Or I will make th'advantage of this hour
Mine own, and what to come shall threaten me
I fear less than my fortune. Know, weak cousin,   125
I love Emilia, and in that I'll bury
Thee and all crosses else.

ARCITE                         Then come what can come,
Thou shalt know, Palamon, I dare as well
Die as discourse or sleep. Only this fears me,
The law will have the honour of our ends.       130
Have at thy life!

PALAMON          Look to thine own well, Arcite!
 *They fight again.*
 *Horns. Enter Theseus, Hippolyta, Emilia, Pirithous,*
 *and train.* ⌜*Theseus*⌝ *separates Palamon and Arcite*

THESEUS
What ignorant and mad malicious traitors
Are you, that 'gainst the tenor of my laws
Are making battle, thus like knights appointed,
Without my leave and officers of arms?          135
By Castor, both shall die.

PALAMON                         Hold thy word, Theseus.
We are certainly both traitors, both despisers
Of thee and of thy goodness. I am Palamon,
That cannot love thee, he that broke thy prison—
Think well what that deserves. And this is Arcite;   140
A bolder traitor never trod thy ground,
A falser ne'er seemed friend. This is the man
Was begged and banished; this is he contemns thee,
And what thou dar'st do; and in this disguise,
Against thine own edict, follows thy sister,    145
That fortunate bright star, the fair Emilia,
Whose servant—if there be a right in seeing
And first bequeathing of the soul to—justly
I am; and, which is more, dares think her his.
This treachery, like a most trusty lover,        150
I called him now to answer. If thou be'st
As thou art spoken, great and virtuous,
The true decider of all injuries,

Say, 'Fight again', and thou shalt see me, Theseus,
Do such a justice thou thyself wilt envy.        155
Then take my life—I'll woo thee to't.

PIRITHOUS                               O heaven,
What more than man is this!

THESEUS                         I have sworn.

ARCITE                                    We seek not
Thy breath of mercy, Theseus. 'Tis to me
A thing as soon to die as thee to say it,
And no more moved. Where this man calls me traitor
Let me say thus much—if in love be treason,    161
In service of so excellent a beauty,
As I love most, and in that faith will perish,
As I have brought my life here to confirm it,
As I have served her truest, worthiest,         165
As I dare kill this cousin that denies it,
So let me be most traitor and ye please me.
For scorning thy edict, Duke, ask that lady
Why she is fair, and why her eyes command me
Stay here to love her, and if she say, 'Traitor',   170
I am a villain fit to lie unburied.

PALAMON
Thou shalt have pity of us both, O Theseus,
If unto neither thou show mercy. Stop,
As thou art just, thy noble ear against us;
As thou art valiant, for thy cousin's soul,      175
Whose twelve strong labours crown his memory,
Let's die together, at one instant, Duke.
Only a little let him fall before me,
That I may tell my soul he shall not have her.

THESEUS
I grant your wish; for to say true, your cousin   180
Has ten times more offended, for I gave him
More mercy than you found, sir, your offences
Being no more than his. None here speak for 'em,
For ere the sun set both shall sleep for ever.

HIPPOLYTA (*to Emilia*)
Alas, the pity! Now or never, sister,            185
Speak, not to be denied. That face of yours
Will bear the curses else of after ages
For these lost cousins.

EMILIA                    In my face, dear sister,
I find no anger to 'em, nor no ruin.
The misadventure of their own eyes kill 'em.    190
Yet that I will be woman and have pity,
 ⌜*She kneels*⌝
My knees shall grow to th' ground, but I'll get mercy.
Help me, dear sister—in a deed so virtuous
The powers of all women will be with us.
 *Hippolyta kneels*
Most royal brother—

HIPPOLYTA          Sir, by our tie of marriage—   195

EMILIA
By your own spotless honour—

HIPPOLYTA                    By that faith,
That fair hand, and that honest heart you gave me—

EMILIA
By that you would have pity in another,
By your own virtues infinite—

HIPPOLYTA                    By valour,
By all the chaste nights I have ever pleased you—   200

THESEUS
These are strange conjurings.

PIRITHOUS                    Nay, then, I'll in too.
 ⌜*He kneels*⌝
By all our friendship, sir, by all our dangers,
By all you love most: wars, and this sweet lady—

EMILIA
By that you would have trembled to deny
A blushing maid—
HIPPOLYTA            By your own eyes, by strength—
In which you swore I went beyond all women,    206
Almost all men—and yet I yielded, Theseus—
PIRITHOUS
To crown all this, by your most noble soul,
Which cannot want due mercy, I beg first—
HIPPOLYTA
Next hear my prayers—
EMILIA                Last let me entreat, sir—    210
PIRITHOUS
For mercy.
HIPPOLYTA    Mercy.
EMILIA                Mercy on these princes.
THESEUS
Ye make my faith reel. Say I felt
Compassion to 'em both, how would you place it?
    ⌈They rise⌉
EMILIA
Upon their lives—but with their banishments.
THESEUS
You are a right woman, sister: you have pity,    215
But want the understanding where to use it.
If you desire their lives, invent a way
Safer than banishment. Can these two live,
And have the agony of love about 'em,
And not kill one another? Every day    220
They'd fight about you, hourly bring your honour
In public question with their swords. Be wise, then,
And here forget 'em. It concerns your credit
And my oath equally. I have said—they die.
Better they fall by th' law than one another.    225
Bow not my honour.
EMILIA                O my noble brother,
That oath was rashly made, and in your anger.
Your reason will not hold it. If such vows
Stand for express will, all the world must perish.
Beside, I have another oath 'gainst yours,    230
Of more authority, I am sure more love—
Not made in passion, neither, but good heed.
THESEUS
What is it, sister?
PIRITHOUS (to Emilia) Urge it home, brave lady.
EMILIA
That you would ne'er deny me anything
Fit for my modest suit and your free granting.    235
I tie you to your word now; if ye fail in't,
Think how you maim your honour—
For now I am set a-begging, sir. I am deaf
To all but your compassion—how their lives
Might breed the ruin of my name, opinion—    240
Shall anything that loves me perish for me?
That were a cruel wisdom: do men prune
The straight young boughs that blush with thousand
    blossoms
Because they may be rotten? O, Duke Theseus,
The goodly mothers that have groaned for these,    245
And all the longing maids that ever loved,
If your vow stand, shall curse me and my beauty,
And in their funeral songs for these two cousins
Despise my cruelty and cry woe worth me,
Till I am nothing but the scorn of women.    250
For heaven's sake, save their lives and banish 'em.

THESEUS
On what conditions?
EMILIA                Swear 'em never more
To make me their contention, or to know me,
To tread upon thy dukedom; and to be,
Wherever they shall travel, ever strangers    255
To one another.
PALAMON            I'll be cut a-pieces
Before I take this oath—forget I love her?
O all ye gods, despise me, then. Thy banishment
I not mislike, so we may fairly carry
Our swords and cause along—else, never trifle,    260
But take our lives, Duke. I must love, and will;
And for that love must and dare kill this cousin
On any piece the earth has.
THESEUS                Will you, Arcite,
Take these conditions?
PALAMON                He's a villain then.
PIRITHOUS                These are men!
ARCITE
No, never, Duke. 'Tis worse to me than begging,    265
To take my life so basely. Though I think
I never shall enjoy her, yet I'll preserve
The honour of affection and die for her,
Make death a devil.
THESEUS
What may be done? For now I feel compassion.    270
PIRITHOUS
Let it not fall again, sir.
THESEUS                Say, Emilia,
If one of them were dead—as one must—are you
Content to take the other to your husband?
They cannot both enjoy you. They are princes
As goodly as your own eyes, and as noble    275
As ever fame yet spoke of. Look upon 'em,
And if you can love, end this difference.
I give consent. (To Palamon and Arcite) Are you
    content too, princes?
PALAMON and ARCITE
With all our souls.
THESEUS                He that she refuses
Must die, then.
PALAMON and ARCITE
                Any death thou canst invent, Duke.
PALAMON
If I fall from that mouth, I fall with favour,    281
And lovers yet unborn shall bless my ashes.
ARCITE
If she refuse me, yet my grave will wed me,
And soldiers sing my epitaph.
THESEUS (to Emilia)                Make choice, then.
EMILIA
I cannot, sir. They are both too excellent.    285
For me, a hair shall never fall of these men.
HIPPOLYTA ⌈to Theseus⌉
What will become of 'em?
THESEUS                Thus I ordain it,
And by mine honour once again it stands,
Or both shall die. (To Palamon and Arcite) You shall
    both to your country,
And each within this month, accompanied    290
With three fair knights, appear again in this place,
In which I'll plant a pyramid; and whether,
Before us that are here, can force his cousin,
By fair and knightly strength, to touch the pillar,
He shall enjoy her; the other lose his head,    295

And all his friends; nor shall he grudge to fall,
Nor think he dies with interest in this lady.
Will this content ye?
PALAMON                    Yes. Here, cousin Arcite,
I am friends again till that hour.
ARCITE                              I embrace ye.
THESEUS (to Emilia)
Are you content, sister?
EMILIA                    Yes, I must, sir,          300
Else both miscarry.
THESEUS (to Palamon and Arcite)
                    Come, shake hands again, then,
And take heed, as you are gentlemen, this quarrel
Sleep till the hour prefixed, and hold your course.
PALAMON
We dare not fail thee, Theseus.
THESEUS                    Come, I'll give ye
Now usage like to princes and to friends.          305
When ye return, who wins I'll settle here,
Who loses, yet I'll weep upon his bier.
                    Exeunt. [In the act-time the bush is removed]

✿

4.1     Enter the Jailer and his Friend
JAILER
Hear you no more? Was nothing said of me
Concerning the escape of Palamon?
Good sir, remember.
FRIEND                    Nothing that I heard,
For I came home before the business
Was fully ended. Yet I might perceive,          5
Ere I departed, a great likelihood
Of both their pardons: for Hippolyta
And fair-eyed Emily upon their knees
Begged with such handsome pity that the Duke,
Methought, stood staggering whether he should
        follow                                   10
His rash oath or the sweet compassion
Of those two ladies; and to second them
That truly noble prince, Pirithous—
Half his own heart—set in too, that I hope
All shall be well. Neither heard I one question          15
Of your name or his scape.
            Enter the Second Friend
JAILER                    Pray heaven it hold so.
SECOND FRIEND
Be of good comfort, man. I bring you news,
Good news.
JAILER        They are welcome.
SECOND FRIEND                    Palamon has cleared you,
And got your pardon, and discovered how
And by whose means he scaped—which was your
        daughter's,                              20
Whose pardon is procured too; and the prisoner,
Not to be held ungrateful to her goodness,
Has given a sum of money to her marriage—
A large one, I'll assure you.
JAILER                    Ye are a good man,
And ever bring good news.
FIRST FRIEND                    How was it ended?          25
SECOND FRIEND
Why, as it should be: they that ne'er begged,
But they prevailed, had their suits fairly granted—
The prisoners have their lives.
FIRST FRIEND                    I knew 'twould be so.

SECOND FRIEND
But there be new conditions which you'll hear of          29
At better time.
JAILER        I hope they are good.
SECOND FRIEND                    They are honourable—
How good they'll prove I know not.
        Enter the Wooer
FIRST FRIEND                    'Twill be known.
WOOER
Alas, sir, where's your daughter?
JAILER                    Why do you ask?
WOOER
O, sir, when did you see her?
SECOND FRIEND                    How he looks!
JAILER
This morning.
WOOER        Was she well? Was she in health?          34
Sir, when did she sleep?
FIRST FRIEND                    These are strange questions.
JAILER
I do not think she was very well: for now
You make me mind her, but this very day
I asked her questions and she answered me
So far from what she was, so childishly,
So sillily, as if she were a fool,          40
An innocent—and I was very angry.
But what of her, sir?
WOOER                    Nothing, but my pity—
But you must know it, and as good by me
As by another that less loves her—
JAILER
Well, sir?
FIRST FRIEND Not right?
WOOER                    No, sir, not well.
SECOND FRIEND                    Not well?          45
WOOER
'Tis too true—she is mad.
FIRST FRIEND                    It cannot be.
WOOER
Believe, you'll find it so.
JAILER                    I half suspected
What you told me—the gods comfort her!
Either this was her love to Palamon,
Or fear of my miscarrying on his scape,          50
Or both.
WOOER        'Tis likely.
JAILER                    But why all this haste, sir?
WOOER
I'll tell you quickly. As I late was angling
In the great lake that lies behind the palace,
From the far shore, thick set with reeds and sedges,
As patiently I was attending sport,          55
I heard a voice—a shrill one—and attentive
I gave my ear, when I might well perceive
'Twas one that sung, and by the smallness of it
A boy or woman. I then left my angle
To his own skill, came near, but yet perceived not          60
Who made the sound, the rushes and the reeds
Had so encompassed it. I laid me down
And listened to the words she sung, for then,
Through a small glade cut by the fishermen,
I saw it was your daughter.
JAILER                    Pray go on, sir.          65
WOOER
She sung much, but no sense; only I heard her
Repeat this often—'Palamon is gone,

Is gone to th' wood to gather mulberries;
I'll find him out tomorrow.'
FIRST FRIEND                          Pretty soul!
WOOER
'His shackles will betray him—he'll be taken,          70
And what shall I do then? I'll bring a bevy,
A hundred black-eyed maids that love as I do,
With chaplets on their heads of daffodillies,
With cherry lips and cheeks of damask roses,
And all we'll dance an antic fore the Duke          75
And beg his pardon.' Then she talked of you, sir—
That you must lose your head tomorrow morning,
And she must gather flowers to bury you,
And see the house made handsome. Then she sung
Nothing but 'willow, willow, willow', and between   80
Ever was 'Palamon, fair Palamon',
And 'Palamon was a tall young man'. The place
Was knee-deep where she sat; her careless tresses
A wreath of bull-rush rounded; about her stuck
Thousand freshwater flowers of several colours—      85
That she appeared, methought, like the fair nymph
That feeds the lake with waters, or as Iris
Newly dropped down from heaven. Rings she made
Of rushes that grew by, and to 'em spoke
The prettiest posies—'Thus our true love's tied',    90
'This you may lose, not me', and many a one.
And then she wept, and sung again, and sighed—
And with the same breath smiled and kissed her
    hand.
SECOND FRIEND
Alas, what pity it is!
WOOER                    I made in to her:
She saw me and straight sought the flood—I saved
    her,                                              95
And set her safe to land, when presently
She slipped away and to the city made,
With such a cry and swiftness that, believe me,
She left me far behind her. Three or four
I saw from far off cross her—one of 'em             100
I knew to be your brother, where she stayed
And fell, scarce to be got away. I left them with her,
    Enter the Jailer's Brother, the Jailer's Daughter, and
    others
And hither came to tell you—here they are.
JAILER'S DAUGHTER (sings)
'May you never more enjoy the light . . .'—
Is not this a fine song?
JAILER'S BROTHER          O, a very fine one.          105
JAILER'S DAUGHTER
I can sing twenty more.
JAILER'S BROTHER          I think you can.
JAILER'S DAUGHTER
Yes, truly can I—I can sing 'The Broom'
And 'Bonny Robin'—are not you a tailor?
JAILER'S BROTHER
Yes.
JAILER'S DAUGHTER Where's my wedding gown?
JAILER'S BROTHER                 I'll bring it tomorrow.
JAILER'S DAUGHTER
Do, very rarely—I must be abroad else,               110
To call the maids and pay the minstrels,
For I must lose my maidenhead by cocklight,
'Twill never thrive else. (Sings) 'O fair, O sweet . . .'
JAILER'S BROTHER ⌈to the Jailer⌉
You must e'en take it patiently.
JAILER                              'Tis true.

JAILER'S DAUGHTER
Good ev'n, good men. Pray, did you ever hear       115
Of one young Palamon?
JAILER                        Yes, wench, we know him.
JAILER'S DAUGHTER
Is't not a fine young gentleman?
JAILER                                  'Tis, love.
JAILER'S BROTHER
By no mean cross her, she is then distempered
Far worse than now she shows.
FIRST FRIEND (to the Jailer's Daughter)
                              Yes, he's a fine man.
JAILER'S DAUGHTER
O, is he so? You have a sister.
FIRST FRIEND                    Yes.                  120
JAILER'S DAUGHTER
But she shall never have him, tell her so,
For a trick that I know. You'd best look to her,
For if she see him once, she's gone—she's done
And undone in an hour. All the young maids
Of our town are in love with him, but I laugh at
    'em                                              125
And let 'em all alone. Is't not a wise course?
FIRST FRIEND                                Yes.
JAILER'S DAUGHTER
There is at least two hundred now with child by him,
There must be four; yet I keep close for all this,
Close as a cockle; and all these must be boys—
He has the trick on't—and at ten years old          130
They must be all gelt for musicians
And sing the wars of Theseus.
SECOND FRIEND                      This is strange.
⌈JAILER'S BROTHER⌉
As ever you heard, but say nothing.
FIRST FRIEND                          No.
JAILER'S DAUGHTER
They come from all parts of the dukedom to him.
I'll warrant ye, he had not so few last night        135
As twenty to dispatch. He'll tickle't up
In two hours, if his hand be in.
JAILER                              She's lost
Past all cure.
JAILER'S BROTHER Heaven forbid, man!
JAILER'S DAUGHTER (to the Jailer)
Come hither—you are a wise man.
FIRST FRIEND                        Does she know him?
SECOND FRIEND
No—would she did.
JAILER'S DAUGHTER    You are master of a ship?       140
JAILER
Yes.
JAILER'S DAUGHTER Where's your compass?
JAILER                                    Here.
JAILER'S DAUGHTER                    Set it to th' north.
And now direct your course to th' wood where
    Palamon
Lies longing for me. For the tackling,
Let me alone. Come, weigh, my hearts, cheerly all.
Uff, uff, uff! 'Tis up. The wind's fair. Top the bowline.
Out with the mainsail. Where's your whistle, master?
JAILER'S BROTHER Let's get her in.                   147
JAILER
Up to the top, boy!
JAILER'S BROTHER     Where's the pilot?
FIRST FRIEND                              Here.

JAILER'S DAUGHTER
  What kenn'st thou?
SECOND FRIEND          A fair wood.
JAILER'S DAUGHTER                Bear for it, master.
  Tack about!                                    150
  (*Sings*) 'When Cynthia with her borrowed light . . .'
                                                 *Exeunt*

4.2   ⌈*Enter Emilia, with two pictures*⌉
EMILIA
  Yet I may bind those wounds up that must open
  And bleed to death for my sake else—I'll choose,
  And end their strife. Two such young handsome men
  Shall never fall for me; their weeping mothers
  Following the dead cold ashes of their sons,     5
  Shall never curse my cruelty. Good heaven,
  What a sweet face has Arcite! If wise nature,
  With all her best endowments, all those beauties
  She sows into the births of noble bodies,
  Were here a mortal woman and had in her         10
  The coy denials of young maids, yet doubtless
  She would run mad for this man. What an eye,
  Of what a fiery sparkle and quick sweetness
  Has this young prince! Here love himself sits smiling!
  Just such another wanton Ganymede              15
  Set Jove afire once, and enforced the god
  Snatch up the goodly boy and set him by him,
  A shining constellation. What a brow,
  Of what a spacious majesty, he carries!
  Arched like the great-eyed Juno's, but far sweeter, 20
  Smoother than Pelops' shoulder! Fame and honour,
  Methinks, from hence, as from a promontory
  Pointed in heaven, should clap their wings and sing
  To all the under world the loves and fights
  Of gods, and such men near 'em. Palamon        25
  Is but his foil; to him a mere dull shadow;
  He's swart and meagre, of an eye as heavy
  As if he had lost his mother; a still temper,
  No stirring in him, no alacrity,
  Of all this sprightly sharpness, not a smile.   30
  Yet these that we count errors may become him:
  Narcissus was a sad boy, but a heavenly.
  O, who can find the bent of woman's fancy?
  I am a fool, my reason is lost in me,
  I have no choice, and I have lied so lewdly     35
  That women ought to beat me. On my knees
  I ask thy pardon, Palamon, thou art alone
  And only beautiful, and these the eyes,
  These the bright lamps of beauty, that command
  And threaten love—and what young maid dare cross
    'em?                                          40
  What a bold gravity, and yet inviting,
  Has this brown manly face? O, love, this only
  From this hour is complexion. Lie there, Arcite,
  Thou art a changeling to him, a mere gypsy,
  And this the noble body. I am sotted,           45
  Utterly lost—my virgin's faith has fled me.
  For if my brother, but even now, had asked me
  Whether I loved, I had run mad for Arcite;
  Now if my sister, more for Palamon.
  Stand both together. Now come ask me, brother—  50
  Alas, I know not; ask me now, sweet sister—
  I may go look. What a mere child is fancy,
  That having two fair gauds of equal sweetness,
  Cannot distinguish, but must cry for both!

⌈*Enter a Gentleman*⌉
  How now, sir?
GENTLEMAN      From the noble Duke your brother,  55
  Madam, I bring you news. The knights are come.
EMILIA
  To end the quarrel?
GENTLEMAN            Yes.
EMILIA                    Would I might end first!
  What sins have I committed, chaste Diana,
  That my unspotted youth must now be soiled
  With blood of princes, and my chastity          60
  Be made the altar where the lives of lovers—
  Two greater and two better never yet
  Made mothers joy—must be the sacrifice
  To my unhappy beauty?
    *Enter Theseus, Hippolyta, Pirithous, and attendants*
THESEUS                Bring 'em in
  Quickly, by any means, I long to see 'em.       65
                          *Exit one or more*
  (*To Emilia*) Your two contending lovers are returned,
  And with them their fair knights. Now, my fair sister,
  You must love one of them.
EMILIA                      I had rather both,
  So neither for my sake should fall untimely.
    *Enter a Messenger*
THESEUS
  Who saw 'em?
PIRITHOUS       I a while.
GENTLEMAN                  And I.                  70
THESEUS (*to the Messenger*)
  From whence come you, sir?
MESSENGER                    From the knights.
THESEUS                                    Pray speak,
  You that have seen them, what they are.
MESSENGER                                I will, sir,
  And truly what I think. Six braver spirits
  Than these they have brought, if we judge by the
    outside,
  I never saw nor read of. He that stands         75
  In the first place with Arcite, by his seeming,
  Should be a stout man; by his face, a prince.
  His very looks so say him: his complexion,
  Nearer a brown than black, stern and yet noble,
  Which shows him hardy, fearless, proud of dangers.
  The circles of his eyes show fire within him,   81
  And, as a heated lion, so he looks.
  His hair hangs long behind him, black and shining,
  Like ravens' wings. His shoulders, broad and strong;
  Armed long and round; and on his thigh a sword  85
  Hung by a curious baldric, when he frowns
  To seal his will with. Better, o' my conscience,
  Was never soldier's friend.
THESEUS Thou hast well described him.
PIRITHOUS Yet a great deal short,                 90
  Methinks, of him that's first with Palamon.
THESEUS
  Pray speak him, friend.
PIRITHOUS                 I guess he is a prince too,
  And, if it may be, greater—for his show
  Has all the ornament of honour in't.
  He's somewhat bigger than the knight he spoke of, 95
  But of a face far sweeter. His complexion
  Is as a ripe grape, ruddy. He has felt,
  Without doubt, what he fights for, and so apter
  To make this cause his own. In's face appears
  All the fair hopes of what he undertakes,       100

And when he's angry, then a settled valour,
Not tainted with extremes, runs through his body
And guides his arm to brave things. Fear he cannot—
He shows no such soft temper. His head's yellow,
Hard-haired and curled, thick twined: like ivy tods,      106
Not to undo with thunder. In his face
The livery of the warlike maid appears,
Pure red and white—for yet no beard has blessed
    him—
And in his rolling eyes sits victory,
As if she ever meant to court his valour.                 110
His nose stands high, a character of honour;
His red lips, after fights, are fit for ladies.

EMILIA
Must these men die too?

PIRITHOUS                      When he speaks, his tongue
Sounds like a trumpet. All his lineaments
Are as a man would wish 'em—strong and clean.            115
He wears a well-steeled axe, the staff of gold.
His age, some five-and-twenty.

MESSENGER                           There's another—
A little man, but of a tough soul, seeming
As great as any. Fairer promises
In such a body yet I never looked on.                     120

PIRITHOUS
O, he that's freckle-faced?

MESSENGER                      The same, my lord.
Are they not sweet ones?

PIRITHOUS                      Yes, they are well.

MESSENGER                                      Methinks,
Being so few and well disposed, they show
Great and fine art in nature. He's white-haired—
Not wanton white, but such a manly colour                125
Next to an auburn, tough and nimble set,
Which shows an active soul. His arms are brawny,
Lined with strong sinews—to the shoulder piece
Gently they swell, like women new-conceived,
Which speaks him prone to labour, never fainting         130
Under the weight of arms; stout-hearted, still,
But when he stirs, a tiger. He's grey-eyed,
Which yields compassion where he conquers; sharp
To spy advantages, and where he finds 'em,
He's swift to make 'em his. He does no wrongs,           135
Nor takes none. He's round-faced, and when he smiles
He shows a lover; when he frowns, a soldier.
About his head he wears the winner's oak,
And in it stuck the favour of his lady.
His age, some six-and-thirty. In his hand                140
He bears a charging staff embossed with silver.

THESEUS
Are they all thus?

PIRITHOUS              They are all the sons of honour.

THESEUS
Now as I have a soul, I long to see 'em.
(To Hippolyta) Lady, you shall see men fight now.

HIPPOLYTA                                      I wish it,
But not the cause, my lord. They would show              145
Bravely about the titles of two kingdoms—
'Tis pity love should be so tyrannous.
(To Emilia) O my soft-hearted sister, what think you?
Weep not till they weep blood. Wench, it must be.

THESEUS (to Emilia)
You have steeled 'em with your beauty.
(To Pirithous)                          Honoured friend,
To you I give the field: pray order it                   151
Fitting the persons that must use it.

PIRITHOUS                              Yes, sir.

THESEUS
Come, I'll go visit 'em—I cannot stay,
Their fame has fired me so. Till they appear,
Good friend, be royal.

PIRITHOUS                  There shall want no bravery.

EMILIA ⌈aside⌉
Poor wench, go weep—for whosoever wins              156
Loses a noble cousin for thy sins.              Exeunt

**4.3**    *Enter the Jailer, the Wooer, and the Doctor*

DOCTOR Her distraction is more at some time of the moon
    than at other some, is it not?

JAILER She is continually in a harmless distemper: sleeps
    little; altogether without appetite, save often drinking;
    dreaming of another world, and a better; and what
    broken piece of matter soe'er she's about, the name
    'Palamon' lards it, that she farces every business
        *Enter the Jailer's Daughter*
    withal, fits it to every question. Look where she comes—
    you shall perceive her behaviour.                    9
        *They stand apart*

JAILER'S DAUGHTER I have forgot it quite—the burden on't
    was 'Down-a, down-a', and penned by no worse man
    than Giraldo, Emilia's schoolmaster. He's as fantastical,
    too, as ever he may go upon's legs—for in the next
    world will Dido see Palamon, and then will she be out
    of love with Aeneas.                                 15

DOCTOR What stuff's here? Poor soul.

JAILER E'en thus all day long.

JAILER'S DAUGHTER Now for this charm that I told you
    of—you must bring a piece of silver on the tip of your
    tongue, or no ferry: then, if it be your chance to come
    where the blessed spirits are—there's a sight now! We
    maids that have our livers perished, cracked to pieces
    with love, we shall come there and do nothing all day
    long but pick flowers with Proserpine. Then will I make
    Palamon a nosegay, then let him mark me, then—

DOCTOR How prettily she's amiss! Note her a little further.

JAILER'S DAUGHTER Faith, I'll tell you: sometime we go to
    barley-break, we of the blessed. Alas, 'tis a sore life
    they have i'th' other place—such burning, frying,
    boiling, hissing, howling, chattering, cursing—O they
    have shrewd measure—take heed! If one be mad or
    hang or drown themselves, thither they go, Jupiter
    bless us, and there shall we be put in a cauldron of
    lead and usurers' grease, amongst a whole million of
    cutpurses, and there boil like a gammon of bacon that
    will never be enough.                                36

DOCTOR How her brain coins!

JAILER'S DAUGHTER Lords and courtiers that have got
    maids with child—they are in this place. They shall
    stand in fire up to the navel and in ice up to th' heart,
    and there th'offending part burns, and the deceiving
    part freezes—in truth a very grievous punishment as
    one would think for such a trifle. Believe me, one would
    marry a leprous witch to be rid on't, I'll assure you.

DOCTOR How she continues this fancy! 'Tis not an
    engrafted madness, but a most thick and profound
    melancholy.                                          47

JAILER'S DAUGHTER To hear there a proud lady and a
    proud city wife howl together! I were a beast an I'd
    call it good sport. One cries, 'O this smoke!', th'other,
    'This fire!'; one cries, 'O that ever I did it behind the
    arras!', and then howls—th'other curses a suing fellow
    and her garden-house.
    (*Sings*) 'I will be true, my stars, my fate . . .'
                                            *Exit Daughter*

JAILER (*to the Doctor*) What think you of her, sir?    55
DOCTOR I think she has a perturbed mind, which I cannot
   minister to.
JAILER Alas, what then?
DOCTOR Understand you she ever affected any man ere
   she beheld Palamon?    60
JAILER I was once, sir, in great hope she had fixed her
   liking on this gentleman, my friend.
WOOER I did think so too, and would account I had a
   great penn'orth on't to give half my state that both
   she and I, at this present, stood unfeignedly on the
   same terms.    66
DOCTOR That intemperate surfeit of her eye hath distempered the other senses. They may return and settle
   again to execute their preordained faculties, but they
   are now in a most extravagant vagary. This you must
   do: confine her to a place where the light may rather
   seem to steal in than be permitted; take upon you,
   young sir her friend, the name of Palamon; say you
   come to eat with her and to commune of love. This
   will catch her attention, for this her mind beats upon—
   other objects that are inserted 'tween her mind and
   eye become the pranks and friskins of her madness.
   Sing to her such green songs of love as she says
   Palamon hath sung in prison; come to her stuck in as
   sweet flowers as the season is mistress of, and thereto
   make an addition of some other compounded odours
   which are grateful to the sense. All this shall become
   Palamon, for Palamon can sing, and Palamon is sweet
   and every good thing. Desire to eat with her, carve
   her, drink to her, and still among intermingle your
   petition of grace and acceptance into her favour. Learn
   what maids have been her companions and playferes,
   and let them repair to her, with Palamon in their
   mouths, and appear with tokens as if they suggested
   for him. It is a falsehood she is in, which is with
   falsehoods to be combated. This may bring her to eat,
   to sleep, and reduce what's now out of square in her
   into their former law and regiment. I have seen it
   approved, how many times I know not, but to make
   the number more I have great hope in this. I will
   between the passages of this project come in with my
   appliance. Let us put it in execution, and hasten the
   success, which doubt not will bring forth comfort.    98
                                          *Exeunt*

                            ✪

**5.1** ⌈*An altar prepared.*⌉ *Flourish. Enter Theseus,*
      *Pirithous, Hippolyta, attendants*
THESEUS
Now let 'em enter and before the gods
Tender their holy prayers. Let the temples
Burn bright with sacred fires, and the altars
In hallowed clouds commend their swelling incense
To those above us. Let no due be wanting.    5
      *Flourish of cornetts*
They have a noble work in hand, will honour
The very powers that love 'em.
      *Enter Palamon with his three Knights* ⌈*at*
      *one door*⌉, *and Arcite with his three Knights*
      ⌈*at the other door*⌉
PIRITHOUS                          Sir, they enter.
THESEUS
You valiant and strong-hearted enemies,
You royal german foes that this day come
To blow that nearness out that flames between ye,    10
Lay by your anger for an hour and, dove-like,

Before the holy altars of your helpers,
The all-feared gods, bow down your stubborn bodies.
Your ire is more than mortal—so your help be;
And as the gods regard ye, fight with justice.    15
I'll leave you to your prayers, and betwixt ye
I part my wishes.
PIRITHOUS              Honour crown the worthiest.
                              *Exit Theseus and his train*
PALAMON (*to Arcite*)
The glass is running now that cannot finish
Till one of us expire. Think you but thus,
That were there aught in me which strove to show    20
Mine enemy in this business, were't one eye
Against another, arm oppressed by arm,
I would destroy th'offender—coz, I would,
Though parcel of myself. Then from this gather
How I should tender you.
ARCITE              I am in labour    25
To push your name, your ancient love, our kindred,
Out of my memory, and i'th' selfsame place
To seat something I would confound. So hoist we
The sails that must these vessels port even where
The heavenly limiter pleases.
PALAMON              You speak well.    30
Before I turn, let me embrace thee, cousin—
This I shall never do again.
ARCITE              One farewell.
PALAMON
Why, let it be so—farewell, coz.
ARCITE                          Farewell, sir.
                  *Exeunt Palamon and his three Knights*
Knights, kinsmen, lovers—yea, my sacrifices,
True worshippers of Mars, whose spirit in you    35
Expels the seeds of fear and th'apprehension
Which still is father of it, go with me
Before the god of our profession. There
Require of him the hearts of lions and
The breath of tigers, yea, the fierceness too,    40
Yea, the speed also—to go on, I mean,
Else wish we to be snails. You know my prize
Must be dragged out of blood—force and great feat
Must put my garland on me, where she sticks,
The queen of flowers. Our intercession, then,    45
Must be to him that makes the camp a cistern
Brimmed with the blood of men—give me your aid,
And bend your spirits towards him.
      *They kneel before the altar,* ⌈*fall on their faces, then*
      *on their knees again*⌉
(*Praying to Mars*)              Thou mighty one,
That with thy power hast turned green Neptune into
   purple;
Whose havoc in vast field comets prewarn,    50
Unearthèd skulls proclaim; whose breath blows down
The teeming Ceres' foison; who dost pluck
With hand armipotent from forth blue clouds
The masoned turrets, that both mak'st and break'st
The stony girths of cities; me thy pupil,    55
Youngest follower of thy drum, instruct this day
With military skill, that to thy laud
I may advance my streamer, and by thee
Be styled the lord o'th' day. Give me, great Mars,
Some token of thy pleasure.    60
      *Here they fall on their faces, as formerly, and there*
      *is heard clanging of armour, with a short thunder,*
      *as the burst of a battle, whereupon they all rise and*
      *bow to the altar*
O great corrector of enormous times,

Shaker of o'er-rank states, thou grand decider
Of dusty and old titles, that heal'st with blood
The earth when it is sick, and cur'st the world
O'th' plurisy of people, I do take     65
Thy signs auspiciously, and in thy name,
To my design, march boldly. (*To his Knights*) Let us go.
                                      *Exeunt*

**5.2**     *Enter Palamon and his Knights with the former
          observance*
PALAMON (*to his Knights*)
Our stars must glister with new fire, or be
Today extinct. Our argument is love,
Which if the goddess of it grant, she gives
Victory too. Then blend your spirits with mine,
You whose free nobleness do make my cause     5
Your personal hazard. To the goddess Venus
Commend we our proceeding, and implore
Her power unto our party.
         *Here they kneel before the altar, ⌈fall on⌉ their faces
         then on their knees again⌉*
(*Praying to Venus*) Hail, sovereign queen of secrets,
         who hast power
To call the fiercest tyrant from his rage     10
And weep unto a girl; that hast the might,
Even with an eye-glance, to choke Mars's drum
And turn th'alarum to whispers; that canst make
A cripple flourish with his crutch, and cure him
Before Apollo; that mayst force the king     15
To be his subject's vassal, and induce
Stale gravity to dance; the polled bachelor
Whose youth, like wanton boys through bonfires,
Have skipped thy flame, at seventy thou canst catch
And make him to the scorn of his hoarse throat     20
Abuse young lays of love. What godlike power
Hast thou not power upon? To Phoebus thou
Add'st flames hotter than his—the heavenly fires
Did scorch his mortal son, thine him. The huntress,
All moist and cold, some say, began to throw     25
Her bow away and sigh. Take to thy grace
Me, thy vowed soldier, who do bear thy yoke
As 'twere a wreath of roses, yet is heavier
Than lead itself, stings more than nettles.
I have never been foul-mouthed against thy law;     30
Ne'er revealed secret, for I knew none; would not,
Had I kenned all that were. I never practised
Upon man's wife, nor would the libels read
Of liberal wits. I never at great feasts
Sought to betray a beauty, but have blushed     35
At simp'ring sirs that did. I have been harsh
To large confessors, and have hotly asked them
If they had mothers—I had one, a woman,
And women 'twere they wronged. I knew a man
Of eighty winters, this I told them, who     40
A lass of fourteen brided—'twas thy power
To put life into dust. The agèd cramp
Had screwed his square foot round,
The gout had knit his fingers into knots,
Torturing convulsions from his globy eyes     45
Had almost drawn their spheres, that what was life
In him seemed torture. This anatomy
Had by his young fair fere a boy, and I
Believed it was his, for she swore it was,
And who would not believe her? Brief—I am     50
To those that prate and have done, no companion;
To those that boast and have not, a defier;

To those that would and cannot, a rejoicer.
Yea, him I do not love that tells close offices
The foulest way, nor names concealments in     55
The boldest language. Such a one I am,
And vow that lover never yet made sigh
Truer than I. O, then, most soft sweet goddess,
Give me the victory of this question, which
Is true love's merit, and bless me with a sign     60
Of thy great pleasure.
         *Here music is heard, doves are seen to flutter. They
         fall again upon their faces, then on their knees*
O thou that from eleven to ninety reign'st
In mortal bosoms, whose chase is this world
And we in herds thy game, I give thee thanks
For this fair token, which, being laid unto     65
Mine innocent true heart, arms in assurance
My body to this business. (*To his Knights*) Let us rise
And bow before the goddess.
         *They rise and bow*
                        *Time comes on.*     *Exeunt*

**5.3**     *Still music of recorders. Enter Emilia in white, her
          hair about her shoulders, with a wheaten wreath;
          one in white holding up her train, her hair stuck
          with flowers; one before her carrying a silver hind
          in which is conveyed incense and sweet odours,
          which being set upon the altar, her maids standing
          apart, she sets fire to it. Then they curtsy and kneel*
EMILIA (*praying to Diana*)
O sacred, shadowy, cold, and constant queen,
Abandoner of revels, mute contemplative,
Sweet, solitary, white as chaste, and pure
As wind-fanned snow, who to thy female knights
Allow'st no more blood than will make a blush,     5
Which is their order's robe: I here, thy priest,
Am humbled fore thine altar. O, vouchsafe
With that thy rare green eye, which never yet
Beheld thing maculate, look on thy virgin;
And, sacred silver mistress, lend thine ear—     10
Which ne'er heard scurril term, into whose port
Ne'er entered wanton sound—to my petition,
Seasoned with holy fear. This is my last
Of vestal office. I am bride-habited,
But maiden-hearted. A husband I have 'pointed,     15
But do not know him. Out of two, I should
Choose one and pray for his success, but I
Am guiltless of election. Of mine eyes
Were I to lose one, they are equal precious—
I could doom neither: that which perished should     20
Go to't unsentenced. Therefore, most modest queen,
He of the two pretenders that best loves me
And has the truest title in't, let him
Take off my wheaten garland, or else grant
The file and quality I hold I may     25
Continue in thy band.
         *Here the hind vanishes under the altar and in the
         place ascends a rose tree having one rose upon it*
(*To her women*) See what our general of ebbs and flows
Out from the bowels of her holy altar,
With sacred act, advances—but one rose!
If well inspired, this battle shall confound     30
Both these brave knights, and I a virgin flower
Must grow alone, unplucked.
         *Here is heard a sudden twang of instruments and
         the rose falls from the tree*
The flower is fall'n, the tree descends. (*To Diana*) O
   mistress,

Thou here dischargest me—I shall be gathered.
I think so, but I know not thine own will.    35
Unclasp thy mystery. ⌜*To her women*⌝ I hope she's
pleased;
Her signs were gracious.

                        *They curtsy and exeunt*

**5.4**    *Enter the Doctor, the Jailer, and the Wooer in the
habit of Palamon*

DOCTOR Has this advice I told you done any good upon
her?

WOOER O, very much. The maids that kept her company
have half persuaded her that I am Palamon. Within
this half-hour she came smiling to me, and asked me
what I would eat, and when I would kiss her.    6
I told her presently, and kissed her twice.

DOCTOR
'Twas well done—twenty times had been far better,
For there the cure lies mainly.

WOOER                Then she told me
She would watch with me tonight, for well she knew
What hour my fit would take me.

DOCTOR              Let her do so,    11
And when your fit comes, fit her home,
And presently.

WOOER          She would have me sing.

DOCTOR
You did so?

WOOER     No.

DOCTOR         'Twas very ill done, then.
You should observe her every way.

WOOER               Alas,    15
I have no voice, sir, to confirm her that way.

DOCTOR
That's all one, if ye make a noise.
If she entreat again, do anything—
Lie with her if she ask you.

JAILER            Ho there, Doctor.

DOCTOR
Yes, in the way of cure.

JAILER           But first, by your leave,    20
I'th' way of honesty.

DOCTOR            That's but a niceness—
Ne'er cast your child away for honesty.
Cure her first this way, then if she will be honest,
She has the path before her.

JAILER           Thank ye, Doctor.

DOCTOR
Pray bring her in and let's see how she is.    25

JAILER
I will, and tell her her Palamon stays for her.
But, Doctor, methinks you are i'th' wrong still.

                             *Exit Jailer*

DOCTOR
Go, go. You fathers are fine fools—her honesty?
An we should give her physic till we find that—

WOOER
Why, do you think she is not honest, sir?    30

DOCTOR
How old is she?

WOOER         She's eighteen.

DOCTOR              She may be—
But that's all one. 'Tis nothing to our purpose.
Whate'er her father says, if you perceive
Her mood inclining that way that I spoke of,
*Videlicet*, the way of flesh—you have me?    35

WOOER
Yes, very well, sir.

DOCTOR            Please her appetite,
And do it home—it cures her, *ipso facto*,
The melancholy humour that infects her.

WOOER I am of your mind, Doctor.

          *Enter the Jailer and his Daughter, ⌜mad⌝*

DOCTOR
You'll find it so—she comes: pray humour her.    40
    ⌜*The Doctor and the Wooer stand apart*⌝

JAILER (*to his Daughter*)
Come, your love Palamon stays for you, child,
And has done this long hour, to visit you.

JAILER'S DAUGHTER
I thank him for his gentle patience.
He's a kind gentleman, and I am much bound to
him.
Did you ne'er see the horse he gave me?

JAILER                       Yes.    45

JAILER'S DAUGHTER
How do you like him?

JAILER              He's a very fair one.

JAILER'S DAUGHTER
You never saw him dance?

JAILER              No.

JAILER'S DAUGHTER           I have, often.
He dances very finely, very comely,
And, for a jig, come cut and long-tail to him,
He turns ye like a top.

JAILER             That's fine, indeed.    50

JAILER'S DAUGHTER
He'll dance the morris twenty mile an hour,
And that will founder the best hobbyhorse,
If I have any skill, in all the parish—
And gallops to the tune of 'Light o' love'.
What think you of this horse?

JAILER             Having these virtues    55
I think he might be brought to play at tennis.

JAILER'S DAUGHTER
Alas, that's nothing.

JAILER            Can he write and read too?

JAILER'S DAUGHTER
A very fair hand, and casts himself th'accounts
Of all his hay and provender. That ostler
Must rise betime that cozens him. You know    60
The chestnut mare the Duke has?

JAILER               Very well.

JAILER'S DAUGHTER
She is horribly in love with him, poor beast,
But he is like his master—coy and scornful.

JAILER
What dowry has she?

JAILER'S DAUGHTER         Some two hundred bottles
And twenty strike of oats, but he'll ne'er have her.  65
He lisps in's neighing, able to entice
A miller's mare. He'll be the death of her.

DOCTOR What stuff she utters!

JAILER Make curtsy—here your love comes.

WOOER (*coming forward*) Pretty soul,    70
How do ye?
    *She curtsies*
              That's a fine maid, there's a curtsy.

JAILER'S DAUGHTER
Yours to command, i'th' way of honesty—
How far is't now to th' end o'th' world, my masters?

DOCTOR
Why, a day's journey, wench.
JAILER'S DAUGHTER (to Wooer)      Will you go with me?
WOOER
What shall we do there, wench?
JAILER'S DAUGHTER            Why, play at stool-ball—
What is there else to do?
WOOER                  I am content          76
If we shall keep our wedding there.
JAILER'S DAUGHTER               'Tis true—
For there, I will assure you, we shall find
Some blind priest for the purpose that will venture
To marry us, for here they are nice, and foolish.     80
Besides, my father must be hanged tomorrow,
And that would be a blot i'th' business.
Are not you Palamon?
WOOER               Do not you know me?
JAILER'S DAUGHTER
Yes, but you care not for me. I have nothing
But this poor petticoat and two coarse smocks.        85
WOOER
That's all one—I will have you.
JAILER'S DAUGHTER            Will you surely?
WOOER
Yes, by this fair hand, will I.
JAILER'S DAUGHTER            We'll to bed then.
WOOER
E'en when you will.
     He kisses her
JAILER'S DAUGHTER (rubbing off the kiss)
               O, sir, you would fain be nibbling.
WOOER
Why do you rub my kiss off?
JAILER'S DAUGHTER            'Tis a sweet one,
And will perfume me finely against the wedding.    90
(Indicating the Doctor) Is not this your cousin Arcite?
DOCTOR                 Yes, sweetheart,
And I am glad my cousin Palamon
Has made so fair a choice.
JAILER'S DAUGHTER           Do you think he'll have me?
DOCTOR
Yes, without doubt.
JAILER'S DAUGHTER (to the Jailer) Do you think so too?
JAILER                                        Yes.
JAILER'S DAUGHTER
We shall have many children. ⌜To the Doctor⌝ Lord,
     how you're grown!                         95
My Palamon, I hope, will grow too, finely,
Now he's at liberty. Alas, poor chicken,
He was kept down with hard meat and ill lodging,
But I'll kiss him up again.
     Enter a Messenger
MESSENGER
What do you here? You'll lose the noblest sight    100
That e'er was seen.
JAILER              Are they i'th' field?
MESSENGER                          They are—
You bear a charge there too.
JAILER               I'll away straight.
⌜To the others⌝ I must e'en leave you here.
DOCTOR              Nay, we'll go with you—
I will not lose the sight.
JAILER             How did you like her?
DOCTOR
I'll warrant you, within these three or four days    105

I'll make her right again.
               ⌜Exit the Jailer with the Messenger⌝
(To the Wooer)          You must not from her,
But still preserve her in this way.
WOOER                         I will.
DOCTOR
Let's get her in.
WOOER (to the Jailer's Daughter)
               Come, sweet, we'll go to dinner,
And then we'll play at cards.
JAILER'S DAUGHTER            And shall we kiss too?
WOOER
A hundred times.
JAILER'S DAUGHTER   And twenty.
WOOER                          Ay, and twenty.    110
JAILER'S DAUGHTER
And then we'll sleep together.
DOCTOR (to the Wooer)           Take her offer.
WOOER (to the Jailer's Daughter)
Yes, marry, will we.
JAILER'S DAUGHTER       But you shall not hurt me.
WOOER
I will not, sweet.
JAILER'S DAUGHTER If you do, love, I'll cry.      Exeunt

**5.5**    *Flourish. Enter Theseus, Hippolyta, Emilia,*
          *Pirithous, and some attendants*
EMILIA
I'll no step further.
PIRITHOUS             Will you lose this sight?
EMILIA
I had rather see a wren hawk at a fly
Than this decision. Every blow that falls
Threats a brave life; each stroke laments
The place whereon it falls, and sounds more like    5
A bell than blade. I will stay here.
It is enough my hearing shall be punished
With what shall happen, 'gainst the which there is
No deafing, but to hear; not taint mine eye
With dread sights it may shun.
PIRITHOUS (to Theseus)         Sir, my good lord,   10
Your sister will no further.
THESEUS                 O, she must.
She shall see deeds of honour in their kind,
Which sometime show well pencilled. Nature now
Shall make and act the story, the belief
Both sealed with eye and ear. (To Emilia) You must be
     present—                                  15
You are the victor's meed, the price and garland
To crown the question's title.
EMILIA                       Pardon me,
If I were there I'd wink.
THESEUS               You must be there—
This trial is, as 'twere, i'th' night, and you
The only star to shine.
EMILIA              I am extinct.              20
There is but envy in that light which shows
The one the other. Darkness, which ever was
The dam of horror, who does stand accursed
Of many mortal millions, may even now,
By casting her black mantle over both,            25
That neither could find other, get herself
Some part of a good name, and many a murder
Set off whereto she's guilty.
HIPPOLYTA                   You must go.

EMILIA
In faith, I will not.
THESEUS      Why, the knights must kindle
Their valour at your eye. Know, of this war     30
You are the treasure, and must needs be by
To give the service pay.
EMILIA      Sir, pardon me—
The title of a kingdom may be tried
Out of itself.
THESEUS      Well, well—then at your pleasure.
Those that remain with you could wish their office     35
To any of their enemies.
HIPPOLYTA      Farewell, sister.
I am like to know your husband fore yourself,
By some small start of time. He whom the gods
Do of the two know best, I pray them he
Be made your lot.      *Exeunt all but Emilia*
⌈*Emilia takes out two pictures, one from her right
side, and one from her left*⌉
EMILIA
Arcite is gently visaged, yet his eye     41
Is like an engine bent or a sharp weapon
In a soft sheath. Mercy and manly courage
Are bedfellows in his visage. Palamon
Has a most menacing aspect. His brow     45
Is graved and seems to bury what it frowns on,
Yet sometime 'tis not so, but alters to
The quality of his thoughts. Long time his eye
Will dwell upon his object. Melancholy
Becomes him nobly—so does Arcite's mirth.     50
But Palamon's sadness is a kind of mirth,
So mingled as if mirth did make him sad
And sadness merry. Those darker humours that
Stick misbecomingly on others, on them
Live in fair dwelling.     55
*Cornetts. Trumpets sound as to a charge*
Hark, how yon spurs to spirit do incite
The princes to their proof. Arcite may win me,
And yet may Palamon wound Arcite to
The spoiling of his figure. O, what pity
Enough for such a chance! If I were by     60
I might do hurt, for they would glance their eyes
Toward my seat, and in that motion might
Omit a ward or forfeit an offence
Which craved that very time. It is much better
*Cornetts. A great cry and noise within, crying, 'A
Palamon'*
I am not there. O better never born,     65
Than minister to such harm.
*Enter Servant*
     What is the chance?
SERVANT The cry's 'A Palamon'.
EMILIA
Then he has won. 'Twas ever likely—
He looked all grace and success, and he is
Doubtless the prim'st of men. I prithee run     70
And tell me how it goes.
*Shout and cornetts, crying, 'A Palamon'*
SERVANT      Still 'Palamon'.
EMILIA
Run and enquire.      *Exit Servant*
⌈*She speaks to the picture in her right hand*⌉
     Poor servant, thou hast lost.
Upon my right side still I wore thy picture,
Palamon's on the left. Why so, I know not.
I had no end in't, else chance would have it so.     75

*Another cry and shout within and cornetts*
On the sinister side the heart lies—Palamon
Had the best-boding chance. This burst of clamour
Is sure the end o'th' combat.
*Enter Servant*
SERVANT
They said that Palamon had Arcite's body
Within an inch o'th' pyramid—that the cry     80
Was general 'A Palamon'. But anon
Th'assistants made a brave redemption, and
The two bold titlers at this instant are
Hand to hand at it.
EMILIA      Were they metamorphosed
Both into one! O why? There were no woman     85
Worth so composed a man: their single share,
Their nobleness peculiar to them, gives
The prejudice of disparity, value's shortness,
To any lady breathing—
*Cornetts. Cry within, 'Arcite, Arcite'*
     More exulting?
'Palamon' still?
SERVANT      Nay, now the sound is 'Arcite'.     90
EMILIA
I prithee, lay attention to the cry.
*Cornetts. A great shout and cry, 'Arcite, victory!'*
Set both thine ears to th' business.
SERVANT      The cry is
'Arcite' and 'Victory'—hark, 'Arcite, victory!'
The combat's consummation is proclaimed
By the wind instruments.
EMILIA      Half sights saw     95
That Arcite was no babe. God's lid, his richness
And costliness of spirit looked through him—it could
No more be hid in him than fire in flax,
Than humble banks can go to law with waters
That drift winds force to raging. I did think     100
Good Palamon would miscarry, yet I knew not
Why I did think so. Our reasons are not prophets
When oft our fancies are. They are coming off—
Alas, poor Palamon.
⌈*She puts away the pictures.*⌉
*Cornetts. Enter Theseus, Hippolyta, Pirithous,
Arcite as victor, and attendants*
THESEUS
Lo, where our sister is in expectation,     105
Yet quaking and unsettled. Fairest Emily,
The gods by their divine arbitrament
Have given you this knight. He is a good one
As ever struck at head. ⌈*To Arcite and Emilia*⌉ Give me
your hands.
(*To Arcite*) Receive you her, (*to Emilia*) you him: (*to
both*) be plighted with     110
A love that grows as you decay.
ARCITE      Emilia,
To buy you I have lost what's dearest to me
Save what is bought, and yet I purchase cheaply
As I do rate your value.
THESEUS (*to Emilia*)      O lovèd sister,
He speaks now of as brave a knight as e'er     115
Did spur a noble steed. Surely the gods
Would have him die a bachelor lest his race
Should show i'th' world too godlike. His behaviour
So charmed me that, methought, Alcides was
To him a sow of lead. If I could praise     120
Each part of him to th'all I have spoke, your Arcite
Did not lose by't; for he that was thus good,

Encountered yet his better. I have heard
Two emulous Philomels beat the ear o'th' night
With their contentious throats, now one the higher,
Anon the other, then again the first,                     126
And by and by out-breasted, that the sense
Could not be judge between 'em—so it fared
Good space between these kinsmen, till heavens did
Make hardly one the winner. (*To Arcite*) Wear the
    garland                                                130
With joy that you have won.—For the subdued,
Give them our present justice, since I know
Their lives but pinch 'em. Let it here be done.
The scene's not for our seeing; go we hence
Right joyful, with some sorrow. (*To Arcite*) Arm your
    prize;                                                 135
I know you will not lose her. Hippolyta,
I see one eye of yours conceives a tear,
The which it will deliver.
        *Flourish*
EMILIA                          Is this winning?
O all you heavenly powers, where is your mercy?
But that your wills have said it must be so,               140
And charge me live to comfort this unfriended,
This miserable prince, that cuts away
A life more worthy from him than all women,
I should and would die too.
HIPPOLYTA                       Infinite pity
That four such eyes should be so fixed on one             145
That two must needs be blind for't.
THESEUS                    So it is.        *Exeunt*

**5.6**    *Enter, guarded, Palamon and his three Knights
        pinioned; enter with them the Jailer and an
        executioner with block and axe*
PALAMON
There's many a man alive that hath outlived
The love o'th' people; yea, i'th' selfsame state
Stands many a father with his child: some comfort
We have by so considering. We expire,
And not without men's pity; to live still,                  5
Have their good wishes. We prevent
The loathsome misery of age, beguile
The gout and rheum that in lag hours attend
For grey approachers; we come towards the gods
Young and unwappered, not halting under crimes             10
Many and stale—that sure shall please the gods
Sooner than such, to give us nectar with 'em,
For we are more clear spirits. My dear kinsmen,
Whose lives for this poor comfort are laid down,
You have sold 'em too too cheap.
FIRST KNIGHT                    What ending could be
Of more content? O'er us the victors have                  16
Fortune, whose title is as momentary
As to us death is certain—a grain of honour
They not o'erweigh us.
SECOND KNIGHT              Let us bid farewell,
And with our patience anger tott'ring fortune,             20
Who at her certain'st reels.
THIRD KNIGHT              Come, who begins?
PALAMON
E'en he that led you to this banquet shall
Taste to you all. (*To the Jailer*) Aha, my friend, my
    friend,
Your gentle daughter gave me freedom once;
You'll see't done now for ever. Pray, how does she?

I heard she was not well; her kind of ill                   26
Gave me some sorrow.
JAILER                Sir, she's well restored
And to be married shortly.
PALAMON                By my short life,
I am most glad on't. 'Tis the latest thing
I shall be glad of. Prithee, tell her so;                   30
Commend me to her, and to piece her portion
Tender her this.
        *He gives his purse*
FIRST KNIGHT        Nay, let's be offerers all.
SECOND KNIGHT
Is it a maid?
PALAMON        Verily, I think so—
A right good creature more to me deserving
Than I can quit or speak of.
ALL THREE KNIGHTS            Commend us to her.       35
        *They give their purses*
JAILER
The gods requite you all, and make her thankful.
PALAMON
Adieu, and let my life be now as short
As my leave-taking.
        *He lies on the block*
FIRST KNIGHT        Lead, courageous cousin.
SECOND *and* THIRD KNIGHTS We'll follow cheerfully.
        *A great noise within: crying, 'Run! Save! Hold!'*
        *Enter in haste a Messenger*
MESSENGER Hold! Hold! O, hold! Hold! Hold!               40
        *Enter Pirithous in haste*
PIRITHOUS
Hold, ho! It is a cursèd haste you made
If you have done so quickly! Noble Palamon,
The gods will show their glory in a life
That thou art yet to lead.
PALAMON                    Can that be,
When Venus, I have said, is false? How do things
    fare?                                                   45
PIRITHOUS
Arise, great sir, and give the tidings ear
That are most rarely sweet and bitter.
PALAMON                          What
Hath waked us from our dream?
PIRITHOUS                  List, then: your cousin,
Mounted upon a steed that Emily
Did first bestow on him, a black one owing                  50
Not a hair-worth of white—which some will say
Weakens his price and many will not buy
His goodness with this note; which superstition
Here finds allowance—on this horse is Arcite
Trotting the stones of Athens, which the calkins           55
Did rather tell than trample; for the horse
Would make his length a mile, if't pleased his rider
To put pride in him. As he thus went counting
The flinty pavement, dancing, as 'twere, to th' music
His own hooves made—for, as they say, from iron            60
Came music's origin—what envious flint,
Cold as old Saturn and like him possessed
With fire malevolent, darted a spark,
Or what fierce sulphur else, to this end made,
I comment not—the hot horse, hot as fire,                   65
Took toy at this and fell to what disorder
His power could give his will; bounds; comes on end;
Forgets school-doing, being therein trained
And of kind manège; pig-like he whines
At the sharp rowel, which he frets at rather                70

Than any jot obeys; seeks all foul means
Of boist'rous and rough jad'ry to disseat
His lord, that kept it bravely. When naught served,
When neither curb would crack, girth break, nor
   diff'ring plunges
Disroot his rider whence he grew, but that    75
He kept him 'tween his legs, on his hind hooves—
On end he stands—
That Arcite's legs, being higher than his head,
Seemed with strange art to hang. His victor's wreath
Even then fell off his head; and presently    80
Backward the jade comes o'er and his full poise
Becomes the rider's load. Yet is he living;
But such a vessel 'tis that floats but for
The surge that next approaches. He much desires
To have some speech with you—lo, he appears.    85
   *Enter Theseus, Hippolyta, Emilia, and Arcite in a*
   *chair borne by attendants*

PALAMON
O miserable end of our alliance!
The gods are mighty. Arcite, if thy heart,
Thy worthy manly heart, be yet unbroken,
Give me thy last words. I am Palamon,
One that yet loves thee dying.

ARCITE               Take Emilia,    90
And with her all the world's joy. Reach thy hand—
Farewell—I have told my last hour. I was false,
Yet never treacherous. Forgive me, cousin—
One kiss from fair Emilia—(*they kiss*) 'tis done.
Take her; I die.                *He dies*

PALAMON      Thy brave soul seek Elysium.    95

EMILIA (*to Arcite's body*)
I'll close thine eyes, Prince. Blessèd souls be with thee.
Thou art a right good man, and, while I live,
This day I give to tears.

PALAMON          And I to honour.

THESEUS
In this place first you fought, e'en very here
I sundered you. Acknowledge to the gods    100
Our thanks that you are living.
His part is played, and, though it were too short,
He did it well. Your day is lengthened and
The blissful dew of heaven does arrouse you.
The powerful Venus well hath graced her altar,    105
And given you your love; our master, Mars,
Hath vouched his oracle, and to Arcite gave
The grace of the contention. So the deities
Have showed due justice.—Bear this hence.
   ⌈*Exeunt attendants with Arcite's body*⌉

PALAMON             O cousin,
That we should things desire which do cost us    110

The loss of our desire! That naught could buy
Dear love, but loss of dear love!

THESEUS           Never fortune
Did play a subtler game—the conquered triumphs,
The victor has the loss. Yet in the passage
The gods have been most equal. Palamon,    115
Your kinsman hath confessed the right o'th' lady
Did lie in you, for you first saw her and
Even then proclaimed your fancy. He restored her
As your stol'n jewel, and desired your spirit
To send him hence forgiven. The gods my justice    120
Take from my hand, and·they themselves become
The executioners. Lead your lady off,
And call your lovers from the stage of death,
Whom I adopt my friends. A day or two
Let us look sadly and give grace unto    125
The funeral of Arcite, in whose end
The visages of bridegrooms we'll put on
And smile with Palamon, for whom an hour,
But one hour since, I was as dearly sorry
As glad of Arcite, and am now as glad    130
As for him sorry. O you heavenly charmers,
What things you make of us! For what we lack
We laugh, for what we have, are sorry; still
Are children in some kind. Let us be thankful
For that which is, and with you leave dispute    135
That are above our question. Let's go off
And bear us like the time.      *Flourish. Exeunt*

**Epilogue** *Enter Epilogue*

EPILOGUE
I would now ask ye how ye like the play,
But, as it is with schoolboys, cannot say.
I am cruel fearful. Pray yet stay awhile,
And let me look upon ye. No man smile?
Then it goes hard, I see. He that has    5
Loved a young handsome wench, then, show his
   face—
'Tis strange if none be here—and, if he will,
Against his conscience let him hiss and kill
Our market. 'Tis in vain, I see, to stay ye.
Have at the worst can come, then! Now, what say ye?
And yet mistake me not—I am not bold—    11
We have no such cause. If the tale we have told—
For 'tis no other—any way content ye,
For to that honest purpose it was meant ye,
We have our end; and ye shall have ere long,    15
I dare say, many a better to prolong
Your old loves to us. We and all our might
Rest at your service. Gentlemen, good night.
                 *Flourish. Exit*

# A SELECT GLOSSARY

**a,** (as pronoun) familiar, unstressed form of 'he'

**abate,** to shorten, take from, deprive, except, blunt

**abatement,** reduction, depreciation

**abhor,** disgust, protest against

**abide,** await the issue of, pay the penalty for

**able,** to vouch for

**abode,** delay, stay; to foretell

**about,** irregularly, indirectly; be on the move

**abram,** auburn

**abridgement,** reduction, pastime

**abroad,** away, apart, on foot, current

**abrogate,** abstain from

**abruption,** breaking off

**absolute,** complete, certain, positive, beyond doubt

**Absyrtus,** see MEDEA

**abuse,** wrong, ill-usage, deception; to deceive, dishonour

**aby,** pay the penalty for

**accident,** occurrence, event, incident

**accite,** summon

**accommodate,** equip, adapt itself to

**accommodation,** comfort, entertainment

**accomplish,** equip, obtain

**accountant,** accountable

**accoutred,** dressed, equipped

**acerb,** bitter

**Acheron,** river of the underworld

**achieve,** make an end, finish, win, obtain

**Achilles' spear,** a mythical spear: rust scraped from it cured a wound that it had inflicted

**acknow,** acknowledge

**acknown,** *be acknown,* confess knowledge

**aconitum,** poison

**acquit,** atone for, repay, release

**Actaeon.** Diana turned him into a stag because he saw her bathing; he was torn to pieces by his dogs.

**acture,** action

**adamant,** impenetrably hard stone; magnet

**addition,** mark of distinction, title

**admiration,** wonder, astonishment, marvel

**admire,** wonder, marvel

**admittance,** fashion, reception

**adoptious christendoms,** fond nicknames

**advantage,** opportunity, interest on money; to profit

**adventure,** chance, hazard, to risk

**advertise,** inform

**advertisement,** information, advice

**advice,** consideration, forethought

**advised,** cautious, aware, carefully considered

**Aeneas,** a Trojan prince who carried his father, Anchises, from the blazing city. Dido, Queen of Carthage, received him and his son, Ascanius. She fell in love with him, but he left Carthage at the gods' command, and Dido committed suicide.

**Aeolus,** god of the winds

**Aesculapius,** god of medicine

**affect,** affection, tendency, disposition; love, like, imitate

**affected,** disposed, in love

**affection,** passion, desire, disposition, affectation

**affeer,** confirm

**affiance,** confidence

**affined,** related, obliged

**affront,** meet, confront

**affy,** trust, betroth

**after,** according to, at the rate of

**against,** in expectation of, in preparation for the time when, in time for

**Agenor,** father of Europa

**aglet-baby,** tag in shape of a tiny figure

**agnize,** confess, acknowledge

**aim,** target, guess

**Ajax,** a strong, dim-witted Greek hero; mad with anger at not being given the arms of the dead Achilles he slaughtered a flock of sheep as if they were human enemies and killed himself

**alarum,** call to arms, assault

**Alcides,** Hercules

**alder-liefest,** dearest of all

**Alecto,** one of the three fates; her head was wreathed with serpents

**allay,** relief; to qualify

**All-hallond eve,** Halloween, the eve of All Saints' Day

**Allhallowmas,** All Saints' Day (1 November)

**All-hallown summer,** fine weather in late autumn

**allowance,** admission of a claim, reputation

**alter,** exchange

**Althaea.** Her son, Meleager, was fated to live until a brand of fire burned away. After he killed her brothers she burned it.

**amerce,** punish with a fine

**ames ace,** two aces, lowest possible throw at dice

**amort,** spiritless, dejected

**an,** if, though, whether, as if

**anatomize,** dissect, lay bare

**anatomy,** skeleton

**Anchises,** see AENEAS

**anchor,** hermit

**Anna,** sister of Dido of Carthage

**anon,** soon, 'coming'

**Anthropophagi,** cannibals

**antic,** grotesque pageant, clown; fantastic

**antre,** cave

**ape, to lead apes in hell,** an old maid's function

**Apollo,** god of the sun, music and poetry. Daphne, escaping from his pursuit, was changed to a laurel.

**appaid,** contented, satisfied

**apparently,** openly

**appeach,** inform against

**appeal,** accusation; to accuse

**apple-john,** apple eaten when shrivelled

**appliance,** service, remedy, treatment

**appointment,** equipment, instruction

**apprehensive,** lively, quick-witted

**approof,** proof, trial, approval

**approve,** prove, show to be true, confirm, put to the proof, test, convict

**apt,** willing, impressionable

**Aquilon,** north wind

**Arabian bird,** phoenix

**argosy,** large merchant ship

**argue,** prove, show

**argument,** proof, subject of debate, subject-matter, summary

**Ariadne,** deserted by her lover, Theseus

**Arion,** a singer carried ashore by a dolphin

**arm,** reach, take in one's arms

**arm-jaunced,** jolted by armour

**armipotent,** powerful in arms

**aroint thee,** be gone

**arras,** wall-tapestry

**articulate,** arrange terms, specify

**artificial,** made by art, skilled, skilful

**artist,** scholar, doctor

**Ascanius,** see AENEAS

**asnico,** ass

**aspect,** look, glance, position and influence of a planet, sight

**aspersion,** sprinkling

**assay,** trial, attempt

**assubjugate,** debase

**assurance,** pledge, deed of conveyance, guarantee

**assure,** betroth, convey property

**Astraea,** goddess of justice

**Atalanta,** maiden huntress who killed the suitors she outraced

**Ate,** goddess of mischief and destruction

**atomy,** atom, mote

**atone,** reconcile, unite, agree

**atonement,** reconciliation

**Atropos,** one of the three fates; her duty was to cut the thread of life

**attach,** arrest, seize

**attachment,** arrest, stop

**attaint,** conviction, infection; infect, convict of treason, disgrace

**attribute,** reputation, credit

**attribution,** praise

**aught**, anything

**aunt**, old woman, bawd, girl friend

**austringer**, falconer

**avoid**, get rid of, get out of

**awkward**, oblique, not straightforward

**back friend**, pretended friend

**baffle**, disgrace

**bait**, set dogs on, worry, persecute, entice with bait, feed, feast

**balk**, let slip, quibble, heap

**ballow**, cudgel

**ban**, to curse

**Banbury cheese**, proverbially thin

**bandog**, fierce chained dog

**banquet**, dessert, light meal of fruit and sweetmeats

**bases**, skirt-like garment worn by a knight

**Basilisco-like**, like Basilisco, a braggart knight in the play *Soliman and Perseda*

**basilisk**, fabulous reptile whose look was fatal, large cannon

**basta**, enough!

**bastard**, sweet Spanish wine

**bastinado**, beating

**bate**, trouble; beat wings ready for flight, blunt, reduce, grow less, deduct

**bateless**, not to be blunted

**bat-fowling**, catching birds at night

**batlet**, bat used in washing clothes

**bauble**, jester's stick

**bavin**, brush-wood

**bawd**, procurer (male or female)

**beadle**, parish officer with power to punish

**beadsman**, one who prays for another

**bearherd**, bear-keeper

**bearing cloth**, christening garment

**beaver**, visor, helmet

**bedlam**, lunatic hospital, lunatic

**beetle**, heavy hammer-like tool; overhang

**beldam**, grandmother, hag

**bell-wether** , castrated ram with a bell round its neck

**be-mete**, measure

**bemoiled**, covered with mud

**bench**, raise to authority, sit as judge

**bench-hole**, privy

**bend**, look, glance; to turn, incline, direct, strain, submit

**bent**, inclination, direction, tension, force, range, aim

**berayed**, defiled

**bergamask**, a rustic dance

**besonian**, beggar, scoundrel

**besort**, suitable company; to suit

**beteem**, pour over, grant

**bewray**, reveal

**bias**, in bowls, weight which makes a bowl swerve; natural bent, inclination, compelling influence

**bigamy**, marriage with widow(er)

**biggin**, nightcap

**bilbo**, finely-tempered sword

**bilboes**, shackles

**bill**, halberd, pike, note, catalogue, label

**birdbolt**, blunt-headed arrow for shooting birds

**bisson**, partly blind, blinding

**blank**, blank page or charter, white mark in centre of target, aim, range; to make pale

**blazon**, coat of arms, description, proclamation; to proclaim, praise

**blood-baltered**, stained with clots of blood

**blow**, swell, blossom, (of flies) deposit eggs (on), defile

**blowse**, chubby girl

**bluecap**, Scotsman

**blurt**, make light of

**board**, to address, make advances to, mount sexually

**bob**, taunt, mock, cheat, get by trickery, pummel

**bodkin**, dagger, hair-pin or ornament

**boll'n**, swollen

**bolt**, broad-headed arrow, shackle; to sieve, fetter

**bolter**, sieve

**bolting-hutch**, sifting-bin

**bombard**, leather jug or bottle

**bombast**, cotton-wool padding for clothes, bombastic

**bona-roba**, well dressed prostitute

**bone-ache**, syphilis

**boot**, booty, profit, advantage, help, use, avail, addition; to be of use, profit, present in addition

**Boreas**, north wind

**borrow**, receive, assume, counterfeit

**bosky**, wooded

**botchy**, ulcerous

**bots**, disease of horses caused by worms

**bottom**, ship, valley, bobbin; to wind on a bobbin

**bounce**, bang

**brabble**, brawl

**brace**, suit of armour, readiness

**brach**, hound, bitch

**braid**, deceitful; to reproach

**branched**, patterned as with branches

**brave**, finely dressed, splendid, excellent; bravado or threat; to adorn, challenge, defy, swagger, taunt

**bravery**, bravado, finery, splendour, ostentation, defiance

**brawl**, French dance; quarrel

**break-neck**, ruinous course of action

**breast**, voice

**breathe**, speak, exercise, rest

**breathed**, exercised, valiant, inspired

**breese**, gadfly

**Briareus**, hundred-handed giant

**brief**, letter, summary

**broke**, bargain

**broken**, (of music) in parts, scored for different instruments

**broker**, agent, go-between

**Brownist**, member of a Puritan sect founded by Robert Browne

**bruit**, rumour, report; to announce

**bubuncle** , facial eruption

**buck**, washing, dirty clothes for washing

**buckler**, shield

**Bucklersbury**, street of apothecaries' and druggists' shops off Cheapside

**budget**, wallet, bag

**buff**, strong leather used for coats of bailiffs and legal officers

**bug**, bogey, terror

**bugle**, bead, hunting-horn

**bulk**, body, stall in front of a shop

**bully**, friend, fine fellow

**bung**, pickpocket

**burgonet**, helmet

**burn**, infect with venereal disease

**buss**, kiss

**cabin**, den

**cabinet**, dwelling

**cacodemon**, evil spirit

**caddis**, woollen tape

**cade**, barrel

**caduceus**, Mercury's magic wand entwined by two serpents

**caitiff**, wretch, miserable person

**caliver**, light musket

**callet**, whore

**Cambyses**, hero of a bombastic tragedy

**can**, to know, be skilled in

**canary**, lively dance, light sweet wine

**cantle**, segment

**canvass**, toss (as in a blanket)

**capable**, able to receive, feel, or understand

**cap-à-pie**, from head to foot

**capitulate**, specify terms

**captious**, capacious

**carbonado**, meat scored across for cooking

**carcanet**, necklace

**card**, playing card, compass card; mix, debase

**carl, carlot**, peasant

**carnation**, flesh-colour

**carrack**, galleon

**carriage**, ability to bear

**carve**, cut, shape, invite with look and gesture

**case**, vagina

**cast**, throw of dice, tinge, founding; to throw, vomit, reckon, add

**casual**, accidental, subject to accident

**cataplasm**, poultice

**catastrophe**, outcome, end, rear

**caterpillar**, extortioner, parasite

**cates**, food, delicacies

**catling**, catgut string

**cautel**, trick, deceit

**cautelous**, deceitful

**censure**, judgement, blame; judge, estimate

**centre**, centre of the earth or the universe

**Cerberus**, three-headed dog of the underworld

**cerecloth, cerements**, winding-sheet

**Ceres**, goddess of agriculture

**certain**, fixed

**cess**, death; *out of all cess*, beyond all measure

**challenge**, claim, accuse

**chamber-lye**, urine

**chamblet**, light fabric

**champaign**, open country

**changeling**, waverer

**chape**, sheath

**chapman**, merchant, customer

**chaps**, jaws

**character**, writing, hand-writing; to write

**charactery**, writing

**charneco**, Portuguese wine

**Charon**, ferryman of the underworld

**chaudron**, entrails

**cheapen**, bargain, bid for

**cheater**, officer appointed to look after property forfeited to the King

**cherry-pit,** game of throwing cherry stones into a little hole

**cheverel,** kid leather, pliant and easily stretched

**child,** baby girl, youth of noble birth

**childing,** fertile

**chopine,** shoe with high platform-sole

**cicatrice,** scar, impression

**cinquepace,** lively dance

**cipher,** zero, nought; to express, decipher

**cittern,** wire-stringed instrument

**civil,** of the city, well-ordered

**clack-dish,** begging bowl

**clapperclaw,** maul, thrash

**clepe,** call

**clerk,** scholar

**clew,** ball of thread

**climate,** region, dwell

**cling,** shrivel

**clinquant,** glittering

**clip,** embrace

**closure,** bound, enclosure, conclusion

**clown,** rustic, jester

**clyster,** enema

**cockney,** milksop, squeamish or affected woman

**cod,** testicle

**codpiece,** bag-shaped flap on breeches, covering the genitals, tied with laces, often embroidered and padded

**coffin,** pastry case

**cog,** cheat, flatter

**coign,** corner-stone, corner

**coil,** noisy disturbance, fuss, trouble

**Colbrand,** legendary Danish giant

**collection,** inference, understanding

**collied,** darkened

**collop,** slice, offspring

**colour,** pretext, excuse

**colours,** military ensigns

**colt,** young fool; to make a fool of; to have sexual intercourse with

**co-mart,** agreement, bargain

**commodity,** commercial privileges, expediency, advantage, consignment

**companion,** knave

**comparative,** proportionate, full of comparisons; one who assumes equality

**competitor,** associate, partner

**complexion,** bodily habit or constitution, temperament, appearance, colour

**complot,** conspiracy

**composition,** consistency, agreement

**con,** learn by heart

**conceit,** idea, device, apprehension, understanding, opinion, judgement, fancy, imagination, fancy trifle; to think, estimate

**conceited,** full of imagination, ingenious, having a certain opinion

**conclusion,** experiment, riddle

**condolement,** mourning

**conference,** conversation, talk, discussion

**confiner,** inhabitant

**congree,** harmonize

**conscience,** knowledge, understanding, scruple

**conscionable,** ruled by conscience

**consign,** agree, yield up possession

**consist,** insist

**conspectuity,** sight

**contain,** keep

**continent,** container, sum; restrained, temperate

**controller,** detractor

**convenience,** fitness, advantage

**convent,** convene, summon

**conversation,** intercourse, behaviour

**convert,** turn, change

**convey,** lead away, carry

**conveyance,** underhand dealing

**convince,** overcome

**convive,** feast together

**convoy,** means of conveyance

**cony,** rabbit

**cony-catch,** cheat

**copatain,** highcrowned

**cope,** sky; have to do with, encounter, recompense

**copped,** peaked

**copy,** original; subject-matter, copyhold-tenure

**coranto,** a dance

**Corinthian,** reveller

**corky,** withered

**Cornelia,** mother of the Gracchi, model of Roman motherhood

**cornett,** a brass instrument, capable of great brilliance

**cornuto,** cuckold, deceived husband

**corollary,** surplus

**costard,** an apple; the head

**cote,** cottage; pass by

**cotquean,** 'old woman', man who interferes in housekeeping

**couch,** hide, lie hidden, make crouch

**counsel,** secret, secret purpose or thought

**Counter,** debtors' prison

**counterpoint,** quilt, counterpane

**countervail,** equal, counter-balance

**court-cupboard,** sideboard

**cousin,** nephew, kinsman, relative

**cowl-staff,** pole on which a 'cowl' (or basket) is carried

**coy,** scorn, stroke

**cozen,** cheat

**cozier,** cobbler

**crackhemp,** gallows-bird, one born (or deserving) to be hanged

**crank,** twist; wind

**crant,** wreath

**crare,** small trading vessel

**craze,** break

**credent,** believing, credible

**credit,** credibility, reputation, report

**cresset,** fire-basket, torch

**crow-keeper,** one employed to drive away crows

**crusado,** Portuguese coin

**cry,** pack of hounds; yelp in following scent

**cubiculo,** bedroom

**cullion,** testicle (term of abuse)

**culverin,** large cannon

**cunning,** knowledge, skill; ingenious

**Cupid,** god of love, son of Venus and Mars (or Mercury), usually thought of as a boy armed with bow and arrows

**curate,** parish priest

**curiosity,** exactness, over-scrupulousness, delicacy

**curious,** anxious, needing care, fastidious,

difficult to please, delicate, beautifully made; delicately

**curst,** shrewish, cross, cantankerous, malignant, fierce

**curtal,** having the tail docked

**curtal-axe,** cutlass

**customer,** prostitute

**cut,** docked or gelded horse; vulva

**Cyclops,** one of a race of one-eyed giants, workmen for Vulcan the smith.

**cyme,** medicinal plant

**Cynthia,** goddess of the moon

**cypress,** fine lawn fabric

**Cytherea,** Venus, goddess of love

**Daedalus,** with his son Icarus, escaped from imprisonment on home-made wings. Icarus flew too high, the sun melted the wax, and he was drowned. Daedalus escaped.

**danger,** harm, injury, power to harm, range (of a weapon), debt

**Daphne,** see APOLLO

**dare,** dazzle

**date,** time, term, term of life, end

**daub,** cover with false show; *daub it,* pretend

**dear,** important, energetic, dire

**debile,** weak

**Deborah,** prophetess who inspired Israel to victory

**decimation,** execution of every tenth man

**decoct,** heat

**defeat,** destruction; to ruin, destroy, disfigure, defraud

**defeature,** disfigurement

**defend,** forbid

**defunction,** death

**defunctive,** funereal

**delightsome,** delightful, delighted

**demerit,** merit, sin

**denier,** copper coin of little value

**denunciation,** formal declaration

**deplore,** tell with grief

**depose,** swear, examine on oath

**deprave,** defame

**deputation,** office of deputy

**deracinate,** uproot

**derive,** inherit, descend, bring down on, draw

**dern,** dark, drear

**determinate,** fix; ended, decisive, intended

**determination,** ending, decision, intention

**determine,** end, settle, decide

**Deucalion,** the Greek Noah

**dexter,** right

**dey-woman,** dairy-woman

**Diana,** goddess of hunting, the moon, and chastity

**diaper,** table napkin

**dich,** attach to

**Dido,** see AENEAS

**diffidence,** distrust, suspicion

**diffused,** confused, disordered

**dilate,** relate, express at length

**dildo,** penis, phallus; used in ballad refrains

**dime,** tenth man

**Dis,** god of the underworld

**disable,** impair, disparage

**disappointed,** unprepared

**discase,** undress

**discourse**, reasoning, talk, conversational power, familiarity

**discover**, uncover, reveal, make known, recognize, spy out, reconnoitre

**disease**, trouble, annoyance; to disturb

**disedge**, dull the appetite, sate

**disgracious**, disliked, out of favour

**dishabit**, dislodge

**dishonest**, dishonourable, unchaste, immoral

**dishonesty**, dishonour, immorality

**dislike**, disagreement, disapproval; to displease

**disliken**, disguise

**dismiss**, forgive

**dismission**, dismissal

**dismount**, lower, draw from sheath

**dispiteous**, pitiless

**dispose**, disposal, control, disposition, temperament, manner; to control, direct, incline, come to terms

**disputable**, argumentative

**distain**, stain, defile

**distemper**, ill humour, illness of mind or body, intoxication; disturb, disorder

**distemperature**, intemperateness of weather, illness, ailment, disturbance of mind

**distinction**, discrimination

**distinctly**, separately, individually

**distract**, divide, perplex, drive mad

**distrain**, seize

**distressful**, hard-earned

**dive-dapper**, dabchick, little grebe

**division**, variation, modulation, disposition

**divulge**, proclaim

**do**, copulate (with)

**doctrine**, lesson, learning

**document**, lesson

**dogged**, cruel

**doit**, coin worth half a farthing, a minute sum

**dole**, portion, share, grief, sorrow

**domineer**, feast riotously

**dominical**, red-printed letter in calendar marking the Sundays

**doom**, judgement

**double**, false, deceitful; wraith

**doubt**, suspicion, fear; to suspect, fear

**dout**, extinguish

**dowlas**, coarse linen

**dowle**, downy feather

**doxy**, beggar's wench

**draff**, pigwash

**draught**, cesspool, privy

**draw**, withdraw, empty, search for game, track by scent

**drawer**, tapster

**dressing**, trimming

**dribbling**, falling short or wide of the mark

**drift**, purpose, plot, shower

**drollery**, puppet-show, comic picture

**drumble**, move slowly

**drybeat**, beat soundly

**dudgeon**, hilt of a dagger

**duello**, duelling code

**dump**, mournful tune or song

**dun**, dark; dun horse

**dup**, open

**durance**, durability, strong and durable cloth, imprisonment

**eager**, sour, bitter

**ean**, to bring forth (lambs)

**eanling**, young lamb

**ear**, plough

**ecstasy**, excitement, trance, madness

**edge**, appetite, desire

**effectual**, pertinent, to the point

**eftest**, easiest

**eftsoons**, soon

**egg**, epitome of worthlessness; *take eggs for money*, accept injury tamely

**eisel**, vinegar

**eld**, old age, ancient time

**elder gun**, pop-gun made from elder wood

**element**, sky, *pl.* atmospheric powers

**elf**, tangle

**embarquement**, embargo, prohibition

**emboss**, to drive (a hunted animal) to extremity

**embossed**, swollen, foaming at the mouth

**embowel**, disembowel

**empiric**, quack

**emulation**, ambitious rivalry, grudge, envy

**emulator**, disparager

**encompassment**, winding course, 'talking round' a subject

**enew**, drive into water

**engine**, artifice, plot, mechanical contrivance, rack

**engross**, write out in a fair hand; collect, monopolize, fatten

**enlarge**, set free

**enlargement**, release, liberty

**enormous**, disordered, irregular

**ensconce**, shelter, hide

**enseamed**, defiled with sweat

**ensear**, dry up

**enshield**, concealed, emblazoned

**ensteeped**, lying under water

**entreat**, treat, negotiate, intercede

**entreatment**, entering into negotiation

**envy**, malice, enmity; show malice towards

**Ephesian**, boon companion

**epicurism**, luxury, excess

**Erebus**, place of darkness, hell

**ergo**, therefore

**eringo**, aphrodisiac sweetmeat

**erne**, grieve

**erst**, formerly

**escot**, support financially

**espial**, spy

**estimable**, appraising

**estimate**, valuation, value, repute

**estimation**, value, thing of value, esteem, reputation, conjecture

**estridge**, goshawk

**Europa**, carried away by Jupiter who had taken the form of a bull

**event**, outcome, issue, result

**evitate**, avoid

**exception**, objection, dissatisfaction

**exclamation**, loud reproach

**excrement**, growth (of hair)

**excursion**, rush, passage of arms

**exempt**, cut off

**exhalation**, meteor

**exhale**, draw forth

**exhaust**, draw out

**exhibit**, to submit, present

**exhibition**, allowance of money, gift

**expectancy**, hope

**expedience**, speed, expedition

**expedient**, speedy, direct

**experimental**, of experience

**expiate**, end

**extend**, be lavish in praise, exaggerate the worth of, take by force

**extent**, seizure of property in execution of a writ, attack

**extirp**, root out

**extracting**, distracting

**extravagancy**, wandering

**extravagant**, straying, vagrant

**eyas**, young hawk

**eyas-musket**, young sparrow-hawk

**eye-glass**, retina, eye's lens

**eyestrings**, muscles or nerves of eye

**face**, appearance, appearance of right; to put on a false appearance, brave, bully, brazen, trim

**facinorous**, vile

**fact**, deed, crime

**factious**, seditious

**factor**, agent

**faculty**, disposition, quality

**fadge**, be suitable, succeed

**fading**, refrain of a song

**fail**, failure, fault; to offend, die

**fain**, glad, obliged

**fairing**, gift

**faithed**, believed

**faitor**, cheat, rogue

**falchion**, sword, scimitar

**falling sickness**, epilepsy

**falsing**, deceptive

**fame**, rumour, report, reputation; make famous

**familiar**, attendant spirit

**fancy**, love, whimsicality; to love, fall in love with

**fang**, seize

**fangled**, foppish

**fantastic**, imaginary, fanciful, extravagant

**fantasy**, delusion, imagination, fancy, whim

**fap**, drunk

**farce**, to cram, stuff

**farced**, stuffed out

**fardel**, bundle, burden

**farm**, lease

**farthingale**, hooped petticoat

**fashions**, disease of horses

**fault**, lack, (in hunting) break in the scent

**favour**, leniency, something given as a mark of favour, badge, charm, appearance, look, face, feature

**feat**, dexterous, graceful; show to advantage; deed

**feature**, shape, form, comeliness

**fedary**, **federary**, confederate, accomplice

**feed**, pasture

**feeder**, servant, parasite, shepherd

**fee-farm**, fixed rent for perpetual tenancy

**fee-grief**, individual sorrow

**fee-simple**, estate belonging to owner and his heirs for ever, absolute possession

**felicitate**, happy

**fell**, fierce, cruel, enraged; skin, covering of hair or wool, fleece

**felly**, section of rim of wooden wheel

**fence**, art of fencing, defence; to defend

**fere,** spouse

**fern seed,** believed to be invisible and to confer invisibility

**ferret,** worry

**festinate,** hasty

**fetch,** stratagem, trick; to draw, derive, strike a blow

**fettle,** make ready

**fig,** to insult with a 'figo'

**fig of Spain, figo, fico,** scornful gesture made by thrusting the thumb between two of the closed fingers or into the mouth.

**fight,** screen for protection of crew in sea battle

**file,** catalogue, list, roll, rank, number; to smooth, polish, defile

**fill,** fulfil

**fill-horse,** shaft horse

**fine,** end; to end, pay, fix as sum payable, punish

**fire-drake,** fiery dragon, meteor

**firk,** thrash

**fitchew,** polecat

**fitment,** fit equipment, fitting office

**fives,** strangles, a disease of the horse

**flap-dragon** a raisin in burning brandy; to swallow as a flap-dragon

**flapjack,** pancake

**flatness,** completeness

**flaw,** flake of snow, gust, fragment, fault, outburst; to crack, break

**fleckled,** dappled

**fledge,** fledged, covered with down

**fleer,** mock, sneer; to gibe

**flesh,** initiate in bloodshed, inflame, gratify

**fleshment,** excitement resulting from a first success

**flewed,** having large chaps

**flirt-jill,** woman of loose character

**float,** sea

**flourish,** gloss, embellishment, florid decoration, fanfare of trumpets

**flush,** full, lusty

**flux,** discharge, flowing, stream

**fob,** cheat, trick

**foin,** thrust

**foison,** harvest, plentiful crop

**fond,** foolish, silly, trivial, eager, desirous

**fool,** professional jester, term of endearment or pity, plaything

**foot,** see FOUTRE

**footcloth,** saddle cloth

**fop,** to dupe, fool

**foppery,** folly, deceit

**foppish,** foolish

**forage,** preying

**forbid,** cursed

**fordo,** kill, destroy

**fordone,** exhausted

**forecast,** foresight

**forehorse,** first in team

**forespent,** previously spent, past

**forestall,** condemn in advance

**forgetive,** inventive

**fork,** forked tongue, barbed arrow-head

**forked,** horned

**formal,** traditional, dignified, sane

**former,** foremost

**forsake,** refuse, reject, renounce

**forslow,** delay

**forspent,** exhausted

**fortitude,** strength

**forwhy,** because

**foutre,** strong expression of contempt, a fuck (French)

**fox,** kind of sword

**fracted,** broken

**fraction,** discord, quarrel, fragment

**frame,** contriving, structure, plan; to prepare, go, bring to pass, perform

**frampold,** disagreeable

**franchise,** freedom, privilege

**franchised,** free

**frank,** unrestrained, generous, free; sty; to pen up in a sty

**franklin,** yeoman

**fray,** frighten

**free,** generous, magnanimous, innocent, untroubled; to absolve, banish

**French crown,** French coin, baldness produced by venereal disease

**frequent,** addicted, familiar

**friend,** lover, mistress

**frieze,** coarse woollen cloth

**frippery,** old-clothes' shop

**front,** forehead, face, foremost line of battle, beginning; to confront, oppose

**frontlet,** band worn on forehead

**froward,** perverse, wilful, rebellious

**frush,** smash

**fullam,** false dice

**fulsome,** pregnant, loathsome, filthy

**furnishings,** externals

**furniture,** equipment, harness

**fury,** rage, passion, poetic passion, goddess of vengeance

**fustian,** coarse cloth, bombastic gibberish

**gaberdine,** loose-fitting coat or cloak

**gage,** pledge; to stake, bind, engage

**gain-giving,** misgiving

**Galen,** famous physician of second century

**gall,** resentment, bitterness; to rub sore, chafe, graze, wound, harass, scoff

**galliard,** lively dance

**galliass,** large, heavy ship

**galloglasses,** heavily-armed foot-soldiers of Ireland or the Western Isles

**gallow,** frighten

**gallows,** gallows-bird, one born (or deserving) to be hanged

**gamut,** musical scale

**garboil,** disturbance, quarrel

**Gargantua,** large-mouthed giant in Rabelais

**garland,** royal crown, glory

**garnish,** dress; to adorn, equip

**gaskins,** breeches

**gaud,** plaything, showy ornament; to ornament

**gear,** stuff, talk, matter, business

**geck,** dupe

**gender,** kind, sort, offspring

**generosity,** high birth

**generous,** high-born

**genius,** spirit, good or bad angel, embodied spirit

**gentle,** of noble birth; to ennoble

**gentry,** rank by birth, good breeding, courtesy

**George,** jewel bearing figure of the saint, part of insignia of Order of the Garter

**germen,** seed

**gest,** deed, time allotted to stage of journey

**gesture,** bearing

**ghasted,** frightened

**ghastness,** terror

**ghostly,** spiritual

**gig,** whipping-top

**giglet,** wanton

**gimmaled,** jointed, hinged

**ging,** gang

**gird,** gibe; taunt, besiege

**glance** (*at*), hint at, cast a slight on

**glass,** mirror, sand-glass

**glaze,** stare

**gleek,** gibe; to gibe, jest

**glib,** castrate

**glut,** swallow

**go,** walk

**go to!** expression of disapproval, protest, or disbelief

**goatish,** lustful

**good,** financially sound, rich

**good-brother,** brother-in-law

**goodman,** husband, yeoman, master

**goose,** smoothing iron

**goose of Winchester,** prostitute

**gorbellied,** fat-bellied

**gore,** to defile, wound

**gorge,** what has been swallowed

**gorget,** piece of armour for the throat

**Gorgon,** woman whose look turned the beholder to stone

**gossip,** god-father or -mother, sponsor; make merry

**gourd,** kind of false dice

**gout,** drop

**government,** control, self-command, evenness of temper

**grained,** ingrained, showing grain of the wood, lined, forked

**gramercy,** thank you

**grange,** outlying farmhouse

**gratify,** thank, reward, pay, do honour to

**gratulate,** pleasing; greet, express joy at

**greasily,** obscenely

**grece,** step, stair, degree

**gree,** agree

**groat,** fourpenny piece

**groundlings,** those who stood in the cheapest part of the theatre

**grow,** be or become due

**guard,** caution, border, trimming; to ornament

**guidon,** pennant

**guiled,** treacherous

**guinea-hen,** prostitute

**guise,** custom, habit

**gules,** red (heraldry)

**gull,** unfledged bird, dupe, fool, trick; to cheat

**gummed,** stiffened with gum

**gust,** taste, relish

**Guy,** Guy of Warwick, slayer of the giant Colbrand

**gyve,** fetter

**habit,** dress, appearance

**habitude,** temperament

**hackney,** prostitute

**haggard,** wild hawk

**haggle,** to hack, gash

**hag-seed,** child of a witch

**hai,** home-thrust in fencing

**hair,** kind, character

**halcyon,** kingfisher

**half,** partner

**half-blooded,** of noble blood by one parent only

**half-cheek,** profile

**half-cheeked,** with a piece missing or broken on one side

**half-face,** thin face

**half-faced,** showing half the face

**half-sword,** half a sword's length

**halidom,** holy relic

**Hallowmas,** All Saints' Day, 1 November

**handfast,** firm hold, marriage contract

**handy-dandy,** choose which you please (in a children's game)

**haply,** perhaps, by chance

**happiness,** handsomeness, appropriateness, opportunity

**hardiment,** bold exploit

**hardness,** difficult, hardship

**harlot,** man or woman of promiscuous life

**harlotry,** harlot, silly wench

**hatch,** engrave, inlay, lower half of a divided door

**hatchment,** memorial tablet with coat of arms

**hautboy,** wood-wind instrument, ancestor of the oboe

**havoc,** devastation; *cry havoc,* give the signal to an army to plunder

**hazard,** game at dice, chance, venture

**head,** headland, topic, army

**head borough,** parish officer

**heap,** crowd

**heavy,** important, dull, sluggish, sleepy, grievous

**hectic,** wasting fever

**Hector,** Trojan hero

**Hecuba,** Queen of Troy, wife of Priam and mother of Hector

**hedge-pig,** hedgehog

**Helen,** most beautiful woman of her world, wife of the Greek Menelaus, carried off to Troy by Paris

**Helicon,** mountain of Greece sacred to the Muses

**hemp-seed,** gallows-bird, one born (or deserving) to be hanged

**Hercules,** as a baby strangled two serpents; performed twelve great labours, including the obtaining of the golden apples of the Hesperides and the overcoming of Cerberus, the three-headed dog of the underworld

**Hero,** see LEANDER

**Hesperus,** the evening star

**hest,** command

**heyday,** excitement

**high-lone,** quite alone, without support

**high-proof,** in the highest degree

**high-stomached,** haughty

**hight,** called

**hilding,** contemptible, good-for-nothing, baggage

**him,** male (dog)

**hint,** occasion, reason, opportunity

**hipped,** lame in the hip

**hit,** succeed, agree

**hive,** hive-shaped hat

**hoar,** grow mouldy

**hobby-horse,** figure of a horse used in morris dances, etc., buffoon, prostitute

**holding,** consistency, burden of a song

**holp,** helped

**honest,** worthy, virtuous, chaste

**honesty,** honour, decency, chastity

**honour,** chastity

**hoodman-blind,** blind man's buff

**horn,** the mark of a cuckold

**horn-book,** child's first reader

**horn-mad,** ready to gore the enemy, enraged at being cuckolded

**hose,** stockings, breeches

**host,** lodge

**hot-house,** brothel

**housekeeper,** householder, watch-dog, stay-at-home

**hox,** hamstring

**hoy,** a small vessel

**hugger-mugger,** secrecy

**hull,** float, drift with sails furled

**humorous,** moist, capricious, moody

**humour,** moisture; bodily fluid supposedly composed of blood, phlegm, choler, and melancholy, the proportions determining personal temperament; temperament, mood, whim, caprice, inclination

**Hungarian,** hungry, needy

**hunt's-up,** morning-song to arouse huntsmen

**hurricano,** waterspout

**hurry,** commotion, disorder

**husband,** one who keeps house; to farm, till

**husbandry,** management, thrift

**Hybla,** mountain in Sicily famous for fragrant flowers and honey

**Hydra,** many-headed snake whose heads grew again as they were cut off

**Hymen,** god of marriage

**Hyperion,** sun god

**Icarus,** see DAEDALUS

**idea,** image

**Ides of March,** according to the old Roman calendar, 15 March

**idle,** empty, trifling, worthless, useless, foolish, out of one's mind

**ignis fatuus,** will-o'-the-wisp

**Ilion, Ilium,** citadel of Troy

**image,** likeness, copy, representation, sign, embodiment, idea

**imbecility,** weakness

**imbrue,** pierce, shed blood of

**immanity,** cruelty

**imminence,** impending evil

**immodest,** arrogant, immoderate

**immoment,** insignificant

**imp,** young shoot, child; to engraft feathers into a bird's wing

**impart,** afford, make known

**impartment,** communication

**impeach,** accusation, reproach; to accuse, challenge, discredit

**impeachment,** accusation, detriment, hindrance

**impertinency,** irrelevancy, ramblings

**impertinent,** irrelevant

**imply,** involve

**import,** involve, imply, express, be important, concern, portend

**importance,** matter, meaning, importunity

**importancy,** significance

**important,** importunate, urgent

**importless,** meaningless, trivial

**imposition,** imputation, accusation, command

**impostume,** abscess

**impress,** impression, to stamp

**imputation,** reputation

**incapable,** unable to receive or realize

**inclining,** compliant; party, inclination

**incomprehensible,** boundless

**incontinently,** immediately

**incorpsed,** made one body

**incorrect,** unchastened, rebellious

**incredulous,** incredible

**index,** table of contents, preface

**indifferency,** impartiality, moderate size

**indifferent,** impartial, ordinary; tolerably, fairly

**indigest,** unformed, unformed mass

**indign,** unworthy

**indirection,** roundabout method, dishonest practice

**indirectly,** wrongfully, evasively, by suggestion, inattentively

**indiscreet,** lacking judgement

**indiscretion,** want of judgement

**indisposition,** disinclination

**indistinguishable,** mongrel

**indrenched,** waterlogged

**infect,** affect with some feeling

**influence,** supposed flowing from the heavens of an ethereal fluid acting on human character and destiny, inspiration

**inform,** take shape, inspire, instruct

**infuse,** shed, imbue

**ingenious,** talented, intelligent, discerning, skilfully contrived

**ingling,** engaging in sexual play

**inhabitable,** uninhabitable

**inherit,** possess

**inhibit,** prohibit

**inhibition,** prohibition

**injurious,** insulting, malicious

**inkle,** linen tape, thread, or yarn

**inland,** of the central, more cultured, part of a country

**innocent,** idiot, half-wit

**inoculate,** engraft by budding

**insensible,** imperceptible to senses

**insinuate,** ingratiate, suggest

**insinuation,** ingratiation, hint

**instance,** cause, detail, proof, mark, presence

**intellect,** meaning, content

**intelligence,** communication, information, news, obtaining of secret information, spy; to pass information

**intelligencer,** agent

**intelligencing,** informative

**intelligent,** giving information, open, informative

**intemperature,** wildness, intemperance

**intenable,** unretentive

intendment, purpose, intention

intercept, interrupt

interest, right, title, share

interressed, invested right or share

interlude, entertainment, play

intrenchant, invulnerable

intrinse, intrinsicate, closely entwined

investments, vestments, clothes

irreconciled, unexpiated

irregular, irregulous, lawless

iterance, repetition

iwis, indeed, certainly

Jack, jack, fellow, scoundrel, figure striking the bell on a clock; key of a virginal, smaller bowl aimed at, quarter of a pint

Jack-a-Lent, puppet set up as target during Lent

jade, horse of poor condition or vicious temper, term of contempt; to wear out, make a fool of

jakes, privy

jaunce, prance, trudge up and down

jay, a flashy whore

jealous, suspicious, afraid, apprehensive, doubtful

jealousy, suspicion, apprehension, mistrust

jennet, small Spanish horse

jess, strap attached to the leg of a trained hawk

jet, strut, encroach

jig, quick lively dance, short lively comic entertainment

jointress, dowager

jollity, finery

jolthead, block-head

jordan, chamber-pot

journal, daily

journey-bated, exhausted with travel

Jove, poetic form of JUPITER

jowl, strike, knock

jump, just, precisely; hazard; to hazard, agree, coincide

junket, sweetmeat, delicacy

Juno, queen of the gods and wife of Jupiter; goddess of marriage

Jupiter, ruler of the gods. He was thought to hurl thunderbolts at mortals who displeased him, but otherwise was best known for his many amorous adventures.

just, true, honourable, exact

justicer, judge

justify, maintain the innocence of, vindicate, prove, corroborate

justly, with good reason

jut, encroach

keech, fat of slaughtered animal rolled into a lump

keel, skim

ken, range of sight; to see, recognize, know

kennel, pack, gutter

kern, light-armed Irish foot-soldier

kersey, homely; coarse woollen material

kibe, chilblain

kind, natural, tender, courteous, affectionate; nature, way, race, sort

kindle, bring forth

kindly, naturally, properly, exactly

knack, trifle, knick-knack

knap, nibble, strike

knot, fancifully laid-out flower bed or garden plot

laager, camp

label, slip of paper, strip of paper or parchment by which a seal is attached; to add as a codicil

laboured, worn out, highly finished

laboursome, elaborate

lace, to ornament

lag, last, late

lampass, disease of the horse in which flesh swells behind front teeth

land-damn, damn in this world

lank, become thin

lantern, window-turret

lard, fatten, garnish

large, generous, lavish, free, improper

latch, strike, catch, receive, bewitch

laund, glade

lavolt, lavolta, high-leaping dance

lay, wager

lazar, leper, sick beggar

leading, command, direction, generalship

leaguer, camp

Leander, lover of Hero of Sestos, he swam the Hellespont nightly to visit her in her tower, and was drowned in a storm

learn, teach

leasing, lying

leather-coat, russet apple

lecture, lesson, instruction

leer, appearance, complexion

leese, lose

leet, court held by lord of the manor

leman, sweetheart, lover

lengthen, delay, postpone

lenity, gentleness

leno, pimp, pander

lenten, meagre

let, hindrance; to hinder, forbear, cause

Lethe, river in Hades; to drink its waters gave forgetfulness

level, aim, line of fire, range; to aim, guess

liable, subject, suitable

liberal, accomplished, humane, abundant, free in speech, unrestrained, gross

lie, to lodge, stay, be still, in prison or in defensive posture

lief, dear

liege, sovereign lord

light, frivolous, unchaste, swift, easy, merry, trivial, delirious

like, please, be in good condition

liking, bodily condition

limbeck, distilling vessel

limber, flexible

lime, cement, catch with birdlime

limit, prescribed time, time of rest after childbearing, region; to appoint

limitation, allotted time

line, rank, Equator, cord for taking measurements; copulate with

linger, prolong, defer

link, torch, lamp-black

linsey-woolsey, woven material of wool and flax, hence medley, nonsense

linstock, forked stick holding a gunner's match

list, limit, bound, barriers enclosing tilting ground, desire; to please, choose

lither, yielding

livelihood, liveliness, animation

liver, supposedly the seat of love and strong emotion

lob, bumpkin

lockram, linen cloth

lodge, flatten, beat down

loggats, game of aiming small logs at a fixed stake

long, belong

loose, unattached, negligent; moment of arrow's discharge, last moment

lose, to ruin, forget

lover, friend, mistress

luce, pike as heraldic device

luggage, baggage of an army

lune, fit of temper, of frenzy

lurch, lurk, rob

lure, dummy bird for recalling hawk

lust, pleasure, desire

lustihood, vigour

lusty, merry, lustful

luxurious, lustful, lecherous

luxury, lust

maculate, stained, impure

maculation, stain, impurity

mainly, with force, greatly, perfectly

majority, pre-eminence

make, mate, husband or wife

making, form, appearance

malapert, saucy

malkin, servant wench, slut

mammer, hesitate

mammock, tear to bits

manège, art of horsemanship

mandrake, poisonous plant believed to shriek when pulled up

mankind, resembling a man, violent, ferocious

manner, stolen goods found on a thief

mansionry, dwelling-place

map, picture, embodiment

mappery, map-making

marches, border country next to Scotland or Wales

marry, (as an exclamation) by (the Virgin) Mary

Mars, god of war and patron of soldiers

mate, rival, checkmate, destroy, stupefy

material, important, forming the substance, full of sense

maugre, in spite of

maund, basket

mazzard, head

meacock, coward, weakling

meal, spot, stain

mean, something between or intervening, middle, medium position, tenor, alto; to lament

measurable, suitable

measure, a stately dance; tune; to measure

mechanic, labourer

mechanical, mean, vulgar; labourer

Medea, escaping with Jason she tore to pieces her brother Absyrtus, scattering his limbs in her father's path to delay him; restored the youth of Jason's father, Aeson

**medicinable,** healing, medicinal

**medlar,** fruit eaten soft and pulpy; prostitute

**meed,** reward, gift, merit

**meiny,** company of retainers, multitude

**mell,** to meddle

**memorize,** make memorable

**memory,** memorial, memento, remembrance

**Mercury,** messenger of the gods

**mere,** sure, absolute, unqualified, only

**merely,** simply, entirely

**merit,** reward

**mess,** dish, portion, group originally of four persons eating together, set of four

**metaphysical,** supernatural

**mete,** measure, aim

**metheglin,** spiced mead

**methinks,** it seems to me

**method,** table of contents

**micher,** truant

**mickle,** great

**microcosm,** little world, man considered as epitome of the universe

**milch,** in milk, tearful

**mince,** to extenuate, moderate, affect

**mineral,** mine

**Minerva,** goddess of wisdom

**minimus,** creature of tiniest size

**minion,** favourite, darling, harlot, saucy creature

**Minotaur,** devouring monster dwelling in labyrinth of Crete

**mirable,** wonderful

**mirror,** model, pattern

**mischief,** misfortune, injury, disease

**misdread,** fear of evil

**miser,** miserable wretch

**misgoverned,** unruly

**misgovernment,** misconduct

**misgrafted,** badly matched

**misprision,** contempt, mistake, misunderstanding

**misproud,** arrogant

**missive,** messenger

**mistake,** take, undertake or deliver wrongly, misjudge, blunder, feel misgiving about

**misthink,** think evil of

**mistress,** the jack at bowls

**mobled,** muffled

**mockery,** imitation, futile action

**model,** architect's plan, mould, copy

**modern,** everyday, commonplace

**modest,** moderate, satisfactory, becoming

**modestly,** without exaggeration

**modesty,** moderation, avoidance of exaggeration

**moiety,** half, share, small part

**moldwarp,** mole

**mome,** blockhead

**moment,** cause

**monster,** make a monster of, show as monstrous

**monument,** sepulchre, effigy, portent

**monumental,** memorial, commemorating

**mood,** anger, outward appearance, mode

**mooncalf,** misshapen creature

**moonish,** changeable

**mop,** grimace

**mope,** move blindly and stupidly, be bewildered

**moralize,** interpret, explain

**mort,** note on a horn at the death of the deer

**mortified,** dead to the world, deadened, destroyed

**mortifying,** mortal

**mose in the chine,** suffer from glanders

**mot,** motto

**mother,** hysteria

**motion,** motive, puppet show, puppet; to propose

**motive,** cause, instigator, instrument, moving limb or organ

**mould,** earth

**mouse,** tear, bite

**mouse-hunt,** woman-chaser

**mow,** grimace

**mulled,** thick, heavy

**mummy,** dead flesh, medicinal or magical preparation from this

**muniments,** furnishings

**mure,** wall

**murrain,** plague

**muse,** wonder

**muset,** gap

**muss,** scramble

**mutine,** mutineer; rebel

**mutiny,** strife, quarrel

**mutual,** common, intimate

**mutually,** in return, together

**mystery,** trade, profession, skill

**naked,** unarmed, plain

**Naso,** family name of the poet Ovid

**native,** source, origin; natural, kindred, closely related, rightful

**natural,** that is so by birth, related by blood, kind, tender; half-wit

**naturalize,** familiarize

**naught,** wickedness, wicked, ruined, ruin

**naughty,** bad, wicked, good-for-nothing

**nave,** nub, navel

**neaf,** fist

**neat,** animal of the ox kind, cattle

**neb,** mouth

**Nebuchadnezzar,** king of Babylon, driven out to eat grass like cattle

**neglect,** cause of neglect

**nephew,** cousin, grandson

**Neptune,** god of the sea

**Nereides,** sea-nymphs

**Nero,** Roman emperor, responsible for the assassination of his mother; believed to have played on the lyre and recited while watching the burning of the city set on fire by his orders

**nerve,** sinew

**nervy,** sinewy

**Nessus,** centaur killed by Hercules for trying to rape Deianira; a tunic dyed with his blood poisoned Hercules

**Nestor,** oldest and wisest of the Greeks at the siege of Troy

**next,** nearest, quickest

**nice,** wanton, delicate, shy, difficult to please, fastidious, scrupulous, subtle, needing precision, delicately balanced, intricate, exact, skilful, trivial

**nicely,** elegantly, scrupulously, sophistically, exactly

**niceness,** coyness, fastidiousness

**nicety,** coyness

**Nicholas,** patron saint of scholars; *Saint Nicholas' clerks,* highwaymen

**nickname,** name wrongly

**niggard,** to act in a miserly way, supply sparingly

**night-gown,** dressing gown

**night-rule,** disorder by night

**nine-men's morris,** cutting in turf for game played with nine pegs or discs

**Niobe,** overboastful of her children, who were slain; she was herself turned to stone.

**noise,** rumour, music, band of musicians; clamour, spread by rumour

**noll,** head

**nonsuit,** to refuse to listen to or grant the suit of

**nook-shotten,** having a very uneven outline

**notedly,** precisely

**nothing,** vulva

**notion,** understanding, mind

**novum,** dice-game in which chief throws are nine and five

**noyance,** injury, harm

**nuncio,** messenger

**nursery,** nursing, care

**nut-hook,** beadle, constable

**oathable,** fit to take an oath

**ob.,** abbreviation of 'obolus', halfpenny

**objection,** charge, accusation

**obligation,** contract

**obliged,** pledged

**obsequious,** dutiful, dutiful in funeral rites

**observant,** obsequious servant

**observe,** humour

**occasion,** opportunity, pretext, cause, course of events

**occulted,** hidden

**occupation,** business, handicraft, trade

**occupy,** have to do with sexually

**occurrent,** event

**oddly,** unevenly

**oeillade,** inviting glance, ogling

**o'erblow,** blow away

**o'ercount,** outnumber

**o'ercrow,** triumph over

**o'erdyed,** dyed with a second colour

**o'ereaten,** left after most has been eaten

**o'erflourished,** decorated on the outside

**o'erlook,** examine, bewitch, despise

**o'ermaster,** possess

**o'erparted,** having too difficult a part

**o'er-peer,** rise above

**o'erpost,** get over easily

**o'er-raught,** overtook

**o'er-teemed,** worn out by child-bearing

**o'er-watch,** stay awake too long

**o'erwhelm,** overhang

**o'erwrested,** strained

**offend,** harm, hurt

**offer,** act on the offensive, venture

**office,** function, service

**offices,** parts of a house devoted to household matters

**omit,** neglect, disregard, lay aside

omittance, postponement
open-arse, medlar
operant, effecting, effective
operation, effect, efficacy
opinion, censure, public judgement, self-conceit, self-confidence
opposeless, irresistible
opposite, antagonist, opponent; hostile, adverse
oppress, suppress, distress
oppression, burden, distress
oppugnancy, conflict
orb, circle, sphere, sphere in which a star moves, heavenly body, earth
ordinance, what is ordained, established rule, decree, rank
ordinary, fixed-price tavern meal, 'ordinary run'
orgulous, proud
orifex, aperture
orison, prayer
ort, fragment, scrap
ostent, appearance, show
ostentation, appearance, show, spectacle
othergates, in another way
otherwhiles, at times
ouch, jewel
ounce, lynx
ouph, elf
ousel, blackbird
outbrave, to surpass in beauty or valour
outlandish, foreign
outlive, survive
outpeer, surpass
outsell, exceed in value
outward, outward appearance
overhear, hear over again
overhold, overestimate
overlook, overtop, look down on from above, read through
overpeer, rise above, look down on
overscutched, made haggard with beating, worn out
overture, disclosure; formal opening, first indication
Ovid, Roman poet who was sent into exile
owe, own, possess

pace, train (a horse) to pace
pack, gang; conspire, shuffle (cards), to cheat, be off
packing, plotting
paddock, toad
pageant, show, spectacle
pain, trouble, punishment
painful, laborious, toilsome
pale, fence, enclosure; enclose, encircle
palisado, staked fence
pall, fail
palliament, white robe of a candidate for the Roman consulship
palmer, pilgrim
palter, equivocate, use trickery
pander, go-between in a love affair, pimp
Pandion, see PHILOMEL
pantaloon, foolish old man
pantler, servant in charge of the pantry
paper, notice specifying offence committed; write down
paragon, compare, excel, show as a model

Parca, one of the three fates who prepared and cut the thread of human life
parcel, part, item, group
parcel-gilt, partly gilded
pard, panther or leopard
pardie, certainly, indeed
Paris ball, tennis ball
paritor, official who summoned offenders to an ecclesiastical court
parlous, perilous, cunning, dreadful
part, action, side
partage, freight, cargo
partake, impart, take sides
partaker, confederate, supporter
parted, divided, gifted
partialize, make partial
particular, detail, individual, personal interest, intimacy; private, personal
partisan, footsoldier's weapon, a long-handled spear with a blade
pash, head; to strike violently, smash
passado, a forward thrust in fencing
passant, heraldic term of beast stepping; surpassing
passenger, traveller on foot
passion, suffering, affliction, fit of disease, overpowering emotion, passionate speech, sorrow; feel deep emotion
passionate, express with passion; compassionate, sorrowful
patch, fool
patchery, trickery
paten, thin circular metal plate
patent, title, privilege, authority
patience, permission, leave
patronage, uphold, defend
pattern, precedent, model; to give an example, be a pattern or precedent for
pax, tablet kissed by priest and congregation in celebration of mass
peach, denounce, turn informer
peat, spoilt girl
peck, pitch
peculiar, individual, private, belonging to one person
pedant, schoolmaster
peeled, tonsured
Pegasus, winged horse
peise, balance, weight, suspend
pelf, property, possessions
pelican, believed to feed its young with its own blood
pelt, rage, scold
pelting, paltry, worthless
pencil, paintbrush
pendulous, impending, suspended
pensioners, royal bodyguard
Penthesilea, Amazon queen
peradventure, by chance, perhaps
perdition, ruin, loss, damnation
perdurable, everlasting
peregrinate, travelled, foreign in style
perfect, fully informed, equipped, ready; accomplish, instruct
perfection, performance, completion
periapt, amulet
period, end, goal, highest point, full pause, full stop; to end
perjure, perjurer; corrupt
perniciously, to destruction
perpend, consider

persistive, steadfast
perspective, optical device for producing fantastic images; picture of figure producing a distorted or unexpected effect
pert, lively
pervert, turn aside
pester, throng, obstruct
pestiferous, pernicious
phantasim, fantastic being
Philomel, Philomela, daughter of Pandion, raped by Tereus (husband of her sister Progne), who cut out her tongue; she wove her story into a tapestry. Progne feasted Tereus on their murdered son; Philomela was changed into a nightingale
Phoebe, Diana, goddess of the moon
Phoebus, Apollo, god of the sun
phoenix, unique Arabian bird which, dying, is recreated from its own ashes
phraseless, inexpressible
phthisic, consumptive cough
physical, beneficial to health
pia mater, brain
picked, refined, fastidious
Pickt-hatch, London district noted for brothels
pie, magpie
piece, cask of liquor, masterpiece; add to, augment
pied, parti-coloured
pike, spike in centre of buckler
pill, plunder, rob, strip (of bark), make bald
pillicock, penis
pin and web, disease of the eye
pinch, bite, pang; to bite, harass, distress
pismire, ant
placket, petticoat, opening, slit
planched, made of boards
plant, sole of the foot
plantage, plants
plash, pool
plausible, pleasing
plausibly, with applause
plausive, deserving of applause
Plautus, Roman writer of comedy
pleached, formed by intertwined over-arched boughs, folded
pleasant, merry, jokey
pleurisy, abundance, excess
plight, pledge
Pluto, ruler of the underworld
Plutus, god of riches
poach, to thrust, stab
point, highest point, conclusion, lace with tags (for attaching hose to doublet, etc.), full stop
poking-stick, rod used for pleating ruffs
polecat, prostitute
policy, government, administration, prudence in managing public or private affairs, cunning, craftiness, trick
politic, dealing with government and administration, crafty
politician, schemer, scheming statesman
pomewater, large juicy kind of apple
pomp, procession, pageant
Pontic sea, Black Sea
poop, swamp, overwhelm
poor-john, salted fish
popular, plebeian, vulgar

**popularity**, common pledge

**porridge, porrage**, soup

**porringer**, basin

**port**, gate, bearing, style of living

**portable**, bearable, endurable

**portage**, portholes

**portance**, behaviour

**posied**, inscribed with a motto

**possess**, inform, acquaint

**posset**, drink of hot milk curdled with ale or wine; curdle, thicken

**post**, post set up for notices, etc., doorpost on which tavern reckoning was kept, courier, messenger, post-horse; to hasten, carry swiftly

**poster**, swift traveller

**posy**, motto inscribed inside a ring

**potable**, able to be drunk

**potato**, sweet potato, supposedly aphrodisiac

**potency**, power, authority

**potent**, potentate

**potential**, powerful

**potting**, tippling

**pouncet-box**, small box for perfumes

**powder**, preserve meat with salt or spice

**powdering-tub**, pickling-tub, sweating-tub used in cure of venereal disease

**power**, army

**pox**, syphilis, venereal disease

**practice**, trickery, conspiracy, plot

**practisant**, one who carries out a trick

**practise**, plot, conspire

**praemunire**, writ for maintaining papal jurisdiction in England

**praetor**, Roman magistrate

**praise**, appraise, value

**prank**, adorn

**precedent**, former; original from which a copy is made, sign

**precisian**, puritan

**precurrer**, forerunner

**precurse**, heralding

**predicament**, situation

**prefer**, present, advance, introduce, recommend

**pregnancy**, quickness of wit

**pregnant**, clear, fertile, compelling, resourceful, receptive

**prejudicate**, influence beforehand

**prejudice**, injury; injure

**premised**, sent before

**preparation**, accomplishment

**preposterous**, inverting the natural order of things, monstrous

**prerogative**, precedence, pre-eminence

**presage**, omen, prognostication, foreboding

**presence**, presence-chamber, company, person

**present**, immediate, instant; ready money, to show, represent, bring a charge against

**presently**, immediately

**presentment**, dedication

**press**, crowd, crowding, printing-press, cupboard, authority to enlist men compulsorily; to crowd, oppress, force into military service

**pressure**, impression, character impressed

**pretence**, expressed aim, intention, pretext

**prevent**, anticipate, escape, avoid

**prick**, mark made by pricking, dot, point, spot in centre of target, prickle, penis; to mark by making a dot, etc., pierce, fix, point, spur

**pricket**, buck in its second year

**prick-song**, music sung from notes

**pride**, magnificence, splendid adornment, highest state, mettle, sexual desire

**prig**, thief

**prime**, first, chief, sexually excited; springtime

**primogenitive**, first-born's right to inherit

**principal**, superior, abettor, principal rafter

**princox**, pert youth

**Priscian**, Roman grammarian

**prithee**, please

**private**, private person

**privates**, sexual organs

**prize**, contest; to value, esteem

**prizer**, prize-fighter, wrestler

**probal**, *to*, able to bear the probe or examination of

**probation**, trial, proof

**proceeder**, one proceeding to a university degree

**procurator**, proxy

**prodigious**, ominous, portentous, monstrous, abnormal

**proditor**, traitor

**proface**, 'may it do you good', formula of welcome before a meal

**profited**, proficient

**progeny**, lineage, race

**project**, notion, idea; set forth, exhibit

**projection**, plan

**prolixious**, prolonged

**prolong**, postpone

**Prometheus**, stole fire from heaven and was chained to Mount Caucasus

**prone**, ready, eager

**proof**, test, trial, experience, issue, result

**propend**, incline

**proper**, (one's) own, private, peculiar, excellent, handsome

**property**, identity, particular quality, take possession of, endow with qualities, make a tool of

**proportion**, portion, division, relative size, proportioning, rhythm

**propose**, put forward, set before one's mind, suppose, converse

**propriety**, identity, proper state

**propugnation**, defence

**prorogue**, prolong, postpone

**prosecution**, pursuit

**Proserpina**, daughter of Ceres, carried away by Pluto to become queen of hell

**prosperous**, favourable

**Proteus**, a sea-god, able to assume different shapes

**protractive**, consisting in delay

**proud**, elated, giving cause for pride, lofty, splendid, spirited, swollen, over-luxuriant, sexually excited

**prove**, try, test, find by experience, experience

**publish**, make known, proclaim, denounce

**pucelle**, virgin, maid, slut

**puck**, pixy, mischievous spirit

**pudding**, sausage, stuffing

**pudency**, modesty

**puissance**, strength, armed force

**puissant**, strong, powerful

**pumpion**, pumpkin

**punk**, prostitute, harlot

**punto reverso**, back-handed thrust in fencing

**purblind**, quite blind, partially blind

**purchase**, booty, prize, annual rent from land; to strive, gain, acquire otherwise than by inheritance

**purgation**, clearing from accusation or suspicion

**purlieu**, tract of land bordering on a forest

**purpose**, proposal, conversation

**pursuivant**, junior attendant on heralds, herald, messenger

**pursy**, short-winded, fat

**purveyor**, steward going ahead to make provision

**putter on**, instigator

**puttock**, bird of prey, kite, buzzard

**quail**, prostitute; to fail, faint, overpower

**quaint**, skilful, clever, dainty, fine, beautiful, elaborate

**qualification**, calm and controlled condition

**qualify**, moderate, mitigate, appease, control, dilute

**quality**, accomplishment, rank, profession, party, side, manner, cause

**quarry**, heap of slaughtered game

**quarter**, part, watch, relations with and conduct towards another

**quat**, pimple

**quatch**, word of unknown meaning

**quean**, hussy, whore

**queasy**, hazardous, squeamish

**quell**, slaughter

**quest**, jury, inquest, search party

**questant**, seeker

**questionable**, inviting question

**questrist**, seeker

**quick**, flowing, fresh, impatient

**quiddity**, subtlety, equivocation

**quietus**, clearing of accounts

**quintain**, object or figure to practise tilting at

**quirk**, quibble, clever expression, peculiarity of behaviour, fit, start

**quit**, set free, rid, acquit, acquit oneself of, revenge, repay, requite

**quittal**, requital

**quittance**, discharge from debt, requital; requite

**quoif**, close-fitting cap

**quondam**, former

**quote**, give marginal references to, set down in writing, note, observe, regard

**quoth**, said

**quotidian**, (of an intermittent fever) returning every day

**race**, root (of ginger), lineage, breed, natural disposition

**rack**, driving cloud; to torture, extend, stretch, strain

**rage**, madness, angry disposition, sexual passion; to enrage

**ramp**, loose woman

**rampallian**, riotous woman

**rampire,** barricade

**rank,** growing too luxuriantly, swollen, grown too fat, rebellious, high, full, lustful, in heat, coarse, festering; closely

**rankle,** cause a festering wound

**rankness,** overflowing, insolence

**rascal,** inferior deer

**rate,** estimate, value, expense

**ravel,** become entangled, disentangle

**raven,** devour greedily

**rayed,** bespattered, fouled

**read,** teach, discover the meaning of, expound a riddle

**re-answer,** compensate for

**reason,** speech, remark; to talk, discuss, explain

**reasonable,** needing the use of reason

**reave,** deprive, take away

**rebate,** make dull, blunt

**rebato,** stiff collar

**receipt,** what is received, receptacle, receiving, capacity, recipe

**receiving,** understanding, reception

**recheat,** notes sounded on a horn to call hounds together in stag-hunting

**reck,** to care (for), mind

**reclaim,** tame, subdue

**recognizance,** bond, token

**recomfort,** console

**record,** witness, memory; to witness, sing

**recordation,** memorial, impression on the memory

**recourse,** access, admission, flowing

**recover,** reconcile, reach, rescue

**recovery,** process by which entailed estate was transferred from one party to another

**rector,** ruler, head

**rectorship,** rule

**recure,** restore, heal

**reduce,** bring, bring back

**reechy,** smoky, dirty, stinking

**reek,** smoke, steam

**refel,** refute

**reflect,** shine

**regiment,** authority, rule

**region,** upper air, air

**regreet,** greeting, contract

**reguerdon,** reward

**rehearsal,** account

**rehearse,** describe, tell

**reins,** kidneys

**rejoindure,** reunion

**rejourn,** adjourn

**relative,** able to be related and believed

**religion,** strict fidelity, religious duty

**religious,** exact, conscientious, strict

**remember,** mention, commemorate, remind

**remit,** pardon, give up

**remonstrance,** demonstration

**remorse,** pity, tenderness, moderation, qualification, mitigation

**remorseful,** compassionate

**remotion,** departing

**removed,** remote, secluded, separated by time or space

**removedness,** absence

**remover,** one who changes

**render,** surrender, rendering of an account, statement; to give back, represent, describe as being, declare, state, surrender

**rendezvous,** refuge, last resort

**renege,** deny, renounce

**repair,** restoring, coming; go, come, return

**repairing,** able to renew the attack

**repasture,** food

**repeal,** recall from exile; recall as from banishment, call back into favour

**repetition,** recital, mention

**repine,** vexation

**replenished,** perfect

**replication,** reply, echo

**reprisal,** prize

**reprobance,** damnation, rejection by God

**reprobate,** depraved

**reproof,** disgrace, disproof, refutation

**reprove,** disprove, refute

**repugn,** oppose

**repugnancy,** resistance

**repugnant,** refractory

**rescue,** freeing from legal custody by force

**resemblance,** appearance, likelihood

**resist,** affect with distaste

**resolve,** dissolve, answer (a question), solve (a problem), convince, inform

**resolvedly,** answering all questions

**respect,** relationship, discrimination, consideration, esteem; to regard, care for, esteem

**respective,** careful, worthy of being cared for, discriminating

**respectively,** respectfully

**respite,** delay, date to which something is postponed, limit

**responsive,** suited

**rest,** place to rest, restored strength, resolution, stakes kept in reserve

**restrain,** draw tight, withhold

**resty,** restive, sluggish

**retention,** detention, reserve, power to hold or retain

**retentive,** confining, restraining

**retire,** return

**retort,** to reflect (heat), reject (an appeal)

**retreat,** recall of pursuing force

**retrograde,** contrary, seeming to move in a backward course

**reverb,** resound

**revolt,** revulsion, change; rebel

**revolution,** alteration, change produced by time

**revolve,** consider

**rhapsody,** confused medley

**rheum,** mucus from nose or throat, cold in the head, rheumatism

**riggish,** wanton

**rigol,** ring, circle

**rim,** belly, membrane lining the abdomen

**ring-carrier,** go-between

**ringlet,** little circle

**riot,** loose living, debauchery

**rivage,** shore

**rival,** partner

**rive,** split

**rivelled,** wrinkled

**road,** riding, period of riding, stage (of journey), roadstead, highway

**robustious,** boisterous

**roguing,** roaming

**roguish,** vagrant

**roisting,** blustering, bullying

**rondure,** circle

**ropery,** knavery

**Roscius,** famous Roman actor

**rote,** fix by memory

**round,** plain, plain-spoken, severe; round dance, roundabout way, rung of a ladder; surround, become round

**roundel,** round dance

**rouse,** full draught of liquor, drinking bout

**royal,** gold coin of the value of ten shillings (fifty pence)

**royalty,** prerogative enjoyed, or granted, by the sovereign

**roynish,** scurvy

**rub,** in bowls an obstacle hindering or turning aside the bowl, obstacle, hindrance, roughness, unevenness; to turn aside, hinder

**rubious,** ruby-red

**ruddock,** robin

**rude,** ignorant, barbarous, violent, rough

**rudeness,** violence, roughness

**rudesby,** rough unmannerly brute

**ruffle,** bustle; swagger, bear oneself proudly, bluster, be turbulent

**rugged,** shaggy, frowning

**rug-headed,** having shaggy hair

**ruinous,** brought to ruin

**rumour,** fame, tumult, uproar

**runagate,** deserter, vagabond

**runnion,** wretch (a term of abuse)

**russet,** homely, simple

**sable,** black

**sack,** white wine

**sackbut,** trumpet resembling a trombone

**sad,** steadfast, grave, serious, dismal

**sadly,** gravely, seriously

**sadness,** seriousness

**safe,** make safe

**safety,** custody, safeguard

**sain,** said

**salamander,** lizard-like animal supposed to live in fire

**sale-work,** ready-made goods not of the highest quality

**sallet,** light helmet, something mixed or savoury

**salt,** lecherous

**salute,** touch, affect

**sample,** example

**sanctimony,** holiness, sanctity

**sanctuarize,** give sanctuary to

**sand-blind,** half-blind

**sanguine,** red

**sans,** without

**sapient,** wise

**sarcenet,** fine soft silk material

**sauce,** over-charge, rebuke

**saucy,** insolent, presumptuous, wanton

**savagery,** wild growth

**savour,** smell, perfume, style, character

**saw,** saying, maxim, proverb

**say,** finely woven cloth, taste, saying

**scab,** rascal, scurvy fellow

**scald,** scabby, mean

**scale,** weigh

**scandalize,** disgrace

**scantling,** specimen, sample

**scantly,** slightingly

**scape,** escapade, transgression; escape

**scarf,** officer's sash, sling, streamer; blindfold, deck with streamers

**scathe,** harm; to injure

**schedule,** document

**school,** university, instruction, learning; reprimand, discipline

**science,** knowledge

**scion,** shoot, slip for grafting

**sconce,** head, small fort or earthwork

**scope,** object, aim, purpose, theme, liberty

**score,** notch cut in stock or tally in keeping accounts, account; to notch

**scorn,** taunt, insult, object of contempt

**scotch,** cut, gash

**scour,** hasten

**scrimer,** fencer

**scrip,** small bag, piece of paper with writing on

**scrippage,** contents of scrip

**scripture,** writing

**scrubbed,** stunted

**scruple,** tiny part, doubt, difficulty

**scurril,** coarsely abusive

**scut,** deer's tail

**scutcheon,** shield with coat of arms; tablet showing armorial bearings of a dead man

**seam,** fat, grease

**sectary,** dissenter, one who pursues a particular study

**secure,** free from care, confident, unsuspicious; make confident or overconfident

**securely,** confidently, without suspicion

**security,** confidence, overconfidence

**seed,** mature, run to seed

**seedness,** sowing

**seeing,** appearance

**seel,** make blind, close (a person's eyes)

**seely,** foolish, innocent, harmless

**seen,** skilled

**seld,** seldom

**self,** one's own, same

**semblable,** similar, like, equal

**semblative,** like, resembling

**Seneca,** Roman tragic dramatist

**sennight,** week

**sense,** physical feeling, sensuality, sexual desire, mental apprehension, mind, opinion

**senseless,** without feeling or perception, free from sensual sin

**sensible,** perceptible, tangible, substantial, having sensation, sensitive, endowed with feeling

**sensibly,** having sensation, feelingly

**sentence,** sententious saying, maxim

**septentrion,** north

**sequel,** series

**sequent,** succeeding, following

**sequester,** separation; to separate

**sequestration,** separation, seclusion

**sere,** dry, withered

**serpigo,** a skin disease

**servanted,** subject

**service,** what is placed on the table for a meal

**several,** distinct, different, individual, respective, various

**severally,** separately, at different entrances or exits

**sewer,** attendant in charge of service at a meal

**shadow,** shade; to hide, shelter

**shame,** to be ashamed

**shard,** piece of broken pottery, patch of cow-dung

**shark,** gather hastily together

**shearman,** one who shears woollen cloth

**sheaved,** made of straw

**sheep-biter,** sneaking rascal

**sheep-biting,** sneaking

**shent,** blamed, rebuked, reproved

**shift,** expedient, resource, trick; for (a) shift, to serve a turn; make (a) shift, contrive, manage

**ship-tire,** kind of head-dress

**shive,** slice

**shoal,** shallow, sand-bar

**shock,** to meet with force

**shog,** go, shift

**shotten,** that has spawned, lean

**shoulder-shotten,** having a dislocated shoulder

**shove-groat,** shovel-board

**shrewd,** wicked, mischievous, bad-tempered, dangerous, evil, difficult

**shrewdly,** sharply, grievously, intensely, very much

**shrieve,** sheriff

**shrift,** confession and absolution, penance

**shrill-gorged,** shrill-throated

**shrive,** to hear confession and absolve

**shroud,** protection, sailropes; shelter, hide

**shuffle,** use trickery, smuggle, shift

**siege,** seat, rank, turd

**sight,** visor

**sightless,** unseen, unsightly

**simple,** medicinal herb, single ingredient in a compound

**simpleness,** innocence, simplicity, foolishness

**simplicity,** ignorance, silliness

**simular,** pretended, plausible; counterfeiter

**sinfully,** with sins unatoned for

**single,** slight, trivial, sincere, simple

**singularity,** eccentricity, own person

**sinister,** left (hand), unfavourable, unjust

**sink,** cause to fall, ruin; sewer, drain

**Sinon,** a Greek who by guile persuaded the Trojans to take the Grecians' wooden horse into Troy

**sirrah,** form of address used mainly to inferiors

**sith, sithence,** since

**size,** allowance

**skill,** judgement, reason, ability; to make a difference, matter

**skilless,** ignorant

**skimble-skamble,** confused, rubbishy

**skipper,** flighty youth

**skirr,** fly, scour

**slab,** half-solid

**slander,** ill-repute, disgrace, discredit; bring into disgrace, reproach

**slanderous,** shameful, disgraceful

**slave,** enslave, make subservient to oneself

**sleave,** floss-silk, silk untwisted into fine threads

**sleeve-hand,** cuff

**sleided,** (of silk) floss, untwisted into fine threads

**sliding,** slip

**slipper,** shifty

**slippery,** unstable, fickle

**slipshod,** in slippers

**slobbery,** sloppy, slovenly

**slops,** wide breeches

**slovenly,** nasty, disgusting

**slubber,** sully, hurry over

**smatch,** taste

**smatter,** chatter

**smock,** woman's undergarment, woman

**smooth,** flatter, gloss over

**smother,** suffocating smoke

**smug,** trim, spruce

**sneap,** snub, pinch with cold

**snuff,** resentment

**sob,** opportunity for a horse to recover his wind, respite

**sod,** steeped, boiled

**soiled,** fed with fresh-cut green fodder

**soilure,** sullying, defilement

**sole,** unique, mere, alone

**solely,** alone, entirely

**solemnity,** ceremony, festivity

**solicit,** urging, entreaty; move, stir, bring something about

**sometime,** at some time, sometimes, once, formerly

**sonance,** sounding, signal

**sonties,** saints

**sooth,** truth, flattery

**soothe,** humour, flatter

**soother,** flatterer

**sop,** bread or wafer dipped in wine, etc

**sophister,** one using false arguments

**sore,** buck, deer, in its fourth year

**sorel,** buck, deer, in its third year

**sort,** lot, rank, company, group, way, state; allot, ordain, come about, turn out, be suitable, correspond, adapt, fit, classify, choose, contrive, go in company

**sortance,** agreement, suitableness

**sound,** utter, proclaim, keep sound

**souse,** swoop on, pickle

**span-counter,** game in which counters are thrown to lie within a hand's span

**spavin,** disease of horses consisting of swelling of joints, producing lameness

**specialty,** special characteristic, possession, special contract under seal

**spectacles,** organs of sight

**speculation,** watcher, power of seeing, sight, looking on

**speculative,** seeing

**speed,** fortune, outcome, protection, assistance; to fare (well or ill), turn out, be successful, assist, favour

**speeding,** successful, fruitful; lot, outcome, success

**spend,** consume, exhaust

**sphere,** orbit of a planet, one of the concentric globes supposed to revolve round the earth with a harmonious sound

**spherical,** planetary

**spial,** spy

**spigot,** peg in faucet of a barrel

**spill,** destroy

**spilth,** spilling

**spin,** spurt

**spinner,** spider

**spinster,** spinner

**spital,** hospital

**spleen,** bodily organ regarded as the seat of emotions

**spoil,** plundering, prey, ruin; plunder, seize, destroy

**spot,** stain, disgrace, embroidered pattern

**spring,** fountain, source, shoot of a plant

**springe,** snare for birds

**springhalt,** disease of horse characterized by twitching and lameness

**spurn,** kick, insult, blow; to kick, oppose scornfully

**square,** fair, just; carpenter's set-square, measure, rule; body of troops in square formation, square of material in bosom of a dress; to regulate, quarrel (among), be at variance

**squarer,** quarreller

**squash,** unripe peapod

**squiny,** look asquint

**staff,** shaft of lance, lance, stanza

**stagger,** hesitate, waver

**staggers,** giddiness

**stain,** tinge, to eclipse, dim, be obscured

**stair-work,** furtive love-making

**stale,** decoy, bait, prostitute, laughing-stock, urine (of horses)

**stall,** install, keep, dwell, bring (a hunted animal) to a stand

**stamp,** stamping tool, coin, medal, distinguishing mark, imprint; to impress, mark with an impression, give approval to

**stanch,** satisfy

**stanchless,** insatiable

**stand,** place where one stands in ambush or in hiding; confront, oppose, stand firm; *stand at a guard with,* be fully protected against; *stand on, upon,* insist on, persist in, depend on, rely on, concern, be the duty or interest of; *stand to,* have an erection, support, maintain, be firm in, persist in; *stand to it,* maintain a cause, take a stand

**standard,** standard-bearer

**staniel,** kestrel

**stare,** (of hair) stand on end

**start,** sudden invasion, sudden flight, impulse; to startle, rush

**starting-hole,** refuge, loop-hole

**starve,** die, die of cold, kill or benumb with cold

**state,** condition, condition of health or prosperity, rank, dignity, chair of state, throne, nobles, ruling body, government

**station,** way of standing, attitude

**statist,** statesman

**statute,** bond, mortgage

**statute-cap,** woollen cap ordered in 1571 to be worn on Sundays and holy days by all below a certain rank

**stay,** obstacle; detain, stand, stand firm, wait (for), attend on

**stead,** to be of use to, help

**stell,** to fix, portray

**stern,** rudder

**stew,** brothel, stewpan

**stick,** stab, be fastened or fixed, hesitate

**stickler-like,** like an umpire

**stigmatic,** deformed person

**stigmatical,** deformed

**still,** always, continually

**stillitory,** apparatus for distillation

**stilly,** softly

**sting,** sexual desire

**stint,** stop

**stithy,** forge

**stoccado,** thrust with a rapier

**stock,** block of wood, person without feeling, stocking, dowry; set in the stocks

**stockfish,** dried cod or other fish, beaten before cooking

**stockish,** blockish, unfeeling

**stomach,** appetite, inclination, temper, courage, pride, anger; resent

**stomacher,** ornamental front over which the bodice was laced

**stone,** mirror, thunderbolt, testicle; turn to stone

**stone-bow,** cross-bow used for shooting stones

**stoop,** (of a bird of prey) swoop

**stop,** hole in a wind-instrument, stopped to produce a difference in pitch, fingering of musical chords, fret of a lute; staunch, heal

**store,** breeding, increase; to populate

**stout,** bold, strong, proud

**stoutness,** stubborn pride

**stover,** fodder for cattle

**strain,** race, character, kind, class, tune; clasp, force, constrain, urge

**strait,** narrow, tight-fitting, strict, niggardly; narrow place, difficulty

**strange,** foreign, new, not knowing, unfriendly, cold, shy

**strange-achieved,** gained at a distance and for others

**strange-disposed,** of unusual character

**strangely,** coldly, without greeting, as a stranger, to an extraordinary degree, in an unusual way

**strangeness,** behaving as a stranger, aloofness, reserve

**strappado,** torture in which victim was hoisted up by his arms, which were tied behind his back, then let down halfway with a jerk

**stratagem,** deed of violence

**straw,** *wisp of straw,* mark of disgrace for a scolding woman

**stray,** body of stragglers; lead astray

**strength,** authority, legal power, body of troops

**strewment,** strewing of flowers

**stricture,** strictness

**strike,** to lower (sail), blast, destroy, tap (a cask)

**strossers,** trousers

**stubborn,** inflexible, stiff, rude, harsh, ruthless

**stuff,** semen

**style,** title

**subduement,** conquest

**subjection,** duty as, or of, a subject

**submission,** confession of error

**subscribe,** sign, write down, assent, acknowledge, admit, submit, yield up, answer

**subscription,** submission

**substractor,** slanderer

**subtle,** thin, fine, cunning, treacherous, tricky

**subtlety,** illusion

**success,** outcome, result, good or bad fortune, succession

**successfully,** likely to succeed

**succession,** following on, successors, success

**successive,** hereditary, descending by succession

**successively,** by inheritance

**sufferance,** suffering, damage, patient endurance

**suffice,** content, satisfy

**sufficient,** able, able to meet liabilities, solvent

**suggest,** tempt, prompt to evil, insinuate to

**suggestion,** prompting to evil, temptation

**sumless,** incalculable

**summoner,** officer who summoned offenders to an ecclesiastical court

**sumpter,** driver of pack-horse, pack-horse

**superflux,** superabundance, surplus

**supernal,** heavenly

**superscript, superscription,** address or direction on letter

**superstitious,** idolatrously devoted

**supervise,** perusal; look over

**suppliance,** pastime

**supply,** help, reinforcements

**supposal,** opinion

**suppose,** supposition, expectation; believe, imagine, guess

**supposition,** doubt

**surcease,** cessation; cease

**sure,** safe, beyond power of doing harm, reliable, united

**surety,** confidence of safety, certainty, stability, warrant

**sur-reined,** overridden

**suspect,** suspicion

**suspire,** breathe, draw breath

**sutler,** one who sells provisions to soldiers

**swagger,** rant, bluster, quarrel

**swaggerer,** blusterer, quarreller

**swart,** dark, swarthy

**swasher,** blustering ruffian

**swashing,** swaggering, slashing

**sway,** direction, control, sovereignty; to rule, move

**swayed,** curved in

**swear,** *swear out,* forswear

**sweat,** sweating sickness, sweating cure; take a sweating cure

**sweet,** scented; scent

**sweeting,** sweet variety of apple, term of endearment

**swim,** float

**swinge,** thrash

**swinge-buckler,** roisterer

**sworder,** cut-throat, gladiator

**'Swounds,** by God's wounds (a strong oath)

**sympathize,** to be similarly affected, agree (with), match

**table,** tablet for an inscription, writing-tablet, flat surface on which a picture is painted, quadrangular space between chief lines on palm of hand

**table-book,** notebook

**tables,** backgammon

**tabor,** small drum

**tackled stair,** rope ladder

**tag, tag-rag,** rabble

**tailor,** ? sex organ

**take,** strike, strike with disease or enchantment, catch, take effect, reckon, measure, write down, accept as true, catch fire, perceive, understand, esteem, take away, conclude; *take head,* deviate, run off its course; *take in,* capture; *take me with you,* speak so that I can understand you; *take it on,* assume authority; *take on,* rage, show great distress, pretend; *take out,* make a copy of; *take up* lift, enlist, arrest, buy on credit, rebuke, reprimand, oppose, encounter, make up (a quarrel)

**taking,** blasting; state of great excitement or alarm, malignant influence

**tale,** numbering one after another, talk, story, falsehood

**talent,** measure for a large sum of money

**tall,** long, lofty, goodly, fine, brave

**tally,** stick marked with notches for keeping accounts

**'tame,** broach (a cask)

**tang,** ring out

**tardy,** delay

**targe,** light shield

**tarre,** urge, incite

**tarriance,** delay, waiting

**tarry,** wait, await, remain, delay

**task,** tax, impose a task on, occupy, put a strain on, put to proof

**tasking,** challenge

**tassel-gentle,** tercel, male hawk

**tawdry-lace,** silk lace or ribbon for the neck

**tax,** accusation; blame, accuse, charge

**taxation,** claim, demand, slander

**teem,** conceive, bring forth, bear children, be fruitful

**teen,** affliction, grief, sorrow

**tell,** count

**temper,** disposition, temperament, mental balance, hardness and elasticity imparted to steel; to compound, mix, persuade

**temperance,** mildness of temperature, calmness, chastity

**temperate,** chaste, secular

**temporary,** secular

**tempt,** test, risk

**tenant,** vassal

**tend,** listen, watch over, attend on, wait (for)

**tendance,** attending to, service, people in attendance

**tender,** offer, thing offered, care; to exhibit, pay down, care for; sensitive, compassionate

**tender-hafted, tender-hested,** gently framed

**tent,** probe; to probe, cure (a wound), lodge

**Tereus,** see PHILOMEL

**Termagant,** violent character, supposedly god of the Mohammedans, in old miracle plays

**termination,** expression, term

**termless,** indescribable

**tertian,** fever returning every third day

**tester,** sixpenny piece

**tetter,** skin disease

**text,** capital letter

**thane,** Scottish title or rank, somewhat lower than earl

**thankful,** deserving thanks

**theoric,** theory

**therefor,** for that

**Thetis,** a sea-goddess, the sea

**thews,** bodily parts, strength

**thick,** rapid, dim

**thills,** shafts of a cart

**thing,** sexual organ

**think,** seem

**thought,** anxiety, sorrow

**thrasonical,** boastful

**three-farthings,** thin silver coin, having a profile of Queen Elizabeth with a rose behind the ear

**three-nooked,** three-cornered

**threne, threnos,** dirge

**thrift,** profit, gain, advantage

**thriftless,** profitless

**thrum,** thread left at end of a warp

**thunder-stone,** thunderbolt

**tickle,** unstable, precarious; please, provoke

**tick-tack,** game in which pegs were put into holes; fornication

**tide,** time, course

**tidy,** plump

**tight,** sound, able

**tilt,** thrust, fight, encounter

**timeless,** untimely, premature

**timely,** early, in due season

**tinct,** colour, elixir of the alchemists

**tire,** equipment, dress, head-dress; attire, prey (on), feed greedily

**tis,** this (dialectal)

**tisick,** consumptive cough

**Titan,** god of the sun, the sun

**tithe,** tenth; to take the tenth part

**tithing,** district

**tittle,** dot

**to,** in addition to, against, appropriate to, in comparison with, in respect of, as to

**tod,** 28 lb. weight of wool; to yield this amount of wool

**tofore,** previously

**toge,** Roman toga

**toil,** net, snare; to cause to work hard, weary with work

**token,** spot of infection, plague-spot

**toll,** exact toll or tribute

**tool,** weapon, penis

**top,** head, forelock, highest point; to surpass, copulate with

**top-gallant,** highest mast, summit

**topless,** supreme

**tortive,** twisted

**toss,** impale, toss on a pike

**tottering,** ragged

**touch,** touchstone, taint; sound, test, wound

**touse,** pull out of joint

**trace,** follow, pass (through)

**tract,** trace, track, course

**trade,** coming and going, path, habit, business

**traded,** practised

**trade-fallen,** out of work

**traditional,** bound by tradition

**train,** tail, troop, bait; draw, entice, lead astray

**traject,** ferry

**trammel,** bind with the corpse, entangle

**transfix,** remove

**translate,** change, transform

**transpose,** change

**trash,** check, hold in leash

**traverse,** march, move from side to side

**treatise,** tale, talk

**treble-dated,** living three times as long as man

**trench,** cut

**trencher,** wooden plate

**trencher-friend,** parasite

**trenchering,** plates

**trencher-knight,** hanger-on

**tribunal,** dais

**trick,** custom, way, knack, touch (of a disease), toy; to adorn, blazon

**tricking,** adornment

**trill,** trickle

**triple,** third

**triple-turned,** three times faithless

**tristful,** sorrowful

**triumph,** public festivity, tournament, trump card

**triumphant,** triumphal, celebrating a triumph

**triumphantly,** in celebration

**Trojan,** good fellow

**troll,** sing

**troll-my-dames, troll-madams,** game resembling bagatelle

**trophy,** token of victory, memorial, monument

**tropically,** figuratively

**trot,** bawd

**troth,** truth, faith, word of honour

**trow,** believe, think, know

**truepenny,** honest fellow

**trull,** prostitute, wench

**truncheon,** staff carried as a symbol of office

**try,** test; purify, refine, prove

**tub,** sweating-tub used in the treatment of venereal disease

**tuck,** rapier

**tucket,** signal, flourish on a trumpet

**tun-dish,** funnel

**turtle,** turtle-dove, symbol of chaste and faithful love

**tushes,** tusks

**twire,** twinkle

**type,** distinguishing sign, stamp

**tyrannically,** vehemently

**tyranny,** violence, outrage

**tyrant,** usurper

**umbrage,** shadow

**unaccommodated,** unprovided

**unacquainted,** unfamiliar

**unadvised,** rash, without consideration or knowledge

**unagreeable,** unsuited

**unaneled,** not having received extreme unction

**unapproved,** unconfirmed by proof

**unaptness,** disinclination

**unattainted,** without infection, unprejudiced

**unattempted,** untempted
**unavoided,** inevitable
**unbarbed,** unarmed
**unbated,** unabated, not blunted (with a button)
**unbent,** with bow unbent, unprepared, unfrowning
**unbid,** unexpected
**unbitted,** unbridled
**unbolted,** unsifted, coarse
**unbookish,** unskilled, inexperienced
**unbreathed,** unpractised
**uncase,** strip, lay bare
**uncharge,** acquit of guilt
**uncharged,** unattacked
**unchary,** carelessly
**unchecked,** not contradicted
**unclasp,** reveal
**unclew,** undo, ruin
**uncomprehensive,** unplumbable
**unconfirmed,** uninstructed, ignorant
**uncovered,** open, bare-headed
**uncrossed,** not crossed through because bill still unpaid
**unction,** salve, ointment
**uncurrent,** not current, not permisible
**undercrest,** wear as a device of honour
**undergo,** undertake, come under, support, carry, bear
**underhand,** unobtrusive
**underskinker,** under-tapster, assistant barman
**undertake,** take charge of, assume, have to do with, venture
**undertaker,** one who takes on himself another's quarrel, one who settles with
**underwrite,** subscribe, submit to
**uneath,** with difficulty, scarcely
**unexperient,** inexperienced
**unexpressive,** inexpressible
**unfair,** rob of beauty
**unfashionable,** badly formed
**unfenced,** defenceless
**unfold,** disclose, display, reveal, release from the fold
**unfurnished,** unprovided, unprepared
**ungalled,** uninjured
**ungenitured,** impotent
**ungored,** unwounded
**unhandsome,** unskilled, nasty
**unhappily,** unfavourably, evilly
**unhappiness,** evil nature
**unhappy,** ill-fated, wretched, mischievous
**unhouseled,** not having received the sacrament
**union,** pearl
**unkind,** unnatural
**unluckily,** with foreboding
**unmanned,** not trained, not broken in
**unowed,** having no owner
**unpaved,** without testicles, castrated
**unpitied,** unmerciful
**unplausive,** unapproving
**unpregnant,** unapt
**unprevailing,** unavailing
**unprizable,** without value, invaluable
**unprized,** not valued, beyond value
**unproper,** not belonging to one person, indecent
**unprovide,** make unprepared
**unquestionable,** not inviting conversation

**unraked,** not covered with fuel to keep it burning
**unrecalling,** past recall
**unreclaimed,** untamed
**unrecuring,** past cure
**unrespective,** heedless, held back by no consideration, undiscriminating
**unrolled,** struck off the register
**unscanned,** unconsidered
**unseasoned,** ill-timed, inexperienced
**unseminared,** without seed
**unsifted,** untried
**unsorted,** ill-chosen
**unspeakable,** indescribable, inexpressible
**unspeaking,** unable to speak
**unsquared,** inappropriate
**unstanched,** unquenchable, freely menstruating
**unstate,** deprive of rank and state
**untainted,** unaccused
**untempering,** unwinning
**untented,** unable to be treated, incurable
**untoward,** perverse, unruly
**untowardly,** unluckily
**untraded,** not customary
**untrimmed,** undressed
**untrussing,** undoing hose, undressing
**unwarily,** unexpectedly
**unweighed,** hasty, unconsidered
**unwrung,** not pinched or rubbed
**up,** in arms, in rebellion, in prison; *up and down,* completely, exactly
**upon,** because of, in consequence of
**urchin,** hedgehog, goblin
**urinal,** glass to hold urine
**usance,** interest on money
**use,** habit, custom, usual experience, advantage, profit, lending at interest, interest on something lent, need; to be accustomed, continue, make a practice of, deal with, treat, go often
**usurp,** take or hold what belongs to another, supplant, assume
**utterance,** utmost extremity
**uttermost,** latest

**vail,** gratuity, tip, setting (of the sun); lower, let fall, do homage
**vain,** foolish, silly, unreal
**valanced,** fringed with hair
**validity,** strength, value
**value,** estimate, be worth
**valued,** denoting the value
**vantage,** advantage, gain, superiority, vantage-ground, opportunity
**varletry,** rabble
**vastidity,** immensity
**vaulty,** arched, hollow
**vaunt,** beginning
**vaunt-courier,** herald, harbinger
**vengeance,** harm, injury
**vent,** emission, utterance, outlet for energy; let out, emit, utter, make known
**ventage,** aperture, finger-hole
**ventricle,** cavity of the brain
**Venus,** goddess of love and beauty; wife of Vulcan, the smith-god, but more often associated with her lover Mars
**verge,** compass, circle
**vexation,** agitation, torment, grief
**via,** away

**vice,** character in a morality play representing a vice, jester, buffoon, gripping tool; to screw
**vicegerent,** deputy, representative
**vicious,** blameworthy, blemished
**vie,** add one to another
**villein,** peasant, servant
**vinewed'st,** most mouldy
**violent,** to rage, storm
**virginalling,** fingering as on the virginals— a keyed musical instrument
**virtue,** courage, merit, accomplishment, power, efficacy, essence, essential characteristic
**virtuous,** powerful, beneficial
**visitor,** one who visits to offer spiritual comfort
**voice,** speech, words, common talk, rumour, report, expressed opinion, judgement, vote, approval, authority to be heard; to acclaim, vote
**voiding lobby,** anteroom
**voluntary,** volunteer
**votaress,** woman who has taken a vow
**votarist, votary** one who has taken a vow
**vouch,** assertion, testimony; affirm, guarantee, bear witness
**Vulcan,** the smith-god, whose wife, Venus, was unfaithful
**vulgar,** of the common people, commonly known, common, public, mean; common people, vernacular

**waft,** to convey by water, beckon, turn
**waftage,** passage by water
**wafture,** wave
**wage,** to wager, hazard, attempt, carry on war, pay
**waist,** girdle
**wake,** remain awake, be up late for revelry or on guard, wear out with lack of sleep, arouse
**walk,** tract of forest
**wall-eyed,** white-eyed, having glaring eyes
**want,** lack, miss
**wanton,** frolicsome, lawless, capricious, luxurious, luxuriant, lustful, unchaste; spoilt child, pampered darling, roguish, sportful, unruly or lustful creature
**wappered,** worn out
**ward,** guard, custody, prison-cell, defensive position in fencing; guard, protect
**warden,** kind of cooking pear
**warp,** distort, deviate
**war-proof,** war-tested, courage
**warrant,** guarantee, assure
**warrantize,** surety, authorization
**warranty,** sanction, authorization
**warren,** game enclosure
**warrener,** gamekeeper
**wassail,** drinking, revelling, feast
**waste,** spend, consume
**watch,** wakefulness, sleeplessness, watchfulness; be awake, keep from sleep, catch in the act
**watchful,** sleepless
**water,** lustre
**water-gall,** secondary rainbow
**watering,** drinking
**water-standing,** tearful
**wave,** waver

waxen, increase

weal, welfare, commonwealth

wear, fashion; carry, possess, be fashionable, weary

weather, storm

weed, garment, dress

week, *to be in by the*, be caught, ensnared, deeply in love

ween, expect, hope

weet, know

welkin, sky

well-liking, plump

well-respected, well-considered

well to live, prosperous

wharf, river bank

wheel, spinning wheel, tread-wheel on which a dog was harnessed to turn a roasting-spit

whelk, pimple

where, whereas

whiffler, officer who keeps the way clear for a procession

while, whiles, whilst, till

whipstock, whip-handle

whist, become silent

whiting-time, bleaching time

whitster, bleacher

whittle, small clasp-knife

wide, missing the mark, astray

widgeon, to cheat

wight, creature

wild, reckless, distracted, (of sea) open

wilderness, wildness of character, licentiousness

wildly, without cultivation, naturally

wildness, madness

will, sexual desire, sexual organ (male or female)

wimpled, hooded, blinkered

wince, kick

Winchester goose, sufferer from syphilis, prostitute

windgall, soft tumour on horse's leg

windlass, circuit made to intercept game, crafty device

wink, sleep, close one's eyes

wintered, used in winter

wipe, scar

wistly, intently, closely

wit, mental power, mind, sense, wisdom, imagination, one who has such qualities; know

withal, with this, with it, as well, at the same time, with

without, beyond

wittol, a man aware of and tolerating his wife's adultery

witty, wise, cunning

woman-tired, henpecked

wonder, admiration; admire, marvel

wondered, performing wonders

wondering, admiration

wood, mad

woodcock, dupe

woodman, hunter

woolward, wearing wool next to the skin

world, *to go to the*, marry; *a woman of the*, married woman

worm, serpent, snake

worn, exhausted, past

worship, dignity, honour, authority; to honour

wort, vegetable, unfermented beer

worthy, excellent, valuable, deserved, well-founded, fitting

wot, know

wrack, ruin, destruction

wreak, vengeance, revenge

wrest, tuning-key; take by force

wring, wrest, force, writhe, press painfully on

writ, document, writing, mandate, written command, scripture

writhled, wrinkled

wry, to swerve

Xantippe, scolding wife of the philosopher Socrates

yard, yard measure, penis

yare, ready, quick, moving lightly

yaw, sail out of course, lose direction

yellowness, jealousy

yellows, jaundice

yerk, thrust suddenly

younker, fine young man, novice, greenhorn

zany, comic performer awkwardly imitating a clown or mountebank

# INDEX OF FIRST LINES OF
# SONNETS

# INDEX OF FIRST LINES OF SONNETS